中国工程院院士文集系列

钱七虎院士文集 上册

Qian Qihu Yuanshi Wenji

人民交通出版社股份有限公司
北京

内 容 提 要

本书是"中国工程院院士文集系列"之一,主要包括学术论文、学术报告、思想与观点、吾家吾国四个部分。本书汇集了钱七虎院士在防护工程与结构抗爆、岩石力学与工程、隧道及地下工程、绿色与智能建造等领域的学术研究成果,记录了院士在相关学科的前沿技术发展、重大工程建设方面的理念与观点,呈现了院士勤勉严谨治学、矢志不渝报国、不遗余力为民的精神追求。

本书可供防护工程、岩石力学与工程、隧道及地下工程等相关专业领域的技术人员学习参考。

图书在版编目(CIP)数据

钱七虎院士文集 / 钱七虎著. — 北京:人民交通出版社股份有限公司,2023.8
ISBN 978-7-114-18925-8

Ⅰ.①钱… Ⅱ.①钱… Ⅲ.①土木工程—文集 Ⅳ.①TU-53

中国国家版本馆 CIP 数据核字(2023)第 145897 号

Qian Qihu Yuanshi Wenji

书　名	钱七虎院士文集(上册)
著作者	钱七虎
责任编辑	李学会　谢海龙　吴燕伶
责任校对	赵媛媛　魏佳宁　刘　璇
责任印制	张　凯
出版发行	人民交通出版社股份有限公司
地　址	(100011)北京市朝阳区安定门外外馆斜街 3 号
网　址	http://www.ccpcl.com.cn
销售电话	(010)59757973
总 经 销	人民交通出版社股份有限公司发行部
经　销	各地新华书店
印　刷	北京印匠彩色印刷有限公司
开　本	880×1230　1/16
印　张	78.5
字　数	2372 千
版　次	2023 年 8 月　第 1 版
印　次	2023 年 8 月　第 1 次印刷
书　号	ISBN 978-7-114-18925-8
定　价	498.00 元(含上、下册)

(有印刷、装订质量问题的图书,由本公司负责调换)

爱国、创新、求实、奉献、协同、育人

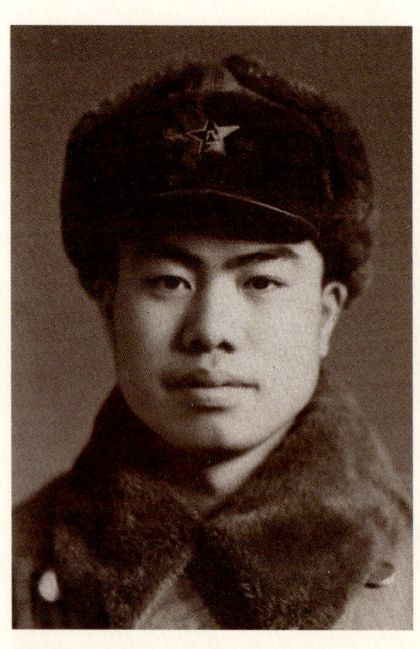

1954 年

钱七虎院士就读哈尔滨军事工程学院
防护工程专业

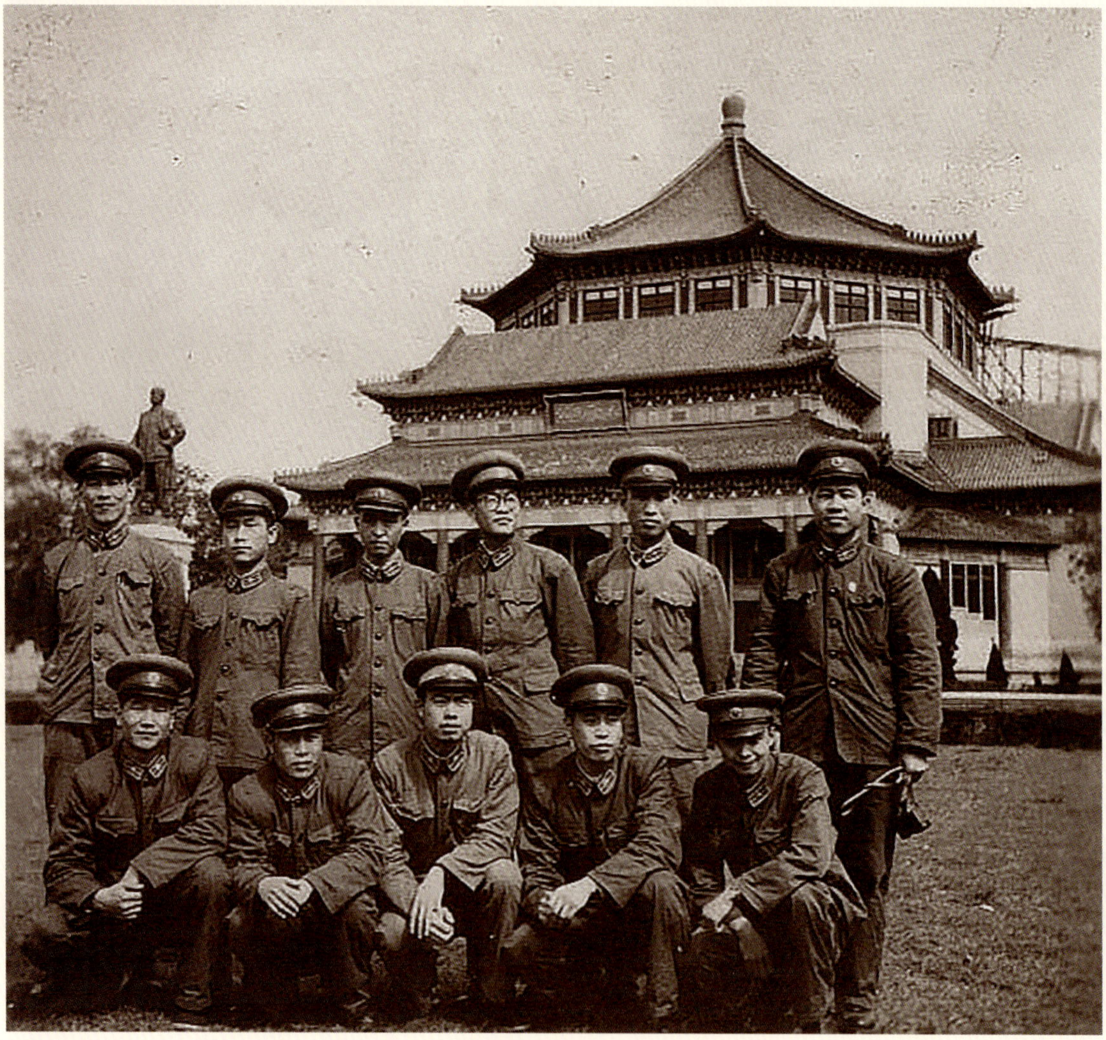

1959 年 钱七虎院士（前排中）赴广州实习

1965 年
钱七虎院士在西安工程兵学院实验照

1969 年
钱七虎院士在西安工程兵工程学院图书馆查阅资料

1984 年
钱七虎院士全家照

1998 年 钱七虎院士在古比雪夫军事工程学院

2003 年 钱七虎院士出席全国政协会议

2010 年 钱七虎院士作《未来的绿色、低碳的城市和城际交通》专题报告

2010 年　钱七虎院士在南京长江路隧道工程

2015 年

钱七虎院士在兰州地铁工程

2017 年　钱七虎院士在世界首例公铁合建盾构法隧道——武汉三阳路长江隧道工程

2018 钱七虎院士在京张高铁工程

2018 钱七虎院士在苏通GIL综合管廊工程

2018 钱七虎院士在指导学生

2019 钱七虎院士在北京东六环改造工程

2019 钱七虎院士在中国岩石力学与工程学会水下隧道工程技术分会成立大会上

2020 钱七虎院士在穿江越海超大断面盾构隧道建造技术高端论坛作报告

2020
钱七虎院士在胶州湾第二海底隧道工程专家咨询会上

2020
钱七虎院士在济南黄河隧道工程

2021 年　钱七虎院士在白鹤滩水电站工程

2021年
钱七虎院士在深中通道工程

钱七虎院士文集  序　PREFACE

　　回想自己的人生道路，首先是感恩，我一生的成长、进步完全是国家、人民和党培养教育的结果。我是在1937年淞沪会战爆发后出生的，民族灾难深重，母亲在逃难途中的小船上生下了我。我七岁时，父亲因贫病离世。没有1949年江苏的解放，我就会和我兄、姐一样失学、失业。是人民的助学金支持我念完了中学；是组织的保送，使我进入了哈军工的大门；又是组织的选拔，我得以到苏联古比雪夫军事工程学院留学，成为一名技术科学副博士。是党、团的教育，使我不断提高自己的革命觉悟，我十四岁加入了中国共青团，十八岁加入了中国共产党，在革命的队伍里，我树立了为人民服务的人生观，才有了不断进步的动力。所以我在哈军工，五年本科成绩全优，年年被评为社会主义建设积极分子和优秀学员，四十五岁时总参总政通过调研、考察，选拔我担任工程兵工程学院院长，成为我军的高级干部。回顾自己的进步，归结到一点：国家、人民和党对我的恩情说不完。

　　熟悉我的同志，都说我的进步道路非常顺利，这和我一生中周围同志对我的支持分不开，对此我衷心感谢。是同志们在我入团后选我担任团支部委员、书记，又是同志们在我入党后选我担任党支部委员。在哈军工，是群众的支持，年年评选我为社会主义建设积极分子。在我担任工程兵工程学院院长期间，当参政胡主任带领总参工作组在学院调研时，是由于学院广大干部、教员的积极反映，总参党委研究决定准备把我们学院党委树为总参勤政廉政的典型，并在北京召开了全国新闻媒体参加的新闻通气会，宣传我院党委的勤政廉政事迹。我本人也被总参党委选为总参唯一

本文节选自"在中央军委给钱七虎记一等功庆功大会上的讲话"，2013年。

的代表参加全军爱国奉献优秀干部事迹报告团在全军巡回宣讲。我能当上教授、中国工程院首届院士，是我国已故著名科学家张维、李国豪推荐提携的结果。我能获得国家科技进步三等奖、二等奖和一等奖，其科研成果是我团队和合作团队集体奋斗的结晶，是大家的支持把我排在第一。我能连续三届担任中国岩石力学与工程学会理事长，是岩土工程领域同志们信任、支持和选举的结果。总结我一生，我深深体会：一个人没有大家的支持，就不可能有什么成就和进步，而要得到别人的支持，就要支持别人；一个领导要得到群众的支持，就要树立和实践领导就是服务的理念。国家最高科学技术奖获得者王选教授就说过："一个人做事处世既要想到自己，更要想到别人，而我的老师季羡林说，做事处世首先要想到别人，其次才想到自己。"这些话是我的座右铭，也是我经常教育我的学生的。在今天的庆功会上，我要再一次衷心感谢我的老前辈、老领导、老同事，感谢我的专业领域内的同人，感谢我学术团队的每位成员和我的学生对我的支持。

最后，请允许我表达我的感奋之情，国家和人民给予我的荣誉是对我的有力鞭策，在我有生之年，我要始终秉持"位卑未敢忘忧国""不待扬鞭自奋蹄"的信念，要学习 2012 年"感动中国"人物林俊德院士，至死攻坚不放松。"老骥伏枥，志在千里"，我时刻感到还有很多东西需要我学习，头脑中还有大量课题需要我研究，还有很多社会焦点和技术攻关，我可以贡献我的学习心得和见解。特别对于我国深地下工程如何可靠地防护，包括应对不断发展的钻地核弹的打击，是我们超高抗力工程建设必须持续研究的、关系国家战略安全的关键课题，也是我有生之年为之奉献全部精力的目标。

作为一名老科学家，培养更多的优秀中青年学术骨干，后继有人，是我义不容辞的责任，我要责无旁贷地把他们带向更高层次，引领防护工程学科发展，推动我国岩石力学与工程领域技术水平的持续提升。为此奋斗，是我的幸福所在。

2023 年 6 月

目录 CONTENTS

钱七虎院士文集

上册

学术论文

爆破荷载作用下岩石破坏特性的"共轭键"基近场动力学数值模拟研究
……………………………………………………………… 周小平,王允腾,钱七虎(003)
事故型冲击荷载对结构作用研究总论 ………………………………… 钱七虎(015)
准脆性材料断裂模拟方法研究 ……………… 燕秀发,钱七虎,赵跃堂,周寅智(026)
基于广义粒子动力学的巷道围岩弹塑性分析 …………… 赵 毅,周小平,钱七虎(033)
单轴压缩条件下岩石破坏的光滑粒子流体动力学数值模拟 … 周小平,赵 毅,钱七虎(046)
克里金算法与多重分形理论在岩土参数随机场分析中的应用 … 王长虹,朱合华,钱七虎(058)
深部硐室围岩分区破裂化现象数值模拟研究 ……………… 苏仲杰,钱七虎(065)
岩爆、冲击地压的定义、机制、分类及其定量预测模型 ……………… 钱七虎(071)
准脆材料动力强度的本质和侧向惯性约束作用 …… 戚承志,钱七虎,陈灿寿,陈剑杰(077)
岩体非协调变形对围岩中的应力和破坏的影响 ……………… 钱七虎,周小平(083)
非协调变形下深部岩体破坏的非欧模型 ……………………… 周小平,钱七虎(091)
Effects of the axial in situ stresses on the zonal disintegration phenomenon in the
　surrounding rock masses around a deep circular tunnel ……… Q. Qian,X. Zhou,E. Xia(099)
深隧道围岩分区破裂的数学模拟 …………… 戚承志,钱七虎,王明洋,陈剑杰(108)
分区破裂化现象的研究进展 ………………… 戚承志,钱七虎,王明洋,罗 健(116)
基于线法的功能梯度材料断裂分析 …… 燕秀发,钱七虎,方国强,赵敏福,郭延宝(124)
基于突变理论的滑坡时间预测模型 …………… 周小平,钱七虎,张永兴,杨海清(132)

Quantitative analysis of rockburst for surrounding rocks and zonal disintegration

  mechanism in deep tunnels ················ Qihu Qian, Xiaoping Zhou（142）

浅埋地下结构顶板在竖向地震作用下的动力响应 ············ 陈灿寿,戚承志,钱七虎,李信桥（155）

功能梯度材料结构分析的半解析数值方法研究 ············ 燕秀发,钱七虎,王 玮,孙 翱,戴 耀（162）

岩质高边坡稳定性分析与评价中的四个准则 ······················· 李 宁,钱七虎（169）

中国岩石工程技术的新进展 ································· 钱七虎（175）

泥岩损伤特性试验研究 ···························· 许宝田,钱七虎,阎长虹,许宏发（186）

深部非均匀岩体卸载拉裂的时间效应和主要影响因素 ············ 范鹏贤,王明洋,钱七虎（191）

大型抛掷爆破中的重力影响 ································· 钱七虎（199）

Effect of loading rate on fracture characteristics of rock

  ················ Xiaoping Zhou, Qihu qian, Haiqing Yang（202）

深埋球形洞室围岩分区破裂化机理 ················ 周小平,钱七虎,张伯虎,张永兴（208）

岩石在过临界荷载作用下变形异常现象的模拟 ············ 戚承志,钱七虎,王明洋,吴 徽（216）

岩爆机理的简化分析和决定性参数的思考 ································· 钱七虎（224）

深部裂隙岩体岩爆定量预测模型 ································· 周小平,钱七虎（233）

Crack tip higher order stress fields for functionally graded materials with generalized form of gradation

  ················ Xiufa Yan, Qihu Qian, Hongbiao Lu, Wei Wang, Ao Sun（243）

岩体结构面对应力波传播规律影响的研究进展 ················ 俞 缙,钱七虎,赵晓豹（251）

岩石爆炸动力学的若干进展 ································· 钱七虎（260）

纵波在改进的弹性非线性法向变形行为单节理处的传播特性研究

  ················ 俞 缙,钱七虎,林从谋,赵晓豹（284）

爆炸作用下岩石破裂块度分布特点及其物理机理 ············ 戚承志,王明洋,钱七虎,罗 健（293）

电阻率法在深部巷道分区破裂探测中的应用 ············ 许宏发,钱七虎,王发军,李术才,袁 亮（297）

多层软弱夹层边坡岩体稳定性及加固分析 ············ 许宝田,钱七虎,阎长虹,许宏发（306）

点滴化学注浆技术加固土遗址工程实例 ············ 柴新军,钱七虎,杨泽平,林重德,松永和也（312）

Dynamic strength of rocks and physical nature of rock strength

  ················ Qihu Qian, Chengzhi Qi, Mingyang Wang（318）

Zonal disintegration of surrounding rock mass around the diversion tunnels

  in Jinping II Hydropower Station, Southwestern China

  ················ Q. H. Qian, X. P. Zhou, H. Q. Yang, Y. X. Zhang, X. H. Li（335）

Bifurcation condition of crack pattern in the periodic rectangular array

  ················ X. P. Zhou, Q. H. Qian, H. Q. Yang（345）

岩石、岩体的动力强度与动力破坏准则 ································· 钱七虎,戚承志（352）

深部巷道围岩分区破裂化现象现场监测研究
 ………………………… 李术才,王汉鹏,钱七虎,李树忱,范庆忠,袁　亮,薛俊华,张庆松(359)
深部岩体工程围岩分区破裂化现象研究综述 ………………………………… 钱七虎,李树忱(368)
岩石爆破的破碎块体大小控制 ………………………………………………… 戴　俊,钱七虎(375)
深部岩体强度准则 …………………………………………………… 周小平,钱七虎,杨海清(378)
深部巷道围岩变形破坏的时间过程及支护 …………………………… 戚承志,钱七虎,王明洋(385)
高放废物地质处置的成本估算 …………… 罗嗣海,钱七虎,王　驹,刘晓东,赖敏慧,杨普济(394)
微型土钉微型化学注浆技术加固土质古窑 ……………………… 柴新军,钱七虎,罗嗣海,林重德(401)
我国高放废物地质处置战略研究 ……………………………………………… 潘自强,钱七虎(408)
深埋巷道分区破裂化机制 ……………………………………………………… 周小平,钱七虎(413)
基于滑面正应力假设的土压力计算方法 ……………………… 刘华丽,钱七虎,朱大勇,周先华(422)
深部隧道围岩的流变 …………………………………………… 戚承志,钱七虎,王明洋,丁常树(427)
深部岩体力学研究进展 ………………………………………………………… 何满潮,钱七虎(432)
高放废物深地质处置中的多场耦合与核素迁移 ……………… 罗嗣海,钱七虎,李金轩,周文斌(446)
深部坑道围岩的变形与承载能力问题 ………………………………… 陈士林,钱七虎,王明洋(453)
深部岩体块系介质变形与运动特性研究 ……………………………… 王明洋,戚承志,钱七虎(462)
岩石中侵彻与爆炸作用的近区问题研究 ……………………………… 王明洋,邓宏见,钱七虎(468)
岩体的构造层次及其成因 ……………………………………… 戚承志,钱七虎,王明洋,董　军(473)
岩体的构造层次粘性及动力强度 ……………………………………… 戚承志,钱七虎,王明洋(482)
弹粘塑性孔隙介质在冲击荷载作用下的一种本构关系——第二部分:弹粘塑性孔隙
　介质的畸变行为 ……………………………………………………… 戚承志,王明洋,钱七虎(491)
强动载作用下的岩石动力学 …………………………………………… 钱七虎,戚承志,王明洋(495)
块体结构岩体中超低摩擦效应的理论研究 …………………………… 戴　俊,钱七虎,王明洋(515)
锦屏二级水电站引水隧道围岩分区破裂化研究 ……………………………… 钱七虎,周小平(522)
大直径盾构掘进风险分析及对特大直径盾构挑战的思考 …………………… 钱七虎,陈　健(537)
建设城市地下综合管廊,转变城市发展方式 ……………………………………………… 钱七虎(545)
隧道工程建设地质预报及信息化技术的主要进展及发展方向 …………………………… 钱七虎(553)
推进城市地下空间规划建设的思考 ………………………………………………………… 钱七虎(566)
公路隧道大断面改扩建施工开挖方案研究 …………………………………… 黄伦海,钱七虎(572)
Safety risk management of underground engineering in China: progress, challenges and strategies
　………………………………………………………………………………… Qihu Qian,Peng Lin(581)
水下隧道工程实践面临的挑战、对策及思考 ……………………………………………… 钱七虎(615)
地铁工程建设安全风险管理研究 ……………………………………… 解东升,钱七虎,戎晓力(620)

地下工程建设安全面临的挑战与对策 ………………………………………………………… 钱七虎(627)
一种岩溶地质条件下的城市地铁超前预报方法
　　……………………………… 苏茂鑫,钱七虎,李术才,薛翊国,张庆松,邱道宏,聂利超(644)
地下磁悬浮交通设计研究的若干问题 …………………………………………………… 钱七虎(651)
基于地下集装箱运输的城市地下环境物流系统建设 ………………………… 范益群,钱七虎(658)
地下洞室连续围岩岩爆的定量分析及其与分区破裂化之间的关系 …………………… 钱七虎(663)
我国城市地下空间综合管理的探讨 …………………………………………… 陈晓强,钱七虎(675)
隧道工程动力响应特性与汶川地震隧道震害分析及启示 …………… 钱七虎,何　川,晏启祥(681)
深埋隧道开挖过程动态及破裂形态分析 ………………… 李树忱,钱七虎,张敦福,李术才(692)
中国地下工程安全风险管理的现状、问题及相关建议 ……………………… 钱七虎,戎晓力(701)
从河床冲淤分析沉管法修建长江水下隧道问题 ………………………………………… 钱七虎(708)
国内外地下综合管线廊道发展的现状、问题及对策 ………………………… 钱七虎,陈晓强(712)
地下空间作为城市空间结构的社会学内涵 …………………………………… 奚江琳,钱七虎(716)
现代地下货物运输系统的研究与进展 …………………………………… 张明聚,钱七虎,唐　劼(720)
大盾构工程关键技术的新进展 …………………………………………………………… 钱七虎(725)
Mechanism and classification and quantitative prediction model of strain rockburst in the
　　surrounding rock masses around a deep circular tunnel ……………… Qihu Qian,Xiaoping Zhou(734)
Evaluation of the status and outlook of the urban underground space development and
　　utilization in China …………………………………………………………… Qihu Qian(749)
关于绿色发展与智能建造的若干思考 …………………………………………………… 钱七虎(755)
工程建设领域要向智慧建造迈进 ………………………………………………………… 钱七虎(757)
利用地下空间助力发展绿色建筑与绿色城市 …………………………………………… 钱七虎(759)
科学利用城市地下空间,建设和谐宜居、美丽城市 ……………………………………… 钱七虎(770)

学术报告

动力扰动(爆破或顶板塌落)诱发冲击地压事故的机理 ……………………………………………(779)
抗爆结构的膜力(in-plane-force)效应 ………………………………………………………………(783)
重要目标防爆抗爆的主要技术措施 …………………………………………………………………(787)
关于空腔爆炸的隔震技术 ……………………………………………………………………………(798)

高地应力岩体地下工程围岩支护设计计算的问题、原因和对策 ……………………………（808）

锦屏一级电站开挖过程地下洞室群围岩变形破坏数值模拟和监测及其分析和启示 …………（816）

深部岩体分区破裂化机理的研究进展 ……………………………………………………………（824）

非线性岩石力学（非传统）研究前沿导论 ………………………………………………………（835）

应变型岩爆的机理、分类及其定量预测模型 ……………………………………………………（845）

非协调变形与岩爆的机理和预测 …………………………………………………………………（855）

Deformation and failure mechanism of surrounding rock mass around underground caverns in Jinping I hydropower station ……………………………………………………………（863）

超深盐膏层地应力与井筒完整性 …………………………………………………………………（870）

Recent advances in the zonal disintegrarion phenomenon in the deep rock mass engineering ………（881）

Dynamic strength and it's physical nature and dynamic fracture criteria of rock ………………（885）

大盾构掘进的事故及对特大盾构工程的思考 ……………………………………………………（893）

地下空间开发利用、防治城市病及地下物流系统的发展方向 …………………………………（898）

21世纪是地下空间开发利用的世纪 ………………………………………………………………（912）

建设城市地下综合管廊转变城市发展方式 ………………………………………………………（944）

隧道岩爆监测预警 …………………………………………………………………………………（956）

地下工程建设安全面临的挑战与对策 ……………………………………………………………（961）

世界地下物流系统研究动态与新进展 ……………………………………………………………（970）

The evaluation of status quo and outlook for the underground logistics system (ULS) development in China …………………………………………………………………………（981）

The present situation and future prospect in application of tunneling machine in china underground engineering construction ………………………………………………………（987）

21世纪前期我国城市地下空间开发利用的战略及对策 …………………………………………（1000）

"双碳"目标下的城市建设 …………………………………………………………………………（1006）

"双碳"目标下的能源地下工程 ……………………………………………………………………（1014）

审读大百科轨道交通文稿时的思考 ………………………………………………………………（1017）

利用地下空间,助力发展绿色建筑与绿色城市 …………………………………………………（1026）

绿色城镇与绿色城镇基础设施 ……………………………………………………………………（1032）

数字隧道和智慧隧道——隧道建设信息化技术的发展方向 ……………………………………（1051）

若干重要建设工程中的哲学思考 …………………………………………………………………（1057）

中国岩石工程进展和规划 …………………………………………………………………………（1060）

思想与观点

建设科技强国迫切需要科学家精神 …………………………………………………………………… (1081)
白鹤滩工程所取得的经验对地下工程建设有重要借鉴意义 ……………………………………… (1084)
又好又快稳步推进城市地下综合管廊建设的思考 ………………………………………………… (1085)
21世纪,让我们向地下要空间 ………………………………………………………………………… (1088)
高度重视地下工程安全——专访中国工程院钱七虎院士 ………………………………………… (1092)
关于北京地下道路建设和地下空间开发相关问题的访谈 ………………………………………… (1094)
通过数字化向智慧建造迈进 …………………………………………………………………………… (1097)
特大城市解决交通拥堵问题的思路与出路 …………………………………………………………… (1100)
又快又好地推进地下空间建设——对话中国工程院钱七虎院士 ………………………………… (1103)
迎接气候变化的挑战,推进绿色建设、实施生态大保护 …………………………………………… (1107)
依托中国的独特优势,加速迈向科技强国的伟大目标 ……………………………………………… (1109)
隧道与地下开发实现历史性"穿越"——访中国工程院院士钱七虎 ……………………………… (1111)
"岩爆"可测时代即将到来 …………………………………………………………………………… (1115)
利用地下空间建设"花园城市" ……………………………………………………………………… (1116)
春风化雨　润物无声——深切缅怀潘家铮院士 …………………………………………………… (1120)
城市交通拥堵、空气污染以及雨洪内涝的治本之策 ………………………………………………… (1122)
城市化发展呼唤积极和科学开发利用城市地下空间 ………………………………………………… (1124)
在苏州地铁11号线开通仪式上的发言 ……………………………………………………………… (1126)
在中国城镇供热协会地下综合管廊运营维护专业委员会成立大会上的致辞 …………………… (1127)
在2021年中国国际服务贸易交易会智慧建造与绿色发展高峰论坛的发言 …………………… (1128)
永远跟党走　永葆革命青春——在中国工程院学习习近平总书记"七一"
　　重要讲话精神报告会上的报告 …………………………………………………………………… (1130)
在中国科协学会党建工作指导委员会成立大会暨学会党建工作先进表彰大会上的发言 ……… (1132)
在"国土空间规划契机下地下空间的机遇与挑战学术研讨会暨《2020中国城市地下空间
　　发展蓝皮书》发布会"上的致辞 ………………………………………………………………… (1134)
在江苏省高层次人才主题学习会上的发言——让生命在科技报国中闪光 ……………………… (1135)
在习近平总书记关于乌东德水电站首批水轮机组发电亲切祝贺和重要指示的
　　重大意义的认识座谈会上的发言 ………………………………………………………………… (1139)
在中国岩石力学与工程学会"弘扬科学家精神,加强作风和学风建设"主题宣讲会上的发言 …… (1140)

在超大直径全断面竖井掘进机下线仪式上的发言 （1143）
在江苏省过江通道建设技术专家委员会成立大会上的发言 （1144）
在纪念于学馥先生百年诞辰大会上的发言 （1145）
在中国岩石力学与工程学会水下隧道工程技术分会成立大会上的发言 （1146）
在2019年盾构与掘进关键技术暨盾构再制造技术国际峰会的致辞 （1148）
在国家科学技术奖励大会上的发言 （1150）
在第三届全国工程安全与防护学术会议上的发言 （1151）
在中国岩石力学与工程学会工程安全与防护分会成立大会上的发言 （1152）
在第七届中俄深部岩石动力学高层论坛开幕式上的发言 （1153）
在中国城市建设科学发展论坛上的发言 （1154）
在第十四次全国岩石力学与工程学术大会开幕式上的发言 （1156）
在苏通GIL综合管廊工程专家聘任座谈会上的发言 （1157）
在东华理工大学建校60周年庆典大会上的致辞 （1158）
在第四届GeoChina国际会议上的致辞 （1160）
在全国隧道及地下工程不良地质超前预报与突水突泥灾害防治学术会议上的发言 （1161）
在"精细爆破学术研讨会"上的致辞 （1163）
在汕头市苏埃通道工程专家委员会成立暨技术咨询会上的发言 （1164）
在"中国矿业科学协同创新联盟"成立大会上的发言 （1165）
中国科协第80期新观点新学说学术沙龙开幕词 （1166）
在人民交通出版社创建六十周年纪念活动上的发言 （1167）
在中国工程科技论坛上的发言 （1168）
在中国人民解放军理工大学溯源碑揭幕仪式上的发言 （1169）
在中国岩石力学与工程学会岩爆机理探索沙龙上的发言 （1170）
在第三届全国水工岩石力学学术会议上的致辞 （1171）

吾家吾国

淡泊名利品自高 （1175）
耿耿丹心　为国铸盾 （1176）
科技强军、为国铸盾的防护工程专家 （1177）
铸盾一生 （1179）

科研工作者要永远跟党走 ……………………………………………………………………… (1183)
科学家精神的核心是追求真理和献身科学 …………………………………………………… (1186)
淡泊名利是科学家精神的重要内核 …………………………………………………………… (1189)
为武汉捐赠650万元,钱七虎:烈士献出了生命,我有什么不能贡献？ ………………… (1190)
老党员履新职 …………………………………………………………………………………… (1193)
中国智慧建造必将走在世界前列 ……………………………………………………………… (1195)
和平年代要树立忧患意识科学家更要有责任担当 …………………………………………… (1197)
一生一事,为国为民——采访1954届校友钱七虎院士 ……………………………………… (1199)
铸就共和国"地下钢铁长城" …………………………………………………………………… (1205)
建设科技强国迫切需要工匠精神 ……………………………………………………………… (1207)

钱七虎院士大事记 ……………………………………………………………………………… (1213)

学术论文

吾志所向，一往无前

Qian Qihu
Yuanshi
Wenji

| 学 术 论 文——防护工程与结构抗爆

爆破荷载作用下岩石破坏特性的"共轭键"基近场动力学数值模拟研究

周小平[1*], 王允腾[1], 钱七虎[2]

1. 武汉大学土木建筑工程学院, 武汉 430072;
2. 陆军工程大学爆炸冲击防灾减灾国家重点实验室, 南京 210007
* 联系人, E-mail: xiao_ping_zhou@126.com

收稿日期: 2019-05-23; 接受日期: 2019-06-19; 网络出版日期: 2019-11-20
国家自然科学基金(编号: 51679017和51839009)和重庆市院士牵头科技创新引导专项基金(编号: cstc2017jcyj-yszx0014)资助项目

摘要 近场动力学是一种基于非局部弹性理论的积分形式的无网格数值方法. 本文为克服传统近场动力学模型中的固定泊松比问题, 建立了"共轭键"基近场动力学模型, 实现了岩石爆破破坏特性的近场动力学数值模拟. 通过引入"共轭键"转动角度及建立微观和宏观变形能的等效关系, 推导了"共轭键"基近场动力学模型中的法向刚度参数及切向刚度参数与宏观力学参数之间的关系. 另外, 通过对影响域中每根"键"所储存的能量密度与临界"键"能量密度进行对比, 判断近场动力学模型中的"键"是否断裂, 从而实现裂纹起裂、扩展及连接过程的数值模拟. 三个数值算例说明: 该模型能有效地模拟岩石爆破破坏特性. 数值算例与试验结果对比表明, 本文所提出的数值方法可以预测岩石材料的爆破破坏模式及特性.

关键词 "共轭键"基近场动力学, 脆性岩石材料, 爆破冲击荷载

PACS: 45.20.-d, 46.05.+b, 46.15.-x, 46.50.+a, 62.20.Mk

1 引言

随着计算机水平的发展, 应用数值算法模拟岩体在复杂荷载作用(尤其是爆破冲击荷载作用)下的破坏特性, 是研究岩石或岩体动力学特性的一种有效手段[1]. 目前, 模拟岩石或岩体破坏特性的数值方法主要分为连续介质力学数值方法、非连续介质力学数值方法和连续-非连续介质力学耦合方法三大类. 连续介质力学方法主要适用于分析连续体的小变形及损伤断裂; 非连续介质力学方法适用于分析非连续体的破坏和运动; 连续-非连续介质力学耦合方法可以分析岩体从连续到非连续的全过程[2]. 在连续介质力学方法中, 有限元方法(Finite Element Method, FEM)是目前广泛应用于各领域的一种数值计算方法[3-5]. 但在处理裂纹扩展问题时, 需对裂纹面和裂纹尖端的网格进行重划分, 这导致计算过程复杂化和低效化. 在非连续介质

力学方法中, 非连续变形分析法(Discontinuous Deformation Analysis, DDA)能有效地模拟岩石中的裂纹扩展和连接过程[6-9]. DDA是一种基于离散块体刚体运动的数值计算方法[6]. 但是, DDA存在裂纹扩展路径依赖于离散块体大小的缺陷. 在连续-非连续介质力学耦合方法中, 有限元-离散元耦合方法(Finite Element/Discrete Element Method, FDEM)是一种将有限元和离散元结合起来的数值方法[10-12], 它在模拟裂纹扩展方面展现了巨大的潜力, 并且将其应用于水压致裂模拟, 取得了很好的效果[11-13]. 然而, 其输入参数如何与岩石的宏观力学性质更好地对应还需要进一步研究.

以上三类数值方法各有优缺点, 但是上述数值方法均属于局部力学理论. 然而, 为了解决传统连续介质力学不能模拟不连续问题的缺陷, 美国Sandia国家试验室的Silling[14]于2000年基于非局部作用的思想提出了近场动力学理论(Peridynamics), 该理论后来被称为基于键作用的近场动力学(Bond-Based Peridynamics)[15]. 近场动力学理论将物质离散为有质量的物质点(Material Point), 物质点之间通过对点力相互作用, 用空间积分的方式计算由于物体变形而使物质点产生的作用力, 从而有效避免了基于连续性假设建模和求解空间微分方程的传统连续介质力学方法在面临不连续问题时的奇异性[16-18]. 近年来, 近场动力学模型被应用到岩土工程领域和用于揭示其相关的力学和物理机制[19-25].

然而, 在"键"基近场动力学中, 由于仅有一个近场动力学模型参数用于描述两个相互作用的物质点运动及变形导致其固定泊松比的问题. 在二维平面应力条件下, "键"基近场动力学模型的泊松比固定为1/3; 在三维条件或二维平面应变条件下, "键"基近场动力学模型的泊松比固定为1/4. 为了克服固定泊松比的问题, Zhou和Shou[23], Zhu和Ni[26]在传统近场动力学数值模型中引入切向键的概念, 通过判断"键"与参考坐标轴的夹角变化度量其相应的转动变形值. 本文根据Stillinger-Weber势能函数引入"共轭键"的概念[27-29], 通过对两个相互作用"键"之间的"共轭键"夹角变化度量其剪切变形. 同时, 根据传统断裂力学理论, 建立"键"能相关的断裂准则, 模拟拉伸和剪切裂纹的起裂、扩展过程. 本文通过3个算例分析, 说明了"共轭键"基近场动力学模型可以有效地模拟脆性岩石材料在爆破冲击荷载作用下的破坏特性.

2 近场动力学理论

在传统连续介质力学中, 单元或物质点的线性动量守恒控制方程可以表示为二阶偏微分方程形式:

$$\rho \ddot{\mathbf{u}} = \nabla \cdot \boldsymbol{\sigma} + \mathbf{b}, \tag{1}$$

式中, ρ是材料物质密度, $\ddot{\mathbf{u}}$为物质点加速度矢量, $\boldsymbol{\sigma}$为单元或物质点上应力, $\mathbf{b}$为外力密度.

在近场动力学基本理论中, 计算域$\mathfrak{B}_0$内的研究对象可以离散为具有相关物理变量的物质点, 每个物质点$\mathbf{x}_i$与其周围具有一定距离范围内的其他物质点$\mathbf{x}_j$具有相互作用, 即非局部长程作用力密度$\mathbf{f}(\mathbf{x}_i, \mathbf{x}_j, t)$. 如图1所示, 在近场动力学模型中, 非局部长程作用力$\mathbf{f}(\mathbf{x}_i, \mathbf{x}_j, t)$与参考坐标系$\mathfrak{B}_0$和变形坐标系$\mathfrak{B}$中的相对位置矢量$\boldsymbol{\xi}$和相对位移矢量$\boldsymbol{\eta}$相关. 在图1中的近场动力学原理中, δ表示一定大小的相互作用域的半径; $\mathbf{u}_i$和$\mathbf{u}_j$是两个相互作用物质点$\mathbf{x}_i$和$\mathbf{x}_j$的位移矢量; $\mathbf{y}_i$和$\mathbf{y}_j$表示在变形坐标系下两个相互作用物质点$\mathbf{x}_i$和$\mathbf{x}_j$的位置矢量. 在参考坐标系下, 两个相互作用物质点$\mathbf{x}_i$和$\mathbf{x}_j$的相对位置矢量为$\boldsymbol{\xi}_{ij}=\mathbf{x}_j-\mathbf{x}_i$; 在变形坐标系下, 两个相互作用物质点$\mathbf{x}_i$和$\mathbf{x}_j$的相对位置矢量为$\boldsymbol{\xi}_{ij}+\boldsymbol{\eta}_{ij}=\mathbf{y}_j-\mathbf{y}_i$; 两个相互作用物质点$\mathbf{x}_i$和$\mathbf{x}_j$的相对位移矢量为$\boldsymbol{\eta}_{ij}=\mathbf{u}_j-\mathbf{u}_i$.

在近场动力学理论中, 物质点$\mathbf{x}_i$的运动控制方程可以表示为[14-16]

$$\rho \ddot{\mathbf{u}}(\mathbf{x}_i, t) = \int_{H_{\mathbf{x}_i}} \mathbf{f}(\mathbf{x}_i, \mathbf{x}_j, t) \mathrm{d}V_{\mathbf{x}_i} + \mathbf{b}(\mathbf{x}_i, t), \tag{2}$$

式中, $H_{\mathbf{x}_i}$表示物质点$\mathbf{x}_i$的影响域, $\mathbf{b}$表示外力密度.

在"键"基近场动力学模型中, 两个物质点之间的相互作用力密度可以表示为[14,15]

$$\mathbf{f}(\boldsymbol{\xi}_{ij}, \boldsymbol{\eta}_{ij}, t) = c \frac{|\boldsymbol{\xi}_{ij}+\boldsymbol{\eta}_{ij}|-|\boldsymbol{\xi}_{ij}|}{|\boldsymbol{\xi}_{ij}|} \cdot \frac{\boldsymbol{\xi}_{ij}+\boldsymbol{\eta}_{ij}}{|\boldsymbol{\xi}_{ij}+\boldsymbol{\eta}_{ij}|}, \tag{3}$$

式中, c表示近场动力学刚度参数.

3 "共轭键"基近场动力学原理

由于仅有一个近场动力学刚度参数c, 导致"键"基近场动力学模型中泊松比固定的缺陷[26,30-33]. 为了克服"键"基近场动力学模型中泊松比固定的缺陷, 本文建立了"共轭键"基近场动力学数值模型.

在"共轭键"基近场动力学数值模型中, 每个物质

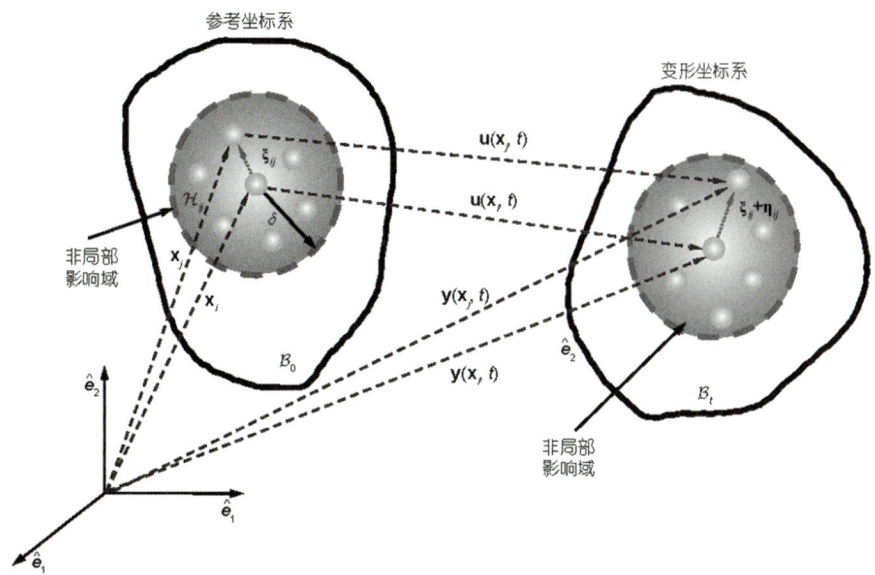

图 1 (网络版彩图)近场动力学物质点运动及变形原理图
Figure 1 (Color online) Schematics of the peridynamic material points' kinematics in the reference and the current configurations.

点的变形能W^{PD}可以分解为膨胀变形能W_l^{PD}和剪切变形能W_β^{PD}, 其相应的计算公式如下:

$$W^{PD} = W_l^{PD} + W_\beta^{PD}, \tag{4}$$

式中, W_l^{PD}为物质点的膨胀变形能; W_β^{PD}为物质点的剪切变形能.

每个物质点的膨胀变形能W_l^{PD}和剪切变形能W_β^{PD}可以分别由其影响域内的每根"键"的膨胀变形能密度w_l和"共轭键"的转动变形能密度w_β积分确定, 其相应的计算公式如下:

$$W_l^{PD} = \frac{1}{2}\int_{H_{x_i}} w_l(\eta_{ij}, \xi_{ij}) dV_{\xi_{ij}}, \tag{5}$$

$$W_\beta^{PD} = \frac{1}{2}\int_{H_{x_i}}\int_{H_{x_j}} w_\beta(\eta_{ij}, \xi_{ij}, \eta_{ik}, \xi_{ik}) dV_{\xi_{ik}} dV_{\xi_{ij}}, \tag{6}$$

式中, w_l为"键"的膨胀变形能密度; w_β为"共轭键"的转动变形能密度.

"共轭键"的转动变形能密度w_β是由一对"共轭键"ξ_{ij}和ξ_{ik}所形成的夹角变形程度进行度量的. 在参考坐标系下, 一对"共轭键"可以表示为ξ_{ij}和ξ_{ik}. 相应地, 在变形坐标系下, 一对"共轭键"则可以表示为$\xi_{ij}+\eta_{ij}$和$\xi_{ik}+\eta_{ik}$. 如图2所示, 当一个物质点所包含的相互作用影响域中含有N个物质点时, 其中表示膨胀变形能的传

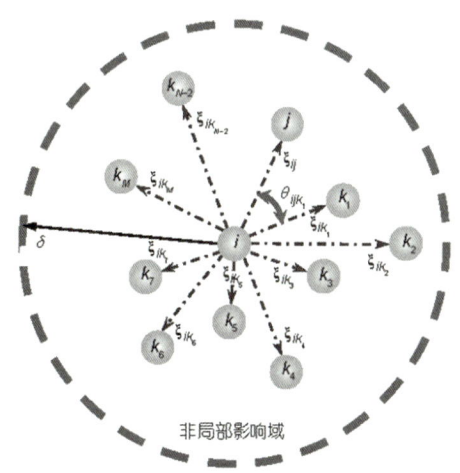

图 2 (网络版彩图)"共轭键"基近场动力学模型理论图
Figure 2 (Color online) Schematic of the conjugated bond-pair-based peridynamic model.

统"键"个数为$N-1$, 表示剪切变形能的"共轭键"个数为$N-2$, 基于小变形假设条件下, 在变形坐标系下, 表示两个相互作用物质点x_i和x_j的"键"长度为

$$|\xi_{ij}+\eta_{ij}| = |\xi_{ij}|\sqrt{\chi^T F^T F \chi} = |\xi_{ij}|\sqrt{\chi^T(2E+I)\chi}, \tag{7}$$

式中, $|\xi_{ij}|$表示参考坐标系下连接两个相互作用物质点x_i和x_j的"键"长; F表示物体变形梯度; E表示物体的应变张量; χ表示在参考坐标系下的"键"ξ_{ij}的单位方向向量.

式(7)可以进一步写为

$$\left|\boldsymbol{\xi}_{ij}+\boldsymbol{\eta}_{ij}\right|=\left|\boldsymbol{\xi}_{ij}\right|\chi_m\varepsilon_{mn}\chi_n+\left|\boldsymbol{\xi}_{ij}\right|, \tag{8}$$

式中, ε_{mn} 为"键"的局部应变分量.

在变形过程中, 两个相互作用物质点 $\mathbf{x}_i$ 和 $\mathbf{x}_j$ 之间的"键"的伸长量可以表达为

$$s_{ij}=\frac{\left|\boldsymbol{\xi}_{ij}+\boldsymbol{\eta}_{ij}\right|-\left|\boldsymbol{\xi}_{ij}\right|}{\left|\boldsymbol{\xi}_{ij}\right|}=\chi_m\varepsilon_{mn}\chi_n, \tag{9}$$

式中, $|\boldsymbol{\xi}_{ij}+\boldsymbol{\eta}_{ij}|$ 为变形坐标系下的"键"长; $|\boldsymbol{\xi}_{ij}|$ 为参考坐标系下的"键"长.

根据参考文献[14–16], 由于"键"的伸长储存的微观变形势能密度可以表示为

$$w_l(\boldsymbol{\eta}_{ij},\boldsymbol{\xi}_{ij})=\frac{1}{2}c_n s_{ij}^2\left|\boldsymbol{\xi}_{ij}\right|=\frac{1}{2}c_n(\chi_m\varepsilon_{mn}\chi_n)^2\left|\boldsymbol{\xi}_{ij}\right|, \tag{10}$$

式中, c_n 为近场动力学"键"的伸长变形刚度参数.

将式(10)代入式(5)可以得到物质点上的膨胀变形能:

$$W_l^{\mathrm{PD}}=\frac{1}{2}\int_0^\delta\int_0^{2\pi}\frac{1}{2}c_n(\chi_m\varepsilon_{mn}\chi_n)^2\left|\boldsymbol{\xi}_{ij}\right|^2\mathrm{d}\varphi\mathrm{d}\left|\boldsymbol{\xi}_{ij}\right|. \tag{11}$$

类似地, 由于"共轭键"转动引起的微观变形势能密度可以表示为

$$w_\beta(\boldsymbol{\eta}_{ij},\boldsymbol{\xi}_{ij},\boldsymbol{\eta}_{ik},\boldsymbol{\xi}_{ik})=\frac{1}{2}\lambda_t\gamma^2\approx\frac{1}{2}\lambda_t(\theta_{jik}-\theta_{jik0})^2, \tag{12}$$

式中, λ_t 为"共轭键"近场动力学转动变形刚度参数; γ 为"共轭键"上的剪切应变; θ_{jik} 为变形坐标系下的"共轭键" $\boldsymbol{\xi}_{ij}+\boldsymbol{\eta}_{ij}$ 和 $\boldsymbol{\xi}_{ik}+\boldsymbol{\eta}_{ik}$ 的夹角; θ_{jik0} 为参考坐标系下的"共轭键" $\boldsymbol{\xi}_{ij}$ 和 $\boldsymbol{\xi}_{ik}$ 的夹角.

根据小变形假设及相应的几何条件[27–29], "共轭键"转动引起的微观变形势能密度 $w_\beta(\boldsymbol{\eta}_{ij},\boldsymbol{\xi}_{ij},\boldsymbol{\xi}_{ik},\boldsymbol{\eta}_{ik})$ 可以表达为

$$w_\beta(\boldsymbol{\eta}_{ij},\boldsymbol{\xi}_{ij},\boldsymbol{\eta}_{ik},\boldsymbol{\xi}_{ik})=\frac{1}{2}\lambda_t(\Theta_{ij}\varepsilon_{ij})^2=\frac{1}{2}\lambda_t \\ \times\left[\frac{1}{\sqrt{1-(\boldsymbol{\chi}^{\mathrm{T}}\boldsymbol{\psi})^2}}\cdot\left(2\chi_i\psi_j-\boldsymbol{\chi}^{\mathrm{T}}\boldsymbol{\psi}\cdot\chi_i\chi_j-\boldsymbol{\chi}^{\mathrm{T}}\boldsymbol{\psi}\cdot\psi_i\psi_j\right)\varepsilon_{ij}\right]^2, \tag{13}$$

式中, $\boldsymbol{\chi}$ 为在参考坐标系下的"键" $\boldsymbol{\xi}_{ij}$ 的单位方向向量; $\boldsymbol{\psi}$ 为在参考坐标系下的"键" $\boldsymbol{\xi}_{ik}$ 的单位方向向量.

将式(13)代入式(6)中, 可以得到物质点上的剪切变形能[31,32].

$$W_\beta^{\mathrm{PD}}=\frac{1}{2}\int_0^{2\pi}\left\{\int_0^\delta\int_0^{2\pi}\frac{1}{2}\lambda_t\frac{1}{1-(\boldsymbol{\chi}^{\mathrm{T}}\boldsymbol{\psi})} \\ \cdot\left[\left(2\chi_i\psi_j-\boldsymbol{\chi}^{\mathrm{T}}\boldsymbol{\psi}\cdot\chi_i\chi_j-\boldsymbol{\chi}^{\mathrm{T}}\boldsymbol{\psi}\cdot\psi_i\psi_j\right)\right]^2 \\ \cdot\left|\boldsymbol{\xi}_{ij}\right|\mathrm{d}\varphi\mathrm{d}\left|\boldsymbol{\xi}_{ij}\right|\right\}\mathrm{d}\overline{\varphi}. \tag{14}$$

将式(11)和(14)代入式(4)中, 并进行相应的积分计算, 可以得到每个物质点上的总变形能:

$$W^{\mathrm{PD}}=W_l^{\mathrm{PD}}+W_\beta^{\mathrm{PD}} \\ =\left(\frac{\pi\delta^3 hc_n}{16}+\frac{\pi^2\delta^2 h\lambda_t}{16}\right)\cdot(\varepsilon_{11}^2+\varepsilon_{22}^2) \\ +\left(\frac{\pi\delta^3 hc_n}{24}-\frac{\pi^2\delta^2 h\lambda_t}{4}\right)\cdot\varepsilon_{11}\varepsilon_{22} \\ +\left(\frac{\pi\delta^3 hc_n}{12}+\frac{\pi^2\delta^2 h\lambda_t}{2}\right)\cdot\varepsilon_{12}^2. \tag{15}$$

根据传统连续介质力学(Classical Continuum Mechanics), 相应宏观变形能为

$$W^{\mathrm{CCM}}=\left(\frac{1}{2}K+\frac{2}{3}G\right)\cdot(\varepsilon_{11}^2+\varepsilon_{22}^2) \\ +\left(K-\frac{2}{3}G\right)\cdot\varepsilon_{11}\varepsilon_{22}+2G\varepsilon_{12}^2. \tag{16}$$

根据能量守恒原理, 建立近场动力学变形能和传统连续介质力学变形能等效表达式, 可以得到相应的"共轭键"基近场动力学微观刚度参数表达式:

$$c_n=\frac{6E}{(1+v)(1-2v)\pi\delta^3 h}, \tag{17}$$

$$\lambda_t=\frac{E(1-4v)}{(1+v)(1-2v)\pi^2\delta^2 h}, \tag{18}$$

式中, E 为材料杨氏弹性模量, v 为材料泊松比.

根据式(17)和(18)可知, "共轭键"基近场动力学模型可以克服传统近场动力学模型中的泊松比固定的缺陷.

相互作用影响域内的膨胀力密度 $\mathbf{f}_n$ 可以表示为

$$\mathbf{f}_n=c_n\cdot s_{ij}\cdot\frac{\boldsymbol{\xi}_{ij}+\boldsymbol{\eta}_{ij}}{\left|\boldsymbol{\xi}_{ij}+\boldsymbol{\eta}_{ij}\right|}=c_n\cdot s_{ij}\cdot\hat{\mathbf{e}}_n \\ =\frac{6E}{(1+v)(1-2v)\pi\delta^3 h} \\ \cdot\frac{\left|\boldsymbol{\xi}_{ij}+\boldsymbol{\eta}_{ij}\right|-\left|\boldsymbol{\xi}_{ij}\right|}{\left|\boldsymbol{\xi}_{ij}\right|}\cdot\frac{\boldsymbol{\xi}_{ij}+\boldsymbol{\eta}_{ij}}{\left|\boldsymbol{\xi}_{ij}+\boldsymbol{\eta}_{ij}\right|}, \tag{19}$$

式中, δ 为影响域半径; h 为模型厚度.

如图3(a)可得, 膨胀力密度 $\mathbf{f}_n$ 沿着两个物质点在变形后的相应"键"方向.

相应地, 相互作用影响域内的剪切力密度 $\mathbf{f}_t$ 可以表示为

$$\begin{aligned}\mathbf{f}_t &= \frac{m_{jik}}{|\boldsymbol{\xi}_{ij}+\boldsymbol{\eta}_{ij}|}\cdot\hat{\mathbf{e}}_t \\ &= \frac{\lambda_t(\theta_{jik}-\theta_{jik0})}{|\boldsymbol{\xi}_{ij}+\boldsymbol{\eta}_{ij}|}\cdot\frac{\boldsymbol{\xi}_{ij}+\boldsymbol{\eta}_{ij}}{|\boldsymbol{\xi}_{ij}+\boldsymbol{\eta}_{ij}|} \\ &\times\left[\frac{\boldsymbol{\xi}_{ij}+\boldsymbol{\eta}_{ij}}{|\boldsymbol{\xi}_{ij}+\boldsymbol{\eta}_{ij}|}\times\frac{\boldsymbol{\xi}_{ik}+\boldsymbol{\eta}_{ik}}{|\boldsymbol{\xi}_{ik}+\boldsymbol{\eta}_{ik}|}\right],\end{aligned} \quad (20)$$

式中, $\dfrac{\boldsymbol{\xi}_{ij}+\boldsymbol{\eta}_{ij}}{|\boldsymbol{\xi}_{ij}+\boldsymbol{\eta}_{ij}|}\times\left[\dfrac{\boldsymbol{\xi}_{ij}+\boldsymbol{\eta}_{ij}}{|\boldsymbol{\xi}_{ij}+\boldsymbol{\eta}_{ij}|}\times\dfrac{\boldsymbol{\xi}_{ik}+\boldsymbol{\eta}_{ik}}{|\boldsymbol{\xi}_{ik}+\boldsymbol{\eta}_{ik}|}\right]$ 表示沿着变形后"共轭键"形成的剪切力密度的单位方向向量, 如图3(b)和(c)所示.

根据近场动力学理论, 引入物质点损伤变量 $D=(\mathbf{x}_i, t)$ 描述裂纹扩展路径, 其表达式为

$$D(\mathbf{x}_i,t) = 1 - \frac{\int_{H_{\mathbf{x}_i}}\mu(\mathbf{x}_i,\boldsymbol{\xi}_{ij},t)\mathrm{d}V_{\mathbf{x}_j}}{\int_{H_{\mathbf{x}_i}}1\mathrm{d}V_{\mathbf{x}_j}}, \quad (21)$$

式中, $\mu(\mathbf{x}_i, \boldsymbol{\xi}_{ij}, t)$ 为"键"的历史标量, 其表达式为

$$\mu(\mathbf{x}_i,\boldsymbol{\xi}_{ij},t) = \begin{cases} 1, & \text{完整"键"}, \\ 0, & \text{破坏"键"}, \end{cases} \quad (22)$$

式中, 当 $\mu(\mathbf{x}_i, \boldsymbol{\xi}_{ij}, t)=1$ 时, 表示物质点之间的"键"为完整的; 当 $\mu(\mathbf{x}_i, \boldsymbol{\xi}_{ij}, t)=0$ 时, 表示物质点之间的"键"为破裂的.

在"共轭键"基近场动力学模型中, 每根"键"的变形能力密度由其相应的"键"膨胀变形能密度和"键"的剪切变形能密度组成[31,32]

$$w_{\boldsymbol{\xi}_{ij}} = \frac{1}{2}c_n s_{ij}^2|\boldsymbol{\xi}_{ij}| + \sum_{k}^{N_t-2}\frac{1}{2}\lambda_t(\Delta\theta_{jik})^2. \quad (23)$$

根据线弹性断裂力学理论和近场动力学理论, 能量释放率可以表示为在影响域内断裂"键"释放的变形能密度之和. 如图4所示, 其相应的计算表达式为

$$G_c = \int_0^\delta\int_0^{2\pi}\int_0^\delta\int_0^{\arccos(z/\xi)} w_c|\boldsymbol{\xi}|^2\sin\theta\mathrm{d}\theta\mathrm{d}|\boldsymbol{\xi}|\mathrm{d}\varphi\mathrm{d}z. \quad (24)$$

通过对式(24)进行积分, 可以得到材料中"键"的临界断裂能密度为

$$w_c = \frac{4G_c}{\pi\delta^4}, \quad (25)$$

式中, G_c 为线弹性断裂力学理论中材料的能量释放率.

在变形过程中, 对每根"键"的变形能密度 $w_{\boldsymbol{\xi}_{ij}}$ 与材料中"键"的临界断裂能密度 w_c 的大小进行比较, 可以得到"共轭键"基近场动力学模型中的"键"的历史标量:

$$\mu(w_{\boldsymbol{\xi}_{ij}},t) = \begin{cases} 1, & w_{\boldsymbol{\xi}_{ij}} \le w_c, \\ 0, & w_{\boldsymbol{\xi}_{ij}} > w_c. \end{cases} \quad (26)$$

4 数值模拟结果及分析

4.1 算例I: 脆性材料动力断裂

为了验证"共轭键"基近场动力学模型模拟动荷载作用下固体材料的破裂特性, 本文首先模拟了动荷载

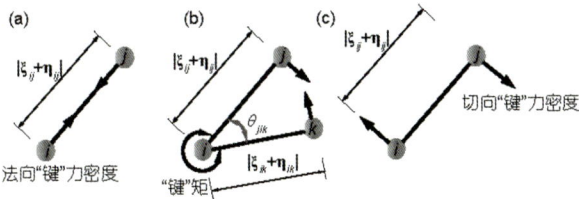

图 3 (网络版彩图)法向"键"(a)、"键"矩(b)及切向"键"(c)原理图

Figure 3 (Color online) Schematics of normal bond force (a), bond moment (b), and tangent bond force (c).

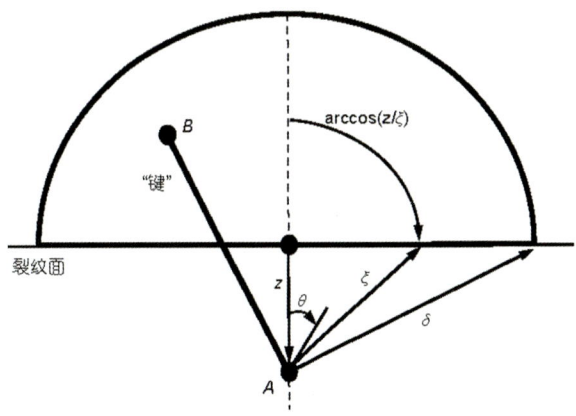

图 4 "共轭键"基近场动力学模型中I型及II型断裂原理图

Figure 4 Illustrations of Mode-I and Mode-II fracture in the conjugated bond-based peridynamics.

作用下PMMA材料的破坏过程[34,35], 并将"共轭键"基近场动力学模型模拟结果与试验结果和其他传统数值结果相对比. 如图5所示, 一块含有水平预制裂纹的PMMA板受到竖向动荷载作用. 根据经典试验[34,35], PMMA材料参数为: 密度ρ=2450 kg/m³; 杨氏弹性模量E=32.0 GPa; 泊松比υ=0.2; 能量释放率G_c=3.0 N/m; 膨胀弹性波速c_d=3809.5 m/s; 剪切弹性波速c_s=2332.8 m/s; 瑞利波波速c_R=2119.0 m/s. 如图5所示, 动荷载施加在PMMA版的上下两个边界, 其具体表达式如下:

$$\sigma_y(t) = \begin{cases} \sigma_0 t/t_0, & \text{for } t \le t_0, \\ \sigma_0, & \text{for } t > t_0, \end{cases} \tag{27}$$

式中, σ_0=1.0 MPa; t_0=50.0 μs.

在"共轭键"基近场动力学模型中, 采用17200个近场动力学物质点表示PMMA板. 近场动力学物质点间

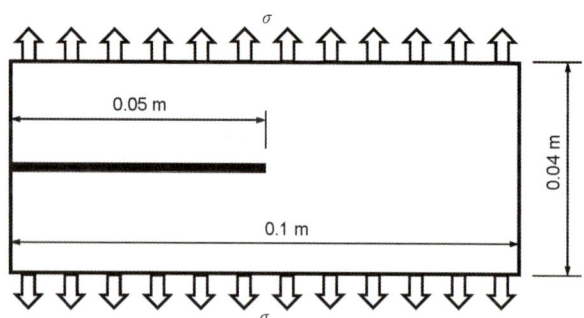

图 5 含预制水平裂纹的PMMA板在动荷载作用下的几何及边界条件

Figure 5 Geometric and boundary conditions of the PMMA plate containing a single pre-existing flaw under dynamic loads.

距为Δx=0.5 mm, 影响域尺寸为δ=1.5 mm; 时间步长度为Δt=0.05 μs.

图6表示动荷载作用下裂纹起裂、扩展和分叉的

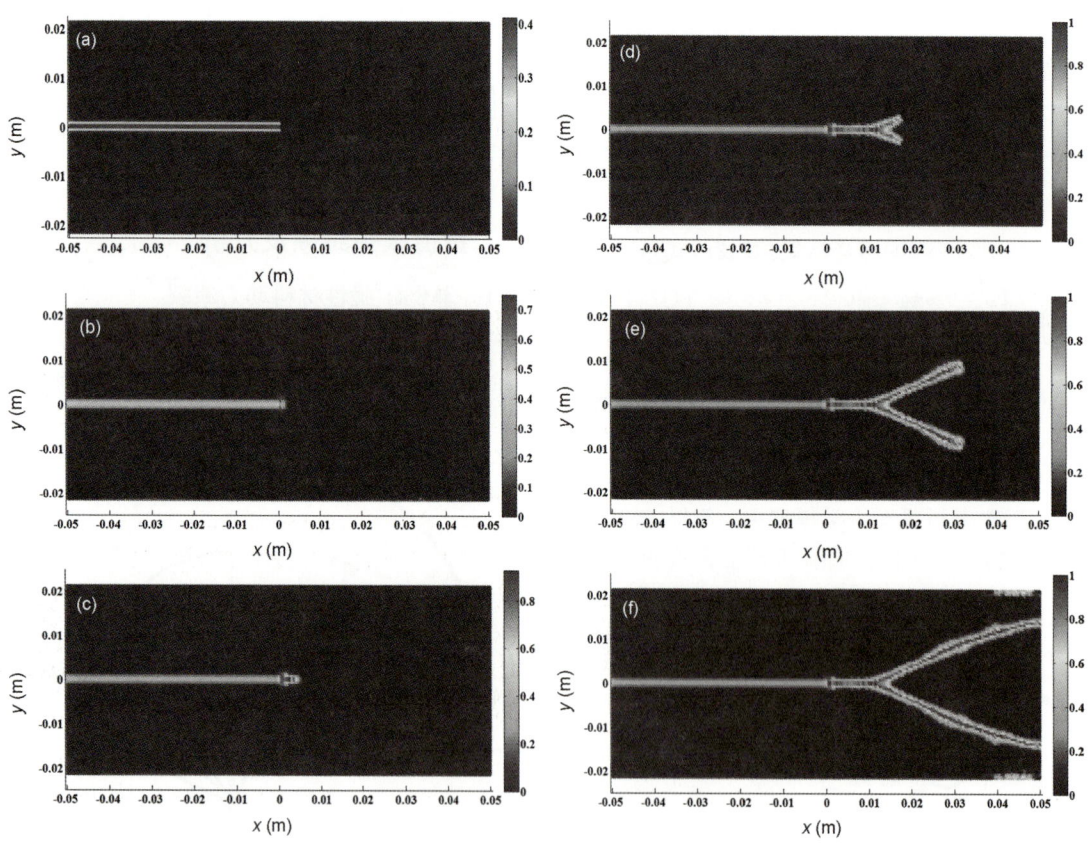

图 6 (网络版彩图)PMMA板在动荷载作用下的破坏过程损伤图. (a) Δt=1.0 μs; (b) Δt=18.0 μs; (c) Δt=30.0 μs; (d) Δt=40.0 μs; (e) Δt=50.0 μs; (f) Δt=62.0 μs

Figure 6 (Color online) The process of crack growth in brittle PMMA plate under dynamic loads. (a) Δt=1.0 μs; (b) Δt=18.0 μs; (c) Δt=30.0 μs; (d) Δt=40.0 μs; (e) Δt=50.0 μs; (f) Δt=62.0 μs.

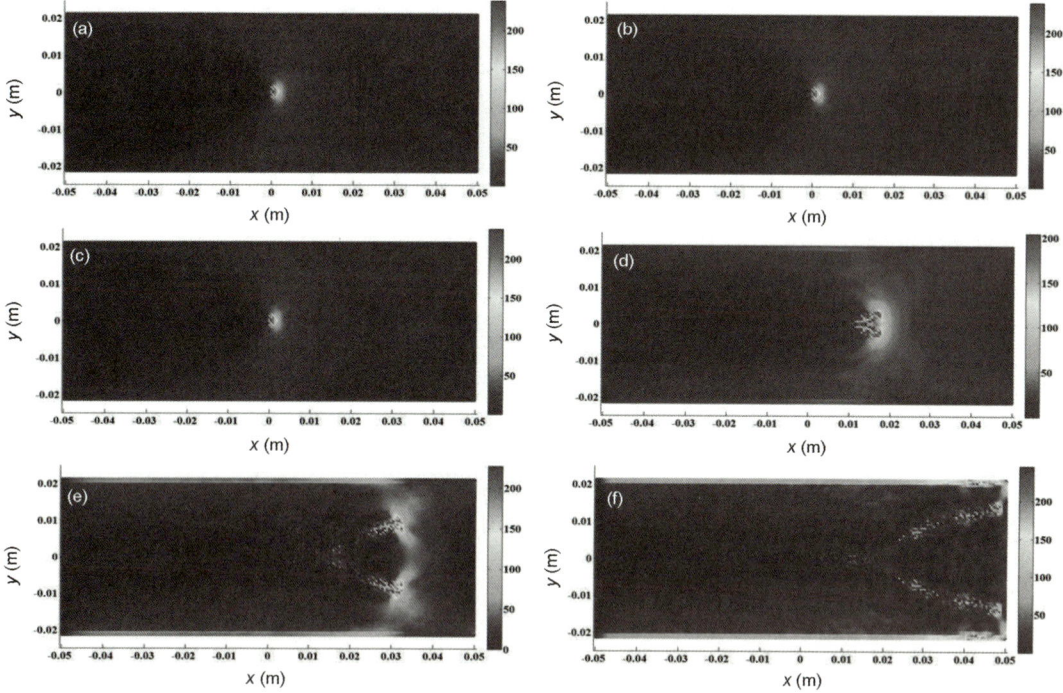

图 7 (网络版彩图)动荷载作用下PMMA板破坏过程的应变能密度演化过程图. (a) Δt=1.0 μs; (b) Δt=18.0 μs; (c) Δt=30.0 μs; (d) Δt=40.0 μs; (e) Δt=50.0 μs; (f) Δt=62.0 μs

Figure 7 (Color online) Strain energy density evolution of the fracturing process in the PMMA plate containing a single pre-existing flaw under dynamic loads. (a) Δt=1.0 μs; (b) Δt=18.0 μs; (c) Δt=30.0 μs; (d) Δt=40.0 μs; (e) Δt=50.0 μs; (f) Δt=62.0 μs.

演化过程. 图7表示应变能密度演化过程. 从图6(a)和图7(a)可知, 当时间为1.0 μs时, 应变能密度在预制裂纹尖端集中; 如图6(b)和7(b)所示, 当时间为18.0 μs时, 随着动荷载不断增加, 新裂纹从预制裂纹尖端起裂, 应变能密度集中于新裂纹尖端. 如图6(c)和7(c)所示, 随着动荷载继续增加, 裂纹沿着水平方向扩展. 如图6(d)和7(d)所示, 当时间为40.0 μs左右时, 两个分叉裂纹从主裂纹尖端出现, 并沿着一定角度向右边界扩展. 如图6(e)和7(e)所示, 两个分叉裂纹继续沿着一定角度向右边界扩展. 如图6(f)和7(f)所示, 当时间为62.0 μs时, PMMA板完全破坏.

将"共轭键"基近场动力学数值结果(图8)与试验结果[35]、扩展有限元数值模拟结果[36]和离散元数值模拟结果[37]进行对比, 可知: "共轭键"基近场动力学数值预测的PMMA板在动荷载作用下的最终破坏模式与试验结果[35]、扩展有限元模拟结果[36]和离散元模拟结果[37]相吻合, 说明"共轭键"基近场动力学数值模型可以准确模拟脆性固体在动荷载作用下的破坏模式.

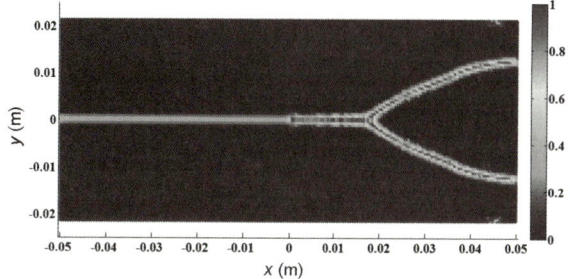

图 8 (网络版彩图)"共轭键"基近场动力学模拟PMMA板在动荷载作用下的最终破裂模式

Figure 8 (Color online) Ultimate failure pattern of PMMA plate with a pre-existing crack subjected to dynamic loads obtained by the conjugated bond-pair-based peridynamics.

图9表示"共轭键"基近场动力学数值预测的裂纹尖端扩展速度与其他传统数值方法预测的裂纹尖端扩展速度的对比结果. 从图9可得, 当裂纹分叉产生前, 裂纹尖端扩展速度不断增大, 但没有超过瑞利波速; 当裂纹动力分叉发生后, 裂纹扩展速度在低于瑞利波速度的

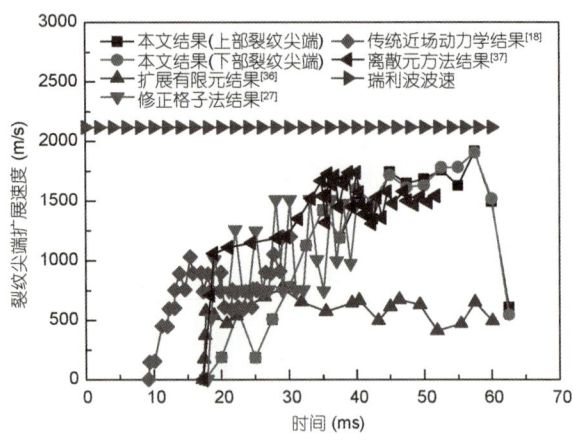

图 9 (网络版彩图)PMMA板在动力荷载作用下的裂纹扩展速度演化图

Figure 9 (Color online) Comparison of crack growth velocity predicted by different numerical methods.

一定范围内波动. 另外, 通过与其他传统数值模拟结果对比, 说明"共轭键"基近场动力学数值模型可以定量地模拟脆性固体材料在动荷载作用下的破坏特性.

4.2 算例II: 爆炸荷载作用下脆性岩石材料的破裂特性

根据Rabczuk和Belytschko[38]的研究, 为验证"共轭键"基近场动力学数值模型, 模拟了岩石材料受爆炸荷载作用下的破裂特性. 圆环内部半径为80 mm; 外部半径150 mm. 岩石材料力学参数为: 杨氏弹性模量E=200.0 GPa; 泊松比υ=0.2; 材料密度ρ=2450 kg/m^3; 断裂韧度K_I^c=6.0 GPa m$^{1/2}$. 圆柱内部爆炸压力荷载为

$$p = p_0 e^{-t/t_0}, \quad (28)$$

式中, p_0=10000 MPa; t_0=0.1 ms.

在"共轭键"基近场动力学数值模型中, 圆环由均匀分布的12668个离散的近场动力学物质点组成. 为验证"共轭键"基近场动力学数值计算结果的稳定性, 采用4种不同旋转角度、均匀分布的近场动力学物质点, 4种不同的物质点旋转角度分别为α=0°, α=15°, α=30°和α=45°, 其中物质点旋转角度为均匀分布物质点沿水平轴旋转的角度.

图10显示了物质点旋转角度α=0°的"共轭键"基近场动力学数值模型受到爆炸荷载作用下的破裂过程. 如图10(a)所示, 当t=50.0 μs时, 径向裂纹从圆环内部起裂. 如图10(b)所示, 当t=60.0 μs时, 径向裂纹向圆环外

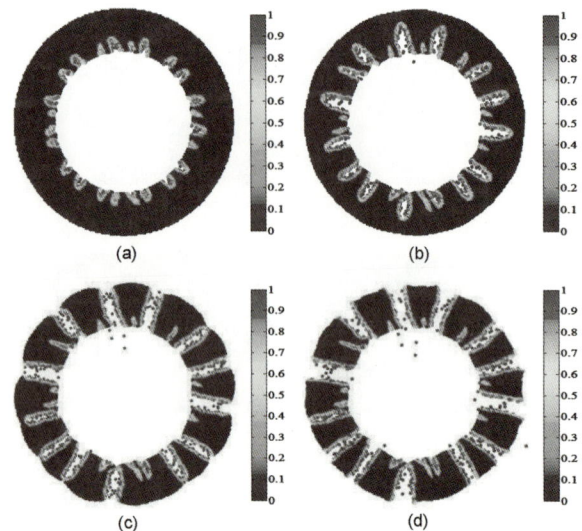

图 10 (网络版彩图)爆炸荷载作用下岩石破坏过程的"共轭键"基近场动力学模型数值模拟结果. (a) t=50.0 μs; (b) t=60.0 μs; (c) t=70.0 μs; (d) t=75.0 μs

Figure 10 (Color online) Fracturing process of rock cylinder under internal blasting pressure at different times. (a) t=50.0 μs; (b) t=60.0 μs; (c) t=70.0 μs; (d) t=75.0 μs.

部扩展, 且有岩石碎屑产生. 如图10(c)所示, 当t=70.0 μs时, 爆炸荷载作用产生的径向裂纹导致圆环整体破裂. 如图10(d)所示, 当t=75.0 μs时, 爆炸荷载作用下产生的岩石破裂体沿着径向方向运动.

4种不同分布的近场动力学物质点数值模型预测的岩石圆环最终破坏模式绘制在图11中. 如图11所示, 4种不同数值模型预测的结果相似. 另外, 岩石碎块数量统计在表1中. 如表1所示, 4种不同数值模型模拟的岩石碎块数量相似. 以上分析表明, 从定性和定量角度说明"共轭键"基近场动力学数值模型的稳定性.

4.3 算例III: 岩石爆破试验数值模拟

根据以前的岩石爆破试验[39], 圆柱状岩石试样的直径为144 mm, 高度为150 mm. 如图12所示, 圆柱状岩石试样中心含有直径为6.0 mm的圆孔作为爆破孔. 相关的岩石材料力学参数[39,40]: 密度ρ=2700 kg/m^3, 体积模量K=37.0 GPa, 泊松比υ=0.25, 断裂能量释放率G_c=298.0 J/m^2. 爆破荷载如图12(c)所示.

在近场动力学数值模型中, 66890个近场动力学物质点代表圆柱状岩石试样, 相应的非局部系数为4.0. 图13表示圆柱状岩石试样在爆破荷载作用下的破裂过

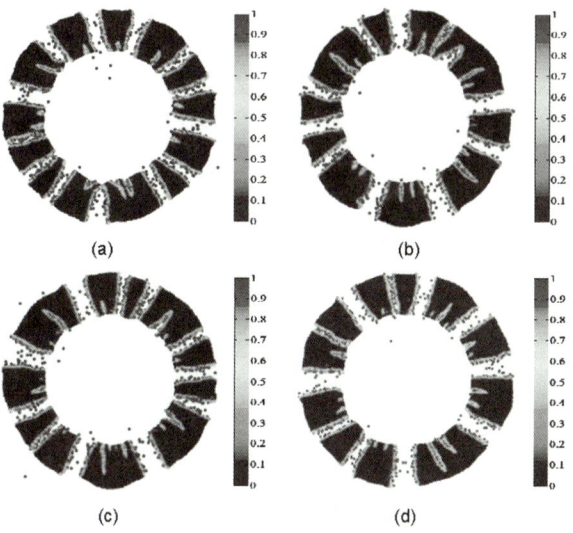

图 11 (网络版彩图)4种不同分布的近场动力学物质点数值模型的破坏模式. (a) $\alpha=0°$; (b) $\alpha=15°$; (c) $\alpha=30°$; (d) $\alpha=45°$

Figure 11 (Color online) Final failure patterns of brittle rock specimen under the internal blasting loads predicted by conjugated bond-pair-based peridynamic model with four different material point orientations. (a) $\alpha=0°$; (b) $\alpha=15°$; (c) $\alpha=30°$; (d) $\alpha=45°$.

表 1 岩石圆环受到内部爆破荷载作用下的碎块数量

Table 1 The number of rock fragmentations under internally blasting pressure for four different numerical samples with various orientations of peridynamics material points

近场动力学粒子分布角度	岩石碎块数量(块)
$\alpha=0°$	16
$\alpha=15°$	12
$\alpha=30°$	12
$\alpha=45°$	11

程图. 如图13(a)所示, 当$t=0.1$ ms时, 圆柱状岩石试样的爆破孔周围产生损伤破坏区, 即裂纹从爆破孔周边起裂. 如图13(b)~(d)所示, 当$t=0.1$~0.6 ms时, 部分裂纹沿着径向方向向外边界扩展, 且部分径向裂纹在扩展过程中出现分叉现象. 如图13(e)和(f)所示, 当$t=0.8$~1.0 ms时, 径向裂纹继续向圆柱状岩石试样外边界扩展, 同时, 少量环向裂纹出现.

图14表示"共轭键"基近场动力学数值模型预测岩石试样在爆破荷载作用下的最终破坏模式. 与已有试验结果[39]和数值模拟结果[41,42]对比可知, "共轭键"基近场动力学数值模型预测的岩石破坏模式与其他数值模拟方法预测的结果以及试验结果相似. 以上分析表明, "共轭键"基近场动力学数值方法可以有效地模拟爆破冲击荷载作用下岩石的破坏特性.

本文数值模拟结果表明爆炸荷载引起的裂纹扩展速度大约为100 m/s, 裂纹传播速度小于爆炸产生的应力波传播速度($V_R \approx 2000$ m/s). 由于本文的数值模拟结果基于弹脆性本构模型的"共轭键"基近场动力学模型, 爆炸荷载作用产生的环向裂纹主要集中于爆破孔附近, 在远离爆破孔的位置仅出现少量环向裂纹(图14), 这可能是由于不同的本构模型及边界条件导致的.

5 结论

本文提出的"共轭键"基近场动力学数值模型克服了传统"键"基近场动力学数值模型的泊松比固定的缺

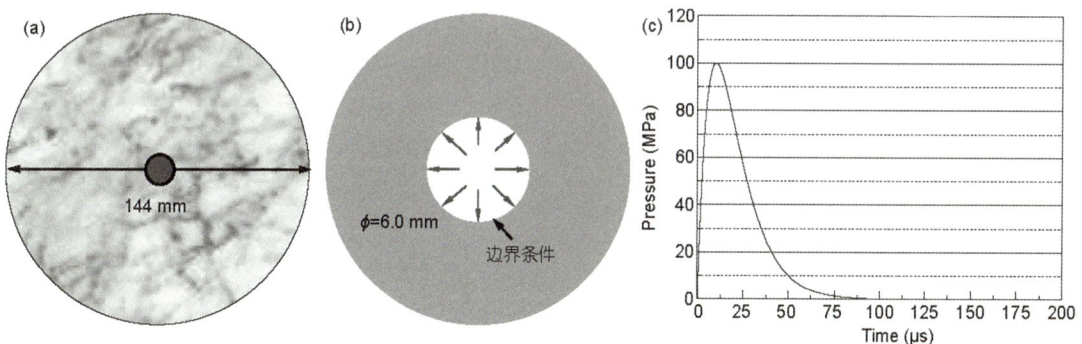

图 12 (网络版彩图)岩石爆破荷载试验几何及边界条件. (a) 数值模型几何条件; (b) 数值模型边界荷载条件; (c) 爆破荷载演化图

Figure 12 (Color online) Numerical model of the rock blasting experimental tests. (a) Geometrical conditions; (b) boundary conditions; (c) blasting loading evolution.

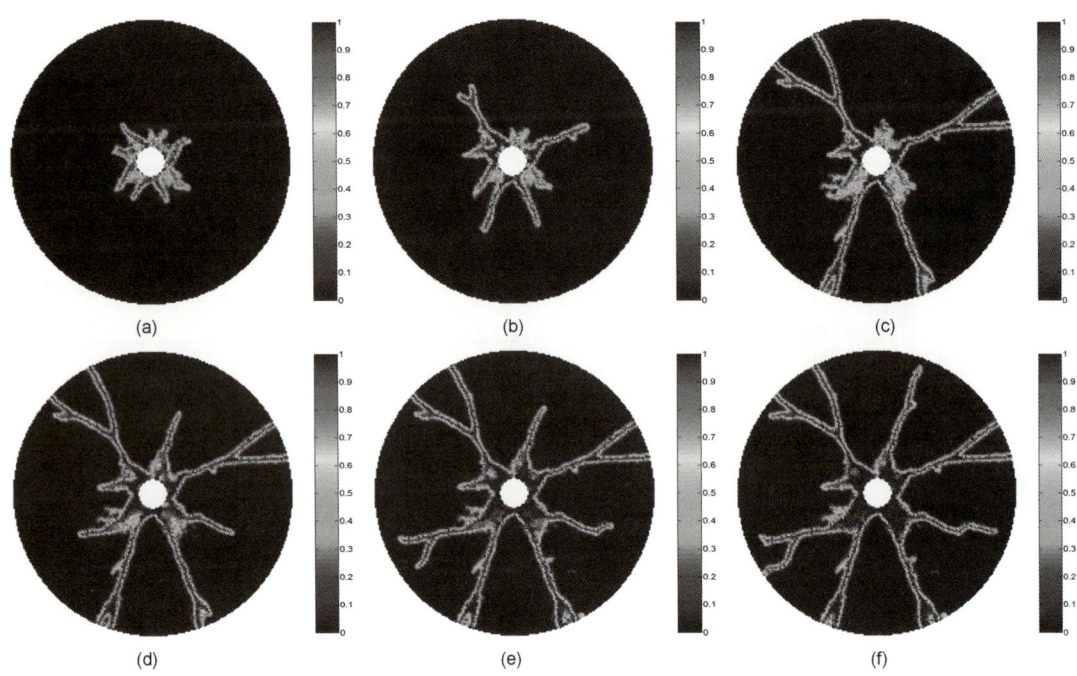

图 13 (网络版彩图)岩石爆破试验中裂纹起裂及扩展过程图. (a) t=0.1 ms; (b) t=0.2 ms; (c) t=0.4 ms; (d) t=0.6 ms; (e) t=0.8 ms; (f) t=1.0 ms

Figure 13 (Color online) Crack initiation and propagation in the rock specimen under blasting loads. (a) t=0.1 ms; (b) t=0.2 ms; (c) t=0.4 ms; (d) t=0.6 ms; (e) t=0.8 ms; (f) t=1.0 ms.

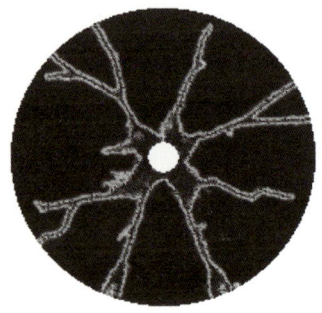

图 14 (网络版彩图)"共轭键"基近场动力学数值模型预测岩石试样在爆破荷载作用下的最终破坏模式

Figure 14 (Color online) Ultimate failure patterns of rocks subjected to blasting loads obtained by the conjugated bond-pair-based peridynamics.

陷. "共轭键"基近场动力学数值模型可以广泛地、有效地应用到岩石材料的破坏特性研究. 在"共轭键"基近场动力学数值方法中, 通过引入"共轭键"转动角度及建立微观和宏观变形能的等效关系, 推导了"共轭键"基近场动力学模型中的法向刚度参数及切向刚度参数与宏观力学参数之间的关系. 另外, 通过建立与线弹性断裂力学的能量释放率等效关系, 推导了临界微观"键"能密度的表达式. 然后, 通过对影响域中每根"键"所储存的能量密度与临界"键"能密度进行对比, 判断近场动力学模型中的"键"是否断裂, 从而实现固体材料中裂纹的起裂、扩展及分叉现象的准确预测. 本文通过3个数值算例说明"共轭键"基近场动力学数值方法的有效性和准确性. 另外, 本文在"共轭键"近场动力学数值模型中仅考虑了弹脆性本构模型, 在未来的研究工作中, 将发展基于J-H和RTH等弹塑性本构模型的"共轭键"近场动力学数值模型及相应的非局部弹塑性边界条件.

参考文献

1 Zhou S, Zhuang X, Zhu H, et al. Phase field modelling of crack propagation, branching and coalescence in rocks. Theor Appl Fract Mech, 2018, 96: 174–192

2 Li S H, Feng C, Zhou D, et al. Mechanical Methods in Landslide Research (in Chinese). Beijing: Science Press, 2018 [李世海, 冯春, 周东, 等. 滑坡研究中的力学方法. 北京: 科学出版社, 2018]

3 Tang C. Numerical simulation of progressive rock failure and associated seismicity. Int J Rock Mech Min Sci, 1997, 34: 249–261

4 Areias P, Rabczuk T. Steiner-point free edge cutting of tetrahedral meshes with applications in fracture. Finite Elem Anal Des, 2017, 132: 27–41

5 Areias P, Reinoso J, Camanho P P, et al. Effective 2D and 3D crack propagation with local mesh refinement and the screened Poisson equation. Eng Fract Mech, 2018, 189: 339–360

6 Shi G H, Goodman R E. Generalization of two-dimensional discontinuous deformation analysis for forward modelling. Int J Numer Anal Methods Geomech, 1989, 13: 359–380

7 Ning Y, Yang J, An X, et al. Modelling rock fracturing and blast-induced rock mass failure via advanced discretisation within the discontinuous deformation analysis framework. Comput Geotech, 2011, 38: 40–49

8 Shi G H. Manifold method of material analysis. In: Transactions of the 9th Army Conference on Applied Mathematics and Computing. Minneapolis, 1991

9 Wu Z J, Wong L N Y. Frictional crack initiation and propagation analysis using the numerical manifold method. Comput Geotech, 2012, 39: 38–53

10 Yan C Z, Zheng H, Sun G H, et al. Polygon characterization of coarse aggregate and two-dimensional combined finite discrete element method analysis (in Chinese). Rock Soil Mech, 2015, 36(S2): 95–103 [严成增, 郑宏, 孙冠华, 等. 粗粒料多边形表征及二维FEM/DEM分析. 岩土力学, 2015, 36(S2): 95–103]

11 Yan C Z, Zheng H, Sun G H, et al. Combined finite-discrete element method for simulation of hydraulic fracturing. Rock Mech Rock Eng, 2016, 49: 1389–1410

12 Yan C Z, Zheng H. Three-dimensional hydromechanical model of hydraulic fracturing with arbitrarily discrete fracture networks using finite-discrete element method. Int J Geomech, 2016, 17: 04016133

13 Yan C Z, Jiao Y Y, Zheng H. A fully coupled three-dimensional hydro-mechanical finite discrete element approach with real porous seepage for simulating 3D hydraulic fracturing. Comput Geotech, 2018, 96: 73–89

14 Silling S A. Reformulation of elasticity theory for discontinuities and long-range forces. J Mech Phys Solids, 2000, 48: 175–209

15 Silling S A, Askari E. A meshfree method based on the peridynamic model of solid mechanics. Comput Struct, 2005, 83: 1526–1535

16 Huang D, Zhang Q, Qiao P Z, et al. A review on peridynamics (PD) method and its applications. Adv Mech, 2010, 40: 448–459 [黄丹, 章青, 乔丕忠, 等. 近场动力学方法及其应用. 力学进展, 2010, 40: 448–459]

17 Huang D, Zhang Q, Qiao P Z. Damage and progressive failure of concrete structures using non-local peridynamic modeling. Sci China Tech Sci, 2011, 54: 591–596

18 Madenci E, Oterkus E. Peridynamic Theory and Its Applications. Boston: Springer, 2014

19 Zhou X P, Wang Y T. Numerical simulation of crack propagation and coalescence in pre-cracked rock-like Brazilian disks using the non-ordinary state-based peridynamics. Int J Rock Mech Min Sci, 2016, 89: 235–249

20 Wang Y T, Zhou X P. Peridynamic simulation of thermal failure behaviors in rocks subjected to heating from boreholes. Int J Rock Mech Min Sci, 2019, 117: 31–48

21 Wang L, Xu J, Wang J, et al. A mechanism-based spatiotemporal non-local constitutive formulation for elastodynamics of composites. Mech Mater, 2019, 128: 105–116

22 Shojaei A, Mudric T, Zaccariotto M, et al. A coupled meshless finite point/Peridynamic method for 2D dynamic fracture analysis. Int J Mech Sci, 2016, 119: 419–431

23 Zhou X P, Shou Y D. Numerical simulation of failure of rock-like material subjected to compressive loads using improved peridynamic method. Int J Geomech, 2017, 17: 04016086

24 Ren H L, Zhuang X Y, Rabczuk T. Dual-horizon peridynamics: A stable solution to varying horizons. Comput Methods Appl Mech Eng, 2017, 318: 762–782

25 Oterkus S, Madenci E. Peridynamic modeling of fuel pellet cracking. Eng Fract Mech, 2017, 176: 23–37

26 Zhu Q Z, Ni T. Peridynamic formulations enriched with bond rotation effects. Int J Eng Sci, 2017, 121: 118–129

27 Zhang Z N, Chen Y X. Modeling nonlinear elastic solid with correlated lattice bond cell for dynamic fracture simulation. Comput Methods Appl Mech Eng, 2014, 279: 325–347

28 Zhang Z N, Chen Y X, Zheng H. A modified Stillinger-Weber potential-based hyperelastic constitutive model for nonlinear elasticity. Int J Solids Struct, 2014, 51: 1542–1554

29 Zhang Z N, Yao Y, Mao X B. Modeling wave propagation induced fracture in rock with correlated lattice bond cell. Int J Rock Mech Min Sci, 2015, 78: 262–270

30 Huang D, Lu G, Qiao P. An improved peridynamic approach for quasi-static elastic deformation and brittle fracture analysis. Int J Mech Sci, 2015, 94-95: 111–122

31 Wang Y, Zhou X, Shou Y. The modeling of crack propagation and coalescence in rocks under uniaxial compression using the novel conjugated bond-based peridynamics. Int J Mech Sci, 2017, 128-129: 614–643

32 Wang Y, Zhou X, Wang Y, et al. A 3D conjugated bond-pair-based peridynamic formulation for initiation and propagation of cracks in brittle solids. Int J Solids Struct, 2018, 134: 89–115

33 Diana V, Casolo S. A bond-based micropolar peridynamic model with shear deformability: Elasticity, failure properties and initial yield domains. Int J Solids Struct, 2019, 160: 201–231

34 Ravi-Chandar K, Knauss W G. An experimental investigation into dynamic fracture: II. Microstructural aspects. Int J Fract, 1984, 26: 65–80

35 Ravi-Chandar K, Knauss W G. An experimental investigation into dynamic fracture: III. On steady-state crack propagation and crack branching. Int J Fract, 1984, 26: 141–154

36 Song J H, Wang H, Belytschko T. A comparative study on finite element methods for dynamic fracture. Comput Mech, 2008, 42: 239–250

37 Braun M, Fernández-Sáez J. A new 2D discrete model applied to dynamic crack propagation in brittle materials. Int J Solids Struct, 2014, 51: 3787–3797

38 Rabczuk T, Belytschko T. Cracking particles: A simplified meshfree method for arbitrary evolving cracks. Int J Numer Meth Engng, 2004, 61: 2316–2343

39 Banadaki M M D, Mohanty B. Numerical simulation of stress wave induced fractures in rock. Int J Impact Eng, 2012, 40-41: 16–25

40 Panchadhara R, Gordon P A, Parks M L. Modeling propellant-based stimulation of a borehole with peridynamics. Int J Rock Mech Min Sci, 2017, 93: 330–343

41 Xie L X, Lu W B, Zhang Q B, et al. Analysis of damage mechanisms and optimization of cut blasting design under high *in-situ* stresses. Tunn Undergr Space Technol, 2017, 66: 19–33

42 Ma G W, An X M. Numerical simulation of blasting-induced rock fractures. Int J Rock Mech Min Sci, 2008, 45: 966–975

事故型冲击荷载对结构作用研究总论

钱七虎

(解放军理工大学,江苏 南京,210007)

一、事故型冲击荷载发生领域和冲击量级

事故型冲击荷载在很多建筑结构领域中都可以遇见。

在工业建筑领域可以发生于重物掉落于楼板上,重物重量可达若干吨,速度可达 15~20m/s。

可以发生于工业爆炸事故(煤气、粉尘爆炸)中,由于设备部件飞散以及结构坍塌引起的冲击,其速度前者达 100m/s,后者可达 15~20m/s。

在能源设施中,在空难中由于飞机堕落在核反应堆的防护壳上,飓风携物件的冲击,反应堆事故中设备部件的内部冲击等,飞机重量可达 90t,速度可达 250~300m/s。

交通建筑中由于运输工具汽车的撞击,运输工具重量可达数吨,速度数十米/秒。

水工结构领域中河岸、海岸建筑的支座由于舰船的挤压碰撞,塔架设备的摔落于站台上,速度达 30m/s,重量达 20t。

防护工程中炮、炸弹最大可达 1000m/s 多速度和 30000b 的重量(MOP 钻地弹)。

二、冲击的概念、冲击荷载的类型和总特点

1. 冲击概念

发生在力学系统中由于冲击体与系统的动力接触,由于十分剧烈的很短时间的力学作用引起冲击体和力学系统作用点的速度剧烈改变的现象。

2. 类型

冲击荷载有两大类型:

(1)作用于防护结构的冲击荷载——炮、炸弹及其他军事技术装置对防护工程产生的冲击荷载。

(2)产生于民用及工业、水工建筑以及能源设施建筑中的冲击荷载,它又可分两类:

①使用型冲击荷载:通常产生于支撑特殊设备(锤击、柱夯设备)的结构中,它的特点表现为多次冲击作用。所以这些结构中承受冲击时只允许结构材料在弹性阶段工作。

②事故型冲击荷载:特点是高强度、一次性,因此允许结构产生局部破坏和显著的塑性变形。

3. 冲击荷载的总特点

冲击荷载区别于爆炸荷载的基本特点是冲击体与结构的相互作用、冲击荷载施加面积的有限性,作用时间的短暂性,以及在结构中产生的波动过程。与此相应,在结构中要研究冲击对结构的局部作用和整体作用和相应的破坏。

局部作用的特点是在结构中冲击区域附近产生的局部变形和破坏,包括形成漏斗坑的侵彻、结构背面的混凝土的震塌以及贯穿和冲切。

整体作用是伴随着可以导致结构坍塌的整体变形。

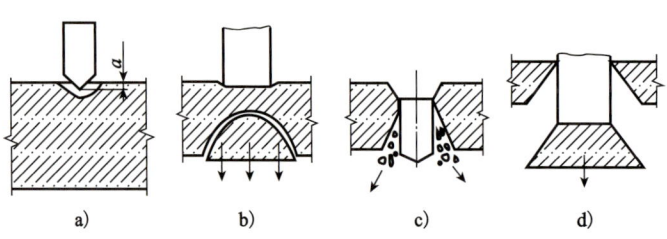

图1 事故型冲击产生的局部破坏

三、冲击荷载按作用性质的分类

1. 按冲击速度可区分为低速、高速、超高速冲击

前两类属于力学冲击荷载,而超高速冲击由于材料转入另一类集聚态,因此属于热动力学范畴。

2. 按冲击能量的吸收可区分为剧烈的、中度剧烈的、非剧烈的冲击

剧烈冲击指的是冲击时结构吸收的能量大大超过冲击体吸收的能量,因此研究剧烈冲击时可以忽略冲击体的变形,认为它不变形。

在中度剧烈冲击中,结构与冲击体在冲击时吸收的能量相当或相可比较。

在非剧烈(柔和)冲击中,冲击体吸收的能量显著超过结构吸收的能量。显然,在前两种情况中必须考虑冲击体与结构的相互作用,而在后者,认为冲击时结构是不变形的,冲击作用可以用不取决于结构变形的局部冲量荷载 $P(t)$ 来代替。

事故型冲击荷载大都属于低速剧烈冲击或低速中度剧烈冲击,因此,须考虑冲体与结构在冲击时的相互作用。

四、事故型冲击荷载的局部作用计算

1. 事故型冲击区别于军事弹体冲击的特点

事故型冲击区别于军事弹体冲击在于其冲击体质量大(达数吨、数十吨)、冲击速度低(数米至数十米/秒,小于250m/s)、径厚比 d/δ 大(冲击体等效直径与被冲击结构厚度的比值),一般其 $d/\delta > 1$,而军事弹体冲击其 d/δ 远小于1。因此,军事弹体型冲击局部作用的计算公式(包括侵彻、震塌和贯穿的计算)推广到事故型冲击局部作用的计算很不成功。

试验表明:径厚比 $d/\delta > 2$ 时,计算弹体侵彻深度的公式用以计算事故型冲击体侵入结构深度时,与试验结果相比,误差达 8~9 倍,计算震塌与贯穿厚度的误差也急剧增大。这是不难理解的,因为在大的 d/δ 时,侵彻实际上不发生,只有压入,而其贯穿则带有动力冲切的特性,在 V_0 稍大时,$\delta_p \approx \delta_s$(震塌厚度约等于贯穿厚度),所以事故型冲击局部破坏的特性与弹体冲击侵彻相比具有相异的特性。

由于与冲击局部作用相伴随的波动过程,具有很大的复杂性,所以冲击局部作用的计算目前基本上是采用试验公式,而事故型冲击由于冲击体的不规则性和随机性,呈现出更加复杂,它更加依赖试验研究成果。近年来,开始进行了一些冲击局部作用的理论研究探索,其运用波动理论的成果主要集中于弹体冲击的局部作用,涉及事故型冲击局部作用的则基于视为简化的假设仅确定结构变形的极限变形状态,假设的简化裕度则作为结构的安全储备。

2. 事故型冲击局部作用计算

(1)俄罗斯古比雪夫军事工程大学关于钢筋混凝土板的试验公式

$75\text{m/s} \leqslant V_0 \leqslant 150\text{m/s}$ 时,

震塌厚度 $\dfrac{h_s}{d} = 1.429 NQ^{-0.5673}$

贯穿(冲切)厚度 $\dfrac{h_p}{d} = 0.9432 NQ^{-0.5530}$

$V_0 \leq 75\text{m/s}$ 时,

$\dfrac{h_s}{d} = \dfrac{h_p}{d} = 2.019 NQ^{-0.3439}$

式中,d 为冲击体等效直径。N 为考虑冲击体头部形状的系数,平头 $N = 1.0$;大角度锥体头 $N = 1.07$;小角度锥体头 $N = 1.28$;球面 $N = 1.18$。V_0 为冲击速度。Q 为无因次量,$Q = 10^4 R_c d^3 / M_s V_0^2$。$R_c$ 为混凝土立方体强度。M_s 为冲击体质量。

该公式试验范围 $d/\delta = 1.0,\cdots,3.7$

试验结果示于图2。

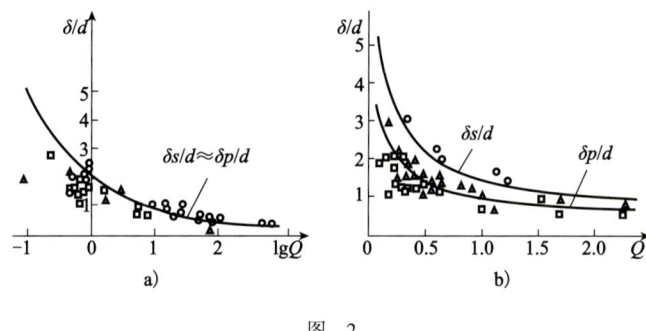

图 2

(2) 钢筋混凝土板贯穿临界冲击速度和贯穿后的冲击体残余速度

$$V_{cr.p} = 0.616 R_c^{\frac{1}{2}} \rho^{\frac{1}{6}} \left(\dfrac{d\delta^2}{F} \right)^{\frac{2}{3}}$$

式中,F 为冲击体重量;ρ 为混凝土密度;

$$V_{res} = \left[\dfrac{1}{1 + \dfrac{F_t}{F}} (V_0^2 - V_{cr.p}^2) \right]^{\frac{1}{2}}$$

式中,F_t 为混凝土冲切体的重量。

3. 事故型冲击作用下核反应堆防护壳的防冲击(贯穿)计算

(1) 计算简图(图3)

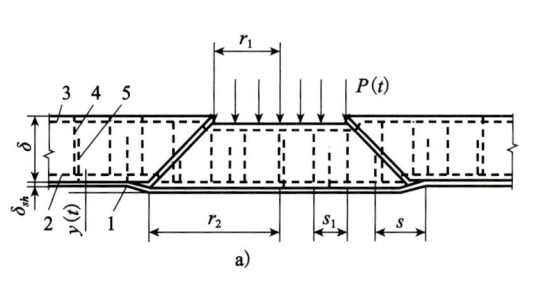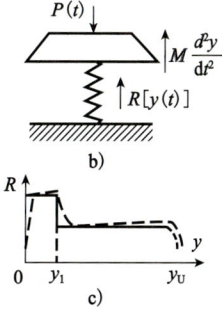

图 3

(2) 计算模型:刚塑化模型

核反应堆防护壳事故型冲击包括自外(包括飞机堕落时发动机和飓风携带的重物等)和自内(事故发生后透平的断裂叶片),前者 V_0 达 250m/s,F 达 20t;后者 $M_s \approx 300$kg,$V_0 \leq 150$m/s。计算的前提是:$\delta < \delta_p$,$V_0 > V_{cr.p}$,计算出 V_{res}。带有初速 V_{res} 的冲切体的运动受到混凝土壳中的横向钢筋、纵向钢筋,金属壳板以及固定螺栓的阻力。

极限支护能力的确定是所有钢构件的破坏。

冲切体运动方程 $M_s \dfrac{\partial^2 y}{\partial t^2} + F(y) = 0$,求出 $y = y(t)$

式中,y 为冲击体相对于残余结构的垂直位移;F 为钢构件的支撑力。

初始条件:$t = 0, y = 0, \dot{y} = V_{res}$

$$F = F_{sw} + F_{s.ex} + F_{s.in} + F_a$$

式中,F_{sw} 为横向钢筋的支撑力;$F_{s.ex}$、$F_{s.in}$ 为壳体的外层和内层内的纵向钢筋的支撑力;F_a 为金属壳板的支撑力。

(3) 结构极限状态—壳体连续性的破坏

冲击体停止运动瞬间的最大变形 y_{max} 超过钢材料的极限变形值 $\varepsilon_{s.u}$,则壳体破坏,若 $y_{max} < \varepsilon_{s.u}$,则壳体安全。

t_{max}、y_{max} 由下列方程解求出:

$$\dot{y}(t_{max}) = 0 \quad y_{max} = y \cdot (t_{max})$$

五、事故型冲击的结构整体作用计算

1. 计算特点

(1) 要考虑冲击体与结构的相互作用,因此局部作用与整体作用是耦合的;

(2) 冲击局部破坏对结构截面质量,刚度(弯、剪)都有影响;

(3) 整体作用要考虑冲击的波动过程的影响;

(4) 冲击作用下钢筋混凝土结构的变形破坏特点。

2. 冲击力

(1) 冲击力的试验曲线与理论曲线

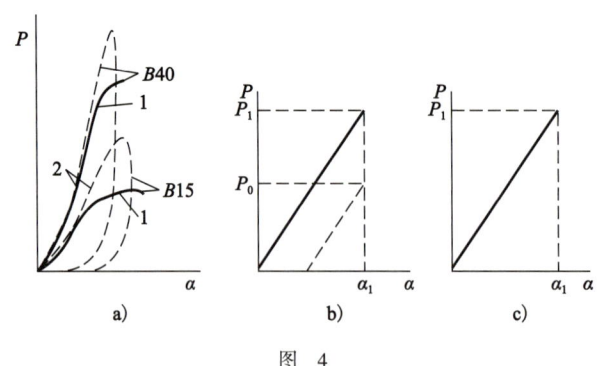

图 4

(2) 冲击力 $P(t)$ 的试验公式

对于钢筋混凝土梁的低速冲击,试验表明可将上式写为如下:

$$P(t) = \begin{cases} k_c \alpha & P \leq P_r \\ 0 & P > P_r \end{cases}$$

式中，P_r 为接触力的极限值；k_c 为表征接触区的刚度系数。

接触力增加的主要原因归结于混凝土的应变速率效应，应变速率的影响目前可由主应力极限值乘以相应的动力强化系数来体现。确定 p_r 的最简单而可靠的公式是：

$$p_r = k_R R_{d,loc} A_{loc}$$

式中，k_R 为考虑上部钢筋和横向钢筋对局部挤压阻抗贡献的系数，对于标准冲击头和 $\mu\% = 0.5 \sim 0.8\%$，$\mu\% = 0.25 \sim 0.5\%$，可取 $k_R = 1.15$，对于没有横向钢筋的构件 $k_R = 1$，在中间情况（$0 \leqslant \mu\% < 0.25\%$），用线性内插法求；$R_{d,loc} = k_d R_{s,loc}$，为无筋混凝土对局部挤压的动阻抗；$k_d$ 为混凝土的动力强化系数；$R_{s,loc}$ 为局部受压时混凝土的静力强度，$R_{s,loc} = R_b \gamma_{loc}$，$\gamma_{loc} = \sqrt[3]{A/A_{loc}}$，$A$ 为计算挤压面积，R_b 为混凝土的立方强度，对于平冲击头 $A = r_2 l_p$，$l_p = 2r_1 + 2r_2$，$A_{loc} = 2r_1 r_2$，r_1，r_2 为分别为平头冲击体截面纵横向一半的尺寸。

试验表明，对于平面的绝对刚性冲击体低速冲击时，k_c 采取如下形式：

$$k_c = \frac{\pi r_2 E_b \psi}{\ln 2\lambda - 0.527 - 0.716/\lambda^2 - 0.678/\lambda^4} \qquad \lambda < 2$$

$$k_c = \pi r_2 E_b \psi / [\ln 2\lambda + 0.527] \qquad \lambda > 8$$

式中，$\lambda = \frac{h}{r_1}$，$\psi = 0.875\phi(1+\nu)^2/(1+2\nu)$，没有上部纵向钢筋和横向钢筋时 $\phi = 0.15$，有纵向钢筋和横向钢筋时 $\phi = 0.26$；ν 为泊松系数；E_b 为混凝土的初始弹性模量。

3. 事故型冲击钢筋混凝土结构的变形与破坏特征

1）梁的变形与破坏特征

在冲击荷载作用下，钢筋混凝土梁的应变过程与静载条件下一样，主要经历三个阶段：裂缝形成以前阶段；裂缝形成到纵向受拉钢筋屈服的阶段；屈服后阶段。但其裂缝形成与集中静载以及分布冲量爆炸荷载作用下所观察到的不同。对梁的冲击不同于静载或分布冲量荷载作用的一系列特点，首先取决于结构的动力特性（通常用最低的自振频率 ω_0 来表征）、冲击体与梁的质量比 M_s/M_b、冲击速度 v_0、冲击时间 t，其次是梁纵向钢筋含量、横向钢筋含量及其类型。

试验表明，在足够宽的 ω_0、M_s/M_b、v_0 的变化范围内，钢筋混凝土梁呈现出一致的状态特性。在低速冲击下梁的裂缝主要集中在接触点附近很短的区段内，而不是像集中静载以及分布冲量爆炸荷载下所观察到的其分布遍及整个梁段上（图5）。

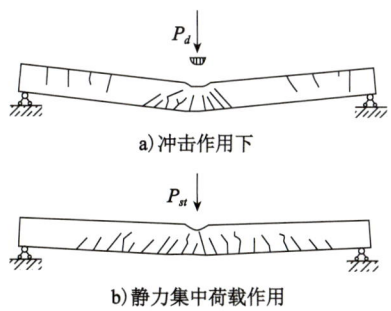

图5　钢筋混凝土梁的裂缝形成特征

（1）冲击速度的影响

在冲击速度较高（$v_0 > 15$ 米/秒）的情况下，有横向钢筋的梁，因受拉箍筋的作用阻碍了混凝土碎块的崩落，首先产生马蹄形的震塌裂缝，裂缝的两端方向与梁轴线大约成45°，其后在震塌裂缝区内出现垂直的弯曲裂缝（图6a）。此时，在震塌裂缝的水平投影范围内，其结果是有若干根横向杆把上下结构连接了起来。因此，在大多数情况下可以不考虑梁的震塌和贯穿的局部问题（没有横向钢筋的梁可能导致被冲击物贯穿的情况,如图6c)。

在冲击速度很小($v_0 < 15$ 米/秒)的情况下,首先形成垂直的弯曲裂缝,其后形成由弯-剪联合作用产生的倾斜裂缝(图6b)。

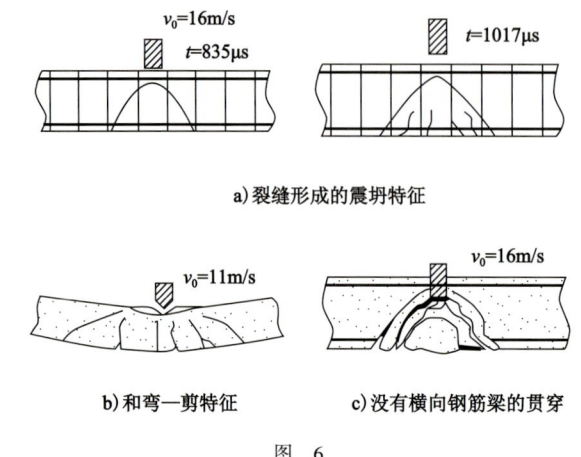

a) 裂缝形成的震坍特征

b) 和弯—剪特征

c) 没有横向钢筋梁的贯穿

图 6

(2) 钢筋的影响

尽管裂缝产生的性质有明显的不同,但它们对梁以后的变形有着相同的影响。考虑到这种情况以及裂缝形成的时间与结构变形的总时间相比很小,可以近似认为,垂直和倾斜的裂缝都是在结构整体变形开始之前形成的。裂缝对接触处下方区域的初始削弱在很大程度上决定了梁以后的状态,也就是说变形集中在倾斜裂缝之间的区域。在随后的运动中,裂缝张开,钢筋达到屈服极限。因此,把整体变形条件下纵向钢筋达到动力屈服极限之前的结构工作阶段,称为准弹性阶段,以后进入塑性工作阶段,最终导致破坏。

箍筋的类型也是一个重要的因素。与上下纵向钢筋固定连接的封闭箍筋尤为可取,这样的钢筋栅格有助于把冲击接触作用分布到较大的混凝土体积中去,并依靠对横向应变的束缚作用改善压缩区混凝土的工作特性。箍筋的"捆束"可以使混凝土的包括下降段全程应力应变曲线得到实现,并保证钢筋的潜力得到充分的利用,该潜力是与更完整地实现下部纵向钢筋的应变图形相联系的(直到下部纵向钢筋断裂)。但不封闭的箍筋效果比较差。

(3) 冲击位置的影响

当梁承受跨中冲击时,将会由于冲击位置下方的垂直截面和倾斜截面的转动而发生破坏(图7);当冲击位置接近支座时,混凝土可能从短跨的一方沿斜裂缝剪坏(图8)。在后一情形中,裂缝对梁轴线斜度不是45°(45°是跨中冲击的特征),而大体与梁跨的两段长度比成比例。

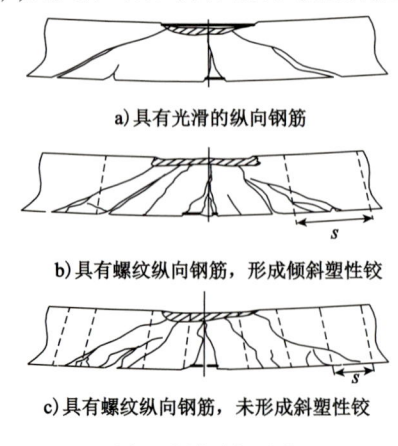

a) 具有光滑的纵向钢筋

b) 具有螺纹纵向钢筋,形成倾斜塑性铰

c) 具有螺纹纵向钢筋,未形成斜塑性铰

图 7 梁的破坏图形

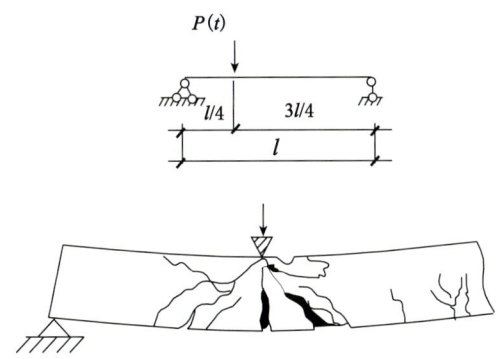

图 8　在 1/4 跨度处承受冲击荷载梁的剪切破坏

实验还表明,也有可能在开始加载时,在 1/3 跨度处由于混凝土剪坏而沿着斜截面发生梁的破坏情形(图 9)。因为相应的动力横向力曲线的最大值正好是发生在 1/3 跨度处。以这种形式破坏的梁,常是其跨中部分的箍筋间距是在假设仅有分布荷载,而未考虑冲击荷载特点的条件下按构造要求确定的。

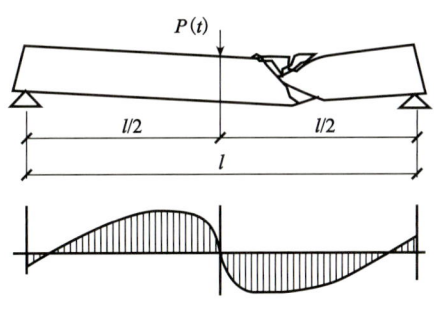

图 9　梁沿斜截面的破坏

(4) 冲击体头部形状的影响

在低速冲击下,冲击体头部的形状也是一个不容忽视的重要因素。楔形冲击体侵入梁的深度通常不大,因为与箍筋连接的上部纵向钢筋形成一种独特的阻尼器作用,阻滞了冲击体的侵彻,这使得临界截面的工作高度没有太大减少。平头冲击体的动力效应似乎明显地高于楔形冲击体。图 10 表示了在其它条件相同时,上述两种冲击体的接触力和时间的试验曲线。由图 10 可见,与楔形冲击体(头部呈 90°)相比,平头冲击的荷载强度较高,而作用时间较短。其他形状的冲击头(圆柱体等)的作用介于上述二者之间。因此,平头冲击体的冲击是最不利的情况,计算时应把它作为最不利情况加以考虑计算。

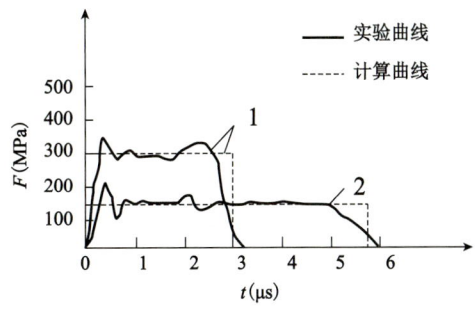

图 10　平头冲击体
(1) 和楔形头冲击体(2) 冲击梁构件的 $F-t$ 曲线

动载通过一个构件(例如金属横梁,图11a)传递到梁试件上的试验表明,没有出现局部的压碎现象,梁的破坏特性接近于在冲量(爆炸)荷载下的破坏,即在最大弯矩区有一个**塑性铰**(图11b),而没有出现在直接冲击中由高振型引起的梁上部的垂直裂缝(图11c)。

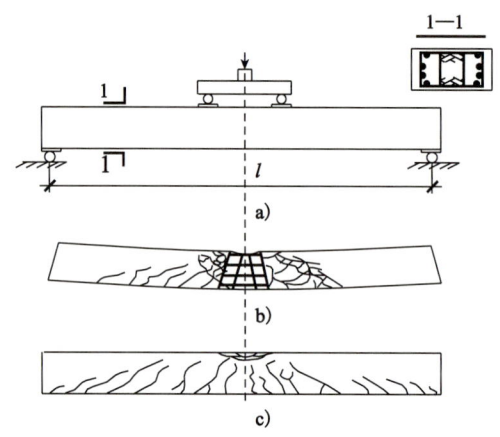

图11 冲击荷载通过一个结构传递给梁的裂缝形成和破坏

综上试验资料可以归纳出一系列建立钢筋混凝土梁在低速冲击下物理力学模型所必需的基本变形特征。

①由于梁具有横向钢筋,在低速冲击下梁不会发生震塌和贯穿的局部破坏,只须计算梁整体响应。此时要考虑冲击体侵入梁内一定深度从而造成的局部变形对梁整体受力及变形的影响。

②冲击体的头部形状对梁的响应会有重要的影响,计算中应给于考虑。通过对比知平头冲击体在其他条件相同时,其消耗在侵彻和局部变形的能量最小,其最大部分能量都传给了梁因而是最不利的。

③低速冲击产生的沿纵向传播的波动过程对弹塑性挠度值的影响很小而可以忽略不计。

④梁的应变一般要经历三个阶段:裂缝形成以前;裂缝形成后到纵向钢筋达到屈服;纵向钢筋屈服后直至破坏。

2)柱的变形与破坏特征

冲击作用下钢筋混凝土柱的破坏类型主要有:发生于冲击点在柱高度中央时的弯曲破坏,与梁体系中的破坏特征相似;当支座在冲击方向内具有一定柔度时,则冲击点接近支座时也发生该类破坏;破坏发生在对不可移动的支座冲击(如交通工具冲击)时的剪切破坏,产生从接触面向支座倾斜的斜裂缝,并沿所形成的倾斜受压带发生破坏,如图12所示。

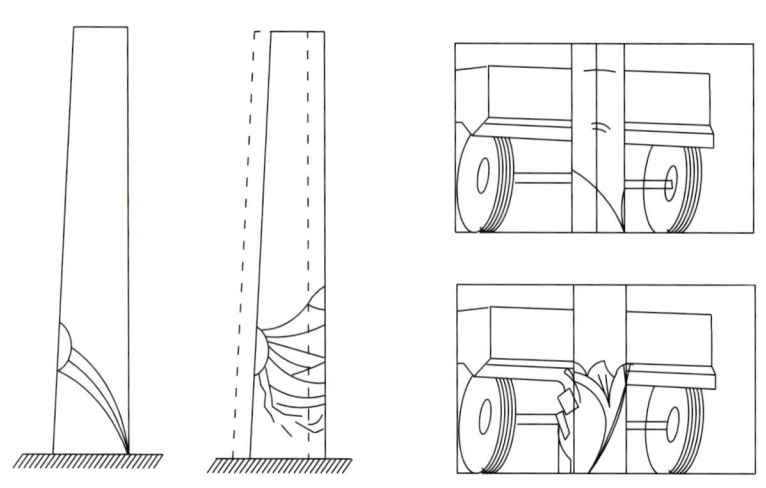

图12 圆截面钢筋混凝土悬臂灯柱的破坏图

4. 计算理论

1) 基于铁木辛柯梁的计算理论（以梁为例）

(1) 要点

分布参数；考虑转动惯量和剪切变形的影响（波动过程）；局部破坏影响；弹性阶段和塑性阶段工作；极限状态条件 $\varphi_{\max} \leqslant \varphi_u$（塑性铰展开角）。

(2) 运动方程

利用拉格朗日方程和冲击体的运动的方程，得到：

$$\begin{cases} \sum_{n=1}^{\infty} \ddot{\bar{T}}_n (H_{1n} + \xi_{1n}) + \sum_{n=1}^{\infty} \ddot{\bar{\bar{T}}}_n H_{1n} + \sum_{n=1}^{\infty} \bar{T}_n \eta_{1n} - k_1 \Delta_1 \alpha(t) = 0 \\ \cdots\cdots\cdots\cdots\cdots\cdots\cdots\cdots\cdots\cdots\cdots\cdots\cdots\cdots \\ \sum_{n=1}^{\infty} \ddot{\bar{T}}_n (H_{nn} + \xi_{nn}) + \sum_{n=1}^{\infty} \ddot{\bar{\bar{T}}}_n H_{nn} + \sum_{n=1}^{\infty} \bar{T}_n \eta_{nn} - k_1 \Delta_n \alpha(t) = 0 \\ \sum_{n=1}^{\infty} \ddot{\bar{T}}_n H_{1n} + \sum_{n=1}^{\infty} \ddot{\bar{\bar{T}}}_n (H_{1n} + \xi_{1n}) + \sum_{n=1}^{\infty} \bar{\bar{T}}_n \psi_{1n} - k_1 \Delta_1 \alpha(t) = 0 \\ \cdots\cdots\cdots\cdots\cdots\cdots\cdots\cdots\cdots\cdots\cdots\cdots\cdots\cdots \\ \sum_{n=1}^{\infty} \ddot{\bar{T}}_n H_{nn} + \sum_{n=1}^{\infty} \ddot{\bar{\bar{T}}}_n (H_{nn} + \xi_{nn}) + \sum_{n=1}^{\infty} \bar{\bar{T}}_n \psi_{nn} - k_1 \Delta_n \alpha(t) = 0 \\ \sum_{n=1}^{\infty} \ddot{\bar{T}}_n X_n(\bar{x}) + \sum_{n=1}^{\infty} \ddot{\bar{\bar{T}}}_n X_n(\bar{x}) + \ddot{\alpha}(t) + \omega_0^2 \alpha(t) = 0 \end{cases}$$

(3) 计算结果的讨论

当 $V_0 \leqslant 30 \text{m/s}$，考虑局部作用与局部破坏的铁木辛柯梁冲击计算结果与工程弯曲理论计算结果（不考虑转动惯量与剪切影响）对比表明：波动过程的影响仅表现在梁运动的初始阶段，而在梁纵向筋达到屈服后的塑性阶段中，对梁的挠度与内力的影响区别不大；极限状态条件没有充分考虑冲击下钢筋混凝土结构变形特点，因而过低地估价了结构的强度，需要研究另外的模型。

2) 基于片式相连的钢筋混凝土梁的计算理论

(1) 计算模型

在该模型（图13）中，梁简化为由三个元件（刚性单元）和五个连系件模拟钢筋和混凝土的连系件连接在一起的体系。无裂缝工作阶段影响可以忽略不计，各计算单元的几何参数取决于荷载的作用位置和接触面的长度，也取决于临界截面的应力—应变状态。当集中作用时（楔形冲击头），接触面的长度趋近于0。对计算界面做平截面假设，即应变沿截面高度成线性分布。

(2) 运动过程

$$\begin{cases} g_1(h_0 - x_0)\varphi_1 + g_2\ddot{\varphi}_1 + g_3\ddot{\varphi}_2 = (\bar{y} - a)f_1 k_1 \alpha/2 \\ (q_1 + q_2 x_0 + q_3 x_0^2)\varphi_1 + (q_4 x_0 + q_3 x_0^2)\varphi_2 + q_5\ddot{\varphi}_1 + q_6\ddot{\varphi}_2 = (1 - f_1) C k_1 \alpha/2 \\ (b_1 + b_2 x_0 + b_3 x_0^2 + b_4 x_0^3)\varphi_1 + b_5\ddot{\varphi}_1 + b_5\ddot{\varphi}_2 = -k_1 a \alpha/2 \\ M_{st}[(\ddot{\varphi}_1 + \ddot{\varphi}_2)(\bar{y} - a) + \ddot{\alpha}] = -k_1 \alpha \end{cases}$$

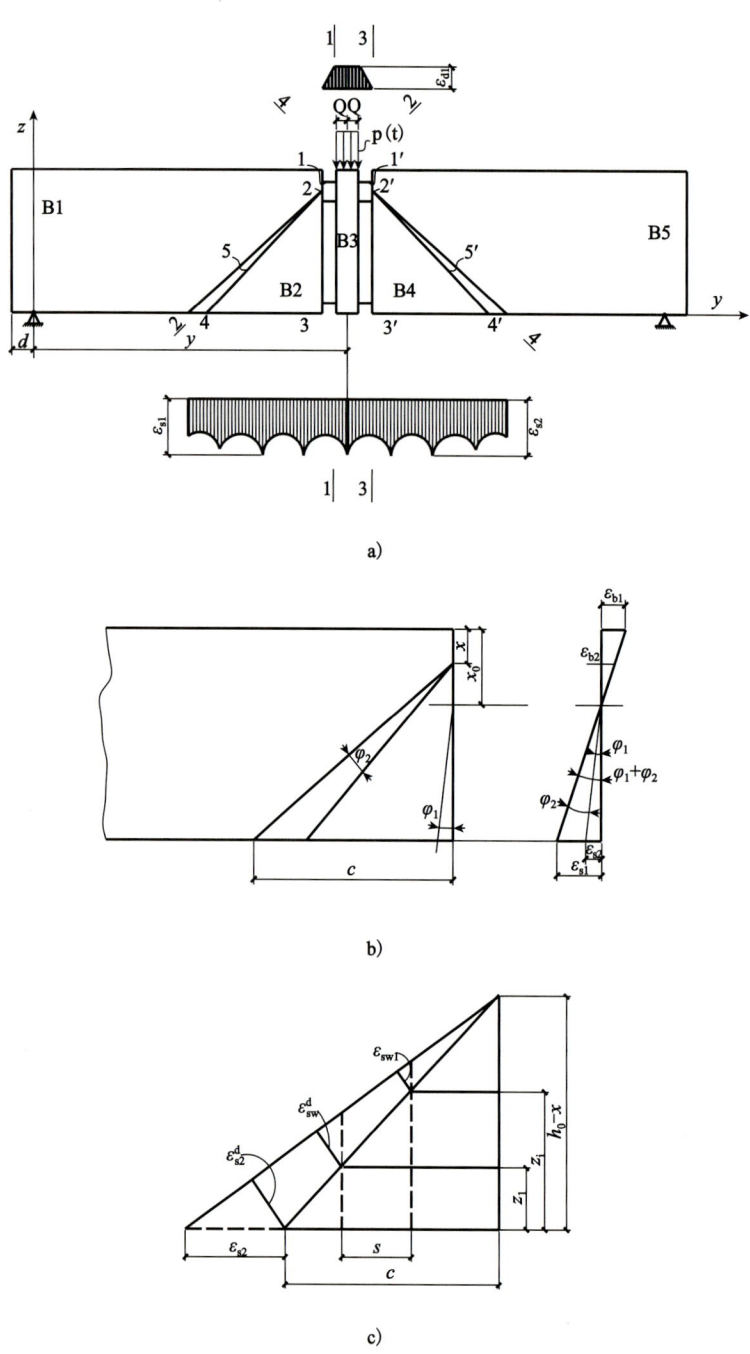

图 13　钢筋混凝土梁的刚片—连系件模型

(3) 计算方法

初始条件是为：$t=0$ 时，$\varphi_1=\varphi_2=\alpha=0$；$x_0=h/2$；$\dot{\varphi}_1=\dot{\varphi}_2=0$；$\dot{\alpha}=v_0$。

为了进行下一步的计算，将梁沿截面高度分成若干相等的层，得到应力应变沿计算截面高度的分布，然后得到连系件内力。利用该模型可以研究梁应变的全部基本特征：当具有屈服极限的纵向钢筋达到极限弹性应变时，相应层内的应力取为常数且等于 R_{sd}；用相同的方法可以考虑与斜裂缝相交的箍筋，逐步达到动力屈服极限的情况。还可以考虑到箍筋对受压区混凝土的束缚作用。考虑到这些因素的影响，则混凝土的应变曲线示于图 14。在边缘层内达到最大应力 R_{sd} 后，其应力开始下降，而在所有较低的层内达到最大值，由实验可以确定，在向下分支（$\approx 0.2R_{bd}$）的低应力条件下，约束混凝土能长时间在很大的非弹性应变条件下工作,从而使纵向钢筋的强度特性在很大程度上得以实现。该方法能恰当地评估横向钢筋的作用,并选择其最佳含量。

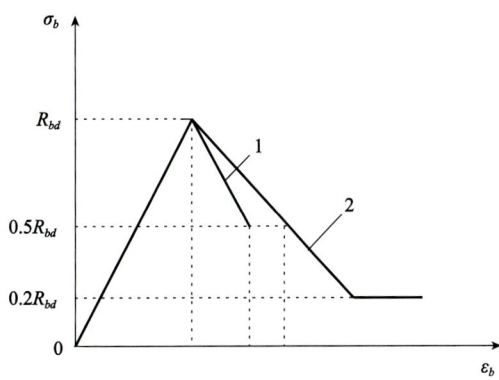

图 14　考虑约束变形的受压混凝土应变曲线
1-非约束混凝土;2-约束混凝土

准脆性材料断裂模拟方法研究

燕秀发[1,2]，钱七虎[1]，赵跃堂[1]，周寅智[1]

（1. 陆军工程大学 爆炸冲击防灾减灾国家重点试验室，江苏 南京 210007；2. 解放军 91550 部队，辽宁 大连 116023）

摘　要：针对岩石、混凝土类准脆性材料的断裂过程模拟，发展了基于黏聚裂纹模型的虚节点扩展有限元法，并给出了该法的数值原理和控制方程。通过三点弯曲梁拉伸断裂、单边缺口试件 I–II 复合型断裂和 Nooru-Mohammed 试验多裂纹断裂等典型算例，并与已有解或试验结果对比，表明该法适合于模拟准脆性材料由张开型裂纹支配的断裂过程。相对于节点分离有限元法，该法无需预设开裂路径；相对于塑性损伤有限元法，该法能够可靠模拟多裂纹曲线扩展；相对于标准扩展有限元法，该法无需引入裂尖单元，避免了应力强度因子的计算；相对于高阶富集扩展有限元法，该法具有良好的适用性，便于得到收敛的计算结果。此外，基于单元的位移场描述使其更易于嵌入常用有限元软件，从而利用后者良好的非线性计算功能求解复杂问题。

关　键　词：扩展有限元法；虚节点；黏聚裂纹模型；准脆性材料；裂纹扩展
中图分类号：TU 457　　　　**文献识别码**：A　　　　**文章编号**：1000－7598 (2017) 12－3462－07

A method for simulating fracture in quasi-brittle materials

YAN Xiu-fa[1,2], QIAN Qi-hu[1], ZHAO Yue-tang[1], ZHOU Yin-zhi[1]

(1. State Key Laboratory of Disaster Prevention & Mitigation of Explosion & Impact, Army Engineering University of PLA, Nanjing, Jiangsu 210007, China; 2. PLA 91550 Unit, Dalian, Liaoning 116023, China)

Abstract: Based on the cohesive crack model, a virtual node extended finite element method (XFEM) was developed to simulate the fracturing process of quasi-brittle materials, such as rock and concrete. Numerical principles and governing equations of this method were also proposed. Typical numerical examples were presented, including tension fracture of the three-point bending beam, I–II mixed mode fracture of a single edge notched specimen and fracture of multiple cracks in the Nooru-Mohammed experiment. Moreover, these results were compared with existing solutions or experimental results. It is found that this method is suitable to simulate fracturing process of quasi-brittle materials dominated by opening-mode cracks. Pre-assignment of crack-growth paths is not necessary for the proposed method, in comparison with the node-separation finite element method (FEM). Contrary to the plastic-damage FEM, this method also can reliably simulate the propagation of multiple cracks. Meanwhile, it is not necessary to introduce crack-tip elements and calculate stress intensity factors in comparison with the normal XFEM. Lastly, it is particularly applicable to the acquirement of convergent computational results by contrast with the XFEM with higher-order enriched elements. Moreover, the developed method can be easily embedded into conventional finite element software to solve complex problems by applying powerful nonlinear computational functions of the latter for its displacement description based on elements.

Keywords: extended finite element method (XFEM); virtual node; cohesive crack model; quasi-brittle materials; crack propagation

1　引　言

断裂是岩石、混凝土等准脆性材料的主要破坏形式，数值模拟是揭示其断裂机制的重要手段。常规有限元法求解裂纹类强间断问题仍存在一定的局限性，如裂尖区域单元局部细化、裂纹面与单元边界一致以及随裂纹扩展进行的网格重构等，给断裂过程模拟带来极大的困难。

收稿日期：2015-12-16。
基金项目：中国博士后科学基金特别资助项目（No. 201003768）；国家自然科学基金创新研究群体科学基金（No. 51321064）；爆炸冲击防灾减灾国家重点试验室开放课题资助项目（No. DPMEIKF201308）。
本文原载于《岩土力学》（2017 年第 38 卷第 12 期）。

扩展有限元法[1]的提出，显著发展了裂纹扩展模拟技术。该法通过在位移模式中加入跳跃函数和裂尖渐近位移场函数，实现了网格剖分与裂纹构形的相互独立，避免了裂纹扩展过程中的网格重构，从而克服了常规有限元法的局限性，受到了广泛关注。然而该法仍处于完善之中，如基于节点影响域的自由度富集判断方法存在缺陷，对单元类型的支持还不够丰富以及难以模拟多裂纹扩展等问题。与上述标准扩展有限元法基于单位分解的特殊函数插值方法不同，Hansbo 等[2]采用等参元的位移插值模式，提出了虚节点扩展有限元法的思想，Song 等[3]进一步发展了该法的断裂力学计算形式。Pan 等[4]应用虚节点扩展有限元法模拟了混凝土拱坝的开裂过程，结合试验并对比其他方法评估了计算结果。但这种仅针对特别实例进行的研究缺乏代表性，虚节点扩展有限元法仍需要深入的数值试验来确定其模拟准脆性材料断裂的适用性。

本文发展了基于黏聚裂纹模型的虚节点扩展有限元法，结合典型算例，通过与节点分离有限元法、塑性损伤有限元法和标准扩展有限元法以及高阶富集法的模拟结果相互对比，评估了该法模拟准脆性材料断裂过程的适用性，并给出了改善计算收敛性的数值技巧。在已有研究[5]对商业有限元软件扩展有限元法功能模块的可靠性和实用性提出质疑的背景下，本文工作具有特殊的意义。

2 基于黏聚裂纹模型的虚节点扩展有限元法

2.1 虚节点扩展有限元法

假定裂纹沿单元逐一扩展，裂纹尖总是终止在单元边界上，如图 1 所示，虚节点扩展有限元法的数值原理是，在真实单元各节点上设置相应的虚拟节点，单元一旦开裂，则该单元上的虚拟节点即被激活，真实单元分解成为两个重叠的虚拟单元 1 和 2，每个虚拟单元由真实节点和虚拟节点构成，则裂纹单元的位移场可以看作系数为 Heaviside 阶跃函数的两虚拟单元的加权叠加

$$u(X) = \sum_{I \in S_1} u_I^1 N_I(X) H(-\Phi) + \sum_{J \in S_2} u_J^2 N_J(X) H(\Phi)$$
(1)

式中：$N_I(X)$ 和 $N_J(X)$ 为常规有限元形函数；I、J 分别表示虚拟单元 1、2 的节点编号；Φ 为符号距离函数，如图中所示，在裂纹面上 $\Phi = 0$，在裂纹一侧为正，另一侧为负；$H(x)$ 为 Heaviside 阶跃函数：当 $x > 0$ 时，$H(x) = 1$，当 $x \leq 0$ 时，$H(x) = 0$；S_1、S_2 分别为虚拟单元 1 和 2 的节点集合；当 $\Phi < 0$，$u_I^1 = u_I$，$u_J^2 = u_J + a_J$；当 $\Phi > 0$，$u_I^1 = u_I - a_I$ 和 $u_J^2 = u_J$；其中 u_I 和 u_J 为常规有限元节点自由度，a_I 和 a_J 为引入 Heaviside 函数产生的附加自由度。由虚拟单元在裂纹面上的位移差可得裂纹两侧的位移跳跃为

$$[u] = \sum_{I \in S_1} u_I^1 N_I(X) - \sum_{J \in S_2} u_J^2 N_J(X) \Big|_{\Phi(X)=0}$$
(2)

裂纹法向张开和切向滑动位移分别为

$$\delta_n = n[u], \quad \delta_t = \|[u] - n\delta_n\|$$
(3)

式中：n 为裂纹面的单位外法线方向矢量。

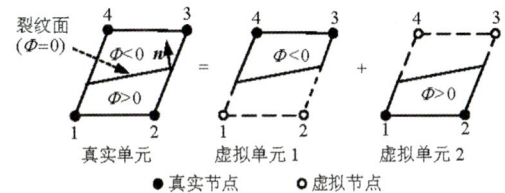

图 1 裂纹单元分解示意图
Fig.1 Decomposition of a cracked element

岩石、混凝土类准脆性材料的断裂过程应用黏聚裂纹模型描述，考虑求解区域 Ω，Γ_t 和 Γ_c 分别为该区域的 Neumann 边界与裂纹面，并记 ρ 为密度，σ 为应力，b 为体力，t 为 Neumann 边界作用力，τ 为裂纹面黏聚力，则对于准静态问题，由动量定理等效积分的弱形式可得

$$\int_\Omega \frac{\partial \delta u}{\partial X} : \sigma \mathrm{d}\Omega = \int_{\Gamma_t} \delta u \cdot \rho b \mathrm{d}\Omega + \int_{\Gamma_t} \delta u \cdot t \mathrm{d}\Gamma_t + \int_{\Gamma_c} \delta[u] \cdot \tau \mathrm{d}\Gamma_c$$
(4)

根据变分原理，可得离散方程为

$$\sum_{e=1}^{2} \int_{\Omega_e} B^T \sigma_e H((-1)^e \Phi) \mathrm{d}\Omega_e - \sum_{e=1}^{2} [\int_{\Omega_e} \rho N^T b \cdot H((-1)^e \Phi) \mathrm{d}\Omega_e + \int_{\Gamma_t^e} N^T t H((-1)^e \Phi) \mathrm{d}\Gamma_t^e] + \sum_{e=1}^{2} (-1)^e \int_{\Gamma_c^e} N^T \tau \mathrm{d}\Gamma_c^e = 0$$
(5)

式中：e 为虚拟单元编号；N 和 B 分别为形函数矩阵与应变矩阵。

2.2 黏聚裂纹模型

黏聚裂纹模型[6]准脆性材料断裂过程中裂纹面上的黏聚力 $\tau = \{\tau_n, \tau_t\}$ 可应用如图 2 所示的双线性本构关系计算，假设起裂前材料的初始刚度为 K_0，当最大主应力超过材料的拉伸强度 τ_0 时材料基体开始起裂，裂纹扩展方向与最大主应力方向垂直。

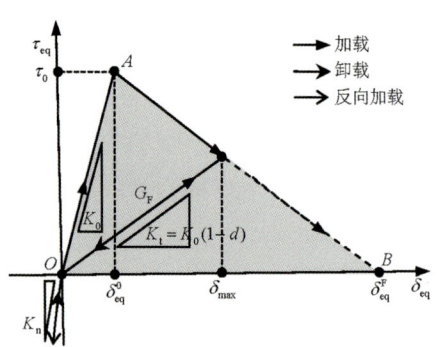

图 2 黏聚裂纹双线性本构关系
Fig.2 Bilinear traction-separation law for cohesive cracks

定义裂纹面等效开度为

$$\delta_{eq} = \sqrt{\delta_n^2 + \delta_t^2} \quad (6)$$

由起裂准则有裂纹初始开度：

$$\delta_{eq}^0 = \frac{\tau_0}{K_0} \quad (7)$$

随着 δ_{eq} 增加，材料基体逐渐相互分离，当裂纹张开消耗的能量达到材料的断裂能 G_F，即图 2 中三角形 OAB 包围区域面积时，基体完全断裂分离，此时裂纹临界开度为

$$\delta_{eq}^F = \frac{2G_F}{\tau_0} \quad (8)$$

材料完全断裂后，为避免（反向加载）裂纹闭合时裂纹面相互嵌入引入罚刚度 K_n。

注意到加载过程中的卸载-再加载情形，令 $\delta_{max} = \mathrm{MAX}(\delta_{eq})$ 为起裂后裂纹的最大开度，与其对应的材料刚度为 K_t，定义损伤变量为

$$d = \frac{K_0 - K_t}{K_0} = \frac{\delta_{eq}^F(\delta_{max} - \delta_{eq}^0)}{\delta_{max}(\delta_{eq}^F - \delta_{eq}^0)} \quad (9)$$

由 $K_t = (1-d)K_0$ 进一步可得裂纹面等效黏聚力为

$$\tau_{eq} = (1-d)K_0\delta_{eq} \quad (10)$$

对于势函数：

$$\Psi(\delta_n, \delta_t) = \int_0^{\delta_{eq}} \tau_{eq} d\delta \quad (11)$$

则有裂纹面法向和切向黏聚力为

$$\left.\begin{array}{l} \tau_n = \dfrac{\partial \Psi(\delta_n, \delta_t)}{\partial \delta_n} = \dfrac{\delta_n}{\delta_{eq}}\tau_{eq} \\ \tau_t = \dfrac{\partial \Psi(\delta_n, \delta_t)}{\partial \delta_t} = \dfrac{\delta_t}{\delta_{eq}}\tau_{eq} \end{array}\right\} \quad (12)$$

3 数值算例

3.1 三点弯曲梁的拉伸断裂

如图 3 所示的混凝土三点弯曲梁问题，梁高度 $T = 0.15$ m，长度 $L = 4T$，宽度 $B = T$，位移加载，加载区域宽度 $l = 0.01$ m。材料参数为弹性模量 $E = 36.5$ GPa，泊松比 $\nu = 0.1$，拉伸强度 $\tau_0 = 3.19$ MPa，断裂能 $G_F = 50$ N/m。考虑平面应变情况，应用弧长法求解。数值模拟得到的梁断裂变形以及为与已有结果对比无量纲化后的加载区域 P（载荷）-D（位移）曲线分别如图 4、5 所示。由图 4 可见，本文方法得到的裂纹扩展途径完全反映了问题的拉伸断裂特点，与裂纹实际扩展途径基本一致。对比图 5 所示的节点分离有限元法解[7]以及图 6 所示的标准扩展有限元法解（图中 r 为以裂尖为中心的富集区域半径）[8]，本文解与这两种方法的计算结果都符合得非常好。

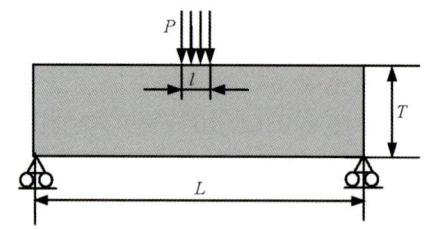

图 3 三点弯曲混凝土梁
Fig.3 Three-point bending concrete beam

图 4 三点弯曲梁的断裂变形
Fig.4 Fracture deformation of three-point bending beam

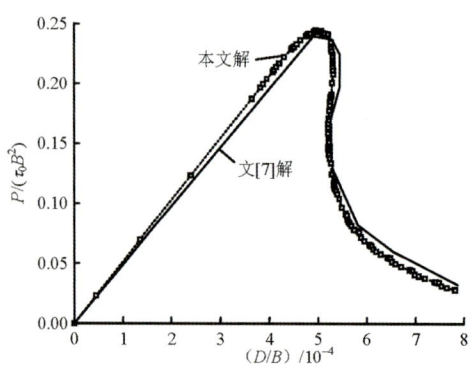

图 5 P（载荷）-D（位移）无量纲曲线
Fig.5 Curves of non-dimensional P(Load)-D(Displacement)

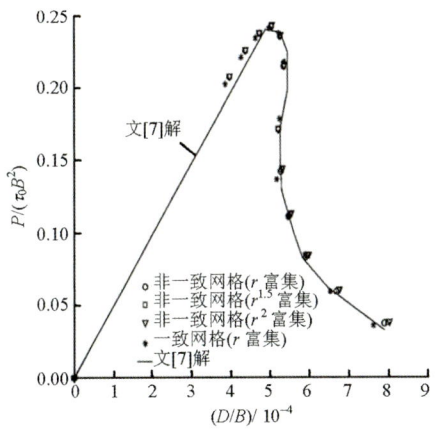

图6 P-D 无量纲曲线文献[8]解
Fig.6 Curves of non-dimensional P-D given by Ref. [8]

3.2 单边缺口试件（SEN）I-II 复合型断裂

如图 7 所示为预制偏置裂纹的三点弯曲试件，裂纹长度 $H=19$ mm，试件长度 $L=230$ mm，高度 $T=75$ mm，宽度 $B=25$ mm，支撑跨距 $2S=205$ mm，偏置裂纹位置为 χS，其中 χ 为偏置系数。材料参数为弹性模量 $E=31.37$ GPa，泊松比 $\nu=0.2$，拉伸强度 $\tau_0=4.4$ MPa，断裂能为 $G_F=0.17$ N/mm。令 $\chi=0.25$，计算结果如图 8、9 所示。为便于与已知解[9]（加载位移约为 0.1 mm）对比，图 9 中同时给出了标定在未变形网格上的相同加载位移时的裂纹扩展路径。由图 8 试件加载至濒临折断时（加载位移约 0.145 mm）的变形可见，本文解明确反映了问题的 I-II 复合型断裂性质以及 I-II 复合型断裂逐渐向 I 型断裂转变的断裂特点。如图 9 中与已知解加载点 P-D 响应曲线及裂纹扩展路径的对比表明：本文解的裂纹扩展方向与文献[9]解基本一致，裂纹扩展路径范围相差很小（本例中裂纹扩展区域单元的横向和纵向尺寸为 1.437 5 mm×1 mm），且 P-D 响应曲线具有相同的数值趋势和较小的误差。

图8 试件断裂变形
Fig.8 Specimen fracture deformation

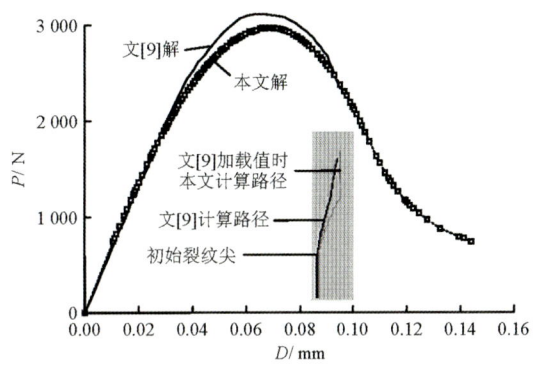

图9 $\chi=0.25$ 时 P-D 曲线和裂纹扩展路径
Fig.9 P-D curses and crack-growth paths for $\chi=0.25$

3.3 多裂纹曲线断裂

著名的 Nooru-Mohammed 试验[10]可以作为检验准脆性材料断裂模拟方法是否有效的基准之一。该试验如图 10 所示，预制双边中心缺口水泥砂浆试样通过固定其上的 L 形钢架水平和垂直加载。水泥砂浆试样尺寸为 200 mm×200 mm×50 mm，各边中心缺口尺寸为 25 mm× 5 mm。先施加水平方向剪力 F_s 至指定值，然后保持 F_s 恒定，再垂直方向位移加载（U_n），加载过程中与 U_n 对应的载荷为 F_n。试样材料参数为弹性模量 $E=30$ GPa，泊松比 $\nu=0.2$，拉伸强度 $\tau_0=3.0$ MPa，断裂能 $G_F=110$ N/m。当 $F_s=10$ kN 时，计算结果如图 11 和图 12 所示。

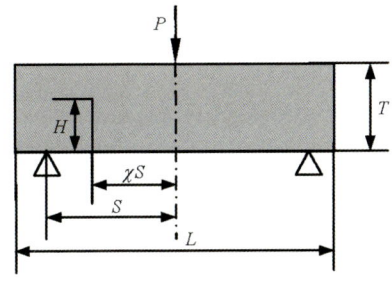

图7 I-II 复合型断裂三点弯曲试验
Fig.7 Three-point bending test on I-II mixed mode fracture

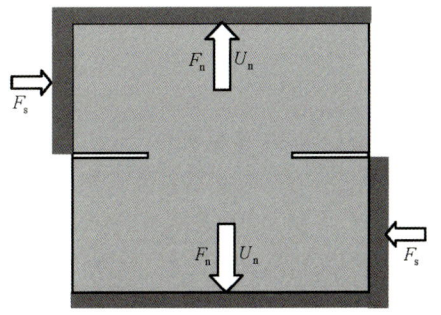

图10 Nooru-Mohammed 试验示意图
Fig.10 Sketch of Nooru-Mohammed experiment

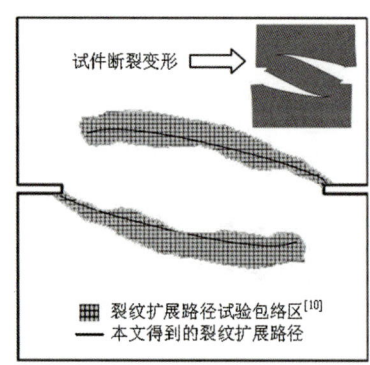

图 11 裂纹扩展路径和试件断裂变形
Fig.11 Crack-growth paths and specimen fracture deformation

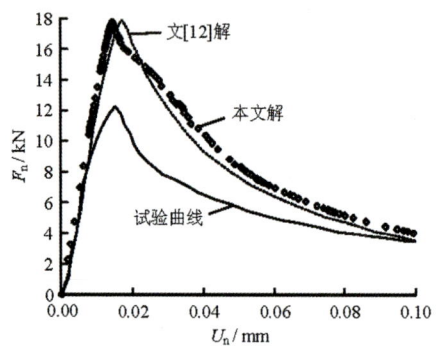

图 12 F_n-U_n 曲线
Fig.12 F_n-U_n curves

如图 11 所示,本文方法得到的裂纹扩展路径完全位于试验包络区内,不但与包络区的伸展方向一致,而且与包络区轮廓线有相似的几何形状。图 11 中同时给出了试件的断裂变形图,对比图 13 所示[11]的裂纹近于平直扩展且交汇的塑性损伤有限元法预测结果,本文得到的裂纹扩展路径充分体现了试验特有的多裂纹并发、最大主应力方向偏转裂纹弯曲扩展的特点。如图 12 所示,本文方法得到的 F_n(载荷)-U_n(位移)曲线与文献[12]应用指数型黏聚本

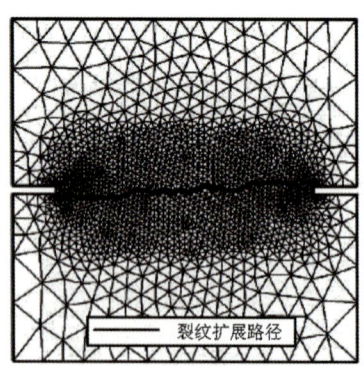

图 13 裂纹扩展路径的塑性损伤有限元法解[11]
Fig.13 Crack-growth paths obtained by the plastic-damage finite method[11]

构关系曲线的扩展有限法解符合得较好,然而相对试验曲线虽然变化趋势相同,但存在数值误差。事实上,数值误差产生的原因是原试验给出的拉伸强度和断裂能参数并非直接来自于实测,而是被高估的推测值。因此,综合前面得到裂纹扩展路径结果,此处对 Nooru-Mohammed 试验的模拟是可靠的。

4 讨 论

4.1 与标准扩展有限元法的比较

3.1 节问题的标准扩展有限元法解[8]如图 6 所示,但与本文算法相比,文献[8]中的方法要复杂得多。该法的复杂性首先在于,为计及裂纹扩展过程中裂尖位于单元内部的情况,相应引入了裂尖单元,并以裂纹尖端位移渐近解作为裂尖单元位移场的特定富集函数,但为了消除裂尖位置对计算精度的影响,不得不对以裂尖为中心半径 r 区域内的节点都应用富集函数,但 r 的取值目前仍具有经验性的特点;其次裂纹扩展方向应用线弹性断裂力学的最大周向应力准则确定,即

$$\theta = 2\arctan\frac{1}{4}[\frac{K_\mathrm{I}}{K_\mathrm{II}} \pm \sqrt{(\frac{K_\mathrm{I}}{K_\mathrm{II}})^2+8}] \quad (13)$$

式中:K_I 和 K_II 为 I 型和 II 型应力强度因子。

由 2.1、2.2 节可见,虚节点扩展有限元法事实上是标准扩展有限元法的简化:通过假设裂纹沿单元逐一扩展,裂尖终止在单元边上,无需引入裂尖单元,相应也克服了富集区域半径选择的经验性;黏聚裂纹模型的应用,移去了裂纹尖端应力奇异性,避免了应力强度因子的计算;令裂纹开裂方向与最大主应力方向垂直,实际上是对最大周向应力断裂准则的简化,降低了问题的复杂性,有利于提高计算效率。本文算例表明,当单元尺寸足够精细时,上述处理方法对分析结果影响不大,仍能得到裂纹正确和光滑的扩展路径。

4.2 与高阶富集元法的比较

由 3.2 节问题的 I-II 复合型断裂特点,分析图 9 所示的裂纹扩展路径能够发现,文献[9]解应当比本文解更接近试验结果,这是因为该文算法应用了高阶富集函数,其富集单元的位移模式为

$$\boldsymbol{u}(X) = \sum_{i\in \boldsymbol{I}}\boldsymbol{u}_i N_i(X) + \sum_{j\in \boldsymbol{J}}N_j(X)H(X)[\boldsymbol{u}_{0j} +$$
$$\boldsymbol{u}_{1j}(\frac{x-x_j}{h_j})^2 + \boldsymbol{u}_{2j}(\frac{y-y_j}{h_j})^2 +$$
$$\boldsymbol{u}_{3j}(\frac{x-x_j}{h_j})(\frac{y-y_j}{h_j})] \quad (14)$$

式中：u_i 为常规有限元节点自由度；I 为单元常规节点集合；J 为单元富集节点集合；i、j 为节点编号；$u_{kj}(k=0\sim3)$为富集节点附加自由度；h_j 为与节点相关的特征长度[9]。

由式（14）可见，二次多项式被引入节点富集函数，而本文应用的节点富集函数仅是保留其 u_{0j} 项的低阶情况。从理论上讲，高阶富集元的应用能够改进计算精度，但对于本例——存在极值临界点，具有急速跳过（snap-through)特征的高度非线性问题，除增加了计算量外，还可能引起收敛困难。由图 9 的 P-D 曲线可见，本文方法的加载量达 0.145 mm，此时试件接近折断，结构濒临完全失稳，而文献[9]仅加载至约 0.1 mm，远小于本文算例加载量。结构越接近完全失稳，问题的非线性响应越强烈，因此，不排除该文算法之后遭遇收敛困难的可能。对于高度非线性问题，数值模拟的首要任务是得到收敛的计算结果，其次是提高解的精度。可见，本文方法具有良好的适用性，便于得到收敛的计算结果。

4.3 改善计算收敛性的方法

岩石、混凝土类不稳定材料的断裂过程模拟是伴随着损伤软化与刚度衰减的高度非线性问题，如何获得收敛的数值结果是始终需要关注的问题。由数值原理可见，不同于一般扩展有限元法基于节点影响域的位移场描述，虚节点扩展有限元法同常规有限元法一样对位移场的描述都是基于单元的，因此，更易于嵌入常用有限元软件，从而利用其良好的非线性功能求解复杂问题，如选择 Newton 法或弧长法求解非线性方程组以及引入黏性正则化。如图 14 所示，当求解具有急速跳回（snap-back）特点的问题 3.1 时，Newton 法因迭代发散导致求解过程中断，改用弧长法后收敛困难被克服，得到了与已知解符合得较好的数值结果。然而，若令问题 3.2 中 $\chi=0.5$，则如图 15 所示，按 Duvaut-Lions 正则化原理[13]引入黏性（黏性系数 $c=0.000\,1$）后才获得收敛解。分析此时出现收敛困难的原因，首先是虚节点扩展有限元法建立在单元水平上的裂纹扩展模式使结构表现出过高的脆性响应，合理引入黏性可以起到抵减作用；其次，对比图 15 和图 9 得到的裂纹扩展路径和 P-D 曲线能够发现，$\chi=0.5$ 时问题本身具有更显著的剪切断裂特点和较弱的拉伸断裂性质，进而反映出虚节点扩展有限元法更适合模拟其中张开型成分较大的断裂模式的数值特性。

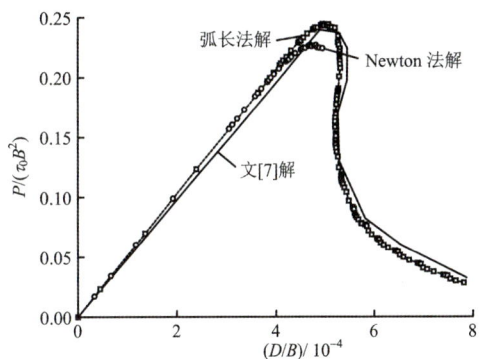

图 14　Newton 法和弧长法的解
Fig.14　Solutions by Newton method and arc-length method

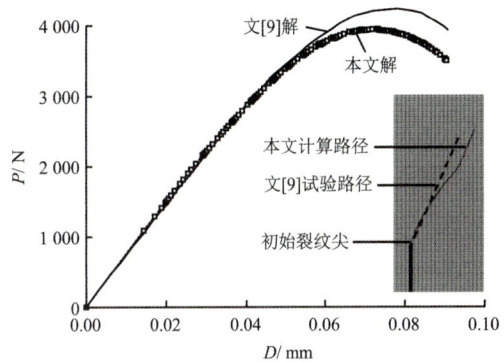

图 15　P-D 曲线和裂纹扩展路径($\chi=0.5$，$c=0.000\,1$)
Fig.15　P-D curves and crack-growth paths ($\chi=0.5$，$c=0.000\,1$)

5　结　论

（1）针对岩石、混凝土类准脆性材料的断裂过程模拟，发展了基于黏聚裂纹模型的虚节点扩展有限元法，给出了存在和不存在初始裂纹、单一拉伸型断裂和I-II复合型断裂、单裂纹断裂和多裂纹断裂、平直扩展裂纹和弯曲扩展裂纹等涵盖该类材料主要断裂形式的典型算例。与已有解或试验结果的对比表明，该法适合于模拟准脆性材料由张开型裂纹支配的断裂过程。

（2）本文方法与其他断裂模拟方法的比较表明，相对于节点分离有限元法，该法不需预设开裂路径；相对于塑性损伤有限元法，该法能够正确模拟多裂纹曲线扩展；相对于标准扩展有限元法，该法无需引入裂尖单元，避免了应力强度因子的计算；相对于高阶富集扩展有限元法，该法具有良好的适用性，便于得到收敛的计算结果。此外，基于单元的位移场描述使其更易于嵌入常用有限元软件，从而利用后者良好的非线性计算功能求解复杂问题。

（3）此处的计算经验表明，岩石、混凝土类不稳定材料的断裂过程模拟是伴随着损伤软化和刚度衰减的高度非线性问题，首要的是得到收敛的计算结果，因此，隐含在本文算例细节中的提高收敛性的数值技巧同样需要给予关注。

参 考 文 献

[1] MOËS N, DOLBOW J, BELYTSCHKO T. A finite element method for crack growth without remeshing[J]. **International Journal for Numerical Methods in Engineering**, 1999, 46(1): 131—150.

[2] HANSBO A, HANSBO P. A finite element method for the simulation of strong and weak discontinuities in solid mechanics[J]. **Computer Methods in Applied Mechanics and Engineering**, 2004,193(33–35): 3523—3540.

[3] SONG J H, AREIAS P M A, BELYTSCHKO T. A method for dynamic crack and shear band propagation with phantom nodes[J]. **International Journal for Numerical Methods in Engineering**, 2006, 67(6): 868—893.

[4] PAN J W, ZHANG C H, XU Y J, et al. Comparative study of the different procedures for seismic cracking analysis of concrete dams[J]. **Soil Dynamics and Earthquake Engineering**, 2011, 31(10): 1594—1606.

[5] GIGLIOTTI L. Assessment of the applicability of XFEM in Abaqus for modeling crack growth in rubber[D]. Sweden Stockholm: Royal Institute of Technology, 2012.

[6] HILLERBORG A, MODÉER M, PETERSSON P E. Analysis of crack formation and crack growth in concrete by means of fracture mechanics and finite elements[J]. **Cement and Concrete Research**, 1976, 6(6): 773—782.

[7] CARPINTERI A, COLOMBO G. Numerical analysis of catastrophic softening behavior (snap-back instability)[J]. **Computers and Structures**, 1989, 31(4): 607—636.

[8] MOËS N, BELYTSCHKO T. Extended finite element method for cohesive crack growth[J]. **Engineering Fracture Mechanics**, 2002, 69(7): 813—833.

[9] STEFANO M, UMBERTO P. Extended fininite element method for quasi-brittle fracture[J]. **International Journal for Numerical Methods in Engineering**, 2003, 58(1): 103—126.

[10] NOORU-MOHAMED M B. Mixed-mode fracture of concrete: An experimental approach[Ph. D. Thesis D]. Netherlands: Delft University of Technology, 1992.

[11] FEIST C, HOFSTETTER G. An embedded strong discontinuity model for cracking of plain concrete[J]. **Computer Methods in Applied Mechanics and Engineering**, 2006, 195(61–63): 7115—7138.

[12] XU Y, YUAN H. Applications of normal stress dominated cohesive zone models for mixed-mode crack simulation based on extended finite element methods[J]. **Engineering Fracture Mechanics**, 2011, 78(5): 544—558.

[13] HIBBITT, KARLSSON & SORENSEN Inc. ABAQUS /Standard User's Manual[M]. Pawtucket Rhode Island: HKS Co., 2010.

基于广义粒子动力学的巷道围岩弹塑性分析

赵 毅[1]，周小平[1]，钱七虎[2]

(1. 重庆大学土木工程学院，重庆 400045；2. 解放军理工大学国防工程学院，江苏 南京 210007)

摘 要：提出了广义粒子动力学数值分析方法，该方法是一种无网格数值分析方法，可以考虑关联塑性流动法则和非关联塑性流动法则对岩石材料塑性变形的影响。将广义粒子动力学数值分析方法应用于巷道围岩的弹塑性分析，确定了巷道围岩的应力场、位移场和塑性区。该数值模拟结果与有限元结果吻合较好，表明将考虑岩石材料剪胀特性的弹塑性本构理论引入到广义粒子动力学数值分析方法，不失为模拟岩石类材料弹塑性破坏的一种有效数值手段，研究结果为更好地理解岩石材料的屈服破坏过程提供重要的参考。

关键词：巷道围岩；剪胀角；弹塑性；广义粒子动力学；位移；塑性区

中图分类号：TU43   **文献标识码**：A   **文章编号**：1000-4548(2015)06-1104-13

作者简介：赵 毅(1983-)，男，博士研究生，主要从事岩土工程的科研工作。E-mail: zhaoyi0622@163.com。

Elastoplastic analysis of surrounding rock masses around tunnels using general particle dynamics method

ZHAO Yi[1], ZHOU Xiao-ping[1], QIAN Qi-hu[2]

(1. School of Civil Engineering, Chongqing University, Chongqing 400045, China; 2. PLA University of Science and Technology, Nanjing 210007, China)

Abstract: The novel meshless numerical method, which is known as general particle dynamics (GPD) method, is proposed. The non-associated flow law and the associated flow law can be employed to analyze the plastic deformation of the surrounding rock masses around tunnels using the GPD method. The stability of the surrounding rock masses around tunnels are also determined using the GPD method as well as the stress fields, displacement fields and plastic zone. The numerical results by the proposed method are in good agreement with the FEM results. It is proved that the GPD method is efficient to predict the elastic-plastic properties of the surrounding rock masses around tunnels.

Key words: surrounding rock mass around tunnel; dilatancy angle; elastoplasticity; general particle dynamics; displacement; plastic zone

0 引 言

对于地下洞室围岩稳定性分析，众多学者已经做了大量的研究[1-3]。理论研究主要基于理想弹塑性、弹脆性力学模型研究围岩的应力场和位移场，如 Park 等[4]给出了弹脆性模型下洞室围岩的应力和变形解析解；Sharan[5-6]给出了 Hoek-Brown 弹脆性模型的围岩应力和变形的解析解，算例表明 Hoek-Brown 弹脆性模型计算误差是满足工程精度要求的；蒋斌松等[7]通过考虑深部岩体的塑性承载特性，分析了弹塑脆性模型围岩应力和变形的规律。

数值分析法也应用于研究巷道围岩的应力及变形规律。如 Brown 等[8]认为峰后岩体的弹性变形为一定值，通过计算得到了应力和变形场；Wang 等[9]基于数值分析给出了脆塑性逼近形式的数值解。Lee 等[10]采用有限差分方法给出了圆形洞室围岩的弹塑性数值解；Park 等[11]分析了不同剪胀特性的圆形洞室围岩的变形规律。

对于岩石、硬黏土及混凝土等材料，随着应变的增大，体积先是稍微减小，随后逐渐增大，称为剪胀。这一现象由著名学者 Prevost 等[12]最早提出。在巷道围岩稳定性理论和数值研究中，考虑围岩受剪体积变化是导致巷道围岩大变形的主要原因。

基于传统网格方法（如有限元）模拟时，需要将材料不连续面设置成单元的边，面临网格重划分等繁琐的前处理，因此预测巷道围岩的塑性区乃至松动区时受网格划分精度的影响。而无网格 GPD 方法在处理

收稿日期：2015-02-26。

基金项目：国家重点基础研究发展计划("973"计划)项目(2014CB046903)；国家自然科学基金项目(51325903,51279218)；重庆市自然科学基金院士专项项目(cstc2013jcyjys30002)。

本文原载于《岩土工程学报》(2016年第38卷第6期)。

大变形和不连续问题上具有优势[13-14]。

在模拟岩石类材料塑性变形时，本文提出的无网格 general particle dynamics 方法（以下简称 GPD 法）相对于基于连续介质力学的传统数值方法有诸多优点[13-17]。在无网格 GPD 法的框架下，引入考虑中间主应力影响的 Drucker-Prager 屈服准则（以下简称 D-P 准则），并考虑材料剪胀性的非关联流动法则，对圆形巷道围岩进行弹塑性力学分析，能给出更为准确的数值模拟解答。

1 GPD 法的基本控制方程和考虑材料剪胀的的弹塑性本构模型

1.1 GPD 法的基本控制方程

GPD 法是一种纯拉格朗日形式的无网格粒子法，用于获取偏微分控制方程解的数值方法。在 GPD 方法中，当前计算步的支持域内的离散化粒子将场函数及其导数的连续积分表达式近似转化为粒子离散化求和形式。该方法用于解决固体大变形问题具有优势。粒子 j 为粒子 i 支持域内的所有粒子（如图1），粒子 i 处的离散化粒子场量函数 及其梯度形式表达式为[17-18]

$$f(x)_{x=x_i} = \sum_{j=1}^{N} \frac{m_j}{\rho_j} f(x_j) W_{x_i}(x_i-x_j, h) \quad , \quad (1)$$

$$\nabla f(x)_{x=x_i} = -\sum_{j=1}^{N} \frac{m_j}{\rho_j} [f(x_i)-f(x_j)] \nabla W_{x_i}(x_i-x_j, h), \quad (2)$$

式中，m，ρ 分别为粒子质量和密度，$W(x_i-x_j, h)$ 为光滑核函数，x_i-x_j 为粒子 j 和粒子 i 的距离，h 为光滑长度，N 为粒子 i 的支持域内的 j 粒子总数。

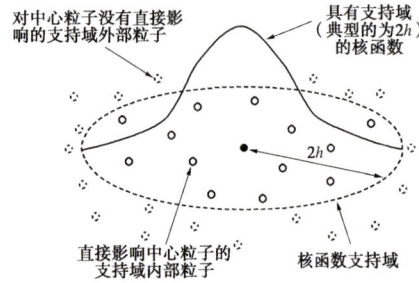

图 1 核函数和支持域示意图

Fig. 1 Kernel function and supporting domain

光滑核函数的表达式为[13-14, 16, 19]

$$W(R,h) = \begin{cases} \left(\frac{2}{3}-R^2+\frac{1}{2}R^3\right)\alpha_d & (0 \leq R<1) \\ \frac{1}{4}(2-R)^3 \cdot \alpha_d & (1 \leq R<2) \\ 0 & (R \geq 2) \end{cases}, \quad (3)$$

式中，在一维、二维、三维情况下分别有 $\alpha_d=1/h$，$15/(7\pi h^2)$，$3/(2\pi h^2)$。

连续介质力学中的质量和动量守恒方程经过 GPD 法离散处理后为[17-18]

$$\frac{d\rho_i}{dt} = \rho_i \sum_{j=1}^{N} \frac{m_j}{\rho_j}[(v_\alpha)_i-(v_\alpha)_j]\frac{\partial W_{ij}}{\partial (x_\alpha)_i} \quad , \quad (4)$$

$$\frac{d(v_\alpha)_i}{dt} = \sum_{j=1}^{N} m_j\left[\frac{(\sigma_{\alpha\beta})_i}{\rho_i^2}+\frac{(\sigma_{\alpha\beta})_j}{\rho_j^2}-\Pi_{ij}\delta_{\alpha\beta}\right]\frac{\partial W_{ij}}{\partial (x_\beta)_i}+f_i^\alpha, \quad (5)$$

式中，密度 ρ，速度 v_α 和应力 $\sigma_{\alpha\beta}$ 为因变量，空间坐标 x_α 和时间 t_α 为自变量。移动拉格朗日框架下的时间导数算子为 (d/dt)，下标 α，β 表示坐标方向。

1.2 考虑岩石材料剪胀的弹塑性本构模型

（1）弹塑性本构方程

为描述岩石类材料的力学性质，在研究中引入理想弹塑性模型。本文将阐述弹塑性模型的一般形式[20]。

首先，应变率张量定义为

$$\dot{\varepsilon}^{\alpha\beta} = \frac{1}{2}\left(\frac{\partial v^\alpha}{\partial x^\beta}+\frac{\partial v^\beta}{\partial x^\alpha}\right) \quad . \quad (6)$$

对于理想弹塑性材料，$\dot{\varepsilon}^{\alpha\beta}$ 由弹性应变率张量 $\dot{\varepsilon}_e^{\alpha\beta}$ 和塑性应变率张量 $\dot{\varepsilon}_p^{\alpha\beta}$ 两部分组成，其表达式为

$$\dot{\varepsilon}^{\alpha\beta} = \dot{\varepsilon}_e^{\alpha\beta}+\dot{\varepsilon}_p^{\alpha\beta} \quad . \quad (7)$$

根据广义胡克定律，弹性应变率张量 $\dot{\varepsilon}_e^{\alpha\beta}$ 的表达式为

$$\dot{\varepsilon}_e^{\alpha\beta} = \frac{\dot{s}^{\alpha\beta}}{2G}+\frac{1-2\nu}{3E}\dot{\sigma}^{\gamma\gamma}\delta^{\alpha\beta} \quad , \quad (8)$$

式中，$\dot{s}_e^{\alpha\beta}$ 是偏应力率张量，ν 为泊松比，E 为杨氏模量，G 为剪切模量，$\dot{\sigma}^{\gamma\gamma}$ 为 3 个应力率之和，即 $\dot{\sigma}^{\gamma\gamma} = \dot{\sigma}^{xx}+\dot{\sigma}^{yy}+\dot{\sigma}^{zz}$。

塑性应变率张量 $\dot{\varepsilon}_p^{\alpha\beta}$ 通过塑性流动法则进行计算，其表达式为

$$\dot{\varepsilon}_p^{\alpha\beta} = \dot{\lambda}\frac{\partial g}{\partial \sigma^{\alpha\beta}} \quad , \quad (9)$$

式中，塑性乘子变化率 $\dot{\lambda}$ 取决于当前应力状态和加载历史，g 为塑性势函数，它决定了塑性应变发展的方向和趋势。

剪胀是岩石类材料区别于金属材料的重要特征，岩石材料具有剪胀性质，不考虑剪胀的关联流动法则并不适用于岩石材料。Vermeer 等[21]从理论上证明了关联流动法则不适用于岩石类材料，它会引起岩石塑性变形时无能量耗散。之后，范文等[22-24]对岩土类材料的剪胀特性进行了大量研究，指出用理论或数值方法分析边坡和隧洞等工程的稳定性时，剪胀角的选取对结果影响很大。现阶段试验数据较少，岩石剪胀特性的研究不够深入，各种岩石材料剪胀角的取值还不明确。在进行岩石类材料数值计算时，对剪胀角的处理往往采用两种做法：①将材料的屈服函数 f 作为材料的塑性势 g（即关联流动法则），这种忽略剪胀角

的做法，其后果是扩大了材料的剪胀性；②采用考虑剪胀角的非关联流动法则，一般取剪胀角为特殊值（0°）。

根据塑性力学理论，塑性乘子λ和屈服函数f有以下的对应关系：

$$\mathrm{d}f = \frac{\partial f}{\partial \sigma^{\alpha\beta}} \mathrm{d}\sigma^{\alpha\beta} = 0 \quad 。 \tag{10}$$

将弹性应变率张量式（7）和塑性应变率式（8）代入式（6），总应变率张量表达式为

$$\dot{\varepsilon}^{\alpha\beta} = \frac{\dot{s}^{\alpha\beta}}{2G} + \frac{1-2\nu}{3E}\dot{\sigma}^{\eta\eta}\delta^{\alpha\beta} + \dot{\lambda}\frac{\partial g}{\partial \sigma^{\alpha\beta}} \quad 。 \tag{11}$$

根据柯西应力张量，偏应变张量和静水应力张量的关系式为

$$\sigma^{\alpha\beta} = s^{\alpha\beta} + \frac{1}{3}\sigma^{\eta\eta}\delta^{\alpha\beta} \quad 。 \tag{12}$$

整理式（11），完全弹塑性材料的全应力应变关系可以表示为

$$\dot{\sigma}_i^{\alpha\beta} = 2G\dot{e}^{\alpha\beta} + K\dot{\varepsilon}^{\eta\eta}\delta^{\alpha\beta} -$$
$$\dot{\lambda}\left[\left(K - \frac{2G}{3}\right)\frac{\partial g}{\partial \sigma^{mn}}\delta^{mn}\delta^{\alpha\beta} + 2G\frac{\partial g}{\partial \sigma^{\alpha\beta}}\right] , \tag{13}$$

式中，α和β为自由指标，m和n为哑指标，应力率偏量为$\dot{e}^{\alpha\beta} = \dot{\varepsilon}^{\alpha\beta} - \frac{1}{3}\dot{\varepsilon}^{\eta\eta}\delta^{\alpha\beta}$，体积模量为$K = \frac{E}{3(1-2\nu)}$，剪切模量为$G = \frac{E}{2(1+\nu)}$。

将全应力应变关系式（13）代入一致性条件（10）可得理想弹塑性乘子的表达式为

$$\dot{\lambda} = \frac{2G\dot{\varepsilon}^{\alpha\beta} + \left(K - \frac{2G}{3}\right)\dot{\varepsilon}^{\eta\eta}\frac{\partial f}{\partial \sigma^{\alpha\beta}}\delta^{\alpha\beta}}{2G\frac{\partial f}{\partial \sigma^{mn}} + \left(K - \frac{2G}{3}\right)\frac{\partial f}{\partial \sigma^{mn}}\delta^{mn}\frac{\partial g}{\partial \sigma^{mn}}\delta^{mn}} \quad 。 \tag{14}$$

将岩石材料的屈服函数f和塑性势函数g代入式（13）和（14），可得全应变率张量$\dot{\varepsilon}_e^{\alpha\beta}$和塑性乘子。将本构关系（13）得到的应力张量代入式（5），GPD法的控制方程即可形成封闭。

（2）Drucker-Prager 模型

通过是否越过 Drucker-Prager 屈服条件来判断岩石是否进入塑性流动区域。Drucker-Prager 屈服条件可表示为

$$f(I_1, J_2) = \sqrt{J_2} + \alpha_\varphi I_1 - k_c = 0 , \tag{15}$$

式中，I_1和J_2分别表示第一应力不变量和第二应力不变偏量，其表达式分别为

$$\left.\begin{array}{l} I_1 = \sigma^{xx} + \sigma^{yy} + \sigma^{zz} \\ J_2 = \frac{1}{2}s^{\alpha\beta}s^{\alpha\beta} \end{array}\right\} \tag{16}$$

α_φ和k_c是 Drucker-Prager 模型的参数，对于平面应变问题，其与莫尔–库仑材料参数的凝聚力c和内摩擦角φ存在以下关系：

$$\left.\begin{array}{l} \alpha_\varphi = \dfrac{\tan\varphi}{\sqrt{9 + 12\tan^2\varphi}} \\ J_2 = \dfrac{3c}{\sqrt{9 + 12\tan^2\varphi}} \end{array}\right\} \tag{17}$$

除了屈服准则，塑性势函数用于准确描述应力应变关系。本文将关联和非关联塑性流动法则分别植入GPD法。

对于关联流动法则，塑性势函数和屈服准则有如下形式：

$$g = \sqrt{J_2} + \alpha_\varphi I_1 - k_c \quad 。 \tag{18}$$

对于非关联流动法则，其塑性势函数有如下形式：

$$g = \sqrt{J_2} + 3I_1 \sin\psi , \tag{19}$$

式中，ψ是膨胀角。

对于关联塑性流动问题，将塑性势能函数式（18）代入应力应变关系式（13），并联合塑性乘子表达式（14），Drucker-Prager 模型下的理想弹塑性应力应变关系可表达式为

$$\dot{\sigma}_i^{\alpha\beta} = 2G\dot{e}^{\alpha\beta} + K\dot{\varepsilon}^{\eta\eta}\delta^{\alpha\beta} -$$
$$\dot{\lambda}\left[3\alpha_\varphi K\delta^{\alpha\beta} + \frac{G}{\sqrt{J_2}}s^{\alpha\beta}\right] 。 \tag{20}$$

式中，塑性乘子的变化率$\dot{\lambda}$的表达式为

$$\dot{\lambda} = \frac{3\alpha_\varphi K\dot{\varepsilon}^{\eta\eta} + (G/\sqrt{J_2})s^{\alpha\beta}\dot{\varepsilon}^{\alpha\beta}}{9\alpha_\varphi^2 K + G} \quad 。 \tag{21}$$

对于非关联塑性流动问题,将式(19)代入式(13),并联合塑性乘子表达式（14），可以得到如下表达式：

$$\dot{\sigma}_i^{\alpha\beta} = 2G\dot{e}^{\alpha\beta} + K\dot{\varepsilon}^{\eta\eta}\delta^{\alpha\beta} -$$
$$\dot{\lambda}\left[9K\sin\psi\delta^{\alpha\beta} + \frac{G}{\sqrt{J_2}}s^{\alpha\beta}\right] , \tag{22}$$

式中，塑性乘子的变化率为

$$\dot{\lambda} = \frac{3\alpha_\varphi K\dot{\varepsilon}^{\eta\eta} + (G/\sqrt{J_2})s^{\alpha\beta}\dot{\varepsilon}^{\alpha\beta}}{27\alpha_\varphi K\sin\psi + G} \quad 。 \tag{23}$$

从上述应力应变关系可看出，关联和非关联塑性流动法则的主要不同在于是否存在剪胀角的影响。剪胀角参与到非关联塑性流动模型中，通过选择不同的值，对非关联塑性流动有不同贡献。在下面的拉格朗日 GPD 算法框架下，用于求解式（20）、（22），而无需像有限元一样构建刚度矩阵。

考虑大变形问题，本构关系中应力率不变式需考虑刚体转动。GPD 法中采用 Jaumann 应力率为

$$\dot{\sigma}^{\alpha\beta} = \dot{\sigma}^{\alpha\beta} - \sigma^{\alpha\gamma}\dot{\omega}^{\beta\gamma} - \sigma^{\gamma\beta}\dot{\omega}^{\alpha\gamma} , \tag{24}$$

式中，旋转张量$\dot{\omega}^{\alpha\beta}$的表达式为

$$\dot{\omega}^{\alpha\beta} = \frac{1}{2}\left(\frac{\partial v^\alpha}{\partial x^\beta} - \frac{\partial v^\beta}{\partial x^\alpha}\right) \quad . \tag{25}$$

最后，关联和非相关联流动模型的应力应变关系可以分别表达为

$$\dot{\sigma}^{\alpha\beta} - \sigma^{\alpha\gamma}\dot{\omega}^{\beta\gamma} - \sigma^{\gamma\beta}\dot{\omega}^{\alpha\gamma}$$
$$= 2G\dot{e}^{\alpha\beta} + K\dot{\varepsilon}^{\gamma\gamma}\delta^{\alpha\beta} - \dot{\lambda}\left[3\alpha_\phi K\delta^{\alpha\beta} + \frac{G}{\sqrt{J_2}}s^{\alpha\beta}\right], \tag{26}$$

$$\dot{\sigma}^{\alpha\beta} - \sigma^{\alpha\gamma}\dot{\omega}^{\beta\gamma} - \sigma^{\gamma\beta}\dot{\omega}^{\alpha\gamma}$$
$$= 2G\dot{e}^{\alpha\beta} + K\dot{\varepsilon}^{\gamma\gamma}\delta^{\alpha\beta} - \dot{\lambda}\left[9K\sin\psi\delta^{\alpha\beta} + \frac{G}{\sqrt{J_2}}s^{\alpha\beta}\right], \tag{27}$$

式中，塑性乘子变化率 $\dot{\lambda}$ 可通过式（21）、（23）分别进行计算。

（3）引入计算塑性力学的数值误差处理

本文采用完全弹塑性模型，需保证岩石的性质和模型的一致性，比如：某个区域塑性变形发生时，该区域应力状态不能超过塑性屈服面。但是，由于计算塑性力学中常出现数值误差，岩石应力状态常常超越塑性面。在这种情况下，需要将应力状态通过数值的方法折减回到屈服面。在 GPD 弹塑性模型中，同样的问题也会出现，采用以下数值处理过程用于将应力状态回归映射到屈服面上。

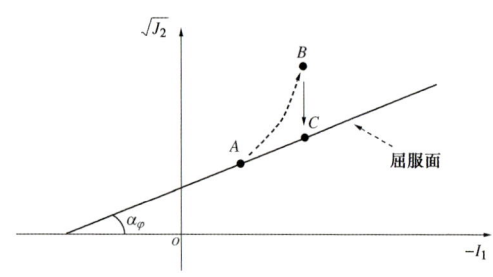

图 2 非物理塑性状态回归处理

Fig. 2 Stress-scaling back procedure

在塑性加载过程中，理想弹塑性材料经历塑性形变，其应力状态也在屈服面上。但是，计算的误差会使应力状态越过塑性屈服面，如图 2 中的路径 AB。在这样的情况下，B 点应力状态是计算出来的非真实的塑性应力状态，应力状态需要进行折减回到屈服面。过程如下：如图 2 中所示，几何上进行了一个应力折减（垂直于 x 轴方向映射到屈服面上得到 C 点）使得 B 点到 C 点，于是应力状态 B 点回归到屈服面的 C 点。数值上引入比例因子 r，对于 Drucker-Prager 屈服条件而言，第 n 步的比例因子为

$$r^n = \frac{-\alpha_\varphi I_1^n + k_c}{\sqrt{J_2^n}} \quad . \tag{28}$$

相应地，如果弹塑性材料为岩石类材料，若应力状态越过屈服面，比如 Drucker-Prager 屈服准则，应满足

$$-\alpha_\varphi I_1 + k_c < \sqrt{J_2} \quad . \tag{29}$$

在静水压力 $I_1^n = \sigma_n^{xx} + \sigma_n^{yy} + \sigma_n^{zz}$ 保持不变的情况下，第 n 步偏剪切应力通过比例因子 r 进行折减，根据式（12）有如下关系：

$$\tilde{\sigma}_n^{xx} = r^n s^{xx} + \frac{1}{3}I_1^n \quad , \tag{30}$$

$$\tilde{\sigma}_n^{yy} = r^n s^{yy} + \frac{1}{3}I_1^n \quad , \tag{31}$$

$$\tilde{\sigma}_n^{zz} = r^n s^{zz} + \frac{1}{3}I_1^n \quad , \tag{32}$$

$$\tilde{\sigma}_n^{xy} = r^n s^{xy} \quad , \tag{33}$$

$$\tilde{\sigma}_n^{xz} = r^n s^{xz} \quad , \tag{34}$$

$$\tilde{\sigma}_n^{yz} = r^n s^{yz} \quad . \tag{35}$$

模型由于采用蛙跳法进行积分，在 $n+0.5$ 步进行判断更新回归。在每个半步的时候，岩石应力状态越过塑性屈服面，按上述映射方法将应力状态回归到屈服面。

2 GPD 中的弹塑性本构方程

2.1 GPD 法的控制方程离散化

连续介质力学中的质量和动量守恒方程经过 GPD 法离散处理后，对于任意质点 i，其密度和速度导数形式可表示为[17-18]

$$\frac{d\rho_i}{dt} = \rho_i \sum_{j=1}^{N}\frac{m_j}{\rho_j}\left[v_i^\alpha - v_j^\alpha\right]\frac{\partial W_{ij}}{\partial x_i^\alpha} \quad , \tag{36}$$

$$\frac{dv_i^\alpha}{dt} = \sum_{j=1}^{N}m_j\left(\frac{\sigma_i^{\alpha\beta}}{\rho_i^2} + \frac{\sigma_j^{\alpha\beta}}{\rho_j^2}\right)\frac{\partial W_{ij}}{\partial x^\beta} + f^\alpha \quad , \tag{37}$$

式中，密度 ρ，速度 v 和应力 σ 为因变量，空间坐标 x 和时间 t 是自变量。移动拉格朗日框架下的时间导数为 (d/dt)，上标 α，β 表示坐标方向。

对于质点 i，在关联塑性流动理论下，由式（26）可得

$$\frac{d\sigma_i^{\alpha\beta}}{dt} = \sigma_i^{\alpha\gamma}\dot{\omega}_i^{\beta\gamma} + \sigma_i^{\gamma\beta}\dot{\omega}_i^{\alpha\gamma} + 2G\dot{e}_i^{\alpha\beta} + K\dot{\varepsilon}_i^{\gamma\gamma}\delta_i^{\alpha\beta} -$$
$$\dot{\lambda}_i\left[3\alpha_\varphi K\delta_i^{\alpha\beta} + \frac{G}{\sqrt{J_2}}s_i^{\alpha\beta}\right] \quad . \tag{38}$$

由式（21）可得塑性乘子变化率为

$$\dot{\lambda}_i = \frac{3\alpha_\varphi K\dot{\varepsilon}_i^{\gamma\gamma} + (G/\sqrt{J_2})s_i^{\alpha\beta}\dot{\varepsilon}_i^{\alpha\beta}}{9\alpha_\varphi^2 K + G} \quad . \tag{39}$$

对于质点 i，在非关联塑性流动理论下，由(26)式可得

$$\frac{d\sigma_i^{\alpha\beta}}{dt} = \sigma_i^{\alpha\gamma}\dot{\omega}_i^{\beta\gamma} + \sigma_i^{\gamma\beta}\dot{\omega}_i^{\alpha\gamma} + 2G\dot{e}_i^{\alpha\beta} + K\dot{\varepsilon}_i^{\gamma\gamma}\delta_i^{\alpha\beta} - \dot{\lambda}\left[9K\sin\psi\delta_i^{\alpha\beta} + \frac{G}{\sqrt{J_2}}s_i^{\alpha\beta}\right], \quad (40)$$

式中，塑性乘子变化率由式（23）可得

$$\dot{\lambda}_i = \frac{3\alpha_\varphi K\dot{\varepsilon}_i^{\gamma\gamma} + (G/\sqrt{J_2})s_i^{\alpha\beta}\dot{\varepsilon}_i^{\alpha\beta}}{27\alpha_\varphi^2 K + G}. \quad (41)$$

由 GPD 法的一个基本数学变换，可得下列表达式

$$\frac{\partial v_i^\alpha}{\partial x^\beta} = \sum_{j=1}^N \frac{m_j}{\rho_j}(v_j^\alpha - v_i^\alpha)\cdot\frac{\partial W_{ij}}{\partial x_i^\beta}. \quad (42)$$

应变率张量式（6）和角应变率张量式（25）离散后，得到应变率 $\dot{\varepsilon}^{\alpha\beta}$ 和 $\dot{\omega}^{\alpha\beta}$ 的 GPD 离散形式为

$$\dot{\varepsilon}^{\alpha\beta} = \frac{1}{2}\left(\frac{\partial v^\alpha}{\partial x^\beta} + \frac{\partial v^\beta}{\partial x^\alpha}\right)$$
$$= \frac{1}{2}\left[\sum_{j=1}^N \frac{m_j}{\rho_j}(v_j^\alpha - v_i^\alpha)\cdot\frac{\partial W_{ij}}{\partial x_i^\beta} + \sum_{j=1}^N \frac{m_j}{\rho_j}(v_j^\beta - v_i^\beta)\cdot\frac{\partial W_{ij}}{\partial x_i^\alpha}\right], \quad (43)$$

$$\dot{\omega}^{\alpha\beta} = \frac{1}{2}\left(\frac{\partial v^\alpha}{\partial x^\beta} - \frac{\partial v^\beta}{\partial x^\alpha}\right)$$
$$= \frac{1}{2}\left[\sum_{j=1}^N \frac{m_j}{\rho_j}(v_j^\alpha - v_i^\alpha)\cdot\frac{\partial W_{ij}}{\partial x_i^\beta} - \sum_{j=1}^N \frac{m_j}{\rho_j}(v_j^\beta - v_i^\beta)\cdot\frac{\partial W_{ij}}{\partial x_i^\alpha}\right]. \quad (44)$$

2.2 人工黏度

为使得算法更稳定，人工黏度 Π_{ij} 引入动量方程用于解决非物理的数值振荡。因此，式（37）修正为[16, 25]

$$\frac{dv_i^\alpha}{dt} = \sum_{j=1}^N m_j\left[\left(\frac{\sigma_i^{\alpha\beta}}{\rho_i^2} + \frac{\sigma_j^{\alpha\beta}}{\rho_j^2}\right) - \Pi_{ij}\delta_i^{\alpha\beta}\right]\frac{\partial W_{ij}}{\partial x^\beta} + f_i^\alpha. \quad (45)$$

式中 当 $\alpha = \beta$ 时，$\delta^{\alpha\beta}=1$；当 $\alpha\neq\beta$ 时，$\delta^{\alpha\beta}=0$。

迄今为止，有许多种人工黏度项被提出，用于改进各种数值算法的稳定性和防止粒子穿透。Monaghan 型的人工黏度 Π_{ij} 是最为广泛使用的人工黏度，本文将采用 Monaghan 型的人工黏度，其具体表达式如下[25]：

$$\Pi_{ij} = \begin{cases} \dfrac{-\alpha_\Pi c_{ij}\phi_{ij} + \beta_\Pi\phi_{ij}^2}{\rho_{ij}} & (v_{ij}\cdot x_{ij}<0) \\ 0 & (v_{ij}\cdot x_{ij}\geq 0) \end{cases}, \quad (46)$$

式中 $\phi_{ij}=\dfrac{h_{ij}v_{ij}\cdot x_{ij}}{|x_{ij}|^2 + 0.01h_{ij}^2}$，$c_{ij}=\dfrac{1}{2}(c_i+c_j)$，$\rho_{ij}=\dfrac{1}{2}(\rho_i+\rho_j)$，$h_{ij}=\dfrac{1}{2}(h_i+h_j)$，$x_{ij}=x_i-x_j$，$v_{ij}=v_i-v_j$，$\alpha_\Pi$ 是体积黏度，而与 β_Π 相关的项是用来防止在高马赫数时粒子的相互穿透，一般取值在 1.0 左右；因子 $0.01h_{ij}^2$ 用于防止粒子相互靠近时产生的数值发散；c 和 v 分别表示固体中声速和粒子的速度。

3 巷道围岩的弹塑性区、应力场及位移场的数值模拟

3.1 数值计算模型

圆形巷道受力模型如图 3 所示。计算时作如下假设：①岩体是各向同性的；②巷道为深埋圆形平巷，承受的水平应力为垂直应力的 λ 倍，考虑到深埋巷道实际情况，忽略所取模型范围内的重力影响；③模型简化为平面应变问题；④塑性区内岩体满足 Drucker-Prager 强度准则。

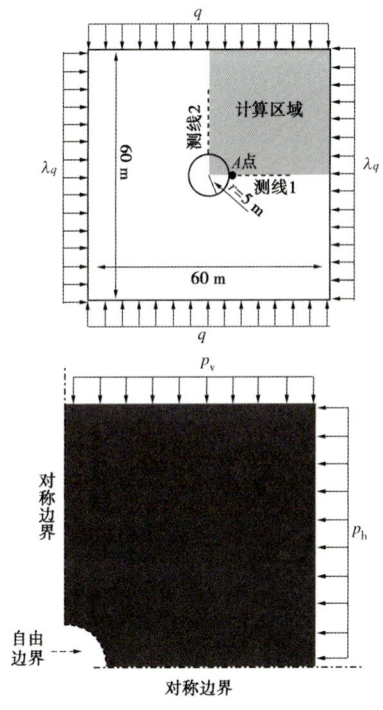

图 3 数值分析模型

Fig. 3 Numerical model

本文将 GPD 自编程序应用于巷道围岩的弹塑性分析，模拟圆形巷道的弹塑性区、应力场及位移场。如图 3 所示，计算模型的宽为 60 m，高为 60 m，巷道的直径为 10 m，采用无 GPD 粒子模拟巷道，类似有限元中空单元模拟巷道。由于研究模型的对称性，本文截取计算模型的四分之一进行模拟。阴影部分为计算区域，该模拟区域的长及高均为 30 m。岩体的密度 ρ 为 2 500 kg/m³、弹性模量 E、泊松比 ν、体积模量 K 为 2 GPa 和剪切模量 G 为 1 GPa，内摩擦角为 30°，黏结力 2 MPa，抗拉强度为 2 MPa。

将图 3 的阴影区模型通过 GPD 法离散为数值计算模型，计算模型的水平方向离散为 200 个 GPD 粒子，垂直方向离散为 200 个 GPD 粒子，挖去洞室 2 480 个 GPD 实粒子，岩石的宏观力学性质（如应力，应变，

· 037

位移）集中在每个 GPD 粒子上，模型共 36520（40000-2480）个 GPD 实粒子。计算模型受垂直方向地应力 $p_v=q$，水平方向地应力 $p_h=\lambda q$ 作用。共 5 个计算方案：①方案 1，p_v 和 p_h 分别为 10 MPa 和 5 MPa；②方案 2，p_v 和 p_h 10, 10 MPa；③方案 3，p_v 和 p_h 分别为 10, 15 MPa；④方案 4，p_v 和 p_h 分别为 10, 20 MPa；⑤方案 5，p_v 和 p_h 分别为 10, 25 MPa。

3.2 侧压系数对巷道围岩塑性区的影响

图 4 给出了 GPD 法模拟的塑性区大小和形状随侧压系数的变化云图。作为对比，图 5 给出了 FEM 法模拟的塑性区大小和形状随侧压系数的变化云图。表 1 分别列出了 GPD 法和 FEM 法洞室顶、底板与两帮塑性区半径与侧压系数 λ 之间的关系。

表 1 巷道围岩塑性区半径 R_p（m）

Table 1 Radii of plastic zones width of roof and floor and side wall around tunnel

侧压系数 λ	GPD		FEM	
	顶底	两帮	顶底	两帮
0.5	0.0	0.8	0.0	0.5
1.0	0.5	0.5	0.0	0.0
1.5	1.8	0.0	1.0	0.0
2.0	3.0	0.0	2.0	0.0
2.5	4.0	0.0	2.8	0.0

从图 4,5 及表 1 可以看出：①对于塑性区的形状和位置，当 $\lambda<1.0$ 时，塑性区对称分布于圆形巷道两帮，单侧帮呈"弦月"形；当 $\lambda=1.0$ 时，塑性区为圆形；当 $1.0<\lambda<2.0$ 时，塑性区主要对称分布于圆形巷道的顶部和底部，呈对称"弦月"形；当 $\lambda>2.0$ 时，塑性区向巷道两肩与两底角转移。②对于塑性区的分布规律，随水平地应力增大，塑性区逐渐由两帮向顶底板转移。顶底板塑性区的宽度随水平地应力的增加而增大。当侧压系数 $\lambda>2.0$ 以后，塑性区向肩部与底角迅速扩展，成为圆形巷道失稳的主要位置。③GPD 法塑性区结果和 FEM 法塑性区结果相比较，GPD 法塑性区形状、位置随侧压系数变化规律更接近实际工程中对围岩塑性区的描述，GPD 法模拟围岩塑性区具有合理性。

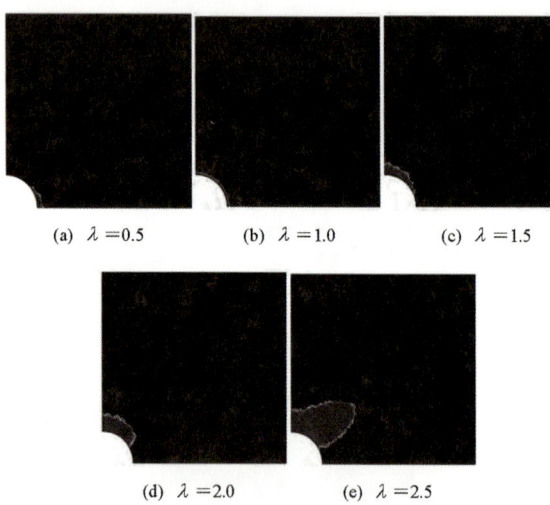

图 4 不同侧压系数下巷道围岩的塑性区分布（GPD 法）

Fig. 4 Plastic zones of surrounding rock around tunnel under different lateral pressure coefficients using GPD

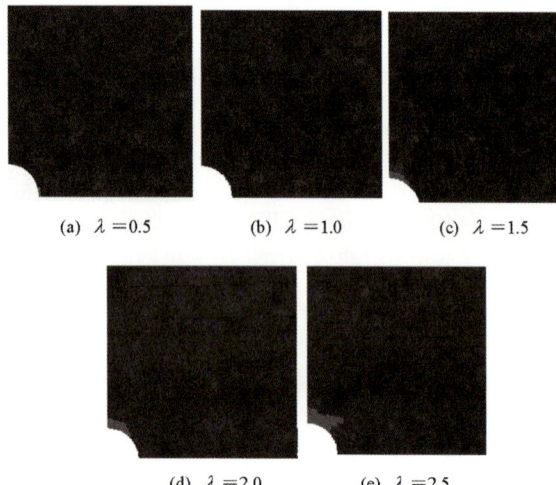

图 5 不同侧压系数下巷道围岩的塑性区分布（FEM 法）

Fig. 5 Plastic zones of surrounding rock around tunnel under different lateral pressure coefficients using FEM

3.3 侧压系数对巷道围岩应力场的影响

（1）不同侧压系数下径向应力和切向应力随半径的变化规律

假设垂直方向地应力 $q=10$ MPa，侧压系数 λ 分别为 0.5, 1.0, 1.5, 2.0, 2.5 时，图 6～10 分别给出巷道在 $\theta=0, \pi/2$ 时的径向应力 σ_r 及切向应力 σ_θ 随半径的变化规律。

如图 6 所示，当侧压系数 $\lambda=0.5$ 时，在巷道边墙范围内，圆形巷道围岩的切向应力 σ_θ 随离洞壁距离的增加而先增后减，在距洞壁 1 m 处，切向应力达到最大值 22.6 MPa；径向应力 σ_r 随离洞壁距离的增加而增大，在距洞壁 15 m 处接近原岩应力。

在巷道顶板范围内，圆形巷道围岩的切向应力 σ_θ 随离洞壁距离增加而先增后减，距洞壁 2 m 处切向应力 σ_θ 达到最大值 6.6 MPa；径向应力 σ_r 随离洞壁距离增加而增大；在距周边 16 m 左右接近原岩应力。从图 6 可以看出：巷道围岩应力场的 GPD 计算结果和有限元计算结果非常吻合。

如图 7 所示，当侧压系数 $\lambda=1$ 时，在巷道边墙、巷道顶板范围内，圆形巷道围岩的切向应力 σ_θ 随离洞壁距离增加而先增后减，在距洞壁 0.9 m 处有最大应力 19.1 MPa；径向应力 σ_r 随离洞壁距离增加而增大，在距洞壁 14 m 左右接近原岩应力。从图 7 可以看出：巷道围岩应力场的 GPD 计算结果和有限元计算结果非常吻合。

如图 8 所示，当侧压系数 $\lambda=1.5$ 时，圆形巷道边墙的切向应力 σ_θ 随离洞壁距离增加而先增后减，距洞壁 1.5 m 处切向应力达最大值为 14.5 MPa；径向应力 σ_r 随离洞壁距离增加而增大；在距洞壁 15 m 左右接近原岩应力。圆形巷道顶拱的切向应力 σ_θ 随离洞壁距离增加而先增后减，距洞壁 1.5 m 处切向应力达最大值为 29.1 MPa；径向应力 σ_r 随离洞壁距离增加而增加；在距洞壁 10 m 左右接近原岩应力。从图 8 可以看出：巷道围岩应力场的 GPD 计算结果和有限元计算结果非常吻合。

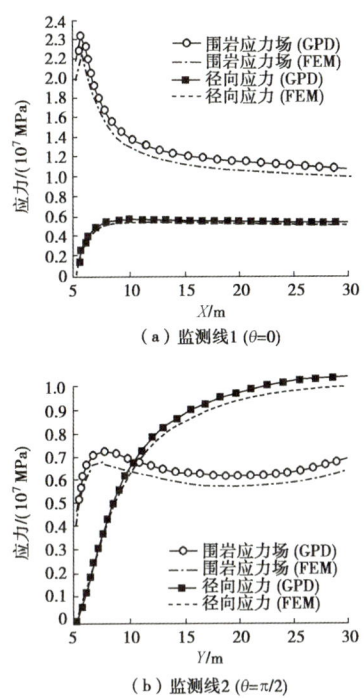

图 6 当 $\lambda=0.5$ 时不同监测线上应力分布规律

Fig. 6 Stress distribution along different monitoring lines when $\lambda=0.5$

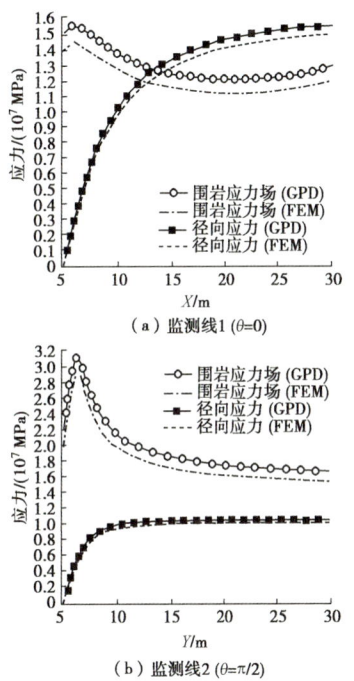

图 8 当 $\lambda=1.5$ 时不同监测线上应力分布规律

Fig. 8 Stress distribution along different monitoring lines when $\lambda=1.5$

如图 9 所示，当侧压系数 $\lambda=2.0$ 时，圆形巷道边墙的切向应力 σ_θ 随距洞壁距离增加而先增后减，距洞壁 2.5 m 处有最大应力 12.8 MPa；径向应力 σ_r 随离洞壁距离增加而增大；在距洞壁 20 m 左右接近原岩应力。圆形巷道顶板围岩的切向应力 σ_θ 随离洞壁距离增加而先增后减，距洞壁 2 m 处有最大应力 37.1 MPa；径向应力 σ_r 随离洞壁距离增加而增大，在距洞壁 12 m 左右接近原岩应力。

如图 10 所示，当侧压系数 $\lambda=2.5$ 时，圆形巷道边墙的切向应力 σ_θ 随离洞壁距离增加而增大；径向应力 σ_r 随离洞壁距离增加而增大；在距洞壁 12 m 左右接近原岩应力。圆形巷道顶板的切向应力 σ_θ 随离洞壁距离增加而先增后减，距洞壁 2.5 m 处切向应力达最大值为 46.5 MPa；径向应力 σ_r 随离洞壁距离增加而增

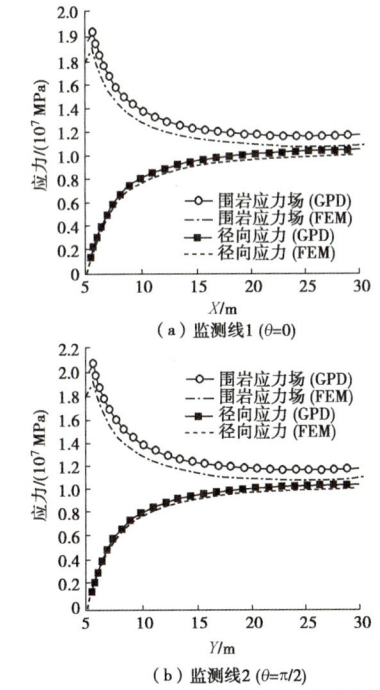

图 7 当 $\lambda=1.0$ 时不同监测线上应力分布规律

Fig. 7 Stress distribution along different monitoring lines when $\lambda=1.0$

大，在距洞壁 12 m 左右接近原岩应力。

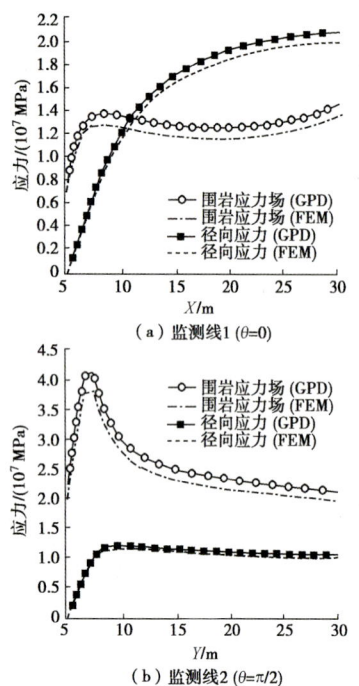

图 9 当 $\lambda = 2.0$ 时不同监测线上的应力分布规律

Fig. 9 Stress distribution along different monitoring lines when $\lambda = 2.0$

综上分析，在圆形巷道监测线 1（$\theta = 0$）及监测线 2（$\theta = \pi/2$）方向上，巷道的切向应力 σ_θ 随离洞壁距离增加而先增后减，最后达到原岩应力状态；径向应力 σ_r 随离洞壁距离的增加而增大，最后达到原岩应力状态。

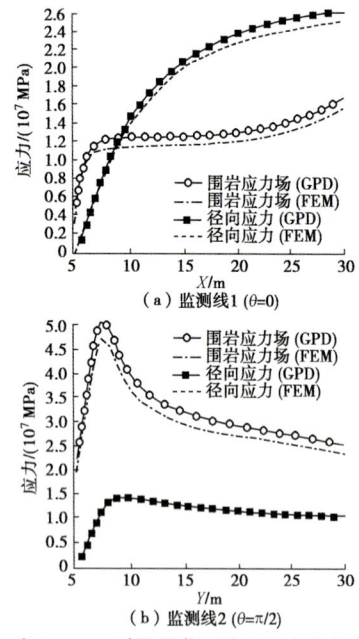

图 10 当 $\lambda = 2.5$ 时不同监测线上的应力分布规律

Fig. 10 Stress distribution along different monitoring lines when $\lambda = 2.5$

（2）不同侧压系数对巷道围岩应力场的影响

图 11～13 给出了不同侧压系数下，GPD 法巷道围岩的最大主应力、最小主应力、最大剪应力分布图和 FEM 法巷道围岩的最大主应力、最小主应力、最大剪应力分布图。表 2 给出了 GPD 法计算的不同侧压系数的最大主应力、最小主应力、最大剪应力的极值及应力集中系数。相应地，图 14，15 分别给出了上述 3 种应力值随侧压系数的变化规律。

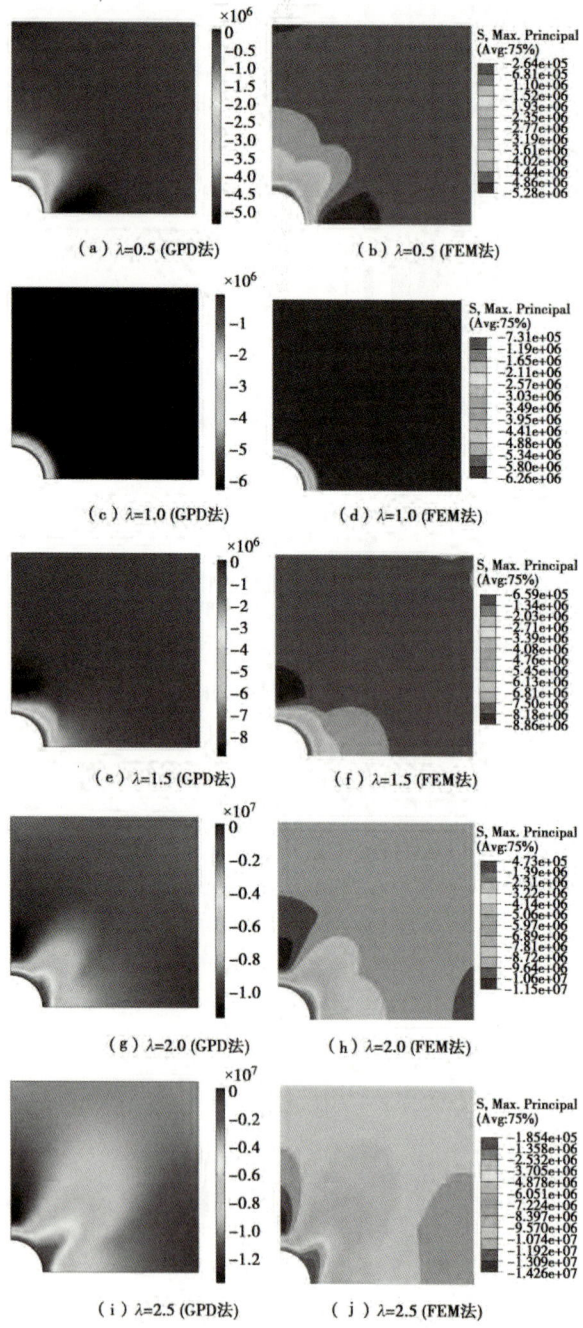

图 11 不同侧压系数下巷道围岩最大主应力 σ_1 分布云图

Fig. 11 Distribution of maximum principal stress of surrounding rock around tunnel at different lateral pressure coefficient

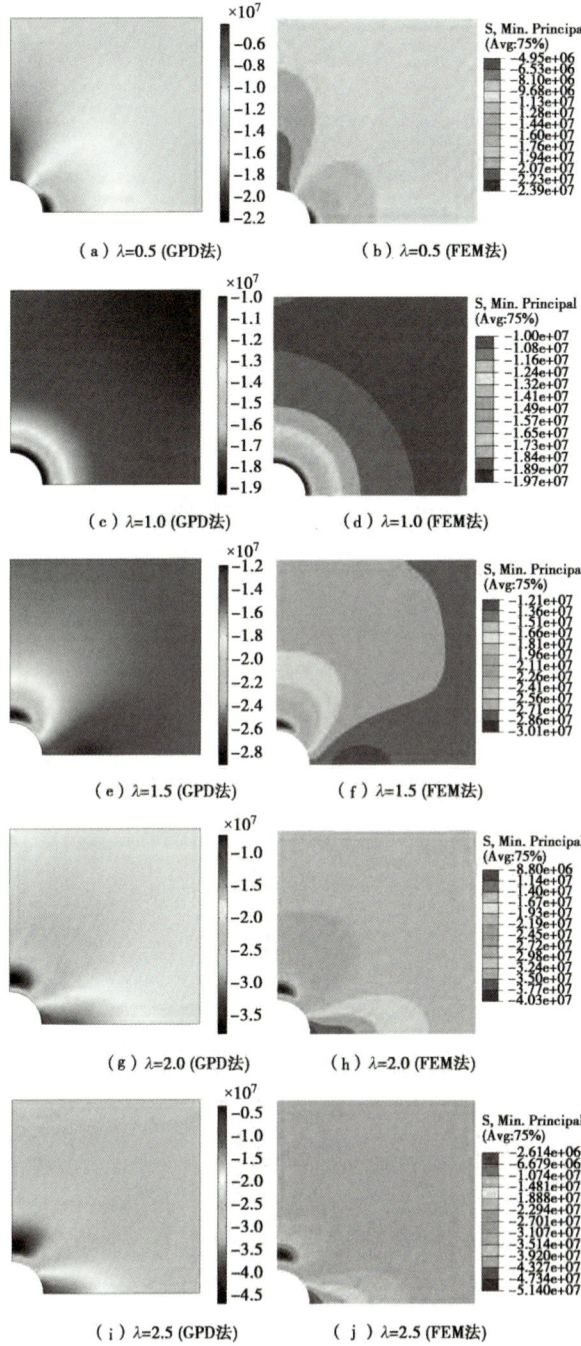

图 12 不同侧压系数下巷道围岩最小主应力 σ_3 分布云图

Fig. 12 Distribution of minimum principal stress of surrounding rock around tunnel under different lateral pressure coefficients

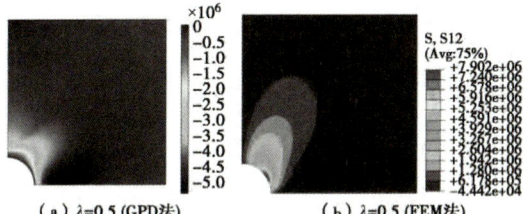

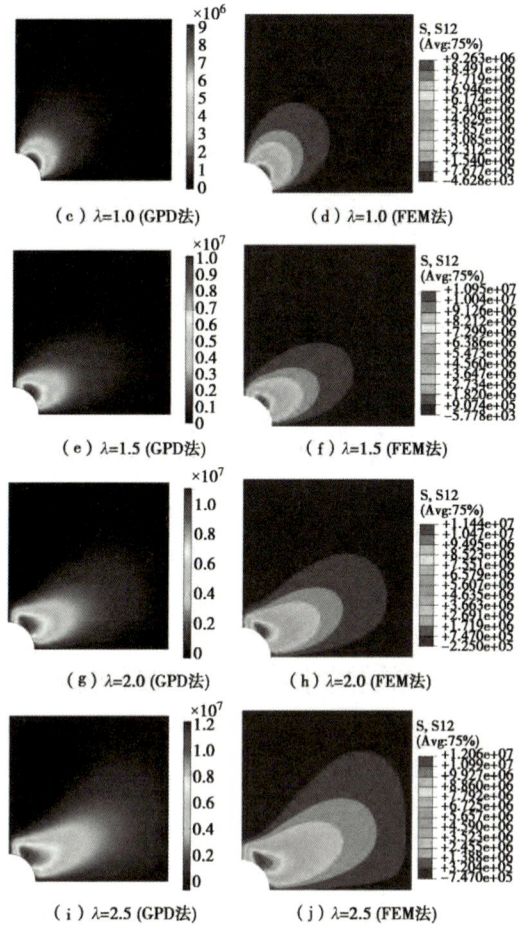

图 13 不同侧压系数下巷道围岩剪应力 σ_{12} 分布云图

Fig. 13 Distribution of shear stress of surrounding rock around tunnel under different lateral pressure coefficients

表 2 应力极值随侧压系数的变化规律

Table 2 Extreme values of principal stresses varying with different lateral pressure coefficients

侧压系数 λ	$\sigma_{1\max}$ /MPa	$\sigma_{3\max}$ /MPa	$\tau_{xy\max}$ /MPa	应力集中系数 β
0.5	−5.3	−24.0	7.9	2.26
1.0	−6.3	−19.7	9.3	1.94
1.5	−8.9	−30.1	11.0	1.89
2.0	−11.5	−40.3	11.4	1.85
2.5	−14.3	−51.4	12.1	1.88

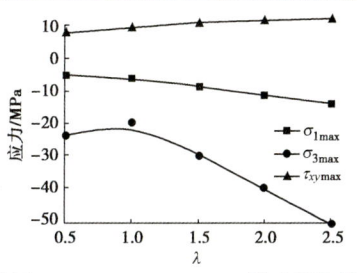

图 14 $\sigma_{1\max}$，$\sigma_{3\max}$，$\tau_{xy\max}$ 随 λ 变化规律

Fig. 14 Relationship between stress and lateral pressure coefficient

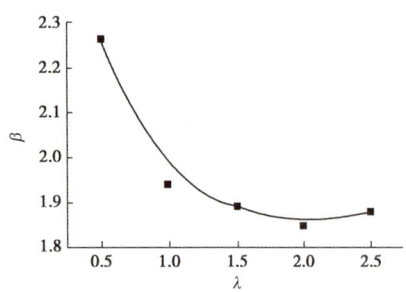

图 15 应力集中系数 β 随 λ 变化规律

Fig. 15 Relationship between stress concentration factor and lateral pressure coefficient

从图 14，15 及表 3 可以看出：①当 $\lambda < 1.0$ 时，最小主应力的极值 $\sigma_{3\max}$ 随侧压系数增大而增大；当 $\lambda > 1.0$ 时，最小主应力的极值 $\sigma_{3\max}$ 随侧压系数增大而减小。②最大主应力的极值 $\sigma_{1\max}$ 随侧压系数的增大而增大；剪应力 $\tau_{xy\max}$ 也随侧压系数的增大而增加。③当 $\lambda < 2.0$ 时，应力集中系数 β 随侧压系数 λ 的增大而减小；当 $\lambda > 2.0$ 时，应力集中系数 β 随侧压系数 λ 的增大而增大。④当 $\lambda = 2.0$ 时，应力集中系数 β 最小。⑤使用广义粒子动力学数值模拟方法计算的巷道围岩应力场分布规律和有限元法一致。

3.4 不同侧压系数对巷道围岩变形特征的影响

图 16～18 给出了不同侧压系数下巷道围岩水平位移、垂直位移及总位移的分布规律。从图 16～18 中可以看出：①随着侧压系数 λ 的增大，巷道顶板和边墙的水平位移增大。当 $\lambda = 0.5$ 时，巷道顶底板的水平位移接近于边墙的水平位移值。②随着侧压系数 λ 的增大，顶部、帮部垂直位移都在均匀增大。当 $\lambda = 2.0$ 时候，巷道顶底板的水平位移接近于边墙的水平位移值。③随着侧压系数 λ 的增大，总位移不断增大。当 $\lambda = 1.0$ 时，巷道顶底板总位移和巷道边墙的总位移值相当。当 $\lambda < 1.0$ 时，巷道顶底板总位移大于巷道边墙；当 $\lambda > 1.0$ 时，巷道边墙总位移大于巷道顶底板。④使用广义粒子动力学数值模拟方法计算的巷道围岩位移分布规律和有限元法一致。

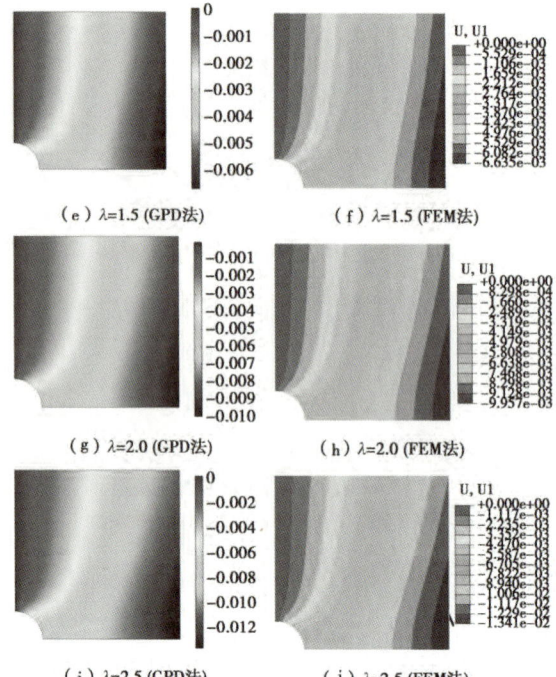

图 16 不同侧压系数下巷道围岩的水平方向位移分布图

Fig. 16 Distribution of horizontal displacement of surrounding rock around tunnel under different lateral pressure coefficient

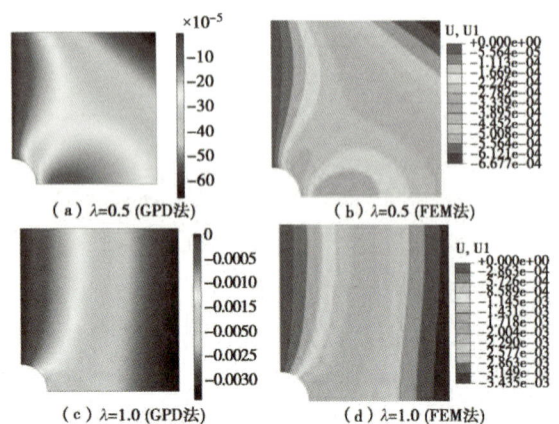

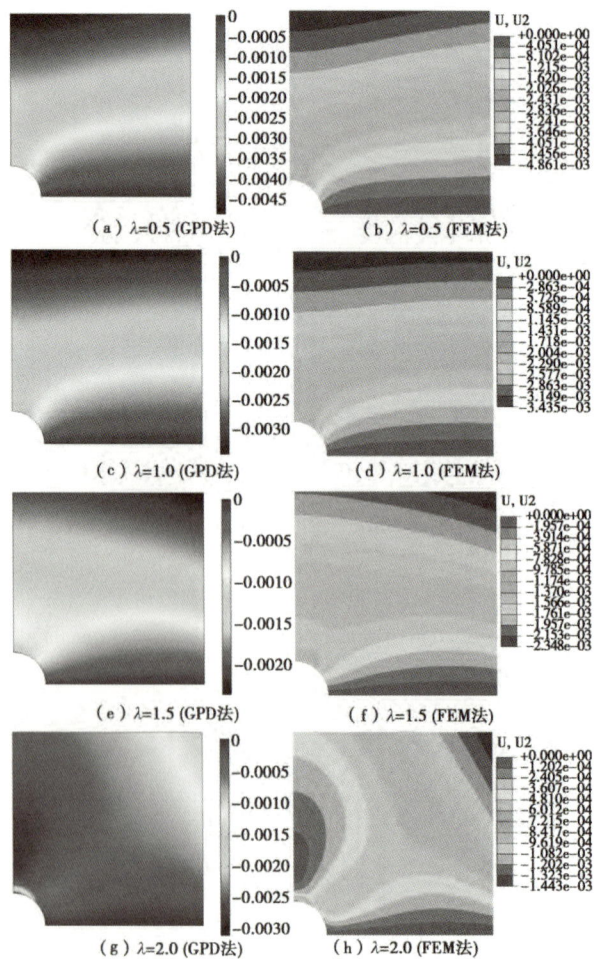

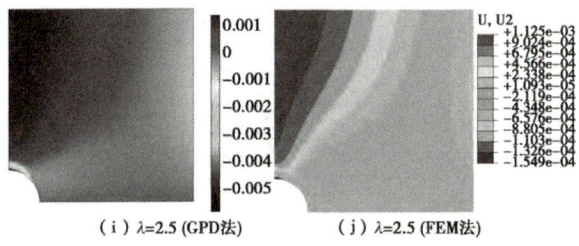

图 17 不同侧压系数下巷道的垂直方向位移云图

Fig. 17 Distribution of vertical displacement of the surrounding rock around tunnel under different lateral pressure coefficients

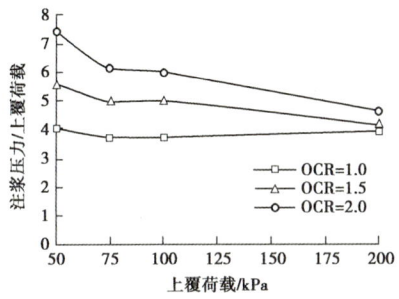

图 18 不同侧压系数下巷道围岩的总位移云图

Fig. 18 Distribution of total displacement of surrounding rock around tunnel under different lateral pressure coefficients

3.5 GPD 法的精度和收敛性研究

本文分别对有限元法（FEM 法）和 GPD 法模拟垂直方向地应力 q=10 MPa，侧压系数 λ 为 1 的巷道围岩等效应力的精确度和收敛性进行研究。采用 FEM 法模拟巷道围岩应力时，采用 8 节点二次减缩积分四边形网格。采用 GPD 法模拟巷道围岩应力时，GPD 粒子与 FEM 法的网格采用同样的尺度（FEM 法的网格取 0.3 m）。对比如下：

图 19 为图 3 中 A 点（图 3 测线 1 上距离洞壁 1 m 处）的瞬态和稳态阶段等效应力曲线图，GPD 法和 FEM 法的数值结果基本一致。在瞬态下（如图 19（a）），FEM 法的数值结果有明显的震荡，而 GPD 法模拟结果更平滑且没有震荡，这表明 GPD 法在捕捉结构瞬态力学响应时更稳定。

由于 GPD 法较为新颖，在结构应力分析方面的研究成果并不多，有必要研究 GPD 法的收敛性。采用 3 种 GPD 粒子精度（0.3，0.375，0.5 m）分别对图 3 中 A 点计算，将模拟的等效应力场结果与 0.3 m 网格精度的 FEM 法模拟结果进行对比分析。由图 20 可见，采用不同粒子精度的 GPD 法模拟巷道围岩应力时，应力和时间是线性关系，均未出现不稳定。

由 GPD 法和 FEM 法模拟巷道围岩应力场的对比可以看出，GPD 法和 FEM 法模拟结果比较一致，且 GPD 法中 GPD 粒子越密，模拟结果越精确。

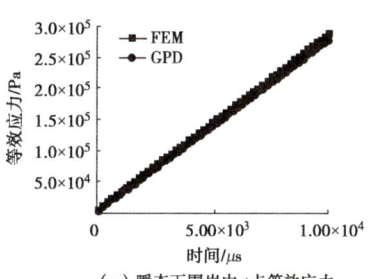

（a）瞬态下围岩中 A 点等效应力

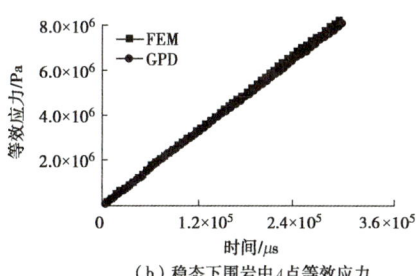

（b）稳态下围岩中 A 点等效应力

图 19 GPD 法模拟结果精度分析

Fig. 19 Accuracy studies by using GPD

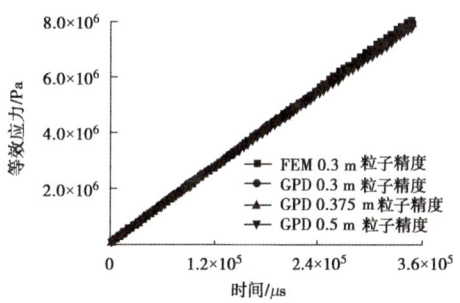

（a）不同粒子精度的GPD法和FEM法等效应力数值结果对比

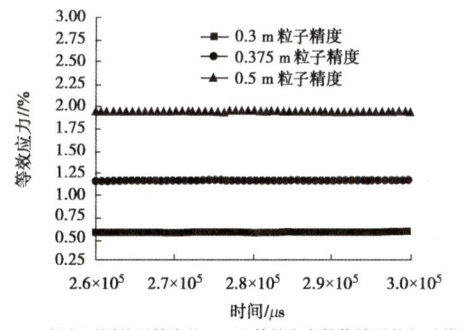

（b）不同粒子精度的GPD法等效应力数值结果的相对偏差

图 20 GPD 法数值结果收敛性的分析

Fig. 20 Convergence studies by using GPD

3.6 考虑剪胀特性的巷道围岩特征

岩石材料受剪应力作用时会出现颗粒错动，往往会产生塑性体积变形，岩石不可避免出现剪胀特性。上文中采用非关联塑性流动模型的 GPD 法来描述岩石材料的剪胀特性，但剪胀角为 0。在非关联塑性流动模型中，剪胀角为 0 的物理意义是该岩石材料具有塑性不可压缩特性。本节中，对不同剪胀角情况下的巷道围岩位移和塑形区的情况分别进行讨论。

在模拟直径 $D=10$ m 的圆形巷道时（图 3），在保持其他条件不变的情况下只改变剪胀角的值。图 21 表示垂直方向地应力 $q=10$ MPa，侧压系数 λ 为 0.5 时圆形巷道围岩发生塑性变形的情况，其最大位移出现在圆形巷道顶部，随着剪胀角 ψ 的增加，巷道围岩的位移也增加。当 $\psi=0°$ 时，最大位移 $d_{max}=0.45$ cm；当 $\psi=10°$ 时，$d_{max}=0.80$ cm；当 $\psi=25°$ 时，$d_{max}=1.2$ cm。同时，由于剪胀角的增大，使围岩塑性区的范围逐渐增大（如图 22（a）～（c）），宏观表现为围岩承载能力的提高。

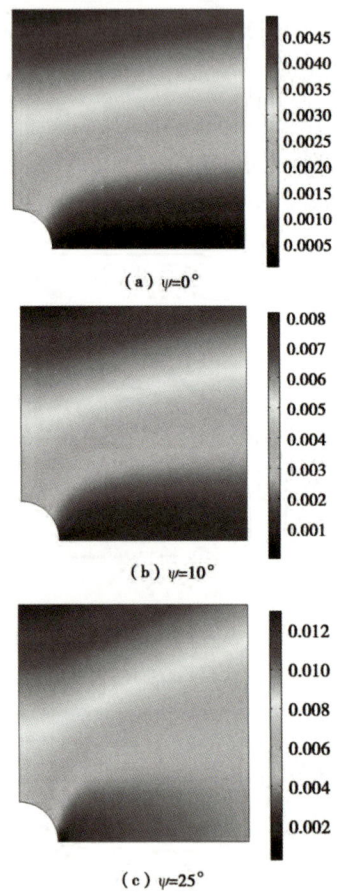

图 21 不同膨胀角下 GPD 法巷道围岩位移的数值结果

Fig. 21 Displacement contours of surrounding rock around tunnel under different dilatancy angles by using GPD

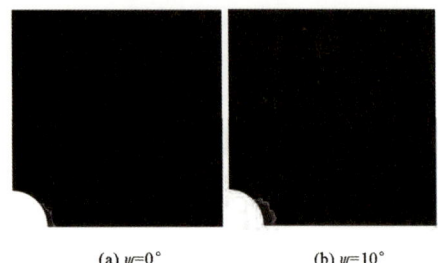

图 22 不同膨胀角下 GPD 法巷道围岩塑形区的数值结果

Fig. 22 Plastic zone of surrounding rock around tunnel under different dilatancy angles by using GPD

4 结 语

本文提出了一种无网格数值模拟方法，即广义粒子动力学数值分析方法，可以考虑剪胀角对岩石材料塑性变形的影响，能模拟岩石材料的弹塑性变形特性。本文将广义粒子动力学数值分析方法应用于不同侧压系数下巷道围岩的弹塑性分析，确定了不同侧压系数下巷道围岩的应力场、位移场和塑性区。本文的数值模拟结果和有限元结果比较吻合。表明将考虑岩石材料剪胀特性的弹塑性本构理论引入到广义粒子动力学数值分析方法，不失为模拟岩石类材料弹塑性破坏的一种有效数值手段，研究结果为更好地理解岩石材料的屈服破坏过程提供重要的参考。

参考文献：

[1] 潘 岳, 王志强. 基于应变非线性软化的圆形硐室围岩弹塑性分析[J]. 岩石力学与工程学报, 2005, **24**(6): 915－920. (PAN Yue, WANG Zhi-qiang. Elastoplastic analysis of surrounding rock of circular chamber based on strain nonlinear softening[J]. Chinese Journal of Rock Mechanics and Engineering, 2005, **24**(6): 915－920. (in Chinese))

[2] CARRANZA-TORRES C, FAIRHURST C. The elastoplastic response of underground excavations in rock masses that satisfy the Hoek-Brown failure criterion[J]. International Journal of Rock Mechanics and Mining Sciences, 1999, **36**(6): 777－809.

[3] WANG Y. Ground response of circular tunnel in poorly consolidated rock[J]. Journal of Geotechnical Engineering, 1996, **122**(9): 703－708.

[4] PARK K H, KIM Y J. Analytical solution for a circular opening in an elastic-brittle-plastic rock[J]. International Journal of Rock Mechanics and Mining Sciences, 2006, **43**(4): 616－622.

[5] SHARAN S K. Elastic-brittle-plastic analysis of circular openings in Hoek-Brown media[J]. International Journal of

Rock Mechanics and Mining Sciences, 2003, **40**(6): 817–824.

[6] SHARAN S K. Analytical solutions for stresses and displacements around a circular opening in a generalized Hoek-Brown rock[J]. International Journal of Rock Mechanics and Mining Sciences, 2008, **40**(1): 78–85.

[7] 蒋斌松, 张 强, 贺永年, 等. 深部圆形巷道破裂围岩的弹塑性分析[J]. 岩石力学与工程学报, 2007, **26**(5): 982–986. (JIANG Bin-song, ZHANG Qiang, HE Yong-nian, et al. Elastoplastic analysis of cracked surrounding rocks in deep circular openings[J]. Chinese Journal of Rock Mechanics and Engineering, 2007, **26**(5): 982–986. (in Chinese))

[8] BROWN E T, BRAY J W, LADANYI B, et al. Ground response curves for rock tunnels[J]. Journal of Geotechnical Engineering, 1983, **109**(1): 15–39.

[9] WANG S L, YIN X T, TANG H, et al. A new approach for analyzing circular tunnel in strain-softening rock masses[J]. International Journal of Rock Mechanics and Mining Sciences, 2010, **47**(1): 170–178.

[10] LEE Y K, PIETRUSZCZAK S. A new numerical procedure for elastoplastic analysis of a circular opening excavated in a strain-softening rock mass[J]. Tunnelling and Underground Space Technology, 2008, **23**(5): 588–599.

[11] PARK K H, TONTAVANICH B, LEE J G. A simple procedure for ground response curve of circular tunnel in elastic-strain softening rock masses[J]. Tunnelling and Underground Space Technology, 2008, **23**(2): 151–159.

[12] PREVOST J H, HUGHES T J R. Finite element solution of elastic-plastic boundary value problems[J]. Journal of Appied Mechanics, 1984, **48**: 69–74.

[13] ZHOU X P, BI J, QIAN Q H. Numerical simulation of crack growth and coalescence in rock-like materials containing multiple pre-existing flaws[J]. Rock Mechanics and Rock Engineering, 2015, **48**(3): 1097–1114.

[14] ZHOU X P, ZHAO Y, QIAN Q H. A novel meshless numerical method for modeling progressive failure processes of slopes[J]. Engineering Geology, 2015, **192**(18): 139–153.

[15] GINGOLD R A, MONAGHAN J J. Smoothed particle hydrodynamics: theory an application to non-spherical stars[J]. Mon Not R Astron Soc, 1977, **181**: 375–389.

[16] MONAGHAN J J, LATTANZIO J C. A refined particle method for astrophysical problems[J]. Astron Astrophys, 1985, **149**(1): 135–143.

[17] LIBERSKY L D, PETSCHEK A G, CARNEY T C, et al. High strain Lagrangian hydrodynamics a three-dimensional SPH code for dynamic material response[J]. J Comput Phys, 1993, **109**(1): 67–75.

[18] LIBERSKY L D, PETSCHEK A G. Smoothed particle hydrodynamics with strength of materials[J]. Advances in the Free Lagrange Method Lecture Notes in Physics, 1990, **395**: 248–257.

[19] MONAGHAN J J. Simulating free surface flows with SPH[J]. J Comput Phys, 1994, **110**(2): 399–406.

[20] YU M H, LI J C. Computatioal plastictiy: with emphasis on the application of the unified strength theory and associated flow rule[M]. Berlia: Springer, 2012.

[21] VERMEER P A, DE B R. Non-associated plasticity for soils, concrete and rock[J]. Heron, 1984, **29**(3): 1–65.

[22] 范 文, 俞茂宏, 陈立伟. 考虑材料剪胀及软化的有压隧洞弹塑性分析的解析解[J]. 工程力学, 2004, **21**(5): 16–24. (FAN Wen, YU Mao-hong, CHEN Li-wei. An analytic solution of elastoplastic pressure tunnel considering material softening and dilatancy[J]. Engineering Mechanics, 2004, **21**(5): 16–24. (in Chinese))

[23] 邓楚键, 郑颖人, 王 凯, 等. 有关岩土材料剪胀的讨论[J]. 岩土工程学报, 2009, **31**(7): 1110–1114. (DENG Chu-jian, ZHENG Ying-ren, WANG Kai, et al. Some discussion on the dilatancy of geotechnical materials[J]. Chinese Journal of Geotechnical Engineering, 2009, **31**(7): 1110–1114. (in Chinese))

[24] 张培文, 陈祖煜. 剪胀角对求解边坡稳定的安全系数的影响[J]. 岩土力学, 2004, **25**(11): 1757–1760. (ZHANG Pei-wen, CHEN Zu-yu. Finite element method for solving safety factor of slope stability[J]. Rock and Soil Mechanics, 2004, **25**(11): 1757–1760. (in Chinese))

[25] MONAGHAN J J. On the problem of penetration in particle methods[J]. Journal of Computational Physics, 1989, **82**: 1–15.

单轴压缩条件下岩石破坏的光滑粒子流体动力学数值模拟

周小平[1, 2, 3]，赵 毅[1]，钱七虎[4]

(1. 重庆大学 土木工程学院，重庆 400045；2. 重庆大学 山地城镇建设新技术教育部重点实验室，重庆 400045；3. 重庆大学 煤矿灾害动力学与控制国家重点实验室，重庆 400044；4. 解放军理工大学 国防工程学院，江苏 南京 210007)

摘要： 岩石破坏问题是非连续问题。采用传统的有限元方法模拟岩石破坏时，计算结果依赖于网格，计算效率低。光滑粒子流体动力学(SPH)法主是一种模拟流体的无网格方法。对 SPH 法进行改进，将 SPH 法中流体的本构关系修正为弹脆性固体的本构关系，采用 Weibull 统计方法描述岩石材料的非均匀性，使 SPH 法能有效地模拟各向异性弹脆性岩石的破坏。改进的 SPH 法克服了有限元的缺点，在模拟裂纹启裂、扩展和连接时，计算结果不依赖网格，计算效率高。通过对单轴压缩条件下岩石破坏的数值计算结果表明：改进的 SPH 法能有效地理解和预测岩石材料的复杂破裂过程。

关键词： 岩石力学；光滑粒子法；单轴压缩；岩石类材料

中图分类号： TU 45　　**文献标识码：** A　　**文章编号：** 1000 – 6915(2015)增 1 – 2647 – 12

NUMERICAL SIMULATION OF ROCK FAILURE PROCESS IN UNIAXIAL COMPRESSION USING SMOOTHED PARTICLE HYDRODYNAMICS

ZHOU Xiaoping[1, 2, 3], ZHAO Yi[1], QIAN Qihu[4]

(1. School of Civil Engineering, Chongqing University, Chongqing 400045, China; 2. Key Laboratory of New Technology for Construction of Cities in Mountain Area, Chongqing University, Chongqing 400045, China; 3. State Key Laboratory of Coal Mine Disaster Dynamics and Control, Chongqing University, Chongqing 400044, China; 4. Engineering Institute of National Defense Engineering, PLA University of Science and Technology, Nanjing, Jiangsu 210007, China)

Abstract: The problem of rock failure is discontinuous one. The numerical results depend on meshes and computational efficiency is low when finite element method is applied to simulate the failure of rock. Standard smoothed particle hydrodynamics(SPH) method is mesh-free numerical one, which is mainly applied to model the problem of fluid. In this paper, a corrected smoothed particle hydrodynamics, in which constitutive relation of fluid is replaced by constitutive relation of elasto-brittle solid and Weibull statistical approach is used to describe the heterogeneity of the rock-like materials, is developed to simulate the failure of heterogeneous elasto-brittle rock. The corrected smoothed particle hydrodynamics overcomes the shortcomings of finite element method. The numerical results is independence of meshes and computational efficiency is high when the corrected smoothed particle hydrodynamics is applied to simulate growth and coalescence of cracks. The corrected smoothed particle

---

收稿日期：2013–09–26；修回日期：2014–04–15。
基金项目：国家重点基础研究发展计划(973)项目(2014CB046903)；教育部博士点基金项目(20130191110037)；重庆市院士基金项目(cstc2013jcyjys0005)。
本文原载于《岩石力学与工程学报》(2015 年第 34 卷增 1 期)。

hydrodynamics is helpful to understanding and predicting complex fracture processes of rock-like materials.

Key words: rock mechanics; smoothed particle hydrodynamics(SPH) method; compressive failure process; rock-like material

1 引 言

岩石是自然界中最复杂的固体材料之一，它一般是由多种矿物晶体、胶结物及孔隙组成的混合体，其结构呈现非连续、非均匀、非线性和各向异性等特征，岩石变形破坏过程的实质是岩石材料中缺陷的萌生、扩展、相互作用和贯通的过程[1]。已有研究表明，岩石的宏观力学性质及变形破坏机制与裂纹的萌生、扩展和贯通息息相关。

虽然岩石损伤断裂机制的理论研究取得了较大进展，但是数值模拟岩石断裂过程依然是研究的难点。基于传统网格方法(如有限元)模拟时，需要将裂纹面设置成单元的边，裂尖设置成单元的节点，网格重划分等繁琐的前处理，因此预测裂纹扩展路径会受网格划分精度的影响；离散元方法(DEM)中人工胶结处理在模拟材料断裂过程也会出现同样的困难。

光滑粒子动力学(SPH)法是拉格朗日公式与粒子近似法的和谐结合，最初是用于解决三维开放空间天体物理学问题[2-3]，然后被广泛地应用于工程和科学的不同领域，如流体动力学，分子动力学以及固体力学等[4-6]。

在模拟岩石类材料变形破坏过程时，SPH法相对于基于连续介质力学的传统数值方法有诸多优点。首先，固体在极限变形的情况下，也会出现类似流体流动的大变形，SPH法自适应的特点易模拟固体大变形、冲击和爆破等。其次，在膨胀应变场的作用下，SPH粒子间距变远失去相互作用，会自适应出现损伤和断裂，断裂带自动产生。最后，作为一种插值算法，微观结构和缺陷可利用粒子的物理量来表征，SPH法对模拟非均匀材料中裂纹的启裂、扩展和连接非常有效[7]。

本文引入SPH法模拟非均质岩石类材料的破坏模式，利用Weibull分布来考虑岩石非均匀性，建立岩石类材料的弹脆性破坏模型。数值计算结果与实验结果对比分析，表明自主开发的SPH程序模拟岩石变形破坏的有效性。

2 SPH法的基本方程

SPH法作为一种纯拉格朗日形式的无网格粒子法，是用于获取偏微分方程解的数值方法，模型通常被离散成占有一定体积的粒子。该方法原本是用于模拟流体以及热学问题，近年来，将该方法用于固体大变形问题，并逐渐成为热点。

SPH法的场函数及其导数的核近似式是在连续域内实现的，应用当前计算步的支持域内的离散化粒子将连续积分表达式近似转化为粒子离散化求和形式。粒子j为粒子i支持域W内的所有粒子(见图1)，粒子i处的离散化粒子场量函数$f(x)$及其梯度形式$\nabla f(x)$表达式为

$$f(x)_{x=x_i} = \sum_{j=1}^{N} \frac{m_j}{\rho_j} f(x_j) W_{x_i}(x_i - x_j, h) \tag{1}$$

$$\nabla f(x)_{x=x_i} = -\sum_{j=1}^{N} \frac{m_j}{\rho_j} [f(x_i) - f(x_j)] \nabla W_{x_i}(x_i - x_j, h) \tag{2}$$

式中：$W_{x_i}(x_i - x_j, h)$为核近似函数，常用的光滑核函数有3次样条函数，高斯型核函数和5次样条函数等；h为光滑长度，是核函数W支持域的度量；N为粒子i的支持域内的粒子总数；m, ρ分别为粒子质量和密度。

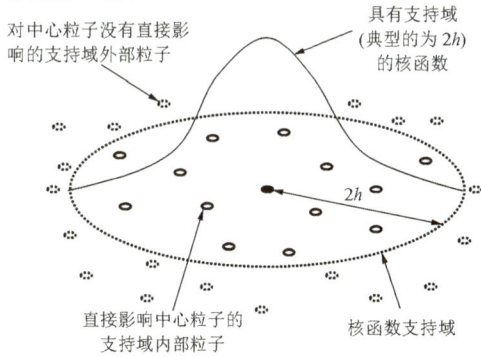

图1 核函数和支持域示意图

Fig.1 Kernel function and the supporting domain

连续介质力学中的质量和动量守恒方程经过SPH法处理后为

$$\frac{d\rho_i}{dt} = \rho_i \sum_{j=1}^{N} \frac{m_j}{\rho_j} [(v_\alpha)_i - (v_\alpha)_j] \frac{\partial W_{ij}}{\partial (x_\alpha)_i} \tag{3}$$

$$\frac{d(v_\alpha)_i}{dt} = \sum_{j=1}^{N} m_j \left[\frac{(\sigma_{\alpha\beta})_i}{\rho_i^2} + \frac{(\sigma_{\alpha\beta})_j}{\rho_j^2} + \Pi_{ij}\delta_{\alpha\beta} \right] \frac{\partial W_{ij}}{\partial (x_\beta)_i} \quad (4)$$

式中：v_α 为速度；$\sigma_{\alpha\beta}$ 为应力，是因变量；x_α 为空间坐标；t_α 为时间。移动拉格朗日框架下的时间导数算子为 (d/dt)，下标 "α"，"β" 表示坐标方向。当 $\alpha = \beta$ 时，$\sigma_{\alpha\beta} = 1$；当 $\alpha \neq \beta$ 时，$\sigma_{\alpha\beta} = 0$。σ 为负时表示受压。人工黏度 Π_{ij} 用于解决非物理的数值振荡[3]。

3 岩石单轴压缩力学模型

3.1 试验模型

如图 2 所示，SPH 法模拟岩石单轴压缩的力学模型，假设该模型为平面应变问题，岩样的几何尺寸为宽 50 mm，高 100 mm，上边界以 1.5 mm/s 匀速位移加载，下边界固定，左右两边为自由边界。试样的密度为 3 200 kg/m³，弹性模量 $E = 10$ GPa，泊松比 $\mu = 0.25$，体积模量 $K = 6.7$ GPa，剪切模量 $G = 4$ GPa。

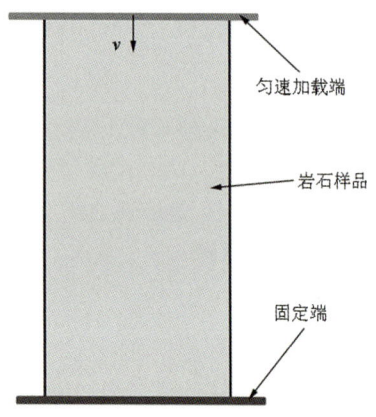

图 2 岩石单轴压缩模型
Fig.2 Rock specimen subjected to uniaxial compression

将力学模型通过 SPH 法离散为数值模型，试样水平方向离散为 50 个粒子，垂直方向离散为 100 个粒子，模型共 5 000 个实粒子。

式(4)中应力张量 σ_{ij} 由两部分组成：一部分是各向同性压力 P，另一部分是黏性剪切应力 τ_{ij}：

$$\sigma_{ij} = -P\delta_{ij} + \tau_{ij} \quad (5)$$

在固体力学中，其基本模型一般允许应力为应变和应变率的函数。对于各向异性剪切应力，若假设为小位移，则应力率与应变率互成比例，比例系数为剪切模量，即

$$\dot{\tau}_{ij} = 2G\bar{\dot{\varepsilon}}_{ij} = 2G(\dot{\varepsilon}_{ij} - \frac{1}{3}\delta_{ij}\dot{\varepsilon}_{kk}) \quad (6)$$

式中：G 为剪切模量，$\dot{\tau}_{ij}$ 为应力率，$\bar{\dot{\varepsilon}}_{ij}$ 为 $\dot{\varepsilon}_{ij}$ 的剪切变形部分，应变率 $\dot{\varepsilon}_{ij}$ 定义为

$$\dot{\varepsilon}_{ij} = \frac{1}{2}\left(\frac{\partial v_i}{\partial x_j} + \frac{\partial v_j}{\partial x_i}\right) \quad (7)$$

在 SPH 法中，空间连续体的变形、运动均是通过离散的质点运动进行描述，采用 Jaumann 应力率，在应力率张量与应变率张量之间建立的合理关系如下：

$$\dot{\tau}_{ij} - \tau_{ik}\dot{\omega}_{jk} - \tau_{kj}\dot{\omega}_{ik} = 2G\bar{\dot{\varepsilon}}_{ij} \quad (8)$$

式中：$\dot{\tau}_{ij}$ 为 Cauchy 应力率张量，$\dot{\tau}_{ij} = d\tau_{ij}/dt$，$\dot{\omega}_{ij}$ 为扭转率张量，定义为

$$\dot{\omega}_{ij} = \frac{1}{2}\left(\frac{\partial v_i}{\partial x_j} - \frac{\partial v_j}{\partial x_i}\right) \quad (9)$$

SPH 法不同光滑核函数对数值模拟结果影响不大，为此本文采用比近似高斯核函数更加稳定的五次样条函数进行数值模拟。同时，采用 Monaghan 型人工黏度修正避免非物理的数值振荡。搜索影响域粒子采用计算效率较高的链表搜索法，利用蛙跳法进行积分，时间步长为 0.3 μs。

3.2 边界处理

因而 SPH 法较其他数值方法在边界条件的处理上有自身的特点，是研究的热点。作为基于插值的 SPH 法，其边界的处理方法主要是虚粒子的补位，现有的虚粒子主要有镜像虚粒子等[8-15]。本文采用镜像虚粒子的办法处理边界。

3.3 von Mises 等效应力稳定性

数值手段要能有效、准确地预测岩石弹性阶段的瞬态应力场，然后才可用来模拟压缩条件下岩石的脆性断裂。通过试样早期等效应力在岩样中的传播考察 SPH 法框架下岩石单轴压缩模型的有效性和稳定性。

采用 von Mises 等效应力[16]分析岩样中的应力场和弹性波的力学响应。图 3 为匀速载荷单轴压缩模型中等效应力波传播情况。在图 4(a)~(c)中，等效应力波从模型上端产生，沿模型向下传播，到达固定的刚性下端面发生反射；在图 4(d)~(f)中，等效应力波在底端反射后沿模型向上传播，与上端新产生的入射等效应力波发生干涉，于是模型中会第一次出现入射应力波和反射应力波相叠加的应力波

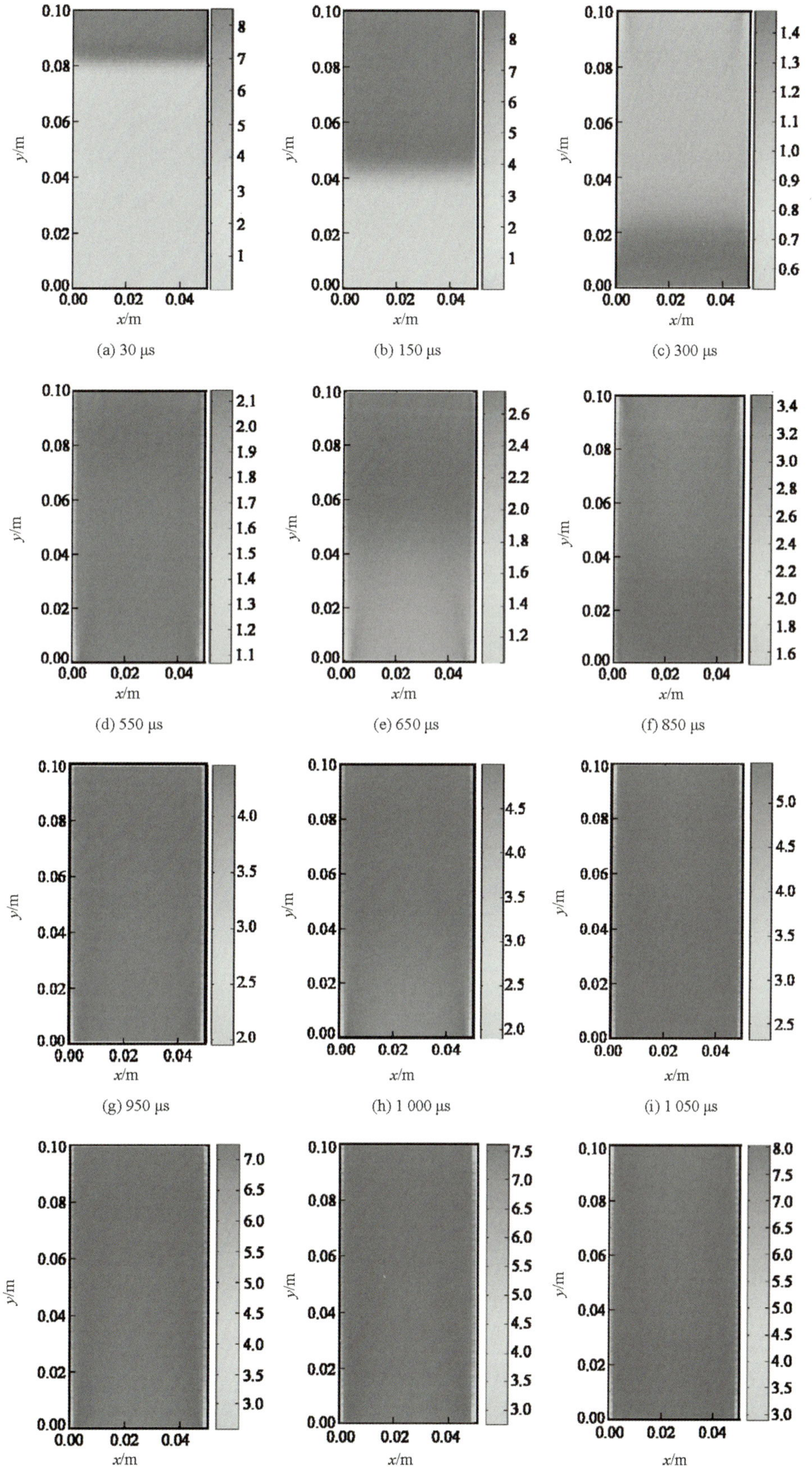

(a) 30 μs (b) 150 μs (c) 300 μs

(d) 550 μs (e) 650 μs (f) 850 μs

(g) 950 μs (h) 1 000 μs (i) 1 050 μs

(j) 1 100 μs (k) 1 200 μs (l) 1 350 μs

图 3　匀速载荷单轴压缩模型中等效应力波传播情况(单位：10^4 Pa)

Fig.3　Stress wave propagation in specimen when uniform velocity loading(unit：10^4 Pa)

模式；在图 4(g)~(i)中，等效应力波在试样中来回振荡，形成复杂的交互作用的等效应力波模式，随着时间的推移，这个时间通常很短暂(1 ms 左右)；如图 4(j)~(l)所示，模型空间各点等效应力的波动越来越小，试样迅速达到较稳定的状态。在试样破坏之前，空间各点等效应力绝对大小随着匀速加载端的载荷增加而线性增加。

4 SPH 法模拟岩石单轴压缩破坏过程

在 SPH 法研究岩石单轴压缩破坏时，利用莫尔-库仑强度准则判断岩石的破坏。对单轴压缩条件下岩石试样一点的应力状态进行分析，随着外载荷的不断增加，当岩石中某个面上的 σ，τ 在莫尔-库仑准则破坏包络线上或者跨越准则包络线时，岩石 SPH 粒子点发生破坏。莫尔-库仑强度准则表示为

$$F_s = \tau - \sigma\tan\varphi - c = 0 \quad (10)$$

式中：φ 为 τ-σ 应力面上莫尔-库仑破坏面的倾斜角，也称为岩石材料的内摩擦角；c 为岩石材料的黏聚力，随粒子点位置不同呈 Weibull 统计分布。当模型内某个 SPH 粒子破坏时，引起周围 SPH 粒子应力的集中，导致裂纹开始扩展，而无网格 SPH 法相较于其他数值方法具有自适应的优势，只需对每个 SPH 粒子应力状态代入莫尔-库仑准则就可判断岩石是否破裂。

通过岩样粒子点抗压强度的 Weibull 统计分布来描述岩石类材料的非均质性，本文的 SPH 粒子点黏聚力 $c(i)$ Weibull 统计分布表示如下：

$$c(i) = c_0(-\ln(x))^{1/m} \quad (11)$$

式中：c_0 为黏聚力期望值，本文取 0.1 MPa；m 为不均度参数，其值越大，SPH 粒子场量离散程度。

Δc 越小，不均度越小(见图 4)。本文算例中，选取 $m=20$ 进行研究，可以看出，即使程序中不均度是相同的，随着时间推移，每次运算的模型，只能保证各粒子点场量离散程度相同，而粒子场量值的空间分布初始值状态是不同的，这跟自然界中相同抗压强度的岩样细观缺陷的分布也存在千差万别相类似。因此，引入 Weibull 统计分布的相同 SPH 程序在不同时间模拟时会出现不同断裂带形式，这和单轴压缩实验中同类岩石多次实验出现不同破裂模式相类似。以往，通过反复的岩石单轴压缩室内实验，归纳岩石有那些破坏模式，而 SPH 数值实验也可预测单轴压缩条件下岩样的破坏模式。

4.1 岩石破坏应力应变分析

图 5~8 模拟了单轴压缩岩样破坏过程中水平、

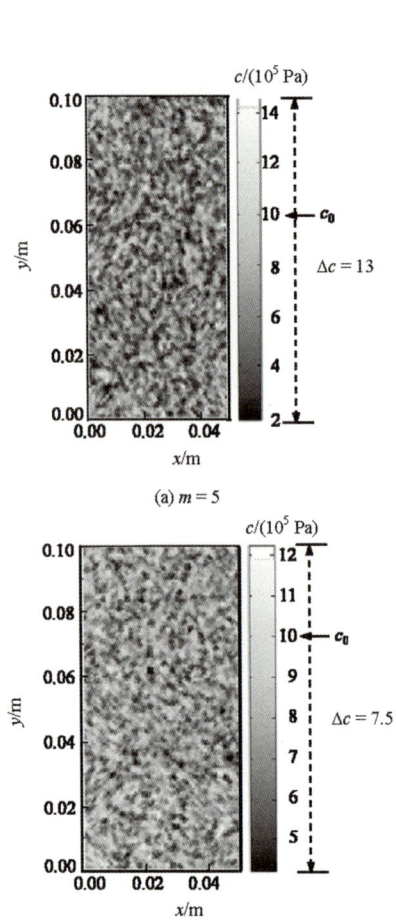

图 4 满足 Weibull 分布的黏聚力在不同均值度 m 下的云图

Fig.4 Random cohesion with Weibull distribution of different homogeneous index m

竖直方向应力和应变的分布规律。图 5~8 中(a)图均为破坏点前的岩样应力或应变分布图；图 5~8 中(b)和(c)图均为出现单一破坏点时的岩样应力或应变的分布，且(b)图对应于图 9 中应力-应变曲线上的点 a；图 5~8 中(d)，(e)和(f)图均为出现多个破坏点时的岩样应力或应变分布图，且(d)图对应于图 10 中应力-应变曲线上点 b；图 5~8 中(g)，(h)和(i)图

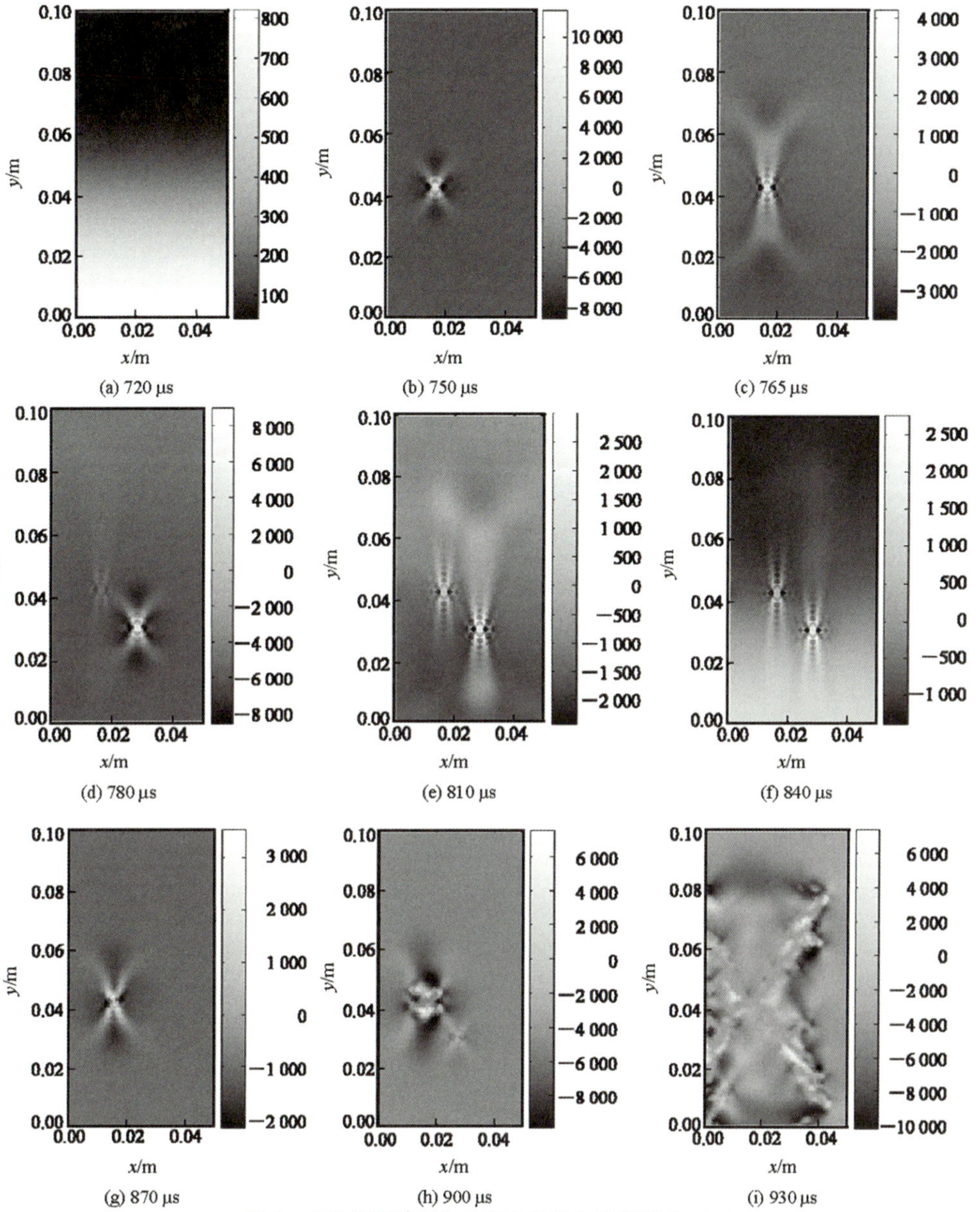

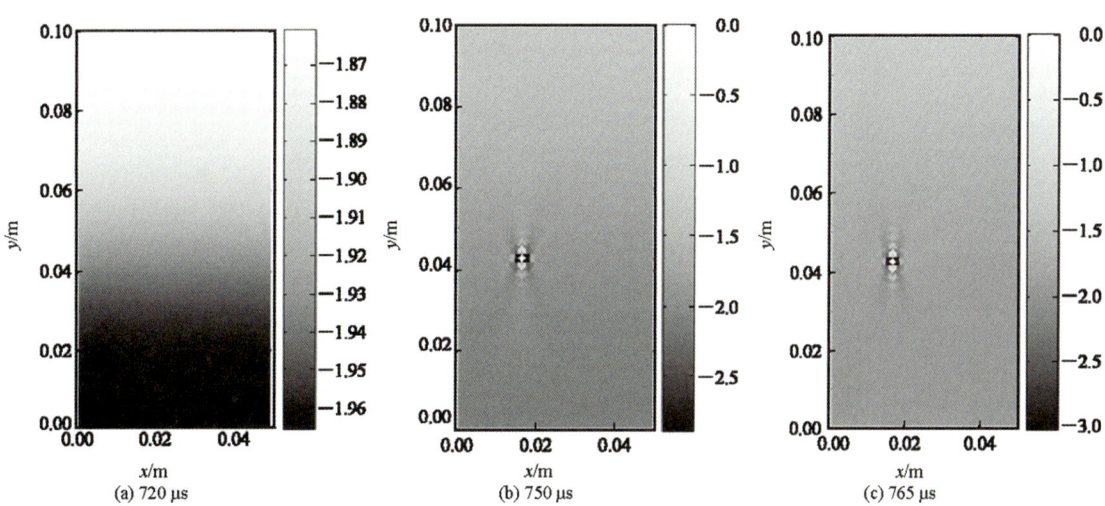

图 5 单轴压缩模型中水平方向应力云图(单位：Pa)

Fig.5 Horizontal stress contour plot of uniaxial compression using SPH(unit：Pa)

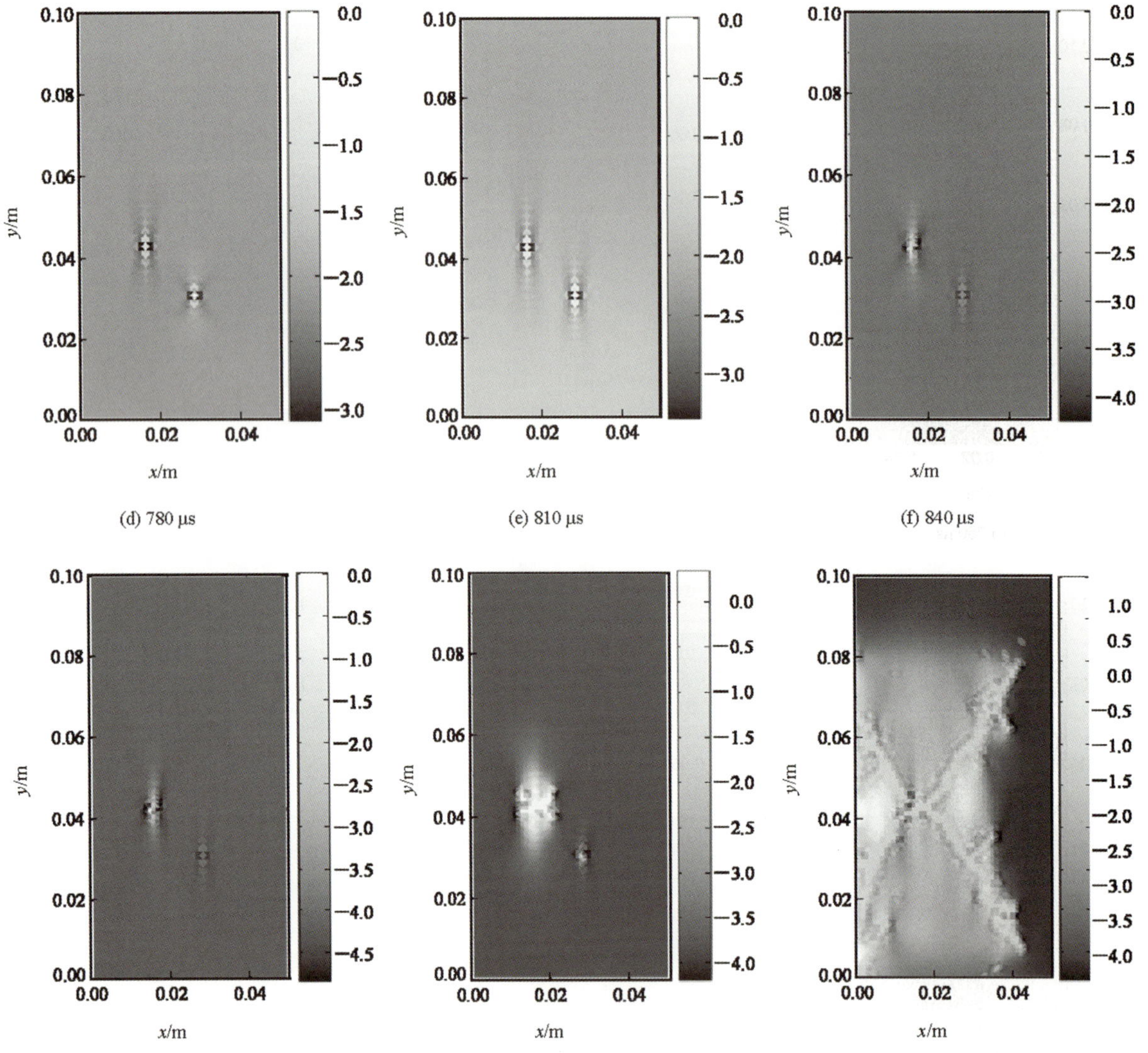

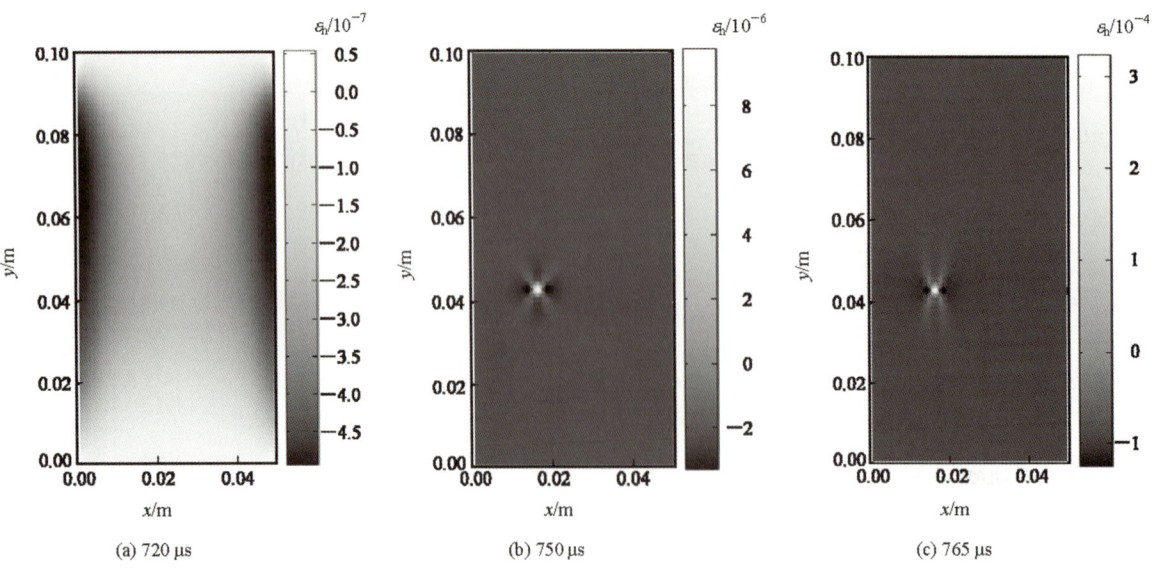

图6 单轴压缩模型中竖直方向应力云图(单位：MPa)

Fig.6 Vertical stress contour plot of uniaxial compression(unit：MPa)

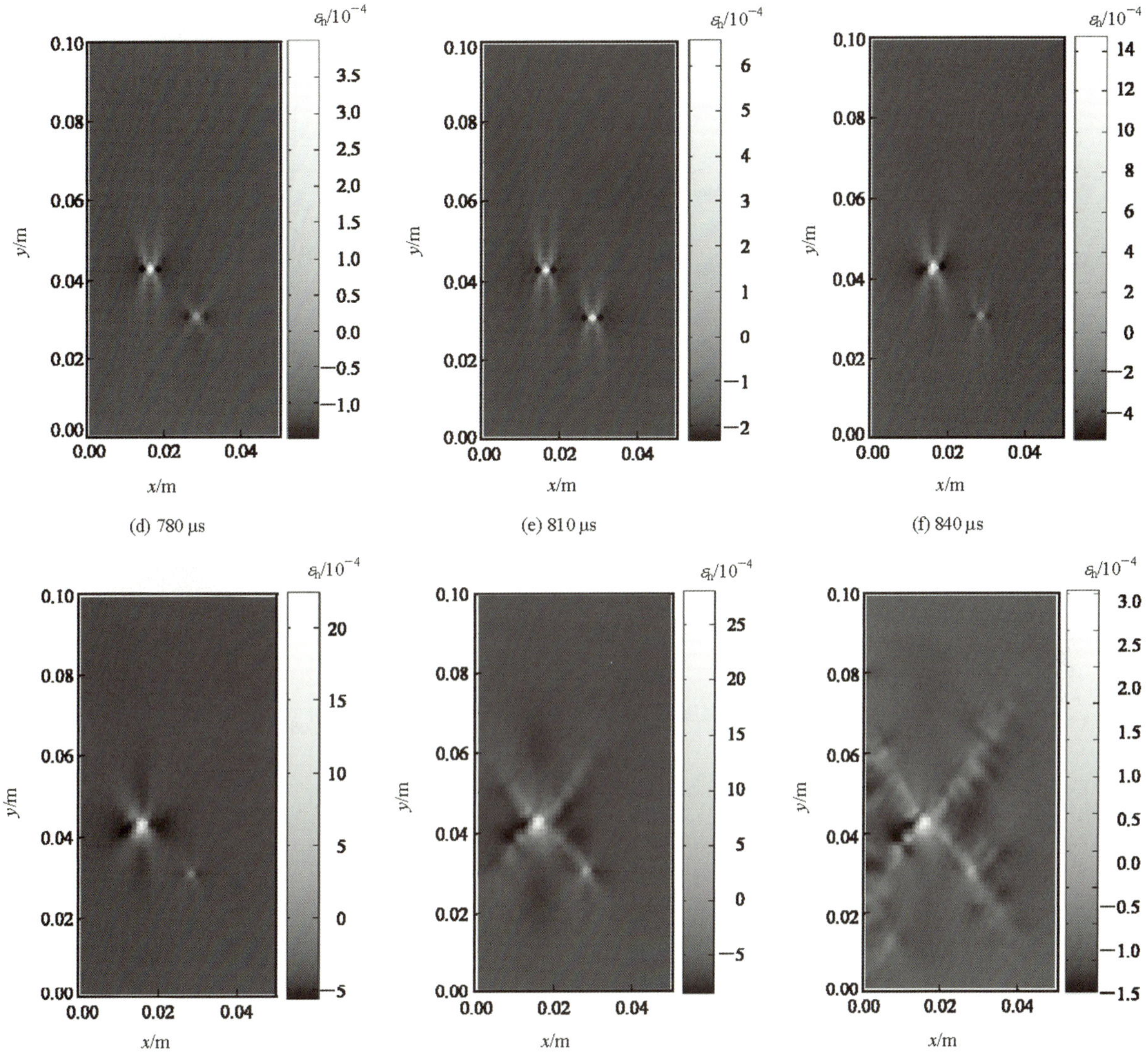

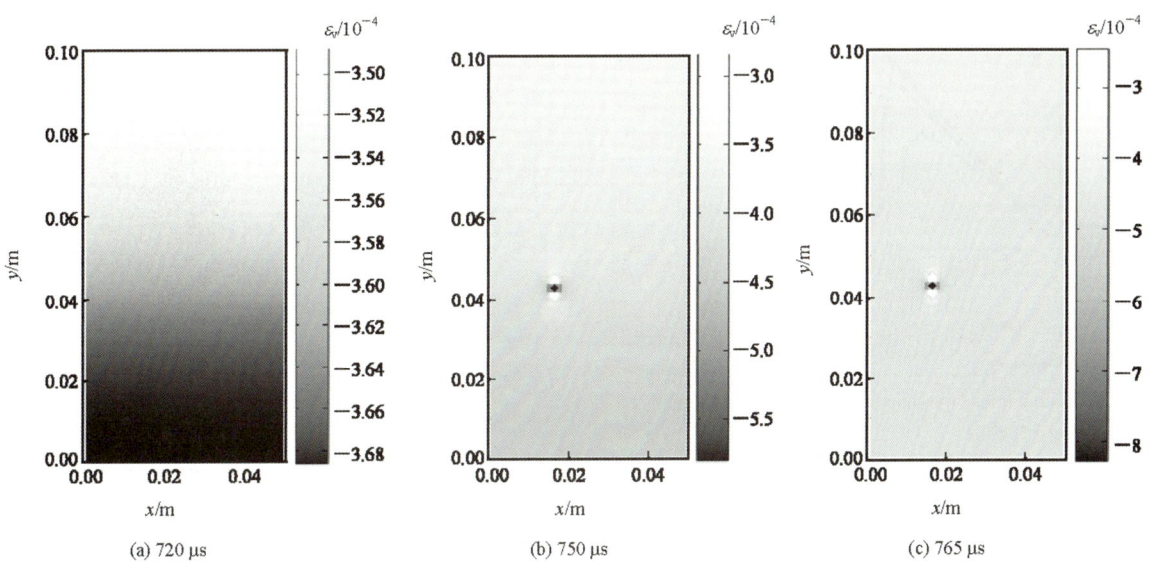

(d) 780 μs (e) 810 μs (f) 840 μs

(g) 870 μs (h) 900 μs (i) 930 μs

图 7 单轴压缩模型中水平方向应变云图
Fig.7 Vertical steers contour plot of uniaxial compression using SPH

(a) 720 μs (b) 750 μs (c) 765 μs

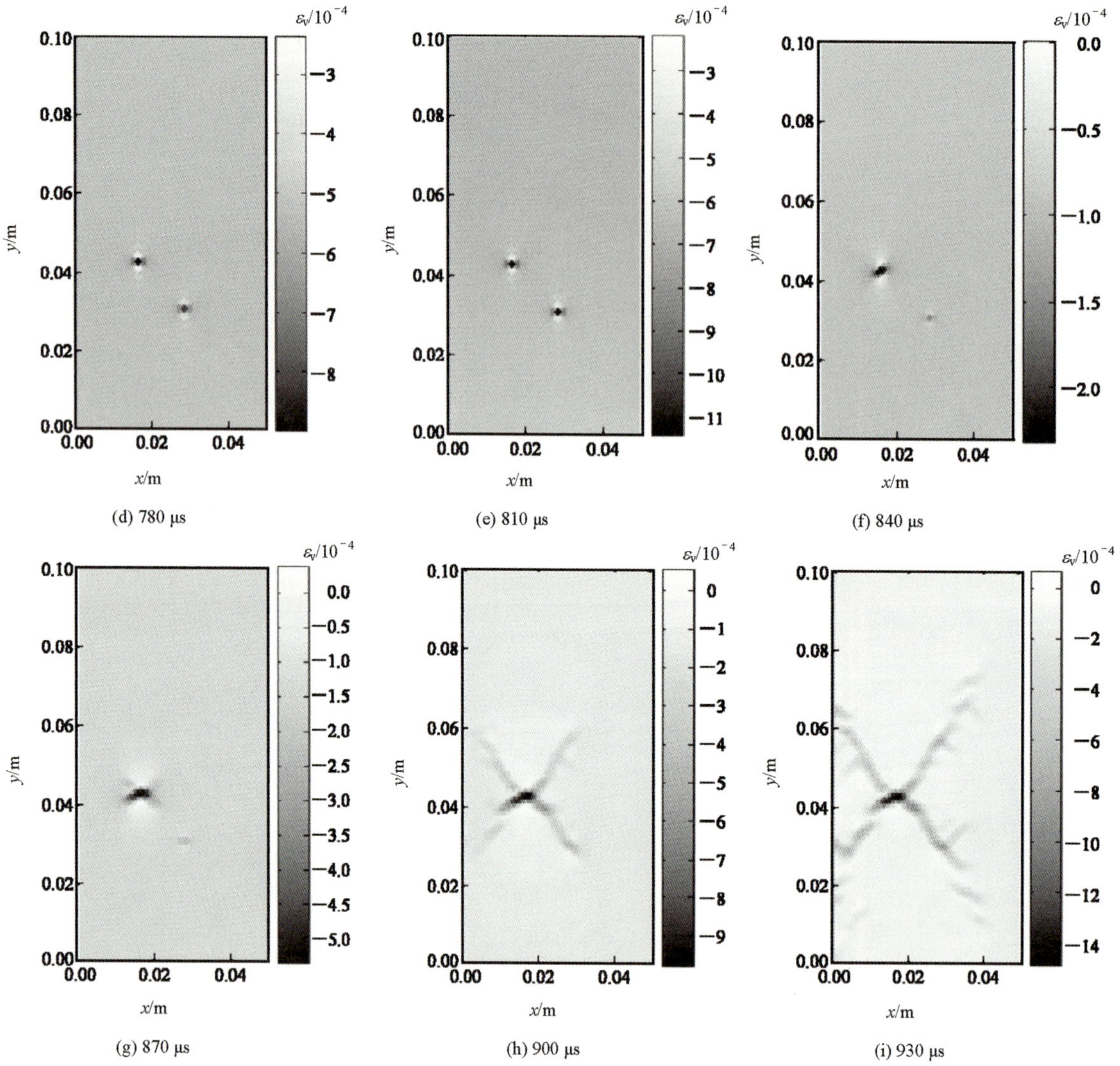

(d) 780 μs
(e) 810 μs
(f) 840 μs
(g) 870 μs
(h) 900 μs
(i) 930 μs
x/m
y/m
εv/10⁻⁴

图 8 单轴压缩模型中竖直方向应变云图

Fig.8 Vertical strain contour plot of uniaxial compression using SPH

m = 20
m = 10
m = 5
轴向应力/(10⁷ Pa)
轴向应变/10⁻⁴

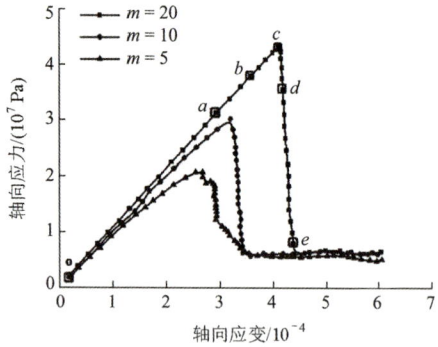

图 9 SPH 模拟的不同均值度岩样的应力－应变曲线

Fig.9 Stress-strain curves for different heterogeneous specimens by uniaxial compression numerical SPH code

均为某一处或者多处破坏点“成团”，破坏从该“成团”区域沿某个角度迅速展开时的应力或应变图，且(g)，(h)和(i)图分别对应于图 9 中应力－应变曲线上点 c，d 和 e。

从图 5～8 中的(a)，(b)和(c)图可以看到，首先失去承载力的粒子周围应力或应变场变化明显，大体上 45°，135°，225°，315°方向的应力、应变较大，这几个方向上的剪应力也是最大的，因此，这几个方向是剪切破坏的发展方向。

从图 5～8 中的(d)，(e)和(f)可以看出，多处粒子出现破坏，新的破坏点不断出现，数量增加，这些点通常都是黏聚力相对小的点。本算例只有 2 个破坏点，出现 2 个破坏点后，破坏点的数量不再增加，所有破坏点周围的应力或应变随着时间而增加。从图 5～8(g)，(h)和(i)可以看出，左侧破坏点发育“成团”，并扩展成破裂带，破裂带周围有明显的应力奇异现象，使破裂带迅速扩展。由于左侧点破裂带的

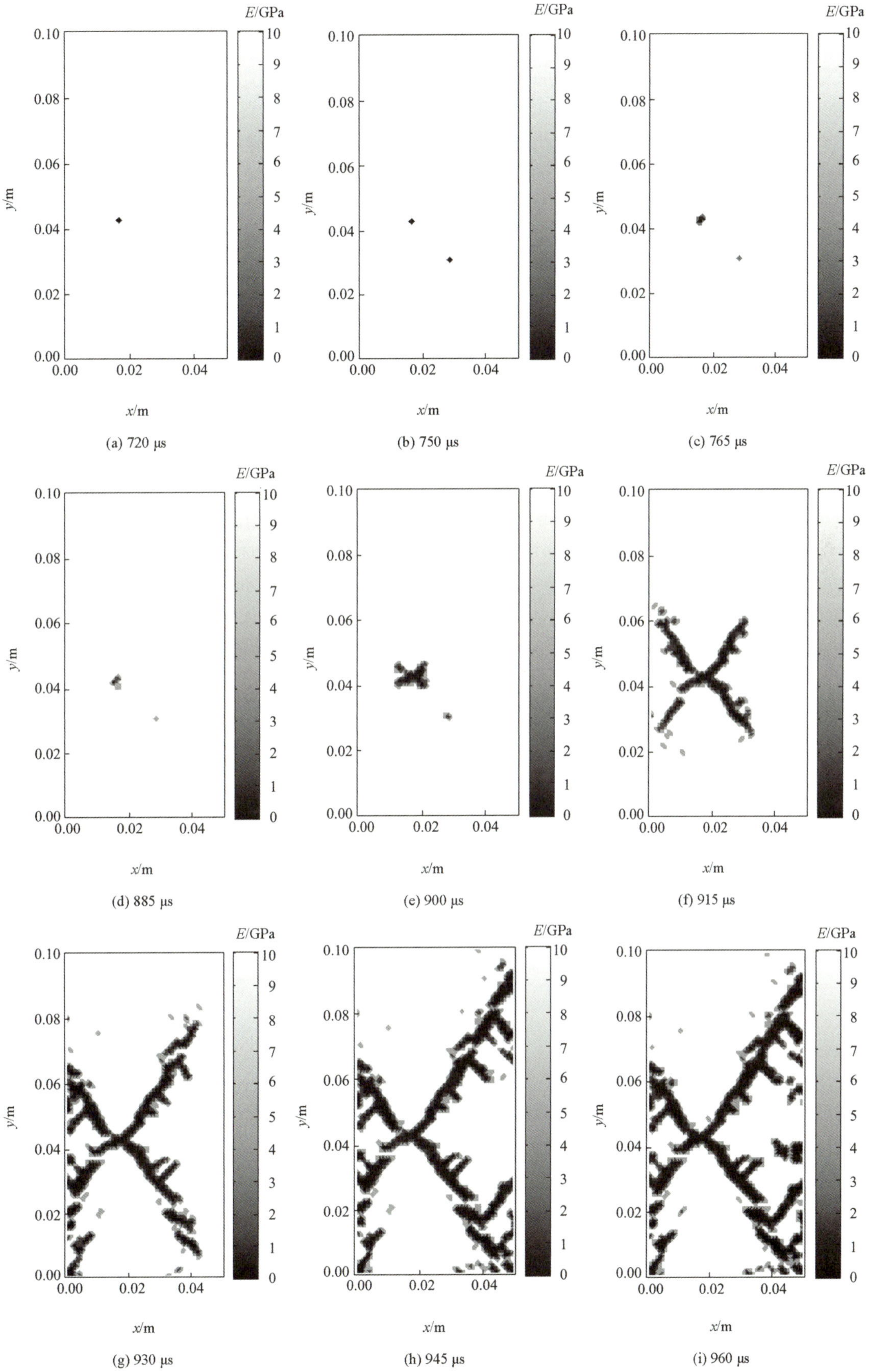

图 10 单轴压缩模型中岩样破坏模拟
Fig.10 Rock specimen failure simulations of uniaxial compression

形成导致应力的释放，右侧点不再发育，其两端应力逐渐减小至正常水平。

在试样破坏过程中，破坏点的分布模式(破坏点的数量和位置)反映了不均质岩样的应力和应变分布，以及破坏过程中的应力-应变重分布机制。

图 9 中 $m = 20$ 的曲线对应于图 5～8 的岩样算例，岩样破裂经历了如下过程：图 9 中点 a 对应于随机出现单个破坏粒子，其周围应力不断增大；点 b 开始，破坏粒子数量增加，当增加到某个时刻，破坏粒子数量不再增加而其周围应力应变增大；曲线顶端 c，某个或几个破坏粒子开始"成团"破坏；点 d 开始，应力跌落对应于破裂区迅速增大；点 e 开始，裂纹相互作用、贯通并形成剪切破裂带。

图 10 模拟了岩石破裂带的形成。单轴压缩条件下岩石的最终破坏模式，是通过模拟破坏粒子点的位置随时间变化而获得的。在图 10(a)中，独立的破坏点随机出现在岩样中。在图 10(b)中，破坏点数量增加，若调节不均质度参数 m，岩样不均质度越大，承载力越低，初期破坏粒子发育更密集。在图 10(c)中，左侧破坏点"成团"，联合近邻粒子，发育成独立裂隙和区域破坏，而其右下侧的破坏点，由于应力重分布机制作用，不再发育，但是岩样依然完整。在图 10(d)和(e)中，该破坏区域迅速扩展。在图 10(f)和(g)中，更多的裂隙相互作用、连接和贯通。在图 10(h)和(i)中，最终形成 X 型共轭剪切破裂带。

4.2 岩石单轴压缩破坏模式

尤明庆和华安增[17]指出，岩石材料的破坏就相邻岩石颗粒间的离合关系来看，可能的形式只有错开或分离 2 种，分别代表材料的剪切破坏和拉伸破坏。本文的岩样单轴压缩的最终宏观破坏表现形式是明显的剪切破坏。在数值模拟中，可以采用类似室内实验中大量统计研究其破坏规律的方法，通过数值"实验"研究可能出现的破裂模式，再用实验结果进行佐证。尤明庆[18]通过大量的单轴压缩实验，总结了几种基本的破坏形式。本文通过 SPH 数值模拟也得到下列岩样破坏模式：

(1) 单一的剪切破坏面形式，并由它贯穿整个岩样，产生一个剪切破坏面(见图 11(a))。

(2) 存在与主剪切面平行的剪切破坏面，或者还存在一些局部剪切破坏面(见图 11(b))。

(3) 2 个相互连接的剪切破坏面(见图 11(c))。

(4) 岩样一端或者两端同时出现破裂圆锥面。

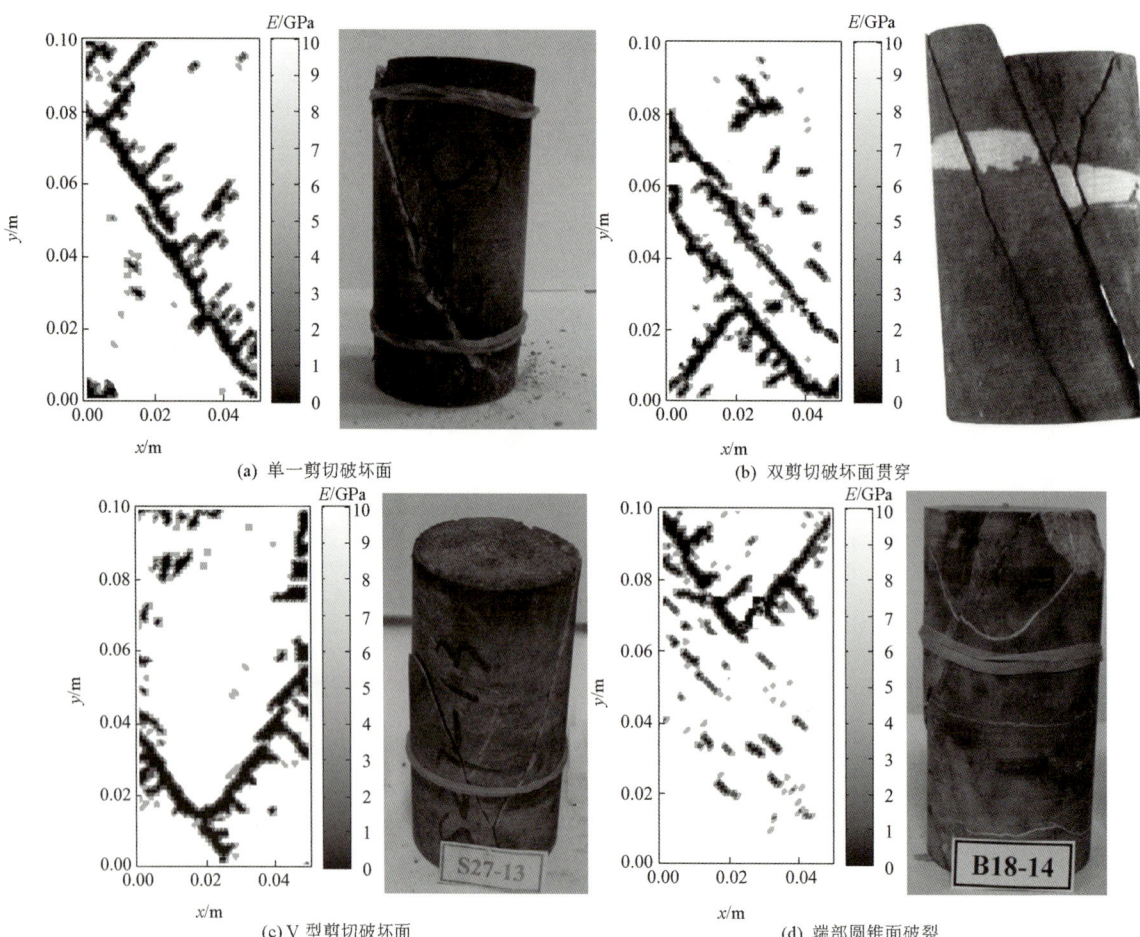

(a) 单一剪切破坏面 (b) 双剪切破坏面贯穿

(c) V 型剪切破坏面 (d) 端部圆锥面破裂

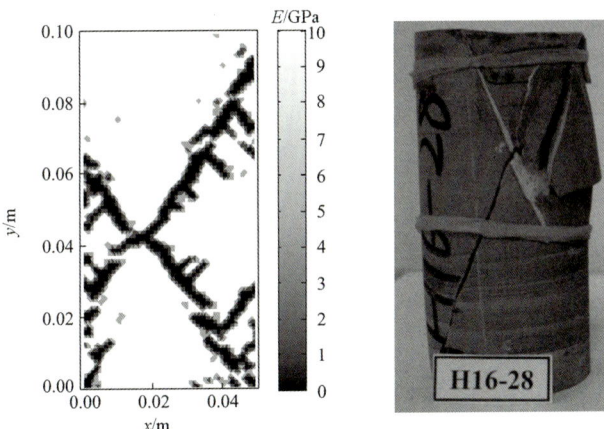

(e) X 状共轭剪切破坏

图 11 单轴压缩岩样最终破坏模式数值结果和试验结果对比

Fig.11 Rock specimen failure simulations of uniaxial compression

在实验中,这种情况较少出现。SPH 法数值模拟同样也会少量出现该情况(见图 11(d))。

(5) X 形共轭剪切破坏(见图 11(e)),最能体现剪切破坏机制在岩石单轴压缩中的作用。本文多采用该破坏模式来诠释岩石单轴压缩的破坏过程,但在实验中,由于岩石材料非连续,非均匀性,该形式出现并不多。

5 结 论

本文建立和发展了有效的无网 SPH 数值方法,该方法能用于模拟不均质弹脆性岩样在单轴压缩下的变形破坏特性。基于拉格朗日方法的 SPH 法适合追踪粒子点场量,易获得固体瞬时的应力-应变状态。SPH 法由于其无网格特性易于模拟岩石类弹脆性材料破裂中的不连续问题,如裂纹的启裂、扩展和贯通问题。

模拟结果与已有的岩石单轴压缩理论和实验结果对比分析,验证了 SPH 法在模拟岩石类材料变形中的力学性质、破裂过程和破裂模式的有效性。

SPH 法在模拟岩石类弹脆性材料的破坏时相较于基于连续介质力学的传统数值模拟方法有明显的优势,预期可应用于更多土木工程数值模拟领域。

参考文献(References):

[1] JAEGER J C, COOK N G W, ZIMMERMAN R W. Fundamentals of rock mechanics[M]. 4th ed. Malden, MA: Blackwell Publishing, 2007: 1 – 98.

[2] LUCY L B. A numerical approach to the testing of the fission hypothesis[J]. The Astronomical Journal, 1977, 82(12): 1 013 – 1 024.

[3] GINGOLD R A, MONAGHAN J J. Smoothed particle hydrodynamics: theory and application to non-spherical stars[J]. Monthly Notices of the Royal Astronomical Society, 1977, 181(2): 375 – 389.

[4] LIBERSKY L D, PETSCHEK A G. Smoothed particle hydrodynamics with strength of materials[C]// Advances in the Free-Lagrange Method Including Contributions on Adaptive Gridding and the Smooth Particle Hydrodynamics Method, Lecture Notes in Physics. [S.l.]: [s.n.], 1991: 248 – 257.

[5] BENZ W, ASPHAUG E. Impact simulations with fracture. I. Method and tests[J]. Icarus, 1994, 107(1): 98 – 116.

[6] MONAGHAN J J. Smoothed particle hydrodynamics[J]. Reports on Progress in Physics, 2005, 68(8): 1 703 – 1 759.

[7] MA G W, WANG X J, LI Q M. Modeling strain rate effect of heterogeneous materials using SPH method[J]. Rock Mechanics and Rock Engineering, 2010, 43(6): 763 – 776.

[8] LIU M B, LIU G R. Smoothed Particle Hydrodynamics(SPH): an overview and recent developments[J]. Archives of Computational Methods in Engineering, 2010, 17(1): 25 – 76.

[9] KOSHIZUKA S, NOBE A, OKA Y. Numerical analysis of breaking waves using the moving particle semi implicit method[J]. International Journal for Numerical Methods in Fluids, 1998, 26(7): 751 – 769.

[10] CUMMINS S J, RUDMAN M. An SPH projection method[J]. Journal of Computational Physics, 1999, 152(2): 584 – 607.

[11] OGER G, DORING M, ALESSANDRINI B, et al. Two-dimensional SPH simulations of wedge water entries[J]. Journal of Computational Physics, 2006, 213(2): 803 – 822.

[12] COLAGROSSI A, LANDRINI M. Numerical simulation of interfacial flows by smoothed particle hydrodynamics[J]. Journal of Computational Physics, 2003, 191(2): 448 – 475.

[13] LIBERSKY L D, PETSCHEK A G, CARNEY T C, et al. High strain Lagrangian hydrodynamics a three-dimensional SPH code for dynamic material response[J]. Journal of Computational Physics, 1993, 109(1): 67 – 75.

[14] TAKEDA H, MIYAMA S M, SEKIYA M. Numerical simulation of viscous flow by smoothed particle hydrodynamics[J]. Progress of Theoretical Physics, 1994, 92(5): 939 – 960.

[15] MORRIS J P, FOX P J, ZHU Y. Modeling low reynolds number incompressible flows using SPH[J]. Journal of Computational Physics, 1997, 136(1): 214 – 226.

[16] TIMOSHENKO S P, GOODIER J N. Theory of elasticity[M]. 3rd ed. [S.l.]: McGraw-Hill, 1970: 11 – 50.

[17] 尤明庆, 华安增. 岩石试样单轴压缩的破坏形式与承载能力的降低[J]. 岩石力学与工程学报, 1998, 17(3): 292 – 296.(YOU Mingqing, HUA Anzeng. Fracture of rock specimen and decrement of bearing capacity in uniaxial compression[J]. Chinese Journal of Rock Mechanics and Engineering, 1998, 17(3): 292 – 296.(in Chinese))

[18] 尤明庆. 岩石的力学性质[M]. 北京: 地质出版社, 2007: 22 – 25.(YOU Mingqing. Geotechnical property[M]. Beijing: Geology Press, 2007: 22 – 25.(in Chinese))

克里金算法与多重分形理论在岩土参数随机场分析中的应用

王长虹[1,3]，朱合华[2]，钱七虎[3]

(1. 上海应用技术学院 轨道工程系，上海 200235；2. 同济大学 地下建筑与工程系，上海 200092；
3. 解放军理工大学 国防工程学院，南京 210007)

摘 要：岩土参数的空间分布特征由于存在取样数据之间自相关和互相关的特性，未知点的岩土参数属性可通过特定的方法内插或外推，经典的数理统计方法难以确定周围的数据样本点以及相应的插值系数。首先介绍地统计学中基于距离加权的普通克里金（ordinary kriging, OK）算法、泛克里金算法（UK）和协克里金算法（CK）。由于基于滑动距离加权的 OK 算法无法度量局部空间的奇异性，将引入多重分形理论弥补该缺陷。以 2010 上海世博会的世博轴区域（长 525 m，宽 80 m）为工程背景，区域内共有 42 个取土钻孔，以典型的粉质黏土层 3 个重要的物理力学指标，即黏聚力、内摩擦角和压缩模量验证以上算法。对于岩土参数黏聚力和内摩擦角，预测精度由高至低为多重分形联合模型（MK）、协克里金模型（CK）、泛克里金模型（UK）、普通克里金模型（OK）；对于岩土参数压缩模量，相应的顺序为泛克里金模型和普通克里金模型位置互换。研究结果证明，在岩土参数空间场的分析中，辅助信息有助于提高数据预测精度，并且多重分形联合模型有助于分析空间局部的奇异性。

关 键 词：普通克里金；泛克里金；协克里金；多重分形；随机场
中图分类号：TU 452   **文献标识码**：A

Application of Kriging methods and multi-fractal theory to estimate of geotechnical parameters spatial distribution

WANG chang-hong[1,3], ZHU He-hua[2], QIAN Qi-hu[3]

(1. Department of Rail Track Engineering, Shanghai Institute of Technology, Shanghai 200235, China;
2. Department of Geotechnical Engineering, Tongji University, Shanghai 200092, China
3. Engineering Institute of National Defense Engineering, PLA University of Science and Technology, Nanjing 210007, China)

Abstract: Due to spatial auto-correlation and inter-correlation among geotechnical observed data, the spatial geotechnical characteristics distribution at unknown location has to be extrapolated or interpolated by some special methods. However, classical statistical methods could not rationally resolve the problems which include selection of sample points, and comparison of spatial estimating weights between bilateral data. The distance-weighted ordinary Kriging (OK), universal Kriging (UK), and co-Kriging (CK) prediction methods for scattered data are introduced at first, which are known as the Kriging family in global geostatistics. Moreover, multi-fractal theory combining with co-Kriging (MK) is presented to depict the local singularity which should be ignored by Kriging of sliding weighted average algorithm. The performance is compared in different typical geotechnical parameters: cohesion coefficient c, friction angle φ and compression modulus E_s. This study is based on the main axis (525 m long, 80 m wide) of Expo 2010 area of Shanghai, geotechnical test samples come from 42 boreholes. The performance of the different model fitness used in this study is MK, CK, OK and UK from the best to poor for parameters c and φ, and for parameter E_s simulation, the sequence is MK, CK, UK and OK. The results prove that, in the most geotechnical occasions, auxiliary information would improve the prediction accuracy, and MK theory is useful tool to measure local singularity.

Key words: ordinary Kriging(OK); universal Kriging(UK); co-Kriging(CK); multi-fractal(MK); random field

---

收稿日期：2014-03-26。
基金项目：国家自然科学基金（No. 51208303）；国家 973 计划（No. 2011CB013800‑G）；上海市教委青年教师培养计划（No. yyy11007）；上海市教委科技创新计划（No. 14ZZ162）。
本文原载于《岩土力学》（2014 年第 35 卷增刊 2）。

1 引 言

地统计学是统计学理论的一个分支,主要用于空间数据的插值和映射,最早用于时间序列数据分析,逐渐深入到空间数据分析领域。空间相关性对于经典统计学理论而言是一个难题,如方差分析和最小二乘法回归分析都假定数据为独立同分布[1]。当数据自相关性存在时,独立分布的假定自然就不成立了,而协方差往往难以准确描述这种趋势[2]。克里金(Kriging)模型通过周边观测值的加权平均值预测未知点的数值,加权系数的计算主要通过研究区域内的半变差函数而确定,具有良好的数理统计基础[3]。

普通克里金(OK)模型[4-5]、泛克里金(UK)模型[6]、协克里金(CK)模型[7]是常用的克里金算法。泛克里金模型通过设定空间分布趋势和分析残差的克里金过程来研究空间场特征。协克里金模型可以将空间互相关的变量信息融合在一起考虑,提高了未知量的预测精度。

原2010上海世博会园区正在面临新的一轮基础建设开发,由于建设场地分布面积巨大的原因,采用岩土参数的统计一阶矩和二阶矩,难以描述岩土参数的空间分布特征,影响岩土工程的设计参数取值,从而需要引入岩土参数随机场计算理论。

研究证明,协克里金模型一般优于普通克里金模型[8-10],但随机变量之间的空间互相关性不高时,优势也随之降低[11-12]。当变量的空间自相关性表现较强时,普通克里金算法表现较好;当变量之间的协相关性较强时,协克里金方法表现更好;当变量存在空间明显趋势时,泛克里金模型表现最好。

克里金算法通过半变差函数提供了一种很好的空间统计和插值算法,但是半变差函数的优劣取决于观测的数据量和尺度,具有空间平滑的缺陷,为克服上述问题,引入多重分形理论,通过刻画测量尺度与观测值之间的关系,可以较好地度量空间局部的奇异性[13]。

2 背景工程

2010上海世博会的世博轴区域(长525 m,宽80 m),共有42个取土钻孔,该研究是原2010年上海世博会数字化地下工程的拓展,岩土参数c、φ、E_S和其他参数信息可以通过研究区域内的42个钻孔统计数据获取。以粉质黏土层为研究对象,经典统计学指标见表1。

表1 粉质黏土层统计数据
Table 1 Statistics parameters of silty clam stratum

参数	钻孔数量/个	最小值	最大值	平均值	标准差
c / kPa	42	4.0	29.0	13.7	5.2
φ / (°)	42	8.4	32.5	21.1	4.8
E_S / MPa	42	2.3	13.9	5.2	2.7

3 理论方法

3.1 普通克里金模型

观测值的半变差函数由式(1)定义。

$$\hat{\gamma}(h) = \frac{1}{2N(h)} \sum_{i=1}^{N(h)} [\hat{Z}(x_i) - \hat{Z}(x_i + h)]^2 \quad (1)$$

式中:$\hat{Z}(x_i)$、$\hat{Z}(x_i + h)$分别为x_i和$x_i + h$的观测值;$N(h)$为滞后距h分割的区间数目,通过作出$\hat{\gamma}(h)-h$图,可以拟合$\hat{\gamma}(h)$成指数模型、高斯模型和球状模型等,最佳模型可以通过最小误差值确定。

假设$Z(x)$为一阶平稳正态分布函数,数学期望为未知的常数μ,具有以下内蕴假定[14-15]:

$$\left.\begin{array}{l} E[Z(x+h) - Z(x)] = 0 \\ Var[Z(x+h) - Z(x)] = 2\gamma(h) \end{array}\right\} \quad (2)$$

普通克里金估值公式为

$$\hat{Z}(x_0) = \sum_{i=1}^{n} \lambda_i Z(x_i) \quad (3)$$

式中:$\hat{Z}(x_0)$为$Z(x_i)$在x_0处的预测值;$Z(x_i)$为点x_i处的观测值;λ_i为插值系数。普通克里金系数计算方程式为

$$\left.\begin{array}{l} \sum_{i=1}^{n} \lambda_i \gamma(x_i - x_j) + \mu = \gamma(x_0 - x_j), \ (j = 1, 2, \cdots, n) \\ \sum_{i=1}^{n} \lambda_i = 1 \end{array}\right\} \quad (4)$$

普通克里金模型估值误差为

$$\sigma_{OK}^2 = E\{[Z(x_0) - \hat{Z}(x_0)]^2\} = \sum_{i=1}^{n} \lambda_i \gamma(x_0 - x_i) + \mu \quad (5)$$

3.2 泛克里金模型

当随机函数$Z(x)$的一阶统计矩存在时,但数学期望$m(x)$不是一个常数或一个常数表达式,这种估值方法可称为泛克里金模型[16]。假定$Z(x)$是一阶稳定的随机函数,残差的协方差函数$cov(x_i, x_j)$存在,在x_0邻域,泛克里金预测式为

$$\hat{Z}(x_0) = \sum_{i=1}^{n} \lambda_i Z(x_i) \quad (6)$$

漂移式 $m(x)$ 定义为

$$m(x)=\sum_{i=0}^{n}\mu_i f^i(x), \quad (f^0(x)=1) \quad (7)$$

泛克里金求解未知权值系数 λ 和 μ 公式为

$$\left.\begin{array}{l}\sum_{jj=1}^{k}\lambda_{jj}cov(x_j,x_{jj})-\sum_{i=0}^{n}\mu_i f^i(x_j)=cov(x_j,x_0),\\ \qquad (j=1,2,\cdots,k)\\ \sum_{j=1}^{k}\lambda_j f^i(x_j)=f^i(x_0),\quad (i=0,1,2,\cdots,n)\end{array}\right\} \quad (8)$$

可以改写为以下的矩阵的形式：

$$HU=V \quad (9)$$

$$H=\begin{bmatrix}cov(x_1,x_1) & \cdots & cov(x_1,x_k) & f^0(x_1) & \cdots & f^n(x_1)\\ cov(x_2,x_1) & \cdots & cov(x_2,x_k) & f^0(x_2) & \cdots & f^n(x_2)\\ \vdots & & \vdots & \vdots & & \vdots\\ cov(x_k,x_1) & \cdots & cov(x_k,x_k) & f^0(x_k) & \cdots & f^0(x_k)\\ f^0(x_1) & \cdots & f^0(x_k) & 0 & \cdots & 0\\ \vdots & & \vdots & \vdots & & \vdots\\ f^n(x_1) & \cdots & f^n(x_k) & 0 & \cdots & 0\end{bmatrix}$$

$$U^T=(\lambda_1,\lambda_2\cdots,\lambda_k,-\mu_0,\cdots,-\mu_n)$$

$$V^T=[cov(x_0-x_1),\cdots,cov(x_0-x_k),f^0(x_0),\cdots,f^n(x_0)]$$

(10)

当 $Z(x)$ 半变差函数存在时，以上泛克里金求解未知权值系数方程式改写为

$$\left.\begin{array}{l}\sum_{jj=1}^{k}\lambda_{jj}\gamma(x_j-x_{jj})+\mu_0+\sum_{i=1}^{n}\mu_i f^i(x_j)=\gamma(x_j-x_0),\\ \qquad (j=1,2,\cdots,k)\\ \sum_{j=1}^{k}\lambda_j=1;\quad \sum_{j=1}^{k}\lambda_j f^i(x_j)=f^i(x_0),\quad (i=1,2,\cdots,n)\end{array}\right\}$$

(11)

泛克里金模型的插值误差为

$$\sigma_{UK}^2=E\{[Z(x_0)-\hat{Z}(x_0)]^2\}=cov(x_0,x_0)-U^TV \quad (12)$$

3.3 协克里金模型

实际的工程中许多变量之间存在一定的相关性，其中一些有较丰富的资料，另一些变量的资料比较少，有的变量很容易测量，一些变量则难以测量或者费时费力。能否利用资料多、容易测量的变量资料去提高资料少、难以测量的变量的估计精度，协克里金法就是这种多变量的估值方法。

假设 $Z_i(x)$，$Z_j(x)$ 为二阶平稳或满足广义的内蕴假设：

$$E\{Z_j(x)\}=m_j(x) \quad (13)$$

$$r_{jj}=\frac{1}{2}E\{[Z_j(x+h)-Z_j(x)]^2\} \quad (14)$$

$$r_{ij}=\frac{1}{2}E\{[Z_i(x+h)-Z_i(x)][Z_j(x+h)-Z_j(x)]\} \quad (15)$$

$Z_1(x)$，$Z_2(x)$ 为区域内 2 个变量，观测值为 n_1、n_2，且 $n_1>n_2$，假定 $Z_1(x)$ 比 $Z_2(x)$ 容易测量，使用 $Z_1(x)$ 辅助信息分析 $Z_2(x)$ 在 x_0 处的值为

$$\hat{Z}_2(x_0)=\sum_{i=1}^{n_1}\lambda_{1i}Z_1(x_{1i})+\sum_{j=1}^{n_2}\lambda_{2j}Z_2(x_{2j}) \quad (16)$$

根据估值无偏及误差方差最小的要求，得到以下计算未知权值 λ 和 μ 的协克里金线性方程组：

$$\left.\begin{array}{l}\sum_{i=1}^{n_1}\lambda_{1i}r_{11}(x_{1i}-x_I)+\sum_{j=1}^{n_2}\lambda_{2j}r_{12}(x_{2j}-x_I)+\mu_1=\\ \qquad r_{12}(x_0-x_I),\quad (I=1,2,\cdots,n_1)\\ \sum_{i=1}^{n_1}\lambda_{1i}r_{21}(x_{1i}-x_J)+\sum_{j=1}^{n_2}\lambda_{2j}r_{22}(x_{2j}-x_J)+\mu_2=\\ \qquad r_{22}(x_0-x_J),\quad (J=1,2,\cdots,n_2)\\ \sum_{i=1}^{n_1}\lambda_{1i}=0;\quad \sum_{j=1}^{n_2}\lambda_{2j}=1\end{array}\right\} \quad (17)$$

协-半变差函数 $r_{12}(h)=r_{21}(h)$ 为

$$r_{12}(h)=\frac{1}{2}E\{[Z_1(x+h)-Z_1(x)][Z_2(x+h)-Z_2(x)]\}$$

(18)

为了计算协变异函数，先定义一个新的变量 $Z_{12}^+(x)=Z_1(x)+Z_2(x)$，然后计算新的变量的变异函数：

$$r_{12}^+(h)=\frac{1}{2}E\{[Z_{12}^+(x+h)-Z_{12}^+(x)]^2\} \quad (19)$$

这一新变量的变异函数与原协变异函数有如下关系：

$$r_{12}^+(h)=\frac{1}{2}E\{\{[Z_1(x+h)+Z_2(x+h)]-[Z_1(x)+Z_2(x)]\}^2\}=r_{11}(h)+r_{22}(h)+2r_{12}(h) \quad (20)$$

由此可以得到

$$r_{12}(h)=\frac{1}{2}[r_{12}^+(h)-r_{11}(h)-r_{22}(h)] \quad (21)$$

2 个变量之间的变异函数和协变异函数需要满足柯西-史克瓦兹不等式：

$$|r_{12}(h)| \leq |r_{11}(h)r_{22}(h)|^{\frac{1}{2}} \quad (22)$$

协克里金的估值误差为

$$\sigma_{ck}^2 = \sum_{i=1}^{n_1} \lambda_{1i} r_{12}(x_{1i} - x_0) + \sum_{j=1}^{n_2} \lambda_{2j} r_{22}(x_{2j} - x_0) + \mu_2 \quad (23)$$

3.4 多重分形理论

多重分形是由 Mandelbrot(1982 年)在研究湍流时提出来的,以后逐渐建立它的数学形式并在各领域得到应用。多重分形是研究物理量和其他量在几何支撑上的奇异性分布而引入的基本概念。用矩方法特征化多重分形,矩方法采用剖分函数计算多重分形维数 α,公式为

$$\alpha = \lim_{\varepsilon \to 0} \frac{\lg \langle \chi(\varepsilon) \rangle}{\lg \varepsilon} = \lim_{\varepsilon \to 0} \frac{\lg \left\langle \sum_{i=1}^{N(\varepsilon)} \mu_i(\varepsilon) \right\rangle}{\lg \varepsilon} \quad (24)$$

式中:$\langle \rangle$ 为定义在剖分数据集合 S 上的测度 $\mu_i(\varepsilon)$ 的统计矩;ε 为测量尺度的大小,本场的值取 0.1 m(钻孔取样直径);$N(\varepsilon)$ 为容量的个数。

众多的 Kriging 空间插值或滤波方法都是基于对场值的某种滑动加权平均,将预测式改写为

$$\hat{Z}(x_0) = \sum_{\Omega(x_0,\varepsilon)} \omega(\|x_0 - x\|) Z(x) \quad (25)$$

式中:$\Omega(x_0, \varepsilon)$ 为围绕中心点 x_0 半径为 ε 的小滑动窗口;$\omega(\|x_0 - x\|)$ 为 $\Omega(x_0, \varepsilon)$ 中与中心点 x_0 相隔距离 $\|x_0 - x\|$ 的任意点 x 的加权函数。它可由 Kriging 方法确定,但不涉及局部奇异性度量。由 Cheng 提出的多重分形方法[17]将滑动平均关系式表达为

$$\hat{Z}(x_0) = \varepsilon^{\alpha-2} \sum_{\Omega(x_0,\varepsilon)} \omega(\|x_0 - x\|) Z(x) \quad (26)$$

式中:α 为 x_0 处的局部多重分形维数。

以上表达式不仅包含了空间相关性的成分,而且具有度量局部奇异性的因子。由此可见,多重分形适合于局部异常的描述,通常的加权平均方法只是该多重分形方法的特殊情况。

3.5 交叉检验

模型交叉检验基于以下定义:正则化平均值绝对误差(normalized mean absolute error,NMAE),正则化平方根误差(normalized root mean square error,NRMSE)。NMAE 定义如下:

$$NMAE = 100\% \frac{\sum_{i=1}^{M} |\hat{z}_i - z_i|}{\sum_{i=1}^{M} z_i} \quad (27)$$

NRMSE 定义如下:

$$NRMSE = 100\% \frac{\sum_{i=1}^{M} [\hat{z}_i - z_i]^2}{\frac{1}{M} \sum_{i=1}^{M} z_i} \quad (28)$$

式中:$\hat{z}_i$ 为变量 z_i 的估计值;z_i 为点 i($i = 1, 2, \cdots, M$)处的观测值。NRMSE 在工程应用更广泛,在某些场合,NMAE 的健壮性更好。

预测模型的效率还可以通过 G 值来定义:

$$G = \left(1 - \left\{\sum_{i=1}^{M} [\hat{z}_i - z_i]^2 / \sum_{i=1}^{M} [\hat{z}_i - \bar{z}]^2 \right\}\right) \quad (29)$$

式中:$\bar{z}$ 为所有观测值的平均值。当 $G = 1$ 时,所采用的估计方法是最好的估计;当 $0<G<1$ 时,所采用的估计方法是良好的估计,愈接近1,效果更好;当 $G = 0$ 时,所采用的估计方法和采用数据的平均值估计没有区别;当 $G<0$ 时,所采用的估计方法比直接采用数据的平均值估计效果要差。

4 结果与讨论

在协方差预分析中,c 和 φ 以及 φ 和 E_s 的互相关关系更高,在协克里金模型中,一个将作为第一变量,另一个将作为第二变量。

在泛克里金模型中,漂移形式取以下形式:

$$\hat{m}(x_0) = \hat{\mu}_0 + \hat{\mu}_1 x + \hat{\mu}_2 y \quad (30)$$

式中:$\hat{m}(x_0)$ 为未知点 x_0 处的平均值;$\hat{\mu}_i$ 为权值系数;x、y 为点 x_0 处的坐标值。

图 1 为采用普通克里金(OK)模型、泛克里金(UK)模型、协克里金(CK)模型,多重分形联合(MK)模型的 c、φ 和 E_s 的观测值与预测值的散点图。散点图是大规模数据统计质量控制的基本工具,能够有反映以下特征:(1)数据强度;(2)形状(线性,非线性等);(3)斜率方向(正向或负向);(4)离散值。数据强度对于 3 个变量的 4 个模型没有区别,观测值域预测值基本接近直线关系,且回归直线的斜率为正值,多重分形联合模型分析的效果最好,协克里金的模型效果次之,对于岩土参数 c、φ,普通克里金模型稍优于泛克里金模型,而对于岩土参数 E_s,泛克里金模型略优于普通克里金模型,4 个模型都存在不同程度的离散值。

图 2 为采用普通克里金(OK)模型、泛克里金(UK)模型、协克里金(CK)模型,多重分形联合(MK)模型的 c、φ 和 E_s 的盒须图。图 2(a)为原始数据图,图 2(b)为去除最大、最小值后的盒须图。

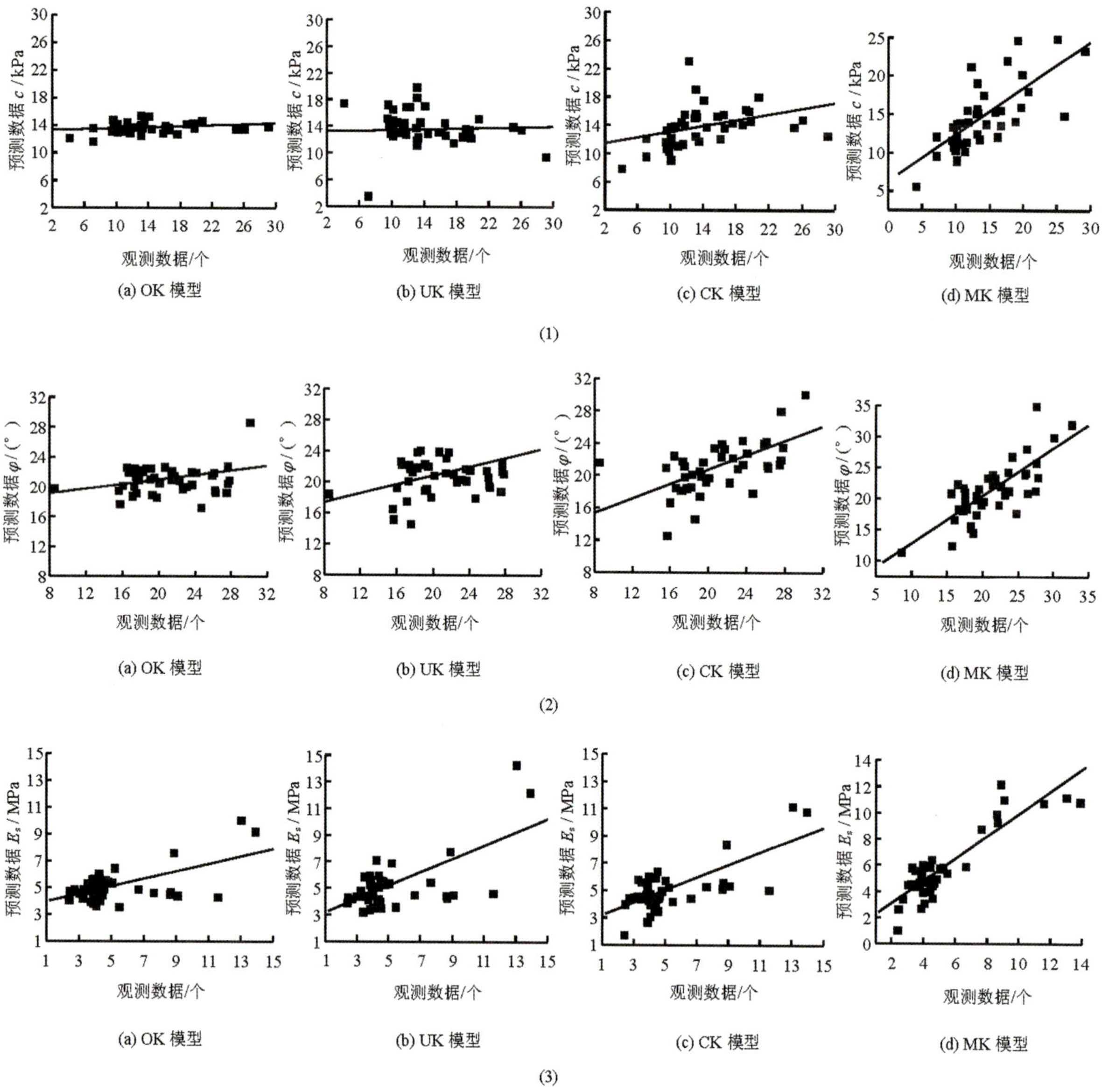

图 1 粉质黏土层 c、φ 和 E_s 采用不同模型的观测数据与预测数据散点图

Fig.1 Scatter plots of estimated data versus observed data for parameters parameters c, φ and E_s using different models of stratum silty clay

盒须图可以提供以下信息：(1) 数据分布在盒须图中的分布；(2) 按照数据偏离中线程度判断偏态性；(3) 50% 的数据集中在方盒内；(4) 25% 和 75% 的分位数为方盒上端和下端的直线；(5) 最大、最下值为直线端点；(6) 数据中位数为方盒中部的刻画线；(7) 数据离散点为须线点。

从图 2(b) 可以看出，25% 至 75% 之间的数据由多重分形联合（MK）模型、协克里金（CK）模型、泛克里金（UK）模型、普通克里金（OK）模型依此模拟较好，4 个模型的偏态性和 50% 的中位数模拟都较好。对于参数 c，除了多重分形联合模型，其余 3 个模型都低估了其最大值，最小值也是 MK

模型模拟最好，OK 模型和 CK 模型低估了其值，UK 模型高估了其值；对于参数 φ，CK 模型较好地模拟了其分布，OK 模型和 UK 模型低估了其分布，MK 联合模型高估了分布范围；对于参数 E_s，MK 联合模型较好的模拟了其分布，OK 模型、UK 模型、CK 模型不同程度缩短了数据分布范围。

图 3 为不同模型的 c、φ 和 E_s 的 $NMAE$、$NRMSE$ 和 G 值。对于 $NMAE$，OK 模型的值最小，CK 模型次之，正因为 MK 模型的局部加强作用，导致该值偏大，需要加入 $NRMSE$ 等参数综合判断；对于 $NRMSE$，MK 联合模型的值最小，CK 模型次之，UK 模型除了参数 E_s，数值最大；对于 G 值，MK

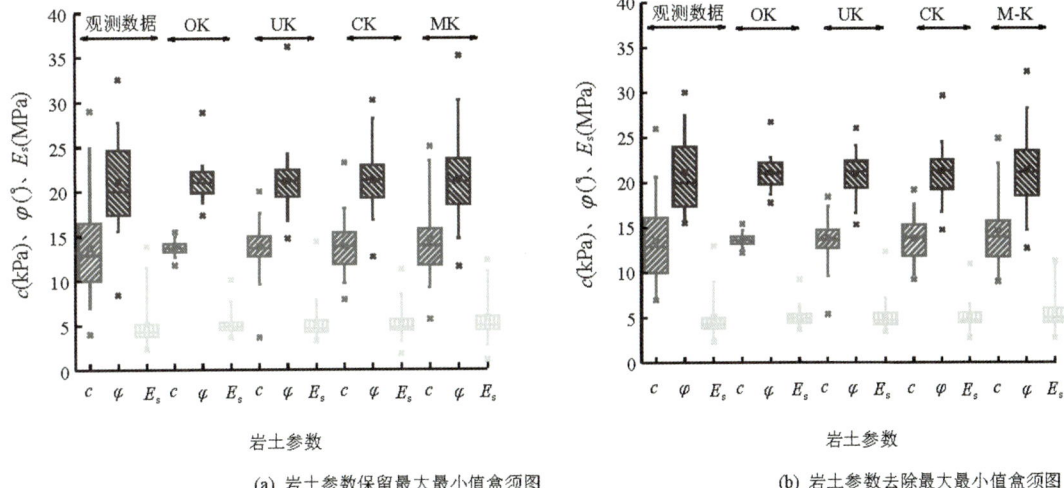

(a) 岩土参数保留最大最小值盒须图   (b) 岩土参数去除最大最小值盒须图

图 2　粉质黏土层 c、φ 和 E_s 采用不同模型的盒须图
Fig.2　Box-and-whisker plots of the observed data and estimated data for parameters parameters c, φ and E_s using different Kriging models

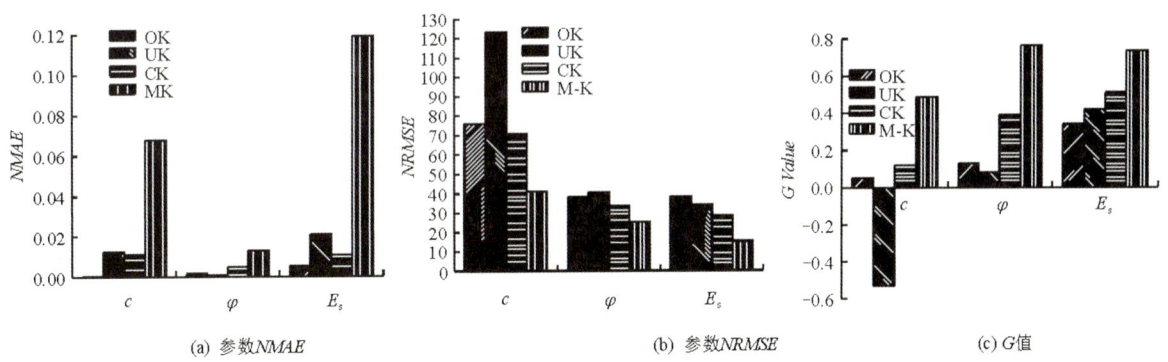

(a) 参数 NMAE　　(b) 参数 NRMSE　　(c) G 值

图 3　粉质黏土层 c、φ 和 E_s 采用不同模型的 NMAE、NRMSE 和 G 值
Fig.3　NMAE, NRMSE, and G-value for parameters c, φ and E_s using different Kriging models

联合模型的值最大，CK 模型次之，UK 模型对于参数 c，还出现了负值。基于以上数据，模型模拟优劣顺序为：MK 联合模型、CK 模型、OK 模型和 UK 模型。

4　结　语

通过介绍原 2010 世博园世博轴区域粉质黏土的岩土参数 c、φ 和 E_s 的 4 个空间模拟模型：普通克里金（OK）模型、泛克里金（UK）模型、协克里金（CK）模型，多重分形联合（MK）模型。基于经典统计学和地统计结果，虽然每一种模型都有其局限性，但多重分形联合模型表现最好，能够捕捉到岩土参数局部的奇异性特征，其次为协克里金模型，由于岩土参数 c、φ 缺乏明显的空间趋势，泛克里金模型仅在参数 E_s 的预测中稍优于普通克里金模型。

多重分形理论能够度量岩土参数空间的局部奇异性，多变量协同分析可以提高岩土参数空间分布的预测精度。由于克里金算法属于确定性算法，在以后的研究中将深入研究岩土参数软、硬数据分布的随机模拟算法，提高对勘察数据的解译能力。

参 考 文 献

[1] LEGENDRE P. Spatial autocorrelation: Trouble or new paradigm[J]. **Ecology**, 1993, 74(6): 1659－1673.

[2] GUMPERTZ M L, GRAHAM J M, RISTAINO J B. Autologistic model of spatial pattern of phytophthora epidemic in bell pepper: Effects of soil variables on disease presence[J]. **Journal of Agricultural, Biological, and Environmental Statistics**, 1997, 2(2): 131－156.

[3] MILLER J, FRANKLIN J, ASPINALL R. Incorporating spatial dependence in predictive vegetation models[J].

Ecological Modelling, 2007, 202(3): 225−242.

[4] TRIANTAFILIS J, ODEH I O A, MCBRATNEY A B. Five geostatistical models to predict soil salinity from electromagnetic induction data across irrigated cotton[J]. **Soil Science Society of America Journal**, 2001, 65(3): 869−878.

[5] ELDEIRY A, GARCIA L A. Detecting soil salinity in alfalfa fields using spatial modeling and remote sensing[J]. **Soil Science Society of America Journal**, 2008, 72(1): 201−211.

[6] ODEH I O A, MCBRATNEY A B, CHITTLEBOROUGH D J. Further results on prediction of soil properties from terrain attributes: Heterotopic cokriging and regression-kriging[J]. **Geoderma**, 1995, 67(3): 215−226.

[7] YATES S R, WARRICK A W. Estimating soil water content using cokriging[J]. **Soil Science Society of America Journal**, 1987, 51(1): 23−30.

[8] STEIN A, DOOREMOLEN W, BOUMA J et al. Cokriging point data on moisture deficit[J]. **Soil Science Society of America Journal**, 1988, 52(5): 1418−1423.

[9] ZHANG R, MYERS D E, WARRICK A W. Estimation of the spatial distribution of the soil chemicals using pseudo-cross-variograms[J]. **Soil Science Society of America Journal**, 1992, 56(5): 1444−1452.

[10] ISTOK J D, SMYTH J D, FLINT A L. Multivariate geostatistical analysis of ground-water contaminant: A case history[J]. **Ground Water**, 1993, 31(1): 63−74.

[11] SHOUSE P J, GERIK T J, RUSSELL W B et al. Spatial distribution of soil particle size and aggregate stability index in a clay soil[J]. **Soil Science**, 1990, 149(6): 351−360.

[12] MARTINEZ C A. Multivariate geostatistical analysis of evapotranspiration and precipitation in mountainous terrain[J]. **Journal of Hydrology**, 1996, 174(1): 19−35.

[13] 李晓军, 王长虹, 朱合华. Kriging 插值方法在地层模型生成中的应用研究[J]. 岩土力学, 2009, 30(1): 157−162.
LI Xiao-jun, WANG Chang-hong, ZHU He-hua. Kriging interpolation and its application to generating stratum model[J]. **Rock and Soil Mechanics**, 2009, 30(1): 157−161.

[14] MILLER J, FRANKLIN J, ASPINALL R. Incorporating spatial dependence in predictive vegetation models[J]. **Ecological Modelling**, 2007, 202(3): 225−242.

[15] PHILLIPS J D. Measuring complexity of environmental gradients[J]. **Vegetatio**, 1986, 64(3): 95−102.

[16] ELDEIRY A, GARCIA L A. Detecting soil salinity in alfalfa fields using spatial modeling and remote sensing[J]. **Soil Science Society of America Journal**, 2008, 72(1): 201−211.

[17] 王长虹, 朱合华. 多重分形与 Kriging 插值在地层模型生成中的应用[J]. 岩土力学, 2011, 32(6): 1864−1868.
WANG Chang-hong, ZHU He-hua. Application of multi-fractal and Kriging interpolation to reconstruction of stratum[J]. **Rock and Soil Mechanics**, 2011, 32(6): 1864−1869.

| 学术论文——岩石力学与工程

深部硐室围岩分区破裂化现象数值模拟研究

苏仲杰[1,2]，钱七虎[2]

(1.辽宁工程技术大学土木与交通学院,阜新 123000；2.解放军理工大学国防工程学院,南京 210007)

摘　要：围绕深部岩体工程的分区破裂化现象,通过数值模拟方法研究了深部硐室围岩分区破裂化现象产生的力学机理及其应力的变化规律。针对深部圆形硐室围岩,分别对应力状态 $P_y > P_x$、$P_y = P_x$、$P_y < P_x$ 进行数值模拟,发现了深部围岩产生的应力集中现象。当应力集中中的应力值超过岩石的强度时,就会产生分区破裂化现象。

关键词：深部硐室围岩；　分区破裂化现象；　数值模拟；　应力集中

中图分类号：TU 45　　　　**文献标识码**：A　　　　**文章编号**：1671-4431(2014)02-0089-06

Numerical Simulation Research on Zonal Disintegration Phenomenon of Rock Mass Around Deep Grotto

SU Zhong-jie[1,2], *QIAN Qi-hu*[2]

(1. School of Civil Engineering and Transportation, Liaoning Technology University, Fuxin 123000, China；
2. Institute of National Defense Engineering, PLA University of Science and Technology, Nanjing 210007, China)

Abstract: The mechanics mechanism and stress changing law of the occurrence of zonal disintegration phenomenon of rock mass around deep grotto was researched through the numerical simulation revolved around this phenomenon of deep rock mass engineering. The stress state of rock mass around deep round grotto was simulated under three stress states ($P_y > P_x, P_y = P_x, P_y < P_x$). Then stress concentration phenomenon which was engendered by the deep surrounding rock was found. It was the zonal disintegration phenomenon that occurred when the concentrated stress value of the deep surrounding rock was more than its strength value.

Key words: rock mass around deep grotto； zonal disintegration phenomenon； numerical simulation； stress concentration

从很多的深部工程实例可以归纳出如下的分区破裂化现象的规律性[1-5]：围岩中的分区破裂化现象大致发生在深部岩体围岩中的初始垂直地应力 $\sigma_{地}$ 大于岩体单轴压缩强度极限 R_c 的情况下；分区破裂化现象中破裂区的数量取决于比值 $\sigma_{地}/R_c$,比值越大,破裂区越多,反之则越少；分区破裂化现象既发生在巷道钻爆法施工时的情况下,也发生在巷道机械化掘进时的情况下；巷道机械法掘进时开始发生分区破裂化现象时的岩体初始地应力 $\sigma_{地}$ 一般高于钻爆法掘进时开始发生分区破裂化现象时的相应地应力 $\sigma_{地}$。这意味着,卸载自由面形成的速度对分区破裂化现象的产生也有一定的影响。

分区破裂化现象和规律与传统的连续介质岩石力学理论不相一致,依据后者,地下硐室和巷道围岩依次出现处于不同应力应变状态的破裂区、塑性区和弹性区。

1　围岩的两类破坏形态与分区破裂化现象的发生机理

依据统计断裂力学和岩体的层次构造理论,岩体可看作为含有微裂纹的块体材料[6-9]。地下巷道的开挖

收稿日期：2013-11-29。
基金项目：国家自然科学基金(2005037222)和中国博士后科学基金(2005037222)。
本文原载于《武汉理工大学学报》(2014 年第 36 卷第 2 期)。

导致应力集中,在低围压荷载作用下,当支座压力荷载达到岩石材料峰值极限应力 1/2～1/3 时,微裂纹很快地在该区域均匀地发展,并密集成羽状排列,发展成微裂缝密集带,当达到一定密度时,然后自发地联合成粗裂纹、宏观裂缝。在产生宏观裂缝——岩石材料的破坏的同时,在其邻域由于裂缝卸荷的作用,停止了微裂纹的产生,从而停止了材料的破坏。

上述围岩破坏的过程需要一个时间准备,即大量微裂的发展、联合和形成宏观裂缝的过程不是瞬时,而是在一个时间段内完成的。

当初地应力水平高时,岩石从高应力处向卸荷区(巷道周边)位移速度急剧提高,超过了微裂纹发生发展的速度,微裂缝来不及发展,就很快地形成宏观裂缝,随之在宏观裂缝邻域,由于其卸荷作用停止了微裂纹的产生。同时,由于应力重分布,应力集中的峰值转移到围岩深处(试验表明,该转移速度小于岩石径向位移速度)。当初应力水平足够高,即岩石中储积的弹性能足够高,上述能量流过程得以继续,则在产生第一批宏观裂缝一定距离处,即其卸荷影响范围外,产生新的宏观裂缝,上述过程一直持续到平衡状态的恢复,即在围岩深处的应力集中系数降低,径向应力增加,从而岩石位移速度大为降低,来临了岩石平衡变形的阶段。综上所述,分区破裂化现象的发生主要是由于较高的初地应力水平和巷道开挖形成的卸荷面,破裂区的数量也取决于初应力水平的高低。

2 深部围岩分区破裂化现象数值模拟研究

选择合适的数值仿真模拟软件,对于模拟对象来说是非常必要的。选择的数值仿真模拟软件,应该既能模拟岩石的连续变形过程又能模拟岩石的破断过程。RFPA^{2D}仿真软件比较符合要求。此次数值仿真模型的原型与物理仿真模拟相同,目的在于验证两种仿真模拟结果是否一致,为后续研究工作提供可靠的依据。

程序 RFPA 假设岩层中每一个微分为均质,各向同性线弹脆性体。不同单元的弹模和强度等力学参数服从韦伯分布。认为当单元强度达到破坏认为节理、裂隙与岩石一样,也是岩体的组成部分,区别仅仅是力学性质的不同,只是具有极低的弹模和强度。因此该程序可以用连续介质力学的方法处理非连续性的问题。

RFPA^{2D}(Rock Failure Process Analysis)岩层破断过程分析系统是由东北大学唐春安等[10]人研制开发的仿真软件。该系统将岩石材料视为非线性材料,从损伤力学的角度,考虑到岩石的损伤过程,提出了连续损伤力学的概念,建立了岩石的损伤模型。

2.1 RFPA^{2D}仿真数值模拟设计

2.1.1 硐室围岩的设计

模拟硐室围岩的岩石重度为 2.5,抗压强度为 60 MPa,弹性模量为 6 000 MPa,泊松比为 0.25;模型的长度 40 m,宽 40 m,硐室为圆形,直径为 6.8 m。

2.1.2 硐室围岩施加载荷设计

圆形硐室围岩的受力状态分为 3 种情况:

1)自重应力状态加上垂直应力大于水平应力状态($P_y > P_x$) 自重应力状态加上垂直应力 P_y 初始载荷为 10 MPa,P_x 初始载荷为 5 MPa,然后 P_y、P_x 每一步增加 2 MPa,直至圆形硐室破坏。

2)自重应力状态加上垂直应力等于水平应力状态($P_y = P_x$) 自重应力状态加上垂直应力 P_y 初始载荷为 5 MPa,P_x 初始载荷为 5 MPa,然后 P_y、P_x 每一步增加 2 MPa,直至圆形硐室破坏。

3)自重应力状态加上垂直应力小于水平应力状态($P_y < P_x$) 自重应力状态加上垂直应力 P_y 初始载荷为 5 MPa,P_x 初始载荷为 10 MPa,然后 P_y、P_x 每一步增加 2 MPa,直至圆形硐室破坏。

圆形硐室围岩的受力状态各剖面情况:确定 0°剖面上、45°剖面上、90°剖面上、135°剖面上的水平应力状态、垂直应力状态、剪应力状态,各剖面及模型如图 1 所示。

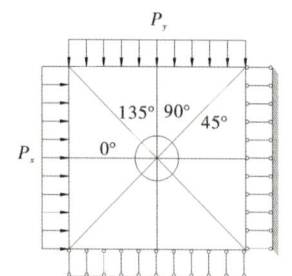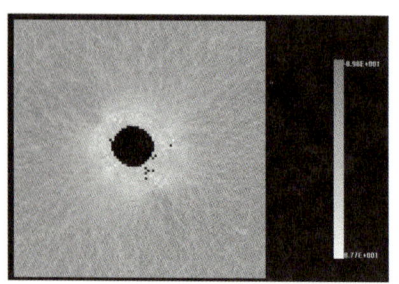

图1 圆形硐室围岩的受力状态、各剖面及模型图

2.2 深部围岩分区破裂化现象数值模拟结果与分析

2.2.1 自重应力状态加上垂直应力等于水平应力状态（$P_y = P_x$）数值模拟结果与分析

由图2、图3可知，$P_x = P_y = 31$ MPa条件下，水平在0°剖面上，在圆形硐室的内壁的应力集中最大；水平应力在45°剖面，在圆形硐室的围岩内与剪应力的分布规律相同，且在数值上基本相等；水平应力在90°剖面上，在圆形硐室的内壁的应力最小，而剪切应力在圆形硐室的内壁上达到最大值；水平应力在135°剖面上与45°剖面上的应力分布规律相同。在45°、135°剖面垂直应力为拉应力。随着$P_x = P_y$的增大，在未达到硐室围岩的单向抗压、抗剪强度情况下，应力的分布规律保持上述的分布规律。

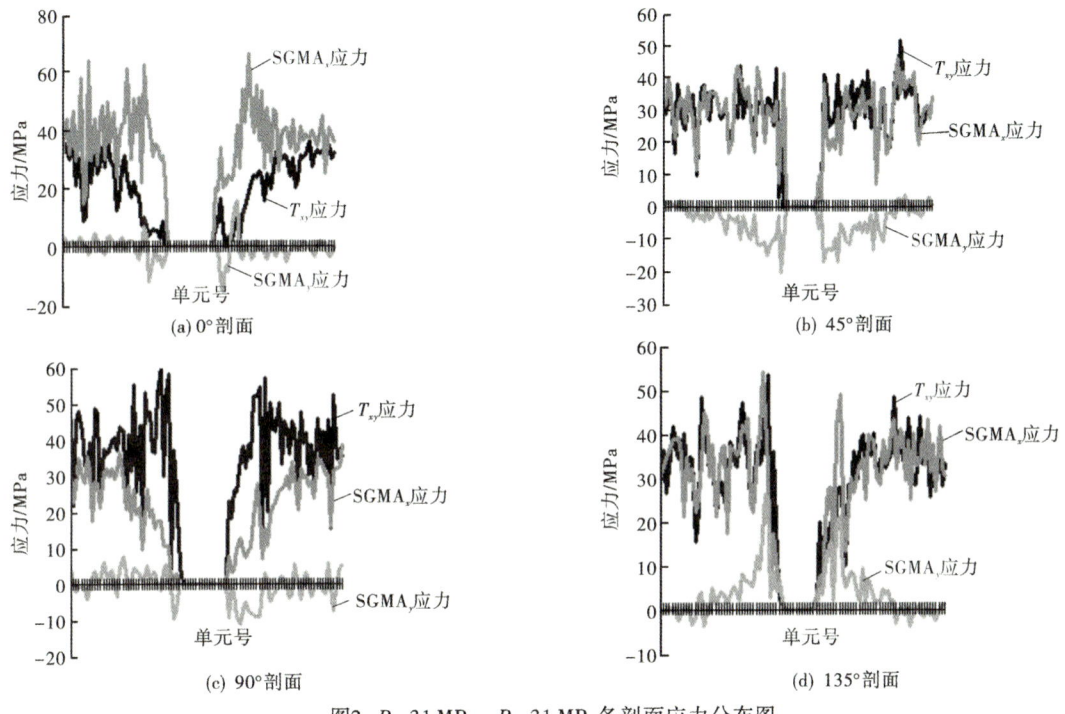

图2 $P_y = 31$ MPa，$P_x = 31$ MPa各剖面应力分布图

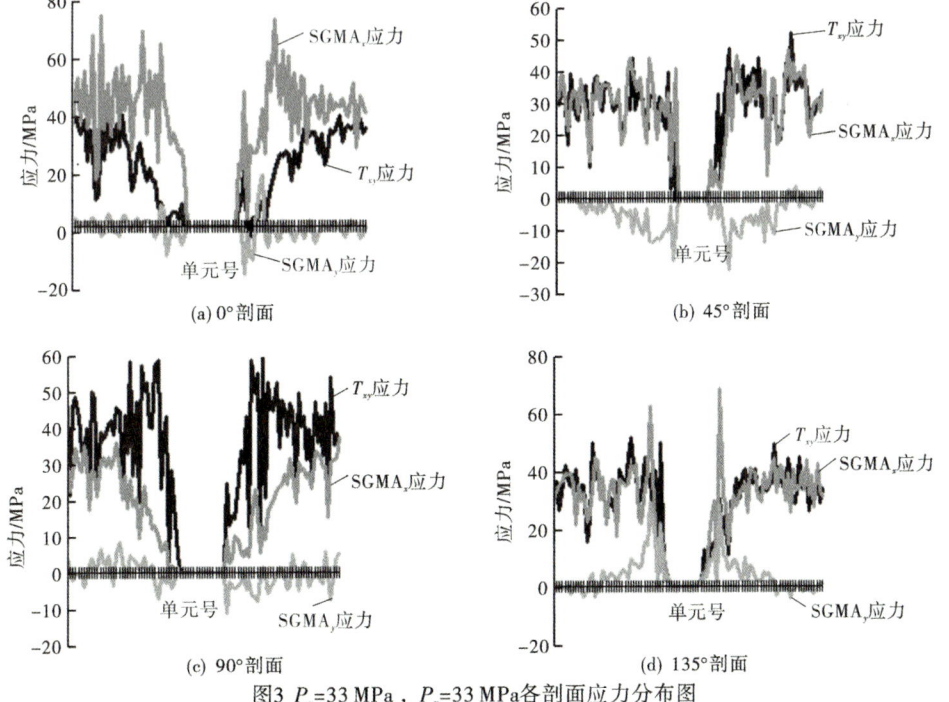

图3 $P_y = 33$ MPa，$P_x = 33$ MPa各剖面应力分布图

2.2.2 自重应力状态加上垂直应力小于水平应力状态（$P_y < P_x$）数值模拟结果与分析

由图4、图5可知，$P_y < P_x$ 条件下，水平在0°剖面上，在圆形硐室的内壁的应力集中最大；水平应力在45°剖面，在圆形硐室的围岩内与剪应力的分布规律相同，且在数值上基本相等；水平应力在90°剖面上，在圆形硐室的内壁的应力最小，而剪切应力在圆形硐室的内壁上达到最大值；水平应力在135°剖面上与45°剖面上的应力分布规律相同。在45°、135°剖面垂直应力为拉应力。

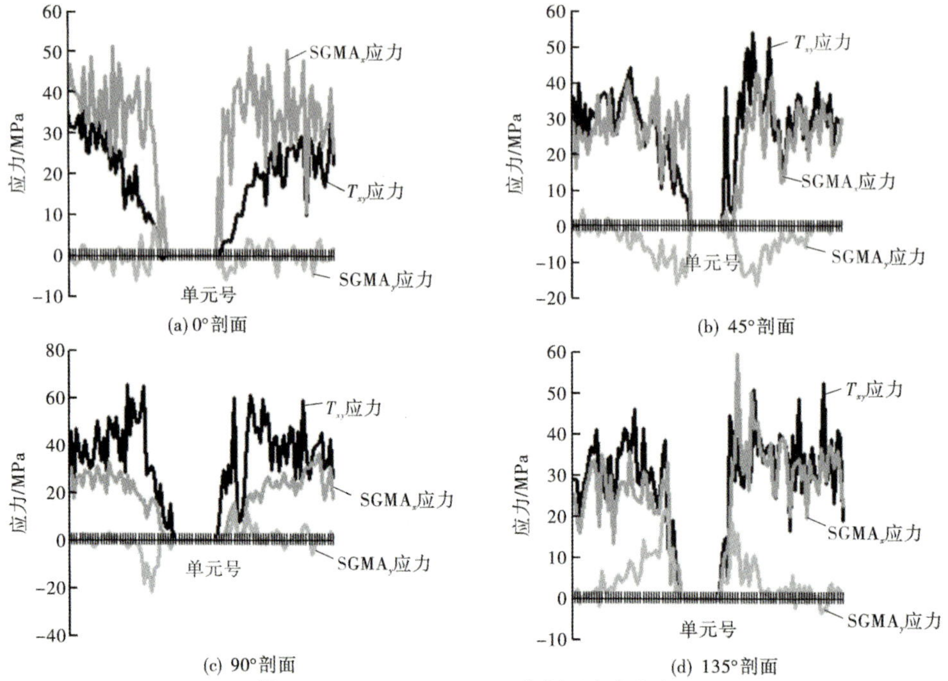

图4 $P_y=27$ MPa，$P_x=32$ MPa各剖面应力分布图

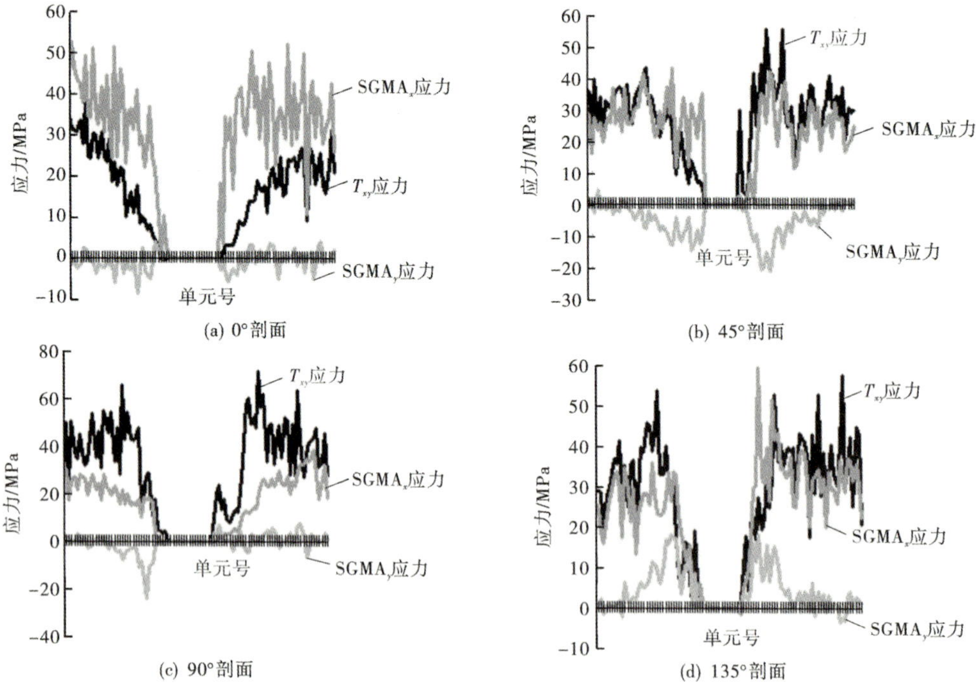

图5 $P_y=29$ MPa，$P_x=34$ MPa各剖面应力分布图

2.2.3 自重应力状态加上垂直应力小于水平应力状态（$P_y > P_x$）数值模拟结果与分析

由图6、图7可知，$P_y > P_x$ 条件下，水平在0°剖面上，在圆形硐室的内壁的应力集中最大；水平应力在

45°剖面,在圆形碉室的围岩内与剪应力的分布规律相同,且在数值上基本相等;水平应力在90°剖面上,在圆形碉室的内壁的应力最小,而剪切应力在圆形碉室的内壁上达到最大值;水平应力在135°剖面上与45°剖面上的应力分布规律相同。在45°、135°剖面垂直应力为拉应力。

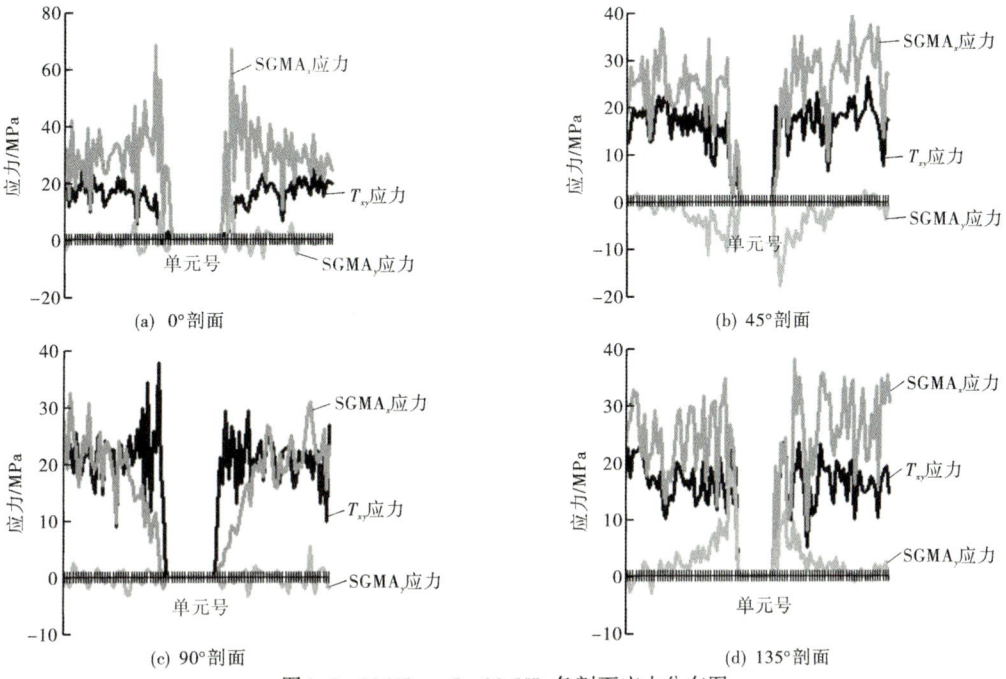

图6 $P_y=28$ MPa,$P_x=23$ MPa各剖面应力分布图

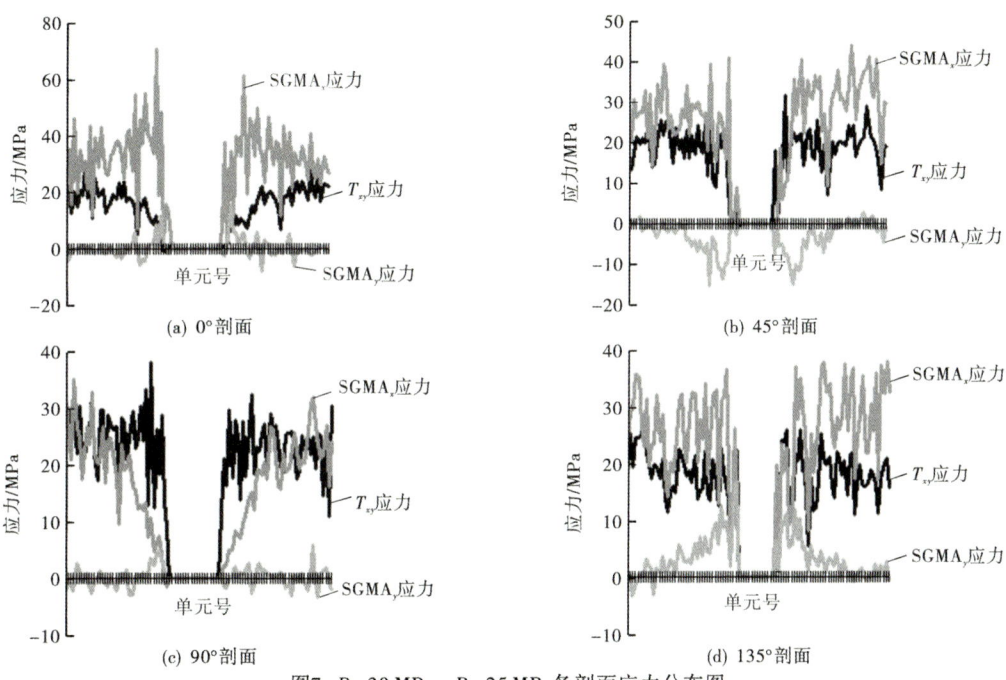

图7 $P_y=30$ MPa,$P_x=25$ MPa各剖面应力分布图

3 结 论

a. 在$P_y > P_x$情况下,圆形碉室围岩的局部会产生破裂化现象。

b. 在$P_y = P_x$情况下,圆形碉室围岩的局部会产生破裂化现象,但相对$P_y > P_x$情况下局部会产生破裂化现象要多一些。

c. 在 $P_y < P_x$ 情况下,圆形硐室围岩的局部会产生围绕硐室的分区破裂化现象。

d. 深部地下硐室的围岩分区破裂化现象产生的条件:$\sigma_{地}/R_c$ 越大,破裂区越多;$P_x > P_y$ 的情况下,会产生闭合分区破裂化现象;$T_{xy} > [T_{xy}]$;在硐室的围岩内 0°、45°、90°、135°剖面内产生交替的应力升高区和应力降低区。随着深度的增加,产生的围岩应力集中现象,应力集中的应力值超过岩石的强度(单向抗压、单向抗剪强度)时,就会产生分区破裂化现象。

参考文献

[1] 钱七虎. 深部地下空间开发中的关键科学问题[R]. 南京:解放军理工大学工程兵工程学院,2004.

[2] 王明洋,周泽平,钱七虎. 深部岩体的构造和变形与破坏问题[J]. 岩石力学与工程学报,2006,25(3):448-455.

[3] 戚承志. 岩石的构造层次及力学性质和研究方法[R]. 南京:解放军理工大学工程兵工程学院,2004.

[4] 王明洋,宋华,郑大亮,等. 深部巷道围岩的分区破裂机制及"深部"界定探讨[J]. 岩石力学与工程学报,2006,25(9):77-82.

[5] 钱七虎. 深部岩体工程响应的特征科学现象及"深部"的界定[J]. 东华理工学院学报,2004,27(1):1-5.

[6] 戚承志,钱七虎. 材料变形及损伤演化的微观物理机制[J]. 固体力学学报,2002,23(3):312-317.

[7] 戚承志,钱七虎. 关于岩石的剥离破坏过程及混合破坏准则[J]. 世界地震工程,2002,18(4):70-76.

[8] 戚承志,王明洋,钱七虎. 弹粘塑性孔隙介质在冲击载作用下的一种本构关系 第二部分:在冲击载作用下弹粘塑性孔隙介质的畸变行为[J]. 岩石力学与工程学报,2003,21(11):1763-1766.

[9] 戚承志,钱七虎. 岩石等脆性材料动力强度依赖应变率的物理机制[J]. 岩石力学与工程学报,2003,21(2):177-181.

[10] 唐春安. 岩石破裂过程中的灾变[M]. 北京:煤炭工业出版社,1993.

学术论文——岩石力学与工程

岩爆、冲击地压的定义、机制、分类及其定量预测模型

钱七虎[1,2]

（1. 解放军理工大学 国防工程学院，南京 210007；2. 解放军理工大学 爆炸冲击防灾减灾国家重点实验室，南京 210007）

摘 要：引述了若干国际权威学者关于岩爆的机制和定义的论述，在此基础上，依据岩爆发生的不同机制，将岩爆分为断层滑移或者剪切断裂所导致的断裂滑移型和岩石破坏导致的应变型岩爆，并结合事故案例，分析了应变型和滑移型两类岩爆及冲击地压的发生机制和特点。在机制分析的基础上，介绍了岩（煤）柱应变型岩爆和围岩应变型岩爆以及断裂滑移型岩爆定量预测的研究成果，其中，包括作者利用非欧几何模型研究非协调变形影响后对应变型岩爆进行了定量预测和数值模拟的新研究成果。

关 键 词：岩爆机制；断裂滑移；应变型岩爆；定量预测

中图分类号：TD 325　　**文献标识码**：A

Definition, mechanism, classification and quantitative forecast model for rockburst and pressure bump

QIAN Qi-hu[1,2]

（1. College of National Defence Engineering, PLA University of Science and Technology, Nanjing 210007, China; 2. State Key Laboratory of Disaster Prevention & Mitigation of Explosion & Impact, PLA University of Science and Technology, Nanjing 210007, China）

Abstract: As a review paper, the discourses on the mechanism and the definition of rockburst of authoritative experts were quoted, based on these discourses and according to the different mechanisms of rockburst, rockbursts are divided into the sliding mode rockburst resulting from fault-slip events and strain mode rockburst resulting from the failure of the rock. Combined with the specific accident cases, the mechanism and characteristics of the two types of rockburst are analyzed. Finally, based on the analysis of the mechanism, the quantitative forecast and numerical simulation of rock (coal) pillar strain mode rockburst, enclosing rock strain mode rockburst and sliding mode rockburst are introduced, in which, the author's latest research of quantitative forecast and numerical simulation for strain mode rockburst after incompatible deformation by using non-Euclidean geometry model is included.

Key words: rock burst mechanism; sliding mode fracture; strain mode rockburst; quantitative forecast

1 引 言

随着交通和经济的发展，在深部地下空间修建了大量的矿井和隧道，矿井开挖深度已达地面下4 000 m，规划深度已达地下 5 000 m，民用隧道最大埋深已超 2 500 m。在修建大深度矿井和隧道时，岩爆已成为普遍问题[1]，发生在南非、智利、加拿大、澳大利亚、俄罗斯等国的矿井以及挪威、美国、中国、瑞典、瑞士等国的隧道工程中。岩爆严重威胁着施工人员的安全，造成了巨大的经济损失。例如 2010 年 11 月 28 日发生于中国锦屏二级水电站 2 500 m 埋深的引水隧洞中的岩爆，导致了 7 人死亡和全断面岩石掘进机的严重损坏。

2 岩爆的机制和定义

虽然国内文献中关于岩爆的机制、定义和分类众说纷纭，不尽一致，但是国际上研究岩爆的权威学者关于岩爆的定义、机制和分类是基本相同的。

岩爆是一个物理现象，所以其定义应是现象的描述，以下引述了若干定义和机制：

收稿日期：2013-11-09。

基金项目：国家自然科学基金资助项目（No. 51325903，No. 51021001，No. 512792）。

本文原载于《岩土力学》（2014 年第 35 卷第 1 期）。

（1）岩爆是突然的岩石破坏，其特征是岩石的破碎和从围岩中突出并伴随着能量的猛烈释放[2]。

（2）开挖一个新地下孔洞或者改变一个已有孔洞造成围岩中的应力变化，这些应力变化能导致孔洞附近岩体的破坏，或者诱发已有断裂的滑移[3]；第1类岩爆定义为已有断裂的滑移，第2类岩爆定义为一定体积的岩石的脆性破坏[3]。

（3）为了发生断裂型岩爆，原生的或感生的应力水平必须足够高，以能激活原生的断层面的运动，或启动地质结构面的活动，或者在岩体中形成断裂（新断层），岩爆是岩体动力破坏的后果[1]。

（4）矿井中动力现象（岩爆）的本质是围岩获得了动能[4]。

（5）采矿诱发的岩爆是与平衡状态的失稳相连的，它可以包括：①已有断裂面的滑移；②岩体的破裂[3]。

3 岩爆的分类和冲击地压

岩爆因其表现形式的不同，形成不同类型的岩爆，但其物理现象的本质是相同的，所以都应称为岩爆。

"地震性事件"、"动力事件"、"动力现象"、"动力失稳"、"突变"、"突变形式的失稳"都是同义的，都意味由一个平衡状态向另一个状态的突变，伴随着剩余能量以应力波形式的释放。它可以发生于不同的尺度：在加载试件中，它呈现于微观尺度和颗粒形式，伴随着声发射；在矿井中，它不仅有声发射，而且有岩爆；在地壳中，除了相对小的突变外，有时还发生十分猛烈的突变，如地震[5]。

两类岩爆为两类突变，即体积性失稳和接触面失稳[5]：Ⅰ型岩爆，由断层滑移事件所导致的；Ⅱ型岩爆，由岩石破坏所导致的，它们包括围岩应变岩爆和岩柱应变岩爆[3]。

岩爆形式的动力破坏，基本可以区分为2类：第1类，通常称为应变型岩爆，是由岩石破坏导致的；第2类是断层滑移或者剪切破裂所导致的。两类岩爆的主要差别是：在第1类中，扰动源（开挖）和岩爆破坏部位是相重合的；在第2类中，其扰动源和所导致的岩爆破坏部位可以分离相当大的距离。与第2类滑移断裂型岩爆相联系的能量通常甚大于应变型岩爆的能量。断裂滑移事件的岩爆破坏通常远比应变型岩爆事件强烈得多，在矿井环境中通常在单一事件中有数十米或者甚至数百米巷道被破坏[1]。

硬岩矿井中大量的地震性事件属于剪切型，或滑移型失稳类[6]。

累积的证据导致的结论是：在南非矿井中，剪切型，也可能是沿已有断裂面或新鲜断裂面滑移型破坏的矿震起了优势地位[7]。

切不可忽视爆炸波对触发煤爆的影响。门头沟矿在700～900 m深处的114起煤爆中，有89起（占78%）是因爆破触发的，在龙凤矿700 m深处因爆破而触发的煤爆也超过总数的50%。由此可以推断，在这样的深度，煤一定是处于准稳定平衡状态，爆炸波的触发作用能导致死亡事故发生。有些矿也出现过顶板爆裂，坚硬岩层顶板更为不利的影响是爆裂会扩展而远远超前于长壁工作面，从而导致失稳和煤爆[8]。

为了分析冲击地压和岩爆的异同，下面再介绍中国煤矿的两个实际冲击地压事故的调查结论。

（1）华丰煤矿冲击地压事故综合剖析：4层煤的多个工作面发生过冲击地压。华丰井田煤系地层以上为500～800 m砾岩层，砾岩层坚硬整体性强，其断裂垮落对下部的煤岩体产生冲击载荷，是4层煤工作面发生冲击地压的主要力源。

（2）义马煤业集团千秋煤矿"11·3"重大冲击地压事故调查报告：10人死亡、64人受伤，本矿区煤层顶板为巨厚砂砾岩（380～600 m），事故发生区域接近落差达50～500 m的F16逆断层，地层局部直立或倒转，构造应力极大，处在强冲击地压危险区域；煤矿开采后，上覆砾岩层诱发下伏F16逆断层活化，瞬间诱发了井下能量巨大的冲击地压事故。

有了上述关于岩爆的定义、机制和分类的描述，特别是陈宗基关于煤爆的认识以及两个煤矿实际冲击地压事故的调查结论，可以得出如下关于冲击地压发生机制的判断：在多层多巷开挖的矿井中，巷道的围岩（煤）由于采矿的多次开挖所积累的大量破裂，使已连续破碎和不连续破碎块系处在准稳定平衡状态，这种准稳定平衡状态主要属于接触面准平衡，即剪切准平衡状态，当爆破时或顶板断裂时，爆炸波的传播与顶板断裂的冲击诱发了该准平衡块系的失稳，导致冲击地压的发生，这种冲击地压究其发生机制，是与断裂滑移型或剪切型岩爆为同一类型，也是广义岩爆的一类。

4 岩爆的定量预测和数值模拟

岩爆造成人员伤亡和设备损坏的后果使得岩爆的预测预报十分必要和紧迫，但是岩爆发生机制的复杂性使得岩爆的预测、预报十分困难。首先对岩爆发生时间的准确预报实际上是不可能的，这是由岩爆发生机制的随机性和复杂性所决定。但是岩爆的发生主要是由地下深部岩体的开挖所引起的地应力变化所决定，因此地质勘察技术、地应力检测技术、基于岩石力学理论和方法以及计算技术的岩石力学数值模拟技术的长足发展，使得岩爆发生处以及等级的定量预测和数值模拟成为可能。

岩爆研究的权威学者认为定量预测的时代已经到来，如此的定量进展需要通过数值模拟和现场观测的精细结合才可实现[1, 4]。

4.1 应变型岩爆的定量预测原理

应变型岩爆再可分为两类，在矿井中常留有岩（煤）柱，岩柱的岩爆和巷道围岩的岩爆其定量预测原理稍有不同，分述如下。

4.1.1 岩（煤）柱应变型岩爆的定量预测[3]

岩柱应变型岩爆的预测是基于矿井中岩柱的动力破坏（岩爆）与岩石试件在柔性试验机上发生猛烈破坏的机制相似。众所周知，为了获得岩石试件单轴抗压全应力-应变曲线，需要在刚性试验机上进行，这是20世纪50～60年代Petukov[9–10]和Cook[11]所指出的，反之在柔性试验机上岩石试件在应力下降段上则发生猛烈破裂，所谓刚、柔是指试验机与试件刚度之比较，试验机刚度大于试件刚度，则为刚，反之则为柔。图1所示即为两者相对刚度的影响。

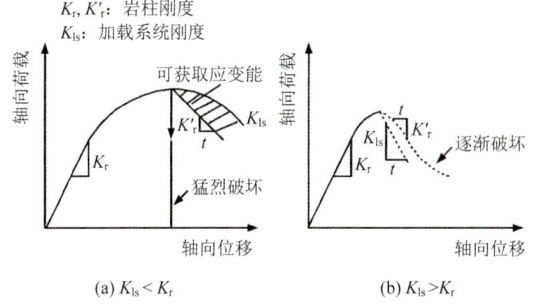

图 1　单轴加载条件下加载体系和岩石试件相对刚度对破坏特性的影响[11]

Fig.1　Impact of relative stiffness between rock specimen and loading system under uniaxial loading on destruction of property

岩柱岩爆定量预测的关键是如何计算岩柱和围岩的刚度，如果岩柱峰后刚度大于围岩刚度，则可能发生岩爆，否则，则发生渐变性破坏（塌方）。

（1）岩柱刚度（K_{pr}、K'_{pr}）计算

①长岩柱（平面应变型）刚度（单位厚度）：

$$K_{pr} = \frac{EB}{H(1-v^2)} \quad \text{（峰值前）} \quad (1)$$

$$K'_{pr} = \frac{E'B}{H(1-v^2)} \quad \text{（峰值后）} \quad (2)$$

式中：E、E'为峰值前、后岩体弹性模量；v为泊松比；B为柱宽；H为柱高。

②相应对于窄岩柱（柱厚为S），则有

$$K_{pr} = \frac{EBS}{H(1-v^2)} \quad (3)$$

$$K'_{pr} = \frac{E'BS}{H(1-v^2)} \quad (4)$$

（2）围岩刚度（K_{ls}）计算

计算模型见图2。

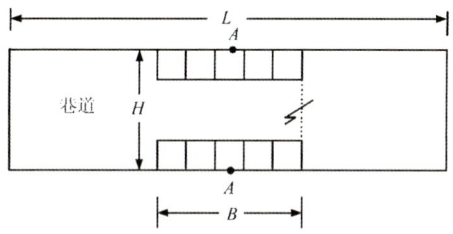

图 2　围岩刚度计算模型

Fig.2　Analytical model for stiffness of surrounding rocks

对长岩柱进行二维计算，对窄岩柱则由二维改为三维，数值计算点A的荷载$BS\sigma_p$与位移D曲线；求其曲线斜率即为围岩的局部刚度K_{ls}：$K_{ls} < K'_{pr}$岩爆（猛烈破坏）；$K_{ls} > K'_{pr}$逐渐破坏（塌方）。

以上判别该岩柱是否发生岩爆的机制、岩柱岩爆等级的定量预测的原理，与下述围岩应变型岩爆的相同，此处不再赘述。

4.1.2 围岩应变型岩爆定量预测

为定量预测围岩应变型岩爆的位置、规模和岩块弹射速度，需要解决以下3个问题[12]：

（1）何处发生围岩破裂：基于非协调变形分析的非欧几何模型确定开挖卸荷后在洞室围岩中形成的应力场[12]。

（2）围岩破裂为何发生：为此需要确定围岩各点储存的应变能和破裂耗散能，后者的确定要研究岩石破裂的起裂准则和裂纹的扩展（包括稳定扩展和失稳扩展）准则[12]。

（3）发生何种类型岩爆：需研究破裂区潜在型岩爆、即时型岩爆、时滞型岩爆发生的判别准则。

为解决第1个问题，岩体含有原生微裂纹以及岩爆的前提是岩体的破裂，所以，此时岩体不应看作连续的，不再满足应变协调方程，需采用建立在非欧几何模型上的非协调变形岩体的力学理论来计

算洞室围岩中的应力场分布。基于非欧模型的解与围岩中的微破裂现象以及应力脉动现象是相一致的,而连续介质力学模型的解与此是不相一致的。

岩石介质为非均匀地质体,岩体中存在着不同尺度的原生微裂纹,其分布函数是随机的,可由岩体细观结构试验来确定。为简单计,采用一阶近似,假定原生微裂纹均匀分布,其尺度采用统计平均值,密度用平均化法获得。

对第2个问题,可用断裂力学理论来解决。围岩的破裂过程包括原生裂纹沿界面和向岩体内部扩展、次生裂纹的稳定扩展和失稳扩展、以及微裂纹合并形成宏观裂隙最终导致岩体破裂。

为解决第3个问题,首先应了解应变型岩爆的孕育和发展过程,它可概括为:第1阶段,开挖卸荷导致围岩的破裂和碎化;第2阶段,在围岩中应力梯度作用下,围岩碎块的运动加速直至弹射,与此同时,围岩的应变能转化为动能。

围岩应变型岩爆分类判据的建立基于能量原理,即动力破坏(岩爆)的本质是剩余能量转化为动能[4−5, 9],因此破碎岩块的动能应近似等于高地应力围岩存储的应变能与耗散于岩石破裂的破坏能之差,然后根据这个原理进行分类:潜在型岩爆、即时型岩爆和时滞型岩爆。

(1)潜在型岩爆的机制:周围围岩已经破裂但没有发生岩爆;原生微裂纹和次生微裂纹扩展消耗的总能量大于或等于岩体中存储的弹性应变能(不存在剩余能量)。

(2)即时型岩爆的机制:次生微裂纹的长度大于临界值,于是发生失稳扩展形成宏观裂隙,所产生的的破裂区和洞壁相交,且次生裂纹扩展消耗的总能量小于围岩体存储的弹性应变能。

(3)时滞型岩爆的机制:围岩内所发生的破裂区与洞壁不相交;次生裂纹扩展消耗的总能量小于围岩体存储的弹性应变能,剩余能量转化为碎块动能;裂区岩块的运动足以导致至洞壁的最短路径上的围岩破裂并飞出。

4.2 断裂滑移型岩爆预测[5]

该岩爆预测基于下列考虑和假设:断裂面为单一的;这类岩爆是与作用在该断裂面上的剪抗力突变相联的;已有断裂和新断裂的区别在于前者的方向和位置独立于应力场的计算,而后者是由应力场计算决定的;断裂面合理地假设为平面,于是简化了计算,岩体是弹性的。

定量预测时首先要进行围岩开挖和爆破(动力作用)引起的应力场计算,即确定垂直于断裂面的正应力σ和断裂面上剪应力τ,然后对断裂面依照滑移发生判别准则$\tau - \tau_r > 0$进行判别:

相应的摩擦抗力τ_r为

$$\tau_r = (\mu_s \sigma + c) \text{ 或 } \tau_r = (\mu_{超} \sigma + c) \quad (5)$$

对开挖卸荷诱发的,则有

$$\tau - (\mu_s \sigma + c) > 0 \quad (6)$$

对动力作用诱发的,则有

$$\tau - (\mu_{超} \sigma + c) > 0 \quad (7)$$

式中:σ为断裂面上的正压应力;μ_s为静摩擦系数;$\mu_{超}$为超低摩擦系数;c为黏结力。

若$\tau - \tau_r > 0$,发生滑移,否则断裂面稳定不发生滑移(图3)。

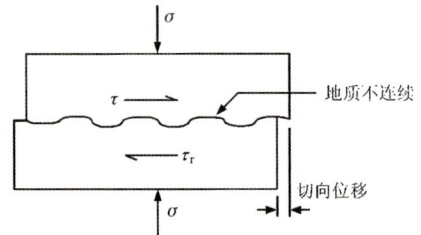

图3 岩体断裂滑移示意图
Fig.3 Slip failures of rocks

为预测岩爆的量级,需进行滑移后岩爆动能的计算:

$$W_k = W_r - W_h \quad (8)$$

式中:W_r为释放能;W_h为消耗于摩擦产生的热能;W_k为转化为动能的剩余能量。

$$\left. \begin{array}{l} W_r = \dfrac{1}{2} \int_A S_t (\tau + \mu\sigma) \mathrm{d}A \\ W_h = \mu \int_A S_t \sigma \mathrm{d}A \\ W_k = \dfrac{1}{2} \int_A S_t (\tau - \mu\sigma) \mathrm{d}A \end{array} \right\} \quad (9)$$

式中:S_t为滑移位移量;A为断裂面面积;μ为断裂面动摩擦系数;若为爆破诱发岩爆,则应为$\mu_{超}$。

进一步简化计算,τ、σ在滑移断裂面空间上假设为常量;断裂面为圆盘;滑移量取平均值S_{AV},则有

$$\left. \begin{array}{l} W_r = \dfrac{1}{2} \int_A (\tau + \mu\sigma) S_{AV} \mathrm{d}A \\ W_h = \mu \sigma A S_{AV} \\ W_k = \dfrac{1}{2} \Delta\tau A S_{AV} \end{array} \right\} \quad (10)$$

滑移时的剪应力降$\Delta\tau$为

$$\Delta\tau = \tau - \mu\sigma \quad (11)$$

滑移量分布函数为

$$S_t = \frac{4(1-v)\Delta\tau}{\pi(1-\nu/2)G}\sqrt{R^2 - r^2} \quad (12)$$

平均滑移量为

$$S_{AV} = \frac{1}{A}\int_A S_t dA = \frac{8(1-v)\Delta\tau R}{3\pi(1-\nu/2)G} \quad (13)$$

式中：R 为断裂面的半径；r 为滑移计算点的坐标；v 泊松比；G 为剪切模量。

4.3 岩石工程岩爆定量预测的流程[3]

具备以上所述的各类岩爆定量预测的原理和方法以后，可以着手进行实际岩石工程的定量预测，其预测的流程见图 4。

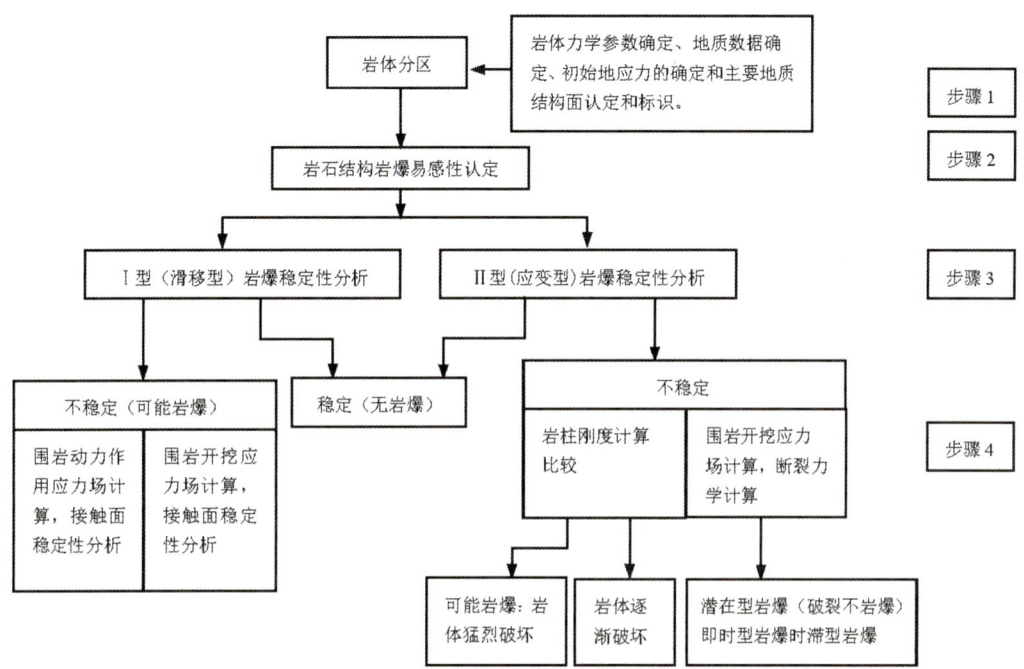

图 4 岩石工程岩爆定量预测流程
Fig.4 Flowchart of quantitative forecasting for rockburst in rock engineering

在该流程中，步骤 1 是基础，只有掌握了该工程区域的地质结构面、初始地应力、岩体力学参数等情况和数据才能实施以下各步骤。步骤 2 是定量预测的关键步骤，只有分析了工程区域各分区岩爆易感性，才能判定各分区岩爆发生的风险大小，区分轻重缓急进行岩爆的定量预测。步骤 3 应根据各工程分区有无地质结构面而定，有的需进行两种类型岩爆校核计算，有的仅需进行一种类型岩爆校核计算。步骤 4 按 4.1 和 4.2 所述内容进行。

4.4 岩体结构易爆易感性判定[3]

下面介绍 3 类岩爆易感的岩体结构，其他的情况应根据岩体开挖时的应力场计算和地质结构面具体部位来确定其岩爆易感性。

（1）图 5(a)所示为巷道接近已有断裂进行开挖。开挖引起的应力变化将增加断裂的切应力，而减少断裂的法向压应力，两个变化都能诱发断裂滑移，故有可能导致断裂滑移型岩爆。

（2）图 5(b)为当巷道开挖通过地质断裂面或岩性分区面的情况，此时若接近时未发生断裂滑移型岩爆，但临近断裂面的某区部分岩体进入破裂状态，取决于两区岩石的相对变形性质，该破坏也可能为猛烈破坏（岩爆）。

（3）图 5(c)为巷道沿断裂面或岩性分区边界开挖，则 C 区岩石为一岩柱，取决于 A 区、B 区岩石与 C 区的变形性质，满足一定要求，C 区岩石（岩柱）也可能为猛烈破坏（岩爆），该类情况也包括孤立的岩体结构的情况，即其呈现与周围岩体岩性不同或呈现几何的不规则性。

5 结 论

（1）从现象学出发对岩爆定义为，矿井或隧道的围岩或岩柱破坏、碎化发生崩出或弹射的现象，伴随能量的猛烈释放。岩爆发生的机制为，因开挖卸荷或动力作用诱发围岩中应力场的变化，或直接导致围岩的破坏碎化和弹射，或通过围岩中的已有断层和结构面滑移（活化）或新结构面滑移引起围岩破坏和弹射。

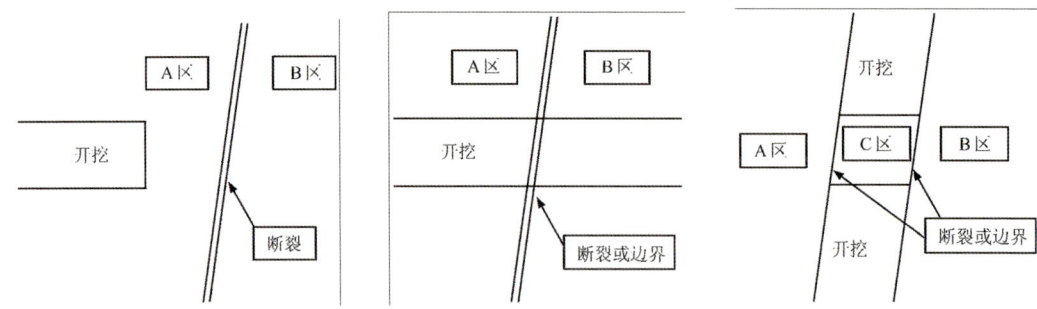

(a) 开挖接近已有主断裂，潜在断裂滑移型岩爆　(b) 开挖通过主断裂或其边界，潜在应变型岩爆　(c) 随主断裂开挖（或边界），潜在应变型岩爆

图 5　岩爆易感的岩体结构
Fig.5　Susceptive rock mass structures for rockburst

（2）岩爆的分类：第 1 类岩爆为应变型岩爆，或称体积不稳定（岩体破坏）导致的岩爆，其特点是扰动源和岩爆发生地一致；第 2 类岩爆为滑移型及剪切型岩爆，或称接触不稳定、通过断层或断裂面滑移导致的岩爆，其特点是扰动源（爆破或顶板断裂）和岩爆（冲击地压）发生地相距一定距离，第 2 类岩爆较第 1 类更普遍、更强烈，其破坏范围甚至达数十和几百米巷道。

（3）矿井中的冲击地压大部分属于第 2 类岩爆，特别是多层多巷矿井中，围岩经过多次开挖卸荷，围岩已成为处于准稳定状态的新老结构面和破裂面所分割的岩块系，在动力作用（爆破或顶板断裂）下，岩块系失稳所导致。在矿井中，这类岩爆习称为冲击地压。

（4）对工程所在部位进行分区并判明各分区的岩爆易感性，分区的基础主要是地质勘察、地质结构面的判定、初始地应力值以及岩体力学参数的确定。有了分区的认识以及相关参数后，就可进行岩爆稳定性分析，稳定性分析的第 1 步是计算开挖或爆破作用产生的应力场，对于断裂滑移型岩爆，接下来是依据判别准则进行结构面（断裂面）稳定性校核和岩爆动能的估算；对于应变型岩爆，接下来是定性预测和按照断裂力学公式进行定量预测。

参 考 文 献

[1] STACEY T R. Dynamic rock failure and its containment[C]//Proceedings of the First International Conference on Rock Dynamics and Applications. Lausanne: CRC Press, 2013: 57－70.

[2] BLAKE W. Rockburst Mechanics[D]. Golden: Colorado School of Mines, 1967, 1: 1－64.

[3] DENIS E GILL, MICHEL AUBERTIN, RICHARD SIMON. A practical engineering approach to the evaluation of rockburst potential[C]//Rockburst and Seismicity in Mines. Rotterdam: A.A. Balkema, 1993: 63－68.

[4] LINKOV A V. Dynamic phenomena in mines and the problem of stability[D]. [S. l.]: University of Minnesota, 1992.

[5] ALEXSANDER M LINKOV. Keynote lecture: New geomechanical approaches to develop quantitative seismicity[C]//Rockburst and Seismicity in Mines. Rotterdam: A. A. Balkema, 1997: 151－166.

[6] SALAMON M D G. Keynote address: Some applications of geomechanical modelling in rockburst and related research[C]//Rockburst and Seismicity in Mines. Rotterdam: A. A. Balkema, 1993: 297－309.

[7] RYDER J A. Excess shear stress in the assessment of geologically hazardous situations[J]. **Journal of South African Institute of Mining and Metallurgy**, 1988, 88: 27－39.

[8] 陈宗基. 岩爆的工程实录、理论与控制[J]. 岩石力学与工程学报, 1987, 6(1): 1－18.
TAN TJONG-KIE. Rockbursts, case records, theory and control[J]. **Chinese Journal Rock Mechanics and Engineering**, 1987, 6(1): 1－18.

[9] PETUKOV I M. Rockbursts in Kizel coalfield mines[M]. Perm: Perm Publ., 1979.

[10] PETUKOV I M, LINKOV A M. The theory of post-failure deformations and the problem of stability in rock mechanics[J]. **International Journal of Rock Mechanics and Mining Sciences & Geomechanics Abstracts,** 1979, 16(5): 57－76.

[11] COOK N G W. A note on rockburst considered as a problem of stability[J]. **Journal of South African Institute of Mining and Metallurgy**, 1965, 65: 437－445.

[12] ZHOU X P, QIAN Q H. The non-Euclidean model of failure of the deep rock masses under the deformation incompatibility condition[J]. **Journal of Mining Science**, 2013, 49(3): 368－375.

准脆材料动力强度的本质和侧向惯性约束作用

戚承志[1],钱七虎[2],陈灿寿[3],陈剑杰[4]

(1. 北京建筑大学,北京市高校工程结构与新材料工程研究中心,北京 100044;2. 解放军理工大学 国防工程学院,南京 210007;3. 解放军理工大学 人防工程设计研究院,南京 210007;4. 西北核技术研究院,西安 710024)

摘 要:从材料内部结构层次的角度,就准脆材料动力强度的本质和侧向惯性效应的作用问题进行了研究。利用岩体的 Maxwell 型松弛模型,得到了试样的强度、应变率、试样尺度和松弛速度之间的关系。利用这一关系解释了实验得到的岩石试样的动力强度与应变率成正比、与试件的尺寸成正比的结果。利用这一关系和实验数据拟合,得到了断裂传播速度随试件尺寸的增大而减小的结论。利用这一关系,结合 Bazant 的静力强度尺寸律,确定了相应的区分动力尺寸效应与静力尺寸效应的特征应变率。进一步的研究表明,岩石试样的动力强度的本质是由于断裂传播速度的有限性,当加载速度足够大时,在试样完全宏观断裂之前所发生的超载。侧向惯性约束的作用可以归结为降低了断裂的传播速度,即降低了松弛速度,从而有更充足的时间保证超载的发生。

关键词:准脆材料;动力强度;应变率效应;惯性约束

中图分类号:TU432　**文献标识码**:A　**文章编号**:1673-0836(2013)05-0975-06

The Essence of Dynamic Strength of Quasi-Brittle Materials and the Role of Lateral Inertia Confinement Effect

Qi Chengzhi[1], Qian Qihu[2], Chen Canshou[3], Chen Jianjie[4]

(1. Beijing High Institution Research Center for Engineering Structures and New Materials, Beijing University of Civil Engineering and Architecture, Beijing 100044, China; 2. Institute of National Defence Engineering, PLA University of Science and Technology, Nanjing 210007, China; 3. Research and Design Institute of Civil Air Defence Engineering, PLA University of Science and Technology, Nanjing 210007, China; 4. Northwestern Institute of Nuclear Technology, Xi'an 710024, China)

Abstract: Dynamic strength of quasi-brittle materials and the role of lateral inertia confinement effect are studied from the viewpoint of internal structural hierarchy of quasi-brittle materials. The relation between strength, strain rate, sample size and relaxation velocity is determined by the Maxwell-type relaxation model, with the help of which we can interpret the experimental results that dynamic strength of rock samples is proportional to the strain rate and to the sample size. With the help of this relation, and by fitting the experimental data, we can conclude that the fracture propagation velocity decreases with the increase of sample size. Combining this relation with Bazant's size effect law of static strength of rock sample we can determine the characteristic strain rate separating the static size effect regime and dynamic size effect regime. Further analysis shows that the essence of dynamic strength is the over-loading that occurs before the macrofracturing of sample because of the lowering of fracture propagation velocity when the loading rate is high enough. The role of lateral inertia confinement is attributed to the decrease of fracture propagation velocity, i. e. , relaxation velocity, in samples that ensures the occurrence of overloading.

Keywords: quasibrittle materials; dynamic strength; strain rate effect; inertia confinement

1 引 言

准脆材料(岩石、混凝土、陶瓷等)和其它材料的强度会随应变率的提高而提高。其一般规律是，首先材料强度随应变率的增加而缓慢地增加(此处称为Ⅰ区)，当应变率继续增加，超过某一值时材料强度随应变率的增加急剧增加(此区称为Ⅱ区)，当应变率进一步增加到冲击爆炸应变率时(Ⅲ区)，材料强度随应变率的增加而增加的速率又变缓，与Ⅰ区情况相当。许多研究者很早就注意到并研究了这一问题，如 Attewell P. B.[1]、Rinehart J. S.[2]、A. Kumar[3]、Lindholm U. S.[4]、Kipp M. E. 和 Grady D. E.[5]、Lankford J.[6]、Perzyna P.[7]、Stavrogin A. N. and Protosenja A. G.[8]、Grady D. E.[9]等从不同的角度研究了这一问题。本文作者在文[10,11]中基于前人提出了热活化与粘性机制并联竞争的材料强度-应变率依赖模型。在强度随应变率增加急剧增加的Ⅱ区，粘性机制占优。而在Ⅰ区和Ⅲ区，热活化机制占优。

在高应变率冲击载作用下，试件的侧向惯性约束变得十分明显。Field J. E. 等认为[12]，随着应变率的提高试件逐渐从一维应力状态向一维应变状态的转变。转变应变率依赖于试件材料的密度和试件的尺寸。试件尺寸越大、密度越高，转变应变率越低[13,14]。Li Q. M. 等展示[15,16,17]，在应变率大约为 $10^2\ s^{-1}$ 量级时，随着侧向约束的增加，混凝土试件的强度会得到显著的增强。他们的结论是，所观测到的动力压缩强度增加的一部分可以归结为侧向惯性约束效应，而不是真正的应变率效应。

现在对于侧向惯性约束效应对于试样动力强度的贡献还缺乏定量的评价。那么侧向惯性约束效应的作用时什么哪？岩石的动力强度的本质是什么哪？本文试图在这些问题上做一些探索。

2 准脆材料变形破坏的松弛模型

理想晶体的强度为其理论强度。但是现实晶体具有复杂的内部结构，所以受力时会产生应力集中和塑性变形，因此其强度比理论强度低 2～3 个数量级。

按照 Radianov 的观点[18]，诸如岩石和混凝土之类的非均匀介质内的应力由两部分组成：一部分由均匀介质的体积变形或者畸变形引起的弹性应力；另一部分是由构造非均匀性所引起的局部应力，这一部分应力引起介质内的不可逆变形。均匀介质的弹性应力与可逆变形成线性关系。至于残余应力(非弹性应力) Δs^l_{ij}，它们在介质变形有限时出现，并随时间自我松弛。构造非均匀性的残余应力可由下列 Maxwell 松弛模型方程来确定：

$$\frac{d\Delta s^l_{ij}}{dt} = 2\rho c_s^2 \dot{e}_{ij} - v\frac{\Delta s^l_{ij}}{l} \quad (1)$$

式中：Δs^l_{ij} 为尺寸为 l 的非均匀构造的残余应力；ρ 为介质密度；$\dot{e}_{ij}$ 为剪切变形速率；v 为一常数，反映了应力的松弛速度；c_s 为弹性横波速度。事实上 $\frac{l}{v}$ 可以解释为松弛时间 $\tau = \frac{l}{v}$。

可以用 Maxwell 松弛模型来描述准脆性材料的变形与破坏的原因是，准脆性材料在变形破坏过程中的塑性变形不显著，准脆性材料的变形破坏过程主要由弹性变形和开裂支配。弹性加载过程由方程(1)右边的第一项来描述；而开裂引起的松弛过程由方程(1)右边的第二项来描述。

Sherman S. I. 等人的现场观测表明[19]，地壳岩石的破裂规律确实遵从 Maxwell 体的破裂规律。

这一方程的主要特点是，构造非均匀单元体的残余应力的松弛速率与残余应力的大小成正比，与非均匀单元体的尺寸成反比。残余应力的增加受到两个相反的过程的控制：加载过程 $2\rho c_s^2 \dot{e}_{ij}$ 和松弛过程 $v\frac{\Delta s^l_{ij}}{l}$。

在常应变率情况下上述方程的解为：

$$\Delta s^l_{ij} = 2\rho c_s^2 \dot{e}_{ij}\frac{l}{v}[1 - e^{-\frac{vt}{l}}] = 2\rho c_s^2 \dot{e}_{ij}\tau[1 - e^{-\frac{t}{\tau}}] \quad (2)$$

如果加载时间短 $t << \tau$，那么松弛过程来不及发展，此时加载过程为主导因素，由方程(2)可得：

$$\Delta s^l_{ij} \approx 2\rho c_s^2 \dot{e}_{ij} t \quad (3)$$

也即残余应力将随时间几乎线性增加。

对于长的加载过程 $t >> \tau$，松弛过程来得及发展，加载过程受限于松弛时间，所以由方程(2)可得：

$$\Delta s_{ij}^l \approx 2\rho c_s^2 \dot{e}_{ij} \tau = 2\rho c_s^2 \dot{e}_{ij} \frac{l}{v} \quad (4)$$

为了能够发生宏观破坏，加载时间需要大于松弛时间 $t > \tau$，所以方程(4)可以用来描述岩石的宏观破坏。

定义残余偏应力强度为 $\Delta \sigma_I = \sqrt{\frac{3\Delta s_{ij}^l \Delta s_{ij}^l}{2}}$，把方程(4)代入得：

$$\Delta \sigma_I = 3\rho c_s^2 \dot{\varepsilon}_I \frac{l}{v} \quad (5)$$

式中：$\dot{\varepsilon}_I = \sqrt{\frac{2\dot{e}_{ij}\dot{e}_{ij}}{3}}$ 为应变率强度。

由(5)式可以看出，如果应变率固定，那么内部单元的尺寸越大，残余应力就越大。这样在描述固体的参数中引入了一个量纲为长度的参数。当 $\Delta \sigma_I$ 达到试样的强度 σ_Y 时，试样开始破坏。所以方程(5)可以写成：

$$\sigma_Y = 3\rho c_s^2 \dot{\varepsilon}_I \frac{l}{v} \quad (6)$$

从方程(6)可以看出，试样的动力强度与试样的尺寸成正比，与松弛速度成反比。这一规律的物理机理如下。实验表明[20]，最大的裂纹扩展速度是有限的，不会超过 Rayleigh 波的速度 C_R。这样试样的尺寸越大，那么宏观断裂需要的时间越长，那么按照方程(6)试件破坏时所加上的荷载越大，外观显现的动力强度越大。所以从断裂发展动力学的角度讲，所谓的动力强度，只不过是由于断裂传播速度的有限性，试样来不及松弛而加上去的荷载，即超载。

3 侧向惯性约束效应的作用

为了研究岩石的动力尺寸效应，洪亮、李夕兵等在 2008 年进行了细致的实验研究[21]。实验结果见图 1 和图 2。实验表明，岩石的动力强度随着应变率的增加呈幂函数增加，这与其他人得到的结果一致(见图 1)。令人感兴趣的结果是，试件尺寸越大，岩石动力强度对于应变率的敏感性越大，也即在同样的应变率条件下，岩石的动力强度随着岩石试样的尺寸增加而增加，这正好与静力加载条件下的尺寸效应率相反(见图 2)。

冯峰等人[22]对三组几何相似，尺寸不同(80，122，155 mm)的中心直裂纹平台巴西圆盘试样，利用霍普金森压杆系统进行径向冲击试验，采用实验-数值方法确定岩石的动态断裂韧度。测试结果表明，岩石的动态断裂韧度同时受到加载率效应和尺寸效应的影响。岩石的动态断裂韧度随着应变率的增加而增加。试件尺寸越大，岩石的动态断裂韧度对于应变率的敏感性越大，也即在同样的应变率条件下，岩石的动力强度随着岩石试样的尺寸增加而增加。考虑到岩石的抗拉强度、抗压强度与岩石的断裂韧度成正比[23-25]，文献[22]的实验结果支持文献[21]的实验结果。

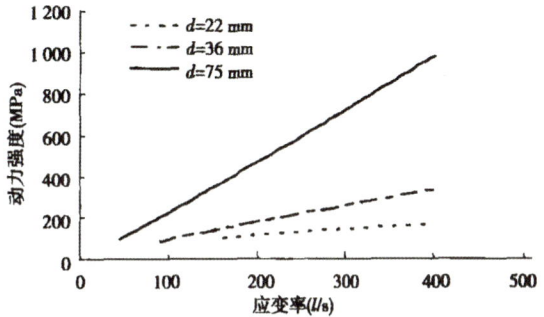

图 1 具有不同尺寸的花岗岩试样的应变率效应[21]

Fig. 1 Strain-rate effect on dynamic strength of rocks with different specimen diameters[21]

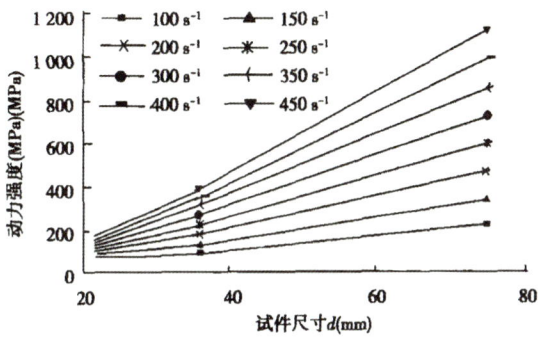

图 2 在不同的应变率下试件尺寸对于动力强度的影响[21]

Fig. 2 Sample size effect on dynamic strength of rocks under various strain-rates[21]

现在我们利用方程(6)来分析如图 1、2 所示的实验数据。对于花岗岩杨氏模量为 $E = 5.5 \times 10^{10}$ Pa，泊淞比为 $\mu = 0.29$，相应的剪切模量为 $G = 2.13 \times 10^{10}$ Pa。对于直径为 $D = 22$ mm $= 0.022$ m、$D = 36$ mm $= 0.036$ m 和 $D = 75$ mm $= 0.075$ m 的试样，按照方程(6)进行拟合实验数据可得对应的有效松弛速度分别为 $v = 3\,515$ m/s、$v = 2\,465$ m/s 和 $v = 1\,867$ m/s。因此可以看到有效松弛速度随着试件的尺寸增加而降低。这种有效松弛速度的降低可以用下述方程来描述：
$$v(D) = 112.59D^2 - 1403D + 6056.6 \quad (7)$$
式中：D 的单位为 cm。

有效松弛速度随着试件的尺寸增加而降低可以归结为动力加载引起的试件侧向惯性约束效应。在高应变率加载时侧向惯性效应非常明显，试件的应力应变状态会从一维应力状态转向一维应变状态[12]。转换的特征应变率依赖于试件的密度和尺寸，试件密度越高、尺寸越大，那么转换的特征应变率越低[14,16]。Li 等的研究展示[15~17]，在应变率大于 10^2 s^{-1} 时试件的侧向惯性约束很大，混凝土的强度得到加强。由高应变率引起的侧向约束效应延迟了材料的宏观破坏，显示的有效松弛速度降低。

因此试件的动力强度可以用下式表示：
$$\sigma_Y = 3\rho c_s^2 \dot{\varepsilon}_I \frac{D}{v(D)} \quad (8)$$

从方程(8)可以看出，侧向约束效应引起的有效松弛速度的降低，使得试件尺寸一定时松弛时间变长，需要较低的应变率就可以达到一定的动力强度。反之，当应变率一定时，更大的试件可以达到更高的动力强度。

应用方程(8)可以很好地拟合图1和2所示实验数据。

利用方程(8)，根据有效松弛速度的变化，可以近似评价侧向约束效应对于试样动力强度增加的贡献份额。由图1可以看出，直径为 22 mm 的试样的应变率效应不明显，因此可以近似认为直径为 22 mm 的试样的有效松弛速度 $v = 3515$ m/s 为无侧向惯性约束时的有效松弛速度。直径为 36 mm 和 75 mm 的试样的应变率效应非常明显，可以认为侧向惯性约束效应方程显著。此时的有效松弛速度分别为为 $v = 2465$ m/s 和 $v = 1867$ m/s。所以，对于直径为 36 mm 的试样侧向惯性效应的相对贡献率为：

$$\eta = \frac{\Delta\sigma_Y}{\sigma_Y}$$
$$= \frac{1}{3\rho c_s^2 \dot{\varepsilon}_I \frac{3.6}{v(3.6)}} \left[3\rho c_s^2 \dot{\varepsilon}_I \frac{3.6}{v(3.6)} - 3\rho c_s^2 \dot{\varepsilon}_I \frac{3.6}{v(2.2)} \right]$$
$$= 1 - \frac{v(3.6)}{v(2.2)} = 1 - \frac{2465}{3515} = 29.8\%$$

对于直径为 75 mm 的试样侧向惯性效应的相对贡献率为：

$$\eta = \frac{\Delta\sigma_Y}{\sigma_Y} = 1 - \frac{v(7.5)}{v(2.2)} = 1 - \frac{1867}{3515} = 46.9\%$$

从上述的简单评价可知，惯性约束的作用还是非常明显的。

那么是否可以从侧向惯性约束效应的观点来解释试件的动力强度呢？作者认为用这种观点解释动力强度存在着一些困难，因为用这种观点来解释试件的动力强度还存在着一些科学逻辑上的矛盾之处。原因在于，科学规律应该具有内在的统一性和广泛性。如果用应力状态的改变来解释动力压缩时强度提高，也应该用应力状态的改变来解释动力拉伸时的强度提高，但是不能。因为在高应变率拉伸情况下，试件内出现三向受拉状态，动力强度不应该提高，但是事实却与之相反，动力强度提高很大，甚至强度提高倍数超过动力压缩时的强度提高倍数（动力压缩时的动力强度提高倍数大约为 3~4，而动力拉伸时的动力强度放大倍数能够达到 5~7 倍）。另外诸如金属之类的材料，其强度几乎不依赖于静水压力，但是它们的动力强度也提高很多。所以从侧向惯性约束效应的观点来解释试件的动力强度不具有内在的统一性和广泛性。

如果我们从断裂动力学的角度，利用方程(6)来解释动力强度，并考虑到侧向惯性约束的作用，那么我们既可以解释动力压缩时的动力强度，也可以解释动力拉伸时的动力强度，以及强度不依赖于静水压力的固体的动力强度，不会遇到利用侧向惯性约束效应的观点来解释试件的动力强度时所遇到的困难。打个比方，如果我们从地球的角度来观察太阳系几大行星的运动，我们看到的将会是杂乱无章的行星运动。而如果我们从太阳的角度来观察太阳系几大行星的运动，我们看到的将会是有规则的圆形或者是椭圆形行星运动。所以研究太阳系行星的运动规律取太阳为基点为正确的选择。同理，以断裂动力学的观点来研究试样的动力强度，解释动力强度的机理会变得很简单。所以作者认为从断裂动力学的角度来研究动力强度为合适的选择。

下面研究试件的动静尺寸效应的转换应变率。试件的动静尺寸效应可以用图3表示。

通常准脆性材料的静力尺寸效应可以用下式表示[26]：
$$\sigma_Y = \sigma_0 \left(1 + \frac{D}{D_0}\right)^{-\frac{1}{2}} \quad (9)$$
式中：σ_0 和 D_0 为常数。

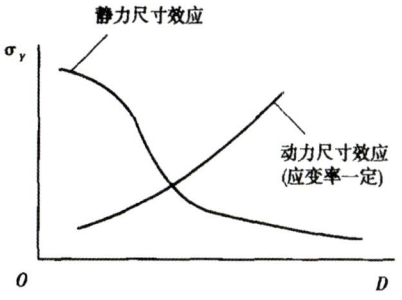

图3 试件的动力和静力尺寸效应
Fig.3 Static and dynamic size effect of samples

对于固定的岩石试样尺寸 D,由下列条件:

$$\sigma_D = \sigma_0\left(1+\frac{D}{D_0}\right)^{-\frac{1}{2}} = \sigma_Y = 3\rho c_s^2 \dot{\varepsilon}_{lc}\frac{D}{v(D)} \quad (10)$$

可得动静尺寸效应转换的特征应变率:

$$\dot{\varepsilon}_{lc} = \frac{\sigma_0 v(D)}{3\rho c_s^2 D\left(1+\frac{D}{D_0}\right)^{\frac{1}{2}}} \quad (11)$$

由方程(11)可以看出,在固定的试样尺寸 D 条件下,当应变率 $\dot{\varepsilon} > \dot{\varepsilon}_{lc}$ 时,动力尺寸效应占据优势;而当 $\dot{\varepsilon} < \dot{\varepsilon}_{lc}$ 时,静力尺寸效应占优。

由方程(11)还可以看出,试样的密度越大,转换特征应变率越低。上述结论与 Forrestal[14]、Zhang et al.[16],与洪亮、李夕兵等[21]的实验结果一致。这说明 Maxwell 模型所得结果是正确的。

现在讨论(11)式的物理意义。

对于具有尺寸为 l 的岩石试样,所谓的静力强度 σ_{st} 是在一定的应变率 $\dot{\varepsilon}_{lc}$ 下测得的。从松弛过程的角度来讲这意味着加载过程与松弛过程平衡,即:

$$\frac{d\Delta s_{ij}^l}{dt} = 2\rho c_s^2 \dot{e}_{ij} - v\frac{\Delta s_{ij}^l}{l} = 0 \quad (12)$$

由此得 $\Delta s_{ij}^l = 2\rho c_s^2 \dot{e}_{ij}\frac{l}{v}$,i.e. $\Delta\sigma_l = 3\rho c_s^2 \dot{\varepsilon}_l \frac{l}{v} = \sigma_{st}$。因此相应的应变率为 $\dot{\varepsilon}_{lc} = \frac{v\sigma_{st}}{3\rho c_s^2 l}$,表明静力强度和应变率存在着一一对应关系。当施加的应变率大于 $\dot{\varepsilon}_{lc}$ 时:$\dot{\varepsilon}_l > \dot{\varepsilon}_{lc}$,$\frac{d\Delta s_{ij}^l}{dt} = 2\rho c_s^2 \dot{e}_{ij} - v\frac{\Delta s_{ij}^l}{l} > 0$,动力加载开始,从而动力尺寸效应占优。

从另一方面来讲,对于固定的应变率 $\dot{\varepsilon}$,相应的满足方程(12)的试样尺寸为 $l_{cr} = \frac{v\sigma_{st}}{3\rho c_s^2 \dot{\varepsilon}_{lc}}$。如果 $l > l_{cr}$,那么 $\frac{d\Delta s_{ij}^l}{dt} = 2\rho c_s^2 \dot{e}_{ij} - v\frac{\Delta s_{ij}^l}{l} > 0$,动力加载开始,从而动力尺寸效应占优。

这样用一个非常简单的松弛模型,结合静力尺寸效应就可以解释冲击试验的主要变形破坏特征,这说明松弛模型抓住了准脆材料动力变形破坏的主要特征,适合描述准脆材料的动力变形与破坏。

4 结 论

岩石具有复杂的结构层次,每一个层次上的力学性质不同,因而材料的变形破坏特性比较复杂。由于断裂扩展速度有限性,岩石力学行为的时间特性反映在每一个结构层次都有自己的特征破坏时间。根据外部动力荷载的时间特性不同,材料会有多个结构层次参加到动力变形破坏过程中来,因而表现出的力学性质会大不相同。本文从材料内部结构层次的角度,利用松弛模型和静力尺寸效应模型,就准脆材料的动力强度的本质、试样侧向惯性约束效应的作用问题进行了研究。研究表明,材料变形破坏的空间尺度和时间尺度之间的关系是由于裂纹传播速度的有限性所决定的。材料的动力强度的实质是,由于裂纹传播速度的有限性,当加载速度足够大时,在物体完全断裂之前所加上去的荷载,即超载。侧向惯性约束的作用可以归结为降低了断裂的传播速度,降低了松弛速度,从而有更充足的时间保证超载的发生。

参考文献(References)

[1] Attewell P B. Dynamic fracturing of rocks, parts I, II, III [J]. Colliery Eng., 1963, 203-10, 248-52, 289-94.

[2] Rinehart J S. Dynamic fracture strength of rock [A]// Proceeding of 7th Symp. on Rock Mech [C]. ALME 1965, 205-08.

[3] Kumar A. The effect of stress rate and temperature on the strength of basalt and granite [J]. Geophysics, 1968, 33(3): 501-10.

[4] Lindholm U S, Yeakley L M, and Nagy A. The dynamic strength and fracture properties of dresser basalt [J]. Int J Rock Mech Mining Sci., 1974, 11: 181-91.

[5] Kipp M E, Grady D E and Chen E P. Strain-rate dependent fracture initiation [J]. Int. J. Fracture, 1980, 16: 471-8.

[6] Lankford J. The role of subcritical microfracture process in compressive failure of ceramics [A]//Fracture mechanics of Ceramics [C]. 1983, Vol .5, 45-54, Plenum Press, New York.

[7] Perzyna P. Constitutive modeling of dissipative solid for localization and fracture [A]//*Localization and fracture*

phenomena in inelastic solids [C]. 99-241, Springer, Wien, New York, 1998.

[8] Stavrogin A N and Protosenja A G. Mechanics of deformation and fracture of rocks [M]. Nedra, Moscow, 1992.

[9] Grady D E. Shock wave properties of brittle solids [A]. in: Shock Compression of Condensed Matters, 9-20, Steve Schmidt (ed.), AIP Press, New York, 1995.

[10] Qi Chengzhi, Wang Mingyang, Qian Qihu. Strain rate effects on the strength and fragmentation size of rocks [J]. Int. J. Impact Eng., 2009, 36(12): 1 355-1 364.

[11] 戚承志, 钱七虎. 岩石等脆性材料动力强度依赖应变率的物理机制[J]. 岩石力学与工程学报, 2003, 21(2):177-181. (Qi Chengzhi, Qian Qihu. Physical mechanism of brittle material strength-strain rate sensitivity[J]. Chinese Journal of Rock Mechanics and Engineering, 2003, 21(2): 177-181. (in Chinese))

[12] Field J E, Walley S M, Proud W G, et al. Review of experimental techniques for high rate deformation and shock studies [J]. Int. J. Impact Eng., 2004, 30(4): 725-775.

[13] Gorham D A. The effect of specimen dimensions on high strain rate compression measurements of copper [J]. J. Phys., D, 1991, 24: 1489-1492.

[14] Forrestal M J, Wright T W, Chen W. The effect of radial inertia on brittle samples during the split Hopkinson pressure bar test[J]. Int. J. Impact Eng., 2007. 34(3), 405-11.

[15] Li Q M., Meng H. About the dynamic strength enhancement of concrete-like materials in a split Hopkinson pressure bar test[J]. Int. J. Solids Struc., 2003, 40(2):343-60.

[16] Zhang M, Wu H J, Li Q M, et al. Further investigation on the dynamic compressive strength enhancement of concrete-like materials based on split Hopkinson pressure bar tests, Part-I: experiments[J]. Int. J. of Impact Engng., 36(12), 1327-1337, 2009.

[17] Li Q M., Lu Y B, Meng H. Further investigation on the dynamic compressive strength enhancement of concrete-like materials based on split Hopkinson pressure bar test, part II: numerical simulations[J]. Int. J. Impact Eng., 2009, 36(12): 1335-1345.

[18] Radionov V N., Sizov I A., Tsvetkov V M. Fundamental of geo-mechanics [M]. Nedra, Moscow, 1986.

[19] Sherman S I, Seminsky K Z, Adamovich A N, et al. Fault formation in lithosphere, zone of tension [M]. Edited by Logachev N. A. Nauka, Novosibirsk, 1992

[20] Fineberg J., Marder M. Instability in dynamic fracture [J]. Phys. Rep., 1999, 313, 101-108.

[21] 洪亮, 李夕兵, 马春德, 等. 岩石动态强度及其应变率灵敏性的尺寸效应研究[J]. 岩石力学与工程学报, 2008, 27(3):526-533. (Hong Liang, Li Xibing, Ma Chunde, et al. Study on size effect of rock dynamic strength and strain rate sensitivity[J]. Chinese Journal of Rock Mechanics and Engineering, 2008, 27(3): 526-533(in Chinese))

[22] 冯峰, 韦重耕, 王启智. 用中心直裂纹平台巴西圆盘测试岩石动态断裂韧度的尺寸效应[J]. 工程力学, 2009, 26(4): 167-173. (Feng Feng, Wei Chenggeng and Wang Qizhi. Size effect for rock dynamic fracture toughness tested with cracked straight through flattened Brazillian disc [J]. Engineering Mechanics. 2009, 26(4): 167-173. (in Chinese))

[23] Bhagat R B. Mode I fracture toughness of coal [J]. Int. J. Min. Eng., 1985;3:229-36.

[24] Whittaker B N, Singh R N, Sun G. Rock fracture mechanics: principles, design and applications [M]. Elsevier, Amsterdam, 1992.

[25] Zhang Z. X. An empirical relation between mode I fracture toughness and the tensile strength of rock [J]. Int. J. Rock Mech. Mining Sci. 2002, 39:401-406.

[26] Bazant Z P. Size effect in blunt fracture: Concrete, rock, cracks[J]. J. Eng. Mech. 1984, 110 (4): 518-535.

| 学术论文——岩石力学与工程

岩体非协调变形对围岩中的应力和破坏的影响

钱七虎[1,2,3]，周小平[2,4,5]

(1. 解放军理工大学 国防工程学院，江苏 南京 210007；2. 重庆大学 土木工程学院，重庆 400045；3. 解放军理工大学 爆炸冲击防灾减灾国家重点实验室，江苏 南京 210007；4. 重庆大学 山地城镇建设新技术教育部重点实验室，重庆 400045；5. 重庆大学 煤矿灾害动力学与控制国家重点实验室，重庆 400044)

摘要：讨论研究岩体非协调变形的重要性，提出岩石非协调变形的物理和数学含义，介绍研究岩体非协调变形的技术思路，阐述岩体非协调变形和协调变形的基本不同点。通过含微裂纹的岩体中开挖圆形洞室的计算实例，具体分析岩体非协调变形的研究对围岩应力和破坏的影响，确定非协调变形产生的自平衡封闭应力，研究微裂纹的密度和长度对自平衡封闭应力和岩体破坏的影响。

关键词：岩石力学；非协调变形；微裂纹；非欧模型；自平衡封闭应力

中图分类号：TU 45　　　**文献标识码**：A　　　**文章编号**：1000 - 6915(2013)04 - 0649 - 08

EFFECTS OF INCOMPATIBLE DEFORMATION ON FAILURE MODE AND STRESS FIELD OF SURROUNDING ROCK MASS

QIAN Qihu[1,2,3], ZHOU Xiaoping[2,4,5]

(1. Engineering Institute of National Defense Engineering, PLA University of Science and Technology, Nanjing, Jiangsu 210007, China; 2. College of Civil Engineering, Chongqing University, Chongqing 400045, China; 3. State Key Laboratory of Disaster Prevention and Mitigation of Explosion and Impact, PLA University of Science and Technology, Nanjing, Jiangsu 210007, China; 4. Key Laboratory of New Technology for Construction of Cities in Mountain Area of Ministry of Education, Chongqing University, Chongqing 400045, China; 5. State Key Laboratory of Coal Mine Disaster Dynamics and Control, Chongqing University, Chongqing 400044, China)

Abstract: The importance of the incompatible deformation of the deep rock masses is discussed. The mathematical and physical meanings of the incompatible deformation of the deep rock masses are proposed. The method to research the incompatible deformation of the deep rock masses is introduced. The difference between the incompatible and compatible deformation of the rock mass is elaborated. The calculation case of a circular tunnel in the rock masses containing microcracks is given based on the non-Euclidean model. The effects of the incompatible deformation on failure mode and the stress field of the surrounding rock mass is analyzed. The self-equilibrated stresses induced by the incompatible deformation are determined. The influences of the density and the length of microcracks on the self-equilibrated stresses and failure mode are investigated.

Key words: rock mechanics; incompatible deformation; microcracks; non-Euclidean model; self-equilibrated stresses

1 引 言

重温陈宗基院士三十多年前的殷切建议[1]："残余应力还没有引起足够的注意，其来源和性质还不明确；残余应力可以分为2种类型：即'封闭应力'和'非封闭应力'，若岩样从岩体中取出时应力消失，则属于后者，反之，尽管岩样的边界是自由的，'封闭应力'仍然可以保存，也即它具有自身平衡的性质；目前，岩石力学仅仅对变形为协调的情况感兴趣，例如，在一般的弹性理论中，非协调张量等于 0。但自身平衡的初始应力在数学上采用

收稿日期：2012-11-15；修回日期：2013-02-05。
基金项目：国家自然科学基金资助项目(51021001，51279218)；重庆市自然科学基金院士专项(CSTC2013JJYS3001)。
本文原载于《岩石力学与工程学报》(2013 年第 32 卷第 4 期)。

非协调张量不等于 0 的假定解析，这个假定将为岩石力学开创一个新的远景。希望我的建议能够为大家所接受，因为这关系到岩石力学的基本概念。"

由上述建议可以认识到：

(1) 当岩体边界是自由时，即没有荷载作用时，岩体中可存在自平衡的"封闭应力"；

(2) 这种自平衡的封闭应力是岩体中的残余应力的一个类型；

(3) 自平衡的"封闭应力"可用非协调张量不等于 0 来分析，即通过分析非协调变形来求得；

(4) 岩石力学以往仅研究协调变形，而非协调变形的研究将开创一个新的远景。

本文的主要研究内容：阐述岩体非协调变形的技术思路，论述岩体的非协调变形和协调变形的基本不同点，分析非协调变形产生的自平衡封闭应力。其研究目的是为了分析深部岩体非协调变形和揭示深部岩体变形和破坏机制提供理论指导，为深部岩体的设计、施工和支护方案的优化提供理论依据。

2 材料的连续性和变形协调性

材料变形的协调性是由材料的连续性所决定的。材料的连续性是弹塑性力学的基本假定，它假定介质无空隙地分布于物体所占的整个空间。这一假定与物质是由不连续的粒子所组成的观点相矛盾。这个矛盾可用统计平均的观点解决，在力学分析中，从物体中取出的任意微小单元，在数学上是一个无穷小量，但它却含有大量的体粒或非晶体的高分子，晶体缺陷与微小单元相比小很多，进而与物体尺寸相比更是小得多，因而连续性假定对于固体材料来说实际上是合理的。

材料连续的结论还必须建立在变形是协调的基础之上：对物体及其微元施加任意变形后，能将这些变形后的微元重新拼合为一个变了形的物体整体，即变形后微元必须能无缝地拼合起来，这样的变形称为协调变形。若变形是不协调的，即不能拼合成整体的，当然也就不能采用基本微元的分析方法。

由此可见，变形的协调性是指物体变形后仍保持其整体性和连续性。从数学的观点来说，要求变形产生的位移函数在其定义域内为单值连续函数。

图 1 给出了岩体的协调和非协调变形[2]。如果物体变形后出现"开裂"和"叠合"，"开裂"和"叠合"显然是变形的不协调现象：它形成物体的非整体和非连续。

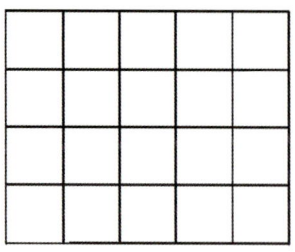

(a) 变形前

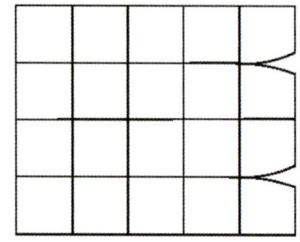

(b) 变形后的"开裂"现象

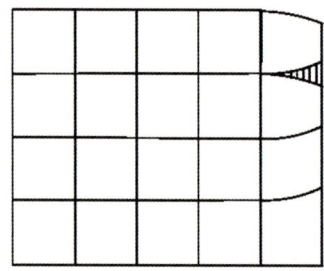

(c) 变形后的"叠合"现象

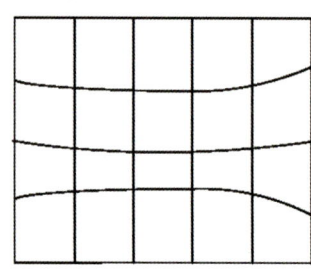

(d) 协调变形状态

图 1 岩体的协调和非协调变形[2]

Fig.1 Compatible and incompatible deformations of rock mass[2]

类似地有封闭应力问题。如果将一个有封闭内应力的物体分割成微元，这些微元被释放为无应力状态。这些释放后的微元一般也不能无缝地拼合成一个处于自然状态(即各点应力为 0)的物体。因此，具有自平衡封闭应力的岩体一般也不满足变形协调条件。

材料变形协调性的条件可以用数学上的变形协调方程来表述。以平面应变为例。平面应变问题有 2 个位移量 u，v，在直角坐标系中 3 个应变量为

$$\varepsilon_x = \frac{\partial u}{\partial x}, \quad \varepsilon_y = \frac{\partial v}{\partial y}, \quad \gamma_{xy} = \frac{\partial u}{\partial y} + \frac{\partial v}{\partial x} \quad (1)$$

3个应变方程对2个位移未知量积分将得出非单值解,这从数学上说明必须满足一个附加条件后才能得出物体变形后单值的位移。这个附加条件可由材料的连续整体性,即位移函数的连续可微性导出[3]:

$$\frac{\partial^2 \varepsilon_x}{\partial y^2}+\frac{\partial^2 \varepsilon_y}{\partial x^2}=\frac{\partial^2 u}{\partial x \partial y}+\frac{\partial^2 u}{\partial y \partial x}=$$

$$\frac{\partial^2}{\partial x \partial y}\left(\frac{\partial u}{\partial y}+\frac{\partial v}{\partial x}\right)=\frac{\partial^2 \gamma_{xy}}{\partial x \partial y} \quad (2)$$

即

$$\frac{\partial^2 \varepsilon_x}{\partial y^2}+\frac{\partial^2 \varepsilon_y}{\partial x^2}-\frac{\partial^2 \gamma_{xy}}{\partial x \partial y}=0 \quad (3)$$

式(3)为二维情况下的应变协调方程,或简称为协调方程或相容方程。

类似地,三维问题的应变协调方程[3]为

$$\frac{\partial^2 \varepsilon_x}{\partial y^2}+\frac{\partial^2 \varepsilon_y}{\partial x^2}=\frac{\partial^2 \gamma_{xy}}{\partial x \partial y} \quad (4)$$

$$\frac{\partial^2 \varepsilon_z}{\partial x^2}+\frac{\partial^2 \varepsilon_x}{\partial z^2}=\frac{\partial^2 \gamma_{xz}}{\partial z \partial x} \quad (5)$$

$$\frac{\partial^2 \varepsilon_y}{\partial z^2}+\frac{\partial^2 \varepsilon_z}{\partial y^2}=\frac{\partial^2 \gamma_{yz}}{\partial y \partial z} \quad (6)$$

$$2\frac{\partial^2 \varepsilon_x}{\partial y \partial x}=\frac{\partial}{\partial x}\left(-\frac{\partial \gamma_{yz}}{\partial x}+\frac{\partial \gamma_{xz}}{\partial y}+\frac{\partial \gamma_{xy}}{\partial z}\right) \quad (7)$$

$$2\frac{\partial^2 \varepsilon_y}{\partial z \partial x}=\frac{\partial}{\partial y}\left(\frac{\partial \gamma_{yz}}{\partial x}-\frac{\partial \gamma_{xz}}{\partial y}+\frac{\partial \gamma_{xy}}{\partial z}\right) \quad (8)$$

$$2\frac{\partial^2 \varepsilon_z}{\partial x \partial y}=\frac{\partial}{\partial z}\left(\frac{\partial \gamma_{yz}}{\partial x}+\frac{\partial \gamma_{xz}}{\partial y}-\frac{\partial \gamma_{xy}}{\partial z}\right) \quad (9)$$

平面应变分量ε_x,ε_y,γ_{xy}满足变形协调方程之后,可保证求出的位移是单值和连续的,从而保证物体在变形后不会出现开裂或叠合现象。

这里必须说明,对于应力未知量求解,也必须满足以应力分量表示的协调方程。例如,对于平面应力问题,3个应力未知分量σ_x,σ_y,τ_{xy}只有2个平衡方程。在忽略体力的情况下,平衡方程可以表示为

$$\left.\begin{array}{l}\dfrac{\partial \sigma_x}{\partial_x}+\dfrac{\partial \tau_{xy}}{\partial_y}=0 \\ \dfrac{\partial \sigma_y}{\partial_y}+\dfrac{\partial \tau_{xy}}{\partial_x}=0\end{array}\right\} \quad (10\text{a})$$

只有满足以应力分量表示的协调方程:

$$\left(\frac{\partial^2}{\partial x^2}+\frac{\partial^2}{\partial y^2}\right)(\sigma_x+\sigma_y)=0 \quad (10\text{b})$$

才能得到包含3个未知函数σ_x,σ_y,τ_{xy}的3个微分方程,求得唯一单值解。实际上,在弹性力学解的唯一性证明过程中,是从应力解都满足平衡方程和协调方程出发的,这是唯一解的前提和基础。

3 岩体的非连续性和变形非协调性

岩体的情况相对于其他变形物体和材料来得复杂。

首先,岩体是非连续的地质体,是由地质构造破碎带、裂隙和节理纵横切割为尺寸大小不同的岩块。岩块中具有细、微观裂纹。上述裂缝和裂纹的延展尺度为$10^6 \sim 10^{-6}$ m。

在不连续岩体力学中,宏观裂隙可以当做各向异性的连续介质来处理,宏观节理可以用节理单元来研究,描述岩体中的裂隙、断层、节理等宏观结构面的不连续位移和运动特性可以采用不连续介质离散模型,如离散元等。但一般微观、亚微观不连续面在分析中忽略不计,当做连续介质来处理,然后用连续介质的弹塑性力学来分析。

岩体一般都具有初始的微观和细观裂纹,例如,根据致密岩体和页岩体的最新研究表明:页岩气储层的孔隙直径为5~200 nm,致密灰岩油储层的孔隙直径为40~500 nm,致密砂岩气储层的孔隙直径为40~700 nm,致密砂岩油储层的孔隙直径为50~900 nm。上述孔隙系统占整个储层空间80%~90%,局部也发育微米到毫米级孔隙和裂缝等储层空间。

由此可见,微、细观裂纹直径为数十纳米到微米以至毫米级,这些初始微裂纹对岩体的连续性和变形协调性的影响是一个值得研究的问题。

其次,地下工程(包括巷道、隧道和洞室)开挖后,开挖卸荷引起围岩的变形和破坏,围岩中不但产生新的微裂纹,原生初始微裂纹也将扩展、连接、合并成粗裂纹甚至宏观裂纹,总之开挖卸荷导致围岩中裂纹的密度和尺寸都将产生变化,相应地对岩体变形非协调性产生新的影响,这也是值得研究的问题。

岩体出现非协调变形后,非协调张量不为0[4-14],对于平面问题,有

$$\frac{\partial^2 \varepsilon_x}{\partial y^2}+\frac{\partial^2 \varepsilon_y}{\partial x^2}-\frac{\partial^2 \gamma_{xy}}{\partial x \partial y} \neq 0 \quad (11)$$

对于变形非协调问题,内度量非欧几何空间的

应变不等于外度量欧氏几何空间的应变。

对于应力，相应有[9]

$$\frac{2(1-\nu)}{E}\Delta\sigma = R \qquad (12)$$

式中：Δ 为拉普拉斯算子，σ 为应力张量第一不变量，E 为内度量非欧几何空间材料的弹性模量，ν 为内度量非欧几何空间材料的泊松比。

4 岩体非协调变形的研究

岩体中的内部缺陷，从微、细观裂纹到构成岩体晶粒的位错(dislocation)、向错(disclination)和螺错(dispiration)导致了岩体变形的非协调性[4-14]，即应变协调方程不再成立和基本微元的概念也破坏。解决这种非协调问题，存在着众多的学派，但主要都是基于非欧模型。人们生活所在的是欧氏几何空间，所谓协调与非协调是相对这个空间而言的，是否可以超脱出这个空间而进入更一般的抽象空间，从另一个角度思考问题，这时非协调性的问题就转化为选择具有适当性质的空间的问题，即选择适当的非欧几何空间。空间的本质由度量张量决定，它定义了空间中的距离等等。利用非欧几何空间所具有的曲率、挠度等几何参量，然后可通过某种方法来描述引起岩体变形非协调性的岩体内部缺陷。这种方法就是建立外部观察者的外度量欧氏几何空间与描述岩体内部缺陷的内度量非欧几何空间的坐标变换关系，这种方法是基于仿射度量的微分几何，是抽象的微分几何在力学问题上的成功运用，是一种漂亮的几何‐拓扑描述，当然它需要复杂的数学知识。

在上述过程中使用的曲率、挠率等几何参量具有明确的物理意义，它们是描述非协调变形的热力学量，所以在研究岩体的非协调变形时，必须借助不可逆热力学理论才能建立求解曲率、挠率等非欧几何参量与岩体内部缺陷之间的关系 $\Delta^2 R = \gamma^2 R$（其中：γ 为与微裂纹的密度、长度等有关的非欧参数），从而建立完整的岩体的非欧模型体系，以拓展传统的弹塑性力学模型解决非连续、非协调问题的能力[9-13]。欧氏模型到非欧模型的转变可表示为

$$\left.\begin{array}{l}\sigma_{ij,j}=0\\ \Delta\sigma=0\end{array}\right\} \xrightarrow{\text{非欧模型}} \begin{array}{l}\sigma_{ij,j}=0\\ \dfrac{2(1-\nu)}{E}\Delta\sigma=R\\ \Delta^2 R=\gamma^2 R\end{array} \qquad (13)$$

5 岩体非协调变形对围岩应力的影响

作为计算实例(见图 2)，取圆形洞室(半径 $r_0 = a$)为平面应变情况中的围岩应力分析问题。假设地层中的初始垂直、水平地应力分别为 σ_v 及 σ_h。众所周知，按连续介质的弹性力学，符合应变协调方程(相容方程)的毛洞围岩中的二次应力场[4]为

$$\left.\begin{array}{l}\sigma_r = \dfrac{1}{2}(\sigma_v + \sigma_h)\left(1 - \dfrac{a^2}{r^2}\right) - \dfrac{1}{2}(\sigma_v - \sigma_h)\cdot\\ \quad\left(1 - \dfrac{4a^2}{r^2} + \dfrac{3a^4}{r^4}\right)\cos(2\theta)\\ \sigma_\theta = \dfrac{1}{2}(\sigma_v + \sigma_h)\left(1 + \dfrac{a^2}{r^2}\right) - \dfrac{1}{2}(\sigma_v - \sigma_h)\cdot\\ \quad\left(1 + 3\dfrac{a^4}{r^4}\right)\cos(2\theta)\\ \tau_{r\theta} = \dfrac{1}{2}(\sigma_v - \sigma_h)\left(1 + \dfrac{2a^2}{r^2} - \dfrac{3a^4}{r^4}\right)\sin(2\theta)\end{array}\right\} \qquad (14)$$

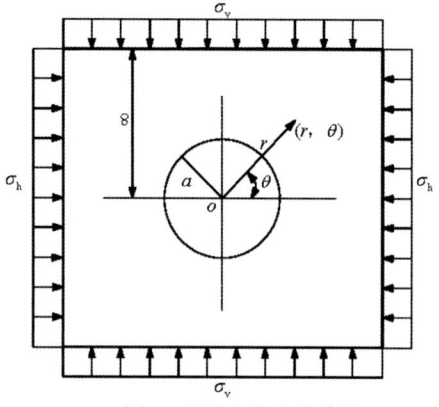

图 2 圆形洞室力学模型
Fig.2 Mechanical model of circular tunnel

对于该计算实例，如果围岩中含有初始微裂纹，或出现次生微裂纹，此时，应变协调条件的假设不再成立，即在非协调变形不等于 0 的情况下，基于非欧模型的围岩中的二次应力场[14]为

$$\left.\begin{array}{l}\sigma_r = \sigma_{r1} + \sigma_{r2}\\ \sigma_\theta = \sigma_{\theta 1} + \sigma_{\theta 2}\\ \tau_{r\theta} = \tau_{r\theta 1} + \tau_{r\theta 2}\end{array}\right\} \qquad (15)$$

式中：σ_{r1}，$\sigma_{\theta 1}$，$\tau_{r\theta 1}$ 均为协调变形时的弹性应力，σ_{r2}，$\sigma_{\theta 2}$，$\tau_{r\theta 2}$ 均为内部缺陷引起的非协调变形产生的自平衡应力。弹性应力和自平衡应力可表示如下：

$$\sigma_{r1} = \dfrac{\sigma_v + \sigma_h}{2}\left(1 - \dfrac{r_0^2}{r^2}\right) + \dfrac{\sigma_h - \sigma_v}{2}\cdot$$
$$\left(1 - 4\dfrac{r_0^2}{r^2} + 3\dfrac{r_0^4}{r^4}\right)\cos(2\theta) \qquad (16a)$$

$$\sigma_{r2} = -\frac{E}{2\gamma^{3/2}(1-v^2)r}[AJ_1(\gamma^{1/2}r) + BN_1(\gamma^{1/2}r) + CK_1(\gamma^{1/2}r)] + \frac{E}{2(1-v^2)r\gamma^{3/2}}[A_1J_1(\gamma^{1/2}r) +$$
$$B_1N_1(\gamma^{1/2}r) + C_1K_1(\gamma^{1/2}r)]\cos(2\theta) - \frac{3E}{(1-v^2)r^2\gamma^2} \cdot$$
$$[A_1J_2(\gamma^{1/2}r) + B_1N_2(\gamma^{1/2}r) + C_1K_2(\gamma^{1/2}r)]\cos(2\theta) \quad (16b)$$

$$\sigma_{\theta 1} = \frac{\sigma_v + \sigma_h}{2}\left(1 + \frac{r_0^2}{r^2}\right) - \frac{\sigma_h - \sigma_v}{2}\left(1 + 3\frac{r_0^4}{r^4}\right)\cos(2\theta) \quad (16c)$$

$$\sigma_{\theta 2} = -\frac{E}{2\gamma(1-v^2)}[AJ_0(\gamma^{1/2}r) + BN_0(\gamma^{1/2}r) - CK_0(\gamma^{1/2}r)] + \frac{E}{2\gamma^{3/2}(1-v^2)r}[AJ_1(\gamma^{1/2}r) +$$
$$BN_1(\gamma^{1/2}r) + CK_1(\gamma^{1/2}r)] + \frac{E}{2(1-v^2)\gamma} \cdot$$
$$\left[A_1J_0\left(r\sqrt{\gamma}\right) + B_1N_0\left(r\sqrt{\gamma}\right) + C_1K_0\left(r\sqrt{\gamma}\right)\right] \cdot$$
$$\cos(2\theta) - \frac{3E}{2r(1-v^2)\gamma^{3/2}}\left[A_1J_1\left(r\sqrt{\gamma}\right) + B_1N_1 \cdot\right.$$
$$\left.\left(r\sqrt{\gamma}\right) + C_1K_1\left(r\sqrt{\gamma}\right)\right]\cos(2\theta) + \frac{3E}{r^2(1-v^2)\gamma^2} \cdot$$
$$\left[A_1J_2\left(r\sqrt{\gamma}\right) + B_1N_2\left(r\sqrt{\gamma}\right) + C_1K_2\left(r\sqrt{\gamma}\right)\right]\cos(2\theta) \quad (16d)$$

$$\tau_{r\theta 1} = \frac{\sigma_v - \sigma_h}{2}\left(1 + 2\frac{r_0^2}{r^2} - 3\frac{r_0^4}{r^4}\right)\sin(2\theta) \quad (16e)$$

$$\tau_{r\theta 2} = \frac{E}{r(1-v^2)\gamma^{3/2}}\left[A_1J_1\left(r\sqrt{\gamma}\right) + B_1N_1\left(r\sqrt{\gamma}\right) + C_1K_1 \cdot\right.$$
$$\left.\left(r\sqrt{\gamma}\right)\right]\sin(2\theta) - \frac{3E}{r^2(1-v^2)\gamma^2}\left[A_1J_2\left(r\sqrt{\gamma}\right) +$$
$$B_1N_2\left(r\sqrt{\gamma}\right) + C_1K_2\left(r\sqrt{\gamma}\right)\right]\sin(2\theta) \quad (16f)$$

$$\gamma^2 = \frac{3\beta[3\lambda_2(1-D_2)^2 + 2\mu_2(1-D_3)^2](1-\alpha)}{2[3\lambda_2(1-D_2)^2(1-\alpha) + 2\mu_2(1-D_3)^2(3-\alpha)]} \quad (16g)$$

$$\beta = \frac{\mu_2(1-D_3)^2}{\mu_3}, \quad \alpha = \frac{3g^2(1-D_0)^2(1-D_2)^2}{\delta} \quad (16h)$$

$$\delta = [\lambda_1(1-D_0)^2 + 2\mu_1(1-D_1)^2] \cdot [3\lambda_2(1-D_2)^2 + 2\mu_2(1-D_3)^2] \quad (16i)$$

$$\left.\begin{array}{l}D_0 = \dfrac{\pi\omega_0}{1+\pi\omega_0}, \quad D_1 = \dfrac{\pi\omega_0}{1+v_0+\pi\omega_0} \\ \omega_0 = \eta_0 c_0^2, \quad v_0 = \dfrac{\lambda_1}{2(\lambda_1+\mu_1)}\end{array}\right\} \quad (16j)$$

$$\left.\begin{array}{l}D_2 = \dfrac{\pi\omega}{1+\pi\omega}, \quad D_3 = \dfrac{\pi\omega}{1+v_1+\pi\omega} \\ \omega = \eta c^2, \quad v_1 = \dfrac{\lambda_2}{2(\lambda_2+\mu_2)}\end{array}\right\} \quad (16k)$$

$$E = \frac{\mu_2[3\lambda_2(1-D_2)^2 + 2\mu_2(1-D_3)^2](1-D_3)^2}{\lambda_2(1-D_2)^2 + \mu_2(1-D_3)^2} \quad (16l)$$

$$v = \frac{\lambda_2(1-D_2)^2}{2[\lambda_2(1-D_2)^2 + \mu_2(1-D_3)^2]} \quad (16m)$$

式中：λ_1，μ_1 均为相应于外度量欧氏几何空间的 Almansi 应变张量的 Lame 常数；λ_2，μ_2 均为相应于内度量非欧几何空间的 Lame 常数；g，μ_3 均为唯象学参数；$r_0 = a$ 为洞室半径；(r, θ) 为极坐标；η 为二次微裂纹密度；c 为二次微裂纹半长；η_0 为初始微裂纹密度；c_0 为初始微裂纹半长；C 为与 Wronskian 相一致的 $J_0\left(\sqrt{\gamma}r\right)$ 和 $N_0\left(\sqrt{\gamma}r\right)$ 的线性独立解；J_0 和 J_1，N_0 和 N_1，K_0 和 K_1 分别为零阶和一阶贝塞尔函数、纽曼函数和麦克唐纳圆柱函数。

参数 A，B，A_1 和 B_1 可由下式得到：

$$\left.\begin{array}{l}A = (C/2)\pi\sqrt{\gamma}r_0\left[K_0\left(\sqrt{\gamma}r_0\right)N_1\left(\sqrt{\gamma}r_0\right) - K_1\left(\sqrt{\gamma}r_0\right)N_0\left(\sqrt{\gamma}r_0\right)\right] \\ B = -(C/2)\pi\sqrt{\gamma}r_0\left[K_0\left(\sqrt{\gamma}r_0\right)J_1\left(\sqrt{\gamma}r_0\right) - K_1\left(\sqrt{\gamma}r_0\right)J_0\left(\sqrt{\gamma}r_0\right)\right] \\ A_1 = (C_1/2)\pi\sqrt{\gamma}r_0\left[K_2\left(\sqrt{\gamma}r_0\right)N_3\left(\sqrt{\gamma}r_0\right) - K_3\left(\sqrt{\gamma}r_0\right)N_2\left(\sqrt{\gamma}r_0\right)\right] \\ B_1 = -(C_1/2)\pi\sqrt{\gamma}r_0\left[K_2\left(\sqrt{\gamma}r_0\right)J_3\left(\sqrt{\gamma}r_0\right) - K_3\left(\sqrt{\gamma}r_0\right)J_2\left(\sqrt{\gamma}r_0\right)\right]\end{array}\right\} \quad (17)$$

式中：C_1 为与 Wronskian 相一致的 $J_2\left(\sqrt{\gamma}r\right)$ 和 $N_2\left(\sqrt{\gamma}r\right)$ 的线性独立解；J_3，N_3 和 K_3 分别为三阶和一阶贝塞尔函数、纽曼函数和麦克唐纳圆柱函数。

比较上述这 2 个解可见，非协调张量不等于 0 的非协调解较之非协调张量等于 0 的协调解(σ_{r1}，$\sigma_{\theta 1}$，$\tau_{r\theta 1}$)增加了与初始微裂纹密度 η_0 和初始微裂纹半长 c_0 以及二次微裂纹密度 η 和二次微裂纹半长 c 有关的部分(σ_{r2}，$\sigma_{\theta 2}$，$\tau_{r\theta 2}$)。众所周知，由于协调解 σ_{r1}，$\sigma_{\theta 1}$，$\tau_{r\theta 1}$ 为与荷载(初始地应力)相平衡及与边界条件相适应的解，所以(σ_{r2}，$\sigma_{\theta 2}$，$\tau_{r\theta 2}$)为自平衡的封闭应力解。可见岩体中的自平衡应力是由岩体中的内部缺陷引起的非协调变形产生的。需要说明的是，岩石的塑性本质是加载过程中二次微裂纹的产生，在计算围岩应力和破坏的非协调变形模型中，岩石塑性不再计入岩石材料的宏观本构中，而是在微观计算中考虑，即计入了二次微裂纹发展的影响。

现在来分析非协调变形产生的岩体中的自平衡应力变化情况，为此，给出相应计算参数如下：$\sigma_v = 40$ MPa，$\sigma_h = 20$ MPa，$v = 0.2$，$r_0 = 7$ m，$g = 90$ MPa，$C = C_1 = 8.25$ m^{-2}，$\lambda_1 = 3.2$ GPa，$\theta = 0°$，$\lambda_2 = 10.417$ GPa，$\mu_1 = 3.2$ GPa，$E = 25$ GPa，$\mu_2 = 6.944\ 4$ GPa。计算结果如图3, 4所示。

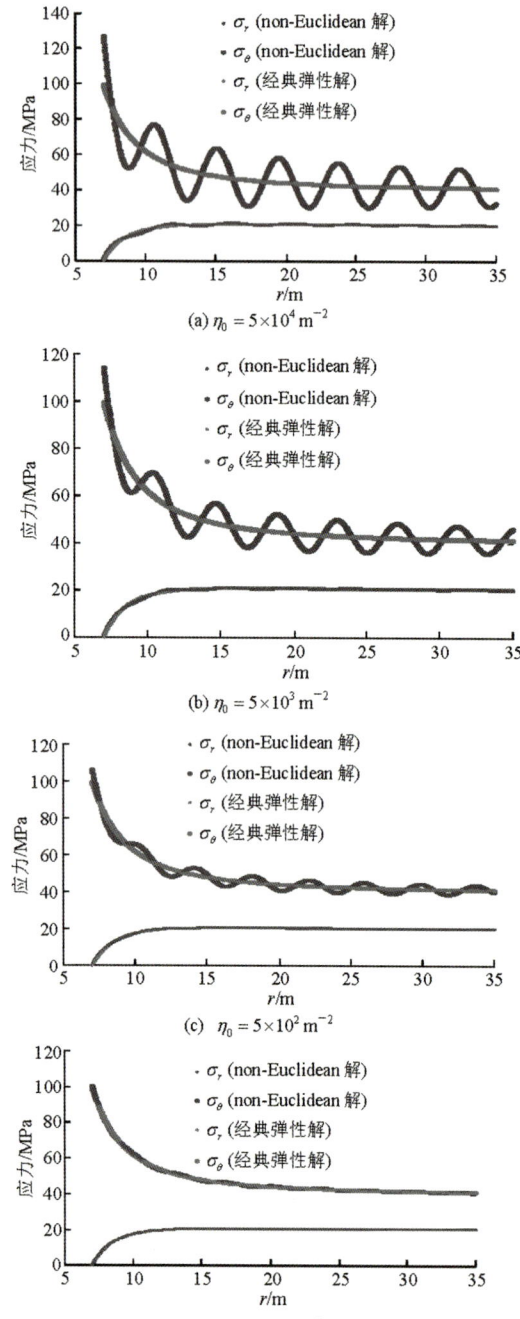

图3　不同初始微裂纹密度情况下隧洞围岩应力变化规律（$c_0 = 0.05$ mm）

Fig.3　Dependence of distribution of stress in surrounding rock mass around circular tunnel on the density of microcracks ($c_0 = 0.05$ mm)

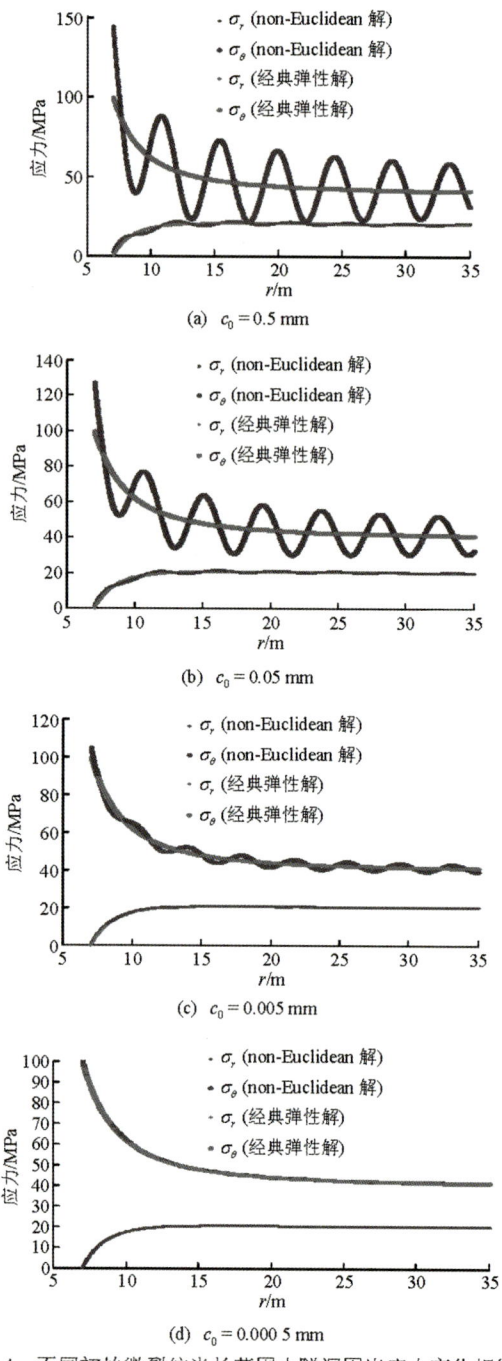

图4　不同初始微裂纹半长范围内隧洞围岩应力变化规律（$\eta_0 = 5×10^4$ m^{-2}）

Fig.4　Dependence of distribution of stress in surrounding rock mass around circular tunnel on the half length of microcracks ($\eta_0 = 5×10^4$ m^{-2})

图3中的计算结果均相应于初始微裂纹半长$c_0 = 0.05$ mm(50 μm)，图3(a)，(b)，(c)和(d)相应于不同初始微裂纹密度，图4中的计算结果均相应于不同的初始微裂纹的半长度，从0.5～500 μm，但相应于同一个初始微裂纹密度$\eta_0 = 5×10^4$ m^{-2}。

由图3，4可见：围岩中非协调变形产生的自平衡应力在径向和切向方向呈波状振荡变化，波幅自洞周往围岩深部逐步减小；这个自平衡应力对总应力的影响与初始微裂纹的密度与长度密切相关：密度越大，影响越大；微裂纹长度越大，影响越大。当初始微裂纹密度较小（$\eta_0 = 50 \text{ m}^{-2}$）时，即使微裂纹半长$c_0 = 50$ μm，其影响可忽略不计，相类似，当微裂纹很小时（$c_0 = 0.5$ μm），即使密度很大（$\eta_0 = 5\times 10^4 \text{ m}^{-2}$），其影响也可忽略不计。

洞室围岩中应力在径向方向呈波状振荡变化的现象30 a前在金川镍矿地下矿井围岩中早有发现，现场测量表明：围岩中松弛区和严密区交替出现[15-16]。

因此，似乎可以得出，初始微裂纹很稀或很小时，是可以忽略不计的，即将不连续岩体看作为连续性介质，可以不研究非协调变形的影响，从物理意义上可推测说，稀疏的很小的微裂纹群所引起岩体变形的内能变化量很小，可以在不可逆热力学方程中忽略它。

6 非协调变形对围岩破坏的影响

由上述算例的协调解与非协调解的对比可见，其主要差别为：协调解的围岩应力是单调函数，其极值在洞周附近，至围岩深部趋近于初始地应力，因此按照协调解，围岩的破坏仅发生在洞周附近；非协调解的围岩应力自洞周至围岩深部呈波状振荡衰减，在洞周附近及离洞周不同距离上有若干个不断减小的极值，如果这些应力极值达到强度破坏准则，就使岩体产生多处破坏。因此按照非协调解，围岩的破坏可以发生在离洞周的不同距离上的几个部位，即分区破裂。这就是围岩分区破裂化和围岩深部产生岩爆的本质原因：围岩的内部缺陷是围岩分区破裂化和围岩深部岩爆的物理本质，这些内部缺陷引起的非协调变形是产生分区破裂化和深部岩爆的力学本质。

7 结 论

(1) 岩体变形的协调性是指岩体变形后仍保持其整体性和协调性，从物理的观点来说，对岩体及其微元施加荷载产生变形后，能将这些变形后的微元重新拼合为一个变了形的岩体整体；从数学的观点来说，岩体变形后产生的位移函数在其定义域内为单值连续函数，即应变(应力)满足应变(应力)协调方程。

(2) 以往岩石力学研究岩体的变形仅仅研究岩体的协调变形，即非协调张量等于0(对平面问题$\partial^2\varepsilon_x/\partial y^2 + \partial^2\varepsilon_y/\partial x^2 - \partial^2\gamma_{xy}/\partial x\partial y = 0$)。无论是位移解和应力解，以往都从满足平衡方程和协调方程出发，这是唯一解的前提和基础。

(3) 岩体变形中出现的"开裂"和"叠合"，形成岩体的非整体和非连续，即呈现变形的不协调性，此时非协调张量不等于0，即

$$\frac{\partial^2\varepsilon_x}{\partial y^2} + \frac{\partial^2\varepsilon_y}{\partial x^2} - \frac{\partial^2\gamma_{xy}}{\partial x\partial y} \neq 0$$

(4) 研究岩体的非协调变形，主要基于引入内度量非欧几何空间，通过曲率、挠度等非欧几何参量来描述引起岩体变形非协调性的岩体内部缺陷，借助于不可逆热力学理论建立求解非欧几何参量与岩体内部缺陷的关系，拓展传统的弹塑性力学模型$\Delta^2 R = \gamma^2 R$。

(5) 岩体是非连续的地质体，在传统连续介质的弹塑性模型中忽略不计的微、细观裂纹，岩体开挖卸荷后产生的新微裂纹以及原生初始微裂纹的扩展、连接和贯通都属于岩体非协调变形范畴。

(6) 岩体的非协调变形是产生岩体中的自平衡封闭应力的重要原因。围岩中的非协调张量不等于0 的非协调应力解较之非协调张量等于0 的协调应力解增加了与岩体内部缺陷相关的自平衡封闭应力解。

(7) 对于地下洞室的围岩应力来说，围岩中非协调变形产生的自平衡应力在半径方向呈波状振荡变化，波幅自洞周往围岩深部逐步减小；自平衡应力的大小与内部缺陷的密度与大小密切相关：密度越大，影响越大；缺陷尺度越大，影响越大，当缺陷尺度与密度很小时，其影响可以忽略不计。

(8) 岩体的非协调变形产生围岩中自平衡封闭应力的振荡变化是围岩分区破裂化和围岩深部产生岩爆的本质原因。

(9) 峰值应力前的岩石塑性变形的实质是少量微裂纹的产生，根据以上分析，忽略不计它的非协调性误差可能不大，但在峰值应力后的应变软化段，此时微裂纹大量产生，有些扩展合并成粗裂纹和主干裂纹，忽略其非协调性，引起的误差可能较大，甚至很大。

参考文献(References)：

[1] 陈宗基. 中国岩石力学学科发展及工程应用[C]// 第四届国际岩石力学大会上的讲话. [S. l.]: [s. n.], 2005: 9 - 10.(CHEN Zongji. The development and engineering application for Chinese rock mechanics[C]// The Fourth International Congress on Rock Mechanics, Montreux. [S. l.]: [s. n.], 2005: 9 - 10 (in Chinese))

[2] 孙 钧, 侯学渊. 地下结构(上册)[M]. 北京：科学出版社, 1987: 152.(SUN Jun, HOU Xueyuan. Underground structure(I)[M]. Beijing: Science Press, 1987: 152.(in Chinese))

[3] 杨桂通. 弹塑性力学[M]. 北京：人民教育出版社, 1980: 49 - 50. (YANG Guitong. Elastoplastic mechanics[M]. Beijing: People's Education Press, 1980: 49 - 50.(in Chinese))

[4] ZHOU X P, CHEN G, QIAN Q H. Zonal disintegration mechanism of cross-anisotropic rock masses around a deep circular tunnel[J]. Theoretical and Applied Fracture Mechanics, 2012, 57(1): 49 - 54.

[5] BENVENSITE Y. On the Mori-Tanaka's method in cracked solids[J]. Mechanics Research Communications, 1986, 13(4): 193 - 201.

[6] GOLSHANI A, ODA M, OKUI Y, et al. Numerical simulation of the excavation damaged zone around an opening in brittle rock[J]. International Journal of Rock Mechanics and Mining Sciences, 2007, 44(6): 835 - 845.

[7] GOLSHANI A, OKUI Y, ODA M, et al. A micromechanical model for brittle failure of rock and its relation to crack growth observed in triaxial compression tests of granite[J]. Mechanics of Materials, 2006, 38(4): 287 - 303.

[8] QIAN Q H, ZHOU X P, XIA E M. Effects of the axial in situ stresses on the zonal disintegration phenomenon in the surrounding rock masses around a deep circular tunnel[J]. Journal of Mining Science, 2012, 48(1): 276 - 285

[9] GUZEV M A, PAROSHIN A A. Non-Euclidean model of the zonal disintegration of rocks around an underground working[J]. Journal of Applied Mechanics and Technical Physics, 2001, 42(1): 131 - 139.

[10] GUZEV M A. Structure of kinematic and force fields in the Riemannian continuum model[J]. Journal of Applied Mechanics and Technical Physics, 2011, 52(5): 709 - 716.

[11] МЯСНИКОВ В П, ГУЗЕВ М А Геометрическая модель внутренних самоуравновешенных напряжений в твердых телах докл[J]. РАН, 2001, 380(5): 627 - 629.

[12] ГУЗЕВ М А, МЯСНИКОВ В П. Термомеханическая модель упругопластического материала с дефектами структуры[J]. Изв. РАН. Механика Твердого Тела, 1998, (4): 156 - 172.

[13] МЯСНИКОВ В П, ГУЭЕВ М А. Неевклидова модель деформирования материалов на различных структурных уровнях[J]. Физ. Мезомеханика, 2000, (3): 5 - 16.

[14] ZHOU X P, QiAN Q H. The non-Euclidean model of failure of the deep rock masses under the deformation incompatibility condition[J]. Journal of Mining Science, 2013, to be pressed.

[15] 方祖烈, 陈新万, 丁延棱, 等. 对金川镍矿不良岩体水平巷道地压活动规律的几点认识[J]. 北京钢铁学院学报, 1982, (4): 9 - 15. (FANG Zulie, CHEN Xinwan, DING Yanling, et al. A few understanding of ground pressure activities around horizontal roadways in weak rock masses in Jinchuan nickel mine[J]. Journal of Beijing University of Iron and Steel Technology, 1982, (4): 9 - 15.(in Chinese))

[16] 何满潮. 中国煤矿软岩巷道支护理论与实践[M]. 北京：中国矿业大学出版社, 1996: 64 - 69.(HE Manchao. Support theory and practice of soft rock roadway in China coal mine[M]. Beijing: China University of Mining and Technology Press, 1996: 64 - 69.(in Chinese))

非协调变形下深部岩体破坏的非欧模型

周小平[1,2,3]，钱七虎[1,4,5]

(1. 重庆大学 土木工程学院，重庆 400045；2. 重庆大学 山地城镇建设新技术教育部重点实验室，重庆 400045；3. 重庆大学 煤矿灾害动力学与控制国家重点实验室，重庆 400044；4. 解放军理工大学 国防工程学院，江苏 南京 210007；5. 解放军理工大学 爆炸冲击防灾减灾国家重点实验室，江苏 南京 210007)

摘要：将岩体看成不含原生宏观裂隙而仅含原生微裂纹的颗粒材料。在开挖卸荷过程中，原生微裂纹启裂、扩展并穿过岩石基质，产生次生裂纹。此时，岩体出现非连续和非协调变形，经典的弹塑性力学理论不再适用于研究岩体的非连续和非协调变形情况。基于自由能密度、平衡方程和变形非协调条件，提出一种新的非欧模型，确定微裂纹半长度和密度对标量曲率和自平衡应力的影响，获得深部圆形洞室围岩应力场，它包括弹性应力和自平衡应力。由于自平衡应力的影响，当微裂纹的密度和半长度较大时，深部圆形洞室围岩的应力场具有明显的振荡特性。但是，当微裂纹的密度和半长度较小时，深部圆形洞室围岩的应力场振荡特性不明显。对于位于波峰附近的微裂纹，其尖端的应力集中将导致次生裂纹的失稳扩展、连接、合并成宏观裂隙，形成破裂区；而对于位于波谷附近的微裂纹，其尖端应力集中不足以导致次生裂纹发生失稳扩展，因此出现非破裂区。深部圆形洞室围岩应力场交替出现波峰和波谷现象导致破裂区和非破裂区的交替出现，即分区破裂化现象。通过数值模拟，详细研究微裂纹半长度和密度对深部岩体分区破裂化现象的影响。

关键词：岩石力学；自平衡应力；分区破裂化现象；非欧模型

中图分类号：TU 45　　**文献标识码**：A　　**文章编号**：1000 – 6915(2013)04 – 0767 – 08

NON-EUCLIDEAN MODEL OF FAILURE OF DEEP ROCK MASSES UNDER INCOMPATIBLE DEFORMATION

ZHOU Xiaoping[1,2,3], QIAN Qihu[1,4,5]

(1. *College of Civil Engineering，Chongqing University，Chongqing* 400045，*China*；2. *Key Laboratory of New Technology for Construction of Cities in Mountain Area，Ministry of Education，Chongqing University，Chongqing* 400045，*China*；3. *State Key Laboratory of Coal Mine Disaster Dynamics and Control，Chongqing University，Chongqing* 400044，*China*；4. *Engineering Institute of National Defense Engineering，PLA University of Science and Technology，Nanjing，Jiangsu* 210007，*China*；5. *State Key Laboratory of Disaster Prevention and Mitigation of Explosion and Impact，PLA University of Science and Technology，Nanjing，Jiangsu* 210007，*China*)

Abstract：Rock masses without pre-existing macrocracks are considered as granular materials with only microcracks. During excavation of tunnels，microcracks may nucleate，grow and propagate through rock matrix；secondary micrcracks may appear，and discontinuous and incompatible deformation of rock masses may occur. The classical continuum elastoplastic theory is not suitable for analyzing discontinuous and incompatible deformation of rock masses any more. A new non-Euclidean model is established based on free energy density，the equilibrium equation and the deformation incompatibility condition，where effects of the half length and density of microcracks on scalar curvature and the self-equilibrated stresses in deep rock mass are investigated. Stress fields in the surrounding rock masses around a deep circular tunnel are determined，which are the sum of elastic stresses and the self-equilibrated stresses determined by the scalar curvature. Due to the self-equilibrated stresses, the distribution of stresses in the surrounding rock masses around deep tunnels is obviously fluctuant or wave-like when the half length and density of microcracks are large，while the distribution of stresses in the surrounding rock

---

收稿日期：2012-11-19；修回日期：2013-01-17。
基金项目：国家自然科学基金资助项目（51078371，51279218，51021001）；重庆大学煤矿灾害动力学与控制国家重点实验室课题（2011DA105287-MS201204）。
本文原载于《岩石力学与工程学报》(2013 年第 32 卷第 4 期)。

masses around deep tunnels is not obviously fluctuant or wave-like when the half length and density of microcracks are small. The stress concentration at the tips of microcracks located in vicinity of stress wave crest is comparatively large, which may lead to the unstable growth and coalescence of secondary microcracks, and consequently the occurrence of fractured zones. On the other hand, the stress concentration at the tips of microcracks located around stress wave trough is relatively small, which may lead to arrest of microcracks, and thus to the non-fractured zones. The alternative appearance of stress wave crest and stress trough thus may induce the alternative occurrence of fractured and non-fractured zones in deep rock masses. The influences of the density and half length of microcracks on zonal disintegration and self-equilibrated stresses are investigated in detail by using numerical simulation.

Key words: rock mechanics; self-equilibrated stresses; zonal disintegration phenomenon; non-Euclidean model

1 引 言

岩体既含有宏观裂隙也含有微观缺陷，在不连续岩体力学中，宏观裂隙可以当做各向异性的连续介质来处理，宏观节理可以用节理单元来研究，描述岩体中的裂隙、断层、节理等宏观结构面的不连续位移和运动特性可以采用不连续介质离散模型，如离散元等。位错对材料非协调变形的影响已开展一些研究[1-5]。而一般微观、亚微观不连续面在分析中忽略不计，当做连续介质来处理，用连续介质的弹塑性力学来分析[1-8]。因此，微观和亚微观裂纹对岩体变形协调性的影响没有研究，是一个值得研究的问题。

类似地，对于存在自平衡封闭应力问题，如果将一个有封闭内应力的物体分割成微元，这些微元被释放为无应力状态。这些释放后的微元一般不能无缝地拼合成一个处于自然状态(即各点应力为 0)的物体。因此，具有自平衡封闭应力的岩体一般也不满足变形协调条件。自平衡封闭应力对变形协调性的影响也没有研究，是一个值得研究的问题。

由于岩石力学界长期以来将含微观和亚微观裂纹的岩体当做连续介质处理，很少研究微观和亚微观裂纹对深部岩体的变形和破坏规律的影响。本文从非协调方程、能量方程和平衡方程出发，利用热力学理论研究了微裂纹密度和长度对深部岩体内度量张量、外度量张量、标量曲率、自平衡应力和深部圆形洞室围岩应力场的影响，研究结果将为揭示深部岩体变形和破坏规律提供理论基础，为深部岩体的设计、施工和支护方案的优化提供理论依据。

2 理论模型

2.1 均匀分布微裂纹对内度量非欧几何空间标量曲率的影响

假设深部岩体含均匀分布的原生微裂纹，原生微裂纹的密度为 η_0，原生微裂纹的半长度为 c_0。

对于不含缺陷或不存在自平衡封闭应力的岩石(体)，从外部观察者的外度量欧氏几何空间中度量的应变 $a_{ij} = \frac{1}{2}(\boldsymbol{\delta}_{ij} - D_{ij}) = \frac{1}{2}\left(\boldsymbol{\delta}_{ij} - \frac{\partial u_i}{\partial x_j}\frac{\partial u_i}{\partial x_j}\right)$ 与从内度量非欧几何空间中度量的应变 $\boldsymbol{\varepsilon}_{ij} = \frac{1}{2}(\boldsymbol{\delta}_{ij} - d_{ij}) = \frac{1}{2}\left(\boldsymbol{\delta}_{ij} - \frac{\partial u_i'}{\partial x_j}\frac{\partial u_i'}{\partial x_j}\right)$ 一致，D_{ij} 和 d_{ij} 分别为外度量张量和内度量张量，且外度量张量等于内度量张量 $D_{ij} = d_{ij}$，$u_i = x_i - U_i(x,t)$ 和 $u_i' = u_i - U_i'(x,t)$，$U_i(x,t)$ 和 $U_i'(x,t)$ 分别为外度量欧氏几何空间的位移矢量和内度量非欧几何空间的位移矢量。此时，岩体变形协调的假设成立，变形协调条件也称为 Saint-Venant 变形协调方程。在直角坐标系下，Saint-Venant 变形协调方程[9-10]可以表示为

$$R = 2\left(\Delta\boldsymbol{\varepsilon}_{kk} - \frac{\partial^2 \boldsymbol{\varepsilon}_{ij}}{\partial x_i \partial x_j}\right) = 0 \quad (1)$$

式中：ε_{kk} 为内度量非欧几何空间的体积应变，$\varepsilon_{kk} = \varepsilon_{11} + \varepsilon_{22} + \varepsilon_{33}$；$\boldsymbol{\varepsilon}_{ij}$ 为内度量非欧几何空间的应变张量；R 为内度量非欧几何空间的标量曲率。

对于含缺陷或存在自平衡封闭应力的岩石(体)，从外部观察者的外度量欧氏几何空间中度量的应变 a_{ij} 不等于从描述岩体内部缺陷的内度量非欧几何空间中度量的应变 $\boldsymbol{\varepsilon}_{ij}$，也即外度量张量不等于内度量张量 $D_{ij} \neq d_{ij}$。

因此，Saint-Venant 变形协调的假设不成立。此时，变形非协调方程[1-8]为

$$R = 2\left(\Delta\boldsymbol{\varepsilon}_{kk} - \frac{\partial^2 \boldsymbol{\varepsilon}_{ij}}{\partial x_i \partial x_j}\right) \neq 0 \quad (2)$$

含微裂纹岩体的自由能密度包括外度量欧氏几

何空间的应变能密度、内度量非欧几何空间的应变能密度、外度量欧氏几何空间和内度量非欧几何空间的应变场相互作用产生的应变能密度以及与内度量非欧几何空间的标量曲率相关的应变能密度。根据能量等效的方法，含微裂纹岩体的自由能密度[11-12]可以表示为

$$F = \frac{\lambda_1(1-D_0)^2 a_{kk}^2}{2} + \mu_1(1-D_1)^2 a_{ij}a_{ij} + \frac{\lambda_2(1-D_2)^2 \varepsilon_{kk}^2}{2} + \mu_2(1-D_3)^2 \varepsilon_{ij}\varepsilon_{ij} + g(1-D_0)(1-D_2)a_{kk}\varepsilon_{ii} - \frac{\mu_3(1-D_3)^2 R^2}{4} \quad (3a)$$

其中，

$$\left.\begin{array}{l} D_0 = \pi\omega_0/(1+\pi\omega_0),\ D_1 = \pi\omega_0/(1+v_0+\pi\omega_0) \\ D_2 = \pi\omega/(1+\pi\omega),\ D_3 = \pi\omega/(1+v_1+\pi\omega) \\ \omega = \eta c^2,\ \omega_0 = \eta_0 c_0^2 \end{array}\right\} \quad (3b)$$

式中：λ_1，μ_1均为相应于外度量欧氏几何空间的Almansi应变张量的Lame常数；λ_2，μ_2均为相应于内度量非欧几何空间的Lame常数；g，μ_3均为唯象学参数；a_{kk}为外度量欧氏几何空间的Almansi体积应变；D_1，D_0均为原生微裂纹引起的损伤变量；D_2，D_3均为次生微裂纹引起的损伤变量；η为次生微裂纹的密度；c为次生微裂纹的半长度；η_0为原生微裂纹的密度；c_0为原生微裂纹的半长度。外度量欧氏几何空间材料的泊松比可表示为：$v_0 = \lambda_1/[2(\lambda_1+\mu_1)]$，内度量非欧几何空间材料的泊松比可表示为：$v_1 = \lambda_2/[2(\lambda_2+\mu_2)]$。

根据热力学理论，对自由能密度求外度量欧氏几何空间的Almansi应变张量的偏导数，可得含微裂纹岩体的应力为

$$\boldsymbol{\sigma}_{ij} = \frac{\partial F}{\partial a_{ij}} = \lambda_1\boldsymbol{\delta}_{ij}a_{kk}(1-D_0)^2 + 2\mu_1 a_{ij} \cdot (1-D_1)^2 + g\boldsymbol{\delta}_{ij}\varepsilon_{kk}(1-D_0)(1-D_2) \quad (4)$$

根据热力学理论[1-8]，自由能与内度量非欧几何空间的标量曲率和内度量非欧几何空间的应变张量之间的关系为

$$\frac{\partial F}{\partial \varepsilon_{ij}} + 2\left(\boldsymbol{\delta}_{ij}\Delta - \frac{\partial^2}{\partial x_i \partial x_j}\right)\frac{\partial F}{\partial R} = 0 \quad (5)$$

式中：Δ为拉普拉斯算子。

将式(3)代入式(5)，可得内度量非欧几何空间的应变张量、内度量非欧几何空间的体积应变、标量曲率和外度量欧氏几何空间的Almansi体积应变之间的关系为

$$\lambda_2\boldsymbol{\delta}_{ij}\varepsilon_{kk}(1-D_2)^2 + 2\mu_2\varepsilon_{ij}(1-D_3)^2 + g\boldsymbol{\delta}_{ij}a_{kk}(1-D_0)\cdot$$
$$(1-D_2) - \mu_3(1-D_3)^2\left(\boldsymbol{\delta}_{ij}\Delta - \frac{\partial^2}{\partial x_i \partial x_j}\right)R = 0 \quad (6)$$

根据式(6)，可得内度量非欧几何空间的体积应变、标量曲率与外度量欧氏几何空间的Almansi体积应变之间的关系为

$$\boldsymbol{\varepsilon}_{kk} = \frac{2\mu_3(1-D_3)^2}{[3\lambda_2(1-D_2)^2 + 2\mu_2(1-D_3)^2]}\Delta R - \frac{3g(1-D_0)(1-D_2)}{3\lambda_2(1-D_2)^2 + 2\mu_2(1-D_3)^2}a_{kk} \quad (7)$$

将式(7)代入式(6)并消去内度量非欧几何空间的体积应变，可得内度量非欧几何空间的应变张量、标量曲率和外度量欧氏几何空间的Almansi体积应变之间的关系为

$$\boldsymbol{\varepsilon}_{ij} = \frac{\mu_3}{2\mu_2}\left(\boldsymbol{\delta}_{ij}\Delta - \frac{\partial^2}{\partial x_i \partial x_j}\right)R - \frac{g(1-D_0)(1-D_2)}{3\lambda_2(1-D_2)^2 + 2\mu_2(1-D_3)^2}\cdot$$
$$a_{kk}\boldsymbol{\delta}_{ij} - \boldsymbol{\delta}_{ij}\frac{\lambda_2\mu_3(1-D_2)^2}{\mu_2[3\lambda_2(1-D_2)^2 + 2\mu_2(1-D_3)^2]}\Delta R \quad (8)$$

将式(8)代入式(2)，可得外度量欧氏几何空间的Almansi体积应变和内度量非欧几何空间的标量曲率之间的关系为

$$\frac{2g(1-D_0)(1-D_2)}{3\lambda_2(1-D_2)^2 + 2\mu_2(1-D_3)^2}\Delta a_{kk} + \frac{R}{2} =$$
$$\frac{\mu_3}{\mu_2}\frac{\lambda_2(1-D_2)^2 + 2\mu_2(1-D_3)^2}{3\lambda_2(1-D_2)^2 + 2\mu_2(1-D_3)^2}\Delta^2 R \quad (9)$$

对于静力问题，其平衡方程可以表示为

$$\frac{\partial \boldsymbol{\sigma}_{ij}}{\partial x_j} = 0 \quad (10)$$

将应力表达式(4)代入式(10)，可得外度量欧氏几何空间的位移、Almansi体积应变和内度量非欧几何空间的体积应变之间的关系为

$$[\lambda_1(1-D_0)^2 + \mu_1(1-D_1)^2]\nabla a_{kk} + \mu_1(1-D_1)^2\Delta U + g(1-D_0)(1-D_2)\nabla \boldsymbol{\varepsilon}_{kk} = 0 \quad (11)$$

式中：U为外度量欧氏几何空间的位移矢量。

对式(11)运用张量分析，可得外度量欧氏几何空间的位移梯度、旋转位移和内度量非欧几何空间的体积应变之间的关系为

$$[\lambda_1(1-D_0)^2 + 2\mu_1(1-D_1)^2]\nabla\text{div}U + g(1-D_0)(1-D_2)\nabla \boldsymbol{\varepsilon}_{kk} = \mu_1(1-D_1)^2\text{rot}(\text{rot}U) \quad (12)$$

式中：$\text{rot}U$为外度量欧氏几何空间的旋转位移，

$\mathrm{div}U$ 为外度量欧氏几何空间的位移梯度。

对于深部岩体问题，外度量欧氏几何空间的旋转位移 $\mathrm{rot}U=0$，因此，式(12)可以表示为

$$[\lambda_1(1-D_0)^2+2\mu_1(1-D_1)^2]\nabla\mathrm{div}U+g(1-D_0)(1-D_2)\nabla\boldsymbol{\varepsilon}_{kk}=0 \quad (13)$$

利用张量分析，式(13)可以表示为

$$\nabla\{[\lambda_1(1-D_0)^2+2\mu_1(1-D_1)^2]\mathrm{div}U+g(1-D_0)(1-D_2)\boldsymbol{\varepsilon}_{kk}\}=0 \quad (14)$$

对式(14)积分，可得外度量欧氏几何空间的位移梯度、内度量非欧几何空间的体积应变之间的关系为

$$[\lambda_1(1-D_0)^2+2\mu_1(1-D_1)^2](1-\alpha)\mathrm{div}U+g(1-D_0)(1-D_2)(1-\alpha)\boldsymbol{\varepsilon}_{kk}=B \quad (15)$$

式中：B 为积分常数。

将式(7)代入式(15)，消去内度量非欧几何空间的体积应变，可得外度量欧氏几何空间的位移梯度、Almansi 体积应变和内度量非欧几何空间的标量曲率之间的关系为

$$[\lambda_1(1-D_0)^2+2\mu_1(1-D_1)^2](1-\alpha)\mathrm{div}U-\frac{2\mu_3g(1-D_3)^2(1-D_0)(1-D_2)(1-\alpha)}{3\lambda_2(1-D_2)^2+2\mu_2(1-D_3)^2}\Delta R=$$
$$B+\frac{3g^2(1-D_0)^2(1-D_2)^2(1-\alpha)}{3\lambda_2(1-D_2)^2+2\mu_2(1-D_3)^2}\boldsymbol{a}_{kk} \quad (16)$$

其中，

$$\alpha=[3g^2(1-D_0)^2(1-D_2)^2]/\{[\lambda_1(1-D_0)^2+2\mu_1(1-D_1)^2][3\lambda_2(1-D_2)^2+2\mu_2(1-D_3)^2]\}$$

外度量欧氏几何空间的位移可以分解为弹性位移 U_0 和微裂纹引起的非弹性位移 W。

根据弹性力学理论[1-2, 9-10]，外度量欧氏几何空间的弹性位移梯度可以表示为

$$\mathrm{div}U_0=\frac{B}{[\lambda_1(1-D_0)^2+2\mu_1(1-D_1)^2](1-\alpha)} \quad (17)$$

根据细观力学理论[1-2]，微裂纹引起的位移梯度可以表示为

$$\mathrm{div}W=\{[2\mu_3g(1-D_3)^2(1-D_0)(1-D_2)(1-2\alpha)]\Delta R\}/\{[3\lambda_2(1-D_2)^2+2\mu_2(1-D_3)^2][\lambda_1(1-D_0)^2+2\mu_1(1-D_1)^2](1-\alpha)\} \quad (18)$$

将式(17)和(18)代入式(16)，可得外度量欧氏几何空间的 Almansi 体积应变和内度量非欧几何空间的标量曲率之间的关系为

$$\boldsymbol{a}_{kk}=-\frac{2\alpha\mu_3(1-D_3)^2}{3g(1-D_0)(1-D_2)(1-\alpha)}\Delta R \quad (19)$$

将式(19)代入式(9)，可得内度量非欧几何空间的标量曲率的微分为

$$\Delta^2 R=\gamma^2 R \quad (20)$$

其中，

$$\gamma^2=\frac{3\beta[3\lambda_2(1-D_2)^2+2\mu_2(1-D_3)^2](1-\alpha)}{2[3\lambda_2(1-D_2)^2(1-\alpha)+2\mu_2(1-D_3)^2(3-\alpha)]}$$

$$\beta=\frac{\mu_2}{\mu_3}$$

2.2 内度量非欧几何空间的标量曲率对深部圆形洞室围岩应力场的影响

如图 1 所示，假设圆形洞室的半径为 $r_0=a$，深部岩体所受的水平方向地应力为 σ_h，垂直方向地应力为 σ_v，对于静水压力问题有 $\sigma_h=\sigma_v=\sigma_\infty$。由于该问题为轴对称问题，因此式(20)可以用极坐标表示为

$$\left(\frac{\partial^2}{\partial r^2}+\frac{1}{r}\frac{\partial}{\partial r}\right)^2 R=\gamma^2 R \quad (21)$$

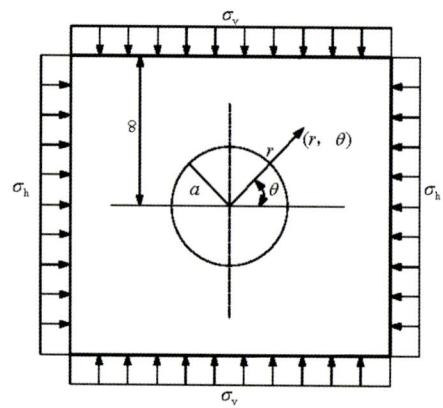

图 1　深部圆形洞室的力学模型

Fig.1　Mechanical model of a deep circular tunnel

对式(21)求解，可得内度量非欧几何空间的标量曲率为 4 项，其中一项 $I_0(\sqrt{\gamma}r)$ 因 $r\to\infty$ 时 R 有限，故该项前系数取 0。经化简后，内度量非欧几何空间的标量曲率为

$$R=AJ_0(\sqrt{\gamma}r)+BN_0(\sqrt{\gamma}r)+CK_0(\sqrt{\gamma}r) \quad (22)$$

式中：A，B 均为积分常数，由边界条件确定。

由热力学理论[5-8]可知，内度量非欧几何空间的标量曲率的边界条件可以表示为

$$R|_{r=r_0}=0,\quad \frac{\partial R}{\partial r}\bigg|_{r=r_0}=0 \quad (23)$$

将式(22)代入式(23)，积分常数 A，B 可以表示为

$$A = (C/2)\pi\sqrt{\gamma}r_0\left[K_0\left(\sqrt{\gamma}r_0\right)N_1\left(\sqrt{\gamma}r_0\right) - K_1\left(\sqrt{\gamma}r_0\right)N_0\left(\sqrt{\gamma}r_0\right)\right] \quad (24)$$

$$B = -(C/2)\pi\sqrt{\gamma}r_0\left[K_0\left(\sqrt{\gamma}r_0\right)J_1\left(\sqrt{\gamma}r_0\right) - K_1\left(\sqrt{\gamma}r_0\right)J_0\left(\sqrt{\gamma}r_0\right)\right] \quad (25)$$

式中：C 为保证解的唯一性而与 J_0，N_0 线性相关的 Wronskian 解；J_0 和 J_1，N_0 和 N_1，K_0 和 K_1 分别为零阶和一阶 Bessel，Neumann 和 Macdonald 柱状函数；r_0 为圆形洞室的半径。

对于含微裂纹的岩体，内度量非欧几何空间的应变可以表示为

$$\boldsymbol{\varepsilon}_{ij} = \frac{(1+v)\boldsymbol{\sigma}_{ij} - \boldsymbol{\delta}_{ij}v\boldsymbol{\sigma}}{E} \quad (26)$$

其中，

$$v = \frac{\lambda_2(1-D_2)^2}{2[\lambda_2(1-D_2)^2 + \mu_2(1-D_3)^2]}$$

$$E = \frac{\mu_2[3\lambda_2(1-D_2)^2 + 2\mu_2(1-D_3)^2](1-D_3)^2}{\lambda_2(1-D_2)^2 + \mu_2(1-D_3)^2}$$

$$\boldsymbol{\sigma} = \boldsymbol{\sigma}_{ii}$$

根据式(2)和(26)，内部度量非欧几何空间的标量曲率和应力张量第一不变量之间的关系可以表示为

$$R = \frac{2(1-v)}{E}\Delta\boldsymbol{\sigma} \quad (27)$$

将式(22)代入式(27)，可得应力张量第一不变量为

$$\boldsymbol{\sigma} = \sigma_r + \sigma_\theta = 2(1+v)\sigma_\infty - \frac{E}{2\gamma(1-v)} \cdot \left[AJ_1\left(\sqrt{\gamma}r\right) + BN_1\left(\sqrt{\gamma}r\right) + CK_1\left(\sqrt{\gamma}r\right)\right] \quad (28)$$

对于静水压力条件下的轴对称问题，其平衡方程式(10)可以表示为

$$\frac{\partial\sigma_r}{\partial r} + \frac{\sigma_r - \sigma_\theta}{r} = 0 \quad (29)$$

根据式(28)，(29)，可得静水压力条件下深部圆形洞室围岩径向应力为

$$\sigma_r = \frac{\sigma_v + \sigma_h}{2}\left(1 - \frac{r_0^2}{r^2}\right) - \frac{E}{2\gamma^{3/2}(1-v^2)r} \cdot \left[AJ_1\left(\sqrt{\gamma}r\right) + BN_1\left(\sqrt{\gamma}r\right) + CK_1\left(\sqrt{\gamma}r\right)\right] \quad (30)$$

根据式(28)，(30)，可得静水压力条件下深部圆形洞室围岩切向应力可以表示为

$$\sigma_\theta = \frac{\sigma_v + \sigma_h}{2}\left(1 + \frac{r_0^2}{r^2}\right) - \frac{E}{2\gamma(1-v^2)}\left[AJ_0\left(\sqrt{\gamma}r\right) + BN_0\left(\sqrt{\gamma}r\right) - CK_0\left(\sqrt{\gamma}r\right)\right] + \frac{E}{2\gamma^{3/2}(1-v^2)r} \cdot \left[AJ_1\left(\sqrt{\gamma}r\right) + BN_1\left(\sqrt{\gamma}r\right) + CK_1\left(\sqrt{\gamma}r\right)\right] \quad (31)$$

同理，对于非静水压力问题，内度量非欧几何空间的标量曲率可以表示为

$$R = \left[A_1J_2\left(r\sqrt{\gamma}\right) + B_1N_2\left(r\sqrt{\gamma}\right) + C_1K_2\left(r\sqrt{\gamma}\right)\right]\cos(2\theta) \quad (32)$$

非静水压力条件下，深部圆形洞室围岩应力场可以表示为

$$\sigma_r = \frac{\sigma_v + \sigma_h}{2}\left(1 - \frac{r_0^2}{r^2}\right) + \frac{\sigma_h - \sigma_v}{2}\left(1 - \frac{4r_0^2}{r^2} + \frac{3r_0^4}{r^4}\right)\cos(2\theta) - \frac{E}{2\gamma^{3/2}(1-v^2)r}\left[AJ_1\left(\sqrt{\gamma}r\right) + BN_1\left(\sqrt{\gamma}r\right) + CK_1\left(\sqrt{\gamma}r\right)\right] - \frac{E}{2\gamma^{3/2}(1-v)r} \cdot \left[A_1J_1\left(r\sqrt{\gamma}\right) + B_1N_1\left(r\sqrt{\gamma}\right) + C_1K_1\left(r\sqrt{\gamma}\right)\right] \cdot \cos(2\theta) - \frac{3E}{\gamma^2(1-v^2)r^2}\left[A_1J_2\left(\sqrt{\gamma}r\right) + B_1N_2\left(\sqrt{\gamma}r\right) + C_1K_2\left(\sqrt{\gamma}r\right)\right]\cos(2\theta) \quad (33a)$$

$$\sigma_\theta = \frac{\sigma_v + \sigma_h}{2}\left(1 + \frac{r_0^2}{r^2}\right) - \frac{\sigma_h - \sigma_v}{2}\left(1 + \frac{3r_0^4}{r^4}\right)\cos(2\theta) - \frac{E\left[AJ_0\left(\sqrt{\gamma}r\right) + BN_0\left(\sqrt{\gamma}r\right) + CK_0\left(\sqrt{\gamma}r\right)\right]}{2\gamma(1-v^2)} + \frac{E\left[AJ_1\left(\sqrt{\gamma}r\right) + BN_1\left(\sqrt{\gamma}r\right) + CK_1\left(\sqrt{\gamma}r\right)\right]}{2\gamma^{3/2}(1-v^2)r} - \frac{3E\left[A_1J_1\left(\sqrt{\gamma}r\right) + B_1N_1\left(\sqrt{\gamma}r\right) + C_1K_1\left(\sqrt{\gamma}r\right)\right]\cos(2\theta)}{2\gamma^{3/2}(1-v^2)r} + \frac{E\left[A_1J_0\left(r\sqrt{\gamma}\right) + B_1N_0\left(r\sqrt{\gamma}\right) + C_1K_0\left(r\sqrt{\gamma}\right)\right]\cos(2\theta)}{2\gamma(1-v^2)} + \frac{3E\left[A_1J_2\left(r\sqrt{\gamma}\right) + B_1N_2\left(r\sqrt{\gamma}\right)C_1K_2\left(r\sqrt{\gamma}\right)\right]\cos(2\theta)}{\gamma^2(1-v^2)r^2} \quad (33b)$$

$$\tau_{r\theta} = \frac{\sigma_v - \sigma_h}{2}\left(1 + \frac{2r_0^2}{r^2} - \frac{3r_0^4}{r^4}\right)\sin(2\theta) + \frac{E}{r(1-v^2)\gamma^{3/2}}\left[A_1J_1\left(r\sqrt{\gamma}\right) + B_1N_1\left(r\sqrt{\gamma}\right) + C_1K_1\left(r\sqrt{\gamma}\right)\right]\sin(2\theta) - \frac{3E\sin(2\theta)}{r^2(1-v^2)\gamma^2}\left[A_1 \cdot J_2\left(r\sqrt{\gamma}\right) + B_1N_2\left(r\sqrt{\gamma}\right) + C_1K_2\left(r\sqrt{\gamma}\right)\right] \quad (33c)$$

式中：J_2，N_2，K_2 分别为二阶 Bessel，Neumann 和 Macdonald 柱状函数。A_1，B_1，C_1 由边界条件确定，C_1 为保证解的唯一性而与 J_2，N_2 线性相关的 Wronskian 解。A_1，B_1 可表示为

$$\left.\begin{array}{l} A_1 = (C_1/2)\pi\sqrt{\gamma}r_0\left[K_2\left(\sqrt{\gamma}r_0\right)N_3\left(\sqrt{\gamma}r_0\right)-\right.\\ \left.K_3\left(\sqrt{\gamma}r_0\right)N_2\left(\sqrt{\gamma}r_0\right)\right] \\ B_1 = -(C_1/2)\pi\sqrt{\gamma}r_0\left[K_2\left(\sqrt{\gamma}r_0\right)J_3\left(\sqrt{\gamma}r_0\right)-\right.\\ \left.K_3\left(\sqrt{\gamma}r_0\right)J_2\left(\sqrt{\gamma}r_0\right)\right] \end{array}\right\} \quad (33d)$$

式中：J_3，N_3 和 K_3 分别为三阶 Bessel，Neumann 和 Macdonald 圆柱函数。

式(33a)仅仅适用于非静水压力条件下深部圆形洞室围岩应力场；对于静水压力状态，深部圆形洞室围岩径向应力采用式(30)、切向应力采用式(31)。

2.3 微裂纹对深部圆形洞室围岩应力场的影响

根据圆形洞室围岩应力分量式(33)，可以确定主应力为

$$\left.\begin{array}{l} \sigma_1' = \dfrac{\sigma_r + \sigma_\theta}{2} + \dfrac{1}{2}\sqrt{(\sigma_r - \sigma_\theta)^2 + 4\tau_{r\theta}^2} \\ \sigma_2' = \dfrac{\sigma_r + \sigma_\theta}{2} - \dfrac{1}{2}\sqrt{(\sigma_r - \sigma_\theta)^2 + 4\tau_{r\theta}^2} \end{array}\right\} \quad (34)$$

式中：σ_r，σ_θ，$\tau_{r\theta}$ 可以按照式(33)计算，只要将 γ 中的次生裂纹半长度和密度替换为 0 即可。

原生裂纹尖端的应力强度因子[13]可以表示为

$$K_I = -[\sigma_2' + f(c_0)S_2']\sqrt{2d_i\tan\left(\dfrac{\pi c_0}{2d_i}\right)} \quad (35)$$

式中：$S_2' = \sigma_2' - (\sigma_1' + \sigma_2')/2$，$f(c_0) = d/c_0$，$d$ 为岩石材料的晶粒直径，σ_1' 和 σ_2' 由式(34)确定，d_i 为微裂纹之间的间距。

原生微裂纹沿界面扩展的临界条件为

$$K_I = K_{IC}' \quad (36)$$

式中：K_{IC}' 为界面的断裂韧度。

当式(35)满足式(36)时，原生微裂纹将扩展，原生微裂纹扩展后的平均半长度为 c_1，平均密度为 η_1。由于原生微裂纹的长度增加，深部圆形洞室围岩应力场将改变。改变后的深部圆形洞室围岩应力场 σ_r'，σ_θ'，$\tau_{r\theta}'$ 仍然可以按照式(33)确定，只要将 γ 中的原生裂纹半长度和密度替换为 c_1 和 η_1 即可。同时，圆形洞室围岩中的主应力也可以按照式(34)确定，只要将深部圆形洞室围岩应力分量替换为 σ_r'，σ_θ'，$\tau_{r\theta}'$ 即可。

当原生微裂纹的半长度沿界面扩展到 c_1 后，原生微裂纹将扩展到岩石基质，产生次生裂纹。此时，次生裂纹尖端的应力强度因子为

$$K_I = -[\sigma_2'' + f(l)S_2'']\sqrt{2d_i\tan\left(\dfrac{\pi l}{2d_i}\right)} \quad (37a)$$

其中，

$$S_2'' = \sigma_2'' - (\sigma_1'' + \sigma_2'')/2 \quad (37b)$$

$$\sigma_1'' = \dfrac{\sigma_r' + \sigma_\theta'}{2} + \dfrac{1}{2}\sqrt{(\sigma_r' - \sigma_\theta')^2 + 4\tau_{r\theta}'^2} \quad (37c)$$

$$\sigma_2'' = \dfrac{\sigma_r' + \sigma_\theta'}{2} - \dfrac{1}{2}\sqrt{(\sigma_r' - \sigma_\theta')^2 + 4\tau_{r\theta}'^2} \quad (37d)$$

式中：$f(l) = d/l$，l 为次生裂纹的扩展长度；σ_r'，σ_θ' 和 $\tau_{r\theta}'$ 均为原生微裂纹扩展后的平均长度为 c_1 时的深部圆形洞室围岩应力场，可由式(33)确定。

次生裂纹稳定扩展的临界条件为

$$K_I = K_{IC} \quad (38)$$

式中：K_{IC} 为岩石的断裂韧度。

当次生裂纹扩展长度 $l = c_2$ 时，深部岩体的承载能力最大，达到峰值强度。此时，深部岩体峰值强度可以表示为

$$\sigma_2'' + f(c_2)S_2'' = -\dfrac{K_{IC}}{\sqrt{2d_i\tan\left(\dfrac{\pi c_2}{2d_i}\right)}} \quad (39)$$

这里，c_2 可以通过迭代计算确定。

由于次生裂纹的扩展，深部圆形洞室围岩应力场将发生改变。此时，深部圆形洞室围岩应力场 σ_r''，σ_θ''，$\tau_{r\theta}''$ 仍然可以按照式(33)确定，只要将 γ 中的原生裂纹半长度和密度替换为 c_1 和 η_1、次生裂纹的平均半长度和密度替换为 l 和 η_2 即可。同时，圆形洞室围岩中的主应力也可以按照式(34)确定，只要将深部圆形洞室围岩应力分量替换为 σ_r''，σ_θ''，$\tau_{r\theta}''$ 即可。

当次生裂纹扩展长度 $l > c_2$ 时，次生裂纹将发生失稳扩展，深部岩体的承载能力下降。当次生裂纹扩展长度 $l = d_i - c_1$ 时，次生裂纹连接、贯通形成宏观裂隙和破裂区。

3 数值计算结果和参数敏感性性分析

3.1 深部圆形洞室围岩应力场的特性

为了研究深部圆形洞室围岩应力场的特性，在数值模拟中，下列计算参数被使用：$\sigma_h = 40$ MPa，$\sigma_v = 20$ MPa，$C = C_1 = 8.25$ m^{-2}，$r_0 = 7$ m，$\theta = 0°$，$\lambda_1 = 10$ GPa，$\mu_1 = 7$ GPa，$\lambda_2 = 8$ GPa，$\mu_2 = 5$ GPa，$\mu_3 = 1.6\times10^6$ kN·m^2，$g = 900$ MPa，$c_1 = 0.2$ mm，$\eta_1 = 1\,000$ m^{-2}，$d = 0.9$ mm，$d_i = 0.69$ mm，$\eta_0 = $

1×10^4 m^{-2},$c_0 = 0.05$ mm,$K_{IC} = 1.5$ MPa·$\sqrt{m}$,$K'_{IC} =$ 0.5 MPa·$\sqrt{m}$。图 2 给出了深部圆形洞室围岩应力场特性,由图可知:与弹性应力的单调特性不同,自平衡应力具有振荡特性。由于自平衡应力的振荡特性导致了深部岩体出现分区破裂化现象。

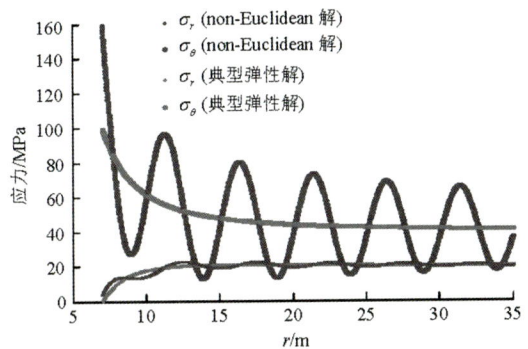

图 2 深部圆形洞室围岩应力场特性

Fig.2 Behaviors of stress field of surrounding rock mass around a deep circular tunnel

3.2 微裂纹密度和半长度对圆形洞室围岩分区破裂化现象的影响

为了研究微裂纹半长度和密度对深部圆形洞室围岩分区破裂化现象的影响,在数值模拟中,下列参数被使用:$r_0 = 7$ m,$g = 90$ MPa,$d = 3.5$ mm,$K_{IC} =$ 1.9 MPa·$\sqrt{m}$,$C = C_1 = 2.25\times10^4$ m^{-2},$d_i = 8.9$ mm,$K'_{IC} = 0.3$ MPa·$\sqrt{m}$,$c_1 = 0.9$ mm,$\lambda_1 = 10$ GPa,$\mu_1 =$ 5 GPa,$\lambda_2 = 8$ GPa,$\mu_2 = 7$ GPa,$\mu_3 = 500\,000$ kN·m^2,$\eta_1 = \eta_2 = 500$ m^{-2},$\sigma_h = 30$ MPa,$\sigma_v = 30$ MPa。

在图 3,4 中,黑色表示破裂区。图 3 表明,当微裂纹的半长度较大时,深部圆形洞室围岩才会出现分区破裂化现象。反之,当微裂纹的半长度较小时,深部圆形洞室围岩可能不会出现分区破裂化现象;同时图 3 也表明,破裂区的数量和宽度随半长度的增大而增大。从图 4 可以看出,当微裂纹密度较大时,深部圆形洞室围岩才出现分区破裂化现象,

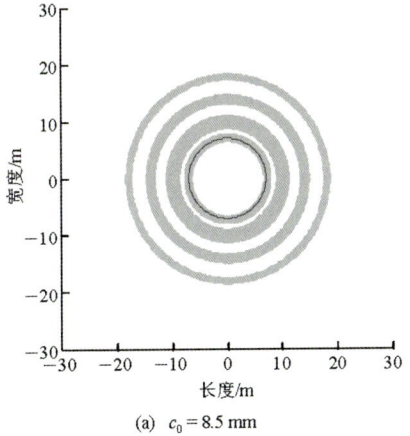

(a) $c_0 = 8.5$ mm

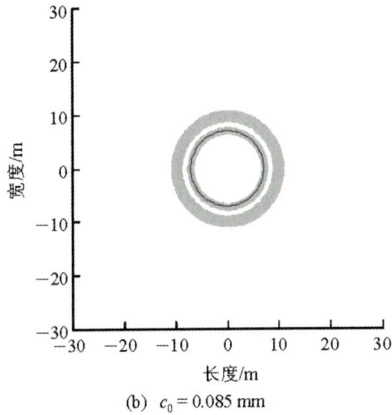

(b) $c_0 = 0.085$ mm

图 3 当原生微裂纹密度 $\eta_0 = 5\,000$ m^{-2}时,原生微裂纹半长度对深部圆形洞室围岩破裂区的影响

Fig.3 Dependence of distribution of fractured zones on the half length of microcracks when $\eta_0 = 5\,000$ m^{-2}

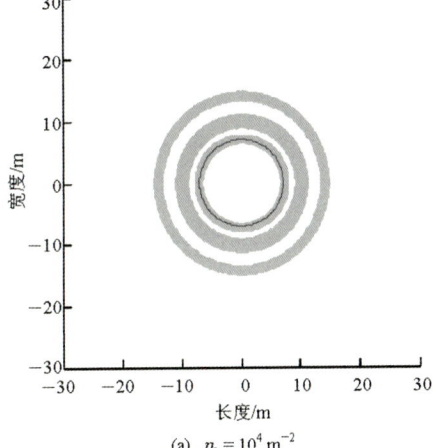

(a) $\eta_0 = 10^4$ m^{-2}

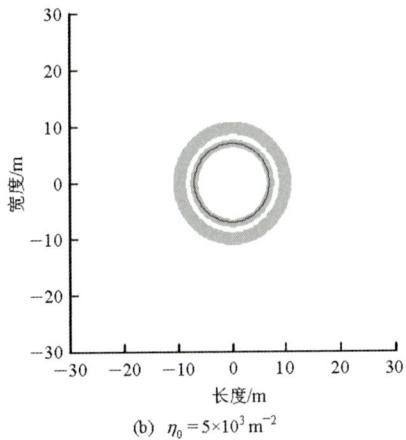

(b) $\eta_0 = 5\times10^3$ m^{-2}

图 4 当原生微裂半长度 $c_0 = 0.085$ mm 时,微裂纹密度对深部圆形洞室围岩破裂区的影响

Fig.4 Dependence of distribution of fractured zones on density of microcracks when $c_0 = 0.085$ mm

而当微裂纹密度较小时,深部圆形洞室围岩可能不会出现分区破裂化现象。图 4 也表明,破裂区的数量和宽度随微裂纹的密度增加而增多。

4 结 论

基于自由能密度、平衡方程和变形非协调条件，本文提出了一种新的非欧模型，该模型确定了静水压力和非静水压力条件下内度量非欧几何空间的标量曲率。通过能量方程、标量曲率和平衡方程，确定了静水压力和非静水压力条件下深部圆形洞室围岩的应力场，该应力场包括自平衡应力和弹性应力。与经典的弹性应力的单调特性不同，自平衡应力具有振荡特性。由于自平衡应力的振荡特性导致了深部圆形洞室围岩应力场也具有振荡特性。对于位于波峰附近的微裂纹，其尖端的应力集中将导致次生裂纹的失稳扩展、连接、汇合成宏观裂隙并形成破裂区；而对于位于波谷附近的微裂纹，其尖端应力集中不足以导致次生裂纹发生失稳扩展，因此出现非破裂区。深部圆形洞室围岩应力场交替出现波峰和波谷现象导致了破裂区和非破裂区交替出现，即分区破裂化现象。当微裂纹的半长度和密度越大，深部岩体分区破裂化现象越明显，当微裂纹的半长度和密度小于临界值时，深部岩体可能不出现分区破裂化现象。同时，微裂纹的半长度和密度越大，破裂区的宽度和数量越大。

参考文献(References)：

[1] GUZEV M A. Structure of kinematic and force fields in the Riemannian continuum model[J]. Journal of Applied Mechanics and Technical Physics，2011，52(5)：709－716.

[2] GUZEV M A, MYASNIKOV V P. Thermomechanical model of an elastic-plastic material with structural defects[J]. Izvestiya Rossijskaya Akademiya Nauk Mekhanika Tverdogo Tela, 1998, 33(4): 156－172.

[3] MYASNIKOV V P, GUZEV M A. Non-Euclidean model of deformation of materials at different structural levels[J]. Physical Mesomechanics，2000，3(1)：5－16.

[4] MYASNIKOV V P, GUZEV M A. Geometric model of internal self-equilibrated stresses in solids[J]. Doklady Rossijskaya Akademiya Nauk，2001，380(5)：627－629.

[5] GUZEV M A, PAROSHIN A A. Non-Euclidean model of the zonal disintegration of rocks around an underground working[J]. Journal of Applied Mechanics and Technical Physics，2001，42(1)：131－139.

[6] QIAN Q H, ZHOU X P. Non-Euclidean continuum model of the zonal disintegration of surrounding rocks around a deep circular tunnel in a non-hydrostatic pressure state[J]. Journal of Mining Science，2011，47(1)：37－46.

[7] QIAN Q H, ZHOU X P. Effects of the axial in situ stresses on the zonal disintegration phenomenon in the surrounding rock masses around a deep circular tunnel[J]. Journal of Mining Science，2012，48(1)：276－285.

[8] ZHOU X P, CHEN G, QIAN Q H. Zonal disintegration mechanism of cross-anisotropic rock masses around a deep circular tunnel[J]. Theoretical and Applied Fracture Mechanics，2012，57(1)：49－54.

[9] A J C B de Saint-Venant. Mémoire sur la torsion des prismes[J]. Mémoires présentés par divers Savants à l'Académie des Sciences de l'Institut Impérial de France，1855，14(1)：233－560.

[10] A J C B de Saint-Venant. Mémoire sur la flexion des prismes[J]. Journal de Mathématiques Pures et Appliquées，1856，1(2)：89－189.

[11] BENVENSITE Y. On the Mori-Tanaka's method in cracked solids[J]. Mechanics Research Communications，1986，13(4)：193－201.

[12] SIDOROFF F. Description of anisotropic damage application to elasticity[C]// Proceedings of IUTAM Colloquium Physical Nonlinearities in Structural Analysis. Berlin：Springer，1981：237－244.

[13] GOLSHANI A, OKUI Y, ODA M, et al. A micromechanical model for brittle failure of rock and its relation to crack growth observed in triaxial compression tests of granite[J]. Mechanics of Materials，2006，38(4)：287－303.

Effects of the axial in situ stresses on the zonal disintegration phenomenon in the surrounding rock masses around a deep circular tunnel

Q. Qian, X. Zhou, and E. Xia

aSchool of Civil Engineering, Chongqing University, Chongqing 400045,China

bKey Laboratory of New Technology for Construction of Cities in Mountain Area,Chongqing University, Ministry of Education of China, Chongqing 400045, PR China

cPLA University of Science and Technology, Nanjing210007, China

Received January 16, 2012

Abstract: A new non-Euclidean model, in which effects of the axial in-situ stress with arbitrary value on the zonal disintegration phenomenon in the surrounding rock masses around a deep circular tunnel is taken into account, is established. The total stress-field distributions of the surrounding rock around a deep circular tunnel including effects of the axial in-situ stress were given out. The strength criterion of the deep rock masses is applied to determine the occurrence of disintegration zones. The numerical computation was carried out. It is found from numerical results that number and size of fractured and nonfractured zones are sensitive to the axial in-situ stress, tangential, radial and axial normal stress, the intermediate principal stress coefficient and the rock mass rating classification RMR.

Keywords: axial in-situ stress, the zonal disintegration phenomenon, deep circular tunnel, fractured zones, non-fractured zones

INTRODUCTION

The zonal disintegration phenomenon occurs in the surrounding rock masses around deep tunnels and has never been observed in shallow rock engineering before. The zonal disintegration phenomenon attracted attentions of many researchers [1-13]. It is observed from experiments that the zonal disintegration phenomenon is sensitive to the axial in-situ stress p_z. In the previous theoretical analyses of the zonal disintegration behavior of the deep rock masses around circular tunnels (as shown in Fig. 1), the plane strain assumption is

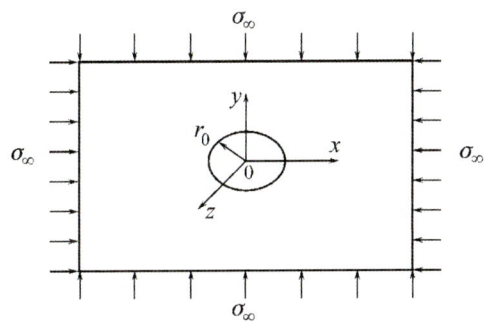

Fig. 1. The deep circular tunnel subjected to hydrostatic in situ stress σ_∞.

本文原载于《Journal of Mining Science》(2012 年)。

adopted. The axial in-situ stress p_z along the axis of the circular tunnel cannot be considered arbitrary, because of the plane strain assumption. Instead, $p_z = 2\nu\sigma_\infty$ since the axial normal strain ε_z is assumed to be 0. As $0 < \nu < 0.5$, p_z is always less than σ_∞. In fact, the horizontal in-situ stresses are sensitive to the direction. The difference between the maximum and minimum horizontal in-situ stress at the same depth can be large, which depends on the local geological condition [14, 15]. Generally, the axial normal strain ε_z is not equal to zero, which gives a non-plane-strain problem. To the author's knowledge, the axisymmetrical theoretical analysis on the mechanism of the zonal disintegration phenomenon with arbitrary value pz as a general case ($p_z = 2\nu\sigma_\infty + E\varepsilon_z$, E is Young's modulus of elasticity of the rock) has not been discussed in the previous studies.

In this paper, the axial normal strain ε_z is assumed to be ε_0 before excavation, which increases slightly after tunnel excavation. For a deep tunnel, it can be considered as an infinite domain problem and the strain increment after excavation, compared to ε_0, can be ignored. If the axial length of the tunnel is long enough, it can be considered as a plane problem with the assumption that ε_0 is a non-zero constant, which is called a quasi-plane strain problem. In the present study, ε_0 is assumed to be non-zero constant, effect of the axial in-situ stress p_z on the stress field distribution of the surrounding rock masses around deep tunnels is researched, as well as magnitude and site of non-fractured zones and fractured zones.

1. the analytical model

It is assumed that a deep circular tunnel, whose radius is r_0, is subjected to the vertical in-situ stress σ_∞, the horizontal in-situ stress σ_∞ at infinity and the axial in-situ stress p_z along the axis of the tunnel. For three-dimensional problem, the compatible condition for shallow rock masses are called the Saint-Venant compatibility condition, they can be expressed as follows:

$$\begin{cases} 6\left[\dfrac{\partial^2 \varepsilon_{11}}{\partial x_3^2} + \dfrac{\partial^2 \varepsilon_{33}}{\partial x_1^2} - \dfrac{\partial^2 \varepsilon_{13}}{\partial x_1 \partial x_3}\right] = R \equiv 0, \\ 6\left[\dfrac{\partial^2 \varepsilon_{11}}{\partial x_2^2} + \dfrac{\partial^2 \varepsilon_{22}}{\partial x_1^2} - \dfrac{\partial^2 \varepsilon_{12}}{\partial x_1 \partial x_2}\right] = R \equiv 0, \\ 6\left[\dfrac{\partial^2 \varepsilon_{22}}{\partial x_3^2} + \dfrac{\partial^2 \varepsilon_{33}}{\partial x_2^2} - \dfrac{\partial^2 \varepsilon_{23}}{\partial x_2 \partial x_3}\right] = R \equiv 0, \end{cases} \quad (1)$$

where R has the meaning of a scalar curvature.

The formation of fractured zones around works in deep rock masses is irreversible. Therefore, together with the elastic-strain tensor ε_{ij}^e, it is necessary to introduce incompatible irreversible strain tensor ε_{ij}^{in} as an additional parameter of the problem [12, 13]. In this case, the equations of state of the deep rock masses are assumed to be the thermodynamic variable. Therefore, the total strain ε_{ij} includes the reversible strain ε_{ij}^e and irreversible strains ε_{ij}^{ir} consisting of compatible irreversible strains ε_{ij}^c and incompatibility irreversible strains ε_{ij}^{in}, then, the following expression can be obtained as:

$$\varepsilon_{ij} = \varepsilon_{ij}^e + \varepsilon_{ij}^c + \varepsilon_{ij}^{in}. \tag{2}$$

Since $\varepsilon_{ij} \neq \varepsilon_{ij}^e + \varepsilon_{ij}^c$ and $\varepsilon_{ij} \neq \varepsilon_{ij}^e$, the function R is not equal to zero. The occurrence of incompatible irreversible strains leads to incompatibility condition in deep rock masses. From mathematical viewpoint, this means that the internal geometrical structure of deep rock masses is non-Euclidean. Here, R has the meaning of a scalar curvature [16].

For three-dimensional problem, incompatible condition in deep rock masses is written as follows:

$$R \equiv 2\left(\frac{\partial^2 \varepsilon_{11}}{\partial x_3^2} + \frac{\partial^2 \varepsilon_{33}}{\partial x_1^2} + \frac{\partial^2 \varepsilon_{11}}{\partial x_2^2} + \frac{\partial^2 \varepsilon_{22}}{\partial x_1^2} + \frac{\partial^2 \varepsilon_{33}}{\partial x_2^2} + \frac{\partial^2 \varepsilon_{22}}{\partial x_3^2} \right.$$
$$\left. - \frac{\partial^2 \varepsilon_{12}}{\partial x_1 \partial x_2} - \frac{\partial^2 \varepsilon_{13}}{\partial x_1 \partial x_3} - \frac{\partial^2 \varepsilon_{23}}{\partial x_2 \partial x_3}\right) = 2\left(\Delta \varepsilon_{cc} - \frac{\partial^2 \varepsilon_{ij}}{\partial x_i \partial x_j}\right), \tag{3}$$

where Δ is the Laplace operator, $\varepsilon_{ll} = \varepsilon_{11} + \varepsilon_{22} + \varepsilon_{33}$. From Eq. (3), it is assumed that the contribution enters additively and the internal energy is square in R, i.e.:

$$\rho_0 U = \frac{E}{1+v}\left[\frac{v}{2(1-2v)}\varepsilon_{ll}^2 + \frac{1}{2}\varepsilon_{ij}\varepsilon_{ij}\right] - \frac{q}{4}R^2, \tag{4}$$

where q is the fitting parameter of the model determined from the experimental data; ρ_0 is the density of deep rock masses; E is Young's modulus; v is Poisson's ratio.

According to [12], the following expressions can be written as:

$$\Delta E_{ll} - \frac{\partial^2 E_{ij}}{\partial x_i \partial x_j} = 0, \tag{5}$$

$$E_{ij} = \xi\left(\sigma_{ij} - q\Delta R \delta_{ij} + q\frac{\partial^2 R}{\partial x_i \partial x_j}\right), \tag{6}$$

From (5) and (6), the following expression can be obtained as:

$$\Delta \sigma - 2q\Delta^2 R = 0. \tag{7}$$

Hooke's law can be written as follows:

$$\varepsilon_{ij} = \frac{(1+v)\sigma_{ij} - \delta_{ij}v\sigma}{E}. \tag{8}$$

Substituting (8) into (3), the following can be expressed as:

$$R = \frac{2(1-v)}{E}\Delta\sigma. \tag{9}$$

From (7) and (9), the following expression can written as:

$$\Delta^2 R - \rho^2 R = 0, \tag{10}$$

where $\rho^2 = \dfrac{E}{2q(1-v)}$.

For the quasi-plane strain problem, Eq. (10) can be rewritten as follows:

$$\left(\frac{\partial^2}{\partial r^2} + \frac{1}{r}\frac{\partial}{\partial r}\right)^2 R - \rho^2 R = 0. \tag{11}$$

This is a fourth-order linear differential equation, and its solution which decreases as $r \to \infty$ is written in the following form:

$$R = AJ_0(\rho^{1/2}r) + BN_0(\rho^{1/2}r) + CK_0(\rho^{1/2}r), \quad (12)$$

where $\rho^2 = \dfrac{E}{2q(1-v)}$; $A = (C/2)\pi\sqrt{\rho}r_0[K_0(\sqrt{\rho}r_0)N_1(\sqrt{\rho}r_0) - K_1(\sqrt{\rho}r_0)N_0(\sqrt{\rho}r_0)]$; $B = (C/2)\pi\sqrt{\rho}r_0[K_0(\sqrt{\rho}r_0)J_1(\sqrt{\rho}r_0) - K_1(\sqrt{\rho}r_0)J_0(\sqrt{\rho}r_0)]$; C coincides with the Wronskian of the linearly independent solutions J_0 and N_0; J_0, N_0 and K_0 are zero-order Bessel, Neumann, and MacDonald cylindrical functions, respectively.

For the case of three-dimensional problem, the axial normal strain components ε_z are not zero, then the following expression can be obtained as:

$$\varepsilon_z = \frac{1}{E}[\sigma_z - v(\sigma_r - \sigma_\theta)]. \quad (13)$$

For a deep tunnel, it is assumed that the axial normal strain components ε_z is equal to ε_0, the axial stress component σ_z is:

$$\sigma_z = E\varepsilon_0 + v(\sigma_r + \sigma_\theta). \quad (14)$$

As $r \to \infty$, $\sigma_r = \sigma_\theta = \sigma_\infty$, hence as $r \to \infty$, the following can be obtained as:

$$\sigma = (\sigma_z + \sigma_r + \sigma_\theta) \to p_z + 2\sigma_\infty. \quad (15)$$

From Eqs. (5) and (9), σ can be obtained as:

$$\sigma = -\frac{E}{2\rho(1-v)}[AJ_0(\rho^{1/2}r) + BN_0(\rho^{1/2}r) - CK_0(\rho^{1/2}r)] + p_z + 2\sigma_\infty. \quad (16)$$

According to $\sigma = (\sigma_z + \sigma_r + \sigma_\theta)$, the stress component σ_θ can be given as follows:

$$\sigma_\theta = \frac{\sigma - E\varepsilon_0}{v+1} - \sigma_r. \quad (17)$$

The equations of equilibrium have the following form:

$$\frac{\partial \sigma_r}{\partial r} + \frac{\sigma_r - \sigma_\theta}{r} = 0. \quad (18)$$

Substituting (17) into (18), the equations of equilibrium (18) can be rewritten as:

$$\frac{\partial \sigma_r}{\partial r} + \frac{2\sigma_r}{r} - \frac{\sigma - E\varepsilon_0}{(1+v)r} = 0, \quad (19)$$

Integrating (19), the radial stress component σ_r can be obtained as follows:

$$\sigma_r = \left[\frac{p_z + 2\sigma_\infty - E\varepsilon_0}{2(1+v)}\right]\left(1 - \frac{r_0^2}{r^2}\right) - \frac{E}{2\rho^{3/2}(1-v^2)r}[AJ_1(\rho^{1/2}r) + BN_1(\rho^{1/2}r) - CK_1(\rho^{1/2}r)]. \quad (20)$$

Substituting (20) into (17), the tangential stress component σ_θ can be expressed as follows:

$$\sigma_\theta = \left[\frac{p_z + 2\sigma_\infty - E\varepsilon_0}{2(1+v)}\right]\left(1 - \frac{r_0^2}{r^2}\right) - \frac{E}{2\rho(1-v^2)}[AJ_0(\rho^{1/2}r) + BN_0(\rho^{1/2}r) - CK_0(\rho^{1/2}r)] + \frac{E}{2\rho^{3/2}(1-v^2)r}[AJ_1(\rho^{1/2}r) + BN_1(\rho^{1/2}r) - CK_1(\rho^{1/2}r)] \quad (21)$$

Substituting (20) and (21) into (14), the axial stress component σ_z can be written as follows:

$$\sigma_z = \left[\frac{v(p_z + 2\sigma_\infty) + E\varepsilon_0}{1+v}\right] - \frac{vE}{2\rho(1-v^2)}[AJ_0(\rho^{1/2}r) + BN_0(\rho^{1/2}r) - CK_0(\rho^{1/2}r)]. \quad (22)$$

From formulas (20)(22), the total stress-field distributions of the surrounding rock masses around a deep circular tunnel including effects of the axial in-situ stress can be determined.

2. the determination of fractured zones

Knowing the stress components, it is necessary to determine the rock regions that correspond to fractured zones and nonfractured zones. Disintegration zones appear when the stresses in deep rock masses reach a certain critical value. From the physical viewpoint, this means that it is necessary to use a strength criterion, whose fulfillment in a selected region corresponds to the occurrence of disintegration zones. In the present paper, the strength criterion of the deep rock masses is applied to determine the occurrence of disintegration zones.

The strength criterion of the deep rock masses can be expressed as follows [17]:

$$\begin{cases} F_1 = \sigma_1 - \dfrac{1}{1+a}(a\sigma_2 + \sigma_3) + W(\sigma_1 + b\sigma_2 + \sigma_3)\dfrac{I_1}{\sigma_c} - \sigma_c\left[\dfrac{b\sigma_1}{\sigma_c} + \dfrac{d}{(1+a)\sigma_c}(a\sigma_2 + \sigma_3) + b^2\right]^n = 0, \quad F_1 \geq F_2, \\ F_2 = \dfrac{1}{1+a}(\sigma_1 + a\sigma_2) - \sigma_3 + W(\sigma_1 + b\sigma_2 + \sigma_3)\dfrac{I_1}{\sigma_c} - \sigma_c\left[\dfrac{b\sigma_1 + a\sigma_2}{(1+a)\sigma_c} + \dfrac{d\sigma_3}{\sigma_c} + b^2\right]^n = 0, \quad F_1 < F_2, \end{cases}$$
(23)

where m_i is strength parameter of the intact rock, a is the intermediate principal stress coefficient: σ_c is the uniaxial compressive strength of the intact rock; RMR is rock mass rating classification; $n = 0.65 - \dfrac{RMR-5}{200}$ ($RMR > 23$), $b = 0.61\mathrm{Exp}\left(\dfrac{RMR-100}{12}\right)$ (for disturbed rock mass); $d = 1.4m_i\mathrm{Exp}\left(\dfrac{RMR-100}{14}\right) - 0.31\sqrt{m_i^2\mathrm{Exp}\left(\dfrac{RMR-100}{7}\right) + 4\mathrm{Exp}\left(\dfrac{RMR-100}{6}\right)}$ (for disturbed rock mass); $b = 0.619\mathrm{Exp}\left(\dfrac{RMR-100}{18}\right)$ (for undisturbed rock mass); $d = 0.9m_i\mathrm{Exp}\left(\dfrac{RMR-100}{28}\right) - -0.31\sqrt{m_i^2\mathrm{Exp}\left(\dfrac{RMR-100}{14}\right) + 4\mathrm{Exp}\left(\dfrac{RMR-100}{9}\right)}$ (for undisturbed rock mass); W is dimensionless parameter.

If the principal stress components satisfy strength criterion of the deep rock masses, the number, size and site of occurrence of the disintegration zones can be obtained.

3. numerical computation results

3.1. Effects of the Axial In-Situ Stress on Distribution of Fractured and Nonfractured Zones in Deep Surrounding Rock Masses

Effects of the axial in-situ stress on the zonal disintegration phenomenon in surrounding rock mass around a deep circular tunnel are taken into account. In Fig. 2, the following computation parameters are used: the strength criterion of rock masses $\sigma_c = 73.4$ MPa, $RMR = 70$, $W = 0.05$, $r_0 = 7$ m, $C = 1.66 \cdot 10^{10}$ m^{-2}, $q = 1.79 \cdot 10^7$, $\sigma_\infty = 35$ MPa, $a = 0.55$, $v = 0.3$, $E = 18$ GPa, $m_i = 7.5$.

In Fig. 3, the following computation parameters are used: the strength criterion of rock mass $\sigma_c = 60.4$ MPa, $RMR = 70$, $W = 0.05$, $E = 18$ GPa, $v = 0.28$; $r_0 = 7$ m, $m_i = 7.5$, $C = 1.46 \cdot 10^{10}$ m^{-2}, $q = 1.79 \cdot 10^7$, $\sigma_\infty = 63$ MPa, $a = 0.55$.

Effect of the axial in-situ stress on distribution of fractured zones and nonfractured zones in deep surrounding rock masses is depicted in Figs. 2 and 3. It can be found from Figs. 2 and 3 that the zonal disintegration phenomenon can occur, not only is the value of the axial in-situ stress larger than that of the horizontal and vertical in-situ stresses, but also is smaller than that of the horizontal and vertical in-situ stresses. It can also be observed from Figs. 2 and 3 that the number and size of fractured zones increase with an increase in the axial in-situ stress when the value of the axial in-situ stress is larger than that of the horizontal and vertical in-situ stresses, while the number of fractured zones decreases with an increase in the axial in-situ stress when the value of the axial in-situ stress is smaller than that of the horizontal and vertical in-situ stresses.

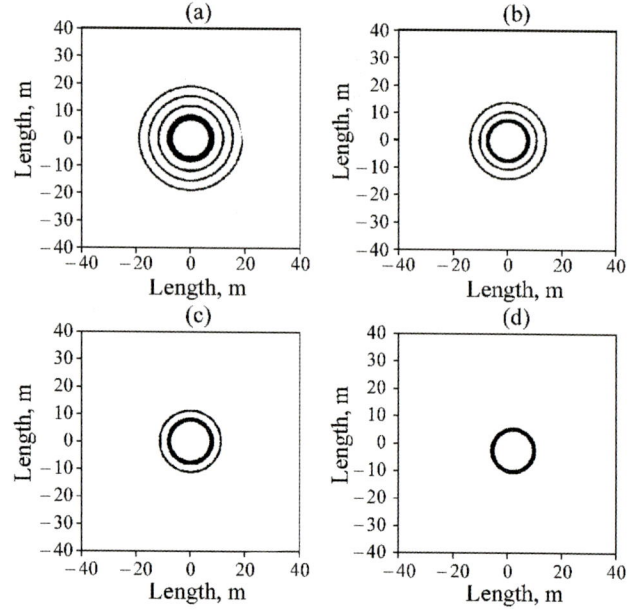

Fig. 2. Effects of the axial in-situ stress on distribution of fractured and nonfractured zones when the value of the axial in-situ stress is larger than that of the horizontal and vertical in-situ stresses: (a) $p_z = 93$ MPa; (b) $p_z = 82.6$ MPa; (c) $p_z = 78$ MPa; (d) $p_z = 73.5$ MPa.

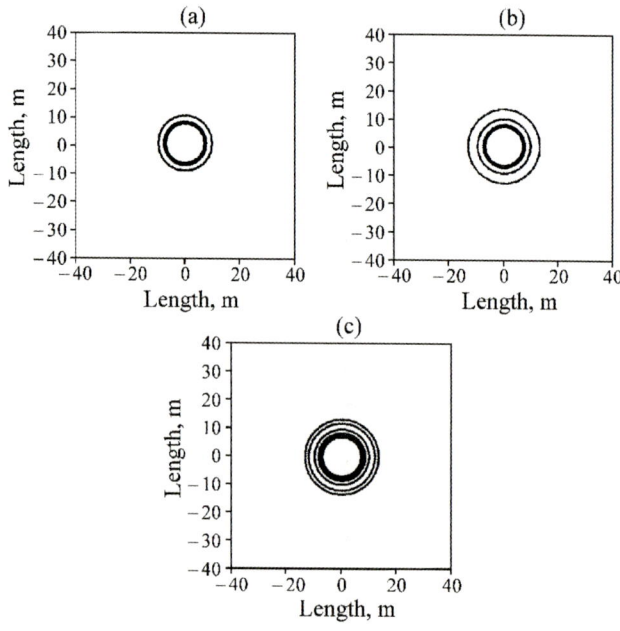

Fig. 3. Effects of the axial in-situ stress on distribution of fractured and nonfractured zones when the value of the axial in-situ stress is smaller than that of the horizontal and vertical in-situ stresses (a) $p_z = 58.8$ MPa; (b) $p_z = 51.6$ MPa; (c) $p_z = 38.5$ MPa.

3.2. Effects of the Rock Mass Rating Classification RMR on Distribution of Fractured and Nonfractured Zones in Deep Surrounding Rock Masses

In simulation, the following computation parameters are used: the strength criterion of intact rock $\sigma_{ci} = 125$ MPa, $W = 0.05$, $E = 18.5$ GPa, $v = 0.28$, $r_0 = 7$ m, $m_i = 7.5$, $a = 0.55$, $p_z = 47.35$ MPa, $q = 1.8 \cdot 10^7$, $\sigma_\infty = 35$ MPa, $a = 0.55$, $\varepsilon_0 = 4.0 \cdot 10^{-3}$, $C = 1.66 \cdot 10^{10}$ m^{-2}.

Effect of RMR on distribution of fractured zones and nonfractured zones in deep surrounding rock masses is shown in Fig. 4. It can be revealed from Fig. 4 that the number and size of fractured zones decreases with increasing the rock mass rating classification RMR.

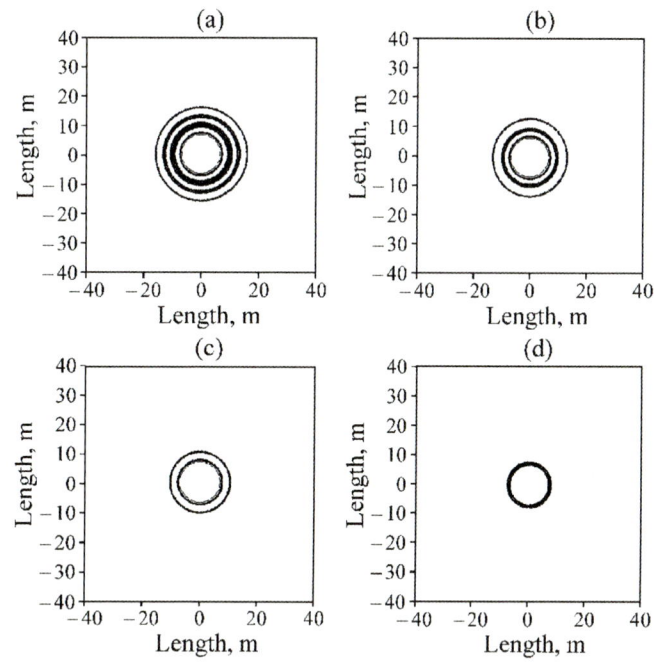

Fig. 4. Effects of RMR on distribution of fractured and nonfractured zones: (a) RMR = 57; (b) RMR = 70; (c) RMR = 81; (d) 89.

3.3. Effects of the Intermediate Principal Stress Coefficient on Distribution of Fractured and Nonfractured Zones in Deep Surrounding Rock Masses

In simulation, the following computation parameters are used: the strength criterion of rock masses $\sigma_c = 73.4$ MPa, $W = 0.05$, $E = 17.6$ GPa, $v = 0.3$, $r_0 = 7$ m, $C = 1.66 \cdot 10^{10}$ m^{-2}, $p_z = 82.6$ MPa, $m_i = 7.5$, $q = 1.79 \cdot 10^7$, $\sigma_\infty = 35$ MPa, $RMR = 70$, $\varepsilon_0 = 3.5 \cdot 10^{-3}$. Effect of the intermediate principal stress coefficient on distribution of fractured zones and nonfractured zones in deep surrounding rock masses is plotted in Fig. 5. It can be seen from Fig. 5 that the number and size of fractured zones increase with decreasing the intermediate principal stress coefficient.

3.4. Effects of the Uniaxial Compressive Strength of Intact Rock on Distribution of Fractured and Nonfractured Zones in Deep Surrounding Rock Masses

In simulation, the following computation parameters are used: $E = 17.6$ GPa, $v = 0.3$, $W = 0.05$, $r_0 = 7$ m, $C = 1.66 \cdot 10^{10}$ m^{-2}, $p_z = 82.6$ MPa, $m_i = 7.5$, $a = 0.55$, $\sigma_\infty = 35$ MPa, $q = 1.79 \cdot 10^7$, $RMR = 70$, $\varepsilon_0 = 3.5 \cdot 10^{-3}$ (refer to Fig. 6). It can be seen from Fig. 6 that the number and size of fractured zones increase with decreasing the uniaxial compressive strength of intact rock.

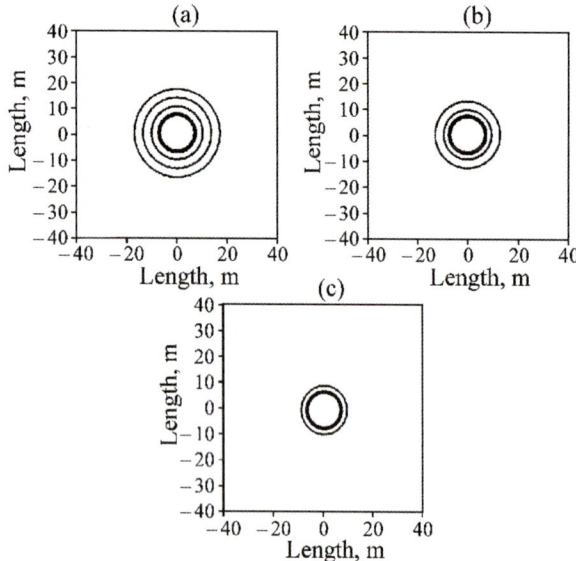

Fig. 5. Effects of the intermediate principal stress coefficient on distribution of fractured and nonfractured zones: (a) a = 0.47; (b) a = 0.55; (c) a = 0.8.

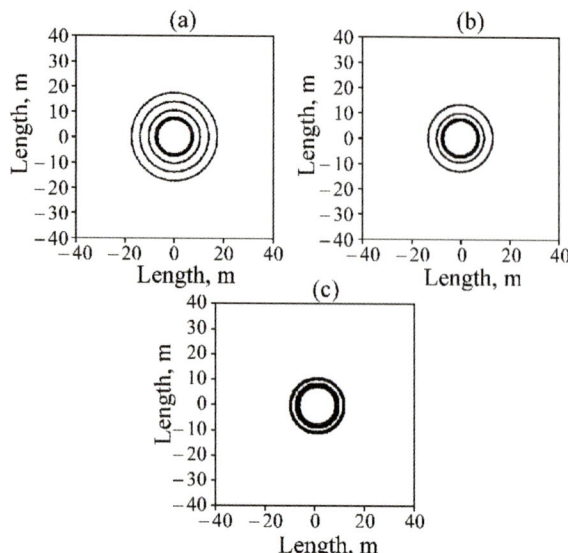

Fig. 6. Effects of the uniaxial compressive strength of intact rock on distribution of fractured and nonfractured zones: (a) σ_c =120 MPa; (b) σ_c =125 MPa; (c) σ_c =135 MPa.

conclusion

A new non-Euclidean model is proposed to investigate effects of the axial in-situ stress with arbitrary value on the zonal disintegration phenomenon, on the total elastic stress-field distributions of the surrounding rock around a deep circular tunnel. The strength criterion of the deep rock masses is applied to determine the number and size of fractured and nonfractured zones. The main conclusions are summarized as follows:

(1) The zonal disintegration phenomenon occurs, not only is the value of the axial in-situ stress larger than that of the horizontal and vertical in-situ stresses, but also is smaller than that of the horizontal and vertical in-situ stresses. The number and size of fractured zones increase with an increase in the axial in-situ stress when the value of the axial in-situ stress is larger than that of the horizontal and vertical in-situ stresses, while the number and size of fractured zones decrease with increasing the axial in-situ stress when the value of the axial in-situ stress is smaller than that of the horizontal and vertical in-situ stresses.

(2) The number and size of fractured zones increase with decreasing the intermediate principal stress coefficient and the rock mass rating classification RMR and the uniaxial compressive strength of intact rock.

acknowledgments

This work was supported by the National Natural Science Foundation of China, projects nos. 51021001 and 51078371, Natural Science Foundation Project of CQ CSTC, projects nos. CSTC and 2009BA4046, and the Fundamental Research Funds for the Central Universities, projects nos. CDJZR10205501.

references

1. Cloete, D.R. and Jager, A.J., "The Nature of the Fracture Zone in Gold Mines as Revealed by Diamond Core Drilling," Association of Mine Managers, Papers and Discussions, 19721973.
2. Adams, G.R. and Jager, A.J., "Petroscopic Observations of Rock Fracturing ahead of Stope Faces in Deep-Level Gold Mine," *Journal of the South African Institute of Mining and Metallurgy*, 1980, vol. 80, no. 6.
3. Shemyakin, E.I., Fisenko, G.L., Kurlenya, M.V., Oparin, V.N., et al., "Disintegration Zone of Rocks around Underground Workings, Part I: Data of Full-Scale Observations," *Fiz. Tekh. Probl. Razrab. Polezn. Iskop.*, 1986, no. 3.
4. Shemyakin, E.I., Fisenko, G.L., Kurlenya, M.V., Oparin, V.N., et al., "Disintegration Zone of Rocks around Underground Workings, Part II: Rock Fracture on Models from Equivalent Materials," *Fiz. Tekh. Probl. Razrab. Polezn.Iskop.*, 1986, no. 4.
5. Shemyakin, E.I., Fisenko, G.L., Kurlenya, M.V., Oparin, V.N., et al., "Disintegration Zone of Rocks around Underground Workings, Part III: Theoretical Concepts," *Fiz. Tekh. Probl. Razrab. Polezn. Iskop.*, 1987, no. 1.
6. Shemyakin, E.I., Kurlenya, M.V., Oparin, V.N., Reva, V.N., et al., "Disintegration Zone of Rocks around Underground Workings, Part IV: Practical Applications," *Fiz. Tekh. Probl. Razrab. Polezn. Iskop.*, 1989, no. 4.
7. Reva, V.N. and Tropp, E.A., "Elastoplastic Model of the Zonal Disintegration of the Neighborhood of an Underground Working," Physics and Mechanics of Rock Fracture as Applied to Prediction of Dynamic Phenomena (Collected Scientific Papers), Mine Surveying Inst., Saint Petersburg, 1995.
8. Qian Qihu, "The Key Problems of Deep Underground Space Development," The Key Technical Problems of Base Research in Deep Underground Space Development, The 230th Xiangshan Science Conference, Beijing, 2004.
9. Qian Qihu, "The Current Development of Nonlinear Rock Mechanics: The Mechanics Problems of Deep Rock Mass," Chinese Society for Rock Mechanics and Engineering, Proc. 8th Conf. Rock Mechanics and Engineering, Beijing: Science Press, 2004.
10. Zhou Xiaoping and Qian Qihu, "Zonal Fracturing Mechanism in Deep Tunnel," *Chinese Journal of Rock Mechanics and Engineering*, 2007, vol. 26, no. 5.
11. Zhou, X.P., Wang, F.H., Qian, Q.H., and Zhang, B.H., "Zonal Fracturing Mechanism in Deep Crack-Weakened Rock Masses," *Theoretical and Applied Fracture Mechanics*, 2008, vol. 50, no. 1.
12. Guzev, M.A. and Paroshin, A.A., "Non-Euclidean Model of the Zonal Disintegration of Rocks around an Underground Working," *Journal of Applied Mechanics and Technical Physics*, 2001, vol. 42, no. 1.
13. Qian Qihu and Zhou Xiaoping, "Non-Euclidean Continuum Model of the Zonal Disintegration of Surrounding Rocks around a Deep Circular Tunnel in a Non-Hydrostatic Pressure State," *Journal of Mining Science*, 2011, vol. 47, no. 1.
14. Cai, M.F., He, M.C., and Liu, Dy., Rock Mechanics and Engineering, Beijing: Science Press, 2002.
15. Lu, A., Xu, G., Sun, F., and Sun, W., "Elasto-Plastic Analysis of a Circular Tunnel Including the Effect of the Axial In Situ Stress," *International Journal of Rock Mechanics and Mining Sciences*, 2010, vol. 47, no. 1.
16. Dubrovin, B.A., Novikov, S.P., and Fomenko, A.T., Modern Geometry: Methods and Applications, Moscow: Nauka, 1986.
17. Zhou Xiaoping, Qian Qihu, and Yang Haiqing, "Strength Criterion of Deep Rock Masses," *Chinese Journal of Rock Mechanics and Engineering*, 2008, vol. 27, no. 1.

深隧道围岩分区破裂的数学模拟

戚承志[1],钱七虎[2],王明洋[2],陈剑杰[3]

(1. 北京建筑工程学院 工程结构与新材料北京市高校工程研究中心,北京 100044;
2. 解放军理工大学 工程兵工程学院,南京 210007;3. 西北核技术研究院,西安 710024)

摘 要:该研究为先前深隧道围岩分区破裂现象内变量梯度塑性模型的进一步发展。利用应变梯度模型研究了深隧道围岩的分区破裂现象。作为额外的状态变量,在此引入应变梯度这一新变量。利用虚功原理得到了深隧道围岩的平衡方程、边界条件和流动准则,利用 Clausius-Duhem 不等式获得了岩体的本构方程。对于圆形深隧道,由上述模型的一般方程得到了弹性变形情况下、具有下降段的弹塑性变形情况下和不考虑弹性变形的塑性变形情况下圆形深隧道围岩的支配方程,得到了解析解,并讨论了解析解的性质。这一模型不仅扩展了隧道围岩的经典弹塑性模型,也为下一步数值研究深隧道围岩的分区破裂现象奠定了理论基础。

关 键 词:深隧道;分区破裂;应变梯度理论

Mathematical modeling of zonal disintegration of surrounding rock near deep tunnels

QI Cheng-zhi[1], QIAN Qi-hu[2], WANG Ming-yang[2], CHEN Jian-jie[3]

(1. Beijing Higher Institution Engineering Research Center for Engineering Structures and New Materials, Beijing University of Civil Engineering and Architecture, Beijing 100044, China; 2. Engineering Institute of Engineering Cops, PLA University of Science and Technology, Nanjing 210007, China; 3. Northwest Institute of Nuclear Technology, Xi'an 710024, China)

Abstract: The present work is the further development of the previous works of the writers on internal variable gradient model for zonal disintegration phenomenon near deep tunnels. Strain gradient model is used to study zonal disintegration phenomenon. As an additional variable, strain gradient, is introduced. The equilibrium equations, boundary conditions and flow rule are obtained by using virtual work principle. Constitutive equations are obtained by using Clausius-Duhem inequality. For deep circular tunnels the governing equations are obtained from the general equations of the proposed model for elastic deformation regime, elastoplastic deformation regime and plastic deformation regime without consideration of elastic deformations. The solutions of the governing equations are obtained; and the behaviour of the solutions is analyzed. The proposed model not only expands the classical elastoplastic model for surrounding rock near tunnels, but also serves as theoretical basis for numerical study of zonal disintegration phenomenon near deep tunnels.

Key words: deep tunnel; zonal disintegration; strain gradient theory

1 引 言

自俄罗斯学者发现深部隧道围岩的分区破裂现象以来[1]30 多年过去了,俄罗斯学者[2-8]、中国学者[9-13]进行了一系列理论和试验研究,这些研究从不同的角度分析了分区破裂现象的机制,促进了对于这一现象的理解。笔者在前期工作中[9,13-14],利用连续相变理论成功地模拟了深部隧道围岩分区破裂化现象的时空结构,也解释了与这一现象密切相关的受损岩石试样临界后变形增量的逆转现象。但上述工作的不足是,没有把解析解与岩石的临界后变形特征参数(例如岩石临界后变形的软化模量等)联系起来。20 世纪 80 年代之后发展起来的梯度塑性理论[15-22]在解释固体的变形局部化和有规律的变形模式空间分布方面取得了很大的进展,这些理论采用应变梯度或者其他内变量梯度作为额外的模型变量,这些梯度变量进入到介质的平衡方程中或者内变量演化方程中,得到了不可逆变形的局部化解,或者空间分布解。在文献[23-24]中作者利用了内变量梯度理论,模拟了深部隧道围岩的分区破裂现象。但这一理论的缺点是梯度项没有进入到平衡方程中去,因此,不能够通过平衡方程把隧道围岩弹性区和塑性区的受力和变形状态统一解出来。本文将致力于把梯度项融进平衡方程中去,以

基金项目:国家自然基金资助项目(No.51174012);国家重点基础研究发展计划(973)项目(No.2010CB732003);北京自然基金资助项目(No. KZ200810016007);非线性动立系统建模与分析团阶(No. PHR201107123)。

本文原载于《岩石力学》(2012 年第 33 卷第 11 期)。

便建立自洽的深部隧道围岩的分区破裂模型。

2 介质变形的非局部变量

位移矢量 $\boldsymbol{u}$ 的分量为 u_i，应变张量 $\boldsymbol{\varepsilon}$ 的分量为 ε_{ij}。

$$\varepsilon_{ij} = \frac{1}{2}(u_{i,j} + u_{j,i}) \qquad (1)$$

应变张量可以分解为弹性部分 ε_{ij}^{e} 与塑性部分 ε_{ij}^{p} 之和

$$\varepsilon_{ij} = \varepsilon_{ij}^{e} + \varepsilon_{ij}^{p} \text{ 或者 } \boldsymbol{\varepsilon} = \boldsymbol{\varepsilon}^{e} + \boldsymbol{\varepsilon}^{p} \qquad (2)$$

其中

$$\varepsilon_{ij}^{e} = \frac{1}{2}(u_{i,j}^{e} + u_{j,i}^{e}),\ \varepsilon_{ij}^{p} = \frac{1}{2}(u_{i,j}^{p} + u_{j,i}^{p}) \qquad (3)$$

梯度理论可以分成两类：应变梯度和内变量梯度理论。这两种理论的根本区别在于，应变梯度理论中，应变梯度被选为状态变量，与应变梯度共轭的高阶应力进入到平衡方程中去；而内变量梯度与某些耗散热力学力相对应，这些热力学力可以进入到内变量的演化方程中去，但不出现在平衡方程或者动量守恒方程中。因此，内变量梯度理论只修正本构关系的描述，而几何方程与平衡方程没变。从热力学的角度来看，内变量梯度模型只修正了自由能函数和耗散势函数，而应变梯度理论需要扩展内力功与外力功的表达式。

在本文中将利用内变量模型来研究岩体的变形。

在描述连续介质的非局部或者梯度本构关系时通常引入下列形式的单位体积的内能[21]：

$$U = U(\boldsymbol{\varepsilon}^{e}, \boldsymbol{q}, s, \boldsymbol{Q}(\boldsymbol{q})) \qquad (4)$$

式中：$\boldsymbol{\varepsilon}^{e}$ 为弹性应变张量；$\boldsymbol{q}$ 为内变量矢量；s 为熵；$\boldsymbol{Q}(\boldsymbol{q})$ 为非局部变量矢量，为内变量矢量 $\boldsymbol{q}$ 的函数。

3 虚功原理

对于给定介质的一个子区域 V，作用在 V 内材料上的外力功等于花费于 V 内材料上的内力功。外部功是由宏观体力 $\boldsymbol{F}_b$、作用于介质表面 Γ 上的宏观面力 $\boldsymbol{F}_s$ 产生：

$$W_{ext} = \int_V \boldsymbol{F}_b \cdot \delta \boldsymbol{u} dV + \int_\Gamma \boldsymbol{F}_s \cdot \delta \boldsymbol{u} d\Gamma \qquad (5)$$

式中：δ 为变分算符；$\delta \boldsymbol{u}$ 为虚位移。

为了描述介质的不可逆变形，内变量 $\boldsymbol{q}$ 可以取为介质的全应变 $\boldsymbol{\varepsilon}$、累积有效塑性应变 $E^p = \sqrt{\varepsilon_{ij}^{p}\varepsilon_{ij}^{p}}$，而非局部变量 $\boldsymbol{Q}(\boldsymbol{q})$ 取为应变梯度 $\nabla \boldsymbol{\varepsilon}$。

外力功由内力功 W_{int} 来平衡，内力功由弹性应力张量 $\boldsymbol{\sigma}$、与塑性应变张量 ε_{ij}^{p} 对应的广义力 $\boldsymbol{\Omega}$、与累积有效塑性应变 E^p 相对应的广义力 $\boldsymbol{B}$，应变梯度 $\nabla \boldsymbol{\varepsilon}$ 对应的广义力 $\boldsymbol{T}$ 来完成：

$$W_{int} = \int_V \boldsymbol{\sigma} \cdot \delta \boldsymbol{\varepsilon}^{e} dV + \int_V \boldsymbol{\Omega} \cdot \delta \boldsymbol{\varepsilon}^{p} dV + \int_V \boldsymbol{B} \cdot \delta E^p dV + \int_V \boldsymbol{T} \cdot \delta \nabla \boldsymbol{\varepsilon} dV \qquad (6)$$

因为

$$\delta E^p = \frac{\varepsilon_{ij}^{p}\delta\varepsilon_{ij}^{p}}{\sqrt{\varepsilon_{mn}^{p}\varepsilon_{mn}^{p}}} = \frac{\varepsilon_{ij}^{p}\delta\varepsilon_{ij}^{p}}{E^p} = p_{ij}\delta\varepsilon_{ij}^{p} \qquad (7)$$

$$p_{ij} = \varepsilon_{ij}^{p}/E^p \qquad (8)$$

式中：p_{ij} 为塑性应变方向张量，为 Ilyushin A A 首次引进[25]。

把 $\boldsymbol{\varepsilon}^{e} = \boldsymbol{\varepsilon} - \boldsymbol{\varepsilon}^{p}$ 代入式（6），得

$$W_{int} = -\int_V (\sigma_{ij,j} - T_{ijk,kj})\delta u_i dV + \int_\Gamma [(\sigma_{ij} - T_{ijk,k})n_j]\delta u_i d\Gamma - \int_V (\sigma_{ij} - \Omega_{ij} - Bp_{ij})\delta\varepsilon_{ij}^{p} dV + \iint_\Gamma T_{ijk}\delta\varepsilon_{ij} n_k d\Gamma \qquad (9)$$

虚功原理可以表示为

$$W_{ext} = W_{int} \qquad (10)$$

将式（5）、（9）代入式（10），得

$$\int_V (\sigma_{ij,j} - T_{ijk,kj} + F_{bi})\delta u_i dV + \int_\Gamma [F_{si} - (\sigma_{ij} - T_{ijk,k})n_j + \kappa_i]\delta u_i d\Gamma + \int_V [(\sigma_{ij} - \Omega_{ij} - Bp_{ij})]\delta\varepsilon_{ij}^{p} dV + \int_\Gamma (T_{ijk}n_k)n_j \frac{\partial \delta u_i}{\partial n} d\Gamma = 0 \qquad (11)$$

其中 $\kappa_i = [kn_j - (\delta_{jm} - n_j n_m)\partial_m](T_{ijk}n_k)$。在方程（11）的推导中除了应用了奥高散度定理外，还应用了表面散度定理[18]。

由于 δu_i、$\delta \varepsilon_{ij}^{p}$ 和 $\partial \delta u_i /\partial n$ 的任意性可得下列方程：

$$\sigma_{ij,j} - T_{ijk,kj} + F_{bi} = 0 \qquad (12)$$

$$F_{si} - (\sigma_{ij} - T_{ijk,k})n_j + \kappa_i = 0 \qquad (13)$$

$$\sigma_{ij} - \Omega_{ij} - Bp_{ij} = 0 \qquad (14)$$

$$T_{ijk}n_k n_j = 0 \qquad (15)$$

式（12）为介质的平衡方程，其中考虑了梯度效应；式（13）为介质的宏观边界条件；式（15）为高阶的应力边界条件；而式（14）定义了介质的屈服条件。对于该方程两边取绝对值得

$$\|\sigma_{ij} - \Omega_{ij}\| = B \qquad (16)$$

式中:Ω_{ij} 为材料随动硬化参数,当材料各向同性硬化时,有 $\Omega_{ij} = 0$,此时式(16)变为

$$\|\sigma_{ij}\| = B \qquad (17)$$

所以 B 具有材料塑性限的物理意义。

4 本构关系的热力学限制

介质的不可逆变形的出现引起了介质内部能量的耗散,所以介质的本构关系必须符合热力学定律的限制。

热力学第一定律可以表示为

$$\int_V \dot{U} \mathrm{d}V = \int_V \boldsymbol{\sigma} : \dot{\boldsymbol{\varepsilon}} \mathrm{d}V \qquad (18)$$

如果要求能量守恒方程对于每一点都成立,那么必须在等式的右边增加一项非局部能量剩余项 R,来表示位于该点的介质单元由于相互作用得到的来自介质中其他单元的能量:

$$\dot{U} = \boldsymbol{\sigma} : \dot{\boldsymbol{\varepsilon}} + R \qquad (19)$$

对于非局部能量剩余项 R,Polizzotto[21-22]引入下列隔离条件来表示体积 V 内没有任何非局部能量流出该体积之外,即

$$\int_V R \mathrm{d}V = 0 \qquad (20)$$

非局部能量剩余项 R 如同 Abu-Alrub 等[20]所指出的那样,除了与塑性梯度效应外,还可能与表面效应和界面效应有关。如果不考虑表面和界面效应、在弹性区域以及不考虑梯度效应时,$R = 0$。在现有的很多梯度塑性模型中忽略了非局部能量剩余项 R,此处将忽略非局部能量剩余项 R。

引入自由能函数 $\psi = U - Ts$,对于等温过程,即 $\dot{T} = 0$,由式(19)可得

$$T\dot{s} = \boldsymbol{\sigma} : \dot{\boldsymbol{\varepsilon}} - \dot{\psi} \geq 0 \qquad (21)$$

自由能密度函数 ψ 可以表示为

$$\psi = \psi(\varepsilon_{ij}^e, \varepsilon_{ij}^p, E^p, \varepsilon_{ij,k}) \qquad (22)$$

对于式(22)关于时间求导得

$$\dot{\psi} = \frac{\partial \psi}{\partial \varepsilon_{ij}^e} \dot{\varepsilon}_{ij}^e + \frac{\partial \psi}{\partial \varepsilon_{ij}^p} \dot{\varepsilon}_{ij}^p + \frac{\partial \psi}{\partial E^p} \dot{E}^p + \frac{\partial \psi}{\partial \varepsilon_{ij,k}} \dot{\varepsilon}_{ij,k} \qquad (23)$$

因为由式(8)有

$$\dot{E}^p = p_{ij} \dot{\varepsilon}_{ij}^p$$

把上式代入到式(23)中得

$$\dot{\psi} = \frac{\partial \psi}{\partial \varepsilon_{ij}^e} \dot{\varepsilon}_{ij}^e + \frac{\partial \psi}{\partial \varepsilon_{ij}^p} \dot{\varepsilon}_{ij}^p + \frac{\partial \psi}{\partial E^p} p_{ij} \dot{\varepsilon}_{ij}^p + \frac{\partial \psi}{\partial \varepsilon_{ij,k}} \dot{\varepsilon}_{ij,k} = \frac{\partial \psi}{\partial \varepsilon_{ij}^e} \dot{\varepsilon}_{ij}^e + \left(\frac{\partial \psi}{\partial \varepsilon_{ij}^p} + \frac{\partial \psi}{\partial E^p} p_{ij} \right) \dot{\varepsilon}_{ij}^p + \frac{\partial \psi}{\partial \varepsilon_{ij,k}} \dot{\varepsilon}_{ij,k} \qquad (24)$$

把式(24)代入到式(21)得

$$T\dot{s} = \left(\sigma_{ij} - \frac{\partial \psi}{\partial \varepsilon_{ij}^e} \right) \dot{\varepsilon}_{ij}^e + \left(\sigma_{ij} - \frac{\partial \psi}{\partial \varepsilon_{ij}^p} - \frac{\partial \psi}{\partial E^p} p_{ij} \right) \dot{\varepsilon}_{ij}^p - \frac{\partial \psi}{\partial \varepsilon_{ij,k}} \dot{\varepsilon}_{ij,k} \geq 0 \qquad (25)$$

由式(25)得

$$\sigma_{ij} = \frac{\partial \psi}{\partial \varepsilon_{ij}^e} \qquad (26)$$

$$\Sigma_{ij} = \sigma_{ij} - \frac{\partial \psi}{\partial \varepsilon_{ij}^p} - \frac{\partial \psi}{\partial E^p} p_{ij} \qquad (27)$$

$$T_{ijk} = \frac{\partial \psi}{\partial \varepsilon_{ij,k}} \qquad (28)$$

可以引入下列广义力

$$\Omega_{ij} = \frac{\partial \psi}{\partial \varepsilon_{ij}^p}, \quad B = Y + \frac{\partial \psi}{\partial E^p} \qquad (29)$$

其中 $B = Y + \frac{\partial \psi}{\partial E^p}$ 包含两部分,$\frac{\partial \psi}{\partial E^p}$ 表示能量(保守)项,当 $\frac{\partial \psi}{\partial E^p} > 0$ 时表示材料的硬化效应;而 Y 表示耗散项,当 $\frac{\partial \psi}{\partial E^p} = 0$ 时,即材料为理想塑性材料时,$Y \dot{\varepsilon}_{ij}^p$ 表示塑性变形引起的能量耗散,所以 Y 有初始屈服限的意义。

这样式(27)可以写为

$$\Sigma_{ij} = \sigma_{ij} - \Omega_{ij} - B p_{ij} + Y p_{ij} \qquad (30)$$

式(25)可简化为

$$T\dot{\eta} = \Sigma_{ij} \dot{\varepsilon}_{ij}^p - T_{ijk} \dot{\varepsilon}_{ij,k} \geq 0 \qquad (31)$$

所以为了满足式(31)必须有如下关系:

$$\dot{\varepsilon}_{ij}^p = \dot{\lambda} \Sigma_{ij}, \quad \dot{\varepsilon}_{ij,k} = -\dot{\eta} T_{ijk} \qquad (32)$$

式中:$\dot{\lambda} \geq 0$ 和 $\dot{\eta} \geq 0$ 为流动系数。这样就保证了满足式(31)。

为了简化分析,在此忽略掉随动硬化效应,即设塑性应变张量不进入到塑性自由能函数中去。设

塑性自由能函数为

$$\psi = \frac{1}{2}\varepsilon_{ij}^e C_{ijkl}\varepsilon_{ij}^e + \frac{1}{2}H(E^p)^2 \quad (33)$$

式中：H 为硬化（软化）模量，当 $H>0$ 时材料具有硬化效应，当 $H<0$ 时材料具有软化效应。

这样除了得到广义虎克定律外还会得到下列广义力：

$$\Omega_{ij} = \frac{\partial \psi^p}{\partial \varepsilon_{ij}^p} = 0, \quad B = Y + \frac{\partial \psi^p}{\partial E^p} = Y + HE^p \quad (34)$$

把式（34）代入到式（16），得

$$\|\sigma_{ij}\| = Y + HE^p \quad (35)$$

5 考虑应变梯度时问题的支配方程

对于平衡方程式（11），当忽略体力时变为

$$\sigma_{ij,j} - T_{ijk,kj} = 0 \quad (36)$$

对于深部圆形隧道围岩，利用极坐标分析较为方便。取圆柱坐标的径向、切向坐标分别为 r 和 θ。高阶应力导数 $T_{ijk,kj}$ 的展开式较复杂，为了简化求解，只保留 T_{rrr} 项，更一般的情况将在以后研究。这样在极坐标里式（36）变为

$$\frac{\partial \sigma_r}{\partial r} + \frac{\sigma_r - \sigma_\theta}{r} - \frac{\partial^2 T_{rrr}}{\partial r^2} - \frac{1}{r}\frac{\partial T_{rrr}}{\partial r} = 0 \quad (37)$$

5.1 弹性情况下的支配方程

几何方程为

$$\varepsilon_r = \frac{du_r}{dr}, \quad \varepsilon_\theta = \frac{u_r}{r} \quad (38)$$

对于 T_{ijk}，在轴对称情况下可以取为下列简单的形式

$$T_{rrr} = c_1\left(\frac{\partial \varepsilon_r}{\partial r} + \frac{\partial \varepsilon_\theta}{\partial r}\right) = c_1\left(\frac{\partial^2 u_r}{\partial r^2} + \frac{\partial u_r}{r\partial r} - \frac{u_r}{r^2}\right) \quad (39)$$

其中 $c_1 > 0$ 为系数。这样平衡方程（37）变为

$$\left[\frac{E(1-\nu)}{(1+\nu)(1-2\nu)} - c_1 \nabla^2\right]\left[\frac{\partial^2 u_r}{\partial r^2} + \frac{\partial u_r}{r\partial r} - \frac{u_r}{r^2}\right] = 0 \quad (40)$$

假定 $\beta^2 = \dfrac{E(1-\nu)}{c_1(1+\nu)(1-2\nu)}$（显然 β 的量纲为 $[m]^{-1}$），则上式变为

$$(\nabla^2 - \beta^2)\left(\frac{d^2}{dr^2} + \frac{d}{rdr} - \frac{1}{r^2}\right)u_r = 0 \quad (41)$$

式（41）的解为

$$u_r = \frac{A}{r} + Br + CI_1(\beta r) + DK_1(\beta r) \quad (42)$$

式中：A、B、C、D 为待定常数；I_1、K_1 分别为第一类和第二类变型 Bessel 函数。

因为在 $r \to \infty$ 时 u_r 应该有限，所以 $B=0$、$C=0$，所以式（26）最终变为

$$u_r = \frac{A}{r} + DK_1(\beta r) \quad (43)$$

所以径向变形 ε_r 和环向变形 ε_θ 为

$$\left.\begin{aligned}\varepsilon_r &= \frac{\partial u_r}{\partial r} = -\frac{A}{r^2} + \eta D\left[K_0(\eta r) - \frac{K_1(\eta r)}{\eta r}\right] \\ \varepsilon_\theta &= \frac{u_r}{r} = \frac{A}{r^2} + \frac{D}{r}K_1(\eta r)\end{aligned}\right\} \quad (44)$$

而应力为

$$\left.\begin{aligned}\sigma_r &= E'\left[\frac{\partial u_r}{\partial r} + \nu'\frac{u_r}{r}\right] = E'\left[-\frac{A}{r^2} + \right. \\ & D\eta\left(K_0(\eta r) - \frac{K_1(\eta r)}{\eta r}\right) + \nu'\left(\frac{A}{r^2} + \frac{DK_1(\eta r)}{r}\right)\right] = \\ & E'\left[-(1-\nu')\frac{A}{r^2} + D\eta K_0(\eta r) + (\nu' - D)\frac{K_1(\eta r)}{r}\right] \\ \sigma_\theta &= E'\left[\frac{u_r}{r} + \nu'\frac{\partial u_r}{\partial r}\right] = E'\left[\frac{A}{r^2} + \frac{DK_1(\eta r)}{r} - \right. \\ & \left. \nu'\frac{A}{r^2} + \nu'D\eta\left(K_0(\eta r) - \frac{K_1(\eta r)}{\eta r}\right)\right] = E'\left[(1-\nu')\frac{A}{r^2} + \right. \\ & \left. \nu'D\eta K_0(\eta r) + D(1-\nu')\frac{K_1(\eta r)}{r}\right]\end{aligned}\right\}$$

$$(45)$$

5.2 具有下降段的弹塑性情况下的支配方程

具有软化段的应力-应变曲线如图1所示[26]。

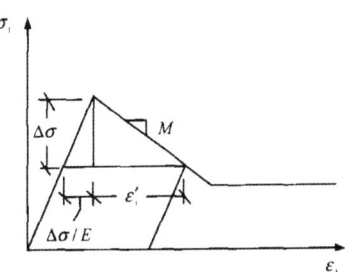

图1 具有软化段的应力-应变曲线
Fig.1 Stress-strain curve with softening

下降段用应力可以表示为

$$\sigma_1 - A\sigma_3 = \sigma_{com} - |M|\varepsilon_1' \quad (46)$$

式中：$|M|$ 为软化模量；ε_1' 为 σ_1 方向的峰值后变形；σ_{com} 为单轴抗压强度。

如果引入侧向变形系数

$$\beta = |\varepsilon_3'/\varepsilon_1'| \quad (47)$$

那么式（46）变为

$$\sigma_1 - A\sigma_3 = \sigma_{com} - |M|\varepsilon_3'/\beta \quad (48)$$

对于弹性变形如果为了简单起见取泊松比为 $\mu = 0.5$，那么弹性变形可以表示为

$$\varepsilon_1^e = \frac{3}{4E}(\sigma_1 - \sigma_3), \quad \varepsilon_3^e = -\varepsilon_1^e \quad (49)$$

如果设 $\xi = |M/E|$，那么在 σ_1 和 σ_2 方向总的可逆变形为

$$\left.\begin{array}{l}\varepsilon_1^n = \varepsilon_1' + |\Delta\sigma|/E = \varepsilon_1' + \varepsilon_1'|M/E| = \varepsilon_1'(1+\xi) = \\ \dfrac{(1+\xi)}{|M|}(-\sigma_1 + A\sigma_3 + \sigma_{com}) \\ \varepsilon_3^n = \varepsilon_3' - \Delta\sigma/E = -\beta\varepsilon_1' - |M|\varepsilon_1'/E = -\varepsilon_1'(\beta+\xi) = \\ -\dfrac{(\beta+\xi)}{|M|}(-\sigma_1 + A\sigma_3 + \sigma_{com})\end{array}\right\} \quad (50)$$

而总的变形为

$$\left.\begin{array}{l}\varepsilon_1 = \varepsilon_1^e + \varepsilon_1^n = \\ \dfrac{3}{4E}(\sigma_1 - \sigma_3) + \dfrac{(1+\xi)}{|M|}(-\sigma_1 + A\sigma_3 + \sigma_{com}) \\ \varepsilon_3 = \varepsilon_3^e + \varepsilon_3^n = \\ -\dfrac{3}{4E}(\sigma_1 - \sigma_3) - \dfrac{(\beta+\xi)}{|M|}(-\sigma_1 + A\sigma_3 + \sigma_{com})\end{array}\right\} \quad (51)$$

式（51）可以变为

$$\left.\begin{array}{l}|M|\varepsilon_1 = -a_1\sigma_1 + b_1\sigma_3 + c_1 \\ |M|\varepsilon_3 = a_2\sigma_1 - b_2\sigma_3 - c_2\end{array}\right\} \quad (52)$$

其中 $a_1 = (1+0.25\xi)$, $b_1 = A + \xi(A-0.75)$, $c_1 = \sigma_{com}(1+\xi)$, $a_2 = \beta + 0.25\xi$, $b_2 = A\beta + \xi(A-0.75)$, $c_2 = \sigma_{com}(\xi+\beta)$。

形式上式（52）可以变为

$$\left.\begin{array}{l}\sigma_1 = -k_1\varepsilon_1 - l_1\varepsilon_3 + m_1 \\ \sigma_3 = -k_2\varepsilon_1 - l_2\varepsilon_3 + m_2\end{array}\right\} \quad (53)$$

其中：$k_1 = \dfrac{|M|b_2}{\Delta}$, $l_1 = \dfrac{|M|b_1}{\Delta}$, $m_1 = \dfrac{b_1c_2 - b_2c_1}{\Delta} = -\dfrac{0.75\xi\sigma_c(\beta-1)}{\Delta} = -\dfrac{\xi\sigma_c}{(1-A)}$, $k_2 = \dfrac{|M|a_2}{\Delta}$, $l_2 = \dfrac{|M|a_1}{\Delta}$, $m_2 = \dfrac{c_2a_1 - c_1a_2}{\Delta} = -\dfrac{0.75\xi\sigma_c(\beta-1)}{\Delta} = m_1$, $\Delta = a_1b_2 - a_2b_1 = 0.75(\beta-1)(1-A) \neq 0$ $(\beta \neq 1)$, σ_c 为岩体单轴抗压强度。

对于弹塑性问题，通常假定式（39）成立[19]。利用公式（53）和方程（31），用 σ_1 和 σ_2 分别代替（31）中的 σ_r 和 σ_θ，最终得到具有圆形横截面的深部隧道围岩的平衡方程

$$\begin{aligned}&\left(k_1 + c_1\nabla^2\right)\left(\frac{d^2u_r}{dr^2} + \frac{du_r}{rdr} - \frac{u_r}{r^2}\right) + \\ &(l_1 - k_2)\frac{du_r}{rdr} + (k_1 - l_2)\frac{u_r}{r^2} = 0\end{aligned} \quad (54)$$

在隧道壁 $r = r_0$ 处由式（13）、（15）可得边界条件

$$(\sigma_r - T_{rrr,r}) - \kappa_r|_{r=r_0} = 0 \quad (55)$$

$$T_{rrr}|_{r=r_0} = 0 \quad (56)$$

由于边界条件展开式复杂，此处不给出展开式。

式（54）~（56）的解析解较难得到。现在考虑一种特殊情况 $l_1 - k_2 = (k_1 - l_2)$，如果忽略体积变形 $\varepsilon_r + \varepsilon_\theta = 0$。那么

$$(l_1 - k_2)\frac{du_r}{rdr} + (k_1 - l_2)\frac{u_r}{r^2} = \frac{(l_1 - k_2)}{r}(\varepsilon_r + \varepsilon_\theta) = 0 \quad (57)$$

这样式（54）变为

$$\left(\xi^2 + \nabla^2\right)\left(\frac{d^2u_r}{dr^2} + \frac{du_r}{rdr} - \frac{u_r}{r^2}\right) = 0 \quad (58)$$

其中 $\xi^2 = k_1/c_1$。

式（58）的解为

$$u_r = \frac{A}{r} + Br + CJ_0(\xi r) + DY_0(\xi r) \quad (59)$$

围岩的应变为

$$\left.\begin{array}{l}\varepsilon_r = \dfrac{\partial u_r}{\partial r} = -\dfrac{A}{r^2} + B - \xi\left[CJ_1(\xi r) + DY_1(\xi r)\right] \\ \varepsilon_\theta = \dfrac{u_r}{r} = \dfrac{A}{r^2} + B + \dfrac{1}{r}\left[CJ_0(\xi r) + DY_0(\xi r)\right]\end{array}\right\} \quad (60)$$

围岩的应力为

$$\left.\begin{aligned}\sigma_r &= k_1\varepsilon_r + l_1\varepsilon_\theta + m_1 = \\ &(l_1+k_1)B + (l_1-k_1)\frac{A}{r^2} - k_1\xi[CJ_1(\xi r) + \\ &DY_1(\xi r)] + l_1\frac{1}{r}[CJ_0(\xi r) + DY_0(\xi r)] + m_1 \\ \sigma_\theta &= k_2\varepsilon_r + l_2\varepsilon_\theta + m_2 = \\ &(k_2+l_2)B + (l_2-k_2)\frac{A}{r^2} - k_2\xi[CJ_1(\xi r) + \\ &DY_1(\xi r)] + l_2\frac{1}{r}[CJ_0(\xi r) + DY_0(\xi r)] + m_2 \end{aligned}\right\} \quad (61)$$

式（59）～（61）中：J_0 和 J_1 分别为第一类 0 阶和 1 阶 Bessel 函数；Y_0 和 Y_1 分别为第二类 0 阶和 1 阶 Bessel 函数。

由式（41）和式（58）的比较可见，由于岩体本构关系中出现了下降段，使得式（41）中的算符 $(\nabla^2-\beta^2)$ 中常数 β 前面的"–"符号变为式（58）中算符 $(\xi^2+\nabla^2)$ 中常数 ξ 前面的"+"符号，从而使得方程的解由单调性的函数式（42）转变为式（59）的具有准周期振荡的解。因此，可见岩体本构关系中下降段的存在是引起分区破裂的必要条件。

由位移、应力和应变表达式（59）～（61）可见，这 3 组解的右边前两项对应着弹塑性理论的解，而后面的 Bessel 函数项代表着对于弹塑性解的修正，这种修正随着不可逆变形的增加而增加。这里的位移、应变和应力表达式与笔者在文献[9, 13]中得到的解的差别在于，此处出现了弹塑性理论解，是在连续介质力学框架内的自洽解。

位移、应变和应力表达式具有随距隧道壁距离增加幅值减小的准周期性，能够反映分区破裂现象的位移、应变和应力变化情况。文献[9, 13]的研究表明，选取适当的 Bessel 函数解的常数可以模拟试验和现场得到的破裂区随距隧道壁距离的分布情况，此处适当选取式（59）～（61）中 Bessel 函数前面的系数可以模拟试验和现场得到的破裂区随距隧道壁距离的分布情况。将在后续研究中研究式（59）～（61）的解。

5.3 忽略弹性变形的塑性情况下的支配方程

对于岩体的极限后变形，在此采用具有下降段的 Tresca 本构模型。为了简化计算，在此采用刚塑性模型，即忽略弹性变形。

对于深部围岩的分区破裂化，一般认为，在地应力达到岩体的单轴抗压强度后开始出现分区破裂化现象，此时按照 Tresca 屈服条件，两个滑移面上的剪应力 T 和 T_1 达到极限状态：

$$\left.\begin{aligned} 2T &= \sigma_\theta - \sigma_r \geq 2\tau_c = \sigma_c \\ 2T_1 &= \sigma_z - \sigma_r \geq 2\tau_c = \sigma_c \end{aligned}\right\} \quad (62)$$

式中：τ_c 为岩体的抗剪强度。

在极限后区域，岩体中的剪力将下降，两个剪应力随着剪应变的下降情况如图 2 所示。可以表示为

$$\left.\begin{aligned} \sigma_\theta - \sigma_r &= \sigma_c - M(\varepsilon_\theta - \varepsilon_r) \\ \sigma_z - \sigma_r &= \sigma_c - M(\varepsilon_z - \varepsilon_r) \end{aligned}\right\} \quad (63)$$

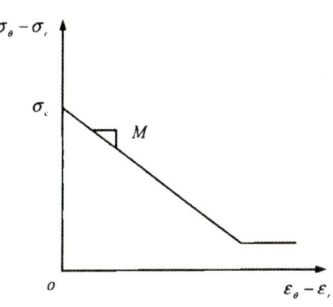

(a) $\sigma_\theta - \sigma_r$ 随 $\varepsilon_\theta - \varepsilon_r$ 的变化曲线

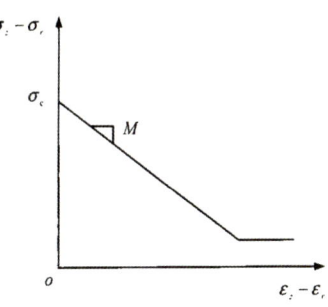

(b) $\sigma_z - \sigma_r$ 随 $\varepsilon_z - \varepsilon_r$ 的变化曲线

图 2 忽略弹性变形时具有软化段的应力-应变曲线
Fig.2 Stress-strain curves with softening without consideration of elastic deformations

对于式（63）第 2 式，σ_z 等于岩体的原始地应力：$\sigma_z = \sigma_\infty$，而 $\varepsilon_z = 0$，所以式（63）第 2 式变为 $\sigma_r = \sigma_\infty - \sigma_c - M\varepsilon_r$，$\mathrm{d}\sigma_r = -M\mathrm{d}\varepsilon_r$。

考虑到几何方程 $\varepsilon_r = \dfrac{\mathrm{d}u_r}{\mathrm{d}r}$，$\varepsilon_\theta = \dfrac{u_r}{r}$，平衡方程变为

$$(M+c_1\nabla^2)\left(\frac{\mathrm{d}^2}{\mathrm{d}r^2} + \frac{\mathrm{d}}{r\mathrm{d}r} - \frac{1}{r^2}\right)u_r + \frac{\sigma_c}{r} = 0 \quad (64)$$

式（64）的解为[27]

$$\begin{aligned} u_r =\ & b_1 J_0(\zeta) + b_2 Y_0(\zeta) + \\ & \frac{\pi}{2M} Y_0(\zeta) \int \eta J_0(\zeta)\left(d_1 r + d_3\frac{1}{r} - \frac{\sigma_c}{2} r\ln r\right)\mathrm{d}\zeta - \\ & \frac{\pi}{2M} J_0(\zeta) \int \eta Y_0(\zeta)\left(d_1 r + d_3\frac{1}{r} - \frac{\sigma_c}{2} r\ln r\right)\mathrm{d}\zeta \end{aligned}$$

$$(65)$$

式中：$\eta = kr$，$k^2 = M/c_1$，其他符号含义详见文献[27]。

将式（65）代入到应变和应力的表达式中，可以得到应变和应力的表达式。

与式（41）中的算符 $(\nabla^2 - \beta^2)$ 中常数 β 前面的"-"符号不同，式（58）中的算符 $(\xi^2 + \nabla^2)$ 中常数 ξ 前面的符号为"+"，从而使得方程（64）的解式（65）具有随距隧道壁距离增加幅值减小的准周期性，能够反映分区破裂现象。方程（64）的解将在后续工作中进行数值研究。

6 结 论

（1）本文利用塑性梯度理论研究了深部隧道围岩的分区破裂现象。作为额外的状态变量，在此引入应变梯度这一新变量。利用虚功原理得到了岩体的平衡方程、边界条件和流动准则。利用非局部的 Clausius-Duhem 不等式获得了岩体的本构方程。

（2）对于圆形深部隧道，由上述理论得到了弹性变形情况下、具有下降段的弹塑性变形情况下和不考虑弹性变形的塑性变形情况下深部隧道围岩的支配方程。

（3）对于隧道围岩的临界后课题来讲，应变梯度的引进能够描述岩体处于临界后状态的自组织行为。

（4）岩体应力-应变曲线下降段的存在，是引起隧道围岩出现随距隧道壁距离增加位移、应变和应力准周期性变化的必要条件。

参 考 文 献

[1] SHEMYAKIN E I, FISENKO G L, KURLENYA M V, et al. Zonal disintegration of rocks around underground workings. Part 1: Data of in situ observations[J]. **Journal of Mining Science**, 1986, 22(3): 157－168.

[2] SHEMYAKIN E I, FISENKO G L, KURLENYA M V, et al. Zonal disintegration of rocks around underground workings. Part II: Rock fracture simulated in equivalent materials[J]. **Journal of Mining Science**, 1986, 22(4): 223－232.

[3] SHEMYAKIN E I, FISENKO G L, KURLENYA M V, et al. Zonal disintegration of rocks around underground mines. Part III: Theoretical concepts[J]. **Journal of Mining Science**, 1987, 23(1): 1－5.

[4] SHEMYAKIN E I, FISENKO G L, KURLENYA M V, et al. Zonal disintegration of rocks around underground workings. Part IV: Practical applications[J]. **Journal of Mining Science**, 1989, 25(4): 297－302.

[5] ODINTSEV V N. Mechanism of the zonal disintegration of a rock mass in the vicinity of deep-level workings[J]. **Journal of Mining Science**, 1994, 30(4): 334－343.

[6] CHANYSHEV A I. To the problem of deformable medium failure. Part I: Basic equations[J]. **Journal of Mining Science**, 2001, 37(3): 273－288.

[7] CHANYSHEV A I. To the problem of deformable medium failure. Part II: Discussion of the results of analytical solutions[J]. **Journal of Mining Science**, 2001, 37(4): 392－400.

[8] GUZEV M A, POROSHIN A A. Non-Euclidean model of the zonal disintegration of rocks around an underground working[J]. **Journal of Applied Mechanics and Technical Physics**, 2001, 42(1): 131－139.

[9] QI Cheng-zhi, QIAN Qi-hu, WANG Ming-yang. Evolution of the deformation and fracturing in rock masses near deep-level tunnels[J]. **Journal of Mining Science**, 2009, 45(2): 112－119.

[10] 潘一山，李英杰，唐鑫，等. 岩石分区破裂化现象研究[J]. 岩石力学与工程学报, 2007, 26(增刊 1): 3335－3341.

PAN Yi-shan, LI Ying-jie, TANG Xin, et al. Study of zonal disintegration of rock[J]. **Chinese Journal of Rock Mechanics and Engineering**, 2007, 26(Supp.1): 3335－3341.

[11] 贺永年，蒋斌松，韩立军，等. 深部巷道围岩间隔性区域断裂研究[J]. 中国矿业大学学报, 2008, 37(3): 300－304.

HE Yong-nian, JIANG Bin-song, HAN Li-jun, et al. Study of intermittent zonal fracturing of surrounding rock in deep roadways[J]. **Journal of China University of Mining & Technology**, 2008, 37(3): 300－304.

[12] 顾金才，顾雷雨，陈安敏，等. 深部开挖洞室围岩分层断裂破坏机制模型试验研究[J]. 岩石力学与工程学报, 2008, 27(3): 433－438.

GU Jin-cai, GU Lei-yu, CHEN An-min, et al. Model test study of mechanism of layered fracture within surrounding rock of tunnels in deep stratum[J]. **Chinese Journal of Rock Mechanics and Engineering**, 2008, 27(3): 433－438.

[13] WANG Ming-yang, QI Cheng-zhi, QIAN Qi-hu, et al. Physical modeling of the deformation increment sign change effect of rock sample under compression[J]. **Journal of Mining Science**, 2010, 46(4): 21－28.

[14] 戚承志, 钱七虎. 岩体动力变形破坏的基本问题[M]. 北京: 科学出版社, 2009.

[15] AIFANTIS E C. On the microstructural origin of certain inelastic models[J]. **Journal of Engineering Materials and Technology**, 1984, 106(2): 326－330.

[16] VARDOULAKIS I, AIFANTIS E C. A gradient flow theory of plasticity for granular materials[J]. **Acta Mechanics**, 1991, 87(3－4): 197－217.

[17] AIFANTIS E C. Pattern formation in plasticity[J]. **International Journal of Engineering Science**, 1995, 33(15): 2161－2178.

[18] FLECK N A, HUTCHINSON J W. Strain gradient plasticity[C]//Advances in Applied Mechanics. New York: Academic Press, 1997: 295－361.

[19] CHAMBON R, CAILERIE D, HASAN N E. One-dimensional localization studied with a second grade model[J]. **European Journal of Mechanics/A: Solids**, 1998, 17(4): 637－656.

[20] ABU AL-RUB R K, VOYIADJIS G Z. A physically based gradient plasticity theory[J]. **International Journal of Plasticity**, 2006, 22(3): 654－684.

[21] POLIZZOTTO C. Unified thermodynamic framework for nonlocal/gradient continuum theories[J]. **European Journal of Mechanics/A: Solids**, 2003, 22(5): 651－668.

[22] POLIZZOTTO C. Interfacial energy effects within the framework of strain gradient plasticity[J]. **International Journal of Solids and Structures**, 2009, 46(7－8): 1685－1694.

[23] QI Cheng-zhi, QIAN Qi-hu, WANG Ming-yang, et al. Mechanical modeling of zonal disintegration of rock mass near deep level tunnels[C]//Proceedings of the 3rd International Conference on Contemporary Problems in Architecture and Construction. Beijing: [s. n.], 2011: 114－119.

[24] WANG Ming-yang, QI Cheng-zhi, QIAN Qi-hu, et al. One plastic gradient model of zonal disintegration of rock mass near deep level tunnels[J]. **Journal of Mining Science**, 2012, 48(1): 46－54.

[25] ILYJUSHIN A A. Plasticity[M]. Moscow & Leningrad: OGIZ Press, 1948.

[26] BAKLASHOV I V, KARTOZIA B A. Mechanics of underground buildings and structures[M]. Moscow: Nedra Press, 1984.

[27] ZAITSEV V F, POLYANIN A D. Hand book for ordinary differential equations[M]. Moscow: Physmatlit Press, 2001.

分区破裂化现象的研究进展

戚承志[1,2]，钱七虎[2]，王明洋[2]，罗健[1]

(1. 北京建筑工程学院 北京市工程结构与新材料工程研究中心，北京 100044；
2. 解放军理工大学 工程兵工程学院，江苏 南京 210007)

摘 要：通过对分区破裂化现象的发现过程进行回顾、对分区破裂化现象的研究现状进行综述，对深部巷道围岩的受力变形的特点有了更好的认识，并利用弹塑性理论、断裂力学理论、非线性科学理论等对分区破裂化现象形成的机理进行了探索，得到了一些有益的结论。在现场观测到了分区破裂化现象，从试验上再现了分区破裂化现象的一些主要特点，在数值模拟方面能够定性地模拟这一现象。但还有很多问题没有解决，包括岩体的力学性质及结构特性对于分区破裂化现象时间演化过程的影响、分区破裂化现象与受损岩石变形增量变符号现象之间的关系、岩爆与分区破裂化现象之间的关系问题等。

关键词：分区破裂化现象；理论研究；试验研究；数值模拟；研究展望

中图分类号：TU45　　**文献标识码**：A　　**文章编号**：1009-3443(2011)05-0472-08

Advance in investigation of zonal disintegration phenomenon

QI Cheng-zhi[1,2], QIAN Qi-hu[2], WANG Ming-yang[2], LUO Jian[1]

(1. Beijing Engeneering Structure and New Material Research Center, Beijing Institute of Civil Engineering and Architecture, Beijing 100044, China; 2. Engineering Institute of Corps of Engineers, PLA Univ. of Sci. & Tech., Nanjing 210007, China)

Abstract: The finding process of the zonal disintegration phenomenon was reviewed, and the state-of-the-art of investigations of zonal disintegration phenomenon overviewed. The review shows that our understanding on the stressing and the deformation of the rock mass near the deep level openings has been greatly improved. This phenomenon was investigated with the help of the elasticity and plasticity theories, fracture mechanics and nonlinear science etc., and many helpful conclusions reached. In situ observations the existence of zonal disintegration phenomenon was observed, laboratory model experiments reproduced the main features of zonal disintegration phenomenon, and numerical modeling qualitatively reproduced zonal disintegration phenomenon. But many problems still exist, including the influence of the mechanical and the structural properties of the rock mass on the temporal evolution process of the zonal disintegration, the relationship between the zonal disintegration and the phenomenon of the sign change of deformation increment after the beginning of the damaging process in rock, the relationship between the zonal disintegration and the rock bursts etc.

Key words: zonal disintegration; theoretical research; experimental research; numerical simulation; research perspective

随着社会经济的发展，很多工程领域，例如采矿工程、大型水利水电建设、交通工程和国防工程等都涉及深部岩体的变形与破坏问题。与浅部岩体的变形破坏不同，深部工程围岩由于处于高的地应力状态，变形与破坏具有显著的非线性特性，并伴随着诸

收稿日期：2009-03-02。
基金项目：国家自然科学基金资助项目(50825403,51174012)；国家973计划资助项目(2010CB732003)；北京市自然科学基金资助项目(KZ200810016007)。
本文原载于《解放军理工大学学报(自然科学版)》(2011年第12卷第5期)。

如岩爆、瓦斯突出、大体积塌方、塌陷等重大工程灾害,对生命安全及财产构成巨大威胁。正因为如此,世界各国学者对深部岩体的变形与破坏很早就给予了高度重视,并进行了深入研究。在最近30年,随着现代科技水平的提高及岩土工程的发展,深部非线性岩体力学作为一个科学方向已获得了飞速发展,研究范围在不断扩大,理论研究、实验研究及数值模拟都取得了丰富的成果。

深部围岩变形破坏具有显著的非线性特征,在深部围岩中出现了与浅部围岩不同的非线性变形与破坏现象。深部围岩的分区破裂化现象就是这些非线性现象之一。该现象自从20世纪70年代被发现以来吸引了许多学者的关注,他们进行了理论、试验与数值分析等方面的研究,并取得了一定的成果。总结通过现场观测及实验室试验发现的分区破裂化现象的新规律,整理研究成果,发现研究中存在的问题,指出今后需要研究的方向,对于今后分区破裂化现象的研究十分重要。

笔者从事分区破裂化方面的研究,近几年来,取得了一定进展,并掌握了国内外分区破裂化现象研究方面较多的资料。希望本综述能够为分区破裂化现象的研究工作者们提供借鉴。

1 分区破裂化现象的发现

巷道围岩物理力学参数随着距巷道壁距离的增加而呈周期性变化,这一规律是俄罗斯学者В. А. Борисовец在1972年首次发现的[1]。他观测到在200 m深度时竖向巷道的砂岩围岩的单轴强度、孔隙度、弹性模量以及密度随着距巷道壁距离的增加发生了周期性变化。后来他又和同事一起在450 m深度处的竖井中,在一系列长达11.5 m深的钻孔中进行测量,证实了这一规律。观测表明,在竖井围岩中出现了周期长度为1.5~2.0 m的3~5个变化周期[2]。

В. Н. Опарин等[3,4]在Норильск的深达1 km的矿井中,利用地球物理方法测量了电阻率、密度等物理参数,围岩的裂隙度通过间接方法确定。在巷道围岩中他们发现了上述周期性变化,且裂纹区及压密区交替出现,并重复洞室的形状,裂纹区的宽度为0.5~2.5 m,而裂纹之间的过渡区宽度为1.5~3.0 m。

М. П. Зборщик等[5,6]研究了在沉积岩中950 m深度处单个准备巷道围岩的电阻率,并用潜望镜研究了裂纹结构。在90 d的时间内观察到了主干裂纹从巷道壁向围岩内发展的过程。观测到围岩中形成了岩块结构之后,环绕巷道的岩体因产生裂纹而被破坏。在径向观测到了由于裂纹的张开及闭合而产生"附加压缩及膨胀纵波"的传播过程。

Ю. С. Кузнецов等[7,8]在Норельск的100~1 100 m的深度,利用地球物理方法、潜望镜观测法及肉眼观测法,在与主巷道交叉的巷道采掘面上观测到在主巷道的周围产生了多达4个裂纹区。裂纹区之间的区域破坏非常弱,并重复巷道的横截面形状。张开的裂纹壁比较光滑,显示了裂纹的张开特性。破裂区总的张开宽度大于巷道半径的减小值,这说明裂纹区之间的过渡区被压密。裂纹区及过渡区的宽度为1~1.5 m。裂纹区的数量取决于地压力水平,最远的裂纹区离巷道壁5~11 m。

А. В. Шмиголь、В. Я. Кириченко等[9,10]研究了地压力水平为$\gamma H/\sigma_c = 2.2 \sim 2.4$时的围岩变形与破坏情况($\gamma$为岩体密度,$H$为巷道埋深,$\sigma_c$为岩石单轴抗压强度)。发现了4个破裂区,宽度为0.2~1.8 m,且随离巷道壁的距离增加而减小。过渡区的宽度为1.0~2.3 m,且随离巷道壁的距离增加而增加。裂纹的张开宽度为1.5~5 mm,它们之间的距离为0.10~0.25 m。在与支护的接触面附近岩体被压密,巷道围岩的剪切位移明显。

О. И. Мельников等[11]在采掘巷道的支撑压力区的周期性变形岩体中观测到相对压缩段及相对拉伸段的运动情况。И. Л. Черняк在采掘掌子面的前方观测到固定的压缩区及张拉区,它们的位置与到采掘掌子面的距离无关[12]。

南非学者G. R. Adams、A. J. Jager[13]于1980年在南非2 300 m深度处金矿采掘面的前方发现了周期性的破裂面现象。他们的工作与俄罗斯学者的工作几乎是同时及相互独立地进行的。

我国在现场测量中也发现了类似的现象。1979年李世平在徐州权台矿的锚杆受力观测中发现锚杆不仅受拉应力作用,而且受压应力作用[14]。1991年贺永年在淮北朱仙庄矿巷道发现围岩超声波速呈波峰—波谷—波峰状态[15]。1996年方祖烈在金川矿观测到巷道围岩中应变随着距巷道壁的距离呈拉—压—拉分布,试验证实了这种分布的普遍性[16]。但是当时这种现象还没有引起人们的足够重视,没有被作为一个课题进行研究。最近在淮南矿区[17]及锦屏水电站隧洞[18]进行的现场观测中也发现了分区破裂化现象的存在。

2 分区破裂化现象的研究进展

2.1 定性研究

分区破裂化现象与通常的连续介质的弹塑性力学解完全不一样,因而引起了人们的兴趣,他们进行了大量的理论、实验与数值模拟研究。A. M. Козел 等[2]认为,巷道围岩的分区破裂化现象是由于爆炸引起的压缩波及拉伸波的不均匀分布造成的。但是这种观点不能解释不采用爆破法施工时也能产生分区破裂化现象的事实。

俄罗斯学者 Е. И. Шемякин、М. В. Курленя 等[19~24]在 20 世纪 80 年代末进行了一系列理论与试验研究工作。在分区破裂化现象的形成机理方面,Е. И. Шемякин 等认为,破裂区的形成是由于支撑压力区在较大的环向压应力作用下沿着环向压应力方向形成劈裂裂纹所致。理论及实验研究表明[24],从岩体的受力角度讲,岩体在纵向压力很大及侧压力较小时,沿最大纵向主压应力方向可以发生劈裂。由于岩块之间存在着缝隙,岩体的变形不满足圣维南变形协调条件,侧压力较小的条件总能在局部得到满足,沿着最大压应力方向形成劈裂裂纹的条件是满足的。当形成劈裂时,劈裂面便可作为新的"假洞室壁","假洞室壁"后面岩体的变形破坏过程重复前述的变形破坏过程,直到围岩中应力状态不再满足劈裂形成的条件。但是这种观点不能解释若干个破裂区同时发展的事实,也不能解释破裂区之间的过渡区破坏很弱的变形特点。

В. Н. Одинцев[25,26]研究了岩体的极限后变形及弹性区岩体与应力极限后岩体组成的系统稳定性问题,考虑到了支护提供的压力 p_0,求解了巷道围岩的边值问题。结果表明,径向压力沿径向分布 $p(r)$ 有极值。当压力达到某个值 $p = p^*$ 时,在无限小的变形上对应的应力所作的无限小功为负,这时系统失稳,出现动力破坏。作者推断:最简单的能够保证岩体稳定状态的裂纹结构为环形开裂裂纹。所形成的环向开裂裂纹又作为"假巷道轮廓",巷道围岩的变形破坏重复上述过程便形成了分区破裂现象。类似地,对于深部巷道开挖面前面的分区破裂化现象,В. Н. Одинцев[27]从地质材料结构的角度进行了分析,认为破碎区的出现是因为岩体在准静变形过程中,达到不稳定状态时发生动力破坏,并形成新的稳定结构,这一结构在后续的一定加载范围内保持稳定。这样的过程重复便形成了分区破裂化现象。

В. Н. Одинцев 在上述研究中首次考虑了大深度巷道围岩在压缩状态下张开裂纹的发展情况,并考虑到岩体的结构,建立了开裂破坏准则。但是他没有考虑在强烈的非均匀压缩状态下岩体裂纹结构的发展情况,没有给出数值结果以及数值结果与试验结果的比较。

2.2 弹塑性理论方法

А. И. Чанышев 首次应用严格的弹塑性分析来解释分区破裂化现象,研究了在无穷远处受静水压力作用下岩体中无支护巷道的变形问题[28]。考虑了岩体的体积膨胀及压缩,通过解边值问题,确定了膨胀区及压缩区的边界。膨胀区靠近未扰动岩体,而压缩区靠近巷道。在膨胀区径向应力增加,在压缩区径向应力减小。膨胀区的数量取决于巷道壁上的剪切变形,围岩的应力状态及岩体的力学参数(内聚力、内摩擦角)。

在后续的工作中,А. И. Чанышев[29~31]在塑性形变理论及流动理论的框架内,考虑了体积的弹性任意压缩性及软化的任意性,建立了求解方程,求出了方程的特征线。安瓦尔·恰内晒夫证明当岩石软化段的下降变形模量趋近于体积的弹性压缩模量时,一组特征线将重复巷道的形状,这就是分区破裂化现象。

В. Н. Рева 和 Э. Н. Тропп[32]解决了在静水压力作用下圆形巷道的弹塑性边值问题。围岩具有 2 个塑性区,一个位于巷道壁附近,另一个位于围岩中,塑性区满足 Mises 条件,在弹性区变形满足变形协调条件。解的结果表明,只能唯一地确定第一个破裂区到巷道壁的距离。如果不引入额外的假定,内部破裂区距巷道壁的距离无法确定。

Л. С. Метров、А. Д. АЛЕКСЕЕВ 等[33,34]利用线性热粘弹性模型及协同学方法来研究岩体的破坏过程。通过数值解法获得了岩体的形变参数(lame 系数)在巷道围岩中的周期性分布。但是此模型没有考虑岩体的破坏特征。

L. S. Metlov 等[35]以非平衡热力学的观点分析了巷道围岩的分区破裂化现象。认为巷道开挖之后巷道围岩处于热力学不平衡状态,围岩中的缺陷逐渐积累,在离巷道壁较远处围岩处于受压状态,将会出现强化,当此处的应力随着时间逐渐变大达到围岩的强度极限时产生脆性破坏。这一过程不断重复便形成分区破裂化现象。但是这种观点不能解释若干个破裂区同时发展的情况。

经典的弹性理论能够很好地描述材料破坏前的

力学行为,而经典的塑性理论不能精确地描述材料屈服后的宏观力学行为,这是因为在不可逆变形后材料中出现了能量的耗散,产生了自我组织现象,形成了耗散结构。耗散结构需要借助于物理学的有关理论进行描述。

2.3 非经典弹塑性理论方法

在应力应变状态改变时岩体的原有裂纹和新生裂纹使得岩体的变形协调条件不再满足。为了更好地描述岩体的变形与破坏,从欧氏几何过渡到非欧几何是必须的。М. А. Гузев 等利用非欧几何的方法来研究含缺陷介质的变形破坏问题[36]及深部巷道周围的分区破裂化现象[37]。除了弹性变形张量外还必须引进不可逆变形张量。这时 Riemann-Christoffel 张量 R_{lijk} 不为 0。由 Riemann-Christoffel 张量确定的 Ricci 张量的迹 R 具有标量曲率的意义。在平面变形状态下标量曲率为唯一的内部构造非欧性的参数。М. А. Гузев 等[37]引入了 R 参数,利用了非平衡热动力学理论,得到了求解 R 的定解方程,求出了非协调参数 R 及巷道围岩中的应力随着距巷道壁的距离的增加而周期性变化的规律。但是这一解没有考虑岩体中存在的初始缺陷,没有给出分区破裂化现象的时间演化过程。

我国在分区破裂化现象方面的研究起步较晚。在 20 世纪 90 年代末,钱七虎、戚承志、王明洋等把俄罗斯学者的岩体结构层次概念、深部岩体力学的非线性力学现象(分区破裂化现象、摆型波、超低摩擦现象等)介绍到我国,引起了我国学者的兴趣。在钱七虎院士的倡导下我国的岩石力学与工程学者们开始了这一领域的研究,取得了一定的成果,在岩体的结构层次及其形成机理、力学性质的关系、深部岩体的变形与运动特性方面进行了卓有成效的工作[38-45]。在深部岩体的分区破裂化现象机理方面也进行了深入探索。

深部巷道围岩在变形及应力超过弹性限之后出现不可逆的变形及能量的耗散,围岩发生结构变化,即围岩中出现了新的缺陷,这种变化可以用渐进相变来描述。能量的耗散及渐进相变伴随着自我组织的发生,这是传统的弹塑性理论无法描述的。戚承志等[43,44]利用连续相变理论研究了巷道围岩的分区破裂化现象。取相对塑性剪切变形作为序参量,利用 Ginzburg-Landau 形式的自由能及不可逆非平衡热动力学理论得到了确定序参量的求解方程,解得了序参量及塑性剪切变形在巷道围岩中的周期性分布,首次得到了序参量及剪切变形随时间演化的方程,得到了若干破裂区同时发展的结构,求得的结果与现场观测结果相符合。岩体物理力学与结构性质的影响、围岩开挖加载卸载过程的影响、岩爆与分区破裂化现象之间的相互关系问题、围岩变形破坏的波动性是需要进一步研究的问题。

周小平等[45]利用弹性力学理论得到了在原岩应力及开挖扰动作用下巷道围岩的弹性应力场及位移场。当弹性应力场满足破裂条件时岩体发生破裂,形成破碎区。他们利用断裂力学理论确定了破裂区岩体的残余强度和产生破裂区的时间,进而确定了破裂区和非破裂区的宽度及数量。研究表明巷道分区破裂化的产生与开挖速度及岩石的强度有关。但是这一模型采用的是岩体的弹性模型及脆性破坏,不能解释若干破裂区同时破坏及破裂区之间的过渡区变形情况。

2.4 试验研究

除了俄罗斯学者进行的试验研究以外,我国学者也进行了一些试验研究。

潘一山等[46]进行了模型试验,部分得到了分区破裂化现象。认为深部岩体裂纹密度大,支撑压力区峰值应力很高,环向应力大,径向应力小,所以此处岩体处于近乎单轴应力状态,岩体将很快进入加速蠕变阶段。在此基础上利用改进的流变模型研究了分区破裂化现象。

贺永年等[47]进行了厚壁圆筒的卸载试验,得到了环形断裂带。认为围岩中能量分布存在极值,极值位于弹塑性边界上。这种能量分布是不稳定的,当有外部扰动时,系统会寻求能量更低的状态以达到新的平衡,于是能量的高峰位置出现断裂。由于深部岩体含有很大的压缩势能,洞室开挖卸压降低了能流的阈值,引起势能从三轴状态向单轴状态流动,为岩石的破坏提供足够的能量。岩石的承载力具有不稳定的特点,峰值后承载力急剧下降,由于流入的能量超过断裂需要的能量,巷道围岩可以形成有规律的破坏,即分区破裂化现象。

顾金才院士[48]认为,分区破裂化现象的产生是有条件的。第一,需要较大的巷道轴向压力,因为只有较大的轴向压力才能引起较大的径向位移,当巷道围岩的径向拉应变超过极限值时就会发生破坏。如果轴向压力足够大,那么就会引起二三次破坏。第二,储存于围岩中的能量须突然释放,产生卸载波,使得围岩受拉破坏。如果卸载波强度很高,那么就会产生二三次破裂,产生分区破裂化现象。基于上述认识,顾院士通过实验再现了分区破裂化现象的主要特点。

张强勇等[49]采用铁晶砂胶结新型岩土相似材料,以及自行研制的高地应力真三维地质力学模型试验系统,再现了巷道围岩破裂区与非破裂区交替出现的规律。试验证实了顾金才院士关于分区破裂化现象需要有较大的巷道轴向压力的观点。

相信随着相似材料制作的不断改进,相似条件的进一步满足,试验设备的不断完善,分区破裂化现象产生的条件及其时间空间演化规律会更真实地被揭示出来。

3 展 望

对于分区破裂化现象的数值模拟依赖于正确的岩体变形破坏模型的建立。这一方面的工作已经开始[18,50],但是模拟效果还不够理想。

对于深部的界定问题也在讨论之中,如何满潮[51]及王明洋[52]等人的研究。

В. В. Макаров 在现场及实验室试验中发现了分区破裂化现象的新特点。В. В. Макаров 观测表明[53,54],巷道开挖之后巷道围岩的分区破裂化现象有 2 个阶段。该现象在第 1 个阶段形成强烈的分区破坏结构,在第 2 个阶段发生相对慢的过程,自发地出现破裂波阵面。这些波阵面在受压区形成,以一定的速度穿越所有的分区破裂化结构,从而引起分区破裂结构随时间的变化。В. В. Макаров 观测还表明[53~55],分区破裂化的形成与岩体的强烈受压有关。根据岩体性质的不同,形成的波动状的周期性分区破裂结构具有不同的性质及表现形式。在坚硬岩体中,受到岩爆作用时形成环形裂纹结构。在弱岩中,永久巷道围岩中岩体的密度发生分区现象,而且分区结构的形成伴随着破坏波阵面的产生及传播。在破坏强烈的岩体中,每一次开挖过程都伴随着破坏波阵面的产生及传播,而且破坏波阵面具有波动状周期性结构。В. В. Макаров 和 М. А. Гузев[55,56]在实验研究中还发现了岩石试件在产生损伤后局部出现变形增量的逆转现象,这种现象与试件中形成的分区破坏现象密切相关。上述这些新现象在目前还没有得到很好地研究。

在研究岩体的变形与破坏时一个不可回避的重要问题是岩体的结构层次。岩体介质的块系构造作为非线性岩体力学的一个基本概念具有非常重要的意义。这一概念最早由 М. А. Садовский 院士提出[57],并已经得到了证实[58,59]。研究表明,岩体块系系统存在着块体尺寸的自相似等级序列关系[57]。

另外通过实际考察还发现[59],某一尺度水平上裂纹的张开尺寸 δ_i 及由裂纹分开的同级块体尺寸 Δ_i 之间存在着稳定的统计关系:$\mu_\Delta(\delta) = \delta_i/\Delta_i = \Theta \cdot 10^{-2}$,其中 Θ 为系数,其变化范围为 $1/2 \sim 2$。而 $\mu_\Delta(\delta)$ 被称作岩石力学"不变量"。这种概念就使得在分析中所使用的基本体元的概念,及连续介质力学数学模型中使用的圣维南变形协调条件受到了置疑。$\mu_\Delta(\delta)$ 可以作为某一构造水平上实际岩体变形或位移的一个不协调尺度,也可以作为一个表征岩块接触面粗糙度的尺度。

介质的非均匀性、变形的不协调性及非线性都要求引入新的思想和新的数学手段来描述岩体的力学行为。В. Е. Панин 院士等创立的结构非均匀介质的物理细观力学观点及方法是很值得借鉴的[60]。物理细观力学从 3 个层次,即微观、细观、宏观来考察材料的变形与破坏,把受载体看作多水平有机联系的自组织系统。在协同学方法的范围内,受载固体被看作是开放的,在应力集中局部区域内的强非平衡系统。在应力集中的局部区域内的加载过程中,发生着非平衡的局部构造转化,这种转化在不同的尺寸水平上发生,其特性、能量、尺寸、速度各不相同。在给定的加载边界条件下,它们的自我组织以耗散结构的形成为前提。耗散结构的演化决定着塑性变形及破坏的特征。

在描述具有非均匀内部构造的介质力学行为时,另一种方法为非标准分析法[61],采用具有构造的函数来研究岩体的变形与破坏问题。这种方法解决了这样的矛盾:一方面,函数应处处非连续;另一方面,能应用连续函数的数学手段。但是总的来讲,到目前为止,对于块系岩体还没有建立统一的数学框架。

由上述事实可见,对深部围岩的分区破裂化现象的研究在定性及定量方面都取得了一定的进展,作者认为还需要对下列问题做进一步研究。

(1) 确立块系岩体在外界作用时协同学原则问题,建立一个考虑变形非协调性、非均匀、不可逆性、体积变形、非平衡性、能量耗散性的数学框架。在这一框架内应考虑岩体的结构特征、几何特征及力学特性等。

(2) 分区破裂化现象裂纹结构的时间空间演化问题。正如观测所示,在强度不同、原始裂隙度不同的岩体中分区破裂化的时间过程是不一样的,这还需要进一步解释。

(3) 观测发现在发生岩爆的巷道中存在分区破裂化现象,岩爆的发生与分区破裂化现象形成的关系需进一步研究。

(4) 岩石试件在产生破坏后局部出现变形增量的逆转现象与分区破裂化现象的关系及其机理需进一步研究。

(5) 相似材料的改进,相似条件的进一步满足,试验设备的不断完善。

弄清上述问题对于研究及有效地利用分区破裂化现象进行巷道的开挖,选择高效的支护形式及其设立时间,进行矿藏的高效率安全开采具有重要的意义。

参考文献:

[1] БОРИСОВЕЦ В А. Неоднородности волнового характера в породах вблизи выработок, сооружаемых буровзрывным способом [J]. Шахтное Строительство, 1972(9):7-11.

[2] КОЗЕЛ А М, БОРИСОВЕЦ В А, Репко А А. Горное давление и способы поддержания вертикальных стволов [M]. Москва: Недра, 1976.

[3] ОПАРИН В Н, ТАПСИЕВ А П. О некоторых закономерностях трещинообразования вокруг горных выработок [C]. Горные удары, методы оценки и контроля удароопасности массива горных пород. Фрунзе: Илим, 1979.

[4] ОПАРИН В Н, ЕЛИСОВЕТСКИЙ И Я. О некоторых закономерностях в напряженно-деформированном состоянии окресности горных пород [C]. Геофизические методы контроля напряжений в горных породах. Новосибирск: ИГД СО АН СССР, 1980.

[5] ЗБОРЩИК М П, МАРЯРЧУК А М, МОРОЗОВ А Ф. Электромагнитный контроль трещиноватости слоистого массива горных пород вокруг выработок [C]. Геофизические методы контроля напряжений в горных породах. Новосибирск: ИГД СО АН СССР, 1980

[6] ЗБОРЩИК М П, МОРОЗОВ А Ф. Механизм разрушения слоистых пород и взаимодействие их с крепью полевых подготовительных выработок [C]. II всесоюзная конференция "Проблемы механики подземных сооружений": Тезиз доклада. Тула: ТПИ, 1982.

[7] КУЗНЕЦОВ Ю С. Устойчивость горных выработок на удароопасных участках рудников "Октябрьский" и "Таймырский" Норильского ГМК [C]. Устойчивость и крепление горных выработок. Ленинград: ЛГИ, 1981.

[8] КУЗНЕЦОВ Ю С, ЕРОФЕЕВ Ю Н, ЕРОФЕЕВ Н П. Сферические зоны влияния с дифференциацией пород в приконтурном массиве вокруг горизонтальной выработки [C]. Совершинствоание технологии производства плавикового шпата и цветных металлов: Материалы конференции. Красноярск: Изд-воКПИ, 1989.

[9] ШМИГОЛЬ А В, КИРИЧЕНКО В Я, РЕВА В Н. Шахтные исследования характера разрушения слабых пород [J]. Шахтное стройтельство, 1987(5):11-12.

[10] КИРИЧЕНКО В Я, ШМИГОЛЬ А В, РЕВА В Н. О механизме пучения выработок, сооруженных в слабых породах [J]. Шахтное стройтельство, 1988(11):3-5.

[11] МЕЛЬНИКОВ О И, МУХИН Н А, ПЕТРЕНКО С И,и другие. Деформирование массива пород вокруг выработок, расположенных в зонах ПГД [C]. Проблемы горных давлений. СПб.: ВНИМИ, 1991.

[12] ЧЕРНЯК И Л. Повышение устойчивости подготовительных выработок [M]. Москва: Недра, 1993.

[13] ADAMS G R, JAGER A J. Peteroscopic observations of rock fracturing ahead of slope face in deep-level gold mines [J]. Journal of the South African Institute of Mining and Metallurgy, 1980, 80(6):204-209.

[14] 李世平. 权台煤矿煤巷锚杆试验观测报告 兼论煤巷锚杆特点与参数选择新观点[J]. 中国矿业学院学报, 1979(4):19-57.
LI Shi-ping. Experimental observation report on anchor test in roadway of Quantai coal mine and discussion on new viewpoint of anchor characteristics and parameter selection [J]. Journal of China Institute of Mining and Technology, 1979(4): 19-57. (in Chinese).

[15] 贺永年. 软岩巷道围岩松动带及其状态分析[J]. 煤炭学报, 1991, 16(2):63-69.
HE Yong-nian. Analysis of loose zone around the roadway in soft rock [J]. Journal of China Coal Society, 1991, 16(2):63-69. (in Chinese).

[16] 方祖烈. 软岩巷道维护原理与控制措施[C]//何满潮. 中国煤矿软岩巷道支护理论与实践. 徐州: 中国矿业大学出版社, 1996.

[17] 李术才. 用钻孔电视观测分区破裂化 [C]//中国科协学术部新观点新学说沙龙文集. 深部岩石工程围岩分区破裂化效应. 北京: 中国科学技术出版社, 2008.

[18] 李树忱. 分区破裂化现象的现场观测分析与数值模拟 [M]//中国科协学术部. 深部岩石工程围岩分区破裂化效应. 北京: 中国科学技术出版社, 2008.

[19] ШЕМЯКИН Е И, ФИСЕНКО Г Л, КУРЛЕНЯ М В,и другие. Эффект зональной дезинтеграции горной породы вокруг подземноых выработок [J]. Доклады АН СССР, 1986, 289(5):1088-1094.

[20] ШЕМЯКИН Е И, ФИСЕНКО Г Л, КУРЛЕНЯ М

В, и другие. Зональная дезинтеграция горных пород вокруг подземных выработок Ч. I [J]. ФТПРПИ, 1986(3):3-15.

[21] ШЕМЯКИН Е И, ФИСЕНКО Г Л, КУРЛЕНЯ М В. и другие. Зональная дезинтеграция горных пород вокруг подземных выработок, Ч. II [J], ФТПРПИ, 1986(4):3-12.

[22] ШЕМЯКИН Е И, ФИСЕНКО Г Л, КУРЛЕНЯ М В, и другие. Зональная дезинтеграция горных пород вокруг подземных выработок, Ч. III [J]. ФТПРПИ, 1987(1):3-8.

[23] ШЕМЯКИН Е И, КУРЛЕНЯ М В, ОПАРИН В Н, и другие. Зональная дезинтеграция горных пород вокруг подземных выработок, Ч. IV [J]. ФТПРПИ, 1989(4):3-9.

[24] КУРЛЕНЯ М В, ОПАРИН В Н, Бобров Г Ф, и другие. О расклинивающем эффекте зон опорного давления [J]. ФТПРПИ, 1995(4):3-11.

[25] ОДИНЦЕВ В Н. Запредельное деформирование и зональная дезинтеграция горных пород вблизи выработок на больших глубинах [C]. Горные давления и техноллогии подземной разработки руд на больших глубинах: доклад всесоюзного совещания "Интенсивные методы разработки руд на больших глубинах", Москва: Издательство ИПКОН, 1990.

[26] ОДИНЦЕВ В Н. О механизме зональной дезинтеграции массива горных пород вблизи глубоких выработок [J]. ФТПРПИ, 1994(4):10-19.

[27] ОДИНЦЕВ В Н. Отрывное разрушение массива горных пород [M]. М.: ИПКОН РАН, 1996.

[28] ЧАНЫШЕВА А И. К исследованию зональной дезинтеграции горных пород [C]. Напряженно-деформированное состояние массива горных пород. Новосибирск: ИГД, 1988.

[29] 安瓦尔·恰内晒夫. 深部围岩的塑性变形及围岩分区破裂化现象[M]//中国科协学术部. 深部岩石工程围岩分区破裂化效应. 北京: 中国科学技术出版社, 2008.

[30] ЧАНЫШЕВ А И. К проблеме разрушения деформируемых сред. Ч. I: основные уравнения [J]. ФТПРПИ, 2001(3):53-67.

[31] ЧАНЫШЕВ А И. К проблеме разрушения деформируемых сред. Ч. II: обсуждение результатов аналитических решений [J]. ФТПРПИ, 2001(4):57-66.

[32] РЕВА В Н, ТРОПП Э Н. Упруго-пластическая модель зональной дезинтеграции окркстности подземной вырвьотки [C]. Физика и механика разрушения горных пород применительны к прогнозу динамических явлений. С-Петербург: ВНИМИ, 1995.

[33] МЕТЛОВ Л С. Механическая модель зональной дезинтеграции. Физика и технология высоких давлений [J]. ФТПРПИ, 1995(1):57-63.

[34] АЛЕКСЕЕВ А Д, МОРОЗОВ А Ф, МЕТЛОВ Л С, и другие. Синегетические модели зональных дезинтеграционных явлений [C]. Материалы международной конференции Эффективная и безопасная подземная добыча угля на базе современных достижений геомеханики. С-Петербург: Изд-воСИбГГУ, 1996.

[35] METLOV L S, MOROZOV A F, ZBORSCHIK M P. Rock failure foundations of mechanism of zonal rock failure in the vicinity of mining work [J]. Journal of Mining Science, 2002,38(2):150-155.

[36] ГУЗЕВ М А, МЯСНИКОВ В П. Термомеханическая модель упругопластического материала с дефектами структуры [J]. Механика твердого тела, 1998(4):156-172.

[37] ГУЗЕВ М А, ПАРОШИН А А. Неевклидова модель зональной дезинтеграции горных пород вокруг подземной выработки [J]. Прикладная механика и техническая физика, 2001(42): 147-156.

[38] 戚承志, 钱七虎, 王明洋. 岩体的构造层次及其成因[J]. 岩石力学与工程学报, 2005,24(16):2838-2846.
QI Cheng-zhi, QIAN Qi-hu, WANG Ming-yang. Structural hierarchy of rock mass and the mechanism of its formation [J]. Chinese Journal of Rock Mechanics and Engineering, 2005,24(16): 2838-2846. (in Chinese).

[39] 戚承志, 钱七虎, 王明洋. 岩体的构造层次粘性及动力强度[J]. 岩石力学与工程学报, 2005(增刊1):4679-4687.
QI Cheng-zhi, QIAN Qi-hu, WANG Ming-yang. The structural hierarchy viscosity and dynamic strength of rock mass [J]. Chinese Journal of Rock Mechanics and Engineering, 2005, 24 (Sup 1):4679-4687. (in Chinese).

[40] 王明洋, 戚承志, 钱七虎. 深部岩体块系介质变形与运动特性研究[J]. 岩石力学与工程学报, 2005,24(16):2825-2830.
WANG Ming-yang, QI Cheng-zhi, QIAN Qi-hu. Study on deformation and motion characteristics of blocks in deep rock mass [J]. Chinese Journal of Rock Mechanics and Engineering, 2005,24(16):2825-2830. (in Chinese).

[41] 戚承志, 钱七虎, 王明洋. 岩体非协调变形的非欧模型[C]. 南京: 第五届全国工程结构安全防护学术会议, 2005.

[42] 王明洋,钱七虎,戚承志.深部岩体的构造和变形与破坏问题[J].岩石力学与工程学报,2006,25(3):448-455.
WANG Ming-yang, QIAN Qi-hu, QI Cheng-zhi. The structure, deformation and fracture of deep level rock mass [J]. Chinese Journal of Rock Mechanics and Engineering, 2006,25(3):448-455. (in Chinese).

[43] 戚承志,钱七虎,王明洋.深部巷道围岩变形破坏的时间过程及支护[C].南京:第一届全国工程安全与防护会, 2008.

[44] QI Cheng-zhi, WANG Ming-yang, QIAN Qi-hu. Evolution of the deformation and fracturing in rock masses near deep-level tunnels [J]. Journal of Mining Science, 2009,45(2): 112-119.

[45] 周小平,钱七虎.深埋巷道分区破裂化机制[J].岩石力学与工程学报, 2007,26(5):877-885.
ZHOU Xiao-ping, QIAN Qi-hu. Zonal fracturing mechanism in deep tunnel [J]. Chinese Journal of Rock Mechanics and Engineering, 2007,26(5):877-885. (in Chinese).

[46] 潘一山,李英杰,唐鑫.岩石分区破裂化研究[J].岩石力学与工程学报, 2007,26(增刊1):3335-3341.
PAN Yi-shan, LI Ying-jie, TANG Xin. Study on rock zonal disintegration [J]. Chinese Journal of Rock Mechanics and Engineering, 2007, 26 (Sup 1): 3335-3341. (in Chinese).

[47] 贺永年,蒋斌松,韩立军.巷道围岩的间隔性区域断裂何耗散结构[J].中国矿业大学学报, 2008,37(3):300-304.
HE Yong-nian, JIANG Bin-song, HAN Li-jun. Study on intermittent zonal fracturing of surrounding rock in deep roadway [J]. Journal of China University of Mining and Technology, 2008,37(3):300-304. (in Chinese).

[48] 顾金才,顾雷雨,陈安敏.深部开挖洞室围岩分层断裂破坏机理模型试验与分析[J].岩石力学与工程学报, 2008,27(5):433-438.
GU Jin-cai, GU Lei-yu, CHEN An-min. Model test study on mechanism of layered fracture within surrounding rock of tunnels in deep stratum [J]. Chinese Journal of Rock Mechanics and Engineering, 2008,27(5):433-4384. (in Chinese).

[49] 张强勇,陈旭光,林波,等.深部开巷道围岩区破裂三维地质力学模型试验研究[J].岩石力学与工程学报, 2009,28(9):1757-1766.
ZHANG Qiang-yong, CHEN Xu-guang, LIN Bo, et al. Study of 3D geomechanical model test of zonal disintegration of surrounding rock of deep tunnel [J]. Chinese Journal of Rock Mechanics and Engineering, 2009,28(9):1757-1766. (in Chinese).

[50] 周小平.锦屏二级水电站引水隧道的数值模拟[M]//中国科协学术部.深部岩石工程围岩分区破裂化效应,北京:中国科学技术出版社, 2008.

[51] 何满潮.深部的概念体系及工程评价指标[J].岩石力学与工程学报, 2005,24(16):2854-2858.
HE Man-chao. Study on rock mechanics in deep mining engineering [J]. Chinese Journal of Rock Mechanics and Engineering, 2005, 24(16): 2854-2858. (in Chinese).

[52] 王明洋,宋华,郑大亮,等.深部巷道围岩的分区破裂机制及深部界定探索[J].岩石力学与工程学报, 2006,25(9):1771-1776.
WANG Ming-yang, SONG Hua, ZHENG Da-liang,et al. On mechanism of zonal disintegration within rock mass around deep tunnel and definition of deep rock engineering [J]. Chinese Journal of Rock Mechanics and Engineering, 2006,25(9):1771-1776. (in Chinese).

[53] МАКАРОВ В В, ЕМЕЛЬЯНОВ Б И, ЗВОНАРЕВ М И. Особенности формирования зон компрессии и дилатансии горных пород вокруг авработок [C]. Вопросы разработки месторождений Дальнего Востока. Владивосток: ДВПИ, 1990.

[54] МАКАРОВ В В. Зональное деформирование горных пород вокруг выработок и в образцах [C]. Тез. Докл. X международ. Конф. По мех. Горных пород. Москва: 1993.

[55] МАКАРОВ В В. О зональном деформировании горных пород вокруг одиночных капитальных выработок [C]// Механика подземных сооружений. Тула: ТПИ, 1995.

[56] ГУЗЕВ М А, МАКАРОВ Н Н. Деформирование и разрушение силино сжатых горных пород вокруг выработок [M]. Владивоток: Дальнаука, 2007.

[57] САДОВСКИЙ М А. Кусковатость горной породы [J]. Доклады АН СССР, 1979,247(4):829-831.

[58] ОПАРИН В Н,ЮШКИН В Ф, АКИНИН А А, и другие. О новой шкале структурно-иерахических представлений как паспортной характеристике объектов геосреды [J]. ФТПРПИ, 1998(5):17-33.

[59] КУРЛЕНЯ М В, ОПАРИН В Н, ЕРЕМЕНКО А А. Об отношении линейных размеров блоков пород к раскрытию трещин в структурной иерархии массивов [J].ФТПРПИ, 1998(2):6-33.

[60] PANIN V E. Physical mesomechanics of heterogeneous media and computer-aided design of materials [M]. Cambridge: Cambridge Intersci Pub, 1998.

[61] РЕВУЖЕНКО А Ф. Механика упругопластических сред и нестандартный анализ [M]. Новосибирск: Издательство Новосибирского Университета, 2000.

基于线法的功能梯度材料断裂分析

燕秀发[1,2]， 钱七虎[1]， 方国强[3]， 赵敏福[4]， 郭延宝[5]

(1. 解放军理工大学 工程兵工程学院，江苏 南京 210007；2. 解放军 91550 部队，辽宁 大连 116023；3. 海军装备部，北京 100841；4. 皖西学院，安徽 六安 237300；5. 装甲兵工程学院，北京 100072)

摘 要：为了克服材料非均匀性引起的数值困难，一种半解析数值方法——线法，被引入功能梯度材料的断裂分析。通过有限差分将问题的控制方程半离散为定义在沿梯度方向离散节线上的常微分方程组，然后应用 B 样条高斯配点法求解该常微分方程组。为了演示线法功能梯度材料断裂分析的方法，给出了指数梯度含裂纹功能梯度材料板分别在恒定位移、弯曲载荷作用下应力强度因子的数值算例，同时给出了节线和配点局部加密、区间映射以及变间距离散等提高计算精度和效率的线法断裂分析技巧。与相关问题理论解的对比分析表明，该法的计算结果具有很高的精度。

关键词：功能梯度材料；线法；非均匀性；断裂分析；半解析法

中图分类号：O346.2　　**文献标识码**：A　　**文章编号**：1009-3443(2010)04-0346-08

Method of lines for fracture analysis of functionally graded materials

YAN Xiu-fa[1,2], QIAN Qi-hu[1], FANG Guo-qiang[3], ZHAO Min-fu[4], GUO Yan-bao[5]

(1. Engineering Institute of Corps of Engineers, PLA Univ. of Sci. & Tech., Nanjing 210007, China; 2. Unit No.91550 of PLA, Dalian 116023, China; 3. Navy Equipment Department, Beijing 100841, China; 4. West Anhui University, Liuan 237300, China; 5. PLA Armoring Force Engineering College, Beijing 100072, China)

Abstract: To overcome the numerical difficulties caused by material nonhomogeneity, a semi-analytical numerical method, method of lines (MOLs), was introduced into the fracture analysis of functionally graded materials (FGMs). The basic idea of that method was to semi-discretize the governing equations into a system of ordinary differential equations (ODEs) by means of finite difference approach. The ODEs were defined on the discrete lines along the gradation direction. By using B spline collocation method at Gaussian points, solutions to the problem were obtained. The computational examples of the stress intensity factors for the cracked plate of FGMs with an exponential gradation under constant displacement and bending loading were presented to demonstrate the fracture analysis method of FGMs. In such examples, some numerical techniques, including local refinement of discrete lines and collocation points, interval map and varying space discretization, were employed to improve the computational accuracy and the efficiency. Numerical results of the examples by MOLs are quite accurate compared with the theoretical solutions to the corresponding problems.

Key words: functionally graded materials; method of lines; nonhomogeneity; facture analysis; semi-analytical method

　　功能梯度材料是新一代非均匀材料，具有随空间坐标变化的材料成分和特性，其独特之处在于可以通过定制材料成分和梯度来满足最终的需要,因此在航空航天（热障涂层）、军事（梯度装甲）、光学

收稿日期：2009-01-08。
基金项目：国家自然科学创新研究群体科学基金资助项目(51021001)；中国博士后科学基金资助项目(20080431344)。
本文原载于《解放军理工大学学报(自然科学版)》(2011 年第 12 卷第 4 期)。

(光纤)、核能(核反应堆第一壁)等领域都得到了广泛的应用[1]。由于制备工艺的局限性和恶劣的工作环境,断裂破坏是功能梯度材料的主要失效模式[2]。因为功能梯度材料具有空间变化的材料属性参数,使得描述该材料断裂问题的控制方程都成为变系数偏微分方程,获得问题的解析解非常困难,所以数值方法对于功能梯度材料的断裂分析具有特别重要的意义。

目前,在功能梯度材料断裂分析中得到广泛应用的是有限单元法和无网格法。根据实现材料物性函数模拟的不同途径,功能梯度材料断裂分析的有限单元法可分为 2 种类型,即均匀元法和等参梯度元法(多重等参元法)。应用均匀元法,张幸红等[3]研究了功能梯度材料裂纹应力强度因子与材料非均匀性参数的关系。G·Anals 等[4]分析了裂纹尖端应力场的适用范围。Li Chun-yu 等[5]应用等参梯度元法计算了该类材料环形裂纹的应力强度因子。J·H·Kim 等[6]进一步将该法推广应用于正交各向异性功能梯度材料结构的断裂分析。文献[7,8]应用无网格法计算了功能梯度材料板边缘裂纹的应力强度因子,与解析解的对比表明该计算结果具有较高的精度。

然而,在随后的研究中发现,上述方法存在一些问题。均匀元法的基本原理是在单元内部采用零次材料属性插值,通过不同单元间材料属性参数的变化以分段近似来模拟功能梯度材料的材料梯度,其根本缺陷是不能反映单元内部材料属性的变化,计算精度依赖于网格细化程度[9]。等参梯度元法扩展了均匀材料有限单元法中等参变换的概念,应用材料属性参数与位移场相同的插值函数进行模拟,因此能够反映单元内部材料属性的变化,从而改善计算精度。但是由于材料属性参数的连续变化可能导致单元特性恶化,等参梯度元法的计算结果具有单元类型敏感性和不确定性,表现为同一问题应用不同的单元类型可能得到差异很大的计算结果[10]。无网格法的问题在于,节点的影响域(支撑域)是重叠的,与有限单元法中求解域被离散成有限个相互连接的单元相比,无网格法的计算量远大于有限元法的计算量,同时对材料属性参数的模拟进一步提高了计算强度。特别是对于涂层、裂纹等属性参数或待求场量具有高梯度性质的问题,计算强度的增大更加明显,不能适应分析各种梯度形式对结构物理特性影响的计算需要。

上述方法在功能梯度材料断裂分析中出现困难的主要原因是,这些方法都是基于结构完全离散的数值方法,计算误差不仅来源于求解区域的离散,也来源于沿梯度方向材料属性参数的离散,连续变化的材料属性参数导致了刚度矩阵数值积分的复杂化以及可能的单元特性恶化。因此,李永等[11]提出了求解功能梯度材料板弯曲问题的半解析法——康托洛维奇宏细观精化法(新康法)。与实验结果的对比表明,该法具有很高的计算精度,且便于分析各种参数对结构物理特性的影响,提供了发展功能梯度材料结构分析半解析数值方法的新思路。然而新康法针对特定问题选取特别的待定函数形式,还不适合推广应用于一般性的功能梯度材料结构分析问题。本文尝试将一种典型的具有普遍意义的半解析数值方法——线法(method of lines)引入功能梯度材料的断裂分析,通过具体的算例演示该法的数值过程和计算技巧,并通过与解析解的对比分析,最终表明线法在功能梯度材料断裂分析中的有效性和适用性。

1 问题描述

应力强度因子是断裂力学的重要参数,该参数的计算是功能梯度材料断裂分析的重要内容。能否克服裂纹尖端应力奇异性及其附近应力和应变的高梯度,方便有效地得到应力强度因子的精确结果,是检验断裂分析方法性能的有效手段。因此,本文要研究的内容是沿 y 向梯度含裂纹功能梯度材料板应力强度因子的求解问题,如图 1 所示。为使问题具有代表性,分别考虑位移(图 1(a))和应力(图 1(b))2 种载荷情况。

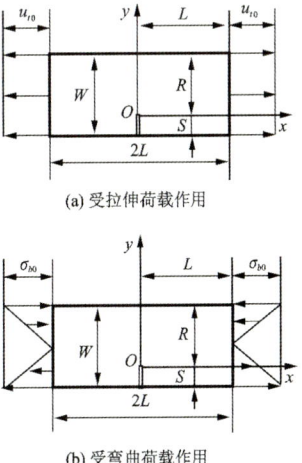

(a) 受拉伸荷载作用

(b) 受弯曲荷载作用

图 1 含裂纹功能梯度材料板

Fig.1 Functionally graded material plate with a crack

(1) 均匀拉伸
$$\bar{u}_t = u_{t0},$$
$\bar{u}_t$ 为作用在板两侧边界处的拉伸位移。

(2) 纯弯曲
$$\bar{\sigma}_b = \sigma_{b0}[1 - \frac{2}{W}(y + S)]_。$$
式中：$\bar{\sigma}_b$ 为作用在板两侧边界处的弯曲应力，σ_{b0} 为弯曲应力的幅值，W 为板的宽度，S 为裂纹长度，R 为裂纹尖至板上端面的距离。

为与解析解比较，设材料弹性属性形式为
$$E = E(y), \quad (1)$$
$$\mu = C_。 \quad (2)$$
式中：E 为弹性模量，μ 为泊松比，C 为常数。考虑问题的对称性，则相应载荷类型下待求问题的等效计算构形如图 2 所示。

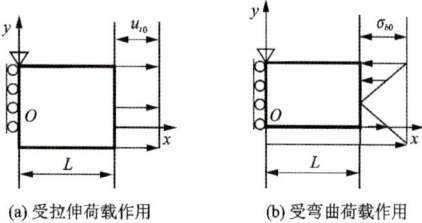

图 2 含裂纹功能梯度材料板的等效构形

Fig.2 Equivalent model of functionally graded material plate with a crack

2 数值实现

2.1 线法功能梯度材料断裂分析的控制方程

类似其他结构分析方法，线法也以位移作为基本求解量，因此必须导出功能梯度材料用位移表达的平衡方程和边界条件。

2.1.1 平衡方程

根据功能梯度材料制备完成后具有连续变化的微结构和材料属性的特点，可将其处理为物性参数是空间坐标连续函数的非均匀材料，而且功能梯度材料的属性一般从一点到另一点逐渐变化，可以假定给定一点的材料性质在任意方向上都是相同的。因此，在连续介质意义上功能梯度材料被认为是各向同性的非均匀实体。

综上所述，考虑功能梯度材料为物性参数是空间坐标连续函数的各向同性非均匀材料，记 $\boldsymbol{x} = x_i = (x\ y\ z), \boldsymbol{u} = u_i = (u\ v\ w), i = 1, 2, 3$。

为不失一般性，设材料参数为 $E = E(\boldsymbol{x}), \mu = \mu(\boldsymbol{x}), G = G(\boldsymbol{x}) = \frac{E}{2(1+\mu)}, H = H(\boldsymbol{x}) = \frac{\mu}{(1-2\mu)}$，则由物理方程、几何方程和应力平衡微分方程可得功能梯度材料位移表示的平衡方程
$$G(u_{i,jj} + u_{j,ij}) + 2GHu_{k,ki} + (\frac{\partial G}{\partial x_j}u_{i,j} + \frac{\partial G}{\partial x_j}u_{j,i}) + (2\frac{\partial G}{\partial x_i}H + 2G\frac{\partial H}{\partial x_i})u_{k,k} = 0, \quad (3)$$
式中，$i, j = 1, 2, 3$。

2.1.2 边界条件

由物理方程、几何方程可得在应力边界 S_σ 上，
$$T_i = [G(u_{i,j} + u_{j,i}) + 2GHu_{k,k}\delta_{ij}]n_j_。 \quad (4)$$
式中：T_i 为作用在 S_σ 上的表面力；$n_j = \cos(n, x_j)$，n 为边界面的外法线方向；δ_{ij} 为 Kronecker 符号。

在位移边界 S_u 上有
$$u_i = \bar{u}_i, \quad (5)$$
式中，$\bar{u}_i$ 为作用在 S_u 上的位移。

由式(3)～(5)，对比均匀材料位移形式的平衡方程和边界条件可以看出，除位移边界条件外，功能材料的控制方程均为包含材料属性函数的变系数偏微分方程。

2.2 求解区域的离散和定义在节线上的控制方程

线法功能梯度材料结构分析的求解目标是定义在沿梯度方向离散节线上的常微分边值问题，首先必须将求解区域用若干沿梯度方向的节线离散。考虑求解问题的数值特点，如图 3 所示，原问题的等效计算构形为疏密相间的 $1\sim m$ 条节线所离散。图中 i 为节线编号，并记各疏密区域分界节线编号为 n_1、n_2、n_5、n_6，求解区域边界节线编号为 1、n_3、n_4、n_7。h_i 为节线间距，前后相邻节线的间距分别为 h_{i+1} 和 h_{i-1}，并记 H_1 和 H_2 分别为等距离散区域(1)(4)和区域(3)(6)的节线间距。图中边界节线外部编号为 0、n_3^0、n_4^0、n_7^0 的节线是引入边界条件所需的虚拟线。

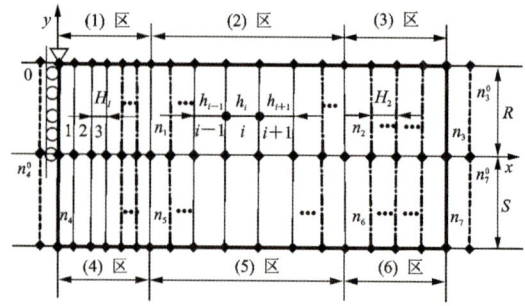

图 3 等效计算构形中求解区域的离散

Fig.3 Domain discretization of the problem in the equivalent model

断裂问题是典型的高梯度问题,裂纹尖端附近应力、应变具有剧烈变化的特点,同时该处还存在材料属性参数的变化。为兼顾计算精度和效率,求解区域采取局部节线细化,节线细化区与粗糙区变间距过渡的离散方法,具体说明如下:

(1) 区别于裂纹边界和非裂纹边界,以裂纹尖端为界,应用上下两排离散线;

(2) 在裂纹尖端附近,如图3中的(1)(4)区,设置高度细化的等间距节线,以克服该处待求场量的高梯度困难;

(3) 设置变间距(过渡)节线区,如图3中的(2)(5)区,但注意避免相邻节线间距比值过大;

(4) 在边界附近区域,如图3中的(3)(6)区,应用相对粗糙的等间距节线,同样注意避免该区域离散节线间距相对于过渡区邻近节线离散间距的较大波动。

2.2.1 定义在离散节线上的控制方程

由式(1)~(3)可得图1、2中待求问题平面应力情况下的控制方程,则相应的定义在节线i上的平衡方程为:

$$(\frac{\partial^2 u}{\partial x^2})_i + a(\frac{\partial^2 u}{\partial y^2})_i + b(\frac{\partial^2 v}{\partial x \partial y})_i + d_1[(\frac{\partial u}{\partial y})_i + (\frac{\partial v}{\partial x})_i] = 0, \quad (6)$$

$$(\frac{\partial^2 v}{\partial y^2})_i + a(\frac{\partial^2 v}{\partial x^2})_i + b(\frac{\partial^2 u}{\partial x \partial y})_i + d_2(\frac{\partial v}{\partial y})_i + d_3(\frac{\partial u}{\partial x})_i = 0。 \quad (7)$$

式中:$a = \frac{1-\mu}{2}$,$b = \frac{1+\mu}{2}$,$d_1 = \frac{aE'}{E}$,$d_2 = \frac{E'}{E}$,$d_3 = \frac{\mu E'}{E}$。其中,$E' = \frac{dE}{dy}$。

2.2.2 离散节线端点处的边界条件

记x_i为节线i的x轴坐标,下标i为节线编号。由式(4)(5)有:

(1) 在节线$i = 1$的端点$(0, R)$处由约束条件(避免结构发生平动)有:

$$u|_{(0,R)} = 0,$$
$$v|_{(0,R)} = 0。$$

(2) 在节线$i \in [2, n_3]$的端点(x_i, R)处有:

$$(\frac{\partial v}{\partial x} + \mu \frac{\partial u}{\partial y})|_{(x_i, R)} = 0, \quad (8)$$

$$(\frac{\partial u}{\partial y} + \frac{\partial v}{\partial x})|_{(x_i, R)} = 0。 \quad (9)$$

(3) 在节线$i \in [1, m]$的端点$(x_i, 0)$处由位移和应力连续性条件有:

$$u|_{(x_i, 0)} = u|_{(x_j, 0)}, \quad v|_{(x_i, 0)} = v|_{(x_j, 0)},$$

$$(\frac{\partial u}{\partial y} + \frac{\partial v}{\partial x})|_{(x_i, 0)} = (\frac{\partial u}{\partial y} + \frac{\partial v}{\partial x})|_{(x_j, 0)}, \quad (10)$$

$$(\frac{\partial v}{\partial y} + \mu \frac{\partial u}{\partial x})|_{(x_i, 0)} = (\frac{\partial v}{\partial y} + \mu \frac{\partial u}{\partial x})|_{(x_j, 0)}, \quad (11)$$

式中,$j = n_3 + i$。

(4) 在节线$i \in [n_4, m]$的端点$(x_i, -S)$处有

$$(\frac{\partial v}{\partial y} + \mu \frac{\partial u}{\partial x})|_{(x_i, -S)} = 0, \quad (12)$$

$$(\frac{\partial u}{\partial y} + \frac{\partial v}{\partial x})|_{(x_i, -S)} = 0。 \quad (13)$$

2.3 常微分边值问题的导出

2.3.1 区间映射

由图3中求解区域的离散方法和§2.2.2中离散节线的端点边界条件可见,沿梯度y方向,上下两组离散节线分别定义在不同的区间,即线$1 \sim n_3$沿y方向的定义区间为$[0, R]$,线$n_4 \sim m$沿该方向的定义区间为$[-S, 0]$。这种不规则的节线定义区间,导致问题成为多点边值问题,非常不利于程序的实现和求解效率的提高。因此需要应用区间映射技术,将定义在不同区间的节线控制方程变换到同一区间,使多点边值问题转变为两点边值问题。采用如下的坐标变换:

$$\left.\begin{array}{l} \xi = \frac{y}{R}, \\ \frac{df(\xi)}{dy} = \frac{1}{R}\frac{df(\xi)}{d\xi}, 0 \leq y \leq R。 \end{array}\right\} \quad (14)$$

$$\left.\begin{array}{l} \xi = -\frac{y}{S}, \\ \frac{df(\xi)}{dy} = -\frac{1}{S}\frac{df(\xi)}{d\xi}, -S \leq y \leq 0。 \end{array}\right\} \quad (15)$$

则上、下两排节线沿梯度方向均被定义于$[0, 1]$区间,各离散节线端点坐标也成为$(x_i, 0)$和$(x_i, 1)$。

2.3.2 节线位移函数对离散坐标偏导数的有限差分近似

为将定义在节线上的控制方程转变为常微分方程,应用由Taylor公式得到的具有二阶精度的非等距三点中心差分近似该方程中待求场量对离散坐标x的偏导数,记节线位移函数$u_i = u_i(y) = u(x_i, y)$,$v_i = v_i(y) = v(x_i, y)$及$\frac{du_i}{dy} = u_i'$,$\frac{dv_i}{dy} = v_i'$,$\frac{d^2 u_i}{dy^2} = u_i''$,$\frac{d^2 v_i}{dy^2} = v_i''$,则对于图3中节线间距为$h_{i+1}$、$h_i$、$h_{i-1}$的区域(2)(5)有:

$$\left.\begin{aligned}
\left(\frac{\partial^2 u}{\partial x^2}\right)_i &= \frac{2[h_i u_{i-1} - (h_i + h_{i-1})u_i + h_{i-1}u_{i+1}]}{h_{i-1}h_i(h_{i-1}+h_i)}, \\
\left(\frac{\partial^2 v}{\partial x^2}\right)_i &= \frac{2[h_i v_{i-1} - (h_i + h_{i-1})v_i + h_{i-1}v_{i+1}]}{h_{i-1}h_i(h_{i-1}+h_i)}, \\
\left(\frac{\partial u}{\partial x}\right)_i &= \frac{h_{i-1}^2 u_{i+1} + (h_i^2 - h_{i-1}^2)u_i - h_i^2 u_{i-1}}{h_i h_{i-1}(h_{i-1}+h_i)}, \\
\left(\frac{\partial v}{\partial x}\right)_i &= \frac{h_{i-1}^2 v_{i+1} + (h_i^2 - h_{i-1}^2)v_i - h_i^2 v_{i-1}}{h_i h_{i-1}(h_{i-1}+h_i)}, \\
\left(\frac{\partial^2 u}{\partial x\partial y}\right)_i &= \frac{h_{i-1}^2 u'_{i+1} + (h_i^2 - h_{i-1}^2)u'_i - h_i^2 u'_{i-1}}{h_i h_{i-1}(h_{i-1}+h_i)}, \\
\left(\frac{\partial^2 v}{\partial x\partial y}\right)_i &= \frac{h_{i-1}^2 v'_{i+1} + (h_i^2 - h_{i-1}^2)v'_i - h_i^2 v'_{i-1}}{h_i h_{i-1}(h_{i-1}+h_i)}.
\end{aligned}\right\} \quad (16)$$

如果式(16)中 $h_{i+1} = h_i = h_{i-1} = H_1$ 或 H_2，则得到适用于图3中的(1)(3)(4)(6)等区域的等距差分公式。当 $i \in (1, n_3)$ 和 $i \in (n_4, n_7)$ 时，将式(16)代入式(6)(7)，则定义在节线上的控制方程由偏微分方程组转变为以节线位移函数为未知量的常微分方程组。显然，对于边界节线即图3中的线1、n_3 和线 n_4、n_7，为了应用式(16)，针对本文问题需要得到虚拟线上的 v_0、v'_0、$v_{n_3}^0$、$v'^0_{n_3}$、$u_{n_4}^0$、$u'^0_{n_4}$、$v^0_{n_4}$、$v'^0_{n_4}$、$u^0_{n_7}$、$u'^0_{n_7}$、$v^0_{n_7}$、$v'^0_{n_7}$ 等辅助量，这些辅助量通过引入该线处的边界条件获得。

2.3.3 边界线上边界条件的引入和虚拟线上的辅助量

(1) 拉伸和弯曲2种载荷情况下虚拟线 $i = 0$ 上的辅助量 v_0、v'_0。

根据问题的对称性，在边界线($i = 1$)处有边界条件：
$$\left.\begin{aligned} u_1 &= 0, \\ G\left[\left(\frac{\partial u}{\partial y}\right)_1 + \left(\frac{\partial v}{\partial x}\right)_1\right] &= 0. \end{aligned}\right\} \quad (17)$$

由式(16)的第4式有 $\left(\frac{\partial v}{\partial x}\right)_1 = \frac{v_2 - v_0}{2H_1}$。

代入式(17)得：
$$\left.\begin{aligned} v_0 &= v_2, \\ v'_0 &= v'_2. \end{aligned}\right\} \quad (18)$$

(2) 拉伸和弯曲2种载荷情况下虚拟线 $i = n_4^0$ 上的辅助量 $u^0_{n_4}$、$u'^0_{n_4}$、$v^0_{n_4}$、$v'^0_{n_4}$。

在 $i = n_4$ 处有边界条件：
$$\left.\begin{aligned} \frac{E}{1-\mu^2}\left[\left(\frac{\partial u}{\partial x}\right)_{n_4} + \mu\left(\frac{\partial v}{\partial y}\right)_{n_4}\right] &= 0, \\ G\left[\left(\frac{\partial u}{\partial y}\right)_{n_4} + \left(\frac{\partial v}{\partial x}\right)_{n_4}\right] &= 0. \end{aligned}\right\} \quad (19)$$

由式(16)第3、4式有：
$$\left(\frac{\partial u}{\partial x}\right)_{n_4} = \frac{u_{n_4+1} - u^0_{n_4}}{2H_1},$$

$$\left(\frac{\partial v}{\partial x}\right)_{n_4} = \frac{v_{n_4+1} - v^0_{n_4}}{2H_1}. \quad (20)$$

将式(20)代入式(19)，并注意式(1)(2)，可得：
$$\left.\begin{aligned}
u^0_{n_4} &= u_{n_4+1} + 2H_1 \mu v'_{n_4}, \\
v^0_{n_4} &= v_{n_4+1} + 2H_1 u'_{n_4}, \\
u'^0_{n_4} &= u'_{n_4+1} + 2H_1 \mu v''_{n_4}, \\
v'^0_{n_4} &= v'_{n_4+1} + 2H_1 u''_{n_4}.
\end{aligned}\right\} \quad (21)$$

类似 v_0、v'_0 的求解过程可得到均匀拉伸载荷下虚拟线 $i = n_3^0$ 和 $i = n_7^0$ 上的辅助量：
$$\left.\begin{aligned}
v^0_{n_3} &= v_{n_3-1}, \\
v'^0_{n_3} &= v'_{n_3-1}, \\
v^0_{n_7} &= v_{n_7-1}, \\
v'^0_{n_7} &= v'_{n_7-1}.
\end{aligned}\right\} \quad (22)$$

类似 $u^0_{n_4}$、$u'^0_{n_4}$、$v^0_{n_4}$、$v'^0_{n_4}$ 的求解过程可得到弯曲载荷下虚拟线 $i = n_3^0$ 和 $i = n_7^0$ 上的辅助量：
$$\left.\begin{aligned}
u^0_{n_3} &= u_{n_3-1} - 2H_2 \mu v'_{n_3} + \frac{2H_2 \sigma_{b0}}{\bar{E}}\left[1 - \frac{2}{W}(y+S)\right], \\
v^0_{n_3} &= v_{n_3-1} - 2H_2 u'_{n_3}, \\
u'^0_{n_3} &= u'_{n_3-1} - 2H_2 \mu v''_{n_3} - \frac{2H_2 \sigma_{b0}}{\bar{E}^2}\bar{E}' - \frac{4H_2 \sigma_{b0}}{\bar{E}W} + \\
&\quad \frac{4H_2 \sigma_{b0}}{\bar{E}^2 W}\bar{E}'(y+S), \\
v'^0_{n_3} &= v'_{n_3-1} - 2H_2 u''_{n_3}.
\end{aligned}\right\} \quad (23)$$

$$\left.\begin{aligned}
u^0_{n_7} &= u_{n_7-1} - 2H_2 \mu v'_{n_7} + \frac{2H_2 \sigma_{b0}}{\bar{E}}\left[1 - \frac{2}{W}(y+S)\right], \\
v^0_{n_7} &= v_{n_7-1} - 2H_2 u'_{n_7}, \\
u'^0_{n_7} &= u'_{n_7-1} - 2H_2 \mu v''_{n_7} - \frac{2H_2 \sigma_{b0}}{\bar{E}^2}\bar{E}' - \frac{4H_2 \sigma_{b0}}{\bar{E}W} + \\
&\quad \frac{4H_2 \sigma_{b0}}{\bar{E}^2 W}\bar{E}'(y+S), \\
v'^0_{n_7} &= v'_{n_7-1} - 2H_2 u''_{n_7}.
\end{aligned}\right\} \quad (24)$$

式中，$\bar{E} = E/(1-\mu^2)$，$\bar{E}' = \frac{d}{dy}(\bar{E}) = \frac{1}{(1-\mu^2)}\frac{dE}{dy}$。

2.3.4 定义在节线上的常微分方程

应用式(16)近似式(6)(7)中位移函数对离散坐标 x 的偏导数，对边界线注意式(18)(21)(22)(拉伸载荷)式(23)(24)(弯曲载荷)，然后利用式(14)进行区间变换，则可得定义在节线 $i \in [1, n_3]$ 上的原问题控制方程的常微分方程近似式：

$$\mathbf{U}'' = R\bar{\mathbf{A}}\mathbf{U}' + R^2\bar{\mathbf{B}}\mathbf{U} + R^2\bar{\mathbf{F}}. \quad (25)$$

式中，$\bar{\mathbf{A}}$、$\bar{\mathbf{B}}$ 和 $\bar{\mathbf{F}}$ 为系数矩阵；$\mathbf{U}' = \frac{d\mathbf{U}}{d\xi}$，对于均匀拉伸载荷待求位移函数 $\mathbf{U} = (v_1 \ u_2 \ v_2 \ \cdots \ u_i \ v_i \ \cdots \ u_{n_3-1}$

$v_{n_3-1}\ v_{n_3})^T$,对于弯曲载荷待求位移函数 $U = (v_1\ u_2\ v_2\ \cdots\ u_i\ v_i\ \cdots\ u_{n_3-1}\ v_{n_3-1}\ v_{n_3})^T$。

利用式(15),经过与式(25)类似的过程可得定义在节线 $i \in [n_4, m]$ 上的原问题控制方程的常微分方程近似式:

$$U'' = -SAU' + S^2 BU + S^2 E,\quad (26)$$

式中:A、B 和 E 为系数矩阵;$U' = \dfrac{dU}{d\xi}$,对于均匀拉伸载荷待求位移函数,$U = (u_{n_4}\ v_{n_4}\ u_{n_4+1}\ v_{n_4+1}\ \cdots\ u_j\ v_j\ \cdots\ u_{n_7-1}\ v_{n_7-1}\ v_{n_7})^T$,对于弯曲载荷待求位移函数 $U = (u_{n_4}\ v_{n_4}\ u_{n_4+1}\ v_{n_4+1}\ \cdots\ u_j\ v_j\ \cdots\ u_{n_7}\ v_{n_7})^T$。

2.3.5 端点边界条件的常微分形式

将式(14)代入式(8)~(13),并注意对于 $i \in [2, n_3-1]$ 和 $i \in [n_4+1, m-1]$ 等内部节线,应用式(16)中的第 3、4 近似式(8)~(13)中的位移函数对离散坐标 x 的偏导数。但对于边界节线,为获得与内部节线相同的近似精度,需应用由 Newton 等距插值公式得到的式(27)来近似各边界节线端点边界条件方程中对位移坐标 x 的偏导数[12]:

$$\left.\begin{array}{l}\left(\dfrac{\partial f}{\partial x}\right)_l = \dfrac{-f_{l+2} + 4f_{l+1} - 3f_l}{2h_l} \\ \left(\dfrac{\partial f}{\partial x}\right)_m = \dfrac{3f_m - 4f_{m-1} + f_{m-2}}{2h_m} \end{array}\right\} \quad (27)$$

式中:f 为待求位移函数 u 或 v;l、m 分别为求解域两侧边界节线的编号,$l = 1, n_4, m = n_3, m$;$h_l = H_1, h_m = H_2$。于是可得 $i \in [1, n_3]$ 各节线端点边界条件的常微分方程近似式

$$C|_{\Gamma_k}U' + RD|_{\Gamma_k}U + RG|_{\Gamma_k} = 0. \quad (28)$$

式中:$k = 0, 1$, $\Gamma_0 = (x, 0)$, $\Gamma_1 = (x, 1)$。

利用式(15),经过与式(28)类似的过程可得 $i \in [n_4, m]$ 各节线端点边界条件的常微分方程近似式

$$C|_{\Gamma_k}U' - SD|_{\Gamma_k}U - SG|_{\Gamma_k} = 0. \quad (29)$$

经过上述区间映射和有限差分过程,原求解域内的偏微分边值问题转变为定义在各离散节线上的由式(25)(26)(28)(29)构成的常微分两点边值问题。

3 常微分边值问题的求解和应力强度因子的计算

3.1 常微分边值问题的求解

将原求解域的偏微分边值问题转化为定义在离散节线上的常微分边值问题后,常微分边值问题的求解成为关键。在求解常微分边值问题的方法中,打靶法和样条函数配点法是较为优越的算法。然而,Jones 等的研究表明,应用打靶法求解线法常微分方程组时存在计算结果不稳定的可能[13]。基于这一考虑,本文选择样条函数配点法中的 B 样条 Gauss 配点法求解线法常微分边值问题。

B 样条 Gauss 配点法求解常微分边值问题的基本原理是:以 B 样条基样条插值函数作为问题的近似解,通过使其在各配点区间的 Gauss 映射点和边界点处满足常微分方程和边界条件确定基函数的系数,由此得到近似解。求解节线常微分方程组得到离散节线位移近似函数后,对位移近似函数求导则可得沿节线方向的应变近似函数,插值可得节线上任意点在该方向上的应变值,由相邻节线对应点位移数值的有限差分,则可得离散方向上的应变数值。将所得应变值代入物理方程,于是进一步得到该点处的应力数值。

3.2 计算应力强度因子的方法

裂纹尖端附近的应力场和位移场取决于应力强度因子 K,对功能梯度材料裂纹尖端场的研究表明,该类材料裂纹尖端处应力场与均匀材料具有相同的函数形式,但应力强度因子由裂纹尖端处的材料参数决定,图 1、2 中所示裂纹应力强度因子的定义[14]为:

$$K_I = \lim_{r \to 0} \dfrac{E_0 u}{4}\sqrt{\dfrac{2\pi}{r}}.$$

式中:u 为裂纹面的张开位移,r 为该张开位移点到裂纹尖端的距离,E_0 为裂纹尖端处的弹性模量。将 E_0 用 $E_0/(1-\mu^2)$ 代替可得平面应变情况下应力强度因子的定义。

基于位移的外推法是断裂问题中计算应力强度因子的基本方法。但是,尽管应用线法可以得到节线上任一点的位移值,但在裂纹尖端 $r = 0$ 处,有限差分不能真正反映该处位移对坐标导数趋于无穷的性质。因此,在靠近裂纹尖端处线法计算结果的误差较大,不能直接应用上述极限过程。基于上述考虑,通过下面的方法计算应力强度因子,定义 K_I^* 为名义应力强度因子,则

$$K_I^* = \dfrac{E_0 u}{4}\sqrt{\dfrac{2\pi}{r}}.$$

以 K_I^* 为纵坐标,r 为横坐标,将有关点绘出,用最小二乘法处理,通过这些点绘出一条最佳拟合直线后,该直线至纵坐标轴的交点即为 k_I 的估算值。

3.3 进一步提高计算精度的方法

令图 1、2 中 $W = 1$, $L = 3W$, $R = 0.8$, $S = $

$0.2, u_{s0}=1, \sigma_{s0}=1$,并记 $E_0=E(0)$,$E_1=E(-S)$,$E_2=E(R)$。为了与已知的解析解[15]对比,考虑式(1)(2)中的材料梯度模型为

$$E=E_0 e^{\beta y},$$
$$\mu=0.3,$$

分别计算各载荷在 $E_2/E_1=0.5$、2 两种梯度水平下的应力强度因子。β 和 E_0 可根据材料内部弹性模量空间变化情况的数据拟合得到。

考虑与已知解析解的前提一致,计算问题的平面应变情况,只需令 $E=E/(1-\mu^2)$ 和 $\mu=\mu/(1-\mu)$,即可方便地将平面应力问题转换为平面应变问题。计算过程中取协调一致的封闭单位体系,各计算参数的量纲为相应的单位力、单位长度、单位弹性模量。

考虑到裂纹尖端附近存在着高梯度的应力和应变并兼有材料属性梯度,为了获得良好的计算精度,不仅需要裂纹尖端附近离散节线高度细化,还可以根据 B 样条 Gauss 配点法求解常微分方程的数值原理,采取进一步提高计算精度的措施。对于二阶常微分方程,B 样条 Gauss 配点法的精度可达到 $O(\Delta_{max}^{2+k})$,其中,Δ_{max} 为配点区间的最大长度,k 为各配点区间的 Gauss 映射点数。因此,可以通过在求解区间内设置较多的配点区间以减小 Δ_{max}(类似于有限单元法中的 H 算法)和较多的 Gauss 映射点以增大 k(类似于有限单元法中的 P 算法),并特别注意配点区间向裂纹尖端局部加密以最大限度降低该处的计算误差,进一步提高断裂分析的计算精度。

4 应力强度因子的数值结果

在裂纹尖端附近位移偏导数趋向于无穷,且具有高梯度变化的特点,所以在该处应用有限差分近似位移偏导数存在较大误差,相应的数值结果在裂纹尖端附近并不准确,这种情况同其他结构分析方法如非奇异元有限单元法都是类似的。因此,应力强度因子应由距裂纹尖端一定距离处的名义应力强度因子通过最小二乘法得到的最佳拟合直线外推获得。图 4 是均匀拉伸载荷情况下用 §3.2 中的方法,由最佳拟合直线获得应力强度因子的外推过程。图中拟合点为各配点区间的 Gauss 映射点,为与解析解对比,各拟合点纵坐标为归一化后的名义应力强度因子 $K_I^*/(\sigma_{s0}\sqrt{\pi S})$,其中,$\sigma_{s0}=E_1 u_{s0}/[L(1-\mu^2)]$。外推直线与纵轴的交点即为归一化后的应力强度因子。用类似的方法可得弯曲载荷情况下的归一化应力强度因子 $K_I/(\sigma_{s0}\sqrt{\pi S})$。

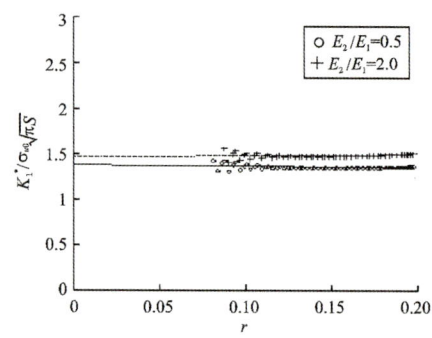

图 4 拉伸载荷下归一化的应力强度因子

Fig. 4 Normalized stress intensity factors under the tension loading

表 1 是图 1、2 中各载荷在两种梯度情况下应力强度因子线法数值解与解析解[15]的对比。比较表明,线法数值解与解析解符合得较好。同时表中结果也说明,材料梯度变化情况直接影响计算精度。因此,上述算例不仅证实了线法功能梯度材料断裂分析的有效性,也进一步表明该法对于材料梯度参数波动具有良好的适应性。分析线法的数值原理可以发现:在非梯度方向,有限差分的离散精度是 $O(h^2)$(h 为节线间距);在梯度方向,节线常微分方程的求解精度是 $O(\Delta_{max}^{2+k})$。因此,加密节线和配点区间或增加 Gauss 映射点数可以进一步提高精度,缩小因材料梯度增大引起的计算误差。

表 1 应力强度因子对比

Tab.1 Contrast of stress intensity factors

载荷类型	E_2/E_1	线法数值解	解析解	误差/%
$K_I/(\sigma_{s0}\sqrt{\pi S})$ 均匀拉伸	0.5	1.3860	1.3336	3.93
	2.0	1.4785	1.4132	4.62
$K_I/(\sigma_{s0}\sqrt{\pi S})$ 纯弯曲	0.5	1.3268	1.2618	5.15
	2.0	0.8258	0.8801	-6.16

5 结 论

本文首先导出了功能梯度材料位移形式的平衡方程和边界条件,然后以拉伸、弯曲等载荷作用下指数梯度材料平面裂纹问题应力强度因子的计算为例,演示了线法功能梯度材料断裂分析的基本过程。该法的基本思想是沿着与材料梯度无关的方向将求解区域半离散化,通过有限差分将问题的控制方程转变成为定义在离散节线上的常微分方程组,应用 B 样条函数 Gauss 配点法求解该常微分方程组得到问题的解答。此外,在算例中同时给出了节线变间距离散、区间映射以及节线和配点区间局部加密等

提高计算精度与效率的数值技巧。线法计算结果与解析解的对比表明了线法在功能梯度材料断裂分析中的有效性。

由线法的求解过程可见，与有限单元法和无网格法等基于结构完全离散并求解节点处代数方程的完全数值方法不同，线法直接求解的是结构半离散后，定义在沿梯度方向节线上保留了材料属性参数变化信息的常微分方程，具有典型的半解析方法的性质。因此，该法能够有效地反映功能梯度材料连续变化的材料属性，不需要像均匀元法分段近似材料属性函数，也不需要类似等参梯度元法进行附加的材料属性参数的插值近似。对比新康法，线法对问题的求解直接从控制方程出发，不需要事先选取满足边界条件的特定待求场函数，求解的问题更加广泛，是一种具有普遍意义的半解析数值方法。

进一步分析线法的数值特点可以看出，线法实质上是在1个或2个方向(三维问题)用有限差分求近似解，而在材料梯度方向用B样条Gauss配点法求高精度解，类似于有限元分析的网格局部加密和采用高阶位移模式，因此非常符合目前具有单向梯度的大多数功能梯度材料的特点。此外，应用B样条Gauss配点法求解常微分边值问题，可以通过缩小配点区间长度或增加配点数，改善局部的计算精度，使线法具有很高的应用灵活性，特别适合求解裂纹、应力集中和材料梯度陡变等高梯度问题。

参考文献：

[1] 黄旭涛，严密. 功能梯度材料：回顾与展望[J]. 材料科学与工程，1997，15(4)：35-38.
 HUANG Xu-tao, YAN Mi. Review and prospects of functional gradient materials [J]. Journal of Materials Science and Engineering, 1997, 15(4): 35-38. (in Chinese).

[2] LANUTTI J J. Functionally graded materials: properties, potential and design guidelines [J]. Composites Engineering, 1994, 24(4): 81-94.

[3] 张幸红，李亚群，韩杰才. TiC-Ni 系功能梯度材料的断裂力学分析[J]. 复合材料学报，2001，11(4)：87-92.
 ZHANG Xing-hong, LI Ya-qun, HAN Jie-cai. Crack problem analysis of TiC-Ni FGM using finite element method [J]. Acta Mechanical Composite Sinica, 2001, 11(4): 87-92. (in Chinese).

[4] ANALS G, LAMBROS J, SANTARE M H. Dominance of asymptotic crack tip fields in elastic functionally materials [J]. International Journal of Fracture, 2003, 115(2): 193-204.

[5] LI Chun-yu, ZOU Zheng-zhu. Stress intensity factor for a functionally graded material cylinder with an external circumferential crack[J]. Fatigue & Fracture of Engineering Materials & Science, 1998, 21(9): 1447-1457.

[6] KIM J H, PAULINO G H. Mixed-mode fracture of orthotropic functionally graded materials using graded finite elements and the modified crack closure method[J]. Engineering Fracture Mechanics, 2002, 69(14-16): 1557-1586.

[7] 陈建，吴林志，杜善义. 采用无单元法计算含边沿裂纹功能梯度材料板的应力强度因子[J]. 工程力学，2000，17(5)：140-144.
 CHEN Jian, WU Lin-zhi, DU Shan-yi. Evaluating SIF of functionally graded plate with an edge crack by element-free method [J]. Engineering Mechanics, 2000, 17(5): 140-144. (in Chinese).

[8] RAO B N, RAHMAN S. Mesh-free analysis of cracks in isotropic functionally graded materials [J]. Engineering Fracture Mechanics, 2003, 70(1): 1-27.

[9] ANALS G, SANTRE M H, LAMBROS J. Numerical calculation of stress intensity factors in functionally graded material [J]. International Journal of Fracture, 2000, 104(2): 131-143.

[10] KIM J H, PAULINO G H. Isoparametric graded finite elements for nonhomogenous isotropic and orthotropic functionally graded materials [J]. Journal of Applied Mechanics, 2002, 69(3): 502-514.

[11] 李永，宋健，张志民. 非均质梯度材料复合结构的 Kantorovich 宏细观精化解法[J]. 中国科学(E 辑)，2003，33(1)：29-41.
 LI Yong, SONG Jian, ZHANG Zhi-min. Kantorovich method of macro or meso-scopic solution for nohomogeneous functionally graded materials composite [J]. Journal of China Science (Series E), 2003, 33(1): 29-41. (in Chinese).

[12] 凌复华，殷学纲，何冶奇. 常微分方程数值方法在力学中应用[M]. 重庆：重庆大学出版社，1990.

[13] JONES D J, SOUTH J C, KLUNKER E B. On the numerical solution of elliptic partial differential equation by the method of lines [J]. Journal of Computational Physics, 1972, 9(2): 496-527.

[14] RAVICHANDRAN K S, BARSOUM I. Determination of stress intensity factor for cracks in finite-width functionally graded material [J]. International Journal of Fracture, 2003, 121(2): 183-203.

[15] ERDOGAN F, WU B H. The surface crack problem for a plate with functionally graded properties [J]. Journal of Applied Mechanics, 1997, 64(3): 449-456.

基于突变理论的滑坡时间预测模型

*周小平¹，钱七虎²，张永兴¹，杨海清¹

(1. 重庆大学土木工程学院，重庆 400045；2. 解放军理工大学，南京 210007)

摘 要：考虑滑面介质的流变特性结合滑坡监测资料，运用尖点突变模型中状态变量的突变来反映滑体的突滑，建立了一个滑坡时间预测模型。基于突变理论，分析了单滑面滑坡的失稳机制，分析结果表明：滑体突滑失稳的必要条件仅取决于滑动面剪切段介质和蠕滑段介质的刚度比。根据尖点突变理论的分叉集方程，得到了滑坡突变时间的计算公式，为滑坡预报提供了初步的理论依据。同时，根据能量转化原理，导出了滑体突滑初速度的计算公式。最后将理论模型应用于某实际的滑坡工程。

关键词：滑坡；突变理论；流变；时间预测；能量转化原理

TIME PREDICTION MODEL OF LANDSLIDES BASED ON CATASTROPHE THEORY

*ZHOU Xiao-ping¹, QIAN Qi-hu², ZHANG Yong-xing¹, YANG Hai-qing¹

(1. School of Civil Engineering, Chongqing University, Chongqing 400045, China; 2. PLA University of Science and Technology, Nanjing 210007, China)

Abstract: Based on catastrophe theory and monitoring data, a time prediction model of landslides is established, in which the rheological behaviors of medium on the sliding surface are taken into account and the catastrophe of a state variable is applied to reflect the instability of landslides. It is found that the catastrophe properties of landslides are controlled by the stiffness ratio of mediums of a shear section to those of rheology section on the sliding surface. According to the bifurcation set of cusp catastrophe theory, the formulae to predict occurrence time of landslides is derived, which can be applied to forecast landslides. Moreover, the formulae of sliding velocity are obtained by using energy transformation theory. Finally, the present model is applied to predict occurrence time of certain landslide.

Key words: landslides; catastrophe theory; rheology; time prediction; energy transformation theory

根据大量的滑坡实例地质剖面图以及实地调研资料，分析发现有许多滑坡滑带可以近似为折线型，也有一些为非折线型滑坡，本文针对折线型滑坡进行讨论。折线型滑动带的形状主要特点：中前部滑带呈缓倾角甚至反倾角直线，后缘滑带呈陡倾角状并接近竖直方向且具有一定张开位移，滑带一般并非全部贯通，而是中部有一定长度的锁固段。

滑坡滑带中往往具有一层厚度较薄、强度低的软弱介质，滑体在降雨和地下水等成灾因素耦合作用下，滑体前缘向临空方向蠕滑变形，同时后缘拉裂段在孔隙水压力的作用下不断张开，水使滑带中部锁固段材料抗剪强度逐渐降低，最终导致滑动面整个贯通，形成"牵引式"的滑坡失稳模式[1-12]。文献[5-12]利用尖点突变模型研究了相关的"牵引

收稿日期：2009-07-30；修改日期：2009-10-29。

基金项目：国家自然科学基金项目(50778184,51078371)；教育部新世纪优秀人才支持计划项目(NECT-07-0911)；重庆市杰出青年基金项目(CSTC,2009BA4046)。

本文原载于《工程力学》(2011年第28卷第2期)。

式"滑坡的失稳机制,但是上述文献都没有研究滑坡的失稳时间。

由于滑坡位移的非均匀性,采用三维模型虽然更能反映滑坡实际位移,但是三维地质模型会使计算变得非常复杂。为了简化分析,本文的时间预测模型建立在二维滑坡模型基础上,只要计算参数准确,同时选择的截面合理同样能较准确预测滑坡失稳时间。

考虑引起滑坡失稳的主导因素,本文建立了图 1 所示的滑坡几何模型。图 1 表示折线型滑动带滑坡二维几何模型,其中 AB 段为后缘拉裂段,BC 段为剪切段,CD 段为蠕滑段。α_i 为滑动带各段与水平面的夹角,剪切段和蠕滑段长度分别为 l_s 和 l_r,滑动带宽度为 h_s,M 为滑体的质量,后缘拉裂段裂隙水高度为 h_w,滑体和水的容重分别为 γ 和 γ_w。

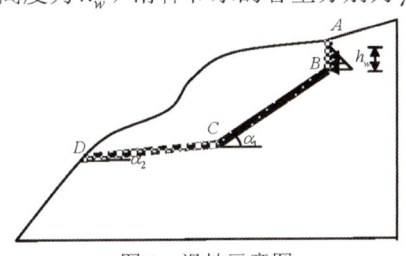

图 1 滑坡示意图
Fig.1 Sketch of landslides

1 滑带岩土材料的本构关系

对于图 1 所示的滑坡力学模型,后缘拉裂段 AB 张开且充填有高度为 h_w 的孔隙水。剪切段和蠕滑段由于岩土材料力学特性不一样,因此本文考虑各段具有不同的本构关系。

1.1 剪切段岩土体的本构关系

剪切段介质具有明显的应变软化性质,如图 2 所示,文献[13-14]的研究表明剪切段介质的本构关系可以表示为:

$$f(u) = \lambda u_s \exp(-u_s / u_c) \quad (1)$$

式中:$\lambda = G_s A_l / h_s$ 为剪切段初始刚度;A_l 为滑带各分段的面积,在二维情况下取单位宽度,因此有 $A_l = l_s$;G_s 为剪切段介质的剪切模量;u_c 为剪切段介质的峰值位移;u_s 为滑体沿滑面的位移。

在降雨条件下,滑带土由于后缘拉裂段的张开,地表水渗入滑床,这些水对滑带岩土介质的侵蚀作用可以导致滑带土抗剪强度的部分丧失,即水致弱化。为了反映地表渗水和地下水对滑带土的弱化作用,根据文献[14-15]的研究,可以在本构方程中引入一个水致弱化函数,该函数可以表示为

$$g(\zeta) = (1-\phi)(1-\zeta)^2 + \phi \quad (2)$$

式中:ζ 为含水量;ϕ 为岩土体的软化系数,可由试验确定。

式(2)是一个单调下降的函数,在干燥情况下 $\zeta = 0$,$g(0) = 1$;在饱和情况下,$\zeta = 1$,$g(1) = \phi < 1$。考虑水致弱化后的剪切段本构方程为:

$$f(u) = \lambda g(\zeta) u_s \exp\left(-\frac{u_s}{u_c}\right) \quad (3)$$

图 2 剪切段介质本构关系
Fig.2 The constitutive relationship of the shear zone

1.2 蠕滑段

为了全面的反映蠕滑段滑带土的粘弹塑性性质,本文选用五元件的西原体本构模型。西原模型实质上是由虎克体(H)、粘弹性体(N//H)和粘塑性体(N//St.V)串联而成,能最全面反映岩土材料的粘弹性、粘弹塑性效应[16]。

图 3 为西原体元件模型,在粘塑性体(N//St.V) 中,摩擦滑块承受的应力为 σ_p,σ_s 为单向屈服应力,σ 为总的外加应力,只有当 $\sigma_p > \sigma_s$ 时滑块才能滑动,超出的应力 $\delta_d = \sigma - \sigma_p$ 则由粘性阻尼器来承担。在粘弹性体(N//H)中,在应力作用下应变不是立即达到弹性应变的终值,而是有一个相对滞后的过程,应力也由最初的全部由粘壶承受而逐渐转移到线性弹簧上。在该模型中,瞬时弹性响应是虎克体(H)提供的,由于阻尼器的存在,使应力值能够在瞬间超过塑性理论的预期值,而当系统达到稳定状态时,就将趋向于这个平衡值。

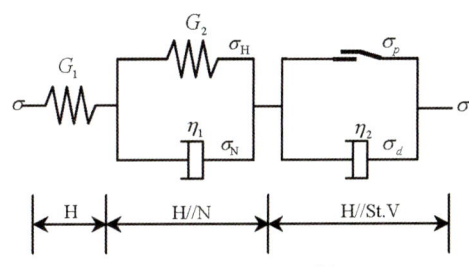

图 3 西原体模型[5]
Fig.3 Visco-elastoplastic model[5]

图 3 表示蠕滑段介质元件本构模型,西原体本构方程可以表示为:

$$\frac{\eta_1}{G_1+G_2}\dot{\sigma}+\sigma=$$
$$\frac{G_1\eta_1}{G_1+G_2}\dot{\varepsilon}+\frac{G_1G_2}{G_1+G_2}\varepsilon,\quad \sigma<\sigma_s \quad (4)$$

$$\ddot{\sigma}+\left(\frac{G_1}{\eta_1}+\frac{G_2}{\eta_2}+\frac{G_2}{\eta_1}\right)\dot{\sigma}+\frac{G_1G_2}{\eta_1\eta_2}(\sigma-\sigma_s)=$$
$$G_1\ddot{\varepsilon}+\frac{G_1G_2}{\eta_1}\dot{\varepsilon},\quad \sigma\geq\sigma_s \quad (5)$$

根据徐曾和的研究[17]，式(4)和式(5)可以转化为荷载位移关系：

$$G_\infty\tau_r\frac{du_r}{dt}+G_\infty u_r=\frac{G_\infty\tau_r}{G_0}\frac{dR_1}{dt}+R_1,\quad R_1<R_f \quad (6)$$

$$\frac{d^2R_2}{dt^2}+\left(\frac{\eta_1+\eta_2}{\tau_r\eta_2}+\frac{G_0}{\eta_1}\right)\frac{dR_2}{dt}+\frac{G_0}{\tau_r\eta_2}(R_2-R_f)=$$
$$\frac{G_0}{\eta_2}\frac{du_r}{dt}+\frac{G_\infty G_0}{G_0-G_\infty}\frac{d^2u_r}{dt^2},\quad R_2\geq R_f \quad (7)$$

式中：u_r 为蠕滑段位移，蠕滑段位移方向始终与滑动面平行；$G_\infty=(G_1G_2)/(G_1+G_2)$；$\tau_r=\eta_1/G_2$；$G_0=G_1$；$G_\infty$ 为蠕滑段介质的长期剪切刚度；G_0 为蠕滑段介质的瞬时剪切刚度；τ_r 为松弛时间；R_1、R_2 为蠕滑段荷载；R_f 为屈服荷载；$R_1<R_f$ 表示蠕滑段荷载小于屈服荷载；$R_2\geq R_f$ 表示蠕滑段荷载大于屈服荷载。

2 力学模型

2.1 监测数据拟合

将滑体视为刚体，即认为滑体滑动过程中剪切段和蠕滑段作为整体沿滑动面下滑。滑体水平位移 u 与时间的关系可通过拟合已有滑坡水平位移观测数据得到，并且可以根据具体的数据选择满足需要的多项式最高阶数。另外，根据实际监测数据分布情况选择最合适函数进行拟合。本文以五次多项式为例，监测数据拟合方程为：

$$u=u(0)+h_1t+h_2t^2+h_3t^3+h_4t^4+h_5t^5 \quad (8)$$

式中：$u(0)$ 为 $t=0$ 时刻的初始水平位移，参数 h_1、h_2、h_3、h_4、h_5 可通过拟合实测数据得到。

2.2 平衡条件

根据静力平衡条件，$t=0$ 时刻的初始条件为：

$$f[u(0)]=\lambda g(\zeta)u(0)\exp\left[-\frac{u(0)}{u_c}\right] \quad (9)$$

$$R(0)=\frac{f[u(0)]\cos\alpha_1-0.5\gamma_w h_w^2}{\cos\alpha_2} \quad (10)$$

式中 $i=1,2$。

1) 当 $R_1<R_f$ 时，对于蠕滑段介质本构模型为广义开尔文模型情况。

沿蠕滑面的位移为 $u_r=u(\cos\alpha_2)^{-1}$，将荷载位移关系式(6)化为：

$$a\frac{du}{\cos\alpha_2 dt}+\frac{bu}{\cos\alpha_2}=c\frac{dR_1}{dt}+R_1 \quad (11)$$

式中：$a=G_\infty\tau_r$；$b=G_\infty$；$c=G_\infty\tau_r/G_0$，移项得：

$$\frac{d(au\cos^{-1}\alpha_2-cR_1)}{dt}=R_1-bu\cos^{-1}\alpha_2 \quad (12)$$

再通过求解关于 t 的非齐次常微分方程可得：

$$R_1=\frac{a}{c}u\cos^{-1}\alpha_2-\frac{1}{c}e^{-\frac{t}{c}}\times$$
$$\left[\int\left(\frac{au\cos^{-1}\alpha_2}{c}-bu\cos^{-1}\alpha_2\right)e^{\frac{t}{c}}dt+B_1\right] \quad (13)$$

式中，B_1 为积分常数。

将式(8)代入式(13)中可得：

$$R_1=\frac{a}{c}u(0)-\left(\frac{a}{c}-b\right)\cos^{-1}\alpha_2 u(0)-\frac{B_1}{c}e^{-\frac{t}{c}}-$$
$$h_1\cos^{-1}\alpha_2\left(\frac{a}{c}-b\right)(t-c)-h_2\cos^{-1}\alpha_2\times$$
$$\left(\frac{a}{c}-b\right)(2c^2-2ct+t^2)-h_3\cos^{-1}\alpha_2\left(\frac{a}{c}-b\right)\times$$
$$(-6c^3+6c^2t-3ct^2+t^3)-h_4\cos^{-1}\alpha_2\left(\frac{a}{c}-b\right)\times$$
$$(24c^4-24c^3t+12c^2t^2-4ct^3+t^4)-h_5\cos^{-1}\alpha_2\times$$
$$\left(\frac{a}{c}-b\right)(-120c^5+120c^4t-60c^3t^2+$$
$$20c^2t^3-5ct^4+t^5) \quad (14)$$

将 $t=0$ 代入式(14)中，根据初始条件式(9)和式(10)可得：

$$B_1=au(0)-(a-bc)\cos^{-1}\alpha_2 u(0)+h_1\cos^{-1}\alpha_2\times$$
$$(ac-bc^2)-2h_2\cos^{-1}\alpha_2(ac^2-bc^3)+$$
$$6h_3\cos^{-1}\alpha_2(ac^3-bc^4)-24h_4\cos^{-1}\alpha_2\times$$
$$(ac^4-bc^5)+120h_5\cos^{-1}\alpha_2(ac^5-bc^6)-$$
$$c\cos\alpha_2\times\left\{\lambda g(\zeta)u(0)\exp\left[\frac{-u(0)}{u_c}\right]-0.5\gamma_w h_w^2\right\}$$
$$(15)$$

考虑平面问题，取单位宽度的滑体，则图1所示系统的总势能为：

$$E_1 = \int_0^u [R_1 + f(u)]\mathrm{d}u -$$
$$Mgu(\tan\alpha_1 + \tan\alpha_2) - \frac{1}{2}\gamma_w h_w^2 u \quad (16)$$

根据 $grad_u E_1 = 0$，得平衡曲面方程：
$$R_1 + f(u) - Mg(\tan\alpha_1 + \tan\alpha_2) - \frac{1}{2}\gamma_w h_w^2 = 0 \quad (17)$$

根据平衡曲面方程的光滑性质，在尖点处有：
$$grad_u[grad_u(grad_u E_1)] = 0 \quad (18)$$
即：
$$\frac{\mathrm{d}^2 R_1}{\mathrm{d}u^2} + \frac{\mathrm{d}^2 f(u)}{\mathrm{d}u^2} = 0 \quad (19)$$

$$\frac{\left(\dfrac{u}{\cos\alpha_1} - 2u_c\right)\lambda g(\zeta)\mathrm{e}^{-\frac{u}{u_c \cos\alpha_1}}}{u_c^2 \cos^2\alpha_1} = 0 \quad (20)$$

于是，在尖点处，有：
$$u = u_1 = 2u_c \cos\alpha_1 \quad (21)$$
式中，u_c 为剪切段介质的峰值荷载对应的位移。

将平衡曲面方程式(17)在尖点 u_1 处展开成幂级数，截取前三项得：
$$T_1 + S_1(u - u_1) + N(u - u_1)^3 = 0 \quad (22)$$
式中：
$$T_1 = \frac{2a}{c}u_c \cos\alpha_1 - \left(\frac{a}{c} - b\right)\cos^{-1}\alpha_2 u(0) - \frac{B_1}{c}\mathrm{e}^{-\frac{t}{c}} -$$
$$h_1 \cos^{-1}\alpha_2\left(\frac{a}{c} - b\right)(t - c) - h_2 \cos^{-1}\alpha_2 \times$$
$$\left(\frac{a}{c} - b\right)(2c^2 - 2ct + t^2) - h_3 \cos^{-1}\alpha_2 \times$$
$$\left(\frac{a}{c} - b\right)(6c^2 t - 6c^3 - 3ct^2 + t^3) - h_4 \cos^{-1}\alpha_2 \times$$
$$\left(\frac{a}{c} - b\right)(24c^4 - 24c^3 t + 12c^2 t^2 - 4ct^3 + t^4) -$$
$$h_5 \cos^{-1}\alpha_2\left(\frac{a}{c} - b\right) \times$$
$$(120c^4 t - 120c^5 - 60c^3 t^2 + 20c^2 t^3 - 5ct^4 + t^5) -$$
$$Mg(\tan\alpha_1 + \tan\alpha_2) - \frac{1}{2}\gamma_w h_w^2 ;$$
$$S_1 = \frac{a}{c} - \frac{\lambda g(\zeta)}{\cos\alpha_1 \mathrm{e}^2}; \quad N = \frac{\lambda \mathrm{e}^{-2} g(\zeta)}{6u_c^2 \cos^3\alpha_1}。$$

引入无量纲的状态变量：
$$x = \frac{u - u_1}{u_1} \quad (23)$$

将式(23)代入式(22)得尖点突变理论标准形式的平衡曲面方程：
$$x^3 + px + q = 0 \quad (24)$$
式中：
$$p = \frac{S}{Nu_1^2} = \frac{3}{2}\left(\frac{\mathrm{e}^2 a \cos\alpha_1}{\lambda g(\zeta) c} - 1\right) = \frac{3}{2}(k - 1) \quad (25)$$
$$k = \frac{\mathrm{e}^2 a \cos\alpha_1}{\lambda g(\zeta) c} \quad (26)$$

$$q(t) = \frac{3\mathrm{e}^2}{4\lambda g(\zeta) u_c}\left[\frac{2a}{c}u_c \cos\alpha_1 - \left(\frac{a}{c} - b\right)\times\right.$$
$$\cos^{-1}\alpha_2 u(0) - \frac{B_1}{c}\mathrm{e}^{-\frac{t}{c}} - h_1 \cos^{-1}\alpha_2 \times$$
$$\left(\frac{a}{c} - b\right)(t - c) - h_2 \cos^{-1}\alpha_2\left(\frac{a}{c} - b\right)\times$$
$$(2c^2 - 2ct + t^2) - h_3 \cos^{-1}\alpha_2\left(\frac{a}{c} - b\right)\times$$
$$(-6c^3 + 6c^2 t - 3ct^2 + t^3) - h_4 \cos^{-1}\alpha_2 \times$$
$$\left(\frac{a}{c} - b\right)(24c^4 - 24c^3 t + 12c^2 t^2 - 4ct^3 + t^4) -$$
$$h_5 \cos^{-1}\alpha_2\left(\frac{a}{c} - b\right)(120c^4 t - 120c^5 -$$
$$60c^3 t^2 + 20c^2 t^3 - 5ct^4 + t^5) -$$
$$\left. Mg(\tan\alpha_1 + \tan\alpha_2) - \frac{1}{2}\gamma_w h_w^2 \right] \quad (27)$$

2) 当 $R_2 \geq R_f$ 时，对于蠕滑段介质本构模型为西原体模型的情况。

沿蠕滑面的位移为 $u_r = u\cos^{-1}\alpha_2$，将荷载位移关系式(7)化为：
$$\frac{\mathrm{d}^2 R_2}{\mathrm{d}t^2} + b_1\frac{\mathrm{d}R_2}{\mathrm{d}t} + b_2(R_2 - R_f) =$$
$$b_3\frac{\mathrm{d}u_r}{\mathrm{d}t} + b_4\frac{\mathrm{d}^2 u_r}{\mathrm{d}t^2}, \quad R_2 \geq R_f \quad (28)$$
式中：$b_1 = \dfrac{\eta_1 + \eta_2}{\tau_r \eta_2} + \dfrac{G_0}{\eta_1}$；$b_2 = \dfrac{G_0}{\tau_r \eta_2}$；$b_3 = \dfrac{G_0}{\eta_2 \cos\alpha_2}$；
$b_4 = \dfrac{G_\infty G_0}{(G_0 - G_\infty)\cos\alpha_2}$。

移项得：
$$\frac{\mathrm{d}^2(R_2 - b_4 u)}{\mathrm{d}t^2} + \frac{b_1 \mathrm{d}(b_1 R_2 - b_3 u)}{\mathrm{d}t} + b_2(R_2 - R_f) = 0 \quad (29)$$

再通过求解关于 t 的二阶非齐次常微分方程可得：
$$R_2 = \mathrm{e}^{\frac{-\left(b_1 + \sqrt{b_1^2 - 4b_2}\right)t}{2}} \times \left\{ C_1 + \right.$$

$$\int \frac{-e^{\frac{\left(b_1+\sqrt{b_1^2-4b_2}\right)t}{2}}[b_2 R_f + b_3 u'(t) + b_4 u''(t)]}{\sqrt{b_1^2-4b_2}} dt + e^{t\sqrt{b_1^2-4b_2}} \cdot$$

$$\left\{ C_2 + \int \frac{e^{\frac{\left(b_1-\sqrt{b_1^2-4b_2}\right)t}{2}}[b_2 R_f + b_3 u'(t) + b_4 u''(t)]}{\sqrt{b_1^2-4b_2}} dt \right\} \quad (30)$$

式中，C_1 和 C_2 为积分常数。令：

$$b_5 = \frac{b_1 + \sqrt{b_1^2-4b_2}}{2} \quad (31)$$

$$b_6 = \frac{b_1 - \sqrt{b_1^2-4b_2}}{2} \quad (32)$$

将式(8)、式(31)和式(32)代入式(30)中可得：

$$R_2 = e^{-b_5 t}\left\{ C_1 - \frac{e^{b_5 t}}{b_5^5(b_5-b_6)} \times \{b_3(120h_5 - 24b_5 n_1 + 6b_5^2 n_2 - 2b_5^3 n_3 + b_5^4 n_4) + b_5[b_2 b_5^3 R_f + 2b_4 \times (b_5^3 n_3 - 60h_5 + 12b_5 n_1 - 3b_5^2 n_2)]\}\right\} +$$

$$e^{(b_5-b_6)t}\left\{ C_2 + \frac{e^{b_6 t}}{b_6^5(b_5-b_6)}\{b_3(120h_5 - 24b_6 n_1 + 6b_6^2 n_2 - 2b_6^3 n_3 + b_6^4 n_4) + b_6 \times [b_2 b_6^3 R_f + 2b_4 \times (-60h_5 + 12b_6 n_1 - 3b_6^2 n_2 + b_6^3 n_3)]\}\right\} \quad (33)$$

式中：

$n_1 = h_4 + 5h_5 t$，$n_2 = h_3 + 4h_4 t + 10h_5 t^2$，
$n_3 = h_2 + 3h_3 t + 6h_4 t^2 + 10h_5 t^3$，
$n_4 = h_1 + 2h_2 t + 3h_3 t^2 + 4h_4 t^3 + 5h_5 t^4$。

当 $t=0$ 时，初始条件为：

$$R_1(0) = C_1 - \frac{1}{b_5^5(b_5-b_6)}\{b_3(120h_5 - 24b_5 h_4 + 6b_5^2 h_3 - 2b_5^3 h_2 + b_5^4 h_1) + b_5[b_2 b_5^3 R_f + 2b_4 \times (12b_5 h_4 - 60h_5 - 3b_5^2 h_3 + b_5^3 h_2)]\} + C_2 + \frac{1}{b_6^5(b_5-b_6)}\{b_3 \times (120h_5 - 24b_6 h_4 + 6b_6^2 h_3 - 2b_6^3 h_2 + b_6^4 h_1) + b_6[b_2 b_6^3 R_f + 2b_4(12b_6 h_4 - 60h_5 - 3b_6^2 h_3 + b_6^3 h_2)]\} = \frac{f[u(0)]\cos\alpha_1 - 0.5\gamma_w h_w^2}{\cos\alpha_2} \quad (34)$$

$$R_2'(0) = \frac{1}{b_5^4 b_6^5(b_5-b_6)}\{b_5^6 b_6^5(C_2-C_1) - 2b_5^3 b_6^5 \times (b_3 h_2 + 3b_4 h_3) + 6h_5^2 b_6^5(b_3 h_3 + 4b_4 h_4) + 120b_3 b_6^5 h_5 - 24b_6^5(b_3 h_4 + 5b_4 h_5) + b_5^4 b_6 \times \{b_6[b_6(b_6^4 C_2 + 2b_3 b_6 h_2 - 6b_3 h_3 + 6b_4 b_6 h_3) + 24(b_3 - b_4 b_6)h_4] - 120(b_3 - b_4 b_6)h_5\} + b_5^5 \times [b_6^6(C_1 - 2C_2) - 2b_6^3(b_3 h_2 + 3b_4 h_3) + 6b_6^2 \times (b_3 h_3 + 4b_4 h_4) + 120b_3 h_5 - (b_3 h_4 + 5b_4 h_5) \times 24b_6 + b_6^4(b_3 h_1 + 2b_4 h_2 + b_2 R_f)]\} = 0 \quad (35)$$

联立式(34)和式(35)可求得积分常数 C_1 和常数 C_2 分别为：

$$C_1 = [n_5 b_5^4(b_5-b_6)^2 b_5^5 - n_6 b_5^4(b_5-b_6)^2 b_6^5 + b_3 b_5^5 b_6^4 h_1 - 2b_3 b_5^5 b_6^3 h_2 + 2b_3 b_5^4 b_6^4 h_2 + 2b_4 b_5^5 b_6^4 h_2 - 2b_3 b_5^3 b_6^5 h_2 + 6b_3 b_5^5 b_6^2 h_3 - 6b_3 b_5^4 b_6^3 h_3 - 6b_4 b_5^5 b_6^3 h_3 + 6b_3 b_5^4 b_6^4 h_3 + 6b_3 b_5^2 b_6^5 h_3 - 6b_4 b_5^3 b_6^5 h_3 - 24b_3 b_5^5 b_6 h_4 + 24b_3 b_5^4 b_6^2 h_4 + 24b_4 b_5^5 b_6^2 h_4 - 24b_4 b_5^4 b_6^3 h_4 - 24b_3 b_5 b_6^5 h_4 + 24b_4 b_5^2 b_6^5 h_4 + 120b_3 b_5^5 h_5 - 120b_3 b_5^4 b_6 h_5 - 120b_4 b_5^5 b_6 h_5 + 120b_3 b_5^4 \times b_6^2 h_5 + 120b_3 b_6^5 h_5 - 120b_4 b_5 b_6^5 h_5 + b_2 b_5^5 b_6^4 R_f]/[b_5^4 b_6^5(2b_5^2 - 3b_5 b_6 + b_6^2)] \quad (36)$$

$$C_2 = [n_5 b_5^5(b_5-b_6)^2 b_5^5 + n_6 b_5^4(b_6-b_5)^2 b_6^5 - b_3 b_5^5 b_6^4 h_1 + 2b_3 b_5^5 b_6^3 h_2 - 2b_3 b_5^4 b_6^4 h_2 - 2b_4 b_5^5 b_6^4 h_2 + 2b_3 b_5^3 b_6^5 h_2 - 6b_3 b_5^5 b_6^3 h_3 + 6b_3 b_5^4 b_6^3 h_3 + 6b_4 b_5^5 b_6^3 h_3 - 6b_3 b_5^4 b_6^4 h_3 + 6b_3 b_5^2 b_6^5 h_3 + 6b_4 b_5^3 b_6^5 h_3 + 24b_3 b_5^5 b_6 h_4 - 24b_3 b_5^4 b_6^2 h_4 - 24b_4 b_5^5 b_6^2 h_4 + 24b_4 b_5^4 b_6^3 h_4 + 24b_3 b_5 b_6^5 h_4 - 24b_4 b_5^2 b_6^5 h_4 - 120b_3 b_5^5 h_5 + 120b_3 b_5^4 b_6 h_5 + 120b_4 b_5^5 b_6 h_5 - 120b_4 b_5^4 \times b_6^2 h_5 - 120b_3 b_6^5 h_5 + 120b_4 b_5 b_6^5 h_5 - b_2 b_5^5 b_6^4 R_f]/[b_5^4 b_6^5(2b_5^2 - 3b_5 b_6 + b_6^2)] \quad (37)$$

式中，$n_5 = \dfrac{f[u(0)]\cos\alpha_1 - 0.5\gamma_w h_w^2}{\cos\alpha_2}$，

$n_6 = \dfrac{1}{b_5^5(b_5-b_6)}\{b_3(120h_5 - 24b_5 h_4 + 6b_5^2 h_3 - 2b_5^3 h_2 + b_5^4 h_1) + b_5[b_2 b_5^3 R_f + 2b_4(-60h_5 + 12b_5 h_4 - 3b_5^2 h_3 + b_5^3 h_2)]\} + \dfrac{1}{b_6^5(b_5-b_6)} \cdot$

$$\{b_3(120h_5 - 24b_6h_4 + 6b_6^2h_3 - 2b_6^3h_2 + b_6^4h_1) + b_6[b_2b_6^3R_f + 2b_4(12b_6h_4 - 60h_5 - 3b_6^2h_3 + b_6^3h_2)]\}。$$

考虑平面问题，取单位宽度的滑体，则图 1 所示系统的总势能为：

$$E_2 = \int_0^u [R_2 + f(u)]\mathrm{d}u - Mgu(\tan\alpha_1 + \tan\alpha_2) - \frac{1}{2}\gamma_w h_w^2 u \quad (38)$$

根据 $grad_u E_2 = 0$，得平衡曲面方程：

$$R_2 + f(u) - Mg(\tan\alpha_1 + \tan\alpha_2) - \frac{1}{2}\gamma_w h_w^2 = 0 \quad (39)$$

根据平衡曲面方程的光滑性质，在尖点处有：

$$grad_u[grad_u(grad_u E_2)] = 0 \quad (40)$$

即：

$$\frac{\mathrm{d}^2 R_2}{\mathrm{d}u^2} + \frac{\mathrm{d}^2 f(u)}{\mathrm{d}u^2} = 0 \quad (41)$$

$$\frac{\left(\dfrac{u}{\cos\alpha_1} - 2u_c\right)\lambda g(\zeta)\mathrm{e}^{-\frac{u}{u_c\cos\alpha_1}}}{u_c^2\cos^2\alpha_1} = 0 \quad (42)$$

于是，在尖点处，有：

$$u = u_1 = 2u_c\cos\alpha_1 \quad (43)$$

式中，u_c 为剪切段介质的峰值荷载对应的位移。

将平衡曲面方程式(39)在尖点 u_1 处展开成幂级数，截取前三项得：

$$T_2 + S_2(u - u_1) + N(u - u_1)^3 = 0 \quad (44)$$

式中：

$$T_2 = \mathrm{e}^{-b_5 t}\left\{C_1 - \frac{\mathrm{e}^{b_5 t}}{b_5^5(b_5 - b_6)}\{b_3(120h_5 - 24b_5 n_1 + 6b_5^2 n_2 - 2b_5^3 n_3 + b_5^4 n_4) + b_5[b_2 b_5^3 R_f + 2b_4 \times (-60h_5 + 12b_5 n_1 - 3b_5^2 n_2 + b_5^3 n_3)]\}\right\} +$$

$$\mathrm{e}^{(b_5 - b_6)t}\left\{C_2 + \frac{\mathrm{e}^{b_6 t}}{b_6^5(b_5 - b_6)}b_3(120h_5 - 24b_6 n_1 + 6b_6^2 n_2 - 2b_6^3 n_3 + b_6^4 n_4) + b_6[b_2 b_6^3 R_f + 2b_4(-60h_5 + 12b_6 n_1 - 3b_6^2 n_2 + b_6^3 n_3)]\}\right\} +$$

$$\frac{2\lambda g(\zeta)u_c}{\mathrm{e}^2} - Mg(\tan\alpha_1 + \tan\alpha_2) - \frac{1}{2}\gamma_w h_w^2 ;$$

$$S_2 = \frac{-\lambda g(\zeta)}{\cos\alpha_1 \mathrm{e}^2}; \quad N = \frac{\lambda g(\zeta)}{6u_c^2 \mathrm{e}^2 \cos^3\alpha_1}。$$

引入无量纲的状态变量 $x = \dfrac{u - u_1}{u_1}$，根据式(44)得尖点理论标准形式得平衡曲面方程：

$$x^3 + px + q = 0 \quad (45)$$

式中：

$$p = \frac{S}{Nu_1^2} = -\frac{3}{2} \quad (46)$$

$$q(t) = \frac{\mathrm{e}^{-b_5 t}}{Nu_1^3}\left\{C_1 - \frac{\mathrm{e}^{b_5 t}}{b_5^5(b_5 - b_6)}\{b_3(120h_5 - 24b_5 n_1 + 6b_5^2 n_2 - 2b_5^3 n_3 + b_5^4 n_4) + b_5[b_2 b_5^3 R_f + 2b_4 \times (12b_5 n_1 - 60h_5 - 3b_5^2 n_2 + b_5^3 n_3)]\}\right\} + \mathrm{e}^{(b_5 - b_6)t} \times$$

$$\left\{C_2 + \frac{\mathrm{e}^{b_6 t}}{b_6^5(b_5 - b_6)}b_3(120h_5 - 24b_6 n_1 + b_6^2 n_2 - 2b_6^3 n_3 + 6b_6^4 n_4) + b_6[b_2 b_6^3 R_f + 2b_4 \times (-60h_5 + 12b_6 n_1 - 3b_6^2 n_2 + b_6^3 n_3)]\}\right\} +$$

$$\frac{2\lambda g(\zeta)u_c}{\mathrm{e}^2} - Mg(\tan\alpha_1 + \tan\alpha_2) - \frac{1}{2}\gamma_w h_w^2 \quad (47)$$

3 滑坡失稳时间预测

图 4 表示式(17)和式(39)对应的带折叠翼的平衡曲面，x 为状态变量，p 和 q 为控制变量。平衡曲面上的双折线在 p-q 平面上的投影即为尖点突变的分叉集。图 4 中，当控制变量随时间的变化达到分叉集左支 O_1D 时，状态变量 x 发生突变，即滑坡突然失稳。O 对应于尖点，在尖点处状态变量也会发生突跳，但是突跳能量差为 0。

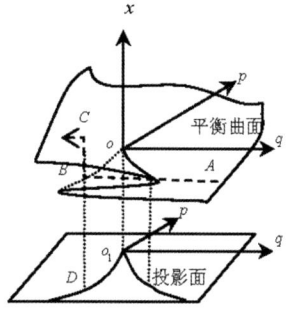

图 4 尖点突变模型的平衡曲面

Fig. 4 Equilibrium surface of cusp catastrophe model

3.1 当 $R < R_f$ 时，对于蠕滑段介质本构模型为广义开尔文模型的情况

式(24)中只有在 $p \leq 0$ 时成立，因而由式(24)可得发生突变的必要条件为：

$$p = \frac{S}{Nu_1^2} = \frac{3}{2}\left(\frac{e^2 a \cos\alpha_1}{\lambda g(\zeta)c} - 1\right) = \frac{3}{2}(k-1) \leq 0 \quad (48)$$

式(48)表明滑体突滑失稳的必要条件仅取决于滑动面剪切段介质和蠕滑段介质的刚度比,即滑体突滑失稳是由系统本身的特性决定的。

由尖点突变理论,式(24)的分叉点集为:
$$D = 4p^3 + 27q^2(t) = 0 \quad (49)$$

$p=0$ 时式(24)有 3 重零根 $x_1 = x_2 = x_3 = 0$; 当 $p<0$ 时,有 3 个实根,它们是:
$$\begin{cases} x_1 = 2\left(\frac{p}{3}\right)^{\frac{1}{3}} = \sqrt{2}(1-k)^{\frac{1}{2}}, \\ x_2 = x_3 = -\left(\frac{p}{3}\right)^{\frac{1}{2}} = -\frac{\sqrt{2}}{2}(1-k)^{\frac{1}{2}}. \end{cases} \quad (50)$$

于是跨越分叉点集的状态 x 发生突跳:
$$\Delta x = x_1 - x_2 = \frac{3\sqrt{2}}{2}(1-k)^{\frac{1}{2}} \quad (51)$$

将式(25)、式(26)和式(27)代入分叉集方程式(49)可得:
$$4p^3 + 27\left\{\frac{3e^2}{4\lambda g(\zeta)u_c}\left[\frac{2a}{c}u_c\cos\alpha_1 - \left(\frac{a}{c}-b\right)\times\right.\right.$$
$$\cos^{-1}\alpha_2 u(0) - \frac{B_1}{c}e^{\frac{t}{c}} - h_1\cos^{-1}\alpha_2\left(\frac{a}{c}-b\right)\times$$
$$(t-c) - h_2\cos^{-1}\alpha_2\left(\frac{a}{c}-b\right)(2c^2 - 2ct + t^2) -$$
$$h_3\cos^{-1}\alpha_2\left(\frac{a}{c}-b\right)(-6c^3 + 6c^2t - 3ct^2 + t^3) -$$
$$h_4\cos^{-1}\alpha_2\left(\frac{a}{c}-b\right)(24c^4 - 24c^3t + 12c^2t^2 -$$
$$4ct^3 + t^4) - h_5\cos^{-1}\alpha_2\left(\frac{a}{c}-b\right)(-120c^5 +$$
$$120c^4t - 60c^3t^2 + 20c^2t^3 - 5ct^4 + t^5) -$$
$$\left.\left.Mg(\tan\alpha_1 + \tan\alpha_2) - \frac{1}{2}\gamma_w h_w^2\right]\right\}^2 = 0 \quad (52)$$

式(52)中只含一个变量 t,且都是关于 t 的初等函数。当滑动面介质参数确定及滑坡实测数据拟合曲线方程已知时,就可根据上式通过数值分析方法确定突滑时间。

3.2 当 $R \geq R_f$ 时,对于蠕滑段介质本构模型为西原体模型的情况

式(45)中只有在 $p \leq 0$ 时成立,因而由式(45)可得发生突变的必要条件为:

$$p = \frac{S}{Nu_1^2} = -\frac{3}{2} < 0 \quad (53)$$

式(53)表明:当蠕滑段荷载大于屈服荷载时,蠕滑段介质处于加速蠕滑阶段,整个系统一定满足突滑的必要条件。

由尖点突变理论,式(45)的分叉点集为:
$$D = 4p^3 + 27q^2(t) = 0 \quad (54)$$

由于 $p<0$,式(45)有 3 个实根,它们是:
$$\begin{cases} x_1 = 2\left(\frac{p}{3}\right)^{\frac{1}{3}} = \sqrt{2}, \\ x_2 = x_3 = -\left(\frac{p}{3}\right)^{\frac{1}{2}} = -\frac{\sqrt{2}}{2}. \end{cases} \quad (55)$$

于是跨越分叉点集的状态 x 发生突跳:
$$\Delta x = x_1 - x_2 = \frac{3\sqrt{2}}{2} \quad (56)$$

将式(46)和式(47)代入分叉集方程式(54)可得:
$$4p^3 + 27\left\{\frac{e^{-b_5 t}}{Nu_1^2}\left\{C_1 - \frac{e^{b_5 t}}{b_5^5(b_5 - b_6)}\{b_3(120h_5 - \right.\right.$$
$$24b_5 n_1 + 6b_5^2 n_2 - 2b_5^3 n_3 + b_5^4 n_4) + b_5[b_2 b_5^3 R_f +$$
$$2b_4(-60h_5 + 12b_5 n_1 - 3b_5^2 n_2 + b_5^3 n_3)]\}\right\} + e^{(b_5-b_6)t} \times$$
$$\left\{C_2 + \frac{e^{b_6 t}}{b_6^5(b_5-b_6)}\{b_3(120h_5 - 24b_6 n_1 + 6b_6^2 n_2 - \right.$$
$$2b_6^3 n_3 + b_6^4 n_4) + b_6[b_2 b_6^3 R_f + 2b_4(12b_6 n_1 -$$
$$\left.60h_5 - 3b_6^2 n_2 + b_6^3 n_3]\}\right\} + \frac{2\lambda g(\zeta)u_c}{e^2} -$$
$$\left.Mg(\tan\alpha_1 + \tan\alpha_2) - \frac{1}{2}\gamma_w h_w^2\right\}^2 = 0 \quad (57)$$

式(57)中,只含一个变量 t,且都是关于 t 的初等函数。当滑动面介质参数确定及滑坡实测数据拟合曲线方程已知时,就可根据上式通过数值分析方法确定突滑时间。

4 滑坡突滑初速度

1) 当 $R < R_f$ 时,对于蠕滑段介质本构模型为广义开尔文模型的情况。

将总势能表达式(16)在尖点 $u = u_1 = 2u_c\cos\alpha_1$ 处展开,截取前四次项有:
$$E_1 = \frac{\lambda g(\zeta)}{24u_c^2 e^2 \cos^3\alpha_1}(u-u_1)^4 + (u-u_1)^2 \times$$

$$\left[\frac{a}{2c} - \frac{\lambda g(\zeta)}{2e^2 \cos\alpha_1}\right] + \left[4u_c \cos\alpha_1 + \frac{2u_c}{e^2} - \frac{1}{2}\gamma_w h_w^2 + m_1 - Mg(\tan\alpha_1 + \tan\alpha_2)\right] \times$$

$$(u - u_1) + 4u_c^2 \cos^2\alpha_1 - \frac{2u_c^2 u_1}{2e^2} + m_1 u_1 - $$

$$\left[\frac{1}{2}\gamma_w h_w^2 + Mg(\tan\alpha_1 + \tan\alpha_2)\right] u_1 \quad (58)$$

式中：

$$m_1 = -\left(\frac{a}{c} - b\right)\cos^{-1}\alpha_2 u(0) - \frac{B}{c}e^{\frac{t}{c}} - h_1(t-c) \times$$

$$\cos^{-1}\alpha_2\left(\frac{a}{c} - b\right) - h_2 \cos^{-1}\alpha_2\left(\frac{a}{c} - b\right) \times$$

$$(2c^2 - 2ct + t^2) - h_3 \cos^{-1}\alpha_2\left(\frac{a}{c} - b\right) \times$$

$$(-6c^3 + 6c^2 t - 3ct^2 + t^3) - h_4 \cos^{-1}\alpha_2 \times$$

$$\left(\frac{a}{c} - b\right)(24c^4 - 24c^3 t + 12c^2 t^2 - 4ct^3 + t^4) - h_5 \cos^{-1}\alpha_2\left(\frac{a}{c} - b\right)(120c^4 t + t^5 - $$

$$120c^5 - 60c^3 t^2 + 20c^2 t^3 - 5ct^4)$$

引入无量纲量 x，将式(58)简化为：

$$E_1 = \frac{\lambda g(\zeta) u_1^2}{24e^2 \cos\alpha_1}(x^4 + A_1 x^2 + Q_1 x + F_1) \quad (59)$$

式中：

$$A_1 = \frac{24e^2 \cos\alpha_1}{\lambda g(\zeta)}\left(\frac{a}{2c} - \frac{\lambda g(\zeta)}{2e^2 \cos\alpha_1}\right)$$

$$Q_1 = \frac{24e^2 \cos\alpha_1}{\lambda g(\zeta) u_1}\left[4u_c \cos\alpha_1 + \frac{2u_c}{e^2} - Mg(\tan\alpha_1 + \tan\alpha_2) - \frac{1}{2}\gamma_w h_w^2\right]$$

$$F_1 = \frac{24e^2 \cos\alpha_1}{\lambda g(\zeta) u_1^2} u_1 \left[4u_c^2 \cos^2\alpha_1 - \frac{2u_c^2 u_1}{2e^2} + m_1 - \left[Mg(\tan\alpha_1 + \tan\alpha_2) + \frac{1}{2}\gamma_w h_w^2\right]\right]$$

引入无量纲能量为：

$$V_1 = \frac{24e^2 \cos\alpha_1}{\lambda g(\zeta) u_1^2} E_1 \quad (60)$$

式中，$V_1 = x^4 + A_1 x^2 + Q_1 x + F_1$。

由式(50)知 $x_1 = -2x_2$，突滑时无量纲能量差为：

$$\Delta V_1 = V_1(x_1) - V_1(x_2) = $$

$$x_1^4 - x_2^4 + A_1(x_1^2 - x_2^2) + Q_1(x_1 - x_2) \quad (61)$$

即：

$$\Delta V_1 = \frac{15}{4}(1-k)^2 + \frac{3A_1}{2}(1-k) + \frac{3\sqrt{2}Q_1}{2}\sqrt{(1-k)} \quad (62)$$

突滑释放的能量为：

$$\Delta E_1 = \frac{\lambda g(\zeta) u_1^2}{24e^2 \cos\alpha_1}\left[\frac{15}{4}(1-k)^2 + \frac{3A_1}{2}(1-k) + \frac{3\sqrt{2}Q_1}{2}\sqrt{(1-k)}\right] \quad (63)$$

根据能量转化原理，突滑初速度为：

$$V_0 = \sqrt{\frac{2\Delta E_1}{M}} \quad (64)$$

2) 当 $R \geq R_f$ 时，对于蠕滑段介质本构模型为西原体模型的情况。

将总势能表达式(38)在尖点 $u = u_1 = 2u_c \cos\alpha_1$ 处展开，截取前四次项有：

$$E_2 = \frac{\lambda g(\zeta)}{24u_c^2 e^2 \cos^3\alpha_1}(u - u_1)^4 + \left[-\frac{\lambda g(\zeta)}{2e^2 \cos\alpha_1}\right] \times$$

$$(u - u_1)^2 + \left[4u_c \cos\alpha_1 + \frac{2u_c}{e^2} - \frac{1}{2}\gamma_w h_w^2 + m_2 - Mg(\tan\alpha_1 + \tan\alpha_2)\right](u - u_1) + 4u_c^2 \cos^2\alpha_1 - $$

$$\left[Mg(\tan\alpha_1 + \tan\alpha_2) + \frac{1}{2}\gamma_w h_w^2\right]u_1 - \frac{2u_c^2 u_1}{2e^2} + m_2 u_1 \quad (65)$$

式中：

$$m_2 = e^{-b_5 t}\left\{C_1 - \frac{e^{b_5 t}}{b_5^5(b_5 - b_6)}\{b_3(120h_5 - 24b_5 n_1 + 6b_5^2 n_2 - 2b_5^3 n_3 + b_5^4 n_4) + b_5 \times [b_2 b_5^3 R_f + 2b_4(12b_5 n_1 - 60h_5 - 3b_5^2 n_2 + b_5^3 n_3)]\}\right\} + e^{(b_5 - b_6)t}\left\{C_2 + \frac{e^{b_6 t}}{b_6^5(b_5 - b_6)} \times \{b_3(120h_5 - 24b_6 n_1 + 6b_6^2 n_2 - 2b_6^3 n_3 + b_6^4 n_4) + b_6[b_2 b_6^3 R_f + 2b_4(12b_6 n_1 - 60h_5 - 3b_6^2 n_2 + b_6^3 n_3)]\}\right\}$$

引入无量纲量 x，将式(65)简化为：

$$E_2 = \frac{\lambda g(\zeta) u_1^2}{24e^2 \cos\alpha_1}(x^4 + A_2 x^2 + Q_2 x + F_2) \quad (66)$$

式中：

$$A_2 = \frac{24e^2 \cos\alpha_1}{\lambda g(\zeta)}\left(-\frac{\lambda g(\zeta)}{2e^2 \cos\alpha_1}\right);$$

$$Q_2 = \frac{24e^2 \cos\alpha_1}{\lambda g(\zeta)u_1}\left[\frac{2u_c}{e^2} - \frac{1}{2}\gamma_w h_w^2 + m_2 + 4u_c\cos\alpha_1 - Mg(\tan\alpha_1 + \tan\alpha_2)\right];$$

$$F_2 = \frac{24e^2 \cos\alpha_1}{\lambda g(\zeta)u_1^2}\left[4u_c^2\cos^2\alpha_1 - \frac{2u_c^2 u_1}{e^2} + m_2 u_1 - \left[Mg(\tan\alpha_1 + \tan\alpha_2) + \frac{1}{2}\gamma_w h_w^2\right]u_1\right].$$

引入无量纲能量为：

$$V_2 = \frac{24e^2\cos\alpha_1}{\lambda g(\zeta)u_1^2}E_2 \tag{67}$$

式中，$V = x^4 + A_2 x^2 + Q_2 x + F_2$。

由式(55)可知 $x_1 = -2x_2 = \sqrt{2}$，突滑时无量纲能量差为：

$$\Delta V = V(x_1) - V(x_2) = x_1^4 - x_2^4 + A_2(x_1^2 - x_2^2) + Q_2(x_1 - x_2) \tag{68}$$

即：

$$\Delta V_2 = \frac{15}{4} + \frac{3A_2}{2} + \frac{3\sqrt{2}Q_2}{2} \tag{69}$$

突滑释放的能量为：

$$\Delta E_2 = \frac{\lambda g(\zeta)u_1^2}{24e^2\cos\alpha_1}\left[\frac{15}{4} + \frac{3A_2}{2} + \frac{3\sqrt{2}Q_2}{2}\right] \tag{70}$$

根据能量转化原理，突滑初速度为：

$$V_0 = \sqrt{\frac{2\Delta E_2}{M}} \tag{71}$$

5 工程实例

图 5 表示重庆某滑坡剖面图，滑坡区总体为斜坡地形。滑体主要为粉质粘土夹碎块石组成，下部为粉砂质泥岩。滑动面分为蠕滑段和剪切段，各段介质的物理力学参数如表 1 所示，根据勘测资料，滑坡角 $\beta=60°$，取单位宽度计算，滑坡岩土体的重量为 $M=10^4$ kg，滑体中含水的重量为 $M=10^2$ kg，剪切段介质的峰值位移 $u_c = 2$ m，$t=0$ 时刻的初始位移为 $u(0)=0.025$ m，监测点处于蠕滑段的前端，便于真实地反映滑坡时间-位移关系，根据图 6 所示的位移监测数据拟合得到位移随时间变化的关系为：

$$u = 2\times10^{-37}t^5 - 1\times10^{-29}t^4 + 3\times10^{-22}t^3 - 3\times10^{-15}t^2 + 2\times10^{-8}t + 0.025 \tag{72}$$

式中：长度单位为 m；时间单位为 s。

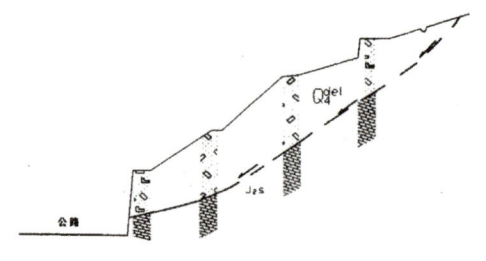

图 5　某滑坡剖面图
Fig.5　Profile of landslides

表 1　滑动面各段介质参数表
Table 1　Parameters of slip surface

剪切段	蠕滑段		
刚度 λ/MPa	弹性模量 G_1/MPa	粘弹性模量 G_2/MPa	粘滞系数 η/(MPa·s)
3200	430	373.9	3.739×10⁶

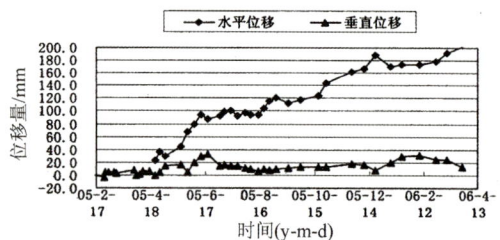

图 6　位移监测数据[15]
Fig.6　Monitoring data of displacement[15]

经计算：$p = 3(k-1)/2 = -0.0106434 < 0$，满足发生突变的必要条件。根据式(27)得到突滑时间的数值解为 $t = 501.86$ d，根据式：

$$\Delta E = \frac{\lambda\mu_1^2}{6e^2}\left[\frac{15}{4}(1-k)^2 + \frac{3A}{2}(1-k) + \frac{3\sqrt{2}Q}{2}\sqrt{(1-k)}\right] \tag{73}$$

可得滑体突滑瞬时释放能量为 $\Delta E = 43608.2$ N·m，根据式：

$$V_0 = \sqrt{\frac{2\Delta E}{M + M_W}} \tag{74}$$

可得突滑初速度 $V_0 = 2.93858$ m/s。

根据我们的建议，该边坡将要滑动，因此相关建设部门进行了抗滑处理，现在抗滑处理效果良好。

6 结论

根据滑面介质流变特性利用尖点突变模型和

实测资料，建立了滑坡的时间预测模型。通过分析表明：滑坡突滑失稳的必要条件仅取决于滑动面剪切段介质和蠕滑段介质的刚度比，即滑体突滑失稳是由系统本身特性决定的。根据尖点突变理论的分叉集方程，得到了滑坡突滑时间的计算公式，同时根据能量转化原理，导出了滑体突滑初速度的计算公式。最后将理论模型应用于某实际的滑坡工程，证明了理论模型的有效性和可行性。

参考文献：

[1] 文宝萍. 黄土地区典型滑坡预测预报及减灾对策研究[M]. 北京: 地质出版社, 1997.
Wen Baoping, Research on the forecast and disaster reduction of typical landslides in loess area [M]. Beijing: Geology Press,1997. (in Chinese)

[2] 潘家铮. 建筑物的抗滑稳定和滑坡分析[M]. 北京: 水利出版社, 1980: 120-132.
Pan Jiazheng. Analysis of stability-improving and Landslides [M]. Beijing: China Waterpower Press, 1980: 120-132. (in Chinese)

[3] 刘忠玉, 马崇武, 苗天德. 高速滑坡运程预测的块体运动模型[J]. 岩石力学与工程学报, 2000, 19(6): 742-746.
Liu Zhongyu, Ma Congwu, Miao Tiande. Kinematic block model of long run-out prediction for high-speed landslides [J]. Chinese Journal of Rock Mechanics and Engineering, 2000, 19(6): 742-746. (in Chinese)

[4] 汪洋, 刘波, 汪为. 滑坡速度计算的改进条分法[J]. 安全与环境工程, 2004, 11(3): 68-70.
Wang Yang, Liu Bo, Wang Wei. Improved slice method for landslide speed calculation [J]. Ground Engineering & Environmental Services, 2004, 11(3): 68-70. (in Chinese)

[5] 肖盛燮, 周小平, 杨海清. 二维高速滑坡力学模型[J]. 岩石力学与工程学报, 2006, 25(3): 456-461.
Xiao Shengxian, Zhou Xiaoping, Yang Haiqing. Two-dimensional mechanical model of high-speed landslides [J]. Chinese Journal of Rock Mechanics and Engineering, 2006, 25(3): 456-461. (in Chinese)

[6] 秦四清. 斜坡失稳的突变模型与混沌机制[J]. 岩石力学与工程学报, 2000, 19(4): 486-492.
Qin Siqing. Nonlinear catastrophic model of slope instability and chaotic dynamics mechanism of slope evolution process [J]. Chinese Journal of Rock Mechanics and Engineering, 2000, 19(4): 486-492. (in Chinese)

[7] 秦四清. 斜坡失稳过程的非线性演化机制与物理预报[J]. 岩土工程学报, 2005, 27(11): 1241-1248.
Qin Siqing. Nonlinear evolutionary mechanisms and physical prediction of instability of planar-slip slope [J]. Chinese Journal of Geotechnical Engineering, 2005, 27(11): 1241-1248. (in Chinese)

[8] van Asch Th W J, Malet J P, van Beek L P H. Influence of landslide geometry and kinematic deformation to describe the liquefaction of landslides: Some theoretical considerations [J]. Engineering Geology, 2006, 88(1-2): 59-69.

[9] 熊传祥, 龚晓南, 王成华. 高速滑坡临滑变形能突变模型的研究[J]. 浙江大学学报(工学版), 2000, 34(4): 443-447.
Xiong Chuanxiang, Gong Xiaonan, Wang Chenghua. A deformation energy catastrophic model of high-speed landslide before sliding [J]. Journal of Zhejiang University (Engineering Science), 2000, 34(4): 443-447. (in Chinese)

[10] 周利杰, 方云. 降雨作用下反倾岩质边坡尖点突变模型研究[J]. 水利与建筑工程学报, 2008, 6(4): 130-131.
Zhou Lijie, Fang Yun. Study on cusp-catastrophic model of anti-dip rock slope under rainfall action [J]. Journal of Water Resources and Architectural Engineering, 2008, 6(4): 130-131. (in Chinese)

[11] 许建聪, 尚岳全, 郑束宁, 张天宝. 强降雨作用下浅层滑坡尖点突变模型研究[J]. 浙江大学学报(工学版), 2005, 39(11): 1676-1679.
Xu Jiancong, Shang Yuequan, Zheng Shuning, Zhang Tianbao. Cusp-catastrophic model of shallow landslide under intensive rainfall [J]. Journal of Zhejiang University (Engineering Science), 2005, 39(11): 1676-1679. (in Chinese)

[12] 龙辉, 秦四清, 万志清. 降雨触发滑坡的尖点突变型[J]. 岩石力学与工程学报, 2002, 21(4): 502-508.
Long Hui, Qin Siqing, Wan Zhiqing. Catastrophe analysis of rainfall-induced landslides [J]. Chinese Journal of Rock Mechanics and Engineering, 2002, 21(4): 502-508. (in Chinese)

[13] 唐春安. 岩石破裂过程的灾变[M]. 北京: 煤炭工业出版社, 1993.
Tang Chun'an. Catastrophe in rock unstable failure [M]. Beijing: China Coal industry Publishing House, 1993. (in Chinese)

[14] 殷有泉, 杜静. 地震过程的燕尾型突变模型[J]. 地震学报, 1994, 16(4): 416-422.
Yin Youquan, Du Jing. Coattail catastrophe model in the process of earthquake [J]. Acta Seismologica Sinica, 1994, 16(4): 416-422. (in Chinese)

[15] 姜永东. 三峡库区边坡岩土体蠕滑与控制的现代非线性科学研究[D]. 重庆: 重庆大学, 2006: 122-123.
Jiang Yongdong. Study on Creep caused landslide and its control in the reservoir area of the Three Gorges based on the modern non-linear sciences gorges [D]. Chongqing: Chongqing University, 2006: 122-123. (in Chinese)

[16] 孙钧. 岩土材料流变及其工程应用[M]. 北京: 中国建筑工业出版社, 1999.
Sun Jun. Rock rheological and engineering applications [M]. Beijing: China Architecture & Building Press, 1999. (in Chinese)

[17] 徐曾和, 徐小荷. 考虑围岩流变特性的地震尖点型突变模型[J]. 岩土力学, 2000, 21(1): 24-27.
Xu Zenghe, Xu Xiaohe. Instability of fault earthquake in rheological media and cusp catastrophe [J]. Rock and Soil Mechanics, 2000, 21(1): 24-27. (in Chinese)

Quantitative analysis of rockburst for surrounding rocks and zonal disintegration mechanism in deep tunnels

Qihu Qian[1] *, Xiaoping Zhou[2]

[1] Engineering Institute of Engineering Corps, PLA University of Science and Technology, Nanjing, 210007, China

[2] School of Civil Engineering, Chongqing University, Chongqing, 400045, China

Received 21 December 2010; received in revised form 25 January 2011; accepted 5 February 2011

Abstract: Rock masses without preexisting macrocracks can usually be considered as granular materials with only microcracks. During the excavation of the tunnels, microcracks may nucleate, grow and propagate through the rock matrix; secondary microcracks may appear, and discontinuous and incompatible deformation of rock masses may occur. The classical continuum elastoplastic theory is not suitable for analyzing discontinuous and incompatible deformation of rock masses. Based on nonEuclidean model of the discontinuous and incompatible deformation of rock masses, the distribution of stresses in the surrounding rock masses in deep tunnels is fluctuant or wavelike. The stress concentration at the tips of microcracks located in vicinity of stress wave crest is comparatively large, which may lead to the unstable growth and coalescence of secondary microcracks, and consequently the occurrence of fractured zones. On the other hand, the stress concentration at the tips of microcracks located around stress wave trough is relatively small, which may lead to the arrest of microcracks, and thus the nonfractured zones. The alternate appearance of stress wave crest and trough thus may induce the alternate occurrence of fractured and nonfractured zones in deep rock masses. For brittle rocks, the dissipated energy of microcrack growth is small, but the elastic strain energy stored in rock masses may be larger than the dissipated energy growths of preexisting microcracks and secondary microcracks. The sudden release of the residual elastic strain energy may lead to rockburst. Based on this understanding, the criteria of rockburst are established. Furthermore, the relationship between rockbursts and zonal disintegration in the surrounding rock masses around deep tunnels is studied. The influences of the insitu stresses and the physicomechanical parameters on the distribution of rockburst zones and the ejection velocity of rock fragments are investigated in detail.

Key words: underground tunnel; rockburst; zonal disintegration; nonEuclidean model

1 Introduction

With the development of underground rock engineering structures at great depth, rock mass failure becomes an increasingly challenging issue for underground engineers. Unlike shallowburied rock masses, deep rock masses in complex geological condition with high insitu stress, high earth temperature, high water pressure and engineering disturbance, are characterized by discontinuous, incompatible and large deformations. Rockbursts and zonal disintegration are typical failure modes of deep rock masses. During the excavation of a tunnel in the deep rock masses, the fractured and nonfractured zones occur alternately around tunnels, which was referred as the zonal disintegration phenomenon in some related publications [14] and was never reported in shallow rock engineering before. Rockburst is a rock failure phenomenon associated with ejection, acoustic emission or microseismic events induced by a sudden release of elastic strain energy stored in the rock masses.

Great efforts have been made to understand rockburst and zonal disintegration phenomena as two main types of failure modes of deep rock masses. Under the hypothesis of linear damage of rock after its peak stresses, rockburst around circular tunnels was analyzed by Pan and Xu [5]. Based on crack propagation in rock masses, the rockburst mechanism was studied and the concept of stress intensity isogram for surrounding rock masses was proposed by Wang et al. [6]. Insitu investigation on zonal disintegration in surrounding rock masses around tunnels was carried out, and reliable evidences to prove the existence of zonal disintegration phenomenon were obtained by Shemyakin et al. [7]. Based on the incompatible deformation of rock masses, a nonEuclidean model of surrounding rock masses around circular tunnels under hydrostatic pressure condition was presented by Guzev et al. [8, 9]. Moreover, on the basis of the discontinuous and incompatible deformation of rock masses, a new nonEuclidean continuum model of the zonal disintegration of surrounding rock masses around a deep circular tunnel in a nonhydrostatic pressure state was established by Qian and Zhou [10]. In addition, the mechanism of zonal disintegration phenomenon of surrounding rock masses was analyzed by using energy criterion [11]. However, the efforts are mainly made to study rockburst and zonal disintegration separately. The studies on the relationship between rockburst and zonal disintegration are very limited.

Relationship between zonal disintegration and rockburst is studied in the paper. The mechanism of the nucleation of secondary microcracks induced by microcracks, the mechanism of the transition of the secondary microcracks from stable to unstable propagation, and the mechanism of the coalescence of secondary microcracks and formation of macrocracks (which lead to rockbursts) are investigated. A new method to analyze stability of deep rock masses is also established.

2 Stress field obtained by nonEuclidean model

In the following discussions, the discontinuous and incompatible deformation of rocklike material was considered using an incompatible strain tensor in the kinematic equations. The internal space of rocklike material after deformation was treated as a nonEuclidean one [10]. Moreover, based on the evolution equation of incompatible tensor, a new nonEuclidean model suitable for nonhydrostatic stress condition was established, and the elastic stress field of surrounding rock mass around deep circular tunnels under unloading was obtained [10]. The expression can be written as follows:

$$\tau_{r\theta} = \frac{\sigma_v - \sigma_h}{2}\left(1 + 2\frac{r_0^2}{r^2} - 3\frac{r_0^4}{r^4}\right)\sin(2\theta)$$
$$+ \frac{E}{r(1-v)\gamma^{3/2}}\left[A_1 J_1(\gamma^{1/2} r) + B_1 N_1(\gamma^{1/2} r) + C_1 K_1(\gamma^{1/2} r)\right]\sin(2\theta) \quad (3)$$
$$- \frac{3E}{r^2(1-v)\gamma^2}\left[A_1 J_2(\gamma^{1/2} r) + B_1 N_2(\gamma^{1/2} r) + C_1 K_2(\gamma^{1/2} r)\right]\sin(2\theta)$$

where r_0 is the radius of tunnels; (r,θ) is the polar coordinates; $\gamma^2 = E/[4q(1-v)]$, q is a nonEuclidean parameter that can be determined by experiments; E is the Young's modulus; v is the Poisson's ratio; σ_v is the vertical insitu stress; σ_h is the horizontal insitustress; and C is determined by the function that coincides with the Wronskian of the linearly independent solutions $J_0(\gamma^{1/2} r)$ and $N_0(\gamma^{1/2} r)$, J_0, N_0 and K_0 are zero order Bessel, Neumann and Macdonald cylindrical functions, respectively. The parameters A, B, A_1 and B_1 can be written as follows:

$$A = (C/2)\pi\gamma^{1/2} r_0\left[K_0(\gamma^{1/2} r_0) N_1(\gamma^{1/2} r_0) - K_1(\gamma^{1/2} r_0) N_0(\gamma^{1/2} r_0)\right] \quad (4)$$

$$B = -(C/2)\pi\gamma^{1/2} r_0 \cdot \left[K_0(\gamma^{1/2} r_0) J_1(\gamma^{1/2} r_0) - K_1(\gamma^{1/2} r_0) J_0(\gamma^{1/2} r_0)\right] \quad (5)$$

$$A_1 = (C_1/2)\pi\gamma^{1/2} r_0\left[K_2(\gamma^{1/2} r_0) N_3(\gamma^{1/2} r_0) - K_3(\gamma^{1/2} r_0) N_2(\gamma^{1/2} r_0)\right] \quad (6)$$

$$B_1 = -(C_1/2)\pi\gamma^{1/2} r_0 \cdot \left[K_2(\gamma^{1/2} r_0) J_3(\gamma^{1/2} r_0) - K_3(\gamma^{1/2} r_0) J_2(\gamma^{1/2} r_0)\right] \quad (7)$$

According to Eqs.(1)(3), the distribution of stresses is plotted in Fig.1. It is obvious, from Fig.1, that the distribution of stresses of surrounding rock masses is fluctuant. The following parameters are used in Figs.1 and 2: $\sigma_v = 0.4$ MPa, $\sigma_h = 0.1$ MPa, $v = 0.2$, $q = 1.448$, $C = 18620\text{m}^{-2}$, $C_1 = 18620\text{m}^{-2}$, $E = 450$ MPa, $r_0 = 0.07$ m, $\theta = 0°$, $c_0 = 4$ mm (initial length of microcrack), $c_1 = 8$ mm (final length of microcrack after growth), $K'_{IC} = 0.03\text{MPa}\cdot\text{m}^{1/2}$ (fracture toughness of rock interface), $K_{IC} = 0.12\text{MPa}\cdot\text{m}^{1/2}$ (fracture toughness of rock). The distribution of fractured zones in surrounding rock masses around tunnel is shown in Fig.2. Unlike that predicted by traditional continuum theory, the fractured and nonfractured zones shown in Fig.2 alternately occur, and the width of fractured zones decreases with increasing distance away from tunnel wall.

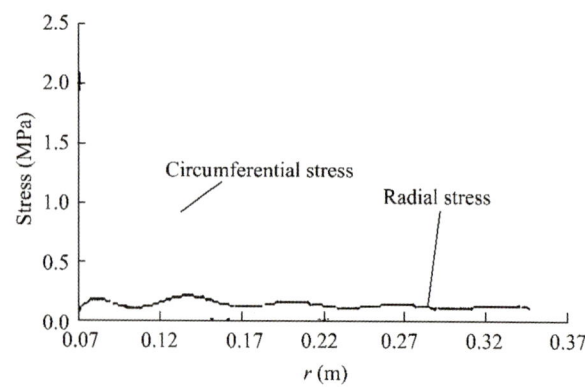

Fig. 1 Distribution of stress in surrounding rock mass around tunnel.

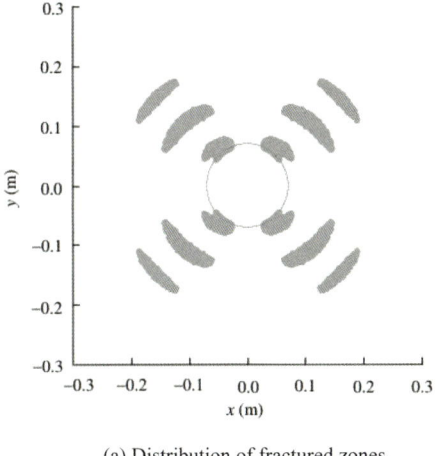
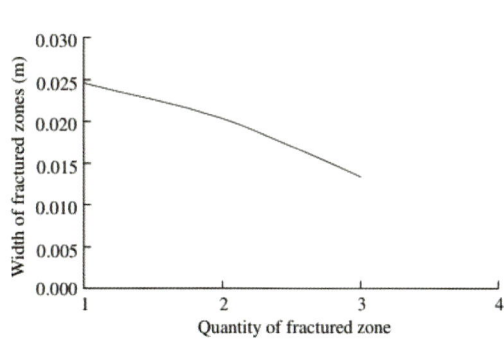

(a) Distribution of fractured zones. (b) Dependence of width of fractured zones on its quantity.

Fig. 2 Fractured zone in surrounding rock mass around a tunnel.

3　Characteristics of fractured zonesand rockburst zones

Relationship between the evolution of microcracksand the evolution of damage of rock was established by Golshani et al.[12,15]. The stage before microcracks growth is the elastic phase of rocks. The stages of growth of microcracks and stable growth of secondary microcracks are the nonlinear hardening phases of rocks. The stage of unstable propagation of the secondary microcracks corresponds to the strain softening phase of rocks [13,15]. Based on the above studies, the developments of rockbursts in continuum surrounding rock mass are categorized into three stages, including microcrack growth, stable propagation of secondary microcracks, and unstable growth and coalescence of secondary microcracks to form macrocracks.

3.1　Microcrack growth

It is assumed that the distribution of microcracks is homogeneous, and the initial length of microcracks is regarded as c_0, which can be determined by mesoscopic experiment. The final length of microcracks after growth is c_1 and the fracture toughness of the rock interface is K_{IC}'. After the excavation of tunnel in the deep rock masses, microcracks will grow along the rock interface. The mechanical behaviors of deep rock masses are characterized by discontinuous and incompatible deformation. The classical elastoplastic theory is not applicable anymore. The nonEuclidean model can be used to analyze the distribution of stresses of surrounding rock mass after the microcrack growth. Based on the nonEuclidean theory, the principal stresses can be written as

$$\left.\begin{aligned}\sigma_1' &= \frac{\sigma_r + \sigma_\theta}{2} + \frac{1}{2}\sqrt{(\sigma_r - \sigma_\theta)^2 + 4\tau_{r\theta}^2} \\ \sigma_2' &= \frac{\sigma_r + \sigma_\theta}{2} - \frac{1}{2}\sqrt{(\sigma_r - \sigma_\theta)^2 + 4\tau_{r\theta}^2}\end{aligned}\right\} \quad (8)$$

The tensile stresses on the surfaces of microcracks are obtained as

$$\sigma'_t = \sigma'_2 + f(c_0)S'_2 \qquad (9)$$

$$S'_2 = \sigma'_2 - (\sigma'_1 + \sigma'_2)/2 \qquad (10)$$

where $f(c_0) = d/c_0$, d is the diameter of rock grain; and σ'_t is the tensile stress on the surfaces of microcracks.

The mode I stress intensity factor at the tips of microcracks is expressed as

$$K_I = -\left[\sigma'_2 + f(c_0)S'_2\right]\sqrt{2d_i \tan\frac{\pi c_0}{2d_i}} \qquad (11)$$

where d_i is the spacing between microcracks that can be obtained by mesoscopic experiments.

The critical condition for the growth of microcracks along the rock interface is

$$K_I = K'_{IC} \qquad (12)$$

When the stress intensity factor at the tips of microcracks satisfies the critical condition (Eq.(12)), the microcracks will grow. The dissipated energy density of microcrack growth can be determined by the integral of energy release rate G along the length of microcracks $c = c_1 - c_0$:

$$U_1 = \frac{N_1(\kappa+1)(1+\nu)}{2E} \cdot \int_{c_0}^{c_1} \left\{\sqrt{2d_i \tan\frac{\pi c}{2d_i}}\left[\sigma'_2 + f(c)S'_2\right]\right\}^2 dc \qquad (13)$$

where $f(c) = d/c$, and N_1 is the density of microcracks that can be obtained by mesoscopic experiment.

3.2 Stable propagation of secondary microcracks

It is observed from the experiments that microcracks will propagate through rock matrix under certain stress conditions, leading to the appearance of microcracks [15, 16]. Based on the length of microcracks and fracture toughness of rock, the critical condition for secondary microcrack nucleation can be written as

$$\sigma'_2 + f(c_1)S'_2 = -\frac{K_{IC}}{\sqrt{2d_i \tan\frac{\pi c_1}{2d_i}}} \qquad (14)$$

where K_{IC} can be determined by experiments.

The stress intensity factor at the tips of the secondary microcracks can be expressed as

$$K_I = -\left[\sigma'_2 + f(l)S'_2\right]\sqrt{2d_i \tan\frac{\pi l}{2d_i}} \qquad (15)$$

where $f(l) = d/l$, l is the growth length of secondary microcracks.

The critical condition for stable growth of secondary microcracks is

$$K_I = K_{IC} \qquad (16)$$

When the growth length of secondary microcracks l reaches c_2, the load capacity of the rock reaches its maximum potential and the rock is fractured.

The critical condition for the rock fracture can be defined as

$$\sigma_2' + f(c_2)S_2' = -\frac{K_{IC}}{\sqrt{2d_i \tan\dfrac{\pi c_2}{2d_i}}} \tag{17}$$

Replacing c_0 with c_2 in Eq.(9), the uniaxial tensile strength of rock $\sigma_{\max}' = \sigma_2' + f(c_2)S_2'$ and the stable growth length of secondary microcrack c_2 can be determined by iteratively using Eqs.(9) and (17).

The dissipated energy density of stable growth of the secondary microcracks can be determined by integration of the energy release rate G along the growth length of secondary microcracks $l = c_2 - c_1$:

$$U_2 = \frac{N_2(\kappa+1)(1+v)}{2E} \cdot \int_{c_1}^{c_2}\left\{\sqrt{2d_i \tan\frac{\pi l}{2d_i}}\left[\sigma_2' + f(c)S_2'\right]\right\}^2 dl \tag{18}$$

where N_2 is the density of secondary microcracks of stable growth, which can be obtained by the numerical simulation.

3.3 Unstable propagation, coalescence of secondary microcracks and formation of macrocracks

When the length of secondary microcracks is larger than c_2, the secondary microcracks will grow unstably and the load capacity of rock decreases, leading to the damage localization of rocks. The critical condition for damage localization is expressed as Eq.(17). Unstable growth of secondary microcracks will lead to the coalescence of secondary microcracks, and further the appearance of macrocracks and the failure of rock masses.

Growth length from nucleation to coalescence of secondary microcracks is

$$l = d_i - c_1 \tag{19}$$

The growth length from unstable propagation to coalescence of secondary microcracks can be written as

$$l = d_i - c_2 \tag{20}$$

It is assumed that the postpeak deformation modulus of rock is E_1, and the dissipated energy density of unstable growth of secondary microcracks can be determined as

$$U_3 = \frac{N_3(\kappa+1)(1+v)}{2E_1} K_{IC}^2 (d_i - c_2) \tag{21}$$

where N_3 is the density of secondary microcracks of unstable propagation, and it can be determined by the numerical computation; and E_1 can be determined by the methods suggested in Ref.[15].

3.4 Criteria of rockburst based on energy analysis

The strain energy stored in rock masses can be defined as

$$U_e = \frac{1}{2}\sigma_{ij}\varepsilon_{ij} \tag{22}$$

For brittle rock, the elastic strain energy stored in the rock masses can be approximately replaced by the strain energy density. Substituting Hooke's law into Eq.(22), the strain energy density stored in rock masses is

$$U_e = \frac{1-\upsilon^2}{2E}\left(\sigma_r^2 + \sigma_\theta^2\right) - \frac{\upsilon(1+\upsilon)}{E}\sigma_r\sigma_\theta + \frac{1+\upsilon}{E}\tau_{r\theta}^2 \tag{23}$$

Neglecting thermal dissipation density during the growth of microcracks and secondary microcracks, the total dissipated energy density can be written as

$$U = U_1 + U_2 + U_3 \tag{24}$$

The occurrence of rockburst should satisfy the following two conditions: (1) the coalescence of secondary microcracks and occurrence of macrocracks; and (2) the total dissipated energy density should be smaller than the elastic strain energy density:

$$\left.\begin{array}{l} l = d_i - c_1 \\ U < U_e \end{array}\right\} \tag{25}$$

If surrounding rock masses around tunnels satisfy the conditions (Eq.(25)), rockburst tends to occur. If the location of rockburst is close to the tunnel wall, rockburst will more likely occur. If the first condition in Eq.(25) is satisfied, only fractured zones will occur.

The present model is suitable for continuum surrounding rock masses that only contain microcracks. Because the discontinuous and incompatible deformation of rock masses is taken into account, the present results are precise.

According to the energy conservation law, the kinetic energy density can be expressed as

$$W = U_e - U \geqslant 0 \tag{26}$$

The ejection velocity of rock fragments is

$$V = \sqrt{2W/\rho} \tag{27}$$

where ρ is the density of rock masses.

The location of fractured zones and rockburst zones, the ejection velocity of rock fragments and the kinetic energy density can all be determined by the above equations.

4 Numerical simulations

Fractured zones and rockburst zones in surrounding rock masses around circular tunnels are analyzed. In the simulations, the following material parameters are used: $r_0 = 7\,\text{m}$, $v = 0.15$, $q = 1.460$, $C = 4.599 \times 10^5\,\text{m}^{-2}$, $C_1 = 4.599 \times 10^5\,\text{m}^{-2}$, $K_{IC}' = 0.13\,\text{MPa}\cdot\text{m}^{1/2}$, $N_1 = 1650$, $N_2 = 1650$, $N_3 = 300$, $d_i = 7.5$ mm, $c_0 = 0.4\,\text{mm}$, $c_1 = 0.8\,\text{mm}$, $\rho = 2200\,\text{kg/m}^3$.

The distribution of fractured zones and rockburst zones in surrounding rock masses under different stress conditions is plotted in Fig.3. For brittle rock, the value of Young's modulus is smaller than that of postpeak modulus. The following parameters are used in the simulation: $d = 8\,\text{mm}$, $E = 28\,\text{GPa}$, $E_1 = 280\,\text{GPa}$, $K_{IC} = 1\,\text{MPa}\cdot\text{m}^{1/2}$. It can be observed from Fig.3 that the distribution of fractured zones and rockburst zones is sensitive to the difference between the horizontal and vertical stresses. When the difference between the horizontal and vertical stresses is large enough, rockburst will occur in places far from the tunnel wall.

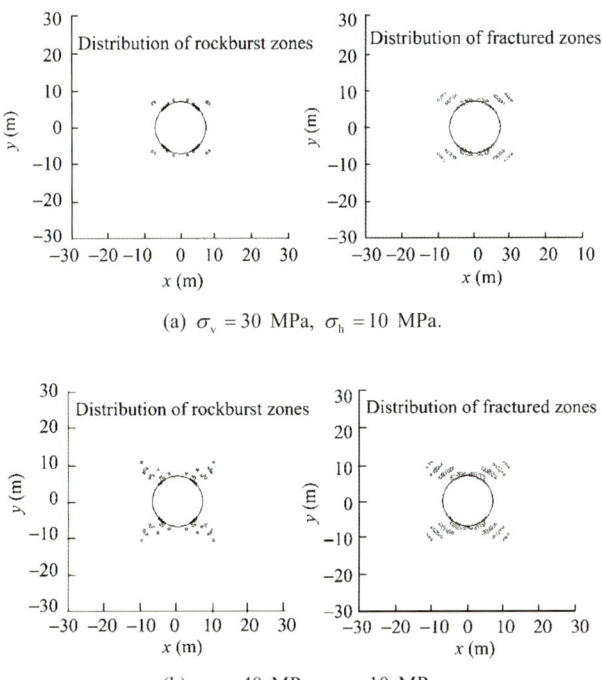

Fig. 3 Distribution of fractured zones and rockburst zones under different stress conditions.

The dependence of area of rockburst zones in surrounding rock masses on the lateral pressure coefficient is plotted in Fig.4. It can be found from Fig.4(a) that the area of rockburst zones increases with the increasing lateral pressure coefficient when it is larger than 1. It can also be found from Fig.4(b) that the area of rockburst zones decreases with the increasing lateral pressure coefficient when it is smaller than 1.

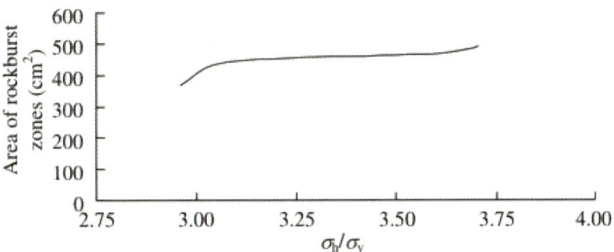

(a) When the lateral pressure coefficient is larger than 1.

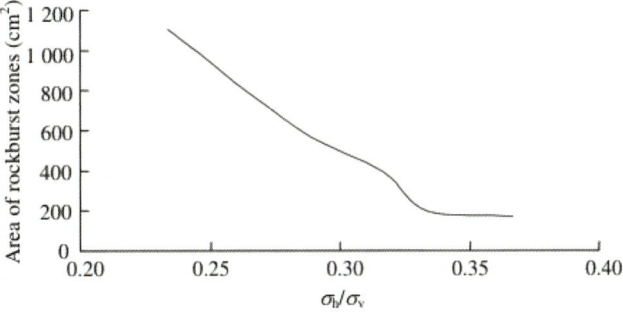

(b) When the lateral pressure coefficient is smaller than 1.

Fig. 4 Area of rockburst zones under different stress conditions.

The dependence of average ejection velocity of rockburst in surrounding rock masses on the lateral pressure coefficient is shown in Fig.5. It is observed from Fig.5(a) that the average ejection velocity of rockburst decreases with the increasing lateral pressure coefficient when it is larger than 1. It is seen from Fig.5(b) that the average ejection velocity of rockburst roughly increases with the increasing lateral pressure coefficient when it is smaller than 1.

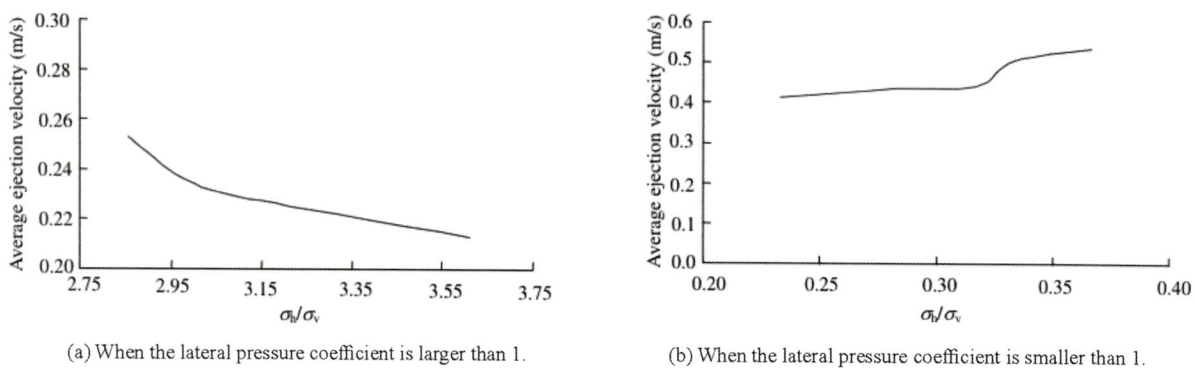

(a) When the lateral pressure coefficient is larger than 1. (b) When the lateral pressure coefficient is smaller than 1.

Fig. 5 The average ejection velocity of rock fragments under different stress states.

The dependence of rockburst zones in surrounding rock mass on fracture toughness is plotted in Fig.6. In the simulations, the following parameters are used: $\sigma_v = 30\text{MPa}$, $\sigma_h = 10\text{MPa}$, $E = 28\text{ GPa}$, $E_1 = 280\text{ GPa}$, $d = 8\text{ mm}$. It is found from Fig.6 that the distribution of rockburst zones is sensitive to the fracture toughness.

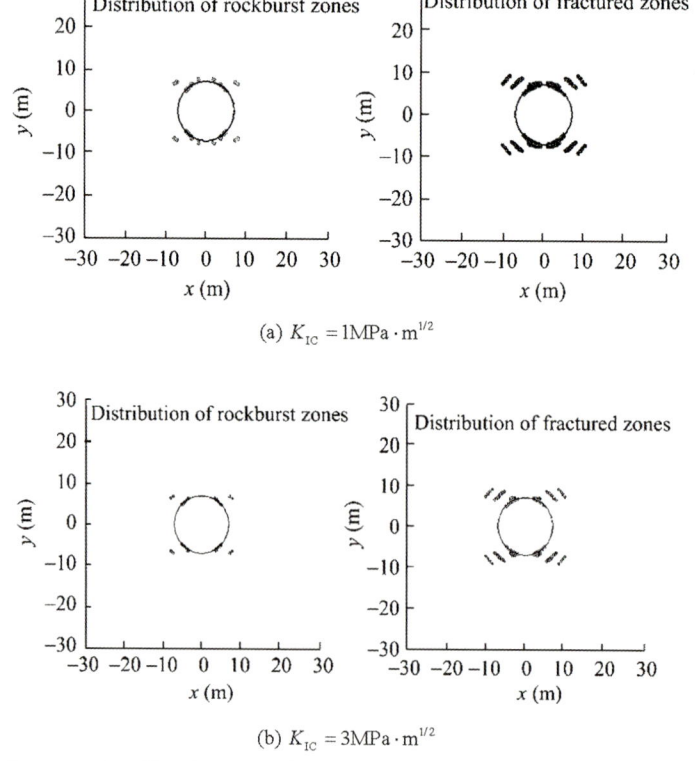

(a) $K_{IC} = 1\text{MPa}\cdot\text{m}^{1/2}$

(b) $K_{IC} = 3\text{MPa}\cdot\text{m}^{1/2}$

Fig. 6 Distribution of fractured zones and rockburst zones under different fracture toughnesses.

The dependence of area of rockburst zones on fracture toughness is shown in Fig.7. It can be seen from Fig.7 that the area of rockburst zones decreases as the fracture toughness increases.

The dependence of average ejection velocity of rockburst zones on fracture toughness is plotted in Fig.8. It is observed from Fig.8 that the average ejection velocity of rockburst zones increases as the fracture toughness increases.

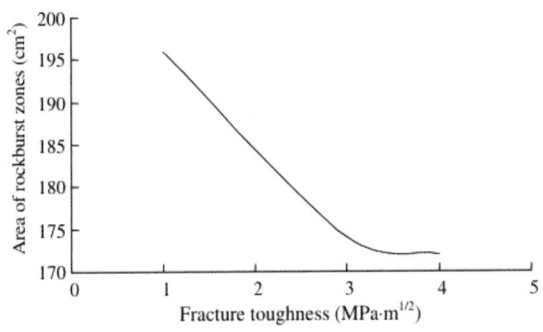
Fig. 7 Area of rockburst zones under different fracture toughnesses.

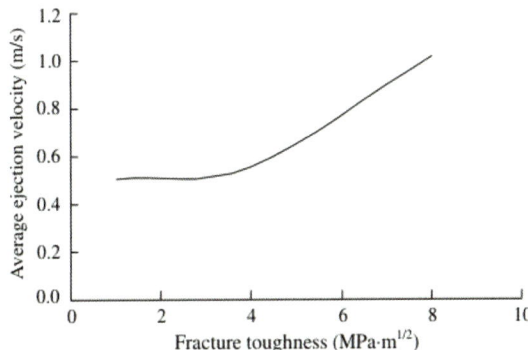
Fig. 8 Average ejection velocity of rock fragments under different fracture toughnesses.

The dependence of fractured zones and rockburst zones in surrounding rock mass on the postpeak modulus of rock is plotted in Fig.9. In the simulations, the following parameters are used: $\sigma_v = 30$ MPa, $\sigma_h = 10$ MPa, $E = 28$ GPa, $d = 8$ mm. It is found from Fig.9 that the distribution of both fractured zones and rockburst zones is sensitive to the postpeak modulus of rock.

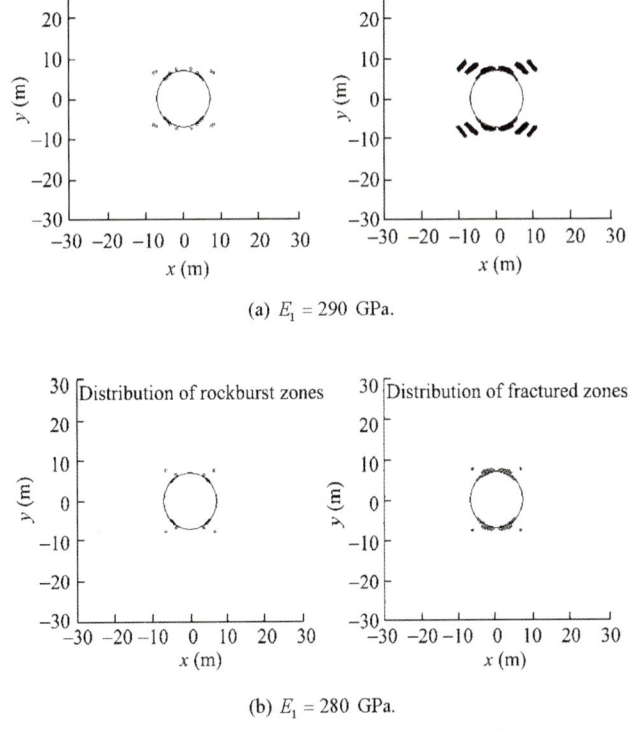

Fig. 9 Distribution of fractured zones and rockburst zones under different postpeak moduli.

The dependence of area of rockburst zones on postpeak modulus of rock is shown in Fig.10. It can be seen from Fig.10 that the area of rockburst zones increases as the postpeak modulus of rock increases.

The dependence of the average ejection velocity of rock fragments on postpeak modulus of rock is shown in Fig.11. It can be seen from Fig.11 that the average ejection velocity of rock fragments increases as the postpeak modulus of rock increases.

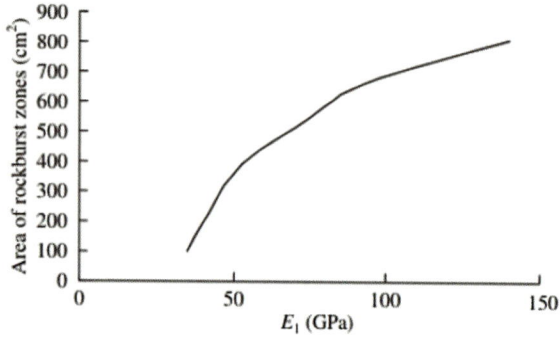

Fig. 10 Area of rockburst zones under different postpeak moduli.

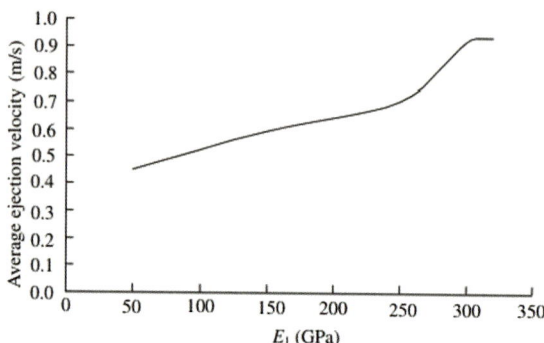

Fig. 11 Average ejection velocity of rock fragments under different postpeak moduli.

5 Discussions and conclusions

Rockburst and zonal disintegration possibly occur in rock masses with high insitu stresses. They are of two different types of failure modes for deep rock masses. The mechanisms of zonal disintegration and rockburst in surrounding rock mass around circular tunnel are analyzed above and discussed below.

The physical process of zonal disintegration can be summarised as follows. During the excavation of tunnels in deep rock masses, the microcracks propagate through the rock matrix, secondary microcracks then appear, and the discontinuous and incompatible deformation of rock masses occurs. Based on nonEuclidean model of the discontinuous and incompatible deformation of deep rock masses, the distribution of stresses of surrounding rock masses is fluctuant. As a result, the stress concentration at the tips of microcracks located around stress wave crest is comparatively large, which leads to the arrest of microcracks, and the occurrence of macrocracks and fractured zones. However, the stress concentration at the tips of microcracks located nearby stress wave trough is relatively small, which leads to the stop of growth of microcracks, and the occurrence of nonfractured zones. The alternate appearance of stress wave crests and troughs induces the alternate occurrence of fractured and formation of nonfractured zones in deep surrounding rock masses. The deep rock masses don't obey the rule of successive condition of fractured zones and nonfractured zones interpreted by classical continuum theory. Meanwhile, the classical continuum theory that is suitable for the shallow rock masses is not valid for the deep rock masses.

The mechanism of rockburst can be summarised as follows. When a rockburst occurs during the excavation of tunnels, it experiences three stages: (1) microcrack growth; (2) stable growth of secondary microcracks; and (3) unstable growth and coalescence of secondary microcracks to form macrocracks. If the dissipated energy to grow microcracks and secondary microcracks is smaller than the elastic strain energy stored in rock masses, the residual strain energy will be released suddenly in the form of the kinetic energy of rock fragments, resulting in rockburst.

It can be concluded from the mechanisms of rockburst and zonal disintegration that both of them are induced by the unstable growth and coalescence of secondary microcracks to form macrocracks. The elastic strain energy stored in rock masses can be either smaller or larger than the dissipated energy to grow microcracks and secondary microcracks. If the elastic strain energy stored in rock masses is larger than the dissipated energy to grow microcracks and secondary microcracks, the residual elastic strain energy will transform into the kinetic energy of rock fragments, leading to the occurrence of rockburst. Otherwise, zonal disintegration will occur.

A prediction model for rockburst, taking into account the zonal disintegration under nonhydrostatic stress condition, has been established. The numerical analysis is made based on the prediction model. The main conclusions from the numerical results are drawn as follows:

（1）Rockburst occurs not only at the tunnel wall but also at the location far from the tunnel wall. The probability of occurrence of rockburst at the location far from the tunnel wall increases as the difference between horizontal and vertical stresses increases.

（2）When the lateral pressure coefficient is smaller than 1, the area of rockburst zones decreases, while the average ejection velocity of rockburst zones increases as the lateral pressure coefficient increases. When the lateral pressure coefficient is larger than 1, the area of rockburst zones increases, while the average ejection velocity of rockburst zones decreases as the lateral pressure coefficient increases.

（3）The area of rockburst zones decreases as the fracture toughness and the postpeak modulus of rock increase. However, the average ejection velocity of rock fragment increases as the fracture toughness increases.

（4）The distribution of fractured zones and rockburst zones depends on the postpeak modulus of rock. The area of rockburst zones and the average ejection velocity of rock fragment increase as the postpeak modulus of rock increases.

References

[1] Qian Qihu. The new development of nonlinear rock mechanics—many key problems of deep rock mass mechanics. In: The Eighth National Rock Mechanics and Engineering Academic Memoir. Beijing: Science Press, 2004 (in Chinese).

[2] Qian Qihu, Li Shuchen. A review of research on zonal disintegration phenomenon in deep rock mass engineering. Chinese Journal of Rock Mechanics and Engineering, 2008, 27 (6): 1 2781 284 (in Chinese).

[3] Qian Q H, Zhou X P, Yang H Q, et al. Zonal disintegration of surrounding rock mass around the diversion tunnels in Jinping II hydropower station, Southwestern China. Theoretical and Applied Fracture Mechanics, 2009, 51 (2): 129138.

[4] Zhou Xiaoping, Qian Qihu. Zonal fracturing mechanism in deep tunnel.Chinese Journal of Rock Mechanics and Engineering, 2007, 26 (5): 877885 (in Chinese).

[5] Pan Yishan, Xu Bingye. The rockburst analysis of circular chamber under consideration of rock damage. Chinese Journal of

Rock Mechanics and Engineering, 1999, 18 (2): 152156 (in Chinese).

[6] Wang Guiyao, Sun Zongqi, Qing Dugan. Fracture mechanics analysis of rock burst mechanism and prediction. The Chinese Journal of Nonferrous Metals, 1999, 9 (4): 841845 (in Chinese).

[7] Shemyakin I, Kyrlenya M V, Reva V N, et al. USSR discovery No.400, phenomenon of zonal disintegration of rocks around underground workings. Byull. Izobret., 1992, (1): 715.

[8] Guzev M A, Paroshin A A. NonEuclidean model of the zonal disintegration of rocks around an underground working. Journal of Applied Mechanics and Technical Physics, 2001, 42 (1): 131139.

[9] Myasnikov V P, Guzev M A. Thermomechanical model of elasticplastic materials with defect structures. Theoretical and Applied Fracture Mechanics, 2000, 33 (3): 165171.

[10] Qian Qihu, Zhou Xiaoping. NonEuclidean continuum model of the zonal disintegration of surrounding rocks around a deep circular tunnel in a nonhydrostatic pressure state. Journal of Mining Science, 2011 (in press).

[11] Reva V N. Stability criteria of underground workings under zonal disintegration of rocks. Fiz. Tekh. Probl. Razrab. Polezn. Iskop., 2002, 38 (1): 3538.

[12] Golshani A, Okui Y, Oda M, et al. A micromechanical model for brittle failure of rock and its relation to crack growth observed in triaxial compression tests of granite. Mechanics of Materials, 2006, 38 (4): 287303.

[13] Li Chunrui, Kang, Lijun, Qi Qingxin, et al. Probe into relationship between zonal fracturing and rock burst in deep tunnel. Journal of China Coal Society, 2010, 35 (2): 185190 (in Chinese).

[14] Golshani A, Oda M, Okui Y, et al. Numerical simulation of the excavation damaged zone around an opening in brittle rock. International Journal of Rock Mechanics and Mining Sciences, 2007, 44 (6): 835845.

[15] Zhou Xiaoping. Analysis of the localization of deformation and the complete stressstrain relation for mesoscopic heterogeneous brittle rock under dynamic uniaxial tensile loading. International Journal of Solids and Structures, 2004, 41 (5/6): 1 7251 738.

[16] Sahouryeh E, Dyskin A V, Germanovich L N. Crack growth under biaxial compression. Engineering Fracture Mechanics, 2002, 69 (18): 2 1872 198.

浅埋地下结构顶板在竖向地震作用下的动力响应

陈灿寿[1], 戚承志[2], 钱七虎[1], 李信桥[2]

(1.解放军理工大学工程兵工程学院,南京210007;2.北京建筑工程学院北京市工程结构与新材料工程研究中心,北京100044)

摘　要：采用结构动力学的方法研究了浅埋地下箱形结构在竖向地震作用下的动力响应。鉴于土与结构动力相互作用分析的复杂性,为了简化分析,整个分析过程分为2步。第1步把结构看作是刚体,利用刚体与地基的相互作用分析求得刚体在竖向地震分量作用下的动力响应;第2步首先考虑到了侧墙对于顶板的抗弯约束作用,求得了顶板的固有频率及振型,并把第1步刚体的动力响应作为输入求解顶板梁的受迫振动,进而求得了顶板弯矩。

关键词：竖向地震分量；顶板；土与结构相互作用；动力响应
中图分类号：P315.96; TU43;U448.22+5　　**文献标志码**：A

Dynamic response of the roof slab of a shallowly buried underground structure under vertical seismic excitation

CHEN Canshou, QI Chengzhi, QIAN Qhu, LI Xinqiao

(1 Engineering Institite of Cops of Engiheers PLA University of Science and Technobgy Nanjing210007, Ching
2. Beijing Engineering Stncture and New Maleral Research Center Beijing Instinte of Civil Engineering and Achitecture Beijng100044, China)

Abstract: Structural dynamic method is applied to study the dynamic response of a shalbwly buried underground stuctire under vertical seismic excitation In view of the camplexity of dynamic soilunderground stuctire in lteraction analysis and in order to sinplify the analysis bwo sleps are adopted in the analysis In the first step the structure is looked at as a rigil body and the dynamic response of the structure under vertical seimic excitation is obtained on the basis of soil-stucture dynamic interaction analysis In the second slep at first the natural frequencies and mode shapes of the structure are obtained with considering the bending restraint of the sidle walls on roof slab and then the dynamic response of roof slab is detemined by taking the dynamic response of the rigid body in the first step as input and firthemore the bending mament of the roof shb is obtained

Key words vertical seismic excitation roof slab; soil-stucture in teractions dynamic response

引言

在早期的地震工程中一个被忽略的问题是地震竖向分量的作用问题。造成这一状况的原因有2个, 一个是人们怀疑是否存在振幅大的竖向地震分量;另一个是竖向地震分量引起的结构动力效应是否强烈。上述第1个原因,主要是由于国际上强震数据库的主要成分是远场地震数据,而强烈震中附近的地震数据较少。还因为地震竖向分量的衰减比水平分量要快,所以回归分析得到的结果是竖向加速度幅值较小。至于结构的动力效应,因为结构抵抗竖向力(重力)的安全系数较高,因此观测到的结果是由竖向地震分量引起

收稿日期:2010-06-11；修订日期:2010-07-02。
基金项目:北京市自然科学基金项目(KZ200810016007);国家自然科学基金项目(50825403);国家973重大项目(2010CB732003)。
本文原载于《世界地震工程》(2010年第26卷第4期)。

的结构内力不大。只是最近30多年,人们才从工程地震学及地震工程学的角度来重新审视上述观点。

Newmark等是最早研究竖向及水平地震峰值的学者之一[1],他们确定了位移、速度和角速度的放大系数。他们基于有限的、主要是美国本土的33组地震数据进行了分析,研究表明,竖向地震加速度峰值可以比较安全地取为水平分量的2/3。但是大量的地震记录表明,许多地震产生的竖向加速度不仅相对于水平地震分量来讲很高,而且就绝对值大小来讲也很大。例如在日本的阪神地震中竖向地震分量与水平地震分量的比值达到了1.69,最大值达到了0.57g。竖向地震分量被认为是阪神地震中许多结构破坏的原因。例如某一大型桁架结构的钢柱的水平截面裂纹被认为是倾覆效应及竖向反应联合作用引起的。在地下结构中,中柱水平截面的环向周边裂纹以及中柱屈曲都表明地震的竖向分量对于结构的破坏具有显著的影响。在阪神地震中,在某些地铁的中柱中观测到中间部分有竖向裂纹,或者中间部分严重压碎。因为对于对称结构来讲,在水平地震动作用下,中柱中间部分的弯矩基本上为零,因此中柱中间部分的破坏不能够认为是水平地震动引起的。

对于地震动记录的分析表明,与水平地震动相比竖向地震动分量到达场地要快,持时要短,强度较高,卓越周期短(在0.2~0.8 s之间)。如果持时短,地震效应在地震持时过后发生,那么可以利用冲量理论来分析结构的动力响应。例如在文献[2]中作者利用冲量力量来分析地下结构在竖向地震动作用下的动力效应。分析结果表明,箱型双层框架结构在竖向地震动作用下,第1层中柱的竖向地震力及重力之和大于第2层中柱的竖向地震力及重力之和。分析还表明结构埋深越大,竖向地震力与重力之比越小。因此竖向地震动必须在地下结构的计算中考虑。

文献[3]利用波动理论得到了中柱的运动方程,他的结论是,中柱的破坏与上覆土的重量密切相关,某一频率的地震波能够引起中柱的共振,并进而引起中柱的破坏。

在文献[4]中利用了数值分析的方法得到了地铁结构中柱和顶板的应力分布,分析表明,地下结构中柱底部到顶部竖向应力依次减小,并且应力反向,使中柱受到拉应力作用,会降低中柱的抵抗力。

文献[5]利用数值分析法分析了大开地铁站在地震作用下的动力响应,他们的计算结果表明:大开地铁站的坍塌主要是由顶板和中柱的破坏引起的,顶板的破坏使得上覆土压力完全由中柱承担,由于中柱本身也存在很大的动力损伤,从而引起中柱和顶板的坍塌,进而引起地面的大沉降量。因此,在现有的地铁车站结构设计中,有必要加强对顶板和中柱的抗压和抗冲剪强度的设计。

总的来讲,竖向地震分量对于地下结构的作用是显著的,在某些情况下会非常严重的,所以在地下结构的抗震设计中应该考虑竖向地震分量的作用。虽然当前数值分析法得到了很大的发展,但是简化的解析分析法仍然具有很大的理论及实际意义,因为简化的解析方法能够直观地表达出结构动力响应的依赖因素和沿着结构的分布。顶板和中柱是地下结构地震时的薄弱部分,而中柱在竖向地震作用时反应的确定有赖于顶板动力反应的确定,所以文中研究浅埋地下箱形结构顶板在竖向地震作用下的动力响应。为了阐述分析方法方便起见,只考虑单跨结构,带中柱的双跨结构将在后续工作中研究。因为总体地考虑土与结构的动力相互作用会非常复杂,不便于分析,所以为了简化分析,整个分析过程分为两步。第1步把结构看作是刚体,利用刚体与地基的相互作用分析求得刚体在竖向地震分量作用下的动力响应。第2步考虑到了侧墙对于顶板的抗弯约束作用求得了顶板的固有频率及振型,并把第1步刚体的动力响应作为输入求解顶板梁的动力响应,求得顶板的内力。

1 结构侧墙处运动规律的近似求解

首先假定地基为温克尔弹性地基,侧墙与地基之间的相互作用系数为 k_1 及 k_2,土体对侧墙的阻尼为 C_1,地基对于基底的弹簧系数为 K_1,阻尼为 C_2(见图1)。

为了在求顶板的动力响应时给出顶板的边界条件,必须知道结构两个侧墙处的运动规律。为了确定结构侧墙处的运动规律,必须求解结构与土的相互作用。为了简化计算,我们此节假定结构为刚性的,即侧墙运动规律的求解问题归结为求解刚性体与地基的相互作用问题。质量块的位移应该与底板在 $x=0$ 及 $x=a$ 处的位移一致。

设刚体块的绝对位移为 $u(t)$,相对位移为 $\tilde{u}(t)$,地基的位移为 $u_0(t)$。这样相对位移与绝对位移和地基的位移之间的关系为:

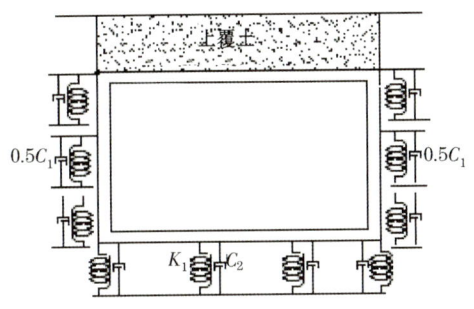

图 1 地下箱形结构的计算模型

Fig 1　Analytical model of an underground box-type structure

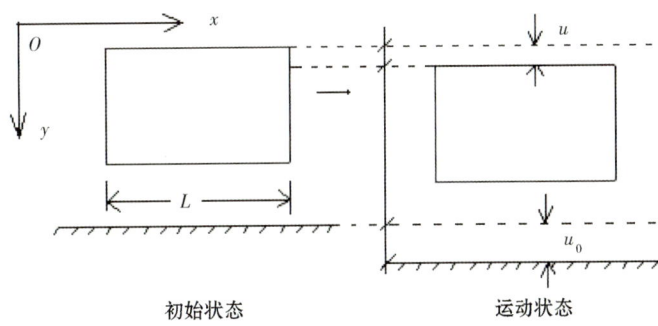

图 2　土与结构的位移图

Fig 2　The displacements of soil and structure

$$\tilde{u}(t) = u_0(t) - u(t) \tag{1}$$

侧墙与地基之间的相互作用系数为 k_{s1} 及 k_{s2}。那么单位面积上的相互作用力可以写为：

$$\sigma_1(t) = k_{s1}(u_0 - u) = k_{s1}\tilde{u}(t),$$
$$\sigma_2(t) = k_{s2}(u_0 - u) = k_{s2}\tilde{u}(t) \tag{2}$$

假设此时土体的位移为 u_0，结构的位移为 u，两侧土体对于结构的粘性阻尼系数简化为每侧 $0.5C_1$，结构与上覆土看作整体，质量为 M，底板文克尔地基模型弹簧系数为 K_1，地基对于结构的粘性阻尼系数简化为 C_2，则根据达朗贝尔原理，结构的运动方程为：

$$|M - M^r|\ddot{u} + (C_1 + C_2)(\dot{u_0} - \dot{u}) + (K_1 + k_{s1} + k_{s2})(u_0 - u) = 0 \tag{3}$$

式中：M^r 为结构所占体积置换的土体质量。引入 M^r 可以使我们在去掉土中结构，回复原状土时得到结构所占体积处的土与周围土一致运动的结果（图 2）。

利用式（1），式（3）重写为：

$$-|M - M^r|\ddot{u} + (C_1 + C_2)\dot{u} + (K_1 + k_{s1} + k_{s2})u = -|M - M^r|\ddot{u_0} \tag{4}$$

当 $|M - M^r| \neq 0$ 时，上式可以简化为：

$$\ddot{u} - 2\xi\omega\dot{u} - \omega^2 u = \ddot{u_0} \tag{5}$$

式中：

$$2\omega\xi = (C_1 + C_2)/|M - M^r|$$
$$\omega^2 = (K_1 + k_{s1} + k_{s2})/|M - M^r| \tag{6}$$

方程（5）的解为：

$$u = \frac{1}{\omega}\int_0^t \ddot{u_0}(\tau) e^{-\xi\omega(t-\tau)}\sin\omega(t-\tau)d\tau \tag{7}$$

式中：$\omega = \omega\sqrt{1-\xi^2}$ 为有阻尼振动频率。

如果取地基的振动为简谐振动 $\ddot{u_0} = \alpha\sin\theta t$ 即 $u_0 = (\alpha/\theta^2)\sin\theta t$ 则在经过一端时间以后自由振动部分将会衰减掉，只剩下稳态受迫振动部分：

$$u = \frac{\alpha}{(\omega^2 - \theta^2)^2 + 4\xi^2\omega^2\theta^2}[(\omega^2 - \theta^2)\cos\theta t - 2\xi\theta\sin\theta t] = A\sin(\theta t - \phi) \tag{8}$$

式中：振幅 A 及相位差 ϕ 分布由下面两式确定：

$$A = \frac{\alpha}{\sqrt{(\omega^2 - \theta^2)^2 + 4\xi^2\omega^2\theta^2}}; \quad \phi = \tan^{-1}\left(\frac{2\xi\omega\theta}{\omega^2 - \theta^2}\right)$$

利用式（1）可以得到刚体块的绝对位移：

$$u(t) = u_0(t) - u(t) = (\alpha/\theta^2)\sin\theta t - A\sin(\theta t - \phi) \tag{9}$$

分析式（6）我们可以看到 刚性质量块的频率主要依赖于地基力学参数及其埋深 因为 M 依赖于埋深。

2 顶板的运动方程及求解

为了简化顶板的动力分析,我们假定底板为刚体,侧墙刚性联结在底板上,并考虑侧墙的柔性。设顶板为弹性的,将顶板在轴线长度取单位长度,简化为梁,假定顶板梁与侧墙为刚性联结。

顶板的绝对位移和相对于侧墙的相对位移分别表示为:

$$w = w(x,t), \quad \bar{w} = \bar{w}(x,t) \tag{10}$$

那么顶板的相对位移 $\bar{w}$ 与结构为刚体时的绝对位移 $u(t)$ 之间的关系为:

$$\bar{w}(x,t) = u(t) - w(x,t) \tag{11}$$

由于结构是浅埋,为了简化计算,我们假定顶板及其上面的覆土一起运动。假定横截面单位长度梁及上覆土的质量为 m,那么顶板运动方程可以写为:

$$D\frac{\partial^4 \bar{w}}{\partial x^4} + m\frac{\partial^2 \bar{w}}{\partial t^2} + C\frac{\partial \bar{w}}{\partial t} = m\frac{\partial^2 u}{\partial t^2} \tag{12}$$

式中:$D = Eh^3/[12(1-\nu^2)]$ 为顶板的抗弯刚度,h 为顶板的厚度。

式(12)为线性偏微分方程,可采用可以分离变量法来求解[6]。为此首先讨论下列齐次方程的解:

$$D\frac{\partial^4 \bar{w}}{\partial x^4} + m\frac{\partial^2 \bar{w}}{\partial t^2} + C\frac{\partial \bar{w}}{\partial t} = 0 \tag{13}$$

采用变量分离法求解,即设式(13)的解为:

$$\bar{w}(x,t) = \sum_{i=1}^{\infty} Y_i(x)\phi_i(t) \tag{14}$$

式中:$Y_i(x)$ 为梁的振型;$\phi_i(t)$ 为依赖于时间的函数,称为广义坐标。

把式(14)代入到式(13)得到 2 个独立的方程:

$$\frac{\partial^2 \phi_i}{\partial t^2} + \frac{C}{m}\frac{\partial \phi_i}{\partial t} + \omega^2 \phi_i = 0 \tag{15}$$

$$\frac{\partial^4 Y_i}{\partial x^4} - k^4 Y_i = 0 \tag{16}$$

式中:$k^4 = \dfrac{m\omega^2}{D}$,由式(16)可以解出振型 $Y(x)$:

$$Y(x) = A_1\cos kx + A_2\sin kx + A_3\cosh kx + A_4\sinh kx \tag{17}$$

对于顶板与侧墙刚结的情况,求解结构的动力响应严格地讲需要求解顶板与侧墙的联合振动问题。此处为了简化分析,我们假定侧墙对于顶板的抗弯作用用抗弯弹簧来代替,抗弯弹簧系数为 k_ϕ,即侧墙顶端转动单位角度需要施加的力矩。

顶板此时的边界条件为:

$$\begin{aligned}&x=0,\ Y(0)=y_0=0,\ DY''(0)=M_0=k_\phi Y'(0)\neq 0,\ F_{s0}\neq 0,\ Y'(0)=y_0'\neq 0;\\ &x=L,\ Y(L)=0,\ DY''(L)=M_L=-k_\phi Y'(L)\end{aligned} \tag{18}$$

把式(18)代入到式(17)整理后得到:

$$\begin{aligned}&A_3 = -A_1\\ &2DkA_1 + k_\phi A_2 + k_\phi A_4 = 0\\ &(\cos kL - \cosh kL)A_1 + \sin kL\cdot A_2 + \sinh kL\cdot A_4 = 0\\ &(Dk\cos kL + Dk\cosh kL + k_\phi \sin kL + k_\phi \sinh kL)A_1\\ &+ (Dk\sin kL - k_\phi \cos kL)A_2 - (Dk\sinh kL + k_\phi \cos kL)A_4 = 0\end{aligned} \tag{19}$$

用矩阵形式表示为:

$$\begin{bmatrix} 2Dk & k_\phi & k_\phi \\ (\cos kL - \cosh kL) & \sin kL & \sinh kL \\ (Dk\cos kL + Dk\cosh kL & (Dk\sin kL & -(Dk\sinh kL \\ + k_\phi \sin kL + k_\phi \sinh kL) & -k_\phi \cos kL) & +k_\phi \cosh kL) \end{bmatrix} \begin{Bmatrix} A_1 \\ A_2 \\ A_4 \end{Bmatrix} = 0 \tag{20}$$

上式为关于 $A_1 \sim A_4$ 的线性方程?因为 $A_1 \sim A_4$ 不为 0,故式(20)的系数行列式应该为 0。

$$\begin{bmatrix} 2Dk(\cos kL - \cosh kL) & k_\phi \sin kL & k_\phi \sinh kL \\ (Dk\cos kL + Dk\cosh kL + k_\phi \sin kL + k_\phi \sinh kL) & (Dk\sin kL - k_\phi \cos kL) & -(Dk\sinh kL + k_\phi \cosh kL) \end{bmatrix} = 0 \quad (21)$$

将上行列式展开并整理,最终得到问题的频率方程:

$$\sin kL \cdot \sinh kL + \frac{k_\phi}{Dk}(\sin kL \cdot \cosh kL - \cos kL \sinh kL) + \frac{k_\phi^2}{2D^2k^2}(1 - \cos kL \cosh kL) = 0 \quad (22)$$

式(22)在 $k_\phi = 0$ 时得到两端简支梁的特征方程:$\sinh kL = 0$;而在 $k_\phi = \infty$ 时得到两端固定梁的特征方程:$\sinh kL \cdot \cos kL = 1$。所以特征方程(22)介于两端简支梁和两端固定梁之间的梁的特征方程。因此可以定性地判断,对于两端刚结的梁,kL 的取值范围介于两端简支和两端固定梁的 kL 值之间:$kL \in (i\pi, (i+0.5)\pi)$。下面就通过例子来验证。

3 算例

对于侧墙,设墙厚为 $\delta_w = 0.7$ m,从底板内侧板面到顶板下侧板面的高度为 $h_w = 5.5$ m,混凝土的杨氏模量为 $E = 2.35 \times 10^{10}$ N/m^2,泊淞比 $v = 0.16$,混凝土的密度为 $\rho = 2400$ kg/m^3,则对于沿结构轴向取 1m 得到的板梁结构,截面惯性矩 $I_w = 1 \times 0.7^3/12 = 0.0286$ m^4。变形为平面应变,侧墙抗弯刚度为 $D_w = Eh^3/[12(1-v^2)] = 6.9 \times 10^8$ N·m^2,地基系数 $k_g = 2.0 \times 10^8$ N/m^3,参数 $\beta_w = \sqrt[4]{k_g \cdot 1m/(4D_w)} = 0.519$/m,$\beta_w h = 0.519 \times 5.5 = 2.85$。对于动力问题来讲,侧墙顶端的转动刚度依赖于振动频率,此处为简化计算,作为初级近似,取墙顶端的转动刚度为弹性地基梁远端固定时的转动刚度[7]:

$$k_\phi = 2D\beta \frac{\sinh \beta h \cosh \beta h - \sin \beta h \cos \beta h}{\sinh^2 \beta h - \sin^2 \beta h} = 7.244 \times 10^8 \text{ N} \cdot \text{m} \quad (23)$$

设顶板梁的跨度为 $L = 7.0$ m,顶板厚为 $\delta_c = 0.7$ m,抗弯刚度为:$D_c = Eh^3/[12(1-v^2)] = 6.9 \times 10^8$ N·m^2,设上覆土的厚度为 4m,土密度为 $\rho_s = 1800$ kg/m^3,则梁及上覆土单位长度的质量为:$m = 0.7 \times 1 \times 2400 + 4 \times 1 \times 1800 = 8880$ kg/m,通过试算法可得梁振动前3阶振动模式的特征频率和振型。

对于第1振动模态,特征值为:

$$(kL)_1 = 1.287\pi = 4.042$$

特征频率为:

$$\omega_1 = \frac{(kL)_1^2}{L^2}\sqrt{\frac{D}{m}} = \frac{4.042^2}{7^2}\sqrt{\frac{6.9 \times 10^8}{8880}} = 92.94 \text{ Hz}$$

振型为:

$$Y_1(x) = \cos kx - 2.065\sin kx - \cosh kx + 0.965\sinh kx \quad (24)$$

对于第2振动模态,特征值为:

$$kL = 2.212\pi = 6.949$$

特征频率为:

$$\omega_2 = \frac{(kL)_2^2}{L^2}\sqrt{\frac{D}{m}} = \frac{6.949^2}{7^2}\sqrt{\frac{6.9 \times 10^8}{8880}} = 274.72 \text{ Hz}$$

振型为:

$$Y_2(x) = \cos kx - 2.8916\sin kx - \cosh kx + 1.002\sinh kx \quad (25)$$

对于第3振动模态,特征值为:

$$(kL)_3 = 3.1674\pi = 9.951$$

特征频率为:

$$\omega_3 = \frac{(kL)_3^2}{L^2}\sqrt{\frac{D}{m}} = \frac{9.951^2}{7^2}\sqrt{\frac{6.9 \times 10^8}{8880}} = 560.9 \text{ Hz}$$

振型为:

$$Y_3(x) = \cos kx - 3.7082\sin kx - \cosh kx + \sinh kx \quad (26)$$

表 1 给出对于简支梁、两端固定梁[8]及本例梁的特征值的比较。

从比较可以看出,确实本例的两端刚结梁的固有频率介于简支梁情况及两端固定梁情况的频率之间。

下面研究结构的受迫振动。

取底板地基系数为 $k_s = 2.0 \times 10^8 \text{ N/m}^3$,侧墙抗剪系数:$k_{x10} = k_{x20} = 0.7 \times 10^8 \text{ N/m}^3$,单位宽度底板总面积为 $S_b = 8.4 \times 1 = 8.4 \text{ m}^2$,两侧侧墙总面积为 $S_w = 2 \times 6.9 \times 1 = 13.8 \text{ m}^2$,所以 $K_1 + k_1 + k_2 = 26.46 \times 10^8 \text{ N/m}$。梁及上覆土单位长度的质量为:$m = 0.7 \times 1 \times 2400 + 4 \times 1 \times 1800 = 8880 \text{ kg/m}$,土密度为 $\rho_s = 1800 \text{ kg/m}^3$,结构与上覆土的质量为 $M = 19.46 \times 2400 + 8.4 \times 1800 = 1.072 \times 10^5 \text{ kg}$。结构的体积取代的土的质量为 $M^r = 57.96 \times 1800 = 1.0433 \times 10^5 \text{ kg}$,所以

$$\omega^2 = (K_1 + k_1 + k_2)/|M - M^r| = \frac{26.46 \times 10^8}{(1.072 - 1.0433) \times 10^5} = 8.522 \times 10^5$$

表 1 3种梁端支撑情况下特征值的比较
Table 1 The comparison of the characteristic values of beam for 3 support conditions

支撑情况	基频 $(kL)_1$	2阶频率 $(kL)_2$	3阶频率 $(kL)_3$
两端简支	π	2π	3π
两端固定	1.5π	2.5π	3.5π
本例梁	1.287π	2.21π	3.167π

如果取地基振动为谐振动 $\ddot{u}_0 = \alpha \sin\theta t = 5\sin 50 t \text{ m/s}^2$,阻尼比 $\xi = 0.2$,则在经过一段时间以后自由振动部分将会衰减掉,只剩下稳态受迫振动部分:

$$\tilde{u} = A\sin(\theta t - \phi) \tag{27}$$

式中:振幅 A 及相位差 ϕ 分布由下两式确定:

$$A = \frac{\alpha}{\sqrt{(\omega^2 - \theta^2)^2 + 4\xi^2\omega^2\theta^2}} = 5.86 \times 10^{-6} \text{m} \approx 0; \quad \phi = \tan^{-1}\left(\frac{2\xi\omega\theta}{\omega^2 - \theta^2}\right) \approx 0$$

结构的绝对运动为:

$$\ddot{u}(t) = \ddot{u}_0(t) - \ddot{\tilde{u}}(t) \approx \ddot{u}_0(t) = 5\sin 25 t \text{ m/s}^2 \tag{28}$$

所以结构基本上按照土的运动规律运动。下面利用振型分解法求解顶板的运动方程。

为了能够利用振型的正交性来解方程(12),我们设单位长度上的阻尼系数与该处的质量密度成正比:

$$C(x) = \alpha_0 m(x) \tag{29}$$

式中:α_0 为系数。

把式(14)代入到方程式(12)中去,再把得到的方程两边乘以第 n 振型 $Y_n(x)$,并进行积分,根据振型的正交性可得解耦的振型方程:

$$M_n \ddot{\phi}_n(t) + \alpha_0 M_n \dot{\phi}_n(t) + \omega_n^2 M_n \phi_n(t) = P_n(t) \tag{30}$$

设 $\xi_n = \frac{\alpha_0}{2\omega_n}$,则上式可以写为:

$$\ddot{\phi}_n(t) + 2\xi_n\omega_n\dot{\phi}_n(t) + \omega_n^2\phi_n(t) = P_n(t)/M_n, \quad n = 1, 2, \ldots \tag{31}$$

式中:广义质量 M_n 及广义力 $P_n(t)$ 由下式确定:

$$M_n = \int_0^L Y_n(x)^2 m(x) dx \quad P_n(t) = \int_0^L Y_n(x) m(x) \ddot{u}(t) dx \tag{32}$$

对于第 1 振型

$Y_1(x) = \cos kx - 2.065\sin kx - \cosh kx + 0.9656\sinh kx$ 利用式(32)可以求得等效质量及等效荷载

$$M_1 = 21.3964 m; \quad P_1(t) = 50.465 \cdot m\sin 50 t$$

所以广义坐标方程为:

$$\ddot{\phi}_1(t) + 2\xi_1\omega_1\dot{\phi}_1(t) + \omega_1^2\phi_1(t) = P_1(t)/M_1 = 2.452\sin 50 t \tag{33}$$

取 $\xi_1 = 0.1$,得到的稳态解为:

$$\phi_1 = A_1\sin(50 t - \phi_1) \tag{34}$$

式中:

$$A_1 = \frac{-2.452}{\omega_1^2\sqrt{(1 - \theta^2/\omega_1^2)^2 + 4\xi_1^2\theta^2/\omega_1^2}} = 3.95 \times 10^{-4} \text{m}$$

$$\phi_1 = \tan^{-1}\left(\frac{2\xi\omega\theta}{\omega^2 - \theta^2}\right) = 8.61°$$

由于外载为对称的,第 2 振型是反对称的,振动能量几乎不会分配到该振型上的。第 3 振型频率与外载频率相比非常大,故这一振型的贡献也不大。因此可以只取第 1 振型:

$$\tilde{w} = Y_1(x)\phi_1(t) = A_1(\cos kx - 2.065\sin kx - \cosh kx + 0.9656\sinh kx)\sin(50t - \phi_1)$$

峰值 $|\tilde{w}| \approx 0.001$ m。

所以板的绝对运动位移为:

$$w(x,t) = u(t) - \tilde{w}(x,t)$$

弯矩为:

$$M(x,t) = D\tilde{w}'' = Y_1''(x)\phi_1(t)$$
$$= DA_1 k^2(-\cos kx + 2.065\sin kx - \cosh kx + 0.9656\sinh kx)\sin(50t - \phi_1) \tag{35}$$

顶板与竖墙连接处的弯矩为:

$$M(0,t) = 6.9 \times 10^8 \times 3.94 \times 10^{-4} \times 0.5774^2 \times (-1 + 0 - 1 + 0)\sin(50t - \phi_1)$$
$$= -1.8 \times 10^5 \sin(50t - \phi_1) \text{ N·m}$$

顶板跨中的弯矩为:

$$M(3.5,t) = 6.9 \times 10^8 \times 3.94 \times 10^{-4} \times 0.5774^2 \times (0.435 + 1.8585 - 3.8392 + 3.5792)\sin(50t - \phi_1)$$
$$= 1.84 \times 10^5 \sin(50t - \phi_1) \text{ N·m}$$

要求总的弯矩,还需要叠加上静力荷载引起的弯矩。

4 结论

为了评价在竖向地震作用下顶板的受力及变形,文中采用结构动力学的方法研究浅埋地下箱形结构在竖向地震作用下的动力响应。为了简化分析,整个分析过程分为两步。第 1 步把结构看作是刚体,利用刚体与地基的相互作用分析求得刚体在竖向地震分量作用下的动力响应。第 2 步首先考虑到了侧墙对于顶板的抗弯约束作用求得了顶板的固有频率及振型,并把第 1 步刚体的动力响应作为输入求解顶板梁的动力响应,得到了顶板的位移及内力。

参考文献

[1] Newmark N M, Blume J A, Kapur K K. Seismic design spectra for nuclear power plants[J]. Journal of Power Division, 1973, 99(2): 287–303.

[2] 杨春田. 日本阪神地震地铁工程的震害分析[J]. 工程抗震, 1996, 17(2): 40–42.
Yang Chuntian. Damage analysis of subway during Kobe earthquake[J]. Earthquake Resistant Engineering, 1996, 17(2): 40–42.

[3] 于翔,钱七虎,赵跃堂,等. 地铁工程结构破坏的竖向地震力影响分析[J]. 解放军理工大学学报(自然科学版), 2001, 2(3): 30–35.
Yu Xiang, Qian Qihu, Zhao Yuetang, et al. Influence of vertical seismic excitation on subway structure damage[J]. Journal of PLA university of Science and Technology (Natural science edition), 2001, 2(3): 30–35.

[4] 国胜兵,赵毅,赵跃堂,等. 地下结构在竖向和水平地震荷载作用下的动力分析[J]. 地下空间, 2002, 22(4): 314–319.
Guo Shengbing, Zhao Yi, Zhao Yuetang, et al. Dynamic analysis of underground structures under vertical horizontal seismic excitations[J]. Underground Space, 2002, 22(4): 314–319.

[5] 庄海洋,程绍革,陈国兴. 阪神地震中大开地铁车站震害机制数值仿真分析[J]. 岩土力学, 2008, 29(1): 246–250.
Zhuang Haiyang, Cheng Shaoge, Chen Guoxing. Numerical simulation and analysis of earthquake damages of Dakai metro station caused by Kobe earthquake[J]. Rock and Soil Mechanics, 2008, 29(1): 246–250.

[6] 克拉夫 R,彭津 J. 结构动力学(第 2 版(修订本))[M]. 王光远,等译校. 北京: 高等教育出版社, 2006.
Clough R W, Penzien J. Dynamics of Structures[M]. Wang Guangyuan, translated. Beijing: High Educational Press, 2006.

[7] 龙驭球. 弹性地基梁的计算[M]. 北京: 人民教育出版社, 1981.
Long Yuqiu. Computation of Beams on Elastic Foundation[M]. Beijing: People's Educational Publisher, 1981.

[8] Thomson W T, Dahleh M D. Theory of Vibration with Application (Fifth edition)[M]. Prentice-Hall Inc, 1998.

功能梯度材料结构分析的半解析数值方法研究

燕秀发[*1,2], 钱七虎[1], 王 玮[2], 孙 翱[2], 戴 耀[3]

(1. 解放军理工大学 工程兵工程学院, 南京 210007; 2. 解放军 91550 部队, 大连 116023;
3. 解放军装甲兵工程学院, 北京 100072)

摘 要: 一种典型的半解析数值方法——线法被引入功能梯度材料的结构分析。首先推导了功能梯度材料位移形式的平衡方程和边界条件, 然后阐述了线法功能梯度材料结构分析的基本步骤和数值原理。该方法的基本思想是通过有限差分将问题的控制方程半离散为定义在沿梯度方向离散节线上的常微分方程组, 然后应用 B 样条函数 Gauss 配点法求解该常微分方程组得到问题的解答。为演示线法在功能梯度材料结构分析中的应用, 给出了线性梯度和指数梯度功能梯度材料板分别受恒定位移、均匀拉伸载荷和弯曲载荷作用的数值算例。与相应问题解析解和其他数值方法的比较表明, 线法的计算结果具有很高的精度, 而且不需要任何特殊的考虑就能够有效模拟材料内部物性参数的连续变化, 也无需事先选取满足特定条件的待定场函数, 是一种非常适合功能梯度材料结构形式和材料特点的半解析数值方法。

关键词: 线法; 功能梯度材料; 有限差分; B 样条函数 Gauss 配点法; 半解析法
中图分类号: O34.2   **文献标识码**: A

1 引 言

功能梯度材料是新一代非均匀材料, 具有空间变化的微观结构和热传导率、密度、弹性模量及泊松比等连续变化的材料宏观属性。该类材料的独特之处在于可以通过定制材料的成分和梯度满足最终的需要, 提供了材料设计的新思想。因此, 在众多工程领域都得到了广泛应用。但是由于材料内部连续变化的物性参数导致该类材料结构分析的控制方程都成为变系数微分方程(组), 并具有多场相互耦合的特点, 引起了数学处理上的极大困难。因此, 数值方法在功能梯度材料研究中具有重要的地位。

目前, 有限单元法和无网格法在功能梯度材料结构分析中得到了深入研究和广泛应用。Anals等[2]应用基于复合材料层板模型的均匀元法, 研究了功能梯度材料结构的断裂特性。Kim[3] 和 Zou[4] 扩展了均匀材料有限单元法中等参变换的概念, 提出了等参梯度元法。Rao[5]等应用无网格法计算了含裂纹功能梯度材料板的应力强度因子, 得到了较高精度的数值结果。然而在这些研究中也发现了上述方法存在的问题, 层板模型均匀元法的主要问题是计算结果严重依赖于子层细化程度, 层间材料属性的剧烈变化会带来很大的计算误差[2]。等参梯度元法的计算实践表明, 该法的计算结果具有单元类型敏感性和不确定性[3], 表现为同一问题应用不同的单元类型可能得到差异很大的计算结果。无网格法的问题在于, 该法中节点的影响域(支撑域)是重叠的, 与有限单元法中求解域被离散成相互联接的单元相比, 其计算量远大于有限单元法, 不便于分析各种梯度形式对结构物理特性的影响。

因此, 李永等[6]提出了求解功能梯度材料板弯曲问题的半解析数值方法——新康法。与实验结果的对比表明该法具有很高的计算精度, 提供了发展功能梯度材料结构分析半解析数值方法的新思路。然而, 该法针对特定问题特别选取的待定函数形式使其还不便推广应用于一般性的功能梯度材料结构分析问题。本文尝试将一种典型的具有普遍意义的半解析数值方法——线法引入功能梯度

收稿日期: 2009-01-15; 修改稿收到日期: 2009-06-02。
基金项目: 中国博士后科学基金面上资助(20080431344); 国家自然科学创新研究团体基金(51021001)资助项目。
本文原载于《计算力学学报》(2010 年第 27 卷第 6 期)。

材料的结构分析。首先推导了线法功能梯度材料结构分析的控制方程,然后阐述了求解步骤和数值原理,最后通过具体算例演示了该法的实现过程,并与解析解以及其他数值方法的计算结果进行对比,进一步表明了线法对于功能梯度材料结构分析的有效性和适用性。

2 控制方程

功能梯度材料具有连续变化的微结构和材料属性。在连续介质意义上该类材料可以被考虑为物性参数是空间坐标连续函数的各向同性的非均匀实体。

2.1 平衡方程

类似于其他常见结构分析方法,线法也是以位移作为基本求解量。记 $x_i = \{x, y, z\}$,不失一般性,定义材料参数为弹性模量 $E = E(x_i)$,泊松比 $\mu = \mu(x_i)$,剪切模量 $G = G(x_i) = E/[2(1+\mu)]$,并记 $H = H(x_i) = \mu/[(1-2\mu)]$。不计体力,将几何方程代入物理方程,由应力平衡微分方程可得

$$G(u_{i,jj} + u_{j,ij}) + 2GHu_{k,ki} + (G_{,j}u_{i,j} + G_{,j}u_{j,i}) + (2G_{,i}H + 2GH_{,i})u_{k,k} = 0 \quad (1)$$

式中 $i, j, k = 1, 2, 3$;下标 $,j$ 代表 $\partial/\partial x_j$。

2.2 边界条件

类似式(1)的过程,在应力边界 S_σ 上有

$$[G(u_{i,j} + u_{j,i}) + 2GHu_{k,k}\delta_{ij}]n_j = T_i \quad (2)$$

式中 T_i 为作用在 S_σ 上的表面力;$n_j = \cos(n, x_j)$,n 为边界面的外法线方向。在位移边界 S_u 上有

$$u_i = \bar{u}_i \quad (3)$$

式中 $\bar{u}_i$ 为作用在 S_u 上的位移。

3 求解步骤和数值原理

3.1 离散求解区域

当前制备和应用的功能梯度材料结构普遍具有单向梯度的特点。因此,在线法功能梯度材料分析中,仅应用若干节线将求解区域沿非梯度方向离散。例如对于图1所示由边界 Γ_1,Γ_2,Γ_3 和 Γ_4 构成的具有沿 y 方向材料梯度的求解区域 Ω,该区域被沿 x 方向的 n 条节线离散,但保持梯度方向 y 不被离散。显然,这种离散方式与功能梯度材料结构的特点相适应并具有半离散的性质。

求解区域离散后,原问题转化为求解定义在节线上的待求场量的控制方程(简称为节线控制方程)。

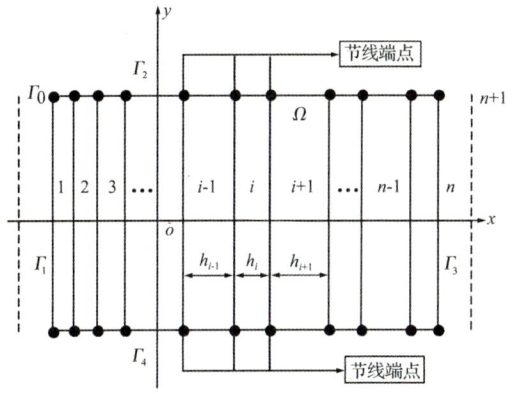

图 1 求解区域的离散
Fig.1 The partition of problem domain by discrete lines

3.2 建立定义在节线上的常微分方程组

将与式(1)相对应的节线平衡方程中待求位移函数对离散坐标的偏导数用由 Taylor 公式得到的具有二阶精度的三点中心差分近似,如对图1中离散坐标为 x 的平面问题,设 $f = f(x,y)$ 为待求场函数,记 $f_i = f(x_i,y)$(此处下标 i 为节线编号),$f'_i = df_i/dy$,并注意 $x = x_i$ 处的相邻节线间距 h_i,h_{i+1} 和 h_{i-1},则该处的偏导数近似公式为

$$\begin{cases} \left(\dfrac{\partial^2 f}{\partial x^2}\right)_i = \dfrac{2[h_i f_{i-1} - (h_i + h_{i-1})f_i + h_{i-1}f_{i+1}]}{h_{i-1}h_i(h_{i-1} + h_i)} \\ \left(\dfrac{\partial f}{\partial x}\right)_i = \dfrac{h_{i-1}^2 f_{i+1} + (h_i^2 - h_{i-1}^2)f_i - h_i^2 f_{i-1}}{h_i h_{i-1}(h_{i-1} + h_i)} \\ \left(\dfrac{\partial^2 f}{\partial x \partial y}\right)_i = \dfrac{h_{i-1}^2 f'_{i+1} + (h_i^2 - h_{i-1}^2)f'_i - h_i^2 f'_{i-1}}{h_i h_{i-1}(h_{i-1} + h_i)} \end{cases} \quad (4)$$

于是节线平衡方程由偏微分方程组转变为以定义在节线上的场函数 f_i 为未知量的常微分方程组。但对于边界节线,例如图1中对应边界 Γ_1 和 Γ_3 的线1和线 n,为应用上式,需要引入边界外部的虚拟线,即图中所示的线0和线 $n+1$。基本方法是将该处应力边界条件方程中(对于位移边界条件,节线控制方程中相应位移分量的偏导数已知无需处理)关于离散坐标的导数应用式(4)近似,解得 f_0,f'_0,f_{n+1} 和 f'_{n+1} 后,再经该式代入节线平衡方程,则在得到边界节线常微分方程的同时,也引入了该处的应力边界条件。

3.3 引入求解区域非边界节线处的边界条件

除通过边界节线引入求解区域的部分边界条件外,其他边界处的边界条件通过各离散节线的端点边界条件(如图1中各节线的上下端点,分别对应边界 Γ_2 和 Γ_4)引入。对于内部节线,该线端点边界条件中对离散坐标的一阶偏导数应用式(4)第

二式近似，但对于边界节线，其端点边界条件中的一阶偏导数一般可用后差分或前差分近似，即

$$\left(\frac{\partial f}{\partial x}\right)_1 = \frac{f_2 - f_1}{h_1}, \quad \left(\frac{\partial f}{\partial x}\right)_n = \frac{f_n - f_{n-1}}{h_{n-1}} \quad (5)$$

但精度较低。更精巧的差分格式是，在边界区域设置等距离散节线，设节线间距为h，由Newton等距插值公式有[7]：

$$\begin{cases} \left(\dfrac{\partial f}{\partial x}\right)_1 = \dfrac{-f_3 + 4f_2 - 3f_1}{2h} \\ \left(\dfrac{\partial f}{\partial x}\right)_n = \dfrac{3f_n - 4f_{n-1} + f_{n-2}}{2h} \end{cases} \quad (6)$$

应用上式近似边界节线端点边界条件方程中对离散坐标的偏导数，能够获得与内部节线相同的差分精度。于是通过有限差分近似，将节线端点边界条件由偏微分方程转变为常微分方程，结合前面得到的定义在节线上的常微分方程，原问题由求解偏微分边值问题转变为求解常微分边值问题。

3.4 求解常微分边值问题和应力场的获得

原求解问题转化为常微分边值问题后，常微分边值问题的求解成为关键。本文选择B样条函数Gauss配点法求解该常微分边值问题[8]。B样条函数Gauss配点法求解常微分边值问题的基本原理是：以B样条基样条插值函数作为常微分边值问题的近似解，基函数的系数为解中的待定系数。将求解区间划分为若干个配点区间，通过使近似解在各配点区间的Gauss映射点和边界点处满足常微分方程和边界条件建立以近似解中待定系数为未知量的代数方程组，求解该方程组确定基函数的系数，由此得到问题的近似解。研究表明，用B样条函数Gauss配点法求解常微分边值问题能够获得很高的精度和较好的数值稳定性[9]，对于r阶常微分方程组，若k为配点区间的Gauss映射点数，$\Delta_{\max}$为配点区间的最大长度，则B样条函数Gauss配点法的精度可达到$O(\Delta_{\max}^{r+k})$。

求解常微分边值问题得到离散节线位移近似函数后，对该函数求导则可得沿节线方向位移导数的近似函数，对其插值可得节线上任意点在梯度方向上的位移导数值；由相邻节线对应点位移数值的有限差分，则可得离散方向上的位移导数值；将上述导数值代入物理方程，进而得到该点处的应力数值。

4 数值算例

4.1 问题描述和平衡方程

如图2所示，计算功能梯度材料板分别在下列

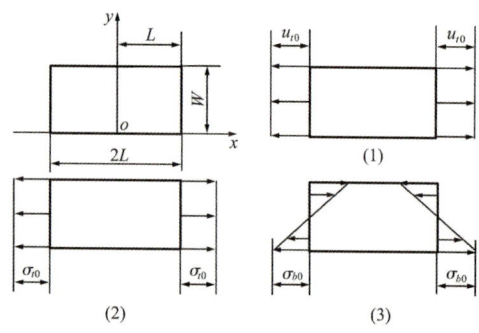

图2 拉伸或弯曲载荷作用下的功能梯度材料板
Fig.2 Functionally graded material plate under tension or bending loading

三种载荷作用下的应力分布，考虑问题为平面应力情况。

(1) 均匀拉伸：$\bar{u}_t = u_{t0}$ (7)

(2) 均布拉力：$\bar{\sigma}_t = \sigma_{t0}$ (8)

(3) 纯弯曲：$\bar{\sigma}_b = \sigma_{b0}\left(1 - \dfrac{2}{W}y\right)$ (9)

为便于与解析解比较，设功能梯度材料板的材料属性形式为

$$E = E(y), \quad \mu = C \quad (10)$$

由式(10)和式(1)可得问题的平衡方程：

$$\begin{cases} \dfrac{\partial^2 u}{\partial x^2} + a\dfrac{\partial^2 u}{\partial y^2} + b\dfrac{\partial^2 v}{\partial x \partial y} + d_1\left(\dfrac{\partial u}{\partial y} + \dfrac{\partial v}{\partial x}\right) = 0 \\ \dfrac{\partial^2 v}{\partial y^2} + a\dfrac{\partial^2 v}{\partial x^2} + b\dfrac{\partial^2 u}{\partial x \partial y} + d_2\dfrac{\partial v}{\partial y} + d_3\dfrac{\partial u}{\partial x} = 0 \end{cases} \quad (11)$$

式中 $a = \dfrac{(1-\mu)}{2}$，$b = \dfrac{(1+\mu)}{2}$，$d_1 = \dfrac{aE'}{E}$

$d_2 = \dfrac{E'}{E}$，$d_3 = \dfrac{\mu E'}{E}$，$E' = \dfrac{\mathrm{d}E}{\mathrm{d}y}$

4.2 求解区域的离散

利用求解问题的对称性，原问题的等效计算构形如图3所示。考虑到材料梯度沿y轴方向，将求解区域沿x轴方向用间距为$h = L/(n-1)$的n条等距垂直节线半离散化，节线编号如图3所示，记x_i为节线i的x坐标，节线位移函数为

$$u_i = u_i(y) = u(x_i, y), \quad v_i = v_i(y) = v(x_i, y)$$

以及

$$\frac{\mathrm{d}u_i}{\mathrm{d}y} = u_i', \quad \frac{\mathrm{d}v_i}{\mathrm{d}y} = v_i', \quad \frac{\mathrm{d}^2 u_i}{\mathrm{d}y^2} = u_i'', \quad \frac{\mathrm{d}^2 v_i}{\mathrm{d}y^2} = v_i''$$

图中各边界编号为Γ_j，$j = 1, 2, 3, 4$。注意此处$h_i = h_{i-1} = h_{i+1}$，由式(4)可得节线i处具有$O(h^2)$精度的中心差分公式，即

$$\left(\frac{\partial u}{\partial x}\right)_i = \frac{u_{i+1} - u_{i-1}}{2h}, \quad \left(\frac{\partial v}{\partial x}\right)_i = \frac{v_{i+1} - v_{i-1}}{2h}$$

$$\left(\frac{\partial^2 u}{\partial x \partial y}\right)_i = \frac{u'_{i+1} - u'_{i-1}}{2h}, \quad \left(\frac{\partial^2 v}{\partial x \partial y}\right)_i = \frac{v'_{i+1} - v'_{i-1}}{2h}$$

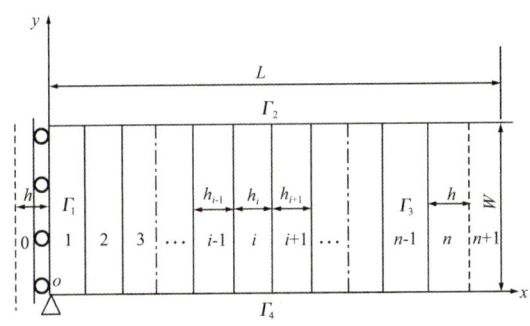

图 3 等效计算构形和离散节线
Fig.3 Equivalent computation model and discrete lines

$$\left\{\begin{array}{l}\left(\dfrac{\partial^2 u}{\partial x^2}\right)_i = \dfrac{u_{i-1}-2u_i+u_{i+1}}{h^2} \\ \left(\dfrac{\partial^2 v}{\partial x^2}\right)_i = \dfrac{v_{i-1}-2v_i+v_{i+1}}{h^2}\end{array}\right. \quad (12)$$

4.3 节线平衡方程的常微分形式转化

4.3.1 节线平衡方程

由式(11)可得离散节线 i 上的平衡方程为

$$\left\{\begin{array}{l}\left(\dfrac{\partial^2 u}{\partial x^2}\right)_i + a\left(\dfrac{\partial^2 u}{\partial y^2}\right)_i + b\left(\dfrac{\partial^2 v}{\partial x \partial y}\right)_i + d_1\left[\left(\dfrac{\partial u}{\partial y}\right)_i + \left(\dfrac{\partial v}{\partial x}\right)_i\right] = 0 \\ \left(\dfrac{\partial^2 v}{\partial y^2}\right)_i + a\left(\dfrac{\partial^2 v}{\partial x^2}\right)_i + b\left(\dfrac{\partial^2 u}{\partial x \partial y}\right)_i + d_2\left(\dfrac{\partial v}{\partial y}\right)_i + d_3\left(\dfrac{\partial u}{\partial x}\right)_i = 0\end{array}\right. \quad (13)$$

将式(12)代入上式,则上式转化为具有二阶精度的常微分方程,但要获得定义在边界线 $i=1$ 和 $i=n$ 上的平衡方程的常微分方程形式,需要借助虚拟线上的辅助量。如下所述这些辅助量通过引入与边界线间距均为 h 的虚拟线 $i=0$ 和 $i=n+1$ 并结合边界线处的边界条件求得。

4.3.2 虚拟线上的辅助量

(1) 虚拟线 $i=0$ 上的辅助量

如图3所示,对于上述三种载荷,在边界 Γ_1(线 $i=1$)处有边界条件:

$$u_1=0,\; G\left[\left(\dfrac{\partial u}{\partial y}\right)_1+\left(\dfrac{\partial v}{\partial x}\right)_1\right]=0 \quad (14)$$

令式(12)第二式中 $i=1$,代入上式得

$$v_0=v_2,\; v_0'=v_2' \quad (15)$$

(2) 虚拟线 $i=n+1$ 上的辅助量

对于均匀拉伸载荷,在边界 Γ_3(线 $i=n$)处有边界条件:

$$u_n=\bar{u},\; G\left[\left(\dfrac{\partial u}{\partial y}\right)_n+\left(\dfrac{\partial v}{\partial x}\right)_n\right]=0 \quad (16)$$

令式(12)第二式中 $i=n$,代入上式得

$$v_{n+1}=v_{n-1},\; v_{n+1}'=v_{n-1}' \quad (17)$$

类似上述过程,对于均布拉力载荷情况有

$$u_{n+1}=u_{n-1}-2hv_n'+\dfrac{2h}{\mathrm{E}}\sigma_{t0},\; v_{n+1}=v_{n-1}-2hu_n'$$

$$u_{n+1}'=u_{n-1}'-2hv_n''-\dfrac{2h}{\mathrm{E}^2}\sigma_{t0}\mathrm{E}',\; v_{n+1}'=v_{n-1}'-2hu_n'' \quad (18)$$

式中 $\mathrm{E}=\dfrac{E}{(1-\mu^2)}$,$\mathrm{E}'=\dfrac{\mathrm{d}}{\mathrm{d}y}\left(\dfrac{E}{1-\mu^2}\right)=\dfrac{1}{(1-\mu^2)}\dfrac{\mathrm{d}E}{\mathrm{d}y}$。

对于纯弯曲载荷情况有

$$u_{n+1}=u_{n-1}-2hv_n'+\dfrac{2h\sigma_{b0}}{\mathrm{E}}\left(1-\dfrac{2}{W}y\right)$$

$$v_{n+1}=v_{n-1}-2hu_n',\; v_{n+1}'=v_{n-1}'-2hu_n''$$

$$u_{n+1}'=u_{n-1}'-2hv_n''-\dfrac{2h\sigma_{b0}}{\mathrm{E}^2}\mathrm{E}'-\dfrac{4h\sigma_{b0}}{\mathrm{E}W}+\dfrac{4h\sigma_{b0}}{\mathrm{E}^2W}\mathrm{E}'y \quad (19)$$

4.3.3 节线平衡方程的常微分方程近似

设式(13)经有限差分后得到的定义在各离散节线上的常微分方程为

$$\boldsymbol{U}''=\boldsymbol{A}\boldsymbol{U}'+\boldsymbol{B}\boldsymbol{U}+\boldsymbol{F} \quad (20)$$

式中 $\boldsymbol{U}$ 为待求节线位移函数向量,$\boldsymbol{A},\boldsymbol{B}$ 和 $\boldsymbol{F}$ 为系数矩阵。

对于均匀拉伸载荷,式(20)中 $\boldsymbol{U}=\{v_1,u_2,v_2,\cdots,u_i,v_i,\cdots,u_{n-1},v_{n-1},u_n\}^{\mathrm{T}}$,将式(12)代入式(13),在边界线处应用式(15,17),并注意已知 $u_1=0,u_n=\bar{u}$,则可得载荷(1)情况下节线平衡方程的常微分方程近似式,并确定式(20)中的各系数矩阵。

对于均布拉力或纯弯曲载荷,式(20)中 $\boldsymbol{U}=\{v_1,u_2,v_2,\cdots,u_i,v_i,\cdots,u_{n-1},v_{n-1},u_n,v_n\}^{\mathrm{T}}$,类似上面的过程,但注意在边界线处应用式(15,18)或式(19),且已知 $u_1=0$,则可分别得到载荷(2)和(3)情况下定义在离散节线上的关于节线未知位移函数的常微分方程组。

4.4 节线端点边界条件的常微分形式转化

4.4.1 节线端点边界条件

(1) 在边界 Γ_2 上:记 $\mathrm{E}=E/(1-\mu^2)$,在 $i\in[1,n]$ 的各节线端点 (x_i,W) 处有

$$G\left(\dfrac{\partial u}{\partial y}+\dfrac{\partial v}{\partial x}\right)\bigg|_{(x_i,W)}=0,\; \mathrm{E}\left(\dfrac{\partial v}{\partial y}+\mu\dfrac{\partial u}{\partial x}\right)\bigg|_{(x_i,W)}=0 \quad (21)$$

(2) 在边界 Γ_4 上:在边界线 $i=1$ 的端点 $(0,0)$ 处由约束条件(避免结构发生平动)有

$$u(0)=u|_{(0,0)}=0,\; v_1(0)=v|_{(0,0)}=0 \quad (22)$$

在 $i\in[2,n]$ 的各节线端点 $(x_i,0)$ 处有

$$G\left(\dfrac{\partial u}{\partial y}+\dfrac{\partial v}{\partial x}\right)\bigg|_{(x_i,0)}=0,\; \mathrm{E}\left(\dfrac{\partial v}{\partial y}+\mu\dfrac{\partial u}{\partial x}\right)\bigg|_{(x_i,0)}=0 \quad (23)$$

4.4.2 节线端点边界条件的常微分方程近似

设节线端点边界条件的常微分近似式为

$$C|_{\Gamma_j}U' + D|_{\Gamma_j}U + G|_{\Gamma_j} = 0 \qquad (24)$$

式中 C,D 和 G 为系数矩阵，Γ_j 中 $j=2,4$。对于内部节线 $i\in[2,n-1]$ 的端点边界条件方程，应用式(12)前两式近似方程中位移对坐标 x 的偏导数，但对于边界节线端点边界条件方程中的位移对坐标 x 的偏导数，应用式(6)来近似。则由式(21~23)，可得各载荷情况下离散节线端点处边界条件的常微分方程近似式与式中的系数矩阵。

由式(20,24)，将原问题从偏微分边值问题转化为求解定义在各离散节线上的关于未知节线位移函数的常微分边值问题，可应用 3.4 节中所述的 B 样条函数 Gauss 配点法求解。

4.5 数值结果

令式(7~9)和图2中 $u_0=1,\sigma_0=1,q_0=1$，$W=9,L=9$，材料参数 $\mu=0.3$，设弹性模量分别为指数梯度形式和线性梯度形式，即

$$E(y)=E_0 e^{\beta y},\quad E(y)=E_0+\beta y \qquad(25,26)$$

式中 $E_0=E(0)=1$，β 为非均匀参数，并记 $E_1=E(W)$。

计算过程中取协调一致的封闭单位体系，上述各计算参数的量纲为相应的单位力、单位长度及单位弹性模量。计算结果如图4～图9所示，各梯度形式下的梯度变化情况即 E_1/E_0 如图中标注。各图中的实线为解析解[3]，圆点°为线法数值结果。对比表明，线法数值解与解析解匹配得非常好，对于不同的梯度形式和较大的梯度变化范围均具有很好的适应性。

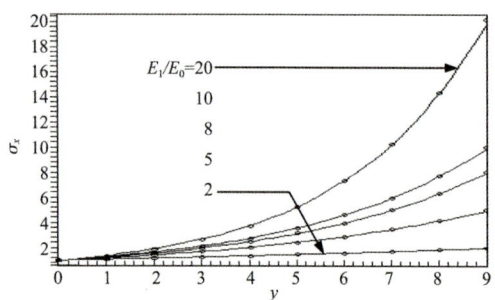

图4 均匀拉伸载荷作用下的应力分布(指数梯度)
Fig.4 Stress under constant displacement for exponential gradation

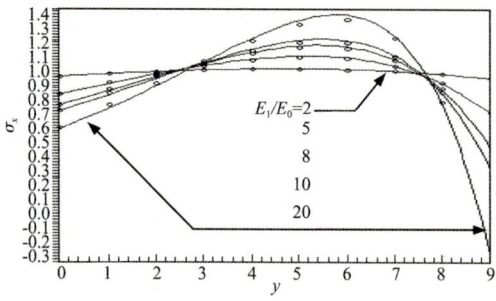

图5 均布拉力载荷作用下的应力分布(指数梯度)
Fig.5 Stress under tension loading for exponential gradation

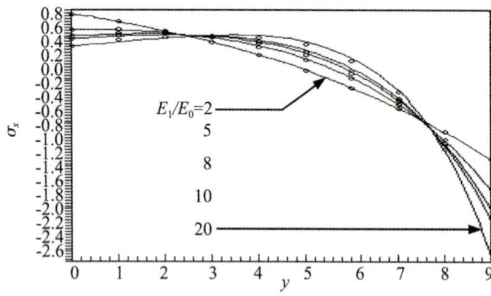

图6 弯曲载荷作用下的应力分布(指数梯度)
Fig.6 Stress under bending loading for exponential gradation

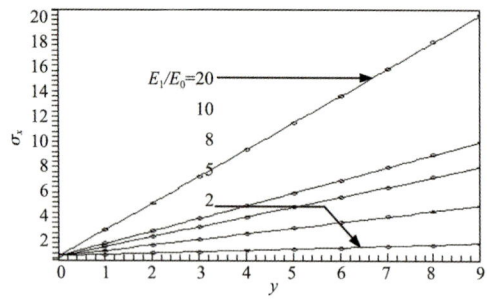

图7 均匀拉伸载荷作用下的应力分布(线性梯度)
Fig.7 Stress under constant displacement for linear gradation

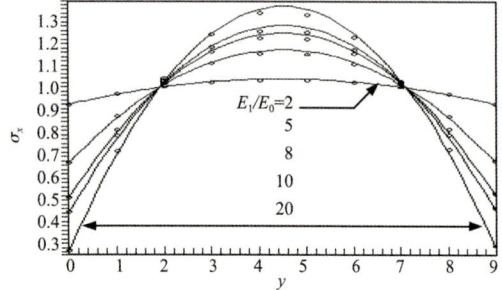

图8 均布拉力载荷作用下的应力分布(线性梯度)
Fig.8 Stress under tension loading for linear gradation

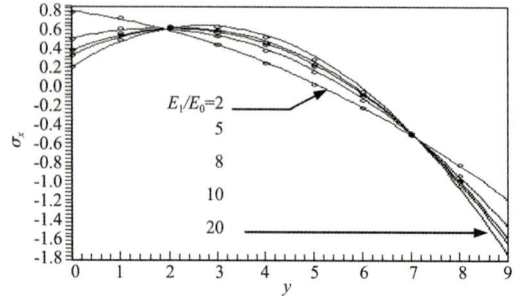

图9 弯曲载荷作用下的应力分布(线性梯度)
Fig.9 Stress under bending loading for linear gradation

5 分析和讨论

下面通过与其他数值方法计算结果的比较，研究线法的数值特性。

5.1 线法与均匀元法的比较

图 10 所示是文献[3]中分别应用层板模型均匀元法(四节点单元)和等参梯度元法(四节点单元)对图 2 中载荷(1)、指数梯度形式、$E_1/E_0=8$ 情况得到的单元节点处的数值结果(为了与本文构形一致原文符号与标注作了调整)。分析图 10 可以看出，均匀元法得到的结果具有分段连续的特点，其根本原因是均匀元法对材料属性函数的模拟是建立在单元尺度以上的，不能反映单元尺度以下即单元内部各点的材料属性变化情况。图 10 和图 11 中等参梯度元法和线法的结果都与解析解匹配得很好，表明这两种方法非常有效地模拟了结构的材料梯度属性，能够反映单元内部或节线上各点材料属性的连续变化。

5.2 线法与等参梯度元法的比较

为与等参梯度元法对比，考虑如图 12 所示，泊松比 $\mu=0$，沿梯度方向受单位力 $\bar{\sigma}$ 作用的功能梯度材料板。

基于复合材料层板模型的四节点均匀元和四节点等参梯度元的计算结果如图 13 所示[3]。线法对该问题的计算结果如图 14 所示。由图 13 可见，等参梯度元法的计算结果是分段且不连续的曲线，计算精度极低，相反分层均匀元法的结果却与解析解匹配得很好。分析这一现象后发现，分层均匀法与解析解匹配良好的原因是节点处材料属性离散值与该处应变值恰好互为倒数，且四节点单元为常应力单元；等参梯度元法的计算结果较差是由于单元位移模式选择不当，材料属性的连续变化导致单元特性恶化造成的。线法数值结果如图 14 所示，可以看出线法数值结果与解析解符合得非常好，这是因为线法特别选用的 B 样条基样条插值函数相当于有限单元法 $k+r-1$ 阶多项式位移函数的精度，对各种梯度形式和结构变形情况的适应性较强，可以有效避免因位移模式选择不当导致的计算不确定性。

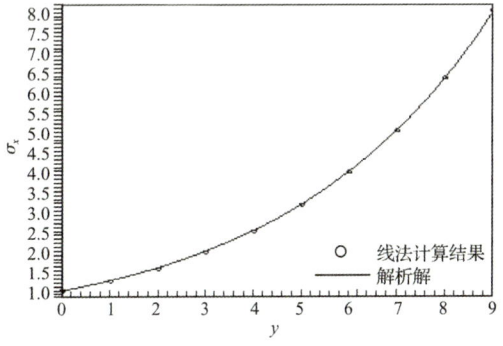

图 11 线法的计算结果与解析解

Fig.11 Computational results of MOLs and analytical solution

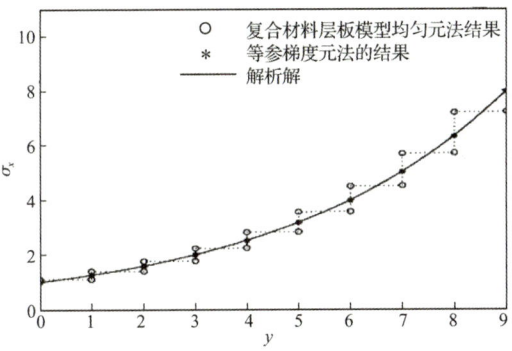

图 10 均匀元法和等参梯度元法的数值结果[3]

Fig.10 Numerical results of homogenous element method and isoparametric graded element method[3]

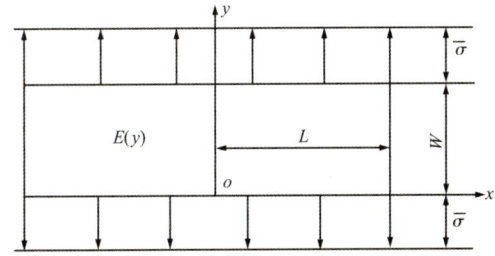

图 12 沿梯度方向拉伸的功能梯度材料板

Fig.12 FGM plate under tension loading along the gradation

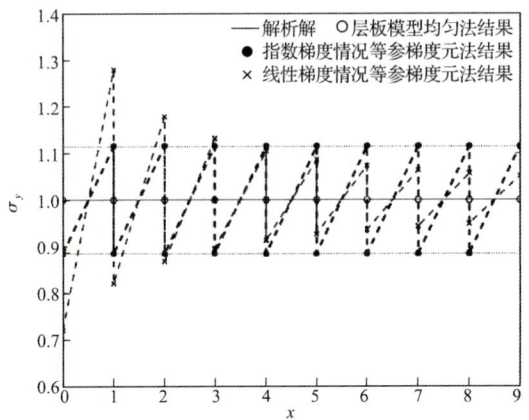

图 13 沿梯度拉伸时均匀元法和等参梯度元法结果[3]

Fig.13 Computational results of homogenous element and isoparametric graded element for FGM plate under tension loading along the gradation[3]

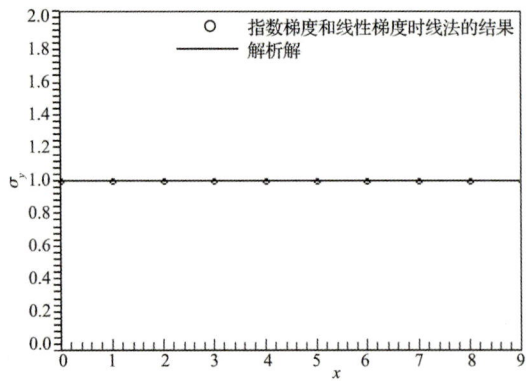

图 14 沿梯度方向拉伸时线法的计算结果
Fig. 14 Computational results of MOLs for FGM plate under tension loading along the gradation

6 结 论

本文首先阐述了线法功能梯度材料结构分析的求解步骤和数值原理，然后以均匀拉伸、均布拉力和纯弯曲等载荷作用下线性梯度与指数梯度材料板的平面问题为例，演示了线法功能梯度材料结构分析的实现过程。线法的基本思想是沿着与材料梯度无关的方向将求解区域半离散化，应用有限差分将问题的控制方程转变成为定义在离散节线上的常微分边值问题，然后应用 B 样条函数 Gauss 配点法求解该常微分边值问题得到问题的解答。与相应问题解析解和其他数值方法的比较表明，线法非常适合于功能梯度材料结构分析。

由线法的求解过程可见，线法直接求解的是结构半离散后定义在沿梯度方向节线上保留了材料属性参数变化信息的常微分方程，不需要像均匀元法分段近似材料属性函数，也不需要类似等参梯度元法进行附加的材料属性参数的插值近似。通过对相应问题不同数值方法计算结果的分析，线法克服了均匀元法不能模拟单元尺度以下材料属性梯度的缺点，能够反映功能梯度材料连续变化的材料属性，避免了等参梯度元法位移模式选择不当引起的计算不确定性。此外，对比新康法，线法对问题的求解直接从控制方程出发，无需事先选取满足特定条件的待求场函数，因此是具有一定普遍意义的半解析数值方法。

根据线法的数值原理可以看出，线法的实质是在非材料梯度方向用有限差分求近似解，而在研究关注的材料梯度方向用 B 样条函数 Gauss 配点法求高精度解，类似于有限元分析的 H 方法和 P 方法，因此非常符合目前具有单向梯度的大多数功能梯度材料结构的特点。特别是对于常见的涂层/基体和三明治夹层结构中薄膜形式的梯度材料层，由于梯度方向的尺寸远远小于其他方向的尺寸，线法的这一数值特点使其相对于基于结构完全离散的数值方法具有明显的优势。

参考文献（References）:

[1] Lanutti J J. Functionally graded materials: properties, potential and design guidelines[J]. Composites Engineering, 1994, 24(4): 81-94.

[2] Anals G, Santre M H, Lambros J. Numerical calculation of stress intensity factors in functionally graded material[J]. International Journal of Fracture, 2000, 104(2): 131-143.

[3] Kim J H, Paulino G H. Isoparametric graded finite elements for nonhomogenous isotropic and orthotropic functionally graded materials[J]. Journal of Applied Mechanics, 2002, 69(3): 502-514.

[4] Zou Z Z, Wu X S, Li C Y. On the multiple isoparametric finite element method and computation of stress intensity factor for cracks in FGMs[J]. Key Engineering Materials, 2000, 16(183-187): 511-516.

[5] Rao B N, Rahman S. Mesh-Free analysis of cracks in isotropic functionally graded materials[J]. Engineering Fracture Mechanics, 2003, 70(1): 1-27.

[6] 李 永, 宋 健, 张志民. 非均质梯度材料复合结构的 Kantorovich 宏细观精化解法[J]. 中国科学, E 辑, 2003, 33(1): 29-41. (LI Yong, SONG Jian, ZHANG Zhi-min, Kantorovich method of macro or mesoscopic solution for nohomogeneous functionally graded materials composite[J]. Journal of China Science, E, 2003, 33(1): 29-41. (in Chinese))

[7] 凌复华, 殷学纲, 何冶奇. 常微分方程数值方法在力学中应用[M]. 重庆: 重庆大学出版社, 1990. (LIN Fu-hua, YIN Xue-gang, HE Ye-qi. Application of Numerical Method of Ordinary Differential Equations to Mechanics[M]. Chongqing: Chongqing University Press, 1990. (in Chinese))

[8] Ascher U, Christiansen J, Russell R D. A collocation solver for mixed order systems of boundary value problems[J]. Mathematics of Computation, 1979, 33(146): 659-679.

[9] Ascher U, Bader G. Stability of collocation at Gaussian points[J]. SIAM Journal of Numerical Analysis, 1986, 23(4): 412-422.

岩质高边坡稳定性分析与评价中的四个准则

李 宁[1,2]，钱七虎[3]

(1. 西安理工大学 岩土工程研究所，陕西 西安 710048；2. 中国科学院寒区旱区环境与工程研究所 冻土工程国家重点试验室，
甘肃 兰州 730000；3. 解放军理工大学 工程兵工程学院，江苏 南京 210007)

摘要：对于岩质高边坡而言，其潜在滑动面上力学参数的合理性、边坡开挖后需要的整体加固力以及计算的边坡安全系数与实际边坡安全储备的接近程度将直接影响岩质高边坡工程的安全性与经济性。从现有的确定边坡潜在滑动面力学参数的反演分析方法入手，指出其存在的局限性，提出边坡潜在滑动面力学参数最小取值准则；在此基础上，根据开挖岩体的总压力充其量是诱发边坡失稳的全部不利荷载的思路，提出边坡最大主动加固力准则。将潘家铮的上、下限原理应用于边坡的数值分析与稳定性评价，提出数值法进行稳定性分析评价的安全系数上、下限准则，并将其在工程评价中进行验证与推论。该研究可为工程设计人员进行有效经济地选择岩质高边坡加固方案提供宏观判据与基本原则。

关键词：边坡工程；岩质边坡；稳定分析；宏观判据；滑动面

中图分类号：P 642   **文献标识码**：A   **文章编号**：1000 - 6915(2010)09 - 1754 - 06

FOUR CRITERIA OF STABILITY ANALYSIS AND ASSESSMENT OF HIGH ROCK SLOPE

LI Ning[1,2], QIAN Qihu[3]

(1. Institute of Geotechnical Engineering, Xi'an University of Technology, Xi'an, Shaanxi 710048, China; 2. State Key Laboratory of Frozen Soil Engineering, Cold and Arid Regions Environmental and Engineering Research Institute, Chinese Academy of Sciences, Lanzhou, Gansu 730000, China; 3. Engineering Institute of Engineering Corps, PLA University of Science and Technology, Nanjing, Jiangsu 210007, China)

Abstract: For high rock slopes, the mechanical parameters on the potential sliding surface, the reinforcement forces needed after slope excavation, and the correlation between the calculated factor of safety and actual safety margin will directly influence engineering safety and economy. Based on back analysis methods for determining mechanical parameters of potential sliding plane and their shortcomings, the minimum value criterion of mechanical parameter selection is proposed; and the maximum reinforcement force criterion is then proposed on the basis of the equivalence of the largest reinforcement force needed and the weight of excavated rock body. Finally, the upper and lower limit criteria of the numerical results used for stability analysis and assessment are put forward and applied to engineering design based on PAN Jiazheng's upper and lower limit theory. It will helpfully provide designers with criteria of slope stability analysis and assessment on reinforcement scheme selection of high rock slope.

Key words: slope engineering; rock slope; stability analysis; analysis and assessment criteria; sliding plane

1 引 言

岩质高边坡稳定分析与评价问题，是水利枢纽和交通铁路工程建设中经常遇到的工程问题。李宁和张鹏[1]在对前人的研究成果与工程经验总结的基础上，对岩质边坡最危险滑动面的确定思路与方法等5个主要问题进行了讨论与分析。对于工程设

收稿日期：2010-02-08；修回日期：2010-05-05。
基金项目：国家自然科学基金资助项目(50879068)；国家自然科学基金创新群体(40821001)。
本文原载于《岩石力学与工程学报》(2010 年第 29 卷第 9 期)。

计人员而言，对边坡稳定性的判断及最终所采取的工程整治措施合理与否，主要取决于以下 3 个方面：(1) 边坡潜在滑动面上力学参数的可靠性；(2) 计算的安全系数与实际情况的接近程度；(3) 整体加固力设计的合理性。

潜在滑动面上力学参数主要由地质人员确定。尽管有多种方法确定参数，如室内和现场试验法、经验公式法、神经网络等智能方法，但由于现有勘探水平的局限，即使进行较为细致的地质工作，由于岩质高边坡的地形特点及复杂的组成，能相对准确地确定出潜在滑动面已属不易，准确地确定能够真实反映潜在滑动面实际力学特性的参数值则更加困难。同时，由于确定方法自身的局限，地质工程师在取舍参数时也犹豫再三，最终的结果往往是从安全的角度出发，选取的参数过于保守。

对边坡的稳定性进行宏观评价，当前仍是以经验方法为主，将不论怎样复杂的分析计算成果归结为一个安全系数，再按规范将该安全系数与规范给出的标准作比较。这是一个令研究人员、设计人员无奈的、粗糙的、落后的、不得已的评价方法。笔者在 2004～2005 年完成的马来西亚巴贡水电站厂房边坡的稳定性分析与评价中，早已不用单纯的安全系数，而是对坡体的应力场、变形场特征及变形趋势进行分析。如何评价复杂的、大型有限元方法对边坡的分析结果是目前尚在探索的研究课题。潘家铮[2]在 20 世纪 80 年代就从宏观角度提出过滑动面的最小值原理与抗滑力最大值原理：滑坡失稳时将沿抵抗力最小的一个滑面滑动破坏(最小值原理)；滑面确定后，则滑面上的反力能够自行调整以发挥最大的抗滑力(最大值原理)。陈祖煜等[3,4]在此基础上，经多年研究，从塑性理论角度严谨推导出了土坡滑动问题的极限上、下限解，为边坡稳定性分析宏观评价提供了新思路与新视角。此外，崔政权等[5~13]在这一领域也进行了诸多探索。

针对近年来笔者承担的锦屏、龙滩、拉西瓦、三峡船闸和紫坪铺等水电工程大型岩体边坡稳定性分析与评价研究工作中所遇到的问题，从设计人员最关心的稳定性分析结果的评价入手，对决定边坡稳定性的潜在滑动面强度参数的合理确定、边坡加固力的宏观判断及边坡稳定性数值分析结果的合理性与安全性等 3 个方面进行探讨，通过对以上水电工程岩质边坡的分析研究成果以及李 宁等[1~19]研究的分析讨论，总结出几点体会。

2 潜在滑动面最小参数取值准则

当前，工程设计人员已经普遍接受了反分析方法，本文将其总结为最小参数准则：对于一个长期稳定的边坡，在其所经历的荷载工况下，其潜在滑动面上的最低强度参数不应使它的安全系数小于 1.0。这个准则已得到较广泛的认可，有一定的合理性与实际应用性。但是对于复杂的岩质高边坡，其潜在滑动面上的强度参数选取过于复杂，地质人员往往从安全的角度出发，过于保守地选取强度参数，使边坡在经受降雨或地震时，往往认为需要大量的加固措施。笔者认为，如果该边坡经历了降雨、地震等工况仍处于稳定状态，那么对于设计的使用期限通常在一百年的边坡来说，也应该根据未开挖边坡在地震、降雨等荷载下，其安全系数不小于 1.0 为条件进行反演，确定边坡的最低强度参数。这样确定出的参数值才能有效地剔除设计参数偏低致使分析结果过于保守的偏差，从而反映坡体的实际情况，给工程设计人员提供合理有效的潜在滑动面力学参数的下限值。

图 1 为猴子岩右岸坝肩边坡计算剖面和潜在滑动面示意图。表 1 为在原设计参数下天然边坡的安全系数。

由表 1 可知，采用二维有限元模型进行边坡安全系数计算，在原设计参数下，天然状态边坡不能自稳，这说明所提供的潜在滑动面强度参数过于保守。按天然状态边坡安全系数为 1.05，地震工况边坡安全系数为 1.0 分别反演各滑面的最小强度参数，结果如表 2 所示。潜在滑动面强度参数相应得到提高，采用该参数可大大节省加固工程量。同时，该参数的反演是建立在边坡长期宏观稳定的实践基础

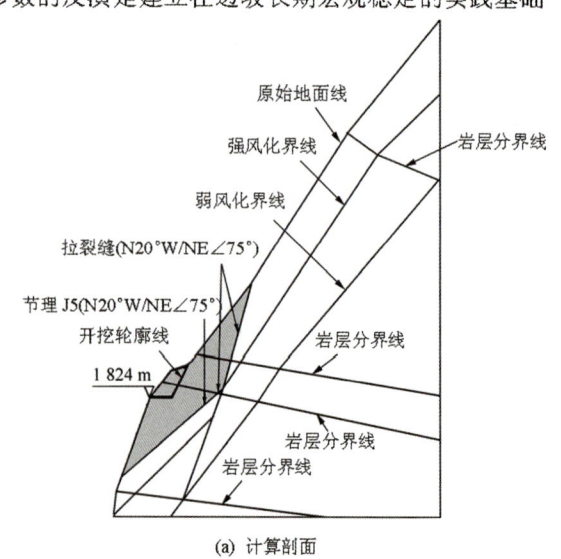

(a) 计算剖面

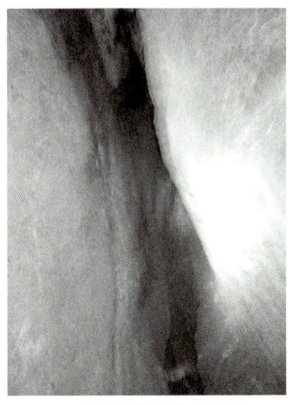

(b) 后缘拉裂缝(潜在滑面)

图1 猴子岩右岸坝肩边坡计算剖面和潜在滑动面示意图

Fig.1 Sketches of calculating profile and potential sliding plane of right abutment slope

表1 原设计参数下天然状态边坡的安全系数

Table 1 Factors of safety of natural slope using original design parameters

结构面名称	滑动面强度参数		边坡安全系数	
	c'/MPa	f'	自重工况	自重+地震工况
潜在滑动面节理 J5	0.11	0.595	0.69	0.59

表2 反演后建议的滑动面强度参数

Table 2 Strength parameters of sliding plane derived from back analysis

结构面名称	天然状态反演		地震工况反演	
	c'/MPa	f'	c'/MPa	f'
潜在滑动面节理 J5	0.24	0.79	0.31	0.90

上的，潜在滑动面强度参数的可靠性是有保障的。因此，使得加固方案达到了合理有效且更加经济的要求。

3 边坡主动加固力上限准则

在边坡加固设计中，往往由于前期地质工作不充分，过分依赖于锚索或抗滑桩等加固措施，使设计的加固量常常偏大，缺少加固量的宏观判断依据。加固量的合理确定，是一个非常复杂的、高度非线性的问题，很难直接计算得到。但不妨换个角度思考：假设边坡未开挖之前是稳定的(或处于临界稳定状态)，则开挖后，坡体的法向压力减小，使抗滑能力减小，加固力最多恢复至开挖前的法向力，故可认为，边坡开挖后，其最大加固力在开挖面的法向分量不应超过被挖掉的岩体质量在开挖表面的法向分量。该准则只能用于一般情况。图2 为某边

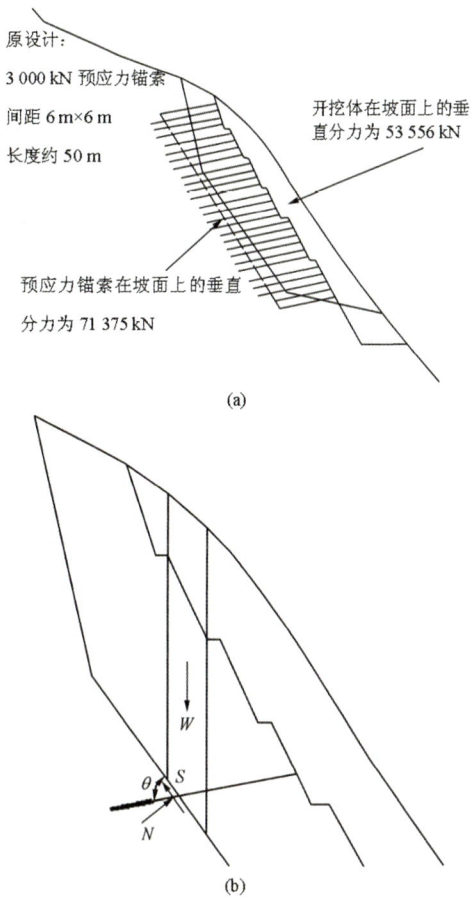

图2 某边坡开挖体和加固锚索在滑动面上的垂直分力

Fig.2 Vertical components of excavated rock body and anchor cable of a slope

坡开挖体和加固锚索在滑动面上的垂直分力。表3为某边坡原加固设计安全系数表。表4为不同锚索加固量滑面垂直分量及安全系数对比表。

表3 某边坡原加固设计的安全系数

Table 3 Factors of safety of a slope with original enforcement design

工况	安全系数
开挖前	1.25
开挖后	1.09
27 排预应力锚索加固后	4.70

表4 不同锚索加固量坡面垂直分量及安全系数对比

Table 4 Comparison of vertical component of enforcement force and factor of safety

方案	锚索数量/排	垂直分力/kN	安全系数
原设计方案	27	7.1×10^4	4.7
修正方案1	19	5.4×10^4	2.7
修正方案2	7	1.8×10^4	1.3

由图2可知,锚索沿滑面垂直分力远大于开挖体在滑面上的垂直分力,边坡安全系数达到4.7,远大于原自然边坡,说明边坡的加固方案太过保守。当开挖体在滑面上的垂直分力与锚索在滑面垂直分力相近时,边坡的安全系数为2.7,依然相差较大。当锚索在滑面的垂直分力约为开挖体在滑面上的垂直分力的1/3时,边坡的安全系数基本接近天然状态下的边坡安全系数。综上所述,可以将最大加固力准则表述为:对于一个处于稳定状态的边坡,边坡开挖后,其最大加固力在开挖面法向上的分量不应超过被挖掉的岩体质量在开挖面上的法向分量。

应该注意的是,在使用这个判据时,如果不太稳定的边坡开挖后长时间得不到支护与加固,则较大的潜在滑动面上可能出现较多的塑性应力调整,使滑面上已调整的岩体部分的峰值强度降为残余强度,则可能使所需加固力远大于原挖掉的岩体质量。所以本准则具体应用条件为:(1) 边坡在开挖后及时支护;(2) 边坡采用加固形式应为主动加固,例如预应力锚索等,且加固区位于滑动面中部;(3) 滑坡类型为推移式滑坡。

4 滑动面滑动的下限准则

对已选定的滑动面,实际边坡的破坏是一个渐进的自组织调整过程,潜在滑面上超限的拉、剪应力充分转移释放使滑面达到峰值强度,从而使得滑面抗滑力充分发挥。而在有限元模拟分析过程中,由于单元网格总有一定的尺寸,且对超限应力的非线性迭代逼近总有循环次数的限制,故潜在滑动面最终由有一定尺寸的单元网格的节点来模拟其超限应力的有限次数的转移与释放,滑面上的应力难以自行充分调整,从而发挥其最大的抗滑能力,也就是说,有限元法(包括强度折减法)很难模拟出实际坡面的峰值抗滑能力。基于此,提出的滑面滑动的安全系数下限准则可表述为:对于已选定的滑动面,其潜在滑面上超限的拉、剪应力的转移释放所达到的滑面强度总是小于该滑面上实际的最大抗滑能力,从而使计算所得的安全系数总是小于该滑面实际的安全系数。图3为边坡破坏时潜在滑动面上的应力转移示意图。其中,第一单元的超限拉、剪应力为

$$\sigma_i - [\sigma_i] = \Delta \sigma_i \quad (1)$$

通过节点荷载增量为

$$\Delta F_i = \int [B]^T \Delta \sigma_i \mathrm{d}A \quad (2)$$

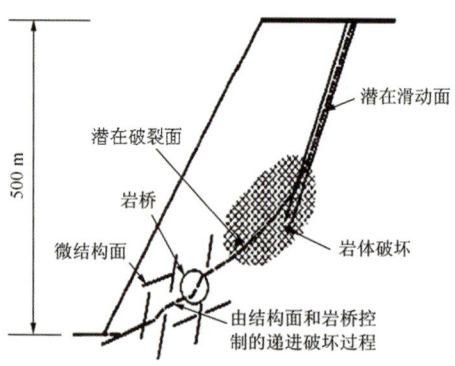

(a) 地质结构示意图

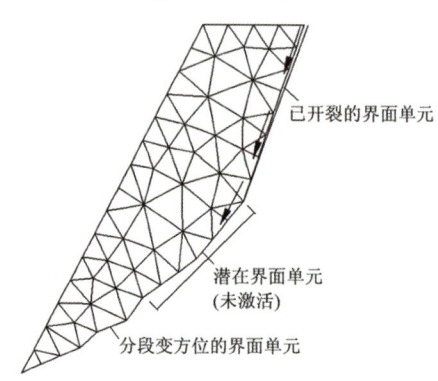

(b) 有限元数值模型

图 3 边坡破坏时潜在滑动面上的应力转移示意图

Fig.3 Sketches of stress transfer at potential sliding plane

在相邻单元产生转移来的应力增量 $\Delta \sigma_{i+1}$ 为

$$\Delta \sigma_{i+1} = [S] \Delta u_i \quad (3a)$$

其中,

$$\Delta u_i = [K]^{-1} \Delta F_i \quad (3b)$$

被释放单元的新一轮应力为

$$\{\sigma\}_{i+1} = \{\sigma\}_i + \Delta \sigma_{i+1} \quad (4)$$

式(1)~(4)中:σ_i 为第 i 次非线性循环应力,$[\sigma_i]$ 为单元抗拉或抗压强度,$\Delta \sigma_i$ 为第 i 次应力增量,A 为单元面积,Δu_i 为第 i 次的位移,$[B]^T$ 为超限单元的几何矩阵,$[K]$ 为刚度矩阵,$[S]$ 为应力矩阵。

经过 N 次迭代后,超限的拉、剪应力永远不可能完全转移释放掉,故单元的真实抗力不可能被全部模拟出。本准则也是对潘家铮的最大抵抗力原理在有限元法等数值分析结果评价中的诠释。

5 潜在滑动面选取的上限准则

在当前应用有限元法分析岩质边坡稳定性的问题中,最关键的是如何较准确选取潜在滑动面的位

置，岩质边坡不像均匀的土质边坡，其潜在滑动面可以发生在任何部位，有限元法也不能如同极限平衡法那样可以自动寻找所有可能的滑动面，进而选出安全系数最小的那个滑动面。

岩质滑坡体若由单个滑面组成，所选定模型刚好选准该滑面，则所选择的滑面的抗滑能力等于实际滑体的抗滑能力；当滑坡体有多个滑面组成，滑体失稳时它将沿抵抗力最小的一个滑面滑动。对于这种情况，边坡分析人员所选择的有限滑面的破坏很难准确地描述实际的潜在滑面，很可能会遗漏部分滑面而选择较好的岩体为破坏界面。由最小抵抗力原理[2]可知，滑坡失稳时将沿其抗滑力最小的滑动面破坏，因此，人为选择的潜在破坏面的抗滑能力永远大于或等于实际滑体的抗滑能力。图4为铜湾右岸4#山头边坡楔形体滑移图，图中为开挖完成后的坡体，人为选择的滑坡体的滑面不是实际的滑坡面，由于人为破坏面抗滑力高于实际破坏面的抗滑力，滑体没能沿人为滑面破坏，而是沿低于人为破坏面的地方破坏。图5为猴子岩水电站右岸坝址工程地质剖面图，由于存在多个滑面，很难准确地确定出实际滑体边界，人为选择的有限元组合滑动面的抗滑力总是大于实际滑体的抗滑能力。

图4 铜湾右岸4#山头边坡楔形体滑移图
Fig.4 Photograph of wedge sliding of right bank slope No.4

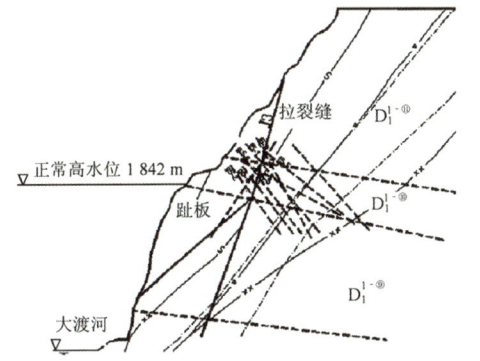

图5 猴子岩水电站右岸坝址工程地质剖面图
Fig.5 Geological profile of right abutment in Houziyan hydropower station

综上所述，根据"最小抵抗力原理"可得出其在边坡稳定分析评价中的推论：设计分析所选择的破坏面的实际抗滑能力大于或等于实际滑体的抗滑能力。该推论等价于下述推论：各种设计分析所选择的边坡的安全系数(包括有限元法在潜在滑动面上的拉剪、应力充分释放转移后所得的安全系数)大于或等于实际边坡的安全系数。本文将这一现象称为"潜在滑动面选取的上限准则"。

6 结 论

在利用潜在滑动面最小参数取值准则进行反演确定滑面强度参数时，反演分析方法应与稳定性分析方法相一致、相协调，即对未开挖边坡应用极限平衡法、传统有限元法或有限元强度折减法等不同方法反演得出的强度参数，在后续边坡开挖分析中，也应用相同分析方法。

对于主动加固力上限准则，如果不太稳定的边坡开挖后长时间得不到支护与加固，则潜在滑动面上可能出现较多的塑性应力调整，使滑面上已调整的岩体部分的峰值强度降为残余强度，则可能使所需加固力远大于挖掉的岩体质量。应用时应该注意这种现象。其具体的应用条件为：(1) 边坡在开挖前是稳定的；(2) 开挖后潜在滑坡类型为推移式滑坡；(3) 边坡采用的加固形式应为主动式加固，例如预应力锚索等，且加固区位于滑动面中部。

应用上下限准则进行边坡稳定性评判时，若坡体地质条件简单，块体组合明显，最危险滑面的选取接近实际情况，则上限准则影响较小，稳定性评价基于下限准则，数值计算结果可能偏于安全；当岩体破碎，地质条件复杂时，难以确定最危险滑面，则上限准则的影响更为显著。因此，计算的安全系数有可能偏于危险。

进一步对上、下限准则进行有限元分析结果的评价与推论：

(1) 一般情况下，前期地质工作越粗糙，潜在滑动面的确定越不准，则分析的安全系数一般较实际偏大，除非分析人员人为过度降低力学参数。

(2) 相同参数下，应用有限元分析的安全系数一般较刚体极限平衡分析大，且更接近实际。

(3) 有限元弹塑性分析较纯弹性分析时的安全系数大，且更接近于实际。

(4) 强度折减法得到的安全系数较一般的弹塑性分析得到的大，更加接近于实际。

(5) 三维分析的安全系数较二维分析的安全系数大，在相同单元精度下，三维分析结果更接近于实际。

以上推论是基于纯客观、理性分析得出的，不考虑设计或分析人员根据个人经验人为地调整参数或模型所产生的影响。

致谢 丁卫华博士、姚显春讲师对论文提出了不少有意义的修改，特此致谢!

参考文献(References)：

[1] 李 宁，张 鹏. 岩质边坡稳定分析与设计中的几个基本问题[C]// 中国岩石力学与工程学会第七次学术大会论文集. 北京：中国科学技术出版社，2002：395 – 397.(LI Ning, ZHANG Peng. Discusses on the fundamental problems in rock slope engineering[C]// Proceedings of 7th Academic Conference of CSRME. Beijing: China Science and Technology Press, 2002: 395 – 397.(in Chinese))

[2] 潘家铮. 建筑物的抗滑稳定和滑坡分析[M]. 北京：水利出版社，1980.(PAN Jiazheng. The sliding stability and landslide analysis of structure[M]. Beijing: China Water Conservancy Press, 1980.(in Chinese))

[3] 陈祖煜，汪晓刚，杨 健，等. 岩质边坡稳定分析——原理·方法·程序[M]. 北京：中国水利水电出版社，2005.(CHEN Zuyu, WANG Xiaogang, YANG Jian, et al. The stability analysis of rock slope——theory, method and process[M]. Beijing: China Water Power Press, 2005.(in Chinese))

[4] 陈祖煜. 土力学经典问题的极限分析上、下限解[J]. 岩土工程学报，2002，24(1)：1 – 11.(CHEN Zuyu . Limit analysis for the classic problems of soil mechanics[J]. Chinese Journal of Geotechnical Engineering, 2002, 24(1): 1 – 11.(in Chinese))

[5] 崔政权，李 宁. 边坡工程——理论与实践最新发展[M]. 北京：中国水利水电出版社，2004.(CUI Zhengquan, LI Ning. Slope engineering——the development of theory and practice[M]. Beijing: China Water Power Press, 2004.(in Chinese))

[6] DUNCAN J M. State of the art: limit equilibrium and finite-element analysis of slopes[J]. Journal of Geotechnical Engineering, 1996, 122(7): 577 – 596.

[7] 孙君实. 条分法的提法及数值计算的最优化方法[J]. 水力发电学报，1983，(1)：52 – 64.(SUN Junshi. The formulation and optimization method for numerical evaluation of slice method[J]. Journal of Hydroelectric Engineering, 1983, (1): 52 – 64.(in Chinese))

[8] 孙君实. 条分法的数值分析[J]. 岩土工程学报，1984，6(2)：1 – 12. (SUN Junshi. The numerical analysis of strip partition method[J]. Chinese Journal of Geotechnical Engineering, 1984, 6(2): 1 – 12.(in Chinese))

[9] 朱大勇，钱七虎. 严格极限平衡条分法框架下的边坡临界滑动场[J]. 土木工程学报，2000，33(5)：67 – 73.(ZHU Dayong, QIAN Qihu. Critical slip fields of slopes satisfied all conditions of limit equilibrium for slices[J]. China Civil Engineering Journal, 2000, 33(5): 67 – 73.(in Chinese))

[10] 张鲁渝，郑颖人，赵尚毅，等. 有限元强度折减系数法计算土坡稳定安全系数的精度研究[J]. 水利学报，2003，(1)：21 – 27.(ZHANG Luyu, ZHENG Yingren, ZHAO Shangyi, et al. The feasibility study of strength-reduction method with FEM for calculating safety factors of soil slope stability[J]. Journal of Hydraulic Engineering, 2003, (1): 21 – 27.(in Chinese))

[11] 郑 宏，刘德富，罗先启. 基于变形分析的边坡潜在滑面的确定[J]. 岩石力学与工程学报，2004，23(5)：709 – 716.(ZHENG Hong, LIU Defu, LUO Xianqi. Determination of potential slide line of slopes based on deformation analysis[J]. Chinese Journal of Rock Mechanics and Engineering, 2004, 23(5): 709 – 716.(in Chinese))

[12] 郑颖人，赵尚毅. 有限元强度折减法在土坡与岩坡中的应用[J]. 岩石力学与工程学报，2004，23(19)：3 381 – 3 388.(ZHENG Yingren, ZHAO Shangyi. Application of strength reduction FEM in soil and rock slope[J]. Chinese Journal of Rock Mechanics and Engineering, 2004, 23(19): 3 381 – 3 388.(in Chinese))

[13] 郑 宏，刘德福. 弹塑性矩阵 D_{ep} 的特性和有限元边坡稳定性分析中的极限状态标准[J]. 岩石力学与工程学报，2005，24(7)：1 099 – 1 105.(ZHENG Hong, LIU Defu. Properties of elasto-plastic matrix D_{ep} and a criterion on limiting state of slope stability by FEM[J]. Chinese Journal of Rock Mechanics and Engineering, 2005, 24(7): 1 099 – 1 105.(in Chinese))

[14] 李 宁，张 平，李国玉. 岩质边坡预应力锚固的设计原则与方法探讨[J]. 岩石力学与工程学报，2004，23(17)：2 972 – 2 976.(LI Ning, ZHANG Ping, LI Guoyu . Discussion on design principle and method of prestressed cable for support of rock slopes[J]. Chinese Journal of Rock Mechanics and Engineering, 2004, 23(17): 2 972 – 2 976.(in Chinese))

[15] 李 宁，张 鹏，曲 星. 岩质边坡稳定性分析中的几个关键问题[C]// 西部矿山建设工程理论与实践会议论文集. 徐州：中国矿业大学出版社，2009：51 – 60.(LI Ning, ZHANG Peng, QU Xing. Several key problems of analysis on rock slope stability[C]// Proceedings of Theory and Application in Western Mine Construction. Xuzhou: China University of Mining and Technology Press, 2009: 51 – 60.(in Chinese))

[16] 朱大勇，钱七虎，周早生，等. 基于余推力法的边坡临界滑动场[J]. 岩石力学与工程学报，1999，18(6)：667 – 670.(ZHU Dayong, QIAN Qihu, ZHOU Zaosheng, et al. Critical slip field of slope based on the assumption of unbalanced thrust method[J]. Chinese Journal of Rock Mechanics and Engineering, 1999, 18(6): 667 – 670.(in Chinese))

[17] 朱大勇，钱七虎，周早生，等. 岩体边坡临界滑动场计算方法及其在露天矿边坡设计中的应用[J]. 岩石力学与工程学报，1999，18(5)：497 – 502.(ZHU Dayong, QIAN Qihu, ZHOU Zaosheng, et al. Technique for computing critical slip field of rock slope and its application to design open pit slope[J]. Chinese Journal of Rock Mechanics and Engineering, 1999, 18(5): 497 – 502.(in Chinese))

[18] 刘华丽，朱大勇，钱七虎，等. 滑面正应力分布对边坡安全系数的影响[J]. 岩石力学与工程学报，2006，25(7)：1 323 – 1 330.(LIU Huali, ZHU Dayong, QIAN Qihu, et al. Effect of normal stress distribution on factor of safety of a slope[J]. Chinese Journal of Rock Mechanics and Engineering, 2006, 25(7): 1 323 – 1 330.(in Chinese))

[19] 朱大勇，钱七虎. 三维边坡严格与准严格极限平衡解答及工程应用[J]. 岩石力学与工程学报，2007，26(8)：1 513 – 1 528.(ZHU Dayong, QIAN Qihu. Rigorous and quasi-rigorous limit equilibrium solutions of 3D slope stability and application to engineering[J]. Chinese Journal of Rock Mechanics and Engineering, 2007, 26(8): 1 513 – 1 528.(in Chinese))

中国岩石工程技术的新进展

钱七虎[1,2]

(1. 解放军总参科技委,北京 100857;2. 解放军理工大学,南京 210007)

[摘要] 在水电建设的岩石工程技术方面,介绍了三峡船闸高边坡稳定性及监控施工技术;深切河谷水电工程高边坡稳定性和支护技术;大跨度高边墙地下洞室群围岩稳定技术。在公路、铁路建设中的岩石工程技术方面,介绍了青藏铁路——穿越长年冻土层的施工技术;公路、铁路隧道中乌鞘岭隧道挤压大变形支护技术,高寒高海拔的风火山隧道、二郎山隧道施工技术;岩溶地区隧道超前地质预报技术。在矿业工程的岩石工程技术方面,介绍了煤矿巷道支护成套技术创新体系;低透气性煤层群无煤柱煤与瓦斯共采技术;深凹露天矿安全高效开采技术。另外,还介绍了岩石工程锚固新技术和精细爆破技术。

[关键词] 岩石工程;水电工程;支护技术;地质预报;锚固技术;爆破技术

[中图分类号] TU45 [文献标识码] A [文章编号] 1009-1742(2010)08-0037-12

1 前言

"岩石力学与岩石工程"是一门与国民经济建设、国防建设有着极其密切关系的应用学科。人类为了开发能源、防灾减灾、发展农业,要兴修大量的水利工程;为了修公路、筑铁路,要劈山越岭、开凿隧道(包括水下隧道);为了国民经济建设的需要,要从地下索取大量的矿产资源;为了修筑大型桥梁,要解决最关键的桥墩岩基问题;为了缓解地面居住及城市交通的紧张状况,势将更大规模地开发和利用地下空间;为了国防建设的需要,要修筑地下洞(井)库工程等。人类这些重要的工程活动,实际上都要依据"岩石力学与岩石工程"学科的理论原理作为指导,才能保证正确进行。没有"岩石力学与岩石工程"学科知识的武装,这些重要工程的设计、施工和成功修建是不可能的。我国的岩石工程,无论是地面的,还是地下的,近些年来其规模之大、难度之高、数量之多,已居世界之前列。例如,世人瞩目的三峡工程以及金沙江、雅砻江、大渡河上已建和在建的大型水电工程;南水北调水利工程;成昆(成都—昆明)、南昆(南宁—昆明)、京九(北京—九龙)、青藏(青海—西藏)、滇藏(云南—西藏)等铁路工程;"7918"(7条射线、9条纵线、18条横线)高速公路工程;大冶、攀枝花、鞍本、金川等矿山工程;抚顺、大同、两淮、兖州等煤炭工程;大庆、胜利、克拉玛依等石油工程;秦山、大亚湾、岭澳等核电工程;北京、上海、广州、深圳等城市地铁工程;一些在建或蓄势待建的石油战略储存、核废料储存工程、国防工程等。这些工程相当一部分在我国西部地区,不仅地势险峻、地质环境复杂,而且还处于地质灾害高风险条件下,带来的技术难题也是空前的。几年来中国岩石力学与岩石工程科技人员为解决这些技术难题,奋力拼搏,开拓进取,取得了一大批开创性成果。

2 水电建设中的岩石工程技术

2.1 三峡船闸高边坡稳定性及其监控施工技术

水电工程最具代表性的是三峡工程。三峡岩石工程建设重大成果之一是三峡船闸高边坡稳定性及其监控施工技术。永久船闸位于三峡大坝左岸山体中,是在山林中深切开挖修建的双线连续 5 级船闸。船闸线路总长 6 442 m,其中船闸主体段(闸室段)

本文原载于《中国工程科学》(2010 年第 12 卷第 8 期)。

长1 621 m。船闸沿线形成W形(见图1)。

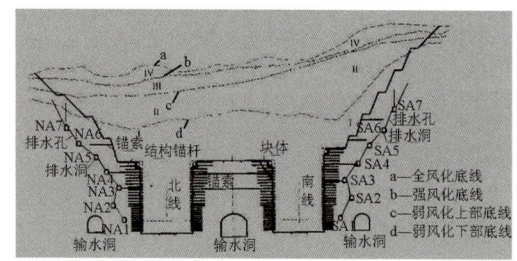

图1 船闸高边坡典型锚固断面
Fig.1 Typical anchoring section of high-rock slope of ship lock

人工开挖岩质高边坡,最大开挖高度173 m,最大坡高160 m,边坡高度连续超过120 m范围长约460 m,土石方明挖4 196×10⁴ m³,土石方洞挖9.8×10⁵ m³,两线船闸中间保留有50～60 m中隔墩。船闸最大总水头113 m,地下输水系统反弧门单级最大工作水头45.2 m,是目前世界上规模最大、级数最多、总设计水头最高的船闸。

和一般高边坡相比,船闸高边坡具有以下特点:a 它是在山体中深切出来的陡高边坡,高度大、形态复杂、范围广、应力释放充分,呈现出明显的卸荷和非均质特征;b 为确保黄金水道畅通和满足船闸人字门的正常运行,对边坡稳定和变形特性有严格的特殊要求;c 施工难度大、干扰多、工期紧,不仅地面施工强度高,窄、深且陡的闸室直立边墙开挖更为困难,而且与地下隧洞与竖井开挖同步进行,如何解决开挖与爆破的相互影响,最大限度地减少岩体损伤和确保施工安全都是施工中极具挑战性的难题。

三峡船闸20世纪50年代末开始设计研究,1994年开工建设,2003年6月建成投产,它是多门类、多学科先进技术与方法综合运用和研究成果的汇萃。几十年来,数百名高校、科研、设计、施工等部门科技人员进行联合攻关,经过长期研究,所有技术难题都得到妥善解决。它涉及高边坡工程地质、岩体力学特性、地下渗流及排水措施、施工方法优化与开挖爆破技术、高边坡锚固技术、高边坡稳定性分析等方面。研究获得了有关三峡船闸高陡边坡工程稳定大量定性和定量的信息,各种二维、三维数值分析方法得出大量计算成果,为设计、施工和科学评价稳定性提供了可靠依据。

三峡船闸高边坡稳定性及其监控施工技术成果是多方面的,文章主要介绍它的综合锚固技术。三峡船闸高边坡共安装普通钢筋锚杆近3.6万根、高强结构锚杆10万根、1 000 kN级锚索229束、3 000 kN级锚索3 975束,块体支护1 054块,其边坡锚固量和块体支护量均居世界之最,并研究成功了多形式的锚索结构。三峡船闸高边坡锚固工程浩大,锚索结构形式及施工工艺直接关系到工程进度和造价。为充分利用各种锚索的性能,根据船闸边坡的结构特点,在船闸不同部位采用不同结构形式的锚索。

1)船闸高边坡使用的支护锚索有:1 000 kN、3 000 kN级端头锚和3 000 kN级对穿锚,除113束监测锚索和121束闸首混凝土结构加固锚索为无黏结结构外,其他均为全长黏结锚索。

2)设计吨位1 000 kN锚索,超张拉吨位1 150 kN,采用7根φ15.24 mm的1 860 MPa级钢绞线,设计强度利用系数为0.55,超张拉强度系数为0.63,孔深30～40 m,孔径115 mm,内锚段长5 m,内锚段采用枣核状结构。

3)3 000 kN级端头锚索,锚索孔深30～60 m,内锚段长8 m,孔径165 mm。锚索设计吨位3 000 kN,超张拉吨位为3 450 kN,采用19根φ15.24 mm的1 860 MPa级钢绞线,设计强度利用系数为0.606,超张拉强度系数为0.697,其他与1 000 kN级端头锚索相同。

4)3 000 kN级对穿锚索,主要用于中隔墩及南北坡的系统支护,南北坡与岩体深部的排水洞对穿,中隔墩两侧对穿。对穿锚索两端均设锚墩,只有张拉段无内锚段。锚索孔深40～60 m,孔径165 mm,其他与3 000 kN级端头锚索相同。闸首支持体加固采用专门设计的能适应混凝土与岩体间可能产生变形的无黏结锚索,并考虑闸首闸门交变荷载的作用,外锚头还剥皮8 m形成黏结段,使外锚端有锚具和黏结段的双重保险。

5)另外,根据水电工程的特点,对锚索的耐久性也作了系统研究,并采取了相应的增强和耐久性措施,确保耐久、实用。

工程完工后,截止2007年7月20日,测得南北坡岩体向闸室中心线方向的最大累计位移分别为72.07 mm和52.96 mm,位移月变化在1.64 mm和1.96 mm之间;南北坡直立墙最大位移分别为36.93 mm和30.08 mm;中隔墩南、北侧最大累计位移分别为23.16 mm和31.50 mm,闸首顶部向闸室方向最大位移 4.73 mm,闸首顶部向上游方向最大位移 2.75 mm,闸首顶部向下游方向最大位移 2.84 mm,变形基本结束。船闸投入运用后,闸室充水过程中船闸首的位移不大于 0.5 mm,完全满足船闸人字门正常运用的要求,均控制在设计预测范围之内。

2.2 深切河谷水电工程高边坡稳定性和支护技术

除三峡工程外,中国还有大批大型和超大型水电站正在和即将建设中。西电东送的规模将超过 $1.5×10^8$ kW。如长江支流金沙江干流上,规划建设 20 个梯级电站,其中 $100×10^4$ kW 以上的有 17 座,如溪洛渡、白鹤滩、乌东德和向家坝的规模是在 $500×10^4$ kW 以上;雅砻江干流规划建设 21 个梯级电站,$100×10^4$ kW 以上的有 10 座,其中锦屏一级、二级总装机容量为 $840×10^4$ kW;大渡河干流规划建设 22 个梯级电站,其中 $100×10^4$ kW 以上电站有 7 个,例如,瀑布沟电站为 $360×10^4$ kW。另外,还有澜沧江上的小湾、糯扎渡、乌江上的构皮滩、红水河上的龙滩,黄河上游的拉瓦西等。

这些电站大多处在我国青藏高原向四川盆地过渡地带,由于晚近期以来地质构造的作用,青藏高原快速隆升,形成了一系列的大江大河以及高差达 2 000~3 000 m 的深切河谷,同时也发育了我国著名的西部深大断裂带和地震带。因此,地质条件十分复杂,在开发水力资源的过程中,存在着复杂的大型工程高边坡问题。这类高边坡具有如下一些显著特征:a.自然谷坡陡峻,坡角一般在 45°以上,大多在 70°~90°,谷深多在千米以上;b.工程边坡高,大多在 300 m 以上,最高可达到 600 m 以上;c.边坡的地质条件复杂,断层发育,岩体地应力高,岩体卸荷强烈,卸荷深度大。锦屏一级水电站就是一个典型的例子。

锦屏一级水电站枢纽区为典型的深切"V"型峡谷,相对高差 1 500~1 700 m。左岸为反向坡,1 820~1 900 m 高程以下为大理岩,坡度 55°~70°;以上为砂板岩,坡度 40°~50°,呈山梁与浅沟相间的微地貌特征。该工程将建世界最高拱坝,左岸肩拱肩槽边坡开挖高度达到 530~540 m,国内外少有,而且地质条件十分复杂,设计和实施难度很大,引起国内外工程界高度关注。

成都勘测设计研究院联合国内多家科研单位和大专院校对坝址区开展了深入细致的研究,基本查清了该区的工程地质条件和主要的工程地质和岩石力学问题。针对结构面不利组合情况,分别采用三维极限平衡法、有限元强度折减法和三维非线性有限元法等方法,考虑边坡不同工况,分析了边坡整体滑移破坏模式和局部滑移破坏模式下的稳定状况,并采取相应确保稳定的措施。

1)边坡截、排水系统:在开挖边坡周边设置截水沟、排水沟等地表排水系统和地下排水孔、排水平洞系统,避免地表水、地下水入渗开挖边坡,引起边坡失稳。

2)锚杆、锚筋加固系统:采用系统锚杆、锚筋束对开挖边坡浅表松动岩体进行系统处理,确保岩体稳定,施工安全。

3)锚索布置:左岸开挖边坡 1 820 m 高程以上的砂板岩卸荷岩体及 1 800 m 高程附近 f_{42-9} 断层剪出口部位岩体,采用 2 000 kN(必要时 3 000 kN)级预应力锚索加固,锚索间排距 4 m×4 m,锚索长度 40~80 m。大理岩卸荷岩体采用 2 000~1 000 kN 级锚索加固,随机布置。开口线附近浅表危岩体采用 2 000 kN 级预应力锚索加固,锚索间排距 5 m×5 m。

4)抗剪置换洞:f_{42-9} 断层为边坡潜在整体滑动的滑移面,采用抗剪置换洞方式予以加固。分别在 1 834 m,1 860 m 和 1 885 m 高程设置 3 层抗剪置换洞,采用 9 m×10 m 断面,挖除断层及破碎带,周围岩体固结灌浆,回填混凝土。

5)灌浆处理:利用现有锚固洞对卸荷裂隙和卸荷岩体进行灌浆处理,采用马道锚筋束孔及马道间的预灌浆孔对边坡进行预灌浆处理。

通过采取以上措施处理的边坡工程,经过几年的监测,边坡变形逐步收敛,证明边坡加固实施效果较好。该边坡成功实施,表明我国边坡工程实践取得新的跨越,在边坡工程地质调查、稳定性分析、加固技术、施工开挖技术等方面提高到一个新水平。

2.3 大跨度高边墙地下洞室群围岩稳定技术

随着我国国民经济的不断发展和对地下空间建设的需求,大跨度高边墙地下洞室(群)在我国的能源、交通、采掘、国防等行业得到快速发展。以水电行业为例,近 10 年来,在我国西南地区兴建或将要建设一大批地下式水力发电站,其洞室的跨度、高度和规模位居世界前列。表 1 是西南地区部分水电工程规模统计表。

表 1 西南地区部分水电工程规模统计表

Table 1 The statistics of scales about hydropower engineering in southwest China

名称	流域	装机容量/MW	坝型/坝高/m	地下厂房(宽/m×高/m×长/m)	调压室(宽/m×高/m×长/m)	引水(泄洪)隧洞(宽/m×高/m×长/m)	工程边坡高/m	开挖工程量(石/$10^4 m^3$土)	建设状况	特性说明
二滩	雅砻江	3 300	拱坝/240	25.5×65.4×280.3	19.5×69.8×217	16.5×16.5×876.6	240	1 151.5	已建,2000年竣工	国内已建最高双曲拱坝
溪洛渡	金沙江	12 600	拱坝/278	28.4×75.1×381	23×94×300	φ10×183	>300	3 543.5	在建,2005年开工	尾水洞长1 802 m
锦屏一级	雅砻江	3 600	拱坝/305	29.6×68.8×277	φ32×92	15×16×—	>350	1 853.8	在建,2005年开工	在建最高拱坝
大岗山	大渡河	2 400	拱坝/210	31.2×72.4×277.9	20.5×72.3×130	14.5×16×—	>260	912.3	筹建	
官地	雅砻江	2 400	重力坝/168	31.9×77.5×243	21.5×18×96	18.8×27×—	>120	1 231	筹建	
双江口	雅砻江	2 000	堆石坝/314	29.3×64×—	18×52×120	9×10×73	>200	1 052	拟建	在建最高堆石坝
两河口	大渡河	3 000	堆石坝/295	28.5×63×273	20.5×83.7×215	12×15×215	>200	2 120	拟建	295 m堆石坝
瀑布沟	大渡河	3 300	堆石坝/186	26.8×70.1×294.1		φ9.5×533.7	>150	1 529.5	在建,2003年开工	75.4 m深厚覆盖层
构皮滩	乌江	3 000	拱坝/232.5	27×73.2×230.5	24×46×110.2	φ9.5×—	>280	1 609.3	在建,2003年开工	
小湾	澜沧江	4 200	拱坝/292	30.6×79.2×298	φ32×87	φ18×932	>400	2 468	在建,2002年开工	大圆筒型调压室

近 10年来,我国水电站地下厂房特别是大型地下洞室群的设计、施工和管理的理论、方法、技术和措施已取得长足的发展,地下工程的设计理念不断创新和突破,新技术和新方法不断应用和完善,积累了丰富的设计、施工经验,也产生了显著的经济效益和社会效益。然而,大型水电工程地下厂房往往地处高山峡谷,洞室规模巨大,其主体洞室(发电厂房、主变室、尾水调压室或尾水闸门室等)跨度接近或超出 30 m,高度可达到 80~90 m,长度可达 400~500 m,主体洞室与压力管洞、母线洞、尾水管洞以及通向地面的出线洞(井)、通风洞(井)、交通洞、尾水隧洞、排水洞等附属洞室一起,形成规模宏大、纵横交错的地下洞室群(见图 2)。对于各种类型地下厂房,不仅需要解决好洞室群在开挖过程中的洞室围岩稳定问题,更要确保地下厂房能长期安全稳定运行。

大跨度高边墙地下洞室(群)围岩稳定面临以下主要关键技术问题,包括地下洞室群合理布置、地

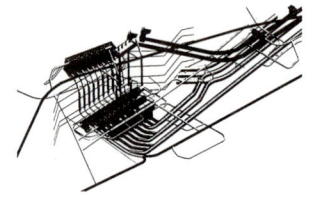

图 2 某大型地下厂房洞室群布置示意图
Fig. 2 The sketch of caverns group in a big-scale underground workshop

下洞室围岩稳定性与评判标准、地下洞室群支护设计与优化、高地应力地下洞室群岩爆的预防和处置、地下洞室群施工期快速监测与反馈分析和地下洞室开挖程序等。文章仅就开挖顺序和监测与反馈分析作简要介绍。

地下洞室群的施工采取不同的开挖顺序将产生不同的应力变化途径,研究不同的施工顺序以寻求岩体的最佳稳定效果是大型地下洞室群围岩稳定与支护方式研究的重要方面。目前,在地下洞室群施工开挖顺序研究方面,国内外已从平面分析发展到

反映大型地下洞室群的三维非线性特征的定量化分析,并与施工组织优化相配合,使施工力学分析与施工系统工程紧密结合,并进行定量化评估。

总结已建和在建的地下洞室群施工情况,我国科技工作者提出的开挖顺序是:"先拱后墙、自上而下、逐层开挖、逐层支护,平面多工序,竖向多层次,多工作面交叉作业"的施工方案。图3是某工程主厂房及主变室分层开挖示意图。

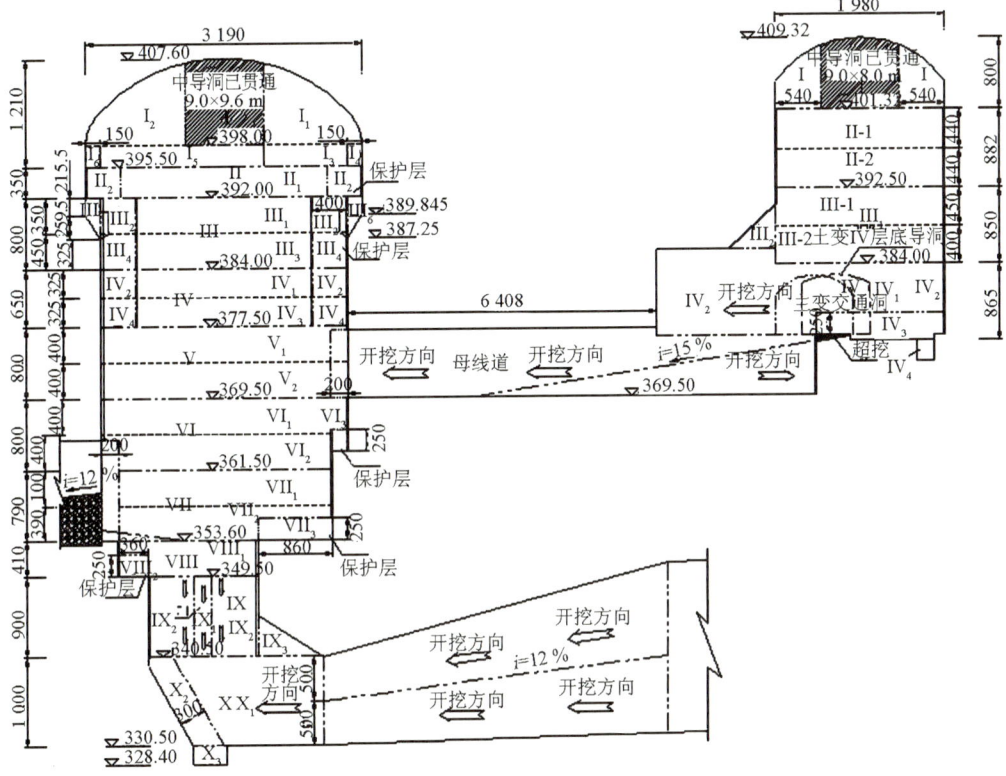

图 3 某工程主厂房及主变室开挖分层示意图 (单位: cm)

Fig. 3 Sketch of stratified excavation on main workshop and the primary chamber (unit: cm)

由于岩体是一个包含各种复杂因素和本构关系的模糊系统,为弥补预可行研究和设计中结论和实际工程之间差距,在施工方案确定以后,还要在施工过程中建立快速监测和反馈分析系统,分析研究在一定工程措施下(开挖顺序、爆破参数、支护措施)的输出信息(应力变形等),以不断修正施工方案中结构和支护参数,这个过程要不断循环反复,也就是说整个施工要不断调整和优化,确保地下洞室群施工的经济与安全。图4是溪洛渡水电站地下厂房典型监测剖面多点位移计布置及监测反馈分析成果。从图4可清楚看出,洞室各关键部位变形量的大小,图4中虚线为实测数据,实线为反馈分析计算数据,两者总的趋势吻合较好。

通过溪洛渡地下厂房施工期监测反馈分析,逐层进行围岩稳定性评价和后续开挖对围岩影响的预测,为整个开挖过程中的设计支护参数的调整、施工开挖方案的优化提供了重要的依据。

3 公路、铁路建设中的岩石工程技术

根据规划,2020年以前中国将建成布局为"7918"的高速公路网络,即7条射线、9条纵线、18条横线,总里程约 8.5×10^4 km。其中北京至各省会城市的7条射线总里程为 1.8×10^4 km。该高速公路网将连接所有现在人口在20万人以上的319个城市。除高速公路外,其他各种等级的公路也在大规模建设,到2010年全国公路总里程将达到 230×10^4 km,到2020年将达到 300×10^4 km。

中国铁路营运里程到2007年底为 7.8×10^4 km,居亚洲第一位。根据规划,2020年前铁路营运里程将达到 10×10^4 km。中国铁路工程除已建成的4纵4横线路外,还有一大批重大的待建和在建项目,其中举世瞩目的是目前正在施工的京沪高速铁路,总投资为2 209亿元,是一次性投资最多的铁路建设项目,我国铁路建设的速度在东部、铁

路建设的难点在西南。例如,滇藏铁路是规划线路之一,自云南大理站西,经澜沧江、怒江、雅鲁藏布江到拉萨,路经滇西北横断山脉、藏东南高山峡谷地区及藏南谷地,全长 1 594 km。沿线依次通过唐古拉—三江折褶带、拉萨—波密折褶带和雅鲁藏布江缝合带等三大构造带。铁路沿线地区新构造运动十分强烈,据我国 30 多年精密复测水准点资料,年平均上升速度达 12 mm,珠穆朗玛峰地区达 12~50 mm。据国家地震局 1990 年《中国地震烈度区划图》(1:400 万),滇藏铁路沿线工程设防烈度≥Ⅸ度区线路累计长 127 km,占 8%。全线设特大、大、重型桥梁 392 座,总长 108.9 km;隧道 419 座,最长者达 12.59 km,总长 491.8 km。全线桥隧总长 600.7 km,占线路总长的 7.7%。可以说,其技术难度是世界罕见的。

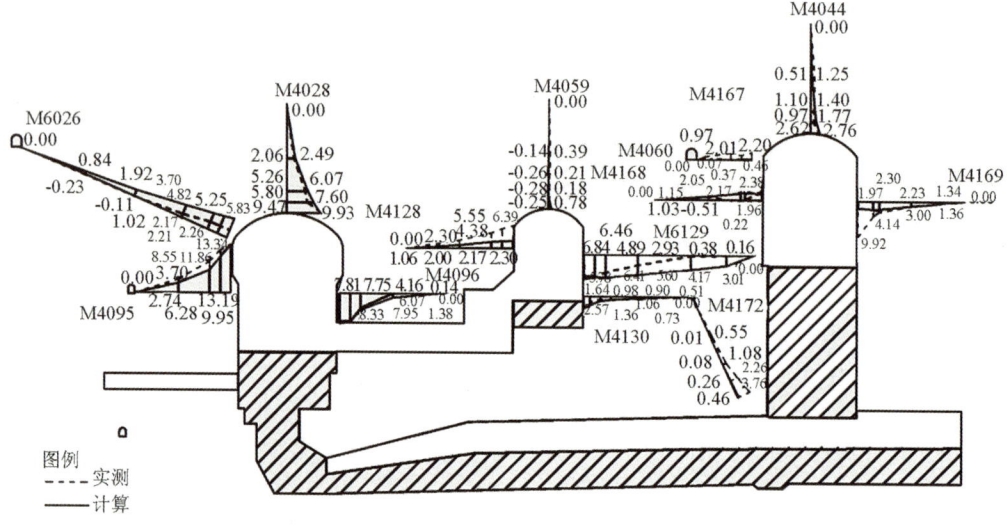

图 4 溪洛渡地下厂房典型剖面多点位移计布置及监测反馈分析成果

Fig. 4 Typical location of multiple point displaymentmeter of the underground powerhouse in Xiluodu and its monitoring feedback

3.1 青藏铁路——穿越长年冻土层的施工新技术

最近几年铁路建设最突出的成就是建成青藏铁路。青藏铁路,从格尔木至拉萨,全长 1 142 km,海拔高于 4 000 m 地段长达 960 km,最高海拔 5 072 m,创造了世界高原铁路的建设奇迹。

工程建设成功克服了冻土、高寒缺氧、生态脆弱三大世界性工程难题,实现了多项技术创新。在关键的多年冻土工程技术方面,中铁第一勘察设计院系统研究了气候变化的影响,制订了高原多年冻土地区铁路勘察、设计和施工的国家技术标准;确立了主动降温、冷却地基、保护冻土的设计思想并确定了关键参数;创造性地综合采用片石气冷、热棒路基、以桥梁跨越特殊不良冻土地段,并研究成功防冻胀隧道衬砌结构等成套冻土路基结构施工工艺和施工技术。

青藏铁路沿线自然条件恶劣,属生命禁区。建设中,针对大群体、高海拔、作业时间长的特点,创建三级医疗保障救治体系,应用高原病综合救治等技术和创鼠疫预防监控体系,实现高原病零死亡和人间鼠疫零感染。青藏高原生态环境脆弱,线路穿越三江源等自然保护区,环境保护任务艰巨。建设者开展野生动物、高寒植被、多年冻土、江河源水质保护等方面的综合研究和创新实践,在大规模建设中保护了生态环境,实现了工程建设与自然环境相和谐。

技术创新、管理创新为青藏铁路建设提供了可靠支撑,青藏铁路建设形成行业标准、部级和国家级工法多项,获专利数十项,发表论文千余篇,极大地推动了多年冻土工程、高原医学和环境保护等领域的科技进步,总体技术达到国际领先水平,是中国铁路建设史上标志性的重大成就。

3.2 '隧道'施工技术新进展

在公路、铁路建设中,隧道工程施工技术近几年也取得了巨大成就。中国公路、铁路等行业已投入使用的各类隧道 1.2 万座以上,延长总计超过 7 000 km。其中铁路隧道 7 000 多座,总长超过 4 000 km,公路隧道 4 000 多座,总长超过 2 000 km。中国建成的水下隧道主要集中在上海地区,已建 8 座跨黄浦江隧道,另外待建和在建的还有 5 座。其他在建的有南水北调的跨黄河跨长江隧道和城市有关隧道;在建的跨海隧道有胶州湾湾口海底隧道、广州生物岛——大学城隧道及狮子洋海底铁路隧道等;拟建的有琼州海峡跨海工程、渤海湾(大连—蓬莱)跨海工程、杭州湾外海工程(上海—宁波)以及大连湾水下隧道等。

在特长隧道建设方面,10 km 以上特长铁路隧道有 21 座,其中最长的铁路隧道为 20.06 km 的乌鞘岭隧道,另外,还有大瑶山隧道(14.3 km)、秦岭一线隧道(18.46 km)、二线隧道(18.46 km)、东秦岭隧道(12.27 km)、圆梁山隧道(11.07 km)。公路隧道方面有秦岭终南山隧道(18 km),规划在建中的 10 km 以上公路隧道有甘肃大坪里隧道、陕西包家山隧道、山西宝塔山隧道等。中国目前规模最大的特长公路隧道群——沪蓉西高速公路全线有隧道 44 座共 155.712 km,隧线比 48.66%,其中特长隧道 10 座共 92.134 km。水工隧洞中,锦屏二级水电站共 4 条引水隧洞处于高地应力和岩爆、岩溶、高地温、有害气体等不良地质地带,开挖洞径 12 m,平均长度 16.625 km,最大埋深达 2 525 m。另外,引大入秦工程是跨流域调度的大型水利工程,全长 884.3 km,有隧洞 77 座,总长 110 km。其中总干渠全长 87 km,有隧洞 33 座,总长 75.14 km,堪称"地下运河"。2009 年 4 月 15 日,大伙房水库输水工程输水隧道全线贯通,贯通误差仅为 3 cm,隧道长 85.3 km,超过目前世界公认最长的 57.67 km 的瑞士戈特哈尔德隧道,成为世界上最长的水工隧道。

3.2.1 乌鞘岭隧道挤压大变形的支护技术

兰新铁路线上的乌鞘岭隧道是中国目前最长的铁路隧道(全长 20.06 km),岭脊地段埋深 500～1 100 m,隧道穿越多条区域性大断层组成的"挤压构造带",高地应力软弱围岩挤压大变形显著,变形控制十分困难,对设计、施工技术都是一次空前的挑战。中铁第一勘察设计院集团有限公司等单位科技人员采取了"短开挖、快封闭、强支护、速成环、二衬适时紧跟"的原则,成功解决了通过挤压大变形的难关,具体措施有:

1)采取辅助工法,对软弱断层围岩进行超前预支护,施工中采用严格控制施工爆破参数,尽量减少对围岩的扰动。

2)优化施工参数,采取超短台阶法施工,使围岩和初期支护系统及早闭合成环。

3)采用"多重支护"的原则,开挖初期及时喷混凝土(厚 20 cm)并施作长系统锚杆,使围岩和支护系统达到变形协调,并释放部分位移。随即施作钢支撑,并复喷混凝土(厚 15 cm)以部分限制围岩挤压变形发展。

4)根据实例变形曲线采取合适的预留变形量和适当的二次衬砌施作时机。

5)加强现场监控量测,动态控制围岩挤压大变形,并重视地质超前预报。

3.2.2 高寒高海拔隧道施工技术

中国隧道工程在机械化程度、修建速度、长大隧道修建能力等方面和国际水平相当,但在穿过高寒、高海拔复杂地质条件地区施工技术已处于世界领先水平。如以青藏铁路风火山隧道为依托而形成的高原多年冻土隧道防排水及保温施工技术、低温早强混凝土施工技术、高原多年冻土隧道光面爆破技术、高原多年冻土隧道施工卫生保障技术和环境保护技术等,填补了国内在该领域的空白。以川藏公路二郎山隧道和鹧鸪山隧道为依托,解决了高海拔大型公路隧道地形地质条件复杂、气候条件恶劣、营运条件特殊带来的三大类关键技术难题,形成了一批创造性综合技术成果。

1)在国内外开创了单洞双向行驶条件下,平导压入半横向式和纵向式通风的特长公路隧道建设模式并解决了相应的关键理论和技术问题。

2)在国内外首次建立了季节性冻胀冻融地区特长公路隧道结构抗防冻设计的技术体系。

3)运用岩体浅表生改造理论、卸荷变形破裂理论和软岩力学理论,在国内外首次建立了高地应力场的相应地质力学模式、岩爆和大变形的力学机制及其预测预报综合集成技术。

4)在国内首次测试了 400～5 000 m 海拔范围考虑烟雾和 CO 的海拔高度系数,填补了我国现行《公路隧道通风照明设计规范》的空白。

5)建立了基于 GIS 的隧道机电设备营运智能监控及维护管理一体化技术系统,实现了高寒区大型公路隧道机电系统监控管理的智能化、网络化、组态化和综合化。

6)根据公路隧道横断面特征,针对高烈度地震区公路隧道,提出了加设减震层、采用聚合物钢筋或钢纤维混凝土衬砌等抗减震措施。

二郎山隧道经历了汶川地震的考验,基本完好无损,证明施工技术措施是正确的。以上核心技术均属于世界领先水平。

3.2.3 岩溶地区隧道超前地质预报技术

在隧道施工中,对不良地质条件的超前预报方面近几年也取得了重大进展。山东大学科技人员针对沪蓉西高速公路上20余条岩溶隧道进行的综合超前预报研究,预报准确率达到70%。主要的技术进展有:

1)在大量分析已有地质勘察资料基础上,分析了隧道岩溶发育的主控因素、岩溶垂直分带和水平展布特点,建立了岩溶隧道灾害风险定量评价方法,提出了相应的隧道施工灾害风险分级标准,并对高风险岩溶隧道各工程段进行了科学的风险评估。

2)针对物探多解性和地质情况的复杂性,研究了不同超前地质预报方法的特点和适用范围;研发了突水预报的瞬变电磁仪和地质雷达专用后处理软件;研究了雷达、瞬变电磁的正反演理论和提高预报精度的方法;提出了综合超前地质预报的原则和技术体系。按照综合预报"合理搭配、科学管理、贯穿全程、因地制宜"的思想,将研究成果成功应用于20余条岩溶隧道的超前地质预报工程实践中,指导了施工,避免了突水和突泥等灾害事故的发生,保证了施工安全。

3)研究了岩溶裂隙水突出的影响因素与其基本力学机理;研究了动力扰动诱发突水的灾变机制、岩溶裂隙水作用下岩石裂纹扩展机制、岩溶水突出耦合突变机制与非线性特点;给出了相应的灾变演化路径和突水判据;分析研究了岩溶裂隙水作用下单裂纹扩展机理、爆破启裂扩展机制和相邻裂纹的贯通机制,并进行了数值模拟,为灾变控制方法的研究奠定了理论基础。

4)提出了岩溶隧道灾害四色预警机制和相应的应急预案;研究了预警信息的发布依据和发布流程,制订了主要高风险岩溶隧道应急救援方案和逃生路线,并在乌池坝隧道进行了突水灾害应急救援演练。

5)研究了高风险岩溶隧道突水灾害的处理原则与标准;在分析不同涌水类型和特点的基础上,建立了"先探后堵、综合治理"的岩溶水处理原则;研究了不同注浆材料的适用性,进行了注浆材料的性能试验;探讨了常见突水灾害的处理方案,应用不同的注浆方案和注浆材料,成功地在现场进行了不同岩溶突水类型的注浆封堵试验,为岩溶突水灾害治理积累了宝贵的经验。

4 矿业工程的岩石工程技术

中国煤矿资源丰富,目前在能源结构中占主导地位,2008年煤炭产量已达27.16×10^8 t,与之配套的每年巷道掘进量大约在12 000 km以上。中国黑色和有色金属矿开采规模巨大,2008年铁矿石原矿产量已达8.4×10^8 t,10种有色金属2008年总产量为$2 520.28\times10^4$ t,相应的每年黑色和有色金属地下矿山巷道掘进量也十分巨大。这是国民经济各行业中最大的岩石工程。

4.1 煤矿巷道支护成套技术创新体系

地下开掘规模巨大的巷道工程,巷道支护是直接影响矿井安全、产量与效益的关键技术。煤炭科学研究总院康红普研究团队针对我国煤矿巷道特点,开发集理论、方法、材料、机具、工艺、仪器和技术规范于一体的巷道支护成套技术创新体系,形成了有中国特色和自主知识产权的巷道支护技术,解决复杂困难巷道支护难题,加快我国高产高效矿井建设,大幅度提高煤炭产量与效益,明显改善了巷道安全状况。

主要研究内容有:煤矿巷道锚杆支护理论、锚杆支护动态信息设计法与软件、高强度树脂锚杆与锚索支护材料、锚杆钻机与快速施工技术、矿压监测仪器及技术、锚杆与注浆联合加固技术及综采放顶煤、深部开采等复杂困难条件巷道支护技术。通过16年连续进行巷道支护技术科技攻关,取得主要创新性研究成果有:

1)提出锚杆支护扩容—稳定理论。认为锚杆支护主要作用在于控制锚固区围岩的离层、滑动、张开裂隙等扩容变形与破坏,在锚固区内形成次生承载层,最大限度地保持锚固区围岩的完整性,避免有害变形的出现,减小锚固区围岩强度的降低。为此,应采用高强度、高刚度锚杆组合支护系统。高强度要求锚杆具有较大的破断力,高刚度要求具有较大的预紧力并实施加长或全长锚固,组合支护要求采用钢带、金属网等护表构件,应尽量一次支护有效控制围岩变形。

2)提出锚杆支护动态信息设计法,具有两大特点:a 设计不是一次完成,而是一个动态过程;

b．设计充分利用每个过程中提供的信息,实时进行信息收集、分析与反馈。

3) 开发出注浆锚杆、注浆锚索及钻锚注一体化等锚固与注浆联合加固技术。注浆锚索是在小孔径树脂锚固锚索的基础上,开发的小孔径树脂与注浆联合锚固锚索,兼有树脂锚固和注浆锚固锚索的优点。先对锚索进行树脂端部锚固,施加预紧力,使锚索及时承载,然后水泥注浆全长锚固。通过控制注浆参数,使浆液渗透到钻孔周围的煤岩体内,达到注浆加固的目的。引进与改进了钻锚注锚杆,兼有钻进、锚固、注浆加固的功能。这些技术将锚固与注浆加固有机结合,为松软破碎围岩巷道提供了有效加固手段。

4) 开发出巷道矿压和安全监测成套仪器,包括测力锚杆,锚索测力计,顶板离层指示仪,多点位移计等。这些仪器已大面积推广应用于煤矿井下,对巷道支护设计优化、评价巷道支护效果、保证巷道安全起到重要作用。

5) 开发出系列锚杆钻机,并经过不断完善和提高,产品已经根据我国煤矿巷道条件形成系列,性能指标达到国际先进水平,满足了煤矿锚杆支护技术的要求,结束了锚杆钻机主要靠进口的历史。

该项成果已推广应用于20个省(自治区)的58个大中型矿区,使我国煤矿巷道支护技术水平实现了跨越式提升。它深刻地改变了矿井的开拓部署与巷道布置方式,对我国高产高效矿井建设、煤炭产量与效益的大幅度提高及安全状况改善起到重要作用。

4.2 低透气性煤层群无煤柱煤与瓦斯共采技术

中国煤炭资源70%以上为低透气性、高瓦斯、高吸附性、煤岩松软的地质条件,瓦斯事故频发,已成为煤矿安全生产中一个突出问题。淮南矿业集团袁亮研究团队研究成功的"低透气性煤层群无煤柱煤与瓦斯共采技术",为低透气性、高瓦斯煤层安全高效开采开辟新的方向,其主要成果:

1) 采取卸压开采,增加煤层透气性,在采动区预先布置巷道和钻孔抽采瓦斯,实现卸压开采抽采瓦斯、煤与瓦斯共采构想。

2) 针对煤岩松软、极易离层,卸压开采采动区巷道围岩控制难题,提出锚索网注围岩的整体控制技术,并在工程实践中成功应用,保证了生产安全。

3) 变传统的U型通风方式为沿空留巷(替代预先布置的专用瓦斯抽采岩巷) Y型通风方式,工作面瓦斯抽采率达70%,确保被卸压高瓦斯煤层安全高效开采。

世界采矿大会国际组委会主席,波兰科学院院士杜宾斯基对此成果的评价是:"采用无煤柱沿空留巷Y型通风卸压开采,解决深井低透气性,高瓦斯,高地应力等构造复杂地质条件矿区煤与瓦斯共采技术难题,是无煤柱煤与瓦斯共采关键技术重大突破,成果达到国际领先水平"。

4.3 深凹露天矿安全高效开采技术

我国80%以上的铁矿石来自于露天开采。目前,很多大中型露天矿山已由山坡露天开采进入深部凹陷开采。矿山转入深凹开采后,一方面,随着开采深度的增加,边坡不断加高加陡,形成名副其实的高陡边坡,如首钢水厂铁矿的最终边坡垂直高度达到670 m,凹陷开采深度430 m,边坡的加高加陡导致边坡的稳定性和开采的安全性越来越差。另一方面,提高边坡角又是露天矿减少剥岩量、降低生产成本的重要手段,如一座年产千万吨级的露天矿,边坡角提高$1°$,就可减少剥岩量$1\,000×10^4$ t以上,经济效益可达亿元。所以,这是一把双刃剑。为此必须系统地进行边坡优化设计和稳定性综合监测与控制的研究,才能在保证开采安全的前提下,最大限度地提高边坡角,减少剥岩量,降低成本,增加效益。

中国大型露天矿的总体边坡角与国外同类矿山相比,普遍偏缓。这主要是由传统的经验类比和二维极限平衡分析的边坡设计方法所决定的。北京科技大学蔡美峰研究团队在大量系统的工程地质、水文地质勘查和试验研究、矿区地应力场测量和矿岩物理力学特性测试的基础上,首次采用大型三维非线性有限差分法、离散单元法和基于GIS的三维极限平衡分析相结合的方法进行边坡稳定性分析和设计优化(其中,基于GIS的三维极限平衡分析方法为国内外首创),将水厂铁矿各区的总体边坡角分别提高了$1°\sim 6°$,平均提高了$3°\sim 4°$,使水厂铁矿边坡设计和开采优化达到并超过世界先进国家水平。同时采用了GPS全站仪等多种手段,结合网络理论,在水厂铁矿建立了边坡位移监测网,对边坡稳定性进行动态分析和预测预报,为及时采取必要的防控措施、保证边坡稳定提供了依据。同时,制定并采取了一系列保障深部高陡边坡强化开采安全的技术措施,包括:根据边坡潜在破坏模式调查和预测,制定了相应的边坡稳定性控制技术;针对水厂铁矿位于地震危险区,基于边坡动力失稳机制研究,制定了高陡边坡开采的防震减灾技术;针对水厂铁矿边界紧邻滦河,制定了邻水开采高陡边坡采场渗透破坏的监测与防治措施等,从而保证了深凹露天高陡边坡开采的安全。为了降低运输成本、提高生产效率,还在该矿建立了新型高效的汽车—胶带半连续运输系统和基于 GPS 的自动化实时生产调度系统。

通过 4 年的攻关研究,使水厂铁矿生产成本显著下降,生产效率大幅度提高。2001—2004 年,在开采深度和提升高度不断增加的情况下,矿石成本不但没有上升,反而逐年下降,下降幅度达 32%;4 年中全员劳动生产效率增长了 2.3 倍,当年该矿的全员劳动生产率是全国同行业重点矿山平均水平的 6 倍以上,位列全国第一。创造经济效益 1.39 亿元/年。

通过攻关研究,不但将水厂铁矿的生产和管理整体提升到同期国际先进水平,而且解决了中国大型露天矿深部开采带有共性的关键技术问题,有力地推动了中国露天矿的科技进步、生产发展。

5 岩石工程锚固新技术

近年来,为了适应我国土木、水利水电、矿山和建筑等工程建设高速发展的需要,加强了对岩石锚固综合技术的研究,中冶建筑研究总院程良奎研究团队在岩土锚杆的荷载传递机制、设计、结构形式、灌浆工艺以及长期性能与安全评价等方面取得了一系列创新成果,使我国岩土锚固的综合技术水平得到了提升和跨越,主要成果有:

5.1 开发了可重复高压灌浆型和旋喷灌浆扩体型锚杆

针对复杂地层锚杆承载力低、蠕变变形大无法满足工程使用要求的突出难题,以及工程经济性对高承载力锚杆的需要,相继开发了可重复高压灌浆型及旋喷灌浆扩体型锚杆。

可重复高压灌浆型锚杆的技术关键是采用独特的注浆套管和注浆枪对锚杆锚固段圆柱形注浆体实施一次或多次高压劈裂灌浆,从而有效地提高锚固体的粘结强度和锚杆承载力。旋喷灌浆扩体锚杆即采用高压喷射原理在锚杆锚固段范围内对岩土体进行水力切割扩孔并置换充填水泥浆,形成一个圆柱状的扩大头,从而充分发挥扩大头的端承作用,极大地提高锚杆承载能力,通常扩大头直径为 0.7～0.8 m,最大可达 1.2 m。

5.2 开发了能显著提高承载力和防腐特性的荷载分散型锚固(单孔复合锚固)体系

此锚固体系是在同一钻孔中布设若干一定间距配置的单元锚杆,每个单元锚杆有其独立的自由段和锚固段,而且承受的荷载也是通过各自的张拉千斤顶施加的,并通过预先补偿张拉(补偿各单元锚杆在同等荷载下因自由段长度不等而引起的位移差),而使所有单元锚杆承受相同的荷载。荷载分散型锚杆锚固段长度粘结应力分布均匀,其承载力能随锚固段长度的增加而线性增加。

特别是由无粘结钢绞线绕承载体弯曲成"U"型的单元锚杆组合而成的压力分散型锚杆体系,除能形成双层防腐外,由于灌浆体受压,不易开裂,大大提高了锚杆的耐久性。若用于临时工程,在其使用功能完成后,可方便地拆除芯体,对周边地下工程的开发不构成障碍。

压力分散型锚固体系具有粘结应力分布均匀、较强的防腐能力、高承载力、蠕变变形小的显著特点。目前,以其为主的荷载分散型锚固方法已在我国城市深基坑支挡、复杂地层高边坡加固、地下室与低注结构抗浮及运河船闸抗倾结构、水利水电重力坝中得到广泛应用,并展示了广阔的发展前景。

5.3 建立了锚杆长期性能与安全评价模式

目前,岩土锚杆、隧道、地下洞室、混凝土坝及抗浮抗倾结构等永久性工程的广泛应用,岩土锚固工程长期性能与安全评价逐步成为岩土工程科技工作者关注的热点。

中冶集团建筑研究总院提出了包括锚杆锁定荷载(初始预应力)变化量、锚杆现有承载力降低率、被锚固的岩土体与结构物变形速率以及锚杆的腐蚀损伤程度为主的安全控制指标;建立了包括风险源识别、长期性能检测、监测项目与方法、安全评价的临界技术指标以及安全度不足锚固工程的处治方法等项内容的安全评价模式。并对所收集到国内外 17 项被检验的岩土锚固工程长期性能状况进行了分析研究,研究结果表明:具有足够安全度的锚杆设计、锚杆全长完善的防腐措施,采用能改善力学与化学稳定性的锚固结构、规范的锚杆验收试验、完善系统的长期性能监测与维护管理体系是提高岩土锚固的长期性能、确保锚固工程的长期安全工作的主要途径和方法。

5.4 建立了我国岩土锚固技术标准体系

随着我国岩土锚固技术的广泛应用,岩土锚固的标准化建设也得到发展。至今,已建立了较完整的岩土锚固技术标准体系。国家标准《锚杆喷射混凝土支护技术规范》(GB 50086—2001)及中国建设标准化协会标准《岩土锚杆（索）的技术规程》(CECS22：2005)等技术标准对岩土锚杆的设计、材料、防腐、施工、试验、监测与验收都作了明确的规定，为我国岩土锚杆的设计、施工沿着安全可靠、技术先进、经济合理和有利环保的轨道发展发挥了重要作用。

6 岩石工程的精细爆破技术

进入21世纪尤其是最近5年，随着爆炸力学、岩石动力学、工程力学和工程爆破技术等基础理论研究的不断进展，借助计算机技术、爆破实验和测量技术的进步，使得定量化的爆破设计成为可能；优良的便携式和遥控式爆破有害效应监测系统的出现，使爆破实时监控成为可能；爆破器材的进步、施工机械化水平和自动化水平的提高，为现代矿山和水利水电工程精细化的爆破施工提供了技术支持。精细爆破的萌芽和发展促进了岩石工程爆破技术脱胎换骨的变化。

精细爆破，即通过定量化的爆破设计和精心的爆破施工，实现炸药能量释放与爆破作用过程的精确控制，既达到爆破目的，又实现对爆破有害效应的精确控制，最终实现安全可靠、绿色环保及经济合理的爆破作业。精细爆破秉承了传统控制爆破的理念，但与传统控制爆破又有着明显的区别。精细爆破的目标与传统控制爆破一样，既要达到预期的爆破效果，又要实现对爆破效果和爆破危害的双重控制。精细爆破更注重利用爆炸力学、岩石动力学、结构力学、材料力学和工程爆破等相关学科的最新研究成果，并充分利用飞速发展的计算机技术、数值分析技术，采用定量化爆破设计计算理论、方法和试验手段，对爆破方案和参数进行优化，实现对爆破效果及有害效应的精确控制。精细爆破更注重根据爆破对象的力学特性、爆破条件及工程要求，依赖性能优良的爆破器材及先进可靠的起爆技术，辅以精心施工和严格管理，实现爆破全过程的精密控制。与传统控制爆破相比，精细爆破在定量化的爆破试验、爆破设计、炸药能量释放和爆破作用过程控制、爆破效果的定量评价等方面，均提出了更高的要求。精细爆破不仅仅局限于传统控制爆破，其概念适用于岩石、拆除及特种爆破等工程爆破的方方面面，它不是一种爆破方法，而是含义颇广的概念。

精细爆破的支撑条件包括：爆破岩与设计理论、爆破数值模拟及计算机辅助设计、高可靠性和安全性的爆破器材、爆破测试与检测技术、现代信息及控制技术等，其核心包括"定量设计，精心施工，实时监控，科学管理"。中国工程院冯叔瑜院士认为：精细爆破是我国工程爆破技术发展的第三个里程碑。

在溪洛渡水电站大坝拱肩槽开挖中，贯彻精细爆破理念，爆破形成的建基面光滑平整，平整度、半孔率整体达到优秀水平，爆破对建基面岩石的损伤得到有效控制。建基面法线方向的平均超欠挖、平整度、残孔率的整体合格率分别为97.2%，98.8%，99.8%，钻孔声波法检测平均爆破影响深度均在1.0 m以内，质量等级全部达到优良标准。

在溪洛渡、向家坝水电站超大规模地下厂房洞室群开挖中，为了克服不利地质缺陷的影响，在开挖中贯彻精细爆破理念，采用质点振动速度、岩石声波等多种手段对爆破效果进行定量评估的评价体系。采用整体集成样架高精度自动施工量角器等措施，实现了精确定位、精确钻进的施工程序。建立了一整套精细爆破管理体系，形成的保留面光滑平整，平整度、半孔率整体达到优秀水平。

我国已将建设"资源节约型"和"环境友好型"社会作为21世纪的重要战略，精细爆破符合时代需求，有望作为引领中国工程爆破行业科技创新的重要手段与发展方向之一，将在实现爆破行业的可持续发展的过程中发挥重要作用，对我国工程爆破的发展产生深远的影响。

7 结语

以上几个方面说明我国是岩石工程世界第一大国，不仅工程量极大，而且地质条件复杂，技术难度也是空前的。近年来，岩石工程科学技术已取得了一系列创新性成果，在勘测设计、施工、监测等方面，也积累了丰富的经验，当然也有许多教训，这些都是岩石工程领域的宝贵财富。以上介绍的只是笔者认为比较突出的几个方面，不一定全面，文章只作为抛砖引玉，希望全体岩石力学与岩石工程科技工作者都来总结已取得的新成果，共同推进岩石力学与岩石工程科学技术的发展，开辟岩石工程的新纪元。

致谢：衷心感谢戴会超、宋胜武、李术才、何满潮、程良奎、邬爱清、何川、蔡美峰、康红普提供相关资料；感谢方祖烈教授对资料的整理、归纳。

泥岩损伤特性试验研究

许宝田[1,2]　钱七虎[1]　阎长虹[2]　许宏发[1]

(1.解放军理工大学　南京　210007；2.南京大学地球科学与工程学院　南京　210093)

摘　要　以南京长江三桥地基中的泥岩为对象，对泥岩进行三轴试验。试验结果表明：随着侧压的增大，破坏荷载增大，塑性变形明显增大，岩石破坏后，残余强度随侧压增大而提高。在此基础上研究分析了泥岩微元强度服从 Weibull 分布，泥岩微元体破坏服从莫尔－库仑岩石强度判据时的损伤软化参数与围压的关系特征。结合岩石破裂过程应力-应变全过程曲线，讨论了初始损伤特性，分析结果表明：泥岩初始损伤时的主应力差对数随围压增大而增大，两者呈线性关系；分析了泥岩损伤变量随主应力差变化关系，结果表明泥岩损伤变量与主应力差呈双曲线数学关系，通过对双曲线模型作线性化处理，结合试验数据采用回归分析法确定模型参数，分析结果发现 F_0 随围压的增大而增大，而 m 则随压的增大而减小，反映泥岩随围压的增大，脆性度降低。

关键词　泥岩　损伤特性　试验研究
中图分类号：TU451　　**文献标识码**：A

TRIAXIAL TESTING STUDY ON DAMAGE CHARACTERISTICS OF MUDSTONE

XU Baotian[1,2]　QIAN Qihu[1]　YAN Changhong[2]　XU Hongfa[1]

(①PLA University of Science and Technology, Nanjing　210007)
(②School of Earth Sciences and Engineering, Nanjing University, Nanjing　210093)

Abstract　This paper presents the triaxial test results of mudstone from the foundation of the Nanjing third bridge on Yangtse River. It is found that the failure pressure, the plastic deformation and the residual strength of the mudstone after subversion increase as the confining pressure increases. Then, it studies and analyses the relation between the damage soften parameters and the confining pressure, when the strength of rock's micro—unit is of the Weibull distribution and the strength of rock's micro—unit conformed to the Mohr-Coulomb strength criterion. Connecting with the stress-strain full procedure curves, the initial damage characteristic is discussed. The results indicate that the relation of logarithm of pressure and confining pressure of initial damage is linear. By studying on the relation between damage variable and main stress, it is found the relation of damage variable and main stress submits to a hyperbola model. The hyperbola model can be transformed into a linear equation. So the model parameters can be gotten by regressive analysis based on the test results. The results indicate 'F_0' increases as the confining pressure increases, but 'm' declines, which reflects the brittleness tolerance of the mudstone declines as the confining pressure increases.

Key words　Mudstone, Damage characteristic, Triaxial test, Rock mechanics

---

收稿日期：2009-08-31；修回日期：2010-01-12。
本文原载于《工程地质学报》(2010 年第 18 卷第 4 期)。

1 引言

岩石损伤扩展是岩体损伤力学研究中的一个重要问题[1]。从岩石材料内部所含缺陷分布的随机性出发，将连续损伤理论和统计强度理论有机地结合起来，从岩石微元强度服从一定概率分布的角度出发，使岩石本构模型的研究取得了重大进展[2]。岩体材料中大小不一、形状各异的裂隙，在受力时裂隙不断扩展演化，从而使其表现出复杂的应力-应变关系，而损伤力学就是从这些缺陷的不断劣化着手解决其应力-应变关系的一种有效手段[3]。

以往研究成果表明，对泥岩初始损伤门槛值和损伤变量随应力、应变变化的特征问题国内研究较少。本文以南京长江三桥地基中的泥岩为对象，基于三轴试验，对其损伤特征作了研究分析。

2 泥岩三轴应力-应变特征

南京长江三桥地基中的白垩系浦口组泥岩天然密度为 2.39 g·cm^{-3}，天然单轴抗压强度 1.84～4.68 MPa，含水率 6%～8%[4]。试验岩样尺寸：φ50.10 mm×100.34 mm，本次试验过程中对样品在围压 σ_3 分别为 0.5、1.0、1.5、2.0 MPa 下进行三轴压缩试验(图1)。

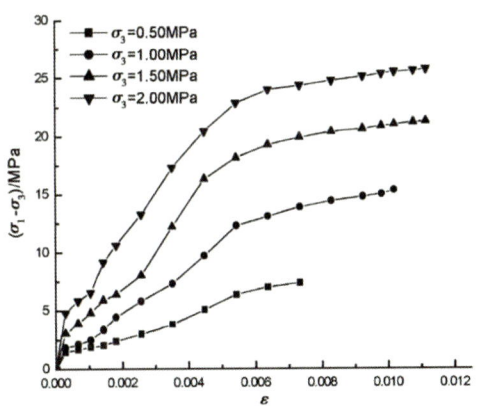

图 1 不同侧压下泥岩主应力差-应变曲线

Fig. 1 The principle stress-strain curves of mud stone under various confining pressure

3 泥岩初始损伤特性分析

由图1可以看出，应力-应变曲线中，存在转折点。加荷初期，应变变化较缓，随应力增大，到达转折点后，应变变化较块，在对数坐标中尤其明显(图2)，随着围压增大，转折点对应的主应力差逐渐增大。

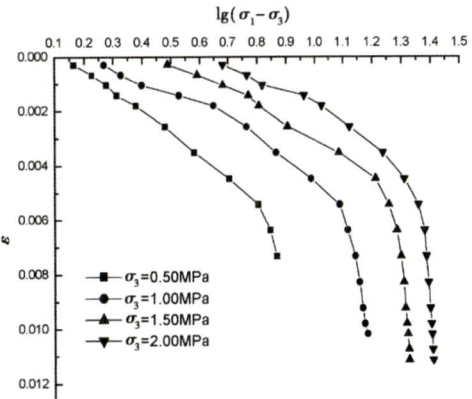

图 2 不同侧压下泥岩主应力差对数-应变曲线

Fig. 2 The principle stress logarithm-strain curves of mud stone under various confining pressure

赵锡宏教授[5]通过对土的三轴损伤特性研究提出了确定初始损伤的方法。本文应用该方法确定泥岩的初始损伤点，研究表明，该方法对泥岩是适用的。该方法将试验结果绘成 $\varepsilon - \lg(\sigma_1 - \sigma_3)$（主应变-主应力差对数）曲线(图3)，它由一段缓变的曲线和一条陡降直线组成。在缓变的曲线上找出曲率半径最小的点 O，过 O 点作直线与直线段的延长线交于 C 点，作 ∠OCB 的角平分线 CD，D 点为初始损伤点，相应的应力和应变称为初始损伤应力、应变门槛值。

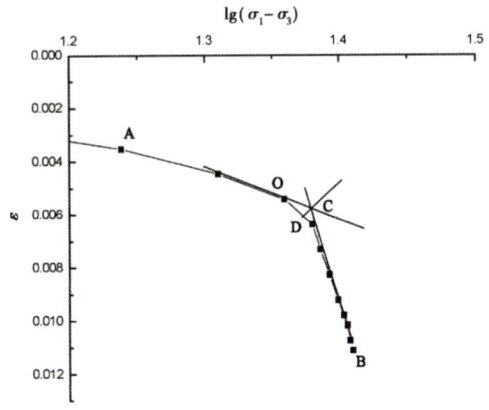

图 3 初始损伤点的确定

Fig. 3 Initial damage threshold

根据以上方法，分析试验结果发现初始损伤是主应力差对数随围压增大而增大，两者呈线性关系

(图4)。

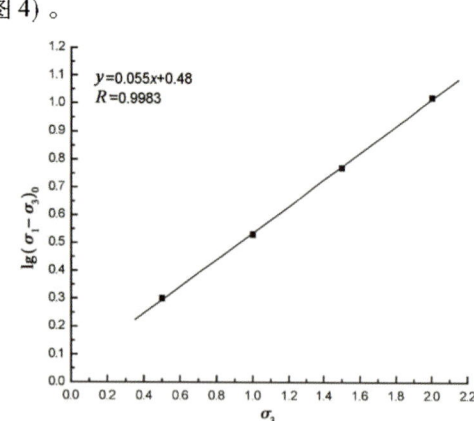

图4 初始损伤主应力差对数-围压关系曲线

Fig. 4 The principle stress logarithm-confining pressure curves of initial damage

4 泥岩三轴损伤特性

根据等效应变假定,有[6~8]
$$\sigma = \sigma^*(1-D) \quad (1)$$
式中:σ 为名义应力,σ^* 为有效应力,D 为损伤变量。

假定岩石微元强度 F^* 服从 Weibull 分布[6~8],则其概率密度函数可以表示为:
$$P[F^*] = \frac{m}{F_0}\left(\frac{F^*}{F_0}\right)^{m-1}\exp\left[-\left(\frac{F^*}{F_0}\right)^m\right] \quad (2)$$
式中:F^* 为微元破坏 Weibull 分布的分布变量;m、F_0 为模型参数。则损伤变量 D 可表示为:
$$D = \int_0^{F^*} P(y)\,\mathrm{d}y = 1 - \exp\left[-(F^*/F_0)^m\right] \quad (3)$$
故,要计算损伤变量 D 则首先要确定微元强度 F^*。

曹文贵[6]结合岩石的破坏模式与判据,提出的岩石微元强度表示方法,假设岩石的破坏准则为:
$$F^* = f(\sigma^*) \quad (4)$$
上式为与岩石强度参数有关的函数。假设岩石微元体破坏服从莫尔-库仑强度判据:
$$\sigma_1^* - \frac{1+\sin\varphi}{1-\sin\varphi}\sigma_3^* = \frac{2c\cos\varphi}{1-\sin\varphi} \quad (5)$$
c、φ 分别为岩石的内聚力和内摩擦角。F^* 全面反映了岩石微元破坏的危险程度,可作为岩石微元的强度:
$$F^* = \sigma_1^* - \alpha\sigma_3^* \quad (6)$$
式中:$\alpha = (1-\sin\varphi)/(1+\sin\varphi)$,$\sigma_1^*$、$\sigma_3^*$ 为有效应力:
$$\sigma_1^* = \sigma_1/(1-D) \quad (7)$$

$$\sigma_3^* = \sigma_3/(1-D) \quad (8)$$

采用文[8]和[9]中的方法确定参数 m、F_0,根据文[9]研究结果,对于模型参数 m、F_0,F_0 反映岩石宏观平均强度的大小,m 则反映岩石的脆性度。
$$F_0 = F_c \cdot m^{1/m} \quad (9)$$
式中:F_c 为峰值下的 F^* 值,
$$m = 1/\ln[E\varepsilon_c/(\sigma_c - 2\nu\sigma_3)] \quad (10)$$
E 为弹性模量,ε_c、σ_c 为峰值下的应变、应力,ν 为泊松比。

由此得到不同围压的参数值(表1),损伤变量-主应力差关系曲线如图5。由此可以看出泥岩损伤变形具有如下特性:

表1 不同围压下的模型参数

Table 1 Model parameters of different confining pressure

σ_3/MPa	F_c/MPa	F_0/MPa	m	E/MPa	ν
0.5	7.24	6.8	0.95	1483	0.3
1.0	12.82	10.4	0.84	3208	0.3
1.5	18.88	14.3	0.80	5028	0.3
2.0	23.48	16.4	0.76	6632	0.3

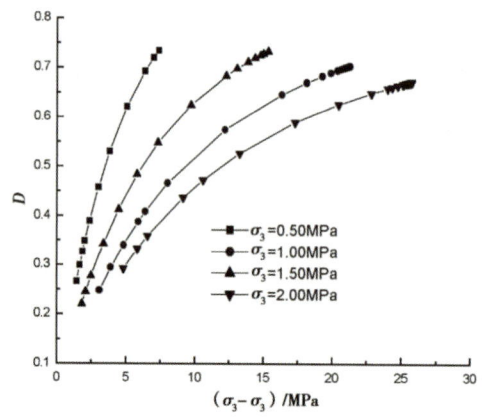

图5 损伤变量-主应力差关系曲线

Fig. 5 Damage variable-principle stress curves

(1) F_0 随围压的增大而增大,而 m 则随围压的增大而减小,反映泥岩随围压的增大,脆性度降低;

(2) 损伤变量随主应力差增大而增大,两者呈非线性关系。

根据以上分析,泥岩损伤变量 D 是描述岩石损伤特性最为重要的参数,该参数除与岩石力学特性有关外,跟三轴试验中的主应力大小有直接关系,如能直接建立应力跟损伤变量之间的数学关系,将能大大简化损伤变量的计算过程,根据图5曲线特征,假设损伤变量-主应力差成双曲线函数关系,建立的

模型方程为:
$$1 - D = \frac{a}{\sigma_1 - \sigma_3} + b \quad (11)$$
式中 a、b 为试验常数。

$D = 1$ 时，$\sigma_1 = \sigma_0$（σ_0 为初始损伤时的最大主应力）；$D = 0$ 时，$\sigma_1 = \sigma_c$。

所以，得：
$$a = \frac{(\sigma_0 - \sigma_3)(\sigma_c - \sigma_3)}{\sigma_c - \sigma_0}, \quad b = -\frac{\sigma_0 - \sigma_3}{\sigma_c - \sigma_0} \quad (12)$$

根据以上分析，得到不同围压下的损伤变量-主应力差关系（表2），计算结果表明模型结果与实验结果相关系数均接近于1。

表 2 不围压下的模型方程
Tab. 2 Model result under different confining pressure

σ_3 MPa	拟合方程	相关系数(R)
0.5	$y = 1.28 - 1.91/(\sigma_1 - \sigma_3)$	0.99973
1.0	$y = 1.25 - 3.00/(\sigma_1 - \sigma_3)$	0.99994
1.5	$y = 1.22 - 4.50/(\sigma_1 - \sigma_3)$	0.99999
2.0	$y = 1.33 - 6.0/(\sigma_1 - \sigma_3)$	0.99999

以围压为0.5MPa为例，根据模型计算的结果和试验结果如图6。两者符合得较好，所以本文提出的损伤变量表达式可以简单、较好地描述泥岩损伤演化过程。

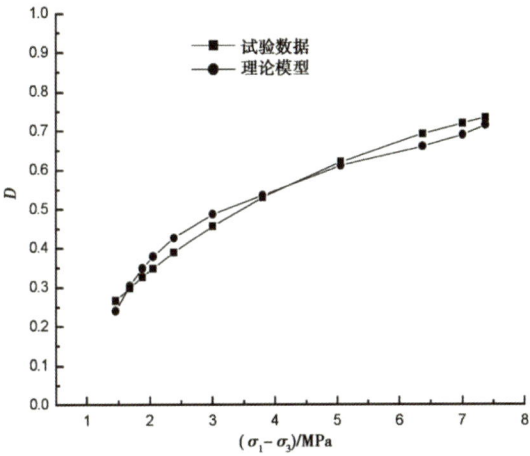

图 6 理论模型与实测结果
Fig. 6 Curves of theoretical model and test result

5 结 论

本文在相关研究成果的基础上对泥岩损伤力模型参数进行了研究，同时对泥岩的损伤变形特性进行分析，分析结果表明：

(1) 初始损伤主应力差对数随围压增大而增大，两者呈线性关系；

(2) 泥岩微元强度概率分布模型参数 F_0 随围压的增大而增大，而 m 则随围压的增大而减小，反映泥岩随围压的增大，脆性度降低；

(3) 损伤变量与主应力差呈双曲线函数关系，拟合结果证明该函数关系非常接近实际情况，假设的模型是合理的。

参 考 文 献

[1] 刘成禹,何满潮,王树仁. 花岗岩低温冻融损伤特性的实验研究[J]. 湖南科技大学学报,2005,(1): 37~40.
Liu Chengyu, He Manchao, Wang Shuren. Experimental investigation freeze-thawing damage characteristics of granite at low temperature. Journal of Hunan University of Science & Technology, 2005, (1): 37~40.

[2] 曹文贵,赵明华,刘成学. 基于Weibull分布的岩石损伤软化模型及其修正方法研究[J]. 岩石力学与工程学报,2004,(19): 3226~3231.
Can Wengui, Zhao Minghua, Liu Chengxue. Study on the model and its modifying method for rock softening and damage based on weibull random distribution. Chinese Journal of Rock Mechanics and Engineerin, 2004, (19): 3226~3231.

[3] 李杭州,廖红建,盛谦. 基于统一强度理论的软岩损伤统计本构模型研究[J]. 岩石力学与工程学报,2006,(7): 1331~1336.
Li Hangzhou, Liao Hongjian, Sheng Qian. Study on statistical damage constitutive model of soft rock based on unified strength theory. Chinese Journal of Rock Mechanics and Engineerin, 2006, (7): 1331~1336.

[4] 许宝田,阎长虹,许宏发. 基于三轴试验的泥岩应力-应变特征分析[J]. 岩土工程学报,2004,(6): 863~865.
Xu Baotian, Yan Changhong, Xu Hongfa. Study of stress-strain for mud stone based on triaxial test. Chinese Journal of Geotechnical Engineering, 2004, (6): 863~865.

[5] 赵锡宏,孙红,罗冠威. 损伤土力学[M]. 上海: 同济大学出版社, 2000.
Zhao Xihong, Sun Hong, Luo Guanwei. Damage Soil Mechanics. Shanghai: Tongji University Press, 2000.

[6] 曹文贵,赵明华,刘成学. 基于统计损伤理论的莫尔—库仑岩石强度判据修正方法之研究[J]. 岩石力学与工程学报, 2006, (7): 2403~2408.
Cao Wengui, Zhao Minghua, Liu Chengxue. Study on rectified method of Mohr-coulomb strength criterion for rock based on statistical damage theory. Chinese Journal of Rock Mechanics and Engineerin, 2006, (7): 2403~2408.

[7] 杨明辉,赵明华,曹文贵. 岩石损伤软化统计本构模型参数的确定方法[J]. 水利学报,2005,(3): 345~349.
Yang Minghui, Zhao Minghua, Cao Wengui. Method for determining the parameters of statistical damage softening constitutive model for rock. Journal of Hydro-science, 2005,(3): 345~349.

[8] 李杭州,廖红建,盛谦. 基于统一强度理论的软岩损伤统计本构模型研究[J]. 岩石力学与工程学报,2006,(7): 1331~1336.
Li Hangzhou, Liao Hongjian, Sheng Qian. Study on statistical damage constitutive model of soft rock based on unified strength theory. Chinese Journal of Rock Mechanics and Engineerin, 2006, (7): 1331~1336.

[9] 杨明辉,赵明华,曹文贵. 岩石损伤软化统计本构模型参数的确定方法[J]. 水利学报,2005,(3): 345~350.
Yang Minghui, Zhao Minghua, Cao Wengui. Method for determining the parameters of statistical damage softening constitutive model for rock. Journal of Hydraulic Engineering, 2005,(3): 345~350.

深部非均匀岩体卸载拉裂的时间效应和主要影响因素

范鹏贤[1,2]，王明洋[1]，钱七虎[1]

(1. 解放军理工大学 工程兵工程学院，江苏 南京 210007；2. 中国矿业大学 深部岩土力学与地下工程国家重点实验室，江苏 徐州 221008)

摘要： 深部地下工程围岩的拉伸破坏是一个常见的但没有得到充分研究的科学现象。岩石内部包含的缺陷和缺陷处的应力集中被认为与岩石在压应力作用下的拉伸破坏密切相关。基于应力集中的非均匀岩石拉伸破坏模型的理论框架，系统的研究 3 个影响岩石拉伸破坏的主要因素：加卸载速率(偏应变率)、缺陷尺度和初始地应力，对岩石内部缺陷处拉应力的产生和发展过程的影响。研究表明：(1) 偏应变速率对缺陷处附加拉应力具有显著的影响，偏应变率越高，缺陷处的附加拉应力越大；(2) 不同层次的缺陷对应于不同大小的偏应变率，偏应变率越高，对应缺陷的尺寸就越小；(3) 缺陷处的附加拉应力与初始应力成正比，缺陷较大时，岩体是否发生卸载破坏由初始应力、应力集中系数和抗拉强度间的关系决定。引用模拟试验和现场观测的结果和现象对理论分析结果进行讨论，试验现象和现场观测数据均与理论分析基本吻合。

关键词： 岩石力学；非均匀岩石；卸载；拉破坏；影响因素

中图分类号： TU 45　　**文献标识码：** A　　**文章编号：** 1000 - 6915(2010)07 - 1389 - 08

TIME EFFECT AND MAIN INFLUENCE FACTORS OF UNLOADING SPLITTING OF DEEP-SEATED ROCK WITH NONUNIFORMITIES

FAN Pengxian[1,2], WANG Mingyang[1], QIAN Qihu[1]

(1. *Engineering Institute of Engineering Corps, PLA University of Science and Technology, Nanjing, Jiangsu* 210007, *China*;
2. *State Key Laboratory for Geomechanics and Deep Underground Engineering, China University of Mining and Technology, Xuzhou, Jiangsu* 221008, *China*)

Abstract: The splitting of stressed rock in the neighborhood of underground openings is a common phenomenon that is not well explained theoretically up to the present. The stress concentration at nonuniformities is believed to be the crucial one among the factors influencing the tensile failure process under pressure conditions. Based on the model of rock with nonuniformities, three main factors of the splitting fracture of rock, including unloading duration(or deviator strain rate), scale of nonuniformities and initial geostress, are detailed investigated in order to reveal the mechanism behind the phenomenon of splitting. The results of theoretical analysis show that: (1) The excess tension stress of nonuniformities is affected by deviator strain rate significantly; and it increases nonlinearly as the unloading duration shrinks. (2) There is a nonlinear corresponding relationship between the scale of nonuniformities and the magnitude of deviator strain rate; and higher deviator strain rate tends to excite

---

收稿日期：2010-01-18；修回日期：2010-03-23。
基金项目：国家自然科学杰出青年基金(50825403)；国家重点基础研究发展计划(973)项目(2010CB732003)；中国矿业大学深部岩土力学与地下工程国家重点实验室开放基金项目(SKLGDUEK0907)。
本文原载于《岩石力学与工程学报》(2010 年第 29 卷第 7 期)。

smaller nonuniformities. (3) If there exist nonuniformities with big scale enough, the maximum excess tension stress is proportional to the initial geostress. Therefore, the possibility of unloading splitting is determined by the relation among the three parameters: initial geostress, stress concentration coefficient and tensile strength. Results of an unloading test, in which the axial stress of hydraulic stressed rock samples is unloaded with different control modes, and data of in-situ core disking observation, are cited to verify the analytical analysis results. The main phenomena and critical stress condition of core disking predicted by the theoretical model coincide well with the results of the cited experiments and in-situ observation data.

Key words: rock mechanics; rock with nonuniformities; unloading; splitting; influence factor

1 引言

在岩石力学领域，长期以来对地下坑道周围岩石应力应变状态和破坏的研究主要是利用弹塑性力学的知识，在连续介质力学框架内进行，但是岩石内部包含了大量的节理、微裂纹、包含物、结晶缺陷等尺度不同的构造缺陷。岩石的破坏与均匀连续介质具有本质的区别。大量研究[1~6]表明，岩石的破坏往往是从局部发展起来的，导致这种现象的原因可归结为2个方面：一是由于岩石材料的非均匀性导致的抗力大小差异，二是由于内部缺陷处变形不协调造成的应力起伏。

近十几年来，前一方面的研究取得了重大的进展，在对岩石材料强度不均匀性分布进行假设的研究基础上，结合有限元程序，形成了较成熟理论体系和有限元软件[7]。但是对后一方面的研究一直困难重重。以 Griffith 理论为代表的微观力学对微裂纹的大小、形状、倾角等进行假设，研究单个裂纹周边的应力分布和扩展准则，而宏观力学则在试验和观测的基础上对岩石进行唯像和经验性的描述。两类理论在近乎平行的道路上各自向前发展。但是，研究人员一直没有放弃沟通两类理论。岩石细观力学的提出和兴起即反映了这种趋势。

在研究细观力学的过程中，M. A. Grinfeld 等[8,9]指出：即使是单个的缺陷也是极端复杂的，具体的考虑种类繁多、性状各异的缺陷的形状和特性既无可能也无必要。很多情况下，一些简化描述方法是必要的。基于这种实用主义的思想，一些新方法被提出并付诸应用，具有较大影响的有统计分形强度理论[10]，代表性体元法[9,11]等。基于缺陷处应力集中的非均匀岩体拉破坏模型[12]也是建立岩体细观非均匀性与宏观力学性质之间联系的有益尝试。

大量试验结果[1,4,13,14]表明，岩石的拉伸破坏在所有破坏形式中是最基本和普遍的，甚至在宏观压应力作用下，拉伸破坏在脆性断裂的起始阶段也是主要的破坏形式[1,13,14]。岩石的宏观压应力条件和拉破坏模式之间的悖论引起了学界广泛的讨论。先后有等效拉应力理论[15]、极限拉应变理论[16]、剪胀角理论[17]等被提出，用于分析岩石在宏观压应力作用下的拉伸破坏机制和现象。部分理论由于其简明性已经得到许多学者和工程师的认同，但是在解释一些深地下岩石力学现象时，这些理论显得过于简化，甚至与某些试验现象相抵触[14]。

非均匀岩体拉破坏模型[12]认为岩石在压应力下发生拉伸破坏的主要原因是缺陷处的应力集中，建立了缺陷处应力状态的微分方程，推导了岩石在轴向加载和卸载情况下缺陷处局部拉应力的解析表达式，为解释完整岩块的劈裂和卸载张拉破坏提供了新的理论视角。

需要说明的是，王明洋等[12]和本文中的缺陷并不是特指某种具体形状的微裂纹，而是一个不均匀材料占据和影响的区域。为简化分析，假设岩块内部缺陷稀疏分布，缺陷应力场之间不发生相互作用[5]。变形过程中，材料不均匀性将引起一个局部的自平衡附加应力场，该应力场有可能导致较高的局部应力，从而引起材料的微观动力断裂。同时，应力集中系数 K 也并不对应某种几何形状的缺陷，而应看作是缺陷的整体效应，是一种宏观的材料参数。

本文在王明洋等[12]的基础上对具有初始应力的岩块卸载时发生拉伸破坏的 3 个主要影响因素(卸载速率、缺陷尺度和初始应力)进行了进一步探讨。

2 非均匀岩体模型及其主要结果

2.1 基本公式

具有内部缺陷的脆性岩体宏观上的应力－应变关系服从胡克定律，但缺陷附近有应力的起伏分布，缺陷处的不协调变形和应力集中导致的拉应力是岩石发生拉伸破坏的主要原因[12]。

岩体内部的局部应力 σ_{ij}^{loc} 可分为 2 个部分：一是弹性应力 σ_{ij}^e，遵循胡克定律，在整个区域内均匀分布；二是缺陷引起的附加应力 $\Delta \sigma_{ij}$，仅在缺陷区域内不为 0。忽略脆性岩石拉伸破坏过程中的塑性变形，局部应力 σ_{ij}^{loc} 可表示为两部分之和：

$$\sigma_{ij}^{\text{loc}} = \sigma_{ij}^e + \Delta \sigma_{ij} \tag{1}$$

在准静态加卸载或较低应变率情况下，缺陷处应力集中的演化方程[12]可近似为

$$\frac{\mathrm{d}\Delta\sigma_{ij}}{\mathrm{d}t} = K\rho c_s^2 \dot{e}_{ij} - \eta \frac{\Delta\sigma_{ij}}{l} \tag{2}$$

式中：e_{ij} 为偏应变，且 $e_{ij} = u_{ij} - (1/3)\delta_{ij}u_{ij}$，$\delta_{ij}$ 为 Kronecher 符号，$\dot{e}_{ij}$ 为 e_{ij} 随时间的变化率；$\Delta\sigma_{ij}$ 和 K 分别为特定尺寸的缺陷上的附加应力和材料的应力集中系数；ρ 为介质的密度；c_s 为剪切波波速；η 为影响附加应力松弛速率的材料参数；l 为缺陷尺度。

式(2)右侧第一项代表剪切应变导致的附加应力，其形式类似于胡克定律；第二项描述了附加应力的松弛，假设附加应力松弛速率 $\mathrm{d}\Delta\sigma_{ij}/\mathrm{d}t$ 的大小与缺陷尺寸成反比。该模型考虑了缺陷处的应力集中和松弛对材料应力场的影响，也考虑了加卸载速率或应变速率对附加应力的影响。

2.2 应力集中产生机制及相关参数说明

陈宗基[18]很早就注意到深部岩体与一般岩石不同，并提出了"封闭应力"的概念。他认为，由于岩石是非均匀材料，在变形过程中，晶体间存在摩擦力，局部变形将受到阻碍，从而引起应力积累。因此岩体中有部分应力被封存着，并处于平衡状态。岩爆和岩芯饼化即是封闭应力释放的结果。

缺陷处的应力集中和起伏与封闭应力类似，也是由于材料的非均匀性引起的。由于材料的非均匀性，不同部位的变形是不协调的，这就不可避免的会在缺陷内部和边界产生应力集中和次生应力场。由于变形不协调而导致的应力集中不仅和缺陷的尺寸有关，还与应力的变化速率有关，不同尺度的缺陷处附加应力的大小和消散的快慢也必定是不一样的。岩石内部包含了大量的缺陷，其大小形状各异，既无可能也无必要对其一一考察。充分利用岩石内部缺陷的自相似构造特性，在细观机制和宏观力学表现之间建立简洁但基本符合实际的定量联系，通过尽可能少的参数来反映尽可能多的特性是必要且可行的。

图 1 概略描绘了不均匀材料中缺陷处应力集中的一种特殊情况：在 z 轴应力卸载过程中，由于不同材料之间变形的不协调，将在 z 轴方向引起自平衡局部附加应力场 $\Delta\sigma_{zz}(x, y, z, t)$。如果附加应力和平均应力的合力超出了材料的局部抗拉强度，则将发生微观裂纹的生成和扩展。$\Delta\sigma_{zz}(t)$ 则特指 t 时刻应力集中区 z 方向附加应力的最大值。

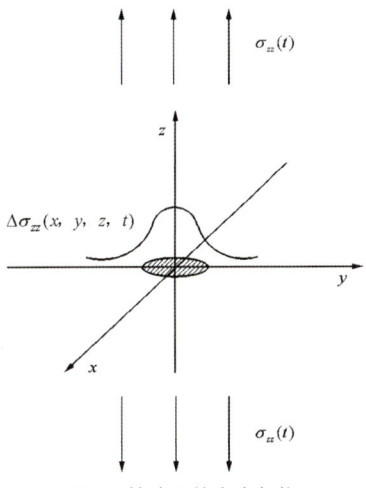

图 1　缺陷处的应力起伏
Fig.1　Stress fluctuation of nonuniformity

式(2)中出现的岩石参数有 5 个，分别是密度、剪切波速、材料应力集中系数、材料应力松弛系数和缺陷尺度。其中材料应力集中系数 K 和材料应力松弛系数 η 可以通过宏观试验确定，如轴压变形试验和弹性后效试验等。

缺陷尺度的物理意义不同于断裂力学的微裂纹尺寸，而应该看作是一种材料脆性断裂过程的特征参数。在不同的变形速率下，起到决定作用的缺陷具有不同的尺度。

岩石具有复杂的构造层次，这种内部构造层次显著影响岩石的物理力学性质[19]。不同的构造层次对应着不同的强度、黏性及应变率。通常随着变形速度的增大，发生变形破坏的优势构造层次将逐渐由宏观向微观层次过度，并表现在破坏后块体的尺寸上[20]。岩石结构的层次结构至少在 5～6 个数量级上具有自相似性，其分布可以通过分析弹性波在岩石中的衰减规律确定[21]。由于确定岩石中缺陷的分布方法及其原理比较复杂，本文中假设缺陷尺度及其分布是已知的。

2.3 围岩的卸载拉裂

假设岩体受静水压力 $-\sigma_0$ 的作用，应力作用的

时间足够长，岩体中的缺陷处的应力集中已经完全松弛，在其中一个方向上卸载(见图2)，考察岩石的应力－应变状态。试验研究表明，卸载引起的破坏一般为垂直于卸载方向的拉裂，因此仅考虑 z 方向的附加应力。

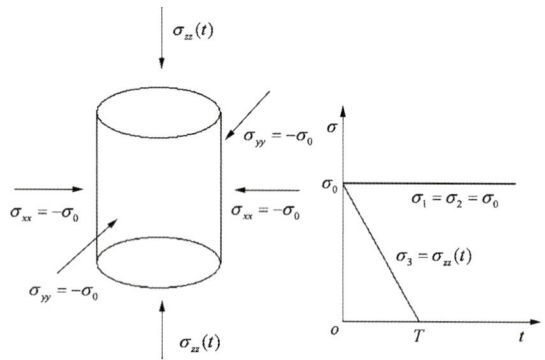

图 2　卸载示意图
Fig.2　Sketch of unloading

令岩体 x，y 轴方向应力保持不变，z 轴方向线性卸载，卸载公式为

$$\sigma_{zz}(t) = \begin{cases} -\sigma_0\left(1-\dfrac{t}{T}\right) & (0<t \leqslant T) \\ 0 & (t>T) \end{cases} \quad (3)$$

式中：T 为卸载持续时间。

缺陷处 z 方向的局部附加拉应力[12]为

$$\Delta\sigma_{zz} = \begin{cases} \dfrac{K}{3}\dfrac{\sigma_0}{T}\dfrac{l}{\eta}\left(1-\mathrm{e}^{-\frac{\eta}{l}t}\right) & (0<t \leqslant T) \\ \dfrac{K}{3}\dfrac{\sigma_0}{T}\dfrac{l}{\eta}\left(1-\mathrm{e}^{-\frac{\eta}{l}T}\right)\mathrm{e}^{-\frac{\eta}{l}(t-T)} & (t>T) \end{cases} \quad (4)$$

缺陷处的局部应力 $\sigma_{zz}^{\mathrm{loc}}$ 是由 σ_{zz}^{e} 和 $\Delta\sigma_{zz}$ 组成的。考虑到 $\sigma_{zz}^{\mathrm{e}} \approx \sigma_{zz}(t)$，可得

$$\sigma_{zz}^{\mathrm{loc}} = \sigma_{zz}^{\mathrm{e}} + \Delta\sigma_{zz} \cong \sigma_{zz}(t) + \Delta\sigma_{zz} =$$

$$-\sigma_0\left(1-\dfrac{t}{T}\right) + \dfrac{K}{3}\dfrac{\sigma_0}{T}\dfrac{l}{\eta}\left(1-\mathrm{e}^{-\frac{\eta}{l}t}\right) \quad (5)$$

3　深部围岩卸载拉伸破坏的影响因素

依据节 2 的模型结果，岩体中缺陷处的局部拉应力与很多因素有关，主要包括缺陷处的应力集中系数、缺陷的尺寸、缺陷处应力松弛的速度、初始应力和加卸载时间等。下面分别对 3 个影响岩体卸载破坏的主要因素：卸载速率、缺陷尺寸分布和初始地应力进行分析。

3.1　卸载速率

设岩体参数为：$\eta = 2\times 10^{-8}$ m/s，$K=1$，$l=1\times 10^{-6}$ m，初始应力 $\sigma_0 = 50$ MPa，卸载模式采用图 2 所示形式，持续时间分别为 50，100，200 和 400 s。

图 3 为不同卸载时间卸载时缺陷处轴向附加拉应力的时程曲线。图 3 表明：卸载速率越快，缺陷处的附加应力增长越快，附加拉应力峰值越高，卸载速率对缺陷处的附加应力具有显著影响。在快速卸载的情况下($T=50$ s)，平行于卸载方向的拉应力峰值可达近 11 MPa，超过一般岩石的抗拉强度，将导致一定尺度范围内的缺陷被激活，触发岩石内部微裂纹的不稳定发展过程，进而可能导致岩石在快速卸载情况下的拉伸破坏。卸载结束后，附加应力逐步松弛，松弛的速度随附加应力峰值的增大而增大。

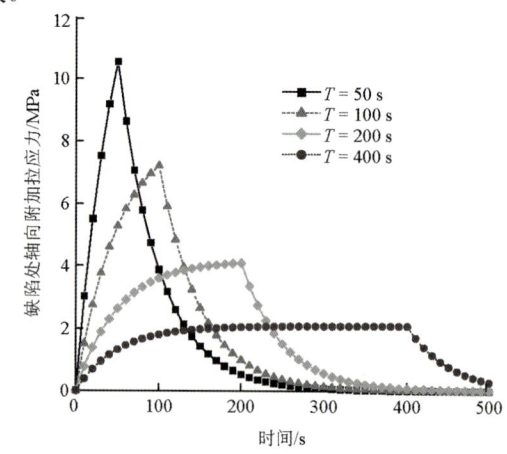

图 3　不同卸载时间卸载时缺陷处轴向附加拉应力的时程曲线
Fig.3　Time-history curves of excess tension stress with different unloading durations

图 4 为具有初始应力的岩石试件卸载时缺陷处轴向局部应力的时程曲线。从图 4 可以看出，卸载速率较快时(如 $T=50$ s)，缺陷处的局部应力迅速从压应力转变为拉应力，并在卸载结束时达最大值，之后逐步松弛。如果局部应力超过缺陷处材料的抗拉强度，将使岩体在缺陷处发生破坏，进而导致裂纹的扩展。而卸载速率较慢时(如 $T=400$ s)，缺陷处的附加拉应力较小，并在卸载结束后逐渐松弛，较难引起岩体的破坏。

3.2　缺陷尺度

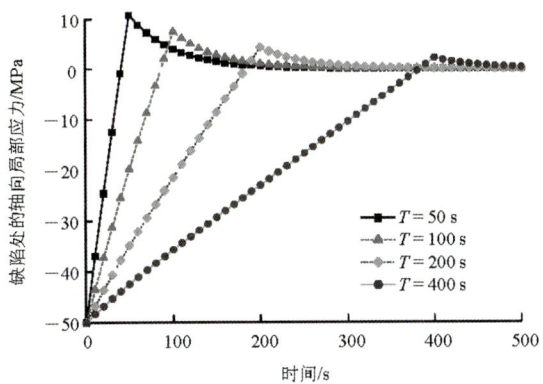

图 4 卸载时缺陷处轴向局部应力的时程曲线

Fig.4 Time-history curves of local stress of nonuniformities

设定应力松弛速率 $\eta = 2 \times 10^{-8}$ m/s，应力集中系数 $K = 1$，初始应力 $\sigma_0 = 50$ MPa，变化加卸载速率和缺陷尺寸大小，分别考察附加拉应力与卸载速率及最大附加拉应力与缺陷尺寸之间的关系曲线。卸载模式仍采用图 2 所示形式。

轴向卸载时不同卸载周期的岩石内部不同尺寸缺陷处附加拉应力的时程曲线见图 5。从图 5 可以得出如下结论：

(1) 若卸载速率相同，则卸载持续过程中，较大尺寸缺陷处的局部附加应力与时间基本成正比，而较小尺寸缺陷处的局部附加应力与时间的关系则是非线性的；

(2) 卸载速率相同的情况下，缺陷的尺寸越大，则同一时刻缺陷处局部附加拉应力的数值就越大；

(3) 缺陷尺寸较大时，卸载过程完成后其局部应力集中衰减较慢；

(4) 缺陷尺寸大小与局部附加拉应力之间的关系是非线性的，缺陷尺寸很小或很大时，局部附加拉应力对缺陷尺寸的变化不敏感。

图 6 为轴向卸载情况下卸载结束时的最大局部附加拉应力和缺陷尺寸之间的关系。

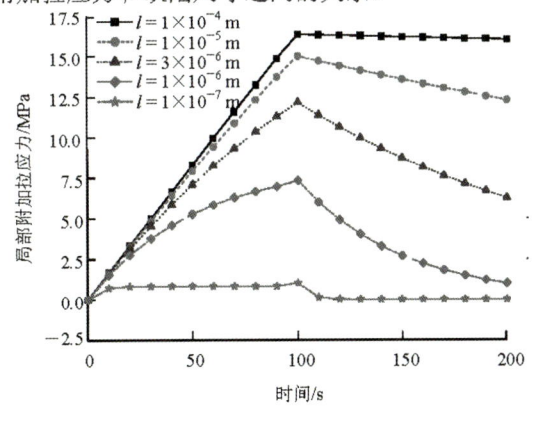

(a) $T = 100$ s

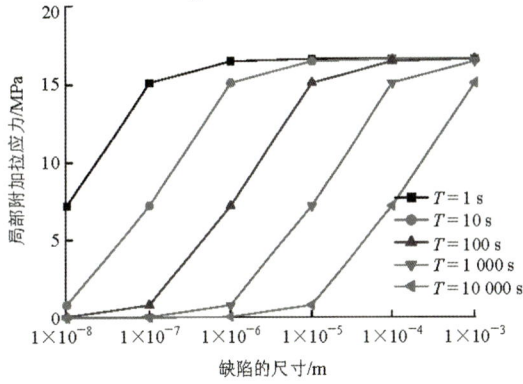

(b) $T = 400$ s

图 5 卸载时不同尺寸缺陷处附加拉应力的时程曲线

Fig.5 Time-history curves of access tension stress of nonuniformities with different scales during unloading

图 6 卸载时最大局部附加拉应力与缺陷尺寸之间的关系

Fig.6 Relationships between the maximum access tension stress and the scale of nonuniformities during unloading

从图 6 可以看出，不同加卸载速率激活的缺陷尺寸不同；对于一定的卸载速率，只有大于一定尺寸的缺陷处才会产生较大的附加拉应力；卸载速率越快，被激活的缺陷的临界尺寸越小。也就是说，当偏应变率较高时，细观和微观层面上的缺陷被同时激活，均产生较大的局部拉应力；而在偏应变率较小的情况下，只有尺寸较大的缺陷才会被激活。

依据王明洋等[12]模型的公式，假设岩石的剪切波速为 3 000 m/s，密度为 2 500 kg/m³，则当 $T = 1$ s 时，岩石的偏应变率约为 10^{-3} 量级。对照图 6 中的曲线和偏应变量级，可得偏应变率量级与被激活缺陷的临界尺寸的量级之间的大致对应关系(见表 1)。

3.3 初始地应力

若缺陷尺寸较大，卸载过程中始终有 $l \gg \eta t$，则可将式(4)中的第一式作迈克劳林展开：

表1 偏应变率和被激活的缺陷尺寸量级对应关系

Table 1 Relationship between the scale of excited nonuniformities and the deviator strain rates

卸载时间/s	偏应变率量级/s^{-1}	被激活缺陷的尺寸量级/m
10^0	10^{-3}	$>10^{-7}$
10^1	10^{-4}	$>10^{-6}$
10^2	10^{-5}	$>10^{-5}$
10^3	10^{-6}	$>10^{-4}$
10^4	10^{-7}	$>10^{-3}$

$$\Delta\sigma_{zz} = \frac{1}{3T}\frac{l}{\eta}K\sigma_0\left(1-e^{-\frac{\eta}{l}t}\right) =$$

$$\frac{1}{3T}\frac{l}{\eta}K\sigma_0\left(1-1+\frac{\eta}{l}t+R_n\right) \cong \frac{1}{3}K\sigma_0\frac{t}{T} \quad (6)$$

当 $t=T$ 时，缺陷处的局部拉应力最大，岩体在缺陷处发生破坏的临界条件可表示为

$$\sigma_{zz\;max}^{loc} = \sigma_{zz}(T) + \Delta\sigma_{zz}(T) = \frac{K}{3}\sigma_0 = \sigma_t \quad (7)$$

初始静水压力的临界值为

$$\sigma_{0c} = \frac{3}{K}\sigma_t \quad (8)$$

由式(8)可以看出，如果岩体中包含有较大尺度的缺陷，则岩体是否发生破坏主要是由初始应力的大小、应力集中系数和岩石的抗拉强度3个参数之间的相对关系决定的。式(8)可以用来估计能够发生岩芯饼化现象的临界地应力条件和临界深度。

对于特定的岩石材料，应力集中系数和抗拉强度是固定的，在初始静水压力的作用下某一方向的快速卸载能否导致岩体在垂直于卸载方向的拉伸破坏的决定性因素是初始压力的大小。

4 讨 论

4.1 卸载试验现象及其理论解释

岩石在卸载条件下的破坏是一个采矿和其他地下工程中经常遇到的现象，产生的拉裂缝往往平行于地下工程的边界。卸载破坏现象虽引起了学术界的注意，但相关研究还不够充分，对卸载拉裂机制认识不深。为检验本文理论的适用性，下面讨论一个国外报道的具有初始应力岩石的轴向卸载试验，并给出试验现象的定性解释。

为模拟岩石的卸载拉裂，E. Bauch 和 C. Lemmp[14]利用直径为 70 mm，长径比为 2∶1 的德国红砂岩试件进行了一系列加卸载试验。试验岩石的力学参数见表2。

表2 试验岩石的力学参数

Table 2 Mechanical parameters of sample sets used

试件组别	抗压强度/MPa	抗拉强度/MPa
1	54	3.9
2	88	4.9

所有的试件均在静水压力($\sigma_1=\sigma_2=\sigma_3$)的基础上开始加载或卸载。为比较不同影响因素对岩石拉伸破坏模式的影响，设计了3种卸载方式，它们分别是：(1) 围压不变，卸载速率或应变率控制下的轴压缓慢卸载；(2) 围压不变，轴压的突然卸载；(3) 轴压不变，围压的逐步加载。

第一种卸载方式下，卸载以准静态的方式进行。虽然尝试了不同的围压和卸载速率或应变率的组合，但始终未观察到岩石试件的张性破坏。在卸载过程中，试件的主应力差最大达到 75 MPa，平均轴向拉应变达到 0.7%，最大轴向应变则达到 1.5%。

第二种卸载方式下，围压保持不变，轴向压力快速卸载到 0。在这种情况下，2 组岩石试件均发生了快速的拉裂破坏，其中储存的能量以爆炸性的形式快速释放。2 种组别的岩石试件，发生拉裂破坏的静水压力强度分别大于 20 和 25 MPa。一般情况下，0.1%的应变足以导致试件的破坏。试件的破裂面为单个或数个平行于端面的拉伸裂缝，破坏后的试件如图 7 所示。

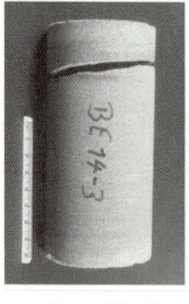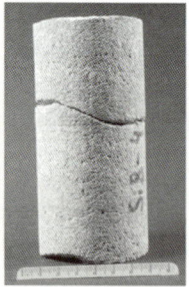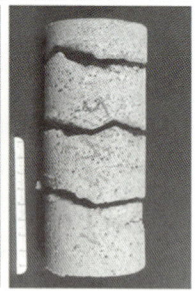

图7 围压恒定时轴压突然卸载后试件的拉伸破坏

Fig.7 Tensile fracture of samples due to abrupt unloading of axial stress with constant radial stress

第三种试验方式是保持轴向压力不变，缓慢的增加围压，试验中试件的应变与第一种加载条件基本相当，同样没有观察到试件的拉伸破坏。

E. Bauch 和 C. Lemmp[14]的试验向现有的理论提出了难题。岩石试件在较小的拉应变时拉裂，却

在 10 倍以上的拉应变情况下保持完好。试验表明，拉应变、最大最小主应力差值均不是影响岩石试件拉伸破坏的主要因素，而卸载速度和抗拉强度则具有显著的影响。

如果采用王明洋等[12]基于应力集中的拉裂准则和本文对卸载拉破坏影响因素的探讨，试验中发现的这些现象则能在很大程度上得到解释。

首先考察第一种试验条件——缓慢卸载。试验中，正应变率范围为 $1.2×10^{-7}\sim 2.4×10^{-6}\ s^{-1}$。对于试验中试件的应力状态，偏应变在数值上比正应变稍小。试验中最大偏应变约为 $10^{-6}\ s^{-1}$ 量级，根据节 3.2 的分析，卸载过程中只有远大于 10^{-4} m 的缺陷会被激活产生显著的拉应力。而试验所用试件均匀完整，无肉眼可见的裂纹和缺陷，可以认为其中包含的缺陷尺寸达不到 10^{-3} m 量级。试验采用的应变率不足以激活试件中的缺陷，也就不会出现拉伸破坏。可以推测，如果进一步增加卸载速率，也就是增加应变率，则岩石将会发生卸载拉伸破坏。

在轴压的突然卸载条件下，试件毫无例外的发生了拉伸破坏。试验没有给出突然卸载的具体速率，估计其应变率可达 $10^{-4}\ s^{-1}$ 量级。也就是说，试件中尺寸大于 10^{-6} m 的缺陷都将被激活，从而产生拉应力。这样大小的缺陷在试件中是必然存在的。由于卸载持续时间很短，卸载结束时，被激活的缺陷处的拉应力可达 $\sigma_{zz\ \max}^{loc} = K\sigma_0/3$。

岩石的应力集中系数可以估算[12]为

$$K = \frac{6\sigma_t}{\sigma_{ci}} \qquad (9)$$

式中：σ_t，σ_{ci} 分别为岩石的抗拉强度和起裂应力。

假设 2 种岩石的起裂应力约为其抗压强度的 70%，那么两种岩石的应力集中系数分别为 0.61 和 0.47。设局部拉伸破坏发生临界条件为 $\sigma_{zz\ \max}^{loc} = \sigma_t$，可得能够导致试件产生拉伸破坏的临界应力条件分别为 σ_0/σ_t 大于 4.92 和 6.38。试验测得 2 种岩石的 σ_0/σ_t 分别大于 5.12 和 5.10，与理论计算值相差不大。造成偏差的可能原因有：(1) 由于资料匮乏，对红砂岩起裂应力的判断偏差；(2) 试验应力级差过大，试验结果精度不够。

4.2 现场观测成果及理论对比

S. S. Lim 和 C. D. Martin[22]报道了对取自加拿大 Underground Research Laboratory 深地下巷道岩芯的观测统计成果。他们着重研究了岩芯饼化程度与最大主应力的关系及岩芯饼化的临界应力条件，并将其观测统计成果与 N. Kaga 等[23]的理论进行了比较，发现后者的理论严重高估了岩性饼化现象发生的临界应力条件。

S. S. Lim 和 C. D. Martin[22]统计得到的主要成果见图 8。其中 σ_1 为最大主应力，d 为饼化厚度，D 为岩芯直径。

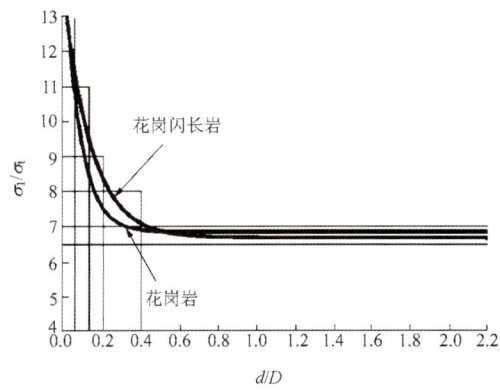

图 8 σ_1/σ_t 与 d/D 的关系[22]

Fig.8 Relationships between σ_1/σ_t and d/D[22]

从图 8 可以看出，岩芯饼化现象出现的临界应力条件为 $\sigma_1/\sigma_t \approx 6.5$，岩石的粒径组成对卸载拉裂稍有影响(粒径较粗的岩石临界 σ_1/σ_t 稍低)。

根据 E. Eberhardt 等[24]对取自 URL 同类岩石破坏过程的研究，URL 产花岗岩抗拉强度 $\sigma_t \approx 9$ MPa，起裂应力 σ_{ci} 约为 104 MPa，其应力集中系数 $K = 6\sigma_t/\sigma_{ci} \approx 0.52$，进而可以估计发生卸载拉裂的临界应力条件约为 $\sigma_{0c} = 3\sigma_t/K = 5.77\sigma_t$。

相比于 N. Kaga 等[23]的理论，本文理论预测结果大为接近，稍小于基于观测数据的条件。但考虑到：(1) 现场地应力条件不规则，最大最小主应力相差较大，统计结果采用最大主应力；(2) 岩芯套取时轴向应力的卸载是一个渐进的过程；(3) 现场取芯难以保证岩芯轴向与最大主应力平行，因此，有理由相信，观测结果高估了岩芯发生饼化的临界应力条件，理论预测的结果基本可信。

5 结 论

利用基于应力集中的非均匀岩体模型对卸载速率(偏应变率)、缺陷尺寸和初始应力对岩石卸载过程中缺陷处的应力集中的影响进行了研究，分析结果表明：(1) 偏应变速率对岩石内部缺陷处附加拉应力的大小具有显著的影响，偏应变率越高，缺陷处的附加拉应力越大；(2) 不同层次的缺陷对应于不同大小的偏应变率，一定的应变率条件下，只有大于对应层次的缺陷被激活,产生较明显的拉应力,加卸载速率或偏应变率越高,其所激活的缺陷的尺寸就越小;(3) 岩石包含较大缺陷时,缺陷处的局部拉应力与初始应力成正比,岩体是否能发生卸载破坏主要由初始应力、应力集中系数和抗拉强度 3 个参数之间的相对关系决定的。

卸载试验[14]和现场观测[22]的结果和现象初步验证了本文模型和分析的正确性。不过,由于试验设计缺乏明确的理论指导,以及现场地应力条件的复杂性和不确定性,其结果对确定模型参数的作用有限。为充分探讨本文模型适用范围,确定不同岩性岩石的相关参数,并进一步研究几个主要因素之间的相互关系,笔者设计了针对性的具有初始应力岩石的卸载破坏试验。试验研究已获得深部岩土力学与地下工程国家重点实验室开放基金项目的资助,将于近期开展。

参考文献(References):

[1] FAIRHURST C. Experimental physics and rock mechanics: results of laboratory study[M]. Tokyo: A. A. Balkema, 2001.

[2] BAHAT D, RABINOVITCH A, FRID V. Tensile fracturing in rocks[M]. Berlin: Springer-Verlag, 2005.

[3] BAZANT Z. Scaling laws in mechanics of failure[J]. Journal of Engineering Mechanics, 1993, 119(9): 1 828–1 844.

[4] PATERSON M S, WONG T F. Experimental rock deformation—the brittle field[M]. Berlin: Springer-Verlag, 2005.

[5] DYSKIN A V. On the role of stress fluctuations in brittle fracture[J]. International Journal of Fracture, 1999, 100(1): 29–53.

[6] RODIONOV V N, SIZOV I A. Model of a rigid body with dissipative structure for geomechanics[J]. Journal of Mining Sciences, 1989, 25(6): 491–501.

[7] ZHU W C, TANG C A. Micromechanical model for simulating the fracture process of rock[J]. Rock Mechanics and Rock Engineering, 2004, 37(1): 25–56.

[8] GRINFELD M A, WRIGHT T W. Morphology of fractured domains in brittle fracture[J]. Metallurgical and Materials Transactions A – Physical Metallurgy and Material, 2004, 35(9): 2 651–2 661.

[9] DYSKIN A V, VANVLIET M, VANMIER J. Size effect in tensile strength caused by stress fluctuations[J]. International Journal of Fracture, 2001, 108(1): 43–61.

[10] XIE H P, GAO F. The mechanics of cracks and a statistical strength theory for rocks[J]. International Journal of Rock Mechanics and Mining Sciences, 2000, 37(3): 477–488.

[11] 朱珍德,渠文平,蒋志坚. 岩石细观结构量化试验研究[J]. 岩石力学与工程学报,2007,26(7):1 313–1 324.(ZHU Zhende, QU Wenping, JIANG Zhijian. Quantitative test study on mesostructure of rock[J]. Chinese Journal of Rock Mechanics and Engineering, 2007, 26(7): 1 313–1 324.(in Chinese))

[12] 王明洋,范鹏贤,李文培. 岩石的劈裂和卸载破坏机制[J]. 岩石力学与工程学报,2010,29(2):234–241.(WANG Mingyang, FAN Pengxian, LI Wenpei. The mechanism of splitting and unloading failure of rock[J]. Chinese Journal of Rock Mechanics and Engineering, 2010, 29(2): 234–241.(in Chinese))

[13] CHEN Y Q. Observation of microcracks patterns in westerly granite specimens stressed immediately before failure by uniaxial compressive loading[J]. Chinese Journal of Rock Mechanics and Engineering, 2008, 27(12): 2 440–2 448.

[14] BAUCH E, LEMMP C. Rock splitting in the surrounds of underground openings: an experimental approach using triaxial extension test[C]// HACK R, AZZAM R, CHARLIER R ed. Engineering Geology for Infrastructure Planning in Europe. Berlin: Springer-Verlag, 2004: 244–254.

[15] BROWN E T, TROLLOPE D H. The failure of linear brittle materials under effective tensile stress[J]. Felsmechanik and Ingenieurgeologie, 1967, 6(1): 229–241.

[16] STACEY T R. A simple extension strain criterion for fracture of brittle rock[J]. International Journal of Rock Mechanics and Mining Sciences and Geomechanics Abstracts. 1981, 18(6): 469–474.

[17] LEMPP C, MÜHLHAUS H B. Splitting and core disking in deep boreholes[C]// Proceedings of 2nd International Symposium on Observation of the Continental Crust through Drilling. [S.l.]: [s.n.], 1985: 94.

[18] 陈宗基. 地下巷道长期稳定性的力学问题[J]. 岩石力学与工程学报,1982,1(1):1–20.(TAN Tjong-kie. The mechanical problems for the long-term stability of underground galleries[J]. Chinese Journal of Rock Mechanics and Engineering, 1982, 1(1): 1–20.(in Chinese))

[19] 戚承志,钱七虎,王明洋. 岩体的构造层次及其成因[J]. 岩石力学与工程学报,2005,24(16):2 838–2 846.(QI Chengzhi, QIAN Qihu, WANG Mingyang. Structural hierarchy of rock massif and mechanics of its formation[J]. Chinese Journal of Rock Mechanics and Engineering, 2005, 24(16): 2 838–2 846.(in Chinese))

[20] QI C Z, WANG M Y, QIAN Q H. Strain-rate effects on the strength and fragmentation size of rocks[J]. International Journal of Impact Engineering, 2009, 36(12): 1 355–1 364.

[21] 戚承志,钱七虎. 岩体动力变形与破坏的基本问题[M]. 北京:科学出版社,2009.(QI Chengzhi, QIAN Qihu. Basic problems of dynamic deformation and fracture of rock mass[M]. Beijing: Science Press, 2009.(in Chinese))

[22] LIM S S, MARTIN C D. Core disking and its relationship with stress magnitude for Lac du bonnet granite[J]. International Journal of Rock Mechanics and Mining Sciences, 2010, 47(2): 254–264.

[23] KAGA N, MATSUKI K, SAKAGUCHI K. The in-situ stress states associated with core disking estimated by analysis of principal tensile stress[J]. International Journal of Rock Mechanics and Mining Sciences, 2003, 40(5): 653–665.

[24] EBERHARDT E, STEAD D, STIMPSON B. Quantifying progressive pre-peak brittle fracture damage in rock during uniaxial compression[J]. International Journal of Rock Mechanics and Mining Sciences, 1999, 36(3): 361–380.

大型抛掷爆破中的重力影响

钱七虎

(解放军理工大学 工程兵工程学院,江苏 南京 210007)

摘 要:通过对抛掷爆破中漏斗坑形成的机理进行分析,确定了重力和岩土介质粘结力在抛掷爆破过程中的作用。基于对抛掷爆破相似理论量纲关系式的分析,提出了重力为决定性作用情况下的抛掷爆破量纲关系式,并进行了模拟试验。比较了模拟试验、原型大型TNT试验和核抛掷爆破试验的结果,表明了重力作用在大型爆破时,特别是在软弱岩层、漏斗坑形成过程中的决定性作用,并在此基础上介绍了基于几何相似原则下 Боресков 公式的修正公式及其在大型爆破中的应用。

关键词: 大型抛掷爆破;重力影响;漏斗坑;几何相似

中图分类号: TJ01   **文献标识码**: A   **文章编号**: 1009-3443(2010)02-0103-03

Influence of gravity in large-scale throwblasting

QIAN Qi-hu[1,2]

(Engineering Institute of Corps of Engineers, PLA Univ. of Sci. & Tech., Nanjing 210007, China)

Abstract: By the mechanism analysis of the funnel pit formation in throwblasting, the influence of gravity and geotechnical media adhesion was determined respectively. Based on the similarity theory, the dimensionless relationship of throwblasting was proposed when the gravity played a decisive role. In addition, the simulation experiment was also conducted. Besides, the determining factor of gravity on the formation of funnel pit, especially in the soft rock, was also declared by comparing the results of the simulation test, the prototypical large TNT test and the nuclear throwblasting test. On this basis, a correction formula for the Боресков formula and its application were introduced when the blasting was a large-scale one.

Key words: large-scale throwblasting; influence of gravity; funnel pit; geometric similarity

爆破计算大多是基于几何相似原则的试验公式,以抛掷爆破为例,其装药量 ε 的计算基于 Боресков 试验公式[1]。

$$\varepsilon = Kh^3 f(n), \quad (1)$$

式中: K 为取决于炸药与岩土特性的系数;h 为装药的埋深;n 为抛掷指数,$n=B/H$,B、H 为抛掷漏斗坑的半径和深度。

式(1)表明,爆破规模尺度之间的转换符合几何相似原则:在相同的抛掷指数 n 下,产生的爆破抛掷漏斗坑相似,其炸药量正比于漏斗坑的体积,即正比于 h^3。

但是大量的爆破实践表明,在爆破规模增大时,特别对于核爆炸的弹坑计算,几何相似不完全适用,必须引入修正,即所需炸药量的增大要比漏斗坑体积增加更快,即比式(1)计算的要多。这可从抛掷漏斗坑形成的物理本质来理解:岩土介质被爆炸所破碎,然后被破碎的岩土由爆炸产生的气状产物所推出。因此,爆炸的能量首先耗于克服岩土介质间的粘结力 σ,其次是耗于重力场中的岩土介质的抛掷位移上。前者与抛掷体积,即 h^3 成正比,而后者则与 $\rho g h^4$ 成正比。对于漏斗坑小(h 小)的爆破,σh^3 远大于 $\rho g h^4$,可以不考虑重力作用的影响,符合几何相似的原则,式(1)适用。对于 h 较大的爆破,σh^3 与 $\rho g h^4$ 相比,不能视后者为微量,即重力作用影响不能忽略,式(1)不完全适用。相反,当爆破规模很大,当重力作用为决定性时,则计算药量将正比于 h^4。为了证实重

收稿日期:2010-03-20。

本文原载于《解放军理工大学学报(自然科学版)》(2010 年第 11 卷第 2 期)。

力相似的重要性,曾进行如下的实验室模拟,在该模拟实验中,代替爆炸气体的为压缩空气[2,3]。

从上述的抛掷爆破的物理过程出发,参与的物理量有:岩土介质密度ρ,内摩擦系数K以及岩土介质强度σ.过程的决定性参数为空腔中的能量ε;空腔直径a_m;气体绝热参数γ,装药埋深h;重力加速度g;以及地面大气压力p_a.漏斗坑的基本参数为其半径B。

根据相似理论,抛掷漏斗坑的量纲关系式应为

$$\frac{B}{h} = f\left(\frac{\varepsilon}{\rho g h^4}; \frac{\varepsilon}{\sigma h^3}; \frac{\varepsilon}{p_a h^3}; \frac{a_m}{h}; \gamma, K\right). \quad (2)$$

由式(2)得出,如果仅考虑几何相似,完全忽略重力相似,即排除参数$\varepsilon/\rho g h^4$,其充要条件是$\rho g h$与p_a及σ相比足够小。相反,如只考虑重力相似,忽略几何相似,即排除参数$\varepsilon/\sigma h^3$及$\varepsilon/p_a h^3$,其充要条件是$\rho g h$比p_a及σ大很多。对于中间情况的埋深与炸药量,如果介质不变,则参数$\varepsilon/\rho g h^4$的作用随着爆破规模增大而越发显著。为了模拟重力作用显著的大规模爆破,必须要足够地减小p_a和σ使得与$\rho g h$相比可以忽略它们,即在量纲关系式中忽略无因次参数$\varepsilon/\sigma h^3$和$\varepsilon/p_a h^3$的影响,为此在模拟试验中采用真空以减小p_a,采用干砂以减小介质间的粘结力σ。

排除参数$\varepsilon/\sigma h^3$,摩擦力还存在,还要消耗能量,所以要保留参数K和γ,考虑到空腔中并非是绝对真空,为了更准确考虑抛掷阻力,代替$\varepsilon/\rho g h^4$以参数$\varepsilon/(\rho g h + p_a)h^3$,即总阻力为$\rho g h + p_a$。

于是,在重力作用为决定性的情况下,式(2)的新形式为

$$\frac{B}{h} = f\left(\frac{\varepsilon}{(\rho g h + p_a)h^3}, \frac{a_m}{h}\right), \quad (3)$$

为了书写简单,令

$$\bar{\varepsilon} = \frac{\varepsilon}{(\rho g h + p_a)h^3}.$$

现在来比较模拟试验和原型的TNT以及核大型抛掷爆破试验的结果[4]。

图1显示了模拟试验与原型试验中地表投影中心土体隆起速度最大值的数据比较。模拟试验的经验公式为$v_0 = 0.6 h_0^{1.85}$,无量纲v_0及h_0为下式表示:

$$v_0 = \frac{v_m}{\sqrt{gh + \frac{p_a}{\rho}}}; \quad h_0 = \frac{h}{\sqrt[3]{\frac{\varepsilon}{(\rho g h + p_a)}}}.$$

在图1上注明了模拟试验和原型的TNT及核试验的曲线,其相应数据为:曲线2为黄土中TNT

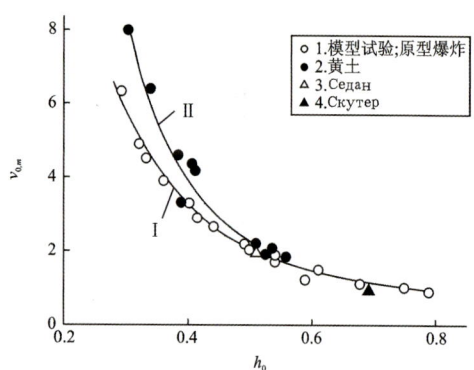

图1 模拟试验(Ⅰ)与原型爆炸(Ⅱ)地面投影点隆起速度最大值的比较

Fig.1 Comparison of the maximum uplift velocity at projection center between the model blasting and the prototype blasting

爆炸,药量1 t,$h = 7$ m;曲线3为冲积土中TNT爆炸[4],药量454 t,$h = 38$ m;曲线4为冲积土中核爆炸[4],药量100 kt,$h = 194$ m。由该图曲线的比较,可以看出模拟试验与原型试验的一致性较好,这表明,对于大型爆破,形成抛掷漏斗坑时,重力作用是决定性的。

由图1还可见,在$h_0 < 0.5$时,原型中的抛掷速度超过模拟试验中的速度,这是由于随着炸药埋深的减少,在自由面附近原型试验中爆炸压缩波对抛掷漏斗坑形成的作用增强了。而在模拟试验中压缩空气是没有爆炸压缩波的。

图2上显示了3个原型试验的地表面投影中心隆起速度时程曲线与模拟试验的相应时程曲线的比较[4]。

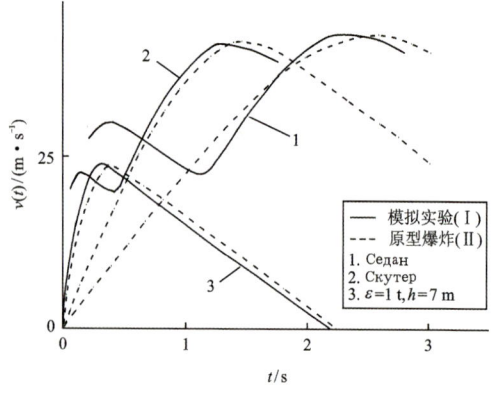

图2 模拟试验(Ⅰ)与原型爆炸(Ⅱ)的速度曲线$v(t)$的比较

Fig.2 Comparison of velocity between the model blasting(Ⅰ) and the prototype blasting(Ⅱ)

由图 2 可见,两者 $v(t)$ 时程曲线的区别仅仅在初始阶段,这时在原型试验中,隆起速度是由爆炸压缩波的作用决定的,因此这个区别是自然可理解的。因为在模拟试验中没有爆炸压缩波,在后续的气相加速阶段,图上曲线显示了两者一致性的满意结果。而在试验曲线 3 中,因为原型试验规模稍小,高速摄影机不能区分出爆炸压缩波所产生的土体运动,所以两者也就完全一致了。

原型试验与模型试验抛掷爆破的数据一致性充分证明了在岩石中,特别是弱岩中形成抛掷漏斗坑时,重力的作用是决定性的。

为了弄清大埋深情况下抛掷爆破中的尺度效应,在前苏联和美国进行了大量的现场试验,试验结果确实显示了重力影响对于漏斗坑尺寸的重要性,即显示了几何相似的偏差。

为了修正基于几何相似的 Боресков 公式,Садовский 和 Покровский 建议如下的应用于 $h > 25$ m 情况下计算爆破药量的公式[1]

$$\varepsilon = Kh^3\ \frac{h}{25}(0.4 + 0.6n^3)。 \quad (4)$$

近年来,按抛掷漏斗坑尺寸的试验成果整理的另一公式也得到广泛应用[5,6],即

$$B/\varepsilon^{1/N} = f_1(h/\varepsilon^{1/N}),\ H/\varepsilon^{1/N} = f_2(h/\varepsilon^{1/N})。 \quad (5)$$

在炸药量 $0.116 \sim 454$ t 的大小爆破范围内,在冲积土中,式(5)以 $N = 3.4$ 计算的结果与试验结果相比的偏差最小(图 3、4)[4],上述结论也证明了式(4)的正确性,在式(5)中,N 实际为 3.5。

抛掷爆破过程的物理本质为岩土介质的破碎及破碎后的岩土介质的被抛出,前者能量耗于克服岩土介质的强度,与破碎体积,即与 h^3 成正比,符合

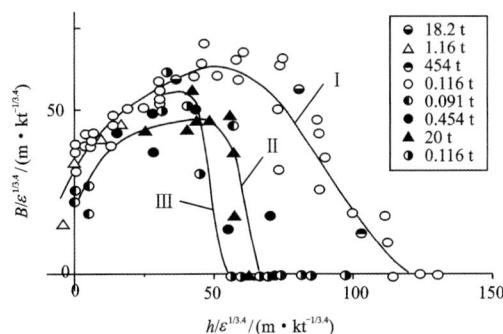

图 3 爆炸漏斗坑比例半径与装药比例埋深的关系
Ⅰ—冲积土;Ⅱ—玄武岩;Ⅲ—凝灰岩

Fig.3 Relationship between the ratio radius and the ratio depth in the funnel pit
Ⅰ—Alluvial soil;Ⅱ—basalt;Ⅲ—tuff

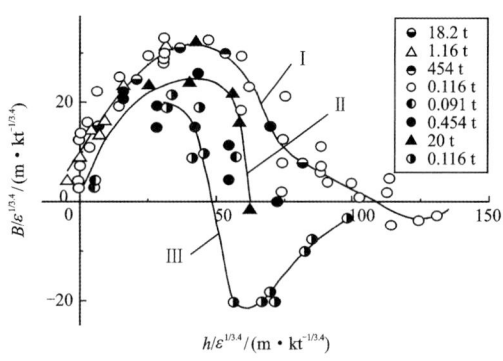

图 4 爆炸漏斗坑比例深度与装药比例埋深的关系
Ⅰ—冲积土;Ⅱ—玄武岩;Ⅲ—凝灰岩

Fig.4 Relationship between the ratio radius and the ratio depth in the funnel pit
Ⅰ—Alluvial soil;Ⅱ—basalt;Ⅲ—tuff

几何相似原则;后者能量耗于重力场中岩土介质在抛掷位移上所作的功,即与 $\rho g h^4$ 成正比,重力相似不符合几何相似原则。对于小埋深爆破,h 小,重力影响小,可以忽略,基于几何相似的 Боресков 试验公式可以适用;对于大埋深(大药量)爆破,重力影响很大,甚至是决定性的,必须采用修正几何相似的试验公式,因此,大埋深抛掷爆破或弹坑爆炸的模拟试验必须在离心机上进行,以得出符合重力相似的试验公式。

参考文献:

[1] 钱七虎. 岩石爆炸动力学的若干进展[J]. 岩石力学与工程学报,2009,28(10):1945-1968.
QIAN Qi-hu. Some advances in rock blasting dynamics [J]. Chinese Journal of Rocks and Engineering, 2009,28(10):1945-1968. (in Chinese).

[2] ВОВК А А,КРАВЕЦ В Г,ЛУУНКО И А,и другие. Геодинамика взрыва и ее приложение[R]. Киев:Ваукова Думка,1981.

[3] БЕЛЯЕВ А Ф. О расчете зарядов при взрывена выброс [R].[S.l.]:Горный Журнал,1957.

[4] САДОВСКИЙ М А. Механический эффект подземного взрыва. Издательство Москва,Недра,1971.

[5] NORDYKE M D. An analysis of cratering data from desert alluvium[J]. Journal of Geophysical Research, 1961,66(10):3389-3404.

[6] САДОВСКИЙ М А,АДУШКИН В В,СПИВАК А А. О размера зон необратимого деформирования при взрыве в блочной среде[J]. Физики Земли,1989,(9):9-15.

Effect of loading rate on fracture characteristics of rock

Xiaoping Zhou[1], Qihu Qian[2], Haiqing Yang[1]

1. School of Civil Engineering, Chongqing University, Chongqing 400045, China;
2. PLA University of Science and Technology, Nanjing 210007, China

© Central South University Press and Springer-Verlag Berlin Heidelberg 2010

Abstract: The three-point bending experiments were applied to investigating effects of loading rates on fracture toughness of Huanglong limestone. The fracture toughness of Huanglong limestone was measured over a wide range of loading rates from 9×10^{-4} to 1.537 MPa·m$^{1/2}$/s. According to the approximate relationship between static and dynamic fracture toughness of Huanglong limestone, relationship between the growth velocity of crack and dynamic fracture toughness was obtained. The main conclusions are summarized as follows. (1) When the loading rate is higher than 0.027 MPa·m$^{1/2}$/s, the fracture toughness of Huanglong limestone increases markedly with increasing loading rate. However, when loading rate is lower than 0.027 MPa·m$^{1/2}$/s, fracture toughness slightly increases with an increase in loading rate. (2) It is found from experimental results that fracture toughness is linearly proportional to the logarithmic expression of loading rate. (3) For Huanglong limestone, when the growth velocity of crack is lower than 100 m/s, the energy release rate slightly decreases with increasing the growth velocity of crack. However, when the growth velocity of crack is higher than 1 000 m/s, the energy release rate dramatically decreases with an increase in the crack growth velocity.

Key words: fracture toughness of rock; three-point bending round bar; loading rate; velocity of crack growth

1 Introduction

Rock material encountered in engineering practice is inhomogeneous and discontinuous containing joints, cracks etc. It is a well-known fact that all forms of rock breakage are caused by the extension of one or more cracks. In many practical applications of rock engineering, such as rock excavation, tunneling, rock cutting process and rock blasting, fracture toughness of rock material is an important parameter for rock failure analysis or for increasing the performance of rock cutting tools. The dynamic mechanical properties of rock are basic information in assessing the stability of rock structures under dynamic loads. Moreover, dynamic fracture toughness of rock is considered to be a fundamental parameter in rock fracture mechanics, an intrinsic material property related to resistance to crack initiation and propagation. Analysis of properties of fracture resistance is essential to understand rock dynamic fragmentation process. It is widely recognized that rock engineering design and/or assessment of critical rock infrastructures need to allow for particularly severe dynamic scenarios which could occur in rock engineering during their lifetime (i.e. strong earthquakes and man-made blast events). Since the dynamic mechanical properties of rock materials differ greatly from those exhibited in quasi-static conditions, specific investigations in such dynamic ranges are necessary in order to fully understand the behavior of rock engineering subjected to high dynamic loading conditions. Therefore, investigating the dynamic fracture toughness of rock materials is a crucial matter for rock engineering.

Studies on dynamic fracture toughness of rock were conducted primarily by laboratory tests. Four methods for measuring rock fracture toughness were suggested by the ISRM [1−2], such as three-point chevron bending (CB) specimens, short rod (SR) specimens, straight-through cracked three-point bend round beam (SB) specimens and cracked chevron notched Brazilian (CCNB) disc specimens. Great efforts have already been made to thoroughly investigate the dynamic fracture toughness of rock. Generally, fracture toughness is dependent on the fracture velocity. At low velocities, fracture toughness, K_{IC}, remains almost constant. Exceeding a threshold of fracture velocity, fracture toughness increases significantly. Fracture toughness also increases with increasing loading rate. For example, XIA et al [3] demonstrated that the fracture toughness of Barre granite depended on the direction of these weak planes. NASSERI et al [4−5] investigated the relationship between the fracture toughness of granite and the crack length. BACKERS et al [6] pointed out that the fracture toughness stays constant irrespective of crack opening rate between 9.8×10^{-6} and 1.1×10^{-2} MPa·m$^{1/2}$/s in the three-point CB tests. The dynamic fracture toughness of marble, which increases remarkably with the increase of loading rate, is obtained by using Split Hopkinson Pressure Bar (SHPB) [7]. With the aid of the SHPB testing system and the scanning electron microscope (SEM), ZHANG et al [8] measured the dynamic fracture toughness and fracture characteristics of Fangshan marble and Fangshan gabbro. The experimental results show that the fracture toughness is nearly a constant when the loading rate is lower than 1.0×10^4 MPa·m$^{1/2}$/s, but the fracture toughness increases with increasing the loading rate when the loading rate is higher than 1.0×10^4 MPa·m$^{1/2}$/s. JIN and YANG [9] measured the fracture toughness of andesite and found that fracture toughness increases with increasing the loading rate. However, they did not obtain a quantitative relationship between the fracture toughness and the loading rate due to the limitation of the testing conditions.

In this work, by means of the three-point bending experiments, the fracture toughness of Huanglong limestone was measured over a wide range of loading rates from 9×10^{-4} to 1.537 MPa·m$^{1/2}$/s.

2 Experimental

The Huanglong limestone test specimen is from Lead-Zinc Mine in Nanjing, China. The values of mechanical parameters of Huanglong limestone are summarized as follows: elastic modulus E=26.911 GPa, Poisson ratio v=0.151, and mass density ρ=2 730 kg/m^3. The specimens in this work, cylinders d50 mm × 200 mm, were machined strictly according to Ref.[1]. Moreover, using metal milling cutter, a chevron notch was prefabricated at the center of round bar. As depicted in Fig.1, the specimen dimensions were as follows: D= 50 mm, L= 200 mm, S=166.6 mm, θ=90°, a_0=7.5 mm, t=2 mm, and h=12.6 mm.

All the tests were performed on an INSTRON1342 servo-control machine. Load was applied at six different loading rates, which was controlled by load point displacement (LPD) δ_F. The load point displacement rates were 0.1, 0.04, 0.01, 0.004, 0.000 4 and 0.000 1 mm/s, respectively. A total of 30 specimens were tested in this survey. The tests were conducted in six groups. There were five sets of tests in each group. During the process of loading, crack opening displacement was recorded by clip gauge at the same time, as shown in Fig.2.

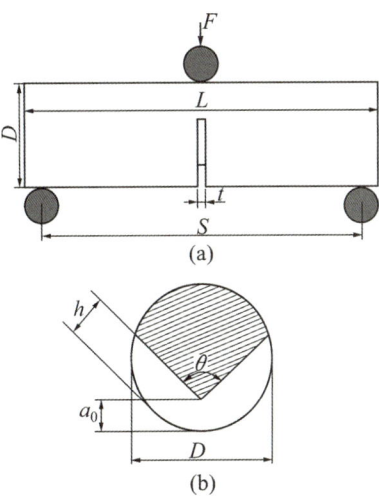

Fig.1 Sketch of three-point CB specimens: (a) Edge crack beam; (b) Cross section

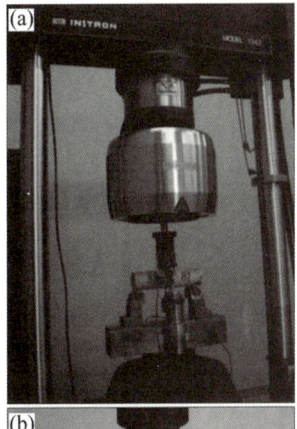

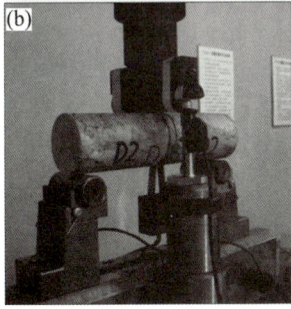

Fig.2 Photographs of three-point bending test: (a) Beam specimen; (b) Near view

3 Effect of loading rate on fracture toughness

Fig.3 illuminates the relationship between load and crack opening displacement. Initially, specimens follow a period of linear ascending, and the load reaches its maximum, which corresponds to the initiation of a crack. Then the load must go down since the dimensionless stress intensity factor rises, and the crack propagates unstably at this stage. The load soon reaches the local minimum value P_{min}, which corresponds to the critical crack opening displacement. Moreover, it can be observed from Fig.3 that the maximum load is higher as loading rate is higher.

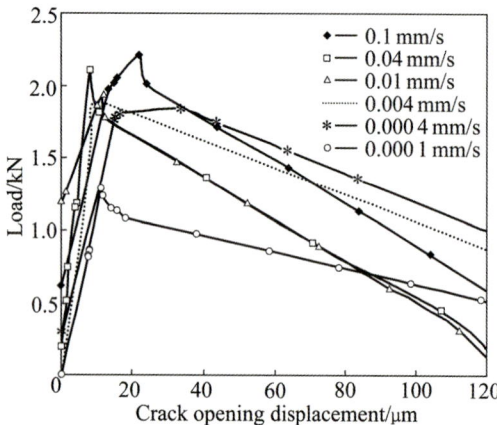

Fig.3 Test records of load−displacement for three-point CB specimens of Huanglong limestone at different load point displacement rates

The maximum load, critical time at the peak, crack opening displacement and load point displacement are shown in Table 1. From Table 1, the main conclusions are summarized as follows. (1) When the load point displacement rate is higher than 0.004 mm/s, the maximum load increases markedly with increasing the loading rate. However, when the load point displacement rate is lower than 0.004 mm/s, the maximum load slightly increases with increasing the loading rate. (2) Crack opening displacement and load point displacement at the peak slightly vary with loading rate, which seem to be dispersive at different loading rates.

According to Ref.[10], fracture toughness of three-point CB specimen can be determined as follows:

$$K_{IC} = \frac{A_{min} P_{max}}{D^{1.5}} \quad (1)$$

$$A_{min} = \frac{S}{D}[1.835 + 7.15(\frac{a_0}{D}) + 9.85(\frac{a_0}{D})^2] \quad (2)$$

where a_0 is the chevron tip distance from specimen surface, P_{max} is the maximum load, D is the diameter of CB specimen, and S is the distance between support points.

According to the work of OUCHTERLONY [1, 10] and ZHANG et al [8], the relationships between loading rate k' of three point bending test and average velocity v_c of crack growth are defined as

$$k' = K_{IC}/t_m \quad (3)$$

$$v_c = \frac{E\sqrt{D}}{6.4 K_{IC}} \delta'_F \quad (4)$$

where t_m is the time at the peak load, $\delta'_F = \delta_F / t_m$.

Table 2 illuminates the dependence of fracture toughness on loading rate and average velocity of crack growth. The range of loading rate is from 9×10^{-4} to 1.537 MPa·m$^{1/2}$/s, and the range of corresponding load point displacement rates is from 0.000 1 to 0.1 mm/s. It can be observed from Table 2 that the average velocity of crack growth increases with increasing the loading rate.

Due to dispersion of the test specimens, the fracture toughness at the load point displacement rate of 0.000 1 mm/s is apparently lower than that in other conditions. The fracture toughness at the load point displacement rate of 0.000 1 mm/s will be deleted in this work. According to the experimental data in Table 2, the relationship between fracture toughness K_{IC} and critical time t_m is presented in Fig.4, and the relationship between fracture toughness K_{IC} and loading rate k' is shown in Fig.5. It is shown from Figs.4 and 5 that when the loading rate is higher than 0.027 MPa·m$^{1/2}$/s, the fracture toughness of rock increases markedly with increasing loading rate; when the loading rate is lower than 0.027 MPa·m$^{1/2}$/s, fracture toughness slightly increases with an increase in loading rate.

From Fig.4 and Fig.5, the following quantitative expressions are initially determined as

$$K_{IC} = 0.485\,31\exp(\frac{-t_m}{3.568\,25}) + 1.706\,18, R = 0.991\,5 \quad (5)$$

$$K_{IC} = 0.054\,7\ln k' + 1.948\,5, R = 0.856\,5 \quad (6)$$

where R is the correlative coefficient.

Table 1 Experimental results of maximum load and crack opening displacement and load point displacement

Load point displacement rate/(mm·s^{-1})	Load point displacement at peak/mm	Crack opening displacement at peak/μm	Critical time at peak/s	Maximum load/kN
0.1	0.131	22.2	1.327 8	2.201 244
0.04	0.123	8.0	3.000 0	2.103 850
0.01	0.051	12.1	5.160 8	1.940 191
0.004	0.251	9.8	63.000 0	1.883 000
0.000 4	0.156	33.4	390.960 8	1.842 096
0.000 1	0.134	11.4	1 347.000 0	1.296 461

Table 2 Effect of loading rate on fracture toughness of Huanglong limestone

Load point displacement rate/(mm·s^{-1})	K_{IC}/(MPa·m$^{1/2}$)	k'/(MPa·m$^{1/2}$·s^{-1})	v_c/(m·s^{-1})
0.1	2.040 57	1.537 00	4.63×10^{-2}
0.04	1.915 87	0.639 00	2.02×10^{-2}
0.01	1.820 17	0.353 00	5.10×10^{-3}
0.004	1.724 94	0.027 00	2.18×10^{-3}
0.000 4	1.687 47	0.004 32	2.23×10^{-4}
0.000 1	1.216 26	0.000 90	7.68×10^{-5}

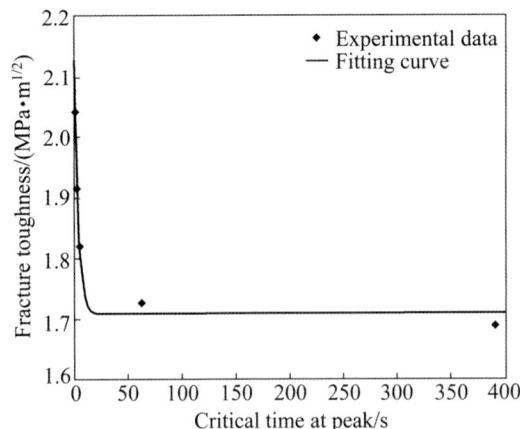

Fig.4 Relationship between fracture toughness and critical time at peak

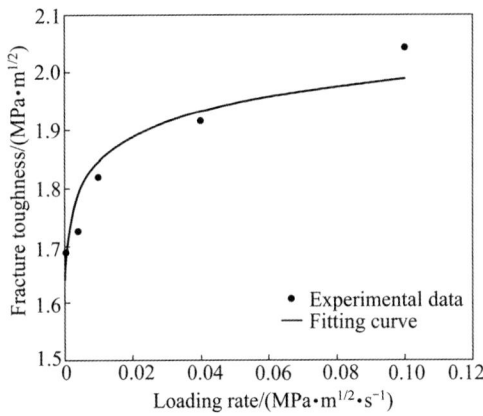

Fig.5 Relationship between fracture toughness and loading rate

4 Dynamic fracture mechanism of rock

Fig.6 illuminates growth of the chevron notch, where a is the height of crack tip, and b is the width of crack tip. It can be seen from Fig.6 that width b is varied with the height of crack tip a. The critical height a_c is obtained as follows:

$$a_c = \frac{D}{2} + \sqrt{\frac{D^2}{4} - \frac{1}{2}[\sqrt{\frac{D^2}{4} - \frac{1}{2}(\frac{D}{2}-a_0)^2} + \frac{\sqrt{2}}{2}(\frac{D}{2}-a_0)]^2} \quad (7)$$

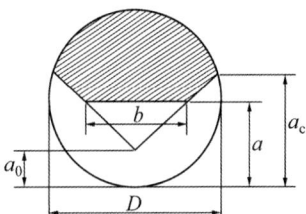

Fig.6 Sketch of crack growth

According to Refs.[1, 11], for three-point bending round bar, stress intensity factor at the crack tip can be expressed as follows:

$$K_I = \frac{PL\alpha^{0.5}}{D^{2.5}(1-\alpha)^2}[3.75 - 11.98\alpha + 24.4\alpha^2 - 25.69\alpha^3 + 10.02\alpha^4] \times [\frac{\alpha^{0.5}(1-\alpha)^{0.5}}{\alpha-\alpha_0}]^{0.5}, \quad a < a_c \quad (8)$$

$$K_I = \frac{PL\alpha^{0.5}}{D^{2.5}(1-\alpha)^2}[3.75 - 11.98\alpha + 24.4\alpha^2 - 25.69\alpha^3 + 10.02\alpha^4], \quad a_c < a < D \quad (9)$$

where $\alpha = a/D$.

Dynamic critical stress intensity factor K_{ID} at the crack tip is related to the velocity of crack growth. Then, the following expression can be obtained as [11–15]

$$K_{ID} = k(v_a)K_I \quad (10)$$

where v_a is the velocity of crack growth, and $k(v_a) = (c_R - v_a)/(c_R - 0.75v_a)$ is a function of crack growth velocity. The function $k(v_a)=1.0$, when $v_a=0$, and $k(v_a)=0$ when v_a reaches the critical velocity of the crack growth, which is normally regarded as the velocity of a Rayleigh wave (c_R) in the rock material.

The relationship between stress intensity factor of mode I and the energy release rate is given by [13–15]

$$G_I = \frac{1-v^2}{E}A_I K_{ID}^2 \quad (11)$$

$$A_I = \frac{v_a^2 \alpha_d}{(1-v)c_s^2[4\alpha_d - (1+\alpha_s^2)^2]} \quad (12)$$

$$c_d^2 = \frac{E(1-v)}{\rho(1+v)(1-2v)} \quad (13)$$

$$c_s^2 = \frac{E}{2\rho(1+v)} \quad (14)$$

where $\alpha_d = \sqrt{1-v_a^2/c_d^2}$, $\alpha_s = \sqrt{1-v_a^2/c_s^2}$, G_I is the energy release rate, c_d is dilatational wave velocity, c_s is the velocity of a shear wave, and K_{ID} is dynamic critical stress intensity factor of mode I.

The relationship between the energy release rate and stress intensity factor of mode I can be obtained from Eq.(11). Eq.(11) can be rewritten as [13–15]

$$G_{\mathrm{I}} = \frac{A_{\mathrm{I}}[k(v_{\mathrm{a}})]^2}{E}(1-v^2)K_{\mathrm{I}}^2 \qquad (15)$$

Fig.7 shows the relationship between energy release rate and dynamic fracture toughness at different growth velocities of crack. The velocity of Rayleigh wave of Huanglong limestone (c_{R}) is 2 km/s. It is indicated from Fig.7 that (1) the energy release rate nonlinearly increases as dynamic fracture toughness increases; (2) the energy release rate is lower as the velocity of crack growth is faster; (3) the energy release rate is nearly a constant when the velocity of crack growth is lower than 100 m/s; and (4) the energy release rate dramatically decreases with increasing the crack growth velocity when the growth velocity of crack is higher than 1 000 m/s.

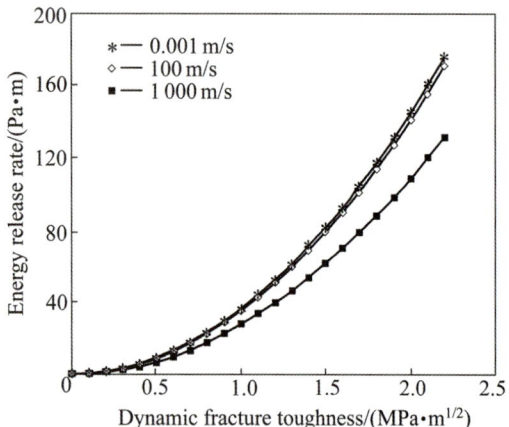

Fig.7 Relationship between energy release rate and dynamic fracture toughness at different growth velocities of crack

5 Comparison with results of some rock and metal materials

ZHANG et al [8] measured the dynamic fracture toughness and fracture characteristics of the SR Fangshan marble and Fangshan gabbro specimens by SHPB. The relationship between the fracture toughness of Fangshan marble and Fangshan gabbro and the loading rate is close to the results for the Huanglong limestone in Fig.5. Similar to the results for Huanglong limestone, the dynamic fracture toughness of some metal materials is also greater than their static fracture toughness. For example, the relationship between the fracture toughness of PMMA (polymethyl methacrylate) and the loading rate [16−17] is extremely close to the results for Huanglong limestone in Fig.5. In addition, experimental results [18] for Si_3N_4 (silicon nitride) and PSZ (partially stabilised zirconia) show that the fracture toughness of both materials has the same tendency to vary with the loading rate as the results of Huanglong limestone in Fig.5. It can be inferred that some common properties probably exist between the above metal materials and the rocks. That is, the above metal materials and the rocks show an increase in fracture toughness at high loading rates.

6 Conclusions

(1) It is found from experimental results that loading rate significantly affects the fracture toughness of Huanglong limestone. When the loading rate is higher than 0.027 MPa·m$^{1/2}$/s, the fracture toughness increases markedly with an increase in loading rate. However, when the loading rate is lower than 0.027 MPa·m$^{1/2}$/s, fracture toughness slightly increases with an increase in the loading rate.

(2) On the basis of experimental results, the quantitative relationship between fracture toughness and loading rate is determined. The fracture toughness of Huanglong limestone and the loading rate can be related to each other by the equation, $K_{\mathrm{IC}}=0.054\,7\ln k'+1.948\,5$.

(3) The fracture toughness of Huanglong limestone and the critical time at the peak can be related to each other by the equation, $K_{\mathrm{IC}}=0.485\,31\exp(-t_{\mathrm{m}}/3.568\,25)+1.706\,18$.

References

[1] OUCHTERLONY F. Suggested methods for determining the fracture toughness of rock [J]. International Journal of Rock Mechanics and Mining Sciences and Geomechanics Abstracts, 1988, 25(2): 71−96.

[2] FOWELL R J. Suggested methods for determining mode I fracture toughness using cracked chevron notched Brazilian disc (CCNBD) specimens [J]. International Journal of Rock Mechanics and Mining Sciences and Geomechanics Abstracts, 1995, 32(1): 57−64.

[3] XIA K, NASSERI M, MOHANTY B, LU F. Effects of microstructures on dynamic compression of Barre granite [J]. International Journal of Rock Mechanics and Mining Sciences, 2008, 45(6): 879−887.

[4] NASSERI M, MOHANTY B, YOUNG R P. Fracture toughness measurements and acoustic emission activity in brittle rocks [J]. Pure Applied Geophysics, 2006, 163(5/6): 917−945.

[5] NASSERI M, MOHANTY B. Fracture toughness anisotropy in granitic rocks [J]. International Journal of Rock Mechanics and Mining Sciences, 2008, 45(2): 167−193.

[6] BACKERS T, FARDIN N, DRESEN G, STEPHANSSON O. Effect of loading rate on mode I fracture toughness, roughness and micromechanics of sandstone [J]. International Journal of Rock Mechanics and Mining Sciences, 2003, 40(3): 425−433.

[7] LI Zhan-lu, WANG Qi-zhi. Experimental research on effect of loading rate for dynamic fracture toughness of rock [J]. Chinese Journal of Geotechnical Engineering, 2006, 28(12): 2116−2120. (in Chinese)

[8] ZHANG Z X, YU Y, ZHAO Q. Influences of loading rates on fracture toughness of gabbro and marble [C]// ROSSMANITH H P. Fracture and Damage of Concrete and Rock FDCR−2. London: E & FN Spon, 1993: 418−425.

[9] JIN Feng-nian, YANG Hai-jie. Loading-rate effect of rock [J].

Chinese Journal of Rock Mechanics and Engineering, 1998, 17(6): 711–717. (in Chinese)

[10] OUCHTERLONY F. On the background to the formulae and accuracy of rock fracture toughness measurements using ISRM standard core specimens [J]. International Journal of Rock Mechanics and Mining Sciences, 1989, 26(1): 13–23.

[11] ZHAO J, LI H B, ZHAO Y H. Dynamic strength tests of the Bukit Timah granite [R]. Singapore: Nanyang Technological University, 1998.

[12] ZHAI Yue, MA Wei-guo, ZHAO Jun-hai, HU Chang-ming. Dynamic failure analysis on granite under uniaxial impact compressive load [J]. Chinese Journal of Geotechnical Engineering, 2007, 29(3): 385–390. (in Chinese)

[13] ZHOU Xiao-ping, YANG Hai-qing. Micromechanical modeling of dynamic compressive responses of mesoscopic heterogeneous brittle rock [J]. Theoretical and Applied Fracture Mechanics, 2007, 48(1): 1–20.

[14] ZHOU Xiao-ping. Analysis of the localization of deformation and the complete stress–strain relation for mesoscopic heterogeneous brittle rock under dynamic uniaxial tensile loading [J]. International Journal of Solids and Structures, 2004, 41(5/6): 1725–1738.

[15] FREUND L B. Dynamic fracture mechanics [M]. Cambridge: Cambridge University Press, 1990: 49–51.

[16] RITTEL D, MAIGRE H. An investigation of dynamic crack initiation in PMMA [J]. Mechanics of Materials, 1996, 23(3): 229–239.

[17] WADA H. Determination of dynamic fracture toughness for PMMA [J]. Engineering Fracture Mechanics, 1992, 41(6): 821–831.

[18] KOBAYASHI T, KAZINO T, KAMIMURA M, IKAWA H. Basic principle of dynamic fracture toughness evaluation by computer aided instrumented impact testing (CAI) system [C]// SALAMA K, RAVICHANDAR K, TAPLIN D M R, RAMA R P. Advances in Fracture Research. Oxford: Pergamon Press, 1989: 651–658.

深埋球形洞室围岩分区破裂化机理

*周小平[1]，钱七虎[2]，张伯虎[1]，张永兴[1]

(1. 重庆大学土木工程学院，重庆 400045；2. 解放军理工大学，南京 210007)

摘 要：在高应力和复杂的地质环境中，深部球形洞室围岩在开挖扰动下会出现破裂区和非破裂区交替出现的分区破裂化现象，研究分区破裂化现象可以深化深部岩体的力学性能研究，同时对深埋洞室的开挖和支护设计提供理论基础。该文研究的深部球形洞室外部受到静水压力的作用，洞室内壁受到一个随时间变化的开挖荷载的作用，其运动方程用位移势函数来表示。通过Laplace变换简化计算，获得势函数的通解，从而获得了在开挖扰动下洞室围岩的应力场和位移场。当围岩应力场满足深部岩体强度准则时，岩体破裂，从而产生变形局部化。岩体破坏产生的应力重分布可能导致新的围岩破坏，从而产生二次破裂区；以此类推，直到应力释放后不能产生破裂区为止。根据断裂力学确定破裂区的残余强度，并确定破裂区和非破裂区的宽度和数量。破裂区的形成受到岩体力学性能、洞室开挖方式和速度等较大的影响。

关键词：球形巷道；深部岩体工程；分区破裂化；深部岩体强度准则；动力问题

中图分类号：TU452　　**文献标识码**：A

THE MECHANISM OF THE ZONAL DISINTEGRATION PHENOMENON AROUND DEEP SPHERICAL TUNNELS

*ZHOU Xiao-ping[1], QIAN Qi-hu[2], ZHANG Bo-hu[1], ZHANG Yong-xing[1]

(1. School of Civil Engineering, Chongqing University, Chongqing 400045, China; 2. PLA University of Science and Technology, Nanjing 210007, China)

Abstract: High stress, high temperature and complicated environment are encountered in deep geology. The mechanical behavior of the deep rock mass is different from that of shallow rock mass. In shallow rock mass engineering, the excavation-affected rock around tunnel contains a loose zone, a plastic zone and an elastic zone. In deep rock mass engineering, the surrounding rock mass around spherical tunnels is divided into fractured zone and non-fractured zone, which occurs alternatively. The mechanism of the zonal disintegration phenomenon around deep spherical tunnels is analyzed. The present model is helpful to understand the failure and deformation of the deep rock mass. The present theory model contains a spherical tunnel which is subjected to an in-situ far-field hydrostatic stress and a changing internal pressure during tunnel excavation. The dynamic problem can be expressed by displacement potential function and the general solution of motion equations can be obtained by using Laplace transform. The stress and displacement field around spherical opening are obtained. The fractured zone in the surrounding rock mass around deep spherical tunnel can be determined by using the deep rock mass strength criterion. Size and quantity of fractured zone and non-fractured zone in the surrounding rock mass around deep spherical tunnel are given. The stress field in the fractured zone is obtained based on the theory of deformation localization. The influencing factors of the zonal disintegration phenomenon include the mechanical properties, excavation method and the excavation velocity, etc.

收稿日期：2008-08-29；修回日期：2008-10-06。
基金项目：国家自然科学基金项目(50490275,50778184)；教育部新世纪优秀人才支持计划项目(NCET-07-0911)。
本文原载于《工程力学》(2010 年第 27 卷第 1 期)。

Key words: spherical tunnel; deep rock mass engineering; the alternative occurrence of fractured zone and non-fractured zone; strength criterion of deep rock masses; dynamic problem

近年来，地下球形结构日渐增多。如水工埋藏式球形岔管、矿山竖井球形马头门、地铁球形穹顶、地下球形贮仓乃至球形掩体等。它们在施工前先将围岩开挖成球形洞室，然后再喷一薄层混凝土支护以利围岩稳定，故弄清球形洞室围岩应力、位移以及围岩的破坏特征就显得十分重要。

深部岩体的力学特性和浅部岩体的力学特性有很大的不同[1-9]。浅部球形洞室围岩状态通常可以分为松动区、塑性区和弹性区，其本构关系可以采用弹塑性力学理论进行推导求解。深部岩体主要特点是地应力高，在高应力和复杂的地质环境中，球形洞室围岩产生膨胀带和压缩带，或称为破裂区和未破裂区交替出现的情形，这一现象被称为分区破裂化现象。文献[1-2]认为分区破裂化现象是深部岩体的一个重要特征。文献[2-5]认为深部岩体会出现大变形和强流变等特性。深部岩体分区破裂化现象于 20 世纪 70 年代在南非 2073m 深的Witwatersrand金矿中首次被发现[6]。之后，在南非深部金矿的巷道中被系统地观察到了围岩中存在分区破裂化现象[7-8]。俄罗斯在Taimyrskii和маяк矿山中也出现了类似的现象[9]。研究深部岩体的分区破裂化现象有助于我们了解深部岩体的力学性能和破坏特性，同时对深部岩体的支护设计提供理论基础。对于深部岩体分区破裂化现象的研究，俄罗斯学者Shemyakin在 80 年代作了大量开创性的工作，在现场观测、实验室模拟和理论研究方面都取得了一些成就，但是都还没有达到机理清晰程度[10-13]。文献[14]提出利用分区破裂化现象界定深部岩体。文献[15]研究了深埋圆形洞室围岩分区破裂化现象的机理。文献[16]从试验的角度研究了圆形洞室围岩的分区破裂化现象。从现有文献看，研究深埋圆形洞室围岩的分区破裂化现象较多，而对深埋球形洞室围岩的分区破裂化现象研究较少。因此，有必要对球形洞室围岩分区破裂化的机理进行研究，以便利用分区破裂化的相关参数，设计经济合理的洞室尺寸和支护方式等。现阶段对深部岩体的分区破裂化现象主要从实验角度进行现象的描述，而对机理的研究较少，为此本文将重点研究深埋球形洞室围岩的分区破裂化现象的机理，并利用深部岩体强度准则研究深部岩体的破坏[17]。

1 开挖卸荷引起的围岩弹性应力场和弹性位移场

如图 1(a)所示，在深部岩体中开挖一个半径为 R 的球形洞室，假设岩体处于静水压力状态，静水压力为 q。在初始地应力场中，由洞室开挖卸荷引起围岩应力场的重分布，并产生相应的位移场。如果扰动后的二次应力场为弹性应力场，则它可以分解为两个问题求解：

1) 在开挖过程中，由开挖卸荷 $p(t)$ 引起的应力场和位移场(图(1b))。

2) 在开挖过程中，由原岩应力 q 引起的应力场和位移场(图(1c))。

两者之和为总的弹性区次生应力场和位移场。

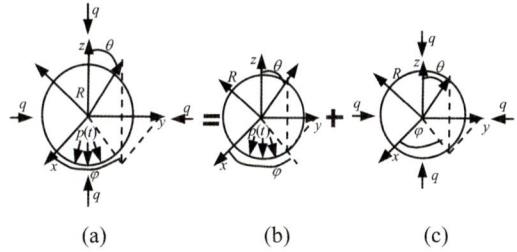

图 1 球形洞室开挖可以分解为两个问题

Fig.1 The original problem is divided into two sub-problems

1.1 开挖卸荷引起的次生应力场

如图 1(a)，洞室内壁从初始时刻 $t = 0$ 开始受到均匀法向压力 $p(t)$ 的作用，如图 1(a)所示采用球坐标系。如图 1(b)所示，对于开挖卸荷条件下的洞室围岩，非零的位移量为 $u_r = u(r,t)$，并且有 $\partial u_r / \partial \theta = \partial u_r / \partial \varphi = 0$，从而位移场的Lame分解简化为：

$$u = \partial \phi / \partial r \tag{1}$$

式中：ϕ 为标量势函数；$r、\theta、\varphi$ 为 3 个坐标轴方向参数。

用位移势表达的运动方程简化为：

$$\frac{\partial^2 \phi}{\partial r^2} + \frac{2}{r}\frac{\partial \phi}{\partial r} = \frac{1}{c_d^2}\frac{\partial^2 \phi}{\partial t^2} \tag{2}$$

式中：$c_d^2 = (\lambda + 2\mu)/\rho$，$\lambda$和$\mu$为Lame常数，$\rho$为岩石的密度。

开挖卸荷问题引起的次生应力场可以归结为：

$$\frac{\partial^2 \phi}{\partial r^2} + \frac{2}{r}\frac{\partial \phi}{\partial r} = \frac{\ddot{\phi}}{c_d^2}, \quad r > R, t > 0 \tag{3}$$

$$\phi(r,0) = \dot{\phi}(r,0) = 0, \quad r \geq R \tag{4}$$

$$\sigma_r(R,t) = p(t) \quad (5)$$

$$\lim_{r \to \infty} \phi(r,t) = 0, \quad t > 0 \quad (6)$$

对式(3)取关于 t 的Laplace变换，可得：

$$\frac{d^2\bar{\phi}}{dr^2} + \frac{2}{r}\frac{d\bar{\phi}}{dr} = k_d^2 \bar{\phi} \quad (7)$$

式中：$k_d = s/c_d$，s 为Laplace变换参数；$\bar{\phi}$ 是 ϕ 的变换。

根据弹性动力学知识，有：

$$\bar{\phi}(r,s) = \frac{A(s)\exp(-k_d r)}{r} \quad (8)$$

式中 $A(s)$ 由 $r = R$ 处的边界条件确定。

式(5)的Laplace变换为：

$$\bar{\sigma}_r(R,s) = \bar{p}(s) \quad (9)$$

根据弹性动力学知识，在 $r = R$ 处的应力为：

$$\bar{\sigma}_r(R,s) = \frac{\mu}{Rc_s^2}\left[s^2 + \frac{4c_s^2}{Rc_d}s + \frac{4c_s^2}{R^2}\right]A(s)e^{-Rk_d} \quad (10)$$

式中 $c_s^2 = \mu/\rho$。

由式(9)和式(10)可得：

$$A(s) = \frac{Rc_s^2 \bar{p}(s) e^{-Rk_d}}{\mu\left[s^2 + \frac{4c_s^2}{Rc_d}s + \frac{4c_s^2}{R^2}\right]} \quad (11)$$

开挖卸荷应力 $p(t)$ 随时间的变化函数可假设为(见图2)：

$$p(t) = f_1 t^2 + g_1 t + h_1 \quad (12)$$

式中：$f_1 = -q/t_0^2$；$g_1 = 0$；$h_1 = q$，t_0 为开挖完成时间，q 为洞室开挖前洞室内边缘岩体受的初始地应力值。

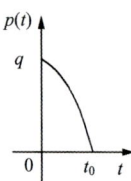

图2 开挖过程中巷道内边缘荷载与时间的函数

Fig.2 The relationship between the internal pressure and time during shield excavation

将式(12)进行Laplace变换，有：

$$\bar{p}(s) = \frac{2f_1}{s^3} + \frac{g_1}{s^2} + \frac{h_1}{s} \quad (13)$$

根据式(8)、式(10)和式(13)，有：

$$\varphi(r,\tau) = \frac{c_s^2 R}{r\mu} \frac{1}{2\pi i} \int_{B_r} \frac{\left(\frac{2f_1}{s^3} + \frac{g_1}{s^2} + \frac{h_1}{s}\right)e^{s\tau}}{s^2 + js + k} ds \quad (14)$$

式中：

$\tau = t - (r-R)/c_d$，$j = 4c_s^2/(Rc_d)$，$k = 4c_s^2/R^2$。

此处要求 $\tau > 0$，即 $r < R + tc_d$，要求所研究的离洞室内表面的最大距离为破坏应力波达到的距离。这样问题就可以归结为求 $(2f_1/s^3 + g_1/s^2 + h_1/s)/(s^2 + js + k)$ 的Laplace逆变换，进而可得：

$$\varphi(r,t) = \frac{Rc_s^2}{\mu r k^3}\left\{\frac{1}{\alpha}e^{\frac{j\delta}{2}}\left[\alpha(-2j^2 f_1 + jkg_1 + 2kf_1 - k^2 h_1)\cdot\right.\right.$$
$$\cos\left(\frac{\delta\alpha}{2}\right) + 2j^2 f_1 - jk[g_1 + 2f_1\delta] +$$
$$(-2j^3 f_1 + j^2 kg_1 - 2k^2 g_1 + 6jkf_1 - jk^2 h_1)\cdot$$
$$\left.\left.\sin\left(\frac{\delta\alpha}{2}\right)\right] + k^2[h_1 + g_1\delta] + kf_1(k\delta^2 - 2)\right\} \quad (15)$$

式中：$\alpha = \sqrt{4k - j^2}$；$\delta = (R - r + tc_d)/c_d$。

根据式(15)，可得洞室围岩径向应力为：

$$\sigma_{r1}(r,t) = \frac{Rc_s^2}{\mu c_d^2 r^3 k^3 \alpha}e^{-\frac{j\delta}{2}}\left\{2\alpha e^{\frac{j\delta}{2}}[R^2 k^2 f_1(\lambda + \mu) +\right.$$
$$2j^2 k_d^2 f_1(\lambda + \mu) - jkc_d^2(\lambda + \mu)(g_1 + 2f_1 t) +$$
$$Rkc_d(kg_1 + 2f_1 kt - 2jf_1)(\lambda + \mu) +$$
$$k(kf_1 r^2 \mu + c_d^2 ch_1 + c_d^2 g_1 kt + f_1 kt^2 c_d^2 - 2f_1 c_d^2)\cdot$$
$$(\lambda + \mu)] - \alpha[4j^2 c_d^2 f_1(\lambda + \mu) - 2jkc_d(\lambda + \mu)\cdot$$
$$(g_1 c_d + 2f_1 r) - 2kc_d^2(\lambda + \mu)(2f_1 - kh_1) +$$
$$2k^2 g_1 rc_d(\lambda + \mu) + k^2 r^2(\lambda + \mu)\cdot$$
$$(2f_1 - kh_1)]\cos\left(\frac{\delta\alpha}{2}\right) - \{4j^3 c_d^2 f_1(\lambda + \mu) -$$
$$2j^2 kc_d(c_d g_1 + 2f_1 r)(\lambda + \mu) - 2k^2[-2g_1 c_d^2\cdot$$
$$(\lambda + \mu) - 2c_d r(2f_1 - kh_1)(\lambda + \mu) +$$
$$kg_1 r^2(\lambda + 2\mu)] + jk[-2c_d^2(\lambda + \mu)\cdot$$
$$(6f_1 - kh_1) + 2kc_d g_1 r(\lambda + \mu) + kr^2(\lambda + 2\mu)\cdot$$
$$\left.(2f_1 + kh_1)]\}\sin\left(\frac{\delta\alpha}{2}\right)\right\} \quad (16)$$

根据式(15)，可得洞室围岩切向应力为：

$$\sigma_{\theta 1}(r,t) = \frac{Rc_s^2}{\mu c_d^2 r^3 k^3 \alpha}e^{-\frac{j\delta}{2}}\left\{-2\alpha e^{\frac{j\delta}{2}}[2j^2 c_d^2 f_1 \mu -\right.$$
$$kc_d(2Rjf_1 + 2f_1 c_d + jc_d g_1 + 2jc_d f_1 t) +$$
$$k^2 \mu(Rc_d g_1 + c_d^2 h_1 + c_d^2 g_1 t) +$$
$$f_1 \mu k^2(R + c_d t)^2 - f_1 k^2 r^2(\lambda + \mu)] +$$
$$\alpha[k^2 r^2 \lambda(kh_1 - 2f_1) + 2c_d(2j^2 c_d f_1 -$$

$$\begin{aligned}&jkc_dg_1-2jkf_1r)+k\mu(kc_dh_1-2c_df_1+\\&kg_1r)]\cos\left(\frac{\delta\alpha}{2}\right)+[-k^2r^2\lambda(2jf_1-2kg_1+\\&jkh_1)+2c_d\mu(2j^3c_df_1-j^2kc_dg_1-\\&2j^2kf_1r)+2c_d\mu jk(kc_dh_1-6c_df_1+kg_1r)+\\&4k^2c_d\mu(c_dg_1+2f_1r-kh_1r)]\sin\left(\frac{\delta\alpha}{2}\right)\}\end{aligned} \quad (17)$$

1.2 开挖卸荷引起的弹性位移场

根据式(1)和式(15)可得位移场为：

$$\begin{aligned}u_1(r,t)=&\frac{Rc_s^2}{\mu c_d^2 k^3 r^2 \alpha}\mathrm{e}^{-\frac{j\delta}{2}}\{-\alpha\mathrm{e}^{\frac{j\delta}{2}}[R^2k^2f_1+\\&2j^2c_d^2f_1-jkc_d^2(g_1+2f_1t)+Rkc_d(-2jf_1+\\&kg_1+2f_1kt)-k^2f_1r^2+kc_d^2(kh_1-2f_1+\\&kg_1t+kf_1t^2)]+\alpha c_d[2j^2c_df_1-jk(c_dg_1+\\&2f_1r)+k(-2c_df_1+kc_dh_1+kg_1r)]\cos\left(\frac{\delta\alpha}{2}\right)+\\&c_d[2j^3c_df_1-j^2kc_dg_1-2j^2kf_1r+\\&jk(-6c_df_1+kc_dh_1+kg_1r)+2k^2(c_dg_1+\\&2f_1r-kh_1r)]\sin\left(\frac{\delta\alpha}{2}\right)\}\end{aligned}\quad (18)$$

2 原岩应力引起的围岩弹性应力场和弹性位移场

2.1 弹性应力场

如图1(c)所示，边界条件为：

$$\begin{cases}r=R,\quad \sigma_r=0\\r\to\infty,\quad \sigma_r=q\end{cases} \quad (19)$$

根据弹性力学知识，可得在开挖过程中由原岩应力q引起的次生应力场为：

$$\begin{cases}\sigma_{r2}=q\left[1-\left(\dfrac{R}{r}\right)^3\right]\\\sigma_{\theta 2}=\sigma_{\varphi 2}=q\left[1+\dfrac{1}{2}\left(\dfrac{R}{r}\right)^3\right]\end{cases} \quad (20)$$

2.2 弹性位移场

根据弹性力学知识，可得在开挖过程中由q和支护力p_i引起的洞室周边$r=R$处弹性位移为：

$$u_{r2}=u_R-u_0=\frac{1+v_0}{2E_0}R(q-p_i) \quad (21)$$

总的次生应力场和位移场分别为：

$$\sigma_r^e=\sigma_{r1}+\sigma_{r2},\quad \sigma_\theta^e=\sigma_{\theta 1}+\sigma_{\theta 2} \quad (22)$$

$$u^e=u_{r1}(r,t)+u_{r2} \quad (23)$$

3 分区破裂化的形成

3.1 分区破裂化的形成机理

在高地应力状态下，深埋球形洞室围岩的应力场在开挖卸荷条件下会出现应力集中现象，洞室周围岩体将会破裂。岩体破裂产生能量释放，并以应力波的形式向远离巷道表面的方向传递，在洞室围岩中形成新的弹性应力场。深部岩体破裂区的外边界相当于新的开挖边界，当重分布应力场满足深部岩体强度准则时，应力再一次释放，产生第二破裂区；依此类推，直到应力释放后不能再产生破裂区为止。从应力分析和数值计算可以看出，每次应力释放产生的破裂区一般远离新开挖边界一定距离，这个区域形成非破裂区。随着破裂区的位置远离洞室内边界，应力场逐渐减小，围岩的破坏也就逐渐停止，岩体逐步趋于稳定，从而出现了破裂区和非破裂区多次交替出现的现象。

一般来说，对于浅部岩体，由于地应力低，洞室围岩在第一次破坏后，产生的破坏应力波与原岩应力场的叠加应力场无法使岩体再次发生破坏，因此，无法形成第二破裂区，故浅部岩体中不会出现分区破裂化现象，而在浅部洞室围岩中出现松动区、塑性区和弹性区。对于深部岩体，洞室围岩受到高地应力作用，也即原岩应力较大。当洞室受到开挖扰动后，开挖卸荷应力场与原岩应力场共同作用导致岩体破坏，岩体破坏产生的应力波完全可能继续使岩体产生新的破坏。因此在高地应力条件下，洞室开挖卸荷导致岩体的分区破坏是深埋洞室围岩分区破裂化现象产生的根本原因。

3.2 破裂区和非破裂区的半径

深部岩体由于构造运动和围岩自重的作用一般处于高地应力状态下，其强度特性明显区别于浅部岩体。采用现有的岩体强度理论研究深部岩体的破坏特性，往往与实际情况不符。因此，2008年周小平和钱七虎[17]等提出了一个适用于深部岩体的强度准则。该准则考虑了岩体拉压不同强度特性和中间主应力的影响，表达式简单，参数具有明确的物理意义，并与RMR岩体地质力学分类指标建立了联系，能反应深部岩体特有的力学特性和破坏模式。本文采用深部岩体强度准则来研究深部岩体的破坏模式。深部岩体强度准则表达式为[17]：

$$\begin{cases} F_1 = \sigma_1 - \dfrac{1}{1+a}(a\sigma_2 + \sigma_3) + W(\sigma_1 + \sigma_3 + b\sigma_2) \\ \qquad \dfrac{I_1}{\sigma_c} - \sigma_c \left[\dfrac{b\sigma_1}{\sigma_c} + \dfrac{d}{(1+a)\sigma_c}(a\sigma_2 + \sigma_3) + b^2 \right]^n = 0 \\ \qquad (\text{当}\, F_1 \geqslant F_2\,\text{时}) \\ F_2 = \dfrac{1}{1+a}(\sigma_1 + a\sigma_2) - \sigma_3 + W(\sigma_1 + \sigma_3 + b\sigma_2) \\ \qquad \dfrac{I_1}{\sigma_c} - \sigma_c \left[\dfrac{b\sigma_1 + a\sigma_2}{\sigma_c(1+a)} + \dfrac{d\sigma_3}{\sigma_c} + b^2 \right]^n = 0 \\ \qquad (\text{当}\, F_1 < F_2\,\text{时}) \end{cases}$$

(24)

式中：a 为中间主应力影响系数；σ_c 为完整岩石的单轴抗压强度；b 是岩石完整性参数；d 与岩性和裂隙分布有关的参数；n 和深部岩体的结构面分布有关，一般取 $n = 1/2$；W 为剪缩参数，可以通过室内三轴实验确定，b、d 的取值详见文献[17]。

一般来说，在球形洞室中，主应力与围岩应力的关系为：

$$\begin{cases} \sigma_1 = \sigma_2 = \sigma_\theta^e \\ \sigma_3 = \sigma_r^e \end{cases} \text{或} \begin{cases} \sigma_1 = \sigma_r^e \\ \sigma_2 = \sigma_3 = \sigma_\theta^e \end{cases} \tag{25}$$

对于式(25)，应根据应力的大小判断选取第一项或第二项。

将式(25)代入式(24)，结合式(23)解方程可得破裂区的宽度 $R_1 - R_0$，其中 R_0 为破裂区的内边界，R_1 为破裂区的外边界。同时可以得到产生破裂区的时间 t_0。

当岩体发生破坏，产生破裂区后，岩体中应力场重新调整，破裂区的外边界成为新的弹性区的内边界，在该边界上的应力 $p(t)$ 函数可假定为(图3)：

$$p(t) = f_2 t^2 + g_2 t + h_2 \tag{26}$$

式中：$f_2 = \left[\sigma_0\left(1 - 2m - 2\sqrt{m^2 - m}\right)\right]/t_1^2$；$g_2 = \left[2\sigma_0\left(m - 1 + \sqrt{m^2 - m}\right)\right]/t_1$；$h_2 = \sigma_0$；$m$ 为非破裂区内边界最大应力与初始应力 σ_0 比值。其中初始应力 σ_0 为新的弹性内边界形成初始时刻由原岩应力引起的应力场。其边界条件为：$t = 0$，$p(t) = \sigma_0$；$t = t_1$，$p(t) = \sigma_R$；$t = -g_2/(2f_2)$，$p(t) = m\sigma_0$。

在产生第一个破裂区后，围岩内的弹性应力场同样分解为破裂区外边界应力 $p(t)$ 产生的应力场和原岩应力场。$p(t)$ 产生的应力场同样按照式(16)和式(17)计算，只是将其中的 f_1、g_1、h_1 变为 f_2、g_2、h_2 即可。按照式(22)、式(24)和式(25)可以计算新的破裂区的位置和半径，以此类推，从而可以计算每个破裂区的位置和半径。而每个破裂区之间为非破裂区，其半径和位置也可以确定。

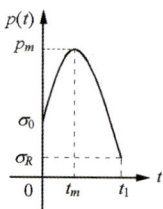

图3　新的破裂区区内边界荷载与时间的函数

Fig.3　The relationship between the internal pressure and time when the first fractured zone appears

3.3 破裂区应力场

以前研究岩体变形破坏特性，利用弹塑性理论进行分析。弹塑性理论仅仅适用于浅部岩体峰值荷载前的非线性强化阶段，对应变软化阶段并不适用。深部岩体由于承受的地应力大，岩体的变形破坏将进入软化阶段，深部岩体的分区破裂化现象和岩体的软化阶段密切相关，研究深部岩体的分区破裂化现象必须研究深部岩体软化阶段的力学特性。当岩体的应力状态满足深部岩体强度准则时，岩体发生破坏并产生变形局部化，同时其应力下降到残余应力 σ_R 水平。在变形局部化区域，位移将不连续，因此不能采用弹塑性的厚壁球理论求解[18]，必须利用变形局部化理论。利用变形局部化理论，可以确定岩体发生变形局部化的位置和时间，因而可以确定破裂区的位置和时间。

岩体变形局部化区域的本构关系为[19]：

$$\sigma_{ij} = D_{ijkl}\varepsilon_{kl} \tag{27}$$

式中：ε_{kl} 为变形局部化区域的应变；$\boldsymbol{D}$ 为变形局部化时岩石的刚度矩阵。

变形局部化情况下岩体的刚度矩阵为：

$$D_{ijkl} = D_{ijkl}^e - \frac{1}{A}D_{ijrs}^e Q_{rs} P_{mn} D_{mnkl}^e \tag{28}$$

式中：$\boldsymbol{D}_{ijkl}^e$ 为弹性刚度矩阵；G 为剪切模量。

$$P_{ij} = \frac{\partial \phi_\beta}{\partial \sigma_{ij}}, \quad Q_{ij} = \frac{\partial \phi_\varphi}{\partial \sigma_{ij}},$$

$$\phi_\beta = \frac{1}{2}(\sigma_1 - \sigma_3) + \frac{1}{2}(\sigma_1 + \sigma_3)\sin\beta_d,$$

$$\phi_\varphi = \frac{1}{2}(\sigma_1 - \sigma_3) + \frac{1}{2}(\sigma_1 + \sigma_3)\sin\varphi_d.$$

其中：ϕ_β 为屈服函数；ϕ_φ 为塑性势函数；φ_d 为动荷载下的剪胀摩擦角；β_d 为动荷载下岩石的内摩擦角。

式(28)中的参数 A 表达式为[19]:
$$A = H + P_{ij}D^e_{ijkl}Q_{kl} \quad (29)$$

其中 H 为塑性软化(硬化)模量,其值为:

$$\frac{H}{2G} = \left\{2(P_1Q_1 - P_3Q_3) - \frac{1}{1-\nu_0}[(P_1-P_3)Q_3 + (Q_1-Q_3)P_3] + \frac{\nu_0}{1-\nu_0}[trQ(P_1-P_3) + trP(Q_1-Q_3)]\right\} - \frac{n_1^4}{1-\nu_0}(P_1-P_3)(Q_1-Q_3) - P_1Q_1 - P_2Q_2 - \frac{\nu_0}{1-\nu_0}(Q_1+Q_2)(P_1+P_2)$$

变形局部化区域的应变张量为:
$$\boldsymbol{\varepsilon} = 0.5(\boldsymbol{g}\otimes\boldsymbol{n} + \boldsymbol{n}\otimes\boldsymbol{g}) \quad (30)$$

其中:
$$g_1 = n_1\left\{2P_1 - \frac{1}{1-\nu_0}[n_1^2(P_1-P_3) + p_3] + \frac{\nu_0}{1-\nu_0}trP\right\},$$
$$g_3 = n_3\left\{2P_3 - \frac{1}{1-\nu_0}[n_1^2(P_1-P_3) + p_3] + \frac{\nu_0}{1-\nu_0}trP\right\},$$
$$n_1^2 = (1-\nu_0)\frac{P_1Q_1 - P_2Q_1}{(P_1-P_3)(Q_1-Q_3)} - \frac{Q_1}{2(Q_1-Q_3)} - \frac{P_1}{2(P_1-P_3)} + \frac{\nu_0(Q_1+Q_2+Q_3)}{2(Q_1-Q_3)} + \frac{\nu_0(P_1+P_2+P_3)}{2(P_1-P_3)},$$
$$n_3^2 = 1 - n_1^2, \quad n_2 = 0, \quad g_2 = 0。$$

此处取式(25)的第二项,由此可以解出 σ_R,从而得到破裂区的应力场:

$$\sigma_R = E_0[A_{18} + A_{19} + A_{20} + A_{21} + \sin\beta_d(A_{22} + A_{23} + A_{24}) + \sin^2\beta_d(A_{25} + A_{26})]/8A_{27}(-1+\nu_0^2) \quad (31)$$

式中,
$$A_{18} = \sin^3\beta_d(7 + 11\nu_0 - 100\nu_0^2 + 72\nu_0^3 + 96\nu_0^4 - 80\nu_0^5) + \sin^3\varphi_d(-1+\nu_0+10\nu_0^2-16\nu_0^3),$$
$$A_{19} = -4\nu_0(-1+\nu_0+2\nu_0^2)(-4+4\nu_0+2\nu_0^2 - \nu_0 A_{30} + A_{30}^2),$$
$$A_{20} = (-1+2\nu_0)\sin^2\varphi_d(-2\nu_0 - 18\nu_0^2 + 32\nu_0^3 + A_{30} - 9\nu_0^2 A_{30} + 8\nu_0^3 A_{30}),$$
$$A_{21} = -4\sin\varphi_d(2 - 4\nu_0 - 3\nu_0^2 + 13\nu_0^3 - 22\nu_0^4 + 16\nu_0^5 - 2\nu_0 A_{30} - 2\nu_0^2 A_{30} + 14\nu_0^3 A_{30} - 20\nu_0^4 A_{30} + 8\nu_0^5 A_{30}),$$
$$A_{22} = -8 - 32\nu_0 - 4\nu_0^2 + 14\nu_0^3 + 16\nu_0^4 - 112\nu_0^5 + \sin\varphi_d(28\nu_0 - 52\nu_0^2 + 48\nu_0^3 - 112\nu_0^4 + 128\nu_0^5),$$
$$A_{23} = \sin^2\varphi_d(-3 + 5\nu_0 - 4\nu_0^2 + 52\nu_0^3 - 64\nu_0^4) + A_{30}(-8 + 48\nu_0 - 80\nu_0^2 + 88\nu_0^4 - 48\nu_0^5),$$
$$A_{24} = A_{30}\sin^2\varphi_d(-10 + 20\nu_0 - 14\nu_0^2 + 20\nu_0^3 - 48\nu_0^4 + 32\nu_0^5),$$
$$A_{25} = \sin\varphi_d(5 - 33\nu_0 + 62\nu_0^2 + 20\nu_0^3 + 24\nu_0^4 - 64\nu_0^5),$$
$$A_{26} = (1+\nu_0)[(10\nu_0 + 148\nu_0^2 - 328\nu_0^3 + 176\nu_0^4) + A_{30}(7 - 45\nu_0 + 106\nu_0^2 - 108\nu_0^3 + 40\nu_0^4)],$$
$$A_{27} = 5 - 5\nu_0 + 2\nu_0^2 + \sin^2\beta_d(1-2\nu_0^2)(-7 + 3\nu_0 + 10\nu_0^2) + 2\sin\beta_d[4(1 - 5\nu_0 + 5\nu_0^2 + 5\nu_0^3 - 64\nu_0^4) + A_{28}],$$
$$A_{28} = -8\nu_0^3 + 16\nu_0^4\sin\varphi_d - (1+2\nu_0)[-4(-2+3\nu_0^2+\nu_0^3) + 16\nu_0^2\sin\varphi_d(-1+\nu_0)] + \sin^2\varphi_d(1+\nu_0-8\nu_0^2),$$
$$A_{29} = -4\nu_0\sin\varphi_d(-3+4\nu_0) + \sin^2\varphi_d(-3+4\nu_0) + 2\sin\beta_d(5 - 14\nu_0 + 8\nu_0^2)(-2\nu_0 + \sin\varphi_d),$$
$$A_{30} = [16 - 32\nu_0 + 4\nu_0^2 + 16\nu_0^3 + \sin^2\beta_d(16 + 32\nu_0 - 44\nu_0^2 + 16\nu_0^3 + A_{29})]^{0.5}/(1-\nu_0)。$$

根据应力的大小,若取式(25)的第一项,利用相同的方法可以解得 σ_r 从峰值应力下降到残余应力 σ_R 的时间 t_1。由于篇幅限制,这里不再详述,可参见文献[15]。

4 算例分析

设球形洞室的半径为 $R = 4\text{m}$, $q = 100\text{MPa}$,岩石的泊松比为 $\nu_0 = 0.15$,弹性模量 $E_0 = 18.9\text{GPa}$,岩石的密度 $\rho = 2720\text{kg/m}^3$,深部岩体强度准则参数为:$RMR = 55$,$m = 5$。破裂区宽度、数量随单轴抗压强度变化的情况见表1。

表1 开挖持续时间为100s时破裂区宽度和数量
Table 1 The width and quantity of fractured zone when excavation time is 100s

单轴抗压强度/MPa	半径/m	破裂区圈数			
		1	2	3	4
50	R_0	4.000	5.088	6.151	7.619
	R_1	4.805	5.843	6.900	7.619
	R_1-R_0	0.805	0.755	0.749	0.000
60	R_0	4.000	4.958	5.943	7.198
	R_1	4.605	5.542	6.490	7.198
	R_1-R_0	0.605	0.584	0.547	0.000
70	R_0	4.000	4.873	5.847	6.641
	R_1	4.417	5.219	5.923	6.641
	R_1-R_0	0.417	0.346	0.076	0.000
80	R_0	4.000	4.841		
	R_1	4.228	4.841		
	R_1-R_0	0.228	0.000		
90	无				

从表 1 可以看出，岩体单轴抗压强度对深埋球形洞室围岩分区破裂化数量和尺寸有较大的影响。岩体单轴抗压强度将直接影响岩体的破裂状态，在同样的情况下，岩体单轴抗压强度越小，破裂区的数量越多。岩体单轴抗压强度越大，破裂区数量越少。当岩体的单轴抗压强度接近或达到地应力大小时，岩体不会产生分区破裂化现象。上述结果和俄罗斯学者 Shemyakin 的结果[10-12]吻合。

5 结论

(1) 深埋洞室开挖卸荷使围岩应力集中，造成岩体破坏。岩体破坏产生的应力波向远离洞室表面方向传递，并在围岩中产生新的应力场，与原岩应力场叠加后可能使岩体重新发生破裂，形成新的破裂区和非破裂区。当岩体破坏产生的应力波减小到无法使岩体破坏时，深埋洞室围岩分区破裂化停止。

(2) 对于一定的开挖时间和开挖方式，深埋洞室围岩会产生固定数量的破裂区和非破裂区，而破裂区的宽度呈等差数列由洞口向外递减。

(3) 岩体单轴抗压强度也会影响破裂区的数量和宽度。当岩体单轴抗压强度小于地应力时，深部岩体可能会发生分区破裂化现象，反之则不会产生。在同等条件下，岩体单轴抗压强度越低，则破裂区数量越多，破裂区宽度越大。

(4) 当深部岩体满足深部岩体强度准则时，岩体将破坏并产生变形局部化现象，从而出现位移不连续，此时可以用断裂力学相关知识确定破裂区的应力场和破裂区形成时间。

参考文献：

[1] 钱七虎. 非线性岩石力学的新进展——深部岩体力学的若干问题[C]// 第八次全国岩石力学与工程学术大会论文集. 北京：科学出版社，2004：10－17.
Qian Qihu. The current development of nonlinear rock mechanics: The mechanics problems of deep rock mass [C]// Proceedings of the 8th Rock Mechanics and Engineering Conference. Beijing: Science Press, 2004: 10－17. (in Chinese)

[2] 钱七虎. 深部地下工空间开发中的关键科学问题[C]// 深部地下空间开发中的基础研究关键技术问题. 北京：科学出版社，2004.
Qian Qihu. The key problems of deep underground space development [C]// Key Technical Problems of Base Research in Deep Underground Space Development. Beijing: Science Press, 2004. (in Chinese)

[3] 古德生. 金属矿床深部开采中的科学问题[C]// 科学前沿与未来(第六集). 北京：中国环境科学出版社，2002：192－201.
Gu Desheng. The science problems in deep mining of metal deposit [C]// Science Foreland and Future (Volume VI). Beijing: China Environment Science Press, 2002: 192－201. (in Chinese)

[4] 谢和平. 深部高应力下的资源开采——现状、基础科学问题与展望[C]// 科学前沿与未来(第六集). 北京：中国环境科学出版社，2002：179－191.
Xie Heping. Resources development under high ground stress; present state, base science problems and perspective [C]//Science Foreland and Future (Volume VI). Beijing: China Environment Science Press, 2002: 179－191. (in Chinese)

[5] 何满潮，谢和平，彭苏萍，姜耀东. 深部开采岩体力学研究[J]. 岩石力学与工程学报，2005，24(16)：2803－2813.
He Manchao, Xie Heping, Peng Suping, Jiang Yaodong. Study on rock mechanics in deep mining engineering [J]. Chinese Journal of Rock Mechanics and Engineering, 2005, 24(16): 2803－2813. (in Chinese)

[6] Adams G R, Jager A J. Petroscopic observations of rock fracturing ahead of stop faces in deep-level gold mines [J]. Journal of the South African Institute of Mining and Metallurgy, 1980, 80(6): 204－209.

[7] Fairhurst C. Deformation, yield, rupture and stability of excavations at great depth [C]// Fairhurst C. Rockburst and Seismacity in Mines. Rotterdam: A.A. Balkema, 1990: 1103－1114.

[8] Sellers E J, Klerck P. Modeling of the effect of discontinuities on the extent of the fracture zone surrounding deep tunnels [J]. Tunneling and Underground Space Technology, 2000, 15(4): 463－469.

[9] Shemyakin I, Kyrlenya M V, Reva V N. Effect of zonal disintegration of rocks around underground workings [J]. Doklady Akademii Nauk USSR, 1986, 289(5): 1088－1094.

[10] Shemyakin E I, Fisenko G L, Kurlenya M V. Zone disintegration of rocks around underground workings. I. Data of in-site observation [J]. Journal of Mining Science, 1986, 22(3): 157－168.

[11] Shemyakin E I, Fisenko G L, Kurlenya M V. Zone disintegration of rocks around underground workings. II. Disintegration of rocks on models of equivalent materials [J]. Journal of Mining Science, 1986, 22(4): 223－232.

[12] Shemyakin E I, Fisenko G L, Kurlenya M V. Zone disintegration of rocks around underground workings. III. Thoeretical notions [J]. Journal of Mining Science, 1987, 23(1): 1－6.

[13] Shemyakin E I, Fisenko G L, Kurlenya M V. Zone disintegration of rocks around underground workings. IV. Practical applications [J]. Journal of Mining Science, 1988, 24(3): 238−241.

[14] 钱七虎. 深部岩体工程响应的特征科学现象及"深部"的界定[J]. 东华理工学院学报, 2004, 27(1): 1−5.
Qian Qihu. The characteristic scientific phenomena of engineering response to deep rock mass and the implication of deepness [J]. Journal of East China Institute of Technology, 2004, 27(1): 1−5. (in Chinese)

[15] Zhou X P, Wang F H, Qian Q H. Zonal fracturing mechanism in deep crack-weakened rock masses [J]. Theoretical and Applied Fracture Mechanics, 2008, 50(1): 57−65.

[16] 顾金才, 顾雷雨, 陈安敏. 深部开挖洞室围岩分层断裂破坏机制模型试验研究[J]. 岩石力学与工程学报, 2008, 27(3): 433−438.
Gu Jincai, Gu Leiyu, Chen Anmin. Model test study on mechanism of layered fracture within surrounding rock of tunnels in deep stratum [J]. Chinese Journal of Rock Mechanics and Engineering, 2008, 27(3): 433−438. (in Chinese)

[17] 周小平, 钱七虎, 杨海清. 深部岩体强度准则[J]. 岩石力学与工程学报, 2008, 27(1): 117−123.
Zhou Xiaoping, Qian Qihu, Yang Haiqing. Strength criteria of deep rock mass [J]. Chinese Journal of Rock Mechanics and Engineering, 2008, 27(1): 117−123. (in Chinese)

[18] Nikolaevshij V N. Mechanical of porous and fractured media [M]. Moscow: Scientific Press, 1990.

[19] Davide Bigont, Tomasz Hueckel. Uniqueness and localization associative and non-associative elasto-plasticity [J]. International Journal Solids and Structures, 1991, 28(2): 197−213.

岩石在过临界荷载作用下变形异常现象的模拟

戚承志[1]　钱七虎[2]　王明洋[2]　吴 徽[1]

(1. 北京建筑工程学院土木交通学院，北京，100044；2.解放军理工大学，南京，210007)

摘 要：对于受损伤圆柱状岩石试件的试验表明，当荷载超过一定临界值时，会出现岩石试件变形增量符号改变的现象，且变形增量分布依赖于环向坐标及轴向坐标。试验表明，这种临界后变形的异常现象发生在破坏之前，在外载超过岩体的体积扩容极限之后，也即在强烈受压状态才会发生。本文利用连续相变理论研究了这种变形异常现象，得到了描述临界后岩石试验变形的支配方程及其解析解。计算结果表明，求得的解能够很好地描述变形增量依赖于试件的环向坐标及轴向坐标的现象。

关键词：强烈受压，变形增量变符号现象，连续相变理论

MODELING OF THE ANOMALOUS DEFORMATION OF ROCK UNDER POST-CRITICAL LOADING

Qi Chengzhi,[1] Qian Qihu,[2] Wang Mingyang,[2] ,Wu Hui[1]

(1. School of Civil and Communication Engineering, Beijing Institute of Civil Engineering and Architecture,

Beijing, P.R.China,100044; 2. PLA University of Science and Technology, 210007, Nanijing, P.R.China)

Abstract: Experiments on preliminarily damaged cylindrical rock samples show that deformation increment sign change effect will occur when the loads exceed critical values, and the deformation increment depends on axial and circumferential coordinates. Experiments also show that this kind of deformation anomaly takes places in pre-failure loading stage when the load exceeds critical value for dilatancy, i.e. takes place in highly stressed state. In the present paper continuous phase transition theory is used to study this kind of deformation anomaly; the governing equation and its solution are obtained. Computation results show that the obtained solution describes the deformation anomaly alone axial and circumferential coordinates very well.

Key words: highly stressed states, deformation increment sign change effect, continuous phase transition theory

1 引言

在上个世纪七八十年代俄罗斯学者在深地下矿山巷道周围观测发现[1-4]，环绕巷道出现了重复巷道形状的破坏区与弱破坏区交互出现的分区破裂现象。这一现象与利用弹塑性理论得到的解不同，因而引起了人们的兴趣。出现分区破裂的深部巷道围岩处于强度极限之后，因此引起学者们对于岩石临界后变形破坏特性的研究。Makarov V.V.在岩石试件的加载试验研究中发现，在强烈受压岩石中出现了变形增量异常现象[5]。Makarov V.V.等对于从坚硬的火成岩到软弱的沉积岩的多种岩石进行了压缩试验。他发现，对于预先受损的岩石试样，当轴向加载应力σ_z超过一定的临界值σ^*时，随着应力的增加，纵向及横向变形增量符号出现了逆转。典型的变形曲线如图 1（b）所示。

基金项目：国家 973 重大项目（2010CB732003），北京市自然科学基金项目（KZ200810016007）及国家自然基金项目（No. 50825403）。

本文原载于《第 2 届全国工程安全与防护学术会议论文集》（2010 年）。

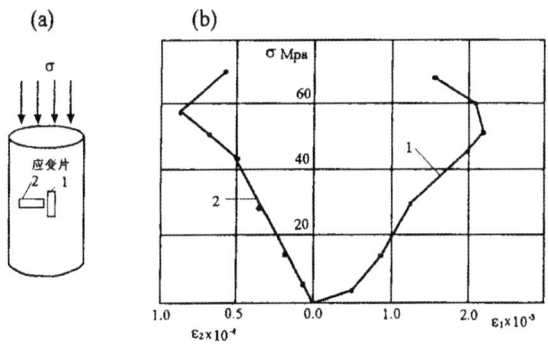

图1 试件受压(a)及变形增量符号改变效应(b)

Fig.1 Sample under compression (a) and the deformation sign change effect (b)

在对泥板岩试件的压缩实验中，在加载至 0.6-$0.8\sigma_c$ 时出现变形异常现象，其中 σ_c 为单轴抗压强度。在循环加载实验中也出现变形异常现象。

对于完整的高强度花岗岩试件，加载至 $0.8\sigma_c$ 时没有出现变形异常现象。只有在预先受损的试件中在加载至 0.6-$0.8\sigma_c$ 时出现变形异常现象。得到的结果间接证明裂纹的形成过程为岩石出现变形异常现象的原因。

对于体积变形的分析表明[5]，在纵向变形增量改变符号之前的时刻，试件的体积开始增加。在横向变形异常现象出现之后，试件加载增加的状态被破坏。在试件的相反面也同时观测到了变形的异常现象。

Makarov V.V. 研究了若干变形异常的假设：残余应力假设、岩石的变形性质非均匀性假设等。但是上述假定都不能令人满意地解释变形增量符号改变现象。因此可以断定，变形的异常现象发生在破坏之前，在外载超过岩体的体积扩容限之后，也即在强烈受压状态才会发生。因此可以断定，纵向及横向变形增量的变符号现象与试件中裂纹开始形成密切相关。

对于试件应变的测量表明[5]，存在着两种变形异常：即上面讨论的"负的变形异常"，以及"正的变形异常"。"正的变形异常"的实质在于：伴随着在试件某一段变形的负增长，在相邻区段出现变形的显著正增长，即比在通常情况下的平均变形高出很多，一般为通常情况的1.5-3倍。

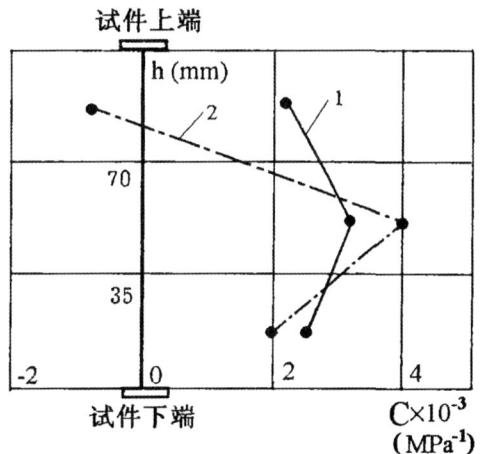

图2 沿着花刚斑岩试件高度坐标三点测量的变形增量的变符号分布(曲线1对应着应力范围为136-144 Mpa, 曲线2对应的应力范围为144 -150Mpa)（其中横坐标为变形增量除以相应的应力增量，即为柔度）

Fig.2 The distribution of the measured deformation increment at 3 points alone sample height (curve 1 corresponds to the stress change range 136- 144 MPa, and curve 2 corresponds to the stress change range 144 -150MPa)（the abscissa is the strain increment divided by corresponding stress increment, i.e. the flexibility）

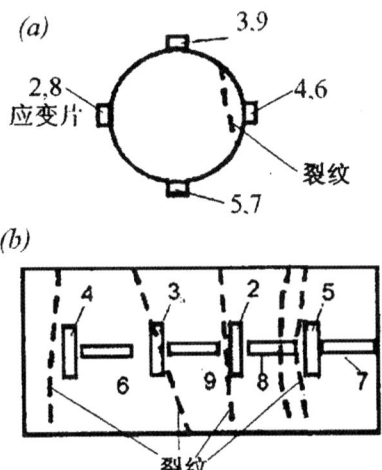

图3 在单轴压缩情况下试件中部周边变形增量的周期性变化(图(a)为应变片布置图，图(b)为裂纹沿着周边的分布)。

Fig.3 The periodical distribution of deformation increment alone periphery of the middle part of the sample under uniaxial compression (panel (a) - the arrangement of strain gauges; panel (b) - the distribution of cracks alone periphery of the sample)

图 2 (b)给出了沿着预先受损的花刚岩试件高度上变形增量的变符号分布及其重分布。试验表明，在出现变形异常的区段，沿着试件的周边也出现了周期

性的变形异常现象及变形正常区段交替出现的现象。图3给出了沿着试件周边变形增量的周期性分布。图4给出了体积变形增量沿着岩石试件周边变化的情况。

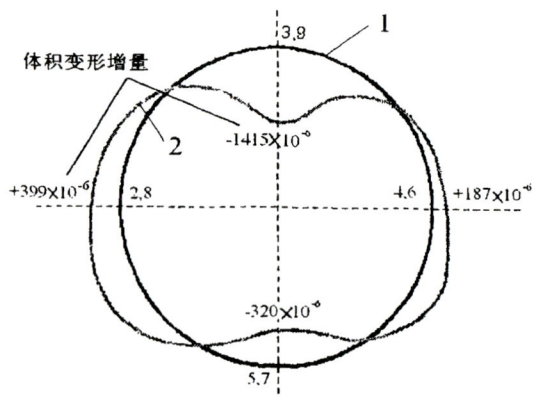

图4 体积变形增量(曲线2)沿着岩石试件周边(圆1)的变化
Fig.4 Volumetric deformation increment (curve 2) alone periphery of the sample (circle 1)

在文[6]中建议了一种力学模型来模拟这种变形异常现象。这种模型假定极限荷载后的变形增量与应力增量成正比。这种模型能够定性地描述试样的力学行为，而在量的方面误差较大。试验表明[7]，岩石的极限后变形破坏过程为连续相变过程。这种相变过程伴随着能量的耗散和自我组织现象的发生。在描述自我组织方面，连续相变理论是一个强有力的工具。这种途径的优点在于，不必追究材料的微细观结构的具体变化情况，而从宏观上把握材料变形破坏的基本规律，因而这种途径的适用范围较微细观理论的适用范围更广，是对于非线性宏观连续介质力学的有力补充。在文[8]中作者利用连续相变理论成功地模拟了深部隧道围岩分区破裂化现象的时空结构。在本文中作者试图利用连续相变理论模拟岩石极限后的变形异常现象。从本文后面得到的支配方程的解析解的计算结果可以看出，这种途径能够非常精确地模拟极限后变形的异常现象。

2 过临界荷载作用下变形异常现象的支配方程

因为在做圆柱形试件的试验中，试件的端面被抛光，并加以砂，这样试件的两端可以认为是没有摩擦的。在圆柱坐标里面，如果用 r、θ、z 分别代表径向、环向及轴向坐标，那么在试件的两端 $z=0$、h 处，除了正应力外，其余应力分量均为零，而侧面 $r=a$（a 为试件半径）为自由边界，即边界条件为：

$$\sigma_z|_{z=0,h} = -\sigma, \quad \tau_{zr}|_{z=0,h} = 0, \quad \tau_{z\theta}|_{z=0,h} = 0;$$

$$\sigma_r|_{r=a} = 0, \quad \tau_{r\theta}|_{r=a} = 0, \quad \tau_{rz}|_{r=a} = 0;$$

(1)

柱坐标中介质的平衡方程为：

$$\left.\begin{array}{l}\dfrac{\partial \sigma_r}{\partial r} + \dfrac{\partial \tau_{r\theta}}{r\partial \theta} + \dfrac{\partial \tau_{rz}}{\partial z} + \dfrac{\sigma_r - \sigma_\theta}{r} = 0 \\ \dfrac{\partial \tau_{\theta r}}{\partial r} + \dfrac{\partial \sigma_\theta}{r\partial \theta} + \dfrac{\partial \tau_{\theta z}}{\partial z} + \dfrac{2\tau_{r\theta}}{r} = 0 \\ \dfrac{\partial \tau_{zr}}{\partial z} + \dfrac{\partial \tau_{z\theta}}{r\partial \theta} + \dfrac{\partial \sigma_z}{\partial z} + \dfrac{\tau_{rz}}{r} = 0\end{array}\right\}$$

(2)

满足上述平衡方程及边界条件的应力分量表达式为：

$$\sigma_z = -\sigma, \sigma_r = \sigma_\theta = \tau_{r\theta} = \tau_{rz} = \tau_{z\theta} = 0 \quad (3)$$

当 σ_z 小于临界值 σ^* 时，岩石试件材料服从胡克定律：

$$\sigma_{ij} = \dfrac{E}{1+\nu}\left(\dfrac{\nu}{1-2\nu}\vartheta\delta_{ij} + \varepsilon_{ij}\right) \quad (4)$$

其中 E 为杨氏模量；ν 为泊淞比；$\vartheta = \varepsilon_{ii}$ 为体积应变。

那么满足条件（4）的应变分量为：

$$\varepsilon_z = -\dfrac{\sigma}{E}, \varepsilon_\theta = \nu\dfrac{\sigma}{E} \quad \varepsilon_r = \nu\dfrac{\sigma}{E} \quad (5)$$

这样2个最大剪切应力为：

$$T = \dfrac{\sigma_\theta - \sigma_z}{2} = \dfrac{\sigma}{2} = T_{23} = \dfrac{\sigma_r - \sigma_z}{2} \quad (6)$$

2个最大剪切应变为：

$$\Gamma = \varepsilon_\theta - \varepsilon_z = (1+\nu)\dfrac{\sigma}{E} = \Gamma_{23} = \varepsilon_r - \varepsilon_z \quad (7)$$

由方程（6）、（7）可以看出，两个最大剪应力及两个最大剪应变大小都相等。随着加载值的增加，这两个最大剪应力同时达到临界值。

岩石极限后变形增量异常现象出现在强烈受压

状态，是与岩体的不可逆变形紧密相关的。在不可逆变形的产生过程中伴随着缺陷的大量产生和岩体构造的变化，即伴随能量的耗散及耗散结构的形成。要描述岩体在出现不可逆变形后的内部构造变化，必须引入新的内变量。

典型的最大剪应力 T（曲线 1）、相对的体积变形 θ（曲线2）与最大剪切变形 Γ 之间的关系曲线如图 5 所示[7]。在弹性阶段（$\Gamma < \Gamma_e, T < T_e$）之后是强化阶段，强化阶段在 $\Gamma = \Gamma_0, T = T_0$ 之后进入软化阶段。在达到极限剪切变形 T_C 后破坏。岩石的体积变形在达到弹性极限前为减小的，在此之后体积开始增加，在达到强度极限时体积应变为零。之后绝对体积变形为增加。这是典型的塑性材料行为。

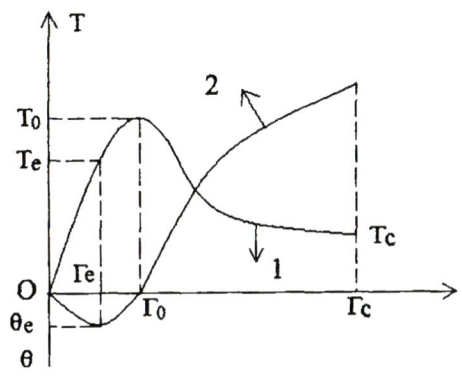

图 5 岩石的应力-变形图

Fig.5 The stress-strain diagram of rock

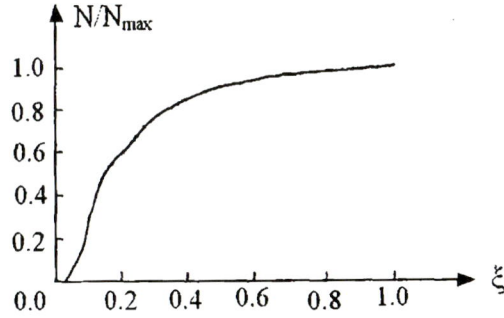

图 6 岩石的微裂纹形成与变形之间的关系（$\xi = \Gamma/\Gamma_C$）

Fig.6 The cracking-deformation dependence of rock ($\xi = \Gamma/\Gamma_C$)

图 6 给出了在岩石试样加载时微裂纹形成的动力过程[7]。图中 N/N_{max} 为在岩石试样中当前的裂纹数量与对应于破坏时的最大微裂纹数量之比。

N/N_{max} 可以看作是岩石的损伤量 ω。在弹性区 ω 的增长非常缓慢，在进入塑性状态之后迅速增长。在硬化阶段，变形局域化区也出现了，在这一区中材料粒子变碎。在新形成的结构中产生了更复杂的结构，向这种复杂结构的转换伴随着稳定性类型的变换。损伤量 ω 在应力极限之后快速增加，在出现了主干裂纹之后其趋向于1。微裂纹的动力发展过程可以看作是多阶段的渐进相变。从图6可以看出渐进相变的一些特点。在压力极限之后伴随着稳定性类型的改变，在图中出现了与材料的强度极限相对应的拐点。这一结论可以使我们利用相变的理论来研究深部坑道周围的分区破裂化现象。

因为两个最大剪应变大小都相等，同时达到临界值，所以可以采用 Γ 来描述岩石的变形过程。对于描述岩石的不可逆变形规律来讲可以引入下列无量纲变形参数：

$$\psi = \frac{\Gamma - \Gamma_e}{\Gamma_C - \Gamma_e} = \frac{\Delta\Gamma}{\Delta\Gamma_{max}} \quad (8)$$

岩体的不可逆变形的出现伴随着能量耗散的产生及耗散结构的形成。因此根据物理学的概念 ψ 可以称作序参量。我们可以首先把变形体看作是静力保守系统，其总能量为其势能 H。

在变形过程中系统经历一系列状态，它们在临界点处转换。在 $\psi = 0$ 时，开始塑性变形，开始形成耗散结构。在 $\psi > \psi_0$ 时，形成新的耗散结构，在破坏的材料中出现了协同学效应。在岩体进入到极限状态以后，介质的粒子之间产生了长程的相互作用。类似于 Ginzburg-Landau 展开式[9]，在势能 H 公式中保留 4 次项，增加梯度项，则有

$$H = -\frac{1}{2}V_2\psi^2 + \frac{1}{4}V_4\psi^4 + \frac{1}{2}C(\nabla\eta)^2 \quad (9)$$

式中 V_2、V_4、C 为系数，且 $V_2 > 0$、$V_4 > 0$、$C > 0$。

这样在整个体积 Ω 中的总势能为：

$$\Phi = \iiint_\Omega \left[-\frac{1}{2}V_2\psi^2 + \frac{1}{4}V_4\psi^4 + \frac{1}{2}C(\nabla\psi)^2 \right] d\Omega \quad (10)$$

根据物理学的基本定律，对于稳定的状态，介质总会选择使自由能为最小的参数，因此在介质的稳定状态时上式所表示的自由能应该为最小。对于上式的 ψ 参数取变分有

$$\delta \Phi = \int_\Omega \delta\psi \left[-V_2\psi + V_4\psi^3\right] d\Omega + \int_\Omega C\nabla\psi\nabla(\delta\psi) d\Omega =$$
$$= \int_\Omega \delta\psi \left[-V_2\psi + V_4\psi^3 - \nabla(C\nabla\psi)\right] d\Omega$$
$$+ \int_{\partial\Omega} C\delta\psi[(\nabla\psi)\vec{n}] dS = 0 \quad (11)$$

这样就得问题的求解方程

$$\nabla(C\nabla\psi) + V_2\psi - V_4\psi^3 = 0 \quad (12)$$

及边界条件为

$$(\nabla\psi)\cdot\vec{n}\big|_{r=R} = 0, \quad (\nabla\psi)\cdot\vec{n}\big|_{z=0,h} = 0 \quad (13)$$

在圆柱坐标里面，如果用 r、θ、z 分别代表径向、环向及竖向坐标，那么方程 (12) 可以写为：

$$C\left(\frac{\partial^2\psi}{\partial r^2} + \frac{\partial\psi}{r\partial r}\right) + \frac{1}{r^2}\frac{\partial^2\psi}{\partial\theta^2} + \frac{\partial^2\psi}{\partial z^2} + V_2\psi - V_3\psi^2 = 0$$
(14)

对于小变形来讲，可以忽略掉方程（14）中的序参量的三次项，那么方程（14）变为

$$\frac{\partial^2\psi}{\partial r^2} + \frac{\partial\psi}{r\partial r} + \frac{1}{r^2}\frac{\partial^2\psi}{\partial\theta^2} + \frac{\partial^2\psi}{\partial z^2} + D^2\psi = 0 \quad (15)$$

其中：$D^2 = V_2/C$。

方程（15）即为问题的支配方程。

3 支配方程的解

利用分离变量法求解支配方程（8）的解[10]。把序参量表示为分别依赖于 3 个坐标函数的乘积的形式：$\psi(r,\theta,z) = R(r)\cdot\Theta(\theta)\cdot Z(z)$，并代入到方程（15），可得 3 个独立方程

$$\frac{d^2Z}{dz^2} + \lambda Z = 0 \quad (16)$$

$$\frac{\partial^2 R}{\partial r^2} + \frac{\partial R}{r\partial r} + \left(D^2 - \lambda - \frac{\mu}{r^2}\right)R = 0 \quad (17)$$

$$\frac{d^2\Theta}{d\theta^2} + \mu\Theta = 0 \quad (18)$$

其中 λ、μ 为常数，不依赖于坐标 r、θ、z。

方程（18）的解为：

$$\Theta(\theta) = A\sin\sqrt{\mu}\theta + B\cos\sqrt{\mu}\theta \quad (19)$$

由周期性条件 $\Theta(0) = \Theta(2\pi)$，$\Theta'(0) = \Theta'(2\pi)$ 可得

$$\mu = m^2, m = 0,1,2,\cdots \quad (20)$$

所以(19)式变为

$$\Theta(\theta) = \sum_{m=0}^{\infty}(A_m\sin m\theta + B_m\cos m\theta) \quad (21)$$

其中：A_m、B_m 为系数。

由方程（16）可得

$$Z(z) = P\sin\sqrt{\lambda}z + Q\cos\sqrt{\lambda}z \quad (22)$$

其中 P、Q 由边界条件 $Z'(0) = 0$，$Z'(h) = 0$ 确定

$$P = 0, \quad \lambda = \left(\frac{n\pi}{h}\right)^2, n = 0,1,2,\cdots \quad (23)$$

所以式（22）变为

$$Z(z) = \sum_{n=0}^{\infty} Q_n \cos\frac{n z\pi}{h} \quad (24)$$

其中：Q_n 为系数。

对于方程（17），可以做变量替换：$\rho = \sqrt{D^2 - \lambda}\cdot r = kr$（其中 $k = \sqrt{D^2 - \lambda}$），$y(\rho) = R(r)$，从而得到下列方程

$$\frac{\partial^2 y_m}{\partial\rho^2} + \frac{\partial y_m}{\rho\partial\rho} + \left(1 - \frac{m^2}{\rho^2}\right)y_m = 0 \quad (25)$$

其中 y_m 为与 m 相对应的 $y(\rho)$ 值。

方程（25）为 m 阶 Bessel 方程，其解为：

$$y_m(\rho) = c_{m1}J_m(\rho) + c_{m2}N_m(\rho) \quad (26)$$

其中 J_m 和 N_m 分别为 m 阶 Bessel 函数及 m 阶

Neumann 函数。

根据在 $\rho = 0$ 时 $y(\rho)$ 值为有界的，可得

$$c_{m2} = 0 \tag{27}$$

所以方程 $y(\rho)$ 的解为所有 m 阶 Bessel 方程的解的线性组合

$$y(\rho) = \sum_{m=0}^{\infty} y_m(kr) = \sum_{m=0}^{\infty} c_{m1} J_m(kr) = R(r) \tag{28}$$

同时根据在试件侧面的边界条件：

$\rho = \sqrt{D^2 - \lambda} \cdot a = \rho_a$ 时，$y'(\rho)\big|_{\rho = \rho_a} = 0$，可得

$$y'(\rho_a) = \sum_{m=0}^{\infty} c_{m1} J'_m(\rho_a) =$$
$$= \sum_{m=0}^{\infty} c_{m1} \left[J_{m-1}(\rho_a) - \frac{m}{\rho^{-1}} J_m(\rho_a) \right] = 0 \tag{29}$$

这样就可以确定 ρ_a 大小，并进而确定 D 的大小。

所以方程（15）的解为：

$$\psi(r,\theta,z) = \sum_{m=0}^{\infty} \left[c_{m1}(A_m \sin m\theta + B_m \cos m\theta) J_m(kr) \right]$$
$$\cdot \sum_{n=0}^{n=\infty} Q_n \cos \frac{n z \pi}{h}$$
$$= \sum_{m=0}^{\infty} \left[(A'_m \sin m\theta + B'_m \cos m\theta) J_m(kr) \right] \cdot \sum_{n=0}^{n=\infty} Q_n \cos \frac{n z \pi}{h} \tag{30}$$

其中 A'_m、B'_m 为系数。

对于岩体的体积膨胀来讲，最简单的模型为体积膨胀与剪切变形成正比的模型[11]：

$$\varepsilon_V = \Lambda \psi \tag{31}$$

其中 Λ 为比例系数。

对于固定的 z 值及圆柱侧边界 $r = a$ 来讲，如果以图 4 中竖轴作为极坐标的角度起算点，则可以取下列角坐标的谐函数项来模拟如图 4 所示的体积变形随角度的变化：

$$\varepsilon_V = \Lambda \psi = -(348\cos\theta - 106\sin\theta + 580\cos 2\theta + 200\cos 3\theta + 287\cos 4\theta) \times 10^{-3} \tag{32}$$

具体的计算结果见表 1。计算结果标示于图 7 之中（见曲线 3）。由图可以看出，在取五项谐函数的情况下模拟的结果已经相当精确。

表 1 一系列角坐标情况下体积变形值

角度 $\theta(°)$	体积变形增量 $\varepsilon_V \times 10^{-3}$	角度 $\theta(°)$	体积变形增量 $\varepsilon_V \times 10^{-3}$
0	-1415	210	8
30	-394	240	316
45	257	270	187
90	399	315	107
120	500	330	-500
150	208	345	-1150
180	-320	360	-1415

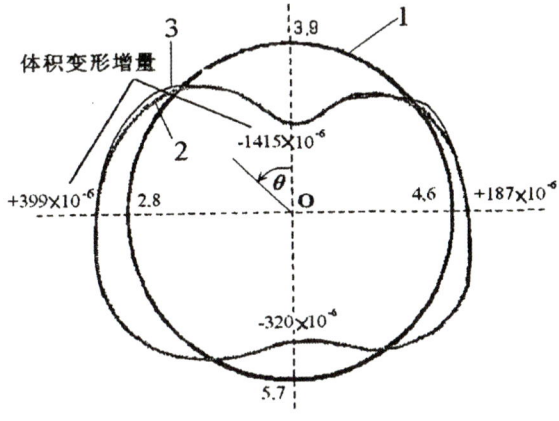

图 7 公式(31)的模拟结果(曲线 3)与试验结果(曲线 2)的比较

Fig.7 Comparison between the modeling by Eq.(31) (curve 3) with the experimental results (curve 2)

现在来考察对于变形增量沿着试件高度分布的模拟。由于此文选择的序参量为相对剪切变形，而试验数据为竖向变形分量 ε_z 及环向变形分量 ε_θ。为了能够模拟试验数据，需要做一些近似。从图 1(b)可以看出，对于岩石试件同一个地方极限后竖向变形分量的增量 $\Delta\varepsilon_z$ 及环向变形分量的增量 $\Delta\varepsilon_\theta$ 符号相反，大小近似成正比。所以可以近似取

$$\Delta\varepsilon_\theta = -\eta\Delta\varepsilon_z \tag{33}$$

其中 η 为比例系数。

由于 $\Delta\Gamma = \Delta\varepsilon_\theta - \Delta\varepsilon_z$ 所以有

$$\Delta\varepsilon_z = k\Delta\Gamma \tag{34}$$

其中 k 为比例系数。

对于固定的角坐标 θ，根据解（30）对于图 2 曲线 1 和 2 可以用下列三项来模拟：

$$\frac{\Delta\varepsilon_z(z)}{\Delta\sigma_z} = \left(H_0 + H_1\cos\frac{z\pi}{h} + H_2\cos\frac{2z\pi}{h}\right)\cdot 10^{-3} \tag{35}$$

对于曲线 1 得到的具体曲线表达式为：

$$\frac{\Delta\varepsilon_z(z)}{\Delta\sigma_z} = \left(2.63 + 0.173\cos\frac{z\pi}{h} - 0.57\cos\frac{2z\pi}{h}\right)\cdot 10^{-3} \tag{36}$$

计算结果如图 8 曲线 3 所示。

对于曲线 2 得到的具体曲线表达式为：

$$\frac{\Delta\varepsilon_z(z)}{\Delta\sigma_z} = \left(1.733 + 1.617\cdot\cos\frac{z\pi}{h} - 2.267\cdot\cos\frac{2z\pi}{h}\right)\cdot 10^{-3} \tag{37}$$

计算结果如图 8 曲线 4 所示。

从以上的模拟结果可以看出，由连续相变理论得到的方程能够很好地描述岩石试样过临界变形异常现象。

相变理论得到的结果与力学分析的结果是相符的。由方程（6）、（7）可知，在轴力作用下，岩石试件在两个相互垂直的方向应力和应变同时达到临界状态，在超过临界状态后岩石试件在两个相互垂直方向产生自我组织现象，这一结果与本文的模拟结果一致。所以相变理论是对于非线性连续介质力学的很好的补充。两者结合对于描述岩石在强烈受压状态下的非线性变形与破坏能够取得更好的效果。

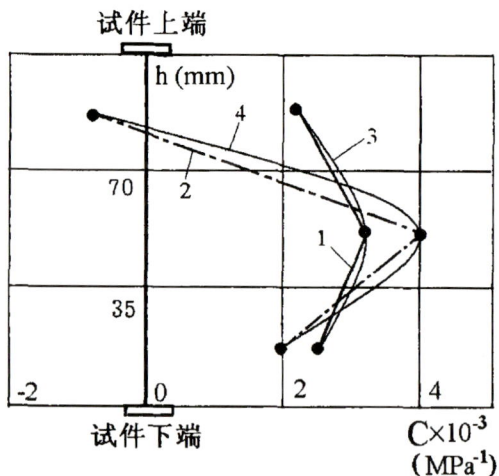

图 8 公式(36)(曲线 3)和(37)(曲线 4)模拟结果与相应的的试验结果（曲线 1 和 2）的比较

Fig.8 Comparison between the modeling by Eq.(37) (curve 3) and Eq. 38 (curve 4) with the corresponding experimental results (curve 1 and 2)

4 结 论

受损岩石的过临界变形为不可逆的变形过程，并伴随着材料内部结构的变化及能量的耗散。这一过程为一个耗散过程，伴随着有规则的空间结构的形成。在进行分析时除了分析应力应变状态外，还必须考虑材料的自我组织现象。试验表明，岩石的过临界变形损伤过程为连续相变过程，这可以使我们利用连续相变的理论研究岩石的过临界变形与损伤过程。本文利用连续相变理论对于过临界荷载作用时变形异常现象进行了模拟，得到了描述临界后岩石试样变形的支配方程，得到了这一方程的解析解，这一解能够同时描述两个相互垂直方向的不可逆变形的周期性空间分布。计算结果表明，求得的解析解能够很好地描述变形增量依赖于圆形试件的环向坐标及轴向坐标现象。

参考文献(References):

[1] SHEMYAKIN E I, FISENKO G L, KURLENYA M V, OPARIN V N, REVA V N, GLUSHIKHIN F P, ROZENBAUM M A, TROPP T A and KUZNETSOV Yu S. Zonal disintegration of rocks around underground workings, Part 1: Data of in situ observations[J], Journal of Mining Science, 1986, 22(3):157-168.

[2] SHEMYAKIN E I, FISENKO G L, KURLENYA M V,

OPARIN V N, REVA V N, GLUSHIKHIN F P, ROZENBAUM M A, TROPP T A and KUZNETSOV Yu S. Zonal disintegration of rocks around underground workings. Part II: Rock fracture simulated in equivalent materials[J]. Journal of Mining Science, 1986, 22(4):223-232.

[3] SHEMYAKIN E I, FISENKO G L, KURLENYA M V, OPARIN V N, REVA V N, GLUSHIKHIN F P, ROZENBAUM M A, TROPP T A and KUZNETSOV Yu S. Zonal disintegration of rocks around underground mines, part III: Theoretical concepts [J], Journal of Mining Science, 1987, 23(1):1-6.

[4] SHEMYAKIN E I, FISENKO G L, KURLENYA M V, OPARIN V N, REVA V N, GLUSHIKHIN F P, ROZENBAUM M A, TROPP T A. Zonal disintegration of rocks around underground workings. IV: Practical applications [J]. 1989, Journal of Mining Science, 25(4): 297-302.

[5] GUZEV M F, MAKAROV V V. Deformation and fracture of the highly stressed rocks around openings [M]. Vladivostok: Dalnauka, 2007.

[6] GUZEV M F, MAKAROV V V and USHAKOV F F. Modeling elastic behavior of compressed rock samples in the pre-failure zone [J], Journal of Mining Science, 41(6): 497-506.

[7] ADIGAMOV N S, RUDAYEV Ya I. Equation of state allowing for loss strength of material [J]. Journal of Mining Science, 35(4): 333-340, 1999.

[8] QI CHENGZHI, QIAN QIHU, WANG MINGYANG. Evolution of the deformation and fracturing in rock masses near deep-level tunnels [J]. Journal of mining science, 2009, 45(2):112-119.

[9] LANDAU L.D., LIFSHITS E.M. Statistical physics, part I [M]. Beijing World Publishing Corporation, Beijing, 1999.

[10] ASMAR N H. Partial differential equations with Fourier series and boundary value problems, second edition [M], Pearson Education, 2005.

[11] NIKOLAEVSKY V N. Mechanics of porous and fractured media [M]. Singapore: World Scientific, 1990.

岩爆机理的简化分析和决定性参数的思考

钱七虎

(解放军理工大学,江苏 南京,210007)

1. 引言

岩爆,指的是岩石的爆炸,俄文中称为岩石的冲击,叫做冲击地压。大量的研究文献表明,岩爆在多个领域均有发生。俄罗斯对岩爆的研究主要是集中于矿山开采领域,而我国在交通工程、水利工程开发中,也对岩爆现象进行了研究。以前,我国的岩爆主要发生在矿山开采领域,特别是煤矿中经常发生岩爆。欧洲曾经发生岩爆的阿尔卑斯山水道埋深约2100m,我国现在部分水利工程的埋深达到了2500m,也出现了岩爆现象。

岩爆按其发生机理上可以分为三大类,一类叫应变型岩爆,另一类叫构造型岩爆,还有一类是应变和构造混合型岩爆。依据岩体类别来考虑,可以分为均匀岩体岩爆和非均匀岩体岩爆两类。

2. 均匀岩体岩爆机理的简化宏观分析

发生岩爆的均匀岩体要具备以下的特征:岩体一定是强力压缩的岩体,岩体中的地应力如果不高,就不可能会发生岩爆;岩体要经过卸载或者开挖卸载,矿井里叫做采动。从能量的观点来分析,卸载就是变形能转化为动能的过程,转化的关键是要有卸载过程,这是最基本的,卸载的情况,一部分由于开挖引起的,一部分是由于顶板破坏引起的;岩爆的动能等于总变形能减去破坏能。

均匀岩体岩爆的机理,简单地说就类似于一个强力压缩的弹簧突然卸载的情况。

(1)一维弹脆性介质均匀压缩杆件受压破坏模型。

以一维弹脆性介质均匀压缩杆件为例(图 1),在$t = 0$时刻,移除σ_0。根据应力波的理论,在杆件里面传播应力波,应力波所到之处应力变成零,向外的速度$v = \frac{\sigma_0}{\rho c}$。此时,岩块的动能为$\frac{1}{2}\rho v^2 = \frac{\sigma_0^2}{2\rho c^2} = \frac{\sigma_0^2}{2E}$,岩石破坏(拉断)的破坏能为$\frac{1}{2}E\varepsilon_p^2 = \frac{\sigma_p^2}{2E}$,这样,可以得到岩爆块的动能(岩爆的强烈程度)为$\frac{1}{2E}(\sigma_0^2 - \sigma_p^2)$岩块的动能就等于变形能减掉破坏能。

从能量里看,E是卸载后的模量,也就是下降段的模量。下降段模量或者卸后模量越大,破坏能就越小,岩爆块的动能就越大。如果E很小,假如下降段水平接近零,破坏就无穷大,所以动能就会很小。σ_0是压缩的高应力,应力越高越有可能发生岩爆。一维杆件,有压缩、卸载就有可能破坏,就会发生岩爆。

本文原载于《岩爆机理探索》(2010 年)。

由上述分析可以得知，均匀岩体中发生岩爆的几个关键参数主要为压缩、卸载、模量等。

(2) 多维强力压缩岩块受压破坏分析。

二维强力压缩岩块的卸载问题(图 2)，由于应力还要转到支柱，不仅仅是一个σ_0的问题，要比一维卸载复杂很多。三维卸载应力更复杂。

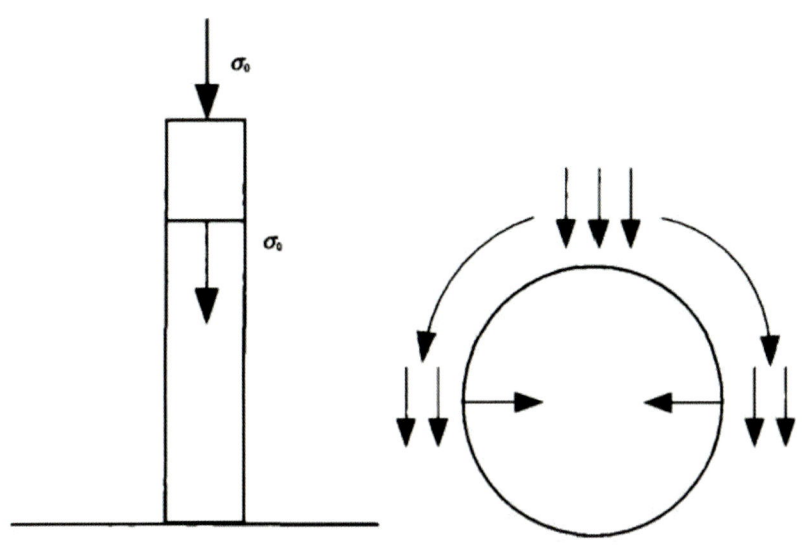

图 1 岩爆原理示意图　　图 2 二维情况下岩爆原理示意图

一维的杆件就是一个卸载，二维采动除了卸载以外还有一个应力，情况就复杂一些。三维的更复杂，多个采动面的岩爆问题就是三维卸载情况下的岩爆问题。

3. 岩体非均匀性影响及局部细观应力分析

实际上的岩体都是非均匀性的，构造就很重要。岩体存在微裂纹、结构面、软弱面等缺陷。构造性岩爆的发生就是结构面、软弱面的存在。岩体中的缺陷决定了各种各样的岩爆。分析研究微裂纹时如何破裂，要确定哪些因素决定破坏能，因为要分析岩爆就要分析原来变形能，变形能比较简单，均匀的地应力，就是破坏分析比较复杂。

(1) 基本模型和公式。

非均匀岩石拉破坏模型认为，具有内部缺陷的脆性岩体宏观上应力与应变服从胡克定律，但缺陷附近有应力的起伏分布，缺陷处的不协调变形和应力集中是岩石发生拉伸破坏的主要原因。

岩体内部的局部应力σ_{ij}^{loc}可分为两个部分：一是弹性应力σ_{ij}^{e},遵循胡克定律，在整个区域内均匀分布；二是缺陷引起的附加应力，仅在缺陷区域内不为零。忽略脆性岩石拉伸破坏过程中的塑性变形，局部应力σ_{ij}^{loc}可表示为两部分之和：

$$\sigma_{ij}^{loc} = \sigma_{ij}^{e} + \Delta\sigma_{ij} \tag{1}$$

在准静态加卸载或较低应变率($< 10^0 s^{-1}$)情况下，缺陷处应力集中的演化方程可用下式近似：

$$\frac{\mathrm{d}\Delta\sigma_{ij}}{\mathrm{d}t} = K\rho c_s^2 \dot{e}_{ij} - \eta\frac{\Delta\sigma_{ij}}{l} \tag{2}$$

其中e_{ij},为偏应变, $e_{ij} = u_{ij} - (1/3)\delta u_{ij}$, δ为克朗内克符号, $\dot{e}_{ij}$上的点代表其随时间的变化率, $\Delta\sigma_{ij}$和K分别为特定尺寸的缺陷上的附加应力和材料的应力集中系数, ρ为介质的密度, c为剪切波波速, η为影响附加应力松弛速率的材料参数, l为缺陷的尺度。

式(2)右侧第一项代表剪切应变导致的附加应力,其形式类似于胡克定律;第二项描述了附加应力的松弛,假设附加应力松弛速率$\mathrm{d}\Delta\sigma_{ij}/\mathrm{d}t$的大小与缺陷尺寸成反比。

该模型考虑了缺陷处的应力集中和松弛对材料应力场的影响,也考虑了加载速率或应变速率对附加应力的影响。

(2)应力集中产生机制。

图3概略描绘了不均匀材料中缺陷处应力集中的一种特殊情况。在z轴应力卸载过程中,由于不同材料之间变形的不协调性,将在z轴方向引起局部应力场。$\Delta\sigma_{zz}(t)$特指t时刻应力集中区z方向附加应力的最大值。缺陷的尺度则定义为应力集中区在垂直于考察方向的跨度。

缺陷处的应力集中既可能是由于微裂纹导致的,也可能是由于晶核或其他包含物与其周围材料的变形特性差异导致的。

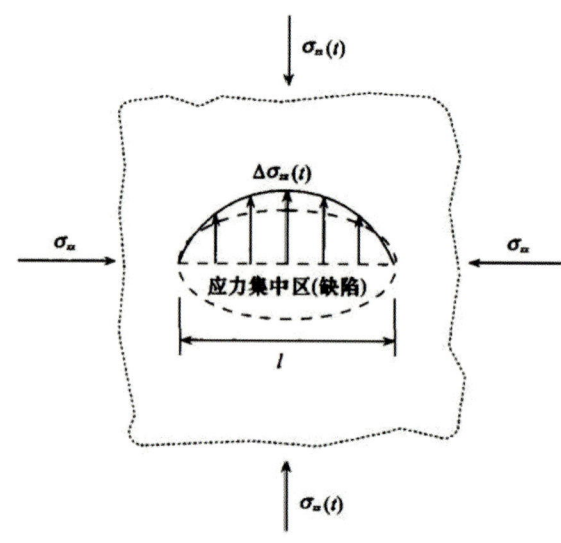

图3 缺陷应力集中区示意图

为简要说明岩体中缺陷处拉应力集中的产生机制,假设一平面应力状态的岩体,其侧向应力保持不变,而按一定速率卸载。由于竖向应力的卸载,岩体中必然产生剪切应力和偏应变。若材料均匀连续,则岩体中微小单元将在应力的作用下发生畸变,占据新位置,但是实际的岩体中,由于材料的非均匀性,不同部位的变形是不同步的,这就不可避免地会在缺陷内部和边界产生应力集中。

可以预见,由于变形不协调而导致的应力集中不仅和缺陷的尺寸有关,还必将与应力的变化速率有关。根据缺陷的形状和缺陷内外材料变形特性的差异,其应力集中的程度和消散的快慢必定是不一样的。

但正如前面提到的,岩石内部包含了大量的缺陷,其大小形状各异,既无可能也无必要对其一一考察。因而充分利用岩石内部缺陷的自相似构造特性,在细观机理和宏观力学表现之间建立简洁但基本符合实际的定量联系,通过尽可能少的参数来反映尽可能多的特性是必要且可行的。

式(2)中出现的岩石参数有 5 个,分别是密度、剪切波速、材料应力集中系数、材料应力松弛系数和缺陷尺度。前四个参数可以通过宏观试验确定,缺陷尺度的确定较复杂,拟另文讨论,本文中假设缺陷尺度及其分布是已知的。

(3)围岩的卸载拉裂。

假设岩体受静水压力的作用,应力作用的时间足够长,岩体中的缺陷处的应力集中已经完全松弛。在其中一个方向上卸载(图4),来考察岩石的应力应变状态。

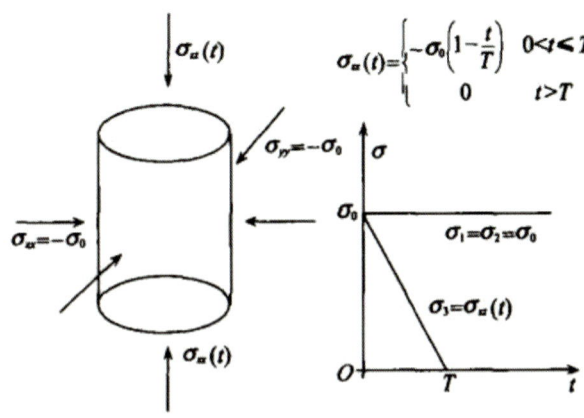

图4 卸载示意图

假设岩体x轴、y轴方向应力保持不变,z轴方向线性卸载,卸载公式为:

$$\sigma_x = \begin{cases} -\sigma_0\left(1-\dfrac{t}{T}\right) & 0 < t \leqslant T \\ 0 & t > T \end{cases} \tag{3}$$

其中T为卸载持续时间。

缺陷处z方向的局部拉应力的大小为:

$$\Delta\sigma_{zz} = \begin{cases} \dfrac{K}{3}\dfrac{\sigma_0}{T}\dfrac{l}{\eta}\left(1-e^{-\frac{\eta}{l}t}\right) & 0 < t \leqslant T \\ \dfrac{K\sigma_0}{3T}\dfrac{l}{\eta}(1-e^{-\frac{\eta}{l}T})e^{-\frac{\eta}{l}(t-T)} & t > T \end{cases} \tag{4}$$

缺陷处的局部应力σ_{zz}^{loc}是由σ_{zz}^{e}和$\Delta\sigma_{zz}$组成的。考虑到$\sigma_{zz}^{e} \approx \Delta\sigma_{zz}(t)$可得:$\sigma_{zz}^{loc} = \sigma_{zz}^{e} + \Delta\sigma_{zz} \cong \Delta\sigma_{zz}(t) + \Delta\sigma_{zz} = -\sigma_0(1-\dfrac{t}{T}) + \dfrac{K}{3}\dfrac{\sigma_0}{T}\dfrac{l}{\eta}(1-e^{-\frac{\eta}{l}t})$ (5)

4. 深部围岩卸载拉伸破坏的影响因素

随着对地下空间和矿产资源的开发逐步深入深部，深部围岩的破坏问题引起了越来越多的关注。研究深部围岩在卸载条件(绝大多数地下工程涉及卸载过程)下的拉伸破坏模式(包括剥落、劈裂、岩爆和分区破裂化等)对于加深复杂应力条件下岩石的力学性质和提高地下空间、资源利用的安全性具有重要的意义。

依据前述模型结果，岩体中缺陷处的局部拉应力与很多因素有关，主要包括缺陷处的应力集中系数、缺陷的尺寸、缺陷处应力松弛的速度、初始应力和加卸载时间等。下面分别对三个影响岩体卸载破坏的主要因素：卸载速率、缺陷尺寸分布和初始地应力进行分析。

(1) 卸载速率。

设岩体参数为：$\eta = 2 \times \frac{10^{-8} \text{m}}{\text{sec}}, K = 1, l = 1 \times 10^{-6}\text{m}$，初始应力 $\sigma_0 = 50\text{MPa}$，卸载模式采用图5所示形式，持续时间分别为50s、100s、200s和400s。

从图5中可以看出，卸载速率越快，缺陷处的附加应力增长就越快，附加拉应力峰值就越高。依据本例的参数，在快速卸载的情况下($T = 50\text{s}$)，平行于卸载方向的拉应力峰值可达11MPa，超过一般岩石的抗拉强度，将导致一定尺度范围内的缺陷被激活，触发岩石内部微裂纹的不稳定发展过程，进而可能导致岩石在快速卸载情况下的拉伸破坏。卸载结束后，附加应力逐步松弛，松弛的速度随附加应力峰值的增大而增大。

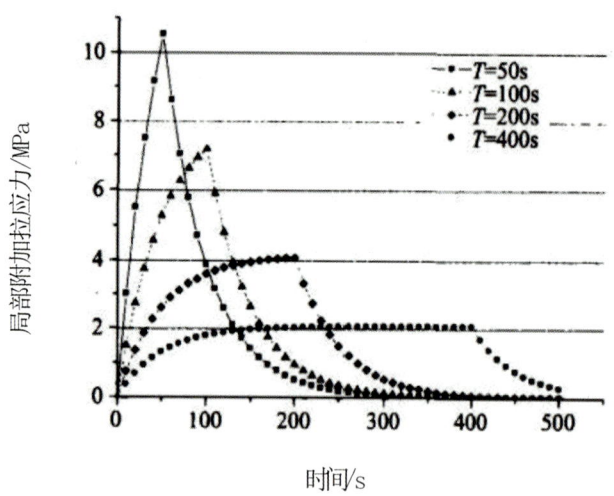

图5 卸载时缺陷处附加拉应力的时程曲线

从图6中可以看出，卸载速率较快时(如$T = 50\text{s}$)，缺陷处的局部应力迅速从压应力转变为拉应力，并在卸载结束时达到最大值，之后逐步松弛。如果局部应力超过缺陷处材料的抗拉强度，将使岩体在缺陷处发生破坏，进而导致裂纹的扩展。而卸载速率较慢时(如$T = 400\text{s}$)，缺陷处的附加拉应力较小，并在卸载结束后逐渐松弛，较难引起岩体的破坏。

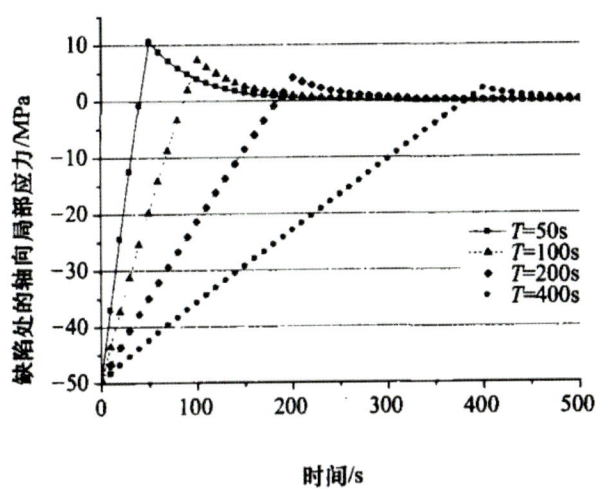

图6 卸载时缺陷处局部应力的时程曲线

(2)缺陷尺度。

设定应力松弛速率 $\eta = 2 \times 10^{-8}$ m/sec，应力集中系数 $K = 1$，初始应力 $\sigma_0 = 50$ MPa，改变加卸载速率和缺陷尺寸大小，分别考察附加拉应力与卸载速率及最大附加拉应力与缺陷尺寸之间的关系曲线。卸载模式仍采用图2所示形式。轴向卸载时，不同卸载周期的岩石内部缺陷处，局部附加应力大小变化情况参见图7和图8。

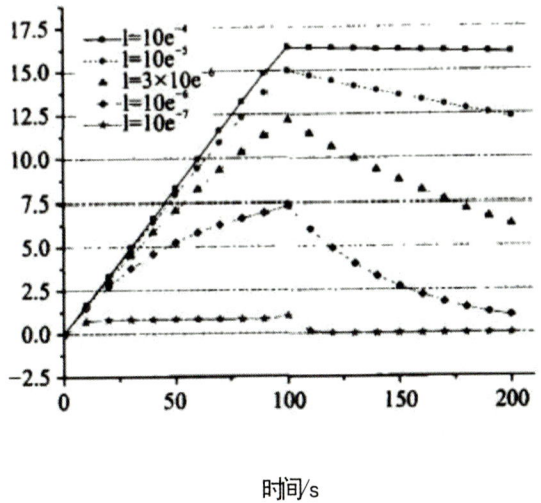

图7 卸载时缺陷处附加拉应力的时程曲线（T=100s）

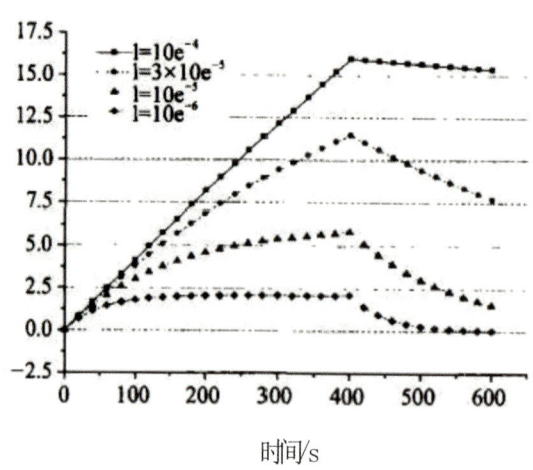

图8 卸载时缺陷处附加拉应力的时程曲线（T=400s）

从图7、图8可以得出如下结论：①对于同样的卸载速率，卸载持续过程中，较大尺寸缺陷处的局部附加应力与时间基本成正比，而较小尺寸缺陷处的局部附加应力与时间的关系则是非线性的；②卸载速率相同的情况下，缺陷的尺寸越大，则同一时刻缺陷处局部附加拉应力的数值就越大；③缺陷尺寸较大时，卸载过程完成后其局部应力集中衰减较慢；④缺陷尺寸大小与局部附加拉应力之间的关系是非线性的，缺陷尺寸很小或很大时，局部附加拉应力对缺陷尺寸的变化不敏感。

图9为轴向卸载情况下卸载结束时的最大局部附加拉应力和缺陷尺寸之间的关系。从图9可以看出，不同的加卸载速率激活的缺陷尺寸不同，对于一定的卸载速率，只有大于一定尺寸的缺陷处才会产生较大的拉应力集中，加卸载速率越快，被激活的缺陷的临界尺寸越小。也就是说，当偏应变率较高时，细观层面上和微观层面上的缺陷被同时激活，均产生较大的局部拉应力；而在偏应变率较小的情况下，只有尺寸较大的缺陷才会被激活。

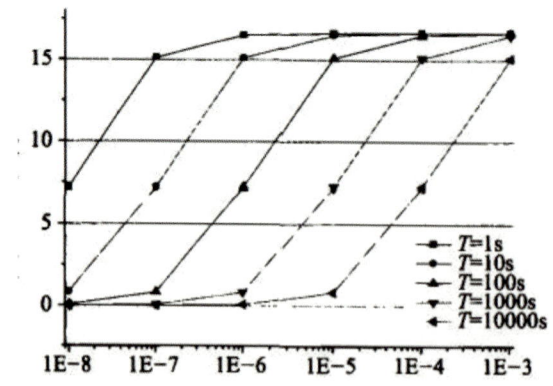

图9 卸载时最大附加拉应力与缺陷尺度之间的关系

依据模型公式，假设岩石的剪切波速为3000m/s，密度为2500kg/m³，则当 $T=1$s 时，岩石的偏应变率约为 10^{-3} 量级。对照图9中的曲线和偏应变量级，可得偏应变率量级与被激活缺陷的临界尺寸的量级之间的大致对应关系 如下表。

偏应变率和被激活的缺陷尺寸量级对应关系表

卸载时间(s)	偏应变率量级(s^{-1})	被激活缺陷的尺寸量级(m)
10^0	10^{-3}	$>10^{-7}$
10^1	10^{-4}	$>10^{-6}$
10^2	10^{-5}	$>10^{-5}$
10^3	10^{-6}	$>10^{-4}$
10^4	10^{-7}	$>10^{-3}$

(3)初始地应力。

若缺陷尺寸较大，卸载过程中始终有 $l \gg \eta t$，则可将式(4)中的第一式作迈克劳林展开：

$$\Delta\sigma_{zz} = \frac{1}{3T}\frac{l}{\eta}K\sigma_0(1-e^{-\frac{\eta}{l}t}) = \frac{1}{3T}\frac{l}{\eta}K\sigma_0(1-1+\frac{\eta}{l}t+R_n) \cong \frac{1}{3}K\sigma_0\frac{t}{T} \quad (6)$$

当 $t=T$ 时，缺陷处的局部拉应力最大，岩体在缺陷处发生破坏的临界条件可表示为：

$$\sigma_{zz\max}^{loc} = \sigma_{zz}(T) + \Delta\sigma_{zz}(T) = \frac{K}{3}\sigma_0 = \sigma_p \quad (7)$$

也就是初始的静水压力大于其临界值：

$$\sigma_{0c} = \frac{3}{K}\sigma_p \quad (8)$$

由上式可以看出，如果岩体中包含有较大尺度的缺陷，则岩体是否发生破坏主要是由初始应力的大小、缺陷处的应力集中系数和岩石的抗拉强度之间的关系决定的。

对于特定的岩石材料，应力集中系数和抗拉强度是一定的，可以预测，在初始静水压力的作用下某一方向的快速卸载将导致岩体在垂直于卸载方向的拉伸破坏，该破坏形式的决定性因素是卸载前初始静水压力的大小。

5. 结论

岩爆的决定性参数主要包括初始地应力、卸载速率(开挖速率)、非均匀性(微缺陷、结构面、软弱面)。对卸载速率(偏应变率)、缺陷尺寸和初始应力对岩石卸载过程中缺陷处的应力集中的影响的研究结果表明：①缺陷的尺寸越大，局部附加拉应力的数值就越大；②缺陷尺寸很小或很大时，局部附加拉应力对缺陷尺寸的变化不敏感；③一定的应变率条件下，只有大于一定尺寸的缺陷被激活，产生较明显的拉应力，加卸载速率或偏应变率越高，其所激活的缺陷的尺寸就越小；④缺陷处的局部拉应力与初始应力成正比。

深部裂隙岩体岩爆定量预测模型

周小平　钱七虎

(解放军理工大学, 江苏　南京, 210007)

1. 引言

岩爆是高应力条件下地下岩体工程开挖过程中, 因开挖卸载引起围岩内应力场重新分布, 导致储存于硬脆性围岩中的弹性应变能突然释放, 产生爆裂、松脱、剥离、弹射甚至抛掷等现象的一种动力失稳的地质灾害, 岩爆具有突发性, 在地下工程中对施工人员和施工设备的威胁最为严重。如果处理不当, 就会给施工安全、岩体及建筑物的稳定带来很多困难, 甚至造成重大的安全事故。

人们首次发现岩爆是18世纪30年代末英国锡矿岩爆, 此后, 在美国、苏联、中国、南非、瑞典等国家关于岩爆的记录不断地增多, 据不完全统计, 从1949年到1985年间, 在我国的32个重要的煤矿中, 至少发生过1842起煤爆和岩爆, 发生地点一般在200～1000m深处的地质构造复杂、煤层突然变化、水平突然弯曲变成陡倾这样一些部位, 我国水电工程的一些地下硐室中也曾发生过岩爆, 地点大多位于高地应力地带的结晶岩和灰岩中, 或位于河谷近地表处, 在高地应力区开挖隧道, 当岩层较为完整、坚硬时, 也常发生岩爆现象。

关于岩爆的发生, 由于其发生机理的复杂性, 到目前为止对岩爆的形成条件及机理还没有形成统一的认识, 学术界提出了若干假说, 归纳起来, 主要包括强度理论、能量理论、刚度理论以及冲击倾向理论等。有的学者认为岩爆是受剪破裂, 也有的学者根据自己的观察和试验结果得出张破裂的结论。还有一种观点把岩爆的岩体破坏过程分为: 劈裂成板条、剪断成块、块片弹射三个阶段式破坏。从国内外的规定和研究成果来看关于岩爆预测判据都是大同小异, 主要包括应力判据、能量判据、岩芯判据、临界深度判据等。虽然岩爆现象的研究取得一定的进展, 但世界范围内频繁发生的岩爆灾害, 使我们不得不承认现有的研究成果还远没有从根本上解决岩爆预测及有效防治等问题。本文假设岩体含有节理和裂隙, 首先计算裂隙尖端应力强度因子, 并利用断裂准则判断裂隙的扩展, 然后基于能量分析确定岩爆判据, 最后建立深部裂隙岩体岩爆定量预测模型。

本文原载于《岩爆机理探索》(2010年)。

2. 非静水压力条件下围岩应力场

考虑岩石材料变形的非连续和非协调特点，在几何方程中采用应变不协调 张量，认为变形后岩石材料内部为非欧几何空间，根据不协调量的演化方程，建 立了非静水压力条件下的非欧几何模型，获得了卸载条件下深部圆形洞室围岩弹性应力场，其表达式可以表示为：

$$\sigma_r = \frac{\sigma_v + \sigma_h}{2}(1 - \frac{r_0^2}{r^2}) + \frac{\sigma_h - \sigma_v}{2}(1 - 4\frac{r_0^2}{r^2} + 3\frac{r_0^4}{r^4})\cos(2\theta)$$

$$- \frac{E}{2\gamma^{3/2}(1-v)r}[AJ_1(\gamma^{1/2}r) + BN_1(\gamma^{1/2}r) + CK_1(\gamma^{1/2}r)] \quad (1)$$

$$+ \frac{E}{2(1-v)r\gamma^{\frac{3}{2}}}[A_1J_1(\gamma^{1/2}r) + B_1N_1(\gamma^{1/2}r) + C_1K_1(\gamma^{1/2}r)]\cos(2\theta)$$

$$- \frac{3E}{(1-v)r^2\gamma^2}[A_1J_2(\gamma^{1/2}r) + B_1N_2(\gamma^{1/2}r) + C_1K_2(\gamma^{1/2}r)]\cos(2\theta)$$

$$\sigma_\theta = \frac{\sigma_v + \sigma_h}{2}(1 + \frac{r_0^2}{r^2}) - \frac{\sigma_h - \sigma_v}{2}(1 + 3\frac{r_0^4}{r^4})\cos(2\theta)$$

$$- \frac{E}{2\gamma(1-v)}[AJ_0(\gamma^{1/2}r) + BN_0(\gamma^{1/2}r) - CK_0(\gamma^{1/2}r)]$$

$$+ \frac{E}{2\gamma^{3/2}(1-v)r}[AJ_1(\gamma^{1/2}r) + BN_1(\gamma^{1/2}r) + CK_1(\gamma^{1/2}r)] \quad (2)$$

$$+ \frac{E}{2(1-v)\gamma}[A_1J_0(r\sqrt{\gamma}) + B_1N_0(r\sqrt{\gamma}) + C_1K_0(r\sqrt{\gamma})]\cos(2\theta)$$

$$- \frac{3E}{2r(1-v)\gamma^{3/2}}[A_1J_1(r\sqrt{\gamma}) + B_1N_1(r\sqrt{\gamma}) + C_1K_1(r\sqrt{\gamma})]\cos(2\theta)$$

$$+ \frac{3E}{r^2(1-v)\gamma^2}[A_1J_2(r\sqrt{\gamma}) + B_1N_2(r\sqrt{\gamma}) + C_1K_2(r\sqrt{\gamma})]\cos(2\theta)$$

$$\tau_{r\theta} = \frac{\sigma_v - \sigma_h}{2}(1 + 2\frac{r_0^2}{r^2} - 3\frac{r_0^4}{r^4})\sin(2\theta)$$

$$+ \frac{E}{r(1-v)\gamma^{\frac{3}{2}}}[A_1J_1(r\sqrt{\gamma}) + B_1N_1(r\sqrt{\gamma}) + C_1K_1(r\sqrt{\gamma})]\sin(2\theta) \quad (3)$$

$$- \frac{3E}{r^2(1-v)\gamma^2}[A_1J_2(r\sqrt{\gamma}) + B_1N_2(r\sqrt{\gamma}) + C_1K_2(r\sqrt{\gamma})]\sin(2\theta)$$

其中，$\gamma^2 = \frac{E}{[4q(1-v)]}$；$E$ 为弹性模量；q 为可由实验确定的系数；v 为泊松比。

$$A = (\frac{C}{2})\pi\sqrt{\gamma}r_0[K_0(\sqrt{\gamma}r_0)N_1(\sqrt{\gamma}r_0) - K_1(\sqrt{\gamma}r_0)N_0(\sqrt{\gamma}r_0)];$$

$$B = -(\frac{C}{2})\pi\sqrt{\gamma}r_0 \times [K_0(\sqrt{\gamma}r_0)J_1(\sqrt{\gamma}r_0) - K_1(\sqrt{\gamma}r_0)J_0(\sqrt{\gamma}r_0)];$$

$$A_1 = (\frac{C_1}{2})\pi\sqrt{\gamma}r_0[K_2(\sqrt{\gamma}r_0)N_3(\sqrt{\gamma}r_0) - K_3(\sqrt{\gamma}r_0)N_2(\sqrt{\gamma}r_0)];$$

$$B_1 = -(\frac{C_1}{2})\pi\sqrt{\gamma}r_0[K_2(\sqrt{\gamma}r_0)J_3(\sqrt{\gamma}r_0) - K_3(\sqrt{\gamma}r_0)J_2(\sqrt{\gamma}r_0)];$$

3. 围岩岩爆特征分析

周小平、Aliakbar 和 Golshani 等建立了裂隙演化过程和岩石损伤演化规律之间的关系。裂隙扩展前对应于岩石的弹性阶段，裂隙扩展阶段和次生裂纹稳定扩展阶段对应于岩石的非线性强化阶段，次生裂纹失稳扩展阶段对应于岩石的应变软化阶段。为此，将深部裂隙岩体岩爆的孕育过程分为三个阶段：裂隙扩展阶段、次生裂纹稳定扩展阶段和次生裂纹失稳扩展、连接并汇合成宏观裂隙阶段。

3.1 破裂区出现的条件

如图1所示，假设有一组裂纹分布于地下硐室围岩，裂隙初始长度为2a，裂纹倾角为α，裂纹间距为2d，地下硐室半径为r_0，垂直方向地应力为α_v，水平方向地应力为α_h，次生裂纹扩展长度为l，岩石的断裂韧度为K_{IC}。地下硐室开挖后，裂隙将发生扩展，岩体将发生非连续和非协调变形，传统的弹塑性理论获得的应力场不再适用于此时岩体的非连续和非协调变形。非欧几何模型的优点是能准确地分析裂隙扩展后的深部围岩应力场。

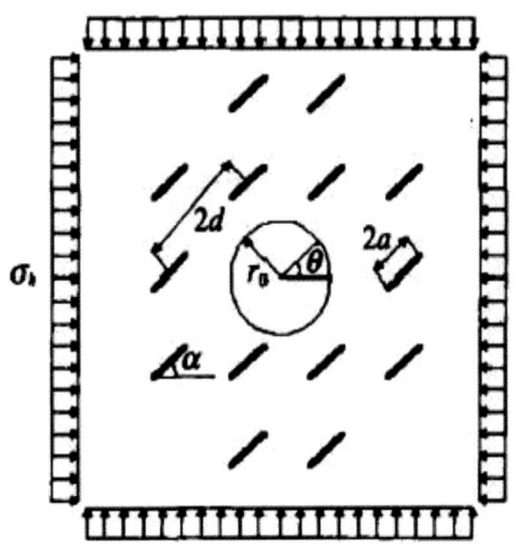

图1 含初始原生裂隙围岩体力学模型

根据非欧几何理论，深部围岩的主应力可以准确地表示为：

$$\begin{cases} \sigma_1' = \frac{\sigma_r+\sigma_\theta}{2} + \frac{1}{2}\sqrt{(\sigma_r-\sigma_\theta)^2 + 4\tau_{r\theta}^2} \\ \sigma_2' = \frac{\sigma_r+\sigma_\theta}{2} - \frac{1}{2}\sqrt{(\sigma_r-\sigma_\theta)^2 + 4\tau_{r\theta}^2} \end{cases} \quad (4)$$

忽略初始应力分布随裂隙扩展变化的影响，则在围岩中距圆形巷道中心为 r 的裂隙表面所受的应力可以表示为：

$$\begin{cases} \sigma_n = \sigma_r \sin^2(\theta-\alpha) + \sigma_\varphi \cos^2(\theta-\alpha) + \tau_{r\theta}\sin(\theta-\alpha)\cos(\theta-\alpha) \\ \tau_n = (\sigma_\varphi - \sigma_r)\sin(\theta-\alpha)\cos(\theta-\alpha) - \tau_{r\theta}\cos^2(\theta-\alpha) \end{cases} \quad (5)$$

如果裂隙表面满足 Mohr-Coulomb 准则，裂隙发生摩擦滑动的有效剪应力为：

$$\tau_{eff} = |\tau_n| - \mu\sigma_n \quad (6)$$

式中，μ 为裂隙表面摩擦系数。

原生裂隙启裂并出现次生裂纹的临界条件为：

$$(|\tau_n| - \mu\sigma_n)\sqrt{\pi a} = \frac{\sqrt{3}}{2}K_{IC} \quad (7)$$

次生裂纹尖端的 I 型应力强度因子可以表示为：

(8)

$$K_I = \frac{F\cos(\theta-\alpha-\alpha_0)}{\sqrt{\pi l}} - \sigma_2'\sqrt{\pi l}$$

式中，$F = 2a\tau_{eff}$；a 为原生裂纹长度；l 为弯折裂隙扩展长度；最大主应力和 σ_r 的夹角为 $\alpha_0 = \arctan(\frac{2\tau_{r\theta}}{\sigma_r - \sigma_\theta})$。

当即次生裂纹尖端应力强度因子 K_I，等于岩石断裂韧度 K_{IC} 时，次生裂纹将发生弯折扩展

$$K_I = K_{IC} \quad (9)$$

根据式（8）和式（9）可以求出 l 的长度。

次生裂纹汇合成宏观裂隙并出现破裂区的条件为：

$$l = d \quad (10)$$

式中，d 为裂纹之间的间距。

3.2 基于能量分析的岩爆判据

原生裂隙摩擦滑动的位移为：

$$b = \frac{\pi a \tau_{eff}(1-v^2)}{E} \tag{11}$$

如果发生摩擦滑动的裂隙密度为 ω_1，则原生裂隙滑移所耗散的总能量密度可以表示为：

$$\chi_{f1} = 2\omega_1 a \tau_{\text{eff}} b \tag{12}$$

如果发生弯折扩展的次生裂纹密度为 ω_2，则次生裂纹发生弯折扩展所耗散的总能量密度可以表示为：

$$\xi_1(l) = \frac{\omega_2(\kappa+1)(1+v)}{2E} \int_0^l \{[\frac{F\cos(\theta-\alpha-\alpha_0)}{\sqrt{\pi l}} - \sigma'_2\sqrt{\pi l}]\}^2 \, dl \tag{13}$$

忽略裂隙扩展过程热量损失，总耗散能可以表示为摩擦滑移耗散能与弯折扩展耗散能之和。

$$U_1 = \chi_{f1} + \xi_1(l) \tag{14}$$

岩体储存的弹性应变能为：

$$U_e = \frac{1}{2}\sigma_{ij}\varepsilon_{ij} \tag{15}$$

将胡克定律代入式（15）可得岩体储存的弹性应变能为：

$$U_e = \frac{1-v^2}{2E}(\sigma_r^2 + \sigma_\varphi^2) - \frac{v(1+v)}{E}\sigma_r\sigma_\varphi + \frac{1+v}{E}\tau_{r\theta}^2 \tag{16}$$

岩爆判据为：

$$\begin{cases} l = d \\ U_1 < U_e \end{cases} \tag{17}$$

岩爆产生的动能为：

$$W = U_e - U_1 \geqslant 0 \tag{18}$$

利用式（10）可以确定破裂区的位置和范围。根据破裂区的位置和范围并结合式（18）可以确定发生岩爆的岩体面积 A。

岩石碎块的弹射速度为：

$$V = \sqrt{2W/\rho} \tag{19}$$

4. 算例与分析

4.1 裂隙倾角对岩爆的影响

计算参数为：弹性模量 E = 25GPa，泊松比 v = 0.15，裂隙表面摩擦系数 μ = 0.38，竖向地应力 σ_v = 30MPa，水平地应力 σ_h = 10MPa，岩石断裂韧度 K_{IC} = 1MPa·$m^{1/2}$，原生

裂隙长度 a = 0.006m，裂隙间距 d = 0.3m，非欧几何参数 C = 2150000，C_1 = 4950000，q = 2.05×10^8，硐室半径 r = 7m。

图2表明：岩爆位置随裂隙倾角变化而变化。图3表明：岩爆面积随裂隙倾角变化而变化，且裂隙倾角为45°或135°时岩爆的面积最小。

4.2 裂隙间距对岩爆的影响

计算参数为：弹性模量 E = 25GPa，泊松比 v = 0.15，裂隙倾角 α = 60°，裂隙表面摩擦系数 μ = 0.38，竖向地应力 σ_v = 30MPa，水平地应力 σ_h = 10MPa，岩石断裂韧度 K_{IC} = 1MPa·m$^{1/2}$，原生裂隙长度 a = 0.006m，非欧几何参数 C = 2150000，C_1 = 4950000，q = 2.05×10^8，硐室半径 r_0 = 7m。

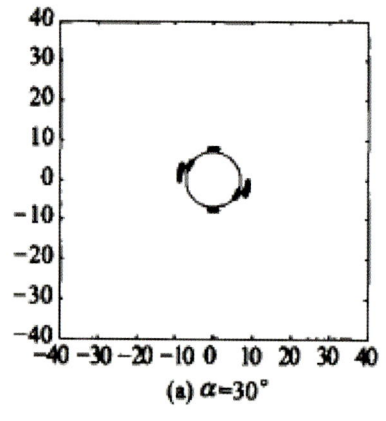

(a) α=30°

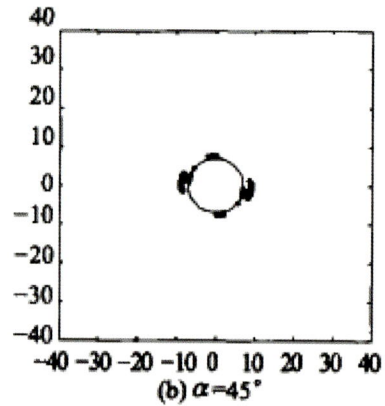

(b) α=45°

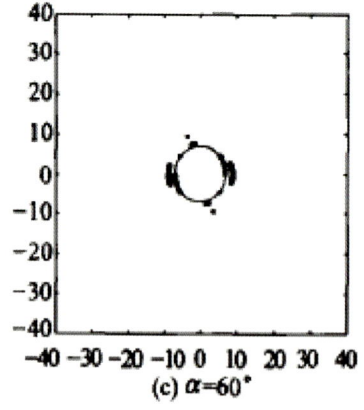

(c) α=60°

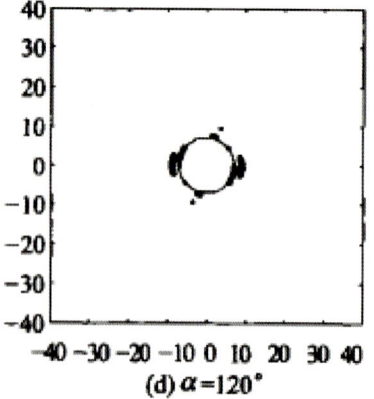

(d) α=120°

图2

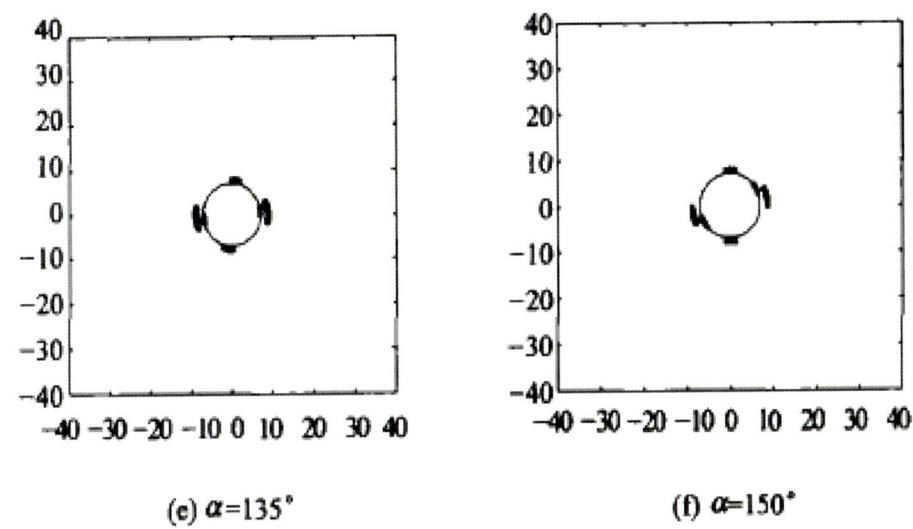

图2 岩爆位置随裂隙倾角的变化规律

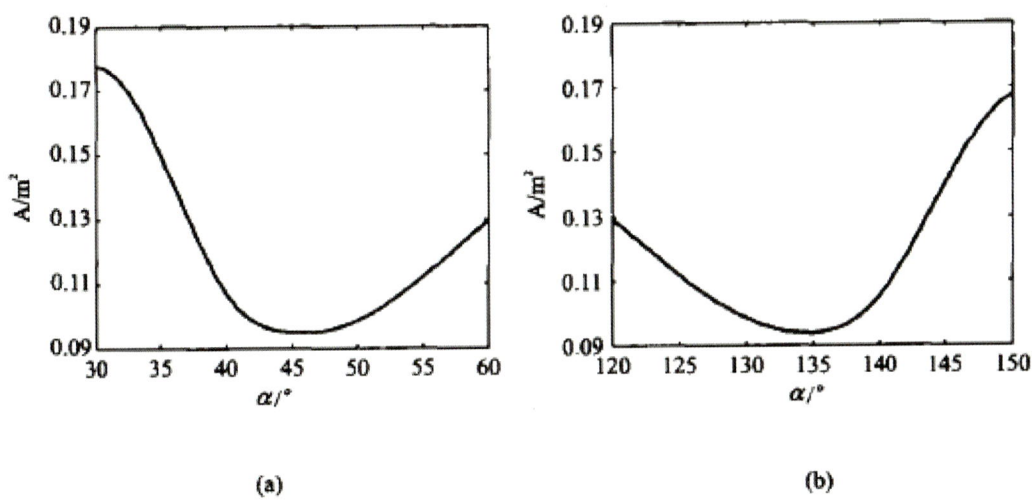

图3 岩爆面积随裂隙倾角的变化规律

图4表明：岩爆位置随裂隙间距变化而变化。图5表明：岩爆面积随裂隙间距增大而减小。

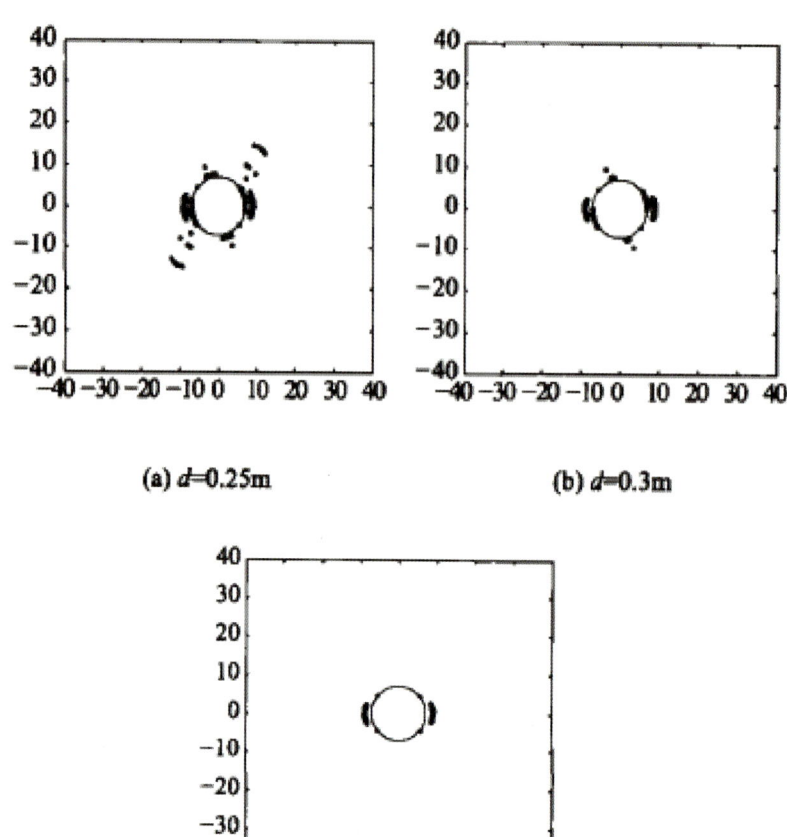

(a) $d=0.25$m

(b) $d=0.3$m

(c) $d=0.35$m

图4 岩爆位置随裂隙间距的变化规律

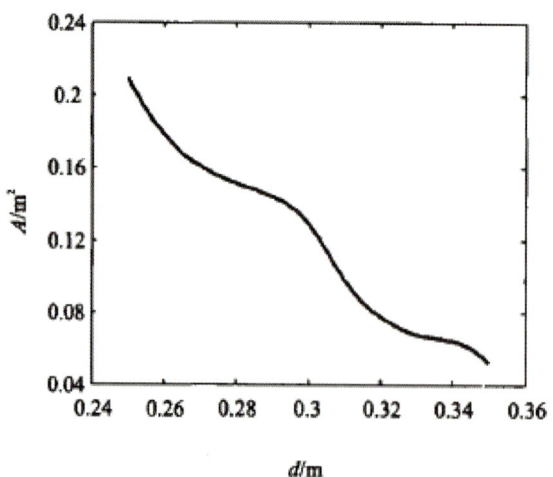

图5 岩爆面积随裂隙间距的变化规律

4.3 原生裂隙长度对岩爆的影响

弹性模量 $E = 25\text{GPa}$,泊松比 $v = 0.2$,裂纹倾角 $\alpha = 30°$,裂隙表面摩擦系数 $\mu = 0.2$,竖向地应力 $\sigma_v = 30\text{MPa}$,水平地应力 $\sigma_h = 10\text{MPa}$,岩石断裂韧度 $K_{IC} = 1\text{MPa}\cdot\text{m}^{1/2}$,裂隙间距 $d = 0.6\text{m}$,非欧几何参数 $C = 1862000$,$C_1 = 1862000$,$q = 3.2\times10^8$,硐室半径 $r_0 = 7\text{m}$。

图6表明:岩爆位置随原生裂隙长度的变化而变化。图7表明:岩爆面积随原生裂隙长度的增加而增加。

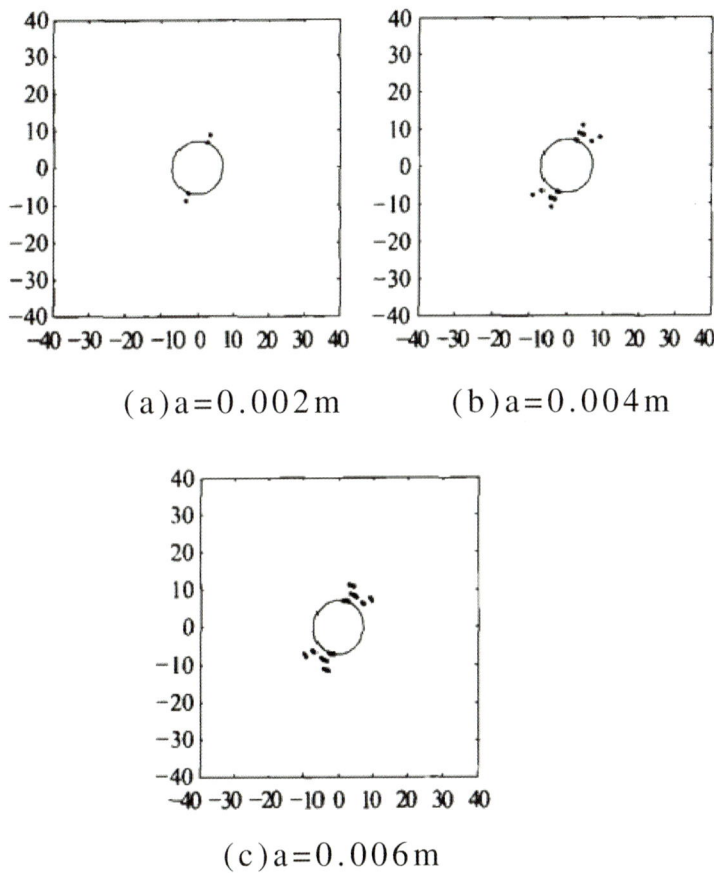

(a) $a = 0.002\text{m}$ (b) $a = 0.004\text{m}$

(c) $a = 0.006\text{m}$

图 6 岩爆位置随原生裂隙长度的变化规律

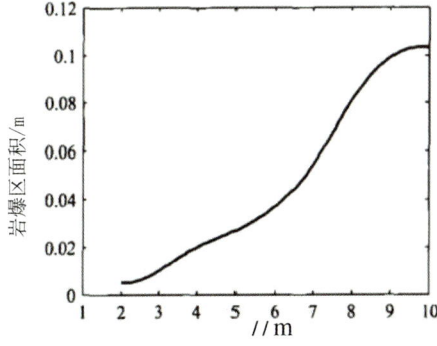

图 7 岩爆面积随原生裂隙长度的变化规律

5. 结论

将岩体视为含初始原生裂隙的缺陷体。当裂纹扩展消耗的能量较小，破裂区岩体储存的弹性应变能大于原生裂隙和次生裂纹扩展消耗的总能量时，剩余的弹性应变能会突然释放，产生岩爆。根据宏观裂隙出现的条件和次生裂纹扩展过程的耗散能小于岩体储存的弹性应变能的条件，建立了岩爆的判据。数值计算表明：

(1) 岩爆位置随原生裂隙倾角、长度和间距的变化而变化。

(2) 岩爆面积随裂隙倾角变化而变化，且裂隙倾角为45°或135°时岩爆的面积最小。

(3) 岩爆面积随裂隙间距增大而减小，岩爆面积随原生裂隙的长度增加而增加。

Crack tip higher order stress fields for functionally graded materials with generalized form of gradation

Xiufa Yan[1,2], Qihu Qin[1], Hongbiao Lu[1],
Wei Wang[2], Ao Sun[2]

1. Engineering Institute of Engineering Corps, People's Liberation Army University of Science and Technology, Nanjing 210007, China;
2. People's Liberation Army 91550 Unit, Dalian 116023, China

© Central South University Press and Springer-Verlag Berlin Heidelberg 2010

Abstract: A generalized form of material gradation applicable to a more broad range of functionally graded materials (FGMs) was presented. With the material model, analytical expressions of crack tip higher order stress fields in a series form for opening mode and shear mode cracks under quasi-static loading were developed through the approach of asymptotic analysis. Then, a numerical experiment was conducted to verify the accuracy of the developed expressions for representing crack tip stress fields and their validity in full field data analysis by using them to extract the stress intensity factors from the results of a finite element analysis by local collocation and then comparing the estimations with the existing solution. The expressions show that nonhomogeneity parameters are embedded in the angular functions associated with higher terms in a recursive manner and at least the first three terms in the expansions must be considered to explicitly account for material nonhomogeneity effects on crack tip stress fields in the case of FGMs. The numerical experiment further confirms that the addition of the nonhomogeneity specific terms in the expressions not only improves estimates of stress intensity factor, but also gives consistent estimates as the distance away from the crack tip increases. Hence, the analytical expressions are suitable for the representation of crack tip stress fields and the analysis of full field data.

Key words: functionally graded materials; crack tip; nonhomogeneity; asymptotic analysis; higher order stress field

1 Introduction

Recent advances in materials processing and engineering have led to a new class of materials called functionally graded materials (FGMs) that possess spatially varying composition and mechanical properties. FGMs are unique in that they offer the possibility of tailoring their constituents and gradation to match the end use [1]. Therefore, they are currently considered for many demanding engineering applications including military armor, earth moving equipment, thermal barrier coatings for turbine blades and internal combustion engines, machine tools and so on. Even though the initial research on FGMs is largely motivated by the practical applications of the concept in a wide variety of thermal shielding problems, materials with graded physical properties have almost unlimited potential in many other technological applications.

Experimental results show that fracture remains a key failure mode of these materials [2]. Consequently, FGMs pose new challenges in terms of modeling, characterization and optimization from the point of view of fracture behavior [3]. So far, considerable researches on various aspects of fracture in the FGM have been carried out. In the early work, EISCHEN [4], using a series representation of variation of elastic modulus, showed that the leading term in the crack tip stress field was of the inverse square root form for any functional form of elastic modulus variation. SHUKLA et al [5] reviewed the elementary concepts of fracture mechanics of FGMs and then presented a comprehensive experimental study of dynamic crack growth using the optical method of reflection photoelasticity and high-speed photography. Assuming that the shear modulus varied exponentially, the behavior of two parallel cracks in FGMs subjected to a tensile stress loading was investigated using Schmidt method by LIANG [6] and it was also revealed that the crack shielding effect was present in FGMs. MA et al [7] considered the plane strain problem of a crack in a functionally graded strip with a power form shear modulus by means of Fourier transform and obtained both mode I and mode II stress intensity factors (SIFs) at the crack tip of FGMs for a pure normal loading or a pure shear loading. CHEN and CHUE [8] studied antiplane fracture problems of two bonded functionally graded strips with internal cracks by using Gauss–Chebyshev integration formula for numerically solving a system of singular integral equations and further discussed the effects of material nonhomogeneous parameters, crack locations and boundary conditions on the stress intensity factor. Subsequently, CHEN et al [9] studied an antiplane collinear crack problem for a strip of FGMs with mixed boundary conditions by the similar method. The primary conclusions of the above investigations are that the classical inverse square root singular nature of the crack tip stress field is preserved in FGMs, but the stress intensity factor is influenced by material nonhomogeneity and the structure of the stress field away from crack tip is significantly altered. In addition, the SIFs for cracks in FGMs for different geometry and loading conditions are provided.

However, because of the mathematical difficulties arising from the fact that the properties of FGMs can vary arbitrarily in space, for the gradation form in a general sense, crack tip higher order stress fields in FGMs embodying nonhomogeneity specific terms for individual stress components have not been developed. Such stress fields are necessary in the analysis of full field experimental data obtained through techniques such as photoelasticity and coherent gradient and sensing (CGS) or computational results obtained by numerical methods. Most of stress fields currently available are obtained through integral transform method and the inversion has to be carried out numerically, making these stress field expressions not feasible for the extraction of fracture parameters from experimental data or numerical results [10–11]. PARAMESWARAN and SHUKLA [12] developed the structure of the first stress invariant and the out of plane displacement to bring out the effects of nonhomogeneity. However, nonhomogeneity specific terms for individual stress components were not derived. LEE [13] provided three-term asymptotic expansions of crack tip stress fields for a crack in an FGM. But, universal differential equations to solve arbitrary order expansions were not developed, and then characteristics of crack tip higher order stress fields for FGMs were not explicitly revealed. It is also important to point out that, owing to complexity of the problem, all the existing studies on crack tip stress fields have to be confined to exponential or linear variation of material properties. Moreover, so far, the validity of asymptotic expansions available for representing crack tip fields has not been investigated and verified.

In this work, to overcome the disadvantages of the works before-mentioned, a generalized form of material gradation applicable to a more broad range of FGMs was presented. The crack tip higher order stress fields for opening mode and shear mode cracks embodying nonhomogeneity specific terms for individual stress components under quasi-static loading were developed under the material model through the approach of asymptotic analysis. Not only were the six-term expansions of the stress fields obtained, but also universal differential equations for solving arbitrary term expansions were derived. On the basis of the analytical expressions, characteristics and structure of crack tip higher order stress fields for FGMs were completely revealed. In addition, in order to verify the accuracy of the higher order stress fields above-developed for representing crack tip stress fields and their validity in the full field data analysis, a numerical experiment was conducted by using them to extract the stress intensity factor from full field data obtained from a finite element analysis.

2 Analytical solutions of crack tip higher order stress fields for FGMs with generalized form of gradation

2.1 Theoretical formulation

The physical properties of FGMs varied from point to point in the material. At a given point in the material, the properties could be assumed to be the same in all directions, and hence at a continuum level FGMs were isotropic nonhomogeneous solids. For instance, large bulk FGMs produced by spark plasma sintering (SPS) technique might be modeled as isotropic materials. Therefore, a nonhomogeneous isotropic model might be appropriate in studying the mechanics of FGMs. In this work, crack tip higher order stress fields were investigated under the model.

The plane elasticity problem of a finite crack lying in a medium of FGMs was considered here. Without losing generality, it was assumed that the material properties, such as elastic modulus E and Poisson ratio μ, varied according to

$$\begin{cases} E = E(x, y) \\ \mu = \mu(x, y) \end{cases} \quad (1)$$

where E and μ are continuous, bounded and at least piecewise differentiable functions. The in-plane stress components (σ_{ij}, $i,j \in \{x,y\}$) can be defined in terms of Airy's stress function $F=F(x, y)$, as given by

$$\begin{cases} \sigma_{xx} = \dfrac{\partial^2 F}{\partial y^2} \\ \sigma_{yy} = \dfrac{\partial^2 F}{\partial x^2} \\ \sigma_{xy} = -\dfrac{\partial^2 F}{\partial x \partial y} \end{cases} \quad (2)$$

Substituting Eq.(2) into the compatibility equation through Hooke's law, the compatibility equation for the plane problem can be expressed as

$$\frac{1}{E}\nabla^4 F + 2[\frac{\partial}{\partial x}(\frac{1}{E})\frac{\partial \nabla^2 F}{\partial x} + \frac{\partial}{\partial y}(\frac{1}{E})\frac{\partial \nabla^2 F}{\partial y}] +$$
$$\nabla^2(\frac{1}{E})(\frac{\partial^2 F}{\partial x^2} + \frac{\partial^2 F}{\partial y^2}) - \frac{\partial^2}{\partial y^2}(\frac{1+\mu}{E})\frac{\partial^2 F}{\partial x^2} -$$
$$\frac{\partial^2}{\partial x^2}(\frac{1+\mu}{E})\frac{\partial^2 F}{\partial y^2} + 2\frac{\partial^2}{\partial x \partial y}(\frac{1+\mu}{E})\frac{\partial^2 F}{\partial x \partial y} = 0 \quad (3)$$

where $\nabla^2 = \frac{\partial^2}{\partial x^2} + \frac{\partial^2}{\partial y^2}$, $\nabla^4 = \frac{\partial^4}{\partial x^4} + 2\frac{\partial^4}{\partial x^2 \partial y^2} + \frac{\partial^4}{\partial y^4}$. Eq.(3) is for the generalized plane stress. The corresponding equation for plane strain is obtained by replacing E and μ by $E/(1-\mu^2)$ and $\mu/(1-\mu)$, respectively.

2.2 Generalized gradation form and governing equations

Consider the plane elasticity problem containing a finite crack with length of 2S on y=0 plane as shown in Fig.1. In the previous study [14], it was shown that the effect of Poisson ratio μ on crack tip stress fields was rather negligible. Consequently, μ might be assumed to be constant in the study of fracture mechanics of FGMs. In addition, the variation of the elastic and physical properties of FGMs is in general limited to a single direction. Hence, in order to describe more extensive variation of material property of FGMs, a generalized form of gradation is presented as follows:

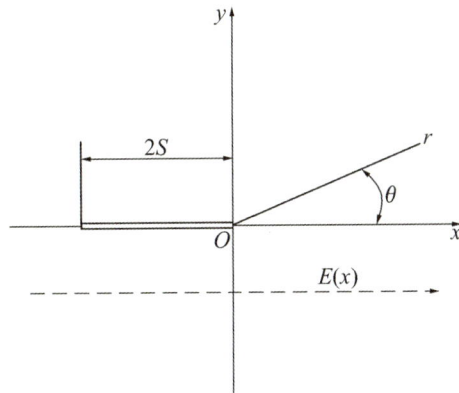

Fig.1 Crack along gradation direction in FGM medium and reference coordinate

$$\begin{cases} E = E_0(1+\beta x)^k \\ \mu = C \end{cases} \quad (4)$$

where E_0 is the elastic modulus at the crack tip (x=y=0); and β and k are nonhomogeneity parameters.

Substituting Eq.(4) into Eq.(3), under the gradation model, the equation to be satisfied by the Airy's stress function $F=F(x, y)$ is

$$\nabla^4 F + \beta x(2+\beta x)\nabla^4 F - 2\beta k(1+\beta x)\frac{\partial}{\partial x}\nabla^2 F +$$
$$k\beta^2(1+k)\frac{\partial^2 F}{\partial x^2} - \mu k \beta^2(1+k)\frac{\partial^2 F}{\partial y^2} = 0 \quad (5)$$

For FGMs, the stress fields near the crack tip maintained the classical inverse square root singularity [4]. Therefore, expressing $F(x, y)$ in terms of polar coordinates centered at the crack tip shown in Fig.1 as $F(r, \theta)$, it can be assumed that the stress function $F(r, \theta)$ can be expanded by a power series as follows:

$$F(r, \theta) = \sum_{n=1}^{\infty} F_n(r, \theta) = \sum_{n=1}^{\infty} r^{\frac{n}{2}+1} f_n(\theta) \quad (6)$$

where $F_n = r^{\frac{n}{2}+1} f_n(r, \theta)$. Switching Eq.(5) into polar coordinates, the above series on substitution into Eq.(5) taking the form of polar coordinates leads to an infinite series involving differential equations associated with each power of r ($r^{-5/2}, r^{-2}, r^{-3/2}, r^{-1}, \cdots, r^{\frac{n}{2}-3}$). For Eq.(5) is valid for any arbitrary r, the differential equations corresponding to each power of r ($r^{-5/2}, r^{-2}, r^{-3/2}, r^{-1}, \cdots, r^{\frac{n}{2}-3}$) should be identically zero. This leads to a set of differential equations for $f_n(\theta)$ (n=1, 2, 3, $\cdots$) as follows:

$$f_1^{(4)}(\theta) + \frac{5}{2}f_1^{(2)}(\theta) + \frac{9}{16}f_1(\theta) = 0 \quad (7)$$

$$f_2^{(4)}(\theta) + 4f_2^{(2)}(\theta) = 0 \quad (8)$$

$$f_3^{(4)}(\theta) + \frac{13}{2}f_3^{(2)}(\theta) + \frac{25}{16}f_3(\theta) + 2\beta\cos\theta f_1^{(4)}(\theta) +$$
$$2k\beta\sin\theta f_1^{(3)}(\theta) + (5+k)\beta\cos\theta f_1^{(2)}(\theta) +$$
$$\frac{9}{2}k\beta\sin\theta f_1'(\theta) + (\frac{9}{8} + \frac{9}{4}k)\beta\cos\theta f_1(\theta) = 0 \quad (9)$$

$$f_4^{(4)}(\theta) + 10f_4^{(2)}(\theta) + 9f_4(\theta) + 2\beta\cos\theta f_2^{(4)}(\theta) +$$
$$2k\beta\sin\theta f_2^{(3)}(\theta) + 8\beta\cos\theta f_2^{(2)}(\theta) +$$
$$8k\beta\sin\theta f_2'(\theta) = 0 \quad (10)$$

$$\vdots$$

$$f_n^{(4)}(\theta) + (\frac{n^2}{2}+2)f_n^{(2)}(\theta) + (\frac{n}{2}+1)^2(\frac{n}{2}-1)^2 f_n(\theta) +$$
$$2\beta\cos\theta f_{n-2}^{(4)}(\theta) + 2k\beta\sin\theta f_{n-2}^{(3)}(\theta) + (n^2 - 4n -$$
$$nk + 4k + 8)\beta\cos\theta f_{n-2}^{(2)}(\theta) + \frac{1}{2}k\beta n^2 \sin\theta f_{n-2}'(\theta) +$$
$$\frac{1}{8}(n^2 - 2nk - 8n + 8k + 16)\beta n^2 \cos\theta f_{n-2}(\theta) +$$
$$\beta^2 \cos^2\theta f_{n-4}^{(4)}(\theta) + k\beta^2 \sin 2\theta f_{n-4}^{(3)}(\theta) + \frac{1}{2}[(n^2 - 2kn -$$
$$8n - 2\mu k^2 - 2\mu k + 10k + 20)\cos^2\theta + 2k^2\sin^2\theta +$$

$2k]\beta^2 f_{n-4}^{(2)}(\theta) + \frac{1}{4}(n^2 - 6n + 8\mu k - 2n\mu + 8\mu - 2\mu nk -$

$2nk + 8k + 12)k\beta^2 \sin(2\theta) f'_{n-4}(\theta) + \frac{1}{32}[(n^4 - 16n^3 +$

$88n^2 - 192n - 4n^3 k + 44n^2 k + 4\mu n^2 k + 4\mu n^2 k^2 +$

$4n^2 k^2 - 144nk - 32\mu nk + 144k + 48\mu k + 48k^2 +$

$48\mu k^2 - 32\mu nk^2 - 32nk^2 + 144)\cos(2\theta) + (n^4 - 16n^3 +$

$88n^2 - 192n - 4n^3 k + 44n^2 k - 4\mu n^2 k - 4\mu n^2 k^2 +$

$4n^2 k^2 - 128nk + 16\mu nk + 112k - 16\mu k + 16k^2 +$

$16\mu nk^2 - 16\mu k^2 - 16nk^2 + 144)]\beta^2 f_{n-4}(\theta) = 0 \quad (11)$

In general, crack faces are supposed to be traction free, which correspond to the following boundary conditions:

$$\begin{cases} \sigma_{\theta\theta}(r, \pm\pi) = \left.\frac{\partial^2 F}{\partial r^2}\right|_{\theta=\pm\pi} = 0 \\ \sigma_{r\theta}(r, \pm\pi) = -\frac{1}{r}\frac{\partial^2 F}{\partial r \partial \theta} + \frac{1}{r^2}\frac{\partial F}{\partial \theta}\bigg|_{\theta=\pm\pi} = 0 \end{cases} \quad (12)$$

Substituting Eq.(6) into Eq.(12), Eq.(12) reduces to

$$\begin{cases} f_n(\pm\pi) = 0 \\ f'_n(\pm\pi) = 0 \end{cases} \quad (13)$$

Solving the above-developed set of differential equations for $f_n(\theta)$ ($n=1, 2, 3, \cdots$) under boundary conditions (Eq.(13)), it should be noticed from Eqs.(9)–(11) that they are coupled to the lower order unknown functions such as $f_1(\theta)$, $f_2(\theta)$, $f_{n-2}(\theta)$ and $f_{n-4}(\theta)$, and Eqs.(7)–(8) can be solved easily. Hence $f_n(\theta)$ ($n \geq 3$) can be obtained in a recursive manner. It is worth noting that Eqs.(11) and (13) are universal differential equations for solving arbitrary unknown functions $f_n(\theta)$ ($n \geq 5$). The solutions of crack tip high order stress fields for opening mode and shear mode are provided in the following sections.

2.3 Crack tip higher order stress fields for opening mode loading

After obtaining $F(r, \theta)$, by substituting $F(r, \theta)$ for $n \leq 6$ into Eq.(2) taking the form of polar coordinates (r, θ), and considering the symmetry of the normal stress components about the line of the crack, a six-term expansion of the crack tip stress field for opening mode loading is obtained as follows:

$\sigma_{yy} = r^{-\frac{1}{2}}\{\frac{1}{4}[5A_{11}\cos(\frac{\theta}{2}) - A_{11}\cos(\frac{5\theta}{2})]\} + r^{\frac{1}{2}}[(\frac{9A_{31}}{4} +$

$\frac{17k\beta A_{11}}{16})\cos(\frac{\theta}{2}) + (\frac{3A_{31}}{4} + \frac{k\beta A_{11}}{2})\cos(\frac{3\theta}{2}) -$

$\frac{k\beta A_{11}}{16}\cos(\frac{7\theta}{2})] + r^{\frac{3}{2}}[(\frac{15A_{51}}{4} + \frac{3\beta kA_{31}}{2} +$

$\frac{3k^2\beta^2 A_{11}}{8} - \frac{3k\beta^2 A_{11}}{8} + \frac{\mu k\beta^2 A_{11}}{8} +$

$\frac{\mu k\beta^2 A_{11}}{8})\cos(\frac{\theta}{2}) + (\frac{5A_{51}}{4} + \frac{13k\beta A_{31}}{16} +$

$\frac{19k^2\beta^2 A_{11}}{64} - \frac{23k\beta^2 A_{11}}{64} + \frac{7\mu k^2\beta^2 A_{11}}{192} +$

$\frac{7\mu k\beta^2 A_{11}}{192})\cos(\frac{3\theta}{2}) + (\frac{3k\beta A_{31}}{16} + \frac{15k^2\beta^2 A_{11}}{128} -$

$\frac{21k\beta^2 A_{11}}{128} - \frac{\mu k^2\beta^2 A_{11}}{128} - \frac{\mu k\beta^2 A_{11}}{128})\cos(\frac{5\theta}{2}) +$

$(\frac{5k\beta^2 A_{11}}{384} - \frac{k^2\beta^2 A_{11}}{128} + \frac{\mu k\beta^2 A_{11}}{384} +$

$\frac{\mu k^2\beta^2 A_{11}}{384})\cos(\frac{9\theta}{2})] + r^2[(-6A_{61} - 2k\beta A_{41} -$

$\frac{\mu k\beta^2 A_{21}}{2} - \frac{\mu k^2\beta^2 A_{21}}{2})\cos(2\theta) + 6A_{61} +$

$2k\beta A_{41} + \frac{\mu k\beta^2 A_{21}}{2} + \frac{\mu k^2\beta^2 A_{21}}{2}] \quad (14)$

$\sigma_{xx} = r^{-\frac{1}{2}}\{\frac{1}{4}[3A_{11}\cos(\frac{\theta}{2}) + A_{11}\cos(\frac{5\theta}{2})]\} + 4A_{21} +$

$r^{\frac{1}{2}}[(\frac{15A_{31}}{4} - \frac{17k\beta A_{11}}{16})\cos(\frac{\theta}{2}) - (\frac{3A_{31}}{4} -$

$\frac{k\beta A_{11}}{2})\cos(\frac{3\theta}{2}) + \frac{k\beta A_{11}}{16}\cos(\frac{7\theta}{2})] +$

$8rA_{41}\cos\theta + r^{\frac{3}{2}}[(-\frac{15A_{51}}{4} + \frac{3k\beta A_{31}}{2} +$

$\frac{\mu k\beta^2 A_{11}}{4} + \frac{\mu k^2\beta^2 A_{11}}{4} + \frac{k\beta^2 A_{11}}{4})\cos(\frac{\theta}{2}) +$

$(\frac{35A_{51}}{4} - \frac{13k\beta A_{31}}{16} + \frac{23k^2\beta^2 A_{11}}{64} - \frac{19k\beta^2 A_{11}}{64} -$

$\frac{7k\mu\beta^2 A_{11}}{192} - \frac{7k^2\mu\beta^2 A_{11}}{192})\cos(\frac{3\theta}{2}) + (-\frac{3k\beta A_{31}}{16} -$

$\frac{19k\beta^2 A_{11}}{128} + \frac{9k^2\beta^2 A_{11}}{128} - \frac{7\mu k\beta^2 A_{11}}{128} +$

$\frac{7\mu k^2\beta^2 A_{11}}{128})\cos(\frac{5\theta}{2}) + (\frac{k^2\beta^2 A_{11}}{128} - \frac{5k\beta^2 A_{11}}{384} -$

$\frac{\mu k\beta^2 A_{11}}{384} - \frac{\mu k^2\beta^2 A_{11}}{384})\cos(\frac{9\theta}{2})] + r^2[(18A_{61} +$

$2k\beta A_{41} + \frac{\mu k\beta^2 A_{21}}{2} + \frac{\mu k^2\beta^2 A_{21}}{2})\cos(2\theta) -$

$6A_{61} + 2k\beta A_{41} + \frac{\mu k\beta^2 A_{21}}{2} + \frac{\mu k^2\beta^2 A_{21}}{2}] \quad (15)$

$\sigma_{xy} = r^{-\frac{1}{2}}\{\frac{1}{4}[-A_{11}\sin(\frac{\theta}{2}) + A_{11}\sin(\frac{5\theta}{2})]\} + r^{\frac{1}{2}}[(-\frac{3A_{31}}{4} +$

$\frac{k\beta A_{11}}{16})\sin(\frac{\theta}{2}) - \frac{3A_{31}}{4}\sin(\frac{3\theta}{2}) + \frac{\beta kA_{11}}{16}\sin(\frac{7\theta}{2})] -$

$8rA_{41}\sin\theta + r^{\frac{3}{2}}[(-\frac{15A_{51}}{4} - \frac{5k^2\beta^2 A_{11}}{16} + \frac{3k^2\beta^2 A_{11}}{16}$

$$\frac{\mu k \beta^2 A_{11}}{16} - \frac{\mu k^2 \beta^2 A_{11}}{16}) \sin(\frac{\theta}{2}) - (\frac{15 A_{51}}{4} + \frac{3 \beta k A_{31}}{16} +$$
$$\frac{61 k \beta^2 A_{11}}{192} - \frac{11 k^2 \beta^2 A_{11}}{64} + \frac{17 \mu k \beta^2 A_{11}}{192} +$$
$$\frac{17 \mu k^2 \beta^2 A_{11}}{192}) \sin(\frac{3\theta}{2}) - (\frac{3 \beta k A_{31}}{16} - \frac{k \beta^2 A_{11}}{128} +$$
$$\frac{3 k^2 \beta^2 A_{11}}{128} + \frac{3 k \mu \beta^2 A_{11}}{128} + \frac{3 k^2 \mu \beta^2 A_{11}}{128}) \sin(\frac{5\theta}{2}) +$$
$$(\frac{k^2 \beta^2 A_{11}}{128} - \frac{5 k \beta^2 A_{11}}{384} - \frac{\mu k \beta^2 A_{11}}{384} -$$
$$\frac{\mu k \beta^2 A_{11}}{384}) \sin(\frac{9\theta}{2})] - r^2 [(12 A_{61} + 4 k \beta A_{41} +$$
$$\mu k \beta^2 A_{21} + \mu k^2 \beta^2 A_{21}) \sin(2\theta)] \quad (16)$$

where A_{11}, A_{21}, A_{31}, A_{41}, A_{51} and A_{61} are unknown constants that can be determined by fitting the stress field to experimental or numerical data.

2.4 Crack tip higher order stress fields for shear mode loading

Following the same procedure and keeping in view the dissymmetry nature of the shear mode problem, the first six terms of the expansion for the stress field are obtained as

$$\sigma_{yy} = r^{-\frac{1}{2}} \{ \frac{1}{4} [A_{12} \sin(\frac{\theta}{2}) - A_{12} \sin(\frac{5\theta}{2})] \} +$$
$$r^{\frac{1}{2}} [(-\frac{3 A_{32}}{4} + \frac{7 k \beta A_{12}}{16}) \sin(\frac{\theta}{2}) + (-\frac{3 A_{32}}{4} +$$
$$\frac{k \beta A_{12}}{2}) \sin(\frac{3\theta}{2}) - \frac{k \beta A_{12}}{16} \sin(\frac{7\theta}{2})] + r^{\frac{3}{2}} [(-\frac{15 A_{52}}{4} -$$
$$\frac{3 \beta k A_{32}}{2} - \frac{k \beta^2 A_{12}}{8} + \frac{5 k^2 \beta^2 A_{12}}{8} + \frac{3 \mu \beta^2 A_{12}}{8} +$$
$$\frac{3 \mu k^2 \beta^2 A_{12}}{8}) \sin(\frac{\theta}{2}) + (-\frac{15 A_{52}}{4} - \frac{27 k \beta A_{32}}{16} -$$
$$\frac{47 k \beta^2 A_{12}}{192} + \frac{49 k^2 \beta^2 A_{12}}{64} + \frac{77 \mu k \beta^2 A_{12}}{192} +$$
$$\frac{77 \mu k^2 \beta^2 A_{12}}{192}) \sin(\frac{3\theta}{2}) - (\frac{3 \beta k A_{32}}{16} + \frac{17 k \beta^2 A_{12}}{128} -$$
$$\frac{19 k^2 \beta^2 A_{12}}{128} - \frac{3 \mu k \beta^2 A_{12}}{128} - \frac{3 \mu k^2 \beta^2 A_{12}}{128}) \sin(\frac{5\theta}{2}) +$$
$$(\frac{5 k \beta^2 A_{12}}{384} - \frac{k^2 \beta^2 A_{12}}{128} + \frac{\mu k \beta^2 A_{12}}{384} +$$
$$\frac{\mu k^2 \beta^2 A_{12}}{384}) \sin(\frac{9\theta}{2})] \quad (17)$$

$$\sigma_{xx} = r^{-\frac{1}{2}} \{ \frac{1}{4} [7 A_{12} \sin(\frac{\theta}{2}) + A_{12} \sin(\frac{5\theta}{2})] \} +$$
$$r^{\frac{1}{2}} [(\frac{27 A_{32}}{4} - \frac{7 k \beta A_{12}}{16}) \sin(\frac{\theta}{2}) + (\frac{3 A_{32}}{4} -$$
$$\frac{k \beta A_{12}}{2}) \sin(\frac{3\theta}{2}) + \frac{k \beta A_{12}}{16} \sin(\frac{7\theta}{2})] + 8 r A_{42} \sin\theta +$$
$$r^{\frac{3}{2}} [(\frac{15 A_{52}}{4} - \frac{3 \beta k A_{32}}{2} + \frac{k \beta^2 A_{12}}{2} + \frac{k^2 \beta^2 A_{12}}{4} +$$
$$\frac{\mu k \beta^2 A_{12}}{2} + \frac{\mu k^2 \beta^2 A_{12}}{2}) \sin(\frac{\theta}{2}) + (\frac{55 A_{52}}{4} +$$
$$\frac{27 \beta k A_{32}}{16} - \frac{49 k^2 \beta^2 A_{12}}{64} + \frac{47 k \beta^2 A_{12}}{192} -$$
$$\frac{77 k \mu \beta^2 A_{12}}{192} - \frac{77 k^2 \mu \beta^2 A_{12}}{192}) \sin(\frac{3\theta}{2}) + (\frac{3 \beta k A_{32}}{16} -$$
$$\frac{23 k \beta^2 A_{12}}{128} + \frac{5 k^2 \beta^2 A_{12}}{128} - \frac{11 k \mu \beta^2 A_{12}}{128} -$$
$$\frac{11 k^2 \mu \beta^2 A_{12}}{128}) \sin(\frac{5\theta}{2}) + (\frac{k^2 \beta^2 A_{12}}{128} - \frac{5 k \beta^2 A_{12}}{384} -$$
$$\frac{\mu k \beta^2 A_{12}}{384} - \frac{\mu k^2 \beta^2 A_{12}}{384}) \sin(\frac{9\theta}{2})] + r^2 [12 A_{62} \sin(2\theta)] \quad (18)$$

$$\sigma_{xy} = r^{-\frac{1}{2}} \{ \frac{1}{4} [-3 A_{12} \cos(\frac{\theta}{2}) - A_{12} \cos(\frac{5\theta}{2})] \} +$$
$$r^{\frac{1}{2}} [(\frac{15 A_{32}}{4} - \frac{23 k \beta A_{12}}{16}) \cos(\frac{\theta}{2}) - \frac{3 A_{32}}{4} \cos(\frac{3\theta}{2}) -$$
$$\frac{k \beta A_{12}}{16} \cos(\frac{7\theta}{2})] + r^{\frac{3}{2}} [(-\frac{15 A_{52}}{4} + \frac{5 k \beta^2 A_{12}}{16} -$$
$$\frac{3 k^2 \beta^2 A_{12}}{16} + \frac{\mu k \beta^2 A_{12}}{16} + \frac{\mu k^2 \beta^2 A_{12}}{16}) \cos(\frac{\theta}{2}) +$$
$$(\frac{35 A_{52}}{4} + \frac{43 \beta k A_{32}}{16} + \frac{23 k \beta^2 A_{12}}{192} - \frac{203 k^2 \beta^2 A_{12}}{192} -$$
$$\frac{133 \mu k \beta^2 A_{12}}{192} - \frac{133 \mu k^2 \beta^2 A_{12}}{192}) \cos(\frac{3\theta}{2}) + (-\frac{3 \beta k A_{32}}{16} +$$
$$\frac{3 k \beta^2 A_{12}}{128} + \frac{7 k^2 \beta^2 A_{12}}{128} + \frac{7 \mu k \beta^2 A_{12}}{128} +$$
$$\frac{7 \mu k^2 \beta^2 A_{12}}{128}) \cos(\frac{5\theta}{2}) + (\frac{5 k \beta^2 A_{12}}{384} - \frac{k^2 \beta^2 A_{12}}{128} +$$
$$\frac{\mu k \beta^2 A_{12}}{384} + \frac{\mu k^2 \beta^2 A_{12}}{384}) \cos(\frac{9\theta}{2})] +$$
$$r^2 [6 A_{62} \cos(2\theta) - 6 A_{62}] \quad (19)$$

where A_{12}, A_{32}, A_{42}, A_{52} and A_{62} are unknown constants that can be determined by fitting the stress field to experimental or numerical data.

2.5 Characteristics of crack tip higher order stress fields for FGMs

It can be seen from the above-developed analytical expressions of crack tip higher order stress fields for FGMs, Eqs.(14)–(19), that they consist of two parts: the classical solution for homogenous materials and additional terms due to nonhomogeneity. By setting β or k to zero in the expressions, the stresses collapse to their homogeneous counterparts. However, unlike homogeneous materials, the stress fields contain terms having Poisson ratio for the presence of the Poisson ratio-dependent coefficient in the governing equation (Eq.(5)) as opposed to the bi-harmonic equation for homogeneous materials. The expressions further indicate that at least the first three terms in the expansions should be considered in the case of FGMs in order to explicitly account for nonhomogeneity effect on the structure of crack tip stress fields. However, the fourth term in the expansions is not affected by material nonhomogeneity. Thus, the first five terms of the expansions ought to be considered to further account for the effects of material nonhomogeneity on crack tip stress fields. Then, in order to entirely reveal the structure of crack tip higher order stress fields for FGMs, the six-term expansions above should be obtained. It can also be noted that the unknown coefficients of the angular functions associated with lower powers of r along with nonhomogeneity parameters, β and k, are embedded in the angular functions associated with higher powers of r in a recursive manner. Material nonhomogeneity shows that more evident influences on crack tip stress fields as power of r increases. A conclusion can be drawn that the angular functions corresponding to powers of r greater than 2 in the expansions all contain the terms whose coefficient includes nonhomogeneity parameters. In fact, the essential discrepancy between the crack tip higher order stress fields for FGMs presented here with ones of homogeneous materials is that the former will not satisfy the governing equations term-by-term, but only as a whole series in a more rigorous mathematical sense. Hence, the analytical expressions are of much more importance for representing the crack tip field for FGMs.

3 Verification of accuracy of crack tip higher order stress fields in representing crack tip stress fields and their validity in full field data analysis

3.1 Numerical example

Consider the beam sample of FGMs with the linear gradation form of elastic modulus for $k=1$ in Eq.(4) as shown in Fig.2, which is loaded in four-point bending under the plane stress condition as shown in Fig.3. In Figs.2–3, $E=E(x)$, $E_1=E(0)$, $E_2=E(W)$, $\mu=C$, a is the crack length, P is the loading, and W and L are the width and length of the beam, respectively.

Particular cases of sample and gradation form are chosen so that the results of the numerical experiment presented here can be compared with the existing solution [15]. The problem described above is a crack problem for opening mode loading. Then, the expansions

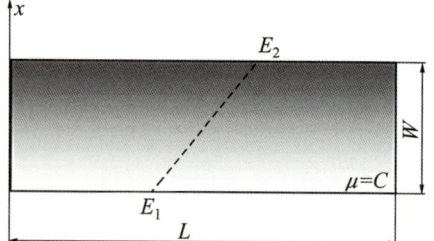

Fig.2 FGM medium with linear variation of elastic modulus

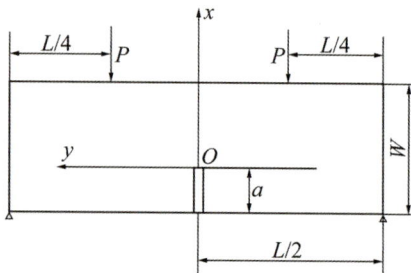

Fig.3 Beam sample of FGMs with linear gradation loaded in four-point bending

above-developed can be verified through Eq.(14). By setting $k=1$ in Eq.(14), stress σ_{yy} of crack tip fields for FGMs with the linear gradation of elastic modulus can be written as

$$\sigma_{yy} = r^{-\frac{1}{2}}\{\frac{1}{4}[5A_{11}\cos(\frac{\theta}{2}) - A_{11}\cos(\frac{5\theta}{2})]\} +$$
$$r^{\frac{1}{2}}[(\frac{9A_{31}}{4} + \frac{17\beta A_{11}}{16})\cos(\frac{\theta}{2}) + (\frac{3A_{31}}{4} +$$
$$\frac{\beta A_{11}}{2})\cos(\frac{3\theta}{2}) - \frac{\beta A_{11}}{16}\cos(\frac{7\theta}{2})] + r^{\frac{3}{2}}[(\frac{15A_{51}}{4} +$$
$$\frac{3\beta A_{31}}{2} + \frac{\mu\beta^2 A_{11}}{4})\cos(\frac{\theta}{2}) + (\frac{5A_{51}}{4} + \frac{13\beta A_{31}}{16} -$$
$$\frac{\beta^2 A_{11}}{16} + \frac{7\mu\beta^2 A_{11}}{96})\cos(\frac{3\theta}{2}) + (\frac{3\beta A_{31}}{16} -$$
$$\frac{3\beta^2 A_{11}}{64} - \frac{\mu\beta^2 A_{11}}{64})\cos(\frac{5\theta}{2}) + (\frac{\beta^2 A_{11}}{192} +$$
$$\frac{\mu\beta^2 A_{11}}{192})\cos(\frac{9\theta}{2})] + r^2[(-6A_{61} - 2\beta A_{41} -$$
$$\mu\beta^2 A_{21})\cos(2\theta) + 6A_{61} + 2\beta A_{41} + \mu\beta^2 A_{21}] \quad (20)$$

By setting β to zero in the above equation, stress σ_{yy} of crack tip fields for homogeneous materials can be obtained as

$$\sigma_{yy} = r^{-\frac{1}{2}}\{\frac{1}{4}[5A_{11}\cos(\frac{\theta}{2}) - A_{11}\cos(\frac{5\theta}{2})]\} +$$
$$r^{\frac{1}{2}}[(\frac{9A_{31}}{4})\cos(\frac{\theta}{2}) + (\frac{3A_{31}}{4})\cos(\frac{3\theta}{2})] +$$
$$r^{\frac{3}{2}}[(\frac{15A_{51}}{4})\cos(\frac{\theta}{2}) + (\frac{5A_{51}}{4})\cos(\frac{3\theta}{2})] +$$

$$r^2[(-6A_{61})\cos(2\theta)+6A_{61}] \qquad (21)$$

The accuracy of the asymptotic expansions in representing crack tip stress fields is evaluated by using Eq.(20) to extract the stress intensity factors from the results of a finite element analysis by local collocation and comparing them with the known solution provided by ROUSSEAU and TIPPUR [15]. Their validity in full field data analysis is evaluated by comparing the stress intensity factors obtained by using Eq.(20) for FGMs with those obtained by using its homogeneous counterpart, Eq.(21).

3.2 Numerical implement and computational results

Numerical simulations of FGMs require implementing the required property variation, i.e., elastic modulus in the present case, in the problem domain. In the finite element analysis, this is done by discretizing the rectangular domain into rows of elements and imposing a constant value of elastic modulus to each row of elements. Then, a stepwise change in modulus is realized. For a mesh consisting of eight-node plane stress elements of size, $a/200$ is employed around the crack tip to model the edge crack ($a/W=0.3$, $a/L=0.05$) in the present numerical experiment, as shown in Fig.4. The level of nonhomogeneity is considered, which will result in the elastic modulus varying by a factor (E_2/E_1) of 2.23 over a width of W. Due to the symmetry of the problem, only the right half of the specimen is considered in the analysis. The nodes in the region, $-\pi \leqslant \theta \leqslant 0$ and $0.1a \leqslant r \leqslant a$, are used as the local collocation points. Stress σ_{yy} at the each collocation node is calculated. Using σ_{yy} field, the coefficients of Eqs.(20)–(21) ($A_{11}, A_{21}, \cdots, A_{61}$) are obtained through an over deterministic least squares method, and then stress intensity factor K_I can be obtained by the relation, $K_I = A_{11}\sqrt{2\pi}$.

Fig.5 shows the estimates of stress intensity factor K_I extracted from the finite element stress σ_{yy} field. These estimates are obtained from the individual regions, $R_{i-1} < r < R_i$ ($R_i = 0.1a_i$, $i=2, 3, \cdots, 10$) around the crack tip, where R_i is the boundary of the data points, so that the accuracy and validity of crack tip higher order stress fields can be verified in different regions, respectively. The circles in Fig.5 indicate values of the stress intensity factor extracted using Eq.(20) for FGMs. It can be noticed that these values are closer to the results provided by ROUSSEAU and TIPPUR [15] than the estimates obtained by using Eq.(21) for homogeneous materials in all the regions of local collocation points. Moreover, these values are very consistent with the existing results as the distance away from the crack tip increases. On the contrary, the crosses in Fig.5, representing the estimates using Eq.(21), show that higher and higher errors are introduced as the distance away from the crack tip increases. It is not worthy to surprise that stress intensity factors obtained by both approaches in the regions very close to the crack tip are relatively coarse for numerical results of the standard finite element method are relatively inaccurate in the immediate vicinity of crack tip under the effect of singularity at the crack tip [16]. Contrarily, it is further confirmed that the addition of the nonhomogeneity specific terms in the above-developed crack tip higher order stress fields not only improves the estimates, but also gives consistent estimates of the stress intensity factor as the distance away from the crack tip increases. Hence, a conclusion can be drawn that the analytical expressions of crack tip higher order stress fields are suitable for the representation of crack tip stress fields and the analysis of full field data.

4 Conclusions

(1) Under a generalized form of material gradation applicable to a more broad range of FGMs, analytical expressions of crack tip higher order stress fields in a series form for opening mode and shear mode cracks are developed under quasi-static loading. It can be noted from the expressions that the unknown coefficients of the angular functions associated with lower order terms along with nonhomogeneity parameters are embedded in the angular functions associated with higher terms in a recursive manner. The essential characteristic of crack tip

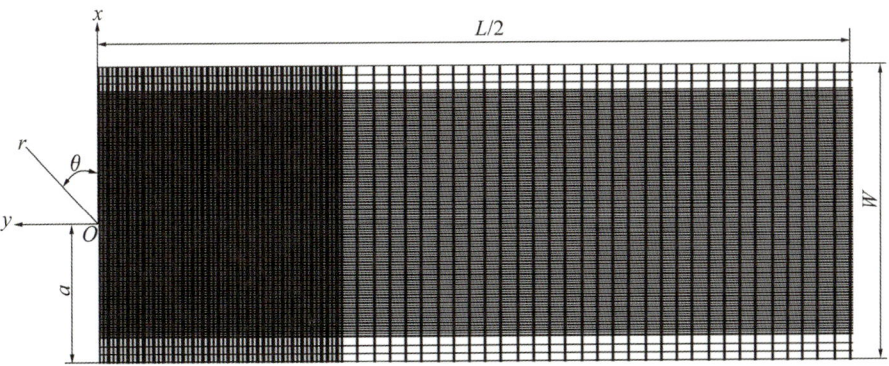

Fig.4 Finite element model for numerical experiment

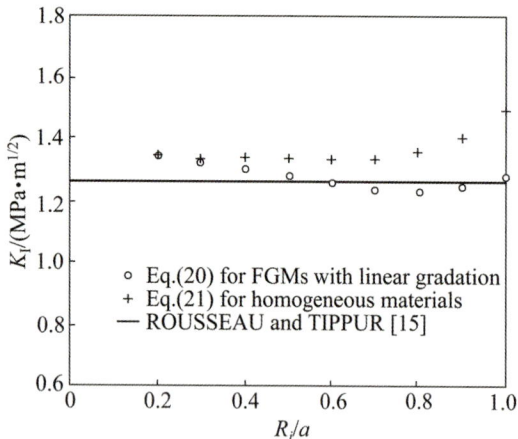

Fig.5 Local collocation estimates of stress intensity factor from finite element results

stress fields for FGMs is that the stress fields will not satisfy the governing equation term-by-term, but only as a whole series in a more rigorous mathematical sense.

(2) A numerical experiment is conducted to verify the accuracy of the above-developed analytical expressions in representing crack tip stress fields and their validity for full field data analysis. It is confirmed that the addition of the nonhomogeneity specific terms in the crack tip higher order stress fields not only improves the estimates, but also gives consistent estimates of stress intensity factor as the distance away from the crack tip increases. Hence, the analytical expressions of crack tip higher order stress fields are suitable for the representation of crack tip stress fields and the analysis of full field data. They can further be the base of numerical methods for fracture analysis of FGMs such as the boundary collocation method and the higher order approximate method.

References

[1] DIAS C M, SAVASTANO H, JOHN V M. Exploring the potential of functionally graded materials concept for the development of fiber cement [J]. Construction and Building Materials, 2010, 24(2): 140−146.

[2] GUO L C, NODA N. Fracture mechanics analysis of functionally graded layered structures with a crack crossing the interface [J]. Mechanics of Materials, 2008, 40(1): 81−99.

[3] AHANKARI S S, KAR K K. Processing and characterization of functionally graded materials through mechanical properties and glass transition temperature [J]. Materials Letters, 2008, 62(32): 3398−3400.

[4] EISCHEN J W. Fracture of nonhomogeneous materials [J]. International Journal of Fracture, 1987, 34(1): 3−22.

[5] SHUKLA A, JAIN N, CHONA A. A review of dynamic fracture studies in functionally graded materials [J]. Stain, 2007, 43(2): 76−95.

[6] LIANG Jun. Basic solution of two parallel mode-I cracks in functionally graded materials [J]. Science in China Series E: Technological Sciences, 2008, 51(9): 1380−1393.

[7] MA Jin-ju, ZHONG Zheng, ZHANG Chuan-zeng. Analysis of a crack in a functionally graded strip with a power form shear modulus [J]. Acta Mechanica Solida Sinica, 2009, 22(5): 465−473.

[8] CHEN Y J, CHUE C H. Mode III crack problems of two bonded functionally graded strips with internal cracks [J]. International Journal of Solids and Structures, 2009, 46(3): 331−343.

[9] CHEN Y Z, LIN X Y, WANG Z X. Antiplane elasticity crack problem for a strip of functionally graded materials with mixed boundary condition [J]. Mechanics Research Communications, 2010, 37(1): 50−53.

[10] DING Sheng-hu, LI Xing. Anti-plane problem of periodic interface cracks in a functionally graded coating-substrate structure [J]. International Journal of Fracture, 2008, 153(1): 53−62.

[11] ZHONG Zheng, CHENG Zhan-qi. Fracture analysis of a functionally graded strip with arbitrary distributed material properties [J]. International Journal of Solids and Structures, 2008, 45(28): 3711−3725.

[12] PARAMESWARAN V, SHUKLA A. Asymptotic stress fields for stationary cracks along the gradient in functionally graded materials [J]. Journal of Applied Mechanics, 2002, 69(3): 240−243.

[13] LEE K H. Characteristics of a crack propagating along the gradient in functionally gradient materials [J]. International Journal of Solids and Structures, 2004, 41(21): 2879−2898.

[14] DELALE F, ERDOGAN F. The crack problem for a nonhomogeneous plane [J]. Journal of Applied Mechanics, 1983, 50(3): 609−614.

[15] ROUSSEAU C E, TIPPUR H V. Evaluation of crack tip fields and stress intensity factors in functionally graded elastic materials: Cracks parallel to elastic gradation [J]. International Journal of Fracture, 2002, 114(1): 87−111.

[16] HILTON P D, SIH G C. Applications of the finite element method to calculation of stress intensity factors [C] // Mechanics of Fracture. Leyden: Noordhoff International, 1973: 426−483.

岩体结构面对应力波传播规律影响的研究进展

俞缙[1,2]，钱七虎[2]，赵晓豹[3]

(1. 华侨大学 岩土工程研究所，福建 厦门 361021；2. 中国人民解放军理工大学 工程兵工程学院，江苏 南京 210007；
3. 南京大学 地球科学与工程学院，江苏 南京 210093)

摘要：天然岩体中广泛存在着各种结构面，当应力波穿越岩体时，由于结构面的存在会产生波速下降和振幅衰减等现象。各种结构面条件下的应力波传播规律问题有着重要的理论与实际意义。主要针对结构面模型的建立、波传播规律的研究方法及主要结论，综述了微裂隙、宏观节理等复杂结构面条件下弹性波传播问题的研究现状，并提出了值得进一步研究的问题。

关键词：岩体；结构面；应力波；传播规律
中图分类号：O347　　**文献标志码**：A　　**文章编号**：1000-1093(2009)S2-0308-09

Research Progress on Effects of Structural Planes of Rock Mass on Stress Wave Propagation Law

YU Jin[1,2], QIAN Qi-hu[2], ZHAO Xiao-bao[3]

(1. Institute of Geotechnical Engineering, Huaqiao University, Xiamen 361021, Fujian, China;
2. Engineering Institute of Engineering Corps, PLA University of Science and Technology, Nanjing 210007, Jiangsu, China;
3. School of Earth Sciences and Engineering, Nanjing University, Nanjing 210093, Jiangsu, China)

Abstract: When a stress wave propagates in the rock mass, it is greatly slowed and attenuated due to the structural planes in the rock mass. The propagation law of the stress wave under the various structural plane conditions has important theoretical and practical significance. Taking aim at the establishment of the structural plane models, the methodology and the main results of wave propagation, the research status quo of the problem of elastic wave propagation under the conditions of microcracks and macrojoint was reviewed, and the problems investigated further were suggested.

Key words: rock mass; structural planes; stress wave; propagation law

0 引言

节理、裂隙对岩体中应力波的传播过程及规律影响很大。通常认为节理间的反射与透射可导致应力波振幅衰减及波速降低，且已被众多研究者的实验所验证。由于岩体结构破坏准则往往由波振幅门槛值(如质点位移、速度或加速度峰值)确定，因此在研究防护工程及岩体动力学问题时，节理裂隙对应力波的影响成为关键课题[1]。

Seinov and Chevkin(1968)[2]很早就指出应力波衰减取决于裂隙的数量、宽度以及充填物的波阻抗。

Morris 等(1964)[3]通过现场及室内实验发现横截有单节理的钻孔墙的声波测井信号振幅下降。Yu and Telford(1973)[4]发现单节理在受载荷情况下仍能反射 60 Hz～1 kHz 入射波 96 %的能量。Kleinberg 等(1982)[5]发现应力波穿越单节理时振幅下降并伴有波形转换。King 等(1986)[6]进行了跨孔(节理间距 0.2～0.5 m)纵波(波长约 0.1 m)测试，发现与平行节理方向应力波相比，穿越节理方向的应力波波速、波幅更小、波频更低。在诸多实验研究的同时，相关理论研究也层出不穷，主要可分为 2 类：1) 微裂隙的影响；2) 节理(众多共线微裂隙、微空隙和微接触体的集合体)

收稿日期：2009-10-20。
基金项目：福建省自然科学基金资助项目(2009J05125)。
本文原载于《兵工学报》(2009 年第 30 卷增刊 2)。

的影响。

1 微裂隙对应力波传播的影响

微裂隙对波的散射作用已通过透反射系数的测定所证实。Mal(1970)[7], Martin 和 Wickham(1981)[8], Achenbach 等(1982)[9], Martin 和 Wickham(1983)[10]分析了硬币状(penny-shaped)单一微裂隙的散射作用。Boström 和 Eriksson(1993)[11]用弹簧界面验证了双微裂隙的散射作用, Angle 和 Achenbach(1985a, 1985b)[12-13]研究了共列裂隙对法向及斜向入射波的散射。此外, Achenbach 和 Kitahara(1986)[14]对共列球形孔隙、Angle 和 Achenbach(1987)[15]对双列裂隙、Achenbach 和 Li(1986a)[16]对多列裂隙、Achenbach 和 Li(1986b)[17]对单列斜向掩蔽物情况、Mikata 和 Achenbach(1988)[18]对单列斜向裂隙情况、Piau(1979)[19]、Sotiropoulos 和 Achenbach(1988a, 1988b)[20-21]、Zhang 和 Gross(1993)[22]、Eriksson 等(1995)[23]对随机分布裂隙等情况进行了详细的研究。其中解析方法被一定程度的采用, 如 Rinehart(1975)[24]用弹性波在平面界面上的透反射理论研究裂隙长度比间距大很多的贯穿裂隙, 而对于较复杂的情况却很难得到解析解, 因此数值方法(如边界元法)被大量运用。Morris 等(1979)[25]对非线性裂隙的影响进行了计算研究并发现了高频谐振(higher harmonics)现象。Hirose 和 Achenbach(1993)[26]采用时间域边界元法对非线性变形单裂隙的散射作用进行了计算发现高频谐振现象在远场中仍有体现。Capuani 和 Willis(1997)[27]还对一维瞬态波入射线性与非线性变形平面界面进行了比较研究, 得到一些有益的结论。

除解析与数值研究以外, 等效介质理论(equivalent medium theories)也被广泛的运用在波在两相介质和裂隙(裂隙尺寸远小于入射波长)介质的传播问题中。该理论中岩块与裂隙被等效的看作为连续、均匀、各向异性介质, 通过波动方程, 利用有效弹模(effective elastic moduli)建立与波速、衰减的联系, 揭示不同的衰减机制。White(1983)[28]对于散射介质采用了复合模量(complex moduli)法(两个独立复合模量要求各向同性介质, 介质各向异性程度决定模量个数)。

Biot(1956a, 1956b)[29-30]较早研究了波在流体饱和(fluid-saturated)介质中的传播问题, 他对饱和平行圆柱孔隙对波衰减的影响提出了许多重要的假设。他利用液体与固体骨架间的粘性界面来体现非弹性特性, 导出了依赖频率的衰减系数并发现其在低频条件下与f^2、在高频条件下与$f^{1/2}$成正比。之后, McCann 等(1985)[31]将 Biot 理论扩展到实际大小分布孔隙的液体饱和(liquid-saturated)岩体中, 并发现衰减系数在 10 Hz~2.25 MHz 范围内与频率成线性关系。Eshelby(1957)[32]基于能量考虑建立了有效弹模(effective elastic moduli), 研究了含椭圆体的各向同性介质内部及外部的应力场、应变场。该方法后来被 Nur 和 Simmon(1969)[33]和 Nur(1971)[34]推广到各向异性及裂隙分布的岩石介质中去, 其中波速的各向异性通过考虑随压力变化的裂隙参数的方法来评价。Walsh(1965a, 1965b)[35-36]假设裂隙处无应力场相互作用, 建立了一种残余应变能法计算有效弹模, 该法仅适合于小裂隙密度。随后, O'Connell 和 Budianski(1974)[37]提出了一种自相容方法来计算含随机方向性的椭圆裂隙固体的有效弹模, 此时考虑了高度集中裂隙的应力场相互作用, 且是在干性或部分饱和情况下计算波速。O'Connell 和 Budianski(1977)[38]运用同样方法对流体饱和裂隙介质的波衰减作用作了进一步研究。Garbin 和 Knoff(1973, 1975a, 1975b)[39-41]提出在随机分布圆形空裂隙及液体饱和裂隙介质中, 考虑入射角及波偏振现象的波速变化时的有效模量计算方法。Chatterjee 等(1980)[42]运用同样方法研究了充填粘性液体的平行硬币状裂隙对波速、波散射及粘性衰减的影响, 并指出在低频情况下, 液态粘性比散射对波的影响更为显著。Hudson(1980, 1981)[43-44]提出一种滤波法(method of smoothing)计算有效弹模, 进而研究裂隙尺寸与分布集中程度都较小的介质的波速及衰减问题。研究表明波散射造成的衰减不仅与波频有关, 还与裂隙密度、裂隙半径与波长比值的三次方成正比。同时 Hudson(1981)[44]认为裂隙对波的散射及依赖频率的衰减系数与 Chatterjee 等(1980)[42]推导出的结果相同。Hudson(1986, 1990)[45-46]、Hudson 和 Knopoff(1989)[47]、Peacock 和 Hudson(1990)[48]将滤波法推广到裂隙密度无序且含多方向裂隙列的介质中。Hudson(1988)[49]修正了解析表达式来评价部分饱和椭圆裂隙介质的波衰减, 结果显示该情况因液体的流动而不同于完全干性和完全饱和裂隙情况。后来 Xu 和 King(1989, 1990)[50-51]的实验研究证实了 Hudson 理论, 并指出品质因子比波速对裂隙介质更为敏感。Ass'ad 等(1993)[52]进行了随机裂隙分布介质中 S 波散射实验, 与 Hudson 理论计算结果比较后发现在裂隙密度为 7%时吻合良好, 但在裂隙密度在 10%以上时偏差很大。

Mavko 和 Nur(1975)[53]和 O'Connell 和 Budianski(1977)[54]提出液体喷射理论来解释波在完全饱和裂隙岩体中的衰减现象。Mavko 和 Nur(1979)[55]又提出气泡在不完全饱和裂隙岩体中的运动机制。Johnston(1978)[56]对干性及饱和裂隙岩体中波的衰减(归结于裂隙面和颗粒边界的滑移摩擦)作了深入研究。Miksis(1988)[57]基于孔隙与所含液体的接触线运动(contact line movement)研究波的衰减。Murphy(1982a,1982b,1984)[58-60]通过对颗粒及裂隙边界的滑移实验验证了波的衰减,之后 Mochizuki(1982)[61]又将 Murphy 的实验结果与 Biot 模型的理论预测进行比较,并阐明其中的差异是由于实验频率范围内不同的衰减机制。Johnston 等(1979)[62]和 Crampin(1981)[63]详细讨论了其衰减机制,并得出摩擦消散主要存在于超声频域,流体喷射流动主要存在于低频范围,显然衰减机制的类型还与裂隙特性及饱和条件有关(例如摩擦更容易发生于窄裂隙,而气泡运动只发生在部分饱和裂隙中,喷射流只发生于完全饱和裂隙中)。Spencer(1981)[64]、Jones 和 Nur(1983)[65]、Winkler(1983)[66]的实验结果均体现出不同饱和程度下孔隙岩体的波速差异(1 Hz~1 MHz)。Spencer(1981)[64]和 Winkler(1985)[67]又揭示了部分饱和及完全饱和裂隙岩体中波的品质因子对频率的依赖性。Green 等(1993)[68]、Green 和 Wang(1994)[69]对不同约束应力、饱和程度的 Berea 砂岩和熔融玻璃样品在 0.03 Hz 到 0.1 kHz、0.6 kHz 到 1 MHz 波条件下的衰减及品质因子进行测试发现品质因子均随约束应力与波频率的变化而改变。

国内,刘斌等(1998)[70]对不同围压下孔隙度不同的干燥及水饱和岩样中的纵横波速度及衰减特性进行研究。赵明阶(1998)[71]、赵明阶,吴德伦(1999a,1999b)[72-73]进行了裂隙岩体在受荷条件下的声学特性研究。张光莹(2003)[74]对含分布裂缝岩石的弹性本构及波传播特性作了研究。李晓昭等(2003)[75]针对岩芯卸荷后裂隙张开导致的声学敏感性进行试验研究,俞缙等(2007)[76]基于小波理论对其作了进一步探讨。

2 节理对应力波传播的影响

有关应力波在裂隙岩体中的传播理论对节理(相对于波长而言长度很大但厚度很小)岩体是不适用的。当节理间含液体或固体充填物时常被看作在高波速介质中的低波速夹层处理,既层状介质(Ewing 等 1957,Brekhovskikh 1980,Bedford 和 Drumheller 1994)[77-79]。当 P、S 波射入介质的界面,除了发生反射与透射外,还会出现局部波(localized wave)。自由界面会产生 Rayleigh 波(固—固界面上为 Stoneley 波,固—液界面上为 Scholte 波)。局部波在自由边界临近处无散射传播,衰减比 P 波、S 波稍慢,因此造成破碎的潜力大,而在成层介质中,局部波是在自由边界处有散射(由多重反射引起)的传播。Thomson(1950)[80]、Haskell(1953)[81]首先推导出波在成层介质中传播的矩阵表达式,此时单一界面上的位移与应力与其它界面的有关。Aki 和 Richard(1980)[82]、Kennett(1983)[83]用传播矩阵法对多层介质中一维波传播及质点共振进行解析与数值研究。Watanabe 和 Sassa(1995)[84]对一维 P 波穿越节理区时的波速及振幅变化进行试验研究,此时节理带被看作低波速夹层介质,主要研究了层数、层厚、分布规律及其物理力学参数对应力波传播的影响,所得试验结果与传播矩阵法所得计算结果吻合较好。Rytov(1956)[85]、Helbig(1984)[86]推导了周期性层状介质(periodically layered medium)中波的散射方程,发现透射波频域中有明显的高能与低能透射带宽。Cetinkaya 和 Vakakis(1996)[87]采用双重积分变换法研究轴对称加载条件下有限双周期性层状介质的瞬态响应。

当节理厚度远小于入射波波长时,节理处应力场是连续的,而位移场是非连续的,一些学者提出位移不连续理论(节理被视为非粘结界面、不完全粘结界面、滑移界面、)来研究此类问题。Goodman(1976)[88]、Swan(1981)[89]针对拉张节理,Bandis 等(1983)[90]、Zhao(1987)[91]针对天然节理受法向或切向应力造成的张开、闭合、滑移现象进行研究,发现含节理岩样变形要大于完整岩样,这启发人们用位移不连续体对其进行描述,用节理比刚度(fracture specific stiffness)来描述节理面上应力与不连续位移的关系。Jones 和 Whitter(1967)[92]用位移不连续理论研究波传过弹性粘结的两个相异空间。Schoenberg(1980)[93]、Myer 等(1985,1990,1997)[94-96]、Myer(1991,1998)[97-98]、Pyrak-Nolte(1988,1996)[99-100]、Pyrak-Nolte 等(1990a)[101]、Suárez-Rivera(1992)[102]、Haugen 等(2000)[103]研究了 P、S 波穿越率线性变形单节理的传播特性。其中,Schoenberg(1980)[93]、Pyrak-Nolte(1988)[99]推导出线弹性位移不连续模型下,一维情况时以任意入射角射入干性(线性弹簧模型)单节理应力波(P、SV、SH 波)的透、反射系数精确解析解,仅适合小振幅波情况;Myer 等(1985)[94]通过声波传过单一且部份接触节理(通过调节节理接触面积调整刚度)的室内试验,试验结果与位移不连续模型理论解吻合较好;Suárez-Rivera(1992)[102]对 S 波传过充填粘土或流体的节理(被表示为质点速度不连续边界)时,做了解析计算与试验验证且结果良好。White(1965)[104]、Miller(1977,1978)[105-106]、Chen 等(1993)[107]研究了 S 波穿越率相关滑移节理,Pyrak-Nolte 和 Cook(1987)[108]、Pyrak-Nolte 等(1992,1996)[109,110]、Gu(1994)[111]、Pyrak-Nolte 和 Nolte(1995)[112]、Roy 和 Pyrak-Nolte(1995,1997)[113-114]研究了平面波穿越线性变形单节理问题。其中,White(1965)[104]、Miller

(1977,1978)[105-106]、Chen 等（1993）[107]建立了节理滑移率相关的位移不连续模型来研究 S 波在节理中的传播，Miller（1977,1978）[105-106]还提出了吸收系数（absorption coefficient）来反映由节理面滑移引起的能量损失；Pyrak-Nolte 等（1996）[110]推导出用流变模型（Kelvin 模型或 Maxwell 模型）描述饱和含水节理的透、反射系数解；Pyrak-Nolte and Cook（1987）[108]、Pyrak-Nolte 等（1992,1996）[109-110]、Roy 和 Pyrak-Nolte（1995,1997）[113-114]对 P、S 波在单一闭合天然节理中的传播效应进行了可控室内试验验证，他们认为节理对波传播的主要影响表现在信号延迟、信号衰减、高频滤波三方面。此外 Pyrak-Nolte（1995）[113]还对试验数据进行小波分析并发现速度频散现象与位移不连续理论预测结果吻合。Cai（2001）[115]、Zhao 和 Cai（2001）[116]在不考虑剪切波的情况下研究了正向入射大振幅 P 波（如临近震源冲击波及地下开凿和采矿中的爆炸波）穿越具有弹性非线性法向变形本构关系节理时的传播特性，将传统的线性位移不连续理论模型发展为非线性模型（BB 模型），获得了节理透射和反射系数的数值解并进行一系列的参数研究，同时对线性和非线性位移不连续理论模型作了比较。结果表明线性模型得到的透射和反射系数解是非线性解的特例，即当入射波的振幅很弱以至于在波传播中产生的最大节理闭合量相对于最大允许闭合量足够小的条件下，非线性与线性解等同，还发现节理非线性变形行为会引起一种高频波现象。此外，Frazer（1995）[117]在不假设大波长情况下对 SH 波传过平行多节理时衰减及波速降低现象进行了理论研究。

数值模拟方面，基于连续位移假设，Goodman 等（1968）[118]、Ghaboussi 等（1973）[119]在有限元法（FEM）计算中将节理视为特殊单元，Crotty 等（1985）[120]、Pande 等（1990）[121]的边界元法（BEM）计算中节理被视为边界面，Gu 等（1995,1996）[122-123]、Coates and Schoenberg（1995）[124]的有限差分（FDM）计算中节理被视为滑移线。其中 Gu 等（1995,1996）[122-123]对简谐波斜入射单节理时的透射、反射和波形转换（wave conversion）研究时发现当 SV 波以临界角（critical angle）入射时会出现 Head 波和不均匀表面纵波，同时通过解析与数值法研究界面波传过单节理问题。因受连续位移假设的节理位移与岩块旋转均过小的限制，基于位移不连续假设的离散元法（DEM）被较多采用。Lemos（1987）[125]用离散元法对坝基与断裂带处节理岩体进行动态分析，Dowding（1983a,1983b）[126-127]用离散元程序 UDEC 对节理岩体中的井巷工程进行计算分析，Chen 和 Zhao（1998）[128]、Chen 等（2000）[129]用UDEC 模拟爆炸波在节理岩体中的传播，Fan 等（2004）[130]针对应力波在节理岩体中传播时入射边界条件对模拟结果的影响进行探讨，提出了将 UDEC 与有限元程序 AUTODYN^{2D}耦合的方法。Zukas 等（2000）[131]、Scheffler（2000）[132]研究了动态建模中单元网格划分、材料本构和参数对模拟结果的影响。Nakagawa（1998）[133]、Nakagawa 等（2000a）[134]发现 P、S 波法向入射具有切向变形节理时仍会发生波形转换。Wu 等（1998）[135]、Hao 等（2001）[136]通过现场实验验证了节理对爆炸应力波传播的影响。

Nihei 等（1999）[137]指出波沿节理面方向传播时没有多重反射发生，但当应力波斜入射平行多节理时会发生多重反射与透射，而进行精确的叠加分析是十分困难的。Pyrak-Nolte 等（1990b）[138]、Myer 等（1995）[139]提出用 $|T_N|=|T_1|^N$（$|T_1|$ 为单节理透射系数、N 为节理数）来计算透射系数的简便方法。但 Pyrak-Nolte 等（1990b）[138]、Myer 等（1995）[139]、Hopkins 等（1988）[140]、Nakagawa 等（2000b）[141]的实验表明该公式只适用于不含多重反射现象的初至波情况，而且 Hopkins 等（1988）[140]、Myer 等（1995）[139]还发现节理间距相对波长较小情况下，$|T_N|$ 大于 $|T_1|^N$。Cai（2001）[115]、Cai 和 Zhao（2000）[142]、Zhao 等（2006a）[143]考虑多重反射，用特征线法与位移不连续模型结合的方式对线性变形平行多节理处垂直入射 P、S 波的传播进行理论研究，对节理间距与波长相对大小 ζ 不作限制。他们发现了节理间距门槛值 ζ_{thr} 与临界值 ζ_{cri}（$\zeta_{thr} > \zeta_{cri}$）两个重要指标，并依此将节理无量纲间距划分为 3 个部分，$|T_N|$ 在这 3 部分中有不同的变化规律。Zhao（2004）[144]、Zhao 等（2006b,2008）[145-146]延续上述思路研究了具有 BB 非线性变形和 Coulomb 滑移变形本构关系的平行多节理处纵波的传播和衰减特性，并且通过与 UDEC 模拟结果的对比得到了一些新的结论。

国内，张奇（1986）[147]认为当岩体纵波波速、节理内充填物纵波波速、波长和节理宽度满足式 $\alpha_{岩体}/\alpha_{节理} \cdot \lambda/\Delta r = n$ 时，应力波在节理内部能够发生 n 次反射，节理内将存在 $n+1$ 个波相互作用，并认为如果节理宽度在应力波波长的 10% 以下时，应力波幅值的 80% 可以通过节理，当 $\Delta r/\lambda \to 0$ 时，充填物对应力波传播的影响没有作用。李夕兵等（1992）[148]针对波斜向入射节理裂隙可能产生的滑移现象，引用库仑摩擦边界条件给出爆炸波通过节理裂隙带的透、反射关系。王明洋、钱七虎（1995a）[149]研究了爆炸波斜向通过闭合 n 条平行节理裂隙带的衰减规律，之后王明洋、钱七虎（1995b,1996）[150-151]通过微结构连续力学和多刚体系统动力学研究应力波作用下断层中颗粒体状透镜体的动力特性,同时进行了试验研究。王明洋等(1999)[152]还特别针对缓倾角断裂隔震效应的机理进行了定量计算。李宁等(1994)[153]进行了岩体节理在动载作用下的有限元分析。王卫华等(2004)[154]借鉴Lemaitre 假设思想建立节理非线性位移不连续模型,获得了垂直入射纵波在非线性法向变形节理时的透、反射系数的近似解析解,并与数值计算结果进行比较。雷卫东等(2006)[155]研究了二维波穿过单节理的透射率特性及其隐含意义。鞠杨等(2006)[156]研究了节理岩石的应力波动与能量耗散。郭易圆,李世海(2002)[157]对有限长岩柱中纵波传播规律与爆破振动波在节理岩体中进行离散元数值分析。俞缙等(2007)[158]分别利用 BB 模型和经典指数模型研究了不同非线性节理变形行为下弹性纵波的传播规律,之后俞缙等(2008,2009)[159-160]又在改进的节理非线性法向变形本构关系基础上,将原有的节理变形非线性程度固定情况发展到可调情况,对弹性纵波在单节理处的传播特性和衰减规律作了进一步研究。

3 展望

1) 目前对于应力波在具有单一或平行结构面岩体内传播的理论研究较为成熟,但当波在具有成组交叉节理岩体中传播时,会发生更复杂的多重透、反射及波的转换,该问题有待进一步研究。

2) 当应力波能量大到不但会使节理发生非线性变形,导致透射波在时间域上发生畸变而产生高频谐波,而且会使完整岩石本身发生非线性变形,导致波在空间域上发生变而产生孤立波和冲击波的情况,这两者间的耦合效应很值得深入研究。

3) 目前已有的各种数值分析方法在分析复杂动力学问题时都具有一定缺陷,采用多种计算方法的耦合计算可能是未来的一个发展方向。

参考文献

[1] Mohanty B. Physics of exploration hazard[M]. London: Taylor and Francis Group, 1998.

[2] Seinov N P, Chevkin A I. Effect of fissure on the fragmentation of a medium by blasting[J]. Journal of Mining Science, 1968, 4(3): 254-259.

[3] Morris R L, Grine D R, Arkfeld T E. Using compressional and shear acoustic amplitude for the location of fractures[J]. Journal of Petroleum Technology, 1964, 16(6): 623-632.

[4] Yu T R, Telford W M. An ultrasonic system for fracture detection in rock faces[J]. Canadian Mining and Metallurgical Bulletin, 1973, 66(7): 96-106.

[5] Kleinberg R L, Chow E Y, Plona T J, et al. Sensitivity and reliability of two fracture detection techniques for borehole application[J]. Journal of Petroleum Technology, 1984, 36(4): 657-663.

[6] King M S, Myer L R, Rezowalli J J. Experimental studies of elastic wave propagation in a columnar-jointed rock mass[J]. Geophysical Prospecting, 1986, 34(8): 1185-1199.

[7] Mal A K. Interaction of elastic waves with a penny-shaped crack[J]. International Journal of Engineering Science, 1970(8): 381-388.

[8] Martin P A, Wickham. Diffraction of elastic waves by a penny-shaped crack[D]. London: Mathematical and Physical Sciences, 1981, 378: 263-285.

[9] Achenbach J D, Gautesen A K, Mcmaken H. Ray methods for waves in elastic solids[M]. Boston: Pitman Publishing Inc, 1982.

[10] Martin P A, Wickham. Diffraction of elastic waves by a penny-shaped crack: Analytical and numerical results[D]. London: Mathematical and Physical Sciences, 1983, 390: 91-129.

[11] Boström A, Eriksson A S. Scattering by two penny-shaped cracks with spring boundary conditions[D]. London: Mathematical and Physical Sciences, 1993, 443(1917): 183-201.

[12] Angel Y C, Achenbach J D. Reflection and transmission of elastic waves by a periodic array of cracks[J]. Journal of Applied Mechanics, 1985a, 52(1): 33-46.

[13] Angel Y C, Achenbach J D. Reflection and transmission of elastic waves by a periodic array of cracks: oblique incidence[J]. Wave Motion, 1985b, 9(3): 375-388.

[14] Achenbach J D, Kitahara M. Reflection and transmission of an obliquely incident wave by an array of spherical cavities[J]. Journal of Acoustic Society of America, 1986, 80(4): 1209-1214.

[15] Angel Y C, Achenbach J D. Harmonic waves in an elastic solid containing a doubly periodic array of cracks[J]. Wave Motion, 1987, 9(5): 375-382.

[16] Achenbach J D, Li Z L. Reflection and transmission of scalar waves by a periodic array of screens[J]. Wave Motion, 1986a, 8(3): 225-234.

[17] Achenbach J D, Li Z L. Propagation of horizontally polarized transverse waves in a solid with a periodic distribution of cracks[J]. Wave Motion, 1986b, 8(4): 371-379.

[18] Mikata Y, Achenbach J D. Interaction of harmonic waves with a periodic array of inclined cracks[J]. Wave Motion, 1988, 10(1): 59-78.

[19] Piau M. Attenuation of a plane compressional wave by a random distribution of thin circular cracks[J]. International Journal of Engineering Science, 1979, 17: 151-167.

[20] Sotiropoulos D A, Achenbach J D. Reflection of elastic waves by a distribution of coplanar cracks[J]. Journal of Acoustic Society of America, 1988a, 84(2): 752-762.

[21] Sotiropoulos D A, Achenbach J D. Ultrasonic reflection by a planar distribution of cracks[J]. Journal of Nondestructive Evaluation, 1988b, 7(3/4): 123-135.

[22] Zhang C, Gross D. Wave Attenuation and dispersion in randomly cracked solids · Ⅱ: Penny-shaped cracks[J]. International Journal of Engineering Science, 1993, 31(6): 859-872.

[23] Eriksson A S, Boström A, Datta S K. Ultrasonic wave propagation through a cracked solid[J]. Wave Motion, 1995, 22(3): 297-310.

[24] Rinehart J S. Stress transients in solids[M]. Santa Fe: Hyperdynamics, 1975.

[25] Morris W L, Buck O, Inman R V. Acoustic harmonic generation due to fatigue damage in high-strength aluminium[J]. Journal of Applied Physics, 1979, 50(11): 6737-6741.

[26] Hirose S, Achenbach J D. Higher harmonics in the far field due to dynamic crack-face contacting[J]. Journal of Acoustic Society of America, 1993, 93(1): 142—147.

[27] Capuani D, Willis J R. Wave propagation in elastic media with cracks. Part I: transient nonlinear response of a single crack[J]. European Journal of Mechanics A·Solids, 1997, 16(3): 377—408.

[28] White J E. Underground sound[M]. New York: Elsevier, 1983.

[29] Biot M A. Theory of propagation of elastic waves in a fluid-saturated, porous solid I: low-frequency range[J]. Journal of Acoustic society of America, 1956a, 28(2): 168—178.

[30] Biot M A. Theory of propagation of elastic waves in a fluid-saturated, porous solid II: higher-frequency range[J]. Journal of Acoustic Society of America, 1956b, 28(2): 179—191.

[31] McCann C, McCann D M. A theory of compressional wave attenuation in non-cohesive sediments [J]. Geophysics, 1985, 50(8): 1311—1317.

[32] Eshelby J D. The determination of the elastic field of an ellipsoidal inclusion and related problems[C] // Proceedings of the Royal Society of London, London: Series A, Mathematical and Physical Sciences, 1957, 241: 376—396.

[33] Nur A, Simmons G. Stress-induced velocity anisotropy in rock: An experimental study[J]. Journal of Geophysical Research, 1969, 74(27): 6667—6674.

[34] Nur A. Effects of stress on velocity anisotropy in rock with cracks [J]. Journal of Geophysical Research, 1971, 76(8): 2022—2034.

[35] Walsh J B. The effect of cracks on the compressibility of rocks[J]. Journal of Geophysical Research, 1965a, 70(2): 381—389.

[36] Walsh J B. The effect of cracks on the uniaxial elastic compression of rocks[J]. Journal of Geophysical Research, 1965b, 70(2): 399—411.

[37] O'Connell R J, Budiansky B. Seismic velocities in dry and saturated cracked solids[J]. Journal of Geophysical Research, 1974, 79(35): 5412—5425.

[38] O'Connell R J, Budiansky B. Viscoelastic properties of fluid-saturated cracked solids[J]. Journal of Geophysical Research, 1977, 82(36): 5719—5730.

[39] Garbin H D, Knopoff L. The compressional modulus of a material permeated by a random distribution of circular cracks[J]. Quarterly of Applied Mathematics, 1973, 30: 453—464.

[40] Garbin H D, Knopoff L. The shear modulus of a material permeated by a random distribution of free circular cracks[J]. Quarterly of Applied Mathematics, 1975a, 33: 296—300.

[41] Garbin H D, Knopoff L. Elastic moduli of a medium with liquid-filled cracks[J]. Quarterly of Applied Mathematics, 1975b, 33: 301—303.

[42] Chatterjee A K, Mall A K, Knopoff M L, et al. Attenuation of elastic waves in a cracked, fluid-saturated solid[J]. Mathematical Proceedings of the Cambridge Philosophical Society, 1980, 88(3): 547—561.

[43] Hudson J A. Overall properties of a cracked solid[J]. Mathematical Proceedings of the Cambridge Philosophical Society, 1980, 88(2): 371—384.

[44] Hudson J A. Wave speeds and attenuation of elastic waves in material containing cracks[J]. Geophysical Journal of the Royal Astronomical Society, 1981, 64(1): 133—150.

[45] Hudson J A. A high order approximation to the wave propagation constants for a cracked solid [J]. Geophysical Journal of the Royal Astronomical Society, 1986, 87(1): 265—274.

[46] Hudson J A. Attenuation due to second-order scattering in material containing cracks[J]. Geophysical Journal International, 1990, 102(2): 485—490.

[47] Hudson J A, Knopoff L. Predicting the overall properties of composite materials with small-scale inclusion or cracks[J]. Pure and Applied Geophysics, 1989, 131(4): 551—576.

[48] Peacock S, Hudson J A. Seismic properties of rocks with distributions of small cracks[J]. Geophysical Journal International, 1990, 102(2): 471—484.

[49] Hudson J A. Seismic wave propagation through material containing partially saturated cracks[J]. Geophysical Journal, 1988, 92(1): 33—37.

[50] Xu S, King M S. Shear wave birefringence and directional permeability in fractured Rock[J]. Scientific Drilling, 1989, 1(1): 27—33.

[51] Xu S, King M S. Attenuation of elastic waves in a cracked solid[J]. Geophysical Journal International, 1990, 101(1): 169—180.

[52] Ass'ad J M, Tatham R H, Mcdonald J A, et al. A physical model study of scattering of waves by aligned cracks: Comparison between experiment and theory[J]. Geophysical Prospecting, 1993, 41(3): 323—339.

[53] Mavko G, Nur A. Melt squirt in the asthenosphere[J]. Journal of Geophysical Research, 1975, 80(11): 1444—1448.

[54] O'Connell R J, Budiansky B. Viscoelstic properties of fluid-saturated cracked solids[J]. Journal of Geophysical Research, 1977, 82(B36): 5719—5730.

[55] Mavko G, Nur A. Wave attenuation in partially saturated rocks[J]. Geophysics, 1979, 44(2): 161—178.

[56] Johnston D H. The attenuation of seismic waves in dry and saturated rock [D]. Cambridge: Massachusetts Institute of Technology, 1978.

[57] Miksis M J. Effects of contact line movement on the dissipation of waves in partially saturated rocks[J]. Journal of Geophysical Research, 1988, 93(6): 6624—6634.

[58] Murphy W F. Effects of microstructure and pore fluids on the acoustic properties of granular sedimentary materials[D]. Stanford: Stanford University, 1982.

[59] Murphy W F. Effects of partial water saturation on attenuation in Massilon sandstone and Vycor porous class[J]. Journal of Acoustic Society of America, 1982b, 71(6): 1458—1468.

[60] Murphy W F. Acoustic measures of partial gas saturation in tight sandstones[J]. Journal of Geophysical Research, 1984, 89(13): 11549—11559.

[61] Mochizuki S. Attenuation in partially saturated rocks[J]. Journal of Geophysical Research, 1982, 87(10): 8598—8604.

[62] Johnston D H, Toksöz M N, Timur A. Attenuation of seismic waves in dry and saturated rocks II: mechanisms[J]. Geophysics, 1979, 44(4): 691—711.

[63] Crampin S. A review of wave motion in anisotropic and cracked elastic media[J]. Wave Motion, 1981, 3(4): 342—391.

[64] Spencer J W. Stress Relaxations at low frequencies in fluid saturated rocks: Attenuation and modulus dispersion[J]. Journal of Geophysical Research, 1981, 86(3): 1803—1812.

[65] Jones T, Nur A. Velocity and attenuation in sandstone at elevated temperature and pressures[J]. Geophysical Research Letters, 1983, 10(2): 140-143.

[66] Winkler K W. Frequency dependent ultrasonic properties of high-porosity sandstones[J]. Journal of Geophysical Research, 1983, 88(11): 9493–9499.

[67] Winkler K W. Dispersion analysis of velocity and attenuation in Berea sandstone[J]. Journal of Geophysical Research, 1985, 90(8): 6793–6800.

[68] Green D H, Wang H F, Bonner B P. Shear wave attenuation in dry and saturated sandstone at seismic to ultrasonic frequencies[J]. International Journal of Rock Mechanics and mining Science and Geomechanics Abstracts, 1993, 30(7): 755–761.

[69] Green D H, Wang H F. Shear wave velocity and attenuation from pulse-echo studies of Berea sandstone[J]. Journal of Geophysical Research, 1994, 99(6): 11755–11763.

[70] 刘斌, Kern H, Popp T. 不同围压下孔隙不同的干燥及水饱和岩样中的纵横波速度及衰减[J]. 地球物理学报, 1998, 41(4): 537–546.
LIU Bin, Kern H, Popp T. Velocities and attenuation of p-and s-waves in dry and wet rocks with different porosities under different confining pressures[J]. Chinese Journal of Geophysics, 1998, 41(4): 537–546. (in Chinese)

[71] 赵明阶. 裂隙岩体在受荷条件下的声学特性研究[D]. 重庆: 重庆建筑大学, 1998.
ZHAO Ming-jie. The study on the acoustic characteristics of rock with crevices in loaded conditions [D]. Chongqing: Chongqing Jianzhu University, 1998. (in Chinese)

[72] 赵明阶, 吴德伦. 单轴受荷条件下岩石的声学特性模型与实验研究[J]. 岩土工程学报, 1999a, 21(5): 540–545.
ZHAO Ming-jie, WU De-lun. Ultrasonic properties of rock under loading and unloading: theoretical model and experimental research[J]. Chinese Journal of Geotechnical Engineering, 1999a, 21(5): 540–545. (in Chinese)

[73] 赵明阶, 吴德伦. 单轴加载条件下岩石声学参数与应力的关系研究[J]. 岩石力学与工程学报, 1999b, 18(1): 50–54.
ZHAO Ming-jie, WU De-lun. Ultrasonic velocity and attenuation of rock under uniaxial loading[J]. Chinese Journal of Rock Mechanics and Engineering, 1999b, 18(1): 50–54. (in Chinese)

[74] 张光莹. 含分布裂缝岩石的弹性本构及波传播特性研究[D]. 长沙: 国防科学技术大学, 2003.
ZHANG Guang-ying. Research on the elastic constitutive model and wave propagation properties of rocks with distributed fractures[D]. Changsha: National University of Defense Technology, 2003. (in Chinese)

[75] 李晓昭, 安英杰, 俞缙, 等. 岩芯卸荷扰动的声学反应与卸荷敏感岩体[J]. 岩石力学与工程学报, 2003, 22(12): 2086–2092.
LI Xiao-zhao, AN Ying-jie, YU Jin, et al. Acoustic responses to rock core unloading-disturbance and unloading-sensitive rock mass[J]. Chinese Journal of Rock Mechanics and Engineering, 2003, 22(12): 2086–2092. (in Chinese)

[76] 俞缙, 赵维炳, 李晓昭, 等. 基于小波变换的岩芯卸荷扰动声学反应分析[J]. 岩石力学与工程学报, 2007, 26(增1): 3558–3564.
YU Jin, ZHAO Wei-bing, LI Xiao-zhao, et al. Acoustic responses to rock core unloading-disturbance based on wavelet transformation[J]. Chinese Journal of Rock Mechanics and Engineering, 2007, 26(Suppl1): 3558–3564. (in Chinese)

[77] Ewing W M, Jardetzky W S, Press F. Elastic waves in layered media[M]. New York: McGraw-Hill, 1957.

[78] Brekhovskikh L M. Waves in layered media [M]. New York: Academic Press, 1980.

[79] Bedford A, Drumheller D S. Introduction to elastic wave propagation[M]. Chichester: John Wiley and Sons, 1994.

[80] Thomson W T. Transmission of elastic waves through a stratified solid media[J]. Journal of Applied Physics, 1950, 21(1): 89–93.

[81] Haskell N A. The dispersion of surface waves on multilayered media[J]. Bulletin of the Seismological Society of America, 1953, 43(1): 17–34.

[82] Aki K, Richards P G. Quantitative seismology, vol. 1: theory and methods[M]. New York: W H Freeman and Company, 1980.

[83] Kennett B L N. Seismic wave propagation in stratified media[M]. London: Cambridge University Press, Cambridge, 1983.

[84] Watanabe T, Sassa K. Velocity and amplitude of p-waves transmitted through fractured zones composed of multiple thin low-velocity layers[J]. International Journal of Rock Mechanics and Mining Science and Geomechanics Abstract, 1995, 32(4): 31–324.

[85] Rytove S M. Acoustical properties of a thinly laminated medium[J]. Soviet Physical Acoustic, 1956, 2: 68–80.

[86] Helbig K. Anisotropy and dispersion in periodically layered media[J]. Geophysics, 46940: 364–373.

[87] Cetinkaya C, Vakakis A F. Transient axisymmetric stress wave propagation in weakly coupled layered structures[J]. Journal of Sound and Vibration, 1996, 194(3): 389–416.

[88] Goodman R E. Methods of geological engineering in discontinuous rocks[M]. New York: West, 1976.

[89] Swan G. Tribology and the characterization of rock joints[C]// Proceeding of 22nd US Symposium on Rock Mechanics, Massachusetts, 1981: 402–407.

[90] Bandis S C, Lumsden A C, Barton N R. Fundamentals of rock fracture deformation[J]. International Journal of Rock Mechanics and Mining Sciences and Geomechnics Abstracts, 1983, 20(6): 249–268.

[91] Zhao J. Experimental studies of the hydro-thermo-mechanical behaviour of joints in granite[D]. London: Imperial College, University of London, 1987.

[92] Jones J P, Whittier J S. Waves at a flexibly bonded interface[J]. Journal of Applied Mechanics, 1967, 40(9): 905–909.

[93] Schoenberg M. Elastic wave behavior across linear slip interfaces[J]. Journal of Acoustic Society of America, 1980, 68(5): 1516–1521.

[94] Myer L R, Hopkins D, Cook N G W. Effects of contact area of an interface on acoustic wave transmission characteristics[C]// Proceedings of the 26th U. S. Rock Mechanics Symposium, Boston, 1985, 1: 565–572.

[95] Myer L R, Pyrak-Nolte L J, Cook N G W. Effects of single fracture on seismic wave propagation[A]. Proceedings of ISRM Symposium on Rock Fractures, Loen: 1990: 467–473.

[96] Myer L R, Nihei K T, Nakagawa S. Dynamic properties of interfaces[A]. Proceedings of the 1st International Conference on Damage and Failure of Interface, Vienna, 1997: 47–56.

[97] Myer L R. Hydromechanical and seismic properties of fracture[C]// Proceeding of the 7th International Congress on Rock Mechanics, Aachen: 1991, 1: 397–404.

[98] Myer L R. Seismic wave propagation in fractured rock[C]// Proceeding of the 3rd International Conference on Mechanics of Jointed and Faulted Rock, Vienna, 1998: 29–38.

[99] Pyrak-Nolte L J. Seismic visibility of fractures [D]. Berkeley: Uni-

versity of California, 1988.

[100] Pyrak-Nolte L J. The seismic response of fractures and the interrelations among fracture properties[J]. International Journal of Mechanics and mining Sciences, 1996, 33(8): 787−802.

[101] Pyrak-Nolte L J, Myer L R, Cook N G W. Transmission of seismic waves across single natural fractures[J]. Journal of Geophysical Research, 1990a, 95(B6): 8617−8638.

[102] Suárez-Rivera R. The influence of thin clay layers containing liquids on the propagation of shear waves[D]. Berkeley: University of California, 1992.

[103] Haugen G U, Schoenberg M A. The echo of a fault or fracture[J]. Geophysics, 1984, 49(4): 364−373.

[104] White J E. Seismic waves: Radiation, transmission and attenuation[M]. New York: McGraw-Hill, 1965.

[105] Miller R K. An approximate method of analysis of the transmission of elastic waves through a Frictional boundary[J]. Journal of Applied Mechanics, 1977, 44: 652−656.

[106] Miller R K. The effects of boundary friction on the propagation of elastic waves[J]. Bulletin of Seismological Society of America, 1978, 68(4): 987−998.

[107] Chen W Y, Lovell C W, Haley G M, et al. Variation of shear wave amplitude during frictional sliding[J]. International Journal of Rock Mechanics and Mining Science and Geomechanics Abstracts, 1993, 30(7): 779−784.

[108] Pyrak-Nolte L J, Cook N G W. Elastic interface waves along a fracture[J]. Geophysical Research Letters, 1987, 14(11): 1107−1110.

[109] Pyrak-Nolte L J, Xu J, Haley G M. Elastic interface waves propagating in a fracture[J]. Physical Review Letters, 1992, 68(24): 3650−3653.

[110] Pyrak-Nolte L J, Roy S, Mullenbach B L. Interface waves propagated along a fracture[J]. Journal of Applied Geophysics, 1996, 35(2/3): 79−87.

[111] Gu B. Interface waves on a france in rock[D]. Berkeley: University of California, 1994.

[112] Pyrak-Nolte L J, Nolte D D. Wavelet analysis of velocity dispersion of elastic interface waves propagating along a fracture[J]. Geophysical Research Letters, 1995, 22(11): 1329−1332.

[113] Roy S, Pyrak-Nolte L J. Interface waves propagating along tensile fracture in dolomite[J]. Geophysical Research Letters, 1995, 22(20): 2773−2777.

[114] Roy S, Pyrak-Nolte L J. Observation of a distinct compressional-mode interface wave on a single fracture[J]. Geophysical Research Letters, 1997, 24(2): 173−176.

[115] Cai J G. Effects of parallel fractures on wave attenuation in rock masses[Ph. D. Thesis][M]. Singapore: Nanyang Technological University, 2001.

[116] Zhao J, Cai J G. Transmission of elastic p-waves across single fractures with a nonlinear normal deformational behavior[J]. Rock Mechanics and Rock Engineering, 2001, 34(1): 3−22.

[117] Frazer L N. SH propagation in rocks with planner fractures[J]. Geophysical Journal International, 1995, 122(1): 33−62.

[118] Goodman R E, Taylor L, Brekke T L. A model for the mechanics of jointed rock[J]. Journal of Soil Mechanics and Foundations Division, 1968, 94(3): 637−659.

[119] Ghaboussi J, Wilson E L, Isenberg J. Finite elements for rock joints and interfaces[J]. Journal of Soil Mechanics and Foundation Division, 1973, 99(10): 833−848.

[120] Crotty J M, Wardle L J. Boundary integral analysis of piecewise homogenous media with structural discontinuities[J]. International Journal of Rock Mechanics and mining Science and Geomechanics Abstracts, 1985, 22: 419−427.

[121] Pande G N, Beer G, Williams J R. Numerical methods in rock mechanics[M]. York: Wiley, New, 1990.

[122] Gu B, Nihei K T, Myer L R, Pyrak-Nolte L J. Fracture interface waves[J]. Journal of Geophysical Research, 1995, 101(1): 827−835.

[123] Gu B, Suáre-Rivera R, Nihei K T, et al. Incidence of plane waves upon a fracture[J]. Journal of Geophysical Research, 1996, 101(11): 25337−25346.

[124] Coates R T, Schoenberg M. Finite difference Modelling of Faults and Fractures[J]. Geophysics, 1995, 60(5): 1514−1526.

[125] Lemos J V. A distinct element model for dynamic analysis of jointed rock with application to dam foundation and fault motion[D]. Minneapolis: University of Minnesota, 1987.

[126] Dowding C H, Belytschko T B, Yen H J. A coupled finite element-rigid block method for transient analysis of rock caverns[J]. International Journal for Numerical and Analytical Methods in Geomechanics, 1983a, 7(1): 117−127.

[127] Dowding C H, Belytschko T B, Yen H J. Dynamic computational analysis of openning in jointed rock[J]. Journal of Geotechnical Engineering Division, 1983b, 109(12): 1551−1566.

[128] Chen S G, Zhao J. A study of UDEC modeling for blast wave propagation in jointed rock masses[J]. International Journal of Rock Mechanics and Mining Sciences, 1998, 35(1): 93−99.

[129] Chen S G, Cai J G, Zhao J, et al. Discrete element modeling of underground explosions in jointed rock mass[J]. Geological and Geotechnical Engineering, 2000, 18(1): 59−78.

[130] Fan S C, Jiao Y Y, Zhao J. On modeling of incident boundary for wave propagation in jointed rock masses using discrete element method[J]. Computers and Geotechnics, 2004, 31(1): 57−66.

[131] Zukas J A, Scheffler D R. Practical aspects of numerical simulations of dynamic events: effects of meshing[J]. Impact Engineering, 2000, 24(9): 925−945.

[132] Scheffler D R, Zukas J A. Practical aspects of numerical simulations of dynamic events: Material interfaces[J]. Impact Engineering, 2000, 24(8): 821−842.

[133] Nakagawa S. Acoustic resonance characteristics of rock and concrete containing fractures[D]. Berkeley: University of California, 1998.

[134] Nakagawa S, Nihei K T, Myer L R. Shear-induced conversion of seismic waves across single fractures[J]. International Journal of Rock Mechanics and Mining Sciences, 2000a, 37(1/2): 203−218.

[135] Wu Y K, Hao H, Zhou Y X, et al. Propagation characteristics of blast-induced shock waves in a jointed rock mass[J]. Soil Dynamics and Earthquake Engineering, 1998, 17(6): 407−412.

[136] Hao H, Wu Y K, Ma G W, et al. Characteristics of surface ground motions induced by blasts in jointed rock mass[J]. Soil Dynamics and Earthquake Engineering, 2001, 20(2): 85−98.

[137] Nihei K T, Yi W, Myer L R, et al. Fracture channel waves[J]. Journal of Geophysical Research, 1999, 104(B3): 4769−4781.

[138] Pyrak-Nolte L J, Myer L R, Cook N G W. Anisotropy in seismic velocities and amplitudes from multiple parallel fractures[J]. Journal of Geophysical Research, 1990b, 95 (B7): 11345−11358.

[139] Myer L R, Hopkins D, Peterson J E, et al. Seismic wave propagation across multiple fractures, in fractured and jointed rock masses [M]. Balkema, 1995.

[140] Hopkins D L, Myer L R, Cook N G W. Seismic wave attenuation across parallel fractures as a function of fracture stiffness and spacing [J]. Eos, Transactions, American Geophysical Union, 1988, 68(44): 1427.

[141] Nakagawa S, Nihei K T, Myer L R. Stop-pass behaviour of acoustic waves in a 1D fractured system [J]. Journal of Acoustics Society of America, 2000b, 107(1): 40−50.

[142] Cai J G, Zhao J. Effects of multiple parallel fractures on apparent attenuation of stress waves in rock mass [J]. International Journal of Rock Mechanics and Mining Sciences, 2000, 37(4): 661−682.

[143] Zhao J, Zhao X B, Cai J G. A further study of P-wave attenuation across parallel fractures with linear deformational behaviour [J]. International Journal of Rock Mechanics and Mining Science, 2006a, 43(5): 776−788.

[144] Zhao X B. Theoretical and numerical studies of wave attenuation across parallel fractures [D]. Singapore: Nanyang Technological University, 2004.

[145] Zhao X B, Zhao J, HEFNY A M, et al. Normal transmission of s-wave across parallel fractures with coulomb slip behavior [J]. Journal of engineering mechanics, 2006b, 132(6): 641−650.

[146] Zhao X B, Zhao J, Cai J G, et al. UDEC modeling on wave propagation across fractured rock masses [J]. Computers and Geotechnics, 2008, 35(1): 97−104.

[147] 张奇. 应力波在节理处的传递过程 [J]. 岩土工程学报, 1986, 8(6): 99−105.
ZHANG Qi. The transfer process of stress wave at joint [J]. Chinese Journal of Geotechnical Engineering, 1986, 8(6): 99−105. (in Chinese)

[148] 李夕兵, 赖海辉, 古德生. 爆炸应力波斜入射岩体软弱结构面的透反射关系和滑移准则 [J]. 中国有色金属学报, 1992, 2(1): 9−14.
LI Xi-bing, LAI Hai-hui, GU De-sheng. The transmission and reflection relationship and slip rule of the obliquely incident blast wave across rock mass structure plane [J]. The Chinese Journal of Nonferrous Metals, 1992, 2(1): 9−14. (in Chinese)

[149] 王明洋, 钱七虎. 爆炸应力波通过节理裂隙带的衰减规律 [J]. 岩土工程学报, 1995a, 17(2): 42−46.
WANG Ming-yang, QIAN Qi-hu. Attenuation law of explosive wave propagation in cracks [J]. Chinese Journal of Geotechnical Engineering, 1995a, 17(2): 42−46. (in Chinese)

[150] 王明洋, 钱七虎. 颗粒介质的弹塑性动态本构关系研究 [J]. 固体力学学报, 1995b, 16(2): 213−219.
WANG Ming-yang, QIAN Qi-hu. A Study on the elastoplastic dynamic constitutive law of granular medium [J]. Acta Mechanica Solida Sinica, 1995b, 16(2): 213−219. (in Chinese)

[151] 王明洋, 钱七虎. 应力波作用下颗粒介质的动力学特性研究 [J]. 爆炸与冲击, 1996, 16(1): 11−20.
WANG Ming-yang, QIAN Qi-hu. Studies on the dynamic properties for granular medium under stress wave [J]. Explosion and Shock Waves, 1996, 16(1): 11−20. (in Chinese)

[152] 王明洋, 赵跃堂, 钱七虎. 缓倾角断裂隔震效应的机理及定量研究 [J]. 岩石力学与工程学报, 1999, 18(1): 60−64.
WANG Ming-yang, ZHAO Yue-tang, QIAN Qi-hu. Studies on mechanism and quantization of isolation effect of slow angle fault [J]. Chinese Journal of Rock Mechanics and Engineering, 1999, 18(1): 60−64. (in Chinese)

[153] 李宁, Swoboda G, 葛修润. 岩体节理在动载作用下的有限元分析 [J]. 岩土工程学报, 1994, 16(1): 29−38.
LI Ning, Swoboda G, GE Xiu-run. Numerical modelling of rock joints under dynamic loading [J]. Chinese Journal of Geotechnical Engineering, 1994, 16(1): 29−38. (in Chinese)

[154] 王卫华, 李夕兵, 左宇军. 非线性法向变形节理对弹性纵波传播的影响 [J]. 岩石力学与工程学报, 2006, 25(6): 1218−1225.
WANG Wei-hua, LI Xi-bing, ZUO Yu-jun. Effects of single joint with nonlinear normal deformation on P-wave propagation [J]. Chinese Journal of Rock Mechanics and Engineering, 2006, 25(6): 1218−1225. (in Chinese)

[155] 雷卫东, Ashraf M H, 滕军, 等. 二维波穿过单节理的透射率特性及其隐含意义 [J]. 中国矿业大学学报, 2006, 35(4): 492−497.
LEI Wei-dong, Ashraf M H, TENG Jun, et al. Characteristics of transmittivity for a single joint in 2D wave propagation and implications [J]. Journal of China University of Mining and Technology, 2006, 35(4): 492−497. (in Chinese)

[156] 鞠杨, 李业学, 谢和平, 等. 节理岩石的应力波动与能量耗散 [J]. 岩石力学与工程学报, 2006, 25(12): 2426−2434.
JU Yang, LI Ye-xue, XIE He-ping, et al. Stress wave propagation and energy dissipation in jointed rocks [J]. Chinese Journal of Rock Mechanics and Engineering, 2006, 25(12): 2426−2434. (in Chinese)

[157] 郭易圆, 李世海. 有限长岩柱中纵波传播规律的离散元数值分析 [J]. 岩石力学与工程学报, 2002, 21(8): 1124−1129.
GUO Yi-yuan, LI Shi-hai. Distinct element analysis on propagation characteristics of p-wave in rock pillar with finite length [J]. Chinese Journal of Rock Mechanics and Engineering, 2002, 21(8): 1124−1129. (in Chinese)

[158] 俞缙, 关云飞, 肖琳, 等. 弹性纵波在不同非线性法向变形行为节理处的传播 [J]. 解放军理工大学学报: 自然科学版, 2007, 8(6): 589−594.
YU Jin, GUAN Yun-fei, XIAO Lin, et al. Transmission of elastic P-wave across single fracture with different nonlinear normal deformation behaviors [J]. Journal of PLA University of Science and Technology: Natural Science Edition, 2007, 8(6): 589−594. (in Chinese)

[159] 俞缙, 赵晓豹, 赵维炳, 等. 改进的岩石节理弹性非线性法向变形本构模型研究 [J]. 岩土工程学报, 2008, 30(9): 1316−1321.
YU Jin, ZHAO Xiao-bao, ZHAO Wei-bing, et al. Improved nonlinear elastic constitutive model for normal deformaton of rock fractures [J]. Chinese Journal of Geotechnical Engineering, 2008, 30(9): 1316−1321. (in Chinese)

[160] 俞缙, 钱七虎, 林从谋, 等. 纵波在改进的弹性非线性法向变形行为单节理处的传播特性研究 [J]. 岩土工程学报, 2009, 31(8): 1156−1164.
YU Jin, QIAN Qi-hu, LIN Cong-mou, et al. Transmission of elastic P-wave across one fracture with improved nonlinear normal deformational behavior [J]. Chinese Journal of Geotechnical Engineering, 2009, 31(8): 1156−1164. (in Chinese)

岩石爆炸动力学的若干进展

钱七虎

(解放军理工大学 工程兵工程学院，江苏 南京 210007)

摘要：论述岩石爆炸动力学原理及其工程应用研究近年来的若干进展，主要内容包括爆炸空腔范围以及各类破坏区范围的理论确定方法，地下爆炸近区的"短波"和"弱波"理论，爆炸远区——弹性区的运动和力学参数以及地下爆炸时岩石破碎等理论研究成果。同时还介绍了实验室条件下均匀介质中爆炸效应的规律和地应力、裂隙、浅埋时 等不均匀、不连续性影响因素的实验室模拟爆炸试验研究。在相关研究的基础上，给出实际岩体中的爆炸效应试验，包括近区破坏效应、远区地震效应、不可逆变形区以及地下爆炸和浅埋(抛掷和定向)爆炸中相似关系的最新成果。此外，简要介绍岩石爆炸动力学在不同领域中的工程应用，并就今后岩石爆炸动力学研究的展望阐述一点认识。

关键词：岩石力学；爆炸动力学；爆炸效应；破坏区；不可逆变形区
中图分类号：TU 45　　　　　**文献标识码**：A　　　　　**文章编号**：1000 - 6915(2009)10 - 1945 - 24

SOME ADVANCES IN ROCK BLASTING DYNAMICS

QIAN Qihu

(*Engineering Institute of Engineering Corps*，*PLA University of Science and Technology*，*Nanjing*，*Jiangsu* 210007，*China*)

Abstract：The principles of rock blasting dynamics and related engineering applications in the recent years are reviewed. The main issues are covered in blasting cave range，various determination methods for failure zones，theories of short-wave and weak-wave in near-field of cave blasting, the movement in plastic regions and mechanical parameters in far-field，and rock fracture under blasting for underground works. The laws of blasting effects in uniform media in indoor test are also introduced，which include heterogeneous and homogeneous characteristics of geostress，fissure，and shllow-buried cave in laboratory simulation or experimental test. The actual rock blasting tests are touched in failure effects of near-field，seismic effects in far-field，and irreversible deformation regions. The latest similarity research results in underground blasting，shallow-buried blasting (throwing and directional blastings) are also presented. In addition，engineering applications of rock blasting dynamics in various fields are briefly introduced. Filnally，some ideas and prospects are proposed for later research in rock blasting dynamics.

Key words：rock mechanics；blating dynamics；blating effects；failure zone；irreversible deformation region

1 引 言

岩石爆炸动力学研究岩石和岩体中爆炸的力学效应，其特点应从与空气中、水下爆炸效应的对比才能理解更深刻。空气中和水下爆炸时的物理景象和介质的运动已经研究得很全面和很详尽，相比之下，岩石中地下爆炸的物理机制和介质的运动和破

收稿日期：2009-07-22；修回日期：2009-09-09。
基金项目：国家自然科学基金重大项目(50490275)。
本文原载于《岩石力学与工程学报》(2009 年第 28 卷第 10 期)。

坏的研究还很不透彻。

无论空气中爆炸,水下爆炸或岩、土中的爆炸,就爆炸现象的物理本质来说,都是在有一个限空间内快速地释放出巨大能量,从而导致周围介质的非定常运动过程。如果爆炸发生在空气中,传播扰动的波阵面是随距离衰减的冲击波,爆炸能量通过冲击波转递给周围介质,同时也通过冲击波的几何扩展和波阵面处的阻尼耗损爆炸能量。从本质上说,空气中的爆炸作用理论可以归纳为爆炸产生的空气冲击波的传播与衰减规律。在水下爆炸中,情况稍有不同,除了冲击波外,还产生水中气泡的脉动,冲击波所携带和耗散的能量和气泡脉动所耗散的能量各占整个爆炸能量一半左右。

岩体中地下爆炸的现象远比空气和水中复杂得多,在其整个景象中,各个过程相互紧密相连:从炸药的爆轰开始,到在冲击波压缩下的岩、土介质相变,再到周围岩体的破坏,最后至地震波的激发。炸药爆轰完成后,冲击波在岩石介质中开始传播,并很快蜕变成压缩波(在 $2 \sim 5$ 倍装药半径 R_z 范围内),这类压缩波的压力上升梯度很大,其应力上升区间长度 s_1 与整个应力波传播的距离 s_2 相比甚短($s_1 / s_2 = \Delta_1 \ll 0.1$),因此被称为"短波"。由于短波的应力 σ_r 幅值量级小于 10^4 MPa,且小于岩石压缩模量 K_1,其所引起的岩石变形是一个微量 $\sigma_r / K_1 = M_0 \ll 1$,在更远的距离上则更小,所以它又被称为"弱波"。试验结果表明,"弱波"的传播速度接近于岩石中纵波的传播波速 a_0。另在岩体中地下爆炸时,爆炸能主要耗散于岩石介质的不可逆变形和破坏以及推动周围岩石介质的运动。

根据上面所阐明的岩石爆炸动力学特点,本文根据爆炸时岩石动力变形与破坏研究现状及研究趋向[1, 2],针对岩石变形与破坏的若干重要问题[3~24],力求介绍简单实用的岩石爆炸动力学理论研究成果,突出其主要影响因素,以使理论模型概念清晰,避免繁琐复杂的数学表达式;另一方面介绍经实践检验的岩石爆炸动力学的试验研究成果及其相关的工程应用。

2 岩石爆炸动力学的理论研究

本节重点阐述包括爆炸空腔范围、各类破坏区范围的确定以及在地下爆炸近区——破坏区的"短波"和"弱波"方面的研究成果。同时还将介绍应用弹性动力学理论研究爆炸远区的运动和力学参数的研究成果,特别是爆炸对自由地表面的影响,即装药埋深对地表地震动参数确定的影响;其次还将介绍岩体非均匀、非连续性构造——块系和裂隙对地下爆炸效应影响的理论研究成果;最后还将介绍应用不可压缩流体模型研究地下爆炸时岩石破碎的理论方面的研究成果[25~28]。

2.1 岩石中地下封闭爆炸时爆炸空腔及各类破坏范围的半径

地下爆炸时介质的运动、变形和破坏与空腔的扩张密切相关,如图1所示。空腔的扩张以及介质的运动、变形和破坏的研究可以分为几个阶段:当空腔中的爆炸产物压力大于 $0.1 \rho a_0^2$ 量级(ρa_0^2 为岩石侧限变形模量,ρ 为岩石的密度),此时可以忽略岩石中主应力分量的差值,即剪应力的影响,空腔的扩张可以看作为在不可压缩流体中的扩张;第二阶段为冲击压碎阶段,此时岩石中应力超过岩石的压碎强度;在破坏波与空腔之间岩石的运动为仅具内摩擦力的破碎岩块的运动;再之后阶段为空腔的动力无波扩张阶段,此时破坏波速小于先导的弹性波速,在破坏波与空腔之间介质仍看作为具有内摩擦力的破碎材料;最后一个阶段是弹性波传播阶段。弹性波开始于破坏区停止发展的瞬间,弹性介质的运动仅保持于外部的弹性区内。

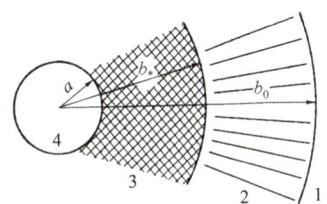

1—弹性变形区;2—径向破裂区;3—破碎区;4—空腔

图 1 岩石破坏示意图

Fig.1 Sketch of rock failure

理论计算时,假定爆炸当量 Q 全部转化为爆炸产物的压缩能,爆轰产物的初始体积为半径为 a_* 的球腔。

相应于球形空腔气体绝热膨胀,爆炸气体对周围介质所作的功为

$$A = Q\left[1 - \left(\frac{a_*}{a_t}\right)^{3(\gamma-1)}\right] \tag{1}$$

式中:γ 为气体的绝热指数,a_t 为 t 时刻空腔半径。

理想不可压缩流体运动阶段：理想不可压缩流体的空腔周围介质在 t_1 时刻获得的动能 W 近似为

$$W \approx 4\pi a_1^3 \rho_0 \frac{[\dot{a}(t_1)]^2}{2} \\ a_1 = a(t_1) \quad (2)$$

式中：$\dot{a}(t_1)$ 为在 t_1 时刻的空腔壁速度，ρ_0 为介质初始密度。

不可压缩流体模型适用条件为空腔中的气体压力大于岩石的晶格理论强度 Y ($Y = -0.1\rho a_0^2$)。由绝热扩张条件可得

$$\frac{a_1}{a_*} = \left(\frac{P_0}{Y}\right)^{\frac{1}{3\gamma}} = Z \quad (3)$$

式中：P_0 为爆炸气体产物在 $t=0$ 时空腔中的压力。由式(1)~(3)可求得流体模型时的空腔壁运动速度的关系式：

$$\dot{a}_1^2 = \frac{Q}{2\pi a_*^3 \rho_0 Z^3}\left[1 - \frac{1}{Z^{3(\gamma-1)}}\right] \quad (4)$$

冲击压碎阶段：在综合分析冲击破坏波阵面上质量守恒、动量守恒条件，冲击破坏后介质的剪胀以及空腔运动与冲击破坏波对介质压缩的相互关系，并在保持无黏聚力仅有摩擦力时的库仑条件下的应力关系后，得到该阶段的关于空腔壁面的运动速度关系式：

$$\dot{a}^2 \frac{\rho_0}{P_0} = \left\{\frac{2Z^{\alpha_1-3}}{3(\gamma-1)}[1 - Z^{3(1-\gamma)}] - \frac{\beta_1 Z^{\alpha_1-3\gamma}}{\alpha_1 - 3\gamma}\right\} \cdot \\ \left(\frac{a_*}{a}\right)^{\alpha_1} + \frac{\beta_1}{\alpha_1 - 3\gamma}\left(\frac{a_*}{a}\right)^{3\gamma} \quad (5a)$$

其中，

$$\alpha_1 = -\frac{2}{\ln \varepsilon} - \frac{4\varepsilon}{\ln \varepsilon} + 4, \quad \beta_1 = -\frac{6}{\ln \varepsilon} \quad (5b)$$

式中：ε 为破坏波阵面上介质的压缩变形。

根据冲击压碎的物理含义，有

$$\sigma_r = -\rho_0 \varepsilon^{1/3} \dot{a}_2^2 = -\sigma_* \\ a_2 = a(t_2) \quad (6)$$

式中：σ_* 为岩石介质的压碎应力极限，σ_r 为径向应力。

在冲击压碎阶段，冲击破坏波与弹性前驱波尚未分离，所以 $\varepsilon = \sigma_*/(\rho a_0^2)$。由式(6)求出相应于冲击压碎区运动速度极限值 $\dot{a}_2^2$，再求得 a_2。

空腔的无(冲击)波动力扩张阶段：自 t_2 瞬间，破坏波与弹性前驱波开始分离，在这个阶段介质开始了无冲击波仅有压缩波的地下动力运动，空腔半径从 a_2 扩张到最大值 a_m，在这个阶段，介质形成径向破裂区(其半径为 b_*)和径向破碎区(其半径为 b_0)。在压碎区的区域中和径向裂纹区边界 $r = b_*$ 上条件为

$$\sigma_r = -\sigma_* \quad (7)$$

在径向破碎区的内边界和外边界上相应地有

$$\sigma_r = -\sigma_* \quad (r = b_*) \\ \sigma_r = -2\sigma_0 \quad (r = b_0) \quad (8)$$

式中：σ_0 为岩石介质的拉裂应力极限。

与冲击压碎阶段所作的分析相类似，无波动力扩张阶段破坏介质的运动方程也相同，唯一不同的是边界条件。对该阶段空腔壁运动速度的表达式 $a(t)$ 求导，在 $\dot{a}(t) = 0$ 时，求出最大空腔半径 a_m 的表达式：

$$\frac{a_m}{a_2} = \left[1 + \frac{\alpha_2}{\beta_2 \chi}\frac{\rho_0 \dot{a}_2^2}{\beta_2 \chi \sigma_*} - \frac{\alpha_2}{\alpha_2 - 3\mu}\frac{P_0}{\chi \sigma_*}\left(\frac{a_*}{a_2}\right)^{3\gamma}\beta_3\right]^{\frac{1}{\alpha_2}} \quad (9a)$$

其中，

$$\beta_2 = \frac{2}{\ln \chi}, \quad \alpha_2 = 4 - \frac{4}{3\ln \chi} + \frac{4}{3\chi^3 \ln \chi} \\ \chi = \frac{b_*}{a_m}, \quad \beta_3 = 1 - \left(\frac{a_2}{a_m}\right)^{3\gamma - \alpha_2} \quad (9b)$$

为了便于工程计算，引入爆炸当量 Q，并且对于比较宽泛的爆炸条件 $10^{-3} < \varepsilon < 10^{-2}$，$1.2 < \gamma < 1.67$，$1 < P_0^{-1}(\rho a_0^2) < 10$，可以近似来代替式(9a)，则有

$$\frac{4}{3}\pi a_m^3 \frac{\rho a_0^2}{Q} = 38\left(\frac{\rho a_0^2}{250\sigma_*}\right)^{\frac{2}{3}} \quad (10)$$

最大空腔半径为

$$a_m = \frac{0.61 Q^{\frac{1}{3}}}{(\rho a_0^2 \sigma_*)^{\frac{1}{9}}} \quad (11)$$

压碎破坏区半径为

$$b_* = \left(\frac{E_0}{3\sigma_*}\right)^{\frac{1}{3}} a_m \quad (12)$$

式中：E_0 为介质的杨氏模量。

在径裂区成立条件 $\sigma_\varphi = \sigma_\theta = 0$，因此粗略地有 $\sigma_r = E_0 e_r$，由边界条件式(8)在径裂区内成立下列关

系式：

$$\sigma_r = -\sigma_* \left(\frac{b_*}{r}\right)^2, \quad e_r = -\frac{\sigma_*}{E_0}\left(\frac{b_*}{r}\right)^2 \quad (13)$$

由式(8)可得径裂区半径为

$$b_0 = b_* \left(\frac{\sigma_*}{2\sigma_0}\right)^{\frac{1}{2}} = \left[\frac{(\rho a_0^2)^2 \sigma_*}{2^7 \sigma_0^3}\right]^{\frac{1}{6}} a_m \quad (14)$$

式(14)所给出的计算结果与试验结果吻合较好。

2.2 岩石中地下爆炸时破坏区内运动参数的衰减规律

众所周知，弹性体模型适用于弹性波的传播，具有小变形的弹塑性模型(形变模型)对于应力幅值超过弹性限不多的情况证明是适用的，而理想塑性体模型正好相反，适用于非常高的压力区，此时岩石在动力载作用下的行为接近于流体动力行为。在中间过渡区域岩石的描述问题是一个更值得关注的问题，因为这一过渡区的范围对于岩石来讲不大于(10～15) GPa。岩石中地下爆炸的$(2～5)R_z$至$(100～120)R_z$的实际工作属于该范围，该范围最大应力和应力上升段的应力随距离的增加大致按比例r^{-n}下降，其中$n = 1.6～1.8$。根据小变形弹塑性模型并考虑到卸载情况，应力波一般按照$n = 1.1～1.2$的衰减律进行衰减。在该压力范围内应力波传播时幅值快速衰减的原因在于介质受限变形的内摩擦力。用试验确定包括内摩擦角(内摩擦角随压力变化)在内的连续介质参数后，可以评估给定岩石的应力波波幅和质点运动速度随距离增大的衰减规律。

在受限变形条件下，地下球腔爆炸的破坏区的岩石颗粒之间的黏聚力可以忽略，但是其摩擦力不可忽略，根据库仑准则可知：$\sigma_\varphi / \sigma_r = \alpha^* < 1$，$\alpha^* = (1-\sin\varphi)/(1+\sin\varphi)$；$\tan\varphi$为摩擦因数，$\alpha^* = \alpha^*(\sigma)$为非线性的表达式。

下面以球对称一维问题研究在破坏区中"短波"的传播与衰减规律，但利用参数改变得出的最终推论也可用于柱对称和平面波的情况。采用拉格朗日变量的运动方程和连续方程有以下形式：

$$\left. \begin{array}{l} \rho \dfrac{\partial v}{\partial t} + \dfrac{1}{\partial r / \partial r_0} \dfrac{\partial \sigma_r}{\partial r_0} + \dfrac{2(\sigma_r - \sigma_\theta)}{r} \dfrac{r_0}{r} = 0 \\[2mm] \dfrac{\partial v}{\partial r_0} = \left(\dfrac{r_0}{r}\right)^2 \dfrac{\partial \varepsilon}{\partial t} - \dfrac{r_0}{r}\dfrac{\partial v}{\partial r_0}\dfrac{2v}{r_0} \\[2mm] v = \dfrac{\partial r}{\partial t} = \dfrac{\partial}{\partial t}(r - r_0), \quad \varepsilon = \dfrac{\rho_0}{\rho} - 1 \end{array} \right\} \quad (15)$$

式中：r_0为拉格朗日坐标系统中质点的初始位置，$r(r_0, t)$为t时刻质点的坐标，v为r方向上的质点移动速度。

质点位置$r(r_0, t)$和质点位移$w(t)$可分别表示为

$$r(r_0, t) = \int_0^t v \mathrm{d}t + r_0, \quad w(t) = \int_0^t v \mathrm{d}t \quad (16)$$

基于试验结果的分析，引入新的自变量δ，τ和新的未知函数m和e：

$$\left. \begin{array}{l} r = a_0 t(1 + \Delta_0 \delta), \quad \tau = \ln t \\ v = a_0 M_0 m(\delta, \tau), \quad \varepsilon = \varepsilon_0 e(\delta, \tau) \end{array} \right\} \quad (17)$$

基于"弱波"理论，M_0和ε_0为微量。基于"短波"理论，Δ_0同样为微量。从式(17)和s_1的计算可以看出，由于Δ_0是微量，有$\Delta_0 \delta \ll 1$。应力上升区的尺度$s_1 = a_0 t \Delta_0$远远小于距爆心的波的经过距离$s_2 \approx a_0 t$或$s_1 \ll s_2$。而对于短波($\Delta_0 \ll 1$)和弱波($M_0 \ll 1$；$\varepsilon_0 \ll 1$)而言，球对称波的环向变形和径向变形分别为

$$\varepsilon_\theta \propto \Delta_0 M_0, \quad \varepsilon_r = \frac{\partial w}{\partial r_0} \propto M_0 \quad (18)$$

从式(18)可看出，ε_r为M_0数量级的微量，而ε_θ为更高数量级的小量，比值$\varepsilon_\theta / \varepsilon_r$是微量$\Delta_0$，上述特性正是短应力波的主要特征。

将内摩擦介质的应力-应变关系代入式(15)，经过运算可得内摩擦介质运动方程：

$$\frac{1}{(1+\varepsilon_\theta)^2}\frac{\partial v}{\partial t} - a_{10}^2 \frac{\partial \varepsilon}{\partial r_0} - \frac{2a_{10}^2(1-\alpha^*)\varepsilon}{r_0}\frac{1+\varepsilon_r}{1+\varepsilon_\theta} = 0 \quad (19)$$

式中：a_{10}为内摩擦状态区域内的当地声速，且有

$$a_{10}^2 = \frac{3K}{\rho_0(1 + 2\alpha^*)} \quad (20)$$

式中：K为岩体的体积模量。

将弹性介质应力-应变关系代入式(15)后，也可得到弹性介质的运动方程。利用上述变换公式就可以对弹性区介质与内摩擦状态的介质的运动方程进行变换，求解短波的传播与衰减问题。经过必要的代换和考虑到对应的M_0和Δ_0为一阶微量，可以得到弹性介质的短波方程：

$$\frac{\partial m}{\partial \delta} + \frac{\partial e}{\partial \delta} = 0, \quad \frac{\partial m}{\partial \tau} - \delta\frac{\partial m}{\partial \delta} + m = 0, \quad \varepsilon_0 = M_0 \quad (21)$$

同样，可得出内摩擦状态的介质的短波方程为

$$\frac{\partial m}{\partial \delta} + \frac{\partial e}{\partial \delta} = 0, \quad \frac{\partial m}{\partial \tau} - \delta\frac{\partial m}{\partial \delta} + (2-\alpha^*)m = 0 \quad (22)$$

由式(22)第一个方程，可得

$$m = -e + \psi(\tau) \quad (23)$$

式中：$\psi(\tau)$为任意函数，若为波前的静止部分，它可以取为0。

对式(22)的第二个方程积分，就可以得到特征方程的通解：$C_1 = mt^{2-\alpha^*}$，$\delta = C_2 m^{\frac{1}{2-\alpha^*}}$（$C_1$，$C_2$均为常量），且有

$$m = \frac{1}{t^{2-\alpha^*}}\psi(\delta t) \quad (24)$$

当取内摩擦介质的极限状态时，即$\alpha^* = \alpha = \frac{\nu}{1-\nu}$，$\nu$为泊松比），且$a_{10} = a_0$。

当换为物理量和利用ψ的任意性，可得

$$\frac{\upsilon}{a_0} = \frac{1}{r^{2-\alpha}}\psi(\xi) \quad (25)$$

式中：ξ为与r和α相关的变量表达式。

短弹性波的通解为

$$\frac{\upsilon}{a_0} = \frac{1}{r}\Phi(r - a_0 t) \quad (26)$$

式中：$\Phi(\cdot)$为任意函数。

比较式(25)和(26)后可知，在内摩擦极限状态区中，运动参数随着距离增加，运动参数波幅的减小与$r^{2-\alpha^*}$成反比。而在弹性区中，幅值的减小与r成反比。因此，在强烈的动载下，岩石介质中爆炸近区将会达到加载极限状态，即内摩擦状态，而在较远处岩石介质仍处于弹性状态，短波在这2种状态中传播但主要在内摩擦状态中衰减。

2.3 岩石中地下爆炸时弹性区运动参数及地表运动参数

当炸药在地下爆炸时，会对周围介质产生破坏和震动效应。浅埋地下爆炸条件下属于弹性区内的爆炸地震效应的地表运动，可以用弹性动力学理论进行研究。整个非弹性区(破坏区)看成爆炸的震动源，简化后的爆炸地震波传播如图2所示。

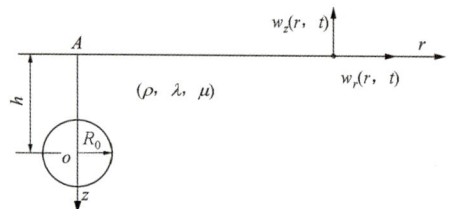

图2 浅埋地下的爆炸简化模型

Fig.2 Simplified model for blasting of shallow-buried underground works

设弹性半空间介质的材料特性用密度ρ，λ和μ表示，纵波速度用C_P，横波速度用C_S表示。震源位于自由表面下深h处，以o为球心，R_0为破坏区半径。震动源的表示形式通常有2种：第一种由破坏区表面的径向粒子速度$\upsilon_R = \upsilon_R(t)$来描述；第二种由空腔表面径向正应力$\sigma_R = \sigma_R(t)$来描述。取第一种表示形式即速度形式作为震源参数：

$$\upsilon_R(t) = \upsilon_0 f(t) \quad (27)$$

式中：$f(t)$为速度时程的函数形式。

首先求解球面波的传播，依据震源周围介质产生的空间球对称运动，可用球坐标$r (r > R_0)$和θ（$0 \leq \theta < \pi$）表示，其初始状态与球形震源中心重合。介质质点的一维径向位移满足线性波动方程：

$$\left.\begin{array}{l}\dfrac{\partial^2 u}{\partial r^2} + \dfrac{2}{r}\dfrac{\partial u}{\partial r} - \dfrac{2u}{r} = \dfrac{1}{C_P^2}\dfrac{\partial^2 u}{\partial t^2} \\ C_P = [(\lambda + 2\mu)/\rho]^{1/2}\end{array}\right\} \quad (r > R, \ 0 < t < t') \quad (28)$$

式中：u为质点的径向位移。

根据初始及边界条件解得速度的近似表达式为

$$\upsilon \approx \frac{\upsilon_0 R_0}{r}\mathrm{e}^{-n\xi}\sin\left(m\xi - \arctan\sqrt{1-2\mu}\right) \quad (29\mathrm{a})$$

式(29)表明：介质的空间(径向)运动具有使振动衰减的特点。振幅的衰减参数和振动的周期(与土壤的性质和震源的半径，即破坏区的半径有关)分别为

$$\left.\begin{array}{l}n = [(1-2\mu)C_P]/[(1-\mu)R_0] \\ T = 2\pi/m = 2\pi(1-\mu)R_0/\left(C_P\sqrt{1-2\mu}\right)\end{array}\right\} \quad (29\mathrm{b})$$

图3计算了3个给定时刻$\bar{t} = t/t_0 = 0.3$，1.0和2.0的剪应力$\tau_{\max} = (\sigma_\theta - \sigma_r)/\sigma_0$沿时间坐标$r/R_0$传播的情况。相对时间$t$按与时间间隔$t_0 = R_0/C_P$的比值进行量纲一化。可见，剪应力$\tau_{\max}$在震源$r = R_0$和弹性波阵$r = R_0 + C_P t$的区域内，是坐标$r/R_0$的单调递减函数。在震源表面上将达到最大值，而在波前上则达到最小值。波阵上各点处的振幅，由于

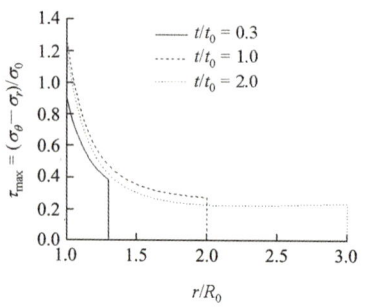

图3 最大剪应力沿半径的分布

Fig.3 The maximum shear stress distribution along radial direction

弹性压缩波的传播有一个突降值，因此剪切力的振幅也将有一个突变。

图 4 计算给出了观察点 $r/R_0=14$ 处相对环向应力 $\bar{\sigma}_\theta$ 随时间 $\bar{t}$ 变化的历程($\bar{\sigma}_\theta=\sigma_\theta/\sigma_0$)。在介质运动的初始阶段，环向应力是拉应力($\bar{\sigma}_\theta \geq 0$)，然后拉应力变成压应力($\bar{\sigma}_\theta < 0$)。拉应力将引起(在许多情况下都会引起)产生径向裂纹破坏。当 $t\to\infty$ 时，$\bar{\sigma}_\theta$ 趋近于静力解。

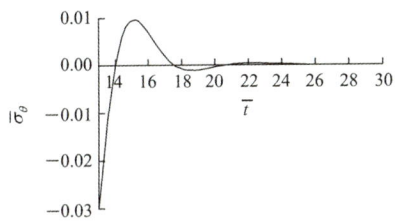

图 4 环向正应力的变化曲线

Fig.4 Variation curve of ring positive stress

再次求解地表自由面振动问题解，压缩波到达自由表面后，与自由表面相互作用，这一过程的特点是形成纵向和横向的拉伸柱面波，这些波会在自由表面附近引起开裂现象，并在土壤的地表层内产生垂直向和水平向的大位移等。

根据弹性动力学中经典的 Lama 问题解，可以构建弹性半空间内自由表面振动问题的精确解。基本脉冲扰动情况下水平和垂直方向的地表位移的精确值[29~45]可分别表示为

$$\left.\begin{array}{l}w_{r\delta}=\dfrac{k}{t^*}(U_{r0}+U_{rR}+U_{r\lambda})\\ w_{z\delta}=\dfrac{k}{t^*}(U_{z0}+U_{zR}+U_{z\lambda})\end{array}\right\} \quad (30a)$$

其中，

$$\left.\begin{array}{l}k=\upsilon_0 R_0/C_S,\ t^*=tC_P/R_0\\ \eta_{rt}=r/(tC_P),\ \eta_{ht}=h/(tC_P)\end{array}\right\} \quad (30b)$$

式中：U_{r0}，U_{rR}，$U_{r\lambda}$，U_{z0}，U_{zR}，$U_{z\lambda}$ 含义参见有关研究成果[14]。

任意脉冲形式作用下，地表运动位移可写为

$$w(r,\ t)=\int_0^t f_*(t-\tau)w_\delta(r,\ t)\mathrm{d}\tau \quad (31)$$

$f_*(t-\tau)$ 按由基本型集中震源向其他源过渡公式 $f_*(t)=f(t)-\int_0^\tau f(s)\exp[-(t-s)]\mathrm{d}s$ 转换完成。

上述理论公式可由如下算例得以验证：设岩石密度为 $2.7\times 10^3\ \text{kg/m}^3$，纵波速度 $C_P=4\ 500\ \text{m/s}$，泊松比为 $\mu=0.25$，可以导出横波和纵波的波速比为 $\gamma=1/\sqrt{3}$。设装药的埋置深度为 h，当球形装药半径为 R_z 时，可取非弹性变形区半径为 $R_0=25R_z$。以 100 kg TNT 为例，装药半径为 $R_z=0.25\ \text{m}$，边界荷载采取速度荷载形式：

$$\upsilon(t^*)\big|_{r=R_0}=\upsilon_0 f(t),\ f(t)=\upsilon_0 \mathrm{e}^{-ct}\sin(dt) \quad (32)$$

式中：c 为衰减系数，且 $c=(1-2\mu)C_P/[(1-\mu)R_0]=480$；$d$ 为振动频率，且 $d=\sqrt{1-2\mu}C_P/[(1-\mu)R_0]=680$，$\upsilon_0=10\ \text{m/s}$；计算与分析均采用量纲一的坐标进行，量纲一的时间参量 $t^*=(tC_P)/R_0$，量纲一的水平距离为 r/R_z，量纲一的装药埋深 h/R_z。边界上速度荷载波形如图 5 所示。

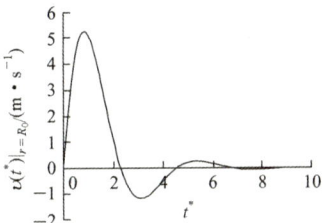

图 5 边界上速度荷载波形

Fig.5 Velocity loading waveform in boundary

计算结果分析按水平方向和垂直方向分别进行。取 $h=30R_z$，根据式(30a)，(30b)进行计算可以得到地表振动的位移和速度波形。图 6 中给出了比例距离为 $r/R_z=50$，125，250，500 处的水平位移和速度波形图(量纲一的时间坐标为 $t^*=tC_P/R_0$，量

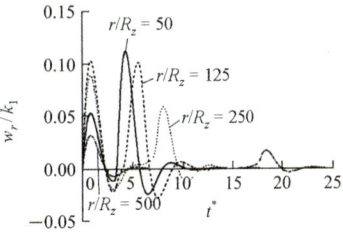

(a) 地表水平方向位移

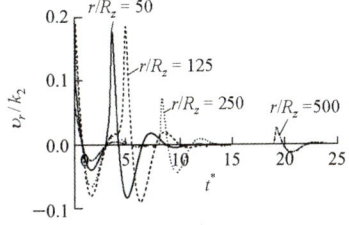

(b) 地表水平方向速度

图 6 地表水平方向位移和速度($h=30R_z$)

Fig.6 Horizontal dispancements and velocies of ground($h=30R_z$)

纲一的位移坐标为 w_r/k_1，量纲一的速度坐标为 v_r/k_2；其中 $k_1=v_0R_0/C_s$，$k_2=v_0/\gamma$）。图7给出了地表比例距离 r/R_z=50，125，250和500位置处的垂直方向位移波形和速度波形图。

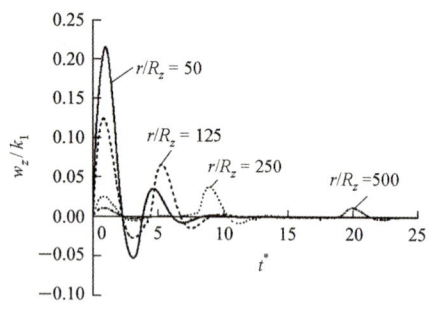

(a) 地表垂直方向位移

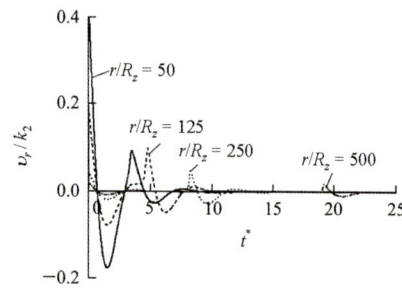

(b) 地表垂直方向速度

图7　地表垂直方向位移速度($h=30R_z$)

Fig.7　Vertical dispancements and velocities of ground($h=30R_z$)

水平方向上的运动存在2种不同形式的波。第一种是纵波，它的粒子振动频率与球腔内的荷载频率一致，也就是与破碎区的自振频率基本一致。第二种是瑞利波，它的周期比纵波周期要大，大概是纵波周期的1.5～2.0倍。也就是说，地表纵波震动的频率约等于瑞利波振动频率的1.5～2.0倍。纵波传播的速度快，而瑞利波传播的速度慢，因此随着传播距离的加大，纵波波峰与瑞利波波峰之间的间隔不断加大，当取水平距离为500倍装药半径的点观察时，瑞利波落后纵波近20倍的比例时间。

根据纵波和瑞利波在地表不同距离上的运动特点，可以将地下爆炸的地表划分为近区和远区。随着地震波传播距离的加大，纵波和瑞利波均发生衰减。纵波的衰减要比瑞利波快得多，当纵波所引的地表震动大于瑞利波引起的地表震动时，可以认为这属于浅埋爆炸的近区。而纵波衰减至与瑞利波幅值相近或小于瑞利波的幅值时，即认为这属于浅埋爆炸的远区。近区与远区的分界点用地表上距爆心的水平比例距离来表示，设 $R^*=r/R_z$。在比例埋深 $h=30R_z$ 的情况下，$R^*\approx 500$。

由式(30a)和计算结果可以看出，地表水平方向位移的尺度系数为 $k_1=v_0R_0/C_s$，即地表水平方向位移与边界荷载和空腔半径之积成正比，与介质中的横波速度成反比；而速度的尺度系数为 $k_2=v_0/\gamma$，即地表水平方向振动速度与空腔边界荷载的大小成正比，与波速比成反比。不同装药情况下，在相同比例埋深和相同比例距离上有相同的速度，但不具有相同的位移。

与水平方向相同，近区垂直方向上地面振动的主要载体是纵波，而在远区的主要载体是瑞利波。在弹性条件的假设下，纵波与瑞利波的震动频率几乎不随距离发生变化，纵波的震动频率约等于爆炸破碎区空腔的自振频率，而表面瑞利波的频率大约为纵波频率的一半。

垂直方向和水平方向位移幅值随比例距离增大以不同的方式衰减，地表最大位移随距离的变化曲线如图8所示。

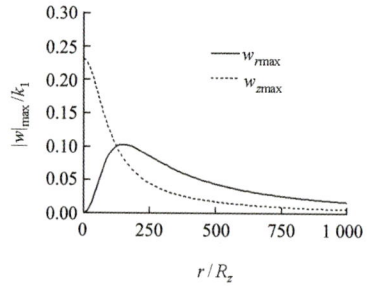

图8　地表最大位移随距离的变化曲线

Fig.8　Variation curves of ground maximum displacements

由图8可知，在距离爆心较近的位置上，地表的垂直振动占主要部分，随着距离爆心水平距离的增加，地表垂直振动位移幅值急剧衰减，而地表水平振动位移幅值先增加而后衰减，垂直振动的衰减速度要快于水平振动的衰减速度。$r\approx 130R_z$ 处是个转折点，在此位置以内垂直方向的振动大于水平方向的振动，而此位置以外水平方向振动大于垂直方向振动。地下爆炸地表近区与远区动力学特征参数的规律使爆炸地震效应的安全评估选用何种评判准则变得更有物理基础。

2.4　地下爆炸时岩石破碎的计算原理

(1) 计算模型

爆炸气体膨胀做功的同时，将能量传给周围岩体介质，由于爆炸应力波阵面传播速度很快，因而可认为能量传播过程瞬间完成，爆炸气体做功转换成破坏岩体介质的压缩变形能和岩体介质的动能，由于岩石破坏的压缩变形能较其动能小很多，因此可近似认为爆炸能全部转换成周围岩石介质的动能。这样一来，爆炸对周围岩体破碎的作用过程可近似简化为2个阶段：第一阶段相应于爆炸应力波传播过程，在此过程中炸药将其爆炸能转交给周围岩石介质，而且近似地假定全部转变成介质的动能，这个过程持续时间极短，由于岩石中应力波可假设瞬间完成。在此阶段内，介质没有任何变形发生；第二阶段介质以其在第一阶段获得的速度开始运动，由于周围介质获得的速度为非均匀的球对称扩展，因此介质随即产生变形，最终导致岩石的破坏或振动，这个阶段持续时间将由运动速度决定。

第一阶段内介质在高压力作用下没有变形发生的情况仅在下述条件下才有可能，即各向均匀压缩，此时3个主应力相等，没有剪应力。这种情况仅相应于不可压缩的流体，所以第一阶段的计算模型又可称为不可压缩流体模型。这种模型简单、直观，实践表明对高爆压力下的岩石爆炸问题它是可以接受的，而且可通过不太复杂的计算就可获得结果。

(2) 速度场

如上所述，爆炸能量传递给周围介质的过程可以归纳为爆炸瞬间理想不可压缩流体中在一个比冲量作用下的状态的确定，这个比冲量$S(S = p\Delta t)$的值在流体的各点处是不同的，它是所研究的介质点坐标的函数。S的大小可通过下述的能量平衡原理来求得。爆炸时形成的速度场具有势，为此引入势函数$\phi = S/\rho$，则直角坐标3个速度场可表示为

$$u_\phi = -\frac{\partial \phi}{\partial x}, \quad v_\phi = -\frac{\partial \phi}{\partial y}, \quad w_\phi = -\frac{\partial \phi}{\partial z} \quad (33)$$

对于无限介质中的爆炸情况，根据有限装(炸)药其爆炸作用仅传播到有限的范围，因此离开爆炸中心无限远距离上，爆炸产生的作用应该无穷小，也即$S = 0$。在自由面上，因为不可能有反力作用，根据牛顿第三定律，比冲量和速度势也应等于0。最后，在装药与岩石介质的分界面上，爆炸压力为常值，等于爆轰产物(气体)爆炸后的压力，因此可认为在整个装药面上比冲量，因而速度势为常数。

由于流体不可压缩，根据运动连续性条件可知：

$$\frac{\partial^2 \phi}{\partial x^2} + \frac{\partial^2 \phi}{\partial y^2} + \frac{\partial^2 \phi}{\partial z^2} = 0 \quad (34)$$

最后一个重要条件是介质在爆炸瞬间获得的动能应该等于炸药的爆炸当量Q，新的关系式为

$$Q = \frac{\rho}{2}\iiint (u_\phi^2 + v_\phi^2 + w_\phi^2) \mathrm{d}x\mathrm{d}y\mathrm{d}z =$$

$$\frac{\rho}{2}\iiint \left[\left(\frac{\partial \phi}{\partial x}\right)^2 + \left(\frac{\partial \phi}{\partial y}\right)^2 + \left(\frac{\partial \phi}{\partial z}\right)^2\right] \mathrm{d}x\mathrm{d}y\mathrm{d}z \quad (35)$$

求出$\phi(x, y, z)$后，很容易求得速度场u_ϕ，v_ϕ，w_ϕ。因此，由于爆炸能量传播给周围介质的结果，爆炸后瞬时，介质没有任何变形和位移，仅仅获得一个初始速度，这个速度场在各点具有不同的方向和量值。

(3) 介质的变形

当介质受到压缩变形时，引起破坏的变形能为$A_s = \rho u_s^2 / 2$，而岩石介质所引起的局部断裂为$A_z = \sigma_z^2/(2E_0)$，相应于比断裂能，其相应的临界速度分别为

$$\left.\begin{array}{l} u_s = \sigma_s / \sqrt{E_0 \rho} \\ u_z = \sigma_z / \sqrt{E_0 \rho} \end{array}\right\} \quad (36)$$

式中：σ_s为屈服流限，σ_z为脆性断裂时的断裂拉应力。

A_s是确定不可逆塑性变形的开始，A_z是描述脆性拉断破坏。因此，式(36)是个补充要求，其条件是体变形必须是正的，即$\varepsilon_1 + \varepsilon_2 + \varepsilon_3 > 0$。如果在某个区域发生介质因爆炸压力产生的压缩，则破坏将在该区域发生于$A \geq A_s$条件下。但是因为振动，在其他区域则可能产生拉伸，其可能的破坏，则由条件$A \geq A_z$确定。

(4) 破坏准则

把速度场分解成2个，一个是相应于分出来的区域运动的速度场，其运动速度为该区域中心的速度u_0，v_0和w_0；另一个是引起变形的速度场。引起变形的速度场即为相对速度(速度差)的速度场，若坐标中心点取为分离区域中心，则相应该区域变形的全部动能：

$$E = \frac{\rho}{2}\iiint \left[(u_\phi - u_0)^2 + (v_\phi - v_0)^2 + (w_\phi - w_0)^2\right] \mathrm{d}x\mathrm{d}y\mathrm{d}z \quad (37)$$

将速度与势函数的关系代入式(37)，把速度场按马克洛林级数分解，仅取一阶微量，可得边长$2a$的立方体分离区域的变形能为

$$E = \frac{4}{3}a^5 \rho D \quad (38)$$

$$D = \left(\frac{\partial^2 \phi}{\partial x^2}\right)^2 + \left(\frac{\partial^2 \phi}{\partial y^2}\right)^2 + \left(\frac{\partial^2 \phi}{\partial z^2}\right)^2 + 2\left(\frac{\partial^2 \phi}{\partial x \partial y}\right)^2 +$$

$$2\left(\frac{\partial^2 \phi}{\partial x \partial z}\right)^2 + 2\left(\frac{\partial^2 z}{\partial x \partial y}\right)^2 \quad (39)$$

式中：D 为破碎性判据。

由 D 的表达式可知，破碎性判据量值为点坐标的函数，因此介质的破碎特性一般情况下随各点坐标而异，这已为爆破作业实践所证实。

由式(38)可知，分离区域的变形能不是和立方体体积成正比，而是和立方体半棱长的五次方成正比。这意味着，对于较大的 a，按式(38)计算的能量 E 将会大大超过耗散于塑性条件的能量，即 $E_s = 8a^3 A_s = 4a^3 \rho u_s^2$。

如果分离区域很小，按式(38)计算的变形能很小，有可能在其中没有任何不可逆变形过程发生。因此保持完整的没有随后的破碎的分离立方体区域将由下述条件确定：

$$E = E_s, \quad \frac{4}{3} a^5 \rho D = 4 a^3 \rho u_s^2 \quad (40)$$

由此可以得

$$a = \sqrt{3} u_s / \sqrt{D} \quad (41)$$

式(41)用以计算碎块大小，函数 D 的破碎判据的名称由此得到证实。作为例子，现计算一个半径为 R、爆炸当量为 Q 的球形药包爆炸后的碎块分布。按式(41)计算出不含裂缝的岩体碎块的平均概率尺寸为

$$\bar{a} = \frac{\sqrt{3} u_s}{\sqrt{\dfrac{3RQ}{\pi \rho r^6}}} = u_s r^3 \sqrt{\frac{\pi \rho}{RQ}} \quad (42)$$

由式(42)可得，在离爆源相同距离 r 上，装药半径 R 或爆炸能 Q 越大，临界速度 u_s 或岩石屈服极限越小，则碎块尺寸 a 越小。

(5) 破碎区

由式(42)可见，最小的碎块位于装药面附近。显然，邻近装药面的块体尺寸为

$$a = r - R \quad (43)$$

为了确定碎块尺寸，应联立求解式(42)和(43)，图 9 给出了相应的图解法。由图 9 可见，在 a, r 为正的象限内，直线与曲线可能有 2 个交点、一个交点或没有交点的 2 种情况。没有交点的情况相应于爆炸不产生破碎。一个交点的情况相应于式(43)为曲线式(42)的切线。此时，易得 $r = 3a$，因此有

$$Q = \frac{729}{16} \pi \rho u_s^2 R^3 \quad (44)$$

式(44)表明，在确定的爆能情况下，其碎块尺寸较大，但破碎块数不多，而且碎块尺寸 $a = R/2$，

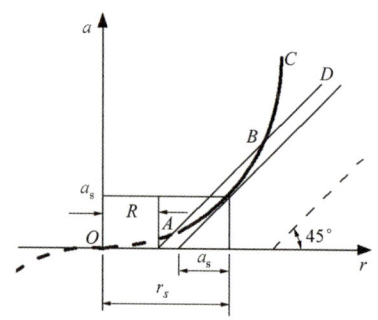

图 9 破碎区的计算图

Fig.9 Calculation scheme of broken region

如果爆能值较按式(44)所求出的较大者，即

$$Q > \frac{729}{16} \pi \rho u_s^2 R^3 \quad (45)$$

则爆炸破碎将形成较多的小块，即 $a < R/2$。

爆炸形成最大碎块的情况相应于该曲线的切线，见图 9。此时 r 由式(42)可知：

$$r = (3 u_s)^{-\frac{1}{2}} \left(\frac{RQ}{\pi \rho} \right)^{\frac{1}{4}} \quad (46)$$

将式(46)代入(42)，得

$$a_s = 3^{-\frac{3}{2}} u_s^{-\frac{1}{2}} \left(\frac{RQ}{\pi \rho} \right)^{\frac{1}{4}} \quad (47)$$

同时求得破碎区的边界为

$$r_s = r + a_s = \frac{4}{\sqrt{27 u_s}} \left(\frac{RQ}{\pi \rho} \right)^{\frac{1}{4}} \quad (48)$$

上述研究介质破碎的方法可以推广到其他的爆炸情况。为此在每一具体情况中，确定最大破碎块尺寸以及破碎区边界时，仅需求作破碎曲线 $a(r)$ 的切线，即 $\partial a(r) / \partial n = 1$。

由此可知，一般情况下可能存在 2 个破碎区域，第一个破碎区域由 u_s 确定，它是由爆炸气体压力直接扩张作用下产生的，第二个区域是按脆性拉裂的临界速度 u_z 所确定的，由爆炸引起的震动产生，按照物理原理，在这些区域点，第三不变量 I_3 应保持正值。

(6) 块度组成

为了阐明确定碎块块度总体组成的图解过程，仅研究无限均匀介质中球形装药爆炸的例子。最简单的极限情况相应于压碎区的半径 $r_s < R$，此时将不存在压缩引起的剪切屈服，只存在脆性破裂区，相应的计算公式为

$$r_z = \frac{4}{\sqrt{27u_z}}\left(\frac{RQ}{\pi\rho}\right)^{\frac{1}{4}} \quad (49)$$

在该区域内，碎块尺寸可由式(43)确定，但需以 u_z 置换 u_s，即

$$\overline{a} = u_z r^3 \sqrt{\frac{\pi\rho}{RQ}} \quad (50)$$

显然，尺寸 $a < \overline{a}$ 的碎块总体积为

$$V = \frac{4}{3}\pi(r^3 - R^3) \quad (51)$$

由式(50)，(51)消除 r^3 后，求出总体积量 V 和碎块尺寸 $a = \overline{a}$ 的线性关系：

$$V = \frac{4a}{3u_z}\sqrt{\frac{\pi RQ}{\rho}} - V_0 \quad (52)$$

式中：V_0 为装药体积。

对于较复杂的情况，即存在爆炸压力压碎的情况 ($r_s > R$)，此时存在 2 个破碎区，其边界相应成比例为 $r_z/r_s = \sqrt{u_s/u_z}$。在每一个破碎区中，其块度总体组成图在 (a, r) 平面上都表示为通过原点的直线。块度按体积分布曲线一般形式见图10。由图10可知，在位于半径 r_s 和 r_z 之间的第二个区内，碎块的块度由上述的 a_z 变到 a_z'：

$$a_z' = a_z\left(\frac{r_s}{r_z}\right)^3 \quad (53)$$

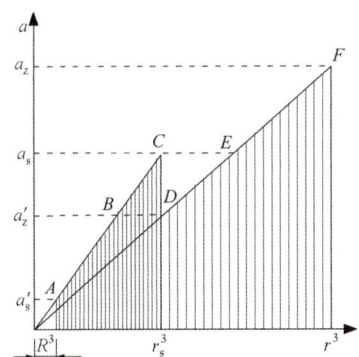

图10 存在 2 个破碎区的球形装药爆炸时破碎块度分布

Fig.10 Broken block distribution of spherical charge blasting when existing two broken regions

而转到位于半径 r_s 和装药半径 R 之间的第一区时，碎块块度按式(42)，相应于临界速度的改变增加至 a_s，即

$$a_s = a_z' \frac{u_s}{u_z} \quad (54)$$

在第一区范围内，块体块度重新减小到量 a_s'：

$$a_s' = a_s\left(\frac{R}{r_s}\right)^3 \quad (55)$$

按照每一个区域内的块度总量分布曲线图，可以构成碎块块度总体组成图。但是应指出，处于直线段 AB 上靠近装药的最小块度在第二个区域中不再出现，但是在总图上这个线段还将画出来，第二区中线段 EF 上的那些最大的块度仍用那个线段表示。在 a_z' 和 a_s 之间的块度在第二个区域中都将出现，应在筛网分析中将其联起来，因此 BE 应联成直线，最终曲线的一般形式见图11。

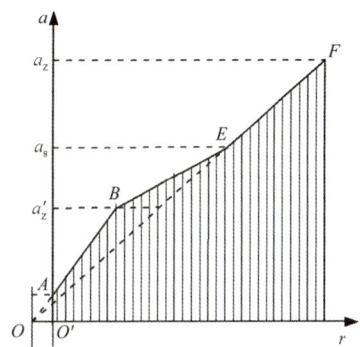

图11 岩体破碎碎块块度总体组成图

Fig.11 General scheme of rock broken block regions

(7) 均匀破碎

上述理论也给出了解反问题的可能性，即按预定的碎块块度寻找装药配置。根据式(41)，显然破碎判据此时应保持常值，即 $D = \text{const}$。根据势函数的连续方程及边界条件可以得

$$\phi = Axz \quad (56)$$

该情况下炸药应该配置在双曲柱面系的表面上，如图12所示。在相互垂直的平面偶 $x = z = 0$ 上，自由面的条件 $\phi = 0$ 中止了。由此可见，理想的均匀破碎仅在下述情况下可能实现，即工作面端头不对结果产生影响，并在 $x = z = 0$ 面组成的直角内部适当配置装药可以达到岩体的最均匀破碎。必须指出，按双曲面配置的装药密度应该是变化的。

图12 炸药的配置表面示意图

Fig.12 Surface scheme of blasting charge

以上在条件 $z=0$ 下求得的特解并不穷尽实践中其他方案,重要的是,上述内容是从理论上说明在进行爆破作业中获得经济上有益的均匀破碎是可能的。

(8) 破碎的持续时间

破碎过程实际上是的一类特殊波(称为破坏波)的传播过程,这个破坏波的波阵面把介质的未破碎部分区分开,在这个破坏波后,介质的联系被破坏,介质破成碎块,破坏波前面,介质完全保持黏结。破坏波的速度和介质最大运动速度相接近,因为介质的破坏是由介质最大速度导致介质的变形所引起的。由此可见,在某个距离 r 上介质的破碎时间为

$$t \leqslant \int_R^r \frac{\mathrm{d}r}{u} \tag{57}$$

对于无限均匀介质中球形装药爆炸的情况,半径 r 的介质破碎的最大可能时间为

$$t = \int_R^r \frac{\mathrm{d}r}{\dfrac{1}{r^2}\sqrt{\dfrac{RQ}{2\pi\rho}}} = \sqrt{\dfrac{2\pi\rho}{RQ}}\int_R^r r^2 \mathrm{d}r = \dfrac{1}{3}\sqrt{\dfrac{2\pi\rho}{RQ}}(r^3 - R^3) \tag{58}$$

如前所述,破碎过程在 $r = r_z$ 时将结束,因此整个无限介质的破碎过程结束时间为

$$t_{\max} = \dfrac{1}{3}\sqrt{\dfrac{2\pi\rho}{RQ}}(r_z^3 - R^3) \tag{59}$$

式(59)适用于无限介质球对称情况,在其余情况,仍然可以利用式(57),但此时需预先知道速度场参数及流线。

确定破碎的延续时间对控制装药体系的爆炸有重要意义。在某个炸药爆炸时发生了介质的某种破碎情况;随后炸药的爆炸将不对已破碎部分的介质产生作用。如果所有炸药同时爆炸,则破碎作用将加强并且所有的碎块都将变小。为了不出现过大的碎块,同时爆炸将是合理的,但是同时又增加了小碎块的量,消耗了爆炸能量。因此在很多情况下,在某个中间瞬间采用随后装药的爆炸是有利的,这时前面的一些装药炸药产生的小碎块已经产生,而大碎块还未形成。此时,联合爆炸的效果仅仅使得大碎块变小,而不会使已碎的小碎块变小,于是最终得到的岩石碎块比较均匀。

2.5 大规模地下爆炸作用下岩体中的岩块运动

岩体为由断裂构造、裂隙、层理和节理所分割的地质体。按照 M. A. Сабовский 院士的观点,岩体是一个复杂的层次构造系统,即岩体是块体的集合体,这些块体被软弱层所分开,这些软弱层包括填充的或部分填充的断层和裂隙。大规模的地下爆炸试验表明,在爆炸作用下,这些岩块在不同层次的等级上发生转动和移动,且主要是转动。由于岩块的运动,岩块产生不可逆的相对线位移和角位移。由于爆炸作用下岩块运动涉及到非常复杂的力学过程,在这节中只能介绍最可能的也是最概略的岩块的块间运动的机制和模型,以能估算块体运动参数和爆炸威力(当量)、与爆炸源的距离、软弱层的特性、应力状态等关系。

(1) 岩系中岩块的惯性移动(平动)模型

爆炸作用下块系单元的模型如图13所示[45~47]。

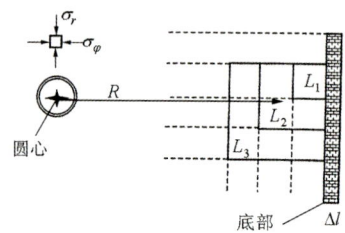

图 13 块系单元的模型[45~47]

Fig.13 Model of block element[45~47]

假定压缩波沿裂缝平面的法线方向传播,块间摩擦力按库仑定律确定:

$$\left. \begin{array}{l} F_{fi} = (\sigma_n \mu_0 + c_R)4L_i^2 \\ \sigma_n = \sigma_0 + \sigma_f \end{array} \right\} \tag{60}$$

式中:F_{fi} 为作用于 i 岩块边侧面上摩擦力;σ_n 为作用于 i 岩块边侧面上的法向应力;σ_0 为初始地应力;σ_f 为爆炸作用产生的自由场应力,且有 $\sigma_f = \rho C \dot{W}$;$\mu_0$ 为摩擦因数;c_R 为裂缝软弱层黏聚力。

为简化计算作如下假定:不同大小的裂缝具有相同的 c_R 值;边长为 L_i 的立方体岩块(包括较低构造层次的小尺度岩块)都可被认为是一个节理,波在节理中传播时没有形状和幅值的改变,波的运动参数在块体中不随时间改变,并且仅由从爆源到块体中心的距离确定。

波在岩体中传播,当波速 $\dot{W}(R, t)$ 达到最大值以前,所有尺度的岩块将会一起运动。当波速 $\dot{W}(R, t)$ 开始衰减时,块体受到侧表面上惯性力和摩擦力的影响。如果下列条件得到满足,单个块体将与周围块体或整个岩体作相对运动:

$$\left| \dfrac{\partial^2 W(R, t)}{\partial t^2} \right| \geqslant \left| \dfrac{F_{fi}}{M_i} \right| = a_{fi} \tag{61}$$

式中:M_i 为第 i 个岩块的质量,t 为从波到达岩块边界开始算起的时间。

当满足式(61)时，在 $t=t_*$ 时刻，块体与整个岩体开始有相对运动，而在此之前，块体与整个岩体一起运动。在 $t>t_*$ 时刻，块体 L_i 的绝对速度为

$$V_b(t) = V(t_*) - \int_{t_*}^{t} a_{fi}(\tau)\mathrm{d}\tau \quad (62)$$

块体与整个岩体的相对速度为

$$\left.\begin{array}{l} V_{\mathrm{brel}}(t) = V(t_*) - V(t) - \int_{t_*}^{t} a_{fi}(\tau)\mathrm{d}\tau \\ V(t) = \partial W(R,\ t)/\partial t \end{array}\right\} \quad (63)$$

当 $\sigma_0 + \sigma_f < 0$ 时为压应力，说明块体之间有接触摩擦作用；当 $\sigma_0 + \sigma_f > 0$ 时块体之间的作用力仅为黏聚力 c_R。块体 1 和块体 2 运动速度示意图见图 14。

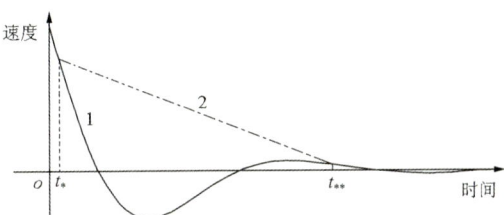

图 14 块体 1 和块体 2 运动速度示意图

Fig.14 Sketch of velocities of blocks 1 and 2

岩块位移将持续增加，直到满足条件 $V_b(t_{**}) = V(t_{**})$，即 $V_{\mathrm{brel}}(t_{**}) = 0$。该时刻在图 14 表示为 t_{**}。如果条件式(63)得到满足，岩块将再一次发生分离破坏。然而，通常规律是，对于强爆炸产生的应力波而言，在 t_{**} 时刻的加速度已经不是特别大。这样，块体的相对位移为

$$\Delta W = \int_{t_*}^{t_{**}} V_{\mathrm{brel}}(\tau)\mathrm{d}\tau = \int_{t_*}^{t_{**}} \left[V(t_*) - V(\tau) - \int_{t_*}^{t} a_{fi}\mathrm{d}t\right]\mathrm{d}\tau \quad (64)$$

根据输入的爆炸地震波波形及相关参数，运用模型(式(61)~(64))，便可以研究块体运动参数与初始地应力、波形参数及节理物理特征之间的关系。

(2) 块系中岩块的转动模型

为了定量描写紧密毗连条件下块系中岩块的变形过程，可以应用复合体变形理论原理以及弹性力学的矩理论，从而克服在块系介质中缺乏尺度因素的困难。按照矩理论，旋转矢量 $\boldsymbol{\omega}$ 和位移矢量 $\boldsymbol{u}$ 的旋度有下列关系：

$$\boldsymbol{\omega} = \frac{1}{2}\mathrm{rot}\boldsymbol{u} \quad (65)$$

这与经典的弹性关系一致(因此，广义应变张量是对称的并与经典的应变张量一致)。在这一阶段，介质的变形过程将在具有受限转动的矩理论的框架内描述。不均匀构造介质进一步变形时，第二阶段的内应力增加从而产生了力矩，限制了不同等价构造岩体的独立旋转，这在很大程度上表明在某一等级水平上存在着介质的非连续。

假定构造单元在受限条件下发生了旋转，介质中的宏观位移场 $\vec{U}$ 可表达成如下形式：

$$|\vec{U}_n| = \frac{1}{2}L_n^2 |\nabla \times \boldsymbol{\omega}_{n-1}| \quad (66)$$

式中：n 为构造等级，L_n 为该构造等级岩块的线性尺寸，$\boldsymbol{\omega}_{n-1}$ 为较高等级岩块的旋转矢量。

现根据上所概念，对于具体的爆炸波的形式评价构造介质的变形，选择持续时间为 τ 的准谐波：

$$\upsilon(r,\ t) = \begin{cases} 0 & (t<0) \\ \upsilon_0(r)\phi(t) & (0 \leqslant t \leqslant \tau) \\ 0 & (t>\tau) \end{cases} \quad (67)$$

假定变形过程产生转动向量的最优势方向，而且单元的转动为不可逆，考虑到式(66)和(67)，在每一个构造单元的界限内岩块的最大旋转为

$$|\boldsymbol{\omega}_{n-1}| = \frac{2\upsilon_0(r)\tau(r)}{\pi L_n}\left(1 - \cos\frac{\pi t}{\tau}\right) \quad (68)$$

式(68)的成立基于如下假定：

$$\frac{\partial \boldsymbol{\omega}_{n-1}^i}{\partial x_k} = \mathrm{const} \quad (69)$$

式中：x_k ($k=1,\ 2,\ 3$) 为笛卡尔坐标，i 为矢量 $\boldsymbol{\omega}_{n-1}$ 的相应分量。

块体旋转方向是由在扰动从一个块体传到另一个块体时作用于具体构造块上的力矩，以及变形构造单元的初始空间走向所决定的。考虑到围绕垂直于 $\vec{U}_n$ 方向的 2 个轴的转角等价的，因此可允许评估岩块的不同方向旋转。岩块的运动将会导致块间间隙的有效宽度的变化，其值为

$$\Delta a = \pm \frac{\upsilon_0(r)\tau(r)}{\pi}\left(1 - \cos\frac{\pi t}{\tau}\right) \quad (70)$$

函数 $\boldsymbol{\omega}(x) = |\boldsymbol{\omega}|(x)$ 可以表示为在 $(x_j,\ x_{j+1})$ 区间上的分段光滑函数，其中构造单元的边界用 x_m 来表示(坐标轴 x 是沿着岩块表面方向)。在 $t \to 0$ 的特殊情况下，构造单元沿着 x 轴的转角对于每一个区间 $(x_j,\ x_{j+1})$ 可以表示为

$$|\Delta \boldsymbol{\omega}(x_1)| = \frac{4\upsilon_0(x_j)\tau(x_j)}{\pi L_j}\left(1 + \frac{x_j - x}{L_j}\right) \quad (71)$$

从式(71)可以看出，残余转角为锯齿形的，对

于每一个区间 (x_j, x_{j+1})，函数在 $x = x_j$ 时达到最大值：

$$|\Delta \boldsymbol{\omega}_{\max}(x)| = \frac{4\upsilon_0(x_j)\tau(x_j)}{\pi L_j} \qquad (72)$$

如果块体尺寸已知，式(72)可以用来确定具体块的棱面的转角，也可以反过来：根据棱面的转角来确定块体的线性尺寸。爆炸荷载作用下激活块体的尺度体现了爆炸能量的大小，块体尺度的估算已显得非常重要。

为此研究岩块间隙的最大可能位移。假定岩石介质为不同等级构造单元相互嵌入的复合体，每一个构造等级都用空间尺寸来表征 L_i。用单个块体的广义位移 U 作为表征介质受到动载作用时的响应尺度，当 $t \to \tau$ 时有

$$|U| \to \frac{\Delta \omega_*(x_j)}{2}L_j + \frac{\Delta \omega(x_j)}{2}L_{j+1} \qquad (73)$$

式中：$\Delta \omega_*(x_j)$ 为第 j 水平的构造单元的最大可能转角。

第 i 水平的构造单元之间的间隙宽度由与之接邻的构造单元的特征尺度来决定。为了简单起见假设：$\varepsilon_i = k_i L_i$。在这种条件下，有

$$\Delta \omega(x_j) \approx 2\varepsilon_i / L_i \qquad (74)$$

因此广义位移 U 为

$$U \approx k_1 L_i + k_1 L_{i+1} = \frac{k_1 L_*}{k_2} + k_1 L_* \qquad (75)$$

$$k_2 = L_{i+1}/L_i \approx \text{const} = 3\sim5 \qquad (76)$$

式中：L_* 为激活块体尺度。

考虑到岩块转动应符合块系层次构造，即式(73)和(75)应相等，就可以得到激活块体的尺度：

$$L_* = \frac{\upsilon_0 \tau_0 k_2}{\pi k_1 (1+k_2)} \qquad (77)$$

其中，

$$\left.\begin{array}{l} \upsilon_0 = A_L r^{-n} Q^{n/3} \\ \tau_0 = B_L q^{1/3} (r/Q^{1/3})^m = B_L r^m Q^{(1-m)/3} \end{array}\right\} \qquad (78)$$

由此可得到激活块体的尺度与爆炸当量及爆心距的关系：

$$L_* = \frac{A_L B_L k_2}{\pi k_1 (1+k_2)} r^{m-n} Q^{(m-n+1)/3} \qquad (79)$$

式中：A_L，B_L，n，m 均为常数，可以通过核爆试验资料来得到。

激活块体的尺度按式(79)确定后，就可以根据式(71)确定岩块的最大转角。把介质表示为不同尺寸的遵从构造层次的构造单元的集合的概念基本相应于岩体的实际构造。从确定岩体的变形特性来讲，以及从外载作用下块体的尺寸的确定角度来讲，块体获得了受限的运动的可能性(依赖于块体间隙及填充材料的特性)。

3 岩石爆炸动力学的试验研究

3.1 岩石中爆炸波的基本参数及其相似关系

岩土中爆炸作用的相似律可表述为：对于同一种爆炸装药，在与装药尺寸(装药半径、埋深或其他线性尺寸)比例相同的比例距离上，在相同介质中的爆炸作用参数相等。其条件是可忽略重力作用的相似，否则几何相似律不能采用为岩土中爆炸作用的相似律，而需予以修正。由于装药尺寸是装药质量的1/3次方，所以几何相似律又可表述为广泛应用的能量相似律。

能量相似破坏的根本原因是在降低爆源的装填密度的情况下，爆轰完成后爆轰产物和周围介质之间的能量分配比例有了改变，随着装填密度的降低，传递给周围介质中的能量将减少。于是，精确的能量相似律应表述成：在相同的比距离 $r/e^{1/3}$ 上的爆炸作用参数相同，e 为爆炸时传递给周围介质的能量值(或炸药当量 Q_e)。对于不同的炸药装填密度 ρ_1，有

$$e/E_0 = 0.16 \rho_1^{3/4} \qquad (80)$$

因此，按照爆炸时传递给周围介质的能量，速度场(及加速度场)的几何相似原则，即运动相似的原则是正确的。在浅埋爆炸试验中，由于浅埋爆炸条件下所得质点速度变化曲线与地下封闭爆炸条件下无本质区别，质点最大速度的变化仍然分为2个具有不同规律的区域：

$$v_m = \begin{cases} a_0 r_0^{-n_0} & (r_0 \leqslant r_*) \\ b_0 r_0^{-p_0} & (r_0 > r_*) \end{cases} \qquad (81)$$

式中：a_0，b_0，n_0，p_0 以及 r_* 均为参数，且破坏半径与装药半径的比值($r_* = R_*/R_z$)在封闭爆炸中随不同介质而变化；而在浅埋爆炸中不但随介质不同，随相对埋深 η 不同而变。η 可表示为

$$\eta = h/R_z \qquad (82)$$

式中：h 为爆心距离自由面的距离。

当自由面卸载波接近爆心时，与 $\eta \to \infty$ 的封闭

爆炸相比,将导致压缩波作用时间缩短,且爆心离自由面越近,对爆炸波作用时间影响就越大,因此在离爆心近区的不同范围内,在不同埋深的爆炸条件下,会形成幅值一样但作用时间完全不同的波动。

3.2 岩体中爆炸效应若干问题的室内试验研究

(1) 地应力对岩体中爆炸破坏的影响

在测量精度的范围内,图 15 给出了外部压力 σ_d 作用下,爆炸的爆炸波阵面、爆炸空腔和破坏波的时空发展的情况。曲线 1 表明冲击波阵面传播速度在有、无初始压力情况下是相同的;曲线 2,2′ 分别表明有压情况下的空腔(2)和小于无压情况下的空腔(2);曲线 3 所示的是有机玻璃块中爆炸时破坏波传播的情况。当试件所受的压力为 50 MPa 时,材料没有破坏,爆炸空腔的尺寸比不受压的试件要小,空腔最大体积约小 17%(曲线 2 与 2′)。在此结果的基础上,图 16 给出了模拟材料(有机玻璃)中空腔最大体积与静水压力的关系曲线,近似地表达为

$$\frac{V_p}{V_m} = 1 - 0.77 \frac{\sigma_d}{\sigma_*} \quad (83)$$

式中:V_p 和 V_m 为分别为受压和非受压试件中的空腔最大体积;σ_* 为材料极限单轴抗压强度,在试验情况中,设定 $\sigma_d / \sigma_* = 0.42$。

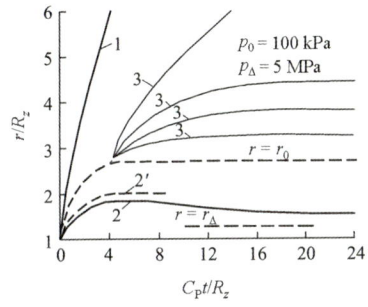

图 15 爆炸波阵面、爆炸空腔和破坏波的时空发展

Fig.15 Spatial developments of blasting wave front, blasting cave and fauilre wave

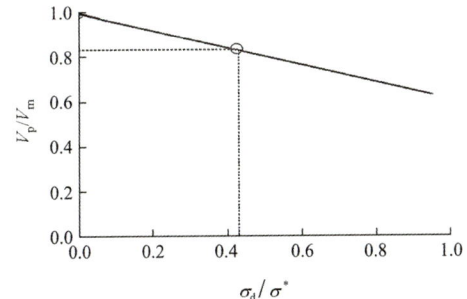

图 16 不同压力时爆炸空腔的变化曲线

Fig.16 Curves of blasting cave under different pressures

由此可见,球形装药爆炸时,爆炸压力形成空腔所作的功减小了,而空腔中爆炸产物中所含的剩余能量增加了。

径向裂缝区与抗拉强度和地压总和的不同相关性相应于空腔周围裂缝发展的两个阶段:在 $r \leq 4R_z$ 区域,爆炸压缩波内,处于拉应力作用下大量裂缝的产生的阶段,$r > 4R_z$ 的区域内,基本是在空腔准静态压力作用下个别裂缝产生的阶段。为了改变 $r > 4R_z$ 区域内个别径向裂缝的扩展范围,由于地压存在,只增加较小的抗拉强度就足够了。而在弱相关区间 $r \leq 4R_z$,要增加很大的地压力才能压制住群生的径向裂缝的产生。

因此,在不超过岩体的强度范围内,即使增加不大的初始地应力量值,也可能显著地改变岩体爆炸破坏的特性。在深部岩体条件下,岩体的三向爆炸的破坏效应可以减小,直至当地压很大时破坏消失。当然,模拟试验成果仅表明定性的特征,在实际岩体的爆炸中,地应力对爆炸破坏的影响在定量上可能还将有所不同。

(2) 岩体中的裂隙对爆炸波运动参数的影响

在岩体中存在有或大或小、或长或短的断裂构造、裂隙、节理和裂纹,统称为裂隙。裂隙的存在将使爆炸效应的分析和预测复杂化,当爆炸压缩波通过裂隙时,爆炸作用的强度会减弱。通过对试验(具有不同的 Δ 和 l 值的裂缝)得出的运动时间参数的分析,得到如下关系式:

$$\Delta t_0 = \Delta / (2 v_0 (l)) \quad (84)$$

式中:Δ 为裂缝宽度(m),l 为裂缝离爆点距离(m)。

式(84)可以较精确地获得波通过裂隙时的延迟时间。在抛掷试验中,裂隙对质点速度的上升时间有显著的影响。裂缝开度 Δ 对 t_1 值的影响可描述($\Delta / Q^{1/3}$ 单位为 $m/kg^{1/3}$,下同)为

$$\left.\begin{array}{l} t_1 / t_0 (l) = 55 (\Delta / Q^{1/3}) \\ 0.014 < \Delta / Q^{1/3} < 0.050 \end{array}\right\} \quad (85)$$

在试验中值得注意的是裂隙的设置位置对上升时间不产生影响。图 17 说明当压缩波通过整个裂隙时其幅值的变化是平稳的,从图 17 可看出,在距离 $r > l + \Delta$ 处裂隙的存在将对固体介质中爆炸作用的强度有重要影响。

对于压缩波幅值,介质中爆炸作用的屏蔽系数为

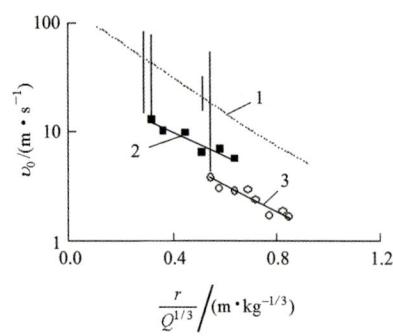

1—无裂隙爆炸；2—$l/Q^{1/3}=0.32$ m/kg$^{1/3}$；3—0.49 m/kg$^{1/3}$

图 17 裂隙后的质点最大速度

Fig.17 The maximum particle velocities after cracking

$$k = \upsilon_0 / \upsilon_1 \tag{86}$$

类似地，k 有以下形式：

$$\left. \begin{array}{l} k = 80\left(\dfrac{\varDelta}{Q^{1/3}}\right)^{0.6}\left(\dfrac{l}{Q^{1/3}}\right)^{0.9} \\ 0.014 < \varDelta/Q^{1/3} < 0.050 \\ 0.25 < l/Q^{1/3} < 0.70 \end{array} \right\} \tag{87}$$

$$\left. \begin{array}{l} k = 22(\varDelta/Q^{1/3})^{0.6} \\ l/Q^{1/3} < 0.2 \end{array} \right\} \tag{88}$$

由此可见，爆炸源的屏蔽在本质上改变了爆炸信号，除了直接衰减压缩波的幅值外，重要的是减轻了爆炸地震效应，即裂缝是对高频的有效过滤。质点速度幅值爆炸作用的屏蔽系数与裂隙距离的关系，计算值和试验值分别见表 1。

表 1 质点速度的屏蔽系数与裂隙距离的关系

Table 1 Shielding coefficients of particle velocity and cracking space

$(l/Q^{1/3})$/(m·kg$^{-1/3}$)	υ_0/(m·s^{-1})	屏蔽系数	
		计算值	试验值
0.22	90	2.70	2.7
0.33	38	3.44	3.4
0.49	19	4.77	4.9
0.65	10	7.30	6.8

必须指出，在工程爆破条件下，模拟试验结果与真实岩块运动得到的数据非常一致。对于上述的爆炸($l/Q^{1/3} \approx 0.44$ m/kg$^{1/3}$)，关于最大质点速度的裂隙屏蔽系数值在 4～6 的范围内。

(3) 岩体中裂隙对爆炸岩石破坏的影响

以裂隙形式存在的初始未填充空间将保证爆炸破坏岩石可能的附加疏松。破坏岩石的疏松程度首先会对渗透性产生影响，并且疏松程度越高渗透性就越强。图 18 给出了在等效威力为 2 kt 的工程爆破条件下(曲线 2)，装药和裂隙之间的空间内岩体的渗透性的试验结果。同时也给出了相同爆炸条件下破坏无裂隙岩体渗透性的试验结果(曲线 1)。由此可知，岩体区域内破坏介质的渗透特性得到改善外，裂隙还减轻了裂隙后区域内爆炸作用的强度。裂隙远壁像是一种边界，边界后的渗透性要小于整体介质中爆炸条件下的值。此时裂隙是限制爆炸时单个裂缝扩展进程中的良好障碍。此时，若在液体和气体物质渗流的优势方向上存在裂隙(大裂缝、构造地质端口、破坏性和渗透性增强的区域)，裂隙的隔离作用会得到加强。

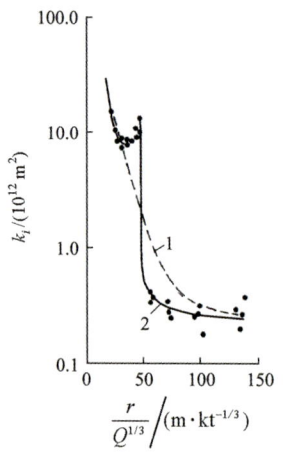

图 18 装药和垂直裂隙之间的空间内及裂隙后岩体的渗透性

Fig.18 Permeability of charge-vertical fissure space and fissured rock

图 19 标出了碎块平均尺寸 $\langle x \rangle$ 的试验数据。从图中可看出，随着至爆心距离的增加，开始破坏介质的碎块平均尺寸增大，然后减小，而在裂缝的位置处达到局部最小值。由此可知，裂缝位置处存在更细小的粒度可以解释为卸载和碎块撞击静止障碍物时发生的介质材料的附加破坏。

裂隙的衰减作用可导致破坏区尺寸的减小。例如，对于开度 $\varDelta/Q^{1/3}=0.027$ m/kg$^{1/3}$ 的裂缝，在裂隙方向上的破坏区尺寸可表达为以下形式：

$$R_\delta = R'_*(l/R'_*)^{1/3} \tag{89}$$

式中：R'_* 为整块介质中爆炸时的破坏区尺寸，并有 $R'_*/Q^{1/3}=1.36$ m/kg$^{1/3}$。

图 20 可用于确定位于装药和裂隙所包含空间内材料碎块的平均尺寸。图中 $\langle x_2 \rangle$ 为装药和裂隙之间区域内碎块的平均尺寸，$\langle x_0 \rangle$ 为无裂隙介质中爆炸时相同范围区域内碎块的平均尺寸。图 20 的试验

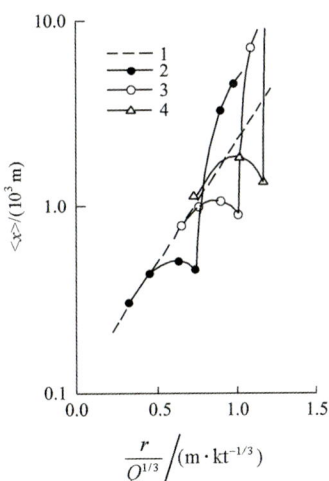
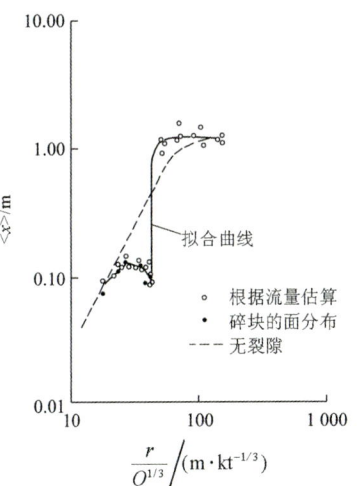

1—整块介质中爆炸时的数据；2—0.65m/kg$^{1/3}$；3—1.0 m/kg$^{1/3}$；4—1.3 m/kg$^{1/3}$

图 19 在裂隙$\it{l}/Q^{1/3}$ = 0.027m/kg$^{1/3}$的爆炸条件下碎块的平均尺寸

Fig.19 Averege sizes of broken block under blasting when $\it{l}/Q^{1/3}$ = 0.027 m/kg$^{1/3}$

图 21 碎块平均尺寸与$r/Q^{1/3}$的关系

Fig.21 Relationship between average size of broken block and $r/Q^{1/3}$

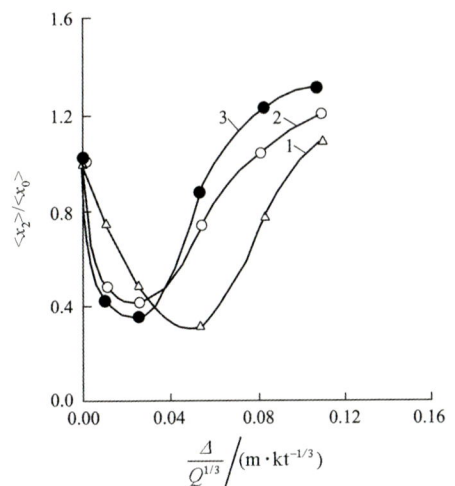

1—$l/\it{\Delta}^{1/3}$ = 0.44 m/kg$^{1/3}$；2—0.65 m/kg$^{1/3}$；3—0.98 m/kg$^{1/3}$

图 20 碎块的相对尺寸与$\it{\Delta}/Q^{1/3}$关系

Fig.20 Relationship between relative size of block and $\it{\Delta}/Q^{1/3}$

结果表明，采用裂隙可明显提高该空间内破碎的均匀度，并且提高破碎程度，并且对一定的$l/Q^{1/3}$值，存在一个裂隙宽度以实现最佳破碎。

图 21 给出了等效威力为 2 kt 的工程爆破破坏区域内介质的试验结果。由此可知，在有裂隙的爆炸条件下，介质的附加疏松导致破碎岩石总体积的明显增大，岩石破碎成更小的粒度级。分析表明，与无裂隙介质中的爆炸相比，由于裂隙作用，破碎体中其中 50%的尺寸小于⟨x⟩的碎块的相对体积非常大。从岩石破碎的观点出发，若将爆炸产生的碎块表面积作为爆炸效能的指标，则可得出四面限定爆源的裂隙可将爆炸效能提高 1.35 倍。如果裂缝从一个方向限制爆炸，预期的效能增长也足够高，约提高了爆炸能量的 10%。

3.3 实际岩体中爆炸效应的试验研究

限于篇幅，仅介绍关于岩体在地下强爆炸时的爆炸效应局部不可逆位移。

(1) 地下强爆炸条件下岩体变形

对于工程观点来说，最为关注的问题集中在破坏区外的变形问题，特别是不可逆位移区的范围及不可逆变形的大小。因为工程可能配置在破坏区内，即破碎区和径向裂纹区内。是否有可能处在不可逆位移区，即非弹性变形区是主要关心的问题。

俄罗斯根据试验资料整理，局部不可逆变形区的半径(m)为(800～1 100)$Q^{1/3}$。美国 Benham 地下封闭核试验的爆心埋深为 1 380 m，因此换算到断面层发生位移的爆心距为 $R = 1\ 410 Q^{1/3}$，美、俄公式是在比较接近的同一个数量级。

俄罗斯学者[46, 47]对于连续均匀介质模型数值计算数据与地下强爆炸试验实测数据进行了比较，如图 22 所示。图 22 实线所示的岩体实测最大比位移大大超过了按连续均匀介质模型的理论计算所得的最大比位移(虚线)和永久比位移(点划线)。其差值随着比距离($R/Q^{1/3}$)增大而减小，当$R/Q^{1/3}$由 30 增至 180 时，误差由 6.6 倍减为 0.0。距离越远，误差越大；在很远的距离上，岩块不发生相对的移动和转动时，连续介质模型计算的成果就比较准确了。

图 23 实线表示为均匀介质中最大变形的均值关系曲线，虚线表示均匀介质中残余变形的关系曲

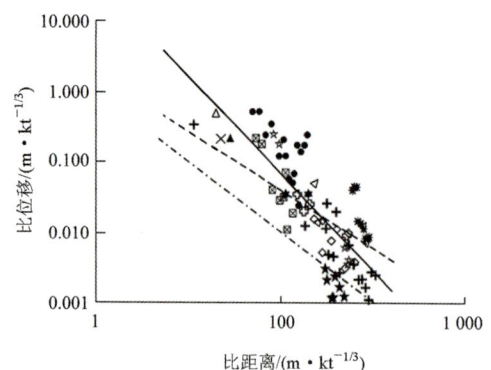

图 22 实测与计算最大比位移、比距离的比较图

Fig.22 Comparison between of measured and calculation diaplacements and distances

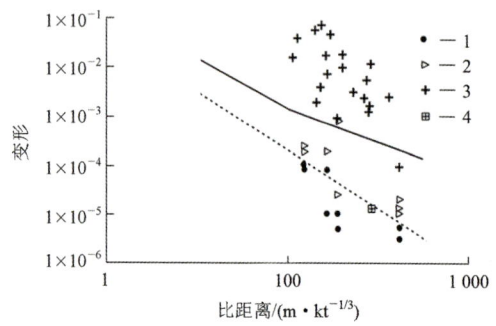

图 23 实测与计算最大岩体变形(残余变形)的比较图

Fig.23 Comparison between of measured and calculation maximum diaplacements and residual diaplacements

线。这表明在岩体中地下核爆炸时,在断裂、裂隙附近,产生了显著的岩体变形局部化现象。当岩体中存在断裂、裂隙、节理时,岩体中的岩块发生了旋转和相互间的滑移,如图 24 所示。可见岩体在裂隙、节理两侧的位移速度相差很大,甚至符号(方向)相反,这是连续介质模型的计算所无法得到的。

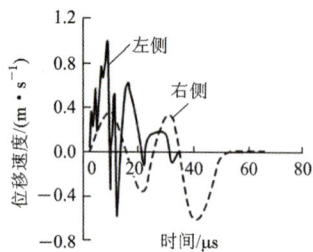

图 24 当量 1.4 kt 爆炸中离爆心 160 m 处与巷道相交的裂隙两侧的速度测量结果

Fig.24 Measuring results of both sides of fissures in interaction of roadway and the place 160 m away from blasting center when charge is 1.4 kt

(2) 连续介质力学模型不适用性产生的物理本质

在地下强爆炸作用下,岩体沿连续性破坏的结构面处发生了显著的相差位移,该显著的岩块相差位移以及岩体变形的强烈局部化现象表明了强地下爆炸时发生了岩块的刚体位移,主要是岩块的整体转动。强地下爆炸时岩体的不可逆位移和岩体连续性破坏处的变号反应的主要源于岩块间的相互转动和相互滑移。

岩体的"块体性",在岩体中地下大爆炸试验时由"岩块激活现象"所充分证明。某岩块被"激活"意味着该岩块相对于周围岩块发生了转动或移动。俄罗斯地下大爆炸试验表明,岩体"块体性"的激活程度由爆炸当量决定,其"激活"岩块的边界基本上就是观测所得到的岩体角位移和线位移的最大梯度处。爆炸当量越大,则"激活"岩块的尺度越大。图 25 为"激活"岩块尺寸在历次爆炸中的测量结果,大多数情况下,激活块体的尺寸可归纳为 $L = (10 \sim 30) Q^{1/3}$ m/kt$^{1/3}$。

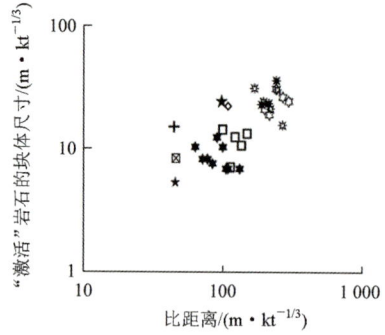

图 25 "激活"岩块尺寸在历次爆炸中的测量结果

Fig.25 Measuring results size of reactived block in various blastings

连续介质模型远离实测结果的另一个物理原因在于该模型忽略岩体作为地质体的另一个构造变形特性——岩体的含能特性。岩体作为地壳岩石圈的一部分,由于地质构造作用和重力作用引起的变形,积累了变形能。俄罗斯地下强爆炸试验数据表明,实际岩体在地下强爆炸时,强爆炸解除了岩体中岩块变形的约束后,岩体变形能得到了释放,转变为岩体的动能,使得岩体获得更大的运动和位移。在这种情况下岩块发生了最大位移,所以在某种意义上说岩体中的岩块的运动也是由岩体在地质构造运动中长期积累的变形能的重分布所决定的。根据同样的道理,俄罗斯科学家解释了地下爆炸诱发了地震的总能量大于其炸药总能量的工程性地震现象。

(3) 地下强爆炸条件下岩体局部不可逆变形对地下工程的危害及其估算

美国在进行地下强爆炸时,沿着已有的地质断层记录了很大的局部不可逆位移,同时在地表也观测到断层竖向位移为 1.0～1.2 m,沿断层为 0.15 m,发生永久位移的断层长度为 0.3～8.0 km。工程对于过大的不可逆位移水平,俄罗斯资料认为是不可接受的;"不可逆变形的尺度相应于破坏地下工程结构的水平","与地下巷道相交的岩块位移可以导致地下工程结构的强烈破坏","小型岩块(尺度为 10 cm 级或米级)的稳定性可以由工程措施保证;大型岩块(尺度为 10 m 级或百米级)的稳定性无法由工程措施来保证"。

因此估算地下强爆炸时出现局部不可逆位移区 R_{**} 的范围以及不可逆位移的大小成为十分必要的。实测资料表明,地下强爆炸时 R_{**}(m)及 $R = R_{**}$ 时的岩体质点速度 v_{**}(m/s)分别为

$$\left.\begin{array}{l}R_{**} = (650\sim1\,400)Q^{1/3}\\ v_{**} = 0.05\sim 0.15\end{array}\right\} \quad (90)$$

根据 R_{**} 的实测数据,俄罗斯学者推荐 R_{**}(m),且 $R_{**} = 1\,000Q^{1/3}$。有了 v_{**} 值,根据爆炸规模估算的爆炸压缩波持续时间 τ 就可估算出不可逆位移值,此时可假定波速波形为抛物形和三角形。实测岩块位移最大幅值和实测岩块残余(永久)位移为分别为

$$\left.\begin{array}{l}\overline{w}_m = 2.5\times 10^4 \overline{R}^{-2.78}\\ \overline{w}_H = 100\overline{R}^{-2}\end{array}\right\} \quad (91)$$

3.4 爆炸抛掷效应的试验研究和效应中的相似关系

爆炸抛掷,又称抛掷爆破,是地下爆炸的一种类型。其典型特点是形成漏斗坑,其大小由装药能量、装药的埋置深度以及岩土特性决定。对于每一种岩土,给定了炸药量后,存在一个最佳埋深,在该埋深爆炸,该药量形成的漏斗坑最大。

典型岩体中的漏斗坑断面如图 26 所示。漏斗坑的特征是其半径 B、深度 H、抛掷指数 $n = B/H$,以及体积 V_B。由此可见,漏斗坑的体积并非等于其抛掷土的体积,由于可见漏斗坑是由于松散土的填充,部分是由于坑沿土的塌落,部分是由于被抛掷土的回落所形成。

当埋深增加,爆炸气体能量不足以产生抛掷时,仅能形成陷落坑。图 27 显示了冲积土中爆炸形成的塌陷坑的横断面。

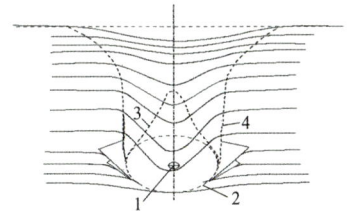

1—爆炸源;2—初始空腔;3—塌落柱;4—沉陷土的边界

图 27 软土中塌陷坑的横断面

Fig.27 Cross-section of collapse crater in soft soil

按照爆炸能量传递给被抛掷岩土的力学机制,抛掷过程可以区分为 3 个基本阶段:第一阶段为球形空腔对称发展的封闭阶段;第二阶段为气相加速阶段——爆炸气体产物的能量主要耗费于加速指向自由表面的土体运动上,一部分能量耗费于克服被抛掷岩土与相邻土体的黏聚力,形成土体隆起的穹顶,穹顶继续运动,再形成飞散块的蜂窝时开始破坏。第二阶段结束时,穹顶完全破坏,爆炸产物的能量实际上已完全耗尽;第三阶段是结束阶段——土体在重力场中惯性飞散的阶段,其飞散远度由碎块所拥有的动能大小、飞散角度以及空气阻力所决定。一般来说,抛掷爆破漏斗坑的尺寸主要是由第二阶段的发展所决定,在埋深不大的爆炸至最佳埋深爆炸的很大范围内,第三阶段的土体运动对于漏斗坑的尺寸影响较小。

(1) 土的运动与飞散

图 28 提供了装药埋深中心投影点处抛掷的第二阶段,即运动阶段的试验结果,结果表明,在所爆炸当量(0.1～10 t)范围内,对于黏土和黄土,以及和不同的爆破规模都没有偏离几何相似。

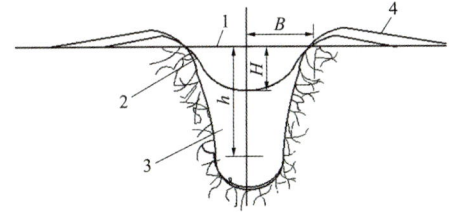

1—自由地表面;2—可见漏斗坑周边;3—回落土;4—土堆

图 26 典型抛掷漏斗坑的横断面

Fig.26 Cross-section of typical throwing funnel

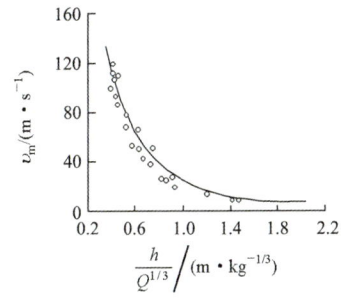

图 28 土体隆起最大速度与 $h/Q^{1/3}$ 的关系

Fig.28 Relationship between uplift maximum velocity and $h/Q^{1/3}$

由图29可见，随炸药埋设深度增加，爆炸抛掷系数 η 略为减少，系数 η 取决于土性。隆起土的动能在黏土中比在黄土中大3~4倍。

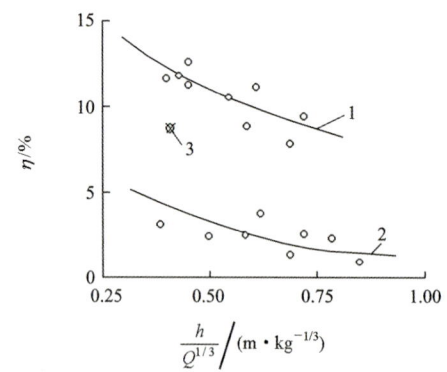

1—黏土中 0.1~10 t；2—黄土中 0.01~1 t；3—黏土中 1 000 t

图29 爆炸抛掷系数 η 和 $h/Q^{1/3}$ 的关系

Fig.29 Relationship between uplift coefficient η and $h/Q^{1/3}$

图30比较了对于炸药当量显著不同的爆破试验结果。表明在同一个比例深度值 $h/Q^{1/3}$，1 000 t 爆炸的最大隆起速度比10 t爆炸以及更少炸药爆炸的相应速度低。

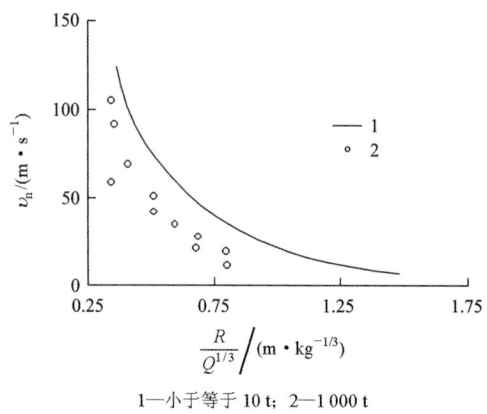

1—小于等于 10 t；2—1 000 t

图30 自由面隆起法向速度和炸药中心至投影点距离关系

Fig.30 Relationship between free-uplift normal velocity and projection point diatance

由此可见，1 000 t 爆炸时，被抛掷土体获得的能量显著地小于按几何相似原则计算的值，这个结果也相应于按单位炸药量计算的被抛掷的土体体积，由试验数据可见。1 000 t 黏土中的爆炸，其抛掷的效果明显的低于按几何相似原则计算值。

在土体惯性飞散的第三阶段中，进行定量研究存在很大困难，在试验中采用回落于离开爆炸投影点不同距离的单位面积上的土量作为土飞散的特征量度。试验表明，在比例埋深 $h/Q^{1/3} \leqslant 0.7$ 时，位于漏斗坑内的土被抛掷至漏斗坑界限外，即使松散土的量常常超过可见漏斗坑的体积。当抛掷指数 $n \geqslant 1.5$ 时，土体返回坠落于漏斗坑内的现象没有被观察到，证明了在大多数情况下，抛掷的第三阶段在形成可见抛掷漏斗坑的过程中不起重要作用。

(2) 抛掷漏斗坑的基本参数和重力相似

多年的爆破作业试验表明，抛掷漏斗坑的尺寸不仅取决于上述的基本指标——爆炸能、炸药埋深、岩土类型，而且还应考虑其他细节，包括爆轰气体产物的特性，地面的地形以及岩体的构造(裂隙性、层理性等等)。如果仅考虑基本指标，则会导致实际结果与预报结果的"偶然"偏差。

通常抛掷爆破装药量的计算是基于试验公式，对于不大装药量的集团装药爆破，装药量为

$$Q_1 = Kh^3 f(n) \tag{92}$$

式中：K 为取决于炸药与岩土特性的系数；$f(n)$ 为抛掷指数 n 的函数，它也取决于岩土特性，不同的作者建议函数 $f(n)$ 取不同的类型。

由式(92)得出，爆破规模尺度之间的转换相应于几何相似原则：在相同的抛掷指数 n 下，产生相似的爆破抛掷漏斗坑的炸药量正比于漏斗坑的体积，即正比于 h^3。但是抛掷爆破实践表明，在爆破规模增大的情况下，必须修正，即所需炸药量的增大要比漏斗坑体积增加更快。在该条件下要实现漏斗坑的相似，必须保持无因次量 $Q_1/(\rho g h^4)$ 为常数，因此在巨大的抛掷爆破中，爆炸药量应正比于埋设深度的四次方：

$$Q_1 = K_1 \rho g h^4 \tag{93}$$

为了修正基于几何相似的 Боресков 公式，俄罗斯学者 Садовский 和 Покровский 建议当 $h > 25$ m 时，有

$$Q_1 = Kh^3 \sqrt{\frac{h}{25}}(0.4 + 0.6n^3) \tag{94}$$

近年来，按抛掷漏斗坑尺寸的试验成果整理的公式得到广泛应用，即

$$H/Q_1^{1/N} = f_2(h/Q_1^{1/N}) \tag{95}$$

在炸药量从 0.116~454 t 的大小爆破范围内，在冲积土中，式(94)以 $n=3.4$ 计算的结果与试验结果相比的偏差最小，在实际工程中一般 $n=3.5$。

关于多种类型岩土的最佳深度和极限深度问

题，所谓极限深度是指开始不形成可见漏斗坑的深度。对于玄武岩、凝灰岩和黏土，装药埋设的最佳深度实际上是重合的，在 $40\sim45$ m/kt$^{1/3.4}$ 时达到。当超过最佳装药埋设深度后，这些岩土介质很快达到极限装药埋设深度，其值为 $55\sim65$ m/kt$^{1/3.4}$，此时可见漏斗坑实际上没形成。而在黄土和冲积土中，装药埋设最佳深度为 $50\sim60$ m/kt$^{1/3.4}$；极限装药埋设深度 2 倍，即为 $110\sim120$ m/kt$^{1/3.4}$。

对于非岩性土，最大的抛掷体积是在抛掷指数为 $n=1.5\sim2.0$ 时记录到，最大漏斗坑体积可按下式计算：

$$V_B = 1.4B^2 H \tag{96}$$

对于硬岩，最大的抛掷体积是在抛掷指数 $n=1.2\sim1.5$ 时记录到，而其漏斗坑体积比式(96)计算的要小 $1.5\sim2.0$ 倍。

(3) 抛掷大爆破中重力相似实验室模拟

抛掷漏斗坑的形成可看作为破碎的岩土介质被爆炸产生的气状产物所推出的过程。因为该过程比较缓慢，所以岩土介质的压缩性可以忽略，但应考虑介质的密度 ρ，内摩擦因数 μ_0 以及岩土介质的黏聚力 c_R。过程的决定性参数主要为空腔中的能量 Q_π，压力 p_k（或空腔半径 a_m），气体绝热指数 γ，以及装药埋深 h、重力加速度 g，以及地面大气压力 p_a。漏斗坑的基本参数为其半径 B。

抛掷爆破漏斗坑取决于初始条件及被抛掷介质性质的量纲关系的一般形式为

$$\frac{B}{h} = f\left(\frac{Q_\pi}{\rho g h^4}, \frac{Q_\pi}{\sigma h^3}, \frac{Q_\pi}{p_a h^3}, \frac{a_m}{h}, \gamma, \mu_0\right) \tag{97}$$

由式(97)可见，为了模拟大爆破，试验中必须不仅要减小 Q_π，而且要足够地减小 p_a 和 c_R，使得与 $\rho g h$ 相比可以忽略它们，即对于参数 $Q_\pi/(p_a h^3)$ 和 $Q_\pi/(c_R h^3)$ 的影响很小，为此在模型中采用真空以减小 p_a，采用干砂以减少黏聚力 c_R。在模型试验中代替爆炸气体产物的是压缩空气。

模拟试验的基本任务在于确定式(97)的具体表达式，首先要排除参数 $Q_\pi/(p_a h^3)$，由于摩擦力的存在，还要消耗能量，所以要保留函数 μ_0 和 γ，其次在排除参数 $Q_\pi/(p_a h^3)$ 的同时，要考虑到并非是绝对真空，为了更准确考虑抛掷阻力，用参数 $Q_\pi/[(\rho g h + p_a)h^3]$ 代替 $Q_\pi/(\rho g h^4)$，即总阻力为 $\rho g h + p_a$。由此可得 Q_π：

$$\left.\begin{array}{l}Q_\pi = \dfrac{p_k V_B}{\gamma - 1} \\[2mm] \dfrac{B}{h} = f\left(Q_\pi^0, \dfrac{a_m}{h}\right) \\[2mm] Q_\pi^0 = \dfrac{Q_\pi}{(\rho g h + p_a)h^3}\end{array}\right\} \tag{98}$$

在模拟试验中，用压缩空气代替高压爆炸气状产物。同时记录漏斗坑尺寸，岩土表面向上的运动速度。图 31 显示了抛掷过程中地表面投影中心升高运动速度的曲线，几何参数 a_m/h 的值为 0.42 和 0.21，p_k/p_a 均为 100。可见地面隆起速度的改变有两种类型：一种是 Q_π^0 足够大，速度单调的增大到最大值，然后岩土块按惯性在重力场中飞散；另一种是 Q_π^0 值不大，$Q_\pi^0 \approx 1$，地面的隆起在定性上有改变，其速度增长有个拐点，图上的标注该拐点时间为 t_*，这个标志二次加速的拐点的本质是爆炸气体由空腔穿透土穹顶而溢出，测量数据表明，此时土穹顶部分运动速度可以以十至百倍大于穿透前的土面隆起速度。

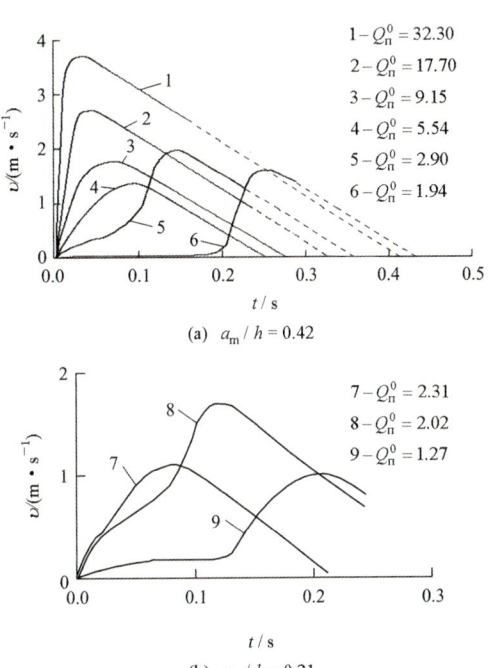

图 31 在不同 Q_π^0 值情况下地表面投影中心升高速度随时间变化曲线

Fig.31 Variation curves of height of ground projection center rising velocity vs. time under different Q_π^0 values

按漏斗坑形成过程中地表面的运动特征，可区分成 3 种类型漏斗坑：在 $Q_\pi^0>4$ 时，形成抛掷漏斗坑,此时空腔中的气体有足够能量很快的推出整个空腔上面的岩土形成 $n>1$ 漏斗坑;在 $4 \geq Q_\textrm{п}^0 >1$ 时,即使地表面的初始运动是向上的,但土穹顶的运动及抛掷有个迟滞,其特征是有个转折时间 t_* 以及部分土塌落于空腔中,此时形成由抛掷漏斗坑向塌陷漏斗坑过渡的过渡型漏斗坑;在 $Q_\textrm{п}^0<1$ 时,可能仅形成塌陷漏斗坑。在此情况下,地表在土穹顶破坏后立刻下沉,塌陷漏斗坑的尺寸明显地取决于几何尺寸 $a_\textrm{m}/h$ 和 $\lg Q_\textrm{п}^0$。

图32显示了3个原型试验的地表面投影中心隆起速度时程曲线与模拟试验的相应时程曲线的比较:核试验《Седан》取当量100 kt,埋深 $h=194$ m,在冲积土中;TNT 爆炸:《Скутер》取当量454 t,$h=38$ m,在冲积土中;1 t NT 爆炸,$h=7$ m,在黄土中。

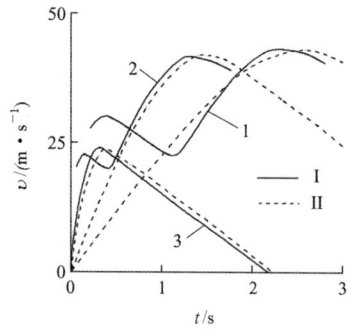

1—《Седан》;2—《Скутер》;3—$\varepsilon=1$ t,$h=7$ m

图32 原型爆炸(I)与模拟试验(II)的速度曲线 $v(t)$ 的比较

Fig.32 Velocity curves comparison between prototype blasting (I) and model blasting(II)

由图32可知,原型和模拟速度时程曲线的区别仅仅在初始阶段,这时在原型试验中,隆起速度是由冲击波的作用决定的,在后续气相加速阶段,图32的曲线显示了两者一致性的满意结果。由此可知,原型试验与模型试验抛掷爆破的数据较为一致,特别是弱岩中形成抛掷漏斗坑时,重力作用是决定性的。

4 岩石爆炸动力学的工程应用

岩石爆炸动力学的工程应用是基于岩石和岩体中爆炸的力学效应:形成空腔;破碎岩石;压密或松散岩体;抛掷岩石介质,形成抛掷漏斗坑;岩体中传播压缩波与地震波。这些效应常常在矿山、水工建筑、水土改良、工业和民用建筑领域找到广泛的应用。具体来说,有下列几个方面:

(1) 采矿工程领域:岩体中的剥离爆破——矿藏的露天开采被实践证明是经济有效的,特别适用于山区和未开发地区。根据爆破技术水平,目前在深度200 m左右进行爆破剥离上覆岩层是可行的,在可望的未来,剥离更大埋深的岩层也是可能的。矿场越大,爆破法剥离的经济性越好。

(2) 水利建筑工程领域:抛掷爆破用以开挖路线通过分水岭的河渠;抛掷筑坝和构筑水库,即利用爆破形成的抛掷或塌陷漏斗坑构筑水库已有成功的先例。在形成塌陷漏斗坑时,在其漏斗边沿形成土堆,但又未破坏其表层,得到的斜坡较抛掷爆破时更陡;在山区构筑堰塞坝,此时爆破可发挥精细的局部化地震作用,造成陡坡的滑坡,形成山峡的堰塞。

(3) 建设地下储库工程领域:地下爆炸时形成的空腔体积($\textrm{m}^3$)大约为 $10^4\ \xi^4$(ξ 为爆炸能,单位为kt TNT 当量)。在黏土型介质或盐岩中进行化爆时,所形成的空腔有很好的稳定性。这样的空腔用以储水和油、气以及储存有毒废料。如在美国《萨尔蒙》爆炸中,构筑了直径为 $\phi 35$ m 的球形空腔。爆炸法构筑油、气储库的造价较常规方法降低2倍。在采用核装药爆破时,其体积可以达到 $1\times10^8\ \textrm{m}^3$ 或更大。

(4) 油、气开采工程领域:爆炸法应用于油气采掘工业主要体现在切割井筒和设备的夹具,以及为了建立油气的流入通道,还采用爆炸特别是聚能爆炸在井管和钢筋混凝土环上穿孔。此外还可利用爆炸法改进油气岩层工作面附近的渗透性,需要利用专门的炸雷或者其他专门设备,此时爆炸法常和水力压裂法联合使用。除力学效应外,爆炸还对周围介质产生热效应,这也会对油气开采产生正面影响。

(5) 民用建筑工程领域:主要用以构筑爆扩桩,即采用爆炸方法形成柱状的带大球状端部的空腔,然后浇筑混凝土或钢筋混凝土。另外就是爆炸法处理深层软弱地基,利用爆炸在软弱土层内压缩形成空腔,空腔周围的土体被压密,空腔内则回填砂形成砂桩,它和周围被压密土体共同组成复合地基,提高地基承载力,并降低工后沉降量。

(6) 交通工程领域:爆炸法在该领域主要应用于构筑铁路和公路的路基,根据不同的地形和地质条件采用抛掷爆破,扬弃爆破,崩坍爆破和定向爆破。爆破法构筑路基优点在于缩短施工工期，节省工程造价，不受气候和地理条件的限制。

岩石爆炸动力学在上述各领域主要用以计算爆炸空腔体积、各类破坏区范围、岩石破碎程度和碎块大小、爆炸漏斗坑的大小以及爆炸所引起岩石介质的运动参数与力学参数和所辐射的地震弹性波的参数，最终来确定工程爆炸方案中的装药量及其布置方案和各种安全范围。

5 展 望

岩石爆炸动力学的研究是岩石力学领域一个崭新的研究方向和富有挑战性的课题。本文综述了国内外岩石爆炸动力学的试验和理论研究进展，归纳了岩石爆炸动力学的理论研究、试验研究及其工程应用。该领域下一步研究的主要方向有：

（1）岩石爆炸动力学理论研究的深入很大程度上依赖于高压高温下岩石状态方程和岩石动力强度理论和动力破坏准则须深入研究。这方面的任务任重道远。

（2）地下爆炸时介质力学运动的分析缺乏强动载下岩石和岩体变形的详尽信息，特别是关于促使岩石破坏的高应力区域中的岩石状态方程(动压力－比热容特性)的冲击绝热特性外，动应力－应变特性往往呈现出应变速率效应，此时岩石的变形和破坏与加载的时间因素有关，因此考虑岩石介质的动力黏性对于研究岩石中的爆炸作用是必要的，即研究强动载作用下的岩石动力本构模型。

（3）还应考虑爆炸作用于周围岩体后，介质并未因运动停止而应力消失，岩体内还可能积累弹性变形能；岩体并非是均匀、连续的岩石介质，而是包含断层、裂隙、层理和节理的非均匀非连续的地质体。综上所述的各种复杂性，导致现阶段水平的岩石力学尚不能作为地下爆炸时分析计算岩石介质运动的基础，所以在研究岩石介质中爆炸作用过程，建立爆炸作用下岩石运动、变形和破坏理论，解决各种实际爆炸工程应用问题时，考虑所有影响介质运动、变形和破坏的因素是不尽合理的。应当力求依赖于最少的岩石和岩体介质的信息和基本的数学工具研制出比较简明的计算模型和方案，以能完成爆炸作用的力学参数计算，并利于其实际应用。

虽然如此，但至今很多工程爆破方案的作业参数计算，包括复杂的装药配置方案的拟定，往往还是利用在总结实际试验数据的基础上得到的试验公式。

参考文献(References)：

[1] 钱七虎，王明洋，赵跃堂. 岩石中爆炸与冲击作用研究进展[C]// 中国岩石力学与工程学会第四次学术大会文集. 北京：中国科学技术文献出版社，1996：53 - 65.(QIAN Qihu, WANG Mingyang, ZHAO Yuetang. Research progress of rock dynamics and impact effects[C]// Proceedings of the fourth academic meeting of Chinese Society for Rock Mechanics and Engineering. Beijing: Scientific and Technical Documents Publishing House, 1996: 53 - 65.(in Chinese))

[2] 钱七虎，戚承志，王明洋. 岩石爆炸动力学[M]. 北京：科学出版社，2006.(QIAN Qihu, QI Chengzhi, WANG Mingyang. Rock blasting dynamics[M]. Beijing: Science Press, 2006.(in Chinese))

[3] 王明洋. 爆炸应力波通过地质构造断层的动力模型理论与试验研究[博士学位论文][D]. 南京：解放军工程兵工程学院，1994. (WANG Mingyang. Dynamic model theory and test research of blasting stress through geological structure faults[Ph. D. Thesis][D]. Nanjing: PLA University of Science and Technology, 1994.(in Chinese))

[4] 王明洋，唐 廷，周泽平. 断裂构造对爆炸地震波传播规律的影响[J]. 岩石力学与工程学报，2005，24(增1)：4 635 - 4 641.(WANG Mingyang, TANG Ting, ZHOU Zeping. Effects of geological structural faults on blasting seismic wave propagation[J]. Chinese Journal of Rock Mechanics and Engineering, 2005, 24(Supp.1): 4 635 - 4 641.(in Chinese))

[5] 戚承志，王明洋，赵跃堂，等. 关于地质材料内部构造层次的若干模型[J]. 世界地震工程，2003，19(1)：70 - 76.(QI Chengzhi, WANG Mingyang, ZHAO Yuetang. On some models of internal structural hierarchy of geomaterials[J]. World Earthquake Engineering, 2003, 19(1): 70 - 76.(in Chinese))

[6] 王明洋，杨林德，钱七虎. 缓倾角断裂隔震效应的定量研究[J]. 同济大学学报(自然科学版)，1999，27(1)：34 - 37.(WANG Mingyang, YANG Linden, QIAN Qihu. Studies on mechanism and quantization of isolation effect of slow angle fault[J]. Journal of Tongji University (Natural Science), 1999, 27(1): 34 - 37.(in Chinese))

[7] 王明洋，钱七虎. 爆炸应力波通过节理裂隙带的衰减规律[J]. 岩土工程学报，1995，17(2)：42 - 46.(WANG Mingyang, QIANQihu. Propagation characteristics of blast-induced shock waves in jointed rock mass[J]. Chinese Journal of Geotechnical Engineering, 1995, 17(2): 42 - 46.(in Chinese))

[8] 戚承志，钱七虎. 材料变形及损伤演化的微观物理动力机制[J].

固体力学学报,2002,23(3):312-317.(QI Chengzhi, QIAN Qihu. Physical kinetic mechanism of evolution of deformation and damage of materials[J]. Acta Mechanica Solida Sinica, 2002, 23(3): 312-317.(in Chinese))

[9] 王明洋,戚承志,钱七虎. 岩体中爆炸与冲击下的破坏研究[J]. 辽宁工程技术大学学报(自然科学版),2001,20(4):385-389.(WANG Mingyang, QI Chengzhi, QIAN Qihu. Studies on failure of rock under explosion and impact[J]. Journal of Liaoning Technical University (Natural Science), 2001, 20(4): 385-389.(in Chinese))

[10] 戚承志,钱七虎. 考虑到时效的一维剥离破坏及损伤破坏机制[J]. 解放军理工大学学报(自然科学版),2000,1(5):1-6.(QI Chengzhi, QIAN Qihu. One-dimensional spallation with consideration of time characteristic and the mechanism of damage and fracture[J]. Journal of PLA University of Science and Technology(Natural Science), 2000, 1(5): 1-6.(in Chinese))

[11] 王明洋,国胜兵,邓国强. 岩石覆盖层抗震塌机制研究[J]. 解放军理工大学学报(自然科学版),2001,2(3):12-19.(WANG Mingyang, GUO Shengbing, DENG Guoqiang. Studies on mechanism of spalling of rock cover[J]. Journal of PLA University of Science and Technology(Natural Science), 2001, 2(3): 12-19.(in Chinese))

[12] 戚承志,钱七虎,王明洋. 固体材料的结构层次及其动力变形与破坏[J]. 解放军理工大学学报(自然科学版),2007,8(5):454-463.(QI Chengzhi, QIAN Qihu, WANG Mingyang. Structural hierarchy and dynamic deformation and fracture of solids[J]. Journal of PLA University of Science and Technology(Natural Science), 2007, 8(5): 454-463.(in Chinese))

[13] 葛涛,王明洋. 坚硬岩石在强冲击荷载作用近区的性状研究[J]. 爆炸与冲击,2007,27(4):306-311.(GE Tao, WANG Mingyang. Characters near strong impact loading zone in hard rock[J]. Explosion and Shock Waves, 2007, 27(4): 306-311.(in Chinese))

[14] 唐廷,王明洋,葛涛. 地下爆炸的地表运动研究[J]. 岩石力学与工程学报,2007,26(增1):3528-3532.(TANG Ting, WANG Mingyang, GE Tao. Study of ground movement of underground explosions[J]. Chinese Journal of Rock Mechanics and Engineering, 2007, 26(Supp.1): 3528-3532.(in Chinese))

[15] 戚承志,王明洋,钱七虎,等. 冲击载作用下岩石变形破坏的细观结构特性[J]. 岩石力学与工程学报,2007,26(增1):3367-3372.(QI Chengzhi, WANG Mingyang, QIAN Qihu, et al. Mesostructural aspects of deformation and fracture of rock under shock loading[J]. Chinese Journal of Rock Mechanics and Engineering, 2007, 26(Supp.1): 3367-3372.(in Chinese))

[16] QI C Z, WANG M Y. Structural hierarchy and mechanical properties of rocks. part I. structural hierarchy and viscosity[J]. Physical Meso-mechanics, 2006, 96(1): 29-39.

[17] QI C Z, WANG M Y. Structural hierarchy and mechanical properties of rocks. part II. structural hierarchy and Size effect and strength[J]. Physical Meso-mechanics, 2006, 96: 41-53.

[18] 戚承志,钱七虎,王明洋,等. 岩体的构造层次及其成因[J]. 岩石力学与工程学报,2005,24(16):2838-2846.(QI Chengzhi, QIAN Qihu, WANG Mingyang, et al. Structural hierarchy of rock massif and mechanism of its formation[J]. Chinese Journal of Rock Mechanics and Engineering, 2005, 24(16): 2838-2846.(in Chinese))

[19] QI C Z, WANG M Y, QIAN Q H. Strain-rate effects on the strength and fragmentation size of rocks[J]. International Journal of Impact Engineering, 2009, 36(12): 1355-1364.

[20] 陈士海,王明洋,赵跃堂,等. 岩石爆破破坏界面上的应力时程研究[J]. 岩石力学与工程学报,2003,22(11):1784-1788.(CHEN Shihai, WANG Mingyang, ZHAO Yuetang, et al. Time-stress history on interface between cracked and uncracked zones under rock blasting[J]. Chinese Journal of Rock Mechanics and Engineering, 2003, 22(11): 1784-1788. (in Chinese))

[21] 戚承志,王明洋,钱七虎. 弹黏塑性孔隙介质在冲击荷载作用下的一种本构关系——第二部分:弹黏塑性孔隙介质的畸变行为[J]. 岩石力学与工程学报,2003,22(11):1763-1766.(QI Chengzhi, WANG Mingyang, QIAN Qihu. Constitutive model of porous elastoviscoplastic materials subjected to shock—part II: distorsional behavior of porous elastoviscoplastic materials[J]. Chinese Journal of Rock Mechanics and Engineering, 2003, 22(11): 1763-1766.(in Chinese))

[22] 戚承志,王明洋,钱七虎. 弹黏塑性孔隙介质在冲击荷载作用下的一种本构关系——第一部分:状态方程[J]. 岩石力学与工程学报,2003,22(9):1405-1410.(QI Chengzhi, WANG Mingyang, QIAN Qihu. Constitutive relation of elastoviscoplastic porous media subjected to shock—part I: state equation[J]. Chinese Journal of Rock Mechanics and Engineering, 2003, 22(9): 1405-1410.(in Chinese))

[23] КОЧАРЯН Г Г, СПИВАК А А. Движение блоков горной при крупномасштабных подземных взрывах. Ч.1: экспериментальные данные[R]. [S.l.]: [s.n.], 2001.

[24] ВОВК А А, КРАВЕЦ В Г, ЛУЧКО И А, и другие. Геодинамика взрыва и ее приложение[R]. Киев: Наукова Думка, 1981.

[25] АДУШКИН В В, СПИВАК А А. Необратимые проявления крупномасштабного подземного взрыва в неоднородной среде[M]. [S. l.]: ИФЗ АН, СССР, 1989.

[26] РОДИНОВ В Н, АДУШКИН В В, КОСТЮЧЕНКО В Н.

Механический эффект подземного взрыва[M]. [S. l.]: Недра, 1971.

[27] РОДИОНОВ В Н, СИЗОВ И А, СПИРАК А А, и другие. О поведении среды в зоне разрушения при камуфлетном взрыве[M]. [S. l.]: Взрывное Дело, 1976: 24 – 29.

[28] 舍米亚金 Е И. 弹塑性理论的动力学问题[M]. 戚承志译. 北京: 科学出版社, 2009.(SHEMYAKIN E I. Dynamical problems of elasto plastic theory[M]. Translated by QI Chengzhi. Beijing: Science Press, 2009.(in Chinese))

[29] ВЛАСОВ О Е, СМИРНОВ С А. Основы расчета дробления горных пород при взрыве[M]. Москва: Издательство Академии Наук СССР, 1962.

[30] ВЛАСОВ О Е. Основы теории действия взрыва[M]. [S.l.]: М. Издание ВИА, 1957.

[31] МАКСИМОВА Е П. Моделирование процесса взрывного разрушения вопросы горного дела[M]. [S.l.]: Углетехиздат, 1958.

[32] БЕЛЯЕВ А Ф. О расчете зарядов при взрывена выброс[R]. [S.l.]: Горный Журнал, 1957.

[33] BOARDMAN C R, RABB D D, MCARTHUR R D. Contained nuclear detonations in four media geological factors in cavity and chimney formation[C]// Proceedings of the 3rd Plowshare Symposium. Engineering with Nuclear Explosives. Livermore, CA: [s. n.], 1964: 106 – 119.

[34] University of California. Engineering with nuclear explosives[C]// Proceedings of the 3rd Plowshare Symposium. Engineering with Nuclear Explosives. Livermore, CA: [s. n.], 11964: 26 – 37.

[35] MARSAL R J. Large scale testing of rock fill materials[J]. Journal of the Soil Mechanics and Foundations Division, 1967, 93(2): 27 – 43.

[36] MURPHAY B F, VORTMAN L J. High-explosive craters in desert alluvium, tuff and basalt[J]. Journal of Geophysical Research, 1961, 66(10): 3 389 – 3 404.

[37] NORDYKE M D. An analysis of cratering data from desert alluvium[J]. Journal of Geophysical Research,1962,67(5):1 965 – 1 974.

[38] САДОВСКИЙ М А, АДУШКИН В В, СПИВАК А А. О размере зон необратимого деформирования при взрыве в блочной среде[J]. Физики Земли, 1989, (9): 9 – 15.

[39] 王洪亮, 葛涛, 王德荣, 等. 块系岩体动力特性理论与试验对比分析[J]. 岩石力学与工程学报, 2007, 26(5): 951 – 958.(WANG Hongliang, GE Tao, WANG Derong, et al. Comparison of theoretical and experimental analyses of dynamic characteristics of block rock mass[J]. Chinese Journal of Rock Mechanics and Engineering, 2007, 26(5): 951 – 958.(in Chinese))

[40] 范新, 王明洋, 谭可可. 爆炸荷载作用下深部块体变形运动规律研究[J]. 岩石力学与工程学报, 2007, 26(5): 1 019 – 1 025.(FAN Xin, WANG Mingyang, TAN Keke. Study on rule of block deformation and movement under explosion loads[J]. Chinese Journal of Rock Mechanics and Engineering, 2007, 26(5): 1 019 – 1 025.(in Chinese))

[41] 范新. 深部岩体的变形与动力破坏效应研究[博士学位论文][D]. 南京:解放军理工大学, 2006.(FAN Xin. Research on deformation of deep rocks and dynamic failure effects[Ph. D. Thesis]. Nanjing: PLA University of Science and Technology, 2006.(in Chinese))

[42] 王明洋, 周泽平, 钱七虎. 深部岩体的构造和变形与破坏问题[J]. 岩石力学与工程学报, 2006, 25(3): 448 – 455.(WANG Mingyang, ZHOU Zeping, QIAN Qihu. Tectonic, deformation and failure problems of deep rock mass[J]. Chinese Journal of Rock Mechanics and Engineering, 2006, 25(3): 448 – 455.(in Chinese))

[43] 王明洋, 戚承志, 钱七虎. 深部岩体块系介质变形与运动特性研究[J]. 岩石力学与工程学报, 2005, 24(16): 2 825 – 2 830.(WANG Mingyang, QI Chengzhi, QIAN Qihu. Study on deformation and motion characteristics of blocks in deep rock mass[J]. Chinese Journal of Rock Mechanics and Engineering, 2005, 24(16): 2 825 – 2 830.(in Chinese))

[44] 钱七虎. 战略防护工程面临的核钻地弹威胁及连续介质力学模型的不适用性[J]. 防护工程, 2005, 26(5): 1 – 10.(QIAN Qihu. Threat of nuclear missile faced in strategic protection works and inapplicability of continuum mechanical model[J]. Protective Engineering, 2005, 26(5): 1 – 10.(in Chinese))

[45] АДУШКИН В В, СПИВАК А А. Геомеханика крупномасштабных взрывов[M]. Москва: Недра, 1993.

[46] KOCHARYAN G G, SPIVAK A A. Movement of rock blocks during large-scale underground explosions. part I: experimental data[J]. Journal of Mining Science, 2001, 37(1): 1999, 20(3): 12 – 26.

[47] KOCHARYAN G G, SPIVAK A A, BUDKOV A M. Movement of rock blocks during large-scale underground explosions. part II: estimates by analytical models, numerical calculations, and comparative analysis of theoretical and experimental data[J]. Journal of Mining Science, 2001, 37(2): 149 – 168.

纵波在改进的弹性非线性法向变形行为单节理处的传播特性研究

俞 缙[1,2]，钱七虎[2]，林从谋[1]，赵晓豹[3]

(1. 华侨大学岩土工程研究所，福建 厦门 361021；2. 解放军理工大学工程兵工程学院，江苏 南京 210007；3. 南京大学地球科学与工程学院，江苏 南京 210093)

摘 要：将传统的 BB 模型与经典指数模型统一起来，提出的准静态条件下改进的节理弹性非线性法向变形本构关系可定量化描述节理变形的这类差异。在不考虑加载速率对节理变形行为影响的情况下，将该本构关系推广至动态条件，建立了法向入射纵波在弹性半无限空间中干性单节理处传播的位移不连续模型，基于 Lemaitre 假设获得了节理透、反射系数 $|T_{IMP}|$ 和 $|R_{IMP}|$ 的近似解析解；同时结合一维波动方程特征线法推导了节理透、反射波质点速度时域数值差分格式并自编了计算程序，进而得到 $|T_{IMP}|$、$|R_{IMP}|$、透、反射波能量 e_{tra} 和 e_{ref}、延迟时间 T_{del} 的半数值解，依此研究弹性纵波在单节理处的传播过程及特征。在针对节理法向初始切线刚度 $K_{n,i}$、节理闭合量与最大允许闭合量比值 $\gamma=d_n/d_{max}$、入射波频率 f 等因素对 $|T_{IMP}|$、$|R_{IMP}|$ 的影响进行探讨的同时，着重分析模型非线性程度 ξ 及入射波最大振幅 $|v_{inc}|$ 的变化对透射波振幅及能量衰减、波形及频谱畸变、时间延迟所产生的影响，并解释了一种"超越"现象。

关键词：岩石节理；纵波；位移不连续模型；非线性程度

中图分类号：TU45 **文献标识码**：A **文章编号**：1000-4548(2009)08-1156-09

作者简介：俞 缙(1978-)，男，江苏苏州人，博士后，讲师，硕士生导师，主要从事岩土力学、岩土工程测试技术及地基处理等方面的研究工作。E-mail: bugyu0717@hqu.edu.cn。

Transmission of elastic P-wave across one fracture with improved nonlinear normal deformation behaviors

YU Jin[1,2], QIAN Qi-hu[2], LIN Cong-mou[1], ZHAO Xiao-bao[3]

(1. Geotechnical Engineering Institute, Huaqiao University, Xiamen 361021, China; 2. Engineering Institute of Engineering Corps, PLA University of Science and Technology, Nanjing 210007, China; 3. School of Earth Sciences and Engineering, Nanjing University, Nanjing 210093, China)

Abstract: The improved elastic nonlinear normal deformation constitutive model under the quasi-static condition which generalizes the BB model and the classical exponential model can describe the extent of nonlinearity quantitatively. Without the loading-rate effect considered, the improved constitutive relationship is extended to be under the dynamic condition, and then a nonlinear displacement discontinuity model for normally incident P-wave propagation across a dry fracture is established in an elastic half-space. By introducing the Lemaitre hypothesis, the approximate analytical solutions of transmission and reflection coefficients $|T_{IMP}|$ and $|R_{IMP}|$ are obtained. Furthermore, using one-dimensional wave equation characteristic method, the time-domain numerical difference scheme of reflected and transmitted particle velocities is proposed, and computational programs are compiled to obtain semi-numerical solutions of $|T_{IMP}|$ and $|R_{IMP}|$, transmission and reflection energy e_{tra} and e_{ref}, and delay time T_{del}. Parameter studies are conducted to get an insight into the effects of the extent of nonlinearity ξ and the incident wave maximum amplitude $|v_{inc}|$ on transmitted wave amplitude and energy attenuation, waveform and spectral aberration, and time delays, in terms of fracture initial stiffness $K_{n,i}$, the ratio of given closure to the maximum allowable closure $\gamma=d_n/d_{max}$, and frequency f. In addition, the phenomenon of exceeding is explained.

Key words: rock fracture; P-wave; displacement discontinuity model; extent of nonlinearity

0 引 言

节理、裂隙对岩体中应力波的传播过程及规律影响很大。通常认为节理间的反射与透射可导致应力波

收稿日期：2008-10-20。

基金项目：国家自然科学基金项目(40702046)；福建省自然科学基金项目(2009J05125)；建设部"科学技术项目计划"(06-K1-23)。

本文原载于《岩土工程学报》(2009 年第 31 卷第 8 期)。

振幅衰减及波速降低。由于岩体结构破坏准则往往由波振幅门槛值（如质点位移、速度或加速度峰值）确定，因此在研究岩体动力学及防护工程问题时，节理裂隙对应力波的影响成为关键课题[1]。

早期的研究主要是考虑应力波穿越含黏结界面或界面层介质时界面对波影响的理论计算及试验验证[2-4]。随后的研究又建立于波散射模型基础之上，集中在相对于波长具有较小尺度节理（如微裂隙、微空隙）的影响方面，并经历了从考虑线性接触条件[5-8]到考虑非线性接触条件[9-11]的发展过程。然而多数天然节理岩体中大尺度平面节理（相对于波长伸展尺寸较大而厚度较小，也称宏观节理，以下简称为节理）的影响占主导地位[12-14]。从物理角度上节理可被看作众多共线微裂隙、微空隙和粗糙面上微接触体的集合体，它们的变形产生节理总体变形（如张开、闭合、滑移），这会使岩体结构质点的位移和速度变化在空间上不连续，故节理又被视为非黏结界面[15]、滑移界面[16]或位移不连续体[17]。当波穿过它们时应力场连续，位移场却是非连续的，显然此时采用黏结界面假设或波散射模型均不适宜，因此一些学者提出了位移不连续理论（把节理变形本构关系看作波动方程中的位移不连续边界条件）来解释节理对波的影响[15-21]。此类研究同样经历了从考虑适用于小振幅波的线性位移不连续模型[16-20]到考虑适用于大振幅波的非线性位移不连续模型的发展过程[12-14, 21-22]。

关于非线性模型，目前仅局限于研究单一类型节理变形关系，如具有双曲型本构关系（BB模型）的节理对纵波的影响[12-14]，而根据不同的节理变形关系可建立不同的位移不连续模型。笔者曾分别利用BB模型和经典指数模型来探讨不同类型非线性节理变形行为下纵波的透、反射规律[21]。该研究虽在某种程度上弥补了前人采用单一模型分析的不足，但两类模型在数学上都存在半值应力（half-closure stress）不可调、模型非线性程度（节理闭合量发展速度）固定不变的缺陷，因此该研究不能反映节理闭合量与应力关系非线性程度的变化对应力波传播的影响。为此笔者随后提出了改进的弹性非线性法向变形本构模型[23]，该模型可定量化描述节理变形非线性程度的差异。本文将该模型推广至动态条件，建立法向入射纵波（不考虑剪切变形）在弹性半空间中干性单节理处传播的位移不连续模型，基于Lemaitre假设[23]获得了节理透、反射系数近似解析解；同时结合一维波动方程特征线法[12-13]推导了节理处透、反射波质点速度数值差分格式并利用Fortran语言编制了计算程序，获得了质点速度时域半数值半解析解，从而得到透、反射系数、透、反射波能量、延迟时间的半数值解，依此探讨纵波在单节理处的传播过程及特征。在针对节理法向初始切线刚度、节理闭合量与最大允许闭合量比值、频率等因素对透、反射系数的影响进行探讨的同时，着重分析模型非线性程度和入射波振幅的改变对透射波振幅及能量衰减、波形及频谱畸变和时间延迟现象所产生的影响，并解释了入射波振幅较高时出现的一种"超越"现象。

1 改进的节理非线性模型及动态推广

1.1 改进的节理弹性非线性法向变形本构模型简述

岩石节理准静态法向单调加载及循环加、卸载条件下代表性模型较多，如Shehata[24]、Goodman[25]、Kulhaway[26]、Bandis等[27]、Barton等[28]、Malama等[29]提出的模型，而动态模型较少且基本是从准静态模型中移植而来，如Cai等[30]、Yang等[31]分别对天然及人工制备节理进行动态单轴加载试验，将BB（Barton and Bandis）模型推广至动态，Wang等[32]用Instron1342电液伺服试验机对不同表面形态人工节理试验建立的考虑率效应的幂函数形式节理动态经验模型。笔者等[23]已探讨了上述各模型非线性程度不可调的数学缺陷，基于前人准静态法向单调加载试验结果，将传统的BB模型与经典指数模型统一起来，建立了改进的节理弹性非线性法向变形本构模型，并在数学上严格证明了二者是新模型的两个特例。模型方程及节理法向切线刚度K_n的表达式为

$$d_n = \frac{\xi d_{max}\left\{\exp\left[\frac{(1-\xi)\sigma_n}{\xi K_{n,i}d_{max}}\right]-1\right\}}{\exp\left[\frac{(1-\xi)\sigma_n}{\xi K_{n,i}d_{max}}\right]-\xi}, \quad (1)$$

$$K_n = \frac{\partial \sigma_n}{\partial d_n} = \frac{\xi K_{n,i}d_{max}^2}{(d_{max}-d_n)(\xi d_{max}-d_n)}, \quad (2)$$

式中，σ_n为节理法向压应力，d_n为节理法向闭合量，$K_{n,i}$为节理法向初始切线刚度，d_{max}为节理法向最大允许闭合量。这里规定σ_n，d_n以压缩为正。ξ为用于对σ_n-d_n关系非线性程度进行修正的系数（具体确定方法见文献[22]）。模型σ_n-d_n关系曲线见图1，其中$\xi_1 < \xi_2 < \xi_3$。新模型以BB模型（$\xi \to 1$）和经典指数模型（$\xi \to \infty$）为边界，随着ξ值的增大，模型曲线的非线性程度随之增大。该模型克服了传统节理弹性非线性模型半值应力不可调、非线性程度固定不变的缺点。

1.2 模型的动态推广

Zhao等[12]、Zhao等[13-14]及笔者等[21]认为连续的循环加卸载对节理的刚化作用可使其σ_n-d_n关系成为无滞回环的弹性关系，而天然岩石节理在漫长的地质历史中经历了多次变形，故起初的循环加卸载过程

中的滞回现象可被忽略，同时在不考虑加载速率对节理变形特性影响情况下，解释了准静态条件下 BB 模型与经典指数模型在纵波传播问题中运用的合理性。这里沿用该思想，将改进的节理弹性非线性法向变形本构模型推广至动态条件，进行计算分析研究。

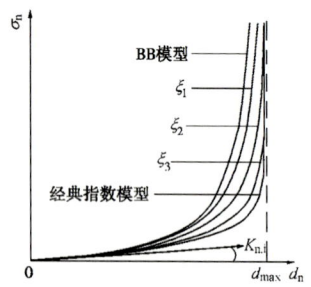

图 1 改进的岩石节理法向变形本构模型示意图

Fig. 1 Improved normal deformation constitutive model of rock fractures

2 近似解析计算

2.1 透、反射系数近似解析解

Schoenberg[16]、Pyrak-Nolte 等[17]已得出线性位移不连续模型下，任意入射角射入干性单节理应力波（P，SV，SH 波）的透、反射系数精确解析解，

$$|T_{\text{lin}}| = \sqrt{\frac{4(K/z\omega)^2}{4(K/z\omega)^2+1}},$$
$$|R_{\text{lin}}| = \sqrt{\frac{1}{4(K/z\omega)^2+1}}, \quad (3)$$

式中，$|T_{\text{lin}}|$，$|R_{\text{lin}}|$ 分别为线性变形节理的透、反射系数，z 为波阻抗，K 为节理法向刚度，ω 为入射波角频率。

为得到非线性位移不连续模型下单节理处纵波透、反射系数解析解，王卫华等[23]借鉴岩石损伤力学领域中的 Lemaitre 等效应变假设思想，利用等效刚度将非线性节理刚度用线性变形本构关系式表示，进而得到 BB 模型情况下节理透、反射系数解析解。所得的计算结果虽并非如该文中所阐述的与 Zhao 等[12]的数值解完全一致，而只是与其存在一定误差的近似解（笔者[21]曾详细分析其原因，此处不再赘述），却不失为一种简便计算方法。本文依此法求得具有改进的弹性非线性法向变形本构关系单节理处的纵波透、反射系数近似解析解，

$$|T_{\text{IMP}}| = \frac{1}{\sqrt{\left[\frac{z\omega(1-d_n/d_{\max})(\xi-d_n/d_{\max})}{2\xi K_{n,i}}\right]^2+1}}, \quad (4)$$

$$|R_{\text{IMP}}| = \frac{1}{\sqrt{\left[\frac{2\xi K_{n,i}}{z\omega(1-d_n/d_{\max})(\xi-d_n/d_{\max})}\right]^2+1}}, \quad (5)$$

式中，$|T_{\text{IMP}}|$，$|R_{\text{IMP}}|$ 分别为改进后的单节理透、反射系数。由式（4）、（5）可知 $|T_{\text{IMP}}|^2+|R_{\text{IMP}}|^2=1$，即由于应用弹性模型假设，透、反射波能量之和遵守能量守恒定律。

2.2 参数研究

依据不同修正系数 $\xi\to1$，$\xi=1.5$，$\xi=5.5$，$\xi\to\infty$，针对 $K_{n,i}$、节理闭合量与最大允许闭合量比值 $\gamma=d_n/d_{\max}$（反映入射波振幅大小）以及入射波频率 f 等参数进行研究。为了与前人研究结果进行对比，参数均来自文献[12，23]。岩石密度为 2.4×10^3 kg/m^3，波频 $f=\omega/2\pi$ 分别为 50 Hz、150 Hz，纵波波速为 4500 m/s，波阻抗=密度×波速=1.08×10^7 kg/(m$^2\cdot$s)，$K_{n,i}$ 与 $d_{\max}$ 的 5 组参数组合见表 1，计算结果见图 2、3。

表 1 $d_{\max}$ 和 $K_{n,i}$ 参数值的组合[12, 23]

Table 1 Various combinations of $d_{\max}$ and $K_{n,i}$

参数组合	节理最大允许闭合量 $d_{\max}$/mm	节理初始切线刚度 $K_{n,i}$/(GPa·m^{-1})
1	0.61	1.25
2	0.57	2.00
3	0.53	3.00
4	0.50	3.80
5	0.40	5.50

由图 2 可知，$|T_{\text{IMP}}|$ 不仅随 $K_{n,i}$，γ 的增大而增大，随 $d_{\max}$，f 的增大而减小，还较强地依赖于修正系数 ξ。在其他参数不变的情况下，ξ 越大，$|T_{\text{IMP}}|$ 越小，这意味着节理位移不连续模型非线性程度越大，对入射波的阻碍也越大，致使 $|T_{\text{IMP}}|$ 下降。从图 3 上也可以看出 $|R_{\text{IMP}}|$ 所具有的与其相反的规律。

3 半数值半解析计算

3.1 一维波动方程特征线法

弹性纵波法向穿越具有 BB 变形本构关系的单节理及平行多节理时的透、反射系数数值解可通过一维波动方程特征线法得到[12-13]。这里借鉴该思路推导改进的弹性非线性法向本构模型单节理处（篇幅所限，平行多节理问题另文撰述）一维波动方程的特征线解。

弹性波的一维波动方程为

$$\frac{\partial^2 u(x,t)}{\partial t^2}=\alpha_{\text{P}}^2\frac{\partial^2 u(x,t)}{\partial x^2}, \quad (6)$$

式中，$u(x,t)$ 为质点位移，α_{P} 为波速，且 $\alpha_{\text{P}}^2=E/\rho$，其中 E 为杨氏模量，ρ 为密度。质点速度 $v(x,t)=\partial u(x,t)/\partial t$，应变 $\varepsilon(x,t)=\partial u(x,t)/\partial x$，式（6）可写为

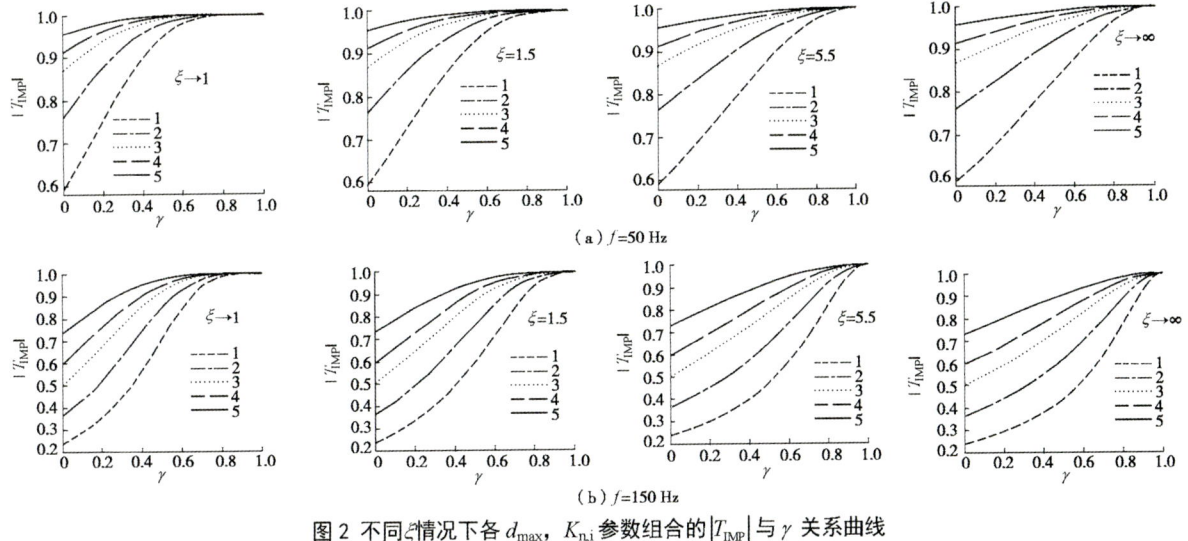

图 2 不同 ξ 情况下各 d_{max}，$K_{n,i}$ 参数组合的 $|T_{IMP}|$ 与 γ 关系曲线

Fig. 2 $|T_{IMP}|$ as a function of γ for various combinations of d_{max} and $K_{n,i}$ under different ξ conditions

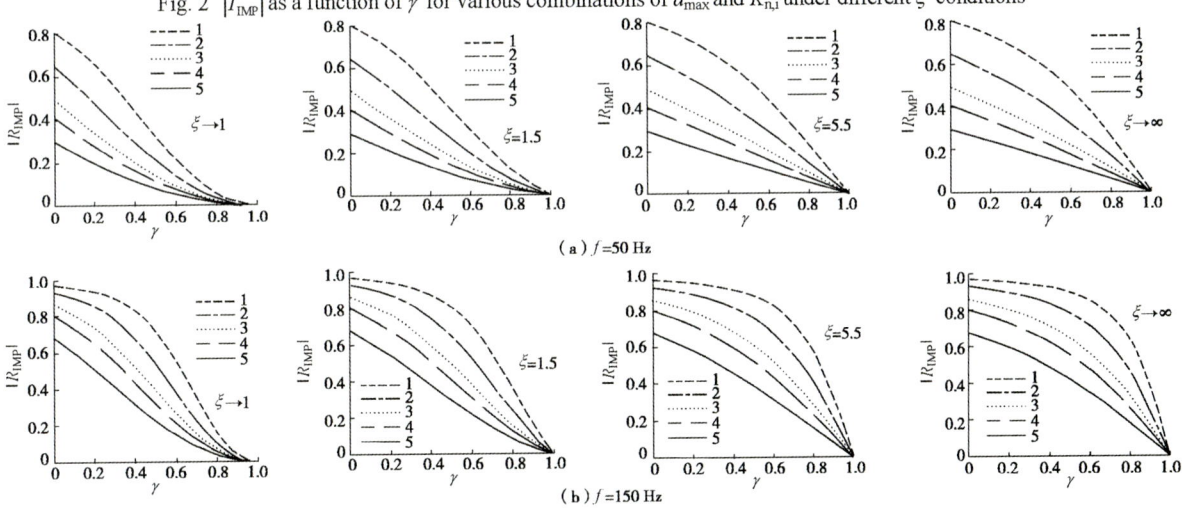

图 3 不同 ξ 情况下各 d_{max}，$K_{n,i}$ 参数组合的 $|R_{IMP}|$ 与 γ 关系曲线

Fig. 3 $|R_{IMP}|$ as a function of γ for various combinations of d_{max} and $K_{n,i}$ under different ξ

$$\left.\begin{array}{l}\dfrac{\partial v(x,t)}{\partial t}=\alpha_p^2\dfrac{\partial \varepsilon(x,t)}{\partial x},\\ \dfrac{\partial v(x,t)}{\partial x}=\dfrac{\partial \varepsilon(x,t)}{\partial t}\end{array}\right\} \quad (7)$$

假设线弹性半空间内位于 $x=x_1$ 处有一干性节理，空间一侧受到质点速度函数为 $p(t)$ 的法向入射平面纵波，根据特征线法，x-t 平面中函数 $v(x,t)-\alpha_p\varepsilon(x,t)$ 的增量为

$$d[v(x,t)-\alpha_p\varepsilon(x,t)]=\left[\dfrac{\partial v(x,t)}{\partial t}-\alpha_p\dfrac{\partial \varepsilon(x,t)}{\partial t}\right]dt+\left[\dfrac{\partial v(x,t)}{\partial x}-\alpha_p\dfrac{\partial \varepsilon(x,t)}{\partial x}\right]dx。 \quad (8)$$

由式（7），可将式（8）改写为

$$d[v(x,t)-\alpha_p\varepsilon(x,t)]=\left[\dfrac{\partial v(x,t)}{\partial x}-\dfrac{1}{\alpha_p}\dfrac{\partial v(x,t)}{\partial t}\right](dx-\alpha_p dt)。 \quad (9)$$

当 $dx/dt=\alpha_p$ 时，$d[v(x,t)-\alpha_p\varepsilon(x,t)]=0$。既沿着 x-t 平面中斜率为 $1/\alpha_p$ 的任何直线，有特征方程：

$$v(x,t)-\alpha_p\varepsilon(x,t)=\text{constant}， \quad (10)$$

由波阻抗 $z=\rho\alpha_p$，$\alpha_p^2=E/\rho$ 容易得到：

$$z\alpha_p\varepsilon(x,t)=-\sigma(x,t)， \quad (11)$$

式中，$\sigma(x,t)$ 为质点应力（定义压为正，拉为负），将式（10）两边同乘以 z，引入式（11）得一维波特征方程：

$$z[v(x,t)-\alpha_p\varepsilon(x,t)]=zv(x,t)+\sigma(x,t)=\text{constant}。 \quad (12)$$

同理，当 $dx/dt=-\alpha_p$ 时，既沿着 x-t 平面中斜率为 $-1/\alpha_p$ 的任何直线，有另一特征方程：

$$z[v(x,t)+\alpha_P \varepsilon(x,t)] = zv(x,t) - \sigma(x,t) = \text{constant}. \tag{13}$$

如图4所示，x-t平面中始于节理处并交于x轴，斜率为$-1/\alpha_P$的直线AB是波动方程的左行特征线；始于节理处并交于t轴上点$(0, t-x_1/\alpha_P)$，斜率为$1/\alpha_P$的直线AC是右行特征线；自点$(0, t-x_1/\alpha_P)$，交于x轴的直线CD，为另一斜率为$-1/\alpha_P$的左行特征线。上标"−"、"+"分别表示参数属于节理前、后弹性波场。

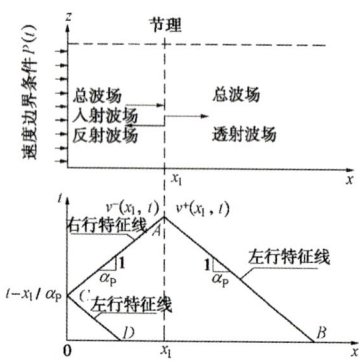

图4 半无限空间中纵波法向穿越单节理波场示意图与对应 x-t 平面中左、右行特征线示意图

Fig. 4 Normal incidence of P-wave through a fracture in a half-space and the corresponding left- and right-running characteristics in the x-t plane

左行特征线AB与x轴交于B点，当$t=0$时，研究域未受扰动，因此有：

$$\left.\begin{array}{l}v^+(x,0)=0,\\ \sigma(x,0)=0,\end{array}\right\} \tag{14}$$

代入式（13）可得节理后波场中左行特征线AB上有：

$$zv^+(x_1,t) - \sigma(x_1,t) = v^+(x,0) + \sigma(x,0) = 0. \tag{15}$$

右行特征线AC与t轴交于C点，在$t-x_1/\alpha_P$时刻，研究域左边界质点速度 $v^-(0,t-x_1/\alpha_P) = p(t-x_1/\alpha_P)$，再由式（12）可得节理前波场中右行特征线$AC$上有：

$$zv^-(x_1,t) + \sigma(x_1,t) = zp(t-x_1/\alpha_P) + \sigma(0,t-x_1/\alpha_P), \tag{16}$$

与特征线AB相同，另一左行特征线CD上有：

$$zp(t-x_1/\alpha_P) - \sigma(0,t-x_1/\alpha_P) = 0. \tag{17}$$

由式（16）、（17）可得：

$$zv^-(x_1,t) + \sigma(x_1,t) = 2zp(t-x_1/\alpha_P). \tag{18}$$

将式（16）、（18）相加得

$$v^-(x_1,t) + v^+(x_1,t) = 2p(t-x_1/\alpha_P), \tag{19}$$

式（19）表达了在节理前后质点速度间的关系。

3.2 透、反射波质点速度数值差分格式

波场在$x=x_1$处的位移不连续边界条件：

$$\sigma^-(x_1,t) = \sigma^+(x_1,t) = \sigma(x_1,t), \tag{20}$$

$$u^-(x_1,t) - u^+(x_1,t) = \frac{\xi d_{\max}\left\{\exp\left[\dfrac{(1-\xi)\sigma(x_1,t)}{\xi K_{n,i} d_{\max}}\right]-1\right\}}{\exp\left[\dfrac{(1-\xi)\sigma(x_1,t)}{\xi K_{n,i} d_{\max}}\right]-\xi}. \tag{21}$$

为简化计算过程，令 $(1-\xi)/(\xi K_{n,i} d_{\max}) = \psi$，则式（21）可简写为

$$u^-(x_1,t) - u^+(x_1,t) = \frac{\xi d_{\max}\{\exp[\psi\sigma(x_1,t)]-1\}}{\exp[\psi\sigma(x_1,t)]-\xi}, \tag{22}$$

对式（22）在时间域求导得：

$$v^-(x_1,t) - v^+(x_1,t)$$
$$= \frac{\xi d_{\max}\psi(1-\xi)\exp[\psi\sigma(x_1,t)]}{\{\exp[\psi\sigma(x_1,t)]-\xi\}^2}\frac{\partial\sigma(x_1,t)}{\partial t}. \tag{23}$$

将式（15）代入式（23）可得：

$$v^-(x_1,t) - v^+(x_1,t)$$
$$= \frac{\xi d_{\max}\psi(1-\xi)\exp[\psi zv^+(x_1,t)]}{\{\exp[\psi zv^+(x_1,t)]-\xi\}^2}\frac{z\partial v^+(x_1,t)}{\partial t}. \tag{24}$$

根据式（19）、（24），得到关于$v^+(x_1,t)$的微分方程：

$$\frac{\partial v^+(x_1,t)}{\partial t}$$
$$= \frac{2[p(t-x_1/\alpha_P) - v^+(x_1,t)]\{\exp[\psi zv^+(x_1,t)]-\xi\}^2}{z\xi d_{\max}\psi(1-\xi)\exp[\psi zv^+(x_1,t)]}. \tag{25}$$

将时间域$[0, t]$划分为J个等时段$(j=1,2,\cdots,J)$，每个时段时间增加值Δt（取为$\Delta t = T_e/m$）。$v^+(x_1,t)$在每个时段有一阶向前差商：

$$\frac{\partial v^+(x_1,t_j)}{\partial t} = \frac{v^+(x_1,t_{j+1}) - v^+(x_1,t_j)}{\Delta t}$$
$$= \frac{m[v^+(x_1,t_{j+1}) - v^+(x_1,t_j)]}{T_e}, \tag{26}$$

式中，T_e为入射波周期，m为在入射波一周期内时间的段个数。可以得到计算$v^+(x_1,t)$的差分递归方程：

$$v^+(x_1,t_{j+1}) = v^+(x_1,t_j) +$$
$$\frac{2T_e[p(t_j - x_1/\alpha_P) - v^+(x_1,t_j)]\{\exp[\psi zv^+(x_1,t_j)]-\xi\}^2}{mz\xi d_{\max}(1-\xi)\psi\exp[\psi zv^+(x_1,t_j)]}, \tag{27}$$

代入式（22）得到计算$v^+(x_1,t)$完整形式差分递归方程：

$$v^+(x_1,t_{j+1}) = v^+(x_1,t_j) +$$
$$\frac{2T_e K_{n,i}[p(t_j-x_1/\alpha_P) - v^+(x_1,t_j)]}{mz(1-\xi)^2} \rightarrow$$

$$\leftarrow \frac{\left\{\exp\left[\dfrac{z(1-\xi)v^+(x_1,t_j)}{\xi K_{n,i}d_{max}}\right]-\xi\right\}^2}{\exp\left[\dfrac{z(1-\xi)v^+(x_1,t_j)}{\xi K_{n,i}d_{max}}\right]} \quad (28)$$

如果入射波的质点速度波形函数 $p(t)$ 和初始条件 $v^+(x_1,0)$ 已知，利用上式可迭代计算得出 $v^+(x_1,t)$，即透射波质点速度 $v_{tra}(x_1,t)$。在计算中将 m 取得较大（m=1000），可以得到一个足够小的 Δt，使得计算得到具有足够精度的 $v^+(x_1,t)$ 差分数值解。再借助式（19）易求出 $v^-(x_1,t)$，从而求出节理处反射波的质点速度：

$$v_{ref}(x_1,t) = v^-(x_1,t) - v_{inc}(x_1,t) , \quad (29)$$

式中，$v_{inc}(x_1,t)$ 为节理处入射波质点速度。节理透、反射系数半数值解 $|T_{IMP}|$、$|R_{IMP}|$ 由透、反射波速度振幅与入射波速度振幅比表示：

$$\left.\begin{aligned}|T_{IMP}| &= \frac{v_{tra}(x_1,t)|_{max}}{v_{inc}(x_1,t)|_{max}} , \\ |R_{IMP}| &= \frac{v_{ref}(x_1,t)|_{max}}{v_{inc}(x_1,t)|_{max}} .\end{aligned}\right\} \quad (30)$$

为检验入射波在节理处的透、反射过程中是否有能量消耗，通过对 $v_{tra}(x_1,t)$ 和 $v_{ref}(x_1,t)$ 的数值积分，计算归一化透、反射波能量 e_{tra} 和 e_{ref}：

$$e_{tra} = \frac{E_{tra}}{E_{inc}} = \frac{\int_{t_{tra}^0}^{t_{tra}^1} z[v_{tra}(x_1,t)]^2 \mathrm{d}t}{\int_{t_{inc}^0}^{t_{inc}^1} z[v_{inc}(x_1,t)]^2 \mathrm{d}t} = \frac{\sum_{j=t_{tra}^0}^{j=t_{tra}^1}[v_{tra}(x_1,t_j)]^2 \Delta t}{\sum_{j=t_{inc}^0}^{j=t_{inc}^1}[v_{inc}(x_1,t_j)]^2 \Delta t} , \quad (31)$$

$$e_{ref} = \frac{E_{ref}}{E_{inc}} = \frac{\int_{t_{ref}^0}^{t_{ref}^1} z[v_{ref}(x_1,t)]^2 \mathrm{d}t}{\int_{t_{inc}^0}^{t_{inc}^1} z[v_{inc}(x_1,t)]^2 \mathrm{d}t} = \frac{\sum_{j=t_{ref}^0}^{j=t_{ref}^1}[v_{ref}(x_1,t_j)]^2 \Delta t}{\sum_{j=t_{inc}^0}^{j=t_{inc}^1}[v_{inc}(x_1,t_j)]^2 \Delta t} . \quad (32)$$

式中 E_{inc}，E_{tra}，E_{ref} 分别为入、透、反射波能量；t_{inc}^0，t_{tra}^0，t_{ref}^0 分别为入、透、反射波初始时刻；t_{inc}^1，t_{tra}^1，t_{ref}^1 为终了时刻。

4 参数研究

利用 Fortran 语言编制计算程序，采用前述近似解析计算中使用的 α_p，ρ，z 参数值，选择表 1 中参数组合 1，且垂直于左边界（$x=0$）的入射纵波为周期是 20 ms 的半正弦波，应力最大振幅 $|\sigma_{inc}| = \rho\alpha_p|v_{inc}|$ 分别定为 0.22，0.54，1.1，2.2，4.4 MPa，其中 $|v_{inc}|$ 为入射波时域最大振幅，分别为 0.02，0.05，0.10，0.20，0.40 m/s，模型修正系数选择 $\xi \to 1$，$\xi=1.8$，$\xi \to \infty$ 三种情况进行计算。入、透射波波形及频谱曲线见图 5。

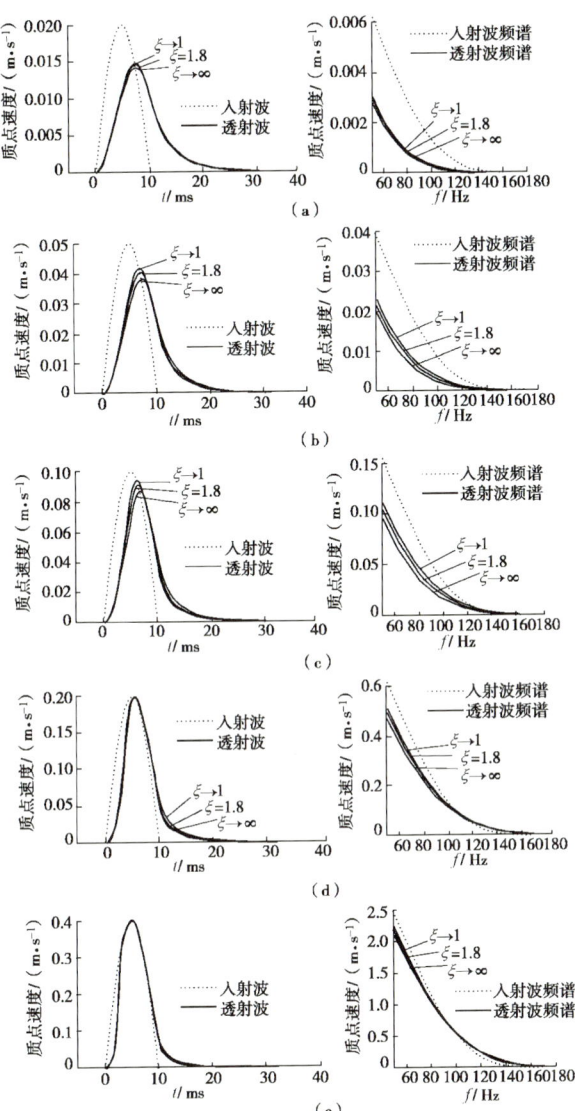

图 5 不同入射波振幅下透射波波形图及频谱曲线图

Fig. 5 Incident P-wave pulses with different amplitudes and the resulting transmitted pulses and the corresponding spectrum

纵波传播过程中，节理首先受压应力作用达到一定闭合量，然后恢复至初始状态，即随 σ_n 的增大，d_n 以非线性方式增加到一个最大值后再减小为 0。由图 5 可见，$|v_{inc}|$ 增大，透射波时域最大振幅 $|v_{tra}|$ 与频域最大振幅相应增大，$|v_{inc}|$ 很大时，$|v_{tra}| \to |v_{inc}|$。同时，透射波波形相对于入射波而言均不同程度地在时域上发生畸变，波起跳阶段体现出"由缓到陡"的变化趋势，到达波峰后又不同程度地经历"由陡至缓"的趋势。$|v_{inc}|$ 较小时，不同 ξ 值的透射波波形及频谱图差别不大，这是因为 ξ 在节理变形初期对节理闭合程度影响不大，对透射波的影响相应也较小；$|v_{inc}|$ 较大时，不同 ξ 值的透射波波形及频谱图差别增大，ξ 较大的节理其 $|v_{tra}|$ 明显较小，这反映了变形非线性程度较大的

节理对纵波的透射能力较低,也意味着非线性程度较小的节理中有更多的波传过,这是因为此时节理变形发展至中期阶段,ξ 较大的节理相对于较小 ξ 值的节理更易闭合,对纵波穿越起到一定阻碍作用,而不利于其传播;$|v_{inc}|$ 很大时,不同 ξ 值的透射波波形及频谱图差别又变小,甚至接近重合,这是由于此时节理变形发展至后期阶段,闭合趋于完全,波可顺利传过节理,这也可以从 $|v_{inc}|=0.4$ m/s 时,透射波与入射波"先分离后重合"的现象上体现出来。值得注意的是,$|v_{inc}|$ 很大时频谱图显示相对于入射波而言,透射波产生振幅更大的高频分量,这与 Zhao 等[12]的认识一致,且出现了 ξ 越大高频部分振幅越大的情况。此外,无论 ξ 取值多大,透射波波峰总是与入射波波峰后部相交。

进一步对 $|T_{IMP}|$,$|R_{IMP}|$,e_{tra},e_{ref} 进行计算,$|v_{inc}|$ 分别为 0.001,0.01,0.05,0.08,0.10,0.15,0.20,0.40 m/s,整个计算结果显示对任何情形有 $e_{tra}+e_{ref}=1$,即透、反射波的能量之和总符合能量守恒定律,这也从侧面体现了计算的正确性。$|T_{IMP}|$,e_{tra} 与 ξ 关系见图 6,7。

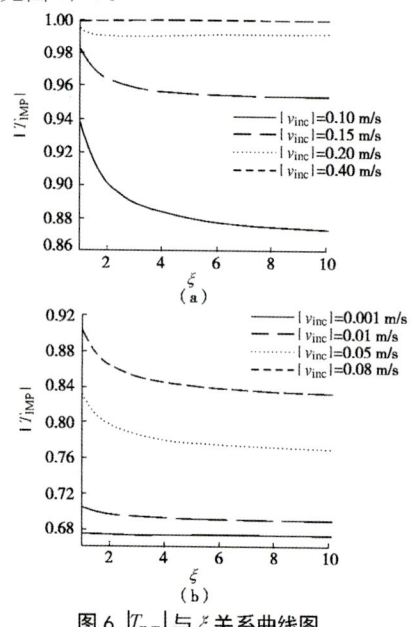

图 6 $|T_{IMP}|$ 与 ξ 关系曲线图

Fig. 6 $|T_{IMP}|$-ξ curves

由图 6,7 可见:①$|T_{IMP}|$,e_{tra} 随 $|v_{inc}|$ 的增加而增大;②$|T_{IMP}|$,e_{tra} 总体上随 ξ 的增加而降低,并在 $|v_{inc}|$ 处于中等水平时($|v_{inc}|=0.10$ m/s)该趋势较为显著,随着 $|v_{inc}|$ 的增大或减小,该趋势均趋于缓和,这与图 5 显示的结果一致;③$|v_{inc}|$ 很小($|v_{inc}|=0.001$m/s)时,$|T_{IMP}|$、e_{tra} 与 ξ 间的关系均为一水平直线,表示 $|T_{IMP}|$,e_{tra} 与 ξ 无关,这是由于此时节理 σ_n-d_n 关系接近线性情况,故 $|T_{IMP}| \to T_{lin}$,e_{tra} 亦趋近于定值;④$|v_{inc}|$ 很大($|v_{inc}|=0.4$ m/s)时,节理接近完全闭合,$K_n \to \infty$,应力波不受阻碍,故 $|T_{IMP}| \to 1$。但此时 $e_{tra}<1$,且随 ξ 的增加而略有降低,这是因为 $|v_{tra}|$ 达到最大值时,节理必将经历非线性变形、即刚度不断增大的硬化过程,也就必然会发生反射现象($e_{ref}>0$),从而导致能量无法完全透射,e_{tra} 也就相对变小。

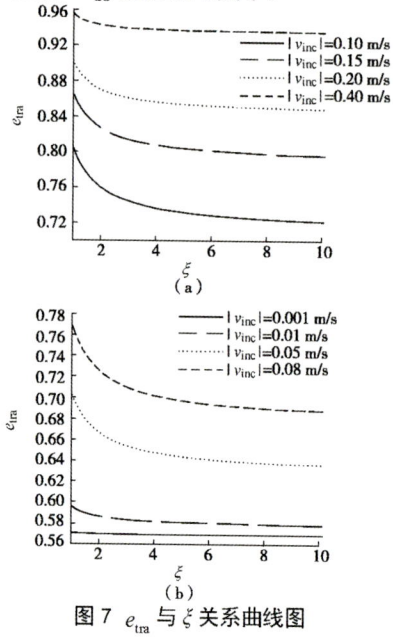

图 7 e_{tra} 与 ξ 关系曲线图

Fig. 7 e_{tra}-ξ curves

纵波穿越节理除了振幅下降和能量衰减外还会出现时间延迟现象。由于起跳点难以识别,入、透射波时间间隔常通过两波波峰时刻差计算,即峰—峰间隔,这里称为延迟时间,记作 T_{del},计算结果见图 8。

由图 8 可见,①T_{del} 随 $|v_{inc}|$ 增大而减小;②T_{del} 总体上随 ξ 的增加而增加,并在 $|v_{inc}|$ 处于中等水平时($|v_{inc}|=0.10$ m/s)该趋势较为显著,随着 $|v_{inc}|$ 的增大或减小,该趋势均趋于缓和,这亦与图 5 显示的结果一致;③$|v_{inc}|$ 很小($|v_{inc}|=0.001$ m/s)时,T_{del} 与 ξ 值无关,$|v_{inc}|$ 较大($|v_{inc}|=0.4$ m/s)时,T_{del} 随 ξ 的增加反而减小,最终趋近于 0,这里不妨称之为"超越"现象。对上述现象的解释是:随着 $|v_{inc}|$ 的增加,节理被一定程度的压缩,其刚度随之增大,即发生硬化作用,较大 ξ 值的节理由于变形的非线性程度较大,在节理闭合发展的初期刚度比小 ξ 值情况增长得慢,因此 T_{del} 较大。$|v_{inc}|$ 继续增加使节理被紧密压缩,在节理闭合发展到一定程度后,较大 ξ 值的节理刚度比小 ξ 值情况增长得快,硬化作用更为显著,其透射波开始"追赶"小 ξ 值的透射波,最终导致其 T_{del} 更小,在这个节理"先软后硬"的发展过程中,是否发生"超越"现象主要看软化和硬化哪一方在竞争中占主要地位。"超越"现象见图 9(a)显示了 $\xi \to \infty$ 情况透射波超越 $\xi \to 1$ 情况的过程,图 9(b)从小尺度上体现了 $\xi \to \infty$ 情况延迟时间的降低,由于 $|T_{IMP}| \to 1$,因此 $|T_{IMP}|$ 的"超越"现象不显著。

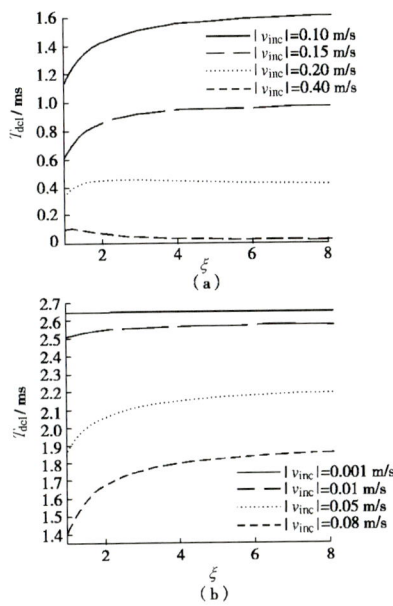

图 8 T_{del} 与 ξ 关系曲线图

Fig. 8 T_{del}–ξ curves

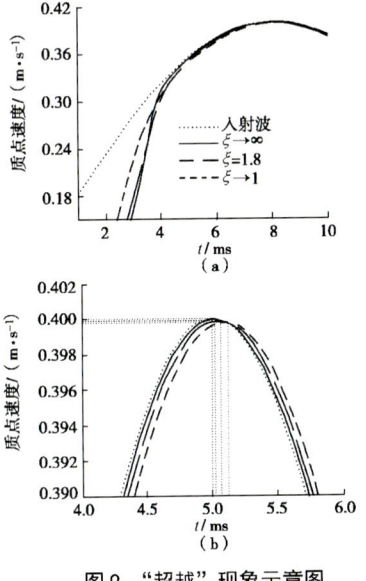

图 9 "超越"现象示意图

Fig. 9 Illustration of the phenomenon of exceeding

5 结　论

节理对岩体中纵波传播规律影响的理论研究经历了从考虑微裂隙到宏观节理、从位移连续到位移不连续、从线性模型到非线性模型的发展过程，本文沿着其发展脉络，将模型非线性程度固定的情况发展到可调情况，通过解析和数值计算得到了一些有益的结论。

（1）基于 Lemaitre 假设获得了弹性纵波法向穿越具有改进的非线性变形行为单节理处 $|T_{IMP}|$，$|R_{IMP}|$ 的近似解析解。计算结果显示 $|T_{IMP}|$ 随 $K_{n,i}$，γ 的增大而增大，d_{max}，f 的增大而减小，同时较强地依赖于修正系数 ξ。其他参数不变时 ξ 越大，$|T_{IMP}|$ 越小，这意味着节理变形非线性程度越大，对入射波的阻碍也越大。$|R_{IMP}|$ 具有的与其相反的规律。

（2）结合一维波动方程特征线法推导了节理处前、后波场质点速度数值差分格式并编制了计算程序，获得了 $v_{tra}(x_1,t)$，$v_{ref}(x_1,t)$ 半数值解，进而得到 $|T_{IMP}|$，$|R_{IMP}|$，e_{tra}，e_{ref}，T_{del} 解。

（3）计算结果显示：$|T_{IMP}|$ 不仅像线性变形节理那样，依赖于 $K_{n,i}$，d_{max}，f，还较强的依赖于修正系数 ξ。对 ξ 的依赖性，反映了节理变形本构关系的非线性程度与 $|v_{inc}|$ 产生的综合影响。$|v_{inc}|$ 很小时，d_n 远小于 d_{max}，无论 ξ 值多大均有：$|T_{IMP}| \to |T_{lin}|$，即 $|T_{lin}|$ 是 $|T_{IMP}|$ 的特例，此时波形畸变、频谱、e_{tra}，T_{del} 也均不受 ξ 影响；$|v_{inc}|$ 较大时，节理闭合进一步发生，非线性特性程度的差异开始体现，$|T_{IMP}|$，e_{tra} 随 ξ 的增加而降低，波形畸变程度、频谱差异、T_{del} 随 ξ 的增加而增大；$|v_{inc}|$ 很大时，节理变形发展至后期阶段，节理闭合趋于完全，$|T_{IMP}|=1$（与 ξ 无关），$e_{tra}<1$ 且随 ξ 的增加略有降低，频谱中出现高频区域振幅增大现象。此外，由于 ξ 较大的节理 K_n 会迅速提高，还可能产生一种"超越"现象。

（4）以上结论有助于进一步了解节理岩体中纵波的传播规律，实际情况有待通过试验进一步验证。

参考文献：

[1] MOHANTY B. Physics of exploration hazard[M]. London: Taylor and Francis Group, 1998.

[2] SCHOENBERGER M, KEVIN F K. Apparent attenuation due to intrabed multiples[J]. Geophysics, 1974, **39**(3): 278 – 291.

[3] SPENCER J W, EDWARDS C M, SONNAD J R. Seismic wave attenuation in nonresolvable cyclic stratification[J]. Geophysics, 1977, **42**(5): 939 – 949.

[4] BANIK N C, LERCHE I, SHUEY R T. Stratigraphic filtering based on derivation of the O'Doherty-Anstey formula[J]. Geophysics, 1985, **50**(12): 2768 – 2774.

[5] HUDSON J A. Wave speeds and attenuation of elastic waves in material containing cracks[J]. Geophysical Journal of the Royal Astronomical Society, 1981, **64**(1): 133 – 150.

[6] ANGLE Y C, ACHENBACH J D. Reflection and transmission of elastic waves by a periodic array of cracks[J]. Journal of Applied Mechanics, 1985, **52**(1): 33 – 46.

[7] ANGLE Y C, ACHENBACH J D. Reflection and transmission of elastic waves by a periodic array of cracks: oblique incidence[J]. Wave Motion, 1985, **9**(3): 375 – 388.

[8] HIROSE S, KITAHARA M. Scattering of elastic waves by a crack with spring-mass contact[J]. International Journal of Numerical Method in Engineering, 1991, **31**(7): 789 – 801.

[9] ACHENBACH J D, NORRIS A N. Loss of specular reflection due to nonlinear crack-face interaction[J]. Journal of Nondestructive Evaluation, 1982, **3**(4): 229 - 239.

[10] SMYSHLYAEV V P, Willis J R. Linear and nonlinear scattering of elastic waves by microcracks[J]. Journal of Mechanics and Physics of Solids, 1994, **42**(4): 585 - 610.

[11] CAPUANI D, WILLIS J R. Wave propagation in elastic media with cracks. Part I: transient nonlinear response of a single crack[J]. European Journal of Mechanics A/Solids, 1977, **16**(3): 377 - 408.

[12] ZHAO J, CAI J G. Transmission of elastic P-waves across single fractures with a nonlinear normal deformational behavior[J]. Rock Mechanics and Rock Engineering, 2001, **34**(1): 3 - 22.

[13] ZHAO X B. Theoretical and numerical studies of wave attenuation across parallel fractures[D]. Singapore: Nanyang Technological University, 2004.

[14] ZHAO X B, ZHAO J, CAI J G, et al. UDEC modelling on wave propagation across fractured rock masses[J]. Computers and Geotechnics, 2008, **35**(1): 97 - 104.

[15] JONES J P, WHITTIER J S. Waves at a flexibly bonded interface[J]. Journal of Applied Mechanics, 1967, **40**: 905 - 909.

[16] SCHOENBERG M. Elastic Wave behavior across linear slip interfaces[J]. Journal of Acoustic Society of America, 1980, **68**(5): 1516 - 1521.

[17] PYRAK-NOLTE L J, MYER L R, COOK N G W. Transmission of seismic waves across single natural fractures[J]. Journal of Geophysical Research, 1990, **95**(6): 8617 - 8638.

[18] KITSUNRZAKI C. Behavior of plane elastic waves across a plane crack[J]. Journal of Mining College of Akita University, 1983, **6**(3): 173 - 187.

[19] CAI J G, ZHAO J. Effects of multiple parallel fractures on apparent attenuation of stress waves in rock mass[J]. International Journal of Rock Mechanics and Mining Sciences, 2000, **37**: 661 - 682.

[20] ZHAO J, ZHAO X B, CAI J G. A further study of P-wave attenuation across parallel fractures with linear deformational behaviour[J]. International Journal of Rock Mechanics and Mining Science, 2006, **44**: 776 - 788.

[21] YU J. Effects of single joint with different nonlinear normal deformational behaviors on P-wave propagation[C]// Proceedings of the 2nd International Conference on Geotechnical Engineering for Disaster Mitigation and Rehabilitation. 2008: 458 - 465.

[22] 王卫华, 李夕兵, 左宇军. 非线性法向变形节理对弹性纵波传播的影响[J]. 岩石力学与工程学报, 2006, **25**(6): 1218 - 1225. (WANG Wei-hua, LI Xi-bing, ZUO Yu-jun. Effects of single joint with nonlinear normal deformation on P-wave propagation[J]. Chinese Journal of Rock Mechanics and Engineering, 2006, **25**(6): 1218 - 1225. (in Chinese))

[23] 俞缙, 赵晓豹, 赵维炳, 等. 改进的岩体节理弹性非线性法向变形本构模型研究[J]. 岩土工程学报, 2008, **30**(9): 1316 - 1321. (YU Jin, ZHAO Xiao-bao, ZHAO Wei-bing, et al. Study on improved elastic nonlinear deformation constitutive model of rock fractures[J]. Chinese Journal of Geotechnical Engineering, 2008, **30**(9): 1316 - 1321. (in Chinese))

[24] SHEHATA W M. Geohydrology of mount vernon canyon area[D]. Golden: Colorado School of Mines, 1971.

[25] GOODMAN R E. Methods of Geological Engineering in discontinuous rocks[M]. New York: West, 1976: 472 - 490.

[26] KULHAWAY F H. Stress-deformation properties of rock and rock discontinuities[J]. Engineering Geology, 1975, **8**: 327 - 350.

[27] BANDIS S C, LUMSDEN A C, BARTON N R. Fundamentals of rock fracture deformation[J]. International Journal of Rock Mechanics and Mining Sciences and Geomechnics Abstracts, 1983, **20**(6): 249 - 268.

[28] BARTON N R, BANDIS S C, BAKHTAR K. Strength, deformation and conductivity coupling of rock joints[J]. International Journal of Rock Mechanics and Mining Sciences and Geomechnics Abstracts, 1985, **22**(3): 121 - 140.

[29] MALAMA B, KULATILAKE P H S W. Models for normal fracture deformation under compressive loading[J]. International Journal of Rock Mechanics and Mining Sciences, 2003, **40**(6): 893 - 901.

[30] CAI J G, ZHAO J, LI H B. Effects of loading rate on fracture normal behaviour[C]// Proceedings of the 2nd Asian Rock Mechanics Symposium, Beijing, 2001: 197 - 200.

[31] YANG W Y, KONG G Y, CAI J G. Dynamic model of normal behavior of rock fractures[J]. Journal of Coal Science & Engineering (China), 2005, **11**(2): 24 - 28.

[32] WANG W H, LI X B, ZHANG Y P, et al. Closure behavior of rock joint under dynamic loading[J]. Journal of Central South University of Technology (China), 2007, **14**(3): 408 - 412.

爆炸作用下岩石破裂块度分布特点及其物理机理

戚承志[1,2]，王明洋[2]，钱七虎[2]，罗 健[1]

(1. 北京建筑工程学院 土木交通学院，北京 100044；2. 解放军理工大学 工程兵工程学院，南京 210007)

摘 要：就岩石在爆炸载作用下破坏块度分布的物理机理进行了分析。从分析可以看出，岩石破坏块度的对数正态分布与材料的多重破坏有关。在封闭爆炸情况下这种分布描述爆心附近岩石破坏块度的分布，此处材料处于静水压力状态，应变率很高，材料的破坏为多重破坏。而 Rosin-Rammler 分布主要描述离爆心较远处岩石破坏的块度分布，此处岩石的破坏主要是由环向拉力引起的径向裂纹所致，以单重破坏为主。

关 键 词：块度分布；单重破坏；多重破坏
中图分类号：O 212.2；O 383　　**文献标识码**：A

Features and physical mechanism of fragmentation distribution of rock under explosion

QI Cheng-zhi[1,2], WANG Ming-yang[2], QIAN Qi-hu[2], LUO Jian[1]

(1. School of Civil and Transportation Engineering, Beijing University of Civil Engineering and Architecture, Beijing 100044, China;
2. Engineering Institute of Cops of Engineers, PLA University of Science and Technology, Nanjing, 210007, China)

Abstract: Physical mechanism of fragmentation distribution of rock mass under explosion is investigated. It is demonstrated that lognormal distribution is closely related to the multi-fold fracture of the material. In the case of contained explosion, lognormal distribution describes the fragmentation distribution of rock in the vicinity of explosion center where the material is under hydrostatic compression, the strain rate is high, and the fracture is multi-fold. While Rosin-Rammler distribution mainly describes the fragmentation distribution of rock mass beyond the vicinity of explosion center where the fracture is induced mainly by the tangential tensile stress, and the fracture is one-fold.

Key words: fragmentation distribution; onefold fracture; multi-fold fracture

1 引 言

在许多实际工程中，例如采矿、磨矿等领域，需要知道地质材料在动力破坏后的块体尺寸及其分布。在常见工程问题中固体动力破坏块体的尺寸分布范围很广，从微米级到米级，涉及的工程问题也非常广泛，从粉尘程度的减少，到大尺寸块体份额的降低等。除此之外，研究固体的动力破坏块度也具有重要的理论意义，因为固体动力破坏块度的研究是固体动力变形破坏研究的延伸，固体的动力破碎块度及其分布不仅能反映固体动力破坏条件的影响，而且也能反映固体的重要性质，例如硬度、塑性性质、粘性性质等等。通过对于固体动力破坏块度的分析能够加深我们对于固体中裂纹群体的非线性相互作用的了解，帮助我们正确解释试验结果，准确地评价固体的物理力学性质。

研究固体动力破碎过程的方法之一是建立数学模型[1-2]，这样可以不用解出固体的应力应变状态而得到固体动力破碎块度分布的信息。这是目前变形体非线性力学最复杂的领域之一。要建立这种数学基础需要知道裂纹的扩展速度、缺陷连接的临界密度及连接条件。虽然在这一领域取得了很大的进展，但是由于这一问题非常复杂，因而现有的数学模型都采用一些简化假定。为了使模型能够描述实际需要解决的许多问题，比如说在冲击载作用下裂纹结构演化的群体性及非线性的描述，有许多问

收稿日期：2008-04-18。
基金项目：国家自然基金项目（No. 50825403）；北京市教委及北京自然基金项目（No. KZ200810016007）。
本文原载于《岩土力学》（2009 年第 30 卷增刊）。

题，比如说在裂纹群体相互作用条件下新的裂纹的产生及扩展条件、分岔的条件等等，还研究的很不够。除此之外在材料构造层次水平上固体的动力破碎规律也研究的不够，尤其破坏块度随应变率变化的物理机理方面。因此本文将对爆炸作用下岩石动力破坏成块的特点及其物理机理进行研究。

2 爆炸作用下岩石破坏成块的特点

在描述固体的动力破坏块度分布时常用的分布函数有泊淞分布、正态分布、伽马分布、Rozin-Rammler 分布及对数正态分布[3–6]。在地下封闭爆炸情况下常用的分布函数有 Rosin-Rammler 分布及对数正态分布[3]。

Rosin-Rammler 分布的数学表达式为

$$m_+(d) = m_0 \exp[-(d/d_{0RR})^n] \qquad (1)$$

式中：m_0 为破坏材料的总的重量；d 为块体的特征尺寸；$m_+(d)$ 为尺寸大于 d 的所有块体的重量；d_{0RR}，n 为常数。通常 d 为相应块体的等效球体的直径。

在 $\ln\ln(m_0/m_+)$ 及 $\ln d$ 坐标里 Rosin-Rammler 分布为一条直线。但是试验证明不是所有的试验点都位于直线上[3–4]。这种偏差发生在爆炸近区，这时可以用对数正态分布函数来描述破坏块度分布

$$\Phi(y) = \frac{1}{\sqrt{2\pi}} \int_{-\infty}^{y} \exp(-t^2/2)\mathrm{d}t \qquad (2)$$

式中：$y = (\ln d - \ln d_{LN})/\sigma_{LN}$，其中 d_{LN} 为分布的中数，σ_{LN} 为块体尺寸的对数对于其平均值的均平方偏差；$\Phi(y)$ 为尺寸小于 d 的所有块体的重量份额。

图 1 为在松脂中所做的爆炸试验的结果[3]。从图中可以看出，随着离爆心距离的增加，破坏块度的分布律逐渐由对数正态分布律转换为 Rosin-Rammler 分布律，这种转换是逐渐进行的。为了描述这种分布，文献[7]建议了如下的破坏块尺度分布公式：

$$\Omega(d,R) = \frac{\alpha(R)}{\sqrt{2\pi}} \int_{-\infty}^{y} \exp(-t^2/2)\mathrm{d}t + [1-\alpha(R)]f_{RR}(d,R) \qquad (3)$$

式中：d 为块体的特征尺寸；Ω 为尺寸小于 d 的所有块体的重量；$\alpha(R)$ 为服从对数正态分布的块体的份额；$y = (\ln d - \ln d_{LN}(R))/\sigma_{LN}(R)$，其中 $d_{LN}(R)$ 为对数分布的块体尺寸中数；$\sigma_{LN}(R)$ 为块体尺寸的对数对于其平均值的均平方偏差；$f_{RR}(d,R)$ 为具有平移的 Rosin-Rammler 分布函数，具有如下形式：

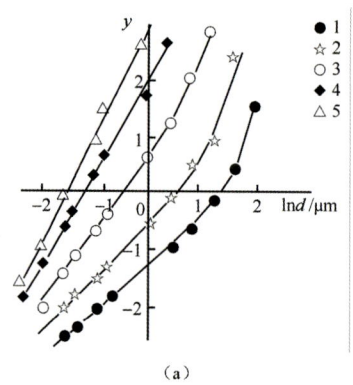

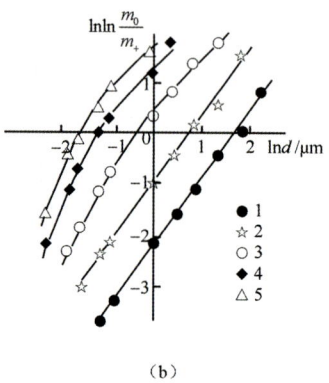

图 1 在不同的距爆心距离上破碎块度分布的变化情况（符号标示 1：$r = 22R_0$，2：$r = 15R_0$；3：$r = 10.5R_0$；4：$r = 7.5R_0$；5：$r = 6.5R_0$。其中 R_0 为装药半径）

Fig.1 Particle size distribution with the distance to explosion center (1：$r = 22R_0$, 2：$r = 15R_0$; 3：$r = 10.5R_0$; 4：$r = 7.5R_0$; 5：$r = 6.5R_0$. Where R_0 - radius of charge)

$$f_{RR}(d,R) = \begin{cases} 0, & d \leq d_k \\ 1-\exp\{-[(d-d_k)/d_{0RR}]^n\} \end{cases} \qquad (4)$$

式中：n 为确定分布形状的指数，表征破裂的均匀程度，它的值越大，材料的破裂均匀度越高；d_k 为块度平移大小，物理意义为细观及微观水平之间的分界线；d_{0RR} 为某一特征尺度，它与平均块体尺寸 d_{RR} 之间具有如下关系：

$$d_{0RR} = \frac{d_{RR} - d_k}{\Gamma(1+1/n)} \qquad (5)$$

式中：Γ 为伽马函数。

形成这种双模态分布的物理机理是什么呢？下面就分析双模态分布的物理意义。

2 爆炸作用下岩石破坏成块分布的物理机理

在动载作用下材料强度对于应变率的依赖关系的一般规律见图 2。首先材料强度随应变率的增加

而缓慢地增加（此处称为Ⅰ区），当应变率继续增加，超过某一值时，材料强度随应变率的增加急剧增加（此区称为Ⅱ区），当应变率进一步增加到冲击爆炸应变率时（Ⅲ区），材料强度随应变率的增加而增加的速率又变缓，与Ⅰ区情况相当。

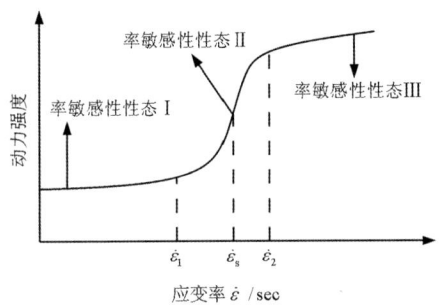

图 2　材料动力强度对于应变率的依赖规律
($\dot{\varepsilon}_1 \approx 10^0 \sim 10^2 \text{ sec}^{-1}, \dot{\varepsilon}_s \approx 10^3 \text{ sec}^{-1}, \dot{\varepsilon}_2 \approx 10^4 \text{ sec}^{-1}$)
Fig.2　Dependence of dynamical strength on strain rates of brittle materials
($\dot{\varepsilon}_1 \approx 10^0 \sim 10^2 \text{ sec}^{-1}, \dot{\varepsilon}_s \approx 10^3 \text{ sec}^{-1}, \dot{\varepsilon}_2 \approx 10^4 \text{ sec}^{-1}$)

作者的分析表明[8-9]，在小的应变率范围内（Ⅰ区），材料的变形及破坏受热活化机制控制，这时材料的变形与破坏主要集中于材料单元之间的界面上，材料变形类似蠕变，具有塑性特性。由于界面破坏后产生应力释放，材料的其它地方难以发生破坏，即这时的破坏具有局部特性及单重特性；当应变率进一步增加时，材料的宏观粘性阻尼机制逐渐加入到材料强度的应变率依赖性中来，并逐渐占据主导地位，材料的破坏靠裂纹的产生、扩展及连接来实现，材料的变形与破坏具有脆性性质，材料变形与破坏的区域逐渐由材料单元的界面向材料内部扩展，也即材料变形破坏逐渐失去局部特性；当应变率继续增加而进入到高应变率区（Ⅲ区）时，不同尺寸的缺陷的增长将同时启动，在没有缺陷的区域，材料的热活化启动，并产生原子键的断裂，形成裂纹产生的源泉，这时的破坏丧失了局部性，破坏变的均匀，破坏源独立发展，受其它破坏源的影响小，变形与破坏具有塑性性质，破坏具有多重破坏的特征。

Колмогоров А.Н.（Kolmogorov A.N.）证明[2]，如果每个粒子的破裂概率不依赖于其绝对尺寸及其先前的破裂，不依赖于其它粒子，那么在所有粒子的多重破裂时，无论它们的初始尺寸分布如何，最终破裂块体的分布函数逐渐趋向于对数正态分布函数。也就是说在多重破坏区，材料的破坏块度服从对数正态分布。

那么封闭爆炸的试验结果怎样哪？试验表明[10]，对于每一种材料都存在着某一个临界应力值σ^*，当爆炸载引起的应力大于该值时就会发生自发的破坏，即材料丧失力学稳定性而快速释放能量。应力值σ^*通常高于材料的强度极限。在材料破坏时表征动载对于材料动力作用的一个重要参数为应变率。试验表明[10]当岩石中的应变率达到$(0.1\sim3)\times10^2 \text{ s}^{-1}$时会发生自相持破坏。在松脂材料中所做的封闭爆炸试验表明，应变率随着距爆心的相对距离r/R_0（其中R_0为装药半径，r为实际距离爆心距离）的增加而减小，具体数据见下表 1。

表 1　球形装药情况下松脂中应变率随距离的变化数据
Table 1　Dependence of strain rate on the distance from spherical charge in rosin

应变率 $\dot{\varepsilon}$	$5\times10^4 \text{ s}^{-1}$	$8\times10^3 \text{ s}^{-1}$	$2\times10^3 \text{ s}^{-1}$	$6\times10^2 \text{ s}^{-1}$
相对距离 r/R_0	2.50	5.00	8.75	13.75

文献[3]对于地下封闭爆炸时岩石破坏块度的研究表明（见图 3），当距爆心相对距离$r/R_0<7.7$时，破坏块尺寸对于距离的依赖关系偏离直线，破坏块变小。在剔除了多重破坏因素后，平均块度的分布规律接近于远区的线性依赖关系，见图中虚线所示。这说明$r/R_0<7.7$的范围内多重破坏的影响是造成平均块度偏离较远处直线关系的原因。$r/R_0=7.7$这一数据与图 2 及表 1 所示数据符合。

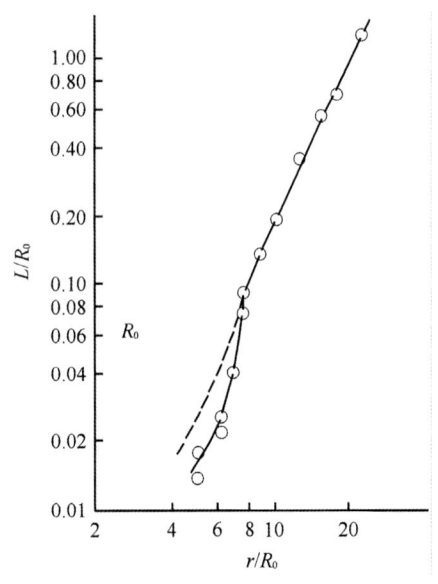

图 3　在松脂中爆炸时破坏块的尺寸与距爆心距离之间的依赖关系[3]
Fig.3　Dependence of fragmentation size on distance under explosion of TEN in pine rosin[3]

在松脂中做的试验表明[4]，在$r \leq 2.7R_0$的范围

内应力大小为 $\sigma_r = 1.1 \cdot 10^3$ MPa，松脂的透明度没有丧失，这说明在这一范围内材料的变形是塑性的，这与图2及表1所示结果相符合。在这一范围内材料主要受到静水压力的作用，而且材料的应变率很高。在 $2.7R_0 < r \leq 4R_0$ 的范围内，破裂波阵面与爆炸波阵面重合，介质的应力状态逐渐从静水压力状态向剪压受力状态过渡。在 $r = 4R_0$ 处破裂波阵面的速度突然降低，这应该与剪切破坏的开始有关。在 $r > 4R_0$ 时破裂波阵面逐渐落后于爆炸波阵面，速度逐渐降低，在 $r = 10R_0$ 时趋于稳定，达到 0.96 km/s 时，破裂波速度接近于瑞利波速（$c_R = 1.0$ km/s）。在这一范围内剪切应力与静水压力共存，这一距离范围内的材料破坏主要是以剪切破坏为主，随着静水压力的降低，材料的破坏逐渐从多重破坏向单重破坏过渡。在 $10R_0 < r \leq 22R_0$ 的范围内破裂波稳定传播，这一范围内的应变率为 $10^2 \sim 10^3$ /s，这一段距离上材料的破坏就是前面所说的自相持破坏，直到 $r = 25R_0$ 时破裂波失稳并停止。在 $10R_0 < r \leq 22R_0$ 范围内，材料的破坏主要是由环向拉力引起的径向裂纹所致，由于受拉破坏形成径向裂纹后，裂纹附近的应力迅速释放，附近不可能再出现裂纹，因此在此范围内破坏是单重的，这时创造径向裂纹需要的能耗少，裂纹能够长距离以稳定的速度传播。

所以由上面的分析可以看出，材料破坏块度的对数正态分布主要集中于爆心附近，此处的破坏是以材料的多重破坏为主。而 Rosin-Rammler 分布主要发生于离爆心较远处，此处材料的破坏以单重破坏为主。由于 Rosin-Rammler 分布所涉及的区域比对数正态分布所涉及的区域大一个数量级，所以总的块度分布还是服从 Rosin-Rammler 分布。

3 结 论

本文就岩石在爆炸载作用下破坏块度分布的物理机理进行了分析。从分析可以看出，岩石破坏块度的对数正态分布与材料的多重破坏有关。在封闭爆炸情况下这种分布描述爆心附近岩石破裂块度的分布，此处材料处于静水压力状态，应变率很高。而 Rosin-Rammler 分布主要描述离爆心较远处岩石破坏的块度分布，此处材料的破坏主要是由环向拉力引起的径向裂纹所致，以单重破坏为主。

参 考 文 献

[1] Кузнецов В М. Математические модели взрывного дела[M]. Новосибирск: Наука, 1977.

[2] Колмогоров А Н. О логрифмически нормальном законе распределения размеров частиц при драблении [J]. **ДАН СССР**, 1941, 31(2): 99－101.

[3] Радионов В Н, Сизов И А, Цветков В М. Основы геомеханики[M]. Москва: Недро, 1986.

[4] Сизов И А, Цветков В М. О механизме образования осколков при камуфлетном взрыве [J]. **Физика горения и взрыва**, 1979, 15(5): 108－113.

[5] Кошелев Э А, Кузнецов В М, Софронов С Т. и др. Статистика осколков, образующихся при разрушении твердых тел взрывом[J]. **ПМТФ**, 1971, 2: 87－100.

[6] Барон Л И, Сиротюк Г Н. Проверка применимости уравнения Розина-Заммлера для исчисления диаметра среднего куска при взрывной отбойке горных пород[J]. Взрывное дело №62/19: 111－121. Москва: Недра, 1979.

[7] Замышляев Б В, Евтерев Л С, Лоборев В М, Чернейкин В А. Локальный двухмодальный закон распределения обломков по размерам при взрывном разрушении горных пород [J]. **ДАН СССР**, 1987, 293(2): 326－329.

[8] 戚承志, 钱七虎. 岩石等脆性材料动力强度依赖应变率的物理机制[J]. 岩石力学与工程学报, 2003, 21(2): 177－181.

[9] QI C Z, WANG M Y, QIAN Q H, et al. Structural hierarchy and mechanical properties of rock mass. Part II, Structural hierarchy, size effect and strength of rock mass[J]. **Physical Mesomechanics**, 2006, 9(6): 41－53.

[10] Спивак А А. Поведение среды при самопроизвольном разрушении[J]. **ФТПРПИ**, 1982, (4): 51－56.

[11] Адушкин В В, Спивак А А. Разрушающее действие взрыва в предварительно напряженной среде[J]. **ФТПРПИ**, 2002, (4): 61－69.

电阻率法在深部巷道分区破裂探测中的应用

许宏发[1]，钱七虎[1]，王发军[1]，李术才[2]，袁 亮[3]

(1. 解放军理工大学 工程兵工程学院，江苏 南京 210007；2. 山东大学 岩土与结构工程研究中心，山东 济南 250061；
3. 淮南矿业集团，安徽 淮南 232001)

摘要：淮南丁集煤矿西部采区南运输大巷埋深达 955 m，已显现出深部开挖的特征。为研究深部巷道围岩破裂情况，在该巷道选取 2 个监测断面(宽 5.0 m，高 3.8 m)，每个断面布置 5 个钻孔，测量围岩电阻率沿钻孔深度的变化。电阻率是岩石的重要电性参数，岩体的破碎程度对电阻率的影响较大，一般有裂纹的地方，电阻率产生突变。采用 ResiTest–4000 电阻率测试仪和研制的孔内探头，对钻孔内岩体电阻率进行测试。岩石破碎区电阻率基准值根据每一钻孔电阻率平均值(剔除特异点)确定，大于基准值的为破碎区。根据测试结果，绘制巷道围岩分区破裂图，与钻孔电视观测结果较吻合。结果表明，该巷道围岩有 4 个破裂分区；破裂分区带的半径与巷道半径基本呈线性关系；巷道周边破裂区宽度最大，平均达到 3.12 m，依次分区破裂带的宽度有递减趋势。

关键词：采矿工程；电阻率；深部巷道；深部岩体；分区破裂

中图分类号：TD 35　　**文献标识码**：A　　**文章编号**：1000–6915(2009)01–0111–09

APPLICATION OF ELECTRIC RESISTIVITY METHOD TO ZONAL DISINTEGRATION EXPLORATION OF DEEP ROADWAY

XU Hongfa[1]，QIAN Qihu[1]，WANG Fajun[1]，LI Shucai[2]，YUAN Liang[3]

(1. Engineering Institute of Engineering Corps，PLA University of Science and Technology，Nanjing，Jiangsu 210007，China；
2. Research Center of Geotechnical and Structural Engineering，Shandong University，Jinan，Shandong 250061，China；
3. Huainan Coal Mining Group，Huainan，Anhui 232001，China)

Abstract：South transportation roadway of depth of 955 m in Huainan Dingji Mine western coal area has shown the characteristics of the deep excavation. In order to investigate fracturing behavior of the surrounding rock in the deep roadway，two sections(the width is 5.0 m，and the height is 3.8 m) are selected. Five boreholes are drilled as fan array in the arch of the every section. The variation of electrical resistivity of surrounding rock with borehole depth can be measured. Electrical resistivity is one of the important electric parameters for rocks. The degree of rock mass fracturing has great influence on electric resistivity；and generally electrical resistivity changes suddenly in the fractural location. Using ResiTest–4000 resistivity tester and self-developed downhole probe，the electric resistivity of rock mass in the boreholes is tested. A reference value of electrical resistivity on disintegrated zones of rock mass can be determined according to average electrical resistivity(to reject abnormal data) in every hole；and the ones greater than the reference values are disintegrated rock mass. According to the test results，the diagrams of zonal disintegration distribution of surrounding rock of the roadway are drawn. In addition，the diagrams are in agreement with ones obtained from borehole TV video. The results show that there are four disintegration zones in the surrounding rock of the roadway; the radius of the disintegrated zones has a linear relationship with the radius of the roadway; the thickness of the disintegrated zones near the roadway boundary is maximal and up to 3.12 m; and the thickness of disintegrated zones decreases in turn with borehole depth.

Key words: mining engineering; electrical resistivity; deep roadway; deep rock mass; zonal disintegration

1 引 言

通过近几年来的研究，深部岩体的分区破裂化现象得到越来越多的学者认可。钱七虎[1]定义深部岩体分区破裂化现象为："在深部岩体中开挖洞室或巷道时，其两侧和工作面前的围岩中会产生交替的破裂区和不破裂区，称这种现象为分区破裂化现象"。顾金才等[2]通过模型试验，发现在高的平行于洞室轴向的水平压力作用下，洞室围岩出现多条裂缝，裂缝之间存在未破坏区域，证明分区破裂存在的事实。М. В. Курленя 和 В. Н. Опарин[3]在对大量试验数据的分析和理论研究基础上，发现各破裂区的半径服从某种规律，并和地质力学不变量有关。坑道围岩分区破裂的最大范围，是工程实际中较重要的参数，E. I. Shemyakin 等[4]建立了用相同材料制成的未加固坑道模型，通过试验分析，得到了分区破裂最大半径的计算经验公式。L. S. Metlov 等[5]利用非平衡热动力学方程，揭示了围岩分区破裂的物理基础，并进行了计算机模拟。周小平和钱七虎[6]把深部巷道的开挖看作动力问题，运动方程采用位移势函数，运用弹性力学和断裂力学，确定了破裂区岩体的残余强度和产生破裂区的时间，进而确定了破裂区和非破裂区的宽度和数量。李英杰等[7]对现场观测和相似材料模拟试验中岩石分区碎裂化的时间效应进行了总结，得出了岩石分区碎裂化现象是与岩体蠕变有密切关系，求出了岩石分区破裂化发生时破裂带半径的公式。

目前，围岩的分区破裂测试的手段很多，主要有钻孔潜望镜法、电阻率法、超声波法和γ射线法等。E. I. Shemyakin 等[8]在 1 050 m 深度的 TALNAKH-OKTYARBSKIG 矿进行了围岩破裂测试，使用 REP－451 潜望镜、直流低频率电流电测(电阻率)、γ射线、超声波和局部光学等方法，清楚地揭示了地下开采围岩的状态。

钻孔潜望镜法比较直观，通过录像资料，肉眼判别钻孔内孔壁围岩的裂隙分布情况，但由于孔内雾气、灰尘等原因，往往单靠录像资料很难准确判别破碎程度。尤其对于细小裂隙，由于电视分辨率的限制，也很难判别。因此，本文利用电阻率法测试钻孔内围岩电阻率的变化，进而判定围岩的破裂范围，为更准确确定围岩破裂区，提供了新的尝试，这是对钻孔潜望镜法测试结果的有效补充。

目前，深部巷道分区破裂化实测资料很少。本文在淮南丁集煤矿近千米水平大巷进行了围岩电阻率探测，给出了围岩破裂分区，总结得到了破裂分区带中线半径 r_{cn} 随巷道半径和分区号 n 的变化规律。

2 试验现场概述和深部条件评价

2.1 巷道断面与地质条件

试验点位于淮南丁集煤矿西部 11－2 采区南运输大巷，水平标高−910 m，地面标高约 45 m，埋深约 955 m。巷道附近岩层柱状图如图 1 所示。巷道所处岩层近似水平，运输轨道大巷位于 11－2 煤底板，岩性为砂质泥岩、中砂岩和粉细砂岩，根据附近的 29－8，29－12 钻孔资料分析，11－2 煤厚 2.3～3.1 m，平均厚 2.7 m。在 11－2 煤大巷内选取 2 个监测断面 B 和 C，2 个监测断面相距约 95 m。

(1) 地质构造情况：本次测试区域南部地层平缓，为近水平状态，构造简单；向北倾角逐渐增大，为 0°～6°，并发育有 SF82，F84，F151 等断层。其中，F117，F84，F151 为井田内大中型断层，对掘进影响重大。

(2) 水文地质情况：该掘进区段巷道施工过程中，水文地质条件较简单，巷道内水量不大，钻孔岩壁无水迹。

(3) 地温：该区属地温异常区，地温梯度为 2.3～4.0 ℃/(100 m)，平均地温梯度为 3.21 ℃/(100 m)，根据 11－2 煤底板温度与底板标高关系回归方程推算，−910 m 水平地温约为 45 ℃，为二级

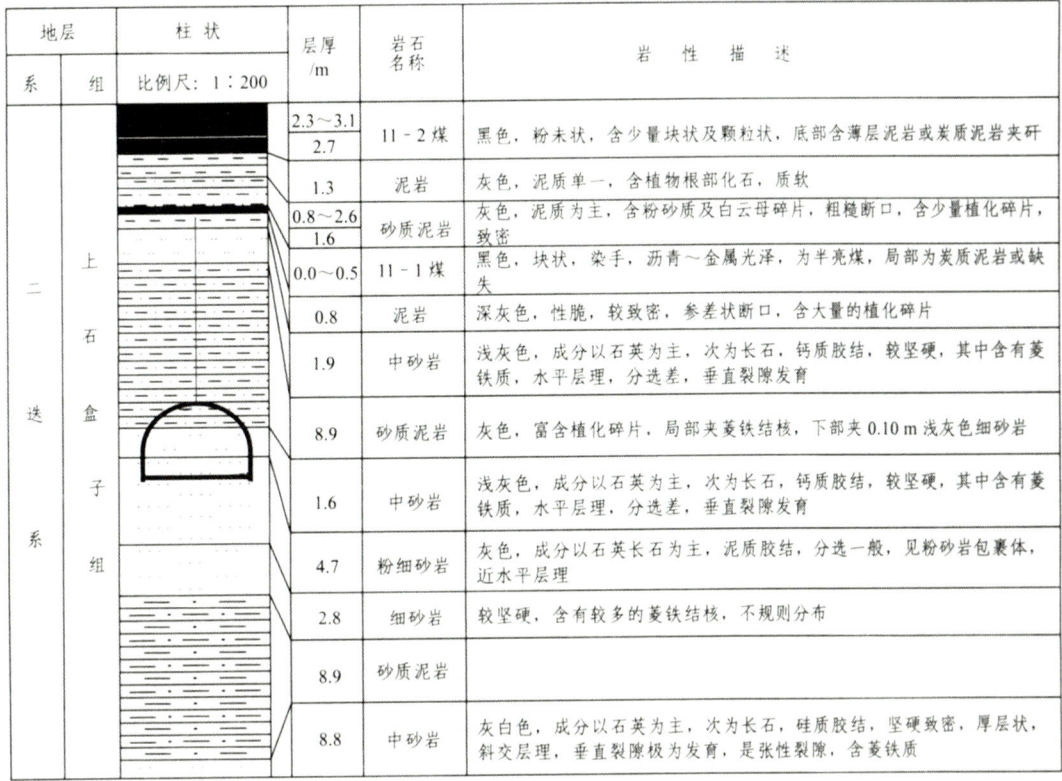

图 1 巷道附近岩层柱状图

Fig.1 Terrane histogram round roadway

热害区。

(4) 巷道断面设计：轨道大巷断面形状为直墙半圆拱形，宽 5.0 m，高 3.8 m。采用锚网喷+锚索支护，锚杆的规格是 ϕ20 mm，L2 200 mm 的高强度螺纹钢，间排距为 800 mm；钢筋网的规格为 ϕ6 mm；喷层采用 C20 强度等级混凝土，厚度为 150 mm。锚索规格为 ϕ15.24 mm，长 6 200 mm，间排距 2.4 m，按 2-3-2 布置。每个断面共有 5 个探测孔(见图 2)，采用地质钻机钻孔，钻孔直径为 73 mm，深度 10 m，2 个孔水平布置，1 个孔垂直布置，另外 2 个孔与水平面的夹角为 45°。

2.2 深部力学条件

(1) 围岩强度：巷道顶板主要围岩为砂质泥岩，由于现场未进行强度试验，可参考淮南潘谢矿区和淮南谢一矿的试验结果。淮南潘谢矿区砂质泥岩单轴抗压强度 σ_c = 30.9 MPa(标高-376.4 m)[9]。淮南谢一矿区，在-780 m 水平(埋深 810 m)处，自然含水状态泥岩单轴抗压强度 σ_c=28.0～30.3 MPa[10]。可以估计，淮南丁集矿区砂质泥岩的单轴抗压强度约为 30 MPa。

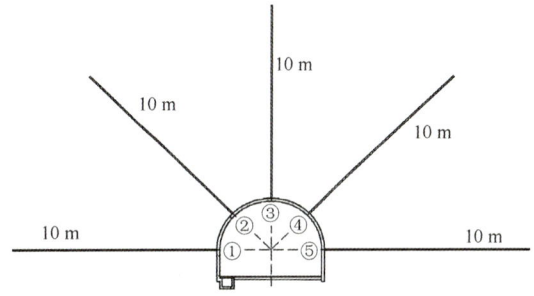

图 2 钻孔分布图

Fig.2 Borehole arrangement plan

(2) 原岩应力：根据刘泉声等[10]淮南地区地应力现场测量结果(-780 m 水平，埋深 810 m)，垂直于巷道轴线方向的水平原岩应力高达约 30 MPa，垂直应力高达约 25 MPa。侧向压力系数 λ=1.2。可以估计出丁集矿 955 m 深度，垂直原岩应力约为 28 MPa，水平原岩应力约为 34 MPa。该原岩应力大于砂质泥岩的抗压强度。在测试期间曾发生两起顶板小块岩石爆落的现象，说明现场围岩应力是很大的，已接近岩爆发生的条件。根据钱七虎[1]对深部界定分析所得的结论，一般当原岩应力大于岩石单轴抗压强度(即 $\sigma_地 > \sigma_c$)时，围岩中可发生分区破裂化现象。因此，本文所选测量巷道已满足深部界定的条件，有发生分区破裂的可能。

3 试验原理与方法

电阻率是岩石一个重要的电性参数。电阻率通常采用四电极法测量(见图3)，2 个电极(A，B)接地，将电流通入地下，由另 2 个电极(M，N)测量所产生的电位差，再通过计算获得电阻率值。4 个电极与中心点对称布置，对称等距的装置为 Wenner 装置，对称不等距的装置为 Schlumberger 装置。对于 Wenner 装置，其电阻率计算式为

$$\rho = \frac{2\pi a V}{I} \qquad (1)$$

式中：V 为电极电压，I 为电极电流，ρ 为电阻率，a 为电极间距。

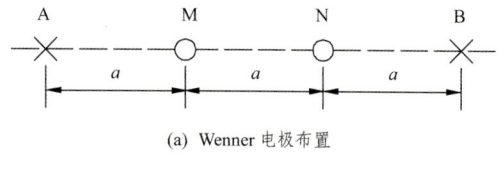

(a) Wenner 电极布置

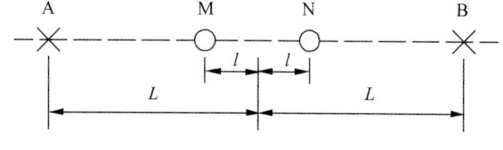

(b) Schlumberger 电极布置

图 3 电阻率测量电极布置

Fig.3 Electrode arrangement in electrical resistivity measurement

本次现场电阻率测试采用的 ResiTest–4000 电阻率测试仪，包括主机、Wenner 探头和标准块。Wenner 探头的长度为 15 cm，其上一排 4 个电极，每个电极的间距是 50 mm；通过外面 2 个电极产生电流，通过里面 2 个电极测试所产生的电压；标准块的作用是检验仪器是否工作正常；ResiTest–4000 主机上显示的结果为电阻率($k\Omega \cdot cm$)。

为了能在钻孔内测试围岩的电阻率，试制了带滚轮的孔内探头。孔内探头与 Wenner 探头的原理一致，也有 4 个电极，电极间距 50 mm，每个电极安装了导电性能极好的泡沫尖端，确保能与不规则表面充分接触。每个电极与 Wenner 上的电极用屏蔽电缆对应连接。用试制的探头测试标准电阻率块，误差在 1%以内，满足测量要求。实际测量时，把自制探头放入孔内，每步前进 0.2 m 测量一次，直至孔底。测试时先向孔内吹入空气，排出孔内瓦斯，以防瓦斯爆炸。

很多学者研究发现，岩性[11]、含水量[12, 13]、孔隙率[11]、温度[14]、破裂[15]对岩石电阻率的影响较大。在前四项条件变化不大的情况下，破裂程度是影响电阻率的主要因素。从岩石单轴压缩过程中电阻率的变化就能反映破裂程度对电阻率的影响。陆阳泉等[16]进行了大型天然灰岩样品从微裂到大破裂的电阻率试验，在整个加载过程中电阻率变化有明显的"趋势"异常，岩石大破裂前观测到了"短期"前兆和"临震"突变。王云刚等[17]采用具有冲击性倾向的煤样，在压力试验机上进行了单轴压缩试验，电阻率变化随压力的变化曲线大致成"凹"型曲线，即煤样电阻率随压力增加逐渐减少，当达到破裂应力值的一半左右时，电阻率达到最小值；继续加载电阻率将升高，直到出现第一次主破裂；煤样出现主破裂后在继续加载的情况下，电阻率会继续升高，甚至会超出电阻率测试仪的量程。这充分说明，受载煤岩的电阻率变化与其破裂过程中的裂隙形成、发育和扩展密不可分。

因此根据电阻率的突变情况，判断钻孔内岩石的破裂和完整状态是可行。

4 试验结果

根据现场量测的孔内岩壁电阻率值，分别绘制出 2 个断面 5 个钻孔的岩体电阻率随孔深变化的曲线，如图 4，5 所示(图中，x 为钻孔深度坐标)。

由以上分析可知，岩石破碎区的电阻率有突变现象。由于每孔的条件不同，各孔中完整岩石的电阻率变化范围有差异，但就某单个孔来说，变化范围不大。因此，必须确定岩石破碎区电阻率的基准值。E. I. Shemyakin 等[8]提出了根据平均值确定破碎岩体电阻率基准值的方法。根据这一观点，本文采用下式进行计算：

$$\left.\begin{array}{l}\rho_0^{(k)} = \dfrac{1}{N_k}\sum_{i=1}^{N_k}\rho_k(x_i) \\ (0 < x_i \leq L;\ k = 1, 2, \cdots, 5)\end{array}\right\} \qquad (2)$$

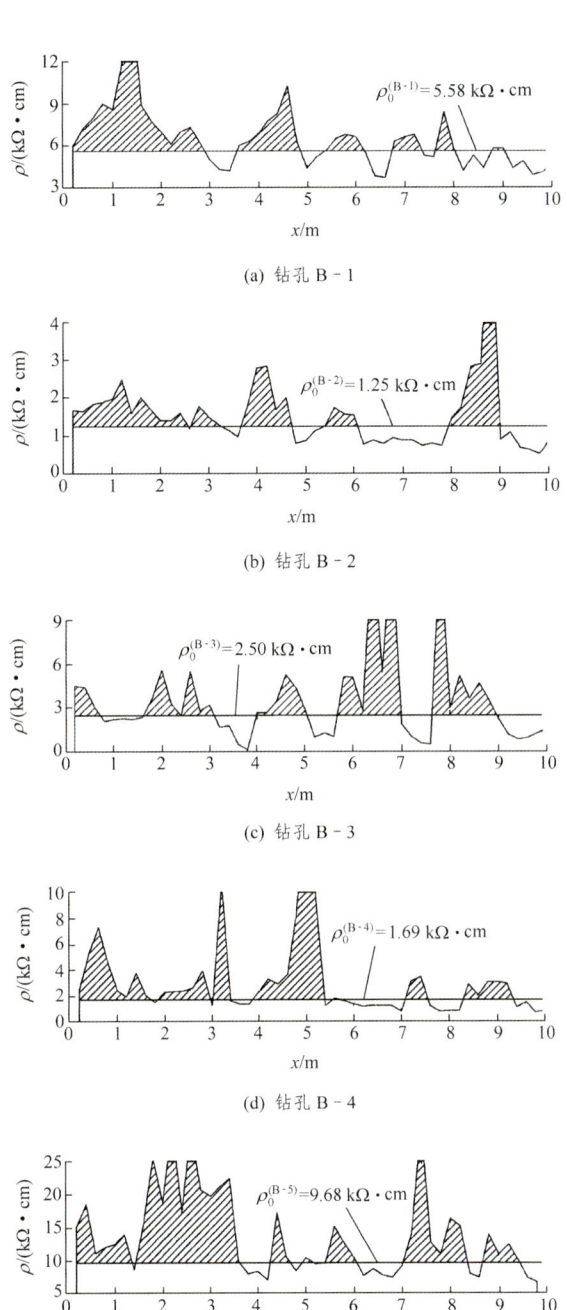

(a) 钻孔 B‑1

(b) 钻孔 B‑2

(c) 钻孔 B‑3

(d) 钻孔 B‑4

(e) 钻孔 B‑5

图 4　断面 B 各钻孔围岩电阻率变化曲线

Fig.4　Variable curves of electrical resistivity of surrounding rock in boreholes of section B

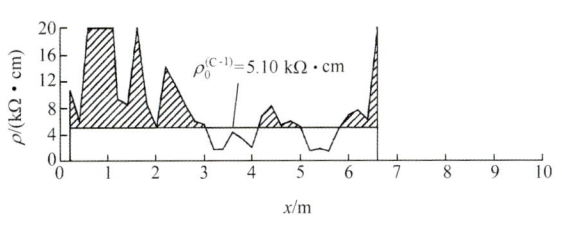

(a) 钻孔 C‑1

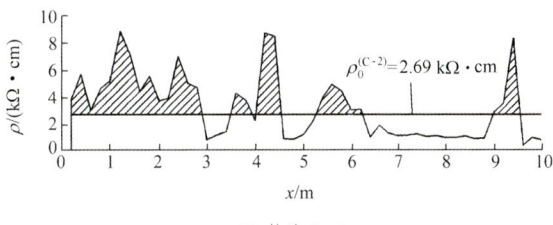

(b) 钻孔 C‑2

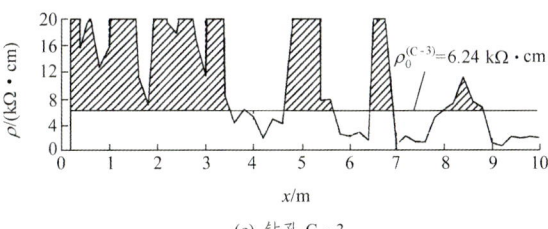

(c) 钻孔 C‑3

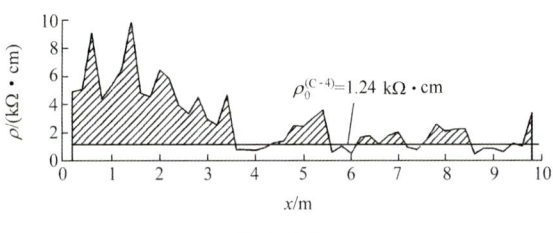

(d) 钻孔 C‑4

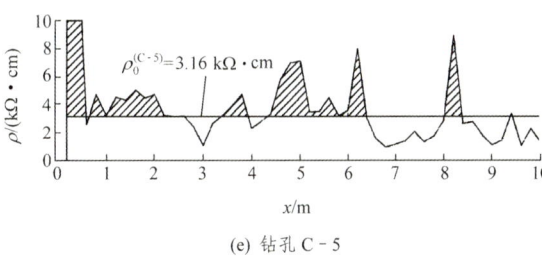

(e) 钻孔 C‑5

图 5　断面 C 各钻孔围岩电阻率变化曲线

Fig.5　Variable curves of electrical resistivity of surrounding rock in boreholes of section C

式中：k 为钻孔号；i 为测量点号；$\rho_0^{(k)}$ 为 k 钻孔岩石破碎区电阻率基准值；N_k 为 k 钻孔测量总数；$\rho_k(x_i)$ 为 k 钻孔电阻率分布函数，x_i 为测量点 i 到孔口的距离；L 为孔深。

根据式(2)计算过程中，对电阻率值明显偏离的特异点不参加计算，例如有些电阻率值为正常值的十几倍，是由于岩石破碎引起的，计算时剔除。表 1 中列出了各钻孔岩石破碎区电阻率基准值。岩石破碎区电阻率基准值见图 4，5 中的横线及表 1。这样，电阻率位于 $\rho_0^{(k)}$ 之上的为破碎区，在图 4，5 中用斜线填充表示。将每个钻孔内的破坏区(电阻率大于基准值)的部分依次描绘在断面图上，把相邻孔关联的破碎区用线连接，且用斜十字线填充，形成不同厚度的破碎带。各破碎带之间未填充的空白

表 1 各钻孔岩石破碎区电阻率基准值

Table 1 Electrical resistivity reference values of fracturing area for all boreholes

钻孔号	电阻率计算范围/(kΩ·cm)	电阻率基准值 $\rho_0^{(k)}$/(kΩ·cm)
B-1	3.70~9.00	5.58
B-2	0.45~2.50	1.25
B-3	0.20~5.60	2.50
B-4	0.75~3.94	1.69
B-5	6.80~19.80	9.68
C-1	1.14~10.65	5.10
C-2	0.29~4.70	2.69
C-3	0.40~8.40	6.24
C-4	0.56~2.13	1.24
C-5	0.97~5.80	3.16

为完整区。图6，7分别给出了断面B，C围岩分区破裂化分布情况。

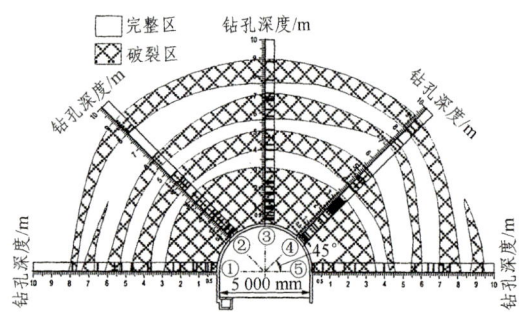

图 6 断面B围岩分区破裂化分布

Fig.6 Zonal disintegration distribution of section B

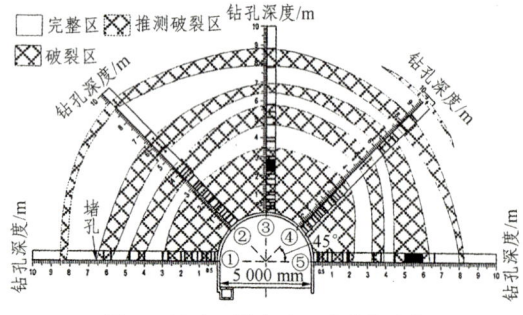

图 7 断面C围岩分区破裂化分布

Fig.7 Zonal disintegration distribution of section C

5 比较和分析

5.1 与钻孔电视观测比较

本文进行围岩电阻率测试的过程中，李术才等[18]也进行了钻孔电视成像观测。在图6，7中，垂直钻孔轴线的小粗线即为钻孔电视观察得到的岩石裂缝。

从图6看出，断面B围岩破碎区可分为4个，用电阻率法确定的断面B各孔围岩破裂区分布与钻孔电视观测结果十分类似，只是在个别地方可能由于2种方法距离量测误差，对位略有差别，总体趋势十分相合。由于在裂隙两侧可能存在细小裂隙，用电阻率法标定的破碎区宽度略大于直接按电视钻孔观测得到的宽度。断面B各钻孔破裂分区范围(内外半径)见表2。

表 2 断面B各钻孔破裂分区范围

Table 2 Disintegrated zones for bores in section B m

钻孔号	第1破碎区		第2破碎区		第3破碎区		第4破碎区	
	内径	外径	内径	外径	内径	外径	内径	外径
B-1	2.5	5.4	6.1	7.3	7.90	8.60	10.1	10.55
B-2	2.5	5.7	6.2	7.2	8.00	8.60	10.5	11.50
B-3	2.5	5.5	6.5	7.5	8.20	9.50	10.1	11.40
B-4	2.5	5.9	6.4	7.9	9.60	10.00	10.8	11.80
B-5	2.5	6.1	6.8	7.1	7.90	8.60	9.5	10.90
平均	2.5	5.7	6.4	7.4	8.32	9.06	10.2	11.23

从图7可以看出，断面C围岩破碎也可分为4个区，各破碎区的具体内外半径见表3所示。

表 3 C断面各钻孔破裂分区范围

Table 3 Disintegrated zones for bores in section C m

钻孔号	第1破碎区		第2破碎区		第3破碎区		第4破碎区	
	内径	外径	内径	外径	内径	外径	内径	外径
C-1	2.5	5.5	6.6	7.5	8.3	9.1	(10.5)	(11.4)
C-2	2.5	5.4	6.0	7.0	7.8	8.9	[11.5]	[12.0]
C-3	2.5	6.0	7.2	8.2	[8.9]	[9.4]	10.5	11.3
C-4	2.5	6.1	[7.1]	[8.0]	8.6	9.6	10.0	11.0
C-5	2.5	4.7	5.9	6.4	6.9	8.9	[10.5]	[10.7]
平均	2.50	5.54	6.56	7.42	8.10	9.18	10.60	11.28

注："( )"内数值表示推测值，"[ ]"内数值表示与钻孔电视观测结果有差异。

用电阻率确定的断面C各钻孔围岩破裂区分布

与钻孔电视观测结果大部分类似,少部分有差异。对差异分析如下:

(1) 钻孔 C‑1 在深 6.5 m 附近堵孔,第 4 破碎分区依据相邻孔破碎分区推测得到。

(2) 钻孔 C‑2 在深度 9.0~9.5 m 范围出现了电阻率突变,而钻孔电视观测见图 8 所示。可以看到表面粗糙,加上电阻率增大,判断围岩内部存在小的裂隙。

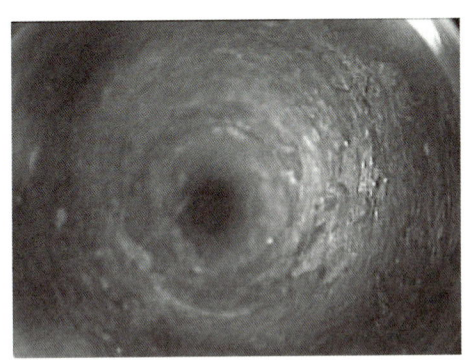

图 8 钻孔 C‑2 在 9.0 m 深附近电视观测图
Fig.8 TV video around x = 9.0 m in borehole C‑2

(3) 钻孔 C‑3 在 6.5~7.0 m 深处出现电阻率突变,其钻孔电视观测情况见图 9。可以看出,这一段画面恰好比较模糊,孔内雾气重,粉尘多,说明从孔壁冒出了温度较高的气体或水汽,影响了观测效果,从钻孔电视很难判别围岩是否破裂,但根据电阻率突变,可以判断存在细小裂隙。

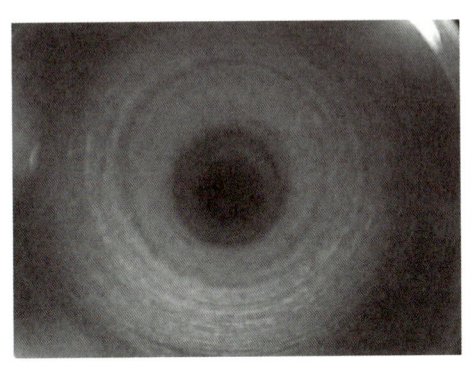

图 9 钻孔 C‑3 在 6.5 m 深附近钻孔电视观测图
Fig.9 TV video around x = 6.5 m in borehole C‑3

(4) 钻孔 C‑4 在 4.5~5.5 m 深处出现电阻率突变,其钻孔电视观测情况见图 10。但这一段画面十分模糊,雾蒙蒙,说明从孔壁冒出了大量的温度较高的气体或水汽,影响了观测所致。从钻孔电视很难判别围岩是否破裂,但根据电阻率突变,也可以

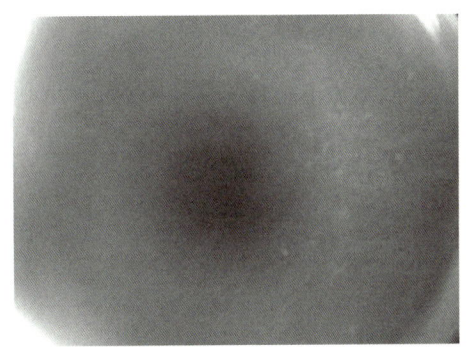

图 10 钻孔 C‑4 在 5.0 m 深附近电视观测图
Fig.10 TV video around x = 5.0 m in borehole C‑4

判断此处存在小裂隙。

(5) 钻孔 C‑5 在 8.0~8.5 m 深处出现电阻率突变。由于钻孔电视测量时,在深度 7 m 左右堵塞了一个小石块,加上钻孔电视探头较大和脆弱,没有通过,所以没有观测到钻孔端部的裂隙。在电阻率测量时,将该石块推入到了孔底,测量获知 8.0~8.5 m 范围电阻率突变,判断存在裂隙。

由上述分析可知,大部分测试结果与钻孔电视基本吻合,由于钻孔电视的固有缺陷和围岩裂隙对电阻率的敏感性大的特点,对少部分不相吻合之处建议遵从电阻率法的测试结果。用电阻率探测法进行围岩破裂分区,是对钻孔电视观测方法的有力补充,弥补了钻孔电视在观测过程中的不足,对提高分析结果的准确性十分有利。

5.2 综合分析

综合 2 个断面各破裂分区的数据,得到了该巷道围岩破裂带的平均内外半径、平均中心半径等参数(见表 4)。根据表 4 可绘制出监测断面围岩平均分区破裂分布图(见图 11)。

表 4 中数据显示,监测断面围岩分区破裂带中,第 1 个破裂分区带平均宽度最大,达到 3.12 m,其他破裂带的宽度有较小的趋势,第 2,3,4 破裂分区带的平均宽度分别为 0.93,0.91 和 0.86 m。根据每个破裂分区中线半径 r_{cn} 与巷道半径 r_0(本文取 r_0 = 2.5 m)的比值,作出该比值与分区破裂号之间的关系曲线,如图 12 所示。通过数学回归,可以获得如下关系式:

$$r_{cn} = 0.86(1+n)r_0 \quad (n = 1, 2, 3, 4) \quad (3)$$

表 4 中列出了利用式(3)计算的结果与试验观测结果的比较(表中误差=[(式(3)计算值−中心半径均值)/中心半径均值]×100%),最大误差为 7.13%。由

表 4 监测断面围岩破裂分区带平均半径

Table 4 Observation sections disintegrated zones average radius   m

取值类型	第1破碎区		第2破碎区		第3破碎区		第4破碎区	
	内径	外径	内径	外径	内径	外径	内径	外径
断面 B 均值	2.50	5.70	6.40	7.40	8.32	9.06	10.20	11.23
断面 C 均值	2.50	5.54	6.56	7.42	8.10	9.18	10.60	11.28
中心半径均值	4.06		6.95		8.67		10.83	
式(3)	4.30		6.45		8.60		10.75	
误差/%	5.91		−7.13		−0.75		−0.74	

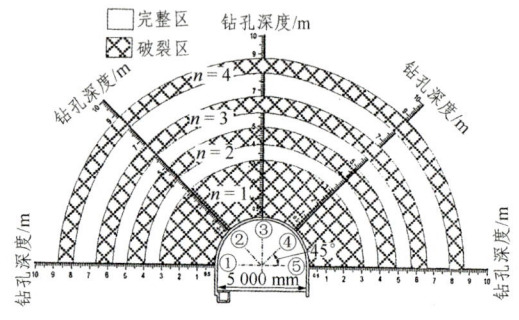

图 11 监测断面围岩分区破裂平均分布图

Fig.11 Zonal disintegration average distribution of different sections

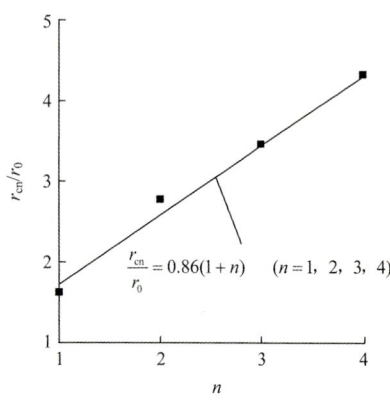

图 12 破裂带中线半径 r_{cn} 与巷道半径 r_0 的比值

Fig.12 Ratio of disintegrated zones' midline radius r_{cn} to roadway radius r_0

此可见，利用式(3)估计试验点分区破裂带半径是可行的。在今后大量的试验结果累积之后，可对式(3)进行修改，进一步加以推广使用。

6 结 论

淮南丁集煤矿西部采区南运输大巷埋深已达 955 m，已显现出深部开挖的特征。本文在近千米埋深的巷道中选取了2个监测断面，采用围岩电阻率测试方法，得到了围岩分区破裂分区图，并与钻孔电视观测结果进行了对比分析，其结果较吻合，可以得出以下结论：

(1) 在岩石破裂处，岩石的电阻率明显偏高，完整岩石的电阻率值较低，用围岩电阻率法进行围岩破裂分区测试，是可行的。围岩破裂区电阻率基准值，可根据每一钻孔测量点的平均值确定，计算时注意剔除特异点，大于基准值的定为破裂区。

(2) 用围岩电阻率划定的破裂分区与钻孔电视观测结果进行了对比，重点分析了几个有差异的地方。由分析可知，围岩电阻率法是对钻孔电视观测方法的有力补充，弥补了钻孔电视在观测过程中的不足，能有效提高测试结果的准确性。

(3) 淮南丁集煤矿 955 m 深的西部采区南运输大巷，围岩呈分区破裂状态，有4个破裂分区带。第1个破裂分区带平均宽度最大，达到3.12 m，第2~4个破裂分区带的宽度有减小的趋势，平均宽度分别为 0.93，0.91 和 0.86 m。

(4) 围岩分区破裂带的半径与巷道半径基本呈线性关系。进一步分析获得了破裂分区带中线半径 r_{cn} 随巷道半径和分区号 n 的变化规律，可用表达式 $r_{cn} = 0.86(1+n)r_0$ ($n = 1, 2, 3, 4$) 来表达。

(5) 本文推导的分区破裂带中线半径经验公式，最大误差小于 7.13%，在今后大量的围岩分区破裂测试结果累积之后，可对该经验公式进行修改，进一步推广使用。

参考文献(References)：

[1] 钱七虎. 深部岩体工程响应的特征科学现象及"深部"的界定[J]. 东华理工学院学报，2004，27(1)：1 - 5.(QIAN Qihu. The characteristic scientific phenomena of engineering response to deep rock mass and the implication of deepness[J]. Journal of East China Institute of Technology，2004，27(1)：1 - 5.(in Chinese))

[2] 顾金才，顾雷雨，陈安敏，等. 深部开挖洞室围岩分层断裂破坏机制模型试验与分析[J]. 岩石力学与工程学报，2008，27(3)：433 - 438.(GU Jincai，GU Leiyu，CHEN Anmin，et al. Model test study on mechanism of layered fracture within surrounding rock of tunnels in deep stratum[J]. Chinese Journal of Rock Mechanics and Engineering，2008，27(3)：433 - 438.(in Chinese))

[3] КУРЛЕНЯ М В，ОПАРИН В Н. Лроблемы нелинейной геомеханики[J]. Ч. I. Фтпрпи，1999，(3)：12－23.

[4] SHEMYAKIN E I，KURLENYA M V，OPARIN V N，et al. Zonal disintegration of rocks around underground workings，part IV：practical applications[J]. Journal of Mining Science，1989，25(4)：297－302.

[5] METLOV L S，MOROZOV A F，ZBORSHCHIK M P. Physical foundations of mechanism of zonal rock failure in the vicinity of mine working[J]. Journal of Mining Science，2002，38(2)：150－155.

[6] 周小平，钱七虎. 深埋巷道分区破裂化机制[J]. 岩石力学与工程学报，2007，26(5)：877－885.(ZHOU Xiaoping，QIAN Qihu. Zonal fracturing mechanism in deep tunnel[J]. Chinese Journal of Rock Mechanics and Engineering，2007，26(5)：877－885.(in Chinese))

[7] 李英杰，潘一山，章梦涛. 深部岩体分区碎裂化进程的时间效应研究[J]. 中国地质灾害与防治学报，2006，17(4)：119－122.(LI Yingjie，PAN Yishan，ZHANG Mengtao. Time effect on zonal disintegration process of deep rock mass[J]. The Chinese Journal of Geological Hazard and Control，2006，17(4)：119－122.(in Chinese))

[8] SHEMYAKIN E I，FISENKO G L，KURLENYA M V，et al. Zonal disintegration of rocks around underground workings，part I：data of in-situ observations[J]. Journal of Mining Sciences，1986，22(3)：157－168.

[9] 涂敏. 潘谢矿区采动岩体裂隙发育高度的研究[J]. 煤炭学报，2004，29(6)：641－645.(TU Min. Study on growth height of separation fracture of mining rock in Panxie Area[J]. Journal of China Coal Society，2004，29(6)：641－645.(in Chinese))

[10] 刘泉声，张华，林涛. 煤矿深部岩巷围岩稳定与支护对策[J]. 岩石力学与工程学报，2004，23(21)：3 732－3 737.(LIU Quansheng，ZHANG Hua，LIN Tao. Study on stability of deep rock roadways in coal mines and their support measures[J]. Chinese Journal of Rock Mechanics and Engineering，2004，23(21)：3 732－3 737.(in Chinese))

[11] 嵇艳鞠，林君，朱凯光，等. 利用瞬变电磁技术进行地下水资源勘察[J]. 工程勘察，2005，(3)：64－67.(JI Yanju，LIN Jun，ZHU Kaiguang，et al. Underground water prospecting by transient electromagnetic method[J]. Geotechnical Investigation and Surveying，2005，(3)：65－67.(in Chinese))

[12] 张流，黄建国，高平. 水对岩石变形过程中电阻率变化的影响[J]. 地震，2003，23(1)：8－14.(ZHANG Liu，HUANG Jianguo，GAO Ping. Influence of water on electric resistivity of deforming rock samples[J]. Earthquake，2003，23(1)：8－14.(in Chinese))

[13] GLOVER P W J. Modeling the stress-strain behavior of saturated rocks undergoing triaxial deformation using complex electrical conductivity measurements[J]. Surveys in Geophysics，1996，17(3)：307－330.

[14] 邓少贵，范宜仁，段兆芳，等. 多温度多矿化度岩石电阻率试验研究[J]. 石油地球物理勘探，2000，35(6)：763－767.(DENG Shaogui，FAN Yiren，DUAN Zhaofang，et al. Experiment study of rock resistivity with multi-temperature and multi-salinity[J]. Oil Geophysical Prospecting，2000，35(6)：763－767.(in Chinese))

[15] TOMECKA-SUCHON S，RUMMEL F. Fracture-induced resistivity changes in granite[J]. International Journal of Rock Mechanics and Mining Sciences，1992，29(6)：583－588.

[16] 陆阳泉，刘建毅，梁子斌. 受载条件下大型灰岩样品的电阻率前兆特征[J]. 华南地震，1998，18(3)：21－27.(LU Yangquan，LIU Jianyi，LIANG Zibin. The resistivity precursor features of the large-scale limestone sample under loading condition[J]. South China Journal of Seismology，1998，18(3)：21－27.(in Chinese))

[17] 王云刚，王恩元，李忠辉，等. 受载煤岩电阻率变化的试验研究[OL]. http：//www.paper.edu.cn，2007.(WANG Yungang，WANG Enyuan，LI Zhonghui，et al. Experimental study on the electric resistivity of loaded coal samples[OL]. http：//www.paper.edu.cn，2007.(in Chinese))

[18] 李术才，王汉鹏，钱七虎，等. 深部巷道围岩分区破裂化现象现场监测研究[J]. 岩石力学与工程学报，2008，27(8)：1 545－1 553.(LI Shucai，WANG Hanpeng，QIAN Qihu，et al. In-situ monitoring research on zonal disintegration of surrounding rock mass in deep mine roadways[J]. Chinese Journal of Rock Mechanics and Engineering，2008，27(8)：1 545－1 553.(in Chinese))

多层软弱夹层边坡岩体稳定性及加固分析

许宝田[1,2]，钱七虎[1]，阎长虹[2]，许宏发[1]

(1. 解放军理工大学，江苏 南京 210007；2. 南京大学 地球科学与工程学院，江苏 南京 210093)

摘要：野外地质调查发现九顶山边坡岩体内存在3条规模较大的软弱夹层控制着岩体的稳定性。采用数值法对该边坡岩体的变形特征进行研究，发现直接开挖后沿软弱夹层将发生大的相对滑动，岩体最大水平位移位于 P_2 软弱夹层坡面附近，达 1.25 cm，相对滑动造成软弱夹层强度降低为残余强度，易造成边坡失稳。P_1，P_2 软弱夹层上剪应力最大值分别为 239.0，172.4 kPa，位于滑面中前部。当采用锚喷加固边坡时，锚杆穿过软弱夹层时轴力突然增大，表明3条软弱夹层均发生较大的变形，对边坡稳定较为不利，但锚杆加固效果明显，能较大地提高边坡稳定性，采用强度折减法计算得到加固后边坡稳定性系数为 1.65。结果表明，应用数值模拟技术，可以直观形象地反映出边坡变形及应力变化的全过程，从而对工程措施优化、信息化设计和施工起超前预报与辅助决策起到一定的作用。

关键词：岩石力学；软弱夹层；边坡；岩体；稳定性

中图分类号：TU 45；P 642.22　　**文献标识码**：A　　**文章编号**：1000 - 6915(2009)增2 - 3959 - 06

STABILITY AND STRENGTHENING ANALYSES OF SLOPE ROCK MASS CONTAINING MULTI-WEAK INTERLAYERS

XU Baotian[1,2], QIAN Qihu[1], YAN Changhong[2], XU Hongfa[1]

(1. PLA University of Science and Technology, Nanjing, Jiangsu 210007, China; 2. School of Earth Sciences and Engineering, Nanjing University, Nanjing, Jiangsu 210093, China)

Abstract: Three weak interlayers were found in the rock mass of Jiudingshan slope by outdoor investigation, which control the stability of the slope. The deformation characteristics of the slope rock mass are studied using numerical simulation. The results indicate that: (1) large relative slide will be induced by the direct excavation, and the maximum horizontal displacement, 1.25 cm, happens on the slope surface where the weak interlayer P_2 emerges; and (2) the relative deformation makes the strength of the weak interlayers reduce to be the residual one, which makes the slope more instable. The maximum shear stresses at the middle and front of the weak interlayers P_1 and P_2 are 239.0, 172.4 kPa respectively. When the slope is strengthened by anchors, the axial force increases rapidly near the weak interlayers, which indicates that the anchors make the stability of the slope be improved. The stability coefficient of the slope after reinforcement calculated by strength reduction method is 1.65. It is shown that the numerical simulation can give the variation process of deformation and stress in the slope, and it is helpful for engineering optimization, informational design and construction.

Key words: rock mechanics; weak interlayers; slope; rock mass; stability

1 引言

含软弱夹层岩体的破坏形式及其发展过程取决于组合系统的稳定性，系统的稳定性则与系统中夹层与围岩间的相互作用密切相关，而系统中岩层间的相互作用又影响着岩体破坏的发展。因此，岩层及其结构的破坏与系统的失稳是相互影响和相互制约的[1]。目前已有不少学者[1~9]对软弱夹层对边坡岩体稳定性的影响进行了研究。软弱夹层是岩体中的不连续面，由于其物理力学性质差，不论厚薄，都会给工程建设带来一系列问题，常成为地下洞室、边坡稳定、坝基、坝肩抗滑稳定等的控制性弱面。历史上许多工程的失事、失稳，究其原因，大多是由于沿着软弱夹层或软弱结构面发生位移量很大的滑动而造成的[10]。目前考虑软弱夹层对岩体稳定性影响的研究成果已经很多，但对岩体内部含多条软弱夹层时，岩体与软弱夹层相互作用导致的变形和稳定性问题则研究较少。本文针对具体边坡工程，运用 FLAC 数值模拟方法，对含多条软弱夹层岩体的应力分布和变形进行计算，并在此基础上，选取适当的边坡加固方法，最后采用强度折减法对含多条软弱夹层情况下加固后边坡岩体的稳定性进行计算，验证加固方案的效果。

2 工程概况

九顶山人工边坡是莱芜钢铁股份有限公司特钢厂扩建开坯车间征地需要而开挖形成的。车间长度方向与边坡走向一致，两者水平距离为 6.0 m。原自然坡度小于 30°，坡面未发现有软弱夹层出露，开挖后边坡最高约 35.0 m，坡角一般为 70°。边坡开挖后、厂房施工前发现该边坡存在较大安全隐患，需要对其进行稳定性评价，以便采取相应的加固措施，以策安全，且为下一阶段开山放坡工作提供指导。

坡体内主要地层为上古生界寒武系凤山组的石灰岩，局部覆盖松散第四系土薄层，自上而下具体为：

(1) 强风化石灰岩：青灰色，泥晶质结构，中厚层、薄层及板状构造。主要矿物成分为方解石，岩层产状为 28°∠22°。分布于边坡的顶部。

(2) 中风化石灰岩：青灰色，泥晶质结构，中厚～薄层。主要矿物成分为方解石，岩层产状为 28°∠22°，$RQD = 50\sim60$。

(3) 风化软弱夹层：主要成分为全风化泥质灰岩，一般厚为 5～10 cm，产状与岩层一致，并有东厚西薄、局部不连续的特征。本工程中具代表性的强～中风化泥质灰岩软弱层，经调查发现，在坡面共出露 P_1，P_2，P_3 三层(见图 1)。

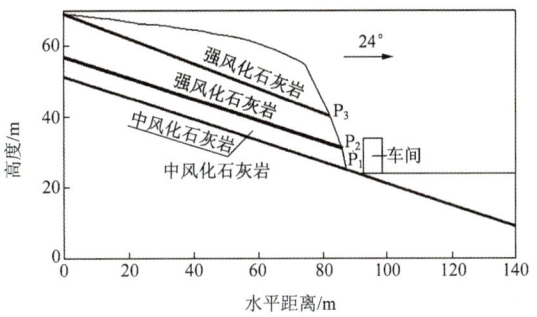

图 1 边坡工程地质剖面图

Fig.1 Geological cross-section of slope engineering

此外，在岩体内还发育有产状为 200°∠70°，128°∠70°的节理，坡面岩体破碎。

可以确定软弱层和节理面为控制边坡稳定性的主要结构面。在进行勘察工作前，边坡西侧坡顶岩体沿 P_3 软弱夹层发生块体滑动破坏，滑体体积达上千方，表明边坡岩体不稳定，处于临界状态。同时，根据勘察结果发现，软弱夹层强度参数虽然不高，但层面倾角不大，仅 20°左右，小于软弱夹层的内摩擦角，再加上软弱夹层本身有一定的黏聚力，从通常的极限平衡角度来看，边坡是稳定的，但实际上并非如此，而且坡顶也未增加荷载，故造成不稳定的原因主要包括：(1) 滑动面抗剪强度随岩体变形增大而下降；(2) 软弱夹层上的应力分布有变化。说明边坡岩体的变形在不断发展，塑性区不断扩大，若任其发展将可能导致岩体破坏的规模和程度不断加大。所以，要分析边坡的稳定性就应对岩体内的变形特征和滑动面上的应力分布状况作详细了解，以便尽快对岩体稳定性作出准确判断，及时采取加固措施，为加固方案提供依据。

3 计算方法和模型

在边坡稳定性分析方法中，极限平衡分析法固然有它的优点，但不能解决分析边坡应力和应变的问题。有限元在解决小变形方面有其优越性，但通常的边坡破坏多具弹塑性体特征，为大变形问题，而有限元在解决大变形方面不十分方便。FLAC程序在计算过程中允许材料发生屈服及大塑性变形，可以模拟岩土的力学性能，尤其在弹塑性分析、大变形分析方面有其独到的优点[11~13]。

从该边坡区域出露的地层岩性特征分析来看，软弱夹层的产状与岩层一致，力学强度低，为潜在滑动面和分离体边界。采用 FLAC 软件对图 1 所示的边坡剖面建立二维 FLAC 计算模型，模拟的边坡岩体范围为：水平向为 140 m，竖向为 75 m；采用四边形四节点单元进行剖分，水平向最大网格数 162 个，竖向最大网格数 138 个，软弱夹层处加密。岩体采用横观各向同性材料的节理化本构模型。由于软弱夹层比较薄，采用接触单元模拟，岩体和接触单元的层面切向均采用 Mohr-Coulomb 屈服准则，层面法向不抗拉。

软弱夹层力学参数根据现场原位大型剪切试验确定。勘察发现，3 条软弱夹层物质成分相同，加上现场条件限制，试验时假设 3 条软弱夹层力学性质相同，具体试验原理和试验方法见作者[14]的研究，该剪切试验设备在"马鞍山马三峰边坡'十一五'技改新建 500 万吨钢厂开山形成的人工边坡工程"边坡稳定性和加固方案设计研究中使用效果较好，测得的试验数据可靠，故在本工程中再次使用该设备对软弱夹层的力学参数进行测试，根据作者[14]研究结果，强度参数取屈服值，石灰岩力学参数在现场勘察和室内试验的基础上确定。岩体物理力学参数见表 1。

表 1 岩体的物理力学参数
Table 1 Physico-mechanical parameters of rock masses

地层	剪切模量/GPa	体积模量/GPa	天然容重/(kN·m^{-3})	黏聚力/kPa	内摩擦角/(°)	剪切刚度/(MPa·m^{-1})	法向刚度/(MPa·m^{-1})	抗拉强度/MPa
强风化石灰岩	0.37	1.1	25.0	100	27			0.0
中风化石灰岩	2.30	5.0	25.0	200	30			1.0
软弱夹层				40	24	65	400	0.0

模型结构除对地质原型进行了必要的简化处理外，基本上保持了与地质原型的一致性。模型边界采用铰支约束，即左右边界无水平位移，底边界无竖向位移。

4 边坡稳定性计算结果

首先假设不考虑软弱夹层的影响，只考虑边坡岩体完全为石灰岩体组成的情况。采用强度折减法[15]计算得到边坡的稳定性系数为 1.51(计算时对各地层和软弱夹层按同比例折减)，表明边坡稳定，并且有足够的安全储备，而实际情况是边坡已经发生破坏，并不稳定，说明软弱夹层的存在对边坡稳定性影响大，岩体整体性差，各向异性特征显著。

因此，必须对边坡的稳定性做出全面评价，以便为加固方案提供必要的参考依据，杜绝重大事故的发生。

因此，这里先对边坡岩体的位移和软弱夹层的应力分布特征进行计算，准确评价软弱夹层对边坡稳定性的影响程度，根据计算结果确定具体加固方案。

4.1 未加固岩体稳定性分析

根据计算结果得岩体水平变形等值线见图 2，位移速度矢量图见图 3，在顺层岩体中变形从上部岩层开始逐步向深部岩层传递，当穿越软弱夹层时向下层的传递被削弱，其中 P_3 软弱夹层以上水平位移达 1 cm 的范围约为 P_2，P_3 软弱夹层之间水平位移变化幅值的 2 倍，P_1 软弱夹层由于接近坡脚，受坡前岩体的支撑作用，变形相对较小；边坡的变形以坡顶最为显著，随着边坡内变形的不断积累和向深层及坡脚的传递，在软弱夹层两侧水平位移等值线为折线，表明在该处水平位移值有突变，沿 3 条软弱夹层两侧岩体均产生不同程度的层间错动。

图 2 水平位移等值线(单位：m)
Fig.2 Isolines of horizontal displacement(unit：m)

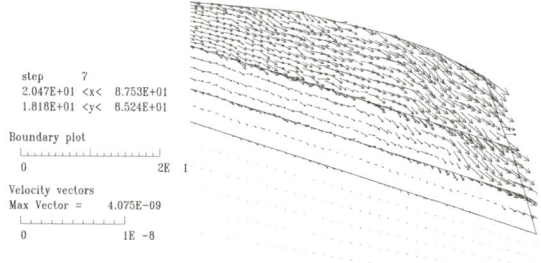

图 3 位移速度矢量图(单位：m/s)
Fig.3 Vectogram of displacement velocity(unit：m/s)

坡面最大水平位移位置在 P_3 软弱夹层坡面位置，最大值达 1.25 cm；P_1，P_2 软弱夹层坡面最大水平位移分别为 1.00，0.50 cm，说明软弱夹层控制岩体的水平位移，决定着岩体的稳定性状况。

以上计算结果表明，沿软弱夹层两侧岩体发生了较大的相对滑动，表明边坡岩体的稳定性决定于变形是否进一步发展和变形后的软弱夹层力学参数是否发生变化。

根据作者[14]研究结果及现场剪切试验,软弱夹层力学参数随剪切变形变化的同时,参数的取值也发生变化。根据试验结果,该软弱夹层的黏聚力 $c = 20\sim 60$ kPa(峰值为 60 kPa,屈服值为 40 kPa,残余强度为 20 kPa),内摩擦角 $\varphi = 20°\sim 26°$(峰值为 26°,屈服值为 24°,残余值为 20°)。为了更准确地分析软弱夹层对边坡的稳定性的影响程度,采用强度折减法[15]计算软弱夹层在不同强度参数下的边坡稳定性系数,见表 2。

表 2 软弱夹层在不同强度参数下的边坡稳定性系数
Table 2 Stability coefficients of slopes containing weak interlayers with different strength parameters

$\varphi /(°)$	边坡稳定性系数				
	$c=20$ kPa	$c=30$ kPa	$c=40$ kPa	$c=50$ kPa	$c=60$ kPa
20	0.92	0.96	0.99	1.02	1.04
22	0.97	1.01	1.03	1.05	1.06
24	1.01	1.03	1.05	1.06	1.07
26	1.04	1.05	1.06	1.07	1.08

根据稳定性系数计算结果,尽管强度参数在较大范围内变化,但稳定性系数仅为 0.92~1.08,非常接近临界状态,表明边坡不稳定或缺少很大的安全储备,一旦受不利因素影响,其破坏的可能性大。

图 3 表明边坡的变形仍在继续,岩体运动的方向平行于软弱夹层,由坡底向上,位移速度逐渐增加。软弱夹层现场大型剪切试验结果表明,软弱夹层发生错动之后,其强度降低为残余强度,黏聚力接近 20~30 kPa,内摩擦角为 20°~22°,边坡岩体处于临界状态,这对边坡稳定极为不利。若受降雨或外部动荷载等不利因素影响,则必然造成岩体的滑动破坏。

从图 4 可以看出,P_1,P_2 软弱夹层上的剪应力从前缘开始向滑面中部由小到大分布,从滑面中部到滑面后缘剪应力从大到小分布,滑面中部剪应力最大,剪应力最大位置分别位于(68.9 m, 27.1 m),(59.8 m, 36.0 m)处,最大值分别为 239.0,172.4 kPa,说明滑面中前部(此处与前文所提到的滑动破坏岩体的位置一致)剪应力较大,产生了应力集中现象,使该处剪应力更为接近软弱夹层的抗剪强度,一旦受不利因素影响,将从剪应力最大处首先发生屈服破坏。

图 4 软弱夹层上剪应力分布图(单位:Pa)
Fig.4 Distribution of shear stress on weak inter layers (unit:Pa)

图 5 的塑性区分布特征也证明了以上结论,图中"*"表示剪切屈服,"o"表示拉伸屈服。剪切屈服位置均位于软弱夹层附近,其中 P_3 软弱夹层附近也局部有拉伸屈服现象。由于边坡岩体中分布有走向与边坡走向一致的延伸性好的节理,若软弱夹层的错动进一步发展,沿塑性区首先破坏后,破坏面贯穿形成滑动面时岩体就可能沿软弱面发生顺层滑脱,并在软弱夹层两侧将岩石拉裂。因此,从坡脚先发生屈服后形成的破坏面逐渐向后、向上延伸,一旦屈服带贯通至地表便成为滑动面,导致边坡失稳。

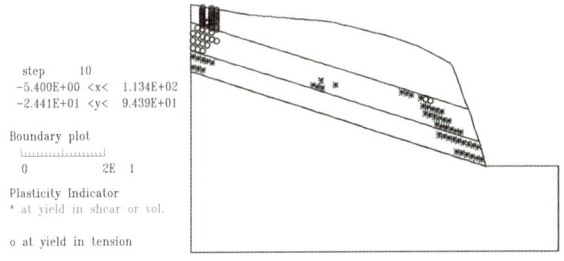

图 5 岩体塑性区分布特征
Fig.5 Distribution character of plastic zone in rock mass

以上分析表明,受软弱夹层影响,边坡水平变形大,岩体沿软弱夹层发生层间错动,导致层面抗剪强度下降,同时,由于软弱夹层中前部的应力集中现象,均导致了边坡稳定性的降低,所以,边坡的破坏过程是沿着软弱夹层的局部首先发生屈服并导致屈服区附近岩石拉裂,随着岩体位移的增大,破坏区也逐渐加大,最终导致边坡失稳,所以边坡实际上不安全,必须要采取加固措施。

4.2 岩体加固效果分析

根据边坡实际状况,拟采用锚喷法加固,即采用全长黏结型锚杆+表面挂网喷射素混凝土加固法。锚杆孔径 110 mm,弹性模量 210 GPa。锚杆锚固段必须深入到软弱夹层下的岩体内并有足够的长度,以满足承载力要求。因此，锚杆总长度应达到 16～24 m，边坡上部抗滑锚杆长度应在 16 m 以上，坡底锚杆长度应为 14～22 m。确定锚杆间距的主要因素是锚固强度需求和锚杆的相互作用。通过对锚杆的相互作用和锚固强度的研究，锚杆间距为 2.0～2.5 m 时，锚杆的锚固效果最佳[16]。

计算时不考虑表面钢筋网和混凝土的加固作用(即钢筋网只考虑起护面作用)，根据多次反复试算，确定锚杆水平、垂直间距均为 2 m，整个边坡体内由上到下共布置 14 道锚杆，锚杆下倾 15°安放，计算得到的锚杆长度和轴力图见图 6。根据计算结果，考虑 2.0 的安全系数，杆体采用 $2\phi 28$ mm 可满足抗拉断条件，在锚杆穿过软弱夹层时轴力突然增大，锚杆加固效果明显，能较大地提高边坡稳定性，也说明岩体在软弱夹层附近有较大的变形，导致锚杆轴力增大。

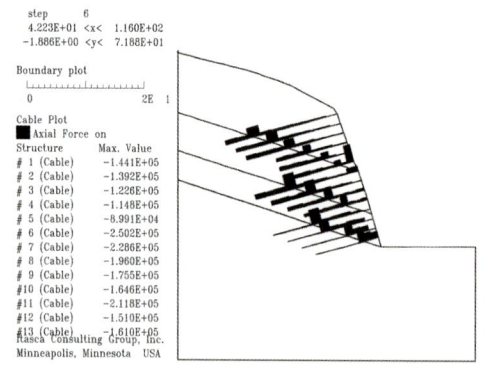

图 6 锚杆轴力分布图(单位：N)

Fig.6 Diagram of axial forces distribution in anchors(unit：N)

图 7 表明，加固后边坡岩体塑性区范围很小。在近坡面位置，仅在 P_2 软弱夹层附近有小范围达到剪切塑性状态，可能引起小型块体崩塌，但对边坡的整体稳定性并不构成影响。根据经验，坡面钢筋网和喷射混凝土完全可以保障其不发生破坏。

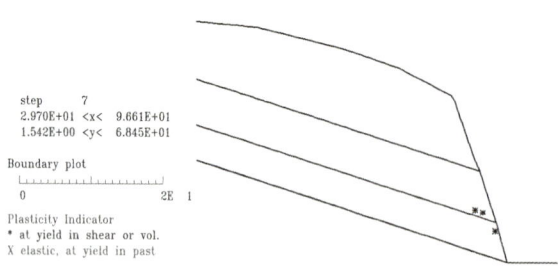

图 7 锚固后岩体塑性区分布

Fig.7 Distribution of plastic zone after reinforcement by anchors

锚喷加固提高了岩体的整体性，故采用强度折减法计算加固后边坡整体稳定性系数为 1.65，表明采取锚喷加固措施后边坡整体稳定并有足够的安全储备。

5 结 论

经过以上分析，得出如下结论：

(1) 对含多条软弱夹层的岩石边坡，直接开挖后沿软弱夹层将发生大的相对滑动变形。当边坡稳定性接近临界状态时，相对滑动造成软弱夹层强度降低为残余强度，易造成边坡失稳。

(2) 由于软弱夹层发生较大的剪切变形，造成抗剪强度发生变化，根据试验得到的软弱夹层强度参数变化范围计算得到的边坡稳定性系数均接近于临界状态，证明受 3 条软弱夹层影响边坡不稳定或缺乏足够的安全储备。

(3) 软弱夹层上的剪应力分布具有中前部高、前后低的特征，剪切屈服位置均位于软弱夹层附近，从坡脚首先发生屈服后形成的破坏面逐渐向后、向上延伸，一旦屈服带贯通至地表便成为滑动面，导致边坡失稳。

(4) 采用锚喷加固后，在软弱夹层附近锚杆轴力迅速增大，表明锚杆具有较好的加固作用，能提高岩体的整体性，对提高边坡的稳定性效果明显，同时也进一步证明了几条软弱夹层对岩体稳定性的影响是很大的。

(5) 模拟结果表明，应用数值模拟技术研究边坡的变形和破坏特征，可以直观形象地反映出边坡变形及应力变化的全过程。各种状态下的仿真可以为设计提供较好的反馈，从而对工程措施优化、信息化设计和施工起超前预报与辅助决策的作用。

参考文献(References)：

[1] 张顶立，王悦汉，曲天智. 夹层对层状岩体稳定性的影响分析[J]. 岩石力学与工程学报，2000，19(2)：140－144.(ZHANG Dingli, WANG Yuehan, QU Tianzhi. Influence analysis of interband on stability of stratified rock mass[J]. Chinese Journal of Rock Mechanics and Engineering, 2000, 19(2): 140–144.(in Chinese))

[2] 刘文方，隋严春，周菊芳，等. 含软弱夹层岩体边坡的突变模式分析[J]. 岩石力学与工程学报，2006，25(增 1)：2 663－2 669.(LIU Wenfang, SUI Yanchun, ZHOU Jufang, et al. Catastrophe analysis of rock mass slope with weak intercalated layers[J]. Chinese Journal of Rock Mechanics and Engineering, 2006, 25(Supp.1): 2 663–2 669.(in

Chinese))

[3] 赵永红，杨振涛. 含软弱夹层岩石材料的损伤破坏过程[J]. 岩石力学与工程学报，2005，24(13)：2 350－2 355.(ZHAO Yonghong，YANG Zhentao. Research on fracturing around cemented slot in rock specimen[J]. Chinese Journal of Rock Mechanics and Engineering，2005，24(13)：2 350－2 355.(in Chinese))

[4] 金 峰，邵 伟，张立翔，等. 模拟软弱夹层动力特性的薄层单元及其工程应用[J]. 工程力学，2002，19(2)：36－40.(JIN Feng，SHAO Wei，ZHANG Lixiang，et al. A thin-layer element for simulation of static and dynamic characteristics of soft interlayer and its application[J]. Engineering Mechanics，2002，19(2)：36－40.(in Chinese))

[5] 黎剑华，颜荣贵，陈寿如，等. 高等级公路缓斜陡节理边坡的复合破坏机制与治理对策[J]. 岩石力学与工程学报，2001，20(3)：415－418.(LI Jianhua，YAN Ronggui，CHEN Shouru，et al. Mechanism and control method of complex landslide of steep cut slope with brandy-inclined joints in high-grade highway[J]. Chinese Journal of Rock Mechanics and Engineering，2001，20(3)：415－418.(in Chinese))

[6] 陈静曦，章 光，袁从华，等. 顺层滑移路堑边坡的分析和治理[J]. 岩石力学与工程学报，2002，21(1)：48－51.(CHEN Jingxi，ZHANG Guang，YUAN Conghua，et al. Analysis and treatment of bedding-slip cut slope[J]. Chinese Journal of Rock Mechanics and Engineering，2002，21(1)：48－51.(in Chinese))

[7] OKUBO C H. Rock mass strength and slope stability of the Hilina slump，KIlauea volcano，Hawai'i[J] Journal of Volcanology and Geothermal Research，2004，138(1/2)：43－76.

[8] 任光明，聂德新，米德才，等. 软弱层带夹泥物理力学特征的仿真研究[J]. 工程地质学报，1999，7(1)：65－71.(REN Guangming，NIE Dexin，MI Decai，et al. A simulation study of physico-mechanical properties of intercalated gouge in layered weak zones[J]. Journal of Engineering Geology，1999，7(1)：65－71.(in Chinese))

[9] 陈 镕，陈竹昌，薛松涛，等. 夹有软弱土层的层状场地对入射SH波的响应分析[J]. 岩石力学与工程学报，1999，18(2)：161－165.(CHEN Rong，CHEN Zhuchang，XUE Songtao，et al. The response analysis of strata with extremely soft layer to incident SH waves[J]. Chinese Journal of Rock Mechanics and Engineering，1999，18(2)：161－165.(in Chinese))

[10] 李克钢，许 江，李树春. 三峡库区岩体天然结构面抗剪性能试验研究[J]. 岩土力学，2005，26(7)：1 063－1 067.(LI Kegang，XU Jiang，LI Shuchun. Study on property of rock mass discontinuity shear strength in the Three Gorges reservoir area[J]. Rock and Soil Mechanics，2005，26(7)：1 063－1 067.(in Chinese))

[11] 郭军辉，阎长虹，许宝田，等. 马三峰土质边坡稳定性分析与评价[J]. 防灾减灾工程学报，2007，27(1)：101－106.(GUO Junhui，YAN Changhong，XU Baotian，et al. Analysis and evaluation on stability of Masanfeng soil slope[J]. Journal of Disaster Prevention and Mitigation Engineering，2007，27(1)：101－106.(in Chinese))

[12] Itasca Consulting Group Inc.. FLAC users manual(version 5.0)[R]. Minneapolis：Itasca Consulting Group Inc.，2005.

[13] 刘 波，韩彦辉. FLAC 原理、实例与应用指南[M]. 北京：人民交通出版社，2005.(LIU Bo，HAN Yanhui. Guideline for principle，example and application of FLAC[M]. Beijing：China Communications Press，2005.(in Chinese))

[14] 许宝田，阎长虹，陈汉永，等. 边坡岩体软弱夹层力学特性试验研究[J]. 岩土力学，2008，29(11)：3 077－3 081.(XU Baotian，YAN Changhong，CHEN Hanyong，et al. Experiment study of mechanical property of weak intercalated layers in slope rock mass[J]. Rock and Soil Mechanics，2008，29(11)：3 077－3 081.(in Chinese))

[15] 郑颖人，赵尚毅. 有限元强度折减法在土坡与岩坡中的应用[J]. 岩石力学与工程学报，2004，23(19)：3 381－3 388.(ZHENG Yingren，ZHAO Shangyi. Application of strength reduction FEM to soil and rock slope[J]. Chinese Journal of Rock Mechanics and Engineering，2004，23(19)：3 381－3 388.(in Chinese))

[16] 吴顺川，高永涛，王金安. 坡间路基挡土墙"双锚"建设方案评价及参数优化数值模拟[J]. 岩土工程学报，2006，28(3)：332－336.(WU Shunchuan，GAO Yongtao，WANG Jin'an. Assessment on double anchor construction project of subgrade retaining wall on slope and its parameters optimization with numerical simulation[J]. Chinese Journal of Geotechnical Engineering，2006，28(3)：332－336.(in Chinese))

点滴化学注浆技术加固土遗址工程实例

柴新军[1]，钱七虎[2]，杨泽平[1]，林重德[3]，松永和也[3]

(1. 东华理工大学 土木与环境工程学院，江西 抚州 344000；2. 解放军理工大学，江苏 南京 210007；
3. 佐贺大学 低平地研究中心，日本 佐贺 840-8502)

摘要：表面防风化问题一直是土遗址保护研究的重点和难点。将开发研制的点滴化学注浆装置应用于工程实际，对堂加 2# 土遗址表层进行化学加固，内容包括：室内模拟场地点滴注浆试验、浆液固化体三轴试验、微观结构扫描电镜(SEM)观察、现场模拟土遗址的点滴注浆试验、点滴化学注浆技术加固堂加 2# 土质古窑。试验结果表明：(1) 点滴化学注浆装置采用的气球密封注入前端构造，可有效解决注入过程中浆液外溢问题，并可在地下遗址顶部及其附近实现由下而上的化学注浆加固；(2) 化学加固剂(硅酸乙酯)通过改善土粒团块之间的胶结状态，可有效提高遗址土的黏聚力，对内摩擦角的影响不甚明显；(3) 点滴化学注浆技术可对预定厚度的遗址土进行有效加固，克服了喷涂工艺加固深度浅且易出现两张皮的现象；(4) 点滴化学注浆技术对堂加 2# 土质古窑的成功加固，验证了该技术的现场适宜性和可行性，对类似的土遗址固化处理具有借鉴和参考意义。

关键词：土力学；点滴注浆；化学加固；土遗址；防风化
中图分类号：TU 44；TV 543　　**文献标识码**：A　　**文章编号**：1000-6915(2009)增1-2980-06

CASE STUDY OF DRIP INJECTION OF CHEMICAL GROUTS IN EARTHEN RUINS REINFORCEMENT

CHAI Xinjun[1], QIAN Qihu[2], YANG Zeping[1], HAYASHI S[3], MATSUNAGA K[3]

(1. *School of Civil and Environmental Engineering*, *East China Institute of Technology*, *Fuzhou*, *Jiangxi* 344000, *China*;
2. *PLA University of Science and Technology*, *Nanjing*, *Jiangsu* 210007, *China*; 3. *Institute of Lowland Technology*,
Saga University, *Saga* 840-8502, *Japan*)

Abstract: Weathering prevention is one of the difficulties in earthen ruins conservation. The self-developed drip injection apparatus of chemical grouts is first applied to engineering practice to perform chemical reinforcement on the surface of Dougaeri earthen ruins No.2. The main contents include drip injection of chemical grouts in laboratory model ground, the strength characteristics of chemical grouted soil by triaxial tests, microstructures investigated by SEM observation, drip injection in field model earthen kiln, and reinforcement of Dougaeri earthen kiln No.2 by drip injection of chemical grouts. The test results show that: (1) the balloon cover injection tip specially developed for the drip injection tube can prohibits chemical grout flowing upward to the ground surface, in particular it can implement down-to-up chemical grouting on the top of underground earthen ruins due to its soft and tight attachment with the drill hole sides；(2) chemical grouts(ethyl silicate) can increase cohesion effectively, but it has little influence on the internal friction angle；(3) the thickness of chemical grouted soil by the drip injection apparatus is much larger than that of spraying chemical grouts on the surface of earthen sites；and (4) the field feasibility and adaptability of the drip injection apparatus has been verified by its successful application on strengthening Dougaeri earthen kiln No.2，which will benefit to preservation of other similar earthen ruins.

Key words：soil mechanics；drip injection；chemical grouting；earthen ruins；weathering prevention

1 引 言

土遗址是指以土作为主要建筑材料的人类历史上生产、生活等各种活动遗留下来的遗迹，是一种重要的文物资源。土遗址本身病害主要表现为土遗址表层土的风化酥粉、开裂、块状剥落和坑壁坍塌等病害。土遗址的保护加固是一个世界性的难题[1~3]。

土遗址的加固保护主要是解决两个方面的问题，一是表面防风化，二是整体稳定性。对稳定性的加固主要采用砌补、灌浆、锚杆锚固。锚杆材料主要有钢筋[4]、木锚杆[5]、和微型土钉[6]等。锚固技术在石质文物(如石窟、石像、摩崖危岩等)加固工程中的应用已渐趋成熟[7~11]，但在土遗址加固中的应用仍处于探索和研究阶段[1, 3~6, 12]。

土遗址表面防风化问题一直是土遗址保护研究的重点和难题[3, 13]。文物工作者尝试研制了多种防风化材料：无机材料、有机高分子材料及无机－有机复合材料，并取得了一定进展[14, 15]。如李最雄等[16~18]研制出一种特别适用于干旱地区土遗址保护的 PS 材料(即高模数的硅酸钾溶液)，并在西北地区大面积推广使用，成效显著。

这些防风化材料主要采用喷涂工艺进行表面防风化处理，但化学浆液在土遗址表层的渗透深度一般只有几个厘米，且在土遗址表层易形成一薄层硬壳而出现两张皮的现象[16~19]。传统的化学注浆技术可解决加固深度问题，但应用于土遗址表层加固时，注浆管存在一些缺陷和不足，限制了化学注浆技术的应用，主要表现在：其一，易出现浆液沿孔壁和注入管之间的间隙外溢至地表，降低了注入效果，且污损外观，常用的做法是用水泥浆和膨润土密封孔口间隙，但同样会在遗址表面留下污痕；其二，在地下遗址顶部难以进行由下而上的注浆固化[20]。

鉴于传统化学注浆管的不足，X. J. Chai 等[21~23]通过室内试验，开发研制出了点滴化学注浆技术装置。点滴化学注浆装置的关键部位是注入管采用了气球密封注入前端(BCIT)构造，可有效解决注入过程中浆液外溢的问题，更重要的是可实现在地下遗址顶部及其附近由下而上的注浆加固。本文首次将该点滴化学注浆装置应用于工程实际，对堂加 2#土遗址(日本九州地区)[24]表层进行化学加固。

2 点滴化学注浆装置及室内试验简介

2.1 点滴化学注浆装置

点滴化学注浆装置如图 1 所示，主要组成为气球密封注入前端(BCIT)、三叉管和点滴观察器。气球密封注入前端是该装置的核心部位。注浆管加气球密封结构的直径为 4 mm，注入孔直径设定为 5 mm。化学注浆时，将注浆管注入前端插入场地指定深度处，向气球密封结构注入空气，气球膨胀后可密封注入管与注入孔壁之间的间隙。注浆完成后，排出气球内空气可方便抽出气球密封注入前端，在遗址表面不留明显痕迹。注入孔可用遗址土混合适量的化学浆液填塞处理。室内多次试验表明，气球密封注入前端可有效密封间隙，解决浆液外溢问题。在地下遗址顶部注浆时，通过调节空气注入量可调节附着力的大小，使之能足够支撑注入前端结构的重量，实现由下而上化学注浆加固的目的。详细构造见相关研究[21~23]。

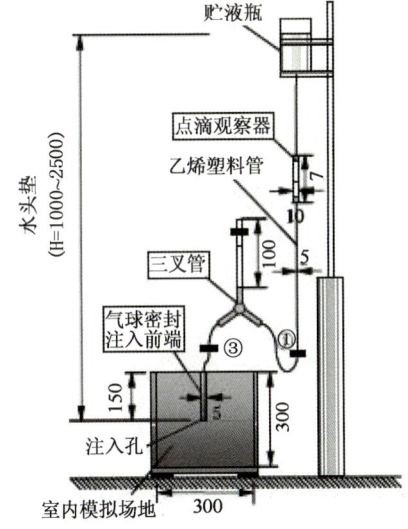

图1 点滴化学注浆装置示意图(单位：mm)

Fig.1 Schematic diagram of drip injection apparatus for chemical grouting(unit：mm)

2.2 室内模拟场地点滴注浆试验

室内模拟场地所用试样取自堂加 2#土质古窑(日本九州地区)附近的现场模拟窑迹。古窑位于丘陵地带，土质为花岗岩风化物，黏粒成分较多。按照 ASTM(美国材料与试验学会)D422 分类可定为 MH(高塑性粉土)，其物理力学性质指标见表1。

表1 堂加2#古窑遗迹土物理力学性质指标

Table 1 Physico-mechanical properties of Dougaeri No.2 earthen kiln

土粒相对密度 G_s	土粒百分含量/%			塑限 w_P/%	液限 w_L/%	现场密度 /(g·cm⁻³)
	黏粒 (<5 μm)	粉粒 (5~75 μm)	砂粒 (0.075~4.75 mm)			
2.71	34	12	54	37.4	54.8	1.21~1.31

室内模拟场地在PVC圆柱筒(高300 mm,直径300 mm)中通过击实成型。含水量为25%,干密度 ρ_d = 1.30 g/cm³,饱和度为50%~60%,模拟现场状态。根据岩土遗址加固保存的经验和日本地盘工学会的规定,通过系列室内试验和评价,硅酸乙酯被优先选定进行室内场地及现场模拟窑迹的点滴注入试验。

注入孔直径为5 mm,孔深150 mm。选定浆液在两个室内模拟场地中进行了点滴注入试验。注浆后模拟场地在室内标准条件下养护7 d(温度23 ℃,相对湿度65%);然后将试样浸入水中,流水冲掉未固化土体。化学浆液固化土体照片如图2所示,近似为直径200 mm的球体。经计算固结效率(注入浆液体积与固化土体体积之比)为1∶10~1∶11。浆液填充率(注入浆液体积与固化土体中总间隙体积之比)约为18%。

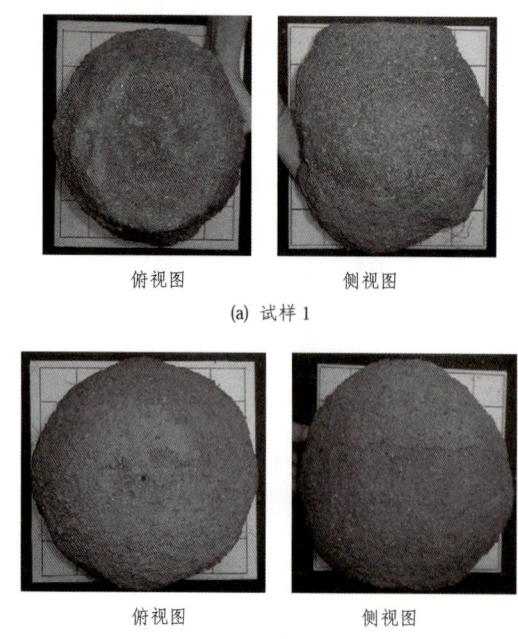

(a) 试样1

(b) 试样2

图2 室内模拟场地浆液固化土体照片

Fig.2 Photos of chemical grouted soil in laboratory model ground

3 浆液固化体特征

3.1 浆液固化体强度特征

进行三轴试验测定固化土体的强度参数(总应力法)。试样取自现场模拟遗址,用标准模型成型(ϕ 5 cm×10 cm)。试样分为两类。一类是试样中掺入填充率为18%的浆液(土A),另一类为未掺加浆液的天然遗迹土(土B)。所有试样静压成型为含水量26%,干密度1.30 g/cm³,近似模拟现场状况。

试样在试验室养护28 d后进行三轴试验,试验时测定的含水量约为4%。三轴试验选定的围压为30,60,100和200 kPa,轴向应变率为 10^{-2} min⁻¹。两种试样的偏应力-轴向应变关系如图3所示。图3表明在相同的围压下,土A比土B具有更高的抗剪强度。

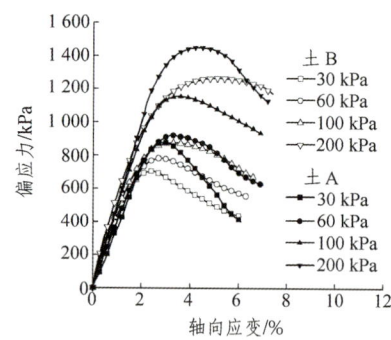

图3 试样(土A,B)偏应力-轴向应变关系

Fig.3 Deviator stress-axial strain relations of soils A and B

2种试样在 $t\text{-}s$ ($t=(\sigma_1-\sigma_3)/2$, $s=(\sigma_1+\sigma_3)/2$) 应力空间的强度包络线和计算的强度参数见图4。试验结果表明化学加固剂(硅酸乙酯)可以有效提高遗址土体的黏聚力,但对内摩擦角的影响不甚明显。

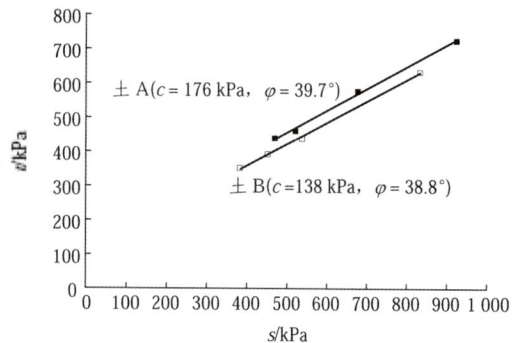

图4 试样(土A,B)在 $t\text{-}s$ 应力空间强度包络线

Fig.4 Failure envelope of total stress in $t\text{-}s$ stress space for soils A and B

3.2 浆液固化体微观结构特征

取浆液固化土(土A)和天然遗址土(土B)样品进行扫描电镜(SEM)观察，观察结果照片见图5，其中图5(b)为硅酸乙酯在空气中风干后晶体的扫描电镜照片。对比图5中的3组照片可以看出，浆液固化土体中土粒团块较小且分布较均匀，土粒团块表面有硅酸乙酯晶体黏结，这是浆液固化体黏聚力提高的微观机制。

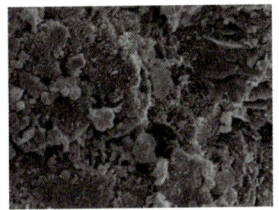

(a) 天然遗迹土(土B, ×2000, 加速电压 15 kV, 宽 66.0 μm)

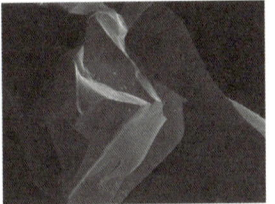

(b) 硅酸乙酯结晶体(×75, 加速电压 15 kV, 宽 1.76 mm)

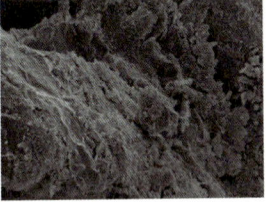

(c) 浆液固化土(土A, ×2000, 加速电压 15 kV, 宽 66.0 μm)

图 5 　扫描电镜(SEM)照片(土 A 和 B)
Fig.5　SEM photos for soils A and B

4　现场模拟窑迹点滴注入试验

为验证开发的点滴化学注浆装置的现场有效性和可行性，在堂加 2# 土质古窑迹附近的斜坡上开挖了长约 4.6 m，深、宽各 0.8 m 的土槽模拟窑迹。采用点滴注浆装置在指定部位实施化学注浆(硅酸乙酯)。注入孔呈正三角形布置，注入孔中心间距为 300 mm。注入孔设置为直径 5 mm，孔深 200 mm，注入压力设定为 20 kPa，最大注入时间设定为 6 h。

注入点总孔数为 18 个，注入量 - 时间关系如图 6 所示。根据注入关系曲线，注入效果可以分为

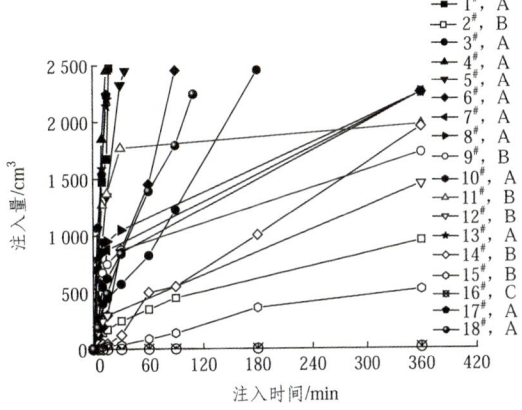

图 6 　现场模拟窑迹注入量 - 注入时间关系
Fig.6　Injection volume-injection time relation in field model kiln

3 个等级：A，B 和 C。在设定注入时间内，注入量超过 2 000 cm^3 定为等级 A，为注入效果良好；注入量为 200～2 000 cm^3 定为等级 B，为注入效果一般；注入量小于 200 cm^3 定为等级 C，为注入效果差。根据室内注入固结效率，预期固结体形象图及评定等级如图 7 所示。

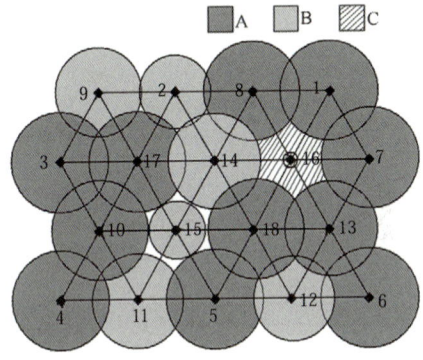

图 7 　现场模拟窑迹预期浆液固化土示意图
Fig.7　Imagine of expected solidified soil in field model kiln

现场模拟窑迹实施化学注浆一年后，对固化部位开挖评定注入效果。浆液固结土体表面硬度指数明显高于周边未固结土体。固结土体近似呈球状，见图8，同预期固结形象示意图(图7)基本相似。

5　点滴化学注浆装置加固堂加 2# 古窑

2004 年 9 月～2005 年 11 月，应用开发的点滴化学注浆装置及选定的化学浆液(硅酸乙酯)对堂加 2# 古窑迹表面实施了点滴化学注浆加固。注入孔设定为孔径 5 mm，孔深 200～350 mm。注入孔呈正

图 8 现场模拟窑迹浆液固化土照片
Fig.8 Photo of chemical grouted soil in field model kiln

三角形布置,中心间距为 300 mm。注入压力设定为 15~25 kPa。最大注入时间设定为 4 h。

注入孔总数目为 593 孔,注入总量为 1 302 680 cm³。现场点滴注入照片见图 9。依据现场模拟窑迹注入效果等级分类,注入效果为等级 A,B 和 C 的注入孔数目分别为 278,256 和 49,平均每孔注入量约为 2 197 cm³。典型的注入量 - 注入时间关系如图 10 所示。对比图 10 和 6 注浆曲线可见,点滴化学注浆加固堂加 2#窑效果良好。化学注浆完

图 9 微型化学注浆技术固化堂加 2 号古窑
Fig.9 Solidification of Dougaeri kiln No.2 with drip injection of chemical grouts

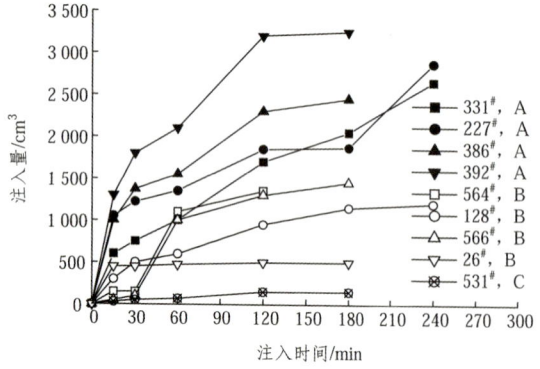

图 10 堂加 2#古窑典型的注入量 - 注入时间关系
Fig.10 Typical injection volume-time relation for Dougaeri kiln No.2

成后,注入孔采用浆液混合遗址细黏土填堵,遗址表面无明显痕迹(图 11),达到了做旧如旧的目的,至今防风化效果良好。

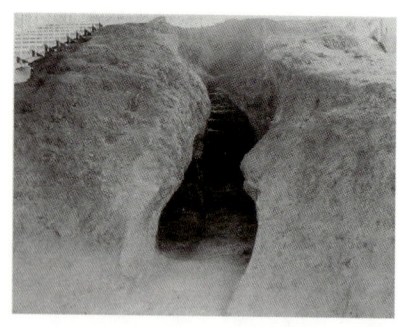

图 11 化学注浆表层加固后堂加 2#古窑外观照片
Fig.11 Photo of Dougaeri kiln No.2 after fulfillment of chemical grouting with drip injection apparatus

6 结 论

本文首次将自主开发研制的点滴化学注浆技术应用于工程实际,对堂加 2#土质古窑表层进行了化学加固,结论归纳如下:

(1) 点滴化学注浆装置的主要组成为:气球密封注入前端、三叉管和点滴注入观察器。气球密封注入前端构造可有效解决注浆过程中浆液外溢问题,特别是在地下遗址顶部及附近可实现由下而上的注浆加固。

(2) 化学浆液(硅酸乙酯)通过改善土粒团块之间的胶结状态,可有效提高遗址土的黏聚力,但对内摩擦角的影响不明显。

(3) 点滴化学注浆技术可对预定厚度的遗址土进行有效加固,克服了喷涂工艺加固深度浅且易出现两张皮的现象。

(4) 点滴化学注浆技术对堂加 2#土质古窑的成功加固,验证了该技术的现场适宜性和可行性,对类似的土遗址固化处理具有借鉴和参考意义。

参考文献(References):

[1] 黄克忠. 走向 21 世纪的中国文物科技保护[J]. 敦煌研究, 2000, (1): 5 - 9.(HUANG Kezhong. Technological protection of Chinese cultural relics tending towards 21st century[J]. Dunhuang Research, 2000, (1): 5 - 9.(in Chinese))

[2] 黄克忠. 任重而道远的莫高窟文化遗产保护[J]. 敦煌研究, 2006, (6): 200 - 202.(HUANG Kezhong. The important conservation of the cultural heritage site of Mogao Grottoes: the duty is sacred and heavy while the way ahead is long[J]. Dunhuang Research, 2006, (6): 200 - 202.(in Chinese))

[3] 孙满丽. 土遗址保护研究现状与进展[J]. 文物保护与考古科学, 2007, 19(4): 64 – 70.(SUN Manli. Research status and development of the conservation of earthen sites[J]. Sciences of Conservation and Archaeology, 2007, 19(4): 64 – 70.(in Chinese))

[4] 李最雄, 王旭东. 古代土建筑遗址保护加固研究的新进展[J]. 敦煌研究, 1997, (4): 167 – 172.(LI Zuixiong, WANG Xudong. The new progress of protection and reinforcement for the ancient earthen architecture site[J]. Dunhuang Research, 1997, (4): 167 – 172.(in Chinese))

[5] 孙满丽, 王旭东, 李最雄, 等. 木质锚杆加固生土遗址研究[J]. 岩土工程学报, 2006, 28(12): 2 156 – 2 159.(SUN Manli, WANG Xudong, LI Zuixiong, et al. Study on immature earthen sites reinforced with wood anchor[J]. Chinese Journal of Geotechnical Engineering, 2006, 28(12): 2 156 – 2 159.(in Chinese))

[6] 柴新军, 钱七虎, 罗嗣海, 等. 微型土顶微型化学注浆技术加固土质古窑[J]. 岩石力学与工程学报, 2007, 27(2): 347 – 353.(CHAI Xinjun, QIAN Qihu, LUO Sihai, et al. Historical earthen kiln reinforcement with micro-soil nailing and mini-chemical grouting techniques[J]. Chinese Journal of Rock Mechanics and Engineering, 2007, 27(2): 347 – 353.(in Chinese))

[7] 何燕, 李智毅. 关于河南灵泉寺石窟地质病害及整治方法的研究[J]. 岩土力学, 2000, 21(1): 56 – 59.(HE Yan, LI Zhiyi. A study of geological disease analyses and renovation methods in Lingquan Temple Grotto of Henan[J]. Rock and Soil Mechanics, 2000, 21(1): 56 – 59.(in Chinese))

[8] 李丽慧, 杨志法, 岳中琦, 等. 龙游大型古洞室群变形破坏方式及加固方法研究[J]. 岩石力学与工程学报, 2005, 24(12): 2 018 – 2 028.(LI Lihui, YANG Zhifa, YUE Z Q, et al. Deformation and failure modes and reinforcement methods of ancient cavern group in Longyou County[J]. Chinese Journal of Rock Mechanics and Engineering, 2005, 24(12): 2 018 – 2 028.(in Chinese))

[9] 张旭辉. 土钉与锚杆的关系探讨[J]. 岩石力学与工程学报, 2002, 21(11): 1 744 – 1 746.(ZHANG Xuhui. Relations between soil nailing and bolting[J]. Chinese Journal of Rock Mechanics and Engineering, 2002, 21(11): 1 744 – 1 746.(in Chinese))

[10] 朱浮声, 郑雨天. 全长黏结式锚杆的加固作用分析[J]. 岩石力学与工程学报, 1996, 15(4): 333 – 337.(ZHU Fusheng, ZHENG Yutian. Analysis of reinforcement of full-length grouted bolts[J]. Chinese Journal of Rock Mechanics and Engineering, 1996, 15(4): 333 – 337.(in Chinese))

[11] 程良奎. 岩土锚固的现状与发展[J]. 土木工程学报, 2001, 34(3): 7 – 12.(CHENG Liangkui. Present status and development of ground anchorages[J]. China Civil Engineering Journal, 2001, 34(3): 7 – 12.(in Chinese))

[12] 李最雄. 丝绸之路古遗址保护[M]. 北京: 科学出版社, 2003.(LI Zuixiong. Conservation of ancient sites on the silk road [M]. Beijing: Science Press, 2003.(in Chinese))

[13] 周双林. 土遗址防风化保护概括[J]. 中原文物, 2003, (6): 78 – 83. (ZHOU Shuanglin. The protection general situation of anti-weathering of the earthen sites[J]. Culture Relics of Central China, 2003, (6): 78 – 83.(in Chinese))

[14] 单玮, 张康生, 刘世勋. 秦俑一号坑碳化遗迹的加固[C]// 秦始皇兵马俑博物馆编. 秦俑学研究. 西安: 山西人民教育出版社, 1996: 1 385 – 1 387.(SHAN Wei, ZHANG Kangsheng, LIU Shixun. The reinforcement of carbonization relic of Emperor Qin's Terrn-cotta Warriors and Horse 1 pit[C]// Emperor Qin's Terra-cotta Warriors and Horse Museum ed. Studies on Chin Yung. Xi'an: Shanxi People's Education Press, 1996: 1 385 – 1 387.(in Chinese))

[15] 张宗仁, 樊北平. 几处商秦土遗迹的保护[C]// 秦始皇兵马俑博物馆编. 秦俑学研究. 西安: 山西人民教育出版社, 1996: 1 379 – 1 383.(ZHANG Zongren, FAN Beiping. The conservation of some earthen sites of Shang and Qin dynasty[C]// Emperor Qin's Terrn-cotta Warriors and Horse Museum ed. Studies on Chin Yung. Xi'an: Shanxi People's Education Press, 1996: 1 379 – 1 383. (in Chinese))

[16] 李最雄, 王旭东, 张志军, 等. 秦俑坑土遗址的加固试验[J]. 敦煌研究, 1998, (4): 151 – 158.(LI Zuixiong, WANG Xudong, ZHANG Zhijun, et al. A reinforcement test of the earthen sites of Qin's Terra-cotta Army pit[J]. Dunhuang Research, 1998, (4): 151 – 158.(in Chinese))

[17] 李最雄, 王旭东, 田琳. 交河故城土建筑遗址的加固试验[J]. 敦煌研究, 1997, (3): 171 – 181.(LI Zuixiong, WANG Xudong, TIAN Lin. Experimentation of chemical consolidation on ancient earth-structure sites of Jiaohe[J]. Dunhuang Research, 1997, (3): 171 – 181. (in Chinese))

[18] 李最雄, 王旭东, 郝利民. 室内土建筑遗址的加固试验——半坡土建筑遗址的加固试验[J]. 敦煌研究, 1998, (4): 144 – 149.(LI Zuixiong, WANG Xudong, HAO Liming. A reinforcement test of indoor ancient earth—structure sites, Banpo sites[J]. Dunhuang Research, 1998, (4): 144 – 149.(in Chinese))

[19] 周双林, 王雪莹, 胡原, 等. 辽宁牛河梁红山文化遗址土体加固保护材料的筛选[J]. 岩土工程学报, 2005, 27(5): 567 – 570.(ZHOU Shuanglin, WANG Xueying, HU Yuan, et al. Selection of strengthening medium for archaeological site of Hongshan Culture at Niuheliang, Liaoning Province[J]. Chinese Journal of Geotechnical Engineering, 2005, 27(5): 567 – 570.(in Chinese))

[20] 坪井直道. 薬液注入工法の実際[M]. 东京: 鹿島出版会, 1977.

[21] CHAI X J. Development of earth sewing technique and its application to reinforcing Funasako historical kiln site[Ph. D. Thesis][D]. Saga, Japan: Saga University, 2005.

[22] HAYASHI S, CHAI X J, MATSUNAGA K, et al. Drip injection of chemical grouts: a new apparatus[J]. Geotechnical Testing Journal, American Society for Testing and Materials, 2006, 29(2): 108 – 116.

[23] MATSUNAGA K. Fundamental study on the techniques of soil remains by chemical grouting and evaluation method[M. S. Thesis][D]. Saga, Japan: Saga University, 2002.

[24] 築城町教育委員会. 船迫窯跡群保存整備基本計画報告書[R]. 福岡县: 築城町教育委員会, 1998.

Dynamic strength of rocks and physical nature of rock strength

Qihu Qian[1], Chengzhi Qi[2], Mingyang Wang[1]

Engineering Institute of Engineering Crops, PLA University of Science and Technology, Nanjing, 210007, China
School of Civil and Communication Engineering, Beijing Institute of Civil Engineering and Architecture, Beijing, 100044, China Received 18 December 2008; received in revised form 19 April 2009; accepted 26 May 2009

Abstract: Time-dependence of rock deformation and fracturing is often ignored. However, the consideration of the time-dependence is essential to the study of the deformation and fracturing processes of materials, especially for those subject to strong dynamic loadings. In this paper, we investigate the deformation and fracturing of rocks, its physical origin at the microscopic scale, as well as the mechanisms of the time-dependence of rock strength. Using the thermo-activated and macro-viscous mechanisms, we explained the sensitivity of rock strength to strain rate. These mechanisms dominate the rock strength in different ranges of strain rates. It is also shown that a strain-rate dependent Mohr-Coulomb-type constitutive relationship can be used to describe the influence of strain rate on dynamic rock fragmentation. A relationship between the particle sizes of fractured rocks and the strain rate is also proposed. Several time-dependent fracture criteria are discussed, and their intrinsic relations are discussed. Finally, the application of dynamic strength theories is discussed.

Key words: rock dynamics; deformation and fracturing; time-dependence; dynamic strength; criteria of fracturing

1 Introduction

Traditional strength theories are mainly concerned with the macroscopic deformation and fracturing of continuum materials. The time-dependence of the material strength is usually neglected. In those theories, the failure takes place when the combination of stresses or strains at one point in a solid reaches a limit value. The selection of stress and strain combinations and the determination of their limit values are the basis of particular strength theories.

Actually, failure processes of material generally take place over some time. As the failure of rock is

本文原载于《Journal of Rock Mechanics and Geotechnical Engineering》(2009 年)。

resulted from the nucleation, growth and coalescence of inherent microcracks progressively at a limited velocity, the macroscopic deformation and fracturing of materials are time-dependent. The strain rate sensitivity of strength and the incubation time of fracturing for rock material when the strength limit is reached are the typical examples of the time-dependence of material response.

Hence, the careful consideration of the time-dependence is essential to study the deformation and fracturing processes of materials, especially for those subject to strong dynamic loadings. Therefore, the time-dependence of the deformation and fracturing of rocks, its origin at the microscopic scale, and the mechanisms of rock dynamic strength are investigated in this paper.

2 Traditional strength theories

Traditional strength theories (or criteria) may be divided into five classes: (1) the maximum normal stress theory, (2) the maximum normal strain theory, (3) the maximum shear stress theory, (4) the maximum specific strain energy theory (Von Mises criterion), and (5) the Mohr-Coulomb (M-C) criterion. Among these strength theories, the M-C criterion, modified from the shear stress theory, is widely used in geotechnical engineering practice. It is noted that the Hoek-Brown criterion and the Drucker-Prager (D-P) criterion, which are also widely used in geotechnical engineering, are the modifications of the M-C criterion and the Von Mises criterion [1-3].

The above-mentioned criteria are applicable to special failure modes under a certain stress state. For example, the M-C criterion mathematically does not consider the influence of intermediate principal stress on the strength of material. It takes only shear and normal stresses into account on one shear plane. Therefore, it may be called a single-shear stress theory.

Further development of single-shear stress theory produces the twin-shear stress theory, which in turn is the basis of the unified stress theory [4, 5]. The single-shear stress theory, the twin-shear stress theory and other strength theories apply to particular cases, or linear approximations of the unified strength theory.

The unified strength theory represents advancement in the development of a more general strength theory. However, the above-mentioned strength theories are far from being perfect or mature. The main shortcoming of these strength theories includes the neglect of time-dependence of the deformation and the internal structure of solids.

3 Kinetic nature of solid strength

Investigations of the microscopic physical nature and failure mechanisms of solids fall into two categories: static methods and kinetic methods.

Static methods are characterized by the transition from viewing solids as elastic or viscoelastic media to viewing solids as atom or molecule systems. In these systems, atoms or molecules are connected by cohesion, and the external forces applied to the solids are distributed on the links between atoms or molecules. In this way, the internal forces are induced. Therefore, the stability of solids before failure is determined by the relationship between (1) the cohesions between atoms or molecules and (2) the internal forces in bonds induced by external forces. If the internal forces are less than cohesions, elastic deformation will be induced, otherwise irreversible deformation and fracturing will occur.

In microscopic static theories, the strength property of solids is described by the concept of limit strength, and the failure of materials is considered to be a critical event that takes place instantaneously when the internal force in any bond of atoms reaches its critical value. Based on an understanding of atomic structure of solids, a theoretical strength of a solid can be determined.

However, there are two contradictions between the static microscopic failure mechanism and experimental observations for materials. The first is that the actual strength of materials is much less (1-3 orders lower) than the theoretical strength($\sigma_{th} \approx 0.1E$, where E is the Young's modulus). According to previous investigations, the remarkable difference between the theoretical and the actual strengths may be attributed to the existence of defects near which significant stress concentration takes place.

The second contradiction is that the static microscopic failure concept assumes that the failure is of instantaneous event, but experiments show that the failure of materials is a time-dependent process. The duration of failure may be determined by Zhurkov's formula.

The attempts to solve the second contradiction give rise to kinetic theories, the second category of theories describing the deformation and fracturing of materials. In kinetic theories, the atomic system is under thermal vibration, and it interacts with the external loads. The atom vibration changes the distances between atoms and consequently changes the forces in the bonds of atoms. Rough estimations show that the frequency of thermal vibration of molecules is approximately 10^{12}–$10^{13} s^{-1}$, and the average kinetic energy distributed to every degree of freedom for an atom is $KT/2$ (where K is Boltzmann's constant and T is absolute temperature). When $T = 300K$, the resultant average force in atomic bonds is of the order of 9 800 MPa, and the force for the breakage of atomic bonds has the same order, 14 700–29 400 MPa. The difference between the two energies is called the energy barrier.

The problem is also related to the non-uniformity of atom vibration, called the thermodynamic energy fluctuation, resulting from the chaotic thermal motions of atoms. This means that the kinetic energy distributed to individual degrees of freedom within individual atom may be much higher than the average vibration energy of the atom. As a result, the forces in atomic bonds in individual atom may exceed the limit forces for the breakage of atomic bonds. The breakage of atomic links thus will occur, leading to fracturing.

It is clear from the foregoing analysis that thermal fluctuation plays a fundamental role in the breakage of atomic bonds.

The roles of external forces applied to solids are two-fold. First, the external forces are smaller than the energy barrier U for breakage of atomic bonds defined by $\Delta U(f) = f\Delta r$, where f is the force induced by external forces in every atomic bond and Δr is the change in distance between atoms induced by external forces.

Second, the force f reduces the probability of the restoration of broken atomic bonds because the action of f increases the distance between atoms. Therefore, a mutual compensation between external forces and thermal fluctuation exists: the thermodynamic energy fluctuation makes the breakage of atomic bonds possible, but external forces exclude the possibility of the restoration of broken atomic bonds (some chemical processes may restore broken atomic bonds, e.g. by sealing micro-fractures in clays).

The foregoing discussion deals with the kinetic nature of the breakage of bonds at atomic scales. However, the development of fracturing in a material should be treated as the accumulation of breakages of atomic bonds in a solid, leading to the initiation of fractures (micro-cracks and micro-voids). This process is called fracturing localization.

Thermal fluctuation is a time-dependent stochastic process. Furthermore, the force f needs a certain time to overcome the resistance provided by the energy barrier and to increase the distances between atoms. The process of fracturing localization also needs a certain time to activate and develop. All these facts indicate that the failure of a material founded on thermal fluctuation at the atomic level is a time-dependent process, which presumably needs time to be activated and to develope. The larger the external force is, the shorter the time needed for overcoming the energy barrier will be, i.e. fracturing will happen more quickly.

There are still many problems to be solved regarding kinetic theories of deformation and fracturing in solids. Such theories are under development.

4 Dynamic strength theories

From the above discussion, we can conclude that material strengths are not physical constants and fracturing of solids needs time to be activated, to develop and to complete. These conclusions are also based on experimental data. Indeed, many solids show the strain rate sensitivity of strength. In this case, new parameters, e.g. strain rate or stress rate, should be taken into account in the description of deformation and fracturing of solids. Dynamic strength theories expand upon traditional and kinetic strength theories of solids by considering the dynamic effects induced by high strain rate loading.

4.1 Experimental observation

The fracturing strengths of rocks increase significantly under intensive dynamic loading. Some experimental data are presented in Table 1 [6].

Figures 1 and 2 represent laboratory experimental data collected under a constant loading rate [7], where τ is the loading time from initial application of the load to failure, σ_f is the failure stress, $\dot{\varepsilon}$ is the strain rate, and σ_{dyn} and σ_{st} are the dynamic and static failure stresses.

Table 1 The fracturing strengths of rocks [6].

Rocks	Static strength (MPa)	Dynamic strength (MPa)	The ratio of dynamic strength to static strength
Limestone	42.56	276.62	6.5
Marble (normal to the deposit)	21.28	191.50	9.0
Marble (parallel to the deposit)	63.84	496.49	7.8
Granite	70.93	405.30	5.7

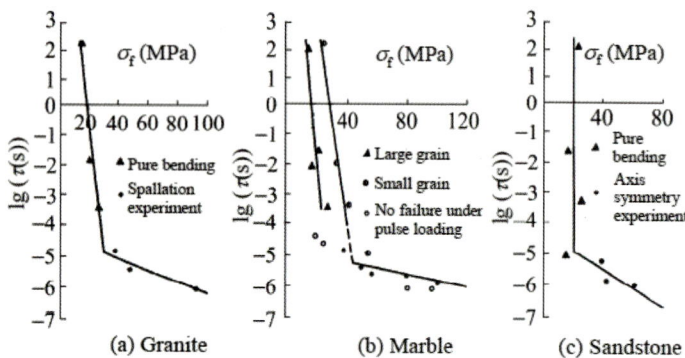

Fig. 1 Relationship between fracture time and load amplitudes [7].

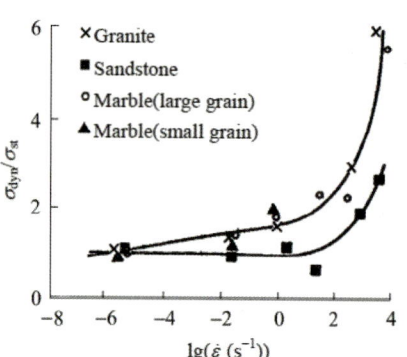

Fig. 2 Strain rate dependence of rock strength [8].

It can be observed from Fig.1 that when $\tau > 10^{-5}$ s, the loading is quasi-static and a weak time-dependence of failure stress is observed. When $\tau < 10^{-5}$ s, failure stress increases remarkably with a decrease in loading time.

It can be observed from Figs.2 and 3 that dynamic strength rapidly increases with the strain rate when strain rate $\dot{\varepsilon} > 10 \text{s}^{-5}$ [8].

To establish deformation and fracturing models for surrounding rocks of tunnels, it is necessary to apply dynamic strength theories and failure criteria.

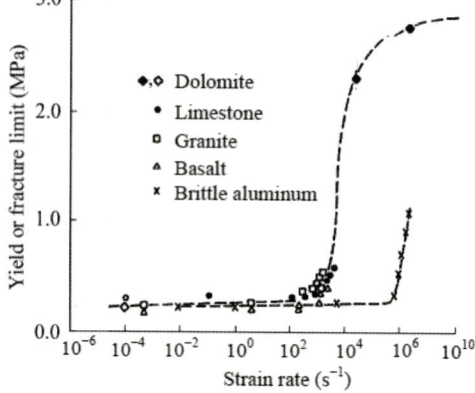

Fig. 3 Strain rate dependence of yield or fracture limit [8].

4.2 Stain rate sensitivity of rock strength

Under moderate uniaxial tension, the expected life time (instigation of loading to failure) τ may be determined by Zhurkov's formula:

$$\tau = \tau_0 \exp\left(\frac{U_0 - \gamma \sigma_t}{KT}\right) \quad (1)$$

where U_0 is the activation energy, σ_t is the uniaxial tensile stress, γ is the activation volume, and τ_0 is a temporal parameter which is in the order of the thermal vibration period of atoms [9].

Zhurkov's formula shows the thermo-activated nature of deformation and fracturing of solids, which gives the dependence of the strength on the life time as

$$\sigma = Y = \frac{1}{\gamma}\left(U_0 + KT \ln \frac{\tau_0}{\tau}\right) \quad (2)$$

where ε_0 is the limit deformation at failure, and $\dot{\varepsilon}$ is the constant strain rate of loading process. When $\tau = \varepsilon_0 / \dot{\varepsilon}$, the Eq.(2) becomes

$$\sigma = Y = \frac{1}{\gamma}\left[U_0 + KT\left(\ln \dot{\varepsilon} - \ln \frac{\varepsilon_0}{\tau_0}\right)\right] \quad (3)$$

i.e.

$$\sigma = Y = \frac{1}{\gamma}\left(U_0 + KT \ln \frac{\dot{\varepsilon}}{\dot{\varepsilon}_0}\right) \quad (4)$$

where $\dot{\varepsilon}_0 = \varepsilon_0 / \tau_0$ represents the maximum possible tensile strain rate in the material.

A similar formula holds true for the dynamic shear strength Y_τ:

$$Y_\tau = \frac{1}{\gamma_\tau}\left(G_0 + KT \ln \frac{\dot{\gamma}}{\dot{\gamma}_0}\right) \quad (5)$$

where γ_τ is the activation volume under shear deformation, G_0 is the activation energy, $\dot{\gamma}$ is the shear strain rate, and $\dot{\gamma}_0$ is the limit shear strain, $\dot{\gamma}_0 = \gamma_0 / \tau_0$.

In principle, the dependence of the compressive strengths of solids on strain rate is similar to that of the tensile strengths, but the values of the parameters in the formulae are different. Only compressive and shear strengths will be examined in this paper. Qi and Qian[10] have re-derived Zhurkov's formula on the basis of microscopic theory.

Experiments by Stavrogin and Protosenja [11] showed that Eqs.(4) and (5) can describe the strain rate sensitivity of compressive, shear and tensile strengths of solids at relatively low strain rates. Their results indicate that a thermo-activated mechanism dominates the strain rate sensitivity of strength. When the strain rate exceeds a certain threshold value, the strain rate sensitivity of strength moves into a new regime, where the strength increases rapidly with increases in the strain rate, and the deformation and fracturing of solids are more adiabatic. In this case, according to current knowledge, phonon damping (macroscopic viscosity) plays a predominant role.

The investigation shows the general features of dynamic strength of solids as illustrated in Fig.4 [11]. In a low strain rate regime, the strength of materials increases slowly as the strain rate increases. This regime is provisionally named Regime 1. When the strain rate exceeds a threshold value, the strength increases rapidly with the increase of the strain rate. This regime is named Regime 2. When the strain rate is very high, the dependence of strength on strain rate becomes weak again and is somewhat similar to that in Regime 1. This regime is named Regime 3 (Fig.4).

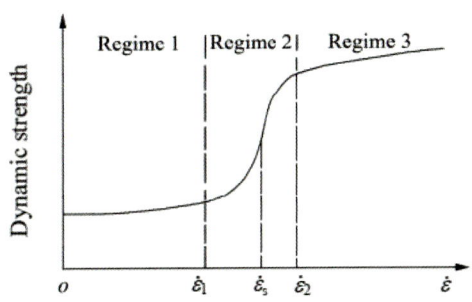

Fig. 4 Dependence of dynamic strength on strain rates of brittle materials ($\dot{\varepsilon}_1 \approx 10^0 - 10^2\ \mathrm{s}^{-1}$, $\dot{\varepsilon}_s \approx 10^3\ \mathrm{s}^{-1}$, $\dot{\varepsilon}_2 \approx 10^4\ \mathrm{s}^{-1}$).

The smooth transition from Regime 1 to Regime 2 represents a gradual change of the deformation and failure mechanisms during the transition, i.e. the thermo-activated mechanism gradually loses its predominance and phonon damping (macroscopic viscosity) gradually emerges as the dominant mechanism. However, the two mechanisms do coexist.

In Regime 2, the material behavior is closely related to its viscosity. Generally, viscosity can be defined as the transport of momentum along a velocity gradient. In a steady shock wave process, viscosity can be viewed as the diffusion of momentum along the axis of wave propagation [12]. The viscosity is commonly considered to be a material property, which describes proportionality between a viscous stress component and the velocity gradient or strain rate, and depends on temperature. However, more complex constitutive behaviors can appear under shock loading.

The transition from Regime 2 to Regime 3 is accompanied by weaker dependence of strength on strain rate. Kipp et al. [13] determined the fracturing stresses of penny-shaped cracks at different strain rates. They showed that when the strain rate grows, the fracturing stress of cracks increases in magnitude and becomes effectively independent of crack size at very high strain rates. At very high strain rates, a wide range of crack sizes is initiated simultaneously and the failure grows by multiple cracks growth and coalescence. An experimental demonstration of this effect was reported by Kalthoff and Shockey [14] using short pulse loading on cracks with finite lengths. The results imply that in the transition process from Regime 2 to Regime 3, the effect of locality (or localization) of deformation and fracturing is reduced gradually, and homogeneity of deformation and fracturing gradually emerges.

At high strain rates, the number of crack nuclei grows rapidly due to the thermo-fluctuation rupture of inter-molecular bonds in intact regions, in addition to the athermal growth of cracks. In other words, the thermo-activated mechanism is reactivated in the absence of a significant stress concentration. The presence of defects only results in an increase in the rate of these ruptures due to the specific local features of energy dissipation in the deformation and fracturing process of solids. The last situation should cause a weaker dependence on temperature of solid strength in sub-microsecond lifetime intervals. At very high stain rates, the material fragments after fracturing are very small due to the simultaneous initiation of damage throughout the solid volume.

Therefore, based on analysis of the available experimental data, another conclusion about the dependence of solid strength on strain rate can be drawn. At low strain rates, the deformation and fracturing of rocks are controlled by the thermo-activated mechanism and the strength sensitivity to strain rate can be expressed by Eqs.(4) and (5). When the strain rate increases, the phonon damping (macroscopic viscosity) mechanism emerges, and gradually plays the dominant role. Since the crack propagation velocity in a solid is limited by the Rayleigh wave velocity, the viscosity coefficient decreases with the strain rate. At the structural level, the decrease of viscosity with the strain rate activates internal degrees of freedom and the related motion of meso-particles. At very high strain rates, the stress attained in solids approaches the theoretical limit of the strengths. In this case, a wide range of crack sizes is simultaneously initiated. In intact areas, the inter-molecular bonds are broken. These broken bonds serve as the growing athermal nuclei of damage, and the thermo-activated mechanism is reactivated. This means that the localization of the deformation and damage is gradually lost. Thus, the thermo-activated mechanism again emerges as the dominant mechanism of deformation and fracture at high strain rates. Also, the strength sensitivity to strain rate can be considered as the result of competition between the thermo-activated and the macroscopic viscous mechanisms. The viscous mechanism and its mathematic formula are examined below.

The divergence of viscosity within a rock, even one that is experiencing a constant deformation rate, is very large. This is obviously related to the fact that deformation and fracturing take place at different scales.

Rocks have multi-level structures. This observation is critical for the determination of their physical and mechanical properties. For example, the viscosity of a rock is directly related to its multi-level structure. In engineering practice, viscosity can be grouped into three scales, i.e. macro-, meso-, and microscopic levels.

Mathematically, viscosity η is expressed by the following equation:

$$\eta = G\tau' \tag{6}$$

where G is the shear modulus and τ' is the relaxation time [15].

The relaxation of materials is due to not only the relative sliding between structural elements, but also the reorganization of these elements and their internal structural changes. Thus, rock relaxation is accompanied by dilatancy. When a rock structure is fractured, stress concentrations arise, which then diminish with time. This relaxation time is proportional to the structural element size and inversely proportional to the growth velocity of the induced flaws. During the process of dilatancy, structural flaws tend to occur somewhat uniformly in the rock. The growth velocity of the induced flaws (e.g. dislocation and micro- and macro- cracks) is limited. Also, it depends on the applied external stresses and therefore on stress relaxation.

From a phenomenological point of view, the growth velocity of flaws is assumed to be a function of strain rate, i.e.

$$v = v(\dot{\varepsilon}) \tag{7}$$

Expanding Eq.(7) into Taylor's series, we obtain

$$v(\dot{\varepsilon}) = v_0 + \alpha\dot{\varepsilon} + \cdots \tag{8}$$

where $v_0 = v_0(0)$ may be regarded as the growth velocity of flaws at a fixed magnitude of deformation.

It is necessary here to point out that the thermo- activated mechanism also contributes to the velocity of the growth of flaws. Experiments show that the propagation velocity of cracks (growth velocity of flaws) is limited to the range from 0.2 to 0.5 of the shear wave velocity. Taking this into consideration, we choose the following formula to approximate the change of growth velocity with strain rate due only to macro-viscosity:

$$v = b\left(\frac{1+\lambda\dot{\varepsilon}^n}{\xi\dot{\varepsilon}^n}\right) \tag{9}$$

where b, ξ, λ and n are experimental constants.

On the other hand, according to the Maxwell model of Radionov et al. [23], a rock with an internal element size L can not withstand the deformation when the deformation rate is more than $\dot{\varepsilon}^* = \sigma_f \upsilon/(GL)$, where G is the shear modulus and $\upsilon \approx 2\times 10^{-6}$ cm/s is a parameter characterizing the relaxation rate of stress, because of the stress concentration due to the heterogeneity of the rock. Therefore, the resulted strain rate is inversely proportional to the size L, i.e. $\dot{\varepsilon} \propto 1/L$ [23]. Hence, the viscosity can be given as

$$\eta = A\frac{L}{v} = Ab\left(\frac{\xi\dot{\varepsilon}^n}{1+\lambda\dot{\varepsilon}^n}\right)\frac{1}{\dot{\varepsilon}} = \frac{C_0\dot{\varepsilon}^{n-1}}{B+B_1\dot{\varepsilon}^N} \tag{10}$$

where A, B, B_1 and C_0 are experimental constants.

Equation (10) can be re-written as

$$\eta = C_1\frac{\dot{\varepsilon}^{n-1}}{1+\dot{\varepsilon}^n} \tag{11}$$

where C_1 is an experimental constant.

Equation (11) is similar to the second term of the following expression:

$$\eta = A\frac{\ln(\dot{\varepsilon}/\dot{\varepsilon}_0)}{\dot{\varepsilon}} + C\frac{(\dot{\varepsilon}/\dot{\varepsilon}_s)^{n-1}}{\left[(\dot{\varepsilon}/\dot{\varepsilon}_s)^n+1\right]} \quad (n \geqslant 1) \tag{12}$$

where $\dot{\varepsilon}_0$ is a deformation rate on the order of 10^{12}s^{-1}, $\dot{\varepsilon}_s$ is an approximation parameter, and C is an experimental constant[16].

In this case, the first term on the right-hand side of Eq.(12) may be regarded as the contribution of the thermo-activated mechanism of deformation[16], and the second term is the contribution of the macro-viscosity mechanism.

Hence, the macro-viscosity can be expressed as

$$\eta_{\text{macv}} = C\frac{(\dot{\varepsilon}/\dot{\varepsilon}_s)^{n-1}}{\left[(\dot{\varepsilon}/\dot{\varepsilon}_s)^n+1\right]} \tag{13}$$

From the above analysis, it is shown that an increase of the deformation rate leads to a decrease of viscosity, which means that the deformation and fracturing of rocks gradually converge at the macro- and micro- scopic scales.

The compressive strength sensitivity to the strain rate then can be expressed as the summation of the following two terms (Y_D is the compressive strength, and $Y_{\tau D}$ is the shear strength):

$$Y_D = Y_T(\dot{\varepsilon}) + Y_V(\dot{\varepsilon}) \tag{14}$$

$$Y_{\tau D} = Y_{\tau T}(\dot{\gamma}) + Y_{\tau V}(\dot{\gamma}) \tag{15}$$

The first term on the right-hand side of Eqs.(14) and (15) represents the contributions of the thermo-activated mechanism, and the second term represents those of the viscosity mechanisms.

According to Eq.(13), the contributions of the macro-viscosity may be expressed as

$$Y_V(\dot{\varepsilon}) = \dot{\varepsilon}\eta = \frac{b(\dot{\varepsilon}/\dot{\varepsilon}_s)^n}{(\dot{\varepsilon}/\dot{\varepsilon}_s)^n+1} \tag{16}$$

and

$$Y_{\tau V}(\dot{\gamma}) = \dot{\gamma}\eta = \frac{b_1(\dot{\gamma}/\gamma_s)^n}{(\dot{\gamma}/\gamma_s)^n + 1} \tag{17}$$

where b and b_1 may be interpreted as the maximum values of the contributions from the macro-viscosity mechanism, and $\dot{\gamma}_s$ is the experimental parameter. The effect of temperature is implicitly included in these formulae.

Finally, a unified relationship between strength and strain rate is obtained. It includes the thermo-activation and the viscosity mechanisms as two coexisting and competing mechanisms, i.e.

$$Y_D = \frac{1}{\gamma}\left(U_0 + KT\ln\frac{\dot{\varepsilon}}{\dot{\varepsilon}_0}\right) + \frac{b(\dot{\varepsilon}/\dot{\varepsilon}_s)^n}{(\dot{\varepsilon}/\dot{\varepsilon}_s)^n + 1} \tag{18}$$

$$Y_{\tau D} = \frac{1}{\gamma_\tau}\left(G_0 + KT\ln\frac{\dot{\gamma}}{\dot{\gamma}_0}\right) + \frac{b_1(\dot{\gamma}/\gamma_s)^n}{(\dot{\gamma}/\gamma_s)^n + 1} \tag{19}$$

The temperature rise caused by impact affects the strength of a solid. Generally, the strength of a metal decreases significantly when the temperature reaches 85%–90% of its melting temperature, and is different from the results calculated by Eqs.(18) and (19). However, the melting temperatures of rocks are significantly higher than those of metals. Generally the temperature caused by impact in rock is not very close to the melting temperature. Therefore, Eqs.(18) and (19) are applicable to rocks.

The thermo-activated mechanism acts more significantly near stress concentration areas and along grain boundaries at low strain rates. At very high strain rates, the thermo-activated mechanism is activated again in the intact areas of rocks, but the parameters in Eqs.(4) and (5) in these regimes should be different, with γ, γ_τ being less at high strain rates.

The experimental data for silicon carbide, aluminum oxide, granodiorite and dolomite are shown in Fig.5. The left parts of the experimental curves are almost straight horizontal lines (Fig.5(a)). Thus, it is very easy to determine items U_0/γ, K/γ, G/γ_s and K/γ_s in Eqs.(18) and (19) by data fitting.

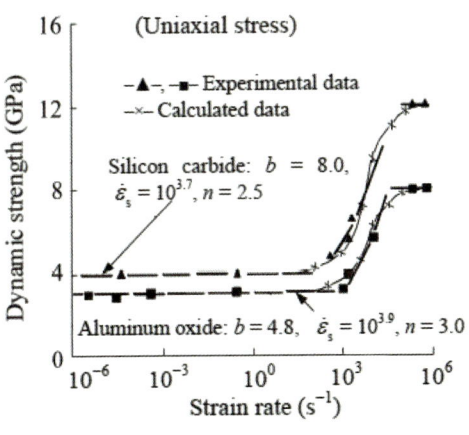

(a) Aluminum oxide and silicon carbide.

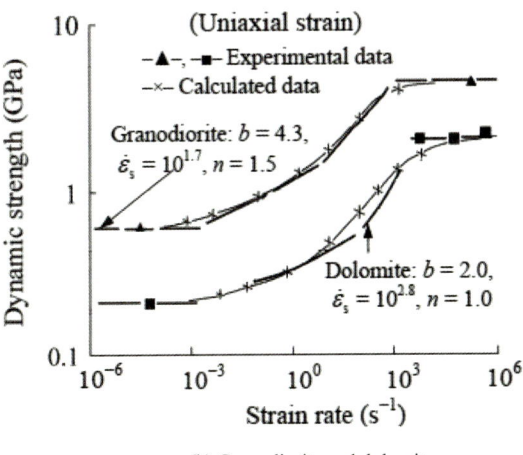

(b) Granodiorite and dolomite.

Fig. 5 Comparison of Eq.(18) with the experimental data.

Considering that the contribution of macro-viscosity is very small at low strain rates, it may be argued that the contribution of the thermo-activation mechanism calculated by Eqs.(4) and (5) is also small, and that the material strength is weakly dependent on the strain rate. On the other hand, at high strain rates, the contribution of macro-viscosity, i.e. the second terms of the right-hand side of Eqs.(18) and (19), is dominant. For convenience, the left (horizontal) parts of the curves in Fig.5 may be prolonged to the right. The straight lines are chosen as the basis for superimposing the contribution of macro-viscosity.

According to the above descriptions, example calculations have been carried out using Eqs.(18) and (19). The results of these calculations are compared with the experimental results reported by Grady [12]. The calculated and experimental results agree well (Fig.5). This indicates that the given model has a sound physical basis, and it is applicable to a wide range of strain rates, and is simple and convenient for practical use.

For step-type loading and continuously changing load (t), Bailey's damage accumulation criterion can be used:

$$\sum \frac{\Delta t_i}{\tau(\sigma_i)} = 1 \quad \text{or} \quad \int_0^{t_p} \frac{dt}{\tau[\sigma(t)]} = 1 \tag{20}$$

where t is time, the subscript "i" denotes the loading ordering, t_p is the loading time to failure, and $\tau(\sigma_i)$ is the life time of material under stress σ_i.

4.3 Other strength theories that consider temporal factors

4.3.1 Nikiforovsky-Shemyakin impulse criterion

According to the Nikiforovsky-Shemyakin impulse criterion [17], when the total pulse J_0 reaches its limit value, i.e. when $\int_0^{t_p} \sigma(t) dt = J_0$, failure will take place. In a one-dimensional case, the relationship between the stress in solids σ and the particle velocity V can be expressed as $= \rho D V$, where ρ is the solid's density and D is the shock wave propagation velocity. Substituting this relationship into the Nikiforovsky-Shemyakin impulse criterion gives

$$\int_0^{t_p} \sigma(t) dt = \int_0^{t_p} \rho D V dt = \rho D u = J_0 \tag{21}$$

where u is the displacement at failure.

The impulse criterion indicates the damage accumulation nature of the fracturing processes, which coincides with Zhukov's criterion.

On the other hand, if the characteristic length of the shock wave is $'$, then $u = L'\varepsilon$, and Eq.(21) becomes

$$\int_0^{t_p} \sigma(t) dt = \rho D L' \varepsilon = J_0 \tag{22}$$

which shows that fracturing takes place when strain reaches a critical value. Therefore, the second strength theory can be applied to dynamic fracturing problems. The creep phenomenon and strength-strain rate sensitivity show the temporal effect of fracturing; their physical origin is identical. By multiplying Zhukov's formula with the Aleksandro creep formula $\dot{\varepsilon} = \dot{\varepsilon}_0 \exp\left[-(U_0 - \gamma\sigma)/KT\right]$, then $\tau\dot{\varepsilon} = \tau_0\dot{\varepsilon}_0 = \dot{\varepsilon}_{st}$ is obtained. Thus, the same conclusion is drawn, i.e. the critical failure strain is the same no matter what strain rate is applied.

Experiments show that, under shear, triaxial compression and other complex loading conditions, in a wide range of strain rates covering 9–10 orders of magnitude, the critical strain rate $\dot{\varepsilon}_{st}$ is only weakly dependent on temperature, stresses and strain rate. Thus, it can be considered as a constant [18]. This situation indicates a close relationship between deformation and fracturing. Therefore, the second strength theory may be considered as a quasi-temporal criterion.

4.3.2 Failure criterion based on damage evolution

According to Kachanov [19], the evolution of the damage parameter ψ may be described by the following equation:

$$\frac{d\psi}{dt} = f(\sigma,\psi) = \begin{cases} A\left(\dfrac{\sigma}{1-\psi}\right)^n & (\sigma \geqslant 0) \\ 0 & (\sigma \leqslant 0) \end{cases} \quad (23)$$

Failure takes place when the damage parameter reaches its critical value.

Integrating Eq.(23) yields

$$\int_0^{t_p} \sigma^n dt = \frac{1-(1-\psi_p)^{n+1}}{A(n+1)} = J_0 \quad (24)$$

which coincides with the criterion of Eq.(22) when $n=1$.

4.3.3 Structural-temporal criterion

According to the principles of the fracture mechanics of solids, when the average stress $\sigma(t,x)$ over spatial-temporal cells $[t-\tau,t]\times[0,d]$ reaches its static strength σ_c, failure takes place, i.e.

$$\frac{1}{\tau}\int_{t-\tau}^{t} dt' \int_0^d \sigma(t',r)dr \leqslant \sigma_c \quad (25)$$

where r is the spatial coordinate. This criterion is called Morozov-Petrov's structural-temporal criterion [20].

If we introduce a new parameter $J_c = \sigma_c \tau_c$, then Eq.(25) becomes

$$\int_{t-\tau}^{t} dt' \int_0^d \sigma(t',r')dr' = J_c(t) \quad (26)$$

where τ_c is the fracture incubation time corresponding o the characteristic time for the energy transfer between two neighboring cells d/v', where v' is the elastic wave velocity and d is the structural element size.

Therefore, Morozov-Petrov's structural-temporal criterion is physically manifested as a critical structural impulse.

4.3.4 Mohr-Coulomb-type constitutive models of strain rate dependence

The M-C criterion is a simple and practical criterion for geological materials. The strengths of geological materials show a significant strain rate dependency (sensitivity). Therefore, when analyzing geomechanical problems, it is necessary to consider the dependence of strength on strain rates. Under general stress states expressed in terms of principal stresses, the Mohr- Coulomb failure criterion can be written

$$\frac{\sigma_1 - \sigma_3}{2} = \frac{\sigma_1 + \sigma_3}{2}\sin\phi + c\cos\phi \tag{27}$$

where σ_1 and σ_3 are the maximum and minimum principal stresses, respectively, ϕ is the internal friction angle, and c is the internal cohesion of the material.

With a uniaxial compression test, the internal cohesion c can be determined by

$$c = \frac{\sigma_Y^C(1-\sin\phi)}{2\cos\phi} \tag{28}$$

where σ_Y^C is the uniaxial compressive strength. Substituting Eq.(18) into Eq.(28), we obtain

$$c = \frac{1-\sin\phi}{2\cos\phi}\left[\frac{1}{\gamma}\left(U_0 + KT\ln\frac{\dot{\varepsilon}}{\dot{\varepsilon}_0}\right) + \frac{b(\dot{\varepsilon}/\dot{\varepsilon}_s)^n}{(\dot{\varepsilon}/\dot{\varepsilon}_s)^n + 1}\right]e^{AC_0} \tag{29}$$

The last term in Eq.(29), e^{AC_0}, expresses the influence of strain on internal cohesion, and $C_0 = \sigma_1/\sigma_3$ is a parameter of stress state.

Substituting Eq.(29) into Eq.(27), a Mohr-Coulomb- type failure (strength) criterion with strain rate dependence can be obtained.

For loading conditions with high strain rates, the thermo-activated term may be replaced by static uniaxial compressional strength σ_{YS}^C because of the weak influence of the thermo-activated term on strength:

$$\sigma_1 - \sigma_3 = (\sigma_1 + \sigma_3)\sin\phi + (1-\sin\phi)\left[\sigma_{YS}^C + \frac{b(\dot{\varepsilon}/\dot{\varepsilon}_s)^n}{(\dot{\varepsilon}/\dot{\varepsilon}_s)^n + 1}\right]e^{AC_0} \tag{30}$$

or

$$\dot{\varepsilon}_s = \frac{1+\sin\phi}{1-\sin\phi}\sigma_3 + \left[\sigma_{YS}^C + \frac{b(\dot{\varepsilon}/\dot{\varepsilon}_s)^n}{(\dot{\varepsilon}/\dot{\varepsilon}_s)^n + 1}\right]e^{AC_0} \tag{31}$$

For underground explosions, explosion-induced fractures occur in proximity to the center of the explosion by a shear mechanism. The problem may be simplified furthermore, because $\varepsilon_r \gg \varepsilon_\theta$, where ε_r is the radial strain and ε_θ is the tangential strain. Therefore, the shear strain $\varepsilon_r = \varepsilon_r - \varepsilon_\theta \approx \varepsilon_r$, and the volumetric strain are approximated as $\varepsilon_v = \varepsilon_r + 2\varepsilon_\theta \approx \varepsilon_r$. Furthermore, it can be taken $\dot{\varepsilon} = \dot{\varepsilon}_r$. The relationship between the two principal stresses is $\sigma_1 = \alpha\sigma_3$, where $\alpha = C_0 = \mu/(1-\mu)$ and μ is Poisson's ratio.

In this case, the M-C criterion in the vicinity of explosion may be written as

$$\sigma_1 = \frac{1+\sin\phi}{1-\sin\phi}\sigma_3 + \left[\sigma_{YS}^C + \frac{b(\dot{\varepsilon}/\dot{\varepsilon}_s)^n}{(\dot{\varepsilon}/\dot{\varepsilon}_s)^n + 1}\right]e^{A\alpha} \tag{32}$$

4.3.5 Fragment size of fractured rock mass under dynamic loading

The strength of a fractured rock mass depends on the sample size. Generally, the compressive strength of materials σ_D can be expressed as a function of the sample size D as follows [21, 22]:

$$\sigma_D = \sigma_0(1 + D/D_0)^{-1/2} \tag{33}$$

where σ_0 and D_0 are constan.

If D_0 in Eq.(33) is replaced by Δ_i/D_0, where Δ_i is the size of blocks of i-th rank, then Eq.(33) becomes

$$\sigma_D = \sigma_0 \left(1 + \Delta_i / D_0\right)^{-1/2} \tag{34}$$

which can be rewritten as

$$\Delta_i = D_0 \left[\left(\sigma_0 / \sigma\right)^2 - 1\right] \tag{35}$$

where parameter σ_D is replaced by σ representing the applied load.

Replacing σ in Eq.(35) by the strength of rock mass, the following formula is obtained to determine the fragment size of fractured rock mass:

$$\Delta_i = D_0 \left[\left(\frac{\sigma_0}{Y_\tau}\right)^2 - 1\right] \tag{36}$$

Equation (36) shows that the mean fragment sizes of fractured rock mass decrease with the growth of external loads.

This conclusion is confirmed by quasi-static and dynamic experiments. In the case of a one-fold fracture, under both dynamic and quasi-static conditions, Fig.6 shows the relationship between the specific shear deformation energy E_τ and mean fragment size D given by the same curve [23]. This relationship applies to both shear fracture and cleavage fracture, which can be approximated by the following equation:

$$D \sim \frac{1}{E_\tau} = \frac{3}{1+\mu} \frac{E}{\left(\sigma_r - \sigma_\phi\right)^2} \tag{37}$$

where σ and σ_ϕ are the radial and tangential stresses respectively.

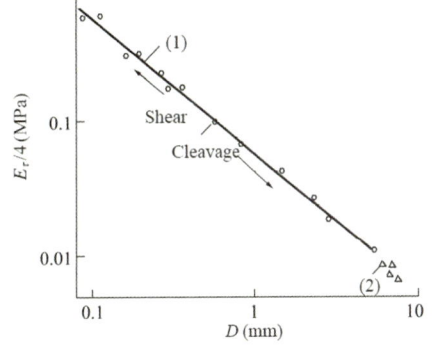

Fig. 6 Dependence of specific shear deformation energy on rock fragment size based on the results of (1) a chemical explosion blast (0.4g) and (2) quasi-static experiments when $\sigma_{22} = 0$ MPa.

Replacing σr − σφ in Eq.(37) by $2Y\tau$, the following result is obtained:

$$D \sim \frac{1}{E_\tau} = \frac{3}{4(1+\mu)} \frac{E}{Y_\tau^2} \tag{38}$$

which has the same D-Y_τ dependence as Eq.(36).

To predict the mean fragment size under uniaxial dynamic loading, Y_τ should be replaced in Eq.(36) by Y_D or $Y_{\tau D}$, which are determined by Eq.(18) and (19) respectively.

For predicting the mean fragment size near the center of the explosion, Eq.(36) is used. But, the dynamic shear strength of rock mass should be determined by the following formula:

$$Y_{\tau D} = \left[\frac{1}{\gamma_\tau}\left(G_0 + KT \ln \frac{\dot{\gamma}}{\dot{\gamma}_0}\right) + \frac{b_1\left(\dot{\gamma}/\dot{\gamma}_s\right)^n}{\left(\dot{\gamma}/\dot{\gamma}_s\right)^n + 1}\right] e^{AC} \tag{39}$$

Under external loads, fracturing takes place on the structural surfaces of the largest scale fragments in the medium. The fragment size is the characteristic size of the structural elements at this scale. A further increase in the stress intensity results in fracturing at the next lower scale, and the size of fragments is the characteristic size of the structural elements at that scale. The increases of strain rate, confining pressure and plastic strain hardening may enhance the strength of the medium. Consequently, the deformation and fracturing may cover the small-scale levels of a rock mass and the fragment size will decrease.

5 Application of modern strength theories

5.1 Internal structure of a spallation plane

The introduction of temporal factors into strength criteria improves our understanding of failure mechanisms, and may produce results that are significantly different from those of traditional strength theories. As an example, the spallation problem for the propagation of a triangular impulse stress wave without rise time is considered below.

According to the traditional strength theories, tensile stress occurs when a stress wave reflects from free surfaces. When the resultant tensile stresses at some distance from the free surfaces reach a critical value σ_t, spalling takes place at that distance $x = \sigma_t / (2\sigma_m)$, where σ_m is the amplitude of the stress wave. However, according to the new strength theories, every point in the rock covered by the reflected wave is under tensile stress and prone to tensile fracturing, i.e. spalling.

In a rock section where tensile stresses are small at a particular time, it is reasonable to assume that the rock will need a longer time to be fractured. Such a section may fracture simultaneously with other sections where tensile stress is greater and incubation time for fracture is longer, i.e. connected rock within a definite width may fracture simultaneously. Experiments validate this hypothesis. Therefore, spalling zones generally are defined to have certain widths (or thickness), which in turn means that spalling zones have internal structures. It is apparent that the simulation of such an event will not be realistic if we use traditional static strength theories to simulate dynamic fracture.

5.2 Safety thresholds for explosive detonations

Ground vibration induced by explosions may damage surface infrastructure and underground facilities. The decisive parameters for assessing the likely degree of damage and the safety of proximal infrastructure are the seismic vibration parameters: acceleration, particle velocity, and displacement. At present, consensus on the issue of which parameters should be used for safety evaluations of structures by explosion-induced seismicity has not been completely resolved. Most jurisdictions around the world take the ground surface velocity as the control parameter. The use of such a parameter agrees with in-situ investigation, i.e. it is the ground surface velocity or displacement, not internal forces, that controls the damage to buildings and facilities. This also agrees with the modern strength theories.

According to the modern strength theories, the damage to infrastructure under explosion-induced ground vibration is caused by dynamic failures. The control parameter for dynamic failure is displacement or particle velocity. Considering that displacement is the integration of velocity over time, the introduction of failure criteria involving vibration velocity and frequency as controlling parameters, as proposed by American Mining Bureau, and authorities in Germany and Finland, is more reasonable. Large numbers of observations show that, under the same geological conditions, at the same site and for the same type of structures, the degree of damage to the structures is the same when the vibration velocity exceeds a characteristic value for the particular kind of construction.

6 Conclusions

Usually, the attention is focused on the spatial aspect of rock engineering problems, and the time-dependence is often ignored. However, the dependence of deformation and fracturing processes on time lies in the fact that the fracturing of rocks requires time to be mobilized, to develope and to complete, and rock strengths depend on strain rates. The consideration of the time-dependence of material deformation improves our understanding of material deformation and fracturing.

At low strain rates, the deformation and fracturing of rock are controlled by the thermally activated mechanisms. With an increase of strain rate, the phonon damping (macroscopic viscosity) mechanism emerges and gradually dominates. At very high strain rates, deformation and fracturing occur gradually at the microscopic scale, under which conditions the thermally activated mechanism is reactivated. In this case, a wide range of crack sizes is initiated simultaneously in the rock and inter-molecular bonds in previously intact regions are broken. These broken bonds serve as the nuclei from which damage (micro-cracks) grows. This means that the localization of deformation and damage will be gradually reduced and finally disappear.

At high strain rates, thermal activation emerges as the dominant mechanism of deformation and fracturing. Thus, the dependence of strength on strain rate may be considered as the result of competition between two coexisting mechanisms, thermally activated and macro-viscous mechanisms, which take turns playing the leading role over different ranges of strain rate. The dependence of rock strength on strain rate may be expressed as the summation of the contributions from these two mechanisms. A comparison between experimental and calculated data has shown that the proposed model describes the strength dependence (sensitivity) on strain rate very well over a wide range of strain rates. The proposed model has a sound physical foundation, is applicable to a wide range of strain rates, and is simple and convenient for practical applications.

The influence of dynamic loading on the fragment size of rock has also been shown to depend on the accumulation of increased shear deformation energy at the moment of fracturing due to the strength enhancement originating from the change in stress states, the accumulation of plastic deformation, and strain rate. The suggested relationship describes the fragmentation size well.

A Mohr-Coulomb-type constitutive relationship has been proposed, and the intrinsic relations between different temporal failure criteria have been expressed and explained. The applicability of these modern strength theories has been shown by explaining some unusual phenomena that cannot be easily explained with the traditional strength theories.

References

[1] Hoek E, Brown E T. Empirical strength criterion for rock masses. Journal of Geotechnical and Geoenvironment Engineering, ASCE, 1980, 106 (9): 1 013–1 019.

[2] Hoek E, Brown E T. The Hoek-Brown failure criteriona 1988 update. In: Proc. 15th Can. Rock Mech. Symp.. Toronto: Press of University of Toronto, 1988: 31–36.

[3] Drucker D C, Prager W. Soil mechanics and plastic analysis or limit design. Quart of Applied Mathematics, 1952, 10: 157–162.

[4] Yu M H, He L N, Song L Y. Twin shear stress strength theory and its generalization. Science in China (serials A), 1985, 28 (11): 1 113–1 120.

[5] Yu M H, He L N. A new model and theory on yield and failure of materials under complex stress state. In: Mechanical Behavior of Materials (VI). Oxford: Pergamon Press, 1991: 851–856.

[6] Rinehart J S, Dynamic fracture strength of rocks. In: Proceedings of the 7th Symposium on Rock Mechanics. New York: American Institute of Mining, 1965: 205–208.

[7] Bellendir E N, Klyadchenko V F, Kazachuk A I. Strength of rock mass under the intervals of loading of 102–106s. Journal Mining Science, 1991, 27 (2): 46–49 (in Russian).

[8] Nikolaevsky V N. Mechanics of porous and fractured media. Singapore: World Scientific, 1990.

[9] Kuksenko V S, Betechtin V I, Ryskin V S, et al. Nucleation of submicroscopic cracks in stressed solid. International Journal Fracture, 1975, 11 (5): 829–840.

[10] Qi Chengzhi, Qian Qihu. The microscopic physical mechanism of deformation and damage of materials. Chinese Journal of Solid Mechanics, 2002, 23 (3): 312–317 (in Chinese).

[11] Stavrogin A N, Protosenja A G. Mechanics of deformation and fracture of rocks. Moscow: Nedra, 1992 (in Russian).

[12] Grady D E. Shock wave properties of brittle solids, in: Shock Compression of Condensed Matters. New York: AIP Press, 1995: 9–20.

[13] Kipp M E, Grady D E, Chen E P. Strain-rate dependent fracture initiation. International Journal Fracture 1980, 16 (3): 471–478.

[14] Kalthoff J F, Shockey D A. Instability of cracks under impulse loading. Journal of Applied Physics, 1977, 48 (3): 986–993.

[15] Landau L D, Lifshits E M. Theory of elasticity. New York: Pergamon Press, 1959.

[16] Qi Chengzhi, Qian Qihu. Physical mechanism of brittle material strength-strain rate sensitivity. Chinese Journal of Rock Mechnics and Engineering, 2003, 21 (2): 177–181 (in Chinese).

[17] Nikiforovsky V S, Shemyakin E I. Dynamic fracture of solid. Nauka: Novosibirck, 1979 (in Russian).

[18] Regel V P, Slutsker A N, Tomashevsky E E. Kinetic nature of strength of solids. Moscow: Nedra, 1974 (in Russian).

[19] Kachanov L M. Fundamentals of fracture mechanics. Moscow: Nedra, 1974 (in Russian).

[20] Morozov N, Petrov Y. Dynamics of fracture. Berlin: Springer-Verlag, 2000.

[21] Bažant Z P, Ožbolt J, Eligehausen R. Fracture size effect: review of evidence for concrete structures. Journal of Structural Engineering, 1994, 120: 2 377–2 398.

[22] Bažant Z P. Scaling in nonlinear fracture mechanics. In: IUTAM symposium on nonlinear analysis of fracture. Dordrecht: Kluwer Academic Publishers, 1997: 1–12.

[23] Radionov V N, Sizov I A, Tsvetkov V M. Fundamentals of geomechanics. Moscow: Nedra, 1986 (in Russian).

Zonal disintegration of surrounding rock mass around the diversion tunnels in Jinping II Hydropower Station, Southwestern China

Q.H. Qian [a,b], X.P. Zhou [a,*], H.Q. Yang [a], Y.X. Zhang [a], X.H. Li [c]

[a] School of Civil Engineering, Chongqing University, Chongqing 400045, China
[b] PLA University of Science and Technology, Nanjing 210007, China
[c] Key Lab for the Exploitation of Southwestern Resources and the Environmental Disaster Control Engineering, Ministry of Education, Chongqing University, Chongqing 400044, China

ARTICLE INFO

Article history:
Available online 10 April 2009

Keywords:
Jinping II Hydropower Station
Deep rock mass
Zonal disintegration
Numerical simulation

ABSTRACT

By means of numerical simulation, the special phenomenon of zonal disintegration of surrounding rock mass around the diversion tunnels of Jinping II Hydropower Station is analyzed in this paper. In order to model the growth and coalescence of cracks within rock mass in Jinping II Hydropower Station, the weak-element is adopted. When cracks coalesce, failure of deep crack-weakened rock masses occurs and fractured zone is formed. The present result is different from the one obtained by the traditional elasto-plastic theory. The numerical results show that the slip-line zonal fracture is created within rock mass around the diversion tunnels in Jinping II Hydropower Station. Meanwhile, the magnitude and distributions of fractured zones are determined by numerical simulation. It is shown that the present results are in good agreement with the one observed by model tests. Through sensitivity analysis, the effects of stress condition, cohesion and the angle of internal friction on the phenomenon of zonal disintegration is determined.

© 2009 Elsevier Ltd. All rights reserved.

1. Introduction

In shallow rock mass engineering, the excavation-affected rock around tunnel contains an excavation-damage zone and an excavation disturbed zone [1]. The excavation-damage zone (EDZ) refers to irreversible damage in the shallow crack-weakened rock resulting from the excavation of a tunnel. Damage occurs when the energy is dissipated in the frictional sliding along preexisting flaw and the growth of wing cracks. The excavation-damage zone can be further divided into loose zone and plastic zone. The loose zone is close to the tunnel wall and is delineated by a more rapid decrease in measured acoustic velocity and more rapid increase in hydraulic transmissivity than in the plastic zone of the EDZ. It is found from visual observations that macroscopic cracks exists in the loose zone. The loose zone is divided into failed and a non-failed zones [2,3]. The failed zone occurs where cracks coalesce and rock has completely detached from the surrounding rock mass. The non-failed part of the loose zone, in contrast to the failed zone, may have visible macro-cracking but is not detached. The plastic zone displays a more gradual change in acoustic velocity and hydraulic transmissivity, which eventually return to background levels [1]. Beyond the EDZ itself, in situ stresses are disturbed in a range of several excavation radii [4] but not to such a degree that irreversible stress-induced changes to the rock properties have occurred. Therefore, the volume of stress-disturbed rock is referred to as the elastic zone, in which no microcrack propagates.

The mechanical behaviors of deep rock mass are different from those of shallow rock mass. Zonal disintegration refers to the alternative occurrence of disintegration zone and non-disintegration zone around tunnel in deep rock mass engineering [5]. With the development of deep rock mass engineering, the special phenomenon of zonal disintegration, which is different from the traditional elasto-plastic theory, is observed in deep underground engineering. However, the mechanism of zonal disintegration is still not explicitly revealed. Better understanding of the mechanism of zonal fracture within deep rock mass promises benefit in many areas from rock mechanics to deep underground engineering and earthquake prediction. It is essential and important to understand how zonal fracture occurs under high geostress in order to provide better understanding of fracturing process of deep rock mass that occur in the deep engineering fields.

At present, many approaches, such as in situ observations, model test and theoretical analysis, are applied to study the special phenomenon of zonal disintegration in deep rock mass. The phenomenon of zonal disintegration is initially observed in the mines of Talnakh-Oktyarb' skiy deposit at depths up to 1050 m, in Russia. It was demonstrated that around the tunnel there are zones of fissured and non-fissured, propagating discretely into the depths of surrounding rock mass [6–9]. Similarly, zonal disintegration are observed by the in situ velocity tests of ultrasonic wave in the mine of Jinchuan, China [10]. Many model tests have also been

本文原载于《Theoretical and Applied Fracture Mechanics》(2009 年)。

conducted. For example, the results of layered fracture within surrounding rock have been obtained by Gu et al. [11]. It is obvious that both in situ test and model test are expensive and confined in certain condition. In the author's previous work, the mechanism of zonal disintegration in deep rock mass is investigated, and the size and distributions of zonal disintegration are determined from theoretical point of view [12]. The theoretical results are more accurate, but the phenomenon of zonal disintegration is so complex that not every problem has explicit theoretical solution now. Usually, numerical simulation is effective in solving these problems. However, no numerical simulation of zonal disintegration is available now.

In this paper, in order to take the growth and coalescence of cracks within rock mass into account, a new numerical simulation method, in which the weak-element is adopted, is proposed. It is shown that the present results are in good agreement with the one observed by model tests. Finally, the size and distributions of zonal disintegration in Jinping II Hydropower Station is obtained. Through sensitivity analysis, the effect of stress condition, cohesion and the angle of internal friction on the phenomenon of zonal disintegration are determined.

2. Comparison with the experimental observation

To validate the weak-element model the weak-element method, the experiment observation on facture and failure around underground openings is chosen. Laboratory experiments were performed on the thick-walled hollow cylinders of Berea sandstone with a hole diameter of 25.4 mm, an external diameter of 89 mm, and a length of 152 mm [13,14]. The samples were subjected to axisymmetric pressures on outer diameter, internal pressure is zero. The sample were constrained to near zero axial deformation

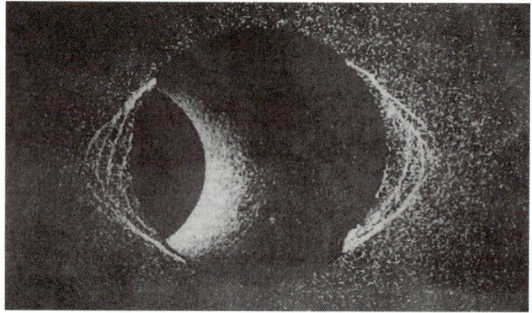

Fig. 1a. Typical failure around the thick-walled hollow cylinders of Berea sandston [13].

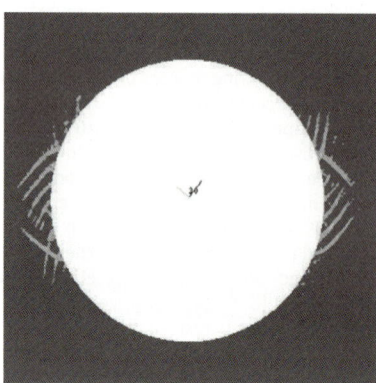

Fig. 1b. The numerical result of failure in Berea sandston.

(plane strain). Failure around the hole was caused by loading to a state of high external pressure (75 MPa). Uniaxial compressive strengths reported in the literature rang from 44–74 MPa for Berea sandstone. For Berea sandstone, Young's modulus is 17 GPa, Poisson's ratio is 0.32. The above parameters were applied to model facture and failure around the thick-walled hollow cylinders of Berea sandstone. The boundary conditions of numerical model is the same as those of experimental model. The experimental observation on the experiment observation on facture and failure around the thick-walled hollow cylinders of Berea sandstone is shown in Fig. 1a. The numerical result is depicted in Fig. 1b.

Comparison between numerical and experimental results, it is found that the numerical simulation result is in good agreement with the experimental one on failure modes around the hole of Berea sandstone. It is implied that the present weak-element method can be applied to model failure modes around underground openings.

3. Background of Jinping II Hydropower Project

Jinping II Hydropower Project at the upriver of the Yalong River is located in Sichuan Province, Southwest of China. The installed capacity of the project is 4800 MW. Four diversion tunnels with a diameter of 12–13 m and a total length of 16.67 km are constructed at depth of 1500–2000 m. In somewhere, the depth is up to 2525 m. There exist a series of difficulties, such as high geo-stress (max = 70 MPa), rock burst, karst, water flow, and instability of the surrounding rock mass [15,16]. Experience in dealing with some of these problems is still lacking.

The main strata outcropping in this area consist of marble, limestone, sandstone and so on. There are three main sets of fractures within rock mass. One set of fractures is dominant. Dip angle of the dominant discontinuities is 60°, the length and spacing of the dominant discontinuities are 1 m and 0.5 m, respectively. The dominant discontinuities are filled with chip of rock whose thicknesses is equal to 2 mm.

Both TBM and NATW are adopted in the construction of diversion tunnels. The No. 1 and No. 3 tunnels with a diameter of 13 m are constructed by TBM. However, the No. 2 and No. 4 tunnels are the four-arcs ones, which are constructed by NATW. The tests show that the maximum principal stress is about 70 MPa and the minimum principal stress is about 31 MPa.

4. Numerical simulation of zonal disintegration around diversion tunnel in Jinping II Hydropower Station

In the middle of diversion tunnels, a cross-section, which is mainly composed of marble of the Zagu'nao stratum (T_{2b}), is chosen as a representative section. The uniaxial compressive strength

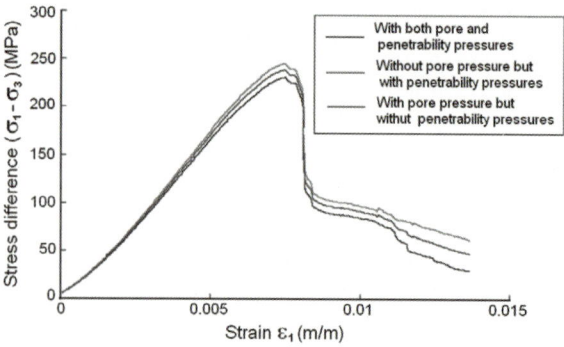

Fig. 2. The complete stress–strain curves of marble with the confining pressure of 30 MPa.

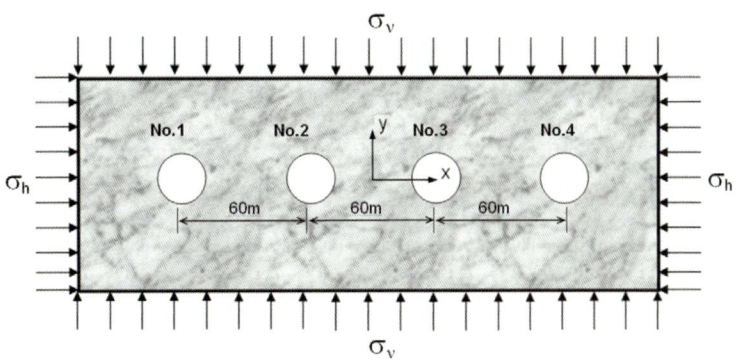

Fig. 3. Sketch of diversion tunnel in Jinping II Hydropower Station (plain strain).

Table 1
Parameters of rock mass by tests.

	Cohesion (MPa)	Angle of internal friction (°)	Poisson's ratio	Young's modulus (GPa)	Weight density (kN m^{-3})
Intact rock element (marble)	10.6	34	0.23	18.9	27.20
Weak-element (faults)	2.1	34	0.23	10.9	25.16

of marble is about 75–85 MPa. The complete stress–strain curves of marble are shown in Fig. 2.

4.1. The numerical model

A cross-section (300 m × 90 m) with four diversion tunnels was analyzed by using multiscale simulation, as shown in Fig. 3. In order to model the growth and coalescence of cracks within rock

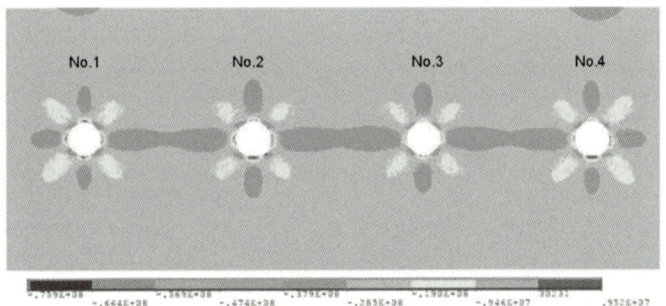

(a) Distributions of horizontal stress around four diversion tunnels (Unit: Pa)

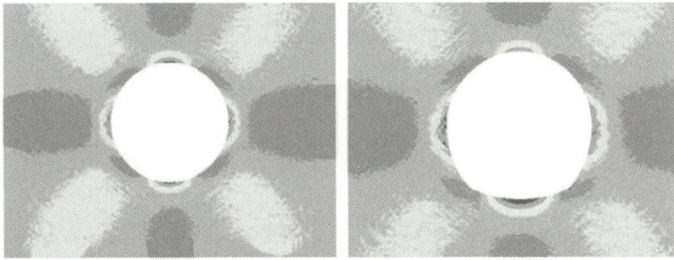

(b) Distributions of horizontal stress around No.1 tunnel   (c) Distributions of horizontal stress around No.2 tunnel

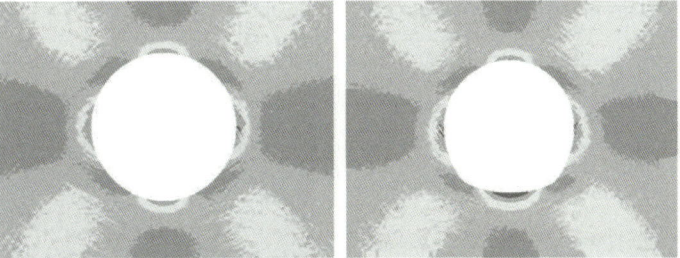

(d) Distributions of horizontal stress around No.3 tunnel   (e) Distributions of horizontal stress around No.4 tunnel

Fig. 4. The distributions of horizontal stress in the surrounding rock mass around diversion tunnel.

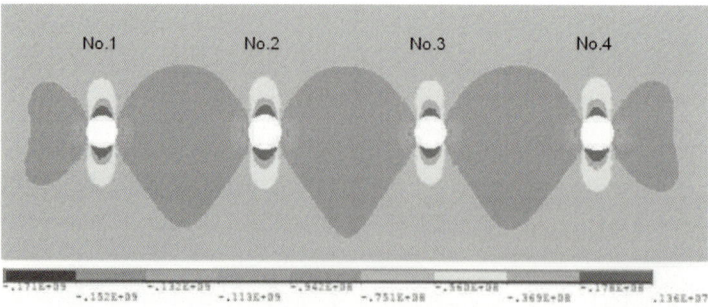
(a) Distributions of vertical stress around four diversion tunnels

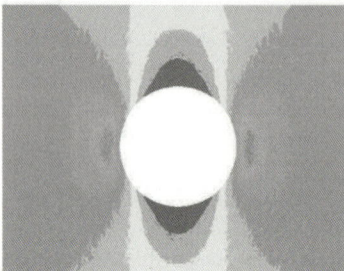
(b) Distributions of vertical stress around No.1 tunnel

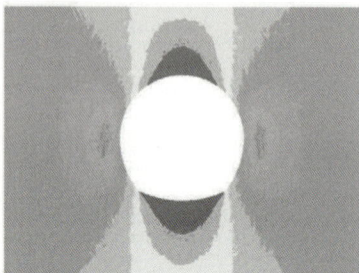
(c) Distributions of vertical stress around No.2 tunnel

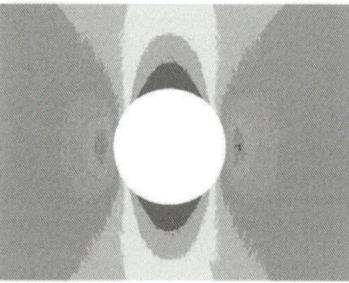
(d) Distributions of vertical stress around No.3 tunnel

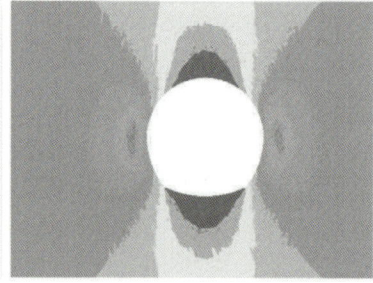
(e) Distributions of vertical stress around No.4 tunnel

Fig. 5. The distributions of vertical stress in the surrounding rock mass around diversion tunnel.

mass in Jinping II Hydropower Station, the weak-element method is applied to multiscale simulation. The present method is different from the traditional continuum mechanics-based approaches. The present method tries to consider the pre-existing crack from the microstructure level rather than the continuous level. The constitutive relation of the weak-element intersected by the pre-existing cracks, in which the nucleation, growth and coalescence of cracks are analyzed, is derived from mesomechanical theory [17]. This model is investigated under plane strain condition, in which there are 449,600 quadrangular elements.

The cross-section in Fig. 3 is subjected to geostresses of $\sigma_v = 69.5$ MPa and $\sigma_h = 23$ MPa. It should be noted that the compressive stress is negative in this paper. The vertical displacement components of the bottom are restricted, and both left and right boundaries are loaded by geostress. All the material parameters are listed in Table 1.

4.2. Analysis of zonal disintegration around diversion tunnel in Jinping II Hydropower Station

Horizontal and vertical stress are adopted, since horizontal and vertical stress can be monitored in the future.

The distribution of horizontal stress around the diversion tunnel is depicted in Fig. 4. It is noted that the surrounding rock mass around arch crown and sidewalls is subjected to tensile stress, which distributes in few region. It also shown in Fig. 4a that the horizontal stress is decreased at the distance of 2.2–4.2 m from the tunnel. This phenomenon of decrease of the horizontal stress can be observed more clearly in the enlarged photos, as shown in Fig. 4b–e.

Fig. 5 illustrates the distributions of vertical stress around the diversion tunnel. The stress distributions obtained by the present numerical analysis are different from those obtained by traditional elasto-plastic theory. It is obvious that the stress field obtained by the present numerical analysis is not continuous, while the stress field obtained by traditional elasto-plastic theory is continuous.

Fig. 6 shows the zonal disintegration around the diversion tunnel in Jinping II Hydropower Station. It is shown from the

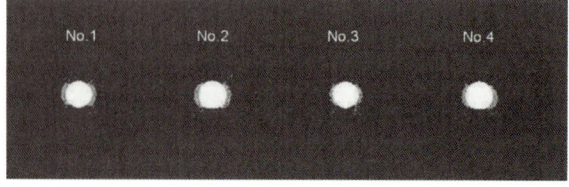

Fig. 6. Distributions of zonal disintegration around four diversion tunnels.

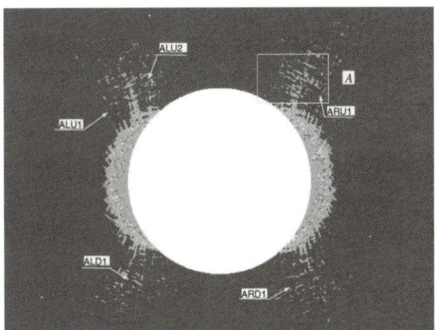
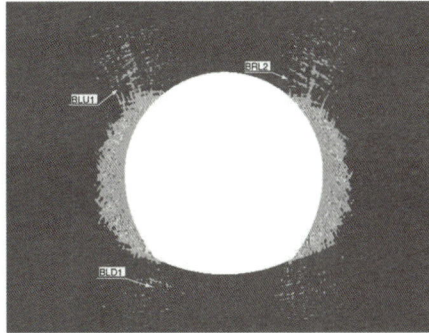
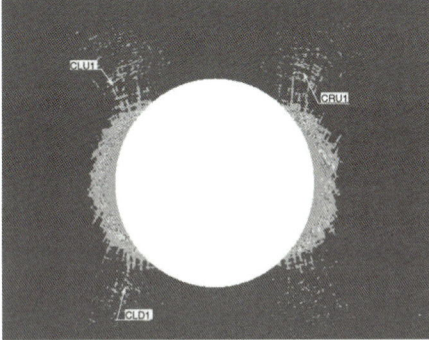
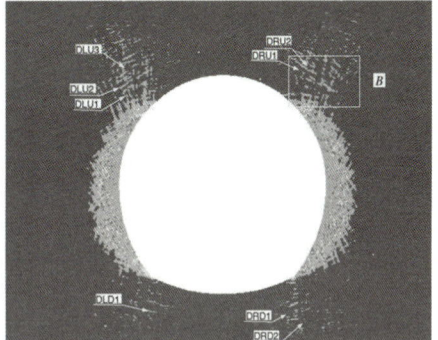

Fig. 7. The enlarged photo of the distributions of disintegration zone within the surrounding rock mass around diversion tunnel.

numerical results that there exist zones of fissured and non-fissured, extending discretely into surrounding rock mass around

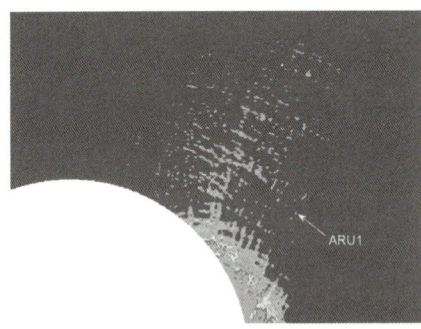

(a) Enlarged photo of zonal disintegration in area A around No.4 tunnel

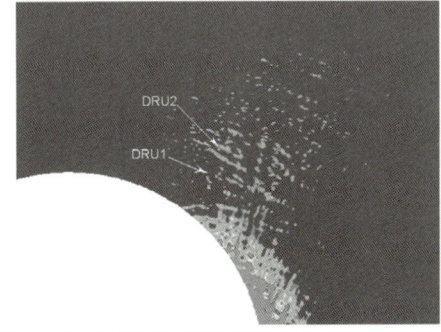

(b) Enlarged photo of zonal disintegration in area B around No.4 tunnel

Fig. 8. Enlarged photo of disintegration zone.

the tunnel. Zonal disintegration, whose width is about 2 m, exists around sidewalls of the tunnel, but also some disintegration zones discretely distribute far from the tunnel. It is shown from Fig. 7 that the fractured zones look like slip-line. The enlarged photo of Fig. 6 is depicted in Figs. 7 and 8.

Table 2 illustrates the sizes of disintegration zone which is far from diversion tunnel. It is suggested that the zonal disintegration is sensitive to tunnel shape. In this model, the magnitude of disintegration zones around circular tunnels is less than that of disintegration zones around the four-arcs ones.

5. Comparisons between presented method and traditional elasto-plastic theory

Figs. 9 and 10 illustrate the stress distributions in surrounding rock mass around diversion tunnels. Material parameters applied in the elasto-plastic analysis are as follows: weight density $\gamma = 27.2$ kN m^{-3}, Young's modulus $E = 18.9$ GPa, Poisson's ratio $\upsilon = 0.23$, cohesion $C = 10.6$ MPa, the angle of internal friction $\varphi = 34°$, vertical loading $\sigma_v = 69.5$ MPa, horizontal loading $\sigma_h = 23$ MPa.

It is noted from Fig. 9 that the surrounding rock mass around the arch crown and sidewalls are subjected to tensile stress, which distributes in a little region. It is shown from Fig. 10 that the vertical stress is increased around diversion tunnel. Compared with Figs. 4 and 5, the maximum horizontal stress is larger in Fig. 4 than that in Fig. 9 and the maximum vertical stress in Fig. 5 is smaller than that in Fig. 10. Moreover, the stress distributions in Fig. 9 are more homogeneous than those in Fig. 4.

It can be seen from single tunnel that the stress distributions obtained by elasto-plastic theory and by present method are distinct. Take No. 1 tunnel for example, it is revealed from Fig. 11a that the horizontal stress distribution is homogeneous

Table 2
Magnitude and distribution of disintegration zones far from diversion tunnel.

Diversion tunnel	Number of fracture	Location of fracture	Orientation of fracture	Distance from tunnel (m)	Length of fracture (m)	Width of fracture (cm)
No. 1 tunnel	ALU1	Top of left corner	Tangential	1.30	1.87	8–10
	ALU2	Top of left corner	Tangential	2.69	2.44	7–8
	ARU1	Top of right corner	Tangential	2.58	3.22	6–10
	ALD1	Bottom of left corner	Radial	–	3.56	6–8
	ARD1	Bottom of right corner	Tangential	1.93	1.50	6–7
No. 2 tunnel	BLU1	Top of left corner	Tangential	1.96	1.91	6–7
	BRU1	Top of right corner	Tangential	2.63	1.82	6–8
	BLD1	Bottom of left corner	Tangential	1.67	1.69	8–16
No. 3 tunnel	CLU1	Top of left corner	Tangential	2.75	2.08	6–10
	CLD1	Bottom of left corner	Radial	–	2.87	10–16
	CRU1	Top of right corner	Tangential	2.96	2.12	8–10
No. 4 tunnel	DLU1	Top of left corner	Tangential	1.93	1.19	5–6
	DLU2	Top of left corner	Tangential	2.26	2.82	6–10
	DLU3	Top of left corner	Tangential	3.03	1.89	5–8
	DRU1	Top of right corner	Tangential	2.11	2.03	8–11
	DRU2	Top of right corner	Tangential	3.28	2.23	10–12
	DLD1	Bottom of left corner	Tangential	1.28	2.83	<6
	DRD1	Bottom of right corner	Tangential	1.48	1.69	<8
	DRD2	Bottom of right corner	Tangential	2.20	2.43	8–11

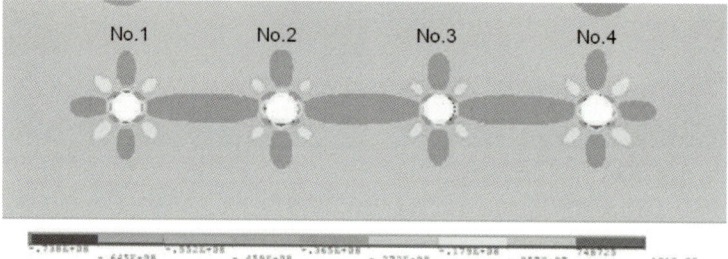

Fig. 9. The distributions of horizontal stress obtained by elasto-plastic theory.

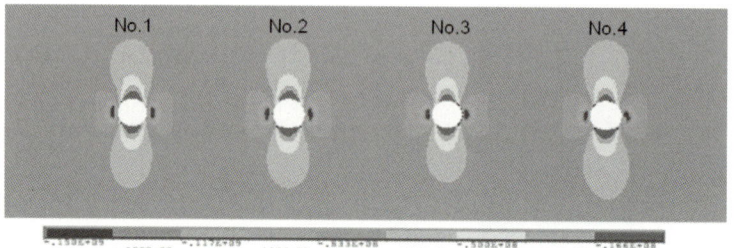

Fig. 10. The distributions of vertical stress obtained by using elasto-plastic theory.

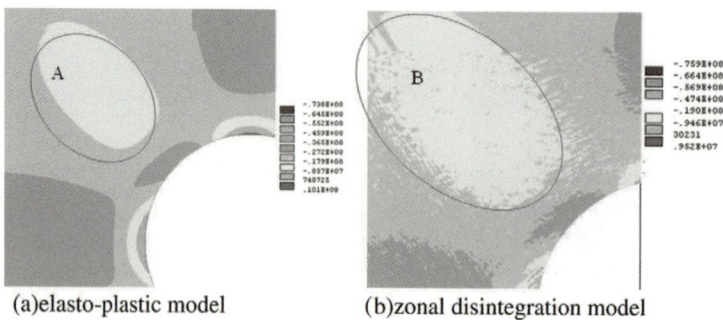

Fig. 11. Comparison of the distributions of horizontal stress near No. 1 tunnel between two different models.

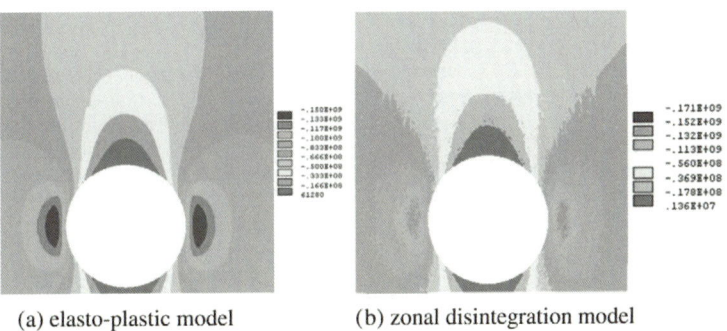

Fig. 12. Comparison of the distributions of vertical stress around No. 1 tunnel between two different models.

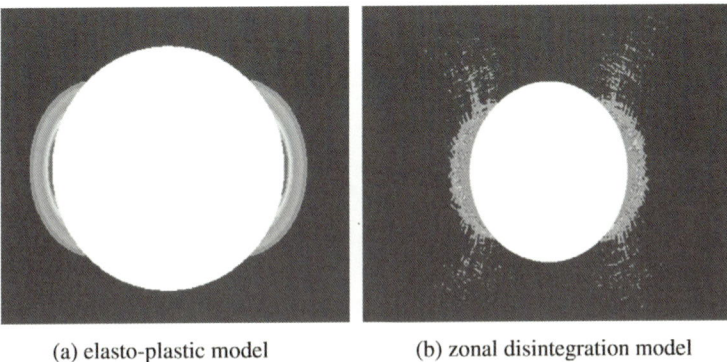

Fig. 13. Enlarged photo of disintegration zone around No. 3 tunnel.

and continuous, such as part A. However, in the zonal disintegration model in Fig. 11b, there exist some areas around No. 1 tunnel, in which the horizontal stress is decreased and nonhomogeneous and discontinuous after excavation, such as part B. The same distribution rule of vertical stress has been observed in Fig. 12a and b.

The phenomenon of zonal disintegration only occur when the geostress and strength of rock mass satisfy certain relationship. According to experimental observation and theoretical analysis [13,14,17], under plane strain condition, when value of geostress is less than that of uniaxial compressive strength of rock mass, the phenomenon of zonal disintegration may not occur. The present numerical results also show that (1) when value of geostress is less than that of uniaxial compressive strength of rock mass, the phenomenon of zonal disintegration may not occur, as depicted in Fig. 13a; (2) when value of geostress is more than that of uniaxial compressive strength of rock mass, the phenomenon of zonal disintegration may occur, as shown in Fig. 13b.

6. Comparisons between presented method and model tests

In order to compare with model tests, a four-arcs tunnel is chosen, as shown in Fig. 14. In Fig. 14a, material parameters applied in numerical simulation are as follows: for the intact rock, weight density $\gamma = 27.2$ kN m^{-3}, Young's modulus $E = 18.9$ GPa, Poisson's

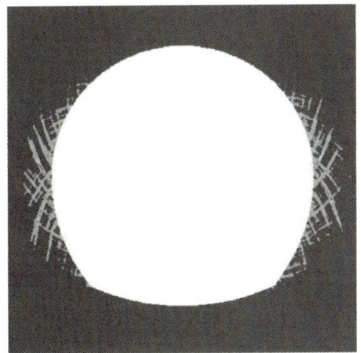

Fig. 14. Disintegration zone around four-arcs tunnel.

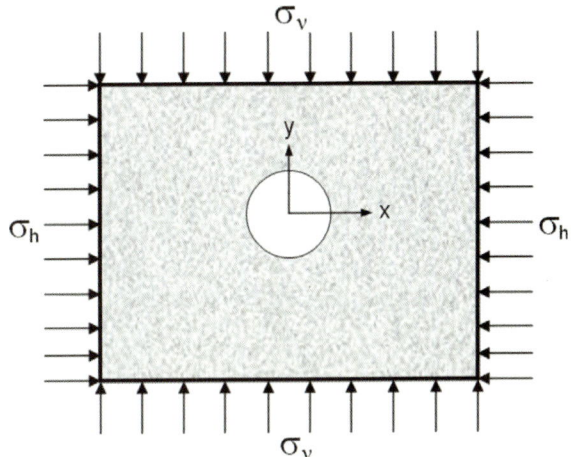

Fig. 15. Model of an included four-arcs tunnel.

ratio $v = 0.23$, cohesion $C = 10.6$ MPa, the angle of internal friction $\varphi = 30°$. For the weak-element, weight density $\gamma = 25.16$ kN m^{-3}, Young's modulus $E = 10.9$ GPa, Poisson's ratio $v = 0.23$, cohesion $C = 4$ MPa, the angle of internal friction $\varphi = 30°$, vertical geostress $\sigma_v = 70$ MPa, horizontal geostress $\sigma_h = 40$ MPa, The uniaxial compressive strength of rock mass is 31 MPa.

It can be seen from Fig. 14 that the disintegration zones look like slip-line. The present results are in good agreement with model tests by Gu et al. [11], as shown in Figs. 1a and b.

Table 3
Material properties.

Intact rock			Weak-element		
Young's modulus (GPa)	Poisson's ratio	Weight density (kN m^{-3})	Young's modulus (GPa)	Poisson's ratio	Weight density (kN m^{-3})
18.9	0.23	27.2	10.9	0.23	25.16

7. Sensitivity analysis

The mechanism of zonal disintegration is not clearly revealed in the previous analysis. In order to reveal the mechanism of zonal disintegration in depth rock masses. The numerical analysis is preformed. The mechanism of zonal disintegration in depth rock masses is revealed by using the present numerical analysis. Size of the fractured zone under different condition is also obtained from the present numerical analysis.

In order to study the parameter sensitivity, a single tunnel is modeled under plain strain condition, as shown in Fig. 15. A cross-section (60 m × 60 m) with an included diversion tunnel of 12–13 m in diameter was analyzed and the total number of elements is 122,240. Material properties of intact rock and weak-element are listed in Table 3.

Fig. 16 illustrates influence of value of geostress on the fractured zone. Material parameters applied to numerical simulation are as follows: for intact rock, cohesion $C = 16$ MPa, the angle of internal friction $\varphi = 30°$. For weak-element, cohesion $C = 4$ MPa,

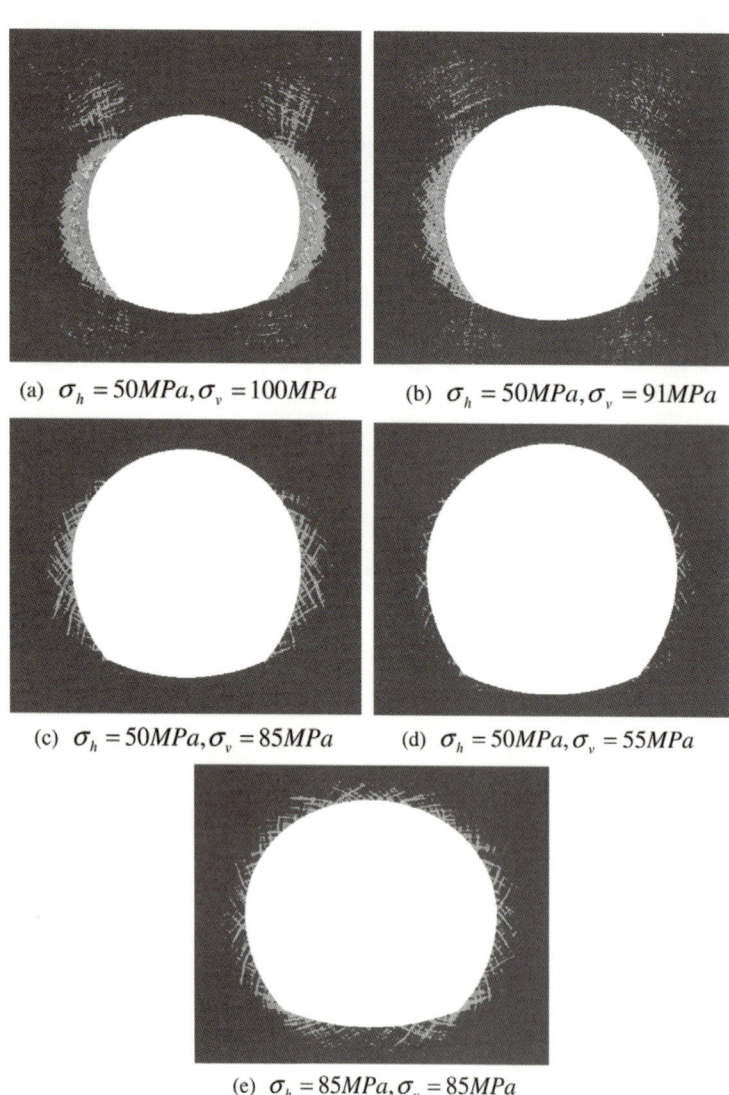

(a) $\sigma_h = 50 MPa, \sigma_v = 100 MPa$

(b) $\sigma_h = 50 MPa, \sigma_v = 91 MPa$

(c) $\sigma_h = 50 MPa, \sigma_v = 85 MPa$

(d) $\sigma_h = 50 MPa, \sigma_v = 55 MPa$

(e) $\sigma_h = 85 MPa, \sigma_v = 85 MPa$

Fig. 16. Comparison of disintegration zone under different stress condition.

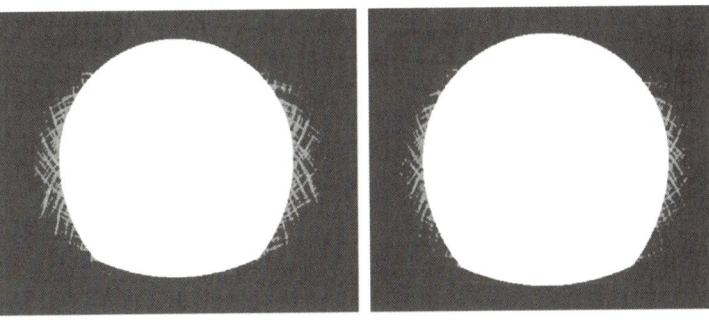

(a) The angle of internal friction $\beta = 30^0$ (b) The angle of internal friction $\varphi = 33^0$

Fig. 17. Comparison of disintegration zones in surrounding rock mass between different angle of internal friction.

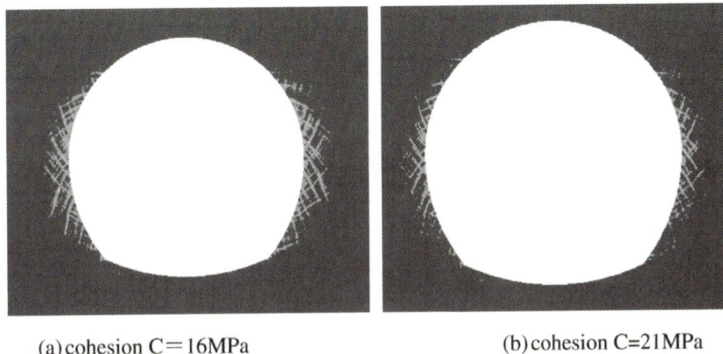

(a) cohesion C＝16MPa (b) cohesion C=21MPa

Fig. 18. Comparison of disintegration zones in surrounding rock mass between different cohesion values.

the angle of internal friction $\varphi = 30°$, the uniaxial compressive strength of rock mass is about 55 MPa.

It is observed from Fig. 16a and b that fractured zone, which causes apparent failure of rock mass around tunnel, exists around the sidewalls of tunnel, but also fractured zone discretely distributes far from the tunnel when value of the geostress is large enough. It is shown from Fig. 16c that zonal disintegration only exists around the sidewalls of tunnel since value of geostress is not large enough to lead to apparent failure of rock mass around tunnel. It is found from Fig. 16d that zonal disintegration does not occur when value of geostress is less than the uniaxial compressive strength of rock mass. It is suggested from Fig. 16 that zonal disintegration only occurs within rock mass around tunnel when value of geostress is larger than the uniaxial compressive strength of rock mass. In Fig. 16a and b, since the value of geostress is more than that of the uniaxial compressive strength of rock mass, zonal disintegration causes apparent failure of rock mass around tunnel. As a result, the phenomenon of disintegration zones transfers to the left and right corner far from the tunnel, which is founded first by authors. By comparison Fig. 16a with Fig. 16b, it is found that magnitude and size of fractured zone increases with increasing the vertical stress when value of the horizontal stress is smaller than that of the vertical stress. It is observed from Fig. 16e that zonal disintegration exists around the sidewalls and arch crown of tunnel when the vertical and horizontal stresses are nearly the same.

Fig. 17 illustrates effect of the angle of internal friction on the fractured zone. Material parameters applied to numerical simulation are as follows: for intact rock, cohesion $C = 16$ MPa. For weak-element, cohesion $C = 4$ MPa. The vertical and horizontal geostress are 85 MPa and 50 MPa, respectively.

It can be concluded from Fig. 17 that zonal disintegration depends on the angle of internal friction of rock mass. Magnitude and size of fractured zone increase with decreasing the angle of internal friction of rock mass.

Fig. 18 shows effect of magnitude of cohesion on disintegration zone. Material parameters applied to numerical simulation are as follows: for intact rock, the angle of internal friction of rock mass $\varphi = 30°$. For weak-element, the angle of internal friction $\varphi = 30°$, cohesion $C = 4$ MPa. The values of vertical and horizontal geostress are 85 MPa and 50 MPa, respectively.

It is found from Fig. 18 that zonal disintegration is sensitive to the value of cohesion. It can be concluded from Fig. 18 that magnitude and size of the disintegration zones increase with decreasing the value of cohesion.

In order to study the parameter sensitivity, an included circular tunnel is modeled under plain strain condition, as shown in Fig. 19. A cross-section (60 m × 60 m) with an included circular tunnel of 13 m in diameter was analyzed. In numerical simulation, material parameters are as follows: for intact rock, the angle of internal friction of rock mass $\varphi = 30°$, $C = 10$ MPa. For weak-element, the angle of internal friction $\varphi = 30°$, cohesion $C = 4$ MPa.

It is shown from Fig. 19 that zonal disintegration apparently exists around circular tunnel, but also fractured zone apparently distributes far from the tunnel when value of the geostress is large enough. In Fig. 19a, magnitude of fractured zone around left and right sidewall tunnel is about 3, respectively. magnitude of fractured zone around left and right upper corner of tunnel is about 2, respectively. In Fig. 19b, since the value of geostress is more than that of the uniaxial compressive strength of rock mass, zonal disintegration causes failure of parts of rock mass around circular tunnel. Magnitude of fractured zone around left and right sidewall

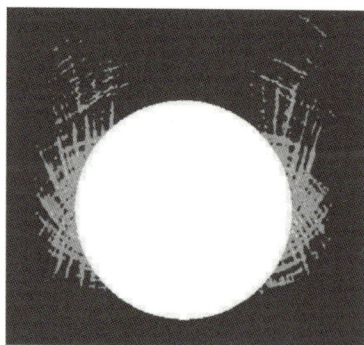

(a) $\sigma_h = 28MPa, \sigma_v = 75MPa$

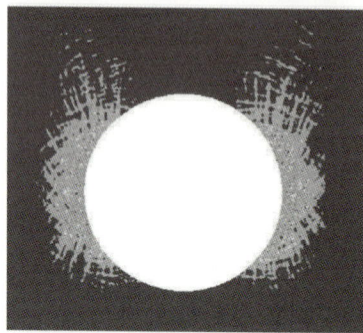

(b) $\sigma_h = 28MPa, \sigma_v = 85MPa$

Fig. 19. Comparison of disintegration zone around circular tunnel under different stress condition.

tunnel is about 4, respectively. Magnitude of fractured zone around left and right upper corner of tunnel is about 3, respectively.

8. Conclusion

By means of numerical simulation, the special phenomenon of zonal disintegration in surrounding rock mass around the diversion tunnels of Jinping II Hydropower Station is discussed in this paper. The conclusions are summarized as follows:

(1) In order to model the growth and coalescence of cracks within rock mass in Jinping II Hydropower Station, the weak-element method is adopted. Meanwhile, the sizes and magnitudes of fractured zone around diversion tunnels are determined.
(2) The present numerical results are different from the one obtained by the elasto-plastic theory. Zonal disintegration, which is in the slip-line pattern, is observed in the numerical simulation. The present numerical results are in good agreement with the model tests.
(3) The effect of geostress condition, cohesion and the angle of internal friction on the phenomenon of zonal disintegration is determined. The phenomenon of zonal fracture may occur when value of geostress is larger than the uniaxial compressive strength of rock mass. Magnitude and size of the disintegration zones increase with decreasing the value of cohesion and the angle of internal friction.

Acknowledgement

The authors would like to express their sincere thank to Professor G.C. Sih for his kind help and remarks. This work is supported by the National Natural Science Foundation of China (Nos. 50490275, 50679097, 50621403, 50778184).

References

[1] J.B. Martion, N.A. Chandler, Excavation-induced damage studies at the underground research laboratory, International Journal of Rock Mechanics and Mining Sciences 41 (8) (2004) 1413–1426.
[2] R.S. Read, Characterizing excavation damage in highly-stressed granite at AECL's underground research laboratory, in: J.B. Martino, C.D. Martin (Eds.), Proceedings of the EDZ Workshop, Designing the Excavation Disturbed Zone for a Nuclear Repository in Hard Rock held at the International Conference on Deep Geological Disposal of Radioactive Waste, Winnipeg, MB, September 20, 1996.
[3] C.D. Martino, R.S. Read, AECL's mine-by experiment: a test tunnel in brittle rock, in: M. Aubertin, F. Hassani, H. Mitri (Eds.), Proceedings of the Second North American Rock Mechanics Symposium: Narms '96, Rock Mechanics Tools and Techniques, June 19–21, Rotterdam, Balkema, 1996.
[4] B.H.G. Brady, E.T. Brown, Rock Mechanics for Underground Mining, George Allen and Unwin, 1985.
[5] Q.H. Qian, The characteristic scientific phenomena of engineering response to deep rock mass and the implication of deepness, Journal of East China Institute of Technology 27 (1) (2004) 1–5 (in Chinese).
[6] E.I. Shemyakin, G.L. Fisenko, M.V. Kurlenya, et al., Zonal disintegration of rocks around underground workings, part I: data of in-situ observations, Journal of Mining Science 22 (3) (1986) 157–168.
[7] E.I. Shemyakin, G.L. Fisenko, M.V. Kurlenya, et al., Zonal disintegration of rocks around underground workings, part II: rock fracture simulated in equivalent materials, Journal of Mining Science 22 (4) (1986) 223–232.
[8] E.I. Shemyakin, G.L. Fisenko, M.V. Kurlenya, et al., Zonal disintegration of rocks around underground workings, part III: theoretical concepts, Journal of Mining Science 23 (1) (1987) 1–6.
[9] E.I. Shemyakin, G.L. Fisenko, M.V. Kurlenya, et al., Journal of Mining Science 25 (4) (1989) 297–302.
[10] Y.S. Pan, Y.J. Li, X. Tang, Z.H. Zhang, Study on zonal disintegration of rock, Chinese Journal of Rock Mechanics and Engineering 26 (s1) (2007) 3335–3341 (in Chinese).
[11] J.C. Gu, L.Y. Gu, A.M. Chen, J.M. Xu, W. Chen, Model test study on mechanism of layered fracture within surrounding rock of tunnels in deep stratum, Chinese Journal of Rock Mechanics and Engineering 27 (3) (2008) 433–438 (in Chinese).
[12] X.P. Zhou, Q.H. Qian, Zonal fracturing mechanism in deep tunnel, Chinese Journal of Rock Mechanics and Engineering 26 (5) (2007) 877–895 (in Chinese).
[13] R.T. Ewy, N.G.W. Cook, Deformation fracture around cylindrical openings in rock. I. Observations analysis of deformations, International Journal of Rock Mechanics and Mining Sciences & Geomechanics Abstracts 27 (5) (1990) 387–407.
[14] R.T. Ewy, N.G.W. Cook, Deformation fracture around cylindrical openings in rock. II. Initiation growth and interaction of fractures, International Journal of Rock Mechanics and Mining Sciences & Geomechanics Abstracts 27 (5) (1990) 409–427.
[15] G.Q. Chen, X.T. Feng, H. Zhou, B.R. Chen, S.L. Huang, C.Q. Zhang, Numerical analysis of the long-term stability of the seepage tunnel in Jinping II Hydropower Station, Rock and Soil Mechanics 28 (s2) (2007) 417–422 (in Chinese).
[16] S.Y. Wu, X.H. Ren, X.R. Chen, et al., Stability analysis and supporting design of surrounding rocks of diversion tunnel for Jinping Hydropower Station, Chinese Journal of Rock Mechanics and Engineering 24 (20) (2005) 3777–3782 (in Chinese).
[17] X.P. Zhou, F.H. Wang, Q.H. Qian, B.H. Zhang, Zonal fracturing mechanism in deep crack-weakened rock masses, Theoretical and Applied Fracture Mechanics 50 (1) (2008) 57–65.

Bifurcation condition of crack pattern in the periodic rectangular array

X.P. Zhou*, Q.H. Qian, H.Q. Yang

School of Civil Engineering, Chongqing University, Chongqing 40045, PR China

Available online 8 December 2007

Abstract

Bifurcation condition of crack pattern in the periodic rectangular array plays an important role in determining the final failure pattern of rock material. An approximation for the critical crack size/spacing ratio is established for a uniformly growing periodic rectangular array yields to a non-uniform growing pattern of crack growth. Numerical results show that the critical crack size/spacing ratio λ_{cr} depends on the number of cracks, the crack spacing, the perpendicular distance between two adjacent rows, as well as the loading conditions. In general, λ_{cr} increases with the number of lines. It is observed that the critical crack size/spacing ratio λ_{cr} for the periodic rectangular array decreases with an increase in the perpendicular distance between two adjacent rows. It is clear that the critical crack size/spacing ratio λ_{cr} for the periodic rectangular array under shear stress increases with increasing the crack spacing.
© 2007 Elsevier Ltd. All rights reserved.

Keywords: The periodic rectangular array; Bifurcation; Crack interaction

1. Introduction

The periodic rectangular array of cracks always manifest themselves in the form of rock joints [1,2]. The evolution pattern of crack array plays a very important role in determining the final failure pattern of rock material. To reveal the evolution pattern of crack array, some experimental methods are developed. Among these experimental studies, bifurcation in growth pattern of cracks captured by using CCD in Riedel shear test for a transparent gelatin sample seems most interesting [3]. According to the experimental result, at a certain loads, uniform growth of crack arrays stops while periodic growth of every alternate crack occurs. Then, this alternating evolution of crack growth continues before the coalescence of neighbouring cracks appears. On the other hand, some theoretical consideration on the alternating pattern of crack growth were provided.

On the basis of group theory, it is shown that bifurcation of periodicity of $n = 2$ should occur [3]. The theoretical s [3] agree with their observation of alternating evolution pattern. Based on the crack interaction, stability of the growth patterns in crack arrays has theoretically been researched [4–14]. It is found that the crack interaction affects mainly the bifurcation in the growth pattern of crack array. The previous studies concentrated primarily on thermally induced edge cracks and echelon cracks.

To the authors' knowledge, the condition for onset of bifurcation in the growth pattern of the periodic rectangular array has not been investigated previously. In this paper, the possible bifurcation of crack growth pattern in the periodic rectangular array is analyzed. The main objective of the present study is to derive the critical ratio λ_{cr} (size/spacing ratio $\lambda_{cr} = l/w$, l is the half crack length and w is the distance between the center of two adjacent cracks) for the problem of the periodic rectangular array under tension and shear stress. In particular, the possible bifurcation in two situations is considered. Established is the periodic rectangular array subjected to far-field tension applying perpendicular to the crack faces (see Fig. 1). The periodic rectangular array is then subjected to far-field shear applied in parallel with the crack faces depicted in Fig. 2.

To formulate the nonlinear system for the growth of the periodic rectangular array and to examine the possibility of

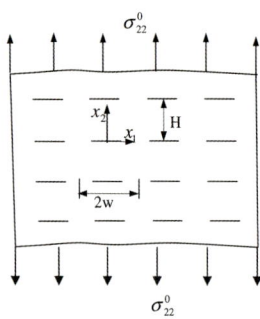

Fig. 1. The periodic rectangular array under far-field tension stress perpendicular to the crack face.

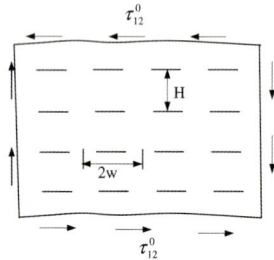

Fig. 2. The periodic rectangular array under far-field shear acting parallel to the crack face.

bifurcation of such a system, crack interaction must be taken into account. For two-dimensional problem of the periodic rectangular array, the asymptotic analysis is applied to estimate the effect of crack interaction, as this method of analysis has been used and found useful for crack interactions. Although strictly speaking the asymptotic analysis is valid only when the cracks are far apart, numerical calculations [15–17] showed that the method is accurate even when the neighbouring cracks are very close to each other. More specifically, both shear and tensile pseudo-tractions are first formulated using asymptotic method. A nonlinear system for the incremental crack growth of the periodic rectangular array are established. A uniformly growing crack array is always one of the possible solutions. It is found the possibility that at a certain crack size, non-uniformly growing crack array also becomes an alternate solution. Perturbations are applied to both the crack sizes and pseudo-tractions, then linear stability analysis is employed. The final eigenvalue equation is obtained by setting the determinant of the coefficient matrix for the homogeneous system of the increments in crack length to zero. Numerical results will then be obtained for λ_{cr} under various loading conditions.

2. Asymptotic analysis of crack interaction

For the periodic rectangular array, the asymptotic analysis is applied to estimate the effect of the crack interaction. In particular, when the interaction of neighbouring cracks cannot be neglected, the presence of one crack inevitably affects the growth of the others. The problem can be solved by decomposing the problem into two subsidiary problems. The first one is the solution of the uncrack solid under external far-field traction, and the second is a solid with an isolated crack loaded by an internal pressure, which will balance the stress field on the crack face when the two solutions are superimposed. Due to crack interactions, the local internal pressure on each crack face needs to be adjusted because of the presence of all the other cracks.

Consider an infinite elastic solid containing N parallel cracks of equal crack length. By superposition of the subsidiary problems, the traction consistency condition on each crack in a subsidiary problem can be written as [1,16,17]

$$\sigma_{ij}^{pj}(x_1) - 2\sum_{j=1}^{N}\int_0^l K_{ijkl}(x_1,x_1^j)\sigma_{kl}^{pj}(x_1^j)\mathrm{d}x_1^j = \sigma_{ij}^0,$$
$$x_1^j \in [0,l] \tag{1}$$

where $\sigma_{ij}^p(x_1)$ is the pseudo-traction on the crack faces, σ_{ij}^0 is the applied stress.

Now examine the variations of $K_{ijkl}(x_1,x_1^j)$ in order to characterize the crack interactions. Firstly, let the body be subjected to $\sigma_{22}^0 \neq 0$ or $\tau_{12}^0 \neq 0$. Under these loading conditions, we only study the nontrivial component $K_{2222}(x_1,x_1^j)$ and $K_{1212}(x_1,x_1^j)$. The expression for $K_{2222}(x_1,x_1^j)$ and $K_{1212}(x_1,x_1^j)$ was given, respectively, [15]

$$K_{2222}(x_1,x_1^j) = \frac{2}{w}\mathrm{Re}\left\{\frac{\cos(\pi x_1^j/(2w))\sqrt{(\sin(\pi l/2w))^2-(\sin(\pi x_1^j/2w))^2}}{\left[(\sin(\pi z/2w))^2-(\sin(\pi x_1^j/2w))^2\right]\sqrt{1-(\sin(\pi l/2w)/\sin(\pi z/2w))^2}}\right\}$$

$$-\frac{x_2}{w}\mathrm{Im}\left\{\frac{\cos(\pi x_1^j/(2w))\sqrt{(\sin(\pi l/2w))^2-(\sin(\pi x_1^j/2w))^2}}{\left[(\sin(\pi z/2w))^2-(\sin(\pi x_1^j/2w))^2\right]^2\left[1-(\sin(\pi l/2w)/\sin(\pi z/2w))^2\right]}\left[\frac{\pi}{w}\sin\frac{\pi z}{2w}\cos\frac{\pi z}{2w}\sqrt{1-(\sin(\pi l/2w)/\sin(\pi z/2w))^2}\right.\right.$$

$$\left.\left.+\frac{\pi}{2w}\frac{(\sin(\pi l/2w))^2\cos(\pi z/2w)\left[(\sin(\pi l/2w))^2-(\sin(\pi x_1^j/2w))^2\right]}{(\sin(\pi z/2w))^3\sqrt{1-(\sin(\pi l/2w)/\sin(\pi z/2w))^2}}\right]\right\} \tag{2}$$

$$K_{1212}(x_1, x_1^j) = \frac{1}{2w}\mathrm{Re}\left\{\frac{\cos(\pi x_1^j/(2w))\sqrt{(\sin(\pi l/2w))^2 - (\sin(\pi x_1^j/2w))^2}}{\left[(\sin(\pi z/2w))^2 - (\sin(\pi x_1^j/2w))^2\right]\sqrt{1 - (\sin(\pi l/2w)/\sin(\pi z/2w))^2}}\right\}$$

$$-\frac{x_2}{2w}\mathrm{Im}\left\{\frac{\cos(\pi x_1^j/(2w))\sqrt{(\sin(\pi l/2w))^2 - (\sin(\pi x_1^j/2w))^2}}{\left[(\sin(\pi z/2w))^2 - (\sin(\pi x_1^j/2w))^2\right]^2\left[1 - (\sin(\pi l/2w)/\sin(\pi z/2w))^2\right]}\right.$$

$$\times\left[\frac{\pi}{w}\sin\frac{\pi z}{2w}\cos\frac{\pi z}{2w}\sqrt{1 - (\sin(\pi l/2w)/\sin(\pi z/2w))^2}\right.$$

$$\left.\left.+\frac{\pi}{2w}\frac{(\sin(\pi l/2w))^2\cos(\pi z/2w)\left[(\sin(\pi l/2w))^2 - (\sin(\pi x_1^j/2w))^2\right]}{(\sin(\pi z/2w))^3\sqrt{1 - (\sin(\pi l/2w)/\sin(\pi z/2w))^2}}\right]\right\} \quad (3)$$

where $z = x_1 + ix_2 = x_1 + i(jH), i = \sqrt{-1}, j = 1, 2, \ldots, +\infty$.

An approximate closed-form solution is obtained for the integral equation (1) for the periodic rectangular array arrays in Figs. 1 and 2. To this end, we assume that the cracks are so distributed that the higher-order terms (in comparison with terms of order 1) containing $e^{-(H/2w)\pi}$ ($1 \geq 4j$), $e^{-n(H/2w)\pi}\sin^m(\pi l/2w)$ and $e^{-n(H/2w)\pi}\sin^m(\pi x_1/2w)$ ($n \geq 2j$ and $m \geq 2$) can be neglected. The asymptotic expressions for $K_{2222}(x_1, x_1^j)$ and $K_{1212}(x_1, x_1^j)$ for periodic rectangular array of Figs. 1 and 2 are [1,16,17]

$$\begin{Bmatrix} K_{2222}^r(x_1, x_1^j) \\ K_{1212}^r(x_1, x_1^j) \end{Bmatrix} = \begin{Bmatrix} -2\left[1 + 2j\frac{H}{2w}\pi\right] \\ -\left[1 - 2j\frac{H}{2w}\pi\right] \end{Bmatrix}\frac{2}{w}e^{-j(H/w)\pi}$$

$$\times \cos\frac{\pi x_1^j}{2w}\sqrt{\sin^2\frac{\pi l}{2w} - \sin^2\frac{\pi x_1^j}{2w}} \quad (4)$$

Substituting Eq. (4) into Eq. (1) gives

$$\sigma_{22}^{pj}(x_1) - 2\sum_{j=1}^{N}\int_0^l K_{2222}^r(x_1, x_1^j)\sigma_{22}^{pj}(x_1^j)\mathrm{d}x_1^j = \sigma_{22}^0 \quad (5)$$

$$\tau_{12}^{pj}(x_1) - 2\sum_{j=1}^{N}\int_0^l K_{1212}^r(x_1, x_1^j)\tau_{12}^{pj}(x_1^j)\mathrm{d}x_1^j = \tau_{12}^0 \quad (6)$$

For the sake of conciseness, Eqs. (5) and (6) is first rewritten in matrix form

$$\{\sigma\} = \{\sigma^\infty\} + [\lambda]\{\sigma\} \quad (7)$$

where $\{\sigma^\infty\} = \{\sigma_{22}^0, \sigma_{22}^0, \sigma_{22}^0, \ldots, \sigma_{22}^0\}$, $\{\sigma\} = \{\sigma_{22}^{p1}, \sigma_{22}^{p2}, \sigma_{22}^{p3}, \ldots, \sigma_{22}^{pN}\}^T$ can be computed from Eq. (5) (for far-field tensile stress); $\{\sigma\} = \{\tau_{12}^{p1}, \tau_{12}^{p2}, \tau_{12}^{p3}, \ldots, \tau_{12}^{pN}\}^T$ can be computed from Eq. (6), $\{\sigma^\infty\} = \{\tau_{12}^0, \tau_{12}^0, \tau_{12}^0, \ldots, \tau_{12}^0\}$ (for far-field shear stress). The superscript T means the transpose of the vector.

$$[\lambda] = \begin{bmatrix} \lambda_{12} & \lambda_{22} & \cdots & \lambda_{1N} \\ \lambda_{21} & \lambda_{22} & \cdots & \lambda_{2N} \\ \vdots & \vdots & & \vdots \\ \lambda_{N1} & \lambda_{N2} & \cdots & \lambda_{NN} \end{bmatrix} \text{ is a } N \times N \text{ matrix}$$

obtained from Eqs. (5) and (6) for far-field tensile stress and far-field shear stress, respectively.

Rearranging Eq. (7) leads to

$$\{\sigma\} = [M]^{-1}\{\sigma^\infty\} \quad (8)$$

where $[M] = [I] - [\lambda]$, $[I]$ is a unit matrix of size $N \times N$.

3. Bifurcation in crack pattern

A mixed mode crack growth criterion has to be adopted for the formulation of crack growth and the subsequent stability analysis. Although there are numerous criteria for mixed mode crack growth, there is still considerable debate on which theory the most suitable for the propagation of crack under mixed mode conditions. The criteria for mixed mode crack propagation includes the maximum tangential stress criterion [18], the maximum energy release rate criterion [19], the strain energy density criterion [20] and the maximum dilatational strain density criterion [21]. The investigation on which theory to use is clearly out of scope of the present study. Instead, we quote the strain energy density criterion, which is commonly used in fracture mechanics.

The strain energy density factor of rock material can be written as [20]

$$S = a_{11}K_\mathrm{I}^2 + 2a_{12}K_\mathrm{I}K_{\mathrm{II}} + a_{22}K_{\mathrm{II}}^2 + a_{33}K_{\mathrm{III}}^2 \quad (9)$$

where K_I, K_{II} and K_{III} are the mode I and Mode II and Mode III stress intensity factors, respectively.

$$\begin{cases} a_{11} = \frac{1+v_0}{8\pi E_0}[(3 - 4v_0 - \cos\varphi)(1 + \cos\varphi)] \\ a_{12} = \frac{1+v_0}{8\pi E_0}(2\sin\varphi)[\cos\varphi - (1 - 2v_0)] \\ a_{22} = \frac{1+v_0}{8\pi E_0}[4(1 - \cos\varphi)(1 - v_0) \\ \qquad + (1 + \cos\varphi)(3\cos\varphi - 1)] \\ a_{33} = \frac{1+v_0}{2\pi E_0} \end{cases}$$

For the mixed mode I and II problem, Eq. (9) can be rewritten as

$$S = a_{11}K_\mathrm{I}^2 + 2a_{12}K_\mathrm{I}K_{\mathrm{II}} + a_{22}K_{\mathrm{II}}^2 \quad (10)$$

The onset of rapid crack propagation is assumed to start when the strain energy density $S_{\min}$ associated with $(dW/dV)_{\min}^{\max}$ reaches a critical value, i.e.

$$S_{\min} = S_c \tag{11}$$

where $S_c = r_c\left(\frac{dW}{dV}\right)_c = \frac{(1+v_0)(1-2v_0)}{2\pi E_0}K_{\mathrm{ICC}}^2$ is the critical strain energy density factor, K_{ICC} is mode I critical stress intensity factor.

Mode I and mode II stress intensity factors can be expressed as $K_\mathrm{I} = \sigma_{22}^p\sqrt{\pi l}$ and $K_\mathrm{II} = \tau_{12}^p\sqrt{\pi l}$. Then, take variation (11) leads to

$$\delta l = -\frac{2\pi l^2}{S_c}[a_{11}\sigma_{22}^p\delta\sigma_{22}^p + a_{12}\sigma_{22}^p\tau_{12}^p(\delta\tau_{12}^p + \delta\sigma_{22}^p) + a_{22}\tau_{12}^p\delta\tau_{12}^p] \tag{12}$$

Therefore, crack length increment for the jth crack in the periodic rectangular array becomes

$$\delta l_j = -\frac{2\pi l_j^2}{S_c}[a_{11}\sigma_{22}^{pj}\delta\sigma_{22}^{pj} + a_{12}\sigma_{22}^{pj}\tau_{12}^{pj}(\delta\tau_{12}^{pj} + \delta\sigma_{22}^{pj})$$
$$+ a_{22}\tau_{12}^{pj}\delta\tau_{12}^{pj}] \tag{13}$$

Take the variation of equation for pseudo-traction (8), we have

$$\frac{\partial}{\partial l_j}[M]^{-1}\{\sigma_{22}^0\}\delta l_j = -\sum_{j=1}^N [M]^{-1}\frac{\partial}{\partial l_j}[M][\sigma_{22}^p]\delta l_j \tag{14}$$

$$\{\delta\tau_{12}^p\} = \sum_{j=1}^N \frac{\partial}{\partial l_j}[M]^{-1}\{\tau_{12}^0\}\delta l_j$$
$$= -\sum_{j=1}^N [M]^{-1}\frac{\partial}{\partial l_j}[M][\tau_{12}^p]\delta l_j \tag{15}$$

Consequently, for the jth crack, Eqs. (14) and (15) can be specialized as, respectively

$$\{\delta\sigma_{22}^{pj}\} = \sum_{j=1}^N \frac{\partial}{\partial l_j}[M]^{-1}\{\sigma_{22}^0\}\delta l_j$$
$$= -\sum_{j=1}^N [M]^{-1}\frac{\partial}{\partial l_j}[M][\sigma_{22}^{pj}]\delta l_j \tag{16}$$

$$\{\delta\tau_{12}^{pj}\} = \sum_{j=1}^N \frac{\partial}{\partial l_j}[M]^{-1}\{\tau_{12}^0\}\delta l_j$$
$$= -\sum_{j=1}^N [M]^{-1}\frac{\partial}{\partial l_j}[M][\tau_{12}^{pj}]\delta l_j \tag{17}$$

At the onset of bifurcation, and σ_{22}^{pj} and τ_{12}^{pj} must satisfy the following growth criterion:

$$\sigma_{22}^{pj} = \sqrt{\frac{S_C}{\pi a_{11} l_j}} \tag{18}$$

$$\tau_{12}^{pj} = \sqrt{\frac{S_C}{\pi a_{22} l_j}} \tag{19}$$

Since before the onset of bifurcation in the crack growth pattern all cracks have the same size, we must have at the onset bifurcation $l_1 = l_2 = \cdots = l_N = l$. Substitution of (18) and (16) into (13) yields

$$\{\delta l_n\} = 2l\sum_{j=1}^N \left[[[M]^{-1}]^\mathrm{T}\frac{\partial}{\partial l_j}[M]\right]\delta l_j \tag{20}$$

Substitution of Eqs. (19) and (17) into Eq. (13) yields

$$\{\delta l_n\} = 2l\sum_{j=1}^N \left[[[M]^{-1}]^\mathrm{T}\frac{\partial}{\partial l_j}[M]\right]\delta l_j \tag{21}$$

where $j = 1,2,\ldots,N$. It is straightforward to see that $[[M]^{-1}]^\mathrm{T}$ and $\frac{\partial}{\partial l_j}[M]$ are function of l/w. Thus, it is advantageous to introduce a new parameter $\lambda = l/w$. Finally, Eqs. (20) and (21) can be simplified to

$$\sum_{j=1}^N A_{nj}(\lambda)\delta l_j = 0, \quad j = 1,2,\ldots,N \tag{22}$$

or expressed in matrix form as

$$[A(\lambda)]\{\delta l\} = \{0\} \tag{23}$$

For nontrivial solution $\delta l_j (j=1,2,\ldots,N)$, we must have the determinant of the coefficient matrix being zero. Therefore, we have the following eigenvalue equation for the determination of λ_{cr}:

$$\det\begin{bmatrix} A_{11}(\lambda) & A_{12}(\lambda) & \cdots & A_{1N}(\lambda) \\ A_{21}(\lambda) & A_{22}(\lambda) & \cdots & A_{2N}(\lambda) \\ \vdots & \vdots & & \vdots \\ A_{N1}(\lambda) & A_{N2}(\lambda) & \cdots & A_{NN}(\lambda) \end{bmatrix} = 0 \tag{24}$$

The numerical solution of this eigenvalue equation is considered next.

4. Numerical results and discussion

4.1. Bifurcation of growth pattern of the periodic rectangular array under uniaxial tension

The numerical solution for the eigenvalue equation can be solved by using standard technique of root searching. The critical crack size/spacing ratio λ_{cr} for the periodic rectangular array under uniaxial tension are summarized in Figs. 3–7. It is seen from Figs. 3–7 that the critical crack size/spacing ratio λ_{cr} depends on the crack spacing and the perpendicular distance between two adjacent rows. It is obvious from Figs. 3–7 that the critical crack size/spacing ratio λ_{cr} increases with the number of rows. It is observed from Figs. 3–7 that the critical crack size/spacing ratio λ_{cr} for the periodic rectangular array under uniaxial tension increases with decreasing the perpendicular distance between two adjacent rows.

4.2. Bifurcation of growth pattern of the periodic rectangular array under shear

The critical crack size/spacing ratio λ_{cr} for the periodic rectangular array under shear are summarized in Figs. 8–12. It is seen from Figs. 8–12 that the critical

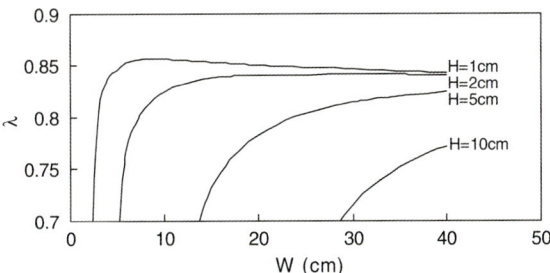

Fig. 3. The critical half crack length/spacing ratio λ_{cr} at the onset of the bifurcation of growth patterns of the periodic rectangular array under uniaxial tension when $j = 2$.

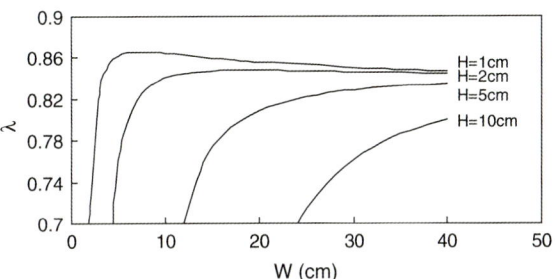

Fig. 4. The critical half crack length/spacing ratio λ_{cr} at the onset of the bifurcation of growth patterns of the periodic rectangular array under uniaxial tension when $j = 3$.

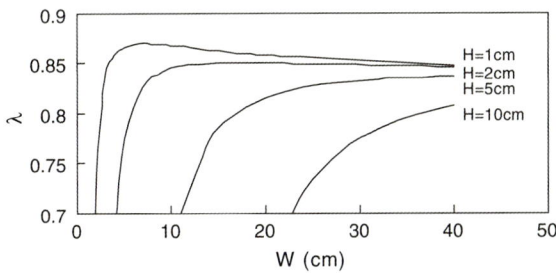

Fig. 5. The critical half crack length/spacing ratio λ_{cr} at the onset of the bifurcation of growth patterns of the periodic rectangular array under uniaxial tension when $j = 4$.

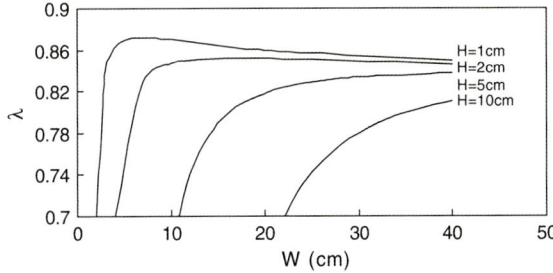

Fig. 6. The critical half crack length/spacing ratio λ_{cr} at the onset of the bifurcation of growth patterns of the periodic rectangular array under uniaxial tension when $j = 5$.

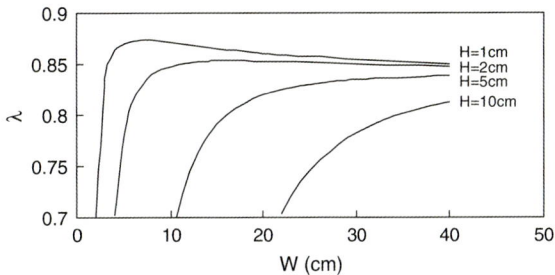

Fig. 7. The critical half crack length/spacing ratio λ_{cr} at the onset of the bifurcation of growth patterns of the periodic rectangular array under uniaxial tension when $j = 6$.

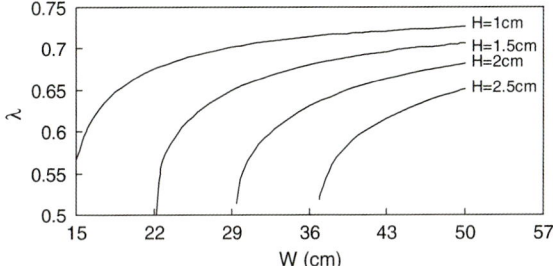

Fig. 8. The critical half crack length/spacing ratio λ_{cr} at the onset of the bifurcation of growth patterns of the periodic rectangular array under shear when $j = 3$.

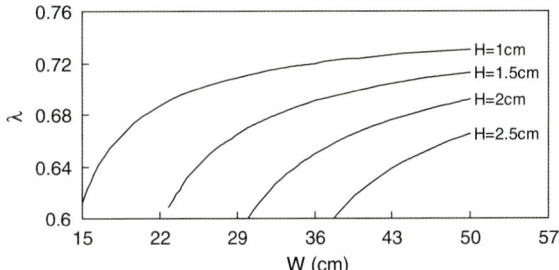

Fig. 9. The critical half crack length/spacing ratio λ_{cr} at the onset of the bifurcation of growth patterns of the periodic rectangular array under shear when $j = 4$.

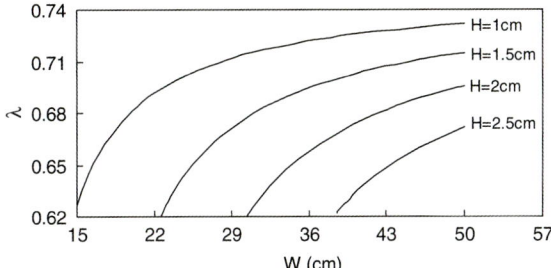

Fig. 10. The critical half crack length/spacing ratio λ_{cr} at the onset of the bifurcation of growth patterns of the periodic rectangular array under shear when $j = 5$.

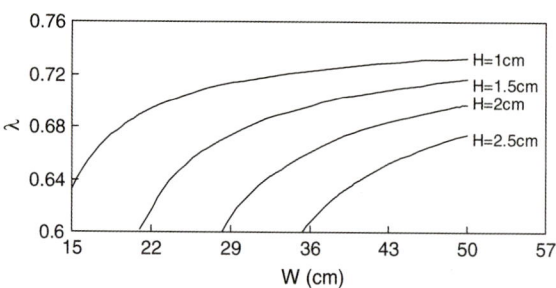

Fig. 11. The critical half crack length/spacing ratio λ_{cr} at the onset of the bifurcation of growth patterns of the periodic rectangular array under shear when $j = 6$.

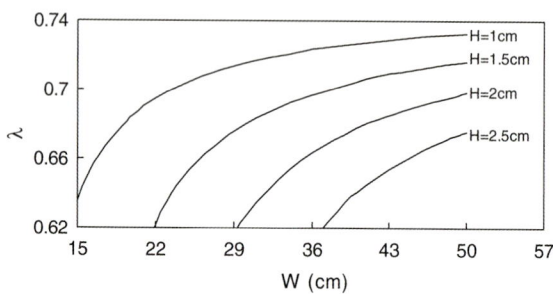

Fig. 12. The critical half crack length/spacing ratio λ_{cr} at the onset of the bifurcation of growth patterns of the periodic rectangular array under shear when $j = 7$.

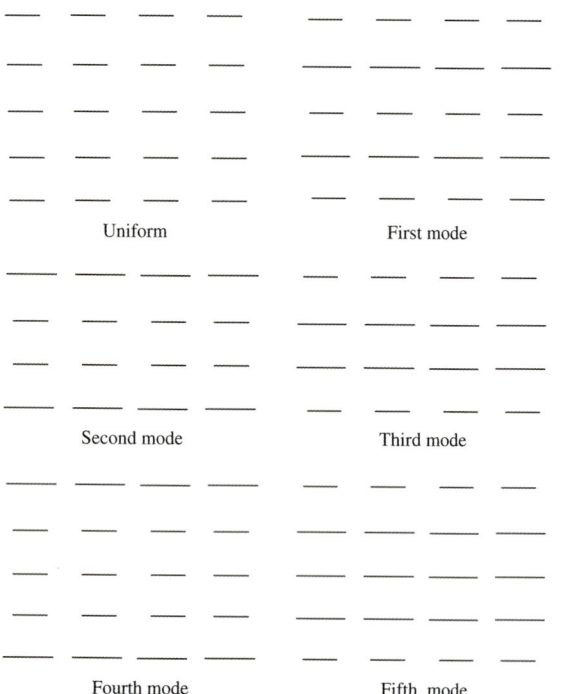

Fig. 13. A typical sketch for the bifurcations in growth pattern corresponding to the first five eigenvalue of homogeneous system of crack increments at the onset of bifurcation.

crack size/spacing ratio λ_{cr} depends on the crack spacing and the perpendicular distance between two adjacent rows. It is obvious from Figs. 8–12 that the critical crack size/spacing ratio λ_{cr} increases with the number of lines. It is observed from Figs. 8–12 that the critical crack size/spacing ratio λ_{cr} for the periodic rectangular array increases with decreasing the perpendicular distance between two adjacent rows. It is clear that the critical crack size/spacing ratio λ_{cr} for the periodic rectangular array under shear increases with increasing the crack spacing.

4.3. Growth pattern of the periodic rectangular array

Based on the "group theoretical bifurcation analysis", Oguni et al. found that the alternative pattern of adjacent long and short crack is expected [3]. No eigenvector has been calculated to examine the actual growth pattern in his works. In view of this, in the present study the eigenvector corresponding to each eigenvalue of homogeneous system of the incremental crack vectors is computed. The eigenvector or the growth pattern depends on the actual loading configuration as well as crack geometry. But, in general, for smaller eigenvalue the components of the eigenvectors are always alternating with one plus and one minus value. It agrees with Oguni's theoretical results [3]. We also observe that for higher eigenvalues eigenvector components can alternate with two minus and one plus, and with two plus and one minus. It is consistent with Chau's theoretical results [13]. However, we also find that for higher eigenvalues eigenvector components can alternate with three minus and one plus, and with three plus and one minus. The first five most likely modes of non-uniform growth pattern of the periodic rectangular array are depicted in Fig. 13.

5. Conclusion

Proposed is a bifurcation analysis for the possible change in growth pattern of the periodic rectangular array under far-field tension acting along the direction perpendicular to the crack faces, far-field shear acting parallel to the crack faces. The method of "pseudo-traction" or "asymptotic analysis" is using to estimate the interaction, which together with the growth criterion forms a nonlinear system for the crack length and the pseudo-traction. The critical crack size to crack spacing ratio or λ_{cr} is viewed as the nonlinear parameter for this nonlinear system. When λ increases to λ_{cr}, non-uniform crack growth will become possible. The critical crack size/spacing ratio λ_{cr} depends on the crack spacing and the perpendicular distance between two adjacent rows.

The fundamental eigenvector of the homogeneous system at the onset of bifurcation of growth pattern is found corresponding to the alternating growth pattern.

Acknowledgments

The work was supported by the National Natural Science Foundation of China (Nos. 50490275, 50778184).

References

[1] X.P. Zhou, Y.X. Zhang, Q.L. Ha, K.S. Zhu, Bounds on the complete stress–strain relation for a crack-weakened rock mass under compressive loads, International Journal of Solids and Structures 41 (22–23) (2004) 6173–6196.

[2] X.P. Zhou, Q.L. Ha, Y.X. Zhang, K.S. Zhu, Analysis of the localization of deformation and study on the complete stress–strain relation for brittle rock subjected to dynamic compressive loads, International Journal of Rock Mechanics and Mining Sciences 41 (2) (2004) 311–319.

[3] K. Oguni, M. Hori, K. Ikeda, Analysis on evolution pattern of periodically distributed defects, International Journal of Solids and Structures 34 (1997) 3259–3272.

[4] Z.P. Bazant, H. Ohtsubo, Stability condition for propagation of a system of cracks in a brittle solid, Mechanics Research Communications 4 (1977) 353–366.

[5] L.M. Keer, S. Nemat-Nasser, A. Oranratnachai, Unstable growth of thermally induced interacting cracks in brittle solids: further results, International Journal of Solids and Structures 15 (1978) 111–126.

[6] S. Nemat-Nasser, L.M. Keer, K.S. Parihar, Unstable growth of thermally induced interacting cracks in brittle solids, International Journal of Solids and Structures 14 (1978) 409–430.

[7] Z.P. Bazant, A.B. Wahab, Instability and spacing of cooling or shrinkage cracks, Journal of Engineering Mechanics, ASCE 105 (EM5) (1979) 873–889.

[8] Z.P. Bazant, Instability of parallel thermal cracks and its consequences for hot-dry rock thermal generation, in: D.P.H. Hasselman, R.A. Heller (Eds.), Thermal Stresses in Severe Environments, Plenum Press, New York, 1980, pp. 169–181.

[9] A.V. Dyskin, On the possibility of bifurcation in linear periodic arrays of 2-D cracks, International Journal of Fracture 67 (1994) R31–R36.

[10] H.B. Muhlhaus, K.T. Chau, A. Ord, Bifurcation pf crack pattern in arrays of two-dimensional cracks, International Journal of Fracture 77 (1996) 1–14.

[11] K.T. Chau, H.B. Muhlhaus, A. Ord, Bifurcation in growth patterns for arrays of parallel Griffith, edge and sliding cracks, Key Engineering Materials 145–149 (1998) 71–76.

[12] A.P. Parker, Stability of arrays of multiple edge cracks, Engineering Fracture Mechanics 62 (1998) 577–591.

[13] K.T. Chau, G.S. Wang, Condition for the onset of bifurcation in en echelon crack arrays, International Journal for Numerical and Analytical Methods in Geomechanics 25 (3) (2001) 289–306.

[14] X.P. Zhou, Onset of bifurcation in crack pattern in two-dimensional intermittent cracks under far field shear stress, International Journal of Nonlinear Sciences and Numerical Simulation 8 (1) (2007) 55–58.

[15] B.L. Karihaloo, J. Wang, On the solution of doubly periodic array of cracks, Mechanics of Materials 26 (1997) 209–212.

[16] J. Wang, J. Fang, B.L. Karihaloo, Asymptotics of multiple crack interactions and prediction of effective modulus, International Journal of Solids and Structures 37 (2000) 4261–4273.

[17] J. Wang, J. Fang, B.L. Karihaloo, Asymptotic bounds on overall moduli of cracked bodies, International Journal of Solids and Structures 37 (2000) 6221–6237.

[18] F. Erdogan, G.C. Sih, On the crack extension path in plates under plane loading and transverse shear, Journal of Basic Engineering, ASME 85 (1963) 519–527.

[19] M.A. Hussain, E.L. Pu, J.H. Underwood, Strain energy release rate for a crack under combined Mode I and II, ASTM STP 560 (1974) 2–28.

[20] G.C. Sih, Strain energy density factor applied to mixed mode crack problems, International Journal of Fracture 10 (1974) 305–321.

[21] N.A. Yehia, On the use of the T-criterion in fracture mechanics, Engineering Fracture Mechanics 22 (1985) 189–199.

岩石、岩体的动力强度与动力破坏准则

钱七虎[1]，戚承志[2]

(1.解放军理工大学 工程兵工程学院，南京 210007; 2.北京建筑工程学院 土木交通学院，北京 100044)

摘要：就岩石与岩体变形与破坏的时间特性进行了理论研究.研究了岩石及岩体的强度对于应变率的依赖关系及其机理,给出了考虑强度对于应变率依赖关系的莫尔库仑准则,确定了岩体的破坏尺寸与应变率之间的关系.讨论了若干时间性破坏准则,揭示了它们之间的内在联系.最后讨论了动力强度理论的应用.

关键词：岩石与岩体；变形与破坏；时间因素；动力强度；破坏准则
中图分类号：TU 451　　**文献标识码**：A　　**文章编号**：0253-374X(2008)12-1599-07

Dynamic Strength and Dynamic Fracture Criteria of Rock and Rock Mass

QIAN Qihu[1], QI Chengzhi[2]

(1. Institute of Engineering Corps, PLA University of Science and Technology, Nanjing 210007, China;
2. School of Civil and Communication Engineering, Beijing Institute of Civil Engineering and Architecture, Beijing 100044, China)

Abstract: The paper presents a theoretical study of the temporal features of deformation and fracture of rock and rock mass and an investigation of the strain rate dependence of strength and its mechanism. The paper also presents the Mohr-Coulomb constitutive relations with a consideration of strain rate dependence and the relationship between the size of blocks of fractured rock mass and the strain rate. Several temporal fracture criteria are discussed, and their intrinsic relationship is revealed. Finally, the application of dynamic strength theories is discussed.

Key words: rock and rock mass; deformation and fracture; temporal factor; dynamic strength; criteria of fracture

岩石作为固体材料，其破坏定义一直是唯象的，即当固体中形成了主干裂纹时，就认为岩石出现了破坏.岩体的破坏过程非常复杂，包括微裂纹的产生、扩展、合并形成主干裂纹.工程对象所处的岩体中都存在有断裂、节理、裂隙，不能按照上述定义判定岩体已受破坏，因此岩体的破坏必然要涉及到空间的概念.

固体从连续状态转变到破坏状态的描述称为破坏模型，联系破坏模型参数之间关系的方程称为破坏准则.破坏模型和破坏准则的研究和表述构成了强度理论的内容.

1 经典强度理论

经典强度理论是现代动力强度理论的基础，现代动力强度理论是经典强度理论的发展和完善.

收稿日期：2008-01-20。
基金项目：国家自然科学基金资助项目(50490275)；北京市自然科学基金及北京市教委联合资助项目(KZ200810016007)。
本文原载于《同济大学学报(自然科学版)》(2008年第36卷第12期)。

经典强度理论中,假设当固体中某点应力及应变张量的某种组合达到极限值时就产生破坏,极限值在主应力空间上组成了一个表面,超出此表面意味着固体的破坏.应力应变的组合及其极限值的确定构成了各个强度理论的基本内容.

经典强度理论有5种:第一强度理论为最大正应力强度理论;第二强度理论为最大正应变强度理论;第三强度理论为最大剪应力强度理论;第四强度理论为最大比应变能理论;第五种强度理论为第三强度理论的变体,即莫尔-库伦理论.莫尔-库伦强度理论目前广泛应用于岩石工程实践.莫尔理论的一个变体为Hoek-Brown准则[1-2].

由于莫尔-库伦准则(C-M准则)没有反映中间主应力的影响,不能解释岩土在静水压力下也能被破坏的现象,所以在岩石工程实践中还广泛应用德鲁克-普拉格准则(D-P准则)[3].D-P准则克服了C-M准则的主要弱点,但实际上D-P准则是在C-M准则和米赛斯准则(第四强度理论)基础上扩展而得.

上述强度理论分别适用于固体的不同应力状态的不同破坏模式.

莫尔-库伦理论的数学表达式中没有中间主应力,它只考虑了一个面上的剪应力及其正应力,因此称之为单剪理论.

单剪理论的进一步发展为双剪理论[4],而双剪理论的进一步发展为统一强度理论[5].单剪、双剪理论及介于二者之间的其它破坏准则都是统一强度理论的特例或线性逼近.因此可以说,统一强度理论在强度理论的发展史上具有突出的贡献.

但是目前现代强度理论尚不完善成熟,还处于不断发展的进程中.主要是在唯象上还没有反映时间因素对固体破坏的影响以及没有考虑固体破坏的物理构造本质和机理.而前者则是动力破坏准则和强度理论的主要内容.

2 动力强度理论

动力强度理论在反映破坏的时间因素以及破坏的物理机理上发展了传统理论,但是动力强度理论还不成熟,尚处于发展中.

2.1 动力强度的试验数据

岩石工程中,岩石的破坏时间从钻探、冲击(钻)到爆炸大致为秒、毫秒、微秒3个量级.在强烈动载作用下岩石的强度会提高很多,一些数据见表1.

表1 岩石断裂强度[6]

Fig.1 The fracture strength of rocks

岩石	静力强度/atm	动力强度/atm	动力、静力强度之比
石灰岩	420	2 730	6.5
大理岩(垂直沉积方向)	210	1 890	9.0
大理岩(平行沉积方向)	630	4 900	7.8
花岗岩	700	4 000	5.7

图1和图2是相应于实验室中动力等速加载试验[7],其中,τ为加载至破坏的时间,s;σ为破坏应力;σ_{dyn},σ_{st}为相应的动力与静力加载下的破坏应力;$\dot{\varepsilon}(t)$为应变率;$\dot{\varepsilon}_1$为Ⅰ区与Ⅱ区之间的分界线;$\dot{\varepsilon}_2$为Ⅱ区与Ⅲ区的分界线.

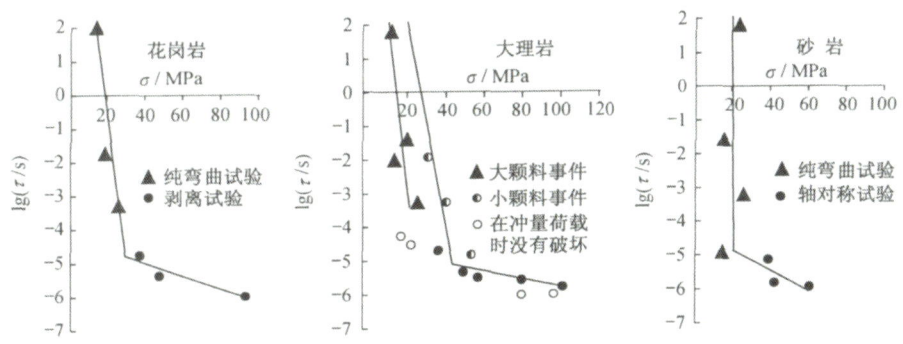

图1 破坏时间与加载之间的关系

Fig.1 Relationship between fracture time and loading

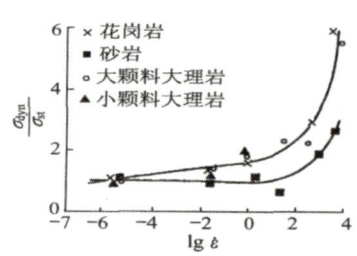

图 2 强度对于应变率的依赖关系

Fig.2 The strain rate dependence of strength

由图 1 试验曲线可见,当 $\tau > 10^{-5}$ s 时为准静载加荷条件,强度与破坏时间弱相关;当 $\tau < 10^{-5}$ s 时,破坏应力随加载时间减少而迅速增加.

由图 2 试验曲线可见,当应变速率 $\dot{\varepsilon} > 10$ s^{-1} 时,随着变形速度增加,动力强度急剧增大.

在抗爆炸的岩石工程中,按围岩的最大变形速率,选取相应的岩石动力强度数值,就可以完成相应的工程设计.但是若要分析围岩的变形破坏,进行动力数值仿真,必须要了解岩石的动力强度理论及其动力破坏准则.

2.2 考虑时间因素和构造的主要强度理论

2.2.1 Журков С Н 准则及强度对于应变率的依赖关系

对单轴适中拉应力,材料在静力外载作用下的破坏时间 τ 服从 Журков С Н 公式[8]

$$\tau = \tau_0 \exp\left[\frac{U_0 - \gamma\sigma}{KT}\right] \quad (1)$$

式(1)还可用以建立强度 Y 对时间和温度的公式

$$\sigma = Y = \frac{1}{\gamma}\left(U_0 + KT\ln\frac{\tau_0}{\tau}\right) \quad (2)$$

式中:τ_0 为原子的 Debye 振动周期数量级别的参数,约为 10^{-12} s;U_0 为活化能;γ 为活化体积,通常为 $10\sim 1\,000$ 原子体积;σ 为单轴拉应力;K 为玻尔兹曼常数;T 为绝对温度.

文献[9]基于微观理论推导了 Журков С Н 公式.

实验表明,式(2)在动载的作用下也成立,它描述了包括岩石在内的多种材料在应变率较低时的抗拉及抗压强度对应变率的依赖性.

如果材料在常应变速率 $\dot{\varepsilon}$ 下达到材料的极限应变 ε_0,则 Журков С Н 公式可变为

$$\sigma = Y = \frac{1}{\gamma}\left[U_0 + KT\left(\ln\dot{\varepsilon} - \ln\frac{\varepsilon_0}{\tau_0}\right)\right] \quad (3)$$

令 $\dot{\varepsilon}_0 = \dfrac{\varepsilon_0}{\tau_0}$,则有

$$\sigma = Y = \frac{1}{\gamma}\left(U_0 + KT\ln\frac{\dot{\varepsilon}}{\dot{\varepsilon}_0}\right) \quad (4)$$

对材料受剪,类似公式为

$$Y_\tau = \frac{1}{\gamma_\tau}\left(G_0 + KT\ln\frac{\dot{\gamma}}{\dot{\gamma}_0}\right) \quad (5)$$

式中:Y_τ 为动力剪切强度;γ_τ 为剪切变形情况下的活化体积;G_0 为剪切情况下的活化能;$\dot{\gamma}$ 为剪切应变率;$\dot{\gamma}_0 = \gamma_0/\tau_0$,其中 γ_0 为材料的极限剪切应变.

研究表明[10],在不同的应变率区段,不同的机制起主导作用.在应变率较低阶段,变形的热活化机制起主导作用;当应变率大于某一值时,材料强度随应变率的增加而急剧增加,此时材料的变形和破坏具有绝热性质,粘性阻尼机制起主导作用;当应变率很大时,粘性系数随应变率增加而减少,热活化机制又重新出现,此时,裂纹的临界应力不依赖裂纹尺寸,这样在广泛的裂纹尺寸范围内,裂纹增长同时启动,多裂纹的增长和连接使得破坏产生.岩石等脆性材料随应变率变化实验曲线的定性一般规律如图 3 所示.

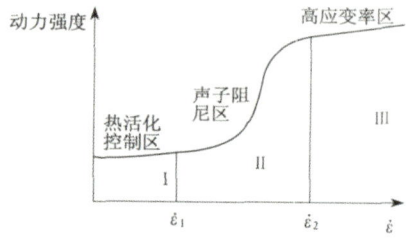

图 3 强度对于应变率依赖机理

Fig.3 The mechanism of strain rate dependence of strength

图 3 的进一步说明是:当高应变率达到爆炸冲击应变率时,激活了新的破坏机制,需要进一步探讨.研究表明,此时不同尺寸的材料缺陷的增长同时启动,而在没有缺陷的区域,材料的热活化启动,产生原子键的断裂,形成了新裂纹产生的源泉,所以材料的热活化机制又重新呈现.而当应变率很高时,材料宏观粘性机制(声子阻尼)由于材料中的裂纹扩展速度受瑞利波速的限制,而表现为粘性系数随着应变率的增加而减少,即粘性机制弱化,热活化机制又居主导.这样可以将材料动力强度对于应变率依赖关系看作是热活化机制与宏观粘性机制并行存在而在不同的应变率区相应占据主导地位的过程.于是

可推得联合的热活化与粘性机制并存竞争的材料动力强度对于应变率的依赖机制[10]

$$Y = \frac{1}{\gamma}\left[U_0 + KT\ln\frac{\dot{\varepsilon}}{\dot{\varepsilon}_0}\right] + \frac{b(\dot{\varepsilon}/\dot{\varepsilon}_s)^n}{(\dot{\varepsilon}/\dot{\varepsilon}_s)^n + 1} \quad (6)$$

$$Y_\tau = \frac{1}{\gamma_\tau}\left[G_0 + KT\ln\frac{\dot{\gamma}}{\dot{\gamma}_0}\right] + \frac{b_1(\dot{\gamma}/\dot{\gamma}_s)^n}{(\dot{\gamma}/\dot{\gamma}_s)^n + 1} \quad (7)$$

式(6)及式(7)中右端第一项反映热活化机制,第二项反映粘性机制,其系数 $b,b_1,\dot{\varepsilon}_s,\dot{\gamma}_s$ 及 n 由试验拟合确定.

为了检验上述动力破坏强度模型的有效性,将式(6)计算结果与前人研究的脆性材料的抗压强度实验数据进行了比较(图4),可见计算结果与试验数据符合得较好.

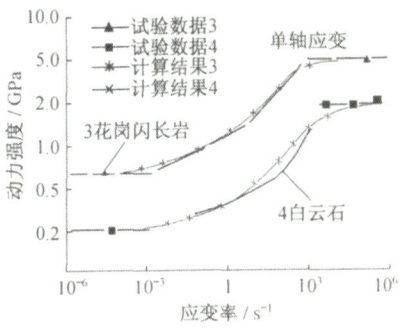

图 4 计算结果与试验数据的比较

Fig.4 Comparison of the calculation result and the experimental data

对于象岩体这样的多层次材料,岩体的粘性系数及动力强度与岩体的结构层次有着紧密的联系.通常尺寸大的结构层次的特征应变率低于尺寸小的结构层次的应变率,大尺寸的结构层次上的强度小于小尺寸结构层次上的强度.随着应变率的增加,岩体的变形与破坏逐渐深入到小尺寸层次上.文献[11]首次将岩体的粘性系数及动力强度与不同的结构层次联系了起来.

对非常应变速率加载,即对阶梯变化力和连续变化力加载 $\sigma(t)$,可应用贝利损伤累积法则,其完全破坏的条件是

$$\sum\frac{\Delta t_i}{\tau(\sigma_i)} = 1 \text{ 及 } \int_0^{t_p}\frac{dt}{\tau[\sigma(t)]} = 1 \quad (8)$$

式中:t 为时间;下标 i 为加载时间段序号;$\tau(\sigma)$ 为加载应力为 σ 时的材料寿命;t_p 为加载至破坏的时间.

2.2.2 Nikiforovsky-Shemyakin 冲量准则

按照 Nikiforovsky-Shemyakin 破坏准则[12],当应力冲量达到临界值 J_0,即 $\int_0^{t_p}\sigma(t)dt = J_0$ 时,材料破坏.一维冲击波作用下,材料内应力 σ 与粒子的运动速度 v 之间的关系为 $\sigma = \rho Dv$,其中 ρ 为介质的密度;D 为介质中的冲击波速度.代入 Nikiforovsky-Shemyakin 破坏准则得

$$\int_0^{t_p}\sigma(t)dt = \int_0^{t_p}\rho Dv\,dt = \rho Du = J_0 \quad (9)$$

式中:u 为破坏位移值.

冲量准则反映了材料的破裂是损伤的累积过程,或者是裂纹的发展过程,即是微裂纹的增多与连接合并产生宏观裂纹(主干裂纹)的过程,这与Журков С Н 破坏时间准则一致.因此,破坏的时间因素表明了动力强度提高的物理本质.

如果冲击波所传播的长度为 L,则 $u = L\varepsilon$,于是有

$$\int_0^{t_p}\sigma(t)dt = \rho DL\varepsilon = J_0 \quad (10)$$

这表明,冲量准则反映了在冲击波作用下,材料在达到临界应变值时破坏.这从而表明第二强度理论可以直接应用于动力破坏.

实际上,因为材料的蠕变现象和动力强度对应变率的依赖现象都反映时间因素对破坏的影响,其物理本质基本相同,所以可将著名的 Aleksandrov 蠕变公式 $\dot{\varepsilon} = \dot{\varepsilon}_0\exp\left[-\dfrac{U_0 - \gamma\sigma}{KT}\right]$ 与 Журков С Н 公式相乘,有 $\tau\dot{\varepsilon} = \tau_0\dot{\varepsilon}_0 = \varepsilon_{st} = $ 常数,即得到与上述相同的结论:不同应变率荷载作用下,其临界破坏应变值不变.

实验表明,在剪切及三轴压缩情况下,在复杂应力及变形情况下,包括在破坏时间偏离Журков С Н 公式的应变率情况下,极限应变 ε_{st} 在比较大的变化范围内(9~10 个量级内)对于温度应力及加载率

(应变率)的依赖性很弱,基本上是一个常数.这表明固体的变形与破坏存在着紧密的联系.因此,可以认为第二强度理论也是一个准时间准则.

2.2.3 Kachanov 损伤演化方程导出的动力破坏准则[13]

在损伤力学理论中,材料的破坏演化过程用唯象的损伤参量 ψ 描述,它的演化方程为

$$\frac{d\psi}{dt} = f(\sigma,\psi) = \begin{cases} A\left(\dfrac{\sigma}{1-\psi}\right)^n & \sigma \geq 0 \\ 0 & \sigma \leq 0 \end{cases} \quad (11)$$

式中:A 及 n 为材料常数.

当损伤参量达到临界值时,材料破坏,$\psi \leq 1$.

积分式(11),可得到

$$\int_0^{t_p} \sigma^n dt = \frac{1-(1-\psi_p)^{n+1}}{A(n+1)} = J_0 \quad (12)$$

当 $n=1$ 时,即得与式(10)相同的冲量准则.

这表明,损伤积累的唯象理论与强度时间准则一致,因而与最大临界应变理论是一致的.

2.2.4 断裂力学导出的动力破坏准则

按断裂力学理论,当裂尖处的最大应力 $\sigma(t,x)$ 在时间-空间 $[t-\tau,t] \times [0,d]$ 上的平均值达到材料的静力强度 σ_c 时,材料就会破坏,这就是 Morozov-Petrov 的时间-空间破坏准则[14]:

$$\frac{1}{\tau}\int_{t-\tau}^{t} dt' \int_0^d \sigma(t',r) dr \leq \sigma_c \quad (13)$$

通过引进 $J_c = \sigma_c \tau$,式(13)可以写成

$$\int_{t-\tau}^{t} dt' \int_0^d \sigma(t',r) dr = J_c(t) \quad (14)$$

式中:τ 为一个与断裂孕育时间相对应的特征时间,它相应于破坏相邻构造单元之间相互作用能量传播的特征时间 d/c,其中 d 为断裂构造的尺度,由带裂纹试件的静力实验确定,c 为最大波速;r 为空间坐标.

因此,由断裂力学导出的破坏准则同样具有构造临界冲量的物理意义.此时,式(14)理解为裂纹处范围的平均冲量达到其临界值 $\sigma_c \tau$ 时材料破坏,此时的时间 τ 可视为破坏时间.

3 岩体的破坏尺寸

材料的破坏强度 Y 与试件尺寸 D 之间存在一定的关系,试件越大,强度越小.不同的学者建议了不同的关系式.

对于受压材料,通常有

$$Y = C_1 D^{-\frac{1}{K}} + C_0 \quad (15)$$

式中:C_1, K, C_0 为正常数.

反之可以利用式(15)讨论施加的荷载(应力 σ)与固体破坏的块体的平均尺寸 D 之间的关系,以 σ 代 Y,将式(15)改写为

$$D = \left(\frac{C_1}{\sigma - C_0}\right)^K \quad (16)$$

式(16)表明,当施加的荷载增加时,破裂的块体尺寸会逐渐减少,进行的静力及动力试验证实了这一点.

动力强度 Y 与应变率及应力状态有关.将式(16)中 σ 用岩体破坏时的动力强度 Y 替代,D 用岩体破坏时的块体尺寸 Δ_i 替代,即得

$$\Delta_i = \left(\frac{C_1}{Y - C_0}\right)^K \quad (17)$$

由于动力强度 Y 依赖于应变率,因此,式(17)表明了破坏块体尺寸对应变率的依赖规律.即应变率也影响着破坏块体的尺寸,应变率越高,动力强度越高,破坏块体尺寸越小.高应变率及其应力状态将决定材料的最终承载能力和最小破坏尺寸.联系到岩体的构造层次,则随着荷载的施加,对应于地质构造运动中不同应变率荷载,岩体首先在大尺寸的构造单元的间隙之间发生破坏,破坏岩体的尺寸为这一水平的构造尺寸.当荷载继续增加至下一级构造单元的破坏强度时,下一级的构造单元间隙之间发生破坏,破坏块体的尺寸为这一级别的构造尺寸.随着荷载的增加,达到破坏的构造单元(岩体的破坏块体)的尺寸逐渐减少,即岩体的节理、裂隙间距逐渐加密.研究表明,裂隙的宽度与岩块(岩体的破坏块体)线性尺寸之间为不变量.因此,随着岩体节理、裂隙的不断加密,其裂隙宽度逐渐减少.

试验表明[15],K 的取值通常为 $K=2$,那么式(17)表示岩体破坏的尺寸与应变能成反比.实验表明从静力到强动力加载范围内材料的破坏尺寸与材料的畸变能成反比,因此应该将式(17)的动力强度理解为剪切动力强度.所以动力加载对于岩体破坏尺度的影响在于由于动力加载所引起的强度升高所导致的畸变能的增加[16].

4 复杂应力状态下考虑强度-应变率效应的莫尔-库伦本构模型

对于岩石类材料,莫尔-库伦准则是一个简单适

用的准则,但是岩石在强度方面表现出对应变率的明显依赖效应,所以分析岩石动力学问题时,应用莫尔-库伦准则必须考虑应变率效应.对于复杂应力状态,用主应力表示的库伦准则可表示为

$$\frac{\sigma_1 - \sigma_3}{2} = \frac{\sigma_1 + \sigma_3}{2}\sin\varphi + C\cos\varphi \quad (18)$$

式中:σ_1,σ_3 分别为最大主应力与最小主应力;φ 为介质的内摩擦角;C 为介质的内聚力.由此得单轴压缩情况下内聚力 C 表达式为

$$C = \frac{\sigma_Y^C}{2} \cdot \frac{1 - \sin\varphi}{\cos\varphi} \quad (19)$$

式中:σ_Y^C 为单轴压缩时的强度极限.

考虑到前述的单轴抗压时强度极限与应变率之间的关系如式(6),式(19)可写为

$$C = \frac{1 - \sin\varphi}{\cos\varphi}\left[\frac{1}{\gamma}\left(U_0 + KT\ln\frac{\dot\varepsilon}{\dot\varepsilon_0}\right) + \frac{b(\dot\varepsilon/\dot\varepsilon_s)^n}{(\dot\varepsilon/\dot\varepsilon_s)^n + 1}\right]e^{AC_0} \quad (20)$$

式中:A 为系数;$C_0 = \sigma_1/\sigma_3$ 表示应力状态;e^{AC_0} 表示应力状态对内聚力的影响.

将式(20)代入用主应力表示的库伦准则,即可得复杂应力状态下考虑强度-应变率效应的莫尔-库伦准则.

对于应变率高的强动载,若考虑到应变率较低时的热活化项对强度的影响不大,可用单轴压缩静力强度 σ_{YS}^C 代替,有

$$\sigma_1 - \sigma_3 = (\sigma_1 + \sigma_3)\sin\varphi + (1 - \sin\varphi) \cdot \left[\sigma_{YS}^C + \frac{b(\dot\varepsilon/\dot\varepsilon_s)^n}{(\dot\varepsilon/\dot\varepsilon_s)^n + 1}\right]e^{AC_0} \quad (21)$$

或 $\sigma_1 = \frac{1 + \sin\varphi}{1 - \sin\varphi}\sigma_3 + \left[\sigma_{YS}^C + \frac{b(\dot\varepsilon/\dot\varepsilon_s)^n}{(\dot\varepsilon/\dot\varepsilon_s)^n + 1}\right]e^{AC_0}$

$$(22)$$

而对于地下封闭爆炸近区在强动载作用下的问题,可进一步简化,因 $\varepsilon_r \gg \varepsilon_\theta$,其剪切变形 $\varepsilon_\gamma = \varepsilon_r - \varepsilon_\theta \approx \varepsilon_r$,体积变形 $\varepsilon_v = \varepsilon_\gamma + 2\varepsilon_\theta \approx \varepsilon_r$,所以可取 $\dot\varepsilon = \dot\varepsilon_r$,这时径向应力与环向应力关系为 $\sigma_1 = \alpha\sigma_3$,其中 ε_r 为径向应变;ε_θ 为环向应变;$\alpha = C_0 = \frac{\mu}{1 - \mu}$,$\mu$ 为泊松比.

于是爆炸近区的莫尔-库伦准则可写为

$$\sigma_1 = \frac{1 + \sin\varphi}{1 - \sin\varphi}\sigma_3 + \left[\sigma_{YS}^C + \frac{b(\dot\varepsilon/\dot\varepsilon_s)^n}{(\dot\varepsilon/\dot\varepsilon_s)^n + 1}\right]e^{A\alpha} \quad (23)$$

5 现代动力强度理论的应用

5.1 剥离面的内部结构

现代动力强度理论将时间因素引入强度概念中,不将破坏作为瞬时的,而将破坏时间作为材料变形及破坏的主要因素.时间因素的引入,加深了对破坏机理的认识,得到了一些与传统强度理论不同的结果.

兹举应力波传播引起的一维剥离破坏(无升压的三角形脉冲)为例.

按照传统的强度理论,入射压应力波反射形成拉应力波后,当某点处的拉应力达到极限拉应力 σ_p 时,材料破坏,此时材料的破坏应在一个面,离自由面的距离 x,$x = \sigma_p/2\sigma_m$,其中 σ_m 为应力波的峰值.而按照现代动力强度理论,在破坏之前,由于反射波产生拉应力,材料各点都孕育着破坏,但并非立即破坏.可以设想,拉应力小的截面,其破坏孕育时间长.因此,它可能与拉应力大、破坏孕育时间短的截面同时破坏.计算结果证实了这个设想,材料的破坏(断裂)具有一定的厚度,即材料的断裂具有构造性,某厚度内的材料同时破坏.

由此可见,按传统强度理论进行动力数值仿真时,其仿真的结果"不真".

5.2 爆破地震动的安全阈值

爆破时产生的地震动会对建筑物及设施造成破坏.衡量地震动对建筑物或设施的破坏程度以及地震动时建筑物或设施安全与否,决定性参量是建筑物或设施所处位置的地震动参数.取地震动加速度、速度和位移中哪一个参数作为爆破地震动效应的安全判据尚无统一意见,但世界各国普遍认同以地表动速度为判据比较可靠稳定,这与以往地震破坏危害的调查分析是一致的,即对结构破坏起控制作用的是速度或位移而不是内力.这可以从现代动力强度理论得到解释.按照传统强度理论,引起结构材料破坏的是材料的应力达到最大临界应力,因此对结构破坏起控制作用的是结构的内力、惯性力和加速度,所以安全判据是地震的加速度.而按现代动力强度理论,结构的爆震破坏是动力破坏,引起结构材料破坏的是材料的应变和位移,当结构材料位移在地震动时达到临界值时材料破坏,所以地震动时的安

全判据应是地震动速度. 考虑到位移是速度在时间上的累积, 所以美国矿业局、德国和芬兰的判据引入震动速度和频率 2 个指标更为合理. 不同的研究者的大量观测结果表明, 在同样的地基条件下, 同一地点、同一类型的建筑物在地基的震动速度超过这一具体建筑物的某个速度阈值时, 其损坏程度相同.

6 结论

自然界中的一切过程都发生于空间及时间之中, 材料的变形与破坏也不例外. 通常研究材料的变形与破坏时, 将注意力主要集中于材料变形与破坏的空间特性方面, 而对于其时间特性注意不足. 材料的变形与破坏具有时间特性, 也即材料变形及破坏依赖于时间因素. 材料对于时间的依赖关系表现为材料的变形与破坏需要一定的时间来完成, 材料的强度依赖于加载时间(应变率). 时间因素的引入加深了人们对材料变形及破坏现象的认识, 使时间因素显现出来, 并成为材料变形与破坏的决定性因素之一.

本文就岩石与岩体变形与破坏的时间特性进行了研究. 研究了岩石及岩体的强度对应变率的依赖关系及这种依赖关系的物理机理. 研究表明, 在不同的应变率区段, 不同的机制起主导作用: 在应变率较低阶段, 变形的热活化机制起主导作用; 当应变率大于某一值时, 材料强度随应变率的增加而急剧增加, 此时材料的变形和破坏具有绝热性质, 粘性阻尼机制起主导作用; 当应变率很大时, 粘性系数随应变率增加而减少, 热活化机制又重新出现. 给出了考虑强度对于应变率依赖关系的莫尔库仑准则, 其中材料的内聚力依赖于应变率. 岩体具有内部结构层次, 大的尺寸水平对应于小的强度, 而小的尺寸水平对应于大的强度. 本文基于岩体强度的尺寸效应及强度的应变率依赖性确定了岩体的破坏尺寸与应变率之间的关系. 探究了若干时间性破坏准则之间的内在联系, 从分析可以发现, 这些准则都反映了材料的变形与破坏需要一定的时间, 以便使材料的内部结构发生变化, 产生破坏. 借助于强度的动力概念, 可以更好地理解各种经典强度理论不能解释的各种现象.

致谢: 本文是在同济大学孙钧讲座报告的基础上整理而成的, 作者对于邀请参加讲座表示感谢!

参考文献:

[1] Hoek E, Brown E T. Empirical strength criterion for rock masses [J]. J Geotech Engrg Div, ASCE, 1980, 106(9): 1013.

[2] Hoek E, Brown E T. The Hoek-Brown failure criterion——a 1988 update[C]// Proc 15th Can Rock Mech Symp. Toronto: Press of University of Toronto, 1988, 31−38.

[3] Drucker D C, Prager W. Soil mechanics and plastic analysis or limit design[J]. Quart of Applied Mathematics, 1952, 10(2): 157.

[4] YU M H, HE L N, SONG L Y. Twin shear stress strength theory and its generalization[J]. Scientia Sinica, Serials A, 1985, 28(11): 1174.

[5] YU M H, HE L N. A new model and theory on yield and failure of materials under complex stress state[C]// Mechanical Behavior of Materials-VI. Oxford: Pergamon Press, 1991, 851−856.

[6] Николаевский В Н. Динамическая прочность и скорость разрушения[C]// Удар, взрыв и разрушение. Москва: Мир, 1981, 166−203.

[7] Беллендир Э Н, Клядченко В Ф, Козачук А И, et al. Сопротивление разрушению горных пород привременнах нагружения 10^2-10^6 С[J]. ФТПРПИ, 1991, (2): 46.

[8] Регель В Р, Слуцкер, Томашевский Э Е. Кинетическая природа прочности твердого дела[M]. Москва, Наука, 1974.

[9] 戚承志, 钱七虎. 材料变形及损伤演化的微观物理机制[J]. 固体力学学报, 2002, 23(3): 312.
QI Chengzhi, QIAN Qihu. The microscopic physical mechanism of deformation and damage of materials[J]. Chinese Journal of Solid Mechanics, 2002, 23(3): 312.

[10] 戚承志, 钱七虎. 岩石等脆性材料动力强度依赖应变率的物理机制[J]. 岩石力学与工程学报, 2003, 21(2): 177.
QI Chengzhi, QIAN Qihu. The physical mechanism of the rate sensitivity of rock-like brittle materials[J]. Chinese Journal of Rock Mechanics and Engineering, 2003, 21(2): 177.

[11] 戚承志, 钱七虎, 王明洋. 岩体的构造层次粘性与动力强度[J]. 岩石力学与工程学报, 2005, 24(增刊): 4679.
QI Chengzhi, QIAN Qihu, WANG Mingyang. The structural hierarchy, viscosity and dynamic strength of rock mass[J]. Chinese Journal of Rock Mechanics and Engineering, 2005, 24(suppl): 4679.

[12] Никифровский В С, Шемякин Е И. Динамическое разрушение тведого тела[M]. Новосибирск: Наука, 1979.

[13] Качанов Л М. Основы механики разрушения[M]. Москва: Наука, 1974.

[14] Morozov N, Petrov Y. Dynamics of fracture[M]. Berlin: Springer-Verlag, 2000.

[15] Bažant Z P, Ožbolt J, Eligehausen R. Fracture size effect: review of evidence for concrete structures[J]. J of Struc Engrg, 1994, 120: 2377.

[16] QI Chengzhi. Dynamic deformation and fracture of geomedium [D]. Moscow: Lomonosov Moscow State University. Faculty of Mechanics and Mathematics, 2006.

深部巷道围岩分区破裂化现象现场监测研究

李术才[1]，王汉鹏[1]，钱七虎[2]，李树忱[1]，范庆忠[1]，袁 亮[3]，薛俊华[3]，张庆松[1]

(1. 山东大学 岩土与结构工程研究中心，山东 济南 250061；2. 解放军理工大学 工程兵工程学院，江苏 南京 210007；
3. 淮南矿业集团公司，安徽 淮南 232001)

摘要：随着矿产资源的开发，深部巷道围岩中出现了破裂区和完整区相间的分区破裂化现象，这是一种新发现的特殊工程地质现象，引起了国内外岩石力学界的关注，但相关研究还不深入。为在现场监测深部巷道围岩的分区破裂化现象，在淮南矿区近千米深井半径不同的巷道选择了 4 个监测断面，每个断面布置了 3～5 个钻孔，采用矿井钻孔电视成像仪对巷道断面围岩不同钻孔内的破裂情况进行了监测，监测到了围岩内的分区破裂化现象，并给出了 4 个断面围岩的分区破裂分布图。通过分析巷道所在岩层、地质资料及钻孔内裂隙的形状，说明裂隙是由巷道开挖引起的。监测结果表明，3 个大断面巷道围岩内的相同破裂分区的半径和厚度相差不大，小断面巷道围岩内的破裂分区与大断面相似，只是相同各破裂分区的半径和厚度相应减小，表明各破裂分区的半径是巷道半径 r 的关系式为 $(\sqrt{2})^{i-1}r$ (i = 1，2，3，4)。4 个监测断面围岩内均产生了 4 个破裂分区，第 1 个破裂分区厚度与巷道半径相当，破裂程度最严重，第 2 个破裂分区其后各破裂分区厚度和破裂程度均依次减小。研究结果对于认识深部巷道围岩的破裂模式及其稳定性支护具有重要意义。

关键词：采矿工程；深部巷道；围岩分区破裂化；现场监测
中图分类号：TD 313 **文献标识码**：A **文章编号**：1000–6915(2008)08–1545–09

IN-SITU MONITORING RESEARCH ON ZONAL DISINTEGRATION OF SURROUNDING ROCK MASS IN DEEP MINE ROADWAYS

LI Shucai[1], WANG Hanpeng[1], QIAN Qihu[2], LI Shuchen[1], FAN Qingzhong[1],
YUAN Liang[3], XUE Junhua[3], ZHANG Qingsong[1]

(1. Research Center of Geotechnical and Structural Engineering, Shandong University, Jinan, Shandong 250061, China；
2. Engineering Institute of Engineering Corps, PLA University of Science and Technology, Nanjing, Jiangsu 210007, China；
3. Huainan Coal Mining Group, Huainan, Anhui 232001, China)

Abstract: With the mine resources development, an alternation of fracturing area and intact area phenomenon so-called zonal disintegration appears in surrounding rock mass of deep mine roadways. This is a newly-discovered special engineering geological phenomenon and it attracts scholars' attentions in rock mechanics fields, but related research is not clear. In order to observe zonal disintegration phenomenon, four sections in roadways with different dimensions under about 1 000 m deep mine are considered; three or five bores being arranged in each section. The fracturing through bores of surrounding rock mass is observed by mine bore TV imager; zonal disintegration phenomenon is observed in surrounding rock mass and zonal disintegration distributions in four sections are obtained. Analyses of roadway location, geological data and fracturing shape in bore show that fracturing is caused by excavation. Observation results also show that the radius and thickness of three large sections' fracturing areas are almost the same. Small sections' fracturing areas are similar with those of

---

收稿日期：2008–03–17；修回日期：2008–05–15。
基金项目：国家杰出青年科学基金项目(A 类)(50625927)；国家自然科学基金面上项目(50574053)；国家自然科学基金资助项目(50744044)。
本文原载于《岩石力学与工程学报》(2008 年第 27 卷第 8 期)。

large sections, but radius and thickness of fracturing areas reduce correspondingly. Analyses display that the relation between each fracturing area radius and roadway radius r is nearly expressed by $(\sqrt{2})^{i-1}r$ ($i = 1, 2, 3, 4$). There are four fracturing areas in surrounding rock of four observation sections. The thickness of first fracturing area is similar to roadway radius, and the first fracturing area is the most seriously fractured; and the other fracturing areas thickness and fracturing extension reduce consequently. The research results are important for surrounding rock fracturing mode of deep mine roadway and the corresponding stability reinforcement.

Key words: mining engineering; deep mine roadway; zonal disintegration of surrounding rock mass; in-situ monitoring

1 引 言

随着对能源需求量的增加和开采强度的不断加大,浅部资源日益减少,国内外矿山都相继进入深部资源开采状态。据不完全统计,国外开采超千米深的金属矿山有 80 多座,其中大多分布在南非和俄罗斯。煤炭占我国能源结构的 70%以上,而 90%的煤炭产量来自井工开采。目前已探明的储量中,约 53%的矿山埋深超过 1 000 m。根据目前资源开采状况,中国煤矿开采深度以每年 8~12 m 的速度增加,东部矿井正以每年 10~25 m 的速度发展。目前我国的新汶、开滦、淮南、兖州等淄博等多数或部分矿井的开采深度均已超过 800 m,部分矿井开采深度达到 1 000~1 300 m。可以预计在未来 20 a 我国很多煤矿将进入到 1 000~1 500 m 深度。在今后 10~20 a 内,我国的金属和有色金属矿山将进入 1 000~2 000 m 深度开采。随着开采深度的不断增加,工程灾害日趋增多,如矿井冲击地压、瓦斯爆炸、矿压显现加剧、巷道围岩大变形、流变和地温升高等,对深部资源的安全高效开采造成了巨大威胁。特别是深部巷道的变形破坏现象与浅部相比有很大的不同,用传统的连续介质力学理论无法科学地解释,引起了国际上岩石力学工程领域专家学者的极大关注,成为近几年该领域研究的热点[1~4],分区破裂化现象即是其中之一。

按照传统的连续介质弹塑性力学理论,浅部巷道开挖后围岩从内到外分别为破裂区、塑性区和弹性区,如图 1 所示。

而在深部巷道围岩中则出现了图 2 所示的破裂区和完整区多次交替的现象,即分区破裂化。钱七虎[5]将分区破裂化定义为"在深部岩体中开挖洞室或者巷道时,在其两侧和工作面前的围岩中,会产生交替的破裂区和不破裂区,称这种现象为分区破裂化"。

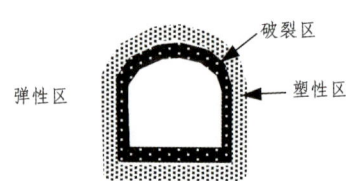

图 1 浅部巷道围岩破裂分布
Fig.1 Surrounding rock fracturing distribution in shallow roadway

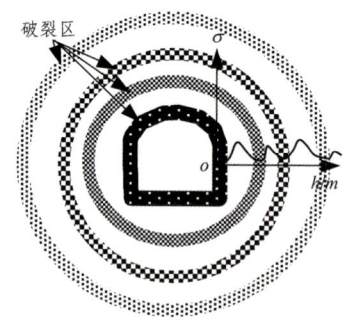

图 2 深部巷道围岩分区破裂化
Fig.2 Surrounding rock zonal disintegration distribution in deep roadway

20 世纪 80 年代,I. Shemyakin 等[6~9]在深部矿山 Маяк 开采现场采用电阻率仪发现了分区破裂化现象(见图 3),并且进一步通过试验验证了该现象的存在。根据实际巷道和模型观察,发现围岩分区破裂时,其承受荷载变化相当缓慢,可以当作静态看待,因此认为间隔破裂现象是在外部条件不变或缓慢变化时形成的,而且延续时间较长。G. D. Adams 和 A. J. Jager[10]在南非 Witwatersrand 金矿 2 000~3 000 m 深处采场采用钻孔潜望镜监测到顶板间隔破裂情形。D. F. Malan 和 S. M. Spottiswoode[11]利用已有的相关监测资料,分析了采矿矿场顶板岩层间隔破裂随时间和采矿活动的发展和形成,同时探

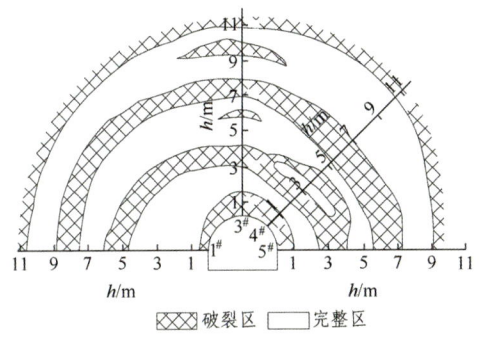

图 3 俄罗斯Маяк矿分区破裂化现象

Fig.3 Zonal disintegration of Маяк mine in Russia

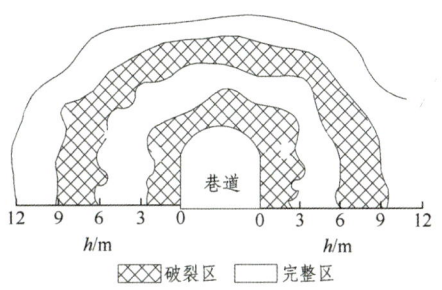

图 4 金川镍矿区深部巷道分区破裂现象[14]

Fig.4 Zonal disintegration of deep roadway in Jinchuan nickel mine area[14]

讨了矿震与矿场顶板围岩分区破坏的关联性。E. J. Sellers 和 P. Klerck[12]通过试验研究了深埋隧洞围岩不连续面对间隔破裂的影响作用，发现在满足一定要求的情况下，不连续面可能成为隧洞围岩间隔破裂的起源之一。М. В. Курленя 和 В. Н. Опарин[13]在对大量试验资料数据库的分析和理论研究的基础上，给出了各破裂区的半径和厚度表达公式，但该公式仅适用于特定矿区，对于我国深部巷道还需要深入研究。方祖烈[14]在我国金川镍矿区某深部巷道采用多点位移计监测围岩变形，得到如图4所示的围岩分区破裂现象。王明洋等[15~18]研究了深部巷道围岩地质力学能的"量子化"效应，指出深埋隧洞围岩间隔破裂的出现，对于隧洞的支护形式、掘进方法、支护范围都需要重新考虑。贺永年等[19]根据隧道围岩分区破裂资料研究，确认分区破裂是深部高应力隧洞围岩的一种广泛的规律性行为，是一种新的工程地质力学现象，揭示了深部隧洞围岩的另一种平衡过程及其新的平衡稳定形式。潘一山等[20~22]通过模型试验发现岩石环形裂纹的出现与围压、加载方式和岩石力学特性有关，分别从巴西劈裂和蠕变的角度研究了分区破裂化。周小平和钱七虎[23]把深部巷道的开挖看作动力问题，运动方程采用位移势函数，运用弹性力学和断裂力学，确定了破裂区岩体的残余强度和产生破裂区的时间，进而确定了破裂区和非破裂区的宽度和数量。顾金才等[24]通过模型试验，发现了在平行于洞室轴向的高水平压力作用下，洞室围岩出现多条裂缝，裂缝之间存在未破坏区域，证明了分区破裂存在的事实。

分区破裂化现象已经引起了许多专家学者的关注，是目前深部工程建设中急需解决的理论和实际问题之一。上述学者分别从现场监测、模型试验和理论方面分别取得了一些成就，但目前关于深部巷道围岩分区破裂化的研究尚处于初级阶段，其机制尚不清楚。由于现场所监测到的分区破裂化结果不多，我国对这一特殊现象的认识还不足。本文在淮南矿区近千米深井巷道进行了现场监测，并给出了破裂分区的预测模型。

2 工程概述

为研究巷道断面尺寸与围岩分区破裂范围的关系，选择淮南矿区丁集矿−910 m西部11-2采区南轨道大巷和采区水仓内仓作为研究对象，分别选取断面尺寸不同的巷道进行监测。轨道大巷和内仓同处一个水平，该矿地面高程45 m左右，因此巷道的埋深为955 m左右，地应力高。

2.1 地质条件

巷道所处岩层近似水平，轨道大巷和水仓内仓位于11-2煤底板，巷顶距11-2煤法距10~15 m，岩性为砂质泥岩、中砂岩和粉细砂岩。根据附近的29-8，29-12钻孔资料分析可知，11-2煤厚2.3~3.1 m，平均2.7 m。表1为岩层柱状与巷道位置。

2.2 巷道断面及支护设计

轨道大巷和水仓内仓断面形状均为半圆拱形，断面尺寸分别为5 000 mm×3 880 mm 和 2 800 mm×2 400 mm。均采用锚网喷＋锚索支护，锚杆的规格是ϕ20 mm×2 200 mm的高强度螺纹钢，间排距均为800 mm×800 mm；钢筋网的规格是ϕ6 mm；喷层采用C20混凝土，厚度为150 mm。锚索规格为ϕ15.24 mm×6 200 mm，间、排距为2.4 m。图5为轨道大巷断面尺寸及支护设计。

表 1 岩层柱状与巷道位置
Table 1 Terrane histogram and roadway location

地层系	地层组	柱状(1∶200)	层厚/m	地层名称	岩性描述
二迭系	上石盒子组		2.3~3.1 / 2.7	11-2 煤	黑色，粉末状，含少量块状及颗粒状，底部含薄层泥岩或灰质泥岩夹矸
			1.3	泥岩	灰色，泥质单一，含植物根部化石，质软
			0.8~2.6 / 1.6	砂质泥岩	灰色，泥质为主，含粉砂质及白云母碎片，粗糙断口，含少量植化碎片，致密
			0.0~0.5	11-1 煤	黑色，块状，染手，沥青～金属光泽，为半亮煤，局部为炭质泥岩或缺失
			0.8	泥岩	深灰色，性脆，较致密，参差状断口，含大量的植化碎片
			1.9	中砂岩	浅灰色，成份以石英为主，次为长石，钙质胶结，较坚硬，其中含菱铁质，水平层理，分选差，垂直裂隙发育
			8.9	砂质泥岩	灰色，富含植化碎片，局部夹菱铁结核，下部夹 0.10 m 浅灰色细砂岩
			1.6	中砂岩	浅灰色，成份以石英为主，次为长石，钙质胶结，较坚硬，其中含菱铁质，水平层理，分选差，垂直裂隙发育
			4.7	粉细砂岩	灰色，成份以石英为主，泥质胶结，分选一般，见粉砂岩包裹体，近水平层理
			2.8	细砂岩	较坚硬，含有较多的菱铁结核，不规则分布
			8.9	砂质泥岩	灰色，块状，砂泥质结构，滑面发育，含植化碎片，参差状断口
			8.8	中砂岩	灰白色，成份以石英为主，次为长石，硅质胶结，坚硬致密，厚层状，斜交层理，垂直裂隙极为发育，是张性裂隙，含菱铁质

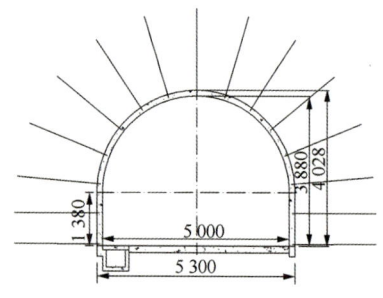

图 5 轨道大巷断面尺寸及支护设计(单位：mm)
Fig.5 Railway large roadway section dimensions and support design(unit：mm)

2.3 监测断面位置

轨道大巷共布置 3 个断面，监测断面 A 距离水仓 80 m，监测断面 A 与 B 相距 60 m，监测断面 B 与 C 相距 95 m。水仓内仓布置了 1 个监测断面 D，其位置选择在起坡处。图 6 为监测巷道位置平面及剖面图。

监测断面 A 测试的时间是 2007 年 11 月 25 日，当时监测断面 A 距离巷道掘进头约 100 m；监测断面 B～D 测试的时间为 2008 年 1 月 28 日，监测断面距离巷道掘进迎头约 100 m。

3 监测仪器及监测方案

3.1 监测仪器及原理

监测巷道围岩内破坏的仪器采用中矿华泰生产的 KDVJ‑400 矿井钻孔电视成像仪。它采用高分辨率彩色电视摄像头进行监测，在液晶显示屏幕上显示钻孔内壁构造，图像清晰(分辨率可达 0.1 mm)、颜色逼真，能监测孔内整体情况，并可录像，具有防爆功能，且体积小、质量轻，操作简便。矿井钻孔电视成像仪包括井下全景摄像探头、系统控制器、专用电视电缆和彩色监视器等部分组成。

3.2 监测方案

监测断面 A 采用锚索机钻孔，钻孔直径 $\phi 32$ mm，深度 10 m。由于锚索钻机在钻水平孔时存在困难，因此，共钻 3 个钻孔，分别位于顶板(90°)和左右两拱肩(与水平的夹角 30°)。

监测断面 B～D 采用地质钻机钻孔，钻孔直径 $\phi 73$ mm，深 10 m。每个断面共钻 5 个钻孔，如图 7 所示，分别位于顶板(90°)、左右两拱肩(45°)和左右两帮(0°)。

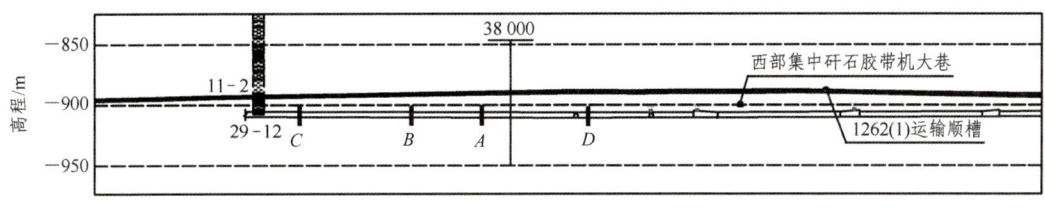

(a) 剖面图

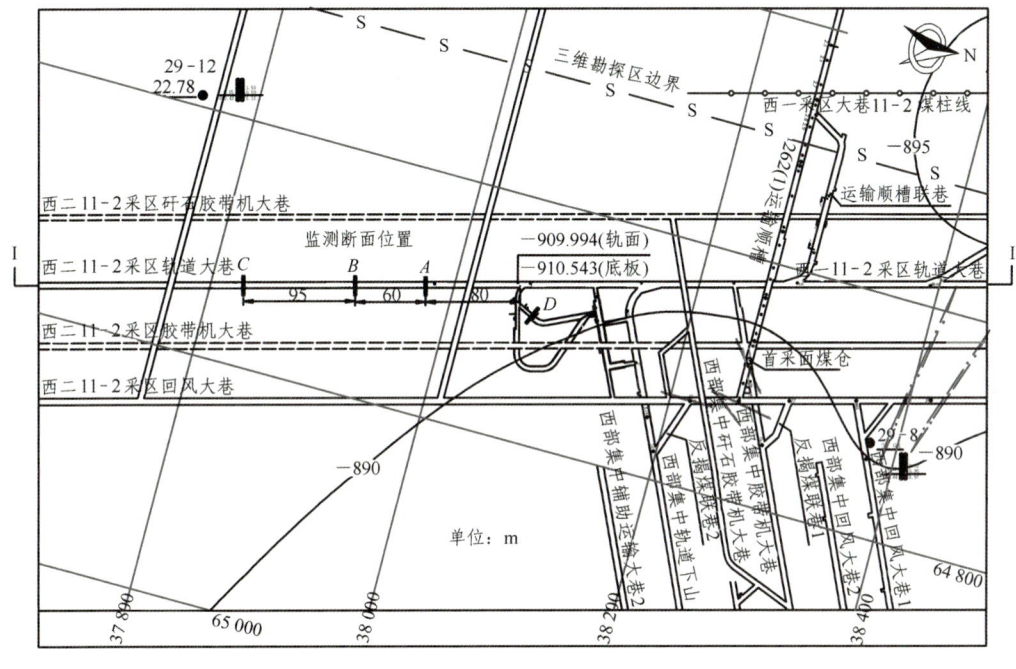

(b) 平面图

图 6 监测巷道位置平面及剖面图

Fig.6 Observation roadway plane location and geological section plan

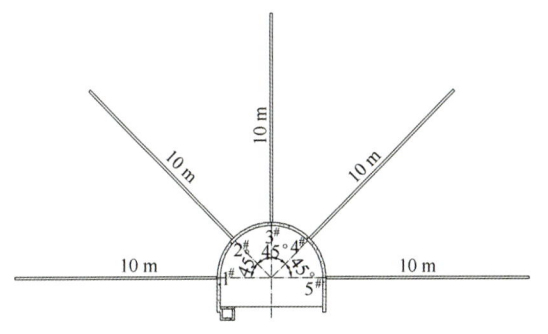

图 7 监测钻孔布置图

Fig.7 Plan of observation bore arrangement

应用钻孔电视监测围岩内部破坏情况时，采用前进式，即一边慢慢推进摄像头，一边记录围岩的破坏情况，直到钻孔底部。当监测到孔内围岩的破坏或裂隙时，记录下围岩破裂的深度、破坏程度和视频录像时间，将围岩破坏的深度、破坏程度和形式与记录的视频文件对应起来。

4 监测结果分析

4.1 钻孔图像

采用钻孔电视对每个断面的所有钻孔进行监测录像，图 8 为提取的部分孔内视频图。由图 8 可知，钻孔电视能清晰地分辨不同岩层、完整岩石与破坏情况，根据破坏程度分为严重破碎、破碎、裂缝和裂隙，在裂隙处可监测到白色气体突出。

4.2 监测结果

根据钻孔电视监测的围岩内部的破坏资料，将每个钻孔内围岩的破坏沿钻孔由内向外依次描绘在图上，并且将不同孔内围岩破裂区间隔超过 0.5 m 的区域用多义线连接起来，并采用十字网格充填起来，形成不同深度的破裂区，各破裂区之间为间隔的完整区。

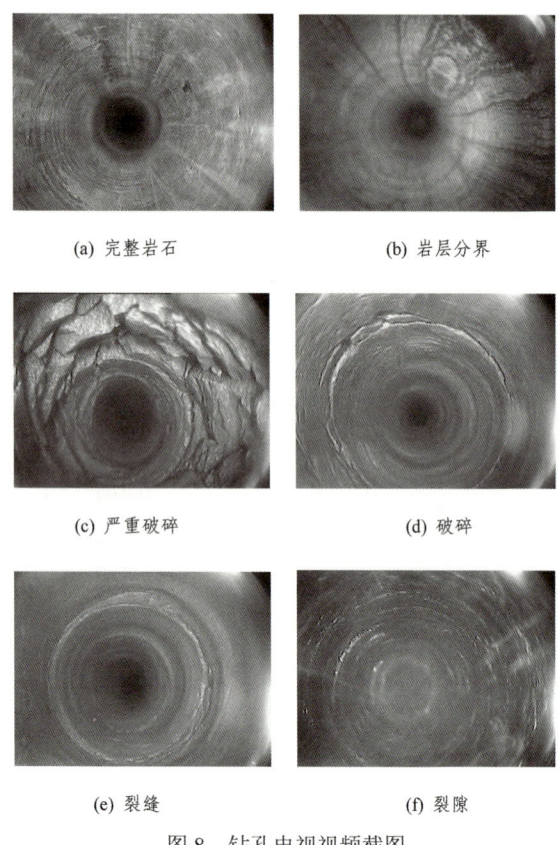

(a) 完整岩石　　　　(b) 岩层分界

(c) 严重破碎　　　　(d) 破碎

(e) 裂缝　　　　　　(f) 裂隙

图 8　钻孔电视视频截图

Fig.8　Bore TV video screenshots

图 9~11 分别为监测断面 A~C 围岩分区破裂分布情况。其中监测断面 C 的 $1^{\#}$，$5^{\#}$ 钻孔由于塌孔，分别只监测到 6.3 和 6.8 m，在图 11 绘出了预测破裂带。

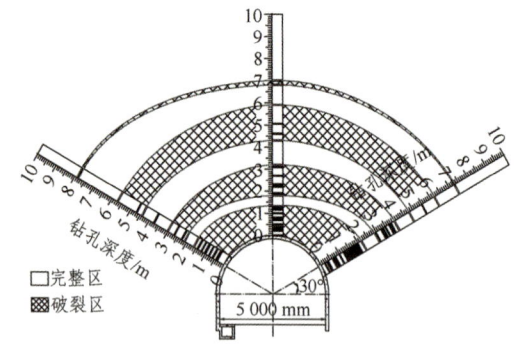

图 9　监测断面 A 围岩分区破裂分布

Fig.9　Zonal disintegration distribution of monitoring section A

由图可以看出，大断面巷道围岩内部共分为 4 个破裂区，巷道表面 1.5~3.0 m 范围内围岩的破坏最为严重，以严重破碎和破碎为主，这个区域可认为是传统意义上的巷道围岩松动圈。表面破坏区之

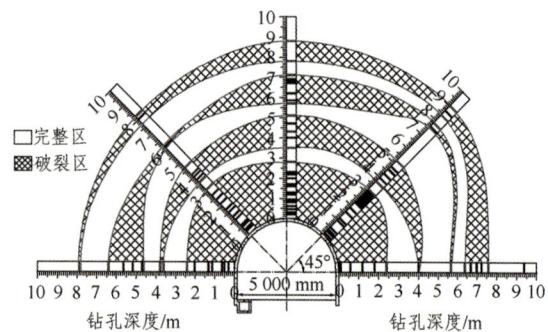

图 10　监测断面 B 围岩分区破裂分布

Fig.10　Zonal disintegration distribution of monitoring section B

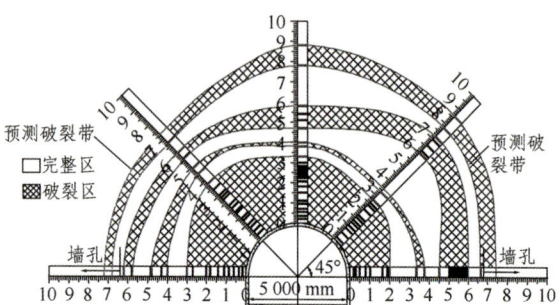

图 11　监测断面 C 围岩分区破裂分布

Fig.11　Zonal disintegration distribution of monitoring section C

外是完整区，再往外进入第二破裂圈，平均深度为 3.5~4.0 m，这一破裂分区围岩的破坏也比较大，以破碎和裂隙为主。然后又是完整区，接着又是第三破裂区，平均深度为 5~6 m，这一破裂分区围岩的破坏比较小，以裂缝为主。间隔完整区，进入第四破裂区，平均深度为 7~8 m，这一分区围岩破裂以裂隙为主。监测断面 D 围岩分区破裂分布见图 12。

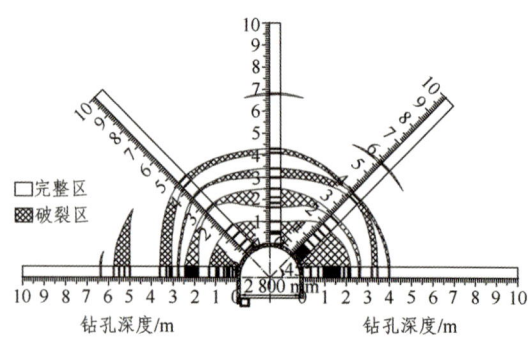

图 12　监测断面 D 围岩分区破裂分布

Fig.12　Zonal disintegration distribution of monitoring section D

由图 12 可以看出，巷道围岩内部大致分为 4 个破裂区，巷道表面 1.0~2.0 m 范围内围岩的破坏最为严重，其中 $2^{\#}$ 钻孔 2.6 m 范围内围岩的破坏也

非常大,因此,这个区域可认为是传统意义上的巷道围岩松动圈。5 个钻孔内的围岩均存在分区破裂化现象,并且破裂区距巷道表面的深度大致相同,将 5 个钻孔内处于同一破裂圈的深度平均后得到每个破裂分区的范围:

(1) 第 1 个破裂区范围为 0.0~1.5 m;
(2) 第 2 个破裂区范围为 2.1~2.6 m;
(3) 第 3 个破裂区范围为 3.2~3.4 m;
(4) 第 4 个破裂区范围为 3.8~4.1 m。

其中 1# 钻孔在 5.00~5.75 m 范围内有一个破裂区,在 6.4 m 处有一个裂隙;4# 钻孔在 5.8 m 处也有一个裂隙。从破坏深度看,小断面巷道较大断面巷道小。由此可见,在相同的地质条件下,巷道断面缩小时,巷道围岩内每个分区破裂的半径也相应的缩小了。

4.3 讨论与分析

(1) 裂隙成因分析

根据图 5,监测断面 A 巷道顶板 10 m 钻孔穿越 7.5 m 厚的砂质泥岩层、1.9 m 厚的中砂岩和 0.8 m 厚的泥岩,从监测钻孔内最大破坏深度来看,均没有超过 7.5 m,因此不存在天然岩层分层。监测断面 D 的破坏范围也在砂质泥岩层。

虽然监测断面 B,C 巷道顶板距离煤层较近,穿过的岩层较多,但天然岩层之间胶结较好,加工好的岩样不少都为两种岩层交界处的。由于岩层近似水平,2 个水平孔穿过同一种岩层,不会存在天然人岩层分层。

若为天然裂隙两倾斜,钻孔内裂隙的形态应为椭圆,不会与钻孔轴向垂直,但孔内裂隙形态均与钻孔轴向垂直,说明岩层裂隙是巷道开挖造成的。左右两帮的水平孔内的裂隙也近似圆形,与钻孔轴向垂直,更进一步说明监测到的裂隙是巷道开挖引起的。

(2) 各个破裂分区的范围

将巷道每个监测断面围岩各钻孔破裂分区的范围平均后的值列于表 2。表中数值分别加上半径 2.5 和 1.4 m 的监测断面;内径和外径分别表示靠近巷道的一侧和远离巷道的一侧。

根据表 2 绘制了 2 个不同断面的分区破裂分布图,如图 13 所示。表 2 数据显示,较大断面破裂分区的最大外径为 10.80 m,破坏深度为 8.30 m;较小断面破裂分区最大外径为 5.5 m,破坏深度为 4.1 m。

表 2 监测断面各破裂分区范围

Table 2 Fracturing scopes for various observation sections

m

分区		监测断面 A	监测断面 B	监测断面 C	监测断面 D	平均值
I	内径	2.50	2.50	2.50	1.40	2.50
	外径	4.23	5.28	5.47	2.87	4.99
II	内径	4.77	6.18	6.29	3.54	5.75
	外径	5.90	7.05	6.69	3.95	6.55
III	内径	7.00	8.39	7.82	4.64	7.74
	外径	8.43	9.14	8.76	4.83	8.78
IV	内径	9.83	10.12	9.66	5.24	9.87
	外径	10.03	10.80	10.37	5.50	10.40

注:平均值为监测断面 A~C 对应值的平均值。

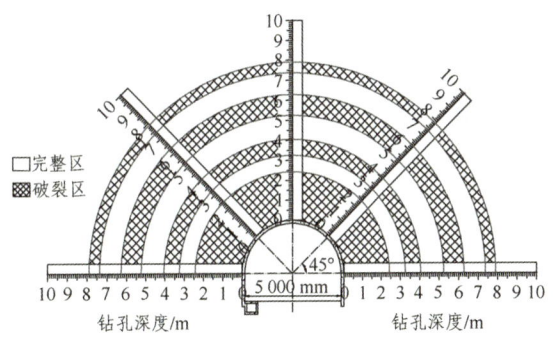

(a) 半径 2.5 m 的巷道

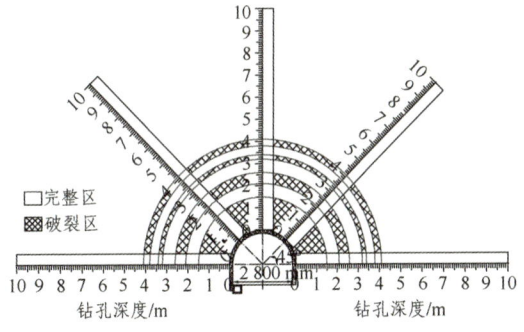

(b) 半径 1.4 m 的巷道

图 13 不同断面围岩分区破裂平均分布

Fig.13 Zonal disintegration of average fracture distribution of different sections

对于同一破裂分区,小断面各个破裂分区的半径均为大断面的 53%~62%,平均为 57%,而小、大断面半径之比为 0.56,二者相差不大。因此,本文认为巷道围岩分区破裂的半径与巷道的直径成正比。

М. В. Курленя 和 В. Н. Опарин[13]在对大量试验资料数据库的分析和理论研究的基础上,发现各破裂区的半径可以用模数 $(\sqrt{2})^{i-1}$(i = 1,2,3,4)来

描述,它与巷道半径 r 有关。根据监测结果分析发现各破裂分区半径与巷道半径存在如下关系:

$$r_i = (\sqrt{2})^{i-1} r \quad (i = 1, 2, 3, 4) \quad (1)$$

式中:i 为破裂区域的编号。

5 结 论

深部巷道围岩分区破裂化现象是目前逐渐引起关注的问题,为研究深部巷道围岩的分区破裂化现象,选择近千米埋深的不同断面尺寸巷道,采用矿井钻孔电视成像仪监测了巷道围岩的内部破裂情况,监测到了分区破裂化现象,得出如下结论:

(1) 对于监测巷道围岩内部的破坏情况,包括裂隙的大小、破坏形状均非常有效,钻孔电视可以非常直观地了解巷道围岩内部的破坏状态和岩层分布。

(2) 在丁集矿 955 m 埋深西二采区南轨道大巷 3 个不同断面和水仓 1 个断面,均监测到巷道围岩内部存在分区破裂化现象,都为 4 个破裂分区。

(3) 4 个监测断面围岩内均产生了 4 个破裂分区,第 1 个破裂分区厚度与巷道半径相当,破裂程度最严重,第 2 个破裂分区其后各破裂分区厚度和破裂程度均依次减小。

(4) 3 个大断面巷道围岩的破裂分区的半径和厚度相差不大,小断面巷道围岩的破裂分区与大断面相似。围岩分区破裂的半径与巷道半径基本呈线性关系。分析表明各破裂分区的半径与巷道半径 r 的关系为 $(\sqrt{2})^{i-1} r (i = 1, 2, 3, 4)$。

本文所获得的研究结果对于认识深部巷道围岩的破裂模式及其稳定性支护具有重要意义。

参考文献(References):

[1] 钱七虎. 深部岩体工程响应的特征科学现象及"深部"的界定[J]. 东华理工学院学报,2004,27(1):1 - 5.(QIAN Qihu. The characteristic scientific phenomena of engineering response to deep rock mass and the implication of deepness[J]. Journal of East China Institute of Technology,2004,27(1):1 - 5.(in Chinese))

[2] 周宏伟,谢和平,左建平. 深部高地应力下岩石力学行为研究进展[J]. 力学进展,2005,35(1):91 - 99.(ZHOU Hongwei,XIE Heping,ZUO Jianping. Developments in researches on mechanical behaviors of rocks under the condition of high ground pressure in depths[J]. Advances in Mechanics,2005,35(1):91 - 99.(in Chinese))

[3] 何满潮,谢和平,彭苏萍,等. 深部开采岩体力学研究[J]. 岩石力学与工程学报,2005,24(16):2 803 - 2 813.(HE Manchao,XIE Heping,PENG Suping,et al. Study on rock mechanics in deep mining engineering[J]. Chinese Journal of Rock Mechanics and Engineering,2005,24(16):2 803 - 2 813.(in Chinese))

[4] 李全生. 我国井下煤矿开采技术现状和发展展望[J]. 煤矿开采,2002,7(3):1 - 5.(LI Quansheng. Present situation of underground coal mining technology and its prospects in China[J]. Coal Mining Technology,2002,7(3):1 - 5.(in Chinese))

[5] 钱七虎. 非线性岩石力学的新进展——深部岩体力学的若干问题[C]// 中国岩石力学与工程学会编. 第八次全国岩石力学与工程学术大会论文集. 北京:科学出版社,2004:10 - 17.(QIAN Qihu. The current development of nonlinear rock mechanics:the mechanical problems of deep rock mass[C]// Chinese Society for Rock Mechanics and Engineering ed. Proceedings of the 8th Rock Mechanics and Engineering Conference. Beijing:Science Press,2004:10 - 17.(in Chinese))

[6] SHEMYAKIN I,FISENKO G L,KURLENYA M V,et al. Zonal disintegration of rocks around underground workings. I. data of in-situ observations[J]. Soviet Mining Science,1986,22(3):157 - 168.

[7] SHEMYAKIN I,FISENKO G L,KURLENYA M V,et al. Zonal disintegration of rocks around underground workings. II. disintegration of rocks on models of equivalent materials[J]. Soviet Mining Science,1986,22(4):223 - 232.

[8] SHEMYAKIN I,FISENKO G L,KURLENYA M V,et al. Zonal disintegration of rocks around underground workings. III. theoretical notions[J]. Soviet Mining Science,1987,23(1):1 - 6.

[9] SHEMYAKIN I,FISENKO G L,KURLENYA M V,et al. Zonal disintegration of rocks around underground workings. IV. practical applications[J]. Soviet Mining Science,1988,24(3):297 - 302.

[10] ADAMS G D,JAGER A J. Etroscopic observations of rock fracturing ahead of the stope faces in deep-level gold mines[J]. Journal of the South Africa Institute of Mining and Metallurgy,1980,(2):115 - 127.

[11] MALAN D F,SPOTTISWOODE S M. Time-dependent fracture zone behavior and seismicity surrounding deep level stopping operations[C]// GIBOWICZ S J,LASOCKI S ed. Rockbursts and Seismicity in Mines. Rotterdam:A. A. Balkema,1997:173 - 177.

[12] SELLERS E J,KLERCK P. Modeling of the effect of discontinuities on the extent of the fracture zone surrounding deep tunnels[J].

Tunneling and Underground Space Technology,2000,15(4):463–469.

[13] КУРЛЕНЯ М В,ОПАРИН В Н. Проблемы нелинейной геомеханики Ч. 1[J] ФТПРПИ,1999,(3):12–23.

[14] 方祖烈. 软岩巷道维护原理与控制措施[C]// 何满潮编. 中国煤矿软岩巷道支护理论与实践. 北京:煤炭工业出版社,1996:64–70.(FANG Zulie. Support principles for roadway in soft rock and its controlling measures[C]// HE Manchao ed. Soft Rock Tunnel Support in Chinese Mines:Theory and Practice. Beijing:China Coal Industry Publishing House,1996:64–70.(in Chinese))

[15] 王明洋,周泽平,钱七虎. 深部岩体的构造和变形与破坏问题[J]. 岩石力学与工程学报,2006,25(3):448–455.(WANG Mingyang,ZHOU Zeping,QIAN Qihu. Tectonic,deformation and failure problems of deep rock mass[J]. Chinese Journal of Rock Mechanics and Engineering,2006,25(3):448–455.(in Chinese))

[16] 陈士林,钱七虎,王明洋. 深部坑道围岩的变形与承载能力问题[J]. 岩石力学与工程学报,2005,24(13):2 203–2 212(CHEN Shilin,QIAN Qihu,WANG Mingyang. Problems of deformation and bearing capacity of rock mass around deep-buried tunnels[J]. Chinese Journal of Rock Mechanics and Engineering,2005,24(13):2 203–2 212.(in Chinese))

[17] 王明洋,戚承志,钱七虎. 深部岩体块系介质变形与运动特性研究[J]. 岩石力学与工程学报,2005,24(16):2 825–2 830.(WANG Mingyang,QI Chengzhi,QIAN Qihu. Study on deformation and motion characteristics of blocks in deep rock mass[J]. Chinese Journal of Rock Mechanics and Engineering,2005,24(16):2 825–2 830.(in Chinese))

[18] 王明洋,宋 华,郑大亮,等. 深部巷道围岩的分区破裂机制及深部界定探讨[J]. 岩石力学与工程学报,2006,25(9):1 771–1 776.(WANG Mingyang,SONG Hua,ZHENG Daliang,et al. On mechanism of zonal disintegration within rock mass around deep tunnel and definition of "deep rock engineering"[J]. Chinese Journal of Rock Mechanics and Engineering,2006,25(9):1 771–1 776.(in Chinese))

[19] 贺永年,韩立军,邵 鹏,等. 深部巷道稳定的若干岩石力学问题[J]. 中国矿业大学学报,2006,35(3):288–296.(HE Yongnian,HAN Lijun,SHAO Peng,et al. Some problems of rock mechanics for roadways stability in depth[J]. Journal of China University of Mining and Technology,2006,35(3):288–296.(in Chinese))

[20] 潘一山,李英杰,唐 鑫,等. 岩石分区破裂化现象研究[J]. 岩石力学与工程学报,2007,26(增1):3 335–3 341.(PAN Yishan,LI Yingjie,TANG Xin,et al. Study on zonal desintegration of rock[J]. Chinese Journal of Rock Mechanics and Engineering,2007,26(Supp.1):3 335–3 341.(in Chinese))

[21] 李英杰,潘一山,李忠华. 岩体产生分区碎裂化现象机制分析[J]. 岩土工程学报,2006,28(9):1 124–1 128.(LI Yingjie,PAN Yishan,LI Zhonghua. Analysis of mechanism of zonal disintegration of rocks[J]. Chinese Journal of Geotechnical Engineering,2006,28(9):1 124–1 128.(in Chinese))

[22] 唐 鑫,潘一山,章梦涛. 深部巷道区域化交替破碎现象的机制分析[J]. 地质灾害与环境保护,2006,17(4):80–84.(TANG Xin,PAN Yishan,ZHANG Mengtao. Mechanism analysis of zonal disintegration in deep level tunnel[J]. Journal of Geological Hazards and Environment Preservation,2006,17(4):80–84.(in Chinese))

[23] 周小平,钱七虎. 深埋巷道分区破裂化机制[J]. 岩石力学与工程学报,2007,26(5):877–885.(ZHOU Xiaoping,QIAN Qihu. Zonal fracturing mechanism in deep tunnel[J]. Chinese Journal of Rock Mechanics and Engineering,2007,26(5):877–885.(in Chinese))

[24] 顾金才,顾雷雨,陈安敏,等. 深部开挖洞室围岩分层断裂破坏机制模型试验与分析[J]. 岩石力学与工程学报,2008,27(3):433–438.(GU Jincai,GU Leiyu,CHEN Anmin,et al. Model test study on mechanism of layered fracture within surrounding rock of tunnels in deep stratum[J]. Chinese Journal of Rock Mechanics and Engineering,2008,27(3):433–438.(in Chinese))

深部岩体工程围岩分区破裂化现象研究综述

钱七虎[1]，李树忱[1,2]

(1. 解放军理工大学 工程兵工程学院，江苏 南京 210007；2. 山东大学 土建与水利学院，山东 济南 250061)

摘要：随着经济建设与国防建设的不断发展，深部岩体工程越来越多，如逾千米乃至数千米的矿山(如金川镍矿和南非金矿等)、锦屏二级引水隧洞及辅助洞、核废料的深层地下存储、深部地下防护工程等。深部岩体工程在开挖洞室或巷道时，围岩变形和破坏等出现了一系列新的科学现象。除了岩爆和围岩挤压大变形以外，围岩的分区破裂化现象也吸引了很多岩石力学工作者的关注。基于国外对分区破裂化现象的实验和理论研究，归纳出分区破裂化现象的主要特征参数及其变化规律，揭示分区破裂化现象产生的条件；提出这一领域的研究方向；同时介绍国内在该领域实验和理论方面的研究进展。

关键词：岩石力学；深部岩体；分区破裂化；非线性岩石力学；动力问题；岩爆

中图分类号：TU 45 **文献标识码**：A **文章编号**：1000 – 6915(2008)06 – 1278 – 07

A REVIEW OF RESEARCH ON ZONAL DISINTEGRATION PHENOMENON IN DEEP ROCK MASS ENGINEERING

QIAN Qihu[1], LI Shuchen[1,2]

(1. Engineering Institute of Engineering Corps, PLA University of Science and Technology, Nanjing, Jiangsu 210007, China; 2. School of Civil and Hydraulic Engineering, Shandong University, Jinan, Shandong 250061, China)

Abstract: With the development of national economy and defense works, there are more and more deep underground rock wass engineering, such as the mines with depth of thousand to several thousands meters(Jinchuan nickel mines and gold mines in South Africa, etc.), water diversion tunnels and auxiliary tunnel of Jinping II hydropower station, deep geological deposition of nuclear waste and deep underground protection engineering. While the deep rock mass are excavated by the blast or other methods, the deformation and fracture of the surrounding rock show several new scientific characteristic phenomena. Besides the deep rockburst and large deformation of tunnel in squeezing ground, the phenomenon of zonal disintegration attracts the attentions of scholars and engineers in the fields of geotechnical engineering and rock mechanics in the world. Based on the experimental and theoretical researches of zonal disintegration phenomenon at home and abroad, the main characteristics and corresponding changing laws of zonal disintegration are summarized. It is revealed the mechanism for the occurrence of zonal disintegration; and the emphasis in this field is proposed. Also the domestic experimental and theoretical investigations of zonal disintegration phenomenon are introduced.

Key words: rock mechanics; deep rock masses; zonal disintegration; nonlinear rock mechanics; dynamic problem; rockburst

---

收稿日期：2008–04–14；修回日期：2008–05–12。
基金项目：国家自然科学基金重大项目"深部岩石力学基础及应用"(50490270)。
本文原载于《岩石力学与工程学报》(2008 年第 27 卷第 6 期)。

1 引言

深部岩石力学之所以在当代发展成岩石力学的一个热点研究方向,是因适应了深部岩体工程围岩变形和破坏的一系列新科学现象研究的需要。除了岩爆和围岩挤压大变形以外,围岩的分区破裂化现象也吸引了很多岩石力学工作者的关注。在深部岩体工程中开挖洞室或坑道时,在其洞室围岩中会产生交替的破裂区和非破裂区的现象,这种现象在相关文献中被称之为分区破裂化现象(zonal disintegration)。这些科学现象的"新"在于浅部岩体工程中未曾发现过这些现象,并且这些现象用传统的连续介质弹塑性力学不能完全解释清楚。解释这些新现象发生的机制,定性以及定量地分析这些现象及其规律,数值仿真出这些科学现象正孕育形成新的岩石力学分支学科——一些学者将其称为深部非线性岩石力学。

本文总结了国内外分区破裂化现象研究领域的实验和理论研究成果和进展,归纳出分区破裂化现象的主要特征参数及其变化规律,定性分析了分区破裂化现象产生的机制,指出了这一领域下一步的研究方向。

2 现场观测与实验研究

深部岩体分区破裂化现象于 20 世纪 70 年代在南非 2 073 m 深的金矿中首次被发现[1]。之后,在南非深部金矿的巷道中系统地观察到了围岩中存在的分区破裂化现象,如图 1 所示[2]。

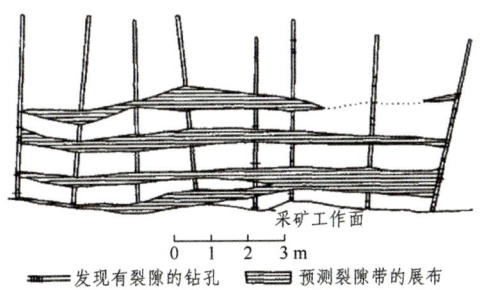

图 1 南非 Witwatersrand 金矿巷道顶板分区破裂化现象[2]

Fig.1 Zonal disintegration phenomenon of the tunnel roof in Witwatersrand gold deposit in South Africa[2]

这些现象是采用岩石潜望镜进行观测的。岩石潜望镜带有电光源和倾斜反射镜,可深入钻孔观察;同时,岩石潜望镜带有照相机,可以把围岩表面裂缝照下来。通过采用岩石潜望镜对南非深部金矿巷道的观察,其结果表明,所有的钻孔都不止有一个比较集中的破裂区,这些破裂区由间隔 5~150 mm 的裂缝组成,这些裂缝把岩体分割成片状的破坏区,这些破坏区被相对完好的、约为 1 m 厚的未破裂区所分割开,其方向垂直于工作面,破裂区范围至少有 5 m,破裂区平行于工作面可延伸至 12 m。在机械化开挖的 Doornfontein 金矿 Chamber 矿井的顶板上也显示出与钻爆法开挖的矿井中相似的分区破裂化现象,所以在 G. R. Adams 和 A. J. Jager[2]的研究中排除了分区破裂化现象的产生是由于爆破引起的,即不论是在钻爆法施工的巷道还是采用机械化施工的巷道中,只要条件满足,巷道围岩都会产生分区破裂化现象。他们还根据围岩中裂缝的走向是否平行于工作面,区分了岩体中原有的地质节理和采掘扰动引起的破坏裂缝,在图上标示以斜线和垂直线。G. R. Adams 和 A. J. Jager[2]指出,裂缝形成于分离的区域中的原因尚未得到理论解释。

俄罗斯科学院(原苏联科学院)西伯利亚分院对分区破裂化现象于 20 世纪 80~90 年代进行了深部矿井现场实验研究、实验室模拟实验研究、理论分析以及现象的应用研究[3~13]。现场实验研究[4]是在诺里尔斯基金属矿山联合企业塔拉娜哈十月矿区的若干矿井中进行的,其深度分别为 957~1 050,800~900,500~550 以及 110~140 m。为了进行实验观察,研制了一系列探测装置及其相应的综合分析方法:常电流和低频电流的电测法、地下电测法、超声透射法、γ 射线的核物理探测法,并且采用钻孔型岩石潜望镜进行选择性的目力检查。把这些钻孔探测的特征测量结果绘制于图上,得到了这些巷道围岩的分区破裂化现象,如图 2 所示。各种方法探测的结果相互印证,通过结果的重复性保证了探测的可靠性。

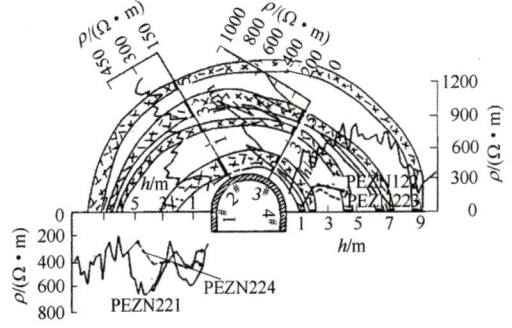

图 2 Taimyrskii 矿山巷道围岩的分区破裂化现象

Fig.2 Zonal disintegration phenomenon of tunnel rockmass in the Taimyrskii Mine

通过基于相对于柱状巷道的对称性和裂缝走向规律的统计规则,在所有测站的钻孔中所观察到的总裂缝中,排除了原生的地质节理,最后把每一测站的所有钻孔中的裂缝探测结果连成综合断面图。Е. И. Шемякин 等[4]在结论中指出,地下巷道周围分区破裂化效应的发现是基于深部巷道围岩现场钻孔探测的数据;这个效应的实质在于岩体中准柱面状的裂缝区和非裂缝区的交替,这与地下巷道周围变形和破坏状态的已知理论在概念和原则上是不同的;巷道围岩中存在 1.0~1.5 m 宽的裂缝区,相互间被宽度为 1.0~1.5 m 的未破坏岩体所间隔开,这些裂缝区的表面大体上重复巷道表面的几何轮廓形状;裂缝区的数量取决于岩石压力的量级,在 1 000 m 深度的地下工程中通常不少于 3 个分区破裂带。

为了校核深部巷道现场实验发现的分区破裂化效应,研究其产生的条件和规律,Е. И. Шемякин 等[5]在实验室进行了等效材料模型实验,包括平面应变和三维立体模型实验,巷道形状分别为圆形和拱形断面,分别考虑了支护和不支护的(锚固和喷混凝土)作用。岩体破坏的记录在平面应变模型上通过照相,在立体模型上通过拆解模型后再照相的方式获得。在模型实验的巷道中同时进行了位移量测。加载方式,平面应变模型从 3 个面上进行加载,三维立体模型从 5 个面上进行加载。围岩由连续介质和裂缝介质组成。巷道通过钻孔来建立。模型的相似比分别为 1∶50 和 1∶100。巷道与模型的尺寸比是 0.12~0.17。加载的强度(初始地应力)为考虑构造而削弱后的岩体单轴抗压强度的 1.1~1.9 倍。模型和巷道的尺寸比例以及实验台的研制,排除了加载装置对巷道围岩变形的影响,巷道是在预加应力的模型材料中开挖(如图 3 所示),得到了围岩分区破裂化现象。

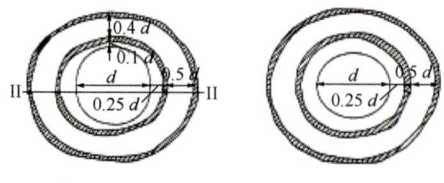

图 3 模型实验得到的分区破裂化现象

Fig.3 Zonal disintegration phenomena obtained by model test

Е. И. Шемякин 等[5]根据模型实验结果再一次做出如下结论:巷道围岩中强破碎区和弱破碎区交替产生的基本原因,不是由于巷道掘进中钻爆施工,而是在于巷道周围应力场的改变;分区破裂化的发现表明,现有关于巷道围岩中应力-应变状态的一般原理与深部巷道围岩中的变形破坏实际不相适应,鉴于分区破裂化效应的明显重要性,迫切要求对其进行理论思考和研究。

3 机制研究

Е. И. Шемякин 等[6]从理论方面研究了分区破裂化现象产生的机制,如图 4 所示。

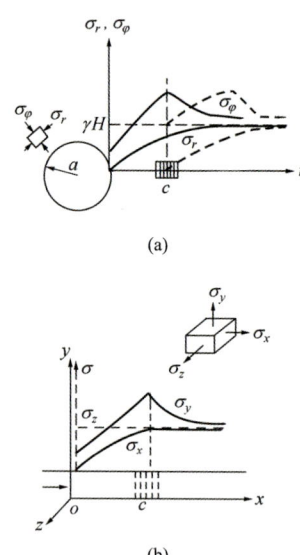

图 4 分区破裂化现象产生的机制

Fig.4 Mechanism of zonal disintegration phenomenon

Е. И. Шемякин 等[6]首先分析了深部巷道围岩在卸荷条件下裂缝的产生,与实验室试件在小侧向压力条件下竖向加载时穿透裂缝产生具有相似性,该相似性表明第一个裂缝区形成的机制,是在应力集中导致的最大支撑压力作用处,由于巷道自由面的影响,产生径向的拉伸变形并达到围岩的极限应变,导致第一个破坏区中裂缝的张开,以及围岩向着巷道自由面位移的发展导致了"伪掌子面",即"伪自由表面"的形成,从而引起围岩再次发生应力重分布。应力重分布将导致第二个"伪自由表面"的形成,从而进一步引起围岩内应力的重分布。上述过程不断发展,一直延续到围岩重分布的应力达不到围岩的破坏条件为止。这就是深部巷道围岩中若干个破坏区与非破坏区交替发生的机制。显然,上述分析仅是定性的,不能定量确定围岩中各个破坏区和非破坏区的厚度。定量的理论分析必须研究主干裂缝形成与展开的时间、"破坏波"从自由表面向围岩深部运动的时间，以及两个时间的相互关系[8]。显然，分区破裂化效应的研究必将为岩爆的研究提供很大兴趣[14]。

Е. И. Шемякин 等[7]研究了分区破裂化效应的实际工程应用。为此，首先要确定在什么深度上可能发生巷道围岩的分区破裂化效应，在该研究中，根据围岩最大支撑压力作用处径向拉伸变形达到极限值的条件，推得产生第一个裂缝区的深度确定公式。但是推导该公式时，初始地应力假定等于 γH 值，也就是没有考虑构造地应力，而且也没有考虑动力因素。因此，该公式应用于实际工程还需要适当修正。关于分区破裂化效应的实际工程应用，Е. И. Шемякин 等[7]列出了用以减少深度岩体工程的岩体坍塌、减少巷道的开挖量、提高钻爆法的效率、提高锚杆支护的效果以及节省爆破材料等方面的工程应用。

Курленя 教授研究了分区破裂化效应中的时间因素。他根据模型实验得出如下结论：初始应力值对巷道周边位移速度具有重要影响，该速度是围岩卸荷产生的位移速度，当 σ/R_c 由 1.75 增大至 3.00 时，巷道周边位移速度由 43 m/h 增至 468 m/h，而支座压力处波阵面的运动速度自巷道周边指向围岩深部，由 12 km/h 增至 22 km/h，前者增大为 10 倍，后者为 1.8 倍。他在引用了其他学者的研究成果后指出：分区破裂化现象的机制是由时间过程所确定的，该过程是由巷道围岩卸荷产生，并由初应力水平所决定；分区破裂化效应既可以由机械化开挖、即由慢速卸荷模型产生，也可以由钻爆法开挖、即由快速卸荷产生，但是慢速卸荷下产生的分区破裂化效应较之快速卸荷下需要更高的应力水平。Курленя 教授提出了围岩变形与破坏的两类过程：一类是连续的逐步发展的过程，在围岩介质中初始产生均匀的微裂纹，然后微裂纹随机地合并成粗裂纹，同时在粗裂纹附近由于应力释放停止微裂纹的发展，再后产生宏观主干裂缝，形成岩石材料的破坏，并在围岩主干裂缝附近停止材料的破坏过程；另一类过程是基于时间因素的影响，在高地应力水平下，围岩介质将由高应力区向卸荷区快速运动，该能量流使得围岩介质来不及产生许多微裂纹后再形成宏观裂纹，即来不及顺序地经过破坏的准备过程，而是介质位移速度超过了微裂纹的积累与聚合速度，宏观粗裂纹立即突跃产生，即材料突然地破坏，同时在其周围由于应力释放，不再产生微裂纹。如果应力水平足够高，围岩积累的变形能足够大，高速能量流过程将继续，得以在第一批裂缝的卸荷影响范围之外，继续突跃地产生第二批粗裂缝，过程将延续到应力水平不再高的情况下，此时则恢复连续平稳的变形过程。

在 Ф. П. Глушихин 等[13]的研究中，前苏联煤炭部全苏矿山力学与矿山测量科研院(ВНИМИ)总结了 20 世纪 90 年代以前 20 a 中在模型实验方面取得的经验，包括模拟理论、模拟方法以及模拟研究的结果，其中也包括"大深度巷道围岩破坏"下的实验结果和新规律，其实验结果总结于表 1。

表 1 大深度巷道围岩破坏实验结果表
Table 1 Results of failure test of surrounding rock in deep tunnels

岩体巷道断面和模型特点	巷道周边破裂区		第 1 个破裂区		第 2 个破裂区		第 3 个破裂区	
	$\gamma H/R_{CЖ}$	R/r	$\gamma H/R_{CЖ}$	R/r	$\gamma H/R_{CЖ}$	R/r	$\gamma H/R_{CЖ}$	R/r
节理岩体，圆型，纵长和立体模型	0.7	1	1.0	1.56	2.0	2.0	2.4	3.1
节理岩体，平巷，平面应变模型	0.5	1	0.9	1.80	1.5	2.3	1.9	3.2
连续岩体，圆型，立体模型	0.5～0.8	1	1.0～1.2	1.30～1.50	2.0	1.7～1.9	2.7	2.6

Ф. П. Глушихин 等[13]特别明确地指出，通过对平面和三维模型研究结果的分析，得出如下结论：对节理岩体巷道周围介质的破坏，第二应力分量 σ_2 将产生重要影响；不论上述特征如何改变，围绕巷道的分区破裂化效应(即破坏与非破坏交替的现象)是确定的；传统的围绕巷道的 3 个区(围岩破坏区、塑性区和弹性区)的围岩变形和破坏概念是相应于埋深较浅的岩体工程的荷载条件。但 Ф. П. Глушихин 等[13]关于分区破裂化效应的理论分析基本上还是定性的，而且和 Е. И. Шемякин 等的分析观点基本一致。

ВНИМИ 还出版了《等效材料平面应变模型上大深度模拟的方法指南》，该指南指出了大深度模拟的困难。首先在于模型实验不能严格满足几何相似，因此在施加荷载的深度上不能保证必须的力的平衡，这就不能满足在研究矿山压力表现时的边界条件；其次在于如何严格保证大深度模拟时的平面应变状态。在多年理论和实验经验基础上，该指南提出了大深度模拟的一般原则、模型应力 - 应变状态的相似条件、加载的方法、加载装置的结构、模型参数的计算实例等所有的必需的内容[14～17]。

4 近期研究及新进展

自 2003 年以来,随着深部工程的不断增加,深部一些新的岩石力学现象不断出现,中国的学者开始关注并开展了分区破裂化现象的研究[18~34]。以钱七虎院士为首的学术团队在国内率先介绍了国外学者关于分区破裂化现象研究的成果,提出了在国外已有研究成果基础上今后开展研究的方向及其关键问题。钱七虎[18,19]除了系统介绍国外学者在深部岩石力学(包括分区破裂化效应及其他关键科学问题)的研究成就外,提出了深部围岩分区破裂化现象是一个与空间、时间效应密切相关的科学现象的新观点,认为分区破裂化效应的产生,一方面是由于高地应力和开挖卸荷导致围岩的"劈裂"效应,另一方面是由于围岩深部高地应力和开挖面应力释放所形成的应力梯度而产生的能量流;提出了分区破裂化的定性规律(影响因素)中应该考虑巷道洞室开挖的速度(卸荷速度);提出了分区破裂化与应变型岩爆是一个问题的两个侧面,都决定于岩体开挖后岩石积聚的变形势能转变为动能和破坏能的分配比例;指出了岩石延性随深度增加而增加,超过临界深度后,岩石都转变为非脆性(延性)破坏,即"硬岩变软岩"的结论不准确,认为该结论仅适应于实验室中双向围压下加载实验的结论,不适用于巷道洞室表面围岩存在一面卸载的情况,即围岩不可能都由硬岩变软岩,否则无法解释岩爆的发生。

2006 年以来,中国学者在一些学报上发表了分区破裂化效应研究的成果,这里应指出:特别重要的是在淮南矿区的深部矿井巷道围岩钻孔中通过用钻孔窥测仪探测并观察到了分区破裂化效应的存在,并以录像的形式记录下来,肯定了深部岩体巷道围岩分区破裂化效应的存在[34];其次是通过实验室模拟实验也获得了分区破裂化现象[28];再次是在国内首次通过数值计算和模拟再现了均匀介质岩体中的巷道围岩分区破裂化现象[26,29,30,32];最后是在国内外首次在节理岩体巷道围岩中数值仿真了分区破裂化效应[30,32]。

在这些研究中,有的并未得到明显的分区破裂化效应,其原因作者认为可能在于仅考虑了基于莫尔-库仑强度理论的剪切滑移破坏[22],当第一批剪切滑移裂缝产生后,不能像拉伸裂缝那样张开,不能形成具有卸荷效应的新自由表面;也可能有的研究仅基于平面应力模型,而不是采用俄罗斯实验以及俄、中现场巷道中的平面应变条件:垂直的以及巷道轴线方向上的高初始压应力将在巷道断面径向产生拉伸变形。而在平面应力模型中,巷道断面径向变形小,形不成拉伸裂缝。Ф. П. Глушихин 等[13]、顾金才等[28,29]均正确地指出了第二主应力,即轴线方向应力,对产生分区破裂化的重要影响,但是决不能因此认为轴线方向的应力必须是"最大主应力"。这只要认真了解俄罗斯学者实验研究的实验参数数据就可得知,实际上从三维应力分析也很易知道,对于巷道断面径向拉伸来说,轴向压应力和垂直压应力都是等价的。有些没有得到分区破坏化效应的数值模拟研究结果,还可能在于没有采用动力过程模拟。因为巷道开挖的卸荷,必在径向上巷道表面与围岩深部间产生应力梯度,从而导致径向加速度、速度和位移的产生,该径向位移叠加"劈裂"产生的拉伸变形,较易形成拉伸裂缝。

有的学者不同意"分区破裂化现象及其规律与传统的连续介质弹塑性力学理论不相一致"的结论。在这里必须指出,正如前文以及 Е. И. Шемякин 等[4,6,13]所指出的,这个不一致首先指的是分区破裂化效应与传统的连续介质弹塑性力学关于巷道围岩破碎区、塑性区、弹性区依次分布的现象不一致。其次,更重要的在于,传统连续介质弹塑性力学不能研究岩石峰值强度后的变形和破坏(即软化段),仅能研究断裂的产生,不能研究裂缝的发展,包括裂缝的合并,即传统的连续介质弹塑性力学不能正确、全面地研究围岩中的局部化变形,当然包括研究节理介质中的分区破裂化效应。周维垣和杨 强[33]正确地指出"传统模型仅能够描述峰值点以前的材料特性,对峰值点后的描述无能为力","传统连续模型在考虑局部化分岔问题时,将实验中力和位移的关系简单映射为应力-应变曲线,而没有考虑内部微结构的变化以及应变梯度的控制作用",而"应变梯度在局部化分岔破坏阶段起到了主导和控制作用。这一点在传统模型中难于得到体现"。其结果是导致在采用连续传统模型模拟局部化破坏过程中,出现局部化分析结果对网格疏密或走向的强烈依赖性,即网格依赖性问题。以有限元法为例,基于传统连续理论的有限元法对局部化破坏问题模拟的网格依赖性问题是其始终无法解决的问题,此外还有零能耗散问题。具体地说,裂缝的产生与网格疏密度有关,由于网格疏密不同,将出现几种不同的裂缝产生的情况,即出现数值仿真不"真"的问题。限于篇幅,不再细述,可参阅相关研究[35~37]。

5 结论

深部岩体分区破裂化现象产生机制的研究是岩

石力学领域一个崭新的研究方向和富有挑战性的课题。本文综述了国内外深部岩体分区破裂化现象的实验和理论研究进展,归纳了分区破裂化现象产生的条件及其变化规律。该领域下一步研究的主要方向有:

(1) 分区破裂化现象和规律是一个与空间、时间效应密切相关的现象,所以分区破裂化现象定性规律的成果必须重新研究,才能进入定量规律的研究。

(2) 深部巷道、洞室在发生分区破裂化现象下围岩变形、稳定性及其支护方法的研究。

围岩破裂区和非破裂区交替的情况和浅部围岩塑性区和弹性区依次排列的情况,其变形和稳定性有很大的不同,前者需要考虑岩石峰值后的残余强度(岩块间的摩擦力)。支护时间和支护变形的大小影响巷道周边岩石位移速度,因此与分区破裂化现象的发展有很大相关性。这样支护的类型也就决定了围岩控制理论,关键是根据分区破裂化现象的原理确定围岩变形速率。反过来,利用围岩变形速率来控制分区破裂化现象的发展。在具有分区破坏现象时锚杆的布置和长度的设计计算方法将与浅部有很大的不同。

(3) 研究分区破裂化效应的发展与开挖卸荷周期的关系(该周期与岩石弹性势能、动能和耗散能的分配比例相关),即不同施工方法(钻爆法、TBM 法)及施工进度对分区破裂现象的影响。

(4) 不同支护方法(钻爆法和 TBM 法相应的支护时机、支护参数等)对分区破裂效应和自承体系的影响,提出合理的支护方式和支护时机。

(5) 从能量耗散的角度,建立分区破裂化效应、应变型岩爆与深部围岩挤压大变形间发生发展的统一理论。

参考文献(References):

[1] CLOETE D R, JAGER A J. The nature of the fracture zone in gold mines as revealed by diamond core drilling[R]. [S. l.]: Association of Mine Managers,1972.

[2] ADAMS G R, JAGER A J. Petroscopic observations of rock fracturing ahead of stope faces in deep-level gold mine[J]. Journal of the South African Institute of Mining and Metallurgy,1980,80(6):204–209.

[3] ШЕМЯКИН Е И, ФИСЕНКО Г Л, КУРЛЕНЯ М В, и др. Эффект зональной дезинтеграции горных пород вокруг подземных выработок[J]. Доклады АН СССР, 1986, 289(5): 1 088–1 094.

[4] ШЕМЯКИН Е И, КУРЛЕНЯ М В, ОПАРИН В Н, и др. Зональная дезинтеграция горных пород вокруг подземных выработок. часть I: данные натурных наблюдений[J]. ФТПРПИ, 1986, (3): 3–15.

[5] ШЕМЯКИН Е И, КУРЛЕНЯ М В, ОПАРИН В Н, и др. Зональная дезинтеграция горных пород вокруг подземных выработок. часть II: разрушение горных пород на моделях из эквивалентных материалов[J]. ФТПРПИ, 1986, (4): 3–13.

[6] ШЕМЯКИН Е И, КУРЛЕНЯ М В, ОПАРИН В Н, и др. Зональная дезинтеграция горных пород вокруг подземных выработок. часть III: теоретические представления[J]. ФТПРПИ, 1987, (1): 3–8.

[7] ШЕМЯКИН Е И, КУРЛЕНЯ М В, ОПАРИН В Н, и др. Зональная дезинтеграция горных пород вокруг подземных выработок. часть IV: практические приложения[J]. ФТПРПИ, 1989, (4): 3–9.

[8] КУРЛЕНЯ М В, ОПАРИН В Н. К вопросу о факторе времени при разрушении горных пород[J]. ФТПРПИ, 1993, (2): 6–33.

[9] SHEMYAKIN E I. New problems in rock mechanics[C]// Proceedings of the 10th Plenary Scientific Session of the International Bureau Strata Mechanics. [S.l.]: A. A. Balkema, 1988: 17–24.

[10] РОДИОНОВ В Н, СИЗОВ И А, ЦВЕТКОВ В М. Основы геомеханики[M]. Москва: Москва Недра, 1986.

[11] КУРКОВ С Н, КУКССЕНКО В С, ПЕТРОВ В А. Фифеские основы прогрозирования механическово разрушения[J]. Доклады АН СССР, 1981, 259(6): 1 350–1 352.

[12] КУРЛЕНЯ М В, ОПАРИН В Н. Проблемы нелинейной геомеханики. часть I[J]. ФТПРПИ, 1999, (3): 310–313.

[13] ГЛУШИХИН Ф П, КУЗНЕЦОВ Г Н, ШКЛЯРСКИЙ М Ф, и др. Моделирование в геомеханике[M]. Москва: Москва Недра, 1991.

[14] ФИЛИНКОВ А А, ЛАЗАРЕВИЧ Л М, НИЛИЦ А И. Некоторые особенности проявления горных ударов на таштагольском руднике[C]// Свойства Горного Массива и Управление его Состоянием. Москва: Недра, 1991.

[15] Методифеские Указания по Воспроизведению Больших Глубин на Плоских Моделях из Эквивалентных Материалов. ВНИМИ—Всесоюзный ордена трудового красного энамени научно-исследовательский институт горной геомеханики и маркшейдерского дела[R]. Ленинград: [s. n.], 1980.

[16] ТРОПП Э А, РОЗАНБАУМ М А, РЕВА В Н, и др. Зональная дезинтеграция породы вокруг горных выработок на больших глубинах[M]. Докл: АН СССР, 1985: 976.

[17] ЛИНЬКОВ А М, ПЕТУХОВ И М. Акселерация разрушения горных пород[C]// Свойства Горного Массива и Управление его Состоянием. Москва: Недра, 1991.

[18] 钱七虎. 深部地下空间开发中的关键科学问题[R]. 南京: 解放军理工大学工程兵工程学院, 2004.(QIAN Qihu. Key scientific problems for deep underground space excavation[R]. Nanjing: Engineering Institute of Engineering Corps, PLA University of Science and Technology, 2004.(in Chinese))

[19] 钱七虎. 非线性岩石力学的新进展——深部岩石力学的若干关键问题[C]// 第八次全国岩石力学与工程学术大会论文集. 北京：科学出版社, 2004: 10 – 17.(QIAN Qihu. The current development of nonlinear rock mechanics: the mechanics problems of deep rock mass[C]// Proceedings of the 8th National Conference on Rock Mechanics and Engineering. Beijing: Science Press, 2004: 10 – 17.(in Chinese))

[20] 王明洋, 宋华, 郑大亮, 等. 深部巷道围岩的分区破裂机制及"深部"界定探讨[J]. 岩石力学与工程学报, 2006, 25(9): 1 771 – 1 776.(WANG Mingyang, SONG Hua, ZHENG Daliang, et al. On mechanism of zonal disintegration within rock mass around deep tunnel and definition of "deep rock engineering"[J]. Chinese Journal of Rock Mechanics and Engineering, 2006, 25(9): 1 771 – 1 776.(in Chinese))

[21] 钱七虎. 深部岩体工程响应的特征科学现象及"深部"的界定[J]. 东华理工学院学报, 2004, 27(1): 1 – 5.(QIAN Qihu. The characteristic scientific phenomena of engineering response to deep rock mass and the implication of deepness[J]. Journal of East China Institute of Technology, 2004, 27(1): 1 – 5.(in Chinese))

[22] 陈士林, 钱七虎, 王明洋. 深部坑道围岩的变形与承载能力问题[J]. 岩石力学与工程学报, 2005, 24(13): 2 203 – 2 211.(CHEN Shilin, QIAN Qihu, WANG Mingyang. Problems of deformation and bearing capability of rock mass around deep buried tunnels[J]. Chinese Journal of Rock Mechanics and Engineering, 2005, 24(13): 2 203 – 2 211.(in Chinese))

[23] 李英杰, 潘一山, 李忠华, 等. 岩体产生分区破裂化现象机制分析[J]. 岩土工程学报, 2006, 28(9): 1 124 – 1 128.(LI Yingjie, PAN Yishan, LI Zhonghua, et al. Analysis of mechanism of zonal disintegration of rocks[J]. Chinese Journal of Geotechnical Engineering, 2006, 28(9): 1 124 – 1 128.(in Chinese))

[24] 贺永年, 韩立军, 邵鹏, 等. 深部巷道稳定的若干岩石力学问题[J]. 中国矿业大学学报, 2006, 35(3): 288 – 295.(HE Yongnian, HAN Lijun, SHAO Peng, et al. Some problems of rock mechanics for roadways stability in depth[J]. Journal of China University of Mining and Technology, 2006, 35(3): 288 – 295.(in Chinese))

[25] 唐鑫, 潘一山, 章梦涛. 深部巷道区域化交替破碎现象的机制分析[J]. 地质灾害与环境保护, 2006, 17(4): 80 – 84.(TANG Xin, PAN Yishan, ZHANG Mengtao. Mechanism analysis of zonal disintegration in deep level tunnel[J]. Jounal of Geological Hazards and Environment Preservation, 2006, 17(4): 80 – 84.(in Chinese))

[26] 周小平, 钱七虎. 深埋巷道分区破裂化机制[J]. 岩石力学与工程学报, 2007, 26(5): 877 – 885.(ZHOU Xiaoping, QIAN Qihu. Zonal fracturing mechanism in deep tunnel[J]. Chinese Journal of Rock Mechanics and Engineering, 2007, 26(5): 877 – 885(in Chinese))

[27] 陈伟. 深埋洞室围岩分层断裂现象模型实验解析. 解放军理工大学学报(自然科学版), 2008, 9(8)(待刊).(CHEN Wei. Model test analysis of layered fracture within surrounding rock of tunnels in deep stratum[J]. Journal of PLA University of Science and Technology (Natural Science), 2008, 9(8)(to be pressed).(in Chinese))

[28] 顾金才, 顾雷雨, 陈安敏, 等. 深部开挖洞室围岩分层断裂破坏机制模型实验研究[J]. 岩石力学与工程学报, 2008, 27(3): 433 – 438.(GU Jincai, GU Leiyu, CHEN Anmin, et al. Model test study on mechanism of layered fracture within surrounding rock of tunnels in deep stratum[J]. Chinese Journal of Rock Mechanics and Engineering, 2008, 27(3): 433 – 438.(in Chinese))

[29] 唐春安, 张勇兵. 深部巷道围岩间隔破裂现象的 RFPA 数值实验研究[J]. 岩石力学与工程学报, 2008, 27(7)(待刊).(TANG Chun'an, ZHANG Yongbing. Numerical model test RFPA of layered fracture within surrounding rock of tunnels in deep stratum[J]. Chinese Journal of Rock Mechanics and Engineering, 2008, 27(7)(to be pressed).(in Chinese))

[30] ZHOU X P, QIAN Q H, ZHANG B H. Zonal disintegration mechanism of deep crack-weakened rock masses under dynamic loads[J]. Journal of Central South University of Technology.(to be pressed)

[31] 于学馥, 宋存义. 不确定性科学决策方法[M]. 北京: 冶金工业出版社, 2003.(YU Xuefu, SONG Cunyi. Uncertain science decision-making method[M]. Beijing: China Metallurgical Industry Press, 2003.(in Chinese))

[32] 李树忱, 钱七虎, 李术才, 等. 深部岩体分区破裂化现象数值实现[J]. 岩石力学与工程学报, 2008, 27(8)(待刊).(LI Shuchen, QIAN Qihu, LI Shucai, et al. Numerical model of the zonal disintegration for the rock mass around deep underground works[J]. Chinese Journal of Rock Mechanics and Engineering, 2008, 27(8)(to be pressed).(in Chinese))

[33] 周维垣, 杨强. 岩石力学数值计算方法[M]. 北京: 中国电力出版社, 2005.(ZHOU Weiyuan, YANG Qiang. Numerical method of the rock mechanics[M]. Beijing: China Electric Power Press, 2005.(in Chinese))

[34] 李术才, 李树忱, 王汉鹏, 等. 淮南煤矿深部矿井围岩中的分区破裂化现象研究[R]. 济南: 山东大学, 2006.(LI Shucai, LI Shuchen, WANG Hanpeng, et al. Study on the zonal disintegration phenomenon of the deep mining roadways in the Huainan[R]. Jinan: Shandong University, 2006.(in Chinese))

[35] BAZANT Z P, PLANANS J. Fracture and size effect in concrete and other quasibrittle materials[M]. Boca Raton: CRC Press LLC, 1998: 9 – 10.

[36] OLIVER J. Modeling strong discontinuities in solid mechanics via strain softening constitutive equation. part 1: fundamentals[J]. International Journal for Numerical Methods in Engineering, 1998, 39(21): 3 575 – 3 600.

[37] OLIVER J. Modeling strong discontinuities in solid mechanics via strain softening constitutive equation. part 2: numerical simulation[J]. International Journal for Numerical Methods in Engineering, 1998, 39(21): 3 601 – 3 623.

岩石爆破的破碎块体大小控制

戴 俊[1,2]，钱七虎[1]

（1.解放军理工大学 工程兵工程学院，江苏 南京 210007；2.西安科技大学 建筑与土木工程学院，陕西 西安 710054）

摘 要：基于岩石组成的层次结构及其力学特性，分析爆炸载荷下岩石破坏的物理本质；探讨岩石爆破中的炸药爆速及装药量对岩石破碎块体尺寸及爆破引起围岩损伤区变化的影响；指出岩石爆破中正确选定炸药品种和确定单位体积炸药消耗量是控制岩石破碎块体大小的有效手段；同时也指出根据爆破目的合理控制岩石破碎块体大小有助于减少爆破引起的围岩损伤。对岩石爆破工程中正确进行装药设计具有重要意义。

关键词：岩石；爆破；破碎块度；控制

中图分类号：TD 235.4；TD 231.1　　　　**文献标识码**：A

Control of size of rock fragmentation by blasting

DAI Jun[1,2], QIAN Qihu[1]

(1. Engineering Institute of Engineer, PLA University of Science and Technology, Nanjing 210007, China; 2.College of Architecture and Civil Engineering, Xi'an University of Science and Technology, Xi'an 710054, China)

Abstract：Based on the hierarchy of rock and their mechanic characteristic, the physical essence on rock fragmentation by explosion loading, the effects of the explosive velocity and the charge on the size of rock fragmentation and the damage of remaining rock by blasting are researched. It is pointed out that it is an effective measure by which the rock fragmentation is controlled to select the variety of explosive and to determine the specific charge accurately. It is also pointed out that the damage area in surrounding rock will be reduced greatly if the rock fragmentation sizes are controlled reasonably according to blasting purpose. The achievements in the present paper are significant for designing the accurate charge in blasting engineering.

Key words：rock；blasting；fragmentation size；control

0 引 言

目前，各类工程中的岩石开挖仍以爆破方法为主。以合理的炸药爆炸能量对开挖范围内的岩石进行适度的破碎，并尽量减少爆破作业对周围保留岩石强度和稳定性的不利影响，是从事爆破理论与技术研究的专家学者追寻的重要目标之一。

近年来，不同学者对岩石爆破破碎的块体尺寸问题进行了研究，提出了岩石爆破破碎块度的分布描述方法和预测模型，对工程实践起到良好的指导作用。但这些方法和模型存在局限性，仍需要进一步完善[1-3]。印度学者 A.K.CHAKRABORTY 等建立了用隧道爆破指数预测爆破效果的模型，但其对爆破的岩石破碎块度重视不够，没有建立起隧道爆破指数与岩石破碎块度的关系[4]。

岩石爆破破碎块体尺寸受岩石性质、炸药性质和爆破设计参数等多方面因素的影响，比较复杂[5]。本文仅打算从岩石爆破破坏的物理本质出发，分析爆破装药参数对岩石破碎块体大小的影响，以及爆破条件的改变对爆破引起开挖范围以外保留岩石损伤区大小的影响关系，进而指出，工程爆破中根据工程目的的不同控制合理的岩石破碎块体尺寸的重要性。

1 岩石爆破破碎的物理本质

研究表明，岩石是成分多变、构造非均匀的介质，岩石中存在着复杂的构造层次系统，这种层次中不同级别的块体尺寸 Δ_i（i=1，2，3，…，）存在自相似关系[6]

$$\Delta_i = 2^{-i/2} \cdot \Delta_0 \qquad (1)$$

式中，Δ_i 为第 i 级别的块体尺寸；i 为自然数；

收稿日期：2006-05-08。

基金项目：国家自然科学基金重大资助项目(50490275)；中国博士后科学基金资助项目(2005037221)；陕西省教育厅专项科研基金资助项目(04JK221)；西安科技大学博士基金资助项目(2004-09)。

本文原载于《辽宁工程技术大学学报(自然科学版)》(2008 年第 27 卷第 1 期)。

$\Delta_0 = 2.5 \times 10^6$ m，为地核的直径。

同级块体之间由张开度为 δ_i 的裂纹分开，这种裂纹的张开度 δ_i 与块体尺寸之间存在稳定的统计关系

$$\mu_\Delta(\delta) = \delta_i / \Delta_i = \Theta \times 10^{-2} \quad (2)$$

式中，Θ 为系数，在 1/2~2 之间取值；$\mu_\Delta(\delta)$ 称为岩石力学不变量。

这些块体层次的分隔裂纹是岩石材料的薄弱连结，岩石材料受外载后破坏的物理本质就是这些不同层次裂纹起裂与扩展。根据 Griffith 的理论，使较长的裂纹起裂、扩展只需要较小的载荷；反之，使较短的裂纹起裂、扩展则需要较大的载荷。实践中，在静态载荷作用下，材料的断裂破坏往往是由其中潜在的较长裂纹起裂、扩展引起的，这时材料的强度较低，而破碎块度较大，破坏后的岩石块体数量少；而在岩石爆破过程中，炸药的爆炸加载具有较高的加载速率，这时不仅较大的块体层次分隔裂纹起裂、扩展，而且较小的块体层次分隔裂纹也起裂、扩展，因此岩石破碎成较小的破碎块度，块体数量也多，相应地表现出较高的强度。这便是对上述岩石破坏物理本质的有效证明，目前已得到普遍认同。

据此，通过不同加载速率加载，控制爆破载荷大小，可以有效达到获得适度破碎块度的目的。

2 炸药爆速对破碎块度的影响

美国 Sandia 国家实验室的 Grady 对岩石动态破碎过程进行了研究，得出脆性岩石动态破碎平均块度与加载率之间的如下关系[7]

$$d = \left[\sqrt{20} K_{IC} / (\rho c \dot{\varepsilon})\right]^{2/3} \quad (3)$$

式中，d 为岩石破碎块度尺寸；K_{IC} 为岩石的断裂韧度；ρ 为岩石容重；c 为岩石的弹性波速度；$\dot{\varepsilon}$ 为加载应变率。

根据前节所述岩石爆破破坏的物理本质，有

$$d \in \{\Delta_i | i = 1, 2, \cdots,\}$$

根据文献[8]，不同加载率下的岩石强度可以近似表示为

$$\hat{\sigma} = k \dot{\varepsilon}^{1/3} \quad (4)$$

式中，$\hat{\sigma}$ 为岩石的强度；k 为系数，与岩石性质有关。

将式（4）代入式（3），有岩石破碎块度尺寸 $d = \Delta_i$ 与岩石强度 $\hat{\sigma}$ 的关系

$$d = \left[\sqrt{20} K_{IC} / (\rho c)\right]^{2/3} \cdot (k/\hat{\sigma})^2 \quad (5)$$

岩石爆破条件下，炮孔中装药爆炸后在岩石中引起的爆炸载荷可以表示为

$$\sigma = p_0 (r/r_b)^{-\alpha} \quad (6)$$

$$p_0 = \rho_e D^2 (V_c / V_b)^\gamma / 8 \quad (7)$$

以上两式中 σ 为炸药爆炸载岩石中引起的压应力；p_0 为作用于炮孔壁的爆炸载荷；r 为计算点到装药中心的距离；α 为岩石中爆炸应力波衰减指数；r_b 为炮孔半径；ρ_e 为炸药密度；D 为炸药爆速；V_b 为炮孔体积；V_c 为装药体积；γ 为爆炸产物压力膨胀衰减指数，当炮孔内压力大于或等于 100 MPa 时，$\gamma = 3$，当炮孔内压力小于 100 MPa 时，$\gamma = 1.4$。

取 $\sigma = \hat{\sigma}$，并将式（6）、式（7）代入式（5），有

$$d = A \cdot k^2 (B \cdot D^2)^{-2} \quad (8)$$

$$A = \left[\sqrt{20} K_{IC} / (\rho c)\right]^{2/3}$$
$$B = \rho_e (V_c / V_b)^\gamma (r/r_b)^{-\alpha} / 8$$

于是，可得到炸药爆速与岩石破碎块度尺寸的关系

$$d \propto D^{-4} \quad (9)$$

根据式（9），要使岩石破碎块度尺寸减小两个层次，即 $d_1 = \Delta_{i+2} = \Delta_i \cdot 2^{-2/2} = d/2$，则

$$(1/2) D^{-4} = D_1^{-4}$$
$$D_1 = \sqrt[4]{2} D = 1.19 D \quad (10)$$

由式（10）知，当炸药爆速提高 1.19 倍时，岩石的破碎块度尺寸减小一半。可见，岩石的破碎块度尺寸对炸药爆破是敏感的。

3 单位体积炸药消耗量与破碎块度的关系

工程爆破实践中，通过改变炮孔内的装药结构可以比较容易地改变爆破作用于岩石的爆炸载荷，从而很好地满足不同工程目的对爆破结果的要求。在炮孔直径不变的条件下，采取不同的装药不耦合值，可以实现不同的爆破装药量，进而便可使岩石受到不等的爆炸载荷值。

利用式（5）~式（7），有

$$d = A \cdot k^2 \cdot (B_1 V_c^\gamma)^{-2} \quad (11)$$

式中，$B_1 = \rho_e D^2 (1/V_b)^\gamma (r/r_b)^{-\alpha} / 8$

于是，得到炮孔装药量与岩石破碎块度尺寸的关系

$$d \propto V_c^{-2\gamma} = (q/\rho_e)^{-2\gamma} = \rho_e^{2\gamma} \cdot q^{-2\gamma} \quad (12)$$

进一步，有
$$d \propto q^{-2\gamma} \quad (13)$$

以上两式中 q 为炮孔装药量。

式（12）也可理解为爆破岩石破碎块度尺寸与爆破单位体积耗药量的关系。由式（12）知，如果爆破岩石破碎块度尺寸减小两个层次，即 $d_1=d/2$，则有

$$q_1 = q^{-2\gamma} \cdot 1/2 = \left(2^{1/2\gamma} \cdot q\right)^{-2\gamma} \quad (14)$$

如果取 $\gamma = 3$，则 $q_1=1.12q$；如果取 1.4，$q_1=1.28q$。

即岩石爆破破碎块度尺寸减小一半，则炸药单位耗药量应增加 12%~28%。在周边爆破条件下，为了降低爆破对围岩的损伤，对岩石的破碎应严格要求，不宜追求过小的破碎块体。

4 爆破装药与爆破损伤区大小的关系

岩石爆破中，周边爆破都是采用多钻眼、少装药的方式进行的，目的在于尽可能减少爆破引起的围岩损伤，尽可能保持围岩的原有强度和稳定性。周边眼的装药量是决定围岩受爆破损伤程度的重要因素，不同的炮孔装药量，炮孔壁受到的爆炸载荷不同，进而起裂、扩展的裂纹层次不同，破碎块度尺寸不同。

由式（5）、式（6）知，如果岩石的破碎块体降低两个层次，大小由 $d=\Delta_i$ 变为 $\Delta_{i+2}=\Delta_i \cdot 2^{-2/2}=d/2$，则炮孔内的爆炸压力变化可依据

$$d \propto p_0^{-2} \quad (15)$$

推得
$$d/2 \propto p_0^{-2}/2 = \left(\sqrt{2}p_0\right)^{-2} \quad (16)$$

即炮孔内的爆炸压力将增大 $\sqrt{2}$ 倍。

如果认为岩石中的爆炸应力波衰减规律不变，且引起岩石损伤的临界应力值为 σ_0，则炮孔内压力为 p_0 时，岩石受到爆破损伤区半径 r 可表示为

$$r = (p_0/\sigma_0)^{1/\alpha} \cdot r_b \quad (17)$$

当炮孔内爆炸压力增大 $\sqrt{2}$ 倍后，岩石受到爆破损伤区半径 r_1 可表示为

$$r_1 = \left(\sqrt{2}p_0/\sigma_0\right)^{1/\alpha} \cdot r_b = 2^{1/2\alpha} \cdot r \quad (18)$$

对于常见岩石，其泊松比 $\mu = 0.2 \sim 0.5$，岩石受到爆破损伤区半径增大为

$$r_1 = 2^{1/2\alpha} \cdot r = (1.2 \sim 1.4)r$$

即，当周边爆破的岩石爆破破碎块体尺度降低两个层次，$d_1=d/2$ 时，炮孔内的爆炸压力需增大 $\sqrt{2}$ 倍，而爆破引起的围岩损伤区半径将增大 1.2~1.4 倍。在围岩损伤范围增大的同时，炮孔近区围岩受到的损伤程度必然增大，甚至出现宏观破坏，引起爆破超挖。

5 结 论

（1）岩石是具有不同结构层次的系统。外载荷作用下，岩石破坏的物理本质是不同结构层次块体的解体，爆炸载荷作用时，由于加在率很高，载荷作用值高，岩石破碎后的块体尺寸小。

（2）爆破采用高爆速炸药时，岩石破碎块体尺寸小；反之，岩石破碎块体尺寸大。

（3）当岩石爆破的单位体积耗药量增大 12%~28%时，岩石的破碎块体尺寸将减小一半。在周边爆破时，通过严格控制炮孔装药量，可以达到在爆破岩石的同时，最大限度降低爆破引起的围岩损伤。

（4）周边爆破时，如果岩石爆破破碎块体尺寸降低两个层次，则炮孔内的爆炸压力需增大 $\sqrt{2}$ 倍，而爆破引起的围岩损伤区半径将增大 1.2~1.4 倍。

参考文献:

[1] 林大泽. 爆破块度评价方法研究的进展[J]. 中国安全科学学报. 2003,13(9):9-13.

[2] 孙保平,徐全军,单海波,等. 深孔爆破岩石破碎块度的控制研究[J]. 爆破,2004,21(3):29-31.

[3] 刘 慧,冯叔瑜.爆破块度分布预测的分形损伤模型[J]. 铁道工程学报,1997,14(1):112-118.

[4] Chakraborty A K,Raina A K, Choudhury P B, et al. Development of rational models for tunnel blast prediction based on a parametric study[J]. Geotechnical and Geological Engineering,2004(22):477-496.

[5] 王明洋,戚承志,钱七虎. 岩石中爆炸与冲击下的破坏研究[J]. 辽宁工程技术大学学报,2001,20(4):385-389.

[6] 戚承志,王明洋,赵跃堂,等.关于地质材料内部构造层次的若干模型[J].世界地震工程,2003,19(1):70-76.

[7] 杨 军,金乾坤,黄风雷. 岩石爆破理论模型及数值计算[M].北京:科学出版社,1999:28-30.

[8] 李夕兵,古德生. 岩石冲击动力学[M]. 长沙:中南工业大学出版社,1994:51-56.

深部岩体强度准则

周小平[1]，钱七虎[2]，杨海清[1]

(1. 重庆大学 土木工程学院，重庆 400045；2. 解放军理工大学，江苏 南京 210007)

摘要：提出一种新的适用于深部岩体的强度准则，该准则考虑深部岩体的拉伸破坏，同时考虑深部岩体的剪胀和剪缩破坏，其破坏面可以是闭合的也可以是张开的，而且与 RMR 岩体地质力学分类指标建立联系。深部岩体准则中的参数均有明确的物理意义，可以很容易通过实验确定或通过 RMR 岩体地质力学分类确定。该准则不仅可考虑所有应力分量对材料破坏的影响，而且可反映深部岩体的受力特点。最后，将新准则与实验结果进行比较，从而验证了新准则的有效性。

关键词：岩石力学；深部岩体；非线性强度准则；岩体地质力学分类

中图分类号：TU 45　　　　**文献标识码**：A　　　　**文章编号**：1000－6915(2008)01－0117－07

STRENGTH CRITERIA OF DEEP ROCK MASS

ZHOU Xiaoping[1]，QIAN Qihu[2]，YANG Haiqing[1]

(1. *College of Civil Engineering*，*Chongqing University*，*Chongqing* 400045，*China*；
2. *PLA University of Science and Technology*，*Nanjing*，*Jiangsu* 210007，*China*)

Abstract：The most distinguished difference between deep and shallow levels is the so-called high geostress. The failure mode of the shallow rock mass is brittle failure. The failure modes of the deep rock mass may be brittle failure or ductile failure that depend on the property of rock mass and the geostress state. Mohr-Coulomb criterion，Hoek-Brown criterion，Druck-Prager criterion and unified strength criterion are suitable for brittle failure of the shallow rock mass. A new strength criterion，which is applicable to deep rock mass，is proposed. The new strength criterion，in which the effect of intermediate principal stress is considered，can consider the brittle failure and ductile failure of the deep rock mass. Moreover，the new strength criterion that is relative to the rock mass rating(RMR) classification can be applied to the research of the tensile failure of deep rock mass. All the parameters have clear physical concept，which can be conveniently determined by experiments or RMR in the presented criterion. Finally，in order to examine the proposed strength criterion，the comparison between the present strength criterion and the experimental data is performed. It is shown that the presented strength criterion is in good agreement with the experimental results.

Key words：rock mechanics；deep rock mass；nonlinear strength criterion；rock mass rating(RMR) classification

1 引 言

近年来，随着水电开发、矿山开采和公路等工程建设的迅速发展，深部岩石工程越来越常见。岩体强度准则是岩石工程结构强度计算和设计的基础理论，研究岩体强度准则在理论创新和工程应用方面都具有重要意义[1~3]。岩体强度准则的研究已有

收稿日期：2007-04-19；修回日期：2007-07-23。
基金项目：国家自然科学基金重大项目(50490275)，国家自然科学基金面上项目(50778184)。
本文原载于《岩石力学与工程学报》(2008 年第 27 卷第 1 期)。

数十年的历史，到目前为止提出过各种各样的岩体强度准则，但它们大都有一定的适用范围。例如：Mohr-Coulomb 和 Hoek-Brown 准则都没有考虑中间主应力的作用，因而较为保守；Drucker-Prager 准则由于不能反映π平面上的拉伸子午线和压缩子午线的不同，而与实际不同；双剪强度理论虽然考虑了中间主应力的影响，但只能适用于剪切和拉压强度满足一定关系的材料；统一强度理论仅仅适合于 $\sigma_3 < \sigma_c$ 时的脆性破坏。此外，虽然提出了众多的岩土类材料强度理论，但是对于深部岩体至今尚未有合适的强度准则。深部岩体由于构造运动和围岩自重的作用处于高地应力状态，其强度特性明显区别于浅部岩体。例如，脆性岩石在高地应力状态下开挖卸荷有可能发生岩爆，而另外一些岩石有可能发生延性破坏或剪缩破坏。浅部岩体的破坏形式随岩性的不同而不同，通常岩石的破坏主要表现为剪胀破坏。采用现有的岩体强度理论计算深部岩体强度，往往与实际情况不符，因为现有的岩体强度理论不能正确反映深部岩体的力学特性，多数强度理论都是基于岩石的剪胀破坏提出的，因而迫切需要一个适用于深部岩体的强度准则。

本文基于断裂力学知识和大量实验结果推导了新的深部岩体强度准则，该准则与 RMR 岩体地质力学分类指标建立了联系。该准则不仅考虑到了所有应力分量对材料破坏的影响，而且反映了深部岩体的受力特点。最后将新准则与目前常用的强度准则进行了比较，验证了新准则的可行性。

2 深部岩体强度准则

在低应力状态下，脆性岩石的破坏主要表现为脆性破坏；而在高应力状态下，岩石的破坏既有可能是脆性破坏也有可能是延性破坏，其破坏特性与岩性和受力状态有关[2~7]。E. Hoek 和 E. T. Brown[4] 根据印第安石灰岩的三轴实验结果认为，从脆性到韧性过渡点大约在主应力比 $\sigma_1/\sigma_3 = 4.3$ 处。K. Mogi[5] 研究了一些岩石的破坏特性，认为大多数岩石的脆性-韧性过渡点在平均应力比 $\sigma_1/\sigma_3 = 3.4$ 处。E. Hoek 和 E. T. Brown[6] 提出一个粗略的标准：$\sigma_3 < \sigma_c$ 时的破坏为脆性破坏，$\sigma_3 > \sigma_c$ 时的破坏为韧性破坏。浅部岩体主要特点是地应力水平较低，通常岩石在开挖卸荷时主要表现为剪胀破坏，其破坏面一般为张开型，因此浅部岩体的屈服和破坏可以用 Mohr-Coulomb 准则、Hoek-Brown 准则和非线性统一强度理论表示。深部岩体的主要特点表现为地应力水平高，岩石在开挖卸荷时可能表现为拉伸破坏，也可能表现为剪缩和剪胀破坏[8]，其破坏形式与岩性和地应力状态有关。为此，迫切需要发展一种既能反映深部岩体的拉伸破坏，也能反映深部岩体的剪胀和剪缩破坏的深部岩体强度准则。基于脆性断裂力学知识[9, 10]，A. A. Griffith 提出了 Griffith 强度理论，但是 Griffith 强度理论没有考虑裂纹间的相互作用，因而和实际不符。为了考虑裂纹间的相互作用，根据作者[11]的相关研究结果，有

$$(\sigma_1' - \sigma_3')^2 = m(\sigma_1' + \sigma_3') \tag{1a}$$

其中，

$$\left. \begin{array}{l} m = 4\xi\sigma_{t\max} \\ \sigma_{t\max} = -\dfrac{(\sigma_1' - \sigma_3')^2}{4(\sigma_1' + \sigma_3')\xi} \end{array} \right\} \tag{1b}$$

式中：ξ 为常数；σ_1'，σ_3' 均为相互作用应力。

考虑裂纹间的相互作用时，远场应力 σ_1，σ_3 和相互作用应力 σ_1'，σ_3' 有如下关系[11]：

$$\sigma_1' = \left(\frac{b\sigma_1}{2\sigma_c} + \frac{d\sigma_3}{2\sigma_c} + \frac{k^2}{2} \right)(\sigma_1 - \sigma_3)^{1/n} \tag{2a}$$

$$\sigma_3' = \left(\frac{b\sigma_1}{2\sigma_c} + \frac{d\sigma_3}{2\sigma_c} + \frac{k^2}{4} \right)(\sigma_1 - \sigma_3)^{1/n} \tag{2b}$$

式中：k 为与岩石的完整性系数有关的参数。

根据式(1)，(2)，可得

$$\sigma_1 - \sigma_3 = \sigma_c \left(\frac{b\sigma_1}{\sigma_c} + \frac{d\sigma_3}{\sigma_c} + b^2 \right)^n \tag{3}$$

根据作者[11]的相关研究结果可以确定 n，b，d 与 RMR 之间的关系为

$$n = 0.65 - \frac{RMR - 5}{200} \quad (RMR > 23) \tag{4a}$$

对于扰动岩体，有

$$b = 0.618 \exp\left(\frac{RMR - 100}{12} \right) \tag{4b}$$

$$d = 1.4 m_i \exp\left(\frac{RMR - 100}{14} \right) -$$

$$0.31\sqrt{m_i^2\exp\left(\frac{RMR-100}{7}\right)+4\exp\left(\frac{RMR-100}{6}\right)} \quad (4c)$$

对于未扰动岩体，有

$$b = 0.618\exp\left(\frac{RMR-100}{18}\right) \quad (5a)$$

$$d = 0.9m_i\exp\left(\frac{RMR-100}{28}\right) -$$

$$0.31\sqrt{m_i^2\exp\left(\frac{RMR-100}{14}\right)+4\exp\left(\frac{RMR-100}{9}\right)} \quad (5b)$$

式中：m_i 为完整岩石的 Hoek-Brown 参数，RMR 为岩体地质力学分类指数。

根据 M. H. Yu 等[7, 12]考虑中间主应力的影响方法，本文提出的深部岩体破坏准则为

$$\left.\begin{aligned}
F_1 &= \sigma_1 - \frac{1}{1+a}(a\sigma_2+\sigma_3) + W(\sigma_1+\sigma_3+b\sigma_2)\frac{I_1}{\sigma_c} - \\
&\quad \sigma_c\left[\frac{b\sigma_1}{\sigma_c}+\frac{d}{(1+a)\sigma_c}(a\sigma_2+\sigma_3)+b^2\right]^n = 0 \\
&\quad\quad\quad\quad\quad\quad\quad (F_1 \geq F_2) \\
F_2 &= \frac{1}{1+a}(\sigma_1+a\sigma_2) - \sigma_3 + W(\sigma_1+\sigma_3+b\sigma_2)\frac{I_1}{\sigma_c} - \\
&\quad \sigma_c\left[\frac{b\sigma_1+a\sigma_2}{\sigma_c(1+a)}+\frac{d\sigma_3}{\sigma_c}+b^2\right]^n = 0 \\
&\quad\quad\quad\quad\quad\quad\quad (F_1 < F_2)
\end{aligned}\right\}$$

$$(6)$$

式中：a 为中间主应力影响系数；σ_c 为完整岩石的单轴抗压强度；b 为岩石完整性参数；d 为与岩性和裂隙分布有关的参数；n 为与深部岩体的结构面分布有关的参数；W 为剪缩参数，可以通过室内三轴实验确定，当 $W=0$ 时，岩体表现为剪胀破坏。

本文应力角定义如下：

$$\theta = \arctan\left[\frac{2\sigma_2-(\sigma_1+\sigma_3)}{\sqrt{3}(\sigma_1+\sigma_3)}\right] \quad (-30°\leq\theta\leq 30°) \quad (7)$$

主应力与应力第一不变量 I_1、偏应力第二不变量 J_2 以及应力角之间的关系为

$$\sigma_1 = \frac{I_1}{3} + \frac{2}{\sqrt{3}}\sqrt{J_2}\sin\left(\theta+\frac{2\pi}{3}\right) \quad (8)$$

$$\sigma_2 = \frac{I_1}{3} + \frac{2}{\sqrt{3}}\sqrt{J_2}\sin\theta \quad (9)$$

$$\sigma_3 = \frac{I_1}{3} + \frac{2}{\sqrt{3}}\sqrt{J_2}\sin\left(\theta-\frac{2\pi}{3}\right) \quad (10)$$

深部岩体强度准则可以用应力第一不变量和偏应力第二不变量表示，则式(6)可表示为

$$F_1 = \frac{1}{3\sigma_c(1+a)}\left\{9(1+a)(2+a)p^2W + \right.$$
$$\sqrt{3}(2+a)q\sigma_c\cos\theta + 3q[2(a^2-1)pW - $$
$$a\sigma_c]\sin\theta - 3(1+a)\sigma_c^2\left[b^2+\frac{(b+d)p}{\sigma_c}+\right.$$
$$\frac{(2a-1)qd\sin\theta-\sqrt{3}qd\cos\theta}{3(1+a)\sigma_c}+$$
$$\left.\left.\frac{2(1+a)qb\sin\left(\theta+\frac{2\pi}{3}\right)}{3(1+a)\sigma_c}\right]^n\right\} \quad (11)$$

$$F_2 = \frac{1}{3\sigma_c(1+a)}\left\{9(1+a)(2+a)p^2W + \right.$$
$$\sqrt{3}(2+a)q\sigma_c\cos\theta + 3q[2(a^2-1)pW + $$
$$a\sigma_c]\sin\theta - 3(1+a)\sigma_c^2\left[b^2+\right.$$
$$\frac{dp-\frac{2}{3}dq\sin\left(\frac{2\pi}{3}-\theta\right)}{\sigma_c}+\frac{3(a+b)p}{3(1+a)\sigma_c}+$$
$$\left.\left.\frac{2aq\sin\theta}{3(1+a)\sigma_c}+\frac{32bq\sin\left(\frac{2\pi}{3}+\theta\right)}{3(1+a)\sigma_c}\right]^n\right\} \quad (12)$$

其中，

$$p = I_1/3, \quad q = \sqrt{3J_2}$$

根据式(6)可以知道岩体的单轴抗拉强度 σ_t 和单轴抗压强度 σ_r 分别为

$$\left.\begin{array}{l} \sigma_t = \left(\dfrac{d - \sqrt{d^2 + 4b^2}}{2}\right)\sigma_c \\[6pt] \sigma_r = \dfrac{2}{1+\sqrt{5}}b\sigma_c \end{array}\right\} \quad (13)$$

因此，该准则考虑了岩体拉压强度的不同。当 $a \neq 0$ 时，该准则考虑了中间主应力的影响；当 $W = 0$ 时，该准则考虑了浅部岩体的剪胀破坏，其破坏面是张开的；当 $W \neq 0$ 时，该准则考虑了深部岩体的剪缩破坏，其破坏面是闭合的。

3 理论值与高应力状态下深部岩石的实验值对比分析

目前，常用的岩体强度准则主要有 Hoek-Brown 准则、统一强度理论[7]和 Druck-Prager 准则等。E. Hoek 和 E. T. Brown[13]提出的经验准则为

$$\sigma_1 - \sigma_3 = \sqrt{m\sigma_c\sigma_3 + s\sigma_c^2} \quad (14)$$

式中：s 为岩体完整性系数，对于完整岩石取 $s = 1$。

Hoek-Brown 准则最大的缺点是没有考虑中间主应力的影响，因而不能很好地符合真三轴实验结果。国内外大量实验结果表明，中间主应力效应是岩土类材料的一个基本特性。B. Singh 等[14]为了考虑中间主应力的影响，将 Hoek-Brown 准则修正为

$$\sigma_1 - \sigma_3 = \sigma_c\left[\frac{m(\sigma_2 + \sigma_3)}{2\sigma_c} + s\right]^n \quad (15)$$

本文用深部岩体强度准则预测 q 值，并与实验值和其他强度准则计算的 q 值进行了对比分析。作者参考了不同岩石的 4 组真三轴实验数据，这 4 组实验的围压均比较高，能较好地模拟深部岩体的高地应力状态。

图 1 给出了 Shirahama 砂岩实验值与理论值比较[15]。修正的 Hoek-Brown 准则计算参数为：$m = 18.2$，$s = 1$；统一强度理论计算参数为：$b = 0.4$，$m = 18.2$，$s = 1$；本文深部强度准则的计算参数为：$\sigma_c = 65$ MPa，$d = 10.7$，$b = 0.618$，$a = 0.4$，$W = 0.04$。

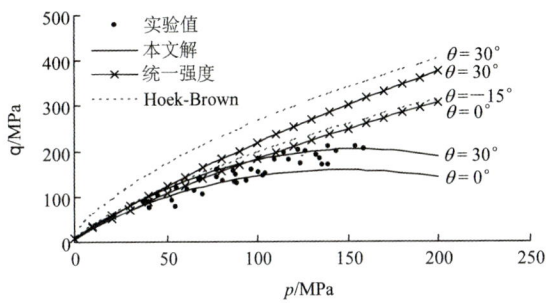

图 1 Shirahama 砂岩实验值与理论值比较[15]

Fig.1 Comparison between theoretical and experimental results of the Shirahama sandstone[15]

图 2 给出了 Yuubari 页岩实验值与理论值比较[15]。修正的 Hoek-Brown 准则计算参数为：$m = 6.5$，$s = 1$；统一强度理论计算参数为：$b = 0.5$，$m = 6.5$，$s = 1$；本文深部强度准则计算参数为：$\sigma_c = 100$ MPa，$d = 3.740$，$b = 0.618$，$a = 0.5$，$W = 0.05$。

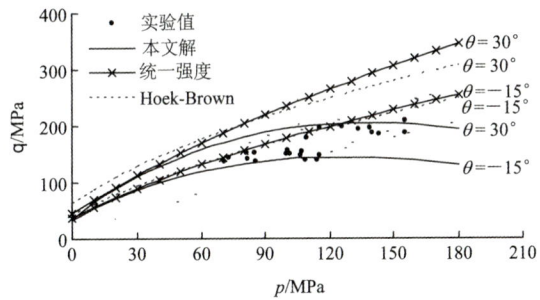

图 2 Yuubari 页岩实验值与理论值比较[15]

Fig.2 Comparison of theoretical and experimental results of the Yuubari shale[15]

图 3 给出了 Dunham 白云岩实验值与理论值比较[16]。修正的 Hoek-Brown 准则计算参数为：$m = 8$，$s = 1$；统一强度理论计算参数为：$b = 0.3$，$m = 8$，$s = 1$；本文深部强度准则计算参数为：$\sigma_c = 400$ MPa，$d = 4.640$，$b = 0.618$，$a = 0.3$，$W = 0.04$。

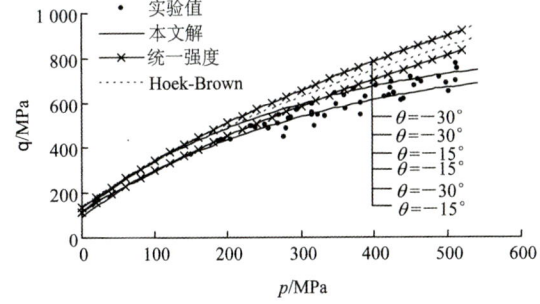

图 3 Dunham 白云岩实验值与理论值比较[16]

Fig.3 Comparison between theoretical and experimental results of the Dunham dolomite[16]

图 4 给出了 KTB Amphibolite 岩实验值与理论值比较[17]。修正的 Hoek-Brown 准则计算参数为：$m=30$，$s=1$；统一强度理论计算参数为：$b=0.3$，$m=30$，$s=1$；本文深部准则计算参数为：$\sigma_c=250$ MPa，$d=17.680$，$b=0.618$，$a=0.3$，$W=0.00$。

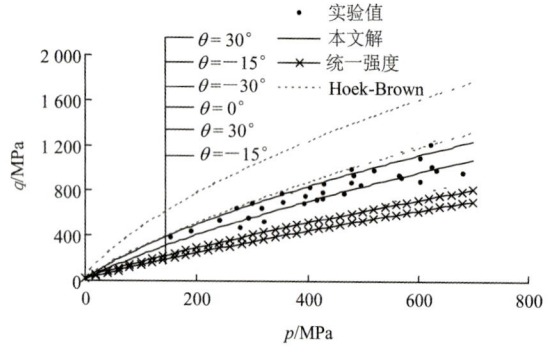

图 4　KTB Amphibolite 岩实验值与理论值比较[17]

Fig.4　Comparison between theoretical and experimental results of the KTB Amphibolite[17]

图 1～4 表明：本文提出的强度理论在高围压时预测的 q 值和实验值偏差很小，大部分偏差小于 10%，因此可以作为深部岩体的强度理论。而 Hoek-Brown 准则和统一强度理论值偏离实验值较大，因此不能作为深部岩体的强度理论。

图 5 给出了 Westerly 花岗岩实验值与理论值比较[18]。Drucker-Prager 准则计算参数为：$\alpha=0.254$，$k=3.34$；Mohr-Coulomb 准则计算参数为：$c=7$ MPa；$\varphi=58°$；本文深部强度准则计算参数：$\sigma_c=300$ MPa，$d=13.129$，$b=0.618$，$a=0.5$，$W=0.030$。

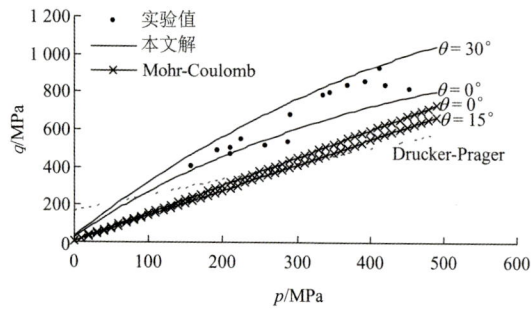

图 5　Westerly 花岗岩实验值与理论值比较[18]

Fig.5　Comparison between theoretical and experimental results of the Westerly granite[18]

图 6 给出了 Carrara 大理岩实验值与理论值比

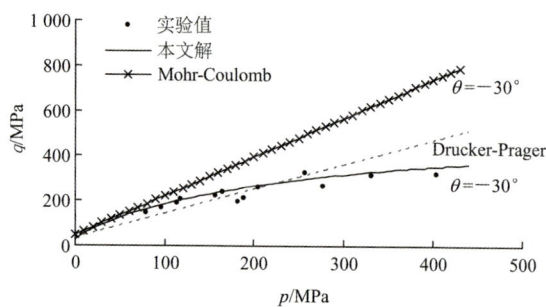

图 6　Carrara 大理岩实验值与理论值比较[19]

Fig.6　Comparison between theoretical and experimental results of the Carrara marble[19]

较[19]。该实验为常规三轴实验，$\sigma_2=\sigma_3$。Drucker-Prager 准则计算采用参数为：$\alpha=0.211$，$k=17.54$；Mohr-Coulomb 准则计算参数为：$c=25$ MPa；$\varphi=42°$；本文准则计算参数为：$\sigma_c=300$ MPa，$d=4.900$，$b=0.618$，$a=0.3$，$W=0.005$。

图 5，6 表明：本文提出的强度理论在高围压时预测的 q 值和实验值偏差很小，大部分偏差小于 10%，因此可以作为深部岩体的强度理论。而 Mohr-Coulomb 准则及 Drucker-Prager 准则偏离实验值较大。

4　理论值与低应力状态下岩石的实验值对比分析

图 7 给出了灰岩不同强度准则比较[20]。统一强度理论计算参数为：$b=0.5$，$m=21.77$，$s=1$；Hoek-Brown 准则计算参数为：$m=21.77$，$s=1$；本文深部强度准则计算参数为：$\sigma_c=78.7$ MPa，$d=14.560$，$b=0.618$，$a=0.5$，$W=0.00$。

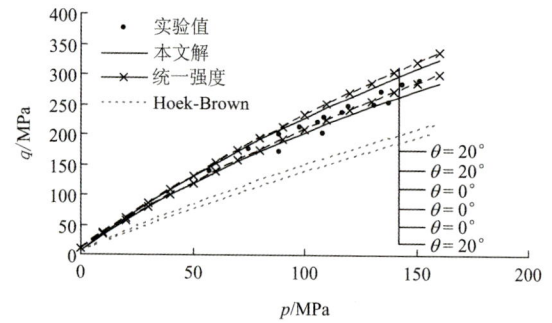

图 7　灰岩不同强度准则比较[20]

Fig.7　Comparison between theoretical and experimental results of limestone[20]

图 8 给出了粗颗粒大理岩不同强度准则比较[21, 22]。统一强度理论计算参数为：$b=1$，$m=18$，$s=1$；Hoek-Brown 准则计算参数为：$m=18$，$s=1$；本文深部强度准则计算参数为：$\sigma_c = 36.454$ MPa，$d=19.415$，$b=0.618$，$a=1.0$，$W=0.00$。

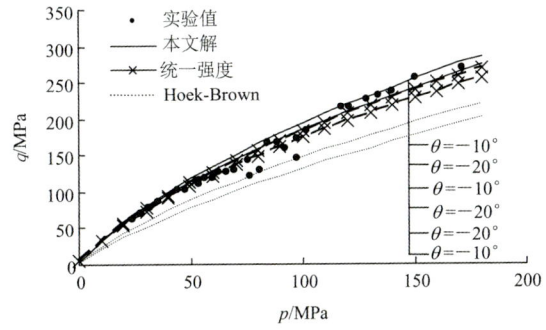

图 8　粗颗粒大理岩不同强度准则比较[21, 22]
Fig.8　Comparison between theoretical and experimental results of coarse grained dense marble[21, 22]

图 7，8 表明，本文深部岩体强度准则和统一强度理论在低应力状态下的计算结果和实验值很接近，因此本文深部岩体强度准则同样适用于浅部岩体；Hoek-Brown 准则由于没有考虑中间主应力的作用，因而较为保守。

5　本文深部岩体强度准则反映岩体的剪胀和剪缩破坏情况

图 9 给出了 Trachyte 岩剪缩破坏面[5]。图 9 表明，本文深部岩体强度准则可以根据剪缩参数 W 的不同反映岩体的剪胀和剪缩破坏，破坏面既可以为张开型($W=0.00$)，也可以为闭合型($W=1.00$)。

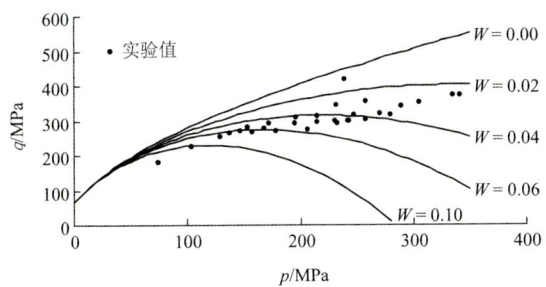

图 9　Trachyte 岩剪缩破坏面[5]
Fig.9　Limit loci in p-q plane of Trachyte[5]

因此，本文的深部岩体强度准则既适用于深部岩体，也适用于浅部岩体。图 9 中的计算参数为：$\sigma_c = 147.866$ MPa，$d=3.580$，$b=0.618$，$\theta = 10°$，$a=1.0$。

6　结　论

(1) 本文提出了深部岩体强度准则，该准则可以考虑岩体拉压不同强度特性和中间主应力的影响，表达式简单，具有明确的物理意义。

(2) 本文提出的深部岩体破坏准则，能反映深部岩体高应力状态下特有的韧性(或称延性)破坏或剪缩破坏，其破坏面可以是闭合的。

(3) 通过将本文深部岩体强度准则的理论值与实验值的对比分析表明，该准则在高应力状态下比统一强度理论和 Hoek-Brown 准则更接近实验值，因而说明该准则既适用于浅部岩体的脆性破坏，也适用于深部岩体的延性或脆性破坏。

(4) 本文提出的深部岩体强度准则中的参数有明确的物理意义，并且与 RMR 岩体地质力学分类指标建立了联系。

参考文献(References)：

[1]　王明洋，周泽平，钱七虎. 深部岩体的构造和变形与破坏问题[J]. 岩石力学与工程学报，2006，25(3)：448－455.(WANG Mingyang，ZHOU Zeping，QIAN Qihu. Tectonic，deformation and failure problems of deep rock mass[J]. Chinese Journal of Rock Mechanics and Engineering，2006，25(3)：448－455.(in Chinese))

[2]　何满潮，谢和平，彭苏萍，等. 深部开采岩体力学研究[J]. 岩石力学与工程学报，2005，24(16)：2 803－2 813.(HE Manchao，XIE Heping，PENG Suping，et al. Study on rock mechanics in deep mining engineering[J]. Chinese Journal of Rock Mechanics and Engineering，2005，24(16)：2 803－2 813.(in Chinese))

[3]　周宏伟，谢和平，左建平. 深部高地应力下岩石力学行为研究进展[J]. 力学进展，2005，35(1)：91－99.(ZHOU Hongwei，XIE Heping，ZUO Jianping. Developments in researches on mechanical behaviors of rocks under the condition of high ground pressure in the depths[J]. Advances in Mechanics，2005，35(1)：91－99.(in Chinese))

[4]　HOEK E，BROWN E T. Strength of jointed rock masses[J]. Geotechnique，1983，33(3)：157－223.

[5]　MOGI K. Pressure dependence of rock strength and transition from brittle fracture to ductile flow[J]. Bulletin of the Earthquake Research Institute，1966，44(1)：215－232.

[6] HOEK E, BROWN E T. Practical estimates of rock masses strength[J]. International Journal of Rock Mechanics and Mining Sciences, 1997, 34(8): 1 165 - 1 186.

[7] YU M H, ZAN Y W, ZHAO J, et al. A unified strength criterion for rock material[J]. International Journal of Rock Mechanics and Mining Sciences, 2002, 39(8): 975 - 989.

[8] AUBERTIN M, LI L, SIMON R. A multiaxial stress criterion for short-and long-term strength of isotropic rock media[J]. International Journal of Rock Mechanics and Mining Sciences, 2000, 37(8): 1 169 - 1 193.

[9] GRIFFITH A A. The phenomena of rupture and flow in solids[J]. Phil Trans, 1920, 221(A): 163 - 198.

[10] 周维垣. 高等岩石力学[M]. 北京:水利电力出版社, 1989.(ZHOU Weiyuan. Advanced rock mechanics[M]. Beijing: Water Resources and Electric Power Press, 1989.(in Chinese))

[11] 周小平,哈秋聆,张永兴. 考虑裂隙间相互作用情况下围压卸荷过程应力应变关系[J]. 力学季刊, 2002, 23(2): 227 - 235.(ZHOU Xiaoping, HA Qiuling, ZHANG Yongxing. Constitutive model for crack-weakened rock mass under confining pressure unloading with crack interaction effect[J]. Chinese Quarterly of Mechanics, 2002, 23(2): 227 - 235.(in Chinese))

[12] 高延法,陶振宇. 岩石的强度准则的真三轴压力实验检验与分析[J]. 岩土工程学报, 1993, 15(4): 26 - 32.(GAO Yanfa, TAO Zhenyu. Examination and analysis of true triaxial compression testing of strength criteria of rock[J]. Chinese Journal of Geotechnical Engineering, 1993, 15(4): 26 - 32.(in Chinese))

[13] HOEK E, BROWN E T. Empirical strength criterion for rock masses[J]. Journal of Geotechnical Engineering Division, ASCE, 1980, 106(GT9): 1 013 - 1 035.

[14] SINGH B, GOEL R K, MEHROTRA V K, et al. Effect of intermediate principal stress on strength of anisotropic rock mass[J]. Tunnelling and Underground Space Technology, 1998, 13(1): 71 - 79.

[15] TAKAHASHI M, KOIDE H. Effect of the intermediate principal stress on strength and deformation behavior of sedimentary rocks at the depth shallower than 2 000 m[C]// MAURY V, FOURMAINTRAUX D ed. Rock at Great Depth. Rotterdam: A. A. Balkema, 1989: 19 - 26.

[16] MOGI K. Fracture and flow of rocks under high triaxial compression[J]. Journal of Geophysical Research, 1971, 76(6): 1 255 - 1 269.

[17] CHANG C, HAIMSON B. True triaxial strength and deformability of the German continental deep drilling program(KTB) deep hole amphibolites[J]. Journal of Geophysical Research, 2000, 105(8): 18 999 - 19 013.

[18] HAIMSON B, CHANG C. A new true triaxial cell for testing mechanical properties of rock, and its use to determine rock strength and deformability of Westerly granite[J]. International Journal of Rock Mechanics and Mining Sciences, 2000, 37(1/2): 285 - 296.

[19] FREDRICH J T, EVANS B, WONG T W. Micromechanics of brittle to plastic transition in Carrara marble[J]. Journal of Geophysical Research, 1989, 94(B4): 4 129 - 4 145.

[20] 尹光志,李贺,鲜学福,等. 工程应力变化对岩石强度特性影响的实验研究[J]. 岩土工程学报, 1987, 9(2): 20 - 28.(YIN Guangzhi, LI He, XIAN Xuefu, et al. Experimental study on the influence of engineering stress changes on strength characteristics of rocks[J]. Chinese Journal of Geotechnical Engineering, 1987, 9(2): 20 - 28.(in Chinese))

[21] MICHELIS P. Polyaxial yielding of granular rock[J]. Journal of Engineering Mechanics, ASCE, 1985, 111(8): 1 049 - 1 066.

[22] MICHELIS P. True triaxial cyclic behavior of concrete and rock in compression[J]. International Journal of Plasticity, 1987, 3(3): 249 - 270.

深部巷道围岩变形破坏的时间过程及支护

戚承志¹ 钱七虎² 王明洋²

(1. 北京建筑工程学院土木系, 北京, 100044; 2. 解放军理工大学工程兵 工程学院, 210007, 南京)

摘要: 在外载作用下岩体的变形破坏过程是一个时间过程。深部围岩变形破坏的一个重要特征是在围岩中出现了分区破裂化现象。分区破裂化现象不是巷道开挖完之后马上出现的, 而是需要一定的时间来完成。弄清分区破裂化现象的时间演化过程对于支护来讲具有重要的意义。本文利用不可逆热力学理论、连续相变理论及弹塑性理论研究了深部隧道围岩变形破坏的时间演化问题, 给出了深部巷道围岩分区破裂化现象的空间及时间解析解, 揭示了分区破裂化现象时间演化过程的一些规律。基于深部巷道围岩分区破裂化现象的时间及空间演化规律本文建议了出现分区破裂化时巷道的一些支护方法。

关键词: 深部围岩, 变形破坏, 时间过程, 巷道支护

中图分类号: TU50, TD32   **文献标识符**: A   **文章编号**:

The Temporal Process of the Deformation and Fracture of Rock Mass near Deep   Level Tunnels and Their Supporting

Qi Chengzhi1 Qian Qihu2 Wang Mingyang2

(1 Department of Civil Engineering of Beijing Institute of Civil Engineering and Architecture, Beijing, 100044; 2.PLA University of Science and Technology, Nanjing, 210007, P.R.China)

Abstract: The process of deformation and fracture of rock mass under external loadings is a temporal process. One of the significant features of the deformation and fracture of rock mass near deep level tunnels is the occurrence of the phenomenon of zonal disintegration. The zonal disintegration phenomenon does not occur immediately after the excavation of tunnels. but needs some time for its' formation. The better understanding of the temporal process of zonal disintegration is very important for supporting of tunnels. In the present paper the problem of the temporal process of zonal disintegration is studied on the basis of irreversible thermal dynamics, theory of continuous phase transitions and elasticity-plasticity theories. The spatial and temporal analytical solution of the zonal disintegration near deep level tunnels is given, some laws of the evoluation of zonal disintegration are revealed. Based on the laws of the spatial and temporal evolution of zonal disintegration some supporting methods are suggested for the strengthening of the tunnels in the case of the occurrence of zonal disintegration.

Key words: Deep level rock mass, deformation and fracture, temporal process, supporting of tunnels

0 引 言

在深部岩体中当地压力超过岩体的单轴抗压极限时, 环绕巷道会出现分区破裂现象。自从上世纪 80 年代这一现象被发现以来学者们一直试图解释这一现象。俄罗斯学者 Шемякин Е.И., Курленя М.В., Опарин В.Н.等认为从岩体的受力角度讲, 岩体的分区破裂现象是由于岩体在压力作用下, 侧压力较小时在沿最大主压应力方向产生岩体劈裂所引起的[1, 2, 3]。实验还证明[4], 由于岩块之间存在着缝隙, 岩体的变形不满足圣维南变形协调条件, 侧压力较小的条件总能局部满足, 劈裂现象总能够发生。对于深部巷道开挖面前面的分区破裂现象, Одинцев В.Н.从地质结构的角度进行了分析, 认为破碎区的出现是岩体在准静变形过程中, 达到不稳定状态时, 发生动力破坏, 并形成新的稳定结构, 这一结构在后续的一定加载范围内保持稳定, 这样的过程重复便形成分区破裂现象[5, 6]。但是目前来讲对于这一现象的理论描述还没有解决。

众所周知, 经典的弹性理论能够很好地描述材料破坏前的力学行为, 而经典的塑性理论能够大

基金项目: 国家自然科学基金重大项目(50490275)。
本文原载于《第一届全国工程安全与防护学术会议论文集》(2008 年)。

致地而不能精确地描述材料屈服后的宏观力学行为，这是因为在出现不可逆变形后材料中出现了能量的耗散，因而出现了自我组织现象，形成了耗散结构。耗散结构需要借助于物理学的有关理论进行描述。实验表明，岩石受力变形的过程可以看作是多阶段渐近相变[7]。在物理学中 Landau 的二级相变理论[8]成功地解释了超导、超流等现象[9]，Abrikosov 利用二级相变理论揭示了超导体中的周期性钉扎结构现象[10]。既然岩体受力变形的过程可以看作是多阶段渐进相变，那么其演化规律也应该遵循相变的规律。因为理想弹塑性没有考虑到岩体在极限状态下的自我组织的协调学过程，故其为隧道应力应变状态的初级近似。因此可以联合利用物理学的有关理论及弹塑性理论来研究分区破裂化现象。

岩石的应力及应变状态随着时间而变化，也即岩石具有流变特性。对于深部隧道，围岩的流变特性更加显著，对于隧道的稳定性的影响更大。观测表明，分区破裂化现象并不是巷道开挖后马上出现的，而是经过一定的时间演化后形成的。深部巷道围岩中分区破裂化现象的时间演化问题是一个还没有解决的问题。分区破裂化现象与某些物理-化学过程的空间结构极其相似，从这些物理-化学过程的时空参数的关系看，分区破裂现象的时空关系似乎是与量子力学方程类似的方程的解答。目前我们只有一些关于深部巷道围岩变形及应力时间演化的观测资料，但是关于深部岩体的分区破裂化现象时间演化的机理问题及理论描述问题还没有得到解决。搞清楚深部围岩分区破裂化现象的时间演化规律对于确定支护的形式及支护次序，达到最佳的支护效果，提高矿山及深部地下工程的安全性具有十分重要的意义。目前对于这一问题的理论研究文献几乎没有。因此本文将基于相变理论、不可逆热力学理论及传统的弹塑性理论研究分区破裂化现象的时间演化问题，并提出支护方面的一些建议。

1 问题的基本方程

典型的最大剪应力 T、相对的体积变形 θ 与最大剪切变形 Γ 之间的关系曲线如图 1 所示[7]。在弹性阶段 ($\Gamma < \Gamma_e, T < T_e$) 之后是强化阶段，强化阶段在 $\Gamma = \Gamma_0, T = T_0$ 之后进入软化阶段。在达到极限剪切变形 T_C 后破坏。岩石的体积变形在达到弹性极限前为减小，在此之后体积开始增加，在达到强度极限时体积应变为零。之后绝对体积变形为增加。这是典型的塑性材料行为。

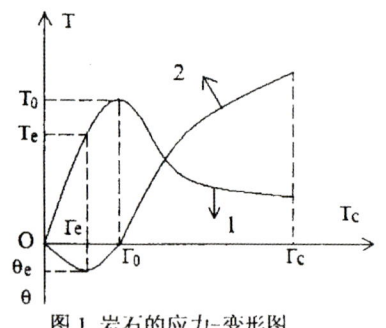
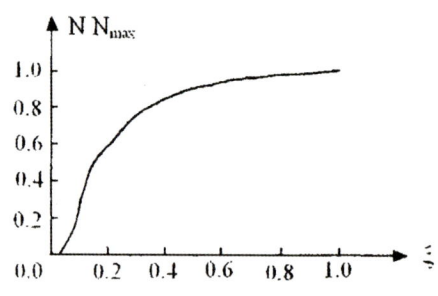

图 1 岩石的应力-变形图　　图 2 岩石的微裂纹形成与变形之间的关系 $(\xi = \Gamma/\Gamma_C)$

Fig.1 The stress-deformation diagram of rock　Fig.2 The cracking-deformation dependence of rock $(\xi = \Gamma/\Gamma_C)$

图 2 给出了在岩石试样加载时微裂纹形成的动力过程[7]。图中 N/N_{max} 为在岩石试样中当前的裂纹数量与对应于破坏时的最大微裂纹数量之比。

N/N_{max} 可以看作是岩石的损伤量 ω。在弹性区 ω 的增长非常缓慢，在进入塑性状态之后迅速增长。在硬化阶段，变形局域化区也出现了，在这一区中材料粒子变碎。在新形成的结构中产生了更复杂的结构，向这种复杂结构的转换伴随着稳定性类型的变换。损伤量 ω 在应力极限之后快速增加，在出现了主干裂纹之后其趋向于 1。微裂纹的动力发展过程可以看作是多阶段的渐进相变。从图 2 可以看出渐进相变的一些特点。在压力极限之后伴随着稳定性类型的改变，在图中出现了与材料的强度极限相对应的拐点。这一结论可以使我们利用相变的理论来研究深部坑道周围的分区破裂化现象。

对于描述岩石的不可逆变形规律来讲可以引入下来无量刚变形参数：

$$\psi = \frac{\Gamma - \Gamma_e}{\Gamma_C - \Gamma_e} \tag{1}$$

岩体的不可逆变形的出现伴随着能量耗散的产生，及耗散结构的形成。因此根据物理学的概念 ψ 可以称作序参量。我们可以首先把变形体看作是静力保守系统，其总能量为其势能 H。

在文中考虑到材料中出现缺陷时的热力学势能函数取为下式：

$$H = V_0 + V_1\psi - \frac{1}{2}V_2\psi^2 + \frac{1}{4}V_4\psi^4 + \cdots \tag{2}$$

根据渐进相变特点 $V_2 > 0$，$V_4 > 0$。

函数（2）应当满足在 $\psi = 0$ 时完整系统的平衡方程 $\nabla H = 0$，即

$$\frac{dH}{d\psi} = \left[V_1 - V_2\psi + V_4\psi^3 + \cdots\right]_{\psi=0} = 0 \tag{3}$$

由此得 $V_1 = 0$。不失一般性可以设 $V_0 = 0$

此处为了研究方便势能函数可以只取 ψ 的四次项：

$$H = -\frac{1}{2}V_2\psi^2 + \frac{1}{4}V_4\psi^4 \tag{4}$$

在变形过程中系统经历一系列状态，它们在临界点处转换。在 $\psi = 0$ 时，开始塑性变形，形成耗散结构。在 $\psi > \psi_0$ 时，新的耗散结构形成。在破坏的材料中出现了协同学效应。在岩体进入到极限状态以后，介质的粒子之间产生了长程的相互作用。类似于 Ginzburg-Landau 展开式[8]，在公式（4）中增加梯度项，则有

$$H = -\frac{1}{2}V_2\psi^2 + \frac{1}{4}V_4\psi^4 + \frac{1}{2}C(\nabla\eta)^2 \tag{5}$$

式中 C 为系数，且 C>0。

这样在整个体积 Ω 中的总势能为：

$$\Phi = \int_\Omega \left[-\frac{1}{2}V_2\psi^2 + \frac{1}{4}V_4\psi^4 + \frac{1}{2}C(\nabla\psi)^2\right]d\Omega \tag{6}$$

根据物理学的基本定律，对于稳定的状态介质总会选择使自由能为最小的参数，因此在介质的稳定状态时上式所表示的自由能应该为最小。对于上式的 ψ 参数取变分有

$$\begin{aligned}\delta\Phi &= \int_\Omega \delta\psi\left[-V_2\psi + V_4\psi^3\right]d\Omega + \int_\Omega C\nabla\psi\nabla(\delta\psi)d\Omega = \\&= \int_\Omega \delta\psi\left[-V_2\psi + V_4\psi^3 - C\nabla^2\psi\right]d\Omega + \\&+ \int_{d\Omega} C\delta\psi[(\nabla\psi)\vec{n}]dS = 0\end{aligned} \tag{7}$$

这样自由能随着时间的变化为：

$$\frac{\delta\Phi}{\delta t} = \int_\Omega \frac{\delta\psi}{\delta t}\left[-V_2\psi + V_4\psi^3 - C\nabla^2\psi\right]d\Omega + \int_{d\Omega}\frac{\delta\psi}{\delta t}(C\nabla\psi)\vec{n}dS \tag{8}$$

对于有能量耗散的系统来讲，其演化方程应该由下列不等式来得到

$$\frac{\delta\Phi}{\delta t} = \frac{\delta\Phi}{\delta\psi}\cdot\frac{\delta\psi}{\delta t} \leq 0$$

为了满足上述关系下式必须成立

$$\frac{\delta \psi}{\delta t} = -K \frac{\delta \Phi}{\delta \psi} = -K\left(-V_2\psi + V_4\psi^3 - C\nabla^2\psi\right), K > 0 \qquad (9)$$

及边界条件为

$$(\nabla \psi)\cdot \vec{n}\big|_{r=a} = 0 \qquad (10)$$

对于圆形洞室,在应力应变状态为轴对称时,如果忽略掉ψ的三次项,则式(32)可以写为

$$\left(\frac{\partial^2 \psi}{\partial r^2} + \frac{\partial \psi}{r \partial r}\right) + D^2\psi = \lambda \frac{d\psi}{dt} \qquad (11)$$

其中λ为系数。

2 问题的解

首先给出地下巷道围岩应力应变状态的弹塑性解。

设地下坑道半径为a,岩体的原始应力为σ_∞,支护受到的径向压力为q。对于出现破坏或塑性变形的极限状态情况,设介质不可压缩,即$\varepsilon_r + \varepsilon_\theta + \varepsilon_z = 0$。对于平面应变情况$\varepsilon_z = 0$,则有$\varepsilon_r = -\varepsilon_\theta$。代入到变形协调方程

$$\frac{d\varepsilon_\theta}{dr} + \frac{\varepsilon_\theta - \varepsilon_r}{r} = 0 \qquad (12)$$

得

$$\varepsilon_\theta = C/r^2 \qquad (13)$$

通常岩石的塑性变形条件为莫尔库仑条件

$$T = k + \lambda \sigma_r \qquad (14)$$

其中$k = K\dfrac{\cos\rho}{1-\sin\rho}, \lambda = \dfrac{\sin\rho}{1-\sin\rho}$,$\rho$为摩擦角。塑性区的半径由下式确定:

$$b(q) = a \cdot \left[\frac{\sigma_r(R) + k/\lambda}{q + k/\lambda}\right]^{\frac{1}{2\lambda}} \qquad (15)$$

在$\lambda = 0$时有

$$b(q) = a \cdot e^{\frac{\sigma_\infty - q - k}{2k}} \qquad (16)$$

对于$\lambda = 0$的情况塑性区的变形为[11]:

$$\Gamma = \Gamma_e\left[-1 + 2\nu + 2(1-\nu)(c/r)^2\right] \qquad (17)$$

岩体为流变介质,应当考虑时间因素。这时(13)式应当写为

$$\varepsilon_\theta = C(t)/r^2 \qquad (18)$$

观测表明,对于深部某些岩体的流变,$C(t)$可以取下列表达式[12]:$C(t) = \alpha(1 - \beta e^{-t/t_0})$,其中$\alpha$及$\beta$为系数。所以$d\psi/dt$可以取下式($\eta$为系数):

$$\frac{d\psi}{dt} = \eta \frac{e^{-t/t_0}}{r^2} \qquad (19)$$

所以(11)式可以写为(χ为系数):

$$\left(\frac{\partial^2 \psi}{\partial r^2} + \frac{\partial \psi}{r \partial r}\right) + D^2 \psi = \chi \frac{e^{-t/t_0}}{r^2}$$

即

$$r^2 \frac{\partial^2 \psi}{\partial r^2} + r \frac{\partial \psi}{\partial r} + D^2 r^2 \psi = \chi e^{-t/t_0} \tag{20}$$

方程（20）为非齐次的 Bessel 方程，其解为[14]

$$\psi = \left[\frac{\pi}{2} N_0(Dr) \int Dr J_0(Dr) d(Dr) - \frac{\pi}{2} J_0(Dr) \int Dr N_0(Dr) d(Dr)\right] \cdot \chi e^{-t/t_0} + C_1 J_0(Dr) + C_2 N_0(Dr) \tag{21}$$

上式中 J_0 及 N_0 分别为 0 阶 Bessel 函数及 0 阶 Neumann 函数，它们都为柱函数。

利用下列任意阶 ν 柱函数 C_ν 的递推关系：

$$\frac{d}{dx}\left[x^\nu C_\nu(x)\right] = x^\nu C_{\nu-1}(x)$$

可以将 ψ 表达式中柱函数的积分进行积分，从而得到

$$\psi = \left[\frac{\pi}{2} N_0(Dr) Dr J_1(Dr) - \frac{\pi}{2} J_0(Dr) Dr N_1(Dr) + C_3 N_0(Dr) + C_4 J_0(Dr)\right] \cdot \chi e^{-t/t_0} + C_1 J_0(Dr) + C_2 N_0(Dr) \tag{22}$$

当 $t \to \infty$，ψ 趋向于稳态解

$$\psi = C_1 J_0(Dr) + C_2 N_0(Dr) \tag{23}$$

利用坐标变换可以把边界条件 $(\nabla \psi) \cdot \vec{n}|_{r=a} = 0$ 化为

$$\partial \psi / \partial r|_{r=a} = 0 \tag{24}$$

利用下列微分关系

$$\frac{d}{Ddr}[J_0(Dr)] = -J_1(Dr), \quad \frac{d}{Ddr}[N_0(Dr)] = -N_1(Dr)$$

可得：

$$\left.\frac{\partial \psi}{\partial r}\right|_{r=a} = -D[C_1 J_1(Da) + C_2 N_1(Da)] = 0$$

即

$$C_2 = -C_1 \frac{J_1(Da)}{N_1(Da)} \tag{25}$$

所以问题的稳态解为：

$$\psi = C_1 \left[J_0(Dr) - \frac{J_1(Da)}{N_1(Da)} \cdot N_0(Dr)\right] \tag{26}$$

由于剪切变形不分正负，所以 ψ 可以取绝对值：

$$\psi = \left|C_1\left[J_0(Dr) - \frac{J_1(Da)}{N_1(Da)} \cdot N_0(Dr)\right]\right|$$

由 $\psi = (\Gamma - \Gamma_e)/(\Gamma_c - \Gamma_e)$ 得巷道围岩的变形

$$\Gamma = \psi(\Gamma_c - \Gamma_e) + \Gamma_e \tag{27}$$

参数 C_1 的选择应该使隧道壁的应变与经典解一致。参数 D 依赖于硐室半径，其选择应该使 ψ 的周期等于硐室壁到第一个破碎区的距离。

取 $Da = 3.2$，则 $D = 3.2/a$。设 $\nu = 0.3$，$\Gamma_C/\Gamma_e = 10$，$k = T_0$，$\sigma_\infty = 3T_0 = 1.5\sigma_0$，则利用

$$\Gamma(a) = \left| C_1 \left[J_0(Dr) - \frac{J_1(Da)}{N_1(Da)} \cdot N_0(Dr) \right] \right| (\Gamma_C - \Gamma_e) + \Gamma_e$$

$$= \Gamma_e \left[-1 + 2\nu + 2(1-\nu) e^{\frac{\sigma_\infty - T_0}{T_0}} \right] \tag{28}$$

可得 $C_1 \approx 1.862$。

现在研究巷道围岩变形的时间演化问题。对于方程（22），为了使其在 $t \to \infty$ 时得到 ψ 的稳态解，可以把其写成如下的形式（ζ 为系数）：

$$\psi = [N_0(Dr)J_1(Dr) - J_0(Dr)N_1(Dr)] \frac{\pi}{2} \chi \cdot Dr \cdot e^{-t/t_0} +$$

$$+ C_1 \left[J_0(Dr) - \frac{J_1(Da)}{N_1(Da)} N_0(Dr) \right] (1 - \zeta \cdot e^{-t/t_0}) \tag{29}$$

为了确定（29）式中的系数 χ、ζ，我们在此假定在初始时刻 $t = 0$ 在围岩无穷远处瞬间加上静水地应力 $\sigma = \sigma_\infty$，则会出现初始屈服区域。我们假定在塑性区边界上 $\psi = 0$。塑性区的半径可以由（15）或者（16）来确定。而 $J_0(Dr) - \frac{J_1(Da)}{N_1(Da)} N_0(Dr)$ 的包罗线为 $\sqrt{2/(\pi Dr)}$。

所以在 $t = 0$ 时刻，在塑性区的边界 $r = b$ 上，由（29）式得

$$\psi = [N_0(Db)J_1(Db) - J_0(Db)N_1(Db)] \frac{\pi}{2} \chi \cdot Db +$$

$$+ C_1 \sqrt{\frac{2}{\pi Db}} \cdot (1 - \zeta) = 0$$

我们还假定在巷道壁上由 $\Gamma = |\psi|(\Gamma_C - \Gamma_e) + \Gamma_e$ 公式计算得到的变形值与（17）式计算的相等。这样就完全可以确定 χ 及 ζ 的值。

在计算初始塑性区半径时应该考虑到此时的岩体强度应该为岩体的动力强度，而动力强度要比静力强度及长期强度要高。例如取 $k = 1.3T_0$，$\nu = 0.3$，$\sigma_\infty = 3T_0 = 1.5\sigma_0$，巷道壁的压力 $q = 0$，围岩初始塑性半径取为

$$b = a \cdot e^{\frac{\sigma_\infty - q - k}{2k}} = a \cdot e^{\frac{2T_0 - 0 - 1.3T_0}{2 \cdot 1.3T_0}} = 1.3a$$

而巷道壁的剪切变形为

$$\Gamma = \Gamma_e \left[-1 + 2\nu + 2(1-\nu)\left(\frac{c}{r}\right)^2 \right] = 1.97\Gamma_e$$

又 $Db = 4.16$，所以可求得

$$\zeta = 0.937；\quad \chi = -0.046。$$

所以序参量的计算公式为：

$$\psi = [N_0(Dr)J_1(Dr) - J_0(Dr)N_1(Dr)] \frac{\pi}{2} \cdot Dr \cdot (-0.046) \cdot e^{-t/t_0} +$$

$$+ 1.862 \cdot [J_0(Dr) - 0.705 N_0(Dr)] [1 - 0.937 \cdot e^{-t/t_0}] \tag{30}$$

计算结果如图 3 所示。

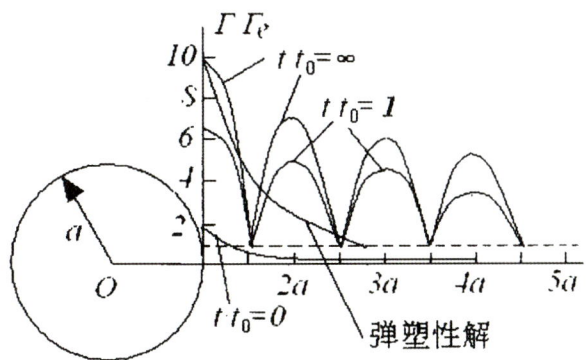

图 3 隧道围岩变形随时间的演化

Fig. 3 Evolution of the deformation of rock near deep level tunnel with time

必须指出的是，本文的计算结果是基于理想的模型而得到的。实际的情况是，由于地应力有限，分区破裂化现象的区域也是有限的。分区破裂化区域随时间逐渐扩大，以类似于波阵面的形式展开。作为初级近似，可以取弹塑性解的塑性区作为分区破裂化发生的区域。图中给出了地应力 $\sigma_\infty = 1.5\sigma_0$，$\Gamma_C/\Gamma_e = 10$ 的弹塑性解。由图可以看出此时除了隧道壁处的破坏区外还产生了一个破裂区，这一结果与实验结果吻合[1]。由图 3 可以可以看出，破坏区之间的区域破坏非常小，这也与试验结果符合。从对于试验模型的刨析来看，破坏区之间的区域破坏很弱，裂纹随机分布，而且这些裂纹似乎出现的较晚，是在围岩向隧道内侧产生位移时产生的。

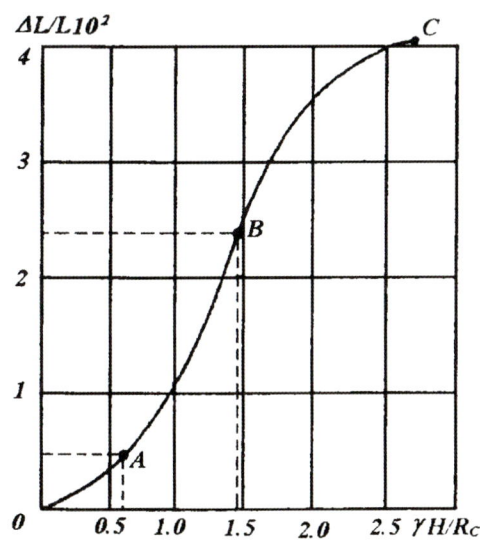

图 4 巷道相对收缩对于相对施加载的依赖关系

Fig.4 The dependence of relative convergence of tunnel on relative loading

分区破裂化现象的产生伴随着巷道围岩向巷道内位移的急剧增加。等效材料模型试验的结果如图 4 所示[1]。图 4 中横标为地应力与岩石单轴抗压强度的比值，而竖标为巷道围岩向巷道内方向的相对位移。A 点表示巷道壁开始出现破坏的时刻，之后巷道壁的位移急剧增加，在 B 点第一个破裂区出现。后续破裂区的出现对于巷道壁位移的影响非常小。因此第一个破裂区对于支护的影响最为显著，在支护时必须认真考虑。

3 深部巷道的支护问题

深部巷道围岩的变形很大。在弹性区域变形 $\varepsilon < 10^{-3}$；在脆塑性变形区孔洞及裂纹闭合，材料密实，一般变形为 $0.5 \cdot 10^{-3} < \varepsilon < 3.5 \cdot 10^{-3}$；在裂纹区及破坏的发展区变形一般为 $\varepsilon > 3.5 \cdot 10^{-3}$。

众多的观测表明,非弹性变形区的半径b与巷道壁位移U之间存在如下关系[13]:

$$\frac{b}{a} = 12\sqrt[3]{U^2/a^2}, \quad U = \left(\frac{b}{12a}\right)^{3/2} a \tag{31}$$

围岩向隧道内部的位移由其状态确定。众多的仪器观测表明,对于中等硬度的围岩来讲,如果隧道直径为4-4.5m,深度为800-1000m,那么围岩的位移大约为100-300mm。

在非弹性变形区积累在围岩中的弹性能会有所降低。其中一部分能量消耗于围岩的破坏及向隧道内的位移,其余部分以势能的形式储存于压密区及弹性区。实验表明在800-1000米处砂岩中的弹性变形能大约为$(0.5-4)10^4 N \cdot m/m^3$,而煤炭中约为$(1.5-5)10^5 N \cdot m/m^3$。

隧道开挖之后在隧道壁上产生最大的应力集中。如果岩体的强度足够高,岩体不破坏,那么应力峰值会长时间保持于隧道壁上。对于泥板岩、片岩、软砂岩等在800-1200m深度处,在双向应力状态时岩体的强度极限低于隧道壁上的应力,岩体逐渐从隧道壁向围岩内部破坏。因为没有发生岩体的完全破坏,岩体还有残余的内聚力,那么隧道周围破坏的一层岩体为围岩内部岩体建立了侧向压力。如果进行支护,那么在隧道壁处$\sigma_r > 0$。这一支撑压力随着离隧道壁的距离而增加,这样就提高了非弹性变形区的岩体承载能力。

在实验机上做$\sigma_1 > \sigma_2 > \sigma_3$的不等分量压力的研究研究表明,在$\sigma_3 \leq 1 \sim 2 \cdot 10^6 pa$时(对应于支护压力为100T/m²至200T/m²)岩体的受力状态与双轴受压状态区别不大。在这种情况下破坏压力σ_1为1.4-1.5倍的短时加载时的抗压强度σ_{comp}。如果考虑到岩体的长期强度(在隧道的使用期内)为0.65-$0.5 \sigma_{comp}$,那么在计算时应该取$\sigma_0 = \sigma_{comp}$。这时特别注意的是,设立抗力为$q = 1000 T/m^2$的强大支护时对于非弹性变形区的大小及围岩的承载力影响不大。因此在硬岩中($\sigma_0 > \sigma_\varphi$)支护的作用是为隧道提供支撑板,以防止岩块坠落到隧道中去。

在弱岩中隧道的支护起着完全不同的作用。这样的岩体只有在未扰动及处于三向加载状态下才能保持承载能力及整体性。随着隧道的开挖及围岩进入到平面应力状态,岩体连续性开始受到破坏,发生着准脆破坏。在高湿度情况下,诸如泥板岩等岩会发生蠕变。这样在隧道附近的围岩中非弹性区会扩展。巷道围岩破坏层类似于支护为其余的围岩建立侧向支撑压力,从而提高它们的承载能力。隧道支撑的效率取决于于承载能力(支撑压力)及支护最大柔度的选择。

在支护之前在隧道开挖面之后岩体的弹性回复(反向变形)发生的非常快。为了能够承受这种非弹性位移而围岩不破坏,根据前面所述,支护的柔度必须是有限的。支护的柔度应当与围岩向隧道内发生的位移相适应。非常明显,柔性支护比刚性支护更能防止低强度岩体的破坏。必须指出,支护也不能过渡柔软,否则破坏难以避免。

在出现分区破裂化现象时岩体的位移会增加很大,这时必须增加支护的柔度。在这种条件下决定性的因素为非弹性区岩体的性质、支护的柔度及承载能力,它们决定了隧道围岩中应力的分布。

在分区破裂产生条件下除了保证支护具有足够的柔性之外,还可以通过在巷道壁附近的岩体中建立柔性层来获得额外的柔性。为了保证在巷道的整个使用期内不需要对于支护进行额外修复,需要计算柔性层的厚度。在永久支护周围建立柔性层时可以利用岩体材料。为了增加岩体的柔度,可以让岩体边缘部分含有空洞及在这些空洞中填以诸如泡沫聚合物这样的柔性材料。

在支护方法方面应该对于围岩的破裂区进行有选择的加固,以便使支护在使用期内可靠工作。加固的时间应该由蓬松区裂纹的最大张开来决定。注浆的凝固时间不应该超过第一个压缩区内的岩体产生破坏所需要的时间。在注浆的时候压力应该保持的足够高,以便使浆能够注入到孔隙及裂纹中,这样可以起到加固的作用。这样在巷道围岩中形成了一个强度大于初始强度区域,从而形成一个保护层,保护支护。在采用锚杆加固的时候应该使得锚杆的端部处于岩体的弱破坏区域。

总之在进行支护时必须充分考虑到分区破裂化现象演化的空间及时间的规律,以便使得支护更加合理及可靠。

4 结论

深部围岩的一个重要特征是在围岩中出现了分区破裂化现象。分区破裂化现象不是巷道开挖完之后马上出现的,而是需要一定的时间来完成。隧道围岩的变形破坏演化过程对于支护来讲非常重要,尤其对于深部隧道来讲更是如此。因此本文利用不可逆热力学理论、相变理论及传统的弹塑性理论研究了隧道围岩变形破坏的时间演化问题，揭示了分区破裂化现象空间及时间演化过程的一些规律。并建议了在出现分区破裂化情况下巷道的一些支护方法。这些方法包括应该保证支护及与支护临近的岩体区域具有足够的柔性，对于围岩要进行有选择的加固，注浆的凝固时间应该小于第一压缩区的岩体产生破坏的时间等等。这样可以使得支护更加合理及可靠。

参考文献：

[1] Шемякин Е.И.,Фисенко Г.Л., Курленя М.В., Опарин В.Н. и другие. Зональная дезинтеграция горных пород вокруг подземных выработок [J], Ч.II, ФТПРПИ,1986, №4, 3-13.

[2] Шемякин Е.И,ФисенкоГ.Л., Курленя М.В., Опарин В.Н. Зональная дезинтеграция горных пород вокруг подземных выработок [J], Ч.III, ФТПРПИ,1987, №1, 3-8.

[3] Шемякин Е.И, ФисенкоГ.Л., Курленя М.В., Опарин В.Н. Зональная дезинтеграция горных пород вокруг подземных выработок [J],Ч.IV ФТПРПИ,1989, №4, 3-9.

[4] Курленя М.В., Опарин В.Н. и другие. О расклинивающем эффекте зон опорного давления [J], ФТПРПИ, №4, 3-11,1995.

[5] Одинцев В.Н. О механизме зональной дезинтеграции массива горных пород вблизи глубоких выработкок[J]. ФТПРПИ, №4, 10-19, 1994.

[6] Одинцев В.Н. Отрывное разрушение массива горных пород [M]. М.: ИПКОН РАН，1996.

[7] Адигамов Н.С., Рудаев Я.И. Уравнение состояния, учитывающие разупрочнение материала [J]. ФТПРПИ, №4, 24-32, 1999.

[8] Landau L.D., Lifshits E.M. Statistical physics, part I [M]. Beijing World Publishing Corporation, Beijing, 1999.

[9] Ginzburg V.L. Nobel lecture: On superconductivity and superfluidity as well as on the "physical minimum" at the beginning of the XXI century [J]. Review of Modern Physics, V. 76, 981-998, July, 2004.

[10] Abrikosov A.A. Nobel lecture: Type II superconductor and the vortex lattice [J]. Review of Modern Physics, V. 76, 975-979, July, 2004.

[11] Шемякин Е.И. Две задачи механики горных пород, связанные с освоением глубоких месторождений руды и угля. ФТПРПИ, №6, 29-45, 1975.

[12] Зборщик М.П., Назимко В.В. Охрана выработок глубоких шахт в зонах разгрузки [M]. Киев: Тэхник, 1991.

[13] Алексеев А.Д., Недодаев Н.В. Предельное состояние горных пород [M]. Киев: Наукова Думка, 1982.-200с.

[14] Зайцев В.Ф., Полянин А.Д. Справочник по нелинейным дифференциальным уравнениям[M]. Москва: Физматлит, 1993。

高放废物地质处置的成本估算

罗嗣海[1,2]，钱七虎[3]，王　驹[4]，刘晓东[2]，赖敏慧[2]，杨普济[2]

(1. 江西理工大学，江西　赣州　341000；2. 东华理工大学，江西　抚州　344000；
3. 解放军理工大学，江苏　南京　210007；4. 核工业北京地质研究院，北京　100029)

[摘要] 文章介绍和分析了国外部分国家处置库成本估算结果和成本筹措方式，据此对我国处置库的建造成本进行了粗略的估算，对成本筹措和当前的投入提出了初步建议。

[关键词] 高放废物；地质处置；成本；成本筹措

[文章编号] 1000-0658(2008)03-0181-07　　[中图分类号] X771　　　[文献标识码] A

1　引　言

高放废物深地质处置工程是一项系统、复杂的巨型工程，其投资额度大、投资时间长、不确定性因素多，部分投资可能在废物产生甚至资金承担方停止运营后数年至数十年后发生。因此，需要有充足的专项经费和专门的经费筹措机制。欧美等有核国家都对处置库的建造成本进行了分析；制定了专门的法律法规，明确和规范资金的筹措办法和机制。笔者介绍和分析了国外部分国家处置库成本估算结果和成本筹措方式；据此对我国处置库的建造成本进行粗略的概念性的估算；对成本筹措和当前的投入进行了初步探讨。

2　美国和日本高放废物地质处置成本估算

2.1　美国对尤卡山处置库的成本估算

美国高放废物的处置对象主要为乏燃料，处置库将建于尤卡山的凝灰岩中。美国对高放废物地质处置的成本进行了系统分析[1]。1998年按照可行性报告中的设计方案，对处置库的成本组成及投资时间进行了估算。2000年对1998年可行性报告设计方案进行了一些修改，2001年完成了与新的设计相应的成本分析报告"民用放射性废物管理总系统生命周期的成本"（Total System Life Cycle Cost（TSLCC）for the Civilian Radioactive Waste Management System）。估算时所作的主要假定包括：废物处置开始后的100年内完成处置库的关闭；处置库在高温模式下运行；处置库完成后，将可以处置83800 t（重金属）商业废核燃料（包括混合氧化废料）、政府所属的大约2500 t（重金属）废核燃料（包括海军舰艇的废核燃料）以及大约22000个高放废物罐（包括含有钚元素的玻璃固化高放废物的废物罐）等。

高放废物处置系统成本的组成分为处置库、废物接收储存和运输、计划的整合、法规四大部分。处置库的成本包括处置库研究

收稿日期：2007-04-22；修回日期：2007-08-06。
本文原载于《铀矿地质》（2008年第24卷第3期）。

和开发、地面设施、地下设施、废物包与金属防水罩、监管和基础设施及管理支持等；废物接收储存和运输成本包括相关的研发成本、废物接收与运输的准备和获准、内华达运输专用线的施工成本、废物接收与运输设备及其退役的成本、废物接收与运输、废物接收与运输的运行成本等，但不包括运输前的暂存成本。计划的整合（Program Integation）成本包括质量保证、计划管理与集成、NRC和NWTRB的费用及曾发生的核废物协调办公室的费用。法规（Institutional）成本包括核废物政策法规定的相应费用，主要有：等量纳税（PETT）、补助金、180（c）条款的费用及财政与技术援助。估算的总成本约为575亿美元（表1）。

表1 高放废物处置系统成本概况表
（以2000年美元物价计算）
Table 1 Summary of cost components for HLW system

花费项目	需资金数额（$\times 10^6$ 美元）	比例（%）
处置库（合计）	42070	73.1
含：研究和开发	6580	11.4
地面设施	7700	13.4
地下设施	8980	15.6
性能确认	2270	3.9
废物包装和防水罩	13290	23.1
监管和基础设施及管理支持	3250	5.7
废物接收、存储及运输	6800	11.8
计划的整合	4070	7.1
法 规	4580	8.0
总 计	57520	100

费用的主要组成部分是处置库部分，需要的投资总额是420.7亿美元。用于处置库研发的经费占项目总成本的11.4%。

处置成本可分为3类，分别是可指定的直接成本、可指定的公共可变成本以及不可分配的公共成本，这3类成本可按不同的原则进行分摊。

2.2 日本对高放废物地质处置的成本估算

日本未来处置库的处置对象为经后处理（指对乏燃料进行溶解和处理，以便将铀和钚分离出来）的玻璃固化体。场址未定，围岩按软质岩石和硬质岩石两大类考虑。处置成本估算时将人员、材料、机械设备等直接成本和设施管理与行政等非直接成本两大类成本相加。成本估算依据的前提条件有：（1）岩类、衬砌和支护：选择何种处置围岩，如选择花岗岩还是诸如沉积岩之类的软岩；（2）深度和进入地下设施的方法：由于处置库选择的围岩类型不同，所以处置库深度的选取也不相同，而进入地下的方法也主要考虑斜井、竖井、或复合方法；（3）工程屏障的技术参数：估算时需考虑缓冲材料的厚度和形状及包装容器的材料和厚度；（4）选址过程：选址的成本取决于拟调查的潜在处置场地的个数，计算时分3个阶段考虑，分别是"潜在场地"（Potential sites）早期调查阶段、"技术可行场地"（Technically feasible sites）的初步地质调查阶段以及为设计和施工的场地特性评价阶段，各阶段拟调查的场地数量随选址过程而定。费用估算时，如果依据的前提条件改变，估算的成本数额也将相应改变。据此，日本共做了11种典型情形的费用估算（表2）[2]，结果及统计见表3；平均成本组成见图1，不同情形下各分项成本的最大变化见图2。

由此可见，处置库的设计建造、运营和项目管理是日本处置库成本的主要组成部分，如果记入勘察和部分设计成本，研发总成本占处置成本的比例应高于10%。不同设计间成本的最大变化项目依次是地下设施、地下设备、项目管理、处置库运营、勘察与土地征用、地面设施，其余项目基本不变。

3 部分国家的单位处置成本及初步分析

各国对HLW/SNF（高放废物/乏燃料）管理成本的测算方法不一样，因此，对其精确的比较和分析有一定困难。各国成本估算的主要差异是：（1）在成本构成的要素上不一样（即研究与开发、贮存、运输、准备、处置）；（2）在计算废物管理成本时的假设前提和边界范围不同（即假设的核电站运行寿

表 2　日本高放废物地质处置费用估算情形

Table 2　Cases for cost estimate of HLW repository in Japan

情 形	1	2	3	4	5	6	7	8	9	10	11
岩 类	软岩						硬岩			软岩	硬岩
深 度（m）	500						1000		1100	500	1000
支 护	混凝土						无			混凝土	无
缓冲方法	砌块				整块		砌块				整块
外包装材料	碳钢			钛-碳钢			碳钢				
缓冲层厚度（cm）	40	70	40				70	40		70	40
包装材料厚度（cm）	18	19	18	7	18		19	18		19	18
进入地下方法	斜井加竖井	竖井	斜井加竖井							竖井	斜井加竖井
选址过程	3 阶段考虑的场址个数：5-2-1					10-5-1	5-2-1			10-5-1	5-2-1

表 3　日本高放废物地质处置费用估算（单位：10 亿日元）

Table 3　Results of repository cost estimate in Japan (in billions of yen)

情 形	1	2	3	4	5	6	7	8	9	10	11
技术开发	113.7	113.7	113.7	113.7	113.7	113.7	113.7	113.7	113.7	113.7	113.7
勘察及土地征收	175.1	195.9	175.1	175.1	175.1	220.3	201.8	183.0	184.2	240.3	183.0
设计及建设	951.7	1108.6	952.6	946.3	942.2	954.8	928.5	878.5	900.1	1112.6	869.0
地面设施	28.7	32.8	28.5	28.7	28.6	28.8	25.4	23.8	23.8	32.8	23.7
地下设施	534.9	680.4	531.6	527.7	534.9	535.0	251.9	203.9	204.9	677.3	203.9
地面设备	249.7	249.7	254.4	251.5	247.1	251.8	304.3	304.3	304.5	256.5	301.7
地下设备	97.7	105.0	97.4	97.7	90.8	97.9	306.2	305.8	326.2	104.8	298.9
其它	40.7	40.7	40.7	40.7	40.7	41.3	40.7	40.7	40.7	41.3	40.7
处置库运营	741.9	813.1	745.4	796.4	736.5	741.9	880.0	843.4	852.7	816.5	833.8
处置库退役及关闭	83.7	87.2	83.3	83.7	83.6	83.7	87.4	85.2	86.6	86.9	85.4
处置库监测	125.8	125.8	125.8	125.8	125.8	125.8	125.8	125.8	125.8	125.8	125.8
项目管理	555.8	617.1	554.8	555.9	554.8	571.2	547.5	484.5	483.9	631.5	484.2
总成本	2747.6	3061.4	2750.6	2797.0	2731.7	2811.4	2884.6	2714.1	2746.9	3127.3	2649.9
统计值	全部范围值 2714.1~3127.3		全部平均值 2820			标准差 148					
	硬岩范围值 2747.6~3061.4		硬岩平均值 2749			标准差 99					
	软岩范围值 2649.9~3127.3		软岩平均值 2861			标准差 163					

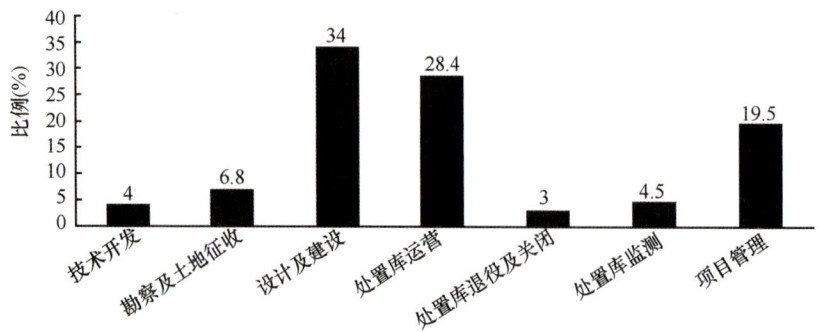

图 1　日本高放废物地质处置的成本组成

Fig. 1　The cost component for Japanese repository

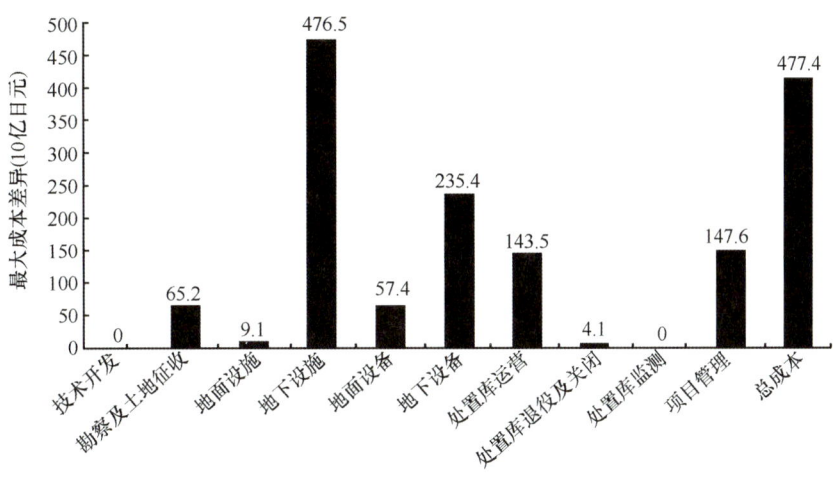

图 2　不同情形间估算的各分项成本最大差异值

Fig. 2　The maximum difference for each component among the estimated cases

期。处置库的贮存及运行寿期，关闭和监督活动的程度）；有的国家在总的测算中包括低、中放废物管理成本，核电站退役成本，而其它国家则不包括；（3）在估价的基点和对通货膨胀及时间范围所作的考虑也不同；（4）对其研究开发初期成本的考虑，即处理其开发各自 HLW/SNF 长期管理计划早期阶段成本的方法也不一样；在总的计划成本里，有的包括开发初期成本，有的则不包括。同时，同一国家对其高放废物的地质处置成本的估算也在不断变化。

表 4 列举了一些国家对高放废物处置成本估算的结果[3~10]，表中数据一般排除了废物暂存成本。将估算的总成本换算成美元并除以拟处置的废物量（拟后处理时用后处理前的乏燃料重量）得到处置每吨乏燃料的成本。

表 4　部分国家高放废物地质处置成本估算结果及单位成本

Table 4　Results of cost estimates and unit cost for HLW/SNF repository in some countries

国　家	处置对象	数　量	围　岩	估算的总成本	单位成本（万美元/tHM）
捷克	乏燃料	3724 tHM	花岗岩	469 亿捷克克郎	47.8
芬兰	乏燃料	5600 tHM	花岗岩	25 亿欧元	53.0
瑞典	乏燃料	7800 tHM	结晶岩	320 亿瑞典克郎	47.5
美国	乏燃料为主	97000 tHM	凝灰岩	575 亿美元	59.3
斯洛伐克	乏燃料	2500 tHM	结晶岩沉积岩	752 亿斯洛伐克克郎	86.4
日本	后处理的固化体	32000 tHM	硬质岩石软质岩石	27489 亿日元 28610 亿日元	73.0 76.0
瑞士	后处理的固化体或固化体＋乏燃料	共 3000 tHM，1200 tHM 已确定后处理，其余待定	粘土	29 亿瑞士法郎	74.6
英国	后处理的固化体＋乏燃料	约 10620 tHM，4700 t 乏燃料，7400 罐固化体	按坚硬岩石估算	50 亿英镑	78.8
法国	后处理的固化体	80000 m³ (B)，6000 m³ (C)	粘土岩或花岗岩	150 亿欧元	未计算
加拿大	CANDU 乏燃料	73000 tHM (3.7×10⁶ 束)	结晶岩	162 亿加元	16.7
单位成本统计值（法，加未参与统计）	全部：	平均值 66.3	标准差 14.5		
	乏燃料：	平均值 58.8	标准差 16.2		
	固化体：	平均值 75.6	标准差 2.5		

结果表明,由于各国处置的规模不同、概念设计不同、处置条件不同、估算假设不同,所得的单位处置成本结果有较大的差别。列入的国家其单位处置成本数值在16.7~86.4万美元/tHM,但在50~80万美元/tHM之间的居多;加拿大CANDU堆乏燃料的处置成本似明显较低,其余各国平均成本为65.3万美元/tHM;表中几个国家的乏燃料平均处置成本为57.3万美元/tHM,这与文献[9]得出的变化295~920美元/kg(U),参考值为560美元/kg(U)相近。表中所列采用后处理时处置成本似高一些,平均为75.6万美元/tHM,但法国的估算认为直接处置乏燃料时成本更高。处置库的规模与单位成本的相关性似不密切。从日本的估算结果看,围岩类别、处置深度对处置成本的影响并不明显。

4 高放废物管理与处置成本的筹措

4.1 筹措方式

由于 HLW/SNF 长期管理的许多活动将要持续几十年或更长时间(可能在废物生产者们已脱离业务之后),因此,明智的做法是在废物生产者仍在运行时就筹集经费,以满足将来废物长期管理和最终处置的需要。基金和储备金(funds and reserves)是两种最常见的财务系统。

基金广泛采用收取年费的办法,其计算和确定的基础是在某年的发电量或产生的废物量(即根据在该年度产生的废物相关的未来责任来确定)。主要有两种收集资金的办法:一是在电价中征收,二是由废物生产者分摊(废物生产者再从电价中收钱),分摊的数量通常是由国家政府机构计算和确定的。在多数情况下电价征收的资金出自核电站发电的收入,但西班牙征收的资金来自所有售电者的收入(包括非核电)。

财务系统设立之前所产生废物的费用收取,在芬兰和瑞典是在成立财务系统时收取一次性费用,在日本和瑞士则是全过程连续收取,而在美国则是两种方法结合起来收取。

4.2 筹措费率

各国对核废物管理和处置估算的成本不一样,基金涵盖的范围不一样,收取的费率也不一样,一些国家从核电中征收的核废物管理费见表5[10]。从表中可见,对于不同涵盖范围的费率变化在0.001~0.0077美元/kWh之间,大多小于0.003美元/kWh,当只考虑HLW/SNF时的费率最低,约为0.001美元/kWh。

表5 国外核废物管理的费率
Table 5 The fee rate for nuclear waste management in some countries

国家	费率(kWh)
只包含HLW/SNF处置的费用	
捷克	0.05CZK(约0.0019美元)
日本	约0.13日元(约0.0011美元)
美国	0.001美元
包含SNF贮存和处置	
芬兰	0.0023EURO(约0.0027美元)
包含退役,HLW/SNF贮存和处置,以及其他废物的处置	
保加利亚	电价销售的3%(近似0.9美元/MW-hr)
匈牙利	1.18HUF(约0.0053美元)
立陶宛	全部售电价的6%(近似0.0049立陶宛塔,约0.0017美元)
斯洛伐克	全部售电价的6.8%(约0.13SK,0.0037美元)
西班牙	所有电站发电零售价的0.8%
瑞典	0.01SEK(约0.0012美元)
瑞士	0.01CHF(约0.0077美元)

5 中国高放废物地质处置成本的粗略分析

随着我国核电事业的迅速发展,将积累大量的高放废物,安全处置这些高放废物需要较大资金,并需要固定的筹措渠道。但我国高放废物数量的预测尚有许多不确定性,高放废物处置的概念设计尚未提出,因此,对中国高放废物处置的成本全面和合理的估算尚有不少困难。现根据国外成本估算的成果，对我国高放废物地质处置成本作一粗略分析，其结果是概念性的。

5.1 总成本的直接类比估算

按中国核电的发展规划推算[11]，到2020年产生的乏燃料累计将达到2000 t；此后，每年将产生约1000 t的乏燃料，假设2050年中国高放废物处置库投入使用，届时高放废物将达约32000 t。初步确定我国将对除CANDU堆外的乏燃料进行后处理[11]；因此，分别按照表4的平均单位成本66.3万美元/tHM和后处理路线平均成本75.6万美元/tHM估算，处置32000 t乏燃料或后处理的固化体需要的总资金将分别约为212亿和242亿美元，约合人民币1700亿和2000亿元。

5.2 研发成本及其投入

研发经费是高放废物处置中的重要组成部分，法国在1992～2005年共投入研发资金9.8亿欧元，年均0.7亿欧元。国外研发经费一般占处置库总经费的10%～15%。如按10%估算，中国处置库研发成本应在170～200亿元人民币，按20年研发期，年平均投入的开发资金至少应为8亿人民币以上。我国的高放废物处置研究工作起步较晚，总投入的研发经费不大，处置库的研发工作进展较慢。为保证在本世纪中叶建成高放废物处置库，中国当前应加大处置库研发的投资力度。

5.3 资金的筹措

高放废物地质处置耗资巨大，建设周期长，相当部分投资可能发生在废物生产者们已脱离业务之后，"谁污染谁付费"（polluters pay）是国际上普遍认同的原则，在高放废物产生的同时从生产者或电费中收取高放废物管理基金并由独立机构管理是国际通行的做法，国外对基金设立之前产生的高放废物也进行某种方式的补征。因此，在我国核电加速发展伊始，宜尽早考虑和研究建立相应的废物管理与处置基金筹资机制。筹资费率以管理和处置成本估算为依据。如果参考并直接类比国外的数据，HLW/SNF处置的筹资费率一般为0.001美元/kW·h按直接汇率换算，我国的处置库筹措费率应为0.008元/kWh。如果包含暂存和核电站退役，其筹措费率应更高。可见，加征废物管理和处置基金并不会对电价产生太大的影响。

6 结束语

本文简要介绍和分析了国外（特别是美国和日本）高放废物地质处置成本估算的结果，计算和比较了不同情况下的单位处置成本；介绍了高放废物管理和处置的经费筹措方法和费率；最后，类比计算和粗略分析了我国高放处置的总成本，对当前处置库的研发投入和筹资机制提出了初步建议。

成本估算是高放废物处置研发中的一项复杂工作，是确定高放废物管理费率的依据，国外一般均有专人承担并不断进行更新，因此，文中对我国高放废物地质处置的经济分析仅是初步的和粗略的。

[参考文献]

[1] U·S· Department of Energy· Analysis of the Total System Life Cycle Cost of the Civilian Radioactive Waste Management Program [R]·Washington，D·C·20585；Office of the Civilian Radioactive Waste Management，May 2001.

[2] JNC·H12：Project to Establish the Scientific and Technical Basis for HLW Disposal in Japan (Supplementary Report) [R]·1-1-2 Marunouchi Chiyoda-ward Tokyo 100～8245 Japan；Japan Nuclear Cycle Development Institute，April 2000.

[3] SKB·Costs for management of the radioactive waste products from nuclear power production(Plan 2003) [R]·Stockholm Sweden；Swedish Nuclear Fuel and Waste Management Company·June 2003.

[4] J·Vogt：Swedish spent fuel management：system，facilities and operating experiences [C]·In：High level radioactive waste management，proceedings of the third international conference·Vol·1，33～40，Las Veegas，Nevada，1992. Edited by James S·Tulenko etc·Published by American Society of Civil Engineers·

[5] A·Ulibarri，A·Veganzones：Spanish high level

radioactive waste management system issues [C]. In: Proceedings of the 1993 international conference on nuclear waste management and environmental remediation. Vol. 1, 101~104, Prague, 1993. Edited by D. Alexandre etc. Published American Society of Mechanical Engineers.

[6] C. McCombie. Swiss high level radioactive waste management system issues [C] . In: High level radioactive waste management, proceedings of the third international conference. Vol. 1, 25~27, Las Veegas, Nevada, 1992.

[7] Nirex: Technic notes on cost profiles for CoRWM option 7 (Deep Geological Disposal) and Option 9 (Phased Deep Geological Disposal) [R], 2006.

[8] NWMO: Choosing a way forward-the future management of Canada's used fuel, final study [R], NWMO, 2005

[9] F. G. Long, R. D. Ward et al. Assessment and comparison of waste management costs for nuclear and fossil energy soureces [C] . In: Proceedings of the 1993 international conference on nuclear waste management and environmental remediation. Vol. 2, 815~824, Prague, 1993. Edited by P.-E. Ahlstroem etc. Published American Society of Mechanical Engineers.

[10] 郑华铃译. 高放废物和/或乏燃料长期管理工程的组织机构框架结构综述 [C]，见：高放废物地质处置资料汇编，国防科学技术工业委员会，北京，2005.

[11] Wang Ju, Su Rui et al. Deep geological disposal of high level radioactive wastes in china [J] . 岩石力学与工程学报, 2006. 25 (4): 649~658.

| 学术论文——岩石力学与工程

微型土钉微型化学注浆技术加固土质古窑

柴新军[1]，钱七虎[2]，罗嗣海[3]，林重德[4]

(1. 东华理工大学 土木与环境工程学院，江西 抚州 344000；2. 中国人民解放军理工大学，江苏 南京 210007；
3. 江西理工大学，江西 赣州 341000；4. 佐贺大学 低平地研究中心，日本 佐贺 840 - 8502)

摘要：土遗址的保护是一个世界性的难题，其本身的病害主要表现为遗址表层土的分化剥落、土体开裂和坑壁坍塌等。传统的土钉锚固技术和化学注浆技术应用于土遗址表层的加固保存时存在一些缺陷：(1) 钻孔直径较大，对遗址易造成较大的扰动，且在遗址表面易留下较明显的痕迹；(2) 传统的化学注浆技术注入压力大，易导致土遗址内部产生微裂纹破坏，常出现化学加固剂沿孔壁和注入管之间的间隙外溢至地表，污损遗迹外观。直接喷淋化学加固剂于遗址表层，渗透量和加固效果则难以评价。以加固堂加 2# 土质古窑址为目标，开发出微型土钉技术和微型化学注浆装置，以克服上述不足。根据具体的工况特征，二者可结合使用，也可单独应用。开发的微型装置主要包括水泥浆注入器、气球密封注入前端等，其显著特征是适合于微型钻孔($\phi 5 \sim 7$ mm)中进行注浆锚固和化学加固，具有一定的创新性。该微型加固技术的有效性和适宜性在现场模拟土窑的加固试验中得到检验。应用有限差分程序 FLAC 对 2# 土质古窑进行稳定分析，根据不同部位的稳定性和可能的破坏模式，进行微型加固方案的设计，并成功地应用于 2# 土质古窑的加固保存。该微型加固技术以其便携、微扰动和现场适应性强等特点在土遗址的加固补强中具有良好的应用前景。

关键词：土力学；土遗址；微型土钉；微型化学注浆；数值分析；补强设计
中图分类号：TU 44；TV 543   **文献标识码**：A   **文章编号**：1000 - 6915(2008)02 - 347 - 07

HISTORICAL EARTHEN KILN REINFORCEMENT WITH MICRO-SOIL NAILING AND MINI-CHEMICAL GROUTING TECHNIQUES

CHAI Xinjun[1]，QIAN Qihu[2]，LUO Sihai[3]，HAYASHI Shigenori[4]

(1. School of Civil and Environmental Engineering，East China Institute of Technology，Fuzhou，Jiangxi 344000，China；
2. PLA University of Science and Technology，Nanjing，Jiangsu 210007，China；3. Jiangxi University of Science and Technology，
Ganzhou，Jiangxi 341000，China；4. Institute of Lowland Technology，Saga University，Saga 840 - 8502，Japan)

Abstract：Preservation of historical earthen ruins against damage is one of the difficult problems. The main geological issues of such earthen ruins are weathering，cracking and falling of the surface soil. Conventionally used soil nailing and chemical grouting techniques have the following limitations when applied to strengthen these ruins. One is that the large diameter of drill hole will disturb the soil ruins and leave evident spots of appearance. The second one is that chemical grouts flow towards ground surface and will also spoil the appearance of the soil ruins. And also the high injection pressure may lead to micro-cracks inside the earthen ruins. Further spraying chemical grouts on the surface is frequently used in practice，but the permeation quantity and solidification effect are difficult to evaluate. From the objectives of strengthening and preserving Dougaeri historical earthen kiln

---

收稿日期：2007-04-26；修回日期：2007-06-22。
基金项目：东华理工大学博士基金资助项目(DHB0602)。
本文原载于《岩石力学与工程学报》(2008 年第 27 卷第 2 期)。

No.2, the micro-soil nailing and mini-chemical grouting techniques have been developed to overcome the above limitations. The two techniques can be used together or individually according to the specific requirements. The important micro-equipments developed are mortar injector and balloon cover injection tip, which are suitable for nailing and chemical grouting in micro drill holes(ϕ 5－7 mm). The feasibility and adaptability of the techniques have been evaluated in a field model kiln test. A fast Lagrangian analysis of continua(FLAC) code has been employed to perform stability analysis. The reinforcement design with the micro-techniques has been presented considering the different stabilities and possible failure modes at different locations of the kiln. The recommended design has been successfully applied to the practical works. It is envisaged that the micro techniques will have broad applications to strengthening and preserving historical earthen ruins due to its portability, little disturbance and well-field adaptability.

Key words: soil mechanics; earthen ruins; micro-soil nailing; mini-chemical grouting; numerical analysis; preservation design

1 引 言

近年来，岩土文物遗址的加固保护受到了我国的重视。2004 年全国文物科技保护工作会议及成果展，昭示了近年来文物科技工作的显著成绩。锚杆技术[1~4]以其原位加固的特点，在岩土工程中得到广泛应用；在石窟、摩岩造像等石质遗址锚固方面的应用也渐趋成熟[5~9]。然而，土遗址的保护仍然是一个世界性的难题，我国《文物事业"十五"发展规划和 2015 年远景目标(纲要)》将土遗址的加固技术研究作为重要加强的基础工作，突现了土遗址加固技术开发研究的急迫性。

调查结果表明，土遗址本身的病害类型主要表现为遗址表层土质的分化酥粉、土体干裂和坑壁坍塌等。造成上述病害的原因，一方面是土遗址本身质地为天然地质土体，为多孔的含黏性成分较多的颗粒集合体，对水分的变化较为敏感；另一方面是自然因素和人为因素等外部因素的影响。从土遗址保存的一些实例来看，锚杆技术和化学注浆技术在土遗址表层加固处理方面有所应用[10~12]，同时也面临一些问题。传统的土钉锚杆技术其钻孔直径(ϕ 40～120 mm)较大，钻孔过程扰动较大且在遗址表面易留下明显的痕迹，其深层锚固的优势在浅层加固中难以发挥。传统的化学注浆技术注入管直径较大，同样易在土遗址表面留下明显痕迹，影响外观；较大的注入压力易导致土遗址内产生微裂纹，且常出现化学加固剂外溢的现象。常用喷淋的方法进行表面基质强化，缺点是化学加固剂渗透深度和渗入量难以控制和定量管理，固化效果难以评价。因此，怎样将岩土锚固技术和化学注浆技术合理地应用于土质遗址表层的加固强化是当前岩土工程师和研究者所面临的一个新课题。

本文以加固强化船迫 2#土质古窑(简称 2#窑，位于日本福冈县北九州市船迫町)表层土为目标，针对传统锚杆技术和化学注浆技术的不足，开发出了微型土钉技术和微型化学注浆装置(简称微型加固技术)[12~14]。开发的微型装置主要有：水泥浆注入器和气球密封注入前端，其显著特征是非常适合在微型钻孔(ϕ 5～7 mm)中进行注浆锚固和化学加固，具有一定的创新性。特别是气球密封前端结构，可有效地解决注入过程中化学加固剂的外溢问题，可实现定点位注入过程的定量控制，具有独创性。首先简要介绍了微型加固技术的概念及相应开发的微型装置。微型加固技术的适宜性和可行性在现场模拟土窑的加固试验中得到验证。最后，采用数值模拟的方法对 2#窑进行稳定性评价，根据不同部位稳定性和破坏模式的不同，提出了相应的微型加固方案，并成功地应用于 2#窑的加固保存。可以预期该微型加固技术以其便携、微扰动、现场适应性强等特点，在土遗址表层的加固保存方面具有良好的应用前景。

2 微型加固技术及其仪器开发

船迫窑迹群位于日本福冈县北九州市船迫町，是茶白山东窑迹群、宇土窑迹群、堂加窑迹群的总称。其中堂加窑迹群 2#窑大约修建于公元 8 世纪中叶，主要用于烧制修建丰前国分寺(建于公元 756 年)所需屋瓦[15]。2#窑发掘于 1996 年，其内部形状照片见图 1，为长约 10 m 的台阶形烧瓦窑。该古窑位于风化花冈岩丘陵地带的斜坡位置，断面形状

图 1 堂加窑迹群 2#窑照片

Fig.1 Photo of Dougaeri historical earthen kiln No.2

较复杂，部分烧结部位有细微裂缝，表层土有风化剥落现象，土质属于含黏性成分较多的砂质黏土。为公开展示的需要，当地行政管理部门要求在尽量减小对窑址扰动损坏的前提下，采取必要的岩土工程措施，进行内部外部一体化加固和表层固化处理，以防止和减弱风化劣化的影响。

针对堂加窑迹群 2#窑的特征和加固要求，加固方案确定为微型土钉和微型化学注浆技术相结合。微型土钉用以实现土遗址表层烧结部和内部一体化加固，微型化学注浆技术用以遗址表层土的固化处理。土钉钻孔直径和化学注浆孔直径分别设定为$\phi 7$和 5 mm，以减小对土遗址的扰动破坏，且表面不留明显痕迹。根据具体的工况特征，微型土钉技术和微型化学注浆技术即可结合使用，也可单独使用[12~14]。

2.1 微型土钉技术

微型土钉技术中，芯材为直径$\phi 2.5$ mm 的全长螺纹细钢杆，见图 2，极限张拉力为 6.95 kN。钻孔直径设置为$\phi 7$ mm。开发使用的设备有：手提式调速电动钻、水泥浆液注入装置以及便携式拉拔试验仪等。

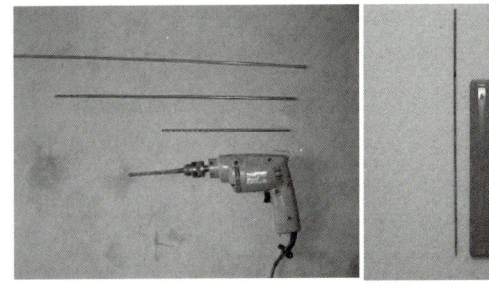

(a) 调速电动钻和钻杆　　(b) 芯材

图 2 手提式调速电动钻、钻杆和芯材

Fig.2 Portable driller，drill rod and steel bar

手提式调速电动钻钻杆直径为$\phi 7$ mm，可通过调整转速以满足不同土质中的稳定钻进。电动钻及钻杆照片见图 2。水泥浆注入器如图 3 所示，由注入管、储液筒和推进器三部分组成。注入管外径设定为$\phi 6$ mm。钻孔完成后，通过水泥浆注入器进行注浆，最后插入芯材锚固。

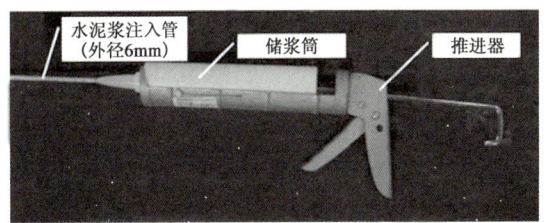

图 3 水泥浆注入器

Fig.3 Developed mortar injector for application

2.2 微型化学注浆技术

传统的化学注浆技术中，注入管注入前端大致可以分为 2 种类型：直管式注入前端和滤网式注入前端。传统的注入管加固土遗址时存在以下一些缺陷：(1) 注入管直径较大，注入孔易在土遗址表面留下较明显的痕迹。(2) 常出现化学加固剂沿孔壁和注入管之间的间隙外溢至地表，降低了注入精度，且污损外观。常用的做法是用水泥浆和膨润土密封孔口间隙，但同样易在遗迹表面留下污痕。(3) 常用直接喷淋化学加固剂于遗址表层进行基质强化的方法，但渗透量和固化效果难以定量评价。

本研究针对传统化学注浆技术应用于土遗址表层固化处理时的不足，开发出了一种新的点滴化学注浆装置，称之为微型化学注浆装置，主要由气球密封注入前端、三叉管和点滴观测器三部分组成。气球密封注入前端是该点滴注入装置的关键部位，如图 4 所示。为适应小孔径化学加固剂注入的要求，注入管直径设定为$\phi 2$ mm。橡皮气球用以密封注入管与钻孔之间的间隙，注入管加密封气球后直径为$\phi 4$ mm。通过调整气球中的气压，可有效地封堵间隙，并有效地解决药液外溢的问题，具体情况可参考有关研究[12~14]。

3 现场模拟窑迹加固试验

对 2#窑进行正式加固补强前，在其附近陡坡位置开挖了一座现场模拟土窑，在指定部位进行微型化学注浆和微型土钉加固原形试验，以检验开发微型加固技术的现场可行性并选择适宜的化学加固

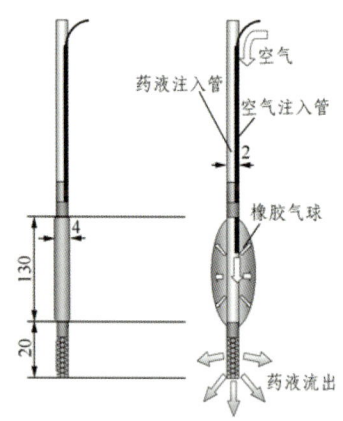

图 4　气球密封注入前端(单位：mm)

Fig.4　Balloon cover injection tip(unit：mm)

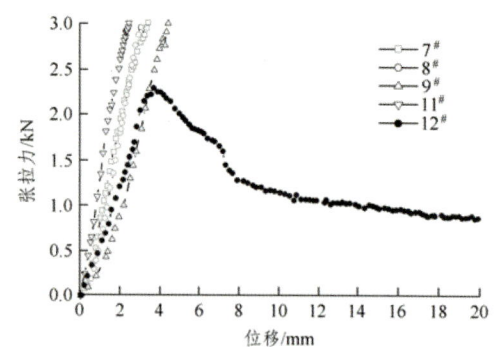

图 6　现场化学加固剂固化土体中拉拔试验(土 A)

Fig.6　Field pullout tests in chemically grouted soil(soil A)

剂。模拟窑长 4.6 m，宽 0.8 m，深 0.8 m。首先根据室内点滴注入固化效果，在指定部位进行化学加固剂点滴注入试验。然后在固化土体和自然土体部位分别进行施作微型土钉锚固，一个月之后进行抗拔试验。

3.1　化学加固剂注入试验

点滴注入所用化学加固剂为通过室内试验推荐的硅酸乙酯。注入孔呈正三角形布置，间距为 300 mm，注入孔长度为 300 mm，注入点共计 28 孔。1 a 后，对化学加固剂固化部位进行开挖，固化体呈球状，如图 5 所示，固结效率(药液注入量/固结土体体积)估算为 1∶(10～11)。

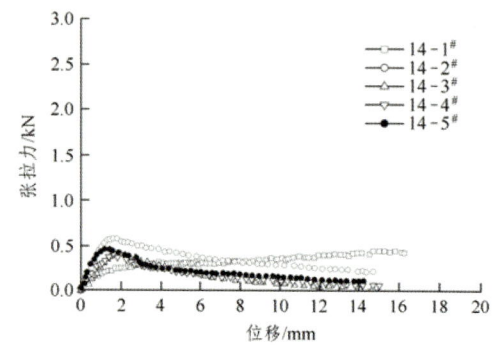

图 7　现场天然土中拉拔试验(土 B)

Fig.7　Field pullout tests of natural soil(soil B)

微型土钉的抗拔力明显高于未处理土中的抗拔力(土 B)。

室内拉拔试验进一步表明，微型土钉对土体的含水量比较敏感，土体的含水量不但影响最终的极限抗拔力，同时也影响土钉的破坏模式[12]。

室内三轴试验结果表明，选定的化学加固剂(硅酸乙酯)可有效提高土体的黏聚力，而对内摩擦角的影响不甚明显[16]。

4　稳定性分析及补强设计

4.1　稳定性分析

应用有限差分程序(FLAC)对堂加窑迹群 2# 窑进行稳定分析和补强设计。莫尔－库仑弹塑性模型用来模拟该砂黏土。现场试验和室内试验表明含水量是影响强度参数的一个重要因素，因此选定含水量 $w = 33\%$ 所对应的状态为标准状态，以对比分析不同部位的稳定性。该状态时的土质参数是根据现场拉拔试验反算，结合室内三轴试验并进行适当的调整后而得出，如表 1 所示。表 1 中土的张拉强度为非饱和土的表观抗张拉强度。

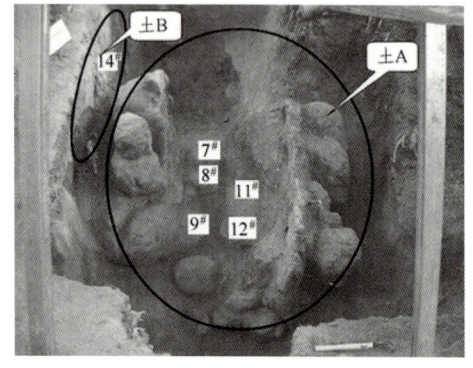

图 5　现场模拟土窑化学固化土体

Fig.5　Chemically grouted body in field model kiln

3.2　微型土钉锚固试验

在现场模拟窑迹化学加固处理部位(土 A)与未处理部位(土 B)均进行微型土钉锚固试验。钻孔直径为 ϕ 7 mm，长 300 mm。水泥浆水灰比确定为 0.45。土 A 和 B 中进行的对比试验点位置如图 5 所示。土 A 和 B 中抗拔试验张拉力－位移关系分别如图 6 和 7 所示。试验结果表明，化学加固剂固化土中(土 A)

表1 模拟分析中所用土质参数
Table 1 Soil parameters used in simulation analysis

密度ρ_d /(g·cm^{-3})	剪切模量 G/MPa	体积模量 K/MPa	黏聚力 c/kPa	内摩擦角 φ/(°)	张拉强度 /kPa
1.30	3.0	5.0	40	20	4,20

注：4 和 20 kPa 分别代表土体在选定标准状态(w = 33%)时可能的较低和较高表观张拉强度。

经现场调查，堂加 2$^{\#}$窑横断面形状可简化为 3 种平面：U 型、凹型和凸型，如图 8 所示。稳定分析中采用强度折减法确定安全系数(FS)。3 种类型断面对应于低、高张拉强度时的破坏模式和安全系数(FS)列于表 2。

表2 3 种类型断面安全系数(FS)和破坏模式汇总
Table 2 Summary of factor of safety(FS) and failure modes for three idealized geometries

张拉强度/kPa	U 型	凹型	凸型
4	3.45(T + S)	1.45(T)	3.85(T + S)
20	4.15(S)	3.65(S)	4.45(S)

注：T 和 S 分别表示模拟结果中单元的主要破坏模式为张拉破坏和剪切破坏模式。

模拟分析结果表明 U 型和凸型断面在低、高张拉强度时均具有相对较高的安全系数，均可保持自身稳定。凹型断面是最薄弱断面，在低、高张拉强度时具有明显不同的破坏模式。最不利的情况是在低张拉强度时的张拉破坏，如图 9 所示，相应的安全系数为 1.45，明显低于其他几种情形。

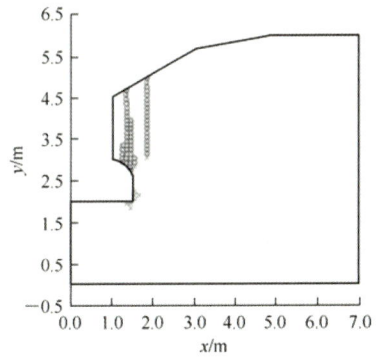

图 9 凹型断面张拉破坏模式(4 kPa，FS = 1.45)
Fig.9 Tensile failure mode of cave-shaped geometry (4 kPa，FS = 1.45)

4.2 补强方案设计

根据稳定分析结果，加固补强方案因断面形状而异。对于 U 型和凸型断面，在低、高张拉强度时均具有相对较好的稳定性，其补强方案设计为：化学加固剂固化表面 0.3～0.5 m 厚的表层土，根据构造需要，仅设置少量微型土钉(间距 S = 1.0 m，长度 L = 0.5～1.0 m)，分别如图 10，11 所示。

对于凹型断面，因其相对为最薄弱部位，必须设置微型土钉锚固以提高抗张拉破坏的稳定性。本研究采用有限差分程序 FLAC 中的桩单元模型模拟土钉，可考虑轴力、剪力和弯矩的作用。主要评价要素有：(1) 土钉长度(L)；(2) 土钉设置角度(α)；(3) 土钉设置间隔(S)。根据模拟分析结果，对凹型断面部位，加固方案设计为：面层 0.3～0.5 m 厚度采用化学注浆固化处理。建议土钉设置为：$L \geq$ 2.0 m，$\alpha \leq 30°$，$S \leq 0.3$ m，如图 12 所示，则相应

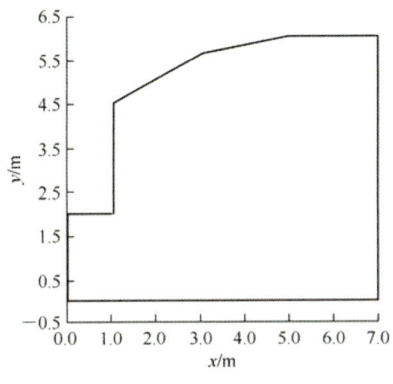

(a) U 型

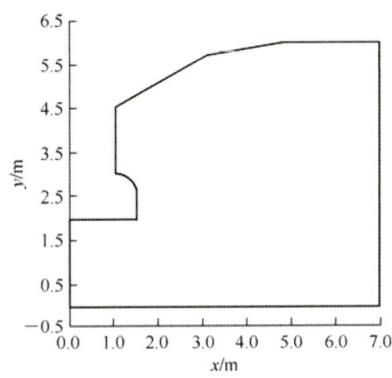

(b) 凹型

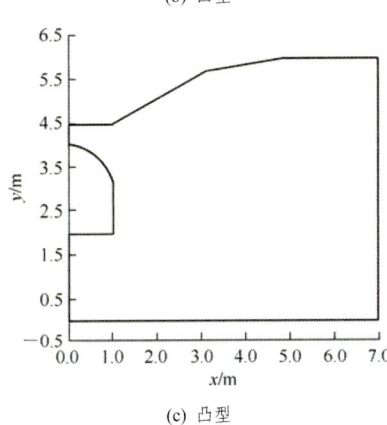

(c) 凸型

图 8 堂加窑迹群 2$^{\#}$古窑断面三种类型
Fig.8 Three idealized geometries of the Dougaeri historical earthen kiln No.2

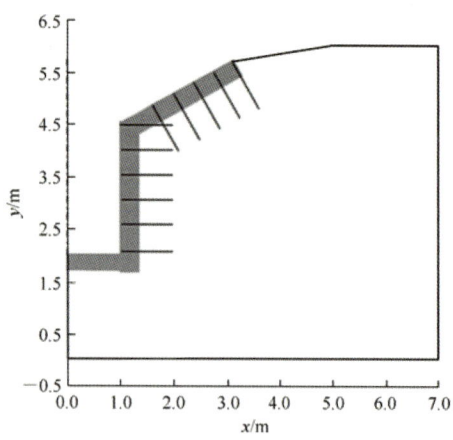

图 10 U 型断面补强设计示意图

Fig.10 Schematic diagram of reinforcement design for U-shaped geometry

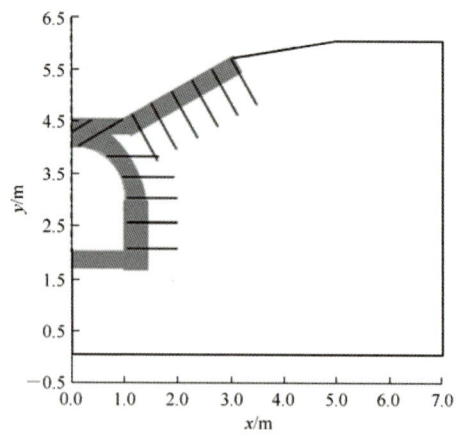

图 11 凸型断面补强设计示意图

Fig.11 Schematic diagram of reinforcement design for arch-shaped geometry

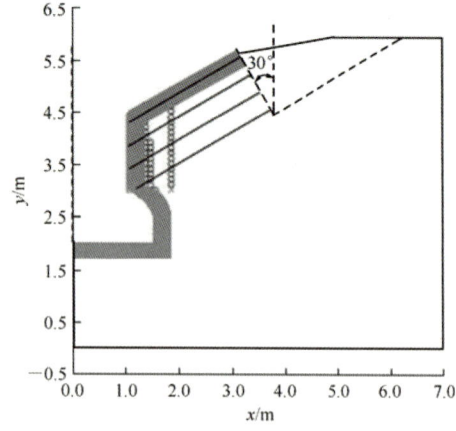

图 12 凹型断面补强设计示意图($FS = 2.55$)

Fig.12 Schematic diagram of reinforcement design for cave-shaped geometry($FS = 2.55$)

的安全系数提高为 2.55。该加固补强方案已成功地应用于 2$^{\#}$窑的加固保存,效果良好。

5 结 论

本研究以船迫窑迹群 2$^{\#}$土质古窑址的加固保存为目标,并开发出了微型加固补强技术,其结论归纳如下:

(1) 微型加固技术包括微型土钉技术和微型化学注浆技术,根据具体工况特征和要求,二者可以结合应用,也可以单独使用。

(2) 微型加固技术中开发使用的微型仪器设备有:便携式变速电动钻、水泥浆注入器、便携式拉拔设备和气球注入前端等,其显著特征为适合于微型钻孔(直径ϕ 5~7 mm)中进行注浆锚固和化学加固,具有一定的创性新。

(3) 微型化学注浆技术中气球密封注入前端构造可实现定点位化学加固剂的定量控制,有效地解决了注入过程药液外溢的问题,也克服了喷淋方法注入效果难以评价的缺点,具有独创性。

(4) 微型加固技术的有效性和可行性在现场模拟土窑的加固试验中得到了检验。现场抗拔试验表明,化学加固土体中微型土钉的抗拔力明显高于未处理土中的抗拔力。

(5) 数值模拟分析表明,2$^{\#}$窑不同断面部位具有不同的稳定性和最不利的破坏模式。在稳定分析的基础上提出了合理的微型加固补强设计方案,并已成功地应用于 2$^{\#}$窑的加固保存。

(6) 可以预期该微型加固技术以便携、微扰动、现场适应性强等优点,在土遗址的加固保存中具有良好的应用前景。

参考文献(References):

[1] 张旭辉. 土钉与锚杆的关系探讨[J]. 岩石力学与工程学报, 2002, 21(11): 1 744 - 1 746.(ZHANG Xuhui. Relations between soil nailing and bolting[J]. Chinese Journal of Rock Mechanics and Engineering, 2002, 21(11): 1 744 - 1 746.(in Chinese))

[2] 朱浮声, 郑雨天. 全长黏结式锚杆的加固作用分析[J]. 岩石力学与工程学报, 1996, 15(4): 333 - 337.(ZHU Fusheng, ZHENG Yutian. Analysis of reinforcement of full-length grouted bolts[J]. Chinese Journal of Rock Mechanics and Engineering, 1996, 15(4): 333 - 337. (in Chinese))

[3] 张乐文, 汪 稔. 岩土锚固理论研究之现状[J]. 岩土力学, 2002, 23(5): 627 - 631.(ZHANG Lewen, WANG Ren. Research on status of anchorage theory of rock and soil[J]. Rock and Soil Mechanics,

2002，23(5)：627 – 631.(in Chinese))

[4] 程良奎. 岩土锚固的现状与发展[J]. 土木工程学报，2001，34(3)：7 – 12.(CHENG Liangkui. Present status and development of ground anchorages[J]. China Civil Engineering Journal，2001，34(3)：7 – 12.(in Chinese))

[5] 黄克忠. 任重而道远的莫高窟文化遗产保护[J]. 敦煌研究，2006，(6)：200 – 202.(HUANG Kezhong. Protection of Mogaoku ashes with shouldering heavy responsibilities[J]. Duhuang Research，2006，(6)：200 – 202.(in Chinese))

[6] 黄克忠. 文物建筑材质的研究与保存[J]. 东南文化，2003，(9)：93 – 96.(HUANG Kezhong. Research and protection on materials for cultural relics and architecture[J]. Southeast Culture，2003，(9)：93 – 96.(in Chinese))

[7] 黄克忠. 走向21世纪的中国文物科技保护[J]. 敦煌研究，2000，(1)：5 – 9.(HUANG Kezhong. Technological protection of Chinese cultural relics tending towards 21st century[J]. Dunhuang Research，2000，(1)：5 – 9.(in Chinese))

[8] 何 燕，李智毅. 关于河南灵泉寺石窟地质病害及整治方法的研究[J]. 岩土力学，2000，21(1)：56 – 59.(HE Yan，LI Zhiyi. A study on geological defect analysis and renovation methods in Lingquan Temple Grotto of Henan[J]. Rock and Soil Mechanics，2000，21(1)：56 – 59.(in Chinese))

[9] 李丽慧，杨志法，岳中琦，等. 龙游大型古洞室群变形破坏方式及加固方法研究[J]. 岩石力学与工程学报，2005，24(12)：2 018 – 2 028.(LI Lihui，YANG Zhifa，YUE Z Q，et al. Deformation and failure modes and reinforcement methods of ancient cavern group in Longyou County[J]. Chinese Journal of Rock Mechanics and Engineering，2005，24(12)：2 018 – 2 028.(in Chinese))

[10] 孙满丽，王旭东，李最雄，等. 木质锚杆加固生土遗址研究[J]. 岩土工程学报，2006，28(12)：2 156 – 2 159.(SUN Manli，WANG Xudong，LI Zuixiong，et al. Study on immature earthen sites reinforced with wood anchor[J]. Chinese Journal of Geotechnical Engineering，2006，28(12)：2 156 – 2 159.(in Chinese))

[11] 周双林，王雪莹，胡 原，等. 辽宁牛河梁红山文化遗址土体加固保护材料的筛选[J]. 岩土工程学报，2005，27(5)：567 – 570.(ZHOU Shuanglin，WANG Xueying，HU Yuan，et al. Selection of strengthening medium for archaeological site of Hongshan Culture at Niuheliang，Liaoning Province[J]. Chinese Journal of Geotechnical Engineering，2005，27(5)：567 – 570.(in Chinese))

[12] CHAI X J. Development of earth sewing technique and its application to reinforcing Funasako historical kiln site[Ph. D. Thesis][D]. Saga，Japan：Saga University，2005.

[13] MATSUNAGA K. Fundamental study on the techniques of soil remains by chemical grouting and evaluation method[M. S. Thesis][D]. Saga，Japan：Saga University，2002.

[14] HAYASHI S，CHAI X J，MATSUNAGA K，et al. Drip injection of chemical grouts：a new apparatus[J]. Geotechnical Testing Journal，2006，29(2)：108 – 116.

[15] 福冈县筑城町教育委员会：船迫窑迹群调查报告[R]. 福冈：福冈县筑城町教育委员会，1998.(Education Committee of Tsuikimachi，Fukuokaken，Japan. Investigation report of Funasako historical kiln sites[R]. Tsuikimachi：Education Committee of Tsuikimachi，Fukuokakcn，Japan，1998.(in Japanese))

[16] CHAI X J，HAYASHI S. Effect of constrained dilatancy on pull-out resistance of nails in sandy clay[J]. Ground Improvement，2005，9(3)：127 – 135.

我国高放废物地质处置战略研究

潘自强　钱七虎

中国工程院

随着我国核能事业的飞速发展，高放废物的处理和处置，已成为公众关心的一个重大安全和环保问题。这体现在如何最终安全处置核电站乏燃料后处理产生的高放废物、核武器研制和生产过程中业已产生的高放废物、以及我国存在的某些可能不宜后处理的乏燃料。

对高放废物的安全处置，是一个与核安全同等重要的问题，是落实科学发展观、建设和谐社会、确保我国核能工业可持续发展和环境保护的重大问题。在研究和开发方面，高放废物安全处置还存在一系列科学技术难题，需要几十年坚持不懈的努力加以解决。在公众接受方面，则存在一些需要认真解决的重大社会学难题。西方国家的核能开发情况表明，安全处置核废物，尤其是高放废物，已成为制约核能工业可持续发展的最关键因素之一。为推进我国高放废物地质处置工作，中国工程院开展了"高放废物地质处置战略研究"咨询项目。本文反映了该项目的最终成果。

1 高放废物安全处置是核能可持续发展的重要保障

高放废物是一种放射性强、毒性大、半衰期长并且发热的特殊废物，对其进行安全处置难度极大，面临一系列科学、技术、工程、人文和社会学的挑战，其最大难点在于使高放废物与生物圈进行充分、可靠的永久隔离，且隔离时间超过一万年。目前公认的安全可靠，且技术上可行的方法为地质处置方法，即在地表以下500～1000米深建造"矿山式"处置库，通过工程屏障和天然屏障永久隔离高放废物。

高放废物的安全处置受到国际组织和世界各国的高度关注。国际原子能机构成员国大会于1997年通过了"国际乏燃料安全与放射性废物安全公约"，明确条约签字国安全处理处置乏燃料和放射性废物的责任。各有核国家也均在国家层面上高度重视高放废物安全处置的工作。他们大部分通过制定国家政策、颁布法律法规、成立专门机构、筹措专门经费、建立专门的地下研究设施和开展长期研究等方式，从政策、法规、机构、经费、

设施和科研等方面确保高放废物的安全处置。例如，美国1982年颁布"核废物法案"，并在能源部成立"民用放射性废物管理局"，专门负责安全处置美国的7,000吨军工高放废物和63,000吨民用核电站乏燃料。美国还制定了长期的研究开发计划，并在内华达州尤卡山建造地下研究设施，开展现场工程研究。经过45年多的基础研究和场址评价工作，尤卡山场址于2002年获美国总统布什批准，预计将于2018年建成高放废物处置库。

高放废物安全处置的研究开发具有长期性的特点。需要进行长期的基础研究、技术开发和工程研究，方可实现安全处置的目标。美国于1957年提出高放废物地质处置的设想并开始研究和技术开发，到2018年才能建成处置库，历经62年。芬兰于1976年开始研究，到2020年建成处置库，将历经45年，足见其工作的长期性。

高放废物地质处置还具有成本高、投资大、投资周期长的特点。国际上每吨乏燃料处置的平均成本为66.3万美元。例如，美国整个处置计划耗资575亿美元，日本的整个处置计划耗资3万亿日元。

高放废物处置经费一般来自政府投资和核电站中一定比例的经费（一般按核电站电费收入的1%左右收取高放废物地质处置基金，美国每年约能收取6亿美元）。前者用于处置军工设施的高放废物，后

本文原载于《第二届废物地下处置学术研讨会论文集》（2008年）。

者用于处置民用核电站的高放废物。研发资金一般占总投资的 10~15%，如美国的总研发经费为 65 亿美元，前期的年度研发经费达 1.5 亿~2 亿美元。

西方国家多年以来的经验表明，安全处置高放废物是核能可持续发展的重要保障。高放废物是核能工业的必然产物，对其安全处置是核能工业界义不容辞的任务。对于此项工作，社会各界均广泛关注。关注之深，在某种程度上足以影响政府对核能发展的政策。芬兰是一个成功的实例，由于其乏燃料安全处置扎实推进，卓有成效，民众支持核电建设，欧洲的第一台 EPR 机组已在芬兰开始建造。

2 我国高放废物地质处置进展和存在问题

据估计，我国的核军工设施已暂存了一定量的高放废液，急需进行玻璃固化和最终地质处置。我国目前运行的 11 个核电机组每年约产生 470 吨乏燃料。

根据 2007 年 10 月国务院批准的《国家核电发展专题规划(2005—2020 年)》中的核电规模，我国到 2020 年投入运行的核电装机容量将达到 4000 万千瓦，在建装机容量 1800 万千瓦。以此为基础计算，到 2020 年我国将累积有约 10,300 tHM 乏燃料（其中压水堆乏燃料约 7,000 tHM 和重水堆乏燃料约 3,300 tHM）。《国家核电发展专题规划(2005-2020 年)》中于 2020 年建成的反应堆，加上届时在建的 18 个反应堆，最终共将产生 82,630 吨乏燃料。关于 2020 年以后的乏燃料数量，每增加一座百万级千瓦的核电站，每年将多产生 22 tHM 乏燃料，每个堆全寿期共产生约 1320 吨乏燃料。对这些军工高放废物和核电站产生的高放废物进行最终安全处置，是确保我国的环境安全和核工业可持续发展的必然要求。

我国高放废物地质处置研究工作于上世纪八十年代中起步，20 多年来，在选址和场址评价、核素迁移、处置工程和安全评价等方面均取得了不同程度的进展。核工业北京地质研究院等单位开展了高放废物处置库场址预选研究，在对华东、华南、西南、内蒙和西北等 5 个预选区进行初步比较的基础上，重点研究了西北甘肃北山地区，在地质调查和水文及工程地质条件、地震地质特征等研究基础上，施工了四口深钻孔，获得了深部岩样、水样和相关资料，初步掌握了场址特性评价方法。在工程方面，研究了内蒙古高庙子膨润土作为缓冲/回填材料的性能，以及低碳钢、钛及钛钼合金等材料在模拟条件下的腐蚀行为。在核素迁移方面，建立了模拟研究试验装置及分析方法；研究了镎、钚、锝在特定条件下的某些行为。在安全评价方面，初步进行了一些调研。总的说来，我国高放废物地质处置研究工作，在经费十分有限，条件很困难的情况下，做了不少工作，特别是在选址和场址特性评价方面取得了一定进展，但从总体上说还处于研究工作的前期阶段，距完成地质处置任务的阶段目标任务还相差甚远。

2003 年我国发布《中华人民共和国放射性污染防治法》，其第四十三条中明确规定"高水平放射性固体废物实行集中的深地质处置"，这从国家层次明确了深地质处置的地位。2006 年原国防科工委、科技部和国家环保总局联合发布《高放废物地质处置研究开发规划指南》，明确了深地质处置开发的主要技术路线和开发的总体设想。2007 年，国务院批准《国家核电发展专题规划(2005-2020 年)》，明确提出 2020 年建成我国高放废物地质处置地下试验室的目标，从而使高放废物地质处置进入了新的阶段。

但是，我国目前的高放废物地质处置研究也面临一些问题：

没有国家级高放废物地质处置专项规划。目前有关高放废物地质处置的政府行为只停留在部委一级层面上，没有达到国家级层面（如全国人大和国务院），因此一些必须在国家级层面决策的事项（如政策和技术路线、政府部门分工、决策机制等）难以进行；对高放废物地质处置如此重大的高难项目，目前还没有国家级高放废物地质处置专项规划，也没有列入国家重大科技工程、973 计划和国家自然科学基金重大研究计划等。目前我国的高放废物地质处置项目仅在原国防科工委"核设施退役和放射性废物处理处置"专项中予以支持，其力度小，远远不能满足需求。

政府法规和标准基本上还是空白。《中华人民共和国放射性污染防治法》明确了高放废物实行集中的深地质处置的原则，但至今尚没有制定相关的法规和技术标准，如《高放废物地质处置规定》、《高放废物处置库选址标准》、《高放废物处置安全评价标准》等，这就严重影响了选址、场址评价、安全评价和工程设计等工作的推进。

尚没有明确实施高放废物地质处置工程的责任主体。高放废物地质处置是国家行为，应由政府总体负责。但具体实施，还需要由政府的专门部门或政府授权的独立的机构负责。目前，我国尚未有这样的实施主体单位，已对当前工作的推进产生严重影响。

决策机制不健全。高放废物地质处置时间跨度长、技术难度大、影响面广，是关系到子孙后代的万年大计，必须科学决策、民主决策，让公众和利益相关者广泛参与决策。这就需要设计一个好的决策机制。但是，我国目前决策机制和决策程序不明确，尤其是没有明确国家层次的决策机制。

经费投入极少。目前，高放废物地质处置工程科研仅有国防科工委"军工核设施退役和三废治理专项"这一个经费渠道，但它要解决的既是军工高放废物问题，又是民用核电站的高放废物问题，而后者的废物量却占绝大部分。我国十五高放废物地质处置的平均年度经费仅为400万元左右。十一五虽有增加，预期可达到年均1000万元的强度，但仍然很低，远不能满足高放废物地质处置的各项需求。更为重要的是，还没有建立从核电电费中收取高放废物地质处置所需资金的筹资机制，相关基础研究也未列入国家计划。

研究开发力量薄弱，缺乏研究平台。目前我国从事高放废物地质处置的专职科研人员约为40人左右，涵盖地质、工程、化学和安全评价等领域，但与高放废物地质处置的艰巨任务相比，这支队伍力量极为薄弱。另外，还严重缺乏高放废物地质处置的研究平台，一些重大科学问题还没有解决、大量工程尺度课题根本就无法开展，距完成地质处置任务的阶段目标还相差甚远。

3 我国高放废物地质处置规划设想及经费预测

3.1 规划设想

我国高放废物地质处置的总目标是：在我国领土内选择地质稳定、社会经济环境适宜的场址，在本世纪中叶建成高放废物地质处置库，通过工程屏障和地质屏障的包容、阻滞，保障国土环境和公众健康不会受到高放废物的不可接受的危害。

研究开发和处置库工程建设可分为三个阶段：

(1) 实验室研究开发和处置库选址阶段（2006—2020）。其目标是，完成各学科领域实验室研究开发任务，初步选出处置库场址并完成初步场址评价。确定地下实验室场址，完成地下实验室的可行性研究，并建成地下实验室。

(2) 地下现场试验阶段（2021—2040）。其目标是，完成地下实验室现场试验，完成场址详细评价，并最终确认处置库场址。掌握处置库建造技术，完成处置库设计和可行性研究，并开始建造处置库。

(3) 处置库建设阶段（2041—本世纪中叶）。其目标是，2050年前后，建成处置库，开展示范处置，并开始接受高放废物。

3.2 研究内容(2006～2020)

（1）战略、规划、法规、标准研究：开展战略、策略和规划方案研究，提出我国高放废物地质处置研究和开发的中长期规划。开展管理体系、法规和标准研究。制定高放废物地质处置的安全目标、环保要求、审批程序和责任制度；建立高放废物地质处置技术标准，规定高放废物地质处置的选址、设计、建造、运行、关闭和监护的工程技术与安全防护技术要求。制定玻璃固化体性能标准、高放废物处置库选址标准和安全评价标准等的制/修订。

（2）选址和场址评价研究。开展处置库场址和地下实验室场址的选址和场址评价。确定地下实验室场址。开展甘肃北山地区预选地段评价和对比研究，推荐出 1～2 处处置库预选地段，提交初步评价报告；对甘肃北山的场址评价研究成果开展回顾性审评；在十一五之后，确定处置库预选区，初步选出处置库场址。主攻甘肃北山花岗岩场址，兼顾其他地区场址和其他围岩类型。对通过国家主管和审管机构审批的预选区和预选地段开展综合研究。开展区域构造研究、地震安全性评价、未来气候和地质变化趋势研究、第四纪地质特征和环境演化研究。通过地面调查和钻孔施工等手段，开展地质研究、水文地质研究、工程地质研究、地球物理测量、综合场址评价方法研究、岩体质量评价技术研究。开展场址建模技术研究、建立处置库预选场址地学信息库。

（3）处置工程研究。开展地下实验室工程设计，建成地下实验室。开展处置库概念设计。根据目前我国高放废物的现状和今后我国核能发展规划，预测拟处置废物的来源、类型、数量、总活度、核素组成和其它物化特性等。开展工程屏障系统研究，开展包装容器、缓冲材料等工程屏障系统的材料筛选、结构及性能验证；开展地下硐室稳定性研究和水-热-力等多场耦合条件下工程屏障特性研究。建立处置工程信息库。

（4）安全评价研究。开展废物源项调查和源项预测。开展安全评价方法学研究，构建技术体系框架，建立安全和环境评价信息系统；开展高放废物地质处置系统的总体安全目标和辅助安全指标，开展情景分析和后果分析方法、模式和参数体系、灵敏度分析和不确定性分析方法研究。以预选出的场址和处置工程概念设计为基础，开展安全评价研究，完成本阶段高放废物处置库安全评价报告。反馈处置系统安全评价结果。

（5）核素迁移行为研究。研究高放废物、乏燃料及 α 废物在处置条件下的性能，开展核素迁移研究，建立核素迁移模型、数据库和高放废物玻璃固化体性能标准；开展关键核素在地下水中的化学反应行为实验室研究，掌握相应的测试技术和方法；完成关键核素在近场处置条件下的化学形态及胶体行为研究；掌握关键核素在近场屏障体系的化学反应及迁移机理，完成关键核素在近场围岩及混合回填材料中的吸附、扩散等迁移参数测定；系统获取安全评价所需的核素迁移数据。初步掌握现场核素迁移试验技术和方法。研究乏燃料及高放废物的内、外包装材料及其处置条件下的腐蚀行为等长期稳定性。

3.3 经费预测

经费预测。全部处置我国《国家核电发展专题规划(2005-2020 年)》中所有核电站（58 个）全寿期产生的 82，630 吨乏燃料所需的成本约为 1343 亿人民币（不含后处理成本），这一数值仅占所有核电站总电费收入的 1.25%。在 2020 年规划之后建成的每一个百万千瓦级核电站，按运行 60 年，共产生 1320 吨乏燃料来计算，所需的处置费用为 21.4 亿元人民币。研发经费是高放废物处置中的重要组成部分，一般占处置库总经费的 10～15%，因此，我国高放废物处置的研发经费至少应为 13.4 亿人民币。关于 2020 年之前的经费测算，选址和场址评价费用约为 5 亿人民币，研究开发的费用约为 3—5 亿人民币，建造地下实验室的费用约为 4—5 亿人民币。若以建成地下实验室为工程目标，从现在起至 2020 年，我国每年应当保证 1 亿～1.5 亿人民币的研究开发资金投入，以解决建造地下实验室之前的各项工作需要。

经费筹措渠道。我国高放废物地质处置的资金筹措应当采取 2 种渠道。一是从核电的电费中提取，这部分费用主要是用于核电站产生的高放废物的最终处置；二是由政府财政支出，其主要用途是处置核军工产生的高放废物以及公益性单位产生的高放固体废物。国外从核电电费中收取的高放废物处置基金的筹资费率一般核电电价的 1%，如美国为 0.001 美元/千瓦时，日本为 0.13 日元/千瓦时。按真实比价理论和实际汇率的平均值估算（即 1 美元=4.9 元人民币），我国的地质处置筹措费率可初步定为 0.005 元/千瓦时，仅相当于电价的 1.25%。

经费管理方式。国家应当建立乏燃料和高放废物安全处置基金，并交由国家授权的单位管理，以确保执行单位的运行、研究开发、地下实验室和处置库建造、运行和关闭等的资金需求。

4 推进我国高放废物地质处置工作的几点建议

建立高放废物地质处置法规和标准体系。在正在制定的《原子能法》中应明确高放废物地质处置的原则要求和经费来源。建议国务院制定"高放废物地质处置条例",条例应当包括地质处置的安全要求、技术路线、工作进度、审管和主管单位的职责、经费来源和实施主体等。加速《核电站后端基金管理条例》的编制,综合解决乏燃料后处理、高放废物处置和核电站退役所需基金的筹措和监督管理问题。为了满足当前工作的需要,在正在制定的《放射性废物安全监督管理条例》中,明确规定选址和处置工程的要求和审批程序,并尽快制定"放射性废物地质处置"、"放射性废物处置设施安全评价"和"地质处置设施选址"等标准。

尽快明确实施高放废物地质处置的责任主体。高放废物安全处置是以建成地质处置库为明确工程目标的,完成这一长期任务需要有组织管理和实施机构的保证。鉴于中国核工业集团公司多年以来已有一支从事高放废物地质处置的专业队伍和设计院所、较早地开展了研发工作,积累了较丰富的经验,因此,为了保持工作的延续性,并充分发挥中核集团所属各科研设计院所的技术优势,建议当前可将实施高放废物地质处置的责任主体置于中核集团公司之下。

开展顶层设计。为了更有效地组织当前高放废物地质处置科研工作,建议尽快制定"高放废物处置科研项目指南"。与此同时,全面开展顶层设计,包括法规体系、管理模式、技术路线、规划目标和筹资机制等。

尽快开展对现已进行的选址工作进行回顾性安全审评。建议中国核工业集团公司组织核工业北京地质研究院等单位对已有选址工作进行总结并向主管部门和审管部门提出报告,审管和主管部门对报告进行回顾性审评,以鉴明现有工作的成果,明确下一步工作的方向。在进一步全面深入开展预选区工作之前,应该完成回顾性安全审评。

建立国家高放废物地质处置研究平台。在2020年前建立以地下实验室为核心的国家高放废物地质处置研究平台。研究平台包括地下实验室以及相关的,包括处置工程、场址评价、核素迁移和安全评价等在内的实验研究平台。

增加研究费用强度和渠道。高放废物地质处置研究开发和工程建造需要较多经费(2020年之前选址和场址评价费用约为5亿人民币,研究开发的费用约为3—5亿人民币,建造地下实验室的费用约为4—5亿人民币)。但当前经费与需求很不适应。建议:(1) 政府增加投入,包括增大军工三废专项对高放废物地质处置的投资,以及在核能科研专项中明确列入高放废物地质处置项目等;(2) 在正在制定的"从核能发电电费中提取核能后端费用"规定中明确地质处置费用比例;(3) 建议在国家发改委设立"高放废物地质处置"国家重大科技开发专项;(4) 在国家自然科学基金和科技部"国家科技支撑计划"和"973计划"中设立高放废物地质处置专项。

加强国际合作。高放废物处置研究工作在国际上是透明、公开的。我国高放废物处置研究工作一直得到了国际原子能机构的支持,取得了良好的效果。今后,在继续争取国际原子能机构支持的同时,有必要开拓或加强双边合作。

深埋巷道分区破裂化机制

周小平[1],钱七虎[2]

(1. 重庆大学 土木工程学院,重庆 400045;2. 解放军理工大学,江苏 南京 210007)

摘要: 深部巷道外部受到远场原岩应力的作用,而内壁受到一个随时间变化的内压作用,开挖过程是动力问题,其运动方程可以用位移势函数来表达。通过对运动方程进行 Laplace 变换,进而求得其通解。根据弹性力学知识和边界条件得到巷道围岩由于开挖扰动和原岩应力作用引起的弹性应力场和位移场。当该弹性应力场满足破裂条件时,岩体发生破裂,位移不连续,形成破裂区。结合断裂力学知识,确定破裂区岩体的残余强度和产生破裂区的时间,进而确定破裂区和非破裂区的宽度和数量。数值分析结果表明,巷道分区破裂化的产生跟开挖速度与岩石强度有关。该研究可为深部岩体的开挖和支护设计提供初步的理论基础。

关键词: 岩石力学;深部岩体;深埋巷道;分区破裂化;动力问题

中图分类号: TD 32   **文献标识码:** A   **文章编号:** 1000 – 6915(2007)05 – 0877 – 09

ZONAL FRACTURING MECHANISM IN DEEP TUNNEL

ZHOU Xiaoping[1], QIAN Qihu[2]

(1. *College of Civil Engineering,Chongqing University,Chongqing* 400045,*China*;
2. *PLA University of Science and Technology,Nanjing,Jiangsu* 210007,*China*)

Abstract: The mechanical behaviors of rock mass in deep tunnel are different from those in shallow tunnel. The surrounding rock in shallow tunnel is classified into loose zone, plastic zone and elastic zone;while the surrounding rock in deep tunnel is classified into fractured zone and non-fractured zone, which occur alternatively. The mechanism of the alternative occurrence of fractured zone and non-fractured zone in deep tunnel is studied. It is assumed that the outer boundary of the deep tunnel is subjected to an in-situ far-field stress field, whose inner surface is subjected to an internal pressure which changes during tunnel excavation. As a result, the tunnel excavation process is related to dynamic problem. The motion equation expressed by displacement potential function is established, which determines the release of pre-existing stress upon excavation of the opening. The general solution of motion equations is obtained by using Laplace transform. Based on the elastic theory and boundary condition, the near-field stress redistribution and displacement field around circular opening induced by excavation are determined. If the elastic stress fields satisfy the failure condition of rock mass, failure of rock mass occurs. It results in the discontinuous displacement, and the fractured zone is formed, and the deformation localization occurs. On the basis of the deformation localization theory, the support reaction acted on the elastic zone is defined. Then, the fracture mechanics theory is applied to analyzing residual strength of the rock mass and the time of onset of fractured zone. Width and quantity of fractured zone and non-fractured zone are given. It is found from numerical results that the width and quantity of fractured zone and non-fractured zone depend on the strength of rock mass and the velocity of tunnel excavation. The quantity of the fractured zone increases with increase of tunnel excavation velocity. The width of the fractured zone decreases with increase of tunnel

---

收稿日期:2007-01-12;修回日期:2007-01-26。
基金项目:国家自然科学基金重大项目(50490275)。
本文原载于《岩石力学与工程学报》(2007 年第 26 卷第 5 期)。

excavation velocity. The quantity and width of the fractured zone increase with decrease of the rock mass strength.

Key words：rock mechanics；deep rock masses；deep tunnel；zonal fracturing；dynamic problem

1 引 言

深部岩体的力学特性与浅部岩体的有很大的不同。浅部巷道围岩状态通常可以分为松动区、塑性区和弹性区，其本构关系可以采用弹塑性力学理论进行推导求解。而深部巷道围岩产生膨胀带和压缩带，或称为破裂区和未破裂区交替出现的情形，这一现象被称为区域破裂现象[1~11]。钱七虎[1~3]研究了深部岩体工程响应的特征并且界定了"深部"的范围。M. V. Kurlenya 等[4~11]在深部开采现场发现了分区破裂化现象，并且进一步通过实验验证了分区破裂化现象的存在，但是没有进行理论研究。V. N. Nikolaevshij 等[12~17]给定了深部岩体的主要特点是地应力高，并且界定了"深部"的范围，但是其研究主要是定性描述，定量研究并不多见。

深埋巷道开挖产生应力重分布，当次生应力场满足岩体破坏条件时，应力释放，深部岩体产生第一次破裂区。对于浅部岩体由于地应力水平低，在应力释放后不可能再产生第二次破裂区。对于深部岩体，由于其主要特点是地应力高[12~17]，因此，应力释放后产生的第一次破裂区的外边界相当于新的开挖边界，这样应力再一次重分布，并且当重分布应力场满足岩体破坏条件时，应力再一次释放，产生第二次破裂区。依次类推，直到应力释放后不能再产生破裂区为止。破裂区是原生共线裂纹在次生应力场作用下贯通、汇合后的结果，而且破裂区的位移不再连续，因而不能继续采用弹塑性力学知识进行求解。为了确定破裂区岩体的残余强度和破裂区形成的时间，将岩体视为存在许多原生裂纹的复合体，并将岩石母体视为均匀介质，采用断裂力学知识确定破裂区岩体在原生共线裂纹贯通、汇合后的残余强度和时间。

研究深部岩体的区域破裂化现象有助于了解深部巷道围岩的破坏机制，同时可为深部岩体的支护设计提供理论基础。因此，研究深部岩体的区域破裂现象具有重要的理论和实际意义。现阶段对深部岩体的分区破裂化现象的研究主要是从实验角度进行定性描述，而对分区破裂化机制的研究较少，为此本文从理论上研究深部巷道围岩分区破裂化现象的机制。

2 开挖过程中，由 $p(t)$ 引起的次生应力场和弹性位移场

如图 1 所示，在深部岩体中开挖一个半径 $r = a$ 的圆形巷道，则在初始地应力场中，由巷道开挖卸荷引起围岩应力场的重分布，并产生相应的位移场。如果该扰动后的二次应力场为弹性的，则可分解为 2 个部分来求解：

(1) 开挖过程中，由 $p(t)$ 引起的应力场和位移场的求解，其中 $p(t)$ 在 $t = 0$ 时刻等于地应力，开挖完成时刻 $t = t_0$ 时 $p(t) = 0$。

(2) 开挖过程中，由原岩应力引起的应力场和位移场的求解。

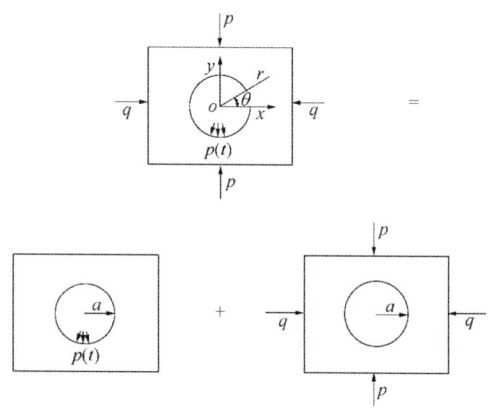

图 1 圆形巷道开挖过程的力学模型分解为 2 个部分

Fig.1 Two parts from mechanical model of circular tunnel during excavation process

上述 2 个部分之和为总的弹性区的次生应力场和位移场。在开挖前，可认为与初始原岩应力场相应的是零位移场，所以所求的第二个位移场应减去在开挖初始时刻由原岩应力引起的初始位移值。

2.1 开挖过程中，由 $p(t)$ 引起的次生应力场

如图 1 所示，非零位移量有 $u_r = u(r, t)$，同时有 $\dfrac{\partial u_r}{\partial \theta} = 0$，从而位移场可简化为

$$u = \frac{\partial \phi}{\partial r} \tag{1}$$

式中：ϕ 为标量势函数。

用位移势表达的运动方程可简化为

$$\frac{\partial^2 \phi}{\partial r^2}+\frac{1}{r}\frac{\partial \phi}{\partial r}=\frac{1}{c_d^2}\frac{\partial^2 \phi}{\partial t^2} \qquad (2)$$

式中：$c_d^2=\dfrac{\lambda+2\mu}{\rho}$，$\lambda$，$\mu$ 为 Lame 常数。于是问题可归结为

$$\frac{\partial^2 \phi}{\partial r^2}+\frac{1}{r}\frac{\partial \phi}{\partial r}=\frac{\ddot{\phi}}{c_d^2} \quad (r>a,\ t>0) \qquad (3)$$

$$\phi(r,\ 0)=\dot{\phi}(r,\ 0)=0 \quad (r\geqslant a) \qquad (4)$$

$$\sigma_r(a,\ t)=p(t) \qquad (5)$$

$$\lim_{r\to\infty}\phi(r,\ t)=0 \quad (t>0) \qquad (6)$$

对式(3)取关于 t 的 Laplace 变换，可得

$$\frac{d^2 \bar{\phi}(r,\ s)}{dr^2}+\frac{1}{r}\frac{d\bar{\phi}}{dr}=k_d^2 \bar{\phi} \qquad (7)$$

式中：s 为 Laplace 变换参数，$k_d=s/c_d$，$\bar{\phi}$ 为 ϕ 的 Laplace 变换。

式(7)的通解为

$$\bar{\phi}(r,\ s)=A(s)I_0(k_d r)+B(s)k_0(k_d r) \qquad (8)$$

式中：$I_0(k_d r)$，$k_0(k_d r)$ 分别为零阶第一类和第二类 Bessel 函数，且有

$$\left. \begin{array}{l} I_0(k_d r) \sim (2\pi k_d r)^{-\frac{1}{2}}\exp(k_d r) \\ k_0(k_d r) \sim \left(\dfrac{\pi}{2k_d r}\right)^{\frac{1}{2}}\exp(-k_d r) \end{array} \right\} \qquad (9)$$

式(6)的 Laplace 变换为

$$\lim_{r\to\infty}\bar{\phi}(r,\ s)=0 \qquad (10)$$

由式(9)和(10)可知，式(8)中的 $A(s)=0$，于是式(8)可变为

$$\bar{\phi}(r,\ s)=B(s)k_0(k_d r) \qquad (11)$$

利用应力 - 势函数之间的关系，应力表达式的 Laplace 变换为

$$\bar{\sigma}_r(r,\ s)=\frac{\lambda}{r}\frac{\partial \bar{\phi}}{\partial r}+(\lambda+2\mu)\frac{\partial^2 \bar{\phi}}{\partial r^2} \qquad (12)$$

$$\bar{\sigma}_\theta(r,\ s)=\frac{\lambda+2\mu}{r}\frac{\partial \bar{\phi}}{\partial r}+\lambda\frac{\partial^2 \bar{\phi}}{\partial r^2} \qquad (13)$$

将式(11)中 Bessel 函数取式(9)中的函数，则有

$$\bar{\phi}(r,\ s)=B(s)\left(\frac{\pi}{2k_d r}\right)^{\frac{1}{2}}\exp(-k_d r) \qquad (14)$$

根据式(12)和(14)，对于圆形巷道有

$$\bar{\sigma}_r(r,\ s)=\frac{B(s)e^{\frac{-rs}{c_d}}\sqrt{\dfrac{\pi}{2}}N_r}{4r^4\left(\dfrac{c_d}{rs}\right)^{\frac{3}{2}}s^2} \qquad (15)$$

其中，

$$N_r=8sc_d r\mu+4r^2 s^2(\lambda+2\mu)+c_d^2(\lambda+6\mu)$$

将 $r=a$ 代入式(15)，有

$$\bar{\sigma}_r(a,\ s)=\frac{B(s)e^{\frac{-as}{c_d}}\sqrt{\dfrac{\pi}{2}}N_a}{4a^4\left(\dfrac{c_d}{as}\right)^{\frac{3}{2}}s^2} \qquad (16)$$

其中，

$$N_a=8c_d as\mu+4a^2 s^2(\lambda+2\mu)+c_d^2(\lambda+6\mu)$$

开挖过程中荷载与时间的函数关系如图 2 所示，其 $p(t)$ 函数关系可假设为

$$p(t)=f_1 t^2+g_1 t+h_1 \qquad (17)$$

式中：$f_1=-\dfrac{q}{t_0^2}$，其中，t_0 为开挖完成时间；$g_1=0$；$h_1=q$。

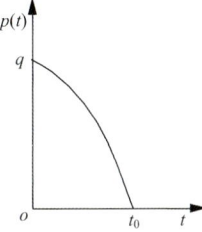

图 2 开挖过程中荷载与时间的函数关系

Fig.2 Function relationship between load and time during excavation process

对式(17)进行 Laplace 变换可得

$$\bar{p}(s)=\frac{2f_1}{s^3}+\frac{g_1}{s^2}+\frac{h_1}{s} \qquad (18)$$

由式(5)和(16)可得

$$B(s)=\frac{4a^3 c_d e^{\frac{as}{c_d}}\sqrt{\dfrac{\pi c_d}{2as}}(2f_1+g_1 s+h_1 s^2)}{s^2 N_0} \qquad (19)$$

其中，

$$N_0=c_d^2\lambda+4a^2 s^2\lambda+6c_d^2\mu+8ac_d s\mu+8a^2 s^2\mu$$

将式(19)代入式(14)有

$$\bar{\phi}(r,\ s)=\frac{4a^3c_d^2\mathrm{e}^{\frac{(a-r)s}{c_d}}(2f_1+g_1s+h_1s^2)}{s^3N_0\sqrt{ar}} \quad (20)$$

对式(20)进行 Laplace 逆变换可得

$$\phi(r,\ t)=\frac{2a^3}{(ar)^{1/2}(\lambda+6\mu)^3}(A_1+A_2+A_3) \quad (21)$$

式中：A_1，A_2，A_3 均为参数，其取值情况详见附录(下同)。

将式(20)代入式(12)，并作 Laplace 逆变换可得圆形巷道的径向应力为

$$\sigma_{r_1}(r,\ t)=\frac{1}{(ar)^{3/2}(\lambda+6\mu)^3}\left\{[-a^4(A_1+A_2+A_3)+2a^4r(A_4+A_5+A_6)]\frac{\lambda}{r}+\frac{\lambda+2\mu}{r}\left[\frac{3a^4}{2r}(A_1+A_2+A_3)-2a^4(A_4+A_5+A_6)+2a^4r(A_7+A_8+A_9)\right]\right\} \quad (22)$$

将式(20)代入式(13)，并作 Laplace 逆变换可得圆形巷道切向应力为

$$\sigma_{\theta_1}(r,\ t)=\frac{1}{(ar)^{3/2}(\lambda+6\mu)^3}\left\{[-a^4(A_1+A_2+A_3)+2a^4r(A_4+A_5+A_6)]\frac{\lambda+2\mu}{r}+\frac{\lambda}{r}\left[\frac{3a^4}{2r}(A_1+A_2+A_3)-2a^4(A_4+A_5+A_6)+2a^4r(A_7+A_8+A_9)\right]\right\} \quad (23)$$

2.2 开挖过程中，由 $p(t)$ 引起的次生位移场

根据式(1)可得，由 $p(t)$ 引起的次生位移场为

$$u_{r_1}(r,\ t)=-\frac{a^4}{(ar)^{3/2}(\lambda+6\mu)^3}[(A_1+A_2+A_3)-2r(A_4+A_5+A_6)] \quad (24)$$

3 在开挖过程中，初始地应力引起的弹性应力场和位移场

3.1 开挖过程中，初始地应力引起的弹性应力场

本文仅考虑静水压力的情况，即，水平方向的初始地应力等于垂直方向的原岩应力 $p=q$，则在开挖过程中由 q 引起的次生应力场为

$$\left.\begin{array}{l}\sigma_{r_2}=q\left(1-\dfrac{a^2}{r^2}\right)\\[6pt]\sigma_{\theta_2}=q\left(1+\dfrac{a^2}{r^2}\right)\end{array}\right\} \quad (25)$$

3.2 开挖过程中，由初始地应力引起的弹性位移场

开挖后围岩产生的径向位移 u 可表示为

$$u=\frac{(1+v_0)}{E_0r}[a^2(q-p_i)+qr^2(1-2v_0)] \quad (26)$$

式中：p_i 为巷道壁的支反力。

在巷道开挖过程中，由 q 引起的巷道周边 $r=a$ 处的弹性位移场可表示为

$$u_{r_2}=\frac{1+v_0}{E_0}a(q-p_i) \quad (27)$$

总的次生应力场和位移场可分别为

$$\left.\begin{array}{l}\sigma_r^e=\sigma_{r_1}+\sigma_{r_2}\\\sigma_\theta^e=\sigma_{\theta_1}+\sigma_{\theta_2}\end{array}\right\} \quad (28)$$

$$u_r^e=u_{r_1}(r,\ t)+u_{r_2} \quad (29)$$

4 破裂区的半径

假设岩石破坏满足 Mohr-Coulomb 准则，Mohr-Coulomb 准则可表示为

$$\sigma_\theta-\sigma_r=2(c_0\cot\beta+\sigma_r)\frac{\sin\beta}{1-\sin\beta} \quad (30)$$

式中：β 为岩石的内摩擦角，c_0 为岩石的黏聚力。

将式(28)代入式(30)可得

$$\sigma_\theta^e-\sigma_r^e=2(c_0\cot\beta+\sigma_r^e)\frac{\sin\beta}{1-\sin\beta} \quad (31)$$

解式(31)可得破裂区的宽度 R_1-R_0 及破裂区发生的时刻 t_{p_1}，其中 R_0，R_1 分别为破裂区的内、外边界。

5 破裂区的应力场

对于第二个破裂区，第一个破裂区的外边界即是求解弹性区的边界，在该边界上的应力 $p(t)$ 函数关系(见图3)可假定为

$$p(t)=f_2t^2+g_2t+h_2 \quad (32)$$

其中，

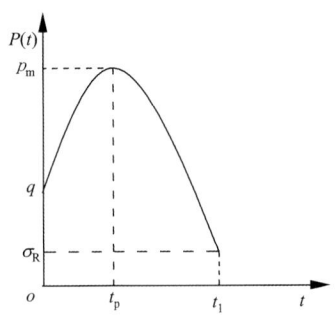

图 3 第二个破裂区的边界荷载与时间的函数关系

Fig.3 Function relationship between boundary load and time in the second fractured zone

$$h_2 = q$$

$$f_2 = \frac{q(1 - 2m - 2\sqrt{m^2 - m})}{t_1^2}$$

$$g_2 = \frac{2q(m - 1 + \sqrt{m^2 - m})}{t_1}$$

图 3 中有 3 个未知数：p_m，σ_R，t_1。将 t_p 代入式(28)可得 p_m。这里，σ_R 为第一个破裂区变形局部化完成时的应力；t_1 为变形局部化完成时间，即裂纹扩展贯通的时间。

当岩石中的应力状态满足 Mohr-Coulomb 准则时，岩石发生破坏并产生变形局部化，同时其应力下降到残余应力 σ_R 水平(见图 4)。在变形局部化区域位移将不连续，因此不能采用厚壁筒理论求解[12]。

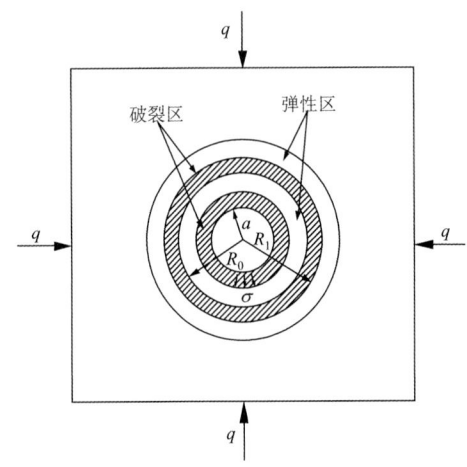

图 4 塑性区外边界对弹性区的支反力

Fig.4 Support reaction acted on elastic zone by outer boundary of plastic zone

在平面应变情况下，变形局部化区域的本构关系为

$$\sigma_{ij} = D_{ijkl}\varepsilon_{kl} \qquad (33)$$

式中：ε_{kl} 为变形局部化区域的应变，D_{ijkl} 为变形局部化时岩石的刚度矩阵。

变形局部化情况下的应变[18]可表示为

$$\varepsilon_{kl} = \frac{1}{2}(c_k n_l + n_k c_l) \quad (l = 1, 2, 3) \qquad (34a)$$

其中，

$$\left. \begin{aligned}
n_1 &= \sqrt{\frac{c_1}{c_1 - c_3}} \\
n_3 &= \sqrt{-\frac{c_3}{c_1 - c_3}} \\
c_1 &= f_{11}(g_{11} - g_{33}) + g_{11}(f_{11} - f_{33}) + v_0 h \\
c_3 &= f_{33}(g_{11} - g_{33}) + g_{33}(f_{11} - f_{33}) + v_0 h \\
h &= f_{22}(g_{11} - g_{33}) + g_{22}(f_{11} - f_{33}) \\
f_{ij} &= \frac{\partial F}{\partial \sigma_{ij}} \quad (i, j = 1, 2, 3) \\
g_{ij} &= \frac{\partial Q}{\partial \sigma_{ij}} \quad (i, j = 1, 2, 3) \\
F &= \frac{1}{2}(\sigma_1 - \sigma_3) + \frac{1}{2}(\sigma_1 + \sigma_3)\sin\beta_d \\
Q &= \frac{1}{2}(\sigma_1 - \sigma_3) + \frac{1}{2}(\sigma_1 + \sigma_3)\sin\varphi_d
\end{aligned} \right\} \qquad (34b)$$

式中：F 为屈服函数；Q 为塑性势函数；v_0 为泊松比；φ_d 为动荷载下的剪胀摩擦角；β_d 为动荷载下岩石的内摩擦角，可通过动荷载实验确定，本文则采用相关联流动法则，因此剪胀摩擦角等于岩石的内摩擦角，当然也可采用非相关联流动法则。

变形局部化情况下岩石的刚度矩阵为

$$D_{ijkl} = D_{ijkl}^e - \frac{k}{A}D_{ijrs}^e g_{rs} f_{mn} D_{mnkl}^e \qquad (35)$$

其中，

$$\left. \begin{aligned}
D_{ijkl}^e &= 2G\left[\frac{1}{2}(\delta_{ik}\delta_{jl} + \delta_{il}\delta_{jk}) + \frac{v_0}{1 - 2v_0}\delta_{ij}\delta_{kl}\right] \\
A &= H + f_{ij}D_{ijkl}^e g_{kl} \\
H &= 2G\left[\frac{v_0^2(1 - \sin\beta_d \sin\varphi_d)^2}{4(1 - v_0)(1 + \sin\beta_d)(1 + \sin\varphi_d)} - \right. \\
&\qquad \left. \frac{(1 - \sin\beta_d)(1 - \sin\varphi_d)}{4(1 - v_0)} \right]
\end{aligned} \right\} \qquad (36)$$

式中：D_{ijkl}^e 为弹性刚度矩阵；G 为剪切模量；k 为

常数，一般取 $k = 1$；H 为塑性软化模量。

变形局部化的方向为

$$\tan^2 \theta = \frac{n_1^2}{n_3^2} = -\frac{c_1}{c_3} \quad (37)$$

对于本文所讨论的二维平面应变情况问题，本构关系即为

$$\begin{Bmatrix} \sigma_{11} \\ \sigma_{33} \end{Bmatrix} = \begin{bmatrix} D_{1111} & D_{1133} \\ D_{3311} & D_{3333} \end{bmatrix} \begin{Bmatrix} \varepsilon_{11} \\ \varepsilon_{33} \end{Bmatrix} \quad (38)$$

$$A = H + f_{11} D_{1111}^e g_{11} + f_{11} D_{1133}^e g_{33} + f_{33} D_{3311}^e g_{11} + f_{33} D_{3333}^e g_{33} \quad (39)$$

其中，

$$\left. \begin{array}{l} D_{1111}^e = D_{3333}^e = \dfrac{2G(1-v_0)}{1-2v_0} \\ D_{1133}^e = D_{3311}^e = \dfrac{2G v_0}{1-2v_0} \end{array} \right\} \quad (40)$$

$$\left. \begin{array}{l} f_{11} = \dfrac{1}{2}(\sin \beta_d + 1), \ f_{33} = \dfrac{1}{2}(\sin \beta_d - 1) \\ g_{11} = \dfrac{1}{2}(\sin \varphi_d + 1), \ g_{33} = \dfrac{1}{2}(\sin \varphi_d - 1) \end{array} \right\} \quad (41)$$

$$c_1 = \frac{1}{2}[1 - v_0 + \sin \beta_d + \sin \varphi_d + (1+v_0)\sin \beta_d \sin \varphi_d] \quad (42)$$

$$c_3 = -\frac{v_0}{2}(1 - \sin \beta_d \sin \varphi_d) \quad (43)$$

本文中，有 $\sigma_{11} = \sigma_r$，$\sigma_{33} = \sigma_\theta$，由此可以解出 σ_r 和 σ_θ，从而得到变形局部化完成后的残余应力场 σ_R 为

$$\sigma_R = \frac{E_0 v_0 (A_{10} + A_{11})}{A_{12}} \quad (44)$$

从峰值应力下降到 σ_R 的时间 t_1 的确定方法如下：

对于共线裂纹，动态应力强度因子[19, 20]为

$$K_{ID} = \frac{v_R - v}{v_R - 0.75v} \frac{F_p \cos \alpha}{\sqrt{w \sin\left(\dfrac{\pi l}{w}\right)}} - \sigma_3 \frac{v_R - v}{v_R - 0.5v} \sqrt{2w \tan\left(\dfrac{\pi l}{2w}\right)} \quad (45)$$

式中：α 为裂纹的方位角；w 为裂纹间距，可以通过扫描电镜实验确定；l 为裂纹的扩展长度；v 为裂纹的扩展速度，且有 $v = l/t$；v_R 为 Rayleigh 波波速；

$F_p = 2b_1 \tau_{eff}$，其中，b_1 为裂纹的半长度，τ_{eff} 可表示为

$$\tau_{eff} = (\sigma_1 - \sigma_3)\cos \alpha \sin \alpha - (\sigma_1 \cos^2 \alpha + \sigma_3 \sin^2 \alpha) \tan \beta_d \quad (46)$$

裂纹扩展准则为

$$K_{ID} = K_{ID}^d \quad (47)$$

式中：K_{ID}^d 为岩石的断裂韧度，可通过三点弯曲实验确定。

当 l 无限接近 w 时，裂纹基本贯通，岩石只有 σ_R，裂纹扩展时间 $t = t_1$。根据式(46)和(47)可以确定卸荷时间 t_1。w，K_{ID}^d 亦可根据已有实验成果从有关资料[19, 20]中查得。

例如，对于花岗岩[20]：$b_1 = 0.75$ mm，$w = 3$ mm，$\alpha = 45°$，$K_{ID}^d = 1.5$ MPa·m$^{1/2}$，$v_R = 2\,000$ m/s。根据计算结果有：$t_1 = 1.51 \times 10^{-6}$ s。

6 数值结果

设圆形巷道的半径 $a = 4$ m，$p = q = 100$ MPa，$p(t=0) = 100$ MPa，岩石的泊松比为 $v_0 = 0.1$，弹性模量 $E_0 = 2\,000$ MPa，岩石的密度 $\rho = 2\,300$ kg/m^3，岩石的动摩擦角假设为定值，即 $\beta = 18°$，不同情况下破裂区数量和宽度见表1～4，位移和速度与 p/σ_c 的关系曲线分别如图5，6所示。

表1 $t_0 = 1$ s，$\sigma_c = 80$ MPa 时破裂区数量和宽度
Table 1 Quantity and width of fractured zone when $t_0 = 1$ s, $\sigma_c = 80$ MPa

破裂区数量 n	R_0/m	R_1/m	$(R_1 - R_0)$/m
1	4.000	4.590	0.586
2	4.796	5.338	0.542
3	5.803	5.930	0.127
4	6.500	6.500	0.000

表2 $t_0 = 1$ s，$\sigma_c = 90$ MPa 时破裂区数量和宽度
Table 2 Quantity and width of fractured zone when $t_0 = 1$ s, $\sigma_c = 90$ MPa

破裂区数量 n	R_0/m	R_1/m	$(R_1 - R_0)$/m
1	4.000	4.476	0.474
2	4.756	5.129	0.373
3	5.610	5.610	0.000

表 3 $t_0 = 100$ s 时破裂区数量及宽度

Table 3 Number and width of fractured zone when $t_0 = 100$ s

破裂区数量 n	$\sigma_c = 100$ MPa	$\sigma_c = 90$ MPa			$\sigma_r = 80$ MPa			$\sigma_c = 60$ MPa		
		R_0/m	R_1/m	(R_1-R_0)/m	R_0/m	R_1/m	(R_1-R_0)/m	R_0/m	R_1/m	(R_1-R_0)/m
1	无	4.000	4.463	0.462	4.000 0	4.650 0	0.646	4.000 0	4.965 0	0.960
2		5.753	6.116	0.363	4.800 0	5.325 0	0.525	4.983 0	5.958 0	0.975
3		7.313	7.313	0.000	5.704 0	5.768 0	0.064	6.041 0	6.109 0	0.068
4					0.602 1	0.602 1	0.000	0.652 3	0.652 3	0.000

表 4 $t_0 = 1\ 000$ s 时破裂区数量及宽度

Table 4 Number and width of fractured zone when $t_0 = 1\ 000$ s

破裂区数量 n	$\sigma_c = 100$ MPa	$\sigma_c = 90$ MPa			$\sigma_r = 80$ MPa			$\sigma_c = 60$ MPa		
		R_0/m	R_1/m	(R_1-R_0)/m	R_0/m	R_1/m	(R_1-R_0)/m	R_0/m	R_1/m	(R_1-R_0)/m
1	无	4.000	5.815	1.814	4.00	5.024	1.022	4.00	5.506	1.501
2		5.964	5.964	0.000	5.107	6.109	1.002	5.653	6.657	1.004
3					0.746	0.746	0.000	0.755	0.755	0.000

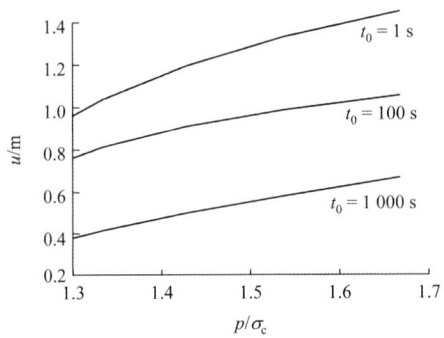

图 5 位移 u 与 p/σ_c 的关系曲线

Fig.5 Relation curves between displacement u and p/σ_c

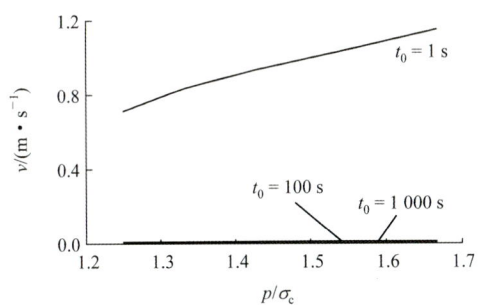

图 6 速度 v 与 p/σ_c 的关系曲线

Fig.6 Relation curves between velocity v and p/σ_c

从表 1～4 中可以看出，岩石的单轴抗压强度越高，则破裂区的数量越少，并且破裂区的宽度越小。比较表 3，4 可知，开挖速度越快，破裂区的数量越多，而破裂区越小。上述结果表明，岩体分区破裂化的产生跟开挖方式与岩石强度有关。

7 结 论

(1) 当开挖卸荷时间较短时，深部岩体可能发生分区破裂化现象，对于给定的开挖方式，其破裂区的数量是确定的，其宽度大致呈等差数列，且由洞口向外递减。

(2) 对于机械开挖，地应力超过岩石单轴抗压强度时，深部岩体才会发生分区破裂化现象。

(3) 在同等条件下，单轴抗压强度越低的深部岩体，破裂区的数量越多，且破裂区宽度越大。

(4) 在同等条件下开挖卸荷时间较长，破裂区的数量越少，但破裂区宽度越大。

(5) 当岩石中的应力状态满足 Mohr-Coulomb 准则时，岩石发生破坏并产生变形局部化，在变形局部化区域，微裂纹扩展合并成粗裂纹，位移将不连续，形成破裂区，可根据断裂力学知识确定破裂区岩体的残余强度和破裂区形成的时间。

(6) 本文假设地应力是定值，水平和垂直方向地应力相等；同时，没有考虑地应力随深度的变化，也没有考虑破裂区的裂纹扩展。因此，本文的解是近似解。

参考文献(References)：

[1] 钱七虎. 深部岩体工程响应的特征科学现象及"深部"的界定[J]. 华东理工学院学报(自然科学版)，2004，27(1)：1－5.(QIAN Qihu. The characteristic scientific phenomena of engineering response to deep

rock mass and the implication of deepness[J]. Journal of East China Institute of Technology(Natural Science), 2004, 27(1): 1 - 5.(in Chinese))

[2] 钱七虎. 非线性岩石力学的新进展——深部岩体力学的若干问题[C]// 中国岩石力学与工程学会编. 第八次全国岩石力学与工程学术大会论文集. 北京：科学出版社，2004：10 - 17.(QIAN Qihu. The current development of nonlinear rock mechanics: the menchanics problems of deep rock mass[C]// Chinese Society for Rock Mechanics and Engineering ed. Proceedings of the 8th Conference on Rock Mechanics and Engineering. Beijing: Science Press, 2004：10 - 17.(in Chinese))

[3] 钱七虎. 深部地下空间开发中的关键科学问题[C]// 第 230 次香山科学会议——深部地下空间开发中的基础研究关键技术问题. 北京：[s. n.]，2004.(QIAN Qihu. The key problems of deep underground space development[C]// The Key Technical Problems of Base Research in Deep Underground Space Development—the 230th Xiangshan Science Conference. Beijing：[s. n.]，2004.(in Chinese))

[4] KURLENYA M V, OPARIN V N. Problems of nonlinear geomechanics, part I[J]. Journal of Mining Science, 1999, 35(3): 216 - 230.

[5] ADAMS C R, JAGER A J. Petroscopic observations of rock fracturing ahead of stope faces in deep-level gold mines[J]. Journal of the South African Institute of Mining and Metallurgy, 1980, 80(6): 204 - 209.

[6] Открытие No.400: Явление зональной дезинтеграции горных пород вокруг подземных выработок[R]. Шемякин Е. И: Курленя М. В, Опарин В. Н, Глущихкн Ф. Н, Розенбаум. М. А –Опубл. ВВИ, 1992.

[7] Шемякин Е И，Курленя М В，Опарин В Н и др. Зоналъная дезинтеграция горных пород вокрут подземных выработок ч .1[J]. ФТПРПИ，1986，(3)：3 - 15.

[8] Шемякин Е И，Курленя М В，Опарин В Н и др. Зоналъная дезинтеграция горных пород вокрут подземных выработок ч .2[J]. ФТПРПИ，1986，(4)：3 - 13.

[9] Шемякин Е И，Курленя М В，Опарин В Н и др. Зоналъная дезинтеграция горных пород вокрут подземных выработок ч .3[J]. ФТПРПИ，1987，(1)：3 - 8.

[10] Шемякин Е И，Курленя М В，Опарин В Н и др. Зоналъная дезинтеграция горных пород вокрут подземных выработок ч .4[J]. ФТПРПИ，1989，(4)：3 - 9.

[11] Шемякин Е И，Фиссико Г Л，Курленя М В，Опарин В Н и др. Эффект зональной дезинтеграции горных пород вокруг подземных выработок[J]. Докл. АН ССР，1986，289(5)：1 088 - 1 094.

[12] NIKOLAEVSHIJ V N. Mechanics of porous and fractured media[M]. New York：[s. n.]，1990.

[13] 何满潮. 深部的概念体系及工程评价指标[J]. 岩石力学与工程学报, 2005, 24(16): 2 854 - 2 858.(HE Manchao. Conception system and evaluation indexes for deep engineering[J]. Chinese Journal of Rock Mechanics and Engineering, 2005, 24(16): 2 854 - 2 858.(in Chinese))

[14] 何满潮，谢和平，彭苏萍，等. 深部开采岩体力学研究[J]. 岩石力学与工程学报，2005，24(16)：2 803 - 2 813.(HE Manchao, XIE Heping, PENG Suping, et al. Study on rock mechanics in deep mining engineering[J]. Chinese Journal of Rock Mechanics and Engineering, 2005, 24(16): 2 803 - 2 813.(in Chinese))

[15] 王明洋，戚承志，钱七虎. 深部岩体块系介质变形与运动特性研究[J]. 岩石力学与工程学，2005，24(16)：2 825 - 2 830.(WANG Mingyang, QI Chengzhi, QIAN Qihu. Study on deformation and motion characteristics of blocks in deep rock mass[J]. Chinese Journal of Rock Mechanics and Engineering, 2005, 24(16): 2 825 - 2 830.(in Chinese))

[16] 王明洋，周泽平，钱七虎. 深部岩体的构造和变形与破坏问题[J]. 岩石力学与工程学报，2006，25(3)：448 - 455.(WANG Mingyang, ZHOU Zeping, QIAN Qihu. Tectonic, deformation and failure problems of deep rock mass[J]. Chinese Journal of Rock Mechanics and Engineering, 2006, 25(3): 448 - 455.(in Chinese))

[17] 王明洋，宋华，郑大亮，等. 深部巷道围岩的分区破裂机制及"深部"界定探讨[J]. 岩石力学与工程学报，2006，25(9)：1 771 - 1 776. (WANG Mingyang, SONG Hua, ZHENG Daliang, et al. On mechanism of zonal disintegration within rock mass around deep tunnel and definition of "deep rock engineering"[J]. Chinese Journal of Rock Mechanics and Engineering, 2006, 25(9): 1 771 - 1 776.(in Chinese))

[18] KENNETH R, NIELS S O, DUNJA P. Discontinuous bifurcations of elastoplastic solutions at plane stress and plane strain[J]. International Journal of Plasticity, 1991, 7(1/2): 99 - 121.

[19] ZHOU X P. Upper and lower bounds for constitutive relation of crack-weakened rock masses under dynamic compressive loads[J]. Theoretical and Applied Fracture Mechanics, 2006, 46(1): 75 - 86.

[20] LI H B, ZHAO J, LI T J. Micromechanical modelling of the mechanical properties of a granite under dynamic uniaxial compressive loads[J]. International Journal of Rock Mechanics and Mining Sciences, 2000, 37(6): 923 - 935.

附 录

本文公式中涉及的部分参数

$$A_1 = h_1(\lambda + 6\mu)^2 \left[2 - E_1\left(1 + \frac{2\mu}{G_1}\right) + E_2\left(-1 + \frac{2\mu}{G_1}\right) \right]$$

$$A_2 = \frac{2g_1(\lambda + 6\mu)}{c} \{-8a\mu + (a - r + c_d t)(\lambda + 6\mu) +$$

$$\frac{4aE_2}{G_1}[\mu G_1(1+E_3)-G_2(-1+E_3)]\}$$

$$A_3=\frac{2f_1}{c^2}\{-16a\mu(\lambda+6\mu)(a-r+c_\mathrm{d}t)+$$

$$(\lambda+6\mu)^2(a-r+c_\mathrm{d}t)^2-8a^2G_3+$$

$$\frac{4a^2E_2}{G_1}[G_1G_3(1+E_3)+2\mu G_4(-1+E_3)]\}$$

$$E_1=\exp\left[\frac{(a-r+c_\mathrm{d}t)(-2\mu+\mathrm{i}G_1)}{2a(\lambda+2\mu)}\right]$$

$$E_2=\exp\left[-\frac{(a-r+c_\mathrm{d}t)(2\mu+\mathrm{i}G_1)}{2a(\lambda+2\mu)}\right]$$

$$E_3=\exp\left[\frac{(a-r+c_\mathrm{d}t)\mathrm{i}G_1}{a(\lambda+2\mu)}\right]$$

$$G_1=\mathrm{i}\sqrt{\lambda^2+8\lambda\mu+8\mu^2}$$

$$G_2=\lambda^2+8\lambda\mu+4\mu^2$$

$$G_3=\lambda^2+8\lambda\mu-4\mu^2$$

$$A_4=\frac{h_1(\lambda+6\mu)^2}{2a(\lambda+2\mu)}\left[E_1(-2\mu+G_1)\left(1+\frac{2\mu}{G_1}\right)+\right.$$

$$\left.E_2(2\mu+G_1)\left(-1+\frac{2\mu}{G_1}\right)\right]$$

$$A_5=\frac{2g_1(\lambda+6\mu)}{c}\left\{-\lambda-6\mu+\frac{E_2(2\mu+G_1)}{2(\lambda+2\mu)G_1}\cdot\right.$$

$$\left[4\mu G_1(1+E_1)-G_2(-1+E_1)+\right.$$

$$\left.\left.\frac{aE_2}{G_1}\left(\frac{-4\mu G_1^2E_3}{a(\lambda+2\mu)}+\frac{G_1G_2E_3}{a(\lambda+2\mu)}\right)\right]\right\}$$

$$A_6=\frac{2f_1}{c^2}\{16a\mu(\lambda+\mu)-2(a-r+c_\mathrm{d}t)(\lambda+6\mu)^2+E_3+$$

$$\frac{2aE_2(2\mu+G_1)}{G_1(\lambda+2\mu)}[G_1G_3(1+2\mu G_4(-1+E_3)]-$$

$$\frac{4E_2a^2}{G_1}\left(\frac{E_3G_3G_1^2}{a(\lambda+2\mu)}+\frac{2\mu E_3G_1G_4}{a(\lambda+2\mu)}\right)\}$$

$$A_7=\frac{h_1(\lambda+6\mu)^2}{4a^2(\lambda+2\mu)^2}\left[E_1(2\mu-G_1)\left(1+\frac{2\mu}{G_1}\right)-\right.$$

$$\left.E_2(2\mu+G_1)\left(-1+\frac{2\mu}{G_1}\right)\right]$$

$$A_8=\frac{2g(\lambda+6\mu)}{c}\left\{\frac{E_2(2\mu+G_1)^2}{4a(\lambda+2\mu)^2G_1}[4\mu G_1(1+E_3)-\right.$$

$$G_2(-1+E_3)]+\frac{E_2(2\mu+G_1)}{a(\lambda+2\mu)^2G_1}[-4\mu E_3G_1^2+G_1G_2(-1+$$

$$\left.E_3)]+\frac{E_2}{G_1a(\lambda+2\mu)^2}(4\mu E_3G_1^{3/2}-G_2E_3G_1^2)\right\}$$

$$A_9=\frac{2f_1}{c^2}\left\{2(\lambda+6\mu)^2+\frac{E_2(2\mu+G_1)^2}{G_1(\lambda+2\mu)^2}[G_1G_3(1+$$

$$E_3)+2\mu G_4(E_3-1)]-\frac{4E_2(2\mu+G_1)}{G_1(\lambda+2\mu)^2}(G_3E_3G_1^2+$$

$$\left.2\mu G_1G_4E_3)+\frac{4E_2}{G_1}\left[\frac{E_3G_3G_1^{3/2}}{(\lambda+2\mu)^2}+\frac{2\mu E_3G_1^2G_4}{(\lambda+2\mu)^2}\right]\right\}$$

$$A_{10}=2c_1A_{14}\nu_0(\nu_0-1)(\sin\beta_\mathrm{d}\sin\varphi_\mathrm{d}-1)^2$$

$$A_{11}=A_{15}(1-\sin\beta_\mathrm{d}\sin\varphi_\mathrm{d})(-1+2\nu_0-\nu_0^3+$$

$$2\nu_0^3\sin\beta_\mathrm{d}\sin\varphi_\mathrm{d}-A_{13})$$

$$A_{12}=2[(1-\nu_0+\sin\beta_\mathrm{d})^2(-1+\nu_0+2\nu_0^2)+A_{16}-A_{17}]$$

$$A_{13}=\sin\varphi_\mathrm{d}(2-2\nu_0^2+\sin\varphi_\mathrm{d})+\sin^2\beta_\mathrm{d}[1-2\nu_0+$$

$$\sin\varphi_\mathrm{d}(2-2\nu_0+\sin\varphi_\mathrm{d}-\nu_0^3\sin\varphi_\mathrm{d})]$$

$$A_{14}=\left(1+\frac{\nu_0\sin\beta_\mathrm{d}}{1+\sin\varphi_\mathrm{d}}-\frac{\nu_0}{1+\sin\varphi_\mathrm{d}}\right)^{\frac{1}{2}}$$

$$A_{15}=\left(-\nu_0+\frac{\nu_0}{1+\sin\varphi_\mathrm{d}}+\frac{\nu_0}{1+\sin\varphi_\mathrm{d}}\right)^{\frac{1}{2}}$$

$$A_{16}=(1+\nu_0)\sin^2\varphi_\mathrm{d}[2\nu_0-1+\sin\beta_\mathrm{d}(\nu_0-1)\cdot$$

$$(2+\sin\beta_\mathrm{d}+\nu_0\sin\beta_\mathrm{d}+2\nu_0^2\sin\beta_\mathrm{d})]$$

$$A_{17}=2(1+\nu_0)\sin\varphi_\mathrm{d}[(2\nu_0-1)(\nu_0-1)]+$$

$$\sin\beta_\mathrm{d}(2-4\nu_0+\nu_0^2+2\nu_0^3+\sin\beta_\mathrm{d}-\nu_0\sin\beta_\mathrm{d})$$

基于滑面正应力假设的土压力计算方法[*]

刘华丽，钱七虎，朱大勇，周先华

(南京解放军理工大学工程兵工程学院,210007)

摘　要：基于滑面正应力假设,提出一种新的极限平衡方法计算挡土结构土压力。首先假定滑裂面上正应力分布为含2个待定参数的三次拉格朗日插值函数；推导出包含主动土压力的水平力、垂直力和绕挡土墙顶点旋转的力矩平衡方程；然后采用优化方法确定最危险滑裂面位置及对应的最大主动土压力。与传统土压力计算方法相比,作者提出的方法可以给出精度较高的土压力分布,且可分析土压力作用点位置对土压力值的影响,其土压力计算方法可应用于工程上。

关键词：主动土压力,极限平衡法,最优化方法

中图分类号：TU 227　　　　　　　　　　**文献标识码**：A

Method for Computation of Earth Pressure Based on Normal Stress Assumpti on

LIU Hua-li, QIAN Qi-hu, ZHU Da-yong, ZHOU Xian-hua

(Engineering Institute of Engineer Corps, PLA University of Science and Technology, Nanjing 210007, China)

Abstract：In this paper, a new li mit equilibrium method is proposed for calculating earth pressures acting upon retaining walls based on assu mptions regarding the nor mal stresses over the potential slip surface. Firstly, the nor mal stress distribution over the slip surface is assu med to be represented by a Lagrangian interpolation function involving two unkno wns. Secondly, the horizontal and vertical force and mo ment equilibriu m equations for the sliding body are derived incorporating the active lateral force. In the end, the most dangerous failure surface and maxi mum active earth pressure can be readily deter mined by opti mization procedure. Compared to conventional met hods of co mputing earth pressure, using the present met hod can yield earth pressure distribution of relatively higher precision and can analyze the relationship bet ween the active earth pressure and the location of the application point. Thus, this new met hod can be applied to practical engineering.

Keywords：active earth pressure; li mit equilibrium met hod; opti mization procedure

1　引言

挡土结构的土压力计算是经典土力学中经常遇到的问题,工程中常采用朗肯理论和库仑理论计算主动土压力[1,2]。但是在一些复杂情况下,例如土体非均质、有外载作用或墙面与土体有摩擦等等,这些经典方法就遇到了难以克服的困难,所假定的直线型滑面与实际也不相符。半个世纪以前,一些学者开始用曲面型滑裂面(圆弧、椭圆、对数螺旋面等)计算土压力[3],这些方法都是建立在极限平衡理论的基础上,计算过程特别繁琐,且精度也得不到保证。后来一些学者利用特征线法分

收稿日期：2006-04-18。
基金项目：国家自然科学基金(NO.40472138)。
本文原载于《地下空间与工程学报》(2006年第2卷第5期)。

析土体的稳定性[4]，来获取土压力问题的塑性力学解，但是由于这种方法经常遇到数值困难，且不容易处理复杂的边界条件，因此在工程应用中受到很大限制。最近几十年来，极限分析法用来确定主动土压力的下限解[5,6]，但是合适的运动许可滑动机制不容易选择，因此也未在工程中普及应用。最近的研究结果表明，直接假设滑面正应力形式，可导出满足严格平衡条件的安全系数显式解[7]。本文对这一方法进行了改进和补充，用于主动土压力计算。且所采用的方法很容易分析主动土压力作用点位置对土压力值的影响。同理这种方法也可适用于被动土压力计算。

2 滑动面与正应力分布构造

如图1所示，为挡土墙后滑动土体。墙体最大主动土压力为 P_a，与墙面摩擦角为 δ，图示为正，其作用点的位置为 (r_p,θ_a)。ω 为墙面与垂直方向夹角(当墙面倾向填土时 ω 为正)，β 为填土表面倾角。挡土墙后方土体沿某一滑面滑动，挡土墙下方的坐标为 (r_a,θ_a)，与坡面交于一点为 (r_b,θ_b)，由图1可知 $\theta_a = \omega - \pi/2$，$\theta_b = \beta$，$r_a = H/\cos\omega$。在 θ_a 和 θ_b 之间均匀选择两个点 θ_1 和 θ_2，即

$$\theta_1 = \theta_a + \frac{\theta_b - \theta_a}{3} \quad (1a)$$

$$\theta_2 = \theta_a + \frac{2(\theta_b - \theta_a)}{3} \quad (1b)$$

构造滑面为三次的拉格朗日插值多项式，其中，r_1，r_2，r_b 为未知变量。

$r(\theta) =$

$$r_a \frac{(\theta-\theta_1)(\theta-\theta_2)(\theta-\theta_b)}{(\theta_a-\theta_1)(\theta_a-\theta_2)(\theta_a-\theta_b)} + r_1 \frac{(\theta-\theta_a)(\theta-\theta_2)(\theta-\theta_b)}{(\theta_1-\theta_a)(\theta_1-\theta_2)(\theta_1-\theta_b)}$$

$$+ r_2 \frac{(\theta-\theta_a)(\theta-\theta_1)(\theta-\theta_b)}{(\theta_2-\theta_a)(\theta_2-\theta_1)(\theta_2-\theta_b)} + r_b \frac{(\theta-\theta_a)(\theta-\theta_1)(\theta-\theta_2)}{(\theta_b-\theta_a)(\theta_b-\theta_1)(\theta_b-\theta_2)}$$

$$(2)$$

如图1所示，土体总体重为 W，水平方向的地震力为 $K_c W$，K_c 是地震影响系数。作用于土体上的正应力与剪切力分别是 $\sigma(\theta)$ 和 $\tau(\theta)$，水压力为 $u(\theta)$，r_p 为土压力合力作用点位置。

假定滑面正应力 $\sigma(\theta)$ 为三次的拉格朗日多项式；两端分别为 σ_a 和 σ_b，在滑面上均匀选择两点 θ_1 和 θ_2 与式(1)相同，引入两个参数 λ_1 和 λ_2。

滑面正应力可由下式表达

$$\sigma(\theta) = \lambda_1 \xi_1(\theta) + \lambda_2 \xi_2(\theta) + \xi_3(\theta) \quad (3)$$

其中

$$\xi_1(\theta) = \frac{(\theta-\theta_a)(\theta-\theta_2)(\theta-\theta_b)}{(\theta_1-\theta_a)(\theta_1-\theta_2)(\theta_1-\theta_b)} \quad (4)$$

$$\xi_2(\theta) = \frac{(\theta-\theta_a)(\theta-\theta_1)(\theta-\theta_b)}{(\theta_2-\theta_a)(\theta_2-\theta_1)(\theta_2-\theta_b)} \quad (5)$$

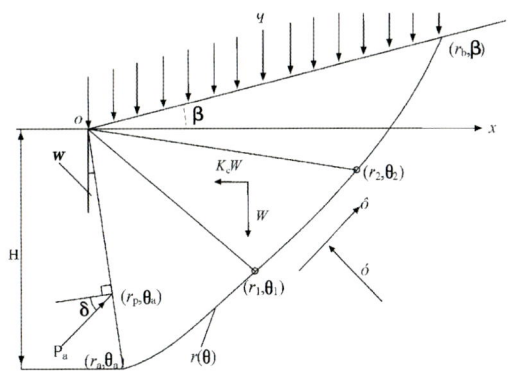

图 1 挡土墙后滑动土体

Fig.1 Potential sliding body behind the retaining wall

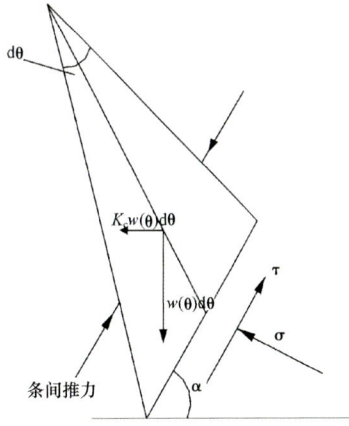

图 2 单个三角形条块受力示意图

Fig.2 Forces acting upon a triangular slice

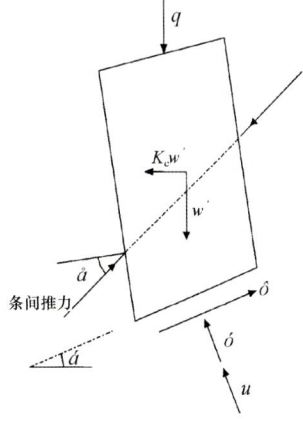

图 3 条块受力分析

Fig.3 Determination of normal stresses

$$\xi_3(\theta) = \sigma_a \frac{(\theta-\theta_1)(\theta-\theta_2)(\theta-\theta_b)}{(\theta_a-\theta_1)(\theta_a-\theta_2)(\theta_a-\theta_b)}$$

$$+ \sigma_b \frac{(\theta - \theta_a)(\theta - \theta_1)(\theta - \theta_2)}{(\theta_b - \theta_a)(\theta_b - \theta_1)(\theta_b - \theta_2)} \quad (6)$$

滑面剪应力 $\tau(\theta)$，由摩尔—库仑强度破坏准则计算

$$\tau(\theta) = [\sigma(\theta) - u(\theta)]\tan\phi(\theta) + c \quad (7)$$

其中，$\Phi(\theta)$ 和 $c(\theta)$ 为有效内摩擦角和凝聚力。

取单位宽度与挡土墙平行的条块，如图(3)所示。假设条间推力与条块面法线方向的倾角为 $\varepsilon(\theta)$，在推力线的法线方向上，根据力的平衡，法向力为

$$\sigma(\theta) = \frac{w' \cdot [\cos(\omega + \varepsilon(\theta)) - K_c\sin(\omega + \varepsilon(\theta))]\cos\alpha(\theta)}{\cos[\alpha(\theta) - \varepsilon(\theta)] + \sin[\alpha(\theta) - \varepsilon(\theta)] \cdot \tan\phi(\theta)} +$$

$$\frac{\alpha\cos(\omega + \varepsilon(\theta)) + [u(\theta)\tan\phi(\theta) - c(\theta)]\sin[\alpha(\theta) - \varepsilon(\theta)]}{\cos[\alpha(\theta) - \varepsilon(\theta)] + \sin[\alpha(\theta) - \varepsilon(\theta)] \cdot \tan\phi(\theta)} \quad (8)$$

σ_a 和 σ_b 可通过(8)式求得。

$\alpha(\theta)$ 是条底面倾角，w' 为图3中单位宽度的条块重量，$w(\theta)$ 是图2中单位弧度夹角对应的扇形体重量，γ 为土体容重。

$$w(\theta) = \frac{1}{2}\gamma \cdot r^2 \quad (9)$$

预先指定 $\varepsilon(a)$ 和 $\varepsilon(b)$ 对整个土压力的求取影响不大，本文中 $\varepsilon(a)$ 为 δ，$\varepsilon(b)$ 为零。

3 平衡方程与土压力计算

土体沿水平方向力平衡，竖直方向力平衡，以及绕极点旋转的力矩平衡分别为

$$\int_{\theta_a}^{\theta_b}(-\sigma \cdot r \cdot r' + \tau \cdot r)d\theta + P_a\cos(\omega + \delta) = \int_{\theta_a}^{\theta_b}K_c w d\theta \quad (10)$$

$$\int_{\theta_a}^{\theta_b}(\sigma \cdot r + \tau \cdot r' \cdot r)d\theta + P_a\sin(\omega + \delta) = \int_{\theta_a}^{\theta_b}w d\theta + \sum q \quad (11)$$

$$-\int_{\theta_a}^{\theta_b}(-\sigma \cdot r \cdot r' + \tau \cdot r) \cdot r \cdot \sin\theta d\theta + \int_{\theta_a}^{\theta_b}(\sigma \cdot r + \tau \cdot r' \cdot r) \cdot r \cdot \cos\theta d\theta$$
$$-P_a\cos(\omega + \delta) \cdot r_p \cdot \sin\theta_p + P_a\sin(\omega + \delta) \cdot r_p \cdot \cos\theta_p =$$
$$-\int_{\theta_a}^{\theta_b}K_c w \cdot \frac{2}{3}r \cdot \sin\theta d\theta + \int_{\theta_a}^{\theta_b}w \cdot \frac{2}{3}r \cdot \cos\theta d\theta + \sum q r_q\cos\theta_q \quad (12)$$

令

$$F_x = \int_{\theta_a}^{\theta_b}K_c w d\theta$$

$$F_y = \int_{\theta_a}^{\theta_b}w d\theta + \sum q$$

$$r_\sigma = r^2(\tan\alpha \cdot \sin\theta + \cos\theta)$$

$$r_\tau = r^2(-\sin\theta + \tan\alpha \cdot \cos\theta)$$

$$M_p = -\cos(\omega + \delta) \cdot r_p \cdot \sin\theta_p + \sin(\omega + \delta) \cdot r_p \cdot \cos\theta_p$$

$$M_c = -\int_{\theta_a}^{\theta_b}K_c w \cdot \frac{2}{3}r \cdot \sin\theta d\theta + \int_{\theta_a}^{\theta_b}w \cdot \frac{2}{3}r \cdot \cos\theta d\theta$$
$$+ \sum q r_q \cdot \cos\theta_q$$

把(7)代入(10)、(11)和(12)式得

$$\int_{\theta_a}^{\theta_b}r \cdot (-\tan\alpha + \tan\phi) \cdot \sigma d\theta + P_a\cos(\omega + \delta)$$
$$= F_x + \int_{\theta_a}^{\theta_b}r \cdot (u\tan\phi - c)d\theta \quad (13)$$

$$\int_{\theta_a}^{\theta_b}r \cdot (1 + \tan\alpha \cdot \tan\phi) \cdot \sigma d\theta + P_a\sin(\omega + \delta)$$
$$= F_y + \int_{\theta_a}^{\theta_b}r \cdot \tan\alpha \cdot (u\tan\phi - c)d\theta \quad (14)$$

$$\int_{\theta_a}^{\theta_b}(r_\sigma + r_\tau \cdot \tan\phi) \cdot \sigma d\theta + P_a \cdot M_p$$
$$= M_c + \int_{\theta_a}^{\theta_b}r_\tau \cdot (u\tan\phi - c)d\theta \quad (15)$$

把(3)代入(13)、(14)和(15)式得

$$\lambda_1\int_{\theta_a}^{\theta_b}r \cdot (-\tan\alpha + \tan\phi) \cdot \xi_1 d\theta +$$
$$\lambda_2\int_{\theta_a}^{\theta_b}r \cdot (-\tan\alpha + \tan\phi) \cdot \xi_2 d\theta +$$
$$P_a\cos(\omega + \delta) + F_x +$$
$$\int_{\theta_a}^{\theta_b}r \cdot (u\tan\phi - c)d\theta - \int_{\theta_a}^{\theta_b}r \cdot (-\tan\alpha + \tan\phi) \cdot \xi_3 d\theta$$
$$\quad (16)$$

$$\lambda_1\int_{\theta_a}^{\theta_b}r \cdot (1 + \tan\alpha \cdot \tan\phi) \cdot \xi_1 d\theta +$$
$$\lambda_2\int_{\theta_a}^{\theta_b}r \cdot (1 + \tan\alpha \cdot \tan\phi) \cdot \xi_2 d\theta +$$
$$P_a\sin(\omega + \delta) = F_y +$$
$$\int_{\theta_a}^{\theta_b}r \cdot \tan\alpha \cdot (u\tan\phi - c)d\theta - \int_{\theta_a}^{\theta_b}r \cdot (1 + \tan\alpha \cdot \tan\phi) \cdot \xi_3 d\theta$$
$$\quad (17)$$

$$\lambda_1\int_{\theta_a}^{\theta_b}(r_\sigma + r_\tau \cdot \tan\phi) \cdot \xi_1 d\theta +$$
$$\lambda_2\int_{\theta_a}^{\theta_b}(r_\sigma + r_\tau \cdot \tan\phi) \cdot \xi_2 d\theta +$$
$$P_a \cdot M_p = M_c + \int_{\theta_a}^{\theta_b}r_\tau \cdot (u\tan\phi - c)$$
$$-\int_{\theta_a}^{\theta_b}(r_\sigma + r_\tau \cdot \tan\phi) \cdot \xi_3 d\theta \quad (18)$$

则式(16)、(17)和(18)可化为如下形式

$$\lambda_1 \cdot A_1 + \lambda_2 \cdot A_2 + P_a \cdot A_3 = A_4 \quad (19)$$
$$\lambda_1 \cdot B_1 + \lambda_2 \cdot B_2 + P_a \cdot B_3 = B_4 \quad (20)$$
$$\lambda_1 \cdot D_1 + \lambda_2 \cdot D_2 + P_a \cdot D_3 = D_4 \quad (21)$$

解方程组(19)、(20)和(21)得

$$P_a = \frac{\Delta_3}{\Delta} \quad (22)$$

$$\Delta = \begin{vmatrix} A_1 & A_2 & A_3 \\ B_1 & B_2 & B_3 \\ D_1 & D_2 & D_3 \end{vmatrix} \quad (23a)$$

$$\Delta_3 = \begin{vmatrix} A_1 & A_2 & A_4 \\ B_1 & B_2 & B_4 \\ D_1 & D_2 & D_4 \end{vmatrix} \quad (23b)$$

采用最优化方法,可搜索出最危险滑面,且对于固定滑面,利用(22)式计算出最大主动土压力。

4 算例与比较

例1:墙面呈负的摩擦,有关参数如下:$\gamma = 20$ kN/m³,$c=0$,$\Phi=30°$,$\delta=-15°$,$\beta=0°$,$H=10$ m。计算得到,最大主动土压力为434.5 kN/m,主动土压力系数为0.435,最危险滑面如图4所示。

例2:填土向下倾斜,墙面光滑。有关参数如下:$\gamma=20$ kN/m³,$c=0$,$\Phi=40°$,$\delta=0°$,$\beta=-30°$,$H=10$ m。计算出最大主动土压力为179.9 kN/m,主动土压力系数为0.180,最危险滑面如图5所示。

本方法与库仑方法和临界滑动场法[8,9]计算出的例1和例2土压力系数比较见表1,由表1可见本方法计算出的结果与库仑方法和临界滑动场法计算出的结果非常接近。又由图4和图5可知本方法优化出的最危险滑面的位置与临界滑动场法比较接近。

表1 主动土压力系数比较

Table 1 Results of computation of active earth coefficient for the examples

方法	例1	例2
库仑	0.420	0.174
临界滑动场法	0.419	0.176
本方法	0.425	0.179

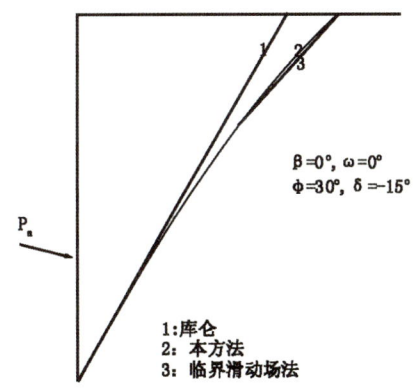

图4 例1最危险滑面

Fig.4 The most dangerous failure surface for the example 1

5 土压力作用点位置对土压力值的分析

传统土压力理论如库仑理论,即只考虑滑体力的平衡,不能反映土压力作用点位置对土压力值的影响。不同的挡土结构,其土压力分布不同,土压

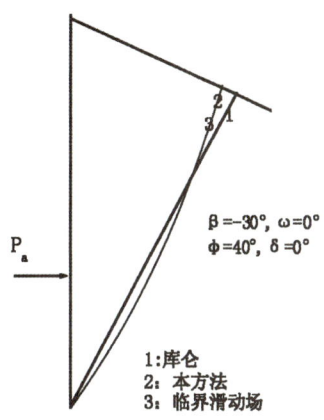

图5 例2最危险滑面

Fig.5 The most dangerous failure surface for the example 2

力合力作用点的位置也不同,而相应的土压力值也不同。分析合力作用点位置,有利于提高土结构设计的合理性。

例3:直立挡墙、水平填土,墙面摩擦角等于土体内摩擦角。有关参数如下:$\gamma=20$ kN/m³,$c=0$,$\Phi=\delta=30°$,$\beta=0°$,$H=10$ m。改变不同的合力作用点位置,分别取合力作用点位于墙面下三分之一点,二分之一点和三分之二点处。利用本文的方法求出不同的合力作用点对应的最危险滑面和最大主动土压力,如图6和7所示。

由图7可知,随着土压力合力作用点位置的上升,土压力合力值逐渐增大,在墙高二分之一处,合力达到最大;随着作用点位置的进一步上升,其土压力合力值又逐渐减小;这与文献[10]得出的结论相吻合。

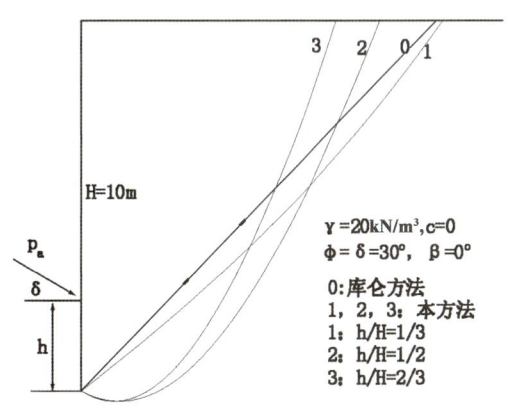

图6 最危险滑面位置比较

Fig.6 Comparison of the most dangerous failure surfaces for the example 3

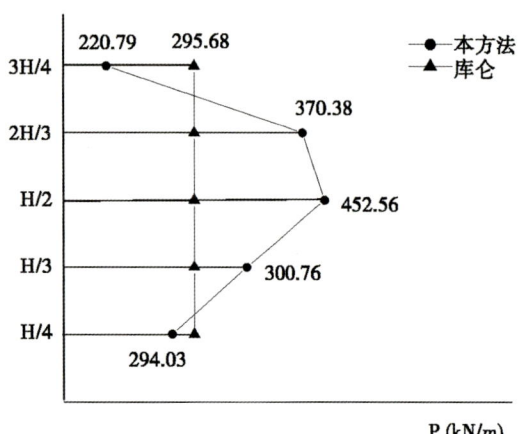

图 7 土压力合力值与作用点位置关系

Fig.7 Relation between the values of lateral earth forces and locations of application point

6 结语

传统的极限平衡条分法计算主动土压力有很多局限性。本文假定滑动面为极坐标下三次的拉格朗日插值函数，假定挡土墙后滑面正应力为三次的拉格朗日插值函数，推导出包含主动土压力的水平力、垂直力和绕挡土墙顶点旋转的力矩平衡方程。采用最优化方法确定最危险滑面和最危险滑面对应的最大主动土压力。同分析主动土压力作用点位置对土压力值的影响。本文方法原理简单，程序易编制，计算精度高，可在实际工程中应用。

参考文献

[1] Terzaghi, K. Theoretical soil mechanics [J]. John Wiley & Sons. New York, N.Y, 1943:35-41

[2] 张天宝. 土坡稳定性分析和土工建筑物的边坡设计 [M]. 成都:成都科技大学出版社,1987

[3] Caquot, A and Kerisel, J. Tables for the calculation of passive pressure, active pressure and bearing capacity of foundations [M]. Gauthier — Villars, Paris. 1948

[4] Lee, I·K and Herington, J·R. A theoretical study of the pressures acting on a rigid wall by a sloping earth or rock fill [J]. Geotechnique. 1972,22(1):1-26

[5] Chen, W·F. Limit Analysis and Soil Plasticity [M]. Elsevier, Amsterdam, 1975

[6] Chen, W·F and Liu, X·L. Limit analysis in soil mechanics [M]. Elsevier, New York. 1990

[7] Zhu, D·Y and Lee C·F. Explicit limit equilibrium solution for slope stability [J]. Int. J. Numer. Anal. Meth Geomech., 2002, 26:1573-1590

[8] 朱大勇,钱七虎等. 极限平衡法计算土压力系数的新途径[J]. 土木工程学报. 2000,1(33):63-68

[9] 朱大勇,钱七虎等. 边坡临界滑动场及其数值模拟 [J]. 岩土工程学报,1997,1(19):63-69

[10] Chen Zuyu. Evaluation of active earth pressure by the generalized method of slices [J]. Can Geotech J. 1998, 35:591-599

深部隧道围岩的流变

戚承志[1]，钱七虎[2]，王明洋[2]，丁常树[3]

(1. 北京建筑工程学院 土木与交通工程学院，北京 100044；2. 解放军理工大学 工程兵工程学院，南京 210007；
3. 二炮营建办公室，北京 100085)

摘　要：隧道围岩的流变对于隧道的长期稳定性影响很大，尤其对于深部隧道来讲更是如此．目前利用线性及非线性粘弹性流变模型对于隧道围岩的研究较多，初始状态多取为弹性状态．但在深部，由于初始地应力很大，在隧道开挖后，可能会马上出现塑性区．因此初始参照状态应该取为弹塑性状态，而且应该考虑塑性区及弹性区力学性质的不同．现利用 Shvedov-Bingham 模型理论研究了隧道围岩在初始时刻出现塑性区情况下的流变情况，并考虑塑性区及弹性区力学性质的不同，给出了流变区半径及应力应变的解析解，并分析了应力分布及流变半径的变化规律．

关键词：深部隧道；流变；塑性软化

中图分类号：TV672.1；U451　　　**文献标识码**：A

Rheology of Rock Around Deep Tunnel

Qi Chengzhi[1], Qian Qihu[2], Wang Mingyang[2]

(1. School of Civil and Traffic Engineering, Beijing 100044; 2. Engrg Institute of Engrg Coeps, University of Science and Technology of the PLA, Nanjing 210007; 3. Construction Office of Second Artillery, Beijing 100085)

Abstract: The rheology of rock around tunnel affects the long-term stability of the tunnel remarkably, especially for tunnels at great depth. At present the majority of the research on rheology is based on the linear or nonlinear visco-elastic models, and the initial state is elastic. But at great depth, because of the great original stress, after excavation of tunnels, plastic zone may arises quickly, therefore elasto-plastic state should be taken as the initial state, and difference of mechanical properties in elastic and plastic zones should be considered. But the study in this aspect is unsufficient. Therefore in this paper, the rheology of rock around tunnels is studied with the use of Shvedov-Bingham model, with the elasto-plastic state being taken as the initial state, and difference of mechanical properties in elastic and plastic zones being considered. Analytic solutions of rheology radius and stress-strain state is obtained, and also the evolution law of them is analyzed.

Key words: tunnel at great depth; rheology; plastic softening

岩石具有流变特性，也即其本身的应力及应变状态随着时间而变化．岩石的流变与岩土工程的长期稳定性密切相关，因而受到岩土工程界的高度关注．对于深部隧道，围岩的流变特性更加显著，对于隧道的稳定性的影响更大．地下隧道围岩中应力及变形的时间过程是当前非线性岩石力学问题(如分区破裂化现象)的研究中一个需要解决的课题．常用的岩石流变模型有粘弹性模型(包括 Maxwell 模

收稿日期：2006-09-06。
基金项目：国家自然科学基金重大项目(50490275)。
本文原载于《北京建筑工程学院学报》(2006 年第 22 卷第 4 期)。

型、Kelvin 模型、Poyting-Thomson 模型及线性后效模型等),以及粘塑性模型(包括粘性液体模型、Shvedov-Bingham 模型等)[1]. 岩石的流变模型已经在分析隧道围岩的流变中得到应用. 如在文[2]中,利用 Shvedov-Bingham 模型研究了隧道围岩在初始时刻未出现塑性区情况下流动半径及应力、坑道壁位移随时间的变化情况,得到了解析解. 在文[3,4]对于隧道的位移及开挖工程给出了一些依赖于时间的分析结果. 在文[5]中,利用非线性粘弹性模型,在考虑变形模量及泊淞比随时间变化时,对于化为二维平面应变问题的圆形隧洞进行了有限元计算,分析了围岩应力、应变及破坏区域随时间变化的情况. 在油井钻孔中基于线性弹粘性模型的钻孔流变在[6]中给出.

但是对于深部隧道,由于初始地应力很大,在隧道开挖后,可能会马上出现塑性区. 因此初始参照状态应该为出现塑性软化区的状态. 但是这一方面的研究还很不够. 因此本文利用 Shvedov-Bingham 模型,并考虑到在塑性区及弹性区岩体性质的不同研究隧道围岩在初始时刻出现塑性区情况下的流变情况,并给出解析解.

1 隧道围岩的弹塑性应力状态

设在深度为 H 的岩体中有半径为 r_0 的隧道. 设在此深度静水压力为 $\sigma_\infty = H\gamma$(γ 为岩体的密度),坑道周边受到的径向压力为 q. 如果此时隧道围岩处于弹性状态,则应力的弹性解为(此处以压为正):

$$\sigma_r = \sigma_\infty - (\sigma_\infty - q) r_0^2 / r^2$$
$$\sigma_\theta = \sigma_\infty + (\sigma_\infty - q) r_0^2 / r^2 \quad (1)$$

设岩体的强度遵从莫尔-库仑准则,即表达式为

$$\tau = k + \lambda \sigma_r \quad (2)$$

其中 $k = K\dfrac{\cos\rho}{1-\sin\rho}$, $\lambda = \dfrac{\sin\rho}{1-\sin\rho}$, ρ 为内摩擦角. 则在坑道周围的岩体中出现塑性区时,把 $\tau_{r\theta} = \tau$ 代入平衡方程

$$\frac{d\sigma_r}{dr} - \frac{2\tau_{r\theta}}{r} = 0 \quad (3)$$

并考虑到边界条件:$r = r_0$ 时 $\sigma_r = q$,得

$$\sigma_r = (q + k/\lambda)(r/r_0)^{2\lambda} - k/\lambda \quad (4)$$

在弹性区应力公式由 Lame 公式表达:

$$\sigma_r = \sigma_\infty - B/r^2; \quad \sigma_\theta = \sigma_\infty + B/r^2 \quad (5)$$

在弹性区与塑性区交界 $r = R$ 两侧的正应力与剪应力强度相等得

$$\frac{B}{R^2} = k + \lambda \sigma_r(R) = \frac{k + \lambda \sigma_\infty}{1+\lambda} \quad (6)$$

所以弹性区的正应力及剪应力的表达式为:

$$\left.\begin{array}{l}\sigma_r \\ \sigma_\theta\end{array}\right\} = \sigma_\infty \mp \frac{B}{r^2} = \sigma_\infty \mp \frac{R^2}{r^2}\frac{k+\lambda\sigma_\infty}{1+\lambda};$$

$$\tau_{r\theta}^e = \frac{R^2}{r^2}\frac{k+\lambda\sigma_\infty}{1+\lambda} \quad (7)$$

弹性区的半径为

$$R^2(q) = r_0^2 \left[\frac{\sigma_r(R) + k/\lambda}{q + k/\lambda}\right]^{1/\lambda} \quad (8)$$

在弹性区与塑性区交界 $r = R$ 处的正应力及剪应力为

$$\sigma_r(R) = \frac{\sigma_\infty - k}{1+\lambda};$$

$$\sigma_\theta(R) = \sigma_\infty + \frac{k+\lambda\sigma_\infty}{1+\lambda} = \frac{k+(2\lambda+1)\sigma_\infty}{1+\lambda}; \quad (9)$$

$$\tau_{r\theta}(R) = \frac{k+\lambda\sigma_\infty}{1+\lambda}$$

对于岩体出现破坏或塑性变形的极限状态的情况,如果作为近似,设介质不可压缩:

$$\varepsilon_r + \varepsilon_\theta + \varepsilon_z = 0, \quad \varepsilon_z = 0$$

代入到变形协调方程

$$\frac{d\varepsilon_\theta}{dr} + \frac{\varepsilon_\theta - \varepsilon_r}{r} = 0 \quad (10)$$

得

$$\varepsilon_\theta = C/r^2 \quad (11)$$

其中 C 为积分常数.

岩体为流变介质,应当考虑时间因素,这时 C 为时间的函数,所以(11)式变为

$$\varepsilon_\theta = C(t)/r^2 \quad (12)$$

2 隧道围岩的流变

岩体的流变按照其变形是否收敛可以分为两类. 第一类岩体初始阶段变形较快,但是随着时间的进行变形会收敛于某一个值. 第二类岩体则没有收敛的变形值. 对于深部隧道来讲,经常遇到的是非收敛变形,因此本文研究第二类岩体.

对于第二类岩体,在坑道壁附近,当剪应力强度超过流动限时发生流动,流动引起变形的增加及应

力的调整. 对于初始时刻出现塑性区的围岩,在塑性软化区发生流动的结果是剪应力趋于长期残余强度 k_r. 而对于塑性软化区外的岩体,由于其流动限低于岩体的短期强度,所以有一部分岩体要发生流变. 如果流变区的半径为 R_1,则在此半径上

$$\tau(R_1,t)=k_r(R_1)=k_1 \quad (13)$$

对于第二类岩体其流变方程可以采用 Shvedov-Bingham 模型[1]:

$$\frac{d\varepsilon_\theta}{dt}=\frac{dT}{dt}+\frac{T-k_r}{t_0} \quad (14)$$

这里引入了无量刚应力,其为原值除以 $2G$(G 为剪切模量),T 为剪应力强度.

在外载作用下实际岩石的典型应力—变形曲线如图 1 所示.

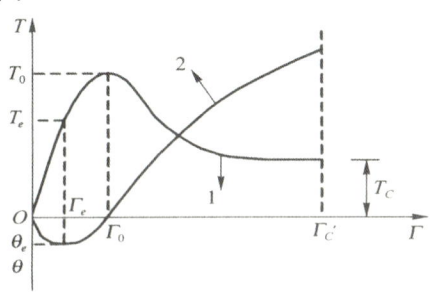

图 1 岩石的受力变形曲线(曲线 1 为剪应力—剪应变曲线,曲线 2 为体变)

由图 1 可以看出,在应力达到应力极限之后岩石开始软化,出现体积膨胀,当剪应力继续增加时体积变形迅速增加. 图 2 为具有软化及残余强度段岩体模型的计算结果. 对于图 2 所示的围岩的应力,围岩的塑性软化半径为 R_1^0,在越靠近坑道壁的地方体积变形越大. $r>R_1^0$ 的范围为弹性区. 在弹性区岩体的流动限为 k_1,k_1 的值小于岩体的强度,当剪应力超过 k_1 时,岩体会发生流变. 因此需要研究在 $r>R_1^0$ 及 $r<R_1^0$ 某个范围内岩体的流变. 下面讨论在这两个区域岩体的流变问题.

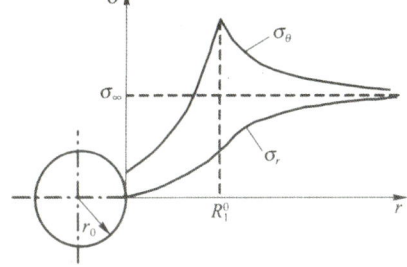

图 2 坑道围岩的应力分布(具软化及残余强度段模型计算结果)

深部岩体中在隧道壁附近的围岩中形成了塑性软化区. 在靠近坑道壁的区域,塑性变形很大,同时体积变形也很大. 由图 2 可知,在塑性变形大的区域剪切变形 Γ 与体积变形 θ 属于同一数量级,即 $\Gamma\sim\theta$

$$\Gamma=\varepsilon_r-\varepsilon_\theta\sim\theta=\varepsilon_\theta+\varepsilon_r \quad (15)$$

因此环向变形接近于零: $\varepsilon_\theta\sim0$. 因此可以把方程(14)左边的 $d\varepsilon_\theta/dt$ 等于 0: $d\varepsilon_\theta/dt=0$. 这就意味着在此区主要的过程为应力松弛的过程. 作为近似,可以把这一结论推广到整个塑性软化区. 因此由方程(14)得到下列方程:

$$dT/dt=-[T-k_r(r)]/t_0 \quad (16)$$

解为

$$T=k_r+(T_0-k_r)e^{-t/t_0}=k_r(1-e^{-t/t_0})+T_0e^{-t/t_0} \quad (17)$$

其中 $T_0=k+\lambda\sigma_r=\lambda(q+k/\lambda)(r/r_0)^{2\lambda}$ 为 $t=0$ 时的剪应力分布.

由(17)式可以看出随着时间的增加,剪应力将逐渐趋于岩体的长期残余强度.

对于围岩来讲隧道壁附近的岩体破坏最为严重,长期残余强度也小. 离坑道壁距离越远,长期残余强度越大. 作为初步近似可以取残余强度从坑道壁处的 k_0 按线性变化到流变区半径 $R_1(t)$ 处的 k_1:

$$k_r(r)=k_0+\frac{r-r_0}{R_1(t)-r_0}(k_1-k_0)=k_0-ar_0+ar \quad (18)$$

其中 $a=(k_1-k_0)/(R_1(t)-r_0)$

把(17)及(18)代入平衡方程

$$\frac{d\sigma_r}{dr}-\frac{2T}{r}=0$$

积分得流变区的径向应力

$$\sigma_r^f=[(k_0-ar_0)\ln(r)^2+2ar](1-e^{-t/t_0})+e^{-t/t_0}(q+k/\lambda)(r/r_0)^{2\lambda}+C$$

考虑到边界条件 $\sigma_r(r_0,t)=q$ 得

$$\sigma_r^f=[(k_0-ar_0)\ln(r/r_0)^2+2a(r-r_0)](1-e^{-t/t_0})+e^{-t/t_0}(q+k/\lambda)[(r/r_0)^{2\lambda}-1]+q \quad (19)$$

随着时间的增加 $t\to\infty$,径向应力趋向于下式

$$\sigma_r^f=(k_0-ar_0)\ln(r/r_0)^2+2a(r-r_0)+q \quad (20)$$

由上式可以看出径向应力由隧道壁的 q 值随着离隧道壁的距离的增加而增加.

环向应力由下式确定

$$\sigma_\theta = 2T + \sigma_r \tag{21}$$

当 $t \to \infty$ 时，由上式及(17)、(20)式有

$$\sigma_\theta = k_r + (k_0 - ar_0)\ln(r/r_0)^2 + 2a(r - r_0) + q \tag{22}$$

其在 $r = R_1^0$ 达到极值.

对于 $r > R_1^0$ 区域内的局部流变区，可近似地认为流动限为一个常数. 取流动限为 $k_r = k_1$，则在 $r = R_1(t)$ 的流变半径上，有

$$T(R_1, t) = k_1 \tag{23}$$

把(12)代入(14)式得

$$\frac{C'}{r^2} = \frac{dT}{dt} + \frac{T - k_1}{t_0}; \quad T \geqslant k_1 \tag{24}$$

把上式改写为

$$\frac{dT}{dt} + \frac{T}{t_0} - \frac{k_1}{t_0} - \frac{C'}{r^2} = 0 \tag{25}$$

对于上式进行积分得

$$T(r, t) = k_1(1 - e^{-t/t_0}) + e^{-t/t_0}\left[T(r, 0) + \frac{1}{r^2}\int_0^t C' e^{t/t_0} dt\right] \tag{26}$$

在初始时刻软化流动区外剪应力可以通过初始软化流动半径来表示. 由(6)式，因为

$$\frac{B}{R^2} = k + \lambda \sigma_r(R) = \frac{k + \lambda \sigma_\infty}{1 + \lambda} = k'$$

所以有

$$B = k'R^2,$$

因此

$$T(r, 0) = \frac{B}{r^2} = k'\frac{R_1^2(0)}{r^2} \tag{27}$$

由(22)式及(23)式得

$$k_1 R_1^2(t) - k' R_1^2(0) = \int_0^t C' e^{t/t_0} dt \tag{28}$$

这样应力软化流动区之外剪应力在任何时刻可以由软化流动区半径 $R_1(t)$ 来表示：

$$T(r, t) = k_1(1 - e^{-t/t_0}) + k_1 e^{-t/t_0} R_1^2(t)/r^2 \tag{29}$$

把(29)式代入平衡方程得

$$\frac{d\sigma_r}{dr} - 2\frac{k_1(1 - e^{-t/t_0})}{r} - 2k_1 e^{-t/t_0}\frac{R_1^2(t)}{r^3} = 0$$

积分后得到弹性区的径向应力表达式：

$$\sigma_r^e(r, t) = k_1(1 - e^{-t/t_0})\ln r^2 - k_1 \frac{1}{r^2} R_1^2(t) e^{-t/t_0} + D \tag{30}$$

为了确定系数 D，利用在流动区与弹性区边界上的径向应力的连续条件

$$\sigma_r^e(R_i, t) = \sigma_r(R_i, t) \tag{31}$$

则得

$$D = [(k_0 - ar_0)\ln(R_1/r_0)^2 + 2(k_1 - k_0)](1 - e^{-t/t_0}) + e^{-t/t_0}(q + k/\lambda)[(R_1/r_0)^{2\lambda} - 1] + q - k_1(1 - e^{-t/t_0})\ln R_1^2 + k_1 e^{-t/t_0} \tag{32}$$

由(30)式可以看出当 $t \to \infty$ 时有

$$\sigma_r = k_1 \ln\left(\frac{r}{R_1}\right)^2 + (k_0 - ar_0)\ln\left(\frac{R_1}{r_0}\right)^2 + 2(k_1 - k_0) + q = k_1 \ln\left(\frac{r}{R_1}\right)^2 + p(R_1) \tag{33}$$

这里

$$p(R_1) = (k_0 - ar_0)\ln(R_1/r_0)^2 + 2(k_1 - k_0) + q.$$

(33)式类似于理想弹塑性时的弹性区应力分布.

在流动区与弹性区边界上有

$$\sigma_r(R_1, t) = \sigma_\infty - k_1 \tag{34}$$

所以有

$$\sigma_\infty - k_1 = [(k_0 - ar_0)\ln(R_1/r_0)^2 + 2(k_1 - k_0)] \cdot (1 - e^{-t/t_0}) + e^{-t/t_0}(q + k/\lambda)[(R_1/r_0)^{2\lambda} - 1] + q \tag{35}$$

由上式可知当 $t \to \infty$ 时，流动区半径 $R_1(\infty)$ 由下式确定：

$$\sigma_\infty - k_1 = (k_0 - ar_0)\ln(R_1(\infty)/r_0)^2 + 2(k_1 - k_0) + q \tag{36}$$

(35)式可以变为如下的形式

$$e^{-t/t_0} = \frac{\sigma_\infty - k_1 - q - [(k_0 - ar_0)\ln(R_1/r_0)^2 + 2(k_1 - k_0)]}{(q + k/\lambda)[(R_1/r_0)^{2\lambda} - 1] - [(k_0 - ar_0)\ln(R_1/r_0)^2 + 2(k_1 - k_0)]} \tag{37}$$

进一步利用(36)上式变为

$$e^{-t/t_0} = \frac{(k_0 - ar_0)\ln(R_1(\infty)/R_1(t))^2}{(q + k/\lambda)[(R_1/r_0)^{2\lambda} - 1] - [(k_0 - ar_0)\ln(R_1/r_0)^2 + 2(k_1 - k_0)]} \tag{38}$$

由上式可以看出，当 $t \to \infty$ 时，流动区半径 $R_t(t) \to R_1(\infty)$.

由上式可以求出每个时刻的流动半径，进而求出围岩中的应力变形状态.

3 结论

流变是岩石的特性之一，隧道围岩的流变对于

隧道的长期稳定性影响很大，尤其对于深部隧道来讲更是如此。目前利用线性及非线性粘弹性流变模型对于隧道围岩的研究较多，初始状态多取为弹性状态。但在深部，由于初始地应力很大，在隧道开挖后，可能会马上出现塑性区。因此初始参照状态应该取为弹塑性状态，而且应该考虑塑性区及弹性区力学性质的不同，目前这一方面的研究还很不够。本文利用Shvedov-Bingham模型研究隧道围岩在初始时刻出现塑性区情况下的流变情况，并考虑塑性区及弹性区力学性质的不同，给出解析解。

参考文献：

[1] Булычев Н. С. Механика подземных сооружений [M]. Москва：Недра，1994.

[2] Рупппенейт К. В и другие. Расчёт крепи шахтных стволов [M]. Москва：Изд-во АН СССР，1962.

[3] Sulem J，Panet M，Guenot A. An analytical solution for time-dependent displacement in a circular tunnel [A]. Int. J. Rock Mech. Min. Sci & Geomech. Abstr.，1987，24：155—164.

[4] Ottosen N S. Viscoelastic-Viscoplastic formulas for analysis in modeling three-dimensional tunnel excavation [A]. Int. J. Rock Mech. Min. Sci & Geomech. Abstr.，1988，25：331—337.

[5] 金丰年. 岩石非线性流变[D]. 解放军理工大学工程兵工程学院(博士后论文)，1997.

[6] 章根德，何鲜，朱维耀. 岩石介质流变学[M]. 北京：科学出版社，1999.

深部岩体力学研究进展

何满潮[1] 钱七虎[2]

(1. 中国矿业大学(北京) 北京 100083；2. 解放军理工大学 南京 210007)

摘　要　深部岩体力学问题是关系到国家财产和人民生命安全的重大课题，也是国内外岩体力学与地下工程领域研究的焦点问题。国家自然科学基金重大项目"深部岩体力学基础研究与应用"(50490270)是我国岩体力学与采矿界截止目前唯一的一项国家自然科学基金委员会资助的重大项目。该项目在 2004~2005 年度研究工作中，在进一步明确"深部"的概念及评价体系、深部高应力场和地质构造精细探测、深部岩体基本力学特性、深部采动覆岩层移动规律及巷道稳定性控制、深部工程灾害发生机理与控制、深部岩体力学实验设备研制、深部工程现场应用研究等方面取得了重要的阶段性研究成果。

关键词　深部，岩体力学，研究进展

Summarize Of Basic Research on Rock Mechanics at Great Depth

He Manchao[1]　Qian Qihu[2]

(1. China University of Mining and Technology, Beijing 100083 China; 2. PLA University of Science and Technology, Nanjing 210007 China)

Abstract　The problem of rock mechanics at great depth that is a closely related to the national property and people's life safety is mot only a very important issue,but also a focus research subject to the field of rock mechanics and underground engineering at home and abroad. The major project of NSF of China, Basic Researches on Mechanics of Rocks at Depth and Their Application (50490270), is so far the only important one financed by Committee of National Natural Science Foundation in the field of rock mechanics and mining in China. For the research period of 2004~2005, some of significant research achievements have been made, such as further definiting the concept of deep and evaluation system, deep high stress field and fine exploration of geological construction, basic mechanics characteristics of deep rock masses, moving rules of mining cover rock at depth and stability control of roadway, deep engineering disaster emergence mechanism and control, experimental equipment manufacture of deep rock mechanics, and research of engineering application.

Key words　Deep, rock mechanics, engineering disaster, research development

1　引言

随着我国地下工程的深度不断加大，地质条件趋于复杂，工程灾害十分严重。另一方面，要保障我国国民经济正常、稳定、持续发展，今后相当长时间内将依赖于深部能源和矿产资源的开发[1~3]。我国政府已经意识到深部资源开采将关系到我国主体能源安全问题，因此，十分重视深部开采岩体力学和工程灾害控制方面的研究。2004年3月20日批准的国家自然科学基金重大项目"深部岩体力学基

基金项目：国家自然科学基金重大项目(50490270)。
本文原载于《第九届全国岩石力学与工程学术大会论文集》(2006年)。

础研究与应用"(50490270)是我国岩体力学与采矿界截止目前唯一的一项国家自然科学基金委员会资助的重大项目。本项目自 2004 年 6 月 6 日正式启动以来,有关专家、教授在深入讨论"深部"概念的基础上,围绕深部高应力场和地质构造精细探测、深部岩体基本力学特性、深部采动覆岩层移动规律及巷道稳定性控制、深部工程灾害发生机理与控制、深部岩体力学实验设备研制、深部工程现场应用研究等方面进行联合攻关,追求科学创新和技术创新,注重新理论、新原理、新发现,目前已取得了一些重要的阶段性成果。

2 深部的基本概念

深部与浅部的明显区别在于深部岩石所处的"三高一扰动"特殊地质力学环境,使深部岩体发生"五个力学特性转化",导致深部开采中以冲击地压(岩爆)、矿压显现剧烈、巷道围岩大变形、突水、地温升高、瓦斯突出(爆炸)等"六大工程灾害"为代表的一系列灾害性事故频繁发生[4]。

浅部开采时所确定的矿井类型,由于进入深部开采之后地质力学环境的改变和力学性质的转化,使得矿井类型也在六个方面发生"转型"(何满潮,2005)[4],即:①硬岩矿井向软岩矿井的转型;②低瓦斯矿井向高瓦斯矿井的转型;③非突矿井向突出矿井的转型;④非冲矿井向冲击矿井的转型;⑤低渗透压矿井向高渗透压矿井的转型;⑥低温矿井向高温矿井的转型。转型期将(已)是事故多发期。

针对原有深部概念以某一开采深度进行深部的界定所具有的局限性,何满潮(2004)结合深部工程所处的特殊地质力学环境,通过对深部工程岩体非线性力学特点的深入研究,提出了"深部"的概念[5],即"深部"是指随着开采深度增加,工程岩体开始出现非线性力学现象的深度及其以下的深度区间。在此概念的基础上,确定了临界深度(H_{cr})的力学模型及公式,建立了深部工程的评价指标[5]。

3 深部高应力场和地质构造精细探测

为了掌握深部资源开采中的地应力场分布特征与变异规律,建立精确地质结构模型,应用高分辨率三维地震勘探等探测技术,探讨了深部高应力场和地质构造精细探测方法。

(1) P-SV 转换波理论和应用。

彭苏萍等[6~8]通过研究,获得了水平单层介质中 P-SV 波转换点的唯一解析解(叠加型和叠前偏移型)和倾斜反射界面 P-SV 转换波转换点坐标的唯一解析解,表达式简捷。应用新的 P-SV 转换波的解析解改进了 P-SV 的动校正、速度分析和叠加算法,提高了计算精度。

(2) 岩层各向异性的地震波响应特征。

苑春芳等[9,10]通过对钻孔岩芯和测井曲线等多属性、多数据体的分析、融合,基本掌握了岩层各向异性的地震波响应特征,为了解深部地震波响应特征奠定了基础。

(3) 水平矿层突变失稳模型及预测技术。

彭苏萍等(2005)将水平矿层简化为两端固支的弹性地基梁模型,导出了矿层应力与位移的解析解。利用李兹(Ritz)能量方法,导出了矿层的近似挠度公式。建立了水平矿层的突变失稳模型,给出了预测方法。

(4) 深部岩体水压致裂法地应力测量。

蔡美峰、乔兰等对深部地应力测量技术进行了应用研究[11~13],在巨野煤田进行了水压致裂法地应力测量,得出了该矿区综合地应力随深度变化规律。在巨厚冲积层土体下进行地应力测量,钻孔最大测点深度在国内首次突破千米大关。

4 深部岩体基本力学特性研究

采用室内岩石力学试验与数值模拟方法，对深部环境下人为工程活动诱发工程灾害的非线性动力学演化机制与成灾机理以及高地应力下岩石破碎机理进行了深入研究。

4.1 深部岩体力学特性及其工程响应

(1) 岩石脆-延转化微观机理。

谢和平、周宏伟等[14,15]通过研究，认为温度和压力在脆性-延性转变过程中起着外因的作用，而岩石内部的微结构的变化起着内部机制的作用，特别是分析了晶体颗粒的平动和旋转、颗粒滑移与生长、键的破裂与接合的相对速率等因素在脆-延转化中的作用。

(2) 温度压力耦合作用下深部岩石流变模型。

谢和平等[16]对三种常用的流变元件(弹性元件、黏性元件和塑性元件)进行了讨论，根据温度和压力作用下的深部岩石的变形特性进行了相应的假设，研究了温度压力作用下西原模型的本构关系，得出了模型在不同条件下的蠕变方程、卸载方程和松弛方程。

(3) 温度压力耦合作用下岩石屈服破坏准则。

左建平等[17]将岩石的屈服破坏过程视为能量释放和能量耗散的过程，当能量耗散到一定程度时岩石即发生破坏失稳。根据最小耗能原理导出了温度压力耦合作用下的深部岩石屈服破坏准则。

(4) 岩石在细观尺度下的热破裂机理。

通过带扫描电镜的高温疲劳实验系统从微细观的角度研究了温度-压力耦合作用下岩石破坏的细观变形机制[18]。实验表明，在温度-压力耦合作用下，岩石的应力-应变关系在应变小于0.3%时，表现出弹性特征；当应变大于0.3%时，表现出非线性的变形特性。且在升温过程中，没有发现明显的热开裂现象，而在降温过程中却出现了裂纹，这是由于在冷却过程中局部矿物颗粒的不均质性导致了冷却过程中的变形不协调。此外，还分析了不同温度下岩石断口的表面形貌特征，以建立断口形貌和温度、压力间的关系，从而搞清岩石温度-压力耦合作用下的破坏机理。

(5) 岩石动态断裂特性。

谢和平等[19,20]利用 Hopkinson 压杆和薄圆形铝片作为波形整形器对大理岩试样进行动态劈裂试验，分析了试样的应变率，破坏时间，破坏模式，以及破坏过程中的载荷-应变关系，得到了关于大理岩在高应变率下拉伸强度、弹性模量、动态破坏应变及裂纹的起裂和扩展特性。

(6) 岩体动力学特性与碎裂诱导机理。

李夕兵等通过力学试验、数值模拟分析等研究得出[21~26]：根据应力波传播理论，在应力脉冲作用时影响近裂纹面动态应力强度因子的主要因素是应力释放区大小以及释放区外的应力分布，由此构造出计算动态应力强度因子的近似表达式，并得到了其相应的起裂角。对垂直、剪切以及斜向等各种冲击载荷作用形式下裂纹的动态响应进行了数值模拟，得到了一系列随时间而变化的动态应力场以及应变场图；根据其定义，计算了相应的动态应力强度因子，进而分析了斜向载荷作用形式下裂纹起裂情况，给出了最优断裂解析式。从波的弹性位移出发，推导了线性位移不连续模型，然后引入等效位移假设，建立了节理非线性位移不连续模型。依据所建立的非线性模型，获得了纵波在非线性法向变形节理处传播的透射和反射系数的解析解。提出了一种适合损伤阶段的新型岩石动态疲劳损伤累积的计算方法。同时，从损伤的实质物理意义出发，提出了损伤参量的抽象分形，得出分形损伤与块度的迭代关系式，由此把应力波作用下的疲劳损伤迭代关系式应用到岩石的破坏后阶段，从而得出冲击能量、岩石损伤、块度分布之间的关系。

4.2 深部巷道围岩峰后破裂演化及失稳对策研究

贺永年等通过对岩石延性变形的实验与研究[27-31]，表明岩石裂隙发育所提供的自由度是岩石延性变形条件。并据此提出软弱岩石开始具有延性转化性质的一般范围。通过岩石细观实验证明，岩石的变形总是与岩石的破裂相伴生，因此，对岩石延性变形的宏观过程以及流变变形而言，本质上具有非线性特点。

根据深部巷道围岩的实测以及试验资料的证明，提出岩石从变形到破坏的过程将使其从连续的材料性质变为具有不连续的结构性质，并且此破坏具有连续和可循环的特点。

对俄罗斯等国有关深部巷道围岩存在有呈间隔分布的断裂形式的非线性动力现象的观点，通过实测资料和现有研究资料分析，提出了国内相应的实证。并独立提出了这种规律形式的断裂和钻孔岩饼的典型高应力现象都有结构有序性的共性，也同时存在有非线性动力破坏规律的一致性。

4.3 组合载荷作用下的岩体特性

古德生、李夕兵等[32-34]利用自行研制的组合加载实验系统，采用低周疲劳加载方法，对红砂岩进行了一维和二维动静组合加载试验，研究了组合加载条件下红砂岩在不同水平静压和不同竖向静压下受不同频率和不同幅值动力扰动作用时的力学响应和破坏特征。

李夕兵等[21,32]分析了一维静加载岩石在动载作用下的损伤破坏机理，建立了受一维静加载岩石在动载作用下的弹塑性损伤本构模型。分析了岩石在不同静载下的损伤应变能释放率变化规律；将突变理论引入受一维静载岩石系统的稳定性分析之中，建立了受一维静载岩石系统在动力扰动下的非线性动力学模型；根据静力预加载结构的冲击屈曲突变理论，建立了静载岩石系统的冲击破坏模型。采用组合模型研究方法，将统计损伤模型和黏弹性模型相结合，建立了一维和多维受静载荷作用的岩石在动载作用下的本构模型——动静组合加载本构模型。

左宇军等[22,33]用应变能密度定义了岩石在动静组合载荷作用下的破坏准则，在动静组合加载本构模型的基础上，求出了受一维静载岩石在动载作用下破坏的应变能密度的临界值；将突变理论应用于分析受静载的岩石因动力扰动而导致的动态断裂的过程，建立了受静载荷作用的岩石内部裂纹在应力谐波扰动下扩展的双尖点突变模型，给出了受静载作用的岩石动态断裂新判据，从几何角度宏观描述了裂纹扩展的复杂过程。建立了动静组合加载的岩石破坏的典型例子——洞室层裂屈曲岩爆的突变模型，得出了洞室层裂屈曲岩爆在准静态破坏条件下的演化规律，建立了动力扰动下的洞室层裂屈曲岩爆的非线性动力学模型。

李夕兵等[34]以单轴动静组合加载实验分析了动静组合加载岩石破坏时所释放的能量转换，并以此为基础，在考虑岩爆损伤能量消耗的前提下，对岩爆岩块弹射速度进行了理论近似确定，其计算方法可供深部有岩爆危险巷道支护设计时参考。

4.4 深井矿岩破坏机理

(1) 随机力学理论与分析方法。

李夕兵等[35-37]建立了采矿工程岩体随开挖与充填过程变化的能量行为函数，该行为函数具有隐式函数和高次非线性功能函数的特点。提出了基于 ANN 的 FORM，SORM 和 MCS 的随机力学理论与分析方法。该方法解决了长期困扰工程随机力学分析的隐式和高次非线性功能函数分析解法的问题。

(2) 小样本条件下岩石参数的统一概率模型。

邓建等[38,39]提出了小样本条件下的地下岩体工程随机参数的统一概率模型和基于样本矩的统一概率模型推断方法，用概率加权矩、正交多项式和信息熵原理建立了小样本条件下的岩石参数的统一概率模型，提高了岩石参数概率模型的精确度。

(3) 岩石的断裂力学参数测试。

陈锋等[40]对大量试样进行了双扭试验，测试了岩石的断裂韧度和亚临界裂纹扩展速率，研究了岩石材料的应力腐蚀机理和裂纹扩展规律，确定了 KI-V 关系，得到了亚临界裂纹扩展参数。采用常位移松弛法所测的四种岩石的的亚临界裂纹扩展速率与裂纹尖端应力强度因子服从幂函数关系。

4.5 深部岩体渗流分析

唐春安等[41~43]在二维 RFPA 渗流—损伤—应力耦合模型的基础上，考虑了温度耦合，建立了三维条件下岩石非均匀性分布的随机统计分布函数，包括正态分布、Weibull 分布、平均分布模型等；建立各种情况下的岩石的细观统计损伤理论模型。编制了深部岩体多场作用模拟软件。

4.6 深部非线性岩体动力学研究

围绕深部岩体的块系构造特点、高地应力及含能特点和非协调变形的特点，钱七虎、潘一山、王明洋等对深部非线性岩体力学进行了深入研究。

(1) 岩体中存在的块系结构的等级关系。

王明洋等利用板块动力学、固体力学及协同学的基本理论研究[44,45]，揭示了来自地球外部的作用对于地球上岩体的块系结构等级形成的关系以及岩体在演化过程中，硬化过程的非平衡特性对于岩体块系结构形成的内部原因。指出岩体块系结构的形成不是偶然的，而是岩体介质在演化过程中外部作用与内部作用共同作用的结果。通过对岩体的构造等级与岩体的力学性质之间的关系进行研究，认为由于岩体的块系等级的存在，在地球的岩体中存在着大规模地质构造等级的慢速过程，一直到微观层次上的快速过程。

(2) 不同的构造层次的黏性及应变率。

研究发现[46]，不同的构造层次对应着不同的黏性及应变率。宏观层次通常对应着低的应变率及高的黏性系数，而高的应变率及低的黏性系数对应着微细观层次。通常随着外载的增加，变形速度的增大，发生变形与破坏的层次逐渐由宏观，经由细观逐步向微观层次过渡，黏性系数逐渐降低。在高应变速率区，材料呈现出流体性质，黏性与变形速率成反比。

(3) 区域破裂化现象。

王明洋等[47]借助光滑函数的组合描述岩块间间断位移场的不光滑性，揭示了岩块界面的局部变形行为以及深部围岩的大变形、稳定性及与峰值后岩体特性的关系进，阐明了深部岩体围岩的变形与破坏分析中连续介质弹塑性力学理论不能适用的物理本质。认为地下坑道周围岩石的破坏形式相似于在小的侧压力时完整试块破坏时产生的裂缝，从而形成类似于坑道外形的破坏区(坑道假外形)。在深部条件下，由于应力向深部转移，形成第二个支承压力区，这样，又出现新的破坏区，产生第二个"假外形"，这就是在深地下矿井里观察到的破坏区与非破坏区轮换的现象。围岩的区域破碎现象必须要满足最小工作深度，它是受岩石在支撑压力峰值区域内的最大主应力方向上产生的劈裂、裂缝的发展以及这个区域内岩石的破坏所制约的，它与岩体所有的最重要的性质有关，因此实践中在相同条件下某个性质发生变化，就可能产生区域破碎与没有破碎互相交替的可能。

(4) 岩块系中超低磨擦效应产生机理和规律。

王明洋等通过研究认为[48]，深部岩体中存在的残留应力大小不仅仅由变形的增降确定，也由所考察的构造集合体的单独区段的变形模量的差值所确定。块体界面的动力变形与稳定性的影响效应规律不仅与块体及界面的体变能力及块体大小密切相关,而且与其动摩擦角超密切相关。揭示了深部岩块系中超低磨擦效应产生的机理、条件和规律。

(5) 岩石的静动力学行为。

钱七虎等(2005)根据颗粒类复合材料的变形物理机制,扩展传统连续介质力学的理论框架,引入间断位移场的概念,描述材料的变形行为,发展建立了非连续局部剪切变形的弹塑性理论。认为不光滑的间断位移场可以借助一些光滑函数的组合来描述,具有二阶精度。在传统质点连续介质力学的基础上,引入了描述间断位移场变形特征的新的自由度和相应的广义力,即旋转自由度和分布力矩应力,通过已有的物理学定律建立相应的控制方程,分析材料的局部变形行为。其中质点连续介质力学描述了宏观尺度上平均的材料变形行为,有限尺度内的间断位移场的变形梯度近似描述了材料的局部变形行为。

(6) 间断面的变形规律。

钱七虎等(2005)根据最小功率耗损原理,采用约束变分方法,建立了局部变形的旋转位移场的演化发展方程,揭示了间断面的主要变形规律。在有限元理论基础上,推导建立了非连续局部剪切变形的弹塑性模型的有限元计算公式,研制了相应的计算程序,为这类问题的数值分析奠定了良好基础。

(7) 煤层变形局部化、分区碎裂化形成机制。

通过理论研究,潘一山等[49]探讨了高压高孔隙压渗流作用下巷道围岩变形局部化和分区碎裂化形成机制,得到了分区碎裂化发生的临界深度,并对其时间效应进行了深入分析。

(8) 岩石等间距破坏现象机理与实验。

潘一山等通过大量研究认为[50,51]:地表岩层在双向应力作用下变形破坏规律的相似材料模拟试验与大陆板块边界推挤作用下的边界条件相类似,实验观测到的现象与岩石中交叉节理有很多相似之处,说明用相似材料模型可以近似模拟地震地质领域中广泛存在着的岩石破裂等间距现象。水平压应力的波动传播,是导致地壳岩层先出现屈曲变形,而后形成类似棋盘格式共轭延性剪胀破裂局部化带现象的力学原因。

5 深部采动覆岩移动规律及巷道稳定性控制

采用工程地质学和现代大变形力学相结合研究方法,在理论研究的基础上,借助物理化学分析、微观结构分析、实验室岩石力学试验、相似材料模拟、数值模拟分析和现场工程地质调查分析等手段,对深部采动覆岩移动规律及巷道稳定性控制进行了深入研究。

5.1 深部采场上覆岩层移动规律与控制

(1) 深部采场覆岩的复合关键层理论。

缪协兴等[52,53]通过研究系统地建立了复合关键层的力学模型;从理论上研究了复合关键层形成的条件及其判别程序,研制开发了相应判别分析软件;用数值模拟方法,研究了复合关键层的破断规律,以及复合关键层对采场矿压、岩层移动及地表沉陷的影响;分析了复合关键层下离层的动态发展规律,为用离层注浆方法控制地表沉陷及离层区瓦斯抽放等提供新的依据。

(2) 深部短壁开采覆岩关键层的黏弹性分析及应用。

缪协兴[54]等系统研究了短壁开采覆岩关键层及保护煤柱的稳定性。通过基于煤(岩)流变的试验测定及短壁开采关键层与保护煤柱的相互作用的时间相关性分析,得到了短壁开采关键层与保护煤柱长期稳定的时间计算公式;基于岩层控制的关键层理论,建立了短壁开采覆岩关键层的力学模型,并对其进行了黏弹塑性分析,得到了关键层及保护煤柱变形随开采参数及时间的变化关系;利用数值模拟和物理模拟方法研究,得到了短壁开采的极限设计参数,并在山西潞安矿区的开采实践中得到了有效应用。

(3) 深部采场覆岩厚关键层的变形、破断规律及其控制。

陈荣华等[55]通过研究认为:厚关键层的破断、垮落规律与长梁(或薄板)矿压理论存在明显的差异,其初次破断与冒落形态多为拱形,周期破断与冒落则呈不等长的短块状。具有厚关键层的采场来压呈现多样性和随机性。不同形式的采场来压对支架的作用差别较大,大块滑落失稳对采场支架的威胁最大。同时,研究了注水软化法对控制厚关键层采场来压的作用,得到了岩层软化系数对注水软化效果的影响规律。

(4) 采动岩体的渗流理论及其应用。

缪协兴等[56~58]以峰后破碎岩体为特殊研究对象,开发了峰后破碎岩体非Darcy流的渗透试验和破碎岩石的渗流试验装置;基于试验研究结果,建立了峰后岩体渗流系统的非线性动力学方程,用谱截断法研究了其非线性动力行为特征;用概率区间分析法建立了破碎岩石渗流分析的随机微分方程,并提出了破碎岩石渗流分析的随机有限元法。

5.2 深部巷道围岩稳定控制理论

(1) 深部巷道围岩稳定控制原则。

深部岩体不仅是局部力学问题,而且是区域地质力学和局部工程力学的耦合作用力学问题。依据上述原理,针对深部开采岩体的非线性力学特性,何满潮(2004)提出了以"大稳定、中稳定、小稳定"为核心的深部巷道支护稳定性控制原则,即:要解决深部巷道围岩稳定性控制问题必须掌握"三个方面",处理好"三个关系"。

"三个方面"主要是指:大稳定、中稳定、小稳定。大稳定是指矿井开采范围内大的构造及区域应力场分布状况;中稳定是指采区范围内采动应力场的三维应力分布状态;小稳定则是指巷道工程岩体结构及工程特性。

"三个关系"则是指大稳定对中稳定的控制关系,大稳定对小稳定的控制关系,中稳定对小稳定之间控制关系。其中:①大稳定对中稳定的控制关系。主要是指矿井地应力场及地质构造应力场与采区布置之间的关系。主要考虑对采区布置有强烈影响的构造应力的大小和方向。②大稳定对小稳定的控制关系。主要是指矿井地应力场及地质构造应力场对巷道布置的控制影响。主要考虑巷道布置的方向和构造应力的方向。③中稳定对小稳定的控制关系。主要是指采动应力场与巷道布置之间的关系。要考虑使巷道布置尽可能少的受到采动应力场的影响。包括:上下关系,即上部采空区对新开巷道的影响;左右关系,即相邻回采工作面之间的关系;前后关系,即巷道受掘进与采动应力场影响的前后稳定性关系。

在处理好上述"三个关系"的基础上,采用耦合支护技术,实现巷道围岩与支护体在强度、刚度以及结构上的耦合,从而才能保证深部巷道围岩的稳定性[59,60]。

(2) 深部高应力软岩巷道支护理论与技术。

随着开采深度的增加,巷道支护技术已从被动支护(以钢架、木支架支护为代表)发展到主动支护(以锚网、锚索支护为代表)。进入深部开采以后,单纯的主动支护已无法保证巷道围岩的稳定性。通过大量的理论与现场试验研究,何满潮等[61~63]提出并建立了深部巷道耦合支护理论与技术。研究表明,深部巷道支护体与围岩实现耦合的特点及标志为:①锚杆与围岩在刚度上实现耦合,调动围岩强度超出锚杆端头范围，从而最大限度地发挥刚性锚杆的支护能力。②采用刚度及强度较高的钢筋网及复合锚杆托盘，从而能够充分转化围岩中膨胀性塑性能并能最大限度的利用围岩的自承能力。③采用锚索在关键部位进行二次支护，从而利用深部围岩强度达到对浅部围岩的控制。④围岩和支护体之间的优化组合，使支护系统达到最佳耦合支护状态，巷道围岩高应力区向低应力区转移，从而实现支护体受力与围岩变形的均匀化。

(3) 巷道围岩锚杆-注浆加固技术及应用。

根据研究，提出巷道围岩破裂后的残余强度是影响围岩稳定的一个关键因素。从破裂围岩而言，控制残余强度的主要因素是提高弱面的连接，实验表明，锚杆-注浆联合的手段在控制围岩残余强度方面有比较有效的作用。工程应用的效果也证明了这一事实[64,65]。

(4) 动载荷作用下深部巷道围岩变形失稳的动力学机制。

王连国等[66~68]针对由某种扰动载荷作用引起或诱发的煤矿巷道围岩失稳破坏，初步研究了扰动应力波在层状岩体界面附近的透射与反射叠加特性，应力波作用下巷道围岩的能量积聚特征，建立了巷道围岩冲击矿压危险性的能量密度因子判据；用非线性结构分析软件 ANSYS/LS-DANY 模拟了巷道围岩层裂屈曲失稳破坏的过程，从一个侧面揭示了煤矿冲击矿压发生的动力学诱发机制。

(5) 深部巷道围岩变形规律。

柏建彪等[69,70]通过研究认为，深部煤层巷道高应力作用于顶板和底板，在两帮相对移近过程中，导致顶板下沉和底板鼓起，两帮相对移近与底臌相互作用，即两帮相对移近促进底臌，底臌又加剧两帮移近。与浅部巷道有着显著的差别，深部巷道不仅顶板下沉，而且两帮相对移近和底臌剧烈，控制两帮相对移近和强烈底臌是深部巷道围岩控制的重要内容。

靖洪文等[71,72]通过研究得到了巷道底板围岩应力转移、上行开采上部岩层应力转移、顶底部掘巷松动围岩应力转移的理论计算式；顶、底部掘巷松动爆破的应力转移规律，巷道松动爆破区域与被保护巷道之间的关系，提出了有关参数的确定方法；掌握了上行开采应力转移的基本原理、覆岩结构状态的分带规律、作用效果以及可行程度的判别方法等；确定了巷道底板松动爆破形成的空腔半径、压碎圈半径、裂隙圈半径、震动圈半径的理论计算式及主要影响因素。

(7) 强度弱化的减冲方法。

陆采平等[73,74]采用弱化煤岩体强度的方法，降低冲击地压的威胁性。该方法一是在冲击危险区域，可以降低煤岩体的冲击倾向性，也即冲击危险性；二是在进行弱化过程中，使应力高峰区向岩体深部转移，并降低应力集中程度；三是在发生冲击矿压时，可以降低冲击的强度。

5.3 深部巷道底臌机理

根据现场观测、物理模拟和数值模拟的研究结果，姜耀东等将深井软岩巷道底臌归结为[75]：挤压流动性底臌、挠曲褶皱性底臌、剪切错动性底臌和遇水膨胀性底臌。同时，研究表明影响巷道底臌最大的因素有底板岩层性质、围岩应力、水理作用和支护强度等。计算表明，底板岩体的破碎程度对底臌量的大小起着决定性的作用，这与现场观测和物理模拟完全一致，同时底臌速度的变化趋势也和现场观测一致。

6 深部工程灾害控制

针对冲击地压、岩爆、煤与瓦斯突出等深部工程灾害，采用理论研究、数值模拟分析、现场试验研究等手段进行了深入研究，取得了如下成果。

6.1 以复合型能量转化为中心的深部巷道防冲支护思想

冲击地压发生的本质就是承受高地应力的煤岩体内，积聚大量弹性能量突然释放的过程。根据冲击地压的能量组成特征，何满潮(2004)按照能量来源不同可将冲击地压分成五类：固体能量诱发型；气体能量诱发型；液体能量诱发型；顶板垮落能量诱发型；构造能量诱发型。上述五种能量源均有可能单一的诱发冲击地压，但更为普遍的是多种成因联合作用的复合型。煤矿冲击地压和岩爆现象是煤岩体系统在变形过程中的一个稳定态积蓄能量向非稳定态释放能量转化的非线性动力学过程。尤其是进入深部开采后，巷道围岩体的变形破坏表现出明显的非线性、大变形特征；从单一能量源分类诱发冲击地压来进行冲击地压的机理研究是很难系统的摸清其发生本质，只有从上述五种能量源的组合及其之间转化特征出发，才能够较好地从能量本源角度认清冲击地压发生的机理，并对其提出行之有效的防治措施。结合以往冲击地压机理研究和实际发生条件的现场调研，可以将冲击发生机理归结为三条定律：能量聚积定律；地质弱面的能量释放定律；工程释放定律。结合具体工程条件，建立了单一重力型工作面冲击地压模型；重力+构造应力型工作面冲击地压模型；特厚坚硬顶板型工作面冲击地压模型；多工作面开采型掘进巷道冲击地压模型；沿空掘进巷道冲击地压模型等 5 种冲击地压力学模型。

6.2 深部复杂开采条件下冲击地压机理

(1) 冲击倾向煤体的细观试验。

以煤岩学、地球化学和断裂力学的观点，姜耀东等(2005)对突出煤的显微组分、微观结构和微观孔隙进行了研究，探讨了突出煤的地质成因；并且通过对 SEM 照片数字图象的分析处理，得到变形过程中形貌变化随载荷变化规律，初步解释突出煤体裂纹损伤演化的细观机理。

(2) 煤岩体结构失稳破坏特征与组合体失稳试验。

在弹塑性损伤理论计算的基础上，姜耀东等[76]提出了煤层突出的组合体力学模型，对深部围岩系统的载荷特性、承载特性、弹性聚能、释能特性、几何结构特性、材料特性等因素的作用机理进行了试验研究和理论分析，认为煤岩体突出过程中存在以能量为基础的结构失稳现象，其特征受多种因素影响。初步提出了煤/岩体突出的结构失稳机理，对深井高应力、大变形采准巷道煤层突出进行了分析。

(3) 深采煤层巷道平动式冲击失稳三维模型。

在 Lippmann H 关于冲击地压基本理论的基础上，姜耀东等[77]建立了深采煤层巷道平动式冲击失稳三维准静态模型，模型适用于顶底板为岩石的煤巷或半煤巷中发生煤层整体冲击失稳的机理分析。该模型考虑了煤层内部滑动摩擦对煤层突出的阻碍作用，分析了煤层整体突出前临界状态下的应力分布情况，得出了更合理的煤层突出范围度量指标及煤层整体突出倾向性的判定方法。结合赵各庄矿实际生产情况，分析了深部开采条件下煤层巷道动力失稳灾害的发生机理以及突出区域相关几何量之间的关系。

(4) 放炮震动诱发煤矿巷道动力失稳机理。

姜耀东等[78]发展了 J. Litwiniszyn 关于震动波诱发巷道动力失稳理论，分析了炮采震动诱发煤层巷道稳定性的影响，从理论上解释了放炮震动诱发冲击地压的根本原因。对比分析了无支护和两种不同柔性支护条件下，巷道受放炮震动影响后的变形破坏情况。

6.3 深部煤与瓦斯突出灾害控制

(1) 基于 Hoek-Brown 强度理论的煤层突出模型。

姜耀东等[79,80]结合 H.Lippmann 的煤层突出基本理论，建立了基于 Hoek-Brown 经验强度理论的煤层整体平移突出模型，计算巷道开挖以后，巷道两侧煤层内应力分布规律，并且探讨有关参数对突出的影响。分析表明，建立的突出模型体现了煤岩材料的内在性质对巷道两侧煤层稳定性的影响。

(2) 突出危险煤层的三点弯曲破坏试验与裂隙扩展的特征。

姜耀东等(2005)通过三点弯曲试验来观察和分析具有冲击倾向的煤体试件细观裂纹的扩展特征，同时分析载荷与变形的关系，研究冲击倾向性煤体在变形机理，从细观层次上研究关键承载区和破碎区的形成与失稳机制。从试验中发现，煤岩体原生裂纹/节理、新裂隙的产生及它们的扩展规律具有非线性自组织和分形混沌的特征，符合非平衡热力学规律的能量耗散结构特征。

(3) 采场围岩能量积聚特征与突出危险性能量判据。

姜耀东等(2005)以采场采动应力变化规律研究为中心，基于 FLAC3D 模拟计算分析了开采参数引起的采场支承压力的变化、岩层结构的运动和弹性应变能的积聚、释放特征，进而监测围岩位移和速度的突出特征，给出采场能量密度因子来评价煤岩突出的判据。

(4) 高地应力低渗透率下瓦斯渗流的滑脱效应、运移理论及数学模型。

潘一山等[81]建立了考虑滑脱效应及孔隙压力的瓦斯渗流模型、考虑滑脱效应及温度场耦合的瓦斯渗流数学模型、考虑滑脱效应的水气耦合瓦斯渗流数学模型。通过数值模拟分析了不同渗透率情况下综合考虑滑脱效应、孔隙压力、温度场和两相耦合作用等深部因素对瓦斯渗流场的影响，为研究非线性瓦斯渗流理论奠定了理论基础，为煤层气大规模工业化开发和产量预测提供了理论指导。唐巨鹏等[82]通过理论分析，研究了煤层储层特征，煤层瓦斯吸附、解吸、扩散、渗流机理，建立了深部煤层瓦斯运移耦合数学模型。

(5) 瓦斯突出与冲击地压统一理论。

通过理论分析和实验室试验，潘一山等[83-86]研究了高孔隙压条件下瓦斯含量与孔隙压力的关系，瓦斯对煤样物理力学性能的影响规律，瓦斯煤层冲击地压与煤和瓦斯突出的区别与联系，瓦斯煤层冲击地压发生机理及其失稳判据，并对阜新矿区深部高瓦斯煤层冲击地压的预测与防治措施进行了深入研究。

6.4 特定环境灾害发生机理与防治

(1) 灾害空区探测及其空区群级联失稳评价。

针对金属矿深部灾害空区赋存的特殊环境，邓建等[87]系统地研究了灾害空区雷达探测信号的解释系统，揭示了各种主要影响因素对雷达探测信号与图像特征的影响规律。建立了一套完整的金属矿山灾害空区雷达精细探测的实验技术。在理论与实验的基础上，解决了涉及地下灾害空区特征对雷达探测信号、图像产生影响的定量精细评判技术和方法，以及金属矿中矿岩含水率、矿石品位对雷达波传播与衰减产生的影响及其关系方程的精细化实验研究关键问题。进行了金属矿井下环境对 GPR 测试的干扰因素分析与研究，获得了雷达探测中各种环境干扰信号识别与压制技术。运用重整化群方法，从系统的角度对空区群稳定性问题进行了整体稳定性研究，给出了特定空区群临界失稳概率，给多空区矿山安全评估及其开采布置方式提供了一种理论指导方法。

(2) 深井特殊环境控制与安全预警。

以声发射技术在岩土工程中的应用为基础，结合国内外的研究成果，李夕兵等(2005)对小铁山矿脉内采准工程中采准和回采巷道的稳定性进行了实时监测，确定了不同岩性的稳固性和开挖后的松动范围和巷道的合理支护方式及支护时间。从岩石应力和声发射次数与能量的关系出发，确定了巷道开挖后破坏形式和声发射的关系，对脉内采准工程合理性和地压控制措施提供了理论指导。王卫华等[88]开展了深井 3D 可视化建模、无线传输微震监测安全预警系统研究。以金属矿深井特殊环境控制与安全预警应用基础研究为核心，建立多相多场耦合岩石动力响应理论与技术与技术及数字矿山与安全预警系统。

7 深部岩体力学实验设备研制

针对深部开采复杂地质力学环境以及深部岩体所表现出的非线性力学特性，研制开发了多套深部岩体力学实验设备，为推进深部岩体基本力学特性研究创造了条件。

(1) 深部岩体非线性力学实验系统研制。

何满潮等针对深部工程岩体所处的复杂地质力学条件及其工程开挖后的复杂受力状态，研制了"深部岩体非线性力学实验系统"[89]。该系统的主要功能为：①单轴、双轴、三轴压缩试验。②单向拉伸试验。③剪切试验。④一向拉一向压、一向拉两向压、一向压一向剪复合试验。⑤可先预设初始应力状态，然后根据施工过程卸载(先三向加载然后两向卸载或一向卸载；先两向加压然后一向卸载)观察施工过程中一点的应力状态变化，获得真正能反映实际情况的工程岩体的力学参数和本构模型。

(2) 深部岩体工程灾害模拟实验系统研制。

为了研究深部复杂开采条件下工程灾害发生过程，何满潮等(2004)研制了"深部岩体工程灾害模拟实验系统"，具有非线性边界条件任意组合和特定条件下可实现重复性模型实验的独特优点，为深部开采条件下工程灾害的发生机理及防治研究提供实验手段。

(3) 新型真三轴巷道模型实验台研制。

为了更好地研究深部高地应力条件下的巷道变形破坏问题，姜耀东等[90]设计研制了一种新型真三轴巷道平面模型试验台。研究表明，新型试验台试验技术具有整体刚度好、柔性均匀加载、成本低和模型试验周期短等特点，更适合于采矿工程中软岩巷道矿压显现的模拟试验。

(4) 与深部开采相适应的动静组合加载与卸载实验系统研制。

在深部岩石工程中，岩石在承受动载荷作用之前，已经处于高静应力或地应力状态之中，岩石的开挖实际上是处于高应力状态的岩石的人为进行的卸载和动力扰动。为了进行静应力与动载荷联合作用下岩石变形及破坏特性的研究，掌握高应力作用下的岩石在动载荷施加后的破碎过程与能量耗散规律，李夕兵等开发研制了两种与深部开采相适应的动静组合加载与卸载实验系统。并进行了水平垂直静压与低频扰动载荷下的岩石性能试验和三向静压与轴向动静组合加载试验。

(5) 核磁共振成像仪的研制及基础试验研究。

采用非磁性聚碳酸酯材料，潘一山等自主成功研制了可改变围压和孔隙压的三轴应力渗透仪。置于核磁共振仪磁体腔中，用此设备可通过直接实验手段在细观水平上观测煤样中水渗流规律，气驱水过程中水气两相流动规律。通过常规试验揭示了煤中瓦斯的吸附、解吸、扩散、渗流的规律和煤体裂隙细观结构。通过自然状态下煤试件的核磁实验、煤试件水渗流核磁成像实验、煤试件气驱水过程核磁成像实验研究了围压、孔隙压力、温度、饱和度对煤层气赋存及运移规律的影响。

(6) 深部高孔隙压下煤变形过程中电荷感应仪的研制。

为了对煤岩体变形破裂时电荷量时空变化以及电荷量与煤岩破裂过程及煤岩物理力学特性(孔隙率、孔隙分布情况，载荷、位移、强度及加载速度等)的关系进行系统研究，潘一山等研制了深部高孔隙压下煤变形过程中电荷感应仪，利用电荷感应方法揭示岩石、煤的变形破裂过程和预测地震、煤和瓦斯突出等煤岩动力灾害机理。

(7) 深部岩体区域化交替破裂现象机理及其时间进程因素的实验装置的研制。

为了揭示深部工程中出现的深部岩体区域化交替破裂现象的机理，钱七虎、王明洋等研制了深部岩体区域化交替破裂现象机理及其时间进程因素的实验装置。

8 结语

上述研究成果已在我国典型深部工程中进行了应用，如：金川镍矿、巨野万福矿、龙口柳海矿第三系软岩矿井、徐州旗山矿古生代千米深井、鹤岗兴安矿中生代矿井、冬瓜山铜矿、贵州开磷矿业总公司等。成果的应用解决现场工程技术难题，同时也极大推动了深部岩体力学的基础理论研究。

深部岩体力学问题是关系到国家财产和人民生命安全的重大课题，也是国内外岩体力学与地下工程领域研究的焦点问题。近年来，在深部资源开采过程中，由于冲击地压、瓦斯爆炸、矿井突水等工程灾害造成的重大安全生产事故在我国频繁发生，造成了巨大的生命、财产损失。其根本原因在于浅部工程中形成的理论、设计和技术体系进入到深部状态已经部分或严重失效。随着国家自然科学基金重大项目"深部岩体力学基础研究与应用"(50490270)研究工作的深入，将会为我国深部工程实践提供更多的新理论、新技术、新设计和新方法，为我国深部地下工程及资源安全、经济、合理开发与利用做出更大的贡献。

参 考 文 献

1 钱七虎. 非线性岩石力学的新进展—深部岩体力学的若干问题[A]. 第八次全国岩石力学与工程学术大会论文集[C], 中国岩石力学与工程学会主编, 北京: 科学出版社, 2004: 10-17.
2 钱七虎. 深部地下工空间开发中的关键科学问题[R]. 第230次香山科学会议, 深部地下空间开发中的基础研究关键技术问题, 2004.6.
3 谢和平. 深部高应力下的资源开采——现状、基础科学问题与展望[A]. 科学前沿与未来(第六集)[C]. 香山科学会议主编. 北京: 中国环境科学出版社, 2002: 179-191.
4 何满潮, 谢和平, 彭苏萍, 姜耀东. 深部开采岩体力学研究. 岩石力学与工程学报, 2005, 24(16): 2803-2813.
5 何满潮. 深部的概念体系及工程评价指标. 岩石力学与工程学报, 2005, 24(16): 2854-2858.
6 彭苏萍, 谢和平, 何满潮等. 沉积相变岩体声波速度特征的试验研究. 岩石力学与工程学报, 2005, 24(16): 2831-2837.
7 Yuan Chunfang, Peng Suping, An exact solution of the coordinates equation of conversion point for P~SV converted waves in a horizontal reflector. Chinese Journal of Geophysics, 2005, 48(5): 1261-1267.
8 苑春方, 彭苏萍. 水平界面上 P~SV 转换波转换点的精确解. 地球物理学报, 2005, 48(5): 1179-1184.
9 苑春方, 彭苏萍, 张中杰等. Kelvin~Voigt 均匀黏弹性介质中传播的地震波. 中国科学, D辑, 2005, 35(10): 957-962.
10 彭晓波, 彭苏萍, 詹阁等. P波方位 AVO 在煤层裂缝探测中的应用. 岩石力学与工程学报, 2005, 24(16): 2960-2965.
11 Meifeng Cai. Techniques for in-situ stress measurement at great depth. Journal University of Science and Technology, 2004, 11(6): 486-488.
12 乔兰, 欧阳振化, 来兴平等. 三山岛金矿采空区地应力测量及其结果分析. 北京科技大学学报, 2004, 26(6): 569-571.
13 Tan Zhuoying, Cai Meifeng. Multi-factor sensitivity study of shallow unsaturated clay slope stability. Journal University of Science and Technology, 2005, 12(3):193-202.
14 左建平, 周宏伟, 鞠杨等. 深部岩石脆性-延性转化机理研究. 深部资源开采基础理论研究与工程实践, 2005: 102-111.
15 周宏伟, 谢和平, 左建平. 深部高地应力下岩石力学行为研究进展. 力学进展, 2005, 35(1): 91-99.
16 谢和平, 左建平, 周宏伟. 温度压力耦合作用下深部岩石流变模型的本构研究. 深部资源开采基础理论研究与工程实践, 2005, 57-65.
17 左建平, 谢和平, 周宏伟. 温度压力耦合作用下岩石的屈服破坏研究. 岩石力学与工程学报, 2005, 24(16): 2917-2921.
18 Wang XS, Wu BS, Wang QY. SEM online investigation of microcrack characterizatics of concretes at various temperatures. Cement and Concrete Research, 2005, 35(7): 1385-1390.
19 谢和平, 左建平. 岩石断裂破坏的声发射机理初探. 深部资源开采基础理论研究与工程实践, 2005: 66-73.
20 李伟, 谢和平, 王启智. 大理岩动态拉伸强度及弹性模量的 SHPB 实验研究. 实验力学, 2005, 20(2): 200-206.
21 赵伏军, 李夕兵. 动静载荷耦合作用下岩石破碎理论及试验研究. 岩石力学与工程学报, 2005, 24(8): 1315-1320.
22 左宇军, 李夕兵. 动静组合载荷作用下岩石失稳破坏的突变理论及模型与试验研究. 岩石力学与工程学报, 2005, 24(5): 741-746.
23 凌同华, 李夕兵. 多段微差爆破振动信号能量分布特征的小波包分析. 岩石力学与工程学报, 2005, 24(7): 1117-1122.
24 周子龙, 李夕兵. 岩石类 SHPB 实验理想加载波形的三维数值分析. 矿冶工程, 2005, 25(3)L 18-20.

25 李夕兵,周子龙,王卫华.运用有限元和神经网络为SHPB装置构造理想冲头.岩石力学与工程学报,2005,24(23):4215-4219.
26 周子龙,李夕兵,龙八军.岩石SHPB试验信号的小波包去噪.岩石力学与工程学报,2005,24(s1):4779-4784.
27 韩立军,贺永年. Experimental Study on Mechanical Characteristics of Cracked Rock Mass Reinforced by Bolting and Grouting. J. China Univ. of Mining & Tech. (English Edition), 2005, 15(3): 177-182.
28 邵鹏,张勇,贺永年等.岩石爆破逾渗断裂行为与块度分布研究.中国矿业大学学报,2004,33(6):638-640.
29 邵鹏,张勇,贺永年等.断续节理岩体动态疲劳损伤研究.岩土工程学报,2005,27(7).
30 周刚.大屯矿区地应力测试与分析.煤炭学报,2005,30(2).
31 窦林名.煤岩体蠕变突变模型.中国煤炭,2005,31(1):37-40.
32 X. Li, C. Ma, et al, Experimental study of dynamic response and failure behavior of rock under coupled static-dynamic loadings, Proceedings of ISRM International Symposium, 3rd ARMS, Ohnishishiv Aoki (eds), Millpress, Rotterdam, 2004: 891-895.
33 左宇军,李夕兵,受静载荷作用的岩石动态断裂的突变模型,煤炭学报,2004,29(6):654-659.
34 李夕兵,左宇军,马春德.动静组合加载下岩石破坏的应变能密度准则及突变理论分析.岩石力学与工程学报,2005,24(16):2814-2824.
35 李夕兵,刘志祥.基于重构相空间充填体变形规律的灰色预测研究,安全与环境学报,2004,4(6):54-57.
36 刘志祥,李夕兵.充填体力学参数的混沌优化反分析研究,湖南科技大学学报,2004,19(4):14-17.
37 刘志祥,李夕兵.充填体变形的混沌时序重构与神经网络预测,矿冶工程,2005,25(1):16-19.
38 Deng Jian, Li Xibing. A distribution-free method using maximum entropy and moments for estimating probability curves of rock variable. International Journal of Rock Mechanics and Mining Sciences, 2004, 41(3): 376-381.
39 邓建等.确定可靠性分析Weibull分布参数的概率加权矩法.计算力学学报,2004,21(3):137-142.
40 Chen Feng, Cao Ping. A mode II fracture analysis to double edge-cracked brazilian disk using the weight function method.Int.J.of Rock Mech. and Min. Sci. 2004, (12): 461-465.
41 L.C. Li, C.A. Tang, T.H. Yang. Numerical approach to fractures saturation behavior in heterogeneous material subjected to thermal loading. The International Conference on Computational Methods, December 15~17, 2004, Singapore, International Journal of computing methods.
42 T.H. Yang, L.C. Li, L.G. Tham, C.A. Tang, Micromechanical model for simulating the hydraulic fractures of rock, ICMM2004, Singapore. International Journal of computing methods.
43 Chunan Tang, Zhengzhao Liang, Yongbin Zhang and Tao Xu Three-Dimensional Material Failure Process Analysis, the Asian Pacific Conference for Fracture and Strength, Key Engineering Material, 2005.
44 王明洋,钱七虎,戚承志.深部岩体的构造和变形与破坏问题,岩石力学与工程学报,2006,25(3).
45 戚承志,钱七虎,王明洋.岩体的构造层次及其成因.岩石力学与工程学报,2005,24(16):2838-2846.
46 戚承志,钱七虎,王明洋.岩体的黏性与其构造层次.岩石力学与工程学报,2005,26(增):4679-4687.
47 王明洋,邓宏见,钱七虎.岩石中侵彻与爆炸作用的近区问题研究.岩石力学与工程学报,2005,24(16):2859-2863.
48 王明洋,戚承志,钱七虎.深部岩体块状介质变形与运动特性研究.岩石力学与工程学报,2005,24(16):2825-2830.
49 Yi-Shan Pan and Zhong-Hua Li. Analysis of rock structure stability in coal mines. International Journal for Numerical and Analytical Methods in Geomechanics, 2005, 29(10): 1045-1063.
50 潘一山,赵扬锋,马瑾.中国矿震受区域应力场影响的探讨.岩石力学与工程学报,2005,24(16):2847-2853.
51 李忠华,潘一山.断层冲击地压的影响因素与震级分析.岩石力学与工程学报,2005,24(S1):5206-5210.
52 缪协兴,陈荣华,浦海等.采动覆岩厚关键层破断与冒落规律分析.岩石力学与工程学报,2005,24(8):1289-1295.
53 缪协兴,茅献彪,孙振武等.采动覆岩中复合关键层的形成条件与判别方法.中国矿业大学学报,2005,34(5):547-550.
54 Feng Meimei, Miao Xiexing, Mao Xianbiao and Xu Jinhai. Study on the Law of Overlaying Key Strata Motion in Short Wall Mining. Progress in Safety Science and Technology (Vol.V) Part B, Proceedings of Asia Pacific Symposium on Safety 2005, November 2~4, 2005, Shaoxing, Zhejiang, China: 1578-1582.
55 陈荣华,钱鸣高,缪协兴.注水软化法控制厚关键层采场来压数值模拟.岩石力学与工程学报,2005,24(13):2266-2271.
56 缪协兴,刘卫群,陈占清.采动岩体渗流理论.北京:科学出版社,2004.
57 W. Q. Liu, X. X. Miao. Experiment testing and simulation for water flow through a crushed rock area with pressure. The 8TH International symposium on fluid control, Measurement and Visualization, China, 2005: 298-1-298-6.
58 W. Q. Liu, X. X. Miao. Gas flow visualization in a gob area with J-type ventilation. The 8TH International symposium on fluid control, Measurement and Visualization, China, 2005: 314-1-314-6.
59 何满潮,郭志飚,任爱武,胡永光.柳海矿运输大巷返修工程深部软岩支护设计研究.岩土工程学报,2005,27(9):977-980.
60 何满潮,胡永光.深部第三系软岩巷道交岔点稳定性及其支护对策研究.建井技术,2005,26(3~4):32-35.
61 孙晓明,何满潮,冯增强.深部松软破碎煤层巷道锚网索支护技术研究.煤炭科学技术,2005,33(3):47-50.
62 孙晓明,何满潮.深部开采软岩巷道耦合支护数值模拟研究.中国矿业大学学报,2005,34(2):166-169.
63 王树仁,何满潮,范新民.JS复合型软岩顶板条件下煤巷锚网支护技术.北京科技大学学报,2005,27(4):390-394.
64 王连国,李明远,王学知.深部高应力极软岩巷道锚注支护技术研究.岩石力学与工程学报,2005,24(16):2889-2893.
65 Mao Xianbiao, Feng Meimei, Yang Jing. Study on the mechanism of stress relief of the underground rooms caused by underground mining. Progress in Safety Science and Technology(Vol.V)Part A, Proceedings of Asia Pacific Symposium on Safety 2005, November 2~4, 2005,

Shaoxing, Zhejiang, China: 781-786.

66　王连国, 缪协兴, 董健涛等. 深部软岩巷道锚注支护数值模拟研究. 岩土力学, 2005, 26(6): 983-985.
67　雷光宇, 卢爱红, 茅献彪. 应力波作用下巷道层裂破坏的数值模拟研究. 岩土力学, 2005, 24(6): 1477-1480.
68　徐金海, 缪协兴, 张晓春. 煤柱稳定性的时间相关性分析. 煤炭学报, 2005, 30(4): 433-437.
69　柏建彪, 侯朝炯. 沿空掘巷窄煤柱稳定性数值模拟研究. 岩石力学与工程学报, 2004, 23(20): 3475-3479.
70　柏建彪. 空巷顶板稳定性原理及支护技术研究. 煤炭学报, 2005, 30(1): 8-11.
71　靖洪文等. 深埋巷道围岩稳定性分析与控制技术研究. 岩土力学, 2005, 26(6): 876-880.
72　许国安, 靖洪文. 煤矿巷道围岩松动圈智能预测研究. 中国矿业大学学报, 2005, 34(2): 152-155.
73　陆采平, 窦林名. 基于能量机理的卸压爆破效果电磁辐射检验法. 岩石力学与工程学报, 2005, 24(6): 1014-1017.
74　陆采平, 窦林名. 煤岩三轴围压钻孔损伤演化冲击实验模拟. 煤炭学报, 2004, 29(6): 659-662.
75　姜耀东, 赵毅鑫, 刘文岗. 深部开采中巷道底臌问题的研究. 岩石力学与工程学报, 2004, 23(14): 2396-2401.
76　姜耀东, 刘文岗, 赵毅鑫. 开滦矿区深部开采高应力围岩稳定性研究. 岩石力学与工程学报, 2005, 24(11): 1857-1862.
77　姜耀东, 赵毅鑫, 刘文岗, 祝捷. 深部煤层巷道平动式冲击失稳三维模型研究. 岩石力学与工程学报, 2005, 24(16): 2864-2869.
78　姜耀东, 赵毅鑫, 宋彦琦等. 放炮震动诱发煤矿巷道动力失稳机理分析. 岩石力学与工程学报, 2005, 24(17): 2968-2973.
79　Y. D. Jiang, Y. X. Zhao & W. G. Liu. Numerical simulation of joint and stiffness effects on coal bumps. Mining Science & Technology, A.A.Balkema. 2004: 459-462.
80　姜耀东, 祝捷, 赵毅鑫. 基于Hoek~Brown强度理论的煤层突出模型研究. 深部资源开采基础理论研究与工程实践, 2005: 222-227.
81　肖晓春, 潘一山. 考虑滑脱效应的煤层气渗流数学模型及数值模拟. 岩石力学与工程学报, 2005, 24(16): 2966-2970.
82　唐巨鹏, 潘一山. 煤层气赋存和运移规律的NMRI研究. 辽宁工程技术大学学报, 2005, 24(5): 674-676.
83　潘一山, 李忠华, 唐鑫. 阜新矿区深部高瓦斯矿井冲击地压研究. 岩石力学与工程学报, 2005, 24(S1): 5202-5205.
84　TANG Ju-peng, PAN Yi-shan, LI Ying-jie. Numerical simulation of deep-level rockburst in Fuxin coalfield. JOURNAL OF COAL SCIENCE & ENGINEERING (CHINA), 2005, 11(1): 13-16.
85　肖晓春, 潘一山, 王秋香. 影响剪切梁层间失效模型的参数分析. 辽宁工程技术大学学报, 2005, 24(3): 354-356.
86　李英杰, 潘一山, 唐巨鹏, 唐鑫. 五龙矿冲击地压危险区划分研究. 矿山压力与顶板管理, 2005, 22(1): 94-96.
87　Jian Deng, et al. Radial basis function network approach for structural reliability analysis. 9th ASCE Joint Specialty Conference on Probabilistic Mechanics and Structural Reliability, July, 2004, Albuquerque, New Mexico, USA: 26-28.
88　王卫华, 李夕兵, 胡盛斌. 模型参数对3DEC动态建模的影响. 岩石力学与工程学报. 2005, 24(s1): 4790-4797.
89　孙晓明, 何满潮, 刘成禹等. 真三轴软岩非线性力学试验系统研制. 岩石力学与工程学报, 2005, 24(16): 2870-2874.
90　姜耀东, 刘文岗, 赵毅鑫. 一种新型真三轴巷道模型试验台的研制. 岩石力学与工程学报, 2004, 23(21): 3727-3731.

高放废物深地质处置中的多场耦合与核素迁移

罗嗣海[1,2]，钱七虎[1,2]，李金轩[1,2]，周文斌[1,3]

(1.东华理工学院，江西 抚州 344000；2.解放军理工大学，南京 210007；3.南昌大学，南昌 330029)

摘　要：概括了深地下工程的深地质处置库的若干特点，简要介绍了多场耦合的机理与类型，论述了高放废物深地质处置多场耦合与核素迁移问题的特点及研究现状，最后提出了高放废物深地质处置多场耦合与核素迁移所需研究的主要问题。

关　键　词：高放废物；深地质处置；多场耦合；核素迁移

中图分类号：TU 473.2；TB 115　　　**文献标识码**：A

Multi-field coupling and nuclide transport in HLW geological disposal repository

LUO Si-hai[1,2], QIAN Qi-hu[1,2], LI Jin-xuan[1,2], ZHOU Wen-bin[1,3]

(1.East China Institute of Technology, Fuzhou 344000, China;
2.PLA University of Science and Technology, Nanjing 210007, China; 3.Nanchang University, Nanchang 330026, China)

Abstract: The features of deep geological disposal repositories as a deep underground engineering are firstly summarized. Then the concept, mechanism and categories of multifeild coupling are introduced; and the characteristics and the state-of-the-art of THMC coupling and nuclide transport in HLW geological disposal repositories are analyzed. Finally, issues related to the THMC coupling and nuclide transport for HLW disposal are stated.

Key words: high level radioactive waste(HLW); deep geological disposal; multi-field coupling; nuclide transport

1　概　述

高放废物的安全处理和处置是制约核工业发展中的重要问题，世界各有核国家都极为重视，从政治、经济、科技与法律等方面进行研究。我国也和其它有核国家一样，已面临如何处置核废料的问题，寻求安全、有效、永久性处置高放废物也是我国应尽早在国家层面重视的重大课题之一。经过近 50 年来的研究，通过多种处置方案的分析和对比，目前，世界各国普遍认为深地质处置是高放废物最现实可行和安全可靠的处置方案，并对此进行了多学科、长时间、强投资的研究。

高放废物的深地质处置是指将固体形式的高放废物埋置在地下 500～1 000 m 的地质体中，即通过地表打竖井至深部、由竖井底部开凿水平坑道，再在水平坑道中打竖井或支坑道，作为废物的存放场所，最后进行封存，从而使之能长时间（至少 1 万年）与人类生存环境隔离。这些坑道和竖井便构成地下处置库。

从形式上看，高放废物深地质处置的处置库也是一项深部大型地下工程，但这类地下工程与一般地下工程相比有其特点，概括地讲，表现在下列几个方面：

（1）从时间跨度上来看，要求处置库的安全期限至少在 1 万年，这样长的时间尺度要求超越了一般意义下社会或技术活动所涉及的时间尺度，也使性能评价中不确定性成为一个重要的必须考虑的因素和研究领域；

（2）从作用因素上看，处置库不仅要经历开挖和运营期间的力学扰动，更重要的还将长时间受放射性辐射和衰变热的作用，因此，温度场成为一个重要的因素；

（3）从评价目标上看出，不仅要评价处置库的区域稳定性和围岩的力学稳定，特别重要的是还要

修改稿收到日期：2005-01-25。
本文原载于《岩土力学》(2005 年第 26 卷增刊)。

保证核素在其有害的年限内不致迁移到生物圈而危害人类生态环境,因此,化学场和核素迁移规律的研究具有特别重要的意义;

(4) 从研究的空间范围看,由于要跟踪核素迁移到生物圈的途径并进行安全性评价,因此,其评价的空间范围不仅限于受机械扰动的围岩,更包括从处置库到核素释放到生物圈的整个地质体;

(5) 从社会影响上看,由于核问题的敏感性和公众的反核情绪,高放废物处置库不仅是一项纯技术性的地下工程,更是一项政治和社会关注的工程,必须向公众和监管机构提供充分的论证;

(6) 从工程数量看,一般一个国家首先考虑 1 个全国性的处置库,因此工程数量少,工程积累的经验和借鉴的可能性相对少。

在开挖和运行阶段,高放废物处置库与一般深部地下工程面临相似的问题,即施工的安全与围岩的力学稳定问题,可用一般的深部岩体力学理论与方法评价和解决。但处置库关闭后长时间尺度下,高放废物深地质处置安全性评价的核心问题是长寿命放射性核素与环境隔离性能的评价。处置库中的核素可能由于下列两种原因进入生物圈:① 由隔离失效引起:比如人类的直接介入;地壳上升、侵蚀作用等因素使处置库直接暴露于地表;处置库周围地震、断裂活动、火山活动等。② 由于地下水的作用引起:地下水对废物固化体进行分解后,核素随地下水在回填材料和天然屏障中迁移,最后进入生物圈。

由隔离失效导致的核素返回生物圈一般可通过选择相对稳定的处置场址来解决。一般认为,处置库安全的关键因素是地下水运动及随地下水运动的核素迁移问题。因此,核素迁移是高放废物深地质处置安全与系统性能评价中特殊而关键的核心问题。同时,处置库的围岩必须能满足处置库长期的力学稳定。

处置库关闭后,由于高放废物的放热引发热传输过程,在开挖扰动的基础上进一步诱发新的围岩应力与变形过程,改变流体渗流过程和近场的地球化学过程,同时,放射性核素随时间在不断衰变,因此,此时不同于一般的深地下工程的围岩稳定和裂隙岩体中溶质迁移情形,关闭后高放废物深地质处置中的围岩稳定与核素迁移,特别是近场的岩体稳定与核素迁移,是长时间尺度、温度不断变化情景中,在温度场、渗流场、应力场和化学场相互耦合的过程中发生的,对此时围岩稳定与核素迁移的正确评价必须研究多场耦合。

2 多场耦合的机理、类型及高放废物深地质处置多场耦合与核素迁移的特点

2.1 多场耦合的基本概念

自然界中存在着不同的物理、力学与化学过程,如果这些过程间存在着相互影响、即一个过程的发生与发展将会受到或(和)影响到另一个过程的发生与发展,则称这些过程为耦合作用过程,这种现象称为多场耦合。多场耦合在数学上表现为过程的控制方程上存在交叉项、或表达某过程的参数受到其它过程机理的影响。

2.2 多场耦合的类型与机理

自然界中存在着多种物理、力学与化学过程,在岩体力学、水资源开发、环境保护中涉及的主要有介质的应力与变形过程(M)、流体(液体与气体)的流动过程(H)、热传输过程(T)和化学过程(化学反应与溶质迁移过程C)。应力与变形过程的主要现象是岩石的应力、变形、损伤、强度与破坏;裂隙的产生、扩展、贯通、损伤及错动;主要的原因是原位应力、地壳运动、重力作用及开挖等。渗流过程的主要现象是:岩石与裂隙中的流体流动;主要形式有:地表水入渗、地下水运动、海水入侵、能源储存中的油/气流,地热田中冷/热水的抽取与注入。热过程的主要表现是由于天然或人为热源引发热传导、热对流;主要原因是:放射性废物的衰变热、地热梯度、冷热水的注汲、永入冻土的冻融等。化学过程主要是反应与非反应性颗粒与溶质迁移,流-固相互作用;主要有污染物迁移(吸附/阻滞、扩散、平流)、固相溶解与沉淀、气体溶解与析出、海水入侵、富氧地表水下渗、材料腐蚀与风化等。

广义而讲,自然界中这些过程总是不同程度地同时存在和相互影响,即总是处在四场全耦合的问题。在许多情形下,有些过程作用不明显,其耦合可忽略不计,因此,只有部分过程产生耦合,从组合上来说,可能存在的各种组合过程共有如下 11 种:

T-H 耦合:典型的例子是地热系统中的热水运移。耦合的机理表现为:一方面,T 对 H 有影响,即温度的变化可引起流体浮力、粘性和渗透系数的变化、流体的相变及引发热扩散;另一方面,H 对 T 有影响,即流体流动将产生热对流,改变温度与热流。

T-M 耦合:如地下开挖中的通风、地下储库中

的霜冻或发热。T 对 M 的影响表现为：温度变化引起介质的机械性质变化，引发岩石的膨胀或收缩并产生热应力，引发裂隙的张闭、损伤及不可恢复的变形。M 对 T 的影响表现为：应力状态变化引起周围介质密实程度变化，从而间接影响传热；变形过程中的机械功转变成热量。

T-C 耦合：温度对化学场的影响表现在由于温度改变流体的密度、粘性及化学势，使反应速率、元素、矿物和反应过程的化学稳定性表现出温度效应；而化学场的变化通过反应的吸热与放热，将改变温度及热流，但这种影响一般相对较小而可忽略。

H-M 耦合：坝基中的渗透与变形、斜坡中的排水与稳定、土的固结、抽取地下水导致的地面沉降与开裂、岩石中的水力压裂等问题可归入此类耦合。渗流场对应力应变场的影响主要表现在固相变形受有效应力控制、裂隙的张开度和刚度与流体压力相关，毛细压力与膨胀压力参与应力平衡且其大小受到饱和度相关；而应力场对渗流场影响主要是应力引起岩石孔隙度、裂隙张开度和连通性的变化并改变岩体的渗透性能。

H-C 耦合：渗流场对化学场的影响表现在流体压力、流速、饱和度及水分变化对固-气溶解、沉淀和溶质阻滞的影响；化学场对渗流场的影响表现在由于气体溶解与析出、化学反应导致的固相溶解与沉淀引起流体粘性和渗透性能的改变。

M-C 耦合：这种耦合实际上是间接的，应力场变化引起的变形、损伤及破裂，可能引起水-岩接触面积变化，引起溶质迁移路径变化及从而影响化学场；而化学场对应力场的影响则是通过化学反应导致固相强度、变形性能的改变而产生的。

T-H-M 耦合：见于深部封闭隧道与采矿、冻结与冻土、储能中的注气、地热开发、核废物地下处置。

T-H-C 耦合：此耦合作用在水热系统中广泛得到研究，主要涉及水热系统中的化学作用问题。

M-H-C 耦合：地基处理时的化学灌浆、地下污染和处理属此种耦合。

T-M-C 耦合：典型的例子是介质的热力变形影响到化学场。

T-H-M-C 耦合：化学采矿、核废物地质处置属于此类耦合。这是一个非常复杂的耦合过程，由于其复杂性，人们有时将其进行分解研究。

2.3 高放废物深地质处置中的主要作用过程和多场耦合与核素迁移的特点

由于地下贮存场地的开挖与建造、地下水的存在以及关闭后的放射性放热效应，高放废物地下处置中涉及以下诸物理-化学过程的相互作用：

(1) 开挖引发介质应力与变形过程及渗流过程，产生 H-M 耦合：由于高放废物处置库的开挖，在围岩中产生变形和附加应力，引起岩石产生区域碎裂和岩体渗透率及裂隙张开度的变化，改变地下水渗流场；流场内压力对岩体应力、变形产生反作用。

(2) 长时间、变强度放射引发热传输过程、热应力变形过程、渗流过程和地球化学反应与溶质迁移过程，产生近场的 T-H-M-C 耦合：处置库关闭后，由于放射性衰变，核废物中随着时间不断放热（几千年至一万年），并在人工屏障和开挖改造后的岩体中传输，改变回填材料和围岩中的温度，引发岩体体积膨胀和热应力，改变流体的性状和运动，破坏水-岩之间及水-缓冲材料间的化学平衡并产生新的地球化学作用；而地球化学作用-渗流-应力变形间产生相互作用并反作用于温度场。

(3) 核素的释放与迁移过程：耦合作用产生新的地球化学环境，引发在多场耦合环境条件下，核素从废物体释放并越过工程屏障随地下水在天然屏障中迁移，最终进入生物圈，其中发生包括对流、扩散、吸附等各种物理化学作用。

(4) 与其它耦合和溶质迁移问题相比，高放废物深地质处置的多场耦合与核素迁移体现下列几个特点：

① 耦合的时间尺度非常长，要求的安全期限至少在 1 万年，因此，不确定性明显；

② 耦合的介质复杂，包括人工回填材料和经历开挖扰动的低渗透性各向异性、非均质性的裂隙岩体；

③ 耦合的空间跨度大，包括处置库和从废物体到生物圈之间人工材料与地质体，空间不同区域，要考虑的作用过程不同，近场是明显的 T-M-H-C 四场耦合，远场主要是溶质运移与地下水运动的 H-C 耦合；

④ 耦合时热源不断衰减，导致温度变化明显；

⑤ 渗流通道组成复杂，既包括地质历史过程中形成的初始结构面，又包括开挖扰动区域化碎裂产生，甚至包括热应力改造的结构面；

⑥ 化学场具有特别重要的作用，表现在因温度和湿度变化导致的近场的化学作用引发的化学成分变化对固化体的溶解与核素的析出具有重要作用，还表现以核素为核心的溶质迁移时处置库性能与安全评价中的核心内容。

⑦ 不同于一般的溶质运移问题,高放废物中的核素的迁移需经过从固化体中析出和近场释放、迁

移是处于复杂的四场耦合环境中和迁移时核素处于不断衰变中。

3 多场耦合与核素迁移问题研究概况

3.1 多场耦合问题研究概况

耦合问题的研究可从 20 世纪 30 年代太沙基发表一维渗透固结理论算起，但在国外引起广泛关注和取得重要进展则是在最近 20 多年，国内则是近十来年的事，这些研究主要源于核废物深地质处置、油/气与地热的开发和能源储存及环境保护的需要。

从研究方法上，多场耦合的研究方法包括理论分析、室内实验、原位实验与监测，研究内容则涉及多场耦合的基本理论、数值方法、计算程序开发及其工程应用等。

室内实验主要有应力与渗流间、温度与应力间、化学与应力间及温度-应力-渗流间的耦合实验研究，通过实验建立了一些相关关系，如温度对岩体力学性状、应力对岩体渗透性状及流体化学成分对岩体的破坏与损伤方面的影响。

理论分析主要是依据固体力学的基本原理和岩石介质的特性，建立各种情形下多场耦合的基本控制方程和定解条件。研究较多的主要是应力-渗流、温度-应力、渗流-化学（溶质运移）、温度-应力-渗流等耦合问题。耦合的类别包括全耦合和部分耦合。方程的类别包括描述各种过程的方程和描述不同过程耦合关系的方程。建立的基本依据包括固相介质的本构方程、几何方程和动力平衡方程，流体的运移规律与质量守恒及热力方面的传输与能量守恒定律。模型类别包括等效连续介质模型、双重介质模型等。

数值方法和程序开发主要是对提出的各类模型建立相应的数值计算方法并开发出相应的计算机程序。提出的主要数值方法有有限元（FEM）、有限差分（FDM）、离散单元（DEM）、离散裂隙网络（DFN）等，开发了多个有较大影响的程序。

在应用方面，多场耦合应用到多个工程领域。比如基础工程、隧道工程、边坡工程、铁路与高速公路工程、油/气储存工程、煤炭开采工程、固结与地面沉降、地热开采、化学采矿及废物处置等。

在现场实验及监测方面，主要通过大规模的现场监测获取资料，作为理论建立的依据及验证理论研究成果的手段。在高放废物深地质处置的地下实验室曾进行大规模和较长时间的加热实验，系统测量了相应的应力、变形和温度分布。如美国能源部的尤卡山（Yucca Mountain）国家高放废物处置库（凝灰岩）、加拿大的白壳（Whiteshell）地下实验室（花岗岩）、法国核废料地质处置实验室（页岩）、瑞典的阿斯泼（Aspo）地下硬岩实验室（花岗岩）、瑞士的格瑞穆塞尔（Grimsel）地下实验室（花岗岩）、德国的高勒本（Corleben）地下实验室（岩盐）、比利时的莫尔（Mol）地下粘土实验室等。

在多场耦合的研究项目方面，开展了几项重大和具有广泛影响的项目，如瑞典核能监察局（SKI）先后组织的三项国际合作项目：INTRACON（1981～1986 年）、HYDROCOIN（1984～1990 年）和 INTRAVAL 及自 1992 年起的国际合作项目 DECOVALEX；美国的尤卡山项目等。

DECOVALEX 是专为高放废物深地质处置的性能与安全评价而开展的一项国际合作项目，该项目始于 1992 年，迄今已有来自 10 个国家的 15 个研究机构参与了该项目。该项合作研究旨在建立 THM 的模型与算法、开发相关的程序并将计算与现场及室内实验结果进行比较，从而设计新的实验来支持程序开发，研究 THM 模拟在性能与安全评价中的作用。DECOVALEX 中研究的问题分为 2 类：一类称为基准试验（benchmark test—BMT），它是假设的一些 THM 问题，目的在于比较不同数学模型与程序的计算结果；另一类称为试验实例（test case—TC），它是原位或现场实验，用它来验证明程序的可靠性。迄今为止，DECOVALEX 已进行了三期，采用离散和连续方法、应用多个程序（MOTIF，THAMES，ADINA-T，ROCMAS，CHEF，HYDREF，VIPLEF，TRIO-EF，CASTEM2000，UDEC 等）研究了多个两场耦合及 THM 三场耦合问题。在今年拟开始的第 4 期任务，拟将水化学场列入，从而以水流-应力-化学或温度-水流-化学耦合过程作为主要研究内容。

3.2 高放废物深地质处置核素迁移的研究概况

作为高放废物深地质处置的围岩中含有不规则的交错裂隙，它们构成溶解于地下水中的放射性核素的主要迁移途径。裂隙岩体中核素迁移研究，包括试验与模拟研究，已成为高放废物地质处置系统中核素迁移研究的最主要内容。

在试验研究方面，作为地质屏障中核素迁移研究最主要内容之一，已开展的裂隙岩体中核素迁移试验研究主要涉及迁移机理研究和迁移参数测定。核素迁移试验研究从规模和形式上大体可分为 3 种类型。

（1）岩样示踪试验：在专门设计的试验装置内对岩石样品进行示踪试验，试件的尺寸从数厘米到数米不等,试验装置由试件主体部分(试件、绝热、密封材料等)、压力传导系统和数据采集系统三部分构成。这类试验的主要目的是:确定裂隙岩体中放射性核素迁移的水动力弥散参数,探求裂隙岩体中放射性核素迁移的时空分布规律,探索裂隙及两侧岩块中核素运移的机理。

(2)钻孔示踪试验:通过钻孔对放射性核素在处置库围岩中的迁移行为进行观测,并用不同的概念模型和数学模型对处置库中核素迁移时空分布规律进行模拟试验研究。

(3)地下实验研究:利用废弃的巷道、矿坑(井)建立地下实验室,对放射性核素在地质屏障中的迁移规律进行试验和模拟研究。美国、日本、瑞典等国家已经或正在建立地下实验室开展高放废物深地质处置研究,如日本正在利用 Tono 铀矿床开展高放废物深地质处置库运营后放射性核素迁移的比拟研究工作。

综上所述,国际上一些西方发达国家已经开展了一系列以高放废物深地质处置为目的的放射性核素迁移试验研究,取得了大量实质性进展,主要表现为:

(1)通过各种试验研究,对放射性核素在裂隙介质中的对流、扩散、吸附和阻滞规律和迁移机理的认识得到了进一步提高,为高放废物深地质处置库围岩中放射性核素迁移概念模型和数学模型的建立奠定了理论基础。

(2)获得了大量试验资料,对以花岗岩、凝灰岩、盐岩、粘土岩等为围岩的处置库,获得了大量放射性核素迁移的水动力弥散参数,为处置库场地的特征评价提供了科学依据。

(3)长时间、大规模的地下实验研究,为长时间尺度下高放废物深地质处置库安全性能评价提供了比拟研究的途径。

(4)裂隙岩体核素迁移试验技术和方法得到了进一步完善和发展,开拓了裂隙岩体核素迁移试验新的途径和方法,丰富了裂隙岩体核素迁移试验理论,研制开发了一些适合于裂隙岩体核素迁移试验的试验装置和测量浓度、温度、压力的传感器。

核素迁移的模拟研究方面,美国、瑞典、日本、法国和加拿大等西方发达国家自 20 世纪 50 年代以来,成立了专门的研究机构就高放废物深地质处置系统中放射性核素的迁移问题进行模拟研究,主要研究内容有:

①深地质处置系统概念模型的建立:研究表明深地质处置系统一般由工程屏障系统(玻璃固化体、包装容器、回填材料)、地质屏障系统和生物圈三部分构成,因此,从模拟对象上高放废物深地质处置系统中放射性核素迁移模拟应包括工程屏障中核素分解和迁移行为模拟、地质屏障中核素迁移模拟和生物圈中核素迁移模拟。

②裂隙岩体核素迁移机理和迁移模型:裂隙岩体核素迁移机理研究主要包括核素在裂隙岩体中的对流、扩散、吸附和阻滞规律的研究以及核素迁移过程中的水岩作用过程和核素存在、迁移形式研究;根据裂隙介质的特征,裂隙岩体核素迁移模型可分为单裂隙核素迁移模型、连续或等效连续介质核素迁移模型、裂隙网络介质核素迁移模型,其中,裂隙网络介质核素迁移模型是高放废物深地质处置系统中放射性核素迁移模拟研究发展趋势。

③多场耦合问题:由于高放废物地质处置系统的复杂性,核素溶解、迁移行为的模拟是一项复杂的系统工程。这里面有热力学、水文地球化学、岩体力学、地下水渗流及溶质迁移等问题。所建立的模型也应该是包括渗流场、化学场、应力场、温度场等在内的多场耦合,或者说是这些场的模型所组成的模型系列。毫无疑问,多场耦合问题是当前高放废物深地质处置系统中放射性核素迁移模拟研究的前沿性基础问题之一。

④数值模拟方法:高放废物地质处置系统中核素迁移介质非均质性和核素迁移条件的复杂性,决定了高放废物深地质处置系统中放射性核素迁移的数学模型难以用解析法求解,核素迁移数学模型数值解法研究和计算软件的开发亦是高放废物深地质处置系统中放射性核素迁移模拟研究的基本问题之一。目前基于随机理论的数值计算方法和基于积分变换的数值逆变换技术在放射性核素迁移模型的求解中越来越受到有关学者的关注。

⑤迁移参数的反演和灵敏性分析:一方面,通过高放废物深地质处置系统中放射性核素迁移模拟研究能模拟放射性核素迁移的时空分布规律,进而达到评价高放废物深地质处置系统安全性能的目的;另一方面,通过核素迁移模拟数据与试验观测数据的拟合可以反演核素迁移的水动力弥散参数。此外,借助核素迁移模拟进行参数灵敏性分析也是高放废物深地质处置系统中放射性核素迁移模拟研究的基本问题之一。

4 高放废物深地质处置多场耦合与核素迁移研究中的主要问题

4.1 深部裂隙岩体的地质、水文地质、水文地球化学环境问题

深部岩体的地应力、节理裂隙及各种构造面是深部岩体在工程扰动及热载作用下响应分析的基础；深部裂隙岩体中天然地下水运动的地质概念模型、物理及数学模型是工程与热作用下渗流分析的基础资料；天然条件下深部水-岩相互作用、地下水的原始化学成分及地球化学环境是受热荷载作用后水-岩作用和核素迁移的环境背景；因此，深部裂隙岩体的地质、水文地质和水文地球化学研究成为高放废物深地质处置多场耦合研究中必须研究的问题和必须获取的基础资料。

4.2 岩块、结构面及回填材料工程性状的温度、应力及时间相关性问题

高放废物深地质处置库中围岩及回填材料在处置库关闭后处于高温、高地应力及高流体压力的长时间作用下，因此，岩块、回填材料、结构面及流体的工程性状与常温、常压和室内实验中测定的数值可能显著不同，研究这些介质与结构面的物理、力学、水力学及热学性质与温度、时间及应力的相关性就成为高放废物深地质处置中多场耦合研究的一个重要基础课题。

4.3 深部节理岩体的应力场、渗流场、温度场及化学场的描述问题

深部节理岩体中的多场耦合是应力场、渗流场、温度场及化学场之间的相互影响，因此，各场自身的地质、力学模型与数学描述，是多场耦合研究的基础，必须进行研究。

4.4 深部岩体开挖扰动时的工程响应及开挖引起的岩体碎裂化与地下水渗流场改变问题

高放废物深地质处置库开挖阶段与一般大型深部地下工程一样，由于开挖的力学扰动，会引发施工中的安全问题、围岩的变形破坏问题，产生的岩体碎裂，改变地下水的渗流场和渗透通道，直接影响了高放废物放置后的性能，因此，必须重视研究这一阶段的应力、变形及流场变化。

4.5 水-乏燃料（玻璃固化体）的相互作用及水-岩、水-缓冲材料间化学作用的机理、化学作用的温度和应力相关性及其力学效应

高放废物处置的核素迁移起源于水对乏燃料或高放废物固化体的溶解，因此研究核素迁移需从源头上研究处置库条件下的水-乏燃料或固化体相互作用与核素的近场释放。

高放废物处置库关闭后，由于核废物的放热，改变废物体外回填材料与围岩中的温度与应力条件，引发水-岩、水-回填材料间的化学作用，产生次生矿物的沉淀与溶解，改变地下水的化学成分、影响孔隙与裂隙率，影响回填材料与岩体的渗透性能、变形性能和强度性能。因此，研究处置库的水-岩、水-回填材料中的地球化学作用类型、机理、温度与压力的影响及地球化学作用的力学效应，成为耦合研究中的一个基础课题。

4.6 多场耦合的基本理论及数值方法

处置库关闭后，处置库的围岩和回填材料中发生热传输、应力与变形、流体流动和地球化学等四种过程，因此建立多场耦合的基本理论及其数值方法，是高废物深地质处置中的最重要课题。

多场耦合基本理论中最重要的部分是建立耦合条件下各场的控制方程及联系各场的关系方程，例如：多场耦合时固相介质中的应力平衡方程、几何方程、本构方程；流体流动的基本规律与耦合条件下流体的质量守恒方程；耦合条件下的能量守恒方程；化学作用的反应方程等。联系各场关系的方程是指各种方程中的参数受其它场结果的影响的相关关系，比如渗透系数与温度和应力的相关关系、岩石的强度与变形参数与温度的相关关系等等。

多场耦合问题的求解必须寻求相应的数值解法，开发相应的计算程序，这成为求解多场耦合问题的技术关键之一。

4.7 高放废物处置库四场耦合条件下放射性核素运移的基本机理、数学描述及其迁移规律

高放废深地质处置的根本目的在于使放射性核素与生物圈的隔离，因此，放射性核素的迁移研究是最核心的课题。

研究的主要内容包括放射性元素在地下水中的传输的化学行为、动力学参数及运移的数学描述与求解的研究，包括：放射性核元素在地下水中的存在形态、运移过程中吸附、扩散与水动力扩散的机理及不同温度与压力条件下各种相关参数的测定；建立放射性核素传输过程的耦合动力学数值模型与求解方法；对核素释放的行为进行环境评价等。

4.8 多场耦合与核素迁移中的不确定性研究及人工类比研究

高放废物深地质处置中多场耦合与核素迁移涉及的时间尺度在一万年以上，这是一个远远跨越

一般工程与技术角度的时间概念，因此，各种过程、模型及参数的不确定性问题是必须研究的课题。

自然界中存在许多与处置库或处置屏障中多场耦合与核素迁移相似的天然类似物和作用过程，研究这些产物或过程是研究高放废物中多场耦合与核素迁移的另一条途径。

5 结 论

高放废物的安全处置是世界各国、也是我国面临的一项重大课题。地质处置是一项被普遍接受的方案。与一般地下工程相比，高放废物深地质处置库有许多特殊性，温度-应力-渗流-化学场的长时间、大范围耦合及耦合条件下的核素迁移是其中最重要的作用过程和关键问题。本文概括了作为深地下工程的深地质处置库的 6 个特点，在简要介绍了多场耦合的机理与类型的基础上，提出了高放废物深地质处置多场耦合与核素迁移问题的特点，简要介绍了研究现状，最后提出了高放废物深地质处置多场耦合与核素迁移所需研究的主要问题。

参 考 文 献

[1] XIA-ting FENG, Jian-jun LIU, Lan-ru JING. Research and application on coupled T-H-M-C processes of geological media in China—a review[A]. **Proceedings of the International Conference on Coulped T-H-M-C Progresses in Geo-system: Fundamental, Modeling, Experimentals & Applications**[C]. Stepansson O, Jing L, Hudson J A, eds., Sweden: Stockholm, 2003. 13－15.

[2] Si-jing WANG, En-zhi WANG, Recent study on coupled processes in geotechnical and geo-envirnomental fields in China[A]. **S Proceedings of the International Conference on Coulped T-H-M-C Progresses in Geo-system: Fundamental, Modeling, Experimentals & Applications** Stepansson O, Jing L, Hudson J A, eds., Sweden: Stockholm, 2003. 13－15.

[3] Willis C, Rubin J. Transport of reactive solutes subject to a moving dissolution boundary: numerical methods and solution[J]. **Water Resource Research**. 1987, 23(8): 1 561–1 574.

[4] Lee Y M, Lee K J. Nuclide transport of decay chain in the fractured rock medium: a model using continuous time Markov process[J]. **Ann. Nucl. Energy**, 1995, 22 (2): 71－84.

[5] 周文斌, 张展适, 李满根. 地球化学程序EQ3/6及其在关键核素的迁移行为研究中的应用[A]. 核资源与环境研究成就与展望[M]. 北京: 原子能出版社, 2001.

[6] 周文斌, 张展适. 元素锝、钚迁移形式的模拟研究[J]. 地球学报, 1999, 20(Supp.): 719－722.

[7] 张展适. 活动水热系统中水-岩作用研究——以墨西哥 Los Azufres 和菲律宾 Tongonan 地热田为例[博士学位论文 D]. 抚州: 华东地质学院, 2000.

[8] 张展适, 周文斌. 锝、钚在处置区的存在形式及影响因素[J]. 辐射防护通讯, 2000, 20(6): 21－25.

[9] 张展适, 周文斌. 锝、钚、锶在黄土地下水中地球化学行为的模拟研究[J]. 辐射防护, 2003, 23(6): 29－35.

[10] 张展适, 周文斌, 李满根. 水力压裂处置中锝、钚迁移行为的模拟研究[J]. 原子能科学技术, 2003, 27(6): 542－546.

[11] 刘亚晨, 吴玉山, 刘泉声. 核废料贮存裂隙岩体耦合分析研究综述[J]. 地质灾害与环境保护, 1999, 10(3): 72－78.

[12] 沈珍瑶. 高放废物处置库耦合过程简介[J]. 辐射防护通讯, 1997, 17(3): 21－23.

深部坑道围岩的变形与承载能力问题

陈士林,钱七虎,王明洋

(解放军理工大学 工程兵工程学院,江苏 南京 210007)

摘要: 通过传统连续介质力学理论描述宏观尺度上的岩块行为,借助光滑函数的组合描述岩块间断位移场的不光滑性,从而揭示了岩块界面的局部变形行为,使得能量在介质中的存储、耗散与通过间断界面边界转移的能得到数学刻画。在该理论框架下,以圆形坑道围岩压力确定为例,通过简化的应力 – 滑移关系,计算出围岩各区所承受的荷载比例,阐明了岩石压力与坑道的埋深、坑道的尺寸和形状、围岩的物理力学性质以及支护的柔度等具有直接的关系。

关键词: 采矿工程;承载能力;滑移间断面;岩石压力

中图分类号: TD 322　　**文献标识码:** A　　**文章编号:** 1000 – 6915(2005)13 – 2203 – 09

PROBLEMS OF DEFORMATION AND BEARING CAPACITY OF ROCK MASS AROUND DEEP BURIED TUNNELS

CHEN Shilin,QIAN Qihu,WANG Ming-yang

(*Engineering Institute of Engineering Corps,PLA University of Science and Technology,Nanjing* 210007,*China*)

Abstract: The behavior of rock block on a macro-scale is described by the classical mechanics of continuum, while the discontinuous displacement field between rock blocks is represented by a combination of smooth functions, which characterizes the deformational behavior of rock block interfaces. Hence, the way is mathematically described in which the energy is stored, dissipated and transferred between rock block interfaces. In the context of the proposed procedure, taking the determination of rock pressures on a circular tunnel for example, the proportion of pressures around the tunnel resisted by each zone is calculated based on a simplified stress-slip displacement relation; the direct relationships among the rock pressures and the buried depth, dimensions and shape of the tunnel, the physico-mechanical characteristics of rock mass, and the flexibility of supports and etc., are illustrated.

Key words: mining engineering;bearing capacity;discontinuous slip plane;rock pressure

1 引 言

M. M. 普罗托奇扬科诺夫第一个注意到岩石与液体介质的不同,岩石具有抗剪切(形状变化)的性质,这种性质对矿业来讲是有利的,能使支护上的荷载不产生深度增长的线性关系[1]。随后的研究中,尤其在能监测固体变形全过程的"刚性"加载实验机出现后,发现岩石即使完全破坏后,也仍然具有强度,并随变形量的增长而呈现强度下降的性质。这个残留强度来源于岩石破碎块体之间的摩擦效应,可以说,在这种状态下残留强度的存在是保证

收稿日期:2004-10-19;修回日期:2004-12-24。
基金项目:国家自然科学基金重大项目(50490275)。
本文原载于《岩石力学与工程学报》(2005 年第 24 卷第 13 期)。

矿山坑道在岩压很高的条件下也具有稳定性的关键[2]。

以往，岩石压力主要在两个学科方向进行发展：一是在矿山和实验室条件下物理模拟方法的发展；另一个是积极地运用连续介质力学模型，采用弹性和塑性理论将岩石的稳定性与其物理 - 力学性质联系起来形成的一些观点[1, 3~6]。这些理论形成的假说与概念，尽管能在某一方面或从某种条件说明岩石压力的形成，但在论述岩石压力形成机理方面还存在与实际较大的分离，尤其是对岩体中的破坏面的运动性状缺乏内在的理解和描画。

岩石力学的自然现象表现在破坏过程中呈现滑移破坏面，其形式有些具有规律性或规则的网状(暴露出岩体的适当边界)或带状形式，这些碎裂表面上粘结力强度降低，存在的是剩留的摩擦强度。此外，观测到滑移面(线)的出现时刻与强度耗尽和在τ - γ (τ 为切应力，γ 为剪切变形)图形中下落的曲线段过渡的时刻是相当的，尽管下落曲线段的实现在很大程度上取决于问题的边界条件。毫无疑问，滑移破坏面的网格大小及形状(曲率)与岩石的性质、地应力水平和边界条件是密切相关的。

岩体变形是其中各个岩块之间沿边界面(结构面)的相对位移与岩块自身的变形发生复杂相互作用的结果[7]，而后者对整个岩体变形起着重要作用。由于这种局部的剪切破坏使其破坏位移矢量经受着很大的断裂(称位移矢量的断裂线为滑移线)，介质的运动是在不光滑的位移场中进行，造成的局部变形运动不仅有平动而且还有转动，导致不可恢复的局部剪切变形(与位移矢量分量滑移线相切的断裂数值就叫做滑动或局部剪切变形)以及应力张量的不对称性。显然，在描述塑性变形时考虑这种界面运动的特性是岩石力学基础研究过程中必须面临的一个重要问题。

本文在用质点连续介质力学描述宏观尺度上的岩块自身变形行为的基础上，借助光滑函数的组合描述间断位移场的不光滑性，从而引出间断位移场变形特征的新的旋转自由度。有限尺度内间断位移场的变形梯度描述了界面的局部变形行为。在该理论框架下以圆形坑道为例阐明了岩石压力的形成机理及基本性质。

2 坑道围岩破坏面变形的数学描述

如前所述，岩体中块体界面上位移矢量场存在着间断。如果采取某些限制，不光滑函数可以用光滑函数的组合来说明，其中一个具有原始函数平均化的意义，其余的是描述原始函数的断裂局部间断信息。将这种描述适用于位移场，就意味着与光滑的(平均化的)位移场一起在引入了平均化时所失去的有关断裂的信息。

讨论平面问题时，矢量自变量 $\bar{r} = x_1\bar{e}_1 + x_2\bar{e}_2$ ($\bar{e}_1$, $\bar{e}_2$ 为标准化正交基底)和矢量函数 $\bar{U} = U_1\bar{e}_1 + U_2\bar{e}_2$ 的情况。令在尺寸为 l ($l << 1$) 的单元中，函数 $\bar{U}$ 是非常光滑的。假设对矢量函数 $\bar{U}$ 来说有一个光滑的平均化 $\bar{u}(\bar{r})$，在单元的中心 $\bar{r}_i$ 处有 $\bar{u}(\bar{r}_i) = \bar{U}(\bar{r}_i)$。在分隔单元(其中心点 $\bar{r}_i$ 和 $\bar{r}_{i+1}$)的边界上，断裂 $\bar{U}(\bar{r})$ 与 $A(\bar{r}_i)(\bar{r}_{i+1} - \bar{r}_i)$ 相等，精度达到 $|\bar{r}_{i+1} - \bar{r}_i|^2$ 以内[8]，其中 A 为具有光滑分量 A_{km} (k, $m = 1, 2$) 的二阶张量。因此，原始场 $\bar{U}(\bar{r})$ 就可转变成相应的光滑矢量场 $\bar{u}(\bar{r})$ 和张量场 $A(\bar{r})$。这里对断裂函数类别的描述中增加了如下限制：如果在断裂线上，单边的导数值 $\partial U_k / \partial x_m$ 彼此相等，则 A_{km} 满足：

$$A_{km} = \frac{\partial u_k}{\partial x_m} - \frac{\partial U_k}{\partial x_m} \tag{1}$$

假定剪切时材料密度不发生变化，而且式(1)中张量 A 的 4 个分量只由 2 个不变量函数 Γ (最大主剪应变)和 Ω 来确定，即

$$\left.\begin{array}{l} A_{11} = \cos 2\theta\Gamma \\ A_{22} = -\cos 2\theta\Gamma \\ A_{21} = -\Omega + \sin 2\theta\Gamma \\ A_{12} = \Omega + \sin 2\theta\Gamma \end{array}\right\} \tag{2}$$

式(2)意味着，没有平均化的位移场垂直于单元边线的分量是连续的。变量 Ω 的力学含义可表示为

$$\Omega = \frac{1}{2}(A_{12} - A_{21}) = -\frac{1}{2}\left(\frac{\partial u_2}{\partial x_1} - \frac{\partial u_1}{\partial x_2}\right) + \frac{1}{2}\left(\frac{\partial U_2}{\partial x_1} - \frac{\partial U_1}{\partial x_2}\right)$$

即变量 Ω 描述了原始位移场和平均化位移场的旋度差。在引入变量 Ω 同时，还可以引入变量 ω：

$$\omega = \Omega + \frac{1}{2}\left(\frac{\partial u_2}{\partial x_1} - \frac{\partial u_1}{\partial x_2}\right) = \frac{1}{2}\left(\frac{\partial U_2}{\partial x_1} - \frac{\partial U_1}{\partial x_2}\right) \tag{3}$$

其含义为原始不光滑位移场旋度的一半。

描述滑移网格采用正交曲线坐标系 λ_1, λ_2 比较方便[9]，即

$$x_1 = x_1(\lambda_1, \lambda_2) \quad x_2 = x_2(\lambda_1, \lambda_2) \tag{4}$$

式中：x_1, x_2 为笛卡尔坐标系(如图 1 所示)。

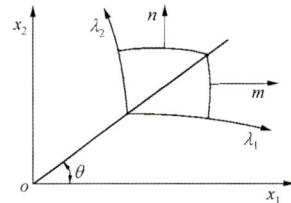

图 1 笛卡尔坐标与曲线坐标

Fig.1 Dikar and curve coordinates

图 1 中 θ 为主应力方向与 ox_1 轴线之间的夹角，λ_1 的切线与 ox_1 轴线之间的夹角为 $\theta - \dfrac{\pi}{4}$；令 $a_1 = \dfrac{\partial l_1}{\partial \lambda_1}$，$a_2 = \dfrac{\partial l_2}{\partial \lambda_2}$（$a_1$，$a_2$ 为正交曲线坐标系与笛卡尔坐标系转换的拉梅常数），则有：$\dfrac{\partial a_1}{\partial \lambda_2} = -a_2 \dfrac{\partial \theta}{\partial \lambda_1}$，$\dfrac{\partial a_2}{\partial \lambda_1} = a_1 \dfrac{\partial \theta}{\partial \lambda_2}$（$l_1$，$l_2$ 为沿相应曲线的弧长）。

描述滑移线网的第 2 个参数是单元的尺寸，选择 λ_1，λ_2 为常数的 2 个坐标线，并把其分成单元段 $\Delta\lambda_1 = f_1(\lambda_1)\chi$，$\Delta\lambda_2 = f_2(\lambda_2)\chi$（$\chi$ 为描述滑移线网格"密度"的参数），将全部塑性区划分成许多单元，其单元的边长在点 (λ_1, λ_2) 处为：$l_1 = f_1(\lambda_1)a_1(\lambda_1, \lambda_2)\chi$，$l_2 = f_2(\lambda_2)a_2(\lambda_1, \lambda_2)\chi$。单元的塑限与其尺寸 l 有关。因而可以假定，作用在单元上的应力也决定着其尺寸，即网的密度取决于应力场。本文研究的是滑移线网形成以后材料的变形(岩体破碎部分占的能量一般份额不大)，可以认为描述网块密度的函数 $f_1\chi$，$f_2\chi$ 是已知的。

介质的总变形是由局部变形(沿滑移线滑过的长度)和普通单元内的变形(由滑移线划分的单元)相互作用而成的。位移场的间断只是发生在滑移线上，也就是说在模型中要描述这个间断信息。下面讨论沿滑移面上的物理方程，U_n，u_n 为原始位移矢量和平均位移矢量在滑移线 λ_2 切线方向(图 1)上的投影，则有

$$\dfrac{\partial u_n}{\partial m} - \dfrac{\partial U_n}{\partial m} = \dfrac{\cos 2\theta}{2}\left(\dfrac{\partial u_1}{\partial x_1} - \dfrac{\partial u_2}{\partial x_2}\right) +$$

$$\dfrac{\sin 2\theta}{2}\left(\dfrac{\partial u_2}{\partial x_1} + \dfrac{\partial u_1}{\partial x_2}\right) + \dfrac{1}{2}\left(\dfrac{\partial u_2}{\partial x_1} - \dfrac{\partial u_1}{\partial x_2}\right) -$$

$$\dfrac{\cos 2\theta}{2}\left(\dfrac{\partial U_1}{\partial x_1} - \dfrac{\partial U_2}{\partial x_2}\right) - \dfrac{\sin 2\theta}{2}\left(\dfrac{\partial U_1}{\partial x_2} + \dfrac{\partial U_2}{\partial x_1}\right) -$$

$$\dfrac{1}{2}\left(\dfrac{\partial U_2}{\partial x_1} - \dfrac{\partial U_1}{\partial x_2}\right) \quad (5)$$

一般情况下，剪切应力 σ_{12}^0，σ_{21}^0 是互不相等的，这里和以后角注"0"都表示相应滑移面上的变量。弹性剪切变形为

$$\dfrac{\cos 2\theta}{2}\left(\dfrac{\partial U_1}{\partial x_1} - \dfrac{\partial U_2}{\partial x_2}\right) + \dfrac{\sin 2\theta}{2}\left(\dfrac{\partial U_2}{\partial x_1} + \dfrac{\partial U_1}{\partial x_2}\right) = \dfrac{\sigma_{12}^0 + \sigma_{21}^0}{4\mu} \quad (6)$$

滑移面上的局部剪切变形为

$$w_{12}^0 = \left(\dfrac{\partial u_n}{\partial m} - \dfrac{\partial U_n}{\partial m}\right)l_1 = \Bigg\{\left[\dfrac{\cos 2\theta}{2}\left(\dfrac{\partial u_1}{\partial x_1} - \dfrac{\partial u_2}{\partial x_2}\right) + \right.$$

$$\dfrac{\sin 2\theta}{2}\left(\dfrac{\partial u_2}{\partial x_1} + \dfrac{\partial u_1}{\partial x_2}\right) - \Omega\Bigg] - \left[\dfrac{\cos 2\theta}{2}\left(\dfrac{\partial U_1}{\partial x_1} - \dfrac{\partial U_2}{\partial x_2}\right) + \right.$$

$$\left.\dfrac{\sin 2\theta}{2}\left(\dfrac{\partial U_1}{\partial x_2} + \dfrac{\partial U_2}{\partial x_1}\right)\right]\Bigg\}l_1 \quad (7a)$$

$$w_{21}^0 = \left(\dfrac{\partial u_m}{\partial n} - \dfrac{\partial U_m}{\partial n}\right)l_2 = \Bigg\{\left[\dfrac{\cos 2\theta}{2}\left(\dfrac{\partial u_1}{\partial x_1} - \dfrac{\partial u_2}{\partial x_2}\right) + \right.$$

$$\dfrac{\sin 2\theta}{2}\left(\dfrac{\partial u_2}{\partial x_1} + \dfrac{\partial u_1}{\partial x_2}\right) + \Omega\Bigg] - \left[\dfrac{\cos 2\theta}{2}\left(\dfrac{\partial U_1}{\partial x_1} - \dfrac{\partial U_2}{\partial x_2}\right) + \right.$$

$$\left.\dfrac{\sin 2\theta}{2}\left(\dfrac{\partial U_1}{\partial x_2} + \dfrac{\partial U_2}{\partial x_1}\right)\right]\Bigg\}l_2 \quad (7b)$$

式(7a)，(7b)确定的局部剪切变形(滑移)与变形张量所确定的描述任意方位角变化的剪切不同，μ 为介质的剪切模量，数值 w_{12}^0，w_{21}^0 只是在相应的滑移面上才具有含义，描述的是单元沿表面上真实的剪切。滑移面上可能发展的切向应力由这些表面上的局部变形值来确定，即

$$w_{12}^0 = w(\sigma_{12}^0) \quad w_{21}^0 = w(\sigma_{21}^0) \quad (8)$$

将式(6)代入式(7a)，(7b)中并结合式(8)，得出滑移面上的物理方程为

$$\dfrac{\cos 2\theta}{2}\left(\dfrac{\partial u_1}{\partial x_1} - \dfrac{\partial u_2}{\partial x_2}\right) + \dfrac{\sin 2\theta}{2}\left(\dfrac{\partial u_2}{\partial x_1} + \dfrac{\partial u_1}{\partial x_2}\right) - \Omega = \dfrac{1}{f_1 a_1 \chi}w(\sigma_{12}^0) + \dfrac{\sigma_{12}^0 + \sigma_{21}^0}{4\mu} \quad (9)$$

$$\frac{\cos 2\theta}{2}\left(\frac{\partial u_1}{\partial x_1}-\frac{\partial u_2}{\partial x_2}\right)+\frac{\sin 2\theta}{2}\left(\frac{\partial u_2}{\partial x_1}+\frac{\partial u_1}{\partial x_2}\right)+\Omega=$$
$$\frac{1}{f_2 a_2 \chi}w(\sigma_{21}^0)+\frac{\sigma_{21}^0+\sigma_{12}^0}{4\mu} \qquad (10)$$

对于圆形坑道来说，可以把坑道围岩的变形作为平面应变问题来考虑，引入以坑道中心为极点的极坐标系(r, θ)，假设介质在向非弹性状态过渡瞬时无剪应力$\sigma_{r\theta}$，这样滑移线将是对数螺旋线[10]，即

$$\left.\begin{array}{l}\lambda_1=\dfrac{1}{\sqrt{2}}\left(\ln\dfrac{r}{r_0}-\theta\right)=\text{const}\\ \lambda_2=\dfrac{1}{\sqrt{2}}\left(\ln\dfrac{r}{r_0}+\theta\right)=\text{const}\end{array}\right\} \qquad (11)$$

式中：r_0为坑道半径。

由对数螺旋滑移线构成的曲线坐标系与极坐标系转化的拉梅常数：$a_1=a_2=r$，将式(9)，(10)转为极坐标系(r, θ)，有

$$\frac{\partial}{\partial r}(u_r+u_\theta)-\frac{1}{r}\frac{\partial}{\partial \theta}(u_r+u_\theta)-\frac{u_r-u_\theta}{r}-2\omega=$$
$$\frac{1}{2\mu}(\sigma_{rr}-\sigma_{\theta\theta})+\frac{2}{f_1 a_1 \chi}S\left(\frac{\sigma_{rr}-\sigma_{\theta\theta}}{2}-\frac{\sigma_{\theta r}-\sigma_{r\theta}}{2}\right) \qquad (12\text{a})$$

$$\frac{\partial}{\partial r}(u_r-u_\theta)+\frac{1}{r}\frac{\partial}{\partial \theta}(u_r-u_\theta)-\frac{u_r+u_\theta}{r}+2\omega=$$
$$\frac{1}{2\mu}(\sigma_{rr}-\sigma_{\theta\theta})+\frac{2}{f_2 a_2 \chi}S\left(\frac{\sigma_{rr}-\sigma_{\theta\theta}}{2}-\frac{\sigma_{\theta r}-\sigma_{r\theta}}{2}\right) \qquad (12\text{b})$$

根据动量守恒条件，极坐标系中的平衡方程[10]为

$$\left.\begin{array}{l}\dfrac{\partial \sigma_{rr}}{\partial r}+\dfrac{1}{r}\dfrac{\partial \sigma_{\theta r}}{\partial \theta}+\dfrac{\sigma_{rr}-\sigma_{\theta\theta}}{r}+X_r=0\\ \dfrac{\partial \sigma_{r\theta}}{\partial r}+\dfrac{1}{r}\dfrac{\partial \sigma_{\theta\theta}}{\partial \theta}+\dfrac{\sigma_{r\theta}+\sigma_{\theta r}}{r}+X_\theta=0\end{array}\right\} \qquad (13)$$

式中：X_r，X_θ分别为体积力矢量在极坐标系(r, θ)上的投影。

根据位移矢量垂直于单元边线分量的连续条件，原始不光滑位移场和平均化的光滑位移场的压缩变形是一致的。因此，在单元体中的相关方程[11]为

$$\left.\begin{array}{l}\dfrac{\partial u_r}{\partial r}+\dfrac{1}{r}\dfrac{\partial u_\theta}{\partial \theta}+\dfrac{u_r}{r}=\dfrac{1-2\nu}{2\mu}(\sigma_{rr}+\sigma_{\theta\theta})\\ \dfrac{\partial u_\theta}{\partial r}+\dfrac{1}{r}\dfrac{\partial u_r}{\partial \theta}-\dfrac{u_\theta}{r}=\dfrac{1}{2\mu}(\sigma_{r\theta}+\sigma_{\theta r})\end{array}\right\} \qquad (14)$$

式中：μ为剪切模量，ν为松比。

不考虑非对称扰动，所有的单元是对称的，其大小与θ无关，即$f_1=f_2=1$，由于$\sigma_{r\theta}\equiv\sigma_{\theta r}=0$，$u_\theta=0$和$X_\theta=0$，在轴对称情况下，式(12)～(14)可简化为

$$\left.\begin{array}{l}\dfrac{\partial \sigma_{rr}}{\partial r}+\dfrac{\sigma_{rr}-\sigma_{\theta\theta}}{r}+X_r=0\\ \dfrac{\partial u_r}{\partial r}+\dfrac{u_r}{r}=\dfrac{1-2\nu}{2\mu}(\sigma_{rr}+\sigma_{\theta\theta})\\ \dfrac{\partial u_r}{\partial r}-\dfrac{u_r}{r}=\dfrac{1}{2\mu}(\sigma_{rr}-\sigma_{\theta\theta})+\dfrac{2}{\chi r}\delta(\tau)\end{array}\right\} \qquad (15)$$

其中，
$$\tau=\frac{\sigma_{rr}-\sigma_{\theta\theta}}{2}$$
$$\delta(\tau)=\frac{w(\sigma_{21}^0)+w(\sigma_{12}^0)}{2}$$

对式(15)进行积分可得

$$\frac{2\mu}{\chi}\delta(\tau)=\frac{2\mu p}{r}-\frac{1-2\nu}{r}Z(r)-2(1-\nu)r\tau \qquad (16)$$

式中：p为积分常数，即载荷参量；$Z(r)=\int r^2 X_r \mathrm{d}r$。

式(15)的通解为

$$\left.\begin{array}{l}\sigma_{rr}=-2\int_{r_0}^{r}\dfrac{\tau}{r}\mathrm{d}r-\int_{r_0}^{r}X_r \mathrm{d}r+\sigma_{rr}(r_0)\\ \sigma_{\theta\theta}=\sigma_{rr}-2\tau\\ u_r=\dfrac{1-2\nu}{2\mu}r\sigma_{rr}+\dfrac{1-2\nu}{2\mu}\dfrac{Z(r)}{r}-\dfrac{p}{r}\end{array}\right\} \qquad (17)$$

确定岩石压力不仅必须知道岩体破坏前的主要变形特征，而且要知道岩体原始应力状态。一般而言，坑道的开挖只影响到周围一定范围内的岩石，这部分岩石简称为围岩，围岩以外的岩石仍保持原始应力状态。在这里围岩的半径用R表示，R通常为坑道半径的3～4倍。由式(17)的第1式即可得到岩石压力的计算公式为

$$Q=-\sigma_{rr}(R)-\int_{r_0}^{R}X_r \mathrm{d}r-2\int_{r_0}^{R}\frac{\tau}{r}\mathrm{d}r \qquad (18)$$

式中：$\sigma_{rr}(R)$为距坑道中心R处的径向应力，即原始地应力。式(18)的前两项相当于坑道的全部上覆载荷，第3项函数是围岩所承担的载荷部分，表明

了围岩自身的承载能力。

函数$\tau(r,p)$可由式(16)确定。式(16)中的$\delta(\tau)$需要通过实验确定,由于研究的是滑移破坏后的情况,因此,试件的变形应在带有监测变形的压力机上进行,另外滑移量δ是有长度量纲的量,在直接确定曲线图时必须测量滑移线"两边"的相对位移,如果材料的弹性性能和滑移线间的距离是已知的,那么滑移量是容易按试件的总变形来计算的。考虑到应力 - 滑移量的变化规律是近似外摩擦的变化规律,因此有相同的表达形式,差别是外摩擦图只是反映预先给定的自由表面在它们互相滑移时的行为,此时应力 - 滑移量图不仅包含了外摩擦的信息,而且包含了沿滑移线的联系破坏过程本身的信息,应用研究外摩擦的方法和实验数据,这些都可以间接地确定应力 - 滑移量图[12]。

在解式(16)中还出现了描述滑移线网格"密度"的参数χ,χ是岩体耗散结构的尺度。在一般情况下,正如对颗粒材料的实验所表明的那样[11],滑移线间距的简单定性概念不是材料常数,而且间距还与荷载条件有关,在这种情况下可以用相关方位上的裂纹密度来决定间距。

3 坑道围岩的承载能力

作为最简单的情况暂不考虑式(16)中体积力对滑移的影响,此时式(16)可简化为

$$\frac{\delta(\tau)}{\chi} = \frac{p}{r} - \frac{1-\nu}{\mu}\tau r \qquad (19)$$

如果$\sigma_{21}^0 = \sigma_{12}^0$,应力$\tau$和滑移量$\delta$是可以通过实验测得的,假设应力 - 滑移量图为图2所示的局部剪切和破坏阶段情况时(τ_4为岩石破坏后的残留强度,δ_4为开始达到残留强度时的滑移值),由式(19)可得

$$\left.\begin{array}{l} r_1 = \sqrt{k_2}\sqrt{p} \\ r_4 = \sqrt{\left(\dfrac{k_1 k_2 \tau_1}{2\tau_4}\right)^2 + \dfrac{k_2 \tau_1 p}{\tau_4} - \dfrac{k_1 k_2 \tau_1}{2\tau_4}} \end{array}\right\} \qquad (20)$$

式中:r_1为坑道中心到围岩弹性区半径,r_4为坑道中心到围岩残留强度区半径,$k_2 = \mu/(1-\nu)\tau_1$,$k_1 = \delta_4/\chi$。

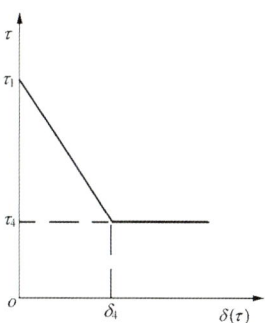

图 2 应力 - 滑移量图

Fig.2 Stress-slip curve

根据式(19),在弹性区:$\delta = 0$,$\tau_e = \dfrac{\mu p}{(1-\nu)r^2}$;在软化区(由图 2 可知):$\delta(\tau) = \dfrac{\delta_4}{\tau_4 - \tau_1}(\tau - \tau_1)$,由此可得$\tau_i = \dfrac{\mu p(\tau_4 - \tau_1) + \mu k_1 r \tau_1}{(1-\nu)(\tau_4-\tau_1)r^2 + \mu k_1 r}$;在强度残留区:$\tau_r \equiv \tau_4$。

为便于讨论,在下面的计算结果中引入无量纲参数,即

$$\left.\begin{array}{l} \lambda^2 = \dfrac{p}{k_1^2 k_2} \\ m^2 = \dfrac{r_0}{k_1 k_2} \\ n^2 = \dfrac{R}{k_1 k_2} \\ \dfrac{m^2}{n^2} = \dfrac{r_0}{R} \end{array}\right\} \qquad (21)$$

式中:λ为荷载参数。

如图 2 所示的状态,设某一试件在平面变形的条件下被压缩,由虎克定律[5]可知,最大的压缩弹性变形$\varepsilon_1 = \dfrac{1-\nu}{2\mu}(\sigma_{rr} - \sigma_{\theta\theta}) = \dfrac{1-\nu}{\mu}\tau = \dfrac{1}{k_2}$,因此,常数$k_2$也有最大压缩弹性变形倒数值的含义;由式(21)可知参数λ的力学含义为:λ的大小相当于材料软化区的无量纲半径$\dfrac{r_4}{r_1}$;常数m也有力学意义,在内边界上滑移线间的距离$l = l_1 = f_1 a_1 \chi$,可得$l = r_0 \chi$,由此,$k_1 = \delta_4/\chi = \delta_4 r_0/l$,则$m^2 = \varepsilon_1/(\delta_4/l)$,假定在试件压缩至弹性极限时在试件上形成滑移线,且滑移量又相同(如果不用专门的方法,则由于不稳定性,完全有可能只有一条滑移线在发展),因为当$\tau = \tau_4$时,每一根线的滑移量等于δ_4,那么,试件

完成局部剪切化时的轴向变形 ε_4 约相当于 δ_4/l，即滑移量变化范围为 $\varepsilon_4 \sim \delta_4/l$，由此可得 $m^2 \sim \varepsilon_1/\varepsilon_4$，亦即 m^2 具有对应于试件轴向最大弹性变形和完成局部剪切化时的轴向变形的比值的含义(也可以近似用弹性介质的压缩模量与软化阶段的压缩模量来表征)，也就是反映了介质存储能量与局部剪切耗损能量的份额比值。可见系统的稳定性与此参数关系极大。

由式(20)可得

$$\left. \begin{array}{l} \dfrac{r_1}{k_1 k_2} = \lambda \\[4pt] \dfrac{r_4}{k_1 k_2} + \dfrac{\tau_4}{\tau_1} \dfrac{r_4^2}{k_1^2 k_2^2} = \lambda^2 \\[4pt] \dfrac{\lambda}{m^2} = \dfrac{r_1}{r_0} \\[4pt] \dfrac{\lambda}{n^2} = \dfrac{r_1}{R} \end{array} \right\} \quad (22)$$

对式(18)进行无量纲化可得

$$Q' = -\dfrac{\sigma_{rr}(R)}{\tau_1} - \dfrac{1}{\tau_1} \int_{r_0}^{R} X_r(r) dr - 2T \quad (23)$$

式中：$-\dfrac{\sigma_{rr}(R)}{\tau_1}$ 为坑道边界上的无量纲岩石压力；T 为围岩的自承能力，且 $T = -\dfrac{1}{\tau_1}\int_{r_0}^{R}\dfrac{\tau}{r}dr$。

如果 $0 \leqslant \dfrac{\lambda}{m^2} < 1$，则有 $r_1 < r_0$，说明围岩是完全弹性的，有

$$T = \dfrac{1}{\tau_1}\int_{r_0}^{R}\dfrac{\tau_e}{r}dr = \dfrac{\lambda^2}{2}\left(\dfrac{1}{m^4} - \dfrac{1}{n^4}\right) \quad (24)$$

当荷载参数继续增大，$\dfrac{\lambda}{m^2} \geqslant 1$ 时，材料的行为将取决于差值 $\left(\dfrac{\tau_1}{\tau_1-\tau_4} - m^2\right)$ 的符号，若 $m^2 > \dfrac{\tau_1}{\tau_1-\tau_4}$，有 $\lambda > \dfrac{\tau_1}{\tau_1-\tau_4}$，由式(22)可得：$r_1 < r_4$。这说明介质存储的弹性能大于耗损能量，系统处于不稳定状态，多余的势能以动能的形式转化，动力学效应将立刻出现在坑道外围(即发生"岩爆")。

如 $m^2 \leqslant \dfrac{\tau_1}{\tau_1-\tau_4}$，当 $1 \leqslant \dfrac{\lambda}{m^2} < \sqrt{\dfrac{1}{m^2}+\dfrac{\tau_4}{\tau_1}}$ 时，则有 $r_4 < r_0 < r_1$，这时还没出现残留强度区，在 $r_0 \leqslant r < r_1$ 区域内材料被软化，而在 $r \geqslant r_1$ 区域材料则处于弹性状态。T 值可由下式确定：

$$T = \dfrac{1}{\tau_1}\int_{r_0}^{r_1}\dfrac{\tau_i}{r}dr + \dfrac{1}{\tau_1}\int_{r_1}^{R}\dfrac{\tau_e}{r}dr = \dfrac{1}{2}\left(1 - \dfrac{\lambda^2}{n^4}\right) +$$

$$\left(\lambda - \dfrac{\lambda^2}{m^2}\right)\dfrac{\tau_1-\tau_4}{\tau_1} + \left[1 - \left(\lambda\dfrac{\tau_1-\tau_4}{\tau_1}\right)^2\right] \cdot$$

$$\left(\ln\dfrac{\lambda}{m^2} - \ln\dfrac{\lambda - \dfrac{\tau_1}{\tau_1-\tau_4}}{m^2 - \dfrac{\tau_1}{\tau_1-\tau_4}}\right) \quad (25)$$

λ 进一步增大，$\sqrt{\dfrac{1}{m^2}+\dfrac{\tau_4}{\tau_1}} \leqslant \dfrac{\lambda}{m^2} < \dfrac{\tau_1}{\tau_1-\tau_4}$ 时，则有 $r_0 \leqslant r_4 < r_1$。这时，$r_0 \leqslant r < r_4$ 区域是残留强度区；$r_4 \leqslant r < r_1$ 区域是软化区；$r_1 \leqslant r$ 区域是弹性区。在上述范围内有

$$T = \dfrac{1}{\tau_1}\int_{r_0}^{r_4}\dfrac{\tau_r}{r}dr + \dfrac{1}{\tau_1}\int_{r_4}^{r_1}\dfrac{\tau_i}{r}dr + \dfrac{1}{\tau_1}\int_{r_1}^{R}\dfrac{\tau_e}{r}dr =$$

$$\dfrac{\tau_4}{\tau_1}\ln\left\{\dfrac{1}{m^2}\left[\sqrt{\left(\dfrac{\tau_1}{2\tau_4}\right)^2 + \dfrac{\lambda^2\tau_1}{\tau_4}} - \dfrac{\tau_1}{2\tau_4}\right]\right\} +$$

$$\left[\lambda - \dfrac{\lambda^2}{\sqrt{\left(\dfrac{\tau_1}{2\tau_4}\right)^2 + \dfrac{\lambda^2\tau_1}{\tau_4}} - \dfrac{\tau_1}{2\tau_4}}\right]\dfrac{\tau_1-\tau_4}{\tau_1} + \dfrac{1}{2}\left(1 - \dfrac{\lambda^2}{n^4}\right) +$$

$$\left[1 - \left(\lambda\dfrac{\tau_1-\tau_4}{\tau_1}\right)^2\right]\ln\dfrac{\lambda}{\sqrt{\left(\dfrac{\tau_1}{2\tau_4}\right)^2 + \dfrac{\lambda^2\tau_1}{\tau_4}} - \dfrac{\tau_1}{2\tau_4}} -$$

$$\ln\dfrac{\lambda - \dfrac{\tau_1}{\tau_1-\tau_4}}{\sqrt{\left(\dfrac{\tau_1}{2\tau_4}\right)^2 + \dfrac{\lambda^2\tau_1}{\tau_4}} - \dfrac{\tau_1}{2\tau_4} - \dfrac{\tau_1}{\tau_1-\tau_4}} \quad (26)$$

当 $\lambda = \dfrac{\tau_1}{\tau_1-\tau_4}$ 时，$r_1 = r_4$，软化区将变化。当 $\lambda > \dfrac{\tau_1}{\tau_1-\tau_4}$ 时，材料内仅保留 2 个区：残留强度区和弹性区。这 2 个区被分出"峰面"，其半径 $r = \lambda k_1 k_2$，且随 λ 的增大而增大。在上述情况下有

$$T = \frac{1}{\tau_1}\int_{r_0}^{r_4}\frac{\tau_r}{r}dr + \frac{1}{\tau_1}\int_{r_4}^{R}\frac{\tau_e}{r}dr = \frac{\tau_4}{\tau_1}\ln\frac{\lambda}{m^2} + \frac{1}{2}\left(1 - \frac{\lambda^2}{n^4}\right)$$

(27)

由式(24)~(27)分别对 λ 求导，可以得出当 λ 为：$1 \leqslant \frac{\lambda}{m^2} < \sqrt{\frac{1}{m^2} + \frac{\tau_4}{\tau_1}}$ 时的某一值 λ^* 时，函数 $T(\lambda)$ 有唯一的最大值 $T(\lambda^*)$，该值对应于岩石的最大承载能力，此时有 $r_4 < r_0$，说明岩体的最大承载能力在出现"残留强度"区前就会降低。

用 $\xi = \sigma_{rr}(R)/\tau_1$ 表示坑道掘进时的无量纲深度，ξ 的大小也表示坑道支柱和已变形材料承受的外荷载。一般情况下，深部岩体的单轴抗压强度 $R_c \approx 2\tau_1$[13, 14]。若 $\xi < 1$，则围岩处于弹性变形，若 $\xi \geqslant 1$，则可出现非相容弹性变形，此时时间效应及体积力对变形影响产生的分区破裂现象[13, 14]拟另文研究。

为了评估围岩各区的承载能力，应讨论岩石具有最大承载能力时的情况，从而得出对实际有用的结果。当围岩处于满载时，即函数 $T(\lambda)$ 为最大值 $T(\lambda^*)$，如果坑道的无量纲深度 ξ 和 $2T(\lambda^*)$ 大小相等，从式(23)中可以看出，此时，岩石压力为 0，所有上覆岩层的荷载正好全部由围岩自己承受，围岩不需要支护，在此把这一无量纲深度 ξ 称作极限深度 ξ_{max}，显然这时的荷载参数为 λ^*，因为其对应于围岩的满载能力，极限深度可以通过式(23)取 $Q = 0$ 时求得。图 3 为坑道稳定时的极限深度与岩体参数的关系。由图 3 可以看出，极限深度随着 m 的增大而变小，且随着 m 的增大，极限深度对 m 的偏差变得更不敏感。由于 λ^* 在 $1 \leqslant \frac{\lambda}{m^2} < \sqrt{\frac{1}{m^2} + \frac{\tau_4}{\tau_1}}$ 的范围内，这样就可以通过式(25)来评估围岩各区的承载能力。

4 算例与应用价值

下面用一个例子来计算在极限深度下围岩各区的承载能力。

已知条件为：$\sigma_{rr}(R) = 25$ MPa(相当于坑道深度为 900~1 000 m 时的地应力)，$\nu = 0.3$，$r_0 = 2$ m，$\tau_1 = 7.5$ MPa，$\tau_4 = 1.5$ MPa，$\mu = 5 \times 10^9$ Pa，$k_1 = 0.05$，$\rho = 2 500$ kg/m³。

由此可得：$k_2 = 950$，$\xi = 3.34$。当 $m = 0.3$ 时，由式(25)可算出弹性区承受了 42%的外力，而非弹性区承受了 58%的外力。在 m 更小时，非弹性区承载的比例更高。如 $m = 0.2$ 时，弹性区与非弹性区承受的外力分别为 33%和 67%；而 $m = 0.1$，弹性区与非弹性区承受的外力分别为 24%和 76%。

由此可见，即使是简单的曲线图(图 2)，也能说明非弹性区包括软化区在岩体承载能力上起着很大的作用。

上述的研究是为了得到解析式便于分析，而按照图 2 所示的简化曲线来计算的，而解式(16)，(17)可以对任何曲线进行数值计算，从而对围岩各区的承载能力进行基本评估。

除 r_0 外，其他已知条件不变，由式(18)计算得到的岩石压力 Q 随坑道半径 r_0 变化的情况见图 4，这个岩石压力 Q 是在对应的坑道半径时，使围岩正好达到满载能力所必须的支护力。从图 4 上可以看出，岩石压力 Q 是随着坑道半径增大而增大，岩石压力为负值，说明在坑道半径小于某个值时，围岩自身提供的承载能力就足够了，不需要支护；此

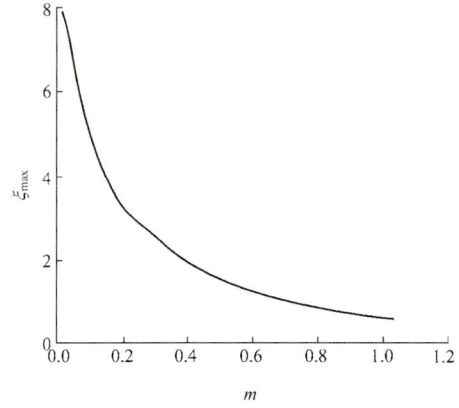

图 3 坑道稳定时的极限深度与岩体参数的关系

Fig.3 Relation between parameter and the limit depth for stability of tunnel

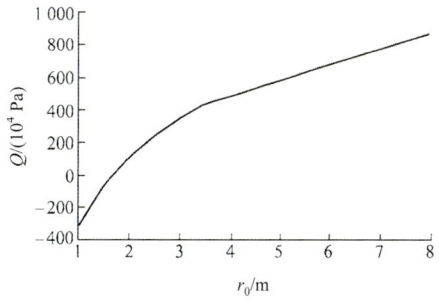

图 4 在不同坑道半径时的岩石压力

Fig.4 Rock pressures with different radii of tunnel

外还可以看出，岩石压力远小于上覆岩层的重量（$\gamma H = 2.45 \times 10^7$ Pa），这说明围岩自身具有承载能力，而且承担了大部分荷载，而支护承受的荷载只是小部分。

由式(17)可以计算坑道内边界在开挖后的位移为

$$-\Delta u_r(r_0) = -\frac{1-2\nu}{\mu} r_0 \int_{r_0}^{R} \frac{\tau}{r} \mathrm{d}r + \frac{\lambda^2 r_0}{m^4 k_2} \quad (28)$$

计算参数除r_0外，其他已知条件不变，在没有支护的情况下，由式(28)计算得到的边界位移(记为Δu_0)随坑道半径r_0变化的情况见图5。在有支护的情况下，由于支护的柔度不同，边界位移受到不同程度的控制，可计算得到相应的岩石压力。

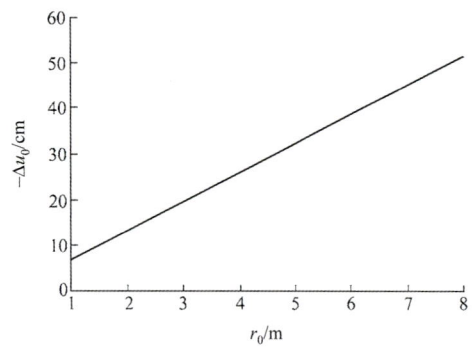

图 5 不同坑道半径时坑道内边界的位移

Fig.5 Displacements of tunnel's inner boundary with different radii of tunnel

图 6～8 分别为坑道内边界位移在 80%Δu_0，50%Δu_0 和 30%Δu_0 时的岩石压力。

图9为不同荷载参数λ时的无量纲压力Q'的变化图，表示的是在不同的参数m下，其他数据按已知条件，无量纲压力Q'随荷载参数λ的变化情况。图10反映了在不同参数m，ν以及其他数据按已知

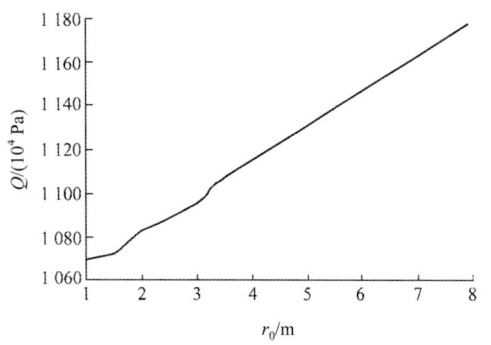

图 6 坑道内边界位移在 80%Δu_0 时的岩石压力

Fig.6 Rock pressures when the displacement of tunnel's inner boundary is 80% Δu_0

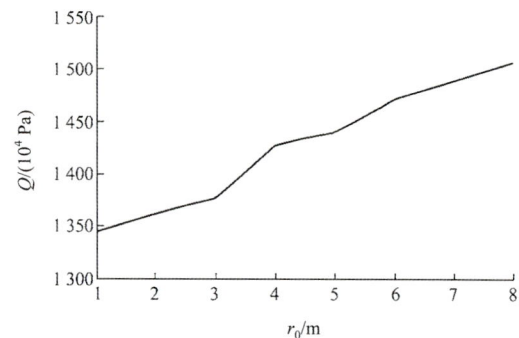

图 7 坑道内边界位移在 50%Δu_0 时的岩石压力

Fig.7 Rock pressures when the displacement of tunnel's inner boundary is 50% Δu_0

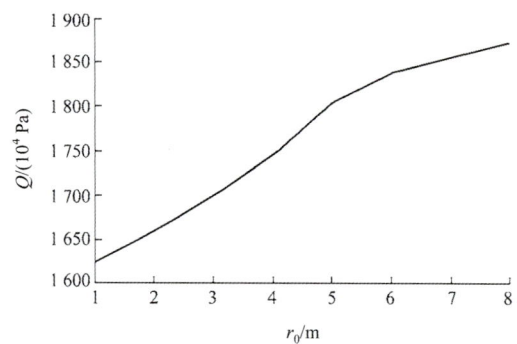

图 8 坑道内边界位移在 30%Δu_0 时的岩石压力

Fig.8 Rock pressures when the displacement of tunnel's inner boundary is 30% Δu_0

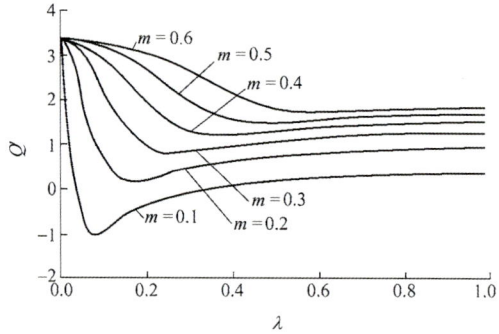

图 9 不同荷载参数λ时的无量纲压力Q'的变化图

Fig.9 Rock pressures Q' with different values of parameter λ

条件时，围岩内边界上荷载与该边界上位移之间的关系。

5 结 论

(1) 本文揭示了深部围岩的大变形、稳定性及与峰值后岩体特性的关系，从而阐明了深部岩体围

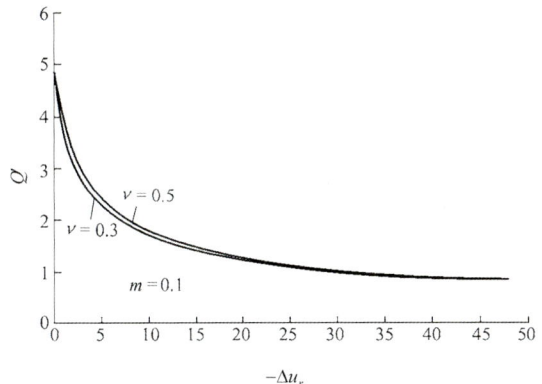

图 10 围岩内边界上荷载与该边界上位移关系

Fig.10 Relation between displacement of tunnel's inner boundary and rock pressure

岩的变形与破坏分析中连续介质弹塑性力学理论不能适用的物理本质。

(2) 计算结果表明，围岩具有很大的自承能力，支护力只占较小部分。岩石压力与坑道的埋深、坑道的尺寸和形状、围岩的物理力学性质以及支护的柔度等具有直接关系。

(3) 参数 m 反映了介质存储能量与局部剪切耗损能量的份额比值。围岩出现不稳定的动力现象(岩爆)与此参数关系极大。如 $\lambda/m^2 > 1$ (λ/m^2 破坏区半径与坑道半径之比 r_1/r_0)，则响应的稳定性只保持到一定的临界状态前。临界状态的参数是与岩体压力、块体接触面上的软化程度、耗散结构的几何尺寸以及弹性模量有关的。在超过临界状态后，岩体将发生弹性势能的动力释放。

(4) 围岩的大变形是由于围岩破坏后滑移所致，故峰值后岩体的特性研究十分重要。

参考文献(References)：

[1] Промобъяконов М М. Давление горных пород и рудничное крепление[R]. [s.l.]: [s.n.], 1931.

[2] Шемякин Е И. Две задачи механики горных пород, связаных с освоением глубоких месторождений угля и руды[J]. Фтпрпи, 1975, (5): 72.

[3] Jaegar J C, Cook N G W. Fundamental of Rock Mechanics[M]. London: Champman and Hall, 1979.

[4] Zheng H T, Guo P J. Rock pressure estimation and application of complete stress-strain curve of rock[A]. In: Proceedings of the Eighth International Conference on Computer Methods and Advances in Geomechanics Morgantown[C]. West Virginia: [s.n.], 1994. 22 – 28.

[5] 董方庭. 巷道围岩松动圈支护理论及应用技术[M]. 北京：煤炭工业出版社，2001.(Dong Fangting. Supporting Theory and Application Technology of Rock Mass around Tunnel[M]. Beijing: China Coal Industry Publishing House, 2001.(in Chinese))

[6] 陈宗基. 地下巷道长期稳定性的力学问题[J]. 岩石力学与工程学报，1982，1(1)：1 – 20.(Tan Tjongkie. Mechanical problems of longterm stability for underground laneway[J]. Chinese Journal of Rock Mechanics and Engineering, 1982, 1(1): 1 – 20.(in Chinese))

[7] 陈宗基，康文法. 岩石的封闭应力、蠕变和扩容及本构方程[J]. 岩石力学与工程学报，1991，10(4)：299 – 312.(Tan Tjongkie, Kang Wenfa. Locked in stresses, creep and dilatancy of rocks and constitutive equations[J]. Chinese Journal of Rock Mechanics and Engineering, 1991, 10(4): 299 – 312.(in Chinese))

[8] Ревуженко А ф, Смажевский С В, Шемякин Е И. О механизме деформирования сыпучего материала при больших сдвигах[J]. Фтпрпи, 1974, (3): 130 – 133.

[9] Ревуженко А ф, Шемякин Е И. Кинематика деформирования сыпучей среды с невязким трением[J]. Прикл. мех и техн.физики, 1974, (4): 119 – 124.

[10] Kachanov L M. Foudationals of Plasticity Theory[M]. Nauka, Moskow: [s.n.], 1969.

[11] Новожилов В В. Две статьи о математических моделях в механике сплошной среды [J]. ИПМ АН СССР, ЛГУ пр. Москва, 1983, 215: 56.

[12] Nikiforovsky V S, Shemyakin Ye I. Dynamic Failure of Solid Bodies[M]. Nauka, Novosbirsk: [s.n.], 1979.

[13] Shemyakin E, Fisenko G L, Kurlenya M V, et al. Effect of zonal disintegration of rocks around the underground-workings[A]. In: Fiz. Tekh. Probl. Razrab. Polezn. Iskop.[C]. [s.l.]: [s.n.], 1986. 1 – 12.

[14] Shemyakin E, Fisenko G L, Kurlenya M V, et al. Effect of zonal disintegration of rocks around the underground workings[J]. Dokl. AN SSSR, 1986, 289(5): 23 – 34.

深部岩体块系介质变形与运动特性研究

王明洋,戚承志,钱七虎

(解放军理工大学 工程兵工程学院,江苏 南京 210007)

摘要: 通过对残留应力的研究表明:其大小不仅仅由变形的增降确定,也由所考察的构造集合体单独区段的变形模量的差值所确定。在具有构造等级的深部岩体介质的变形过程中,储能及返还性状与介质变形的摩擦系数有主要的关系。利用块体动力模型研究表明:块体界面的动力变形与稳定性的影响效应规律与块体及界面的体变能力及块体大小密切相关。

关键词: 井巷工程;深部岩体;变形与运动;超低摩擦现象

中图分类号: TD 322;P 542　　　　**文献标识码:** A　　　　**文章编号:** 1000 - 6915(2005)16 - 2825 - 06

STUDY ON DEFORMATION AND MOTION CHARACTERISTICS OF BLOCKS IN DEEP ROCK MASS

WANG Mingyang,QI Chengzhi,QIAN Qihu

(*Engineering Institute of Engineering Corps*,*PLA University of Science and Technology*,*Nanjing* 210007,*China*)

Abstract: The investigation of rudimental stress indicates that its magnitude has close relationship with the increment of deformation, and the dispersion of deformation module for solitude zone of tectonic groups. The capability of storage and returned energy is related with friction coefficient for the deep rock mass with tectonic level. The study on dynamic model of block indicates that the interface effects of blocks are related with the size of block and the capability of bulk deformation included in block and interface.

Key words: drift engineering;deep rock mass;deformation and motion;anomalous low friction phenomenon

1 引　言

岩体受到地质构造过程及其作用后,呈现裂隙变形应力效应(有张开的和未张开的裂隙),并将其切割为尺寸大小不同的岩块。由于连续体变成碎裂体后,岩块组成的岩体其变形即集中于裂隙处,也就是岩块边界相互关系效应,强度也就取决于裂隙岩体、岩块相互作用产生的强度性质效应问题。我国陈宗基院士早在 20 世纪 60 年代就已充分认识到岩体变形由岩块中的岩石介质变形和岩块在边界面(结构面)附近区域以及岩体的弱化区(裂缝处)的变形组成,并进行了基于流变学观点的理论与实验研究,取得了开创性的成果[1, 2]。因此,岩体介质的基本性质及其变形特性,在解决实际介质在外力作用下的行为特点(能量的储存、耗散、转移)以及稳定性相关联的问题时,特别需要关注介质表现出来的非均匀构造特性,也就是裂隙行为的密度、方向和间隔等关键问题。

岩体介质中存在着不同水平的构造缺陷,其决定了介质的构造等级。在岩体介质的块体中对于由具有任意等级构造组成单元的基本变形特征由以下

收稿日期:2005-03-16;修回日期:2005-04-11。
基金项目:国家自然科学基金重大项目(50490275);国家自然科学基金重点项目(50439030)。
本文原载于《岩石力学与工程学报》(2005 年第 24 卷第 16 期)。

两方面决定：一是其组成部分(块体及其间隙)的变形及强度特征参数；二是集合体的变形特征参数。根据外部作用参数的不同组成具体集合体的构造单元的变形，可以是协调的，也可以是不协调的。由此，集合体的状态在某一时刻可以是密合的，也可以是非密合的。如果在变形的初始阶段所有构造等级的构造单元处于密合状态，那么，根据流变学的概念，可以划分出构造非均匀介质的3个变形阶段。

第1阶段为构造块的可逆变形阶段。在这一阶段中，组成单元的变形在弹性限内是在所有组成部分的协调变形中进行的。由于介质的非均匀性所引起的应力集中完全由外部扰动所决定。在外力撤去后介质返回到初始状态。介质变形的第一阶段行为可以借助于复合介质的弹性理论来描述。

第2阶段为构造块的不可逆的协调变形。此阶段在所有水平上构造单元的变形协调性是由具有不同的力学性质的区段V_i边界上形成变形的塑性非协调区域来达到的。变形发生中是没有破坏介质的连续性，这是因为在强度比较小的单元的某些部位上所形成的塑性非相容性由另外部位的弹性变形来补偿。这时产生的内部应力与在不均匀部位处的应力集中不一样，在外力撤去后仍然存在。这样，在构造非均匀介质中会出现力矩，并作用于单元集合上，或者作用于单独的构造单元上。这一阶段基于流变学对变形参数进行空间上的平均(连续介质近似)基本上不能够描述介质变形时局部特性[3, 4]。

第3阶段为构造块的非协调变形阶段。在这一阶段单元变形时伴随着介质连续性的破坏。目前描述复杂块体系统变形过程的数学方法还没有得到很好的研究[5]。

本文首先研究在具有构造等级的岩体介质的变形过程中的储能性状，然后分析其特点对块系介质动力变形与稳定性的影响效应规律。

2 深部岩块变形的储能性状

深部岩体是位于地壳中的地质体，在地质年代动态构造运动过程中，生成断层、层理、节理、微裂隙等各种缺陷，即张开的和未张开的裂隙。把岩体纵横割分为各种形状不同、大小不等的岩块，具有空间性质和面效应的组合体。根据地质学者的量测和统计，裂缝的宽度可从百米级的构造断裂到10^{-8} m的微裂纹和矿物晶粒的缺陷，裂缝的延展长度为$10^6 \sim 10^{-6}$ m，即从宏观断裂到细观和微观裂纹。按照M. A. Sadovsky院士的观点，岩体中存在着复杂的构造层次系统，该层次系统涉及所有科学研究的尺度，从原子级别到地质构造级别以至行星尺度级别。从众多理论及试验的研究工作，该观点得到了证实[6]。岩体层次构造的概念的定量描述包括：

(1) 构成岩体的不同层次岩石块体尺寸Δ_i的自相似规律，对相邻层次的块体，有

$$\frac{\Delta_i}{\Delta_{i+1}} = \sqrt{2} \qquad (1)$$

(2) 岩体每一构造层次都有自己的结构面，结构面间有一定宽度δ_i的裂缝存在、同一层次的δ_i与Δ_i存在着稳定的统计关系：

$$\mu_\Delta(\delta) = \frac{\delta_i}{\Delta_i} = 0.01\Theta \qquad (2)$$

式中：Θ为系数，其变化范围为0.5～2.0；$\mu_\Delta(\delta)$为岩石力学"不变量"。

δ_i与Δ_i表征了岩体构造等级的尺度，但不决定其力学性质。根据外部作用的量度尺度及特征的不同，同样的构造等级，对于介质的变形过程有不同的影响。与其相连，必须研究划分地质介质构造等级的方法。在其划分方法中，应需引进块间填充材料的具体力学性质，或者利用间接的数据资料，确定不同类型的构造等级、相对变形及强度性质，而地下坑道的地震过滤法是有应用价值的。这种方法是以地震波在非均匀介质中传播时幅值及频率发生变化为依据的。

实际岩体可表示为遵从等级序列的构造单元的集合体(块体)，这些块体由软弱区隔开。作为构造单元边界，可以是张开(未填充)的，也可是部分填充的断层及裂缝的地质构造带。在任何情况下，任何构造水平上岩体介质都可以表示成具有某种体积V_i ($i = 1, 2, 3, \cdots, n$)的块体之集合体，这些块体具有不同的物理力学性质。

通过考虑岩体中具有的间断来对介质构造单元进行等级排列，假设最大的构造块叫做0级块体，组成0级块体的构造单元是由0级块体中最大的间断划分，并被称作1级块体($n = 1$)，逐次类推。这样岩体介质的每一个单元都可以被赋予等级($n = 1, 2, 3, \cdots, N$；其中，N为等级划分的最深层次)。这时$n+1$，$n+2$，$\cdots$，N级构造块体的集合体为服从构造等级的n级块体的独立构造块，同时也是$n-1$级块体的从属单元。

这样形成的实际岩体等级构造模型包括$n = 1$，

2，3，…，N 等级的块体及强度软弱区域(不同等级块体之间的间隔)，而且在每一个集合体中介质是非常不均匀的，并包括具有不同力学性质的区域 V_i，这些区域可以是岩体块、更高级块体的集合、块体之间的间隙。

对于第 n 级的任意集合体的第 i 区段，设 ε_i^e，ε_i^p 为其 i 区段体积内的平均弹性及塑性变形张量，用 ε_i 来描述第 i 区段作为均匀整体的变形，则 i 区段的体积平均变形张量 ε_i 为

$$\varepsilon_i = \varepsilon_i^e + \varepsilon_i^p \tag{3}$$

引入在体积 V 平均意义上的弹性及塑性变形的增降量为

$$\Delta\varepsilon_i^e = \varepsilon_i^e - \varepsilon_V^e, \quad \Delta\varepsilon_i^p = \varepsilon_i^p - \varepsilon_V^p \tag{4}$$

式中：$\varepsilon_V^e = \bar{\varepsilon}^e$，$\varepsilon_V^p = \bar{\varepsilon}^p$（上横杠表示对于体积 V 的平均）。因此，集合体的变形连续性条件为

$$\varepsilon_i^e + \varepsilon_i^p = \varepsilon_V^e + \varepsilon_V^p$$

即

$$\Delta\varepsilon_i^e + \Delta\varepsilon_i^p = 0 \tag{5}$$

式(5)表示在出现塑性非协调性时（$\Delta\varepsilon_i^p \neq 0$），介质的连续性只有在出现弹性变形 $\Delta\varepsilon_i^e = -\Delta\varepsilon_i^p$ 时，方能维持。当式(5)不能满足(弹性变形不足以补偿塑性非协调性)时，那么，介质变形时会失去连续性，该集合体在受到强度削弱的某一水平上发生旋转与断裂。在某一个具体的已被削弱水平上所发生的旋转与断裂过程可以看成在岩体介质中形成耗散结构的最或然机制。这样会形成某个另一级结构单元的耗散结构，这些单元的变形在与外部能量交换中很大程度上决定了介质的总体变形。

应当指出，在 n 级构造水平上非连续性的出现使得其整体变形得以容易，这样导致在更低一级（$n-1$ 级上）块体的变形中塑性非协调性的形成被延迟，意味着大尺度的块体只有在更大的变形时才会参与运动(加速变形)。同样，应该指出的是所引入的变形增降量 $\Delta\varepsilon_i^e$ 及 $\Delta\varepsilon_i^p$ 就集合体的宏观体积变形来讲表征第 i 单元的变形性态，这可以按照增降 $\Delta\varepsilon_i^e$ 及 $\Delta\varepsilon_i^p$ 不依赖于临近单元的变形进行单独考察每个单元的变形。

引入 n 级集合体第 i 区段平均的应力张量及其相对于宏观应力 σ_V 的增降，与张量 ε_i^e，ε_i^p 和 $\Delta\varepsilon_i^e$ 类似得到下式：

$$\left.\begin{array}{l}\Delta\sigma_i = \sigma_i - \sigma_V \\ \sigma_i = k_i\varepsilon_i^e \\ \sigma_V = \bar{\sigma}_i\end{array}\right\} \tag{6}$$

式中：k_i 为在第 i 区段上平均的变形模量，正如前面所述的下标 i 表示参数所属的区段。

用 Δk_i 表示广义变形模量相对于宏观体积上的平均变形模量 $k_0 = \bar{k}_i$ 增降量，利用式(3)~(6)，得到 $\Delta\sigma_i$ 及 σ_V 的表达式为

$$\begin{aligned}\Delta\sigma_i &= \sigma_i - \sigma_V = k_i\varepsilon_i^e - k_0\varepsilon_V^e + \Delta\bar{k}_i\Delta\bar{\varepsilon}_i^p = \\ &\quad k_i\varepsilon_i^e - (k_i - \Delta k_i)\varepsilon_V^e + \Delta\bar{k}_i\Delta\bar{\varepsilon}_i^p = \\ &\quad k_i\Delta\varepsilon_i^e + \Delta k_i\varepsilon_V^e + \Delta\bar{k}_i\Delta\bar{\varepsilon}_i^p\end{aligned} \tag{7a}$$

$$\begin{aligned}\sigma_V &= \bar{\sigma}_i = <k_i\varepsilon_i^e> = <(k_0 + \Delta k_i)(\varepsilon_V^e + \Delta\varepsilon_i^e)> = \\ &\quad <(k_0 + \Delta k_i)(\varepsilon_V^e - \Delta\varepsilon_i^p)> = \\ &\quad <k_0\varepsilon_V^e> - <k_0\Delta\varepsilon_i^p> + <\Delta k_i\varepsilon_V^e> - <\Delta k_i\Delta\varepsilon_i^p> = \\ &\quad <k_0\varepsilon_V^e> - <\Delta k_i\Delta\varepsilon_i^p>\end{aligned} \tag{7b}$$

式中：$<>$ 为平均符号。

利用式(5)，上式可改写为

$$\Delta\sigma_i = -k_i\Delta\varepsilon_i^p + \Delta k_i\varepsilon_V^e + \Delta\bar{k}_i\Delta\bar{\varepsilon}_i^p \tag{8}$$

由此在卸去外力之后：$\sigma_V \to 0$，很容易确定残留内应力值：

$$\Delta\sigma_i = k_i[k_0^{-1}(\Delta\bar{k}\Delta\varepsilon_i^p) - \Delta\varepsilon_i^p] \tag{9}$$

式(9)表明，非均匀介质变形时产生的塑性非协调性，将导致内应力的产生。且这种内应力即便在外载彻底卸去后仍然存在，其大小(包括残留应力值)不仅由变形增降确定，也由所考察构造集合体的单独区段变形模量的差值 Δk_i 所确定，也即与介质变形的摩擦系数有主要的关系(如图 1 所示)。这一性质表明，岩块单元在外载彻底卸去后单元的很大一部分能量还保存下来，在一定条件下，这一储存能量可以释放，既能缓慢释放，也可快速释放。岩体这一特性非常显著影响着岩体变形过程的力学性质和变形稳定性。

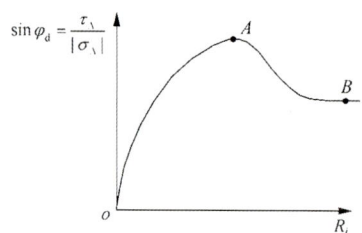

图 1 介质变形的动摩擦系数

Fig.1 Dynamic friction factor of deformation of media

3 块系岩体的变形与运动特性

从力学观点来看,作为能量源的介质储能的存在非常显著地影响了变形的稳定性,并总的来讲影响着变形过程的力学性质。理解深部含能块系介质的变形稳定特性须首先揭示块体界面的法向与切向动力特性。这里研究界面法向的变形及运动特性。

首先研究块系间界面的法向动力特性问题。为了简单起见采用如图2的模型。图2中设可变形块体 A 的长、宽、高特征尺寸分别为 Δ_1,Δ_2,Δ_3,密度为 ρ_A,质量为 $m_A = \rho_A \Delta_1 \Delta_2 \Delta_3$,块体 A 与质量为 m_0 的刚性块体 B 的接触面积为 $S = \Delta_1 \Delta_2$,在块体 A 上作用有冲击动载为 $p(t) = pf(t)$。由于岩块之间界面具有存储和耗散能量的能力,这种能力可通过弹簧与阻尼器来实现,即可通过刚体位移与速度来表征:

$$R_0 = c_0 u(t) + c_1 \dot{u}(t) \tag{10}$$

式中:c_0,c_1 分别为界面的接触刚度系数和波阻抗;$u(t)$ 为块体的刚体位移。但需要注意的是,在模型中弹簧的相对位移不能破坏式(2)的条件,即 $\delta = \Delta_1 \mu(\Delta) \geq u_{\max}(t)$,其中,$\mu(\Delta)$ 为不变量,Δ 为块体的特征尺度。

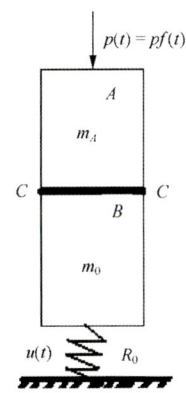

图2 块系法向运动计算简图

Fig.2 Calculation sketch for the vertical motion of rock block displacement

由以上分析,块体 A 的全部位移为

$$Y(x, t) = u(t) + pX(x)y(t) \tag{11}$$

将式(11)代入杆的波动方程式,并按照 Бубнова-Галеркина 方法进行变换,得

$$\ddot{y} + \omega^2 y = \omega^2 \left[f(t) - \frac{m}{p} \ddot{u}(t) \right] \tag{12}$$

式(11),(12)中:m 为块体单位长度质量,且 $m = \dfrac{m_0}{\Delta_3} = \rho_0 A$;$X(x) = \dfrac{1}{EA}\dfrac{x^2}{2}$;$\omega^2 = \dfrac{\int_0^1 X(x)\mathrm{d}x}{X^2(x)\mathrm{d}x} = \dfrac{10}{3}\dfrac{c_t^2}{\Delta_3^2}$;$\omega = 1.83\dfrac{c_t}{\Delta_3}$;$c_t = \sqrt{\dfrac{E}{\rho_0}}$ 块体中的压缩弹性波速(类似一维杆中弹性波速)。

块体界面满足的平衡条件为

$$EAp\left.\dfrac{\partial X}{\partial x}\right|_{x=\Delta_3} y(t) - R_0 - m_0 \ddot{u} = 0 \tag{13a}$$

即

$$p\Delta_{3y} y(t) - R_0 - m_0 \ddot{u} = 0 \tag{13b}$$

令:$r_1 = \dfrac{c_1}{m_0}$,$r_0 = \dfrac{c_0}{m_0} = \omega_0^2$,$\lambda_1 = \dfrac{\omega^2}{\omega_0^2}$,$u(t) = \dfrac{p\Delta_3}{c_0}U(t)$($U(t)$ 为刚体运动的动力函数)。据式(12)和(13)得到动力函数 $U(t)$ 和 $y(t)$ 的方程组:

$$\left.\begin{array}{l}\ddot{U} + r_1 \dot{U} + r_0 U - r_0 y = 0 \\ \lambda_1 \ddot{U} + \ddot{y} + \omega^2 y = \omega^2 f(t)\end{array}\right\} \tag{14}$$

初始条件为:$u|_{t=0} = \dot{u}|_{t=0} = y|_{t=0} = \dot{y}|_{t=0} = 0$。

下面求解相应齐次方程式组的通解。假定 $U_0 = Ce^{st}$,$y_0 = Be^{st}$,则

$$\left.\begin{array}{l}C(s^2 + r_1 s + r_0) - r_0 B = 0 \\ C\lambda_1 s^2 + B(s^2 + \omega^2) = 0\end{array}\right\} \tag{15}$$

由此得

$$(s^2 + r_1 s + r_0)(s^2 + \omega^2) + \omega^2 s^2 = 0 \tag{16}$$

令方程式组的根为:$s_1 = (\alpha_1 + \mathrm{i}\beta_1)\omega$,$s_2 = (\alpha_2 + \mathrm{i}\beta_2)\omega$,计算最终求得

$$\left.\begin{array}{l}U_0 = (C_1 \sin\beta_1\omega t + C_2 \cos\beta_1\omega t)\mathrm{e}^{\alpha_1\omega t} + \\ \quad (C_3 \sin\beta_2\omega t + C_4 \cos\beta_2\omega t)\mathrm{e}^{\alpha_2\omega t} \\ y_0 = (B_1 \sin\beta_1\omega t + B_2 \cos\beta_1\omega t)\mathrm{e}^{\alpha_1\omega t} + \\ \quad (B_3 \sin\beta_2\omega t + B_4 \cos\beta_2\omega t)\mathrm{e}^{\alpha_2\omega t}\end{array}\right\} \tag{17}$$

式中:C_i 为任意常数;B_i 为按式(15)用 C_i 表示;α_i,β_i 为方程式的根。

$$(\alpha^2 - \beta^2 + 1)(n_1 + n_2\alpha + \alpha^2 - \beta^2) - 2\alpha\beta^2(n_2 + 2\alpha) + (\alpha^2 - \beta^2) = 0 \tag{18}$$

$$\beta^2 = \alpha^2 + [n_2 + 2\alpha(2 + n_1 + n_2\alpha)]\dfrac{1}{n_2 + 4\alpha} \tag{19}$$

$$\left.\begin{array}{l}n_1 = \dfrac{r_0}{\omega^2} = \dfrac{\omega_0^2}{\omega^2} \\ n_2 = \dfrac{r_1}{\omega}\end{array}\right\} \tag{20}$$

冲击载荷形式假设为：$f(t) = \left(1 - \dfrac{t}{\theta}\right)(0 \le t \le \theta)$，则 $f(t) = 0$ 其时间段为 $0 < t \le \theta$。因此，式(14)的特解为

$$\left. \begin{aligned} y_1 &= 1 - \dfrac{t}{\theta} \\ U_1 &= 1 + \dfrac{r_1}{r_0 \theta} - \dfrac{t}{\theta} \end{aligned} \right\} \quad (21)$$

同理可以推得 $f(t) = \sin\Omega t$ 或 $f(t) = \mathrm{e}^{-\Omega t}$ 的特解。

在满足初始条件以后可得

$$\left. \begin{aligned} U(t) &= 1 - \dfrac{t}{\theta} + \dfrac{n_2}{n_1 \omega \theta} - n_1 \sum_{i=1,2}(D_i \cos\beta_i\omega t - E_i \sin\beta_i\omega t)\mathrm{e}^{\alpha_i\omega t} \\ y(t) &= 1 - \dfrac{t}{\theta} - \sum_{i=1,2}(A_i \cos\beta_i\omega t - B_i \sin\beta_i\omega t)\mathrm{e}^{\alpha_i\omega t} \end{aligned} \right\} \quad (22)$$

其中，

$$\left. \begin{aligned} D_i &= \dfrac{c_i}{\beta_i(c_i^2 + d_i^2)} - \dfrac{h_i}{\beta_i(h_i^2 + q_i^2)\omega\theta} \\ E_i &= \dfrac{d_i}{\beta_i(c_i^2 + d_i^2)} - \dfrac{q_i}{\beta_i(h_i^2 + q_i^2)\omega\theta} \\ A_i &= \dfrac{g_i}{\beta_i(c_i^2 + d_i^2)} - \dfrac{r_i}{\beta_i(h_i^2 + q_i^2)\omega\theta} \\ B_i &= \dfrac{\overline{g}_i}{\beta_i(c_i^2 + d_i^2)} - \dfrac{\overline{r}_i}{\beta_i(h_i^2 + q_i^2)\omega\theta} \\ c_i &= \beta_i a_i + \alpha_i b_i \\ d_i &= \alpha_i a_i - \beta_i b_i \\ a_1 &= (\alpha_1 - \alpha_2)^2 - \beta_1^2 + \beta_2^2 \\ a_2 &= (\alpha_1 - \alpha_2)^2 + \beta_1^2 - \beta_2^2 \\ b_1 &= 2\beta_1(\alpha_1 - \alpha_2) \\ b_2 &= 2\beta_2(\alpha_2 - \alpha_1) \\ h_i &= b_i(\alpha_i^2 - \beta_i^2) + 2\alpha_i\beta_i a_i \\ q_i &= a_i(\alpha_i^2 - \beta_i^2) - 2\alpha_i\beta_i b_i \\ g_i &= e_i c_i - f_i d_i \\ r_i &= e_i h_i - f_i q_i \\ \overline{g}_i &= e_i d_i + f_i c_i \\ \overline{r}_i &= e_i q_i + f_i h_i \\ e_i &= n_1 + \alpha_i n_2 + \alpha^2 - \beta_i^2 \\ f_i &= \beta_i(n_2 + 2\alpha_i) \end{aligned} \right\} \quad (23)$$

为便于计算，式(18)变换成如下形式：

$$(\alpha^2 - \beta^2)(n_1 + n_2\alpha + 2) + (\alpha^2 - \beta^2)^2 + n_1 + n_2\alpha - 2\alpha\beta^2(n_2 + 2\alpha) \quad (24)$$

由式(19)得

$$(n_2 + 2\alpha)(\alpha^2 + \beta^2)^2 - n_2(\alpha^2 + \beta^2) - 2\alpha n_1 = 0 \quad (25)$$

式(25)可与式(19)联立求解。变换以后的式(19)和(25)具有一个附加解：$\alpha = 0$，$\beta = 1$，在计算中不应予以考虑。式(25)可对其根 α_1 和 α_2 得出很多结论。分析式(25)得到的函数曲线表明(见图3)，必须是 $\alpha_1 < 0$，$\alpha_2 < 0$，在此条件下，耦合的齐次线性系统的平凡解为渐进稳定的。如果式(19)和(25)有一个解，则第 2 个解为

$$\alpha_2 = -0.5 n_2 - \alpha_1 \quad (26)$$

为此，直接从式(18)，(19)或(25)中就足以找到一个解 α_1，β_1。如果 $c_1 = 0$，即 $r_1 = 0$，从式(16)中就可以得出结论：只可能有最小的一个根。这样，$\alpha_i = 0$，并且从含 β 的式(18)中得

$$\beta^4 - \beta^2(2 + n_1) + n_1 = 0 \quad (27)$$

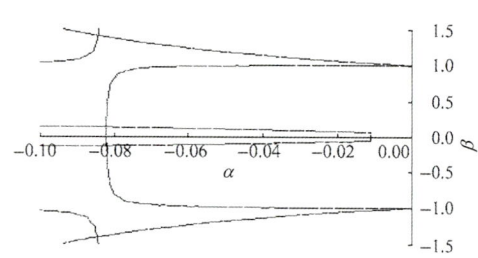

图3　式(25)中 α，β 的曲线图

Fig.3　Curves of α, β in Eq. 25

在这种情况下，振动衰减自然就不存在了。如果块体的界面间均是理想接触(这是不可能的)，也就是不存在刚体运动($m_0 = 0$)，那么，在 $m_0 \to 0$ 时，就得到 $\beta^2(1 + \lambda_1) = 1$。

式(10)给出界面相互作用的内力值，对于下部的块系对块体 B 的阻抗可以根据波传播的相互作用理论得到等效的系数 c_0，c_1，此问题拟另文讨论。作为界面机理性的探讨，设块体 B 放置在半空间均匀的介质上，其系数[7]可取为

$$\left. \begin{aligned} c_0 &= \dfrac{\rho c_\mathrm{p}^2 A}{2\varDelta_1} \\ c_1 &= \rho c_\mathrm{p} A \end{aligned} \right\} \quad (28)$$

式中：ρ，c_p 分别为半空间介质的密度与压缩波的传播速度。

以下列 2 个无量纲参数系数 s_0，s_1 来表达：

$$\left. \begin{aligned} s_0 &= \dfrac{c_\mathrm{p}}{\omega \varDelta_1} \\ s_1 &= \dfrac{4m_0}{\rho A \varDelta_1} = \dfrac{4m_0\mu(\varDelta)}{\rho A \delta} = \dfrac{4m_0\mu(\varDelta)}{m_\mathrm{s}} \\ m_\mathrm{s} &= \rho A \delta \end{aligned} \right\} \quad (29)$$

式(19)的数值用这些参数表示并按下式求得,即

$$\left.\begin{array}{l}n_1 = 2\dfrac{s_0^2}{s_1} \\ n_2 = 4\dfrac{s_0}{s_1}\end{array}\right\} \quad (30)$$

块体界面的纵向力等于$N(Y, t) = -R_0 - m_0\ddot{u}$。计算表明,无论是块体的内力和位移随时间变化的性质,还是其最大值,都与s_0值有关。s_0实际上表征了块体与界面抵抗体积变形的能力(界面的压缩波速与块体中的压缩波速比),当s_0较小时($s_0<1$),界面中最大内力和位移的数值并不是在第1个振动中达到,而是在第2个最大振动以后达到。当s_0值较大时,内力和位移最大值在第一个振动最大值时达到(见图4)。对所有数值来说,随着s_0的增加,动力系数都增大。当s_0值很大时($s_0 \geqslant 4$),块体的刚体运动对界面影响不大,且可以不加考虑。

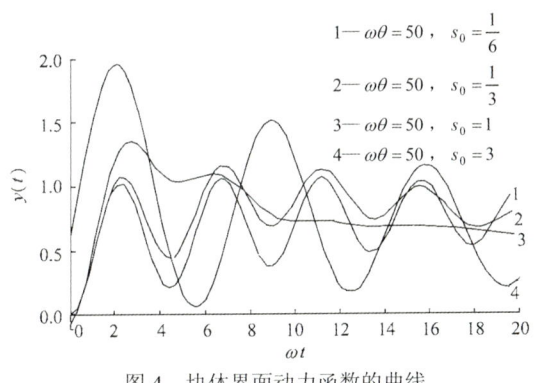

图 4 块体界面动力函数的曲线

Fig.4 Curves of dynamic function for the interface of rock block

以上研究结果可以用来研究文[8,10]指出的超低摩擦现象机理问题。

切向方向运动的界面受摩擦力的影响,摩擦力大小与作用在法向面上的力成正比。在波开始向下传播以后,有效的和附加的惯性力能使动荷载减小;这一点可以用第1个最大振动时$y(t)$减小来解释。在波反射的第2阶段中,当加速度是负的时,惯性力使荷载增加,并使第2个最大振动时$y(t)$的函数值得以增加。这个数值取决于振动的衰减程度以及荷载作用的持续时间。如果衰减没有了($c_1 = 0$),第2个最大振动时的$y(t)$值就会增加到无刚体运动结构的动力系数值。因此,块体在其位移时所产生的内力减少现象主要是由于振动衰减造成的,而振动衰减又受式(10)中的粘性阻抗来约制。要指出的是,荷载作用持续时间对块体刚体位移结构工作的影响,要比无刚体位移的情况更大一些,因为块体的刚体运动使块体中最大内力到达的时间增加。

4 结 论

(1) 深部岩体中存在的残留应力大小不仅由变形的增降确定,也由所考察的构造集合体的单独区段的变形模量差值所确定。

(2) 在具有构造等级的深部岩体介质的变形过程中,储能及返还性状与介质变形的摩擦系数有主要的关系。

(3) 块体界面的动力变形和稳定性的影响效应规律不仅与块体及界面的体变能力及块体大小密切相关,而且超低摩擦现象机理也与此密切相关。

参考文献(References):

[1] 陈宗基,康文法. 岩石的封闭应力、蠕变和扩容及本构方程[J]. 岩石力学与工程学报,1991,10(4):299-312.(Tan Tjongki, Kang Wenfa. On the locked in stress, creep and dilatation of rocks, and the constitutive equations[J]. Chinese Journal of Rock Mechanics and Engineering, 1991, 10(4): 299-312.(in Chinese))

[2] 陈宗基. 地下巷道长期稳定性的力学问题[J]. 岩石力学与工程学报,1982,1(1),1-20.(Tan Tjongkie. The mechanical problems for the long-term stability of ground galleries[J]. Chinese Journal of Rock Mechanics and Engineering, 1982, 1(1): 1-20.(in Chinese))

[3] Sadovsky M A. Mature of lumpiness of rock, Doklady AN SSSR[J]. Geotectonics, 1979, (4): 247-256.

[4] Revuzhenko P. Mechanics of Elastoplastic Media and Nonstandard Analysis[M]. Novosbirsk: Izd NGU, 2000.

[5] Adushkin V V, Spivak A A. Geomechanics of Large-scale Explosions[M]. Moscow: Nedra, 1993.

[6] Kurlenya M V, Adushkin V V, Oparin V N. Alternating reaction of rocks to dynamic action, Doklady AN SSSR[J]. Geotectonics, 1992, (2): 323-329.

[7] Qian Qihu, Wang Mingyang. Comparison studies of loading theory and test results for shallow-buried structures under nuclear explosion[A]. In: International Workshop Blast-resistant Structures[C]. Beijing: Tsinghua University Press, 1992. 76-86.

[8] Kurlenya M V, Oparin V N. Problems of nonlinear geomechanics part I[J]. Fiz.Tekh. Probl. Razrab.Polezn. Iskop., 1999, 3: 435-442.

[9] Kurlenya M V, Oparin V N. Problems of nonlinear geomechanics part II[J]. Fiz. Tekh. Probl. Razrab.Polezn. Iskop., 2000, 4: 689-703.

[10] 钱七虎. 防护结构计算原理[M]. 南京:工程兵工程学院,1981. (Qian Qihu. Calculation Theory of Protective Structures[M]. Nanjing: Engineering Corps Institute Press, 1981.(in Chinese))

岩石中侵彻与爆炸作用的近区问题研究

王明洋，邓宏见，钱七虎

(解放军理工大学 工程兵工程学院，江苏 南京 210007)

摘要：首先，从理论上分析了目前广泛使用的空腔膨胀理论在研究介质中侵彻与爆炸近区过程存在的问题及其产生的原因；然后，从近区应力与变形的实际状态出发，运用动量与质量守恒关系推得了介质近区运动学的关系式，并解决了变形波速确定的问题。运用此关系可以得到揭示侵彻、贯穿比例换算关系和爆炸近区的几何相似关系等简单实用的结果。

关键词：岩石力学；岩石；侵彻与爆炸；近区

中图分类号：TU 452　　　**文献标识码**：A　　　**文章编号**：1000 – 6915(2005)16 – 2859 – 05

STUDY ON PROBLEMS OF NEAR CAVITY OF PENETRATION AND EXPLOSION IN ROCK

WANG Mingyang，DENG Hongjian，QIAN Qihu

(*Engineering Institute of Engineering Corps*，*PLA University of Science and Technology*，*Nanjing* 210007，*China*)

Abstract：The fundamental problems of near cavity of penetration and explosion in rock are discussed；and some development tendencies are presented. The expressions about the kinematics problems near cavity do not agree well with the stress and the deformation states. That are the reasons that considerable errors exist in the results of many scholars studies. According to the stress and deformation states near cavity in practice，the relation formulas for the kinematics of medium are deduced by using the laws of conservations of momentum and mass；and the problem of the speed of the deformation wave is solved. With the proposed relation formulae，some useful results such as the proportional conversion relation of penetration and impenetration，geometrical resemblance relation of explosion near cavity，can be clearly understood. The application results of case study show that the proposed relation formulas are practical.

Key words：rock mechanics；rock；penetration and explosion；near cavity

1 引 言

目前，岩石中侵彻与爆炸问题的研究离实际问题的解决还有相当的距离。除了土中的爆炸荷载确定有了一定的理论基础，其他的计算(围岩动应力、运动参数、侵彻深度、成坑大小和震塌效应等)大都是建立在能量和相似原理的粗略关系式的基础上。

而对近区的研究大多是根据试验归纳拟合的相似系数，往往存在数量级的误差。但是侵彻与爆炸近区性状(空腔及近区破坏半径)是最终决定辐射出来波的基本参数，反映能量的分配份额，揭示侵彻、贯穿、爆炸及爆炸地震动等重要特性关键因素。因此，岩石中侵彻与爆炸近区的研究一直是试验与理论研究的热点与难点。

侵彻与爆炸近区在理论研究方面主要是根据空

收稿日期：2005–05–09；修回日期：2005–06–21。
基金项目：国家自然科学基金重大项目(50490275)。
本文原载于《岩石力学与工程学报》(2005 年第 24 卷第 16 期)。

腔膨胀理论,采用各种或简单或复杂的状态方程及本构关系[1~5]。迄今为止,众多的国内外研究工作中均忽略了一个事实,那就是目前对侵彻与爆炸作用的近区介质运动学关系的表征与近区应力与变形状态不相符合。这也就是为什么许多理论研究成果在近区存在数量级误差的原因。

有鉴于此,本文首先从理论上分析了目前广泛使用的空腔膨胀理论在研究介质中侵彻与爆炸近区过程中存在的问题及其产生的原因,然后从近区应力与变形实际状态出发,运用动量与质量守恒关系推导了介质近区运动学的关系式,运用此结果得到了侵彻、贯穿和爆炸近区的比例换算关系和几何相似关系等简单实用结果,并回答了文[1]需要确定的变形波速问题。

2 目前理论存在的问题

首先分析爆炸与侵彻作用下介质近区的试验结果[6, 7],得到如下定性与定量方面的结论:

(1) 尽管在装药附近有显著的破坏区,但是在质点速度变化图上并没有显示出由于材料破坏所引起的介质不可逆变化的任何征状。这一事实表明,要么是材料的破坏没有损坏运动的连续性,要么是破坏波阵面上的运动参数的变化小到可以忽略不计。因为与近区中波的作用的全部范围相比较,大应力和大速度增长的传播范围比较小,波阵面可以理解成粒子速度达到最大值的面。这样可认为在波阵面上满足下列关系,即

$$\left.\begin{array}{l}\sigma_r = \rho_0 C_p v_r \\ v_r = C_p \varepsilon_r\end{array}\right\} \quad (1)$$

式中:ρ_0 为介质的初始密度,v_r 为粒子速度,C_p 为介质中弹性波速,ε_r 为介质中径向应变,σ_r 为介质中径向应力。

(2) 近区介质的应力波的图形见图 1,可以是 σ_r-t 的图形或是 v_r-t 的图形。图中还列出 w_{max}-t 的图形(即径向位移 - 时间的图形)。

由图 1 比较岩石的应力与变形状态可知,当应力"短"波($t_r/t_+ \approx 0.1$)中应力和速度达到峰值时,介质的变位还很小,即环向应变($\varepsilon_\theta = \varepsilon_\varphi$)很小,体积变化 ε_V 仅由于径向应变 ε_r 产生,而变形 $\varepsilon_\theta = \varepsilon_\varphi$ 的出现应是在变位发展时才应当考虑(如图 2 的应力与应变状态),其全部形成,从 $\sigma_\varphi = \alpha\sigma_r$ 时应力增长阶段开始,在由相邻构件的单轴变形向三轴变形过渡而受到约制条件下发生,那么在按照三轴试验

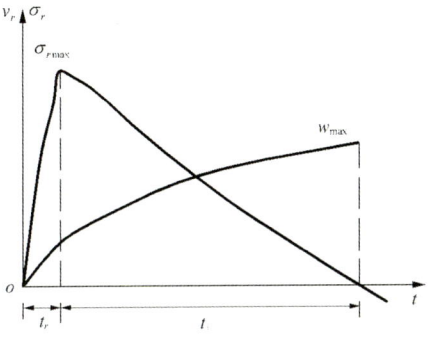

图 1 σ_r-t 曲线图
Fig.1 Stress-time(σ_r-t) curve

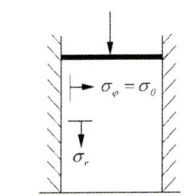

(a) 试件侧面受到约束($\sigma_\varphi = \sigma_\theta = \alpha\sigma_r$)

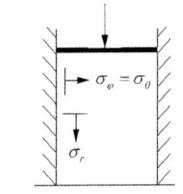

(b) 侧面受到有限约制($\sigma_\varphi = \sigma_\theta \neq \alpha\sigma_r$)

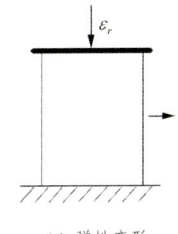

(c) 弹性变形

图 2 压缩时岩石试件变形
Fig.2 Deformation of rock sample under compression

条件下,就会出现另外一种状态:当轴向应力 σ_1 和侧面约束应力($\sigma_2 = \sigma_3$)成比例地变化时,给定的(控制的)变形 ε_1 无论如何也不会限制侧向(法向)变形($\varepsilon_2 = \varepsilon_3$)的发展。这也就是为什么对于球面爆炸波,其弹性波速与塑性波速与一维应变波一样,最多相差 15%左右的原因。即在近区有如下关系:

$$\left.\begin{array}{l}\varepsilon_V \approx \varepsilon_r \approx \gamma_{max} \\ \varepsilon_r \gg \varepsilon_\theta = \varepsilon_\varphi\end{array}\right\} \quad (2)$$

式中:γ_{max} 为最大剪应变。

(3) 图 1 中径向位移达到 w_{max} 以前径向应力下降的阶段,由于径向位移是正的,在应力波中完成

了对摩擦力所作的功，这就需要再耗费一些爆炸产物的能量。因此，近区中对能量耗损最主要的是摩擦运动。

(4) 尽管使用的模型介质其力学性质显著不同，但最大质点速度及其随距离的衰减在装药近区的区别不大，而在远区衰减程度显著不同。在爆炸近区，波动幅值及其衰减程度对介质性质的不敏感性限制了利用运动参数作为评价不同材料破碎特性的信息来源，这也就是为什么在研究岩石中爆炸时不能仅基于介质的变形参数来描述爆炸作用所有性状的原因，但在近区可以运用运动学关系研究确定材料破坏的能量及综合特征参数。

(5) 在图 1 中注意到在点 $\sigma_{r\max}$ 处发生加载(径向压缩增加)到卸载的转变，应力 $\sigma_{r\max}$ 按照卸荷波理论以径向压缩卸载波的速度传播，可是短波和弱波的卸载波速度在加载地段上接近于声速(也就是 C_p)。因此，如果只根据纵波测量到的速度 C_p 的大小来判断介质的状态，那就有可能做出不正确的结论：即近区中的介质处于弹性状态。要对介质状态作出更加精确的结论，就必须对剪切阻抗进行评估。

根据上述试验资料，下面来分析一下在研究侵彻与爆炸等问题时，目前广泛采用的空腔膨胀理论(以球腔理论为例)中的速度场关系[2, 3]为

$$v(r, t) = \dot{a}\left(\frac{a}{r}\right)^K \tag{3}$$

式中：$\dot{a}$ 为空腔的膨胀速度；K 为不同的扩容系数，且 $K = (2-\psi)/(1+\psi) \leqslant 2$，不考虑剪胀时，取 $K = 2$，ψ 为扩容系数。

根据式(3)可计算得到径向应变 $\left(\dot{\varepsilon}_r = \dfrac{\partial v_r}{\partial r}\right)$ 与环向应变 $\left(\dot{\varepsilon}_\theta = \dfrac{v_r}{r}\right)$ 的关系：

$$\varepsilon_r = K\varepsilon_\theta \tag{4}$$

显然，式(4)与近区中的应力与变形真实状态不相符合，这也就是为什么目前理论预测结果误差大的原因。下面根据动量与质量守恒定律对速度场的关系式给予重新表征。

3 侵彻与爆炸近区运动学关系

因为空腔球体由零半径膨胀时，空腔周壁上是有限的位移，变形较大所以不能采用小变形理论。侵彻与爆炸时介质中的能量主要分配在体积压缩存储的势能、耗散的不可恢复变形能和通过边界转移的运动能上。如前所述，近区介质主要以运动能为主，也就是近区破坏介质的体积实际上仍停留在达到峰值应力时的体积大小上，可以认为是不可压缩介质。因此，对于不可压缩介质而言，近区的位移变化可直接采用质量守恒条件($r^3 - a^3 = (r-w)^3$)，即

$$w(r) = r - \sqrt[3]{r^3 - a^3} \tag{5}$$

根据式(5)，径向应变与环向应变分别为

$$\varepsilon_r = \frac{\partial w}{\partial r} = 1 - \left[1\bigg/\sqrt[3]{1-\left(\frac{a}{r}\right)^3}\right]^2 \tag{6}$$

$$\varepsilon_\theta = 1 - \sqrt[3]{1-\left(\frac{a}{r}\right)^3} \tag{7}$$

对于近区 ε_r，ε_θ 随 r/a 的规律曲线示于图 3 中。由图 3 可知，在近区采用式(6)，(7)可满足其应力与变形状态。

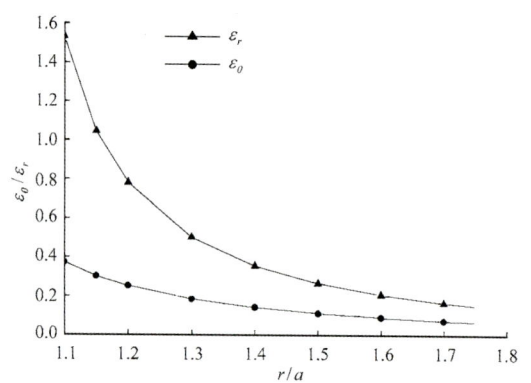

图 3 $\varepsilon_r/\varepsilon_\theta$ 随 r/a 的变化曲线

Fig.3 Variation curves of ε_r and ε_θ with r/a

对于介质的径向运动速度，无论对连续介质还是非连续介质均满足动量守恒定律，根据动量守恒采用式(1)得

$$v_r = C_p \varepsilon_r = C_p\left[1-\left(1\bigg/\sqrt[3]{1-r_*^3}\right)^2\right] \approx$$

$$\eta C_p\left(\frac{a}{r}\right)^K = C_p^*\left(\frac{a}{r}\right)^K \tag{8}$$

式中：η 为等价系数(与近区破碎特性有关)，$C_p^* = \eta C_p$ 为近区破碎介质的等价变形波速，r_* 为近区破碎半径。

对于不同的 K，在不同的近区 r_* 范围根据侵彻

与爆炸时破碎区与空腔半径之比值确定)计算得到的η值列于表1中。

表 1 等价系数η
Table 1 Equivalent coefficient η

K值	r/a				
	1.5~2.2	2.2~2.9	2.9~3.6	3.6~4.3	4.3~5.0
1.6	1/3	1/5	1/8	1/10	1/13
1.8	1/3	1/4	1/6	1/8	1/9
2.0	1/2	1/4	1/5	1/6	1/7

极硬岩石、坚硬岩石、较硬岩石的等价变形波速，根据文[1]和表1可分别取为550，450~550，380~450 m/s。

由于近区介质的运动特性使其变形接近一维平面应变状态，剪切与体积变形的改变属于同一数量级，即式(2)。这一状态导致在坚硬岩体中大幅值纵波的传播速度与小幅值纵波的传播速度差别很小。因此，随着应力σ_r跃迁程度的发展，在应力波中还会产生与σ_φ之间的差值，该差值将达到材料的剪切强度τ_0：

$$\left.\begin{array}{l}\sigma_r-\sigma_\varphi=2\tau_0\\\sigma_r+2\sigma_\varphi=3K\varepsilon\end{array}\right\} \quad (9)$$

一般而言，τ_0和K将由于应力跃迁或者随着体积应变的增大而变化。借助于式(9)，可以计算冲击波的主应力，即

$$\left.\begin{array}{l}\sigma_r=K\varepsilon+\dfrac{4}{3}\tau_0\\\sigma_\varphi=K\varepsilon-\dfrac{2}{3}\tau_0\end{array}\right\} \quad (10)$$

因而，当应变ε为0.1(波中$\sigma_r = 1\times 10^4$ MPa)和$\tau_0 = K\times 10^{-3}$ ($K\varepsilon \gg \tau_0$)时，从力学的观点可知这一状态类似于流体动力学状态：

$$\sigma_r = \sigma_\varphi = \sigma_\theta = -p \quad (11)$$

式中：p为实际的平均压力。

应该指出：在向式(11)的状态变化时，忽略强度与"强度消失"无关，而与消耗在体积变化的能量大大超过消耗在克服形状变化阻力上的能量有关。同时耗费能量的主要部分与不同的范围有关：体积的变化(包括加热和质变)与空腔半径r_c^3成正比，而克服滑移面上的剪应力和破坏(形成新的表面)则与r_c^2成正比。因为这些现象只具有表面的而没有体积的特征，所以计算结果及试验研究[7]表明，爆炸能的绝大部分被消耗在非弹性变形区。在爆炸空腔停止扩张后，地下爆炸的能量分配如下：爆炸产物含总能的10%~20%，介质的熔化占15%左右，耗散于非弹性区(非弹性变形区半径R_d)的能量占60%~70%，主要以摩擦成热的形式耗散，用于介质破裂的能量仅占0.1%~1.0%。在$R<R_d$的区域内，有90%~95%的能量被耗散掉，只有10%~15%的能量进入到弹性区，主要以介质的压缩形式存在。弹性区的能量的一部分用于地震动，而地震动由于几何发散、吸收及散射而逐渐衰减。根据地震动电信号整理结果，地震波能量主要取决于爆心岩体的性质。

因此，描述空腔周围介质运动学关系不能仅仅根据空腔表面的连续条件，而必须按照波阵面上的关系$v_r = C_p\varepsilon_r$和空腔表面上的关系$v_r(a, t) = \dot{a}$来得到。

4 具体应用

4.1 侵彻比例换算关系

侵彻问题的比例换算关系能够把侵彻深度、厚度方向的锥形开裂、震塌破坏和穿透的临界条件与弹丸的动量和能量(撞击速度和质量)、靶板的材料性能(包括强度、韧度和密度)、弹和靶的几何特性如靶板的有限厚度、弹丸的直径和弹丸的弹头形状联系起来并作定量的描述。

侵彻问题的比例换算关系实际上表征了空腔与破坏区的相对大小的关系，揭示了能量的分配关系。这一关系[1]为

$$\frac{a}{b} = \eta_1 \sqrt[4]{\frac{a}{l}} \quad (12)$$

式中：$l = \left(\dfrac{K_c}{\tau_s^e}\right)^2$，$K_c$为介质的抗断裂韧度，$\tau_s^e$为介质的抗剪强度；$a$为弹体半径；$b$为破碎区半径。

穿透的临界条件可以表达为

$$\frac{R_*}{a} = \eta_2 \sqrt{\frac{a}{l}} \quad (13)$$

式中：η_1，η_2分别为与$\sqrt{2}$有关的常系数。

根据式(13)计算到靶板背面的距离，如果在这个距离内时，裂缝超越出背面，贯穿体形成，并且开始对弹体的运动阻抗产生影响，这样就可以进行有限厚度靶板的贯穿过程的研究。

4.2 爆炸几何相似关系

在强烈爆炸作用下，爆炸近区进行着复杂的变化过程，包括压缩波的传播、塑性变形的发生、介质的破坏及爆炸空腔的形成。在孔隙率较低的坚硬岩石中，空腔的形成主要是压缩波向外挤压一定体积的介质所致，被挤压的体积被挤向弹性区并与爆炸能量成正比。空腔的半径 r_c 及非弹性变形区半径 R_d 依赖于爆炸能量及介质的性质，包括可压缩性和强度，根据本文近区速度场的研究结果，其确定公式如下：

$$\left.\begin{array}{l} r_c = \dfrac{0.6 Q^{1/3}}{\left(\rho C_p^2 \sigma_c^2\right)^{1/9}} \\ R_d = \left(\dfrac{\rho C_p^2}{4\sigma_c}\right) r_c \end{array}\right\} \quad (14)$$

式中：ρ，C_p 分别为介质的密度及介质中的声波速；σ_c 为介质的抗压强度；Q 为装药量。

由式(14)可知，空腔半径 r_c 符合几何相似原则，而非弹性变形区半径 R_d 与空腔半径成正比。

5 结 论

(1) 侵彻与爆炸的空腔及近区破坏半径最终决定了辐射出来波的基本参数，反映了能量的分配份额，是揭示侵彻、贯穿和爆炸特性的重要特征量。

(2) 在球腔膨胀条件下近区的应力与应变状态类似一维平面应变。

(3) 近区速度场关系可采用本文相关公式进行计算。

(4) 采用本文的研究结果，可以揭示侵彻问题的比例换算关系、贯穿问题的临界条件和爆炸空腔与非弹性变形区的几何相似关系。

(5) 本文的方法可以推广应用于具有构造特性的介质(如混凝土)。

参考文献(References)：

[1] 王明洋，戎小力，钱七虎，等. 弹体在岩石中侵彻与贯穿计算原理[J]. 岩石力学与工程学报，2003，21(11)：1 811－1 816.(Wang Mingyang, Rong Xiaoli, Qian Qihu, et al. Calculation principle for penetration and perforation of projectile into rock[J]. Chinese Journal of Rock Mechanics and Engineering, 2003, 21(11): 1 811－1 816.(in Chinese))

[2] Forrestal M J, Tzou D Y. A spherical cavity-expansion penetration model for concrete targets[J]. Int. J. Impact Engng., 1997, 34: 4 127－4 146.

[3] Ye N S. Consideration of dilation in determining the dynamics of grinding zone development in an elastoplastic medium with confined explosion of a concentrated charge[J]. Fiz. Tekh. Probl. Razrab. Polezn. Iskop., 1997, (5): 23－34.

[4] Ye N S. Dynamics of breaking zone development during explosion of a concentrated charge in a brittle medium[J]. Journal of Mining Science, 2000, 36(5): 464－475.

[5] Formy B M. High-velocity Interaction of Bodies[M]. Russian: Science Academic Institute of Russian, 1999.

[6] Nikiforovsky V S, Ye I S. Dynamic Failure of Solid Bodies[M]. Novosbirsk: Nauka, 1979.

[7] Adushkin V V, Spivak A A. Geomechanics of Large-scale Explosions[M]. Moscow: Nedra, 1993.

| 学术论文——岩石力学与工程

岩体的构造层次及其成因

戚承志[1]，钱七虎[2]，王明洋[2]，董 军[1]

(1. 北京建筑工程学院 土木工程系，北京 100044；2. 解放军理工大学，江苏 南京 210007)

摘要：岩体具有复杂的构造层次，这种构造层次的范围从微观一致延伸到宏观。但是对于这种构造层次的形成原因到目前为止还没有就内因及外因进行系统的研究。本文基于现有的地质力学研究成果及物理理论对于岩体构造层次形成的原因进行了系统研究。从研究结果可以看出，岩体构造层次的形成既具有岩石初始形成过程中的原因，也有后来的地质运动原因。在岩石形成时，由于不平衡状态的存在及与周围环境进行能量及物质交换，出现了自组织过程及耗散结构，从而形成了相互嵌入的自相似分形结构。地质运动原因是由于外部作用的存在使得地壳处于不平衡状态。不平衡状态为岩体的变形与破坏不断提供能量，从而形成耗散结构。在不同的年代形成不同层次的断层，邻近断层相互成45°角，尺度按$\sqrt{2}$因子变化。由于岩体中存在着自相似分形结构，嵌入系数就是岩体内部构造自相似性的一种外在表现。

关键词：工程地质；构造层次；地质运动；耗散结构；分形结构
中图分类号：TU 435；TU 42   **文献标识码**：A   **文章编号**：1000 - 6915(2005)16 - 2838 - 09

STRUCTURAL HIERARCHY OF ROCK MASSIF AND MECHANISM OF ITS FORMATION

QI Cheng-zhi[1]，QIAN Qi-hu[2]，WANG Ming-yang[2]，DONG Jun[1]

(1. *Department of Civil Engineering*，*Beijing Institute of Civil Engineering and Architecture*，*Beijing* 100044，*China*；
2. *PLA University of Science and Technology*，*Nanjing* 210007，*China*)

Abstract：Rock massif possesses complex structural hierarchy. Such a structural hierarchy involves very wide range of scale levels from microscopic scale level to macroscopic scale level. But until now systematic study on the mechanism of the formation of this kind of structural hierarchy does not exist. Based on the available geomechanical achievements and physical theories，the mechanism for formation of the structural hierarchy of rock massif is studied systematically. It shows that the formation of structural hierarchy is caused by the initial stage of formation of rock massif and the successive tectonic motion processes. In the initial stage of formation of rock massif，because of nonequilibrium nature of the processes and the exchanges of energy with the environment，the self-organization process and dissipative structure arise，and the self-similar fractal structures are formulated. In the view of tectonic，the earth's lithosphere is in nonequilibrium state because of the external actions. The energies of nonequilibrium state are supplied to the deformation and fracture of rock massif and eventually lead to the formation of dissipative structures. At different geological ages，faults of different scales are formed；and the two successively formed faults intersect with about 45°，and the scale ratio is $\sqrt{2}$. Because of the fractal structure in rock massif，coefficient of embedment is a natural expression of this self-similarity.

Key words：engineering geology；structural hierarchy；tectonic motion；dissipative structure；fractal structure

---

收稿日期：2005-03-10；修回日期：2005-04-11。
基金项目：国家自然科学基金重大项目(50490275)。
本文原载于《岩石力学与工程学报》(2005年第24卷第16期)。

1 引 言

岩石的非均匀性、离散性是其最为显著的特点之一。在工程计算中，通常利用连续介质方法来求得所需的参数。连续介质力学采用代表性单元的方法来建立模型。代表性单元的尺寸l必须满足下列条件：$\lambda \ll l \ll L$，其中，λ为介质的天然非均匀性的特征尺寸，而L为所研究物体的宏观尺寸。利用介质平均的特征参数来描述现实离散介质的行为会把介质的特点归结为控制方程的特点上。按照这种方法，任何力学问题的求解都要表示成坐标及时间的连续函数。但是，对于现实材料弹性变形限外行为的观察发现非弹性变形的发展不是连续的，而是具有跳跃性，亦即介质的离散性导致变形的离散性。这种变形的离散性在地壳板块规模上表现为地质板块的大跳跃性运动，从而形成地震扰动源。因此，在岩体介质中这种体积平均法方法受到很大的限制，因为岩体构造的离散性表现在所研究现象的任何线性尺寸上。介质的这种离散性表现为对于任何分离出的体积元在其中具有一定的内部构造，亦即离散介质的任何体积元都具有构造。这就意味着在考察岩体介质中的各种过程中，不能通过选择适当的尺寸来摆脱岩体的离散性，也即不论选择何种线性尺寸，总能找到与尺寸之相仿的单元，使得通过体积平均方法得到的力学值具有与其本身同阶的误差。

在研究介质中波的传播时，波的波形及时间特征参数对于介质的响应来说非常重要。当波长远大于非均匀构造尺寸时，介质可以被认为是连续的。当波长与非均匀性为同一数量级时，介质的离散特性会表现的非常明显，微分波动方程的适应性就值得怀疑了。如果波长小于非均匀性尺寸，那么波的反射就非常明显。这种情况可通过晶体中弹性波的传播来说明。晶格振动的低频部分以声速传播，并由波动方程来描述。而高频部分在介质的非均匀性中进行不断的反射，被认为是热振动，以非常慢的速度传播，并由热传导方程来描述。

岩石的离散块体特性还有一个重要的特点是在一个非常大的尺寸范围内，这种离散性具有相似性，并遵循级串律。研究表明[1, 2]，对于这样的系统存在着下列第i级别的块体尺寸Δ_i的自相似的等级序列公式：

$$\Delta_i = \left(\sqrt{2}\right)^i \Delta_0 \tag{1}$$

式中：Δ_0为地核的直径，且$\Delta_0 = 2.5 \times 10^6$ m；i为负整数。

通过i的降阶，可从地质构造级别一直到微晶体级别确定每一级别上的代表性块体的尺寸。

地壳分离成不同尺寸的单独部分，从大陆板块到细沙构造组成具有嵌入特性，或者层次重复性。在大的部分中嵌入小的部分，而后者又有更小的部分嵌入。令人惊奇的是，这些单独部分的尺寸分布曲线不依赖于这些层次岩石的物理力学性质，而是具有多众数性。较大层次的众数L_{i+1}与下一级众数L_i的比值$\lambda = L_{i+1}/L_i$事实上为常数，被称为嵌入系数。通常λ的变化范围为2.0～5.5[3]。

地壳不断经受着外部的作用，这些外部作用包括宇宙空间中各种物理场的作用、太阳能、月潮、洋流、大气压力的变化、陨石坠落等。在外部作用下地壳内部构造进行重组。这种重组是由局部的破坏及机械能的耗散引起的。这种过程应该由基于统计的动力方程来描绘。由于地外能量的输入，地壳有不断的振动，外部能量转化为内部能，即热流及地质构造运动。由于振动，不断进行着岩块的重组及变向。块体的平动及转动是由于单独晶粒表面摩擦力接触点的位移而产生。在岩体中的转动显然是由块体组成的现实固体在裂纹发展及破坏前所共有的性质。由М. В. Курленя和В. Н. Опарин所预测的非线性摆形波的存在证明在地壳中块体的平动及转动的存在[4]。异常低的岩石块体之间的摩擦力解释了岩块的转动及接触力动力系统中静力平衡的变化[5]。由于构造重组的存在，地质材料的行为就与具有微观结构的一般材料不同，因为对于后者其结构特点一般假定不随时间变化。

按照М. А. Садовский院士的结论，对于产生自我组织来讲，层次构造的离散性及地外能量(宇宙的各种物理场的作用)的不断输入成为耗散结构有序的源泉。地质体有序离散的构造的存在是由其长期的地质运动所造成的。关于上面所述的岩体自相似的等级序列、嵌入系数形成的内在物理原因所在，在大地构造学方面发展了用于解释地球断层构造规律性的地极游移理论[6, 7]，但是对于其他因素的影响，岩体构造层次形成的内部原因目前还缺乏系统的研究。本文就试图就此问题的内在物理与力学机理进行系统的研究，以填补这方面的空白。

2 岩体的构造层次形成的外部原因

2.1 岩体的构造层次形成的地质力学原因

在世界各地的地质构造调查中发现了如下的一些规律[6]：

(1) 块上断层的区域分布不是任意的，而是按照一系列系统的形式来进行分布的；(2) 一个系统的断层走向稳定，各系统之间相互垂直，同一级别断层之间的间距均匀；(3) 通常不同系统断层的地质特点不一样，形成时间不一样，但是形成彼此相似的网格，不同系统之间形成一定的角度；(4) 断层系统及褶皱系统之间有一定的联系。

上述特点在世界各地是一个普遍的现象。这种断层系统保持在一定的方向附近、延展性大、具有一定的继承性，证明这些系统的形成具有全球性的特点。上述那些线性地质构造的分布特点排除了用地壳的局部特点、或者在某个地区地球的内部发生的局部物理地质过程来解释。

在地质学中很久前就确定了具有不同规模及幅值的循环性地质过程。这些过程是在具有发展方向性的背景下进行的，而且还和外部的天文现象相联系。从传统的观点上，例如从板块构造学的观点上不能给出这些过程的满意答案。

利用地球自转速度的减慢来解释地球变形的建议，是在 19 世纪末 20 世纪初由达耳文及雷宾淞提出的。利用地极的移动来解释地球的变形出现的晚一些，在这一方面比较重要的人物是维宁－梅内斯(1944)及沙德格尔(1958)。

随后，获得最大知名度的是旋动说，其基础为地球旋转角速度的改变。在这一领域做出贡献的有：雷宾淞(1955)、察里格拉茨基(1963)、卡特菲尔德(1962)、斯托瓦斯(1975)等。旋转说的实质在于，地球旋转角速度的改变应该导致地球平衡时形状的改变。地球长期的旋转变慢使得两极地区抬高而处于受拉状态，而赤道地区降低却处于受压状态。这 2 个区域的交界位于 $\pm 35°$。这些应力的释放应分阶段间断性地进行。由于应力的释放会产生线性地质构造，特别是走向主要是经向及纬向的断层。实际的地质资料也证明了这种应力的存在。

这种应力并不是产生地壳构造运动的唯一原因。因为仅仅这种应力的释放还不能解释许多得到很好研究的各种具有对角方向的线性地质构造。为了填补这一空白，很多研究者把上述效应与地极的移动结合起来进行解释。其实质在于当同时有旋转速度及旋转轴位置发生变化时，旋转速度的变慢导致地壳中应力的出现，而旋转轴的改变仅仅改变了这些应力的方向。在 2 个旋转参数连续变化时，所形成的应力释放会形成沿着当时处于临界状态的古经向及古纬向断层进行。相对于现代的地理网格，那些构造应有一定的角度。后续的研究表明这种模型太简单。

Д. ХизаношвилиГ(1960)及А. В. Солнцев(1968)建议利用地极的移动来解释地质应力现象。关于这两种原因引起的应力之间的关系问题还没有得到解决。实际的地质资料表明与地极移动相联系的地质构造因素起着很重要的作用。在总结已有的研究成果及自己研究成果的基础上，文[6]发展了新的旋转学说，这一学说的基本要点如下：

(1) 地球地质构造活动的能量源为地球与周围物理场的相互作用力；(2) 地质构造形成的全球性规律是由地幔及上部对于外部作用反应导致的变形规律来决定的；(3) 在地球地质构造历史的初期地壳为均匀的，随着时间的流逝，逐渐被断裂分割；(4) 地壳的变形规律以及与之相连的地壳构造形成过程在整个地质历史中是相似的，可能改变的是地质过程的强度；(5) 每一个断层系统中第一阶的断层应有一定的成矿专门性，该专门性在重复的地质构造活动中随这些断层的某些部分的加入而变得复杂。

从新旋转学说得到的重要结果之一是断层系统分布的全球性特点。在地球旋转轴沿着地球表面平移时(如图 1 所示)[6]在一、三象限产生拉伸，而在二、四象限产生压缩。当旋转角达到一定值时，地壳中某点 K 的应力达到弹性极限。角度的继续增大导致地壳的断裂，发生应力释放，形成深的断层。在其他区域应力得到部分的释放。随着角度的继续增加，会重复上述过程，形成一系列断层。由于地球近似为圆形，形成的断层相互垂直。由旋转轴移动在岩石圈中产生应力的分析表明，应力的聚集是连续的，而应力的释放为分阶段、间断性地进行的。应力释放的离散性是因为在地球的每个象限内至少存在一个区域，其中的应力达到弹性极限，并形成断层。这些断层应该是最大的，因为这些断层的出现使得邻近未破坏部分的地应力部分地释放，但是，在这一区域应力的聚集还是最大的。很明显，这些断层在受拉状态为形成地槽的基体，在受压状态为造山运动的基体。它们把岩石圈分成地质构造运动活性低的区域，称为地台。由于断层的形成部分地减少了临近区域地应力，但不能完全消除这些应力。应力聚集的不断进行导致在同一地质活动年代里依次

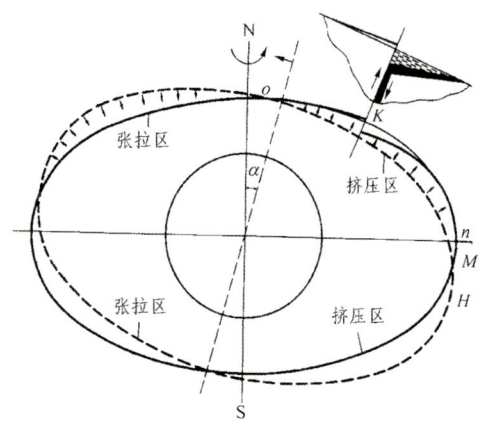

图 1 旋转轴移动时岩石圈中应力的出现及释放

Fig.1 Rising and releasing of stress in lithosphere due to the migration of axial rotation

形成不同级别的断层系统。

由于受外部作用,地壳不止一次的经受地质构造运动。这种运动在不同的地质年代产生不同的断层系统。在每一个地质构造活动年代,断层系统出现的结果是地壳被分割成块。不同级别断层的相互交割形成相应级别的块体系统。这里的块体理解为由断层所分割的地壳部分,且具有相同的地质构造特点及共同的形成历史。在第 2 次及后续的地质运动中,块体的形成是在地壳不均质的各部分中进行。在新的块体中,会有原先形成的块体的一部分进入其中。因此,块体不仅由空间位置来表征,也要由形成时间来表征。由于依次连续的块状构造的形成,地壳中会出现具有一样或者相似的地质状况的部分。该部分为相互重叠的块体的共有部分,并由不同年龄的断层隔开,具有不规则的多边形形状。图 2 表示 2 个邻近的地质构造活动所形成的块体。每一次活动形成自己的块体系统。因此,块体不仅由空间位置,而且也由形成时间来表征。

就方向来讲,在空间中所有的断层形成系统。通常把具有一个方向的断层称为一个系统。也有时方便地把两个相互垂直的断层连成一个系统。除此之外,每一个方向的断层其性质也不一样。对地壳裂隙的研究[7]表明,每一个裂纹系统为从微小裂纹到大断层各种不同层次的裂纹所组成的有规律的裂纹集合体。如果最大断层为第 1 阶断层,那么,在最大的断层之间 1/2,1/4 等处观测到与之平行的断层,被称为第 2 阶、第 3 阶等。图 3 为文[8] 中所提出的 2 个裂缝系统的生成叠加图。粗线为大尺寸块体的裂缝,裂缝越细,对应的块体尺寸越小。在中国情况也是如此[9]。

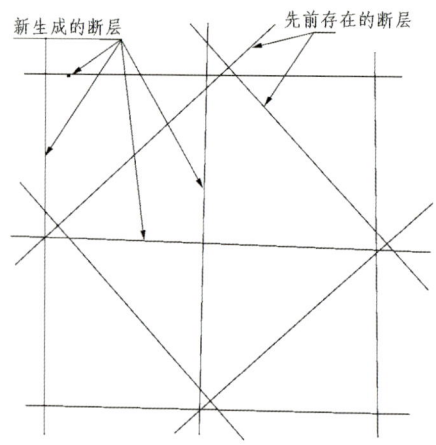

图 2 地质活动对于地壳裂缝形成的影响

Fig.2 Effects of tectonic activity on the formation of cracks of the earth's crust

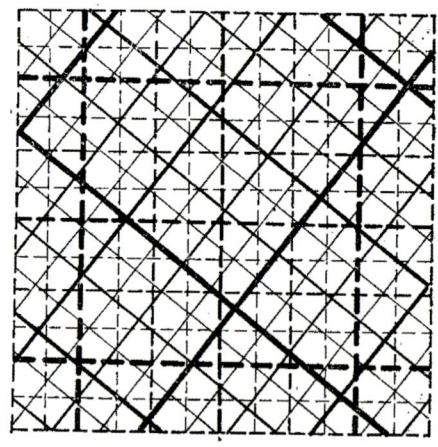

图 3 2 个系统不同等级断层的重叠图

Fig.3 Intersection of faults of different ranks of two systems

在分析岩体的块体构造时需要注意 3 个新的性质。第 1 个性质为裂纹构造及嵌入块体构造中的层次连续性,这在上面已经说明。第 2 个性质为在相邻平行的岩层中块体之间的竖向裂缝(或连接层)不重合(见图 4)[10]。第 3 个性质为先前更早的太古纪岩层的块体对于后来地质时代块体尺寸的影响。

关于所建议的岩层中正则层次序列的证实问题必须解释在广泛的大地构造学的文献中所描述的大地断层之间距离的大的发散性问题。这一问题的解决与块体的第 3 个性质有关,也即更早的太古纪的影响。这里应当从 2 点出发。第 1 点为在地质构造学中关于岩石破坏过程的自成型及相应的自相似性,这些还没有成为实际应用的原则。自成型的提法应该从导致裂缝出现及断层形成的力的全球性转

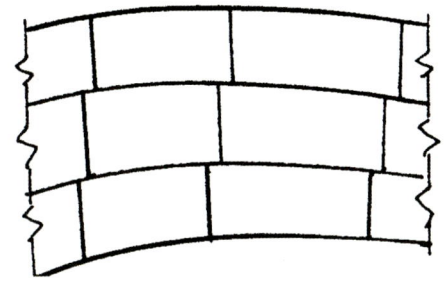

图 4 不同岩层竖向裂缝不重合

Fig.4 Non-coincidence of vertical cracks in different geological strata

动及平动中寻找。由众多的地质构造的研究可以做出这样的结论,即转动及平动运动的普遍性给地壳的块体构造带来规律,而在裂缝形成及块体构造中引入层次序列也是必须的。

第 2 个情况是必须评价先前更早的太古纪的岩层的块体对于后来地质时代块体尺寸的影响。旋转产生不同走向的断层系统见图 5。如果在最近地质年代断层的方向变化了 45°,那么根据基块断层的产生图(见图 3),新的断层在旧的断层的交汇处产生,亦即在旧的断层的正方形网角上产生。如果转动的角度不是 45°,那么,在受压象限所形成的块体的应力状态将是以边界上的压应力为主,剪应力为辅,所形成的破坏面也接近 45°。令人奇怪的是,在所调查的断层系统中,大多数相互成 45°,或者近似成 45°。此时,最新生成的断层之间的距离将比先前的断层之间的距离小 30%(为原断层之间距离的 $1/\sqrt{2}$)左右,如图 2 所示。考虑到断层剪力及拉应力的峰值不重合,新的断层不会准确与旧的断层的网角重合,而实际的新旧之间的距离可以达到 40%,这样断层之间距离就会有大的发散性。М. В. Курленя 和 В. Н. Опарин 等对于岩石块的尺寸进行了测量,在仔细整理块体尺寸的基础上得到了岩石块体尺寸存在着以 $\sqrt{2}$ 为因子的正则关系式(式(1))。那么,通过上面的分析可以知道为什么会有 $\sqrt{2}$ 之一因子。

理论及实验研究表明,$\sqrt{2}$ 因子具有更广泛的适用范围。对于门捷列夫元素周期表中的原子－离子半径的研究也揭示了 $\sqrt{2}$ 因子的存在[11]。深层硐室周围交替破碎区的半径之间的比也存在着 $\sqrt{2}$ 因子[2]。$\sqrt{2}$ 因子在深层硐室周围交替破碎区半径关系中的存在,反映了组成岩石矿物的原子在不同的尺寸水平上其能量结构的复制效应。

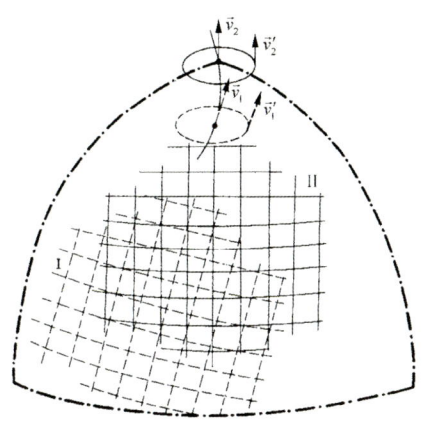

图 5 地球轴旋转时新的裂缝系统的形成

Fig.5 Formation of new crack system under the rotation of the earth axis

在研究所谓的"地球裂缝"时获得了非常重要的结果。结果表明,组成地壳的成分不同、年龄不同的岩石由相互垂直的裂缝分割成级别不同的块体,尺寸从几十厘米到几十公里,甚至几百公里。

舒尔茨及卡木克列里泽对于"地球裂缝"的研究表明,地球裂缝不仅涉及地台,而且也涉及褶皱区域,并保持自己的空间走向,长时间存在;每一个裂缝系统为一个有规律的集合,包含从最微小的裂纹,一直到很深的断层。这一结论也被其他研究者的成果,如尼古拉耶夫对于乌克兰板块的研究所证实。

这样得到的结论是:由于地球旋转状态的变化在岩石圈中产生了应力,应力的释放引起地球的地质构造运动,地球被不同级别的断层所分割,这些断层从微观裂纹一直到参加地槽及造山的深断层。

2.2 嵌入系数的物理实质

正则关系式(式(1))反映了岩石的尺寸存在的关系。岩石的尺寸还存在着成簇的效应。М. А. Садовский(M. A. Sadovsky)等通过对泥炭破碎、石英浸蚀、土壤颗粒的分析,以及地下爆炸和露天采石场中的岩石破碎分析、在建设水电站时利用地震地质方法对岩石非均匀性的分析、利用地质方法对地壳非均匀性、地壳板块和天体尺寸的分析等,发现了块体数在块体尺寸分布中存在着成簇现象,也即存在着优势尺寸,而且相邻的 2 个优势尺寸之比 L_{i+1}/L_i 相对稳定,主要的变化范围为 2.0～5.5[3]。

$\lambda = L_{i+1}/L_i = 2.0～5.5$ 具有深刻的物理含义。文[12]在研究材料的耐久性时发现,在标准的实验条件下,受到低于强度极限的荷载作用时,材料中会出现微裂纹,其数目会随着时间而增加。单位体积内裂纹的密度可以用裂纹的平均长度 r 及裂纹之间的平均距离 d 之比 K 来表示：

$$K = r/d \quad (2)$$

实验表明，当 K 很大时，裂纹之间实际上不相互作用。但当其超过一定的极限值 K_{cr} 时，材料就进入到快速破坏阶段。$K_{cr} = 3 \sim 5$，对材料性质的依赖性非常弱。这一准则被称为材料破坏的密度准则。从物理意义上看，这一准则描述的现象类似于相变。此时，K 可以看作是序参量。当序参量超过临界值时会发生相变。对于具有分布裂纹的物体而言，破坏更类似于逾渗相变。在逾渗相变比拟中，已经张开的裂缝可以看作是打开的通道，而未张开的构造面可以看作是未打通的通道。随着荷载或变形的增加，裂缝不断增加。当裂缝密度达到一定的临界值时，整个裂缝的发展就会发生质的变化，及发生相变。没有连接的裂缝快速连接，最终连起来的裂缝贯串整个物体，从而物体断开。

岩体的变形主要集中在岩块之间的软弱构造面上，裂纹的形成也主要集中在软弱构造面上。那么，对于一定尺寸层次 L_{i+1} 的岩体来讲，它所包含的低一级尺寸 L_i 的岩块之间的边界面即为 L_{i+1} 的岩体中的裂纹源，这时，L_i 相当于式(2)中的 r，而 L_{i+1} 相当于式(2)中的 d。这样一来就不难明白 λ 与 K_{cr} 是如此接近的原因。这表明，介质的破裂在不同尺度水平上应当是具有自相似性。自相似性则存在有其深刻的内在物理原因，在第 3 部分就其物理方面进行了研究。

2.3 地极移动的原因

地球为旋转的层状球体。最上面的固体层为岩石圈。温度及压力随深度的增加而增加。在 $100 \sim 200$ km 深度，多晶体岩石部分熔融，相应的地层叫做软流圈。软流圈具有孔隙(约 10%)，孔隙中填充有较轻的熔体。软流圈下面为中圈，其中发生着循环对流运动。在地球的中心为金属核，金属核由液体与固体两部分组成。两部分之间按照压力温度条件为液体金属的凝结孔隙区。当在外部液核中达到了液体的压力温度条件时，发生强烈的液状金属的流动，起着地球电机的作用，并产生内部地磁场。地球的瞬时旋转轴在地球的地理极附近游移。由于这一游移，岩石圈相对于中圈发生移动，在粘性的软流圈中发生滑移，并具有一定的周期性。这种周期性表现在岩石圈的不同部分(被称作地质构造板块)的相对运动。全球运动构成了大陆的漂移，显然是由地幔中大规模的热重力对流以及岩石圈及地幔地质材料密度的不同引起的。全球性地幔对流分为深部对流及在地幔上部较小规模的格状对流。很显然，上部对流引起了每个板块下面的软流层的流动，并引起了板块的移动及碰撞。

地球为太阳系的单元，而太阳系又为银河系的一部分。因此，研究地球运动规律应该从地球与银河系物理场的相互作用的动力平衡方程开始。地磁体在变化的银河系磁场中运动时应该经受一定的旋转矩。同时，在空间中动量矩应该守恒。这 2 个条件只有当地球的不同部分有相对运动时才能同时满足。大致的计算表明，为了使动量矩守恒，当外核及地幔有角速度变化时，内地核的速度变化比外核及地幔角速度变化最大能够大 3 个数量级。因此，地极移动的可能原因之一是其与周围银河磁场的相互作用。这些磁场由太阳磁场及其他天体的磁场组成。在星际空间中，磁场的强度为几个至几十个毫微特斯拉，而太阳的磁场强度要强 $2 \sim 3$ 个数量级。因此，在地极的漂移中，年移动分量周期的速度比平动的速度大 $1 \sim 2$ 个数量级。

地球旋转的非均匀性由 3 部分组成：(1) 长期性的变化；(2) 非规则变化；(3) 周期性振动，可以分为长周期(周期长于 1 a)、季节性及短周期性振动。从地质学的角度来讲，第 1 部分最重要。另外 2 部分可以用来揭示地球角速度变化的规律性。

天文学及古生物学研究表明，地球的转动系统性地变慢。但是，这种变慢不是线性的，而是受周期性的和非规则性的振动所干扰。天文学、古地磁学及地质年代学的现有资料表明，地极的运动轨迹具有非常复杂的形状，是由地表面的平动及具有不同级别周期的圆周运动叠加而成。

2.4 其他因素的作用

在实际工程中，常见到在形状及尺寸方面偏离前述 $\sqrt{2}$ 正则关系的情况，其中的原因是多方面的。旋转说还不能完全解释地壳的地质构造。地球旋转轴的迁移是形成地质构造的主要影响因素，但不是唯一的因素。因为如果地质构造仅仅是由旋转轴的迁移所引起的，那么，所有的地质构造将会相对于地心对称，但实际情况不是这样，实际情况是地表的地质构造很复杂。

研究表明，地核不处于地球的几何中心，而是相对偏离 436 km。这种偏离毫无疑问会对地壳中的应力分布及应力释放产生影响。其对地壳形成的几何结构分布的影响还不清楚。

在地球内部进行复杂的物理力学化学过程。中国学者陈国达院士提出壳体大地构造学的基本理论[13]，从时间－空间的角度对岩石圈动力源的机制进行了探索。他认为壳体是按照动定转化递进规律演化与运动的，既有内因又有外因，以内因为主。内因中最根本的是地幔由于物质(包括放射性物质元素)及温度(包括由于放射热能积累所致的增温)不均匀性引起的蠕变与流动。在地幔增强的活跃期，有关壳体或相应壳块、壳段获得热能增多，大地热流增高，岩浆活动强烈，同时运动增强，尤以水平运动明显，褶皱逆掩断裂发育，产生造山作用，工作地貌反差增大，剥蚀与沉积增强，形成活动区、蠕地槽区、地洼区等。反之则可转化为稳定区，如地台区。由于地幔蠕动是活跃与宁静交替，依次出现动定转化交替。

上述这些过程引起地球板块的漂移，从而引起相互挤压及拉伸，在板块中产生应力。关于板块的漂移机理及推动力有很多争论。例如文[14，15]建议板块是由于在岩石圈的底部由热点所引起的水平软岩流所导致，这一观点受到文[16]的反对。文[16]的观点是，在这种情况下横向软流各个方向一样，不会在板块中产生任何净作用力。起初，人们认为板块的推动力是由岩石圈下面的软流所施加的粘性力所引起的。但是，这一机理存在若干困难。对流地幔与岩石圈之间的接触区在软流圈的上部，有很低的地震活性及低的运动速度。据计算，为了使岩石圈产生 40 mm/a 的速度，软流圈必须以 200 mm/a 的速度运动。但这一速度高得不太合理，最近地质年代热点小的相对运动表明这是不可能的。另一困难在于，如果这是目前主要的板块移动机理，那么，主要对流环的尺度必须是大洋宽度的一半，而且在岩石圈的大部分区域内运动模式是一样的。但是，这不能解释中大西洋脊线具有形状不规则间隙的板块运动，以及尺寸小的板块运动，如菲律宾及加勒比板块的运动。

另一个得到人们接受的机理认为，板块的运动是由于加在其边缘上的力引起的，软流圈的作用基本上是被动的。这一思想是由文[17～19]发展的。主要的边缘力有下述几种。在洋中脊带，上浮的材料使中脊上升，把新形成的板块推开。在俯冲带岩石圈材料比下面的物质要冷、要重，从而下沉。一部分向下的力被传递到岩石圈板块而成为板块拉力。一部分上面的板块被拖进消减带海沟而处于受拉状态，相应的力为海沟吸力。上述力用于克服地幔阻力、俯冲板块的弯曲阻力以及摩擦力及粘聚力。边缘力机理能够更好地解释板块的运动、所观察到的板块之间的应力模式、板块的运动速度不依赖于板块的面积、依附于向下板块的板运动快、具有大的陆地面积的板块运动慢等现象。因而受到大多数人的接受。

由中国断层现今位移活动分区略图可知，全国具有明显的条块分割现象，径向带断层、维向带断层将中国切割成网格，这一现象与前面所讲述的地极漂移引起的断层网格现象一致。但是由于印度板块的北冲，这一网格不是真正的东西南北方向，而是转动了一个角度。而且因此西部地区的活动速率要高于东部地区可达一个数量级，而且密度也大。因此，可以看出，板块的漂移改变了断层的方向及使得断层密度不均匀。

除此之外，海洋的潮汐、月球引起的岩石圈潮汐等也对地球岩石圈的地质构造的形成产生作用。因此全面综合它们的影响是方程重要的。

3 岩体构造层次形成的内在物理原因

岩石构造体系的形成还有其形成时的原因。岩石形成时成分、温度变化和结晶时速度的不均匀，受外力的作用等，使得岩石形成受许多随机因素影响。影响岩石形成的因素在空间和时间上都是随机的。一些学者(如 Avnie，Malcai 等)的研究表明，分形的形成是由随机性引起得。从热力学的角度讲，在火成岩的形成中，岩浆是由大量的无规则的粒子组成。按照 Ramsey 定理[20]，大数或者众多点(结构单元)的集合一定包含着高度有序的结构。这就意味着任何含有足够大的数量单元的结构可以看作是多重分形体，这种分形体由相互嵌入的自相似构造组成。自然分形结构的重要特点是，其形成需要大的能量输入。从这一点讲，耗散结构可以具有分形特性。从另一方面讲，如果微结构的形成主要是由远离热力学平衡的现象所引起的，那么，这一结构也固有分形特性。岩石形成时成分的不均匀、温度变化的不均匀、结晶时速度的不均匀、受外部作用等，使得岩石形成过程为一个远离平衡的过程。通过与周围环境的能量及物质等的交换，便形成耗散结构。耗散结构的一个重要原则是：不平衡性是有序的源泉[21]。这样就使得岩体从微观结构来讲就具有自相似内部构造的分形体。岩石破坏时裂纹所具有的分形特征、岩石断裂面所具有的分形特征都是岩石具有自相似构造层次的外部表现。

从宏观上讲，正如上面所指出的那样，地壳不断经受着外部作用。由于外部作用的存在，使得地壳处于不平衡状态。不平衡状态为岩体的变形与破坏不断提供能量。地壳中地质体为一个开发系统，由于其不平衡状态的存在及与周围环境进行能量与物质交换，那么，处于不平衡状态的地质体就会形成耗散结构。在这种耗散结构中发生着自组织过程，在分岔点发生着自发的调整。这种调整可看作是一系列动力变换的过程。在这一过程中，随机性、非平衡性、不可逆性是系统秩序的源泉。耗散结构具有一定形状和空间时间特征尺寸，其出现与对称性的破坏及与原先均匀空间状态比对称性低的结构的出现相联。在物理学中，对称定义为空间的各向同性及均匀性。均匀及各向同性的固体具有对称性，这种固体的特点是其性质由两个弹性常数来描述。这种自发的对称性破坏导致固体中各种结构的形成。岩体本身形成时的分形构造面就已经存在。Моисеев(H. H. Moiseev)在考察生物界发展机理的基础上形成了最小能量耗散原则[22]："如果满足守恒律及其他物理限制的稳定运动或状态的集合由一个以上的单元组成，也即它们不能选出唯一的运动或状态，那么，所要实现的运动或状态的最后选择由最少能量耗散(或最小熵增)来决定的"。这一原则决定了岩体在外部作用下的破坏最有可能沿着天然具有的分形构造面来进行。这种情况决定了经过地质运动所形成的块体具有自相似特性。因此，岩石的构造层次性从形成角度讲是由于岩石形成过程中内在的与外在的具有随机特性的因素作用的共同结果。

4 结 语

本文基于现有的地质力学研究成果及物理理论对于岩体构造层次形成的原因进行了研究。可以看出，岩体构造层次的形成既具有内在原因，也具有外在原因。内在原因是在岩石形成时，由于不平衡状态的存在及与周围环境进行能量及物质交换，出现了耗散结构及自组织过程，从而形成了相互嵌入的自相似分形结构。外因是由于外部场作用的存在使得地壳处于不平衡状态。从而引起地球运动状态的改变。其中，地球旋转轴的运动对岩体构造层次的形成起主要作用。不同年代形成不同层次的断层，邻近断层相互成 45°，尺度按 $\sqrt{2}$ 因子变化。由于岩体中存在着相互嵌入的自相似分形结构，嵌入系数就是岩体内部构造自相似性的一种外在表现。其他的原因，如地核不位于地球的几何中心，板块的运动、潮汐作用等也都影响岩体构造的形成。因此，还必须系统研究各种因素对于地质构造形成的影响。

参考文献(References)：

[1] Опарин В Н，Юшкин В Ф，Акинин А А，et al. О новой шкале структурно-иерахических представлений как паспортной характеристике объектов геосреды [J]. ФТПРПИ. 1998，5：17－33.(Oparin V N, Jushkin V F, Akinin A A, et al. On new scale of structural hierarchy presentation as characteristics of geological medium[J]. J. Russian Mining Science[J]. 1998，5：17－33.(in Russian))

[2] Опарин В Н，Курленя М В. О скоростном разрезе Земли по Гутенбергу и возможном его геомеханическом обьяснении.Ч.1 Зональная дезинтеграция и иерархический ряд геоблоков[J]. ФТПРПИ, 1994, 2：14－26. (Oparin V N., Kurlenija M V. On speed cut of earth by Gutenberg and its possible geological interpretation. part I: zone disintegration and hierarchy series of geological blocks[J]. J. Russian Mining Science，1994，2：14－26.(in Russian))

[3] Садовский М А，Волховитинов Л Г，Писаренко В Ф. Деформирование геофизической среды и сейсмический процесс[M]. Москва：Наука，1987.(Sadovsky M A, Volkhovitinov L G, Pisapenko V F. Deformation of Geophysical Medium and Seismic Process[M]. Moscow：Science Press，1987.(in Russian))

[4] Курленя М В，Опарин В Н，Востриков В И. Волны маятникового типа Ч.I[J]. ФТПРПИ，1996，4：3－39. (Kurlenija M V, Oparin V N, Vostrikov N. Waves of pendulum type, part I[J]. J. Russian Mining Science，1996，4：3－39.(in Russian))

[5] Курленя М В，Опарин В Н，Востриков В И. Об эффекте аномально низкого трения в блочных средах [J]. ФТПРПИ, 1997, 1：3－16.(Kurlenija M V, Oparin V N, Vostrikov V I. On anomalous low friction in block medium [J]. J. Russian Mining Science，1997，1：3－16.(in Russian))

[6] Тяпкин К Ф，Кивелюк Т Т. Изучение разломных структур геолого-геофизическими методами[M].

Москва：Недро，1982.(Tiapkin K F，Kiveljuk T T. Study of Fault Structures by Geological and Geophysical Methods[M]. Moscow：Nedro，1982.(in Russian))

[7] Изучение тектоники докремрия геолого-геофизическими методами[M]/под редакциейТяпкина К Ф. Москва：Недро，1972.(Study of Tectonics by Geological and Geophysical Methods[M]. Tiapkin K F ed. Moscow：Nedro，1972.(in Russian))

[8] Тяпкин К Ф. Изучение разломных и складчатых структур докремрия геолого-геофизическими методами[M]. Киев：Наукова Думка，1986.(Tiapkin K F . Study of Fault and Fold Structures by Geological and Geophysical Methods[M]. Kiev：Naukova Dumka，1986.(in Russian))

[9] 陈庆宣，王维襄，孙 叶. 岩石力学与构造应力场分析[M]. 北京：地质出版社，1998.(Chen Qingxuan，Wang Weixiang，Sun Ye. Rock Mechanics and Analysis of Tectonic Stress Field [M]. Beijing：Geological Publishing House，1998.(in Chinese))

[10] Скоробогатов С М. Принцип информационнойэнтропии в механике разрушения инженерных сооружений и горных пластов[M]. Екатеринбург：УрГУПС，2000. (Skorobogatov S M. The Principle of Information Entropy in Fracture Mechanics of Buildings and Rock Seams [M]. Yekaterinburg：UrGUPS，2000.(in Russian))

[11] Курленя М В，Опарин В Н. О масштабном факторе явления зональной дезинтеграции горной породы и канонически ряд атомно-ионных радиусов[J]. ФТПРПИ，1996, 2: 3–14. (Kurlenija M V，Oparin V N. On scale factor of zone disintegration of rock mass and canonic series of atom-ion radius[J]. J. Russian Mining Science，1996，2：3–14. (in Russian))

[12] Журков С Н. Дилатонный механизм прочности твердых тел[J]. ФТТ，1983，25(11)：3 119–3 123.(Zhurkov S N. Dilaton mechanism of strength of solids [J]. Physics of Solids，1983，25(11)：3 119–3 123.(in Russian))

[13] 陈国达. 地质学说的新进展[M]. 北京：科学出版社，1992.(Chen Guoda. New Advances in Geology [M]. Beijing：Sciences Press，1992. (in Chinese))

[14] Morgan W J. Convection plumes in the lower mantle[J]. Nature，1971，1：42–43.

[15] Morgan W J. Deep mantle convection plumes and plate motion[J]. Bull. Am. Assoc. Petroleum Geols.，1972，56：203–213.

[16] Chapple W M，Tullis T E. Evaluation of forces that drive plates[J]. J. Geophys. Res.，1977，82：1 967–1 984.

[17] Orowan E. Convection in a non-Newtonian mantle，continental drift，and mountain building[J]. Phil. Trans. Roy. Soc. Lond.，1965，258(1 088)：284–313.

[18] Forsyth D W，Uyeda S. On the relative importance of the driving forces of plate motion[J]. Geophys. Jour. Roy. Ast，Soc.，1975，43：163–200.

[19] Bott M H P. The Interior of Earth，Its Structure，Constitution and Evolution(Second Edition)[M]. London：Amold，1982.

[20] Иванова В С，Баланкин А С，Бунин И Ж，et al. Синергетика и фракталы в материаловедении[M]. Москва：Наука，1994. (Ivanova V S，Balakin A S，Bunik I G，et al. Synergetics and Fractals in Material Science[M]. Moscow：Science Press，1994.(in Russian))

[21] Haken H. Advanced Synergetics[M]. Berlin：Springer-verlag，1983.

[22] Моисеев Н Н. Алгоритм Развития[M]. Москва：Наука，1987.(Moiseev N N. Algorithm of Growth[M]. Moscow：Science Press，1987.(in Russian))

岩体的构造层次粘性及动力强度

戚承志[1]，钱七虎[2]，王明洋[2]

(1. 北京建工学院 土木系，北京 100044；2. 解放军理工大学，江苏 南京 210007)

摘要： 岩体具有复杂的构造层次，这种构造层次可以从原子尺度水平一直延伸到地质构造水平。这种情况使连续介质力学中使用的基本体元的概念及圣维南变形协调条件受到了置疑。内部构造层次影响岩体的粘性及动力强度，但到目前为止还没有看到连接微细宏观层次上粘性及动力强度的研究。基于已有的资料及物理力学理论对材料的粘性及构造层次之间的关系进行了系统研究。由研究可以看出，不同的构造层次对应着不同的强度、粘性及应变率。宏观层次通常对应着低的强度、低的应变率及高的粘性系数，而微细观层次对应着高的强度、高的应变率及低的粘性系数。通常随着外载的增加，变形速度的增大，发生变形与破坏的层次逐渐由宏观经由细观逐步向微观层次过渡，动力强度增加，粘性系数逐渐降低。在中高应变速率区，岩石的强度增加很大，而粘性与变形速率成反比。因此，强度及粘性系数对于一种介质来讲不是常数，而是在不同的构造层次上具有不同的大小。表现在破坏块体的尺寸上，最小的破坏尺寸与所能达到的强度有直接的关系。通过不同层次上岩体的力学行为的分析，建议了强度、粘性对于构造层次及应变率的依赖关系及岩体破坏块体尺寸的确定公式。

关键词： 岩土力学；构造层次；粘性；松弛；动力强度

中图分类号： TU 435；TU 42　　**文献标识码：** A　　**文章编号：** 1000 - 6915(2005)增 1 - 4679 - 09

THE STRUCTURAL HIERARCHY VISCOSITY AND DYNAMIC STRENGTH OF ROCK MASSIF

QI Cheng-zhi[1], QIAN Qi-hu[2], WANG Ming-yang[2]

(1. Department of Civil Engineering, Beijing Institute of Civil Engineering and Architecture, Beijing 100044, China;
2. PLA University of Science and Technology, Nanjing 210007, China)

Abstract: Rock mass posses complex structural hierarchy. Such structural hierarchy involves a very wide range of scale levels from atomic scale level to tectonic scale level. This situation makes the concept of elementary volume and Saint-Venant's deformation compatibility condition used in continuum mechanics problematic. Internal structural hierarchy influences viscosity and dynamic strength of rock mass, but until now studies haven't been seen to link viscosity and dynamic strength of rock mass on different scale levels. This paper, based on the available investigation data and physical mechanical theories, studied the relationship between the viscosity, dynamic strength and the structure hierarchy of rock mass. It is shown from the study that different hierarchy levels correspond to different strength, viscosity and deformation rate. Macroscopic hierarchy scale level corresponds to low strength, low deformation rate and high viscosity, micro-macroscopic scale levels share high strength, high deformation rate and low viscosity. Generally, with the increase of the external loads and the deformation rate, the deformation and fracture consecutively involve macro-, meso-and microscopic levels, correspondingly dynamic strength increases, and viscosity decreases gradually. In moderate and high deformation rate region, dynamic

---

收稿日期：2005–03–10；修回日期：2005–04–11。
基金项目：国家自然科学基金重大项目(50490275)。
本文原载于《岩石力学与工程学报》(2005 年第 24 卷增 1)。

strength increases substantially; viscosity is inversely proportional to deformation rate. Hence strength and viscosity are not constants for one material, but have different values on different hierarchy levels. As for the size of fractured rock, it is related with the dynamic strength which rock mass can reach. Based on the analysis of mechanical behavior of rock mass on different hierarchy levels, approximation formulae for the dynamic strength and viscosity in dependence on hierarchy levels and deformation rate, formula for the determination of fractured rock mass are suggested.

Key words: rock and soil mechanics; structural hierarchy; viscosity; relaxation; dynamical strength

1 引 言

岩体为天然形成的材料，其内部具有复杂的多层次结构。这种结构的尺寸级别跨度非常大，如岩石可从原子级别直到地质构造级别，这组成了岩石材料的构造层次。按照文[1]的观点，岩体中存在着复杂的构造层次系统，这一层次系统包含所有的科学研究所触及的尺度，从原子级别到地质构造级别及行星尺度级别。这一观点在当时是一个大胆的假设，但目前经过很多的理论及试验工作，这一观点已经得到了令人信服的证实，并有了定量的描述[1,2]。这种概念就使得在分析中所使用的基本体元的概念，以及连续介质力学中的数学模型所使用的圣维南变形协调条件受到了置疑。

这种岩体的构造层次反映在不同层次尺度的关系上，如俄罗斯学者认为对于岩体的构造层次存在着相邻构造级别尺寸之间为 $\sqrt{2}$ 的自相似关系[1,2]。

岩石的尺寸还存在着成簇的效应。Садовский M.A.(M. A. Sadovsky)院士等通过对于泥炭的破碎、石英的侵蚀、土壤颗粒的分析，以及地下爆炸中岩石的破碎分析、露天采石场的岩石破碎分析、在建设水电站时利用地震地质方法对于岩石非均匀性的分析、利用地质方法对于地壳非均匀性的分析、地壳板块的分析、天体尺寸的分析等，发现了块体数随块体尺寸的分布中存在着优势尺寸，而且相邻的 2 个优势尺寸之比 L_{i+1}/L_i 相对稳定，主要的变化范围为 2～5.5[3]。$L_{i+1}/L_i = \lambda$ 被称为岩石的嵌入系数。

由于岩体中存在着不同层次的构造，因此每一构造层次都有自己的构造面。这些构造面是岩体的薄弱面，因此，岩体的变形及破坏主要发生在这些构造面上。这些构造面连接不紧密，有一定的张开宽度，很自然，可以认为，张开宽度越大的构造面，强度越小。

研究表明，裂纹的张开尺寸 δ_i 及由裂纹分开的同级块体的尺寸 Δ_i 之间存在着稳定的统计关系[2]：

$$\mu_\Delta(\delta) = \frac{\delta_i}{\Delta_i} = 10^{-2}\Theta \quad (1)$$

式中：Θ 为系数，其变化范围为 1/2～2；$\mu_\Delta(\delta)$ 为岩石力学"不变量"。

材料的构造影响材料的力学性质，典型的例子为炭原子的不同组合结构形成石墨及金刚石两种力学性质截然不同的材料。同时外部作用也会对介质的内部结构产生影响，并进而对介质的宏观行为产生影响。如很多试验及现场观测发现，在塑性变形局域化过程中、变形波阵面能量耗散过程中、在剥离破坏过程中都观察到了新的结构的形成。基于上述分析，可以看到介质的内部构造及宏观行为是相互作用、相互影响的。这种情况决定了在研究介质的变形与破坏时，应当考虑它们之间的相互作用。岩体构造的多层次性决定了其力学性质的多层次性，其中，岩体的粘性及动力强度是非常重要的力学参数，它们决定着岩体在外载作用下的变形与破坏过程。在深部岩体的非线性力学现象中，深部岩体介质中开挖分区破裂化现象的演化规律(时间因素)是一个急待解决的问题，它正与岩体的应力松弛及粘性密切相关。其他的动力现象，如变符号动力反应、摆型波、超常低磨擦效应，及工程中的岩爆、瓦斯突出、地下工程的防震隔震等都与应力松弛及粘性相关。因此，研究材料的构造层次与材料的动力粘性及强度之间的关系具有重要的意义。在现有的文献中对于材料动力强度及粘性的研究只局限于单独的构造层次上，如文[4]中在地质构造层次上对于岩体的粘性有一些数据。在文[5～8]中给出了材料在细观及微观层次上的粘性。但是它们之间彼此没有联系起来。到目前为止还没有看到关于连接微观细观及宏观层次上粘性及动力强度的研究。正是由于这个原因，基于已有的研究资料及物理力学理论，对材料的粘性、动力强度及构造层次之间的关系进行了系统的研究,并在此基础上建立了强度、粘性对于构造层次及应变率的依赖关系及岩体破坏块体尺寸的确定公式。

2 岩体的构造层次与尺寸效应

材料的力学性能,如强度具有尺寸效应。首次尝试解释尺寸效应的为 Griffith 的关于固体的断裂及流动的著作。他的结论是,具有垂直于荷载方向的长直裂纹的薄玻璃板的拉断强度与裂纹长度的平方根成反比。Wenbull 于 1939 年发展了强度的统计理论。在受拉、受扭及弯曲条件下强度与试件的体积的关系为

$$\sigma_D = \sigma_0 ILV^{-1/\alpha} \quad (2)$$

式中:σ_0 为材料常数;I 为材料的应力状态函数;L 为标准化系数;V 为试件体积;α 为材料常数,表征材料的均匀程度,其越大,表明材料强度对于材料体积的依赖性越低。

Т.А.Контрова及Я.И.Френкель(Kontorova, Frenkel)按照缺陷对于强度危险性的正态分布,建议对于大体积试件采用如下公式:

$$\sigma_D = R_0 - \sqrt{A \lg V - B} \quad (3)$$

对于小体积试件采用如下公式:

$$\sigma_D = a + b/V \quad (4)$$

式(3),(4)中:A,B,a,b,R_0 均为依赖于应力状态及材料性质的常数。

М.М.Протодьяконов (M. M. Protodiakonov) 建立了如下公式:

$$\sigma_D = R_M (d + mb)/(d + b) \quad (5)$$

式中:R_M 为裂隙岩体的强度;m 为裂隙度系数,反映了非裂隙材料强度比裂隙体高几倍;d 为试件直径;b 为裂隙度常数。

В.В.Болотин(V. V. Bolotin)根据变分序列的最小值的第三极限分布及统计强度理论,给出了平均强度极限与试件体积之间的最一般关系:

$$\bar{\sigma}_D = R_{\min} + \beta \left(\frac{V_0}{V}\right)^{1/\alpha} \Gamma\left(1 + \frac{1}{\alpha}\right) \quad (6)$$

式中:α,β,$R_{\min}$ 为分布函数参数;V_0 为某一参照体积;$\Gamma(\cdot)$ 为伽马函数。

文[9]的研究表明,对于诸如岩石、韧化陶瓷、混凝土、灰浆、脆纤维复合材料等,温布尔统计强度理论没有抓住问题的实质。起决定性作用的尺寸效应的源泉不在于统计性,而在于由大的损伤引起的储存于材料中的能量的释放,并通过近似的能量释放分析,导出了准脆性材料的尺寸效应律。自从 Mandelbrot 在 1984 年揭示裂纹表面的分形特性以来,在分形与尺寸效应的联系方面也进行了广泛的研究。但研究表明[10],裂纹的分形特性不是尺寸效应的原因。原因在于在变形损伤过程中,裂纹前端有一个破坏过程区,消耗能量的不是单个裂纹,而是具有众多裂纹的一条带。最终连接而形成单个连续裂纹的是少数微裂纹及滑移面。大多数的能量消耗于破坏过程区中的裂纹中,而不是用于形成具有分形特征的最终裂纹表面。

岩体的构造层次的发现可以为进一步认识岩体强度的体积效应提供帮助。因为每一构造层次都有自己的构造面。这些构造面是岩体的薄弱面,因此岩体的变形及破坏主要发生在这些构造面上。这些构造面连接不紧密,有一定的张开宽度,很自然,张开宽度越大的构造面,强度越小。

通常对于受压材料,强度 σ_D 与试件尺寸 D 之间的关系可以表示为[9]

$$\sigma_D = C_1 D^{-1/k} + C_0 \quad (7)$$

式中:C_1,C_0 为常数,k 为实数。

如果在上式中用 Δ_i 替换 D,则得

$$\sigma_D = C_1 \Delta_i^{-1/k} + C_0 \quad (8)$$

从式(8)可以看出,不同级别的构造单元对应着不同的强度。随着构造单元的尺寸的减小,强度变大。

另一方面,如果把式(1)代入式(8),可得强度与某层次上的单元间隙之间的关系:

$$\sigma_D = C_1' \delta_i^{-1/k} + C_0 \quad (9)$$

上面讨论了构造单元尺寸对于强度的影响,那么反过来讨论施加的应力与岩体破坏的块体的平均尺寸之间的关系。

将式(8)改写为

$$\Delta_i = \left(\frac{C_1}{\sigma - C_0}\right)^k \quad (10)$$

在式(10)中把 σ_D 替换为 σ,以表示施加荷载。

式(10)表明,当施加的荷载增加时,破裂的块

体尺寸会逐渐减小。那么实际情况又是怎样，进行的静力及动力试验均证实了这一点[11]。

对于单轴受拉情况，破坏被看作为串联链中的薄弱环节的破坏，强度对于体积 V 的依赖为

$$\sigma_V = \sigma_{\min} + A/V^{1/n} \quad (11)$$

由于 $V \propto D^3$，所以块体的尺寸与施加的应力之间的关系为

$$D = \left(\frac{\sigma_V - \sigma_{\min}}{A}\right)^{3n} \quad (12)$$

当把式(10)，(12)中的应力替换为破坏强度时便可得破坏块体尺寸。

在节 4 中将利用本节的公式来确定岩体破坏后的尺寸。

3 岩体的构造层次及其粘性

岩体的多层次性决定了发生于岩体中的变形破坏过程的多层次性。通常慢速的变形过程对应着大的构造层次，如在地质构造层次上宏观变形能量的集聚及地震的发生。而快速变形与破坏过程发生在微细观层次上，如在受冲击波作用下材料的变形与破坏。而中等的变形与破坏过程发生在介于宏观地质构造层次及微细观层次之间的宏观层次上，这是接触最多、与工程实际联系最为密切的变形破坏层次。

在实际工程尺度层次上，岩体变形时的粘性系数通常被认为是一个常数。但在地质构造层次上，岩体的变形非常慢，应力松弛非常慢，岩体的粘性系数非常大。而在微细观层次上岩体的粘性系数却要小的多。如在描述冲击波的传播时，在压力高于 100～200 GPa 时，流体动力学近似是合理的，这时材料的强度及粘性效应非常小。在压力为 1～10 GPa 的范围内强度及粘性效应对于冲击波波形的形成具有决定性作用。变形速度的影响由动力粘性来表征。

在现有的文献中，不同材料的粘性系数发散很大，即使在同样的变形速度情况下也如此。这明显是与材料的变形与破坏发生的尺度不同有关。

岩石具有多层次性。岩石的这种内部构造对于岩石的物理力学性质具有决定性的作用。作为岩石重要物理力学性质的粘性与这种构造的层次性具有直接的联系。虽然材料的构造层次有许多种，但是从实际应用方便角度讲，可以分为宏观、细观及微观三种层次。随着现场调查数据的积累，现在可以对于粘性与构造单元的尺度以及与其变形情况的联系进行研究。

材料的粘性 η 的大小可由下式来决定[12]：

$$\eta = G\tau \quad (13)$$

式中：G 为岩体的剪切模量，τ 为松弛时间。

在不同的构造水平上介质的松弛及弹性模量不同，因此引起了介质粘性的不同。现在就不同构造层次上的粘性进行研究。

3.1 宏观层次上的粘性

首先考察在地质构造级别上岩体的应力松弛及粘性。一些地震观察的结果显示[3]，在发生震级 $M \geq 7$ 的强烈地震后，平均需要 84 d 才能基本上恢复地震前的地震学状态。需要几年的时间才能完全恢复。相应的 τ 的范围为：$\tau_1 = 84$ d $\approx 7 \times 10^6$ s，$\tau_2 = 10$ a $\approx 3.1 \times 10^8$ s，$G \approx 45$ GPa，那么粘性的大小为 $\eta \approx 10^{17} \sim 10^{19}$ Pa·s。

粘性值与构造层次之间也存在着联系。根据已有的资料，在地表 $5 \times 5°$ 的梯形面积内，新生代活动断层数 N 与粘性之间的关系如图 1(a)所示。这一关系为非线性关系。

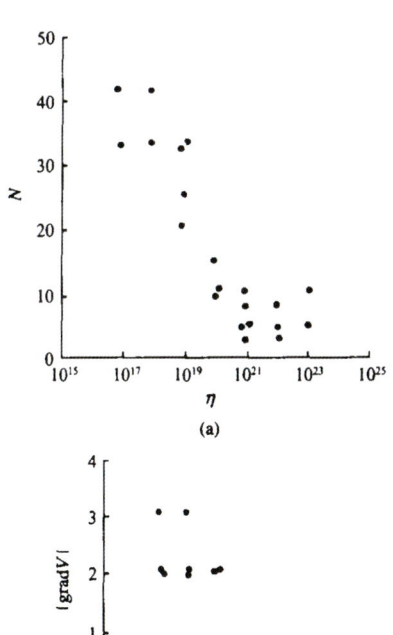

图 1 粘性与断层密度、竖向位移梯度之间的关系[3]

Fig.1 Relations between viscosity, fault density and gradient of vertical movement velocity[3]

统计分析表明，1 km²上的断层密度n与断层尺度L之间存在着紧密的关联关系[8]，即

$$L = \frac{0.78}{n^{0.42}} \quad n = \frac{0.29}{L^{2.2}} \tag{14}$$

不失一般性，式(14)可以写为

$$L = \frac{c}{n^\beta} \quad n = \frac{d}{L^\delta} \tag{15}$$

综合上面的关系就可以得到随着岩体尺度的减小，岩体的粘性值减小的结论。

式(14),(15)反映了在自然的条件下，在地质规模水平上，不论地质构造的发展历史如何、其活动程度如何，岩体的宏观破坏遵循着某种一般的破裂规律。在把实际资料与Maxwell体的试验资料进行比较时发现，Maxwell体在破坏时的破裂曲线与在具有不同发展历史及活动程度的区域中断层的密度长度曲线一样。因此可得出结论：在形成断层网格时，地壳是按照Maxwell体来变形与破坏的。

断层密度是表征岩石圈中断层形成过程的一个参数，它与新地质构造的竖向位移梯度$|\text{grad}V|$紧密相关。这种关联关系不难从图1(a),(b)的对比中看出。在断层密度大的地方，$|\text{grad}V|$值也大；而$|\text{grad}V|$大的地方粘性也小。

令人惊奇的是，作者发现$|\text{grad}V|$是岩体的应变率。这可以从下列推导看出。设竖向坐标为z，位移为w。另外设两个水平方向x,y方向的位移$u=0, v=0$，则沿着x方向w的速度梯度为

$$|\text{grad}V| = \left|\frac{\partial}{\partial x}\left(\frac{\partial w}{\partial t}\right)\right| = \left|\frac{\partial}{\partial t}\left(\frac{\partial w}{\partial x}\right)\right| = \left|\frac{\partial}{\partial t}\left(\frac{\partial w}{\partial x} + \frac{\partial u}{\partial z}\right)\right| = \left|\frac{\partial \gamma_{xz}}{\partial t}\right| \tag{16}$$

因此，可以得出随剪切应变率的增加岩体的粘性减小。

如果把图1(a)中的关系近似用直线代替，则有

$$\ln \eta = b - \alpha|\text{grad}V| = b - \alpha\left|\frac{\partial \gamma_{xz}}{\partial t}\right| = b - \alpha|\dot{\gamma}_{xz}| \tag{17}$$

由式(17)可得

$$\eta = \eta_0' \exp(-\alpha|\dot{\gamma}_{xz}|) = \frac{\eta'}{\exp(\alpha|\dot{\gamma}_{xz}|)} \approx \frac{\eta'}{1+\alpha|\dot{\gamma}_{xz}|} \tag{18}$$

式中：b, α为常数。

由图2可得

$$\dot{\gamma} = \left(\frac{d}{L^\delta}\right)^m \quad (m \leq 1) \tag{19}$$

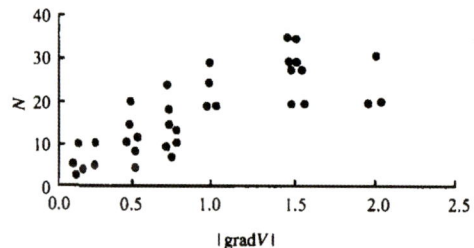

图2 断层密度与位移梯度之间的关系[3]

Fig.2 Relation between fault density and gradient of vertical movement velocity[3]

$$L = \left(\frac{d}{\dot{\gamma}^{1/m}}\right)^{\frac{1}{\delta}} \tag{20}$$

式(20)反映了构造层次与应变率的对应关系。

在实验室尺度的宏观水平上材料的动力粘性系数μ_d由传统的冲击加载方法获得。η的上限为$(10^5 \sim 10^6)$ Pa·s，而η下限由下式确定[9]：

$$\eta \approx \rho_0 u \lambda D t \tag{21}$$

式中：ρ_0为材料的初始密度，D为冲击波的速度，u为粒子速度，λ为$D(u)$冲击压缩律中的系数，t为以时间计的冲击波的波阵面的宽度。对于NaCl的冲击试验表明[10]，根据冲击速度的不同，粘性的变化范围为$\eta \approx (10^3 \sim 10^4)$ Pa·s。

3.2 微观层次上的粘性

微观层次具有原子级别的尺度，在这一尺度上具有点缺陷及线缺陷。这一尺度的动力粘性与位错的动力阻滞有关，可表示为[6]

$$\eta = \alpha \frac{B}{b^2 N_m} \tag{22}$$

式中：B为位错的阻滞粘性系数；b为Burgers矢量；N_m为可动位错密度；α为常数，且$\alpha < 1$。

通常在这一尺度上，粘性很小，主要在30~50 Pa·s范围内变化。

在冲击载作用下，基于位错动力学的材料粘性[7]为

$$\eta = \frac{H}{\dot{\varepsilon}\ln(bN_m v_\infty/\dot{\varepsilon})^2} \tag{23}$$

式中：v_∞为位错的极限速度，H为常数。

由式(23)可知，当$\dot{\varepsilon} \to \infty$时，$\eta \propto \frac{1}{\dot{\varepsilon}}$。

3.3 细观水平上的粘性

从位错动力学直接过渡到宏观塑性是不可能的，这是因为存在有位错群体的集体相互作用以及

有大尺度的变形载体加入到塑性变形中来的缘故。因此，必须研究介于宏观与微观层次之间的细观层次。

介于宏观与微观层次的细观层次起着桥梁的作用，细观层次上的材料行为对于材料的宏观行为起着关键的作用。在岩体受外载作用发生变形与破坏过程中，宏观及细观层次发生着能量的转换，伴随着新的构造的形成。岩石为具有多层次结构的颗粒介质，与金属相比岩石的晶粒之间的夹层的强度比晶粒的强度要小。在强烈的动载作用下，根据冲击波的峰值、上升到最大应力峰值的时间大小及压相持时的不同，在介质的不同细观尺度水平上岩石呈现出流体的特性。流体的粒子之间相互作用，因此必须用统计物理学的理论来处理这一问题。

对于第 i 个粒子，在 t 时刻处于 $\vec{R}_i$ 位置。设有 N 个粒子所组成的系统在 t 时刻的分布函数为 $\psi(\vec{R}_1, \vec{R}_2, \cdots, \vec{R}_N)$。那么在给定的时刻及给定的位置发现某个粒子的概率可由下式来确定[13]：

$$\frac{\partial \psi}{\partial t} = -\sum_N \frac{\partial}{\partial \vec{R}_i}(\vec{R}_i \psi) \quad (24)$$

在文[14]中对于胶体粒子证明将上式变为

$$\frac{\partial \psi}{\partial t} = -\frac{\partial}{\partial \vec{R}}\left(D\frac{\partial}{\partial \vec{R}} - \vec{v}\psi\right) \quad (25)$$

式中：$\vec{v}$ 粒子的相对速度，$D = \frac{kT}{\varsigma}$ 为扩散系数，其中 $\varsigma = 2\pi a\eta$。

那么由给定的流场 $\vec{v}$ 所产生的局部构形平均应力由下式决定：

$$\bar{\sigma}_{\alpha\beta}(\vec{r}) = \int \bar{\sigma}_{\alpha\beta}(\vec{r}, \vec{R})\psi(\vec{R})\mathrm{d}\vec{R} \quad (26)$$

而其空间平均为

$$\langle \bar{\sigma}_{\alpha\beta}(\vec{r})\rangle = \frac{1}{V}\int \bar{\sigma}_{\alpha\beta}(\vec{r})\mathrm{d}\vec{r} = \frac{1}{V}\left[\int_{V_2}\bar{\sigma}_{\alpha\beta}(\vec{r})\mathrm{d}\vec{r} + \sum_N \int_{V_1}\bar{\sigma}_{\alpha\beta}(\vec{r})\mathrm{d}\vec{r}\right] \quad (27)$$

式中：V_1 为粒子的体积，而 V_2 为粒子间介质的体积。有效粘性为

$$\eta_{ij} = \langle \sigma_{ij}\rangle / (2\langle \dot{e}^0_{ij}\rangle) \quad (28)$$

因此

$$\eta = \eta_2(1-\phi) + \frac{\phi\langle\sigma_{\alpha\beta}\rangle_1}{2\dot{e}^0_{\alpha\beta}} \quad (\alpha \neq \beta) \quad (29)$$

式中：$\eta = \eta_{\alpha\beta}$，$\langle\sigma_{\alpha\beta}\rangle_1 = \frac{1}{V_1}\int_{V_1}\bar{\sigma}_{\alpha\beta}(\vec{r})\mathrm{d}\vec{r}$，$\eta_2 = \alpha\eta(\phi)$，$\alpha$ 为系数。

对于二维的情况，设 $\vec{v} = \dot{\gamma}z\vec{j}$
代入式(15)得

$$D\frac{\mathrm{d}^2\bar{\sigma}_{yz}}{\mathrm{d}y^2} + \dot{\gamma}z\frac{\mathrm{d}\bar{\sigma}_{yz}}{\mathrm{d}y} = 0 \quad (30)$$

其解为

$$\bar{\sigma}_{yz}(\vec{r}) = A + b\exp\left(-\frac{\dot{\gamma}\varsigma yz}{kT}\right) \quad (31)$$

代入式(12)得

$$\langle\sigma_{yz}\rangle_1 = A + \frac{B}{V_1}\int_{V_1}\left(1 - \frac{\dot{\gamma}\varsigma yz}{kT} + \cdots\right)\mathrm{d}\vec{r} = A + B\left(1 - \frac{2a^2\dot{\gamma}\varsigma}{5\pi} + \cdots\right) = A + B\exp\left(-\frac{2a^2\dot{\gamma}\varsigma}{5\pi} + \cdots\right) \quad (32)$$

代入式(19)得

$$\eta(\dot{\gamma}) = \frac{A}{\dot{\gamma}} + \frac{B}{\dot{\gamma}}\exp\left[-C\frac{\dot{\gamma}}{kT}\eta(\dot{\gamma})\right] \quad (33)$$

这一方程为非线性的。令 $\eta\dot{\gamma} = X$，则上式变为

$$X = A + B\exp\left[-C\frac{X}{kT}\right] \quad (34)$$

对于同一 T，此方程的解变为 $\eta\dot{\gamma} = X = \mathrm{const}$，如下所示：

$$\eta(\dot{\gamma}) = \frac{\mathrm{const}}{\dot{\gamma}} \quad (35)$$

也即随着应变率的增加粘性与应变率成反比。

3.4 粘性的过渡拟合公式

在本节中作者将尝试把微细宏观上的粘性及动力强度联系起来。由式(2)知，介质的粘性系数决定于介质的松弛时间。介质的松弛除了构造单元之间的相对滑移之外还有介质的构造单元的重组及构造单元内部结构的变化。这种变化包括介质单元之间及内部的断裂及滑移，并伴随着膨胀现象的发生。在介质的结构破坏处产生应力集中，而应力集中现象随着时间而发生松弛。在膨胀现象产生过程中，在介质中结构缺陷的产生大致上是均匀的。介质内产生的缺陷的发展速度，如晶粒中的位错速度、微裂纹及宏观裂纹的扩展速度是有限的，并依赖于所受到的外力的大小。在介质中的各种缺陷的发展速

度与介质中的应力松弛紧密相联，因此与松弛时间成正比。

从唯象角度来讲，假定缺陷的发展速度 v 与应变率 $\dot{\varepsilon}$ 相关，并随着应变率的增加而增加，即
$$v = v(\dot{\varepsilon}) \qquad (36)$$

将上式展开为泰勒级数得
$$v(\dot{\varepsilon}) = v_0(0) + \alpha\dot{\varepsilon} + \cdots \qquad (37)$$

就 $v_0(0)$ 的物理意义来讲，可以理解为在变形大小一定时发生应力松弛的缺陷扩展速度。

对于尺度为 L 的岩体的构造层次，假定松弛时间与缺陷的传播时间成正比，则在取式(37)的线性项时，应力松弛时间为
$$T = \kappa \frac{L}{v_0 + \alpha\dot{\varepsilon}} \qquad (38)$$

虽然上式唯象地推得，但是有足够的证据证明式(38)是正确的。如果把地质体的变形与破坏规律和 Maxwell 体等同起来，那么在分析地震发生时可以把某一震级的地震的重复周期理解为与岩体的松弛时间成正比。按照文[15]的统计，则
$$T \sim L \sim E^{1/3} \qquad (39)$$

这种关系对于里氏 4～8.5 级的地震都适用。作者不排除对于震级 $M<4$ 地震也适用。

非常令人感兴趣的是，按照文[16]研究结果，在从 cm 级至 km 级的尺度上，破坏的时间 t_l 与试件尺寸 l 的关系也成正比：
$$t_l \sim l \qquad (40)$$

这种关系反映了介质的构造层次与时间尺度之间的内在联系。

这样介质的粘性为
$$\eta = G\tau = G\kappa \frac{L}{v_0 + \alpha\dot{\varepsilon}} \qquad (41)$$

则式(41)在小应变率时与式(18)一致，而在 $\dot{\varepsilon} \to \infty$ 时
$$\eta = G\kappa \frac{L}{v_0 + \alpha\dot{\varepsilon}} \propto \frac{1}{\dot{\varepsilon}} \qquad (42)$$

而与式(25)一致。

但是由于材料的多样性，式(41)所含的参数少，因此，其描述能力是有限的。

试验证明，缺陷的发展速度是有限的，例如裂纹的最大扩展速度为剪切波速的 0.2～0.5 倍[7]。考虑到这一情况式(37)可以取如下形式：
$$v = v_0 + b\left(\frac{\varsigma\dot{\varepsilon}^n}{1 + \lambda\dot{\varepsilon}^n}\right) \qquad (43)$$

式中：b，ς，λ，n 为常数。

但是按照 Радионов 的 Maxwell 模型[11]，具有内部结构材料的尺寸 L 由于内部应力集中的存在而不能承受大小超过一定数值的应变率，且有 $L \propto 1/\dot{\varepsilon}$，考虑到在 $\dot{\varepsilon} \to 0$ 时 L 的值是有限的，因此取 $L \propto 1/(\dot{\varepsilon}+d)$，式中 d 为某一常数，这样粘性为
$$\eta = A \frac{1}{v_0 + b\left(\dfrac{\eta\dot{\varepsilon}^n}{1+\lambda\dot{\varepsilon}^n}\right)} \frac{1}{\dot{\varepsilon}+d} =$$
$$\frac{A}{\dot{\varepsilon}+d} + \frac{B + C\dot{\varepsilon}^n}{v_0 + b_1\dot{\varepsilon}^n} \frac{1}{\dot{\varepsilon}+d} \qquad (44)$$

式中：A，B，C，n 为常数。在中高应变率区，考虑到系数 B，d 数值小，式(44)可以近似表示为
$$\eta = \frac{A}{\dot{\varepsilon}} + \frac{C\dot{\varepsilon}^{n-1}}{v_0 + b_1\dot{\varepsilon}^n} \qquad (45)$$

对于 η 与 $\dot{\varepsilon}$ 的关系，式(45)的模拟效果与下式的模拟效果是一样的[17]：
$$\eta = b_1 \frac{\ln(\dot{\varepsilon}/\dot{\varepsilon}_0)}{\dot{\varepsilon}} + b_2 \frac{(\dot{\varepsilon}/\dot{\varepsilon}_s)^{n-1}}{[(\dot{\varepsilon}/\dot{\varepsilon}_s)^n + 1]} \quad n \geq 1 \qquad (46)$$

式中：b_1，b_2 为常数，$\dot{\varepsilon}_0$ 约为 $10^{13} \sim 10^{14}$/s。式(46)的正确性可以通过算例来检验。

由于受动载作用下粘性的影响，材料的动力强度为
$$Y = C_1 + \eta\dot{\varepsilon} = C_1 + b_1 \ln(\dot{\varepsilon}/\dot{\varepsilon}_0) +$$
$$b(\dot{\varepsilon}/\dot{\varepsilon}_s)^n / [(\dot{\varepsilon}/\dot{\varepsilon}_s)^n + 1] \qquad (47)$$

式中：C_1，b_1，b，$\dot{\varepsilon}_0$，$\dot{\varepsilon}_s$ 为材料常数。

对于花岗闪长岩及白云岩，计算结果如图 3 所

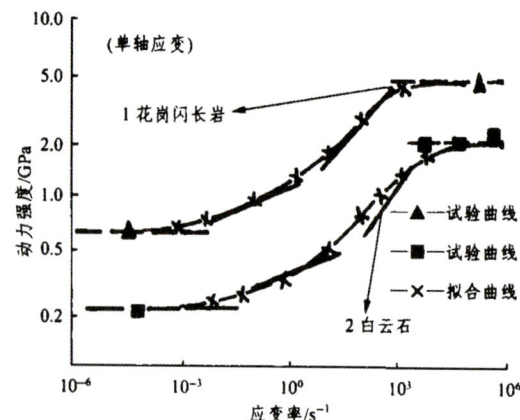

注：拟合参数：曲线 1 (花岗闪长岩)：$b=4.3$，$\dot{\varepsilon}_s=10^{3.9}$，$n=1.5$；
曲线 2 (白云石)：$b=2.0$，$\dot{\varepsilon}_s=10^{2.8}$，$n=1.0$。
试验曲线可参照文[17]

图 3 应用式(39)模拟材料动力强度的结果

Fig.3 Comparison of Eq.(39) with the experimental data

示。在计算中，因为在前 2 项荷载计算的应变率的范围内，变化非常小，故取为常数，其他参数如图 3 所示。

这样得到的粘性拟合式(41)，(44)～(46)，可以根据不同的情况而选用。

从上述不同构造水平上的粘性可以看出，随着外载的增加，变形速度的增大，粘性系数的逐步降低的确意味着变形与破坏的发生层次逐渐由宏观，经由细观向微观层次过渡。

4 岩体的破坏块体尺寸

由式(10)，(12)可以看出岩体破坏时的块体尺寸与破坏时所达到的强度值有关。

在介质出现破坏之后，有侧压静力压缩能使荷载继续增加。在材料屈服之后随着塑性变形的增加材料会出现硬化，材料所能承担的应力继续增加。进行的有侧压静力压缩试验表明，超过弹性限的介质变形导致介质多重体积破裂，不仅使块体体积减小，而且使破裂块体的体积分布变化很大。随着变形增加，分布密度函数极值向小尺寸区移动，而函数本身变窄。试验还表明，块体减小规律与侧压无关，后者仅影响破坏起始时的块体尺寸。

材料变形的速率对于材料的强度提高影响显著。在文[18]中作者对于材料强度随应变率提高的物理机制进行了研究。分析表明，在小应变率的范围内，材料强度 - 应变率依赖性受热活化机制控制；随着应变率进一步增加，材料的宏观粘性阻尼机制(声子阻尼)出现，并逐渐占据主导地位，材料的惯性影响逐渐明显；在高应变率区时，材料惯性的影响非常大，这时不同尺寸的缺陷的增长将同时启动，在材料没有缺陷的地方热活化机制引起原子键的断裂。这样可以把材料强度 - 应变率依赖性看作是热活化机制与宏观粘性机制并行存在，相互竞争的结果。这两种机制在不同的应变率区占据主导地位。强度 Y 与剪应变率 $\dot{\varepsilon}$，塑性变形 ε_p 及应力状态的关系[17]为

$$Y = \left(1 + B_1 \varepsilon_p\right)^\lambda \left[\frac{1}{\gamma}\left(U_0 + kT \ln \frac{\dot{\varepsilon}}{\dot{\varepsilon}_0}\right) + \frac{b(\dot{\varepsilon}/\dot{\varepsilon}_s)^n}{(\dot{\varepsilon}/\dot{\varepsilon}_s)^n + 1}\right] e^{AC} \quad (48)$$

式中：U_0 为活化能，γ 为活化体积，通常为 10～1 000 原子体积，k 为玻尔兹曼常数，T 为绝对温度，b 为宏观粘性对强度增长的贡献的最大幅值；n 为常数，控制材料强度 - 应变率对数曲线的陡峭程度；$\dot{\varepsilon}_s$ 为材料强度 - 应变率半对数曲线的拐点值。$C = \sigma_3/\sigma_1$ 为表示应力状态的参数，$C = -\infty$ 对应于单轴受拉，$C = 1$ 对应于静水受压；A 为常数。应变硬化因子为 $(1 + B_1 \varepsilon_p)^\lambda$ (λ，B_1 为材料常数)。

把式(10)中的应力用式(48)中的强度 Y 替代即得岩体破坏时的尺寸与应变率 $\dot{\varepsilon}$，塑性变形 ε_p 及应力状态的关系：

$$\Delta_i = \left(\frac{C_1}{Y - C_0}\right)^k \quad (49)$$

式(49)在依赖应变率的规律上与试验相符[6]。

这样就可以看出这样的破坏场景：随着荷载的增加，介质在最大尺寸的构造单元的间隙之间发生破坏，破裂块体的尺寸为这一水平的构造尺寸。当荷载继续增加到下一级构造单元的破坏强度时，下一级的构造单元间隙之间发生破坏，破裂块体的尺寸为这一级别的构造尺寸。随着荷载的增加，达到破坏的构造单元的尺寸逐渐减小，破坏块体尺寸逐渐减小。通常应变率 $\dot{\varepsilon}$，塑性变形 ε_p 及应力状态将影响材料的最终承载能力。式(48)，(49)给出了岩体破坏时的尺寸与应变率 $\dot{\varepsilon}$，塑性变形 ε_p 及应力状态的关系。在物理学中，对称是一个非常重要的概念，最常见的一种对称性为空间的各向同性和均匀性，在固体力学中所遇到的各向同性介质就具有对称性的介质，它的一个重要的性质是具有 2 个弹性常数。当介质破坏以后，各向同性消失，也即其对称性被破坏。这里的具有构造层次的岩体的破坏具有连续的性质。在初次破坏时，形成最大等级的块体，这时岩体的整体对称性被破坏了。但是对于新形成的块体来讲，仍然可以看作是均匀的和各向同性的，也就是说对称性已经局部化了。随着外载的增加，块体逐级丧失对称性，对称局部化尺寸逐级减小。因此，岩体的破坏可以看作是一个连续的对称性破坏的过程，及对称性局部化逐级减小的过程。

5 结 论

岩体具有复杂的构造层次，这种构造层次可以从原子尺度水平一直延伸到地质构造水平。这种情况使连续介质力学中使用的基本体元的概念及圣维南变形协调条件受到了置疑。同时，这种内部构造

层次也影响岩体的物理力学性质。本文基于已有的资料及物理力学就材料的粘性及构造层次之间的关系进行了系统研究。由研究可以看出，不同的构造层次对应着不同的强度、粘性及应变率。宏观层次通常对应着低的强度、应变率及高的粘性系数，而微细观层次对应着高的强度、应变率及低的粘性系数。通常随着外载的增加，变形速度的增大，发生变形与破坏的层次逐渐由宏观，经由细观逐步向微观层次过渡，动力强度增加，粘性系数逐渐降低。在高应变速率区，岩石的强度增加很大，而粘性与变形速率成反比。因此强度及粘性系数对于一种介质来讲不是常数，而是在不同的构造层次上具有不同的大小。表现在破坏块体的尺寸上，最小的破坏尺寸与所能达到的强度有直接的关系。本文建议了强度、粘性对于构造层次及应变率的依赖关系。

参考文献(References)：

[1] Опарин В Н，Юшкин В Ф，Акинин А А．Балмашнов Е.Г.О новой шкале структурно-иерахических представлений как паспортной характеристике объектов геосреды[J]. Фтпрпи, 1998, 5: 16 - 33.

[2] Курленя М В，Опарин В Н，Еременко А А. Об отношении линейных размеров блоков пород к раскрытию трещин в структурной иерархии массивов[J]. Фтпрпи, 1993, 3: 3 - 10.

[3] Шерман С И. Разломообразование в литосфере，зон растяжения[M]. Новосибирск: Наука, 1992.

[4] Альтшулер Л В，Доронин Г С，Ким Г Х. Вязкость ударносжатых жидкостей[J]. Пмтф, 1987, 6: 110 - 118.

[5] Белинский И В，Христофоров Б Д. Вязкость N₄Cl при ударном сжатии[J]. Пмтф, 1968, 1: 150 - 151.

[6] Meyers M A. Dynamic deformation and failure[A]. In: Mechanics and Materials: Fundamentals and Linkages[C]. [s. l.]: John Wiley and Sons Inc., 1999. 489 - 594.

[7] Степанов Г В，Харченко В В. Связь напряжений и деформаций в металлах при воздействии импульсной нагрузки[J]. Проблемы Прочности, 1984, 11: 32 - 37.

[8] Садовский М Ф，Болхвитинов Л Г，Писаренко В Ф. Деформирование геофизической среды и сейсмический процесс[M]. Москва: Наука, 1987.

[9] Bažant Z P, Ožbolt J, Eligehausen R. Fracture size effect: review of evidence for concrete structures[J]. J. of Struc. Engrg., 1994, 120: 2 377 - 2 398.

[10] Bažant Z P. Scaling in nonlinear fracture mechanics[A]. In: IUTAM Symposium on Nonlinear Analysis of Fracture[C]. [s. l.]: Kluwer Academic Publishers, 1997. 1 - 12.

[11] Радионов В Н，Сизов И А，Цветков В М. Основы геомеханики[M]. Москва: Недро, 1986.

[12] Landau L D, Lifshits E M. Theory of Elasticity[M]. Beijing: Beijing World Publishing Cooperation, 1999.

[13] Bird R B, Armstrong R C, Hassager O. Dynamics of Polymeric Liquid[M]. New York: [s. n.], 1987.

[14] Chow T S. Mesoscopic physics of complex materials[M]. New York: Springer, 2000.

[15] Содовский М Ф. Еще раз о зависимости объема очага землетрясения от ее энергии[J]. Данссср, Т 275, 1984, 5: 1 087 - 1 088.

[16] Куксенко В С. Физические и методические основы прогнозирования горных ударах[J]. Фтпрпи, 1987, 1: 9 - 22.

[17] Grady D E. Shock wave properties of brittle solids[A]. In: Shock Compression of Condensed Matters[C]. Woodbury, NY: AIP Press, 1995.

[18] 戚承志，钱七虎．岩石等脆性材料动力强度依赖应变率的物理机制[J]. 岩石力学与工程学报, 2003, 21(2): 177 - 181.(Qi Chengzhi, Qian Qihu. Physical mechanism of brittle material strength-strain rate sensitivity[J]. Chinese Journal of Rock Mechanics and Engineering, 2003, 21(2): 177 - 181.(in Chinese))

弹粘塑性孔隙介质在冲击荷载作用下的一种本构关系
——第二部分：弹粘塑性孔隙介质的畸变行为

戚承志¹　　　王明洋²　　　钱七虎³

(¹北京建筑工程学院土木系　北京　100044)　(²解放军理工大学　南京　100038)　(³中国工程院土木水利与建筑学部　北京　100038)

摘要 继发表在本刊 2003 年第 9 期上第一部分的研究，就弹塑性孔隙介质在冲击偏应力张量作用下的行为进行了研究。考虑应变率对材料强度的影响，给出了热活化机理与宏观阻尼机理相互竞争的联合强度-应变率依赖模型。考虑应力状态、应变强化、畸变损伤和与孔隙率相联系的损伤对材料强度的影响，并在此基础上基于热活化原则给出了松弛型的应力偏量与应变偏量之间的关系，且本模型符合热力学第二定律。

关键词 岩石力学，畸变行为，率依赖性，应力状态，应变强化，松弛型关系

分类号 TU 435　　**文献标识码** A　　**文章编号** 1000-6915(2003)11-1763-04

CONSTITUTIVE MODEL OF POROUS ELASTOVISCOPLASTIC MATERIALS SUBJECTED TO SHOCK
——PART II: DISTORSIONAL BEHAVIOR OF POROUS ELASTOVISCOPLASTIC MATERIALS

Qi Chengzhi¹, Wang Mingyang², Qian Qihu³

(¹*Department of Civil Engineering, Beijing Institute of Civil Engineering and Architecture, Beijing 100044 China*)
(²*PLA University of Science and Technology, Nanjing 210007 China*)
(³*Chinese Academy of Engineering, Beijing 100038 China*)

Abstract Continuing the study performed in the first part put in the issue 2003-9 of this journal, the second part of the paper researches the distorsional behavior of porous elastoviscoplastic materials subjected to shock. The influence of strain rate on material strength is considered, and a unified strain rate dependent strength model of competition between thermal activation mechanism and macro-viscosity mechanism is proposed. The influences of stress state, strain hardening, distorsional deformation damage, and damage related with porosity change are considered. And furthermore, on the basis of thermal activation mechanism, a relaxation type relation between distorsional stress tensor and distorsional strain tensor is given, and the model follows the second thermodynamic law.

Key words rock mechanics, distorsional behavior, rate dependence, stress state, strain hardening, relaxation type relation

1 引言

在快速冲击荷载的作用下，物质内部要产生畸变，因而引起物质的损伤，使物质强度减小，同时，孔隙率的增大或减小也会引起物质的损伤，减小物质的强度。因此，必须考虑上述因素，才能得到符合实际的物质强度。此外，当介质同时受到压应力与偏应力时，会发生剪切强化压缩。岩石的屈服强度依赖于压应力、应力状态、材料的损伤度、应变增长率等。因此，模拟介质的弹粘塑性行为十分复杂。本部分基于作者提出的热活化及声子(粘性)阻尼机制相互竞争的强度-应变率依赖模型，并综合考虑应力状态、孔隙率变化和畸变引起的损伤的影响，采用松弛型的应力偏张量与应变偏张量之间的关系来模拟介质的弹粘塑性行为[1~10]。

2 应变率及温度的影响

受单轴拉应力时，材料的应变率服从Журков (Zhurkov)公式[1]：

$$\dot{\varepsilon} = \dot{\varepsilon}_0 \exp\left(-\frac{U_0 - \gamma\sigma}{kT}\right) \quad (1)$$

式中：$\dot{\varepsilon}_0$ 为速度因子，为材料的极限变形速度；U_0 为活化能；σ 为单轴拉应力；γ 为构造敏感系数；k 为玻尔兹曼常数；T 为绝对温度。

式(1)指出了材料变形破坏的热活化本质，同时，它建立了强度对时间和温度的依赖关系：

$$\sigma = Y = \frac{1}{\gamma}\left(U_0 + kT\ln\frac{\dot{\varepsilon}}{\dot{\varepsilon}_0}\right) \quad (2)$$

实验表明[2]，上式足够精确地描述了应变率小于 $10^2\,\text{s}^{-1}$ 时材料强度对应变率的敏感性，在对数坐标里，上述关系为一条直线。当应变率大于某一值时，材料强度随应变率的增加而加速增加，材料的变形具有了更多的绝热性质；而当应变率进一步增加，达到冲击应变率时，材料强度随应变率增加几乎不加速。材料动力强度随应变率变化的规律在对数坐标中如图1所示。上述的材料强度随应变率增加而变化的规律表明，随应变率增加，不同的机制展现了出来，并起主导作用。在应变率较低阶段，变形的热活化机制起主导作用；而在应变率大于某一值时，声子阻尼(粘性)机制起主导作用[3]；当应变率进一步增加，达到冲击应变率时，损伤动力学引起的破坏延迟排除了脆性破坏的可能，并激活了一种新的破坏机制。

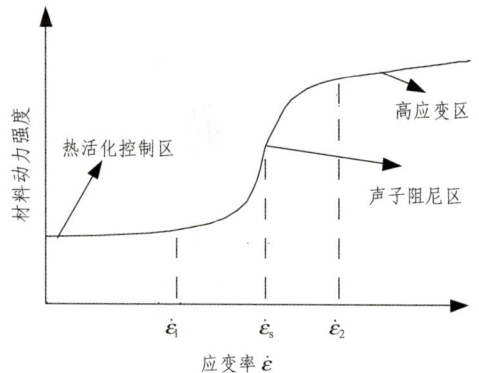

(通常 $\dot{\varepsilon}_1$ 为 $10^1\sim10^2/\text{s}$；$\dot{\varepsilon}_\text{s}$ 为 $10^3/\text{s}$ 左右；$\dot{\varepsilon}_2$ 为 $10^4/\text{s}$ 左右)

图 1 材料动力强度随应变率变化的一般规律

Fig.1 The general law of variation of dynamical strength with strain rate

3 相互竞争的联合强度-应变率依赖模型

虽然可在某个特定的应变率段上只考虑起支配作用的应变率依赖机制，如在应变率较低阶段，采用式(2)，而在应变率大于某一值，如 $100\sim1\,000\,\text{s}^{-1}$ 时，应用线性粘性模型等，这样就能得到折线型的模型。但从材料强度-应变率曲线看，3个应变率区间段是平滑地连接的，这就启发作者假设不同的材料强度-应变率依赖机制是并行存在的，它们之间相互竞争，在不同的阶段某个机制起主导作用，进而提出一种联合的热活化与粘性机制并联的材料强度-应变率依赖模型。这样，就可把式(2)改写为

$$\sigma = Y = \frac{1}{\gamma}\left(U_0 + kT\ln\frac{\dot{\varepsilon}}{\dot{\varepsilon}_0}\right) + \sigma_D \quad (3)$$

式中：σ_D 表示有声子阻尼(粘性)所引起的强度提高，一般依赖于温度，应变率等。为了简化模型，在此不显式地考虑温度的影响，而综合考虑在应变率的影响表达式里边。取 $\sigma_D = \dfrac{b(\dot{\varepsilon}/\dot{\varepsilon}_\text{S})^n}{(\dot{\varepsilon}/\dot{\varepsilon}_\text{S})^n + 1}$，则式(3)变为

$$\sigma = Y = \frac{1}{\gamma}\left(U_0 + kT\ln\frac{\dot{\varepsilon}}{\dot{\varepsilon}_0}\right) + \frac{b(\dot{\varepsilon}/\dot{\varepsilon}_\text{S})^n}{(\dot{\varepsilon}/\dot{\varepsilon}_\text{S})^n + 1} \quad (4)$$

式中：n 为某一指数；b 为常数，表示粘性阻尼贡献

的极限值；$\dot{\varepsilon}_S$ 为某一调整参数，其取值在曲线拐点附近(如图1所示)。

式(4)对粘性模拟的特点是：当 n 取某个数量级值左右，并适当选取 b 值时，在应变率较低阶段，粘性的影响可以忽略不计，而在达到很高的冲击应变率时，粘性影响基本保持不变，并有不大的热活化的材料变形影响。

为了检验上述模型的有效性，利用文[4]所引用的白云岩、花岗闪长岩数据，对白云岩、花岗闪长岩进行了模拟。模拟结果如图 2 所示，与实际数据符合得很好。

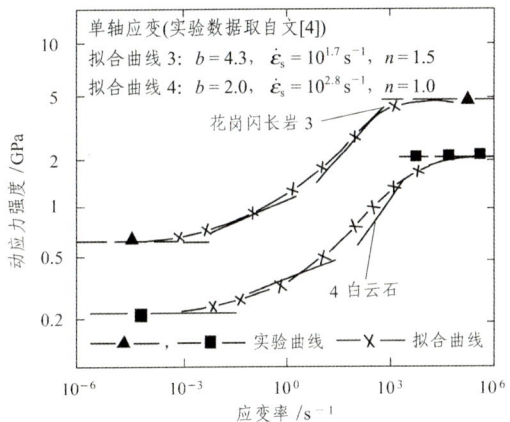

图 2 建议模型与实验数据的比较

Fig.2 Comparison between proposed model and experimental data

对材料受剪，式(4)也成立，只是用剪应力代替拉应力，用剪应变代替拉应变，响应的常数变为受剪时的常数。这时，有公式：

$$\tau_{y0} = \frac{1}{\gamma_\tau}\left(G_0 + kT\ln\frac{\dot{\gamma}}{\dot{\gamma}_0}\right) + \frac{b_1(\dot{\gamma}/\gamma_S)^n}{(\dot{\gamma}/\gamma_S)^n + 1} \quad (5)$$

式中：τ_{y0} 为材料纯剪时的抗剪强度，其余各项符号意义对应于式(4)。

材料的应力状态也对材料的强度产生影响。对于描述地质材料的极限状态，比较常用的是摩尔理论

$$\tau_y = \tau_y(\sigma_m) \quad (6)$$

式中：τ_y 为材料抗剪强度；$\sigma_m = (\sigma_1 + \sigma_3)/2$，$\sigma_1 \geq \sigma_2 \geq \sigma_3$(这里主应力以压为正)。

应力状态为 $\sigma_1 \geq \sigma_2 = \sigma_3$ 时，对于广泛的岩石种类，对极限状态的比较完整的描述为[2]

$$\tau_y = \tau_{y0} e^{AC} \quad (7)$$

式中：$C = \sigma_3/\sigma_1$ 为表示应力状态的参数，$C = -\infty$ 对应于单轴受拉，$C = 1$ 对应于静水受压；A 为常数。

在 $\ln\tau_y - C$ 坐标内，式(7)为一随 C 值增加而增加的斜直线。在 C 的负值区，当其小于某一数值 C_p 时，式(7)可用最大拉应力准则来代替。在 C 的正值区，当 C 大于某一数值 C_y 时，τ_y 保持不变；C_p 和 C_y 的值可由实验获得[2]。

4 畸变形为

材料受压时，产生的畸变会使材料产生损伤。损伤随变形的变化可用温布尔型函数表示[4]

$$D_\phi = 1 - \exp\left[-\frac{1}{2}\left(\frac{\varepsilon_{eq} - \varepsilon_0}{a\varepsilon_0}\right)^2\right], \quad \varepsilon > \varepsilon_0 \quad (8)$$

式中：α 为常数，ε_0 为材料损伤开始的应变值，ε_{eq} 为等效应变。

岩石等孔隙材料在静水压力 P 小于某一值 P_e 时，表现为弹性，P_e 即为孔隙材料的弹性极限，$P_e = (2/3)Y_m\ln(1/\phi_0)$ 为多空材料的弹性极限[5]，Y_m 为基体材料的弹性极限。当大于此值时，孔隙材料会发生"塌缩"，从而孔隙度减少。与"塌缩"现象相联，材料内部会出现损伤。可利用第一部分的式(30)进行评价。这种评价方法与压缩实验数据联系起来，且具有物理意义，因而是合理的。

这样，如果认为孔隙率的变化仅为非弹性变形所致，并假设基体材料塑性不可压缩，则可以把损伤与孔隙率的变化联系起来，并假设为线性关系，则损伤可用下列公式表示[6]：

$$\dot{D}_k = \frac{|\dot{\phi}|}{c(1-\phi)} H(1 - D_k) \quad (9)$$

式中：c 为常数，$H(x)$ 为 Heaviside 函数。

对于受压情况，由式(9)得

$$D_k = \frac{1}{c}\ln\frac{1-\phi}{1-\phi_0} H(1 - D_k) \quad (10)$$

式中：ϕ_0 为初始孔隙率，ϕ 为当前孔隙率。

当前孔隙率 ϕ 可利用第一部分的式(30)进行评价。

对于受拉情况，有

$$D_k = \frac{1}{c}\ln\frac{1-\phi_0}{1-\phi} H(1 - D_k) \quad (11)$$

如果取应变硬化因子为 $(1 + B_1\varepsilon_P)^\lambda$（$\lambda$，$B_1$ 为材料常数)[7, 8]，那么孔隙材料的屈服强度为

$$\tau_y = \tau_{y0}(1+B_1\varepsilon_p)^\lambda \left[\frac{1}{\gamma_\tau}\left(G_0 + kT\ln\frac{\dot\gamma}{\dot\gamma_0}\right) + \frac{b_1(\dot\gamma/\dot\gamma_s)^n}{(\dot\gamma/\dot\gamma_s)^n+1}\right] \cdot$$
$$e^{AC}(1-D_\phi)(1-D_k)\left(1+A_1J_m^{1/2}P_{mV} - A_2(T-T_0)\right) \quad (12)$$

式(12)右边最后一项括号表达式表示基体静水压力及温度的影响。上式为考虑众多因素时计算材料动力强度的一般表达式，在计算具体材料的动力强度时，应根据具体材料的性质确定应考虑那些因素。式(4)为只考虑应变率效应的一种特殊情况。

这样，材料屈服以后，由其变形的热活化性，则可以采用类似于式(1)的下列公式：

$$\dot\varepsilon_{ij}^p = \dot\varepsilon_{ij0}^0 \exp[\gamma_p(\tau_e - \tau_y)] \quad (13)$$

式中：$\tau_e = \sqrt{\frac{3}{2}s_{ij}s_{ij}}$ 为等效应力，且 $\tau_e \geq \tau_y$；$\dot\varepsilon_{ij}^0$ 为应变率因子；γ_p 为常数。

由上式可得等效塑性应变率与等效应力之间的关系为

$$\dot\varepsilon_p = \dot\varepsilon_{p0} \exp[\gamma_p(\tau_e - \tau_y)] \quad (14)$$

式中：$\dot\varepsilon_p = \sqrt{\frac{2}{3}\dot\varepsilon_{ij}\dot\varepsilon_{ij}}$ 为等效塑性应变，$\dot\varepsilon_{p0}$ 为等效塑性应变率因子。

应力偏张量与应变偏张量之间的关系采用松弛型的形式为

$$\dot s_{ij} = 2G(\dot\varepsilon_{ij}^T - \dot\varepsilon_{ij}^p) \quad (15)$$

式中：$\dot\varepsilon_{ij}^T$ 为总的应变偏量速率，$\dot\varepsilon_{ij}^p$ 为塑性应变偏量速率。

这样就闭合了本构方程。

热力学第二定律对上述演化方程施加了限制。对于有损伤的介质，其自由能可写为：

$$\psi = \psi_e^m(1-\phi)(1-D_\phi)(1-D_k) \quad (16)$$

若冲击过程为绝热过程，即不考虑热扩散，则第一部分的式(9)可写为：

$$\sigma_{ij}\dot\varepsilon_{ij}^{(n)} - \beta_i\dot\alpha_i \geq 0 \quad (17)$$

对取式(13)的塑性变形，上式左端的第一项恒为正；对由畸变形和孔隙率变化所引起的损伤，由于第一部分中的式(8)：

$$\beta_i = -\rho_0\frac{\partial\psi}{\partial\alpha_i}$$

及本部分中的式(8),(9)可知，上式左端的第二项也恒为正，因此，式(17)恒满足，本文模型符合热力学第二定律。

5 结 论

继续第一部分的研究，就弹塑性孔隙介质在冲击偏应力张量作用下的行为进行了研究。考虑应变率对材料强度的影响，给出了热活化机理与宏观阻尼机理相互竞争的联合强度-应变率依赖模型。考虑应力状态、应变强化、畸变损伤和与孔隙率相联系的损伤对材料强度的影响，并在此基础上基于热活化原则给出了松弛型的应力偏量与应变偏量之间的关系，且本模型符合热力学第二定律。

参 考 文 献

1 Регель В Р, Слуцкер А И, Томашевский Д Е. Кинетическая Природа Прочности Твердого Тела[M]. Москва：Наука，1974

2 Ставрогин А Н, Протосеня А Г. Пластичность Горных Пород[M]. Москва：Недра，1979

3 Perzyna P. Constitutive modelling of dissipative solid for localization and fracture[A]. In: Localization and Fracture Phenomena in Inelastic Solids[C]. New York：Springer，1998，99～241

4 Grady D E. Shock-wave properties of brittle solids[A]. In: Steve Schmidt ed. Shock Compression of Condensed Matters[C]. New York：AIP Press，1995，9～20

5 Eibl J, Schmidt-Hurtienne B. Strain-rate-sensitive constitutive law for concrete[J]. Journal of Engineering Mechanics，1999，125(12)：1 403～1 410

6 Caroll M M, Holt A C. Static and dynamic pore collapse relations for ductile porous materials[J]. J. Applied Physics，1972，43(2)：1 626～1 635

7 Rubin M B, Vorobiev O Yu, Glenn L A. Mechanical and numerical modeling of porous elastoviscoplastic material with tensile failure[J]. Int. J. Solid and Structures，2000，37(4)：1 841～1 870

8 Steinberg D J, Cochran S G, Guinan M W. A constitutive model for metals applicable athigh-strain rats[J]. J. Applied Physics，1980，5(3)：1 498～1 504

9 戚承志，钱七虎. 岩石等脆性材料动力强度依赖应变率的物理机制[J]. 岩石力学与工程学报，2003，22(2)：177～181

10 戚承志，王明洋，钱七虎. 弹粘塑性孔隙介质在冲击荷载作用下的一种本构关系. 第一部分：状态方程[J]. 岩石力学与工程学报，2003，22(9)：1 401～1 405

强动载作用下的岩石动力学

钱七虎　戚承志　王明洋

(解放军理工大学，210007，南京)

摘要：无论是矿山工程、隧道工程，还是最近一系列高技术常规局部战争中钻地炸弹对地下坚固目标的破坏，都亟待要求对岩石中爆炸引起的应力、变形及其他运动参数和破坏效应作出比较准确的评估。目前对此问题的研究尚只能达到大致的机理及景象的定性认识和描述，定量上还存在着数量级的误差，尤其是爆炸和冲击的近区。其原因在于岩石中爆炸与冲击产生的变形和破坏特征，不仅与作用时发生的相应的物理及力学运动密切相关，而且还强烈地受岩石自身构造缺陷水平及其变化的制约。本文概述了强动载作用下岩石的动力变形与破坏研究现状及研究趋向，并讨论了若干岩石变形与破坏的若干问题。

关键词：爆炸；冲击；动力变形与破坏；岩石的构造水平；物理微细观力学

一、强动载作用下岩石变形与破坏的研究现状

1. 经典弹塑性模型

自20世纪50年代以来，随着试验手段的不断改善，岩石在爆炸作用下基于宏观连续介质力学理论的研究得到了很大的发展[1]。描述介质力学运动性能的基本方程为质量、动量、能量守恒方程在再上介质状态方程式就可构成一组完备的方程。适当选取边值和初始条件后，原则上便可对问题进行数值求解。求解的复杂性在很大程度上取决于选用的介质状态方程式，属于这一类问题的有：线弹性、非线性弹性、弹塑性、流体弹塑性，黏性及其他性质模型等。这些模型的适用性在很大程度上与所研究的过程特性有关，同一种介质在不同的过程中也可能具有不同模型的性质。

岩石在构造水平上呈现的不均匀和相关的应力场，使其在变形过程中与作用的外荷载相适应，这种适应性在宏观上表现为变形的时间相关性，一般呈现非线性弹黏塑性性状。因此，为了考虑岩石变形与加载的时间参数之间的关系常利用 Maxwell、Kelvin-Voigt 等模型进行描述。特别是 Maxwell 模型能够描述岩石的不可逆变形而在工程分析中得到最广泛地应用。

一系列试验结果（上升时间随着距离而增加，准静加载与动力加载、准静模量与动力模量的不同）证明[6]，在非破坏载区岩石的变形伴随着松弛（黏性）机制，而且变形过程具有非线性性质。可是在波的传播过程中使用经典的 Kelvin-Voigt 模型，Maxwell 模型，标准线性模型（又称线性粘弹性模型）表明，这会严重破坏几何相似律。偏离几何相似律的原因分析表明，为了最大限度地达到松弛模型与几何相似的逼近，模型表述方程必须是准齐次的。最后一点意味着这些方程应保证松弛速度与动力变形过程速度近似成正比。

为了更好地反映在非破坏区中波衰减的真实性，在文献[2]中建议了的准弹性模型。这一模型与 Hooke 定律的不同之处在于引入了应力松弛的概念，并且应力松弛的速度与变形进行的速度有关。这样得到的控制方程具有准齐次性，保证了在广泛的爆炸范围内遵守几何相似律，使得在破裂区外波传播时幅值衰减的足够快。

实际数学解的困难使得在工程模型中通常忽略黏性的作用（除非对于多孔岩石，由于受动力压缩时其压力和能量的热分量是急剧增加的，所以用于刻画这种增加程度的黏性效应对变形的影响不能忽略）。通常利用理想弹塑性考虑小压力区域内的弹性和压力较高区域内变形的不可逆性。工程计算时常将一维试验曲线图用折线段来代替真实的弹性变形段和弹塑性变形。

基金项目：国家自然科学基金资助项目（项目编号50179038）。

С.С.Григорян（Grigorian）就广泛的岩石动力学问题进行了研究[3,4]，得出了上述不可压缩的弹塑性介质模型不能在地下爆炸解算时采纳的结论。此后，他较完整地考虑了岩土介质的性质，给出了在任意荷载作用下描述介质行为的一般方程，并进行了详细地解答。假设平均静水压力与介质的密度有一一对应的关系，在加载与卸载时各不相同。对剪切变形，根据与Prandtl-Reuss略不相同的模型得到其解析表达式，方程组由Mises-Schleiher模型来闭合。这一模型的特点为剪切变形为塑性，而体积变形为弹性，也可相反。在非破坏载区，岩石的状态方程变为Hooke定律。试验证明，С.С.Григорян模型对于描述岩土的动力行为来讲是一个可以接受的模型。此模型显然包含了最少的描述介质的变形与破坏机理特征的可能参数，模型的控制方程相对简单是其优点。

利用这些方程的结果是应力及破裂区外上升时间随距离的变化与试验矛盾。矛盾在于衰减系数偏小及上升时间随距离缩小。对于岩石来讲，在破裂区外爆炸波的传播衰减要大的多。

分析表明，С.С.Григорян弹塑性模型有两个缺点：一是模型未考虑破裂波阵面后介质的扩容效应，二是未考虑破裂区介质的松弛性质。

为了克服С.С.Григорян模型的不足，在文献[19]中，建议了推广的弹塑性模型。此模型考虑了在破裂区岩石的变形规律对应力波传播的幅值及时间特征参数有决定性的影响。

弹塑性模型在岩石爆炸动力学分析中得到了广泛和成功的运用。

2. 断裂损伤模型

断裂破碎区和粉碎区对爆炸能量有极大的耗散。岩石爆炸理论的进一步发展是断裂损伤理论的引入。根据断裂力学理论，岩石可看作是含有微裂纹的脆性材料，这些裂纹的稳定性受力学状态的影响，通过统计不同长度的裂纹数及裂纹密度，利用断裂力学即可求得岩石破坏的程度及其分布。这方面研究的代表人物是Curran D. R., Seaman L.,Shockey D.A., Kipp M.E., GradyD.E.等[29,30,31,32]，比较有代表性的模型有NAG-FRAG模型BCM模型。这些模型也有自己的局限。如NAG-FRAG模型用一维荷载作用下的裂缝发展情况来解决三维应力场作用下的爆破过程不太适宜，BCM模型中裂纹均呈水平，仅适用于有层理或沉积类岩石等。要考虑岩石中大量的随机裂纹，引入损伤的概念是非常合宜的。

连续介质损伤力学开始于20世纪50~60年代，从70年代开始应用于岩石力学。岩石破坏损伤模型系统的研究在80年代有KippM.E, GradyD.E.等开展的工作。他们认为原岩中含大量随机分布的原生裂纹，在爆炸和冲击荷载下被激活的裂纹数目服从指数分布，引入损伤参量表示岩石强度的降低，并利用能量平衡理论导出脆性断裂条件下平均破碎块度表达式。随后Chen和Taylor利用O'Connell的成果确立了裂纹密度[34]，有效体积模量及有效Possion比之间的关系。J.S.Kuszmally进一步提出了新的TCK模型。该模型区分了岩石受压及受拉行为的不同，认为岩石处于体积受压时表现为弹塑性性状，而处于体积受拉时则发生与应变率有关的脆性断裂破坏。该模型还利用了KippM.E, GradyD.E.的裂纹激活率及裂纹尺寸的公式，采用了Chen和Taylor的裂纹密度，有效体积模量及有效Possion比之间的关系，在本构关系中引入了损伤随时间的变化率这一参数。模型的材料参数由稳定高应变率条件下岩石的拉伸破坏试验来确定。Thorne又对该模型进行了修正，使其在漏斗深处及以下部位与实际相符，计算的稳定性也增加了。

现代断裂损伤理论的一个不足是，绝大多数断裂损伤理论是建立在唯象的基础上的，缺乏足够的物理基础，因此研究物理上有依据的断裂损伤理论是一个迫切的课题。

岩石等脆性材料在爆炸作用下的断裂破坏问题也是一个非常重要的问题。除了爆炸能产生冲击外，物体的直接撞击也能产生冲击。冲击产生的独特效果是当压应力波在自由面上反射时，当满足一定的应力幅值及时间的联合条件时，会出现剥离现象。起初的断裂准则是众所周知的静力破坏理论的Griffith或Orowan准则，20世纪50年代即通过轻气炮试验发现，发生剥离的最小应力（即剥离门槛）是一个与时间相关的量，而不是一个恒定的量。但当时还缺乏有效的理论与试验手段来认识其机理。在60年代，基于宏观试验，瞬态测量和数值模拟，发展了一些反映时效的剥离准则，比较重要的有：应力梯度准则，应力率准则，积分准则等。60年代，苏联的Zhurkov院士及其小组在广泛的温度及应力范围内，通过大量的试验，发现了破坏的热活化本质，建立了材料的耐久性与所受的应力及温度之间的关系，即Zhurkov公式。后来一些前苏联学者应用了这一准则来研究剥离问题[25]。Zhurkov公式的意义不仅在于确定了材料的耐久性，而且指明了破坏现象的内在物理本质，与物质的微观量联系了起来。

这一领域研究的进一步是微观统计断裂力学理论在岩石破坏研究中得到了迅速的发展。其中引入损伤参数来描述剥离现象，并由此来建立破坏的连续介质力学模型。这种途径最初由Kachanov建议[8]，然后由Rabotnov[9]发展用来描述金属在蠕变条件下的破坏。Rabotnov引入了用于描述材料受损程度的损伤变量ω，这一变量的变化速率依赖于应力水平及受损伤水平，当损伤达到一定程度时，材料就会破坏。

Ильюшин（Iljushin）[10]推广了这一途径，建议了在描述损伤积累时用损伤张量Σ。显然损伤参量的引入一方面可以考虑剥离破坏的时间特征，另一方面也能预测破坏区的位置及尺寸。

很自然，也会假设介质损伤的积累会导致其性质的改变，但Kachanov, Rabotnov，Iljushin没有考虑这一因素。W.Herman[42]首次尝试考虑材料的孔隙度对其力学性质的影响。后来А.И.Глушко[46](Glushko)，Swegle[35]把这一思想应用到非线性粘弹性及弹塑性介质。但是上述对于孔隙率的考虑只是近似的，在大的孔隙率情况下偏差较大。在基本的假设形成之后，下一步就是确定支配关系。

这时的基本前提是：带缺陷的介质的行为可以近似地用连续介质的方程来描述，其状态除用应力，应变张量来描述外，还须增加一个损伤变量。本构关系不应违背热力学第一与第二定律。如果在模型中给出了内能，那么利用热力学定律导出模型的其他关系：应力-应变关系，温度及其他参量，耗散速度函数等关系。如果另外一个热力学量，如温度T作为状态参量的函数给出，那么对这一热力学函数要有热力学限制。

要闭合支配关系，还要增加一个损伤ω的演化过程的描述方程：

$$\dot{\omega} = f(\omega, \sigma, T) \qquad (1)$$

如果说脆性材料在受拉条件下的破坏机理已得到较好的研究，那么其在动力非弹性受压时的情况在很大程度上还不清楚。正是这个原因及实际工程需要，这一课题在最近十年得到了相当的重视。岩石样品受压微裂纹统计分析显示[15]，新的裂纹的走向大多数位于受压方向10°的范围内。众所周知，即使在总的受压状态下，在非均匀性附近，局部应力可能会变成拉应力，可导致裂纹的形成与扩展。现有一种翅膀状的三裂纹开裂模型，虽然试验表明这种形式出现较少，此模型在建造破坏准则及描述膨胀现象时很富成果。

在文献[16]中，利用dilaton模型，理论上证明了介质中微裂纹的产生是自我组织的结果，是一种耗散结构。在这一模型中，裂纹的产生具有动力相变的特性，在这一过程中，稳定态通过不间断的与周围介质的声子交换而维持，产生的裂纹以形成耗散结构的开放系统出现，其独特特征是自由度数比固态相少。这一模型的优点是，得到了裂纹产生的热活化能的表达式，且与试验数据相符。除此之外，这一表达式是通用的，即对任何类型的原子键，超原子及缺陷结构都是适宜的。对破坏现象研究的必然结果是深入到物质的微观层次，而物质的破坏应看作是具有原子，超原子及缺陷结构的受载体之热涨落之热动力学，动力学的统计性质。但总体来讲，就物质破坏的微观动力学来讲，研究较好的是宏观物体在平衡状态下热动力学参数小的可逆的涨落，对足够强的不可逆的破坏性涨落还未得到充分研究，这阻碍了破坏物理的发展，是一个应继续研究的课题。

3. 岩石变形和破坏的构造水平

当前，岩石中变形和破坏的传统描述是依据二种方法来进行的：第一属于连续介质力学，第二个则是位错理论。

利用连续介质力学中的唯象方法以物理和数学为基点的描述进程是完全正确的，应力及应变张量以对称运动效应的塑性变形，通过缺陷受外荷载时的平动来实现的，塑性流动曲线是通过计算应力高于流动限的变形硬化来得到。但是它只适宜去确定宏观均匀岩石介质的整体性质。这一方法用于研究岩石介质时的局限性首先是受制于原有的宏观方法：这里宏观质点实际上是无穷小的点，它不具有常规尺寸，但同时又可以看成相当地大，以致可以平均地反映岩石在宏观水平上出现的力学性质，但这样忽略了一个事实，那就是岩石具有不同的内部构造缺陷水平，其宏观的强度特征在不同的外荷载作用下也是各不相同的。这是因为，在不同形式的荷载作用下，内应力场和变形场在宏观质点本身上是有很大的不同的。一般说来，任何材料都具有复杂的受等级限制的内部构造，它在不同荷载作用下演化成各种各样的形式，不仅导致塑性的各向异性，还导致宏观体积不同形式的破坏。

位错理论用来描述岩石变形的微观性状，根据对塑性变形基本活动的研究，提出了要揭示塑性滑移和裂缝形成机理的任务，以说明总体位错的性状，并且对连续介质力学现象规律在物理上进行解释。在微观上描述岩石流变性状这一方向上，位错理论取得了很大的成就。但是，只是微观分析的应用显而易见对建立宏观的变形与材料的内部构造联系起来的完整理论来说是不够的。这种完整的理论能够把宏观的变形与材料的内部构造联系起来。近些年来，明显地把基本活动的注意力放到了微观水平变形上。非常主要的是，在转向宏观描述时它们的解释不十分正确，而且是用不真实的变形图景来计算的。我们对塑性变形认识的错误在于，对塑性变形的基本行为理解为仅仅是变形缺陷的平动，而把连续性保持条件仅仅理解为变形的平动模式的自我组织，这与不对称运动的特征"切变加转动"现实不符。原则上不能揭示位错源的性质，不能理解位错群体的自组织规律。很明显的是，直接从微观描述过程过渡到宏观现象上来，原则上是不可能的。

为了克服上述缺点，必须考察宏微观之间的尺度水平——细观水平，В.Е.Панин在一般塑性理论及其以后的破坏的研究发展向前迈出了原则性的一步。研究的方法建立在变形和破坏必须至少在三个构造水平（宏观、细观和微观）的思想上，它允许人们把材料当作是等级受限制的体系来加以研究，这种体系演化并能在外荷载作用下自行组织。这种方法使我们能够在唯一的唯象学框架内将变形固体物理学和力学中所得到的结果统一起来。在建议的方法中，中心地位是尝试对细观构造水平的研究[21-24]。

岩石是由尺寸、形状和矿物成分各不相同的许多颗粒牢固地连接在一起而组成的，颗粒又各自具有不同的物理化学和力学性能。根据其复杂的构造运动历史可分为火成岩、沉积岩及变质岩。下面我们以火成岩、变质岩和某些沉积岩的多种矿物和多晶粒结构为基础，在岩石三个构造水平（微观、细观和宏观）上对其基本的物理力学性状进行分析。

岩石中微观上构造缺陷水平的元素既可能是原子，也可能是构造缺陷（包括空缺和位错）。

对于由晶粒组成的岩石，从塑性变形一开始，由于各种晶体具有不同的的力学和热学性能，引起应力微集中，结果就在颗粒内和颗粒边界上造成一些内在的位错。由于应力微集中的作用仅在非常近距离产生影响，因此在应力微集中区域内，位错发生小距离的移动引起部分粒内应力和粒间应力松弛，另一部分由于晶体间的摩擦阻力作用，引起的应力累积并以"封闭应力"的形式存在。"封闭应力"通过"热活化"达到进一步松弛。随着变形过程中位错密度增长，各种物理力学性质之微观不连续性就随之累积，当达到它的某些临界数值时，在比较长的晶格区域内就会丧失剪切稳定性，也就是这些区域中丧失了与作用荷载相适应的能力，进入局部的破坏状态。然而作为传输破坏的位错，只是导致形成与破坏相适应的内部边界，对于任何规模的裂缝扩展都需要积极的旋转。因此，在任意晶相方向上的较大距离上，结构的重新排列就成为可能，它们在中等应力集中区产生，经过许多构造单元扩展较大距离。

岩石中细观上构造缺陷水平的元素既可能是颗粒接头（粒内裂缝、沿过多个颗粒的粒间裂缝和沿粒界的裂缝），也可能是构造上边界碎片的分层、夹杂物等。细观构造元素具有明显的内部边界和特殊的物理-力学特征值，它们比微观尺寸大，但又不是宏观尺寸，包含有足够数量的构造元素以用来平均描述。

在细观水平上的变形过程中，那些早已具有的或者是在塑性变形过程中形成的子构造元素的任何移动和旋转都伴随着子构造内部元素的移动和旋转。因此，在子构造内也应当有适应性的塑性变形发展，以保证子构造元素与相邻者之间的相适应,并以此保持岩石变形的相容性(连续性)。由于在不同尺寸、形成的构造不均匀元素的细观体积中不均匀的塑性变形的发展,形成了不均匀的应力场。在最大切应力的方向上就形成局部剪切带,变形构造元素内局部变形带约束的区域中,子构造元素内部的剪切变形不可能进一步发展,导致非相容变形的产生,进而导致在该处出现不连续性,称之为细观裂纹。同样应力集中可以通过这种方式释放掉一部分,还有一部分则以"封闭应力"的形式保存下来。

细观裂纹是否还会沿细观构造元素边界或在它内部继续发展?取决于细观构造元素边界的状态、元素内部或它的边界上是否存在相应的集中或非均质性等等。细观裂纹在变形子构造元素的边界上形成,并且经常是沿着局部剪切带扩展。当这个过程进入到宏观水平(在宏观水平上形成了局部变形带和宏观裂缝),意味着岩石的整体破坏—整体上丧失掉剪切稳定性。

可见,加载岩石中塑性变形是一个多阶段、多水平的松弛过程,剪切稳定性的丧失是在微观、细观和宏观水平上依次丧失局部剪切稳定性过程中的结果。大量的不同尺度的构造元素的相互作用,无论是原始岩石中具有的,还是塑性变形过程中形成的,都产生大量的试验中观察到的各种各样的材料性质[1,23]。每一个构造水平上塑性变形机制的特征,决定着微观、细观和宏观水平上模拟变形过程的特殊方法。宏观水平上的描述利用泛函力学的方法。微观水平对变形过程的贡献可以通过位错泛函理论给出塑性剪切速度和通过材料对剪切抗力中的贡献在响应中来考虑。细观水平对变形过程的模拟可以在"剪切+旋转"变形图景下通过考虑细观构造元素的演变和新的细观子构造元素的形成来实现。

"连续介质的破坏"意味着介质中形成了断裂,断裂的尺寸超过了作为元素加以研究的材料分子的尺寸。有关介质从连续状态转变到破坏状态这一过程的描述称之为破坏模型。联系模型参数之间关系的方程式就是形式上的破坏准则。

岩石的变形构造水平阶梯的演化能导致塑性流动曲线的阶段性。每一个新的构造水平加入到变形中来都会产生受载变形体新的弹性应力的松弛渠道,这引起变形强化系数的降低及塑性流动速度的增加。岩石的宏观变形特征在达到峰值强度前可以用较确定的力学关系来揭示,达到峰值以后的残余强度是岩石的重要性质,但其大小和走向取决于变形破坏形成的构造尺寸大小,具有非确定的特性,较难用确定的力学关系来描述。因此,连续的与已受破坏的脆性岩石由于构造上的巨大差异,其强度准则存在明显的区别而不能用一个方程描述。这再次证明了在宏观变形特征方面变形构造水平阶梯的决定性作用。连续介质的破坏伴随着剪力的急剧减少并转入破坏材料的强度极限(流限)对应的状态。因此,必须引入两个极限特征参数:未破坏材料的强度$Y_1(I_1,I_2,I_3)$和已破坏材料的强度$Y_2(I_1,I_2,I_3,l)$,而I_1,I_2,I_3为主应力张量的不变量。爆炸波在脆性介质传播时剪切破坏由于受到各个构造缺陷水平的发展限制需要一个迟滞时间,破坏过程中当前极限关系的变化必须以动态演化的形式加以描述,$Y_2(I_1,I_2,I_3,l)$依赖于破坏时在细观水平上所形成的网格形状平均构造尺寸大小及摩擦效应。

目前,尽管细观水平上破坏准则离实际工程应用还存在一段距离,但其对宏观上正确地选择和从物理上揭示作为准则建立基础的各种宏微观关系提供了理论依据。本文采用广义的Mises破坏准则的表述[78]:即在某个静水压力作用下,当剪应力超过了按照该静水压力下屈服条件计算得到的屈服值时,岩石的构造就要发生改变,构造改变的程度则依赖于剪应力超过该屈服值的多少。

在宏观水平上,变形及破坏的描述通常基于连续介质力学,但是必须考虑到来自微观水平及细观水平的贡献。来自微观水平及细观水平的贡献,可以在支配方程中通过给出塑性切变的速度,以及对材料剪切强度的贡献来考虑进来。在微观水平上,材料抗剪强度由位错连续统的演化来决定,而在细观水平上由细观结构的演化及新的细观子结构的形成来决定。因此在模拟大变形应力应变曲线时不仅要考虑材料内部微观结构的演化,而且要考虑各种子结构的形成,以及它们对流动应力的贡献。

内部结构非均匀性对材料力学性质的贡献可以通过著名的Hall-Petch流动应力方程来研究。对剪切流动应力,Hall-Petch方程形式如下[23]

$$\tau_Y = \tau_0 + \tau_{bs} + KD^{-1/2} \tag{2}$$

式中，K 为硬化系数；D 为平均晶粒尺寸；τ_{bs} 为积累的内应力，按塑性物理由下式确定[84]：

$\tau_{bs} = \alpha Gb\sqrt{N}$，其中 G 为剪切模量，N 位错密度，α 为系数，取值范围 0.2~0.5。

在松弛模型中，Hall-Petch 关系反映了在构造非均匀材料中，由于不同部分或点的可动性不同所引起的应力重分布效应。

式（2）的一种情况是[22]

$$\tau_Y = \tau_0 + \tau_{bs} + K_1 D^{-1} + K_2 D^{-1/2} \tag{3}$$

式中各项意义同式（2），K_1 与晶粒内的滑移有关，而 K_2 与晶界内的变形有关，首先是与晶界的滑移有关。因此松弛模型所考虑的是滑移性变形与旋转性变形在不同构造水平上的相互作用。如果在一定条件下旋转变形占支配地位，那么在 Hall-Petch 关系中 K_2 占主要地位。如果转动变形困难，那么 K_1 占主要地位。如果旋转及滑移为同一数量级，那么在 Hall-Petch 关系中 K_1 与 K_2 项都有。

综上所述，揭示地质材料在冲击作用下的变形与破坏过程必须继续研究基于微观物理原理和细观物理力学的理论，并利用现代非线性科学的概念与方法，结合传统的连续介质力学，建立简单的工程实用的介质冲击作用下的统一的分阶段连贯的不同时空尺寸的动力本构模型，并建立动力破坏准则。这一任务是艰巨的，涉及面很广。本文介绍了我们在这方面对于这一课题所做的工作，仅是一些探索，并将在后续的研究工作中继续这方面的探索。

二、在爆炸作用下岩石变形与破坏的若干问题

1. 在爆炸作用下岩石变形与破坏的动态及构造特性

自然界中的岩石，根据其形成过程的不同可分为火成岩、沉积岩及变质岩。一般来讲岩石的物理力学性质取决于岩石的组成，岩石内部结构及所处的热力学环境。但在同一类别内的岩石受外界作用时的反应差别较小，并有许多相似之处。材料的变形及破坏是一个多层次的过程。这一过程既是一个物质粒子在统计物理的意义上翻越能量障碍进行状态转换的过程，也是一个在不同的具体的物质构造水平上发生构造变化的过程。构造水平包括微观水平，细观水平及宏观水平。每一个层次上的变化都有自己的特点，通过对每一个层次上的变形及破坏过程的理解，才能在宏观上更好的理解材料的物理力学行为，也才能正确地对其进行描述，找出其规律。

传统的研究材料破坏的途径为唯象的。按这种方法，在任何有限的体积内，如果应力或应变分量的某些组合超过某些极限值时，材料就会破坏。通常这些极限值在相应的构形空间内构成极限面。应力或应变分量的组合的选取及极限值的确定方法构成了强度理论的主要内容。

现实的固体具有结构，其中含有各种缺陷，空洞裂纹等，使固体具有物理及力学性质的非均匀性。这种固体的非均匀性使试验结果及经典理论具有很大的不同。为了解决这一问题，一些研究者采用了统计物理的方法。采用这种方法的原因是试验中发现了理论强度与实际强度之间的大的差别。为了寻找材料理论强度与实际强度差别原因，断裂力学得到了迅速发展。Griffith（1920）根据弹性理论中的孔洞附近的应力集中的概念首次建议了对这一差距的解释。Griffith 假设在现实固体总会有一个预先存在的连续性缺陷，这种缺陷起着应力集中域的作用，使材料的局部应力增大 n≥1 倍，从而引起材料的破坏。

关于应力集中的概念的发展并没有触及破坏之力学概念的实质：仍然是假设局部应力达到理论强度极限时裂纹产生。但这一假定没有直接的试验证明。

从20世纪50年代开始，在前苏联列宁格勒技术物理研究院的强度物理试验室对在静力作用下的材料破坏进行了系统研究。对单轴适中拉应力，С．Н．Журков等得到了材料在外载作用下的确定破坏时间τ的Журков（Zhurkov）公式[12]。

Stroh的观点也与之非常相近[28]，他把破坏过程与位错的释放、运动及积累联系起来，他认为要释放位错必须要超越活化能$U(\sigma)$，并给出了与Zhurkov公式类似的公式。

Г．М．Бартенев（Bartenev）对高分子材料的联合线形断裂力学及微观动力途径进行了研究，考虑了材料的构造参数，得到了与Журков公式类似的公式。

经典的损伤力学理论研究材料的破坏时，采用唯象的损伤参量Ψ来描述材料的破坏演化过程[8]，当损伤参量Ψ达到一定值时，材料破坏。

在强烈的动载作用下，与在准静力情况不同，材料表现出强烈的动力性质。

对于Zhurkov公式，试验结果表明，在大应力区，Zhurkov公式里面的活化体积要减小。在对数坐标里面，破坏时间对于应力的依赖关系为由两条直线组成的折线，且在小应力区破坏时间对于应力的依赖强烈，而在大应力区破坏时间对于应力的依赖要弱的多。

J．F．Kalthoff和**D.A.Shocky**建议的断裂准则，称为最小时间准则[27]。这一准则的独到之处在于引入了一个具有时间量纲的构造参数，以考虑宏观破坏前的孕育过程。孕育时间被认为是一个与材料特性相关联的常数。按照这一概念，当前应力强度因子$K_I(t)$在一个用于宏观裂纹发展的最短时间内超过材料的动力抗裂韧性时，材料就会发生破坏。这一准则就其物理本质来讲与Журков准则类似。

В．С．Никифоровский和Е．И．Шемякин（Nikiforovsky，Shemyakin）[17]提出了另一个断裂准则——冲量准则。这一断裂准则认为，当局部拉应力在时间上的积分，即整个局部力冲量大于某个临界值时，材料才破坏。

经典的断裂的发展方向之一在于选择一个适当的断裂过程特征参数。H．Neuber和В．В．Новожилов（V.V.Novozhilov）在不同的时间，基于不同的途径，建立了考虑构造度量的破坏准则。基于Neuber-Novozhilov准则，**Morozov-Petrov**考虑了破坏过程的时间因素，从而建立了材料的时间-空间破坏准则[37]。

上述这些准则各自从不同的角度探讨了材料的破坏问题，它们给我们的印象是对于同一种现象可以有不同的描述。那么是不是真的如此哪？这些时间准则之间有没有联系？它们与经典的强度理论又有什么关系哪？在文献[60]中探讨了这一问题。

让我们首先讨论Zhurkov公式。

Zhurkov材料破坏时间τ公式的数学公式为

$$\tau = \tau_0 \exp\left(\frac{U_0 - \gamma\sigma}{kT}\right) \quad (4)$$

式中，U_0为活化能；σ为单轴拉应力；γ为活化体积，通常为10~1000原子体积；k为玻尔兹曼常数；T为绝对温度；τ_0为原子的Debye振动周期数量级别的参数，约为10^{-12}s。

在受剪切应力σ_τ作用下，可取类似于（4）的形式：

$$\tau = \tau_0 \exp\left(\frac{G_0 - \gamma_\tau\sigma_\tau}{kT}\right) \quad (5)$$

上式各项意义与（4）式对应项意义相似，G_0、γ_τ 为材料受剪时的活化能及活化体积。

在工程实际中，常遇到复杂应力的情况，从单轴向复杂应力情况推广困难较大，而且常带主观性，因而试验验证十分重要。对高分子材料，复合材料及钢等，在应力状态变化很大的范围内，通过大量试验，О. Е. Ольховик (O. E. Olkhovik, 1987) 得到如下计算在复合应力作用下活化能的公式：

$$U = U_0 - \gamma_0 R - \alpha_0 L - \beta_0 L^2 \tag{6}$$

式中，β_0，γ_0 为材料常数；

$$L = \frac{1}{\sqrt{3}}(\sigma_1 + \sigma_2 + \sigma_3)$$

$$R = \frac{1}{\sqrt{3}}\left[(\sigma_1 - \sigma_2)^2 + (\sigma_2 - \sigma_3)^2 + (\sigma_3 - \sigma_1)^2\right] \quad \alpha_0 = arctg \frac{2\sigma_2 - \sigma_1 - \sigma_3}{\sqrt{3}(\sigma_1 + \sigma_3)}$$

当 L 值在一个比较小的范围内变化时，可取 $\beta_0 = 0$。

对于材料单轴受拉 变形，有如下变形率的表达式[12]：

$$\dot{\varepsilon} = \dot{\varepsilon}_0 \exp\left(-\frac{U_0 - \gamma\sigma}{kT}\right) \tag{7}$$

式中，$\dot{\varepsilon}_0$ 为极限变形速率；其余各项意义与式(4)中的相同。

上述公式也称 Alexandrov 公式，是 Alexandrov 在 1945 年研究高分子材料时首次得到的。

把式(4) 与式(7)相乘有

$$\tau\dot{\varepsilon} = \tau_0 \dot{\varepsilon}_0 = \varepsilon_{st} = const \tag{8}$$

试验表明[12]，ε_{st} 在比较大的变化范围内（9~10 个数量级内）对于温度、应力及加载率的依赖性很弱，可以认为是一个常数，这证明了变形与破坏这两个宏观过程之间存在着紧密的联系。试验证明[50]，在剪切及三轴压缩情况下，在复杂的应力及变形情况下（包括随时间变化应力的单重加载、循环加载、复杂的应力状态加复杂的加载方式、荷载加辐射及腐蚀情况），在破坏时间偏离 Zhurkov 公式的情况下，在强烈动应力作用下，式(8)仍然成立。因此对于性质不同的材料，及很大的加载变化范围，式(8)都近似成立。

虽然式(8)中没有显形的出现时间变量，但 ε_{st} 却反映了在外力作用下，材料内部结构在时间中发生的变化在破坏时的宏观效果，因此它也可认为是一种准时间准则。这样我们就可以看出，经典的第二强度理论（最大应变理论）包含着材料内部构造在外载作用下变化的时间过程。

В. С Никифоровский 和 Е. И. Шемякин (Nikiforovsky, Shemyakin)[17]提出的断裂准则认为，当局部拉应力在时间上的积分，即整个局部力冲量小于某个临界值 J_c 时，即

$$\int_0^{t_*} \sigma(t)dt \leq J_c \tag{9}$$

这一准则尤其适用于短时外冲量所引起的破坏。它能定量的解释加载率提高时，强度的增加。它的主要的缺陷是不能过渡到静力情况。 这一准则的另一个优点是能直接考虑整个加载历史。在弹脆性模型框架内，能解释固体一系列快速动力破坏的一系列现象。

对于 Nikiforovsky-Shemyakin 冲量准则，由于一维冲击波作用下材料内部 有下列关系：

$$\sigma = \rho D v \tag{10}$$

式中，v 为粒子速度；ρ 为材料密度；D 为冲击波速。所以有

$$\int_0^{t_*} \sigma(t)dt = \int_0^{t_*} \rho D v dt = \rho D u = J_0 \tag{11}$$

式中，u 为位移。

上式表明，当材料粒子的宏观位移达到一定值时，材料破坏。

位移是材料变形的宏观表现，如果应力波所涉及的特征尺寸为 L，则位移可以用应变 ε 表示为：

$$u = L\varepsilon$$

这样式(9)变为

$$\int_0^{t_*} \sigma(t)dt = \rho D L \varepsilon = J_0 \tag{12}$$

也即上式表明，当材料的应变达到一定值时材料破坏。

从上所述 我们就可以看到，Zhurkov 公式，Nikiforovsky-Shemyakin 冲量准则及第二强度理论（最大应变理论）它们是相通的。都反映了材料内部构造在外载作用下变化的时间过程的最终结果。

对于式(11)及式(12)，还需讨论一下结构受地震作用下应变率较低的宏观情况。由地震引起的地震动，及由爆炸引起的地震动都会对地面的建筑物造成破坏。现行的抗震设计是基于承载力或强度进行的，但是通过对以往的地震破坏危害的调查分析发现在一些地震动的某些时段内，对于结构破坏起控制作用的是速度或位移而不是内力。20 世纪 90 年代初期，J.P.Moehle 提出了基于结构性能的抗震设计理论。按照这一理论，结构的塑性变形能力应满足在预定的地震作用下的变形要求，及也即在大震作用下要控制结构的层间位移角限值。这正好与式(11)和 式(12)的结论是一致的。也即基于结构性能的抗震设计理论考虑了结构材料内部构造在地震载作用下变化的时间过程的最终结果。

对于地震动，如果振动的角频率为 ω，则速度 v 与位移 u 的关系为

$$v = u\omega$$

这样我们可以看到速度可以作为控制参数。实际情况也的确如此。不同的研究者研究地震对建筑物作用的大量观测结果表明，在同样的地基条件下，在同一地点，同一类型的建筑物，在地基的速度超过表征这一具体类型的建筑物的某个速度值时，其损坏程度一样。

对于 Kachanov 的损伤积累的唯象理论，其损伤变量 Ψ 演化的方程为：

$$\frac{d\Psi}{dt} = f(\sigma,\Psi) = \begin{cases} A\left(\dfrac{\sigma}{1-\Psi}\right)^n, & \sigma \geq 0 \\ 0 & \sigma \leq 0 \end{cases} \tag{13}$$

积分上式，并假定 $\Psi \leq 1$，可得到

$$A\int_0^{t_*} \sigma^n dt \leq 1 \tag{14}$$

当 $n=1$ 时，就得到公式(9)。

因此 Kachanov 的损伤积累的唯象理论与强度的时间准则是一致的。

对于断裂力学的经典的"力"途径来讲，其意义是要求作用于假定的断裂处的瞬时应力达到临界值。但在动力问题中，必须考虑材料的惯性，因为与断裂面相邻的介质粒子的运动非常快。因此与构造度量相似，很自然地去考虑动力学方面的构造与时间参数。在最简单的情况下，对构造尺寸 d 及最

大波速 c, d/c 是一个相邻的破坏构造单元之间相互作用能量传播的特征时间。这样就可以假定我们有一个与断裂孕育时间相对应的特征时间间隔 τ。一般地，d 与 τ 应看作是相互独立的，因为 τ 由发生在材料结构中的复杂过程决定。这样就可以引入一个空间-时间破坏元胞 $[0,d]\times[t-\tau,t]$，也即在一个给定的尺度水平上，表征破坏过程特性的某种结构是位于空间-时间尺度上的。这样就可有下列 Morozov-Petrov 破坏的空间-构造准则：

$$\frac{1}{\tau}\int_{t-\tau}^{t}dt'\frac{1}{d}\int_{0}^{d}\sigma(t',r)dr \leq \sigma_c \qquad (15)$$

这里 τ—破坏之构造时间；σ_c—材料的静力强度；$\sigma(t,r)$—裂尖拉应力；d—为表征在给定尺度水平上断裂元胞线性尺寸的参数[37]。

按上式，当裂尖处的最大应力在时间-空间 $[0,d]\times[t-\tau,t]$ 上的平均值达到材料强度时，材料就会破坏。通过引进新的符号，上式可以写成

$$J(t)\leq J_c, \quad J(t)=\int_{t-\tau}^{t}dt'\int_{0}^{d}\sigma(t',r)dr, \quad J_c=\sigma_c\tau d \qquad (16)$$

因此这样引进的准则关系就具有了临界构造冲量的物理意义，就其物理本质来讲也与前面所涉及的冲量准则是一致的。动力学的主要问题是确定断裂时间。在所考虑的情况下，自然地把给定的冲量达到其临界值 $J(t_*)=J_c$ 时的时间 t_* 视为破坏时间。

这样我们就完成了对于若干时间性破坏准则的分析。

从前面的分析可以看出，Zhurkov 准则，Nikiforovsky-Shemyakin 最大冲量准则，Kalthoff-Shocky 最小破坏时间准则，Morozov-Petrov 时间-空间破坏准则等时间性破坏准则就其内在的物理机理来讲是相通的。虽然式(8)中没有显形的出现时间变量，但 ε_{st} 却反映了在外力作用下，材料内部结构在时间中发生的动力变化在破坏时的宏观效果，因此它也可认为是一种准时间准则。又其便于测量，因此在应用时间准则时应优先使用。Nikiforovsky-Shemyakin 最大冲量准则考虑了应力波应力随时间的积累过程，也便于测量，所以也可优先应用。唯象的损伤破坏力学的损伤动力演化方程，只有在损伤参量与具体的宏观物理力学量相连，临界损伤值确定之后才可方便应用。Zhurkov 理论对于材料在受静载作用下的破坏时间预测很方便。而 Zhurkov 理论、Kalthoff-Shocky 最小破坏时间准则对于变化的动力荷载较复杂，工程应用复杂。Morozov-Petrov 时间-构造破坏准虽然比经典强度理论更合理，但由于它们涉及微观时间-构造量，这使得它们的利用受到限制。只有在系统的确定了不同材料的这些构造参量之后才可较方便地得到应用。

2. 岩石强度对于应变率的依赖特性

试验表明，在爆炸条件下在不同的加载速度及不同的应力状态下，岩石最重要的特点是，其强度极限，弹性极限，及塑性膨胀依赖于变形速度。因此，必须考虑强度函数与变形过程的速度的依赖行为。岩石材料的率相关的一般规律是：首先材料强度随应变率的增加而缓慢地增加(此处称为 I 区)，当应变率继续增加，超过某一值时，材料强度随应变率的增加急剧增加(此区称为 II 区)，当应变率进一步增加到冲击爆炸应变率时(III 区)，材料强度随应变率的增加而增加的速率又变缓，与 I 区情况相当。

许多研究者很早就注意到并从不同角度研究了这一问题。

A.Kumar 通过对玄武岩及花岗岩在动载作用下的强度，得出结论：应变率的增加对强度的影响类似于温度降低所产生的效应，可以用热活化的观点来描述岩石的动态断裂机制。А.Н.Ставрогин, А.Г.Протосеня 利用了 zhurkov 型的应力-应变率关系式，建立了岩石强度和应变率之间的动力统计理论，并测定了众多岩石的有关参数，应变率范围在 10^2/s 内（I 区），显然这一理论应用范围是有限的。J.D.Cambell 及

W.G.Furguson 研究了低碳钢的动力行为,并进行了区划,指出较低应变率时支配机制为热活化机制,较高应变率时支配机制为宏观黏性机制,在模拟宏观黏性时采用了常黏性系数,但是试验表明,黏性随着应力幅值的增加,也即应变率的增加而减少,因而其模型是不真实的。P.Perzyna 采用了变化的黏性系数,但其模型当应变率无限增大时,材料强度也将无限增加,这显然与实际不符。在文献[46]中指出,从 II 区到 III 区的材料破坏行为可解释为破坏机理上率依赖性从脆性到延性的转换,但其中是何种机理及怎样转换却没有阐述清楚。从上述分析可以看出,虽然已进行了很多研究,但对不同应变率区之间材料强度的应变率依赖性的转换机理还没有一个系统的概念,对其模拟还不够准确。

文献[54,58]对不同应变率区之间脆性材料动力强度的应变率依赖性的转换机理进行了探讨。

本文基于现有的研究资料,对不同应变率区之间脆性材料动力强度的应变率依赖性的转换机理进行的探讨。通过分析可以看到,在应变率小的范围内,材料的变形及破坏受热活化机制控制;当应变率进一步增加时,材料的宏观黏性阻尼机制(声子阻尼)逐渐加入到材料强度的应变率依赖性中来,并逐渐占据主导地位,但是,黏性系数不是一个材料常数,而是随着应力(应变率)的增加而减少;当应变率继续增加而进入到由爆炸及冲击所引起的高应变率区时,材料的惯性显现出来,这时,不同尺寸的缺陷的增长将同时启动,在没有缺陷的地方材料的热活化机制启动,并产生原子键的断裂,这时材料的热活化机制又重新显现出来。这样可以把材料强度随应变率的增加而增加的过程看作是热活化机制与宏观黏性机制并行存在,相互竞争的结果。这两种机制在不同的应变率区占据主导地位。基于上述结论,在文献[54]中给出了联合的热活化与黏性机制并联竞争的材料强度-应变率依赖模型:

$$\tau_Y = \frac{1}{\gamma_\tau}\left(U_0 + kT\ln\frac{\dot{\gamma}}{\dot{\gamma}_0}\right) + \frac{b_1(\dot{\gamma}/\gamma_S)^n}{(\dot{\gamma}/\dot{\gamma}_S)^n + 1} \tag{17}$$

式中,U_0 为活化能;γ_τ 为构造系数,通常为 10-1000 原子体积;k 为玻尔兹曼常数;T 为绝对温度;b_1 表示宏观黏性对强度增长的贡献的最大幅值;n 为常数,控制材料强度-应变率对数曲线的陡峭程度;$\dot{\gamma}_S$ 为材料强度-应变率半对数曲线的拐点值。在这一项中,温度升高的贡献已被隐性地包含在其中。

并对该模型的有效性对照试验资料进行了检验,结果表明与试验符合的很好。本模型物理意义明确,应用的应变率范围广,相对简单,便于实际应用。

3. 强度对于应变率依赖性的弹塑性模型

1) 模型的基本关系

从上面的分析可以看出,Григрян 弹塑性模型[3,4]及推广的弹塑性模型[4]能足够好地再现岩石在爆炸应力波作用下的行为。但在实际工程计算中,需要对于各种岩石确定模型的物理力学参数,但在文献[1,3,4]中未给出,这给本模型的应用设置了障碍。在文献[24]中,给出了地质材料的弹塑性膨胀模型,这一模型考虑了材料的膨胀,与 Григрян 弹塑性模型及推广的弹塑性模型有些相似之处。在文献[24]中,对于弹塑性膨胀模型,基于三轴试验,给出了多种地质材料的弹塑性膨胀模型中的有关参数。因此我们在此综合这些模型,做了一些简化,得到一个实用的模型,以便能利用文献[24]中的参数。

这里对于应力偏张量 S_{ij},不用 Jauman 导数而用其普通导数,又根据文献[1],岩石的膨胀效应在爆炸波的传播过程中影响不大,可不考虑,因此有

$$dS_{ij}/dt = 2G\dot{e}_{ij} + \lambda S_{ij}, \quad dp/dt = -K\varepsilon_V \tag{18}$$

式中,G,K 为剪切及体积模量;ε_V 为体积应变。λ 不采用文献[24]中的形式,而采用以下形式

$$\lambda = (2GW - dJ_2/dt)/2J_2; \quad W = S_{ij}\dot{e}_{ij}$$
$$J_2 = S_{ij}S_{ij}/2 \tag{19}$$

塑性加载条件不考虑膨胀的影响采用下列形式[24]：

$$\tau = \tau_i(p) \tag{20}$$

式中，τ 为剪切应力强度。

$$\frac{4}{3}\tau^2 = \frac{1}{2}S_{ij}S_{ij}, S_{ij} = \sigma_{ij} + p\delta_{ij}, p = -\frac{1}{3}\sigma_{ij} \tag{21}$$

按照弹性极限（$i=1$），最大强度（$i=2$）及残余强度面（$i=3$）的形状，$\tau_i(p)$ 采用下列形式[6]：

$$\tau_i = (Y_i + \alpha_i p)^{S_i} \quad (p < p_i^{(m)})$$
$$\tau_i = \tau_i^{(m)} = const \quad (p > p_i^{(m)}) \tag{22}$$

$$\alpha_i = (Z_i - Y_i)/\tau_i^0, Z_i = (\tau_i^0)^{(1/S_i)}$$
$$\tau_1^0 = \tau_y^0/\tau_f^0, \tau_2^0 = 1, \tau_3^0 = \tau_r^0/\tau_f^0 \tag{23}$$

表 1 一些岩石的模型参数[24]（平均值为这几种岩石的平均值）

参数	闪长岩	大理岩	花岗岩	辉绿岩	粉砂岩	砂岩	石灰岩	平均值
$m_0(\%)$	0.25	0.6	1.2	0.8	1.6	5	9	2.6
τ_f^0	167	45	128	117	98	85	53	99
K_1	210	900	250	430	290	210	510	4.0
G_1	150	450	190	270	180	160	320	2.5
Y_1	0.025	0.014	0.04	0.018	—	0.014	—	22
S_1	0.62	0.59	0.66	0.62	0.47	0.64	0.56	0.59
S_2	0.66	0.64	0.76	0.65	0.61	0.69	0.6	0.66
S_3	—	0.84	0.87	—	—	0.85	—	0.85
τ_1^0	0.75	0.68	0.84	0.87	0.77	0.73	0.65	0.75
τ_3^0	—	0.11	0.04	0.005	0.085	0.32	0.11	0.11
A_1	18	20	39	35	20	24	20	25
M	—	300	1600	4500	—	2700	9000	3600
R_1	0.058	0.05	0.056	—	0.046	—	0.053	0.053

$$p_1^{(m)} \approx (2.3\alpha_1^{S_1})^\varsigma, \varsigma = [1 - S_1]^{-1}$$
$$p_2^{(m)} = p_3^{(m)} \approx (\alpha_2^{S_2}/\alpha_3^{S_3})^\gamma, \gamma = (S_3 - S_2)^{-1} \tag{24}$$

式中，τ_i，p 为无量纲化的参数，分别利用了 τ_f^0 及 $p_f^0 = (2/3)\tau_f^0$ 进行了无量纲化；α_i, S_i 为材料常

数；$\tau_y^0, \tau_f^0, \tau_r^0$ 分别为在单轴压缩时的弹性极限，强度极限及残余强度极限。$p_1^{(m)}$，$p_2^{(m)} = p_3^{(m)}$ 表征在初始膨胀阶段及强度极限以后从库仑塑性状态到 Mises-Tresca 状态的过渡。几种岩石地模型参数如表 1 所示。

上表格中 m_0 为孔隙率，$Y_1 \approx Y_2$，$Y_3 \approx 0$，$K_1 = K/\tau_f^0$，$G_1 = G/\tau_f^0$，$R_1 = R_p/\tau_f^0$，R_p 为拉断强度，$R_c = 2\tau_f^0$。

对于砂岩及粉砂岩，有足够的试验资料来建立某一参数 W_i 与孔隙率 m_0 的关联关系：
$W_i = M_i + N_i m_0$，这里 M_i，N_i 为系数，与关联系数 K 见表 2。

表 2 两种岩石的某些系数[24]

系数	砂岩					粉砂岩
	Y_1	S_1	S_2	τ_1^0	R_1	S_2
M_i	0.02	0.61	0.66	0.8	0.05	0.46
$N_i \cdot 10^9$	-1.1	6.3	5.6	-7.4	-1.4	65
K	--	58	86	--	70	46

2）强度对于应变率依赖效应的考虑

在动力条件下，材料的抗压及抗拉强度与静力条件下数值差别非常大，因此必须考虑强度对于应变率的依赖行为。在文献[53]中，作者发展了一个强度对于应变率依赖性的模型。这一模型不仅解决了岩石强度从低应变率到高应变率的机制转化问题，而且给出了一个统一表达式，从而使这一问题的机制问题及统一表达问题得到了初步解决。

对于地下封闭爆炸这样的球对称应力应变情况，在弹性状态时，可取环向应力 $\sigma_\varphi, \sigma_\theta$ 与径向应力 σ_r 的关系为

$$\sigma_\varphi = \sigma_\theta = \frac{\nu}{1+\nu}\sigma_r \tag{25}$$

式中，ν 为泊淞比。

在塑性状态时，对于孔隙率不大的坚硬岩石，考虑到剪胀的抵消作用，可假定塑性不可压缩，从而有 $\nu = 0.5$，这时可取下列关系

$$\sigma_\varphi = \sigma_\theta = \frac{1}{3}\sigma_r \tag{26}$$

由公式（22）得

$$Y_i = (\tau_i)^{1/S_i} - \alpha_i p \tag{27}$$

对于弹性限，利用式（25）、式(21)可得

$$\tau_1 = \frac{\sigma_r}{2(1+\nu)\tau_f^0}; \quad p = \frac{(1+3\nu)\sigma_r}{2(1+\nu)\tau_f^0} \tag{28}$$

当 σ_r 取屈服应力 σ_Y 时，可得 Y_1 为

$$Y_1 = \left(\frac{\sigma_Y}{2(1+\nu)\tau_f^0}\right)^{1/S_1} - \alpha_1 \frac{(1+3\nu)\sigma_Y}{2(1+\nu)\tau_f^0} \tag{29}$$

其中 σ_Y 对于应变率 $\dot\varepsilon$ 的依赖关系采用文献[53]中的联合的热活化与黏性机制并联与竞争的模型:

$$\sigma_Y = \frac{1}{\gamma_Y}\left(U_{Y0} + kT\ln\frac{\dot\varepsilon}{\dot\varepsilon_0}\right) + \frac{b_Y(\dot\varepsilon/\dot\varepsilon_S)^n}{(\dot\varepsilon/\dot\varepsilon_S)^n + 1} \qquad (30)$$

上式右边第一大项为热活化项,在低应变率区占支配地位,其中 U_{Y0} 为活化能;k 为 Boltzman 常数;T 为绝对温度;γ_Y 为材料活化体积。第二项为黏性贡献项,在中应变率区占支配地位,其中 b_Y 为黏性项贡献地最大幅值;$\dot\varepsilon_0,\dot\varepsilon_S$ 为模型参数。若低应变率区热活化项贡献不大,则热活化项可用静力屈服应力 σ_{Y0} 来代替。即

$$\sigma_Y = \sigma_{Y0} + \frac{b_Y(\dot\varepsilon/\dot\varepsilon_b)^n}{(\dot\varepsilon/\dot\varepsilon_b)^n + 1} \qquad (31)$$

在材料屈服以后,利用(26)式可得

$$\tau_2 = \frac{\sigma_r}{3\tau_f^0}; \qquad p = \frac{5\sigma_r}{6\tau_f^0} \qquad (32)$$

从而有

$$Y_2 = \left(\frac{\sigma_r}{3\tau_f^0}\right)^{1/S_1} - \alpha_2\frac{5\sigma_r}{6\tau_f^0} \qquad (33)$$

当 σ_r 取极限值 σ_U 时,可得

$$Y_2 = \left(\frac{\sigma_U}{3\tau_f^0}\right)^{1/S_2} - \alpha_2\frac{5\sigma_U}{6\tau_f^0} \qquad (34)$$

σ_U 对于应变率的依赖类似于式(30)、式(31),只是其中系数取值不一样:

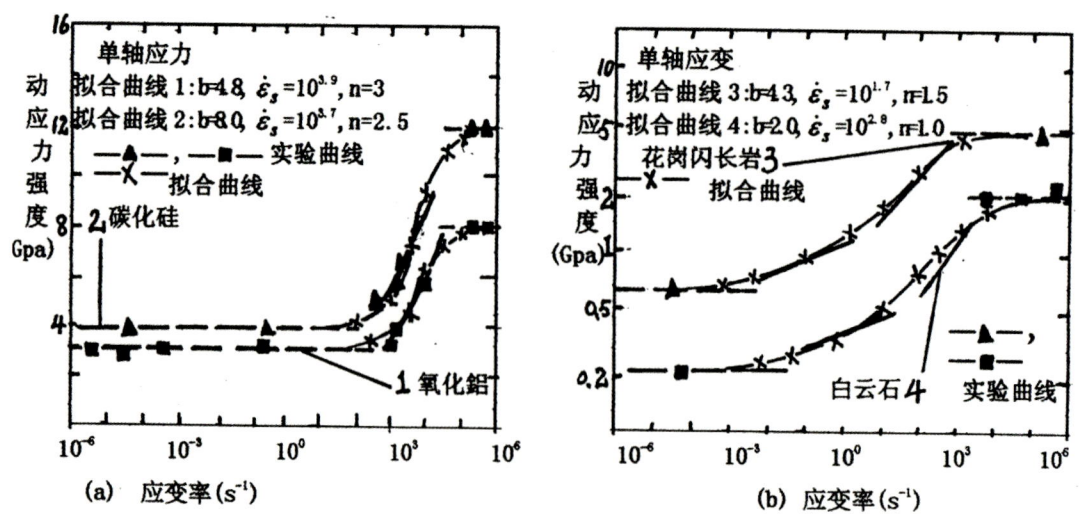

图 1 几种岩石强度对于应变率的依赖性模型参数及拟合情况

$$\sigma_U = \frac{1}{\gamma}\left(U_{U0} + kT\ln\frac{\dot{\varepsilon}}{\dot{\varepsilon}_0}\right) + \frac{b(\dot{\varepsilon}/\dot{\varepsilon}_S)^n}{(\dot{\varepsilon}/\dot{\varepsilon}_S)^n + 1} \tag{35}$$

若低应变率区热活化项贡献不大,则热活化项可用静力屈服应力 σ_{U0} 来代替。即

$$\sigma_U = \sigma_{U0} + \frac{b(\dot{\varepsilon}/\dot{\varepsilon}_S)^n}{(\dot{\varepsilon}/\dot{\varepsilon}_S)^n + 1} \tag{36}$$

几种材料的参数选取如图 1 所示,试验曲线取自文献[46]。

4. 岩石破坏的混合准则

众所周知,微裂纹的产生及增长是一个动力过程。对具有微裂纹的材料的多稳定阶段的考虑,产生了动力变换的思想。动力变换涉及一类非线性现象,不伴有一般意义上的相变。在宏观破坏之前,在爆炸性非稳定性阈值之上,非线性系统发生变换。这一变换可以按照逾渗理论作为动力逾渗问题分析。由逾渗理论所描述的现象属于临界现象类,也即它们由临界点表征,在临界点附近,系统的某一特征参数剧烈变化,系统被划分成块,这是临界现象的显著特征。这些块的尺度在接近临界点时无限增大。因为这些块具有很大的尺度,它们连接成团簇而不依赖于材料的构造。

微裂纹的分布具有空间随机性,其密度可能发生涨落,一些裂纹可能靠的很近,并可能连接,从而形成大的裂纹,这就意味着裂纹从非局部化发展阶段转入到局部化发展阶段。因此令人感兴趣的是把成簇的概率与裂纹的密度联系起来。显而易见,在裂纹密度较小时,裂纹相距太远而无法连接。只有在裂纹密度较大时才有可能连接。

裂纹密度的精确计算受到裂纹的统计概率信息不足的制约,逾渗理论显示,其模型敏感性很大。因此简化模型分析,数值模拟及试验是非常重要的。理论分析及试验证明[7,12],对于一定尺寸 r 的裂纹,当其单位体积内的数目 n 达到一定程度时,会产生裂纹自发地连接成簇的现象,从而产生大一级的裂纹。这一准则称为密度准则,可用下式表示:

$$K_l = n^{-1/3}/r \tag{37}$$

对于岩石,当其达到强度极限时,可取 $K_l = \pi = 3.14$;当岩石完全破裂时,可取 $K_l = 1$[14]。这一准则与 $\omega = \omega_c$ 准则相比,反映了更多的裂纹信息。

对于地质材料,在 20 世纪 70 年代,D.R.Curran,L.Seaman,D.A.Shockey,Kipp,Gradydy 等人利用微观统计本构关系来研究其断裂,并形成 BFRACT 模型[43, 44],其采用的裂纹密度按尺寸的分布关系为指数关系。

$$N_g = N_0 \exp\left(-\frac{R}{R_1}\right) \tag{38}$$

式中,N_0 为单位体积内的裂纹总数;N_g 为半径大于 R 的裂纹数目;R_1 为分布常数,按其物理意义为裂纹的平均尺寸。

采用的形核速率为

$$\dot{N} = \dot{N}_o \exp\left(\frac{\sigma - \sigma_{n0}}{\sigma_1}\right) \tag{39}$$

式中,$\dot{N}_o$ 为门槛形核速率;σ_{n0} 为门槛应力;σ_1 为控制形核速率对于应力大小的敏感性参数。

裂纹的长大由下式决定

$$\frac{dR}{dt}=T_1(\sigma+p_c-\sigma_{g0})R \tag{40}$$

式中，T_1 为长大系数；σ_{g0} 为长大的门槛应力；p_c 为流体或气体对裂纹内表面的压力。

在文献[58]中，结合 BFRACT 模型的有关裂纹的统计信息，讨论材料的破坏，并提出了混合破坏准则。在加载的初始阶段，裂纹处于非局部化阶段，这时可以假设裂纹的数目随尺寸的概率分布保持不变，但其数目按（39）增加，积分得

$$N_t = N + \int_0^t \dot{N}_0 \exp\left(\frac{\sigma-\sigma_{n0}}{\sigma_1}\right)dt \tag{41}$$

而裂纹的尺寸符合式（40），积分得

$$R_t = \mathrm{Re}\,xp\left[\int_0^t \pi(\sigma-\sigma_{g0})dt\right] \tag{42}$$

在 t 时刻，裂纹的平均尺寸为

$$\bar{R}_t = \int R_t \frac{1}{R}\exp\left(-\frac{R}{R_1}\right)dR = R\cdot\exp\left[\int_0^t \pi(\sigma-\sigma_{g0})dt\right] \tag{43}$$

即裂纹的尺寸按同一比例增加。在 t 时刻，裂纹的数目随尺寸的分布概率密度为：

$$n_g = \frac{N_t}{\bar{R}_t}\exp\left(-\frac{R}{\bar{R}_t}\right) \tag{44}$$

这时的裂纹总体积为

$$\begin{aligned}\Delta V(t) = \omega(t) &= \int \frac{N_t}{\bar{R}_t}\exp\left(-\frac{R}{\bar{R}_t}\right)\cdot\pi R^2 d\cdot dR \\ &= 2N_t\pi\bar{R}_t^2 d\end{aligned} \tag{45}$$

式中，d 为裂纹壁间距。

由于裂纹的尺寸是随机分布的，因此这给密度准则的应用带来困难。这时可以对平均的裂纹尺寸利用密度准则，当达到强度极限时有

$$N_t^{-1/3}/(2\bar{R}_t) = \pi \tag{46}$$

也即

$$N_t \bar{R}_t^3 = \frac{1}{8\pi^3} \tag{47}$$

从另一方面，可以把介质的裂纹体积密度 ω 与密度准则结合起来。由式(45)可得

$$\omega(t) = 2N_t\pi\bar{R}_t^2 d = 2(N_t\bar{R}_t^3)\pi d/\bar{R}_t = \frac{\pi d}{4\pi^3\bar{R}_t} \quad ,$$

也即

$$\omega(t)\bar{R}_t/d = \frac{1}{4\pi^2} \tag{48}$$

此公式即为岩石的混合破坏准则，当满足此条件时，就会发生裂纹的连接。此条件包含了裂纹的裂纹

体积密度，及裂纹的尺寸信息。

利用平均尺寸进行评价不能够反映裂纹体系的内部连接情况，不能反映那一个尺寸的裂纹优先连接。为此取裂纹密度分布的单位长度，即 $dR=1$，考察 $\dfrac{d\left(n_g \cdot (2R)^3\right)}{dR}=0$，可得

$$R = 3\overline{R}_t, \quad \left.\dfrac{d^2\left(n_g \cdot (2R)^3\right)}{dR^2}\right|_{R=3\overline{R}_t} \leq 0 \tag{49}$$

也即 $K_l = n_g^{-1/3}/(2R)$ 在 $R=3\overline{R}_t$ 将会取最小值，从而优先满足连接准则，在此尺寸左右的裂纹将优先连接，形成大一级的裂纹。随着大尺度裂纹数目及尺寸的增加，大尺寸的裂纹将会满足密度准则，再连接。随着这一过程的进行最终导致材料的破坏。

到底那一个尺寸的裂纹将优先连接，将会依赖于裂纹数目随其尺寸的分布情况。如果裂纹数目随其尺寸的分布有如下关系 $n_g \sim \exp\left[-(R/\overline{R}_t)^2\right]$，则将会出现另一种情况。在 $R=\sqrt{3/2}\cdot\overline{R}_t$ 处，$K_l = n_g^{-1/3}/(2R)$ 将会取最大值，因此大于或小于此尺寸的裂纹将会优先发展。这将完全不同于前一情况。

四、结论

综上所述，可以得到如下结论：

1. 宏观岩石力学的一个不足是，绝大多数变形与破坏理论是建立在唯象的基础上的，缺乏足够的物理基础。固体物理及非线性科学的概念及方法的应用将会加深我们对岩石变形与破坏的认识，扩展岩石力学的研究范围，丰富岩石力学的以及内容。但研究的最终成果必须化为适宜于工程实际的形式。

2. 岩石是一个具有微细宏观构造层次的复杂的介质。我们对于岩石变形及破坏微观机制及宏观描述方面已经取得了一定的成果，但我们在细观水平上的研究还很薄弱。细观水平在岩石非均匀塑性变形及其随后的破坏中起着很重要的作用，在细观层次上岩石的变形特征是"剪切+旋转"。细观层次上的研究起着连接微观及宏观的桥梁作用。

3. 岩石的变形及破坏是一个动态过程，时间在其中起着关键作用。很多岩石变形损伤演化方程及破坏准则实际上都是具有时间构造性的。进一步完善岩石变形及破坏的时间构造性理论仍是一个重要的课题。

4. 岩石的强度依赖于应变率。在低、中、高应变率区，热活化机制、黏性阻尼机制及热活化机制分别起主导作用，岩石的强度对于应变率依赖的关系式必须考虑这种机制转化。在岩石动力模型中必须考虑强度对于应变率的依赖关系。

5. 岩石的破坏是一个非平衡过程，也是一个岩石微缺陷统计群体的动力演化过程。把岩石微缺陷统计群体的动力演化与非线性物理的概念结合起来的混合破坏准则反映了更多的微缺陷信息，能更好地反映岩石破坏的内部状况。

总之，岩石中强动载研究取得了许多成果，但尚未统一，并留有不少空白。揭示岩石在强动载作用下真实的变形与破坏过程，必须继续在连续介质力学框架内研究基于微、细观物理力学的理论，建立简单工程实用的介质在爆炸作用下，分阶段连贯的不同时空尺度的动力行为；必须研究在冲击与爆炸作用下介质在不同的特征能量尺度及其传输速度下，分阶段连贯的动力破坏过程和动力破坏准则。

参考文献

1. Вовк А. А. и другие. Поведение грунтов при импульсных нагрузок. Киев: Наукова думка, 1984.

2. A. A. Вовк и другие. Геодинамика взрыва и её приложение. Киев: Наукова думка, 1981.

3. Григорян С.С. Об общих уравнениях динамики грунтов. ДАН СССР, 1959, т.124, №2, с.285-287.

4. Григорян С.С. Об основных представлениях динамики грунтов. ПММ, 1960, т.24, вып.6, с.1057-1072.

5. П.В.Макаров. Подход физической мезомеханики к моделированию процессов деформации и разрушения. Физическая Мезомеханика, 1(1):61-81,1998.

6. А.Н.Ставрогин, А.Г.Протосеня. Механика деформирования и разрушения горных пород[M], Москва: Недра, 1985.

7. В.А.Петров, В.И.Башкарев, В.И.Веттегрень. Физические основы прогнози-рования конструкционных материалов. С-Петербург: Политехника, 1994.

8. Л.М.Качанов. О времени разрушения в условии ползучести. Изв. АН СССР. ОТН. 1958, №8, 26-31.

9. Ю.Н. Работнов. О разрушении вследствие ползучести. ПМТФ, 1963, №2, 113-123.

10. А.А.Ильюшин. Об одной теории длительной прочности. МТТ, 1967, №3, 21-35.

11. А.И. Глушко. Исследование откола как прцесс образование микропор. МТТ, 1978, 5, 132-140.

12. В.Р.Регель, А.Е.Слуцкер, Э.Е.Томашевский. Кинетическая природа прочности твердого дела. Москва: Наука, 1974.

13. А.Петров, В.И.Башкарев, В.И.Веттегрень. Физические основы прогнозирования конструкционных материалов. С-Петербург: Политехника, 1994.

14. А.Д.Борулев и другие. Кинетическая модель динамического деформирования и разрушения горных пород. Часть I: Физическое построение. ФТПРПИ, 1985, 3:36-46.

15. Г.И. Канель и другие. Исследование механическх свойств материалов при ударно-волновом нагружении. Механика Твердого Тела, 5:173-188, 1999.

16. В.А. Петров. К дилатонной модели термофлукционного зарождения трещин// Доклады АН СССР.-1988.-Т.301.-№5.-С.1107-1110.

17. В.С.Никифровский, Е.И.Шемякин. Динамическое разрушение тведого тела[M]. Новосибирск: Наука, 1979.
(中译本：固体动力破坏，1981, 煤炭工业出版社)

18. В.Е.ранин, Современные проблемы пластичности и прочности твердых тел[M] Изв. Вуз. Физика, 41(1):7-34,1998.

19. Замышляев Б.В. и другие. Об уравнении состояния горных пород при взрывных нагрузках[J]. ДАН СССР. 1980, Т..25, №2, С.322-326.

20. В.Е.Панин, Ю.В.Гриняев, В.Е.Егорушко. Спектр возвужденных состояний и вехревое механическое поле в деформируемом кристалле[J]. Изв. Вуз. Физика, 1987, 30(1)36-51

21. Е.Панин, В.А.Лихачев, Ю.В.Гриняев, Структурные уровени деформации твердых тел[M] Новосибирск: Наука, 1985.

22. В.Е.Панин, Ю.В.Гриняев и другие. Структурные уровени пластической деформации и разрушения[M]. Новосибирск: Наука, 1990.

23. П.В.макаров, моделирование процесоов деформации и разрушения на мезоуровне, МТТ, №5, 1999, 109-130

24. Капустянский С.М., Николаевский В.Н. Параметры упругопластической дилатансионной модели для геоматериалов[J]. ПМТФ, №6, 1985, С.145-150.

25. Николаевский В. Н. Динамическая прочность и скорость разрушения. В кн: Удар, взрыв и разрушение. М.: Мир, с.166-203.

26. А.Н.Ставрогин,А.Г.Протосеня. Прочность горных пород и устойчивость выработок на больших глубинах[M]. Москва:Недра,1979.

27. J.F.Kalthoff, D.A.Shockey, Instability of cracks under impulse loads. J.Appl. Phys. , 48(5): 986-993 , 1977.

28. A.N.Stroh , Theory of the fracture of metals[J]. Advanc. Physics, 6(24):418-465, 1957.

29. M.E.Kipp , D.E.Grady ,Numerical studies of rock fragmentation,SAND-79-1582,1980

30. .D.E.Grady,M.E.Kipp,Geometric statistics and dynamic fragmentation,J.Appl.Phys.58(3):1210-1222,1985

31. D.Grady,Mechanics of Geomaterial,Edited byZ.Bazant,129-156,1985

32. D.R.Curran,L.Seaman andD.A.Skockey,J.Appl.Phys.1973,44,668

33. D.R.Curran,L.Seaman and D.A.Skockey,Int.J.Impact Engng.13(1),53-83(1993)

34. Chen,W.F.,Pan,A.D.,Proc.of the third Int.Conf. on Constitutive Law of EngineeringMaterials,Tuscon,Arizona(1991)

35. W.Rudnicki, Geomechanics, International Journal of Solids and tructures,37,349-358,2000

35. J.W.Swegle Constitutive equation for porous materials with strength . J. Appl. Phys. , 1980, 51(5): 2574 – 2580.

36. 24A.Kumar, The effect of stress rate and temperature on the strength of basalt and granite [J], Geophysics,33(3):501-510,1968.

37. N. Morozov , Y. Petrov, Dynamics of fracture[M]. Springer-Verlag , Berlin , Heidelberg, 2000.

38. J.D.Cambell, W.G.Furguson, The temperature and stress rate dependence of shear strength of mild steel[J] ,Phil.Mag.,21, 1970.

39. U.S.Lindholm, L.M.Yeakley, and A.Nagy, The dynamic strength and fracture properties of dreser basalt[J], J. Rock Mech. Mining Science, 11:181-191,1974.

40. P.Perzyna, Constitutive modeling of dissipative solid for localization and fracture[A], in: localization and fracture phenomena in inelastic solids, Springer, Wien, New York,1998.

41. J.Lankford, The role of subcritical microfracture process in compressive failure of ceramics[A], in: Fracture Mechanics of Ceramics ,Vol .5, Plenum Press ,N.Y.,1983.

42. W.Herman, Constitutive equation for the dynamic compaction of ductile porous materials, J.Appl.Phys.,1969, 40(6):2490-2499.

43. D.R.Curran , L.Seaman , D.A.Shockey. Computational model for ductile and brittle structures. J. Appl. Phys.,1976, 47(11): 4814 – 4826.

44. D.A.Shockey, L.Seaman , D.R.Curran.The influence of microstructural fractures on dynamic fracture// Metallurgical Effects at High Strain Stress. N.Y.; L.; Plenum Press. 1973, 473-499.

45. Wang Mingyang,Qian Qianhu,Theoretical and Experimental Studies on the Granular Medium under Explosive Loading,Intel. Conf. On Computational methods in Struct, and Geotech. Engrg.1994,Hongkong,1136—1142.

46. Grady D.E. Shock wave properties of brittle solids[C], in: Shock Compression of Condensed Matters, AIP press,1995.

47. 王明洋，钱七虎，郑鸿泰.岩石变形和破坏基本力学问题[C].新世纪岩石力学与工程中数值模拟新进展会议论文集，2001.

48. 王明洋，钱七虎.颗粒介质的弹塑性动态本构关系[J].固体力学学报，1995，Vol.16(2):213-219.

49. 王明洋，赵跃堂，钱七虎.饱和砂土动力特性及数值方法研究[J].岩土工程学报，2002.

50. 王明洋，戚承志，钱七虎.岩石中爆炸与冲击的破坏研究[J].辽宁工程技术大学学报，Vol.20(4):385-389.

51. 戚承志，钱七虎，考虑到时效的一维剥离破坏及损伤破坏机理[J].解放军理工大学学报，2000，1（5），1-6.

52. 王明洋，国胜兵，邓国强.岩石覆盖层抗震塌机理研究，解放军理工大学学报，2001，.2(3),12-19.

53. 戚承志，钱七虎.材料变形及损伤演化的微观物理动力机理[J].固体力学学报，2002.

54. 戚承志，钱七虎.岩石等脆性材料动力强度依赖应变率的物理机制[J].岩石力学与工程学报，2003，22（2）.

55. 戚承志，钱七虎，王明洋，弹塑性孔隙介质的一种本构模型 第一部分：在冲击载作用下弹粘塑性孔隙介质的畸变行为[J].石力学与工程学报，2003，22（11）.

56. 戚承志，钱七虎，王明洋，弹塑性孔隙介质的一种本构模型 第二部分：在冲击载作用下弹粘塑性孔隙介质的体积压缩方程[J].岩石力学与工程学报，2003，22（12）．

57. 钱七虎，王明洋，赵跃堂，岩石中爆炸与冲击作用研究进展[C]//岩石力学与工程学会第四次学术大会文集,中国科学技术文献出版社,53-65.

58. 戚承志.博士后研究报告：岩石介质在强动载作用下的变形与破坏，指导：钱七虎院士。解放军理工大学，2002.

59. 陈宗基，康文法，岩石的封闭应力、蠕变和扩容及本构方程[J]，岩石力学与工程学报，1991，Vol.10(4),299-312．

60. 戚承志，等.时间性破坏准则及其内在联系。世界地震工程，vol.18（2），56-60。

61. 戚承志，等.考虑强度对于应变率依赖性的一种岩石弹塑性动力模型。世界地震工程，vol.18（3）。

62. 王明洋，钱七虎，等.岩石单轴试验全程应力应变曲线讨论，岩石力学与工程学报 1998，Vol.17(1):101-106.

块体结构岩体中超低摩擦效应的理论研究

戴 俊[1,2]，钱七虎[1,3]，王明洋[1]

（1. 解放军理工大学 江苏南京 210007；2. 西安科技大学 陕西西安 710054；3.总参科技委 北京 100857）

摘要：超低摩擦效应是块体结构岩石的一个重要性质方面，有研究认为超低摩擦效应是一定载荷条件下岩石块体界面的暂时脱离造成的。据此，本文利用固体中的应力波理论，分析块体结构岩体中超低摩擦效应产生的机理，探讨影响超低摩擦效应产生的岩体结构组成和外载荷条件，并完成超低摩擦效应引起位移值的计算，结果与文献实验值大体一致。研究发现，岩石组成块体之间存在波阻抗差及施加合理幅值和持续时间的冲击力是决定超低摩擦效应产生的关键因素。研究结论对深部岩体工程结构的设计优化、安全、稳定具有重要理论和实际意义。

关键词：块体岩石，超低摩擦效应，应力波，深部岩体工程

1 概述

岩石是由大小不同的块体构成的，在工程意义上通常称为岩体。岩体依靠各组成块体之间的摩擦力维持稳定，当外力引起的滑动力大于块体之间的摩擦力时，块体之间必产生相对滑动，岩体失去稳定，发生整体破坏。岩体组成块体之间的摩擦力决定于作用于交界面上的正压力和交界面的摩擦系数，交界面上的正压力越大，摩擦力越大，摩擦系数则决定于交界面的性质，不因受载条件的改变而改变。因此，块体之间的摩擦力是随岩体外部受压力的变化而变化的。

然而，在研究深部岩体力学现象中，解释摆锤型波产生机理时，发现[1]：在一定受载条件下，岩体组成块体之间会产生有意义的摩擦暂时"消失"效应，而且这种摩擦暂时"消失"效应是在与冲击压力作用线垂直方向上的岩块交界面上观察到的，这种摩擦暂时"消失"效应称为超低摩擦效应。研究者进一步认为，这种摩擦力"消失"效应是以相互接触的块体表面之间暂时脱开接触为条件的。这种假设在一定程度上已得到了证实。

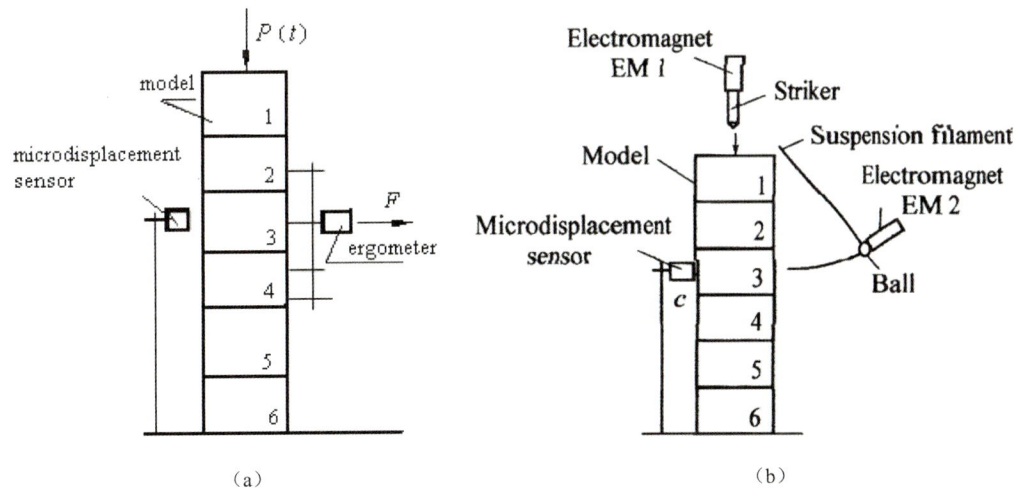

图 1 超低摩擦效应实验研究系统

Fig. 1 test system for studying ultra-low friction effect

a— 水平静载与垂直冲击载荷联合作用；b—水平与垂直冲击载荷共同作用

a—loaded static force horizontally and a impact force vertically; b—loaded impact forces horizontally and vertically

前苏联科学家对超低摩擦效应进行了实验研究[1~3]，其实验系统如图 1 所示。实验材料为有机玻璃

基金项目：国家自然科学基金重大项目资助(50490275)；国家博士后科学基金资助(2005037221)。

块和硅砖块,在垂直于块体交界面的方向上施加冲击载荷,在交界面方向分别施加静态载荷和冲击载荷。实验通过观测块体 3 的水平位移,判定块体之间超低摩擦效应的产生。研究得出的结论归为:

1) 当块体受到水平静力和垂直冲击力,且垂直冲击力能量达到一定值时,块体产生水平位移的水平静力门槛值降低;

2) 对不同材质的实验块体,当水平力一定时,水平位移随垂直冲击力能量的增大而单调递增,当垂直冲击力能量一定值时,水平位移随水平静力增大的关系近似于抛物线函数关系;

3) 在水平和垂直方向均受冲击作用条件下,当水平、垂直冲击力之间的延迟时间间隔按 $\sqrt{2}$ 的幂次关系变化时,块体出现最大水平位移,即块体之间超低摩擦效应的产生具有离散周期;

4) 我们注意到,这种超低摩擦效应是暂时的,只发生在很短的时间内,依冲击载荷情况的不同,可以多次发生。

我们也发现,以上研究没有注意到垂直冲击载荷波型的影响,也没有注意到组成块体力学性质(如:波阻抗)不同时的超低摩擦效应特点。而这几点是非常重要的,因为相同冲击能量的载荷可以由不同的应力波型来实现,此外实际组成岩体的各块体(层)的力学性质往往是不同的。因此,研究超低摩擦效应必须考虑不同冲击载荷波型和岩体组成块体的不同性质的影响,进一步探索超低摩擦效应产生的规律,为解决当前深部矿藏开采、深部岩石隧洞开挖等工程面对的问题提供理论依据和有效方法。

如图 1 所示,在块体系统受到垂直冲击力作用时,将引起应力波在块体中传播。按文献结论,块体之间的超低摩擦效应与块体之间的相互脱离有关,因而断定也与应力波通过块体界面的传递情况有关。为此,本文将利用应力波理论对超低摩擦效应进行研究,进一步分析其规律。

2 应力波通过岩体界面的传递过程

图 2 所示,由 6 块组成的块系结构,左端施加静力载荷 p_0 和冲击载荷 $p(t)$,块体 3 作用有垂直于 $P(t)$ 的水平静力 F_0,各块体的波阻抗为 $(\rho_0 c)_i (i=1,2,\ldots,6)$,假设 F_0 的存在不影响应力波的传播,应力波的传播过程可用图 2 表示。据此,通过分析块体 3 在垂直于冲击力方向的位移,探讨块体(3)两端界面发生超低摩擦效应的情况。

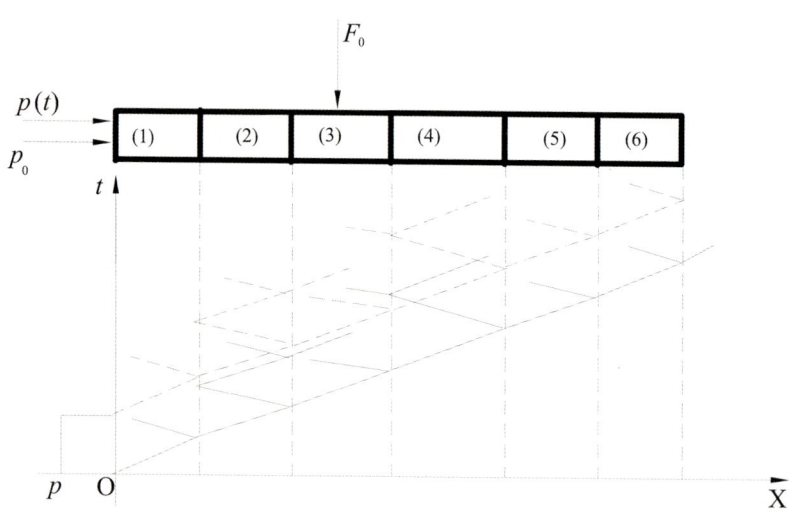

图 2 应力波通过块体界面的传播

Fig. 2 the propagation of a stress wave through the interface of rubbly rock mass

应力波通过块体接触界面的反射、透射情况与界面两侧块体的波阻抗有关,其关系可表示为[4]

$$p_r = Rp, \quad p_t = Tp \tag{1}$$

$$R = \frac{(\rho_0 c)_T - (\rho_0 c)_I}{(\rho_0 c)_T + (\rho_0 c)_I}, \quad T = \frac{2(\rho_0 c)_T}{(\rho_0 c)_T + (\rho_0 c)_I} \tag{2}$$

式中 p, p_r, p_t 分别为入射应力、反射应力和透射应力;R 和 T 分别为反射系数和透射系数;$(\rho_0 c)_I$ 和 $(\rho_0 c)_T$ 分别为入射(反射)块体和透射块体的波阻抗;ρ_0 为块体密度;c 为块体中的弹性波速度。

由式（1）、式（2）可以看出，当$(\rho_0 c)_I < (\rho_0 c)_T$时，$R>0$，反射波与入射波性质相同；反之，当$(\rho_0 c)_I > (\rho_0 c)_T$时，$R<0$，反射波与入射波性质相反，压缩波经过界面将反射回拉伸波。当$(\rho_0 c)_I = (\rho_0 c)_T$时，$R=0$，应力波经过界面时没有反射，透射波与入射波性质和大小相同。

以图2中的块体（3）和（4）之间的界面为例。假定$(\rho_0 c)_3 > (\rho_0 c)_4$，压缩冲击载荷由（3）向（4）传播时，将产生反射拉伸应力波在（3）中反向传播，当这一反射拉伸应力波到达（2）、（3）界面时，如果拉应力足够大，则将引起界面暂时脱离；导致界面之间的摩擦力下降，如果这种摩擦力的降低达到一定值，原来稳定的块体之间将在侧向静力F_0的作用下产生位移。同样透射进入块体（4）中的压缩冲击载荷达到（4）、（5）界面时，也将产生反射和透射，如果$(\rho_0 c)_4 > (\rho_0 c)_5$，则反射波为拉伸波，这一反射拉伸波达到（3）、（4）界面时，仍有引起界面暂时脱离的可能，产生超低摩擦效应。

另一方面，由于块体界面之间的抗拉能力很低，当反射拉应力波达到一定强度时，将会引起块体脱离，于是，这时拉伸应力波不能通过界面。

对于其他的块体波阻抗结合情况，仍然可以进行同样的分析，但不一定导致超低摩擦效应的产生。因此，超低摩擦效应的产生应具备一定的条件。

3 超低摩擦效应的产生条件

当在块体系统左端面仅作用由静力载荷p_0时，块体（3）两端面产生的摩擦力为

$$F_f = 2p_0 A f \qquad (3)$$

式中 F_f为摩擦力，MN；p_0为块体系统左端面上的静力载荷，MPa；A为块体端面面积；m^2；f为接界面的摩擦系数。

图3所示，如果

$$F_0 \geqslant F_f \qquad (4)$$

则块体（3）将会在F_0方向上产生位移；反之，块体（3）保持静止，不会移动。

但当块体系统左端施加冲击载荷后，由于应力波的传播、反射和透射，如果在块体（3）端面引起正压力降低，导致产生超低摩擦效应，则块体将可能会在F_0方向上获得加速度，而移动。

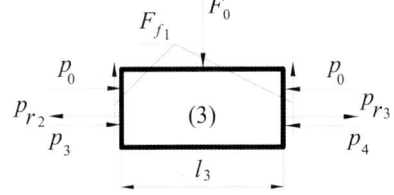

图3 块体（3）的受力分析

Fig. 3 the analysis of forces exerted on the rubbly block No.3

图3中，各应力计算如下：

$$p_3 = T_{12} \cdot T_{23} \cdot p, \quad p_4 = T_{12} \cdot T_{23} \cdot T_{34} \cdot p = p_3 \cdot T_{34}$$

$$p_{r2} = T_{12} \cdot p \cdot R_{23}, \quad p_{r3} = p_3 \cdot R_{34}$$

以上各式中 p_3，p_4分别为投射入块体（3）、（4）中的应力；p_{r2}，p_{r3}分别为反射回块体（2）、（3）中的应力；T，R分别为投射系数和反射系数，下标表示相应的块体界面。参见图2。

类似地，可以得到各界面反射、透射后的各应力。而且可知，如果$(\rho_0 c)_1 > (\rho_0 c)_2 > 0$，则$p_1>p_2>0$，$p_{r1}<0$；如果$(\rho_0 c)_2 > (\rho_0 c)_1 > 0$，则$p_2>p_1>0$，$p_{r1}>0$；

如果$(\rho_0 c)_2 > (\rho_0 c)_3 > 0$，则$p_2>p_3>0$，$p_{r2}<0$；如果$(\rho_0 c)_3 > (\rho_0 c)_2 > 0$，则$p_3>p_2>0$，$p_{r2}>0$。

可见，如果$(\rho_0 c)_1 < (\rho_0 c)_2 < (\rho_0 c)_3$，则压缩冲击载荷将逐渐增强，块体（3）将受到较大的冲击，有利于块体（3）超低摩擦效应的产生。

进一步，如果$(\rho_0 c)_3 > (\rho_0 c)_4 > 0$，则$p_3>p_4>0$，$p_{r3}<0$。当$p_{r3}$传到（2）、（3）界面时，该界面的正压力将降低，甚至降低到零，这样该界面即产生超低摩擦效应。

如果$(\rho_0 c)_4 > (\rho_0 c)_3 > 0$，且$(\rho_0 c)_5 > (\rho_0 c)_4 > 0$，则$p_5>p_4>p_3>0$，$p_{r3}>0$，$p_{r4}>0$，块体（3）不会出现超低摩擦效应。

如果$(\rho_0 c)_4 > (\rho_0 c)_3 > 0$，但$(\rho_0 c)_4 > (\rho_0 c)_5 > 0$，$p_4>p_3$，$p_{r3}>0$，但$p_4>p_5>0$，$p_{r4}<0$。当$p_{r4}$传到传到（3）、（4）界面时，该界面正压力降低，块体（3）产生超低摩擦效应。

由此可知，就块体（3）而言，$(\rho_0 c)_1$，$(\rho_0 c)_2$，$(\rho_0 c)_3$之间的大小关系，对超低摩擦效应的产生

不会有质的改变。但 $(\rho_0 c)_3$ 越大，对超低摩擦的产生越有利。块体（3）产生超低摩擦效应的条件是

$$(\rho_0 c)_3 > (\rho_0 c)_4 \tag{5}$$

或

$$(\rho_0 c)_4 > (\rho_0 c)_3，但 (\rho_0 c)_4 > (\rho_0 c)_5 \tag{6}$$

块体（5）后面块体的 $(\rho_0 c)$ 大小对块体（3）出现超低摩擦效应没有影响。

以此类推，可以分析其他块体的超低摩擦效应。

4 超低摩擦效应引起的位移计算

图3中，当式（4）满足时，块体（3）将失去稳定平衡，而产生侧向移动。当在块体系统左端施加冲击载荷后，由于引起的应力波传播、反射和透射等，将引起块体（3）两端面的正压力降低。冲击载荷作用后，块体（3）两侧的正压力可表示为

$$\bar{p} = (2p_0 - p_{23} - p_{34}) \cdot A \tag{7}$$

$$p_{23} = \langle -p_{r3} \rangle \quad (t_1 < t < t_1 + t_l) \tag{8}$$

$$p_{34} = \langle -p_{r4} \rangle \quad (t_2 < t < t_2 + t_l = \tag{9}$$

式中 $\bar{p}$ 为施加冲击载荷后块体（3）两端面的正压力，MN；A 为块体（3）的截面积，m²；p_{23}，p_{34} 分别为反射波传到界面（2）、（3）与（3）、（4）时的应力，MPa；t_1 为界面（3）、（4）的反射波传到界面（2）、（3）的时间，s；t_2 为应力波由界面（3）、（4）传到界面（4）、（5）并反射回来的时间，s；t_l 为冲击载荷的持续时间，s。这里取冲击应力波传到界面（3）、（4）时刻为时间起点。本文中符号 $\langle \ \rangle$ 的意义为

$$\langle \chi \rangle = \begin{cases} \chi, & \chi \geq 0 \\ 0, & \chi < 0 \end{cases} \tag{10}$$

根据牛顿第二运动定律，块体（3）的侧向移动，可用如下方程描述，

$$a = \frac{1}{m}(F_0 - \bar{p}f) = \frac{1}{m}(F_0 - Af\langle 2p_0 - p_{23} - p_{34}\rangle) \tag{11}$$

$$v = \int_{t_1}^{t_1+x} \langle a \rangle dx \tag{12}$$

$$s' = \int_{t_1}^{t_m} \int_{t_1}^{t_1+x} \langle a \rangle dx dt \tag{13}$$

$$t_m = \max\{t_1 + t_l, t_2 + t_l\} \tag{14}$$

式（11）~（14）中 m 为块体（3）的质量，kg；a 为加速度，m/s²；v 为速度，m/s；s' 为位移，m；x 为任意变量。

考虑到，反射波作用结束时，块体（3）具有最大的移动速度，其后由于惯性效应，块体并不会立刻停止，而是要经历一个减速的过程，才能停止移动，因此块体（3）真正的位移值 s 可取为式（13）计算值的2倍，即

$$s = 2s' = 2\int_{t_1}^{t_m}\int_{t_1}^{t_1+x}\langle a \rangle dx dt \tag{15}$$

由式（15）可以得知，获得一个大于零的加速度 a 是块体（3）产生位移的重要条件。由此得到块体（3）产生超低摩擦效应的另一条件是施加的冲击载荷必须达到一定值，因为如果冲击载荷过小，就不可能获得足够大的 p_{23} 和 p_{34}，以保证加速度 a 大于零。

5 影响超低摩擦效应的因素分析

根据前面的分析，产生超低摩擦效应必须具备一定的条件。只有当载荷条件和块体结构的组成性质满足一定条件时，超低摩擦效应才会产生。这些条件，首先是组成块体系统的各块体应具有不同的波阻抗，以实现冲击载荷传播过程中在界面处的反射；其次是冲击载荷必须具有足够的强度，以使载界面处反射后形成的拉伸应力足以克服静力压缩载荷作用，在界面上形成正压力降低。

为了进一步表述影响超低摩擦效应产生及其程度的因素，这里做出以下的条件限制，即取

$$F_0 = 2p_0 Af \cdot \eta \quad (0 \leqslant \eta \leqslant 1)$$
$$(\rho_0 c)_1 = (\rho_0 c)_2 = (\rho_0 c)_3; (\rho_0 c)_3 = j(\rho_0 c)_4 = j^2(\rho_0 c)_5$$
$$p = np_0; t_l = k(l_3/c_3)$$

这里，j，k 为任意实数；n 为自然数。

于是，利用式（11）~（15），可以推得

$$s = s_1 + s_2 = \frac{Ap_0 f}{m}(l_3/c_c)^2 k^2 \left[\left\langle 2\eta - \left\langle 2 - \frac{j-1}{j+1} \cdot n \right\rangle \right\rangle + \left\langle 2\eta - \left\langle 2 - \frac{j-1}{j+1} \cdot \frac{2}{1+j} \cdot n \right\rangle \right\rangle \right] \quad (16)$$

由式（16）可以看出，影响超低摩擦效应产生的各方面因素。这就是：

1）随冲击载荷作用时间的增加，块体（3）的位移增加，表明冲击载荷作用时间的增加有利于超低摩擦效应的产生；

2）块体介质的波阻抗差值越大，块体（3）的位移越大，表明块体介质的波阻抗差值越大，越有利于超低摩擦效应的产生；

3）侧向静力 F_0 越接近最大摩擦力，越有利于获得较大的侧向位移，但不影响超低摩擦效应的产生；

4）冲击载荷的幅值越大，越有利于超低摩擦效应的产生，但存在一个极值范围，这一范围由保证

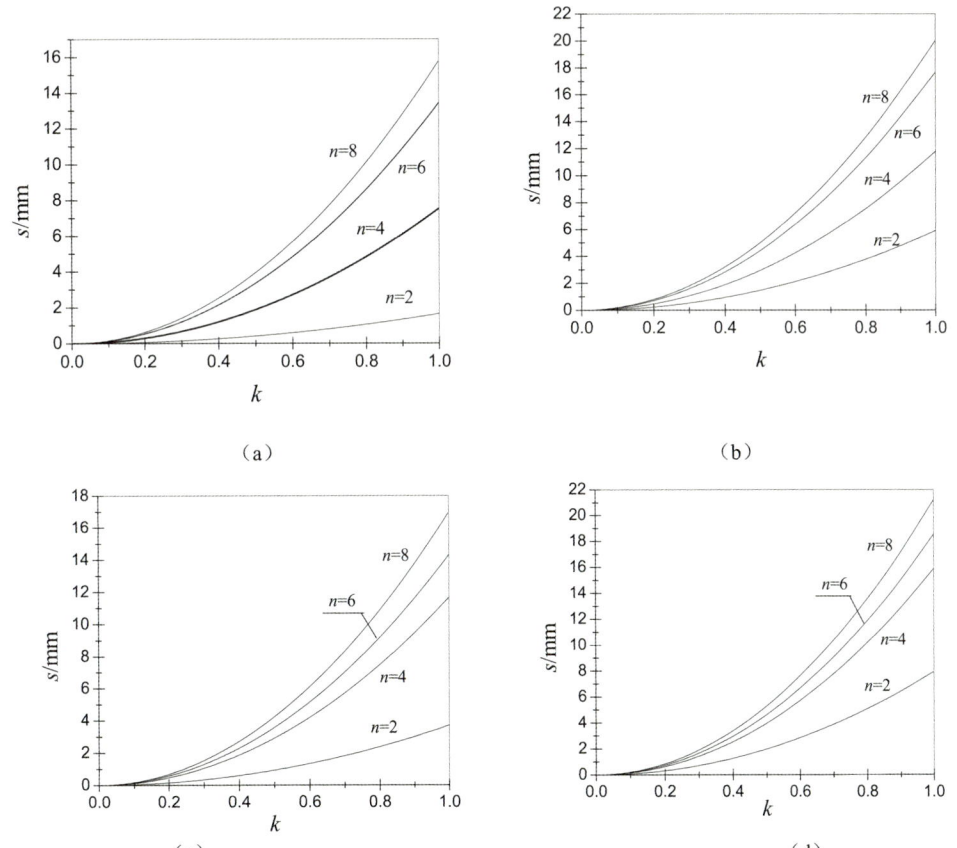

图 4　冲击载荷值与持续时间对块体（3）侧向位移的影响
Fig. 4 the effect of the magnitude and duration of an impact load on the lateral displacement of the rubbly block No.3
a—η =0.8, j=2；b—η =1.0, j=2；c—η =0.8, j=3；d—η =1.0, j=3

所有 $\langle\chi\rangle$ 大于确定。过小时在界面反射引起的拉伸应力不足以克服界面上的压缩应力，不会产生超低摩擦效应；过大时，由于拉伸应力不可能越过界面，因而超出部分对超低摩擦产生不起作用。

5）块体长度、端部静载荷、界面摩擦系数、块体质量等均对超低摩擦效应的产生有一定影响。

此外，这里考虑的矩形冲击载荷波形是最理想的，波形不同效果不同，其他波形时，导致超低摩擦效应的作用将降低。

进一步，根据有关资料[5]，取 p_0=1MPa, f=0.2, ρ_{03}=1200kg/m³, c_3=2800m/s, l_3=0.25m, η=0.8, j=2，经计算得

$$(l_3/c_3)^2 = (0.25/2800)^2 = 8.0 \times 10^{-9} s^2$$

$$Afp_0/m = \frac{Afp_0}{\rho_{03}Al_3} = \frac{0.2 \times 10^6}{1200 \times 0.25} = 666.7 \text{ m/s}^2$$

$$\frac{Afp_0}{m}\left(\frac{l_3}{c_3}\right)^2 = 666.7 \times 8.0 \times 10^{-9} = 5.3 \times 10^{-6} \text{ m}$$

$$s = [(n/3 - 0.4) + (2n/9 - 0.4)]k^2 \times 5.3 \times 10^{-6} \text{ m}$$

由上式，可以看出位移 s 的大小，图 4a 所示（η=0.8, j=2）。类似地，可以得到其他条件下的位移 s 变化，图 4b、4c 和 4d 所示（η=1.0, j=2; η=0.8, j=3; η=1.0, j=3）。计算位移值在 0~21mm 之间，结果与国外资料实验值大体吻合。因而，认为所做分析时可信的。这种方法对一定的外载荷条件下的超低摩擦效应产生的规律分析有理论和实际意义。

6 结束语

超低摩擦效应是块体结构岩石的一个重要性质方面，目前正在受到国内外的日益普遍关注，国外的研究资料认为超低摩擦效应是岩石块体界面的暂时脱离造成的。据此，本文利用固体中的应力波理论，对块体结构岩石的超低摩擦效应产生机理和影响超低摩擦效应外载荷及块体介质等因素进行了研究。研究取得了如下基本结论：

1）超低摩擦效应的产生可认为是冲击载荷引起应力波传播，并在界面发生反射、透射，出现反射拉应力的结果；

2）超低摩擦效应是暂时现象，持续时间很短，但可能会多次发生；

3）产生超低摩擦效应的基本条件是岩石组成块体的波阻抗必须不同，而且受到的冲击载荷必须达到一定值；

4）块体介质的波阻抗比、介质力学性质参数、冲击载荷的幅值和持续时间、静载压力状态等都对超低摩擦效应的产生有影响；

5）导致超低摩擦效应的冲击载荷幅值有一定的极限范围，冲击载荷持续时间越长，对超低摩擦效应的产生越有利，块体介质之间的波阻抗差值越大，越有利于产生超低摩擦效应。

参考文献

1　钱七虎. 非线性岩石力学的新进展——深部岩石力学的若干关键问题. 成都：第八界全国岩石力学与工程学术会议论文集. 2004. 10~17

2　V.N.Oparin, E.G.Balmashnova, and V.L.Vostrikov. On Dynamic Behavior of "Self-Stressed" Block Model, Part 2: Comparison of Theortical and Experimental Data. Journal of Mining Science. Vol. 37, No. 5, 2001：455~461

3　M.V.Kurlenya, V.N.Oparin, E.G.Balmashnova, and V.L.Vostrikov. On Dynamic Behavior of "Self-Stressed" Block Model, Part 1: One-Dimensional Mechanical-Mathematical Model. Journal of Mining Science. Vol. 37, No. 1, 2001：1~9

4　王礼立. 应力波基础. 北京：国防工业出版社.1985

5　М.В.Курленя　В.Н.Опарин　В.И.Востриков. 块状介质中的超低摩擦效应（陆渝生译）．СО РАН физикотехнические проблемы разработки по лезных ископаемых 1997.No.1

THEORETICAL STUDY ON THE ULTRA-LOW FRICTION EFFECT IN NUBBLY ROCK MASS

DAI Jun[1, 2], QIAN Qihu[1, 3], WANG Mingyang[1]

(1. PLA University of Technology   Nanjing, Jiangsu 210007; 2. Xi'an University of Science and Technology   Xi'an, Shanxi 710054; 3. Science and Technology Committee of General PLA brainman board   Beijing   100857)

Abstract: The ultra-low friction effect, an important property of nubbly rock mass, is considered to be from transitory separation between blocks of rock mass under some loading condition. Based on this consideration, the mechanism on that the takes place in nubbly rock mass is analyzed by means of theory on stress wave in solid. The formations of rock mass and load condition affecting ultra-low friction effect taking place are discussed. Finally, the calculation is finished whose result is consistent with that from the finished experiment stated in the reference. The finding that ultra-low friction effect is decided by the key factors, such as the different wave impedance between rock blocks, reasonable load magnitude, and its duration time, is important in theory and practice for the optimization design, safety, and stability of rock projects at depth.

Key words: nubbly rock, ultra-low friction effect, stress wave, rock project at depth

锦屏二级水电站引水隧道围岩分区破裂化研究

钱七虎[1]　周小平[2]

(1. 解放军理工大学，南京 210007；　2. 重庆大学土木工程学院，重庆 400045)

摘要：对岩体分区破裂化进行了数值模拟，探讨了锦屏二级水电站引水隧洞围岩分区破裂化的特点。利用弱单元来考虑锦屏二级水电站引水隧洞围岩中节理的扩展和贯通，得到了与传统弹塑性理论完全不同的洞室围岩分区破裂化形式。同时，通过数值模拟确定了引水隧洞围岩破裂区的数量和分布。计算结果表明洞周围岩破裂带呈"滑移线"型分布，这与室内模型试验结果吻合。通过参数敏感性分析，研究了应力状态、黏聚力和内摩擦角对洞室围岩分区破裂化规律的影响。

关键词：锦屏二级水电站；深部岩体；分区破裂化；数值模拟

中图分类号：　　　**文献标识码**：A

The zonal disintegration of surrounding rock mass in Jinping II hydropower station, Southwestern China

Qihu Qian[1]　　Xiaoping Zhou[2]

(1 PLA University of Science and Technology，Nanjing 210007，China;

2 School of Civil Engineering, Chongqing University, Chongqing400045,China)

Abstract: By means of multiscale simulation, the special phenomenon of zonal disintegration of surrounding rock mass in the diversion tunnels of Jinping II Hydropower Station is analyzed in this paper. In order to model the growth and coalescence of cracks within rock mass in Jinping II Hydropower Station , the weak-element is adopted. When cracks coalesce, failure of deep crack-weakened rock masses occurs, fractured zone is formed. The present result is different from the one obtained by the traditional elasto-plastic theory. The numerical results show that the slip-line zonal fracture is yielded whitin rock mass around the diversion tunnels of Jinping II Hydropower Station.Meanwhile, the magnitude and distributions of fractured zone are determined by numerical simulation. It is shown that the present results are in good agreement with the one observed by model tests. Through sensitivity analyse, the effect of stress condition, cohesion and the angle of internal friction on the phenomenon of zonal disintegration are determined.

Key words: Jinping II Hydropower Station; deep rock mass; zonal disintegration; numerical simulation

1　引言

围岩分区破裂化现象就是指在深部岩体中开挖洞室或巷道时在其两侧或工作面围岩中交替出现破裂区和未破裂区[1]。深部岩体分区破裂化现象是伴随着深部岩体工程的建设出现的特殊工程响应问题，其独特的破裂形式与传统的弹塑性理论解不一致。岩体分区破裂化现象引起了岩石力学界专家和学者的广泛关注，也是当前岩石力学研究的热点和前沿。

目前，对围岩分区破裂化的研究主要集中在现场观测、室内模型试验和理论分析三个方面。在现场观测方面，俄罗斯学者 E.I.Shemyakin 等进行了大量开创性的工作，他们通过现场量测发现，在深部矿井开采面附近围岩中存在呈间隔分布的"环带状碎裂"现象[2~5]；潘一山等根据弹性波速测试结果发现在金川矿区围岩并非全部处于压缩状态，而是呈现结构面张开与闭合相间分布的状态[6]。在室内模型试验研究方面，顾金才等利用混凝土进行室内模型试验得到了洞室围岩"分层断裂"结果，如图 1 所示[7]。在理论研究方面，周小平和钱七虎认为巷道围岩分区破裂化的产生跟开挖速度与岩石强度有关，并给出了破裂区宽度和数量计算公式[8]。综观国内外对围岩分区破裂化问题的研究，现场观测和室内模型试验受到各种测试手段及复杂的现场环境限制,很难揭示围岩分区破裂化的机理,而理论研究往往局限于某些特殊的洞室形式和理性的围岩特性,无法满足工程实际应用的要求。通常,数值模拟手段能帮助我们解决实际中遇到的复杂问题。然而,至今尚未见到国内外关于分区破裂化数值模拟方面的报导。

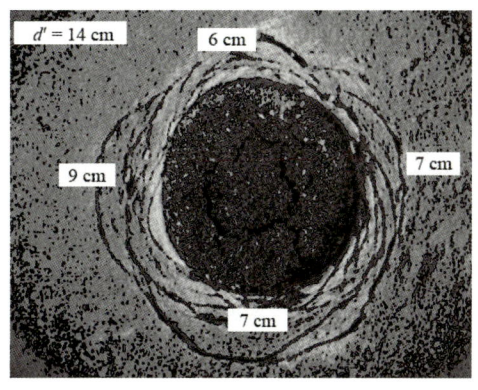

图 1　圆形洞室破坏形态（$d_2 = 160mm, \sigma_z/R_c = 6.83$）[7]

Fig.1　Failure shape of round tunnel($d_2 = 160mm, \sigma_z/R_c = 6.83$)

为此,本文提出了一种新的数值模拟方法,该方法通过弱单元反映岩体的非均质性,将本文提出的数值模拟方法与室内试验结果对比得到了相似的规律。最后,利用本文的数值模拟方法研究锦屏二级水电站引水隧洞围岩分区破裂化特征,并且通过参数敏感性分析,研究了不同应力状态、黏聚力和内摩擦角条件下,四心圆洞室围岩分区破裂化规律。

2　锦屏二级水电站引水隧洞工程概况

工程位于四川省境内的雅砻江锦屏大河弯处,利用雅砻江锦屏 150km 长大河弯的天然落差,截弯取直凿洞引水,电站水头约为 310m,总装机容量为 4,800MW,通过开挖 4 条引水隧洞即可建成引水发电的引水式水电站。4 条引水隧洞,洞线平均长度约为 16.67km,开挖洞径为 13m,衬砌后洞径为 11.8m,上覆岩体一般埋深为 1,500~2,000m,最大埋深约为 2,525m,具有埋深大、洞线长和洞径大的特点,成为锦屏二级水电枢纽的关键控制性工程。

引水隧洞穿越三迭系中上统大理岩、灰岩、结晶灰岩及砂岩、板岩等地层。大部分洞段以 II 类和 II~III 类围岩为主,约占全长的 92%,岩石致密、坚硬、完整。工程区内出露的地层为前泥盆系~第四系的一套浅海~滨海相、海陆交替相地层。引水隧洞所穿越的地层则以三迭系中下统的碳酸盐岩(T_1、T_2、盐塘组 T_{2b}、白山组 T_{2b}、杂谷脑组 T_{2z})为主,其次为三迭系上统 T_3 的砂岩、板岩等。岩层陡倾,其走向与主构造线方向一致。区内节理因构造部位和岩性不同而异。总体而言,以 NNE 向的顺层节理和近 EW 向(NWW 和 NEE)的张扭性节理最为发育,前者多呈闭合状,后者多呈张开。

四条引水隧洞组成的水工洞群采用钻爆法和 TBM 法相结合的施工方案。钻爆法施工洞段为 2#、4# 洞段,其形式为马蹄形断面,开挖直径 13m。采用 TBM 施工的引水隧洞段为 1#、3# 洞段,其形式为圆形断面,开挖直径 12m。锦屏引水隧洞的埋深大且地应力大,仅自重应力已相当可观,再加上构造作用,高地应力的作用非常强烈。工程区实测地应力成果显示,实测最大主应力值约为 70MPa[9,10]。

3　锦屏二级水电站引水隧洞围岩分区破裂化分析

根据有关的试验成果,本文选取引水隧洞中部埋深较大、地应力比较高的白山组(T_{2b})大理岩地层中垂直于引水隧洞轴线的竖直剖面作为计算断面。在大理岩地层中,结构面的走向为 N50~75°W,倾向为 SW，倾角约为 60°，结构面长度约 1m，间距 0.4m，充填物为破碎岩，充填物厚度约为 2 mm。大理岩的单轴抗压强度 75－85MPa，其全过程应力应变曲线如图 2 所示：

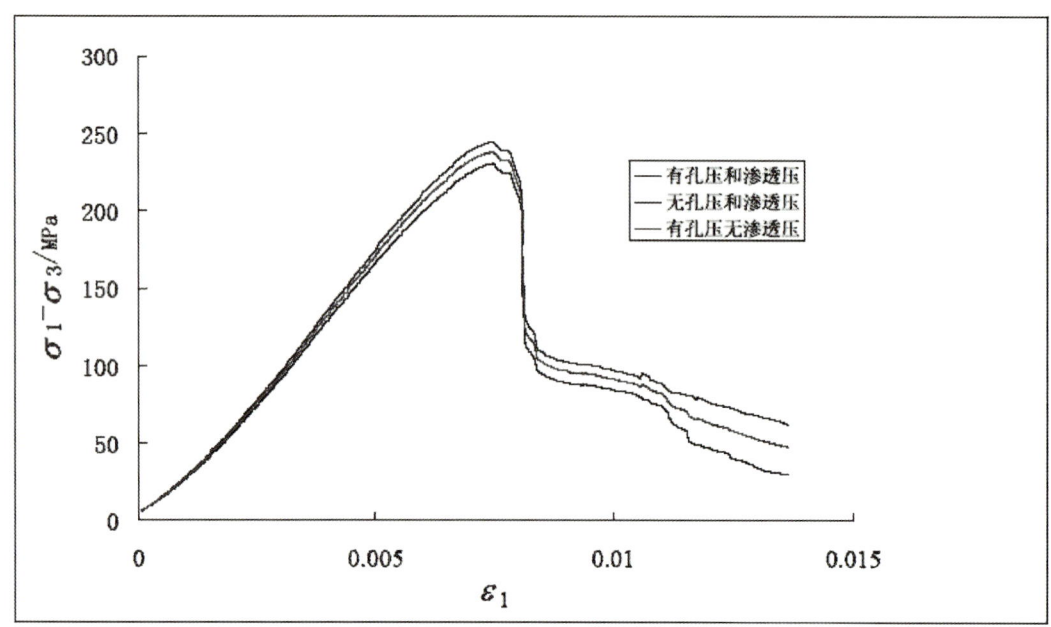

图 2 大理岩的全过程应力应变曲线

Fig.2 Complete stress-strain relation of marble

3.1 计算模型

本文采用多尺度数值模拟方法和多区损伤局部化理论模拟锦屏二级水电站引水隧洞的分区破裂化问题，其本构关系和裂纹的扩展及连接采用细观力学理论进行研究[11]。结构面的扩展和连接采用弱单元进行分析。本文数值计算模型是二维平面应变模型，四条引水隧洞数值计算模型范围为 300m×90m，其中 1 号和 3 号为圆形隧洞，洞径约为 12m；2 号和 4 号为四心圆隧洞，洞径约为 13m，如图 3 所示。模型边界条件为：左右两边施加水平构造应力 23MPa，模型上表面施加垂直应力 69.5 MPa，底边采用法向约束，具体计算参数如表 1 所示。模型的单元数为 449600 个，结点总数为 451917 个。

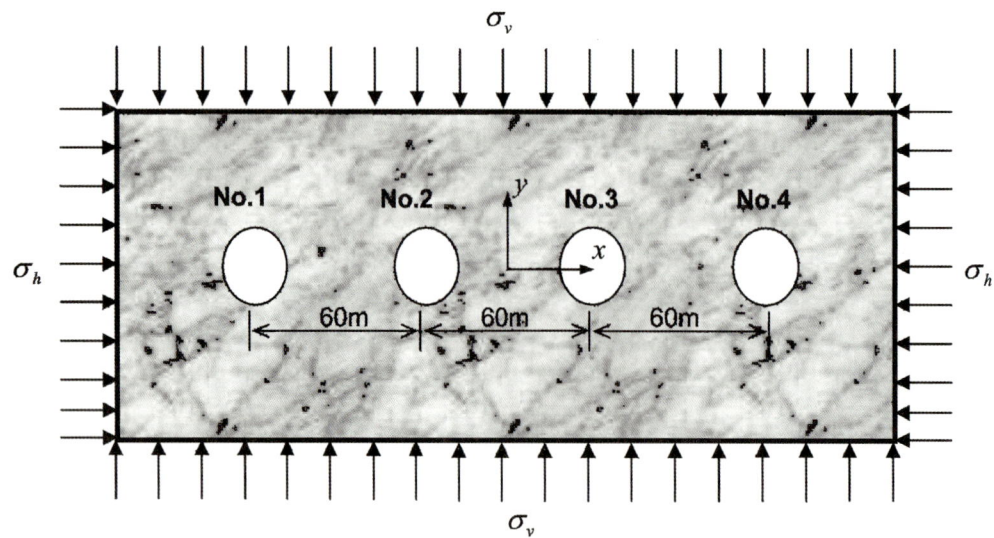

图 3 锦屏二级水电站引水隧洞计算模型简图（平面应变模型）

Fig.3 Sketch of diversion tunnel in Jinping II Hydropower Station(plain strain)

表 1 岩体基本参数取值
Table1 Parameters of rock mass

	黏聚力 C/MPa	内摩擦角/(°)	泊松比	弹性模量/GPa	重度/kN·m^{-3}
岩 石	10.6	34	0.23	18.9	27.20
弱单元	2.1	34	0.23	10.9	25.16

3.2 计算结果分析

图 4 表示引水隧洞水平方向应力分布,图 4(b)~(e)为图 4(a)的局部放大图。由图 4(a)可知在引水隧洞边墙中部和拱顶出现了拉应力,但是岩体受拉应力作用范围很小。图 4(a)中,由于水平构造应力远小于竖直应力,每条引水隧洞四个角上出现了明显的水平应力降低区,尤其是在洞室左右上角离洞周 2.2～4.2m 的范围内,由于节理的存在出现了水平应力下降集中区,局部连成条带状应力下降集中区,如图 4(b)～(e)所示。

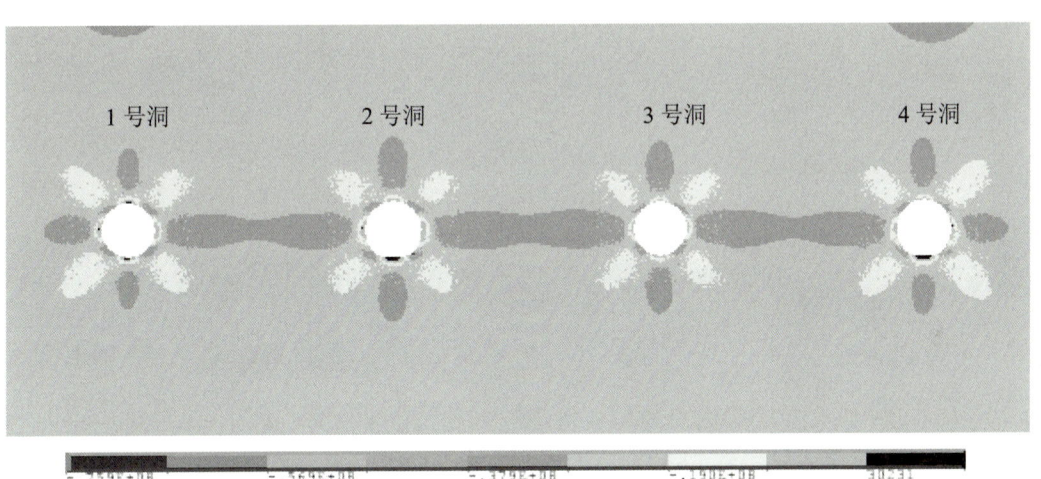

(a)四条引水隧洞水平方向应力总图

(b)1 号洞水平方向应力场

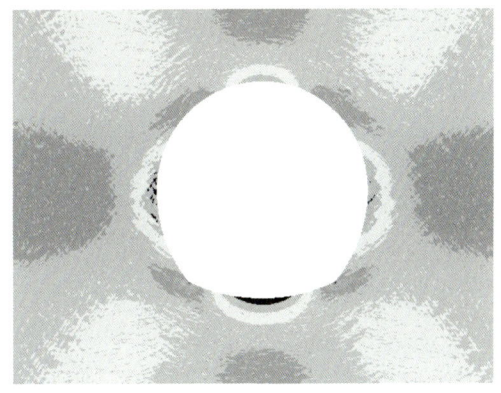

(c)2 号洞水平方向应力场

(d) 3号洞水平方向应力场

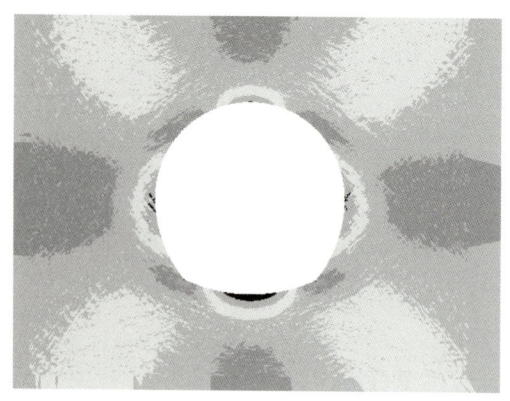
(e) 4号洞水平方向应力场

图 4 引水隧洞围岩水平方向应力图（单位：Pa）

Fig.4 The distributions of horizontal stress in the surrounding rock mass of diversion tunnel

图 5 表示引水隧洞围岩竖直应力分布图，分区破裂化模型计算所得的水平应力场是不连续的，分布也不均匀，出现明显的跳跃。

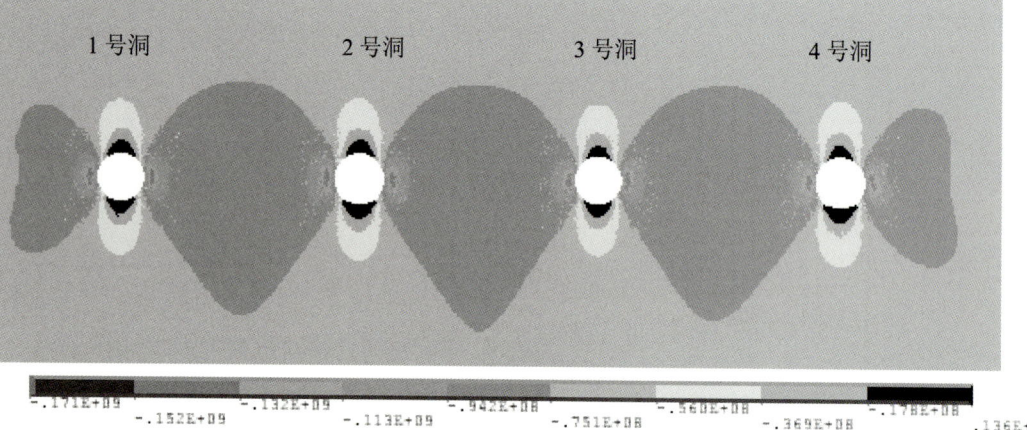
(a) 四条引水隧洞围岩竖直方向应力总图

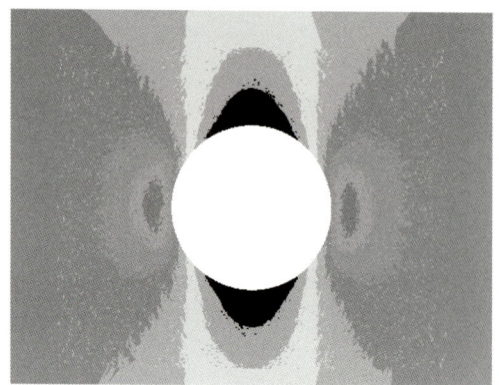
(b) 1号洞竖直方向应力场

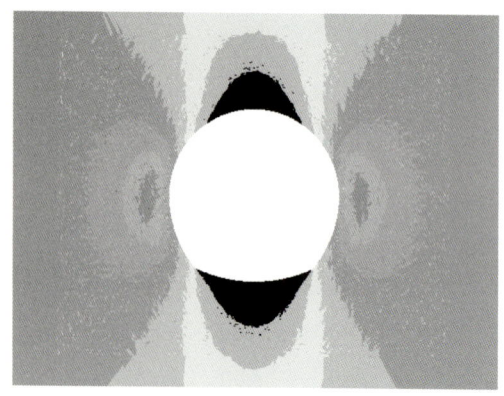
(c) 2号洞竖直方向应力场

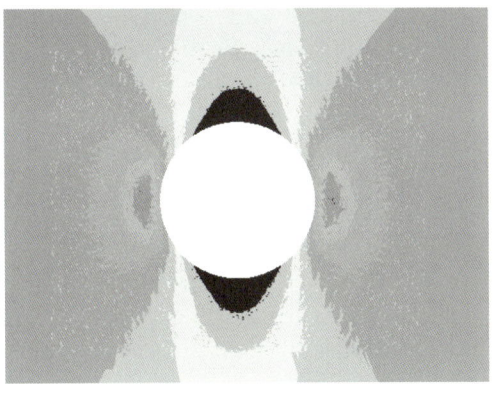
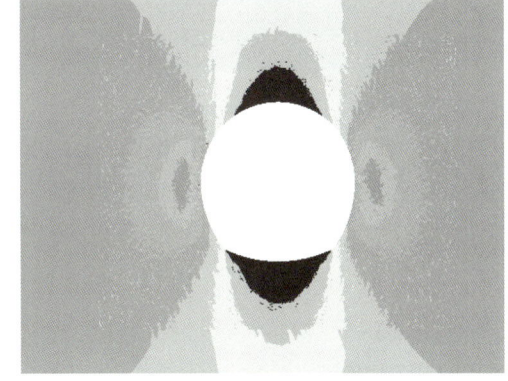

(d) 3号洞竖直方向应力场　　　　　　　(e) 4号洞竖直方向应力场

图 5　引水隧洞围岩竖直方向应力图（单位：Pa）

Fig.5 The distributions of vertical stress in the surrounding rock mass of diversion tunnel

图 6 为引水隧洞围岩分区破裂化分布图，由图 6 可知：每个引水隧洞都出现了较大的破裂范围，在四心拱引水隧道两侧边墙和离洞周一定距离处出现了比较明显的破裂带，在圆拱引水隧道的两侧边墙出现了滑移线型破裂现象，边墙破裂区最大宽度约为 2m。

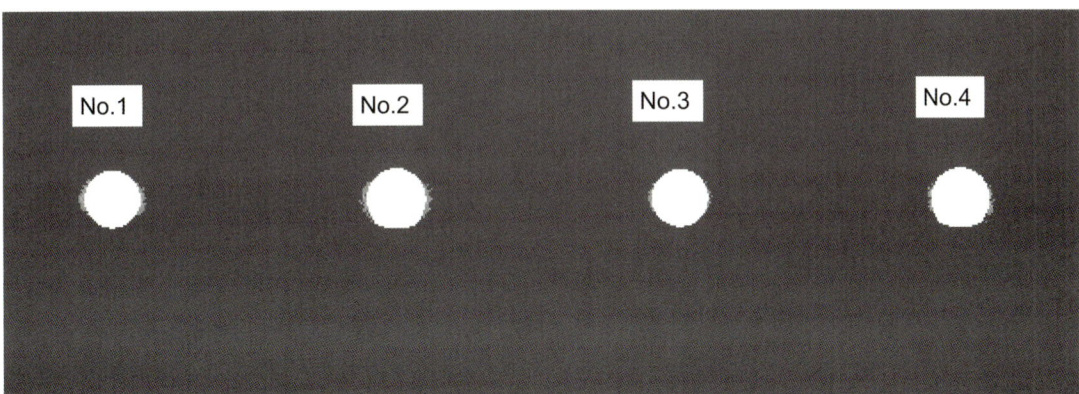

图 6　四条引水隧道的分区破裂化现象

Fig.6 Distributions of zonal disintegration around four diversion tunnels

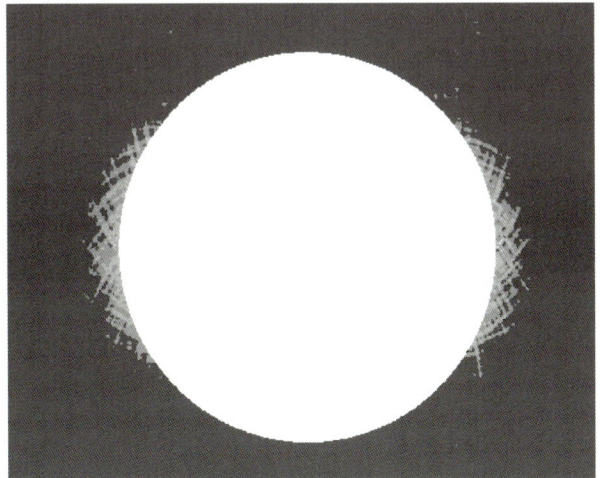

(a) 1号引水隧道分区破裂现象的放大图

(a) Enlarged photo of zonal disintegration around No.1 tunnel

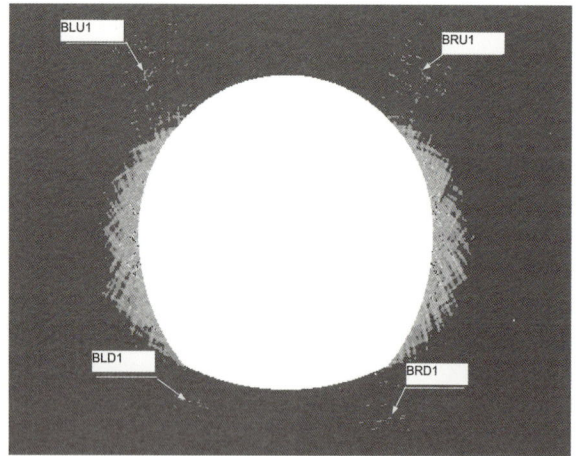
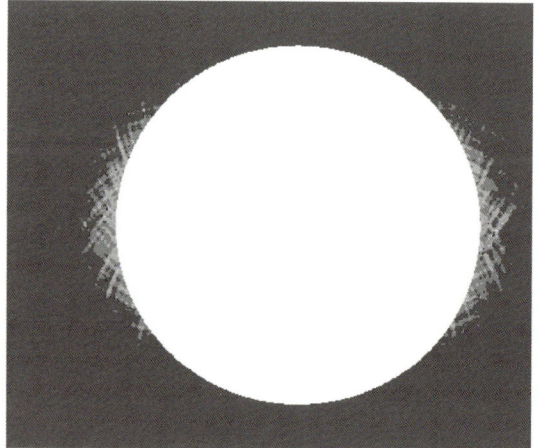

(b) 2号引水隧道分区破裂现象的放大图
(b) Enlarged photo of zonal disintegration around No.2 tunnel

(c) 3号引水隧道分区破裂现象的放大图
(c) Enlarged photo of zonal disintegration around No.3 tunnel

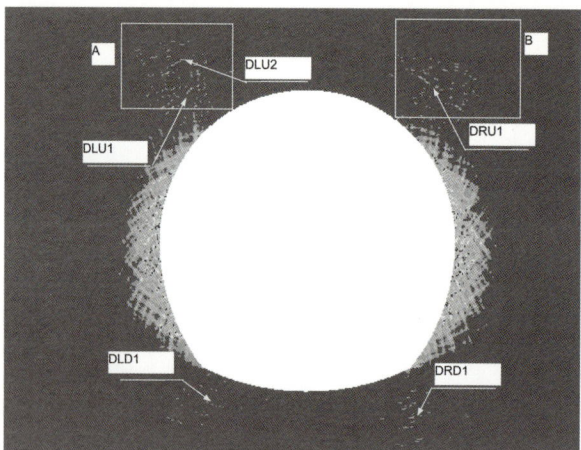

(d) 4号引水隧道分区破裂现象的放大图
(d) Enlarged photo of zonal disintegration around No.4 tunnel

图 7 四条引水隧道分区破裂化现象放大图
Fig.7 The enlarged photo of the distributions of disintegration zone within the surrounding rock mass around diversion tunnel

(a) 在四号引水隧道 A 区的分区破裂化现象放大图
(a) Enlarged photo of zonal disintegration in area A around No.4 tunnel

(b) 在四号引水隧道 B 区的分区破裂化现象放大图
(b) Enlarged photo of zonal disintegration in area B around No.4 tunnel

图 8 在四号引水隧道 A 和 B 区的分区破裂化现象放大图
Fig.8 Enlarged photo of disintegration zone in area B and B around No.4 tunnel

图 8 表示在四号引水隧道 A 和 B 区的分区破裂化现象放大图。表 2 表示 2 号和 4 号引水隧道左右上、下角破裂区的位置、数量和长度。从图 7 和图 8 可以知道隧道形状对分区破裂化现象有明显影响，圆形隧道破裂区数量少而且范围小，而四心圆隧道破裂区数量多而且范围大。

表 2 2 号和 4 号洞左右上、下角破裂区的位置、数量和长度

Table2 Magnitude and distribution of disintegration zones far from diversion tunnel

引水隧道	破裂区数量	破裂区位置	破裂区方位	离洞的距离（m）	破裂区长度（m）	破裂区宽度（cm）
2 号洞	BLU1	左上角	切向	2.03	2.12	6~7
	BRU1	右上角	切向	2.12	3.04	6~8
	BLD1	左下角	切向	1.29	1.69	8~16
	BRD1	右下角	切向	1.95	1.96	8~16
4 号洞	DLU1	左上角	切向	1.65	3.65	5~6
	DLU2	左上角	切向	2.97	1.26	6~10
	DRU1	右上角	切向	2.14	3.02	5~8
	DLD1	左下角	切向	1.21	0.74	5~6
	DRD1	右下角	切向	1.95	0.65	10~12

4 对比分析

岩体在自重及其他因素作用下处于一定的初始应力状态，地下洞室的开挖改变了岩体的初始应力状态，使岩体应力重新分布，洞室围岩就处于二次应力状态当中。分析初始应力与二次应力状态，目前主要采用弹、塑性理论的方法，但运用传统的弹塑性力学方法必将遇到有关岩体介质的假设条件问题。然而，对于特殊的、局部的岩体不连续面，由于其规模大、或产状不利或强度极低等原因，应该将其作为特殊的问题进行研究。因此，研究深部岩体的力学性能，必须考虑岩体的非均质性。本节将本文的计算结果与传统的弹塑性解和室内模型试验结果进行了对比，进一步说明了考虑岩体非均质性的重要性。

4.1 与弹塑性理论计算结果对比

本节在平面应变条件下将建立在传统弹塑性理论上的应力分布与建立在深部岩体分区破裂化模型上的应力分布进行对比分析，并进一步阐释分区破裂化产生的机理和规律。运用锦屏二级水电站引水隧洞围岩的相关力学参数，分别采用弹塑性力学的传统方法和分区破裂化模型的方法进行分析。

传统的弹塑性方法考虑隧洞围岩为完整岩体，无节理发育。传统的弹塑性理论模型计算参数为：容重 $\gamma = 27.2 kN/m^3$，弹性模量 $E = 18.9 GPa$，泊松比 $\upsilon = 0.23$，黏聚力 $C = 10.6 MPa$，内摩擦角 $\beta = 34^0$。

荷载为：垂直应力 $\sigma_v = 69.5 MPa$，水平应力 $\sigma_h = 23 MPa$。

图 9 和图 10 为平面应变弹塑性岩体模型下水平与竖直方向应力分布图。如图 9 所示，从弹塑性方法计算所得的水平方向应力分布来看，洞口附近出现一定范围的拉应力，洞室的四角范围内围岩出现应力减小的情况，而在洞室边墙附近围岩应力出现增加的情况。如图 10 所示，从竖直方向应力分布来看，洞室顶部与底部出现一定的拉应力区，洞室边墙围岩出现较大的压应力。与图 4 和图 5 相比较，破裂前，分区破裂化模型计算所得的水平最大拉应力比弹塑性模型小，压应力值反而增大。竖直方向拉应力与压应力最大值都比弹塑性模型大。

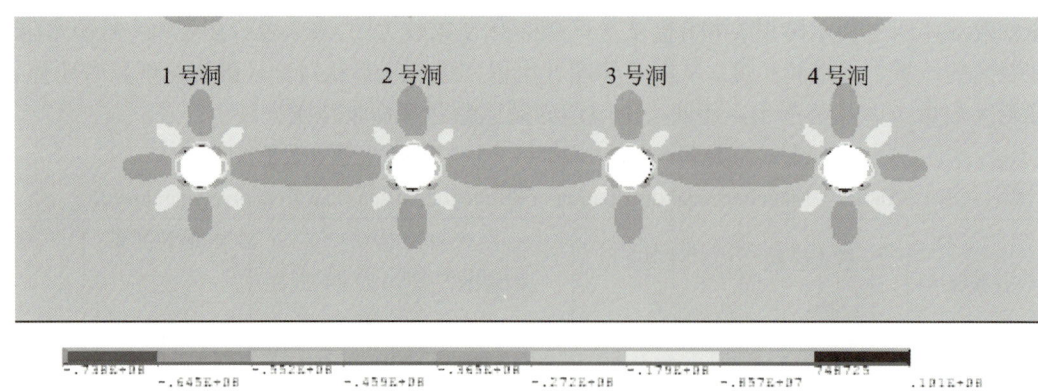

图 9 弹塑性模型水平方向应力分布（单位：Pa）

Fig.9 The distributions of horizontal stress calculated by elasto-plastic theory

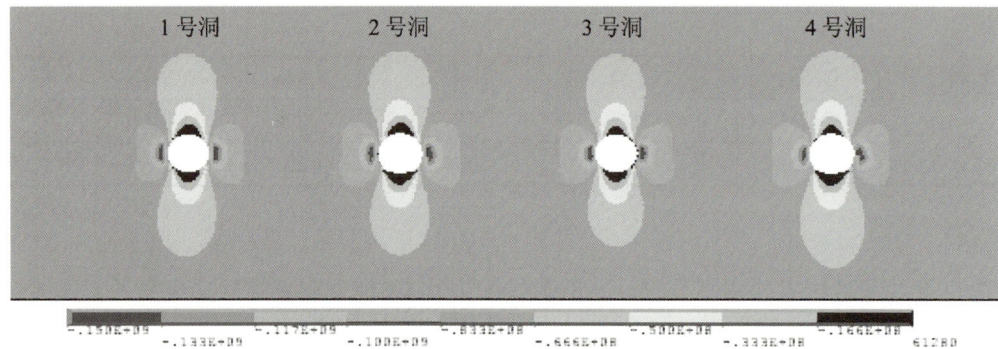

图 10 弹塑性模型竖直方向应力分布 （单位：Pa）

Fig.10 The distributions of vertical stress calculated by elasto-plastic theory

从单个洞室来看，应力场的分布有较大的不同。现取 1 号洞进行对比分析。图 11 和图 12 为 1 号洞在两种模型下洞室左上角水平与竖直方向上的应力分布。从图 11 可以看出，在弹塑性模型中，水平应力场分布是连续的；在分区破裂化模型中，洞室左上角距离洞周表面 2.2～4.2m 的范围内，出现了水平应力下降集中区，局部连成条带状应力下降集中区，整个洞室围岩应力分布不连续，出现明显的跳跃。

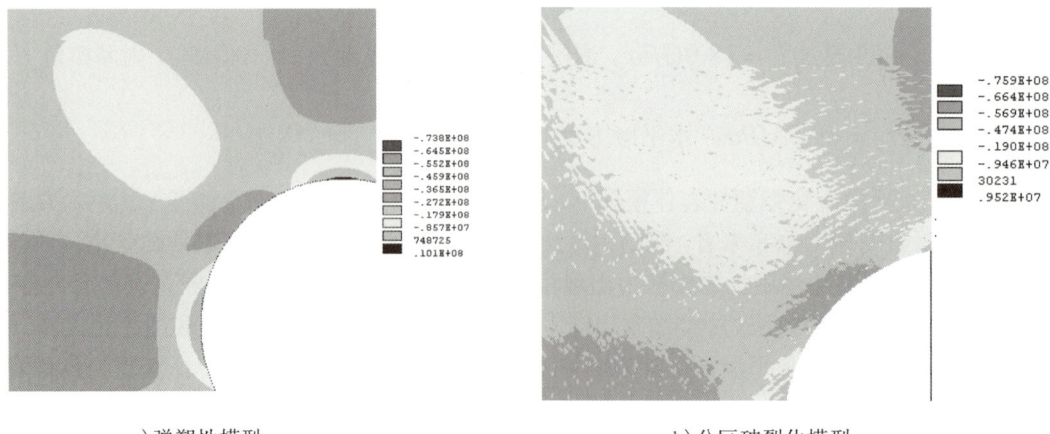

a)弹塑性模型　　　　　　　　　　　　b)分区破裂化模型

图 11 1 号洞两种模型水平方向应力分布对比(单位：Pa)

Fig.11 Comparison of the distributions of horizontal stress near No.1 tunnel between two different models

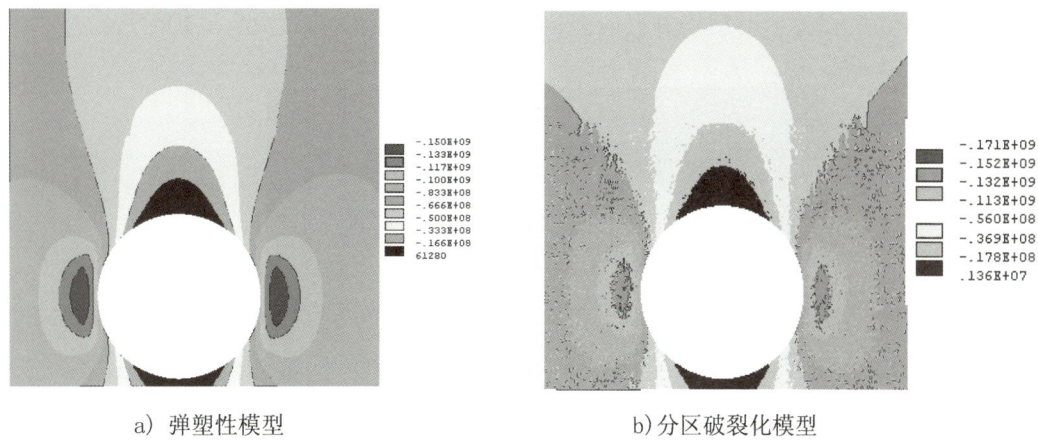

a) 弹塑性模型　　　　　　　　b) 分区破裂化模型

图 12　1 号洞竖直方向应力分布（单位：Pa）

Fig.12　Comparison of the distributions of vertical stress near No.1 tunnel between two different models

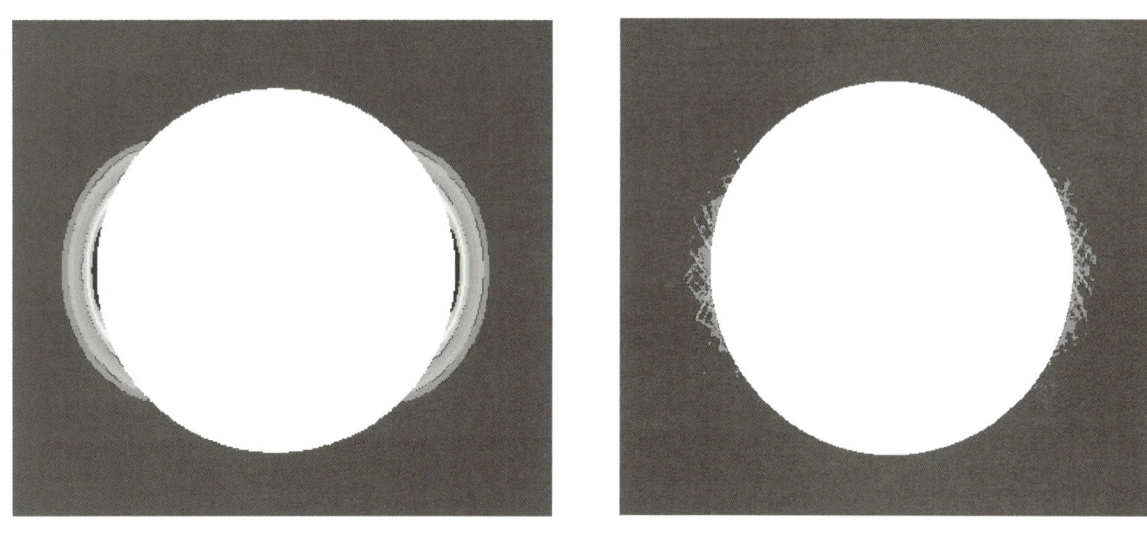

(a) 弹塑性模型　　　　　　　　(b) 分区破裂化模型

图 13　3 号洞围岩破裂区放大图

Fig.13　Enlarged photo of disintegration zone near No.3 tunnel

如图 12 所示，从竖直方向应力分布来看，在弹塑性模型中洞室顶部与边墙的应力场的变化比较均匀，应力场是连续的。而在分区破裂化模型中，由于岩体中节理分布的不均匀性，使洞周围围岩应力场的变化不均匀，应力场是不连续的。

从图 13 可以看出，弹塑性模型不可能出现分区破裂化现象，而分区破裂化模型在一定的应力条件和岩体参数情况下，可能出现分区破裂化现象。从图 13 可知：与传统的弹塑性分区不同，本文计算得到了深部岩体特有的破裂区交替出现的结果。

4.2　与室内模型试验结果对比

本节建立了单洞四心圆模型，得到了四心圆隧洞围岩分区破裂化现象，如图 14 所示。计算参数如下：岩石：容重 $\gamma = 27.2\,kN/m^3$，弹性模量 $E = 18.9\,GPa$，泊松比 $\upsilon = 0.23$，黏聚力 $C = 10.6\,MPa$，内摩擦角 $\varphi = 30^0$；节理：容重 $\gamma = 25.16\,kN/m^3$，弹性模量 $E = 10.9\,GPa$，泊松比 $\upsilon = 0.23$，黏聚

力 C＝4MPa，内摩擦角 $\beta=30^0$；荷载：垂直应力 $\sigma_v=70MPa$，水平应力 $\sigma_h=40MPa$。

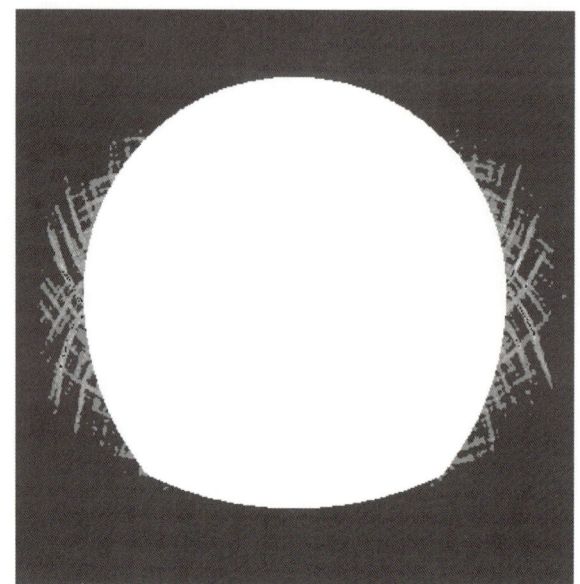

图 14　四心圆隧洞围岩破裂区图

Fig.14　Disintegration zone around four-arcs tunnel

从图 14 可见，围岩破裂区呈"滑移线"形式破裂，这与图 1 所示的顾金才院士室内模型试验得到的结果类似[7]。

5　敏感性分析

参数敏感性分析以单洞四心圆为例，二维平面应变模型范围为 60m×60m，单元总数为 121600 个，结点总数为 122240 个。图 15 表示四心圆单洞计算简图，图 15 中 σ_v 和 σ_h 分别代表竖直方向地应力和水平方向地应力。

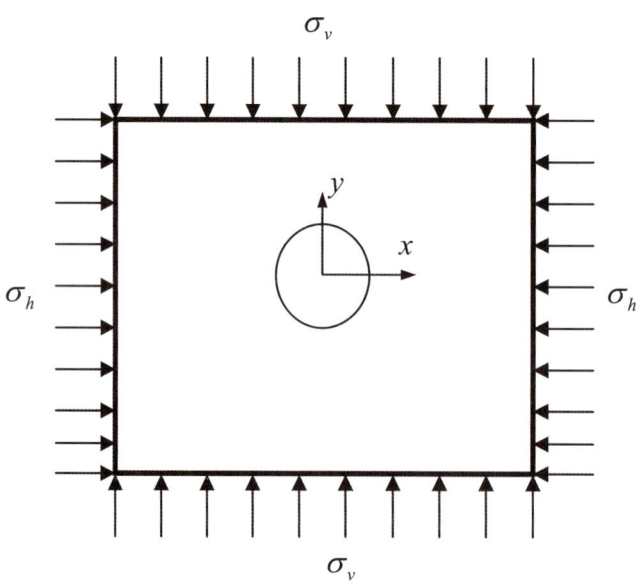

图 15　四心圆单洞计算简图

Fig.15　Model of single four-arcs tunnel

图 16 为不同垂直应力状态下隧洞围岩分区破裂化分析。计算参数为：岩石：容重 $\gamma = 27.2\,kN/m^3$，弹性模量 $E = 18.9\,GPa$，泊松比 $\upsilon = 0.23$，黏聚力 C＝16MPa，内摩擦角 $\varphi = 30^0$；节理：容重 $\gamma = 25.16\,kN/m^3$，弹性模量 $E = 10.9\,GPa$，泊松比 $\upsilon = 0.23$，黏聚力 C＝4MPa，内摩擦角 $\beta = 30^0$。岩石的单轴抗压强度为 55MPa。

(a) $\sigma_h = 50MPa, \sigma_v = 100MPa$

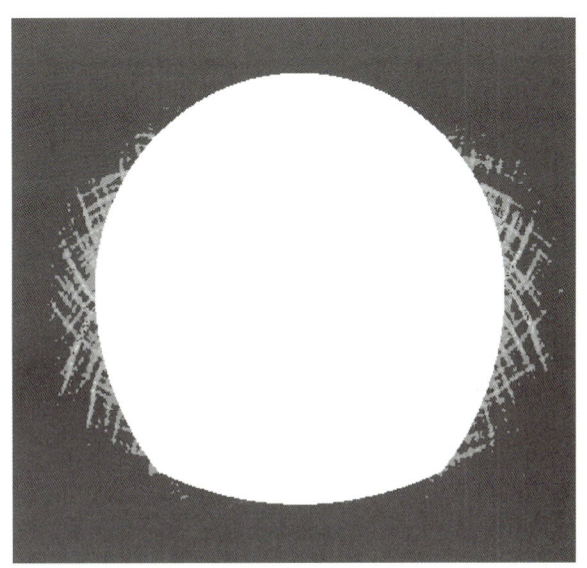

(b) $\sigma_h = 50MPa, \sigma_v = 85MPa$

(c) $\sigma_h = 50MPa, \sigma_v = 65MPa$

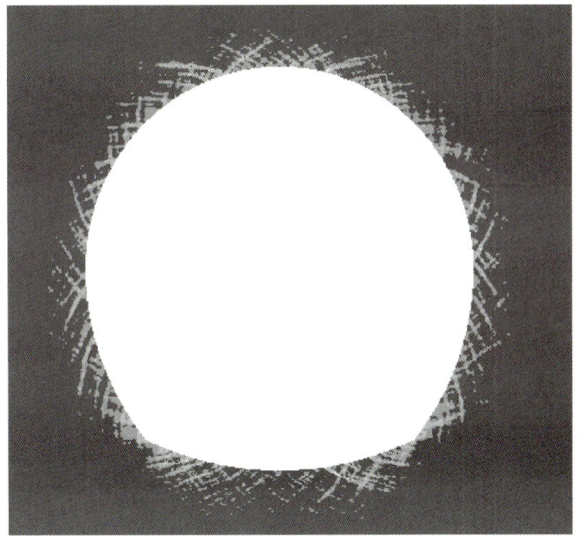

(d) $\sigma_h = 85MPa, \sigma_v = 85MPa$

图 16 不同应力状态下分区破裂化对比

Fig.16 Comparison of disintegration zone in different stress condition

由图 16a 可知，当地应力比岩体的单轴抗压强度大得多时，洞周围岩分区破裂化现象非常严重，以致于洞周围岩已经完全破坏，同时破裂区进一步转移到左右角位置。这是我们发现的最新现象，其他研究者都没有发现这种现象。顾金才院士和 Shemyakin 做实验时由于洞周有支护，所以洞周围岩没有完全破坏，破裂区仅仅出现在洞周围岩。对比图 16(a)和 16(b)，当水平应力小于垂直应力时，随着垂直应力的增加，分区破裂化现象越明显，破裂区的范围和数量越大；当水平应力大于垂直应力时，随着水平应力的增加，分区破裂化现象越明显，破裂区的范围和数量越大。如图 16（c）所示，当水平应力和垂直应力接近时，破裂带集中在洞周附近，洞周围岩呈明显的"滑移线"型破坏。

 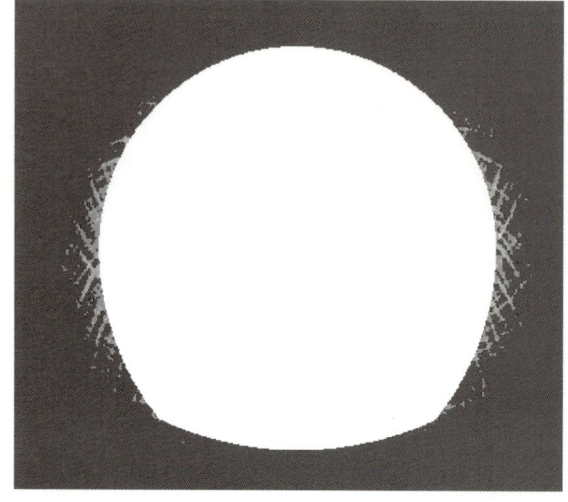

(a) 内摩擦角 $\beta = 30^0$ (b) 内摩擦角 $\varphi = 33^0$

图 17 不同岩体内摩擦角时隧洞围岩破裂区对比

Fig.17 Comparison of disintegration zone in surrounding rock mass between different angle of internal friction

图 17 表示内摩擦角对围岩分区破裂化的影响，计算参数为：岩石：容重 $\gamma = 27.2\,kN/m^3$，弹性模量 $E = 18.9\,GPa$，泊松比 $\upsilon = 0.23$，黏聚力 C=16MPa；节理：容重 $\gamma = 25.16\,kN/m^3$，弹性模量 $E = 10.9\,GPa$，泊松比 $\upsilon = 0.23$，黏聚力 C=4MPa，黏聚力 C=4MPa；荷载：垂直应力 $\sigma_v = 85\,MPa$，水平应力 $\sigma_h = 50\,MPa$。由图 17 可知：岩体内摩擦角对围岩分区破裂化具有重要影响，随着岩体内摩擦角的增加，洞室围岩的破裂区长度、数量和范围都减少。

图 18 表示不同黏聚力条件下围岩分区破裂化对比，计算参数为：岩石：容重 $\gamma = 27.2\,kN/m^3$，弹性模量 $E = 18.9\,GPa$，泊松比 $\upsilon = 0.23$，内摩擦角 $\beta = 30^0$；节理：容重 $\gamma = 25.16\,kN/m^3$，弹性模量 $E = 10.9\,GPa$，泊松比 $\upsilon = 0.23$，黏聚力 C=4MPa，内摩擦角 $\varphi = 30^0$；荷载：垂直应力 $\sigma_v = 85\,MPa$，水平应力 $\sigma_h = 50\,MPa$。

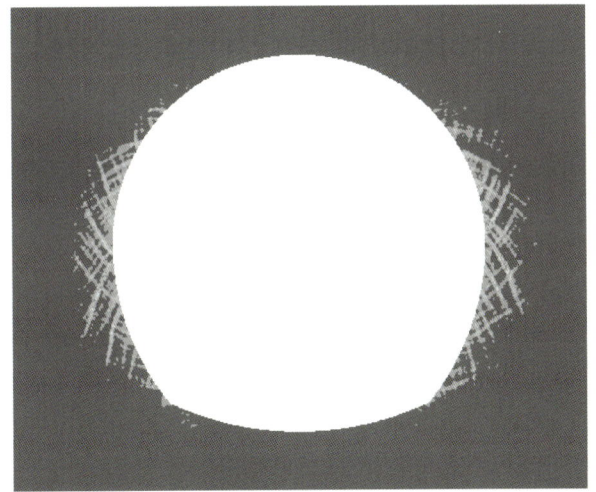 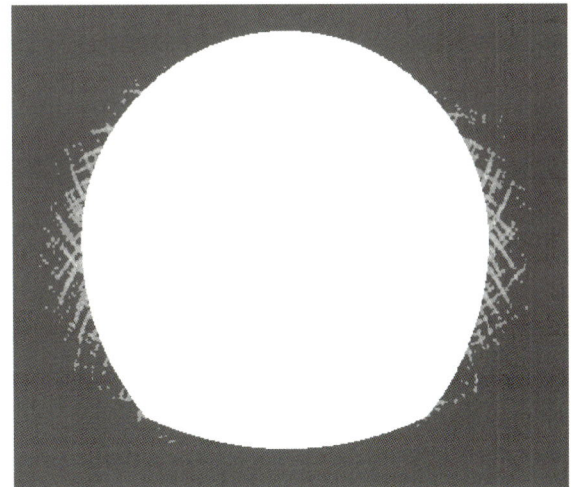

(a) 岩石黏聚力 C=16MPa　　　　　　　　　　(b) 岩石黏聚力 C=21MPa

图 18　不同节理黏聚力时隧洞围岩破裂区对比

Fig.18　Comparison of disintegration zone in surrounding rock mass between different cohesion

由图 18 可知：黏聚力对破裂区长度、数量和范围的影响明显，黏聚力越小，节理之间越容易贯通，破裂区长度越长，破裂区数量越多。

6　结论

本文利用弱单元数值模拟方法，结合锦屏二级水电站实际情况，对引水隧洞围岩分区破裂化进行了初步探讨得到了以下几点结论：

（1）利用若单元法，对锦屏二级水电站引水隧洞围岩分区破裂化进行了模拟，得到了深部岩体分区破裂化的几何特征；

（2）由于岩体的非均质性使得岩体围岩破裂形状与传统的弹塑性理论结果不同，由于考虑了岩体的非均质性，本文得到的应力分布表现为不连续，且洞室围岩的破裂多呈"滑移线"型，这与室内模型试验结果吻合；

（2）通过参数敏感性分析，研究了不同应力状态、黏聚力和内摩擦角条件下，四心圆洞室围岩分区破裂化规律。当水平应力小于垂直应力时，随着垂直应力的增加，分区破裂化现象越明显；当水平应力大于垂直应力时，随着水平应力的增加，分区破裂化现象越明显。随着节理内摩擦角的增加，洞室围岩的破裂区减少，且破裂区范围也减小。节理黏聚力越小，节理之间越容易贯通，破裂区长度越长。

参考文献

[1]钱七虎. 深部岩体工程响应的特征科学现象及"深部"的界定[J]. 华东理工学院学报，2004，27（1）：1-5.

QIAN QIHU. The characteristic scientific phenomena of engineering response to deep rock mass and the implication of deepness [J]. Journal of East China Institute of Technology, 2004,27(1):1-5.

[2] SHEMYAKIN E I, FISENKO G L, KURLENYA M V, et al. Zonal disintegration of rocks around underground workings, part I: data of in-situ observations [J]. Journal of Mining Science, 1986, 22(3):157-168.

[3] SHEMYAKIN E I, FISENKO G L, KURLENYA M V, et al. Zonal disintegration of rocks around underground workings, part II: rock fracture simulated in equivalent materials [J]. Journal of Mining Science, 1986, 22(4):223-232.

[4] SHEMYAKIN E I, FISENKO G L, KURLENYA M V, et al. Zonal disintegration of rocks around underground workings, part III: theoretical concepts [J]. Journal of Mining Science, 1987, 23(1):1-6.

[5] SHEMYAKIN E I, FISENKO G L, KURLENYA M V, et al. Zonal disintegration of rocks around underground workings, part

IV: practical applications [J]. Journal of Mining Science, 1989, 25(4):297-302.

[6] 潘一山，李英杰，唐鑫，张智慧. 岩石分区破裂化现象研究[J]. 岩石力学与工程学报. 2007，26（s1）：3335-3341.
PAN YISHAN, LI YINGJIE, TANG XIN, ZHANG ZHIHUI. Study on zonal disintegration of rock[J]. Chinese Journal of Rock Mechanics and Engineering, 2007,26(s1):3335-3341.

[7] 顾金才，顾雷雨，陈安敏，徐景茂，陈伟. 深部开挖洞室围岩分层断裂破坏机制模型试验研究[J]. 岩石力学与工程学报，2008，27（3）：433-438.
GU JINCAI, GU LEIYU, CHEN ANMIN, XU JINGMAO, CHEN WEI. Model test study on mechanism of layered fracture within surrounding rock of tunnels in deep stratum[J]. Chinese Journal of Rock Mechanics and Engineering, 2008,27(3):433-438.

[8] 周小平，钱七虎. 深埋巷道分区破裂化机制[J]. 岩石力学与工程学报，2007,26(5): 877- 895.
ZHOU XIAOPING, QIAN QIHU. Zonal fracturing mechanism in deep tunnel [J]. Chinese Journal of Rock Mechanics and Engineering, 2007,26(5):877-895.

[9] 陈国庆，冯夏庭，周辉，陈炳瑞，黄书岭，张传庆. 锦屏二级水电站引水隧洞长期稳定性数值分析[J]. 岩土力学，2007，28（s2）：417-422.
CHEN GUOQING, FENG XIATING, ZHOU HUI, CHEN BINGRUI, HUANG SHULING, ZHANG CHUANQING. Numerical analysis of the long-term stability of the seepage tunnel in Jinping II Hydropower Station[J]. Rock and Soil Mechanics, 2007,28(s2):417-422.

[10] 吴世勇，任旭华，陈祥荣，等. 锦屏二级水电站引水隧洞围岩稳定分析及支护设计[J]. 岩石力学与工程学报，2005, 24(20): 3777-3782.
WU SHIYONG, REN XUHUA, CHEN XIANGRONG, et al. Stability analysis and supporting design of surrounding rocks of diversion tunnel for Jinping Hydropower Station[J]. Chinese Journal of Rock Mechanics and Engineering, 2005,24(20): 3777-3782.

[11] ZHOU XIAO-PING, QIAN QI-HU, ZHANG BO-HU, Zonal disintegration mechanism of deep crack-weakened rock masses under dynamic load, Journal of Central South University of Technology(in press)

| 学术论文——隧道及地下工程

大直径盾构掘进风险分析及对特大直径盾构挑战的思考

钱七虎[1]，陈 健[2,3]

(1. 陆军工程大学，江苏 南京 210007；2. 中国海洋大学环境科学与工程学院，山东 青岛 266100；
3. 中铁十四局集团有限公司，山东 济南 250014)

摘要：目前，基于建设规模和设计功能的需求，特大直径盾构已多次应用于铁路、公路、城市轨道交通、地下管廊等领域隧道工程建设。初步统计了国内外 15.5 m 以上的特大直径盾构隧道工程，总结了大直径盾构在掘进过程中的主要风险及常见问题，包括主轴承损坏失效、盾构管片拼装脱出盾尾后上浮、刀具磨损随盾构直径增大而加剧、刀盘前泥饼粘结、渣土滞排、刀盘升温等，并结合案例分析其原因。针对特大直径盾构带来的事故风险，从土压盾构与泥水盾构主轴承密封问题、管片盾尾脱出后上浮风险、常压刀盘与常规刀盘的选择、泥饼粘结和渣土滞排难题、预探前方复杂地质、海中基岩爆破及注浆固结辅助处理等方面提出思考和建议。

关键词：大直径盾构；主轴承密封；管片上浮；刀具磨损；泥饼粘结；渣土滞排；常规刀盘；常压刀盘；地质预探；基岩爆破；注浆固结

DOI: 10.3973/j.issn.2096-4498.2021.02.001　　**文章编号**: 2096-4498(2021)02-0157-08

中图分类号: U 45　　**文献标志码**: A

Analysis of Tunneling Risks of Large-Diameter Shield and Thoughts on Its Challenges

QIAN Qihu[1], CHEN Jian[2,3]

(1. *Army Engineering University, Nanjing* 210007, *Jiangsu, China*; 2. *School of Environmental Science and Engineering, Ocean University of China, Qingdao* 266100, *Shandong, China*;
3. *China Railway 14th Bureau Group Co., Ltd., Jinan* 250014, *Shandong, China*)

Abstract: Presently, owing to the demand of construction scale and design function, many super-large-diameter shields have been widely used in tunnel engineering in railways, highways, urban rail transits, underground utility tunnels, and other areas. In this paper, a preliminary statistic on the super-large-diameter (more than 15.5 m) shield tunnel projects around the world is presented. The main risks of large-diameter shield tunneling, such as main bearing damage and failure, segment floating after segment assembly and removal from shield tail, cutter wear increasing with shield diameter, mud cake bonding on cutterhead, hysteretic discharge of muck, and cutterhead heating, are summarized. Additionally, related countermeasures are provided based on case analysis. Moreover, suggestions are put forward regarding main bearing seal of earth pressure balance shield and slurry shield, floating risk after segment assembly and removal of shield tail, cutterhead selection, mud cake bonding and hysteretic discharge of muck, pre-exploration of complex geology, and blasting and grouting consolidation of sea bedrock.

Keywords: large-diameter shield; main bearing seal; segment up-floating; cutter wear; mud cake bonding; hysteretic discharge of muck; conventional cutterhead; cutterhead with atmospheric pressure; geological pre-exploration; bedrock blasting; grouting consolidation

0 引言

随着隧道施工装备的不断革新，以及新工艺、新技术的不断推广和应用，盾构法作为隧道工程技术的首选引领着隧道工程向大埋深、大断面、长距离的方向发展[1-3]。

收稿日期: 2020-09-26；修回日期: 2020-12-25。
本文原载于《隧道建设(中英文)》(2021 年第 41 卷第 2 期)。

近年来,尽管应用盾构法已成功建成或在建诸如美国西雅图 SR99 隧道、香港屯门隧道、深圳春风路隧道(在建)、武汉三阳路隧道、济南黄河隧道(在建)等一批特大直径的海底隧道[4-5]和城市道路隧道工程,但由于项目多处于环境敏感区域、地质条件复杂、不确定因素多、施工难度大,建设过程中存在隐患和发生事故的可能性较大[6-7]。因此,结合工程实践中大直径盾构掘进出现的常见问题,研判分析工程建设风险,并提出有效的解决措施和建议,具有十分重要的现实意义。

1 国内外 15.5 m 以上特大直径盾构隧道工程初步统计

近年来,国内外 15.5 m 以上特大直径盾构隧道工程见表 1。

表 1 国内外 15.5 m 以上特大直径盾构隧道典型工程实例

Table 1 Tunnel projects using super-large shield with diameter over 15.5 m around world in recent decade

建设年份	隧道名称	盾构掘进长度/km	盾构直径/m	盾构数量/台	盾构类型	主要地质情况	隧道用途
2011—2012	意大利 Sparvo 隧道	2.6+2.5	15.55	1	土压盾构	黏土、泥岩、砂岩	公路隧道
2011—2019	美国西雅图 SR99 隧道	2.8	17.45	1	土压盾构	黏土、粗粒砂岩	公路隧道
2013—2015	香港屯门—赤鱲角隧道	0.8+4.2	17.63/14.00	2	泥水盾构	粉砂层、砂层、花岗岩	公路隧道
2013—2018	武汉三阳路隧道	2.59+2.59	15.76	2	泥水盾构	粉细砂、砾岩、粉砂质泥岩	公轨合建
始建于 2016	意大利圣塔露琪亚隧道	7.5	15.87	1	土压盾构	黏土、砂层、强中风化石灰岩、泥灰岩、粉砂岩等	公路隧道
始建于 2016	上海北横通道	2.76+3.66	15.53	1	泥水盾构	黏土、粉质黏土、粉砂	公路隧道
2017—2021	深圳春风路隧道	3.6	15.80	1	泥水盾构	强中风化花岗岩、石英岩	公路隧道
2017—2021	济南黄河隧道	2.6+2.6	15.74	2	泥水盾构	粉细砂、中粗砂、黏土、钙质结核、黏土夹碎石	公轨合建
始建于 2017	日本东京外环隧道	9.1+9.1+7+7	16.10	4	土压盾构	黏土、砂、砂砾	公路隧道
始建于 2017	武汉和平大道南延隧道	1.4	16.03	1	泥水盾构	石英砂岩、灰质岩、钙质岩、泥岩、黏土	公路+地下管廊
始建于 2018	澳大利亚墨尔本西门隧道	4+2.8	15.60	2	土压盾构	砂层与黏土层、玄武岩	公路隧道
始建于 2019	北京东六环改造工程	7.6	16.07	2	泥水盾构	砂层、粉土夹黏性土互层、局部砾石	公路隧道

2 大直径盾构掘进主要风险及常见问题分析

2.1 主轴承损坏失效

2.1.1 主轴承直接损坏

主轴承直接损坏的形式有以下 4 种:

1)润滑失效造成的轴承疲劳。据调查统计,润滑失效约占轴承损坏成因的 50%,润滑不良是造成轴承过早损坏的最主要原因。

2)选型不当或荷载过大造成的轴承失效。盾构主轴承的不恰当选型、轴承短期严重偏离正常工况工作或长期超负荷运行,都会直接造成轴承失效。

3)微动磨损造成的轴承套圈破坏。即由于轴承滚道面和滚动体接触面间相对微小滑动而产生的磨损,长期反复小振幅的摇摆运动和过盈量不足形成变色磨痕,轴承套圈破坏最终直接导致主轴承的损坏。

4)接触疲劳造成的轴承主要部件失效。轴承表面的摩擦损失几乎都变为热量,主要受力件温度上升会造成运转时的冲击荷载、振动和噪声的加剧,叠加积累的接触疲劳大小决定了主轴承的使用寿命。

通过正确的组合安装和维护保养,轴承的前 3 种失效形式是可以部分或全部避免的。轴承在运转过程中,滚动件与滚道接触面的接触应力作用始终存在,滚动接触疲劳造成的损坏贯穿轴承的整个使用寿命周期,接触疲劳带来的损坏也就成为轴承使用过程中唯一不可避免的失效形式。

盾构选型时应从有效降低应力水平入手,主轴承直径不能太小,直径验算设计应综合考虑负荷承载量;盾构掘进中应控制循环作业时间,盾构推进距离不宜太长,避免或减缓主轴承接触疲劳造成的盾构损坏。前者涉及主轴承直径,后者涉及盾构推进距离(时间)。

2.1.2 密封失效引起的轴承损坏

主驱动密封系统是主轴承的关键部件,特别是位于主轴承前端的主驱动轴承密封,其主要作用是阻止盾构主驱动刀盘内的渣土进入主轴承齿轮箱内部和对主驱动轴承回转滑动机构、密封部位以及与泥砂接触的机构进行冲洗和润滑,密封一旦失效,泥砂进入轴承会造成主轴承损毁并直接导致盾构无法掘进。

2.1.2.1 密封机制

不同形式的主驱动密封系统虽有不同特点,但基本都由多道单唇密封、1道或多道双唇密封和迷宫密封组成,并在唇形密封间设置腔室且充满弹性材料。

2.1.2.2 密封失效案例

1)油脂量不足,油脂溢出压力不足。广深港直径为11.182 m的泥水盾构由于油脂注入不到位,在0.7 MPa水压下,泥浆通过迷宫密封进入轴承密封,轴承滚子及滚道表面在夹杂硬质小颗粒后出现压痕,焊接刀盘时搭铁线未严格按标准放置,致使主轴承滚子和滚道面接触处有大电流通过产生电熔蚀坑,两者在重载作用下造成了主轴承的逐步损坏(轴承直径为4 800 mm)。

2)水土压力过大,超过油脂阻力。在美国西雅图直径17.45 m土压盾构施工中,出现了结泥饼、压力不均、局部土压力增大、刀盘处高温、油脂阻力减小等现象,土舱渣土颗粒进入轴承内部,直接导致主轴承密封失效,较短时间内主轴承发生严重损坏。

2.2 盾构管片拼装脱出盾尾后上浮

盾构管片拼装脱出盾尾后上浮的原因分析[8]如下:

1)盾构隧道直径越大,成型的管片环直径也越大,管片脱出盾尾后,在不能及时填充或浆液凝结速度较慢时,成型密封的管片环在高密度的液态砂浆中受到浮力要更大。

2)盾构直径增大,盾构壳体厚度加大(盾构结构刚度要求),盾尾间隙也适当加大,致使管片外圈与开挖土体间的空隙加大,砂浆填充空间及填充量也在增大,导致管片上浮空间和上浮量也随之增大。

3)隧道直径越大,径向高度越大。由于盾构主机长度基本没有变化,致使盾构主机的高长比加大,更易造成盾构"栽头",使盾尾后方同步注浆浆液流入前舱,脱出盾尾管片上部外圈砂浆填充不密实,加大了成型隧道上浮的趋势。

2.3 刀具磨损随盾构直径增大而加剧

大直径盾构在砂卵砾石地层中推进时,刀具磨损问题格外突出,主要体现在:

1)盾构刀具在同样进尺条件下,其磨损长度与刀具配置部位半径成正比,随着盾构直径的增大,刀具轨迹长度增加,刀具磨损加剧,如南京纬七路直径为14.93 m的大直径盾构的刀具磨损是直径为6.3 m盾构刀具磨损的2.5倍。

2)大直径盾构掘进过程中遇复合地层的可能性更大,刀具配置适应性复杂,更易引起刀具磨损问题,在石英含量高的砂卵石地层中,大直径盾构刀具的磨损可达软土地层中小直径盾构刀具磨损的10倍[1]。

2.4 刀盘前泥饼粘结、渣土滞排、刀盘升温

常规大直径盾构刀盘开口率分布均匀,常压刀盘中心部位一定半径内开口率几乎为零。如果设置开口部位刀盘开口率的均匀度不好(或不合理),将导致刀盘前泥饼粘结、渣土滞排、刀盘升温等风险增大[9]。

典型事例为武汉三阳路长江隧道。越江区间采用直径为15.76 m泥水盾构施工,配有全断面常压可伸缩刀盘,开口率为29%,安装了全断面可更换滚刀,中心区域直径5 m内无开口,在施工过程中刀盘前泥饼粘结严重。三阳路长江隧道盾构见图1,盾构结泥饼见图2[10]。

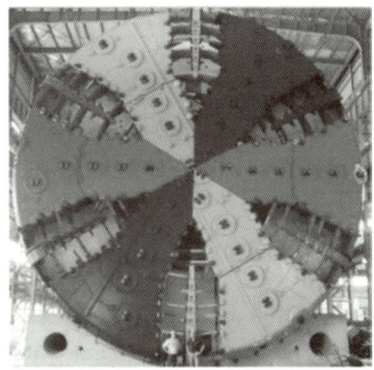

图1 武汉三阳路长江隧道盾构

Fig. 1 Shield used in Wuhan Sanyang Road tunnel

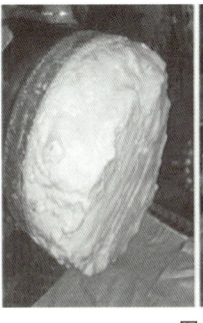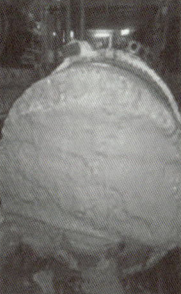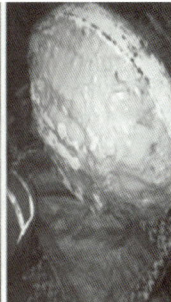

图2 盾构结泥饼现象

Fig. 2 Phenomenon of muck cake

3 对特大直径盾构挑战的思考

3.1 特大直径盾构带来的事故风险

由于工程客观需求,盾构直径面临越来越大的挑战,必须重视由此带来掘进事故风险的防范。特大直径盾构由于盾构直径的增大,主要可能引起的掘进事故风险如下:

1)随着盾构刀盘直径的增加,荷载也随之增大,所配置的主轴承尺寸若不能完全匹配,将导致接触疲劳引起的轴承损坏风险增大;随着盾构直径增大,渣土

舱压力分布不均的概率增大,若主轴承前舱局部压力超出密封和油脂阻力,也会导致主轴承密封失效风险概率增大。

2)随着盾构直径的增大,管片脱出盾尾后受到的浮力比常规盾构隧道管片受到的浮力更大,控制隧道上浮的风险和难度加大。

3)由于盾构直径增大后刀具运行轨迹也随之增大,刀具磨损损坏的概率和风险大大增加。尤其是在复合地层中掘进时,刀具的磨损速度极快,刀具更换次数的增加会造成工期延长、成本增加、安全风险等系列施工问题发生的可能性增大。

4)随着盾构直径的增大,由于受到制造的限制及影响,刀盘中心的开口率也随之降低,在掘进中,使盾构刀盘结泥饼的风险和概率大大增加。

3.2 土压盾构与泥水盾构主轴承密封问题

盾构选型是影响隧道工程成败的重要环节,是盾构施工的关键[2],需要考虑地层条件、地下水位、隧道埋深、开挖面稳定、设计隧道的断面、衬砌类型、工期、工程造价等。伊斯坦布尔海峡公路隧道工程盾构的正确选型和西雅图市 SR99 公路隧道工程盾构的错误选型带来了不同的工程效果[11]。

3.2.1 土压盾构主轴承密封问题

土压盾构适用于稳定性较好、透水性不强的地层以及水土压力较小、盾构驱动功率较大、耐压能力相对较弱的情况。土压盾构结构见图 3。

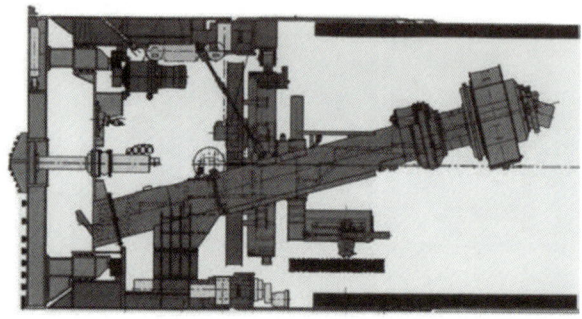

图 3 土压盾构结构

Fig. 3 Structure of earth pressure shield

1)意大利 Sparvo 隧道:双洞双向 6 车道,采用土压盾构施工,盾构直径为 15.55 m,盾构驱动功率达 12 000 kW,刀盘开挖力为 315 000 kN,驱动转矩为 94 793 kN·m。意大利 Sparvo 隧道盾构见图 4。

2)美国西雅图 SR99 隧道:采用土压平衡盾构施工,盾构直径为 17.45 m,主轴承直径为 8.0 m,装机功率为 12 135 kW。美国西雅图 SR99 隧道盾构见图 5[11]。

土压盾构负荷重,驱动功率偏大,主轴承加工尺寸大,特大直径土压盾构主轴承密封以聚氨酯密封结构为主。其主要优缺点如下。

1)优点:采用水冷却降温方式,油脂消耗量小,节约成本和降低能耗。

2)缺点:无法建立备压措施,耐压能力相对偏弱;一旦密封损坏或失效,必须吊出更换修复。如美国西雅图 SR99 土压盾构密封损坏,为吊出更换修复而建设吊出井,付出了较大的代价。

图 4 意大利 Sparvo 隧道盾构

Fig. 4 Shield used in Sparvo tunnel in Italy

图 5 西雅图 SR99 隧道盾构

Fig. 5 Shield used in SR99 tunnel in Seattle

3.2.2 泥水盾构主轴承密封问题

泥水盾构适用于稳定性较差、透水性较强的地层以及水土压力较高、盾构驱动功率较小、耐压能力相对较强的情况。泥水盾构结构见图 6。

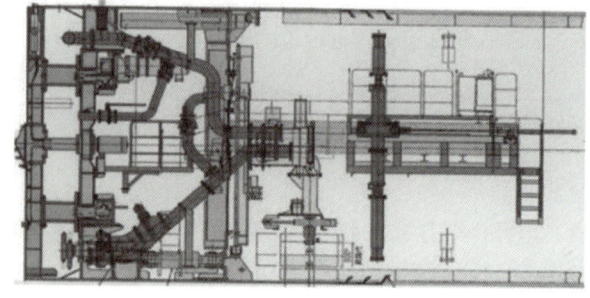

图 6 泥水盾构结构

Fig. 6 Structure of slurry shield

香港屯门隧道：采用泥水盾构施工，盾构直径为17.63 m，主轴承直径为7.6 m，装机功率为8 600 kW。香港屯门隧道盾构见图7。

图7 香港屯门隧道盾构

Fig. 7 Shield used in Tuen Mun Tunnel in Hong Kong, China

泥水盾构负荷小，驱动功率相对较小，主轴承加工尺寸也小。特大直径泥水盾构主轴承密封以唇形密封结构为主。其主要优缺点如下。

1) 优点：采用油脂备压方式，运用成熟高水压条件下的耐压密封体系，能实现原位带压更换作业，且维修操作安全。如济南黄河隧道盾构原位密封修复，已为其他项目提供了借鉴。

2) 缺点：油脂消耗量较大，会相对增大能耗和掘进成本。

3.2.3 对盾构选型的建议

同级别的特大直径土压盾构相对于泥水盾构，渣土舱压力不均匀度及局部土压力更大，轴承承受荷载更大(如美国西雅图SR99土压盾构)，滚动体和滚道表面接触应力更大。因此，土压盾构主轴承密封失效及直接损坏(接触疲劳)概率更高，风险也更大。在特大直径盾构选型方面，尽管泥水盾构比土压盾构造价高，但土压盾构主轴承密封失效风险大，修复造价高，因此，仍建议选择泥水盾构。

3.3 管片盾尾脱出后上浮风险

管片盾尾脱出后上浮会造成管片裂隙、错台、破损、渗漏等施工病害，导致管片拼装质量达不到规范及设计要求，给建成后的运营安全和隧道维护带来巨大挑战。管片上浮病害预防和处理措施应尽早实施。造成管片上浮的原因有多种。

施工中，在管片环与地层之间难以避免地存在盾尾管片拼装空隙，通过加强管理，严格控制空隙量在规范允许范围的前提下，向空隙中泵入浆液，主动控制地层沉降的同时保持管片稳定，是解决以上问题的有效措施[12]。

据统计，管片上浮量的70%发生于安装后的48 h之内。除施工前考虑盾构掘进姿态预留、管片合理选型外，施工中应结合监控量测数据综合分析地质情况及外部影响因素，根据施工信息动态分析研究，适时调整浆液配合比，提高管片拼装精度等，以达到满足规范及设计要求的管片上浮控制效果。

从注浆凝固速度进行对策分析：由于盾构快速推进，必须保证注浆凝固时间小于盾构推进时间，可以通过预先试验来达到并检验。

3.4 常规刀盘与常压刀盘的选择

3.4.1 常规刀盘

当土压盾构地层稳定性较好、透水性不强、水土压力较小时多选用常规刀盘。常规刀盘开口率设计、耐磨设计和刀盘刀具配置应结合工程实际综合考虑。常规刀盘见图8。

图8 常规刀盘

Fig. 8 Conventional cutterhead

3.4.2 常压刀盘

当泥水盾构地层稳定性较差、透水性较强、水土压力较高以及穿越地层为土岩复合地层时，多选用常压刀盘[13-14]。常压刀盘见图9。

图9 常压刀盘

Fig. 9 Cutterhead with atmospheric pressure

在特大盾构工程实践中，为应对高水压、土岩复合地层、减少带压换刀风险，宜配置常压刀盘。常压换刀不需要带压进舱，换刀风险大大降低，但常压刀盘中心开口率小，甚至没有开口，从而增大了结泥饼的风险。

3.5 泥饼粘结、渣土滞排难题

解决泥饼粘结、渣土滞排等难题,多从如下方面入手:

1)盾构配置。针对软塑—硬塑易结泥饼地层,可加大刀盘开口率(缩小中心封闭区域范围),刀具多层次布置,强化切削功能,降低碾磨,尽可能使渣土成块排出;增加、增强刀盘中间结泥饼冲刷系统,创新内循环冲刷;配置刀盘温度自动监测系统和刀盘伸缩功能。

2)盾构掘进。要关注刀盘温度变化预警,加强、加大泥浆循环,保持拼装管片期间的泥渣循环。

3)辅助措施。创新破除泥饼技术,如水刀切割、分散剂(双氧水)化学剥离方法等;在气密性好的围岩下辅助气压作业。

刀盘结饼见图10。高压水刀现场喷射试验见图11。刀盘冲刷系统改造见图12。

图 10 刀盘结饼

Fig. 10 Muck cake on cutterhead

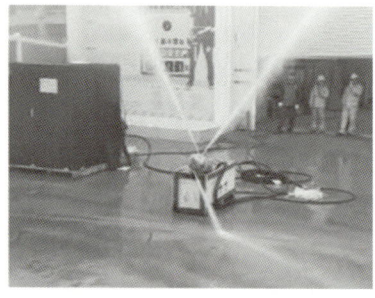

图 11 高压水刀现场喷射试验

Fig. 11 Field injection test of high pressure water knife

在选择冲刷方式时,采取分时集中冲刷和舱底顺流冲刷理念,能较好地解决高黏性地层刀盘粘结的施工难题。

3.6 其他思考

3.6.1 预探前方复杂地质

1)利用泥水盾构SSP(seismic scatter profile)超前探测系统加强预探前方地质分布情况。泥水盾构SSP超前探测系统见图13。

2)利用超前钻机系统超前钻探兼顾预处理措施。超前钻机系统见图14。

3)结合综合掘进参数超前预判前方地质变化。结合预探地质信息,可准确预测孤石基岩、断裂构造和软弱破碎围岩的位置,为后续爆破孤石或注浆固结围岩提供可靠的第一手资料。

(a)刀盘中心冲刷系统

(b)舱底冲刷系统

图 12 刀盘冲刷系统改造

Fig. 12 Retrofit of cutterhead brush system

图 13 泥水盾构 SSP 超前探测系统

Fig. 13 Advanced detection system SSP for slurry shield

图 14 超前钻机系统

Fig. 14 Advanced drilling system

3.6.2 高水压复杂地质特大直径盾构施工辅助处理技术

3.6.2.1 海中基岩爆破辅助处理技术

对施工中遇到的超强基岩、孤石等,采用预裂、微裂或提前爆破技术,实现盾构顺利推进[15]。

珠海横琴三通道马骝洲交通隧道工程,隧道全长2 834.6 m,采用直径为 14.93 m 泥水气压平衡盾构掘进施工,是我国首条海域超大直径复合地层盾构隧道。马骝洲海域岩面变化剧烈,抛石最大直径为 2.2 m,花岗岩层高为 6 m,强度高达 120 MPa。为确保盾构的顺利掘进,采取"物探+爆破"组合的方法进行预处理,对侵入隧道断面内的基岩采用水下爆破,爆破后爆破孔的封堵选用高压旋喷注浆方法并进行河床加固。该方法消除了基岩突起对盾构掘进的威胁,有效降低了盾构刀具的损坏,保证了盾构顺利掘进。珠海横琴三通道马骝洲交通隧道断面见图15。水下爆破封堵见图16。

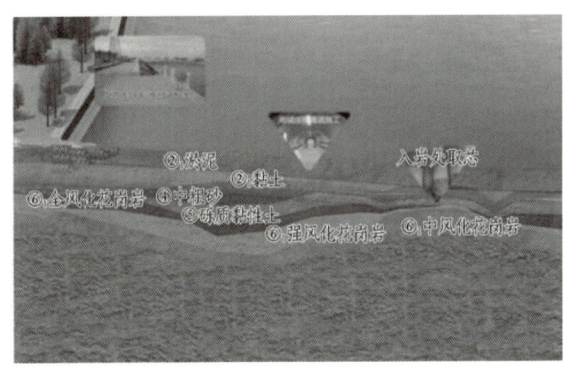

图 15 珠海横琴三通道马骝洲交通隧道断面图

Fig. 15 Cross-sectional view of Maliuzhou traffic tunnel in Hengqin Road, Zhuhai, China

图 16 水下爆破封堵

Fig. 16 Sealing of underwater blasting

3.6.2.2 海中注浆固结辅助处理技术

对施工中遇到的软弱破碎围岩,为防止发生工作面坍塌或卡机等事故,采用地面或水域预注浆固结处理、洞内盾构超前注浆固结等辅助措施,能有效地解决盾构掘进施工难题。

4 结语

盾构法以其明显的技术、经济、安全、环保等优势已成为地下工程首选的施工工法。近 20 年来,随着工程规模和功能的建设需要,盾构有明显向特大直径发展的趋势。

然而,隧道盾构施工是个复杂的系统工程,特大直径盾构往往因隧道掌子面围岩软硬分布不均匀导致施工难度更大,除要突破现有装备水平限制和面对施工技术的挑战外,更需要对施工掘进中常见的问题进行总结分析,及时优化方案,破解各类技术难题。

一系列工程统计数据证明:特大直径盾构不仅是工程建设中的难点、重点,更是业界关注的亮点,特大直径盾构在新技术、新工艺、新材料、新设备的引进、研发和推广应用中发挥了积极的推动作用。应用大直径盾构先进装备发挥其优良性能的同时,要结合工程实践对大直径盾构从整体选型、主轴承密封选择、刀盘设计等方面进行进一步的研究思考,科学研判,研发更多盾构施工的辅助技术,有效化解大直径盾构掘进可能出现的事故和风险,不断迎接新的更大挑战。

参考文献(References):

[1] 钱七虎. 水下隧道工程实践面临的挑战、对策及思考[J]. 隧道建设, 2014, 34(6): 503.

QIAN Qihu. Challenges in construction of underwater tunnels and countermeasures[J]. Tunnel Construction, 2014, 34(6): 503.

[2] 陈健, 黄永亮. 超大直径泥水盾构施工难点与关键技术总结[J]. 地下空间与工程学报, 2015, 11(增刊2): 637.

CHEN Jian, HUANG Yongliang. Summary of key technologies and construction difficulties in large diameter slurry shield tunnel[J]. Chinese Journal of Underground Space and Engineering, 2015, 11(S2): 637.

[3] 王吉云. 近十年来中国超大直径盾构施工经验[J]. 隧道建设, 2017, 37(3): 330.

WANG Jiyun. Super-large diameter shield tunneling technologies in China in recent decade[J]. Tunnel Construction, 2017, 37(3): 330.

[4] 周昆. 复杂地质超大直径越江盾构隧道:和燕路过江通道南段隧道工程A2标项目介绍[J]. 隧道与地下工程灾害防治, 2020, 2(1): 105.

ZHOU Kun. Super-large diameter shield tunnel crossing river in complex geology: Bid A2 of south section tunnel of Heyan Road river-crossing tunnel[J]. Tunnel and Underground Engineering Disaster Prevention and Control, 2020, 2(1): 105.

[5] 张亚果, 姚占虎. 超大直径盾构穿越长江上软下硬地层

刀具配置和优化研究与应用[J]. 现代隧道技术, 2018, 55(2): 180.
ZHANG Yaguo, YAO Zhanhu. On configuration and optimization of cutting tools for the supper-large shield passing through the upper-soft and lower-hard stratum[J]. Modern Tunnelling Technology, 2018, 55(2): 180.

[6] 王发民, 孙振川, 张良辉, 等. 汕头海湾隧道超大直径泥水盾构针对性设计及不良地质施工技术[J]. 隧道建设(中英文), 2020, 40(5): 735.
WANG Famin, SUN Zhenchuan, ZHANG Lianghui, et al. Design of super-large diameter slurry shield used in Shantou Bay tunnel and construction countermeasures for unfavorable geology[J]. Tunnel Construction, 2020, 40(5): 735.

[7] 曾铁梅, 吴贤国, 张立茂, 等. 公路地铁合建越江段大直径盾构隧道工程风险分析[J]. 城市轨道交通研究, 2016, 19(10): 18.
ZENG Tiemei, WU Xianguo, ZHANG Limao, et al. Construction risk analysis for cross-river section of large diameter highway/rail tunnel[J]. Urban Mass Transit, 2016, 19(10): 18.

[8] 牛占威, 张文新, 李云涛, 等. 超大直径海底盾构隧道施工管片上浮与开裂控制技术[J]. 建筑技术, 2019, 50(增刊2): 26.
NIU Zhanwei, ZHANG Wenxin, LI Yuntao, et al. Floating and cracking control technology of segment of super-diameter subsea shield tunneling [J]. Construction Technology, 2019, 50(S2): 26.

[9] 黄大为. 超大直径泥水盾构常压刀盘防结泥饼措施的探讨[J]. 建筑机械化, 2020, 41(3): 30.
HUANG Dawei. Discussion on the measures of super-diameter muddy water shield pressure knife plate to prevent mud pie [J]. Construction Mechanization, 2020, 41(3): 30.

[10] 周文波, 吴惠明, 赵峻. 泥岩地层常压刀盘盾构的掘进策略与分析[J]. 现代隧道技术, 2019, 56(4): 8.
ZHOU Wenbo, WU Huiming, ZHAO Jun. On driving strategy of the shield machine with atmospheric cutterhead in mudstone strata [J]. Modern Tunnelling Technology, 2019, 56(4): 8.

[11] 钟长平, 谢小兵, 刘智, 等. 超大直径盾构施工主要风险与对策[EB/OL]. (2018-10-10)[2020-11-08]. http://zgdgcy.com/index/show/catid/26/id/911.html.
ZHONG Changping, XIE Xiaobing, LIU Zhi, et al. Main risks and Countermeasures of super large diameter shield construction[EB/OL]. (2018-10-10)[2020-11-08]. http://zgdgcy.com/index/show/catid/26/id/911.html.

[12] 赵光, 付士陆, 牛浩. 超大直径泥水盾构隧道抗浮原理及措施综述[J]. 建设监理, 2018(2): 78.
ZHAO Guang, FU Shilu, NIU Hao. Review on anti-floating principle and measures of super-large diameter slurry shield tunnel[J]. Project Management, 2018(2): 78.

[13] 李凤远, 赵海雷, 冯欢欢. 超大直径泥水盾构施工风险防控方法研究[J]. 施工技术, 2019, 48(19): 100.
LI Fengyuan, ZHAO Hailei, FENG Huanhuan. Research on construction risk prevention and control method of super-large diameter slurry shield[J]. Construction Technology, 2019, 48(19): 100.

[14] 李沿宗. 水底隧道上软下硬地层超大直径盾构掘进技术难点与对策探讨[J]. 公路交通技术, 2019, 35(5): 106.
LI Yanzong. Analysis of technical difficulties and countermeasures for the super-large diameter shield tunneling in upper soft and lower hard strata of underwater tunnel[J]. Technology of Highway and Transport, 2019, 35(5): 106.

[15] 张兵, 陈桥, 王国安, 等. 海域含孤石地层超大直径泥水盾构始发关键技术研究[J]. 施工技术, 2019, 48(11): 68.
ZHANG Bing, CHEN Qiao, WANG Guo'an, et al. Research on key launching technology of super large diameter slurry shield machine in boulder stratum of some sea area[J]. Construction Technology, 2019, 48(11): 68.

建设城市地下综合管廊,转变城市发展方式

钱七虎

(解放军理工大学国防工程学院,江苏 南京 210007)

摘要:阐述建设地下综合管廊对推进新型城镇化、补齐城市基础设施短板、实现城市发展方式转变的重大意义;介绍世界发达国家、地区以及我国大陆建设地下综合管廊的情况;简述国务院推进地下综合管廊建设的重要决策部署。针对当前我国地下综合管廊的建设情况,提出以下建议。1)加强顶层设计,积极有序推进。2)完善相关配套政策,包括明确入廊要求、实行有偿使用、加大政府投入、完善融资支持等。3)具体措施:①借鉴国外先进经验,完善法规体系;②系统科学地编制好城市综合管廊规划;③建立统一管理、分工协作、衔接有序的综合管廊管理体制;④地下综合管廊兼顾人防应到位而不越位;⑤推广 PPP 模式,创新投融资机制。

关键词:地下综合管廊;城市发展方式;顶层设计;配套政策;法律法规;规划;管理体制;PPP 模式

DOI: 10.3973/j.issn.1672-741X.2017.06.001

中图分类号: U 45 **文献标志码**: A **文章编号**: 1672-741X(2017)06-0647-08

To Transform Way of Urban Development by Constructing Underground Utility Tunnel

QIAN Qihu

(College of Defense Engineering, PLA University of Science & Technology, Nanjing 210007, Jiangsu, China)

Abstract: The importance of the underground utility tunnel to the new-type urbanization, urban infrastructures and mode transformation of city development is presented. Then the state-of-the-art of construction of underground utility tunnel in mainland China and developed countries are introduced and the key policies provided by the State Council for progress of underground utility tunnel construction are summarized. Some suggestions are given as follows: 1) Strengthening the top-level design positively and orderly. 2) Improving relevant supporting policies, including construction requirements, policies of pay for use, increasing government investment and financing policy, etc. 3) Specific measures, i.e. learning the advanced abroad experiences and improving the regulation system, planning a general project for urban underground utility tunnel systemically and scientifically, establishing a overall management system for underground utility tunnel, giving consideration to civil air defence and promoting the PPP mode and innovating the investment and financing mechanisms, should be implemented.

Keywords: underground utility tunnel; city development mode; top-level design; supporting policies; laws and regulations; planning; management system; PPP mode

0 引言

地下综合管廊是指在城市地下用于集中敷设电力、通信、广播电视、给水、排水、热力、燃气等市政管线的公共隧道[1]。地下综合管廊是"城市地下管线综合体",是保障城市运行的重要基础设施和"生命线"。与发达国家和地区相比,我国地下综合管廊建设起步较晚。目前,我国正处在城镇化快速发展时期,地下基础设施建设相对滞后。进入 2014 年之后,地下综合管廊政策密集出台并不断加码、细化;同时,住建部在全国 36 个大中城市全面启动地下综合管廊试点工程,为后续城市地下综合管廊建设提供案例和参照,有望逐步提高全国建设地下综合管廊的积极性。

本文主要阐述建设地下综合管廊对转变城市发展方式的意义,比较国内外地下综合管廊建设情况,总结

收稿日期:2017-02-15。
本文原载于《隧道建设》(2017 年第 37 卷第 6 期)。

地下综合管廊建设和管理的成功经验,并结合我国当前综合管廊建设情况和相关配套政策,对地下综合管廊的规划设计、法律法规、投融资、建设、运营及管理等方面提出建议措施。

1 建设地下综合管廊对转变城市发展方式的重大意义

20世纪,世界经济快速发展引发了资源巨大消耗、环境污染以及"城市病"剧增。20世纪末,在里约召开的联合国全球环境首脑会议提出了社会和城市的可持续发展问题,讨论并通过了《21世纪议程》。《中国21世纪议程》承诺走可持续发展之路:党的十六大提出建设资源节约型、环境友好型社会;十七大提出建设人与人、人与自然和谐发展的社会;十八大提出生态文明建设以及绿色发展的理念。

城市地下综合管廊建设是我国于2014年启动的一项重大民生工程,也是今后一段时间内我国城市建设的重点工作。推进城市地下综合管廊建设有利于解决路面反复开挖、架空线网密集、管线事故频发以及地下基础设施滞后等问题,同时有利于增加公共产品有效投资、拉动社会资本投入和打造经济发展新动力,对转变城市发展方式具有重大意义。

1.1 建设地下综合管廊是保障城市运营安全的重要环节

传统的市政管线采取直埋方式,导致事故频发。与传统管线直埋方式相比,综合管廊具有较明显的综合优势,见表1。

表1 地下综合管廊相比管线直埋的优势[2]

Table 1 Advantages of underground utility tunnel over the direct burial pipeline[2]

优势	具体内容
合理规划利用城市地下空间	能集约、有效地利用地下空间,改变传统直埋式地下管线杂乱无章的无序状况,并为地下市政管线的远期扩容提前预留空间
管线维护管理更健康	维护、维修人员及机械等在不破坏道路、不影响正常生产生活的情况下直接接触管线,完成维护、维修
明显的经济效益	避免管线破坏、维修、扩容、道路开挖、恢复等造成的直接浪费以及由此导致的商业营业损失、塞车、环境污染、噪声等更大的间接浪费
社会和环境效益更好	具有一定的防灾功能,能够提升城市形象和优化城市环境,有利于城市土地增值

注:由兴业证券研究所整理。

1.2 建设地下综合管廊是我国城市发展方式由粗放发展向集约、绿色、可持续发展模式转变的关键切入点

过去我国城市发展是以注重"面子"为主的粗放发展模式,如管线直埋、城市排水系统设计不足、污水直排、垃圾传统填埋、空气降污减排不控制、公交系统严重滞后等,酿成"里子"——城市地下基础设施建设的短板。

2014年5月,李克强总理考察内蒙古时指出:"面子"是城市的风貌,而"里子"则是城市的良心,只有筑牢"里子",才能撑起"面子",这是城市建设的百年大计。注重"里子"的集约、绿色、可持续发展的举措有科学提高城市交通供给能力(发展大运量快速公交系统、地下快速路和物流系统),垃圾地下集运、卫生填埋和焚烧处理,污水地下集运和处理,雨洪地下储蓄和排洪,建设海绵城市等,而这些都离不开现代化地下综合管廊建设。

1.3 建设地下综合管廊是拉动我国经济增长的一项重要措施

以地下综合管廊建设为龙头的补齐城市地下基础设施短板建设是"供给侧"结构性改革的一个重要方面。

城市地下综合管廊建设的一次性建设成本较高(见图1),但其长远效益十分显著。以我国最大综合管廊——珠海横琴综合管廊为例,管廊全长33.4 km,总投资22亿元,而横琴因地下管廊建设节约土地所产生的直接经济效益就超过80亿元[3]。

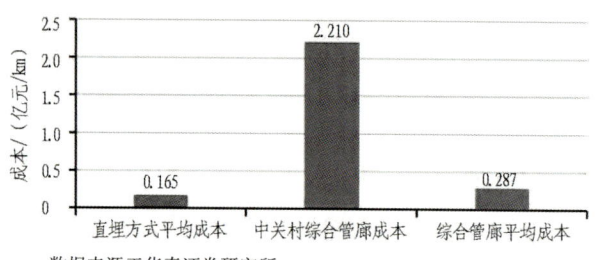

数据来源于华泰证券研究所。

图1 综合管廊与管线直埋成本对比[4]

Fig. 1 Cost comparison between underground utility tunnel and direct burial pipeline[4]

根据住建部测算,按地下综合管廊全寿命周期考虑:50年地下综合管廊与直埋管线相比,所需费用持平;超过50年,综合管廊比直埋管线所需费用低;到100年时,综合管廊与直埋管线相比所需费用低23%[5]。

1.4 建设地下综合管廊是建设中国秀美城市的迫切需要

以台湾地区为例:2001年,台北市挖路13 000次(面积310 000 m²,占台北市面积的1/32),高雄市挖

路 4 000 次(面积 180 000 m²),台北县挖路 5 000 次(面积 230 000 m²),造成交通堵塞、尘土飞扬、噪声不断、空气污染、市民怨声载道。

以城市地下综合管廊为契机的城市地下空间开发利用,可以实现电缆、污水、垃圾等管线的集中敷设,消除了"马路拉链"和"空中蜘蛛网",减少城市污染,对推动秀美城市建设具有重要意义。

1.5 建设和管理地下综合管廊是提升我国城市行政水平的重要抓手

由于给水排水、电力电信、燃气、供热、路灯、垃圾等管线性质各异,且各类管线单位分属不同行政主管,所以,地下综合管廊建设是一个跨行业、跨部门、跨组织的协同工程。

过去城市地下综合管廊之所以存在"建而不入"、"建后难管"的现象,其病根在于分头管理的市政体制与条块分割的部门所有制。在市政体制、机制不改革的条件下,破除分头管理与部门分割的关键在于市政领能否依法执政,克服急功近利意识,处理好"面子"和"里子"的关系,能否秉持"执政为民"的信念,迎难而上,破除部门利益的束缚,处理各种矛盾。例如:欧洲为了破除"马路拉链",市议会通过立法不准马路"开膛";台湾建设"共同管道"的经验是立法先行+"长官魄力与支持"。

2 国内外地下综合管廊建设及成功经验

2.1 国内外地下综合管廊建设比较

地下综合管廊起源于 19 世纪的欧洲。20 世纪,日本和我国台湾地区的综合管廊建设走在了世界前列。相比之下,我国大陆地区综合管廊建设起步晚、发展慢,见表 2。

2.2 有关国家和地区地下综合管廊建设和管理的成功经验

鉴于建设和管理地下综合管廊对转变城市发展方式具有重大意义,各国和一些地区都十分重视地下综合管廊的建设和管理,并取得了一些较为成功的经验。

1) 雨污分流:污水入综合管廊;雨洪不入廊,单独成系统。2) 燃气入廊:大多数国家(英、法、德……)采取燃气入廊;日本允许燃气入廊,设单沟。3) 垃圾入廊:单独设管,真空吹送至垃圾焚烧厂,高温焚烧。4) 预留平台:管廊内保留一些待铺入廊管线的平台。5) 组织先行和稳步推进:组织先行,立法规划紧跟;综合管廊建设稳步推进,质量第一。6) 消费合算,综合效益好。7) 领导高度重视。8) 建设整合:与地铁、城市地下空间、快速路等城市基础设施建设整合。9) 信息化管理:BIM 技术与共同管道作业相结合,采取数字化整合管理。

表 2 国内外地下综合管廊建设历程和运营管理情况

Table 2 Construction progress and running conditions of underground utility tunnel in China and abroad

国家或地区	建设历程和运营管理情况
法国	1832 年,巴黎建造以排水为主的廊道,创造性地布置了供水、煤气、通讯电缆等管线,形成早期的共同沟;1833 年,规划排水网络,兴建地下综合管廊;2001 年,巴黎已有共同沟 100 多 km,收容管线越来越多;目前,巴黎市区及郊区综合管廊里程已达 2 100 km
英国	1861 年,伦敦开始建设综合管廊,至今已有 22 条综合管廊
德国	1893 年始建于汉堡,综合管廊内容纳了自来水、通讯、电力、燃气、污水、热力等市政公用管线
日本	1926 年始建,目前已成为世界上综合管廊建设规模较大、技术最先进的国家之一。在综合管廊的投资、规划设计、建设施工、后期管理等方面,已形成较为成熟完善的系统。东京都的中心城区,已规划建成 200 多 km 的共同沟
前苏联	1933 年,在莫斯科、列宁格勒(今圣彼得堡)、基辅等大城市道路建设时配套建设综合管廊。发展至今,莫斯科的综合管廊长度已超过 130 km
中国大陆	1958 年,北京天安门广场铺设第 1 条长 1 076 m 的地下综合管廊;1994 年,上海浦东新区建成 11.125 km 的综合管廊;2010 年起建设的横琴地下综合管廊,是目前我国大陆建成的里程最长、一次性投资规模最大、服务面积最广、体系最完善的综合管廊;目前,地下综合管廊建设已在全国范围内铺开
中国台湾地区	1991 年台北、淡海及高雄市开始建设共同管道,已建 300 多 km。主管机关和管线单位共同出资建设;维护费用由政府与各管线单位共同分担,管线单位根据其所属管线在综合管廊中所占的面积,来综合确定其所应承担的费用比例

3 国务院推进地下综合管廊建设的重要决策部署

目前,中国正处于城镇化快速发展时期,但地下管廊建设相对滞后。加快这方面的建设,很有必要。为此,国务院及有关部门自 2014 年起密集发布了地下综合管廊相关政策,并提出具体任务目标。目标要求:截至 2015 年底,完成城市地下管线普查,建立综合管理信息系统,编制完成地下管线综合规划(已完成);到 2020 年,完成城市地下老旧管网改造,将管网漏失率控制在国家标准以内,显著降低管网事故,避免重大事故发生;到 2025 年,建成较为完善的城市地下管线体系,使地下管线建设管理水平能够适应经济社会发展需要,应急防灾能力大幅提升[1]。我国已发布的地下综合管廊相关政策见表 3。

表3 我国地下综合管廊相关政策
Table 3 China's related policies for underground utility tunnel

日期	政策或事件	主要内容
2013-09	国务院《关于加强城市基础设施建设的意见》[6]	用3年左右时间,在全国36个大中城市,全面启动地下综合管廊试点工程,中小城市因地制宜建设一批综合管廊项目,新建道路、城市新区和各类园区地下管网应按照综合管廊模式进行开发建设
2014-06	国务院办公厅《关于加强城市地下管线建设管理的指导意见》[7]	2015年底前完成城市地下管线普查,建立综合管理信息系统,编制完成地下管线综合规划,力争用5年时间,完成城市地下老旧管网改造,用10年左右时间,建成较为完善的城市地下管线体系
2014-12	财政部《关于开展中央财政支持地下综合管廊试点工作的通知》[8]	中央财政将对地下综合管廊试点城市给予专项资金补助,直辖市每年5亿元,省会城市每年4亿元,其他城市每年3亿元
2015-01	财政部、住建部《关于组织申报2015年地下综合管廊试点城市的通知》[9]	
2015-04	财政部公布2015年地下综合管廊试点城市[10]	包头、沈阳、哈尔滨、苏州、十堰、长沙、海口、六盘水、白银10个城市入围,计划未来3年将合计建设地下管网389 km,总投资351亿元
2015-05	住建部发布GB 50838—2015《城市综合管廊工程技术规范》国家标准[11]	与2012年的规范相比,增加了雨水、污水管道、燃气管道、热力管道敷设的规定等
2015-05	住建部印发《城市地下综合管廊工程规划编制指引》[12]	指导各地科学合理地规划综合管廊建设区域及布局要求,管廊工程规划期限应与城市总体规划期限一致
2015-06	住建部标准定额司等单位编制完成《城市综合管廊工程投资估算指标》[13]	给出了城市综合管廊工程前期编制投资估算的参考依据,根据舱数不同,每千米投资为0.5亿~3.6亿元
2015-07	国务院常务会议部署推进城市地下综合管廊建设	要求地方政府编制地下综合管廊建设专项计划,在年度建设中优先安排;已建管廊区域,所有管线必须入廊
2015-07	住建部会同财政部开展中央财政支持地下综合管廊试点工作	10个试点城市计划3年内建设389 km,总投资351亿元,其中中央财政投入102亿元,地方政府投入56亿元,拉动社会投资约193亿元
2015-08	国务院办公厅《关于推进城市地下综合管廊建设的指导意见》[1]	提出到2020年建设一批具有国际先进水平的地下综合管廊并投入运营,改善城市地面景观
2015-11	住建部、国家开发银行《关于推进开发性金融支持城市地下综合管廊建设的通知》[14]	入廊费主要用于弥补管廊建设成本,由入廊管线单位向管廊设计、建设、运营单位一次性支付或分期支付;日常维费用主要用于弥补管道日常维护、管理支出,由入廊管线单位按确定的计算周期向管廊运营单位逐期支付
2015-11	国家发展改革委、住建部《关于城市地下综合管廊实行有偿使用制度的指导意见》[15]	把地下综合管廊建设作为信贷支持的重点领域,服务国家战略,各级住建部门要把国家开发银行作为重点合作银行,加强合作,增强地下综合管廊建设项目资金保障,用好用足信贷资金,推进地下综合管廊建设
2016-01	住建部印发《城市综合管廊国家建筑标准设计体系》和《海绵城市建设国家建筑标准设计体系》[16]	《城市综合管廊国家建筑标准设计体系》按照总体设计、结构工程、专项管线、附属设施4部分进行构建,体系中的标准设计项目基本涵盖了城市综合管廊工程设计和施工中各专业的主要工作内容《海绵城市建设国家建筑标准设计体系》主要包括新建、扩建和改建的海绵型建筑与小区、海绵型道路与广场、海绵型公园绿地、城市水系中与保护生态环境相关的技术及相关基础设施的建设、施工验收及运行管理
2016-04	住建部办公厅、财政部办公厅《关于开展地下综合管廊试点年度绩效评价工作的通知》[17]	了解中央财政支持地下综合管廊试点工作进展情况,总结推广试点城市工作经验和做法,查找工作中的不足并提出改进措施,督导各地按计划推进试点工作
2016-05	住建部和财政部组织开展地下综合管廊试点2015年度绩效评价工作	组织专家成立绩效评价小组,对包头等10个第1批中央财政支持地下综合管廊建设试点城市2015年度绩效进行评价
2016-05	住建部、国家能源局《推进电力管线纳入城市地下综合管廊的意见》[18]	电力等管线纳入管廊是城市管线建设发展方式的重大转变,要加强统筹协调、协商合作,认真做好电力管线入廊等相关工作,积极稳妥推进管廊建设
2016-08	住建部《关于提高城市排水防涝能力推进城市地下综合管廊建设的通知》[19]	将城市排水防涝与城市地下综合管廊、海绵城市建设协同推进,构建城市排水防涝工程体系,并与城市防洪规划做好衔接

4 推进地下综合管廊建设相关配套政策及具体措施

4.1 加强顶层设计,积极有序推进

组织先行:明确主管部门,建立协调机制;立法、规划跟进;制定路线图、时间表,积极有序推进工作部署。

4.2 完善相关配套政策

4.2.1 明确入廊要求

城市规划区范围内的各类管线原则上应敷设于地下空间。已建设地下综合管廊的区域,该区域内的所有管线必须入廊。在地下综合管廊以外的位置新建管线的,规划部门不予许可审批,建设部门不予施工许可审批,市政道路部门不予掘路许可审批。既有管线应根据实际情况逐步有序迁移至地下综合管廊,各行业主管部门和有关企业要积极配合城市人民政府做好各自管线入廊工作。

4.2.2 实行有偿使用

入廊管线单位应向地下综合管廊建设运营单位交纳入场费和日常维护费,具体收费标准要统筹考虑建设和运营、成本和收益的关系,由地下综合管廊建设运营单位与入廊管线单位根据市场化原则共同协商确定。入廊费主要根据地下综合管廊本体及附属设施建设成本,以及各入廊管线单独敷设和更新改造成本确定。日常维护费主要根据地下综合管廊本体及附属设施维修、更新等维护成本,以及管线占用地下综合管廊空间比例,对附属设施使用强度等因素合理确定。公益性文化企业的有线电视网入廊,有关收费标准可适当给予优惠,由发改委同住建部制定指导意见,供需双方协调确定地下综合管廊收费标准,形成合理的收费机制。地下综合管廊运营初期不能通过收费弥补成本的,地方人民政府视情况给予必要的财政补贴。

4.2.3 加大政府投入

中央财政要发挥"四两拨千斤"的作用,积极引导地下综合管廊建设,通过现有渠道统筹安排资金予以支持,地方各级人民政府要进一步加大地下综合管廊建设资金投入。省级人民政府要加强地下综合管廊建设资金的统筹,城市人民政府要在年度预算和建设计划中优先安排地下综合管廊项目,并纳入地方政府采购范围。有条件的城市人民政府可对地下综合管廊项目给予贷款贴息。

4.2.4 完善融资支持

将地下综合管廊建设作为国家重点支持的民生工程,充分发挥开发性金融作用,鼓励相关金融机构积极加大对地下综合管廊建设的信贷支持力度。鼓励银行业金融机构在风险可控、商业可持续的前提下,为地下综合管廊项目提供中长期信贷支持,积极开展特许经营权、收费权和购买服务协议预期收益等担保创新类贷款业务,加大对地下综合管廊项目的支持力度。将地下综合管廊建设列入专项金融债支持范围予以长期投资。支持符合条件的地下综合管廊建设运营企业发行企业债券和项目收益票据,专项用于地下综合管廊建设项目。

4.3 落实具体措施

4.3.1 借鉴国内外先进经验,完善法规体系

2015年中央城市工作会议提出:"要保证规划确立后的法律地位,根本点是要提高城市地方人民代表大会的法治权威性"。国外的先进经验是先立法、后建设,这样才能有序、有力推进建设。建设管理综合管廊是一个跨行业、跨部门与单位的协同工程。破除分头管理与部门分割的利器是"法治",依法执政首先要有法可依,光有指导意见不够,需要对综合管廊的所有权、规划权、建设权、管理权、经营权、使用权以及入廊有偿使用费的收取原则等作出明确的、完善的规定。

在这方面,我们可以借鉴德国的《城市建设法典》,日本的《关于建设共同沟的特别措施法》和我国台湾地区的《共同管道法》、《共同管道法施行细则》、《公共建设管线基金收支保管及运用办法》、《共同沟建设及管理经费分摊办法》等现行法规,也可借鉴如《上海市城市道路架空线管理办法》和《重庆市管线工程规划管理办法》等进行完善和补充。

4.3.2 系统科学地编制好城市综合管廊规划

2014年2月25日,习近平总书记在北京市规划展览馆考察时指出:"规划科学是最大的效益,规划失误是最大的浪费,规划折腾是最大的忌讳。"

城市地下综合管廊规划应与城市地下空间规划以及城市建设总体规划同步协调进行。条件不允许时,需要先行或单独规划的综合管廊,必须根据城市总体规划与现有和拟建的城市地下步行道系统、地下轨道交通、地下快速路系统以及为解决城市内涝问题而建设的城市地面与地下海绵设施和贮排雨洪的城市地下河流、蓄排洪设施保持衔接;即通过充分协调,实现编制的统筹规划。综合管廊建设是百年大计,规划时还应充分考虑城市未来发展中对地下物流、垃圾集中处理、中水输送乃至利用地热地冷资源实现城市供热供

冷所需空间的规划要求,同时这也有利于提升综合管廊的经营性。

现在关于燃气管线是否纳入综合管廊还有争议,争论的焦点主要在于燃气管线易燃易爆,危害综合管廊安全。纵观以往的燃气管线事故,如济南和南京等地发生的地下煤气管线爆燃爆炸事故以及黄岛发生的地下输油管线爆炸等,都是由于燃气管线检查维护管理不到位,导致燃气泄漏引发的。将燃气管线统一纳入综合管廊后,可以实现对燃气管线的及时检查、维护和监控,从而能够避免燃气管线泄漏和爆炸事故的发生,因此,将燃气管线纳入到综合管廊当中也是合理可行的。

此外,还有雨洪排泄管道是否入廊的问题。城市要实行雨污分流,则污水管道应纳入到综合管廊中,而雨洪排泄设施不宜入廊,这是因为城市排水系统至少要按5年一遇洪水进行设计,体量较大,导致雨洪排泄设施入廊很难。要彻底解决城市内涝问题,可参照日本和吉隆坡的做法,修建地下河和行车排洪两用隧道。

应大力提倡综合管廊与地铁建设、地下街建设和地下快速路建设相整合,以降低建设成本,减少社会干扰,缩短施工工期,避免重复建设和重复投资。例如:中国台北东西快速道路共同沟的建设,全长6.3 km,其中2.7 km与地铁整合建设,2.5 km与地下街、地下车库整合建设,独立施工的共同沟仅1.1 km,从而大大降低了建设总成本,有效推进了共同沟的发展;武汉地铁3号线宗关站,将电力管、排水管、通讯管等管道集结在一起,与2、3号出入口工程整合建设,既解决了"城市蛛网"埋设随意性较大、分布不合理的问题,也可避免检修时对城市道路重复"开膛破肚";乌鲁木齐市在远景规划城市轨道交通线网中,也考虑了地下综合管廊的同步建设。

编好综合管廊规划的前提和基础是充分掌握各类已有和需建管线的信息和各类地下空间信息,为此,在编制规划前必须开展和完成城市地下管线普查工作。

4.3.3 建立统一管理、分工协作、衔接有序的综合管廊管理体制

2015年中央城市工作会议提出:"推进城市管理机构改革,创新城市工作体制机制。"为了克服条块分割、分头管理的弊端,达到统一协调,当前在尊重现有各部门职能分工的基础上,应尽快明确每个城市的综合管廊管理机构。建议成立城市综合管廊管理委员会,承担国办发[2014]27号《国务院办公厅关于加强城市地下管线建设管理的指导意见》所规定的政府对综合管廊建设的管理责任。由政府主要负责人担任主任,相关政府部门及管线主管部门担任委员,下设综合管理办公室,承担规划编制、项目审批、设计审查、安全和质量监管、地下管线信息共享、项目投融资、专家咨询等综合协调事宜,制定地下综合管廊管理办法、指导运营单位完善管理制度、组织综合管廊建设运营单位与入廊管线单位签订协议,确保地下综合管廊正常运行,指导有关政府部门做好突发事件和应急管理工作。

4.3.4 地下综合管廊兼顾人防应到位而不越位

根据《中华人民共和国人民防空法》第14条规定"城市的地下交通干线以及其他地下工程的建设,应当兼顾人民防空需要。"因此,城市地下综合管廊建设必须兼顾人防需要。什么叫兼顾人防需要,就是设计时,要使管廊在战时具有一定的抗毁能力,保证战时生命线的安全。

到位就是必须兼顾人防,既要提高平时管廊的防灾减灾能力,又要提高战时的防空性能。但到位不能越位,如果兼顾人防的要求过高,使其工程造价提高过大,从而影响综合管廊的建设,即是越位;同时更不能错位,即改变了管廊的功能定位,如将战时兼顾人防要求定位为战时管廊用作人防掩蔽部使用,因按照防护单元的要求需在管廊内施作防爆隔断设施,从而既影响管线敷设,又提高了成本,这就是错位。

4.3.5 推广PPP模式,创新投融资机制

创新投融资机制的关键是吸引社会资本参与综合管廊的建设与运营。除了用足投资补贴、特许经营权、收费权和购置服务与预期收益等担保类贷款贴息和发行专项金融债等财政金融政策外,考虑到社会资本的逐利性是其基本属性之一,要吸引社会资本参与,还必须妥善解决如下问题:

1) 综合管廊集中统一运营,降低运营成本。

2) 严格"风险分担",即政府理应承担的风险不应由社会资本"共同承担"。PPP模式中投资风险不应由项目公司承担,政府更不应让社会资本捆绑打包更大的建设项目。

3) 合理确定项目公司资本金比例(SPV)和政府在项目公司中的持股比例。

4) 采取政府提供购买服务协议的方式解决经营收益不确定性问题。

5) 杜绝地方政府捆绑打包"小马拉大车",增大未来投资风险。

5 结语

目前,我国正处在城镇化快速发展时期,建设城市地下综合管廊对转变城市发展方式意义重大,受到国家和政府的高度重视。与发达国家和地区相比,我国的综合管廊建设相对滞后。全面推进地下综合管廊建设,首先要加强顶层设计,积极有序推进;其次要不断完善相关配套政策,并针对综合管廊的规划设计、法律法规、投融资、建设、运营和管理等方面落实相应的具体措施。通过各方协同努力,共同建设绿色、宜居城市。

参考文献(References):

[1] 中华人民共和国国务院办公厅.关于推进城市地下综合管廊建设的指导意见:国办发[2015]61号[A/OL].[2017-01-11].http://www.gov.cn/zhengce/content/2015-08/10/content_10063.htm.
General Office of the State Council of the People's Republic of China. Guidance on promoting the construction of urban underground utility tunnel: Guo Ban Fa [2015] No. 61 [A/OL]. [2017-01-11]. http://www.gov.cn/zhengce/content/2015-08/10/content_10063.htm.

[2] 兴业证券.地下综合管廊建设研究报告[R/OL].[2017-01-11].http://www.360doc.com/content/15/0902/10/5473201_496410209.shtml.
Industrial Securities. Research report on underground utility tunnel construction [R/OL]. [2017-01-11]. http://www.360doc.com/content/15/0902/10/5473201_496410209.shtml.

[3] 珠海横琴综合管廊告别城市"拉链路"成样板[EB/OL].[2017-01-11].http://www.chinanews.com/gn/2015/08-10/7457058.shtml.
Eliminating the city pull link by underground utility tunnel in Hengqin District of Zhuhai [EB/OL]. [2017-01-11]. http://www.chinanews.com/gn/2015/08-10/7457058.shtml.

[4] 华泰证券.城市综合管廊,从"地下"到"云端"[R/OL].[2017-01-11].http://www.doc88.com/p-6816961492758.html.
Huatai Securities. Urban utility tunnel: From "underground" to "cloud" [R/OL]. [2017-01-11]. http://www.doc88.com/p-6816961492758.html.

[5] 住建部解读地下综合管廊建设相关问题[EB/OL].[2017-01-11].http://www.lyguihua.com/html/zcjd/201581894749428.html.
Interpreting the problems of the construction of underground utility tunnel by MOHURD [EB/OL]. [2017-01-11]. http://www.lyguihua.com/html/zcjd/201581894749428.html.

[6] 中华人民共和国国务院.关于加强城市基础设施建设的意见:国发[2013]36号[A/OL].[2017-01-11].http://www.gov.cn/zwgk/2013-09/16/content_2489070.htm.
The State Council of the People's Republic of China. Opinions on strengthening urban infrastructure construction: Guo Fa [2015] No. 36 [A/OL]. [2017-01-11]. http://www.gov.cn/zwgk/2013-09/16/content_2489070.htm.

[7] 中华人民共和国国务院办公厅.关于加强城市地下管线建设管理的指导意见:国办发[2014]27号[A/OL].[2017-01-11].http://www.gov.cn/zhengce/content/2014-06/14/content_8883.htm.
General Office of the State Council of the People's Republic of China. Guidance on strengthening the management of urban underground pipeline construction: Guo Ban Fa [2014] No. 27 [A/OL]. [2017-01-11]. http://www.gov.cn/zhengce/content/2014-06/14/content_8883.htm.

[8] 中华人民共和国财政部.关于开展中央财政支持地下综合管廊试点工作的通知:财建[2014]839号[A/OL].[2017-01-11].http://jjs.mof.gov.cn/zhengwuxinxi/tongzhigonggao/201501/t20150113_1179393.html.
Ministry of Finance of the People's Republic of China. Notification of the implementation of the central government's support for the underground utility tunnel: Cai Jian [2014] No. 839 [A/OL]. [2017-01-11]. http://jjs.mof.gov.cn/zhengwuxinxi/tongzhigonggao/201501/t20150113_1179393.html.

[9] 中华人民共和国财政部,中华人民共和国住房城乡建设部.关于组织申报2015年地下综合管廊试点城市的通知[A/OL].[2017-01-11].http://www.gov.cn/zhengce/content/2014-06/14/content_8883.htm.
Ministry of Finance of the People's Republic of China, MOHURD. Notification of the organization to declare demonstration cities of the underground utility tunnel in 2015 [A/OL]. [2017-01-11]. http://www.gov.cn/zhengce/content/2014-06/14/content_8883.htm.

[10] 财政部公布2015年地下综合管廊试点城市[EB/OL].[2017-01-11].http://jjs.mof.gov.cn/zhengwuxinxi/tongzhigonggao/201504/t20150409_1214636.html.
Promulgating of demonstration cities of the underground utility tunnel in 2015 by Ministry of Finance of the People's

Republic of China [EB/OL]. [2017 – 01 – 11]. http://jjs.mof.gov.cn/zhengwuxinxi/tongzhigonggao/201504/t20150409_1214636.html.

[11] 城市综合管廊工程技术规范: GB 50838—2015 [S]. 北京: 中国计划出版社, 2015.
Technical code for urban utility tunnel engineering: GB 50838—2015 [S]. Beijing: China Planning Press, 2015.

[12] 中华人民共和国住房城乡建设部. 关于印发《城市地下综合管廊工程规划编制指引》的通知: 建城 [2015] 70 号 [A/OL]. [2017 – 01 – 11]. http://www.mohurd.gov.cn/wjfb/201505/t20150528_221004.html.
MOHURD. Notification of the printing of the guidelines for the planning of the urban underground utility tunnel: Jian Cheng [2015] No. 70 [A/OL]. [2017 – 01 – 11]. http://www.mohurd.gov.cn/wjfb/201505/t20150528_221004.html.

[13] 中华人民共和国住房城乡建设部. 关于印发《城市综合管廊工程投资估算指标》(试行) 的通知: 建标 [2015] 85 号 [A/OL]. [2017 – 01 – 11]. http://www.mohurd.gov.cn/wjfb/201506/t20150629_222705.html.
MOHURD. Notification of the printing of the investment estimate for the urban underground utility tunnel (trial): Jian Biao [2015] No. 85 [A/OL]. [2017 – 01 – 11]. http://www.mohurd.gov.cn/wjfb/201506/t20150629_222705.html.

[14] 中华人民共和国住房城乡建设部, 中国国家开发银行. 关于推进开发性金融支持海绵城市建设的通知: 建城 [2015] 208 号 [A/OL]. [2017 – 01 – 11]. http://www.mohurd.gov.cn/wjfb/201512/t20151230_226176.html.
MOHURD, China Development Bank. Notification of the promotion of development finance to support the construction of sponge cities: Jian Cheng [2015] No. 208 [A/OL]. [2017 – 01 – 11]. http://www.mohurd.gov.cn/wjfb/201512/t20151230_226176.html.

[15] 中华人民共和国国家发展和改革委员会, 中华人民共和国住房和城乡建设部. 关于城市地下综合管廊实行有偿使用制度的指导意见: 发改价格 [2015] 2754 号 [A/OL]. [2017 – 01 – 11]. http://www.mohurd.gov.cn/wjfb/201512/t20151209_225891.html.
National Development and Reform Commission, MOHURD. Guidance on the use of paid-for use systems for urban underground utility tunnel: Fa Gai Jia Ge [2015] No. 2754 [A/OL]. [2017 – 01 – 11]. http://www.mohurd.gov.cn/wjfb/201512/t20151209_225891.html.

[16] 中华人民共和国住房城乡建设部. 关于印发城市综合管廊和海绵城市建设国家建筑标准设计体系的通知 [A/OL]. [2017 – 01 – 11]. http://www.mohurd.gov.cn/wjfb/201602/t20160204_226594.html.
MOHURD. Notification of the design system of the national building standard of the urban underground utility tunnel and the sponge city [A/OL]. [2017 – 01 – 11]. http://www.mohurd.gov.cn/wjfb/201602/t20160204_226594.html.

[17] 中华人民共和国住房城乡建设部办公厅, 中华人民共和国财政部办公厅. 关于开展地下综合管廊试点年度绩效评价工作的通知: 建办城函 [2016] 375 号 [A/OL]. [2017 – 01 – 11]. http://www.mohurd.gov.cn/wjfb/201604/t20160429_227347.html.
General Office of MOHURD, General office of Ministry of Finance of the People's Republic of China. Notification of the annual performance evaluation of the underground utility tunnel: Jian Ban Cheng Han [2016] No. 375 [A/OL]. [2017 – 01 – 11]. http://www.mohurd.gov.cn/wjfb/201604/t20160429_227347.html.

[18] 中华人民共和国住房城乡建设部, 中国国家能源局. 推进电力管线纳入城市地下综合管廊的意见: 建城 [2016] 98 号 [A/OL]. [2017 – 01 – 11]. http://www.mohurd.gov.cn/wjfb/201605/t20160531_227655.html.
MOHURD, National Energy Administration. Promoting the suggestion of moving the electric pipeline into the urban underground utility tunnel: Jian Cheng [2016] No. 98 [A/OL]. [2017 – 01 – 11]. http://www.mohurd.gov.cn/wjfb/201605/t20160531_227655.html.

[19] 中华人民共和国住房城乡建设部. 关于提高城市排水防涝能力推进城市地下综合管廊建设的通知: 建城 [2016] 174 号 [A/OL]. [2017 – 01 – 11]. http://www.mohurd.gov.cn/wjfb/201608/t20160823_228649.html.
MOHURD. Notification of improving drainage capacity in cities to promote the construction of urban underground utility tunnel: Jian Cheng [2016] No. 174 [A/OL]. [2017 – 01 – 11]. http://www.mohurd.gov.cn/wjfb/201608/t20160823_228649.html.

隧道工程建设地质预报及信息化技术的主要进展及发展方向

钱七虎

(解放军理工大学国防工程学院,江苏 南京 210007)

摘要: 复杂的不良地质条件是制约隧道安全高效建设的主要因素,要实现隧道工程的安全高效建设,首先要提高地质预测预报技术水平及其信息化程度。1) 介绍我国复杂不良地质隧道超前预报的方法进展及其应用,包括突水突泥灾害源超前探测方法与设备、断层破碎带超前预报、城市地铁溶洞和孤石等探测的进展及应用等; 2) 介绍我国隧道岩爆监测预警方法及其应用,预报清楚之后就要加强安全风险过程监控; 3) 介绍基于 BIM 技术的建筑物(隧道工程) 安全风险监控最新进展,包括安全风险实时感知系统和实时预警系统; 4) 指出隧道工程建设信息化技术的发展方向,包括开展基于大数据技术的 TBM/盾构施工的分析与控制研究以及数字隧道向智慧隧道(建设和运营维护) 的发展。

关键词: 隧道; 信息化; 不良地质; 超前预报; 岩爆; BIM; 大数据技术; 数字隧道; 智慧隧道

DOI: 10.3973/j.issn.1672 - 741X.2017.03.001

中图分类号: U 455　　　**文献标志码**: A　　　**文章编号**: 1672 - 741X(2017)03 - 0251 - 13

Main Developments and Directions of Geological Prediction and Informatized Technology of Tunnel Construction

QIAN Qihu

(*College of Defense Engineering*, *PLA University of Science & Technology*, *Nanjing* 210007, *Jiangsu*, *China*)

Abstract: The improvements of geological prediction and informatized technologies are the keys to safe and efficient construction of tunnel in complex and bad geological conditions. The technologies used in China are introduced as follows: 1) The development and application of advanced geological prediction technologies of tunnels in complex and bad geological conditions, i.e. methods and equipments for water and mud inrush sources detection, advanced prediction of fault and fracture zones and detection of karst caves and boulders in urban Metro construction. 2) Monitoring and forewarning methods for rockbust in tunnel. 3) Building information modeling (BIM) technology, including real-time perceiving and real-time warning systems. Afterwards, the developing directions of informatized technology of tunnel construction are proposed, including analysis and control of TBM/shield construction based on big data technology and digital tunnel and intellectualized tunnel.

Keywords: tunnel; informatization; bad geological condition; advanced geological prediction; rockburst; BIM; big data technology; digital tunnel; intellectualized tunnel

0 引言

随着经济的持续发展、综合国力的不断提升以及高新技术的不断应用,我国隧道及地下工程建设得到了前所未有的发展。制约隧道安全高效建设的主要因素并公认为隧道建设难点的是极端复杂的不良地质条件。客观复杂的不良地质条件加上施工人员的主观不安全行为酿成了地质灾害和工程事故,对隧道建设的安全、工期和成本造成严重危害。地下地质情况不掌握,事故发生机制不清楚,从信息学的角度来说,就是地下信息采集和传递不及时,信息分析处理不完善,信息共享和利用不充分。随着隧道工程理论和方法的不断进步以及科学技术的不断发展,人们越来越认识到工程地质勘察、设计和施工形成系统和信息化一体的重要性,要想实现隧道安全高效建设就要不断提高隧道工程建设的信息化水平。本文介绍和分析我国隧道地质预报方法及安全监控技术的最新进展,并提出我国隧道工程建设信息化的发展方向,以期对我国隧道技术的发展有一定启示。

收稿日期:2017-02-13。
本文原载于《隧道建设》(2017 年第 37 卷第 3 期)。

1 复杂不良地质超前预报分析方法及其应用

随着我国隧道工程建设规模的不断扩大,隧道工程的数量和长度逐渐增加,部分隧道地质条件异常复杂,施工难度大大增加。暗河、溶洞、断层破碎带、孤石等不良地质条件都会给隧道施工带来严重危害。突水突泥、塌方、卡机、机毁人亡等事故时有发生,如:宜万铁路马鹿箐隧道和野三关隧道、湖北沪蓉西高速公路龙潭隧道、吉莲公路永莲隧道、青海西格铁路关角隧道等突水突泥,甘肃引洮工程、云南那帮水电站工程、辽宁大伙房水库工程、瑞士圣格达铁路隧道工程等遇不良地质掘进机被卡或损坏。这些事故给人民的生命和财产安全造成了极大的危害。为准确查明复杂不良地质的具体情况,及时采取针对性的防治措施,最大限度地减小不良地质对隧道施工与营运的影响,利用地质超前预报技术为不良地质隧道施工提供指导十分必要。

1.1 突水突泥灾害源超前探测方法与设备

对于含水地质构造的超前探测而言,主要任务有: 1) 确定含水构造的具体位置、规模大小和具体形态等特征,即需要实现对含水构造的三维成像; 2) 尽可能准确地测算含水构造内部的含水量和水体特征,突水超前探测的难点是水量的探测,为解决该难题,针对性地提出了基于二电流激发极化半衰时差法的隧道含水构造水量预测方法[1]。半衰时之差包络面积与注入水量的关系见图1,可知含水体静态水量与激发极化半衰时之差呈正相关关系,两者的这种单调线性正相关关系为解决实际工程中含水构造的水量预测奠定了基础。

根据激发极化法等地球物理方法为先导的解决思路,研制出前向三维激发极化探测专用仪器——GEI电法仪,其工作原理见图2,可实现含水构造的三维成像展示。

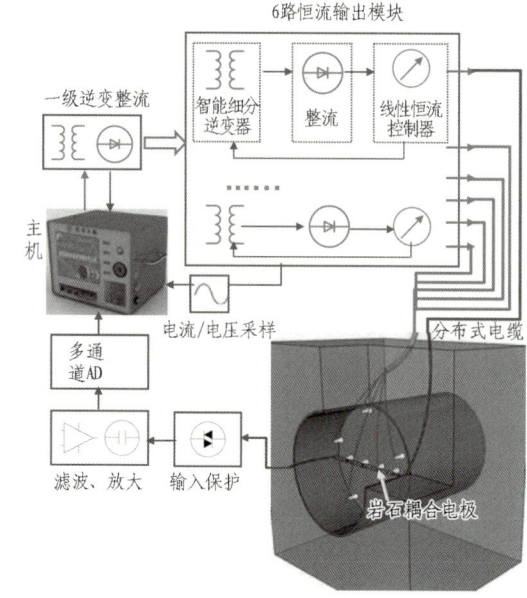

图 2 GEI 电法仪原理

Fig. 2 Working principle of GEI prospecting apparatus

1.1.1 钻爆法突水突泥灾害源超前探测

在成兰铁路跃龙门隧道3#斜井工区采用激发极化法进行超前探水预测预报试验,激发极化探测三维成像见图3。

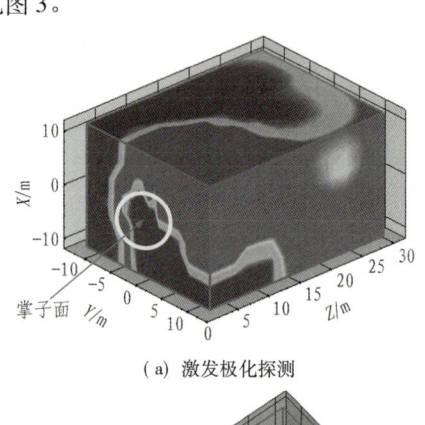

(a) 激发极化探测

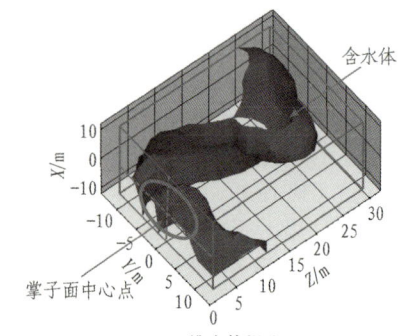

(b) 三维水体提取

图 3 激发极化法超前探水三维成像

Fig. 3 3D images of induced polarization method

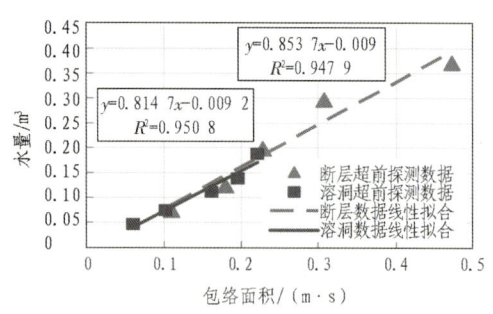

(a) 小型物理模型试验实测数据

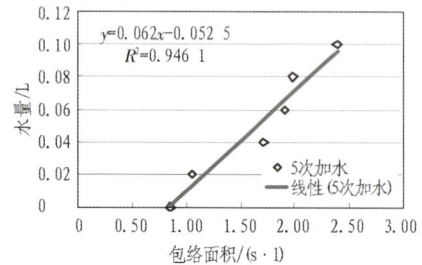

(b) 大型物理模型试验实测数据

图 1 半衰时之差包络面积与注入水量的关系

Fig. 1 Water volumes vs. envelope areas

根据掌子面前方 30 m 范围含导水构造的三维空间展布,综合地质与激发极化分析结果,推断探测区域赋存水量约为 900 m³(静储量,不考虑补给条件),补给条件下总涌水量 1 000 m³/h,实现含水构造的三维成像展示和一定范围内的定性、定位、估量探水,取得了较好的现场验证效果。

1.1.2 TBM 突水突泥灾害源超前探测

在 TBM 突水突泥灾害源超前探测方面,国内外尚没有有效的预报方法。TBM 施工隧道超前地质预报面临的问题见图 4,主要有: 1) TBM 占据大部分隧道,可用观测空间狭小; 2) TBM 掘进电磁环境复杂,干扰严重。

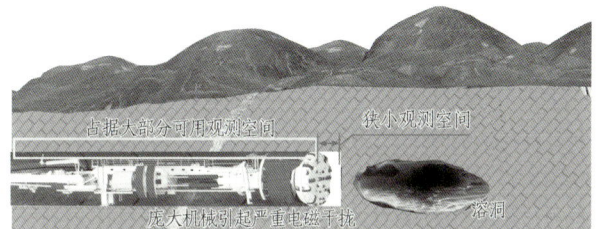

图 4　TBM 施工隧道超前地质预报面临的问题

Fig. 4　Difficulties of advanced geological prediction during TBM tunneling

基于钻爆法超前探水预报方法,提出搭载于 TBM 的激发极化法,搭载方案与总体架构见图 5。

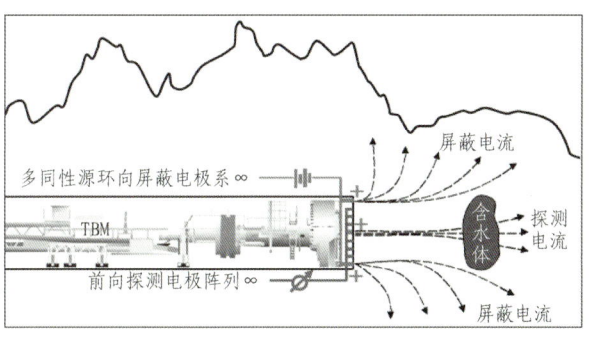

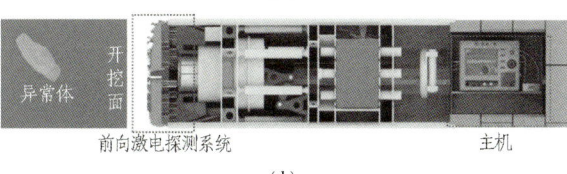

图 5　TBM 激发极化法总体架构

Fig. 5　Structure design of induced polarization method of TBM

将搭载 TBM 的三维激发极化法应用在吉林引松 3 标超前地质探测中,现场应用和部分探测结果分别见图 6 和图 7。引松供水工程从 71+476 始发掘进,累计掘进 1 532 m,激发极化法探测 21 次,准确探测了 2 次岩溶富水区,其他段落为干燥状态或滴水,总体准确度达到 85% 以上。

(a) 搭载激电系统

(b) 主机监测

图 6　TBM 激发极化法现场应用

Fig. 6　Site application of induced polarization method of TBM

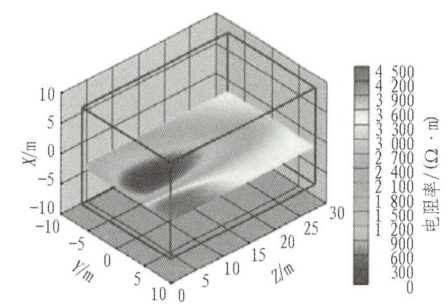

(a) 三维成像

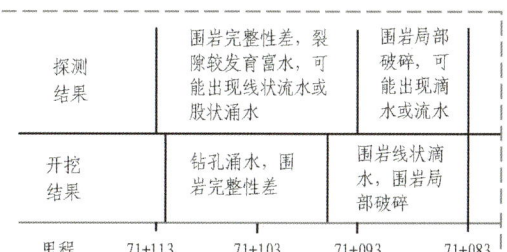

(b) 开挖结果和实测结果对比

图 7　部分探测结果

Fig. 7　Parts of geological prediction results

1.2　隧道前方断层破碎带超前预报

搭载于 TBM 的三维地震法是探测隧道前方断层破碎带的有效方法,其原理见图 8。该方法是在刀盘附近边墙设置 12 个激震点,后方边墙上布置 10 个传感器,利用检修间隙探测,不需对 TBM 机械进行改造,可对断层、溶洞、破碎带等不良地质进行三维定位,探距120 m,断层探测准确率90%,位置误差为探测距离的10%。

图8 地震反射成像法超前探测示意图

Fig. 8 Sketch diagram of seismic reflection imaging method

将搭载于 TBM 的三维地震法应用在辽宁某引水工程中,探测结果见图9。由探测结果可知:在掌子面前方 60～120 m 存在较多强反射界面,且正负交替出现,推测从掌子面前方 60 m 开始进入断层破碎带,围岩强弱交替,这与后期开挖结果相吻合。三维地震探测的实施,保障 TBM 安全穿越 200 m 断层破碎带。

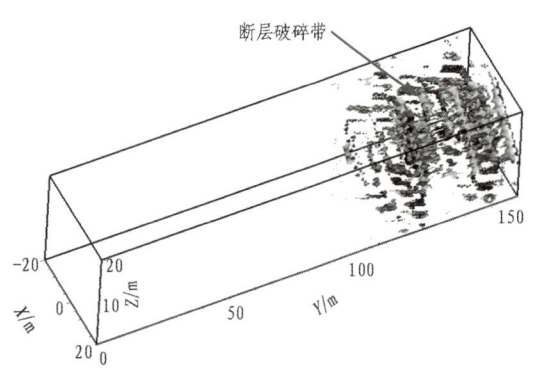

图9 三维探测结果

Fig. 9 3D geological prediction results

1.3 城市地铁溶洞、孤石等探测进展及应用

溶洞、孤石体积小,需要精细化探测,跨孔电阻率 CT 法是一种孔中精细化探测方法,该方法利用不良地质构造与周围介质或者岩层之间的电阻率差异,通过对电阻率的层析成像,来对隐藏在岩体内的不良地质构造和岩层交界面进行识别和定位。该方法的具体工作原理是:在相邻的 2 个地质勘探钻孔中,一个放入供电电极,另一个放入测量电极,利用从钻孔中观测到的电位或电位梯度值进行直接或间接的成像反演,就可以获得这 2 个钻孔间地层的电阻率分布图[2]。跨孔电阻率 CT 法探测原理见图10。

跨孔电阻率 CT 法具有以下优点: 1) 探测电极安装在孔中,深入围岩,可避开各种电磁干扰; 2) 采用跨孔"透视对穿"的观测方式,采集的数据量更多,且更接近勘探目标体; 3) 信号不随深度方向衰减,分辨率更高。

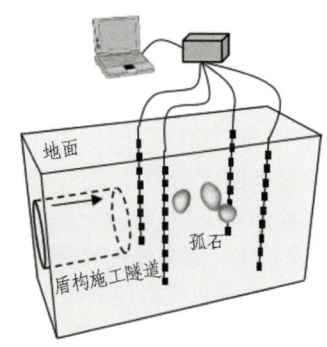

图10 跨孔电阻率 CT 法探测原理

Fig. 10 Detection principle of CT technology by cross-hole resistivity method

1.3.1 厦门市轨道交通 1 号线孤石探测

厦门市轨道交通 1 号线集美大道站—天水路区间站位于厦门市集美区后溪镇,区间起于集美大道站,下穿崎沟村、东宅村民房后到达天水路站,区间穿越残积土、全风化花岗岩、散体状强风化花岗岩等地层,初勘时发现基岩突起及孤石存在,且因区间上部分布大量崎沟村民房,房屋基础薄弱,密集无序,详勘仅钻探 18 个孔位,传统方法无法探明区间孤石、基岩突起具体分布情况。选择地质相近、地势较为开阔的天水路站—厦门北站区间(起讫里程 YDK30 + 095 ～ YDK31 + 035)作为试验段,采用跨孔电阻率 CT 法进行孤石探测试验,部分探测结果见图11。本次跨孔电阻率 CT 法孤石探查试验推断存在 12 处孤石,9 处得到钻孔验证,孤石揭露准确率超过 70%,探测较为成功。

1.3.2 大连地铁 2 号线东春区间溶洞超前探测

跨孔电阻率 CT 法也可对溶洞进行超前探测,通过三维切片成像可推断溶洞的发育位置和规模。采用该方法对大连地铁 2 号线东纬路—春光街区间进行了精细探查,部分探测结果见图12。可知跨孔电阻率 CT 成像结果与实际开挖结果吻合。

1.3.3 南京地铁上元门站基坑涌水探查

超前探测的另一难点是溶管(岩溶裂隙)的探测,南京地铁上元门车站靠近长江,车站基坑开挖时不断涌水,采取注浆措施封堵,因未查明涌水通道,注浆针对性不强,注浆效果不佳。采用跨孔电阻率 CT 法、高密度电法、瞬变电磁法和地质雷达法 4 种方法相结合的综合物探方法对车站基坑进行精细探查,综合物探结果见图13。采用综合物探方法确定了长江水经过基坑的流水通道以及基坑底板富水区的空间分布,根据物探解译结果,推断了长江水源补给通道,合理地设计了注浆孔位,并通过注浆对涌水点进行有效封堵。

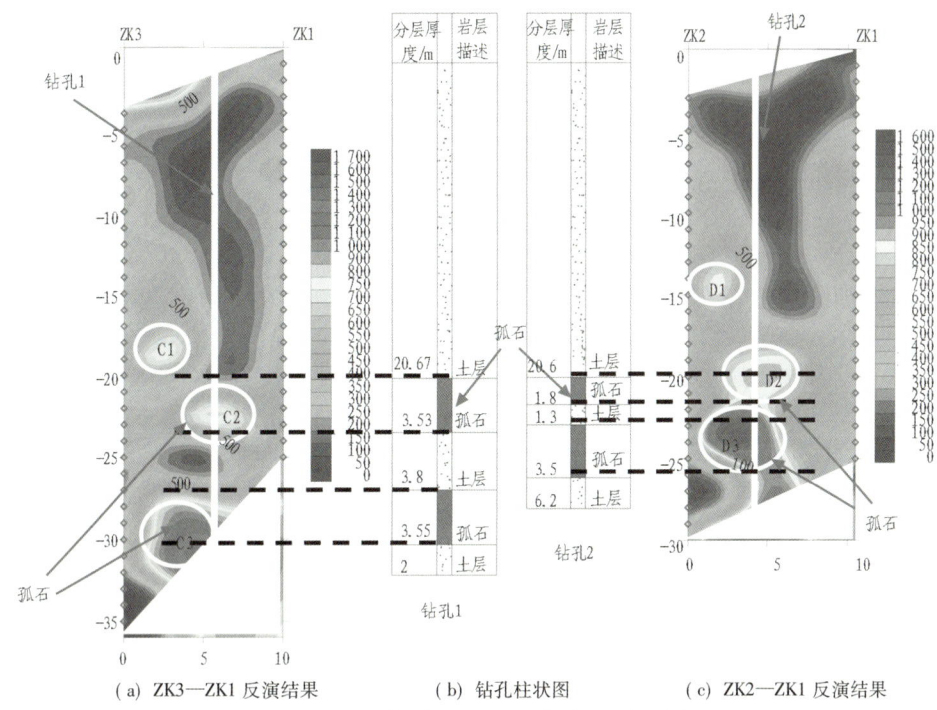

(a) ZK3—ZK1 反演结果　　(b) 钻孔柱状图　　(c) ZK2—ZK1 反演结果

图 11　厦门市轨道交通 1 号线部分孤石探测结果

Fig. 11　Detection results of boulders in Xiamen Rail Transit Line No. 1

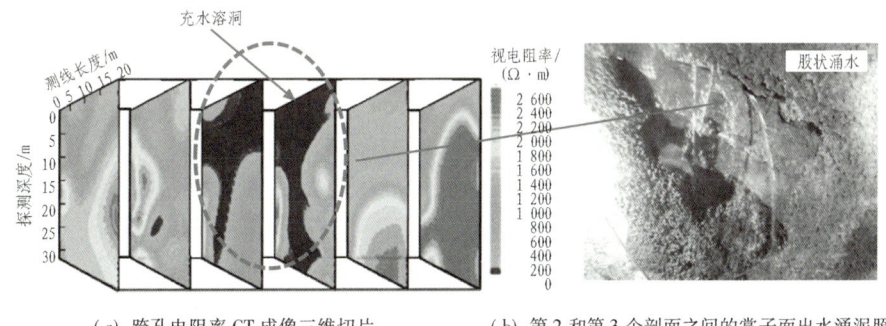

(a) 跨孔电阻率 CT 成像三维切片　　(b) 第 2 和第 3 个剖面之间的掌子面出水涌泥照片

图 12　大连地铁 2 号线东春区间溶洞超前探测结果

Fig. 12　Detection results of karst caves of Dondweilu-Chunguangjie Section on Dalian Metro Line No. 2

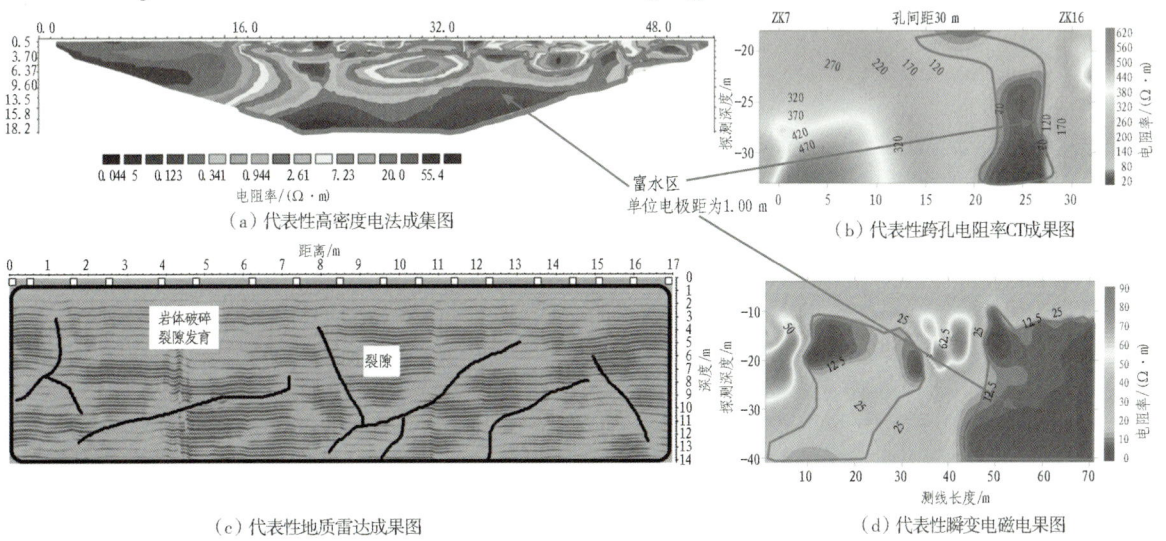

(a) 代表性高密度电法成集图　　(b) 代表性跨孔电阻率CT成果图

(c) 代表性地质雷达成果图　　(d) 代表性瞬变电磁电果图

图 13　南京地铁上元门站基坑涌水综合物探结果

Fig. 13　Comprehensive geophysical prospection results of water inrush of foundation pit of Shangyuanmen Station of Nanjing Metro

2 隧道岩爆监测预警方法及其应用

2.1 隧道岩爆监测预警可行性分析

岩体中一般积聚有弹性变形势能,在一定条件下,这些能量会猛烈释放,岩石发生爆裂并弹射出来的现象,我们称之为岩爆。岩爆的形成一般经历岩体破裂、块体形成、块体弹射3个阶段。岩石是各向异性的非均匀材料,当岩体中的裂纹产生、扩展和摩擦时,岩体内部的能量会以波的形式释放,这就产生了微震事件。我们利用微震监测技术可以接收到岩体内部的微震信息,通过科学反演,就能够得到岩体微破裂发生的时间、具体位置和震裂强度等信息。根据微破裂释放出能量的大小、分布和集中程度,就可以对岩爆的可能性、发生的位置和等级等进行预测预报。现阶段,微震监测技术已经在地下工程中得到了广泛的应用。在多数的岩爆孕育过程中,微震事件及其能量的演化具有自相似性(时间、能量及空间分形特征),且有微震信息前兆特征,这种相似性见图14。可知:在大多数情况下,可以利用已监测到的微震活动性,在基于未来施工不变的情况下,对岩爆的区域和等级进行预警。

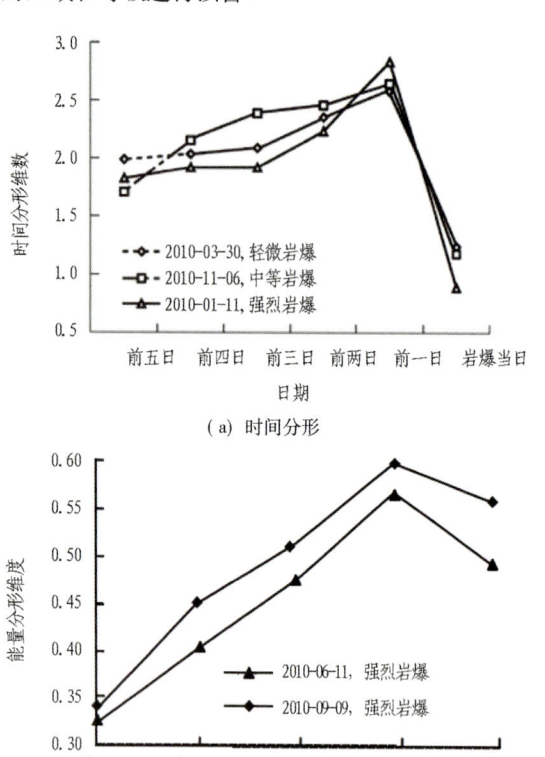

图 14　某工程微震事件及其能量的演化自相似性

Fig. 14　Self-similarity of microseism incidents of a tunnel

2.2 隧道岩爆监测预警技术与方法

2.2.1 按施工方法预警

在隧道 TBM 法和钻爆法施工时,岩爆孕育存在以下显著差异:1) TBM 高等级岩爆孕育伴随低等级岩爆(见图 15);2) TBM 同一区域常发生多次岩爆;3) TBM 岩爆多发生在开挖过程中,钻爆法几小时到几天不等;4) TBM 诱发的微震事件能量一般较大(见图 16)。

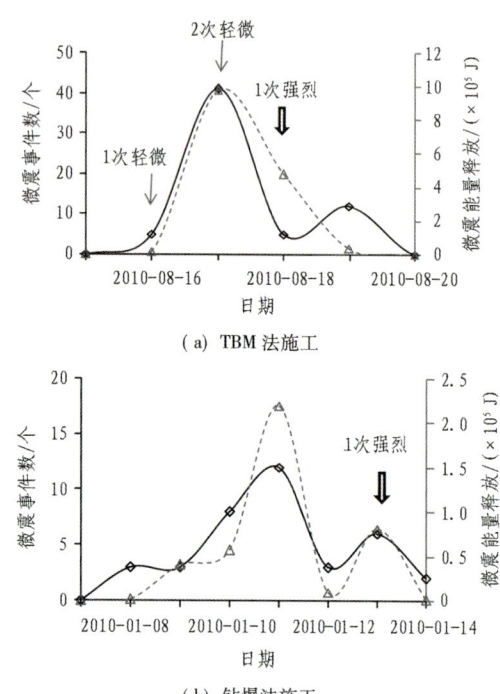

图 15　某工程 TBM 法与钻爆法岩爆等级差异

Fig. 15　Difference between rockburst grades induced by TBM method and those induced by drilling and blasting method

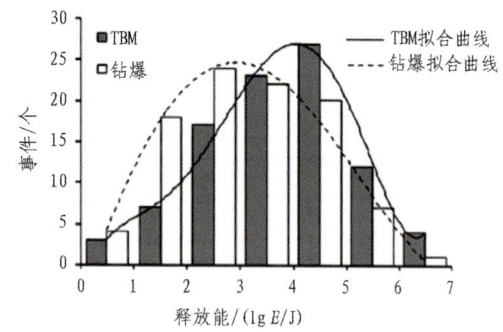

图 16　TBM 法与钻爆法微震事件能量差异

Fig. 16　Difference between seismic energy induced by TBM method and that induced by drilling and blasting method

2.2.2 按岩爆类型预警

岩爆形式的动力破坏基本可以分为 2 类:1) 第 1 类是由岩石破坏导致的,通常称为应变型岩爆;2) 第 2 类是断层滑移或者剪切断裂所导致的,称为应变结构面滑移型和断裂滑移型。这 2 类岩爆的主要区别是:在第 1 类岩爆中,扰动源(开挖)和岩爆发生的部位是相重合的；在第 2 类岩爆中,扰动源和所导致的岩爆发生部位离开一段距离,甚至是相当大的距离。此外,与第 2 类岩爆(滑移断裂型岩爆)相联系的能量通常远远大于应变型岩爆的能量。断裂滑移型岩爆的破坏程度也通常比应变型岩爆强烈得多[3]。因此,建立了隧道不同类型岩爆定量预警公式

$$P_i^{mr} = \sum_{j=1}^{6} w_j^{mr} P_{ji}^{mr} \quad (1)$$

式中：m 表示施工方法,主要有钻爆法和 TBM 法等；r 表示岩爆类型,分为应变型、应变-结构面滑移型和断裂滑移型；i 表示岩爆等级；j 表示用于预警的微震参数；P 表示岩爆发生概率。

基于大量岩爆案例统计分析,锦屏二级水电站钻爆法施工引排水隧道不同类型强烈岩爆的 6 微震参数预警阈值见表 1。

表 1　锦屏二级水电站钻爆法施工引排水隧道不同类型强烈岩爆的预警阈值

Table 1　Warning thresholds of different types of serious rockburst induced by drilling and blasting construction of water diversion and drainage tunnel of Jinping Ⅱ Hydropower Station

岩爆类型	事件数/个	lg E	lg V	事件率/(个/d)	释放能速率 lg(E/t)	视体积率 lg(V/t)
应变型	49.7	6.3	5.0	5.5	5.6	4.1
应变-结构面滑移型	32.1	5.8	4.9	3.0	4.5	3.9

注：E 表示释放能,J；V 表示视体积,m³；t 表示时间,d。

2.3 隧道岩爆孕育过程动态调控

2.3.1 TBM 开挖应变型强烈岩爆预警与调控[4]

通过对某深埋工程 3#TBM 施工隧洞发生的微震事件进行滤噪和定位分析,获得微震事件数量和能量等级随时间的演化规律,见图 17。可知：从 2010 年 9 月 6 日开始至 9 月 8 日,微震活动趋于活跃,微震事件的数量和能量均随时间出现大幅度增加；9 月 6—8 日,一共出现 17 个震级大于 -0.2 的事件,其中震级大于 0.5 的出现 7 个,同时,大事件数量也表现为逐步增加的趋势。3#TBM 隧洞开挖的微震事件累计分布见图 18(图中球体颜色表示微震事件震级大小,颜色与震级大小对应关系如图例所示；球体大小表示微震事件释放能的大小)。由图 18(a) 可看出,该期间共出现 54 个有效微震事件,累计释放能量为 1.1×10^7 J,微震事件比较集中,且释放能量较大。

现场及时采取如下调控措施：1) 降低 TBM 掘进速度,2010 年 9 月 8 日进尺为 16.25 m,9 日进尺降到 9.55 m,10 日再降至 6.25 m,11 日进尺维持在 6.64 m；2) 加强支护措施,增加了 6 m 锚杆的数量。

及时采取调控措施后,效果如下：1) 微震活动趋于平缓,如图 18(b) 所示,微震事件数量和能量都明显降低。2010 年 9 月 6—8 日累计事件数为 54 个,震级大于 -0.2 的事件 27 个,震级大于 0.5 的事件 7 个；9—11 日累计事件数降至 29 个,震级大于 -0.2 的事件降至 10 个,震级大于 0.5 的事件仅有 2 个,微震释放能量也由 1.1×10^7 J 降低至 5.9×10^6 J。微震活动整体趋于平稳。2) 岩爆等级降低,工程现场 9 月 9 日和 10 日分别发生中等岩爆 1 次,随后此洞段安全完成开挖。

2.3.2 钻爆法开挖应变-结构面滑移型中等岩爆预警与调控[4]

通过对某深埋工程 1-1-W 和 2-1-E 施工隧洞发生的微震事件进行滤噪和定位分析,获得的微震事件数量和能量随时间的演化规律见图 19。可知：从 2010 年 12 月 23—26 日,微震活动日渐趋于活跃,微震事件的数量和能量随时间均出现了显著增加,因此怀疑岩体中存在近东西向的硬性结构面。这 4 d 共监测到有效微震事件 35 个,其中震级大于 0.6 的事件有 4 个,且监测到的微震事件大部分集中在 2-1-E 掌子面后方和 2-1-E 与 1-1-W 掌子面中间位置,微震事件累计释放能量 5.4×10^4 J,从中可以看出,微震事件集中而且释放能量较大。

根据以上微震活动的特征和规律,预测到 2010 年 12 月 25 日开挖区域存在高概率中等岩爆发生的风险。因此,建议微震事件集中的 2-1-E 掌子面停止掘进,1-1-W 掌子面单向掘进,及时做好系统支护工作,并对关键部位的支护措施进行加强,同时建议必要时对掌子面进行应力释放。除此之外,在距离掌子面 100 m 处拉上警戒线,禁止车辆和行人进入。26 日

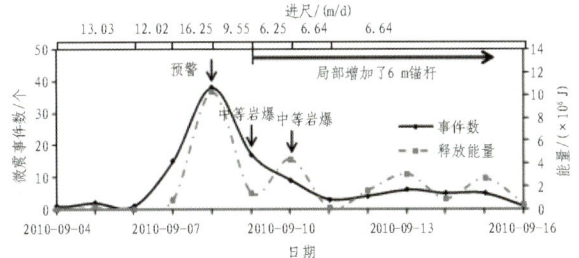

图 17　3#TBM 开挖隧洞时微震事件数和能量随时间演化规律

Fig. 17　Quantities and energy of microseismic incidents vs. time during TBM Tunnel #3

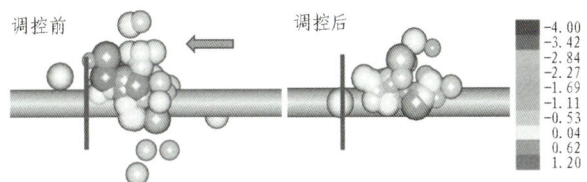

(a) 2010 年 9 月 6—8 日　　(b) 2010 年 9 月 9—11 日

图 18　3#TBM 开挖隧洞累计微震事件分布图

Fig. 18　Distributions of microseismic incidents during construction of TBM Tunnel #3

建议改为由 1-1-W 掌子面单向掘进。根据现场资料显示,26 日 2-1-E 掌子面处实际发生中等岩爆 1 次,证明了岩爆预测的准确性和调控措施的必要性。

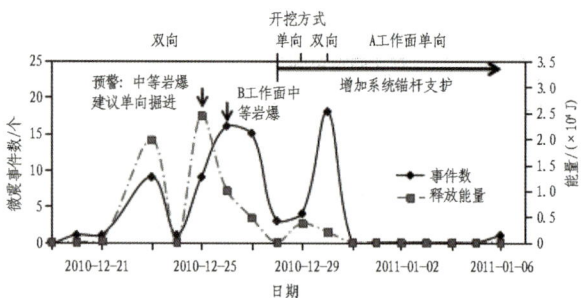

图 19 某深埋工程施工时微震事件数和能量随时间演化规律

Fig. 19 Quantities and energy of microseismic incidents vs. time during a deep tunnel construction

根据建议,2-1-E 掌子面 12 月 28 日暂停掘进,同时增加系统喷锚支护,1-1-W 掌子面单向掘进,掌子面开挖隧洞累计微震事件分布见图 20,可知:采取调控措施前,12 月 23—26 日共发生微震累计事件数量 35 个,能量释放 5.4×10^4 J,震级大于 -0.6 的事件有 4 个,震级为 -2~-1 的事件有 14 个;采取调控措施后,12 月 28—31 日共发生微震累计事件数量 25 个,能量释放 5.8×10^3 J,震级大于 -0.6 的事件 0 个,震级为 -2~-1 的事件 7 个。通过以上对比可以看出,微震事件数和能量都表现为明显的降低,微震活动明显减弱。

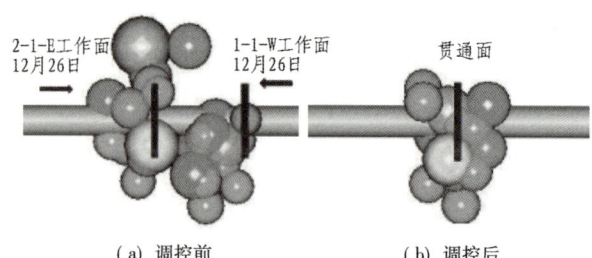

图 20 1-1-W 和 2-1-E 掌子面开挖隧洞累计微震事件分布

Fig. 20 Distributions of microseismic incidents during construction of tunnel face No. 1-1-W and No. 2-1-E

2.4 隧道岩爆监测预警工程实践

2.4.1 锦屏二级水电站深埋隧洞开挖过程岩爆监测预警与防控[5]

锦屏二级水电站位于四川省凉山彝族自治州,该水电站利用雅砻江 150 km 锦屏大河湾处的天然落差,截弯取直开挖隧洞引水发电,水电站总装机规模达 480 万 kW。工程开挖共包括 7 条平行隧洞,即 1#—4# 引水隧洞、施工排水洞和 A、B 辅助洞等,图 21 为其平面布置。其中,4 条平行布置横穿锦屏山的引水隧洞,从进水口至上游调压室的平均洞线长度约 16.67 km,中心距 60 m,洞主轴线方位角为 N58°W,开挖直径 12.40~13.00 m,全线一般埋深为 1 500~2 000 m,最大埋深达 2 525 m。隧洞具有大、长、深等特点。

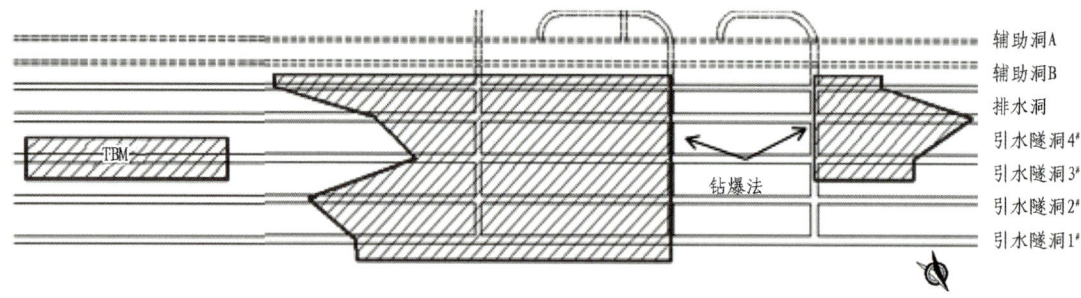

图 21 水电站隧洞布置平面图

Fig. 21 Plan of layout of hydropower station tunnels

在采用岩爆监测预警系统前,2 条辅助洞发生岩爆,造成人员伤亡和严重恐慌,施工队伍被迫更换多次,工期延误 1 年以上。排水洞 2009 年 11 月 28 日发生极强岩爆,TBM 被毁,多人伤亡,停工半年,被迫更改施工方案。

之后采用微震监测技术进行监测预警与预控,通过微震数据采集系统连续进行数据采集、数据远程传输、数据处理与分析,实现对施工隧洞微震事件的连续监测和分析。相邻平行最大埋深洞段岩爆风险控制效果见表 2。

表 2 相邻平行最大埋深洞段岩爆风险控制效果比较

Table 2 Rockburst control effects

对比项	技术应用前	技术应用后
岩爆长度占总长度比	辅助洞 A 为 20.2%;辅助洞 B 为 23.9%	引水隧洞平均 7.3%;排水洞为 5.7%
强烈岩爆次数	辅助洞 A 和辅助洞 B 平均 10.5 次	引水隧洞平均 2.3 次

2.4.2 锦屏地下实验室二期开挖全过程灾害监测预警与防控[6]

锦屏地下实验室二期工程位于锦屏交通洞 A 洞南侧,最大埋深约 2 400 m,是目前世界上埋深最大的实验室,施工过程中潜在的岩爆、片帮、坍塌等硬岩工程灾害发生的频率高、危害严重。根据锦屏深部地下实验室二期的功能设计要求,结合布置区域的地质条件、已有洞室布置和施工条件,地下实验室总体方案采用 4 洞 9 室"错开型"的布置形式,见图 22。目前共有 9 个实验室,其中,1#—6# 为物理实验室,7#—9# 规划为深部岩石力学实验室。1#—8# 实验室长度均为 65 m,城门洞型,隧洞截面 14 m × 14 m,9# 实验室长 60 m(东西两侧各 30 m)。各实验室均采用钻爆法施工,分 3 层开挖,上层 8.0 m,中层 5.0 m,下层 1.0 m。其支护方式主要为锚杆和喷射混凝土。

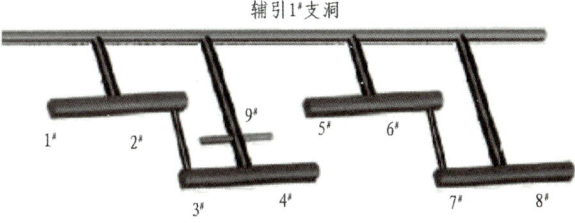

图 22 锦屏地下实验室二期隧洞布置图

Fig. 22 Layout of 2nd phase tunnel projects of Jinping underground laboratory

在 7# 和 8# 实验室施工开挖过程中,于 2015 年 8 月 23 日发生一次极强岩爆。岩爆区域长约 44 m,高 5~6 m,最大爆坑深度 3.1 m,最大爆坑尺寸达 2.4 m × 2.4 m × 1 m,岩块最大弹射距离 7~10 m,爆出岩块体积约 400 m³,岩爆造成 7#、8# 实验室上层南侧边墙已完成的支护系统严重破坏,破坏区的锚杆被拉断和拔出,钢筋网和初喷混凝土被抛出。由于采取全过程灾害监测预警与防控措施,本次"8·23"极强岩爆在发生前 2 h 被成功预警,施工单位根据岩爆预警信息及时通知现场,并撤离了高风险区的施工人员和设备,成功避免了因本次大规模极强岩爆带来的人员伤亡和设备损失,保证了实验室二期工程建设期的施工安全。

3 基于 BIM 技术的建筑物(隧道工程)的安全风险监控

3.1 BIM 技术内涵

BIM(building information modeling)的本质就是把数据转化为信息,通过数字信息仿真模拟建筑物(隧道)具有的真实信息,以三维虚拟现实技术建模,实现可视化的工程数据模型。在建设项目管理过程中,通过运用 BIM 技术,从设计阶段开始就建立互相协调、内部一致的可运算三维信息模型,可以大幅度降低参建各方项目管理的难度,从而解决了许多原来二维平面模型不能解决的问题。从某种程度上说,BIM 不仅仅是一种模型工具,而且是一个协同的工作流程。通过三维模型的可视化演示,可以实现对建设工程项目的碰撞检测、施工进度模拟、工程质量分析和安全风险防控等[7]。

在地铁建设的管线施工中,需要在非常有限的空间内完成通信、信号、综合监控以及通风、暖通、消防、电力等十几个专业系统的安装,各系统之间碰撞冲突、返工修改等问题突出,合理布置这些系统与建筑、结构间的空间关系非常重要。而二维图纸由于无法可视化,往往会导致施工前各专业的冲突问题难以到解决。由于 BIM 技术模型具有数字化、可视化、真实化的特点,采用 BIM 技术模型进行车站的施工碰撞研究,可使项目各参与方进行无间隙共享及无障碍交流,在整个项目周期高效协同工作,从而有效解决上述难题[7]。

在高风险特殊区段地铁施工中,为了降低灾难性事故的发生概率,我们需要实时地对安全风险进行感知,才能及时防范事故发生。因此,进一步将包含岩土地质信息、地下管线信息、周边建筑技术信息、机械人员信息、施工监测信息等在内的工程实体与施工工序信息对应的时间维集成,建立地铁施工的 4D 模型。某地铁车站 4D 模型见图 23。

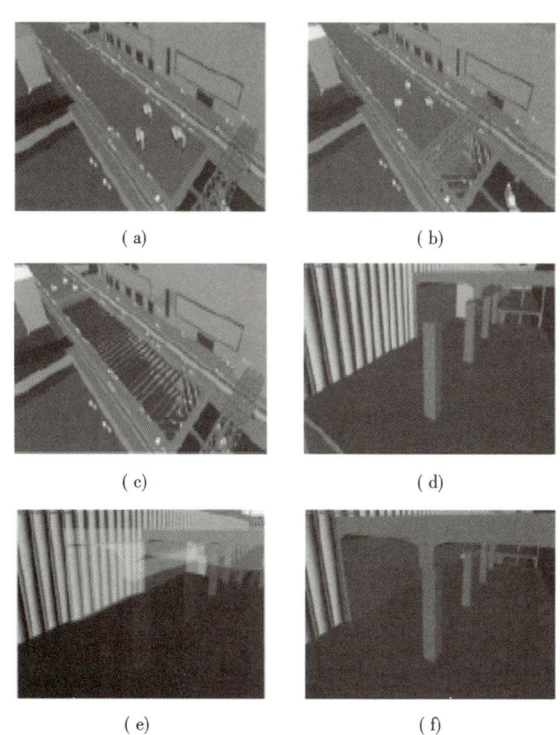

图 23 某地铁车站 4D 模型

Fig. 23 4D models of a Metro station

3.2 隧道施工安全风险实时感知及实时预警系统

下文以武汉某地铁越江隧道联络通道施工过程中的风险控制为例,介绍施工中的安全风险实时感知及实时预警系统。

3.2.1 安全风险实时感知系统

为了实现对工程周边环境与工程结构等多物理量的连续实时感知,提高风险特殊区段地铁施工安全风险信息的采集和传输能力,将光纤光栅传感技术引入到了某地铁隧道联络通道的冻结法施工中。由于光纤光栅传感器具有耦合监测、高精度、自动连续、抗电磁干扰、不受水和潮气影响、远距离传输等诸多优点,工程利用光纤光栅传感器对水平冻土、联络通道初期支护和既有隧道管片分别进行温度-应变耦合监测,构建了联络通道施工多场耦合实时感知系统,从而实现了整个施工周期内对数据的自动连续采集和实时分析与预警[8]。

实时感知系统组成见图24,包括独立供电系统、数据存储分析系统、数据实时采集系统等3部分,并增设防尘防水保护系统。

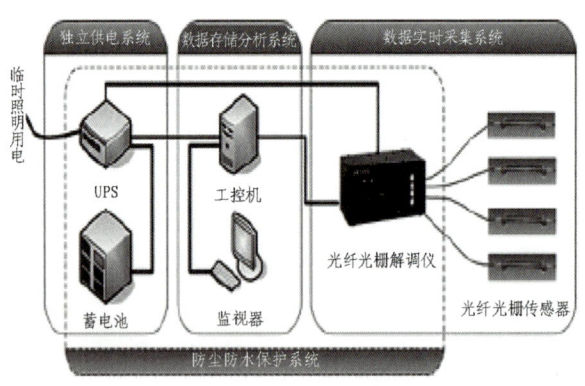

图 24 实时感知系统组成
Fig. 24 Real-time perceiving system

3.2.2 安全风险实时预警系统

从海因里希的事故连锁理论、轨迹交叉理论等事故致因理论中,我们可以得知,施工中人的不安全行为和物的不安全状态会导致安全事故的发生。因此,某地铁隧道联络通道施工过程中,除了建立多场耦合实时感知系统实时获取物的状态外,人的行为对于施工安全风险控制而言更为重要[8]。因此,在多场耦合实时感知系统的基础上,实现了复杂环境下长大隧道中实时跟踪移动目标,并将联络通道施工过程中环境、结构和人的安全信息综合起来进行安全分析判断,并及时有效地发布预警信息,第一时间通知现场作业人员采取应急措施,实现了安全知控一体化和实时化。安全知控一体化和实时化原理见图25。

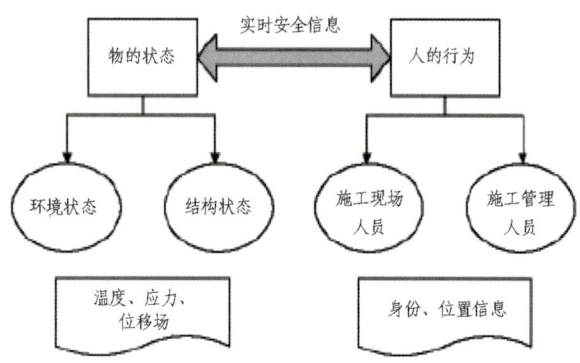

图 25 安全知控一体化和实时化示意图
Fig. 25 Sketch diagram of integration and real-time warning system

地铁施工过程中存在大量的安全信息,必须充分、及时掌握这些安全信息才能对其进行有效的安全控制。因此,BIM、物联网、数据融合等前沿信息技术的应用是提升地铁工程施工安全风险控制水平的重要途径之一,它对于提高地铁施工过程中安全信息的采集、传输、分析和挖掘能力,降低施工安全风险,具有突出作用。隧道实时定位系统架构见图26。

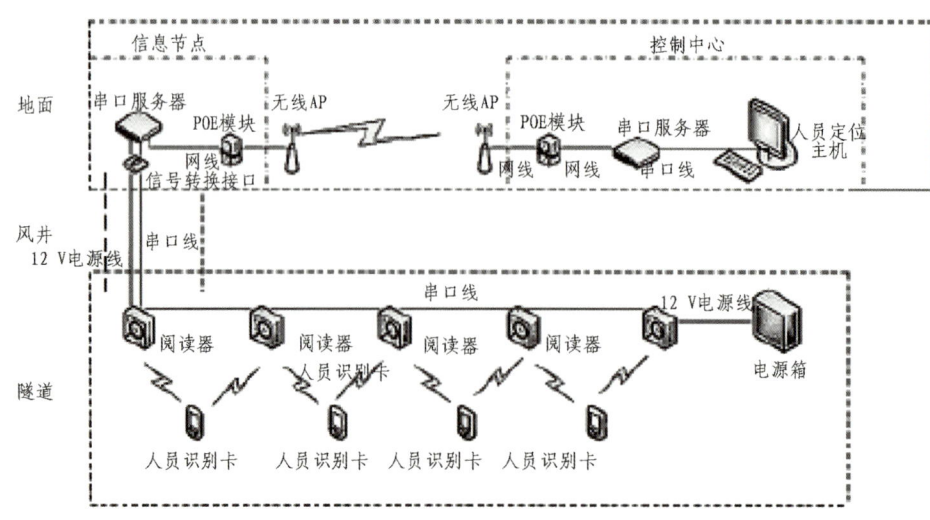

图 26 实时定位系统架构
Fig. 26 Real-time location system

4 隧道工程建设信息化技术的发展方向

4.1 开展基于大数据技术的TBM/盾构施工时的分析与控制研究

隧道建设时的工程事故和TBM/盾构的低运行除了与客观不良地质条件有关,还与TBM/盾构的选型不当以及操控参数选择不合理有关。如何使选型与参数选择从经验上升为科学? 运用大数据技术是一个可行的方向。

4.1.1 大数据技术的定义和特征

大数据技术是指对数据规模大、结构复杂度高、关联度强的数据集进行处理与应用的处理技术。具有以下"4V"特征: 1) Volume(大量),数据量巨大; 2) Variety(多样化),数据类型多且十分复杂; 3) Velocity(快速),处理速度要求快; 4) Value(价值密度低),虽然数据采集量巨大,但有用数据少。

4.1.2 大数据技术内涵

大数据涉及到的技术主要包括数据挖掘与关联分析技术、机器学习、模式识别、预测模型、时序分析以及可视化处理等。

1) 数据挖掘。数据挖掘是一个知识发现的过程,即从大量的数据中自动搜索隐藏在其中的知识或特殊关系信息的过程。

2) 关联分析。关联分析是从大量数据中分析各数据项之间有价值的相关关系。比如20世纪90年代美国沃尔玛超市基于销售数据关联分析,将啤酒与尿布2个看上去没有关系的商品放在一起进行销售,获得了很好的收益。

3) 机器学习。机器学习主要研究如何使用电脑来模拟和实现人类学习时获取知识的过程,重构已有知识,提升自身处理问题能力从而形成创新。机器学习的最终目的是从数据中自动分析获取规律性知识,并用于对未知数据进行预测、判断和评估。机器学习与统计推断学联系密切。

4.1.3 大数据技术应用在TBM/盾构施工中的可行性分析

TBM/盾构施工数据是典型的大数据问题,大数据的概念和TBM/盾构施工数据具有天然的契合性。TBM/盾构在施工掘进过程中连续自动采集数据并存储在数据库中。比如海瑞克盾构每2.5 s会自动采集,每10 s会自动存储一次数据。对于1条长约25 km的隧道来说,单就掘进参数而言,每台盾构每分钟产生包括盾构扭矩、推力、转速、贯入度等掘进参数在内的掘进数据,如果推进2 h,1条线路将产生约4.2×10^5组掘进数据,盾构/TBM掘进数据不仅包括自身的掘进参数(扭矩、推力、转速、贯入度等),还有地层变形数据(包括应力、位移),以及掘进地层地质数据(比如地层类型、土体物理力学参数、地下水位等),加上各监测点获取的数据(包括区间隧道上方、周边布置的监测点以及大量管线、建筑物、道路布置的测点),数据形式多样(包括数值、文字、图片等资料)。总的来说,盾构/TBM施工数据量巨大。盾构掘进参数之间、盾构掘进参数与地层变形之间相互关联,非常复杂,呈现出大量、多态、多源、多维的大数据特征。因此,盾构/TBM施工数据是典型的大数据应用范例。

4.1.4 大数据技术在TBM/盾构施工中的应用前景

目前中国大部分的地铁工程都采用盾构施工,山岭隧道和引水隧道等长大隧道建设中采用TBM施工的也越来越多,并都研发了施工信息管理系统(见图27),系统用以支持TBM/盾构的远程监控,通过多种传输方式将掘进数据发送至中心服务器并自动存储,其采集、翻译、传输和存储分析的过程见图28。因此,TBM/盾构施工管理信息化技术的应用为大数据分析提供了数据基础。

(a) 宁和城际轨道交通一期风险管控系统

(b) 成都地铁建设工程安全风险监控系统

图27 地铁盾构施工信息管理系统

Fig. 27 Construction information management system of Metro shield construction

通过大数据技术有效利用TBM/盾构施工中的大数据,对优化TBM/盾构的设计,正确进行TBM/盾构的选型,提升TBM/盾构的掘进效率和控制施工时的事故发生具有重要意义,其关键是建立数据联盟以及对TBM/盾构施工海量数据的数据挖掘和关联分析技术的研究和提升。

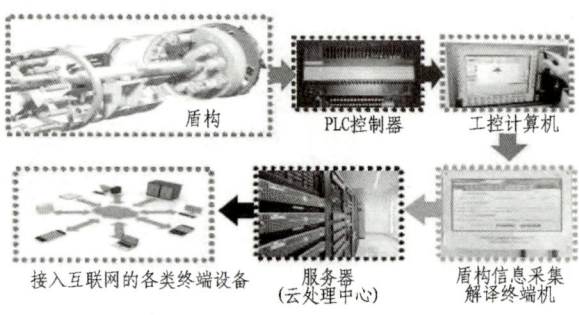

图 28　盾构数据采集过程
Fig. 28　Data acquisition of shield

基于大数据技术开展 TBM/盾构施工中的应用研究在国内刚刚起步,是一个富有探索性和挑战性的课题,尚未有成功的应用实例。例如 TBM/盾构的扭矩是其选型的重要参数,现在依据的是日本公式 $T=aD^2$,式中影响参数单一(仅为盾构外径 D)。盾构外径 D 大时,扭矩偏大,不能充分发挥切削能力;盾构外径 D 小时,扭矩偏小,容易出现扭矩过载。因其复杂性,进行理论研究很难,可以通过大数据技术对以往掘进施工参数进行统计挖掘,分析不同地质条件下不同类型盾构掘进参数的变化规律,得到适用性强的经验公式,为优化国产盾构选型参数提供技术支持。再如何选择与地层特性相适应的掘进参数,从而保证开挖面稳定、减小地层变形位移是盾构施工控制的难点。可以通过对不同地层内掘进时的掘进参数与地层变形参数进行统计来控制掘进,分析盾构掘进参数与地层变形之间复杂的规律性关系,为有效预测复杂条件下地层变形、防治地面隆起或坍塌事故提供依据。

4.2　数字隧道向智慧隧道(建设和运营维护)的发展

智慧工程方面国外已有一些实例,如:2006 年,新加坡启动"智慧国家 2015"计划,建立了针对交通堵塞预报的智慧城市系统;2009 年,韩国仁川实现了房间耗能的智能控制与通过网络监测病人状况的智慧城市应用;昆士兰建立了对桥的智能安全系统,通过装在桥上的传感器,确保桥的安全;爱尔兰"智慧湾"利用浮桥上的传感器及渔民手机,监测水面上的漂浮物、水流,沟通渔民和餐厅的交易;瑞典斯德哥尔摩智慧城市系统使汽车使用量降低 25%,尾气排放降低 8%~14%。隧道工程也将从数字隧道向智慧隧道方向发展。

4.2.1　数字隧道和智慧隧道的概念

数字隧道是隧道工程信息化的初级阶段,是"物理隧道"(实体隧道)的虚拟对照体,以信息化手段对隧道建设过程中的勘察、设计、施工及监测等数据进行集中有效管理。具体体现为数字隧道工程基础平台,该平台是集数据信息存储、查询、三维可视化建模及虚拟浏览为一体的综合系统。其信息包括地形、地理的基础地理数据;包括工程地质和水文地质、环境地质的地质数据(两者共称为地层数据),隧道主体的设计、施工及监测数据等。

智慧隧道是为隧道工程建设和运营服务的隧道工程信息化的高级阶段。包括智慧和智能 2 方面:智慧是对大系统和巨系统而言,例如城市是一个巨系统,包括人、自然和社会等的综合体。隧道是一个大系统,包括隧道本体、隧道环境、隧道建设者和运营者的综合体;智能是对某项技术、某个功能和某种仪器设备而言,如智能手机、智能传感机和智能交通等。

具体来说,智慧隧道就是让作为隧道系统主体的隧道工程建设者和运营者更聪明。首先,它通过互联网把无处不在的被植入隧道本体、周围地层中的智能化传感器、实时跟踪移动目标的 GPS 定位单元以及无线射频识别单元连接起来形成物联网,以此实现对物理隧道(隧道本体和环境)、隧道建设者和设备的全面感知;除此之外,智慧隧道利用云计算技术能对感知信息进行智能处理和分析,实现网上"数字隧道"与物联网的融合;最后,在分析处理后发出对包括超前地质预报、设计方案和修改、施工方案的实施(如盾构推进的操作参数的确定等)、预警信息的发布、应急防治方案的实施等作出智能化响应和决策支持的指令。

4.2.2　智慧隧道和数字隧道的区别和联系

数字隧道是物理隧道在网上的虚拟对照体,两者是分离的;智慧隧道运用物联网可以把数字隧道与物理隧道无缝连接在一起,是物联网与"数字隧道"的融合,智慧隧道是数字隧道功能的延伸、拓展和升华,是数字隧道的智能化。利用云计算对实时感知数据进行快速和协同处理,并在大数据技术所具备的感知能力、逻辑思维能力、自学习与自适应能力和行为决策能力的基础上提供智能化服务。

智慧隧道和数字隧道之间并无绝对界线,是可以过渡的,例如有 4D 数字隧道、N 维数字隧道……,相应有 4D 和 ND BIM 技术。

4.2.3　智慧隧道智的体现

1) 透彻感知。无处不在的智能传感器,对隧道、环境、设备和人及其状态实现全面、综合地感知和对其运行状态的实时感测。

2) 全面互联。通过运用物联网将所有传感器全面连接,通过运用互联网实现感知数据的智能传输和存储。

3) 深度整合。物联网和互联网完全链接和融合,将多源异构数据整合为一致性数据——隧道工程建设和运营全图。

4) 智能服务。在隧道智慧信息(网络、数据)基础上,利用云计算构架一种新的能提供服务的系统结构,基于大数据技术,对海量感知数据进行并行处理、数据挖掘与知识发现,能够为隧道建设和运营提供各种不同层次、不同要求的高效率智能化服务。

4.2.4 智慧隧道建设

建设智慧隧道的关键是做好智慧隧道工程数据交换共享平台建设。这是一个智慧信息基础设施—智慧应用服务技术支持层—智感应用服务层的推进过程。

要有4方面的提升: 1) 三维可视化表达—统一时空基准的四维信息; 2) "静态数据+周期性更新"—"实时获取+动态更新"; 3) "有限服务"—"全面深度服务"; 4) "事后分析+辅助决策"—"实时分析+智能数据挖掘+知识发现+实时决策"。

5 结语

不断提高地质预报技术水平,加强隧道建设、运营维护全过程的信息化、可视化、智慧化研究,逐步实现隧道更加安全、高质、高效、智能的建设与管理,是今后一个时期的发展方向。

参考文献(References):

[1] 李术才,刘斌,李树忱,等.基于激发极化法的隧道含水地质构造超前探测研究[J].岩石力学与工程学报,2011,30(7):1297 – 1309. (LI Shucai,LIU Bin,LI Shuchen,et al. Study of advanced detection for tunnel water-bearing geological structures with induced polarization method [J]. Chinese Journal of Rock Mechanics and Engineering, 2011, 30(7):1297 – 1309. (in Chinese))

[2] 李术才,苏茂鑫,薛翊国,等.城市地铁跨孔电阻率CT超前地质预报方法研究[J].岩石力学与工程学报,2014,33(5):913 – 920. (LI Shucai,SU Maoxin,XUE Yiguo,et al. Study of computed tomography of cross-hole resistivity in urban subway geological prediction [J]. Chinese Journal of Rock Mechanics and Engineering,2014,33(5):913 – 920. (in Chinese))

[3] 钱七虎.岩爆、冲击地压的定义、机制、分类及其定量预测模型[J].岩土力学,2014,35(1):1 – 6. (QIAN Qihu. Definition, mechanism, classification and quantitative forecast model for rockburst and pressure bump [J]. Rock and Soil Mechanics,2014,35(1):1 – 6. (in Chinese))

[4] 冯夏庭,张传庆,陈炳瑞,等.岩爆孕育过程的动态调控[J].岩石力学与工程学报,2012,31(10):1983 – 1997. (FENG Xiating,ZHANG Chuanqing,CHEN Bingrui,et al. Dynamical control of rockburst evolution process [J]. Chinese Journal of Rock Mechanics and Engineering,2012,31(10):1983 – 1997. (in Chinese))

[5] 张文东,马天辉,唐春安,等.锦屏二级水电站引水隧洞岩爆特征及微震监测规律研究[J].岩石力学与工程学报,2014,33(2):339 – 348. (ZHANG Wendong,MA Tianhui,TANG Chun'an,et al. Research on characteristics of rockburst and riles of microseismic monitoring at diversion tunnels in Jinping II hydropower station [J]. Chinese Journal of Rock Mechanics and Engineering, 2014, 33(2):339 – 348. (in Chinese))

[6] 冯夏庭,吴世勇,李邵军,等.中国锦屏地下实验室二期工程安全原位综合监测与分析[J].岩石力学与工程学报,2016,35(4):649 – 657. (FENG Xiating,WU Shiyong,LI Shaojun,et al. Comprehensive field monitoring of deep tunnels at Jinping underground laboratory(CJPL-II) in China [J]. Chinese Journal of Rock Mechanics and Engineering,2016, 35(4):649 – 657. (in Chinese))

[7] 刘卡丁,张永成,陈丽娟.基于BIM技术的地铁车站管线综合安装碰撞分析研究[J].土木工程与管理学报,2015,32(1):53 – 58. (LIU Kading,ZHANG Yongcheng,CHEN Lijuan. Research on subway station pipeline installation collision based on BIM technology [J]. Journal of Civil Engineering and Management,2015,32(1):53 – 58. (in Chinese))

[8] 丁烈云,周诚.复杂环境下地铁施工安全风险自动识别与预警研究[J].中国工程科学,2012,14(12):85 – 93. (DING Lieyun,ZHOU Cheng. Automatic risk identification and safety early-warning of underground construction under complex conditions [J]. Engineering Sciences,2012,14(12):85 – 93. (in Chinese))

推进城市地下空间规划建设的思考

■ 钱七虎

(陆军工程大学,江苏南京210001)

近年来,我国城市地下空间特别是地下综合管廊以及海绵城市的开发建设,正以前所未有的规模和速度蓬勃发展,也带动了城市地下空间综合利用的快速发展。然而,由于工程的规模大、战线长、速度快、时限短,其中暴露的问题值得我们认真地思考。

一、城市地下空间开发利用是城市发展建设主要着力点

以往的城市发展带来许多经验教训:只关注高层建筑、大广场、宽马路等城市"面子"工程,忽视城市的"里子"工程——城市基础设施,尤其是地下空间设施,造成了城市交通拥堵、城市环境恶化、城市内涝、垃圾围城等多种城市病。只有把"里子"工程做好,才能让城市真正有"面子"。充分、科学地进行城市地下空间开发利用是"转变城市发展方式"解决"城市病"的主要着力点,是"建设和谐宜居、美丽城市"的主要途径。

(一)避免看重"面子"的粗放发展模式

我国以往的城市建设过程中,注重"面子"的粗放式发展现象处处可见,如:管线直埋、城市排水系统暴雨重现期设计年限偏短、污水直排、垃圾传统填埋、空气降污减排不控制、公交系统严重滞后。为图一时之快,城市粗放式发展引发了诸多后果:雾霾、交通拥堵、城市内涝、地面地下水系、土壤污染日益严重,甚至出现了垃圾围城、垃圾山泥石流现象,引发了严重性事故。

2009年2月15日,深圳某垃圾填埋场污泥坑上堆填垃圾过程中引发污泥外涌及下游堆体失稳事故,污泥从污泥坑下游垃圾围堰的薄弱部位挤出外涌,在下游面上冲开一道1m多宽的裂缝(见图1)。

2015年12月20日,深圳市光明新区红坳余泥渣土临时收纳场发生渣土堆填体失稳滑坡,造成33栋(间)建筑物被掩埋或不同程度损毁。截至2016年1月6日12时,77

图1 深圳市某垃圾山污泥外涌

人失联,其中58人遇难(见图2)。

除了重大灾害性事件之外,粗放式的发展模式引发了很多的琐碎城市环境问题,比如"马路拉链""空中蜘蛛网"(见图3)等,这些问题小且多,如若不重视,会积少成多产生极大的经济损失。

图2 深圳市某渣土堆填体失稳滑坡　　图3 城市"空中蜘蛛网"

(二)重视"里子"的集约绿色可持续发展模式

重视"里子"实现集约绿色可持续的发展,开发利用城市地下空间是主要途径。通过开发利用城市地下空间,能够有效缓解交通拥堵,减少空气雾霾,治理城市内涝,治理城市水系、土壤污染,缓解水资源短缺等。

地下空间开发利用领域包括:大运量快速公交系统、发展地下快速路和物流系统;垃圾地下集运、卫生填埋和焚烧处理;污水地下集运、建设地下污水处理厂;雨洪地下储蓄和排洪、地下水银行;雨水利用、再生水利用;地下综合管廊;地源热泵系统等。

城市地下综合管廊建设的广泛开展,带动城市地下空间开发利用新的浪潮,进而实现:

本文原载于《城乡建设》(2017年第18期)。

（1）消除"马路拉链"。

（2）电缆入廊，消除"空中蜘蛛网"，有效减少架空高压线和地面变电器。以500万千伏的上海世博地下输变电工程为例，其总建筑面积57615m²，地下55809m²，地面仅有1806m²。

（3）污水入廊，处理并利用污水，解决水污染问题。

（4）垃圾入廊，通过卫生填埋或焚烧的形式处理，不再污染城市土壤和造成事故。

（5）地铁、地下快速路系统、地下物流系统的推进建成。城市内见不到运货的卡车，城市不再交通拥堵，汽车尾气在地下收集处理，雾霾成为历史。

（6）雨洪入渗地下工程和雨洪地下调蓄设施的加大修建，城市内涝现象和"看海"现象得到消除，雨洪资源得到利用。

由此可见，随着地下空间开发利用的全面推进，未来我国城市环境必然会变得越来越秀美。

二、城市地下空间开发利用中的"世界眼光"

（一）进行成本效益分析

但凡地下空间开发利用项目都进行了相关的成本效益分析，进而显现出地下空间开发利用相比于地面建设的优势，以下通过若干案例来进行说明。

1. 荷兰代尔伏特铁路工程

荷兰代尔夫特一条穿越城市的极繁忙的铁路工程的地下和地上方案的评价比较如表1所示。

从表1中可以看出，考虑并量化全生命周期以及对于环境的影响，高架方案同TBM隧道方案相比造价基本相同。由此可见，只考虑直接造价，忽视间接造价的方案比选模式，会对地上、地下方案的选择造成不全面、不客观的负面影响。

表1 荷兰代尔伏特某铁路工程地下和地上方案的评价比较
（单位不详）

	高架方案	隧道（掘开式）	隧道（TBM方法）
施工造价	308	494	536
土地/迁移/破坏	122	146	37
分项目1：直接造价	430	640	573
	100%	148%	133%
运营和维护周期	31	64	47
分项目2：生命造价	461	704	620
	100%	152%	134%
损害（换算成金钱）	133	168	5
分项目3：整个造价和损害	594	872	625
	100%	147%	102%

2. 珠海横琴综合管廊工程

珠海横琴综合管廊全长33.4公里，总投资22亿元，除了供水管网漏损率降低而产生的效益之外，因建设地下管廊而节约的土地产生的间接经济效益就超过了80亿元。

图4 珠海横琴综合管廊

（二）城市地下空间建设质量稳步推进

相比我国大陆地区地下空间建设，港澳台和国际上一些地区、国家更加注意质量，推进速度也相对较为平稳。

以综合管廊为例，台湾从1979年通过建设决议开始，直到1991年才开始建设，整个过程包括立法期、发展期和建设期，过程循序渐进，值得学习（见表2）。

表2 台湾综合管廊开发建设历程

1979/7/17	台北市政府第570次市政会议通过建设共同管道决议
1980/2/15	成立工务局新工处共同管道科
1989/6/14	共同管道法公布实施
1990/12/28	公布共同管道法施行细则
1990/12/19	公布共同管道经费分摊办法
1991/5/1	公布共同管道系统上下空土地使用征收及补偿办法
1992/4/23	公布共同管道设计标准
1991~今	台北、淡海及高雄市开始建设共同管道，至今已建数百公里

同样是共同沟建设，日本1963年通过共同沟措施法，于中央建设省下设26个共同管道科，在全国26个大都市全面推进共同沟工程，全国规划2000公里，电线共同沟15000公里，日本历经1995年阪神大地震，共同沟仍可正常使用，可见其工程质量的高水平。

（三）积极利用污水及雨水

1. 雨污分流，分别处理

积极利用污水及雨水，实现雨污分流，其模式为：污水入综合管廊（见图5）；雨洪不入廊，单独成系统。

各街区的污水依重力汇入污水收集站后，由污水泵打入共同沟内的污水干管，经多次加压后被送至污水处理厂，经二级处理后达到中水水质标准，再由共同沟内的中水管道回

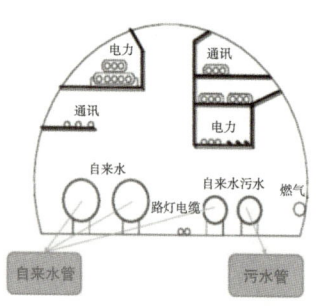

图 5　英国伦敦共同沟（1861）

输到城市各处，作景观、绿化、喷洗道路及冲厕等。

雨洪需要独立成系统，比如日本东京地下河川和日本横滨地下河（见图6和图7）。

图 6　日本东京地下河川　　图 7　日本横滨地下河

2. 建设雨洪地下调蓄措施

建设雨洪地下调蓄措施，防止城市内涝、利用雨洪资源。具有代表性的是马来西亚吉隆坡的"精明隧道"（见图8），是将城市快速路与雨洪调蓄管道合二为一的工程实践。在平时，隧道作为车行的城市快速路；而在暴雨情况下，隧道充作吸纳雨洪和排泄的通道，解决雨涝之害。

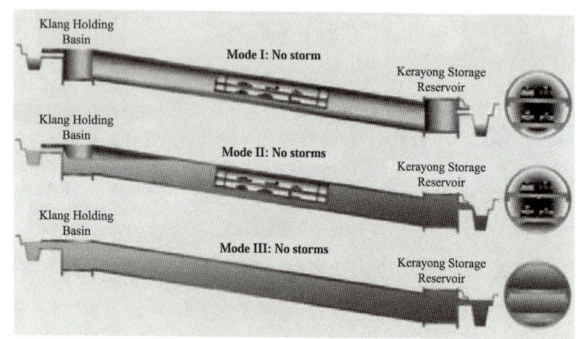

图 8　吉隆坡城市泄洪与公路两用隧道

3. 地下设置雨水收集利用装置（雨水收集和中水利用）

日本、瑞士和牙买加等国利用屋顶收集雨水，并通过管子将雨水输送到地下储水库。

日本要求新建筑物，包括住宅楼、公园、校园、体育场、停车场等处的地下，必须设置雨水贮留设施。东京都、名古屋、大阪、福冈等地的大型建筑物下都设置了雨水利用装置。其中，名古屋体育馆每年积蓄雨水3.6万立方米。1989年开业的东京港区的野鸟公园，园内用水皆来自雨水，形成了湿地、芦苇荡、草地、树林等景点，成为东京地区的著名观光点之一。

上海世博轴利用阳光谷和膜结构收集雨水（见图9）。世博会期间自来水日用量约为2000m³/d。利用雨水时，自来水日用水量降为1100～1200m³/d，回用雨水用水量约800～900m³/d。在可利用雨水时段里，自来水替代率约为40～45%，主要用途为卫生器具冲洗、绿化浇灌等（见图10）。

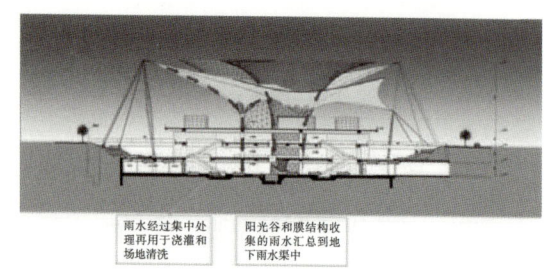

图 9　世博轴雨水收集处理系统

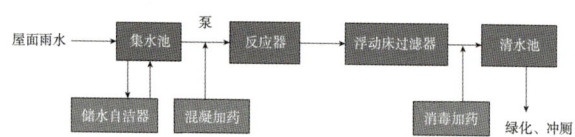

图 10　世博轴雨水处理工艺流程图

4. 利用地下含水层，建立"水银行"

欧美发达国家现在已经基本放弃修建地表水库来储备水资源的传统做法，而是利用地下含水层，建立"水银行"来调节和缓解供水。包括雨洪地表入渗和井灌的人工补给，这是一种可行、费用低廉的解决供水的办法。

建立"水银行"的优势主要体现在：储备地下水，在水短缺的时候提供水量满足用水需求；控制由于地下水位下降引起的海水入侵和地面沉降；提高地下水位，减少地下水的抽取费用；维持河流的基流；通过土壤中的细菌作用、吸附作用和其它物理、化学作用改善水质；通过处理后的污水入渗来实现污水的循环再利用，管理不断增加的大量污水；最后也是最重要的是保护了生态环境。

八十年代初，美国开展了钻孔补给含水层的恢复工程（ASR系统），其中已运行的系统有18个，正在建设之中的系统有40多个。瑞典、荷兰和德国的含水层恢复的人工补给工程（AR工程）在总供水中占20%、15%和10%。

同样，在干旱和半干旱的中东国家，如约旦、科威特、摩洛哥和以色列，都在开展城市污水经处理后补给地下水的工程。阿曼从八十年代初，开始在海岸平原和冲积干谷地区通过地下补给坝截获洪水，进行地下人工水补给工程。

5. 污水处理设施地下化

赫尔辛基污水处理系统将污水处理设施放置在地下，使地面污水处理厂所占的土地得到开发利用，而且使邻厂被污染土地也得到开发（见图11）。

图11 赫尔辛基污水处理系统

（四）燃气入廊

燃气入管廊的目的是为防止因燃气泄漏引起的马路爆炸、马路着火等事故。通常认为燃气一旦在封闭条件下泄露会快速导致浓度增加，从而引发爆燃爆炸事故，但参考国外经验，燃气入廊后可通过检查、检测来防止此类事故的发生。国外大部分国家在管廊建设中，是将燃气管线纳入其中的（见图12）。

日本对于燃气入廊安全性的解决方案是：将燃气管道设置在一个封闭的管沟中，以防燃气泄漏引起的爆炸损害其他管线（见图13）。

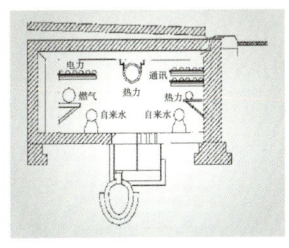 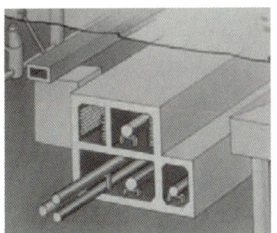

图12 德国汉堡共同沟（1893年）　图13 日本燃气入廊单独设单沟

（五）垃圾入廊

垃圾入廊，单独设管，真空吹送至垃圾焚烧厂，高温焚烧，可明显改善城市环境。

东京在临海副都心共同沟建设中，就专门设置了垃圾管道（图14）。瑞典在二十世纪六十年代在1700人口的居民区内建设并实施采用压缩空气吹运垃圾系统（PWT）。该系统与垃圾收集、处理系统配套，垃圾用于焚烧发电，投资在3～4年内得到了回报。现已在瑞典全国推广，竟出现了垃圾不够焚烧发电用的现象。

发展地下垃圾处理系统可节约城市宝贵的土地资源，控制对环境的噪声和恶臭的污染，能减少对环境的视觉负面影响。

（六）利用低密度地热能源，节能减排

充分开发利用地下空间，建设节能型城市还包括充分利用地下土壤、地下水的低密度天然能源作为冬季热源和夏季冷源，然后再由热泵机组向建筑物供热供冷。地源热泵技术就是这样一种利用地下可再生能源的既可供暖又可供冷的新型中央空调系统，它包括地下埋管式地源热泵和抽取地下水然后回灌的水源热泵。埋管式地源热泵技术目前在国外大面积推广，在欧美更已得到普遍应用，是一种成熟可行、可持续发展的节能新技术。以科罗拉多的地源热泵系统为例，安装地源热泵系统后，每个家庭可降低电力峰值需求、节约能源和减少 CO_2 排放约 10～30%。

该系统主要包括四部分：覆土结构或住宅、地下气热管、地源热泵系统、活化桩（见图15）。

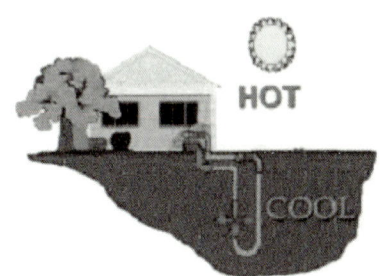

图15 地源热泵系统

阿布扎比在建筑物基坑桩基里面安装地源热泵管道系统，通过采用竖直的地源热泵线圈，来节约成本和能源（见图16）。

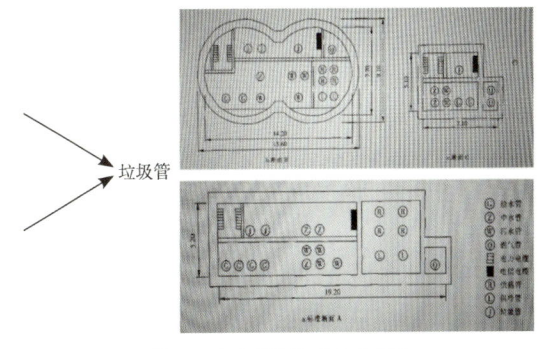

图14 东京临海副都心共同沟

图16 阿布扎比地源热泵管道系统

世博轴的直接式江水源热泵系统和桩基埋管地源热泵系统节能效果明显，利用地表水和地下的可再生能源（冷量和热量），节省用电562.9万千瓦时/年，减少二氧化碳排放

5629吨/年，夏季节能率26.96%，冬季节能率71.03%，全年节能率达到61.4%（见图17）。

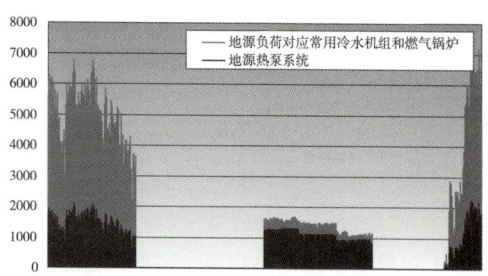

图17 世博轴全年地源承担负荷部分的冷热源功耗比较（kW）

目前，地源热泵利用正进一步与太阳能利用相结合。由于太阳能的辅助供热，可实现系统向地下排热与取热的平衡，从而使得地下温度场保持稳定。既可克服单独使用地源热泵时，土壤温度场不断降低（或升高）后不能有效恢复的局限性，又可克服单独使用太阳能空调系统时，太阳辐射受天候因素制约的局限性。

（七）修建地下快速路及地下物流系统

美国、日本、新加坡等通过修建地下快速路和地下物流系统，进而有效的治理交通拥堵，缓解城市的空气污染问题。

1. 地下快速路

美国波士顿拆除穿过市中心的六车道高架路，建设8～10车道的地下高速路，原有的地面变成林荫路和街心公园。从而使得市区空气的一氧化碳浓度降低了12%；市内栽植了2400棵乔木树，7000多棵灌木树；在海湾的景观岛上栽植了另外的2400棵乔木和26000棵灌木；增加了260英亩新的公园和开敞空间（见图18）。

a. 改建前（地面高架路）　　b. 改建后（地下快速路）

图18 美国波士顿市波士顿中央大道

日本东京中央环状新宿线全长11公里，埋深大于40米，为双向4车道（13米），投资90亿美元（见图19）。该线路穿越池袋、新宿、涩谷3大商业中心，设置6个出入口，于2007年3月实现通车。线路将从池袋到涩谷的行程时间由50分钟缩短到20分钟。通过其除尘及换气技术，每年减少

CO_2 2.5万吨，NO160万吨，悬浮颗粒16吨。值得一提的是，其投资成本低于地面投资约20%。

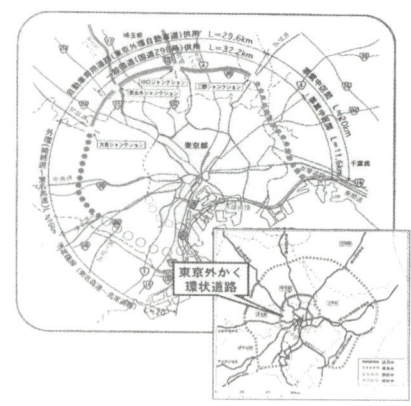

图19 东京中央环装新宿线

东京外环线于1966年宣布为高架路方案，但由于沿线居民以噪声和污染严重为由强烈反对，于1970年宣布冻结。2003年1月日本政府举行新闻发布会，宣布外环线改为建设地下40米深的高速路。由于地下无需动迁和征地费用，可比地面高架路节约建设费用20～30%，工期由15年缩短为8年。地下隧道内设置可将汽车尾气中的烟尘和有害物质进行高科技过滤和分解的设施（见图20），排出的是无污染空气。

新加坡地下道路系统——SURS全长15公里，设四车道，线路为环形。其造价为48亿美元，能够承担城市中心区交通量的40%（见图21）。

2. 地下物流系统

地下货物运输系统，又称地下物流系统（ULS）是基于城内运输和城外运输的结合。城外货物通过各种运输手段到城市边缘的物流园区（CLP），经处理后由CLP通过ULS输送到各个终端。它以集装箱和货盘车为基本单元，以自动导向车（AGV）等为运输工具。

ULS的环境效应显著。世界经济合作组织在2003年的《配送：21世纪城市货运的挑战》报告中指出，发达国家主要城市的货运占城市交通总量的10～15%，而货运车辆对城市环境污染总量的贡献为40～60%。日本东京在建设300公里地下物流系统的评估报告中指出：开发建设地下物流系统可使一氧化氮和二氧化碳分别减少10%和18%，能源消耗减少18%，运输速度提高24%。

三、借鉴先进经验，制定完善的法规体系

地下空间开发利用是一个跨行业、跨部门、跨单位的协同工程。过去我国城市地下空间开发利用综合管理水平低、

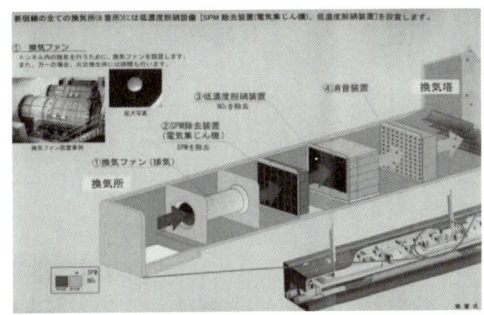
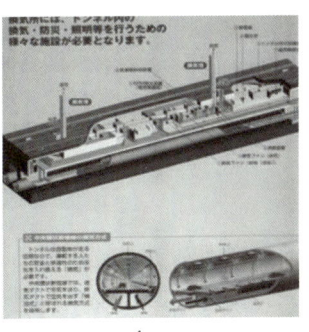
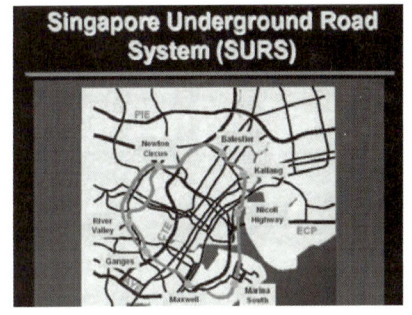

图20 日本东京中央环状新宿线除尘及换气技术

图21 新加坡地下道路系统——SURS

问题多的病根在于分头管理的市政体制与条块分割的部门所有制，破除分头管理与部门分割的利器是"法治"，依法执政首先要有法可依，只有指导意见远远不够，需要对城市地下空间的所有权、规划权、建设权、管理权、经营权、使用权以及有偿使用费的收取原则等作出明确完善的具体规定。

目前我国城市地下空间法规亟待健全。我国现行法律中没有地下空间权的概念和相关规定，只有《物权法》在原则上规定建设用地使用权可以在地表、地上和地下分别设立，地下空间最基础的民事权利并没有得到法律的确认。全国人大和国务院没有制定城市地下空间开发利用的法律法规，原建设部颁布的《城市地下空间开发利用管理规定》属于部门规章，许可权限和法律效力比较低，加之内容不完备、规定不具体，权威性和操作性都不够，在实践中难以有效地贯彻执行。

在这方面，我们可以借鉴德国的《城市建设法典》，日本的《大深度地下公共使用特别措施法》《共同沟法》，台湾地区的"共同管道法""大众捷运法"，美国各州制定的空间法，瑞士的《瑞士民法典》，意大利的《民法典》等。

四、建立统一管理、分工协作、衔接有序的城市地下空间综合管理体制机制

（一）"推进城市管理机构改革、创新城市工作体制机制"

我国城市地下空间管理体制不统一，亟待理顺。城市地下空间开发利用涉及发改委、国土、规划、建设、市政、交通、人防和房产等多个管理部门，而国家在体制上并没有明确一个综合管理机构对其进行统一管理，基本上是各部门依照相关职责对其实施分散管理，相互间缺乏统一协调机制，处于条块分割、各自为战的状态，形成多头管理与无人管理并存的局面。目前，许多城市对地下空间开发利用管理进行了有益探索，分别采取了以规划管理、建设管理、人防管理、安全管理和协调管理为主导的管理模式，但都没有起到统一管理的效果。

（二）必须对管理体制和管理措施进行系统完善

管理法规是实施综合管理的基本依据，在法律法规上明确规定管理体制和管理措施。管理体制是实践综合管理的执行主体，在管理法规的法律授权或行政授权下采取各种管理措施。管理措施是实现综合管理的具体方法，由管理机构依据管理法规，通过有效的管理机制，进行综合运用。

（钱七虎演讲，杨可、李晔整理，范益群校核；发表时略有删节）

公路隧道大断面改扩建施工开挖方案研究

黄伦海[1,3] 钱七虎[1,2]

(1.重庆大学土木工程学院,重庆 400045;2.解放军理工大学工程兵学院,南京 210007;
3.公路隧道建设技术国家工程实验室,招商局重庆交通科研设计院有限公司,重庆 400067)

摘 要 针对大断面公路隧道改扩建施工方案,文章依托国内第一座双洞四车道公路隧道原位扩建为双洞八车道公路隧道的工程实践,采用相似模拟和数值计算相结合的方法,研究软弱围岩条件下隧道改扩建施工开挖方案,得到了大断面原位扩建隧道施工的围岩变形、应力变化规律,提出了先新建再扩建的组合施工方案,可为公路隧道及类似地下工程建设提供参考。

关键词 隧道 改扩建 施工开挖 相似模拟 数值计算

中图分类号:U457⁺.6 **文献标识码:A**

1 引 言

在一些经济发达地区及特大型城市区域,双向四车道乃至双向六车道公路隧道已不能适应交通量日益增长的需求[2],在原有公路或道路标准基础上改建成为双向八车道公路及道路的工程项目越来越多。四车道隧道扩建成八车道隧道的型式主要有[12]:

(1)原位扩建:在原两座两车道隧道基础上扩建成两洞八车道隧道;

(2)并行扩建:在两座两车道隧道旁增建两座两车道隧道,形成四洞八车道;

(3)组合扩建:一座隧道扩建成四车道隧道,另一座增设一个二车道隧道,组合成三洞八车道隧道,或一座隧道扩建成四车道隧道,增设一座单洞四车道隧道,组成双洞八车道隧道。

隧道原位扩建是在原有建成并运营多年的隧道中再进行扩建,需拆除原隧道结构,再开挖形成大断面隧道[7]。从原有隧道开挖形成的小断面隧道到现在拆除原有结构开挖成大断面隧道,隧道围岩经过的应力状态错综复杂,围岩稳定性受开挖方法的影响较大[11],且新扩建的大断面隧道围岩压力与直接在原始地形地貌下开挖成大断面隧道的围岩压力有本质区别[8],因此支护结构的设计和计算不同于普通的大断面隧道[6]。

目前,我国在两车道隧道的基础上扩建成四车道公路隧道的工程实践不多,经验相对缺乏,相关研究工作较少。隧道改扩建施工开挖方案是关系工程建设成败的关键[10],本文采用相似模拟方法并结合数值计算和现场测试的结果综合分析研究洞口V级围岩隧道段原位扩建的施工开挖方案。

2 依托工程概况

福州至厦门高速公路由双向四车道扩建为双向八车道,其中大帽山隧道(长 600 m)采取原位扩建型式,即在原两洞之间新建一座四车道隧道,并将右洞扩建为四车道。各隧道的关系如图 1[5]所示。

隧道围岩为侏罗系南园组凝灰熔岩和燕山早期侵入花岗岩及其风化层。扩建隧道结构按"新奥法"

收稿日期:2015-04-20。
本文原载于《现代隧道技术》(2016 年第 53 卷第 5 期)。

设计[1]，采用复合式衬砌，以喷射混凝土、锚杆、钢筋网为主要支护手段，钢支撑加强，以大管棚、小导管为辅助施工措施，二次衬砌为等厚度C25模筑钢筋混凝土。扩建采用爆破开挖[4]，隧道最大开挖跨度20.77 m，隧道典型支护衬砌断面如图2[5]所示。

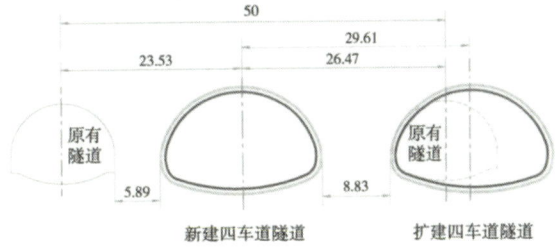

图1 隧道洞室关系
Fig.1 Relationship of tunnel caverns

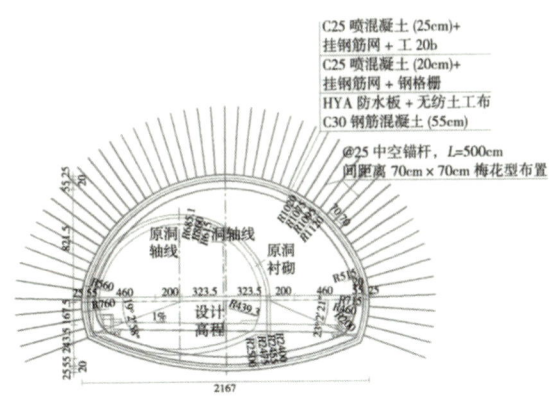

图2 隧道典型衬砌断面
Fig.2 Typical tunnel lining section

3 原位扩建相似模型试验

3.1 试验方案

根据隧道施工开挖方案（图3），制定模型试验方案：(1)试验一方案为先开挖原隧道左侧（包括拆除原洞二次衬砌），再上下台阶扩挖原隧道的右侧；(2)试验二方案为先扩挖原隧道右侧，再开挖左侧（包括拆除原洞二次衬砌）。

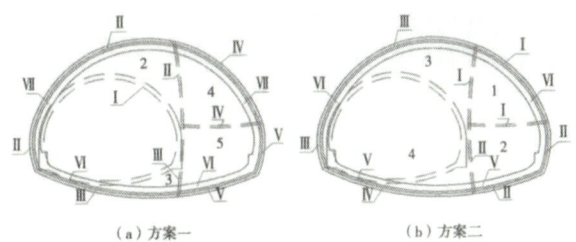

图3 隧道扩建模型试验开挖方案
Fig.3 Excavation schemes for tunnel expansion in the model test

试验采用模型箱内装模型材料，预置隧道支护模型，人工开挖形成无支护隧道。模型几何相似比为1/40，模型箱内空尺寸为2.5 m×1.5 m×1.0 m，模型隧道高0.34 m，以V级围岩隧道段为试验重点，如图4所示。

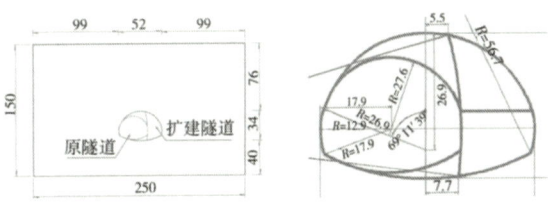

图4 原位扩建模型几何尺寸
Fig.4 Geometries of in-situ expansion model

3.2 试验材料与模型

根据依托工程地质勘察成果，隧道围岩模拟试验材料采用细砂骨料，以石膏粉、膨润土、42.5号普通硅酸盐水泥为胶结料。初选多组配比进行配比实验，测试得到各配比的密度、抗拉强度、试件单轴抗压强度与变形强度、抗剪强度等主要物理、力学性质。假设组成隧道围岩的岩石(体)是均匀的，满足连续性假设且各向同性，则根据平衡微分方程推得 $C_\sigma = C_L \cdot C_\gamma$，根据虎克定律推得 $C_\sigma = C_E$，将各材料的 C_σ 和 C_E 与 $C_L \cdot C_\gamma$ 比较评价其满足相似准则的程度。试验得出砂胶比为3:1，胶结料比为5:3:2，以1:7的拌和水灰比[3]条件下的材料达到96%的相似，相似性最好，其配比参数如表1所示。试验模型采用分层

表1 模型实验的相似系数
Table 1 Similarity factors of the model test

项目	密度相似比	泊松比相似比	弹性模量相似比	应变相似比	应力相似比	位移相似比
相似比	1.24	2.11	14.29	3.47	49.60	138.83

浇注成型法制作，其过程如图5所示。

既有隧道衬砌为C25模筑混凝土，厚h_p=40 cm，E_p=29.5 GPa，μ_p=0.2。试验采用水膏比为1.3∶1的石膏模型，其E_m=2.3 GPa，μ_m=0.163。由于隧道结构弯曲变形是结构安全的控制模式，采用弯曲变形的相似准则，分析得到既有隧道衬砌模型的厚度为h_m=24 mm。既有隧道二次衬砌采用特制的钢模板分段制作，每段长10 cm。

  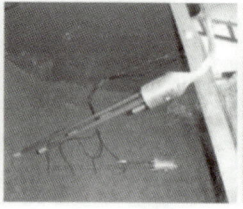 

（a）材料配比　　（b）原隧道二次衬砌埋设　　（c）测试仪器埋设　　（d）模型完毕

图5　原位扩建隧道模型的浇注及制作过程

Fig.5　Casting and manufacturing process of the in-situ tunnel expansion model

3.3　试验测试仪器

试验围岩内部变形采用微型多点位移计、围岩应力采用微型压力盒、地表位移采用百分表进行测试。考虑到边界约束效应，在模型中部位置沿着扩挖隧道的周边布置多点位移计和围岩应力测线，在模型箱的顶部布置地表位移测点。测试点布置如图6所示。

3.4　试验步序

试验需模拟两种不同开挖方案下围岩的开挖力学特性，通过变形、应力等的对比，对两种不同开挖方案进行对比和优选。

步序一（对应方案一）：先开挖A（包括原二次衬砌拆除），待A部分挖通后（分10步），再开挖B（分10步），待B部分挖通后，再开挖C（分10步），每次开挖进尺为10 cm，共30步，如图7所示。

步序二（对应方案二）：先开挖A，待A部分挖通后（分10步），再开挖B（包括原二次衬砌拆除）（分10步），每次开挖进尺为10 cm，共开挖20步，如图8所示。

3.5　围岩变形分析

分析试验数据及变化曲线（图9）可知：

（1）隧道围岩周边变形、内部位移及表面沉降历时曲线总体形状呈S形，变形随开挖步呈台阶式增加，均在当前断面开挖步开挖时变形突增，然后趋于收敛，且隧道同一位置测点位移自开挖轮廓向围岩深部逐渐减小。

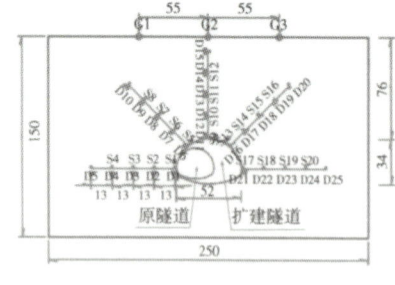 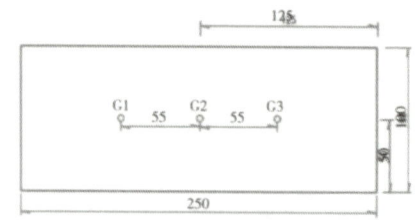

（a）测试剖面布置　　　　　　　（b）地表位移测点布置

（说明：图中D——内部位移测点；S——围岩应力测点；G——地表位移测点）

图6　模型实验测点布置（单位：cm）

Fig.6　Layout of the monitoring points of the model test (Unit: cm)

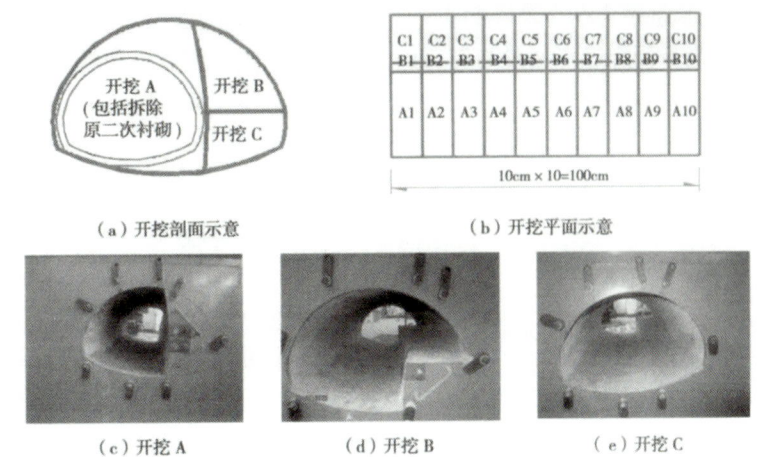

图7 原位扩挖开挖方案一

Fig.7 Scheme 1 for in-situ expanded excavation

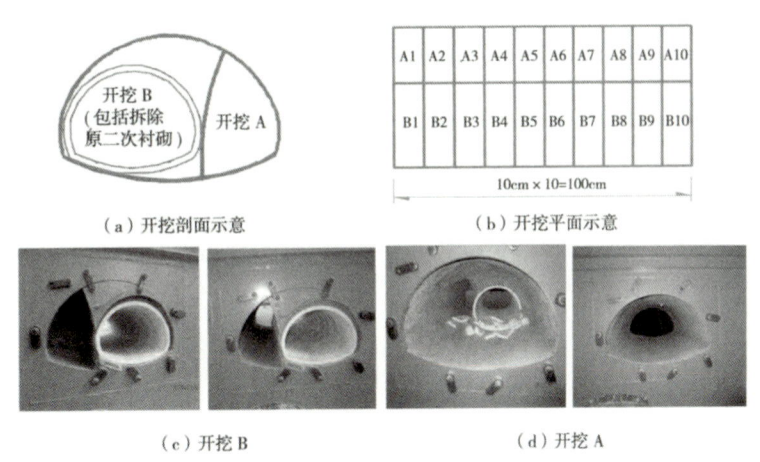

图8 原位扩挖开挖方案二

Fig.8 Scheme 2 for in-situ expanded excavation

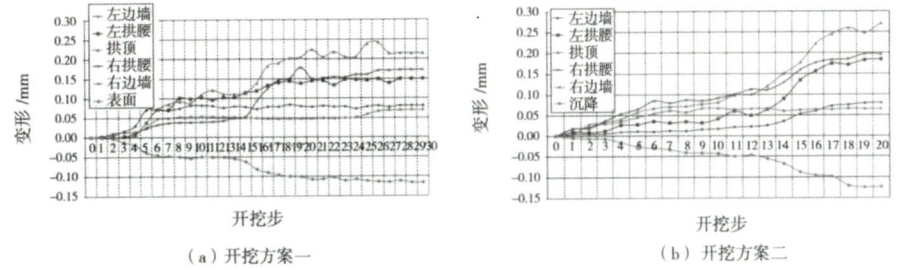

图9 隧道周边变形随开挖步的变化曲线

Fig.9 Curves of tunnel peripheral deformations varying with the excavation steps

(2) 隧道拱顶位移最大,右拱腰次之,左拱腰变形居中,右边墙变形最小。两个方案中变形均表现出一定的偏载,但方案二相应测点的变形总体大于方案一,如图10所示。

（3）隧道拱顶测点变形速率均在当前断面有开挖步时突增，方案一各开挖步变形速率增量均匀，且方案二变形速率大于方案一，如图11所示。

（4）开挖步对隧道变形均有一定影响，拱顶测点最大变形量发生在大断面开挖形成步，超过总变形量的40%，为最不利施工部；拱腰测点变形最大变形量均为当前开挖步，方案二达到67.6%，极为不利。各开挖步变形分配见表2。

3.6 围岩应力分析

分析试验数据及变化曲线（图12）可知：

（1）隧道围岩应力历时曲线总体形状仍呈S形，应力随开挖步呈台阶式变化，表现为受隧道开挖卸载影响；方案1拱顶、边墙部围岩应力均随着开挖进行而减小，左拱腰测点应力先减小后增大，而右拱腰测点应力先增大后减小，变化基本对称；方案2拱顶、左边墙部围岩应力均随着开挖进行而减小，左拱腰测点应力先增大后减小，而右拱腰、右边墙测点应力先减小后增大；

（2）隧道周边应力变化明显，两方案对应测点最终应力基本相同。

（3）各开挖步对隧道围岩应力均有一定影响，两方案测点最大应力释放率均发生在A开挖步，为最不利施工步；方案1在B、C开挖步的应力释放率不大，而方案二在B开挖步的应力释放量与A开挖步相当，同样为不利施工步。各开挖步变形分配见表3。

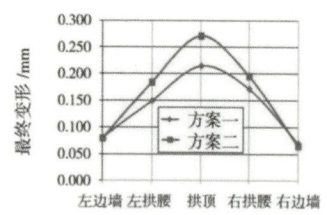

图10 隧道周边变形对比

Fig.10 Comparison of tunnel peripheral deformations

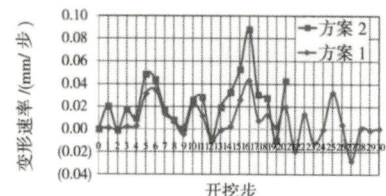

图11 隧道拱顶变形速率对比

Fig.11 Comparison of deformation rates at the tunnel crown

表2 各开挖步的周边变形分配

Table 2 Peripheral deformation distribution of each excavation step

测点位置	拱顶/(%)		左拱腰/(%)		右拱腰/(%)	
	方案一	方案二	方案一	方案二	方案一	方案二
步前	12.5	27.5	25.3	13.5	11.8	44.8
A	35.4	30.0	42.7	18.9	17.6	55.2
B	45.8	42.5	28.7	67.6	58.8	0.0
C	6.3	—	3.3	—	11.8	—

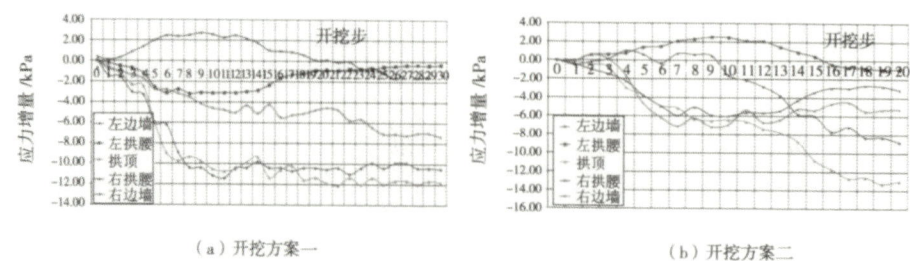

图12 隧道周边围岩应力随开挖步的变化曲线

Fig.12 Curves of surrounding rock stress varying with the excavation steps

学 术 论 文——隧道及地下工程

表3 隧道周边围岩应力释放分配(%)
Table 3 Distribution of rock stress release around the tunnel (%)

方案编号	测点项目	左边墙(S1) 应力/kPa	释放量/kPa	释放率/(%)	拱顶周边(S9) 应力/kPa	释放量/kPa	释放率/(%)	右边墙(S17) 应力/kPa	释放量/kPa	释放率/(%)
方案1	初始值	16.41	–	–	12.34	–	–	12.8	–	–
	初应力	13.1	3.31	31.55	9.89	2.45	20.76	11.07	1.73	23.41
	A	6.6	6.5	61.96	3.01	6.88	58.31	7.67	3.4	46.01
	B	6.5	0.1	0.95	1	2.01	17.03	6.33	1.34	18.13
	C	5.92	0.58	5.53	0.54	0.46	3.90	5.41	0.92	12.45
	残余应力	5.92	10.49	100.00	0.54	11.8	100.00	5.41	7.39	100.00
方案2	初始值	17.10	–	–	13.54	–	–	11.20	–	–
	初应力	15.13	1.97	22.39	10.62	2.92	22.48	9.86	1.34	25.62
	A	11.33	3.8	43.18	4.8	5.82	44.80	7.57	2.29	43.79
	B	8.3	3.03	34.43	0.55	4.25	32.72	5.97	1.6	30.59
	残余应力	8.30	8.8	100.00	0.55	12.99	100.00	5.97	5.23	100.00

3.7 开挖方案分析

对比分析以上实验结果(表4),两方案围岩变形、应力的变化规律相似,围岩变形量、围岩应力大小差异不大,方案一隧道扩挖施工时换算拱顶最大沉降30.54 mm,最大围岩应力释放341.25 kPa,最大残余应力792.6 kPa;方案二隧道扩挖施工时换算拱顶最大沉降37.48 mm,开挖最大围岩应力释放288.67 kPa,最大残余应力785.7 kPa。两方案的变形及应力值均可控。

在最大位移及最大变形速率方面,方案二大于方案一;最大收敛时间方面,方案一大于方案二;初始应力、残余应力及应力释放率方面,方案二总体大于方案一。综合比较并考虑施工组织及扩建施工的特点,采用先进行既有隧道的拆除再进行扩挖施工的施工步序(方案一)更优。

4 改扩建数值模拟

特大断面隧道的开挖方案主要有上下台阶法、单侧壁导坑法、双侧壁导坑法等[12]。每种方法各有优缺点,结合大帽山隧道具体情况,对原位扩建隧道施工开挖方案进行比选分析,拟定研究方案见表5。计算采用MIDAS有限元软件对开挖方案进行数

表4 实验对比分析
Table 4 Contrastive analysis of each scheme

比较项	左边墙 方案一	方案二	左拱腰 方案一	方案二	拱顶 方案一	方案二	右拱腰 方案一	方案二	右边墙 方案一	方案二
最大位移/mm	0.08	0.08	0.15	0.18	0.22	0.27	0.17	0.20	0.07	0.07
最大变形速度/(mm/步)	0.04	0.01	0.04	0.04	0.04	0.05	0.03	0.03	0.01	0.01
最大收敛时间/步	3	3	8	4	10	5	6	3	8	4
初始应力释放/kPa	3.31	1.97	1.19	2.92	2.45	2.92	0.87	2.26	1.73	1.34
最大应力释放/kPa	6.5	3.8	1.65	1.84	6.88	5.82	2.93	2.26	3.40	2.29
残余应力/kPa	5.92	8.3	13.84	15.84	0.54	0.55	15.98	15.09	5.41	5.97

表5 隧道扩建施工开挖方案
Table 5 Excavation scheme of tunnel expansion construction

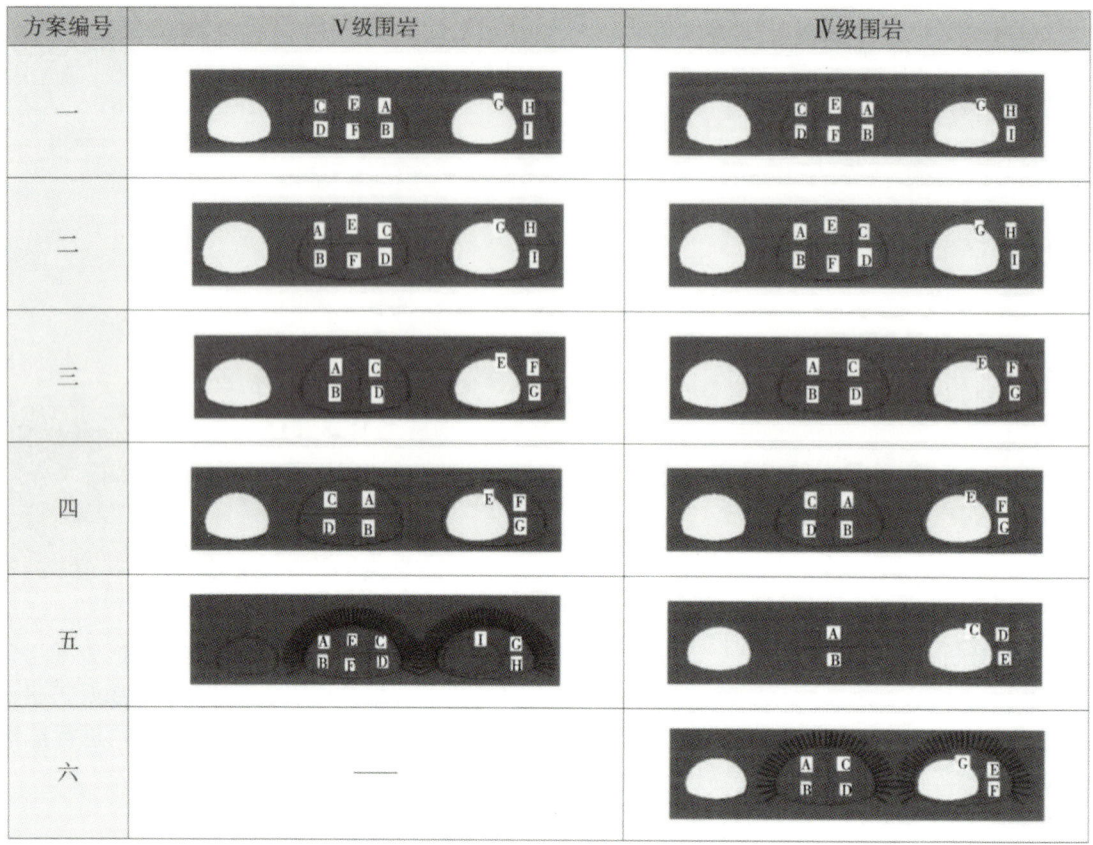

值计算分析,依据工程地质报告和公路隧道设计规范[1]选取围岩及支护衬砌的物理、力学参数,见表6及表7。采用M-C屈服准则,模型左、右边界水平位移约束,下边界竖向位移约束,上边界为地表自由面,主要计算结果见表8。

通过分析得到以下结果:

(1) V级围岩隧道段施工开挖方案一与方案二较接近,方案二优于方案一;方案三与方案四比较接近,方案三略优于方案四;方案五不同于其它几个方案,结合各方案的施工复杂程度,建议采用开挖方案二(对应模型试验研究方案一)。

(2) 对于Ⅳ级围岩隧道段施工开挖方案,方案一与方案二较一致,相比之下方案一优于方案二;方案三、四、六比较接近,相比之下方案四略好;方案五不同于其它几个方案,结合各方案的施工复杂程度,建议采用方案三或方案四。

5 对比分析及现场实施

对比分析模型试验和数值分析结果,开挖方案的围岩内部位移、周边变形、围岩应力、支护衬砌应力等历时曲线总体与模型试验类似,方案结论统一,依托工程隧道扩建施工V级围岩隧道段采用数值分析推荐方案二(对应模型试验方案一),Ⅳ级围岩隧道段采用数值分析方案三。施工中对隧道洞口V级

表6 围岩物理力学参数
Table 6 Physical and mechanical parameters of the surrounding rock

名称	E/kPa	μ	γ/(kN/m³)	c/kPa	φ/(°)	抗拉强度/kPa
坡积亚粘土	50 000	0.35	27.0	30.0	24.8	10.0
强风化花岗岩	1.0×10^6	0.38	24.0	180.0	25.0	80.0
弱风化花岗岩	3.0×10^6	0.30	26.0	600.0	35.0	200.0

表7 支护衬砌物理力学参数
Table 7 Physical and mechanical parameters of the support and lining

名称	E/kPa	μ	$\gamma/(kN/m^3)$	备注
既有隧道二次衬砌	2.85×10^7	0.20	24.0	C20混凝土,厚0.40 m
喷射混凝土	2.85×10^7	0.20	24.0	C25混凝土,厚0.25 m
锚杆	2.0×10^8	0.26	78.0	ϕ25中空注浆锚杆,长5.0 m,间距0.7 m
钢支撑	2.0×10^8	0.26	78.0	20b工字钢,A=0.003 9 m², I_x=1.47×10⁻⁷, I_y=2.49×10⁻⁵, I_z=2.03×10⁻⁶, A_1=0.001 9 m², A_2=0.001 8 m²
扩建隧道二次衬砌	2.85×10^7	0.20	24.0	C25钢筋混凝土,厚0.55 m

表8 主要数值模拟结果对比
Table 8 Comparison of the main numerical simulation results

方案	Ⅴ级围岩隧道段					Ⅳ级围岩隧道段				
	拱顶下沉/mm	围岩最大主应力/MPa	喷射混凝土最大剪应力/MPa	钢支撑受力/kN	锚杆轴力/kN	拱顶下沉/mm	围岩最大主应力/MPa	喷射混凝土最大剪应力/MPa	钢支撑受力/kN	锚杆轴力/kN
一	53.22	1.74	7.58	−473.70	771.80	7.37	2.64	3.40	186.78	38.30
二	53.81	1.73	6.73	−457.35	765.00	7.37	2.65	2.61	162.06	38.31
三	53.81	1.77	6.00	−486.33	686.30	7.46	2.69	1.71	112.65	52.79
四	60.22	1.79	8.09	−480.53	683.60	7.46	2.68	2.96	140.03	55.75
五	57.11	1.65	7.41	−506.84	704.71	7.53	2.69	0.96	60.03	54.56
六	−	−	−	−	−	7.36	2.66	1.80	114.84	65.23

围岩隧道段仰坡地表沉降、水平位移、周边收敛、拱顶沉降、围岩内部位移、混凝土应变、接触压力、爆破震动速度等进行了监测,其结果见表9。由表可以看出,扩建隧道仰坡地表沉降、水平位移、拱顶沉降及围岩收敛均不大,在安全范围内;混凝土应力、接触压力等监测成果也不大,趋势较稳定;爆破振动测点最大振速57.04 cm/s,平均振速21.83 cm/s,除个别测点振速超标外,其余测点振速普遍在控制范围内。

表9 现场监控量测成果
Table 9 Results of site monitoring and measurement

项目	地表沉降/mm	地表水平位移/mm	拱顶沉降/mm	周边收敛/mm	混凝土应力/MPa	接触压力/kPa	爆破震动速度/(cm/s)
最大值	23	25	37	29	2.625	20	57

6 结论及建议

(1)根据既有隧道和新建隧道的断面位置关系,四车道隧道扩建成八车道隧道的形式主要有原位扩建、并行扩建和组合扩建,原位扩建需拆除既有隧道再开挖形成大断面隧道,围岩经多次扰动,应力状态复杂,施工开挖方案是关系工程建设成败的关键;

(2)隧道围岩内部位移、周边变形、围岩应力、支护衬砌应力历时曲线总体呈S形,随开挖步呈台阶式变化,受隧道开挖卸载影响,均在当前断面开挖步开挖时突增,然后趋于收敛,因此应高度重视当前断面开挖步围岩的稳定及支护衬砌的安全;

(3)综合比选并考虑到交通、施工组织等因素,Ⅴ级围岩隧道段改扩建施工推荐先采用双侧壁导坑新建大断面隧道,再超前拆除既有隧道衬砌再扩挖的施工步序；Ⅳ级围岩隧道段施工推荐先采用CD法新建大断面隧道，再超前拆除既有隧道衬砌再扩挖的施工步序；

（4）相似模拟、数值分析和现场监测所得隧道改扩建施工的围岩及支护衬砌的变形和应力均较为接近，三种研究方法相互印证，推荐及实施的开挖方案安全可靠。项目现已建成通车，研究结果对隧道设计规范修订提供了有效的支撑。

参考文献
References

[1] 中华人民共和国交通运输部. 公路隧道设计规范: JTG D70-2004[S]. 北京: 人民交通出版社, 2004.
Ministry of Transport of the People's Republic of China. Code for Design of Road Tunnel: JTG D70-2004[S]. Beijing: China Communications Press, 2004.

[2] 王梦恕. 对21世纪我国隧道工程建设的建议[J]. 现代隧道技术, 2001, 38 (1): 2-4.
WANG Mengshu. Suggestion on China Tunneling Construction Facing the 21st China[J]. Modern Tunnelling Technology, 2001, 38 (1): 2-4.

[3] 蒋树屏, 黄伦海, 等. 利用相似模拟方法研究公路隧道施工力学形态[J]. 岩石力学与工程学报, 2002, 21 (5): 662-666.
JIANG Shuping, HUANG Lunhai, et al. Physical Simulation Analysis on Construction Mechanics Behavior of Road Tunnel[J]. Chinese Journal of Rock Mechanics and Engineering, 2002, 21 (5): 662-666.

[4] 中华人民共和国交通运输部. 公路隧道施工技术规范: JTG F60-2009[S]. 北京: 人民交通出版社, 2009.
Ministry of Transport of the People's Republic of China. Technical Guidelines for Construction of Highway Tunnel: JTG F60-2009 [S]. Beijing: China Communications Press, 2009.

[5] 唐颖. 大帽山大跨度扩建隧道设计要点[J]. 公路, 2009, (10): 263-266.
TANG Ying. Design Points of Damaoshan Extension Tunnel with Large Span[J]. Highway, 2009, (10): 263-266.

[6] 陈正勋, 侯嘉松, 陈福胜. 扩建隧道之设计与施工案例探讨[J]. 中华技术, 2002: 132-150.
CHEN Zhengxun, HOU Jiasong, CHEN Fusheng. Case Study on Design and Construction of Extension Tunnels[J]. China Technology, 2002: 132-150.

[7] 李元福. 关村坝隧道改扩建施工技术[J]. 铁道建筑技术, 1996, (4): 22-26.
LI Yuanfu. Reconstruction and Construction Technology of Guancunba Tunnel[J]. Railway Construction Technology, 1996, (4): 22-26.

[8] 刘正雄, 罗崇篾, 周余奎. 关村坝隧道扩建喇叭口工程控制爆破技术[J]. 爆破, 1999, 16 (3): 40-44.
LIU Zhengxiong, LUO Congchi, ZHOU Yukui. Controlled Blasting of Bell Extension Project of Guancunba Tunnel[J]. Blasting, 1999, 16 (3): 40-44.

[9] 万明富, 海洪, 刘斌. 单洞四车道隧道开挖室内模型试验研究[J]. 东北大学学报(自然科学版), 2007, 28 (2): 266-269.
WAN Mingfu, HAI Hong, LIU Bin. Investigation on Excavating a 4-Lane Road Tunnel through Indoor Model Test[J]. Journal of Northeastern University (Natural Science Edition), 2007, 28 (2): 266-269.

[10] 丁文其, 谢东武, 严宗雪, 等. 动态监测及其在四车道公路隧道施工中的应用[J]. 地下空间与工程学报, 2007, 3 (1): 118-123.
DING Wenqi, XIE Dongwu, YAN Zongxue, et al. Staged Monitoring and Its Application in Construction of Four-Lane Highway Tunnel[J]. Chinese Journal of Underground Space and Engineering, 2007, 3 (1): 118-123.

[11] 杨小礼, 王作伟, 陈杰. 既有隧道扩建工程及衬砌稳定性研究[J]. 交通科学与工程, 2010, 26 (1): 49-52.
YANG Xiaoli, WANG Zuowei, CHEN Jie. Extension Engineering of Tunnel and Stability Analysis of Lining Structures[J]. Journal of Transport Science and Engineering, 2010, 26 (1): 49-52.

[12] 胡居义, 陈礼彪, 黄伦海. 2车道隧道扩建成4车道隧道扩建形式研究[J]. 公路交通技术, 2010, (5): 93-97.
HU Juyi, CHEN Libiao, HUANG Lunhai. Research on Expansion Modes of Two-Lane to Four-Lane Tunnels[J]. Technology of Highway and Transport, 2010, (5): 93-97.

| 学 术 论 文——隧道及地下工程

Safety risk management of underground engineering in China: progress, challenges and strategies

Qihu Qian[a], Peng Lin[b]

a PLA University of Science and Technology, Nanjing, 210014, China
b State Key Laboratory of Hydroscience and Engineering, Tsinghua University, Beijing, 100084, China

ARTICLE INFO

Article history:
Received 4 December 2015 Received in revised form 31 March 2016
Accepted 13 April 2016
Available online 20 May 2016

Keywords:
Underground engineering Safety risk management Safety behaviors
Safety challenges
Safety progress and strategies

ABSTRACT

Underground construction in China is featured by large scale, high speed, long construction period, complex operation and frustrating situations regarding project safety. Various accidents have been reported from time to time, resulting in serious social impact and huge economic loss. This paper presents the main progress in the safety risk management of underground engineering in China over the last decade, i.e. (1) establishment of laws and regulations for safety risk management of underground en-gineering, (2) implementation of the safety risk management plan, (3) establishment of decision support system for risk management and early-warning based on information technology, and (4) strengthening the study on safety risk management, prediction and prevention. Based on the analysis of the typical accidents in China in the last decade, the new challenges in the safety risk management for underground engineering are identified as follows: (1) control of unsafe human behaviors; (2) technological innovation in safety risk management; and (3) design of safety risk management regulations. Finally, the strategies for safety risk management of underground engineering in China are proposed in six aspects, i.e. the safety risk management system and policy, law, administration, economy, education and technology. © 2016 Institute of Rock and Soil Mechanics, Chinese Academy of Sciences. Production and hosting byElsevier B.V. This is an open access article under the CC BY-NC-ND license (http://creativecommons.org/licenses/by-nc-nd/4.0/).

本文原载于《Journal of Rock Mechanics and Geotechnical Engineering》(2016 年)。

1 Introduction

Underground construction is in a great demand in many civil and infrastructure projects all over the world, such as metro (Zhong et al., 2003) and hydropower projects (Lin et al., 2015a). In the last decade, tunnel construction has presented a powerful momentum for rapid economic development. However, owing to various risk factors associated with complex project environments, violations of safety rule occur frequently in tunnel construction, resulting in serious problems in the related project operation (Liu et al., 2005; Qian and Rong, 2008; Qian, 2014; Wu et al., 2015). On 6 July 2010, a tunnel collapse took place in Prague, Czech Republic, causing a 15-m-wide sunken pit on the ground surface (Thomas, 2010). On 23 August 2012, water leakage in metro line caused chaos in Warsaw, Poland (Waltz, 2012). Water flooded into the tunnel at the planned power station, causing considerable transportation problems in the already gridlocked city. Despite the rapid development of construction industry in China, which reached about 2200 billion US dollars in 2012, many problems with regard to the cost and safety risk management are observed (Wu et al., 2015).

In China, the number of construction accidents shows a rising trend in tunnel projects over the past decade. In general, there arises a public concern that tunnel construction may generate ground deformations (Ou et al., 2008; Yoo and Lee, 2008), which may affect the safety of surface buildings and road traffics, and lead to unacceptable damages. Statistics have shown that no significant improvement has been made in safety risk management of tunnel construction from 2008 to 2011. Analysis of accident records in-dicates that the liability accidents account for the majority of ac-cidents in civil engineering, and the reasons for these accidents vary in different projects. With regard to the tunnel engineering, 60%of accidents occur due to both subjective and objective causes, 30%of accidents are liability accidents and 10%are completely triggered by objective causes. The causes for most liability accidents are similar in nature, including poor technologies, management and performance on hazard rectification. The objective causes for accidents in the tunnel engineering include adverse hydro-geological conditions, groundwater or heavy rainfall, and soft soil layers. Collapse is the most dominant accident type in the tunnel engineering, accounting for 60%of the total accident records. Forexample, on 15 November 2008, a fatal tunnel collapse occurred one year after construction of Hangzhou metro line #1, resulting in 12 lives loss. The tunnel section affected by the collapse is 100 m long by 50 m wide, and the depth of the crater is 6 m. Water inrush and object strike are the second and third causes, respectively. In addition, explosion, pipeline damage and rockburst are also the common causes for accidents in underground engineering. The injury and death rates vary among different types of accidents.

Safety risk management has been intensively studied in the USA and Europe (Duddeck, 1996). In 1992, the Council of European Community published the Council Directive 92/57/EEC on the implementation of minimum safety and health requirements at temporary or mobile construction sites (The Council of the European Communities, 1992). In 2004, the working group of In-ternational Tunneling Association proposed its guidelines for tunneling risk management (Eskesen et al., 2004). In 2006, the International Tunneling

Insurance Group issued a code of practice for risk management of tunnel works (ITIG, 2006). The Chinese government has also paid special attention to risk management of underground construction. In 2003, the Ministry of Housing and Urban-Rural Development (MOHURD) of China, together with eight relevant ministries, issued the advice on further enhancement of subway engineering management, proposing detailed re-quirements of safety risk management in the planning, design, construction, and operation of subway engineering. Recently, the MOHURD of China has issued the principle of risk control and the guideline of risk management for construction of underground works, which promote the standardization of safety risk management of underground engineering in China.

To avoid heavy casualties and property losses caused by safety violations, numerous studies have introduced risk-based analysis to safety prevention and control, which can be divided into qualitative and quantitative risk analyses (Smith et al., 2009). The former in- cludes fault tree analysis (FTA), comprehensive fuzzy evaluation method (CFEM), safety check list (SCL) and the others, while the latter includes the job risk analysis method, influence diagrams, neural network (NN), support vector machine (SVM), decision trees and others. The above risk-based analysis methods have made significant contributions to safety risk management in complex engineering projects (Alfredo, 2002; Piniella et al., 2009). However, they are limited to static control and management (Alaeddini and Dogan, 2011). Khakzad et al. (2011) described FTA unsuitable for complex problems due to its limitation in explicitly representing the dependencies of events, updating probabilities, and coping with uncertainties. When the associated parameters, such as geological, design and construction parameters are changed, the aforementioned methods cannot accurately depict the updated feature of dynamic environments as the construction progress continues. Nor can professional supports or suggestions be pro-vided in real time as the parameters are not updated.

In order to address the challenges presented by underground construction projects, studies are also conducted benefitting from new technologies, tools and approaches for construction safety management, including the above-mentioned analytical methods. Structural health monitoring (SHM) system, a comprehensive instrumentation of structures and environment, is widely recog-nized as a crucial element of underground construction safety management (Bhalla et al., 2005). For example, in Italy, the Geo-DATA Company developed an information management system, namely the Geodata Master System (GDMS), which aims at risk management in underground engineering. Based on the geographic information system (GIS) and web technologies, the GDMS pro-vides five sub-systems, including the building condition system (BCS), the building risk assessment (BRA) system, tunnel-boring-machine data management (TDM) system, monitoring data management (MDM) system and document management system (DMS). The GDMS provides complete risk management plans, and has been widely used in subway construction projects in Russia and Italy. However, the GDMS is not suitable for underground con-struction in China, due to the differences in monitoring technolo-gies and construction management regulations.

The SHM system is intended to predict structural and environ-mental instability risks, which are frequently encountered in un-derground construction (Bhalla et al., 2005; Chai et al., 2011; Lin et al., 2014).

By installing durable and robust sensors, it becomes feasible to automatically and intelligently monitor and predict the behaviors of underground structures in real time (Khoury and Kamat, 2009). Substantial work has been carried out on investi-gating the tracking technologies and their applications which can meet various requirements in underground construction practices (Lin et al., 2014). These proactive approaches could be used for dynamic identification and prevention of human error and behavior risks in underground construction. To date, several tracking technologies have been demonstrated as follows: radio frequency identification device (RFID) (Khoury and Kamat, 2009; Tu et al., 2009; Seco et al., 2010; Rao and Chandran, 2013), global positioning system (GPS), wireless local area network (WLAN or WiFi) (Jiang et al., 2015), ultra-wide band (UWB) (Carbonari et al., 2011), ZigBee, and indoor GPS (Ergen et al., 2007). These technol-ogies are able to cover a wide range of area and give relatively ac-curate results (Behzadan et al., 2008).

In this paper, the main progress of the safety risk management of underground engineering in China in the past decade is first presented. Based on the analysis of the major accidents of under- ground engineering in the recent decade, the new challenges of the safety risk control are analyzed. The new strategies of safety risk management in underground construction are proposed in six as- pects, i.e. the management system and policy, the legal, adminis- trative, economic, educational and technical countermeasures. The structure flowchart of this study is illustrated in Fig. 1.

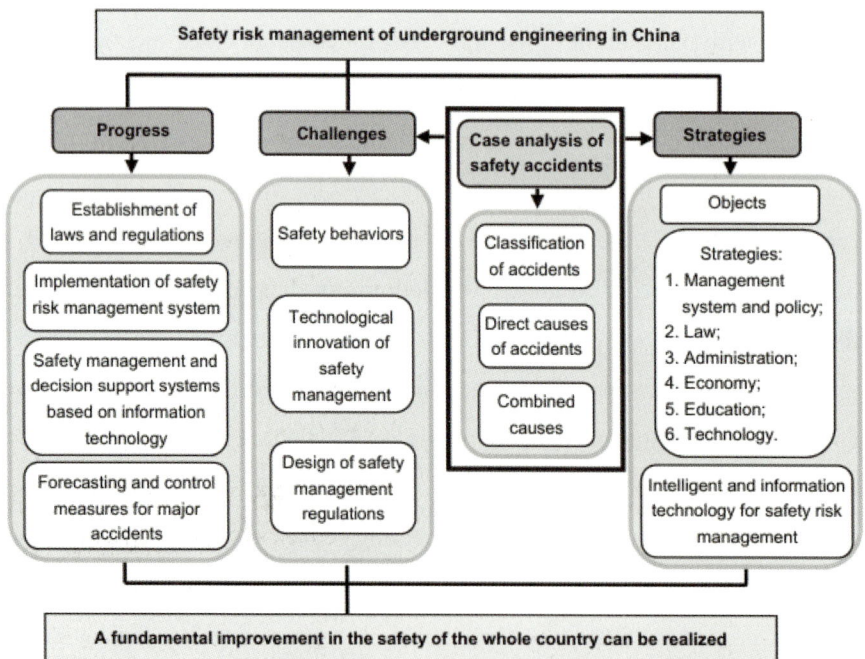

Fig. 1 The structure flowchart of this study.

2 Progress in safety risk management

Qian and Rong (2008) and Qian (2014) proposed four sugges-tions specific to safety risk management of underground engi-neering in China:

（1）The laws and regulations of underground engineering should be improved.

（2）The safety risk management plans should be implemented in construction management of underground projects.

（3）Information technology should be employed to implement early-warning and decision-making support functions for safety risk management.

（4）More resources should be invested into researches on safety risk management, prediction and prevention of major accidents.

Over the past decade, tremendous efforts have been made to the safety risk management of underground engineering in China, including safety risk control regulations, management plans and related technologies. By taking into account the experts' opinions, the guides by researchers and government agencies, a sound safety risk management system has been established on the basis of technology, management, culture, regulations and other means.

2.2 Establishment of laws and regulations

In recent years, Chinese government has paid much attention to the laws and regulations of safety risk management for underground construction. The law and regulation system of safety risk management in China is a combination of several laws, regulations and technical standards (Fig. 2), which can be divided into four levels and nine classes as follows: (1) The law level: the law class; (2) The statue level: the administrative statute class and the local statute class; (3) The regulation level: the ministry regulation class and the local government regulation class; (4) The standard level: the national standard class, the industrial standard class, the local standard class and the com-pany standard class. In general, the law and regulation system of safety risk management for underground construction in China is implemented based on the Constitution, the Construction Law and the Production Safety Law. Issued by administrative regula-tions, including "Regulations on production safety management for construction projects" and "Regulations on safe production license", and supported by ministerial regulations, including "Administrative regulations of safe production for construction projects" and "Regulations of safe production license for con-struction enterprises", the system includes a large number of local statutes, regulations and standards. The main progress in establishment of laws and regulations in the recent decade can be summarized as follows:

（1）With regard to the construction of the urban subway system, the MOHURD of China issued "Temporary regulations on safety and quality management of urban rail transit projects" in January 2010. It provides explicit provisions on risk assessment, risk monitoring and emergency disposal, and stresses that "safety and quality risk management has to be strengthened throughout construction of urban rail transit projects". It clearly defines the responsibilities of all parties participating in urban rail transit construction. For example, the project owner is required to assess the safety and quality risks and organize expert argumentation at the preliminary design stage. Meanwhile, specific evaluations by experts on seismic resistance and wind resistance are also necessary. The costs of risk assessment, field monitoring and environ-ment investigation

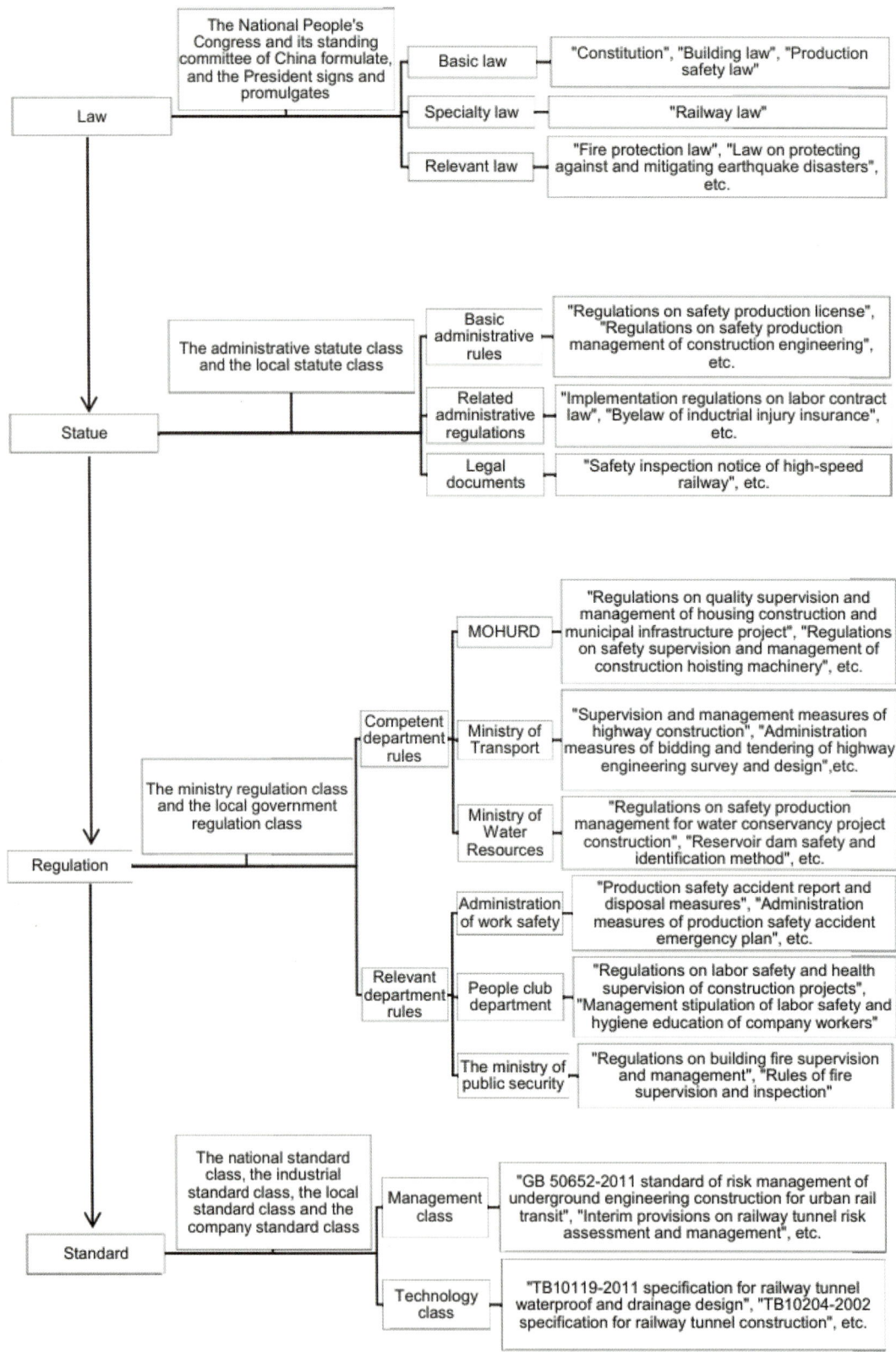

Fig. 2 The legal system for safety risk management of underground construction in China.

should be included in the budget. Geological risks should be specified in the geological survey stage. Guidelines on survey of special geological conditions should be prepared in advance if necessary. The design in-stitutions should organize expert argumentation on the design scheme, monitoring and control standards for the environment, if high risks are expected. Meanwhile, a series of relevant regulations, including "Regulations of safety and quality contingency plan management for urban rail transit", has been issued. The MOHURD of China also issued "Speci-fications on underground construction risk management for urban rail transit" in 2011, in order to regulate the technical details for risk management.

（2）With regard to railway construction, "Temporary regulations of safety risk management for railway construction" was is-sued in September 2010. The regulation extends risk man-agement from tunnel engineering to all kinds of railway projects. In 2014, China Railway Corporation (originated from Ministry of Railways) issued "Technical specifications of risk management for railway construction".

（3）With regard to traffic engineering, the Ministry of Transport of China issued "Guide for safety risk assessment of highway bridge and tunnel design" in 2010 and "Guide for safety risk assessment of highway bridge and tunnel construction" in 2011. On this basis, the policies on risk assessment have been established. The guides provide detailed and feasible methods for evaluating common risks in highway bridge and tunnel construction.

A legal system of safety risk management covering multiple levels of laws, statutes, regulations, and industrial standards has been established in China. However, many challenges are observed in implementation of the legal system.

2.3 Implementation of safety risk management system

Great efforts have been devoted to establish the safety risk management system for underground construction in China in five aspects, including the organization structure system, the safety culture system, the technical management system, the disaster prevention and early-warning system, and the project insurance system.

For construction of urban transit, the safety risk management has been used in construction of metro lines in Beijing, Shanghai, Guangzhou, Shenzhen, Nanjing, Chengdu and other cities in China. The safety risk management systems suitable for local conditions have been gradually established over years. For example, "Safety Risk Management System for Construction of Urban Transit" has been issued in Beijing, which covers engineering survey, design, construction and post-construction phases (Fig. 3). In summary, the safety risk management system of urban transit construction in China has the following characteristics:

（1）Technical management and risk control during the entire construction process: Risks should be identified in the planning and survey stages. Risk sources shall be avoided or mitigated in the design stage. Attention shall be paid to risk control and management during the construction stage. Risk assessment and tracking should be strengthened in the post-construction phase. The third-party supervision should be adopted to reinforce the quality management of technical works at all stages.

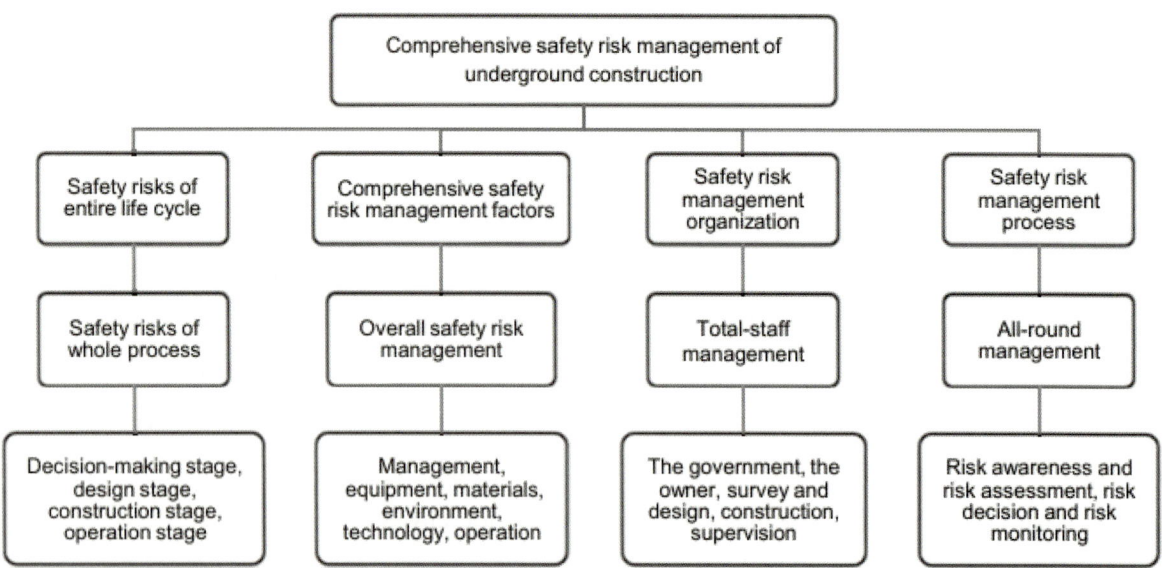

Fig. 3 The safety risk management system of underground construction in China.

(2) Risk assessment, control and prevention of risk sources: Safety risk assessment and hierarchical control system should be generally implemented. Through safety risk identification, risk assessment, hierarchical control and expert checks in advance, safety risk can be mitigated or roughly eliminated. Meanwhile, management measures should be reinforced for potential risks and contingency plans should be drawn to ensure that risks are under control.

(3) Implementation of safety responsibility of each party involved: Proper contracting strategies regarding reward and punishment shall be implemented to reinforce the safety responsibilities of all parties involved in underground construction.

(4) Process monitoring: Managers should pay attention to on-site behaviors during construction. Dynamic management should be strengthened by refined, informationized and programmed means. Process monitoring and control, su-pervision measures and behavior norms shall be imple-mented effectively.

Based on the third-party monitoring, real-time tracking of various hazard sources can be realized. Based on the conditional acceptance of key points before construction, major risks are effectively controlled. Site inspection is used to control safety hazards dynamically. By introducing information technology into monitoring, the efficiency of risk management is improved. Based on the integrated design, construction and management, a linkage mechanism is formed to promote the capability of safety protec-tion. With the hierarchical early-warning management, proper and efficient treatment measures are selected.

2.4 Safety risk management and decision support systems based on information technology

In recent years, significant progress has been made in risk management and early-warning decision support system of un- derground construction in China by adopting information tech- nology. Sun

(1999) studied the intelligent prediction and control of urban underground construction safety and its three- dimensional (3D) simulation system. Lin et al. (2013, 2014) and Jiang et al. (2015) employed several information technologies, including WiFi, 3G, GPS and RFID to establish a risk identification and management system for large-scale hydropower projects. With this system, real-time tracking of site staff and workers can be realized. Together with the construction market and personnel management system, online monitoring of personnel access and activity tracking, early-warning, prediction and evaluation analysis can be implemented. The system promotes site performance analysis and rapid response to quality and safety management, and all-around safety management can be ach-ieved. For instance, a real-time tracking system for personnel safety in the construction area was established for Xiluodu hydropower project on the Jinsha River, China (Fig. 4) (Lin et al., 2014). The 3G technology was adopted for data transmission, GPS auxiliary positioning technique for the deck, and the WiFi-based positioning technique for corridors. Smart phones with multiple sensors were employed as the major devices to monitor site personnel. Highly accurate real-time positioning, safety assessment analysis of site personnel, 3G communication dis-patching, early-warning and messaging were realized. By mining massive data collected by the system, valuable information can be derived. Such information can satisfy the requirements of the owner and the superintendent on labor force consumption and analysis and management of safety behaviors. With the above devices and networks, the system can be extended to safety management, video monitoring, multiple-sensor internet of things and construction machinery management. Scientific allo-cation and integrated management of personnel, resources, environment and regulations can also be achieved.

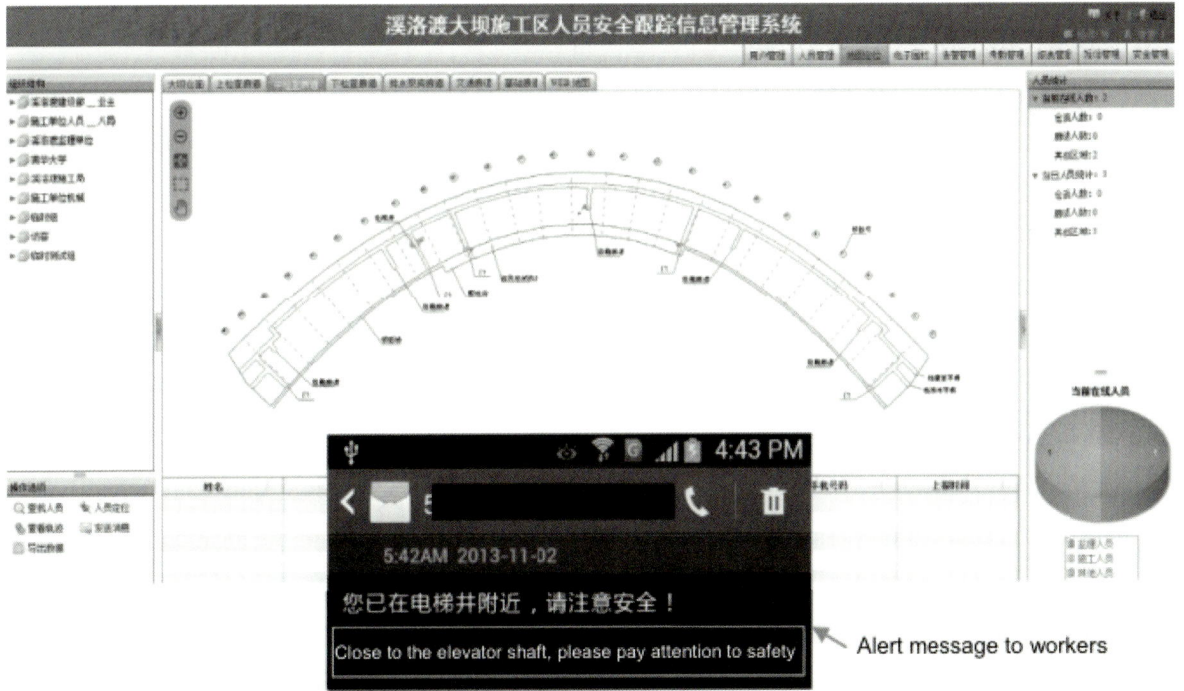

Fig. 4 Location based services (LBS) web client interface on dam construction site (Lin et al., 2014).

2.5 Safety risk management, forecasting and control measures for major accidents in underground engineering

In recent years, in order to enhance the safety risk management in underground construction, great achievements have been made in major accident forecasting and prevention in China, for instance, monitoring and forecasting techniques for water and mud inrush, and rockburst (Li et al., 2013a,b), the risk management and control over the entire construction process of metro projects (Zhong et al., 2008; Shi et al., 2012), and forecasting of rockbursts based on microseismic monitoring (Tang et al., 2011; Ma et al., 2015; Xu et al., 2016).

1) Monitoring and forecasting techniques for water and mud inrush

Water or mud inrush is predictable based on some precursory information. In order to prevent and avoid water or mud inrush, advanced geological forecast has to be performed. Due to the complexity of geological condition and multiple explanations of geophysical detection results, single forecasting method can hardly ensure accurate and reliable results. Different forecasting methods may give different results for the same event and they are sensitive to different events. Therefore, in order to improve the accuracy and precision of forecasting, integrated advanced geological forecasting methods are needed, including macroscopic advanced geological forecast (engineering geological method), long-distance (200-500 m) advanced geological forecast (engi-neering geological method, TSP detection), short-distance (within 50 m) advanced geological forecast (geological radar, infrared detection for groundwater, transient electromagnetic method, advanced drilling, pilot tunnel and empirical methods). The research group led by Professor Li established a four-stage whole-process monitoring and forecasting system for water inrush haz-ards during tunnel construction (Fig. 5), based on a great number of engineering practices and multiple geophysical detection methods (Li et al., 2013a,b). The system stresses the importance of geological analysis and takes advantages of seismic prospecting, transient electromagnetic and induced polarization (IP) methods. The system has greatly improved the prediction precision and ef-ficiency for water inrush hazards, and provides effective measures for problems caused by water inrush. The system has been applied to many tunnel projects in China.

2) Prediction of rockbursts based on microseismic monitoring

The laws of microcrack initiation, propagation and coalescence are the evolution characteristics of macroscopic rockburst, which are also the theoretical basis for monitoring and forecasting of rockburst (Tang et al., 2011; Ma et al., 2015; Xu et al., 2016). The P-and S-waves generated by microcracking can be captured by geo-phones or accelerometers, and these signals are then transformed into digital data signals and processed by data processing software. The time, location and intensity of microseismic events can be determined. Ultimately, the rockburst proneness, and qualitative and quantitative assessment of the location and magnitude of rockbursts can be analyzed according to the evolution of micro-cracks. The microseismic monitoring technique has the following characteristics:

(1) The monitoring scope can be very large. The time, location and magnitude of microcracking events in the rock mass can be determined directly. It overcomes the drawbacks of the traditional "point" monitoring

techniques which are local-ized, discontinuous and labor-intensive. Microseismic monitoring represents the trend of stability monitoring for rock structures.

(2) The monitoring system is automatic and intelligent, which supports remote information transmission. The monitoring instruments are being developed towards highly inte-grated, small-size, multi-channel and highly sensitive devices.

(3) As the monitoring system receives the information of seismic waves, the sensors can be installed in the region far away from the failure-prone area, which is advantageous for ensuring long-term operation of the monitoring system.

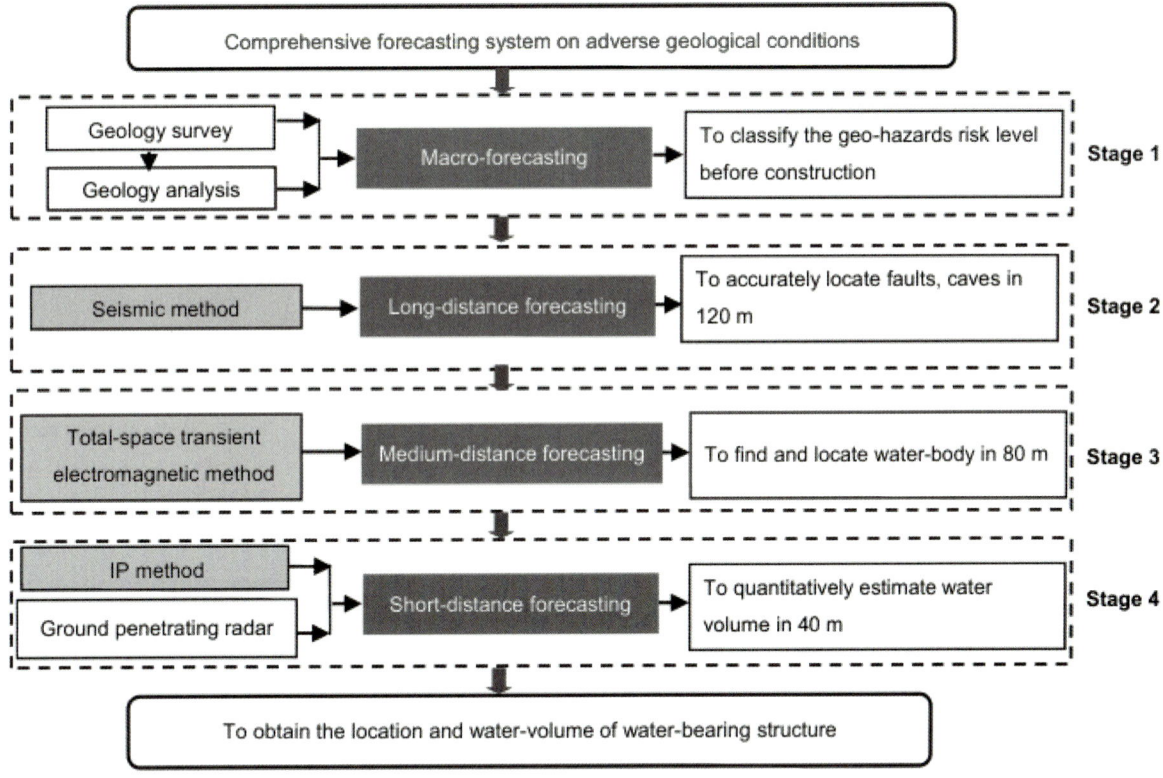

Fig. 5 Schematic diagram for comprehensive forecasting system on adverse geological conditions (Li et al., 2013a,b).

The research team led by Professor Tang carried out micro- seismic monitoring of rockbursts in the headrace tunnels at Jinping II hydropower station (Tang et al., 2011). Rockburst monitoring was performed at 10 working faces. The layout of the microseismic monitoring system is shown in Fig. 6. The moni-toring data showed that the accuracy of rockburst forecasting was as high as 85.5%.

3) Whole-process risk management and control techniques for metro projects.

On account of rapid development of metro projects in China, complex project planning, design, construction and operation, and requirements on safety performance, the research group led by Professor Zhong proposed a system of risk control theory and risk detection method for planning, design, construction, commis-sioning, trial operation and operation stages, by utilizing theoret-ical analyses, small-scale experiments, laboratory tests and full-scale field tests (Zhong et al., 2008; Shi et al., 2012). The system includes identification and assessment of hazard sources, simula-tion of fire hazards, visualized detection of passenger flow, modeling for stability of surrounding rock masses, hot smoke detection, cold slippery detection, simulation of evacuation plans, simulation of large passenger flow, risk early-warning, etc. With this theory, comprehensive risk prevention and control can be realized at the planning, design, construction and operation stages of a metro project. One national standard (GB/T50438—2007) and five industrial standards of production safety (AQ8004—2007, AQ8005—2007, AQ8007—2013, AQ/T9007—2011, AQ/T9002—2006) have been issued. The system has been adopted in design, con-struction and operation stages at more than 300 metro lines in over 30 cities in China.

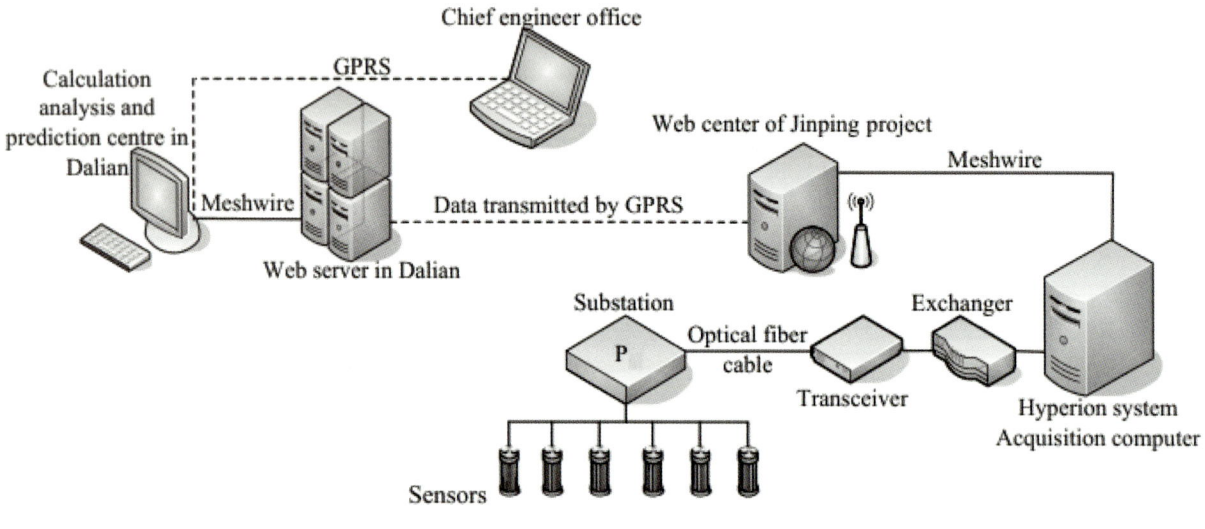

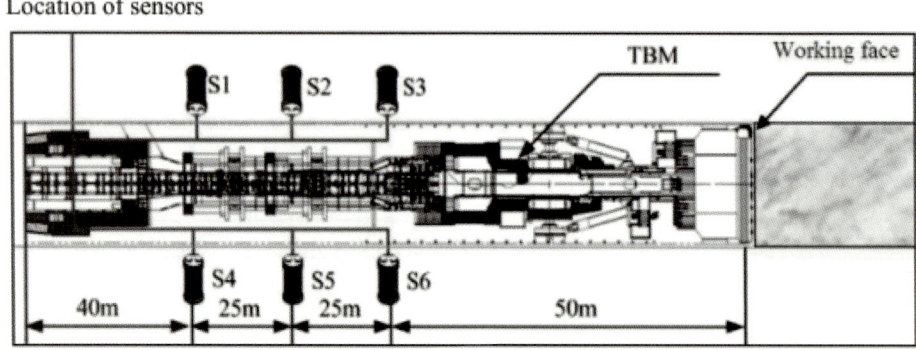

Fig. 6 Monitoring and analysis system for rockbursts during tunnel boring machine (TBM) tunneling for Jinping II hydropower station (Tang et al., 2011).

3 Case analysis of safety accidents

The number of casualties in construction industry in China generally has declined over the past decade. However, the total number of casualties is still large (Fig. 7). According to the analysis in Section 2, despite great progress in four aspects, the safety risk management still faces great challenges with varying economic and engineering scale. Especially for underground construction, a number of safety risks still exist.

3.1 Classification of accidents

Compared to other accidents, the problems, such as safety awareness, safety education, poor safety responsibility, exist in tunnel accidents. In addition, because the hydrogeological condi-tions of underground tunnel projects are generally complex, a number of uncertain factors are encountered. Based on the typical accidents occurred in China in the past decade (Table 1), the acci-dents are classified into collapse, object strike, geological hazard, explosion and toxic gas poisoning (Fig. 8). These accidents have the following features:

(1) Collapse and object strike are the major accident types, accounting for 61% of the total accidents. Collapses may be caused by various factors, such as adverse geological and hydrological conditions, poor management, improper construction, etc. The geological disasters, such as water and mud inrush, cause a number of collapses, mainly the earth collapse and landslide. Collapses often result in a chain reaction. For example, the casualties due to object strike in Xianghu station of Hangzhou metro line #1 was in fact caused by the ground subsidence, due to over excava-tion in the foundation pit, severe defects existed in the support system, delay of erecting the steel bracing and casting the concrete cushion.

(2) The accidents caused by complex geological conditions ac-count for about 21% of the total accidents. In the tunnel projects, soft soil strata, water and mud inrush are the dominant factors. During construction process, complex rock structure is the prominent factor for accidents. Some acci-dents are completely caused by geological disasters without any sign before failure. On the other hand, it also shows that the investigation and prevention of underground disasters need to be further improved. On 28 November 2009, during construction of the drainage tunnel at Jinping II hydropower station by TBM, an extremely strong rockburst occurred. The longitudinal range of rockburst was about 30 m, and the crater depth was around 8 m. A great amount of rock powder was released during the rockburst and spread over the tun-nel for 10 min. The huge amount of energy released during the rockburst destroyed the support system instantaneously, causing a wide range of collapse on the crown and sidewalls of the tunnel. The TBM was partially buried, the main girder was broken, and eight workers were killed.

(3) Toxic gas poisoning, explosion and other accidents account for about 18% of the total accidents. Underground construc-tion is generally carried out in narrow space, leading to higher probability of toxic gas poisoning and explosion.

3.2 Direct causes of accidents

Statistical analyses of the typical accidents (Table 1) indicate that the liability accidents account for about 90% among all the accidents. The direct causes are described as follows.

(1) Poor safety awareness, neglecting of accident precursors, and fluky psychology

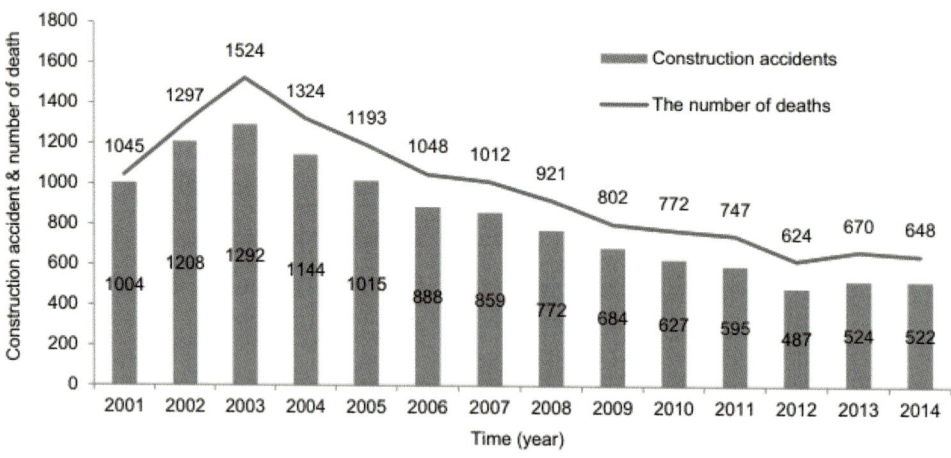

Fig. 7 Statistics of construction accidents and deaths in China from 2001 to 2014.

Table 1 Summary of major accidents in underground engineering in China since 2001.

No.	Accident	Type	Location	Date (MM/DD/YY)	Process	Consequence
1	Landslide of Nanjing metro line #2	Collapse	Nanjing, Jiangsu Province	5/28/2007	At 8:00 am, landslide occurred at east of west foundation pit of Chating station of metro line #2. The accident was located at the cross of Shuiximen street and east road of Jiangdong gate memorial.	2 deaths
2	Collapse accident at Suzhoujie station of Beijing metro line #10	Collapse	Beijing	3/28/2007	At 8:00 am, 1 m3 of soil collapse was found between Section 10-10 and Section 6-6 at the southeast gateway of line #10. The vault was reinforced immediately. At 9:30 am, vault collapsed again during the repairing process.	6 deaths
3	Collapse accident at Shanghai rail transit line #4	Collapse	Shanghai	7/1/2003	At 6:00 am, a large amount of water and sand flooded into the operation face of passageway of Shanghai rail transit line #4, causing local damage of tunnel and ground subsidence in surrounding area. It resulted in severe inclination of 3 buildings and partial collapse of the flood control walls, which caused piping of the cofferdam.	The direct economic loss was 150 million Chinese yuan
4	Ground subsidence at Xianghu station of Hangzhou metro line #1	Collapse	Hangzhou, Zhejiang Province	11/15/2008	Ground subsidence occurred at Xianghu station of Hangzhou metro line #1, forming a cave-in with length of 75 m and depth of 15 m. Eleven cars fell down into the hole.	21 deaths, 24 injuries, and the direct economic loss was 49.62 million Chinese yuan
5	Collapse at Shanghai metro station	Collapse	Shanghai	8/20/2001	At 7:00 pm, 11 workers worked at the platform between axes 14 and 15 of the foundation pit. At 8:00 pm, landslide occurred at axis 16, burying 2 workers immediately. Two more workers were buried up to waist and other 6 rushed away from the pit. At 8:10 pm, the second slide occurred, with earth rushing from axes 18 to 12. The 2 workers were entirely buried and 16 steel bracing were broken.	4 deaths
6	Collapse between Hancheng road and Textile mall at phase I of Xi'an metro line #1	Collapse	Xi'an, Shaanxi Province	8/2/2009	At 1:00 to 5:00 am, the groove in the east section 20-35 m of Guanliang road, on the north side of the Sajinqiao bus station, was excavated. The groove was 4 m wide at the top, 3.5 m at the bottom and 4.5 m deep. At 6:30 am, The pipelines in the groove and slope were cleared. At 9:20 am, the 10 m south pit wall between the pile Nos. 20 and 26 collapsed with a collapsed volume of about 10 m3.	2 deaths

No.	Accident	Type	Location	Date (MM/DD/YY)	Process	Consequence
7	Object strike at Section 10 of Beijing metro line #10	Object strike	Beijing	2/27/2006	The rope of mounted hoist crane was broken suddenly during operation. The drop bucket fell down, smashing 3 workers to death.	3 deaths and 1 injury
8	Object strike at Shunyi station of Beijing metro line #15	Object strike	Beijing	7/14/2010	At 4:30 pm, the steel frame for the wall of deep foundation pit in Shunyi station dropped off, smashing 10 workers.	2 deaths
9	Steel collapse at Chongwenmen station of Beijing metro line #5	Object strike	Beijing	10/8/2003	Groups of workers were assembling reinforcement steel bars. The designed spacing of the main steel bars was only 10 cm and the longitudinal spacing of the scaffold bar was 2 m. It was difficult to assemble reinforcement with bars blocking the stirrup. Workers removed one scaffold bar after asking for instructions from vice-monitor and continued to assemble steel bars at 7:50 pm, when workers were trying to pull the stirrup, the scaffold with reinforcement overturned in the entrance direction of pilot tunnel, pinning 4 workers to the ground.	3 deaths, 1 injury and the direct economic loss was 297,000 Chinese yuan
10	Object strike at Xianghu station of Hangzhou metro line #1	Object strike	Hangzhou, Zhejiang Province	11/25/2010	At 12:30 pm, the landslide in the foundation pit smashed 2 drivers in excavators.	1 death and 1 injury
11	Manual excavation of piles for section 3101 at Shaibu station of Shenzhen metro line #3	Toxic gas poisoning	Shenzhen, Guangdong Province	7/6/2009	At 9:25 am, a worker went down to the well and prepared to cast concrete. Five minutes later, another worker found it was unable to contact the first one and decided to check the condition in the well. Both workers lost contact in the well. The supervisor reported immediately to the project manager. At 9:40 am, the 2 workers were rescued from the well, but died 5 min later.	2 deaths and 8 injuries
12	Collapse at section 3106 of Shenzhen rail transit line #3	Formwork collapse	Shenzhen, Guangdong Province	4/1/2008	Concrete casting was started at 10:50 am. At about 1:00 pm, the workers found that the casting process was too fast and requested to stop casting. The casting was suspended for several times. The construction continued after inspection by workers. At 3:10 pm, the casting was paused again for assembling embedded steel bars at the pier top. Workers from concrete casting were going to take a break. When they were climbing down, the formwork and scaffold overturned suddenly to the southeast. Three workers fell down and died, and 2 workers were injured by the falling formwork.	3 deaths and 2 injuries
13	Gas explosion at the Dongjiashan tunnel of Dujiangyan-Wenchuan highway in Sichuan	Gas explosion	Wenchuan, Sichuan Province	12/22/2005	Tremendous gas explosion took place at the Dongjiashan tunnel of Dujiangyan-Wenchuan highway in Sichuan Province.	44 deaths, 11 injuries and the direct economic loss was 20.35 million Chinese yuan
14	Collapse at the Baiyun tunnel of Nanning-Guangzhou high-speed railway	Geological disasters caused by water gushing and mud inrush	Yun'an County, Guangdong Province	1/18/2010	At 8:00 am, water gushing and mud inrush occurred in the Baiyun tunnel. Collapse occurred at the upper left part of the tunnel. The fault has a height of 80 m, length of 16 m and width of 13 m. Mudflow lasted for 30 s and 2500 m3 mud rushed out by 167 m.	5 deaths and 4 injuries

No.	Accident	Type	Location	Date (MM/DD/YY)	Process	Consequence
15	Collapse at the Gaoyangzhai tunnel of Yichang-Wangzhou railway	Rockfall	Enshi, Hubei Province	11/20/2007	At 8:40 am, rockfall occurred at the Gaoyangzhai tunnel, causing a total landslide volume of 3000 m3. One of the 4 workers was dead, 1 was injured and 2 were missing. Afterwards, 1 bus from Shanghai to Lichuan was reported missing at the accident site.	35 deaths and 1 injury
16	Accident at the Daguishan tunnel of Luoyang-Zhanjiang railway	Explosion	Guangxi Zhuang Autonomous Region	12/10/2006	The 140 kg explosives were left in the tunnel during the day. The night shift workers carried out welding for the shotcrete-blot support.	6 deaths and 1 injury
17	Tunnel collapse at the Xinqixiaying tunnel of Ulanqab-Baotou railway	Collapse	Inner Mongolia Autonomous Region	3/19/2010	At 2:30 pm, the Xinqixiaying tunnel collapsed, killing 10 people.	10 death
18	Water inrush at the Maluqing tunnel of Yichang-Wangzhou railway	Water inrush	Enshi, Hubei Province	4/11/2008	At 5:10 pm, a construction machine collapsed suddenly at the entrance of Maluqing tunnel. When 5 workers entered the tunnel for repairing the machine, water inrush occurred and hundreds of thousands cubic meter water flew into the tunnel.	4 deaths and 1 missing
19	Water and mud inrush at the Yeshanguan tunnel of Yichang-Wangzhou railway	Water and mud inrush	Enshi, Hubei Province	8/5/2007	At 1:00 am, water and mud inrush occurred at the right bottom of the excavation face. The 150,000 m3 water and 54,000 m3 mud rushed into the tunnel. Within 220 m away from the excavation face, the tunnel was filled with mud and rock. The mud in other places was 1-4 m thick. The tunnel passed through the Shimaba anticline and the Erxihe syncline. Five underground rivers and channel flow were identified. After the water inrush, a total of 52 workers at 5 tunnel faces were trapped.	3 deaths and 7 missing
20	Water and mud inrush at the Maluqing tunnel of Yichang-Wangzhou railway	Water and mud inrush	Enshi, Hubei Province	1/21/2006	Water and mud inrush occurred at the exit section of the Maluqing tunnel, with a total water volume of 180,000 m3. Water inrush occurred again for several times during rescue. Dolines, sinkholes and underground rivers were common in the tunnel area. Karst was well developed and the karst water system was complex.	10 deaths and 1 missing
21	Explosion at the Guantoulin tunnel of Wenzhou-Fuzhou railway	Explosion	Lianjiang County, Fujian Province	2/28/2006	At a distance of 50 m from the tunnel entrance, when workers were supporting the formwork of short wall and some were inserting the steel bars while drilling, explosion suddenly occurred in the borehole. Three workers died and one was injured seriously.	3 deaths and 1 injury
22	Collapse at the Binyang tunnel	Collapse	Guangxi Zhuang Autonomous Region	7/11/2010	Vertical joints existed in the surrounding rock masses, which are likely to cause collapses. The rock was weakened in the presence of abundant water in the mountain, causing collapses.	10 deaths
23	Rockburst at the drainage tunnel of Jinping II hydropower station	Rockburst	Liangshan Yi Autonomous Region, Sichuan Province	11/28/2009	At 0:43 am, extremely strong rockburst occurred when TBM excavated at a depth of about 2500 m. The rockburst produced large amounts of dust, which are scattered in the air for 10 min. The huge energy released by rockburst destroyed the support system, causing collapse in large areas in the roof and two sidewalls. About 1000 m3 rock ballast buried the main engine of the TBM and broke the main girder. Eight workers were buried, with 7 dead and 1 injured.	7 deaths and 1 injury
24	Tunnel collapse at Diqing Tibetan Autonomous Region	Collapse	Shangri-La County, Yunnan Province	3/29/2011	At 4:00 pm, due to the unstable geological strata, the Kengu tunnel, which had been supported, collapsed suddenly. Nineteen workers were trapped in the tunnel.	19 injuries

No.	Accident	Type	Location	Date (MM/DD/YY)	Process	Consequence
25	Water and mud inrush at the Taoshuping tunnel of Lanzhou-Chongqing railway	Water and mud inrush	Yuzhong County, Gansu Province	11/4/2014	At 3:10 pm, water and mud inrush occurred at the Taoshuping tunnel of Lanzhou-Chongqing railway. Nine workers were buried and 1 was dead.	1 death and 9 injuries
26	Collapse of the road drainage at the east stretch of Songshan Road	Collapse	Suyu, Jiangsu Province	4/4/2012	At 8:50 am, collapse occurred when 5 workers were constructing the pipeline foundation at the trench bottom. Five workers were buried.	3 deaths and 1 injury
27	Collapse at Xi'an metro line #3	Collapse	Xi'an, Shaanxi Province	5/6/2013	In the early morning, the tunnel section between the Tonghuamen station and the Hujiamiao station of Xi'an metro line #3 was excavated to a depth of 8 m. Tunnel roof collapsed suddenly when 9 workers were working.	5 deaths

The accidents with precursors before failure account for 75% among the total accidents (Fig. 9). The main reason for some accidents with precursors in tunnels is that the precursors are not identified by the superintendent or no countermeasure is taken by the contractor. For instance, before the "12.22" major gas explosion accident in the Dongjiashan tunnel of Dujiangyan-Wenchuan highway in Sichuan (case 13 in Table 1), the tunnel face collapse and abnormal gas gushing led to extremely high

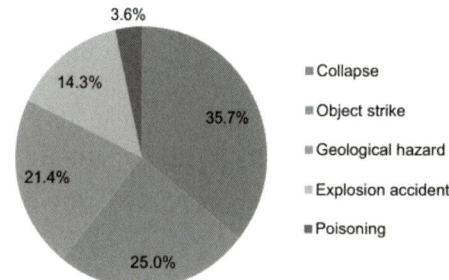

Fig. 8 Classification of major accidents during underground construction in China.

gas content near the platform. The short circuit of three-pin plug near the distribu- tion box of the formwork trolley initiated sparks and caused gas explosion. As the superintendent did not perform the duties properly and non-qualified staff was appointed on the key post, the precursors were not identified. In many cases, no proper counter- measures were taken by the contractor, and little attention was paid by the superintendent and contractor to the precursors. For instance, the surface collapse took place in the Xianghu station of Hangzhou metro line #1 (case 4 in Table 1). A non-qualified chief engineer was appointed for quality control and documentation. During excavation of N2 foundation pit, the cracks on the roads and the displacements in the sidewall of foundation pit were observed. Four alarms were raised by the monitoring data. However, no effective measures were taken and eventually the accident occurred.

（2）Problems associated with the contractor

Firstly, poor safety awareness and unreasonable construction scheme due to deficiency in the existing construction techniques are the main causes (Fig. 10). For example, during construction of Yichang-Wanzhou railway (case 15 in Table 1), geological hazards, such as water and mud inrush, were frequently reported. On 21 January 2006, unprecedented catastrophic flood occurred in the Malujing tunnel. After successful rescue, a tunnel was built for flood discharge in order to eliminate the risks during the construction and operation periods. The flood discharge tunnel project started in August 2006. On 11 April 2008, three days after successful flood discharge, water inrush occurred again, leading to 4 deaths and 1 missing. Violation of operation rules, improper instruction, lack of safety training for workers, and lack of safety technique disclosure were the most important reasons of the accidents, for instance, the object strike accident in Beijing metro line #10 (case 7 in Table 1). In this case, the crane was not checked carefully before the relevant staff signed the acceptance report, and the site manager was lack of law awareness and signed the acceptance report before checking the crane. No measures were taken timely and construction was carried out as schemed. Secondly, illegal subcontracting, improper rectification of hidden dangers, and unqualified appointment are also the common reasons. In addition, improper construction techniques, poor material quality, improper operation of equip-ment, and no warning signs at the dangerous zones are also the common reasons for accidents.

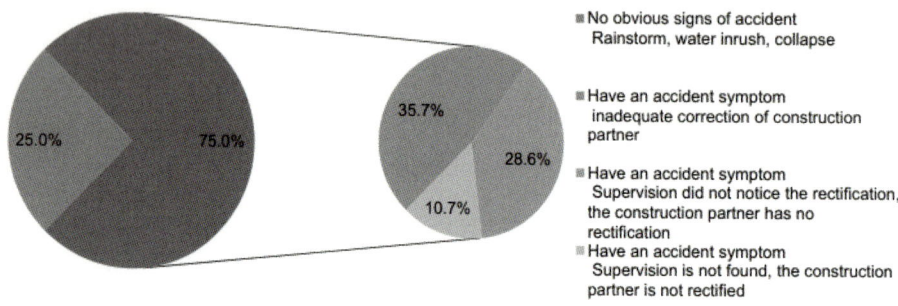

Fig. 9 Proportion of various accident precursors.

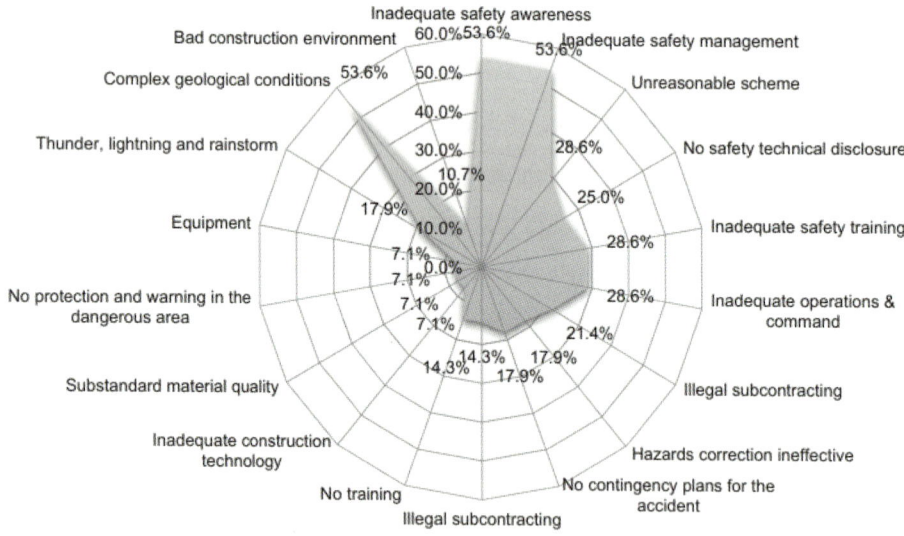

Fig. 10 Accidents due to problems associated with contractor.

(3) Problems related to the superintendent

Nonfeasance and poor supervision ability exist in the tunnel projects. Different from other projects, a higher portion of accidents were caused by poor identification of hidden dangers and lack of professional supervision knowledge, indicating the lack of pro- fessions in superintendent for underground projects. In addition, low responsibilities are taken by the superintendent, such as vio- lations of inspection and acceptance rules and no countermeasure taken after identifying hidden troubles. In the surface collapse of the Xianghu station of Hangzhou metro line #1 (case 4 in Table 1), the superintendents did not perform their duties, for instance, approval of the report for a dangerous construction scheme without careful examination, forceless determent on violations of rules and regulations by the contractor, and delay in reporting to the developer and relevant quality supervision departments. The superintendent did not carry out inspection and acceptance for the project according to stipulations.

(4) Problems related to survey and design

As the tunnel projects are constructed underground, survey data are very important for construction safety. For instance, the "3.19" major collapse in the Xinqixiaying tunnel of Ulanqab- Baotou railway (case 17 in Table 1) was one of these accidents. At 2:30 pm, the Xinqixiaying tunnel collapsed and killed 10 peo-ple. The rock masses along the tunnel alignment are of poor quality and self-stability. The tunnel collapse was caused by insufficient survey depth, untimely and inactive tracking and monitoring of the changes in the surrounding rock masses after excavation, and early commencement of design and construction without inspection and acceptance of geological survey data. Sometimes, accurate survey data cannot be obtained due to complicated geological conditions.

(5) Problems associated with the project owner

One of the problems associated with the project owner is the poor supervision, such as the surface collapse accident in the Xianghu station of Hangzhou metro line #1 (case 4 in Table 1). The owner did not supervise and manage severe hidden dangers properly during the construction process, which was a serious breach of duty. The other is caused by illegal contracting, such as the "3.19" major collapse in the Xinqixiaying tunnel of Ulanqab- Baotou railway (case 17 in Table 1). The standard tendering pro-cedures were not followed by the Mengji Company. The project owner did not supervise the construction strictly. For example, no experts were organized and invited for inspection and acceptance of the survey data in the stipulated time, and the project was started without approval.

3.3　Combined causes of accidents

(1) Individual and departmental interests

The individual and departmental parts mainly refer to some important figures in related units. A few important persons only consider their own achievement and interests, and various important factors are neglected, such as the project scale, con-struction period and cost of scientific research. The construction process and detailed bidding are frequently and randomly inter-fered due to their "power" on project management. Underground projects are becoming the corruption disaster center by power rent-seeking. The owner usually has a tendency to have bidding at a low price, and randomly delays the payment for construction contractor, and shortens the construction period which is out of the schedule. Under the low bid price, the construction contractor has to reduce the cost on safety, design and construction at the same time due to profit margin. Worse, the construction contractor will employ unlicensed workers, regardless of the national and/or in-dustry standards.

(2) Safety inspection just becoming a mere formality

The concept of "massive safety inspection" originates from the political thinking that only focuses on the important person's opinion. The formalism of safety management includes law system and regulations not being efficiently implemented at construction site, which in turn testifies that no one will follow the rules. The government representatives pay attention to the format of exami-nation rather than true inspection, and once an accident occurs, the government representatives only rely on "massive" one-time safety inspection, not on on-site check of the safety management of the enterprises. In this regard, the associated enterprises will be shut down due to the pressures imposed by the superiors and public opinion. Therefore, a scientific engineering safety risk system needs to be established, and the implementation and supervision must be strictly and carefully controlled, while the political thinking and leadership culture in terms of lack in responsibility should be discarded.

(3) Malfunction of governmental supervision

The governmental supervision is usually characterized with multiple and cross leaderships, which is highly dependent on merely using examination and approval mode. This is usually conducted with little attention paid to effective supervision, lead- ing to the lack of independence of inspection. In addition, the government-based investment project implements the notion of "who invests, who manages". At present, the government-based investment project shares the largest percentage in underground construction projects in China. In those projects, the government has a double identity in terms of the supervision and owner. For example, a vice mayor of a city is usually in charge of important urban infrastructure projects. In this circumstance, the multiple and cross leaderships can cause the safety inspection being lack of independence, in the absence of the support of professional in- stitutions and experts team for safety risk management. For an illegal contract and multi-level contract, the government supervi-sion organization has limit capacity to inspect all government-based projects due to the enormous projects under quality con-trol, leading to the occurrence of accidents.

(4) Safe management responsibility

The safe management responsibility is basically not clear, reasonable, and mature, thus the safety management cannot cover the whole process of construction. According to the related regu-lations and rules in China, the construction safety responsibility is almost entirely implemented by the construction enterprises, and the safety management is only focused on construction stage. The owner is the major role that affects safety control, which is responsible for the schedule and construction cost.

(5) Inadequate employment system

The employment system is also not reasonable, in which the untrained workers are vastly employed. Concerning the current labor system in China, a majority of on-site construction employers are migrant workers in order to reduce the construction cost in terms of the migrant employment. This is a common issue. Currently, the population movement of rural migrant in China is significantly large; however, they are basically of poor education. Due to the lack of professional training, the safety consciousness of migrant employment is generally very low. In this regard, they will do at will during construction, not following the construction scheme. Thus brutal construction and operation are a serious problem in field construction.

(6) Unchecked engineering geological data

Engineering geological survey data are basically not examined or checked by the third independent institution. In addition, the construction plan review is missing or inadequate. Based on acci-dent case analyses, the incidents induced by inadequate geological survey data, design and construction plan rank top 2, which ac-count for 83%.

(7) Ambiguous experts' system

The experts' system in engineering evaluation, design and construction plan has the potential defects. Basically, the defects are shown in a form of short-time evaluation, and no detailed and complete review is used for design and construction plan. Experts are usually invited by the owner; therefore the opinions from ex-perts that are not in favor of the owner's intentions would not be adopted. The true suggestions from the experts will not appear in any risk or safety evaluation. Moreover, there is no recourse re-sponsibility mechanism for the experts who propose a false evaluation.

In view of the current major problems, the laws and regulations are not perfect (or mature) for the construction safety manage-ment. Although the government has set up the legal framework for the normal development of construction market (Fig. 2), and the associated regulations and standards for the safety management, the operability in underground construction safety management is weak and, the mandatory regulations considering China's facts do not adopt foreign advanced experiences. The main problems of safety management include the government intervention, lack of responsibility of the owner, artificial construction duration compression, low cost, low investment on safety management, and low number of qualified staff in project site.

4 New challenges in safety risk management

From Sections 2 and 3, although great progress has been made in safety risk control for underground engineering in China, there are still many challenges, particularly safety be-haviors of personnel, technological innovation of safety risk management, and design of regulations and rules for under-ground construction.

4.1 Challenge 1: safety behaviors of personnel

In 2014, the scale of migrant workers reached 274 million in China, among which the migrant workers in construction industry were 61.09 million, accounting for 22.3%(National Bureau of Statistics of China, 2014). The migrant workers in the construc- tion industry generally have low education level, poor safety awareness and knowledge. Frequent accidents have called for higher requirements for safety management. Heinrich et al. (1950) proposed the pyramid theory, which stated that, in a workplace, for every accident that causes a major injury, there are 29 accidents that cause minor injuries and 300 accidents that cause no injuries. Therefore, accidents in underground construction can be effectively prevented by controlling the unsafe behaviors of workers and management staff.

The research group led by Professor Ding carried out studies on recognition, laws and rectification methods for unsafe behaviors during subway construction, based on the visual language (Fig. 11) (Ding and Zhou, 2013; Ding, 2015; Ding and Guo, 2015; Guo et al., 2015). Since 2009, more than 80,000 photos have been collected on the unsafe behaviors of construction workers. First, the vector space model (VSM) (Ding and Guo, 2015; Guo et al., 2015) and the sentence similarity algorithm based on multi-level information fusion are adopted to analyze the unsafe behaviors in the photos and the semantic information is extracted.

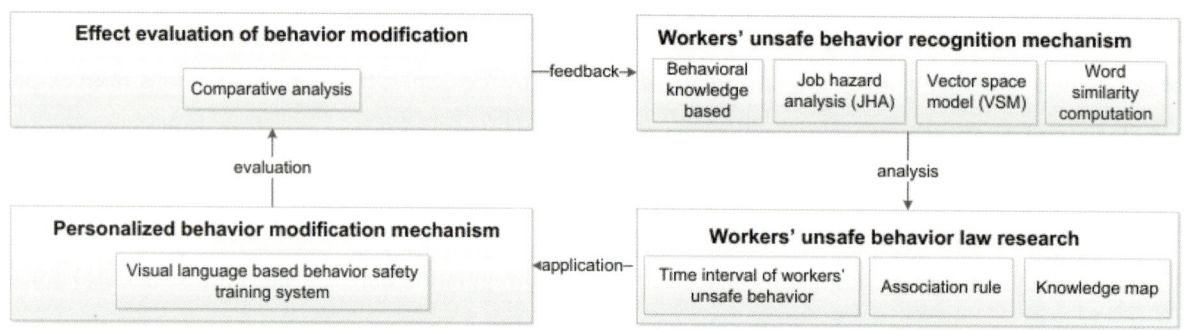

Fig. 11 Flowchart for analysis and extraction of unsafe behaviors of the workers (Ding and Guo, 2015; Guo et al., 2015).

By using the behavior dynamics method, the distribution of time interval between unsafe behaviors is investigated. The results (Ding and Zhou, 2013; Ding, 2015) show that:

（1）The time interval of unsafe behaviors during underground construction is heavy-tailed, i.e. the unsafe behaviors of workers are paroxysmal. One type of unsafe behaviors may not take place in a long period, but may occur frequently in other duration.

（2）The unsafe behaviors of machine operators fluctuate frequently. During excavation, the frequency of such unsafe behavior that "unauthorized workers move around the excavator under operation" reaches 9.6%.

（3）After being rectified by the system, the rate of unsafe be-haviors is reduced by nearly 60%.

With the rapid development of metro lines in China, the daily passenger flow in some cities has reached more than 10 million per day. Large passenger flow in the metro station means a high risk. In view of these problems, Zhong et al. (2003, 2006) proposed a theoretical calculation model with assessment indexes for the maximum passenger capacity of different subway platforms. The individual-based simulation technique for large subway passenger flow is established. The formation and diversion mechanism of large passenger flow is revealed. A complete set of safety analysis models for subway passenger transportation (Fig. 12) is proposed. The monitoring and early-warning system for subway passenger flow is developed to realize real-time warning of unexpected large passenger flow (Zhong et al., 2003, 2006). The calculation model for subway evacuation is put forward, which has been adopted in the design of evacuation passage at subway stations in more than 10 cities, including Beijing, Guangzhou, Shenzhen and so on. With the above theories and techniques, the safety issues in case of large passenger flow are solved successfully in the Dongdan station and Zhichun road station of Beijing metro, and the Huangcun station of Guangzhou metro.

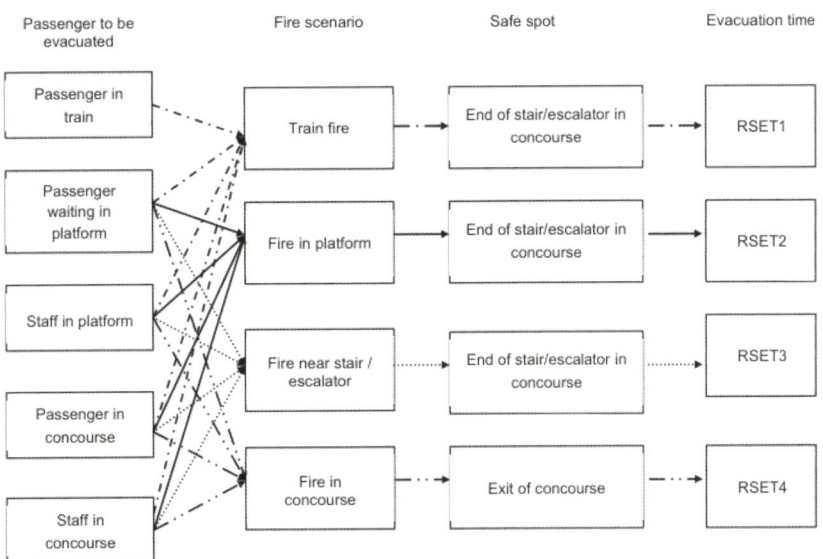

Fig. 12 Passenger evacuation strategy of metro station (Zhong et al., 2008). RSET means the required safety egress/escape time.

4.2 Challenge 2: technological innovation of safety management

Based on statistics of 550,000 accidents, Heinrich et al. (1950) proposed the pyramid theory in 1941: the ratio between death or major injury, minor injury, non-injury and hidden dangers is 1:29:300:1000 among all the accidents. For different production processes or different types of accidents, the above relationship may not be always applicable. However, the statistical law and the theoretical model indicate that, in any project, numerous accidents will inevitably lead to significant casualties. In order to prevent major accidents, attention must be paid to signs or precursors of accidents, and non-injury accidents shall be reduced or eliminated. Otherwise, major accidents will eventu-ally occur.

Due to the impact of external forces (artificial or natural) or intrinsic and extrinsic factors, the balance is broken or some abnormal phenomena or precursors occur before an accident. This imbalance generally implies a process from quantitative change to qualitative change, which usually lasts for a certain period of time. According to the statistics on relevant cases, a large number of precursors exist two or three days to one week before the accident occurrence. However, due to insufficient attention or mishandling of the precursors, the accident finally occurs (Fig. 13). The precursors of accidents or disasters must be

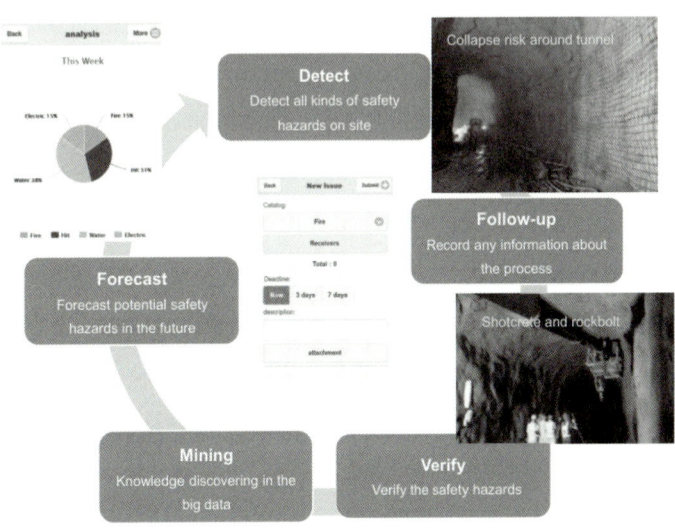

Fig. 13 Safety risk management system based on WeChat.

studied to identify some abnormal phenomena when the balance of an object is broken. Accordingly, people shall take certain precautions to avoid injury or property loss. Thus, in combination with the characteristics of underground construction, timely capture of anomalies or irregularities by instruments, experi-ences and observations, and proper countermeasures can help to eliminate the accident or disaster before it occurs, so as to minimize any losses. Studying the phenomena and applications accident precursors is the most effective way to prevent and reduce accidents or disasters.

1) Integrated safety management and control of metro con-struction based on building information modeling

The building information modeling (BIM) is able to integrate safety management, utilize safety information sources and avoid loss of safety information. Lifecycle safety risk management for metro projects can be achieved by using the BIM method. Safety management covers the entire process of metro construction:

(1) The concept of forward shifting of safety management shall be followed. At the design stage, structural safety analysis and optimization of design schemes shall be performed. Collision detection shall be simulated to identify conflicts between components or equipment in advance so as to avoid safety risks.

(2) During construction, a digitized construction site can be built based on BIM. The real-time safety status can be perceived through the internet of things. The digital site and physical construction site are interconnected so that the digital site serves and guides the physical construction site and dynamic isolation of safety risk energy can be achieved. The results have been applied successfully to construction of the river-cross tunnel (Ding et al., 2014) and station (Chen and Luo, 2014).

(3) In the operation stage, the maintenance plans for equipment and facilities are generated and the evacuation plan in case of emergency is simulated by using the BIM. The BIM technol-ogy is also applied to integrating the management tasks of the metro project, so as to achieve synergy of safety, quality, progress, cost and other operations. The responsibility can be traced back to ensure the project quality and effectively reduce the potential safety risks (Ding et al., 2014).

2) Real-time online safety risk management based on WeChat

In general, the large-scale underground projects are always located in a complex terrain environment. Due to high construction speed, long construction period, and complex interactions between dynamic and complex hidden dangers among workers, machines and environment, the safety risk management is quite difficult. In 2011, Tencent Inc. launched a new instant messaging platform based on mobile internet, WeChat, which has been commonly recognized and widely used nowadays. Lin et al. (2015b) estab-lished a real-time online reporting system for potential safety risks of workers by the following steps (Fig. 13): (i) reporting of safety risks; (ii) analysis of reported data; (iii) distribution of hazard in-formation; (iv) rectification of hidden hazard; (v) analysis of data after rectification; and (vi) closing of hidden hazard. The system includes interactive interfaces and back-end cloud server. The cloud server consists of (i) a data acquisition module, which sends the hazard report by WeChat, (ii) a data identifying module, which analyzes the reported data, identifies the data type and source, and sends the data to the processing center, (iii) a cloud messaging module, which sends information in categories and provides personalized service, and (iv) a data analysis module I, which performs correlation analysis and condition query among the data, spatial location information and attribute data based on the spatial location information. With this system, information of various safety hazards encountered at large-scale construction sites can be collected and reported to the back-end cloud system by the mobile tool WeChat. By data analysis and data mining, enhanced man-agement, personnel and property safety, effective monitoring of the project quality can be achieved eventually. Currently, the system is successfully adopted by Baihetan and Xiluodu construction sites.

4.3 Challenge 3: design of safety management regulations

Heinrich (1941) investigated 75,000 industrial injuries in USA and found that 98% of accidents were preventable, and only 2% of accidents beyond human ability were not preventable. Table 1 shows that most accidents are preventable in the underground construc-tion field. Statistical analysis also shows that precursors occurred two days to one week before the accident (Fig. 10). If attention had been paid to these precursors and proper measures had been taken in time, accidents could have been completely avoided.

The safety science and engineering theory supports the stand-point that the accidents can be prevented, which provides the following methods and principles for accident prevention:

（1）The energy release theory indicates that the accident risk sources can be divided into two categories: static and dy-namic risk sources. With protection technology and treat-ment (such as increasing the design safety factor), the triggering threshold of static risk sources can be enhanced, and the possibility of risk source outbreak and the hazard level after the accident can be reduced. By daily safety risk management, unsafe behaviors of workers and unsafe state of materials can be eliminated. Hence, the accidents trig-gered by dynamic risk sources can be controlled.

（2）The Heinrich's pyramid theory and the causal chain accident theory indicate that as long as the daily unsafe behaviors of people and unsafe state of materials are eliminated, less unsafe factors are accumulated and the direct causes of ac-cidents are cut off. As a result, accidents can be avoided.

（3）The Surry model provides a good idea for accident preven- tion. In order to prevent and control the accidents, technical means should be firstly employed to reveal the dangerous state (precursor), so that the operator can have better awareness on the presence or release of hazards. Secondly, training and education shall be conducted to improve the sensitivity of workers on danger signals, including anti-interference ability and so on. Thirdly, by means of educa- tion and training, the operator can accurately understand the meaning of the warning signal, and know what measures should be taken to avoid the accident or to control its con- sequences. Proper decision can then be made. Finally, the system and its ancillary facilities should be designed in such a way that people have sufficient time and condition for response after proper decision is made. In this manner, the accidents can be controlled to a great extent and good pre-ventive effect can be achieved.

5 New strategies of safety risk management

5.1 Objects of safety risk management for underground engineering

In the background that China is comprehensively deepening the reform and promoting the spirit of administrating the country by law, the relevant laws and regulations are being reformed and enhanced. Correspondingly, actions are taken to achieve the following goals for the safety risk management of underground engineering. The policies and systems of safety risk management for underground engineering should be established and promoted in China. The system and mechanism of safety supervision should be reformed. The manner of supervision should be improved and the supervisory force should be strengthened. The major role of the project owner in safety risk management should be implemented. The market economic means should be introduced to ensure the guarantee policy. The employment policies should be reformed to establish a qualification licensing system. The long-term and effective mechanism of safe production in the construction in-dustry shall be promoted. By 2020, the state of safe production in underground engineering in China shall be significantly enhanced, and the major accidents shall be effectively controlled. By 2030, the state of safety management in underground engineering should be substantially improved.

5.2 Strategies

Valuable lessons have been obtained from bloody accidents in the construction projects, which reveal that safety risk manage- ment must be transformed from passive remediation to proactive planning and prevention. This relies on the combination of tech- nological, managerial, cultural and legal approaches to establish a comprehensive safety risk management system. According to the discussions on progress and challenges in the earlier sections, the safety risk management mode of underground engineering in China is still at a preliminary level or a low level in some regions. The major task of policy makers and researchers is to actively guide the safety risk management mode of underground engineering towards the intermediate level of systematic and scientific safety risk management, and gradually transform toward a high level with safety culture. In this paper, it is suggested to address the problem based on strategies from six perspectives, including management system and policy, law, administration, economy, education and technology, as shown in Fig. 14.

Fig. 14 shows that the theoretical basis of the new strategies is the "3E" theory, i.e. enforcement, education and engineering. The detailed implementation methods include the legal, administrative, economical, educational and technological methods.

The strategy of management system and policy is to establish and promote the safety risk management system and policy for underground engineering which are in line with the international conventions. The safety risk management should be transformed from "rectification after accidents" to "prevention in advance". Meanwhile, the safety risk management mode should be improved from being oriented by experiences and

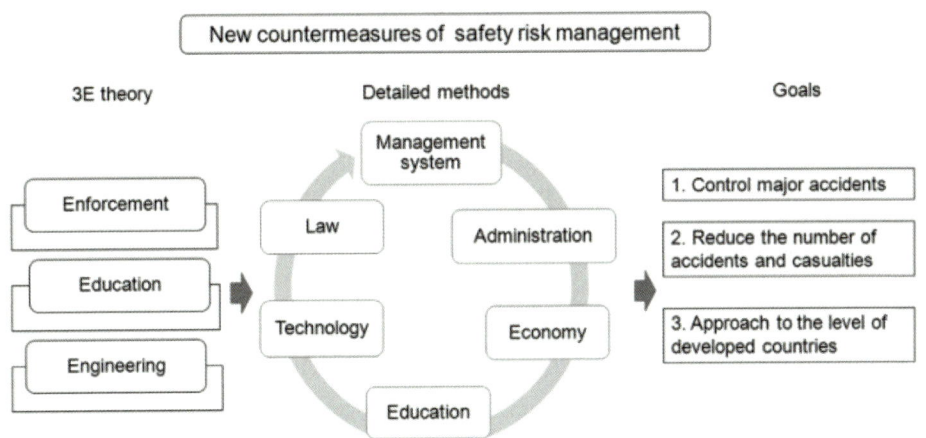

Fig. 14 The goal and new countermeasures of safety risk management of underground engineering in China.

policies to the systematic and scientific stage, and gradually transformed to the stage with safety culture. The safety risk management should be tightly combined with project management and be treated as a mandatory procedure and a core content of project management. Hence, the following four aspects have to be implemented. Firstly, the safety risk management has to be regulated by relevant laws. Particularly, a series of laws and regulations need to be legislated, issued and implemented. For example, "Regulations on safety risk manage- ment of underground construction" can be implemented to enforce the mechanism of risk management and guarantee system. Sec- ondly, an integrated modern safety risk management system, characterized by lifecycle management, various risk-related factors, complete management process and implementation of risk re- sponsibility, should be established. The objective, scientific and dynamic system can be used to identify accident precursors in or- der to control risks and reduce the uncertainties of accidents. Thirdly, the organization management system, technology man- agement system, risk monitoring, emergency management system and safety culture system should be established comprehensively. The safety risk management system should be all-around during the entire construction process. All-around management refers to the involvement of government and market, including the gov-ernment department, the owner, the contractor, the engineering guarantee and insurance company and the consultant. The whole-process management refers to management in various construction stages, including planning, design and construction. Risk identifi-cation, analysis, assessment, monitoring and early-warning should be performed throughout the entire process. Finally, modern technologies should be introduced into the safety risk management system to strengthen the real-time monitoring of safety risks and project quality, and to enhance the capability in identification and treatment of accident precursors.

The legal strategy aims to provide a strong basis for regulating

（1）the safety risk management. Through legislative and contractual methods, the construction industry should promote establishing an engineering guarantee system. On the basis of "Guarantee law", "Contract law", "Construction law", "Bidding law" and "Regulations on quality management of construction engineering", national and regional legislation should be specified in engineering guarantee and insurance. Related laws and regulations should prohibit any bidding with unreasonable low price. The most urgent task is to establish and promote "Regulations on safety risk management of underground construction" in all engineering companies and pro-jects. "Regulations on safety risk pre-evaluation management of underground construction" has defined the supervisory responsibilities of the government and the respective re-sponsibilities of the project owner, design institute, contractor, and companies involved in survey, monitoring and superintendent in the construction process. The so-called "tribute project", "image project" and "special case" should be strictly prohibited. "Regula-tions on comprehensive supervision of safe production of under-ground projects" should regulate the approval process of safety risk management for major engineering projects. The project with extremely high risks should be vetoed by experts. "Regulations on review of investigation, design, and construction scheme of un-derground projects" specifies that geological investigation results, design schemes, construction schemes and the corresponding drawings should be reviewed by the qualified third party. "Regu-lations on guarantee and insurance of underground projects" and "Regulations on qualification licensing of superintendent, design engineer and construction engineer for underground projects" are mandatory to ensure that the safety risk management system and engineering guarantee and insurance system can be implemented in underground construction.

For large-scale tunnel or underground projects, the chief su- pervisor should be a registered geotechnical engineer. In addition, training of practitioners should be strengthened. Particularly, the responsibility of government departments and enterprises in labor training, especially for the migrant workers, should be specified. In order to ensure the safety of products and devices, the third-party authentication policy should be established. When establishing the relevant laws and regulations, punitive measures should be adjusted according to regional economic development levels in China. Stiff punishment should be imposed on illegal activities that may induce serious damage to public, environmental, personal and property safety. Higher cost of illegal activities can help to regulate the behaviors of all parties involved and subsequently reduce safety accidents. Meanwhile, guiding books and manuals shall be pub-lished for better understanding of the laws, regulations, guidelines and technical standards.

The administrative strategy aims to build a harmonious envi-ronment for the construction industry by mobilizing the system and mechanism of market economy. The roles of government, en-terprises, supervisor, guarantee and insurance companies and consultants should be explicated. Detailed suggestions include the following seven aspects. First, the industrial structure of con-struction should be reformed. Leading enterprises are at the up-stream of the industry, in charge of contracts of national major projects. The mainstay of the industry is supported by subsidiary companies of the leading enterprises or medium-scale contractors, in charge of general civil construction projects. The downstream of the industry is composed of small and flexible labor service com-panies, undertaking labor subcontracting. With this kind of in-dustrial structure, companies in different levels have a stable number of employees for the purpose of safety education and training, which can improve the efficiency of safety risk manage-ment and cultivate a safety culture. Second, the construction in-dustry should promote reformation of the migrant labor system and gradually transform the migrant workers into professional workers of labor service companies. Third, reasoning and recording system shall be established for proper project schedule and cost. The associated documents should consider the liability recourse. Fourth, the owner should play a predominant role in safety risk management. The owner should regulate the tendering and bid-ding process and promote the transparency of the tendering and bidding process. Fifth, in addition to strict control of qualification licensing for companies, the access permission policy for technical personnel should be strengthened, particularly the registered en-gineer system. The rights and responsibilities of registered engi-neers in safety risk management should be specified. "One certificate is affiliated with several companies" or "the registered engineer lends his or her certificate to non-qualified companies for bidding" should be prohibited. Sixth, the supervisory management model for safe production should be reformed. The focus should be placed on supervision of establishment and implementation of the engineering guarantee system, the safety risk management system and the safety regulations and standards, rather than on the detailed safety risk management processes of companies in a manner of single-targeted and occasional massive campaigns. Random inspections and surveillances should be carried out, rather than just notifications. The safe production credit system, the reporting and punishment system of major accidents should be established and enhanced. The illegal or improper behaviors of companies should be released to the public on a regular basis, and the public supervisory power should be strengthened. Seventh, the supervisory management system, mechanism and team for safe production in the construction industry should be reformed.

The economic strategy focuses on the following four aspects. First, the engineering guarantee and insurance system should be established and promoted, through legislative and contractual means. Based on "Guarantee law", "Contract law", "Construction law", "Tendering and bidding law" and "Regulations on quality management of construction projects", the content of engineering guarantee and insurance shall be extended to form national laws and local regulations. Second, the guarantor market with proper competition shall be cultivated, and the risk management agency shall be developed as the guarantor. In view of China's facts, inter-company guarantee of the major contractors should be developed with first priority. The third-party guarantee can be provided by contractors with strong ability and good reputation. Third, the risk management consultants can be employed as the brokers and agents for engineering guarantee. The consultants can be commissioned to negotiate with the guarantee company, claim for compensation, develop the risk management techniques, and conduct training and risk consultation. Fourth, the construction industry should rely more on third-party credit investigation companies, establish the databases for recording safety and finan-cial credit of engineering companies, and ensure that the guaran-tors can obtain the credit information of the guarantee companies. Therefore, different players in the construction market will grad-ually become more credible.

The educational strategy focuses on the following three aspects. First, the government and the contractors should put more efforts to the training of rural migrant workers. The development and promotion channels for rural migrant workers shall be solved, so as to realize transformation from migrant workers to professional workers. Second, the quality and professionalization of practi-tioners shall be promoted in the construction industry. Third, more attention shall be paid to safety education of technical professionals in underground engineering, such as continuing education on safety.

The technological strategy includes the following aspects. First, technical standards of investigation, design, construction and acceptance should be enhanced. Higher standards related to safety risk management, including project durability, should be enforced. More resources shall be used in researches of safety risk manage-ment and technology. Second, in order to promote the manage-ment of technical standards, compilation of technical standards shall be implemented by relevant societies and associations. Meanwhile, the technical standards shall serve as technical guides and more emphases shall be put on recommendation of technical methods, in addition to the key technical parameters involved in the safety and quality bottom lines. Third, the leading companies and local governments shall be encouraged to establish company standards and local standards. The national standards are the basic standards with the lowest requirements. The local standards should conform to the economy and society development levels in the region. The company standards should be the standards with the highest requirements. Fourth, during the planning stage, suf-ficient time shall be ensured for investigation, reasoning and design. The time for design cannot be artificially cut down and "design while construction" should be prohibited. Fifth, the expert review mechanism for the key technical works should be reformed. The expert review mechanism is suggested to be replaced by the third-party independent review mechanism. Professional and detailed examination of investigation, design and construction schemes should be conducted by the third-party consultants. The consultants should comment on the examination results and are responsible for the comments proposed. Sixth, a safety risk man-agement intelligent and information system should be established.

5.3 Intelligent and information technology

For improving safety risk management of underground engi-neering, an intelligent system should be developed, which can provide a communication platform for different parties involved in the construction project. Based on the modern intelligent and in-formation technology, monitoring data can be delivered to relevant parties in real time. Activities and behaviors of different parties are transparent in the whole system, which is beneficial to information sharing, mutual supervision and responsibility implementation.

The intelligent underground engineering (IUE) incorporates digital and real underground engineering, which depends on the integration of networking and digital technology. The IUE achieves a unified space-time reference for four-dimensional information show. By employing cloud computing and big data mining method, the IUE can rapidly process and dynamically update the sensor data in real-time, and provide intelligent services based on the percep-tion, logical thinking, self-adaptive and decision-making abilities. The intelligent services include real-time intelligent analysis, data mining, knowledge discovery and real-time decision.

The IUE performance characteristics are described as followings:

（1）Real-time perception: Smart sensors are installed for comprehensive perception and real-time monitoring running state of tunnel, environment, equipment and people.

（2）Comprehensive internet: Using the internet of things, all comprehensive sensors are connected with each other, and intelligent storage and transmission on sensing data are realized.

(3) Deep integration: Combining with the networking and internet of things, integrating multi-source data, which will provide a safety risk management mapping of underground engineering construction and operation.

(4) Intelligent service: Based on the intelligent information infrastructure (network, data), cloud computing, data mining and knowledge discovery, a new system structure can pro-vide intelligent service for the construction and operation periods of the underground engineering.

6 Conclusions

This paper discusses the main progress of the safety risk man-agement of underground engineering in China in the past decade, i.e. (1) establishment of laws and regulations on safety risk man-agement, (2) implementation of the safety risk management plan, (3) risk management and early-warning decision support system for underground engineering based on information technology, and (4) strengthening the studies on safety risk management, prediction and prevention for underground engineering.

In China, the combined cause of safety accidents except for direct causes are summarized as follows: (1) individual and departmental interests, (2) safety inspection merely becoming a mere formality, (3) malfunction of governmental supervision, (4) safe management responsibility, (5) inadequate employment sys-tem, (6) unchecked engineering geological data, and (7) ambiguous experts' system.

The safety risk management of underground construction in China involves hundreds of millions of people, large construction sites, and complex construction processes. The new challenges of the safety risk management of underground engineering include (1) control of unsafe behaviors of workers, (2) technological inno-vation related to safety risk management, and (3) design of safety management regulations. The history of safety risk management in the developed countries shows that with long-term persistence of legal and scientific management, a high level of systematic and scientific safety risk management with safety culture can be achieved.

New strategies of safety risk management for future under-ground construction in China are proposed in six aspects, including the safety risk management system and policy, legal, administra-tive, economic, educational and technical countermeasures. Only in this way, a fundamental improvement in the safety risk manage-ment in China can be realized.

Conflict of interest

We wish to confirm that there are no known conflicts of interest associated with this publication and there has been no significant financial support for this work that could have influenced its outcome.

Acknowledgments

This study is supported by Chinese Academy of Engineering (grant No. 2011-ZD-12), National Natural Science Foundation of China (grant No. 11272178) and National Basic Research Program of China (973 Program) (grant No. 2011CB013502/3). The authors are very grateful to Profs. Shigang She, Songqing Lin, Xiaoli Rong for their critical recommendations, which have greatly helped the authors to improve the paper.

References

[1] Alaeddini A, Dogan I. Using Bayesian networks for root cause analysis in statistical process control. Expert Systems with Applications 2011;38(9):11230-43.

[2] Alfredo DC. Integrated methodology for project risk management. Journal of Con-struction Engineering and Management 2002;128(6):473-85.

[3] Behzadan AH, Aziz Z, Anumba CJ, Kamat VR. Ubiquitous location tracking for context-specific information delivery on construction sites. Automation in Construction 2008;17(6):737-48.

[4] Bhalla S, Yang YW, Zhao J, Soh CK. Structural health monitoring of underground facilities: technological issues and challenges. Tunnelling and Underground Space Technology 2005;20(5):487-500.

[5] Carbonari A, Giretti A, Naticchia B. A proactive system for real-time safety man-agement in construction sites. Automation in Construction 2011;20(6):686-98.

[6] Chai J, Liu JX, Qiu B, Li Y, Zhu L, Wei SM, Wang ZP, Zhang GW, Yang JH. Detecting deformations in uncompacted strata by fiber Bragg grating sensors incorpo- rated into GFRP. Tunnelling and Underground Space Technology 2011;26(1): 92-9.

[7] Chen LJ, Luo H. A BIM-based construction quality management model and its ap-plications. Automation in Construction 2014;46:64-73.

[8] Ding L, Zhou Y, Akinci B. Building information modeling (BIM) application frame-work: the process of expanding from 3D to computable. Automation in Con-struction 2014;46:82-93.

[9] Ding LY, Guo SY. A big data-based workers behavior observation in China's metro construction. In: Proceedings of the creative construction conference, Krakow, Poland; 2015.

[10] Ding LY, Zhou C. Development of web-based system for safety risk early warning in urban metro construction. Automation in Construction 2013;34:45-55.

[11] Ding LY. BIM application and construction. Shanghai, China: Tongji University Press; 2015.

[12] Duddeck H. Challenges to tunneling engineers. Tunnelling and Underground Space Technology 1996;11(1):5-10.

[13] Ergen E, Akinci B, Sacks R. Tracking and locating components in a precast storage yard utilizing radio frequency identification technology and GPS. Automation in Construction 2007;16(3):354-67.

[14] Eskesen SD, Tengborg P, Kampmann J, Veicherts TH. Guidelines for tunneling risk management: international tunneling association working group No. 2. Tunneling and Underground Space Technology 2004;19(3):217-37.

[15] Guo SY, Luo HB, Jiang XY, Xiong CH. Research on metro worker's unsafe behavior knowledge visualization. China Safety Science Journal 2015;25(3):140-5.

[16] Heinrich HW, Petersen D, Roos N. Industrial accident prevention. New York, USA: McGraw-Hill; 1950.

[17] Heinrich HW. Industrial accident prevention: a scientific approach. McGraw-Hill Book Company; 1941.

[18] International Tunnelling Insurance Group (ITIG). A code of practice for risk man-agement of tunnel works. ITIG 2006.

[19] Jiang H, Lin P, Qiang M, Fan QX. A labor consumption measurement system based on real-time tracking technology for

dam construction site. Automation in Con-struction 2015;52:1-15.

[20] Khakzad N, Khan F, Amyotte P. Safety analysis in process facilities: comparison of fault tree and Bayesian network approaches. Reliability Engineering and System Safety 2011;96(8):925-32.

[21] Khoury HM, Kamat VR. Evaluation of position tracking technologies for user localization in indoor construction environments. Automation in Construction 2009;18(4):444-57.

[22] Li SC, Hu C, Li LP, Song SS, Zhou Y, Shi SS. Bidirectional construction process me-chanics for tunnels in dipping layered formation. Tunnelling and Underground Space Technology 2013b;36:57-65.

[23] Li SC, Zhou ZQ, Li LP, Xu ZH, Zhang QQ, Shi SS. Risk assessment of water inrush in karst tunnels based on attribute synthetic evaluation system. Tunnelling and Underground Space Technology 2013a;38:50-8.

[24] Lin P, Li QB, Fan QX, Gao XY, Hu SY. A real time location-based services system using Wi-Fi fingerprinting algorithm for safety risk assessment of workers in tunnels. Mathematical Problems in Engineering 2014. http://dx.doi.org/10.1155/2014/371456.

[25] Lin P, Li QB, Fan QX, Gao XY. Real-time monitoring system for workers' behaviour analysis on a large-dam construction site. International Journal of Distributed Sensor Networks 2013. http://dx.doi.org/10.1155/2013/509423.

[26] Lin P, Liu HY, Zhou WY. Experimental study of failure behavior of deep tunnels under high in-situ stresses. Tunnelling and Underground Space Technology 2015a;46:28-45.

[27] Lin P., Gao X.Y., Wang Z.L. A real-time safety risk management method and system of large civil engineering construction site based on WeChat. Utility model patent, No. 201510845843.6, 2015b.

[28] Liu TM, Zhong MH, Xing JJ. Industrial accidents: challenges for China's economic and social development. Safety Science 2005;43(8):503-22.

[29] Ma TH, Tang CA, Tang LX, Zhang WD, Wang L. Rockburst characteristics and microseismic monitoring of deep-buried tunnels for Jinping II hydro-power station. Tunnelling and Underground Space Technology 2015;49: 345-68.

[30] National Bureau of Statistics of China. Monitoring survey report on China migrant workers. 2014. http://www.stats.gov.cn/tjsj/zxfb/201504/t20150429_797821.html.

[31] Ou C, Teng F, Wang W. Analysis and design of partial ground improvement in deep excavations. Computers and Geotechnics 2008;35(4):576-84.

[32] Piniella F, Fern A, Ndez-Engo MA. Towards system for the management of safety on board artisanal fishing vessels: proposal for check-lists and their application. Safety Science 2009;47(2):265-76.

[33] Qian QH, Rong XL. State, issues and relevant recommendations for security risk management of China's underground engineering. Chinese Journal of Rock Mechanics and Engineering 2008;27(4):649-55 (in Chinese).

[34] Qian QH. Report on the strategy and countermeasure of safety risk management system for civil engineering in China. Consulting Research Project of Chinese Academy of Engineering 2014.

[35] Rao KS, Chandran KR. Mining of customer walking path sequence from RFID su-permarket data. Electronic Government 2013;10(1):34-55.

[36] Seco F, Plagemann C, Jimenez AR, Burgard W. Improving RFID-based indoor posi-tioning accuracy using Gaussian processes. In: Proceedings of the international conference on indoor positioning and indoor navigation (IPIN). IEEE; 2010. p. 1-8.

[37] Shi CL, Zhong MH, Nong XZ, He L, Shi JH, Feng GG. Modeling and safety strategy of passenger evacuation in a metro station in China. Safety Science 2012;50(5): 1319-32.

[38] Smith NJ, Merna T, Jobling P. Managing risk in construction projects. Hoboken, New Jersey, USA: Wiley-Blackwell; 2009.

[39] Sun J. Intelligent prediction, control of construction safety and 3D simulation study on urban underground engineering. Chinese Journal of Rock Mechanics and Engineering 1999;18(Suppl. 1):753-62 (in Chinese).

[40] Tang CA, Wang JM, Zhang JJ. Preliminary engineering application of microseismic monitoring technique to rockburst prediction in tunneling of Jinping II project. Journal of Rock Mechanics and Geotechnical Engineering 2011;2(3):193-208.

[41] The Council of the European Communities. Council directive 92/57/EEC of 24 June 1992 on the implementation of minimum safety and health requirements at temporary or mobile construction sites (eighth individual directive within the meaning of article 16(1) of directive 89/391/EEC). Brussels: The Council of the European Communities; 1992.

[42] Thomas T. Breaking news: Prague ring road tunnel collapse. 2010. http://tunnellingjournal.com/news/breaking-news-prague-ring-road-tunnel-collapse.

[43] Tu YJ, Zhou W, Piramuthu S. Identifying RFID-embedded objects in pervasive healthcare applications. Decision Support Systems 2009;46(2):586-93.

[44] Waltz G. Metro line leak causes chaos in Warsaw. 2012. http://redaktorext.polskieradio.pl/1/9/Artykul/110064,Metro-line-leak-causes-chaos-in-Warsaw.

[45] Wu XG, Liu HT, Zhang LM, Skibniewski MJ, Deng QL, Teng JY. A dynamic Bayesian network based approach to safety decision support in tunnel construction. Reliability Engineering and System Safety 2015;134:157-68.

[46] Xu NW, Li TB, Dai F, Zhang R, Tang CA, Tang LX. Microseismic monitoring of strainburst activities in deep tunnels at the Jinping II hydropower station, China. Rock Mechanics and Rock Engineering 2016;49(3):981-1000.

[47] Yoo C, Lee D. Deep excavation-induced ground surface movement characteristics: a numerical investigation. Computers and Geotechnics 2008;35(2):231-52.

[48] Zhong MH, Zhang XK, Liu TM, Wei X, Fan WC. Safety evaluation of engineering and construction projects in China. Journal of Loss Prevention in the Process In-dustries 2003;16(3):201-7.

[49] Zhong MH, Liu TM, Deng YF, Shi CL, Fu TR. Safety evaluation: important safeguard of work safety for enterprises in China. Journal of Loss Prevention in the Process Industries 2006;19(6):762-8.

[50] Zhong MH, Shi CL, Tu XW, Fu TR, He L. Study of the human evacuation simulation of metro fire safety analysis in China. Journal of Loss Prevention in the Process Industries 2008;21(3):287-98.

水下隧道工程实践面临的挑战、对策及思考

钱七虎[1,2]

(1. 解放军理工大学国防工程学院,江苏 南京 210007;
2. 解放军理工大学爆炸冲击防灾减灾国家重点实验室,江苏 南京 210007)

摘要:首先介绍钻爆法、TBM/盾构法以及沉管法 3 种主要水下隧道建设方法的优点及不足,进而从地质条件、水流速度、水面交通、隧道埋深、通行车道数量、渗漏情况、工程总量、施工工期、施工安全、工程造价等方面对 3 种修建方法进行对比分析。最后重点分析当前水下隧道工程建设过程中存在的主要挑战与相应对策,主要包括:沉管法施工中河床冲刷、管段局部高出河床、埋深增大的挑战及对策,盾构穿越浅覆土层、刀盘刀具磨损、埋深提高、轴承损坏及盾构通过断层和破碎地层施工中的挑战与对策。

关键词:水下隧道;钻爆法;TBM/盾构法;沉管法;盾构施工;工程对策

DOI: 10.3973/j.issn.1672-741X.2014.06.001

中图分类号: U 45　　　**文献标志码**: A　　　**文章编号**: 1672-741X(2014)06-0503-05

Challenges in Construction of Underwater Tunnels and Countermeasures

QIAN Qihu[1,2]

(1. College of National Defense Engineering, PLA University of Science and Technology, Nanjing 210007, Jiangsu, China; 2. State Key Laboratory of Disaster Prevention & Mitigation of Explosion & Impact, PLA University of Science and Technology, Nanjing 210007, Jiangsu, China)

Abstract: The advantages and disadvantages of three major underwater tunnel construction methods, including drill and blast method, TBM/shield method and immersed tube method, are presented. Comparison and contrast is made among the three mentioned construction methods in terms of geological conditions, water flow velocity, water surface transportation, cover depth, number of traffic lanes, water leakage, work quantity, construction period, construction safety and construction cost. Finally, major challenges in the construction of underwater tunnels and their countermeasures are analyzed. In the construction of immersed tunnels, the erosion of the riverbed, the phenomenon of the local parts of tube elements being higher than the riverbed and the increasing cover depth are big challenges; in the TBM/shield method, boring under shallow cover, wearing of cutter heads and cutting tools, increasing cover depth, damage of main bearing and boring in fault and fracture zones are big challenges.

Key words: underwater tunnel; drill and blast method; TBM/shield method; immersed tube method; shield tunneling; countermeasure

0 引言

中国快速的经济发展和城市的不断拓展,需要越江越海交通的配套发展,而作为越江越海的主要手段——桥梁在其数量不断攀升的同时,越来越显现其局限性:在大雪、大风及大雾时节不能保证全天候通行;对沿江、海湾内的航运产生不利影响,包括对港口、码头等航运设施的影响,桥梁净空高度对船舶吨位的影响以及撞桥事故的不断发生,从而对航行船速的限制以及要求船队的解驳等。因而,随着水下隧道修建技术的迅速提高,水下隧道建设步伐就迎势而上并不断推进,工程数量日益增加。黄浦江、甬江、珠江、黄河以及长江等我国大江大河下相继建成并正继续修建不少隧道。继厦门翔安海底隧道、青黄岛胶州湾海底隧道建成后,港珠澳大桥海底隧道也正在修建,而跨越琼州海峡和渤海的海底隧道也正在积极论证中。我国水下隧道建设在不断展现其丰硕成果的同时,面临着不

收稿日期:2014-03-13;修回日期:2014-04-10。
本文原载于《隧道建设》(2014 年第 34 卷第 6 期)。

少的严峻挑战,诸如穿越水下浅覆盖层的挑战、盾构刀具严重磨损的挑战以及 TBM 卡机的挑战等。本文在归纳中国水下隧道工程实践经验教训的基础上拟对面临的挑战与对策做出初步的思考和分析。

1 3 种主要水下隧道修建方法

除在城市湖泊下修建水下隧道可采用筑围堰、排水、明挖方法外,在江、河、海下一般采用暗挖法(钻爆法、TBM 和盾构法)以及沉管法。

1.1 钻爆法

水下隧道钻爆法与陆地隧道的无根本区别,主要关键技术是设计埋深须满足最小覆盖层基岩的要求,以预防坍塌并遏止高压涌水[1-3]。为此,要做好工程地质和水文地质的勘察调查及突水涌水的超前地质预报,并在完全可靠的注浆固结封闭的条件下钻爆开挖。

1.1.1 优点

1) 所需设备投资较小。
2) 对不同地质条件、开挖工艺、方法和断面形状的适应性好,转换场地、移动位置机动、灵活。
3) 动力能源消耗少、要求低,投资小。
4) 设备维修少,利用率高,在不发生事故的前提下施工进度稳定。
5) 比较适合我国当前的国情。

1.1.2 不足

1) 较难提高开挖速度,工期长,特别在发生围岩坍塌及突水涌泥事故情况下,工期难以估计,如:日本青函隧道,因津轻海峡有 9 个断层破碎带,发生 3 次涌水停工,主隧道历时 14 年才贯通。
2) 与 TBM 法相比,隧道断面易发生超挖,围岩破坏大,施工安全差。

1.2 TBM 法和盾构法

TBM 法是隧道全断面掘进机(TBM),利用刀盘刀具破碎岩石和在软土中开挖,边开挖边出渣并进行锚喷支护衬砌或管片支护等连续作业[4-6]。在中国,软土中的 TBM 法称为盾构法。

1.2.1 优点

1) TBM 和盾构法掘进速度快,在均匀的中、硬岩层中的月掘进速度在 600 m 左右。如在英法海峡隧道的英国端,月掘进 764 m,法国端月掘进 685 m(地质较英国端复杂),最高月掘进速度为英国端 1 500 m,法国端也达 1 170 m。最适合长隧道施工。
2) TBM 掘进的隧道具有较平整的临空面,超欠挖极少,围岩松动范围一般都在 200 ~ 500 mm,因受到的破坏很小并能得到及时支护,故围岩稳定,施工安全。
3) 施工安全性大大提高,可最大限度减少施工人员的伤亡。
4) 施工机械化程度达到 95% 以上,降低了劳动强度,改善了劳动环境和条件,实现了隧道施工的快速、高质、高效、安全的目的。
5) 施工中对海底地质不产生任何不利影响,不影响和干扰水面航行;不干扰海洋中保护动物,如白鳍豚等。
6) 具有极其显著的社会效益。

1.2.2 不足

1) 造价昂贵。
2) 遇到复杂多变的地层,如断裂构造、软硬交替或上软下硬的地层时,TBM 和盾构的推进比较困难。
3) 超大直径多车道的 TBM 和盾构还有待研究和开发。

1.3 沉管法

在预制场(船坞或造船厂车间)将隧道整体分段制作好,两端用临时钢封门密封,船坞放水使隧道管段上浮,采用拖运和定位等设备,使管段准确浮运至设计位置;经管段灌水压载,下沉至已开挖好的水底基槽,再依次在水下利用水力压接将管段连接起来,接口使用可靠的密封止水,最后是抛石回填,抽出管段内压载水,开启各段间钢封门,在管段内进行设备安装和装修后,即成水底沉管隧道[7-9]。

1.3.1 优点

1) 建设要求顶面须埋在河流、海流的冲刷线以下,因此能做到沉管断面顶面不侵入航道。
2) 要求地层承载力不大。
3) 隧道各管段可在干邬和工厂车间内预先制备、质量有保证。
4) 管段预制和水底开槽可同时进行,效率高,工期短。
5) 工程造价和投资具有竞争力。

1.3.2 不足

在管节预制、防水、水下开槽、管节浮运、沉放、各管节水下对接以及沉管基础处理的各个工艺流程中出现失误的概率,相对地比其他隧道技术要多。这是因为沉管隧道是埋置或搁置于水下的大型人工构筑物,受河、海地貌、地质、水文、航道,以及设计、不可预见的繁复性施工工艺与方法的直接影响。

在世界上至今已建成 100 多座沉管隧道,其最大沉埋深度(水深加覆盖层厚)已达 50 m,沉管隧道最长的已达 5.825 km。

2 水下隧道的主要修建方法的比较

见表 1。

3 水下隧道建设的挑战与对策思考

3.1 沉管法在河道水流含砂大幅减少的情况下,面临河床冲刷的挑战

如长江中下游建设的水下隧道:武汉、南京及上海长江隧道,由于面临长江三峡水电站、向家坝、溪洛

渡、乌东德及白鹤滩5个水电站(后4个发电功率为三峡电站2.5倍),建成后水流含砂量的大幅减少,从而由中下游河床冲淤不平衡引起的冲刷,难以确保水下隧道在其生命周期内的安全运营。

对策:南京、武汉长江隧道原方案为沉管法,经建议及论证确定后修改为盾构法。

表1 水下隧道主要修建方法比较[10]

Table 1 Comparison and contrast among different construction methods of underwater tunnels

修建方法	地质条件	水流速度	水面交通	隧道埋深	通行车道数量	渗漏情况	工程总量	施工工期	施工安全	工程造价
钻爆法	对地质条件的适应面较宽,灵活性大,进度稳定,水底地质条件差时,需采用掘进辅助工法	无关	无关	一般设置在水下基岩>2.0倍开挖跨度的深度	根据水下基岩情况,可做成多车道隧道	采取有效措施能做到不渗不漏	埋深最大,引道段最长,工程量大	机械化程度不高,可组织快速施工,工期仍较长	水下钻爆施工要严格施工措施,安全施工是个重要问题	单位面积造价较低
掘进机法	对地质条件的适应面较窄,灵活性小,水底地质为多变化的岩体时,不易发挥掘进机优势	无关	无关	一般要设置在水底基岩下>掘进机外径的基岩覆盖深度	TBM当前主要使用两车道,多车道时要增建隧道(直径<12 m)	采取措施能做到不渗不漏	从水下基岩中穿过,两岸引道长,工程量大	机械化程度较高,开挖速度快,工期较有保证	施工作业全部在掘进机中进行,安全性好	单位面积造价高
盾构法	遇到不稳定土层时,需采用土压平衡或泥水加压平衡(气压)盾构施工	无关	无关	一般要设置在水底覆盖层下≥盾构外径的深度	目前已有三车道(15.4 m)的软土型盾构,超大型盾构正在建造或规划中	接缝较多,做到不渗不漏需精心施工	有一定覆盖层厚度,两岸引道较长,工程量相应较大	机械化程度高但推进循环工序较多,工期相对较长	施工作业全部在盾壳掩护下进行,安全性好	单位面积造价较高
沉管法	不怕流砂,基本上不受地质条件限制	水流很急时需用水中作业平台辅助施工	对于繁忙航道,在浮运就位时需采取短期局部航道管理措施	沉管顶面可放置在水底标高处。即与河、海底面齐平,最深记录已达水下50 m(管槽深<65 m,流速≤3 m/s)	一个断面内可设置多条车道,目前已有双向八车道的记录	能实现滴水不漏	可做到隧道长度最短,工程总量最少	基地开槽和管段预制分开同时进行,施工顺利进行条件下管段沉放时间较短,总的工期较短	大量作业在水上,水下作业主要是由潜水员完成,安全性较好	施工顺利的情况下,单位面积造价可以具有竞争力

3.2 沉管顶部深潭部位管段局部顶面高出河床的挑战

河床因冲刷一般呈不对称"V"形,沉管法隧道若在深潭部位按常规设计势必埋深大、工程路线长、造价高。因此,深潭部位管段顶面局部高出河床为具有重大工程实践意义的课题。

国内外已有此类设计成功的实例,如:希腊Preveza Aktio沉管隧道、悉尼港沉管隧道、香港跨港公路沉管隧道西九龙沉管段及上海外滩泰和路隧道。但由于工程条件不同(工程地质、水文地质、河势冲淤变化等),不能套用已建工程经验,必须进行数学模拟和河流模拟试验研究,确定顶面高出河床的幅度及其相应对河流流态、流场以及对河床和附近构筑物(如码头)的影响后,才可确定建设方案。

现介绍上海隧道院对该课题研究的内容和结论以供参考。

3.2.1 数学模型计算研究主要内容

1) 管段不同抬高幅度对流态的影响程度和影响范围(重点是航道和码头)。

2) 基槽开挖对流态的影响程度和影响范围。

3) 临时辅助航道开挖对流态的影响程度和影响范围。

3.2.2 物理模型试验主要研究内容

1) 管段不同抬高方案对流场的影响和对附近码头水域的影响。

2) 分析基槽开挖区的悬砂淤积。

3) 分析不同尺度管段抬高所引起的工程区河床冲淤变化和对附近码头水域的泥砂冲淤影响。

3.2.3 数学模型计算研究结论

抬高的工程实施后,对黄浦江河道内的水位基本没有影响;在隧道工程轴线附近流速有所增加,离开轴线稍远处的上游水域流速略有减少,流速变化范围很小,流场变化无异常现象。

3.2.4 物理模型试验研究结论

流速影响范围在数百米以内,在河床地形调整后流速增大的幅度在 5 cm/s 以内;覆盖区周围发生一定冲刷,冲刷区上下游出现泥砂淤积区,影响范围为数百米;冲刷幅度在 1 m 以内,2 年内,局部地形调整后,趋于稳定;冲刷可影响覆盖层与河床接合处块石稳定;冲淤对工程区附近码头及防洪等工程设施没有产生不利影响。

3.3 沉管埋深增大的挑战

沉管法为浅埋工法,港珠澳大桥的沉管隧道要适应 30 万 t 油轮的通航,因此沉管必然深埋,为此必须应对管槽回淤严重的挑战;此外,深挖管槽导致地基回弹严重和荷载加大共同引起的显著沉降以及对管节段结构的挑战。

对策:边回淤边吸泥排淤,在管段沉放时不断排淤以满足设计要求;设置桩基防止沉降过大;加大管节段结构厚度。应多做同等深度的隧道建设方案比较,深埋隧道应尽量避免采用沉管法。

3.4 大型盾构穿越浅覆土层的挑战

为减小水下隧道的埋深,以减少隧道总长和造价以及隧道在复杂地层中推进的难度,很多隧道常在局部地段面临穿越小于盾构直径的覆盖层深度的挑战,如南京长江隧道 φ14.93 m 盾构,始发段埋深 5.5 m(0.4D);江中冲槽段埋深 10.49 m(0.7D)。具体挑战的风险是盾构姿态稳定性和江底覆土层劈裂穿透涌水。

对策:1) 保证盾构稳定推进的推进参数和推进技术的掌握以及注浆参数和技术的掌握。2) 掘进面泥水压力设定需要满足掘进面的稳定和防止劈裂击穿覆盖层(见图1);为此,需研究稳定掘进面所需泥水压力的确定以及劈裂机制的阐明和泥水击穿压力的判别标准。3) 应进行泥膜形成技术和开舱用气压置换泥浆条件下泥膜稳定时间的研究。

3.5 盾构刀盘刀具磨损的挑战

南京纬七路、纬三路隧道、穿黄隧道、广州、成都隧道在卵砾石层,软硬复合地层中推进时都出现了刀具严重磨损,长时间停工检查修复。大盾构水下砂卵砾石层与上软下硬岩层中推进时检修和更换刀具是施工中必须克服的难题。

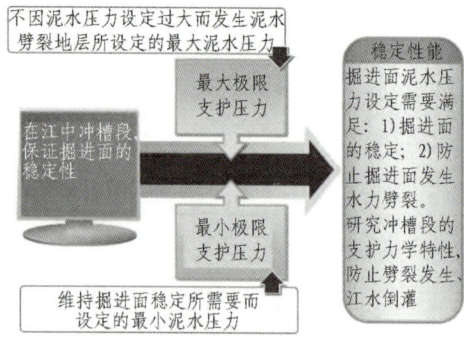

图1 盾构泥水压力平衡示意

Fig. 1 Balance of slurry shield

大型盾构在砂卵砾石层中推进刀具磨损问题更加突出,这是由于:1) 盾构刀具在同样进尺条件下,其磨损工作长度与刀具配置部位半径成正比,例如南京 14.93 m 大盾构掘进刀具的磨损为地铁 6.3 m 盾构磨损的 2.5 倍;2) 在石英含量高的砂卵石层中刀具的磨损可达软土地层中磨损的 10 倍。

南京长江隧道是世界上首次进行高水压条件下常压刀具更换,形成了高水压条件下(0.65 MPa)泥水盾构常压刀具更换技术。常压换刀的成功实施,盾构刀具适时检测和更换技术的建立,部分避免了高压换刀作业的巨大风险,标志着超长隧道的掘进成为可能(见图2)。

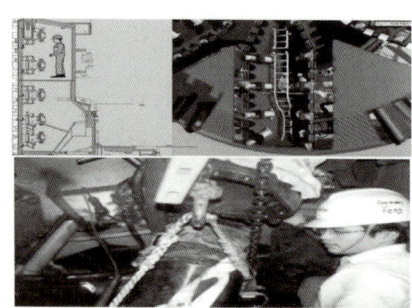

图2 盾构换刀

Fig. 2 Replacement of cutting tools of shield machine

对策:

1) 在盾构选型和配置时能确保实现不在刀盘前高压换刀,而在刀盘后常压下换刀。南京纬七路长江隧道已在中国第 1 次成功实施了高水压条件下常压刀具的更换,降低了高压换刀作业的风险。

2) 研究实行饱和法开舱作业的高压换刀技术。目前的盾构设计和研制水平还不能保证全部刀具在刀盘后常压换刀,高压换刀还难以避免,但常规的压缩空气开舱技术由于每班次作业均须执行加、减压的程序,每班次有效工作时间仅为 25 min,而减压总时间 >180 min,所以工作效率太低。饱和法开舱作业是作业人员一次加压,长期在设定压力的饱和舱内生活和休息,每天乘坐穿梭舱运送至盾构刀盘前的高压舱内从事刀具检修作业,作业时间可达 4 h,每次作业后返回生活舱休息,待全部作业完成后才减压返回至常压下。从而极大地提高了作业效率,显著减少了作业人员减压病发生的概率。

3) 开展砂卵石地层和上软下硬地层中刀具磨损机制和地层适应性研究,目前的初步成果表明,刀具耐磨型的改进有赖于刀具材质的改进和"减小前角,加大后角、增大合金、钝化刀刃"的刀具设计。

3.6 盾构埋深提高的挑战

长江未来的大冲刷量导致长江隧道的设计埋深大,从而引起盾构在基岩中推进的困难。为减小盾构推进难度,必须不按常规设计,即未来覆盖层可小于一个盾构直径,以减少埋深。借鉴沉管隧道高出河床成功的实例,这是可行的,但须解决抗浮和河床覆盖的问题。

3.7 盾构法轴承损坏的挑战

水下长大隧道建设时,盾构轴承有可能损坏,如丹麦大贝尔特(Great Belt) 海峡东隧道,TBM 掘进 500 m 后发生轴承损坏,原因是密封润滑系统钻渣污染。

对策:检测润滑油,查明损坏原因,若为钻渣污染,可采取更换轴承,并增强密封。

3.8 TBM 和盾构通过断层和软弱破碎地层时卡机的挑战

TBM 通过断层和软弱破碎地层时,由于开挖卸荷诱发断层失稳或围岩剪胀扩容导致 TBM 卡机,无法推进。

对策:加强地质超前预报,准确预测断裂构造或软弱破碎围岩位置,超前注浆固结围岩,防止围岩失稳或剪胀扩容。如隧道设计线位通过断裂构造数量较多,可以考虑改变 TBM 或盾构建设工法,并深度比较工程建设方案。

4 讨论与建议

修建桥梁与隧道各有利弊,关键是随工程环境与建设条件而变,因地因时而异,应"宜桥则桥、宜隧则隧、桥隧并举"。

修建水下隧道的各个工法——钻爆法、沉管法与盾构和 TBM 法各有优缺点,关键是修建人员需深刻领会并熟练分析其所适应的工程地质与水文地质环境,只有这样才能发挥其工法的优势,避开其工法的不利,做到趋利避害,不能将某工法应用于不该使用它的环境与条件下。最后要坚持工法的与时俱进,化面临的挑战为发展的机遇,不断实现科技创新,从而发展与完善工法,使我国在科技创新的基础上实现从隧道建设大国走向隧道建设强国。

参考文献(References):

[1] 李树忱,李术才,张京伟,等. 数值方法确定海底隧道最小岩石覆盖厚度研究[J]. 岩土工程学报,2006(10):1304 – 1308. (LI Shuchen, LI Shucai, ZHANG Jingwei, et al. Study on numerical method for the minimum rock covers of subsea tunnels [J]. Chinese Journal of Geotechnical Engineering, 2006(10) : 1304 – 1308. (in Chinese))

[2] 李树忱,张京伟,李术才,等. 海底隧道最小岩石覆盖厚度的位移收敛法[J]. 岩土力学,2007(7):1443 – 1447. (LI Shuchen, ZHANG Jingwei, LI Shucai, et al. Displacement convergence method for minimum rock cover above submarine tunnel [J]. Rock and Soil Mechanics,2007(7) : 1443 – 1447. (in Chinese))

[3] 蔚立元,徐帮树,李术才,等. 确定水下隧道覆盖层厚度的经验公式及其应用[J]. 地下空间与工程学报,2011(3):497 – 503. (YU Liyuan, XU Bangshu, LI Shucai, et al. Empirical formulas for determining overburden thickness of underwater tunnel and their applications [J]. Chinese Journal of Underground Space and Engineer,2011(3) : 497 – 503. (in Chinese))

[4] 郭陕云. 对盾构(TBM) 技术运用及开发的几点认识[J]. 隧道建设,2008,28(6):631 – 637. (GUO Shanyun. Comments on application and development of TBM/shield tunneling technology [J]. Tunnel Construction, 2008, 28 (6) : 631 – 637. (in Chinese))

[5] 李建华,吕瑞虎,李文俊. TBM(盾构) 技术应用及几个主要技术关键的探讨[J]. 隧道建设,2012,32(S2):1 – 5. (LI Jianhua, LV Ruihu, LI Wenjun. Application of TBM tunneling technologies and technical keys [J]. Tunnel Construction, 2012, 32(S2) : 1 – 5. (in Chinese))

[6] Martin Herrenknecht, Karin Bappler,康宝生. 盾构和 TBM 技术的创新与应用[J]. 建筑机械化,2011(8):54 – 57,68,7. (Martin Herrenknecht, Karin Bappler, KANG Baosheng. The technological innovation and application of shield and TBM [J]. Construction Mechanization, 2011(8) : 54 – 57, 68, 7. (in Chinese))

[7] 杜朝伟,王秀英. 水下隧道沉管法设计与施工关键技术[J]. 中国工程科学,2009(7):76 – 80. (DU Chaowei, WANG Xiuying. Key technology of design and construction on immersed tube tunnel [J]. Engineering Science, 2009 (7) : 76 – 80. (in Chinese))

[8] 吴维. 关于沉管法修建长江水下隧道若干问题的研究[J]. 现代隧道技术,2003(2):1 – 4,31. (WU Wei. Study on the problems of building the Yangtze river underwater tunnel by immersed tube method [J]. Modern Tunnelling Technology,2003(2) : 1 – 4, 31. (in Chinese))

[9] 卢普伟,梁邦炎,资利军. 港珠澳大桥隧道工程沉管法与盾构法比选分析[J]. 施工技术,2012(17):89 – 91. (LU Puwei, LIANG Bangyan, ZI Lijun. Scheme Comparison of immersed tunnel and shield tunnel for Hongkong-Zhuhai-Macao bridge [J]. Construction Technology, 2012 (17) : 89 – 91. (in Chinese))

[10] 王梦恕. 中国隧道及地下工程修建技术[M]. 北京:人民交通出版社 2010.

地铁工程建设安全风险管理研究

解东升[1,2]，钱七虎[1]，戎晓力[1]

(1. 解放军理工大学 工程兵工程学院，江苏 南京 210007；
2. 总参工程兵第四设计研究院，北京 100036)

摘 要：地铁工程建设的各种属性决定了其是一项高风险建设项目。由于建设过程中的大规模、高速度以及严峻的安全形势，迫切需要对其进行完善的安全风险管理研究。通过对安全金字塔理论、海恩安全法则、事故变化-失误理论等安全风险管理理论的研究分析，说明通过安全风险管理预防事故发生是完全可行和非常必要的，并在此基础上确立了安全风险管理的理念：管理重点是人；安全风险管理要贯穿整个工程周期；安全风险管理不能完全依靠保险和监测数据；信息化在安全风险管理中具有重要作用。最后，本文还提出了下一步安全风险管理研究所要解决的突出问题。

关键词：地铁工程； 风险管理； 海因里希法则； 海恩法则； 事故预防

中图分类号：U231+.3　**文献标识码**：A　**文章编号**：2095-0985(2012)01-0061-07

Risk Management in Rail Transit Construction

XIE Dong-sheng[1,2], QIAN Qi-hu[1], RONG Xiao-li[1]

(1. Engineering Institute of Engineer Corps, PLA University of Science and Technology, Nanjing 210007, China;
2. The 4th Engineering Design and Research Academy General Staff PLA, Beijing 100036, China)

Abstract: The construction of subway engineering with various internal properties was regarded as a high risk project. Because of the great building size, high construction speed and serious security situation in current rail transport building, it is urgently to do security risk management during the construction. By the analysis of security pyramid theory, Hain rules and variety-fault model, one important conclusion can be obtained that most accidents can be fully avoided by the security risk management. Based on these theories and this conclusion, some crucial ideas on security risk management can be established, which mainly includes: human is the keystone to risk management; risk management must be run through the whole stage of construction; risk management can not only depend on the insurance and the monitoring data, informationization has a significant effect on the risk management. In the end, some outstanding issues about risk management research were mentioned.

Key words: subway project; risk management; Heinrich law; Hain rules; accidents prevention

随着国家经济建设的持续快速稳定发展，城市化水平不断提高，我国的地下空间开发和城市地铁工程建设进入了蓬勃发展的阶段[1]。其中城市地铁工程建设速度较快，而在地铁工程建设中时而突发的工程事故问题，造成了巨大的经济影响和社会影响，引起了人们的极大关注，面向地铁工程建设的安全风险管理专业性研究也越来越得到重视和发展[2~5]。

1 安全风险管理的必要性和紧迫性

1.1 内在属性决定了工程建设的高风险性

地铁工程大多地理位置特殊、质量和安全要求高、涉及工程专业多、工程量巨大、地下和露天作业多、工程和周边环境关系密切、生产的流动性强、生产的周期长，具有几个明显特点[6]：(1) 建设规模大，城市地铁建设线路一般达百余公里甚

收稿日期：2011-07-27　修回日期：2011-12-29。
本文原载于《土木工程与管理学报》(2012年第1期)。

至数百公里;(2)投资大,轨道交通每公里造价达数亿元,在建项目规模普遍达上百亿元;(3)施工周期长,单线建设一般需要 4~5 年;(4)技术要求高、施工工艺复杂,几乎涉及到土木工程、机电设备工程的所有高新技术领域;(5)地质条件复杂,尤其是水文地质条件、地下管线、周围建筑物情况复杂多变等。这些特有属性决定了地铁工程建设是一项高风险建设项目。

1.2 规模大、发展快是安全形势严峻的客观原因

近年来,我国城市轨道交通进入了快速发展阶段。截至目前,全国已有 25 个城市的轨道交通近期建设规划获得国务院批复,到 2015 年前后,我国将建设 87 条轨道交通线路、总里程达 2495 km。未来几年,我国城市轨道交通工程仍将处于大规模、高速度、超常规、跨越式大建设时期。据了解,上海每年轨道交通要完成 60 km,北京轨道交通要完成 40~50 km,而国外大城市每年建设速度一般不超过 10 km,例如莫斯科 200 多公里地铁的建设花费了近 70 年才完成。我国现阶段地铁建设的高速度也为地铁事故频发埋下了很大隐患。使得轨道交通工程建设参与各方处于超负荷工作状态,专业技术人员不足,管理浓度稀释,导致工程建设中出现了不少薄弱环节,尤其是在工程安全风险管理方面,更是造成了地铁工程建设非常严峻的安全形势。

1.3 安全形势是实行安全风险管理的客观需求

2003 年 7 月我国上海地铁 4 号线联络通道施工发生地面沉降,导致一幢 8 层楼房严重倾斜,造成直接经济损失达 1.5 亿元左右,工程延期 3~4 年。2007 年 3 月 28 日北京市地铁十号线苏州街站东南出入口发生严重塌方,导致 6 人死亡。2008 年 11 月 15 日,杭州萧山湘湖段地铁施工现场发生塌陷事故,事故遇难人数达到 21 名(图1)。2010 年 7 月 14 日北京地铁 M15 号线顺义站发生深基坑钢支撑掉落,事故导致 2 死 8 伤(图2)。

接连出现的地铁工程建设事故,不但造成巨大的经济损失和社会影响,有些甚至还付出了生命的惨痛代价。加强工程建设的安全风险管理,控制遏制事故尤其是重大安全事故的发生迫在眉睫。

图 1 杭州地铁 1 号线"11.15"事故现场

图 2 北京地铁 15 号线事故现场

2 典型地铁工程事故案例原因分析

事故的预防要以事故的发生原因为主要导向。已发生事故为我们提供了宝贵的材料,分析已有事故原因,获取有效预防事故的方法,才能针对性地做好事故预防工作。

2.1 北京地铁十号线苏州街站事故原因分析

北京地铁十号线 02 标苏州街站坍塌处周边环境条件十分复杂,工程地质及水文条件极差,坍塌处集隧道爬坡、断面变化及转向、覆土层浅、环境条件和地质条件复杂等多种不利因素,且该暗挖结构本身处于复杂的空间受力状态。施工单位在已发现拱顶裂缝宽度由最初的 1 cm 发展成 10cm,并有少量土方坍塌的情况下,现场管理人员发生严重错误,在没有制定并采取任何保护抢险人员的安全措施的情况下,指挥作业人员实施抢险,造成 6 名作业人员在二次塌方中被埋压。另外,此标段地质勘探以 40 m 为间距设置探孔,事故地点处在探孔间距之间,勘探资料未能显示出事故地点实际地质情况。现场安全生产管理还存在严重漏洞:一是应急预案对施工过程可能出现的风险考虑不全,出现险情后不能按照预案组织抢险;二是对劳务用工管理不严,使用无资质的劳务队伍从事施工作业;三是现场管理人员未严格遵守安全生产有关法律规定。

2.2 杭州地铁一号线萧山湘湖段事故原因分析

杭州事故后的调查分析表明:由于基坑土方开挖过程中基坑超挖,钢管支撑架设不及时,垫层未及时浇筑,钢支撑体系存在薄弱环节等因素,引起局部范围地下连续墙产生过大侧向位移,造成支撑轴力过大及严重偏心。同时基坑监测失效,未采取有效补救措施。致使部分钢管支撑失稳,钢管支撑体系整体破坏,基坑两侧地下连续墙向坑内产生严重位移。

其主要原因有:(1)土方开挖未按照设计工况进行,存在严重超挖现象;土方超挖导致地下连续墙侧向变形、墙身弯矩和支撑轴力增大。(2)钢管支撑体系存在薄弱环节,整体性差,设计不明

确和施工不规范降低了钢支撑体系的整体性和承载力。(3) 监测工作严重失职,监测内容及数量不满足规范要求;测点破坏严重且未修复,造成多处监控盲区;部分监测内容的测试方法存在严重缺陷;提供的监测数据存在伪造现象。(4) 勘察单位未考虑采用薄壁取土器取样对土强度参数的影响,未根据当地软土特点综合判断选用推荐基坑设计参数;设计单位未能根据当地软土特点综合判断、合理选用基坑围护设计参数,力学参数选用偏高降低了基坑围护结构体系的安全储备。(5) 基坑坑底加固措施的改变,降低了基坑围护结构体系的安全储备。(6) 监理工作失职,对施工中出现的不符合设计及规范的严重问题(土方开挖、钢支撑施工、基坑监测等)未能采取有效措施督促整改落实,消除隐患。

3 安全事故风险管理可行性

地铁工程建设的安全事故并非完全不能控制和预防,及时合理地进行风险管理并做好事故预防工作可以大大降低事故发生的概率和事故损失。

3.1 "风险可以监控,事故可以预防"是实行安全风险管理的科学依据

大量实践表明:基本上所有的地铁工程事故都有预兆,它可以反映在监控数据(例如苏州街站事故中拱顶裂缝宽度的变化以及杭州地铁事故发生时监测的实际地表沉降及墙体侧向位移均超过设计报警值)和工程现象中。从事故萌芽出现到险情发生,期间经过事故的缓慢发展期和快速发展期两个阶段,一般有 2~3 天的时间(图 3),因此地铁工程的安全风险是可以监控的,通过风险的分析评估与管理,事故是可以避免的[7,8]。

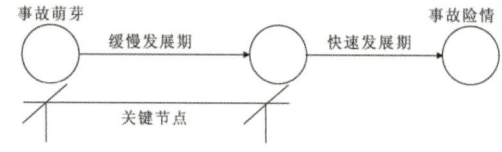

图 3 地铁工程事故发展示意

一般情况下,事故从萌芽状态发展到险情发生要有 2~3 天的时间,而这段时间就是事故预防和规避的关键时期。只要在这个发展期及时合理地进行事故的预防并处置风险事件,就能很好地预防事故或者大大降低事故的损失。

3.2 绝大多数事故是有征兆、可预防的

(1) 海因里希法则

海因里希提出的事故因果连锁理论(Accident Causation Sequence Theory)阐明了导致伤亡事故的各种因素之间以及这些因素与伤害之间的关系。其核心思想是:伤亡事故的发生不是一个孤立事件,而是一系列原因事件相继发生的结果,即伤害与各原因之间具有连锁关系。

在其出版的著作《安全事故预防:一个科学的方法》一书中[9],海因里希提出了安全金字塔法则,通过分析 55 万起工伤事故的发生概率得出,在 1 起重伤害事故背后,有 29 起轻伤害事故,303 起无伤害虚惊事件,以及大量的不安全行为和不安全状态存在,其关系可以形象地用安全金字塔来示例(图 4)。顶峰为严重事故,依次往下为轻伤害事故、无伤害虚惊事件和金字塔最底层的大量的不安全行为和不安全状态。

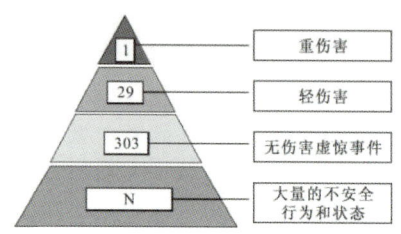

图 4 海因里希安全金字塔理论模型

(2) 海恩安全法则——Hain rules

飞机涡轮机的发明者德国人帕布斯·海恩提出了一个航空界飞行安全的法则,即每一起重大飞行安全事故的背后有 29 个事故征兆,每一个征兆背后又有 303 个事故苗头,每一个事故苗头背后又有 1000 个事故隐患,这就是著名的海恩法则,简记为:1:29:303:1000,这一法则虽然是针对航空界飞行安全而言,但它所揭示的事故背后有征兆,征兆背后有苗头的理论却适用于各行各业。因此,海恩法则也被业界奉为万能法则。

(3) 约翰逊的事故变化-失误理论

美国著名事故分析专家约翰逊的工程事故变化-失误理论认为:事故是由意外的能量释放引起的,这种能量释放的发生是由于管理者或操作者没有适应生产过程中的人或物的因素变化,产生了计划错误或人为失误,从而导致不安全行为或不安全状态,破坏了对能量的屏蔽或控制。这种理论可以用多米诺骨牌来形象解释,即:变化→失误→不安全行为或不安全状态→事故→人的伤害和物的损害。

在工程管理实践中,变化是不可避免的,也并非全部对安全有害,但是,从安全角度考虑,必须及时发现和预测变化发生后的安全隐患,并采取适当的措施或对策,做到顺应有利的变化,克服不利的变化。约翰逊认为,事故的发生一般是多重

因素造成的,应从领导者、计划者、监督者及操作者等四个层面来考虑(图5)。

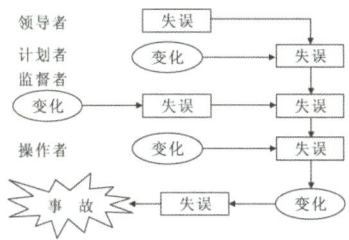

图5　变化-失误模型示意图

约翰逊的工程变化-失误理论、海因里希"安全金字塔"法则和海恩安全法则揭示了一个十分重要的事故预防原理:要预防死亡重伤害事故,必须预防轻伤害事故;预防轻伤害事故,必须预防无伤害虚惊事件;预防无伤害虚惊事件,必须消除日常不安全行为和不安全状态;能否消除日常不安全行为和不安全状态,则取决于日常管理是否到位,这是作为预防死亡重伤害事故的最重要的基础工作。

3.3　安全事故预防关键在人

许多资料显示,由于人的不安全行为导致的事故占事故总数的70%～80%。根据美国50年代的统计数据,在75000件伤亡事故中,自然事故只占2%。海因里奇经过大量研究,通过对55万次事故的统计分析,把事故诱因主要分为人的不安全行为、物的不安全状态和其它不可抗拒因素,认为存在着"88∶10∶2"的规律[9],即100起安全事故中,有88起属于人的失误,有10起是由物的不安全状态造成的,只有2起所谓的"天灾"是难以预防的(图6)。由此可见,要控制事故的发生,控制人的不安全行为是关键。

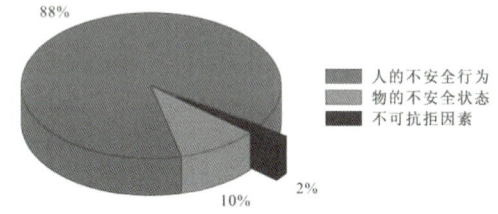

图6　海因里希事故诱因分布模型

前节所列的北京苏州街站事故的最主要原因就是施工现场管理人员的不当指挥和使用无资质的劳务队伍,完全是一件由人的不安全行为导致的事故。

3.4　安全事故具有必然性和偶然性

事故的发生都是一个因素或几个因素共同相互作用产生的结果,单独的事故是偶然的、没有规律的,大量的工程风险进行观察分析之后,就可以发现事故发生其中的规律性,依据规律来预测事故发生概率和导致后果的大小。

事故的偶然性是相对于某一具体事故而言,某一具体事故的发生都是诸多风险因素和其他因素相互作用的结果,是一种随机现象。个别风险事故的发生是偶然的、杂乱无章的,但对大量风险事故资料的观察和统计分析,发现其呈明显的运动规律,表明风险事故的发生又存在着必然性,这就使得人们可以用概率论的方法以及其它现代风险分析方法去计算事故发生的概率和预估损失的程度。

4　安全风险管理应树立的理念

地铁工程的安全管理有着自己的特点和规律,只有把握清楚地铁工程的特点规律,树立正确的管理理念,才能有针对性的做好工程安全风险管理,达到风险管理的高效率和高成果。

4.1　安全风险管理的重点是人

现代工程安全科学认为,工程事故发生的原因主要有两类:人的不安全行为和物的不安全状态。其中人的不安全行为是主要原因,包括管理者的指挥失误和操作者的操作失误,而管理者发生的人因失误是更为危险的人因失误。物的不安全状态包括了施工隧道内部涌水,渗漏水,不均匀沉降和过量沉降,路面塌陷以及支护体系的失稳和破坏等等,它们大都也是由人的不安全行为导致的。

上节中提到的海因里希工程事故诱因调查和在调研分析大量工程事故案例时发现:工程事故的基本模型为:变化带来人的不安全行为,并引起物的不安全状态,往往是一系列的连锁失误造成无法挽回的事故。其中以"人的不安全行为"最难以预测和控制。

4.2　安全风险管理要贯穿于整个工程寿命周期

地铁建设的风险管理时间纵向上应当贯穿于地铁建设的各个阶段,包括勘察、设计、施工及工后运营等。地铁的施工工程,应该有一个足够的安全风险评估,从规划,设计,施工,设备制造、安装到调试、运营,是一个庞大、复杂、多专业、多门类的系统顺序整合过程。

目前的工程合同管理模式中,工程安全风险管理的责任主体主要为施工方,监理负责监督,业主也具有一定的安全风险管理责任。而地铁工程的风险实际并不完全是施工方引起的,国际隧道工程保险集团对施工现场发生安全事故原因的调查结果表明[10],地铁工程发生事故的原因是多方面的,施工方作为工程安全唯一责任主体无法从根本上避免事故的发生。从杭州地铁事故的调查结论中可以看出,设计、施工、监理和监测单位都要对事故负有一定的责任。图7给出了欧共体1993年发布的施工现场安全事故公告中事故原因的调查结果[11],其中认为属于承包商工程管理疏忽的大约占到37%。在隧道工程建设领域,图8给出了国际隧道工程保险集团对于施工现场发生的大量安全事故原因的调查结果[10],其中由于施工缺陷引起的安全事故只占到21%。

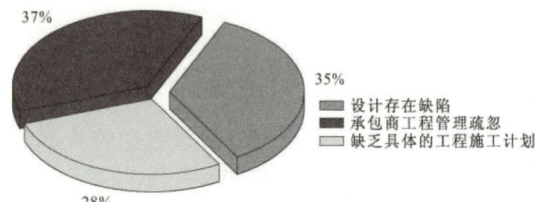

图7 施工现场安全事故原因分析

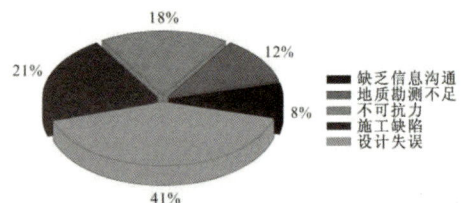

图8 隧道施工现场发生安全事故原因的调查结果

我国的工程勘察受到经费、时间等各方面条件制约,设计施工前不能完全了解工程地质情况和水文地质条件(勘察资料的不完整也是苏州街站事故发生的重要原因),工程设计也存在着设计深度不够、设计计算不精确的问题,而这些问题都会在工程施工期间暴露出来,显然只由施工方完全承担风险是不合理的。

除此之外,对工程的风险管理不能仅仅重视施工阶段,要从工程的立项、规划就考虑风险管理的因素,在工程勘察和设计阶段,更是要严格控制勘察质量,严格审核工程设计,并且在工程施工中,要随着施工状态的变化及时做好应对措施。在必要的情况下,要进行工程的补勘以及设计的更改。另外,还要在工后进行风险的评估,对工程周边的建筑物、地下管线以及工程自身的安全状态进行监控。

4.3 地铁工程安全风险管理具有一定的特殊性

通过对国内2000年后发生的地铁事故进行统计分析,得出其诱因分布如图9所示[12]。

由上图可以看出,诱因分布结果也说明了人的不安全行为是事故发生最重要的诱因。这个结果和海因里希的数据存在一些差异,体现在物的

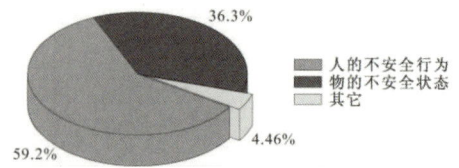

图9 国内地铁工程事故诱因情况分析

不安全状态引发的事故比例明显高于海因里希的统计。这主要是由于地铁工程建设与普通地面工程建设的区别所导致的。地铁工程建造于地下,和周围环境的关系要紧密的多,工程本身和周围环境存在着非常紧密的相互作用,工程施工行为会对周围环境产生影响,同时周围环境也会影响到工程的建设。地铁工程除了常规地面工程中的高空坠落、机械伤害、物体打击等风险外,还具有地面塌陷、基坑倒塌、支撑结构失稳等地铁工程特有的风险。所以地铁工程风险管理和地面工程风险管理存有明显区别,管理过程中应注重加强工程的监测和巡视巡查,控制物的不安全状态。

4.4 依靠保险进行风险管理的局限性

国际上通行采用保险业务来转移工程风险,并由保险公司主导进行风险管理。保险公司在风险分析评价阶段普遍采用的是RPC模式,即:

风险危害(R) = 风险发生概率(P) × 风险引发后果严重程度(C)

该模式是以工程的经济损失为指导性思想的,保险公司对于其利益的判断来源于风险的概率和风险后果两个方面。其指导意义在于给予风险管理单位和保险单位以经济杠杆标准,确定工程投保的方针策略。

我国由于项目管理模式与国外的差别,使得国际上通行的工程保险风险管理不太适合我国国情。国内的大多地铁工程都是政府行为,是与百姓生活密切相关的大型工程项目,事故的发生不但会造成经济损失,还会导致恶劣的社会影响。而这些方面的问题仅仅依靠工程保险的赔偿是不能解决的。所以,我国在地铁工程的风险管理上主要应以全面风险管理为指导思想,就是尽最大可能避免带来人员伤亡和财产损害的风险,而不是以冰冷的经济数字来进行风险分担。

4.5 安全风险管理不能仅仅依靠监测数据

加强工程监测是规避工程风险的重要手段,但是仅仅依靠监测数据来避免风险是远远不够的。首先,我国目前的工程监测仍以人工监测为主,由于经费、仪器、人力投入等原因存在着监测密度不高、监测频率不够等问题。另外,根据测点的监测数据,只能了解到测点的安全状态,而测

点中间是否存在薄弱面却无法得知。其次,有些风险问题仅仅通过监测是很难发现的,单纯的监测数据不能准确反映事物的变化状态。如地表硬路面的突然塌陷等。

从风险管理控制理论来说,工程监测只能监控物的不安全状态。在所有的工程事故中,物的不安全状态引发的事故只占到10%,即使在地铁工程中,物的不安全状态引发的事故也只占到1/3左右,而它们的背后是人的不安全状态。单纯依靠监测来进行风险管理,那将会使大部分的人因事故被忽视,得不到有效控制。所以工程风险管理不能仅仅依靠监测数据。

4.6 信息化在安全风险管理中具有重要作用

随着施工进展,工程地质条件和周围环境会发生变化,传统的施工往往不能根据条件的变化而及时改变施工参数。而信息化施工就可以完全解决此类问题,通过信息化手段,可以随时对施工参数进行调整,对施工进行"迭代式"的动态管理,使得工程施工的风险大大减小。

面对点多、线长、面广的地铁工程管理任务,仅仅还是依靠传统的、落后的管理方法来进行管理,往往容易使得管理者顾此失彼,难免出现疏漏。利用信息化系统可以加快信息传输速度,提高管理的效率和科学水平,增加项目各方的责任。对监测数据和工程事务实行信息化管理,可以大大提高工作效率。既节约了大量的数据处理和数据报告时间,也保证了数据分析的准确性,能够大大改善落后的管理方式。同时,信息化还能对大量的工程数据和处理结果进行备份,随时提供方便的查阅和调用,使风险管理专业化、自动化。

5 安全风险管理需要解决的问题

目前国内在地铁工程安全风险管理方面取得了一些成绩,但是还没有能够达到对工程事故的完全预防和控制。在已有研究成果基础上,安全风险管理还要继续深入研究,以期达到完全控制安全风险事故的目的。

5.1 全面推广安全风险管理制度,将其作为工程项目管理的必要组成部分

安全风险管理活动要涵盖整个工程建设周期,从项目的立项规划,到项目的设计施工,以及工程建设完毕后的工后阶段,都要进行安全风险管理的活动。并且,安全风险管理的活动是一个循环的完整的管理流程,从风险辨识、分析到风险的评估控制,最后再回到新的风险的识别、分析,是一个动态、循环、封闭的过程。

进行全面的安全风险管理,必须建立和推广安全风险管理制度,将其作为工程项目管理的必要组成部分,并且是必不可少的一部分来进行。

5.2 落实风险管理专项经费,逐步建立规范的安全风险管理体系

在工程预算中,关于安全风险管理的费用,如施工监测费用等都没有明确取费标准,在低价中标的管理模式下,安全风险管理有关费用往往被首先挤压,导致一些施工单位在利益的驱使下,冒险施工,"赌"不会发生事故的概率。因此,只有落实安全风险管理专项经费,才能更好地做好风险管理工作。

当前完善的风险管理流程基本为[4,5,13~15]:风险定义、风险辨识、风险估计、风险分析、风险评估、风险处置和风险监控。目前国内各单位从事的隧道及地铁工程风险管理项目上,主要侧重于风险的分析与评估。多数的风险分析和评估是根据经验方法对其进行分析,从风险发生的概率和风险发生后引起的后果两个方面,得出其定性或半定量的可能性大小,予以排序,最后根据量值大小给出建议。

地铁工程的安全风险是一个动态的过程,必须提倡地铁工程进行"迭代"式设计、施工和管理,以期最大可能地规避风险。风险安全管理不仅仅是专项的技术的管理,它还包含管理工作的开拓创新,管理手段的不断进步,甚至于整个建设、施工、监理单位的安全文化的建设以及安全教育的落实、安全意识的培养等等都会对安全风险管理的效果产生重大的影响。因此,逐步建立起适合于我国工程建设实际的完善的安全风险管理体系,并在体系内进行安全风险管理活动,具有重要的意义。

5.3 建立适合地铁工程建设发展现状的安全风险管理技术规范、标准

国家和地方施工规范和标准是进行地铁工程各种工法风险识别和风险评估的基本依据,但目前风险管理相关技术控制规范不够全面。因此,编制地铁工程施工不同工法的技术规范,加强针对地铁工程安全风险管理的法规建设工作,研制不同城市、不同工法的适应不同岩土工程地质、水文地质、环境条件的风险阈值数据库系统已是当务之急。

通过建立安全风险管理体系,明确各方安全风险管理责任;通过立项前进行风险评估,对"严重"等级工程项目一票否决;并对整体项目进行

问卷调查论证；规定安全投入为"专项提取"及其占整个建设费用中的比例，并进行审计、监管；安全风险管理费用专款专用，不得挤压占用；强制实行设计和施工中的风险专项设计和方案。落实各项方案，明确各方安全责任，实现动态施工，动态管理，全面及时高效的管理。

5.4 建立规范的工程安全风险管理专业队伍，突出重大事故预报防治

"第三方监测"近年来成为了各地地铁工程建设风险管理的新举措。实施城市地铁工程施工"第三方监测"是保证施工安全和工程质量的重要手段[16]，但目前针对"第三方监测"没有国家性的法规进行明确规定和管理，各地"第三方监测"基本处于无序状态。国内对工程安全风险管理咨询评估的从业单位和人员没有明确的资质管理。因此，亟需建立规范的专业队伍来保障工程建设中的安全风险管理。

此外，还要加强地铁工程重大事故预测预报和防治技术的研究，对地铁工程安全事故的发生机理，安全监控理论与技术，以及应急处理决策支持技术进行研究。依靠专业化管理水平的提高和技术手段的进步，突出研究重大事故预测预报和防治，主要包括软弱破碎围岩段的塌方、岩溶段突水突泥，岩爆以及水下隧道渗水贯通等。

5.5 建立合适的信息化安全风险管理平台

信息交流不畅也是事故发生的一个重要方面，实现各方操作"透明"，信息共享，责任落实；进行透明式的管理可以集思广益。利用信息化系统可以加快信息传输速度，提高管理效率和水平，增加项目各方责任。在信息化飞速发展的今天，我国安全风险管理的信息化水平还很低，国内还缺乏符合风险管理体系，适合国内地铁工程建设实际的信息化风险管理平台。因此，建立基于现代化信息技术的地铁工程建设安全风险管理信息系统，依靠信息化手段和方法来进行工程安全风险管理必将成为风险管理的发展趋势。

6 结语

尽管地铁工程建设中的风险很大，但是只要合理的进行安全风险管理，依照风险发生发展的规律和特点，突出把握风险管理的重点和难点，运用合理的管理理念和管理方法，建设完善的安全风险管理法规、制度和体系，对工程建设风险进行预防和规避，就可以大大地遏制风险危害的发展，减少工程建设事故特别是重大事故的发生。

参 考 文 献

[1] 周翊民. 中国城市轨道交通的发展策略[J]. 城市轨道交通，2002，24(10)：34-37.

[2] 李元海，朱合华. 岩土工程施工监测信息系统初探[J]. 岩土力学，2002，23(1)：103-106.

[3] 孙 钧. 城市地下工程施工安全的智能控制预测与控制及其三维仿真模拟系统研究[J]. 岩石力学与工程学报，1999，18(s1)：753-762.

[4] 中华人民共和国建设部. 地铁及地下工程建设风险管理指南[M]. 北京：中国建筑工业出版社，2007.

[5] 黄宏伟. 隧道及地下工程建设中的风险管理研究进展[J]. 地下空间与工程学报，2006，2(1)：13-20.

[6] 钱七虎，戎晓力. 中国地下工程安全风险管理的现状、问题及相关建议[J]. 岩石力学与工程学报，2008，27(4)：649-655.

[7] Einstein H H, Vick S G. Geological Model for Tunnel Cost Model [C] // Proceedings of Rapid Excavation and Tunneling Conference. 1974: 1701-1720.

[8] Einstein H H. Risk and risk analysis in rock engineering [J]. Tunneling and Underground Space Technology, 1996, 11(2): 141-155.

[9] Henrich H W. Industrial Accident Prevention: A Scientific Approach [M]. New York: McGraw-Hill Book Company, Inc, 1950.

[10] International Tunneling Insurance Group, A Code Practice For Risk Management of Tunnel Works [S].

[11] The Council of the European Communities. Council Directive92/57/EEC of 24 June 1992 on the Implementation of Minimum Security and Health Requirements at Temporary or Mobile Construction Sites (Eighth Individual Directive Within the Meaning of Article 16(1) of Directive 89/391/EEQ [R]. Brussels: The Council of the European Communities, 1992.

[12] 解东升，王明洋，卢 浩. 地铁工程安全事故统计分析[C] // 第二届全国工程安全与防护学术会议论文集(上册). 北京，2010：205-209.

[13] 路美丽，刘维宁，李兴高. 风险管理在城市地铁工程中的应用初探[J]. 中国安全科学学报，2005，15(5)：96-100.

[14] 路美丽，刘维宁，罗富荣，等. 隧道与地下工程风险评估方法研究进展[J]. 工程地质学报，2006，14(4)：462-469.

[15] 郭 俊. 工程项目风险管理理论与方法研究[D]. 武汉：武汉大学，2005.

[16] 解东升，王明洋，杨效中，等. 轨道交通工程第三方制度研究[C] // 第二届全国工程安全与防护学术会议论文集(上册). 北京，2010：39-45.

| 学术论文——隧道及地下工程

地下工程建设安全面临的挑战与对策

钱七虎

(中国人民解放军理工大学 工程兵工程学院，江苏 南京210007)

摘要：详细介绍我国地下工程建设的安全管理现状和我国地下工程建设安全的严峻形势。从事故发生的类型和原因等对工程建设事故进行分析，指出责任事故及主观原因是构成地下工程建设安全管理巨大挑战的主因。结合海恩安全金字塔、海因里希安全法则、约翰逊的工程变化失误理论模型等工程事故管理理论，提出工程事故预防的可行性和工程建设安全管理实施的对策。对重大地下工程建设中最突出的突水突泥地质灾害和岩爆地质灾害进行详细阐述，主要介绍突水突泥灾害的严重挑战、发生发展的科学规律、事故前兆信息分析以及地质超前预报等事故对策；介绍岩爆地质灾害的定义和分类、岩爆的预测机制、岩爆监测预报的理论基础和工程实践以及岩爆灾害的施工对策等。最后，针对地下工程建设的事故预防提出技术方法和管理手段的一些意见和建议。

关键词：地下工程；安全事故；风险管理；突水突泥；岩爆；地质灾害

中图分类号：TU91　**文献标识码**：A　**文章编号**：10006915（2012）10194512

CHALLENGES FACED BY UNDERGROUND PROJECTS CONSTRUCTION SAFETY AND COUNTERMEASURES

QIAN Qihu

(Engineering Institute of Engineers Corps, PLA University of Science and Technology, Nanjing, Jiangsu 210007, China)

Abstract: The status quo of safety management in China's underground projects construction was introduced. Based on the accidents statistics, the serious security situation of underground projects construction was certificated. The analysis of types and causes for accidents indicated that the responsibility and subjective reasons are the great challenges for security management. Combining with Hains safety pyramid theory, Heinrich's safety rules and Johnson's changeserrors model, the concept that accidents can be prevented by risk management measures was established, and some suggestions for accidents prevention were put forward. Besides, the mechanism and control measures for two typical geological hazards, i. e. water and mud outbursts and rockburst, were represented. For the study of water and mud outbursts, the laws of occurrence and development, the precursory information theory, the geological prediction methods and other incident countermeasures were introduced. For the control measures of rockburst, the definition and classification of rockbursts, the prevention mechanism, theoretical basis and engineering practice for monitoring and prediction, as well as the construction measures, were introduced. Some opinions and suggestions on technical methods and management tools were put forward for the accident prevention for underground projects construction.

Key words: underground engineering; safety accident; risk management; water and mud outbursts; rockburst; geological disaster

1 引言

在我国经济建设迅速发展的背景下，我国现今工程建设速度居于世界前列，尤其是我国的地下工程建设，其工程类型众多、规模之巨大为世界所瞩目。与此同时，由于地质条件的复杂性，地下工程与地下空间建设过程中岩爆、大变形与大面积塌方、突水、地表沉陷等地质与工程灾害事故频发，造成人员伤亡、设备损失、工期延误和工程失效等。地下工程与地下空间的安全建设与风险管理已成为国内外专家关注的重要问题。

2 地下工程建设安全的现状和分析

2.1 地下工程建设安全现状

我国近年来工程事故不断出现，其中不乏一些重特大事故。表 1 为 2008～2011 年我国重特大事故统计[1]。

表 1 2008～2011 年我国重特大事故统计 [1]
Table 1 Statistics of serious accidents happened in China (during 20082011)[1]

年份	事故性质	发生次数	死亡人数
2008	重大事故	86	1304
2008	特大事故	11	667
2009	重特大事故	67	1128
2010	重特大事故	74	1438
2011	重特大事故	59	897

图 1 给出了 2008～2011 年不同领域安全事故数及伤亡人数统计结果。

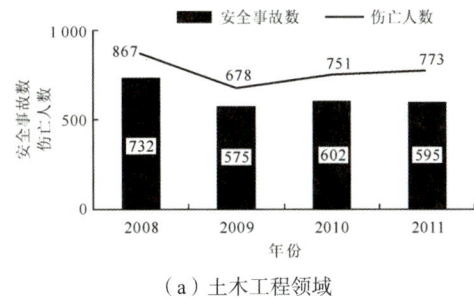
(a)土木工程领域

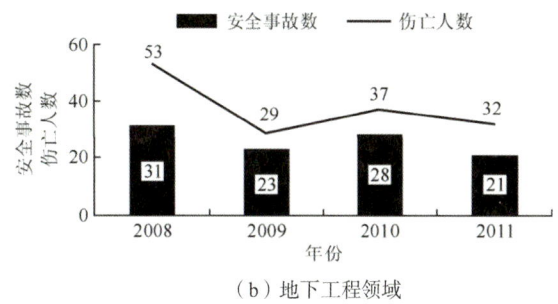
(b)地下工程领域

图1 2008~2011年不同领域安全事故数及伤亡人数
Fig.1 Number of incidents and casualties in different areas (during 20082011)

由表1可见,近年来,安全形势虽有所好转,但事故发生率仍居高不下。历年来,在各行业中我国土木工程领域事故伤亡人数仅低于道路交通事故和煤矿事故,居第三位,在全国重特大事故中,土木工程领域约占10%。

由图1可见,地下工程安全事故数及伤亡人数约占土木工程领域事故数及伤亡人数的10%。在土木工程领域中,由于房建工程数量巨大,地下工程事故数及伤亡数位居第二,图1表明,近4年来地下工程事故数及伤亡数没有明显的下降,总体上安全形势没有明显改善。

2.2 地下工程建设安全事故分析

在地下工程事故案例分析中,共收集50多起矿山法工程事故案例,20多起浅埋暗挖法工程事故案例,40多起盾构与TBM施工事故案例和100多起明挖法基坑及地铁工程事故案例。

2.2.1 事故类型分析

在上述地下工程事故案例中,按各类施工方法区分,其事故类型所占比例如图2所示。

2.2.2 事故原因分析

在地下工程事故直接原因分析中,约1/4为客观原因(22%),1/3多(37%)是主观原因,另外1/3多既有客观原因,又有主观原因(40%)。所有主观原因均可归为责任事故和人为因素,而且所有责任事故都高度近似:政府监管部门监管责任缺失,以罚代管、以包代管;施工方安全管理人员未能履行好安全管理职责,施工指挥人员违规指挥,施工人员违规操作,施工方案不完善;监理不严,责任心不强;业主方招标时违规发包、转包、抢工期;设计方案存在缺陷,勘察资料不全等。通过工程事故的直接主观原因,分析其深层次的原因,主要问题为:

(1)赶工期。这个问题在地下工程建设中相当突出,政府、业主为了献礼或追求政绩,不尊重科学施工规律和相关法规,任意规定工期,造成施工单位忽视安全;任意删减施工中的安全步骤和措施,造成设计方勘察资料不全;边设计边施工,设计方案不完善,酿成重大责任事故。

(2)工程造价低,致使合同价格低,造成安全措施投入的缺少,工程设计和施工方案缺乏严格审查。

(3)用工制度造成的人员技术水平低,安全意识差,无证上岗人员多。这是由于我国基础建设的用工制度中,除管理人员外,现场施工人员很多为农民工为主的流动工人,缺乏培训导致安全意识低,甚至野蛮施工。虽然《安全生产法》要求重要岗位工人有证上岗,但多达36%的案例中存在施工人员无证上岗的现象,12.5%的案例中存在监理人员无证上岗现象。

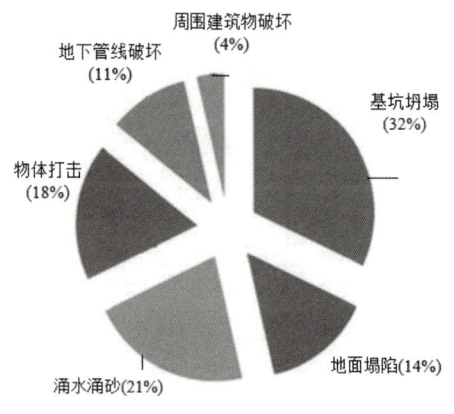
(a)明挖法

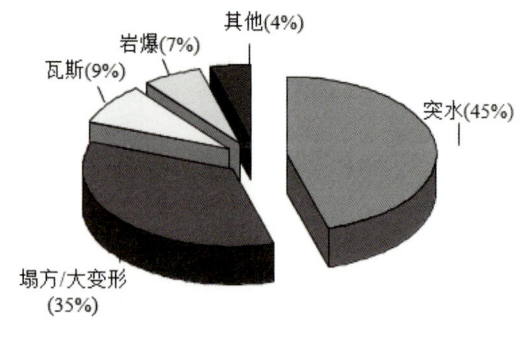
(b)矿山法

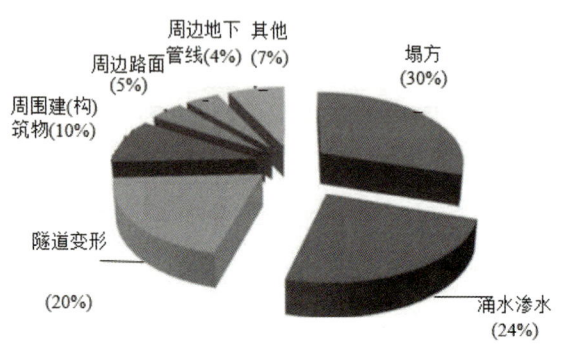
(c)浅埋暗挖法

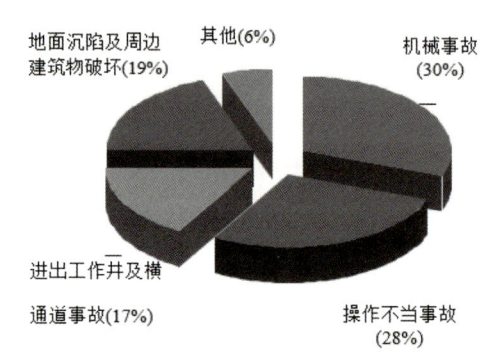
(d)盾构法

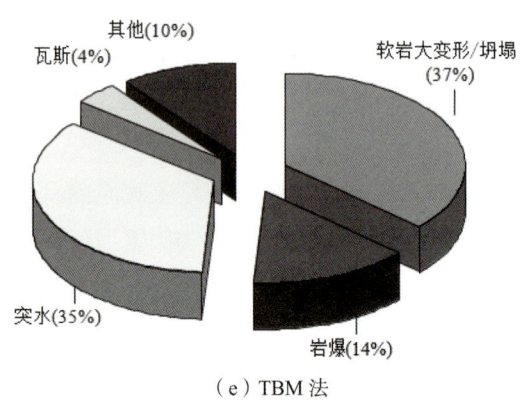
(e)TBM法

图2　不同方法下地下工程安全事故类型及所占比例
Fig.2　Safety accident type and proportion in underground construction by different methods

（4）业主方在工程招投标过程中存在违规现象，违规发包，层层转包，以包代管，以罚代管。招标条件把关不严，将工程发包给不符合相应资质条件的单位，这其中不难发现腐败情节。

上述责任事故及主观原因构成对地下工程建设安全管理的巨大挑战。造成绝大多数事故（坍塌、涌水涌砂、岩爆等）的客观原因是环境和气象，而工程地质环境、水文地质环境复杂是其第一主因，因而查明复杂地质环境并及时防治是地下工程建设面临的又一巨大挑战。特别是突水突泥及岩爆的防治难度最大，挑战最强。

3 地下工程安全管理所面临的挑战和对策

3.1 中国地下工程安全管理所面临的挑战

地下工程建设安全的严峻形势首先是由中国地下工程建设规模大、发展快的客观形势所决定的。中国是目前世界上地下空间开发利用的大国、隧道建设的大国。我国耕地资源已临近18亿亩红线、城市建设用地资源的紧缺、城市交通拥堵的日益突出，促使我国城市地下空间开发利用规模日益广泛拓展，城市轨道交通建设突飞猛进，城市轨道交通建设的速度居世界首位，城市地下快速路建设方兴未艾。以上是造成地下工程建设安全形势严峻的客观原因。由于规模大、发展快，地下工程技术和管理力量难以充分保证。体现在地下工程前期工作（地质勘察和工程可行性研究）不充分；工期偏紧，3~4a建成一条长20~30km的地铁，3~4a建成一条长达数公里的世界级高水压、大直径水下隧道；设计人员、施工管理人员青黄不接，许多年轻工程师甚至助工担任主力。而对于长施工周期、多施工项目、复杂施工技术和不良地质环境的地下工程中诸多不可预见因素的高风险特点认识不足，由上述建设体制、管理体制的深层次原因导致管理体制比较混乱、转包分包现象比较突出；工程招投标制度不规范，项目管理水平参差不齐，官僚主义疏于管理、以包代管、以罚代管经常发生；工程造价偏低造成安全措施投入不足，风险管理的投入不到位等等是地下工程建设安全形势严峻的主要原因，这一切是地下工程建设安全管理面临的严重挑战。

3.2 工程事故预防是可行的

工程安全科学告诉我们：

（1）事故发生发展是有过程的，发生前有预兆，预兆可以反映到监测数据中来。图3给出了工程事故发展示意图。事故萌芽至险情发生在地下工程中一般历时2~3d。

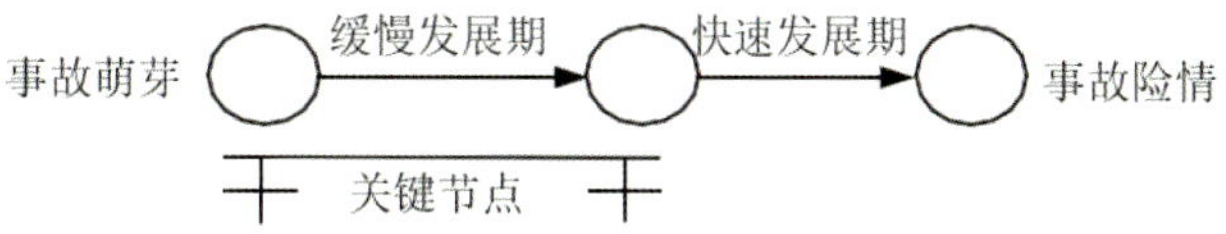

图3 工程事故发展示意图
Fig.3 Sketch of development of engineering accidents

（2）海恩安全金字塔法则

图4给出了海恩安全金字塔理论模型。海恩安全金字塔法则[2]是德国飞机涡轮机的发明者德国人帕布斯·海恩提出的一个在航空界关于飞行安全的法则。海恩法则指出，每1起严重事故的背后，必然有29起轻微事故和300起未遂先兆以及1000起事故隐患。

（3）海因里希安全法则（通过55万次事故的统计分析）指出，存在88∶10∶2的规律，即100起事故中，有88起纯属人为不安全状态引起的（即主观原因），有10起是物（环境）的不安全状态造成的，只有2起是所谓的"天灾"，是难以预防的。

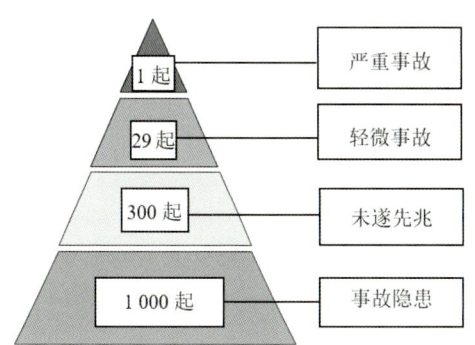

图4 海恩安全金字塔理论模型
Fig.4 Model of Hain's safety pyramid theory

（4）图5给出了约翰逊的变化失误模型示意图。该模型表明，事故酿成是领导者、计划者、监督者和操作者一系列的变化失误链引起的，这些变化和失误往往具有关联性，工程事故的发生大多是由于在错误的时间、地点，采用了错误的技术或管理方法而导致的，并不是一个单独的错误事件，只有切断这个变化失误链条，事故才可以避免。

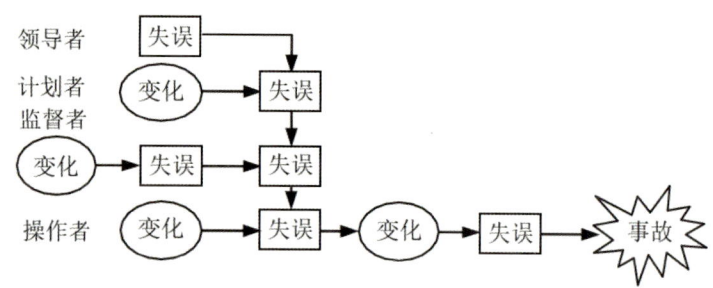

图5　变化失误模型示意图
Fig.5　Sketch of changeserrors model

上述法则、理论和模型揭示了一个十分重要的事故预防原理：事故的发生过程是一个由反应链过程导致的小概率事件，这个反应链上任一环节得到控制，就可以很好地遏制事故的发生。因此，事故是可以预防的，预防的重点就在于对工程中大量的人的不安全行为和物的不安全状态的有效控制。因此，必须不厌其烦地关注施工中的日常安全管理，特别要注重细节管理，才能很好地控制事故隐患，实现重大事故的预防。

3.3　地下工程建设安全管理应实施的对策

面对地下工程建设安全的挑战，根据工程建设安全科学及重特大工程事故原因的分析，其应实施的对策为：

（1）加强针对地下工程安全风险管理的法规建设

针对地下工程高风险工程的特性以及地下工程事故高发态势，需要政府有关部门（城市建设部、交通部、铁道部和水利部等）以及行业协会制订强有力的法规作指导和规范，通过法规明确政府监管部门、业主、监理、设计和施工等各有关单位的安全风险管理责任，建立全面科学的安全风险管理体系；通过法规强制实行重大工程立项前的安全风险分析、评估研究，凡安全风险评估为"严重"等级的工程方案应实行一票否决制，必须进行修改；强制实行重大地下工程的设计和施工方案审核制；通过法规规定在工程预算及定额中安全风险管理的相关费用，及其在整个地下工程建设费用中所占的合理比例，且为"专项提取"，不列入商务造价，并加强其审计和监督，确保这些费用切实落在安全风险管理工作上。

（2）实施地下工程安全风险管理目的重在预防提高地下工程风险意识，全面推广在一切地下工程领域实行地下工程安全风险管理制度，将其作为地下工程项目管理的必要组成部分。

安全风险管理主要包括风险因素辨识和分析、风险工程评估、风险监测和控制，内容涉及探明风险源、分析确定风险因素导致事故发生的概率和事故后果影响的大小、评估风险工程的等级以及建立监测系统，采取预防措施及预警应急方案以规避、转移和减缓风险等。

安全风险管理应贯穿地下工程建设的全过程，包括勘察阶段、方案设计阶段、初步设计阶段、施工设计阶段、施工阶段及工后阶段。

（3）建立地下工程安全监控中心（分中心）和第三方监测、监理制度

安全监控中心是地下工程安全风险管理的信息枢纽，负责组织工程各部分、各阶段的安全风险监控、

综合预警发布、信息管理和相关咨询工作,并进行监督管理。为保证风险管理工作的权威性和专业性,一方面明确安全监控中心是业主的职能部门,同时以合同方式委托专业咨询单位(或第三方监测单位)负责,安全监控中心的所有指令都以业主名义发出。

对于安全风险专业咨询单位和第三方监测、监理单位的资质和监测、监理人员的考核、审查,政府有关部门应建立相应的严格管理制度。

(4)建立基于现代化、信息化技术的地下工程建设安全风险管理信息系统

这种系统能克服目前工程管理中参建各方职责不清、信息不畅的弊端,利用信息技术将工程安全管理的有关信息及时地传递到有关各方,使有关各方在工程中的"操作"相对"透明",有利于提高效率和落实责任。

这种系统包含地理、地质信息系统,为风险管理提供地理、地质、工程和环境等信息支持;包含风险基础信息数据库,为风险管理提供风险工程信息,风险阈值,风险监测数据库(空域和时域),风险处置方案等;包含风险预警管理模块,为风险管理提供风险分析,预警发布,风险处置消警及历史预警查询等。

(5)加强地下工程安全风险管理理论以及重大事故预测预报和防治技术的研究

工程事故的预防主要从创新技术方法和改善管理方法两方面来进行。在管理上,要加强工程建设的安全风险管理,推广安全风险管理制度。在技术方法上,要综合利用现代数值模拟手段和地球物探方法,不断研究新的技术方法,对重大事故进行预测预报。

突水突泥地质灾害和岩爆地质灾害居地下工程事故客观原因中的第一主因,并有时成为不可预防的"天灾"。下面主要对这两类工程地质灾害的机制和防治理论进行阐述。

4 突水突泥地质灾害的挑战和对策

4.1 突水突泥地质灾害对地下工程建设安全的挑战

我国70%的国土面积为山区,且岩溶地区分布广泛,可溶岩分布面积约占国土总面积的1/3,其中以西南部云、贵、桂和川、鄂、湘部分地区的岩溶最为发育。因此在我国地形、地貌及地质条件复杂的地区,特别是西部地区修建地下工程,尤其是长大隧道的建设将会遇到很多的岩溶和不良地质问题。随着我国地下工程建设的大、长、深发展趋势,突水突泥地质灾害逐渐成为制约我国隧道建设发展的瓶颈问题。表2给出了我国的部分长大隧道突水突泥灾害详情[3]。

表2 我国的部分长大隧道突水突泥灾害详情[3]
Table 2 Details of water and mud outburst incidents in China[3]

隧道名称	地质灾害描述
大瑶山铁路隧道	施工期涌水量:4000~15000m³/d,平洞1994+213涌水造成竖井被淹、洞内机具被淹没达数月;正洞DK1994+600涌水淹没隧道200余米,水深1.4m,隧底淤积泥沙厚1m,中断施工长达1a之久;DK1994+636~637处发生涌泥涌沙80m³,淹没轨道,造成短时中断行车;地表斑古坳地区生产生活用水枯竭;地表坍塌约413次
圆梁山铁路隧道	施工期涌水量:110000m³/d(出口DK361+764);DK354+450~510溶洞发生涌水为9.6×10⁴~1.656×10⁵m³/d,伴随涌沙涌泥,淤积长度130m,高度2.5m,涌沙量约1300m³;DK354+879溶洞发生涌水,伴有涌泥涌沙(总量约6000m³),最大涌水量达7.2×10⁴m³/d,造成人员伤亡事故,被迫采用迂回导坑通过;DK360+873涌沙,淹没导洞近200m;DK361+764处发生涌水伴随涌沙涌泥,涌水量为240000m³,涌泥沙覆盖整个掌子面,淤积量约15000m3
歌乐山铁路隧道	DK2+619.6发生涌水,涌水量14400m³/d,涌水泥沙含量达20%~30%

隧道名称	地质灾害描述
宜万马鹿箐隧道	2006年1月位于湖北省恩施市屯堡乡的宜万铁路马鹿箐隧道发生涌水，死亡11人
宜万野三关隧道	2007年8月5日发生突水突泥事故，在30min内，突出15104m³水、泥、石等物，导致大量机械被冲出，并造成人员伤亡，治理时间至少半年，淹没正洞500余米，掌子面附近200m全被大块孤石充满

特别需要指出的，渝怀线铁路圆梁山隧道在毛家坝向斜处最大突水量为 $14.5 \times 10^4 m^3/d$，其中 3# 溶洞曾发生爆喷型涌泥，瞬间涌泥量高达 $4200m^3$，充满下导坑 244m 空间；广渝线华蓥山公路隧道施工中发生多次突水灾害，同时局部伴随大量涌泥，左洞左侧底板突水量高达 $14400m^3/d$，涌泥砂 $463m^3$；渝怀线铁路武隆隧道施工揭露 4 条特大暗河，最大突水量 $718 \times 10^4 m^3/d$。宜万线铁路野三关隧道穿越强岩溶地区，由于连续降雨，地表水与地下溶腔及断层水相通，其 I 线 DK124+602 掌子面在爆破后不久发生大型突水涌泥地质灾害（见图6），突水量约为 $5 \times 10^4 m^3/d$，并伴有大量泥沙涌出，突水部位的水量最终稳定在 $500m^3/h$ 左右，造成 10 人死亡；沪蓉线龙潭隧道左线曾发生大型突泥地质灾害，淹没隧道约 58m，突泥量高达 $1800m^3$，并伴有 $30m^3/d$ 稳定流量的水流，造成施工中断数月之久；宜万线铁路马鹿箐隧道平导施工时上覆溶腔水体压裂拱部围岩，发生特大突水灾害，最大突水量高达 $30 \times 10^4 m^3/h$，造成 11 人死亡；衡广线铁路大瑶山隧道忽略了槽谷地区岩溶现象及岩溶水与断层的连通性，当隧道掘进至 334m 时，发生突水淹井事故，最大突水量达到 $4 \times 10^3 m^3/d$，水中含沙量

（a）突水部位

（b）突水冲出的巨石

图6 宜万线野三关隧道突水 [4]
Fig.6 Water and mud outburst incidents in Yesanguan tunnel on Yichang—Wanzhou railway line[4]

高达 20%，致使水下 6 台高扬程水泵（200D ~ 65×8）因叶轮淤堵磨损而全部失效。后经开挖迂回平行导洞排水清淤才重新开工，由此造成工程停工达 1a 之久。

由表2可见，由于突水突泥等地质灾害造成极大的设备损失、严重的人员伤亡和工期延误。据不完全统计，在水利水电、铁路、公路、矿山的隧（巷）道施工过程中遇到的突水突泥等地质灾害所造成的停工时间约占总工期的 30%。

据相关资料统计，仅20世纪80年代，我国长岩溶隧道共计26座，约占隧道总数的40%，其中西南、中南地区17座，占岩溶隧道数量的65%左右，且在施工过程中均遭受不同程度岩溶水的侵害，其中10座发生了较大岩溶突水灾害。

进入21世纪以来，随着长大隧道等地下工程埋深的增大，施工条件的日趋复杂和生产机械化程度的提高，突水突泥带来的经济损失及人员伤亡呈增长趋势，因此，施工突水突泥灾害对重大地下工程建设安全的挑战相当严峻。

4.2 突水突泥地质灾害的发生是有科学规律可循的

地下工程施工中，不可避免地会破坏隐伏的含水构造，导致导水通道与开挖临空面相连通或处于准连通状态，进一步的扰动会诱发地下水或与导水通道有水力联系的其他水体（地表水、地下暗河以及溶腔水等）突然涌入开挖区，发生突水突泥灾害。因此突水突泥的发生是有条件的：含水构造的能量储存，含水的动水性能和能量释放以及围岩的稳定性。含水构造的能量储存是指含水构造中的岩溶水体由于高水头压力作用相对于开挖中的地下工程具有势能和对围岩具有应变能；岩溶水动力性能与能量释放是指岩溶水压与隔水层厚度的比值，它是突水能量释放与否的关键指标。当开挖过程中，一旦隔水层的防突厚度小于其临界值，能量瞬间释放，形成突水；含水构造围岩的稳定性是阻止突水发生的最后一道屏障，如果地下工程施工中，隔水层被直接挖除或开挖临空面的隔水层围岩，由于工程因素引起变形破坏导致防突厚度减小从而诱发突水。而对于充填型岩溶管道来说，总突水势能达到一定阈值时，充填介质将滑脱，发生渗透失稳突水。

岩溶隧道突水突泥实质上是地下水运移网络或储存条件受外界干扰发生的动力失稳现象，影响因素多且复杂，总体上可概括为工程因素和地质因素。工程因素比较复杂，诸如开挖、爆破、注浆以及其他人为活动，而地质因素则主要包括地下水、岩组、地质结构面、岩层产状、沉积期古剥蚀面、地貌和降雨等。

4.3 突水突泥发生一般有预兆，是可预测预报的

突水突泥地质灾害的前兆包括地质调查前兆、含水构造在地球物理勘探中的前兆以及临突水前的前兆。

4.3.1 突水突泥地质灾害的地质调查前兆特征

通过对岩溶隧道突涌水实例资料的系统收集和归类整理，分析了突涌水地质灾害的前兆信息及临灾特征，总结如下：

（1）在地下工程施工过程中，可通过观测钻孔出水情况或分析开挖揭露围岩的变化情况，获取一定的突水前兆信息，从而避免突水灾害。如：

①当采用超前钻孔探测时，钻进速率突然增大，钻孔开始出现喷水现象，喷距比较大，且呈浑浊状态或时喷时停者。

②在开挖过程中：掌子面附近围岩出现锈状，且开始出现环状滴水或渗水现象者；揭露岩体的岩性发生突变者，如弱可溶岩进入强可溶岩的边界部位；开挖过程中，发现黏土量增多，而涌水量有减小的趋势；或出现大量的剥落碎块，掌子面无涌水现象者；突然遇到断层破碎带、褶曲向斜、裂隙密集带或岩溶管道不良地质，且揭露围岩明显湿化者。

（2）在地下施工中，当接近不良地质或岩溶水时，可以发现一定的突水突泥灾害特征。

①断层破碎带前兆

岩层中伴有牵引褶曲、牵引褶皱、弧形节理组成的小型帚状构造或反倾节理裂隙的出现，节理组数急剧增加，可多达 6~12 组；开始有压碎岩、碎裂岩的出现，岩石强度明显降低；邻近富水断层下盘泥岩、页岩等隔水层明显湿化、软化，或伴有淋水现象和其他水流痕迹。

②临近大型溶洞水体或暗河前兆

钻孔中涌水量剧增，且夹有泥沙或小砾石，或有凉风冒出，洞中有较大的流水声；施工过程中小溶洞出现的频率增加且多有水流、河沙或水流痕迹，裂隙、溶隙间含有较多的铁染锈或夹黏土，岩层明显湿化、软化，或伴有淋水现象。

③岩溶管道前兆

非断层破碎带出现岩层内倾牵引现象或内倾小断层，附近岩层易发生疏松、变暗等氧化现象，且可使岩层潮湿或泥化。

④淤泥带的前兆

频繁出现铁染锈裂隙或有水、无水小溶洞；若淤泥带中含水较多，钻孔中的用水量剧增且浑浊，并夹有大量泥沙和棱角尖锐的小碎石。

⑤邻近人为坑洞积水的前兆

掌子面空气变冷或出现雾气，岩层明显湿化、软化，或伴有淋水现象以及嘶嘶的水声，岩层裂隙有涌水现象；若为煤层老窑积水，则岩层中有暗红色水锈的出现。

4.3.2 突水突泥地质灾害在地球物理勘探中的前兆特征

岩溶水体和不良地质体的存在是突水突泥地质灾害发生的前提，当地下工程开挖邻近它们时，其在电场、磁场和地震波等地球物理信息场上有一定的响应特征，即表现为特殊的前兆信号，通过其辨识，可发现这些前兆信息[5]：地质雷达波通过地下水、含水溶洞及富水介质等含水体后，高频成分被吸收，含水层的反射波的优势频率会降低；电磁波通过含水裂隙或富水破碎带时，其电导率远大于无水岩石，因此雷达波的反射强烈；雷达波遇含水层的反射波相位，相对于入射波，相位差180°。

4.3.3 突水突泥地质灾害在临突水前的前兆信息

模型试验和现场监测表明：

（1）当开挖面接近岩溶含水构造时，隔水层及围岩体经历了内部裂隙萌生、扩展、贯通的细观破裂过程和围岩渗水、局部涌水、整体失稳突水的宏观灾变过程。

（2）临突水前，位移表现为持续增长后的突升，应力表现为持续增长后的突降，渗压表现为持续降低后的突降，隔水层的声发射事件及其能量也在突水前均体现出一定的突升和激增现象。

4.4 应对突水突泥地质灾害的对策

（1）突水突泥地质灾害事故的绝大部分在实行综合超前地质预报与预警机制后是可以避免的

基于突水突泥地质灾害的发生是有科学规律可循的，是有前兆可预测预报的，为了预防和避免突水突泥灾害事故，可以或必须实施超前地质预测预报。由于地质条件的复杂性和物探结果的多解性，单一的预报方法难以保证结果的准确性，而且不同的预报方法所得的结果也有所差异，因此，为了提高预报的准确性和精确度，必须实施综合超前地质预报，从而扬长避短、相互补充、相互印证，多角度、多参数地对施工前方的地质情况进行预报：宏观超前地质预报（工程地质法）、长距离（50~200m）超前地质预报（工程地质法、TSP探测等）、短距离（50m以内）超前地质预报（地质雷达、红外探水、瞬变电磁、超前钻探、超前导洞及经验法等）。

根据综合地质超前预报报告制定相应的预警方案和处理措施。

（2）必须在岩溶地区实施地下工程安全风险管理制度

虽然突水突泥地质灾害是有发生前兆的，是可以预测预报的，绝大部分是可以避免的，但关键是要落实到工程管理中去，因此必须实施地下工程安全风险管理制度。

首先，在重大地下工程可研阶段实行工程安全风险管理立项研究，通过工程风险分析和风险评价决定工程是否批准立项，对于风险等级严重的工程实行一票否决，必须修改方案，特别是修改工程埋深和走向；其次，在立项后工程实施过程中，要通过地质分析确定所要预报工程各段的围岩情况进行风险等级评价和划分，对不同地质灾害风险等级的工程段实行不同的综合预报方案。

实践表明：重大工程区地质分析和不良地质宏观预报是综合超前地质预报的基础和前提，是工程高水平的施工地质灾害综合超前预报的不可或缺的第一道工序。只有在地质分析和宏观预报的指导下，才能准确、有效地实施下一步的地质超前预报、超前钻探、判断及临近警报等后续预报工作。

地下工程地质灾害监测、判断和临近警报技术是在地下工程所在地区不良地质宏观预报和岩溶水体和不良地质体长期和短期超前预报的基础上进行的，包括地质灾害的环境监测、地质灾害发生可能性和位置的判断，以实施地质灾害的临近警报。根据综合超前地质预报基础上风险评价的结果和信息化施工过程中揭示的信息，判断突水的危害程度、发展趋势和可能造成的危害程度。灾害预测预警级别可分为4级：红色（特别严重，I级）、橙色（严重，II级）、黄色（较重，III级）和蓝色（一般，IV级）。预警信息一经发布，必须严格按照预警级别及应急预案制定相应的工程预防措施并按措施进行施工。

5 岩爆地质灾害的挑战和对策

5.1 岩爆地质灾害对地下工程建设安全的挑战

图7给出了我国煤矿历年来出现岩爆矿井数量[6]，该图表明，发生岩爆灾害的煤矿井数量呈指数规律增长。

我国金属矿的开采深度普遍大于煤矿采深，金属矿山的岩爆灾害更为严重突出。如我国年产量超过1.5万吨的冬瓜山铜矿采深超过1000m，自1996年12月5日第一次发生岩爆以来，已记录到岩爆现象超过10余次，多次影响施工进度。表3总结了我国的深部交通及水电地下工程领域岩爆灾害情况[6]。

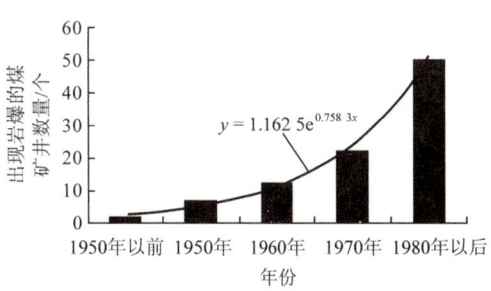

图7 出现岩爆的煤矿井数量[6]

Fig.7 Numbers of coal mine wells where rockburst happened[6]

由表3可见，随着该领域地下工程埋深增加，特别是西部大开发以来，岩爆灾害发生的频次和等级均明显增高。2000年以前，岩爆工程埋深大都为1600m，岩爆等级以轻微岩爆为主，少量为中等岩爆，强烈岩爆极少。2000年以后，岩爆工程最大埋深超2000m，甚达2525m，呈现岩爆等级向中级和强烈等级明显增加的趋势，而像锦屏二级水电站引水隧洞工程更多次发生历史上罕见的极强岩爆（见图8）；2009年11月28日施工排水洞发生极强烈岩爆，该隧洞洞径为7.2m，埋深接近2500m，TBM法开挖，28日0:54极强岩爆位于距掌子面7~20m范围内，爆坑深度达 8～9m，爆方总量达近千立方米，岩爆发生的大量粉尘在洞内弥漫达 10min，正作业的 TBM 设备被埋，主梁断裂，沿轴线约 28m 长原支护系统全部毁损，7 名工人遇难，1 人受伤，严重影响工程进度，采掘设备至今仍未挖出；2010 年 2 月 4 日下午钻爆法施工的 2# 引水隧洞发生极强岩爆，岩爆区段达 37m。岩爆时，停放在该部位的出渣装载车受强烈冲击弹起移位，装载铲车被弹起 3 次，岩爆除造成边坡、拱脚围岩弹出垮塌外，还造成台阶底板出现 3 条裂缝，其中一条横向贯穿隧洞，深约 1m，宽约 10cm。

表 3　我国发生岩爆矿区及隧洞工程汇总（不完全统计）[6]
Table 3　Summary of mine and tunnel projects where rockburst has occurred(incomplete statistics)[6]

工程名称	竣工年份	最大埋深/m	岩爆等级及比例/%			岩爆次数/次	岩爆段长度/m	备注
			轻微	中等	强烈及极强			
成昆铁路关村坝隧道	1966	1650	为主	少量	无			零星岩爆
二滩水电站左岸导流洞	1993	200	为主	少量	无	315		工程区位于深切河谷卸荷集中区域，最大主应力为 26MPa，方位角 N34°E，倾角 23°，因而以水平应力为主
岷江太平驿水电站引水隧洞	1993	600	为主	少量	少量	>400		
天生桥二级水电站引水隧洞	1996	800	70.0	29.5	0.5	30		比例依据岩爆次数统计
秦岭铁路隧道	1998	1615	59.3	34.3	6.4		1894	比例依据岩爆段长度统计
川藏公路二郎山隧道	2001	760	为主	少量	无	>200	1252	
重庆通渝隧道	2002	1050	91.0	7.8	1.2		655	比例依据岩爆段长度统计
重庆陆家岭隧道	2004	600	55.8	39.7	4.5	93		比例依据岩爆次数统计
瀑布沟水电站进厂交通洞	2005	420				183		工程区位于深切河谷卸荷高应力集中区内，地应力方向沿着河谷边坡向与隧洞呈大角度相交
秦岭终南山特长公路隧道	2007	1600	61.7	25.6	12.7		2664	比例依据岩爆段长度统计
锦屏二级水电站引水隧洞	2011	2525	44.9	46.3	8.8	>750	2011	比例依据岩爆次数统计出现数次极强岩爆
江边电站引水隧洞	2012	1678	46.4	50.4	3.2	>300	2012	比例依据岩爆次数统计

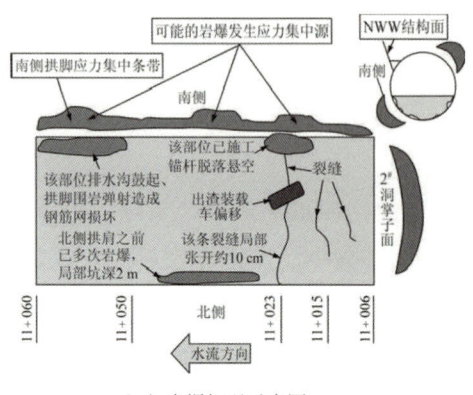

（a）岩爆概况示意图

（b）岩爆导致锚杆脱落

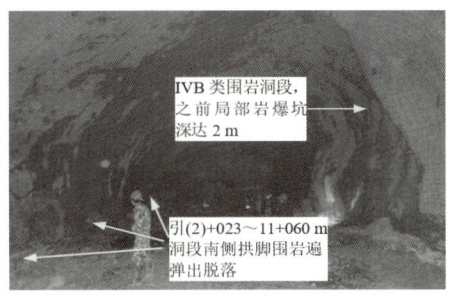
（c）岩爆导致的爆坑

（d）岩爆推翻的出渣装载车

（e）技术人员察看事故现场

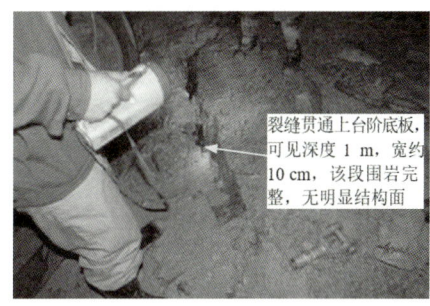
（f）台阶底板的巨大裂缝

图 8 锦屏二级水电站岩爆现场
Fig.8 Accident sites of rockburst in Jinping II hydropower station

由上所述可知，对于深部地下工程，岩爆已成为造成人员伤亡、设备毁损、影响工程施工的突出地质灾害。

5.2 岩爆是可以预测的

岩爆地质灾害发生的主要机制已大致清晰，在一定程度上可以对岩爆进行安全风险分析意义上的预测[7]。

世界上众多岩石力学研究工作者，包括一些知名权威，都认为岩爆的力学机制是一个十分复杂的挑战性的难题。面对其复杂性和难度，我国有的研究专家认为岩爆不能预测预报，只能预警。笔者认为：从目前岩爆研究进展的成果看，其发生的主要（虽非全部）机制已逐渐清晰，因而可以对其进行大致的（虽非准确的）预测；我国已经进行的对岩爆监测预报的大量工程实践[8]表明，对岩爆的发生部位与强度进行大致的预报是可行的；缺乏对岩爆进行预测预报的预警将是盲目的，因而也是事倍功半的。

5.2.1 岩爆的定义和分类

从岩爆的防治角度出发，将岩爆定义为高地应力地区由于地下工程开挖卸荷引起的围岩弹射性破裂的现象，而将无动力弹射现象的围岩脆性破坏归之于围岩静力稳定性丧失现象。从其发生机制出发，基于围岩中有无结构面，可将岩爆类型分为应变型岩爆、地质构造型岩爆以及应变与构造杂交型岩爆。

这样的定义和分类将有利于对岩爆进行预测预报和采取预警措施和工程应急方案。

5.2.2 均质岩体应变型岩爆的预测原理

对于围岩中只存在微缺陷，不存在断层、裂隙、节理等宏观裂纹的围岩称之为均质岩体，在这类岩体中发生的岩爆，一般为应变型岩爆。应变型岩爆的理论机制概括为：工程所在地区的初始高地应力在开挖卸荷引起的应力集中后,使得围岩中的局部地应力水平超过岩体极限强度从而造成围岩的局部破坏,而当破坏岩体相应其围岩高应力水平的围岩弹性应变能超过围岩破坏所耗散的破坏能时,其剩余应变能因围岩的应力梯度转化成破坏岩体的动能,从而破坏岩体向临空面弹射形成岩爆。

现从发生岩爆的冬瓜山铜矿的粉砂岩峰值前加、卸载曲线和荷载变形(见图9[9])全面来说明其预测原理。

由图9可见,当岩石加载至峰值强度后,外载做功 A_1 所产生的能量可分为弹性变形能 Φ_{SP} 和塑性变形能 Φ_{ST} 两部分,前者为岩石储存的变形能 Φ_{SP},后者则在加载过程中以热能等形式耗散掉;而峰值强度后下降段曲线所围的面积则为岩石破坏所耗散的能量 A_2。

对于发生岩爆的岩石来说,其动能即为剩余能量 $\Delta = \Phi_{SP} - A_2$,这只有在 A_2 较小,即 Δ 为正时才可能发生岩爆,即峰值后下降段曲线较陡时才有可能发生岩爆,也即发生脆性破坏时才有可能发生岩爆,所以发生岩爆的岩石一般为弹脆性或弹塑脆性岩石就是这个原因。

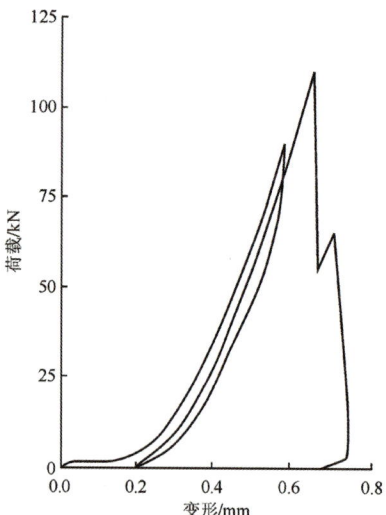

图9 粉砂岩的峰值前加载曲线和荷载变形全图 [9]

Fig.9 Complete curve of cyclic loading and deformation before peak strength of siltstone[9]

因 $\Phi_{SP} = A_1 - \Phi_{ST}$,所以有

$$\Delta = A - \Phi - A \tag{1}$$

式(1)表明,岩爆预测只有在具备下述4个条件下才是可能实现的:

(1)获得相应围岩岩石的加卸载过程的全应力 应变曲线;

(2)计算出高地应力地区洞室开挖卸荷应力重分布后,非连续非协调变形岩体中的应力(应变)场;

(3)计算出相应峰值强度前塑性变形能的微缺陷(裂纹)扩展以及次生裂纹稳定扩展所消耗的能量;

(4)计算出相应于峰值强度下降段曲线岩石破坏能的次生裂纹失稳扩展合并成宏观裂纹的耗散能。

上述4个条件在建立了基于断裂力学和非连续非协调变形介质解析计算模型,或能计算裂纹发生发展的扩散有限元数值计算模型的计算方法体系下是可能的,根据该模型可预测围岩中可能发生岩爆的危险区的部位、面积以及破坏岩块的弹射速度(即岩爆等级)。

5.2.3 围岩中邻近洞室存在单独结构面时应变型岩爆的预测

基于工程实践,当邻近洞室的围岩节理裂隙发育到一定程度,地应力得到松弛,一般不易发生岩爆。但当邻近洞室仅存在单独结构面(裂隙)时,则另当别论,此时构造地应力松弛不充分,而裂隙位于开挖引起的应力重分布的应力集中区,导致原生裂隙滑动、启动出现次生裂隙。当裂隙扩展到一定长度后,就会与洞壁发生相互作用,产生失稳扩展,形成破裂区,对于存在原生裂隙区的脆性岩石,其裂隙扩展、破裂消耗的能量更小,所以与无宏观裂隙的围岩相比,含单条裂隙的围岩洞室更易发生岩爆,岩爆的面积更大,岩爆的强度(弹性速度)更高,这已为预测计算所证实。

5.3 岩爆监测预报在理论上和工程实践上是可行的

由上述可知,岩爆是高地应力地区因开挖卸荷引起的围岩动力破裂现象,由此可以推断(工程实践也表明),岩爆发生前一定是有前兆的,这给岩爆的监测预报提供了可能。如同一切岩体的宏观破坏一样，岩爆的宏观破裂是由大量的微破裂产生、合并、集群导致的失稳破裂发展而成，这就是岩爆的前兆；这些微破裂发生在工程开挖引起的卸荷范围内，因此可以被微震监测器在其有效作用范围内所接收，微破裂产生、合并、集群导致的失稳破裂发展的规律，即是岩爆宏观破坏的规律，也是岩爆监测预报可行性的理论基础。

微震监测系统通过地震检波器或加速度传感器将微破裂产生的 P 波和 S 波接收转化成电信号并转换成数据信号，借助数据处理软件，确定岩体中微震事件的时间、位置、强度，最终依据微破裂稳定或失稳破裂发展的趋势，分别对岩体是否发生岩爆以及岩爆的定位和量级做出定性和定量的评价，也即做出预报。微震监测系统监测范围广，具有监测自动化、信息化、智能化和远程化的特点。微震监测系统的监测范围可以覆盖地下工程开挖卸荷引起的岩爆发生范围。

岩爆监测预报的工程实践（锦屏二级水电站强烈岩爆区域）表明，岩爆监测预报的准确率是相当高的：中国科学院武汉岩土力学研究所的准确率为 88.36%，大连理工大学的准确率为 85.5%。

5.4 对于岩爆地质灾害的对策

综合以上岩爆预测预报的理论研究和工程实践成果，应对岩爆地质灾害的对策归纳如下：

（1）在高地应力地区建设重大地下工程时，必须在原位实测与反演分析的基础上，掌握工程埋深区域的实际初始地应力的分布资料；在室内试验的基础上，获得工程开挖区域岩石的加卸载全过程的应力应变曲线。

（2）进行工程开挖前结构面（断层、裂隙等）的超前地质预报。

（3）进行工程各开挖段的岩爆预测数值研究分析。

（4）选择有经验的科技队伍，进行连续有效的微震监测；为了保证监测的连续有效，必须对现场监测设备及线缆进行有效的保护。

（5）在综合对比分析超前地质预报的结构面信息、微震监测信息、岩爆预测数据的基础上，提出岩爆监测风险预报报告。

（6）建立施工各方领导参加的对岩爆风险预报报告的会商制度（每天、每3d、每周）提出相应岩爆风险等级的防治策略及措施；降低开挖进尺的施工调整方案；增设超前钻孔爆破的应力释放孔；加强抗爆支护措施（增加喷层厚度、系统锚杆密度及长度、加大锚杆垫片的厚度及尺寸，安装等阻大变形抗爆锚杆等）。

6 讨论与展望

重大地下工程的安全建设与风险管理是当前国际岩土力学与工程界最为关注的前沿问题，也是我国未来需要重点研究的关键问题。在我国地下工程建设安全形势没有得到明显改善的情形下，工程建设中安全管理的水平和工程科技水平还有很大的改进空间，针对于地下工程建设的技术方法研究和管理手段研究任重道远。

在管理手段上，城市轨道交通工程建设中已进行了安全风险管理制度的探索，安全风险管理的法规体系建设和制度建设得到了完善，建立了相应的组织管理体系、技术管理体系、管理流程体系、第三方监控体系、预测预警制度及监控预警模型，制订了安全风险管理现场施工指南，开发了信息化的远程监控平台。通过强有效的安全风险管理，重特大事故的发生得到了有效遏制，因此，建议全面推广和加强地下工程安全风险管理制度，加强地下工程全生命周期工作的协调和统筹，加强地质勘查、地应力测试和岩体的非确定性和非均质性研究，加强重大事故预测预报和防治的基础理论、新技术和新材料的研究，建立基于现代化、信息化技术的地下工程安全风险管理信息系统。

在技术方法的创新与发展上，我国经过近些年的努力，取得了一些突出成果，主要包括：发展了有限元极限分析方法技术并在隧道稳定性分析中得到成功应用；开发了光纤光栅传感技术、安全状态实时感知技术、施工便携式智能预警终端；研究了岩爆孕育过程的现场和室内试验方法、特征、规律、机制、分析、预警方法和基于微震信息动态演化规律的动态调控方法；研发了有效控制岩爆的大变形锚杆；建立了陆地声纳、水空间识别、瞬变电磁和复合式激发激化技术为一体的含水构造超前预报技术和系统；探索了水注浆新材料、集地质预报和科学服务为一体的信息化综合治理的决策系统。

为了更好地解决当前和今后重大地下工程建设中的安全问题，建议还应加强以下 3 个方面的研究：

（1）地下工程安全设计理论研究

加强深埋长大隧道动态设计方法和掘进支护技术研究，加强岩石力学建模和岩石工程设计的研究；加强地下工程新方法、新技术、新材料和新工艺的研究；加强地下工程抗震和火灾预防研究；加强工程机械和设备的标准化和产业化。

（2）风险评估理论研究

要研究和推动涵盖地下工程的各个阶段的全生命周期的风险管理理论的研究，研究基于可靠性的风险管理方法和风险量化方法，研究风险评估和决策的新方法和新技术。

（3）岩爆、突水突泥的机制、预防和控制

实施基于实时微震监测的动态控制理论研究；改进前兆规律不明岩爆的预警和动态控制方法；加强微震监测的精确定位研究；加强先进的物理探测，数值分析和微震监测预警和地下工程灾害预测的综合方法的研究与利用；加强快速，准确地应力测量和地球物理勘探新方法和新技术的研究；加强新的数值模拟方法研究，尤其是基于非连续介质力学的数值模拟方法研究；尽快推动建立地下工程灾害数据库。

参考文献（References）：

[1] 中华人民共和国住房和城乡建设部. 2011 年房屋市政工程生产安全事故情况通报 [OL]. 20120227[20120523]. http://www.mohurd.gov.cn/zxydt/201203/t20120314_209111.html. (Ministry of Housing and Urbanrural Development of the People's Republic ofChina. 2011 housing municipal engineering production safety accident report[OL]. 20120227[20120523]. http://www.mohurd.gov.cn/zxydt/201203/t20120314_209111.html. (in Chinese))

[2] 管理科学著名定律连载. NO.39 海恩法则 [J]. 施工企业管理, 2010, (3): 70. (Wellknown Law of Management Science. 39Hain's safety pyramid theory[J]. Construction Enterprise Management, 2010, (3): 70. (in Chinese))

[3] 湖北沪蓉西高速公路建设指挥部. 高风险岩溶隧道地质灾害控制关键技术研究 [R]. 武汉：湖北沪蓉西高速公路建设指挥部, 2009. (Hubei HuRongXi Expressway Construction Headquarters. The research of geological disaster control in high risk karst tunnel[R]. Wuhan:Hubei HuRongXi Expressway Construction Headquarters, 2009. (in Chinese))

[4] 湖北省三峡翻坝高速公路建设指挥部，山东大学. 高风险岩溶隧道不良地质预报与灾害控制研究[R]. 济南：山东大学，2011. (Hubei Three Gorges Fanba Expressway Construction Headquarters, Shandong University. Adverse geological prediction and disaster control in high risk karst tunnel[R]. Jinan:Shandong University, 2011. (in Chinese))

[5] 刘斌，李术才，李树忱，等. 复信号分析技术在地质雷达预报岩溶裂隙水中的应用研究[J]. 岩土力学，2009, 30(7): 2 1912 196. (LIU Bin, LI Shucai, LI Shuchen, et al. Study of application of complex signal analysis to predicting karstfractured ground water with GPR[J]. Rock and Soil Mechanics, 2009, 30(7): 2 1912 196. (in Chinese))

[6] 张晓春. 煤矿岩爆发生机制研究[博士学位论文][D]. 武汉：华中理工大学，1999. (ZHANG Xiaochun. A study on mechanism of rockburst in coal mines[Ph. D. Thesis][D]. Wuhan:Huazhong University of Science and Technology, 1999. (in Chinese))

[7] 王元汉，李卧东，李启光，等. 岩爆预测的"模糊数学"综合评判方法[J]. 岩石力学与工程学报，1998, 17(5): 493501. (WANG Yuanhan, LI Wodong, LI Qiguang, et al. Comprehensive method for estimating rockburst with fuzzy mathematics[J]. Chinese Journal of Rock Mechanics and Engineering, 1998, 17(5): 493501. (in Chinese))

[8] 蔡美峰，王金安，王双红，等. 玲珑金矿深部开采岩体能量分析与岩爆综合预测[J]. 岩石力学与工程学报，2001, 20(1): 3842. (CAI Meifeng, WANG Jin'an, WANG Shuanghong, et al. Analysis of energy distribution and prediction of rockburst during deep mining excavationin Linglong gold mine[J]. Chinese Journal of Rock Mechanics and Engineering, 2001, 20(1): 3842. (in Chinese))

[9] 唐礼忠，王文星. 一种新的岩爆倾向性指标[J]. 岩石力学与工程学报，2002, 21(6): 874878. (TANG Lizhong, WANG Wenxing. A new rock burst proneness index[J]. Chinese Journal of Rock Mechanics and Engineering, 2002, 21(6): 874878. (in Chinese))

一种岩溶地质条件下的城市地铁超前预报方法

苏茂鑫，钱七虎，李术才，薛翊国，张庆松，邱道宏，聂利超

(山东大学 岩土与结构工程研究中心，山东 济南 250061)

摘要：与普通公路或铁路隧道相比，城市地铁一般从城市道路下方通过，因此，城市地铁超前预报(主要指物探类预报)的实施可分洞内和洞外两部分。但由于地铁埋深较浅，干扰因素众多，需要开展综合超前预报。通过分析岩溶地质条件下的城市地铁超前预报特点，利用地质雷达和高密度电法各自优势和互补特点，提出洞内开展地质雷达超前预报和洞外开展高密度电法与地质雷达相结合的综合预报方法。辅以工程应用实例，详细分析该综合预报方法的实施与综合解释过程，证明该综合预报方法对岩溶地质条件下的城市地铁超前预报的有效性。

关键词：工程地质；城市地铁；综合超前预报；地质雷达；高密度电法；岩溶地质

中图分类号：P 642   **文献标识码**：A   **文章编号**：1000 – 6915(2011)07 – 1428 – 07

A METHOD OF METRO ADVANCED GEOLOGICAL PREDICTION UNDER KARST GEOLOGY CONDICTION

SU Maoxin，QIAN Qihu，LI Shucai，XUE Yiguo，ZHANG Qingsong，QIU Daohong，NIE Lichao

(*Geotechnical and Structural Engineering Research Center，Shandong University，Jinan，Shandong 250061，China*)

Abstract：Compared to the normal road or railway tunnels，metro passed through the stratum over the road，so that the metro advanced prediction(mainly geophysical class prediction) implementation can be divided into two parts：prediction inside and outside the tunnel. However，due to shallow depth of the metro and many factors of interference，it′s need to use the comprehensive advanced prediction. By analyzing the metro characters under the karst geological conditions，using the advantages of ground penetrating radar(GPR) and resistivity tomography，a comprehensive prediction method is proposed to carry out GPR forecasts inside and carry out the forecasts of the combination of resistivity tomography and GPR method outside；then through an engineering application，a detailed analysis of the implementation of comprehensive and comprehensive forecasting methods is made to prove that the comprehensive prediction method is effective for metro advanced prediction under the conditions of karst geology.

Key words：engineering geology；metro；comprehensive advanced prediction；ground penetrating radar；high-density resistivity method；karst geology

1 引 言

近十年来，为了加快西部大开发，国家在西部交通基础设施上投入巨大，隧道建设获得了蓬勃发展，隧道超前预报技术(这里主要指物探类的超前预报，以下类同)也得到了相应的重视、发展和提高。当前，随着城市建设的加快发展，我国需要进行地

收稿日期：2011-02-08；修回日期：2011-05-4。
基金项目：国家重点基础研究发展规范(973)项目(2010CB732002)；国家自然科学基金资助项目(50908134，40902084)。
本文原载于《岩石力学与工程学报》(2011 年第 30 卷第 7 期)。

铁建设的城市将越来越多，而地铁在修建的过程往往会遇到许多不良地质情况，严重者会出现突涌水等灾害故事。为了给地铁建设多提供一份决策依据和安全保障，针对城市地铁超前预报方法的研究将越显重要，也很有实际意义。

国内外有关隧道超前预报研究[1-8]的资料较多，但仔细分析这些资料，发现其研究对象绝大多数是针对普通的公路或铁路隧道，而针对城市地铁的却非常少，公开发表的几篇有关城市地铁超前预报方法的文章[9-11]，主要是分析某种物探方法在城市地铁工程中的应用效果，而没有把重点放在如何进行城市地铁超前预报的论述上。

与普通公路或铁路隧道的超前预报相比，由于城市地铁线路很大一部分是从城市道路下方穿过，且埋深较浅(一般为 5～30 m)，而道路上方往往管线密布，车辆来往频繁，各种干扰众多，所以，在进行城市地铁超前预报时，预报方法的选择和实施方式与普通公路或铁路隧道超前预报方法将有所不同。此外，地质条件也是决定城市地铁超前预报方法选择和实施一个重要因素，因此，本文拟对岩溶地质条件下的城市地铁超前预报方法进行探讨和研究，以此抛砖引玉，也为当前和日后的城市地铁超前预报提供参考或借鉴。

2 城市地铁综合超前预报方法

由于每种预报方法都有其适用条件，各有优缺点，加之物探解释具有多解性，因此，开展综合预报尤为必要。综合超前预报[2, 12]是指采用多种预报手段，取长补短，相互补充，相互验证，通过综合解释方法达到预报精度和准确度最大化的过程。

城市地铁超前预报同样需要用到综合预报方法，而综合预报方法涉及到一个预报方法选择与实施的问题。众所周知，普通公路或铁路隧道一般为穿山隧道，埋深较大，其超前预报往往只在洞内开展，而城市地铁上方往往为路面，超前预报除了可以在洞内开展，洞外也可以开展相应的地面探测。因此，城市地铁超前预报方法的选择和实施就分为洞内和洞外两部分。以下将阐述岩溶地质条件下的城市地铁超前预报方法的选择与实施。

2.1 城市地铁超前预报方法的选择

由于城市地铁隧道埋深较浅，不管是洞内还是洞外，应尽量避免选择地震波或弹性波类的探测方法，如 TSP 超前预报、浅层地震勘探等，因为此类方法接收的是震动反馈回来的信号，而路面上来回车辆或者其他施工活动所造成的震动是一个比较大的干扰源，直接影响到此类方法的探测结果。此外，洞外各种电磁干扰较严重，在地面开展探测工作也应尽量不选择无电磁屏蔽处理或电磁屏蔽效果较差的电磁类方法。由于地质雷达与高密度电阻率法是电法勘探中最为成熟的 2 种方法，其中地质雷达精度高，对 20 m 探距范围内的裂隙(包括含水裂隙)、溶洞或破碎岩体探测效果较好。而高密度电阻率法探测深度大，效果好，可以有效地圈出不良地质体的分布范围，也可以为地质雷达的有关水体的解释提供补充验证。因此，对于岩溶地质条件下的城市地铁超前预报方法，提出了洞内采用地质雷达方法，洞外采用高密度电阻率法与地质雷达相结合的方法。

2.2 城市地铁超前预报方法的实施

城市地铁超前预报方法的实施主要分为数据采集和数据解释两大部分。对于数据采集，除了设置好相应的仪器参数，关键要把测线布置好。洞内的测线布置与普通公路或铁路隧道的类似，根据围岩的岩性和构造情况布置好相应测线，这里建议掌子面上的测线布置成井字型，如果条件受限，也应尽可能布置成十字型。洞外的地面高密度电法测线不易在道路中间布置，建议沿着地铁走向，在道路两边各布置一条测线，通过这 2 条测线的探测结果，可以确定是否存在横切地铁隧洞的较大断层、溶沟或溶槽等。对于与地铁走向平行的地质构造，由于地质雷达操作比较灵活，可以在地面上间隔地布置垂直于地铁走向的地质雷达测线进行确定。这里需要说明的是，虽然城市地铁路线设计前一般都会有相应地质勘查工作，但该类地质勘查精度比较粗，很容易漏掉局部一些地质构造，因此，在地面上布置 2 条高密度电阻率法测线对于进一步的详细勘查就显得非常重要。

至于数据解释部分，一定要在相应地质资料分析的基础上，采用洞内与洞外相结合的办法开展综合物探解释。

通过以上分析，城市地铁综合超前预报方法的选择和实施过程可以用图 1 来表示。总体而言，只要方法选择得当，在地质资料分析基础上，通过洞内洞外探测相结合，可大幅度提高地铁超前预报的精确和准确度。

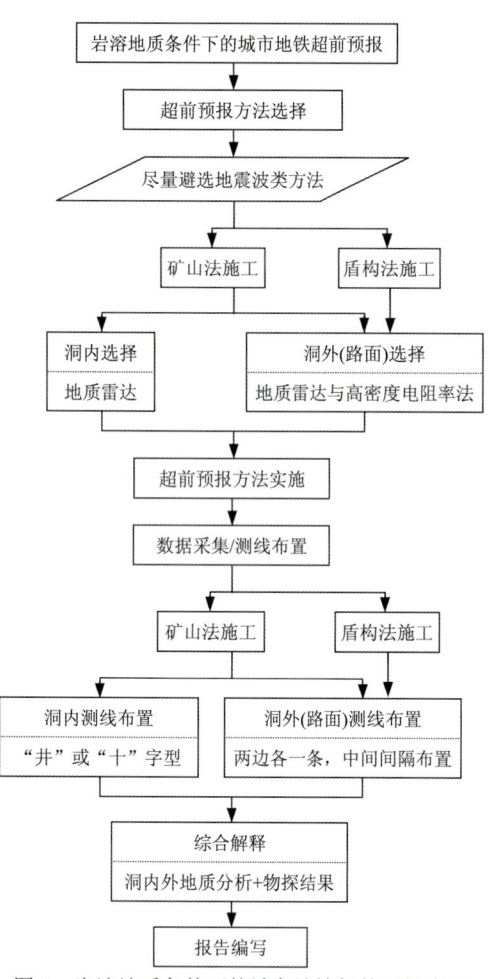

图 1 岩溶地质条件下的城市地铁超前预报流程

Fig.1 Flow chart of metro advanced prediction under karst geology

约 6 m，拱顶埋深约 9 m，采用矿山法施工，针对区间的地质情况，对洞内的超前预报采用地质雷达法，而对洞外(路面)探测则采用高密度电阻率法与地质雷达法相结合的方法，测线布置见图 2，3。

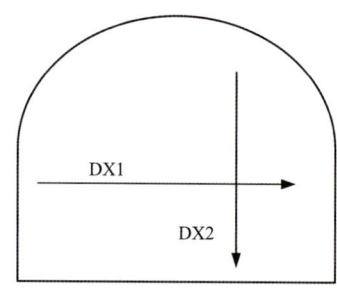

图 2 大连地铁掌子面测线布置

Fig.2 Layout of tunnel face survey lines in Dalian metro

3 城市地铁综合超前预报实例分析

以下将以大连地铁一区间正线的超前预报和广州地铁一车站的超前预报为例来说明岩溶地质条件下的城市地铁超前预报方法。

3.1 大连地铁一区间超前预报实例

探测区间位于城市中心，沿线建筑物众多，管线密集，车流人流非常大。区间范围内上覆第四系人工堆积层(Q_{ml}^4)、第四系中更新统残积层(Q_{el}^2)、第四系中更新统冰碛层(Q_{gl}^2)、下伏震旦系南关岭组泥灰岩层(Z_{hn}^w)、五行山群甘井子组白云质灰岩(Z_{hg}^w)、中生代燕山期辉绿岩($β\mu$)。区间灰岩地段岩溶发育，为了进一步查明区间是否存在较大断层、溶槽、溶洞和破碎带等不良地质体，拟对该区间进行相应的超前预报或探测。

3.1.1 预报方法和测线布置

区间隧道为双洞结构，高约 7 m，每个隧道宽

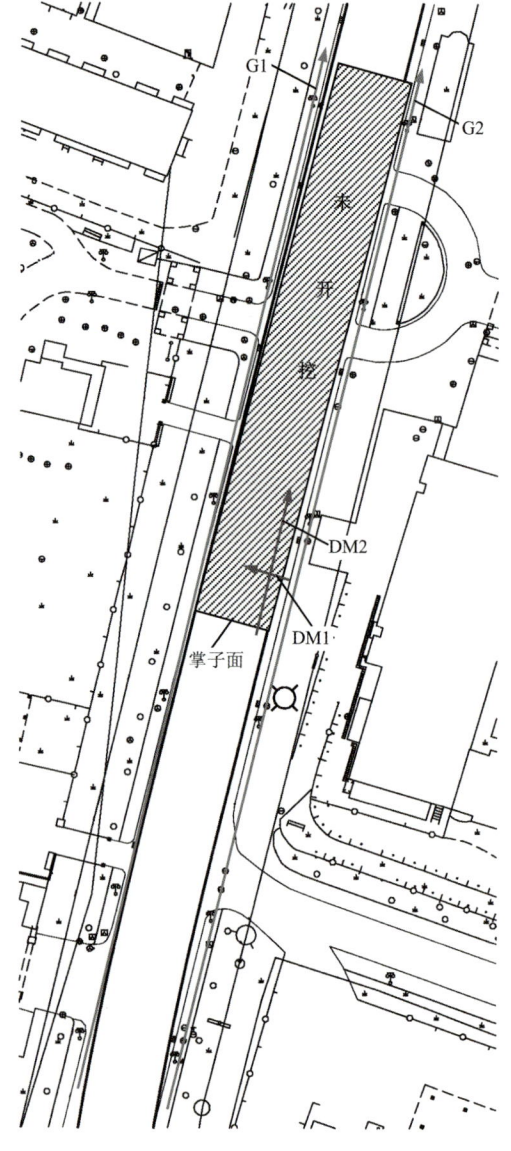

图 3 大连地铁地面测线布置图

Fig.3 Layout of ground survey lines in Dalian metro

洞内洞外的超前探测具体实施过程为：首先在路面上进行高密度电阻率法探测，由于地铁隧道刚好从道路正下方穿过，因此，利用布置在道路边的 2 条测线首先探明区间是否存在横切隧道的比较大的断层、溶沟或溶槽等。待高密度电阻率法探测完后，依据其探测结果，再进行地质雷达的探测。如果高密度电阻率法探测存在较大的横切隧道的断层、溶沟或溶槽等，那么这些不良地质体附近很可能会伴生有岩溶裂隙发育现象，因此，当工作面开挖到其附近处，应采用地质雷达加密测线扫描，这时的雷达测线除了垂直于隧道走向，还需要布置沿着隧道走向的测线，通过交叉网格式的测线布置，可较全面地了解工作面前方的地质情况。

这里需要说明的是，如果高密度电法探测未发现较大的断层、溶沟或溶槽等，针对矿山法施工特点，地质雷达预报只需在洞内发现工作面前方有较大的不良地质体时，才在路面进行加密测线扫描。

3.1.2 干扰因素及其排除或压制方法

在城市环境条件下，高密度电阻率法探测主要受到地下各种管线的干扰影响，特别是金属管道和电力线对高密度电阻率法探测影响较大。对于这种影响，可以在数据采集过程给予一定的排除，如尽可能让电极远离地下管线，如果不能远离，要确保电极不能与金属管线直接接触。此外，在数据处理过程中，通过脊回归和圆滑约束反演等方法可以对该类干扰进行压制。

而对于地质雷达探测，由于其发射和接收的是电磁波信号，受到的干扰更多，不仅受到地下管线的干扰，且受到地面以上的高压线、路灯、来回车辆和道路两边建筑物等干扰影响。在洞内还会受到钢拱架、施工车辆等干扰。这些干扰因素的排除或压制同样可以采取类似于高密度电阻率法的思路，在数据采集过程中尽可能地规避相应的干扰，如不能避免，应记录好相应干扰源的空间分布情况，这对于后面的数据处理和解译非常重要。在数据处理过程中，除了采用常规的滤波、叠加和增益处理外，利用绕射叠加和偏移成像技术可以对高压线、路灯、两边建筑物和洞内钢拱架干扰进行较好的压制。需要指出的是，干扰有些时候是很难压制的，如来往的重型车辆对地质雷达的干扰影响，遇到该类情况，应记录好，以免造成误判。

3.1.3 探测结果分析

高密度电阻率法采用的是偶极布极方式，极距为 3 m，测线 G1 和 G2 的探测结果见图 4。地质雷达采用连续扫描方式，测线 DX1，DX2，DM1 和 DM2 的探测结果见图 5。

从以上高密度电阻率法和地质雷达成果可以看出，高密度电法探测出区间发育有一条较大的充水充泥断层，且断层的底部约 34 m 处岩溶发育，富水。当掌子面开挖到断层附近时，利用地质雷达进

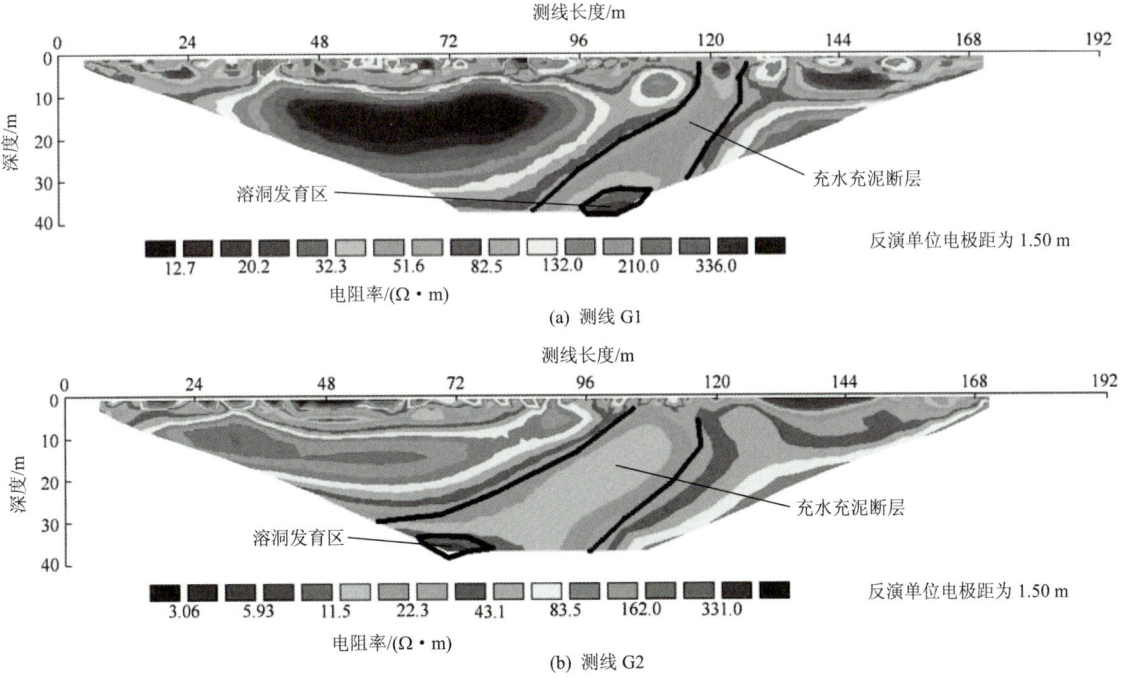

图 4　大连地铁高密度电阻率法视电阻率断面图

Fig.4　Apparent resistivity sections of the high-density resistivity method in Dalian metro

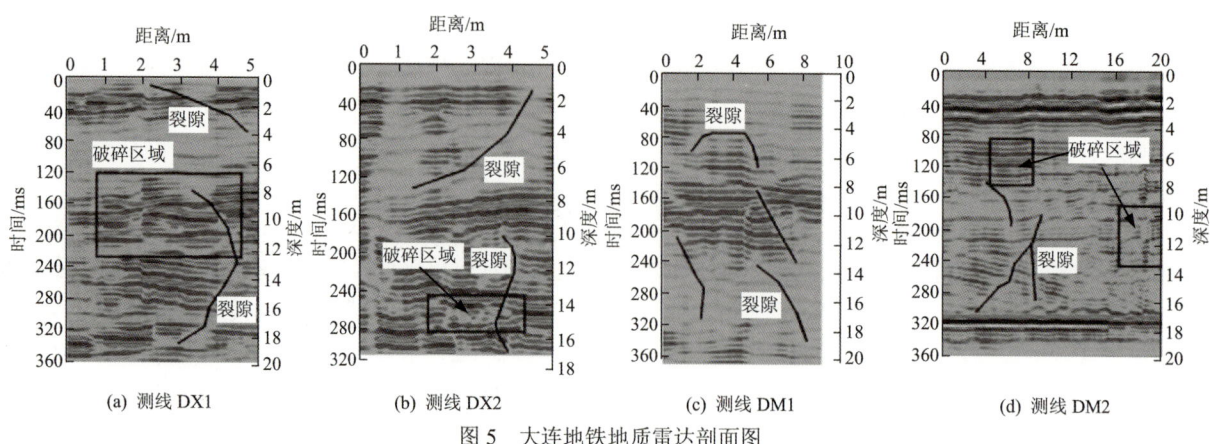

图 5 大连地铁地质雷达剖面图

Fig.5 Sections of ground penetrating radar in Dalian metro

行洞内洞外探测,结果发现,掌子面前方岩体较为破碎,裂隙节理发育,且竖向裂隙比较多,特别是发现了一条位于拱顶上方的弧形裂隙,很容易引起拱顶岩体脱落或垮塌。此外,从裂隙的分布来看,结合高密度电法成果图的富水位置,推断掌子面前方右半部分裂隙水出现,且来自底板以下。

图 6 为工作面开挖 4 m 后的照片,其中拱顶岩体较为破碎,并已进行了挂网及喷锚支护,另外,掌子面右下角有裂隙水流出。可见,开挖结果与预报结果比较吻合。

(a) 拱顶

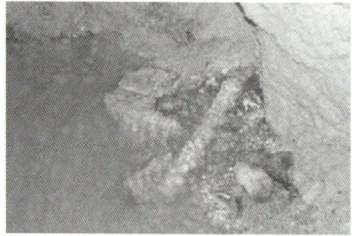

(b) 掌子面

图 6 现场开挖照片

Fig.6 Pictures of in-situ excavation

3.2 广州地铁一车站超前预报实例

车站位于市区,线路沿线地面条件复杂,主要为城市交通干道、居民小区、商业区。沿线城市干道车流量大,住宅、商业区人员密集,建筑物稠密。从区域地质角度出发,沿线揭露的地层按时代、成因和岩性总体上为二元地层,即上部为第四系沉积物,下部为石炭系沉积岩。前期勘察表明,车站附近岩溶发育,而且还存在土洞。由于地质条件复杂,有必要对车站的岩溶和土洞分布情况进行详查。以下为车站竖井附近的探测情况。

本次车站内探测主要是为了了解竖井及其附近地下将要开挖部分(见图 7 右侧虚线框部分)是否存在土洞、溶洞等不良地质体。由于车站已经开始施工,地面存在较多固定建筑和器械,加上各种围挡的阻挡,地质雷达测线布置受到一定的限制,不能和高密度测线一样,只布置了 20 m 长的测线(见图 7)。

本次高密度电阻率法采用的仍为偶极布极方式,极距为 2 m,测线 G 线(108 m)的探测结果见图 8。地质雷达采用连续扫描方式,测线 D 线(20 m)的探测结果见图 9。

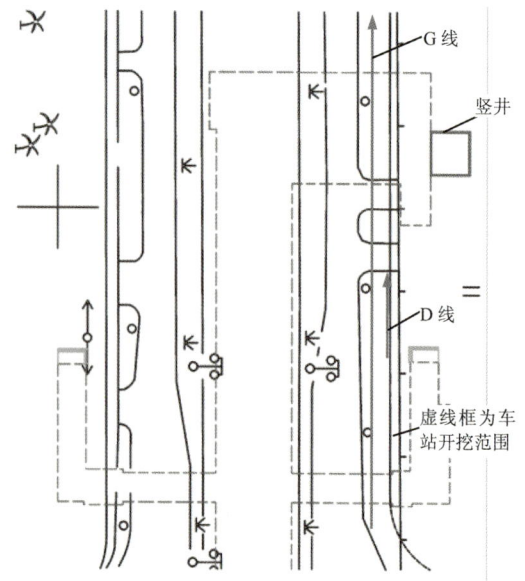

图 7 广州地铁地面测线布置图

Fig.7 Layout of ground survey lines in Guangzhou metro

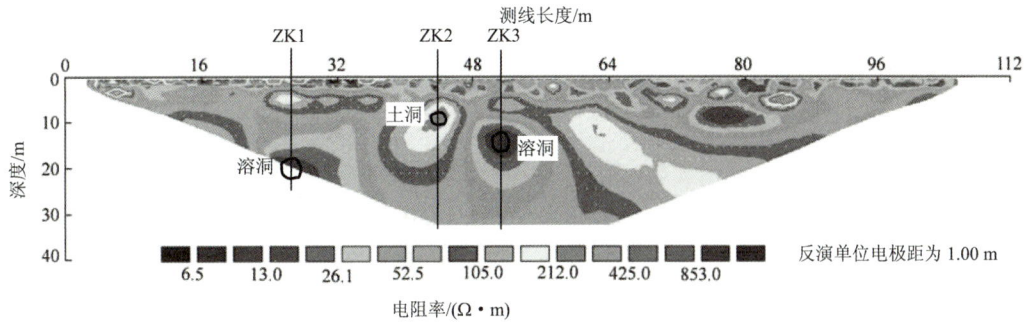

图 8 广州地铁高密度电阻率法视电阻率断面图

Fig.8 Apparent resistivity section of the high-density resistivity method in Guangzhou metro

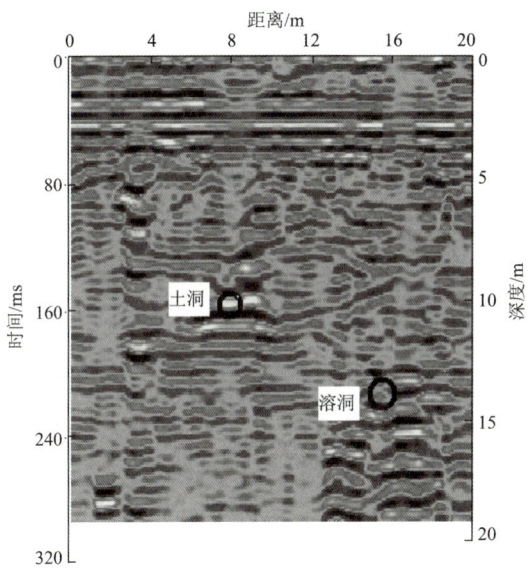

图 9 广州地铁地质雷达剖面图

Fig.9 Section of ground penetrating radar in Guangzhou metro

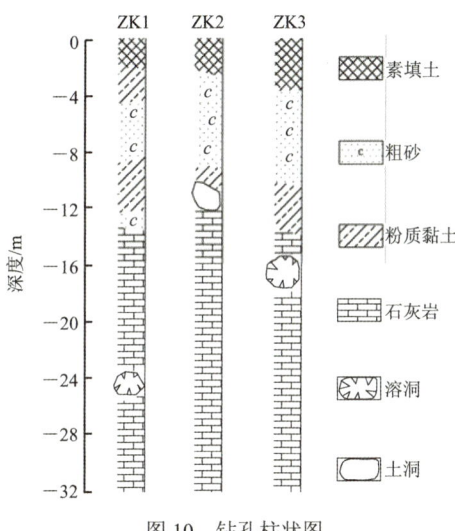

图 10 钻孔柱状图

Fig.10 Bore histograms

虽然地面测线附近存在许多金属器械、电缆干扰,但通过相应的处理(可参考大连地铁实例分析中的干扰分析和压制方法),其最终成像结果是有效的。用高密度电阻率法首先探测到3个较明显的异常体,为了进一步确认该异常是否为假异常,开展了地质雷达探测,其结果与高密度探测结果较为吻合,通过二者解释,可确定一个为土洞,另一个为溶洞(见图 8,9),这与后来的钻孔验证结果(见图 10 中的钻孔 ZK2 和 ZK3)非常吻合。

至于图 8 中最左侧的深部异常体,由于地质雷达探测存在一定困难,该异常体的解释需要结合相应的地质分析。前期勘察表明,该地段的土岩界面基本在地下 13 m 左右,因此,一个发育在石灰岩中的低阻异常体基本为岩溶产物,从该异常体的分布规律和形状而言,判断其为一个溶洞。该解释结果从钻孔柱状图中也得到很好的验证(见图 10 中钻孔 ZK1)。

4 结论与建议

通过以上的分析研究,可以得出以下几点结论及建议:

(1) 理论与实践表明,对于岩溶地质条件下的城市地铁超前预报,在地质分析的基础上,利用高密度电法和地质雷达综合方法,通过洞内洞外相结合可以有效地对掌子面前方围岩情况进行探测。

(2) 与普通公路或铁路隧道不同,城市地铁往往埋深浅,环境干扰因素多,因此,超前预报方法的选择与实施要充分考虑到城市地铁的特点,不能简单照搬普通公路或铁路隧道的超前预报手段。

(3) 城市路面上进行物探工作易受到场地的各种限制,而且各种干扰因素众多,因此,在实际探测过程中要灵活处理测线的布置,包括测线的长度、极距和形状等,并尽可能地减少道路两边的管线、路灯、电线杆和道路来往的大型车辆的影响。

(4) 多种方法的综合解释是城市地铁综合超前预报的最关键环节,要求解释人员既要熟悉每种物探方法的特点,更要密切结合既有钻孔资料和其他地质资料,还有就是现场记录要记详、记全,如此通过洞内洞外相结合的综合解释才能更好地去甄别、排查每一个异常体,确保地质解释成果的有效和精确。

(5) 对于岩溶地质条件下的城市地铁综合超前预报,文中提到的地质雷达与高密度电法相结合方法虽然取得较好的效果,但也存在一些缺点,比如地质雷达探测距离短,探测频次高,会给现场施工造成一定的影响;高密度电法现场需要多人协同工作,操作不便携,因此该综合方法还需要进一步完善和提高。此外,由于三维探测信息量更丰富,日后可以考虑开展地质雷达及高密度电法三维探测,以此不断地提高岩溶地质条件下的城市地铁综合超前预报水平,更好地为城市地铁建设服务。

参考文献(References):

[1] 李术才,李树忱,张庆松,等. 岩溶裂隙水与不良地质情况超前预报研究[J]. 岩石力学与工程学报,2007,26(2):217 – 225.(LI Shucai,LI Shuchen,ZHANG Qingsong,et al. Forecast of karst fractured groundwater and defective geological conditions[J]. Chinese Journal of Rock Mechanics and Engineering,2007,26(2):217 – 225.(in Chinese))

[2] 李术才,薛翊国,张庆松,等. 高风险岩溶地区隧道施工地质灾害综合预报预警关键技术研究[J]. 岩石力学与工程学报,2008,27(7):1 297 – 1 307.(LI Shucai,XUE Yiguo,ZHANG Qingsong,et al. Key technology study on comprehensive prediction andearly-warning of geological hazards during tunnel construction in high-risk karst areas[J].Chinese Journal of Rock Mechanics and Engineering,2008,27(7):1 297 – 1 307.(in Chinese))

[3] 王梦恕. 对岩溶地区隧道施工水文地质超前预报的意见[J]. 铁道勘查,2004,(1):7 – 10.(WANG Mengshu. Hydrogeological and geological forecast of tunnel construction in the karst district[J]. Railroad Survey,2004,(1):7 – 10.(in Chinese))

[4] 钟世航. TSP 作隧道掌子面前方地质预报几例失误原因分析[J]. 岩石力学与工程学报,2003,22(增1):2 443 – 2 446.(ZHONG Shihang. Analysis on the failure reasons in survey forward from working face of tunnels by means of TSP[J]. Chinese Journal of Rock Mechanics and Engineering,2003,22(Supp.1):2 443 – 2 446.(in Chinese))

[5] 薛翊国,李术才,张庆松,等. TSP 超前预报系统探测岩溶隧道的应用研究[J]. 地下空间与工程学报,2007,21(6):1 187 – 1 191.(XUE Yiguo,LI Shucai,ZHANG Qingsong,et al. Application of TSP advanced prediction to tunnel in karst areas[J]. Chinese Journal of Underground Space and Engineering,2007,21(6):1 187 – 1 191.(in Chinese))

[6] 刘传孝. 探地雷达空洞探测机制研究及应用实例分析[J]. 岩石力学与工程学报,2000,19(2):238 – 241.(LIU Chuanxiao. Study on mechanism of cavity detection by GPR and the analysis of its application examples[J]. Chinese Journal of Rock Mechanics and Engineering,2000,19(2):238 – 241.(in Chinese))

[7] 苏茂鑫,李术才,薛翊国,等. 隧道掌子面前方低阻夹层的瞬变电磁探测研究[J]. 岩石力学与工程学报,2010,29(增1):2 645 – 2 650.(SU Maoxin,LI Shucai,XUE Yiguo,et al. The detection of low resistivity interbed in front of the tunnel face by the means of TEM[J]. Chinese Journal of Rock Mechanics and Engineering,2010,29(增1):2 645 – 2 650. (in Chinese))

[8] 雷明锋,彭立敏,施成华. 复杂地质条件下隧道浅埋段全空间综合超前地质预报技术[J]. 岩石力学与工程学报,2009,28(增2):3 571 – 3 576.(LEI Mingfeng,PENG Limin,SHI Chenghua. Whole space comprehensive advanced geological prediction technique for shallow-buried section of tunnel under complicated geological condition[J]. Chinese Journal of Rock Mechanics and Engineering,2009,28(Supp.2):3 571 – 3 576.(in Chinese))

[9] 刘文祥,鲁来玉,石桂,等. 综合物探在地铁隧道工程建设中的应用[J]. 物探与化探,2004,28(2):181 – 183.(LIU Wenxiang,LU Laiyu,SHI Gui,et al. The application of integrated geophysical exploration to the light-rail tunnel construction[J]. Geophysical and Geochemical Exploration,2004,28(2):181 – 183.(in Chinese))

[10] 刘浩杰,白明洲. 探地雷达在城市地铁工程中的应用[J]. 物探与化探,2006,30(3):280 – 282.(LIU Haojie,BAI Minzhou. The application of the ground probing radar to urban subway construction[J]. Geophysical and Geochemical Exploration,2006,30(3):280 – 282. (in Chinese))

[11] 陈荣泰. 地铁浅埋暗挖隧道施工超前地质预报的技术方法[J]. 广州建筑,2007,(1):23 – 27.(CHEN Rongtai. Advanced technical method geological forecast for the construction of shallowly buried and undermined tunnel for metro building[J]. Guangzhou Architecture,2007,(1):23 – 27.(in Chinese))

[12] 李天斌,孟陆波,朱劲,等. 隧道超前地质预报综合分析方法[J]. 岩石力学与工程学报,2009,28(12):2 429 – 2 436.(LI Tianbin,MENG Lubo,ZHU Jin,et al. Comprehensive analysis method for advanced forecast of geology in tunnels[J]. Chinese Journal of Rock Mechanics and Engineering,2009,28(12):2 429 – 2 436.(in Chinese))

| 学术论文——隧道及地下工程

地下磁悬浮交通设计研究的若干问题

钱七虎

(解放军理工大学工程兵工程学院,南京 210007)

摘要:通过分析城市化带来的土地安全问题、环境安全问题、能源安全问题和交通事故引发的公共安全问题,详细论述了发展地下磁悬浮交通的动因。提出未来的城市交通和城际交通方式宜考虑发展地下磁悬浮交通,其中未来的城市客、货运交通宜发展地下中低速磁悬浮交通,而未来的城际交通宜发展成为绿色、低碳的地下高速磁悬浮交通。详细介绍了地下磁悬浮交通和地下物流系统的构造组成,重点介绍了地下磁悬浮系统具有节约土地、环境友好及高速大运量等优越性以及与该系统当前有关的试验研究成果,主要包括隧道的真空压力范围确定、截面设计、空气动力学试验模型、机械电力设备研究、生命支持系统研究以及建设成本估计等。

关键词:地下磁悬浮;绿色低碳交通;地下物流系统;空气动力学

中图分类号:U 237　　文献标志码:A　　文章编号:1672-741X(2011)02-0154-07

Some Issues on Design of Underground Magnetic Levitation Transportation System

QIAN Qihu

(Engineering Institute of Engineer Corps.PLA Univ. of Sci. and Tech., Nanjing 210007,China)

Abstract:The causations for developing underground magnetic levitation are discussed by analyzing the problems of land security,environmental security,energy security and public safety problems caused by traffic accidents,which are brought by the rapid urbanization. The underground magnetic levitation system ought to be taken in to account for the solution of in-city and in tercity transportation problems in future. In this system, the in-city passenger transportation and freight transportation can be realized by low-medium speed underground magnetic levitation system, while the intercity transportation should be realized by the high-speed system, which is more green and more low-carbon. In the description of this new system, the compositions of underground magnetic levitation and underground logistic system are in troduced indetail, the advantages in the underground magnetic levitation system, such as land conservation, environment-friendly, high speed and large capacity,are also presented. Apart from these, some important results of experimental researches are provided, which mainly include the determination of vacuum pressure, design of cross-section, aerodynamics testmodel, machines and electric power equipment research, life support system and construction cost assessment.

Keywords:green and low-carbon transportation;underground magnetic levitation;underground logisticsystem;aerodynamics

0 引言

20世纪全球的社会和经济发展取得了辉煌的成就,但是也出现了很严重的问题。集中表现在社会、经济的发展与资源和环境不相协调,其结果引发了经济和社会发展不可持续的危险。只有建设资源节约型和环境友好型的社会和城市,才能实现可持续发展。于是,绿色经济、低碳经济和绿色交通、低碳交通应运而生,成为关注和研究的热点。在这种背景下,虽然当代中国轮轨高铁引领了风骚,在交通领域,世界有识之士仍然把目光转向地下磁悬浮交通。

1 发展地下磁悬浮交通的动因

发展地下磁悬浮交通的动因主要是为了摆脱21世纪城市交通和城际交通面临的极大困境。具体来说,有下述4个方面。

1.1 城市化凸显的"城市土地"困境

这个困境在中国表现的更加明显和突出。在当前,中国的经济发展速度和城市化速度居世界第一,大量人口涌入城市,其居住、交通、文体、教育、商业等各

收稿日期:2011-02-22;修回日期:2011-03-16。
本文原载于《隧道建设》(2011年第2期)。

方面需要土地供应。据联合国的标准,人均需要 200 m²。按此标准,在中国未来 15 年,按每年平均城市化水平提高 1% 来计算,需供应城市土地约 3 000 km²,即 30 万 hm² 土地。但另一方面,民以食为天,为了养活中国未来最高达 16 亿的人口,必须坚守 1.2 亿 hm² 耕地的红线。据 2007 年国家的统计数据,中国的耕地总量为 1.217 3 亿 hm²,即其余量仅 173 万 hm²,也就是说仅可供 6 年的城市发展之需。如果城市交通再按传统的地面交通发展模式,将面临无地可建公路、铁路、机场的困境。

1.2 环境安全的困境

汽车、喷气式客机、内燃机车组成的传统交通是环境污染的大户。中国的大城市现在已和世界绝大多数城市一样,污染空气的元凶为汽车的尾气,机动车排放的污染物占城市大气污染物的比例高达 40%～60%,并正以每年 2%～3% 的比例增加。汽车排放的 CO 和 NO_x 严重危害市民的健康,不但如此,地面路和高架路交通造成了城市的噪声污染,据统计,中国一些城市主干路的噪声普遍超过 70 dB。交通已成为城市噪声的最大污染源。视觉污染和振动污染等也日益难以被城市居民接受。

传统交通的能源均为化石燃料,据统计,全球交通的碳粒子排放总量约 4.75 亿 t 是形成全球温室效应的重要原因,使得全球气候变暖,极端异常气候频繁。20 世纪,地球平均气温上升 0.6 ℃,致使全球雪域覆盖面积减少 10%;根据联合国气候大会报告《哥本哈根诊断》,预计到 2100 年,地球气温将上升 7 ℃,海平面将上升 1 m,不少岛国和沿海大城市将被淹没,减少温室效应气体排放已变得刻不容缓。必须指出,飞机尾气排放的后果更为严重,因为它排放到同温层,保留时间更长,使得温室反射层更厚。

1.3 能源安全的困境

传统交通是油气消耗的大户,据我国住建部统计,2007 年我国交通能耗已占全社会总能耗 20%,并正逐步上升,未来将如发达国家那样,占全社会总能耗的 1/3。化石能源的储量在不断减少,陆地新增储量已基本没有,其新增储量已基本转为海洋,开采难度和风险大为增加,美墨西哥湾漏油事件即是明证。目前世界石油产量已由峰值不断下降,供不应求趋势难以逆转,当前化石能源价格已在高位运行。据美国能源部统计,20 年后世界石油价格将急剧上涨,因此,几十年后将会出现石油甚至天然气产量难以支撑世界经济和交通运输发展的严峻形势。

1.4 交通事故、交通拥堵和公共安全的困境

据统计,全世界交通事故造成每年死亡 120 万人,其中中国交通事故每年死亡 10 万人。至 2020 年,每年因交通事故死亡人数将比现在上升 92%,世界因交通事故和交通拥堵造成的经济损失每年为 5 000 亿美元,其中美国每年为 800 亿美元。

交通拥堵已成为困扰城市居民和城市当局的主要难题,据调查,美国在上下班高峰期间,没有一个城市不存在道路堵塞情况,在中国 79.3% 的城市居民每天上下班时感觉交通拥堵,33.6% 的人感觉"非常堵"。中国的情况将变得愈来愈严重,中国已经成为全球汽车产销第一大国,近 15 年来,中国汽车拥有量平均增速达 16.74%,北京今年已接近 500 万辆,到 2020 年,中国汽车保有量将达 1.5 亿辆。

除交通事故和飞机坠毁事故造成人员伤亡外,满载汽油的飞机,挂上炸药的汽车(汽车炸弹)是恐怖分子常用的恐怖武器;一场暴风、一场暴雪和一次火山爆发产生的火山灰使地区、国家和世界交通中断和混乱的教训不断重现;大量的城市道路不透水地面已成为城市洪灾和城市热岛效应的重要原因。

2 未来的城市交通和城际交通——地下磁悬浮交通

为了摆脱传统交通所面临的困境,世界各国都依托高科技途径,研究构建资源节约型(节地、节能)和环境友好型(低碳、绿色)的未来城市交通和城际交通:转入地下,实现节地;以电力驱动,摒弃燃料直接驱动,实现绿色环保;以磁悬浮摒弃轮轨摩擦,实现节能低碳;在真空隧道中运行,摒弃车厢高速运行时的空气摩擦,实现进一步的节能低碳[1]。

地下磁悬浮交通是在地下隧道内部铺设磁悬浮轨道,把车体浮起,车体运行在几乎没有摩擦的磁垫上;隧道和车辆协同工作,形成一个"直线电机",以电磁吸力平稳地把车辆加速,以电磁斥力提供刹车力,以确保安全、舒适和精确停车;地下高速磁悬浮交通则是在隧道内沿线布设许多真空泵,从密封的隧道中抽出大部分空气,使空气压力低于航空器在高空飞行时的空气压力或与其相当,所以车体运行时阻力极小,能实现高速或超高速运行。

2.1 低速磁悬浮交通

未来的城市客、货运交通由于站间距短,起、制动频繁,宜发展地下中低速磁悬浮交通[2];而未来的城际交通,为实现绿色、低碳的最大化,宜发展地下高速磁悬浮交通。

城市的货运交通系统宜独立于城市客运交通之外自成城市地下物流系统(ULS-underground logistic system)[3]。在这种系统中,城外货物通过城际货运到城郊的物流园区(CLP-Collective Logistic Park),在 CLP 货物经编组装车后通过 ULS 以集装箱和货盘车为基本单元,以电力自动导向车(AGV)为运输工具运送到

城内的各个终端;反向则由城内多个成品终端通过ULS将产成品运到CLP,再由城际货运系统(地下高速磁悬浮系统)送到其他城市[4]。至于垃圾等废弃物的运送,则可通过气动的地下管道系统进行,这种系统已在瑞典、美国以及中国的上海、广州等地运营[5]。

城市的地下物流系统是无人驾驶的,由程序操控的自动化系统。该系统将以高环境利益、高运送质量指标、高运送效率、高规模效益和标准化突出其竞争性[6-7]。完善的ULS建成后,城区内将没有运货的卡车,城市交通拥堵将大为缓解,城市环境污染将大为降低。据世界经合组织2003年的报告,发达国家城市货运交通占城市交通总量10%~15%,而货运交通尾气污染则占交通诱发的城市空气污染总量的40%~60%;至于发展中国家,由于城市基础设施建设在不断进行,其货运交通比例则更高。日本所进行的东京地下物流系统的评估报告表明:在东京建成了300 km地下物流系统运行后,东京城市的CO及CO_2将分别减少10%和18%,货运能耗减少18%,货运速度提高24%。

2.2 高速磁悬浮系统

该系统建于地下,城市之间直线连接,形成畅通无阻的部分真空的地下隧道网络。隧道采用钢筋混凝土复合衬砌,中间夹一层聚丙烯或钢隔膜,以实现气密和水密,保持隧道内的真空状态。

地下城际高速磁悬浮交通可分2级:区域级和国家级,区域级内呈蜘蛛网状结构,把区域内的城市联系起来;每个区域有一个区域枢纽站,其间由直线隧道相连,构成国家级。2级埋设深度不同,均较城市内的中低速磁悬浮交通深,各级线路之间不交叉,但通过枢纽处的菖蓿叶式立交管道垂直相连。当然,各国家级枢纽之间还可构建洲际级的国际地下高速磁悬浮交通,到那时,就可取代洲际喷气式交通,实现洲际的绿色低碳高速交通。

3 地下磁悬浮交通系统的优越性

3.1 节地

地下磁悬浮系统占用地面土地极少,仅其隧道入口处的车站、机电设备供应点等建在系统隧道的上方地面,因此发展该系统将会大大缓和世界各国特别是我国"耕地安全"的威胁。通过精细的物探和其他勘察技术,可以保证系统避开地下文物、地下矿区和已建成的地下空间,保证不会破坏地下水系。由于系统建设在地下100~200 m,因此也不会涉及较难解决的地热问题,在这个深度内可在多个层面上建成成百上千条隧道,形成庞大的线路网,在这个意义上可以说:系统可在地下三维方向任意地扩展。

3.2 环境友好

由于系统采用电力驱动,因此不污染城市地区的空气,系统不产生温室气体。随着科技的发展,未来可用风能、太阳能、第4代核能甚至地下干热岩等清洁可再生能源发电,从而彻底消除"能源安全"和"环境安全"的威胁,真正实现绿色、低碳以至无碳交通。至于电磁辐射污染的问题,根据中科院电工所对于北京1547 m唐山磁悬浮工程化试验线的实际检测数据和实验室数值仿真数据分析:系统运行时,没有额外的高频(30 KHz~30 MHz)交流电磁辐射;车内及车外5 m以外处低频交流(5~32 Hz)磁场不超过ICNIRP(国际非电离辐射防护委员会)的ICNCRP(电磁辐射公众标准限值);车内和车外1 m外直流磁场不超过ICNIRP的静磁场暴露标准限值;停靠站台时,站台处的直流和低频交流磁场均不超过ICNIRP的电磁辐射公众标准限值。

振动和噪声是所有地面交通工具不可避免的问题,速度越高,噪声和振动越大,但是对于地下磁悬浮系统,不论速度多高,地下噪声和振动都不会通过地层传播,地面上的居民不会察觉到在他们的地下有高速旅行的车辆;此外,由于磁悬浮系统无车轮、无轴承、无旋转电机、无刹车片等电力机械系统中常见的发热、摩擦点,所以维修保养有效减少,污水和固废排放相应明显减少。因此,地下磁悬浮系统在环境友好上具有明显优势。

3.3 高速大运量

有下列因素促使地下磁悬浮系统实现高速行驶,甚至超过飞机800~900 km/h的航速,可达1 000 km/h。

1)显著降低了影响行车速度的因素(接触摩擦和空气摩擦)。该系统通过磁悬浮避免了车辆跟轨道的直接接触,清除了接触摩擦阻力;车辆在隧道中行驶时,空气阻力成指数关系增长,真空是克服空气摩擦的有效办法。

2)系统深埋地下,其行驶路线几乎保持为直线,且无交叉,因此,要求限速行驶的曲线和坡度降低到最低限度。

3)最后系统不需如同飞机起飞和降落时的滑行和爬升。

Swiss Metro研究项目中,列车运行设计速度为400 km/h[8],American Metro研究项目中列车运行速度可达1 000 km/h[9],每次列车可载800名旅客。

3.4 高度安全

近年来,交通隧道内的灾难主要来自于火灾事故。火灾的燃烧条件是燃料、氧气和高温。地下高速磁悬浮隧道,由于电力驱动和真空,因而列车上不带燃料油,隧道内缺乏空气和氧气,火灾无从燃起,即使电气设施由于高温可能引火,但缺乏氧和燃料,火不会蔓延而自动熄灭。

系统在2条管线内双向分开行驶,消除了迎面碰撞的可能性。由于直线电机的电流只能让一个区间内的车辆同向运动,所以磁悬浮列车不可能发生同一行驶线路上前后列车相向碰撞的情况,且刹车和电子信号系统也杜绝了前后车辆发生追尾的可能性。

现代地面和空中交通易遭恐怖袭击。而地下磁浮系统运行在地下,很难遭到系统外的恐怖袭击,系统入口处通过安检系统挡住行为可疑的人和易燃易爆物品,进入系统的任何乘客和货物都会实时显示在控制屏幕上,从而可及时有效地采取反击恐怖的措施。

4 地下高速磁悬浮系统设计试验研究的若干问题

地下高速磁悬浮系统包括2方面主要技术:隧道和真空;悬浮和推进。前者属于空气动力学效应和土建设施研究范畴,后者属于机电设备研究范畴。

4.1 隧道和真空的设计试验研究

4.1.1 真空压力范围的确定

隧道中高速行驶的车体产生的空气动力学效应由阻塞比决定,阻塞比 = 车体截面积/隧道截面积。

列车高速行驶时,当阻塞比超过0.5时,产生声障现象,它限制了空气沿车体侧边的流动,并在车体前面产生一个升压,形成阻力,当速度提高时,空气阻力成指数关系增长。

为了使列车高速行驶时的阻力处在允许的范围内,隧道一般设计成14%~17%的阻塞比,对于等效直径为3.5 m的列车车体,该阻塞比所要求的隧道直径为8.5~9.4 m,这使得地下高速磁悬浮系统在常压隧道下的隧道造价进入不可接受的禁区。

部分真空的理念类似于飞机在万m航行高度时稀薄空气中的情况,该理念使得在建造隧道的造价、真空设备的设施以及能源成本之间找到折衷的可行方案[10]。

隧道内的真空压力确定为10~25 kPa。10 kPa为最低限值,这是由于一方面低于10 kPa的真空形成需要特殊的气泵,更重要的一方面这是由人的生理要求所决定的。在10 kPa的环境下,人的血液开始汽化,不能存活,在10 kPa以上的环境中,人尚能存活1~2 min,给增压或救援留下可能(见图1);25 kPa相当于在高空航行高度时的空气压力,空气阻力较小。

4.1.2 真空压力水平和隧道横截面积的确定

上述参数的确定要从设计车速的技术要求开始,即车速设计为400 km/h还是800 km/h或1 000 km/h。在这个车速下通过真空压力水平和隧道阻塞比值的选择来获得隧道造价、真空设备及设施的造价以及能源消耗费用之间的最佳折衷方案。因为上述3者费用都是真空压力水平和隧道阻塞比(隧道横截面积)2个参数的函数,上述寻优的过程是通过模拟试验来实现的,该模拟试验需要在三维的并在一个时间段上对时程变化灵敏的空气动力模型上进行。

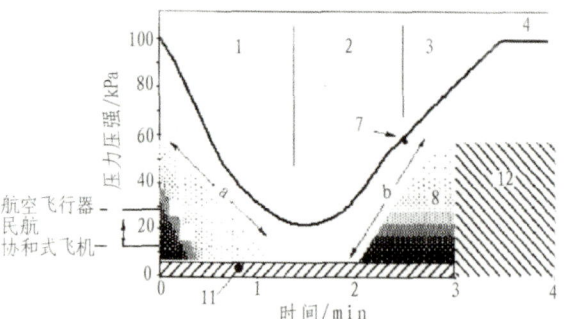

1—减压;2—紧急的压力重建;3—返回正常压力;4—大气压力;5—过压内伤;6—缺氧;7—最小值,2′30″时57 kPa;8—严重不利情况;9—肠道;10—肺部;11—爆炸性减压所造成的严重眩晕;12—不可逆缺氧损伤。

图1 偶然泄压下的人体生理极限值

Fig 1 Limit value of human body under casual pressure releasing

4.1.3 空气动力学模拟实验

一旦真空压力水平和隧道阻塞比确定后,还要通过模拟试验微调隧道几何形状和车体形状,寻求最佳车体形状和隧道几何形状,所谓最佳是指该高速行驶下空气对车体的总阻力最小[11]。

通过模拟试验还可研究确定2个隧道间压力释放管的位置,压力释放管是2个隧道间的空气连通管,用以降低车体在隧道中高速行驶时所受的空气阻力。为了在常压下进行车站和车体之间的换乘,借助2个隧道密闭门组成隔绝车体的空气闸门,在车站的每一端设1个。在该空气闸门的隧管空间内可以加压和泄压,加压至常压后,乘客可在车站与车体间进出,泄压后,车体即可驶入隧道。

4.1.4 保持整个系统真空的措施

为了保证车体在真空环境中高速行驶,对隧道衬砌的密闭性要求很高,哪怕有少量的空气和水分渗漏,都会给运行造成很大问题。隧道衬砌的密闭性能良好、不漏气,真空损失很低,真空泵技术才能可靠地使隧道内保持10 kPa或稍高,从而实现最高可达1 000 km/h的高速运行。隧道衬砌常用的材料是混凝土,其中必然含水分,水分在10 kPa的真空环境中会汽化,而且真空环境具有从围岩中汲取水分的效应,这就对地下高速磁悬浮系统衬砌的防水可靠性要求很高,为此,隧道衬砌应设计成2层或多层的"三明治"衬砌。在Swiss Metro项目研究中,衬砌断面由预制混凝土构件在掘进时逐次安装而成,中间必须嵌入金属或塑料(聚丙烯或其他)隔膜,以实现气密和水密,也有专家认为,钢结构衬砌最可靠,但成本比混凝土衬砌高20%。

在瑞士的 Swiss Metro 项目中为进行上述模拟试验,在洛桑联邦科技大学建立了 Histar(高速列车空气动力学测试平台)和 SETUP(安全、设备与隧道研究平台)[8],见图 2 和图 3。

在 Histar 测试平台上,通过 1∶10 的隧道和车体的小比例尺模型研究高速车体在小隧道内的动态响应。

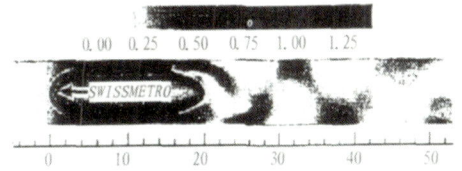

图 2　围绕二维车厢及其后面气流的模拟以及马赫数分布
(100 m/s≈400 km/h)

Fig. 2　Simulation of planar carriage and airflow behind it at the speed of 100 m/s and distribution of Mach number

图 3　Histar 多目标空气动力试验台

Fig. 3　Aerodynamic test bed of Histar multiple targets

在 SETUP 测试平台上,全面研究整个系统在部分真空状态时的效果。这是 1 个 1∶1 的模型,全长 200 m,包括隧道、轨道、车辆、乘降区、空气锁、紧急增压系统、真空泵、隧道衬砌及其他技术方面,对从各项设施到设计乘客安全的所有设备的整个系统进行测试,测试尽量独立进行,以证明技术集成后的可行性,包括用于推进和车辆导向的技术。

4.2 系统机械电力设备的研究

包括车体、电机及其反应部件、磁悬浮及磁导、电力供应及传输系统。

4.2.1 车体

在 Swiss Metro 项目中,车体的设计结构为一整体,由一个变形成流线形曲线的管子组成,低速时曲线的最小曲率半径为 5 km,高速时为 15 km[1]。这样的整体车体结构和质量能保证车体断裂解体的可能性极小。

车体在尺寸和工艺方面相似于麦道-81 型飞机机身,其外径为 3.5 m,所采用的材料和结构原理类似于飞机工业中所采用的。该车体每 m 净质量 0.5 t,200 m 长的车体自质量 100 t,载 800 个旅客时,总质量 180 t。

车体拼接一个电磁悬浮和导向的车架,带有下列部件:线式电机的反应部件、悬浮和导向的电磁部件、能量传递系统的二次绕组线圈、所有空调功能部件。车体断面示意图见图 4。

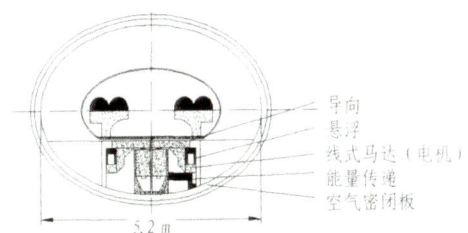

图 4　Swiss Metro 隧道和车体断面

Fig 4　Section of Swiss Metro tunnel and train body

4.2.2 推进电机及其反应部件

车体借助于线式电机推进,并由电磁力悬浮于支撑上,它已成功地在德国 ICE(高速铁路)和日本 MLV(低温超导磁悬浮)开发。

对于 Swiss Metro 项目,设计采用安装于隧道底板上的同极同步电机(见图 5),车体下安装一个反应部件,对一个 200 m 长的车体,每 190 m 必须放 1 个电机以保证连续推动。200 m 长、180 t 质量车体的电机主要参数为[8,12-13]:由静止加速到 160 km/h 的启动加速度为 1.3 m/s^2;以 10.4 MW 的常功率由 160 km/h 提高至 400 km/h;借助线式电机和变频器推进和制动。

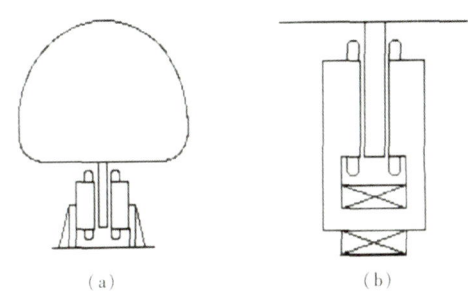

图 5　同极同步电机

Fig 5　Homopolar and synchronous power-driven machine

对 Swiss Metro 项目,其设计运行速度为 400 km/h,平均加速(制动)的距离及时间分别为 10 km、2 min,这和城际设计站间距 50～100 km 是相容的。如果设计运行速度达 1 000 km/h,启动加速度仍为 1.3 m/s^2,则相应的加速(制动)距离和时间近似为 30 km、3.5 min,这将和 50～100 km 站间距不太相容,从而需要相应提高电机的功率以缩短加速和制动距离。

4.2.3 磁悬浮和磁导向

对 Swiss Metro 项目,采用能控制列车稳准行驶的电磁悬浮系统(见图 6),而不采用超导磁悬浮系统,该悬浮系统能在所有速度范围工作,包括车体静止不动时。

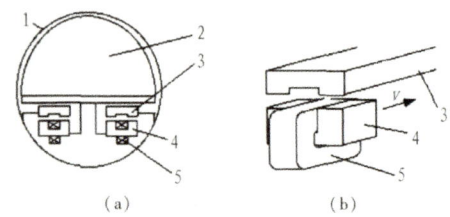

1—隧道;2—客厢;3—轨道;4—轭铁;5—线圈绕组。

图 6 磁悬浮

Fig 6 Magnetic levitation

当对车体的电力供应失效时,电池能接续并满足包括磁悬浮和磁导向的功能,在最不利的不能保证磁悬浮功能的情况下,车体将能下降接触到轮子或滑道上,从而满足制动功能。

设计的悬浮间隙在 20 mm 以内,这是考虑线状布置的约束条件、能耗和装置振动诸因素的一个折衷方案。

4.2.4 车体的电力供应

虽然电机是安装于隧道底板上的,但供应车体一定量的电力还是需要的,它主要应用于空调、电磁悬浮和导向、照明和安全。

Swiss Metro项目不准备采用架空线和弓架的常规电力供应(传输)方式[13],因为它要求车体和隧道间有相当的空间。试验研究采用的是感应传输的方案,它要求的空间很小,该方案包括一个固定于隧道上的 2 根电缆组成的一次绕组线圈和固定于车体上的 10~20 m 的二次绕组线圈(见图 7),专门的模型研究表明,无铁的方案是最好的。用于推进缩尺车体的小功率试验装置,已表明该原理是可行的。图 8 显示了 50 mm 间隙下变压器输出效率与频率的关系。

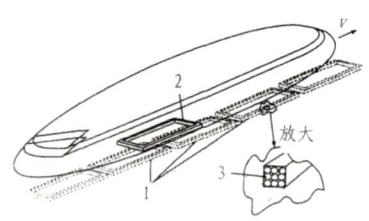

1—安装于隧道的一次绕组;2—安装于车体上的二次绕组;3—里兹电缆。

图 7 感应传输

Fig 7 Induction transfer

4.3 生命环境保障系统

隧道中空气稀少,而车内的乘客在整个旅途中都离不开空气,这就需要一套生命环境保障系统,就像潜水艇和宇宙飞船里一样。

人吸入空气,身体吸收其中的氧气,排出二氧化碳。为保持容器内的空气可供人呼吸,则有 3 件事要做:1)当氧气消耗后,必须补充;2)二氧化碳必须从空气中排出;3)人吸入的水分必须排出。

潜艇和太空飞船上,由高压氧气瓶、制氧机或某种氧气筒提供氧气,另外,有一种燃料电池,在舱内提供电能的同时释放氧。

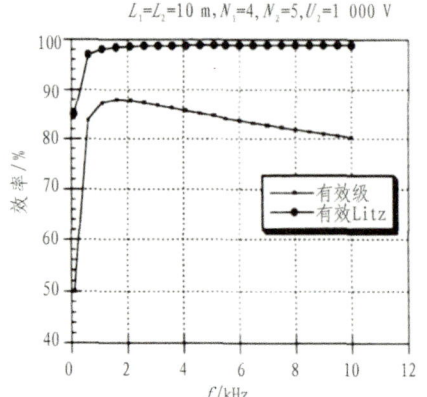

图 8 变压器独立输出

Fig 8 Output of transformer

舱内都安装有二氧化碳清除器,清除二氧化碳通过如下化学反应:当用泵抽舱内的余额空气经过一种叫乙醇胺(MEA)的化合物时,二氧化碳被捕获,这一过程跟真空吸尘器中的过滤器类似,乙醇胺中截取的二氧化碳为气泡形态,经压缩即形成液态;再储存于压力容器中,最后做弃置处理,也可应用于工业;然后乙醇胺恢复为原状态,继续在二氧化碳清除器中使用。美国潜艇舰队长时间、长距离航行时就用这种方法排除二氧化碳。

可用除湿器或化学药品进行除湿,防止水蒸气在车船内部的设备或墙壁上凝结。30多年来,潜艇没发生过一起因生命保障系统故障造成的海员死亡事件。

4.4 造价分析

土建设施(隧道、车站、施工竖井和平巷以及产生、维护真空的设施)占整个地下高速磁悬浮系统设施成本的 70%,这是基于对 Swiss Metro 项目中直径为 4.5~5.5 m 的隧道情况以及砂岩中隧道掘进的已有经验所进行分析的结果。

图 9 给出了取决于隧道所占比例的整个运输设施的成本比较,包括单孔双线铁路、双孔公路、整个建于双孔隧道内的真空磁悬浮系统(Swiss Metro)。

由图 9 可见:如公路隧道比例大于 18% 或者铁路隧道比例大于 42% 的,比建造同样长度的 Swiss Metro 真空隧道的造价要高。

至于 American Metro 项目估算的每 km 总造价为 3 440 万美元(每英里 5 500 万美元),则明显估低了[9]。

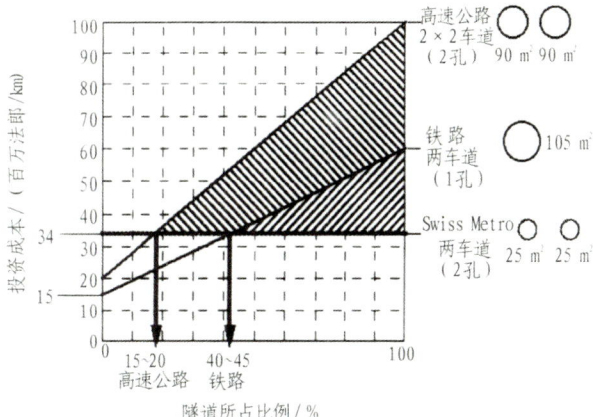

图 9 投资成本的比较
Fig.9 Comparison of the costs

5 结论

地下高速磁悬浮交通是一个节地、节能、低碳、低污染和抗灾安全的资源节约型和环境友好型的未来交通运输方式,是建设投资很大、运行成本较低、客货两用的交通方式,它将会引起全球的交通运输革命。它的发展将会为促进全球温室效应的解决、气候的改善和保障城市的广义安全做出重要贡献。

后记:本文主要是根据 Swiss Metro 以及 American Metro 2个研究项目的成果以及作者本人的思考编写而成,特别是前者。该项目是由瑞士政府科技基金以及瑞士、德国、奥地利 3国 80个工业企业资助完成的。

参考文献(References):

[1] 唐锐,吴俊泉. 中低速磁浮列车在我国城轨交通中的应用前景[J]. 都市快轨交通, 2006, 19(2): 12−16. (TANG Rui, WU Junquan. Application prospects of medium and low speed maglev vehicle in urban rail transit of China[J]. Urban Rapid Rail Transit, 2006, 19(2): 12−16. (in Chinese))

[2] 佟力华,马沂文,胥刃佳. 适用于城市交通的中低速磁悬浮技术[J]. 电力机车与城轨车辆, 2003, 26(5): 4−6. (TONG Lihua, MA Yiwen, XU Renjia. Medium and low speed maglev technology applicable to urban mass transit[J]. Electric Locomotives & Mass Transit Vehicles, 2003, 26(5): 4−6. (in Chinese))

[3] Liu H. Feasibility of underground pneumatic freight transport in New York City[R]. Arlington: ISUFT, 2008.

[4] 郭东军,俞明健. 上海地下集装箱运输系统基础研究[J]. 上海建设科技, 2009(2): 1−3, 14. (GUO Dongjun, YU Mingjian. Fundamental study of Shanghai underground container transport system[J]. Shanghai Construction Science & Technology, 2009(2): 1−3, 14. (in Chinese))

[5] 张耀平. 真空管道运输—真空产业发展的新机遇[J]. 真空, 2006, 43(2): 56−59. (ZHANG Yaoping. Evacuated tube transportation(ETT)— a new opportunity for vacuum industry[J]. Vacuum, 2006, 43(2): 56−59. (in Chinese))

[6] 钱七虎. 建设特大城市地下快速路和地下物流系统—解决中国特大城市交通问题的新思路[J]. 科技导报, 2004, 22(4): 3−6.

[7] 钱七虎,郭东军. 城市地下物流系统导论[M]. 北京:人民交通出版社, 2007.

[8] Marcel Jufer. Swiss Metro— the transport challenge of the next generation[J]. Rail International Schienen der Welt, 1996, 27(11): 4−10.

[9] Brad Swartzwelter. Faster than jets: A solution to America's long-term transportation problems[M]. America: Alder Press, 2003.

[10] 连级三. 磁浮列车原理及技术特征[J]. 电力机车技术, 2001, 24(3): 23−26. (LIAN Jisan. Principle and technology characteristic of maglev vehicle[J]. Technology For Electric Locomotives, 2001, 24(3): 23−26. (in Chinese))

[11] 张瑞华,严陆光,徐善纲,等. 一种新的高速磁悬浮列车—瑞士真空管道高速磁悬浮列车方案[J]. 变流技术与电力牵引, 2004(1): 44−46.

[12] A Cassat, M Rosenmayr, N Macabrey, et al. Swiss Metro and transrapid comparison of the electromechanical components and the power supply in a specified vacuum tunnel environment[J]. International Conference on MAGLEV, 2000(6): 7−10.

[13] M Rosenmayr, A Cassat, H Stemmler. Swiss Metro power supply for a high-power propulsion system with short stator linear motors[C]//International Conference on MAGLEV' 98. Janan: Mt Fuji, 1998.

基于地下集装箱运输的城市地下环境物流系统建设

范益群[1]　钱七虎[2]

(1. 上海市政工程设计研究总院,上海 200092
2. 中国人民解放军理工大学工程兵工程学院,南京 210007)

摘要　城市垃圾危机、土地资源枯竭、城市交通拥堵、生态环境恶化等"城市病"严重阻碍了现代城市的可持续发展。本文通过对城市"垃圾危机"现象的说明,揭示了生活垃圾产生量逐年增加,垃圾处置能力不足以及收运方式不合理,从而导致了"垃圾围城"。解决垃圾危机的有效方法,除了做好垃圾分类以及垃圾"减量化、资源化、无害化"工作外,提高垃圾处置能力和改善收运方式非常重要。进一步,本文通过对上海市建设城市地下环境物流系统的规划与概念设计说明,指出了地下环境物流系统能够为缓解垃圾危机提供一条重要出路。此外,发展地下环境物流系统能够减少交通拥堵、改善城市环境,带动相关高新技术研究,促进城市的可持续发展。

关键词　地下环境物流系统;地下集装箱运输;垃圾危机;交通拥堵;生态环境恶化;城市可持续发展

中图分类号　U412.37　　　　文献标识码　A　　　doi　10.3981/j.issn.1000-7857.2011.07.004

Building Urban Underground Environment Freight Transportation System Based on Underground Container Transportation

FAN Yiqun[1], QIAN Qihu[2]

1. Shanghai Municipal Engineering Design General Institute, Shanghai 200092, China
2. Engineering Institute of Engineer Corps, PLA University of Science and Technology, Nanjing 210007, China

Abstract　City waste crisis, land resource exhaustion, urban traffic congestion, and ecological environment deterioration have seriously hindered the modern urban sustainable development. By describing "waste crisis", it is revealed that the increasing quantity of waste output, the insufficient waste disposal capacity, and the unreasonable waste collection and transportation methods are the cause for the problem of "waste siege" in many big cities. In addition to waste classification and "reduce, reuse, recycle" policy, an effective method for solving waste crisis is to improve the waste disposal capacity as well as the collection and transportation method. Furthermore, the concept of underground environment freight transportation is explained, and plan and conceptual design of urban underground domestic waste transfer logistics system in Shanghai are introduced. It is pointed out that the underground environment freight transportation system provides an important and feasible way to relieve the waste crisis. Besides, underground environment freight transportation system can reduce traffic congestion, improve urban environment, lead to related high-tech researches, and promote the sustainable development of modern cities.

Keywords　underground environment freight transportation system; underground container transportation; waste crisis; traffic congestion; ecological environment deterioration; urban sustainable development

0　引言

所谓城市地下环境物流系统,就是利用地下物流[1-2]进行城市生活垃圾转运[3]的系统,即利用大城市中深层地下空间(地下 40~50 m 以下),通过隧道或大直径的管道连接各主要

收稿日期:2011-01-05;修回日期:2011-02-21。
本文原载于《科技学报》(2011 年第 29 卷第 1 期)。

地下垃圾转运站的运输站点,并连接到垃圾终端处置场所,应用自动化、低能耗的运输方式,进行垃圾转运。

根据美国纽约州能源与发展局(NYSERDA)提供的研究报告[4],纽约市于2002年曾完成了基于气动舱体运输(PCP)技术的地下货物运输系统的技术与经济可行性研究。它运用PCP技术运送9个垃圾周转站的固体垃圾,即将9个分管道(每个管道连接着一个垃圾中转站)连接在一个主管道上,然后通过主管道将各个分管道收集到的垃圾运送到临近的州(如新泽西州)的一个或多个垃圾处理场。通过不间断运转,此系统每天可运送被压缩后的固体垃圾1.8万t。该系统不但可以满足以后20年的垃圾处理需求,而且每吨垃圾运输成本降至原来的1/4。

2009年,上海市政工程设计研究总院立项进行城市地下生活物流研究,提出了采用基于地下集装箱运输的地下物流系统进行上海城市生活垃圾转运的设想[5-7],建议上海地下环境物流系统的网络建设可分近期和远期两步走,其中近期的网络长度为55.0km,较好解决了浦西中心城区约每天1.0万t生活垃圾的环境物流问题,而且它与目前上海市垃圾收运的水陆联运系统相衔接。鉴于当前上海市采用20英尺(外部6.058m)集装箱进行垃圾水陆联运,及上海具备大量地铁盾构,推荐该网络系统采用6.2m直径的地铁隧道和地下集装箱系统[8]进行运输。通过对该系统近期网络的成本效益分析表明,该物流网络采用单管方案可满足运输要求,它的建设成本在50亿~60亿元人民币之间,建成后其运营成本降至原来的1/4~1/10。此外,还获得显著的环境和社会效益。

从中国大型城市的实际情况看,采用不占城市地面空间、无污染的城市地下环境物流系统对城市发展具有极大的现实意义,是解决城市生活垃圾围城、垃圾二次污染、交通堵塞与交通安全、节省土地资源的出路之一。本文通过阐述建设地下环境物流应对垃圾危机构想,及对上海市建设城市地下环境物流系统的规划与概念设计说明,指出地下环境物流系统能够为缓解垃圾危机提供一条重要出路,并希望有关政府部门能重视并推动建设大城市地下环境物流系统的工可研究和试验。

1 建设地下环境物流应对垃圾危机构想

1.1 生活垃圾处理形势严峻

随着城市化进程的加快,中国城市生活垃圾的产生量和堆积量均在逐年增加,2010年以后,垃圾产量将超过2亿t/a。与巨大的城市生活垃圾产生量相对应的,是垃圾处置能力的严重不足。中国现行的生活垃圾处置技术主要有填埋、焚烧和堆肥。截至2008年底,全国城市共建设生活垃圾处理设施579座,处理能力1.03亿t/a,生活垃圾的处理率仅有50%左右,仍有大量的垃圾无法及时处理,只能简单堆放,进一步侵占了土地,加重了污染[9-11]。

统计数据显示,截至2008年,中国城市生活垃圾的历年堆存量已达到70多亿t。许多城市形成了"垃圾围城"的严重污染局面,即城市的生活垃圾收集后统一集中倾倒在郊区,虽然城区较为清洁,但郊区受到了污染,垃圾正逐步将城市包围起来。全国600多座城市,除县城外,已有2/3的大中城市陷入垃圾的包围之中,且有1/4的城市已没有合适场所堆放垃圾。垃圾堆存侵占的土地面积已超过6亿m^2,每年因垃圾造成的经济损失高达300亿元人民币。中国已成为世界上垃圾包围城市最严重的国家之一,"垃圾围城"是每个地方政府都遇到并感到头痛的难题[12]。

1.2 垃圾处置以及收运方式亟待变革

目前国内生活垃圾仍采用人工收集、陆路或水路转运[3],不能承担垃圾长距离外运高昂的成本。此外,在垃圾收运过程中,因人工收拣、垃圾遗落、污水渗漏、恶臭散发等造成的二次环境污染,严重影响了城市的市容环境和居民的身心健康。传统的大型卡车运输方式不可避免地带来尾气污染和噪声污染,并占用了城市的土地资源与道路资源,越来越不适应现代城市高效、节能运转的需要。

解决垃圾危机的有效方法,除了做好垃圾分类以及垃圾"减量化、资源化、无害化"工作[13],提高垃圾处置能力和改善收运方式非常重要。这是由于即使做到垃圾分类和可回收垃圾的循环利用,根据日本环境部统计年鉴提供资料,仍然有80%垃圾需要处理。而由于建设垃圾处理场所受到环境保护要求的限制,在城市周边设置大规模的垃圾填埋场或焚烧厂是不可能的。虽然垃圾焚烧是垃圾处置的有效方法,但焚烧易产生包括一级致癌物二噁英等。因此,比较理想的方式是垃圾处理处置场所着远离城市中心居住区的方向发展,或沿海城市在大海中选择一处荒岛建设一座现代化大规模的垃圾填埋场或焚烧厂,从而提高生活垃圾的处置能力,但这对现有的垃圾收运系统提出一个挑战。

1.3 城市地下环境物流系统建设

根据环卫科技网相关报道提供数据,2007年上海市每吨垃圾的清运处理成本为350~400元人民币,2010年会更高,其中垃圾收运的费用占到了60%~80%。按每天产生生活垃圾2万t计,即每天生活垃圾处理的经济成本将达到700万~800万元人民币,而其中的500万~700万元人民币消耗在垃圾收运过程中。另外,加上输出这些垃圾的直接花费以及诸如燃料消耗、空气污染、事故处理上的间接花费将会更高。如果一个现代化大型垃圾处理场离城市更远,那么垃圾的收运成本会更大,为此,提出了建设大城市地下环境物流系统的设想。

城市地下环境物流系统将地下物流与城市生活垃圾收运系统相结合,即将运输隧道/管道和垃圾转运站布置在地下,可节约地面空间、改善地面卫生环境和交通环境、提高运行效率、降低运输成本;此外,该系统能够提供不受天气影响、全天候、大运量、稳定、高效、节能、环保的运输,可以满足现代化大型垃圾处理场所稳定运行的要求,从而减少甚至取消城市垃圾临时堆放点。另外,它也可节省因建设现代化大型垃圾处理场所修建道路、桥梁、码头等基础设施的投入,这对于缓解日益严重的城市垃圾危机、促进城市节能减排和可持续发展十分有利。

2 上海市建设城市地下环境物流系统的规划和概念设计

2.1 建设城市地下环境物流网络规划构想

建立城市地下环境物流系统网络,主要采取将城市中心环卫规划用地或现有环卫设施用地转化为节点的形式,即根据城市具体环卫设施分布,通过分析选择出具有代表性的地区作为地下环境物流节点。此外,还需要根据城市的具体情况,结合地铁建设规划以及地面物流中心规划等,实现地下物流网络共享。

以上海市为背景的相关研究[5-7]提出整个城市中心区网络可分为近期和远期两步实施。近期的网络建设以静安垃圾中转站为中心,形成一个"+"字型网络路线,见图1。第一期规划网络已可以独立成系统,并较好解决了浦西中心城区的环境物流问题。远期的网络建设,将浦东中心城区的垃圾分流中心相连,构成一条半环线,并与浦西站点相连构成一个大的椭圆线,见图2;同时将浦西地区的东西线延伸到老港垃圾填埋场;构成一个"ø"字型网络路线。总长度近100km。

2.2 上海城市地下环境物流网的几个概念设计

2.2.1 运载工具

上海城市地下环境物流系统,既可选用荷兰AGV(自动导向车)、日本DMT(两用卡车)、德国CargoCap、美国PCP等运载工具[2],也可以选用一种基于磁浮和线性马达泵送式运输的新运载工具。

2.2.2 运输单元和货物处理

由于上海城市生活垃圾在从中转站至远途的焚烧厂、填埋场等终点,多采用8t以上压缩式大型集装垃圾卡车,因此,本项目拟采用5t压缩式的集装箱、1~3辆编为一组进行货物运输,见图3[5]。

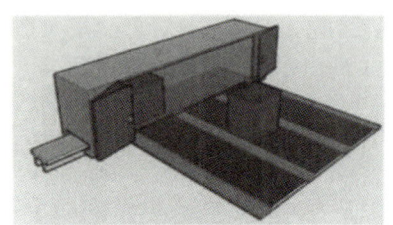

图3 运输舱体设计

Fig. 3 Transport capsule design

2.2.3 运输方式

根据标准箱尺寸,圆形管道直径应为5.5~6.2m,见图4。每个舱体将是8.5英尺(2.59m)宽,10英尺(3.05m)高,42英尺(12.81m)长,以便于能运输一个标准长度的集装箱,其动力推荐利用线性感应电机,为进一步充分利用舱体泵送的特点,也可采用舱体磁浮的方式。每一列由1~3个舱体组成,每5~10min发送一列,见图5[14]。

图1 浦西中心城区地下环境物流网络布局方案

Fig. 1 Layout plan of underground environmental logistics system of Puxi downtown area

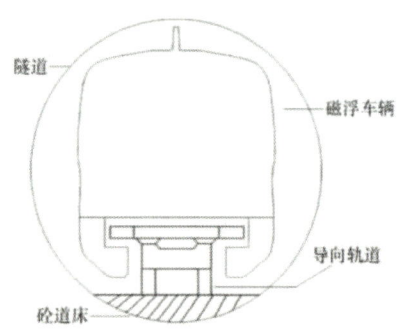

图4 圆形隧道集装箱运输系统

Fig. 4 Round tunnel container transportation system

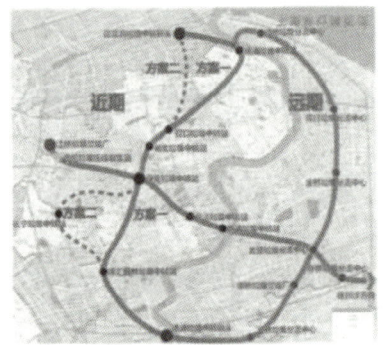

图2 上海市地下环境物流网络布局方案

Fig. 2 Layout plan of underground environmental logistics system in Shanghai

2.2.4 中间站点

城市地下垃圾转运系统的中间站点是一种用于城市地下物流系统的垃圾中转站。该垃圾中转站由地下垃圾压缩站、超深转运竖井和连接区间隧道的地下运输站点三部分组成[6],见图6。

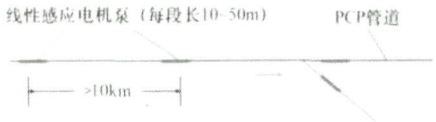

图 5　线性感应电机布置示意

Fig. 5　Schematic diagram of linear induction motor layout

地下垃圾压缩站处于地下浅层空间,地下运输站点位于地下深层空间,两者通过超深转运竖井相衔接。装满垃圾的密封集装箱,通过超深转运竖井运至竖井底部的地下运输站点,并将完成作业任务的空载集装箱由地下运输站点向上运至地下垃圾压缩站的空载集装箱转运点。地下垃圾压缩站平面示意图见图7[6]。

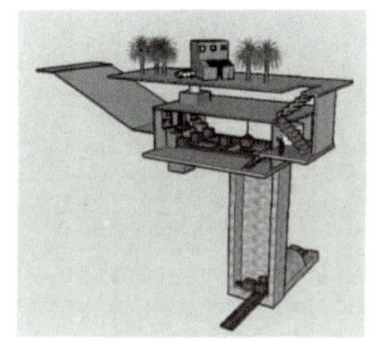

图 6　垃圾中转站立面示意

Fig. 6　Schematic diagram of vertical section of waste transfer station

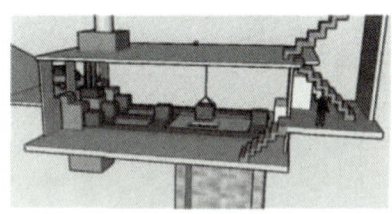

图 7　地下垃圾压缩站平面示意

Fig. 7　Schematic diagram of plan for underground waste compression station

该垃圾中转站具有垃圾压缩、装箱、存放、吊运、地下运输的工艺特点,根据垃圾压缩转运工艺、地下物流运输能力和存储要求确定其布局和规模,从而实现城市生活垃圾收运系统的完全地下集约化。

2.2.5　地下建设

为尽可能避开地下障碍物,地下物流系统拟在−50~−40m以下进行开发,其中,隧道采用盾构机进行施工,隧道结构采用预制管片进行拼装;而地下垃圾中转站则拟采用有支撑围护或压气沉箱法修建[6]。

2.3　经济可行性的考虑

国外研究表明,城市地下物流系统的社会效益、环境效益等间接效益大于其直接效益。在日本的研究成果中,建设城市地下物流系统后,与时间相关的效益(指因节省货物运送时间所获得的社会效益)所占总效益的比例超过了90%。在建设成本方面,根据休斯敦地下物流系统可行性研究[15],建设54英里(86.94km)的地下物流系统,约投资10亿美元,即0.115亿美元/km,相当于当时约1亿元人民币/km。

考虑到对于许多大城市,目前的垃圾转运系统已购置了大量装备,为了实现与现有地面垃圾运输压缩装备衔接,采用了6.2m直径的地铁隧道及地下集装箱系统进行运输是经济可行的。据此,按单管隧道建设网络做成本分析,其建设成本不超过1亿元人民币/km[7]。如不考虑与现有地面垃圾运输压缩装备兼容,地下环境物流系统的管道直径一般为2m左右,其造价会降低更多。

3　结论

通过对城市地下环境物流系统研究,得到以下结论。

(1)建设城市地下环境物流系统可以缓解垃圾危机,释放城市空间,并杜绝垃圾的二次污染、提高运行效率、降低运输成本,从而促进城市节能减排和可持续发展。

(2)地下生活物流系统初期投资较大,但对长期而言,经济、环境和社会的效益明显,政府应给予财政上的支持。

(3)发展城市地下环境物流系统,可在某些特大城市(如北京、上海、广州等)先选择特定物流对象(如生活垃圾),选择适当线路进行可行性研究和试验。

此外,中国当前许多大城市正努力建设成一个创新型的城市,而建设地下环境物流则可带动相关高新产业的发展。

参考文献(References)

[1] 钱七虎. 建设特大城市地下快速路和地下物流系统——解决中国特大城市交通问题的新思路[J]. 科技导报, 2004, 22(4): 3-6.
Qian Qihu. *Science and Technology Review*, 2004, 22(4): 3-6.

[2] 钱七虎, 郭东军. 城市地下物流系统导论 [M]. 北京: 人民交通出版社, 2007.
Qian Qihu, Guo Dongjun. Introduction to urban underground logistics system[M]. Beijing: China Communications Press, 2007.

[3] 王罗春, 赵爱华, 赵由才. 生活垃圾收集与运输[M]. 北京: 化学工业出版社, 2006.
Wang Luochun, Zhao Aihua, Zhao Youcai. Collection and transportation of domestic waste[M]. Beijing: Chemical Industry Press, 2006.

[4] Liu H. Feasibility of underground pneumatic freight transport in New York City: Final report [R]. NYSERDA Agreement No. 7643. New York: New York State Energy Research and Development Authority, 2004.

[5] Yu M J, Fan Y Q. Plan and conceptual design of urban underground domestic waste transfer logistics system [C]. 6th Conference of International Society on Underground Freight Transport (ISUFT2010), Shanghai, Nov 17-18, 2010.

[6] Fan Y Q, Pei Q. Key issues studies in design and construction of transfer station and terminal of municipal underground environmental sanitation

freight transport system: A case study in Shanghai [C]. 6th Conference of International Society on Underground Freight Transport (ISUFT2010), Shanghai, Nov 17–18, 2010.

[7] Fan Y Q, Zhang Z Y, Hu H. Comparative study of domestic waste transfer method in Shanghai Puxi central downtown district [C]. 6th Conference of International Society on Underground Freight Transport (ISUFT2010), Shanghai, Nov 17–18, 2010.

[8] 郭东军. 上海地下集装箱运输系统基础研究[R]. 上海: 同济大学, 上海市政工程设计研究总院, 2008.
Guo Dongjun. Preliminary research on underground container transportation system in Shanghai [R]. Shanghai: Tongji University & Shanghai Municipal Engineering Design and Research Institute, 2008.

[9] 国家统计局. 中国统计年鉴 (2009)[M]. 北京: 中国统计出版社, 2009.
National Bureau of Statistics of China. China statistical yearbook 2009 [M]. Beijing: China Statistics Press, 2009.

[10] 北京市统计局. 北京统计年鉴 (2006)[M]. 北京: 中国统计出版社, 2006.
Beijing Municipal Bureau of Statistics. Beijing statistical yearbook 2006 [M]. Beijing: China Statistics Press, 2006.

[11] 上海市统计局. 上海统计年鉴 (2009) [M]. 北京: 中国统计出版社, 2009.
Shanghai Municipal Statistics Bureau. Shanghai statistical yearbook 2009[M]. Beijing: China Statistics Press, 2009.

[12] 自然之友组织. 中国环境绿皮书——中国环境发展报告(2010)[M]. 北京: 社会科学文献出版社, 2010.
Friends of Nature. Annual report on environment development of China 2010[M]. Beijing: Social Sciences Academic Press, 2010.

[13] 赵由才, 宋玉. 生活垃圾处理与资源化技术手册[M]. 北京: 冶金工业出版社, 2007.
Zhao Youcai, Song Yu. Domestic waste disposal and recycling technical manual[M]. Beijing: Metallurgical Industry Press, 2007.

[14] Liu H, Lenau C. Pneumatic capsule pipeline (PCP): Recent progress[C]. 5th Conference of International Society on Underground Freight Transport (ISUFT2008), Arlington, Texas, USA, March 20–22, 2008.

[15] James A P, Ir Sanders F M, Ir Arends G. Houston projects 2000 and 2001—research leading to the design of a palletized cargo consolidation and distribution ULS for Houston, Texas [C]. 3rd Conference of International Society on Underground Freight Transport (ISUFT2002), Bochum, Sept 19–20, 2002.

地下洞室连续围岩岩爆的定量分析及其与分区破裂化之间的关系

钱七虎

（解放军理工大学 南京，中国）

摘要：将岩体视为不含初始原生裂隙，仅含微缺陷（微裂纹）的颗粒体。深部地下洞室开挖后，微裂纹扩展，穿越岩石基质，产生次生裂纹，岩体也发生非连续和非协调变形，此时传统的弹塑性理论不再适用于岩体的非连续和非协调变形。岩体的非连续和非协调变形导致应力场具有波动特性，处于波峰位置的次生裂纹尖端应力集中较大，发生失稳扩展，连接和汇合成宏观裂隙，产生破裂区；处于波谷位置的次生裂纹尖端应力集中较小，停止扩展，形成非破裂区。由于波峰和波谷交替出现，因此破裂区和非破裂区也交替出现，也即出现分区破裂化现象。对于脆性岩石，裂纹扩展消耗的能量较小，破裂区岩体储存的弹性应变能有可能大于微裂纹和次生裂纹扩展消耗的总能量，此时，剩余的弹性应变能会突然释放，产生岩爆。根据宏观裂隙出现的条件、微裂纹和次生裂纹扩展过程的耗散能小于岩体储存的弹性应变能的条件，建立了岩爆的判据。最后，通过算例探讨了岩爆和分区破裂化现象之间的关系，确定了岩爆发生的位置、岩爆的规模和岩石的弹射速度与应力状态、岩石的物理力学参数之间的关系。

关键词：地下洞室；岩爆；分区破裂化；非欧几何模型

1 引 言

随着各类深部地下工程建设的不断开展，地下洞室埋深不断加深，深部岩体破坏问题也日益突出，深部岩体的破坏对工程建设造成了不良影响。深部岩体在"三高一扰动"的复杂地质力学环境中，表现出与浅部岩体不同的大变形、非连续与非协调的变形特征。岩爆和分区破裂化是深部岩体的典型破坏形式，研究这两种破坏形式对于分析深部洞室围岩稳定性具有重要意义。

所谓分区破裂化现象是指在深部岩体中开挖洞室或巷道时，在其两侧和工作面前的围岩中产生逐次交替的破裂区和非破裂区[1-4]。岩爆是处于高应力状态下的脆性岩石，因弹性应变能突然释放而产生的岩石碎化、弹射、发声甚至地震等破坏现象。针对深部岩体分区破裂化和岩爆这两种特殊工程响应问题，许多学者进行了深入的研究。潘一山等考虑伴随材料变形的渐变损伤，探讨了圆形洞室发生岩爆的可能性[5]。王桂尧等利用断裂力学分析了裂隙扩展特点，讨论了岩爆发生机理，提出了隧道围岩应力强度因子等值线的概念[6]。俄罗斯学者 Shemyakin 等对塔尔赫斯科十月矿区的深部巷道围岩破裂区进行了现场实测，获得了深部岩体分区破裂化的可靠证据[7]。Guzev 等考虑岩体变形的非协调性，建立了静水压力条件下圆形洞室围岩非欧几何模型[8-9]。钱七虎和周小平研

本文原载于《第一届中俄矿山深部开采岩石动力学高层论坛论文集》（2011年）。

究了深部岩体的非连续和非协调变形，建立了非静水压力条件下圆形洞室围岩的非欧几何模型，揭示了深部岩体分区破裂化产生的机制[10]。Reva 等利用能量准则分析了分区破裂化围岩的稳定性问题[11]。目前，研究者只是分别对岩爆和分区破裂化进行研究，而将两者结合分析深部岩体破坏形式还很少[12]。

本文研究的主要目的是结合分区破裂化与岩爆的发生机理，研究二者之间的相互关系并为岩爆预测提供理论依据。本文主要探讨岩体中颗粒间的微裂纹如何演变为次生裂纹，次生裂纹如何由稳定扩展向失稳扩展转化，次生裂纹如何连接和汇合成宏观裂隙并导致岩体产生岩爆的过程。本文的研究可以为深部岩体稳定性分析提供新的思路。

2 基于非欧几何模型的围岩应力场

钱七虎和周小平考虑岩石材料变形的非连续和非协调特点，在几何方程中采用应变不协调张量，认为变形后岩石材料内部为非欧几何空间，根据不协调量的演化方程，建立了非静水压力条件下的非欧几何模型，获得了卸荷条件下深部圆形洞室围岩弹性应力场，其表达式可以表示为[10]：

$$\begin{aligned}\sigma_r =& \frac{\sigma_v+\sigma_h}{2}\left(1-\frac{r_0^2}{r^2}\right)+\frac{\sigma_h-\sigma_v}{2}\times\left(1-4\frac{r_0^2}{r^2}+3\frac{r_0^4}{r^4}\right)\cos(2\theta)-\frac{E}{2\gamma^{3/2}(1-\nu)r}\\
&\times[AJ_1(\gamma^{1/2}r)+BN_1(\gamma^{1/2}r)+CK_1(\gamma^{1/2}r)]+\frac{E}{2(1-\nu)r\gamma^{3/2}}[A_1J_1(\gamma^{1/2}r)+B_1N_1(\gamma^{1/2}r)\\
&+C_1K_1(\gamma^{1/2}r)]\cos(2\theta)-\frac{3E}{(1-\nu)r^2\gamma^2}\times[A_1J_2(\gamma^{1/2}r)+B_1N_2(\gamma^{1/2}r)\\
&+C_1K_2(\gamma^{1/2}r)]\cos(2\theta)\end{aligned} \quad (1)$$

$$\begin{aligned}\sigma_\theta =& \frac{\sigma_v+\sigma_h}{2}\left(1+\frac{r_0^2}{r^2}\right)-\frac{\sigma_h-\sigma_v}{2}\times\left(1+3\frac{r_0^4}{r^4}\right)\cos(2\theta)-\frac{E}{2\gamma(1-\nu)}[AJ_0(\gamma^{1/2}r)\\
&+BN_0(\gamma^{1/2}r)-CK_0(\gamma^{1/2}r)]+\frac{E}{2\gamma^{3/2}(1-\nu)r}\times[AJ_1(\gamma^{1/2}r)+BN_1(\gamma^{1/2}r)+CK_1(\gamma^{1/2}r)]\\
&+\frac{E}{2(1-\nu)\gamma}[A_1J_0(r\sqrt{\gamma})+B_1N_0(r\sqrt{\gamma})+C_1K_0(r\sqrt{\gamma})]\cos(2\theta)\\
&-\frac{3E}{2r(1-\nu)\gamma^{3/2}}[A_1J_1(r\sqrt{\gamma})+B_1N_1(r\sqrt{\gamma})+C_1K_1(r\sqrt{\gamma})]\cos(2\theta)+\frac{3E}{r^2(1-\nu)\gamma^2}\\
&\times[A_1J_2(r\sqrt{\gamma})+B_1N_2(r\sqrt{\gamma})+C_1K_2(r\sqrt{\gamma})]\times\cos(2\theta)\end{aligned} \quad (2)$$

$$\begin{aligned}\tau_{r\theta} =& \frac{\sigma_v-\sigma_h}{2}\left(1+2\frac{r_0^2}{r^2}-3\frac{r_0^4}{r^4}\right)\sin(2\theta)+\frac{E}{r(1-\nu)\gamma^{3/2}}[A_1J_1(r\sqrt{\gamma})+B_1N_1(r\sqrt{\gamma})\\
&+C_1K_1(r\sqrt{\gamma})]\sin(2\theta)-\frac{3E}{r^2(1-\nu)\gamma^2}[A_1J_2(r\sqrt{\gamma})+B_1N_2(r\sqrt{\gamma})+C_1K_2(r\sqrt{\gamma})]\sin(2\theta)\end{aligned} \quad (3)$$

式中，$\gamma^2=E/[4q(1-\nu)]$；q 为非欧几何参数，可由实验确定；E 为弹性模量；ν 为泊松比；σ_v和σ_h分别为垂直和水平方向的地应力；

$A=(C/2)\pi\sqrt{\gamma}r_0[K_0(\sqrt{\gamma}r_0)N_1(\sqrt{\gamma}r_0)-K_1(\sqrt{\gamma}r_0)N_0(\sqrt{\gamma}r_0)]$；

$B=-(C/2)\pi\sqrt{\gamma}r_0\times[K_0(\sqrt{\gamma}r_0)J_1(\sqrt{\gamma}r_0)-K_1(\sqrt{\gamma}r_0)J_0(\sqrt{\gamma}r_0)]$；

$A_1=(C_1/2)\pi\sqrt{\gamma}r_0[K_2(\sqrt{\gamma}r_0)N_3(\sqrt{\gamma}r_0)-K_3(\sqrt{\gamma}r_0)N_2(\sqrt{\gamma}r_0)]$；

$B_1=-(C_1/2)\pi\sqrt{\gamma}r_0[K_2(\sqrt{\gamma}r_0)J_3(\sqrt{\gamma}r_0)-K_3(\sqrt{\gamma}r_0)J_2(\sqrt{\gamma}r_0)]$。

C为由与$J_0(\sqrt{\gamma}r)$和$N_0(\sqrt{\gamma}r)$的线性相关性决定的朗斯基行列式值，J_0和J_1分别为零阶和一阶贝塞尔函数，N_0和N_1分别为零阶和一阶诺依曼函数，K_0和K_1分别为零阶和一阶第二类修正的贝塞尔函数。

由式（1）、（2）和式（3）确定的围岩应力场具有波动特性。其计算参数为：σ_v=0.4MPa，σ_h=0.1MPa，v=0.2，q=1.448，C=18620，C_1=18620，E=450 MPa，r_0=0.07m，θ=0°；c_0=4 mm，c_1=8 mm，K'_{IC}=0.03 MPa·m$^{1/2}$，K_{IC}=0.12 MPa·m$^{1/2}$。该情况下的深部岩体破裂区分布规律，与传统连续介质理论不同，破裂区和非破裂区是交替出现的，且破裂区宽度随离洞壁距离的增加而减小。

3 围岩破裂区及岩爆特征分析

周小平、Aliakbar 和 Golshani 等建立了微裂纹演化过程和岩石损伤演化规律之间的关系。微裂纹扩展前对应于岩石的弹性阶段，微裂纹扩展阶段和次生裂纹稳定扩展阶段对应于岩石的非线性强化阶段，次生裂纹失稳扩展阶段对应于岩石的应变软化阶段[13-15]。为此，将连续围岩岩爆的孕育过程分为三个阶段：微裂纹扩展阶段、次生裂纹稳定扩展阶段和次生裂纹失稳扩展，连接并汇合成宏观裂隙阶段。

3.1 微裂纹扩展阶段

假设微裂纹的分布是均匀的，细观试验确定的微裂纹初始长度为c_0，微裂纹扩展后的最终长度为c_1，界面的断裂韧度为K'_{IC}。深部地下洞室开挖后，微裂纹首先沿颗粒之间的界面扩展。此时岩体将发生非连续和非协调变形，传统的弹塑性理论获得的应力场不再适用于此时岩体的非连续和非协调变形。非欧几何模型的优点是能准确地分析微裂纹扩展后的深部围岩应力场。

根据非欧几何理论，深部围岩的主应力可以准确地表示为：

$$\begin{cases} \sigma_1' = \dfrac{\sigma_r + \sigma_\theta}{2} + \dfrac{1}{2}\sqrt{(\sigma_r - \sigma_\theta)^2 + 4\tau_{r\theta}^2} \\ \sigma_2' = \dfrac{\sigma_r + \sigma_\theta}{2} - \dfrac{1}{2}\sqrt{(\sigma_r - \sigma_\theta)^2 + 4\tau_{r\theta}^2} \end{cases} \quad (4)$$

式中，$\sigma_r, \sigma_\theta, \tau_{r\theta}$ 由式(1)、(2)和(3)的非欧几何模型应力场确定。

微裂纹表面上的拉应力可以表示为：

$$\sigma_t' = \sigma_2' + f(c_0) S_2' \quad (5)$$

$$S_2' = \sigma_2' - (\sigma_1' + \sigma_2')/2 \quad (6)$$

式中，$f(c_0) = d/c_0$，c_0 为微裂纹的初始长度，d 为岩石颗粒直径。

微裂纹尖端的Ⅰ型应力强度因子可以表示为[13, 14]

$$K_I = -\left[\sigma_2' + f(c_0) S_2'\right]\sqrt{2d_i \tan\left(\dfrac{\pi c_0}{2d_i}\right)} \quad (7)$$

式中，d_i 为微裂纹之间的间距，由细观试验确定。

微裂纹沿界面开始扩展（启裂）的临界条件为：

$$K_I = K_{IC}' \quad (8)$$

式中，K_{IC}' 为界面的断裂韧度，可由实验确定。

当微裂纹尖端的应力强度因子满足式(8)时，微裂纹开始扩展，其扩展后的最终长度由界面的范围确定，为统计平均长度 c_1。此时微裂纹扩展所耗散的能量密度可以表示为能量释放率 G 在微裂纹扩展长度 $c = c_1 - c_0$ 上的积分，即

$$U_1 = \dfrac{N_1(\kappa+1)(1+\nu)}{2E} \int_{c_0}^{c_1} \{\sqrt{2d_i \tan\left(\dfrac{\pi c}{2d_i}\right)}\left[\sigma_2' + f(c) S_2'\right]\}^2 dc \quad (9)$$

式中，$f(c) = d/c$，N_1 为微裂纹密度，由细观试验确定。

3.2 次生裂纹稳定扩展阶段

实验结果表明：在一定应力条件下微裂纹将穿越岩石基质，产生次生裂纹[15, 16]。根据微裂纹扩展后的最终长度 c_1 和岩石的断裂韧度，次生裂纹启裂的临界条件可以表示为：

$$\left[\sigma'_2 + f(c_1)S'_2\right] = -\frac{K_{IC}}{\sqrt{2d_i \tan\left(\frac{\pi c_1}{2d_i}\right)}} \quad (10)$$

式中，K_{IC} 为岩石的断裂韧度，可由实验确定。

次生裂纹产生后，次生裂纹尖端的应力强度因子可以表示为：

$$K_I = -\left[\sigma'_2 + f(l)S'_2\right]\sqrt{2d_i \tan\left(\frac{\pi l}{2d_i}\right)} \quad (11)$$

式中，$f(l) = d/l$，l 为次生裂纹扩展的长度。

次生裂纹稳定扩展的条件为：

$$K_I = K_{IC} \quad (12)$$

当次生裂纹扩展长度 l 达到 c_2 时，岩石的承载能力达到最大，岩石开始出现明显的破裂现象。

岩石出现明显破裂的临界应力条件为：

$$\left[\sigma'_2 + f(c_2)S'_2\right] = -\frac{K_{IC}}{\sqrt{2d_i \tan\left(\frac{\pi c_2}{2d_i}\right)}} \quad (13)$$

式中，$S'_2 = \sigma'_2 - (\sigma'_1 + \sigma'_2)/2$。

对式(5)和(13)进行计算，可以确定岩石能承受的最大抗拉强度 $\sigma'_{t\max} = \sigma'_2 + f(c_2)S'_2$ 和次生裂纹稳定扩展的最终长度 c_2。

次生裂纹稳定扩展所耗散的能量密度可以表示为能量释放率 G 在次生裂纹扩展长度 $l = c_2 - c_1$ 上的积分，即

$$U_2 = \frac{N_2(\kappa+1)(1+\nu)}{2E}\int_{c_1}^{c_2}\{\sqrt{2d_i \tan\left(\frac{\pi l}{2d_i}\right)}\left[\sigma'_2 + f(l)S'_2\right]\}^2 dl \quad (14)$$

式中，N_2 为次生裂纹密度，由计算确定。

3.3 次生裂纹失稳扩展，连接并汇合成宏观裂隙阶段

当次生裂纹扩展长度超过 c_2 时，将发生失稳扩展，岩石承载能力开始下降，损伤局

部化开始出现,损伤局部化出现的临界应力条件为式(13)。随着次生裂纹失稳扩展的继续,次生裂纹逐渐连接和汇合。当次生裂纹的尖端相连接时,次生裂纹完全汇合形成宏观裂隙,导致岩体破坏。

次生裂纹启裂至完全汇合形成宏观裂隙的长度为:

$$l = d_i - c_1 \tag{15}$$

次生裂纹从开始失稳扩展到宏观裂隙出现时的扩展长度为:

$$l = d_i - c_2 \tag{16}$$

假设岩石下降段模量 E_1 和岩石的弹性模量 E 之比为 m。次生裂纹失稳扩展所耗散的能量密度可以表示为:

$$U_3 = \frac{N_3 m(\kappa+1)(1+\nu)}{2E} K_{IC}^2 (d_i - c_2) \tag{17}$$

式中,N_3 为失稳扩展的次生裂纹密度,由计算确定。

3.4 基于能量分析的岩爆判据

岩体储存的弹性应变能密度为:

$$U_e = \frac{1}{2} \sigma_{ij} \varepsilon_{ij} \tag{18}$$

将胡克定律代入式(18)可得岩体储存的弹性应变能密度为:

$$U_e = \frac{1-\nu^2}{2E}\left(\sigma_r^2 + \sigma_\varphi^2\right) - \frac{\nu(1+\nu)}{E}\sigma_r\sigma_\varphi + \frac{1+\nu}{E}\tau_{r\theta}^2 \tag{19}$$

忽略微裂纹和次生裂纹扩展过程的热量耗散密度,岩体发生岩爆的总耗散能密度可以表示为:

$$U = U_1 + U_2 + U_3 \tag{20}$$

本文认为岩爆发生的判据应该同时满足以下两个条件:次生裂纹完全汇合形成宏观裂隙,总耗散能密度小于弹性应变能密度,即

$$\begin{cases} l = d_i - c_1 \\ U < U_e \end{cases} \tag{21}$$

如果地下洞室围岩中任意部位同时满足式(21)中的两个条件,该位置将有发生岩爆的趋势,如果该部位临近洞室临空面则更容易发生岩爆。如果仅仅满足式(21)的第1个条件,则岩体仅仅出现破裂区。

本文模型适用于不含初始原生裂隙,仅含微缺陷(微裂纹)的连续围岩。由于本模型考虑了岩体的非连续和非协调变形,因此本文的解是精确解。

根据能量守恒定律,岩爆的动能密度为

$$W = U_e - U \geq 0 \quad (22)$$

发生岩爆的岩石弹射速度为

$$V = \sqrt{2W/\rho} \quad (23)$$

式中,ρ 为岩石的密度。

根据以上的所有公式可以预测围岩中的破裂区和岩爆区的位置、岩爆岩石的弹射速度以及岩爆的等级(动能密度)。

4 算例与分析

本节以算例的形式分析圆形洞室围岩破裂区及岩爆特征。具体计算参数为:r_0=7 m,v=0.15,q=1.46,C=4 599 14 m^{-2},C_1=4 599 14 m^{-2},K'_{IC}= 0.3 MPa·m$^{1/2}$,N_1=1650,N_2=1650,N_3=200,d_i=7.5 mm,c_0=0.4mm,c_1=0.8mm ρ =2200 kg/m^3。

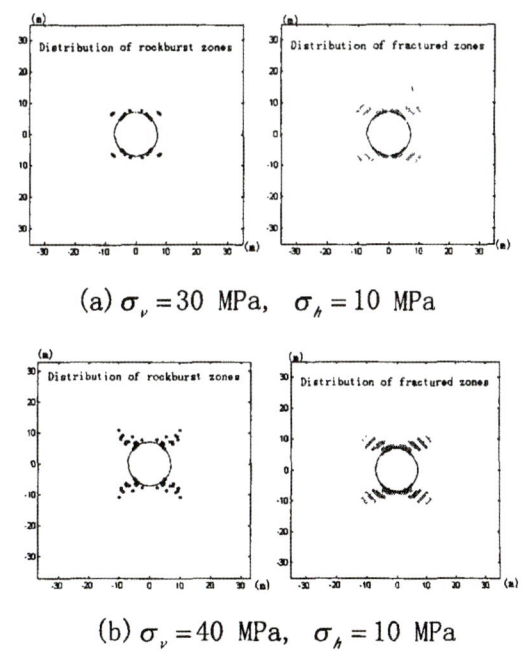

(a) σ_v =30 MPa,σ_h =10 MPa

(b) σ_v =40 MPa,σ_h =10 MPa

图 1 不同应力状态下围岩破裂区及岩爆区分布规律

图 1 表示不同应力状态下围岩破裂区及岩爆区的分布规律,由于岩爆主要发生在脆性岩石材料中,脆性岩石材料峰前模量小于峰后模量,因此,计算参数为 d=8 mm,E=28 GPa,E$_1$=280GPa,K$_{IC}$=1 MPa·m$^{1/2}$。从图 1 中可以看出围岩破裂区和岩爆区分布范围与水平方向应力和竖直方向应力差值有很大关系,当水平方向应力和竖直方向应力差值较大时,远离洞壁也可能会发生岩爆。

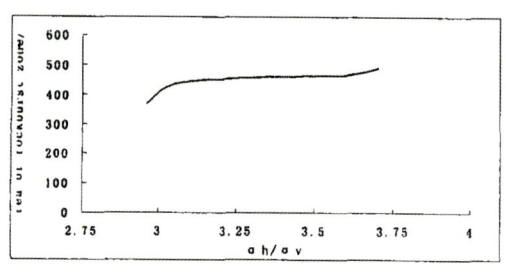

(a)侧压力系数大于1

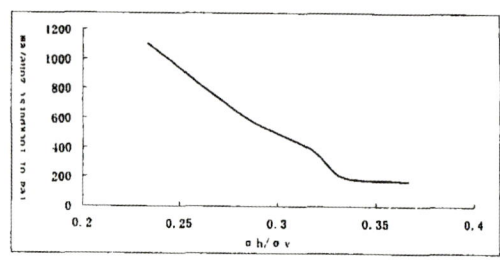

(b)侧压力系数小于1

图2 不同应力状态下围岩岩爆区面积变化规律

图2表示围岩岩爆区面积随侧压力系数的变化规律,从图2中可知:当侧压力系数大于1时,岩爆区面积随侧压力系数的增大而增大;当侧压力系数小于1时,岩爆区面积随侧压力系数的增大而减小。

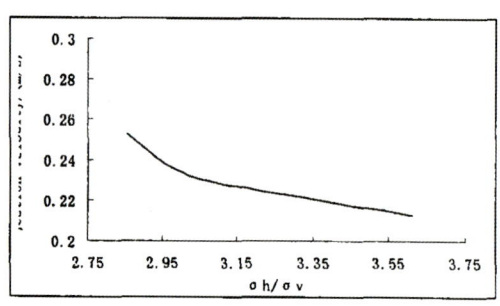

(a)侧压力系数大于1

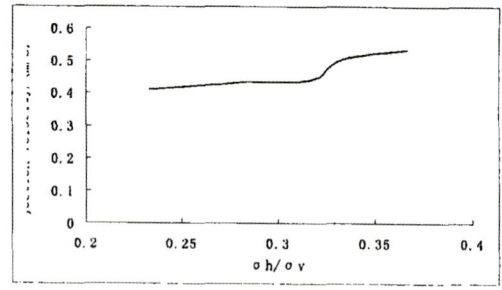

(b)侧压力系数小于1

图3 不同应力状态下碎块平均弹射速度

图3表示碎块平均弹射速度随侧压力系数的变化规律,从图3中可知:当侧压力系数大于1时,碎块平均弹射速度随侧压力系数的增大而减小;当侧压力系数小于1时,碎块平均弹射速度随侧压力系数的增大总体上是增大的。

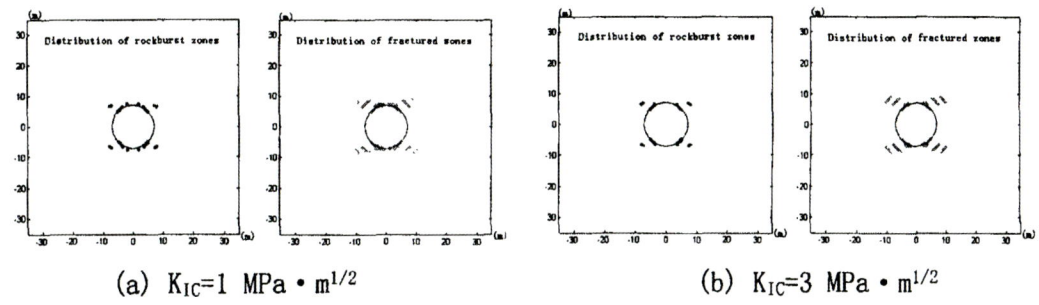

(a) $K_{IC}=1$ MPa·m$^{1/2}$ (b) $K_{IC}=3$ MPa·m$^{1/2}$

图 4 不同岩石断裂韧度条件下围岩岩爆区分布规律

图 4 表示不同岩石断裂韧度条件下围岩岩爆区分布规律，计算参数为：$\sigma_v = 30$ MPa，$\sigma_h = 10$ MPa，$E=28$ GPa，$E_1 = 280$ GPa，$d = 8$ mm。从图 4 可以看出围岩岩爆区分布范围随岩石断裂韧度的增大而减小。

图 5 表示围岩岩爆区面积随岩石断裂韧度变化规律，从图 5 可以看出围岩岩爆区面积随岩石断裂韧度的增大而减小。

图 6 表示碎块平均弹射速度随岩石断裂韧度变化规律，从图 6 可以看出碎块平均弹射速度随岩石断裂韧度的增大而增大。

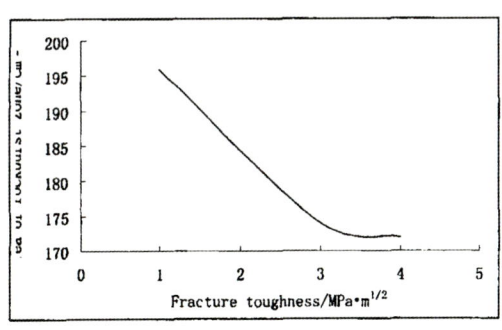

图 5 不同岩石断裂韧度条件下围岩岩爆区面积变化规律

图 6 不同岩石断裂韧度条件下碎块平均弹射速度

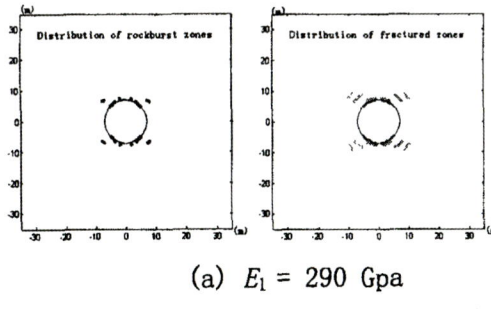
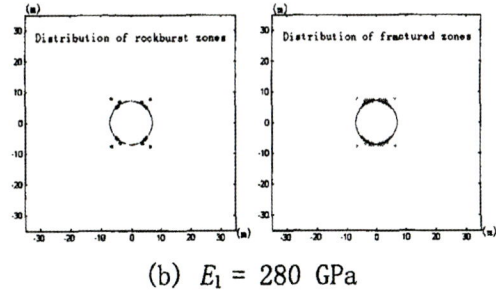

(a) $E_1 = 290$ GPa (b) $E_1 = 280$ GPa

图 7 不同下降段模量下围岩岩爆区分布规律

图 7 表示不同下降段模量条件下围岩岩爆区分布规律，计算参数为：$\sigma_v = 30$ MPa，$\sigma_h = 10$ MPa，$E=28$ GPa，$d=8$ mm。从图 10 可以看出围岩岩爆区分布范围随下降段模量的增大而增大。

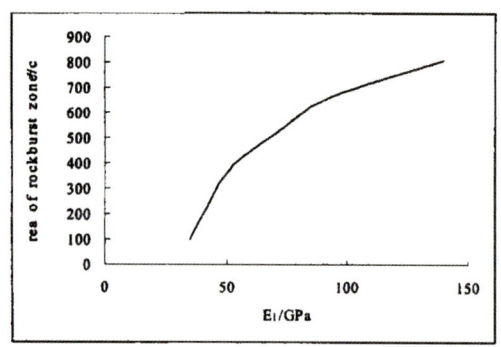

图 8 不同下降段模量条件下围岩
岩爆区面积变化规律

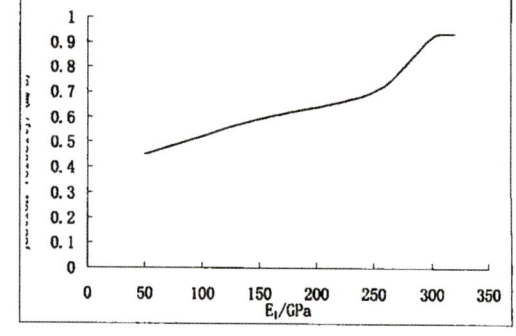

图 9 不同下降段模量条件下碎
块平均弹射速度

图 8 表示围岩岩爆区面积随下降段模量的变化规律，从图 8 可以看出围岩岩爆区面积随下降段模量的增大而增大。

图 9 表示碎块平均弹射速度随下降段模量变化规律，从图 9 可以看出碎块平均弹射速度随下降段模量的增大而总体上增大。

5 结论与讨论

岩爆和分区破裂化是深部岩体破坏的两种不同形式，这两种破坏形式都可能出现在高地应力岩体中。

分区破裂化的形成机理是：深部地下洞室开挖后，微裂纹扩展，穿越岩石基质，产生次生裂纹，岩体表现为非连续和非协调变形。基于非欧几何模型的岩体非连续和非协调变形的分析表明，围岩应力场的分布具有波动特性，处于波峰位置的次生裂纹尖端应力集中较大，并发生失稳扩展，连接和汇合成宏观裂隙，产生破裂区。处于波谷位置的次生裂纹尖端应力集中较小，并停止扩展，形成非破裂区。由于波峰和波谷交替出现，因此破裂区和非破裂区也交替出现。深部岩体不再遵守连续介质理论所解释的破坏区和非破裂区依次出现的规律。深部岩体的分区破裂化现象表明：适用于浅部岩体的连续介质力学理论对深部岩体不再适用。

岩爆的孕育发生机理如下：岩爆的孕育阶段从洞室开挖扰动导致岩体中微裂纹开始扩展到发生岩爆，经历了微裂纹扩展、次生裂纹稳定扩展、次生裂纹失稳扩展和连接并汇合成宏观裂隙 3 个阶段。对于脆性岩石，微裂纹和次生裂纹扩展所消耗的能量将可能小于岩体中储存的弹性应变能，剩余的弹性应变能在次生裂纹失稳扩展后突然释放，转化为破裂区岩体的动能，从而发生岩爆。

从岩爆和分区破裂的发生机理可以看出两者之间具有如下关系：岩爆和分区破裂化现象都是由岩体中的次生裂纹失稳扩展、连接并汇合成宏观裂隙形成的。取决于岩石脆性大小，储存在破裂区岩体中的弹性应变能可能大于或小于微裂纹和次生裂纹扩展消耗的能量，对于前者，剩余的弹性应变能转化为破裂区岩体动能，则发生岩爆；对于后者，破裂区岩体不会发生岩爆,只发生分区破裂化现象。

本文分析了圆形洞室分区破裂化现象和岩爆发生规律,建立了非静水压力条件下考虑岩体分区破裂化的岩爆预测模型,通过数值计算表明:

(1)岩爆不仅发生在洞壁,也可能发生在远离洞壁的地方。水平方向和垂直方向的地应力差越大,越有可能在远离洞壁的地方发生岩爆。

(2)当侧压力系数小于1时,岩爆区面积随侧压力系数的增大而减小,碎块平均弹射速度随侧压力系数的增大而增大;当侧压力系数大于1时,岩爆区面积随侧压力系数的增大而增大,碎块平均弹射速度随侧压力系数的增大而减小。

(3)围岩岩爆区面积随岩石断裂韧度的增大而减小。碎块平均弹射速度随岩石断裂韧度的增大而增大。

(4)围岩岩爆区和破裂区面积随下降段模量的增加而增加,碎块平均弹射速度随下降段模量的增加而总体增加。

(5)本文假设深部巷道开挖卸荷速度很慢,巷道开挖时间很长,深部巷道开挖被认为是静力问题,而实际深部岩体开挖问题是动力问题,因此要研究动力非欧几何模型。

参考文献

[1] 钱七虎.非线性岩石力学的新进展—深部岩体力学的若干关键问题—[C]//第八次全国岩石力学与工程学术会议论文集.北京:科学出版社,2004.(Qian Qi-hu.The new development of nonlinear rock mechanics—many key problem of deep rock mass mechanics[C]//The Eighth National Rock Mechanics and Engineering Academic Memoir.Beijing:Science Press,2004.(in Chinese))

[2] 钱七虎,李树忱.深部岩体工程围岩分区破裂化现象研究综述[J].岩石力学与工程学报,2008,27(6):1278-1284. (Qian Qihu, Li Shuchen. A review of research on zonal disintegration phenomenon in deep rock mass engineering[J]. Chinese Journal of Rock Mechanics and Engineering, 2008,27(6):1278-1284.(in Chinese))

[3]. Q H Qian, X P Zhou, H Q Yang, Y X Zhang, X H Li. Zonal disintegration of surrounding rock mass around the diversion tunnels in Jinping II Hydropower stataion, southwestern china[J]. Theoretical and Applied Fracture Mechanics, 2009,51(2):129-138.

[4]. 周小平,钱七虎.深埋巷道分区破裂化机制[J].岩石力学与工程学报,2007,26(5): 877-885.(Zhou Xiaoping, Qian Qihu. Zonal fracturing mechanism in deep tunnel[J]. Chinese Journal of Rock Mechanics and Engineering, 2007,26(5):877-885.(in Chinese))

[5].潘一山,徐秉业.考虑损伤的圆形洞室岩爆分析[J].岩石力学与工程学报,1999,18(2): 152-156.(Pan Yishan, Xu Bingye. The rockburst analysis of circular chamber under consideration of rock damage[J]. Chinese Journal of Rock Mechanics and Engineering, 1999,18(2): 152-156.(in Chinese))

[6].王桂尧,孙宗颀,卿笃干.隧洞岩爆机理与岩爆预测的断裂力学分析[J].中国有色金属学报,1999,9(4):841-845.(Wang Gui-yao, Sun Zongqi, Qing Dugan. Fracture mechanics analysis of rock burst mechanism and prediction[J]. The Chinese Journal of Nonferrous Metals, 1999,9(4):841-845.(in Chinese))

[7]. Shemyakin I, Kyrlenya M V, Reva V N, et al. USSR discovery No.400, phenomenon of zonal disintegration of Rocks around underground workings[J].Byull.Izobret., 1992,(1):7-15.

[8]. M A Guzev, A A Paroshin. Non-euclidean model of the zonal disintegration of rocks around an underground working[J]. Journal of Applied Mechanics and Technical Physics,2001,42(1):131-139.

[9]. V P Myasnikov, M A Guzev. Thermomechanical model of elastic-plastic materials with defect structures[J]. Theoretical and Applied Fracture Mechanics, 2000,33(3):165-171.

[10]. Qian Qihu Zhou Xiaoping. Non-Euclidean continuum model of the zonal disintegration of surrounding rocks around a deep circular tunnel in a non-hydrostatic pressure state[J]. Journal of Mining Science, 2011(in press).

[11]. Reva V N. Stability criteria of underground workings under zonal disintegration of rocks[J].Fiz.Tekh.Probl.Razrab.Polezn.Iskop., 2002, 38(1): 35-38.

[12]. 李春睿, 康立军, 齐庆新, 等. 深部巷道围岩分区破裂与冲击地压关系初探[J]. 煤炭学报, 2010,35(2): 185-190. (Li Chun-rui, Kang, Li-jun, Qi Qing-xin, et al. Probe into relationship between zonal fracturing and rock burst in deep tunnel[J]. Journal of China Coal Society, 2010,35 (2):185-190.(in Chinese))

[13]. Aliakbar Golshani, Yoshiaki Okui, Masanobu Oda, Takato Takemura. A micromechanical model for brittle failure of rock and its relation to crack growth observed in triaxial compression tests of granite[J]. Mechanics of Materials, 2006,38:287-303.

[14]. A Golshani, M Oda, Y Okui, T Takemura, E Munkhtogoo. Numerical simulation of the excavation damaged zone around an opening in brittle rock[J]. International Journal of Rock Mechanics and Mining Sciences, 2007,44(6):835-845.

[15]. Zhou Xiao-Ping. Analysis of the localization of deformation and the complete stress-strain relation for mesoscopic heterogeneous brittle rock under dynamic uniaxial tensile loading. International Journal of Solids and Structures, 2004, 41 (5-6):1725-1738.

[16]. E Sahouryeh, A V Dyskin, L N Germanovich. Crack growth under biaxial compression[J]. Engineering Fracture Mechanics, 2002,69(18): 2187-2198.

学术论文——隧道及地下工程

我国城市地下空间综合管理的探讨

陈晓强,钱七虎

(解放军理工大学,南京 210007)

摘 要:我国城市地下空间开发利用发展迅速,已成为我国城市建设和改造的重要组成部分。如何提高我国城市地下空间开发利用的层次和水平,已成为亟待解决的问题。通过对目前我国城市地下空间开发利用管理现状的分析,指出了其存在法规不完善、体制不统一和措施不到位等问题。从系统的角度出发,提出了我国城市地下空间综合管理的基本概念,界定了综合管理的定义、内涵和特性,构想了综合管理的组成体系,阐述了综合管理各要素的相互关系,为城市地下空间管理指明了方向。

关键词:地下空间;综合管理;模式

中图分类号:TU99   **文献标识码**:A   **文章编号**:1673-0836(2010)04-0666-06

Study on the Integrated Management of Urban Underground Spacein China

Chen Xiaoqiang Qian Qihu

(PLA University of Science and Technology Nanjing 210007China)

Abstract:The utilization of urban underground space in China is rapidly developing.It has become an important component of urban construction in cities It is urgent to elevate the level of the development and utilization of urban underground space.By analyzing the status of management of urban underground space this paper points out that there are many problems in the area of management legislation management system and managementmeasures.From systematic angles the paper puts forward an integrated management model defines its concept connotation and characteristics conceives its composition system and describes the mutual relations among all elements.The paper shows clearly the future of management of urban underground space

Keywords:underground space;integrated management;model

1 引言

在我国城市化进程中,城市地下空间作为城市发展的空间资源之一,对于解决城市化进程中出现的土地紧张、交通拥堵、环境污染、能源浪费等问题发挥了积极作用。开发利用城市地下空间已成为我国城市建设和改造的重要组成部分,是保证城市可持续发展的必由之路[1]。

进入21世纪以后,我国城市地下空间开发利用迅速发展,特大城市地下空间开发利用的总体规模和发展速度已居世界同类城市的前列,我国已经成为世界城市地下空间开发利用的大国[2]。面对城市地下空间开发利用的发展需求,管理水平不高已成为制约城市地下空间开发利用的"瓶颈",研究和探索适应我国城市地下空间开发利用的管理模式已刻不容缓。

2 我国城市地下空间管理存在的问题

2.1 管理法规不完善,制约了城市地下空间的持续开发利用,亟待健全

收稿日期:2010-03-04(修改稿)。
本文原载于《地下空间与工程学报》(2010 年第6 卷第4 期)。

管理法规是实施城市地下空间开发利用管理的依据。目前,我国在城市地下空间开发利用的立法方面,出台了一些法律法规。在国家层面上,涉及城市地下空间开发利用的主要法规政策有《中华人民共和国物权法》、《中华人民共和国城乡规划法》、《中华人民共和国人民防空法》和《城市地下空间开发利用管理规定》。各省市根据实际需要,也制定了相关的地方性法规、政府规章或规范性文件,如《北京市人防工程和普通地下室安全使用管理办法》、《上海市城市地下空间建设用地审批和房地产登记试行规定》、《天津市地下空间开发利用管理条例》、《重庆市城乡规划地下空间规划导则》、《深圳市地下空间开发利用暂行办法》、《杭州市区地下空间建设用地管理和土地登记暂行规定》等。

虽然我国出台了不少城市地下空间开发利用方面的法律法规、政府规章和规范性文件,但还有许多立法问题没有得到解决。一是没有形成权威的国家立法。国家最高立法机构没有制定关于城市地下空间开发利用的专门法律或国务院的行政法规,建设部的《城市地下空间开发利用管理规定》只是一个政府部门的行政规章,缺乏足够的许可设定权限,法律效力比较低,加上其规定不具体、内容不完备,在实践中难以有效地贯彻执行。二是没有解决基本的权属问题。《中华人民共和国物权法》第136条规定:建设用地使用权可以在土地的地表、地上和地下分别设立[3]。该规定仅停留在宏观层面,并没有从本质上确立地下空间权的概念及物权形态,没有确立地下空间权的取得方式、转让、抵押方法及权属管理等。三是没有明确统一的管理体制。虽然《城市地下空间开发利用管理规定》明确建设行政主管部门负责城市地下空间的开发利用管理工作,但对建设部门的统一管理职责以及如何与土地管理、规划、人防及其他专业管理部门进行协调,则没有明确的规定,无法真正统一起来。四是没有制定专门的技术标准。地下空间建设的标准不统一,没有专业设计、施工、验收方面的标准,只是参照相关技术标准执行[4]。五是没有提出具体的优惠政策。国家对人防工程给予了相应的优惠,而对其他地下空间开发利用在投融资、价格、税收等方面则无具体的优惠政策。

2.2 管理体制不统一,限制了城市地下空间的综合开发利用,亟待理顺

管理体制是实施城市地下空间开发利用管理的基础。目前,我国对城市地下空间的管理由相关的行政管理部门按照职责分别实施相关的管理工作。以上海为例,涉及地下空间管理的部门有19家:发改委、经委、信息委、规划、建设、房地、民防、交通、市政、供电、工商、公安、消防、水务、市容、卫生、通信管理、安监、应急办等[5]。

从我国大多数城市地下空间的管理实践来看,目前主要采取以下管理模式:一是以协调管理为主导。如上海市建立了地下空间管理联席会议这一议事协调工作机制,通过联席会议的跨部门协调平台,协商解决有关地下空间管理中的问题。二是以规划管理为主导。如深圳市建立以规划为龙头的地下空间管理协调机制,通过规划协调相关部门管理的界限,形成以规划管理为核心,以轨道交通沿线为重点,通过详细规划进行控制、协调和平衡的管理体制。三是以建设管理为主导。如根据《城市地下空间开发利用管理规定》,建立以建设行政主管部门为管理主体,以规划控制和建设监管为核心的城市地下空间开发利用管理体制。四是以人防管理为主导。如杭州市人民政府明确"杭州市地下空间的开发、管理归口市人防办,作为市人防办的一项重要职能"。五是以安全管理为主导。如北京市以地下空间使用安全为着力点,以人防和建设(房屋)管理部门为主,分别对人防工程和普通地下室进行管理。

各城市地下空间管理体制的差异性,正反映了目前我国城市地下空间管理体制不统一的尴尬境地。从以上各城市对地下空间开发利用管理体制的不同探索,不难看出,由于地下空间管理部门众多,管理职能分散、交叉,而相互之间又缺乏强有力的统一领导、协调机制和法制保障,无论采用哪种管理模式,由哪个部门牵头管理,都难以得到其他相关部门的有力配合,必然造成多头管理与无人管理并存的局面[6]。

2.3 管理措施不到位,影响了城市地下空间的合理开发利用,亟待规范

管理措施是实施城市地下空间开发利用管理的保证。目前,我国城市地下空间开发利用的管理措施还存在不少值得改进的地方。

在规划管理方面,我国城市地下空间开发利用的规划主要有三类:第一类仅局限于人防工程规划,并作为总体规划的一个专项由人防部门编制;第二类是结合城市整体发展战略,常以地铁规划、共同沟及地下商业街规划为主要内容,地下空间的利用融汇在各个专业规划之中;第三类则较为系统地考虑城市地下空间资源,分层分类地结合城市各

类设施的发展需求,通盘规划地下空间的开发利用。目前,我国城市地下空间规划还是以第一、二类为主,只有少数城市编制了城市地下空间的总体规划[7]。

在建设管理方面,一是建设项目互不连通。不同的开发者根据自己的需要开发建设项目,项目之间缺乏衔接和联系,彼此独立、互不连通,而政府又相应的有效措施。二是建设安全存在隐患。地下施工比地面施工更具危险性和复杂性,技术标准还不成熟,加上建设过程中缺乏统一的管理,形成了管理上的漏洞和盲区。三是建设资金投入不足。地下空间的开发建设投入大、建设周期长、投资回报慢,而由于地下空间产权不清,又缺乏鼓励优惠政策,尚未形成投资收益机制和有偿使用机制,致使地下空间的资金投入不足。

在使用管理方面,一是地下空间使用安全管理主体不清、责任不明。由于地下空间产权不明晰,带来了地下空间使用管理主体不清、责任不明等一系列问题,导致地下空间安全管理无法落到实处。二是地下空间应急资金缺少投入。应急管理资金由谁来投,按什么标准建,由谁来督促验收等还缺乏政策法规的支持。三是缺乏完善的地下空间突发事件应急预案。目前只有重要的地下空间目标制订了相应的应急预案,组织开展了相关演练,而绝大多数的地下空间是既无预案,也无措施,更无演练。

在信息管理方面,一是管理主体不明确。地下空间信息数据资源大都只停留在专业管理功能需要层面,由各相关的专业管理部门分别负责收集、储存和使用,没有明确一个具体部门负责信息资源的整合。二是标准体系不统一。各相关专业管理部门对地下空间数据信息化范围该有多大、如何信息化管理、采用何种形式和标准进行信息化等问题还没有形成统一的认识,都是根据自己的需要制定相关的标准体系。三是信息资源不共享。由于地下空间信息数据量大、形式多、格式杂、分布广,加上地下空间信息化管理系统在负责部门、采用的标准、运用的技术、建设的目标和范围上有所区别,不可避免地导致地下空间信息资源难以实现共享。

3 我国城市地下空间综合管理的基本概念

3.1 城市地下空间综合管理的定义

《中国管理大辞典》对综合管理和集成管理的定义如下:综合管理:专业管理实行横向联系,综合利用各种方法,整体地提高企事业功能的管理。集成管理:对生产要素的集成活动以及集成体的形成、维持及发展的变化进行能动地计划、组织、指挥、协调、控制,以达到整合增效目的的过程[8]。

联合国人类居住中心《关于健全的城市管理规范:建设"包容性城市"的宣言草案》:"城市管理是个人和公私机构用以规划和管理城市公共事务的众多方法的总和。它是一个解决各种冲突或不同利益以及采取合作行动的持续过程,包括正式的制度,也包括非正式的安排和公民社会资本。"我国的学者认为:城市管理是指以城市这个开放的复杂巨系统为对象,以城市基本信息流为基础,运用决策、计划、组织、指挥、协调、控制等一系列机制,采用法律、经济、行政、技术等手段,通过政府、市场与社会的互动,围绕城市运行和发展进行的决策引导、规范协调、服务和经营行为[9]。

根据我国城市地下空间开发利用的特点,结合目前我国城市地下空间开发利用管理的实际,笔者认为:城市地下空间综合管理是指以城市地下空间为客体,从系统的角度,采用法律、行政、技术和经济等手段,统一管理与专业管理相结合,对规划、建设、使用、信息等能动地进行决策、协调和控制,以期发挥城市地下空间开发利用整体效益的过程。

3.2 城市地下空间综合管理的内涵

城市地下空间综合管理的主体:包括综合管理部门和专业管理部门。综合管理部门主要负责从宏观上统筹考虑城市地下空间的各个管理环节,牵头组织综合管理的各项内容,组织协调各专业管理部门的相互关系。专业管理部门在综合管理部门的统一协调下,根据专业特点,各司其职,分工协作,形成管理上的合力。

城市地下空间综合管理的客体:包括所有建于城市地下空间的各类设施,如地下综合体、地下交通设施、地下市政设施、地下商业设施、地下文化娱乐设施、地下防灾设施、地下储存设施、地下生产设施、地下教育科研设施、地下居住设施等[10]。

城市地下空间综合管理的内容:研究制定城市地下空间开发利用的法律法规、行政规章、技术标准和有关政策;监督协调城市地下空间有关行政管理部门的职能履行和相互关系;组织编制城市地下空间开发利用的各项规划;批准城市地下空间开发利用的土地划拨、出让(含转让)、招投标、核价等;进行城市地下空间开发利用工程项目的立项审批、设计审查、施工监管、竣工验收和建档备案;负责城市地下空间的产权登记管理;监督检查城市地下空间的安全使用管理;建设、管理和维护城市地下空间的综合信息;开展城市地下空间开发利用的科学技术研究、学术交流和宣传教育。

3.3 城市地下空间综合管理的特性

3.3.1 系统性

城市地下空间综合管理牵涉到土地、规划、建设等多个管理部门,决策、执行、监督等各个管理阶段,规划、建设、使用、信息等全部管理措施,构成了多主体、多层次、多结构的复杂系统。因此,必须从系统的角度出发,运用系统管理的方法,统筹城市地下空间的开发利用,通过各子系统的有机结合,保证大系统高效、有序地运行。

3.3.2 协调性

一方面是管理主体的协调。综合管理部门与各专业管理部门相互协调,既统筹全局,又不越俎代庖,使综合与专业有机结合;各专业管理部门相互协调,既避免职能相互交叉,又杜绝相互推诿,使分工与协作相结合。另一方面是管理环节的协调。城市地下空间开发利用的规划、建设、使用、信息等各个管理环节要衔接有序,既使城市地下空间与城市发展需求、总体规划、整体功能相协调,也使各个地下空间项目相协调,实现城市地下空间布局合理,开发有序,体系完整,功能综合。

3.3.3 整体性

现代城市将"地面、地下、空中"视为一个完整的"空间整体",城市地下空间的开发利用应与城市的总体发展相适应,成为其整体的有机组成部分。城市地下空间的开发利用不应分散独立,而要相互协调,逐步向布局网络化、开发立体化、利用集约化和功能综合化方向发展。应通过统一管理,分工协作,衔接有序的管理体制,整体筹划,统一组织,使地下空间资源各项开发利用的功能有机联系,保证其综合效益大于单项建设的利益之和,提高城市地下空间开发利用的整体效能。

4 我国城市地下空间综合管理的组成体系

4.1 城市地下空间综合管理

如图1,城市地下空间综合管理包括城市地下空间管理法规、城市地下空间管理体制和城市地下空间管理措施,三者紧密联系。城市地下空间管理法规是实施综合管理的基本依据,在法律法规上明确规定管理体制和管理措施。城市地下空间管理体制是实践综合管理的执行主体,在管理法规的法律授权或行政授权下采取各种管理措施。城市地下空间管理措施是实现综合管理的具体方法,由管理机构依据管理法规,通过有效的管理机制,进行综合运用。

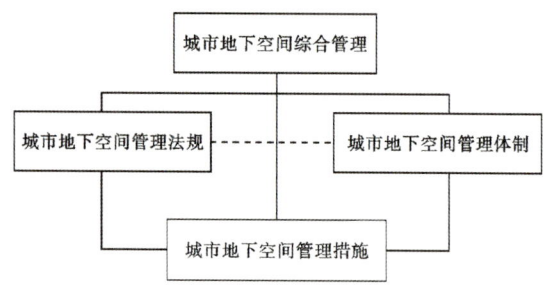

图1 城市地下空间综合管理框图

Fig 1 Integrated management of urban underground space

4.2 城市地下空间管理法规

如图2,城市地下空间管理法规主要包括民事基本法、专项立法或综合立法、配套立法。民事基本法主要是为城市地下空间开发利用提供民事基础权利,解决地下空间权属关系。专项立法主要是针对城市地下空间不同的利用形态进行的立法。然后根据需要,逐步总结经验,对地下空间开发利用进行综合立法。配套立法主要是针对城市地下空间开发利用的相关技术标准、鼓励政策等进行立法。三者统一互补、相辅相成,构成城市地下空间管理法规的完整体系。

4.3 城市地下空间管理体制

如图3,城市地下空间管理体制是指城市地下空间管理系统的组织结构和运行方式,包括管理机构和管理机制。管理机构包括综合管理机构和专业管理机构。各管理机构的职权分配以及相互之间的协调,直接影响到管理的效率和效能。综合管理部门负责牵头组织综合管理的各项内容,各专业管理部门的相互协调。专业管理部门则在综合管理部门的统一协调下,各司其职,分工协作。管理机制是为保证管理目标的实现,各管理机构间相互联系、相互作用的基本准则及相应制度。

4.4 城市地下空间管理措施

如图4,城市地下空间管理措施主要包括规划管理、建设管理、使用管理和信息管理等。规划管理是对城市地下空间的规模、布局和功能进行合理控制和分配。建设管理是为拓展人们的生存空间提供各类地下工程设施。使用管理是保障各类地下工程设施正常运行、安全可靠和提高效率。信息管理是采用信息化方式对城市地下空间的数据资源进行整合。城市地下空间的规划、建设和使用既

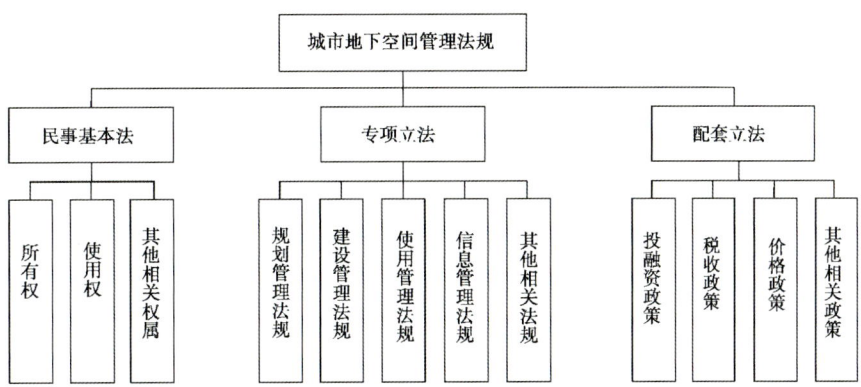

图 2 城市地下空间管理法规框图

Fig 2　The diagram of legislation of integrated management in urban underground space

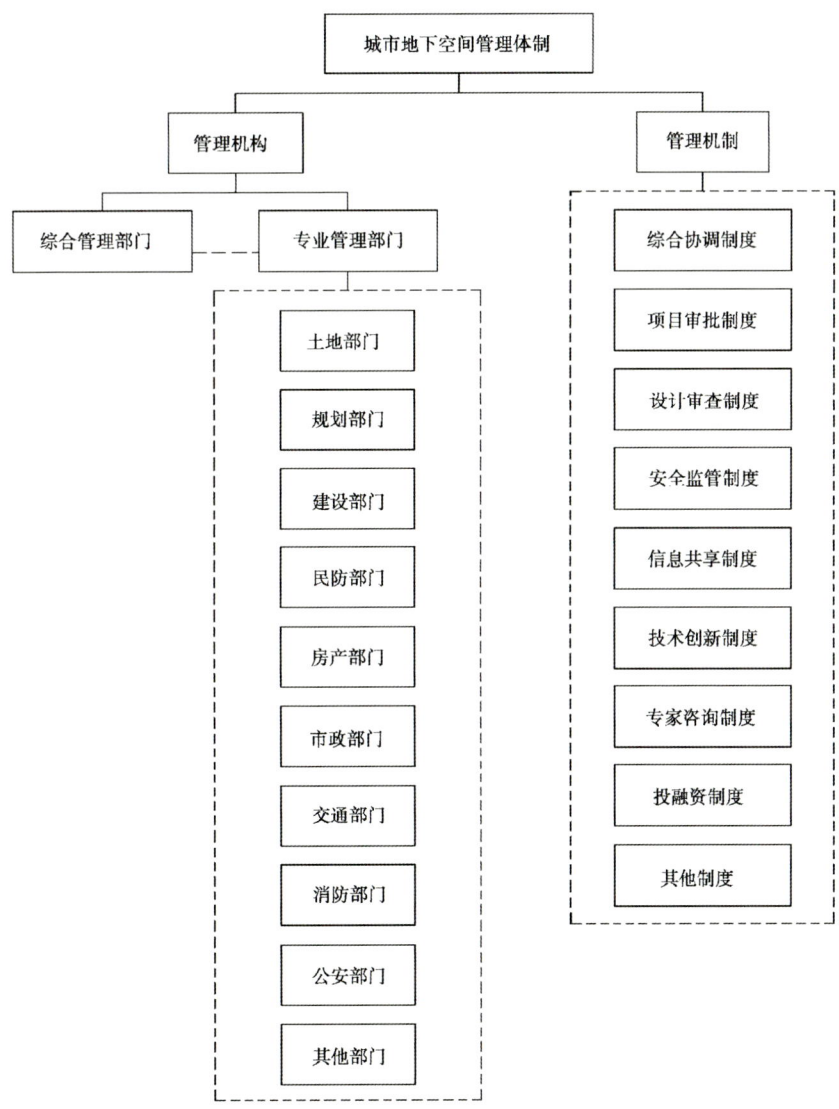

图 3 城市地下空间管理体制框图

Fig 3　The diagram of organization of integrated management in urban underground space

有工作时序上的前后延续性,又有工作性质上的互动互补性。城市地下空间的规划、建设和使用都离不开全面了解和准确掌握城市地下空间的基础数据。城市地下空间信息为合理规划、协调和安全使用提供依据、创造条件。

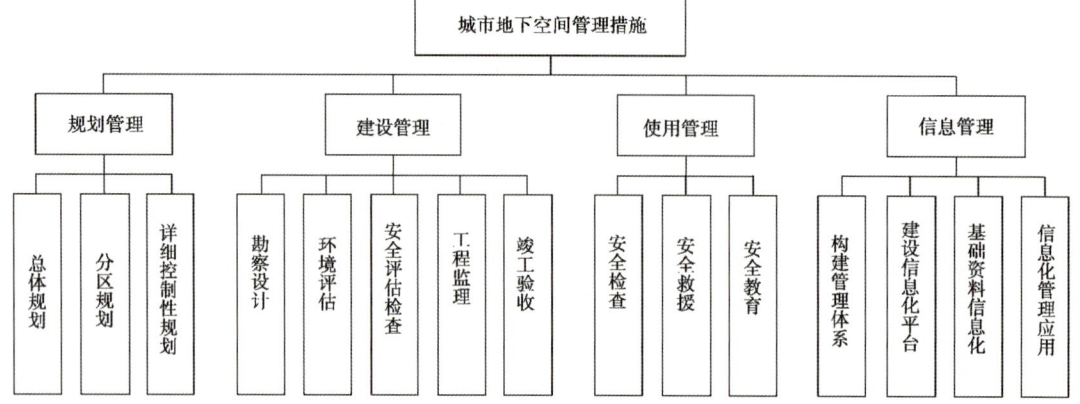

图 4 城市地下空间管理措施框图

Fig.4 The diagram of measures for integrated management of urban underground space

5 结语

实践证明,城市地下空间开发利用的发展需要职责明确、运行高效的城市地下空间管理模式,而良好的管理模式又会规范和促进城市地下空间的开发利用。

为适应城市地下空间"网络化、立体化、集约化、深层化、综合化"的发展需求,达成城市地下空间开发利用的高效率,提升城市地下空间开发利用的层次和水平,使我国从城市地下空间开发利用的大国变为强国,应大力加强城市地下空间开发利用综合管理研究,逐步完善城市地下空间的管理法制、管理体制和管理措施,促进我国城市地下空间开发利用综合管理向法制化、规范化、科学化方向发展。

参考文献(References)

[1] 钱七虎,陈晓强. 充分开发利用地下空间建设资源型和环境友好型城市[J]. 中国人民防空, 2006, (9): 15-18. (Qian Qihu, Chen Xiaoqiang. Full development and utilization of underground space in the building of resource-conserving and environment-friendly cities[J]. Chinese Civil Air Defence, 2006, (9): 15-18. (in Chinese))

[2] 钱七虎. 中国城市地下空间开发利用的现状评价与前景展望[J]. 民防苑, 2006, 88: 1-5. (Qian Qihu. Evaluation of the status and outlook of the urban underground space development and utilization in China[J]. Civil Defense Realm, 2006, 88: 1-5. (in Chinese))

[3] 中华人民共和国物权法[S]. 中国法制出版社, 2007. (Property law of the People's Republic of China [S]. China Legal Publishing House, 2007. (in Chinese))

[4] 钱七虎,陈晓强. 国内外地下综合管线廊道发展的现状、问题及对策[J]. 地下空间与工程学报, 2007, 3(2): 191-194. (Qian Qihu, Chen Xiaoqiang. Situation, problems and countermeasures of utility tunnel's development in China and abroad[J]. Chinese Journal of Underground Space and Engineering, 2007, 3(2): 191-194. (in Chinese))

[5] 上海地下空间开发利用综合管理研究[R]. 2008. (The research of integrated management of development and utilization of underground space in Shanghai[R]. 2008. (in Chinese))

[6] 孙晓波. 城市地下空间实行综合管理的探索与实践[J]. 民防苑, 2006, 88: 26-27. (Sun Xiaobo. The exploration and practice of integrated management of urban underground space[J]. Civil Defense Realm, 2006, 88: 26-27. (in Chinese))

[7] 李迅. 关于城市地下空间规划的若干问题探讨[J]. 民防苑, 2006, 88: 61-67. (Li Xun. The discussion on some problems of the planning of urban underground space[J]. Civil Defense Realm, 2006, 88: 61-67. (in Chinese))

[8] 中国管理科学学会. 中国管理大辞典[M]. 北京: 中央文献出版社, 2008. (Society of Management Science of China. The Chinese dictionary of management. Beijing: The Central Documentary Press, 2008. (in Chinese))

[9] 宋刚,陈锐. 复杂性科学与现代城市管理[J]. 科学对社会的影响, 2006(4): 27-30. (Song Gang, Chen Rui. Complicated science and management of modem city[J]. The Impact of Science on Society, 2006(4): 27-30. (in Chinese))

[10] 陈志龙,王玉北. 城市地下空间规划[M]. 南京: 东南大学出版社, 2005. (Chen Zhilong, Wang Yubei. The planning of city underground space[M]. Nanjing: Southeast University Press, 2005. (in Chinese))

隧道工程动力响应特性与汶川地震隧道震害分析及启示

钱七虎[1] 何 川[2] 晏启祥[2]

(1. 中国人民解放军理工大学，南京 210007；2. 西南交通大学，成都 610031)

我国地处欧亚地震带和环太平洋地震带之间，地震活动频繁，是世界上最大的大陆浅源强震活动区，震害具有分布广、强度高、危害大等特点。从"5·12"汶川大地震及世界范围的工程结构物震害情况看，通常情况下隧道工程震害较房屋、桥梁等地面建筑轻微，隧道工程相对具有较强的抗震性能。但另一方面，在强震条件下，隧道工程震害依然较突出，如"5·12"汶川大地震、台湾集集地震、日本关东大地震、阪神大地震等强震都造成了断层破裂带、缺陷衬砌段隧道结构的破坏，且隧道破坏后修复困难，说明强震条件下的隧道抗减震问题同样需要我们的高度关注。我国现有的山岭隧道有相当部分位于地震区，尤以西南、华北和西北的铁路、公路隧道为甚。而城市地铁区间隧道也有相当数量处于高地震烈度区，据不完全统计，在全国300多个城市中，有一半位于地震基本烈度为Ⅶ度至Ⅶ度以上的地震区，23个百万以上人口的特大城市中，有70%为Ⅶ度和Ⅶ度以上的地震区。20世纪以来，我国已经经历了4个地震活跃期，现在正在经历第5个地震活跃期，作为我国交通命脉的铁路、公路和城市地铁遍及各地，在高烈度地震区修建隧道难以避免，长大隧道作为国家生命线工程的主体结构之一，必须对其地震震害给予高度重视，并从中吸取经验教训，把隧道抗减震设防理念贯穿到各类隧道选线、设计、施工和维修等各个阶段。

1 隧道工程震害

1.1 "5·12"汶川大地震隧道工程震害

近100年以来，全球发生了数次大地震，其中"5·12"汶川大地震、阪神大地震、台湾集集地震当中隧道的震害都比较典型。"5·12"汶川大地震发生在四川龙门山映秀—北川断裂上，龙门山断裂具有十分复杂的结构和构造，主要包括前山断裂（灌县—安县断裂）、中央断裂（北川—映秀断裂）和后山断裂（汶川—茂县断裂），本次地震是龙门山逆冲推覆体向东南方向推挤并伴随顺时针剪切共同作用的结果，见图1。

"5·12"汶川大地震中，都汶公路、国道213线都江堰至汶川段以及剑阁至青川公路等道路上的大量公路隧道都处于高烈度地震区，烈度约Ⅷ~Ⅺ度，这些路段上的隧道都遭受了不同程度的损坏，通过对这些路段上的龙溪隧道、龙洞子隧道、龙池隧道、紫坪铺隧道、友谊隧道、酒家垭隧道以及宝成铁路109号隧道等隧道的震害调查和收集，可把隧道典型病害分为洞口和洞身段病害两大类。隧道洞口区域主要震害现象为：洞口边仰坡垮塌、掩埋洞口；洞口落石、局部边仰坡地面开裂变形；边仰坡防护、截排水沟开裂变形；洞门墙及洞口附近衬砌开裂等，见图2~图6。

本文原载于《汶川大地震工程震害调查分析与研究》（2009年）。

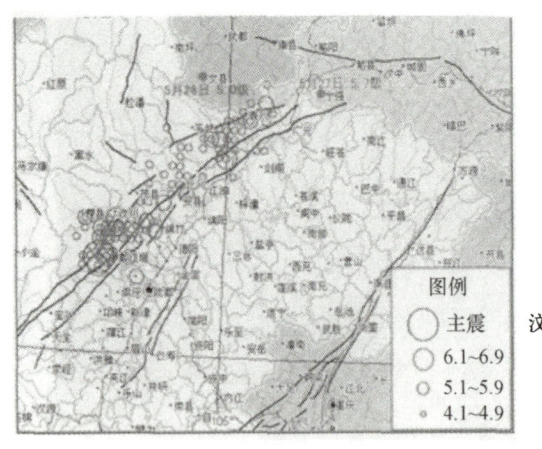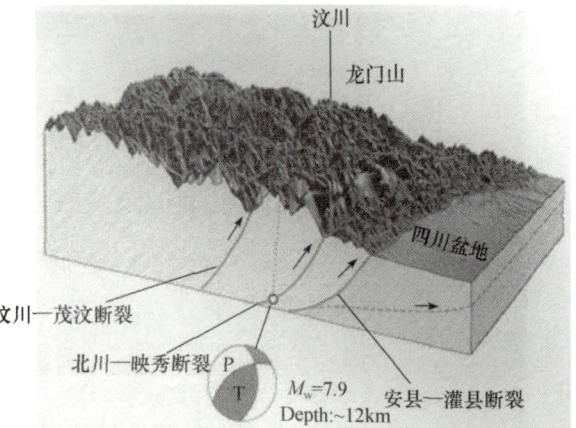

图1 "5·12"汶川大地震成因示意图

图2 洞口边仰坡震害

图3 洞口落石

图4 边仰坡垮塌掩埋洞口

图 5　边仰坡开裂变形

图 6　洞口衬砌开裂

隧道洞身区域主要震害现象有：衬砌环向、纵向、斜向开裂；衬砌错台、钢筋扭曲、局部掉块；衬砌局部垮塌；衬砌上部拱圈整体掉落；路面仰拱隆起；洞体垮塌、瓦斯积聚，见图 7～图 11。除此之外，钢支撑扭曲变形、锚杆垫板脱落等现象也非常普遍。值得指出的是：洞身初期支护和二次衬砌发生严重损坏地段大多处于高地应力区段或者穿越软弱破碎带区域。

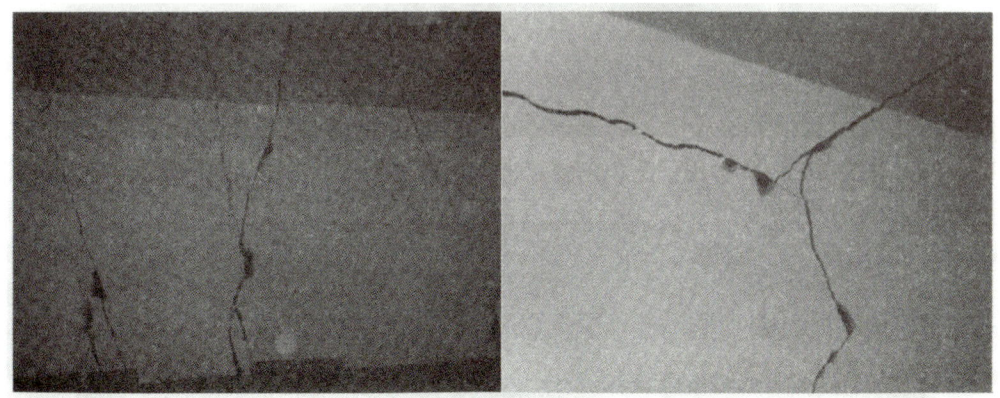

图 7　洞身衬砌开裂

"5·12"汶川地震中，成都地铁工程是距离震中最近的城市地下隧道，烈度约Ⅴ～Ⅶ度，其区间盾构隧道和车站结构局部部位出现了少量较轻微的震害，其中区间盾构隧道的震害特征较为明显，主要表现形式为：管片衬砌裂纹、剥落、错台、渗漏，见图 12。渗漏位置主要发生在横断面 45°方向呈 X 共轭分布，纵向错台主要发生在隧道的侧部。上述震害都没有对隧道的主体安全和建筑限界产生影响。

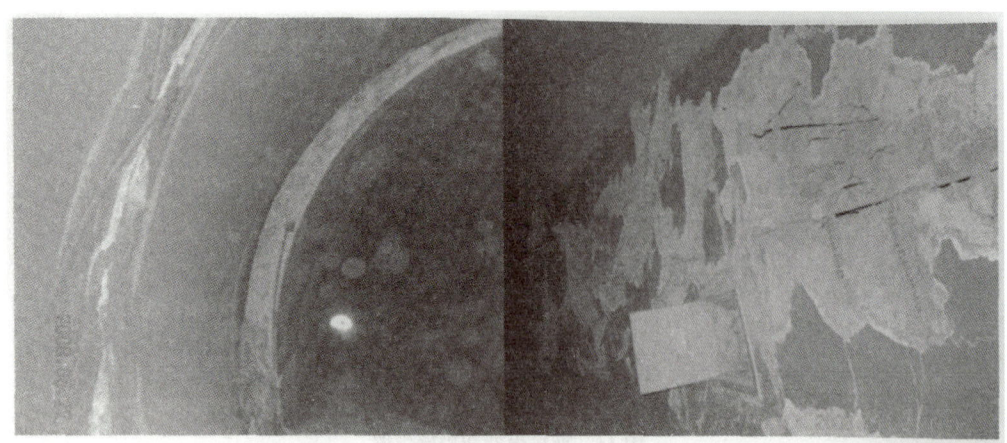

图 8　洞身衬砌错台、剥落和钢筋扭曲

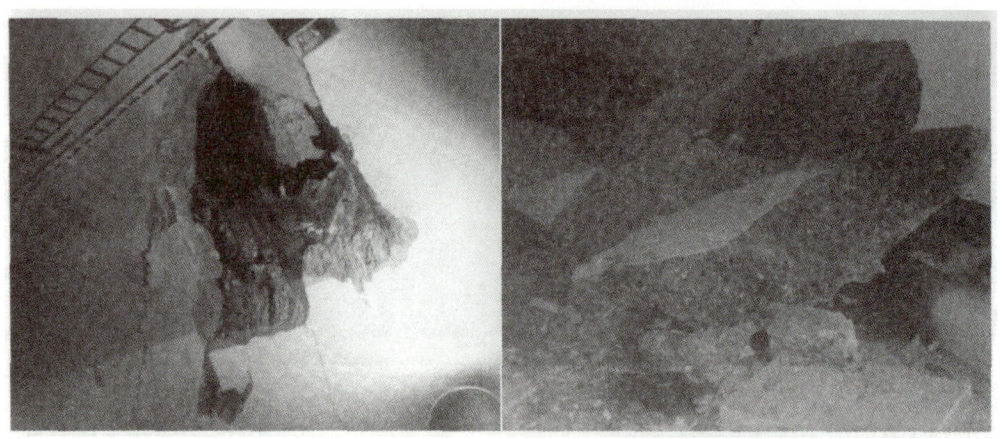

图 9　洞身衬砌局部垮塌

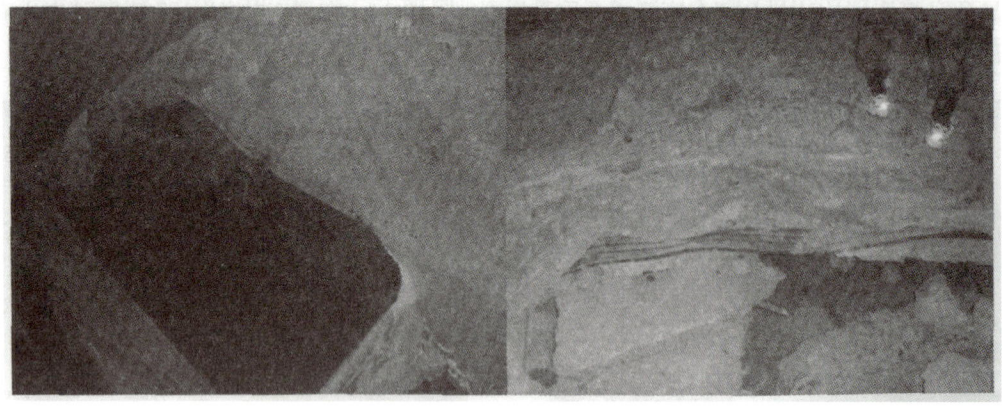

图 10　洞身上部拱圈整体掉落

1.2　国内外其他典型隧道工程震害

我国台湾省和国外的强震同样也造成了大批隧道损坏。如 1923 年日本关东大地震，使震中附近的 25 座隧道受到破坏。1995 年 1 月 17 日在日本发生的阪神大地震，地震规模为里氏 7.3 级，死亡人数约有 6500 人，使一百多座隧道发生不同程度的破坏，有些地方造成了地面塌陷，损坏较为典型的隧道

图 11 仰拱隆起和洞体垮塌

图 12 地铁盾构隧道错台

有神户铁路东山隧道、神户大开地铁车站和六甲隧道等[1],见图13、图15和图16。1999年9月21日台湾发生的集集7.3级地震,震源深度8km,在地表造成了长约100km的破裂带,死亡人数约2300多人,导致台湾8号、14号公路多座隧道严重受损[2],见图14,在调查的总计57座隧道中,有49座隧道衬砌发生混凝土块龟裂、掉落,甚至钢筋弯曲等不同程度的破坏。

图 13 神户铁路东山隧道震害(阪神地震)   图 14 台湾玛陵二号隧道震害(集集地震)

2 隧道工程动力响应特征分析

针对过去国内外隧道工程震害,作者以及国内外相关研究人员开展了大量代表性的研究工作。隧道等地下结构在地震作用下,由于周围岩土介质的存在,会发生不同于地面结构的动力响应[3],地面

图 15 神户地铁大开车站震害（阪神地震）

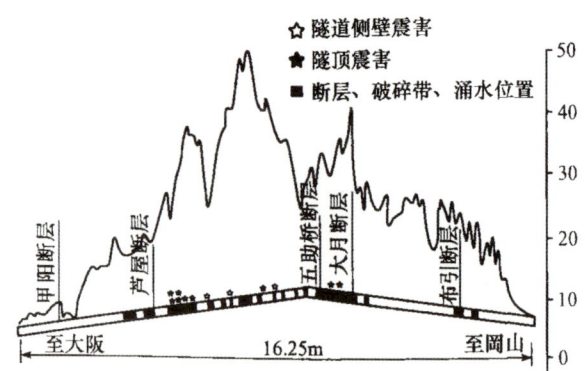
图 16 山阳新干线六甲隧道震害分布（阪神地震）

结构具有明显的加速度放大效应，而地下结构与附近地层的加速度相对比较接近，见图 17。地震以地震波的形式传播能量，当地震波从基岩传入场地时，场地介质在地震波的作用下，将运动传递给隧道等地下结构。通过模型试验和动力分析（见图 18）[4]，揭示了地层中有无隧道时地面和隧道结构不同位置的加速度和相对位移曲线，可以发现隧道的加入对地层的动力响应影响很小，且隧道结构的动力响应很大程度上依赖于地层的振动响应，见图 19 和图 20；进一步的研究还发现，地层加速度与隧道应变时称曲线吻合程度没有地层相对位移与隧道应变时称曲线吻合程度高，说明隧道结构变形对地层变形的依赖性和追随性，见图 21 和图 22。结合目前的相关研究结果，可以总结出隧道具有如下的主

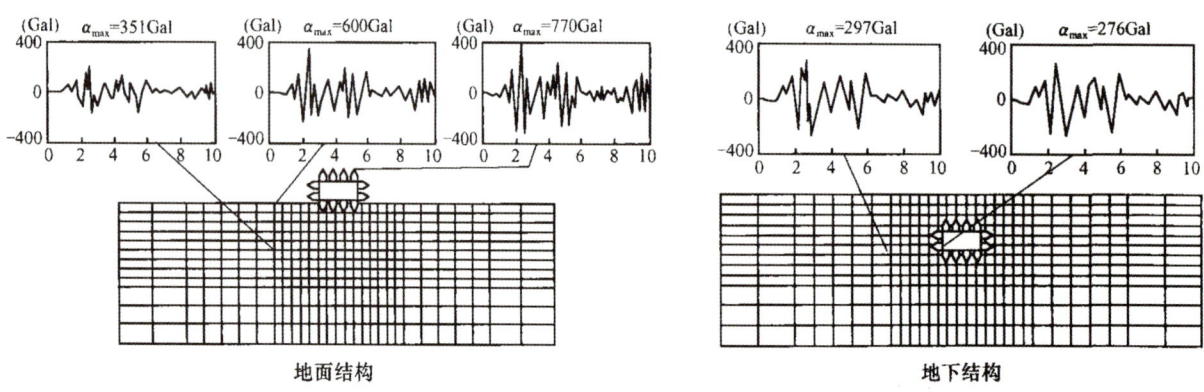

图 17 地面结构与地下结构的动力响应对比

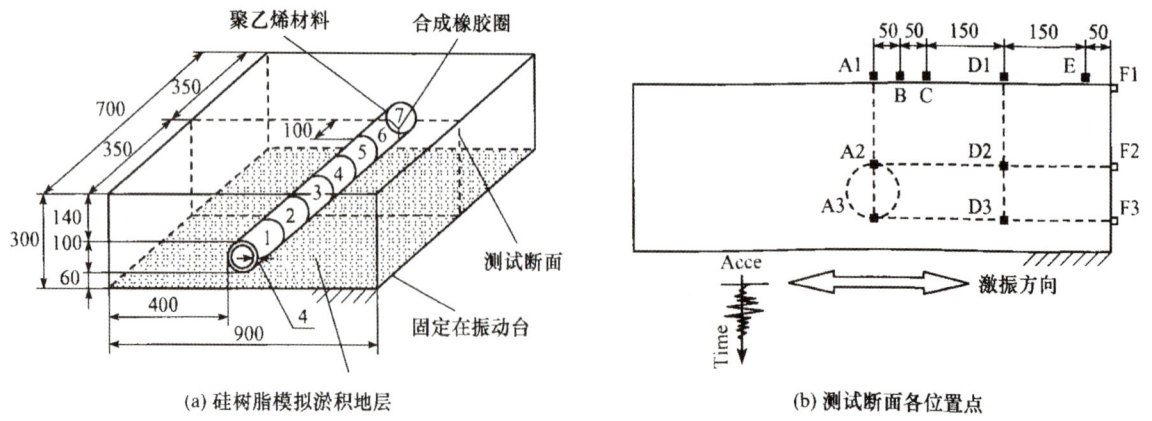

图 18 地铁盾构隧道振动台试验示意图

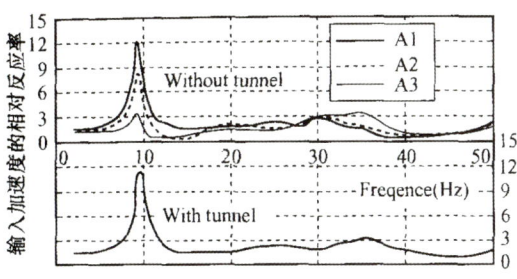

图 19 地层加速度谐振曲线（振动台试验）

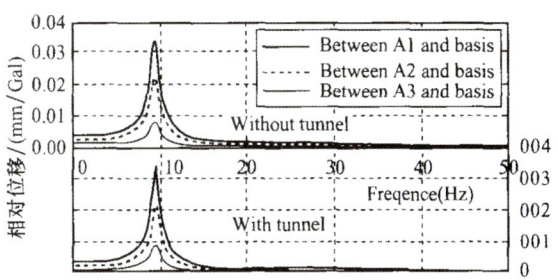

图 20 地层相对位移谐振曲线（动力分析）

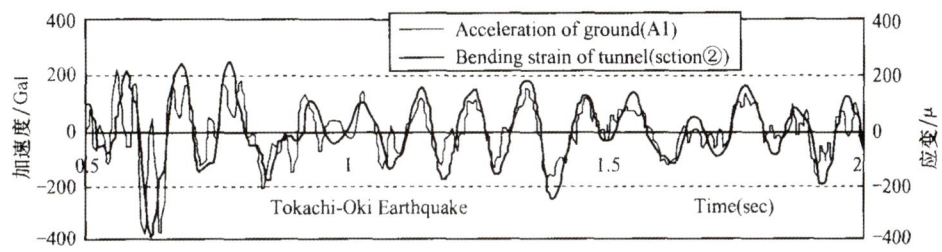

图 21 地面加速度波形和隧道应变波形的对比（振动台试验）

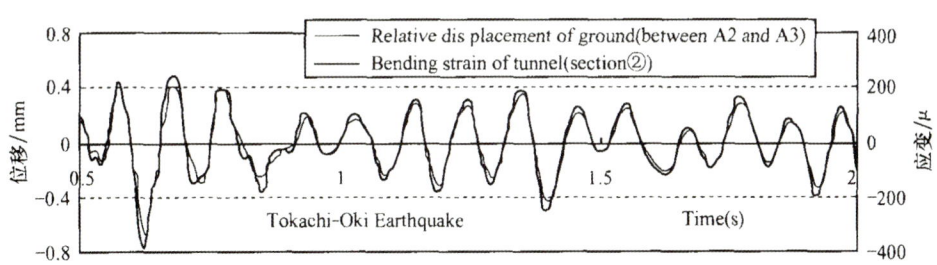

图 22 地层位移波形和隧道应变波形的对比（振动台试验）

要振动特性：与周围地层相比，隧道结构通常具有单位体积重量较小、振动衰减较快、自身惯性力影响较小等特征；地震时隧道结构的振动对地层的振动具有追随性，衬砌所产生的附加应力和变形主要是由地层的相对位移引起的。

另一方面，尽管隧道对地层加速度的追随性不如对其位移的追随性明显，但隧道结构的加速度振幅与围岩是基本一致的，当加速度或速度达到一定量值时，在岩质隧道中是导致衬砌结构出现震害的主要原因。Dowding 和 Rozen 根据 71 座铁路隧道和水工隧洞的震害调查结果发现，当地层加速度大于 $0.5g$ 或速度大于 $0.8m/s$ 时，隧道会发生严重破坏[5]，而"5·12"汶川大地震距震中较近的龙溪公路隧道区域，地层加速度远远超过 $0.5g$，因此采用的普通衬砌结构出现洞体垮塌是必然的（见图 11）。因此，要区别土质地层和岩质地层，正确认识地层加速度和位移各自对隧道结构动力响应的影响。

在隧道纵向抗震方面，当隧道所在地层，上覆地层和下卧地层以及隧道埋深复杂多变，地层围岩性质呈现很大的不均匀性时，在隧道遭受地震等非常时附加荷载情况下，隧道的纵向将发生不协调的三维变形，导致隧道破坏，特别是在隧道穿越活动断层若软弱破碎带时尤其，高地应力条件下也容易出现震害[6~9]，见图 23。"5·12"汶川大地震中洞身衬砌出现严重错台，拱顶衬砌整体塌落和隧道洞体垮塌地段都是处于围岩条件变化很大，尤其是存在软弱破碎带的区域。因此，应当加强对隧道穿越不同地层，穿越大的地质构造等所在区段进行重点设防。

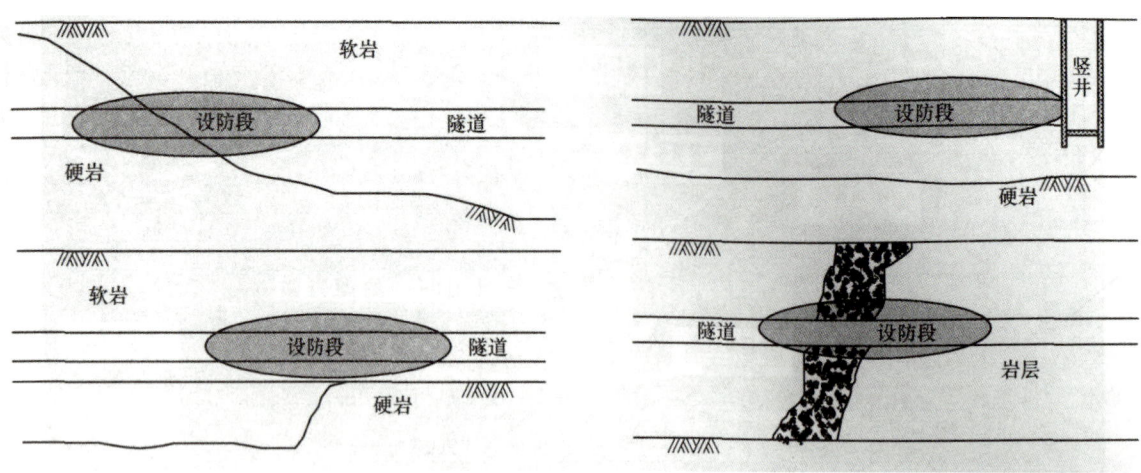

图 23 在纵向方向隧道洞身易出现震害的地段

3 基于隧道工程动力特征和震害的启示

3.1 启示一

隧道工程较其他地面结构具有相对良好的抗震性能，在环境条件复杂的山区，宜多采用隧道方案，它不仅能改善线形提高道路通行能力，同时也利于提高线路整体抵御地震灾害的能力。

我国特别是西部山区，地质构造运动强烈，深大活动断裂发育，高烈度地震频繁，诱发的大、巨型滑坡、崩塌、泥石流等地质灾害频发，若越岭地段线路较多采用明线方案，其潜在危险大，因此越岭地段宜少展线，尽可能以隧道方式通过。在"5·12"汶川地震中，虽然有多座公路铁路隧道发生不同程度的震害，在近震中区甚至出现洞体垮塌，但总体上来看，相对于邻近的桥梁和路基边坡支挡物，隧道的受损状况堪称轻微，岩质隧道工程的良好抗震能力得到证实。二郎山隧道和鹧鸪山隧道由于开展了有针对性的抗震设计，其震害更加轻微，成为震后黄金救援期进入汶川、理县等重灾区唯一"生命线"的关键通道，为抢救人民群众的生命财产作出了贡献。当前，我国正处于扩大内需，加大交通基础设施建设的关键时期，一大批铁路和公路项目纷纷开建。以西南片区特别是四川为例，将建设成（都）九（寨沟）铁路、兰（州）渝（重庆）铁路、大（理）瑞（丽）铁路、贵（阳）广（州）铁路等，另外，成都—马尔康—拉萨铁路正在规划当中。高速公路方面，将建设成都—昆明、成都—拉萨、成都—格尔木、成都—兰州、成都—贵阳等 5 条高速公路。这些铁路和公路工程都必须穿越山区。如大（理）瑞（丽）铁路将穿越高黎贡山越岭地区，成九铁路将穿越龙门山脉[10]，图 24 示出了穿越龙门山脉多种线路方案的隧道布局情况，最后设计部分推荐了龙门山 14km 取直隧道方案，但不管何种方案，其隧道长度所占的比重都相当大。高地震烈度区山区交通道路的建设，必须站在工程安全和线路标准等全局高度正确认识隧道工程在其中发挥的不可或缺的重要作用。

3.2 启示二

要抓紧研究和制定专门针对隧道结构物的抗减震设计规范，尽快改变目前隧道结构抗减震规范条文少，实用性和操作性差的现状。

目前，国内隧道等地下结构的相关的规范基本上采用地面结构的抗震设计方法。《铁路工程抗震设计规范》（GB 50111—2006）针对铁路隧道的条文较少，规范中关于隧道的抗震强度和稳定性验算条文是参照既有设计经验。根据该规范规定：铁路隧道设防目标只需要达到抗震性能要求Ⅱ，在大致为Ⅲ级及以上围岩的洞口、浅埋、偏压隧道和明洞，地震烈度为Ⅶ~Ⅸ度时，需进行抗震验算，验算方法为附加地震力的静力法，并只验算水平地震力对隧道强度和稳定性的影响。但实际工程设计中，铁路

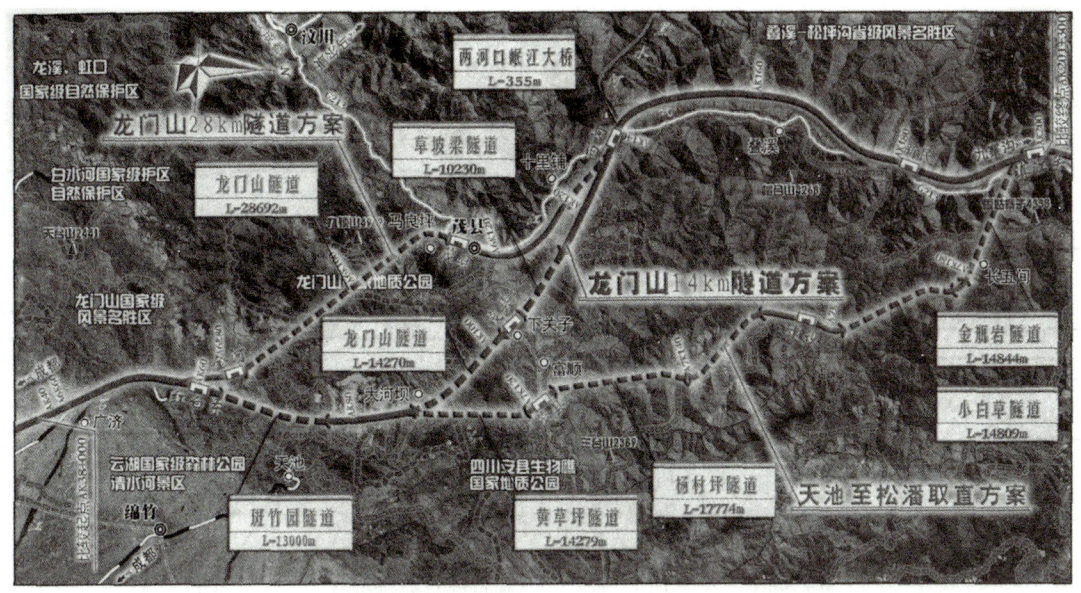

图 24 成九铁路穿越龙门山各方案隧道示意图

隧道稳定性验算范围、洞口设防段长度的决定及其影响因素、活动断层破碎带设防段长度等问题都需要进一步细化并研究其合理程度，对于深埋、浅埋、土层、岩层，其抗震设计方法也应当有所区别；高烈度地震区隧道的抗震设计思路一般是加强衬砌，增大刚度，抵抗地震作用力的方法，但靠一味增大结构刚度来抵抗地震力的设计思路也值得商讨。同时，对于单线、双线、长隧道、短隧道、其洞型型式、结构形态究竟如何决策等问题也一直困扰着隧道工程界，图 25 中通过振动台试验就说明了单双线不同洞型型式条件下衬砌的应变有较大改变[11]。目前作者正在相继开展铁道部重点科技研究开发计划项目"隧道工程结构物抗震设计标准与方法研究"和"高地震烈度地区隧道活动断裂带抗震减震技术研究研究"，正式着力解决上述有关铁路隧道抗震的许多模糊性问题[12,13]。

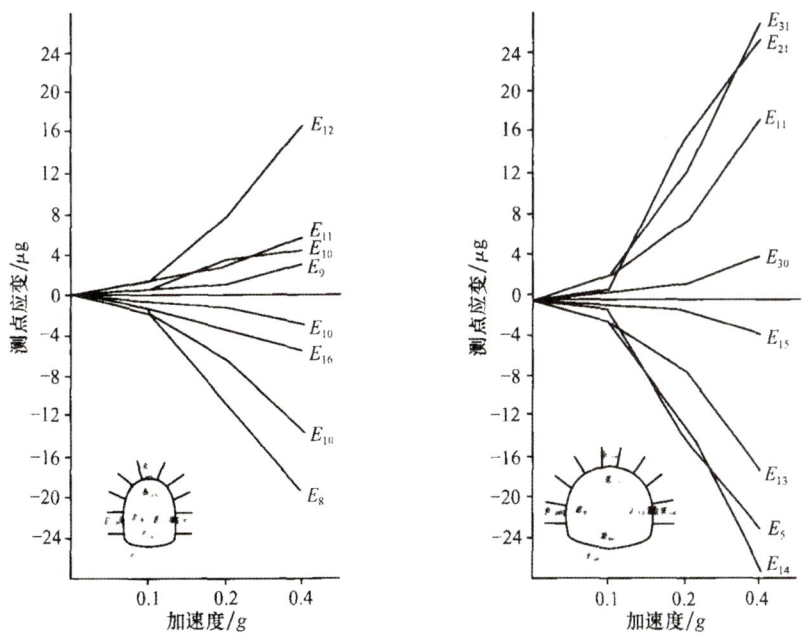

图 25 单线和双线铁路隧道衬砌应变对比（振动台试验）

《公路工程抗震设计规范》(JTG D70—2004) 中对于隧道抗震的规定基本沿用铁路隧道规范的相

关规定，国内也正在研究和对比不同公路隧道衬砌型式的抗震效果[14]。1997年颁布的《核电厂抗震设计规范》（GB50267—97）中规定对于地下结构采用反应位移法进行抗震计算。在我国地下铁道设计中，由于研究工作开展不够，目前规范对抗震设计尚无具体规定。

3.3 启示三

高烈度地震区的隧道工程，应当注意土质和岩质隧道的差异性，采取抗震减震并举的措施抵御地震作用；对于岩质隧道，要把洞口段和洞身周围地质环境复杂地段（围岩急剧变化处、断层破碎带处、结构刚度急剧变化处，高地应力地段）作为重点，前者以抗震措施为主，后者以减震措施为主；对于土质隧道，特别是盾构隧道，重点是进行合理的接头选型设计，增强其适应地震变形的能力，并做好与竖井，联络横通道，地铁车站等连接部位的抗减震设计；建议加大对隧道减震技术的研究开发力度，尽快把有效的减震技术推向实用。

过去，隧道震害问题的解决主要借助抗震手段，如增加衬砌混凝土的刚度、厚度，添加钢纤维形成钢纤维混凝土等措施。但是，仅仅依靠强化衬砌刚度等措施去抗震，效果有时适得其反，有时根本抗不住，且隧道造价也偏高，这时，需要考虑采取以柔克刚的工程措施，减震技术就是在这种条件下提出来的。目前国外减轻隧道地震灾害主要有两种途径：第一是通过改变隧道结构本身的性能，如减小隧道结构的刚性，使之易于追随地层的变形而变形，从而减小隧道的地震响应。第二种途径是在隧道与地层之间设置减震层，使地层的变形难以传递到隧道结构上，从而减小隧道结构的地震响应，减震的基本构想是：在隧道衬砌的外周和地层之间设减震层（某种吸能材料），使原有的衬砌结构围岩系统变为衬砌减震层围岩系统，其目的是通过减震层将衬砌与围岩介质隔开，从而减小和改变地震对结构的作用强度和方式，以此来达到减小结构振动的目的，见图26。

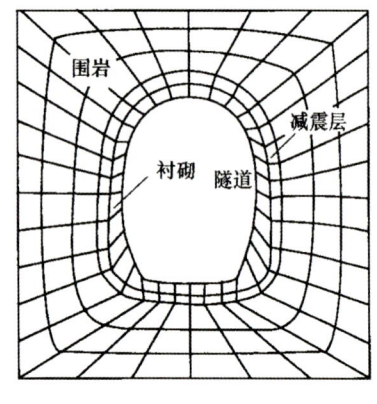
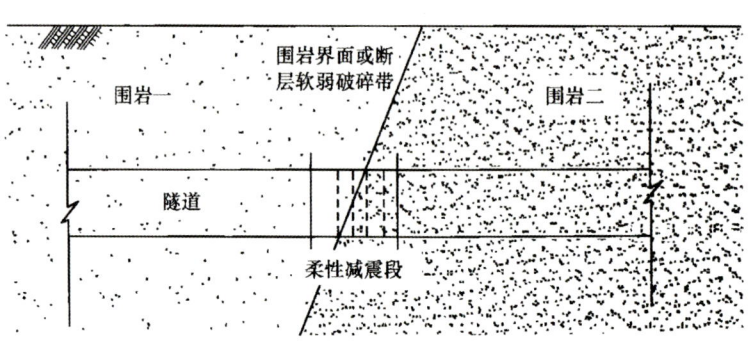

图 26 复合式衬砌设置减震或隔震层示意

国内对山岭隧道复合式衬砌开展了一些模型试验[15]，图 27 就是获得的复合式衬砌有无隔振层时仰拱的加速度对比结果之一，其振动台试验结果表明减震措施对于降低衬砌地震加速度比较有效。目

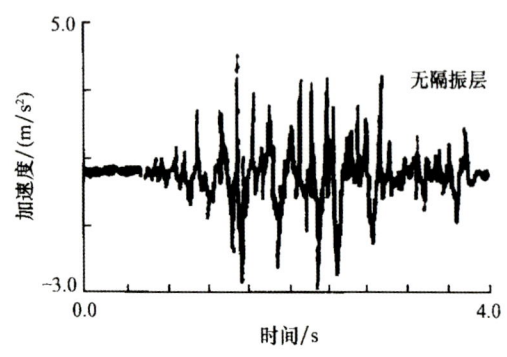
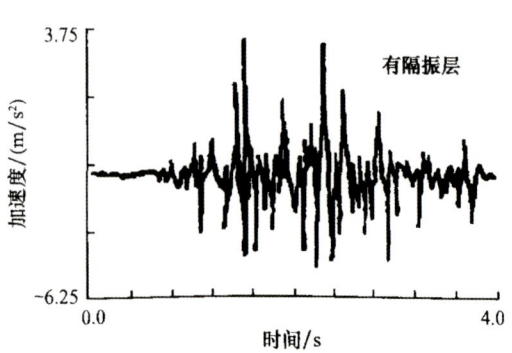

图 27 复合式衬砌有无隔震层时仰拱的加速度对比

本在室内利用盾构隧道模型也开展了大量设置减震层的振动台试验,试验结果表明衬砌的峰值弯矩和峰值轴力都得到了有效衰减[3],如图28所示。

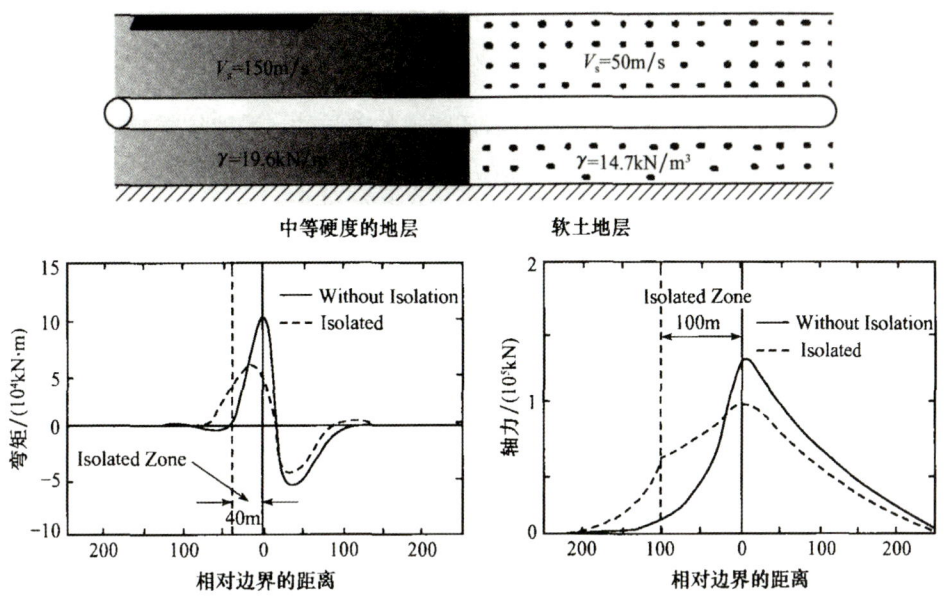

图28 有无减震层的盾构隧道衬砌的弯矩轴力对比

从广义上讲,减震技术属于结构控制技术的范畴。结构控制可以分为主动控制、被动控制、半主动控制和混合控制。对于上部结构,这些技术已在我国被应用于实际工程中,如何在复杂的隧道工程中推广采用结构控制技术,进行隧道减震措施的相关研究是十分必要的。

参 考 文 献

[1] JSCE. The 1995 Hyogoken-Nanbu earthquake. Japan Society of Civil Engineers, 1996, 81 (3): 38—45.

[2] JSCE. The 1999 Ji-Ji earthquake. Japan Society of Civil Engineers, 1999, 81 (3): 38—45.

[3] Kawamata K. Seismic Design of Underground Structure (in Japanese). Tokyo: Kajima Press, 1994: 43—60.

[4] He C, Atsushi K. Study on seismic behavior and seismic design methods in transverse direction of shield tunnels. Structural Engineering and Mechanics, 2001, 11 (6): 651—662.

[5] Dowing C H, Rozen A. Damage to rock tunnels from earthquake shaking. Journal of Geotechnical Engineering Division, 1978, 104: 175—191.

[6] He C, Atsushi K. Seismic behavior in longitudinal direction of shield tunnel located at irregular ground//The 1st International Conference on Advances in Structural Engineering and Mechanics, Seoul, Korea, 1999: 1493—1498.

[7] He C, Atsushi K. A study on seismic behavior of shield tunnels in longitudinal direction//Proceedings of the 4th World Congress on Railway Research, Tokyo, Japan, 1999: 1256—1266.

[8] Atsushi K, He C. Dynamic behavior in longitudinal direction of shield tunnel located at irregular ground with considering effect of secondary lining//Proceedings of the 12th World Conference on Earthquake Engineering, Auckland, New Zealand, 2000: 985—994.

[9] 晏启祥, 何川, 耿萍. 盾构隧道联络通道的地震响应分析. 现代隧道技术, 2008, (增刊): 159—164.

[10] 中铁二院工程集团有限责任公司. 成都至九寨沟线可行性研究报告. 四川成都, 2008.

[11] 邵根大, 骆文海, 李福庭, 等. 强地震作用下铁路隧道衬砌耐震性的研究 [J]. 中国铁道科学, 1992, (13): 93—109.

[12] 西南交通大学. 隧道工程结构物抗震设计标准与方法研究实施大纲. 2008.

[13] 西南交通大学. 高地震烈度地区隧道活动断裂带抗震减震技术研究实施大纲 [R]. 2008.

[14] 申玉生, 高波, 王峥峥, 等. 高烈度地震区山岭隧道模型试验研究. 现代隧道技术, 2008, 45 (5): 38—43.

[15] 王志杰, 高波, 关宝树. 围岩隧道衬砌结构体系的减震研究. 西南交通大学学报, 1996, 31 (6): 590—593.

深埋隧道开挖过程动态及破裂形态分析

李树忱[1,2]，钱七虎[2]，张敦福[1]，李术才[1]

(1. 山东大学 土建与水利学院，山东 济南 250061；2. 解放军理工大学，江苏 南京 210007)

摘要：地下工程开挖过程中，在洞室围岩中会产生拉压交替变化区，当地应力过大时，会产生分区破裂化现象。为了解释洞室围岩拉压交替变化和分区破裂化现象，根据隧道开挖卸荷这一动力学特征，建立隧道开挖过程的动态分析力学模型和计算模式，由此导出由开挖卸荷引起的扰动应力、扰动应变和扰动位移满足的平衡方程、物理方程、几何方程和边界条件。根据实际的位移约束条件，假设位移试函数，利用 Hamilton 时域变分原理，考虑时域变分条件和约束变分条件导出围岩体的积分－变分方程组，建立该方程组的模态矩阵。在给定开挖卸荷路径和零初始条件下采用 Duhamel 积分，得到离散振动方程组的稳态响应。通过矩阵变换，得到隧道围岩体扰动应力、应变和位移的解答函数式。算例分析表明，所给出的理论和方法能正确地反映出隧道开挖引起围岩内的动态过程，并能有效地对开挖引起的围岩的破坏形态进行评价。

关键词：隧道工程；深部岩体；动力问题；时域变分原理；分区破裂化现象
中图分类号：U 45；TU 45　　　**文献标识码**：A　　　**文章编号**：1000 - 6915(2009)10 - 2104 - 09

ANALYSIS OF DYNAMIC AND FRACTURED PHENOMENA FOR EXCAVATION PROCESS OF DEEP TUNNEL

LI Shuchen[1,2], QIAN Qihu[2], ZHANG Dunfu[1], LI Shucai[1]

(1. *School of Civil and Hydraulic Engineering，Shandong University，Jinan，Shandong* 250061；2. *PLA University of Science and Technology，Nanjing，Jiangsu* 210007，*China*)

Abstract：Tension-compression alternating regions can be formed around the rock mass during excavation process. The zonal disintegration phenomena may be brought when the geostress is too large. In order to interpret the tension-compression alternating and zonal disintegration phenomena，the dynamic mechanical model and calculation model of tunnel excavation are presented based on dynamic characteristics of tunnel excavation unloading. And the equilibrium physical equation，geometric equation and boundary condition equation are derived，which are satisfied by the disturbance stress，disturbance strain and disturbance displacement caused by the excavation unloading. Trial function of displacement is assumed considering the practical displacement constraints condition，and integral-variational equation group of the surrounding rock；and its modal matrix is induced by Hamilton time-domain variation principle considering time-domain variation and constraint conditions. The steady-state response of discrete vibration equation group is obtained using Duhamel integral in setting excavation unloading path and zero initial condition. By matrix transformation，the solutions of surrounding rocks disturbance stress，strain and displacement are obtained. The analyses of examples show that the theory and method can correctly reflect the dynamic process caused by excavation；and they effectively simulate the zonal

---

收稿日期：2009-01-14；修回日期：2009-07-18。
基金项目：国家自然科学基金重点项目(50539080，50727904)；国家重点基础研究发展计划(973)项目(2010CB732002)；国家自然科学基金面上项目(5097052)。
本文原载于《岩石力学与工程学报》(2009 年第 28 卷第 10 期)。

disintegration phenomena of the rock mass.

Key words：tunnelling engineering；deep rock masses；dynamic problems；time-domain variation principle；phenomenon of zonal disintegration

1 引 言

1988 年俄罗斯学者 E. И. Shemyakin 从弹塑性应力出发说明了巷道两侧以及工作面前支撑压力的存在，定性地分析了分区破裂化现象产生的机制并在实验室观测到分区破裂化现象，探讨形成机制[1~8]。钱七虎等[9~12]也试图从不同的角度解释岩石分区破裂化产生的机制，取得了一定的成果。周小平等[13~16]将巷道开挖过程视为动力问题，并在此基础上揭示深部岩体分区破裂化的机制。在浅地表地下工程中，洞室在开挖过程中，围岩中也存在拉压交替的不同区域[17]。由于其量值没有超过岩体的强度，没有导致周围岩体的破坏，但围岩中拉压交替出现的现象是客观存在。

通过以上现象可知，分区破裂化现象的前提条件是岩体在开挖过程中，围岩内部存在拉压交替的不同区域。当围岩处于深埋地下工程时，将导致围岩在不同的拉压区产生不同程度的破坏，即分区破裂化现象。分区破裂现象是深部岩体工程特有的现象，而开挖过程中岩体内拉压区交替变化特性不论是深部还是浅部岩石工程都存在的物理现象。为了解释这种现场，本文根据隧道开挖卸荷引起的动力学过程，建立了相应的动力学分析模型和计算模型，推导出了由开挖卸荷引起的扰动应力、扰动应变和扰动位移满足的平衡方程、物理方程、几何方程和边界条件。根据实际的位移约束条件，假设扰动位移试函数，利用 Hamilton 时域变分原理，考虑时域变分条件和约束变分条件导出了围岩体的积分－变分方程组，建立了该方程组的模态矩阵。通过矩阵变换，得到了隧道围岩体的扰动应力、扰动应变和扰动位移的函数式解答。该函数解答能很好的给出地下工程在开挖过程中拉压交替的分区特性。

2 基本方程

假定分析模型满足平面应变问题，由于地应力的存在，开挖区的围岩体的力学模型如图 1 所示。

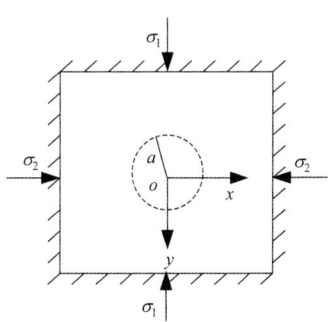

图 1 开挖前力学模型

Fig.1 Mechanical model before tunnel excavation

半径为 a 的虚线为预开挖线。开挖时在半径 a 的周边以约束反力 $f(t)$ 代之，得到开挖卸荷过程的力学模型，如图 2 所示。

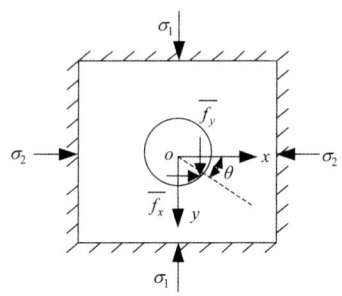

图 2 开挖卸荷过程的力学模型

Fig.2 Mechanical model of unload process by excavation

设开挖前的初始应力分量为 $\sigma_{x0} = -\sigma_2$，$\sigma_{y0} = -\sigma_1$，$\tau_{xy0} = 0$，其对应的初始应变分量为常量。开挖后，总应力分量为 σ_x，σ_y，τ_{xy}。则开挖卸荷引起的扰动应力 σ_{xt}，σ_{yt}，τ_{xyt} 分别为

$$\left.\begin{array}{l}\sigma_{xt} = \sigma_x - \sigma_{x0} \\ \sigma_{yt} = \sigma_y - \sigma_{y0} \\ \tau_{xyt} = \tau_{xy} - \tau_{xy0}\end{array}\right\} \quad (1)$$

开挖引起的扰动应力场满足的平衡方程为

$$\left.\begin{array}{l}\dfrac{\partial \sigma_{xt}}{\partial x} + \dfrac{\partial \tau_{xyt}}{\partial y} - \rho \ddot{u}_t = 0 \\ \dfrac{\partial \tau_{xyt}}{\partial x} + \dfrac{\partial \sigma_{yt}}{\partial y} - \rho \ddot{v}_t = 0\end{array}\right\} \quad (2)$$

式中：ρ 为围岩密度；$\ddot{u}_t$，$\ddot{v}_t$ 分别为水平和竖直向扰动加速度；u_t，v_t 分别为水平和竖直向振动位移

分量，且有

$$\left.\begin{array}{l}u_t = u - u_0 \\ v_t = v - v_0\end{array}\right\} \quad (3)$$

开挖面处扰动力满足的边界条件为

$$\left.\begin{array}{l}l\sigma_{xt} + m\tau_{xyt} = \sigma_2 \cos\theta(f(t)-1) \\ l\tau_{xyt} + m\sigma_{yt} = \sigma_1 \sin\theta(f(t)-1)\end{array}\right\} \quad (4a)$$

其中，

$$l = -\cos\theta, \quad m = -\sin\theta \quad (4b)$$

式中：$f(t)$为开挖面处的约束反力，其卸荷路径假定为$f(t) = 0.5[1 + \cos(\pi t/\tau)]$，$\tau$为卸荷周期。

扰动物理方程为

$$\left.\begin{array}{l}\varepsilon_{xt} = \dfrac{1-\mu^2}{E}\left(\sigma_{xt} - \dfrac{\mu}{1-\mu}\sigma_{yt}\right) \\ \varepsilon_{yt} = \dfrac{1-\mu^2}{E}\left(\sigma_{yt} - \dfrac{\mu}{1-\mu}\sigma_{xt}\right) \\ \gamma_{xyt} = \dfrac{2(1+\mu)}{E}\tau_{xyt}\end{array}\right\} \quad (5)$$

扰动几何方程为

$$\varepsilon_{xt} = \dfrac{\partial u_t}{\partial x}, \quad \varepsilon_{yt} = \dfrac{\partial v_t}{\partial y}, \quad \gamma_{xyt} = \dfrac{\partial u_t}{\partial y} + \dfrac{\partial v_t}{\partial x} \quad (6a)$$

其中，

$$\varepsilon_{xt} = \varepsilon_x - \varepsilon_{x0}, \quad \varepsilon_{yt} = \varepsilon_y - \varepsilon_{y0}, \quad \gamma_{xyt} = \gamma_{xy} - \gamma_{xy0} \quad (6b)$$

3 计算力学模型

由于开挖隧道是圆形隧道（平面应变问题），采用极坐标描述比较方便。

把式(4)用极坐标表示，则开挖面处扰动面力为

$$\left.\begin{array}{l}\overline{f_r} = \left[\dfrac{\sigma_1+\sigma_2}{2} - \dfrac{\sigma_1-\sigma_2}{2}\cos(2\theta)\right](f(t)-1) \\ \overline{f_\theta} = \dfrac{\sigma_1-\sigma_2}{2}\sin(2\theta)(f(t)-1)\end{array}\right\} \quad (7)$$

由于隧道口面力是自成平衡的力系，由圣维南原理可知，这种面力只在隧道口附近产生扰动应力，远处不产生扰动应力，并假定扰动力引起的远场位移边界为固定边界，取半径为b($b \gg a$，且$b \geq \tau_0 c$，τ_0为远场扰动位移衰减为0的时间，c为弹性介质中的波速)的同心大圆，可得隧道开挖动态分析的力学模型如图3所示。

远场（距离b处）的扰动位移边界条件为

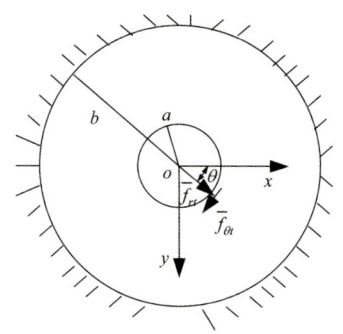

图3 动态分析力学模型
Fig.3 Dynamical analytical mechanical model

$$\left.\begin{array}{l}u_{rt}(b, \theta, t) = 0 \\ u_{\theta t}(b, \theta, t) = 0\end{array}\right\} \quad (8)$$

扰动位移初始条件为

$$\left.\begin{array}{l}u_{rt}(r, \theta, 0) = 0 \\ u_{\theta t}(r, \theta, 0) = 0 \\ \dot{u}_{rt}(r, \theta, 0) = 0 \\ \dot{u}_{\theta t}(r, \theta, 0) = 0\end{array}\right\} \quad (9)$$

式中：上标"·"代表对时间的导数。

4 Hamilton直接法积分–微分方程

通过以上的处理，隧道开挖问题变为卸荷的扰动问题。只要求出扰动位移、扰动应变及扰动应力，即可得到问题的解答。

本文采用Hamilton直接法，考虑到静态解的形式，设扰动位移试函数为

$$\left.\begin{array}{l}u_{rt} = \sum A_m u_{rm} + \sum B_m u_{rm}\cos(2\theta) \\ u_{\theta t} = \sum C_m u_{\theta m}\sin(2\theta)\end{array}\right\} \quad (10)$$

式(10)中，A_m，B_m，$C_m (m=1\sim 6)$分别取前6项计算，即

$$\left.\begin{array}{l}u_{rm} = \cos\left(\dfrac{2m-1}{2}\pi\dfrac{r}{b}\right) \\ u_{\theta m} = 1 - \cos\left(2m\pi\dfrac{r}{b}\right)\end{array}\right\} \quad (11)$$

式(10)能满足式(8)的位移边界条件，扰动系统的变形能为

$$U_t = \dfrac{E}{2(1+\mu)(1-2\mu)}\iint\left[(1-\mu)\varepsilon_{rt}^2 + 2\mu\varepsilon_{rt}\varepsilon_{\theta t} + \right.$$

$$\left.(1-\mu)\varepsilon_{\theta t}^2 + \frac{1-2\mu}{2}\gamma_{r\theta t}^2\right]rdrd\theta \quad (12)$$

扰动系统的动能为

$$T_t = \frac{\rho}{2}\iint(\dot{u}_{rt}^2 + \dot{u}_{\theta t}^2)rdrd\theta \quad (13)$$

扰动面力做的元功为

$$\delta W_t = \oint(\bar{f}_{rt}\delta u_{rt} + \bar{f}_{\theta t}\delta u_{\theta t})d\theta \quad (14)$$

阻尼力做的元功为

$$\delta W_{ct} = -C\oint(\dot{u}_{rt}\delta u_{rt} + \dot{u}_{\theta t}\delta u_{\theta t})d\theta \quad (15)$$

式中：C 为固体阻尼系数。

根据 Hamiltom 时域变分原理，有

$$\int_{t_1}^{t_2}(\delta U_t - \delta T_t - \delta W_t - \delta W_{Ct})dt = 0 \quad (16)$$

式(16)中的变分通过式(10)中的系数 A_m，B_m，C_m 的变分来实现。考虑到时端变分条件和约束变分条件，进行变分运算后，得到围岩体的扰动积分，其微分方程为

$$\frac{\partial U_t}{\partial A_m} + \rho\iint\ddot{u}_{rt}(u_{rm})rd\theta dr +$$

$$C\iint\dot{u}_{rt}(u_{rm})rd\theta dr = \oint\bar{f}_{rt}(u_{rm})ad\theta \quad (17)$$

$$\frac{\partial U_t}{\partial B_m} + \rho\iint\ddot{u}_{rt}[u_{rm}\cos(2\theta)]rd\theta dr +$$

$$C\iint\dot{u}_{rt}[u_{rm}\cos(2\theta)]rd\theta dr = \oint\bar{f}_{rt}[u_{rm}\cos(2\theta)]ad\theta \quad (18)$$

$$\frac{\partial U_t}{\partial C_m} + \rho\iint\ddot{u}_{\theta t}[u_{\theta m}\sin(2\theta)]rd\theta dr +$$

$$C\iint\dot{u}_{\theta t}[u_{\theta m}\sin(2\theta)]rd\theta dr = \oint\bar{f}_{\theta t}[u_{\theta m}\sin(2\theta)]ad\theta \quad (19)$$

将上述积分微分方程组写成矩阵形式为

$$\boldsymbol{M}\ddot{\boldsymbol{X}} + \boldsymbol{C}_n\dot{\boldsymbol{X}} + \boldsymbol{K}\boldsymbol{X} = \boldsymbol{Q}(f(t)-1) \quad (20)$$

式中：$\boldsymbol{X}$ 为系数矩阵，$\boldsymbol{X} = \{A_m, B_m, C_m\}^T$；$\boldsymbol{M}$ 为质量矩阵，围岩的密度和围岩区域大小有关；$\boldsymbol{C}_n$ 为阻尼矩阵，围岩内部阻尼有关；$\boldsymbol{K}$ 为刚度矩阵，与围岩的弹性常数和泊松比有关；$\boldsymbol{Q}$ 为干扰力矩阵，与围岩中初始地应力有关。

式(20)是耦合的系统动力学方程组，可利用系统的模态矩阵 $\boldsymbol{X}_M$ 对式(20)进行解耦运算：

$$\boldsymbol{X}_M^T\boldsymbol{M}\boldsymbol{X}_M\boldsymbol{X}_M^{-1}\ddot{\boldsymbol{X}} + \boldsymbol{X}_M^T\boldsymbol{K}\boldsymbol{X}_M\boldsymbol{X}_M^{-1}\boldsymbol{X} = \boldsymbol{X}_M^T\boldsymbol{Q}(f(t)-1) \quad (21)$$

若记 $\boldsymbol{X}_P = \boldsymbol{X}_M^{-1}\boldsymbol{X}$，$\boldsymbol{K}_P = \boldsymbol{X}_M^T\boldsymbol{K}\boldsymbol{X}_M$，$\boldsymbol{Q}_P = \boldsymbol{X}_M^T\boldsymbol{Q}$，则式(20)解耦后为

$$\ddot{\boldsymbol{X}}_P + 2n\dot{\boldsymbol{X}}_P + P_i^2\boldsymbol{X}_P = \boldsymbol{Q}_P(f(t)-1) \quad (22)$$

式中：n 为阻尼常数，这里取 $n=2.5$；P_i 为第 $i(i=1\sim 18)$ 阶无阻尼固有频率。有阻尼固有频率记为

$$P_{di} = \sqrt{P_i^2 - n^2} \quad (23)$$

干扰力作用下的强迫振动可以通过 Duhamel 积分得到。

(1) 当 $t \leq \tau$ 时，有

$$X_{pi} = Q_{pi}/P_{di}(I_1/4 - I_2/2 + I_3/4) \quad (24)$$

其中，

$$I_1 = \{\{n\sin[(P_{di}-lem_1)t] + lem_1\cos[(P_{di}-lem_1)t]\} - e^{-nt}[n\sin(P_{di}t) + lem_1\cos(P_{di}t)]\}/(n^2+lem_1^2) \quad (25a)$$

$$I_2 = \{\{n\sin[(P_{di}-lem_2)t] + lem_2\cos[(P_{di}-lem_2)t]\} - e^{-nt}[n\sin(P_{di}t) + lem_2\cos(P_{di}t)]\}/(n^2+lem_2^2) \quad (25b)$$

$$I_3 = \{\{n\sin[(P_{di}-lem_3)t] + lem_3\cos[(P_{di}-lem_3)t]\} - e^{-nt}[n\sin(P_{di}t) + lem_3\cos(P_{di}t)]\}/(n^2+lem_3^2) \quad (25c)$$

$$lem_1 = P_{di} - \pi/\tau, \quad lem_2 = P_{di}, \quad lem_3 = P_{di} + \pi/\tau \quad (25d)$$

(2) 当 $t \geq \tau$ 时，有

$$X_{pi} = Q_{pi}/P_{di}(I_1/4 - I_2/2 + I_3/4) \quad (26)$$

其中，

$$I_1 = \{e^{-n(t-\tau)}n\sin[(P_{di}t-lem_1\tau) + lem_1\cos(P_{di}t-lem_1\tau)] - e^{-nt}[n\sin(P_{di}t) + lem_1\cos(P_{di}t)]\}/(n^2+lem_1^2) \quad (27a)$$

$$I_2 = \{e^{-n(t-\tau)}n\sin[(P_{di}t-lem_2\tau) + lem_2\cos(P_{di}t-lem_2\tau)] - e^{-nt}[n\sin(P_{di}t) + lem_2\cos(P_{di}t)]\}/(n^2+lem_2^2) \quad (27b)$$

$$I_3 = \{e^{-n(t-\tau)}n\sin[(P_{di}t-lem_3\tau) + lem_3\cos(P_{di}t-lem_3\tau)] - e^{-nt}[n\sin(P_{di}t) + lem_3\cos(P_{di}t)]\}/(n^2+lem_3^2) \quad (27c)$$

$$lem_1 = P_{di} - \pi/\tau, \quad lem_2 = P_{di}, \quad lem_3 = P_{di} + \pi/\tau \quad (27d)$$

根据变换关系 $\boldsymbol{X}_P = \boldsymbol{X}_M^{-1}\boldsymbol{X}$，变换回去得到耦合的动态稳态响应，即 $\boldsymbol{X} = \boldsymbol{X}_M\boldsymbol{X}_P$。根据假设的扰动

位移试函数以及式(24)或(26)，可得隧道围岩中的扰动位移的近似函数式解答：

$$u_{rt} = \sum_{m=1}^{6} X_m \cos\left(\frac{2m-1}{2}\pi\frac{r}{b}\right) + \sum_{m=1}^{6} X_{m+6} \cos\left(\frac{2m-1}{2}\pi\frac{r}{b}\right) \cos(2\theta) \quad (28)$$

$$u_{\theta t} = \sum_{m=1}^{6} X_{m+12}\left[1 - \cos\left(2m\pi\frac{r}{b}\right)\right]\sin(2\theta) \quad (29)$$

根据极坐标下的几何方程，可以求得围岩中的扰动应变分量：

$$\varepsilon_{rt} = -\sum_{m=1}^{6} X_m \frac{(2m-1)\pi}{2b}\sin\frac{(2m-1)\pi r}{2b} - \sum_{m=1}^{6} X_{m+6}\frac{(2m-1)\pi}{2b}\sin\frac{(2m-1)\pi r}{2b}\cos(2\theta) \quad (30)$$

$$\varepsilon_{\theta t} = \sum_{m=1}^{6} X_m \frac{1}{r}\cos\left[\frac{(2m-1)\pi r}{2b}\right] + \sum_{m=1}^{6} X_{m+6}\frac{1}{r}\cos\left[\frac{(2m-1)\pi r}{2b}\right]\cos(2\theta) + \sum_{m=1}^{6} X_{m+12}\frac{2}{r}\left[1-\cos\left(\frac{2m\pi r}{b}\right)\right]\cos(2\theta) \quad (31)$$

$$\gamma_{r\theta t} = -\sum_{m=1}^{6} X_{m+6}\frac{1}{r}\cos\left[\frac{(2m-1)\pi r}{2b}\right]\sin(2\theta) + \sum_{m=1}^{6} X_{m+12}\frac{2m\pi}{b}\sin\frac{2m\pi r}{b}\sin(2\theta) - \sum_{m=1}^{6} X_{m+12}\frac{1}{r}\left[1-\cos\frac{2m\pi r}{b}\right]\sin(2\theta) \quad (32)$$

根据极坐标下的物理方程，可得围岩中的扰动应力分量：

$$\sigma_{rt} = \frac{1-\mu}{(1+\mu)(1-2\mu)}E\left(\varepsilon_{rt} + \frac{\mu}{1-\mu}\varepsilon_{\theta t}\right) \quad (33)$$

$$\sigma_{\theta t} = \frac{1-\mu}{(1+\mu)(1-2\mu)}E\left(\varepsilon_{\theta t} + \frac{\mu}{1-\mu}\varepsilon_{rt}\right) \quad (34)$$

$$\tau_{r\theta t} = \frac{E}{2(1+\mu)}\gamma_{r\theta t} \quad (35)$$

5 算 例

一隧道开挖半径为 $a = 2.5$ m，取围岩外半径 $b = 20a = 50$ m。围岩弹性模量 $E = 10$ GPa，泊松比 $\mu = 0.25$，密度 $\rho = 2\ 260$ kg/m³。围岩的抗压强度为 $\sigma_c = 22$ MPa，抗拉强度为 $\sigma_t = 2.2$ MPa。初始地应力分别为 $\sigma_1 = 44$ MPa，为单轴抗压强度的 2 倍，$\sigma_2 = 0.8\sigma_1$，卸荷路径为 $f(t)$，周期为 $\tau = 1$ s。

5.1 卸荷过程($t \leqslant \tau$)

在 $t \leqslant \tau$ 时，由计算结果可得围岩中的应变分量。图 4 给出 $\theta = -90°$，$\theta = -45°$ 和 $\theta = 0°$ 处的扰动应变各分量($t = 0.5$ s)。

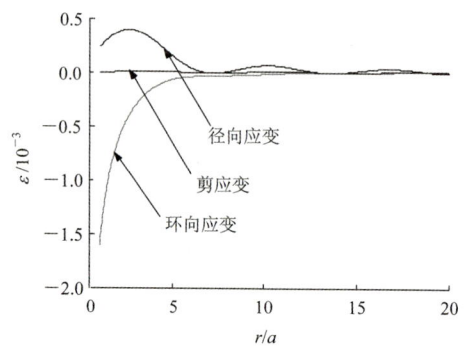

(a) 隧道顶部 $\theta = -90°$ 处

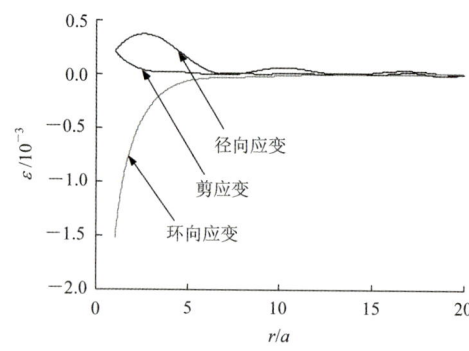

(b) 隧道侧上部 $\theta = -45°$ 处

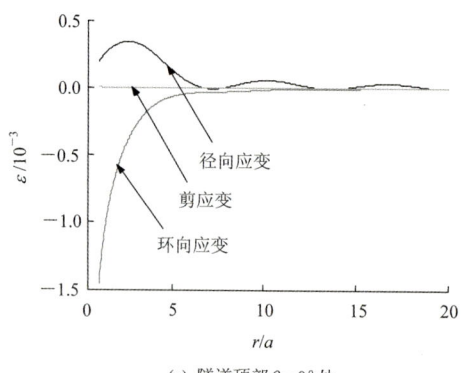

(c) 隧道顶部 $\theta = 0°$ 处

图 4　隧道侧上部各方向的径向、环向应变和剪应变

Fig.4　Radial, circumferential and shear strains at top of tunnel in various directions

由图 4 可知，隧道在开挖周期内，围岩内部各个关键位置在不同时刻环向扰动应变随着距洞口的距离单调递增，而径向扰动应变则随着距离的变化

发生波动，其波动幅值相对较小。

5.2 卸荷完成后($t>\tau$)

为了表述清晰和便于分析，在 $t>\tau$ 时段内，其时间均从卸荷完成后重新计时。隧道顶部 $\theta=-90°$ 处，距洞口表面 $\sqrt{3}a$ 处的径向扰动位移、径向扰动应变和环向扰动应变随时间的变化规律分别如图 5 所示。

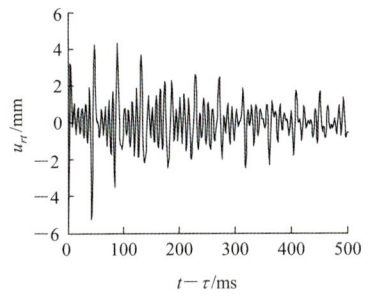

(a) 径向扰动位移

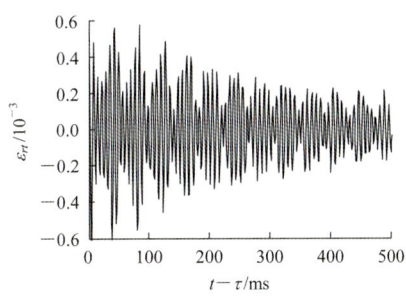

(b) 径向扰动应变

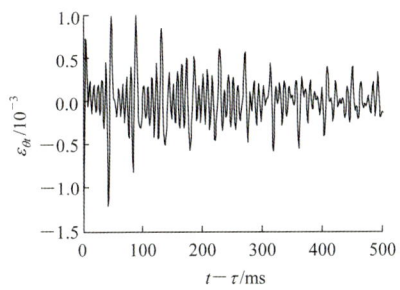

(c) 环向扰动应变

图 5 隧道顶部 $\theta=-90°$ 以及 $r=\sqrt{3}a$ 处径向扰动位移、径向扰动应变和环向扰动应变

Fig.5 Radical disturbing displacements, radical disturbing strains and circumferential disturbing strains when $\theta=-90°$ and $r=\sqrt{3}a$

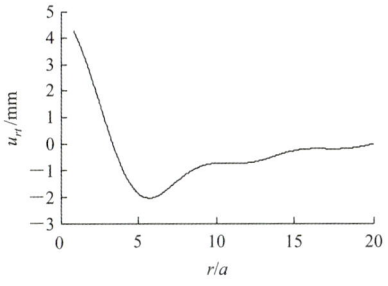

(a) 径向扰动位移($t-\tau=4$ ms)

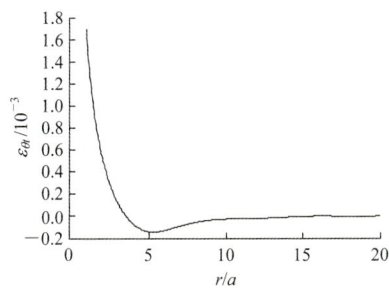

(b) 环向扰动应变($t-\tau=4$ ms)

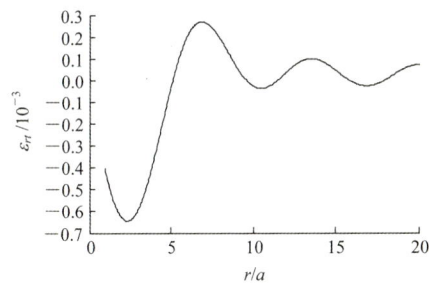

(c) 径向扰动应变($t-\tau=3$ ms)

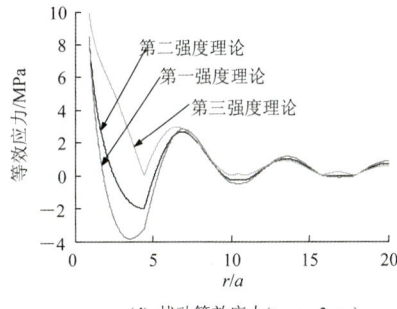

(d) 扰动等效应力($t-\tau=3$ ms)

图 6 隧道顶部 $\theta=-90°$ 的扰动位移、扰动应变和扰动等效应力

Fig.6 Disturbing displacements, disturbing strains and disturbing equivalent stresses at top of tunnel when $\theta=-90°$

由图 5 可知，隧道口附近的径向扰动位移、径向扰动应变和环向扰动应变发生正负变化。

图 6 分别给出了隧道顶部 $\theta=-90°$ 不同时刻的径向扰动位移、环向扰动应变和扰动等效应力分布图。

图 7 分别给出了隧道侧上方 $\theta=-45°$ 不同时刻的径向扰动位移、环向扰动应变、扰动剪应变和扰动等效应力分布图。

图 8 分别给出隧道水平侧边 $\theta = 0°$ 在不同时刻的径向扰动位移、环向扰动应变和扰动等效应力分布图。

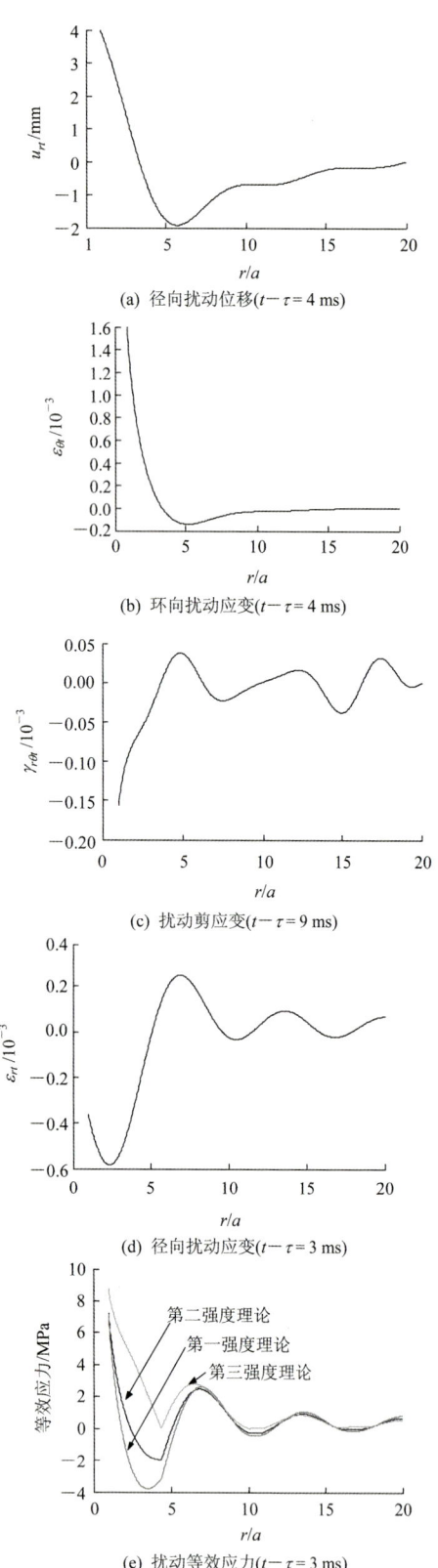

图 7 　隧道侧上方 $\theta = -45°$ 处径向扰动位移、扰动应变和扰动等效应力

Fig.7 　Disturbing displacements, radical disturbing strains and disturbing equivalent stresses at up-side of tunnel when $\theta = -45°$

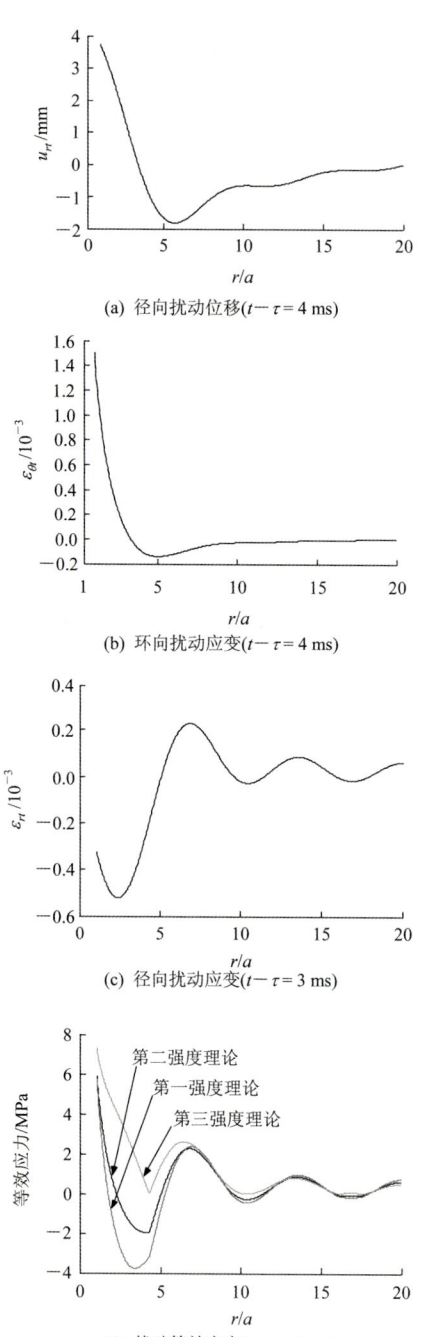

图 8 　隧道侧边 $\theta = -0°$ 处径向扰动位移、扰动应变和扰动等效应力

Fig.8 　Disturbing displacements, radical disturbing strains and disturbing equivalent stresses at side of tunnel when $\theta = -0°$

从图 6～8 可以看出，径向扰动位移和环向扰动应变呈衰减规律变化，最后趋向于 0，而径向扰动

应变和等效应力沿径向呈拉压变化。在卸载后的 3~4 ms 时，洞口附近扰动位移场呈现最大。

图 9 给出了卸载后 $t-\tau=6$ ms 时围岩中的扰动应力分布规律。

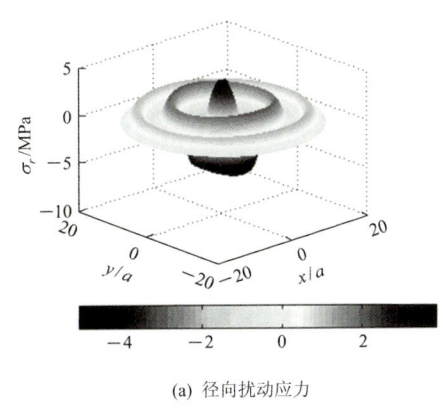

(a) 径向扰动应力

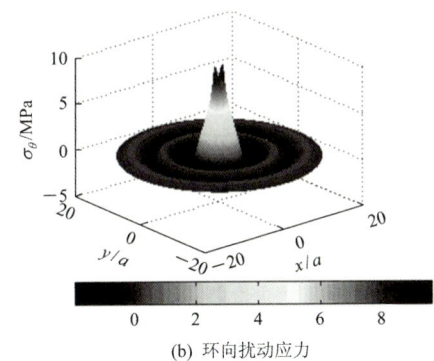

(b) 环向扰动应力

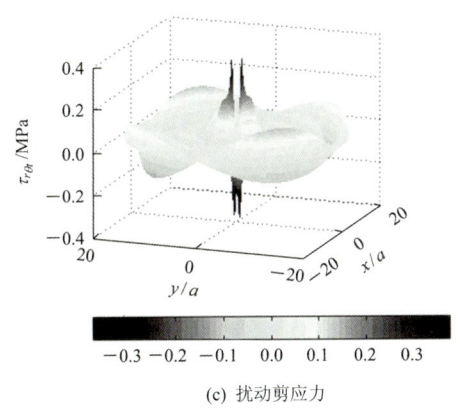

(c) 扰动剪应力

图 9　$t-\tau=6$ ms 时围岩内扰动应力(单位：MPa)

Fig.9　Disturbing stress in the rock-mass at $t-\tau=6$ ms (unit：MPa)

由图 9 可知，在洞室附近围岩的扰动应力场呈拉压变化，在洞室周围较大，随着距离的增大，应力场呈衰减趋势，但依然表现正负交替变化。

综合以上分析可知：在卸载过程中($t\leqslant\tau$)，径向扰动应变为拉应变，沿径向呈衰减性变化，伴有微弱波动。环向扰动应变为压应变，呈递减趋势。

当卸载完成后($t>\tau$)，围岩中各扰动应变呈现正负波动。在卸荷完成后 3~4 ms 时，扰动应变和扰动等效应力的峰值取得最大值，之后波动明显增强。

根据第二强度理论可知，当岩体的临界应变为 0.2×10^{-3} 时，当 $t-\tau=3\sim4$ ms 时，在隧道口附近 $(1\sim\sqrt{3})a$ 的范围内，表现为径向、环向破坏；在 $(\sqrt{3}\sim5)a$ 的范围内，表现为径向破坏；在 $(6.5\sim7.5)a$ 的范围内，表现为环向破坏。

根据第一、三强度理论可知，当岩体的极限拉应力为 2.2 MPa 时，在 $t-\tau=3\sim4$ ms 时，隧道口附近 $(1\sim\sqrt{3})a$ 的范围内，扰动等效应力超过抗拉极限，表现为拉-剪破坏；在 $(\sqrt{3}\sim5)a$ 的范围内，表现为剪切破坏；在 $(6.5\sim7.5)a$ 的范围内，扰动等效应力超过抗拉极限，表现为拉-剪破坏。根据上述分析结果，将围岩破裂分区绘于图 10。

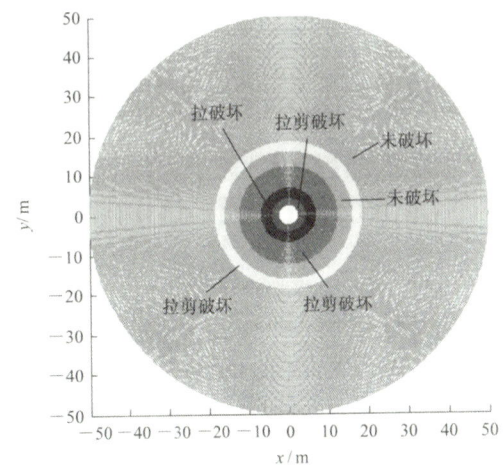

图 10　隧道围岩破坏分区图

Fig.10　Zonal fracture phenomenon around rock mass

6　结　论

(1) 用 Hamilton 时域变分原理的直接法模拟隧道开挖过程，可以得到围岩体内扰动应力场的函数式解答。

(2) 算例结果很好地反映隧道开挖卸荷过程，围岩内扰动应力及应变的变化规律，结果能很好地反映隧道开挖引起的围岩内拉压交替变化特性。

(3) 基于 Hamilton 原理的直接解法可以为隧道开挖模拟提供高效的分析工具，为开挖后围岩内部破坏形态和拉压状态提供可靠的理论依据，为支护设计提供有效的手段。

(4) 本文以线弹性理论为基础，对围岩破坏前扰动应力场引起的动力学行为及破坏形态进行了详细分析，而围岩破坏后的动力学行为及二次破坏形态将在后续工作中展开深入研究。

参考文献(References)：

[1] ADAMS C R，JAGER A J. Petroscopic observations of rock fracturing ahead of stope faces in deep-level gold mines[J]. Journal of South African Institute of Mining and Metallurgy，1980，80(6)：204 - 209.

[2] SHEMYAKIN E I，FISENKO G L，KURLENYA M V，et al. Zonal disintegration of rocks around underground workings，part 1：data of in-situ observations[J]. Journal of Mining Science，1986，22(3)：157 - 168.

[3] SHEMYAKIN E I，FISENKO G L，KURLENYA M V，et al. Zonal disintegration of rocks around underground workings. part II：rock fracture simulated in equivalent materials[J]. Journal of Mining Science，1986，22(4)：223 - 232.

[4] SHEMYAKIN E I，FISENKO G L，KURLENYA M V，et al. Zonal disintegration of rocks around underground mines. part III：theoretical concepts[J]. Journal of Mining Science，1987，23(1)：1 - 6.

[5] SHEMYAKIN E I，KURLENYA M V，OPARIN V N，et al. Zonal disintegration of rocks around underground workings. IV. practical applications[J]. Journal of Mining Science，1989，25(4)：297 - 302.

[6] GUZEV M A，PAROSHIN A A. Non-Euclidean model of the zonal disintegration of rocks around an underground working[J]. Journal of Applied Mechanics and Technical Physics，2001，42(1)：131 - 139.

[7] BORZYKH F. Features of the zonal disintegration of roof rocks and a coal seam around mine workings[J]. Journal of Mining Science，1990，26(5)：418 - 427.

[8] REVA V N. Stability Criteria of Underground Workings under Zonal Disintegration of Rocks[J]. Journal of Mining Science，2002，38(1)：31 - 34.

[9] 钱七虎. 非线性岩石力学新进展——深部岩体力学的若干问题[C]// 中国岩石力学与工程学会编. 第八次全国岩石力学与工程学术大会论文集. 北京：科学出版社，2004：10 - 17.(QIAN Qihu. The current development of nonlinear rock mechanics：the mechanics problems of deep rock mass[C]// Chinese Society for Rock Mechanics and Engineering ed. Proceedings of the 8th Conference on Rock Mechanics and Engineering. Beijing：Science Press，2004：10 - 17.(in Chinese))

[10] 钱七虎. 深部岩体工程响应的特征科学现象及"深部"的界定[J]. 华东理工学院学报(自然科学版)，2004，27(1)：1 - 5.(QIAN Qihu. Characteristic scientific phenomena of engineering response to deep rock mass and the implication of deepness[J]. Journal of East China Institute of Technology(Natural Science)，2004，27(1)：1 - 5.(in Chinese))

[11] 钱七虎. 深部地下空间开发中的关键科学问题[R]. 南京：解放军理工大学工程兵工程学院，2004.(QIAN Qihu. Key scientific problems for deep underground space excavation[R]. Nanjing：Engineering Institute of Engineering Corps，PLA University of Science and Technology，2004.(in Chinese))

[12] 钱七虎，李树忱. 深部岩体工程围岩分区破裂化现场研究综述[J]. 岩石力学与工程学报，2008，27(6)：1 278 - 1 284.(QIAN Qihu，LI Shuchen. A review of research on zonal disintegration phenomenon in deep rock mass engineering[J]. Chinese Journal Rock Mechanics and Engineering，2008，27(6)：1 278 - 1 284.(in Chinese))

[13] 周小平，钱七虎. 深埋巷道分区破裂化机制[J]. 岩石力学与工程学报，2007，26(5)：877 - 885.(ZHOU Xiaoping，QIAN Qihu. Zonal fracturing mechanism in deep tunnel[J]. Chinese Journal Rock Mechanics and Engineering，2007，26(5)：877 - 885.(in Chinese))

[14] 王明洋，周泽平，钱七虎. 深部岩体的构造和变形与破坏问题[J]. 岩石力学与工程学报，2006，25(3)：448 - 455.(WANG Mingyang，ZHOU Zeping，QIAN Qihu. Tectonic，deformation and failure problems of deep rock mass[J]. Chinese Journal Rock Mechanics and Engineering，2006，25(3)：448 - 455.(in Chinese))

[15] 王明洋，戚承志，钱七虎. 深部岩体块系介质变形与运动特性研究[J]. 岩石力学与工程学报，2005，24(16)：2 825 - 2 830.(WANG Mingyang，QI Chengzhi，QIAN Qihu. Study on deformation and motion characteristics of blocks in deep rock mass[J]. Chinese Journal Rock Mechanics and Engineering，2005，24(16)：2 825 - 2 830.(in Chinese))

[16] 王明洋，宋华，郑大亮，等. 深部巷道围岩的分区破裂机制及"深部"界定探讨[J]. 岩石力学与工程学报，2006，25(9)：1 771 - 1 776.(WANG Mingyang，SONG Hua，ZHENG Daliang，et al. On mechanism of zonal disintegration within rock mass around deep tunnel and definition of deep rock engineering[J]. Chinese Journal Rock Mechanics and Engineering，2006，25(9)：1 771 - 1 776.(in Chinese))

[17] 钟世航. 软弱围岩隧道围岩中自承体系及其形成与发展[M]. 武汉：湖北科学技术出版社，1992.(ZHONG Shihang. Formation and developing of self-supporting system in surrounding rock of tunnel[M]. Wuhan：Hubei Science and Technology Press，1992.(in Chinese))

中国地下工程安全风险管理的现状、问题及相关建议

钱七虎，戎晓力

(解放军理工大学 工程兵工程学院，江苏 南京 210007)

摘要：地下工程建设具有投资大、施工周期长、施工项目多、施工技术复杂、不可预见风险因素多和对社会环境影响大等特点，是一项高风险建设工程。建立风险管理制度，对拟建和在建的城市地铁工程项目进行风险评估，继而进行风险控制十分必要，并应扩大到整个地下工程建设领域。主要论述地下工程安全风险管理的现状和地下工程安全风险管理实践中的问题，这些问题包括：(1) 缺乏规范的安全风险管理体系；(2) 工程安全风险管理责任主体不够合理，安全风险管理经费不到位；(3) 工程安全风险管理专业队伍不够规范，专业水平参差不齐；(4) 风险管理相关技术规范、标准不符合目前地下工程发展现状；(5) 缺乏合适的信息化安全风险管理平台。针对我国安全风险管理实践中的问题，提出 4 条建议：(1) 加强针对地下工程安全风险管理的法规建设工作；(2) 推行安全风险管理计划，将安全风险管理作为地下工程建设管理的一个必要组成部分；(3) 安全风险管理要有基于信息化技术的风险管理和预警决策支持系统；(4) 加强地下工程安全风险管理、重大事故预测预报以及防治技术的研究。

关键词：地下工程；安全风险管理；研究现状
中图分类号：TU 94　　**文献标识码**：A　　**文章编号**：1000 - 6915(2008)04 - 0649 - 07

STATE，ISSUES AND RELEVANT RECOMMENDATIONS FOR SECURITY RISK MANAGEMENT OF CHINA′S UNDERGROUND ENGINEERING

QIAN Qihu，RONG Xiaoli

(*Engineering Institute of Engineering Corps*，*PLA University of Science and Technology*，*Nanjing*，*Jiangsu* 210007，*China*)

Abstract：Underground project is a high-risk and complex project with vast construction investment，long-term construction，numerous subprojects，complexity of construction technology，and unpredictability of risk factors and social characteristics of the environmental impact. The establishment of a risk management system，detailed risk assessment for construction of metro project，and risk control，are necessary and should be extended to the overall underground construction areas. This paper mainly describes the present status and issues existing in risk management for safety operation in underground projects，which include：(1) the lack of standardized security risk management system；(2) unreasonable safety risk management，and insufficient funding for security risk management；(3) inadequate or nonprofessional teams or uneven-developed professional level；(4) risk management-related technical specifications，standards are not consistent with the current development of underground works；and (5) the lack of appropriate security risk management information platform. Related to issues existing in risk management in China，four suggestions are proposed as follows：(1) to strengthen codification of risk management for safety in underground projects；(2) to promote risk management as an

---

收稿日期：2007-12-09；修回日期：2007-12-30。
本文原载于《岩石力学与工程学报》(2008 年第 27 卷第 4 期)。

indispensable part of project management; (3) to set up an aid system for warning decision making based on informationization technique; and (4) to strengthen study on major accident prediction and prevention technique for risk management.

Key words: underground engineering; security risk management; research review

1 地下工程安全风险管理的必要性和紧迫性

地下工程安全风险管理的必要性和紧迫性是由中国地下工程建设规模大、发展快的客观事实以及地下工程严峻的安全形势所决定的。中国是目前世界上地下空间开发利用的大国,城市轨道交通建设速度位居世界首位,史无前例。据建设部统计,15个城市近期建设规划了 61 条线路,共长 1 700 km,目前已开工建设 1 000 km 线路,已建成运营 581 km 线路,按规划还有约 1 200 km 线路在今后 10 a 左右内建成。据预测,至 2010 年,我国将有 20 个左右的城市发展以地铁为主的城市轨道交通,到 2020 年,我国将有 30 个左右的城市发展以地铁为主的城市轨道交通。按目前每年开工建设 100~200 km 的速度,到 2020 年,我国城市轨道交通线路规模可达 2 000~2 500 km。

目前,中国城市地下空间的总体规模和总量即将居世界首位。以北京为例,北京地下空间建成面积已达 30 km^2,全市地下空间今后平均每年将增加建筑面积约 3 km^2,占总建筑面积的 10%,到 2020 年,北京地下空间总面积将达 90 km^2。今后为提高土地集约化利用水平,解决城市交通和环境问题,中国许多特大和大城市将结合地铁建设、旧城改造和新区建设,建成众多大型城市地下综合体,除已建的北京中关村西区、北京金融街中心区、大连不夜城、上海龙华南火车站等大型地下综合体外,将建成多个地下城。如:上海世博园区地下城面积达 0.4 km^2,五角场地下城面积约为 0.3 km^2;杭州钱江新城地下城以波浪地下城为骨干,建成地下空间 2 km^2;武汉王家墩商务区地下城总面积超过 3 km^2 等等。特大城市地下快速路的建设,正越来越受到重视,以解决城市交通拥堵问题。上海市政府已批准井字型地下通道方案,该方案全线长 40 km,其中地下 26 km,预计将在 2010 年建成;北京提出在 2020 年以前修建四纵两横地下快速路网方案;深圳建设了约 7 km 长的地下快速通道;南京的内环线地下路约占总长度的 1/2。中国已成为世界上隧道最多,建设发展最快的国家,据规划,中国在 21 世纪前 20 a 要建设 6 000 km 隧道,10 a 内建设 155 km 长的城市公路隧道,其中许多是长、大和深埋隧道。

由于地下工程具有投资大、施工周期长、施工项目多、施工技术复杂、不可预见风险因素多和对社会环境影响大等特点,地下工程建设是一项高风险建设工程。由于规模大、发展快,技术和管理力量难以充分保证的客观原因,加上对地下工程安全风险的认识不客观,风险管理不科学,风险管理的投入不到位的主观原因,所以地下工程建设中,事故频发,形势非常严峻,令人堪忧。建设部领导在 2007 年 9 月召开的地铁建设和运营安全研讨会上说,城市轨道交通和运营的风险正在进一步加大。仅 2006 年至今,不包括矿难,宜万铁路隧道建设就发生两起重大事故,共亡 21 人;2007 年北京、南京地铁施工分别死亡 6 和 2 人;南京地铁 2007 年 3 月份发生地铁施工导致燃气管断裂,引起燃气溢出发生爆炸大火的事件;在经济损失上,以上海地铁 4 号线事故最为严重,损失逾人民币 6 亿元。事实上,业内人员都了解,在长、大隧道施工中几乎每条隧道都有伤亡现象发生。由此可见,在地下工程建设中,实行安全风险管理的必要性和紧迫性十分明显,因此建设部领导提出"建立风险管理制度,对拟建和在建的城市地铁工程项目进行风险评估,继而进行风险控制十分必要"的要求是十分必然的,并应扩大到整个地下工程建设领域。

2 地下工程安全风险管理的现状

2.1 国外安全风险管理实践的借鉴及发展趋势

本文不准备涉及国外安全风险管理研究的评述,仅指出自 2004 年以来,安全、费用与风险已成为国际隧协隧道及地下空间协会每年年会的主题。

地下工程安全风险管理在地下工程中的应用研究在美国、欧洲正在积极开展[1],欧共体行政院于 1992 年发布了"欧共体就在临时或移动施工现场实施最低安全和健康要求的指令"[2],意大利政府于

1996 年由总统签字发布了相应的指令。有些国家已编写出隧道工程风险管理的规范和法规[2~5]。具体包括：国际隧道及地下空间协会 2004 年发布了风险管理的指导方法[3]；英国隧协和保险业协会于 2003 年 9 月联合发布了《英国隧道工程建设风险管理联合规范》[4]；国际隧道工程保险集团(ITIG)于 2006 年 1 月发布了《隧道工程风险管理实践规程》[5]，这个实践规程是基于上述英国联合规范而编制的。

安全风险管理在实际地下工程应用方面，主要由各个岩土工程咨询公司进行。如意大利 GeoDATA 公司针对地下工程施工风险管理推出了名为 GDMS (geodata master system)的信息化管理平台[6]，该系统运用了 GIS 和 WEB 技术，由建筑物状态管理系统(building condition system，BCS)、建筑风险评估系统(building risk assessment，BRA)、盾构数据管理系统(TBM data management，TDM)、监测数据管理系统(monitoring data management，MDM)以及文档管理系统(document management system，DMS)5 个子系统构成，具备完善的风险管理方案，并在俄罗斯圣彼得堡、意大利罗马和圣地亚哥等地铁工程中得到应用，但在监测手段、施工管理等环节上，该系统不完全适合中国国情。由于涉及知识产权问题，故应用该系统必须由该公司承接工程风险管理任务。中国台湾亚新工程顾问股份有限公司针对台湾地下工程施工安全问题开发出 IDEAL——监测资料处理系统，该系统对监测数据有效性检查和处理考虑得比较全面，但监测信息的可视化功能较弱。

在国际上，安全风险管理的以下发展趋势越来越明显：风险管理正成为大型项目发展中的一个例行程序；风险管理与项目管理日趋结合；为风险管理制定强制性的法规，特别是针对施工安全的法规。

2.2 国内安全风险管理的现状和特点

虽然安全风险管理在地下工程中的应用在中国刚刚起步，但是地下工程安全风险及其相关学科的研究自20世纪末中国即已陆续开展。首先，进行较多的是隧道和基坑开挖对环境影响的力学分析；其次，同济大学的丁士昭教授(1992)对我国广州地铁首期工程、上海地铁1号线工程等地铁建设中的风险和保险模式进行了研究。上海隧道建设设计研究院的范益群等[7]以可靠度理论为基础，提出了地下结构的抗风险设计概念；中国科学院地质研究所的刘大安等[8]针对边坡工程开挖而开发的"综合地质信息系统"；同济大学李元海和朱合华[9]开发的"岩土工程施工监测信息系统"，孙钧[10]主持了"城市地下工程施工安全的智能控制预测与控制及其三维仿真模拟系统研究"。特别是黄宏伟等[11~15]在地下工程安全风险研究方面开展了大量的工作，在这些工作基础上，2005年中国土木工程学会召开了中国第一次全国范围的地下工程安全风险分析研讨会，推动了地下工程安全风险研究的全面开展。

中国政府对地下工程的风险管理也相当重视，2003 年建设部等九部委联合印发了《关于进一步加强地铁安全管理工作的意见》，对做好地铁规划、设计、施工、运营的安全工作提出了具体要求。最近，又编发了《地铁与地下工程建设技术风险控制导则》，《地铁及地下工程建设风险管理指南》[16]，对指导中国地铁及地下工程安全风险管理的标准化、程序化和规范化具有促进作用。

安全风险管理的实际应用近两年在我国得到迅速发展，特别是在地铁建设方面，上海、北京新建地铁项目大都进行了风险分析与评估。据了解，北京地铁 5，10 号线等项目的风险评估已取得了具体成果。上海同是工程科技有限公司依托同济大学开发的"安程地铁工程远程监控管理系统"，基于网络传输、无线通讯、网络数据库、数据分析以及自动预测预警等技术，综合了施工、监理、监测、管理以及多媒体等多种信息，已在上海地铁工程中得到应用。针对盾构法隧道施工，上海隧道工程股份有限公司开发了"盾构法隧道施工智能管理系统"，在掌握施工信息的前提下，通过数据分析，对工程施工进行有效管理和技术支持。在大型隧道建设——上海沪崇长江隧道[17]、钱塘江隧道等项目中也进行了风险分析与评估研究，并取得了实际成果。2007年以来，解放军理工大学与意大利 Geodata 公司合作，借鉴意大利先进的风险管理经验和风险管理信息系统，开展了南京地铁建设的安全风险管理的实际工作，并进而与北京市轨道交通建设管理公司合作开展了北京地铁建设的安全风险管理工作。

总体而言，中国地下工程安全风险管理研究与实践已经得到了实质性进展，部分成果已服务于项目的决策，但远远没有达到"风险管理化解地下工程建设之痛"的程度。具体来说，有如下特点：

(1) 发展不平衡。地下工程安全风险管理应用主要集中在地铁建设领域，而在其他地下工程建设领域，开展安全风险管理的仅限于个别大型重点隧道建设项目。这可能和分管地铁建设的建设部召开多次地铁建设风险管理研讨会有关。

(2) 在开展安全风险管理应用的地下工程建设项目中，主要开展的是风险分析和评估，并局限于定性分析或者半定量分析。

(3) 在地下工程建设中，开展和布置较多的是监测系统，但是对风险预警和控制方法没有深入研究，仅实现监测功能，而不能"控制"，没有整合成统一的安全风险管理系统。

(4) 在少数整合了风险辨识、风险分析与评估、风险处置与监控分系统的安全风险管理系统中，往往缺乏坚实的地质地理信息系统和符合实际工程地质、水文地质和环境条件的风险阈值数据库系统。后者需要深入细致的研究基础以及人员和经费的较大投入。

3 地下工程安全风险管理实践中存在的问题

目前，国内许多单位对于风险管理在认识上仍存在许多误区和实施中不完善、不规范的地方，要想以"风险管理化解地下工程建设之痛"，依然有许多工作要做。下面具体阐述当前实施风险管理存在的主要问题。

3.1 缺乏规范的安全风险管理体系

虽然，国内已经发布了一些指导性文件，如建设部：《地铁与地下工程建设技术风险控制导则》，《地铁及地下工程建设风险管理指南》[16]；铁道部：《铁路隧道风险评估暂行规定(报批稿)》[18]。但是目前国家对地下工程项目建设安全风险管理还没有合适的操作性较强的、具有一定强制意义的法规体系，因而风险管理在项目建设中的地位没有明确。项目建设预算中没有明确必须列入风险管理费用，从而造成了风险管理投入不够，风险管理得不到应有重视的状况；项目建设各阶段风险管理责任主体不明，管理程序不清。这些因素使国内地下工程安全风险管理尚处于无序状态，表现为实施安全风险管理的内容和流程不完善、不规范。

随着世界各国研究的深入，尤其是通过意大利、荷兰和澳大利亚等国工程技术人员不懈的研究，当前基本完善了风险管理的流程：即包括风险定义、风险辨识、风险估计、风险分析、风险评估、风险处置和风险监控。该风险管理流程在 2002 年国际隧道及地下空间协会(ITA)起草颁布的《隧道及地下工程风险管理指南》中得到了应用。可以看出，对一个工程项目，应当实施全程的风险管理和完善的流程。

目前国内各单位所从事的隧道及地下工程风险管理项目，主要侧重于风险的分析与评估，而且大多数是应用一种或几种评估方法对工程或工程中的某一部分或某一阶段进行估计，得出风险值，由风险值大小排序。对于其他方面则研究较少，对风险的评价及如何降低风险，只是凭着个人取舍或现有资料来确定使用何种评估方法，各评估方法得出的结果有时也缺少可比性，在评估方法的选取上，尚没有统一的共识。另外，对风险评估的认可也存在很大程度的差异，风险评估的结果本身有多大的风险，有多大的可信度，还存在疑问。

从目前所完成或所进行的一些风险评估项目和有关此方面的成果和论文上可以看出，目前中国的风险分析和评估仅对建设施工中可能发生的事故予以罗列，对其进行简单的分析，得出其定性或半定量的可能性大小，然后予以排序，最后根据风险大小给出建议。事实上这还是片面的，甚至有可能会产生误导。

地下工程的安全风险是一个动态的过程，国外先进的思想是提倡地下工程进行"迭代"式设计、施工和管理，就是为了适应地下工程中地质条件、环境条件等因素与地下工程施工直接的相互影响和变化带来的复杂性，以期最大可能地规避风险。风险管理的很多理论源于金融(保险)风险管理领域，但地下工程的施工安全风险具有独特性，从宏观上说，施工安全事故的发生具有偶然性，但从微观层次上说，大量的工程事故教训已经证明，地下施工的安全事故在发生前往往是有征兆的，是完全可以监控的。因此，仅仅利用普通的风险评估分析方法对地下工程风险进行评估、分级是不够的，从技术上说，所有的风险都是可监控的，而这往往是现阶段工程实践中所忽视的。

随着保险业的发展和管理行业的规范，在隧道及地下工程建设中实行了强制性保险。使得有些建设方、承包和施工方片面地认为买了保险就安全了。事实上，购买保险只是一种转嫁风险的方式，但并不是风险处理的惟一方式。风险管理在风险评估后需要对风险进行处理，可以选择的方式主要包括风险自留、转移和减缓，工程项目的风险事故完全采用保险公司进行风险转移并不一定是最佳选择,而是应该通过科学的决策,作出处理对策。保险公司无法从根本上起到抑制风险事故发生的作用。

3.2 工程安全风险管理责任主体不够合理,安全风险管理经费不到位

在目前中国的工程合同管理模式中,工程安全风险管理的责任主体主要由施工方承担,监理负责监督,业主也具有一定的安全风险管理责任。而地下工程的风险实际不完全是施工引起的,国际隧道工程保险集团对施工现场发生安全事故的原因的调查结果(见图 1)表明,地下工程发生事故的原因是多方面的,施工方作为工程安全唯一责任主体无法根本避免事故的发生,由于目前国内勘察设计失误导致的风险责任界定不清,而这些风险往往到施工时才反映出来,显然由施工方完全承担这些风险显得不够合理。

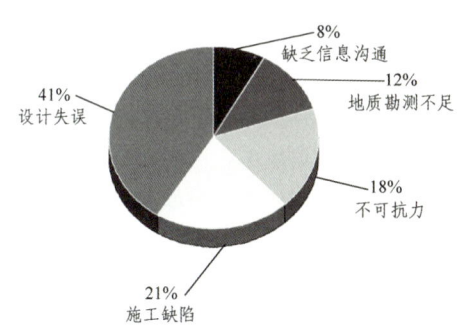

图 1 国际隧道工程保险集团对隧道施工现场发生安全事故原因的调查结果

Fig.1 ITIG on the tunnel construction site safety in the cause of the accident findings

在工程预算中,关于安全风险管理的费用,如施工监测费用等都没有明确取费标准,在低价中标的管理模式下,安全风险管理有关费用往往被首先挤压,导致一些施工单位在利益驱使下,冒险施工,"赌"不会发生事故的概率。国内一些重大事故原因分析中往往发现,在事故发生前已有明显征兆,而施工单位没有足够重视,采取抢救措施不力,最终导致事故发生(如北京地铁 10 号线京广桥事故),这背后深层次原因就是施工方安全经费不足,不愿意加大成本规避风险所致。

3.3 工程安全风险管理专业队伍不够规范,专业水平参差不齐

在工程中加强工程监测,进行必要的安全风险评估咨询工作是规避工程风险的重要手段。实践表明,如果在工程中加强工程监测及安全风险评估咨询工作,可以最大程度的保证施工安全。工程监测在地下工程风险控制管理中起着举足轻重的作用,是预防风险产生的重要手段。但目前,各城市地下开发速度、规模日益加大,专业规范的监测队伍紧缺,使得国内工程监测市场鱼龙混杂,这直接影响监测数据的准确性,以及信息化指导施工的有效性。目前,国内对于监测单位资质、监测人员技术素质没有相应的管理和评价体系,使得监测队伍不够规范,这对项目风险管理十分不利。

当前国内很多地方对实行"第三方监测"进行了探索和实践,实践表明,实施城市地下工程施工"第三方监测"是保证施工安全和工程质量十分重要的举措,有效地避免了施工过程中可能发生的事故。但目前针对"第三方监测"没有国家性的法规进行明确规定和管理,各地"第三方监测"处于无序状态;对"第三方监测"的内容、责任主体、监测指标及管理信息系统数据标准等都没有统一的管理和规定。这不利于有效地通过"第三方监测"进行施工安全风险管理。

国内对工程安全风险管理咨询评估的从业单位和人员没有明确的资质管理,许多工程实践中安全风险评估工作还停留在由院校科研单位以科研项目的形式承担,对于工程安全风险咨询评估工作的内容、质量评价标准、咨询工作的责任认定、从业人员资格认定等都没有统一的管理,使得工程安全风险管理水平参差不齐。但如果进行风险评估的专家队伍水平不够,可以想象,再好的风险监控体系也不能保证全线工程的安全。

3.4 风险管理相关技术规范、标准不符合目前地下工程发展现状

国家和地方施工规范和标准是进行地下工程各种工法风险识别和风险评估的基本依据,但是,目前风险管理相关技术控制规范不够全面,已有的规范有的已落后于时代发展,对实际工程已失去或部分失去指导意义。例如,按现有规范要求地下工程施工地面沉降控制标准不大于 30 mm,实际工程表明,在某些风险等级较高的工程中,这个取值偏松,而有些工程的这个取值又偏严。因此,编制地下工程施工不同工法的技术规范,研制不同城市、不同工法的适应不同岩土岩性的工程地质、水文地质、环境条件的风险阈值数据库系统已是当务之急。

3.5 缺乏合适的信息化安全风险管理平台

利用信息化系统可以加快信息传输速度，提高管理的效率和科学水平，增加项目各方的责任。但是，在信息化飞速发展的今天，国内安全风险管理的信息化水平还很低，还缺乏符合安全风险管理体系，适合地下工程建设实际的信息化风险管理平台。

4 针对中国安全风险管理实践中存在问题的建议

(1) 加强针对地下工程安全风险管理的法规建设工作

加强风险管理意识，遏制工程事故高发态势，需要政府部门(建设部、交通部、铁道部和水利部等)以及行业强有力的法规作指导和规范。国内在这方面的法规还很不健全，亟待加强。特别应在法规中明确规定：在工程设计和施工评标中，除商务造价标和技术标外，必须列入安全风险管理的评定(安全标)，加大安全风险管理的投入。应该在有关预算及定额中明确安全风险管理的相关费用，确保安全风险管理费用在整个工程建设费用中占有合理的比例，且明确为"专项提取"，不列入商务造价标，并且加强审计和监管，确保这些费用切实落在安全风险管理工作上。

目前美国、欧洲在地下工程安全风险管理的应用方面已有较成熟的研究成果，因而在建立自己的有关法规体系时，既要注重吸收国外先进、成熟的经验，使国内在这方面更为迅速且成本更小地发展，又要注重符合中国国情，符合中国的工程建设管理模式和中国工程施工管理水平，使建立的法规体系切实可行。

(2) 推行安全风险管理计划，将安全风险管理作为地下工程建设管理的一个必要组成部分

开展风险管理计划，有利于决策科学化，有利于减少工程事故的发生，有利于提高政府、业主、设计单位、施工单位及运营单位的风险管理意识和风险管理能力，从而达到控制风险、减少损失的目的。每个工程建设项目必须立项开展安全风险管理研究，凡是安全风险等级评价为"严重"的设计和施工方案应实行一票否决制，必须进行修改。

风险管理应明确包括辨识分析、风险评估和风险控制，内容分别涉及探明风险源、分析风险因素、根据事故发生的可能性和影响大小评估工程的风险等级以及建立监测系统、采取预防措施规避、转移和减缓风险等。

(3) 安全风险管理要有基于信息化技术的风险管理及预警决策支持系统

应在风险管理中加强科学的定量研究，加强科学监测、信息化传输和反馈控制。如发展地下工程物探和预报技术、建立地理地质信息系统，利用现代监控技术和信息化手段结合高级专家经验，把进行建设的地下工程周围一定范围内的流沙、软土、管线、岩溶及岩溶水、岩爆以及其他危险情况等地理地质数据储存在计算机里，建立工程管理数据库和决策支持体系。同时，建立施工安全远程监测系统，监测系统通过传感器获取信息后及时反馈到指挥部，这样才能进行数据分析，加强预警，发出警报并及时采取预先储存的应急预案。

(4) 加强地下工程安全风险管理以及重大事故预测预报和防治技术的研究

当前迫切需要对地下工程安全问题的发生机制、安全监测(检测)与监控理论及技术、风险分析技术、灾害预测预报技术、应急处理决策支持技术进行深入系统的研究。评估事故发生概率及灾害后果，提出措施防范依据，发展定量研究，提出和解决其关键科学问题。

鉴于地下工程建设是一项风险较大的建设项目，不可预见的风险因素多和对环境、社会影响大，防治技术难度高，建议科技部将《地下工程安全风险信息化管理以及重大事故(软弱破碎围岩段塌方，岩溶段突水突泥，岩爆区域的预测和防治，河、海底下隧道渗水贯通的预防等)的预测预报和防治技术研究》列为十一五技术支撑研究重大项目，立即开展研究。

参考文献(References)：

[1] DUDDECK H. Challenges to tunnelling engineers[J]. Tunnelling and Underground Space Technology，1996，11(1)：5–10.

[2] The Council of the European Communities. Council directive 92/57/EEC of 24 June 1992 on the implementation of minimum safety and health requirements at temporary or mobile construction sites(eighth individual directive within the meaning of article 16(1) of directive 89/391/EEC)[Z]. Brussels：The Council of the European

Communities, 1992.

[3] ESKESEN S D, TENGBORG P, KAMPMANN J, et al. Guidelines for tunneling risk management: international tunneling association working group No.2[J]. Tunneling and Underground Space Technology, 2004, 19(3): 217–237.

[4] The British Tunnelling Society. The association of British Insurers: Joint code of practice for risk assessment of tunnel works in the UK–09[Z]. London: The British Tunnelling Society, 2003.

[5] The International Tunnelling Insurance Group. A code of practice for risk management of tunnel works[S]. [S. l.]: [s. n.], 2006.

[6] GDMS(Geodata master system): an information system for managing geoengineering projects[OL]. (2006–12–10)[2007–10–10]. http://www.geodata.it.

[7] 范益群,钟万勰,刘建航. 时空效应理论与软土基坑工程现代设计概念[J]. 清华大学学报(自然科学版), 2000, 40(增1): 49–53. (FAN Yiqun, ZHONG Wanxie, LIU Jianhang. Theory of time-space effect and modern design concepts in soft soil pit engineering[J]. Journal of Tsinghua University(Science and Technology), 2000, 40(Supp. 1): 49–53.(in Chinese))

[8] 刘大安,杨志法,柯天河,等. 综合地质信息系统及其应用研究[J]. 岩土工程学报, 2000, 22(2): 182–185.(LIU Da'an, YANG Zhifa, KE Tianhe, et al. Synthetic geology information system and its application to the shiplock slope engineering of a hydropower station[J]. Chinese Journal of Geotechnical Engineering, 2000, 22(2): 182–185.(in Chinese))

[9] 李元海,朱合华. 岩土工程施工监测信息系统初探[J]. 岩土力学, 2002, 23(1): 103–106.(LI Yuanhai, ZHU Hehua. Development of monitoring information system software for geotechnical engineering[J]. Rock and Soil Mechanics, 2002, 23(1): 103–106.(in Chinese))

[10] 孙钧. 城市地下工程施工安全的智能控制预测与控制及其三维仿真模拟系统研究[J]. 岩石力学与工程学报, 1999, 18(增): 753–762.(SUN Jun. Intelligent prediction, control of construction safety and 3D simulation study on urban underground engineering[J]. Chinese Journal of Rock Mechanics and Engineering, 1999, 18(Supp.): 753–762.(in Chinese))

[11] 黄宏伟. 隧道及地下工程建设中的风险管理研究进展[J]. 地下空间与工程学报, 2006, 2(1): 13–20.(HUANG Hongwei. State of the art of the research on risk management in construction of tunnel and underground works[J]. Chinese Journal of Underground Space and Engineering, 2006, 2(1): 13–20.(in Chinese))

[12] 陈桂香,黄宏伟,尤建新. 对地铁项目全寿命周期风险管理的研究[J]. 地下空间与工程学报, 2006, 2(1): 47–51.(CHEN Guixiang, HUANG Hongwei, YOU Jianxin. Study on life cycle risk management of metro[J]. Chinese Journal of Underground Space and Engineering, 2006, 2(1): 47–51.(in Chinese))

[13] 黄宏伟,曾明,陈亮,等. 基于风险数据库的盾构隧道施工风险管理软件(TRM1.0)开发[J]. 地下空间与工程学报, 2006, 2(1): 36–41.(HUANG Hongwei, ZENG Ming, CHEN Liang, et al. Risk management software(TRM1.0) based on risk database for shield tunneling[J]. Chinese Journal of Underground Space and Engineering, 2006, 2(1): 36–41.(in Chinese))

[14] 陈龙,黄宏伟. 上中路隧道工程风险管理的实践[J]. 地下空间与工程学报, 2006, 2(1): 65–69, 73.(CHEN Long, HUANG Hongwei. The practice of risk management in Shangzhong road tunnel engineering[J]. Chinese Journal of Underground Space and Engineering, 2006, 2(1): 65–69, 73.(in Chinese))

[15] 陈龙,黄宏伟. 岩石隧道工程风险浅析[J]. 岩石力学与工程学报, 2005, 24(1): 110–115.(CHEN Long, HUANG Hongwei. Risk analysis of rock tunnel engineering[J]. Chinese Journal of Rock Mechanics and Engineering, 2005, 24(1): 110–115.(in Chinese))

[16] 中华人民共和国建设部. 地铁及地下工程建设风险管理指南[M]. [S. l.]: [s. n.], 2007.(Ministry of Construction, the People's Republic of China. Risk management guide of metro and the mass transit project[M]. [S. l.]: [s. n.], 2007.(in Chinese))

[17] 同济大学. 崇明越江通道工程风险分析研究总报告[R]. 上海: 同济大学, 2003.(Tongji University. The channel's Chongming project risk analysis of the report[R]. Shanghai: Tongji University, 2003.(in Chinese))

[18] 中华人民共和国铁道部. 铁路隧道风险评估暂行规定(报批稿)[S]. [S. l.]: [s. n.], 2007.(Ministry of Railway, the People's Republic of China. Railway tunnel risk assessment interim provisions(draft)[S]. [S. l.]: [s. n.], 2007.(in Chinese))

从河床冲淤分析沉管法修建长江水下隧道问题

钱七虎

(中国人民解放军总参谋部科学技术委员会，北京 100857)

摘 要 文章从三峡水库泥沙入库量的减少以及三峡水库对泥沙的拦截作用两个方面分析了三峡水库对长江中下游河床冲刷的影响，具体剖析了现有冲刷数值模拟的最大冲刷量远较实际冲刷量为小的原因，因此得出了在三峡以下长江中下游不宜采用沉管法修建长江水下隧道的结论。对于重庆地区朝天门两江隧道的修建方案问题，文章在分析了决定修建沉管隧道方案可行性的两个关键参数，即河床水深和水流流速后，指出只要采取相应工程措施，采用沉管法修建两江隧道是可行的。

关键词 水下隧道 沉管法 河床冲刷 三峡水库 河床水深 水流流速

中图分类号：U455.46 **文献标识码**：A

关于沉管法修建水下隧道问题，已有多篇论述谈及其优越性，但是任何事物，包括任何工程技术都有其局限性，包括沉管法修建方案也一样。在一定条件下和适用范围内，可发挥其优越性，若超出其适用条件，就无法发挥其优越性，呈现出其劣势来。

长江水下隧道除上海沪崇隧道外，武汉、南京长江隧道一开始都优选浅埋的沉管方案，但现在实施的方案都是盾构法；西气东输管道穿越长江工程一开始首选也采用定向钻，在南京板桥地区浅穿江底淤泥层，但最后转而采用盾构法在南京三江口地区穿越较深的砂、页岩层修建。修建方案的变更，主要原因在于三峡工程的建成，以及长江上游干、支流上修建大型水库后，长江的泥沙特性和泥沙运动规律都会发生重大变化，从而造成长江中下游的河势演变发生重大变化，不再遵循以往历史演变的规律，不能采用河床深泓历史包络线的办法来确定隧道的埋深。本文主要是归纳三峡工程的泥沙研究成果，讨论沉管法修建长江水下隧道的局限性。

1 三峡水库泥沙的入库量

测量数据表明，20世纪 90年代进入三峡水库地区的泥沙比过去少了，这里有两个数据。

(1)清华大学泥沙专家张仁教授的数据[1]是：嘉陵江的泥沙平均每年比过去减少约 8 000万吨，金沙江的泥沙每年比过去减少约 2 000万吨，总起来减少约 1亿吨，按三峡地区三个入库站至 2000年多年统计，平均入库沙量为 4.531亿吨，则总减少为原入库沙量的 20%强。

(2)中国水利水电科学研究院韩其为院士的数据[2]是"三峡入库泥沙已明显减少"，其数据是1989年以前近 40年的资料，三峡水库入库站多年均输沙量为 4.896亿吨，而 1990年至 2000年 11年间平均输沙量为 3.739亿吨，后者较之前者减少约 1.157亿吨，其中由于径流量减少，导致输沙量减少约 0.188亿吨，其余由于含沙量降低，减少输沙量 0.989亿吨。入库输沙量的减少主要表现为嘉陵江北碚站输沙量的减少，1990年至 2000年较之1956年至 1989年平均年来沙量减少了 69.3%，扣除径流量，含沙量减少 0.64亿吨。

关于泥沙减少的原因，他们的共识是认为主要不是水文资料的波动，而是：①"上游建库和水土保持的拦沙作用"；②"嘉陵江上游建了很多水库，拦沙效果很好"；③水土保持起作用。1989年以后，国家在长江中上游进行水土保持建设，禁伐森林，种草种树，水文部门估算，平均每年可以减沙 1 200万吨；④沿岸城市建设每年挖沙有三四百万吨。

展望未来，可以期待在长江上游干支流上修建水库后将进一步大幅度减少来沙量。根据西部大开发的总部署，金沙江中下段的水电将不断开发，减少三峡水库来沙将不是持续几十年，而是数百年，并且

修回日期：2006-04-05。

本文原载于《现代隧道技术》(2006 年第 43 卷第 4 期)。

在三峡建成后不久就会开始见效,因为其中溪洛渡和向家坝已同时通过立项,并已开工和即将开工。而金沙江是三峡泥沙的主要来源,多年平均输沙量为2.55亿吨,占三峡入库泥沙量(4.531亿吨)的56.2%,其中溪洛渡在100年内可减少113.8亿吨(平均为1.1亿吨/年),向家坝可减少来沙51.9亿吨(平均0.519亿吨/年)。加上金沙江下段三个水电站的开发,较之2000年前4.351亿吨的来沙量,60%以上的大幅度减少来沙量将是完全有根据的。

2 三峡水库对泥沙的拦截和大坝下游河道的冲刷

张仁教授的意见[1]是:"三峡大坝建成后,泥沙大约只有30%~40%能够泄入下游,特别是粗的泥沙都留在库里"。因此,三峡工程的运用改变了进入下游河道的水沙条件,在相当长的时间内,坝下游河道水流挟沙能力处于非饱和状态,河床将发生沿程冲刷,从而可能引起河道的一系列变化。多年来,中国水利水电科学研究院、长江水利委员会及其它有关科研单位作了大量的科研工作,进行了长距离冲刷一维数模和局部河段二维数模计算,各单位成果在定量上存在相当差别,但反映的冲刷过程和特点是基本一致的,计算成果目前仅限于大通以上河段。计算成果表明,宜昌至大通河段,河床冲刷量逐年增加,至水库运用50年达到最大值,累计冲刷量约为43亿吨。冲床冲刷自上游向下游逐年发展,当上游发生强烈冲刷时,下游冲刷很少,或者不冲刷,甚至发生淤积。各河段达到最大冲刷量的时间从上游向下游推进,由于各河段的河道特性不同,河床组成各异,冲刷量及冲刷特点也不同。总的情况是冲刷的距离很长,直至影响三峡坝下1 100 km外的大通河段;河床冲刷自上游向下游逐步发展,各河段经过一定时间冲刷后,冲刷量达到最大,这一过程逐渐向下游发展。大通以下河段并非不冲刷,而是有待研究。

关于长江河床冲刷问题,清华大学张仁教授的意见值得引起隧道工程界的关注。他提及沙市以下细沙河床的冲刷问题,认为"三峡水库建成后,下游河道将会产生长时间长距离的冲刷,造成河床降低,水位下降,这对防洪是有利的。但是,冲刷可能导致护岸底部淘空,进而导致护岸垮塌决口,这是很危险的。"对于埋置于河床表面的沉管隧道来说,会否出现隧道底部淘空的现象,进而重现我国东北埋于河床浅部的油管曾被百年一遇洪水冲断的现象,导致隧管被冲断的事故呢?

3 武汉、南京河段长江隧道隧址附近模型冲刷研究成果

为充分了解隧道工程所在河段的河床演变特性,预测河床变化趋势,有关单位分别建立了反映水沙运动及河床冲淤变化的武汉河段二维水沙数学模型,以及南京河段定床及动床试验模型。现将模型试验的三峡建库前后冲刷研究成果摘引如下(表1,表2)。

三峡建库前,隧址断面最大冲深分别出现在一百年和三百年一遇洪水年中,江汉路隧址断面左深槽最深点高程分别为−10.8 m和−12.9 m,右深槽最深点高程分别为−11.4 m和−13.5 m;南京路隧址断面左深槽最深点高程分别为−8.9 m和−11.1 m,右深槽最深点高程分别为−10.7 m和−12.6 m。

三峡建库后,隧址断面最大冲深处分别出现在一百年和三百年一遇洪水年中。在一百年和三百年设计洪水年中,江汉路隧址断面左深槽最深点分别为−12.6 m和−14.6 m,右深槽最深点高程分别为−13.1 m和−14.9 m;南京路隧址左深槽为−10.6 m和−13.0 m,右深槽为−12.2 m和−14.3 m。

总起来说,三百年一遇洪水年条件下隧址断面最大冲深比一百年一遇洪水年最大冲深大1~2 m左右;由于三峡建库后,计算含沙量较建库前相应减少1/3,故建库后最大冲深较建库前大2 m左右。

南京河段动床试验成果表明,系列年加300年一遇洪水试验条件下,三峡建库前,工程河段隧址断面最低冲刷高程为−28.70 m;建库后,系列年加三百年一遇水文条件下,隧道过江断面最低冲深高程为−30.96 m,即冲深增加−2.26 m。

动床模型试验还表明,在不同水文条件下,工程河段隧址附近河床的最大冲刷点在横向有一定的摆动,最大摆动幅度有160 m;考虑长江上游建坝的影响,300年一遇典型年时−30 m深槽的宽度将达110 m。

对上述长江三峡下游河床的冲刷深度模拟研究成果还持有以下几点认识:

(1)无论是武汉河段还是南京河段的模拟研究成果,作为对比,三峡水库建成前的水沙条件,取自1954年大水以及60年代初的数据,没有考虑到90年代后嘉陵江上游建库和水土保持的拦沙作用导致的含沙量的减少,即入库泥沙数据还是明显偏大的。

(2)河模及数模试验时,三峡水库建成后来水的含沙量为建成前的2/3,但按照大多数单位的泥沙研究成果,大部分的泥沙留在库里,仅少部分,按照张仁教授的意见,即只有30%~40%(即1/3左

表 1 武汉河段三峡建库前、后设计洪水年水沙特征值比较

Table 1 Comparison graph on characteristic values of water and sand in Wuhan channel segment in the year of flood before and after the building of Three Gorges

项目	三百年一遇洪水年			一百年一遇洪水年		
	建库前	建库后	建库前后比值	建库前	建库后	建库前后比值
年径流量 /(亿立方米)	10 958	10 958	1.0	10 257	10 257	1.0
年均含沙量 /(kg/m^3)	0.233	0.155	1.5	0.249	0.166	1.5

表 2 武汉河段隧址断面深槽最低冲刷高程计算成果

Table 2 Calculated graph of minimum scour altitude of deep slot in the section of tunnel site in Wuhan channel segment

水沙条件	江汉路隧址最低冲刷高程				南京路隧址最低冲刷高程			
	左深槽		右深槽		左深槽		右深槽	
	里程	高程 /m	里程	高程 /m	里程	高程 /m	里程	高程 /m
建库前一百年一遇	AK12+776	−10.8	AK13+358	−11.4	ANK12+736	−8.9	ANK13+370	−10.7
建库前三百年一遇	AK12+776	−12.9	AK13+358	−13.5	ANK12+736	−11.1	ANK13+370	−12.6
建库后一百年一遇	AK12+776	−12.6	AK13+358	−13.1	ANK12+736	−10.6	ANK13+370	−12.2
建库后三百年一遇	AK12+776	−14.6	AK13+358	−14.9	ANK12+736	−13.0	ANK13+370	−14.3

右)泄入下游,所以计算时,来沙条件偏大了一倍。

(3) 确定隧道埋深的隧址断面河床深泓包络线,是根据历年的河床横断面测量数据绘制的,而一百年一遇以及三百年一遇洪水时的冲深数据是无法测到的,因此,实际深泓历史包络线的最大冲深较之三百年一遇的最大冲深小得多(图 1)。

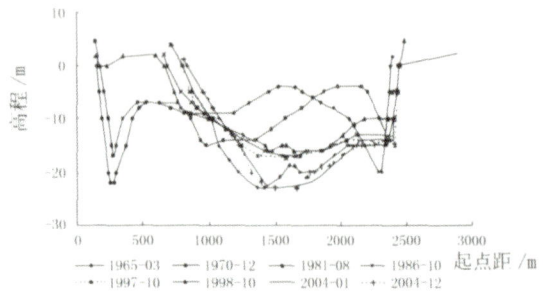

图 1 隧址断面典型年份横断面变化

Fig 1 Graph of the change of cross sections in the section of tunnel site in typical years

(4) 溪洛渡、向家坝等金沙江中、下段大型水电站水库建成后,进一步大幅度减少来沙量的影响在上述模拟研究中并未考虑。

综上所述可见,由于水土保持的拦沙作用以及大量长江中上干、支流大型水库的修建,三峡以下的下游河道将会发生长距离的冲刷,具体冲刷量取决于河道特性和河床组成,目前很难准确预测,但是较之目前河床和数模试验结果将会大很多。因此沉管法修建长江中、下游水下隧道的埋深很难准确确定,若为减少风险,埋深过深,则又导致沉管基槽很难施工,甚至由于吸挖沙困难和回淤的影响而无法实际施工。因此,结论是三峡大坝以下长江水下隧道的修建不宜采用沉管法。对于其它上游修建大型水库的江河,若水量很大、长距离冲刷很严重、河势演变尚未稳定,采用沉管法亦应慎重。

4 重庆朝天门两江隧道采用沉管法修建的难点

连接渝中半岛、江北和弹子石的重庆朝天门两江隧道有望纳入"十一五"规划。由于朝天门位于嘉陵江、长江交汇处,航运繁忙,码头密布,舍桥建隧无疑是十分正确的抉择,具体建隧的方案无论是矿山法、盾构法和沉管法都有不同的难点,各有利弊,应充分论证比较。本文仅就沉管法修建的难点及解决前景简要分析。

重庆位于三峡水库库区,因此不存在河床冲刷问题,只可能发生河床淤积。取决于隧道修建期间三峡水库蓄水位的高低,修建的难点各有不同。若在目前蓄水位处于 135 m 期间修建,其优点是重庆段水深不大,最大只有 10 多米,可充分利用枯水季节两岸裸滩较宽的有利条件,相当长度的管段可采用明挖法施工,从而可节省投资、缩短工期。其主要

难点在长江水流的流速上,一般情况下,管段沉放、对接时对自然条件的要求为:风速<10 m/s波高<0.5 m;流速<0.5~0.6 m/s能见度>1 000 m。但是这样好的沉放施工条件很难满足,而国际上大都为超越此限制自然条件流速的成功先例:比利时斯海尔德隧道水深 25 m,流速 3 m/s;纽约东 63 隧道水深 30 m,流速 2.7 m/s。重庆朝天门地段枯水期流速是否达 3 m/s甚或更多,应仔细调查和预报,高流速下管节的浮运、沉放和对接施工难度较大,应采取专门措施保证施工顺利进行。

为了避开高流速,可选择在三峡水库 156 m 或 175 m 蓄水位期间修建,届时流速降低,但河道宽度和水深加大了,挖泥和吸泥船的挖吸深度有一定的限制。目前国产的抓斗挖泥船最大挖深可达 50 m,适于挖掘粘土和块石,但生产能力较低（320~400 m³/h）;适于绞吸淤泥的链斗绞吸船经吸管加长后挖深可超过 30 m。但是,三峡水库蓄水位达 156 m 或 175 m 后,朝天门长江水位到底抬高多少?是个需要研究的问题,它取决于水库的淤积。必须意识到,朝天门长江水位的抬高大大小于水库蓄水位的增高量,因为长江河床是倾斜的。据韩其为院士的意见[2],三峡水库运用 60 年,流量为 83 421 m³/s(接近百年一遇洪水)时,对于上游无库方案,由于淤积,朝天门水位较之建库前抬高 4.36 m,而建有溪洛渡时,水位仅抬高 0.7 m。因此,朝天门水位的抬高取决于修建的具体时机,即那时水库淤积的多少?而淤积的多少又取决于那个时候上游是否建库?需要

作具体研究。总之,从目前看来,无论是三峡蓄水位在 135 m 或是今后的 175 m,沉管法修建朝天门两江隧道都有难点,但不会存在不可克服的障碍,还是可以着手进行方案可行性研究论证的。

5 结 论

(1) 20 世纪 90 年代以来,三峡地区入库站的年来沙量减少约 1 亿吨,占原来沙量的 20% 强,原因主要是嘉陵江上游建了水库以及水土保持工程起了作用。

(2) 长江上游,特别是金沙江溪洛渡和向家坝水库建成后,来沙量还将大幅减少,大约再可减少 1.5 亿吨。

(3) 目前的河模和数模试验结果表明,三峡水库建成前后,武汉及南京隧道隧址最大冲刷量相比增大 2 m 左右;

(4) 考虑到实际来沙量较河模和数模试验的来沙量小得多,以及河模及数模试验未考虑长江上游一系列大型水库的建成,长江中下游将会发生长距离的冲刷,其最大冲刷量将比试验结果增大很多,因此在三峡以下不宜采用沉管法修建长江水下隧道。

(5) 重庆地区朝天门两江隧道取决于修建时的三峡水库的蓄水位:135 m 时,主要难点在于管节沉放时的水流流速问题;175 m 时,主要难点在于挖吸淤泥的水深问题。根据现有数据,不存在不可克服的障碍,但需研究采取专门的工程措施。

参考文献

[1] 田宗伟.三峡工程为泥沙研究提供了可贵的经验[J].中国三峡建设,2004(5):7-9.
[2] 韩其为.论长江中游防洪的几个问题[J].中国三峡建设,2003(3):8-11.

国内外地下综合管线廊道发展的现状、问题及对策

钱七虎，陈晓强

（解放军理工大学，南京 210007）

摘 要：目前，地下综合管线廊道已成为城市可持续发展的重要方向。作者通过对目前国外（欧洲、日本）国内（北京、杭州、台湾）综合管线廊道发展的现状分析，指出了地下综合管线廊道的优势及其在认识、法规体系、规划管理和投资方面存在的问题，并针对法规体系、发展规划管理、监督机制、投资建设和运营机制几方面提出了相关对策。对地下综合管线廊道建设具有参考价值。

关键词：地下空间；管廊；现状；问题；对策

中图分类号：TU99　　文献标识码：A　　文章编号：1673-0836（2007）02-0191-04

Situation Problems and Countermeasures of Utility Tunnel Development in China and Abroad

QIAN Qi-hu CHEN Xiao-qiang

(PLA University of Science&Technology Nanjing 210007 China)

Abstract: Now Utility Tunnel has become an important direction of sustainable development in cities. Based on the analysis of Utility Tunnel development situation both at home (Beijing, Hangzhou, Tawian) and abroad (Europe, Japan) this paper points out its advantages and problems in understanding legislation planning and investment some Countermeasures are presented. This paper can serve as reference of runderground utility tunnel construction.

Keywords: underground space; utility tunnel; siuation; problem; countermeasure

1 前言

城市中的给水、排水、电力、电信、燃气、热力等市政管线工程，俗称生命线工程，是维持城市功能正常运转的关键。随着城市化水平的不断提高，现代城市对市政管线的需求量越来越大。如北京市在近 20 年，地下市政管线增加了近 10 倍，已达 36000 多公里。

传统中的各种市政管线，一般采用直埋或架空的方式进行铺设，但这种方法已难以满足现代城市发展的需要。目前，世界上比较先进的做法是采用地下综合管线廊道的模式。地下管线综合廊道（日语为"共同沟"，英文为"Utility Tunnel"）就是指将两种以上的城市管线集中设置于同一人工空间中，所形成的一种现代化、集约化的城市基础设施。发展共同沟，已成为城市可持续发展的重要方向。

2 现状分析

2.1 国外情况

欧洲：共同沟的建设最早在欧洲兴起。巴黎在 1832 年建造以排水为主的廊道中，创造性地在其中布置了一些供水管、煤气管和通讯电缆等管线，形成了早期的共同沟。目前，巴黎已建共同沟超过 100 km，且收容的管线也越来越多。西班牙目前有 92 km 长的共同沟，除煤气管外，所有公用设施管

收稿日期：2007-01-13（修改稿）。
本文原载于《地下空间与工程学报》（2007 年第 3 卷第 2 期）。

线均进入廊道,并制定了进一步的规划,准备在马德里主要街道下面继续扩建。俄罗斯莫斯科建有120 km的共同沟,除煤气管外,各种管线均有。瑞典斯德哥尔摩市区街道下有30 km的共同沟,建在岩体中,战时可作为民防工程。前东德在1964年开始修建共同沟,已有15 km建成使用。芬兰将共同沟深埋于地下20 m的岩层中,而不直接建于街道下,其优点是可节省30%的管线长度。

日本:是目前世界上共同沟建设最先进的国家。早在关东大地震以后的东京复兴建设中,就于1926年完成了九段阪和八重洲两处共长1.8 km的共同沟。20世纪六十年代以后,随着城市的恢复和迅速发展,共同沟建设问题再次被提上日程。1963年,日本政府颁布了《关于建设共同沟的特别措施法》,以规范和推动共同沟的建设。到1992年,全国共同沟总长达310 km。因在1995年的阪神大地震中,共同沟发挥了明显的作用,日本计划到21世纪初,在80多个县级中心城市的城市干线道路下建成长约1 100 km的共同沟。

2.2 国内情况

台湾地区:共同沟的建设始于1991年的台北。目前,台湾已建共同沟有300多 km;正在建设淡海及高雄新市镇、南港经贸园区等的共同沟;完成了洲美快速道路、大度路等共同沟工程的设计;制定了台中市、嘉义市、新竹市、台南市、基隆市的共同沟整体规划;进行了配合捷运路网、敦化南北路、新社区、铁路东延等共同沟的规划。台湾共同沟的建设非常重视与地铁、高架道路、道路拓宽等大型城市基础设施的整合建设相结合。如台北东西快速道路共同沟的建设,全长6.3 km,其中2.7 km与地铁整合建设;2.5 km与地下街、地下车库整合建设;独立施工的共同沟仅1.1 km,从而大大地降低了建设总成本,有效地推进了共同沟的发展。

大陆地区:进行共同沟的建设起步较晚。1992年,上海市政府规划建设了大陆第一条规模最大、距离最长的共同沟——浦东新区张杨路共同沟。该共同沟全长11.125 km,共有一条干线共同沟、两条支线共同沟,其中支线共同沟收容了给水、电力、信息与煤气等四种城市管线,为我国其他城市共同沟的发展提供了可供借鉴的经验和教训。目前,上海还建成了松江新城示范性地下共同沟工程(一期)和"一环加一线"总长约6 km的嘉定区安亭新镇共同沟系统。

北京早在1958年就在天安门广场下铺设了1000多米的共同沟。2006年在中关村(西区)建成了我国大陆地区第二条现代化的共同沟。该共同沟主线长2 km,支线长1 km,包括水、电、冷、热、燃气、通讯等市政管线,但为减少施工难度、节省空间、降低工程造价,没有纳入排放雨水和污水的重力流管线。

杭州在站和站前广场改建工程中,为避免站屋和各地块进出管线埋设与维修开挖路面,从而影响车站的运行,将给水管、污水管、电信电缆、电力电缆、铁路特殊电信电缆、有线电视电缆、公交动力线、供热管等置于共同沟内。杭城目前最长的共同沟——钱江新城第一条长达2.16 km的管线共同沟也于2006年初完工。

目前,共同沟还仅在我国一些经济发达的城市和新区有所建设,尚未得到推广和普及。但随着近几年全国掀起的新一轮的城市建设热潮,越来越多的大中城市已开始着手共同沟建设的试验和规划,如重庆、广州、南京、济南、沈阳、福州、郑州、青岛、威海、厦门、大同、嘉兴、衢州、连云港、佳木斯等。

3 存在问题

3.1 思想认识上有误区

共同沟建设的一次性投资远远大于管线独立铺设的成本。据统计,日本、我国台北和上海的共同沟平均造价(按人民币计算)分别是50万元/m、13万元/m和10万元/m。故有人认为其建设成本太高,经济上不合算。其实,这种认识不够全面,没有考虑到共同沟的建设成本与综合效益的关系。台湾曾以信义线6.5 km为例进行过测算,建共同沟比不建只需多投资五亿元台币,但75年间产生的效益却有2337亿元台币(包括堵车、肇事等社会成本的降低、道路及管线维修成本的减少等)。据统计,我国每年因施工产生的地下管线事故造成的直接经济损失约50亿元,间接经济损失约400亿元。同时,交通的破坏、事故的发生使民众怨声载道,"扰民"现象严重。

3.2 法规体系上有空白

为促进和规范共同沟的建设,日本于1963年通过并颁布了《关于建设共同沟的特别措施法》。我国台湾于1989年公布实施了《共同管道法》共34条,并于随后几年陆续公布实施了《共同管道法施行细则》共14条、《共同管道经费分摊办法》共6条、《共同管道系统上下空土地使用征收及补偿办法》共9条和《共同管道设计标准》共18条。目前,在我国大陆地区,对于共同沟的建设,既没有规划、建设、经营、管理方面的法规,也没有设计、施工、验收方面的标准,只是参照相关的政策法规和技术标准进行。

3.3 规划管理上有难度

由于我国市政建设方面的条块分割,各种管线分属不同的单位,报批和施工是各自为政,各权属单位往往不服从规划部门的统一管理,从而造成各类管线相互"打架",不仅影响城市正常的生活和生产,造成重复建设和资源浪费,而且随时间的推移,将使地下空间资源的整合成本越来越大,成为城市可持续发展的瓶颈。因此,为实现地下管线空间开发有序,建设共同沟是一个跨行业、跨组织的协调工程,这既需要政府部门加强集中管理,也需要各管线权属单位进行配合协调,如台北市推动共同沟建设的政府机关与相关管线单位全部算起来有58个之多,其实施难度可想而知。

3.4 资金投入上有不足

据统计,随着我国城市化进程的加快,每年仅支持新增城市人口所需的城市基础设施投资就需4000亿元。而长期以来,我国市政公用设施建设投资一直偏低。1991年前14年占GDP的比重在1%以下,占全社会固定资产投资比重在3%以下。1992年后,其投资比重在稳步上升,至2004年,分别达到了3.48%和6.78%,但与世界银行建议的发展中国家城市基础设施投资占GDP的3%~5%和占全社会固定资产的9%~15%的数据相比,仍显不够。特别是对于我们这样基础设施底子本来就薄的发展中国家来说,更需加大投资力度。

4 主要对策

4.1 借鉴国内外先进经验,制定法规体系

推广和普及共同沟建设的基础是制定相应的法规体系。在这方面,我们既可以借鉴如日本的《关于建设共同沟的特别措施法》和台湾的《共同管道法》、《共同管道法施行细则》、《共同建设管线基金收支保管及运用办法》、《共同沟建设及管理经费分摊办法》等现有法规,也可借鉴如上海市的《上海市城市道路架线管理办法》和重庆市的《重庆市管线工程规划管理办法》等具体规定。

共同沟的法规体系,主要应包括两方面的内容:一是政策立法;二是技术立法。政策立法,必须首先明确:凡是建设共同沟的城市道路,任何单位和部门不得另行开挖道路铺设管线?所有管线必须统一入驻共同沟,并按规定向经营管理企业交纳使用费。并就共同沟的所有权、规划权、建设权、管理权、经营权和使用权等方面做出具体的规定。技术立法,主要是对共同沟的各种设计(如布局设计、结构设计、管线设计、防灾设计等)、施工方法(如施工工艺、施工流程、施工安全等)、检查验收(如验收方法、验收指标等)和材料设备(如建筑材料、监控系统、通风系统、供电系统、排水系统、通信系统、标示系统和地面设施等)制定具体的标准和规范。

4.2 结合城市建设和地下空间利用,统一编制发展规划

发展共同沟,在功能上,要适应好现代城市的全面建设;在建设上,要利用地下空间的有限资源,这就需要通过实施统一的规划,使其与现代城市的全面建设和地下空间的综合利用协调发展。

在统一编制共同沟的发展规划时,一是要发挥政府作用。规划部门作为地下管线工程规划的综合协调机构,要把共同沟的建设纳入到城市的总体规划中,对于有关的新建项目,必须要求其把共同沟作为组成部分规划进去,并提供明确的规划方案,否则不予审批。二是要明确具体内容。规划部门要按照"考虑城市长远需求、结合地下综合开发、完善管线合理配置"的原则,按照城市规划和管线专业规划的要求,组织编制出共同沟的专项规划,具体明确管线布局、建设标准、资金投入和技术要求等内容,对地下管线进行统一规划、统一建设、统一管理。三是要理清发展思路。结合我国城市建设、改造的特点,借鉴有关国家、地区的成功经验,共同沟的发展应立足"先新建,后改造"的指导思想。对于新建的景观区、商务区、开发区和居民区等,应将新建道路与共同沟的建设实行统一规划、同步施工,按道路的等级标准,构建起干线、支线有机结合的共同沟系统。在此基础上,逐步向旧城区展开,结合老道路的整修或拓宽,视情进行共同沟的建设。南宁市就以相思湖10条拟建大道为试点,在该片区实施"地下管线共同沟"项目,并以此为模板向全市铺开。

4.3 成立专门机构和建立信息系统,完善管理、监督机制

目前,在我国的现行管理体制中,市政管委虽是地下管线的管理单位,但其对各地下管线的权属单位并无行政权,在涉及到不同部门或领域的具体利益时,难以做到协调管理。如北京为了建设"北京城市地下管线综合管理信息系统",市政管委向各权属单位要求得到管线数据时?还是经过北京市

委书记会议上讨论通过才得以完成。因此,为推进共同沟的建设,必须打破现有的"谁拥有、谁管理"的各自为政的管理体制,成立专门的管理机构。如日本在中央建设省下设了16个共同管道科;台湾在台北市、高雄市分别成立了共管科与管线科。其主要职责应体现在以下几方面:共同沟始建时,负责相关政策和具体方案的制定;共同沟在建中,负责投资、建设的监控;共同沟建成后,负责工程验收和营运监督等。

目前,地下管线的信息大多还是利用图纸、图表等纸介质进行记录、保存和管理,存在资料不全、查询不便、更新速度慢、利用效率低等问题。为提高地下管线管理的水平,应将各种地下管线信息以数字化的形式显现出来,建立包括有储存、管理、分析、统计、查询、输出、更新等多种功能的地下综合管网信息系统,以满足各种机构、各种应用、各种事件处置的要求。如北京市政管委会在2004年就专门组织开发了"地下管线综合管理系统",全面、系统、准确地反映了地下管线分布现状、运行状态、各类管线间的相关关系等基本信息。

4.4 适应市场经济规律,创新投资、建设和运营机制

根据国内外的做法,目前对共同沟的建设主要有三种投资模式:一是政府投资。如台北市于1991年成立了25亿元(总额度50亿元)的共同沟基金用以推动共同沟的建设。二是企业投资。如北京科技园区建设股份有限公司在中关村西区投资进行了共同沟的建设和地下空间的开发。三是多方投资。如日本规定:共同沟建设投资中的40%由道路部门负担,其余60%由占用共同沟的各管线单位按比例承担。

为改变长期以来我国基础设施领域主要依赖政府财政性资金的传统做法,共同沟的建设应适应市场经济发展的规律,通过"开放市场、引入竞争、公开招标、特许经营"的方式,形成"政府指导、市场运作、社会参与"的投资、建设和运营新型机制(简称BOT),即政府只负责宏观政策的制定和操作程序的规范,如工程的规划论证、项目申请报告的编制、招标方案的设计、投融资方案的研究、特许协议的起草等;而由企业负责投资建设(Build)、运营管理(Operate),并在合同期满后将经营权移交给政府(Transfer),再由政府与其续签合作协议或通过公开招投标确定运营管理企业。这种机制的优点在于:一是转变了政府职能。政府不再费神、费力、费钱直接参与工程的投资和建设管理,只需集中精力担当好"宏观政策制定者"和"具体方案裁判员"的角色。二是拓宽了资金来源。采用公开招标的方式,便于发挥社会资金投融资的积极性,使有实力的多元化经济实体愿意投资参与共同沟的建设。三是保证了项目实施。由于政府与企业签订了具有法律效益的协议,具体规定了企业的责、权、利,企业必定会采取各种有效的办法,保证项目建设和运营的正常进行。四是实现了政企双赢。对政府来说,只需在前期投入少量的启动性资金,既可节约财政投入或实现零投入,又可吸收几十倍甚至上百倍的社会资金投入,从而实现财政性资金的放大效应和倍增效应。对中标企业来说,由于实现了投融资、建设、经营和管理一体化,就可通过对各环节进行系统的规划和控制,提高效率,降低成本,获取较大的收益。需要强调的是,这种投资机制的基础必须建立在共同沟的产权归国有上,这既利于统一规划、协调管理,又可避免地下资源流失或企业垄断。

5 结语

实践证明:建设共同沟,具有实施统一规划,避免重复建设;消除空中网线,美化城市景观;减少道路开挖,改善交通秩序;集中各类管线,减少事故发生;利用地下空间,提高抗灾能力;进行统一管理,方便检查维修等综合效益。因此,认清形势,抓住机遇,通过有效的管理,借助先进的技术,大力发展共同沟,将成为我国建设资源节约型、环境友好型城市的有效途径之一。

参考文献:

[1] 中国工程院课题组. 中国城市地下空间开发利用. 中国建筑工业出版社. 2001. (Project team of China Academy of Engineering. Development and Utilization of Urban Underground Space in China. China Building Industry Press. 2001. (in Chinese))

[2] 陈志龙,王玉北. 城市地下空间规划. 东南大学出版社. 2005. (Chen Zhilong, Wang Yubei. Planning of Urban Underground space. Southeast University Press. 2005. (in Chinese))

[3] 童林旭. 地下空间与城市现代化发展. 中国建筑工业出版社. 2005. (Tong Linxu. Underground Space and Development of Urban Mordenization. China Building Industry Press. 2005. (in Chinese))

[4] 钱维,尤伯军. 政府投资体制的制度创新—项目法人招标制. 中国财政经济出版社. 2006. (Qian Wei, You Bojun. System Creation of Government Investing System— Juristic Person Bidding of the Project. China Finance and Economy Press. 2006. (in Chinese))

[5] 彭保华. 共同管道工程在台湾. 海峡两岸四地21世纪交通与物流学术论坛. 2002. (Peng Baohum. Utility Tunnel in Taiwan. Symposium of 21 Century Transportation and Logistics for Cross—strait. 2002. (in Chinese))

地下空间作为城市空间结构的社会学内涵

奚江琳,钱七虎

(解放军理工大学工程兵工程学院地下空间研究中心,南京 210007)

摘 要:现代城市规划正逐步从注重物质形态规划向多学科渗透的综合科学发展,城市地下空间的规划也应包容更广阔的层面。经济、地理、各种社会因素都将对城市地下空间的功能组成及形态起重要作用。城市空间结构所包含的城市内在各组织的相互作用,必然反映在由各种相互联系所形成的作为一个有机整体的社会之中。文章着重探讨了地下空间作为城市空间结构所具有的社会学内涵,以及地下空间结构对人与人、人与车的社会学关系的反应。

关键词:综合规划;城市地下空间结构;社会学

中图分类号:TU984.11　　**文献标识码**:A

The Socialogical Connotation of Underground Space as Urban Space Structure

XI JIang-lin, QIAN Qi-hu

(*Underground Research Center, PLA University of Science and Technology, Nanjing* 210007, *P. R. China*)

Abstract: The emphasis of modern city planning is put on synthesis of multiple disciplines instead of making substance plan, and the urban underground space plan should be the same as the modern city plan. The Economy, geography and various social factors may affect the function and configuration of urban underground space. The Society is a macrocosm including various relations. Thus the interaction among the inner organizations of the urban space structure must be reflected in the whole society. In this paper, the sociological connotation of the urban space is discussed. The underground space structure will influence the sociological relations, which exist among people vs. peopleor people vs. vehicles and so on.

Keywords: synthetic plan; urban space structure; sociology

1 引言——近现代城市规划科学的演变

1.1 从物质到本质(FROM SUBSTANCE TO ESSENCE)

就近代城市规划科学的自身衍化来看,存在一个从注重城市物质形态到多学科渗透的综合科学的演变过程。城市的空间形态设计确实构成了自工业社会至二十世纪上半叶城市理论的主体也是城市规划中最为发达的一部分。二十世纪初期,以柯布西埃、伊 沙里宁为代表的建筑师用充满激情的建筑师眼光规划城市(包括城市中的社会与经济问题),他们着眼点是城市物质环境,直接以三度图解给予城市发展方向。这种手段一直是当今中国城市规划的主流。

二十世纪中叶以后,国外城市理论的重点逐步从物质环境建设转向了公共政策和社会经济等根本性问题,第 18 版的不列颠百科全书指出,现代城市规划的目的"在于满足城市的社会和经济发展的要求,其意义远超过城市外观的形式和建筑物、街道、公园、公共设施等布局问题,它是政府部门的职责之一,也是一项专门科学。"[1] 这无疑道出了城市

收稿日期:2005-06-07(修改稿)。
本文原载于《地下空间与工程学报》(2005 年第 1 卷第 5 期)。

规划的实质。

1.2 从个体到群体(FOR SINGLE OR FOR GROUP)

以人为本,人是代表狭隘的个人利益,群体的利益,还是人与自然的共同利益?放在不同的命题中就有不同的解答。上世纪70年代社会发展所关注的人文主义,其视角是以具体的人为尺度,主要指个性和生物意义上的人,1990年以来,欧洲城市面对上述的社会变化和挑战,解决深层次社会问题已经成为城市的重点。城市设计和建筑设计的不同不仅体现在空间尺度、内外关系等方面,城市设计所容纳的人群远比建筑复杂,其中的社会学意义也更深远。

《荀子》中说,"人生不能无群","群"的存在乃社会本质,既然有各类人群的客观存在,就有各类群体的城市空间和利益分配问题,在城市环境中确保不同社会成员之间自由交流的平等权利和可能性,这是当代城市环境公共性的核心问题。社会运行的动力蕴藏于人的需要之中,人的需要可在个人与个人之间、群体与群体之间的相互补充中得到满足,各"群体"在社会互动过程中所结成的合力的大小,会影响到社会系统运行的平衡性、协调性以及方向性,这种关系事实上构成城市规划、空间形态重要影响因素。

美国社会学家甘斯 H. Gans 在其著作《人与规划》(People and Plan)(1968)中提出人的生活是由经济、文化和社会权利结构所决定的,而不是规划师们所认为的由物质环境(Physical Environment)及其规划所决定的[2]。随着对规划本身认识的深入,城市规划将冲破功能主义的束缚而进入一个更为广阔的背景之中,目前我国正值大规模和深层次的城市空间演变,对于地下空间规划和发展既是机遇也是挑战,高层次和高起点是必然要求。

2 城市地下空间结构的社会学含义

国内的规划研究近些年在引入国外规划理论方面有所突破,但在实际规划中还存在一些问题。从形态上认知和处理是中国当前规划的现状,缺乏真正多学科的融合。往往是规划设计院依据领导的意图,做范式语言,完整性与科学性均不够充分,城市布局和城市空间结构对经济、产业、社会层次的深入考虑不足,这也是中国特色之一。执政者需要即见效果,而这效果是以目视为准,物质性规划最符合政府胃口,但伴随改革的深入和可持续城市化的继续,城市发展将面临更现实的问题,规划必须包容越来越多的内涵,并且随地下空间包含越来越多的城市功能,通过它折射出一定的社会结构布局。

在地下空间规划中,我们考虑的不仅是功能形态和环境等可视问题,还将包涵文化背景、生活方式、社会层次(共同需求与特殊需求)等内涵,即关注地下空间结构的社会学意义。社会学的研究对象并不存在于社会生活的某一特殊领域中,而是存在于社会生活的各个领域之间的相互联系之中,城市空间结构所包含的城市内在各组织的相互作用,必然反映在由各种相互联系所形成的作为一个有机整体的社会之中。因地下空间所涉及的往往是城市的具体功能,关系到各人群的切身利益和社会公平,那人与人、人与车的社会学关系必然在地下空间结构中有所反应,地下城所容纳的功能内容取决于我们的社会视角(排除经济问题)。

城市大规模地下空间开发将有力地改变城市空间结构,影响城市发展。人与车共同占有城市资源,在交通主导城市形态的现代城市,就拿人车的关系而言,其中包含深刻的社会学意义。

3 人车关系与地下空间结构内涵

城市空间形态首先是一种公共文化,而建立一种真正反映社会各群体的均衡的利益以及平衡机制,将是我们这个时代各种深刻的社会变革的一项重要内容。既然交通方式的改变已成为我们这个时代生活的主导,所以本文的地下空间的社会学以持不同交通工具,占用不同交通空间来分群,这里形成复杂的人(群)与人(群)、人与物的系统关系。

规划需要围绕公共空间(无论地面还是地上公共空间)两种不同的城市生活载体内涵予以表达,其一,即是我们普遍关注的空间内涵的表达,空间是城市公共生活的容器,是规划的物质内涵;其二,社会内涵的表达,社会作为公共生活中市民精神的依存,城市生活与公共空间的互动,社会各阶层的交往,这些是公共空间规划的另一个重要内涵,或许这一层次的表达就暗示了"容器"的未来形态和规模。在今后地下空间规划中,除了经济、容量、布局、分布与发展等研究外,地下空间所影响的人与人、人与车的社会学关系也值得考虑。在2005年1月8日郑州郑东新区地下空间规划评审会上,钱七虎院士指出,要全面系统地对地下空间进行规划,必须面对社会现实,解决城市拥堵。

当中心区的地下空间蕴涵丰富的城市功能,形成相互联系的系统,谓之地下城。巴黎、东京、蒙特利尔等城市的中心均已形成错综复杂的地下系统。地下城的结构从人车分离逐步发展到一定的人车共享。

3.1 人车分离

19世纪末20世纪初,机动化交通出现后,步行与车行的矛盾越来越突出,使得原有的城市结构开始遭到破坏。西方国家开始了道路规划与步行交通的研究。真正意义上的步行中心区开始于20世纪60年代,大规模的道路建设和私人汽车无限制的发展引发了一系列的城市问题,自由经济时期所获得的价值和目标不断受到质疑,为了再现城市中心区的繁荣,西方国家相继实施内城复兴计划,在城市中心区大规模推行步行化策略。交通的立体化也许是有效的办法,空中、地面和地下可以提供三个层次的互不干扰而方便连接的系统。其中往地下发展,重构城市秩序获得了较多的成功实践,尤以巴黎、伦敦等具有悠久历史文化的城市在保留地面环境与历史文脉的同时实现了城市的现代化更新。

人车分离是由密集地铁网络形成的,地铁轨道交通运量大,所形成的城市内聚力更强,使城市呈紧凑的形态,并逐步形成新的城市空间形态与构成。如此轨道交通在城市地下空间规划中不仅具有功能性,同时在地下空间的形态方面起到重要作用。地下城的轨道交通往往交织成网络,并使所辐射区域形成更为密实的团块状发展。巴黎以塞纳河和巴黎圣母院为凝聚力的中心区域创造了世界上最富有魅力的步行环境。

由于受以人为本的思想影响,并且上世纪70年代以前社会发展所关注的人文主义,其视角是以具体的人为尺度,所以在城市设计研究与实践中注重人车分离。这一理论随中国城市轨道交通的兴起受到广泛重视,但这并不是一个最符合现实和需求的境界。

3.2 人车共享

根据近年来的发展,地下机动交通的影响也不可忽视,因为单一的步行化建设并不可能消除私人汽车交通,只是将它们转移到城市的其它地区,同时对于城市中心区而言,步行化削弱了与机动车交通联系紧密的商业的吸引力,所以20世纪80年代以后,在西方发达国家开始了共享理论的研究,人们试图寻求一种合理的规划设计和管理措施,使借助地铁轨道交通成功的步行者与机动车能够平等共存,各类交通方式能够和谐相处,限制交通的人车共存方式更有利于提高城市活力。

向空中发展还是向地下发展(包括人与汽车,两组概念;高架路与天桥;地下机动车道路与地下人行系统)? 一方面,这是个环境命题;另一方面也是一个深刻的社会命题,实际上因为许多汽车的出行的目的地在地下停车场,从地下空间到地下空间,非常直接,人们需要人与车之间便捷的界面,加之地铁的人流也在地下汇集,于是一种积极的功能集合在地下形成,这是人与车、人与人、人与空间的默契。"多样性是大城市的天性"(diversity is nature to big cities)[3]大都市的包容性正是其魅力所在,无论从经济角度,还是从社会角度来看,城市都要尽可能错综复杂并且相互支持的功用的多样性。城市居民的生活方式、文化与价值的角色都包含在内的结构是,它们不仅定义我们在城市环境里的经验,也是在塑造土地利用的形态。

实际上,现代城市既需要人性的步行系统,也需要便捷的快速机动交通在范围广大的不同区域相互联系。现代城市的步行化是结合了机动化之后的步行化,步行系统和机动化交通系统都是现代城市所不可忽视的。二十世纪80年代以来,在西方发达国家开始了共享理论的研究,步行交通与机动车交通之争发生了质的变化,人车平等共存的概念逐渐取代了人车分离的概念。人们试图寻找一种合理的规划设计和管理措施,为所有的道路使用者改善环境,使各种交通方式能够和谐相处。并且这两者并不完全背离,一定程度上可以相互支持,当地面的车辆下地后,更干净更适合人行的地面空间便呈现出来了。

显然,地下城能更好地实现人车共享的理论,在地下连接公共空间和地下交通枢纽,将地面街道空间、地下公共空间、地铁、地下停车场等高效有机地连接在一起。

同时人车共享并不违背以人为本的原则,它不破坏步行创造的以人的适宜尺度和以人的活动和方式的城市结构。北京就存在TOD与AOD协调发展的必要性(即机动车与公共交通共同发展策略)[4],实际上,中国的特大城市中,在以公共交通为主导的发展方向下,人车共享策略更为实际。城市交通不仅局限于道路问题,它与城市空间和人的行为活动紧密联系在一起,通过不同交通方式的叠加与交织,可以优化城市结构,同时增进社会行为的多样性与合理性。但是,多种交通方式的复合带来的是城市空间形态的复杂化,这给地下空间规划带来很大的挑战。

以巴黎为例,巴黎的市中心依托密集的地铁网络构成世界上最宜人的步行区域,但是巴黎在1987年提出过地下机动车交通系统,冠名为"LASER"计划(LASER 与作为巴黎地下城市的挑战与实验,尚未完成),此项计划欲在巴黎构筑两环加放射的地下道路网解决机动车在城市内部的运动问题。

此外,东京首条地下高速路"东京中央环状新宿线"经过池袋、新宿和涩谷三个重要商业区,将于2007年3月通车。开通之后,不仅可以大大缩短通行时间,而且还可以有效地缓解市中心地区的交通拥挤以及减轻市内环境污染问题,可见,轨道交通发达以后,不能消除机动车使用。

2005年1月,南京修改城东干道快速路方案,在密集地段珠江路一段下地,预计到06年9月份,隧道主体完成[5],在此前,南京建设两条过玄武湖的地下机动车干道(其中九华山隧道尚未完工)。

在都市的分中心中,由于所集聚的办公性质的原因,可能对私人机动车交通的需求更为强烈,例如大都市的科技中心和CBD区域。以北京中关村西区规划为例,中关村广场结合我国国情及自身的设计特点,营造了全国最大立体交通网,创立综合管廊+地下空间开发+地下环行车道的三位一体的地下综合构筑物模式,这一方面代表了中国地下空间技术的先进性,另一方面也蕴涵了城市空间结构中的社会学意义。

4 小结

对于都市大规模地下空间的集结方式(地下城)的理解,应该超越传统的偏重物质空间的定义,而被更多地从社会文化角度理解。在整体的城市结构和空间网络中,能够为社会的各种边际要素和人群活动提供场所。

今天的城市规划学科蕴涵深刻的社会学问题,地下空间规划亦不例外。早期的地下空间往往直接到修建的层面上,其设计偏重于物质形态,用建筑师或工程师的眼光来看待地下空间,将钢筋水泥构筑在地下形成人造形式为目的,但当地下空间发展成规模后,其规划可能比地面建筑更为复杂和系统化,不仅需要考虑不同系统的组合、文化保存、环境与生态,包含社会学、经济学等抽象组织形态也将是重要研究内容。地下空间规划中的多目标和系统化的规划设计,将考虑影响城市整体空间的演化结构的社会层域。

参考文献:

[1] 王建国. 现代城市设计理论和方法[M]. 1997:49—50.

[2] 顾朝林. 城市社会学[M]. 东南大学出版社,2002:26.

[3] Jane Jacobs. The Death and Life of Great American Cities. 143.

[4] 陈志龙,奚江琳. 北京市区中心地区地下空间开发利用规划专题研究四—北京地下空间开发利用前景分析,2004:37—38.

[5] 金陵晚报,2005年1月21日 A13版.

现代地下货物运输系统的研究与进展

张明聚[1]，钱七虎[2]，唐 劼[3]

(1. 北京工业大学 建筑工程学院，北京 100022；2. 中国工程院 工程管理学部，北京 100038；
3. 中咨工程建设监理公司，北京 100044)

摘 要：对地下货物运输系统进行研究，以期为我国城市交通运输的发展提供有益的建议．从阐述地下货物运输的概念及分类入手，介绍了地下货物运输系统在国外的研究、应用现状和进展情况，对地下货物运输系统的适用范围、潜在优势和应用前景进行了分析．分析表明，地下货物运输系统是改进城市交通的有效措施，提高运输速度和运行效率；有效减少城市环境污染；提高城市交通的通达性；节约城市用地，促进城市可持续发展；能提高货物配送效率，降低运输和库存成本．在大城市发展现代地下货物运输系统是一个具有战略眼光的研究和发展领域，是解决未来交通难题的发展思路．

关键词：地下空间；隧道；地下货物运输；地下物流
中图分类号：U1；TU91 **文献标识码**：A **文章编号**：0254-0037(2005)06-0580-05

1 概述

从20世纪90年代起，荷兰、日本、德国、美国和英国对地下货运系统非常关注[1-4]．地下运输系统既可以运送乘客，也可以运送货物．地下轨道客运系统，称之为地下铁道(Subway或Metro)．世界大多数发达城市都修建了地下铁道．当然，输送气体和液体的管道也是地下运输的形式，这些运输体系已经发展得很好．在隧道或大直径管道中利用密封舱或车辆运输货物的地下货物运输(underground freight transport，简称UFT)系统却鲜为人知．实际上，在修建地下客运系统的同时，也修建了地下货运系统，只是实际运营的较少．地下货物运输系统的直径、导向系统和推进方法各不相同．

地下货物运输分为密封舱管道系统(Capsule-pipeline Systems)和其它借助隧道运行的地下货物运输系统[1]．密封舱管道系统的直径通常小于1.3 m，没有类似轨道或路面的特别导向系统．形状不同的大直径管道或隧道的其他地下运输系统，具有导向系统．

密封舱管道系统可分为气压或液压密封舱管道(pneumatic or hydraulic capsule-pipeline，简称PCP或HCP)系统．新的进展是在气压和液压系统中使用线性诱导推进力(linear induction propulsion)．

地下货物运输系统虽然能够高效地进行货物配送，又环保．由于财政和组织机构的问题，地下货运系统仍没有得到有效的开发．

目前，在全球范围内，城市化进程的加快使得货物运输需求不断增长、同时又要求减少货运的不利外部影响(如交通事故、噪声危害、空气污染、能量损耗和占用空间)，目前很多大城市的地面货物运输能力已经达到了极限．为此，我国有关专家亦对发展地下货物运输系统提出了初步建议[6-8]．

2 发展历史

2.1 早期的地下货物运输

19世纪初，英国就开发应用了地下货物运输系统[1-3]．第1类管道运输系统是用气压发送(pneumat-

收稿日期：2004-05-20。
基金项目：北京工业大学博士科研启动基金资助(VA00017)。
本文原载于《北京工业大学学报》(2005年第31卷第6期)。

ic-dispatch)的,用来从电报中心向其它邮电局传递电报和信息. 把国际电报公司的办公室与伦敦股票交易所相连,1853 年投入使用. 当时,具有大容量的这种气压发送系统解决了大都市的电报繁忙的问题. 到 1875 年,伦敦约有 53 km 这种管道在运营,英国的利物浦、伯明翰和曼彻斯特等其他 7 个城市也引进了这种系统. 到 1909 年,伦敦有 64 km 这种管道,有 17 个城市建立了这种服务系统. 在欧洲的其他国家也相继建立了这种系统,法国巴黎一直使用到 1984 年.

1859 年伦敦发展了第 1 个带轮气动密封舱管道(Wheeled PCP),在 76 cm×84 cm 的管道中的轨道上运行,轨道的轨距为 611 mm. 单个密封舱质量最多可达 3 048 kg,速度最大为 64 km/h. 1962 年,在德国汉堡的主要邮电局及主要火车站之间,建造了 PCP 系统来运送邮件. 该系统是在直径为 0.45 m 的管道中运行 1.6 m 长的集装箱(密封舱). 密封舱以 30~36 km/h 的速度在长 1.8 km 的管道中运行.

随后,气动快递系统的技术继续发展,到 19 世纪 60 年代,对大直径密封舱管道的关注促使带轮 PCP 系统的发展. 1971 年在佐治亚的 Stocksbridge 及 1973 年在休斯敦建造了管道直径分别为 0.91 m 和 0.45 m 的 2 个试验线,被称为 Tubexpress. 在俄国建立了几个被称为"Transprogress"的系统,在 1971—1983 年进行了商业运营. 第 1 个系统在佐治亚的 Tibilisi 用于运送碎石,其他系统用于运送砾石、砂子、矿物甚至垃圾. 前苏联计划建造 20 多条这种货运系统. 并成立了专门的公司.

日本成功地运用了直径为 1 m 的 PCP 系统,作为永久设施向水泥厂运送石灰石,作为临时施工工程在长大隧道和公路工程中运送土方和施工物资.

在 20 世纪初期,人们研制了另一种货物运输系统. 在芝加哥,从 20 世纪初到 1959 年,运送废料和煤的地下轨道运输系统一直在运行. 在伦敦,被称为"MailRail"的地下运输系统 1927 年投入使用;至今该系统一直在运营,用来运送伦敦中心区内的各邮电局之间的邮件.

2.2 现代地下货物运输的发展

20 世纪 70 年代,在欧洲、日本和美国提出了发展高速运送乘客和货物系统的建议[1]. 除了高速和长距离运输的优点之外,这些系统运行在地下的低压隧道中,有利于减少能源消耗,对环境影响小. 在一些国家已经引起了对地下货物运输的新的关注.

20 世纪 90 年代,荷兰、德国、日本、美国和英国对地下货运系统非常关注[2]. 荷兰在 1987 年提出集成交通运输(integral transport system,简称 ITS)概念,即由长距离地下高速客运及货运网络与局部地区货物短距离集散相结合的运输系统. 1987—1993 年,国家运输部门资助的几个研究项目更加详细地研究了这个运输系统. 短距离低速度地下运输货物受到了关注. 原因是地下货运能够解决城市交通拥挤区域的货物运输问题. 1994 年,荷兰政府可持续发展计划组织(简称 DTO)举办了多次未来交通运输可持续发展的讨论会. 在城市货物运输领域,可持续发展的运输模式除了 UFT 外,别无选择. 在 DTO 的框架之下,接下来的步骤是设计市区内 UFT 的物流规划,并确定实施战略. 这项工作于 1996—1997 年在一个市区货物管道运输的规划项目中展开,称为示范工程. 作为市区内商品运输的新型的、可持续发展的物流概念,应基于小孔径管道网络. 这种地下运输系统将使用小直径隧道(最大直径约为 2.4 m). 这意味着引入该系统需要根本改变目前的货物配送方式. 但是,另一方面,该系统符合小批量高频率随时供货的现代物流趋势. 1995 年,一些私有公司启动了 2 个其他地下货运建设项目. 管道集装运输(unit transport by pipe, 简称 UTP)项目是基于中等距离的地下联结,运送阿姆斯特丹机场和 Antwerp 之间的微型集装箱. 另一个项目,称为地下物流(荷兰语 ondergronds logistiek systeem,简称 OLS)涉及运送航空货物及鲜花的地下运输网络的可行性研究,该系统把 Schiphol 机场和 Aalsmeer 花卉市场与 Hoofddorp 铁路中转站联接起来. 该系统的预设计于 1998 年初启动,1999 年规划了一个 OLS 运输系统试验场. 研究表明,在荷兰 UFT 潜在的货物运输量到 2020 年将占总运输量的 1/3,而且从经济和环境的观点来看,UFT 是一个更具吸引力的发展领域.

德国,1999 年和 2000 年在鲁尔大学(Ruhr University)进行了一项通过地下管道运输普通货物的自动化运输的可行性研究,以及在德国 Ruhrgebiet 经济区域项目的可能经济优点. 在这项研究中,把西南方向的重要经济中心与鲁尔南部联系起来了. 研究表明,直径为 1.6 m 的管道采用无人驾驶密封舱足够用来运输货物,这些密封舱互相之间仅有一个很小的距离、以相对高的速度行驶. 目前的研究合同包括建造足尺中间试验厂的全面规划准备工作. 已研制一个 1:2 的货管(cargo cap)模型来试验密封舱的主要部件. 该模型将用来模拟在密闭系统内的连续运行. 在鲁尔大学的校园内将建造完整的系统.

日本拟建造一个专门设施,借助穿行于隧道中的两用轻型卡车(dual mode truck,简称 DMT)进行货物运输. 这种两用货车在东京可以有人驾驶在公路上运行,也可在地下隧道中无人驾驶运行. 另一项研究是在东京市内建造一个像地铁一样的系统,在地下运送邮件. 由于当时处于经济危机,缺乏研究经费,这 2 个项目被迫于 1994 年暂停. 进入新世纪以来,日本把地下物流网络技术列为 21 世纪 16 项高科技支柱产业之一.

美国研制了称为"Subtrans"的密封舱运输系统,最近对使用该系统运送几个快递分发中心与纽约附近 Newark 机场之间的邮件和包裹的可行性进行了研究.

英国正在进行一项可行性研究,以决定伦敦中心的邮电局之间的地下联络通道是否可以用来运送货物和邮件,即货物地铁(Metrofreight),是一种自动化的货物运输系统. 伦敦中心的购物区有很大的运输量,而且交通又十分拥挤,使得这个系统是可行的.

3 潜在优势及应用前景

3.1 地下货物运输的潜在优势

地下货物运输系统的发展是低污染和环境合理的运输方式. 现代城市交通控制失效及由于交通量的增长不能满足环境保护目标. 同时,由于交通拥堵使得运输经济成本攀升. 特别是接近或市区内的主干道路在白天平均运行速度大幅度下降. 这意味着,运输的成本提高,而车辆和司机的生产率下降.

货物运输在技术上和组织上的变化,为货物运输的可持续发展(从环境、社会和经济观点)提供了机会. 从经济、环境和社会观点而言,地下货物运输具有几方面潜在的优势.

1) 提高运输速度和运行效率. 地下货物运输系统可发展为一种独立有效的交通系统,与传统的运输系统和物流方案相协调. 首先用卡车把货物运送到工业集中或人口密集地区的边沿,然后在地下将货物运送到工业区和城市中心,能够高速、安全运行,迅速、可靠、准时地运输货物.

2) 有效减少城市环境污染. 以地面车辆为主要方式的城市物流配送是造成城市大气污染的主要根源. 地下货物运输系统用于运送货物的专用车辆是由电力这一清洁能源驱动的,不会产生废气,且避免了噪音污染. 通过实施地下货物运输系统,可极大地减少城市环境污染,给人们留下明媚的阳光、清洁的空气和安静的环境.

3) 有助于解决交通拥挤问题,提高城市交通的通达性. 把大量的货物运输转向地下,可极大地缓解交通拥挤问题,降低交通事故率,并给私人小汽车的发展留下巨大的发展空间.

4) 节约城市用地,促进城市可持续发展,是地下空间开发利用的战略选择. 地下货运整个系统建在地面以下一定深度运行,在站点可采用垂直提升的方式集散货物,只占用极少地面空间.

5) 提高货物配送效率,降低运输和库存成本. 城市里的大型超市和企业与地下货物运输系统相连,成为城市地下货物运输系统的终端,企业之间、企业和超市之间便可实现直接配送,配套供货和及时供货,实现企业零库存.

3.2 地下货物运输的应用前景

几乎世界上每个大城市都面临交通拥挤问题. 地下货物运输系统将社会效益与经济效益相结合,把地面的交通运转转移到地下,进行无阻碍的自动化运输. 因有独立的运输环境而受外界影响小,可以保证运输的稳定性. 可以预见,地下货运系统将在 21 世纪的货物运输中起主要作用.

不同的地下货物运输系统有不同的应用领域. 图 1 表示不同的地下货物运输系统、适用范围及其发展状态. PCP 和 HCP 系统的管道直径较小. 它们的应用领域是同种货物的运输. 在这种情况下,意味着只有一种类型的货物,速度相对较低,运距短. 另一方面,大直径系统用以长距离高速度运输不同类型的货物.

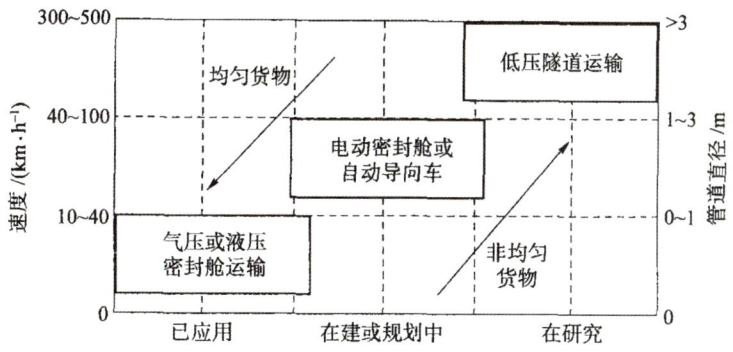

图 1 地下货物运输系统类型、适用范围及其发展状态[1]

Fig.1 Type, scepe of application and development of the underground freight transport system

近些年来,发达国家对地下货物运输系统研究和发展的兴趣激增,发展较快的是荷兰和德国. 现代管道和隧道货物运输系统仍在发展之中,是一个长远发展项目,初期可小规模地应用该系统,并逐步形成城市地下物流网络. 其应用领域有:①特大城市市中心区内部消费物资的配送;②在工业区、港口或机场内的货盘、海运集装箱、空运货盘的运输;③农产品、矿石或固体垃圾的收集和长距离运输;④集装箱在内陆或跨地区的运输.

也有人提出该系统可发展成继公路、铁路、水运和空运之后的第 5 种运输方式. 在国外,一些未来学家甚至考虑得更超前. 将来隧道运输系统将进行长距离的货运和客运. 这些概念基于大直径的隧道(直径大约为 5 m)和更高的运输速度(在 300~500 km/h). 另一方面,水、天然气、暖气和污水处理网已进入家庭、商业和工业企业,以地下货物运输系统为基础形成物流网络的形式,也可把工厂、超市、社区和家庭都连接起来,直接将货物送到千家万户,电子购物的理想就能够变成现实.

现代大城市交通的拥堵、事故和污染的老难题,在我国日益严重. 以北京为例,市区平面交叉道路(除高速以外)高峰出行效率低,平均行车速度仅为 9~14 km/h,其中二环以内不到 10 km/h,全市 60% 的线路满载率在 80% 以上. 除交通拥堵外,交通事故、交通废气排放污染和交通噪声污染也十分严重. 发达国家所经历的"小汽车转换过程"正在我国悄然出现并日益加速. 今后人均出行交通量将持续增加,小汽车的增加也将是持续而明显的. 但是,由于我国特大城市的土地和空间资源已严重短缺,今后地面道路供给的能力将受到严重削弱,不可能持续大幅度扩充道路设施. 地下货物运输由于不占地面土地资源、能隔绝振动和噪声、有利于提高城市空气质量,在工程方案的环境保护指标比较中无疑会处于领先地位,应该受到关注、研究和应用[6]. 在国内的特大城市,如北京、上海、广州,发展现代地下货物运输系统应是一个具有战略眼光的研究和发展领域,是解决未来交通难题的新的发展思路.

4 结束语

1) 在一些发达国家已经引起了对地下货物运输的新的关注,发展较快的是荷兰和德国. 现代管道和隧道货物运输系统仍在发展之中,是一个长远发展项目,初期可小规模地应用该系统,并逐步形成城市地下物流网络.

2) 地下货物运输系统的作用和潜在优势有:提高运输速度和运行效率;有效减少城市环境污染;有助于解决交通拥挤问题,提高城市交通的通达性;节约城市用地,促进城市可持续发展;提高货物配送效率,降低运输和库存成本.

3) 在国内的特大城市发展现代地下货物运输系统应是一个具有战略眼光的研究和发展领域,是解决

未来交通难题的新的发展思路.

参考文献:

[1] TANIGUCHI E, THOMPSON R G. Innovation in Freight Transport[M]. Boston: Witpress Southampton, 2003.

[2] PIELAGE B J. Underground Freight Transportation[A]. A New Development for Automated Freight Transportation Systems in the Netherlands[C]. 2001 IEEE Intelligent Transportation Systems Proceedings, Oakland, 2001. 762-767.

[3] The ASCE task committee on freight pipelines of pipeline division. Freight pipelines: Current status and anticipated future use[J]. Journal of Transportation Engineering, 1998, 124(4): 300-310.

[4] GERARD ARENDS, ENNE DE BOER. Tunnelling of infrastructure: from non-considered to ill considered—lessons from the Netherlands[J]. Tunnelling and Underground Space Technology, 2001, 16(3): 225-234.

[5] 钱七虎. 建设特大城市地下高速路和地下物流系统—解决中国特大城市交通问题的新思路[J]. 科技导报, 2004(4): 3-6.
QIAN Qi-hu. To build a system of underground highwag and underground logistics in metropolises—a novel idea to solve transport problems in metropolises of China[J]. Guide to Science and Technology, 2004(4): 3-6. (in Chinese)

[6] 聂小方, 田丰新. 新兴的城市地下物流系统[J]. 综合运输, 2003(9): 52-53.
NIE Xiao-fang, TIAN Yu-xin. A newly emerging system of city under-ground logistics[J]. Composite Transportation, 2003(9): 52-53. (in Chinese)

[7] 张耀平, 王大庆. 城市地下管道物流发展前景及研究内容初探[J]. 技术经济, 2002(7): 19.
ZHANG Yao-ping, WANG Da-qin. A study on the development prospect and research content of city underground tube logistics[J]. Technical Economics, 2002(7): 19. (in Chinese)

Current Status and Anticipated Future Use of Modern Underground Freight Transport System

ZHANG Ming-ju[1], QIAN Qi-hu[2], TANG Jie[3]

(1. College of Architecture and Civil Engineering, Beijing University of Technology, Beijing 100022, China;
2. Division of Engineering Management, Chinese Academy of Engineering, Beijing 100038, China;
3. CIECC Engineering and Construction Project Management Corp., Beijing 100044, China)

Abstract: A investigation is carried out on the underground freight transport(UFT)system, in which a strong interest grows in several developed country in recently years, so as to provide helpful suggestions to the development of metropolitan transportation in China. The concept and classification of UFT systems are firstly. Secondly, the history of UFT in developed country was introduced and the current status and innovations of UFT systems were given. Lastly, areas of application, potentials and anticipated future use of UFT systems were discussed. In conclusion, UFT systems will be very efficient to the improvement of urban transportation, and the effects and potentials of building UFT systems will be as follows: promotion of travel speed and efficiency; reduction of urban environmental pollution; reduction of traffic congestion and improvement of freight accessibility; and less covering of urban areas and promotion of sustainable development; and promotion of freight distribution productivity and reduction of freight transport and storage cost. It may be a strategy researching and developing area to implement UFT systems in metropolises in China. Building UFT systems will be a new idea to solve future transport problems.

Key words: underground space; underground engineering; tunnel; tube; underground freight transport; underground logistics

大盾构工程关键技术的新进展

钱七虎

（陆军工程大学，南京，210007）

摘　要：本文介绍了大盾构掘进主轴承事故分析、国产主轴承的自主研发、高效掘进渣土适配性改良系列材料及新型盾尾密封油脂的研发与应用等国内关键技术新进展。

关键词：大盾构工程；关键技术；新进展

1 大盾构主轴承事故分析及自主研发

交通强国战略实施的积极推进，迫切需要建设超大直径盾构隧道工程，超大直径盾构机（Φ15m、Φ16m）已能生产，但其核心部件主轴承完全依靠进口，成为"卡脖子"难题。

1.1 大盾构主轴承损坏失效事故的分析

盾构主轴承的主要功能是在盾构机作业过程中承受推力和倾覆力矩将其传递给刀盘，使得盾构能够向前推进、承受刀盘驱动系统的回转力矩（扭矩），将其传递给刀盘使其进行破岩工作，联接刀盘和刀盘支撑并承受其重量。

上述荷载从而对轴承套圈和滚子产生接触应力。接触疲劳造成的轴承部件失效，即滚子和滚道表面剥落。

主轴承损坏失效主要分两类：一类是主轴承直接损坏，另一类是轴承密封失效引起的轴承损坏。

（1）主轴承直接损坏的形式

1）润滑失效工作的轴承疲劳

2）微动磨损造成的轴承套圈破坏

3）安装不当或荷载过大造成的轴承失效

4）接触疲劳造成的轴承部件失效，即滚子和滚道表面剥落

如果加工精良，润滑良好，安装正确，无轴线偏移，无尘埃、水分和腐蚀介质的侵入，且载荷适中（扭矩、推力），前三种失效形式在正确设计和施工及正确选用盾构情况下都是可以避免的。

造成轴承损坏的不可避免的原因只有一个：**接触疲劳**。轴承在运转过程中，接触表面应力的循环作用，滚动件与滚道接触应力区的材料超强度发生变化，金属、微粒从表面剥落，表面此时发生滚动接触疲劳，从而引起主轴承损坏。这种情况发生在主轴承直径偏小及推进距离过长的多个工程中！

若要避免和减缓接触疲劳主要通过降低应力水平和控制循环作用时间。前者涉及主轴承直径，后者涉及盾构推进距离（时间），而后者又与前者正相关，为此，需要增大主轴承直径，降低接触应力水平。

（2）密封失效引起的轴承损坏

主驱动密封系统的作用主要是阻止土仓渣土进入以及冲洗和润滑。

不同形式的主驱动密封系统虽有不同，但基本都由多道单唇密封、一道或多道双唇密封和迷宫密封组成，并在唇形密封间设置腔室且充满弹性材料。在密封腔中注入一定稠度的油脂，直到油脂沿密缝隙溢出，迷宫密封有一定的曲折度，加长油脂溢出通道，增大溢出阻力，密封系统的作用是将泥水和渣土阻挡在外。其条件是土仓或者泥水仓的水压力小于油脂溢出阻力。

密封失效主要原因是油脂量不足，油脂溢出压力不足；或水土压力过大，超过油脂阻力。前者的案例为广深港11.182m泥水盾构主轴承损坏事故，油脂注入不到位，在7bar水压，泥浆通过迷宫进入轴承密封；轴承滚子及滚道表面在夹杂硬质小颗粒后出现压痕；焊接刀盘时搭铁线未严格按标准放置，致使主轴承滚子和滚道面接触有大电流通过产生电熔蚀坑；两者在重载作用下逐步损坏（轴承直径4800m）。后者的案例为美国西雅图17.45m土压盾构及丹麦大贝尔特海峡东通道8.7土压盾构，原因为：结泥饼，压力不均，局部土压力增大，刀盘处出现高温，油脂阻力减小，土仓渣土颗粒进入，主轴承密封失效。

密封失效的另一个原因是产品质量问题。案例为江阴靖江长江隧道大盾构。区间左线采用斯凯孚（SKF）产品。2022年4月29日左线始发，掘进3公里多，于2023年2月8日发生主轴承疲劳严重磨损导致失效停机，停机位置处于深水航道内。判断为主轴承非正常疲劳磨损失效，判断依据为荷载适中，扭

矩、推力在设计范围内,因此排除盾构选型不当;主轴承直径8m,等于盾构直径二分之一,排除主轴承直径偏小不足,盾构距离3公里多不长,在主轴承疲劳寿命内,因此排除主轴承正常疲劳磨损;主轴承零部件破损块的出现证明了此非正常疲劳磨损。失效事故后,对斯凯孚(SKF)的其它新轴承进行的检测结果表明:主轴承材料不合格;装配公差与部件尺寸精度不达标,前者导致疲劳强度不足,后者导致微动磨损。

从以上案例中,我们可以汲取以下教训:

(1)产品制造过程一定要实行监理,并且要进一步研究如何监理。

(2)要研究产品如何验收才能避免上述情况出现。要明确产品的材料、产品参数(加工精度和装配误差)是否符合标准。

(3)失效情况下如何索赔,失效情况引起的工程损失如何包含在合同的索赔条款中。

1.2 大盾构主轴承自主研发

中科院金属研究所Φ8米级盾构主轴承自主研发成功是我国关键科技自立自强重大成果,对于超大直径盾构隧道工程建设具有里程碑意义。

(1)高纯净度、高均质性材料制备技术

1)超纯净材料制备技术【大型套圈】

我国自主研发的大盾构主轴承大型套圈控制元素含量和夹杂物面积分数显著优于进口,表明纯净度超过进口水平,见图1.1,表1.1和表1.2。

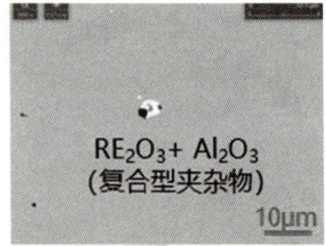

图1.1 自研套圈夹杂物分析图

表1.1 自研与进口夹杂物控制元素含量对比表(ppm)

渠道	五害元素总和	Ti	T.O	N
自研	53	9	8	39
进口	65	13	10	42

表1.2 自研与进口夹杂物尺寸分布表

渠道	扫描面积(mm²)	数量	平均直径	最大直径	数密度	面积百分比
自研	29.12	155	3.50	12.35	5.32	0.016
进口	26.88	769	2.95	28.04	28.61	0.003

2)超纯净材料制备技术【大型滚子】

我国自主研发的大盾构主轴承大型滚子夹杂物(RE_2O_2S和Al_2O_3、MnS)调控和夹杂物面积分数(自主研发0.5‰,进口件1.6‰)显著优于进口。表明纯净度超过进口水平,见图1.2。

3)高均质材料制备技术【大型套圈】

我国自主研发的大盾构主轴承大型套圈全断面解剖表明,硬度261~280HB,Δ≤±10HB,均质性高;自研件拉压高周疲劳极限为487.5MPa(R=-1、10^7),较进口件疲劳极限370MPa(42CrMo4)提升≥100MPa,见图1.3。

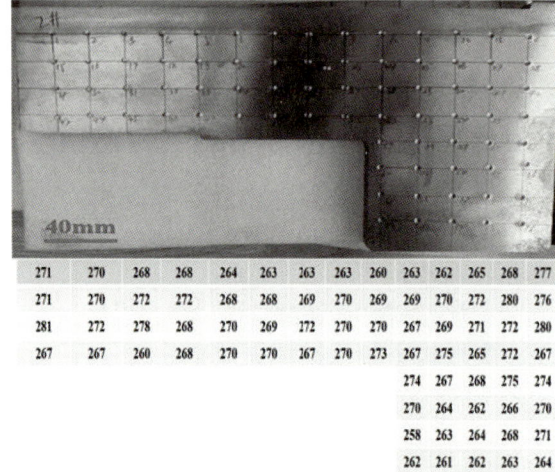

图1.2 自研Φ8m级主轴承套圈硬度及均匀性

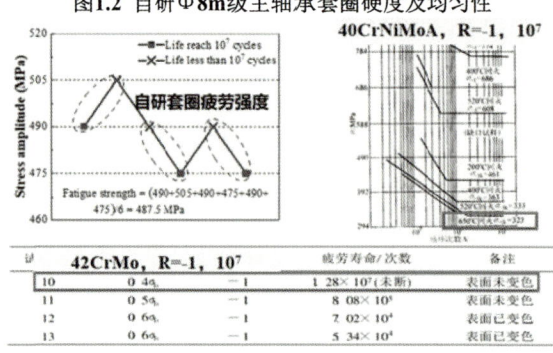

图1.3 自研套圈材料与国外选材疲劳强度对比

4)高均质材料制备技术【大型滚子】

Φ120mm滚子心部到表面硬度59~59.5HRC,硬度均匀性≤±0.5HRC≤±1HRC,超过进口水平【≥±2HRC】;表面、1/2R和心部组织均为马氏体+碳化物,完全淬透,组织均匀,见图1.4~1.8。

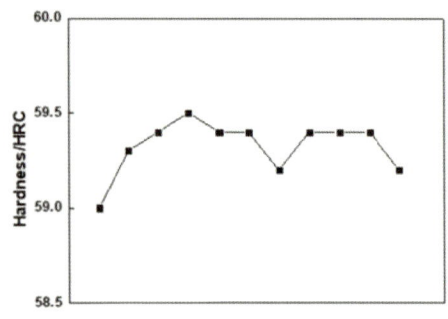

图1.4 Φ120mm滚子心部到表面硬度图

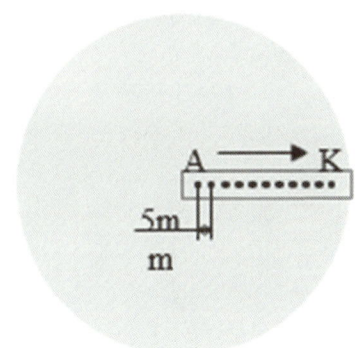

图 1.5 Φ120mm 滚子心部到表面硬度均匀性图

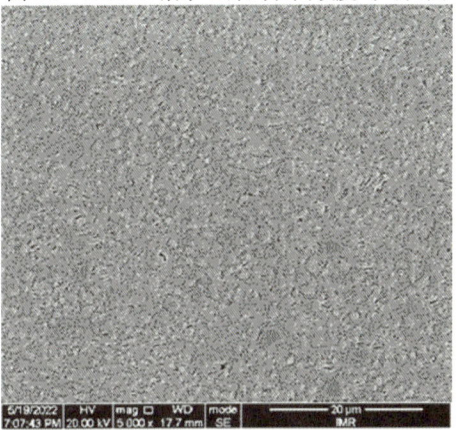

图 1.6 Φ120mm 滚子表面组织情况

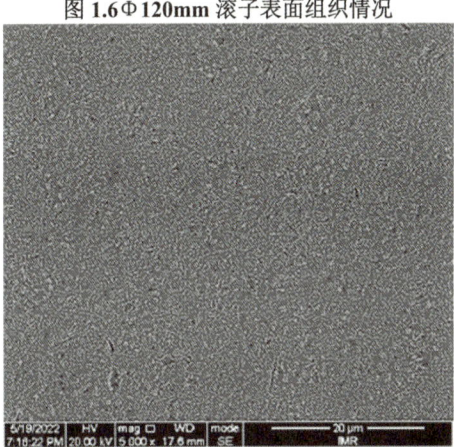

图 1.7 Φ120mm 滚子 1/2R 组织情况

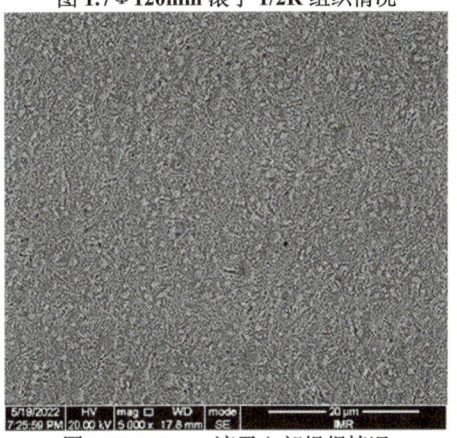

图 1.8 Φ120mm 滚子心部组织情况

（2）大淬深表面淬火技术【大型套圈】

我国自主研发的大盾构主轴承大型套圈材料淬层深度最大可达19mm，远超过进口材料12.5mm；自研套圈样件，淬层深度12~14mm，超过进口水平11mm。表明自研主轴承套圈材料淬透性和淬层深度超过进口水平，见图1.9~图1.11。

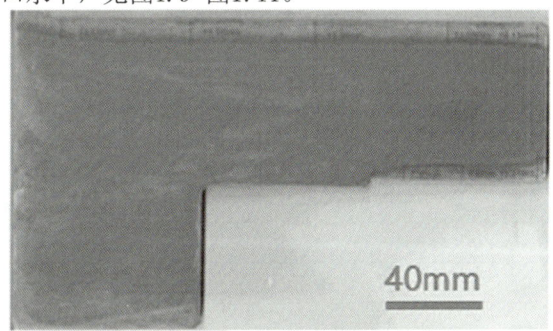

图 1.9 自研Φ8m 级主轴承套圈试验件剖面

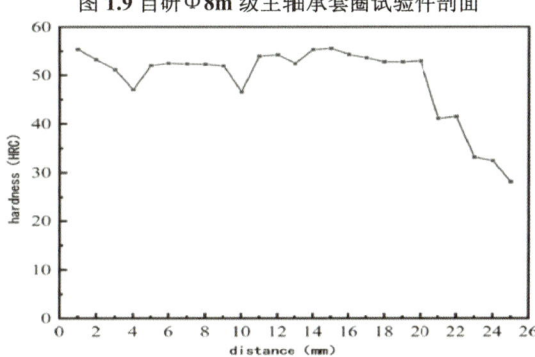

图 1.10 自研套圈滚道面表淬层剖面硬度分布

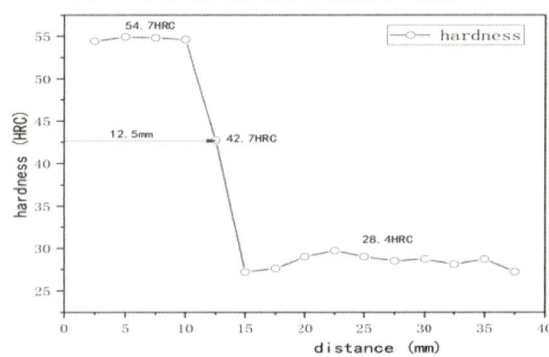

图 1.11 进口套圈滚道面表淬层剖面硬度分布

（3）高性能铜钢复合保持架制备技术

我国自主研发的大盾构主轴承突破了铜钢复合保持架设计与加工技术,开发了铜/钢焊接工艺和检测评价方法。

同时，破坏检验表明自研件拉断力（99kN）远大于进口件拉断力（63kN），自研件焊缝疲劳（拉拉，$F\max=20kN$、应力比0.1、20Hz）为110921次，较进口件23081次，提升了3.8倍，见图1.12,表1.3~表1.6。

图 1.12 自研Φ8m级主轴承铜钢复合保持架

图 1.13 自研Φ8m级主轴承G1级大型滚子(Φ120×120mm)

表 1.3 进口件疲劳寿命

规格	最大力（kN）	周次
主推	20	23081
副推	10	258118

表 1.4 自研件疲劳寿命

规格	最大力（kN）	周次
主推	20	110921
主推	15	763319
副推	20	102485
副推	15	390566

表 1.5 进口件拉断力

规格	平行段（mm）	拉力（kN）
副推	15	55.6
副推	15	56.7
副推	15	25.5
副推	15	58.9
副推	15	63.1
副推	0	28.2

表 1.6 自研件拉断力

规格	平行段（mm）	拉力（kN）
副推	15	99.0
副推	15	92.4
副推	0	64.8

（4）高精密加工和高精度装配技术

1）大型滚子高精密加工技术

我国自主研发的大盾构主轴承大型滚子采用最大许用静负荷设计真实对数曲线构建了真实对数曲线大型滚子设计方法，突破了高精度大型滚子加工技术，见图1.13。

自研Φ120×120mm滚子G1级参数：直径规值≤1μm、端跳≤6μm、表面粗糙度≤0.1μm、圆度误差≤0.5μm（国内之前处于G2A级：直径规值≤1.5μm、端跳≤6μm、表面粗糙度≤0.16μm、圆度误差≤1.3μm）。

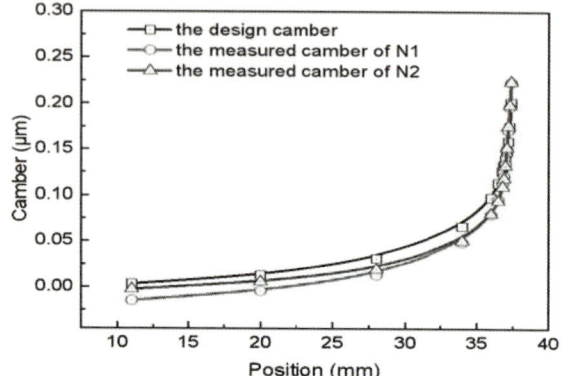

图 1.14 自研Φ8m级主轴承G1级大型滚子素线

2）高精密装配技术

我国自主研发的大盾构主轴承尺寸精度（游隙：径向0.25mm，轴向0.18mm）、形位公差（跳动：内圈轴向0.05mm，外圈轴向0.05mm；内圈径向0.18mm；外圈径向0.16mm，齿0.40mm）、淬层深度等各项指标，均满足设计要求。

该产品经国家轴承质量检验检测中心检验并出具报告，并通过了产品验收、装机应用论证和工程应用，表明应用该大直径主轴承（图1.15）不可能产生接触疲劳强度不足和微动引起的疲劳磨损，完全可采用于超大型隧道工程建设中。

图 1.15 首套自研Φ8m级主轴承成功装配

2 大盾构渣土适配性改良材料及应用

盾构是隧道工程建设的利器，但不是万能的。盾构隧道建设中遇到的工程难题，是盾构机与岩土环境相互作用相互影响的过程中产生的，难题的解决不能只依靠盾构及其高配设施（作用影响的一个方面），还应该从改善岩土环境（作用影响的另一个方面）去思考解决，岩土环境是可以通过针对性的材料注入进行改善。

中国矿大李树忱教授在这个方向上进行了长期系统的研究，取得了成功应用于工程实践中令人关注的成果。

（1）排渣不畅易堵的工程难题

泡沫剂是应用最广泛的渣土改良剂，适用于各种地层，常规地层仅用泡沫剂改良即可。但是离子型普通泡沫剂发泡倍率低，半衰期短，改良效果较差。

1）理论研究

配向密度为表面活性剂分子在泡沫液膜处的分布密度，配向密度越高，泡沫稳定性越强。

由于阴离子电荷互斥导致配向密度低，非离子型表面活性剂的加入有助于提高配向密度。

2）试验研究

开展了泡沫衰变细观测定试验，非离子表面活性剂提高了泡沫的稳定性，衰变时间延长3-5min。

3）材料研发

鉴于普通泡沫剂发泡效果差，发明了非离子聚合型泡沫剂，改善了渣土流塑性，与德国BASF比，发泡倍率提高50%，半衰期提高15%，支撑力提高1倍。显著提升了盾构渣土流动性效果，实现了对国际知名品牌产品的超越。

4）工程应用

新疆EH盾构工程穿越膨胀性泥岩地层，局部夹泥质砂岩，导致排渣堵塞。工程应用了聚合型泡沫剂，3%浓度，30%~50%掺入比，150-170kg/环，解决了现场盾构排渣堵塞难题，掘进效率提高2倍以上。

（2）高黏地层的工程难题

高黏地层常常出现刀盘刀具结泥饼、堵塞、排渣不畅的问题，见图2.1。

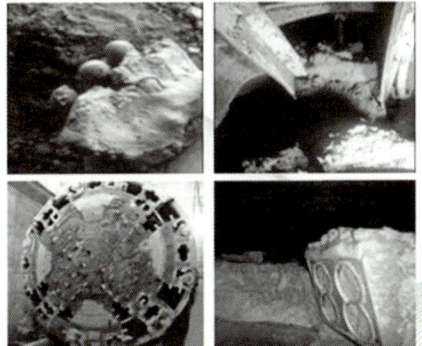

图 2.1 粘土地层盾构结泥饼、堵塞

为解决以上问题，现有解决措施多为增大刀盘开口率（图2.2）、增设中心冲刷装置（图2.3）等机械手段，处理低效、滞后。

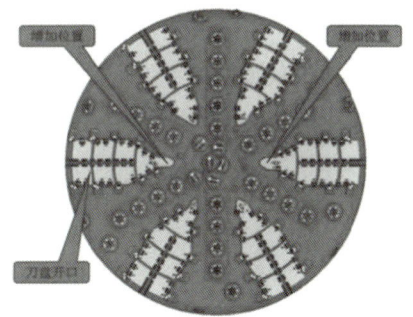

图 2.2 增大刀盘开口率

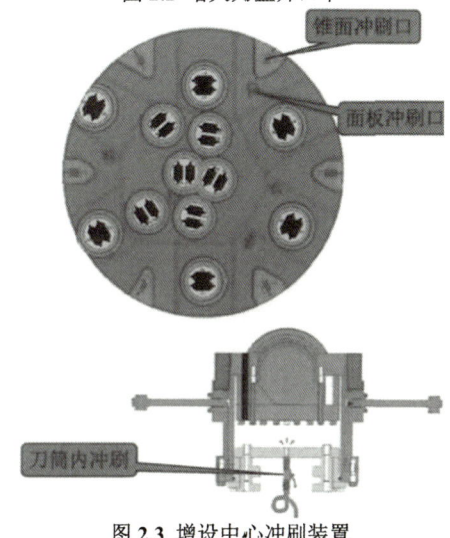

图 2.3 增设中心冲刷装置

1）理论研究

黏性土黏附评价系数α为旋转过程黏土在金属叶片上的黏附比率，用于评价土体黏性大小。

$$\alpha = \frac{G_{MT}}{G_{TOT}}$$

式中：G_{MT} ——叶片黏土黏附质量；

G_{TOT} ——黏土总质量。

研究及工程实践表明，仅用泡沫剂对泥饼的防治作用有限，黏土多带负电荷，可根据静电互斥原理（图2.4）研发高分子抗黏剂。

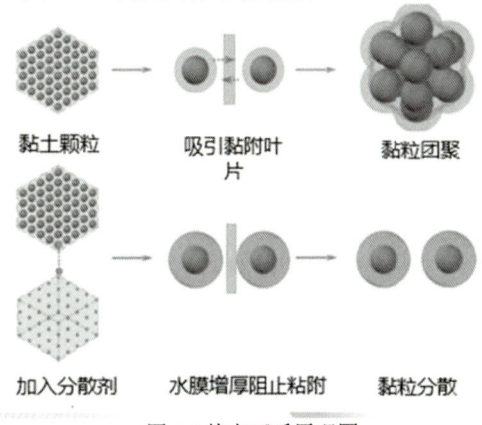

图 2.4 静电互斥原理图

2）材料研发

鉴于传统双氧水浸泡无法根治泥饼，发明了高分子抗粘剂，粘附评价系数可降低60%以上。

3）注入工艺

提出了高黏地层渣土改良定量化注入工艺，根据黏附评价系数 α 动态调控改良剂注入比。

在盾构发泡系统基础上增设抗黏剂注入管路，实现多种改良剂的精确定量化注入，形成了泥饼主动防控技术，从源头上避免了泥饼形成。

4）工程应用

长春地铁2号线西延线盾构隧道穿越强风化泥岩及粉质黏土地层（图2.5），导致结泥饼严重。

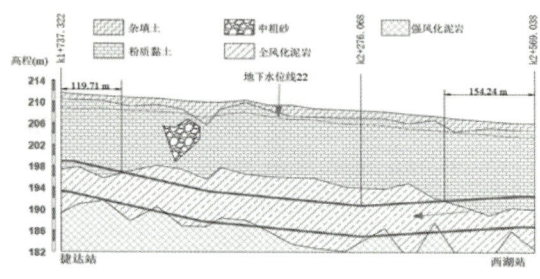

图2.5 盾构隧道地质剖面图

在施工中，采用聚合型泡沫剂、高分子抗黏剂按照体积3:1比例注入，彻底解决了结泥饼难题，掘进效率提高3倍，直至隧道贯通也未出现结泥饼。

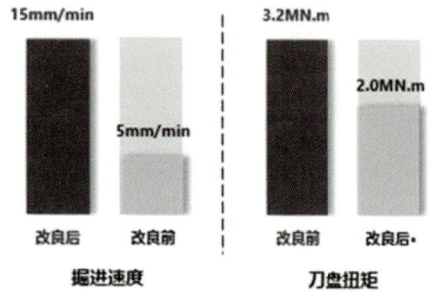

图2.6 改良前后盾构掘进速度和刀盘扭矩变化图

安徽芜湖城南过江隧道采用15.07m超大直径泥水平衡盾构，盾构隧道穿越粉砂、粉细砂、石英砂岩、粉砂岩、泥质粉砂岩、凝灰角砾岩和粉质黏土等地层（图2.7），盾构掘进期间结泥饼严重（图2.8）。现场原状土改良黏附性评价试验表明原状泥浆中黏土黏附评价系数 α 达到95.6%。

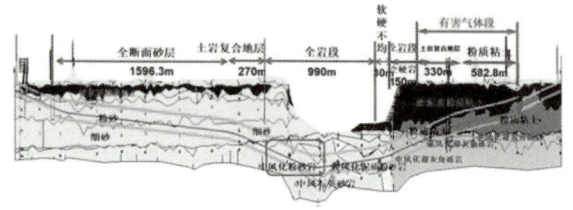

图2.7 盾构隧道地质剖面图

图2.8 刀筒结泥饼图

为了泥饼粘结问题，经试验在泥浆中抗黏剂掺入比>2%时，黏附评价系数 α 降低至2.0%以下，降低了土体对刀具的黏附。

表2.1 不同泥浆的黏附评价系数

泥浆类型	抗黏剂掺入比	黏附评价系数
原状泥浆	/	95.6%
新型抗黏剂泥浆	1%	35.8%
	2%	2.0%
	3%	1.0%
	5%	0.7%

现场根据泥浆比重和掘进参数等反馈信息，按照2%~3%浓度在泥浆调整槽注入抗黏剂，随泥浆循环充分浸润刀具，防止黏土对刀具的黏附，从源头上根治结泥饼问题，已在工程中应用。

（4）富水无黏地层的工程难题

富水无黏地层常常出现螺旋输送机喷涌、排渣困难的问题。

为解决以上问题，现有解决措施常常是采用添加粉状聚合物、膨润土改良，但效果差。

1）理论研究

通过试验提出降低土体渗透性（<10^{-5} cm/s）的喷涌治理定量化指标和通过联合使用多种改良剂（泡沫剂+聚合物+膨润土），减小渗流通道实现盾构喷涌防治的途径。

2）材料研发

鉴于现有粉状防止剂和BASF聚合物改良效果不佳的问题，发明了液态喷涌防止剂，抗渗指标提高24倍，比德国BASF液态防止剂提高3倍（表2.2、表2.3），克服了盾构喷涌-清渣-喷涌恶性循环，解决了高富水地层喷涌难题。

表2.2 防止剂渗透系数对比

名称	渗透系数
喷涌防止剂	6.3×10-6m/s
BASF聚合物	2.4×10-5m/s
粉状防止剂	1.6×10-4m/s

表 2.3 喷涌防止剂指标

形态	液态
颜色	透明
密度（kg/m3）	1035-1045
PH(20℃，5%)	6.5-7.5
水溶性	完全溶解
环保指标	符合OECD指标

3）注入工艺

提出了喷涌定量化防治工艺，在原有管路基础上，增设喷涌防止剂注入管路。待喷涌防止剂（0.2%~1.0%）与泡沫剂（3.0%）混合后，经发泡系统发泡，注入刀盘、土仓、螺旋输送机等部位进行渣土改良。

对于卵砾石等细粒缺失地层，配置15-25%浓度膨润土注入，联合喷涌防止剂共同改良。

4）工程应用

济南地铁2号线盾构隧道穿越富水岩溶地层，地下水丰富，盾构喷涌严重。

在施工中，采用聚合型泡沫剂(3.0%)、喷涌防止剂(0.5%)，通过发泡系统注入，解决了现场喷涌难题，掘进效率提高 2 倍，直至隧道贯通未出现喷涌问题。

深圳地铁12号线盾构隧道穿越富水砂砾地层，渗透性高，盾构喷涌严重。

在施工中，采用聚合型泡沫剂(3.0%)、喷涌防止剂(1.0%)，通过发泡系统注入，解决了喷涌难题，掘进效率提高 3 倍，直至隧道贯通未出现喷涌问题。

（4）卵砾石地层的工程难题

卵砾石地层常常出现地层坚硬，刀盘刀具磨损严重的问题。

为了解决以上问题，现有解决措施常常采用添加泡沫剂、膨润土改良，但效果差。

1）材料研发

鉴于泡沫等改良效果有限的问题，研发了高效耐磨剂，2%浓度改良土体，刀具磨损量降低40%（图2.9、图2.10）。

通过发泡系统注入，起到润滑、冷却作用，有效减少刀仓堵塞，从而减少刀具磨损与偏磨。

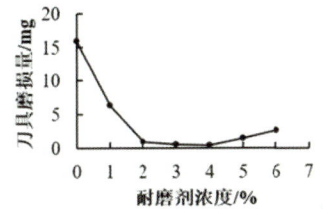

图 2.9 刀具磨损量与耐磨剂浓度变化曲线

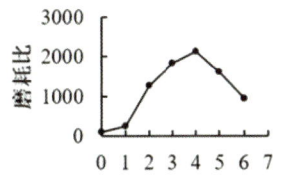

图 2.10 刀具磨耗比与耐磨剂浓度变化曲线

2）工程应用

青岛地铁8号线盾构隧道全断面穿越微风化熔结凝灰岩，刀具磨损十分严重，且出渣困难。使用3%浓度耐磨剂后，刀具使用寿命延长35%，盾构渣土中无自由水，出渣顺畅，日掘进速度提高70%。

（5）高磨蚀地层的TBM工程难题

高磨蚀地层的TBM工程常常出现地层坚硬，刀温过高、刀具偏磨、粉尘过大的问题。

为了解决以上问题，现有解决措施常常采用喷水或泡沫改良，降温、耐磨、抑尘效果差。

1）理论研究

目前研究手段主要有增强盾构刀具材料韧性、强度等性质，金属热处理、刀具表面涂层等。

提出增大刀具与岩土体接触面润滑，降低黏着点区域面积，实现刀具减磨的岩土体改良技术。

针对TBM掘进过程中存在刀温过高、岩粉板结导致受力不均，滚刀偏磨；粉尘过大，损坏设备，影响作业的问题。通过研发高磨蚀地层耐磨抑尘机剂，进一步改良岩土体力学参数。以达到降温、减磨、抑尘防止岩粉板结，提高刀具寿命，改善作业环境的目的。

2）试验研究

开展缩尺滚刀磨损试验，获得滚刀在不同掘进参数下温度演变规律（图2.11~图2.14）。

低温时滚刀磨损主要以微切削为主，温度增高后降低了金属硬度，点蚀和梨沟等破坏模式增加。

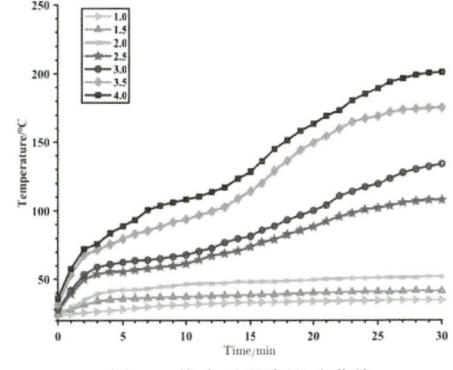

图 2.11 推力对温度影响曲线

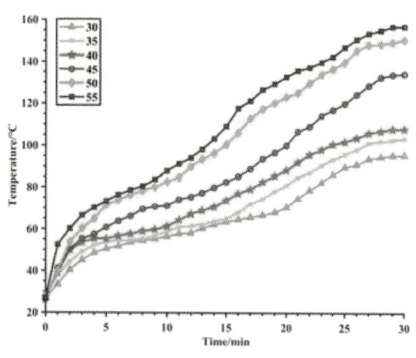

图 2.12 转速对温度影响曲线

图 2.13 刀具表面温度 50℃时磨损模式

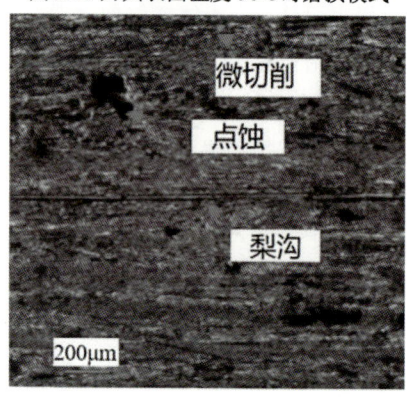

图 2.14 刀具表面温度 210℃时磨损模式

3）材料研发

鉴于喷水和BASF聚合物改良效果差的问题，发明了耐磨抑尘剂，防止岩粉板结，刀具磨损量降低40%，温度降低45%（表2.4）。

表 2.4 刀具减摩材料效果对比表

名称	磨损量	温度
耐磨抑尘剂	31%	47℃
BASF聚合物	42%	62℃
水	51%	86℃

4）注入工艺

搭建TBM耐磨抑尘剂注入系统，包括电控系统、耐磨抑尘剂混合箱体、泵站、空气管路以及泡沫发生管路，根据刀盘直径选择注射口，实现耐磨抑尘剂多路注入功能（表2.5）。

表 2.5 耐磨抑尘剂注射口数量设置原则

TBM直径/m	≤3	3～5	5～7	7～9	≥9
耐磨抑尘剂注射口数量	1	1～2	2～3	3～4	4～5

5）工程应用

青岛地铁1号线TBM隧道主要穿越花岗岩地层，破岩过程粉尘浓度过大，刀具磨损十分严重。使用2.5%浓度耐磨抑尘剂后，TBM主控室粉尘浓度降至<2mg/m³，单把刀具服役寿命提高30%。

3 盾尾密封安全保障技术与材料

（1）工程难题及现有解决措施

在盾构施工过程中，常常出现盾尾密封失效、地层溃塌、地表沉降的问题，见图2.1~图2.3。

图 3.1 盾尾密封失效图

图 3.2 施工现场坍塌图

图 3.3 地表沉降图

为了解决以上问题，现有解决措施常常采用液氮冷冻更换密封系统（图3.4）、改进尾刷形式（图3.5）等手段。

图 3.4 液氮冷冻更换密封系统图

图 3.5 改进尾刷形式

（2）理论研究

基于盾尾密封力学平衡关系（图3.6），开展盾尾击穿数值模拟，通过数值模拟揭示盾尾密封击穿过程浆液渗流规律，模拟多种工况，提出盾尾密封失效临界阈值参数，根据数值计算结果提出盾尾密封油脂密封强度要求。

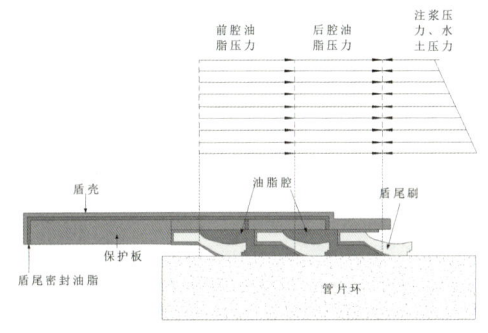

图 3.6 盾尾密封力学平衡关系图

（3）试验研究

基于盾尾密封临界阈值，开展盾尾密封油脂黏温试验，提出温度适应性强的油脂性能指标（图3.7）。

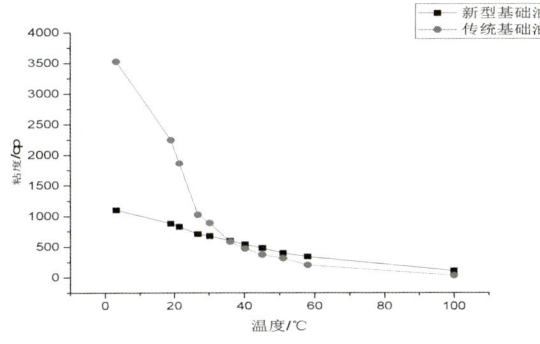

图 3.7 基础油黏温曲线

曲线显示，寒区温度下基础油黏度的温度敏感性显著降低。

（4）材料研发

鉴于普通国产油脂盾尾密封效果不佳的问题，研发了新型盾尾密封油脂，最大抗水压达3.5MPa，整体性能提高1.3倍，优于国外油脂（提高16%），见图3.8、图3.9，打破了国外对高性能盾尾密封油脂合成技术的封锁。

（5）工程应用

广西南崇盾构隧道下穿河流，水压大、盾尾密封保障难。

为了解决以上问题，采用盾尾密封保障技术及新型盾尾密封油脂实现了高水压水下盾构隧道的安全穿越。

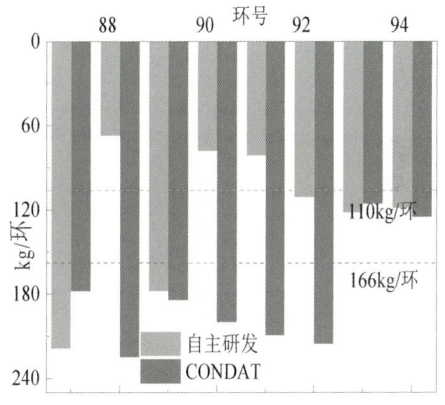

图 3.8 CONDAT 与自主研发盾尾密封油脂环用量对比

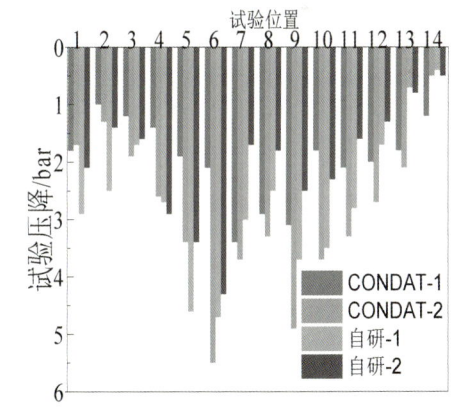

图 3.9 CONDAT 与自主研发盾尾密封油脂试验压降对比

Mechanism and classification and quantitative prediction model of strain rockburst in the surrounding rock masses around a deep circular tunnel

Qihu Qian[a,b,e]   Xiaoping Zhou[a,c,d,*]

(a. *School of Civil Engineering, Chongqing University, Chongqing 400045, China*; b. *Engineering Institute of National Defense Engineering, PLA University of Science and Technology, Nanjing, Jiangsu 210007, China*; c. *Key Laboratory of New Technology for Construction of Cities in Mountain Area, Ministry of Education of Chongqing University, Chongqing 400045, PR China*; d. *State Key Laboratory of Coal Mine Disaster Dynamics and Control, Chongqing University, 400044, China*; e. *State Key Laboratory of Disaster Prevention & Mitigation of Explosion & Impact PLA University of Science and Technology, Nanjing, Jiangsu 210007, China*)

Abstract: Rock masses are considered as geomaterials which contains uniform distribution of pre-existing microcracks. The stress fields in the surrounding rock masses containing uniform distribution of pre-existing microcracks around a deep circular tunnel are obtained based on non-Euclidean model. The critical conditions of initiation of pre-existing microcracks, stable and unstable growth of microcracks are established based on the strain energy density factor approach. The dissipative energy density induced by growth of pre-existing microcracks and secondary microcracks is defined. The mechanism of potential rockburst, immediate rockburst and time delayed rockburst is revealed. Criteria of potential rockburst, immediate rockburst and time delayed rockburst are proposed. Quantitative prediction models of potential rockburst, immediate rockburst and time delayed rockburst are established. Area and location of potential rockburst zones, immediate rockburst zones and time delayed rockburst zones are given out as well as the ejection velocity of rock fragments for immediate rockburst and time delayed rockburst.

Keywords: mechanism of strain rockburst; criteria of potential rockburst; criteria of immediate rockburst; criteria of time delayed rockburst; quantitative prediction models; the strain energy density factor approach

1. Introduction

Rockburst is a phenomenon of rock failure related to ejection of rock fragments, acoustic emission or microseismic events induced by a sudden release of elastic strain energy storied in the rock masses. From the 1st International Symposium on Rockburst and Seismicity in Mines held in Johannesburg in 1982 to the 7th International Symposium on Rockburst held in Dalian in 2009, International Symposium on Rockburst was held 7 times. Great efforts have been made to understand rockburst as a type of the failure of deep rock masses in the past several years. Theories of rockburst, such as the strength theory, the stiffness theory, the energy theory, the theory of impact tendency, the instability theory and the catastrophe theory, were established. In terms of the strength theory, rockburst occurs when stresses in the surrounding rock masses is equal to the ultimate strength of rock masses.However, mechanism of rockburst is simply analyzed by the strength theory, which is lack of researches on the mechanism of rockburst preparation and development. The stiffness theory was proposed by Cook in 1960s which was derived from the theory of rigid press. According to the theory of rigid press, the failure of rock masses is brittle when the stiffness of rock is larger than that of the press machine. Based on the further investigation on mechanism of the failure of rock masses, the stiffness theory treats the surrounding rock masses as the press machine, and treats the pillars as the specimen. Therefore, the essential condition of rockburst is that the stiffness of pillars is larger than that of surrounding rock masses. The stiffness theory can be applied to analyze the mechanism of rockburst in pillars, while it is difficult to give a reasonable explanation of rockburst in surrounding rock masses. The energy theory was proposed by Cook based on the occurrences of rockburst on gold mine in South African in the past several decades. The energy theory deemed that rockburst occurs when the release energy is more than the dissipative energy in the failure process of rock masses. However, the property of equilibrium state of rock masses and condition of the failure of rock masses is not clarified in detail, especially the condition which the energy is released is not clearly clarified. The impact tendency theory of rockburst [1-3] deemed that rockburst occurs when the impact tendency of rock masses is more than the critical value. The indexes of the impact tendency of rockburst mainly include elastic energy index, impacting energy index, brittleness index and so on. The catastrophe theory of rockburst describes a system which saltatorily transforms from a state to another state, or a system evolution during which gradual changes in certain variables lead to a sudden changes in the state of the system. Fracture and damage mechanics hold views that rockburst is induced by microcracks nucleation, propagation through rock matrix during excavation of tunnels. On the basis of the theory of fracture and damage mechanics, Huang [4] investigated effects of structure of rock masses on rockburst intensity from three aspects, such as rock particle arrangement, connection between particles and the distribution and evolution of microcracks. Pan et al. [5] investigated the rockburst mechanism in the surrounding rock masses around a circular tunnel using the creep damage model. In addition, the bifurcation theory, the theory of dissipative structure and the chaos theory are also widely applied to investigate the mechanism of rockburst [6].

According to works by some researchers, the mechanism of rockburst is that the stress concentration of the surrounding rock masses is induced by excavation of tunnels, then the tensile or shear brittle failure with the sound and shake occurs as stress exceeds the strength of rock masses. In this process, parts of the elastic strain energy is dissipated, and the residual elastic strain energy is turned into kinetic energy of rock fragments, which leads to instability of surrounding rock masses, splitting into pieces and ejection of rock fragments. According to works by the other researchers, rockburst is related to the microseismic phenomenon. For example, the mechanism of rockburst from Kaiser [7-8] is that the failure, the fracture and the volume dilatation of rock masses occur when stresses in surrounding rock masses exceed the strength of rock masses. The source of failure of rock masses is the elastic strain energy storied in the surrounding rock masses. The type of the failure of rock mass is called strain rockburst, which is the most common type of rockburst in rock engineering.

Instability theory of rockburst deems that the stress concentration appears during excavation of tunnels. When the strain-softening medium in high stress zone and strain-hardening medium in low stress zone are in quasi-stable equilibrium state, dynamic instability induced by the external disturbances leads to occurrence of rockburst. Tjongkie Tan [10] deemed that *"influence of the explosion wave on rockburst cannot be neglected, ..., 89 out of 114 rockbursts in the coal mine such as Mentougou mine were triggered by the explosion wave, ...,it can be concluded that the coal rock must be in quasi-stable equilibrium state at this depth"*. *"The roof breaking takes place in some mines, ..., resulting in instability and rockburst* of the surrounding rock masses". Linkov deemed that "sudden instability of rock masses leads to rockburst and earthquake".

Although some theories mentioned above can partly explain the mechanism of rockburst, such as the strength theory, the stiffness theory and the impact tendency theory, they can't still clearly reveal the mechanism of rockburst.

In order to determine time of occurrence of rockburst and rockburst intensity, criteria of rockburst, rockburst classification standards and prediction methods have been developed. For instance, Hoek [12] deemed that rockburst intensity can be determined by the ratio of the maximum tangential stress and the uniaxial compressive strength of rock masses. Turchaninov deemed that rockburst intensity can be obtained by the ratio of the sum of both the maximum tangential stress and axial stress of surrounding rock masses and the uniaxial compressive strength of rock masses. Kidybinski deemed that rockburst intensity can be defined by the ratio of the elastic strain energy and the dissipative energy. Russense deemed that rockburst intensity is determined by the ratio of the maximum tangential stress and point load strength of rock masses. According to more than 200 rockburst datum in tunnel of Erlang mountain road, Xu [13] developed the criteria of rockburst in which rockburst intensity is denoted by the ratio of the maximum tangential stress and the uniaxial compressive strength of rock masses. Hou [14] assumed that the maximum principal stress of in-situ stress field is the gravity stress one, then derived the minimum critical depth of rockburst. With the better understanding of rockburst, criteria of rockburst has a tendency from the simple criteria to multi-criteria. For example, Gu [15] proposed the multi-criteria of Qinling Mountains tunnel by synthetically considering the ratio of the elastic strain energy and the dissipative strain energy, the ratio of the maximum tangential stress and the uniaxial compressive strength of rock mass and the integrity index of rock masses. Zhang [16] proposed multi-criteria of rockburst including five factors. With the development of system engineering, some new methods of system engineering have been applied to forecast rockburst, such as neural network algorithm [17], the dynamically weighted grey optimization model [18], the method of fuzzy comprehensive evaluations [19], the probability model [20], the extension judgment method [21], the method of distance discriminant [22] and so on.

Although the criteria of rockburst mentioned above can provide qualitative or limited quantitative analysis of rockburst in the surrounding rock masses around tunnels, the location, dimensions and intensity of rockburst are not quantitatively defined.

2. The mechanism and classification of rockburst

Generalized rockburst is a kind of geological hazard which is characterized by the dynamic failure. When the rock underground engineering is excavated under high stress condition, redistribution of stresses of the surrounding rock masses caused by excavation leads to sudden release of the elastic strain energy which storied in the surrounding rock masses, and to failure, loosening, decohesion and ejection of the surrounding rock masses. The mechanism of generalized rockburst preparation and development process is that: (1)in the first stage, the surrounding rock masses are split into pieces due to excavation of tunnels; (2)in the second stage, ejection of rock fragments is induced by stress gradient in the surrounding rock masses, the elastic strain energy is transformed into kinetic energy of rock fragments, the first type of the failure of rock masses is called strain rockburst. Sudden instability of rock fragments is caused by dynamic disturbances, which may be blasting on a working or roof breaking of tunnels, the second type of the failure of rock masses is called pressure bump.

Therefore, the generalized rockburst can be divided into the strain rockburst and pressure bump. For pressure bump, fragments of the surrounding rock masses, which is in quasi-stable equilibrium state, are accumulatively induced by excavation of tunnels. For the strain rockburst, it can be divided into immediate and time delayed rockburst in terms of the time of occurrence of rockburst after excavation of a working.

In this paper, the strain rockburst is only studied, while the pressure bump will be researched in the other work. For the strain rockburst, the surrounding rock masses, which is assumed to be elastic-brittle medium, is regarded as geomaterial containing uniform distribution of microcracks. Based on non-Euclidean model of the discontinuous and incompatible deformation , the redistribution of stresses of surrounding rock masses around a tunnel is obtained, pre-existing microcracks initiate, propagate along interfaces, and pass through the rock matrix which lead to the appearance of secondary microcracks. On one hand, when the growth length of secondary microcracks is larger than the critical value, secondary microcracks unstably propagate, coalesce and intersect with the tunnel walls, the fractured zones intersecting with the tunnel walls occurs. For brittle rock, if the total dissipative energy induced by propagation of pre-existing and secondary microcracks is less than the elastic strain energy stored in rock masses, the sudden release of the residual elastic strain energy stored in rock pieces may lead to immediate rockburst.On the other hand, the unstable growth of secondary microcracks does not lead to coalescence of microcracks and to the fractured zones intersecting with the tunnel walls. If the total dissipative energy induced by growth of pre-existing and secondary microcracks is less than the elastic strain energy stored in rock masses, the residual elastic strain energy is transformed into kinetic energy of rock fragments. The movement of rock fragments leads to the fracture of the surrounding rock masses along the shortest path to the tunnel walls. As a result, ejection of rock fragments occurs, time delayed rockburst takes place.

3. Quantitative prediction model of strain rockburst

In order to quantitatively predict the location, magnitude and ejection velocity of rock fragments, the following three problems should be solved: ① the stress fields of the surrounding rock masses containing uniform distribution of microcracks around deep circular tunnels after excavation should be determined; ② criterion of initiation and growth of microcracks should be established ; ③ Criteria of potential rockburst, immediate rockburst and time delayed rockburst should be proposed.

For the first problem, the stress fields of the surrounding rock masses containing uniform distribution of pre-existing microcracks cannot be obtained using the theory of elastoplastic mechanics, since rock masses contain the pre-existing microcracks and the prerequisite of rockburst is occurrence of the fracture of rock masses. Therefore, rock masses are discontinuous, and the compatibility condition is not satisfied. The radial stress and the tangential stress have extrema on the working contour and tend monotonically to in-situ stress at infinity within the framework of the classical elastic mechanics, which is different from fluctuation characteristics of stresses in surrounding rock masses after the excavation of tunnels,and is not consistent with the micro-fracture and micro-seismic phenomenon in the surrounding rock masses around deep tunnels[23-24]. In order to solve the first problem, non-Euclidean model in the framework of the discontinuous and incompatible deformation is applied to determine the distribution of stress fields of the surrounding rock masses around tunnels. For the second problem, the fracture mechanics can be applied to solve it . For the third problem, criteria of rockburst can be established based on the theory of energy, that is the kinetic energy is equal to the difference of the elastic strain energy and total dissipative energy induced by propagation of pre-existing microcracks and secondary microcracks.

3.1. The stress field of surrounding rock mass containing uniform distribution of microcracks around the deep circular tunnels

When the surrounding rock masses contain pre-existing microcracks , compatibility conditions of deformation of rock masses is not satisfied. The internal geometry of rock masses does not coincide with the geometry of the observer's Euclidean space. The classical elastic mechanics can not be applied to analyze the stress fields of the surrounding rock masses around a deep circular tunnel. Therefore, non-Euclidean model is used to study the stress fields of the surrounding rock masses around a deep circular tunnel. According to Ref. [25] ,when the density of secondary microcracks $\eta=0$, the stress fields of the surrounding rock masses containing the uniform distribution of pre-existing microcracks around a deep circular tunnel can be expressed as:

$$\sigma_r^i = \frac{\sigma_v + \sigma_h}{2}\left(1 - \frac{r_0^2}{r^2}\right) + \frac{\sigma_h - \sigma_v}{2}\left(1 - 4\frac{r_0^2}{r^2} + 3\frac{r_0^4}{r^4}\right)\cos(2\theta) - \frac{E}{2\gamma^{3/2}(1-v)r} \cdot$$

$$[AJ_1(\gamma^{1/2}r) + BN_1(\gamma^{1/2}r) + CK_1(\gamma^{1/2}r)] + \frac{E}{2(1-v)r\gamma^{3/2}}[A_1J_1(\gamma^{1/2}r) + B_1N_1(\gamma^{1/2}r) + \quad (1)$$

$$C_1K_1(\gamma^{1/2}r)]\cos(2\theta) - \frac{3E}{(1-v)r^2\gamma^2}[A_1J_2(\gamma^{1/2}r) + B_1N_2(\gamma^{1/2}r) + C_1K_2(\gamma^{1/2}r)]\cos(2\theta)$$

$$\sigma_\theta^i = \frac{\sigma_v + \sigma_h}{2}\left(1 + \frac{r_0^2}{r^2}\right) - \frac{\sigma_h - \sigma_v}{2}\left(1 + 3\frac{r_0^4}{r^4}\right)\cos(2\theta) - \frac{E}{2\gamma(1-v)} \cdot$$

$$\left[AJ_0(\gamma^{1/2}r) + BN_0(\gamma^{1/2}r) - CK_0(\gamma^{1/2}r)\right]$$

$$+ \frac{E}{2\gamma^{3/2}(1-v)r} \cdot [AJ_1(\gamma^{1/2}r) + BN_1(\gamma^{1/2}r) + CK_1(\gamma^{1/2}r)] + \frac{E}{2(1-v)\gamma}\left[A_1J_0\left(r\sqrt{\gamma}\right) + B_1N_0\left(r\sqrt{\gamma}\right) \quad (2)\right.$$

$$+ C_1K_0\left(r\sqrt{\gamma}\right)\Big]\cos(2\theta) - \frac{3E}{2r(1-v)\gamma^{3/2}}\left[A_1J_1\left(r\sqrt{\gamma}\right) + B_1N_1\left(r\sqrt{\gamma}\right) + C_1K_1\left(r\sqrt{\gamma}\right)\right]\cos(2\theta)$$

$$+ \frac{3E}{r^2(1-v)\gamma^2}\left[A_1J_2\left(r\sqrt{\gamma}\right) + B_1N_2\left(r\sqrt{\gamma}\right) + C_1K_2\left(r\sqrt{\gamma}\right)\right]\cos(2\theta)$$

$$\tau_{r\theta}^i = \frac{\sigma_v - \sigma_h}{2}\left(1 + 2\frac{r_0^2}{r^2} - 3\frac{r_0^4}{r^4}\right)\sin(2\theta) + \frac{E}{r(1-v)\gamma^{3/2}}\left[A_1J_1\left(r\sqrt{\gamma}\right) + B_1N_1\left(r\sqrt{\gamma}\right) + \right.$$

$$C_1K_1\left(r\sqrt{\gamma}\right)\Big]\sin(2\theta) - \frac{3E}{r^2(1-v)\gamma^2}\left[A_1J_2\left(r\sqrt{\gamma}\right) + B_1N_2\left(r\sqrt{\gamma}\right) + C_1K_2\left(r\sqrt{\gamma}\right)\right]\sin(2\theta) \quad (3)$$

where $\gamma^2 = \dfrac{3\beta\left[3\lambda_2(1-D_0)^2 + 2\mu_2(1-D_1)^2\right](1-\alpha_1)}{2\left[3\lambda_2(1-D_0)^2(1-\alpha_1) + 2\mu_2(1-D_1)^2(3-\alpha_1)\right]}$, $D_0 = \pi\omega_0/(1+\pi\omega_0)$,

$\beta = \dfrac{\mu_2}{\mu_3}$, $D_1 = \pi\omega_0/(1+v_0+\pi\omega_0)$, $\omega_0 = \eta_0 c_0^2$, $v_0 = \dfrac{\lambda_1}{2(\lambda_1+\mu_1)}$, $v_1 = \dfrac{\lambda_2}{2(\lambda_2+\mu_2)}$,

$E = \dfrac{\mu_2\left[3\lambda_2(1-D_0)^2 + 2\mu_2(1-D_1)^2\right](1-D_1)^2}{\lambda_2(1-D_0)^2 + \mu_2(1-D_1)^2}$, $v = \dfrac{\lambda_2(1-D_0)^2}{2\left[\lambda_2(1-D_0)^2 + \mu_2(1-D_1)^2\right]}$,

$\alpha_1 = \dfrac{3g^2(1-D_0)^4}{\left[\lambda_1(1-D_0)^2 + 2\mu_1(1-D_1)^2\right]\left[3\lambda_2(1-D_0)^2 + 2\mu_2(1-D_1)^2\right]}$,

$A = (C/2)\pi\sqrt{\gamma}r_0\left[K_0\left(\sqrt{\gamma}r_0\right)N_1\left(\sqrt{\gamma}r_0\right) - K_1\left(\sqrt{\gamma}r_0\right)N_0\left(\sqrt{\gamma}r_0\right)\right]$,

$B = -(C/2)\pi\sqrt{\gamma}r_0\left[K_0\left(\sqrt{\gamma}r_0\right)J_1\left(\sqrt{\gamma}r_0\right) - K_1\left(\sqrt{\gamma}r_0\right)J_0\left(\sqrt{\gamma}r_0\right)\right]$,

$A_1 = (C_1/2)\pi\sqrt{\gamma}r_0\left[K_2\left(\sqrt{\gamma}r_0\right)N_3\left(\sqrt{\gamma}r_0\right) - K_3\left(\sqrt{\gamma}r_0\right)N_2\left(\sqrt{\gamma}r_0\right)\right]$,

$$B_1 = -(C_1/2)\pi\sqrt{\gamma}r_0\left[K_2\left(\sqrt{\gamma}r_0\right)J_3\left(\sqrt{\gamma}r_0\right) - K_3\left(\sqrt{\gamma}r_0\right)J_2\left(\sqrt{\gamma}r_0\right)\right],$$

where λ_1, μ_1 are Lame parameter corresponding to the geometry of the Euclidean space of the observer, λ_2, μ_2 are Lame parameters related to the internal geometry of the rock material, g and μ_3 are phenomenological parameters, r_0 is the radii of the tunnel, η_0 is the density of pre-existing microcracks, c_0 is statistically-averaged value of the half length of pre-existing microcracks, D_0 and D_1 are the damage variables, determinant of C coincides with the Wronskian of the linearly independent solutions J_0 and N_0, which guarantees its unique solvability, determinant of C_1 coincides with the Wronskian of the linearly independent solutions J_2 and N_2, which guarantees its unique solvability, J_0, N_0, and K_0 are respectively the zero order Bessel, Neumann, and Macdonald cylindrical functions, J_1, N_1, and K_1 are respectively the first-order Bessel, Neumann, and Macdonald cylindrical functions. J_2, N_2, and K_2 are respectively the second-order Bessel, Neumann, and Macdonald cylindrical functions. J_3, N_3, and K_3 are the third-order Bessel, Neumann, and Macdonald cylindrical functions, respectively.

3.2. The stable propagation of pre-existing microcracks and secondary cracks

According to the stress components (1)-(3) of the surrounding rock masses around deep circular tunnels, the principal stresses can be expressed as

$$\left. \begin{array}{l} \sigma'_1 = \dfrac{\sigma_r + \sigma_\theta}{2} + \dfrac{1}{2}\sqrt{(\sigma_r - \sigma_\theta)^2 + 4\tau_{r\theta}^2} \\ \sigma'_2 = \dfrac{\sigma_r + \sigma_\theta}{2} - \dfrac{1}{2}\sqrt{(\sigma_r - \sigma_\theta)^2 + 4\tau_{r\theta}^2} \end{array} \right\} \quad (4)$$

where σ_r, σ_θ and $\tau_{r\theta}$ are determined by Eqs (1)-(3).

The mode I stress intensity factor at tips of pre-existing microcracks is expressed as[26]

$$K_1 = -[\sigma'_2 + f(c_0)S'_2]\sqrt{2d_i \tan\left(\frac{\pi c_0}{2d_i}\right)} \quad (5)$$

where $S'_2 = \sigma'_2 - (\sigma'_1 + \sigma'_2)/2$, $f(c_0) = d/c_0$, σ'_1 and σ'_2 are determined by Eq. (4), d_i is the spacing between microcracks which can be obtained by mesoscopic experiments.

The strain energy density factor of rock material is chosen as the growth condition of pre-existing microcracks along interfaces. The strain energy density factor of rock material can be written as[27-28]

$$S = a_{11}K_I^2 + 2a_{12}K_I K_{II} + a_{22}K_{II}^2 + a_{33}K_{III}^2 \quad (6)$$

where K_I, K_{II} and K_{III} are the mode I and Mode II and Mode III stress intensity factors, respectively.

$$\begin{cases} a_{11} = \dfrac{1+v}{8\pi E}\left[(3-4v-\cos\chi)(1+\cos\chi)\right] \\ a_{12} = \dfrac{1+v}{8\pi E}(2\sin\chi)\left[\cos\chi-(1-2v)\right] \\ a_{22} = \dfrac{1+v}{8\pi E}\left[4(1-\cos\chi)(1-v)+(1+\cos\chi)(3\cos\chi-1)\right] \\ a_{33} = \dfrac{1+v}{2\pi E} \end{cases}$$

The onset of rapid crack propagation is assumed to start when the strain energy density S_{min} associated with $(dW/dV)_{min}^{max}$ reaches a critical value, i.e.

$$S_{min} = S'_C \qquad (7)$$

where $S'_C = r_C \left(\dfrac{dW}{dV}\right)_C = \dfrac{(1+v)(1-2v)}{2\pi E}(K'_{IC})^2$ is the critical strain energy density factor of the rock interface, K'_{IC} is fracture toughness of the rock interface.

When the stress intensity factor at tips of pre-existing microcracks satisfies the critical condition(7), the pre-exisiting microcracks will propagate along interfaces and the growth length of pre-existing microcracks is equal to width of interfaces, which is determined by the mesoscopic experiments. The statistically-averaged value of the growth length of pre-existing microcracks c_1, which depends on the microscopic structures of rock materials, such as size and shapes of grains or aggregates, and the statistically-averaged value of the density of microcracks η_1 can be determined using the homogenization method.

The dissipative energy density induced by pre-existing microcrack growth can be determined by the integral of energy release rate G along the length of microcracks $c = c_1 - c_0$ as[29]:

$$U_1 = \dfrac{\eta_1(\kappa+1)(1+v)}{2E}\int_{c_0}^{c_1}\left\{\sqrt{2d_i \tan\left(\dfrac{\pi c}{2d_i}\right)}\left[\sigma'_2 + f(c)S'_2\right]\right\}^2 dc \qquad (8)$$

where $f(c) = d/c$, $k = 3-4v$.

When the growth length of pre-existing microcracks is equal to c_1, it is observed from experiments that pre-existing microcracks will propagate through rock matrix under certain stress condition, which leads to the nucleation of secondary microcracks.

The critical condition of the occurrence (nucleation) of secondary microcracks is

$$\sigma''_2 + f(c_1)S''_2 = -\dfrac{K_{IC}}{\sqrt{2d_i \tan\left(\dfrac{\pi c_1}{2d_i}\right)}} \qquad (9)$$

Where K_{IC} is the fracture toughness of rocks, $f(c_1) = d/c_1$, $S''_2 = \sigma''_2 - (\sigma''_1 + \sigma''_2)/2$,

$\sigma''_1 = \frac{\sigma'_r + \sigma'_\theta}{2} + \frac{1}{2}\sqrt{(\sigma'_r - \sigma'_\theta)^2 + 4\tau'^2_{r\theta}}$ and $\sigma''_2 = \frac{\sigma'_r + \sigma'_\theta}{2} - \frac{1}{2}\sqrt{(\sigma'_r - \sigma'_\theta)^2 + 4\tau'^2_{r\theta}}$, σ'_r, σ'_θ and $\tau'_{r\theta}$ are the stress components of the surrounding rock mass around deep circular tunnels when the growth length of pre-existing microcracks is equal to c_1. σ'_r, σ'_θ and $\tau'_{r\theta}$ can be determined by replacing c and η with c_1 and η_1 in Ref. [25].

After the nucleation of secondary microcracks, secondary microcracks will propagate. Then the mode I stress intensity factor at tips of secondary microcracks can be expressed as

$$K_1 = -[\sigma'''_2 + f(l)S'''_2]\sqrt{2d_i \tan\left(\frac{\pi l}{2d_i}\right)} \tag{10}$$

where $f(l) = d/l$, l is denoted as the growth length of secondary microcarcks, $S'''_2 = \sigma'''_2 - (\sigma'''_1 + \sigma'''_2)/2$, $\sigma'''_1 = \frac{\sigma''_r + \sigma''_\theta}{2} + \frac{1}{2}\sqrt{(\sigma''_r - \sigma''_\theta)^2 + 4\tau''^2_{r\theta}}$ and $\sigma'''_2 = \frac{\sigma''_r + \sigma''_\theta}{2} - \frac{1}{2}\sqrt{(\sigma''_r - \sigma''_\theta)^2 + 4\tau''^2_{r\theta}}$,

σ''_r, σ''_θ and $\tau''_{r\theta}$ are the stress components of the surrounding rock mass around deep circular tunnels when the growth length of secondary microcracks is equal to l. σ''_r, σ''_θ and $\tau''_{r\theta}$ can be determined by replacing ηc^2 with $\eta_1 c_1^2 + \eta_2 l^2$ in Ref. [25], η_2 is the average density of secondary microcracks of stable growth.

The strain energy density factor of rock material is chosen as the growth condition of pre-existing microcracks along interfaces.

The onset of rapid crack propagation is assumed to start when the strain energy density $S_{\min}$ associated with $(dW/dV)_{\min}^{\max}$ reaches a critical value, i.e.

$$S_{\min} = S_C \tag{11}$$

where $S_C = r_C\left(\frac{dW}{dV}\right)_C = \frac{(1+v)(1-2v)}{2\pi E}(K_{IC})^2$ is the critical strain energy density factor of rock, K_{IC} is fracture toughness of rock.

When the stress intensity factor (10) at tips of secondary microcracks satisfies the critical condition(10), the secondary microcracks will stably propagate. The stable growth length l of secondary microcracks can be determined by (10) and (11).

The dissipative energy density induced by the stable growth of the secondary microcracks can be determined by integration of the energy release rate G along the stable growth length of secondary microcracks.Then we have

$$U_2 = \frac{\eta_2(\kappa+1)(1+v)}{2E}\int_{c_1}^{c_1+l}\left\{\sqrt{2d_i \tan\left(\frac{\pi l}{2d_i}\right)}[\sigma''_2 + f(l)S''_2]\right\}^2 dl \tag{12}$$

When the growth length of secondary microcracks reaches the critical value,parts of secondary microcracks will unstably propagate.

Criterion of unstable growth of secondary microcracks is that the length of secondary microcracks increases, while the stress σ_1^m does not increase. At the beginning of the unstable growth of secondary microcracks, the critical length of secondary microcracks c_2 can be determined by criterion of the unstable growth of secondary microcracks, which can be obtained by the iteratively using Eqs.(10) and (11).

After the unstable growth of secondary microcracks occurs, the load-carrying capacity of rock masses decreases. When the growth length of secondary microcracks is larger than c_2, the unstable growth of secondary microcracks leads to the coalescence of secondary microcracks, and the occurrence of macrocracks and the failure of rock masses.

Growth length from nucleation to coalescence of secondary microcracks is

$$l = d_i - c_2 \tag{13}$$

It is assumed that the post-peak deformation modulus of rock masses is E_1. Then the dissipative energy density of the unstable growth of secondary cracks can be determined as

$$U_3 = \frac{\eta_3(\kappa+1)(1+v)}{2E_1} K_{IC}^2 (d_i - c_2) \tag{14}$$

where η_3 is the density of secondary microcracks of the unstable propagation which can be determined from numerical computation in Ref. [25].

3.3. Criteria of rockburst and kinetic energy of rock fragments
3.3.1. Kinetic energy and the elastic strain energy storied in the rock masses

According to the theory of elastic mechanics, the strain energy density stored in rock masses can be defined by

$$U_e = \sum_{i=0}^{n} A_i \left\{ \frac{1-v^2}{2E} \left[\left(\sigma_{r_i}^e\right)^2 + \left(\sigma_{\theta_i}^e\right)^2 \right] - \frac{v(1+v)}{E} \sigma_{r_i}^e \sigma_{\theta_i}^e + \frac{1+v}{E} \left(\tau_{r_i\theta_i}^e\right)^2 \right\} \tag{15}$$

where A_i is the area of the ith fragment of rocks,

$$\sigma_{r_i}^e = \frac{\sigma_v + \sigma_h}{2}\left(1 - \frac{r_0^2}{r^2}\right) + \frac{\sigma_h - \sigma_v}{2}\left(1 - 4\frac{r_0^2}{r^2} + 3\frac{r_0^4}{r^4}\right)\cos(2\theta),$$

$$\sigma_{\theta_i}^e = \frac{\sigma_v + \sigma_h}{2}\left(1 + \frac{r_0^2}{r^2}\right) - \frac{\sigma_h - \sigma_v}{2}\left(1 + 3\frac{r_0^4}{r^4}\right)\cos(2\theta),$$

$$\tau_{r_i\theta_i}^e = \frac{\sigma_v - \sigma_h}{2}\left(1 + 2\frac{r_0^2}{r^2} - 3\frac{r_0^4}{r^4}\right)\sin(2\theta).$$

3.3.2. The total dissipative energy density induced by the growth of pre-existing microcracks and secondary microcracks

Neglecting the thermal dissipation energy density induced by the growth of pre-existing microcracks and secondary microcracks, the total dissipative energy density can be written as

$$\xi = U_1 + U_2 + U_3 \tag{16}$$

3.3.3. Criteria of rockburst

(1) Criteria of potential rockburst

The occurrence of potential rockburst should satisfy the following two conditions: (a) The total dissipative energy density induced by the propagation of pre-existing and secondary microcracks should be less than the elastic strain energy density; (b) the fractured zones do not intersect with the tunnel walls, and the residual elastic strain energy should be less than the dissipative energy induced by the failure of rock masses along the shortest path to the tunnel walls. Then, criteria of potential rockburst can be expressed as

$$\begin{cases} W = U_e - \xi \geqslant 0 \\ W < U(l_p) \end{cases} \tag{17}$$

where l_p is the shortest path from the fractured zones to the tunnel walls, which can be determined by iterative computation, $U(l_p)$ can be calculated using Eq.(14) by replacing $(d_i - c_2)$ with l_p.

(2) Criteria of immediate rockburst

The occurrence of immediate rockburst should satisfy the following two conditions: (a) The total dissipative energy density induced by the growth of pre-existing and secondary microcracks should be less than the elastic strain energy density; (b) the secondary microcracks unstablely grow to the tunnel walls, and the fractured zones intersect with the tunnel walls. Then, criteria of immediate rockburst is

$$\begin{cases} W = U_e - \xi \geqslant 0 \\ l_p = 0 \end{cases} \tag{18}$$

(3) Criteria of time delayed rockburst

The occurrence of time delayed rockburst should satisfy the following two conditions : (a) The total dissipative energy density induced by the growth of pre-existing and secondary microcracks should be less than the elastic strain energy density; (b) the fractured zones should not intersect with the tunnel walls, and the residual elastic strain energy density should be more than the dissipative energy density induced by the failure of rock masses along the shortest path to the tunnel walls. Then criteria of time delayed rockburst is

$$\begin{cases} W = U_e - \xi \geqslant 0 \\ W \geq U(l_p) \end{cases} \tag{19}$$

3.3.4. The ejection velocity of rock fragments

Areas of the zones of potential, immediate and time delayed rockburst can be determined using criteria of potential, immediate and time delayed rockburst as well as the ejection velocity of rock fragments of the immediate and time delayed rockburst.

According to the energy conservation law, when the total dissipative energy density induced by the growth of pre-existing and secondary microcracks is less than the elastic strain energy density, the residual elastic strain energy density is transformed into the kinetic energy density of rock fragments. Therefore, the ejection velocity of rock fragments for immediate rockburst can be expressed as

$$V = \sqrt{2W/\rho A} \tag{20}$$

where $A = \sum_{i=0}^{n} A_i$ is the area of rockburst.

According to the energy conservation law, the residual elastic strain energy is transformed into the kinetic energy density when the residual elastic strain energy density is more than the dissipative energy density induced by the failure of rock masses along the shortest path to the tunnel walls. Therefore, the ejection velocity of rock fragments for time delayed rockburst can be expressed as

$$V = \sqrt{2[W - U(l_p)]/\rho A} \qquad (21).$$

4. A study case

Taipingyi Hydropower Project at the upriver of the Minjiang River is located in Wenchuan county, Sichuan Province, Southwest of China. The installed capacity of the project is 260MW. The diversion tunnel with a diameter of 9.6 m and a height of 6.8m and a total length of 10.6 km are plotted in Fig.1. The main strata outcropping in this area consist of granite, granodiorite and so on. The in-situ stress is $\sigma_1 = 31.1 MPa$, $\sigma_3 = 10.4 MPa$. Young's modulus of granite is equal to 31GPa, Posson's ratio of granite is equal to 0.20. The uniaxial tensile strength of granite is equal to 10MPa. Cohesion of granite is equal to 35.5MPa. Angle of internal friction of granite is equal to 42.8°. Weight density of granite is 27.2kNm^{-3}. Rockburst zones in diversion tunnel in Taipingyi Hydropower Station obtained from the numerical computation is depicted in Fig.2. By comparison with the numerical results and monitoring results, the location and magnitude of rockburst zone on the left in Fig.2 is in good agreement with the monitoring results in Fig.1.

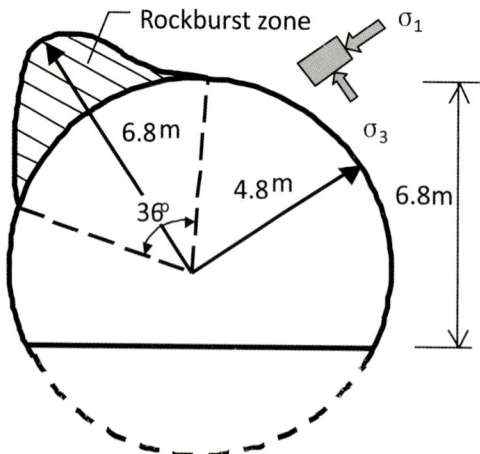

Fig.1 Sketch of diversion tunnel in Taipingyi Hydropower Station

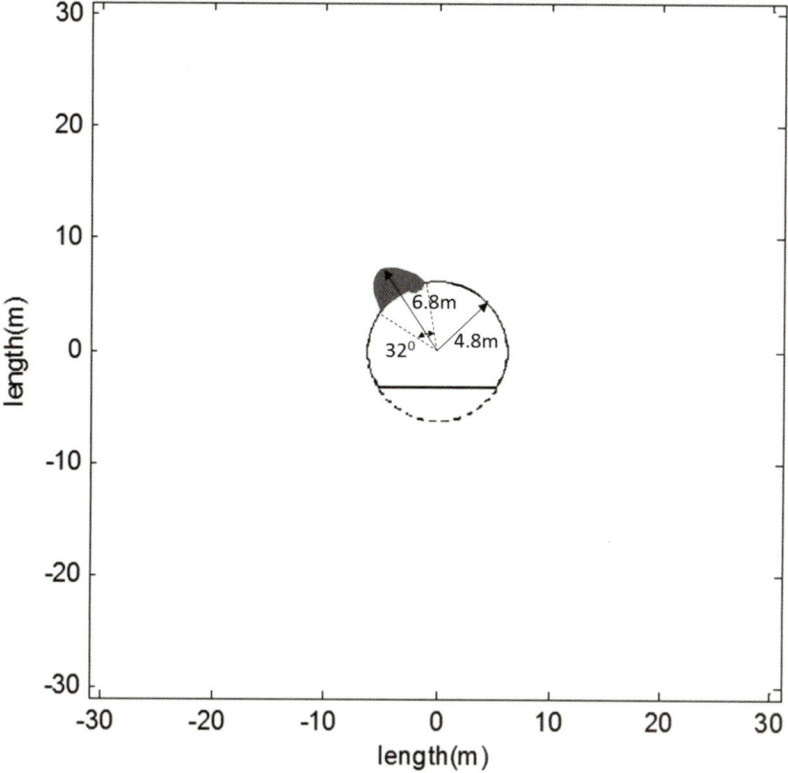

Fig.2 Rockburst zone in diversion tunnel in Taipingyi Hydropower Station

5. Discussions and conclusions

The surrounding rock masses is regarded as geomaterial containing uniform distribution of pre-existing microcracks. Based on non-Euclidean model , stress fields of the surrounding rock masses containing uniform distribution of pre-existing microcracks around the circular tunnel is obtained The mechanism of potential, immediate and time delayed rockburst preparation and development in the surrounding rock masses containing uniform distribution of pre-existing microcracks around the circular tunnel is revealed.

After the excavation of tunnels, redistribution of stress fields of the surrounding rock masses occurs, pre-existing microcracks initiate, propagate along interfaces, and pass through the rock matrix which lead to the appearance of secondary microcracks. On one hand, when the growth length of secondary microcracks is larger than the critical value, secondary microcracks unstablely propagate, coalesce and intersect with the tunnel walls, the fractured zones intersecting with the tunnel walls occurs. For brittle rock, if the total dissipative energy induced by propagation of pre-existing and secondary microcracks is less than the elastic strain energy stored in rock masses, the sudden release of the residual elastic strain energy stored in rock pieces may lead to immediate rockburst.On the other hand, the unstable growth of secondary microcracks does not lead to coalescence of microcracks and to the fractured zones intersecting with the tunnel walls. If the total dissipative energy induced by growth of pre-existing and secondary microcracks is less than the elastic strain energy stored in rock masses, the residual elastic strain energy is transformed into kinetic energy of rock fragments. The movement of rock fragments leads to the fracture of the surrounding rock masses along the shortest path to the tunnel walls. As a result, ejection of rock fragments occurs, time delayed rockburst takes place.

Based on the mechanism of potential, immediate and time delayed rockburst, criteria of potential, immediate and time delayed rockburst is established. The quantitative prediction model of potential, immediate and time delayed rockburst is proposed. Area and location of potential, immediate and time delayed rockburst are determined. The ejection velocity of rock fragments for immediate rockburst and delay rockburst is obtained. The numerical computations are made. It is found from the numerical results that area of potential rockburst zones is maximum; area of time delayed rockburst zones is minimum; and the ejection velocity of rock fragment for immediate rockburst is larger than that for time delayed rockburst.

Strictly, excavation of tunnels and propagation of microcracks cannot be completed immediately, which are related to time, the fracture toughness of rocks is also related to the loading rate. Therefore, analysis and computation of rockburst, in which the inertial effects should be taken into account, are not static problems. In this paper, time of unloading, velocity of crack propagation, the rate dependent fracture toughness of rock and the inertial motion of rock masses are not considered. Moreover, in order to simplify analysis of rockburst, the homogenization method is applied to study the length and density of microcracks, while rock masses in actual engineering is mesoscopic heterogeneous, which distribution of the length and density of microcracks should be described by the probability density function. Therefore, the numerical results from the homogenization method are approximate, in which there is a slight error. In order to eliminate the error, the numerical method will be applied to study effects of the probability density functions describing distributions of the orientations and lengths of microcracks on rockburst in mesoscopic heterogeneous rock masses in the further as well as effects of time on rockburst.

Acknowledgment

This work was supported by project 973 (Grant no. 2014CB046903), the National Natural Science Foundation of China (Grant nos. 51325903,51279218)

References

[1]N.M. Petukhov. Rockburst of Coal Mine, China Coal Industry Publishing House, Beijing, 1980.

[2] H.T Wang, J.Xu, F.S.Wei, X.F.Xian. Evalutaion of tendency indexes of coal or rock burst, Ground Pressure and Strata Control 3(1999) 204-207.

[3] B.J.Zhao, X.J.Teng. Rockburst and its Prevention. China Coal Industry Publishing House, Beijing, 1995.

[4] X.N.Wang, R.Q.Huang. Analysis of deformation and failure features characteristics of rock under unloalding conditions and their effects on rock burst, Mountain Research16(4)(1998)281-285.

[5] Pan Yishan, Xu Bingye. The rockburst analysis of circular chamber under consideration of rock damage, Chinese Journal of Rock Mechanics and Engineering, 1999, 18(2): 152-156.

[6] H.P.Xie, W. G. Pariseau. Fractal character and mechanism of rock burst, Chinese Journal of Rock Mechanics and Engineering12(1)(1993) 28-37.

[7] P.K.Kaiser, D.D.Tannnant, D.R.McCreat. Canadian Rockburst Support, Handbook Geo-mechanics Research Centre, 1996.

[8] F.Pelli, P.K.Kaiser, N.R.Morgenstern. An interpretation of ground movements recorded during construction of Donkin-Morien tunnel. Canadian Geotechnical Journal 28(2)(1991) 239-254.

[9] M.T.Zhang, Z.H.Xu , Y.S.Pan, Y.S. Zhao. A united instability theory on coal (rock) burst and outburst, Journl of China Coal Society 16(4)(1991) 25-31.

[10] T.J Tan. Rockbursts, case records, theory and control, Chinese Journal of Rock Mechanics and Engineering 6(1)(1987) 1-18.

[11] A.M.Linkov. Keynote Lecture: New geomechnical approaches to develop quantitative seismicity, Rockburst and seismicity in Mines, Cibowicz & Lacoski (Vds), Balkema, Rotterdam, 1997.

[12] E.Hoek, P.K.Kaiser, W.F.Bawden. Support of underground excavations in hardrock. A.A.Balkenma, Rotterdam, Brookfield，1995.

[13] L.S. Xu, L.S.Wang. Study on the laws of rockburst and its forecasting in the tunnel of Erlang Mountain road, Chinese Journal of Rock Mechanics and Engineering21(5)(1999) 569-572.

[14] F.L.Hou. The application of rock mechanics in engineering, Knowledge Press, Beijing, 1989.

[15] M.C.Gu, F.L.He, C.Z.Chen. Study of Rockburst in Qingling Tunnel, Chinese Journal of Rock Mechanics and Engineering 21(9)(2002) 1324-1329.

[16] J.J Zhang, B.J.Fu, Z.K.Li, et al. Multi-criteria of rockburst including five factors and classification of rockburst，Chinese Society for Rock Mechanics & Engineering30(1)(2011) 27-41.

[17] X.T.Feng. Adaptive pattern recognition to predict rockbursts in underground openings, Journal of

Northeastern University15(5)(1994) 471-475.

[18] T.Jiang,Z.Q.Huang, Y.Y.Zhao. Dynamically weighted grey optimization model for rockburst risk forecasting and its application to western route of south-north water transfer project, Chinese Journal of Rock Mechanics and Engineering 23(7)(2004) 1104-1108.

[19] Y.H.Wang, W.D. Li, P.K.Lee, et al. Method of fuzzy comprehensive evaluations for rockburst prediction, Chinese Journal of Rock Mechanics and Engineering17(5)(1998) 493-501.

[20]Z.Y. Liang,H.C. Liu, Y.C.Shi, Z.G.Shan. Probability model for rockburst prediction, Chinese Journal of Rock Mechanics and Engineering 23(18)(2004) 3098-3101.

[21] X. Chen,J.Z.Sun, J.K.Zhang, Q.S.Chen. Judgment indexes and classification criteria of rockburst with the extension judgment method, China Civil Engineering Journal 42(9)(2009) 82-88.

[22] J.L Wang,J.P Chen, J.Yang, J.S.Que. Method of distance discriminant analysis for determination of classification of rockburst, Rock and Soil Mechanics 30(7)(2009)2203-2208.

[23] Q.Y. Zhang, X.G. Chen,B.Lin ,D.J. Liu, N.Zhang. Study of 3D geomechanical model test of zonal disintegration of surrounding rock of deep tunnel. Chinese Journal of Rock Mechanics and Engineering 28(9)(2009)1757-1766.

[24] Z.L.Fang, Study on the ground pressure and control method for openings in soft and broken rocks in Jin Chuon Mine No. 2. Journal of Beijing Institute of iron and steel 1(1984) 6-25.

[25] X.P. Zhou, Q.H. Qian. The non-Euclidean model of failure of the deep rock masses under the deformation incompatibility condition. Journal of Mining Science, 2013(in press)

[26] A.Golshani , Y.Okui,M.Oda, et al. A micromechanical model for brittle failure of rock and its relation to crack growth observed in triaxial compression tests of granite. Mechanics of Materials38(4)(2006) 287-303.

[27]G.C.Sih. A special theory of crack propagation: methods of analysis and solutions of crack problems In: G.C.Sih, Editors, Mechanics of Fracture I, Noordhoof, Leyden, 1973.

[28]G.C.Sih. Mechanics of Fracture Initiation and Propagation. Kluwer Academic Publishers, Netherlands, 1991.

[29] Q.H.Qian, X.P.Zhou. Quantitative analysis of rockburst for surrounding rocks and zonal disintegration mechanism in deep tunnels. Journal of Rock Mechanics and Geotechnical Engineering3(1)(2011) 1-9.

Evaluation of the status and outlook of the urban underground space development and utilization in china

Qihu Qian

(Chinese Academy of Engineering)

With the advent of the 21st Century, China has experienced rapid development of urban underground space and constantly improves system of urban underground space that have earned it a leading role in mega cities in the world as regards the overall scale and development speed of the underground space. China has become a major country in the world with respect to the development and utilization of urban underground space which has attracted widespread attention of governments at all levels.

In terms of function, underground traffic plays a major part in the development and utilization of China's urban underground space. The construction rate of China's urban rail transportation ranks first in the world; the construction of urban underground expressways has already kicked off and will be speeding up; urban underground logistics systems are under study; special planning of the underground space is fully underway in the mega and large cities; many cities have compiled or been compiling detailed planning to be aligned with the reconstructing of central districts and the construction of new districts; the large-scale underground complexes have become the highlights in the development and utilization of the underground space in many large cities, and have come up to the internationally advanced level. However, the comprehensive utilization efficiency of the underground spacehas to be improved; the construction of municipal comprehensive pipeline (common tunnel) has just started; thorough development and utilization of underground space has yet to begin; to some extent, there is still a gap between China and the developed countries in terms of legislation, policy, operation and management, underground construction equipment for which the nation has ownership in the core techniques.

Today, in China the development and utilization of urban underground space has become an important means to enlarge urban capacity, relieve urban traffic pressure and improve urban environment, and an important approach to the establishment of a resource-saving and environment-friendly city.

Take Beijing for example. At present, the city has 30 million m^2 underground space. The floor area of underground space in the city is going to increase 3 million m^2 annually, accounting for 10% of the total floor areasof city.

Currently, the construction speed of China'surban subway transportation is top of world. Satistically, China will develop 1200km urban subway transportation till 2010. From now on, China will averagely construct 180km every year.

In recent years, city's subway construction come into a unprecedented developing period in China. Up to June. 2006, there are 7 cities which have subway in mainland China: Beijing-113km; Tianjing-7.4km; Shanghai-123km; Guangzhou-36.6km; Shenzhen-21.8km; Nanjing-16.9km; Total line-318km. And three cities (Chongqing, Dalian, Wuhan) have urban rail transportation. Now 17 subway lines total length 360km construct in other 8 cities.

According construction planning the Beijing's subway line net will compose of 19 lines, total length 561.51km. It will exceed Newyork's totalsubway length and will become the longest in the world. In Shanghai 13 subway lines will be finished till 2012, total length 510km could be used.

Underground complexes in large cities are characterized by their amazing number, large scale and high standard. By taking advantage of the construction of the underground railway and new districts, and urban reconstructing, many cities have established huge-sized complexes that fulfill comprehensive functions and form a complete system of their own, for example: Zhongguancun West district in Beijing underground floor area: 500,000 m^2; Wangfujing district in Beijing available underground floor 600,000 m^2; Central district of financial street in Beijing underground traffic system of financial street is the first large-scale underground traffic system with 30,000 m^2 total floor area. In Shanghai underground syntheses of south station: total exploitation area 90,000 m^2 exploitation depth 15m; Underground syntheses of Renmin Square: total floor area 50,000 m^2. Shenzhen: Luohu Port and railway station underground area of Shenzhen obtained superexcellence award of Asia & Pacific area by international Urban Land Institute in July 2006. Nanjing: Xinjiekou, underground total floor area over 400,000m^2, primarily formed underground city. Underground space of Gongbei Port Square in Zhuhai: underground total floor area over 150,000 m^2. Dalian: Underground Buye City (150,000 m^2), Olympic Park (40,000 m^2). Xi'an underground Gulou Square:50,000 m^2 Ji'nan Spring city Square:40,000 m^2. Other such as Beijing Zhongguancun, Olympic Central Area, Shanghai World Expo Garden, the South Railway Station, Wujiaochang, Guangzhou Pearl River CBD, and the Wave Culture City of Qianjing CBD in Hangzhou, each of these project covers an area of above 100 thousand m^2, built up 3 to 4 storeys, serving simultaneously the purpose of transportation, municipal administration and commerce, providing a superior interior environment and bring on good terms the ground and underground.

Urban underground expressways have started. The finished structures include Xuanwu Lake Underground Expressway, Nanjing East City Underground Trunk Road, West Lake Underground Hubin Road in Hanzhou, Beijing Olympic Center, Zhongguancun, Financial Street Underground Road, the underground sections of Shanghai Middle Ring, the west corridor underground section in Shenzhen, etc. The projects under construction are the Dushu Lake Underground Expressway in Suzhou, etc.

The achievement of tunnel construction attract the world attention. The constructed tunnels have the greatest amount and the most rapid developmentin the world. Cross Yangtze river tunnel from Shanghai to Chongming uses the biggest shield machine in the world. Cross Yangtze river tunnel of Nanjing adopts the second biggest shield machine in the world. The third cross Yangtze river tunnel now is constructed in city Wuhan. The first undersea tunnel of China-Xiangan Tunnel in Xiamen started construction in 2005 with total length of line 9km. Jiaozhouwan undersea tunnel of Qingdao total line 6.17km with undersea tunnel length 5.55km.

Cross Yangtze river tunnel from Shanghai to Chongming

Urban underground space planning is drawing general attention. So far, more than 20 cities, including Beijing, Shanghai, Chongqing, Nanjing, Hangzhou and Qindao, have compiled a special plan for their urban underground space, effectively standardizing its development and utilization. Many cities have composed circumstantial plans for reconstructing the central districts and the construction of the new districts. For instance, Beijing CBD, Zhongguancun West District, Olympic Central Area, Beijing Financial Street, Wangfujing Business District, core district of Qianjiang new city in Hangzhou, Bussiness district of Wangjiadun in Wuhan, etc. effectively control and guide the development and utilization of the underground space along a scientific, reasonable and orderly route. Management regulation on development and utilization of city's underground space' is promulgated in Oct. 27. 1997, providing jural evidence for reasonable development of city's underground space resource.

In the future of 21st century, China will realize sustained and stable growth in economy, and continuous progress in urbanization. Currently, the Chinese government calls for the construction of a resource-saving and environment-friendly society, hence places the increasing emphasis on the development and utilization of China's urban underground space. China's overall scale and area of urban underground space will rank first in the world. Take Beijing for example. Inthe year 2020, the constructed underground space will amount to 90 million m^2, 5 m^2 per capita. According to the underground space plan of Beijing' central urban districts, the underground space of such areas will account for 20% ~ 30% of the total floor area in 2020. The updated design of Beijing Rail Traffic shows that in 2020 the total length of the city's rail line will exceed 1000 km. In the next decade, Beijing's rail transportation will grow 40 km per year; people within Ringroad Five can reach their destinations within 20 minutes by rail; within Ringroad Two they can find a subway station no more than 5 minutes. Public transportation facilities will become the chief means of transportation for 80% of the citizens, half of whom will frequently take the rail.

The following is Underground scope and planning of some cities in China.

City	Planning Range (km²)	Exist Development Quantity (10,000m²)	The time (year)	Prediction Quantity (10,000m²)	Planning Period
Beijing	1085	3000	2006	9000	2004—2020
Shanghai	600	1600	2006	—	—
Nanjing	258	280	2005	730	2002—2010
Shenzhen	2000	1900	2005	—	—
Qingdao	250	200	2004	2544	2004—2020
Wuxi	1662	200	2005	1500	2006—2020

The urban underground space will become a systematic network that develops in line with the ground space. Based on the Management of the Development and Utilization of the Urban Underground Space issued by the Ministry of Construction, cities with over 1 million population will have finished compiling their special plans for urban underground space, bringing designedly and progressively in progress the development and utilization of the underground space of large cities. Many cities will have a systemized and modernized urban underground space system, improving the underground transportation system, the underground lifeline system, the underground air defense and anti-disaster system and the underground public facility system in the key areas of those cities.

A lot of underground cities will be constructed. Core district of Zhujiang new city in Guangzhou underground city floor area 600,000 m² Underground city of business district of Wangjiadun in Wuhan total floor area over 3 million m².

Expos Park district in Shanghai, underground floor area will up to 400,000 m²

Underground city of core district of Qianjiang new city in Hangzhou

The urban underground expressways will also be systemized, so will the underground municipal comprehensive pipeline. Underground logistics systems will be constructed in many cities. Shanghai municipal government have ratified the plan of the-shaped corridor which will extend 40km, 26 km of it being underground. Once completed, it will help the city form a-shaped underground express traffic network. The first session of the project (the south-north corridor) has gone through its design and bidding by the end of Oct. 2006, and the whole project will be completed in 2010. In Beijing underground expressways net will be constructed: system with two longitudinal ways and two transverse ways, every way has 4-6 entrances or exits. The project will be completed in 2020. Gangshen west routeway of Shenzhen constructed about 7km Underground expressway. Nanjing City's underground expressway construction scheme to solve traffic problem is being actively researched.

Gangshen west routeway of Shenzhen

In Shanghai, at the same time, is being studied the feasibility of an underground container transport system. Beijing is also instituting a feasibility plan for its underground systems of expressways and logistics. Many cities are examining the laws and rules of building municipal comprehensive pipeline, exploring its system and mechanism. Now there has been various investment and financing modes of the pipeline. The capital city is launching on a vigorous construction of the municipal comprehensive pipeline in combination with the construction of the subway.

Finally, in a word, by the end of 2020, China is not only a major country but also will become a great power in the world concerning the development and utilization of the urban underground space.

关于绿色发展与智能建造的若干思考

钱七虎

(陆军工程大学,江苏 南京 210001)

2022年是全面实施"十四五"战略规划的关键之年,在实现"双碳"目标的发展过程中,是一个重要的节点和里程碑。"碳达峰""碳中和"的提出是为了应对全球气候变化这一人类面临的最大挑战。中共中央、国务院印发的《国家标准化发展纲要》、国务院发布的《2030年前碳达峰行动方案》应对这个挑战制定了明确的时间表:二氧化碳排放量力争于2030年前实现"碳达峰",努力争取在2060年前实现"碳中和"。而"碳达峰"与"碳中和"主要有两个方面:一是碳排放,二是碳的负排放,也就是碳的吸收,又称"碳汇"。

常规的三个碳排放大户——工业、建筑、交通大约各占1/3。一般来说,建筑全生命期的碳排放总量包含从建材生产、运输到建筑施工、运营、维修、建筑拆除、废弃物处理的7大阶段。

从目前的研究看,在这些阶段中的碳排放,运行阶段占比60%~80%,是最大比重;排名其次的是占比20%~40%左右的建材生产环节的碳排放环节,而施工建造环节过程仅占。5%~10%左右,拆除阶段则占比最低。

面对当前日趋严峻的碳排放形势,优化建造方式,推进绿色建筑发展,是实现降低建筑领域碳排放的重要举措。在建筑领域实现"碳中和"这个目标,除了降低用能外,更需要通过技术创新实现建筑领域的绿色发展,推动以建筑设计为龙头的技术方法创新,推进融合空间节能和设备节能,从而实现大幅降低建筑对空调、供暖、电器、照明等用能需求,促进部分时间、部分空间的低碳用能的理念落实。这些技术创新手段对减少运行阶段的建筑碳排放至关重要。

从长远看,要准确把握智能建造是土木工程产业转型升级高质量发展的关键。这里我结合"智能建造、绿色发展"这个主题,与大家分享一下对绿色发展与智能建造的若干思考"。

首先讲一讲绿色发展在世界和我国的地位和作用,主要有以下几点。

一是世界迎来了绿色发展的时代,绿色发展的思想是习近平新时代中国特色社会主义思想的一个重要组成部分。

二是绿色已成为我国国家建设发展、社会发展的五大理念之一,我国的绿色发展正在进入世界绿色发展的先进行列,未来必定位于世界前列。

目前,气候变暖导致的生态环境恶化是全球正面临的多重挑战,推进绿色发展和绿色建造是应对之策。而在工程建设领域,智能建造是实现绿色建造的必然选择和最佳途径。

1 工程建设应向"绿色建造"转型

1992年,中国承诺将可持续发展纳入国家战略,建设可持续发展的城市和社会,以履行联合国环境

本文原载于《建筑技术》(2022年第53卷第7期)。

与大发展大会通过的《21世纪议程》。当时我们向中国工程院提出了题为《21世纪中国地下空间开发利用的战略和对策》的咨询课题，在这个课题中，建议充分利用地下空间，大力发展以地铁为骨架的轨道交通系统和集约、可持续的城市基础设施，从而提高土地利用效率，节省土地资源，促进地上空间释放用作碳汇的绿色植被和生态空间，支撑可持续城市建设。

建筑业是全球最大的材料与能源消耗产业，全球建筑运营能耗达到30%以上，再加上建设过程中发生能耗，指标接近50%。我国传统粗放建造模式导致资源消耗大、浪费现象多、污染控制难。

为了适应绿色发展的要求，无论是世界还是我国的工程建设，都应该向"绿色建造"转型。

2 智能建造是实现"绿色建造"的必然选择和最佳途径

工程建设要往"绿色建造"的方向发展，需要理念推动、政策驱动、标准引导多方面共同发力。首先是社会各层面理念上的推动，要使大家认识到绿色建造的重要性。其次是利用政策驱动，通过政策鼓励大家节能，也可以通过标准驱动引导，例如制定绿色建筑标准等对建设项目进行准入要求，让高耗能项目有序退出。

除此之外，最重要的还是依靠科技的发展，结合物联网、大数据、5G、人工智能等技术及其在建设场景的落地研究与应用，工程建设才会更加高效节能，实现低碳排放，甚至零碳排放。

这些技术在工程建设中应用，就要涉及到"智能建造"。所以推动和发展"智能建造"是实现"绿色建造"的必然选择与最佳途径。

在工程建设中，"智能建造"首先体现在全面的透彻感知系统。其次是利用先进信息技术手段，通过互联网、物联网的全面互联，实现建设过程感知信息（数据）的高速和实时传输。一定要使建设过程中获取的信息快速传输出去，如果实时获得的信息要过几天才能看到，工程建设就不能实时地反馈和服务。有了互联网、物联网、5G技术后，信息传输速度将显著提升，可以即时地反映认知。

另一个是打造智慧平台，及时汇聚工程建设过程的海量数据，发布安全预警和处理对策预案，并需要技术、管理人员通过平台对采集的数据进行分析、处理、模拟、辅助决策。有了这些智能化的技术赋能工程建设的风险降低，施工人员管理效率提升，同时将最大限度地节约资源，减少环境破坏。

3 未来智慧工程使工程建设向少人化、无人化发展

我国工程建设领域的科技创新与发展经历了从低到高，从局部到全面的历程。改革开放以来，工程建设经历了机械化和信息化的发展。比如地下工程原来都是采取钻爆法、人工打眼、人工放炮，后来可以大量应用机械台钻，多钻台车施工，现在可以采取数字化掘进，这是机械化的进步。

在地下工程的地质探测中信息化的发展也得到了体现，建地下工程，地下有没有水，有没有断层，如何防止地下安全事故发生，需要把情况了解得很清楚，并作出判断，这都是靠信息化设备提升安全性能。

今后，工程建设领域的进步还需通过数字化、智能化向高层次发展迭代——即向智慧化方向迈进，比如传统工程中，做设计是用图纸，但图纸和工程实体是分离的，而在进入BIM时代后，数字工程中BIM技术得到应用，技术人员可以在计算机里建立虚拟可视化的工程模型。

当前科技创新最主要的标志，就是新一代信息技术、数字建筑技术等持续深入地应用到工程建设过程中，即全面地推广数字工程及智能工程。未来，数字工程及智能工程的落地还将促进工程建设领域向少人化、无人化的方向发展。

| 学术论文——绿色建筑与智能建造

工程建设领域要向智慧建造迈进

钱七虎

(解放军理工大学,江苏 南京 210007)

目前全球正面临气候变暖导致的生态环境恶化等多重挑战,应对之策是要推进绿色发展和绿色建造。而在工程建设领域,实现绿色建造的必然选择和最佳途径是智慧建造。信息技术要继续深入地应用到建设工程中去,全面推广数字工程和智慧工程,未来还要向少人化、无人化的方向发展。

工程建设应向"绿色建造"转型

我在上世纪90年代提出开发利用地下空间,可能对当时人均GDP刚超过1000美元的中国来说,是略为超前,但在全世界范围内就不是了。作为一名学者,我参加了一些国际会议,也接触到科学的前沿。当时我看了三本讨论全球生态环境的著作:《寂静的春天》《增长的极限》和《只有一个地球》。《寂静的春天》里讲述了由于大规模采用DDT等剧毒杀虫剂,造成了"没有鸟儿歌唱的树林""死亡的河流"的典型案例,说明"向大自然宣战""征服大自然"的不科学,而应与之和谐相处。《增长的极限》讲的是19世纪到20世纪全球生产力的发展、经济社会的发展取得了巨大成就,但也带来了严重的生态破坏、环境污染、"城市病"等问题。经济社会的发展要有能够适应资源和环境的客观限制,人类不能一味追求增长,不应该去想着战胜自然,要适应自然,与自然和谐相处。

1992年,为了履行联合国环境与发展大会通过的《21世纪议程》,中国承诺将可持续发展纳入国家战略,建设可持续发展的城市和社会。虽然当时中国因为城市化水平和经济发展程度相对其他国家滞后,但是发达国家面临的那些问题在中国也将很快显露出来。当时我们向中国工程院提出了咨询课题《21世纪中国地下空间开发利用的战略和对策》。充分利用地下空间,特别是大力发展以地铁为骨架的轨道交通系统和集约可持续的城市基础设施,从而节省土地资源,提高土地利用效率,释放出大批地上空间用作绿色植被和生态空间,建设可持续城市。可持续发展是一个长久的问题,现在进一步发展到绿色发展。

在当下的经济社会发展背景下,可持续发展、绿色发展已经成为我们全社会包括国家领导人、科学家和大众的普遍共识。在上个世纪,很多人还没有意识到气候变化、碳排放等可能引发的全球灾难。现在我们能够切身感受到了,比如,全球气温上升,洪水、风暴潮、台风、飓风等自然灾害增加。为了适应绿色发展的要求,工程建设应该向"绿色建造"转型。

目前,全球建筑运营能耗已占到总能耗的30%以上,加上建设过程中的能耗,这一指标接近50%。我国传统的粗放建造资源消耗量大、浪费现象严重。许多老旧工程拆除数量规模巨大,许多远未达到使用寿命期限的道路、桥梁、大楼被提前拆除,城市建筑垃圾和建筑物贡献的温室气体,与建筑有关的空气污染、光污染等,都在呼吁"绿色建造"。

本文原载于《建筑》(2020年第18期)。

如何向"绿色建造"方向发展

首先是理念上的推动,大家要认识到绿色建造的重要性,它与人类命运共同体、与未来可能的全球灾难是什么样的关系,要重视并自觉地在实践中贯彻执行绿色发展的理念。其次是政策驱动,比如政府部门对单位和家庭不同等量的用水、用电采取不同的收费价格,鼓励大家节能;也可以通过标准驱动,例如制定绿色建筑标准等对建设项目进行准入,使得高耗能项目退出。当然最重要的,还是要依靠科技的发展,发展大数据技术、人工智能技术、物联网、5G技术等,有了这些科技基础,建设才会更加高效节能、低碳排放,甚至零排放。这些技术应用到工程建设中,其实就是这次要谈到的另一个主题——智慧建造。智慧建造是实现绿色建造的必然选择和最佳途径。

"智慧建造"在工程建设中的体现

首先是全面的透彻感知系统。拿我熟悉的地下工程、隧道工程来讲,地下工程是高风险的工程,因为地下存在很多不确定性,比如地下有没有水、水量多大,地下岩体是破碎的还是坚固的,破碎的岩体有没有可能导致塌方等。这些情况,人都是看不到感受不到的,要通过设备、传感器、信息化的设备去全面感知,摸清情况。第二是通过物联网、互联网的全面互联实现感知信息(数据)的高速和实时传输。只能感知还不行,获取的信息一定要快速传输出去,如果当下获得的信息要过几天才能看到,只能进行事后分析,工程建设就不能实时地反馈和服务。有了互联网、物联网、5G技术后,信息传输非常快。另一个是智慧平台的打造,技术人员要通过这个平台对反馈来的海量数据进行综合分析、处理、模拟,得出决策,从而及时发布安全预警和处理对策预案,这是非常有必要的。有了这些技术,工程建设的风险更低,施工人员更安全,同时也能最大限度地节省材料和减少环境破坏。

未来要向无人化、少人化发展

我国工程领域的科技创新经历了从低到高,从局部到全面的发展历程。改革开放以来,工程建设经历了机械化和信息化的发展,比如地下工程原来都是采取钻爆法、人工打眼、人工放炮,后来可以大量应用机械台钻,多钻台车施工,现在可以采取数字化掘进。再如地铁隧道施工目前大都采用盾构法,使用全断面的TBM掘进机施工。我们已经成功研发生产了直径15米的盾构掘进机,正在设计直径18.2米的,这是机械化的巨大进步。信息化的发展体现在地下工程的地质探测工作中。建地下工程,需要把地下的情况了解得很清楚,地下有没有水,石头有没有节理、裂隙,有没有断层,如何防止地下安全事故发生,这都是靠信息化设备来提升安全性能的。

今后,工程领域的进步要通过数字化向智慧方向迈进,比如数字工程的BIM技术在建筑工程中的应用。原来我们做设计是用图纸的,但图纸和工程实体是分离的,比如不同图纸上画出来的线可能都是交叉的、矛盾的,到建设中就会出现很多问题。进入BIM时代后,技术人员可以在计算机里建立立体的可视化工程模型。另外怎么样安装、怎么样开挖效率最高,通过这个模型都可以模拟得到结论。当前科技创新最主要的标志,就是信息技术继续深入地应用到建设工程中去,也就是全面地推广数字工程和智慧工程。

学术论文——绿色建筑与智能建造

利用地下空间助力发展绿色建筑与绿色城市

钱七虎

（解放军理工大学国防工程学院，江苏 南京 210007）

摘要：当今世界，绿色发展已经成为一个重要趋势，绿色发展理念更是指导我国"十三五"规划的重要理念之一。建设城市地下空间是转变城市发展方式、治理"城市病"并建设绿色城市的主要着力点。首先，说明绿色建筑与绿色城市的概念，以及二者的关系。重点从节约土地、利用地热能、节水、绿色城市基础设施（包括绿色客运城市交通和城际交通，未来城市货运交通，绿色城市污水、雨洪蓄排系统，绿色城市垃圾集运和处理系统，智慧地下综合管廊，城市智慧行车系统）等方面，并结合国内外典型案例分析说明利用地下空间发展绿色建筑和绿色城市的理念、方法、构想。最后，结合实例强调地下空间开发规划的重要性、规划科学的重要性、开发与规划中的问题及规划的具体要求。

关键词：地下空间；绿色建筑；绿色城市；地热能；基础设施；地下物流

DOI：10.3973/j.issn.2096-4498.2019.11.001　　　　**文章编号**：2096-4498(2019)11-1737-11

中图分类号：U 45　　　**文献标志码**：A

Underground Space Utilization Helps Develop Green Buildings and Green Cities

QIAN Qihu

(*College of Defense Engineering, PLA University of Science & Technology, Nanjing 210007, Jiangsu, China*)

Abstract: In the present world, green development has become an important trend, and the concept of green development is one of the important concepts guiding China´s "13th Five-Year Plan". The construction of urban underground space is the main focus of transforming urban development mode, treating "urban disease" and building green cities. Firstly, the concept of green building and green city and the relationship between them are explained. And then the concept, method and conception of using underground space to develop green buildings and green cities are illustrated from the aspects of saving land, utilizing geothermal energy, saving water and green urban infrastructure (including green urban passenger traffic and the intercity traffic, the future urban freight transportation, green city sewage and rain flood storage system, green city garbage pickup and processing system, intelligent underground comprehensive pipe gallery, and urban intelligent traffic system), etc, based on cases study of the typical projects in China and abroad. Finally, the importance of underground space development planning, the importance of planning science, the problems in development and planning and the specific requirements of planning are emphasized.

Keywords: underground space; green buildings; green cities; geothermal energy; infrastructure; underground logistics

0 引言

当今世界迎来了绿色发展时代，习近平总书记提出的绿色发展理念是习近平新时代中国特色社会主义思想的主要组成部分。党的十八大以来，党中央提出了一系列治国理政新理念、新思想、新战略，把生态文明建设和绿色发展提到新的战略高度；党的十八届五中全会更是把绿色列入中国国家建设发展的五大理念之中。中国的绿色发展正在进入世界绿色发展的先进行列。地下空间是一个巨大而丰富的空间资源，对其进行合理开发利用能够促进我国的绿色发展。本文从节约土地、利用地热能、节水、绿色城市基础设施（包括绿色客运城市交通和城际交通，未来城市货运交通，绿色城市污水、雨洪蓄排系统，绿色城市垃圾集运和处理系统，智慧地下综合管廊，城市智慧行车系统）等方面分析说明如何利用地下空间发展绿色建筑与绿色城市，并介绍如何进行地下空间的开发规划。

本文原载于《隧道建设（中英文）》（2019年第39卷第11期）。

1 绿色建筑与绿色城市

绿色建筑和绿色城市的定义在国际上尚无共识。

1.1 绿色建筑

根据《绿色建筑评价标准》[1],绿色建筑是指在建筑的全寿命周期内,最大限度地节约资源(节能、节地、节水、节材),保护环境和减少污染,为人们提供健康、适用和高效的使用空间,与自然和谐共生的建筑。简而言之,即"四节一环保"的建筑。

1.2 绿色城市

绿色城市一方面是"绿色建筑规模化的一个必然结果";另一方面,因为城市要素除了建筑之外,还包括交通等城市基础设施以及自然环境和社会环境等,所以其内涵还应扩展到其他城市要素。绿色城市的实质可概括为:实现更高水平的生态平衡,大幅减少环境污染并使自然资源得到更为合理的配置,同时形成可持续的生态安全保障体系,从而降低城市发展成本,建立起一种自然与社会高度和谐融合、功能高度复合的城市模式。

与绿色建筑的概念相比,绿色城市在节约资源、能源、保护自然之外还加入了社会学范畴的概念,将在西方城市规划领域一直关注的犯罪率、邻里交往、社会和谐发展等命题也一并纳入。二者关系为:绿色城市＝绿色建筑＋绿色城市基础设施＋绿色城市自然环境(蓝天、绿水、青山、净土)＋绿色社会环境。绿色城市的目标是促进城市可持续发展,提高城市的宜居水平。

2 利用地下空间发展绿色建筑与绿色城市

2.1 节约土地

我国现行耕地红线为 1.2 亿 hm^2(18 亿亩),而既有耕地为 1.239 亿 hm^2(18.58 亿亩),城镇化发展面临无地可用的困境,解决土地问题存在必要性与紧迫性。因此,需要推动地上地下 2 个城市建设,破解城市发展空间不足的问题。

城市节约土地的一个主要方面在于宏观上努力实现土地的多重利用。土地的多重利用可沿 2 个方向实施:1)城市无建筑土地的额外利用;2)城市已建成区域的紧密化和功能变化。开发利用地下空间,即把城市交通(地铁和轨道交通、地下快速路、越江和越海湾隧道)尽可能地转入地下,把其他一切可以转入地下的设施(如停车库、污水处理厂、商场、餐饮、休闲、娱乐、健身等)尽可能建于地下,就可实现土地的多重利用,提高土地利用效率,实现节地的要求[2]。

地下空间潜力巨大,能为节约土地提供良好的条件。我国已有许多城市对地下空间进行了开发利用,如南京的玄武湖隧道、九华山隧道,深圳的前海合作区;深圳的福田地下交通枢纽是国内最大的"立体式"交通综合换乘站,汇集了地铁 2、3、11 号线,以及广深港客运专线福田站,是集城市公共交通、地下轨道交通、长途客运、出租小汽车及社会车辆于一体并与地铁竹子林站无缝接驳的立体式交通枢纽换乘中心,地下枢纽空间总建筑面积约 13.73 万 m^2,相当于 192 个足球场的面积;杭州钱江新城核心区地下城以波浪文化城(10 万 m^2)和地铁 1、2 号线换乘站为骨干,地下空间总量达到 200 万 m^2。据初步统计的最新数据,北京、上海、深圳地下空间开发规模分别达到 9 600 万 m^2、9 400 万 m^2、5 200 万 m^2,近 5 年平均增长分别为 410 万、650 万 m^2、680 万 m^2。我国部分重点城市片区地下空间开发规模见图 1,地下空间开发强度见图 2。

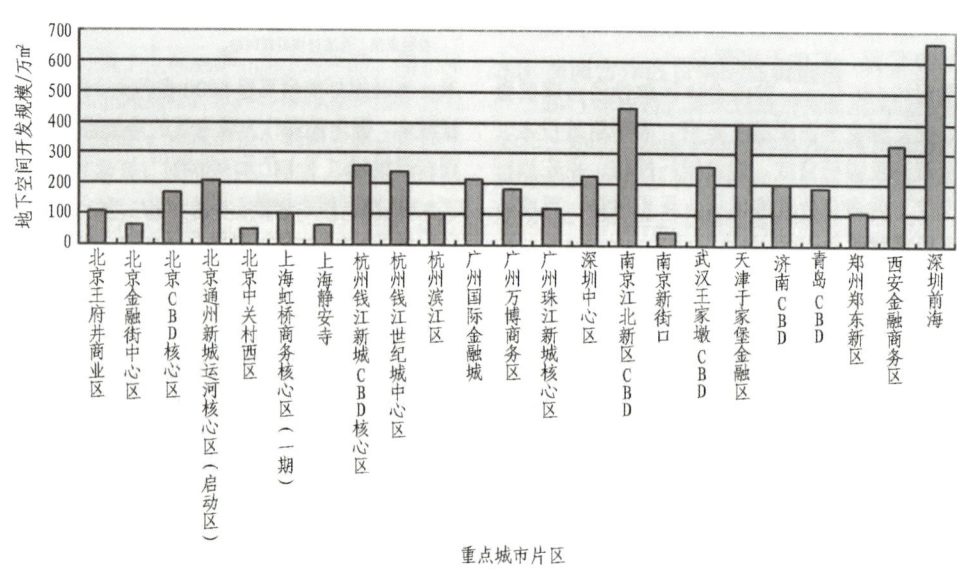

图 1 中国部分重点城市片区地下空间开发规模

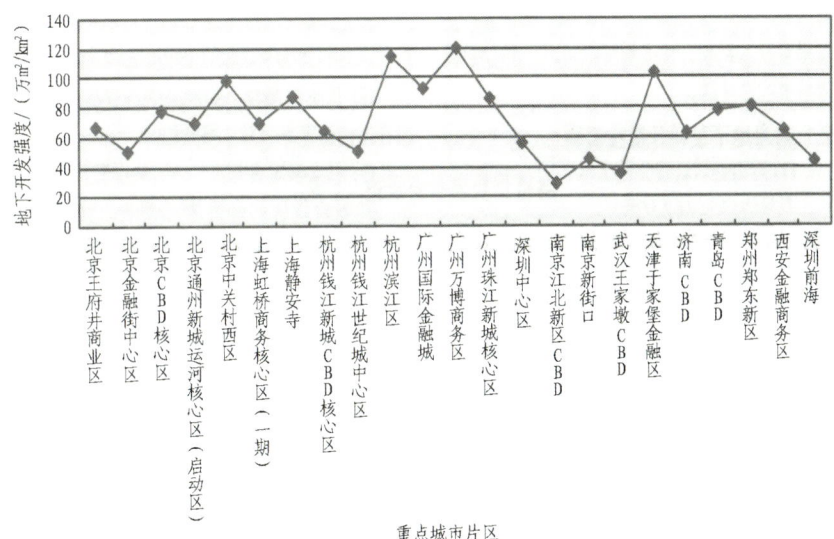

图 2　中国部分重点城市片区地下空间开发强度

Fig. 2　Development intensity of underground space in some key urban areas in China

2.2　利用地热能
2.2.1　低密度地热能源（地温能）
2.2.1.1　地下换热系统

地温能指的是地层中温度小于 25 ℃ 的地层热能。当到达地下一定深度时（5 m 以下），四季的地层温度保持在一稳定值，此时把传统空调器的冷凝器或蒸发器直接埋入地下，利用传热循环介质与大地进行热交换，从而提取地温能，形成地下换热系统。地下换热系统可埋设在地下结构的基坑围护结构（地下连续墙、排桩内）、基础底板下、桩基（钻孔灌注桩、预制桩、PHC 桩等）内；可埋设在新奥法施工的隧道衬砌内或以能源锚杆的形式埋设在其围岩中；也可埋设在地铁区间隧道内、地下输水管道内[3]。隧道结构内埋管换热器工作原理见图 3。

图 3　隧道结构内埋管换热器工作原理示意图

Fig. 3　Sketch of working principle of buried tube heat exchanger in tunnel structure

2.2.1.2　能源隧道

能源隧道是指一种利用隧道衬砌内的热交换管路来提取隧道空气热能或隧道围岩中的地热能，实现隧道附近建筑的供热/制冷服务的技术，见图 4。

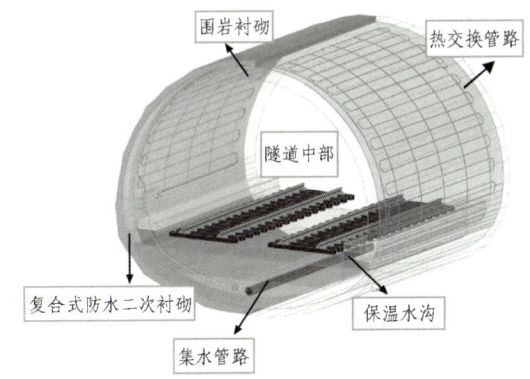

图 4　能源隧道示意图

Fig. 4　Sketch of energy tunnel

能源隧道工作原理是：热交换管由分、集水管与地源热泵前端相连，形成封闭系统，系统内注满循环介质（含防冻液），在水泵的驱动下，热交管内的循环介质在管内循环流动，吸收围岩中的地热能或空气热量，经热泵提升后，用于隧道内部或者周围建筑物的制冷/取热。能源隧道具有以下优势：1) 具有结构和暖通双重功能，比传统空调系统节能 30% 以上；2) 节能环保，无噪音污染，占地少，成本低；3) 能解决寒区隧道的冻胀和结冰等病害问题。

2.2.1.3　工程实例

1) 奥地利政府资助的能源地铁车站示范工程 1984—2004 年桩基埋管数量变化如图 5 所示。该车站位于 Lainzer 隧道 LT24 区，一共有 59 根桩基内埋设热交换管，钻孔灌注桩桩径为 1.2 m，桩长平均为 17.1 m，利用 6 台热泵为附近一所学校供暖。在长期供暖的情况下，车站能提供 150 kW 的热负荷，一个供暖季度可提供 2.14×10^8 kW·h 的能量，天然气的使用量每年减少 34 000 m³,使每年 CO_2 排放量减少 30 t。与传统的靠燃烧天然气供暖的方式比较,可使学校每年用作取暖的费用降低 1 万美元。

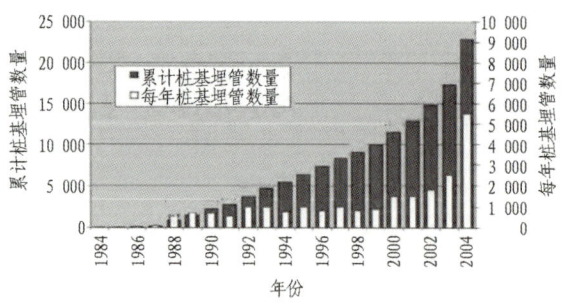

图 5 奥地利能源地铁车站工程桩基埋管数量变化

Fig. 5 Quantity changes of buried pipes in Austrian energy metro station project

2) 瑞士 Grabs 的 PAGO 公司办公楼采用 570 根桩基内埋设热交换管,平均桩长 12 m,以 4 个能源桩为一组,呈方型顶角安装,四边间距为 1.4 m。每延米桩基冬天可获得 35 kW·h 的热量,夏天获得 40 kW·h 的冷量。

3) 上海自然博物馆新馆位于上海市静安区静安雕塑公园内,近北京西路、石门二路,占地面积约为 12 000 m²,总建筑面积约为 45 086 m²,其中地上建筑面积为 12 128 m²,地下建筑面积为 32 958 m²。建筑总高度为 18 m,地上 3 层,地下 2 层,采用地源热泵系统承担建筑部分夏季冷负荷以及冬季热负荷。夏季土壤换热器最大热负荷为 1 639 kW·h,冬季土壤换热器最大热负荷为 1 178 kW·h。

工程采用灌注桩和地下连续墙内埋管 2 种形式。其中,灌注桩内埋管 393 个,平均桩长为 45 m,采用 W 形埋管;地下连续墙内埋管共 452 个,采用 W 形埋管。灌注桩及地下连续墙内埋管布置见图 6。

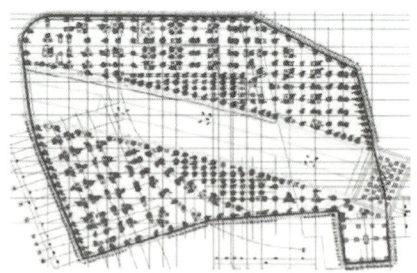

图 6 灌注桩和地下连续墙内埋管布设图

Fig. 6 Layout of cast-in-place piles and buried pipes in diaphragm wall

上海自然博物馆已于 2015 年投入运营,该馆采用了地下结构内埋管热交换系统,初投资比传统的冷水机组+锅炉系统增加 210.2 万元,但该系统利用地温能实现了建筑制冷和供暖,年运行费用可节省 22.3 万元,动态投资回收期为 11.98 年;该系统利用清洁的地温能,每年可节约 117.7 t 标准煤,减排 CO_2 195.5 t。上海中心大厦裙楼、上海世博轴和上海富士康大厦等国内重大工程均采用了地下结构内埋管热交换系统。

4) 某过江隧道在地下连续墙内设置埋管换热管路,来提取地源的地热能,实现隧道附近建筑的供热、制冷。隧道连续墙最大埋深 60 m,连续墙每段长度为 5 m;60 m 深度位置土壤层温度为 16.5~17.5 ℃;采用单 U 型盘管,De32PE 塑料管,埋深按平均 55 m 计算,盘管间距 2.5 m。地下连续墙内埋管直接绑扎在地下连续墙的主筋上,与地下连续墙一起形成换热构件,省去了钻孔费用,且具有传热效果好、稳定性和耐久性好、不占用额外的地下空间等优点。

隧道江北段管理养护中心用房面积约为 4 409.65 m²,利用连续墙埋管及热泵机组进行供暖与制冷,参数如下:1) 制热量 265 kW;2) 制冷量 485 kW;3) 单 U 型盘管,De32PE 塑料管;4) 需要埋管的连续墙长度共 120 m;5) 总投资约 120 万元(含埋管换热器、热泵机组、水泵及管道等费用);6) 节约电能约 23.5 万 kW·h/年(按年运行时间 5 000 h)。隧道江南段参数如下:1) 单 U 型盘管,De32PE 塑料管;2) 可埋管的连续墙长度共 1 300 m;3) 埋管总散热量 2 800 kW;4) 总投资约 850 万元(含埋管换热器、热泵机组、水泵及管道等费用);5) 节约电能约 235.2 万 kW·h/年(按年运行时间 5 000 h)。

5) 科罗拉多安装地源热泵系统后,每个家庭可降低电力峰值需求、节约能源和减少 CO_2 排放 10%~30%。

6) 上海世博轴采用 6 000 根能量桩,是目前世界上单体能量桩用量最大的工程。其节能情况见表 1。

2.2.1.4 未来发展情况

目前,地源热泵正进一步与太阳能结合。通过太阳能的辅助供热,可实现系统向地下排热与取热的平衡,从而使得地下温度场保持稳定,既可以克服单独使用地源热泵时,土壤温度场不断降低(或升高)后不能有效恢复的局限性,又可以克服单独使用太阳能空调系统时,太阳辐射受天候因素制约的局限性[2-3]。

表 1 世博轴节能情况

Table 1 Energy saving of Expo Axis

名称	节能方法	节省用电量/(万 kW·h/年)	减少二氧化碳排放量/(t/年)	节能率/% 夏季	节能率/% 冬季	节能率/% 全年
江水源地源热泵系统	利用地表水和地下的可再生能源(冷量和热量)	562.9	5 629	26.96	71.03	61.40

2.2.2 浅中层地热能(温泉)

河北省地热能资源总量位居全国第二位,2015 年地热资源开采量突破 1.1 亿 m^3,地热供暖面积达到 6 300万 m^2。其中,雄县位于河北省保定地区,是国内首个通过地热供暖实现"无烟城"的县城,拥有享誉全国的"雄县模式"。雄县地热资源分布面积广,出水量大,水温高,现如今地热集中供暖面积已占城区集中供暖面积的85%,覆盖县城80%以上的居民小区,每年可减少 CO_2 排放量12万 t。在收费方面,雄县居民地下水取暖费用为16 元$/m^2$,相较之前燃煤取暖25元$/m^2$的费用更为便宜。

从投入与产出方面分析,地热取暖前期投入较大,后期的年收益率稳定。因此,应推广和提升"雄县模式",在集中成片供暖的基础上,在新兴城镇中打造以地热为主的"绿色热网",解决北方中、小新兴城市和农村冬季供暖的问题,减少燃煤燃烧对空气造成的污染。

2.2.3 深层地热能(干热岩)

我国干热岩资源居世界前列,陆域干热岩资源量为856万亿 t 标准煤,其中青海共和盆地3 705 m深钻获得236 ℃的高温干热岩体。我国已成功在陕西省内进行了干热岩用于供热的商业应用——长安信息大厦2013年共计3.8万 m^2 应用干热岩供热,效果良好。按照2%的可开采资源量计算,我国可开采干热岩量相当于17万亿 t 标准煤,为2016年全国能源消耗量(43.6亿 t)的近4 000倍。至"十三五"末,地热能年利用量相当于替代化石能源7 000万 t 标准煤,减排 CO_2 1.7万亿 t。

国际上有美国、澳大利亚、日本、德国、法国等进行了干热岩发电试验研究项目。美国 Los Alamos 实验室在卡尔德拉的芬登山上建成了一个10 MW 的 HDR(深层干热岩)发电站,该电站主要由2个深度为3 000多米的钻孔及其连通孔组成,冷水由一个钻孔灌入,另一个孔产生200 ℃蒸汽,进入汽轮机发电。我国也已开始干热岩发电的相关研究。

2.3 节水

我国水资源比较短缺,地下空间可以用来解决水资源问题。

2.3.1 雨水

1)世博轴自来水日用量约为2 000 m^3/d,利用雨水时,自来水日用水量降低为1 100~1 200 m^3/d,回用雨水用水量800~900 m^3/d。据此,在可利用时段里,自来水替代率为(800~900)/2 000 = 40%~45%。经处理雨水主要用途为卫生器具冲洗、绿化浇灌等。世博轴雨水收集处理综合利用技术见图7,处理流程见图8。

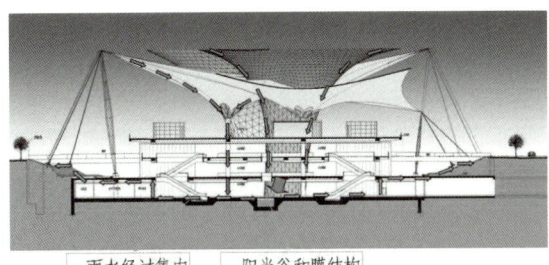

图7 世博轴雨水收集处理综合利用技术

Fig. 7 Comprehensive utilization technology of rainwater collection and treatment on Expo Axis

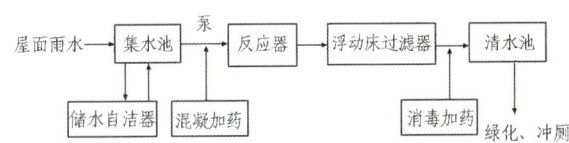

图8 世博轴雨水处理工艺流程图

Fig. 8 Flowchart of Expo Axis rainwater processing

2)西沙群岛上修建了可采集雨水达14万 t 的地下贮水工程。通过技术处理,已达到国家饮用水卫生标准,从而结束了吃水靠大陆船运的历史。

3)北京每年6—9月份降水中,可利用的雨水为2.3亿 m^3,相当于城区全年用水量的1/5多。为积蓄雨水,北京筹划开工修建70个地下小水库,新建公园将首先考虑建设雨水收集利用设施,工程可拦蓄洪水3 559万 m^3。

4)名古屋、大阪、福冈等地的大型建筑物下都设置了雨水利用装置,名古屋体育馆每年利用雨水3.6万 m^3。

2.3.2 再生水(中水)

再生水主要是指城市污水或生活污水经处理后达到一定的水质标准,可在一定范围内重复使用的非饮用水,也称中水。

1)北京2010年再生水利用量达6.8亿 m^3,占总用水量的19%,但利用率仅为60%,"十三五"期间计划将全市再生水用量提升到每年12亿 m^3。

2)天津自2002年正式启动再生水管网建设以来,已铺设了400多千米的再生水管道。目前天津已建成再生水厂8座,全年再生水回用约2 500万 t,大部分用在大型工业项目中,还有河西梅江居住区和南开水上公园周边一些小区也用上了再生水。

3)日本再生水利用量达1.3亿 m^3。

3 绿色城市基础设施建设

建设绿色城市,还要进行绿色城市基础设施的建设。过去由于只注意城市"面子"建设,采用管线直埋、城市排水系统按一年一遇暴雨重现期设计、污水不经处理直排、垃圾传统填埋、空气降污减排不控制、公交系统严重滞后等粗放发展模式，酿成"里子"城市地下基础设施建设短板的恶果：雾霾、交通拥堵、城市内涝、垃圾围城、垃圾山坍塌和泥石流、城市空气污染和雾霾严重、城市地面地下水系和土壤严重污染等"城市病"丛生。

建设城市地下空间是转变城市发展方式、治理"城市病"并建设绿色城市的主要着力点。因此，应以地下综合管廊、海绵城市为主要契机，注意城市"里子"建设的集约绿色可持续发展模式，即：1) 地下综合管廊；2) 海绵城市六字方针(渗、滞、蓄、净、用、排)、雨洪地下储蓄和排洪、地下水银行；3) 污水地下集运和地下污水处理厂；4) 垃圾地下集运、卫生填埋和焚烧处理；5) 雨水、再生水的利用；6) 科学提高城市交通供给能力(发展大运量快速公交系统、发展地下快速路和物流系统)；7) 地下低密度能源利用——地源热泵系统。从而解决交通拥堵、空气雾霾、城市内涝、城市水系和土壤污染、水资源短缺、地下超采漏斗区等问题。

3.1 绿色客运城市交通和城际交通

3.1.1 城市及城际客运交通发展方向

绿色交通要求少占地、低能耗、无污染(电驱动)。未来的城市客运交通以地下轨道交通为主，能够实现少占地、无污染，其中低速磁悬浮由于其能耗低，为城市客运交通最佳选择；未来的城际客运交通应发展地下高速磁悬浮，其速度可达到600 km/h以上(轮轨<400 km/h)，且能耗最低。

3.1.2 地下快速路

地面交通存在以下3个问题：1) 限购限行，利少弊多，不能持久；2) "存量"不减、"增量"攀升，中心城区地面已无地(路)可建；3) 国外曾做过伦敦中心区地面交通供给能力的模拟测算，其结果是即使把伦敦中心区所有建筑架空，其下层地面全变成道路，也不能解决交通拥堵问题。因此，仅大力发展轨道交通不能从根本上解决交通拥堵，即使是轨道交通相当完善的发达国家大城市，如伦敦、东京、芝加哥、马德里、莫斯科等，也困扰于交通拥堵。东京公共交通出行比例达到了60%，地铁建设里程已达300 km(地铁集中于城市中心区，线路里程仅占轨道交通总里程的1/10)，仍然没有解决交通拥堵；伦敦的第一条地铁建成于1863年，至今已建成地铁里程达439 km，但仍然要收取"交通拥堵费"以缓解拥堵之苦。

20世纪末、21世纪初，国际上很多发达国家的特大城市和有识之士把"治堵"目光转向了"地下快速路"和"地下物流系统"。美国波士顿于1994年开始拆除高架路，10年建成了8—10车道的地下快速路；东京正在建设中央环状地下快速线，新宿线已于2007年通车，品川线正在建设；新加坡地下道路系统(SURS)长15 km，为环形4车道，造价48亿美元，承担城市中心区交通量的40%，长9 km的新加坡KPE地下公路目前已建成；马德里已建成36 km的地下快速线；悉尼已建成4条共11.4 km的地下快速线；布里斯班已建成2条6 km地下快速线；巴黎已建成20 km地下快速线，并已完成二代城市地下快速路网设计；莫斯科在三环线中建成了3.6 km地下快速线。

国内大城市的地下快速路建设也已初见端倪，并在缓解交通拥堵问题上发挥了明显作用。南京已建成玄武湖地下快速线、城东干道地下线和内环线地下路；上海正在建设"井"字形通道方案，其中地下线26 km；深圳已建成港深通道的7 km地下快速路。深圳前海的地下空间规划建设在地下车行系统方面，将构建"地下快速路+地下环路+地下车库"的三级地下车行系统，海滨大道、妈湾跨海通道地下道路以分离过境交通为主，以服务区域到发交通为辅；听海路服务于区域到发交通，从而有效改善前海地面交通环境，缓解城市中心区和小街坊路网交通压力。

地下快速路的天然优势是全天候，在暴风、雨雪和大雾等最容易造成地面交通拥堵的天气最能发挥作用，在我国城市建设用地严格控制和大城市人均道路指标普遍偏低的情况下，非常适合我国地少人多的国情[3]。此外，城市地下快速路系统以及地下物流系统的建设也必将为消除汽车尾气对城市空气的污染以及噪声污染作出重要贡献。

3.2 未来的城市货运交通

科技创新转变城市交通运输方式还包括地下物流系统，以替代城市货车限行的行政措施，将城市货运逐步并最终转移至地下。地下货物运输系统，又称地下物流系统(ULS)，是基于城内运输和城外运输的结合。城外货物通过各种运输手段运到城市边缘的物流园区(CLP)，经处理后由CLP通过ULS输送到各个终端。它以集装箱和货盘车为基本单元，以自动导向车(AGV)为运输工具。

根据世界经济合作与发展组织2003年统计，发达国家主要城市的货运占城市交通总量的10%~15%，在"世界工厂"和到处是建设工地的中国则为20%~30%，而在当今电子商务、快递和送快餐发达的中国，这个比例还要高。货运交通转入地下必将对缓解交通拥堵作出重要贡献。

3.2.1 港城地下物流系统

传统的港口集疏运模式以公路集卡运输为主，需要穿越城市，占用城市道路资源，客货交通相互影响，带来了交通拥堵、环境污染等问题。因此,需要更集约化、可持续发展的集疏运新模式——地下集装箱物流系统,以解决港城矛盾,即将原先分散进入港口的货物,首先集中在外围物流综合枢纽,再通过地下专用货运通道集约化转运到港口,实现高效率、规模化运输。从而有效避免集装箱卡车穿越城市,释放地面资源,不影响中心城交通、环境。

3.2.1.1 上海港集装箱集疏运分析

上海港是国际集装箱远洋干线港,连续多年保持集装箱吞吐量世界第一。2009 年以来,上海港集装箱吞吐量增幅基本保持在 20%~30%;2012 年上海港集装箱吞吐量达到 3 253 万标准箱(TEU),位居世界第一;2016 年吞吐总量达到 3 713.3 万 TEU,超过所有欧洲港口集装箱吞吐量总和;根据《上海港总体规划》预测,2020 年上海港集装箱吞吐量将达 4 000 万 TEU[3]。

2015 年,上海港集装箱各集疏运方式中,公路集疏运占 54.8%,水路集疏运占 45.0%,铁路集疏运占 0.15%;2020 年,预测水水中转占 50%,公路集疏运占 48%,铁路集疏运占 2%。分析可知:上海港公路集疏运比例过高,方式单一;外环线和郊环线的集卡交通量较大,特别是外环线的东段和北段,以及郊环线北段,最高断面达 60 000~70 000 pcu/12;北部地区集卡流量集中,高等级道路供给不足;越江通道能力不足[4]。

上海港集装箱集疏运最主要的问题是集疏运比例不合理,公路运输比例过大[4]。首先会引发交通问题:1)公路运输比例大,现有道路不足以提供通行能力,集卡为赶船期,只能提前前往港区附近;2)港区附近停车位不足,提前到达的集卡在路边无序停靠;3)无序停靠的卡车占用路面资源,进一步降低通行能力,造成恶性循环。另外,还会引发环境问题:运输结构失衡,过渡依赖公路运输,带来严重环境污染及能源消耗,使得上海港的快速发展对"节能减排"工作造成巨大的压力。据测算,2016 年上海港集装箱集疏运的能源消费量约 10.5 亿 L 柴油,折合标准煤约 131.25 万 t,排放 CO_2 约 286.47 万 t;最新数据显示,2019 年集装箱吞吐量肯定超过 4 000 万 TEU。

3.2.1.2 国外典型集装箱运输系统

1)阿拉米达走廊(Alameda Corridor)位于美国加州南部,为专用疏港货运的地下铁路,长 32 km,南端连接圣佩罗港湾并延伸到洛杉矶—长滩港区内,北与洛杉矶市内与干线铁路交汇,并与铁路公司的编组站及集装箱枢纽场站连接。铁路线从地下穿越市区,代替了原来包含 200 多个交叉口的通道,缓解了港城交通矛盾,减轻了公路拥堵,减少了卡车和列车停留造成的废气排放,并使铁路运输时间从数小时缩短到 40 min。

2)比利时安特利普港采用地下集装箱运输系统连接港区,承担斯凯尔河(Scheldt)左右两岸集装箱运输。其专用通道运输能力为:单向管,运行速度 7 km/h,全年可处理 200 万 TEU;若运行速度提高至 15 km/h,全年可处理 400 万 TEU。

3.2.2 城市地下立体智慧物流系统

3.2.2.1 背景

诺贝尔经济奖得主斯蒂格利茨的预言正在成为现实:20 世纪信息技术的创新将在 21 世纪引发社会经济最伟大的变革。互联网技术正在改革世界,其改造的第一个传统行业是商贸流通业;通过电子商务进入销售互联网时代,引发第四次零售业革命。电子商务的革命正在循着商业的脉络接口向上下游延伸,向下游延伸的结果是带动了现代物流服务由 toB(面向企业)向 toC(面向客户)变化,新物流使制造端与客户端无缝链接;向上游延伸,正在推动工业 4.0——智慧制造(个性化和定制化)发展。

新零售与新物流的关系如下[5]:1)新零售推动城市物流需求爆发增长。单位与居民物品物流总额近 10 年增长 20 多倍,是增长最快的领域。据罗兰贝格等机构分析,同城 O2O 到 2020 年将达到 1 万亿市场规模。2)新物流是新零售的基础设施。新销售核心是新物流,没有新物流保证,新销售就是空中楼阁。

3.2.2.2 智慧物流

针对城市区域,应建设城市级的系统化、网络化的地下物流运输系统,以解决城市物流配送、垃圾运输等问题。交通需求激增带来的地面车辆、车次数量巨增是导致城市交通拥堵的主要原因,其中对货物物流的需求增长占较大份额。2014 年我国仅快递业务量已达 140 亿件,同比增长 52%;2018 年增为 505 亿件。例如北京,按常规的车辆换算系数,货运车辆占用道路资源的 40%。城市中大气污染物约 60% 来自机动车排放,机动车成为城市 PM2.5 的最大来源,交通也已成为城市噪声的最大污染源。世界经合组织在《配送:21 世纪城市货运挑战》报告中指出:货运对城市污染的占比达 40%~60%,因为 1 辆重卡污染排放量相当于 100 辆小汽车。

智慧物流具有以下特点[5]:1)智慧化平台——数据驱动,智慧布局;2)数字化运营——仓储管理系统、运输管理系统等;3)智能化作业——菜鸟全自动智慧仓储基地;4)智能化作业——菜鸟机器人。(详情扫描二维码)

智慧物流无法突破的瓶颈如下。1)受交通拥堵、交通管制和天气(雨、雪、台风)影响,配送时间和配送效率的要求无法满足;2)由于单位和居民物流总额出现爆发式增长(近10年增长20多倍),同城O2O到2020年将达到1万亿市场规模,城市地面配送使交通拥堵加剧,使已经进入"拥堵时代"的中国城市无法承受;3)在尾气污染已成为城市空气污染元凶的今天,物流的地面货车与摩托车配送对城市雾霾的加剧将使社会无法承受。陈吉宁市长说:"1辆超标重型柴油车相当于200辆小汽车的排放"(2017年1月6日大气污染防治媒体见面会);4)当前最后1 km的快递和外卖主要是摩托车配送,其突出问题是:①由于企业、医院、居住小区等均不让快递员进入,造成门口快递车辆乱停乱放,甚至堆满了物品,社会影响很大;②摩托车配送时不遵守交通规则,违规行驶严重,影响了城市交通秩序,造成了交通事故,社会影响严重。

3.2.2.3 城市地下智慧物流配送系统

地下物流系统是指运用自动导向车为承载工具,通过大直径地下管道、隧道等运输通路,对固体货物实行运输及分拣配送的一种全新概念物流系统。

在城市,物流配送中心与地下物流系统枢纽相结合,或者地下物流系统的物流配送中心和大型零售企业结合在一起,实现网络相互衔接。客户在网上下订单以后,物流中心接到订单,在地下进行客户货物的专业仓储、分拣、加工、包装、分割、组配、配送、交接、信息协同等基础作业或增值作业,通过地下管道物流智能运输系统和分拣配送系统进行运输或配送。也可以与城市商超结合,建立商超地下物流配送,在物流规模爆发式增长情况下,有利于城市"减肥"、节能、货物保存。

地下物流系统末端配送可与居民小区建筑通过运输管道相连,最终发展成一个连接城市各居民楼或生活小区的地下管道物流运输网络,并达到高度智能化。当这一地下物流系统建成后,人们购买任何商品都只需点一下鼠标,所购商品就像自来水一样通过地下管道很快地"流入"家中(或小区的自动提货柜)。

3.2.2.4 地下物流配送系统案例

1)荷兰阿姆斯特丹地下物流系统(OLS-ASH)。荷兰是世界上最大的花卉供应市场,往返在机场和花卉市场的货物供应与配送完全依靠公路,对于一些对时间性要求很高的货物(如空运货物、鲜花、水果等),拥挤的公路交通将是巨大的威胁,供应和配送的滞期会严重影响货物的质量。荷兰1997年规划了连接阿姆斯特丹Schiphol机场、Aalsmeer花卉市场和Hoofddorp铁路中转站的地下物流系统,以缓解鲜花市场与机场铁路中转站往返交通的拥堵。然而,由于政策支持度不足,且因需要巨额投资而难以找到适用的商业融资模式,该项目在2002年不幸搁置。

2)瑞士地下货运系统(CST)。瑞士地区物流存在以下瓶颈:①每日交通拥堵费高达12亿瑞士法郎;②客运和货运车辆在公路和铁路网络上争夺空间;③交通运输网络低效,空载率高。这些问题导致货运的高成本。为解决这些问题,瑞士规划了CST系统(cargo sous terrain,货物地下系统),长80 km,直接连接10个主要物流中心,见图9,橙色部分为现建成试验段。该系统的优势为:①由零售商、邮政、瑞士铁路等共同资助,独立运输设施,全天全自动高效运输,不受地上法规限制;②可与公路、水路、铁路和航空网络高效连接;③采用小型运输单元,直接从生产地点向销售网点交货,无需大面积物流中心;④货物在城市郊区进行集散,降低空载率;⑤中途可搜集废弃物,环境友好,可持续。

图9 瑞士地下货运系统(CST)
Fig. 9 Swiss underground freight system(CST)

3.3 城市绿色污水、雨洪蓄排系统(详情扫描二维码)

3.3.1 雨污分流制城市深层隧道排水系统——新加坡污水输送、集中处理的深层排水隧道

新加坡总面积仅为714.3 km²,年平均降水量为2 400 mm左右。新加坡是全球少数几个采用雨污分流系统的国家之一,新加坡深层隧道能够将整个城市收集的污水输送至处理厂集中处理排放。

3.3.2 雨污合流制城市深层隧道排水系统

3.3.2.1 东京深层排水工程

该工程全长6.3 km,下水道直径约10 m,埋设深度为地下60~100 m,由地下隧道、5座巨型竖井、调压水槽、排水泵房和中控室组成,将东京都18号水路、中川、仓松川、幸松川、大落古利根川与江户川串联在一起,用于超标准暴雨情况下流域内洪水的调蓄和引流排放,调蓄量约为67万 m³,最大排洪量可达200 m³/s,见图10。

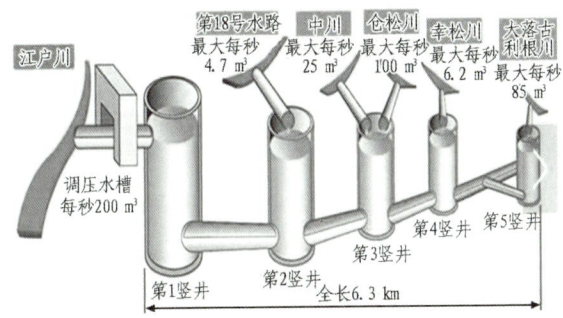

图 10 东京深层排水工程

Fig. 10 Tokyo deep drainage project

3.3.2.2 芝加哥隧道和水库方案

芝加哥年平均降雨量为 910 mm，大部分降雨以强烈夏季暴雨形式发生，每年暴雨约有 100 次，合流制污水最终溢流至芝加哥地区水源地密歇根湖。

1972 年，芝加哥市采取隧道和水库方案（TARP），由 4 个独立隧道和隧道下游 3 座大型水库组成。虽然还没有完全竣工，TARP 已经显著改善了芝加哥河等水道的水质，河岸可供垂钓的鱼种回到之前水平，并带动了旅游业的发展。

3.3.2.3 吉隆坡城市泄洪与公路两用隧道

吉隆坡城市泄洪与公路两用隧道见图 11。在暴雨情况下，吉隆坡市的城市快速路充当吸纳雨洪和排泄的通道，已成功解决雨涝灾害问题。

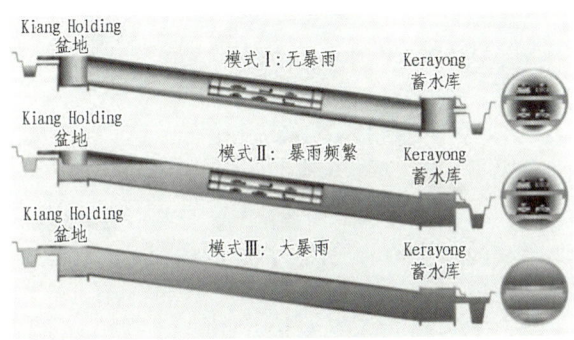

图 11 吉隆坡城市泄洪与公路两用隧道

Fig. 11 Kuala Lumpur urban flood discharge and road dual-use tunnel

3.3.2.4 我国排水系统建设情况

我国上海、广州和武汉正在建设深隧雨洪排水系统。

3.4 城市的绿色垃圾集运和处理系统、智慧地下综合管廊

垃圾通过地下运输，可以做到绿色集运和处理。瑞典是在 20 世纪 60 年代建设并实施压缩空气吹运垃圾系统（PWT）的国家，PWT 系统建在 1 700 人口的居民区内，该系统与收集、处理系统配套，投资在 3~4 年内得到回报。其他案例如设在香港岛西端大维山岩硐中的 RTS 垃圾处理系统。

发展地下垃圾处理系统节约了城市宝贵的土地资源，能得到很大的回报，同时控制了对环境的噪声和恶臭污染，减少了对环境的视觉负面影响。

智能管网管理系统是通过管线系统对水、气、汽压力（传感器）、流量（传感器）、开关量等进行测量、数据通信，实现对供给过程的遥测，再通过反分析遥测数据，对管线有无泄漏及其地点进行报警，及时关闭阀门，实现系统的无人化控制

3.5 城市的智慧行车系统

目前，我国已有许多城市、居民小区建成了智慧行车系统，做到了紧密化和功能变化。如乌镇的智能停车系统、南通的智能地下立体车库（见图 12）。

图 12 江苏南通智能地下立体车库

Fig. 12 Intelligent underground garage of Nantong, Jiangsu

4 地下空间开发规划

4.1 规划的重要性

地下空间是不可逆的空间资源。中共中央、国务院办公厅、住建部文件指出"节约优先、保护优先、自然恢复为主"的方针以及"先规划、后建设"、"规划引领"、"规划先行"的原则，强调了地下空间开发规划的重要性。

4.2 规划科学的重要性

习近平总书记提出："规划科学是最大的效益，规划失误是最大的浪费，规划折腾是最大的忌讳。"说明了规划科学的重要性。

4.3 地下空间开发与规划中的问题

1）城市地上与地下空间规划与开发统筹不足、相互影响，缺少系统性与前瞻性的协同规划。

2）不同功能的地下专项规划彼此缺乏协调，地下占位冲突，地区地下空间开发没有统一规划。

3）地下空间短期需求造成地下空间分布碎片化，导致开发不系统、空间资源浪费严重。

4.4 地下空间规划的具体要求

为了科学系统地编制好地下空间规划,应认真落实"世界眼光、国家标准、中国特色、高点定位"、"以创造历史、追求艺术的精神,把每一寸土地都规划得清清楚楚",并做到以下几点:

1)地下工程地质、水文地质情况要明;2)地下空间开发目标、任务、现状与问题要清;3)地上地下空间规划要统筹,形成一体化,从而发挥地下空间潜力,促进地上地下互动;4)涉及不同功能的多项地下空间规划要相互衔接、协调、尽力促进;5)"成片统一规划、统一设计、统一施工、统一管理"与"多规合一";6)地下空间横向要互联互通,竖向分层立体开发,安排要合理,时序要科学;7)近期建设与中远期发展要预留接口,互相衔接,形成逐步推进格局。

4.5 参考实例

4.5.1 南京高铁南站开发利用地下空间

铁路设计部门以高铁南站地质条件复杂为由,只设计了站房下面负一层地下室。南京地调中心根据地调结果认为高铁南站区域基岩埋深较浅,结构完整,未发现断裂,完全有条件进行地下空间开发,最后说服设计变更优化,见图13。产生的效益为:增加了地面可利用空间约17 hm², 8亿元经济效益;增加了地下可利用空间约4 hm²,若按照20万元/hm²(300万元/亩)的地价简单估算,就是1亿多元的经济效益。

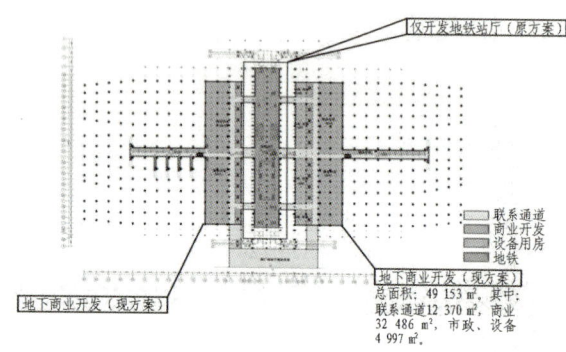

图 13 南京火车南站负一层利用方案
Fig. 13 Utilization plan of B1 floor of Nanjing South Railway Station

4.5.2 基于历史文物保护的地下规划交通设施建设(意大利、希腊)

以保护性开发代替被动保护,主要分为以下2种。
1)空间整合:围绕站点开发城市地下空间,链接地上丰富的历史资源,疏解人流,盘活历史空间,如罗马地铁 Metro Line C,共6个地铁口,分别连接地面历史建筑(见图14)。2)主题展示:结合地上地下历史遗址,围绕历史主题,展现历史魅力,以开发性保护代替被动保护。

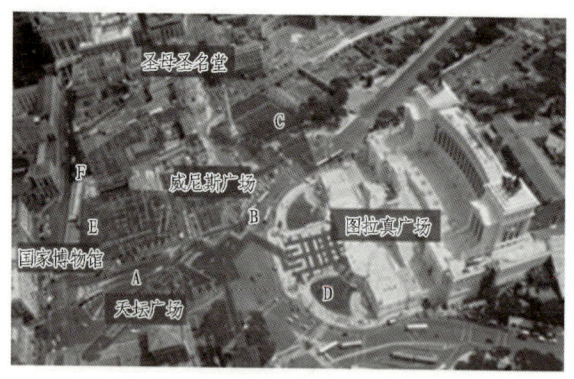

图 14 罗马地铁 Metro Line C 示意图
Fig. 14 Sketch of Metro Line C of Rome Metro

4.5.3 功能设施的复合化与空间立体分层利用(新加坡、日本、香港)

1)功能设施复合化:围绕轨道站点的功能设施布局,注重设施之间及地上地下之间的连通,见图15。

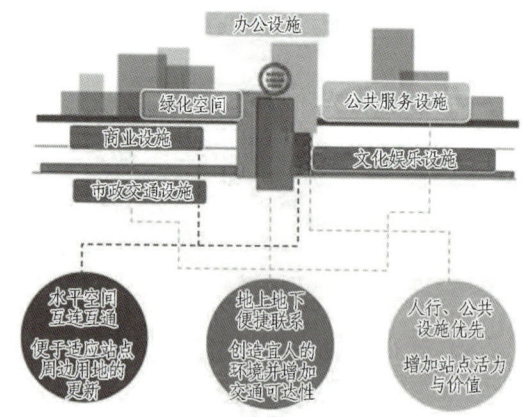

图 15 功能设施复合化
Fig. 15 Compound of functional facilities

2)空间立体分层:商业、交通、公共服务设施、市政、管廊的分层布局,浅层空间优先用于人行空间,见图16。

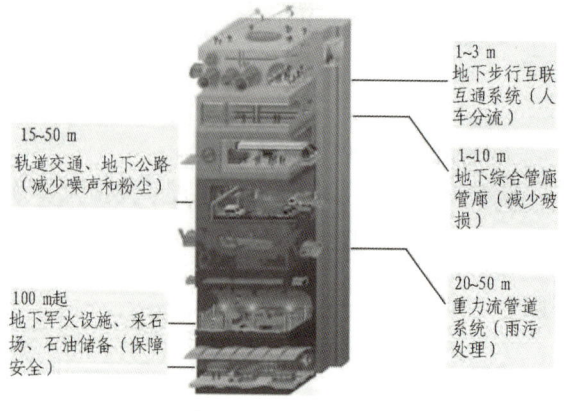

图 16 空间立体分层
Fig. 16 Spatial stratification

4.5.4 基于生态地质评价的地下空间可持续利用(赫尔辛基)

1) 规划思路:地下空间多层次预留,保障地下空间的可持续开发,见图17。

2) 竖向挖掘:预留清晰,分层确权,合理安排地下贮存、市政、管廊设施,见图18。

图17 芬兰赫尔辛基地下空间规划
Fig. 17 Underground space planning of Helsinki, Finland

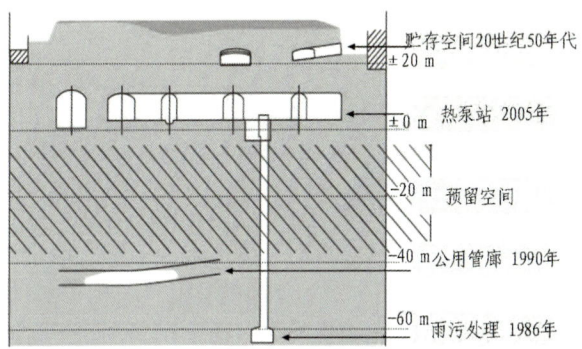

图18 竖向挖掘
Fig. 18 Vertical digging

5 结语

通过若干个五年计划持之以恒的努力,将地上城市中的城市运输与交通系统逐步转移到城市地下去,不但使旅客交通与货物运输转入地下,还包括垃圾、污水等的传输都转入地下,使地上逐步乃至彻底摆脱与传统运输相联系的环境干扰。首先是解决交通拥堵与PM2.5超标的困扰,而释放出的地上空间将用作大片的自然植被和安全的步行,这不是科学幻想,《欧洲地下建设战略研究议程(Strategic Research Agenda For the European Underground Construction Sector)》就已经明确提出,其2030年的远景目标是"将地面交由市民自由使用,将基础设施放入地下"。

参考文献(References):

[1] 绿色建筑评价标准: GB/T 50378—2019 [S]. 北京: 中国建筑工业出版社, 2019.
Assessment standard for green building: GB/T 50378-2019 [S]. Beijing: China Architecture & Building Press, 2019.

[2] 钱七虎, 陈晓强. 充分开发利用地下空间建设资源节约型和环境友好型城市 [M] //钱七虎院士论文选集. 北京: 科学出版社, 2007.
QIAN Qihu, CHEN Xiaoqiang. Make full use of underground space to build resource-saving and environment-friendly cities [M] //Selections from Academician Qian Qihu's Theses. Beijing: Science Press, 2007.

[3] 钱七虎. 科学利用城市地下空间,建设和谐宜居、美丽城市 [J]. 隧道与地下工程灾害防治, 2019, 1(1): 1.
QIAN Qihu. Scientific use of the urban underground space to construction the harmonious livable and beautiful city [J]. Hazard Control in Tunnelling and Underground Engineering, 2019, 1(1): 1.

[4] 范益群, 游克思. 地下集装箱物流系统在港城融合发展中的应用 [J]. 地下空间与工程学报, 2018, 14(增刊1): 49.
FAN Yiqun, YOU Kesi. Application of underground container freight transportation on integration development of city and port [J]. Chinese Journal of Underground Space and Engineering, 2018, 14(S1): 49.

[5] 俞明健. 发展地下物流 打造智慧配送 [J]. 城乡建设, 2019(1): 24.
YU Mingjian. Develop underground logistics and build intelligent distribution [J]. Urban and Rural Development, 2019(1): 24.

科学利用城市地下空间,建设和谐宜居、美丽城市

钱七虎

(解放军陆军工程大学国防工程学院,江苏 南京 210007)

摘要:城市建设发展加速,"城市病"日益突出,地下空间开发利用愈显重要。通过转变城市发展方式,科学规划地下空间开发利用,汲取国际成功经验,注重品质发展,可以实现地下空间开发的多功能利用,实现统筹缓解城市灾害、交通拥堵、空气污染、城市内涝等"城市病"。同时,以地下综合管廊建设和海绵城市建设为契机,及时解决当前存在的法规体系建设和管理体制改革问题,实现人与自然和谐共生,推进城市生态文明建设迈向新高度,推动美丽城市建设。

关键词:城市地下空间;城市发展方式;地下空间灾害;地下基础设施;地下物流;建设法规体系
中图分类号:U45   **文献标志码**:A

引用格式:钱七虎. 科学利用城市地下空间,建设和谐宜居、美丽城市[J]. 隧道与地下工程灾害防治,2019,1(1):1-7. QIAN Qihu. Scientific use of the urban underground space to construction the harmonious livable and beautiful city [J]. Hazard Control in Tunnelling and Underground Engineering, 2019, 1(1):1-7.

Scientific use of the urban underground space to construction the harmonious livable and beautiful city

QIAN Qihu

(College of Defense Engineering, PLA University of Science & Technology, Nanjing 210007, Jiangsu, China)

Abstract: With the development of urban accelerated construction, "urban disease" is becoming more and more prominent, and the development and utilization of underground space is more and more important. By transforming the mode of urban development, scientifically planning the development and utilization of underground space, drawing on international successful experience and paying attention to the development of quality, we can realize the multifunctional utilization of underground space development, realize the "urban diseases" such as urban disaster, traffic congestion, air pollution and urban inland inundation, and take the construction of underground pipe gallery and sponge city construction as an opportunity to resolve the existing problems in the construction of legal system and the reform of management system in a timely manner, realize the harmonious coexistence of man and nature, promote the construction of urban ecological civilization towards a new height, and promote the construction of beautiful cities.
Key words: urban underground space; urban development mode; underground space disaster; underground infrastructure; underground logistics; construction law system

---

收稿日期:2019-01-09。
本文原载于《隧道与地下工程灾害防治》(2019年第1卷第1期)。

0 引言

随着我国城市化进程的快速推进,我国城市建设有了长足发展。但是,以交通拥堵、空气污染和城市内涝为典型特征的"城市病"也日益严重,受到全民关注,已经成为制约环境友好型城市建设的不和谐因素。新时代,我们要坚定不移地建设资源节约型、环境友好型社会,走可持续发展之路,秉持生态文明建设和绿色发展,建设人与人、人与自然和谐发展的社会[1]。在城市建设发展上,需要汲取国内外城市地下空间开发的先进经验,结合目前我国城市的发展情况,实现地下空间多功能利用,发挥地下空间利用在缓解"城市病"中的突出作用,并建立健全城市地下空间建设开发的法律法规与相关配套政策,实现以地下空间开发利用带动城市发展方式的转变。

1 科学规划城市地下空间利用

习总书记指出:城市规划在城市发展中起着重要引领作用,考察一个城市首先看规划,规划科学是最大的效益,规划失误是最大的浪费,规划折腾是最大的忌讳。在城市发展上,必须明确"先规划后建设,先地下后地上",实行"规划引领,规划先行",必须增强规划的科学性、前瞻性、综合性,强化规划的引领与协同促进作用,扎实推进"多规合一"。

1.1 科学系统地编制好城市地下空间开发利用规划

当前我国地下空间建设正处于快速发展阶段,尤其是地下综合管廊建设以及海绵城市建设迅猛发展[2],在全国如火如荼的展开,建设规模和发展速度引领世界,成绩显著。城市地下空间资源有限、宝贵且不可逆,应科学系统地开发利用,规划要坚持高起点、高目标、高瞻远瞩,才能事半功倍,如果急功近利、头疼医头、脚疼医脚,则会事倍功半,要坚决杜绝失误规划与反复规划。要发扬"工匠"精神,精心推进,不留历史遗憾,编制规划时"好"字当头、潜心研究、精心编制,认真规划每一寸土地,不贪功冒进,不贪求建设进度。综合考虑城市总体规划与地下空间开发利用各有关专项建设规划[3],合理确定地下综合管廊建设布局、选择管线种类等,统筹城市建设和发展远景,预留和控制管线扩容空间。

针对开发利用城市地下空间的规划编制,主要应着重解决以下问题:明确开发区域的地质与水文情况;明确区域地下空间开发利用的现状;统筹协调地下、地上空间规划以及涉及地下、地上的各项规划[4],尽力促进"多规合一";横向连通地下空间,合理安排竖向分层立体开发,保持科学时序;互相衔接近期建设与中远期发展规划,形成稳定、逐步推进的格局。

1.2 地下空间建设要提高品质,质量第一,注重"绣花"功夫

在2018年中国城市规划年会上,住建部指出,城市规划建设应突出地方特色和改造人居环境[5-6],注重城市规划和建设的"绣花"功夫,逐步摒弃杜绝急功近利和大拆大建。以往的城市建设,由于粗放式发展,造成了城市宜居性差、人居环境质量品质较低。新时代城市建设应更加注重居民体验,以高品质、高质量标准要求建设城乡人居环境。我们需要探讨更加细致、全面的方法,秉持"绣花"态度和功夫,深入了解社区驻区居民的切实需求与体验后,综合规划、设计和建设城乡的宜居环境。同时,转变过去城市建设的拆迁模式,统筹做好城市建设的更新模式,为城市留住历史沉淀、积累时代变迁的足迹。

在城市规划建设中还应征集参考广大市民的意向,让居民一定程度上介入建设规划的过程,提高市民对建设规划的知悉程度,增加居民的归属感。在城市开发建设中,要以创造优良宜居环境作为核心目标,努力把城市建设成为人与人、人与自然和谐相处的美丽家园。

2 以"世界眼光,国际标准"实现地下空间多功能利用

地下空间的开发,尤其是城市地下空间的利用与城市的经济发展水平密切相关,地下空间的开发是随着城市的发展而逐步建设发展的。我国的地下空间开发利用起步较晚,近年来,逐步为广大人民所认知和熟悉。以前传统的地下空间利用方式主要有地下室及地下人防工程,至20世纪八九十年代,地铁、地下商场与地下停车库快速发展,功能设施逐渐齐备。20世纪末,地下综合体建设开始兴起,地上地下空间布局与开发利用朝着一体化方向发展。

随着高新技术与施工技术的发展,地下空间的利用方式有了新的拓展,在节水、节能以及地下综合体方面有了进一步的深入发展。如国外许多城市在地下空间深部地层建设地下"水银行",实现水资源调节利用。新加坡建设了污水输送、集中处理的深层排水隧道。芬兰赫尔辛基建设了地下污水处理系统。地下空间还可用来建设地下油气储库,建设高危、高辐射医疗废弃物、核废料储库等。

此外,利用地热能的地源热泵近年亦发展迅速。当到达地下一定深度时(<5 m),四季的地层温度基本保持稳定,地下换热系统于地下直埋冷凝器或蒸发器,利用传热循环介质与大地进行热交换,从而汲取地温能[3]。该系统可埋设在地下结构基坑围护结构、基础底板下、桩基内,也可埋设在隧道衬砌内或以能源锚杆的形式布设在隧道围岩中,或埋设在地铁区间隧道内、地下输水管道内。上海世博轴采用江水源热泵系统和桩基埋管地源热泵系统,有效减轻了城市"热岛效应"。国内外已建设多处典型地下换热系统建筑,如奥地利 Lainzer 隧道 LT24 区车站,瑞士 Grabs 的 PAGO 公司大楼,上海自然博物馆新馆。同时,地源热泵可进一步与太阳能结合,由太阳能供能辅热,实现系统与地下的热能交换,从而保持地层热能交换稳定。地热能与太阳能的持续交换,既可克服单独使用地源热泵时,土壤温度场向热传导难以及时有效恢复的局限性,又可摆脱单独使用太阳能系统时太阳能供热交换受天气气候因素制约。

3 城市地下空间建设面临的典型灾害

城市地下空间的开发利用已成为国内外城市建设的关注热点,飞速发展的科学技术与快速完善的施工建设技术为城市地下空间建设提供了不竭的动力,城市地下空间开发利用的方式与建设经验日渐丰富。但在城市地下空间开发利用的过程中,仍面临着一些问题,如地面塌陷、涌水涌砂、管线爆裂、火灾、水灾等。

3.1 地面塌陷

人类活动地区的地表岩、土体由于自然或人为因素作用,下凹陷落,在地面形成塌陷空洞的地质现象,即地面塌陷。常见的地面塌陷可以两大类:岩溶塌陷与非岩溶性塌陷。岩溶塌陷是由于可溶岩中存在的岩溶洞隙而产生,多见"土层塌陷"与"基岩塌陷",在我国 26 个省(区)都有发生。非岩溶性塌陷常见的有采空区塌陷、黄土地区的黄土陷穴塌陷、玄武岩地区通道顶板塌陷。上述塌陷中,以岩溶塌陷分布最广、数量最多、发生频率高、诱发因素最多,且具有较强的隐蔽性和突发性特点,严地威胁到人民群众的生命财产安全。常见的地面塌陷的平面形态有圆形、椭圆形、长条形和不规则形。虽然地面塌陷具有随机、突发的特点,我们仍可采用以下预防措施以避免或减少灾害的损失:减少地表水下渗、预留保护矿柱、科学合理采矿、加强勘察、防治结合。

3.2 涌水涌砂

涌水又称突水,在地下硐室、隧道施工过程中,当穿越岩溶发育地段、临近地下暗河系统,厚层含水砂层以及与地表水连通的断裂破碎带等时,发生短时间内大体量涌水现象,有时伴随流砂涌出现象,称为涌砂。涌水涌砂严重影响城市地下工程安全建设。为有效防治地下空间建设中的涌水涌砂现象,应合理采用排水措施,增设降水排水设施与工序,降低地下水位,对于涌水量较大的溶洞、断层破碎带或砂砾层可采取注浆封堵措施,以封堵地下水的渗流通道。

3.3 管线爆裂

城市地下空间的开发利用过程中,由于管线使用维护不当或工程施工影响,会遭遇各类功能管线爆裂、破裂现象,如供热管道、供水管道、燃气管道以及输油管道爆裂等,对于管线安全、市民正常生活、市政交通带来严重不良影响。常见的管线爆裂的起因有:(1)地质气候条件,如北方寒冷温湿气候交替,表土层的季节性冻土,土层的膨胀反复作用,致使管线错位、侧向位移从而导致管线爆裂发生,管线埋深过浅,产生温度应力致使爆管;(2)管线材质与加工工艺,管线材质以及铸管工艺带来了管线服役性能的优劣,重载、高压、震动、冲击、不均匀沉降的多重环境影响下,造成管线破裂;(3)管线的老龄化、腐蚀,管线的过龄服役,管线运输物与布设环境的腐蚀交互作用,致使管线爆裂。为有效防控城市地下空间管线爆裂灾害,应加强管线(网)的严格监控,设计、施工、选材、质量检验环环把关,建立完善的管线爆裂应急预案,快速高效反应突发事件,也可培养专业的管网技术人员与队伍,提升应对管线爆裂灾害的综合实力。

3.4 其他灾害

管线的失修维护、燃气的泄露以及电力设施的维护不当,使得地下管线、管道面临火灾危害,由于管线的集中敷设,一处火灾往往带来周围管线(网)的严重灾害事故。城市地下空间也会面临雨洪倒灌,管廊严重渗漏水现象,严重威胁管线运营以及人民生命安全。城市地下空间建设应采取必要的工程措施降低失火诱因以及起火、明火时控制火势延燃,设置必要的灭火系统与设施,加强管线的监控排查,设置地面防洪、防汛设施,维护管线以及地下空间的安全运营。

针对近年来城市地下空间开发利用的典型灾害案例,笔者对国内一些典型灾害事故在表 1 中进行了简单梳理,供参考。

表 1 城市地下空间常见灾害
Table 1 Common disasters in urban underground space

灾害类型	典型案例
管线爆裂	2018年7月,深圳地铁在施工时挖断7根电缆,致使大片受影响区域停电;同月施工过程中挖破地下供水主管道,裂口直径约800 mm,致使沿线用户停水 2014年4年,兰州一原油管道发生原油泄漏,致使供水企业自流沟受到污染,自来水苯超标,引起了当地市民抢购矿泉水 2013年11月,青岛黄岛区输油管道破裂原油泄漏流入市政暗渠,在形成密闭空间的暗渠内油气积聚遇火花发生爆炸,造成62人死亡、136人受伤,直接经济损失75 172万元 2013年8月,哈尔滨市连续几场强降雨造成土质沉降,老旧排污管线断裂,泥沙灌入人防工程洞体,最终导致地面塌陷,结果4人落入深坑,两死、两伤 2010年7月,南京市栖霞区一拆除工地发生地下丙烯管道泄漏爆燃事故,共造成22人死亡,120人受伤住院治疗,其中14人重伤,直接经济损失4 784万元 2004年6月,重庆江津市滨江路西段地下污水管网发生爆炸,约100 m的路面被炸得稀烂,一名男子全身重度烧伤
地面塌陷	2018年10月,四川省达州市一处人行道路面塌陷,事故造成2人死亡,2人失踪 2018年8月,南昌市红谷滩新区金融大街由北往南方向、世贸路口附近路面发生大面积塌陷,导致两股车道和人行道全部下陷,旁边工地的施工围墙倒塌 2018年2月,佛山地铁2号线工地突发透水,现场透水面积扩大,导致隧道管片变形及破损,引发地面季华西路三十多米路段坍塌 2013年1月,广州市荔湾区地铁施工工地旁发生地陷后已陆续出现六次坍塌,有2栋共6间商铺坍塌,塌陷面积约690 m^2 2012年1月,湖南省益阳市岳家桥镇发生大面积地面塌陷,因地面岩溶塌陷引起房屋开裂有167户。当地河水也被截流,流入塌陷形成的深坑之中 2012年7月,长沙市紧靠湘江大堤的湘江路中段突然发生塌陷,塌陷坑坑口面积约有30 m^2,一台从此路过的轿车被巨坑吞没,致1人死亡 2009年4月,位于成都市一环路西三段路口一段近30 m的河堤突然塌陷,深达1 m,辆停在河堤上的汽车当场被"吞没" 2008年11月,杭州风情大道地铁施工工地突然发生大面积地面塌陷,正在路面行驶的多辆汽车陷入深坑,多名施工人员被困地下。事故造成了风情大道路面坍塌75 m,下陷15 m,死亡17人,失踪4人,受伤24人
涌水涌砂	2015年1月,西安地铁4号线试验段,隧道洞门出现涌水、涌沙险情,造成地面出现地陷,地下自来水管道、市政排污管道受到不同程度影响 2014年7月,武汉地铁4号线二期工程复兴路站施工时,地下隧道出现涌水涌砂险情,附近人民医院大楼出现开裂现象 2012年11月,南宁地铁施工引发涌水涌沙致使地面陷落,公交车车头陷入塌陷形成的大坑中
火灾	2017年12月,海口市某地下车库突发大火,火灾造成现场大约6辆车受损 2017年2月,厦门轨道2号线海东区间右线盾构现场,在减压舱减压过程中发生起火,导致3人死亡 2016年4月,上海虹桥机场一号航站楼改建工地A楼地下空间发生火灾,火灾造成2人死亡,多人受伤 2016年3月,绵阳金柱圆地下商场起火,导致下面全部近二十家商铺起火 2014年12月,北京地铁7号线广渠门内站工地突然起火,现场冒出大量黑烟,一度造成"浓烟封路"景象
水灾	2017年6月,深圳市暴雨突袭,道路积水严重,部分地铁站点遭水淹,严重影响居民出行 2013年7月,武汉市某小区城下车库被淹,多辆车子被泡在地下车库深水之下,给业主造成重大财产损失 2012年7月,北京房山大雨突袭,某小区地下车库内100多辆轿车惨遭水淹 2010年6月,重庆大渡口区翠园路一地下商场因遭倒灌成"泽国",整体停业 2010年5月,广州暴雨水淹地铁二号线,部分路段停运6 h

4 开发利用城市地下空间,着力解决"城市病"

以往城市建设基本采用注重"面子"建设的粗放发展模式,如修建高层建筑、大广场、宽马路等,酿成了"里子"城市地下基础设施建设的短板,导致了城市病丛生——诸如雾霾严重,交通拥堵,城市内涝,垃圾围城,城市水系、土壤严重污染等。

城市地下空间的开发利用,应以地下综合管廊、海绵城市为主要契机,注重城市"里子"建设,采用集约、绿色可持续发展模式建设地下综合管廊;建设集渗、滞、蓄、净、用、排六种功能于一体的海绵城市，进行雨洪地下储蓄和排洪，建设地下水银行蓄排水；建设地下污水处理厂进行污水地下集运；垃圾地下集运卫生填埋和焚烧处理；雨水利用、再生水利用；发展大运量快速公交系统，地下快速路和物流系统；建立地源热泵系统，利用地下低密度能源[7-8]。通过以上主要途径，解决交通拥堵、空气雾霾、城市内涝等制约城市发展的问题。

4.1 城市地下空间具有防灾减灾的优越性

由于地下结构的固有属性，城市地下空间相比地上空间具有不可比拟的防御优势。由于地下空间受到上覆盖层的防护，能够更好地抵御风、雪等自然灾害。同时地下结构的抗震性能较地上结构更为有效，这对防御恐怖袭击以及地震灾害等具有极大的天然防护能力，极大地提高了城市基础设施的生存能力。

4.2 地下空间开发利用有利于缓解交通拥堵

国外城市建设发展的经历表明，仅大力发展轨道交通也不能从根本上解决交通拥堵问题。21世纪初，众多发达国家的特大城市为解决交通拥堵问题，逐渐开始尝试建设"地下快速路"和"地下物流系统"，如表2所示。

表2 国际典型建设地下快速路与地下物流系统城市
Table 2 International typical constructions of underground expressway and underground logistics system cities

城市名称	建设概况	效果
美国波士顿中央大道	六车道高架路转8~10车道地下高速路，地面变成林荫路和街心公园	空气CO浓度降低12%，市中栽种绿植，公园和开敞空间增加
日本东京中央环状新宿线	双向4车道，长11 km，连贯池袋、新宿、涩谷3大商业中心，6个出口，设除尘换气装置	节约交通时间60%，改善环境，减少160万t NO，降低成本，低于地面成本投资约20%
日本东京外环品川地区联合开发	原高架路方案改建为地下40 m深高速路，隧道内设过滤和分解设施	相比原方案，节约投资成本20%~30%，工期由15 a缩短为8 a
新加坡地下道路系统	长15 km，环形四车道，承担城市中心区交通量的40%	造价48亿美元
美国港城地下物流——阿拉米达走廊	洛杉矶—长滩港，专用疏港货运地下铁路，长32 km，从地下穿越市区	地下走廊代替200多个交叉口的通道，缓解港城交通矛盾，减轻公路拥堵，大幅缩短运输时间
瑞士地下货运系统	长80 km，直接连接10个主要物流中心	全天全自动高效运输，小型运输单元，无需大面积物流中心，货物集散，空载率低

国内部分大城市为缓解城市交通拥堵问题，也逐渐开始建设城市地下快速路，并取得了较为明显的效果。南京市修建了城东干道地下线和内环线地下路、玄武湖地下快速路；上海已在建井字形地下通道方案；深圳建设了港深通道地下快速路，并在前海构建"地下快速路+地上环路+车库"的三级地下车行系统。

为有效应对日益增长的物流规模，缓解城市物流配送运输压力，为城市节能、"减肥"，可将物流配送中心与城市地下物流系统枢纽相结合，在地下保存货物。地下物流系统末端配送可以与居民生活区建筑运输管道相连，最终发展成为高度智能化的连接居民生活区的地下管道物流运输网络。

地下快速路和地下物流系统建设对治理城市交通拥堵、提升空气质量、防治城市空气雾霾问题，增加城市地表绿植覆盖，节省城市土地资源具有显著影响。同时，地下快速路建设和地下物流系统造价较轨道交通系统造价要低，建设经费可以借鉴新加坡的办法解决，即收取城市车辆牌照费以及上下班时间进入市区的执照费。地下快速路的天然优势是全天候，在暴风、雨雪和大雾等最易造成地面交通拥堵的情况下最能发挥作用。这在我国城市建设用地严格控制和大城市人均道路指标普遍偏低的情况下，非常适合我国地少人多的国情。

4.3 地下空间开发利用有利于缓解空气污染

国内外建设实践表明，城市地下快速路系统以及地下物流系统的建设在解决交通拥堵的同时，由于地下快速路系统可以集中收集处理汽车尾气，可以为消除汽车尾气，缓解城市空气污染作出贡献[9]。如波士顿市区修建了地下快速路后，市区一氧化碳浓度降低了12%；东京修建新宿线，每年减少了2.5万t二氧化碳、160 t氮氧化物等，大大改善提升了整体城市视觉面貌[10]。

发展地下垃圾处理系统既能节约城市宝贵的土地资源，带来丰厚经济收益，又可以控制环境噪声和空气恶臭污染，减少环境负面影响。东京临海副都心，将垃圾管道入廊，单独设管，真空吹送至垃圾焚烧厂，高温焚烧[11]。瑞典自1960年开始，建设了压缩空气吹运垃圾系统,该系统建立在1 700人口的居民区内,垃圾吹运系统配套收集、处理系统,最终将垃圾焚烧发电,投资可在3~4 a内得到回报,现已在瑞典全国推广。

以建设城市地下综合管廊为契机,可实现污水入廊集中地下处理再利用,解决地面污水对空气的污染以及城市视觉面貌影响问题[5,10];垃圾入廊,集中卫生填埋或高温焚烧,使其不再污染城市土壤、水源。

通过上述措施多管齐下,可以大大缓解城市空气污染。

4.4 地下空间开发利用有利于缓解城市内涝

雨水是宝贵的淡水资源,而雨洪却是威胁城市安全的不和谐因素。应着力建设地下调蓄措施,对雨洪资源进行储存利用,同时防治城市内涝。首先,要对城市管道实行雨污分流,污水入综合管廊,雨洪不入管廊,单成独立系统;其次,应尽量在地下建设污水集运处理设施与系统以代替地面修建污水处理厂,使原污水厂占地用于其他开发,同时节省地面污水厂邻近土地资源;再次,还可以结合交通隧道建设排洪通道,如马来西亚吉隆坡建设了城市泄洪与公路两用隧道(SMART, Stormwater Management And Road Tunnel 隧道),将城市快速路充作吸纳雨洪和排泄的通道,已成功解决城市雨涝之害。

国际上,部分发达国家采用雨洪地表入渗和井灌的人工补给方式,已成功于地下含水层建立地下"水银行"[2],有效地调节和缓解了供水紧张局面。以地下"水银行"储备地下水,可调节水资源短缺或枯水期补给用水需求,控制海水入侵地下含水层和地面沉降,保持并提高地下水位,减少抽采地下水费用,还能维持河流的基流,并通过土壤多层分解、吸附、过滤等物化作用改善水质,实现污水的循环利用,一举多得,为保护生态环境提供重要保障。

5 地下空间利用要法规先行,完善体制

5.1 建立健全城市地下空间法律法规

当前我国尚无完备的城市地下空间法规,亟待健全,现行法律及规章中缺乏地下空间权的概念和相关规定,且多为地方或部门规章[12-44]。为完善我国地下空间法规体系,我们可以汲取美国各州制定的空间法、瑞士的《瑞士民法典》、日本的《共同沟法》、德国的《城市建设法典》、意大利的《民法典》、中国台湾地区的《共同管道法》等法律法规中的有利经验[2],逐步完善我国城市地下空间开发利用的法律法规,明确所有权、规划权、建设权、管理权、经营权、使用权以及有偿使用费的收取原则等。制定体系完备、可操作性强、具有权威性的法律体系与规章制度[15-46]。

5.2 建立城市地下空间综合管理体制

地下空间开发利用涉及多行业、多部门单位的协同合作。过去由于市政体制的分头管理以及部门条块分割严重,我国城市地下空间开发利用时问题多发,综合管理水平相对低下,因此开发利用城市地下空间应着力推进城市管理机构改革、创新城市工作体制机制,理顺综合管理体制机制。目前,许多城市虽进行了对地下空间开发利用管理的有益探索,如规划管理、建设管理、人防管理、安全管理和协调管理等管理模式,但仍未起到统一管理的作用,必须对管理体制和管理措施进行系统规划、健全与完善。

6 建设地下综合管廊的几点思考

地下综合管廊建设是补足城市基础设施建设短板的重要内容,是城市建设的良心。例如,珠海横琴综合管廊全长33.4 km,建设地下管廊总投资22亿元,节约的土地产生的直接经济效益就超过80亿元;同时,供水管网漏损率也大幅度降低。下面,针对怎么地下综合管廊的建设,提出自己的几点思考。

6.1 燃气管线入廊问题

由于燃气管线易燃易爆,是否将燃气管道纳入综合管廊一直以来争议不断。过往燃气管线事故,均是由于管线管理维护工作疏忽怠慢,以致燃气(油)泄漏引发的。通过加强管理维护和技术升级,可以及时检查、维护和实时监控综合管廊内的燃气管线,避免燃气管线泄漏和杜绝爆炸事故的发生,因此将燃气管线纳入综合管廊是可行的[17]。

6.2 雨洪排泄管道入廊问题

城市需要实行雨污分流,将污水管道纳入到综合管廊中。而雨洪排泄设施体量较大,入廊困难,不宜入廊。为根治城市内涝问题,雨洪的排泄系统应建设雨洪调蓄设施,可参考日本的修建经验建设城市地下河川,或者吉隆坡的SMART隧道经验,整合城市地下快速路系统与暴雨排洪设施。

6.3 地下管廊兼顾人防问题

根据《中华人民共和国人民防空法》第十四条规定[18],建设城市的地下交通干线及其他地下工程时,应兼顾人民防空需要。因此,管廊设计上要确保管廊战时具有一定的抗毁能力[3]。建设城市地下综合管廊时,既要提高平时管廊的防灾减灾能力,也要确保战时防空能力。但兼顾人防要到位而不越

位,更不能改变管廊的功能定位。

7 总结

在地下工程领域,我国工程建设技术水平已日臻成熟,正逐步引领世界,针对地下 100 m 以内的地下工程,我们已经基本掌握开发多层城市地下空间的技术。通过转变传统思维、改革城市开发建设规划与城市交通建设模式,融合科技创新新技术,一定能建成领先国际的城市地下交通系统,彻底摆脱城市交通拥堵和空气污染困扰,真正把中国城市建成青山绿水、交通顺畅、景色宜人的"山水城市",实现伟大的中国梦。

参考文献:

[1] 钱七虎. 建设城市地下综合营廊,转变城市发展方式[J]. 隧道建设, 2017, 37(6): 647-654.
QIAN Qihu. To transform way of urban development by constructing underground utility tunnel [J]. Tunnel Construction, 2017, 37(6): 647-654.

[2] 钱七虎. 推进城市地下些问规划建设的思考[J]. 城乡建设, 2017(18): 60-65.

[3] 钱七虎. 建设城市地下综合管廊转变城市发展方式[J]. 深圳土木&建筑, 2017(1): 6-18.

[4] 陈志龙, 刘宏. 城市地下空间总体规划[M]. 南京: 东南大学出版社, 2011.

[5] 钱七虎. 筑牢城市"里子"才能撑起"面子"[J]. 中华建设, 2018(11): 12-13.

[6] 童彤. 顺应转型用"绣花'功夫做好城市规划建设[N/OL]. 中国经济时报, 2018-11-28 [2019-01-09]. http://sh.qihoo.com/pc/972edd19dfc4e901f? cota=1&refer_scene=so_1&sign=360_e39369d1.

[7] 钱七虎. 城市地下穿越工程的新理论与新技术[J]. 城市轨道交通研究, 2018, 21(11): 7, 20.
QIAN Qihu. New theory and technology of urban underground crossing engineering [J]. Urban Mass Transit, 2018, 21(11): 7, 20.

[8] 钱七虎. 地下空间科学开发与利用[M]. 南京: 江苏科学技术出版社, 2007.

[9] 俞明健, 范益群, 胡昊. 城市地下空间低碳化设计与评估[M]. 上海: 同济大学出版社, 2015.

[10] 钱七虎. 城市交通拥堵和空气污染的治本之策[N]. 科技日报, 2014-04-21(001).

[11] 朱思诚. 东京临海副都心的地下综合管廊[J]. 中国给水排水, 2005(3): 102-103.
ZHU Sicheng. Underground pipe gallery in new Area of Tokyo [J]. China Water&Wastewater, 2005(3): 102-103.

[12] 中华人民共和国住房城乡建设部. 住房城乡建设部关于印发城市综合管廊和海绵城市建设国家建筑标准设计体系的通知: 建质函〔2016〕18号 [EB/OL]. (2016-01-22) [2019-01-09]. http://www.mohurd.gov.cn/wjfb/201602/t20160204_226594.html.

[13] 中华人民共和国国务院办公厅. 国务院办公厅关于推进城市地下综合管廊建设的指导意见: 国办发〔2015〕61号 [EB/OL]. (2015-08-10) [2019-01-09]. http://www.gov.cn/zhengce/content/2015-08/10/content_10063.htm.

[14] 中华人民共和国国务院办公厅. 国务院办公厅关于加强城市地下管线建设管理的指导意见: 国办发〔2014〕27号 [EB/OL]. (2014-06-14) [2019-01-09]. http://www.gov.cn/zhengce/content/2014-06/14/content_8883.htm.

[15] 钱七虎. 开发地下空间要深谋远虑[N]. 中国国防报, 2017-06-20(003).

[16] 顾新, 于文悫. 城市地下空间利用规划编制与管理[M]. 南京: 东南大学出版社, 2014.

[17] 朱合华, 闫治国. 城市地下空间防火与安全[M]. 上海: 同济大学出版社, 2014.

[18] 中华人民共和国政府. 中华人民共和国人民防空法[EB/OL]. (2005-08-05) [2019-01-09]. http://www.gov.cn/ziliao/flfg/2005-08/05/content_20892.htm.

中国工程院院士文集系列

钱七虎院士文集
下册

Qian Qihu Yuanshi Wenji

人民交通出版社股份有限公司
北京

内 容 提 要

本书是"中国工程院院士文集系列"之一，主要包括学术论文、学术报告、思想与观点、吾家吾国四个部分。本书汇集了钱七虎院士在防护工程与结构抗爆、岩石力学与工程、隧道及地下工程、绿色与智能建造等领域的学术研究成果，记录了院士在相关学科的前沿技术发展、重大工程建设方面的理念与观点，呈现了院士勤勉严谨治学、矢志不渝报国、不遗余力为民的精神追求。

本书可供防护工程、岩石力学与工程、隧道及地下工程等相关专业领域的技术人员学习参考。

图书在版编目(CIP)数据

钱七虎院士文集 / 钱七虎著. — 北京：人民交通出版社股份有限公司，2023.8
ISBN 978-7-114-18925-8

Ⅰ.①钱… Ⅱ.①钱… Ⅲ.①土木工程—文集 Ⅳ.①TU-53

中国国家版本馆 CIP 数据核字(2023)第 145897 号

Qian Qihu Yuanshi Wenji
书　　名：钱七虎院士文集（下册）
著 作 者：钱七虎
责任编辑：李学会　谢海龙　吴燕伶
责任校对：赵媛媛　魏佳宁　刘　璇
责任印制：张　凯
出版发行：人民交通出版社股份有限公司
地　　址：(100011)北京市朝阳区安定门外外馆斜街 3 号
网　　址：http://www.ccpcl.com.cn
销售电话：(010)59757973
总 经 销：人民交通出版社股份有限公司发行部
经　　销：各地新华书店
印　　刷：北京印匠彩色印刷有限公司
开　　本：880×1230　1/16
印　　张：78.5
字　　数：2372 千
版　　次：2023 年 8 月　第 1 版
印　　次：2023 年 8 月　第 1 次印刷
书　　号：ISBN 978-7-114-18925-8
定　　价：498.00 元(含上、下册)

(有印刷、装订质量问题的图书，由本公司负责调换)

Qian Qihu
Yuanshi
Wenji

钱七虎 院士 文集

钱七虎院士文集　序　PREFACE

　　回想自己的人生道路，首先是感恩，我一生的成长、进步完全是国家、人民和党培养教育的结果。我是在1937年淞沪会战爆发后出生的，民族灾难深重，母亲在逃难途中的小船上生下了我。我七岁时，父亲因贫病离世。没有1949年江苏的解放，我就会和我兄、姐一样失学、失业。是人民的助学金支持我念完了中学；是组织的保送，使我进入了哈军工的大门；又是组织的选拔，我得以到苏联古比雪夫军事工程学院留学，成为一名技术科学副博士。是党、团的教育，使我不断提高自己的革命觉悟，我十四岁加入了中国共青团，十八岁加入了中国共产党，在革命的队伍里，我树立了为人民服务的人生观，才有了不断进步的动力。所以我在哈军工，五年本科成绩全优，年年被评为社会主义建设积极分子和优秀学员，四十五岁时总参总政通过调研、考察，选拔我担任工程兵工程学院院长，成为我军的高级干部。回顾自己的进步，归结到一点：国家、人民和党对我的恩情说不完。

　　熟悉我的同志，都说我的进步道路非常顺利，这和我一生中周围同志对我的支持分不开，对此我衷心感谢。是同志们在我入团后选我担任团支部委员、书记，又是同志们在我入党后选我担任党支部委员。在哈军工，是群众的支持，年年评选我为社会主义建设积极分子。在我担任工程兵工程学院院长期间，当参政胡主任带领总参工作组在学院调研时，是由于学院广大干部、教员的积极反映，总参党委研究决定准备把我们学院党委树为总参勤政廉政的典型，并在北京召开了全国新闻媒体参加的新闻通气会，宣传我院党委的勤政廉政事迹。我本人也被总参党委选为总参唯一

本文节选自"在中央军委给钱七虎记一等功庆功大会上的讲话"，2013年。

的代表参加全军爱国奉献优秀干部事迹报告团在全军巡回宣讲。我能当上教授、中国工程院首届院士，是我国已故著名科学家张维、李国豪推荐提携的结果。我能获得国家科技进步三等奖、二等奖和一等奖，其科研成果是我团队和合作团队集体奋斗的结晶，是大家的支持把我排在第一。我能连续三届担任中国岩石力学与工程学会理事长，是岩土工程领域同志们信任、支持和选举的结果。总结我一生，我深深体会：一个人没有大家的支持，就不可能有什么成就和进步，而要得到别人的支持，就要支持别人；一个领导要得到群众的支持，就要树立和实践领导就是服务的理念。国家最高科学技术奖获得者王选教授就说过："一个人做事处世既要想到自己，更要想到别人，而我的老师季羡林说，做事处世首先要想到别人，其次才想到自己。"这些话是我的座右铭，也是我经常教育我的学生的。在今天的庆功会上，我要再一次衷心感谢我的老前辈、老领导、老同事，感谢我的专业领域内的同人，感谢我学术团队的每位成员和我的学生对我的支持。

最后，请允许我表达我的感奋之情，国家和人民给予我的荣誉是对我的有力鞭策，在我有生之年，我要始终秉持"位卑未敢忘忧国""不待扬鞭自奋蹄"的信念，要学习 2012 年"感动中国"人物林俊德院士，至死攻坚不放松。"老骥伏枥，志在千里"，我时刻感到还有很多东西需要我学习，头脑中还有大量课题需要我研究，还有很多社会焦点和技术攻关，我可以贡献我的学习心得和见解。特别对于我国深地下工程如何可靠地防护，包括应对不断发展的钻地核弹的打击，是我们超高抗力工程建设必须持续研究的、关系国家战略安全的关键课题，也是我有生之年为之奉献全部精力的目标。

作为一名老科学家，培养更多的优秀中青年学术骨干，后继有人，是我义不容辞的责任，我要责无旁贷地把他们带向更高层次，引领防护工程学科发展，推动我国岩石力学与工程领域技术水平的持续提升。为此奋斗，是我的幸福所在。

2023 年 6 月

钱七虎院士文集 目录 CONTENTS

上册

学术论文

爆破荷载作用下岩石破坏特性的"共轭键"基近场动力学数值模拟研究
......周小平,王允腾,钱七虎(003)
事故型冲击荷载对结构作用研究总论 ...... 钱七虎(015)
准脆性材料断裂模拟方法研究 ...... 燕秀发,钱七虎,赵跃堂,周寅智(026)
基于广义粒子动力学的巷道围岩弹塑性分析 ...... 赵 毅,周小平,钱七虎(033)
单轴压缩条件下岩石破坏的光滑粒子流体动力学数值模拟 ...... 周小平,赵 毅,钱七虎(046)
克里金算法与多重分形理论在岩土参数随机场分析中的应用 ...... 王长虹,朱合华,钱七虎(058)
深部硐室围岩分区破裂化现象数值模拟研究 ...... 苏仲杰,钱七虎(065)
岩爆、冲击地压的定义、机制、分类及其定量预测模型 ...... 钱七虎(071)
准脆材料动力强度的本质和侧向惯性约束作用 ...... 戚承志,钱七虎,陈灿寿,陈剑杰(077)
岩体非协调变形对围岩中的应力和破坏的影响 ...... 钱七虎,周小平(083)
非协调变形下深部岩体破坏的非欧模型 ...... 周小平,钱七虎(091)
Effects of the axial in situ stresses on the zonal disintegration phenomenon in the
  surrounding rock masses around a deep circular tunnel ...... Q. Qian, X. Zhou, E. Xia(099)
深隧道围岩分区破裂的数学模拟 ...... 戚承志,钱七虎,王明洋,陈剑杰(108)
分区破裂化现象的研究进展 ...... 戚承志,钱七虎,王明洋,罗 健(116)
基于线法的功能梯度材料断裂分析 ...... 燕秀发,钱七虎,方国强,赵敏福,郭延宝(124)
基于突变理论的滑坡时间预测模型 ...... 周小平,钱七虎,张永兴,杨海清(132)

Quantitative analysis of rockburst for surrounding rocks and zonal disintegration
　　mechanism in deep tunnels ································· Qihu Qian, Xiaoping Zhou(142)
浅埋地下结构顶板在竖向地震作用下的动力响应 ·············· 陈灿寿,戚承志,钱七虎,李信桥(155)
功能梯度材料结构分析的半解析数值方法研究 ·········· 燕秀发,钱七虎,王　玮,孙　翱,戴　耀(162)
岩质高边坡稳定性分析与评价中的四个准则 ··· 李　宁,钱七虎(169)
中国岩石工程技术的新进展 ·· 钱七虎(175)
泥岩损伤特性试验研究 ··· 许宝田,钱七虎,阎长虹,许宏发(186)
深部非均匀岩体卸载拉裂的时间效应和主要影响因素 ············ 范鹏贤,王明洋,钱七虎(191)
大型抛掷爆破中的重力影响 ·· 钱七虎(199)
Effect of loading rate on fracture characteristics of rock
　　··· Xiaoping Zhou, Qihu qian, Haiqing Yang(202)
深埋球形洞室围岩分区破裂化机理 ···························· 周小平,钱七虎,张伯虎,张永兴(208)
岩石在过临界荷载作用下变形异常现象的模拟 ············ 戚承志,钱七虎,王明洋,吴　徽(216)
岩爆机理的简化分析和决定性参数的思考 ··· 钱七虎(224)
深部裂隙岩体岩爆定量预测模型 ·· 周小平,钱七虎(233)
Crack tip higher order stress fields for functionally graded materials with generalized form of gradation
　　··· Xiufa Yan, Qihu Qian, Hongbiao Lu, Wei Wang, Ao Sun(243)
岩体结构面对应力波传播规律影响的研究进展 ··· 俞　缙,钱七虎,赵晓豹(251)
岩石爆炸动力学的若干进展 ·· 钱七虎(260)
纵波在改进的弹性非线性法向变形行为单节理处的传播特性研究
　　··· 俞　缙,钱七虎,林从谋,赵晓豹(284)
爆炸作用下岩石破裂块度分布特点及其物理机理 ········ 戚承志,王明洋,钱七虎,罗　健(293)
电阻率法在深部巷道分区破裂探测中的应用 ··········· 许宏发,钱七虎,王发军,李术才,袁　亮(297)
多层软弱夹层边坡岩体稳定性及加固分析 ···················· 许宝田,钱七虎,阎长虹,许宏发(306)
点滴化学注浆技术加固土遗址工程实例 ············ 柴新军,钱七虎,杨泽平,林重德,松永和也(312)
Dynamic strength of rocks and physical nature of rock strength
　　··· Qihu Qian, Chengzhi Qi, Mingyang Wang(318)
Zonal disintegration of surrounding rock mass around the diversion tunnels
　　in Jinping II Hydropower Station, Southwestern China
　　··· Q. H. Qian, X. P. Zhou, H. Q. Yang, Y. X. Zhang, X. H. Li(335)
Bifurcation condition of crack pattern in the periodic rectangular array
　　··· X. P. Zhou, Q. H. Qian, H. Q. Yang(345)
岩石、岩体的动力强度与动力破坏准则 ··· 钱七虎,戚承志(352)

深部巷道围岩分区破裂化现象现场监测研究
　　……………………李术才,王汉鹏,钱七虎,李树忱,范庆忠,袁　亮,薛俊华,张庆松(359)
深部岩体工程围岩分区破裂化现象研究综述 …………………………………… 钱七虎,李树忱(368)
岩石爆破的破碎块体大小控制 ………………………………………………………… 戴　俊,钱七虎(375)
深部岩体强度准则 ……………………………………………………… 周小平,钱七虎,杨海清(378)
深部巷道围岩变形破坏的时间过程及支护 ………………………………… 戚承志,钱七虎,王明洋(385)
高放废物地质处置的成本估算 ……………… 罗嗣海,钱七虎,王　驹,刘晓东,赖敏慧,杨普济(394)
微型土钉微型化学注浆技术加固土质古窑 ……………… 柴新军,钱七虎,罗嗣海,林重德(401)
我国高放废物地质处置战略研究 …………………………………………………… 潘自强,钱七虎(408)
深埋巷道分区破裂化机制 …………………………………………………………… 周小平,钱七虎(413)
基于滑面正应力假设的土压力计算方法 ………………………… 刘华丽,钱七虎,朱大勇,周先华(422)
深部隧道围岩的流变 ……………………………………………… 戚承志,钱七虎,王明洋,丁常树(427)
深部岩体力学研究进展 …………………………………………………………… 何满潮,钱七虎(432)
高放废物深地质处置中的多场耦合与核素迁移 ……………… 罗嗣海,钱七虎,李金轩,周文斌(446)
深部坑道围岩的变形与承载能力问题 …………………………………… 陈士林,钱七虎,王明洋(453)
深部岩体块系介质变形与运动特性研究 …………………………………… 王明洋,戚承志,钱七虎(462)
岩石中侵彻与爆炸作用的近区问题研究 …………………………………… 王明洋,邓宏见,钱七虎(468)
岩体的构造层次及其成因 ……………………………………… 戚承志,钱七虎,王明洋,董　军(473)
岩体的构造层次粘性及动力强度 …………………………………… 戚承志,钱七虎,王明洋(482)
弹粘塑性孔隙介质在冲击荷载作用下的一种本构关系——第二部分:弹粘塑性孔隙
　介质的畸变行为 …………………………………………………… 戚承志,王明洋,钱七虎(491)
强动载作用下的岩石动力学 ………………………………………… 钱七虎,戚承志,王明洋(495)
块体结构岩体中超低摩擦效应的理论研究 ………………………… 戴　俊,钱七虎,王明洋(515)
锦屏二级水电站引水隧道围岩分区破裂化研究 ………………………………… 钱七虎,周小平(522)
大直径盾构掘进风险分析及对特大直径盾构挑战的思考 ……………………… 钱七虎,陈　健(537)
建设城市地下综合管廊,转变城市发展方式 ……………………………………………… 钱七虎(545)
隧道工程建设地质预报及信息化技术的主要进展及发展方向 ………………………… 钱七虎(553)
推进城市地下空间规划建设的思考 ………………………………………………………… 钱七虎(566)
公路隧道大断面改扩建施工开挖方案研究 ………………………………………… 黄伦海,钱七虎(572)
Safety risk management of underground engineering in China: progress, challenges and strategies
　………………………………………………………………………………… Qihu Qian,Peng Lin(581)
水下隧道工程实践面临的挑战、对策及思考 …………………………………………… 钱七虎(615)
地铁工程建设安全风险管理研究 …………………………………………… 解东升,钱七虎,戎晓力(620)

地下工程建设安全面临的挑战与对策 ··· 钱七虎（627）

一种岩溶地质条件下的城市地铁超前预报方法
··· 苏茂鑫,钱七虎,李术才,薛翊国,张庆松,邱道宏,聂利超（644）

地下磁悬浮交通设计研究的若干问题 ·· 钱七虎（651）

基于地下集装箱运输的城市地下环境物流系统建设 ······················· 范益群,钱七虎（658）

地下洞室连续围岩岩爆的定量分析及其与分区破裂化之间的关系 ················ 钱七虎（663）

我国城市地下空间综合管理的探讨 ·· 陈晓强,钱七虎（675）

隧道工程动力响应特性与汶川地震隧道震害分析及启示 ············ 钱七虎,何 川,晏启祥（681）

深埋隧道开挖过程动态及破裂形态分析 ···················· 李树忱,钱七虎,张敦福,李术才（692）

中国地下工程安全风险管理的现状、问题及相关建议 ······················· 钱七虎,戎晓力（701）

从河床冲淤分析沉管法修建长江水下隧道问题 ···································· 钱七虎（708）

国内外地下综合管线廊道发展的现状、问题及对策 ························· 钱七虎,陈晓强（712）

地下空间作为城市空间结构的社会学内涵 ··································· 奚江琳,钱七虎（716）

现代地下货物运输系统的研究与进展 ································· 张明聚,钱七虎,唐 劼（720）

大盾构工程关键技术的新进展 ·· 钱七虎（725）

Mechanism and classification and quantitative prediction model of strain rockburst in the
　　surrounding rock masses around a deep circular tunnel ··········· Qihu Qian, Xiaoping Zhou（734）

Evaluation of the status and outlook of the urban underground space development and
　　utilization in China ·· Qihu Qian（749）

关于绿色发展与智能建造的若干思考 ·· 钱七虎（755）

工程建设领域要向智慧建造迈进 ··· 钱七虎（757）

利用地下空间助力发展绿色建筑与绿色城市 ······································ 钱七虎（759）

科学利用城市地下空间,建设和谐宜居、美丽城市 ································· 钱七虎（770）

学术报告

动力扰动（爆破或顶板塌落）诱发冲击地压事故的机理 ·································（779）

抗爆结构的膜力（in-plane-force）效应 ···（783）

重要目标防爆抗爆的主要技术措施 ··（787）

关于空腔爆炸的隔震技术 ···（798）

高地应力岩体地下工程围岩支护设计计算的问题、原因和对策 ……………………… (808)

锦屏一级电站开挖过程地下洞室群围岩变形破坏数值模拟和监测及其分析和启示 …… (816)

深部岩体分区破裂化机理的研究进展 ……………………………………………………… (824)

非线性岩石力学(非传统)研究前沿导论 ………………………………………………… (835)

应变型岩爆的机理、分类及其定量预测模型 ……………………………………………… (845)

非协调变形与岩爆的机理和预测 …………………………………………………………… (855)

Deformation and failure mechanism of surrounding rock mass around underground caverns
 in Jinping I hydropower station …………………………………………………………… (863)

超深盐膏层地应力与井筒完整性 …………………………………………………………… (870)

Recent advances in the zonal disintegrarion phenomenon in the deep rock mass engineering ………… (881)

Dynamic strength and it's physical nature and dynamic fracture criteria of rock ……………… (885)

大盾构掘进的事故及对特大盾构工程的思考 ……………………………………………… (893)

地下空间开发利用、防治城市病及地下物流系统的发展方向 …………………………… (898)

21世纪是地下空间开发利用的世纪 ……………………………………………………… (912)

建设城市地下综合管廊转变城市发展方式 ………………………………………………… (944)

隧道岩爆监测预警 …………………………………………………………………………… (956)

地下工程建设安全面临的挑战与对策 ……………………………………………………… (961)

世界地下物流系统研究动态与新进展 ……………………………………………………… (970)

The evaluation of status quo and outlook for the underground logistics system (ULS)
 development in China ……………………………………………………………………… (981)

The present situation and future prospect in application of tunneling machine in china
 underground engineering construction …………………………………………………… (987)

21世纪前期我国城市地下空间开发利用的战略及对策 ………………………………… (1000)

"双碳"目标下的城市建设 ………………………………………………………………… (1006)

"双碳"目标下的能源地下工程 …………………………………………………………… (1014)

审读大百科轨道交通文稿时的思考 ………………………………………………………… (1017)

利用地下空间,助力发展绿色建筑与绿色城市 …………………………………………… (1026)

绿色城镇与绿色城镇基础设施 ……………………………………………………………… (1032)

数字隧道和智慧隧道——隧道建设信息化技术的发展方向 …………………………… (1051)

若干重要建设工程中的哲学思考 …………………………………………………………… (1057)

中国岩石工程进展和规划 …………………………………………………………………… (1060)

思想与观点

建设科技强国迫切需要科学家精神 ……………………………………………………………… (1081)
白鹤滩工程所取得的经验对地下工程建设有重要借鉴意义 ………………………………… (1084)
又好又快稳步推进城市地下综合管廊建设的思考 …………………………………………… (1085)
21世纪,让我们向地下要空间 …………………………………………………………………… (1088)
高度重视地下工程安全——专访中国工程院钱七虎院士 …………………………………… (1092)
关于北京地下道路建设和地下空间开发相关问题的访谈 …………………………………… (1094)
通过数字化向智慧建造迈进 ……………………………………………………………………… (1097)
特大城市解决交通拥堵问题的思路与出路 ……………………………………………………… (1100)
又快又好地推进地下空间建设——对话中国工程院钱七虎院士 …………………………… (1103)
迎接气候变化的挑战,推进绿色建设、实施生态大保护 ……………………………………… (1107)
依托中国的独特优势,加速迈向科技强国的伟大目标 ………………………………………… (1109)
隧道与地下开发实现历史性"穿越"——访中国工程院院士钱七虎 ………………………… (1111)
"岩爆"可测时代即将到来 ………………………………………………………………………… (1115)
利用地下空间建设"花园城市" …………………………………………………………………… (1116)
春风化雨　润物无声——深切缅怀潘家铮院士 ………………………………………………… (1120)
城市交通拥堵、空气污染以及雨洪内涝的治本之策 …………………………………………… (1122)
城市化发展呼唤积极和科学开发利用城市地下空间 ………………………………………… (1124)
在苏州地铁11号线开通仪式上的发言 ………………………………………………………… (1126)
在中国城镇供热协会地下综合管廊运营维护专业委员会成立大会上的致辞 ……………… (1127)
在2021年中国国际服务贸易交易会智慧建造与绿色发展高峰论坛的发言 ………………… (1128)
永远跟党走　永葆革命青春——在中国工程院学习习近平总书记"七一"
　　重要讲话精神报告会上的报告 ……………………………………………………………… (1130)
在中国科协学会党建工作指导委员会成立大会暨学会党建工作先进表彰大会上的发言 ……… (1132)
在"国土空间规划契机下地下空间的机遇与挑战学术研讨会暨《2020中国城市地下空间
　　发展蓝皮书》发布会"上的致辞 ……………………………………………………………… (1134)
在江苏省高层次人才主题学习会上的发言——让生命在科技报国中闪光 ………………… (1135)
在习近平总书记关于乌东德水电站首批水轮机组发电亲切祝贺和重要指示的
　　重大意义的认识座谈会上的发言 …………………………………………………………… (1139)
在中国岩石力学与工程学会"弘扬科学家精神,加强作风和学风建设"主题宣讲会上的发言 …… (1140)

在超大直径全断面竖井掘进机下线仪式上的发言 ……………………………………（1143）

在江苏省过江通道建设技术专家委员会成立大会上的发言 ……………………………（1144）

在纪念于学馥先生百年诞辰大会上的发言 ………………………………………………（1145）

在中国岩石力学与工程学会水下隧道工程技术分会成立大会上的发言 ………………（1146）

在2019年盾构与掘进关键技术暨盾构再制造技术国际峰会的致辞 …………………（1148）

在国家科学技术奖励大会上的发言 ………………………………………………………（1150）

在第三届全国工程安全与防护学术会议上的发言 ………………………………………（1151）

在中国岩石力学与工程学会工程安全与防护分会成立大会上的发言 …………………（1152）

在第七届中俄深部岩石动力学高层论坛开幕式上的发言 ………………………………（1153）

在中国城市建设科学发展论坛上的发言 …………………………………………………（1154）

在第十四次全国岩石力学与工程学术大会开幕式上的发言 ……………………………（1156）

在苏通GIL综合管廊工程专家聘任座谈会上的发言 ……………………………………（1157）

在东华理工大学建校60周年庆典大会上的致辞 ………………………………………（1158）

在第四届GeoChina国际会议上的致辞 …………………………………………………（1160）

在全国隧道及地下工程不良地质超前预报与突水突泥灾害防治学术会议上的发言 …（1161）

在"精细爆破学术研讨会"上的致辞 ……………………………………………………（1163）

在汕头市苏埃通道工程专家委员会成立暨技术咨询会上的发言 ………………………（1164）

在"中国矿业科学协同创新联盟"成立大会上的发言 …………………………………（1165）

中国科协第80期新观点新学说学术沙龙开幕词 ………………………………………（1166）

在人民交通出版社创建六十周年纪念活动上的发言 ……………………………………（1167）

在中国工程科技论坛上的发言 ……………………………………………………………（1168）

在中国人民解放军理工大学溯源碑揭幕仪式上的发言 …………………………………（1169）

在中国岩石力学与工程学会岩爆机理探索沙龙上的发言 ………………………………（1170）

在第三届全国水工岩石力学学术会议上的致辞 …………………………………………（1171）

吾家吾国

淡泊名利品自高 ……………………………………………………………………………（1175）

耿耿丹心　为国铸盾 ………………………………………………………………………（1176）

科技强军、为国铸盾的防护工程专家 ……………………………………………………（1177）

铸盾一生 ……………………………………………………………………………………（1179）

科研工作者要永远跟党走 ……………………………………………………………（1183）
科学家精神的核心是追求真理和献身科学 …………………………………………（1186）
淡泊名利是科学家精神的重要内核 …………………………………………………（1189）
为武汉捐赠650万元,钱七虎:烈士献出了生命,我有什么不能贡献? ……………（1190）
老党员履新职 …………………………………………………………………………（1193）
中国智慧建造必将走在世界前列 ……………………………………………………（1195）
和平年代要树立忧患意识科学家更要有责任担当 …………………………………（1197）
一生一事,为国为民——采访1954届校友钱七虎院士………………………………（1199）
铸就共和国"地下钢铁长城" …………………………………………………………（1205）
建设科技强国迫切需要工匠精神 ……………………………………………………（1207）

钱七虎院士大事记……………………………………………………………………（1213）

学术报告

学其所用，用其所学

Qian Qihu
Yuanshi
Wenji

钱七虎 院士 文集

动力扰动(爆破或顶板塌落)诱发冲击地压事故的机理

一、引言

欧美文献从现象学出发对岩爆定义为:矿井或隧道的围岩或岩柱破坏、碎化发生弹射、崩出或崩塌的现象,伴随能量的猛烈释放。

岩爆发生的机理:高地应力的地层中,因开挖卸荷或动力作用诱发围岩中应力场的变化,从而或直接导致围岩的破坏碎化、弹射以及崩塌,或通过围岩中的已有断层和结构面滑移(活化)与新结构面滑移引起围岩破坏和弹射、崩塌。

欧美文献关于岩爆的分类如下:

第一类岩爆:应变型岩爆,或称体积失稳,因体应力超过极限引起岩石破裂而导致的岩爆,其特点是扰动源(开挖)和岩爆发生地一致。

第二类岩爆:滑移型岩爆,或称结构面接触失稳,通过断层或断裂面滑移(剪切失稳)导致的岩爆,其特点是扰动源(爆破或顶板断裂)和岩爆(冲击地压)发生地相距一定距离,第二类岩爆较第一类更普遍、更强烈,其破坏范围甚至达数十和几百米巷道。在俄罗斯和中国,矿井中的第二类岩爆,习称为冲击地压事故。

矿井中的冲击地压事故大部分,特别是多层多巷矿井中,其围岩经过多次开挖卸荷,围岩已成为处于准稳定状态的新老结构面和破裂面所分割的岩块系,在动力作用(爆破或顶板断裂)下,岩块系和断裂构造失稳所导致。

冲击地压事故的实例列举如下:

(1)门头沟煤爆分析(陈宗基)

切不可忽视爆炸波对触发煤爆的影响。门头沟矿在 700~900m 深处的 114 起煤爆中,有 89 起(占 78%)是因爆破触发的,在龙凤矿 700m 深处因爆破而触发的煤爆也超过总数的 50%,由此可以推断,在这样的深度,煤一定是处于准稳定平衡状态,爆炸波的触发作用能导致死亡事故发生。

有些矿也出现过顶板爆裂,坚硬岩层顶板更为不利的影响是爆裂会扩展到比长壁工作面遥遥领先,从而导致失稳和煤爆。

(2)华丰煤矿冲击地压事故综合剖析

4层煤的多个工作面发生过冲击地压。华丰井田煤系地层以上为 500~800m 砾岩层,砾岩层坚硬整体性强,其断裂跨落对下部的煤岩体产生冲击载荷,是 4 层煤工作面发生冲击地压的主要力源。

(3)义马煤业集团千秋煤矿"11·3"重大冲击地压事故调查报告

10人死亡、64人受伤。本矿区煤层顶板为巨厚砂砾岩(380~600m),事故发生区域接近落差达 50~500m 的 F16 逆断层,地层局部直立或倒转,构造应力极大,处在强冲击地压危险区域;煤矿开采后,上覆砾岩层诱发下伏 F16 逆断层活化,瞬间诱发了井下能量巨大的冲击地压事故。

以下研究爆破或顶板断裂(动力扰动)是如何诱发冲击地压事故突然发生的?

报告时间:2021 年 11 月 16 日。

二、弱扰动引发残余变形的现场实测信息

(一) 弱扰动引发残余变形的现场实测信息

Science 1993. V. 260 (P. 1617-1623) 上刊登了 Hill. D. D. 等的论文《Seismic ity remoted triggered by the magnitude 7.3. Landers,California. earthquake》介绍了著名的加利福尼亚州兰德斯7.3级地震及在远离震源投影点1250km外的地方在数十小时时段内触发了多次地震事件的研究成果。

上述现象很难用应力场的改变来解释。因为地震诱发的应力随R3很快衰减,根据计算超过200km时,其引起的应力场的改变小于引力场的每天变化量($\sim 2 \cdot 10^{-3}$MΠa)。

这迫使我们设想:地震触发机制的原因是地震波通过时引起的动力变形,这些变形在数十至数百秒内可积累至相当的量,导致后续地震的发生。

下图引用了在远离兰德斯7.3级地震,震源67km外的美国加利福尼亚州PFO试验场处的断层内用激光测距仪实测的变形记录曲线。

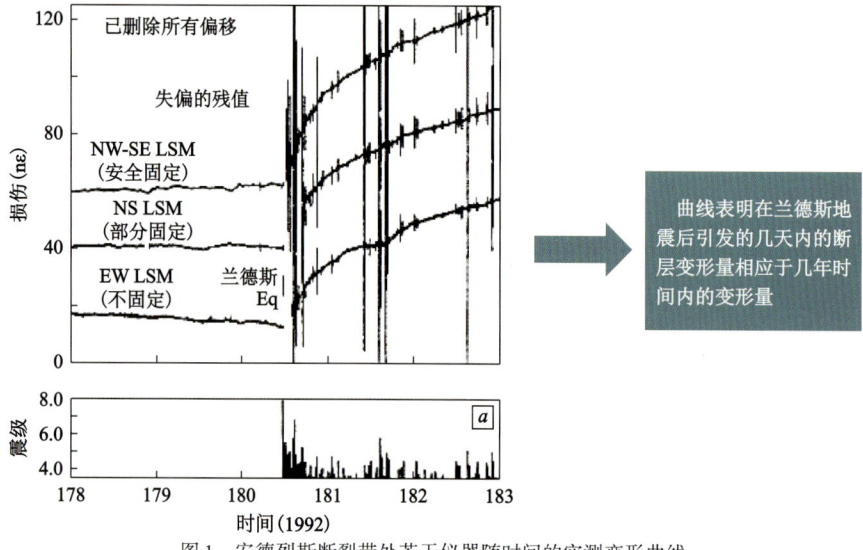

图1 安德烈斯断裂带处若干仪器随时间的实测变形曲线

(二) 微弱的地震波产生断层裂隙的残余变形的原因

这是由于断层与裂隙在加载与卸载时显著的变形不对称性,导致与岩土材料强度相比为很小振幅的动力作用(例如地震波)能产生残余变形。实测的残余变形引用如下:

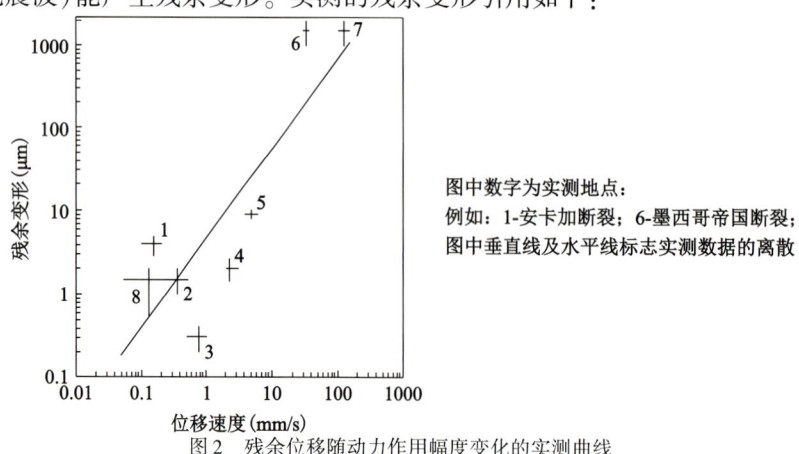

图2 残余位移随动力作用幅度变化的实测曲线

三、弱小扰动诱发的岩块体接触面的滑移变形的试验研究

(一) 弱小扰动诱发的岩块体间接触面滑移变形实验室研究

为了阐明这些动力变形能否诱发灾变,俄学者进行了下述模型试验。
试验装置见图3。

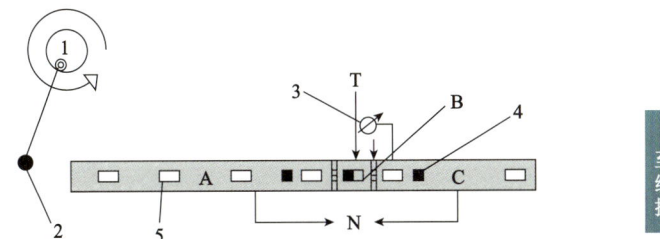

图3 研究微变形积累过程的装置简图

A、C-固定块;B-移动块;N-法向荷载装置;T-切向加荷装置;1-电动机;2-撞击体;3-位移计;4-应变计;5-速度与加速度探头

注:图中充填层中充填2mm厚的石英砂。

试验结果如下:

(1) 没有剪切内力情况下($F_s = 0$),动力影响不引起剩余变形;

(2) 在不大的剪切内力情况下,明显的观察到块间的相对位移,位移方向与施加剪荷载方向一致,与波传播方向无关,在较小剪切内力时,位移时间 15~20MC;在较大剪切内力时,位移时间可达数秒。

(3) 在一定参数组合($F_s/F_p = 0.99$)情况下,位移速度多次增加,并观察到失稳的情况,此时静荷载保持不变,从摩擦离合的观点来说,此时接触面转入快速软化的机制,即接触面剪切抗力逐步降低,失稳过程明显。

接触面剪切变形过程的细观分析:在剪切变形过程中,裂缝(接触面)从强化演化到软化,宏观的强化是与不平整度的破坏相联系,而快速的软化是与局部化的剪切破坏发生有关。动力不稳定性来临在接触面的应力—应变状态达到了流变关系下降段。此时极限状态可以在应力明显小于接触面强度的情况下达到,这可以解释动力事件由地震波引发的很多方面。

试验结果如下:在施加剪切荷载量值不变的情况下,体系失稳状态是通过块间位移幅值的逐渐增加而达到的。临界位移的量值并不取决于脉冲速度的幅值,仅取决于施加剪切荷载的水平。

(二) 扰动诱发的岩块体间初应力接触面滑移变形实验室研究

块间位移积累的力学分析如下:

地震波中,动力振动压缩相通过时,块体界面上的总法向应力将在静应力水平上增加,随后在拉伸相通过时,总法向应力将降低。考虑到当岩块间充填松散体时,其黏结力 c 很小,所以剪切强度基本上正比于作用块面间的法向应力($\tau_p = \mu \cdot \sigma$)。

因此当动应力为压缩时,剪切强度增加,当动应力为拉伸时,剪切强度减小。实际上拉伸相通过时,剪切强度减小更显著,这是因为此时摩擦系数在拉伸波通过时也会减少,这是由于块系介质的超低摩擦效应(effect of anomalously low friction in block media),从而呈现乘积效应。对于每一时间点,计算 τ_s 与 τ_p 的比值,基于上述理由,振动拉伸相通过时,τ_s/τ_p 会明显增加,甚至可能超过 $\tau_s/\tau_p = 1$ 的临界值。

当剪切内力 τ_s 超过剪切强度 τ_p 时,块体沿接触面就滑移。即当 $\tau_s/\tau_p > 1$ 的条件成立,块体相对运动就加速;这个条件不成立,块体相对运动就减速。所以在地震波传播周期内,可以发生若干次间歇性滑

移。这表明,块间位移积分值强烈取决于τ_s/τ_p比值,在低幅值动荷作用时,显著的位移积累可能主要发生于当剪切应力值接近于极限应力时。

(三)不同接触面破坏时的声发射信息及其物理本质

模型试验时,试验了两种充填层:砂充填层和熔接充填层。

不同块体接触面在相对位移和破坏时发出的信息不同:由砂充填层中摩擦时的声发射波形图以及熔接充填层中形成撕裂的声发射图可知:第一种情况中声脉冲的发射是砂粒间以及砂粒与块体不平整面相互作用时发出的,第二种情况中可以见到两个阶段的主动声发射,第一阶段($12.5C,\cdots,18.5C$)可以将其与微裂纹产生相联系,第二阶段($t>18.5C$)开始块体的宏观位移过程,由位移探头记录,即形成宏观裂纹并沿宏观裂纹滑移。

在两个试验中,整个振动都由相当数量的个别的声脉冲所组成。但是在砂充填层接触面中摩擦滑动情况中,在重复试验时,许多脉冲形式几乎都相同,但是在形成裂纹的第二种情况中,冲量按其形成实际上,无论在裂纹形成阶段,还是在宏观裂纹形成后的运动过程中,在重复试验中都不相同。我们注意到,在两种情况中,冲量的振幅至少改变了两个量级。

四、结语

从上述现场监测及模型试验结果及其分析,我们可以得到如下启示:

(1)滑移型岩爆冲击地区的物理本质是接触面变形的失稳。

(2)接触面切向高地应力水平是滑移型岩爆冲击地区发生的必要条件:τ_s接近τ_p,即接触面或围岩的破裂面处于准稳定状态。

(3)触发滑移型岩爆冲击地区机制的充分条件是爆破振动、TBM推进的振动与巷道断裂等动力扰动所诱发的残余变形积累达到临界值。

(4)动力扰动诱发接触面残余滑移变形的物理原因是拉伸波引起的接触面压应力减少以及动摩擦系数值的降低,从而导致剪切强度(抗力)τ_p的降低,使得$\tau_s/\tau_p \geq 1$。

(5)矿井开挖时滑移型岩爆(冲击地区)发生的主要影响因素:围岩中存在结构面或者围岩破裂形成新结构面;接触面的性质;隧道开挖引起的初始地应力的改变;动力扰动(爆破等)的辐射与次数。

(6)滑移型岩爆(冲击地区)发生时接触面的相对位移一定达到临界值,从而发生接触面的失稳。对于泥质充填体的接触面,失稳时并不发生岩体的破裂,因此微震监测不一定十分有效;对于胶结型充填体的接触面,可以发生胶结体的破裂,但其破裂强度低,要求微震监测的灵敏度高。

(7)滑移型岩爆(冲击地区)的发生与优势接触面(结构面)失稳有关,τ_s接近τ_p处于准平衡状态的接触面即为优势结构面,失稳时滑移变形量达到临界值,因此监测预警措施的重点应是找出高剪应力低压应力的结构面,即优势结构面以及监测该结构面的残余滑移变形量。

抗爆结构的膜力(in-plane-force)效应

一、研究必要性

大量试验研究结果表明,抗爆受弯结构的实际承载能力明显高于按目前理论方法计算的值,有时承载能力的提高还非常显著。

Park、Kranthammer 等人研究指出:抗爆结构承载力提高的主要原因是,在爆炸荷载作用过程中结构构件中出现面力(in-plane-force),亦称薄膜力,这种面力使结构构件的抗弯强度提高,从而提高了结构的承载能力。

二、膜力效应的机理

以简支梁为例,如图所示。

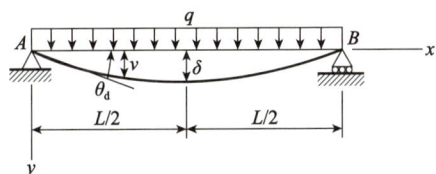

承受均匀荷载的简单梁的挠度曲线

梁在荷载作用下,梁向下挠曲,在小挠度、小变形情况下,梁的纵向伸长可忽略不计。

在大挠度情况下,梁的纵向伸长不可忽略,但支座 B 可自由移动,因此梁在纵长方向不产生拉力。如果支座 B 被约束不能移动,梁在挠曲时其纵向的伸长可产生拉力,这个拉力可提供附加抗力。

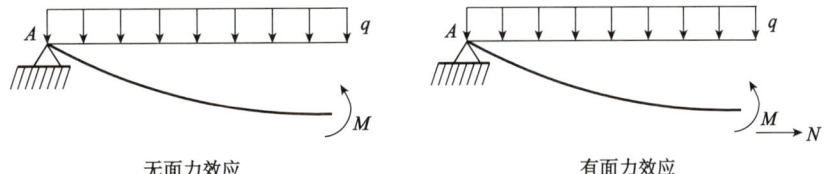

无面力效应　　　　　　有面力效应

由此可见:支座"约束"使其不能自由移动是面力产生的基础,结构发生较大变形是面力效应充分发挥的前提。而抗爆结构允许产生较大变形而不破坏,是因为这种结构主要承受爆炸、冲击荷载的瞬时或短时作用。

核爆结构中约束作用的表现形式有:

(1)箱形结构的环箍作用,它直接对顶板和底板起约束作用。

(2)格构式板盖,划分格构的梁就是它们所包围的板块的约束。

(3)自锁约束作用。钢筋混凝土简支板在大变形情况下会产生自锁作用,当变形很大时,在板的中部混凝土几乎没有作用,只有钢筋受拉,而在板边区域,由于摩擦和双向约束,这就是自锁作用。

承载力提高的程度与板边缘支座(梁端支座)的约束刚度有关,约束刚度越大,产生的面力也越大,因而承载力提高程度也越大;板截面配筋率越低,承载力提高程度越大。

报告时间:2014 年 10 月 25 日。

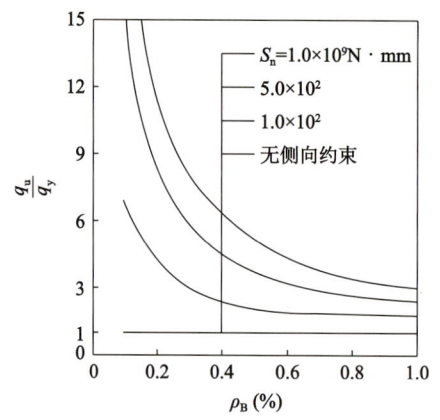

不同约束条件下的结构极限承载力与配筋率的关系

三、试验研究结果

美国陆军工程兵水道实验室和海军土木工程实验室等单位对抗爆结构面力效应的试验结论是：

(1) 快速加载下面力对钢筋混凝土板抗力的影响与静载作用下完全相似，但必须考虑材料的加载速率效应(应变速率效应)。

(2) 考虑面力效应，板的动抗力较之不考虑面力效应，至少可以提高2.5倍，最大动位移则只达到不计面力效应条件下位移的0.1~0.4。

(3) 在因面力效应抗弯能力得到大幅度提高的条件下，剪切破坏是另一种常见的破坏形式，所以必须考虑受弯构件的抗剪强度匹配，提出了几种横向钢筋配量的方案。

(4) 由于在每个方向上，面力沿板全跨均匀分布，因此由外载产生的板平面内轴向变形分布于整个板内，这对于板承受幅值大持续时间极短的爆炸冲击荷载特别有利。

苏联关于抗弯强度的试验研究结论是：对于可能遭受短时动载作用的受弯构件，不允许不配置横向钢筋，否则将可能发生脆性剪切破坏。

四、考虑面力效应的钢筋混凝土结构的动力响应及算例

采用等自由度体系，不考虑面力效应的钢筋混凝土结构的动力方程

$$M\ddot{f} + K \cdot f = F(t)$$

式中，M为等效质量，$M = K_{m \cdot L} m \cdot L$。

$K_{m \cdot L} = \dfrac{K_m}{K_L} K_m$为质量等效系数，$K_L$为荷载等效系数，$m$为单位长度质量，$L$为结构长度，$F(t)$为动荷载考虑面力效应的钢筋混凝土结构的动力方程。

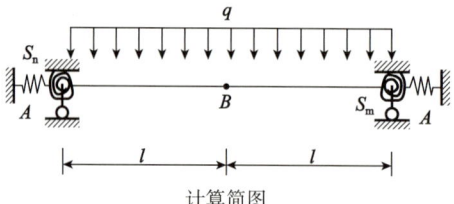

计算简图

塑性铰A及B：

$$M\ddot{f} + R(f) = F(t)$$

$$\frac{l}{2}R(f) = M_R^A + M_R^B + N \cdot f$$

上式中，M_R^A，M_R^B 和面力都是跨度 f 的复杂函数，还应考虑材料的应变速率效应，可以材料动力提高系数 f_d 来考虑。由于 $R(f)$ 是非线性函数，只能通过数值方法求解，对于承受爆炸荷载的结构分析问题，为了保证得到的解能达到满意的精度，必须选择很小的时间步长。

算例：选用 Keenan 试验的两个试件（1 和 2）计算验证，试件是爆炸荷载作用下的钢筋混凝土约束板试验。1 和 2 试件基本参数一致，唯荷载不一样。

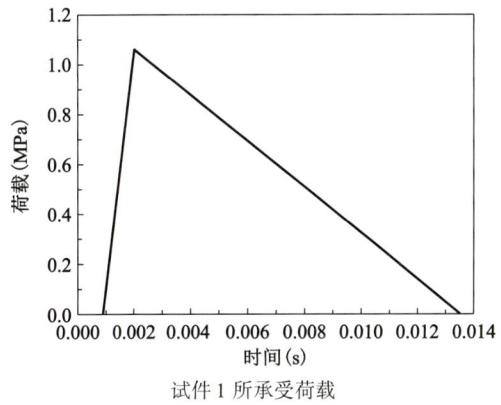

试件 1 所承受荷载

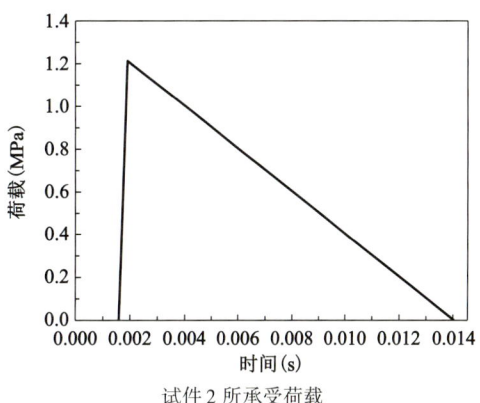

试件 2 所承受荷载

以下两图分别是试验试件在两种不同荷载作用下的跨中位移时程曲线，经比较结果可见：

(1) 不考虑面力效应，计算与试验结果相差甚远；
(2) 考虑面力效应，两者结果基本一致；
(3) 进一步考虑应变速率效应，两者结果更接近。

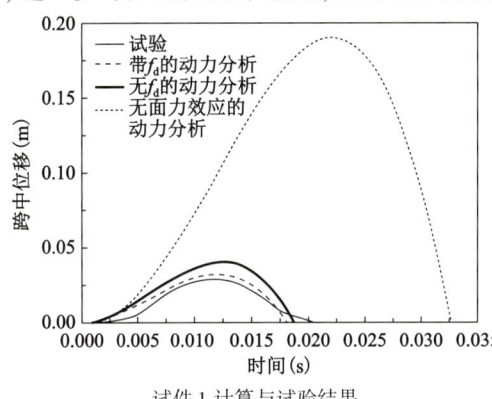

试件 1 计算与试验结果

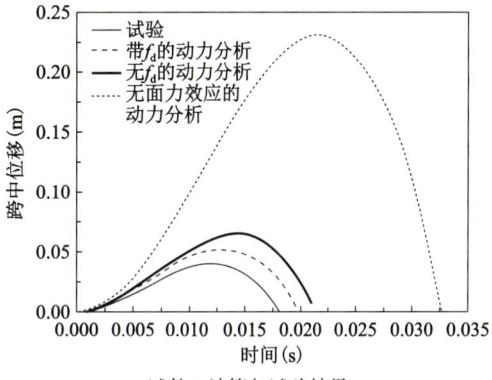

试件 2 计算与试验结果

五、需进一步研究的问题

(1) 面力的发展过程：开始为压面力（AB），后续为拉面力（CD）。
(2) 膜力 $N(f)$ 的计算问题。

绝对约束：膜力 N 只与梁的伸长量有关（铰支座）。

$$N = E \cdot \frac{\Delta L}{L} \cdot S_0 \quad \Delta L = S(f) - L$$

弹性约束：

$$N = S_n \cdot \Delta \quad dN = S_n \cdot d\Delta$$
$$M_R^A = S_m \Theta_s \quad dM_R^A = S_m d\Theta_s$$
$$\Delta L = S(f) - L - 2\Delta$$

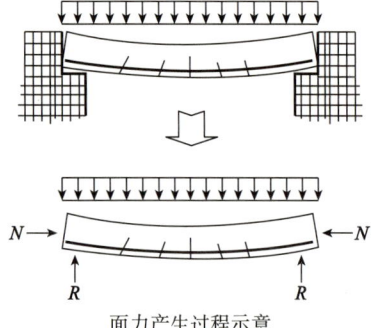

面力产生过程示意

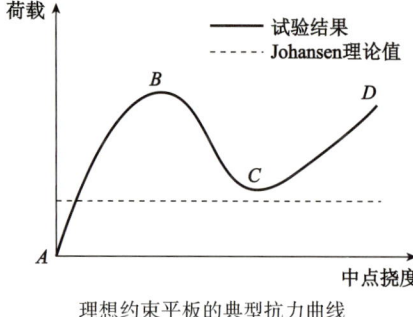

理想约束平板的典型抗力曲线

重要目标防爆抗爆的主要技术措施

一、恐怖爆炸袭击方式、规模及荷载计算

(一) 恐怖爆炸袭击的方式

(1) 汽车炸弹；
(2) 人体炸弹；
(3) 箱包炸弹；
(4) 邮件炸弹。

根据我们对2001年底至2005年4月较大规模恐怖爆炸袭击方式的统计，汽车炸弹袭击约占所有爆炸袭击的78%，人体炸弹约占8%，箱包炸弹和邮件炸弹等其他爆炸物袭击约占14%，如图1所示。从恐怖爆炸袭击的发展趋势看，汽车炸弹袭击愈演愈烈，将成为重要目标的主要防护对象。

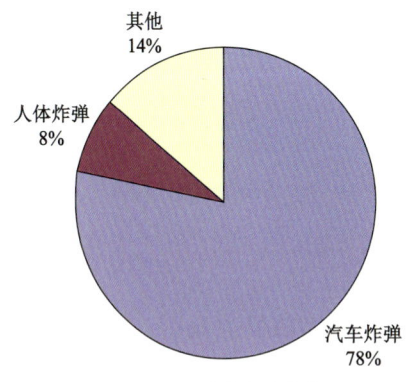

图1 恐怖爆炸袭击方式分布图

(二) 恐怖爆炸袭击的规模

1. 汽车炸弹袭击规模

重要目标的主要防护对象是汽车炸弹，按照我国城市街道常见车辆类型及载重量的不同，通过分析拟将汽车分为五种，如表1所示。由于我国大多数城市对货车在城市行驶都有严格限制，所以表1中只考虑到小型货车。

我国对爆炸物的生产、储存、销售、购买、运输、使用都有严格的管理规定，企图获得大量爆炸物用于非法活动的可能性较小，恐怖分子主要通过自制、购买原材料和盗窃的途径获取炸药，一般多为工业炸药。常见工业炸药TNT当量为0.6~0.8，而自制炸药的爆炸威力更小，约为工业炸药的70%。因此，可取汽车炸弹的TNT当量系数为0.5。

综合汽车类型和炸药的分析结果，拟将汽车炸弹分为五类(表1)。目标设计时，可根据我国各城市道路对车辆管理的规定予以考虑。

汽车炸弹分类及规模　　　　表1

序号	类别	总质量 (kg)	装载质量 (kg)	最高车速 (km/h)	炸药规模 (kgTNT)	备注(车型)
1	基本乘用车型	1500	350~450	175	200	常见的家用轿车
2	交叉乘用车型	2000	500	160	250	微型面包车
3	多功能乘用车型	2300	600~800	160	300	旅行车、SUV、MPV等
4	小客车型	3000	1000	150	500	依维柯、金杯等小型客车
5	厢式货车型	4000以上	2000以上	150	1000以上	小型厢式货车

报告时间：2007年8月6日。

2. 人体炸弹袭击规模

人体炸弹袭击通常是恐怖分子将炸弹隐蔽地缠在身上,由此决定其装药量不会太大。一般成人携带2kg以下的爆炸物不会影响人体的正常活动,且不易被发现。在俄罗斯车臣地区频繁发生的"黑寡妇"人体炸弹事件中,装药量一般为500g~1.5kg,一般与碎铁钉、钢珠等混用,以造成更大的杀伤面积。

3. 邮件炸弹袭击规模

我国邮政规定的允许邮件范围在2~40kg之间,考虑到爆炸物体积以及邮政部门对邮件的检查等因素,邮件炸弹装药量不会太大,一般在2~10kg的TNT炸药。

4. 箱包炸弹袭击规模

箱包炸弹的装药量介于邮件炸弹和汽车炸弹之间,约为25kg的TNT炸药。

(三)恐怖爆炸的荷载计算

通过比较分析并在前人研究和我国规范的基础上,提出了汽车炸弹爆炸荷载经验计算公式:
冲击波入射超压参数:

$$P_m = 1316 \left(\frac{\sqrt[3]{W}}{R}\right)^3 + 369 \left(\frac{\sqrt[3]{W}}{R}\right)^{1.5} \tag{1}$$

式中,P_m为入射冲击波超压峰值(kPa);W为等效TNT装药量(kg);R为结构物距爆炸中心的距离(m)。

冲击波反射超压:

$$P_{rm} = 2P_m + 6P_m^2/(P_m + 700) \tag{2}$$

式中,P_{rm}是正反射冲击波超压峰值(kPa)。

冲击波反射冲量:

$$I_r = 900 \left(\sqrt[3]{W}/R\right)^{1.23} W^{1/3} \tag{3}$$

式中,I_r为冲击波反射冲量(kPa·ms)。

本公式的适用范围是 $0.596 \leq R/\sqrt[3]{W} \leq 9.58$。

二、重要目标界定及分类、分级

(一)重要目标的界定

从反恐怖爆炸角度定义重要目标,既要考察其面临的恐怖爆炸威胁,还要综合考虑其使用功能、人员密度、自身价值等因素。从历史情况看,重要目标主要包括政府建筑、军事设施、外交建筑等,而近期恐怖事件多发生于平民聚集的商业区、居民区、城市交通工具等,而这类目标遭袭造成的后果同样非常严重。

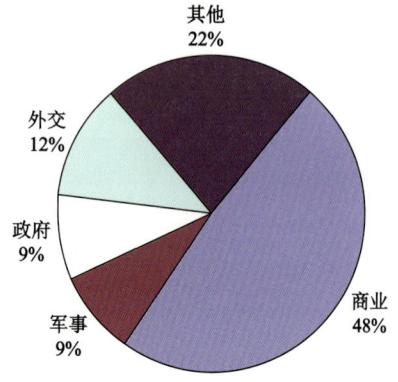

图2 较大规模恐怖爆炸袭击目标分布图

图2是我们对2000年6月至2006年底全球58起较大规模恐怖爆炸袭击事件的统计,其中袭击商业目标共28起,占48%,位居首位。商业目标由于人员密集、防护等级弱且一般地处城市繁华地带,一旦袭击实施成功,会造成巨大的人员伤亡和财产损失,因此成为当前恐怖袭击的主要目标。

因此,我们认为重要目标是指易遭受潜在恐怖爆炸袭击或一旦遭到袭击将会产生严重后果的建筑物或构筑物,其严重后果主要指政治影响、经济损失、社会影响、重大伤亡等方面。如重要军政机关、驻外机构、标志性建筑、核电站、城市交通枢纽、城市地铁、商业建筑、集会建筑和生产易燃易爆、有毒有害物质的企业等。

(二)重要目标的分类

目前恐怖袭击的目标越来越平民化,恐怖分子将原为"等待命令出击预定目标"的方式,改为寻求机会袭击任何可以袭击的目标,因此重要目标的范围越来越广,但防爆抗爆的资源有限,因此有必要对重要目标进行分类分级,突出重点、把握关键,把有限的人力、物力、财力集中在最需要的地方,增强防爆抗爆的针对性和有效性。通过对一些主要目标进行归类分析,我们给出表2所示的分类。

重要目标的分类　　　　　　　　　　　　　　　　　　　　　表2

目标类型	典型建筑物或构筑物
军政目标	党、政、军机关办公楼,标志性建筑等
商业目标	大型商场、超市、酒店、宾馆、地下商场、银行等
交通目标	车站、地铁站、码头、机场、交通枢纽、大型桥梁、隧道等
文体目标	学校、图书馆、体育场馆、电影院(剧院)等
医疗目标	大型医院等
民居目标	大型住宅楼、公寓楼等
外交目标	我国驻外使领馆、外国驻华使领馆等
工业目标	大型企业,尤其是生产易燃易爆、有毒有害物质的企业或核设施

(三)重要目标的分级

遭到恐怖袭击后,不同类型的目标将产生不同的影响,而在同一类型目标中不同等级的目标其影响也各有大小。通过分析,我们拟将重要目标分为极重要的、重要的、一般重要的3个等级,见表3。

重要目标等级分类　　　　　　　　　　　　　　　　　　　　表3

重要目标类型	等级		
	极重要的目标(一级)	重要的目标(二级)	一般重要的目标(三级)
政府目标	国家党、政、军机关办公楼	省(直辖市)党、政、军机关办公楼	市县党、政、军机关办公楼
商业目标	营业面积1.5万 m^2 以上,年销售额10亿元以上	营业面积在1万~1.5万 m^2 ,年销售额达5亿~10亿元	营业面积在0.5万~1万 m^2 ,年销售额达5亿元以下
交通目标	首都火车站、机场、地铁站	各地高峰使用人数或座位数超过1万人(座)车站;或建筑面积5万 m^2 以上,每小时可接纳2000人以上的机场;全国已建的地铁站	各地高峰使用人数或座位数在1万人以下(座)车站;或建筑面积5万 m^2 以上,每小时可接纳2000人以下的机场
文体目标	高峰使用人数超过2万人以上的学校;藏书量1000万册以上的图书馆;占地4万 m^2 以上,座位数在2万以上的体育场馆	高峰使用人数1万~2万人以上的学校;藏书量500万~1000万册的图书馆;占地2万~4万 m^2 ,座位数在1万~2万的体育场馆	高峰使用人数超过1万人以下的学校;藏书量500万册以下的图书馆;占地2万 m^2 以下,座位数在1万以下的体育场馆
医疗目标	三级甲等(含)以上的医院	三级乙等(含)以上的医院	三级乙等以下的医院
民居目标	总建筑面积超过5万 m^2 的住宅楼	总建筑面积1万~5万 m^2 的住宅楼	总建筑面积在1万 m^2 以下的住宅楼
外交目标	联合国驻首都办事处;外国驻首都大使馆	外国驻各省(直辖市)领使馆	外国驻各省市大使馆
工业目标	年营业收入在5000万元以上的工业企业;特大型生产易燃、易爆、有毒、有害化学品的工厂或核化设施	年营业收入在1000万~5000万的工业企业;大型生产易燃、易爆、有毒、有害化学品的工厂或核化工业设施	年营业收入在1000万元以下的工业企业;中型生产易燃、易爆、有毒、有害化学品的工厂或核化工业设施

三、重要目标防爆抗爆的概念设计

（一）重要目标的出入口概念设计

重要目标要提高安全性，最直接、最经济、最有效的办法就是对目标的出入口进行严格控制。在目标出入口附近建立一控制区，安排警卫人员对进出目标的人员及车辆进行严格的检查和管理。其目的是控制外来人员和车辆直接进入目标内部，同时也为重要目标提供了有效防护距离。出入口控制区主要包括：

1. 出入口道路

出入口前应避免采用垂直于目标方向的直线道路，不给汽车获得加速的机会，以防止汽车冲过出入口控制区。在出入口前设置水池、喷泉等景观，同样也是一种降低汽车速度的方法。

出入口前的道路平行于目标正面时，采用高路肩、适当间距的树木或景观等措施，可有效地防止汽车驶离路面。

2. 出入口大门

在满足建筑美观、使用方便的基础上，出入口大门结构主要考虑防汽车冲撞设计，不主张进行抗爆设计，原因是按抗爆设计的大门，门的重量、开启和门框墙的厚度、基础深度等，难以满足平时使用和建筑美观的要求。若非要进行抗爆设计，主张在大门后面一定距离设计防爆屏障来代替。

3. 出入口障碍物

在重要目标出入口设置障碍物，是建立出入口控制区的重要手段。借助于障碍物，可以对进入目标的车辆进行限制和阻滞。一旦发现可疑车辆，可以将其限制在距目标一定距离之外，有效阻止恐怖活动的发生或者减轻爆炸危害。同时，设置障碍物也可以给恐怖分子造成心理上的震慑作用。

4. 出入口警卫室

警卫室主要是警卫人员监视、操作和休息的场所。警卫室通常设有能监视出入口前方道路和进入目标接受检查的窗户或监控设备，也设有操纵大门、障碍物开启和安检设备的按钮等。

（二）重要目标的停车场概念设计

1. 重要目标应尽量避免采用内部停车场

为重要目标设计配套停车场时，应尽可能避免采用地下停车场或目标内部停车场；如必须采用地下停车场时，应对出入口进行严格控制，同时应确保目标主体结构具有较高的抗爆等级。

2. 地面停车场应尽量远离重要目标

地面停车场应远离重要目标且在目标的视线范围之内，从而减轻汽车炸弹爆炸引起的破坏效应。如果条件允许，应将地面停车场设计成单行线，以便更好地观察可疑车辆。

3. 重要目标停车场应有严格的管理措施

重要目标的停车场应安装安检探测仪器、电子监控系统等，严格检查外来车辆。若将前期设计与后期管理结合起来，可大大提高目标的安全性，降低汽车炸弹袭击的潜在危险。

（三）重要目标的建筑概念设计

重要目标防爆抗爆设计，应与抗震和抗风等自然灾害兼顾考虑，这样既可提高目标的综合防灾能力，又能降低工程造价。

1. 平面形状

从冲击波对目标的作用原理和抗侧力结构体系可知：简单、规则、对称、长宽比不大的平面，最好近似于圆形的平面,对防爆抗爆是有利的;而平面突出部分长度过大、形成凹凸状部分过多、有过小开放式平面内折角、折角过多和角部重叠、太狭长和细腰平面等布置,则对防爆抗爆是不利的,如图3所示。

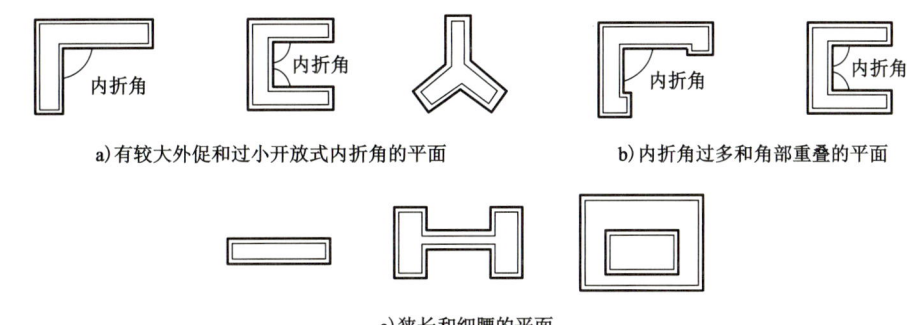

图3 对防爆抗爆不利的目标平面布置

2. 立面布置

①由于冲击波遇到目标及孔洞后会产生反射、绕流及方向改变等作用,因此对防爆抗爆有利的立面应规则、均匀,从上到下外形不变或变化不大,没有过大的外挑和内收,建筑造型总体呈凸状乃至圆形。目标形状对冲击波的影响如图4所示。

②面向街道的立面,禁止采用大面积的玻璃幕墙,以防止爆炸碎片杀伤人员,如非用不可则应采用特殊的防爆玻璃;同时,外墙立面最好采用涂料粉刷,尽量避免粘贴瓷砖。

③面向街道的立面,其窗户开设应尽量小或少,最好不超过立面面积的15%,若有可能开窗户的一面尽量侧向街道。

④面向街道的重要目标底层应高于外地面50cm以上,以防止汽车冲进目标内部爆炸。

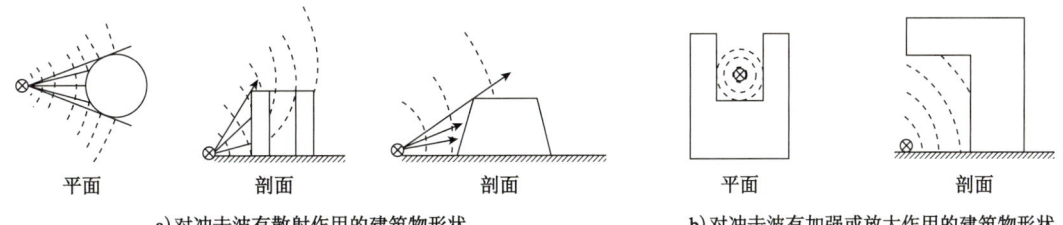

图4 目标形状对冲击波的影响

3. 防护分区

按使用功能要求应对目标室内设计进行必要的防护分区,即将高风险区与低风险区分开,两者之间可设置防爆墙,如图5所示。高风险区设置在目标的外部或边缘,利用目标外侧的裙楼布置,避免目标的主体因爆炸破坏导致坍塌。

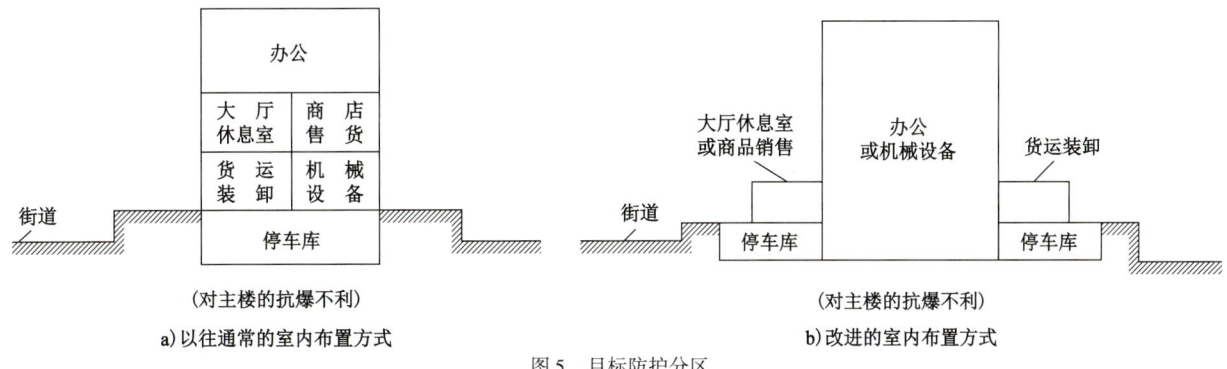

图5 目标防护分区

目标紧急出口所要求的电梯间或楼梯间,应尽量远离爆炸可能发生的区域,如大厅、停车场、货运平台等。

4. 内部装修

重要目标的内部装修应尽量简单,慎用易产生爆炸碎片的装修材料,墙壁上不宜悬挂装饰品,办公家具、计算机显示器等布置在远离临街窗户的部位。

(四) 重要目标的结构概念设计

在满足使用要求的前提下,目标单元的开间和层高应尽量小,这样布置对结构构件抗爆有利。

1. 结构竖向布置

结构竖向布置应做到刚度均匀而连续,避免刚度突变,避免错层,避免软弱层,使结构的侧向刚度呈下大上小、自下而上连续,逐渐减小。图6为几种对防爆抗爆不利的结构竖向布置情况。

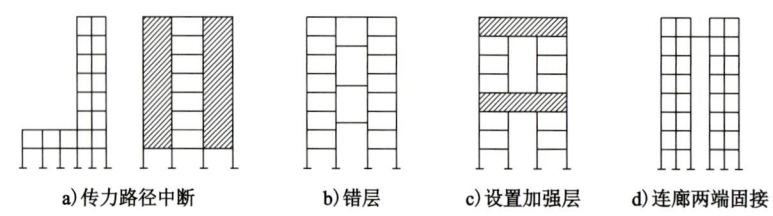

图6 对防爆抗爆不利的结构竖向布置

2. 防止渐进坍塌

防止结构渐进坍塌的关键是结构体系具有整体性和冗余度,以致部分结构局部倒塌或构件严重破坏时,不会导致整个结构丧失承载力。设计时,结构体系应采用超静定结构体系,并设置"骨架结构"或在可能的情况下适当增加一些关键构件,如超静定的水平支撑和垂直的抗压构件等;结构构件应采用延性材料,应考虑抗反复荷载作用,并保证具有足够的抗剪强度;构件相互之间应有可靠的连接和锚固措施。

3. 考虑反向作用

当爆炸冲击波扩散进入结构内部时以及发生内爆炸时,对梁和楼板都会产生向上的推力。因此,重要目标应禁止采用预应力混凝土结构,当采用现浇钢筋混凝土梁、楼板应考虑反向荷载作用,即采用双筋截面配筋;对于外挑的阳台、雨篷等悬挑结构设计,还有结构构件之间的构造连接方式,尤其要考虑爆炸荷载的反向作用。

4. 外墙构造措施

当目标外墙采用轻型混凝土砌块、空心砖等脆性材料时,除砌筑时与柱或墙采用拉结筋连接外,为防止爆炸碎片对目标内部人员的杀伤,外墙粉刷前应在墙体内侧设钢丝网或高强度织物,并与地板和天花板留有一定的锚固;外墙上开设的门窗,其门框、窗框与外墙应有可靠的锚固措施。

当防爆要求很高时,应避免采用脆性材料墙体。

四、重要目标防爆抗爆的工程措施

重要目标的出入口和重要部位应配备有经验的警卫人员,实行严格的人员和车辆进出管理措施,除设置安检探测仪器、电子监控系统外,对重要目标可采用以下防爆抗爆技术措施:

(一) 设置障碍物

障碍物设置应综合考虑目标周围自然环境、潜在的威胁程度、防护等级、目标结构类型等因素,并尽量与目标周围环境融为一体。

1. 障碍物种类

障碍物主要用于防汽车炸弹和防人员入侵,目前国内外障碍物的专利产品很多,我们在总结防汽车炸弹障碍物研究应用现状后,将其归类为两大类,即天然障碍物和人工障碍物。

天然障碍物是借助自然环境设计的、能对汽车起阻碍作用的地貌和地物。如河流、湖泊、排水沟、险峻地形、山脉、植物等,因其难以逾越的特点,可以阻止汽车炸弹接近目标,有效地保护目标的安全。

人工障碍物是通过人工设置的、能对汽车起到减速或阻挡作用的制式设施,通常与景观设计相结合。根据目标周围的环境要求,人工障碍物可预先设置和隐形设置。对周围空旷的重要目标可采用预先设置,如与目标融为一体的景观障碍物;对城市中心的重要目标多采用隐形设置,一旦发现情况能快速展开。

2. 障碍物效能

障碍物效能取决于其对动能的吸收程度,迎头撞击车辆的动能 KE 表示为:

$$KE = \frac{Mv^2}{2}$$

式中:M 是汽车质量,v 是撞击障碍物时的汽车速度。由于公式中 v 是一个平方关系,所以降低汽车冲击速度的重要性远大于降低汽车质量。因此人工障碍物设计思路是:一是降低汽车冲击速度,二是完全拦阻汽车冲击。

3. 人工障碍物

人工障碍物按功能可归类为景观型障碍物、减速型障碍物、破胎型障碍物和拦阻型障碍物,除景观障碍物设于道路两边和目标周围外,其他障碍物主要设于防护目标前方的道路上。

①景观型障碍物

利用道路两边和目标周围的设施和景观设置障碍,如灯柱、公交车站、邮箱、花坛、喷水池、树木、植物、座椅、书报亭、垃圾箱,装饰围栏等,如图7所示。

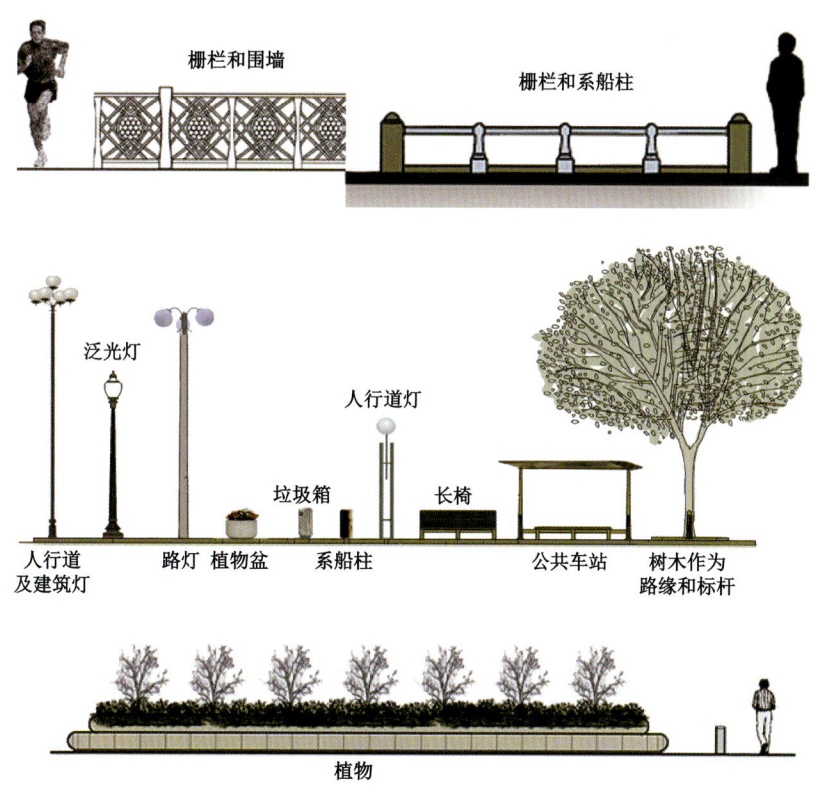

图7 景观型障碍物

②减速型障碍物

主要利用道路凸起或凹凸不平设置障碍,如减速拱(图8)、减速带(图9)和将路面设计成锯齿状等,目的是限速,即控制车辆以一定的速度行驶。

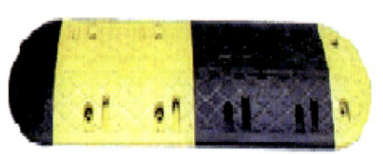

图 8　减速拱

图 9　减速带

③破胎型障碍物

主要利用尖锐的铁刺(钉)扎破充气汽车轮胎的一种设施,此类障碍物不能将冲击的汽车完全阻挡在其设置位置前,即汽车轮胎扎破后仍然可以向前冲击一段距离。因此,其设置距离应考虑该汽车炸弹爆炸后目标的安全距离加上轮胎扎破后的冲击距离,或者与拦阻型障碍物形成障碍物配系设置,主要有地铺式和弹射式两种。

a. 地铺式:ZTT-Y1 型遥控车辆破胎阻截器,如图 10 所示。

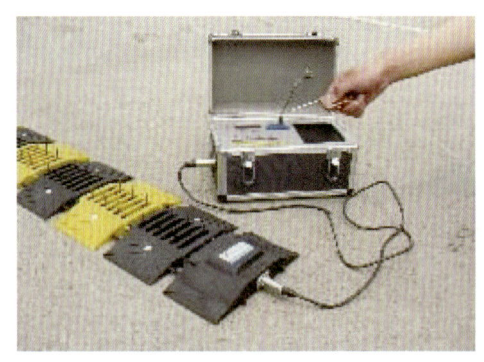

图 10　ZTT-Y1 型遥控车辆破胎阻截器

b. 弹射式:LM-1 便携式遥控阻车器,如图 11 所示。

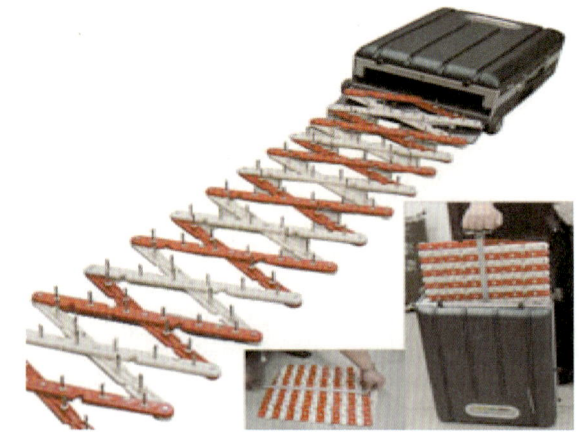

图 11　LM-1 便携式遥控阻车器

④拦阻型障碍物

主要利用自身形状或重量将汽车炸弹完全阻挡在其设置位置前的工程设施,其设置距离主要考虑该汽车炸弹爆炸后目标的安全距离。

　　a. 钢板起落式障碍物(图12)。

　　b. 钢管升降式障碍物。

　　钢管升降式障碍物有全自动液压升降障碍物与半自动液压升降障碍物两种类型,可根据道路通行情况设置。

　　c. 弧板翻转式障碍物。

(二)设置防爆墙

1. 防爆墙的作用

在目标外部一定距离上设置防爆墙,能有效地防护汽车炸弹爆炸冲击波和碎片,保证目标及内部人员的安全。防爆墙可以是目标四周永久性围墙(这些围墙经过抗爆设计),也可以是目标前面或特定位置临时快速设置的装配式防爆墙。防爆墙通常与障碍物联合使用,障碍物设置在防爆墙前一定距离,防爆墙设置在障碍物与目标之间。障碍物用于限制汽车炸弹的起爆位置,防爆墙用于防护爆炸冲击波和碎片,消除爆炸对目标及内部人员的破坏杀伤作用。与目标抗爆加固相比,构筑防爆墙费用低、设置快、效果好,是重要目标防爆抗爆的重要措施之一。

2. 防爆墙的类型

通过对国内外防爆墙归纳总结,防爆墙可分为以下三类:

①刚性防爆墙

这类防爆墙有厚度较大的钢筋混凝土墙、钢纤维钢筋混凝土墙、堆积土防爆墙和加筋土防爆墙等。我国已将钢纤维钢筋混凝土防爆墙用于某驻外使馆防爆加固改造工程。

②柔性防爆墙

这类防爆墙有中等厚度的钢筋混凝土墙、多层复合结构防爆墙、双面钢板复合防爆墙(图13)、配筋砌体防爆墙和柔性织物防爆墙(图14)等。

图12　钢板起落式障碍物

图13　双面钢板复合防爆墙

多层复合结构防爆墙是我国大连理工大学研究发明的,即在混凝土墙体中间添加发泡混凝土、硬质多孔聚氨酯或其他高分子多孔材料。与传统的防爆墙相比,墙体结构"混凝土—多孔材料—混凝土"三层结构或"混凝土—发泡混凝土—多孔材料—发泡混凝土—混凝土"五层结构两种形式。根据防爆墙可以抵抗爆炸冲击能量的设计要求,基于理论计算和实验测试对防爆墙的结构进行优化设计,包括防爆墙层数、各层材料的厚度比以及多孔材料的选择。

 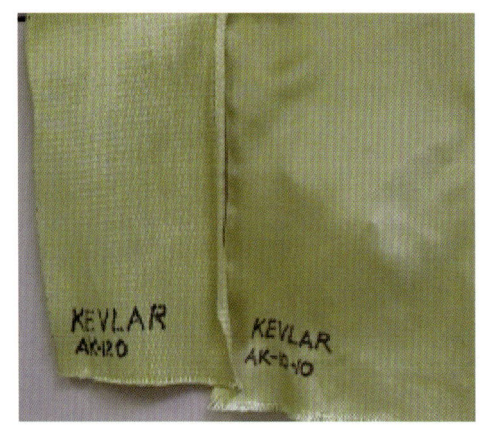

图 14　柔性织物防爆墙

③惯性防爆墙

这类防爆墙有沙袋防爆墙、塑料装配式防爆墙(图 15、图 16)、集装箱式防爆墙、桶体式防爆墙、折叠式防爆墙和水体防爆墙(图 17)等。

图 15　塑料装配式防爆墙之一　　　　　图 16　塑料装配式防爆墙之二

图 17　轻型水体防爆墙

3. 防爆墙的工作机理

①反射与绕流作用

汽车炸弹爆炸产生的冲击波遇到防爆墙发生反射,不能直射到达防爆墙后的目标。实际作用在目标上的空气压力,只是由防爆墙上反射高压区的空气绕流泄压与不规则流动生成,使作用在目标上的空气压力大为衰减。

②变形吸能作用

如果爆炸冲击波作用在柔性防爆墙上,除空气冲击波的反射和绕流作用外(与刚性防爆墙发生的情况有差别),防爆墙产生的塑性大变形(或墙体破碎)要消耗爆炸冲击波能量,从而降低作用在目标上的空气压力荷载。

③动能消耗作用

当大当量汽车炸弹爆炸造成惯性防爆墙破坏时,较大体积的沙土或水体会获得动量飞散开来,沙土或水体在飞散过程中将分散成小颗粒或小水珠,与空气的接触面积增大几个数量级,从而受到很大的空气阻力作用,消耗很大的动能,迅速降低爆炸空气冲击波的压力。动能消耗作用是惯性防爆墙的主要工作机理。

(三)提高结构抗力

不同的结构类型具有不同的受力特点及控制构件,如果控制构件失效可能引发重要目标渐进倒塌,因此应适当提高重要目标控制构件的结构抗力。

1. 新建工程

正如前述,对新建工程结构设计,为防止渐进倒塌,应考虑结构竖向布置、反向荷载作用和相应构造措施等,根据目标防爆抗爆标准对结构控制构件进行加强。

2. 已建工程

对已建工程通常实施结构加固来提高目标的结构抗力,结构加固主要是提高某一特定构件抵抗爆炸荷载破坏的抗力,而不是加强整个目标的总体强度和稳定性。主要构件的加固方法有:

①窗户:主要有封堵、粘贴薄膜或胶带、成层玻璃、抗爆窗、防爆帘(栅栏、百叶窗)等;
②柱:使用沙袋保护柱子、加大柱的截面尺寸、外包钢加固和碳纤维布加固等;
③墙:使用沙袋保护墙体、增加墙体厚度、粘钢加固或碳纤维布加固等;
④梁:梁跨中间加设支撑(跨度大的梁)、粘钢加固或碳纤维布加固等。

关于空腔爆炸的隔震技术

一、本报告的动因

1998年5月13日印度宣布进行核试验,CTBT国际监测系统没有监测到任何信号,很有可能是印度在逃避技术上取得突破:在地震核查逃避技术方面极有可能采用了隔震技术、减震技术措施。

首先我们了解一下爆破地震效应和预裂爆破的隔震效果。

1. 爆破地震效应

在工业爆破中,在给定药量的一定地点的爆震地震动的最大水平加速度值,建议按下式计算:

$$a = K \left(\frac{\sqrt[3]{Q}}{R} \right)^\alpha$$

K和α值,建议按下面不同条件近似采用。

场地为坚硬基岩时:

$$K = 150, \alpha = 1.7$$

场地为基岩时:

$$K = 220, \alpha = 1.67$$

场地为覆盖浅层表土时:

$$K = 300, \alpha = 1.6$$

实际工程时,常常通过小药量试验,确定K和α值。

地震动的径向加速度和垂直加速度在离爆心不远的地点是用同一量级的,在远离爆心的地点,地震动以径向加速度为主,水平向最大加速度作为地震动烈度的物理标准比较适宜。

2. 预裂爆破的降震(隔震)效果

预裂爆破是一项先进的爆破技术。自20世纪50年代末期在水利和矿山工程中应用以来,在提高边坡的稳定性及降低地震效应等方面,显示出一系列优点,引起了国内外研究工作者及工程技术人员的普遍重视。

预裂爆破的降震作用,主要通过爆源和被保护物体间的预裂孔隙面来实现。这一预裂孔隙面垂直于地表,成为爆破地震波的一道屏障。

地震波在传播过程中遇到了波障,就会出现一个地震动强度降低的屏蔽区,这个屏蔽区的大小与顶裂面的深度、长度和缝隙的宽度以及从爆源至顶裂的距离有关。由于顶裂面的深度和长度不可能无限的伸长,所以屏蔽区的范围总是有限的。

试验研究表明:屏蔽区的大小与顶裂面的深度和长度成正比,与爆源至顶裂面的距离成反比;而缝隙的宽度对屏蔽区的大小影响不大。在顶裂面相对于爆源的另一侧,紧靠顶裂面后面降震效果最为显著。

随着距顶裂面距离的增加,降震效果逐渐减弱。当爆源药量一定时,顶裂面至爆源的距离在一定范围内才起屏障作用。目前,关于顶裂爆破降震效果大小的估计,在同一场地和同一爆心距的条件下,通常分别测定顶裂爆破和普通爆破时各测点地震动的强度,然后将这些实际观测数据进行对比分析,以得到

报告时间:2007年5月25日。

降震作用的经验数据。目前还缺乏在不同距离内屏障降震作用的计算公式。(参见:爆破地震效应,张雪亮、黄树棠编著)

对于地下爆室核爆炸解耦技术,不采用隔震技术,则要求较大爆室空腔半径($R \approx 10\text{m}$,能量空腔留有率99.6%),实现起来有困难,需要采用隔震技术以降低要求的爆室空腔半径。

二、大型结构隔震的可行性研究

(一)冲击隔震的设计问题

问题的说明:为一在地下洞室中的大型结构提供冲击隔震,结构质量为700万磅(3178t),其基底面积929.03m²,结构受到突然地冲击,突然向下垂直位移0.762m,相应瞬态地运动峰值速度13.716m/s,结构要求防震到最大加速度为$2g$(包括重力加速度g)。

设计途径:悬置体系振型的解耦Ⅱ;螺旋悬置体系的可行性研究Ⅲ;充气悬置装置Ⅳ;波纹钢风箱用作充气悬置装置以及液体悬置装置Ⅴ;隔震体系中的瞬态波动研究。

结论:

(1)振型的解耦。结构运动的平移和转动模态(振型)的解耦的优点是明显的。上述设施中既包括导弹,又包括水平延展的建筑物。振型的耦合会引起问题的复杂化。

如果(平动时)转动的振型激发的话,结构的某些部位(点)将经受较之其他部位大的运动及其加速度。如果这些点仍处在规定极限以内,则结构的其他部位保护过度。所以,由于非转动的地运动反应能激发转动运动的悬置体系较之振型解耦的体系有效性差一些。

适当地配置悬置体系的隔震元体,耦合项是可以消除的。但是由转动振型对平动振型的耦合可以保留,因为转动振型是不预期的,这一项不重要,如果悬置体系的所有单元(元体)对通过重心的垂直轴已对称的话。

(2)螺旋弹簧。螺旋弹簧作为隔震体系的使用是可行的。一个适当设计的螺旋弹簧悬置体系必须满足三个准则:

①没有沉底或压屈的危险。②最大应力低于允许极限。③适应可利用空间。

研究报告中研制出满足这些准则的弹簧参数选择的图表,可以认为,如此弹簧的生产是可行的,但由于基底的快速下沉可以在弹簧中产生应力波,该应力波可能引起在被支撑结构中的高加速度,这可能是螺旋弹簧的主要缺点。

(3)充气悬置。充气悬置证明是可行的。局限于结构基底下的低压力空气给出了合适的冲击隔震,研制如此体系的主要问题是限制压缩空气的风箱不经受过度的环向应力和剪应力。通过对若干种布置方案的研究,认为在结构下放置相当数量的风箱是最有前景的。

(4)风箱。波纹钢风箱作为充气弹簧显然具有很好的前景。设计方程和图表表明:如此体系在提供合适的冲击隔震时没有产生过大的应力。目前的生产方法能够生产出如此尺寸和复杂类型的风箱。

充液的风箱显然是不可行的,其体系尺寸做不到。如果管道和阀门使用的话,则摩擦损失显得特别大。此外,在液体中传播的压缩波将在结构中产生很大的加速度。

(二)振型的解耦

1. 问题的定义

研究的是三维空间中的大型结构冲击隔震。结构的任何部分的加速度在最不利条件下都不超过$1g$。如果结构的六个振型完全解耦的话,则结构基础的垂直运动只引起结构的垂直运动,基础的水平运动只引起结构的水平运动。如果摇摆运动被消除了,则结构的所有部分将经受同一个加速度。分析中假设振动质量具有足够阻尼,阻尼掉二阶以上使振型畸变的非线性项。

2. 二维振型的解耦

研究图 1 所示的洞室内支撑在弹簧上的结构,以及图 2 所示的结构的运动。

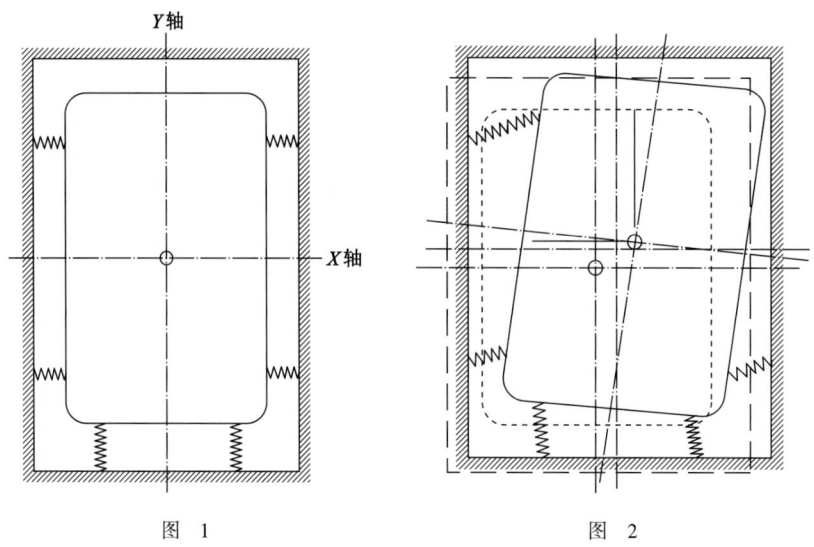

图 1　　　　　　　图 2

结构由垂直弹簧支撑其重量,水平弹簧保持其稳定。当正确配置时,振型可解耦。

假设弹簧对称配置,弹簧是线性的,运动方程:x、y 与 φ 为相对位移。悬置体系(弹簧)由于相对位移施加于质量重心的力、力矩是:

$$F_x = -a_{xx}x - a_{xy}y - a_{x\varphi}\varphi$$
$$F_y = -a_{yx}x - a_{yx}y - a_{y\varphi}\varphi$$
$$M_\varphi = -a_{\varphi x}x - a_{\varphi y}y - a_{\varphi}\varphi$$

式中,a 为挠度系数,例如 a_{xy} 表示 Y 方向单位位移在 X 方向产生的力。

相应的运动方程为(X、Y—洞室基础的运动):

$$F_x = m(\ddot{x} + \ddot{X})$$
$$F_y = m(\ddot{y} + \ddot{Y})$$
$$M_\varphi = I\ddot{\varphi}$$

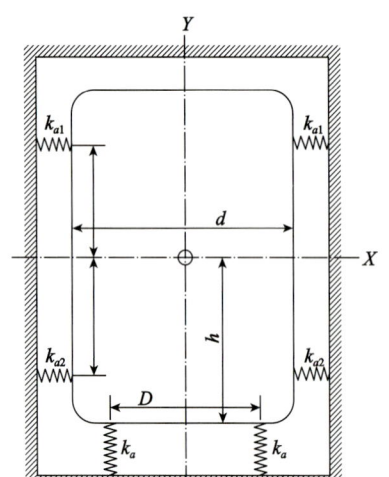

图 3

由上式得:

$$m(\ddot{x} + \ddot{X}) + a_{xx}x + a_{x\varphi}\varphi = 0$$
$$m(\ddot{y} + \ddot{Y}) + a_{yy}y = 0$$
$$I\ddot{\varphi} + a_{\varphi x}x + a_{\varphi\varphi}\varphi = 0$$

上式推导中考虑到体系的对称性,以及弹簧轴线是水平或垂直的。因此,

$$a_{x\varphi} = a_{y\varphi} = 0$$
$$a_{xy} = a_{yx} = 0$$

耦合的消除:体系的各尺寸及各弹簧的刚度系数如图 3 所示。

如果弹簧没有侧向刚度,并忽略二阶项,则各挠度系数如下列方程所示。

$$a_{xx} = 2(K_{a1} + K_{a2})$$

$$a_{yy} = 2K_a$$

$$C_\varphi = 2K_{a1}l_1^2 + 2K_{a2}l_2^2 + \frac{1}{2}K_a D^2 - mgh$$

$$C_{x\varphi} = 2K_{a2}l_2 - 2K_{a1}l_1$$

$$C_{\varphi x} = 2K_{a2}l_2 - 2K_{a1}l_1 + mg$$

a_φ 及 $a_{\varphi x}$ 中最后一项的来源,见图4。

因为基底弹簧初始由自重产生轴向变形,所以系数 $a_{\varphi x}$ 中常忽略了一项 mgx,现解释该情况如下:在静力作用下,两个基底弹簧都施加一个垂直力 $\frac{1}{2}mg$ 于结构。当结构水平位移 x 时,每一个弹簧的变形如图4所示。

对于自由弹簧体来说(图4),在每一垂直弹簧中引起一个弯矩 $\frac{1}{2}mgx$,该弯矩作用在结构上,总共引起 mgx。该弯矩应叠加到没有重力时结构位移产生的弯矩上,并应包括到系数 $a_{\varphi x}$ 中去。

如果结构经受一个对于重心的转角 φ,则其基底将经受一个水平位移 $h\sin\varphi$。φ 很小,则可近似为 $h\varphi$。与上述理由相同,该位移将导致一个负弯矩 $-mgh\varphi$,所以在系数 a_φ 中将包含该项。

两个耦合的系数 $a_{x\varphi}$、$a_{\varphi x}$ 呢?若 $K_{a1}l_1 = K_{a2}l_2 + \frac{mg}{2}$ 则 $a_{\varphi x} = 0$,此时,$a_{x\varphi} = 2K_{a2}l_2 - 2K_{a1}l_1 = -mg$。

所以,代入上述挠度系数于方程中,得

$$I(\ddot{x} + \ddot{X}) + 2(K_{a1} + K_{a2})x\ mg\varphi = 0$$

$$m(\ddot{y} + \ddot{Y}) + 2K_{ar} = 0$$

因为没有初始转动,可得 $\varphi(t) = 0$,所以体系是解耦的,不再推导。对于具有侧向刚度的弹簧(图5),则解耦条件如下:

$$k_a l = k_s h + \frac{mg}{2}$$

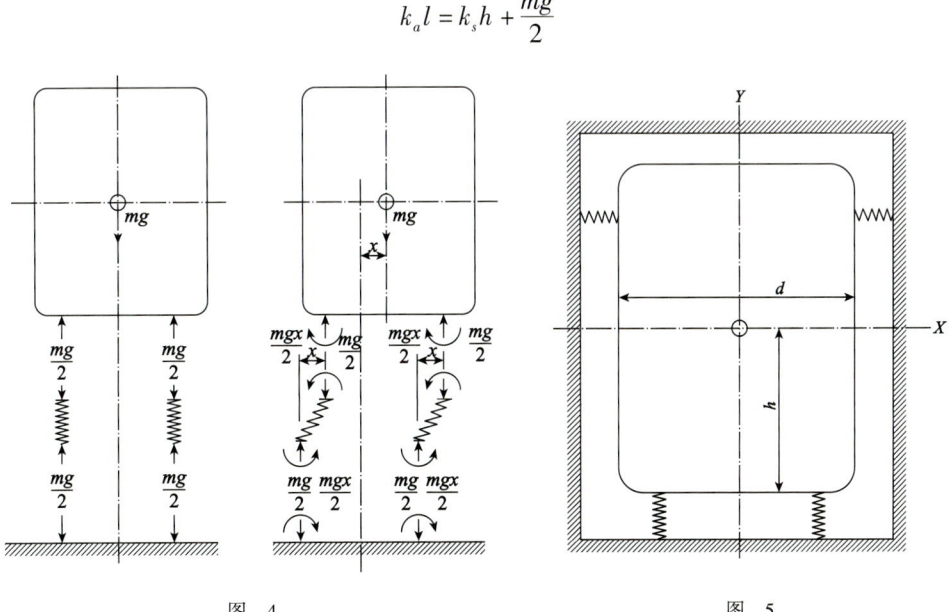

图 4　　　　　　图 5

3. 三维情形下的解耦

不再推导。

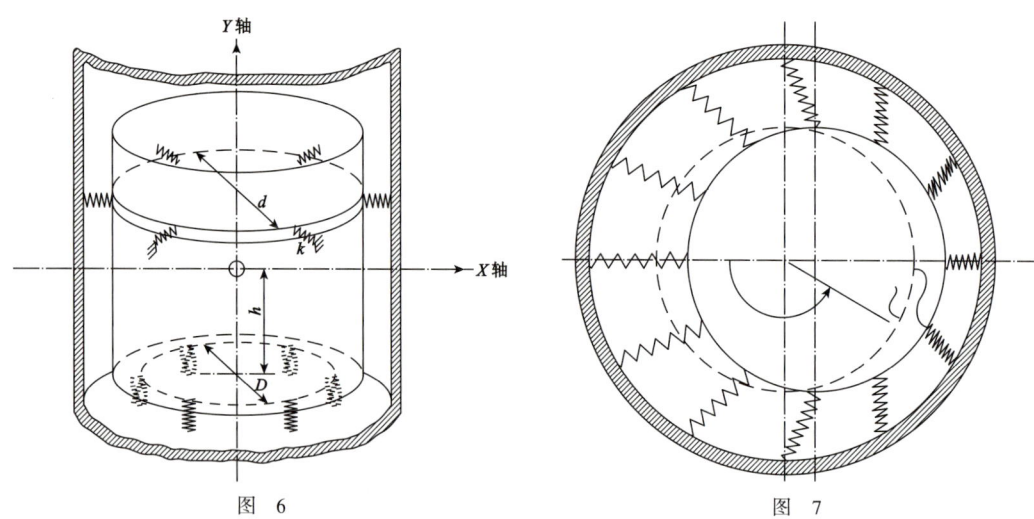

图 6 图 7

其解耦条件为：

$$hNK_s \sim l\frac{n}{2}(k_a + k_s) + mg$$

此时运动方程为：

$$m(\ddot{x} + \ddot{X}) + NK_s + \frac{n}{2}(K_a + K_s)xm\sigma\varphi = 0$$

$$m(\ddot{y} + \ddot{Y}) + NK_A y + nK_s y = 0$$

频率方程为：
垂直运动振型

$$\omega_y = \sqrt{\frac{NK_a + nk_s}{m}}$$

水平运动振型

$$\omega_x = \sqrt{\frac{n_2(k_a + k_s) + NK_s}{m}}$$

转动模型

$$\omega_\varphi = \sqrt{\frac{n_2(k_a + k_s)l^2 + NK_s h^2 + 18(nk_s D^2 + NK_a d^2)}{I}}$$

式中，n 为水平弹簧数量；N 为垂直弹簧数量；K_a、K_s 分别为垂直弹簧的轴向刚度和侧向刚度；k_a、k_s 为水平弹簧的轴向刚度和侧向刚度。

(三) 机械螺旋弹簧的设计

1. 需满足的设计基本要求

(1) 冲击隔震体系有足够低的自振频率，以至其最大动力加速度少于 $1g$，同时应能承担整个结构的重量；

(2) 弹簧中的应力必须小于材料的最大允许剪应力；

(3) 必须有足够空间去构筑弹簧；

(4) 弹簧的面积大小必须不存在屈曲和沉底的危险。

次级的要求有关经济方面和生产方面。例如经济性方面应有少一些的大弹簧，而不是很多小弹簧。弹簧的尺寸使得它是可生产的，例如它应多于半圈，它的直径大于钢材直径的 2 倍。目前最大钢材长度的生产能力限制到 13.208m，如果需要，生产能力可以增加。

2. 设计参数

设计考虑就是在设计过程中变化五个参数（N、D、d、L、n）（弹簧数量、弹簧直径、钢材直径、弹簧自由长度、每一弹簧中圈数）。

3. 设计方程

基座突然下沉后，$t=0$，$x=a$，结构作自由振动：

$$x = a\sin\sqrt{\frac{k}{m}}t$$

$$\ddot{x} = -a\frac{k}{m}\sin\sqrt{\frac{k}{m}}t$$

$$\ddot{x} = a\frac{k}{m} < g$$

$\frac{mg}{a} > a$，$\frac{mg}{k} = \delta_{st}$，即 $\delta_{st} > a$。

考虑到安全系数：$\delta_{ST} = 1.2a$。

最大可能应力发生在向上挠度为 2 倍整个挠度时：

$$\delta_{max} = 2\delta_{st} = 2.4a$$

相应最大剪应力为：

$$\tau_{max} = \frac{\delta_{max}Gd}{\pi D^2 n}$$

$$[\tau_{max}] > \tau_{max} = \frac{2.4Gda}{\pi D^2 n}$$

令材料性质常数 $A = \frac{G}{\pi[\tau_{max}]}$，$[\tau_{max}] = \frac{G}{\pi A}$ 则代入(3.4)，有：

$$\left(\frac{D}{d}\right)^2 nd > 2.4aA$$

此外，对地震加速度而言，对未隔震结构和隔震结构的比较表明：加上隔震系统后，加速度反应谱以及相应的主惯性荷载，可以减少到原来的 1/10~1/5，甚至更多一些。高阶振型可能对未防震结构和非线性隔震结构的震动加速度有较大贡献，但这并不会改变以上反应比较结果。[《工程隔震概论》(新) R. I. Shinner 等]

（四）充气悬置的设计

1. 设计考虑

对于机械弹簧，大结构产生大静挠度，从而导致弹簧中的高应力，并产生屈曲的危险。这个缺点在采用结构支撑在压缩空气垫上时可以被克服。

对于基底面积 929.03m²、3178t 重的结构物，支撑的气压仅为 0.343kg/m² 是一个比较低的压力水平。

2. 空气弹簧性质

图 8 所示的空气垫。初始空气压力 Ψ_0，Ψ_a 大气压力，气垫压缩引程 y 时，压缩空气瞬时压力 Ψ：

空气状态方程：

$$\Psi V^Y = const$$

$$\Psi[(L+y)A]^Y = \Psi_0[LA]^Y$$

$$\frac{\Psi}{\Psi_0} = \left(\frac{1}{1+\frac{Y}{L}}\right)$$

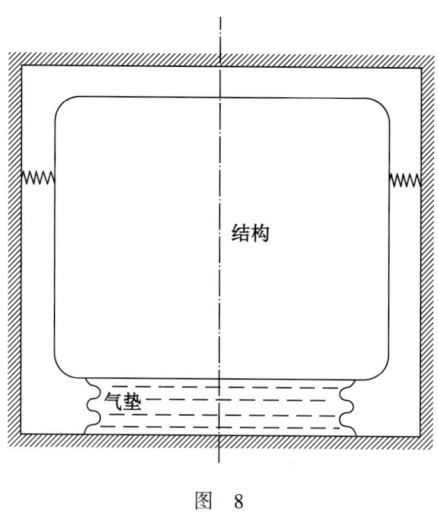

图 8

结构质量的运动方程：
$$\Theta A = mg + m\ddot{y} + \Theta_a A$$

静平衡方程：
$$\Theta_0 A = mg + \Theta_a A$$

联合上式解得：
$$\ddot{y} = g\left(1 + \frac{\varphi_a A}{mg}\right)\left[\frac{1}{\left(1 + \frac{y}{L}\right)^{1.4}} - 1\right]$$

在扰动瞬时产生最大加速度，此时：
$$y = -a$$
$$\ddot{y}_{max} = g\left(1 + \frac{\Theta_a A}{mg}\right)\left[\frac{1}{\left(1 - \frac{a}{L}\right)^{1.4}} - 1\right]$$

令 $\dot{y} = 0$，可求出 y_{max}：
$$\frac{c}{gL}\left[\frac{2.5}{\left(1 + \frac{y_m}{L}\right)^{0.4}} + y_m\right] \quad c = gL\left[\frac{2.5}{\left(1 - \frac{a}{L}\right)^{0.4}} + \frac{a}{L}\right]$$

式中，c 为积分常数。上述方程可画成曲线以求 y_m。

3. 风箱的形状

1) 单一和串联风箱

压缩气垫最可行的方法为一风箱或风箱组，大直径风箱有高环向应力。厚壁风箱可降低环向应力，但引起高的横向应力，并增加风箱刚度。

所以，问题归结为设计一可接受的冲击隔震体系，该体系应不包含高应力水平，并在构造和维护上是可行的。可以采用单一风箱（图 8），但导致高环向应力。采用串联方案（图 10），此时，环应力为一半。

串联风箱体系的冲击隔震特性将不及单一风箱，因为风箱组具有同样的静体积时，串联方案的弹簧系数近似为单一风箱的 1.5 倍。

弹簧系数的比较：

单一风箱的恢复力为：
$$F = (9_0 - 9)A = A9_0\left[1 - \frac{1}{(1+\delta)^\gamma}\right]$$

$\delta = \frac{yA}{V}$ 为相对下沉运动对于位移 $\delta \ll 1$，上式可变成：

$$F = A9_0[1-(1-Y\delta)] = A9_0 Y\delta$$

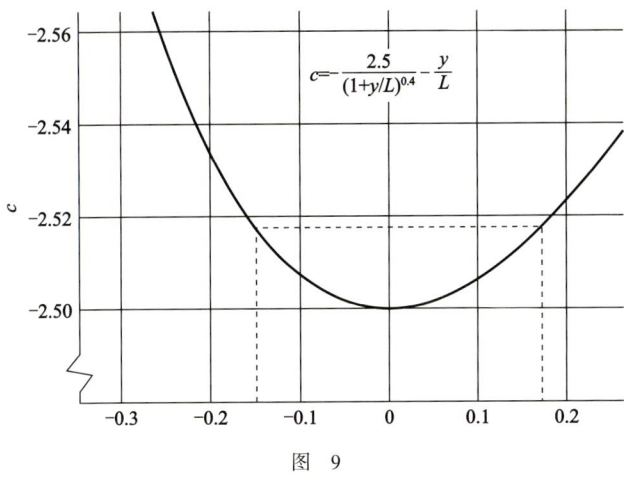

图 9

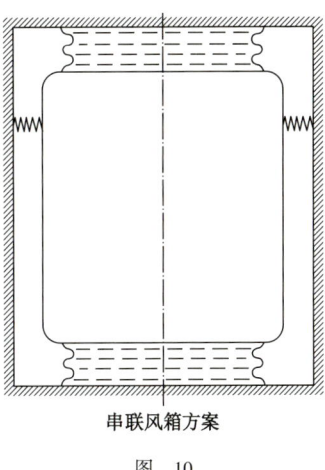

串联风箱方案

图 10

所以,弹簧系数:

$$\frac{F}{y} \cdot \frac{A^2 9_0}{V}\gamma \frac{y}{y} = (A9_a + mg)\frac{A}{V}Y$$

对于串联方案:

$$F = A[(Q_2 - Q_{02}) - (Q_1 - Q_{01})] = A\left\{Q_{02}\left[\frac{1}{(1-\delta)^\gamma} - 1\right] - Q_{01}\left[\frac{1}{(1-\delta)^\gamma} - 1\right]\right\}$$

对于 $\delta \ll 1$:

$$F = A[Q_{02}(1-Y\delta-1) - Q_{01}(1-Y\delta-1)] = A(Q_{02}+Q_{01})Y\delta$$

考虑到一个风箱的压力变化与另一个风箱压力变化相等,符号相反,即

$$Q_{01} = Q_a + \Delta Q$$

$$Q_{02} = Q_a - \Delta Q$$

$$Q_{01} + Q_{02} = 2Q_a$$

$$F = 2AQ_a Y\delta$$

$$\frac{F}{y} = 2\frac{A^2}{V}YQ$$

因而,有:

$$\frac{(F_y)_{串联风箱}}{(F_y)_{单一风箱}} \cdot \frac{2\frac{A^2}{V}YQ_a}{\frac{A^2}{V}YQ_a\left(1+\frac{mg}{Q_a A}\right)} = \frac{2}{1+\frac{mg}{Q_a A}} \approx 1.5$$

(对于我们的结构物, $\frac{Q_a A}{mg} = 3.03$),即 $(F_y)_串 > (F_y)_单$

力-挠度特性:

单一风箱:

$$\frac{F}{mg} = \left(1 + \frac{\varphi_a A}{mg}\right)\left[\frac{1}{(1+\delta)^\gamma} - 1\right]$$

串联风箱:

$$\frac{F}{mg} = \frac{\varphi_a A}{mg}\left[\frac{1}{(1-\delta)^\gamma} - \frac{1}{(1+\delta)^\gamma}\right] + \frac{1}{2}\left[2 - \frac{1}{(1-\delta)^\gamma} - \frac{1}{(1+\delta)^\gamma}\right]$$

当在感兴趣范围($-1 < \frac{F}{mg} < 1$),单一风箱具有较串联风箱更好的冲击隔震特性。

2）盘式风箱

这个方案可以消除高的环向应力，因为它的曲率半径小，所以环向应力低。这个方案允许结构适当位移，但是限制了水平位移。当结构水平运动时，使得风箱边产生剪畸变，在其壁中产生复杂应力状态。横向柔性的缺点压倒了环向应力降低的优点。所以盘式风箱体系将不是一个可行的设计。

3）多个风箱

这种体系降低了风箱半径，所以减少了环向应力。每个风箱能够密封，所以降低了风箱破裂漏气的危险后果。还可将风箱体系与一容器相联通，这将改善冲击隔震性能。如果 N 个风箱支撑结构，其运动方程为：

$$\ddot{y} = g\left(1 + \frac{\Theta_a NA}{mg}\right)\left[\frac{1}{\left(1 + \frac{y}{L}\right)^\varphi} - 1\right]$$

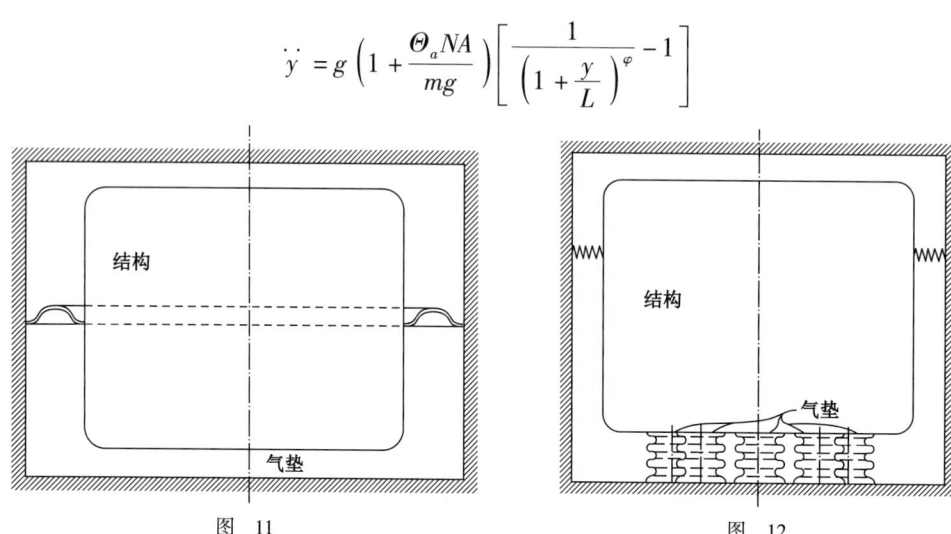

图 11 　　　　　　图 12

在瞬时下沉时，产生最大加速度：

$$\ddot{y}_{max} = g\left(1 + \frac{\Theta_a NA}{mg}\right)\left[\frac{1}{\left(1 - \frac{a}{L}\right)^\Psi} - 1\right]$$

因静平衡方程：
$$mg = (\Psi_0 - \Psi_a)NA$$

有：
$$\frac{\varphi_0}{\varphi_a} = 1 + \frac{mg}{\varphi_a NA}$$

故得：
$$\ddot{y}_{max} = g\left(\frac{1}{1 - \frac{\Psi}{\Psi_0}}\right)\left[\frac{1}{\left(1 - \frac{a}{L}\right)^Y} - 1\right]$$

如果加速度的极限值为 gs，则有：

$$s = \frac{y_{max}}{g} > \left(1 - \frac{\Psi_a NA}{mg}\right)\left[\frac{1}{\left(1 - \frac{a}{L}\right)} - 1\right]$$

或有：

$$s = \frac{y_{max}}{g} > \left(\frac{1}{1 - \frac{\varphi_a}{\varphi_0}}\right)\left[\frac{1}{\left(1 - \frac{a}{L}\right)} - 1\right]$$

对于给定的突然下沉 a_L 和给定的加速度限值，如果风箱的面积小，则要求较高的风箱压力。对于给定的突然下沉，小的风箱面积系统对应最大的力，所以最大的加速度，相对于大的风箱系统为小。

多风箱总面积应小于结构物的总面积为 929.03m²。

故有：
$$NA < 10000\left(\frac{4}{\pi}\right)$$

故 $$\frac{Q_a NA}{mg} < 2.37$$

因此: $$\frac{Q_0}{Q_a} = 1 + \frac{mg}{Q_a NA} \geq 1.422$$

(五) 波纹钢风箱风悬置系统

波纹钢风箱在给定气压下 Q 的环向应力:

$$\sigma_{\text{hoop}} = \frac{YD(Q - Q_a)}{th + tr(\pi - 2)}$$

式中, Y 为波纹半径; D 为风箱平均直径; t 为风箱材厚; h 为波纹高度。

波纹钢中的最大剪应力为:

$$\sigma_{\max} = \frac{3Ph}{\pi Dt^2} \quad \text{或} \quad \sigma_{\max} = \frac{\delta E}{4Jht}\left(\frac{t}{h}\right)^2$$

式中, P 为风箱轴向荷载; δ 为风箱轴向挠度。

(六) 冲击隔震系统中的瞬态现象

瞬态波: 上述各节是基于"准静态"计算, 然而, 十分快的突然下沉将在螺旋弹簧中传播力学波, 而在空气系统中将传播压缩波。压缩波压力 ΔP 的大小取决于速度改变 ΔV 的大小。即: $\Delta P = \pi C \Delta V$ (π 为空气的密度, C 为空气中的声速)。

在本例中, 突然下沉 $a = 0.762\text{m}$ 时, 相应的速度改变 $\Delta V = 13.716\text{m/s}$。

得 $$\Delta p = \frac{0.081 \times 1100 \times 4.5}{32.2 \times 144} = 0.865\text{psi} \ (1\text{psi} = 6894.757\text{Pa})$$

该压缩波传到结构上反射, 在结构表面作用压力为 $2\Delta p$。

但在以前风箱设计中, 结构重量引起的风箱压力为 $14.7\text{psi} = P_s$, 即 $mg = P_s A$, 故 0.865psi 引起的结构加速度为:

$$\ddot{y} = \frac{F}{m} = \frac{2\Delta p A}{mg/g} = \frac{2\Delta p}{P_s} = \frac{2 \times 0.865}{14.7} = 0.117g$$

这个传递的量是可以忽略的。

三、室内爆炸经过隔震系统后的解耦效果

我们通过柱形壳体内空腔爆炸计算已经了解到: 作用在壳体腔壁的爆炸压力 P_m 以及腔壁在爆炸作用下的位移 $y_c(t)$ 及位移速度 $v_c(t) = \dot{y}_c(t)$, 腔型的运动引起风箱传播压缩波, 由此可求得风箱空气柱内压缩波的压力 P_{air} 为:

$$P_{\text{air}} = \pi_{\text{air}} C_{\text{air}} V_c(t)$$

式中, π_{air} 为空气密度; C_{air} 为该压力 (P_{air}) 下压缩波的传播速度 ($C_{\text{air}} = 340\text{m/s}$)。

风箱空气压缩波遇到围岩面反射, 在围岩中传播的压缩波压力 $P_R = 2P_{\text{air}}$。

假设非隔震与隔震情况下, 围岩的加速度与传播的压力成正比, 于是解耦效果可近似表达为:

$$K_{\text{解耦}} = \frac{P_m - P_R}{P_m}$$

相应地, 对于螺旋弹簧的情况, 由于 π、C 很大, 压缩波通过弹簧传递的加速度为 $7.5g$。

高地应力岩体地下工程围岩支护设计计算的问题、原因和对策

一、高地应力岩体洞室支护按连续介质弹塑性力学设计计算的问题

(一) 锦屏一级电站地下洞室群

从 2007 年 1 月到 2009 年初,地下厂房洞室群的开挖到第Ⅷ层;已接近总开挖量的 60%。岩体强度应力比 S 在 2～4 之间,属极高地应力区($S \leqslant 3$)。

1. 变形过大

多点位移计观测到主厂房最大位移达近 87mm,主变室最大位移达 132mm,主变室下游拱肩处最大位移已经达到接近 200mm。

2. 喷混凝土开裂

厂房拱顶滞后开挖约 30 天,开始出现大范围的剥落现象。主厂房开挖到第三层,下游拱肩处开始出现裂缝。

图 1.1　厂房下游拱肩喷混凝土开裂和弯曲的钢筋

厂房拱顶滞后开挖约 30 天,开始出现大范围的剥落现象。主厂房开挖到第三层,下游拱肩处开始出现裂缝。厂房下挖至第四层,厂房下游拱肩裂缝进一步发展,延续到几乎整个下游拱肩。

主变室第三层开挖完成,主变室下游拱肩处也发生与主厂房类似的裂缝,由埋深浅的部位向埋深大的部位扩展。

3. 松动区不断发展

通过声波测试检测到松动区是明显逐步发展的:初期其深度多在 2m 内,随洞室开挖至第三层,其深度增加到 4m,开挖到第四、五层时,某些部位松动区增加到 10m 以上,开挖到六、七层时,某些部位松动区增加到 17m 以上。其次,松动区随深度是间隔的。

报告时间:2018 年 11 月 13 日。

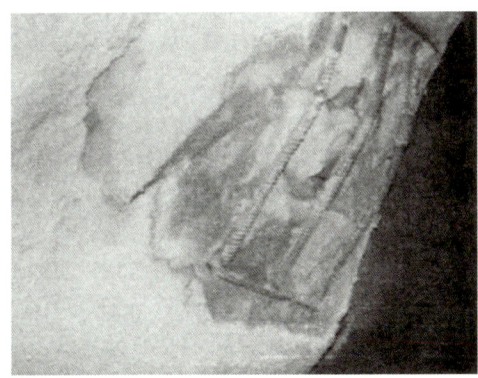

图 1.2　厂房下游喷混凝土的钢筋内鼓及局部裂缝现象

图 1.3　主变室上游侧边墙母线洞间同向张扭性裂缝

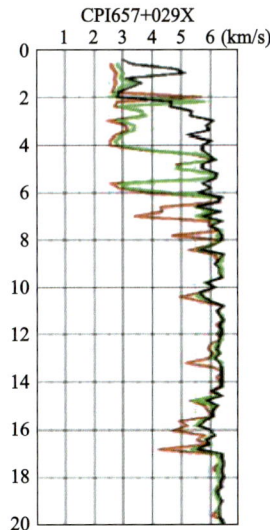
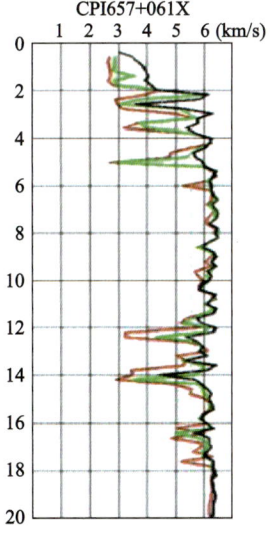

图 1.4　主厂房洞周松动区监测结果

4. 锚索张力超限

现场监测表明,主厂房锚索超出设计吨位的根数,占监测锚索总数的 38.81%,最大超限幅度大于设计吨位的 40% 以上(如设计吨位 200t 的锚索实测值达到 266t,设计吨位 100t 的锚索实测值达到 144t)。

主变室超出设计吨位的锚索根数,占监测锚索总数的43.75%,最大超限幅度约为设计吨位的38%左右(如设计吨位175t的锚索实测值达到241t)。

厂房监测锚杆中已有18.18%的锚杆应力计超过仪器量程。

(二)白鹤滩电站右岸厂房拱顶

围岩强度应力比为2.85~5.09,拱顶深层变形持续发展。

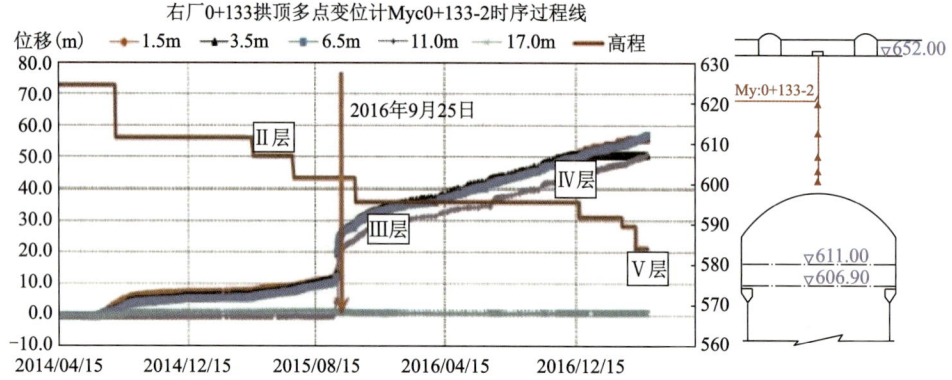

图1.5 右岸厂房顶拱多点位移计监测的顶拱深层变形
(引自白鹤滩监测中心)

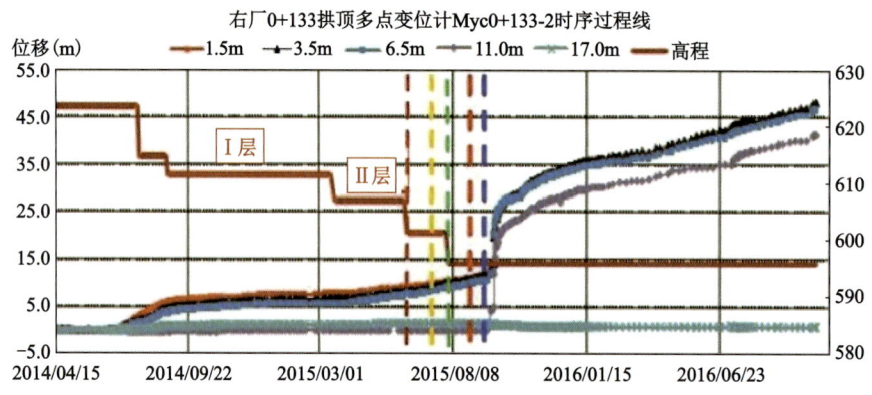

图1.6 右岸厂房顶拱围岩深层变形典型发展过程

白鹤滩电站右岸厂房顶拱0+72洞段,应力转移导致岩体内部8.5m深度位置出现新生破裂,后续1年多时间内破裂面不断扩展。

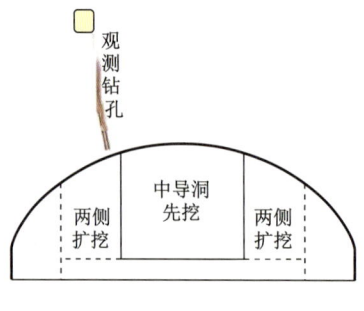

图 1.7

白鹤滩电站右岸厂房顶拱0+133洞段,玄武岩原生裂缝表现时效渐进扩展现象,并促进新的开裂出现。

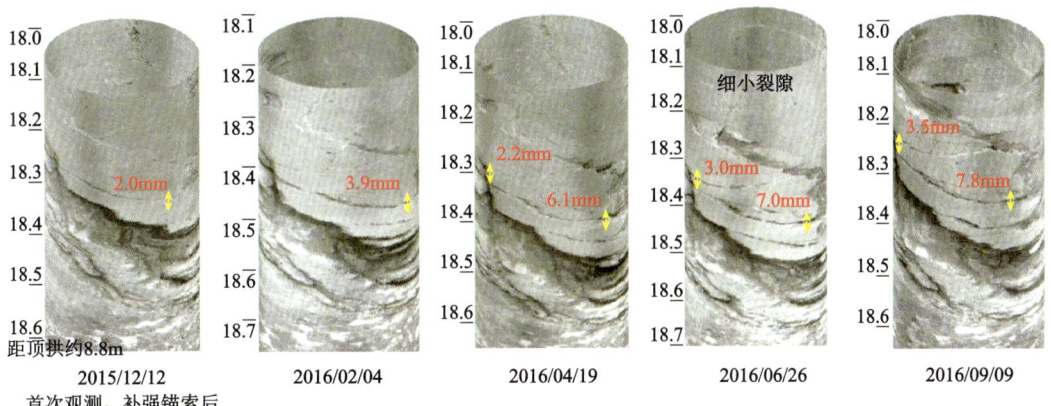

图1.8 R-K0+133-顶拱孔壁岩体破裂时效破裂

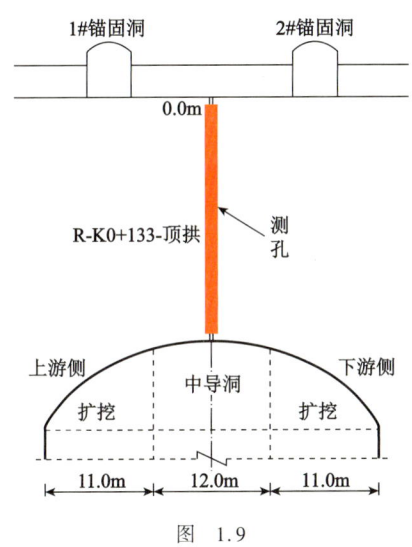

图 1.9

在厂房第3层开挖前,发现距离顶拱轮廓线约11m深度有一条原生裂隙。2015年12月至2016年9月的观测结果显示,后续时效和开挖扰动下该;裂隙面附近出现许多细小破裂,表现为破裂宽度逐渐增加,破裂程度加重。

二、原因

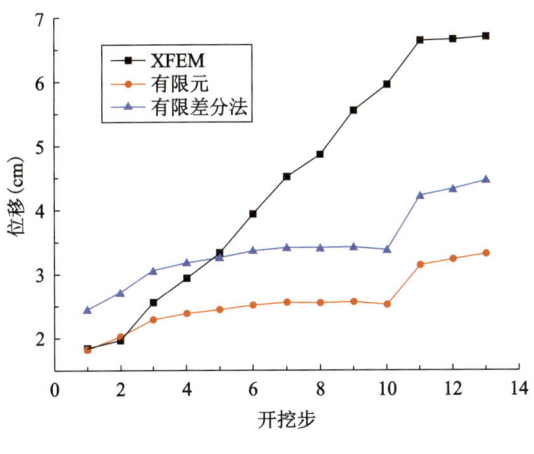

图2.1 主厂房拱顶中点的位移情况对比图

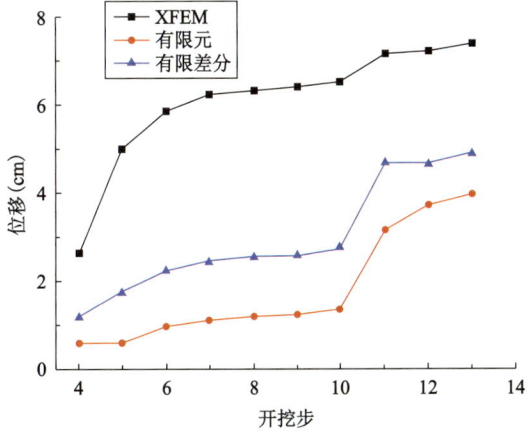

图2.2 主变室锚索位置的位移情况对比图

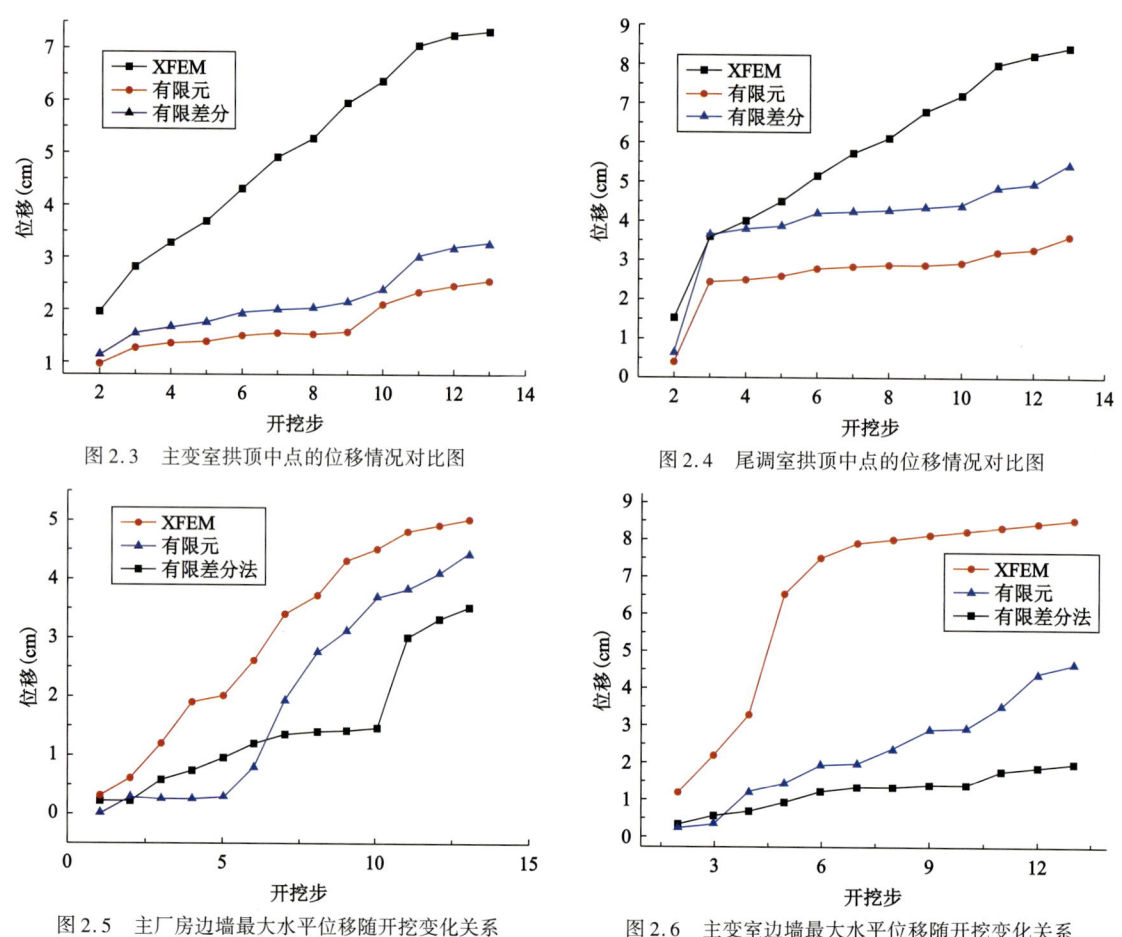

图 2.3　主变室拱顶中点的位移情况对比图　　　　图 2.4　尾调室拱顶中点的位移情况对比图

图 2.5　主厂房边墙最大水平位移随开挖变化关系　　图 2.6　主变室边墙最大水平位移随开挖变化关系

计算结果与锚杆应力监测结果对比　　　　　　　　　　　　　　　表 2.1

位置	锚杆应力监测值(MPa)	XFEM 计算结果(MPa)	有限元计算结果(MPa)	有限差分计算结果(MPa)
主厂房	-38.3 ~ 221.8	0 ~ 259	0 ~ 153.1	0 ~ 139.52
主变室	0.5 ~ 238.0	0 ~ 266	0 ~ 166.94	0 ~ 142.33
1 号尾调室	20.8 ~ 357.7	0 ~ 417	0 ~ 171.7	0 ~ 147.96

计算结果与锚索内力监测结果对比　　　　　　　　　　　　　　　表 2.2

位置		锚索内力监测值(kN)	XFEM 计算结果(kN)	有限元计算结果(kN)	有限差分计算结果(kN)
主厂房	上游边墙 EL1662	1744.46	1800	1385	1242
	下游边墙 EL1662.5	2697.62	2740	2044	1877
主变室	上游边墙 EL1664				
	下游边墙 EL1674.5	2041.77	2055	1874	1726

高地应力岩体工程围岩的变形已发展到破坏阶段：裂缝出现并不断发展，而连续介质弹塑性力学对围岩计算只到屈服为止，屈服不等于破坏，屈服不能反映出现裂缝后的应力释放、应力转移和应力重分布，更不能反映应力重分布后再出现的新裂缝……（剪切破坏只是屈服，不是材料破坏，达到极限应变才是材料破坏-破裂）

所以按该计算方法（以 FLAC 3D 为代表的）计算的围岩变形和松动圈就小，不能反映围岩深层的变形和破裂（应力-应变曲线的下降段）。

从理论上讲，这是建立在变形协调基础上的弹塑性力学不适应非协调变形的情况。

变形协调性:物体变形后仍保持其连续性和整体性。

变形不协调:物体变形后出现"开裂""叠合""封闭内应力"。

对于平面问题,三个应力未知分量为 σ_x、σ_y、τ_{xy},只有两个平衡方程:

$$\frac{\partial \sigma_x}{\partial x} + \frac{\partial \tau_{xy}}{\partial y} + F_x = 0$$

$$\frac{\partial \sigma_y}{\partial y} + \frac{\partial \tau_{xy}}{\partial x} + F_y = 0$$

只有满足以应力分量表示的协调方程:

$$\left(\frac{\partial^2}{\partial x^2} + \frac{\partial^2}{\partial y^2}\right)(\sigma_x + \sigma_y) = -(1+\gamma)\left(\frac{\partial F_x}{\partial x} + \frac{\partial F_y}{\partial y}\right)$$

才能得到包含三个未知函数 σ_x、σ_y、τ_{xy} 的三个微分方程,求得唯一单值解。实际上,在弹性力学解的唯一性证明过程中,是从应力解和位移解都满足平衡方程和协调方程出发的,这是唯一解的前提和基础。

对于位移分量 u、v 表示的平衡方程,似乎不需变形协调方程,即可解得 u、v。这是由于在推导过程中,自动运用位移连续可微条件,即自动代入了变形协调方程。

在有限元求解过程中,单元的位移函数都是连续可微函数!对于围岩出现破坏以后的情况,有限元求得的解已不是真解!

传统有限元在屈服后其单元位移函数仍然是连续函数(多项式),不是裂缝出现后的突跃函数!

所以建立在变形协调解基础上的计算方法不适用于应力-应变曲线上峰值后的下降段,即破裂段。

三、措施

(一)研究变形体的非协调理论

陈宗基院士在第四届国际岩石力学大会的讲话中指出:"岩石力学仅仅对变形为协调的情况感兴趣,例如,在一般的弹性理论中,非协调张量等于零。但自身平衡的初始应力在数学上可用非协调张量不等于零的假定解析,这个假定在岩石力学中将开创一个新的远景。我希望我的建议能够为大家所接受,因为,这关系到岩石力学的基本概念。"

解决所提出问题的措施就是研究发展围岩变形的非协调理论!

1. 岩体非协调变形对围岩应力的影响

岩体的非连续性就是岩体变形的非协调性!

首先来看岩体的非连续性。岩体是非连续的地质体,是由地质构造破碎带、裂隙和节理纵横切割为尺寸大小不同的岩块。岩块中具有细、微观裂纹。上述裂缝和裂纹的延展尺度可以从 10^{-6}m 到 10^6m,宽度可从 10^{-8} 到 10^2m。

在不连续岩体力学中,宏观裂隙可以当作各向异性的连续介质来处理,宏观节理可以用节理单元来研究,描述岩体中的裂隙、断层、节理等宏观结构面的不连续位移和运动特性可以采用不连续介质离散模型,如离散元等。

岩体一般都具有初始的微观和细观裂纹,例如,根据致密岩体和页岩体的最新研究表明:页岩气储层的孔隙直径为 5~200nm,致密灰岩油储层的孔隙直径为 40~500nm,致密砂岩气储层的孔隙直径为 40~700nm,致密砂岩油储层的孔隙直径为 50~900nm。上述孔隙系统占整个储层空间的 80%~90%,局部也发育微米到毫米级孔隙和裂缝等储层空间。

其次，地下工程(包括巷道、隧道和洞室)开挖后，开挖卸荷引起围岩的变形和破坏，围岩中不但产生新的微裂纹，称为次生裂纹，原生初始微裂纹也将扩展、连接、合并成粗裂纹甚至宏观裂纹，总之开挖卸荷导致围岩中裂纹的密度和尺寸都将产生变化，相应地对岩体变形非协调性都将产生新的影响。

他们都将对围岩应力产生影响！

2. 岩体非协调变形对围岩应力的影响

作为计算实例(图3.1)，取圆形洞室(半径a)为平面应变情况中的围岩应力分析问题。假设地层中的初始垂直、水平地应力分别为σ_v、σ_h。

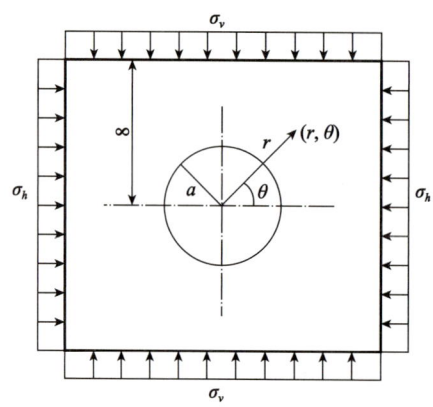

图3.1 圆形洞室力学模型

在岩体存在非协调变形的情况下，围岩应力呈振荡状变化，而且是在单调变化的协调解应力曲线上做非协调应力的振荡变化。

由于协调解应力众所周知为与荷载(初始地应力)相平衡且与边界条件相适应的解，所以非协调解应力即为自平衡(封闭)应力解，这个自平衡应力是由岩体中的缺陷(裂纹)引起的非协调变形(非连续性)产生的。

洞室围岩中应力在径向呈波状振荡变化的现象三十年前在金川镍矿地下矿井围岩中早有发现，现场测量表明围岩中松弛区和严密区交替出现。

因此，我们可以得出结论，初始微裂纹很稀或很小时，是可以忽略不计的，即将不连续岩体看作连续性介质，可以不研究非协调变形的影响，从物理意义上可推测说，稀疏的很小的微裂纹群所引起岩体变形的内能变化量很小，可以在不可逆热力学方程中忽略它。

3. 结论

(1) 岩体变形的协调性是指岩体变形后仍保持其整体性和连续性，其定义为：从物理的观点来说，对岩体及其微元施加荷载产生变形后，能将这些变形后的微元重新拼合为一个变了形的岩体整体；从数学的观点来说，岩体变形后产生的位移函数u、v、w在其定义域内为单值连续函数，即应变和位移满足应变(位移)协调方程。

(2) 以往岩石力学研究岩体的变形仅仅研究岩体的协调变形，即非协调张量等于零(对平面问题$\frac{\partial^2 \varepsilon_x}{\partial y^2} + \frac{\partial^2 \varepsilon_y}{\partial x^2} - \frac{\partial^2 \gamma_{xy}}{\partial x \partial y} = 0$)。无论是位移解和应力解，以往都从满足平衡方程和协调方程出发，这是唯一解的前提和基础。

(3) 岩体变形中出现的"开裂"和"叠合"，形成岩体的非整体和非连续，即呈现变形的不协调性，此时非协调张量不等于零：

$$\frac{\partial \varepsilon_x}{\partial y^2} + \frac{\partial \varepsilon_y}{\partial x^2} - \frac{\partial^2 \gamma_{xy}}{\partial x \partial y} \neq 0$$

（4）研究岩体的非协调变形，主要基于引入内度量非欧几何空间，通过曲率、挠度等非欧几何参量来描述引起岩体变形非协调性的岩体内部缺陷，借助于不可逆热力学理论建立求解非欧几何参量与岩体内部缺陷的关系，拓展传统的弹塑性力学模型：

$$\frac{\partial(1-\nu)}{E}\Delta\sigma = R, \Delta^2 R = \gamma^2 R$$

在非欧空间模型中，R 为缺陷参数——内度量非欧几何空间的标量曲率。

（5）岩体是非连续的地质体，在传统连续介质的弹塑性模型中忽略不计的微、细观裂纹，岩体开挖卸荷后产生的新微裂纹以及原生初始微裂纹的扩展、合并都属于岩体非协调变形范畴。

（6）岩体的非协调变形是产生岩体中的自平衡封闭应力的重要原因。围岩中的非协调张量不等于零的非协调应力解较之非协调张量等于零的协调应力解增加了与岩体内部缺陷相关的自平衡封闭应力解。

（7）对于地下洞室的围岩应力来说，围岩中非协调变形产生的自平衡应力在半径方向呈波状振荡变化，波幅自洞周往围岩深部逐步减小；自平衡应力的大小与内部缺陷的密度与大小密切相关：密度越大，影响越大；缺陷尺度越大，影响越大，当缺陷尺度与密度很小时，其影响可以忽略不计。

因此一般来说，岩体中存在的原生微、细裂纹对围岩应力影响小，都可以忽略不计。

（8）峰值应力前岩石塑性变形的实质是少量微裂纹的产生，根据以上分析，忽略不计它的非协调性误差可能不大，但在峰值应力后的应变软化段，此时微裂纹大量产生，有些扩展合并成粗裂纹和主干裂纹，忽略其非协调性，引起的误差可能较大，甚至很大。所以，高地应力岩体中工程围岩中破坏严重，连续介质弹塑性力学的协调解不适用。

（9）岩体的非协调变形产生围岩中自平衡封闭应力的振荡变化是围岩分区破裂化和围岩深部产生岩爆的本质原因。

（二）研究变形非协调理论

除了上述非欧空间的微分几何描述缺陷的方法外，还有非黎曼几何方法、微分流形方法以及应用近代物理中的规范场论的方法。目前这些方法离开解决实际的岩石力学问题尚有一定距离。但是，研究高地应力岩体工程开挖引起的围岩中裂纹的萌生、发展，从而计算围岩的变形和破坏的数值分析方法发展很快，例如扩展有限元法、DDA 法、RFPA 方法、粒子动力学法等。这些方法正确性的论证以及误差估计目前尚需开展进一步的理论研究。

锦屏一级电站开挖过程地下洞室群围岩变形破坏数值模拟和监测及其分析和启示

一、工程概况及施工中出现的主要问题

(一)厂房布置

地下厂房全长276.99m,吊车梁以下开挖跨度25.60m,以上开挖跨度28.90m,开挖高度68.80m;主变室长197.10m,宽19.30m,总高32.70m。

尾水调压室采用"三机一室一洞"布置形式,设置两个分离圆形调压室,直径分别为34.00m、38.00m(下室),高度分别为80.50m、79.50m。

(二)地质条件

厂区地质条件比较复杂,岩性为大理岩夹绿片岩,围岩类别以Ⅲ1类为主,饱和单轴抗压强度R_b为60~75MPa。地下厂区主要发育有NE向的f13、f14、f18规模较大的三条断层(图),断层走向与厂房轴线约呈45°角;与f18断层相伴还发育有灰绿色云斜煌斑岩脉(X),分布于主厂房空调机房、第一副厂房、主变室及尾水调压室等位置,一般宽2~3m,局部达7m,大多岩性较差,属Ⅳ~Ⅴ类岩体。

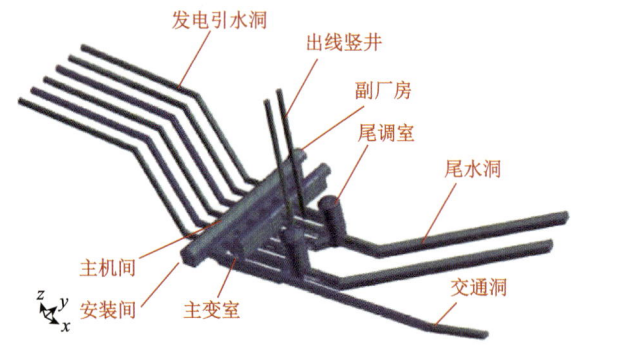
锦屏一级地下厂房洞室群布置示意图

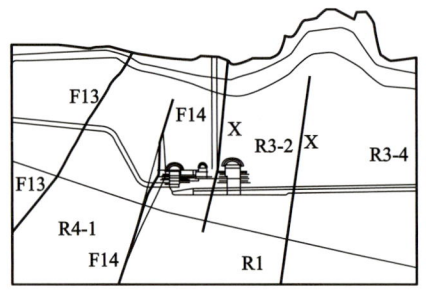
锦屏一级地下厂房地质剖面图

(三)地应力条件

可研阶段在地下厂房范围进行了16个测点的地应力测试实测厂区最大主应力$\sigma_1 = 20 \sim 35.7$MPa,$\sigma_2 = 10 \sim 20$MPa,$\sigma_3 = 4 \sim 12$MPa,σ_1的方向比较一致,介于N28.5°W~N71°W之间,平均N48.7°W,σ_1倾角20°~50°,平均倾角34.2°;属于极高~高地应力区。

报告时间:2016年10月29日。

(四)施工过程中出现的问题

1. 变形过大

从2007年1月到2009年初,地下厂房洞室群的开挖比例接近总开挖量的60%。

多点位移计观测到主厂房最大位移近87mm,主变室最大位移达132mm,明显大于其他地下厂房全部开挖完毕的位移量。

主变室下游拱肩处最大位移接近200mm。

2. 喷射混凝土开裂

主厂房开挖到第三层,开挖高度达到22m时,主厂房5号机组中心线断面附近下游拱肩处开始出现裂缝。

厂房下挖至第四层,主变室下挖至第三层,厂房下游拱肩裂缝进一步发展,并且范围向4号、3号、2号、1号机组方向扩大,延续到几乎整个下游拱肩。

主变室第三层开挖完成,主变室下游拱肩处也发生与主厂房类似的裂缝,由埋深浅的部位向埋深大的部位扩展。

主变室上游侧边墙母线洞间同向张扭性裂缝

厂房下游拱肩喷混凝土开裂和弯曲的钢筋

3. 松动区不断发展

通过物探方法检测到洞周破坏松动区的发展是明显的。统计表明,初期松动区的深度多在2m范围之内,随着洞室继续向下开挖到第三层时,松动区深度增加到2~4m,开挖到第四、五层时,有些部位松动区则增加到10m以上。开挖到第六、七层时,有些部位松动区则增加到17m以上。

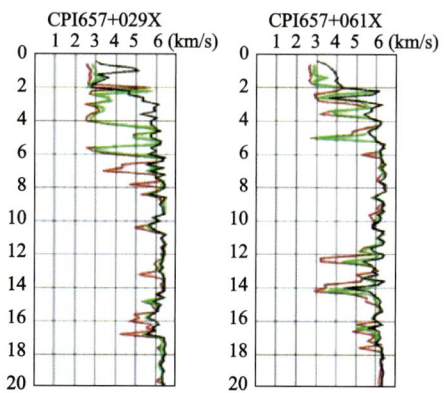

主厂房洞周松动区监测结果

4. 锚索张力超限

现场监测表明,主厂房锚索超出设计吨位的根数,占监测锚索总数的38.81%,最大超限幅度大于设计吨位的40%以上(如设计吨位200t的锚索实测值达到266t,设计吨位100t的锚索实测值达到144t)。

主变室超出设计吨位的锚索根数,占监测锚索总数的43.75%,最大超限幅度约为设计吨位的38%(如设计吨位175t的锚索实测值达到241t)。

二、围岩变形破坏数值模拟和监测结果及其对比

(一)数值模拟内容

(1)采用损伤断裂理论的弱单元扩展有限元(MXFEM)方法,选取纵0+126.8和0+31.7断面,数值模拟Ⅷ分层以上围岩的位移、应力分布规律和锚杆、锚索的内力分布规律,并与现有的监测数据对比分析,以确定计算模型的正确性。

(2)选取纵0+126.8和0+31.7断面,数值模拟Ⅷ分层以下动态施工过程中围岩的位移、应力分布规律。

(3)选取纵0+126.8和0+31.7断面,通过数值模拟Ⅷ分层以下动态施工过程中锚杆的内力分布规律,并预测岩体可能发生的破坏特征。

之所以采用扩展有限元法(ZFEM),是由于它可模拟围岩的变形和破坏,且它不需网格重新剖分就能模拟裂缝扩展。

可以改进单元的形函数以反映裂纹的存在和生长,且改进的形函数在单元内具有单位分解特性,因此扩展有限元法的刚度矩阵具有对阵、稀疏且带状的优点。

(二)计算模型(以纵0+31.7断面为例)

纵0+31.7断面(5号机组断面)计算模型如下图所示,尾调室上游边墙侧靠近f14断层。考虑到锦屏一级水电站地下厂房布置的实际情况,本文计算剖面尺寸取为317m×524m。

模型中按实际支护设计布置了系统锚杆、系统锚索和对穿锚索,以及喷射混凝土层。

纵0+31.7断面计算模型

计算参数表

类别	天然重度(kN/m³)	变形模量(GPa)	泊松比	黏聚力(MPa)	内摩擦角(°)
Ⅲ类围岩	2700	18	0.25	1.5	46.7
f14断层	2700	6.5	0.32	0.5	40
弱单元	2200	3.0	0.35	0.2	37
锚杆	7800	200	0.3	—	—
钢纤维喷射混凝土	2500	23	0.2	—	—

网格中除含 f14 断层外,还随机分布了若干弱单元

(三) 第Ⅰ至第Ⅷ开挖的数值模拟计算结果与监测结果对比

1. 本项目的变形数值模拟计算结果与监测结果(2009 年 11 期安全监测月报)对比

最大水平位移计算结果与变形监测结果对比

位置		多点位移计累计位移(mm)	本项目计算结果(mm)
主厂房	上游边墙 EL1645	10.7	136
	下游边墙 EL1641	59.7	129
主变室	上游边墙 EL1660.8	10.4	59
	下游边墙 EL1664	60.9	140
1号尾调室	上游边墙 EL1666	12.44	39
	下游边墙 EL1651	—	60

从表中可知,本项目的变形计算结果大于监测结果,原因可能是多点位移计安装时围岩已经发生变形,而且监测点位置大都离洞壁 2~6m,不是洞壁的水平位移。但是,从变形规律来看,监测结果和本项目计算都表明主厂房和主变室下游边墙的水平位移较大。

从图中可以看出:

(1) 如果围岩内裂隙扩展并连接形成破裂区,如果围岩内裂隙仅扩展而没有连接则形成扩展区。主厂房上游边墙破裂区的深度约为 11m,主厂房下游边墙破裂区的深度约为 12m,主变室上游边墙破裂区的深度约为 9m,主变室下游边墙破裂区的深度为 8m。

(2) 主厂房上游边墙附近扩展区最大深度为 26m,主厂房和主变室之间围岩的扩展区已经连通。

(3) 主变室下游边墙和尾水调压室上游边墙围岩的扩展区已经贯通。

(4) 主厂房和主变室拱顶扩展区比较小,均小于 3.5m。

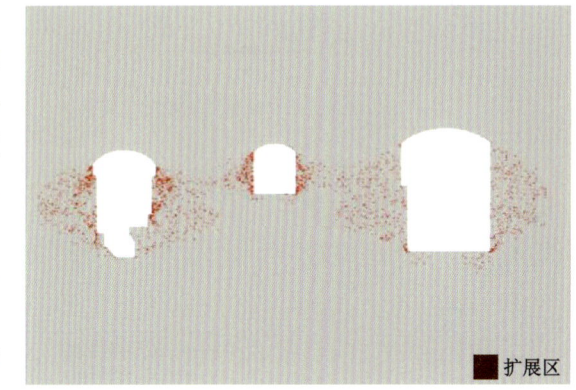

纵 0 + 31.7 断面围岩破裂区与扩展区

(5) 尾水调压室下游边墙围岩的扩展区深度为 23m,无破裂区。

(6) 由于主厂房和主变室的边墙出现了水平方向的拉应力区,因此,主厂房上游边墙 13m 范围内的围岩有劈裂破坏现象,主厂房和主变室之间的岩体基本都出现了劈裂破坏现象,主变室下游边墙 8m 深度范围内存在劈裂破坏现象。

2. 锚杆锚索内力数值模拟计算结果与监测结果对比

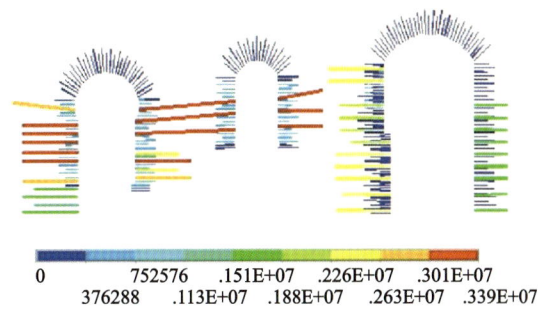

主厂房第Ⅷ分层开挖后锚杆锚索内力总图

(四)第Ⅸ、Ⅹ、Ⅺ分层开挖后的数值模拟计算结果

由左图可知,主厂房第Ⅸ~第Ⅺ分层开挖过程中,主厂房拱顶下沉了约0.2cm,而由于底板宽度随开挖进程不断减小,其上鼓位移也趋于稳定,也就是说主厂房第Ⅸ~第Ⅺ分层开挖对拱顶下沉的影响较小。

由右图可知,主厂房第Ⅸ~第Ⅺ分层开挖过程中,主变室和尾调室底板竖直位移变化不大,也就是说主厂房第Ⅸ~第Ⅺ分层开挖对主变室和尾调室的底板的竖直位移影响较小。此外,主变室拱顶在主厂房第Ⅸ~第Ⅺ分层开挖过程中下沉了1.5cm。

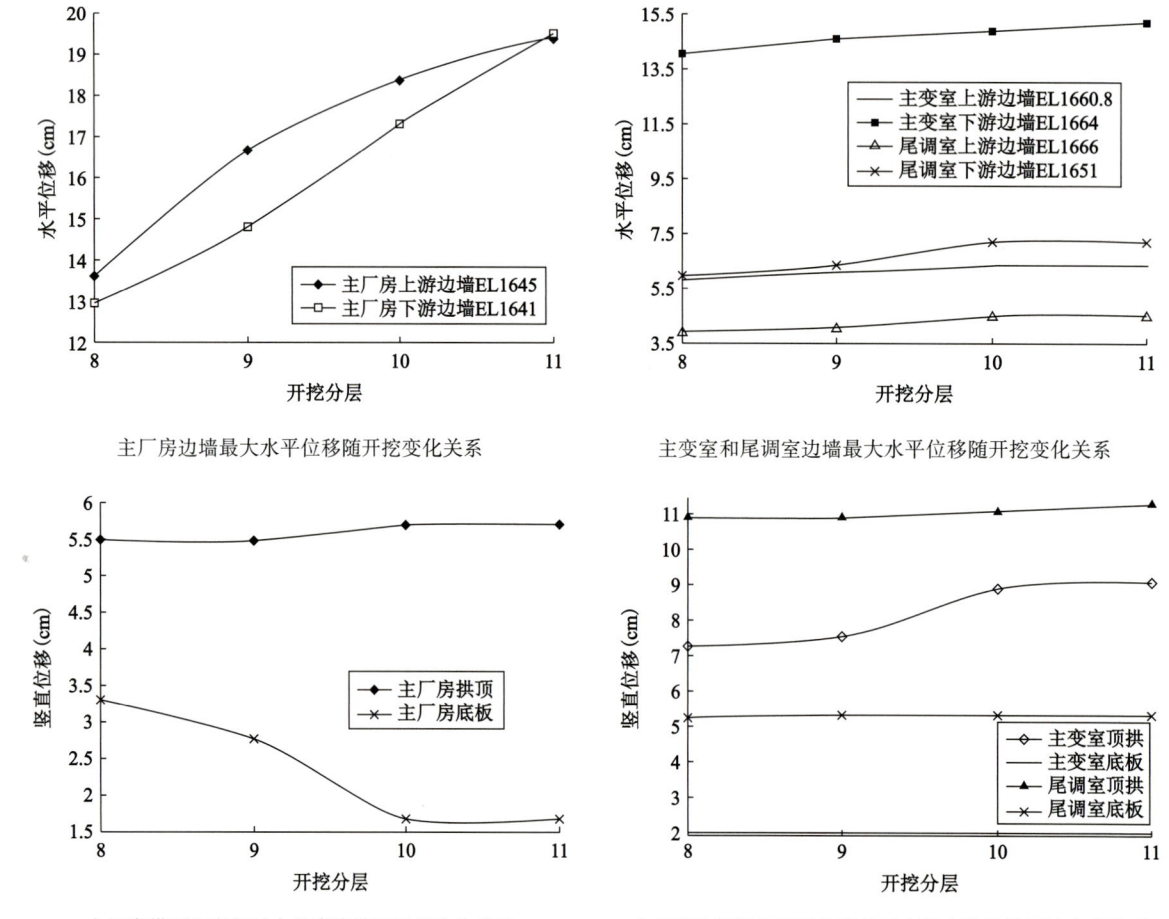

主厂房边墙最大水平位移随开挖变化关系　　主变室和尾调室边墙最大水平位移随开挖变化关系

主厂房拱顶和底板最大竖直位移随开挖变化关系　　主变室和尾调室拱顶和底板最大竖直位移随开挖变化关系

主厂房第Ⅸ~第Ⅺ分层开挖过程中地下洞室群边墙的水平位移增量(单位:cm)

分层开挖	主厂房		主变室		尾调室	
	上游边墙	下游边墙	上游边墙	下游边墙	上游边墙	下游边墙
Ⅸ分层开挖	3.1	1.9	0.2	0.6	0.2	0.4
Ⅹ分层开挖	1.7	2.5	0.3	0.3	0.4	0.8
Ⅺ分层开挖	1.0	2.2	0.0	0.3	0.0	0.0

表中为主厂房第Ⅸ~第Ⅺ分层开挖过程中地下洞室群边墙最大水平位移增量,从表中可以看出,由于地应力高和岩体力学参数低的影响,在主厂房第Ⅸ~第Ⅺ分层开挖过程中主厂房下游边墙最大水平位移变化明显,主变室和尾调室边墙最大水平位移较小。

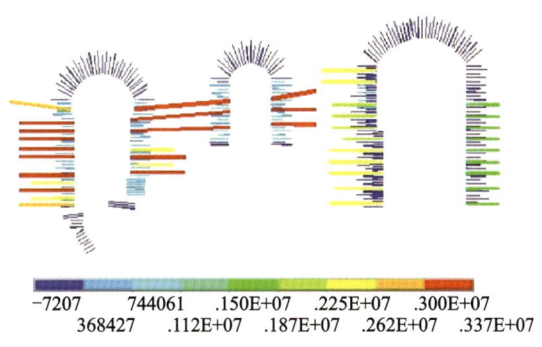

主厂房第Ⅺ分层开挖后锚杆锚索内力总图

(五) 第Ⅰ～第Ⅷ分层开挖的数值模拟计算结果与监测结果对比

(1) 由于高地应力和岩体参数较低的影响,在主厂房第Ⅸ～第Ⅺ分层开挖过程中,主厂房上游边墙的最大水平位移增加了5.8cm,而下游边墙的最大水平位移增加了6.6cm,也就是说,在第Ⅸ～第Ⅺ分层开挖过程中,主厂房下游边墙的水平位移大于对上游边墙的水平位移。但是,在主厂房第Ⅸ～第Ⅺ分层的开挖过程中,拱顶下沉量较小。

(2) 洞室开挖以后主厂房、主变室和尾调室拱脚围岩产生了较大范围的竖直方向压应力集中,最大竖直方向压应力超过65MPa,由于拱脚竖直方向压应力集中,三大洞室拱脚岩体有局部被压碎的可能,尤其是主厂房。由于主厂房和主变室的边墙出现了水平方向的拉应力区,因此,主厂房上游边墙15m范围内的围岩有劈裂破坏现象,主厂房和主变室之间的岩体基本都出现了劈裂破坏现象,主变室下游边墙10m深度范围内存在劈裂破坏现象。

(3) 主厂房和主变室部分预应力锚索超限严重,短锚杆对围岩变形约束作用有限;由于高地应力和岩体参数较低的影响,在主厂房第Ⅸ～第Ⅺ分层开挖过程中,主厂房上游边墙第Ⅷ分层以下的部分锚索轴力明显增加。

三、地下厂房围岩变形破坏严重,部分锚索超限的原因分析

围岩变形、破坏严重,锚索内力超限的根本原因是由强烈压缩岩体开挖卸荷引起的围岩变形、破坏特点所决定的。

地下厂房围岩的强烈压缩是由上覆岩体的自重和构造应力所产生的高地应力所引起的。地应力的高低是相对于岩体的单轴压缩强度而言,锦屏一级电站地下厂房围岩的地应力大都超过其单轴抗压强度的一半,因此属高地应力,即属强烈压缩。

在高地应力岩体中开挖洞室时,在洞室边界上形成高地应力岩体的卸荷,开挖卸荷产生以下三个效应,并自围岩表面向深部逐步转移。

1. 径向拉伸波效应

开挖后向围岩深部转移拉伸效应(拉伸波),因为自由面处径向应力为零,围岩深度处地应力高,因此岩体自围岩深部径向向洞室产生位移和位移速度,形成岩体的拉应变,该值超过岩体的拉伸极限值时,即产生拉伸破坏,如地应力高,则径向应力梯度大,产生的位移速度大,岩体在其破坏后还有剩余速度就形成所谓的"岩爆"。

2. 支座压力转移的支座压力波效应

开挖形成洞室周边围岩的应力集中现象,由于应力集中,洞室周边围岩要承受比未开挖前的地应力更大的周向应力,例如支座一样,所以称为支座压力区,初始强烈压缩(高地应力)围岩中的支座压力区由于周向应力集中形成的高周向压应力和径向拉应变联合极易造成围岩的破坏,破坏后不能承受应力集

中的高周向压应力(又名驼峰压力),支座压力区就要逐步向围岩深部转移,形成所谓向深部转移的支座压力波效应。

3. 伪自由面转移效应

如果洞室开挖的边缘称为自由面,则洞室围岩由径向拉伸和支座压力区联合破坏产生的裂缝将形成新的自由面,产生应力释放和新的卸荷,为区别原生自由面,称为伪自由面,伪自由面将形成新的应力集中区(支座压力区),如果初始地应力很高,则将产生新的围岩破坏区……形成第二个乃至第三个围岩破坏区和伪自由面逐次向深部转移。

综上所述,高地应力岩体开挖卸荷形成的上述三个效应使得围岩的破坏不断向深部转移,不同于浅部低地应力区的岩石工程,其洞室周围仅一个屈服松弛区。而是多区破裂,或称分区破裂化现象。所以高地应力区洞室周围的围岩变形、破坏就较浅部岩石工程低地应力区的严重得多。围岩变形、破坏随时间不断向深部空间的转移表现为围岩变形的流变特点,这就是陈宗基先生所说的残余应力区的围岩流变。

上述分析是仅就单一洞室开挖,二维高地应力卸荷而言,对于目前锦屏一级电站地下厂房的开挖,它是主厂房、主变室和母线洞三维开挖卸荷以及F14、F18、F13多条裂隙分割的情况,更加复杂,其围岩变形、破坏也就更加严重。

四、高地应力围岩开挖卸荷形成的二维分区破裂化现象的一般规律

经过对于高地应力区分区破裂化现象的现场探测研究、实验室的模拟材料试验研究以及数值分析研究,其成果表明:

(1)岩体初始地应力 $\sigma_{地}$ 相对于其单轴抗压强度越高,即 $\sigma_{地}/\sigma_{压}$ 越大,围岩中破裂区数量越多,围岩变形、破坏越严重。

(2)岩体初始水平地应力相对垂直地应力越大,围岩中破裂区数量越多,围岩变形、破坏越严重。

(3)巷道、洞室开挖速度越快,即卸荷越快,围岩中破裂区数量越多,围岩变形、破坏越严重。

(4)岩体原生节理越严重,围岩中破裂区数量越多,围岩变形、破坏越严重。即原生节理对破裂区的形状和分布有重要影响。

五、锦屏一级电站地下厂房开挖及高地应力岩体开挖围岩支护加固措施的建议

针对电站围岩变形破坏严重的原因和机理,现提出下列地下厂房开挖时的围岩支护加固措施建议:

(1)洞室开挖宜分层分段,以降低卸荷速度,缓和应力集中现象,减缓应力集中区及支座压力区向深部的转移。

(2)加固措施紧跟开挖施工,即随分层分段开挖,加固(包括注浆、锚杆、锚索)应逐层逐段跟进,以在围岩破坏前提高围岩强度,减缓支座压力和伪自由面向深部围岩的转移,从而减缓分区破裂现象的产生,以减少围岩的变形、破坏。

(3)锚索长度,即锚固段应达到围岩深部最远破裂区后面的未破裂区;注浆深度应达各个破裂区。并且早注浆、早上锚索,在施工允许条件下越早越好。注浆宜分二次,逐次加压,以适应裂缝逐步扩展的实际情况。对已开挖加固区的变形未收敛区应再次注浆,以提高围岩强度,加速收敛,但注浆压力应控制,边注浆边加强锚索内力观测,防止锚索损伤。

(4)采用压力分散型锚索时,锚固段数量由破裂区数量来确定,几个破裂区采用几个锚固段。锚固段应位于破裂区之间未破裂或微破裂区。

(5) 锚杆长度应位于第一破裂区后面,破裂区可采用钻孔探测或超声波探测。

六、锦屏一级电站地下洞室群开挖变形破坏的启示

(1) 高地应力岩体开挖卸荷的综合效应使得围岩形成破坏并逐次向深部转移,形成分区破裂,或称分区破裂化现象,它一般发生于深部岩体,在高构造地质应力情况下,也可发生于不深部位,例如锦屏一级电站地下洞室群。它不同于浅部低地应力岩体开挖卸荷的围岩,仅在洞室周围形成一个变形较大的屈服区,并不一定导致围岩的破坏。

(2) 传统的弹塑性力学理论,只能分析岩体是否达到屈服强度,不能分析岩体的破坏及其发展:裂缝的产生、扩展和合并。因此只适用于分析浅部低应力岩体开挖时的围岩变形,不适用于分析深部高地应力岩体开挖时的围岩严重变形和破坏,即分析围岩分区破裂化现象,但在传统弹塑性力学理论基础上,郑颖人院士团队发展了数值极限分析法,也可用以分析围岩破裂的产生和发展。

(3) 分析围岩的变形破坏可采用 DDA 分析法、扩展有限单元法、广义粒子动力学和非欧模型分析法等。

深部岩体分区破裂化机理的研究进展

一、研究背景

南非金矿、俄罗斯矿山、中国淮南煤矿、孙村煤矿和梁庄煤矿等深部岩体工程围岩中,采用多种物理手段探测结果表明,分区破裂化现象在高地应力深部巷道围岩中确实存在。

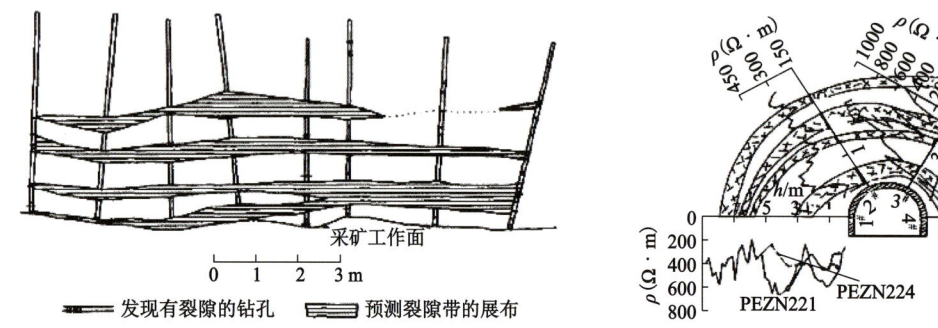

南非 Witwatersrand 金矿巷道顶板分区破裂化现象

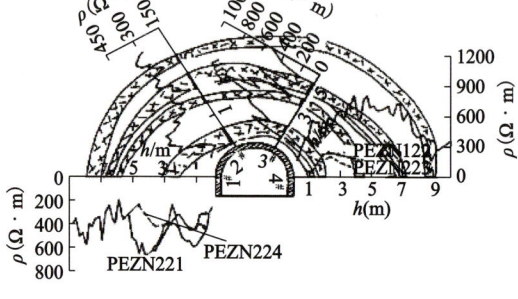

俄罗斯 Taimyrskii 矿山巷道围岩的分区破裂化现象

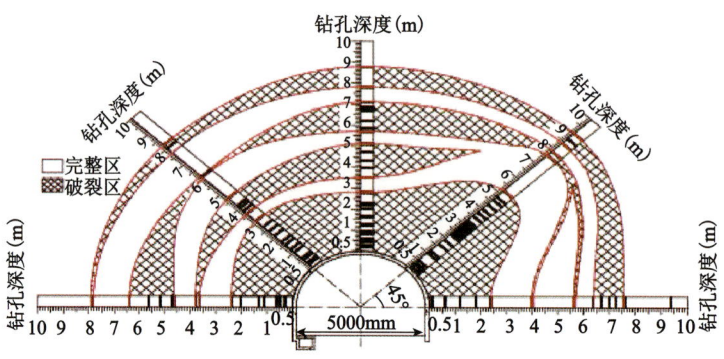

淮南丁集煤矿巷道监测断面 B 围岩分区破裂分布(李术才等)

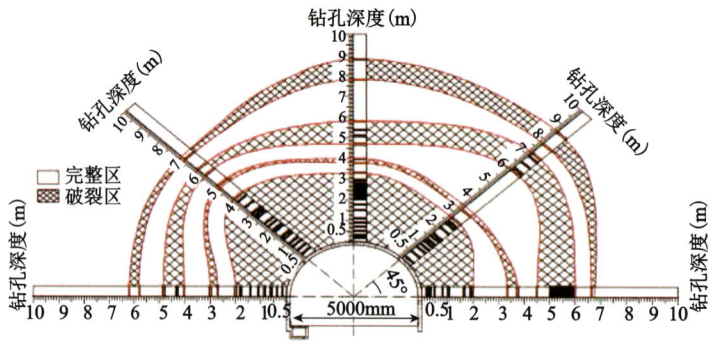

淮南丁集煤矿巷道监测断面 C 围岩分区破裂分布(李术才等)

报告时间:2016 年 6 月 17 日。

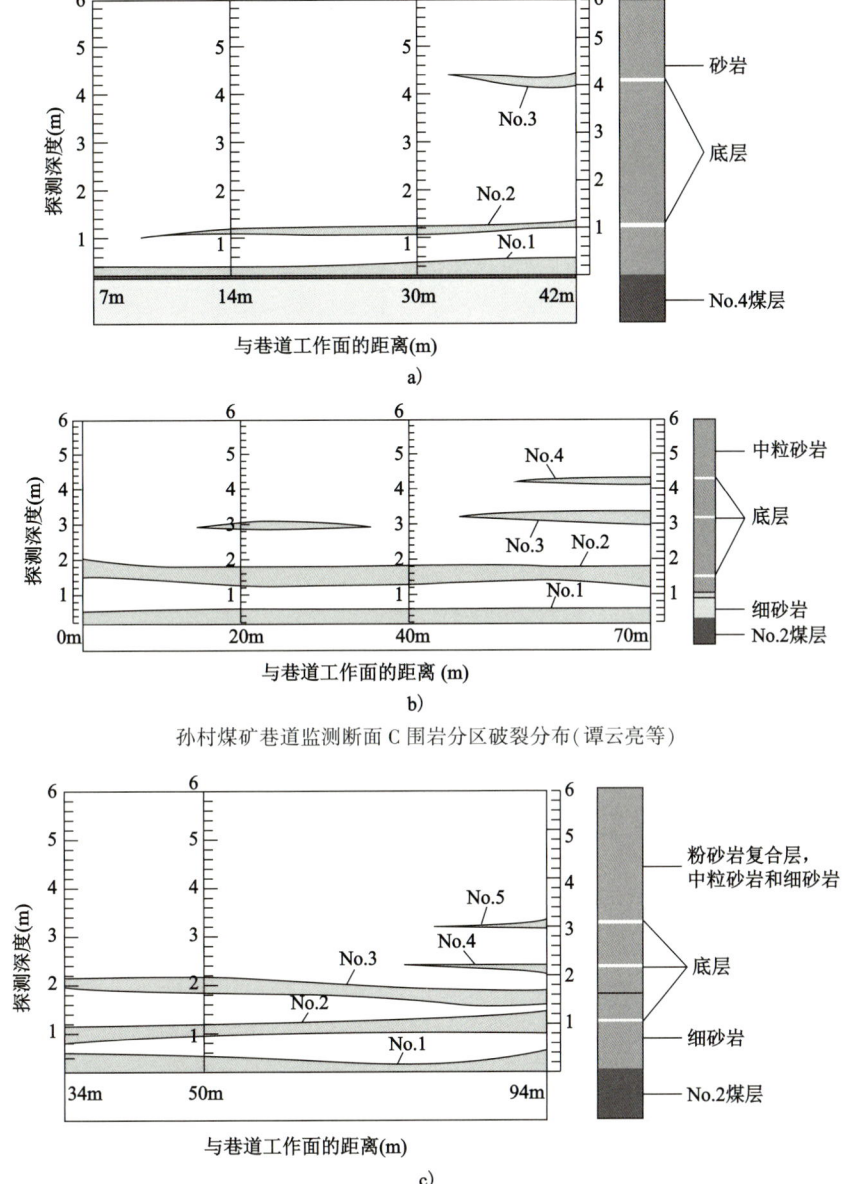

孙村煤矿巷道监测断面 C 围岩分区破裂分布(谭云亮等)

梁庄煤矿巷道监测断面 C 围岩分区破裂分布(谭云亮等)

二、深部岩体分区破裂化现象特征

南非、俄罗斯和中国的现场观测到的分区破裂化现象是相似的:围岩中破裂区和非破裂区交替发生;破裂区是裂缝相对密集的区域,非破裂区又称亚破裂区,是裂缝相对稀疏的区域。

分区破裂化既可在钻爆法矿井中发现,也可在机械化开挖的矿井中发现,因此,可排除分区破裂化是爆破振动引起的结论。

三、深部岩体分区破裂化现象室内模型试验

俄罗斯科学院西伯利亚分院和苏联煤炭部矿岩力学与矿山测量研究院在 20 世纪 80～90 年代,在实验室模拟材料试验中重现了分区破裂化现象,其经验是在平面应变模型中严格保证平面应变条件;巷道

(钻孔)开挖是在预加应力的模型中进行的。

俄罗斯学者格罗斯欣、我国学者张强勇在模型试验中发现:巷道轴线方向的主应力,对产生分区破裂化现象具有重要影响。但不能认为水平方向的主应力必须是"最大主应力",因为在垂直主应力为最大主应力的矿井中以及模型试验中均发现了分区破裂化现象。

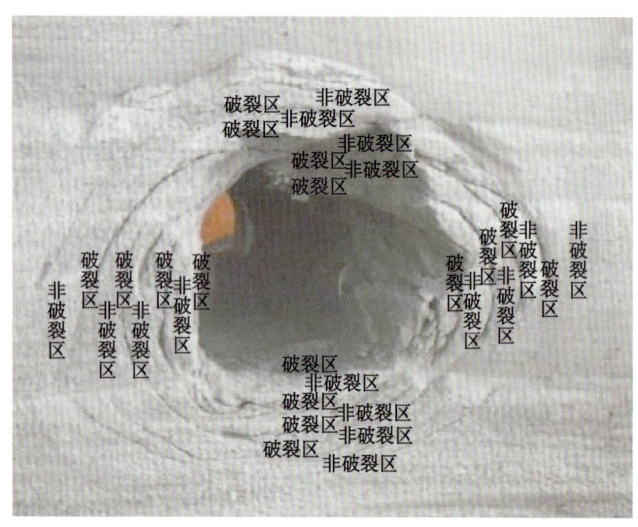

开挖后模型洞周破坏形态(张强勇等)

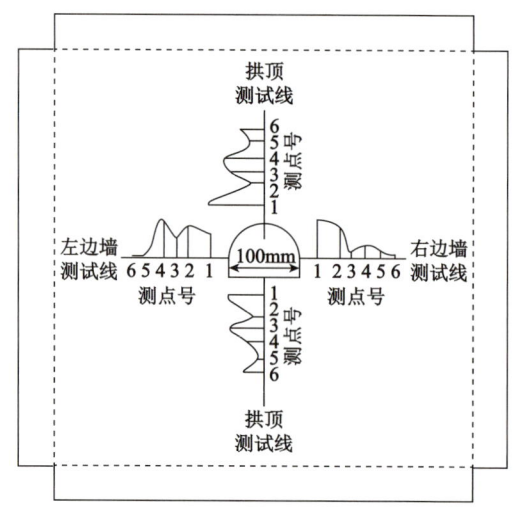

模型洞周测点位移变化规律(张强勇等)

四、深部岩体分区破裂化现象的研究进展

俄罗斯的 Shemyakin 等采用弹塑性模型解释分区破裂化现象,能部分解释分区破裂化现象,但是分区破裂化机理没有完全揭示。弹塑性模型情况下的破裂区数量和位置可以表示为:

$$r_i = n^{i-1} r_1 \tag{1}$$

$$n = [(\sigma_\mu + C\cot\varphi)(1 - \sin\varphi)/(C\cot\varphi)]^{(1-\sin\varphi)/(2\sin\varphi)} \tag{2}$$

Guzev 等在推导深部岩体发生分区破裂化的临界应力条件时,认为洞室边界上的标量曲率(岩石内部结构的非欧参数,或者缺陷参数)为零,其导数也为零,即标量曲率的边界条件可表示为:

$$\begin{cases} R\vert_{r=r_0} = 0 \\ \dfrac{\partial R}{\partial r}\bigg\vert_{r=r_0} = 0 \end{cases} \tag{3}$$

Guzev 推导的深部岩体发生分区破裂化的临界应力条件:

(1)当地应力满足 $\sigma_\infty^* < q\mu_1 = (1+v_1)\mu_1$ 时,不会发生分区破裂化现象,标量曲率没有振荡特性,标量曲率可以表示为:

$$\begin{aligned} R_{1,+} &= c_1 K_0(r\sqrt{Y_-/h}) \\ R_{1,-} &= c_2 K_0(r\sqrt{(2-Y_-)/h}) \\ Y_- &= 1 - \sqrt{P^*/q\mu_1} \end{aligned} \tag{4}$$

应力场可表示为式(2),也没有振荡特性:

$$\begin{cases} \sigma_r = \sigma_伪 \left(1 - \dfrac{r_0^2}{r^2}\right) \\ \sigma_\theta = \sigma_伪 \left(1 + \dfrac{r_0^2}{r^2}\right) \end{cases} \tag{5}$$

(2) 当临界应力满足 $\sigma_\infty^* > q\mu_1 = (1+v_1)\mu_1$ 时，会发生分区破裂化现象，标量曲率具有振荡特性，其标量曲率为：

$$\begin{cases} R_{2,+} = a_1 J_0(r\sqrt{\gamma_+/h}) + b_1 Y_0(r\sqrt{Y_+/h}) \\ R_{2,-} = c_3 K_0[r\sqrt{(2+\gamma_+)/h}] \\ Y_+ = -\sqrt{P_\infty^*/q\mu_1} - 1 \end{cases} \quad (6)$$

同时，获得了静水压力条件下深部圆形洞室围岩的应力场，确定了静水压力条件下深部圆形洞室围岩应力场的波动特性：

$$\begin{cases} \sigma_r = \sigma_父\left(1-\dfrac{r_0^2}{r^2}\right) - \dfrac{E}{2Y^{3/2}(1-V^2)r^4}[aJ_1(\sqrt{\gamma}r) + bN_1(\sqrt{\gamma}r) + cK_1(\sqrt{\gamma}r)] \\ \sigma_\theta = \sigma_父\left(1+\dfrac{r_0^2}{r^2}\right) + \dfrac{E}{2Y^{3/2}(1-V^2)r}[aJ_1(\sqrt{\gamma}r) + bN_1(\sqrt{\gamma}r) + cK_1(\sqrt{\gamma}r)] \\ 2Y(E-V^2)[aJ_0(\sqrt{\gamma}r) + bN_0(\sqrt{\gamma}r) - cK_0(\sqrt{\gamma}r)] \end{cases} \quad (7)$$

式中，J_0 和 J_1 分别为零阶和一阶贝塞尔函数；N_0 和 N_1 分别为零阶和一阶诺依曼函数；K_0 和 K_1 分别为零阶和一阶第二类修正的贝塞尔函数。

Guzev 认为岩体含有缺陷，缺陷会产生附加应力场，即自平衡应力。同时，基于非欧模型获得了静水压力条件下自平衡应力：

$$T_{ij} = \dfrac{4\alpha\mu_1\mu_3}{3(1-\alpha)g}\varepsilon_{ipq}\varepsilon_{jpn}\delta_{pn}\dfrac{\partial^2 R}{\partial x^p \partial x^s} \quad (8)$$

其中：

$$\Delta^2 R + \Gamma R = 0 \quad (9)$$

$$\Gamma = \dfrac{3\beta(3\lambda_2 + 2\mu_2)(1\_C)}{2[3\lambda(1\_C) + 2\mu_2(3\_C)]} \quad (10)$$

Guzev 等采用 Tresca 和 Mises 准则研究深部岩体破裂区的宽度和数量。Tresca 准则可以表示为：

$$K_T = A\sigma_r - \sigma_\theta \quad (11)$$

Mises 准则可以表示为：

$$K_T = B(\sigma_2^r + \sigma_\theta^2 - 2\sigma_r\sigma_\theta)^{1/2} \quad (12)$$

钱七虎院士和周小平在推导深部岩体发生分区破裂化的临界应力条件时，认为洞室边界上的标量曲率为常数且具有极大值，同时该常数依赖于岩体中缺陷的分布情况，标量曲率的边界条件可表示为：

$$R\big|_{r=r_0} = m \quad \begin{cases} \dfrac{\partial R}{\partial r}\big|_{r=r_0} = 0 \\ \dfrac{\partial^2 R}{\partial r^2}\big|_{r=r_0} < 0 \end{cases} \quad (13)$$

其中，m 依赖于岩体中缺陷的分布情况

对平面应变问题，钱七虎院士和周小平推导的深部岩体发生分区破裂化的临界应力条件为：

(1) 当地应力满足 $\sigma_伪 < \dfrac{(\lambda_1 + \mu_1)\mu_1}{2\mu_1 + \lambda_1}$ 时，不会发生分区破裂化现象，标量曲率没有振荡特性：

$$R = 0 \quad (14)$$

应力场可表示为式(2)，也没有振荡特性。

(2) 当地应力满足 $\sigma_伪 \geq \dfrac{(\lambda_1 + \mu_1)\mu_1}{2\mu_1 + \lambda_1}$ 时，会发生分区破裂化现象，标量曲率具有振荡特性，其标量曲率为：

$$R = a_0 J_0(r\sqrt{Y_1}) + b_0 N_0(r\sqrt{\gamma_1}) + c_0 K_0(r\sqrt{\gamma_1}) \tag{15}$$

式中,J_0 为零阶贝塞尔函数;N_0 为零阶诺依曼函数;K_0 为零阶的第二类修正的贝塞尔函数。

对三维问题,钱七虎院士和周小平推导的深部岩体发生分区破裂化的临界应力条件为:

(1)当地应力满足 $\sigma_伪 < \dfrac{(3\lambda_1 + 2\mu_1)\mu_1}{3(2\mu_1 + \lambda_1)}$ 时,不会发生分区破裂化现象,$R=0$,标量曲率没有振荡特性。应力场可表示为式(2),也没有振荡特性。

(2)当地应力满足 $\sigma_父 \geq \dfrac{(3\lambda_1 + 2\mu_1)\mu_1}{3(2\mu_1 + \lambda_1)}$ 时,会发生分区破裂化现象,标量曲率具有振荡特性,其标量曲率为:

$$R = \frac{A_1 \sin(r\sqrt{\chi}) + A_2 \cos(r\sqrt{\chi})}{r} + \frac{A_3[\cosh(r\sqrt{\chi}) - \sinh(r\sqrt{\chi})]}{r}$$

同时,国际上最早确定了轴向应力作用下深部圆形洞室围岩的应力场、破裂区的宽度和数量。

$$\sigma_r = \left[\frac{p_z + 2\sigma_\infty - E\varepsilon_0}{2(1+v)}\right] / \left(1 - \frac{r_0^2}{r^2}\right) - \frac{E}{2\rho^{3/2}(1-v^2)r}[AJ_1(p^{1/2}r) + BN_1(p^{1/2}r) + CK_1(p^{1/2}r)] \tag{16}$$

$$\sigma_\theta = \left[\frac{p_z + 2\sigma_\infty - E\varepsilon_0 - 2(1+v)}{E}\right] / \left(1 + \frac{r_0^2}{r^2}\right) -$$

$$\frac{E}{2p(1-v^2)}[AJ_0(p^{1/2}r) + BN_0(p^{1/2}r) - CK_0(p^{1/2}r)] +$$

$$\frac{E}{2p^{3/2}(1-v^2)r}[AJ_1(p^{1/2}r) + BN_1(p^{1/2}r) + CK_1(p^{1/2}r)] \tag{17}$$

$$\sigma_z = \left[\frac{v(p_z + 2\sigma_父) + E\varepsilon_0}{1+v}\right] - \frac{vE}{2\rho(1v^2)}[AJ_0(p^{1/2}r) + BN_0(p^{1/2}r) - CK_0(p^{1/2}r)] \tag{18}$$

当深部岩体应力满足深部岩体强度准则时,岩体发生破裂。深部岩体强度准则表示如下:

$$\begin{cases} F_1 = \sigma_1 - \dfrac{1}{1+a}(a\sigma_2 + \sigma_3) + W(\sigma_1 + b\sigma_2 + \sigma_3)\dfrac{l_1}{\sigma_c} - \sigma_c\left[\dfrac{b\sigma_1}{\sigma_c} + \dfrac{d}{(1+a)\sigma_c}(a\sigma_2 + \sigma_3) + b^2\right]^n = 0 & F_1 \geq F_2 \\ F_2 = \dfrac{1}{1+a}(\sigma_1 + a\sigma_2) - \sigma_3 + W(\sigma_1 + b\sigma_2 + \sigma_3)\dfrac{l_1}{\sigma_c} - \sigma_c\left[\dfrac{b\sigma_1 + a\sigma_2}{(1+a)\sigma_c} + \dfrac{d\sigma_3}{\sigma_c} + b^2\right]^n = 0 & F_1 < F_2 \end{cases} \tag{19}$$

轴向应力对分区破裂化影响的数值计算结果参数如下:$\sigma_c = 40\text{MPa}$,$r_0 = 7\text{m}$,$V = 0.25$,$GSI = 50$,$m_i = 10$,$\sigma_伪 = 35\text{MPa}$,$p = 2700\text{kg/m}^3$,$E = 18\text{GPa}$。

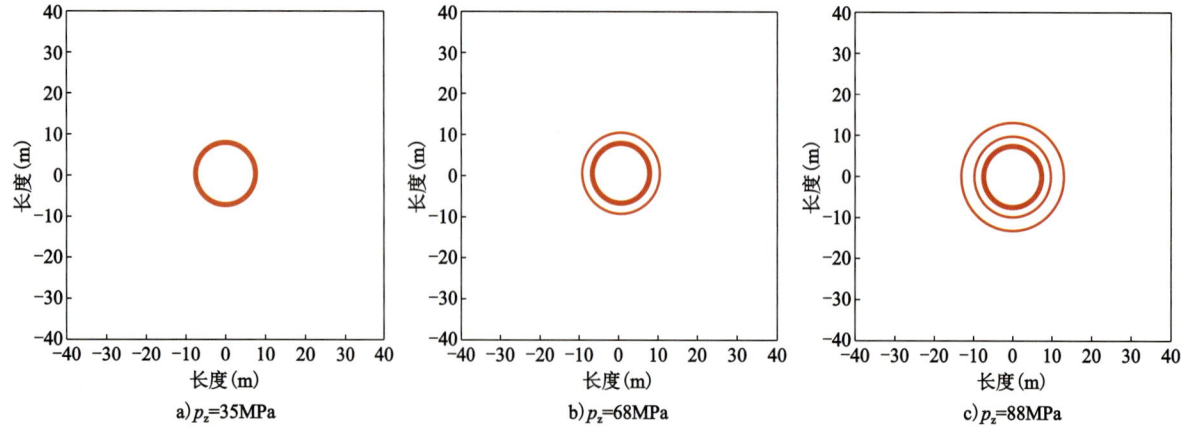

a) p_z=35MPa   b) p_z=68MPa   c) p_z=88MPa

钱七虎院士和周小平国际上最早确定了非静水压力作用下深部圆形洞室围岩的应力场、破裂区的宽度和数量。

$$\sigma_r = \frac{\sigma_h - \sigma_v}{2}\left(\frac{1-4r_0^2}{E\ r^2} + 3\frac{r_0^4}{r^4}\right)\cos(2\theta) +$$

$$2r(1-v)Y^{3/2}[A_1J_1(r\sqrt{\gamma}) + B_1N_1(r\sqrt{Y}) + C_1K_1(r\sqrt{\gamma})]\cos(2\theta) -$$

$$\frac{3E}{r^2(1-v)\gamma^2}[A_1J_2(r\sqrt{\gamma}) + B_1N_2(r\sqrt{\gamma}) + C_1K_2(r\sqrt{\gamma})]\cos(2\theta) \tag{20}$$

$$\sigma_\theta = \frac{\sigma_b - \sigma_v}{2}\left(1 + 3\frac{r_0^2}{r^4}\cos(2\theta) +\right.$$

$$\frac{E}{2(1v)\gamma}[A_1J_0(r\sqrt{\gamma}) + B_1N_0(r\sqrt{\gamma}) + C_1K_0(r\sqrt{\gamma})]\cos(2\theta) -$$

$$\frac{3E}{2r(1v)\gamma^{3/2}}[A_1J_1(r\sqrt{\gamma}) + B_1N_1(r\sqrt{\gamma}) + C_1K_1(r\sqrt{\gamma})]\cos(2\theta) +$$

$$\frac{3E}{r^2(1-v)\gamma^2}[A_1J_2(r\sqrt{\gamma}) + B_1N_2(r\sqrt{\gamma}) + C_1K_2(r\sqrt{\gamma})]\cos(2\theta) \tag{21}$$

$$T_{r\theta} = \frac{\sigma_v - \sigma_h}{2}\left(1 + 2\frac{r_0^2}{r^2} - 3\frac{r_0^2}{r^2} - 3\frac{r_0^4}{r^4}\right)\sin(2\theta) +$$

$$\frac{E}{r(1-v)Y^{3/2}}[A_1J_1(r\sqrt{\gamma}) + B_1N_1(r\sqrt{\gamma}) + C_1K_1(r\sqrt{\gamma})]\sin(2\theta) -$$

$$\frac{3E}{r^2(1-v)\gamma^2}[A_1J_2(r\sqrt{\gamma}) + B_1N_2(r\sqrt{\gamma}) + C_1K_2(r\sqrt{\gamma})]\sin(2\theta) \tag{22}$$

非静水应力对分区破裂化影响的数值计算结果参数如下：$E = 20\text{GPa}, r_0 = 7\text{m}, D = 0.22, p = 2700\text{kg/m}^3$，$Q = 30°, c = 6\text{MPa}, \sigma_h = 30\text{MPa}, \sigma_c = 20.7\text{MPa}$。

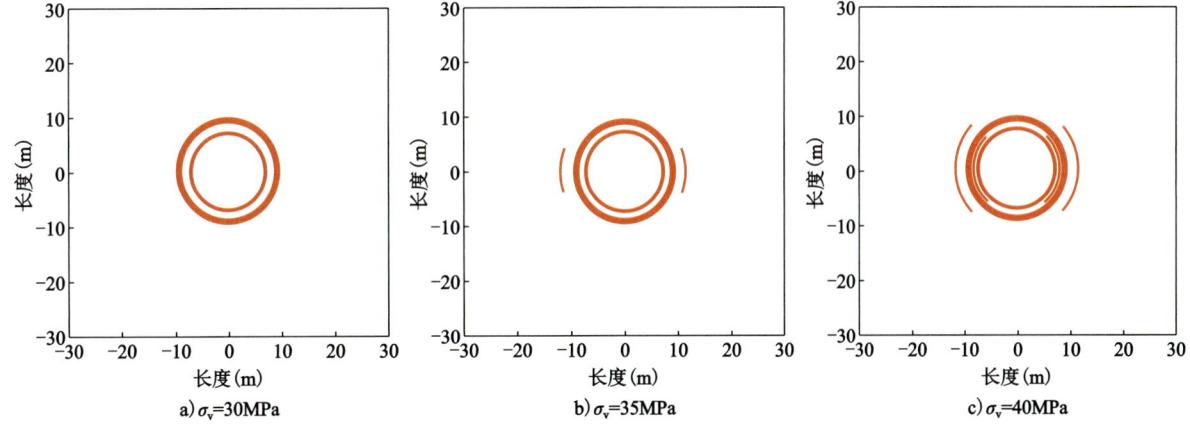

a) σ_v=30MPa   b) σ_v=35MPa   c) σ_v=40MPa

钱七虎院士和周小平国际上最早确定了静水压力作用下深部球形洞室围岩的应力场，其可以表示如下：

$$\sigma_r = \sigma_m\left[1 - \left(\frac{r_0}{r}\right)^3\right] + \frac{p_1 r_0}{2r} + \frac{\beta}{2\beta X^2}[r^2 X\cos(r_0\sqrt{\chi})(A_2 - A_3) + r^2 X\sin(r_0\sqrt{x})(A_1 + A_3)] -$$

$$\frac{\beta}{2r^3\chi^2}\{2[A_1 + A_3 + (A_2 - A_3)r\sqrt{x}]\sin(r\sqrt{x})\} +$$

$$\frac{\beta}{2r^3\chi^2}\{2[A_3 - A_2 + (A_1 + A_3)r\sqrt{\chi}]\cos(r\sqrt{\chi})\} \tag{23}$$

$$\sigma_\theta = \sigma_\psi = \frac{\beta}{4\chi^2 r^3}[r^2 X\cos(r_0\sqrt{\chi})(A_2-A_3)+r^2 X\sin(r_0\sqrt{\chi})(A_1+A_3)] -$$
$$\frac{\beta}{4\chi^2 r^3}\{2[(A_1+A_3)(r^2 X-1)-r\sqrt{\chi}(A_2-A_3)]\sin(r\sqrt{x})\} -$$
$$\frac{\beta}{4\chi^2 r^3}\{2[(A_3-A_2)(1-r^2 X)+(A_1+A_3)r\sqrt{\chi}]\cos(r\sqrt{\chi})\} +$$
$$\sigma_m\left[1+0.5\left(\frac{r_0}{r}\right)^3\right] + \frac{p_1 r_0}{4r} \tag{24}$$

静水压力对深部球形洞室围岩分区破裂化的影响参数如下:$\sigma_c = 30\mathrm{MPa}, r_0 = 7\mathrm{m}, V = 0.25, GSI = 50, m_i = 10, p = 2700\mathrm{kg/m^3}, E = 30\mathrm{GPa}$。

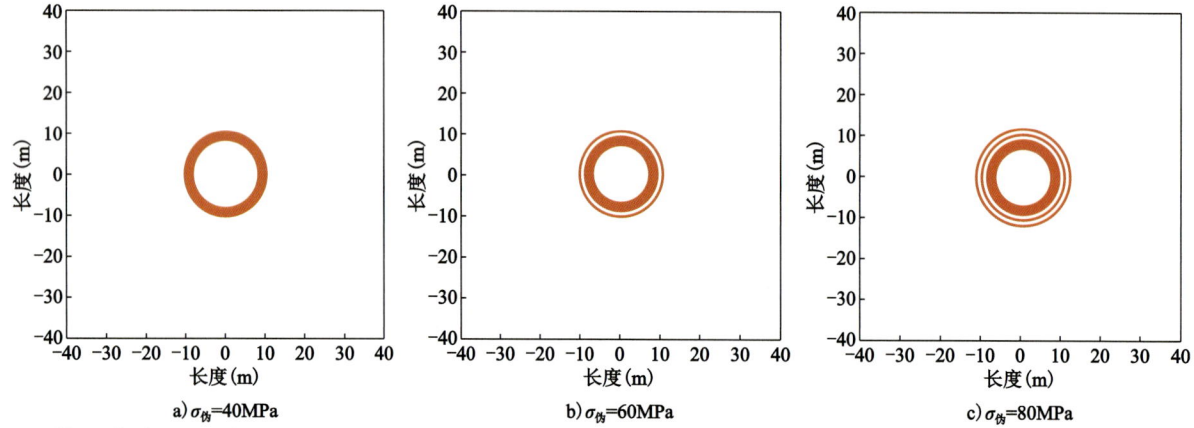

a) $\sigma_{侧}$=40MPa   b) $\sigma_{侧}$=60MPa   c) $\sigma_{侧}$=80MPa

钱七虎院士和周小平国际上最早确定了动态开挖条件下深部圆形洞室围岩应力场,其动态开挖过程可以表示为:

$$\begin{cases} p(r_0,t)=(2-e^{kt})\sigma_{侧} & 0<t<t_0 \\ p(r_0,t)=0 & t>t_0 \end{cases} \tag{25}$$

a) 计算模型   b) 动态开挖过程

式中:
$$t_0 = \frac{0.6931}{k}$$

动态开挖条件下深部圆形洞室围岩应力场可以表示为:

$$\sigma_r = \left\{1 - \frac{r_0^2[A_1(r,0)+A_2(r,0)+A_3(r,0)+A_4(r,0)]}{r^2[A_1(r_0,0)+A_2(r_0,0)+A_3(r_0,0)+A_4(r_0,0)]}\right\}\sigma_m +$$
$$\frac{r_0^2[A_1(r,-k)+A_2(r,-k)+A_3(r,-k)+A_4(r,-k)]}{r^2[A_1(r_0,-k)+A_2(r_0,-k)+A_3(r_0,-k)+A_4(r_0,-k)]} - \sigma_m e^{-kt} +$$
$$\sum_{i=1}^n \frac{r_0^2[A_1(r,p)+A_2(r,p)+A_3(r,p)+A_4(r,p)]}{r^2 \dfrac{a[A_1(r_0,p)+A_2(r_0,p)+A_3(r_0,p)+A_4(r_0,p)]}{as}} \left(\frac{\sigma_m(1-2e^{st_0})}{p+k} - \frac{\sigma_m}{p} - \frac{C_4}{r_0^2}\right)e^{pt}\Bigg|_{p=a_i} \tag{26}$$

$$\sigma = 2(V+1)\sigma_m - \frac{16Er_0^2 y^2(0)(v+1)\sigma_m}{A_1(r_0,0)+A_2(r_0,0)+A_3(r_0,0)+A_4(r_0,0)}T(r,0) +$$

$$\frac{16Er_0^2 y^2(-k)(v+1)\sigma_m}{A_1(r_0,-k)+A_2(r_0,-k)+A_3(r_0,-k)+A_4(r_0,-k)}e^{kt}T(r,p)e^{pt}\bigg|_{p=a_i} \tag{27}$$

$$\sigma_\theta(r,t) = \frac{\sigma}{1+V} - \sigma_r(r,t) \tag{28}$$

卸荷时间对参数如下：$\sigma_c = 30\text{MPa}, r_0 = 7\text{m}, v = 0.25, GSI = 50, m_i = 10, p = 2700\text{kg/m}^3, E = 30\text{GPa}, \sigma_伪 = 40\text{MPa}$。

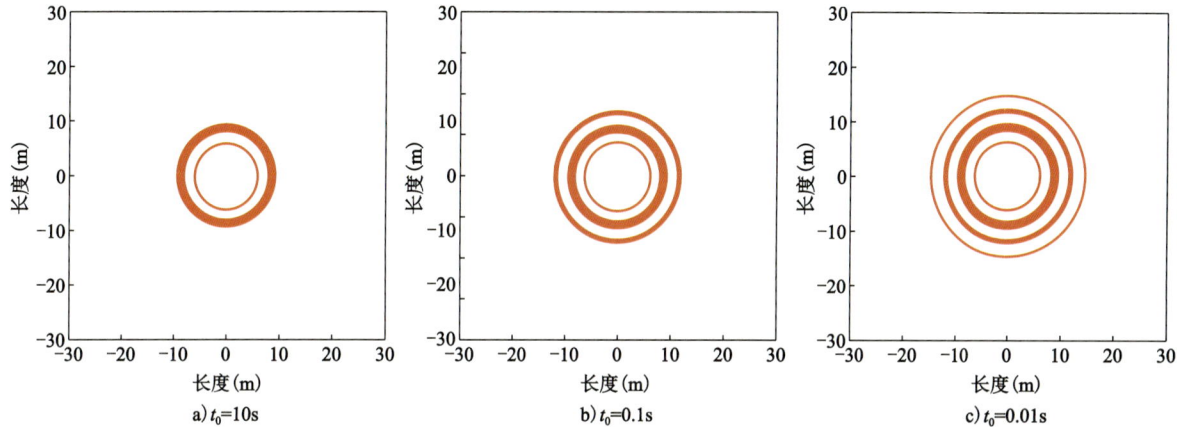

a) $t_0 = 10\text{s}$　　　b) $t_0 = 0.1\text{s}$　　　c) $t_0 = 0.01\text{s}$

钱七虎院士和周小平国际上最早确定了静水压力作用下深部球形洞室围岩的应力场，动态开挖过程可以表达如下：

$$\begin{cases} p(r_0,t) = (2-e^{kt})\sigma_伪 & 0 < t < t_0 \\ p(r_0,t) = 0 & t > t_0 \end{cases} \tag{29}$$

其中：
$$t_0 = \frac{0.6931}{k}$$

a) 计算模型　　　b) 动态开挖过程

动态开挖条件下深部球形洞室围岩应力场可以表示为：

$$\sigma_r(r,t) = -\frac{1}{2}\delta^2\left[\frac{r_0^3 k\sigma_父(1-2e^{-st_0})A(r,s)}{R^3(s-k)B(r_0,s)}e^{st}\right]\delta s^2\bigg|_{s=0} - \delta\left[\frac{C_4 A(r,s)}{r^3 B(r_0,s)}e^{st}\right]\bigg/\delta s\bigg|_{s=0} +$$

$$\sigma_父 - \sum_{i=1}^{n}\frac{r_0^3 A(r,s)}{r^3 s^2 \frac{\delta B(r_0,s)}{\delta s}}\left[\frac{k\sigma_父(1-2e^{-st_0})}{s(s-k)} + \frac{C_4}{r_0^3}\right]e^{st}\bigg|_{s=a_i} \tag{30}$$

$$\sigma(r,t) = -\frac{1}{2}\delta^2\left[\frac{k\sigma_{父}(1-2e^{-st_0})T(r,s)}{(s-k)B(r_0,s)}e^{st}\right]/\delta s^2\bigg|_{s=0} - \delta\left[\frac{C_4 T(r,s)}{r_0^3 B(r_0,s)}e^{st}\right]\delta s\bigg|_{s=0} +$$

$$3\sigma_{父} - \sum_{i=1}^{n}\frac{T(r,s)}{s^2\frac{\delta B(r_0,s)}{\delta s}}\left[\frac{k\sigma_{父}(1-2e^{-st_0})}{s(s-k)} + \frac{C_4}{r_0^3}\right]e^{st}\bigg|_{s=a_i} \tag{31}$$

$$\sigma_Q(r,t) = \sigma_\theta(r,t) = \frac{\sigma(r,t) - \sigma_r(r,t)}{2} \tag{32}$$

卸荷时间对深部球形洞室围岩分区破裂化的影响参数如下：$\sigma_c = 30\text{MPa}, r_0 = 7\text{m}, v = 0.25, GSI = 50, m_i = 10, p = 2700\text{kg/m}^3, E = 30\text{GPa}, \sigma_{侧} = 40\text{MPa}$。

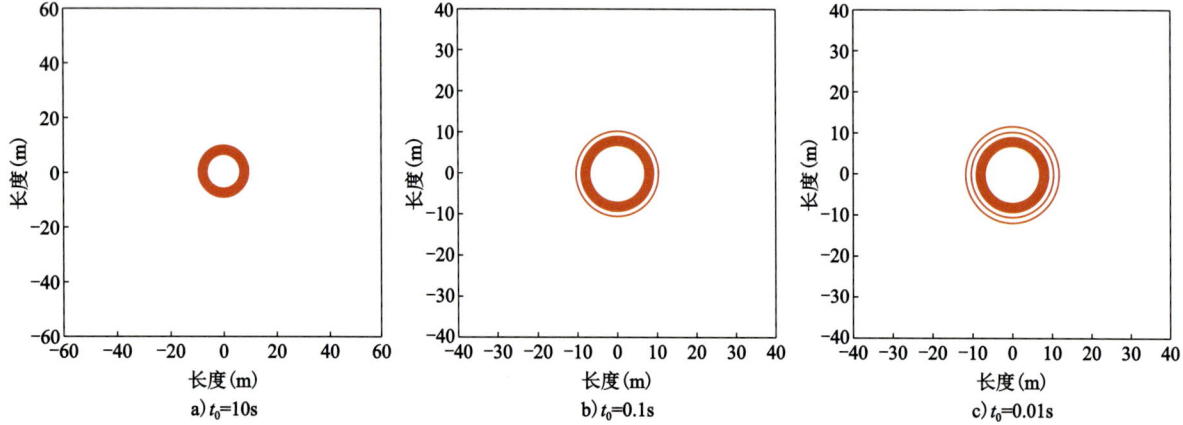

a) $t_0 = 10s$ b) $t_0 = 0.1s$ c) $t_0 = 0.01s$

钱七虎院士和周小平提出了一种能考虑微裂纹影响的非欧几何模型，国际上最早确定了微裂纹对自平衡应力的影响，微裂纹对破裂区的宽度和数量的影响。

$$\sigma_{r2} = \frac{E(1-D)^2}{2(1-v)rX^{3/2}}[A_1 J_1(X^{1/2}r) + B_1 N_1(X^{1/2}r) + C_1 K_1(X^{1/2}r)]\cos(2\theta) -$$

$$\frac{3E(1-D)^2}{(1-v)r^2 X^2}[A_1 J_2(X^{1/2}r) + B_1 N_2(X^{1/2}r) + C_1 K_2(X^{1/2}r)]\cos(2\theta) \tag{33}$$

$$\sigma_{\theta 2} = \frac{E(1-D)^2}{2(1-v)X}[A_1 J_1(r\sqrt{\chi}) + B_1 N_0(r\sqrt{\chi}) + C_1 K_0(r\sqrt{\chi})]\cos(2\theta) -$$

$$\frac{3E}{2r(1v)\chi^{3/2}}[A_1 J_1(r\sqrt{\chi}) + B_1 N_1(r\sqrt{\chi}) + C_1 K_1(r\sqrt{\chi})]\cos(2\theta) +$$

$$\frac{3E}{r^2(1v)\chi^2}[A_1 J_2(r\sqrt{\chi}) + B_1 N_2(r\sqrt{\chi}) + C_1 K_2(r\sqrt{\chi})]\cos(2\theta) \tag{34}$$

$$T_{r\theta 2} = \frac{E(1-D)^2}{2(1v)X^{3/2}}[A_1 J_1(r\sqrt{\chi}) + B_1 N_1(r\sqrt{\chi}) + C_1 K_1(r\sqrt{\chi})]\sin(2\theta) -$$

$$\frac{3E(1-D)^2}{r^2(1v)\chi^2}[A_1 J_1(r\sqrt{\chi}) + B_1 N_2(r\sqrt{\chi}) + C_1 K_2(r\sqrt{\chi})]\sin(2\theta) \tag{35}$$

微裂纹密度对深部圆形洞室围岩应力场的影响参数如下：$\sigma_c = 30\text{MPa}, r_0 = 7\text{m}, v = 0.25, \sigma_h = 20\text{MPa}, p = 2700\text{kg/m}^3, E = 30\text{GPa}, \sigma_v = 40\text{MPa}$。

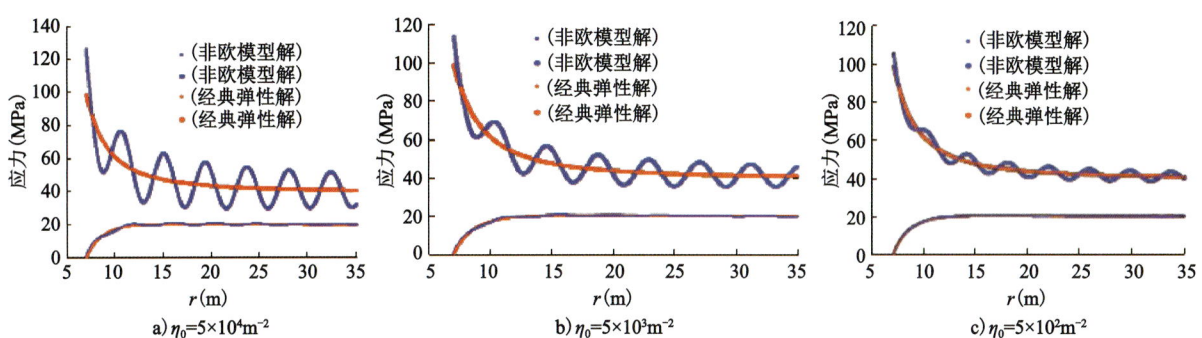

a) $\eta_0 = 5 \times 10^4 \text{m}^{-2}$ b) $\eta_0 = 5 \times 10^3 \text{m}^{-2}$ c) $\eta_0 = 5 \times 10^2 \text{m}^{-2}$

微裂纹长度对深部圆形洞室围岩应力场的影响参数如下：$\sigma_c = 30\text{MPa}$，$r_0 = 7\text{m}$，$v = 0.25$，$\sigma_h = 20\text{MPa}$，$p = 2700\text{kg/m}^3$，$E = 30\text{GPa}$，$\sigma_v = 40\text{MPa}$。

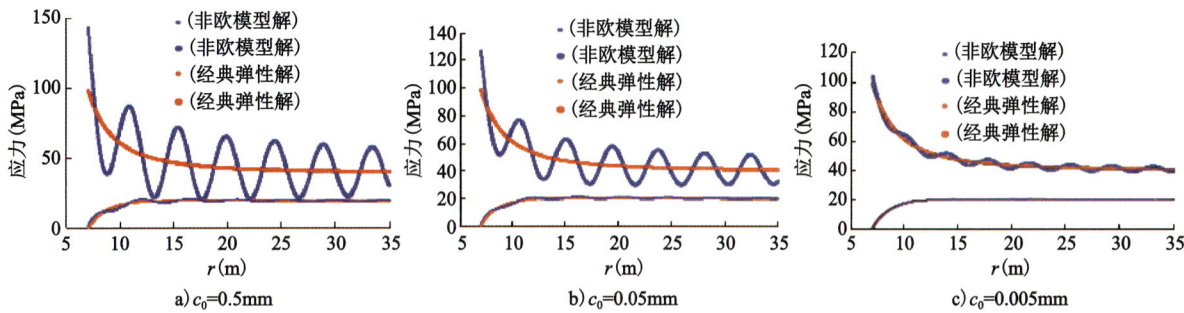

a) $c_0 = 0.5\text{mm}$ b) $c_0 = 0.05\text{mm}$ c) $c_0 = 0.005\text{mm}$

五、理论结果与实验结果对比

计算参数如下：$E = 0.2594\text{GPa}$，$v = 0.268$，$\sigma_伪 = 2.1\text{MPa}$，$p_z = 5.1\text{MPa}$，$c = 0.2\text{MPa}$，$\varphi = 430$，$p = 2620\text{kg/m}^3$，$r_0 = 0.04\text{m}$。

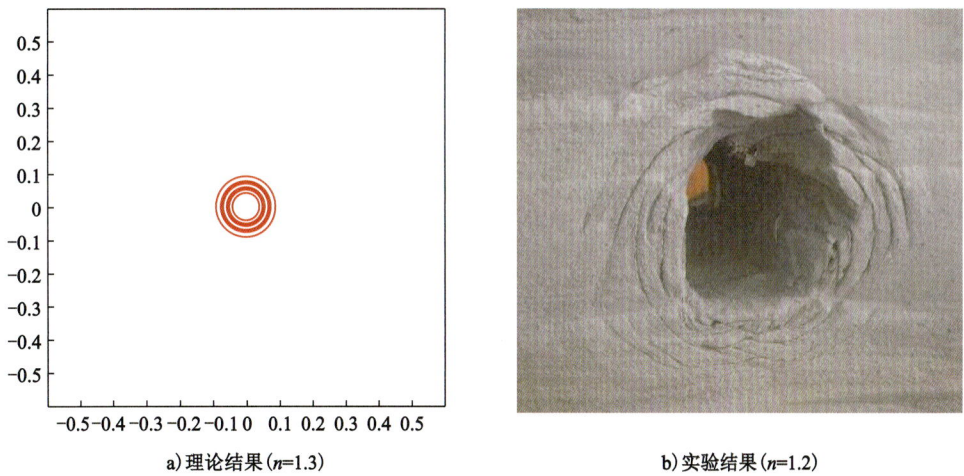

a) 理论结果 ($n=1.3$) b) 实验结果 ($n=1.2$)

理论结果和张强勇的实验结果对比

六、理论结果与广义粒子动力学结果对比

计算参数如下：$E = 17.6\text{GPa}$，$v = 0.3$，$\sigma_c = 50\text{MPa}$，$\sigma_伪 = 65\text{MPa}$，$p = 2750\text{kg/m}^3$，$r_0 = 7\text{m}$。

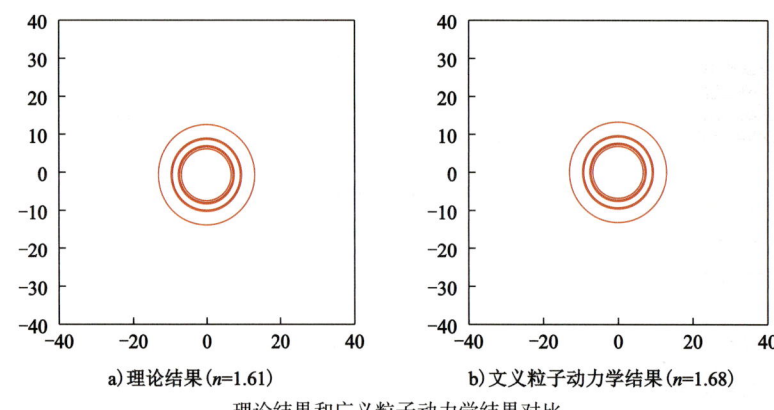

a) 理论结果（$n=1.61$）   b) 广义粒子动力学结果（$n=1.68$）

理论结果和广义粒子动力学结果对比

七、结论

（1）通过标量曲率确定了深部岩体自平衡应力。

（2）自平衡应力和经典的弹性应力组成了深部巷道围岩的应力场，该应力场具有振荡特性。

（3）自平衡应力对深部岩体的分区破裂化现象有重要的影响。

（4）确定了微裂纹对自平衡应力与破裂区的宽度和数量的影响。

（5）提出了深部岩体发生分区破裂化的临界应力条件。

（6）首次提出了非静水压力、轴向应力条件下深部岩体破坏的非欧几何模型，确定了破裂区和非破裂区的数量、位置和宽度。

（7）首次提出了开挖卸荷条件下深部岩体破坏的非欧几何模型，确定了破裂区和非破裂区的数量、位置和宽度。

（8）提出了深部岩体强度准则。

非线性岩石力学(非传统)研究前沿导论

一、引言

传统(经典的)岩石力学的建立是基于连续介质力学的理论框架,在微观与细观层次没有考虑岩石的非连续性;宏观层面上虽然研究岩体的不连续面的影响,但没有考虑岩体在宏观上的块体性,因此无法解释基于岩石微、细观不连续性和宏观块体性产生的科学现象和力学规律。

对于这些内容的研究形成了非线性岩石(静)力学和非线性岩石动力学(非传统)。

这些内容的研究与深部岩体重大工程灾害的孕育演化以及成灾机理密切相关,其成果将成为重大工程灾害时空预测预报和动态调控的理论依据。

二、岩石和岩体的不连续性和岩体的块体性

岩石和岩体的不连续性包括原生(初始)的和次生的。

(一)原生的岩石微细观的不连续性

根据致密岩体和页岩体的最新研究表明:页岩气储层的孔隙直径为5~200nm,致密灰岩油储层的孔隙直径为40~500nm,致密砂岩气储层的孔隙直径为40~700nm,致密砂岩油储层的孔隙直径为50~900nm。上述孔隙系统占整个储层空间的80%~90%。

十种岩样压汞测试结果简表

编号	岩性	孔隙率(%)	平均孔径(nm)	突破半径(μm)	密度(g/cm³)	取样地区
A1	玄武岩	1.7908	28.5	34.4	2.2221	南京
A2	花岗岩	0.8127	260.8	162.9	2.2676	南京
A3	花岗岩	0.6686	480.1	75.1	2.1509	南京
A4	变质岩	1.1222	200.1	144.4	2.1157	南京
A5	变质岩	1.5921	129.7	37.6	2.1355	南京
B1	泥岩	3.1820	20.3	41.3	2.1333	重庆
B2	砂岩	5.9269	63.2	45.9	2.0729	重庆
B3	粉砂岩	10.1074	123.3	6.46	1.9222	南京
B4	粉砂岩	4.4547	37.6	28.3	2.1017	南京
B5	红砂岩	8.9043	38.9	8.34	1.9384	山东临沂

注:突破半径是指汞突破的尺寸最大的第一级喉道的孔隙半径。

次生的岩石不连续性为加、卸载过程萌生的裂纹、裂隙,特别是岩石应力达到峰值后的软化段是大量微裂纹发育并产生宏观主干裂纹的阶段。

报告时间:2014年11月。

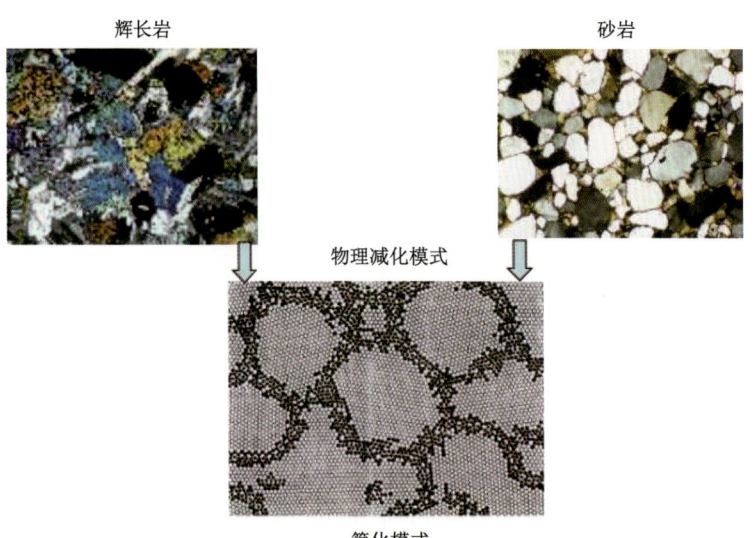

(二)岩体的宏观不连续性

断裂、裂隙的宽度 (m)	裂缝的绝对水平宽度				
	Л Нейштадт(1957)	Е Ромм(1966)	Ж Ферран(1975)	В Жиленков(1975)	С Чернышев(1983)
10^2	非常大的裂隙	宏观裂隙	宏观裂隙	巨大裂隙	区域断裂
10^1					
10^0					裂隙
10^{-1}					宽裂缝
10^{-2}	大裂隙				裂缝 中等裂缝
10^{-3}	中等裂隙				
10^{-4}	小裂隙				
10^{-5}	细裂缝			裂隙	窄裂缝
10^{-6}					
10^{-7}		微观裂纹	细裂纹	小裂纹	毛细裂纹
10^{-8}			微观裂纹		次毛细裂纹

结构性断裂的长度 (m)	裂缝的绝对水平长度				
	по Л Мюллеру (1971)	по В Рацу (1970)	по В Жиленкову (1975)	по Н Красиловой (1979)	по Чернышеву (1983)
10^6	结构性断裂	大构造断裂	Ⅰ级断裂和Ⅱ级断裂	巨大的结构性断裂	主要构造断裂
10^5					
10^4					
10^3		断裂	Ⅲ级断裂		断裂
10^2					
10^1	裂隙 巨大裂隙	宏观裂缝	大裂缝	宏观和中观裂隙	裂缝 长裂缝
10^0	裂隙 大裂隙		裂缝		中等裂缝
10^{-1}	裂隙 小裂隙				短裂缝
10^{-2}	隐性裂纹	微观裂纹	微观裂纹	微观裂纹	微裂纹
10^{-3}					
10^{-4}					
10^{-5}		晶格缺陷		超小裂纹	
10^{-6}					

(三) 科学现象——由岩石不连续性和岩体的块体性决定

(1) Makarov V. V. 等对于从坚硬的火成岩到软弱的沉积岩的多种岩石进行了压缩试验：对于某些试样，在一定条件下，当轴向加载应力超过 $(0.6 \sim 0.8)\sigma_c$ 时在破坏前出现变形异常现象（σ_c 为单轴抗压强度）。典型的变形曲线如下图所示。

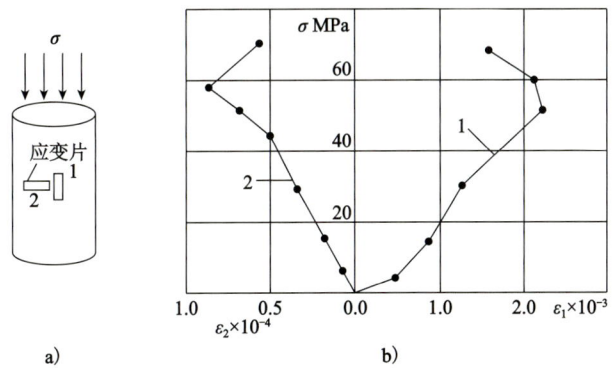

试件受压 a) 及变形增量符号改变效应 b)

(2) 高地应力区洞室围岩出现的应变与位移，随与洞室距离的增加呈波浪形，而非单调变化。

① 山东大学陈旭光和张强勇的试验。

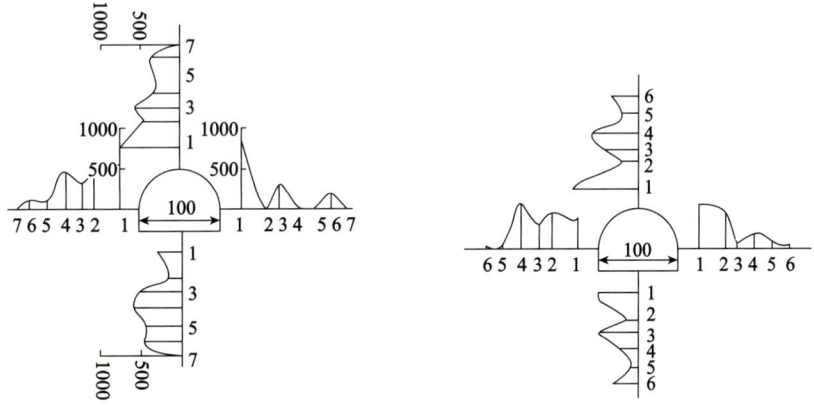

试验实测洞周围岩应变场分布 (单位：μm)　　　型试验洞室开挖后洞周围岩位移场分布

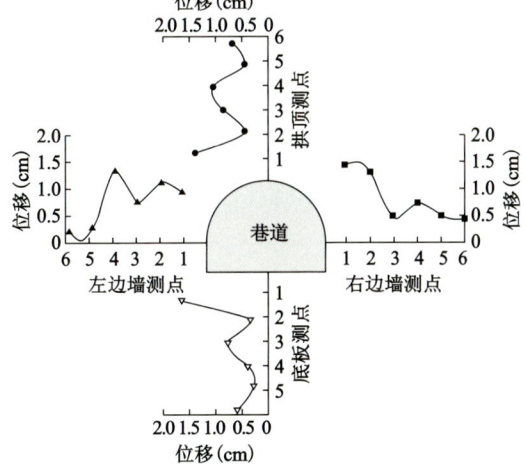

Ⅱ-Ⅱ 断面模型洞周测点位移

② 北京科技大学方祖烈在金川镍矿深井围岩中现场测量表明：围岩中应变程序松弛区和严密区交替现象。

(3)锦屏一级电站地下厂房洞室群开挖过程中和支护后出现的围岩大变形和围岩喷层开裂、岩体劈裂破坏现象。

①变形监测结果。

纵 0+31.70 断面最大水平位移变形监测结果

位置		多点位移计累计位移(mm)
主厂房	上游边墙 EL1648	11.14
	下游拱脚	21.03
主变室	上游边墙 EL1660.8	65.21
	下游边墙 EL1664	61.96
1号尾调室	上游边墙 EL1648	3.38
	下游边墙 EL1674	7.9

纵 0+126.80 断面最大水平位移变形监测结果对比

位置		多点位移计累计位移(mm)
主厂房	上游边墙 EL1648	60.82
	下游拱脚	35.80
主变室	上游边墙 EL1660.8	52.08
	下游边墙 EL1664	233.39
2号尾调室	上游边墙 EL1648	25.18
	下游边墙 EL1674	45.93

②锚索内力监测结果。

主厂房锚索内力超出设计吨位43根,占总数21.83%,设计吨位2000kN。

主变室锚索内力超出设计吨位11根,占总数12.5%(截止到2010年2月)。

③围岩开裂现象。

地下厂房 0+101～0+145m,高程 1670～1672m 范围喷层出现多条裂缝,宽度 0.5～2cm。

下游拱腰至拱座区域内,高程 1667～1672m,0+115～0+145m 段,裂缝张开 2～6cm,伴有混凝土剥落现象,钢筋向洞室内弯曲变形。

乌鞘岭铁路隧洞拱顶及侧墙位移大变形数十厘米至一米多。

(4)地下大当量爆炸数值计算中连续介质力学模型的不适用性。

俄罗斯学者对于连续介质波动力学模型位移数值计算数据与地下核试验实测数据进行了比较研究,结果示于下图。图中实线所示的岩体实测最大比位移大大超过了按连续介质应力波理论计算而得的最大比位移(虚线)和永久比位移(点划线)。其差值随比距离($R/1Q^{\frac{1}{3}}$)增大而减小,当 $R/Q^{\frac{1}{3}}$ 由30增至180时,误差最大可达6.6倍。

当岩体中发生地下核爆炸时,按连续介质力学模型变形计算而得的岩体变形,应是由近而远连续衰减。但俄罗斯在地下核试验中测得的岩体变形数据,却因测点位于岩块(为节理、裂隙所分割的)中间或是位于岩块边缘(位于节理、裂隙面附近)而致其变形值相差达 2～3 个量级。点 1、2、4 为岩块内的变形值(不同应变计基线),点 3 为岩体结构面附近的变形值。下图中的实线表示为均匀介质中最大变形的均值关系曲线,虚线表示均匀介质中残余变形的关系曲线,这表明,在岩体中地下核爆炸时,在断裂、裂隙附近,产生了显著的岩体变形局部化现象。

(5)地下爆炸时地下岩体的位移和倾斜的变号现象。

①在当量1.4kt爆炸中,离爆心160m处与巷道相交的裂隙两侧的位移测量结果。

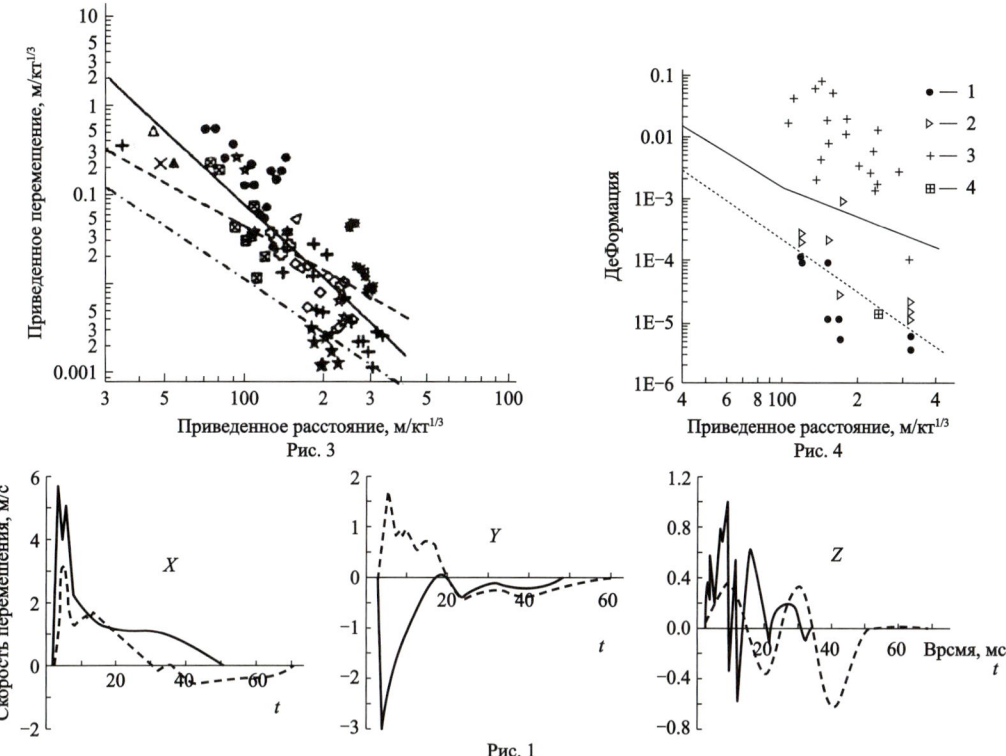

Рис. 3

Рис. 4

Рис. 1

X、Y-断裂面的法、切线方向水平分量；Z-垂直分量

② 巷道墙体的残余（永久）位移和倾斜的测量结果。

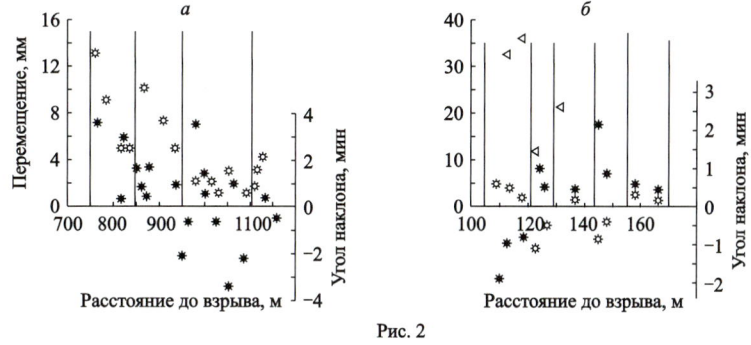

Рис. 2

a -78kt当量；$б$ -1.4kt；*-1巷道内的倾斜；
✿-2横向巷道的倾斜；◁-3巷道内的水准倾斜

（6）分区破裂化现象。

① 南非。

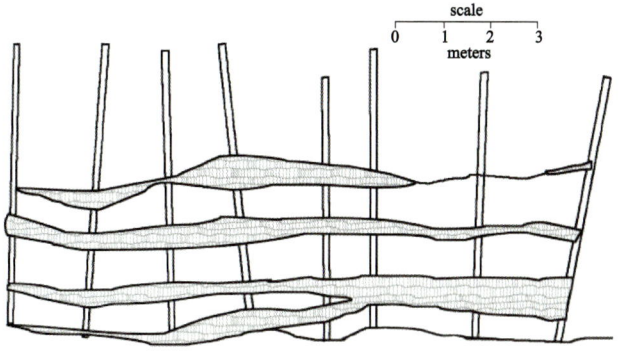

Adams G. R., Jag er A. J. Petroscopic observation of rock fracturing ahead of stop faces in deep-level gold mines //J. South African Inst. Mining and Metallurgy, 1980, vol. 80, № 6, p. 204-209

②俄罗斯。

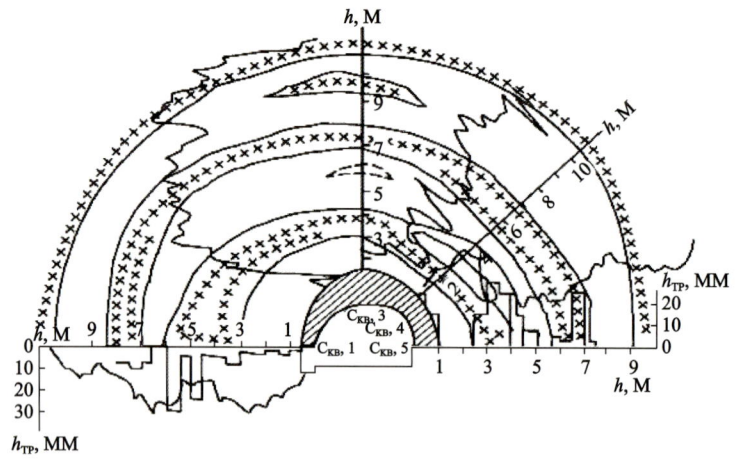

Structure of fractured rock around rock workings
(Kuznetsov Y. S., Ph. D. Th., Russian, 1983)

③中国。

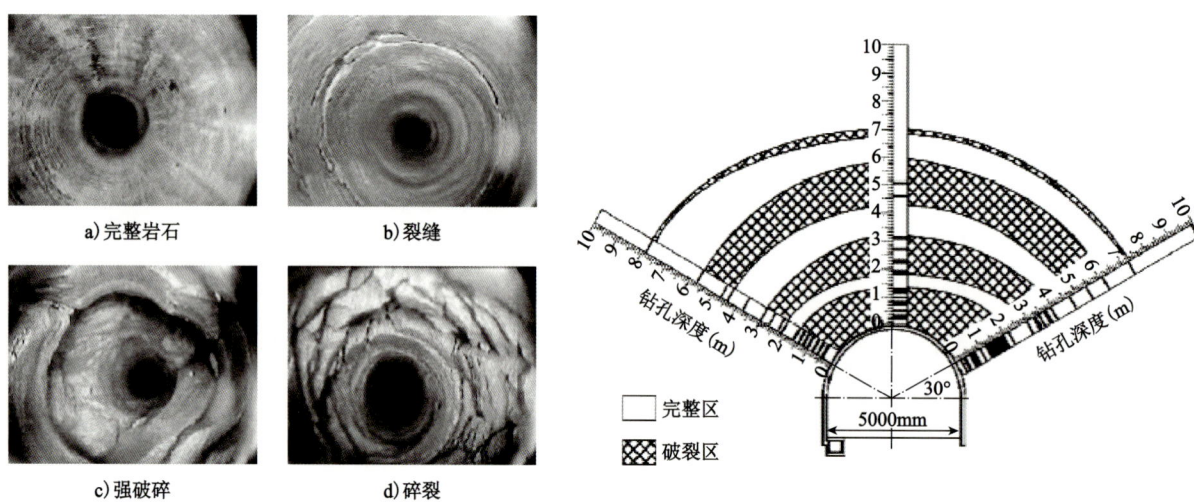

a) 完整岩石　　b) 裂缝

c) 强破碎　　d) 碎裂

淮南矿区丁集煤矿钻孔电视视频截图　　巷道监测断面围岩分区破裂分布

（7）岩石、岩体具有储能和释能特性。

①地下隧道和洞室开挖引起岩爆、矿山开挖诱发冲击地压。

发生在美国大西洋矿井和南非卡尔顿金矿中的岩爆。

大西洋矿井中发生的岩爆

卡尔顿金矿中发生的岩爆

锦屏二级水电站引水隧洞岩爆导致7人死亡和TBM机损坏

②人类大规模工程活动诱发工程性地震（Техногенные землетрясения）。

（8）岩石、岩体中具有残余应力和封闭应力。

陈宗基：

①"残余应力还没有引起适当的注意"。

②"残余应力的来源和性质仍然还不知道"。

③"这种应力可以分为两种类型：即'封闭应力'和'非封闭应力'，若从岩体中取出岩样时其应力便消失，这属于后者，相反地，尽管岩样的边界是自由的，但是'封闭应力'仍然可以保存着，也即它具有自身平衡的性质。"

（9）冲击地压的特征现象。

①冲击地压发生地与扰动源分离。

②矿井中冲击地压事故多发生在回采巷道，较少发生于开采工作面附近？

③矿井中冲击地压事故有时发生在巷道长达数十米甚至上百米纵长上？

④矿井中冲击地压事故的发生迟后于开采时刻，甚至迟后几个小时？

（四）非线性岩石力学（非经典）的研究

1. 岩石、岩体非连续性和块体性的试验研究

（1）由微焦距体工业 CT 实时扫描获取单轴加载不同阶段下的岩石试件的数字化图像，利用数字体散斑法 DVSP 技术计算获得岩石试件不同加载阶段下的内部三维变形场，继而得出三维应变场，揭示出岩石内部变形局部化和破坏，即微裂纹由原生到扩展合并的孕育发展过程。

（2）综合应用超声波钻孔成像系统、三维激光扫描成像系统、数码摄像成像等数字技术进行了岩体结构的识别和描述，直观、精确、快速地获取岩体结构信息。

（3）形成和发展了基于物探、钻探以及开挖面探测相结合的岩体结构特征综合识别方法，弥补了传统勘探技术的不足。

（4）基于"空间结构约束"思想的综合探测联合解释方法成功克服了传统电阻率反演方法定位效果差、界面分辨率低的缺陷，显著提高了岩体结构面的定位精度和界面分辨效果，实现了岩体结构的数字化、精细化识别与解释。

（5）基于钻孔孔壁上某点任意方向的正应变与钻孔围岩远场应力之间的线性关系，提出了一种新的测定深部岩体地应力的新方法——钻孔局部壁面应力全解除法（BWSRM），该方法只要求钻取长度约 4cm 的完整岩芯，因而解决了传统应力解除法要求钻取岩芯 30cm 因而易出现的断芯问题，也克服了水力压裂法必须假定地应力张量的一个主方向与钻孔轴一致的困难。

2. 理论研究

（1）岩体中的缺陷（初始微纹）和开挖卸荷诱发的二次裂纹把岩石连续颗粒分离开，在这些岩石颗粒邻域变形协调条件不成立，所以连续介质弹塑性力学的协调解（经典）不适用于裂缝发育的围岩以及裂纹发育的试件（存在一个阈值）。

（2）岩石中的缺陷和岩体中岩块间的摩擦力、黏结力和剪胀性的存在表明岩石、岩体具有内结构性。

（3）内结构的存在使得岩石和岩体整体变形时，除了整体的变形能外，还存在岩石和岩体内部不协调变形导致的局部的能量耗散，以及该耗散能导致基本微分方程的变化。

（4）岩石与岩体中内结构及非协调变形的存在需要有一个内度量体系及一个隐含参数来描写非协调变形量。

（5）在岩石与岩体的非协调解中出现波动变化的自平衡封闭应力部分，它是由非协调变形产生的。

（6）围岩中波动变化的应力解，是分区破裂化的力学原因和物理本质。

（7）建立在非欧模型基础上的高地应区洞室开挖的应力分析表明：出现围岩的分区破裂化具有一定的高地应力阈值条件，同样的分析表明试件在破坏前的应变逆转现象出现也具有一定的微裂纹阈值。

（8）高地应力地区洞室围岩大变形的主要原因是围岩的变形和破裂逐步向围岩深部转移形成多区破裂现象，即高地应力洞室围岩变形、破裂（非连续性）的时空演变结果。

(五) 非经典的非线性岩石动力学研究

(1) 岩体中由于存在大量的断裂构造、裂隙、节理、裂纹,岩体是一个由不同尺寸等级岩块所组成的岩块系。

$$\Delta_i = (\sqrt{2})^{-i} \cdot \Delta 0 \qquad i = 1,2,\cdots n; \Delta 0 = 2.5 \times 10^6 \text{m}$$

$$\mu_\Delta(\delta) = \frac{\delta_i}{\Delta_t} = \theta \cdot 10^{-2} \qquad (对任何一个 i); \theta \in \frac{1}{2} \sim 2$$

(2) 基于爆炸、爆破效应中存在的变号反应说明在爆炸作用下发生了岩块的移动和转动。

(3) 冲击地压的特征现象表明老巷道围岩已破碎成准平衡(稳定)的岩块系。

①陈宗基:切不可忽视爆炸波对触发煤爆的影响。门头沟矿在 700~900m 深处的 114 起煤爆中,有 89 起(占 78%)是因爆破触发的,在龙凤矿 700m 深处因爆破而触发的煤爆也超过总数的 50%,由此可以推断,在这样的深度,煤一定是处于准稳定平衡状态,爆炸波的触发作用能导致死亡事故发生。

②华丰煤矿冲击地压事故综合剖析:4 层煤的多个工作面发生过冲击地压,华丰井田煤系地层以上为 500~800m 砾岩层,砾岩层坚硬整体性强,其断裂垮落对下部的煤岩体产生冲击载荷,是 4 层煤工作面发生冲击地压的主要力源。

爆破以及顶板断落冲击是如何诱发冲击地压的? 连续介质力学框架的应力波理论存在局限性。

(4) 基于岩体块体性的摆型波研究。

现场和实验室试验表明在岩体中存在传播岩块运动状态的慢"变形"波——摆型波。

①慢波的发生机理。

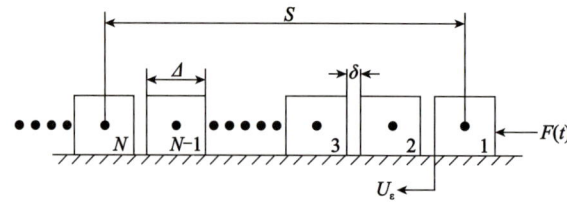

Mechanical model for calculation of velocity characteristic U_v of "deformation" waves

$$U_v = \frac{(1+v)_{upu_c}}{U_\varepsilon + vu_p} \qquad v = \frac{\delta}{\Delta}$$

上述的机理相应于岩块平移运动的慢波,类似还有相应于岩块转动的慢波 $U_{\varepsilon\tau}$、$U_{\varepsilon\Omega}$。

②试验模型。

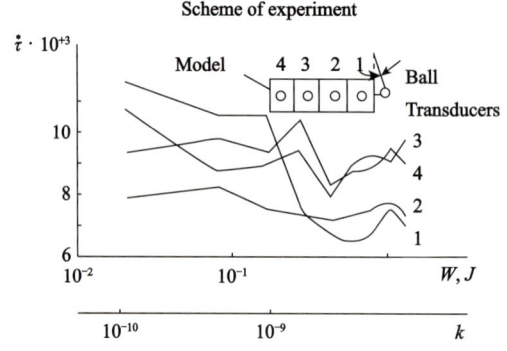

To energy criterion for origin of pendulum-type waves and geomechanical quasi-resonances

③实测的典型摆型波波形。

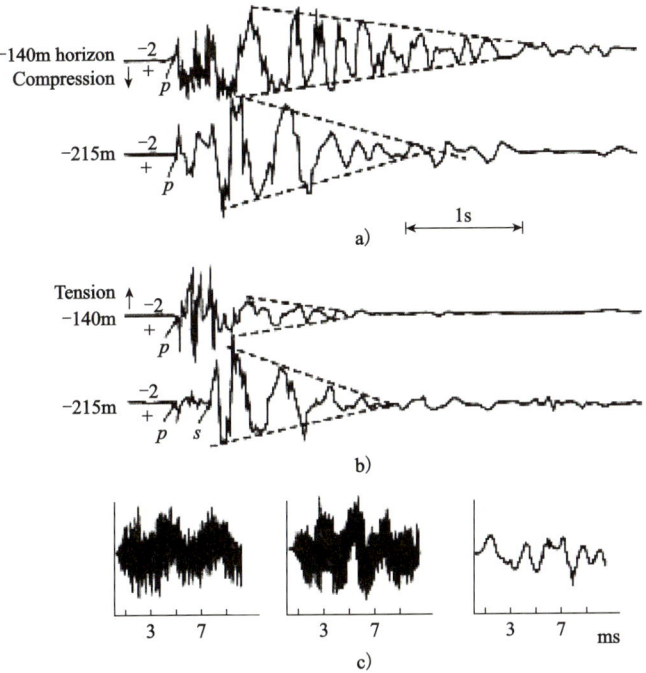

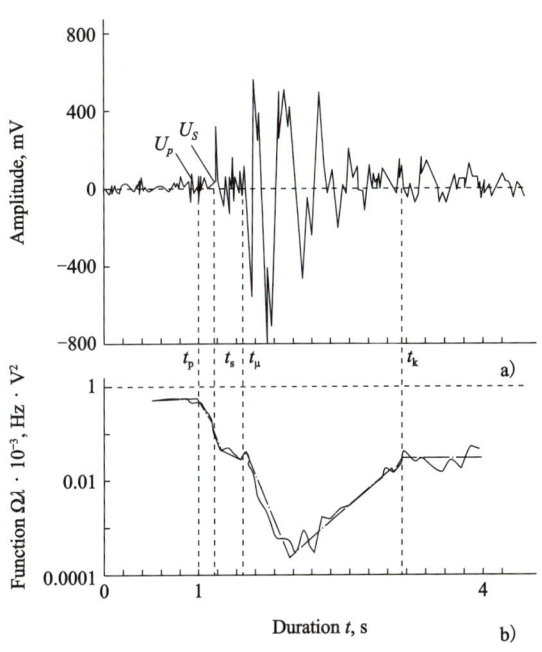

Typical seismograms of pendulum μ-waves（岩爆、爆破、摆锤冲击模型）

Correlation between structures: a) seismograms from explosion; b) plot of scanning function $\Omega_\lambda(t)$ corresponding to seismogram. Times of onset of longitudinal (t_p), transverse (t_s), and pendulum (t_μ) waves and limit of recording of μ-waves packet (t_*) are related to the flexion points on the curve of the scanning function（振幅、摆型波的扫描函数）

④慢"变形"波的实际例子。

Tashtagol 矿井矿体中工程性爆炸在爆炸源区邻域由于摆型波诱发的矿震移动轨迹面。

由于摆型波诱发的余震活动性[10m 岩块的移动速度 $U_\varepsilon \cong 25\text{m/h}(6.9\text{mm/S})$]。

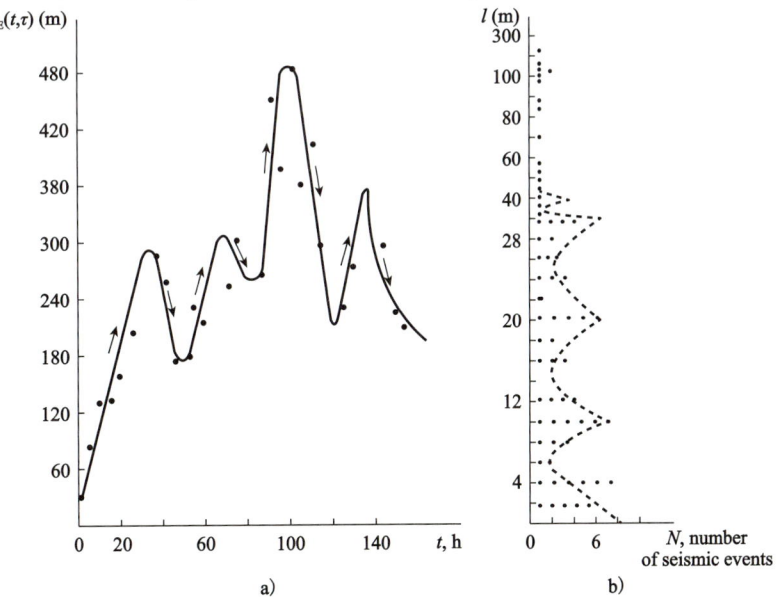

Motion trajectory of sources of seismic energy release versus time for technological explosions in working the Tashtagol deposit: a) generalized plot; b) histogram of distribution of ordinates of experimental points of scanning functions $rE(t,\tau)$ for set of blocks blasted

⑤由大量的工程爆破、岩爆、地震诱发的摆型波的测量数据的分析得出:岩体中发生的动力破坏效应是与摆型波相联系的。

a. 爆腔周围的岩体非线性位移扩展区的尺度范围可由摆型波束的频谱成分改变中估算出来。

b. 由摆型波的波形可以估算出相应尺寸等级水平的岩块载体间的摩擦系数。

c. 冲击地压源区是与动力事件诱发的摆型波频率相联系的。

d. 工程爆破在低岩爆风险岩体中的摆型波摩擦角与在高岩爆风险岩体中的摆型波摩擦角有很大不同(例如前者36°左右,后者22°左右)。

e. 对于Tashtagol矿井潜在的岩爆危险是在矿井区间受到3~4mm/m的绝对应变,冲击危险的边界是4mm/m(深基准点数据)。

(5)基于岩体块体性的动力超低摩擦效应。

超低摩擦效应的本质在于:当在一岩块系正交方向上作用动力冲击时,岩块在力方向上发生位移所需要的力的大小远低于静力下克服摩擦所需要的力的大小。所得出结果对分析冲击地压发生的机理(岩体中贮积能量的释放)以及油气田生产都具有重要意义。

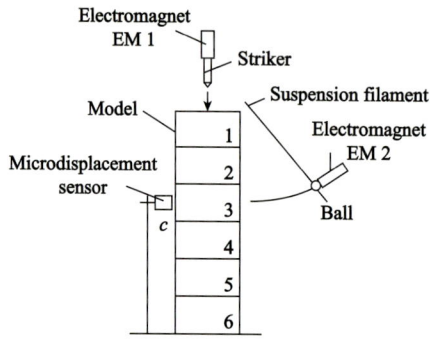

Effect of anomalous low friction in block media

应变型岩爆的机理、分类及其定量预测模型

一、研究背景

随着经济的发展,在深部高地应力地区修建了大量的隧道,这些隧道的施工过程中不时发生岩爆现象。如美国大西洋矿井和南非卡尔顿金矿中发生岩爆,二郎山隧道中发生岩爆(埋深760m),锦屏二级水电站引水隧洞内发生岩爆(埋深2250m)。岩爆严重威胁着施工人员的安全,造成了巨大的经济损失。

二、岩爆的研究进展

总结已发生的岩爆事故和矿山灾害,我们可以把岩爆定义为:处于高地应力下的岩体,因开挖卸荷或与动力作用诱发而产生的围岩岩块碎化、弹射以及工程性地震的现象。

根据岩爆的定义和主要机理的物理本质,岩爆主要可分为两种,一种是由体积不稳定性导致的应变型岩爆,一种是由于接触(剪切)的不稳定性导致的,这包括在中国和俄罗斯往往被称之为冲击地压(Stacey,Linkov)的岩爆。

除物理机理外,上述两种岩爆的主要区别在于:应变型岩爆中,扰动源和岩爆破坏区域是相重合的,冲击地压中扰动源和岩爆破坏区域间会有一定的距离。

对于冲击地压,煤巷围岩经多次开采卸荷的影响累积,处于准平衡状态。在动力作用诱发下,围岩中稳定碎块系的突然失稳导致岩块(煤块)瓦斯的突出,动力作用可以为工作面爆破或为井巷中某处顶板的断裂等。

定量预测岩爆的时代已经到来(Salamon 1993)。本报告的主要目的是说明岩石力学模拟在应变型岩爆研究中的作用,并为减轻和预防岩爆的措施设计而定量预测岩爆的位置、强度和碎块弹射速度。

关于应变型岩爆的机理,一些理论,如强度理论、刚度理论和冲击倾向性理论等,可以部分解释岩爆

报告时间:2013年7月。

发生的机理,但没有从岩爆发生机制和孕育过程上阐明机理。

已有的岩爆判据能为工程现场是否发生岩爆进行一定的定性或岩爆等级的预测,但是仍然没有对岩爆发生的位置、规模和等级进行定量预测。

三、应变型岩爆的定量预测理论

(一) 应变型岩爆的机理和分类

应变型岩爆的孕育和发展过程为:
(1)第一阶段,开挖卸荷导致围岩的破裂和碎化。
(2)第二阶段,在围岩中应力梯度作用下,围岩碎块的运动加速直至弹射,与此同时,围岩的应变能转化为动能。

岩爆根据其孕育发展过程,可分为潜在型岩爆、即时型岩爆和时滞型岩爆

动力现象的本质是剩余能量转化为动能。

1. 潜在型岩爆的机理
(1)周围围岩已经破裂但没有发生岩爆。
(2)原生微裂纹和次生微裂纹扩展消耗的总能量大于或等于岩体中存储的弹性应变能(不存在剩余能量)。

2. 即时型岩爆的机理
(1)次生微裂纹的长度大于临界值,失稳扩展形成宏观裂隙,破裂区和洞壁相交。
(2)次生裂纹扩展消耗的总能量小于围岩体存储的弹性应变能。

3. 时滞型岩爆的机理
(1)围岩内破裂区与洞壁不相交。
(2)次生裂纹扩展消耗的总能量小于围岩体存储的弹性应变能,剩余能量转化为碎块动能。
(3)破裂区岩块的运动导致至洞壁的最短路径上的围岩破裂并飞出。

(二) 岩爆定量预测

为定量预测岩爆的位置、规模和岩块弹射速度,需要解决以下三个问题:
(1)何处发生围岩破裂:基于非欧几何模型确定开挖卸荷后在洞室围岩中形成的应力场。
(2)围岩破裂如何发生:岩石破裂的起裂准则和裂纹的扩展(包括稳定扩展和失稳扩展)准则。
(3)发生何种类型岩爆:破裂区潜在型岩爆、即时型岩爆、时滞型岩爆发生的判别准则。

为解决第一个问题,岩体含有原生微裂纹以及岩爆的前提是岩体的破裂,所以,此时岩体不再是连续的,不再满足应变协调方程,需采用建立在非欧几何模型上的非协调变形岩体的力学理论来计算洞室围岩中的应力场分布。基于非欧模型的解与围岩中的微破裂现象以及应力脉动现象是相一致的,而连续介质力学模型的解与此是不相一致的。

岩石介质为非均匀地质体,岩体中存在着不同尺度的原生微裂纹,其分布函数是随机的,可由岩体细观结构试验来确定。为简单计,采用一阶近似,假定原生微裂纹均匀分布,其尺度采用统计平均值,密度用平均化法获得。

第二个问题可用断裂力学理论来解决。围岩的破裂过程包括原生裂纹沿界面和向岩体内部扩展,次生裂纹的稳定扩展和失稳扩展,微裂纹合并形成宏观裂隙最终导致岩体破裂。

对于第三个问题,基于能量理论可以建立岩爆的判据,即破碎岩块的动能等于高地应力围岩存储的应变能与消耗于岩石破裂的破坏能之差。

1. 第一个问题

深部圆形洞室含均匀分布微裂纹围岩的应力场。

在算例中,围岩中含有均匀分布的原生微裂纹,应变协调条件的假设不再成立。也就意味着,岩体非协调变形张量不等于零。基于非欧几何模型的围岩中的二次应力场可以表示为:

$$\begin{cases} \sigma_r = \sigma_{r1} + \sigma_{r2} \\ \sigma_\theta = \sigma_{\theta 1} + \sigma_{\theta 2} \\ T_{r\theta} = T_{r\theta 1} + T_{r\theta 2} \end{cases} \tag{1}$$

其中:

$\sigma_{r1} = \sigma_{\theta 1} = T_{r\theta 1}$ 为弹性应力分量;

$\sigma_{r2} = \sigma_{\theta 2} = T_{r\theta 2}$ 为自平衡应力,是由原生裂纹非协调变形导致的。

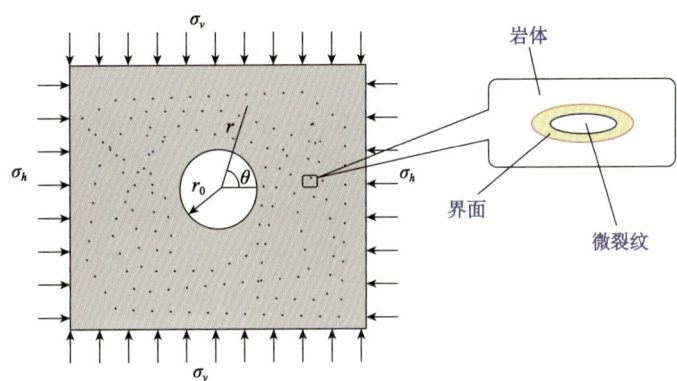

深部圆形洞室计算模型

弹性应力分量为:

$$\begin{cases} \sigma_{r1} = \frac{1}{2}(\sigma_v + \sigma_h)\left(1 - \frac{a^2}{r^2}\right) - \frac{1}{2}(\sigma_v - \sigma_h)\left(1 - \frac{4a^2}{r^2} + \frac{3a^4}{r^4}\right)\cos 2\theta \\ \sigma_{\theta 1} = \frac{1}{2}(\sigma_v + \sigma_h)\left(1 + \frac{a^2}{r^2}\right) - \frac{1}{2}(\sigma_v - \sigma_h)\left(1 + 3\frac{a^4}{r^4}\right)\cos 2\theta \\ T_{r\theta 1} = \frac{1}{2}(\sigma_v - \sigma_h)\left(1 + \frac{2a^2}{r^2} - \frac{3a^4}{r^4}\right)\sin 2\theta \end{cases} \tag{2}$$

其中: σ_h 和 σ_v 分别为初始地应力水平应力分量和垂直应力分量。

自平衡应力分量为:

$$\sigma_{r2} = -\frac{E}{2\gamma^{3/2}(1-v^2)r}[AJ_1(\gamma^{1/2}r) + BN_1(\gamma^{1/2}r) + CK_1(\gamma^{1/2}r)] + \frac{E}{2(1-v^2)r\gamma^{3/2}}[A_1J_1(\gamma^{1/2}r) + B_1N_1(\gamma^{1/2}r) +$$

$$C_1K_1(\gamma^{1/2}r)]\cos(2\theta) - \frac{3E}{(1-v^2)r^2\gamma^2} \times [A_1J_2(\gamma^{1/2}r) + B_1N_2(\gamma^{1/2}r) + C_1K_2(\gamma^{1/2}r)]\cos(2\theta)$$

$$\sigma_{\theta 2} = -\frac{E}{2\gamma(1-v^2)}[AJ_0(\gamma^{1/2}r) + BN_0(\gamma^{1/2}r) - CK_0(\gamma^{1/2}r)] + \frac{E}{2\gamma^{3/2}(1-v^2)r}[AJ_1(\gamma^{1/2}r) + BN_1(\gamma^{1/2}r) + CK_1(\gamma^{1/2}r)] +$$

$$\frac{E}{2(1-v^2)\gamma}[AJ_0(r\sqrt{\gamma}) + B_1N_0(r\sqrt{\gamma}) + C_1K_0(r\sqrt{\gamma})]\cos(2\theta) - \frac{3E}{2r(1-v^2)\gamma^{3/2}}[A_1J_1(r\sqrt{\gamma}) + B_1N_1(r\sqrt{\gamma}) +$$

$$C_1K+(r\sqrt{\gamma})]\cos(2\theta)+\frac{3E}{r^2(1-v^2)\gamma^2}\times[A_1J_2(r\sqrt{\gamma})+B_1N_2(r\sqrt{\gamma})+C_1K_2(r\sqrt{\gamma})]\cos(2\theta)$$

$$T_r\theta_2=\frac{E}{r(1-v^2)\gamma^{3/2}}[A_1J_1(r\sqrt{\gamma})+B_1N_1(r\sqrt{\gamma})+C_1K_1(r\sqrt{\gamma})]\sin(2\theta)-\frac{3E}{r^2(1-v^2)\gamma^2}[A_1J_2(r\sqrt{\gamma})+B_1N_2(r\sqrt{Y})+C_1K_2(r\sqrt{Y})]\sin(2\theta)$$

$$\delta=[\lambda_1(1-D_0)^2+2\mu_1(1-D_1)^2][3\lambda_2(1-D_2)^2+2\mu_2(1-D_3)^2]$$

$$\gamma=\frac{3\beta[3\lambda_2(1-D_2)^2+2\mu_2(1-D_3)^2](1-\alpha)}{2[3\lambda_2(1-D_2)^2(1-\alpha)+2\mu_2(1-D_3)^2(3-\alpha)]}\quad \beta=\frac{\mu_2(1-D_3)^2}{\mu_3},a=\frac{3g^2(1-D_0)^2(1-D_2)^2}{\delta},$$

$$D_0=\frac{\pi\omega}{1+\pi\omega},D_1=\frac{\pi\omega}{1+v+\pi\omega^2},\omega=\eta c^2\quad D_2=\frac{\pi\omega_0}{1+\pi\omega_0},D_3=\frac{\pi\omega_0}{1+v+\pi\omega_0},\omega_0=\eta_0c_0^2$$

其中，η 为次生裂纹密度；c 为次生裂纹统计平均半长度；η_0 为原生裂纹密度；c_0 为原生裂纹统计平均半长度。

与协调解（$\sigma_{r1},\sigma_{\theta1},\tau_{r\theta_1}$）相比，非协调变形量（$\sigma_{r2},\sigma_{\theta2},\tau_{r\theta_2}$）和 η_0,η,c,c_0 有关。众所周知，（$\sigma_{r1},\sigma_{\theta1},\tau_{r\theta_1}$）为与荷载（初始地应力）相平衡及与边界条件相适应的解。所以，（$\sigma_{r2},\sigma_{\theta_2},\tau_{r\theta_2}$）为自平衡应力解，这就容易理解自平衡应力分量是由岩体中的内部缺陷引起的非协调变形产生的。

图 1 和图 2 给出了自平衡应力的数值解。相应计算参数如下：$\sigma_v=40\text{MPa},\sigma_h=20\text{MPa},v=0.2,C=C_1=70.25,g=90\text{MPa},\lambda_1=3.2\text{GPa},\lambda_2=4.1667\text{GPa},\mu_1=3.2\text{GPa},\mu_2=6.25\text{GPa},r_0=7\text{m},\theta=0°,E=15\text{GPa}$。

计算结果均相应于不同的原生微裂纹的半长度，从 1μm 到 500μm，但相应于同一个原生微裂纹密度 $\eta=3\times10^4(1/\text{m}^2)$。

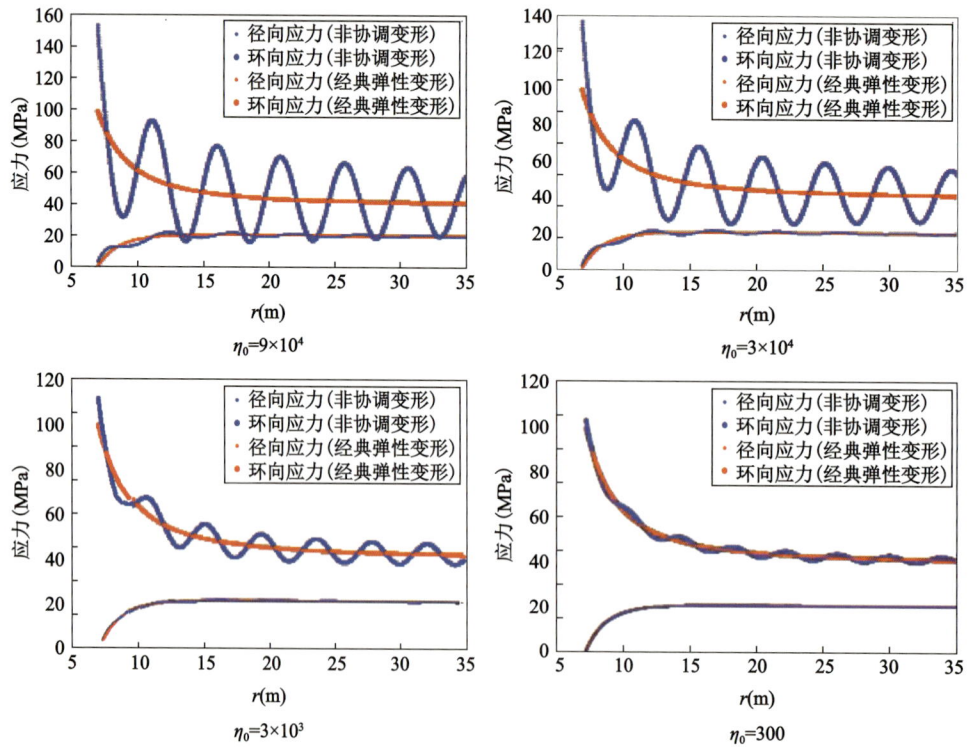

当 $c_0=0.05\text{mm}$ 时，不同原生微裂纹密度情况下隧洞围岩应力变化规律

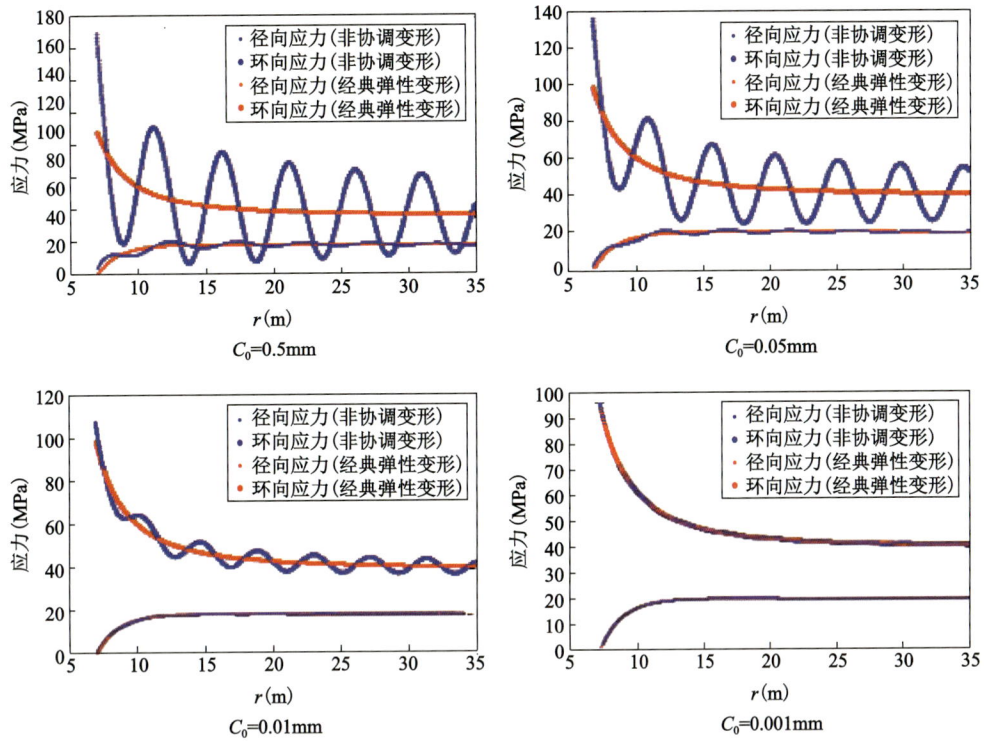

当 $\eta = 3 \times 10^4 (1/m^2)$ 时,不同原生微裂纹半长情况下隧洞围岩应力变化规律

从上两图可以看出,由于非协调变形,围岩的自平衡应力在径向是振荡波动的,其振幅随着离洞壁的距离增加而减小。自平衡应力振荡特性和原生裂纹的密度和尺度有关。

当原生裂纹密度小于临界值时[例如 $\eta_0 = 300(1/m^2)$],振荡特性可以忽略,即使原生裂纹半长达到 0.05mm。同样原生裂纹的半长小于临界值(例如 $c_0 = 0.001$mm)时,振荡特性也可以忽略,即使原生裂纹密度达到 $\eta_0 = 3 \times 10^4 (1/m^2)$。

1983年,在金川矿井中首次发现了围岩应力应变的振荡特性,在山东大学的实验室内,也发现了同样的现象。在淮南煤矿的现场测量中,发现了破裂区和非破裂区交替出现的现象(分区破裂化现象)。

因此,可以得出结论:当原生裂纹的密度或尺度小于某个临界值时,非协调变形效应可以忽略。在此种情况下,岩体可以作为连续介质来考虑。否则,必须要考虑岩体内部缺陷引发的非协调变形的影响。

通过对比协调解和非协调解,我们可以将分区破裂化和围岩内发生岩爆的机理理解为:

(1)其物理本质在于岩体中内部缺陷的存在;
(2)其力学本质在于原生裂纹导致的非协调变形。

2. 第二个问题

基于断裂力学理论的原生微裂纹启裂和次生微裂纹的扩展,原生微裂纹尖端的 1 型应力强度因子可以表示为:

$$K_I = -[\sigma_2' + f(c_0)S_2']\sqrt{2d_i \tan\left(\frac{\pi c_0}{2d_i}\right)} \tag{3}$$

其中:

$$\begin{cases} S_2' = \sigma_2' - (\sigma_1' + \sigma_2')/2 \\ f(c_0) = d/c_0 \\ \sigma_1' = \frac{\sigma_r + \sigma_\theta}{2} + \frac{1}{2}\sqrt{(\sigma_r - \sigma_\theta)^2 + 4\tau_{r\theta}^2} \\ \sigma_2' = \frac{\sigma_r + \sigma_\theta}{2} - \frac{1}{2}\sqrt{(\sigma_r - \sigma_\theta)^2 + 4\tau_{r\theta}^2} \end{cases} \tag{4}$$

$\tau_{r\theta}$、σ_r 和 σ_θ 由公式(1)确定。d_i 是微裂纹之间的间距,可由岩体细观结构试验得到。

原生微裂纹沿界面开始扩展的临界条件为:

$$K_I = K_{IC} \tag{5}$$

其中,K_{IC} 为界面的断裂韧度。

当原生微裂纹尖端的应力强度因子满足判据条件(5),原生裂纹将沿界面扩展扩展至基质,其扩展长度等于界面厚度 c_1,由该岩体细观结构试验决定。

次生裂纹产生的条件为:

$$\sigma_2'' + f(c_1) S_2'' = \frac{K_{IC}}{\sqrt{2d_i \tan\left(\dfrac{\pi c_1}{2d_i}\right)}} \tag{6}$$

其中:

$$f(c_1) = d/c_1 \qquad \sigma_1'' = \frac{\sigma_r' + \sigma_\theta'}{12} + \sqrt{(\sigma_s - \sigma_{s'}')^2 + 4\tau_{r\theta}^2}$$

$$S_2'' = \sigma_2'' - (\sigma_1'' + \sigma_2'')/2 \qquad \sigma_2' = \frac{\sigma_r' + \sigma_\theta'}{12} - \sqrt{(\sigma_r - \sigma_\theta')^2 + 4\tau_{r\theta}^2}$$

σ_r'、σ_θ' 和 $\tau_{r\theta}'$ 是原生微裂纹扩展后的平均长度为 c_1 时,深部圆形洞室围岩应力场分量,其可以用 c_1 替代 c 和 η_1 替代 η 来求得。

次生裂纹尖端的 I 型应力强度因子为:

$$K_I = -[\sigma_2'' + f(l) S_2''']\sqrt{2d_i \tan\left(\frac{\pi l}{2d_i}\right)} \tag{7}$$

其中: $f(l) = d/l;\; \sigma_1''' = \dfrac{\sigma_r'' + \sigma_\theta''}{2} + \dfrac{1}{2}\sqrt{(\sigma_r' - \sigma_\theta'')^2 + 4\tau_{r\theta}''^2}$

$$S_2''' = \sigma_2''' - (\sigma_1''' + \sigma_2''')/2;\; \sigma_2''' = \frac{\sigma_r'' + \sigma_\theta''}{2} - \frac{1}{2}\sqrt{(\sigma_r'' - \sigma_\theta'')^2 + 4\tau_{r\theta}''^2}$$

σ_r''、σ_θ'' 和 $\tau_{r\theta}''$ 是次生裂纹扩展长度为 l 时,深埋圆形洞室围岩应力场分量。可由 $\eta_1 c_1^2 + \eta_2 l^2$ 替代 ηc^2 来求得。η_2 为稳定扩展的次生裂纹平均密度。

次生裂纹在岩石基质稳定扩展的临界条件为:

$$K_I = K_{IC} \tag{8}$$

其中,K_{IC} 为岩石的断裂韧度。

当次生裂纹尖端的应力强度因子即式(7)满足式(8)时,次生裂纹在岩石基质中发生稳定扩展。次生裂纹稳定扩展的长度可由式(9)和式(10)迭代计算确定。

在次生裂纹不稳定扩展初期,次生裂纹的失稳扩展临界长度 c_2 可由次生裂纹失稳扩展的准则确定:当次生裂纹扩展至临界值时,应力不增加,而裂纹长度继续增加。

次生裂纹从失稳扩展发展到连接和汇合成宏观裂隙的临界条件为:

$$l = d_i - c_2 \tag{9}$$

原生微裂纹沿界面扩展消耗的能量密度为:

$$U_1 = \frac{\eta_1(\kappa+1)(1+v)}{2E} \int_{c_0}^{c_1} \left\{ \sqrt{2d_i \tan\left(\frac{\pi c}{2d_i}\right)} [\sigma_2' + f(c) S_2'] \right\}^2 dc \tag{10}$$

次生裂纹稳定扩展消耗的能量密度为：

$$U_2 = \frac{\eta_2(\kappa+1)(1+\nu)}{2E}\int_{c_1}^{c_1+l}\left\{\sqrt{2d_i\tan\left(\frac{\pi l}{2d_i}\right)}\left[\sigma''_2+f(l)S''_2\right]\right\}^2 dl \quad (11)$$

次生裂纹失稳扩展消耗的能量密度为：

$$U_3 = \frac{\eta_3(\kappa+1)(1+\nu)}{2E_1}K_{IC}^2(d_i-c_2) \quad (12)$$

原生裂纹沿界面扩展、次生裂纹稳定扩展和失稳扩展的总耗散能密度为：

$$\xi = U_1 + U_2 + U_3 \quad (13)$$

岩体储存的弹性应变能密度可以表示为：

$$U_e = \sum_{i=0}^{n} A_i\left\{\frac{1-\nu^2}{2E}\left[(\sigma_{r_i}^e)^2+(\sigma_{\theta_i}^e)^2\right]-\frac{\nu(1+\nu)}{E}\sigma_{r_i}^e\sigma_{\theta_i}^e+\frac{1+\nu}{E}(T_{r_i\theta_i}^e)^2\right\} \quad (14)$$

3. 第三个问题

(1) 潜在型岩爆的判据

①原生裂纹和次生裂纹扩展总的耗散能密度小于岩体存储的弹性应变能密度；

②破裂区与洞壁不相交，且剩余的弹性应变能比破坏至洞壁的最短路径所需能量小。

$$\begin{cases} W = U_e - \xi \geq 0 \\ W < U(l_p) \end{cases} \quad (15)$$

l_p 为破裂区至沿洞壁方向的最短破坏路径，可由迭代计算求得。$U(l_p)$ 可由式(12)由 l_p 代替 $(d_i - c_2)$ 求得。

(2) 即时型岩爆的判据

①原生和次生裂纹扩展消耗的总能量小于岩体存储的弹性应变能；

②次生裂隙失稳扩展至洞壁，破裂区和洞壁相交。

$$\begin{cases} W = U_e - \xi \geq 0 \\ l_p = 0 \end{cases} \quad (16)$$

(3) 时滞型岩爆判据

①原生和次生裂纹扩展消耗的总能量小于岩体存储的弹性应变能；

②破裂区与洞壁不相交，但破裂岩体剩余的弹性应变能比破坏至洞壁的最短破坏路径所需能量大，破裂区岩块沿破坏最短路径向洞室方向飞出。

$$\begin{cases} W = U_e - \xi \geq 0 \\ W \geq U(l_p) \end{cases} \quad (17)$$

即时型岩爆岩石碎块的弹射速度为：

$$V = \sqrt{2W/\rho A} \quad (18)$$

其中，$A = \sum_{i=0}^{n} A_i$ 为岩爆区，ρ 为岩体的密度。

时滞型岩爆岩石碎块的弹射速度为：

$$V = \sqrt{2[W-U(l_p)]/\rho A} \quad (19)$$

四、基于数值计算的岩爆分析

(一) 当垂直应力分量大于水平应力分量时的分析岩爆

$r_0 = 6\text{m}, \nu = 0.15, \rho = 2200\text{kg/m}^3, E_1 = 300\text{GPa}, c_0 = 0.01\text{mm}, \eta_0 = 50000\text{m}^{-2}, d = 0.2\text{mm}, d_i = 4\text{mm},$

$\mu_3 = 5.67 \times 10^8 \text{N} \cdot \text{m}^2, E = 30\text{GPa}, C = 3.5378 \times 10^7 \text{m}^{-2}, C_1 = 3.5378 \times 10^7 \text{m}^{-2}, \lambda = 0.33, \lambda_2 = 8\text{GPa}, c_1 = 0.02\text{mm}, \mu_1 = 7\text{GPa}, \mu_2 = 4.9\text{GPa}, g = 2.14\text{MPa}, K_{\text{IC}} = 0.2\text{MPa}, \sigma_v = 40\text{MPa}$。

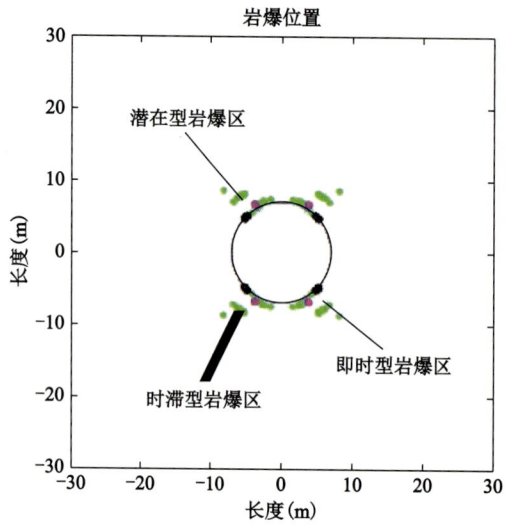

当 $\sigma_v/\sigma_c = 0.375$ 时潜在型岩爆区、即时型岩爆区和时滞型岩爆区的分布

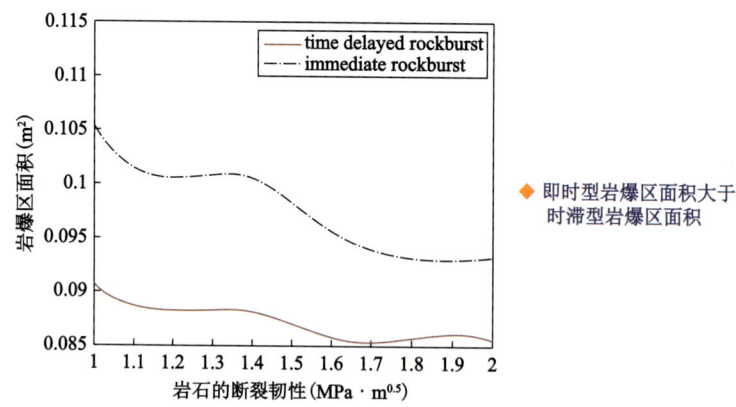

当 $\sigma_v/\sigma_c = 0.375$ 时即时型岩爆区和时滞型岩爆区的面积

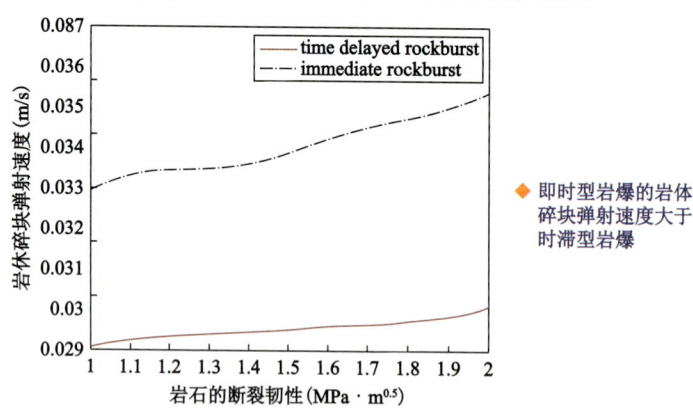

当 $\sigma_v/\sigma_c = 0.375$ 时即时型岩爆和时滞型岩爆的岩体碎块弹射速度

(二)当垂直应力分量小于水平应力分量时的岩爆分析

$r_0 = 6\text{m}, v = 0.15, p = 2200\text{kg/m}^3, E_1 = 300\text{GPa}, c_0 = 0.01\text{mm}, \eta_0 = 50000\text{m}^{-2}, d = 0.2\text{mm}, d_i = 2\text{mm}, \mu_3 = 5.715 \times 10^8 \text{N} \cdot \text{m}^2, E = 30\text{GPa}, C = 5.8653 \times 10^7 \text{m}^{-2}, C_1 = 5.8653 \times 10^7 \text{m}^{-2}, \lambda = 1.2, \lambda_1 = 10\text{GPa}, \lambda_2 = 8\text{GPa}, \mu_2 = 4.9\text{GPa}, g = 2.14\text{MPa}, K'_{\text{IC}} = 0.2\text{MPa}\sqrt{\text{m}}, \sigma_v = 40\text{MPa}, c_1 = 0.05\text{mm}$。

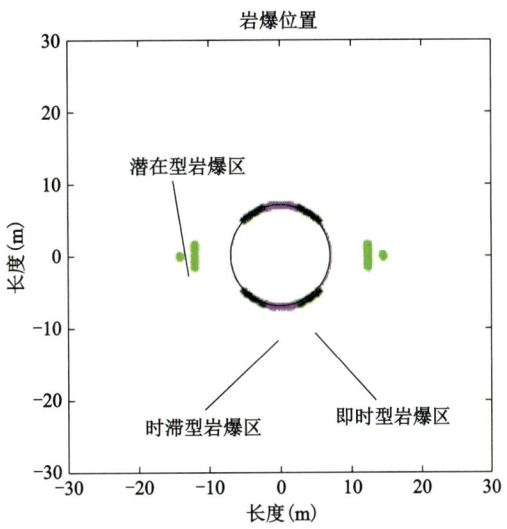

当 $\sigma_h/\sigma_c=0.45$ 时潜在型岩爆区、即时型岩爆区和时滞型岩爆区的分布

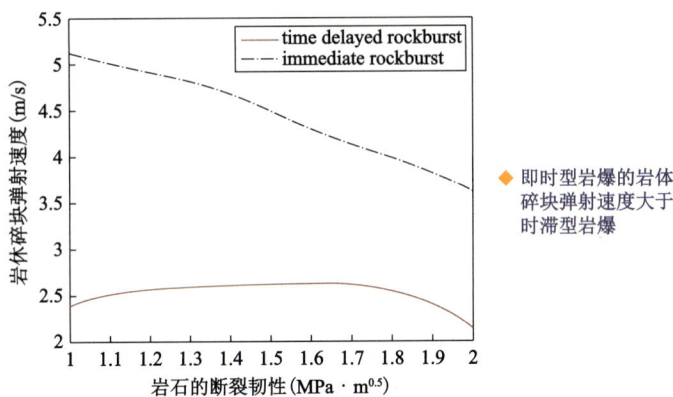

当 $\sigma_h/\sigma_c=0.45$ 时即时型岩爆和时滞型岩爆岩体碎块弹射速度

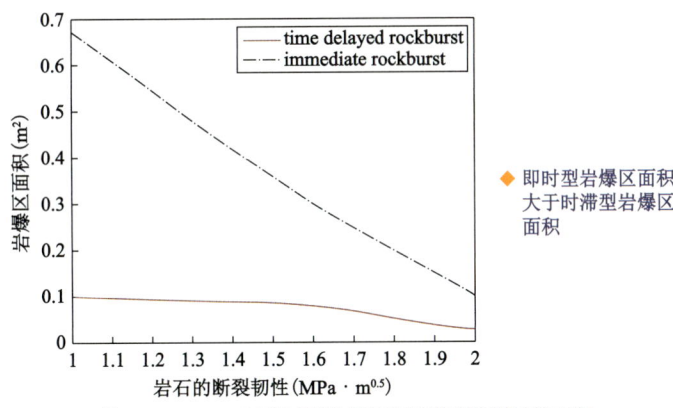

当 $\sigma_h/\sigma_c=0.45$ 时即时型岩爆和时滞型岩爆区的面积

五、结论及讨论

(一)结论

(1)利用非欧几何模型,获得了深部圆形洞室含均匀分布原生微裂纹围岩的应力场。
(2)揭示了含均匀分布微裂纹围岩潜在型岩爆、即时型岩爆和时滞型岩爆发生的机理。

(3)基于潜在岩爆、即时型岩爆和时滞型岩爆发生的机理,建立了潜在型岩爆、即时型岩爆和时滞型岩爆判据。

(4)提出了潜在型岩爆、即时型岩爆和时滞型岩爆的定量预测模型。

(5)确定了潜在型岩爆区、即时型岩爆区和时滞型岩爆区的面积与位置。

(6)获得了即时型岩爆岩石碎块和时滞型岩爆岩石碎块的弹射速度。

(7)潜在型岩爆区分布范围最大,即时型岩爆区面积大于时滞型岩爆区面积;即时型岩爆岩石碎块的弹射速度大于时滞型岩爆岩石碎块的弹射速度。

(二)讨论

(1)开挖卸荷和裂纹的扩展都非瞬时完成,与时间有关。岩石的断裂韧度也与加卸载速率有关,所以岩爆的分析、计算不是静力问题,是动力问题。

(2)卸荷的时间因素、裂纹扩展的速度和断裂韧度的率相关性以及岩石运动惯性没有考虑,今后将进一步考虑时间因素对岩爆的影响。

(3)采用平均化的方法研究微裂纹的长度和密度,而实际岩体(石)中的原生微裂纹的长度和密度是按一定分布规律分布的,平均化会带来一定的误差,今后将采用数值的方法研究按一定规律分布的原生微裂纹以及实际产生的非均匀分布的次生微裂纹对岩爆的影响。

(4)采用平均化的方法研究微裂纹的长度和密度,而实际岩体(石)中的原生微裂纹的长度和密度是按一定分布规律分布的,平均化会带来一定的误差,今后将采用数值的方法研究按一定规律分布的原生微裂纹以及实际产生的非均匀分布的次生微裂纹对岩爆的影响。

(5)初始地应力沿隧道轴长而变化,岩爆的预测是三维问题,基于地应力反演,计算若干断面的预测,可以得到准三维的应变型岩爆预测解。

非协调变形与岩爆的机理和预测

一、研究岩爆必须要研究岩体的破裂

冯夏庭研究团队与唐春安研究团队在锦屏二级水电站利用微震（破裂）监测所进行的岩爆预报预警的成功实践表明，岩爆的发生与岩石的微破裂紧密相关。

钟世航：在锦屏二级电站3号、4号引水隧洞用了陆地声呐法断裂预报，从探查结果表明：
(1) 许多断裂与微震事件密集区吻合；
(2) 许多断裂，特别是交叉断裂的交叉点与发生岩爆的位置十分一致。
岩爆的发生与岩体的大破裂密切相关。

唐春安：卸载岩爆（不管）是在（高应力下或者相对低应力下）岩体开挖后（更易于）沿临空面内裂纹、扩展、贯通直至突然失稳瞬间释放能量而破坏的过程。

二、研究岩体破裂必须要研究岩体的非协调变形

1. 材料变形的协调性

变形产生的位移在其定义域内为单值函数，即变形后的微单元可拼合为一个变了形的物体整体。变形的不协调现象：岩体变形出现"开裂"和"叠合"。

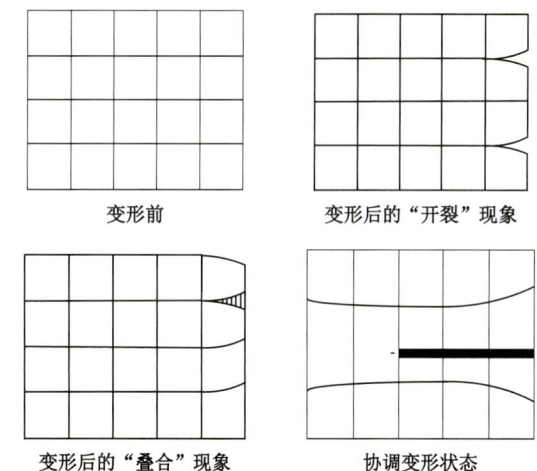

图 1 岩体的协调和非协调变形

2. 变形协调条件（方程）

平面问题：

$$\frac{\partial^2 \varepsilon_x}{\partial y^2} + \frac{\partial^2 \varepsilon_y}{\partial x^2} = \frac{\partial^2}{\partial y^2}\left(\frac{\partial u}{\partial x}\right) + \frac{\partial^2}{\partial x^2}\left(\frac{\partial v}{\partial y}\right) = \frac{\partial^2}{\partial x \partial y}\left(\frac{\partial u}{\partial y} + \frac{\partial v}{\partial x}\right) = \frac{\partial^2 \gamma_{xy}}{\partial x \partial y}$$

位移 u, v（单值→连续可微）

报告时间：2012年11月25日。

3. 变形协调解

满足应变(或应力)平衡方程和应变(或应力)协调方程(非协调张量为零)的解,这个解是单值连续的、唯一的、是不会出现岩体的开裂或套叠。

$$\begin{cases} \sigma_{ij,j} = 0 \\ \Delta \sigma = 0 \end{cases}$$

Δ:二阶拉普拉斯算符。

岩体出现开裂(破裂)或套叠后,其变形就不是协调的,就不满足变形协调方程,在理论上不能采用变形协调解确定应力和位移,在实践上,采用变形协调解就会引起误差,而误差程度未知。

采用变形协调元的有限单元法来研究出现破裂后的岩体变形和破坏在理论上是错误的,在实践上会产生未知的误差。

4. 变形非协调解

岩体出现开裂后,不满足变形协调张量,非协调张量不为零,变形非协调解需满足以下平衡方程和变形非协调方程。

$$\begin{cases} \sigma_{ij,j} = 0 \\ \dfrac{2(1-\nu)}{E}\Delta\sigma = R \\ \Delta^2 R = \gamma^2 R \end{cases}$$

三、非协调变形的实例研究

作为计算实例(图2),取圆形洞室(半径 a)平面应变情况中的围岩应力分析问题,地层中自由场初始垂直、水平地应力为 σ_v 及 σ_h。众所周知,按连续介质弹塑性力学,符合应变协调方程(相容方程)的毛洞围岩中的二次应力场为:

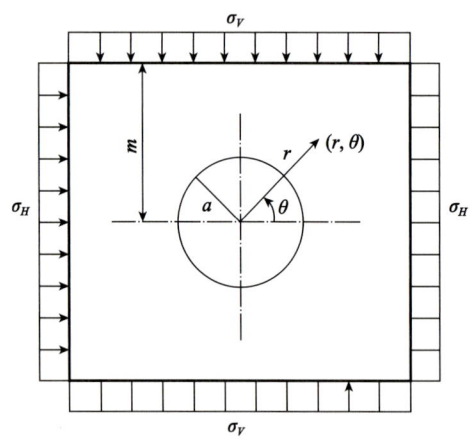

图 2　圆形洞室力学模型

协调解
$$\left\{\begin{aligned} \sigma_r &= \frac{1}{2}(\sigma_v + \sigma_h)\left(1 - \frac{a^2}{r^2}\right) - \frac{1}{2}(\sigma_v - \sigma_h)\left(1 - \frac{4a^2}{r^2} + \frac{3a^4}{r^4}\right)\cos 2\theta \\ \sigma_\theta &= \frac{1}{2}(\sigma_v + \sigma_h)\left(1 + \frac{a^2}{r^2}\right) - \frac{1}{2}(\sigma_v - \sigma_h)\left(1 + 3\frac{a^4}{r^4}\right)\cos 2\theta \\ \tau_{r\theta} &= \frac{1}{2}(\sigma_v - \sigma_h)\left(1 + \frac{2a^2}{r^2} - \frac{3a^4}{r^4}\right)\sin 2\theta \end{aligned}\right.$$

该计算实例,对于含有初始微裂纹和次生微裂纹的围岩,不再保持应变协调条件,即非协调变形不等于零的情况下,基于非欧模型的围岩中的二次应力场为:

$$\left.\begin{array}{l}\sigma_r = \sigma_{r_1} + \sigma_{r_2} \\ \sigma_\theta = \sigma_{\theta 1} + \sigma_{\theta 2} \\ \tau_{r\theta} = \tau_{r\theta_1} + \tau_{r\theta_2}\end{array}\right\}$$

其中:

$\sigma_{r1}, \sigma_{\theta 1}, \tau_{r\theta 1}$ 代表协调变形时的弹性应力,即协调解;

$\sigma_{r2}, \sigma_{\theta 2}, \tau_{r\theta 2}$ 为内部缺陷引起的非协调变形所产生的自平衡应力。

$$\sigma_{r1} = \frac{\sigma_v + \sigma_h}{2}\left(1 - \frac{r_0^2}{r^2}\right) + \frac{\sigma_h - \sigma_v}{2}\left(1 - 4\frac{r_0^2}{r^2} + 3\frac{r_0^4}{r^4}\right)\cos(2\theta)$$

$$\sigma_{r2} = -\frac{E}{2\gamma^{3/2}(1-\nu^2)r}[AJ_1(\gamma^{1/2}r) + BN_1(\gamma^{1/2}r) + CK_1(\gamma^{1/2}r)] + \frac{E}{2(1-\nu^2)r\gamma^{3/2}}[A_1J_1(\gamma^{1/2}r) + B_1N_1(\gamma^{1/2}r) +$$

$$C_1K_1(\gamma^{1/2}\gamma)]\cos(2\theta) - \frac{3E}{(1-\nu^2)\gamma^2 r^2} \times [A_1J_2(\gamma^{1/2}r) + B_1N_2(\gamma^{1/2}r) + C_1K_2(\gamma^{1/2}r)]\cos(2\theta)$$

$$\sigma_{\theta 1} = \frac{\sigma_v + \sigma_h}{2}\left(1 + \frac{r_0^2}{r^2}\right) - \frac{\sigma_h - \sigma_v}{2}\left(1 + 3\frac{r_0^4}{r^4}\right)\cos(2\theta)$$

$$\sigma_{\theta 2} = -\frac{E}{2\gamma(1-\nu^2)}[AJ_0(\gamma^{1/2}r) + BN_0(\gamma^{1/2}r) - CK_0(\gamma^{1/2}r)] + \frac{E}{2\gamma^{3/2}(1-\nu^2)r}[AJ_1(\gamma^{1/2}r) + BN_1(\gamma^{1/2}r) +$$

$$CK_1(\gamma^{1/2}r)] + \frac{E}{2(1-\nu^2)\gamma}[A_1J_0(r\sqrt{\gamma}) + B_1N_0(r\sqrt{\gamma}) + C_1K_0(r\sqrt{\gamma})]\cos(2\theta) -$$

$$\frac{3E}{2r(1-\nu^2)\gamma^{3/2}}[A_1J_1(r\sqrt{\gamma}) + B_1N_1(r\sqrt{\gamma}) + C_1K_1(r\sqrt{\gamma})]\cos(2\theta) + \frac{3E}{r^2(1-\nu^2)\gamma^2} \times$$

$$[A_1J_2(r\sqrt{\gamma}) + B_1N_2(r\sqrt{\gamma}) + C_1K_2(r\sqrt{\gamma})]\cos(2\theta)$$

$$\tau_{r\theta 1} = \frac{\sigma_v - \sigma_h}{2}\left(1 + 2\frac{r_0^2}{r^2} - 3\frac{r_0^4}{r^4}\right)\sin(2\theta)$$

$$\tau_{r\theta 2} = \frac{E}{r(1-\nu^2)\gamma^{3/2}}[A_1J_1(r\sqrt{\gamma}) + B_1N_1(r\sqrt{\gamma}) + C_1K_1(r\sqrt{\gamma})]\sin(2\theta) - \frac{3E}{r^2(1-\nu^2)\gamma^2}[A_1J_2(r\sqrt{\gamma}) +$$

$$B_1N_2(r\sqrt{\gamma}) + C_1K_2(r\sqrt{\gamma})]\sin(2\theta)$$

$$\gamma^2 = \frac{3\beta[3\lambda_2(1-D_2)^2 + 2\mu_2(1-D_3)^2](1-\alpha)}{2[3\lambda_2(1-D_2)^2(1-\alpha) + 2\mu_2(1-D_3)^2(3-\alpha)]}$$

$$\beta = \frac{\mu_2(1-D_3)^2}{\mu_3}, \alpha = \frac{3g^2(1-D_0)^2(1-D_2)^2}{\delta},$$

$$\delta = [\lambda_1(1-D_0)^2 + 2\mu_1(1-D_1)^2] \times [3\lambda_2(1-D_2)^2 + 2\mu_2(1-D_3)^2]$$

$$D_0 = \frac{\pi\omega}{1+\pi\omega}, D_1 = \frac{\pi\omega}{1+\nu+\pi\omega}, \omega = \eta c^2$$

$$D_2 = \frac{\pi\omega_0}{1+\pi\omega_0}, D_3 = \frac{\pi\omega_0}{1+\nu+\pi\omega_0}, \omega_0 = \eta_0 c_0^2$$

其中,T_0 为洞室半径,(T,θ) 是极坐标,E 为杨氏模量,σ_v 为垂直地应力,σ_h 为水平地应力,η 代表次生微裂纹密度,c 为次生微裂纹半长,η_0 是初始微裂纹密度,c_0 为初始微裂纹半长,ν 为泊松比,C 由与 Wronskian 相一致的 $J_0(YT)$ 和 $N_0(YT)$ 的线性独立解确定,J_0、N_0、K_0 相应地分别为零阶贝塞尔函数、纽曼函数和麦克唐纳圆柱函数。

比较上述这两个解可见,非协调张量不等于零的非协调解较之非协调张量等于零的协调解(σ_{r1},$\sigma_{\theta1}$,$\tau_{r\theta1}$)增加了与初始微裂纹密度 η_0 和初始微裂纹半长 c_0 以及次生微裂纹密度 η 和次生微裂纹半长 c 有关的部分(σ_{r2},$\sigma_{\theta2}$,$\tau_{r\theta2}$)。由于协调解 σ_{r1}、$\sigma_{\theta1}$、$\tau_{r\theta1}$,众所周知为与荷载(初始地应力)相平衡及与边界条件相适应的解,所以 σ_{r2}、$\sigma_{\theta2}$、$\tau_{r\theta2}$ 即为自平衡的封闭应力解。可见岩体中的自平衡应力是由岩体中的内部缺陷和次生裂纹引起的非协调变形产生的。

四、非协调变形对围岩应力的影响

协调解和非协调解的比较:现在来分析非协调变形产生的岩体中的自平衡应力变化情况,为此,给出相应参数。

$\sigma_v = 40\text{MPa}, \sigma_h = 20\text{MPa}, \nu = 0.2, C = C_1 = 70.25, g = 90\text{MPa}, \lambda_1 = 3.2\text{GPa}, \lambda_2 = 4.1667\text{GPa}, \mu_1 = 3.2\text{GPa}, \mu_2 = 6.25\text{GPa}, r_0 = 7\text{m}, \theta = 0°, E = 15\text{GPa}$。

(1)协调解(不考虑微裂纹)。
(2)非协调解(考虑均匀初始微裂纹)。
(3)非协调解(考虑均匀初始微裂纹和次生微裂纹)。

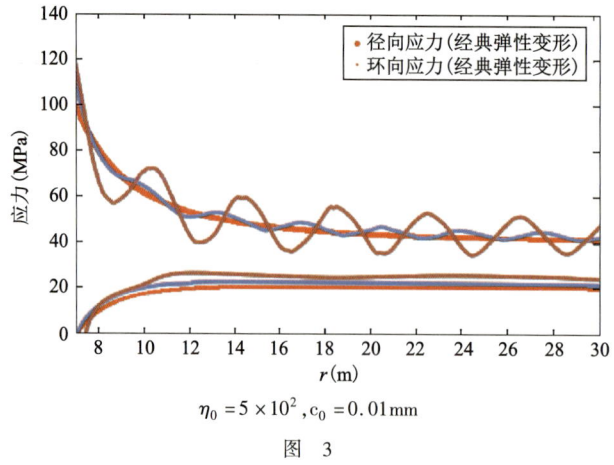

$\eta_0 = 5 \times 10^2, c_0 = 0.01\text{mm}$

图 3

(4)非协调变形的影响因素。
①不同的初始微裂纹密度。

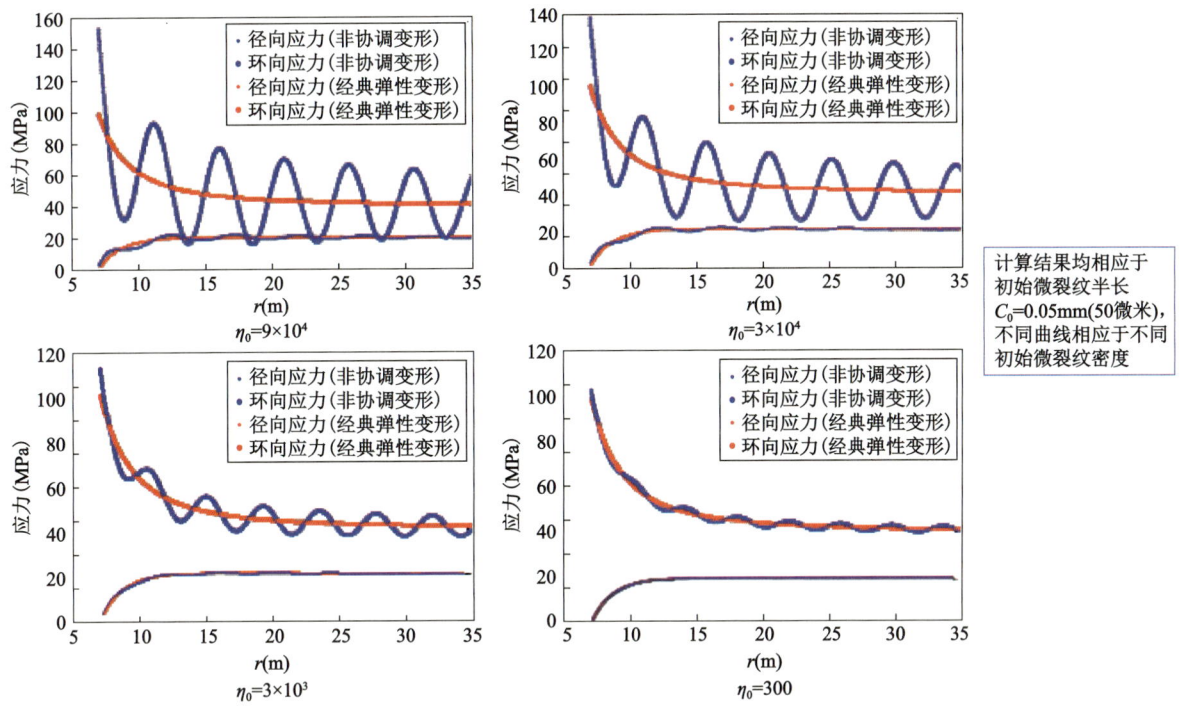

图 4　当 $C_0 = 0.05$ mm 时，不同初始微裂纹密度情况下隧洞围岩应力变化规律

② 不同的初始微裂纹尺度。

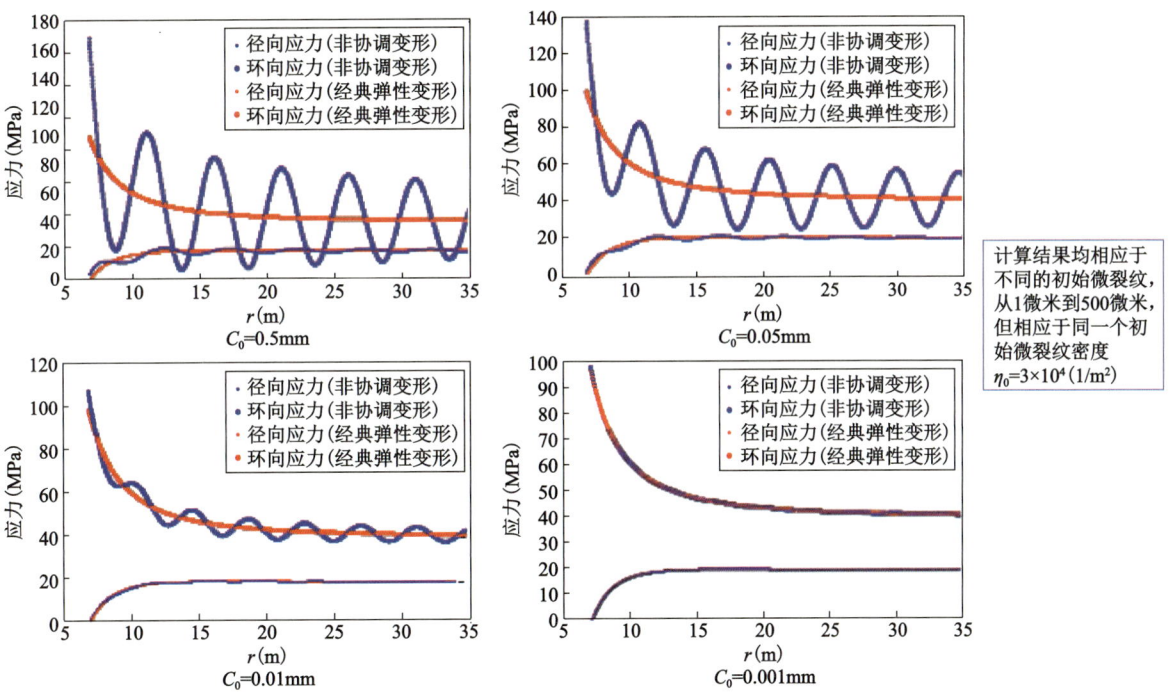

图 5　当 $\eta_0 = 3 \times 10^4$ 时，不同初始微裂纹半长范围内隧洞围岩应力变化规律

③ 除了均匀初始微裂纹，还有一条宏观裂隙。

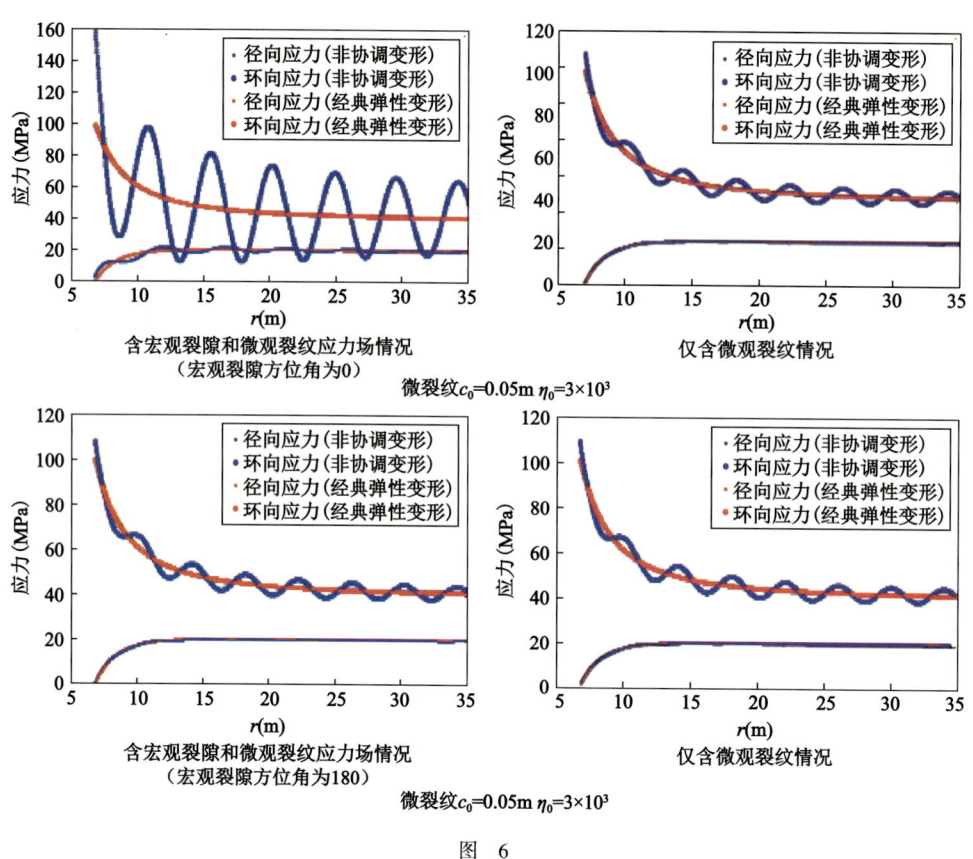

图 6

(5) 小结:非协调变形的影响。

由图可见:围岩中非协调变形产生的自平衡应力在半径方向呈波状振荡变化,波幅自洞周往内逐步减小;它与围岩中的微裂纹的演化密切相关,次生微裂纹的产生与发展增强了非协调变形产生的自平衡应力波状变化的振幅;这个自平衡应力对总应力的影响与初始微裂纹的密度与大小密切相关,密度越大,影响越大,裂纹越大,影响越大,当初始微裂纹密度较小时($\eta_0=300$),即使微裂纹半长c_0为$50\mu m$,其影响可忽略不计,相类似,当微裂纹很小时(c_0为$1\mu m$),即使密度很大($\eta_0=3\times10^4$),其影响也可忽略不计。

宏观裂隙对围岩中自平衡应力的影响,当方位角为 0 时很明显,当方位角为 180°时不明显。

洞室围岩中应力在半径方向呈波荡变化的现象三十年前在金川镍矿地下矿井围岩中早有发现,现场测量表明:围岩中松弛地区和严密区交替出现。

张强勇的实验表明:在高地应力条件下,洞室开挖后洞周径向位移和径向应变出现波峰与波谷间隔分布的拉压交替变化规律。

因此,我们似乎可以得出结论,初始微裂纹很稀或很小时,是可以忽略不计的,即将不连续岩体看作连续性介质,可以不研究非协调变形的影响,从物理意义上可推测说,稀疏的很小的微裂纹群所引起岩体变形的内能变化量很小,可以在不可逆热力学方程中忽略它。

由上述算例的协调解与非协调解的对比可见,其主要差别为:

(1) 协调解的围岩应力极值在洞周附近,然后围岩应力单调衰减,直至围岩深部趋近于初始地应力,因此按照协调解,围岩的破坏仅发生在洞周附近。

(2) 非协调解的围岩应力自洞周至围岩深部呈波形振荡衰减,在洞周附近及离洞周不同距离上有若干个不断减小的极值,如果这些应力极值达到强度破坏准则,就使岩体产生多处破坏。因此按照非协调解,围岩的破坏可以发生在离洞周的不同距离上的几个部位,即分区破裂。

(3) 这就是围岩分区破裂化和围岩深部可能发生岩爆的本质原因:围岩的内部缺陷是围岩分区破裂

化和围岩深部可能发生岩爆的物理本质,这些内部缺陷引起的非协调变形是产生分区破裂化和深部岩爆的力学本质。

五、非协调变形研究对岩爆的影响

1. 协调解和非协调解对动破裂(可能的岩爆)的影响比较

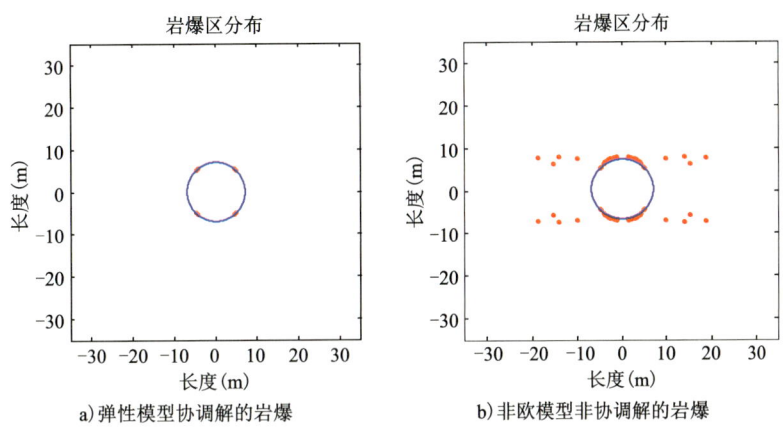

图 7 微裂纹密度 4×10^4,长度 0.05mm

2. 宏观裂隙和微裂纹影响的比较

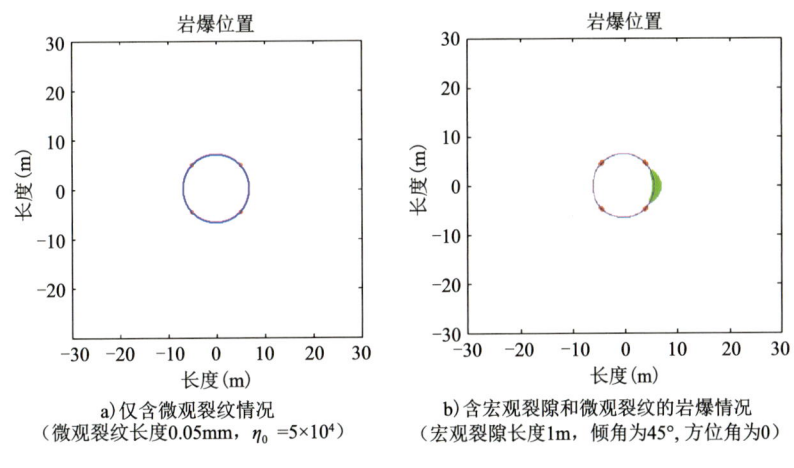

图 8

六、启示和结论

(1)对于地下岩石工程设计而言,围岩中的应力应小于岩石应力应变曲线的峰值应力,处于加载段;峰值应力前岩石塑性变形的实质是围岩中少量次生微裂纹的产生;根据协调解与非协调解中围岩应力的比较可知,当围岩初始微裂纹密度与尺度不大时,忽略变形的非协调性,即协调解的误差不大,这意味连续介质弹塑性力学对工程设计是适用的。

(2)对于岩爆研究而言,围岩中的应力处于岩石应力应变曲线的下降段,次生微裂纹大量萌生,并合并成粗裂纹和宏观裂纹直至突然破裂而弹射。此时协调解误差较大,不能反映围岩中应力自洞壁向围岩深部的波动状分布,因而也不能反映围岩中岩石的破坏以及岩爆可能发生的情况,这意味着不研究岩石破裂的连续介质弹塑性力学不适用岩爆的分析研究。

（3）岩爆机理和预测的研究离不开岩石破裂的研究，因此离不开岩石非协调变形的研究，它可用以确定动破裂区的大小和速度，进而确定岩爆的发生和等级。

（4）为此研究岩爆必须研究围岩的内部缺陷：整体围岩的初始微裂纹以及节理、裂隙等宏观缺陷，它与非协调变形的研究密切相关。非协调变形的研究对于岩爆研究至关重要。

（5）E. T. Brown："甚至在岩爆定义上达到一致意见都是困难的。岩爆这个问题的成功答案，目前正在全世界很多研究中心进行着，它代表着岩石力学这门学科的发展和重大突破"。

（6）陈宗基：以往岩石力学仅仅是对变形为协调的感兴趣，例如，在一般的弹性理论中，非协调张量等于零。但自身平衡的初始应力在数学上可用非协调张量不等于零的假定解析，这个假定将在岩石力学中开创一个新的远景。

目前，非协调变形的研究在国内外尚属初始阶段，因此非协调变形研究任重道远，岩爆机理和预测的研究任重道远。

Deformation and failure mechanism of surrounding rock mass around underground caverns in Jinping I hydropower station

1 Introduction

Jinping I hydropower station underground powerstation mainly consists of the main and auxiliary powerhouse, main transformer chamber, tailrace surge chamber, penstock, diversion tunnels, traffic tunnel and vertical shaft (Fig. 1)

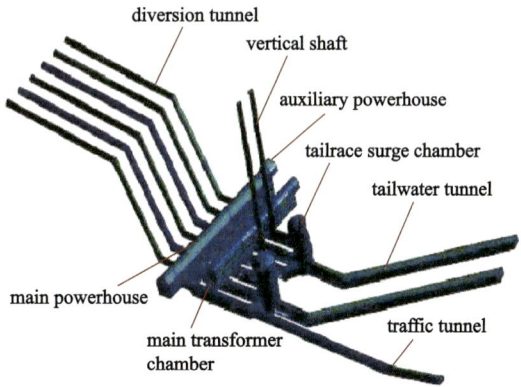

Fig. 1 3D layout of Jinping I underground hydropower station caverns

According to the monitoring data and comprehensive test data of acoustic emission, rock deformation and failure range were significantly greater than other similar depth and scale of the underground powerhouse.

High proportion of rock bolts and anchor cables are overloading than design value.

The deformation and failure mechanism of crack-weakened rock masses around underground caverns in Jinping I hydropower station, which located in a high in-situ stress area, are numerically simulated by the Extended Finite Element Method in this paper.

The deformation and failure mechanism of surrounding rock mass around the underground caverns is investigated.

Finally, comparisons between results from numerical simulation and the monitoring data are made.

2 Background of Jinping I hydropower station

- Principal geostress in situ near the underground powerhouse.

The maximum principal geostress is 20 ~ 35.7MPa;

The intermediate principal geostress is 10 ~ 20MPa;

The minimum principal geostress is 4 ~ 12MPa.

The underground powerhouse is located in high geostress area of rock masses.

Except the f13, f14, f18 and other faults, there are:

- the first bedding fracture set with strike N30 ~ 60°E, dip NW.
- the second joint set with strike N50 ~ 70 °E, dip SE.
- the third joint set with strike N25 ~ 40 °W, dip NE(SW).

The first bedding fracture set develops mainly in second layer of marble with most of the spacing being 20 ~ 50cm.

In the third and fourth marble layers, the spacing of cracks is generally larger than 50cm.

3 Calculation model and parameters

Calculation model of cross-section 0 + 126.8 is shown in Fig. 2. A calculation model of size 317m × 524m for numerical simulation is established.

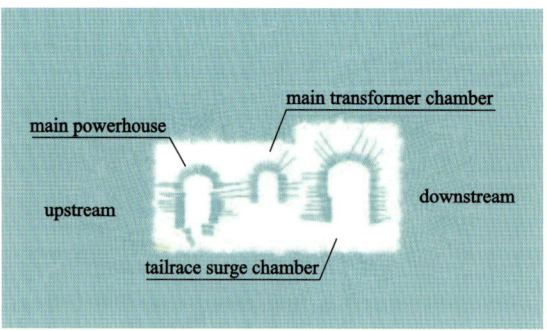

Fig. 2 Calculation model

Based on actual supporting design, system bolt and anchor cable system are applied in the model. The total element number of model is 223,112, the node number is 222,809.

According to the actual geological conditions, the fault f14 are modeled in the grid.

A number of weak elements are applied to model cracks and joints in the rock masses.

The fracture characteristics of crack-weakened rock masses can be modeled by using the Extended Finite Element Method.

All kinds of rock masses and the structure of the physical and mechanical parameters for numerical simulation are listed in Tab. 1.

Calculation parameters Tab. 1

Category	Natural density (kg/m³)	Deformation modulus (GPa)	Poisson's ratio	Cohesion (MPa)	Friction angle (Deg.)
Class II rock	2700	30	0.25	1.7	46.7
f14 Fault	2700	2.5	0.32	0.4	30
Weak element	2200	3.0	0.35	0.5	33
Rock bolt	7800	200	0.3	—	—
Shot concrete	2500	23	0.2	—	—

4 The numerical results

Analysis on deformation of surrounding rock masses around the underground caverns.

The horizontal displacement of surrounding rock masses around main powerhouse during the process of excavation is shown in Fig. 3 ~ Fig. 15.

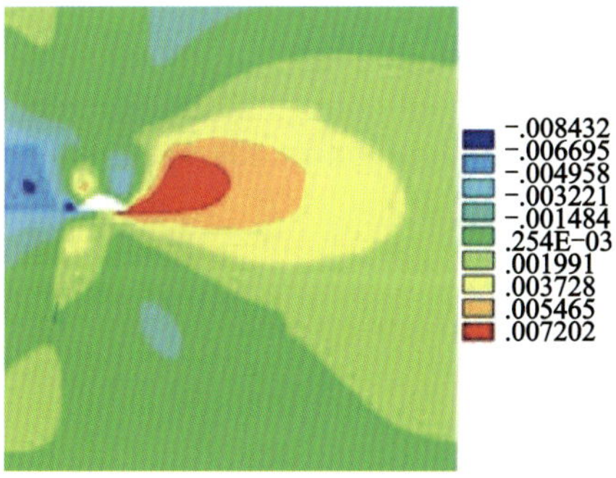

Fig. 3 The first step

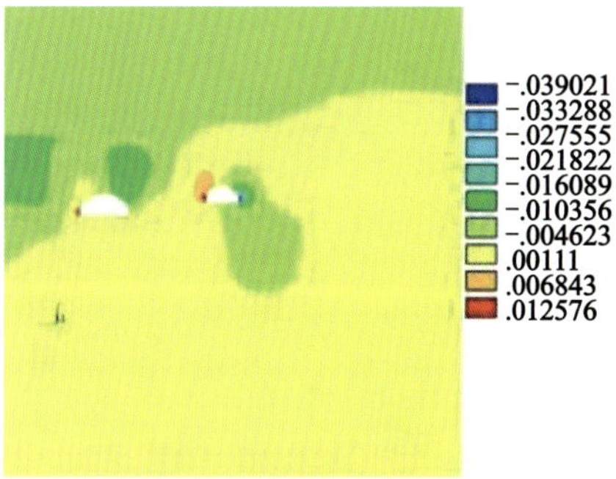

Fig. 4 The second step

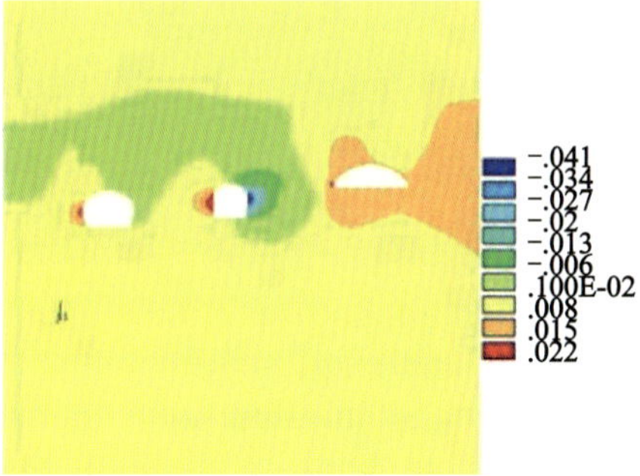

Fig. 5 The third step

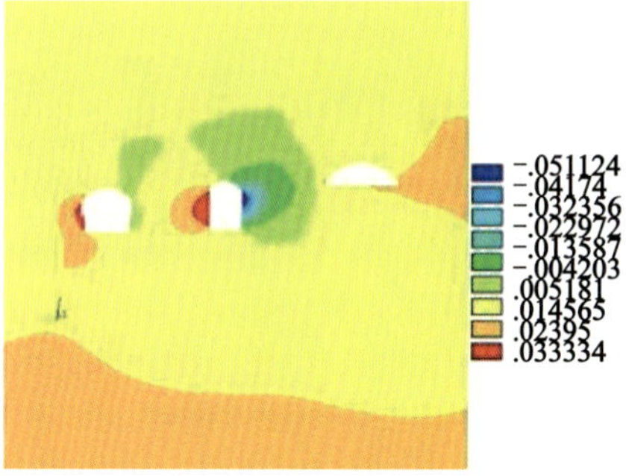

Fig. 6 The fourth step

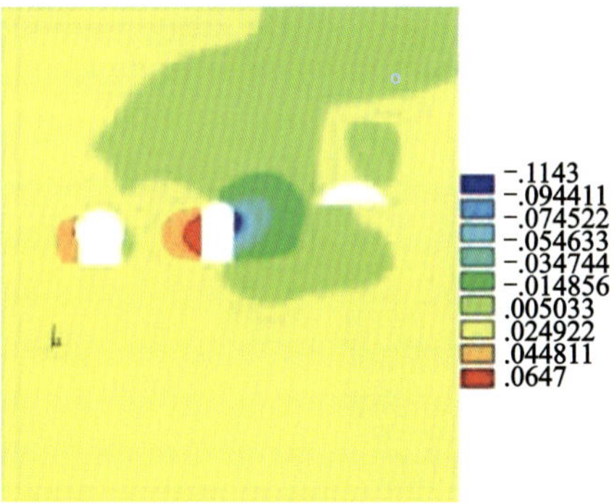

Fig. 7 The fifth step

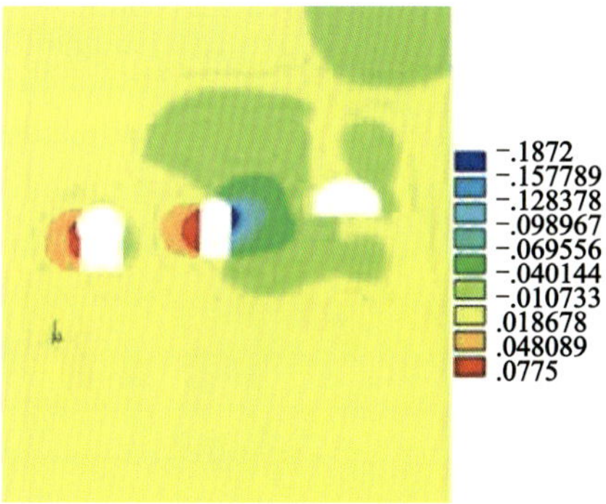

Fig. 8 The sixth step

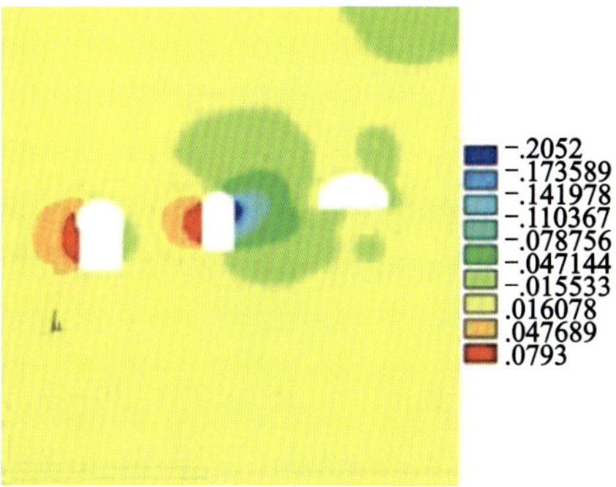

Fig. 9 The seventh step

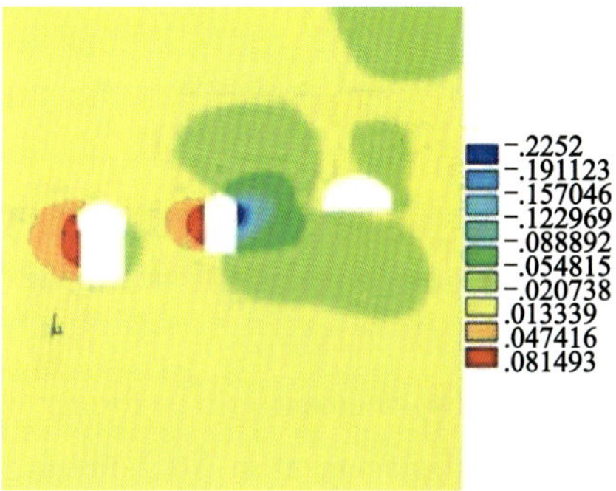

Fig. 10 The eighth step

It is shown from Fig. 12 that after the excavation of the tenth layer of main powerhouse, the maximum horizontal displacement in the upstream sidewall EL1668 of the main transformer chamber reaches 8.2cm, while the maximum horizontal displacement in downstream sidewall EL 1666 of the main transformer chamber is 24.6cm.

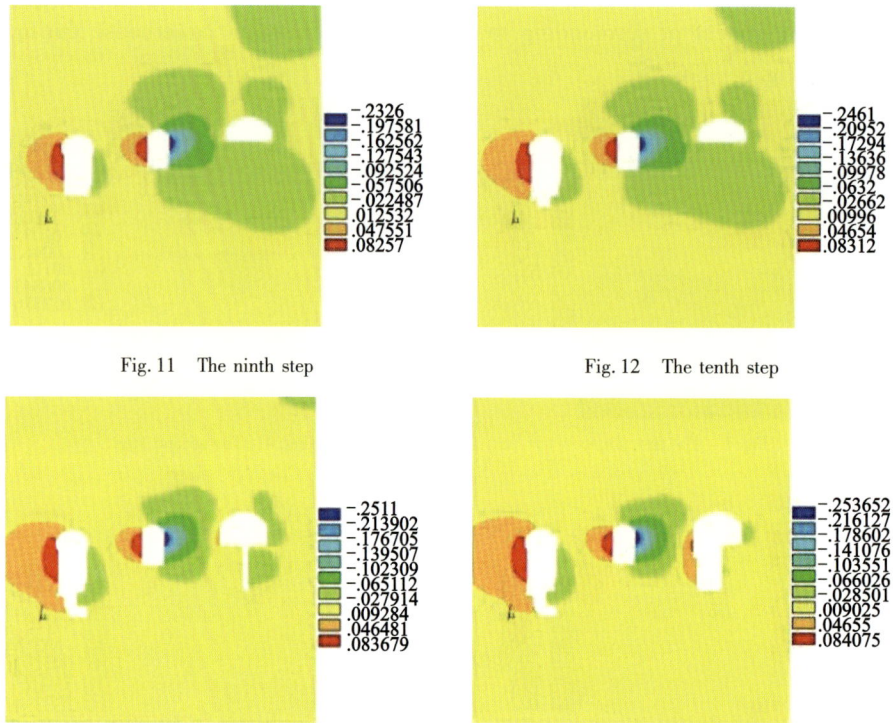

Fig. 11　The ninth step　　　　　　　　Fig. 12　The tenth step

Fig. 13　The eleventh step　　　　　　Fig. 14　The twelfth step

It is found from Fig. 15 that when down stream of tail race surge chamber is excavated, horizontal displacement in the upstream sidewall EL1668m of main transformer chamber is 8.5cm, while horizontal displacement in the downstream sidewall EL 1666m of main transformer chamber is 25.6cm.

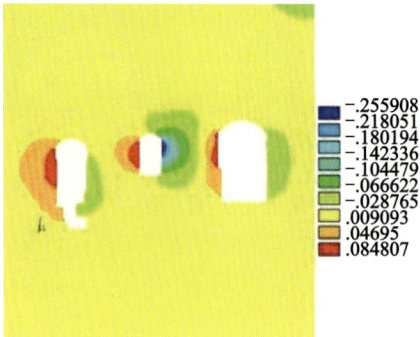

Fig. 15　Horizontal displacement of surrounding rock mass around the powerhouse after the excavation of the downstream part of tailrace surge chamber

According to the weekly records of safety monitoring in 2010 (No. 22).

(1) the horizontal displacement of measuring point around the main powerhouse varies from − 1.07 to 60.82mm.

(2) the horizontal displacement in transformer chamber varies from − 1.75 to 233.39mm.

(3) the horizontal displacement in No. 2 tailrace surge chamber varies from − 23.31 to 88.46mm.

Comparisons between the results from numerical simulation and the monitoring results are shown in Tab. 2. It is shown from Tab. 2 that deformations from numerical results are in good agreement with the monitoring data.

Tab. 2 Comparison between the numerical results of the maximum horizontal displacement and monitoring data.

Location		Monitoring data(mm)	Numerical results(mm)
the main powerhouse	upstream sidewall EL1648	60.82	63
	downstream arch	35.80	42
the main transformer chamber	Upstream sidewall EL1660.8	52.08	59
	Downstream sidewall EL1664	233.39	256
NO.1 tailrace surge chamber	upstream Sidewall EL1648	25.18	32
	Downstream sidewall EL1674	45.93	50

Stress distribution of surrounding rock mass around the underground caverns.

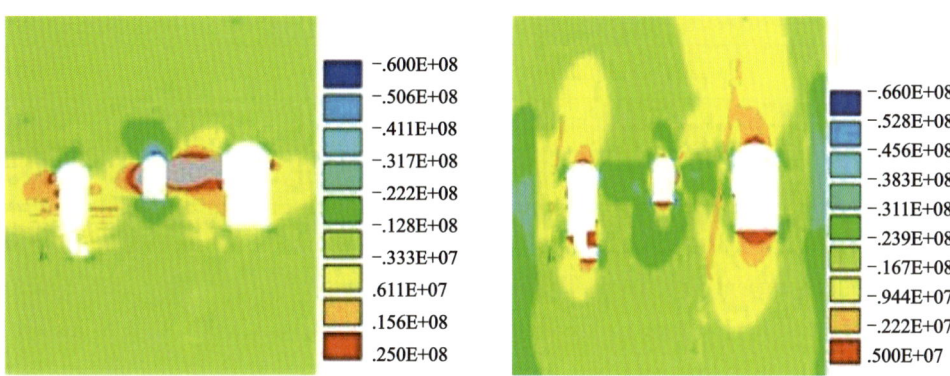

Fig. 16   Horizontal stress field   Fig. 17   Vertical stress field

It can be seen from Fig. 16 and Fig. 17 that there are tensile stress zones around the downstream sidewall of main transformer chamber, skewback of the powerhouse and main transformer chamber in which the value of tensile stress is about 5MPa. It means that splitting failure may occur near the downstream sidewall of main transformer chamber, skewback of the powerhouse and main transformer chamber.

Comparison between the numerical simulation results of the internal force of rock bolt and prestress anchor cable and monitoring data.

The distribution of internal force of rock bolt and presstress anchor cable are shown in Fig. 18 in which the unit is N.

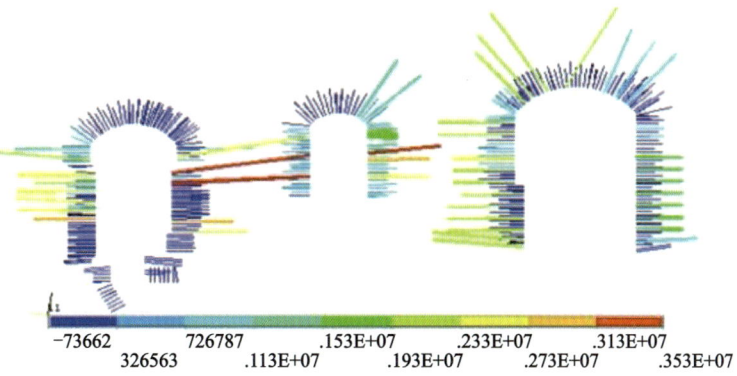

Fig. 18   Internal force of rock bolt and presstress anchor cable after the excavation of downstream part of tailrace surge chamber.

Comparison between the numerical results of the internal force of rock bolt and presstress anchor cable and the monitoring data are listed in Tab. 3 and Tab. 4. The monitoring data comes from the weekly records of safety monitoring (No. 22) in 2010.

Comparison between the numerical results of the internal force of rock bolt and the monitoring data   Tab. 3

Location	Monitoring data (MPa)	Numerical Results (MPa)
Main powerhuose	−38.3 ~ 221.8	0 ~ 259
Main transformer chamber	0.5 ~ 238.0	0 ~ 266
No. 1 tailrace surge chamber	20.8 ~ 357.7	0 ~ 417

Comparison between the numerical results of the internal force of anchor cable and the monitoring data   Tab. 4

Location		Monitoring data (kN)	Numerical Results (kN)
Main powerhouse	Upstream sidewall EL1662	1744.46	1800
	Downstream sidewall EL1662.5	2297.62	2740
Main transformer chamber	Downstream sidewall EL1674.5	1639.77	2055

It is shown from Tables 3 and 4 that the numerical simulation results of the internal force of rock bolt and anchor cable is in good agreement with the monitoring data.

Design value of internal force of long rock bolt and anchor cable is about 2000kN. It is observed from Table 4 that the actual internal force of long rock bolt and anchor cable is bigger than design value. In other words, long rock bolt and anchor cable is more effective to control the deformation of surrounding rock masses than the short one.

In order to enhance the supporting system, the number of long rock bolt and anchor cable must be increased.

Failure mechanism of surrounding rock masses.

During the excavation and supporting process of underground caverns, the fracture behavior in surrounding rock masses is very apparent. During the excavation of the fourth step, it is found that cracks on the profile of caverns obviously propagate. The width of cracks is about 0.5 ~ 2cm, it has a trend to develop into large-scale, and s palling of shot concrete is detected too, as shown in Fig. 19 ~ Fig. 20.

Fig. 19 Splitting failure

Fig. 20 Tension failure

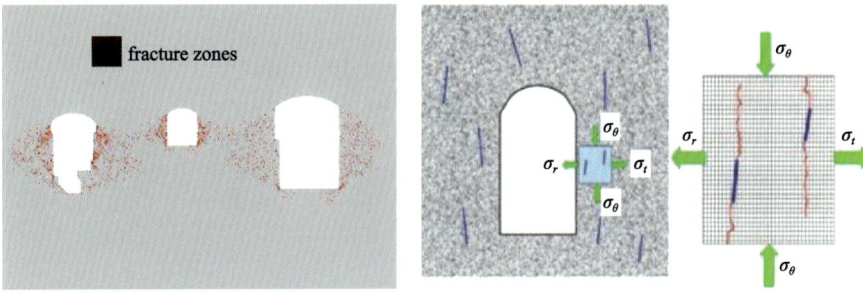

a) The fracture zones　　　　b) Failure mechanism

Fig. 21　Failure mechanism of surrounding rock mass at 0+126.8 cross section.

It is revealed from Fig. 19 ~ Fig. 21 that:

(1) The numerical results of the range and size of fracture zones in skewback of the powerhouse is in good agreement with the actual observed results, the time of failure is in the excavation of the fourth step;

(2) The numerical results of the range and size of the fracture zone in downstream and the arch roof of the main transformer room is in good agreement with the actual observed results, the time of failure is in the excavation of the fourth step;

(3) Splitting failure in downstream sidewall of main transformer chamber is controlled by the fourth joint set, splitting failure in skewback of the powerhouse and the main transformer room is induced by the first bedding fracture set.

5　Conclusions

The scale of underground caverns in Jinping I hydropower station is not the biggest in the giant hydropower projects under construction, but because of the high geostress, relatively low strength of rock masses and the unique complex geological conditions of faults f13, f14, f18 and the lamprophyre veins, the deformation and failure of the surrounding rock masses are relatively intensive. The deformation and failure mechanism of the surrounding rock masses around the underground caverns is investigated by EFEM.

The main conclusions from numerical results are summarized as follows:

(1) Under existing supporting design condition, in the excavation of the cross-section 0+126.8. The numerical results of deformation of the surrounding rock masses are in good agreement with the monitoring data.

(2) The actual internal force of long rock bolt and anchor cable from the numerical results is bigger than design value. Long rock bolt and anchor cable is more effective to control the deformation of surrounding rock masses than the short one. The numerical simulation results of the internal force of rock bolt and anchor cable is in good agreement with the monitoring data.

(3) There are tensile stress zones around the downstream sidewall of main transformer chamber, skewback of the powerhouse and main transformer chamber in which the value of tensile stress is about 5MPa. It means that splitting failure may be occurred near the downstream sidewall of main transformer chamber, skewback of the powerhouse and main transformer chamber. The splitting failure of the downstream sidewall of main transformer chamber is induced by the fourth joint set. The splitting failure of skewback of the powerhouse and main transformer chamber is controlled by the first bedding fracture set. The numerical results of failure of the surrounding rock masses are in good agreement with the actual observed results.

超深盐膏层地应力与井筒完整性

一、盐膏层分布特征

1. 盐膏层的定义

在石油钻井中,对含有以氯化钠为主或其他水溶性无机盐类(如氯化钾、氯化镁、氯化钙、石膏及芒硝等)的地层统称为含盐膏地层。

2. 盐膏层分布

(1)国外:墨西哥湾盆地、哈萨克斯坦和俄罗斯 Precaspian 盆地、加拿大 Scotian 盆地、安哥拉近海的 Kwanza 超深水盆地以及阿尔及利亚、德国、伊朗、西班牙等。

(2)中国:塔里木盆地、江汉平原、四川盆地、华北平原以及青海、胜利、长庆等地区油田。

3. 塔里木盆地库车坳陷盐膏层分布

主要分布在第三系、石炭系和寒武系地层,分属潟湖陆相沉积和滨海相沉积。

吐北2井下第三系盐岩

康2井寒武系盐岩与白云岩混合体

康2井寒武系盐岩被溶解掉的白云岩

克拉苏构造带西段存在两个聚集区:大宛齐、吐北。大宛齐—大北,盐层逐渐减薄;大北—吐北,盐层逐渐增厚;局部盐层厚度大,为盐刺穿。克拉苏构造带东段克深区带盐层最厚,向北部克拉区带减薄。

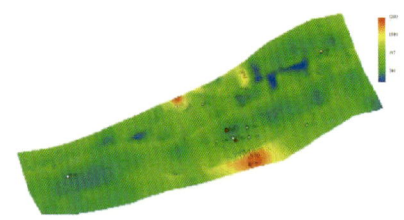
克拉苏西段盐层厚度图(时间域)
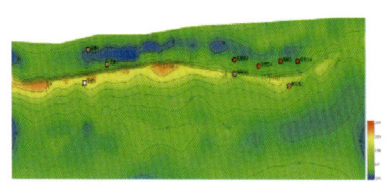
克拉苏东段盐层厚度图(时间域)

纵向:井井不一,埋深差异较大。大北 102 井盐膏层层位最浅,底深为 5338.0m;大北 3 井盐膏层层位最深,底深为 7071m。

二、理化特征

(1)库车 7 口井盐膏层井段矿物 x-射线衍射分析报告。

(2)组分:硬石膏 10%~60%,黏土矿物 10%~30%,石盐 10%~96%。

报告时间:2011 年 1 月 17 日。

(3)黏土矿物:伊利石35%～80%;绿泥石10%～30%;高岭石10%～20%;绿泥石和高岭石混层40%～60%。

井号	井段(m)	矿物种类和含量(%)							黏土(%)
		石英	长石	方解石	白云石	赤铁矿	石盐	硬石膏	
克拉4	6084～6085	9.8	7.5	7.6	—	6.7	—	56.9	11.5
	6101～6102	11.5	6.3	6.8	—	34.7	20.9	11.3	8.5
大北3	6250～6251	19.2	11.8	15.7	3.0	5.1	4.9	11.1	29.2
	6340～6341	3.6	1.2	1.2	1.9	2.1	82.3	2.2	5.5
京深2	6263～6264	15.8	4.1	3.5	—	46.1	9.3	8.4	12.8
却勒4	5975	7.4	1.8	2.3			63.1	19.7	5.2
	5980～5981	3.3	0.8	1.0		0.9	86.5	3.4	4.1
	6000	14.9	5.7	3.5			5.0	58.8	11.7
	6015～6016	28.5	6.8	2.1	11.5	6.4	5.7	22.7	15.3
	6031～6032	30.0	6.7	11.0	7.3	3.2	2.5	19.1	20.2
单塔克502	5200	0.2	2.3	0.1			95.9	1.0	0.5
	5203	0.7	22	0.1			93.5	2.0	1.5
西秋2	6237～6238	13.4	6.4	8.4	—	11.4	1.3	37.3	21.8
	6260～6263	8.3	—	2.2		49.4	27	1.2	11.9
秋参1	6292～6263	13.6	5.6	10.2	1.4	15.7	37.2	8.0	7.9

井号	井段(m)	黏土矿物相对含量(%)							混层比(%S)
		S	I/S	I	K	C	C/S	I/S	C/S
克拉4	6084～6085	—	—	69	8	15	8	—	45
	6101～6102	—	—	66	8	10	16	—	40
大北3	6250～6251	—	—	65	6	8	21	—	50
	6340～6341	—	—	62	14	24	—	—	—
却勒4	5975	—	—	3.1	0.5	0.7	—	0.9	70
	5980～5981	—	—	72	6	10	12	—	55
	6000	—	—	6.7	12	1.6	—	2.2	75
	6015～6016	—	57	34	4	5	—	60	—
	6031～6032	—	56	34	4	6	—	50	—
单塔克502	5200	—	53	25	9	10	3	55	45
	5203	—	50	28	8	11	3	50	40
西秋2	6237～6238	—	—	75	8	12	5	—	45
	6260～6263	—	—	79	13	3	5	—	45
京深2	6263～6264	—	—	63	17	20	—	—	—
秋参1	6293～6263	—	—	57	10	8	25	—	50

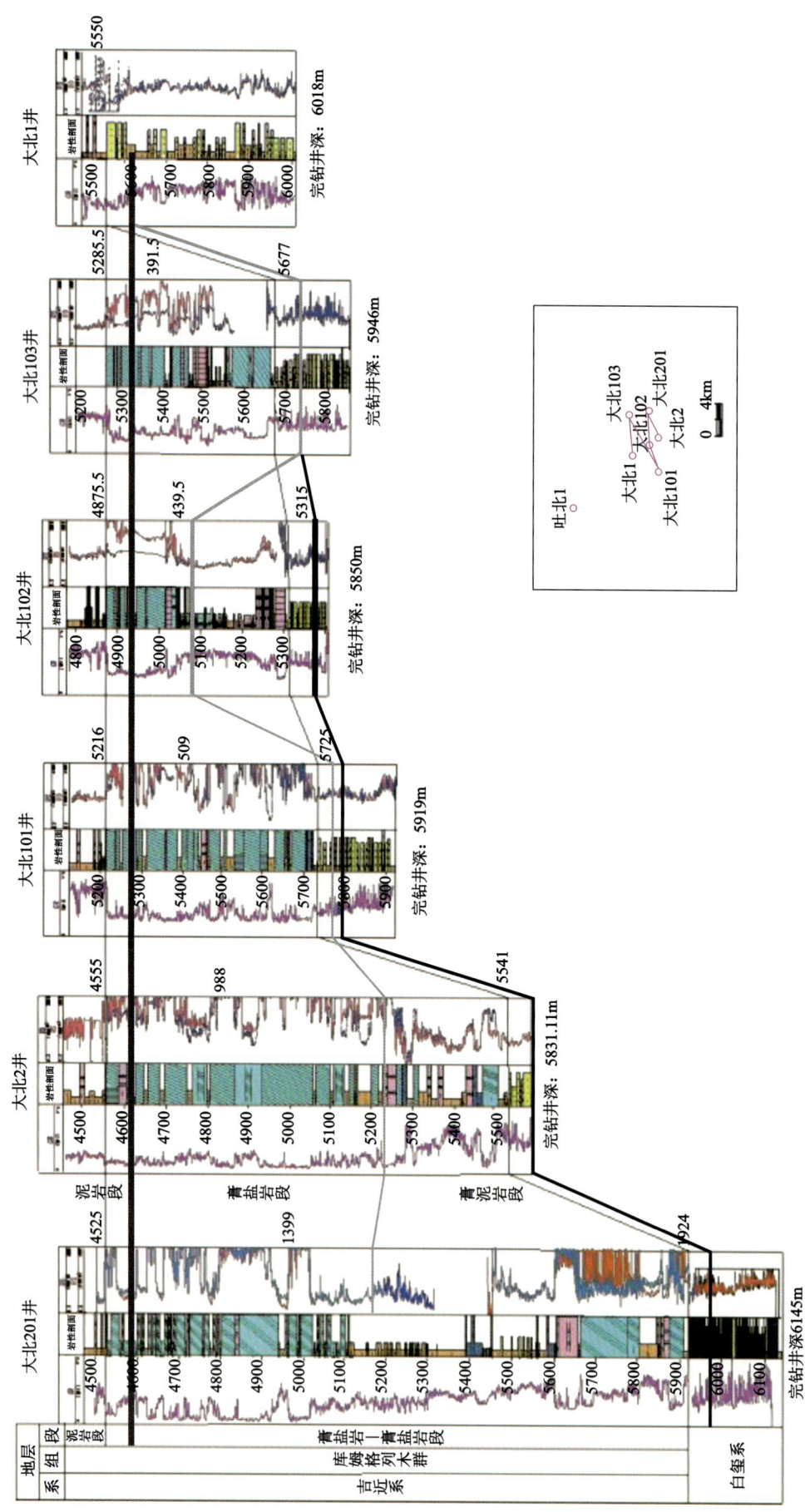

(4)层状复合盐层岩芯。
(5)不同井其矿物成分和含量差异较大;同一口井纵向分布上矿物成分和含量也不同。

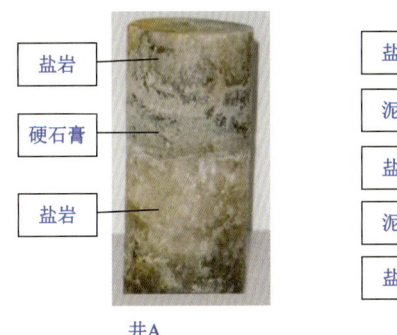

三、弹性、强度特征

1. 常规岩石应力-应变曲线

不同类型应力-应变曲线的内在发生机制,一般可由岩石内存在的裂隙构造、变形来理解。

2. 弹性特征

在低应力状态下,盐岩的变形表现为弹性变形,从试验得到的应力-应变曲线在低应力水平下为直线。

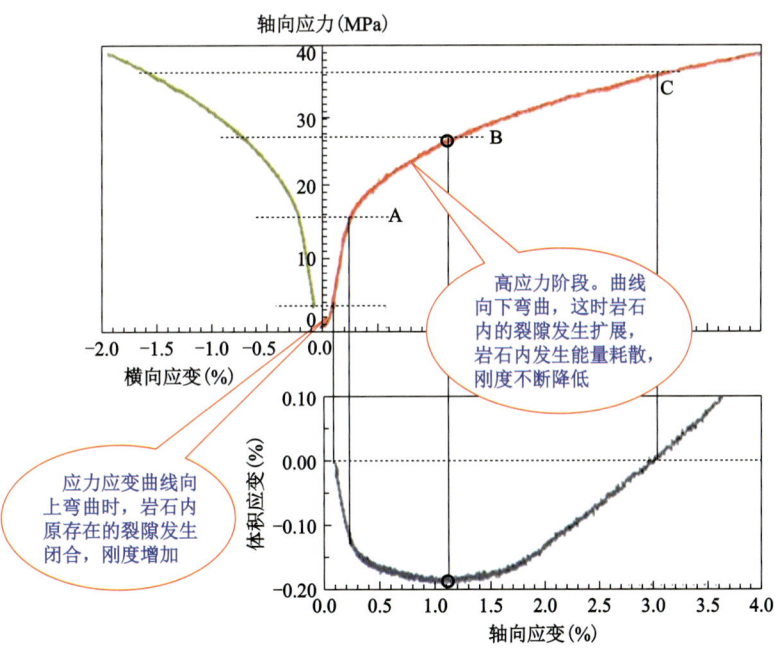

3. 强度特征

强度和弹性模量较低,比花岗岩及普通砂岩小1~2个数量级。
深部地层,强度、弹性模量随岩性、围压而变化。
强度与含盐量、成岩年代、晶体的结构、应力状态有关。

四、弹性、蠕变本构

1. 盐岩损伤扩容边界试验曲线

盐岩损伤扩容边界试验曲线见下图。

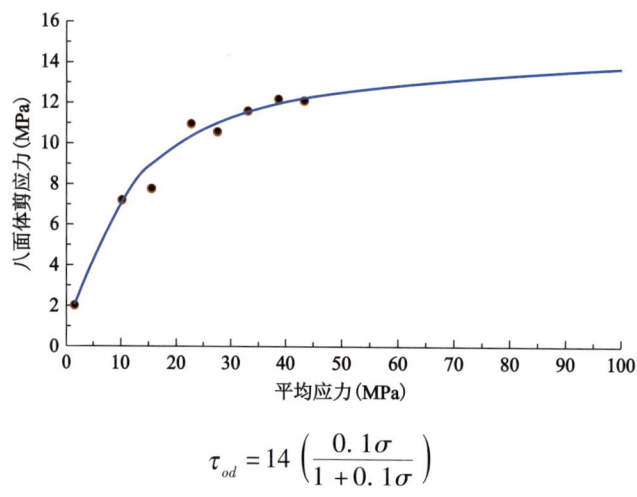

$$\tau_{od} = 14\left(\frac{0.1\sigma}{1+0.1\sigma}\right)$$

2. 盐岩稳态蠕变

稳态蠕变阶段,声发射事件数一直保持在一个较低的水平,说明此阶段盐岩没有明显的损伤,晶界没有产生微裂纹。

3. 弹性本构

盐岩受载以后,首先产生的是弹性变形,当材料处于弹性状态。材料的本构关系就是广义虎克定律。

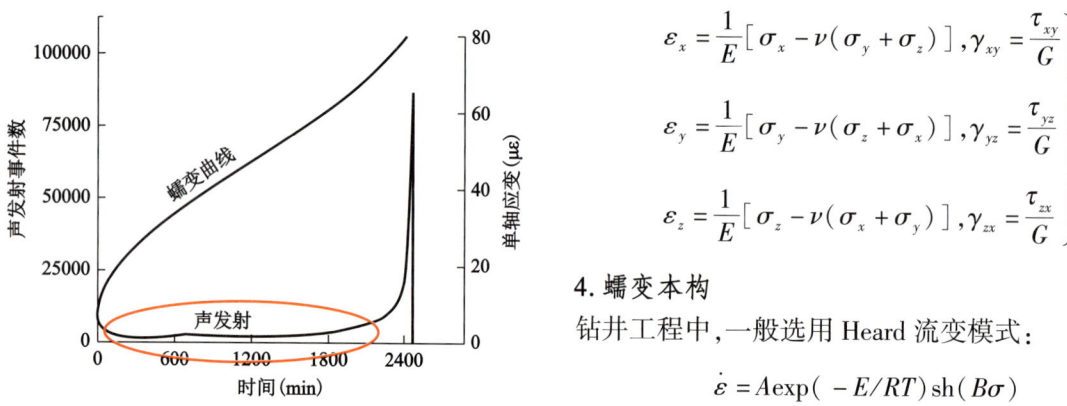

$$\left.\begin{array}{l}\varepsilon_x = \dfrac{1}{E}[\sigma_x - \nu(\sigma_y + \sigma_z)], \gamma_{xy} = \dfrac{\tau_{xy}}{G} \\[4pt] \varepsilon_y = \dfrac{1}{E}[\sigma_y - \nu(\sigma_z + \sigma_x)], \gamma_{yz} = \dfrac{\tau_{yz}}{G} \\[4pt] \varepsilon_z = \dfrac{1}{E}[\sigma_z - \nu(\sigma_x + \sigma_y)], \gamma_{zx} = \dfrac{\tau_{zx}}{G}\end{array}\right\}$$

4. 蠕变本构

钻井工程中,一般选用 Heard 流变模式:

$$\dot{\varepsilon} = A\exp(-E/RT)\operatorname{sh}(B\sigma)$$

由于矿物成分的不同,盐岩进入稳态后蠕变时间、蠕变速率及蠕变量均有所不同,流变模式也有所不同。

5. 存在的问题

(1) 人造盐岩岩心如何做,有何好方法。

(2) 纯盐岩的渗透性可以忽略,但对于含泥、膏的复合盐层,如何考虑渗透性。

(3) 石油钻井中,纯盐岩地层一般采用 Heard 流变模式,在含泥、膏等的复合盐层中,盐岩类型不同,流变模式如何考虑。

(4) 如何考虑盐岩的长期蠕变问题。

五、原始地应力测量方法

1. 各种测量方法对比表

各种测量方法对比见下表。

测量方法	基本思路	特点
水力压裂法	根据井眼的受力状态及其破裂机理来推算地应力	浅部地层,地层易于取芯
井壁崩落法	井壁崩落椭圆的长轴方向总是与最小水平主地应力方向一致,即与最大水平地应力方向垂直,可用井壁崩落椭圆来确定地应力的方向	浅部地层,地层易于取芯
岩石声发射凯塞尔效应	在室内对取自现场的岩芯进行取芯时要在垂直方向取一块,在垂直岩芯轴线平面内相隔取三块,测得四个方向的正应力,利用相应的公式可确定出深部岩石所处的地应力	地层易于取芯
岩盐地应力测量 DRA 法	记录多次单轴压缩应力下每次加载的非弹性应变,两次加载下的应变差曲线的拐点对应以前加载的应力,如果岩芯材料是弹性的那么应变差就是0	根据岩石非弹性变形测地应力
数值拟合	通过建立地质力学模型,借助正演分析所得的结果,确定边界荷载的大小及作用方式,建立反演计算与应力基本影响因素之间的关系	地层取芯较难,且无法通过其他资料获取地应力

2. 岩盐地应力测量 DRA 法

根据岩石非弹性变形测地应力:记录多次单轴压缩应力下每次加载的非弹性应变,两次加载下的应变差曲线的拐点对应以前加载的应力,如果岩芯材料是弹性的那么应变差就是0。

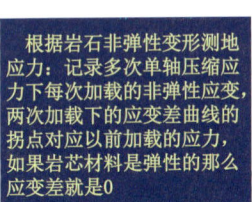
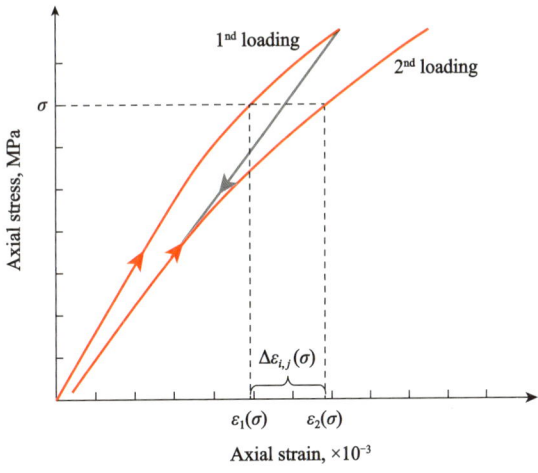

两次加载的应力应变曲线

3. 存在的问题

(1)盐层测量岩盐地应力除了 DRA 法之外,还有什么新方法?
(2)复合盐层原始地应力测量方法有哪些?

六、原始地应力场数值模拟("三明治"理论)

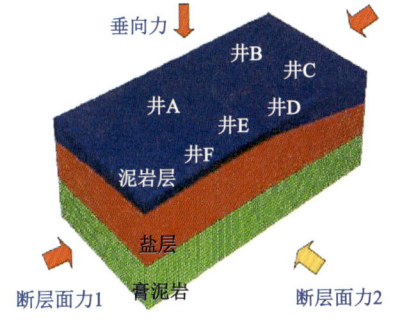
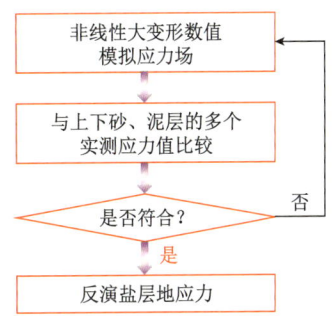

"三明治"理论

1. 基本思路

(1) 边界条件和初始条件的确定。边界条件：前、后、左、右四个表面在垂直于边界的方向上受法向约束；下表面则受垂直于该面的竖向简支约束；上表面受由上覆地层引起的等效重力载荷。

(2) 加载及求解。设上、下部泥岩的测点坐标分别为 (x_1, y_1, z_1) 和 (x_2, y_2, z_2)，测量应力值分别为 $\bar{\sigma}(x_1, y_1, z_1)$ 和 $\bar{\sigma}(x_2, y_2, z_2)$，施加的应力值分别为 $p(x_1, y_1, z_1)$ 和 $p(x_2, y_2, z_2)$，数值模拟计算值为 $\bar{\sigma}(x_1, y_1, z_1)$ 和 $\bar{\sigma}(x_2, y_2, z_2)$。测量值与计算值相近，残差为 ε_i，则有

$$\begin{cases} \sigma(x_1, y_1, z_1) = \varepsilon_1 + \bar{\sigma}(x_1, y_1, z_1) \\ \sigma(x_2, y_2, z_2) = \varepsilon_2 + \bar{\sigma}(x_2, y_2, z_2) \end{cases}$$

(3) 用多元回归方法求 $p(x_1, y_1, z_1)$ 和 $p(x_2, y_2, z_2)$。上述方程组包含有 2 个方程式，含 2 个未知数 $p(x_1, y_1, z_1)$ 和 $p(x_2, y_2, z_2)$，用最小二乘法，使残差取小值，可求得 $p(x_1, y_1, z_1)$ 和 $p(x_2, y_2, z_2)$ 的值，则此时施加的应力边界条件适合于该复合模型。

(4) 利用步骤(3)中的应力边界条件反演得到该地区中间盐层的地应力。

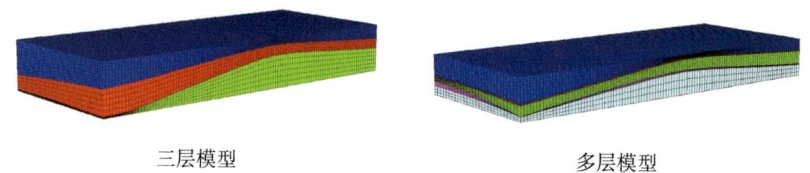

三层模型　　　　　　　　多层模型

2. 存在的问题

(1) 上、下泥岩层层厚较大时，由于测点之间的测距较大，会影响模拟的精度，如何选取测点位置及测点数？

(2) 同一层在空间上的厚度分布不同，导致划分网格时部分区域网格精度不够。

(3) 为提高计算速度，将岩性相近的层位划分为同一层，以减少模型单元数，这将降低模拟的精度。如何解决计算精度的问题，有何等效方法？

七、井眼存在后井周围岩动态应力场

1. 直井

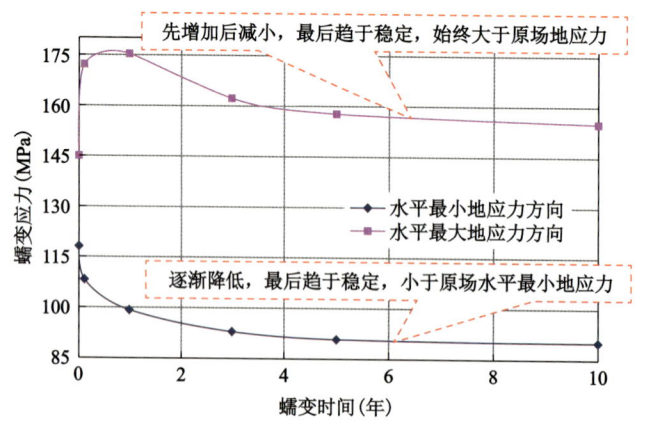

大北 101 井井周应力随蠕变时间变化曲线

2. 弯曲井眼

大北 101-H1 井造斜段 3800～5539m 层段，曲率半径 573m。分析弯曲段 A、B、C 三点的井周应力。

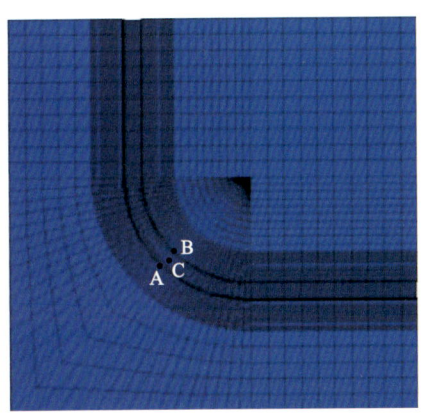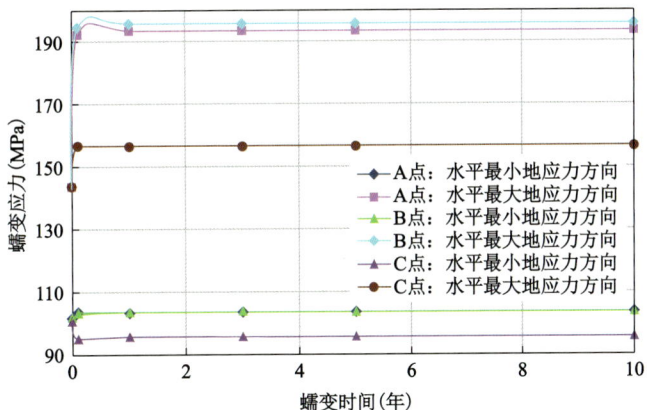

曲率半径573m,井周应力随蠕变时间变化曲线

3. 结果
（1）由于井眼的存在,引起井周应力集中,原场应力发生干扰。
（2）盐岩蠕变的影响,加剧了原场地应力的变化。
（3）水平应力差增大,井壁更容易失稳。

4. 存在的问题
（1）模拟时,施加的边界应力为定值,而实际盐层地应力是变化的,如何考虑边界应力的变化?
（2）施加的井筒内压为定值,而实际由于混入其他流体及温度的影响,井内流体密度是变化的,如何消除内压变化的影响?

八、套管的短期受力与变形

1. 直井短期受力与变形

2006年2月14日下入套管,套管钢级为TP140V,壁厚16mm,套管本身的屈服强度为8453kN。该层位未发生套损情况。

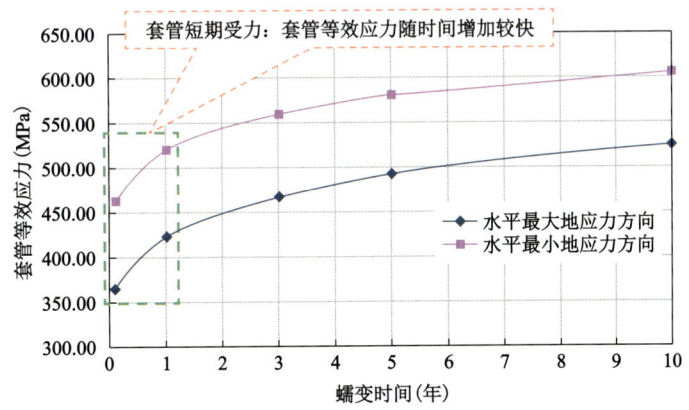

大北101井套管应力随蠕变时间变化曲线

存在的问题:
（1）没有考虑套管的长期力学特性变化。
（2）没有考虑磨损引起的套管力学特性变化。
（3）没有考虑腐蚀引起的套管力学特性变化。

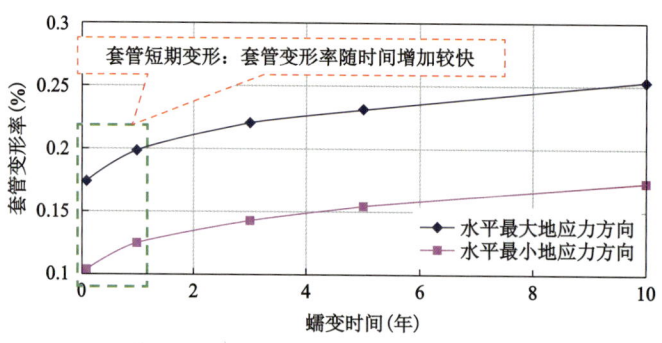

大北 101 井套管变形率随蠕变时间变化曲线

2. 弯曲井眼短期受力与变形

大北 101-H1 井造斜段 3800~5539m 层段采用 10 1/2″TP155 钢级的套管,壁厚 22mm,理想情况下,该型号的套管屈服强度为 965.3MPa。

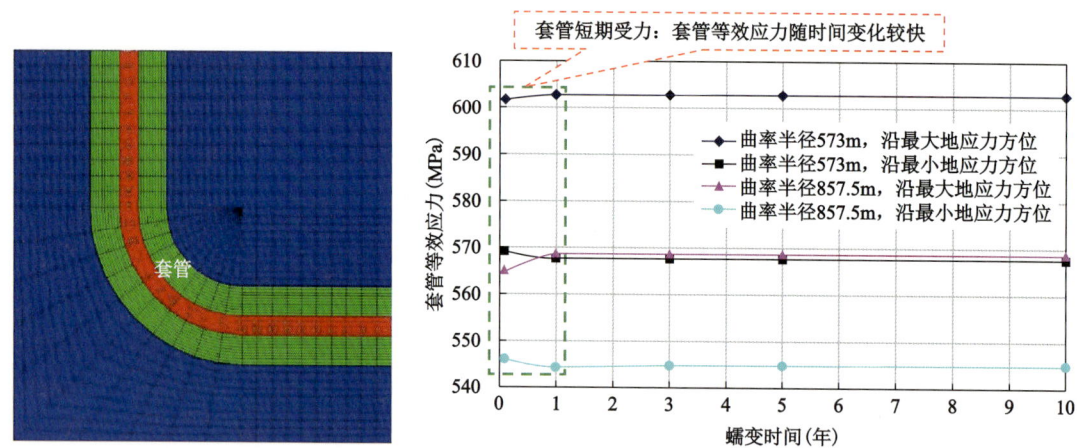

壁厚 22mm,套管等效应力随蠕变时间变化曲线

在大北 101-H1 井造斜段,该钢级的套管临界壁厚 18mm,目前选用壁厚 22mm 的套管,钻井安全。

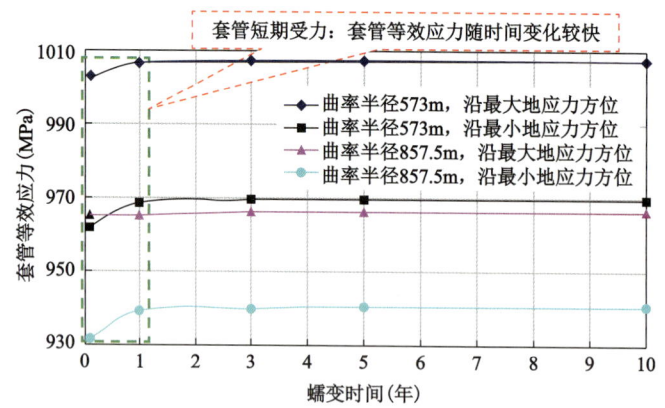

壁厚 12mm,套管等效应力大于套管屈服强度,套管发生屈服

存在的问题:
(1)没有考虑套管的长期力学特性变化。
(2)没有考虑磨损引起的套管力学特性变化。
(3)没有考虑腐蚀引起的套管力学特性变化。
(4)如何消除建模时模型不对称的影响。

九、套管的长期受力与变形

1. 直井受力与变形

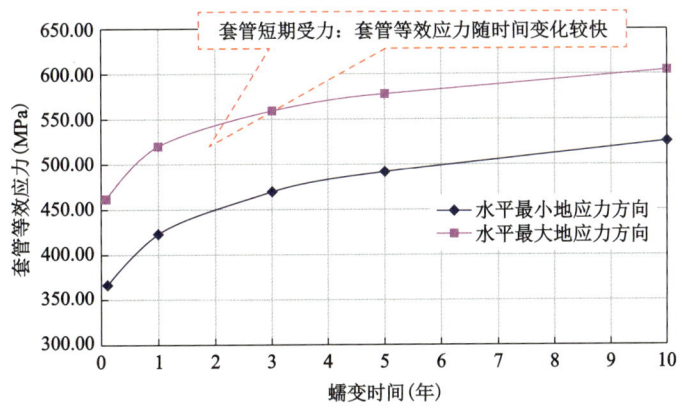

大北 101 井套管应力随蠕变时间变化曲线

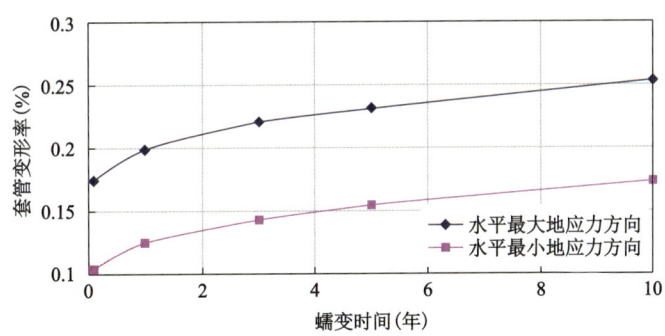

大北 101 井套管变形率随蠕变时间变化曲线

2006 年 2 月 14 日下入套管,套管钢级 TP140V,壁厚 16mm,套管本身的屈服强度为 8453kN,该层位未发生套损情况。

存在的问题:

(1) 没有考虑磨损引起的套管力学特性变化。

(2) 没有考虑套管的长期腐蚀问题。

(3) 没有考虑储层地应力变化对盐膏层的影响。

2. 弯曲井眼受力与变形

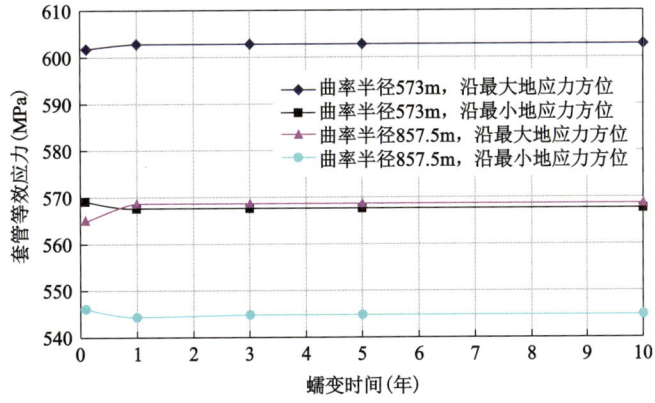

壁厚 22mm,套管等效应力随蠕变时间变化曲线

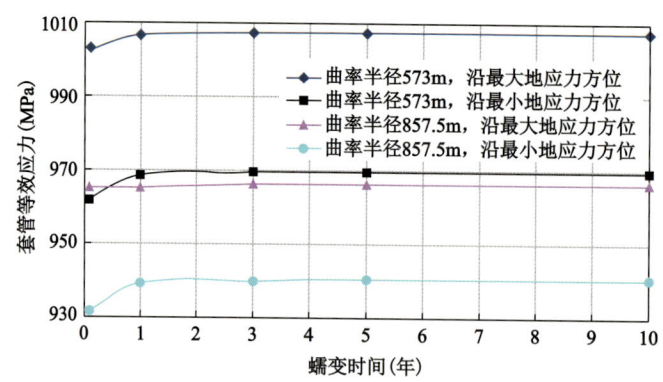

壁厚12mm,套管等效应力大于套管屈服强度,套管发生屈服

在大北101-H1井造斜段,该钢级的套管临界壁厚18mm,目前选用壁厚22mm的套管,安全钻井。
存在的问题:
(1)没有考虑磨损引起的套管力学特性变化。
(2)没有考虑套管的长期腐蚀问题。
(3)没有考虑储层地应力变化对盐膏层的影响。
(4)如何消除建模时模型不对称的影响?

十、咨询的问题

(1)超深盐膏层原始地应力的测试或解释方法。
(2)超深盐膏层的本构与长期强度的测试或解释方法。
(3)超深盐膏层地应力场与井周应力的长期变化规律的确定方法。
(4)盐膏层下部储层压力下降对盐膏层井壁围岩应力的影响。

Recent advances in the zonal disintegration phenomenon in the deep rock mass engineering

Zonal disintegration phenomenon around the tunnels at the 2000~3000 meter depth in Witwatersrand golden mines in South Africa is initially observed with a periscope (Cloete et al. 1972–1973), as shown in Fig. 1.

Similarly, zonal disintegration phenomenon is also discovered in the Taimyrskii and Маяк mines in Russia (Шемякин1986), as depicted in Fig. 2.

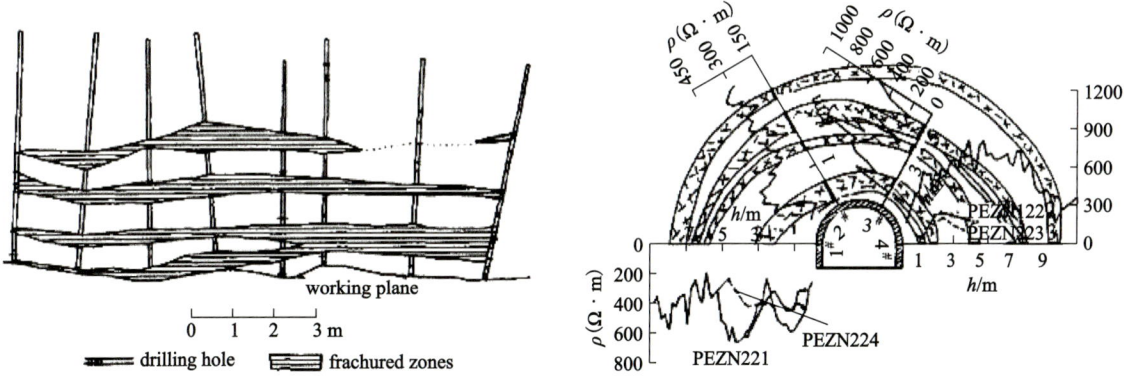

Fig. 1 The situation of fractured zones of tunnel roof in Witsatersrand gold deposit in South Africa

Fig. 2 The situation of fractured zones in Taimyrskii in Russia

During the excavation of a tunnel in the deep rock mass, fractured zone and non-fractured zone occur alternately around deep tunnels, which has been referred as the zonal disintegration phenomenon in some related publications and has never been observed in shallow rock engineering before.

Besides the deep rock burst and large deformation of surrounding rock mass around deep tunnels in squeezing ground, the phenomenon of zonal disintegration attracts the attentions of scholars and engineers in the fields of geotechnical engineering and rock mechanics.

The zonal disintegration phenomenon cannot be clearly explained by conventional elastoplasticity theory.

The progress of experimental and theoretical studies on zonal disintegration phenomenon is concluded in this paper.

1 The stage of qualitative analysis

Shemyakin who initially studied the zonal disintegration phenomenon of deep rock mass in 1986, reached some achievements on in-situ observation, experimental test and theoretical analysis, but the exact mechanism of zonal disintegration phenomenon is not revealed clearly.

Qian presented the new point of view which zonal disintegration phenomenon is a scientific problem related to space and time.

It is suggested in reference (Wang 2006) and (Qian 2004c) that the range of deep rock mass should be determined by zonal disintegration phenomenon.

2 In-situ observation and experimental tests

Zonal disintegration phenomenon is observed and recorded in surrounding rock mass at the 2000 ~ 3000m depth in Huainan mines through the detection machine ofdrilling borehole in 2008 by Li Shucai.

The span of the tunnels is 5 meter. There exist four fractured zones in the surrounding rock mass around each of the tunnels.

Similarly, Li Shuchen observed zonal disintegration phenomenon in surrounding rock mass at the 2500m depth in Jinping II Hydropower Station, southwest of China, as depicted in Fig. 3.

The experimental observation on the experiment observation on facture and failure around the thick-walled hollow cylinders of Berea sandstone is shown in Fig. 4.

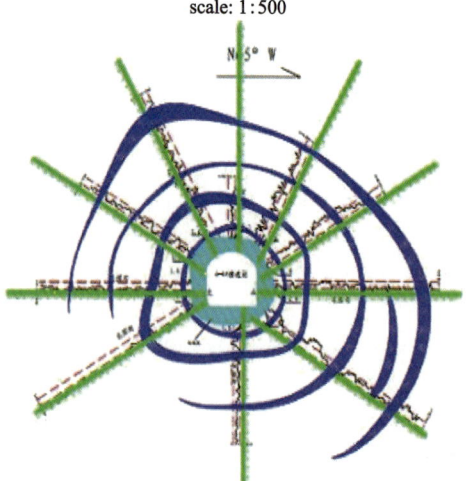

Fig. 3 Zonal disintegration phenomenon in the surrounding rock mass around No. 6-1 auxiliary tunnel in Jinping Hydropower Station

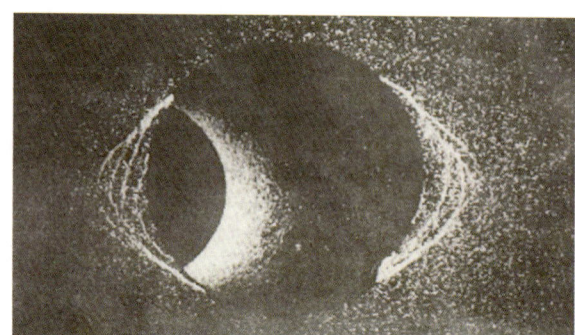

Fig. 4 Typical failure around the thick-walled hollow cylinder of Berea sandstone by Ewy

3 Mechanism research and numerical simulation of zonal disintegration phenomenon

In reference(Wang 2006), the relationships were derived between the condition for the occurrence of zonal disintegration phenomenon in rock masses and the initial in-situ stress, a criterion is proposed for defining the shallow and the deep rock engineering.

In references (Wang 2006, Chen 2005, Wang 2005), effects of zonal disintegration phenomenon on supporting design and excavation of tunnel is discussed.

In reference(Qi), the temporal process of zonal disintegration is studied on the basis of irreversible thermal dynamics, theory of continuous phase transitions and elasticity-plasticity theories, the spatial and temporal analytical solution of zonal disintegration around deep tunnel is given.

In reference(He 2006), it is seen that the zonal disintegration phenomenon of deep rock masses is a kind of natural response caused by the unloading effect of engineering excavation under high in-situ stress.

In the year of 2007, zonal disintegration phenomenon in homogeneous rock mass around deep tunnels is initially numerically simulated (Zhou et al. 2007).

It is found from numerical simulation that zonal disintegration only occur when the geostress is higher than the uniaxial compressive strength of rock mass.

It is shown from numerical results that the quantity and thickness of the fractured zone decreases with increasing the value of geostress.

It is also seen from numerical simulation that the thickness of the first fractured zone is largest, and the thickness of other fractured zone reduce orderly.

In the year of 2008, on the basis of the strength criterion of deep rock mass and the theory of fracture mechanics, Qian initially numerically simulated zonal disintegration phenomenon in crack-weakened rock masses around deep tunnel (Qian et al. 2008).

In the process of numerical simulation, the effects of growth, interaction and coalescence of cracks on zonal disintegration phenomenon are taken into account.

Furthermore, the effects of geometry parameters and array pattern of cracks on zonal disintegration phenomenon in rock mass around deep tunnels are analyzed.

It is shown from numerical results that the quantity and thickness of fractured zone decrease with an increase in the cohesion and internal friction coefficient of rock mass.

The quantity and thickness of fractured zone increase with an increase in tunnel excavation speed, as shown in Fig. 5 and Fig. 6.

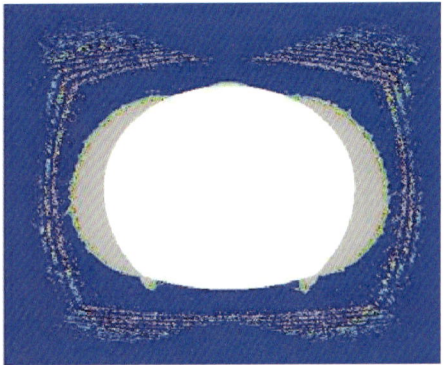 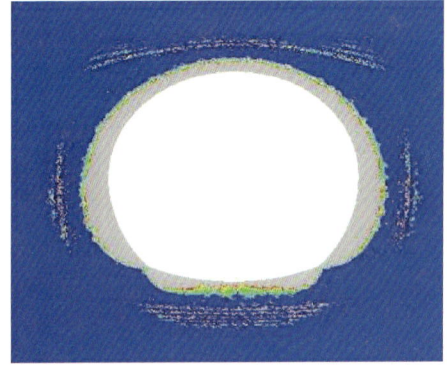

Fig. 5 Zonal disintegration results when the initial geostress is $\sigma_v = 85\text{MPa}$, $\sigma_h = 45\text{MPa}$ and the excavation time is 0.01s.

Fig. 6 Zonal disintegration results when the initial in-situ geostress is $\sigma_v = 85\text{MPa}$, $\sigma_h = 45\text{MPa}$ and the excavation time is 0.1s

The quantity and thickness of fractured zone decrease with decreasing excavation speed, as shown in Fig. 7.

Zonal disintegration phenomenon of deep rock masses is not identical with behaviors of surrounding rock mass containing orderly loose zone, plastic zone and elastic zone which is described in the conventional continuum theory.

Furthermore, the deformation and failure beyond the peak cannot be explained by the conventional continuum theory.

A significant drawback associated with the conventional continuum models is that they lead to mesh-dependent finite element solutions because such model do not contain an intrinsic material length scale needed to define the width of a localized deformation zone and because the resulting governing partial differential equation is ill posed.

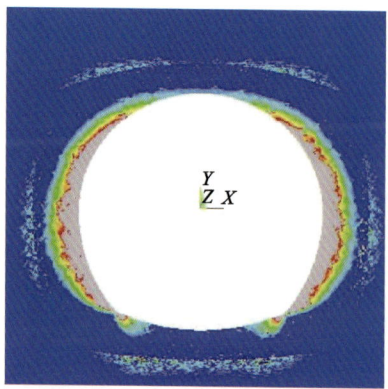

Fig. 7 Zonal disintegration results when the initial in-situ geostress is $\sigma_v = 85\text{MPa}$, $\sigma_h = 50\text{MPa}$ and excavation time is 100s

4 Conclusions

Zonal disintegration phenomenon is closely related to space and time. Only if the qualitative law of zonal disintegration phenomenon is re-studied, the quantitative law of zonal disintegration phenomenon may be obtained.

The mechanical behaviors of deep rock masses is different from those of the shallow rock masses. The surrounding rock massess in shallow rock mass engineering is orderly classified into loose zone, plastic zone and elastic zone, while the surrounding rock masses in deep rock mass engineering is classified into fractured zone and non-fractured zone, which occur alternatively.

The relationship between zonal disintegration and the time of excavation, which is related to the ratio of elastic potential energy, kinetic energy and dissipation energy, needs to be found out.

In order to design reasonable supporting system and determine the supporting time, the effects of supporting system on zonal disintegration phenomenon of deep rock masses needs to be investigated.

At the point view of energy dissipation, a unified theory of zonal disintegration phenomenon, strain-type rock burst and large deformation in squeezing ground in deep rock mass should be established.

Dynamic strength and it's physical nature and dynamic fracture criteria of rock

1 A theoretical study of the temporal features of deformation and fracture of rock

(1) A brief review of classical (static) strength theories Classical (static) strength theories (criteria) may be grouped into 5 classes:

a. maximal normal stress theory;

b. maximal normal strain theory;

c. maximal shear stress theory;

d. maximal specific strain energy theory;

e. modification of third strength theory: the Mohr-Coulomb theory which is widely used in geotechnical engineering practice. The modification of Mohr-Coulomb theory is Hoek-Brown theory.

Mohr-Coulomb criterions (C-M criterion) can not interpretate the failure of rock under hydrostatic pressure because it does not take into account the influence of second principal stress. D-P criterion overcomes the weakness of C-M criterion. Factually, D-P criterion is the further development of C-M criterion and Mises criterion (the fourth strength theory).

Mohr-Coulomb criterion mathematically does not include the intermediate principal stress, i.e. it takes into account shear and normal stresses only on one shear plane, therefore it also may be called single shear stress theory.

The development of single shear stress theory is twin shear stresses theory on which unified stress theory is developed. Single shear stress theory, twin shear stresses theory and other strength theories are particular cases or linear approximation of unified strength theory.

(2) Main shortage of classical (static) strength theories

Above-mentioned modern strength theories are not perfect and mature, and are developing. The main shortage of classical (static) strength theories is that, they don't consider the influence of temporal factor and internal structure of solids. The incorporation of temporal factor constitutes the main content of dynamic strength theories, but now they are not mature and are developing.

(3) Temporal features of deformation and fracture of rock

① Experimental data

Some experimental data of rock strength under intensive dynamic loading

Type	Static strength (atm)	Dynamic strength (atm)	The ratio of dynamic to static strengths
limestone	420	2730	6.5
Marble (normal to the deposit)	210	1890	9.0

Type	Static strength (atm)	Dynamic strength (atm)	The ratio of dynamic to static strengths
Marble (parallel to the deposit)	630	4900	7.8
granite	700	4000	5.7

Experimental data under constant loading rate.

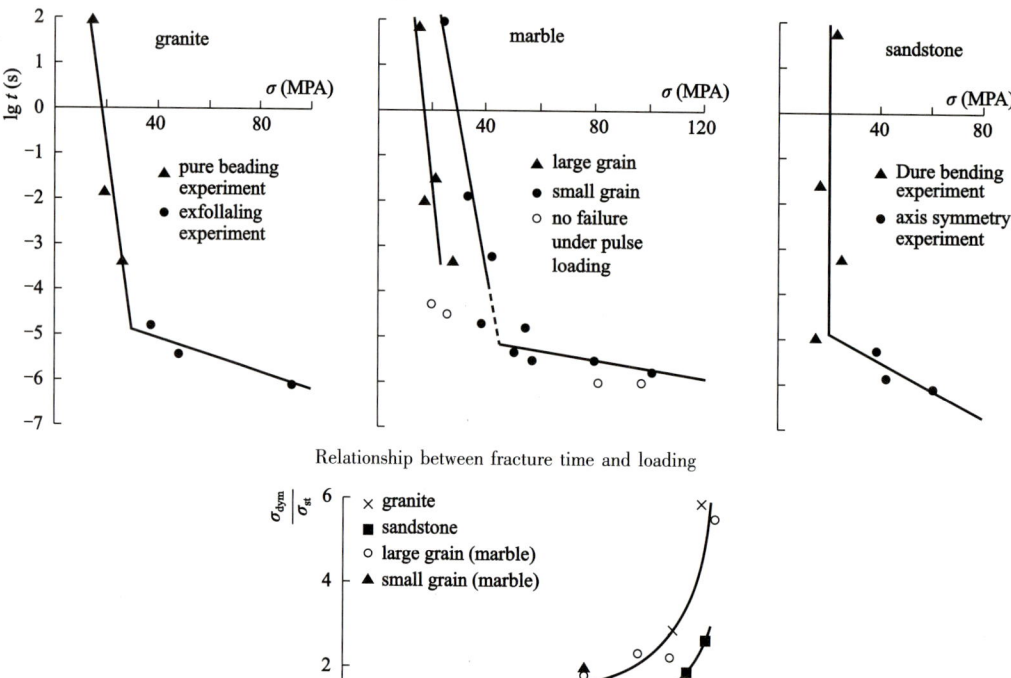

Relationship between fracture time and loading

The strain rate dependence of strength

Failure stress increases with strain rate rapidly when fracture time $\tau < 10^{-5}$s or when strain rate $\dot{\varepsilon} > 10\text{s}^{-1}$

②Theoretical study on considering temporal factor

a. The general features strength theory with of dynamic strength dependence with strain rate is illustrated in following Fig.

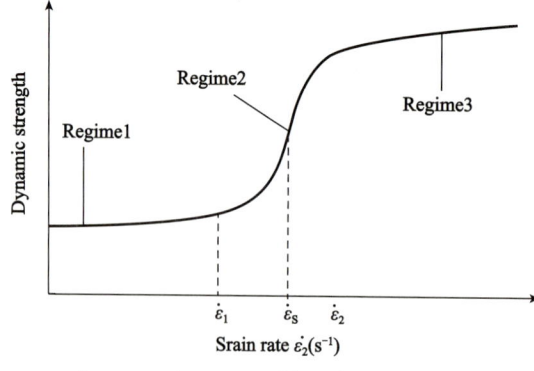

Dependence of dynamical strength on strain rates of brittle materials

$$\dot{\varepsilon}_l \approx 10^0 - 10^2 s^{-1}$$
$$\dot{\varepsilon}_s \approx 10^3 s^{-1}$$
$$\dot{\varepsilon}_2 \approx 10^4 s^{-1}$$

Regime Ⅰ: in the range of low strain rate, the strength of materials increase slowly as the strain rate increases.

Regime Ⅱ: when the strain rate exceeds a definite value the material strength increases rapidly with the strain rate.

Regime Ⅲ: When the strain rate is very high, the dependence of strength on strain rate becomes weak again.

b. Corresponding dominant mechanisms.

Rate sensitivity of the strength of rock at low strain rates (less than 10^2/sec), the thermo-activated mechanism dominates the strength sensitivity at regime Ⅰ.

When the strain rate exceeds a certain threshold value, the strength sensitivity to strain rate transits into a new regime, where the strength increases rapidly with strain rate, phonon damping (macroscopic viscosity) plays a predominant role at regime Ⅱ.

At regime Ⅲ——thermo-fluctuation rapture of the atomic bonds (thermo-activated mechanism is initiated again) in the intact area plays a importment role, material strength approaches the theoretical limit of the material strength ($\sigma_{th} \approx 0.1E$).

However, these two mechanisms (thermo-activated and macroscopic viscosity) coexist, but compete each other.

c. Corresponding mathematic formula.

Thermo-activated mechanism—thermal kinetic influence on fracture of material—Zhurkov's formula.

$$\tau = \tau_0 \exp\left(\frac{U_0 - \gamma\sigma}{kT}\right)$$

or

$$\sigma = Y = \frac{1}{\gamma}(U_0 + kT\ln\frac{\tau_0}{\tau})$$

If

$$\tau = \varepsilon_0/\dot{\varepsilon}$$

ε_0——limit of deformation at failure;

$\dot{\varepsilon}$——constant strain rate of loading.

$$\sigma = Y = \frac{1}{\gamma}[U_0 + kT(\ln\dot{\varepsilon} - \ln\frac{\varepsilon_0}{\tau_0})]$$

or

$$\sigma = Y = \frac{1}{\gamma}\left(U_0 + kT\ln\frac{\dot{\varepsilon}}{\dot{\varepsilon}_0}\right)$$

$\dot{\varepsilon}_0 = \varepsilon_0/\tau_0$——maximum possible strain rate in rocks.

For shear strength Y

$$Y_\tau = \frac{1}{\gamma_\tau}\left(G_0 + kT\ln\frac{\dot{\gamma}}{\dot{\gamma}_0}\right)$$

Macroscopic viscous mechanism

$$\eta_{macv} = C\frac{(\dot{\varepsilon}/\dot{\varepsilon}_S)^{n-1}}{(\dot{\varepsilon}/\dot{\varepsilon}_S)^n + 1}$$

$$Y_V = \dot{\varepsilon}\eta = \frac{b(\dot{\varepsilon}/\dot{\varepsilon}_S)^n}{(\dot{\varepsilon}/\dot{\varepsilon}_S)^n + 1}$$

$$Y_{\tau V} = \dot{\gamma}\eta = \frac{b_1(\dot{\gamma}/\dot{\gamma}_S)^n}{(\dot{\gamma}/\dot{\gamma}_S)^n + 1}$$

unified relation between strength and strain rate

$$Y_D = \frac{1}{\gamma}(U_0 + kT\ln\frac{\dot{\varepsilon}}{\dot{\varepsilon}_0}) + \frac{b(\dot{\varepsilon}/\dot{\varepsilon}_S)^n}{(\dot{\varepsilon}/\dot{\varepsilon}_S)^n + 1}$$

$$Y_{tD} = \frac{1}{\gamma\tau}(G_0 + kT\ln\frac{\dot{\gamma}}{\dot{\gamma}_0}) + \frac{b_1(\dot{\gamma}/\dot{\gamma}_S)^n}{(\dot{\gamma}/\dot{\gamma}_S)^n + 1} \qquad (\Delta)$$

Comparison of Eq. (Δ) with the experimental data.

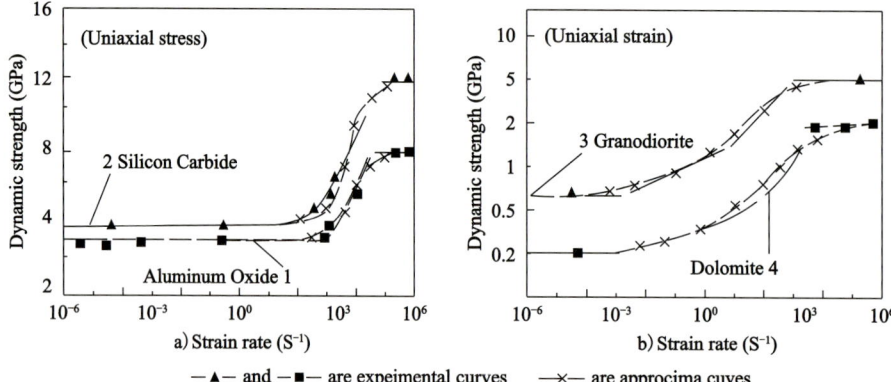

—▲— and —■— are expeimental curves   —×— are approcima cuves

Values of parameters: (a) curve 1 (aluminum oxide): $b = 4.8$, $= 103.9$, $n = 3.0$; curve 2 (silicon carbide): $b = 8.0$, $= 103.7$, $n = 2.5$; (b) curve 3 (granodiorite): $b = 4.3$, $= 101.7$, $n = 1.5$; curve 4 (dolomite): $b = 2.0$, $= 102.8$, $n = 1.0$.

d. Other strength theories with considering temporal factor and connection among these theories.

a) impulse criterion, total pulse reaches limit value J_0.

$$\int_0^{t_p} \sigma(t)\,dt = J_0$$

In one-dimensional case

$$\int_0^{t_p} \sigma(t)\,dt = \int_0^{t_p} \rho Dv\,dt = \rho Du = J_0$$

u——displacement at failure.

Impulse criterion indicates the damage accumulation nature of fracture which coincides with Zhukov's criterion.

If the length of shock wave is L, $u = L\varepsilon$

$$\int_0^{t_p} \sigma(t)\,dt = \rho DL\varepsilon = J_0$$

It indicates that second strength theory can be applied to dynamic fracture problems.

Physical essence of creep phenomenon and strength – strain rate sensitivity is the same thermal kinetic essence of material.

creep formula $\qquad \dot{\varepsilon} = \dot{\varepsilon}_0 \exp\left(-\frac{U_0 - \gamma\sigma}{KT}\right)$

Zhurkov's formula $\qquad \tau = \tau_0 \exp\left(\frac{U_0 - \gamma\sigma}{kT}\right)$

Multiply foregoing two formulas

$\tau\dot{\varepsilon} = \tau_0\dot{\varepsilon}_0 = \varepsilon_{st}$ \qquad critical failure strain

Conclusion: the second strength theory may be considered as quasi-temporal criterion.

b) damage evolution equation from damage mechanics.

$$\frac{d\psi}{dt} = f(\sigma, \psi) = \begin{cases} A\left(\dfrac{\sigma}{1-\psi}\right)^n & \sigma > 0 \\ 0 & \sigma < 0 \end{cases}$$

ψ——damage parameter.

Integrating foregoing Eq yields

$$\int_0^{t_p} \sigma^n \mathrm{d}t = \frac{1-(1-\psi_p)^{n+1}}{A(n^2+1)} = -J_0$$

When $n=1$ coincides with impulse criterion.

c) Structural-temporal criterion from fracture mechanics.

when the average stress $\sigma(t,x)$ over spatial-temporal cell $[t-\tau,t]根[0,d]$ reaches its static strength σ_c, failure takes place. According the fracture mechanics,

$$\frac{1}{\tau}\int_{t-\tau}^{t}\mathrm{d}t'\int_0^d \sigma(t',r)\mathrm{d}r \leq \sigma_c$$

or

$$\int_{t-\tau}^{t}\mathrm{d}t'\int_0^d \sigma(t',r)\mathrm{d}r = J_c(t)$$

τ——corresponds to the characteristic time for the energy; transfer between two neighbouring cells d/c

c——wave velocity;

d——structural element size.

From formula structural-temporal criterion has the meaning of critical structural impulse.

2 Physical nature of dynamic strength of solid

The investigation on the physical microscopic nature of failure of rock may be divided into stage of static methods and stage of Kinetic methods.

(1) Microscopic static theories

In microscopic static theories The stability of solids until failure is related to the relationship between cohesions between atoms and the internal forces in bonds between atoms induced by external forces.

The strength property of solids is described by the conception of limit strength, and the failure of materials is looked at as critical event which takes place instantaneously when the internal force in any of the bonds of atoms reaches its critical value.

There are two contradictions between the static microscopic failure mechanism and experimental observation of materials. The first is that the actual strength of materials is much less (1 – 3 orders lower) than the theoretical strength ($\sigma_{th} \approx 0.1E$, where E is the Young's modulus).

The difference may be attributed to the existence of defects near which significant stress concentration takes place.

The second contradiction is that the static microscopic failure concept assumes that the failure is of instantaneous event, but experiments show that the failure of materials is a time-dependent process. The attempts to solve the second contradiction give rise to kinetic theories.

(2) Microscopic Kinetic theories

In microscopic Kinetic theories:

The atom system is under thermal vibration, the vibration of atoms changes the stress state in the bonds of atoms, the average force resulted from thermal vibration in atomic bonds $\approx 10^3 \mathrm{kg/mm}^2$; the force for the breakage of atomic bonds $\approx (1.5 \sim 3) 10^3 \mathrm{kg/mm}^2$.

The difference between the energies(forces) is called the energy barrier.

The non-uniformity of atomic vibration called the thermodynamical energy fluctuation resulted from the chaotic thermal motion of atoms causes that the kinetic energy distributed to individual degrees of freedom in individual atoms will much higher than the average vibration energy of atoms.

As a result the forces in atomic bonds in individual atoms will exceed the limit forces for the breakage of atomic bonds, and the fracture of atomic links will take place, at the same time the restoration of broken atomic bonds will also take place.

thus, thermal fluctuation plays fundamental role in the breakage of atomic bonds.

The role of external forces is to lower the energy barrier U for the breakage of atomic bonds by $\Delta U(f) = f \cdot \Delta r$, where f is the force induced in every atomic bond by external forces;

and the force f lowers the probability of the restoration of broken atomic bonds, because the action of f increases the distance between atoms.

Therefore external forces raise the possibility of breakage of atomic bonds and reduce the possibility of restoration of broken bonds.

Thermal fluctuation is time-dependent stochastic process. Furthermore internal force f needs some time to overcome energy barrier and to increase the distances between atoms.

So, the development of fracturing in rock should be treated as the accumulation of breakages of atomic bonds in rock.

This process is called fracturing localization, thus localization of fracture also needs some time to be realized.

All these facts indicate that the concept of failure of material is a concept of temporal process, the realization of failure needs some time. The greater the external force, the shorter the time necessary for the overcoming of energy barrier and the increasing the distances between atoms, i.e. the shorter the time for the fracture.

3 Mohr-Coulomb type constitutive model under complex stress state with consideration of strength rate sensitivity

Mohr-Coulomb criterion may be expressed as:
$$\frac{\sigma_1 - \sigma_3}{2} = \frac{\sigma_1 + \sigma_3}{2}\sin Q + C\cos Q$$

In uniaxial compression case, internal cohesion C can be Determined
$$C = \frac{\sigma_Y^C}{2} \cdot \frac{1 - \sin Q}{\cos Q}$$

σ_Y^C——the uni-axial compressional strength.

Considering strength-strain rate sensitivity, we obtain:
$$C = \frac{1 - \sin Q}{2\cos Q}\left[\frac{1}{\gamma}\left(U_0 + KT\ln\frac{\dot{\varepsilon}}{\dot{\varepsilon}_0}\right) + \frac{b(\dot{\varepsilon}/\dot{\varepsilon}_S)^n}{(\dot{\varepsilon}/\dot{\varepsilon}_S)^n + 1}\right]e^{AC_0}$$

e^{AC_0} expresses the influence of stress state on internal cohesion.
$$C_0^2 = \sigma_1^2/\sigma_3$$

The final Mohr-Coulomb type constitutive model with the consideration of strength strain rate sensitivity.
$$\frac{\sigma_1 - \sigma_3}{2} = \frac{\sigma_1 + \sigma_3}{2}\sin\varphi + \frac{1 - \sin\varphi}{2}\left[\frac{1}{\gamma}\left(U_0 + KT\ln\frac{\dot{\varepsilon}}{\dot{\varepsilon}_0}\right) + \frac{b(\dot{\varepsilon}/\dot{\varepsilon}_S)^n}{(\dot{\varepsilon}/\dot{\varepsilon}_S)^n + 1}\right]e^{AC_0}$$

For high strain rate loading (at regime II), because of the weak influence of the thermally activated term on strength, we have
$$\sigma_1 - \sigma_3 = (\sigma_1 + \sigma_3)\sin\varphi + (1 - \sin\varphi)\left[\sigma_{YS}^C + \frac{b(\dot{\varepsilon}/\dot{\varepsilon}_S)^n}{(\dot{\varepsilon}/\dot{\varepsilon}_S)^n + 1}\right]e^{AC_0}$$

In the vicinity of explosion, $\varepsilon_\gamma = \varepsilon_r, \varepsilon_\nu = \varepsilon_r, \sigma_1 = C\sigma_3$

$$C = C_0 = \frac{\mu}{1-\mu}$$

The Mohr-Coulomb criterion may be written as

$$\sigma_1 = \frac{1+\sin Q}{1-\sin Q}\sigma_3 + \left[\sigma_{YS}^C + \frac{b(\varepsilon/\dot{\varepsilon}_S)^n}{(\varepsilon/\dot{\varepsilon}_S)^n + 1}\right]e^{-AC_0}$$

4 Fragment size of fractured rock mass under dynamic loading

As all of us know, the strength of rock mass depends on the sample size.

$$\sigma_D = \sigma_0(1 + D/D_0)^{-1/2}$$

D, D_0——sample size;

σ_D, σ_0——corresponding strength of rock.

It's the same the other way round, the fragment size of fractured rock mass is determined by the strength of rock mass.

$$\Delta_i = D_0[(\sigma_0/\sigma)^2 - 1] \quad \text{or} \quad \Delta_i = D_0\left[\left(\frac{\sigma_0}{Y_\tau}\right)^2 - 1\right]$$

Under uni-axial dynamic loading, Y_τ should be replaced by $Y_{\tau,d}$.

For predicting the mean fragment size near the centre of the explosion, foregoing formulas are used, but Y should be replaced by

$$Y_{\tau,D} = \left[\frac{1}{\gamma_\tau}\left(G_o + KT\ln\frac{\dot{\gamma}}{\dot{\gamma}_o}\right) + \frac{b_1(\dot{\gamma}/\dot{\gamma}_S)^n}{(\dot{\gamma}/\dot{\gamma}_S)^n + 1}\right] \cdot e^{A\alpha}$$

5 Application of modern dynamic strength theories

The introduction of temporal factor into strength may result in some results different from that of traditional strength theories.

(1) Internal structure of spallation plane

According to traditional strength theories, under triangle impulse without rise time spallation take place at the distance from free surface $x = \sigma_t/2\sigma_m$.

According to modern dynamic strength theories, material in a definite width will fracture simultaneously. conclusion. Experiments validate this conclusion.

The reason is that every point in reflected wave under tensile stress prepares to fracture, in section where tensile stress is less, there will need longer time to fracture, and this section may fracture simultaneously with sections where tensile stress is greater and time for fracture is shorter.

(2) Safety threshold of explosion seismic action

Which parameter of acceleration, ground velocity and displacement should be chosen to evaluate building and facility safety?

Most of the countries in the world take ground surface velocity as the control parameter, this agrees with the in-situ investigation, and agrees with the modern dynamic strength theories.

According to modern dynamic strength theories, the damage of buildings under explosion seismic ground shaking is dynamic failure. The control parameters for dynamic failure are displacement or particle velocity.

Considering that displacement is the integration of velocity over time (period of vibration), therefore the failure criterion introducing shaking velocity and frequency as controlling parameters is more reasonable.

6 Conclusions

The dependence of deformation and fracture on temporal factor lies in that fracture of materials needs some time to complete, therefore, material strength depends on strain rate.

The introduction of temporal factor betters our understanding of material deformation and fracture.

The temporal features of fracture of rock are decided by physical nature of rock.

The dependence of strength on strain rate may be considered as the result of competition between two coexisting mechanisms: thermally activated and macro-viscous mechanisms, which play the leading role in different intervals of strain rate. The dependence of rock media strength on strain rate may be expressed as the summation of the contributions from these two mechanisms.

A comparison with the experimental data shows that the proposed model describes the strength sensitivity very well over a wide range of strain rates.

The intrinsic relationship between different temporal failure criteria is revealed.

The strain rate type Mohr-Coulomb criterion is given.

The influence of dynamic loading on the fragment size of rock mass depends on the accumulation of increased shear deformation energy at the moment of fracture due to the strength enhancement originating from the change of stress state, the accumulation of plastic deformation, and strain rate. The suggested relationship describes the fragmentation size well.

大盾构掘进的事故及对特大盾构工程的思考

一、国内外16米级及以上特大盾构工程的初步统计

建设时间	隧道名称	掘进距离（km）		盾构		盾构类型	主要地质情况
		总长	每段掘进长度	直径（m）	数量		
2011—2019	美国西雅图SR99隧道	2.8	2.8	17.45	1台	土压盾构	黏土、粗砾砂岩
2013—2015	香港屯门隧道	4.2	0.65+4.1	17.63/14.0	1/2台	泥水盾构	粉砂层、砂层/砂层、花岗岩
2017至今	日本东京外环隧道	16.2	9.1+7	16.1	4台	土压盾构	黏土、砂、砂砾
2017至今	武汉和平大道南延隧道	1.4	1.4	16.03	1台	泥水盾构	石英砂岩、灰质岩、钙质岩、泥岩、黏土
2016至今	意大利圣塔露琪亚公路隧道	7.5	7.5	15.87	1台	土压平衡盾构	黏土、砂层、强中风化石灰岩、泥灰岩、粉砂岩等
2011—2012	意大利SPARVO隧道	5.1	2.6+2.5	15.55	1台	土压平衡盾构	黏土，泥岩，砂岩
2018至今	墨尔本西门隧道	6.8	4+2.8	15.6	2台	土压盾构	砂层与和黏土层、玄武岩
2013—2018	武汉三阳路隧道	2.59	2.59×2	15.63	2台	泥水盾构	粉细砂、砾岩和粉砂质泥岩
2017—2021	深圳春风路隧道	3.6	3.6	15.8	1台	泥水盾构	强中风化花岗岩、石英岩
2017—2021	济南黄河隧道	5.2	2.6×2	15.76	2台	泥水盾构	粉细砂、中粗砂、黏土、钙质结核、黏土夹碎石
2019至今	北京东六环改造工程	7.6	7.6×2	15.97	2台	泥水盾构	砂层、粉土夹黏性土互层、局部砾岩
2021至今	深圳机荷高速公路荷坳隧道	6.54	3.25+3.29	18.1	2台	泥水盾构	中风化灰岩、砂岩

注：国外超大直径盾构以土压盾构居多；国内超大直径盾构以泥水盾构居多。

二、大盾构掘进主要事故及其初步分析

（一）主轴承损坏失效

盾构主轴承的主要功能：在盾构机作业过程中承受推力和修复力矩将其传递给刀盘，使得盾构能够向前推进；承受刀盘驱动系统的回转力矩（扭矩），将其传递给刀盘使其进行破岩工作；连接刀盘和刀盘支撑并承受其重量。

报告时间：2021年10月。

主轴承损坏失效主要分两类：一类是主轴承直接损坏，另一类是轴承密封失效引起的间接损坏。

1. 主轴承直接损坏的形式

（1）润滑失效工作的轴承疲劳；

（2）微动磨损造成的轴承套圈破坏；

（3）安装不当或荷载过大造成的轴承失效；

（4）接触疲劳造成的轴承部件失效（滚动体、滚柱）和滚道表面剥落。

如果轴承加工精良，润滑良好，安装正确，无轴线偏移，无尘埃、水分和腐蚀介质的侵入，且载荷适中（扭矩、推力），前三种失效形式在正确设计和施工情况下都是可以避免的。造成轴承损坏原因只有一个：接触疲劳。轴承在运转过程中，接触表面应力的循环作用，滚动件与滚道接触应力区的材料超强度发生变化，金属、微粒从表面剥落，表明此时发生滚动接触疲劳，从而引起主轴承损坏。

如何避免和减缓：降低应力水平，控制循环作用时间。

前者涉及主轴承直径，后者涉及盾构推进距离（时间）。

2. 密封失效引起的轴承损坏

主驱动密封系统的作用：阻止土仓渣土进入；冲洗和润滑作用。

不同形式的主驱动密封系统虽有不同点，但基本都由多道单唇密封、一道或多道双唇密封和迷宫密封组成，并在唇形密封间设置腔室且充满弹性材料。在密封腔中注入一定稠度的油脂，直到油脂沿密封缝隙溢出，迷宫密封有一定的曲折度，加长油脂溢出通道，增大溢出阻力，密封系统的作用是将泥水和渣土阻挡在外。其条件是土仓或者泥水仓的水压力小于油脂溢出阻力。

密封失效机理：油脂量不足，油脂溢出压力不足；水土压力过大，超过油脂阻力。

广深港 $\phi 11.182m$ 泥水盾构主轴承损坏事故原因包括：油脂注入不到位，在 7bar（1bar=0.1MPa）水压下，泥浆通过迷宫进入轴承密封；轴承滚子及滚道表面在夹杂硬质小颗粒后出现压痕；焊接刀盘时搭铁线未严格按标准放置，致使主轴承滚子和滚道面接触处有大电流通过产生电熔蚀坑；两者在重载作用下逐步损坏（轴承直径 4800mm）。

美国西雅图 $\phi 17.45m$ 土压盾构及丹麦大贝尔特海峡东通道 $\phi 8.7m$ 土压盾构事故发生的原因为结泥饼，压力不均，局部土压力增大，刀盘处出现高温，油脂阻力减小，主轴承密封失效，土仓渣土颗粒进入。

（二）盾构管片拼装脱出盾尾后上浮

从抗浮系数和上浮空间两方面进行原因分析：

（1）隧道直径越大，管片脱出盾尾后，管片拼装体单位容量减小，抗浮系数减小，受到地层及浆液作用下的浮力会更大。

（2）盾构直径增大，盾构的结构刚度要求，盾构壳体厚度加大，盾尾间隙也适当加大，致使管片外圈的空隙也加大，砂浆填充空间及填充量也再增大，也是导致管片上浮量增大的一个原因。

（3）隧道直径越大，径向高度越大，而盾构机主机长度基本没有变化，致使盾构机主机的高长比加大，更易造成盾尾后方同步注浆浆液流入前舱，造成脱出盾尾管片上部外圈砂浆填充不密实，加大了成型隧道上浮趋势。

（三）刀具磨损随盾构直径增大而加剧

大型盾构在砂卵砾石层推进中刀具磨损问题更加突出，这是由于：

（1）盾构刀具在同样进尺条件下，其磨损工作长度与刀具配置部位半径成正比，例如某直径 14.93m 大盾构掘进刀具的磨损量为地铁直径 6.3m 盾构的 2.5 倍。

（2）在石英含量高的砂卵石层中，刀具的磨损量可达软土地层中磨损的 10 倍。

三、对于特大盾构挑战的思考

（一）对于特大盾构带来的事故风险的思考

由于工程客观的需求，现在出现了盾构直径越来越大的倾向，必须重视由此带来的事故风险。

（1）盾构刀盘直径增加，荷载增大，没有相应增大主轴承尺寸，致使滚动体与滚道接触区应力水平提高，增加接触疲劳引起的轴承损坏的风险。

（2）盾构直径增大，渣土仓压力不均匀度增大，局部大压力的出现，超出油脂阻力，导致主轴承密封失效风险增大。

（3）盾构直径增大，管片盾尾脱出后上浮风险加大。

（4）盾构直径增大，刀具磨损风险增大。

（5）盾构直径增大，刀盘结泥饼风险增大。

（二）对于盾构选型的思考

特大超大盾构：国外土压盾构案例居多；国内泥水盾构案例居多。

土压盾构与泥水盾构选择：

- 地层稳定性较好；
- 地层透水性不强；
- 水土压力较小；
- 盾构驱动功率较大
- 耐压能力相对较小；

→ 土压盾构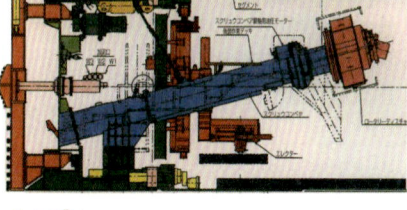

- 地层稳定性较差；
- 地层透水性较强；
- 水土压力较高；
- 盾构驱动功率较小
- 耐压能力相对较强；

→  泥水盾构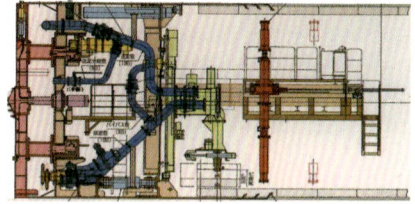

同级别特大超大盾构对比：土压盾构负荷重，驱动功率偏大；泥水盾构负荷小，驱动功率相对较小。

驱动功率对比：

特大土压盾构：盾构驱动功率大，主轴承加工尺寸大

意大利SPARVO隧道：双洞双向六车道，采用一台φ15.55m土压平衡盾构施工。该盾构驱动功率达12000 kW，刀盘最大功率315000 kN，最大扭矩315000 kN·m。2011年8月，北线隧道盾构始发，于2012年7月贯通，盾构调头后施工南线隧道。

美国西雅图SR99隧道：开挖直径17.45m，主轴承8.0m，装机功率12135kW。

特大泥水盾构：盾构驱动功率相对小，主轴承加工尺寸也小

香港屯门隧道：泥水盾构，开挖直径17.63m，主轴承7.6m，装机功率8600kW。

特大超大盾构：高水压、地层差异性大。稳定性差的强透水地质多以唇式橡胶密封为主。

主轴承密封问题

特大土压盾构 — 主轴承密封以聚氨酯密封结构为主
- 优点：采用水冷却降温方式，油脂消耗量小，节约成本和降低能耗。
- 缺点：① 无法建立备压措施，耐压能力相对偏弱；② 一旦密封损坏或失效，必须吊出更换修复。如美国西雅图SR99土压盾构密封损坏，需吊出井在地面修复。

特大泥水盾构 — 主轴承密封以唇式密封结构为主
- 优点：① 采用油脂备压方式，可实现高水压条件的耐压密封体系，作业安全；② 可实现原位带压条件更换。如黄河隧道实现原位密封修复。
- 缺点：油脂消耗量大，能耗和掘进成本高。

土压盾构相对于泥水盾构，渣土舱压力不均匀度及局部土压力更大；轴承承受荷载更大（功率更大），接触表面应力更大。因此主轴承密封失效及直接损坏概率更高。

（三）常压刀盘与常规刀盘选择的思考

对特大盾构，为应对高水压、土岩复合地层、减少带压换刀风险，宜采用常压刀盘配置。

土压平衡盾构；
地层稳定性较好；
地层透水性不强；
水土压力较小
→ 常规刀盘

泥水平衡盾构；
地层稳定性较差；
地层透水性较强；
水土压力较高；
土岩复合地层；
减少带压作业风险
→ 常压刀盘（中心无开口（泥饼））

（1）盾构配置：针对软塑～硬塑易结泥饼地层，加大刀盘开口率（缩小中心封闭区域范围）、刀具多层次布置，强化切削功能，降低碾磨，尽可能使渣土成块排出；增加增强冲刷刀盘中间结泥饼的系统，创新内循环冲刷；配置刀盘温度自动监测系统和刀盘伸缩功能。

（2）盾构掘进：严格刀盘温度变化预警，加强加大泥浆循环，保持拼装管片期间的泥渣循环。

（3）辅助措施：创新破除泥饼的技术，例如水刀切割、分散剂（双氧水）等；在气密性好的围岩下，辅助气压作业。

分时集中冲刷和仓底顺流冲刷理念，解决了高黏性地层刀盘黏结施工难题。

刀盘结饼　　高压水刀现场喷射试验　　刀盘冲刷系统改造

泥饼黏结及滞排

（四）其他思考

（1）利用泥水盾构 SSP 超前探测系统，加强预探前方地质分布情况；
（2）利用超前钻机系统，超前钻探，兼顾预处理措施；
（3）结合综合掘进参数，超前预判前方地质变化。

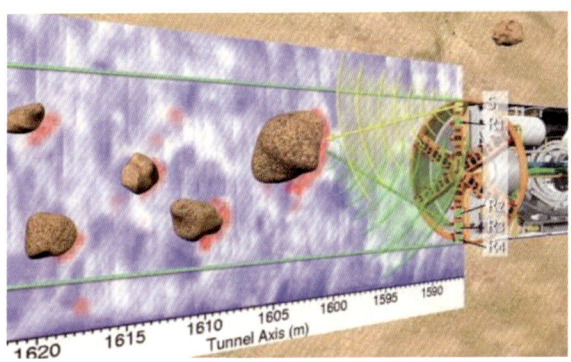

预探前方复杂地质

（4）高水压复杂地质超大直径盾构的其他辅助处理技术。
①海中基岩爆破辅助处理技术——珠海横琴第三通道。

②海中注浆固结辅助处理技术——厦门地铁 2 号线海底隧道。

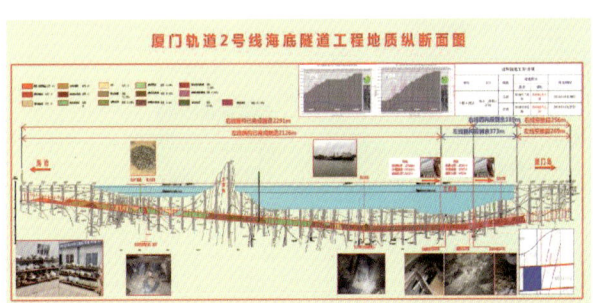

地下空间开发利用、防治城市病及地下物流系统的发展方向

一、引言

以交通拥堵、空气污染和城市内涝为代表的"城市病"全民关注,党和国家领导高度重视。

过去注意高层建筑、大广场、宽马路城市"面子"建设的粗放发展模式酿成"里子"城市地下基础设施建设短板的恶果:"城市病"丛生。

建设城市地下空间是"转变城市发展方式",治理"城市病"的主要着力点。

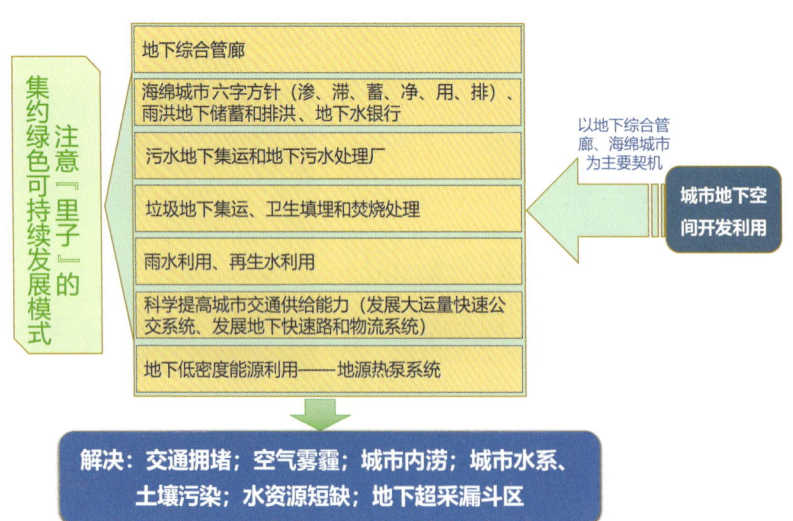

报告时间:2018年10月10日。

二、治理交通拥堵

（1）城市交通拥堵是传统城市化、机动化的必然后果，根本原因在于交通需求与供给的失衡。

（2）解决交通拥堵需要新思维、大手笔：科学抑制交通需求、科学提高交通供给。

（3）限购限行，利少弊多，不能持久。

（4）"存量"不减、"增量"攀升，中心城区地面已无地可建（路）。

（5）国外曾做过伦敦中心区地面交通供给能力的模拟测算，其结果是即使把伦敦中心区所有建筑架空，其下层地面全变成道路，也不能解决交通拥堵问题。

科学提高交通供给能力，出路在地下，具有巨大开发潜力：从国际上看，向地下要交通空间是未来城市交通发展的主要模式之一。以交通功能和改善城市环境功能为主的城市地下空间的开发利用，将是二十一世纪城市地下空间建设的核心内容。

北京"治堵新政"多处提到"地下交通"："地下环廊""建设地下停车场"和"地下快速通道工程"。但远远不够，需要引进"地下城市"的理念。

美国格兰尼教授和日本尾岛俊雄教授提出，地下城市基本理念之一，是将地上城市中的城市运输和传输系统逐步转移到城市地下的不同层次里。是"将地面交由市民自由使用，将基础设施放入地下"。（《欧洲地下建设战略发展议程》）

从目前技术看，地下100m以上都是可以建设的。如果一层交通空间不够，还可以建两层、三层，甚至多层。按上述理念，通过地铁、地下道路以及地下物流系统等方式，将北京地面客、货运交通按规划分阶段逐步转为地下，可以预见，特大城市的交通拥堵将会逐步地，最后从根本上得到改善。

这不是科学幻想，正在逐步转变为现实。

治理交通拥堵应坚持发展以地铁为主的公共交通方式。

截至2017年12月31日，中国内地累计有34个城市建成并投入运营的地铁和轨道交通系统线路5021.7km。

目前正在建设城市地铁1339.5km，车站778个。

预计到2020年底，全国地铁里程数将超过6000km，累计投资4万亿。

（一）国外交通拥堵治理现状

仅大力发展轨道交通不能从根本上解决交通拥堵。即使是轨道交通相当完善的发达国家大城市，如伦敦、东京、芝加哥、马德里、莫斯科等，它们也困扰于交通拥堵。以东京为例，公共交通出行比例达到了60%，地铁建设里程已达300km（地铁集中于城中心区，仅占轨道交通的1/10），仍然没有解决交通拥堵。伦敦也是，它的第一条地铁建成于1863年，至今天已建成439km，但仍然要实行收取"交通拥堵费"以缓解拥堵之苦。

从20世纪末至21世纪初，国际上很多发达国家的特大城市和有识之士把"治堵"目光转向了"地下快速路"和"地下物流系统"。

1. 美国波士顿地下快速路

美国波士顿1994年开始拆除高架路，10年建成8~10车道的地下快速路。

a）改建前（地面高架路）　　b）改建后（地下快速路）

2. 日本东京地下快速线

东京正在建设中央环状地下快速线，新宿线已于 2007 年通车；品川线正在建设。

地下快速线总规模达 11km；埋深 >40m；双向 4 车道（13m）；3 大商业中心：池袋、新宿、涩谷；6 个出入口；投资 90 亿美元；2007 年 3 月通车。

3. 新加坡地下道路系统

新加坡地下道路系统（Singapore Underground Road System，SURS）长 15km，四车道，环形；造价 48 亿美元；承担城市中心区交通量的 40%；新加坡 KPE 地下公路已建成 9km。

4. 其他

（1）马德里已建成 36km 的地下快速线；

（2）悉尼建成 4 条共 11.4km 的地下快速线；

（3）布里斯班建成 2 条 6km 地下快速线；

（4）巴黎建成 20km 地下快速线，并已完成 2 代城市地下快速路网设计；

（5）莫斯科在三环线中建成了 3.6km 地下线。

（二）国内交通拥堵治理现状

国内大城市的地下快速路建设也已初见端倪，并在缓解交通拥堵上发挥了明显作用。

（1）南京已建成玄武湖地下快速路、城东干道地下线和内环线地下路（占总长 1/2）。

（2）上海正在建设井字形通道方案，其中地下线 26km。

（3）深圳已建成港深通道的 7km 地下快速路。

（4）深圳前海的地下空间规划建设在地下车行系统方面，将构建"地下快速路 + 地上环路 + 车库"的三级地下车行系统，海滨大道、妈湾跨海通道地下道路以分离过境交通为主，以服务区域到发交通为辅。听海路服务于区域到发交通，从而有效改善前海地面交通环境，缓解小街坊路网交通压力。

这些地下快速路的天然优势是全天候，在暴风、雨雪和大雾等最易造成地面交通拥堵的情况下最能发挥作用。这在我国城市建设用地严格控制和大城市人均道路指标普遍偏低的情况下，非常适合我国地少人多的国情。

三、治理空气污染

城市地铁与轨道交通系统已为消除城市尾气污染作出重要贡献。城市地下快速路系统以及地下物流系统的建设也必将为消除汽车尾气对城市空气的污染以及噪声污染做出重要贡献。

四、改善城市环境

（1）波士顿拆除穿过市中心的六车道高架路，建设 8~10 车道的地下高速路，原有的地面变成林荫路和街心公园。这样的结果是市区空气的一氧化碳浓度降低了 12%；市中栽植了 2400 棵乔木树，7000

多棵灌木树；在海湾的景观岛上栽植了另外的 2400 棵乔木和 26000 棵灌木；增加了 260 英亩（1 英亩≈4046.86m²）新的公园和开敞空间。

（2）地下货物运输系统（ULS）的环境效应显著。

①世界经济合作组织在 2003 年的《配送：21 世纪城市货运的挑战》报告中指出，发达国家主要城市的货运占城市交通总量的 10%~15%，而货运车辆对城市环境污染总量的贡献为 40%~60%。

②东京 11km 新宿地下线的建设，每年减少 2.5 万 t CO_2，16 万 t 烟尘颗粒和 16 万 t NO；日本东京建设 300km 地下物流系统评估报告的结论是：NO 和 CO_2 将分别减少 10% 和 18%，能耗减少 18%，运输速度提高 24%。

③科技部"十一五"国家的重点科技支撑计划项目"道路隧道空气治理关键技术研究及示范工程应用"已提出一整套综合治理技术，其对人体危害大的 CO、THC（VOC）和 SO_2 的净化率分别达到 80%、65% 和 98%。

（3）城市空气污染的元凶——尾气的污染难题，通过城市地下快速路系统和地下物流系统的修建，也有望得到彻底解决。

（4）城市空气雾霾受农村冬季散煤燃烧的影响。陈吉宁部长："第三个就是农村散煤问题，……，全国大概有 2 亿 t 散煤"。

（一）地热能源利用

利用低密度地热能源，节能减排。

到达地下一定深度时（5m 以下），四季的地层温度保持在一稳定值。

地温能指的是地层中温度小于 25℃ 的地层热能。

将传统空调器的冷凝器或蒸发器直接埋入地下，利用传热循环介质与大地进行热交换，从而提取地温能。

地下换热系统可埋设在地下结构基坑围护结构（地下连续墙、排桩内）、基础底板下、桩基（钻孔灌注桩、预制桩、PHC 桩等）内；可埋设在新奥法施工的隧道衬砌内或以能源锚杆的形式埋设在其围岩中；也可埋设在地铁区间隧道内、地下输水管道内。

奥地利政府资助一下能源地铁车站示范工程，该车站位于 Lainzer 隧道 LT24 区，一共 59 根桩基内埋设热交换管，钻孔灌注桩桩径为 1.2m，桩长平均为 17.1m，利用 6 台热泵为附近一所学校供暖。在长期供暖的情况下，车站能提供 150kW 的热负荷。一个供暖季度可提供 214MkW 的能量，天然气的使用量每年减少 34000m³，使每年二氧化碳排放量减少 30t，与传统的靠燃烧天然气供暖的方式比较，可使学校每年用作取暖的费用降低 1 万美元。

瑞士 Grabs 的 PAGO 公司办公楼采用 570 根桩基内埋设热交换管，平均桩长 12m，以 4 个能源桩为一组，呈方形顶角安装，四边间距为 1.4m。每延米桩基冬天可获得 35kW 时的热量，夏天获得 40kW 时的冷量。

瑞士 Grabs 的 PAGO 公司

上海自然博物馆新馆位于上海市静安区，北京西路、石门二路处在建雕塑公园北部，北邻山海关路。占地面积约12000m²，总建筑面积约为45086m²，其中地上建筑面积12128m²，地下建筑面积32958m²。建筑总高度18m，地上三层，地下两层，采用地源热泵系统承担建筑部分夏季冷负荷以及冬季热负荷。夏季土壤换热器最大算热负荷为1639kW，冬季土壤换热器最大热负荷为1178kW。

上海自然博物馆

工程采用灌注桩与地下连续墙内埋管两种形式，其中，灌注桩埋内埋管393个，平均桩长为45m，采用W形埋管，地下连续墙内埋管共452个，采用W形埋管。灌注桩及地下连续墙内埋管布置见图。

上海自然博物馆已于2015年投入运营，该馆采用了地下结构内埋管热交换系统，其初投资比传统的冷水机组+锅炉系统增加210.2万元，但该系统利用了地温能实现建筑制冷和供暖，年运行费用可节省22.3万元，动态回收期11.98年，该系统利用清洁的地温能，每年可节约117.7t标准煤，减排二氧化碳195.5t。

上海中心大厦裙楼，上海世博轴和上海富士康大厦等国内重大工程均采用了地下结构内埋管热交换系统。

科罗拉多的地源热泵系统：安装地源热泵系统后，每个家庭可降低电力峰值需求、节约能源和减少CO_2排放约10%~30%。

目前，地源热泵正进一步与太阳能结合。由于太阳能的辅助供热，可实现系统向地下排热与取热的平衡，从而使得地下温度场保持稳定。既可克服单独使用地源热泵时，土壤温度场不断降低（或升高）后不能有效恢复的局限性，又可克服单独使用太阳能空调系统时，太阳辐射受天候因素制约的局限性。

河北省地热能资源总量位居全国第二位，2015年地热资源开采量突破了1.1亿m³，地热供暖面积达到了6300万m²。其中，雄县位于河北省保定地区，是国内首个通过地热供暖实现"无烟城"的县城，拥有享誉全国的"雄县模式"。雄县地热资源分布面积广，出水量大，水温高，现如今地热集中供暖面积已占城区集中供暖的85%，覆盖县城80%以上的居民小区，每年可减少二氧化碳排放量12万t。在收费方面，雄县居民地下水取暖收费为16元/m²，相较之前燃煤取暖25元/m²的费用更为便宜。

从投入与产出方面，地热取暖前期投入较大，后期的年收益率稳定。推广和提升"雄县模式"，在集中成片供暖的基础上，在新兴城镇中打造以地热为主的"绿色热网"，解决北方中、小新兴城市和农村冬季供暖的问题，减少燃煤燃烧对空气造成的污染。

（二）干热岩资源利用

全国陆域干热岩资源量为856万亿t标准煤。青海共和盆地3705m深钻获236℃的高温干热岩体。

我国已成功在陕西省内进行了干热岩用于供热的商业应用——长安信息大厦2013年共计3.8万m²中应用干热岩供热，效果良好。

国际上有美国、澳大利亚、日本、德国、法国等进行了干热岩发电试验研究项目。美国Los Alamos实验室在卡尔德拉的芬登山上建成了一个10MW的HDR（深层干热岩）发电站，该电站主要由两个深度为3千多米的钻孔及其连通孔组成，冷水由一个钻孔灌入，另一个孔产生200℃水蒸气，进入汽轮机发电。

五、治理城市内涝

措施:"渗、滞、蓄、净、用、排"。

目标:将 70% 的降雨就地消纳和利用;2020 年,城市建成区 20% 以上面积达到;2030 年,城市建成区 80% 以上面积达到要求。

"海绵城市"六字方针应统筹规划、分工合作、形成合力;突出重点,建设城市深隧排水系统;深层排水系统与浅层排水系统互为补充,兼顾排污与治涝;渗储结合、排蓄结合,达到治涝与雨水利用相兼顾,排洪与治堵相结合。

以下介绍雨污合流制城市的深层隧道排水系统。

1. 东京深层排水工程

东京年平均降雨量为 1800mm,台风和暴雨带来的洪水是最大的威胁。当短历时超常降雨出现时,雨洪超出河道正常排涝能力,引起积水倒灌,城市内涝。分析表明,东京范围内大大小小的河流中,最大的江户川河道宽阔,具有足够的泄洪能力。如何提高其他河道的洪水容纳能力,并及时通过江户川排入东京湾,是解决东京洪水问题的关键,也是深层隧道工程建设的初衷。

该工程全厂 6.3km,下水道直径约 10m,埋设深度为地下 60~100m,由地下隧道、5 座巨型竖井、调压水槽、排水泵房和中控室组成,将东京都十八号水路、中川、仓松川、幸松川、大落古利根川与江户川串联在一起,用于超标准暴雨情况下流域内洪水的调蓄和引流排放,调蓄量约为 67 万 m^3,最大排洪量可达 $200m^3/s$。

在正常状态和普通降雨时,该隧道不必启动,污水及雨水经常规、浅埋的下水道和河道系统排入东京湾,而当诸如台风、超标准暴雨等异常情况出现,并超过上述串联河流的过流能力时,竖井的闸门便会开启,将洪水引入深层下水道系统存储起来,当超过调蓄规模时,排洪泵站自行启动,经江户川将洪水抽排入东京湾。

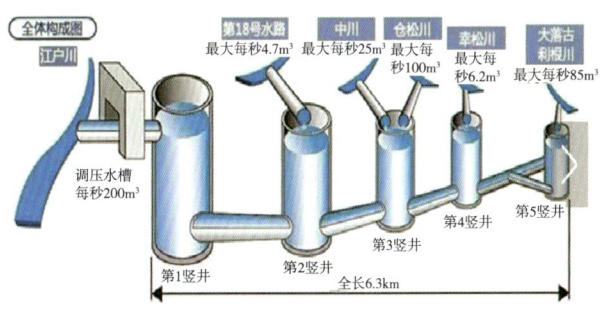

2. 芝加哥隧道和水库方案

芝加哥年平均降雨量为 910mm,大部分降雨以强烈夏季暴雨形式发生,每年暴雨约有 100 次,合流制污水最终溢流至芝加哥地区水源地密歇根湖。

1972 年芝加哥市采取隧道和水库方案(TARP)。TARP 由 4 个独立隧道和隧道下游 3 座大型水库组成。第 1 阶段是 4 条隧道的修建,主要目标是控污,能够把数百个排污口排放的 870 万 m^3 合流制溢流污水收集起来。2006 年 4 条隧道完工并运行。第 2 阶段的主要目的是控制内涝,1998 年已完成奥哈拉水库的修建。另外 2 座水库正在建设中。3 座水库完工后，TARP 的储水量将高达 6600 万 m^3。虽然还没有完全竣工，TARP 已经显著改善了芝加哥河等水道的水质，河岸可供垂钓的鱼种回到之前水平，并带动了旅游业的发展。

3. 新加坡污水输送、集中处理的深层排水隧道

新加坡总面积仅为 714.3km^2，年平均降水量为 2400mm 左右。新加坡是全球少数几个采用雨污分流系统的国家之一。新加坡深层隧道充当的角色是将整个城市收集的污水输送至处理厂集中处理排放。

新加坡淡水资源缺乏，故雨水是其主要淡水资源。雨水通过街道旁方便管理的明渠逐级汇集到 17 个大型蓄水池，经处理后进入配水系统。原来新加坡的污水处理系统由 2450km 管渠、134 座泵站、大型污水处理厂和一些小型处理厂组成。新加坡前瞻性地设计了以污水输送为目的的深层隧道系统，即通过 2 个交错的深层隧道，将覆盖 99% 人口范围的污水集中输送至郊区 2 个大型污水厂，最后将处理过的水排入远离新加坡的深海。此系统将解决新加坡未来 100 年的废水回收、处理和排放需求。

4. 上海苏州河深隧排水系统

该系统是在原有基础上的提升与深化，深隧功能的发挥有赖于深浅结合，只有深浅层之间的管网畅通，深隧的功能才能发挥正常，否则就可能出现"深隧吃不饱，管网顶不住"的现象。

工程服务范围 25 个排水系统共 57.92km^2。

三大建设目标同步实行：①苏州河沿线排水系统设计标准达到 5 年一遇（1h）；②有效应对 100 年一遇降雨（1h，不发生区域性城市运行瘫痪，路中积水深度不超过 15cm）；③22.5mm 以内降雨泵站不溢流（$P=1$ 年），基本消除工程沿线初期雨水污染。

5. 广州东濠涌深隧工程

东濠涌深隧工程为全国首个地下深层隧道项目。建设内容包括新建直径 6m 的东濠涌深层隧道约 1770m，直径 3m 的新河浦涌截污管道约 1390m，以及中山三路竖井、东风东路竖井、玉带濠竖井及江湾竖井（含泵站）。项目主要收集仔鱼岗涌、玉带濠、中山三路的合流污水。深隧建成后可有效减少 70% 雨季溢流污染。

东濠涌流域内排水管网重现期基本处于 1 年以下，洪灾风险较高，5 年一遇较严重水浸点为 28 个。东濠涌深隧建成后，当雨季来临时，深隧将作为东濠涌流域合流溢流污水和初期雨水的调蓄和转输通道，经污水泵组提升后送到污水处理厂处理。如果遇到大型暴雨，深隧将作为雨水排涝通道，行使排涝功能，经尾端排洪泵组提升后排至珠江，提高流域内合流干渠的排水标准到 10 年一遇。浅层排水管道排水标准不低于 5 年一遇。可以有效削减东濠涌流域 70% 雨季溢流污染。深隧建成后，5 年一遇降雨时，较严重水浸点（水浸 20mm 以上）将减少至 17 个。剩下的这 17 个水浸点，将结合工程性改造消除水浸。

6. 吉隆坡城市泄洪与公路两用隧道

Kuala Lumpur – SMART 隧道位于马来西亚吉隆坡市，在暴雨情况下，城市快速路充作吸纳雨洪和排泄的通道，已解决雨涝之害。

7. 雨洪入渗和储水（水银行）相结合

（1）80 年代开始，美国开展了钻孔补给含水层的恢复工程（ASR 系统）。美国已运行的系统有 18 个，正在建设之中的 ASR 系统有 40 多个。

（2）瑞典、荷兰和德国的 AR 工程在总供水中占 20%、15% 和 10%。

（3）荷兰到 1990 年，地下水人工补给量达到 1.8 亿 m^3/年。

（4）伦敦利用地下含水层人工补给后提供水资源 90ML/d。

六、城市地下货物运输系统

科技创新转变城市交通运输方式还包括地下物流系统，以替代城市货车限行的行政措施，将城市货运逐步并最终转移至地下。

地下货物运输系统，又称地下物流系统（ULS）是基于城内运输和城外运输的结合。城外货物通过各种运输手段到城市边缘的物流园区（CLP），经处理后由 CLP 通过 ULS 输送到各个终端。它以集装箱和货盘车为基本单元，以自动导向车（AGV）为运输工具。

根据世界经合组织 2003 年统计，发达国家主要城市的货运占城市交通总量的 10%~15%；在"世界工厂"和到处是建设工地的中国则为 20%~30%；而在当今电子商务、快递和送快餐发达的中国，这个比例还要高。货运交通转入地下必将对缓解交通拥堵作出重要贡献。

港城地下物流系统发展的紧迫性：

传统的港口集疏运模式以公路集卡运输为主，需要穿越城市，占用城市道路资源，客货交通相互影响，带来了交通拥堵、环境污染等问题。需要更集约化、可持续发展的集疏运新模式——地下集装箱物流系统，解决港城矛盾。

将原先分散进入港口的货物，首先集中在外围物流综合枢纽，再通过地下专用货运通道集约化转运到港口，实现高效率、规模化运输。有效避免集装箱货车穿越城市，释放地面资源，不影响中心城交通、环境。

（一）港口地下物流典型案例

1. 上海港

（1）吞吐量分析上海港

① 2009 年以来，上海港集装箱吞吐量基本保持在 20%~30% 的增幅增长。

② 2012 年上海港集装箱吞吐量达到 3253 万标准箱（TEU），位列世界第一，2016 年总量 3713.3 万标准箱，超过所有欧洲港口集装箱吞吐量总和。

③ 根据《上海港总体规划》预测，2020 年上海港集装箱吞吐量将达 4000 万标准箱。

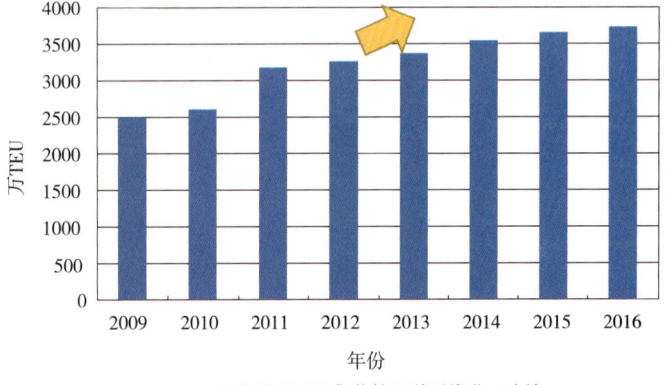

■ 上海港是国际集装箱远洋干线港，连续多年保持集装箱吞吐量世界第一

（2）上海港集装箱集疏运分析

① 2015 年，上海港集装箱各集疏运方式中，公路集疏运占 54.8%，水路集疏运占 45.0%，铁路集疏运占 0.15%。

② 2020 年，预测，水水中转 50%，公路集疏运占 48%，铁路集疏运占 2%。

③ 公路集疏运比例过高，方式单一。

④ 外环线和郊环线的集卡交通量较大，特别是外环线的东段和北段，以及郊环线北段，最高断面达 6 万~7 万 PCU/12。

⑤ 北部地区集卡流量集中，高等级道路供给不足。

⑥ 越江通道能力不足。

（3）集疏运比例不合理，公路运输比例过大引发的交通及环境问题。

① 公路运输比例大，现有道路不足以提供通行能力，集卡为赶船期，只能提前前往港区附近。

② 港区附近停车位不足，提前到达的集卡在路边无序停靠。

③ 无序停靠的货车占用路面资源，进一步降低通行能力，造成恶性循环。

④ 运输结构失衡，过度依赖公路运输，带来严重环境污染及能源消耗，使得上海港的快速发展对"节能减排"工作造成巨大的压力。

⑤ 据测算：2016年，上海公路集装箱集疏运的能源消费量约10.5亿L柴油，折合标准煤约131.25万t，排放二氧化碳约286.47万t。

2. 洛杉矶—长滩港阿拉米达走廊（Alameda Corridor）

（1）阿拉米达走廊位于美国加州南部，专用疏港货运地下铁路，长32 km，南端连接圣佩罗港湾并延伸到洛杉矶–长滩港区内，北至洛杉矶市内与干线铁路交会，并与铁路公司的编组站及集装箱枢纽场站连接。

（2）铁路线从地下穿越市区，代替了原来包含200多个交叉口的通道，缓解了港城交通矛盾，减轻了公路拥堵，减少了货车和列车停留造成的废气排放，并使铁路运输时间从数小时缩短到40 min。

3. 比利时安特利普港地下集装箱运输系统

（1）采用地下集装箱运输系统连接港区，承担斯凯尔河（Scheldt）河左右两岸集装箱运输。

（2）专用通道运输能力：单向管，运行速度7km/h，全年可处理200万TEU，若运行速度提高至15km/h，全年可处理400万TEU。

(二) 发展城市地下物流构建城市立体智慧物流配送系统

1. 引言：新制造与新物流

诺贝尔经济奖得主斯蒂格利茨的预言正在成为现实：二十世纪信息技术的创新将在二十一世纪引发社会经济最伟大的变革。

互联网技术正在改革世界，互联网改造的第一个传统行业是商贸流通业：通过电子商务进入销售互联网时代，引发第四次零售业革命。

零售的本职：成本、效率和体验。

第一次零售革命：百货商店。

第二次零售革命：连锁商店。

第三次零售革命：超级市场。

第四次零售革命：建立在互联网电商基础上的新零售，又是超越互联网的一次革命。

电子商务的革命正在循着商业的脉络接口向上下游延伸。向下游延伸的结果带动了现代物流服务由 toB 向 toC 变化，链接新制造与客户端无缝链条新物流；向上游延伸，正在推动工业4.0——智慧制造发展。智慧制造（新制造）的工业革命实现个性化和定制化的制造；toC（面向客户）的物流服务的物流变革也是革命性的；制造端与客户端的无缝链接。

新零售主要有：互联网＋门店客户端、广告即销售客户端、购物卡客户端、微商客户端、社区与社群客户端、银行商城客户端、保险商城客户端、积分换购客户端等。

新零售推动城市物流需求爆发增长，单位与居民物品物流总额近10年增长20多倍，是增长最快的领域。据罗兰贝格等机构分析，同城O2O到2020年将达到1万亿市场规模。

新物流是新零售的基础设施。新销售核心是新物流，没有新物流保证，新销售就是空中楼阁。

新制造与新物流的创新关键是 CPS（Cyber-physical System），是实体世界与虚拟世界完全融合，是具有了网络、通信、计算、控制等智慧能力的物理网络。CPS首先需要通过物联网技术将物理世界网络化，通过大数据与云计算技术推动物理世界智能化。

2. 智慧物流

1）智慧物流定义与内容

新销售的基础设施——智慧物流。

新物流重点在共配，关键在服务。智慧物流是通过大数据、云计算、智能硬件等智慧化技术与手段，提高物流系统思维、感知、学习、分析决策和智能执行的能力，提升整个物流系统的智能化、自动化水平，降低社会物流成本、提高效率，从而推动物流的发展。

城市配送的定义为：城市配送是指服务于城区以及市近郊的货物配送活动。根据客户的要求对物品进行加工、包装、分割、组配等作业，并按时送达指定地点的物流活动。

城市配送之快递配送特点：小批量、多频次、时效性强。

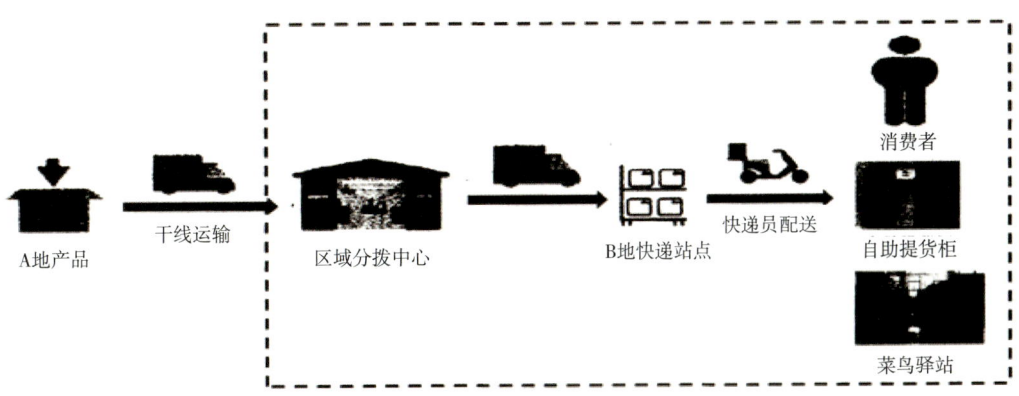

快递末端需要共同配送

收派结构	日均派 1300 票，收 200 票
产品结构	收件几乎全是淘宝件，收件费用 2.8 元 / 件 /1kg 内。散件可以忽略不计
上级网点给的派费	1 元 / 件
给快递员的派费	约 1.05 元 / 件
房租	2160 元 / 月
总体生存状况	生存压力大，亏损严重
对外合作情况	7 家快递企业，4 家电商企业
收取派送费用	1 元 / 件
自提快递费用	0.5 元 / 件
日均派件量	500~600 件之间，峰值 4000 件
雇佣员工	2 个客服、5 个快递员
总体生存状况	生存压力不大，已逐步扭亏为盈

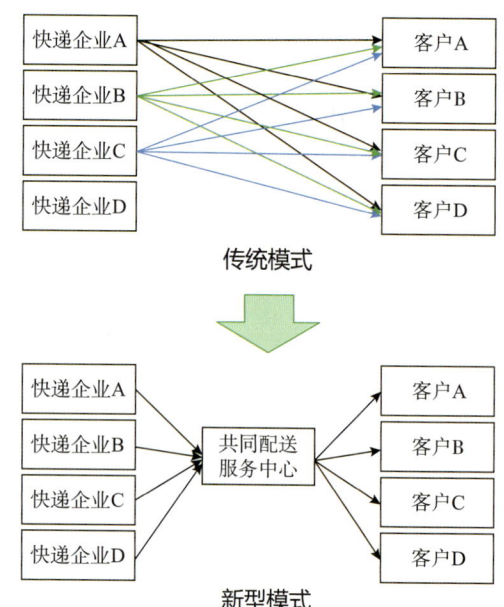

智慧物流的整体架构自上而下体现在三个层面：智慧化平台；数字化运营；智能化作业。

形象地说，如果把智慧物流看作"人"；智慧平台就是"大脑"；数字化运营就是"中枢"；智能化作业就是"四肢"。

（1）智慧化平台——数据驱动，智慧布局

例如通过构建综合评价模型、成本最优模型、站点数量最少模型等多维度模型，基于订单量、路区坐标等输入参数以及传站时间、配送半径等约束条件，采用遗传算法等智能算法进行求解，得出最优的站点数量、每个站点的坐标、平均派送半径等规划决策。在北京市，通过优化已经将 1000km 的距离缩短至 200~500km。

（2）智慧化平台——开放协同，增值共赢

例如菜鸟将加入它平台的物流、仓储等企业的分散资源整合起来，以统一接口面向消费者，使得服务的标准化程度提高。以前消费者、小商家需要面对成千上万个快递公司、仓储企业，现在只需要面对菜鸟即可。

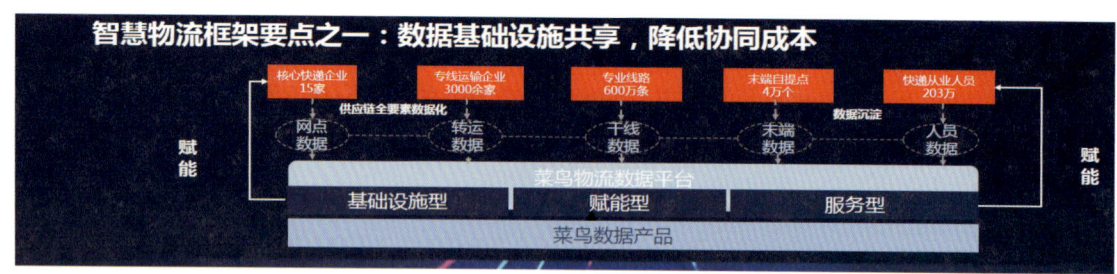

（3）数字化运营——仓储管理系统、运输管理系统等

仓储	智能存储 + 智能拣选 + 智能耗材推荐 +……
	WMS+ 算法 – 最优布局，定位，最短路径，设备调度……
运输	智能调度
	TMS+ 算法 – 动态规划……
配送	智能分拣，路径规划
	配送管理系统 + 算法 – 遗传、蚁群……

（4）智能化作业——菜鸟全自动智慧仓储基地

采用国内目前最好的"货找人"处理系统，消费者在电商平台下单后，仓内同时形成一个贴有货物信息条码的快递箱，扫码后箱子进入自动化轨道"寻找"仓内货物，到达相应货架后，拣货员只需按提示把身后的货物放入箱中即可。

（5）智能化作业——菜鸟机器人

菜鸟网络智慧仓内，单个AGV搬运机器人形同"扫地机器人"。在这里，超过百台这样的机器人。以消费者在猫超下单为例，下单之后，菜鸟网络惠阳智慧仓内的机器人便会接到指令。机器人会自动前往相应的货架，将货架拉到拣货员面前，由拣货员将市民购买的物品放置在购物箱内，随后进行打包配送。目前，机器人与拣货员搭配干活，一个拣货员一小时的拣货数量比传统拣货员多了3倍多。

2）智慧物流无法突破的瓶颈

（1）受交通拥堵、交通管制和气候（雨、雪、台风）影响，配送时间和配送效率的要求无法满足。

（2）由于单位与居民物流总额出现爆发式增长（近10年增长20多倍），同城O2O到2020年将达到1万亿市场规模，城市地面配送使交通拥堵的加剧，使已经进入"拥堵时代"的中国城市无法承受。

（3）在尾气污染已成为城市空气污染元凶的今天，物流的地面货车与摩托车城市配送对城市雾霾的加剧将使得社会无法承受。陈吉宁说，一辆超标重型柴油车相当于200辆小汽车的排放。（2017年1月6日大气污染防治媒体见面会）。

（4）当前最后一公里的快递和外卖主要是摩托车配送，其突出问题一是由于企业、医院。居住小区等均不让快递员进入，在其门口快递车辆乱停乱送，甚至堆满了物品，社会影响很大；二是摩托车快递不遵守交通规则，违规行驶严重，影响了城市交通秩序，造成了交通事故，社会影响严重。

3）城市的地下智慧物流配送系统

地下物流系统是指运用自动导向车和两用卡车等承载工具，通过大直径地下管道、隧道等运输通路，对固体货物实行运输及分拣配送的一种全新概念物流系统。

在城市，地下物流系统与物流配送中心和大型零售企业结合在一起，实现网络相互衔接，客户在网上下订单以后，物流中心接到订单后，迅速在物流中心进行高速分拣，通过地下管道物流智能运输系统和分拣配送系统进行运输或配送。也可以与城市商超结合，建立商超地下物流配送。

地下物流系统示意图

物流配送中心与地下物流系统枢纽相结合，在地下进行客户货物的专业仓储、分拣、加工、包装、分割、组配、配送、交接、信息协同等基础作业或增值作业，在物流规模爆发式增长情况下，有利于城市"减肥"、节能、货物保存。

地下物流系统末端配送可以与居民小区建筑运输管道物相连，最终发展成一个连接城市各居民楼或生活小区的地下管道物流运输网络，并达到高度智能化。当这一地下物流系统建成后，人们购买任何商品都只需点一下鼠标。所购商品就像自来水一样通过地下管道很快地"流入"家中（小区的自动提货柜）。

3. 城市的地下智慧物流配送系统典型案例

1）荷兰阿姆斯特丹地下物流系统（OLS-ASH）

（1）荷兰是世界上最大的花卉供应市场，往返在机场和花卉市场的货物供应与配送完全依靠公路，对于一些时间性很高的货物（如空运货物、鲜花、水果等），拥挤的公路交通将是巨大的威胁，供应和配送的滞期会严重影响货物的质量。

（2）荷兰1997年规划了连接阿姆斯特丹Schiphol机场、Aalsmeer花卉市场和Hoofddorp铁路中转站的地下物流系统，以缓解鲜花市场与机场铁路中转站往返交通的拥堵。然而由于政策支持度不足，且因需要巨额投资而难以找到适用的商业融资模式，该项目在2002年不幸搁置。

2）瑞士地下货运系统（CST）

瑞士地区物流瓶颈导致货物运输的高成本：

（1）每日交通拥堵费高达12亿瑞士法郎；

（2）客运和货运车辆在公路和铁路网络上争夺空间；

（3）交通运输网络低效，空载率高。

瑞士规划了货物地下系统（Cargo Sous Terrain，CST），80km长，直接连接了10个主要物流中心，目前试验段已进入实施阶段。

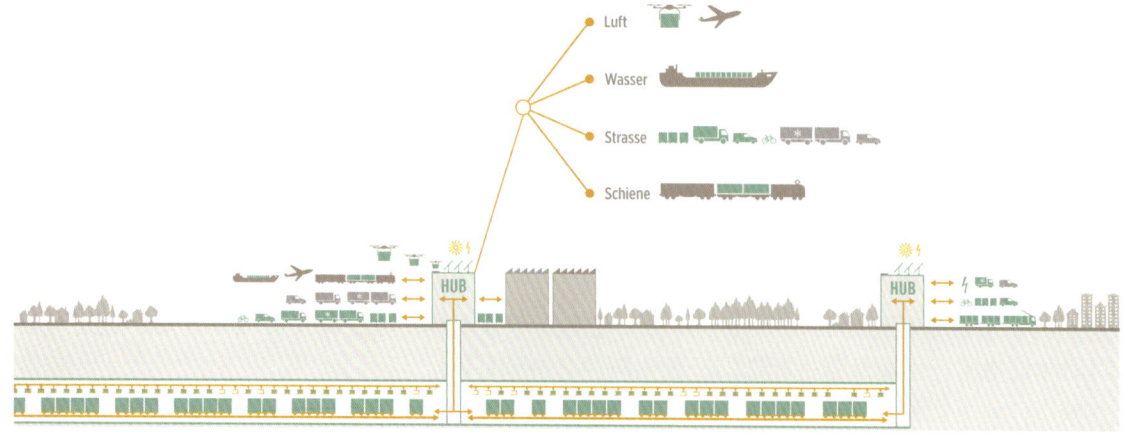

瑞士货物地下系统（CST）

(1)由零售商、邮政、瑞士铁路等共同资助,独立运输设施,全天全自动高效运输,不受地上法规限制;
(2)可与公路,水,铁路和航空网络高效连接;
(3)采用小型运输单元,直接从生产地点向销售网点交货,无需大面积物流中心;
(4)货物在城市郊区进行集散,降低空载率;
(5)中途可搜集废弃物,环境友好,可持续。

七、结语

(1)"规划科学是最大的效益""规划失误是最大的失误""规划折腾是最大的折腾";
(2)"城市工作要把创造优良人居环境作为中心目标,努力把城市建设成为人与人、人与自然和谐相处的美丽家园";
(3)"转变城市发展方式、着力解决城市病,建设和谐宜居、富有活力、各具特色的现代化城市";
(4)世界眼光,国际标准,中国特色,高点定位。

21世纪是地下空间开发利用的世纪

一、引言

20世纪末在巴西的里约热内卢召开了联合国世界可持续发展大会，制定了世界的发展战略——可持续发展。1991年在东京召开的地下空间国际学术会议通过的《东京宣言》指出："二十一世纪是地下空间开发利用的世纪"。2005年党的十六届五中全会明确提出：要加快建设资源节约型、环境友好型社会，促进经济发展与人口、资源、环境相协调。

（1）关于绿色发展理念

党的十八大工作报告提出："坚持节约资源和保护环境的基本国策"。

十八届五中全会提出："推动形成绿色发展方式和生活方式，协同推进人民富裕、国家富强、中国美丽"。

十二届人民代表大会工作报告提出："我们要持之以恒，建设天蓝、地绿、水清的美丽中国"。

（2）可持续发展、绿色发展已成为我国的基本国策

如何建设具有中国特色的资源节约型、环境友好型社会和城市，以及具有高防灾减灾水平的可持续发展的社会和城市，如何实现绿色发展理念，转变城市发展方式，解决城市病，建设天蓝、地绿、水清的美丽中国，成为我国城乡建设面临的重大问题。解决该课题的关键在于运用先进的城乡规划理论、先进适用的科学技术来进行城乡的规划、建设和运营，其中的一个重要方面是必须充分开发利用地下空间。

二、节约土地

1. 紧迫性：土地资源问题

据第六次人口普查的结果，全国总人口为13.397亿人，如此众多的人口需要大量耕地以保证粮食供应问题，国务院提出的我国耕地面积红线是18亿亩。国土资源部数据显示，我国现有耕地为18.18亿亩（2010年）。

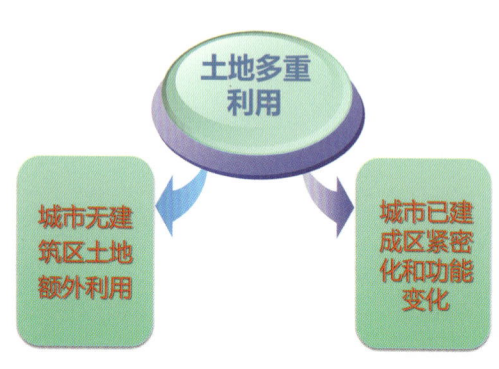

按照联合国的标准，每一个城市人口约需要100平方米的建设土地，中国的城镇化速率约为1%，那么每年约需要使用200万亩土地以满足城市化发展的需求，如果不考虑土地复垦，土地余量仅仅能支撑3~5年的城市化进程，城市建设将面临无地可用的困境。

因此，必须严格控制土地需求，科学合理地利用土地。

2. 节地方向

城市节地的一个重要方面在于宏观上努力实现土地的多重利用。土地的多重利用可沿两个方向实施。

报告时间：2017年7月25日。

开发利用地下空间，把城市交通（地铁和轨道交通、地下快速路、越江和越海湾隧道）尽可能转入地下，把其他一切可以转入地下的设施（如停车库、污水处理厂、商场、餐饮、休闲、娱乐、健身等）尽可能建于地下，实现土地的多重利用，提高土地利用效率，实现节地的要求。

3. 世界眼光

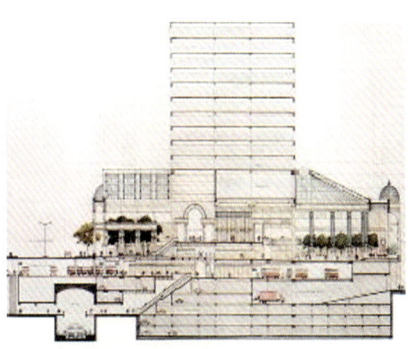

加拿大蒙特利尔和多伦多——地下城

法国双层城市——拉德芳斯新城

法国巴黎——卢浮宫地下空间开发

法国巴黎地下商场　　　　　　德国法兰克福火车站

芬兰赫尔辛基游泳池　　　　　斯德哥尔摩伯尔瓦尔德地下音乐厅

挪威霍姆德亚地下运动厅和游泳池　　　芬兰赫尔辛基地下运动场

芬兰国家技术研究院地下实验室

哈佛大学普塞图书馆

美国明尼苏达大学地下系馆

法国国家图书馆

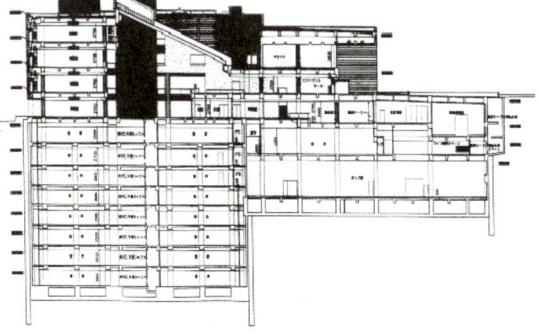

日本地下8层国立国会图书馆

4. 中国特色

1) 修建城市湖下隧道

中国有很多具有城市湖泊的城市，都在湖下修建城市道路，如南京、苏州、无锡、南昌、武汉、杭州等。

（1）南京玄武湖、九华山隧道

南京玄武湖隧道长约 2.68 公里，其中暗埋段长 2.23 公里。九华山隧道全长 2780 米，包括山体隧道和湖底隧道两段。

（2）苏州独墅湖隧道

苏州独墅湖隧桥工程全长 7.37 公里，其中隧道部分长 3.46 公里。

（3）无锡蠡湖隧道

无锡蠡湖隧道全长 1180 米，其中穿越湖中暗埋段长 880 米。

（4）南昌青山湖隧道

工程全长 1965 米，隧道主体工程长 1070 米，其中湖底暗埋段长 550 米。

（5）武汉水果湖隧道

隧道全长 1735 米（含引道），其湖中段长 621 米。

（6）杭州西湖（湖滨）隧道

西湖隧道分东西两线，西线全长 1415 米，东线隧道全长 1335 米。其中湖底部分长约 800 米。

（7）扬州瘦西湖隧道

扬州瘦西湖隧道 3.6 公里，设计为单管双层双向四车道。其中，主体隧道段长 1.28 公里，设计速度为 60 公里/小时。

2) 城市地下立体车库

已建成居民小区紧密化和功能变化的模板，如江苏南通智能地下立体车库。

3) 城市地下变电站

500 千伏静安（世博）输变电工程占地约 13300 平方米，总建筑面积 57615 平方米，其中地下建筑面积 55809 平方米，地上建筑面积 1806 平方米。该变电站是容量最大、电压等级最多、接线最复杂、设备最先进、安全监测系统最完善的中国首座大型超高压地下变电站。

江苏南通智能地下立体车库

逆作结束地上结构施工实景

内部设备实景

500 千伏静安（世博）输变电工程

4）城市地下综合体

（1）上海市人民广场

上海市人民广场的地下，有地铁车站、地下变电站、地下商城、地下车库和香港名品街等市政、商业设施。

其中，上海市人民广场地下商城是中国目前最大的地下商业中心，总面积4万余平方米。地下车库共分为7个区域，拥有600多个车位。

（2）上海世博园地下空间规划

上海世博园区地下空间规划是结合上海世界博览会特点，实现"城市——让生活更美好"的主旨。

地面为广场道路，供游客游览集会；地下一层为地下街，沟通枢纽和主要场馆；地下二层为机动车通道，连接过江通道，同时弥补沿江通道容量的不足。地下空间规划还包含了园内四个集散枢纽、园外三个大型综合型集散换乘枢纽，园内枢纽通过世博大道进行连接。

地下街为结合通道、休闲、购物、展览等多种功能复合的综合型通道模式。

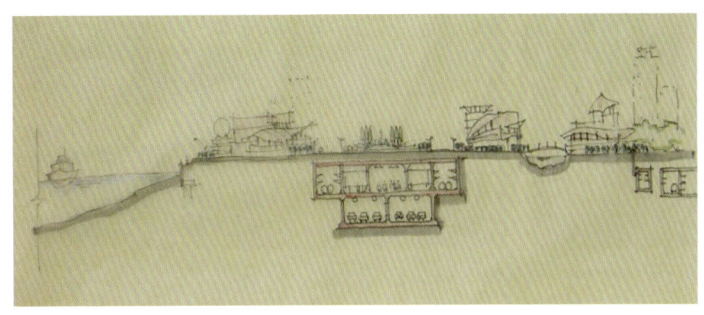

（3）北京国贸中心

国贸中心是典型的建成功能区更新改造的样板。

原本地铁1号线、10号线的两层换乘通道，改建成具备完整功能的地下四层建筑综合体，在这栋地下宫殿中换乘地铁、出入地面、进入国贸，都有宽广空间。

（4）中关村西区地下空间规划

中关村西区采用立体交通系统，实现人车分流，各建筑物地上、地下均可贯通。地下一层的交通环廊全长1.9公里，连接着中关村西区的14座大厦，有10个出入口与地面相连，13个出入口与单体建筑地下车库连通，共规划建设了10000个机动车停车位；地下二层为市政综合管廊的支管廊；地下三层主要是市政综合管廊主管廊。

（5）西安市幸福林带建设工程项目

跨越西安市新城区、雁塔区，规划长度约6公里，平均宽度200米，项目总用地面积约120万平方米。幸福林带地下空间分为三层，总建筑面积72万平方米。其中，地下一层为商业配套，地下二层为地下停车场，地下三层为地铁轨道层。

5)城市综合交通枢纽

立体交通、无缝换乘。

(1)上海虹桥枢纽

虹桥综合交通枢纽是高速铁路、城际和城市轨道交通、公共汽车、出租车及航空港紧密衔接的国际一流的现代化大型综合交通枢纽。建筑面积约100万平方米,其中地下空间面积超过50万平方米。

(2)北京南站

北京南站是集普通铁路、城市轨道交通与公交出租等市政交通设施于一体的大型综合交通枢纽站。地下一层为整个车站的换乘空间以及旅客出站系统,面积119940平方米,是北京南站的枢纽。另外上下两层设停车泊位909个。

(3)深圳福田地下交通枢纽

深圳福田地下交通枢纽国内最大"立体式"交通综合换乘站,汇集了地铁2号线、3号线、11号线,以及广深港客运专线福田站,是集城市公共交通、地下轨道交通、长途客运、出租小汽车及社会车辆于一体的并与地铁竹子林站无缝接驳的立体式交通枢纽换乘中心。地下枢纽空间总建筑面积约13.73万平方米,相当于192个足球场这么大。

 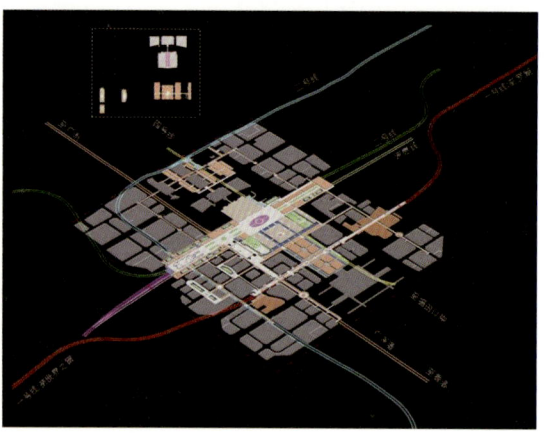

6)地下新城

(1)广州珠江新城

核心区地下城建筑面积60万平方米。

(2)杭州钱江新城

杭州钱江新城核心区地下城,以波浪文化城(10万平方米)和地铁1、2号线换乘站为骨干,形成地下城,地下空间总量达到200万平方米。

核心区地下空间剖面图

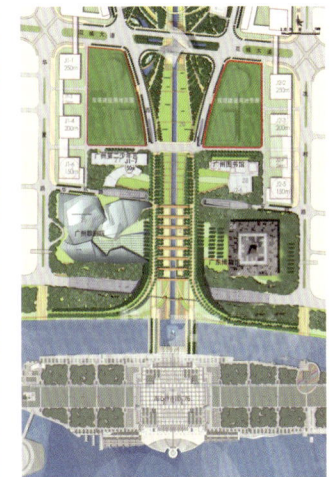
规划图

（3）武汉王家墩商务区地下城

王家墩商务区地下空间建设，包括地下交通、地下市政设施、地下商业和地下防灾等4个部分。其中，地下交通设施面积为196万平方米，包括7.6万个车位以及众多地下道路、地铁站点等；商业面积为46万平方米，包括商业步行街、商场和商铺；地下人防面积为3万平方米；其他各类设施面积为17万平方米。

5. 发展新趋势

成片开发，统一规划、统一设计、统一施工和统一管理。

（1）福州宝龙万象广场

福州宝龙万象广场坐落于福州市万宝商圈核心位置地下，东西长 1100 余米，南北宽 400 余米，地下三层结构，总建筑面积近 20 万平方米（不包括地铁站），开挖深度约 16 米。该工程系国内首座大型融合式人防工程。

综合体内设福州轨道交通 2 号线宁化站（地下三层岛式车站，开挖深度 24 米）、地下双向四车道快速路以及地下车库和地下商业。

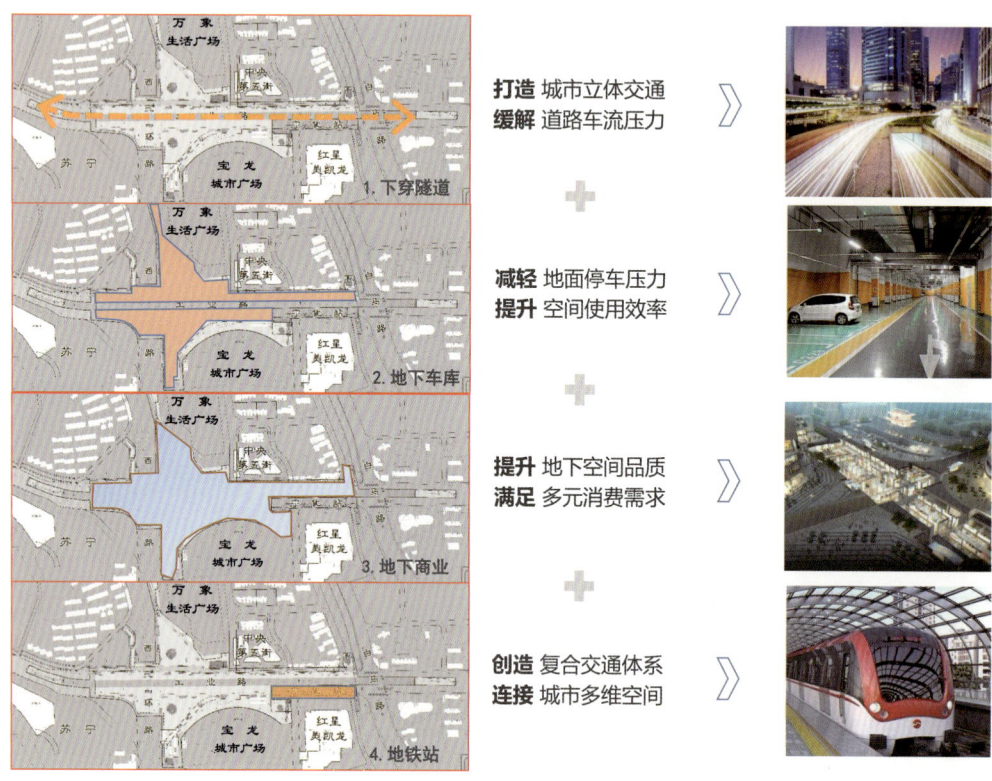

空间规划思路

（2）北京中心商务区（CBD）地下空间

构建"一轴、一区、两点、三线"的 CBD 地下空间结构。

一轴——地下公共空间发展轴（东三环发展轴）。

一区——地下空间核心开发区域（国贸与 CBD 核心区）。

两点——以地铁国贸换乘站、10 号线光华路车站地下空间主要集散点。

三线——地下空间主要公共联络线（长安街、光华路、商务中心区东西街）。

①创造现代时尚的城市活力空间

重点规划区内地下商业相互连通，形成网络。

人行通道两侧设置为商业设施（餐饮、休闲及便利服务设施），且连续设置。从地铁出口可通每一座大楼。

②建立资源共享的地下停车系统

相邻建设用地的地下车库之间尽可能地相互连通。

CBD 核心区地下二层为车行联系层，利用城市道路下的空间建设地下机动车输配环，缓解地面交通的拥堵。

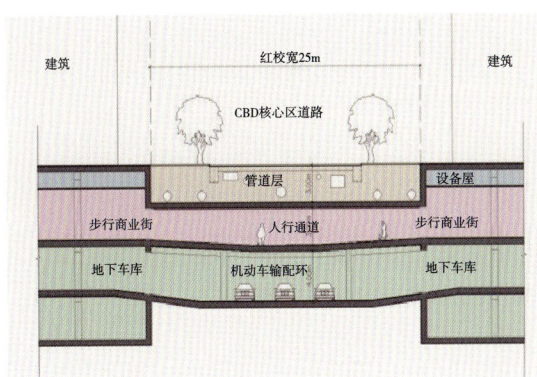

输配环剖面示意

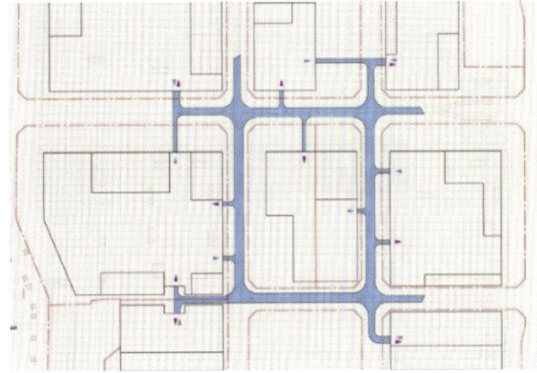

CBD核心区地下机动车输配环系统

（3）上海后世博地下空间开发

①世博园区地下空间开发采取了统一规划、统一设计、统一建设、统一管理的模式（简称"四统一"）。

②已启动建设的A、B片区规划在建建筑面积共170万平方米，其中地下空间77万平方米。

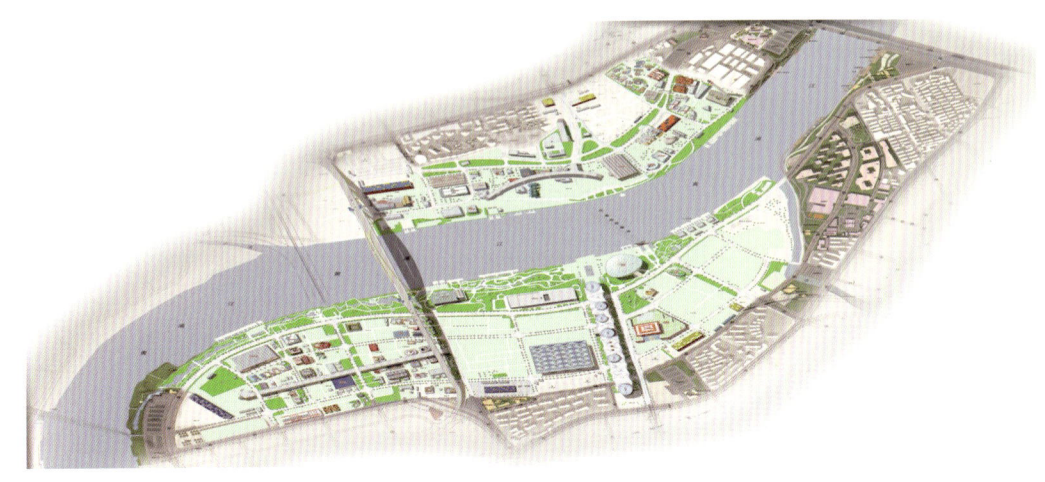

③A、B片区地下空间"四统一"开发建设已经走了三步。

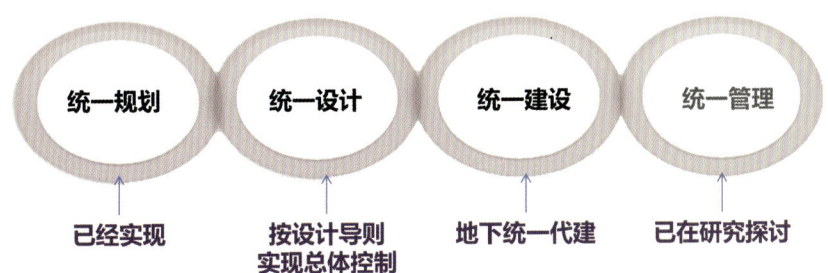

（4）宁波南部商务区地下空间开发

南部商务区在国内首创"统一规划、单体自建、公基统建"的开发建设模式，由政府确定区块规划设计的总体方案，业主单位在同一时间内完成单体建筑方案设计，公共设施由区城投公司统一建设，实行共用共享。

宁波南部商务区"整合建管"模式在国内尚属先例，具有创新性、实用性和示范性，并具有强大的生命力，可能成为未来发展的新趋势。

其核心内容与特色如图所示。

分层出让	• 地下建设用地使用权登记以宗地为基本单位,通过批地水平投影坐标、竖向高程界限确认权属范围,实行分层登记,并在产权证中注明"地下空间",属于人防工程的注明"人防工程"
统一规划	• 引领集中开发片区公共用地和权属用地中的人防工程和普通地下空间设施实现互联互通、成片连网
统一配建	• 地下空间内的人防工程和市政公用设施统一配建,实现地下空间使用功能与工程体系建设最优化、开发成本最低化
统一分配	• 按照地面建筑面积对地下空间开发的投资和地下空间的使用收益尽可能公平地分配出售地下停车场库
统一管理	• 由政府委托一家物业单位对地下空间的使用安全、设备维护、组织协调等全权负责,实现"整合建管"模式综合效益的最大化

(5)南京江北新区中心 CBD 地下空间项目

背景:南京江北新区为第十一个国家级综合开发区。

南京江北新区中心区占地面积约 16 平方公里,其中 CBD 区域占地面积约 7.6 平方公里。

江北新区中心区 CBD 地下空间一期建设项目范围:总用地面积 48 公顷,区域内共计 24 个地块,规划商办用地 21 块,广场绿地 3 块。

江北新区中心区 CBD 地下空间一期建设项目总建筑面积约 100 万平方米。

其中:商业 24 万平方米,停车场 59 万平方米,设备 16 万平方米,市政配套 1 万平方米。

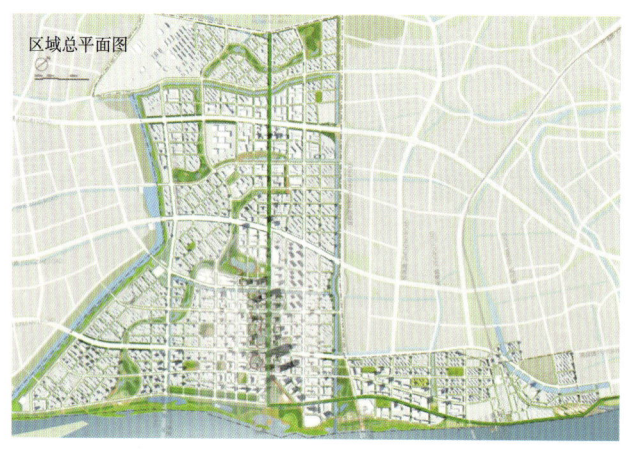

地下空间整体开发有如下配置:

① –6m 市政管廊层;

② –12m 地下步行者空间;

③ –18m 地下车行空间。

其下为三线换乘轨道交通站台及设备空间最深处挖至 –36m。

（6）南京南部新城中片区 EPC 项目

南京市南部新城地下空间成片开发，定位为南京都市新中心、人文绿都窗口。历史文化背景：民国时期大校场机场、美龄宫、明代七桥瓮均位于此处，成为民国文化、明文化、秦淮文化的交汇点，距离南京市高铁南站仅 1 公里。

中片区 EPC 项目内容：地下综合管廊 5 条（8.9 公里）、旧机场跑道历史文化保护（机场跑道长 2600 米，宽 62 米）、跑道下方地下空间综合开发利用（约 19 万平方米）、地下商业面积近 2 万平方米。

三、节能减排

（1）土地的多重利用有助于城市"减肥"。

相对减少城市面积，制止城市超限扩展，建成"紧凑型"的城市结构。国际上提倡的"紧凑型"城市减少了居民的出行距离和机动交通量，相对降低对机动交通，特别是私人轿车的依赖度，相对增加居民的步行和自行车的出行比例，这一切将导致能耗大户——交通能耗的降低，实现城市节能的要求。

（2）掩土住宅、新型窑洞以及地下冷库实现节能。

中国的新式窑洞实现了能源的节约，又改进了阳光和通风。

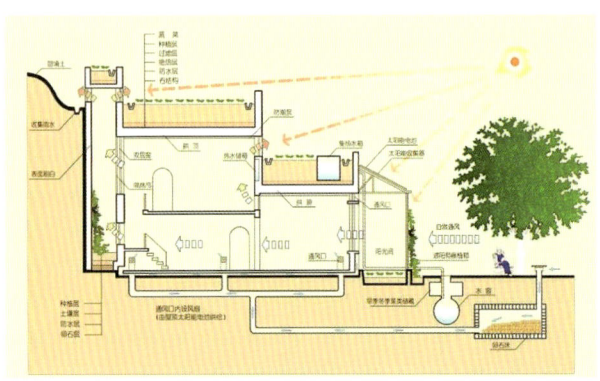

新型窑洞建筑设计原理图　　　　　　　　　　延安大学学生宿舍

（3）地下空间为可再生和洁净能源利用开辟一条有效途径。

地热能包括浅层地温能及中、深层地热能。前者利用地表土壤、地下地表水作为换热介质的节能措施，后者是利用中深层地下热源进行直接供暖或发电的措施。

（一）地温能开发一体化技术

利用低密度地热能源，节能减排。

到达地下一定深度时（5米以下），四季的地层温度保持在一稳定值。

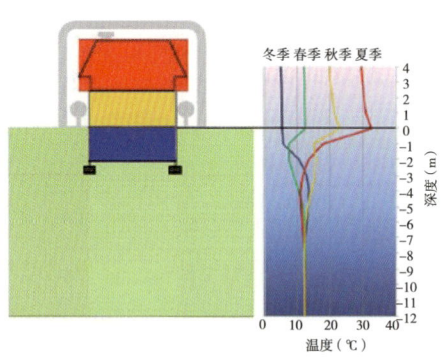
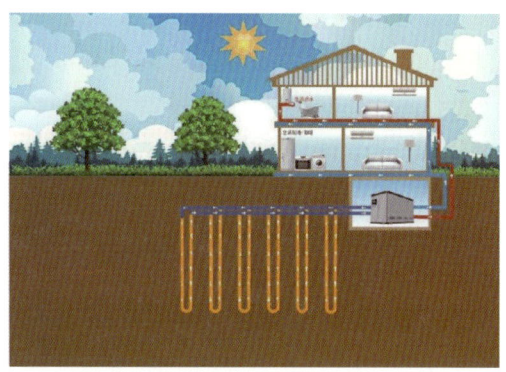

地温能指的是地层中温度小于25℃的地层热能。它把传统空调器的冷凝器或蒸发器直接埋入地下，利用传热循环介质与大地进行热交换，从而提取地温能。

地下换热系统可埋设在地下结构基坑围护结构（地下连续墙、排桩内）、基础底板下、桩基（钻孔灌注桩、预制桩、PHC桩等）内；可埋设在新奥法施工的隧道衬砌内或以能源锚杆的形式埋设在其围岩中；也可埋设在地铁区间隧道内、地下输水管道内。

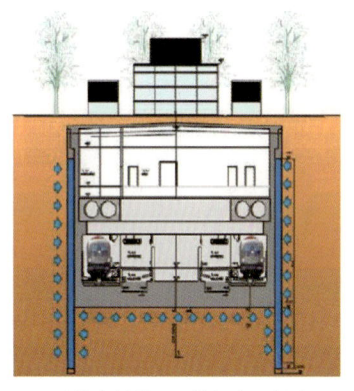

地铁车站的地温能提取示意图
（资料来源：Brandl, 2016）

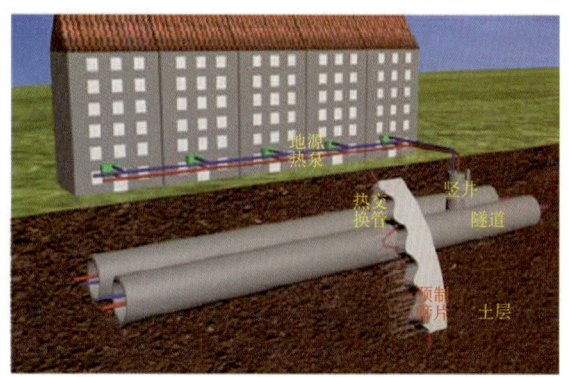

隧道结构内埋管换热器工作示意图

（二）地温能利用的典型案例

1. 阿布扎比的地源热泵系统

通过采用竖直的地源热泵线圈，来节约成本和节能。在建筑物基坑桩基里面安装地源热泵管道系统。

2. 奥地利政府资助的能源地铁车站示范工程

该车站位于 Lainzer 隧道 LT24 区，在一共 59 根桩基内埋设热交换管，钻孔灌注桩桩径为 1.2 米，桩长平均为 17.1 米，利用 6 台热泵为附近一所学校供暖。在长期供暖的情况下，车站能提供 150 千瓦的热负荷。一个供暖季度可提供，214 兆千瓦的能量，天然气的使用量每年减少 34000 立方米，使每年二氧化碳排放量减少 30 吨，与传统的靠燃烧天然气供暖的方式比较，可使学校每年用作取暖的费用降低 1 万美元。

3. 瑞士 Grabs 的 PAGO 公司办公楼

采用 570 根桩基内埋设热交换管，平均桩长 12 米，以 4 个能源桩为一组，呈方形顶角安装，四边间距为 1.4 米。每延米桩基冬天可获得 35 千瓦时的热量，夏天获得 40 千瓦时的冷量。

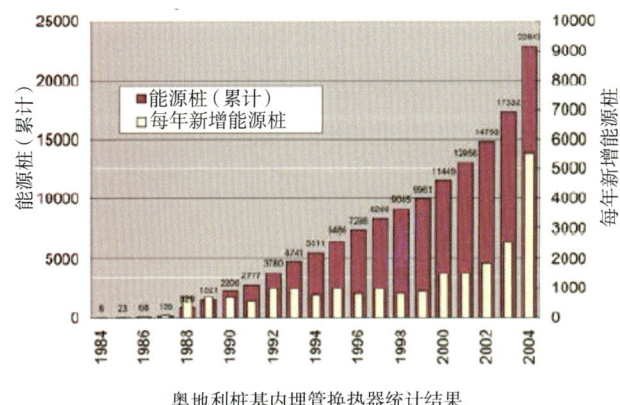

奥地利桩基内埋管换热器统计结果
（资料来源：Brandl，2006）

瑞士 Grabs 的 PAGO 公司办公楼

4. 上海自然博物馆新馆

建筑总高度 18 米，地上三层，地下两层，采用地源热泵系统承担建筑部分夏季冷负荷以及冬季热负荷。工程采用灌注桩与地下连续墙内埋管两种形式，其中，灌注桩埋内埋管 393 个，平均桩长为 45 米，采用 W 形埋管；地下连续墙内埋管共 452 个，采用 W 形埋管。

上海中心大厦裙楼、上海世博轴和上海富士康大厦等国内重大工程均采用了地下结构内埋管热交换系统。

5. 上海世博轴节能技术

采用直接式江水源热泵系统和桩基埋管地源热泵系统。

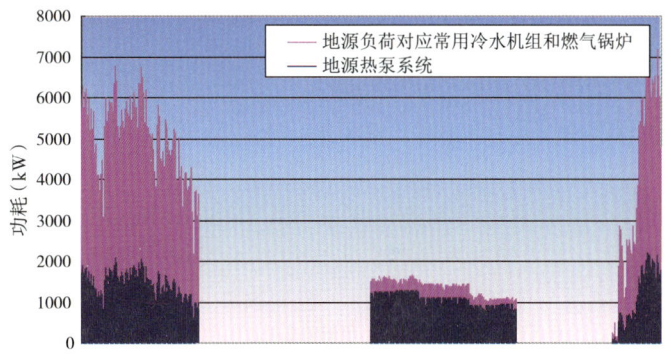

全年地源承担负荷部分的冷热源功耗比较

世博轴节能情况

项目	节能方法	节省用电量	减少二氧化碳排放	节能率（%）		
				夏季	冬季	全年
江水源地源热泵系统	利用地表水和地下的可再生能源（冷量和热量）	562.9 万千瓦时 / 年	5629 吨 / 年	26.96	71.03	61.40

(三)地温能使用总体情况分析

国内外实践证明,利用地源热泵技术开发浅层地温能为建筑物供热、制冷是实现节能减排的有效途径。

1. 国外使用情况

到 2005 年底,已有 33 个国家在推广这项技术。据 2010 年世界地热大会的统计数据,地源热泵的年利用能量达到了 214782TJ,与 2005 年世界地热大会的统计数据相比,5 年内增长了 2.45 倍,平均年增长率达到了 19.7%;地源热泵设备容量为 35236 兆瓦热量,平均年增长率为 18.0%。地源热泵被称为当今社会可利用的最高效的供冷供热技术。

2. 国内使用情况

截至 2012 年底,从全国范围看,现有工程数量已达 23000 多个,总面积 24000 万平方米。项目比较集中的地区有北京、天津、辽宁、河北、河南和山东,78% 的项目集中在华北和东北地区。

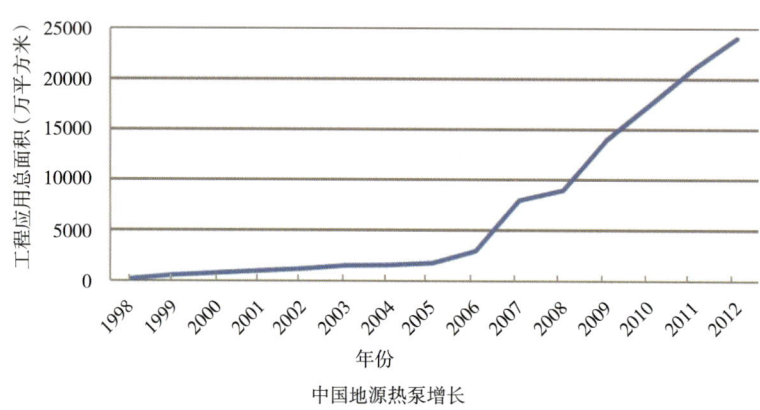

中国地源热泵增长

京津冀规划区 13 个地级以上城市浅层地温能每年可开采热量折合标准煤 9200 万吨,可实现建筑物夏季制冷 35 亿平方米,冬季采暖 29 亿平方米。目前京津冀地区采用浅层地温能供暖制冷面积 8500 万平方米,占全国的 20%。是我国浅层地温能开发程度最高、用于建筑物供暖制冷规模最大的地区之一。

(四)中深层地热能的利用

中、深层地热能同浅层地温能的区别在于,中、深层地热能通常作为冬季供暖的直接热源而不是换热介质。

地下水直接供暖的开凿井深通常要超过地下 1500m。其工程体量庞大,适用于区域集中供暖。直接供暖抽取的地下水使用后,需要进行回灌处理,维持地下水位稳定,维护地质安全。

1. 国外应用情况

德国到 2002 年,有 9 个集中供热站,其地热井深度从 1100~2400m 不等,总供热量 136 兆瓦,用于采暖、理疗、温室等。特点是:多采用热水供暖,大型供热站采用热水供暖系统时,由于热泵用电,引用了"季节特性系数",即供热量与消耗电量之比,一般为 5~7。

美国地热能利用逐年增长:美国不仅利用地热进行供暖,提供 500 兆瓦的热量,同时还利用地热进行发电,约 2,800MW 的电满足美国西部和夏威夷 4000000 人。

2. 国内应用情况

河北省地热能资源总量位居全国第二位,2015 年地热资源开采量突破了 1.1 亿立方米,地热供暖面积达到了 6300 万平方米。其中,雄县位于河北省保定地区,是国内首个通过地热供暖实现"无烟城"的县城,拥有享誉全国的"雄县模式"。雄县地热资源分布面积广,出水量大,水温高,现在地热集中供暖面积已占城区集中供暖的 85%,覆盖县城 80% 以上的居民小区,每年可减少二氧化碳排放量 12 万吨。

在收费方面，雄县居民地下水取暖收费为 16 元 / 平方米，相较之前燃煤取暖 25 元 / 平方米的费用更为便宜。

从投入与产出方面，地热取暖前期投入较大，后期的年收益率稳定。推广和提升"雄县模式"，在集中成片供暖的基础上，在新兴城镇中打造以地热为主的"绿色热网"，解决北方中、小新兴城市和农村冬季供暖的问题，减少燃煤燃烧对空气造成的污染。

（五）干热岩资源利用

中国大陆 3km 至 10km 深处干热岩资源总计为 2.09×10^{17}EJ，合 7.149×10^{14}t 标准煤，若按 2% 可开采资源量计算，相当于中国大陆 2010 年能源消耗总量的 4400 倍。

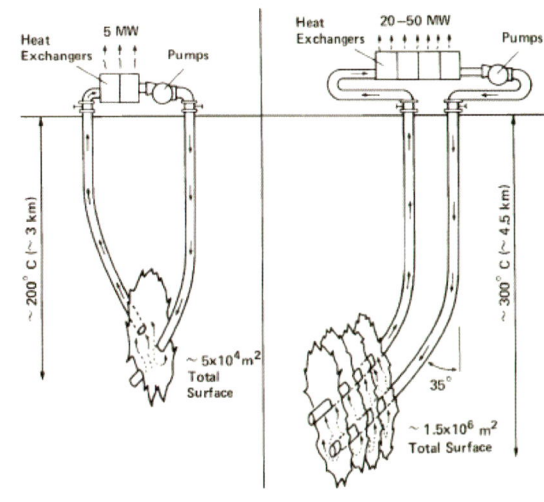

我国已成功在陕西省内进行了干热岩用于供热的商业应用——长安信息大厦 2013 年共计 3.8 万 m^2 中应用干热岩供热，效果良好。

国际上有美国、澳大利亚、日本、德国、法国等进行了干热岩发电试验研究项目。美国 Los Alamos 实验室在卡尔德拉的芬登山上建成了一个 10MW 的 HDR（深层干热岩）发电站，该电站主要由两个深度为 3 千多米的钻孔及其连通孔组成，冷水由一个钻孔灌入，另一个孔产生 200℃蒸汽，进入汽轮机发电。

羊八井地热电站 1975 年 9 月 23 日发电成功，是中国最大的地热能发电站。位于西藏自治区当雄县境内羊八井的地热发电站，海拔 4306m，其地热田地下深 200m，地热蒸汽温度高达 172℃。由 5 眼地热井供水，单井产量为 75～160m^3/h，水温为 145～170℃。每年二、三季度水量丰富时靠水力发电，一、四季度靠水热发电，能源互补。

至 1991 年陆续完成另 8 台 3 兆瓦机组（2.4 万 kW）。此后维持总装机容量看 24.18MW，占拉萨电网总装机容量的 41.5%，在冬季枯水季节，地热发电出力占拉萨电网的 60.0%，成为其主力电网之一。

（六）不稳定可再生能源的储存

太阳能、风能、潮汐能、冰能等不稳定能源需要季节性、昼夜性储存。在地下岩、土介质中修建热水洞库、水库、压缩空气储库和超导磁储电库是最佳的选择。如德国在地下岩层中建成储气压力 8MW 的 29 万 kW 的地下压缩空气库。我国正在江苏金坛地区论证建设压力为 5.6~8MW 的 4×15 万 m^3 的 30 万 kW 的地下压缩空气库供调峰电站用。

四、节约水资源

1. 水资源短缺的形势

中国是世界13个贫水国之一，人均水资源量仅为2251立方米，居世界第110位，是世界平均水平的1/4。据统计，我国每年缺水400亿立方米，其中城市每年缺水60多亿立方米，农村缺水300多亿立方米。在全国666座建制市中，有近400座城市缺水。

建设节水型城市要"开源与节流并重"，关键是要建立城市水系统的良性循环机制，实现水资源的可持续利用。在这方面外国的先进经验是：充分利用雨水、开发污水再生利用和建立地下"水银行"调节和缓解城市供水。这些经验的实施都需要开发利用地下空间。

2. 利用雨水

日本、瑞士和牙买加等国利用屋顶收集雨水，并通过管道将雨水输送到地下储水库。

日本雨水利用的供水量每年为0.05亿立方米。东京都在公园、校园、体育场、停车场等处的地下，修建了大量的雨水储留池。凡是新建筑物，包括住宅楼，都要求设置雨水储留设施。1989年开业的东京港区的野鸟公园，园内用水皆来自雨水，形成了湿地、芦苇荡、草地、树林等景点，成为东京地区的著名观光点之一。名古屋、大阪、福冈等地的大型建筑物下都设置了雨水利用装置其中，名古屋体育馆每年积蓄雨水3.6万立方米。东京江东区南沙地区就建立了雨水调整池，其中最大的一个调整池一次最多可以存储2.5万立方米的雨水。

西沙群岛上修建了可采集雨水达14万吨的地下储水工程。通过技术处理，已达到了国家饮用水卫生标准，从而结束了吃水靠大陆船运的历史。

北京每年6—9月份降水中，可利用的雨水为2.3亿立方米，相当于城区全年用水量的五分之一多。为积蓄雨水，北京将计划开工修建70个地下小水库，在新建公园将首先考虑建设雨水收集利用设施，可工程拦蓄洪水3559万立方米。

香港跑马场地下藏着"大水库"：游乐场下正藏着一座精心设计的宏大蓄洪设施，蓄水量相当于24个标准游泳池。渠务署于2012年在跑马地游乐场展开地下蓄洪计划。整项工程分两期进行，建筑内容包括蓄洪池、雨水泵房、长约1.1公里的接驳渠道以及相关绿化园景工程，预计2018年雨季前全部完成。香港近20年来兴建了约250亿港元防洪工程设施，还有120亿港元的工程正在规划设计和建造当中。

（1）世博轴雨水收集处理综合利用技术研究

世博轴自来水日用量约为2000立方米/天，利用雨水时，自来水日用水量降为1100~1200立方米/天，回用雨水用水量为800~900立方米/天。

据此，在可利用时段里，自来水替代率为（800~900）/2000=40%~45%。主要用途为卫生器具冲洗、绿化浇灌等。

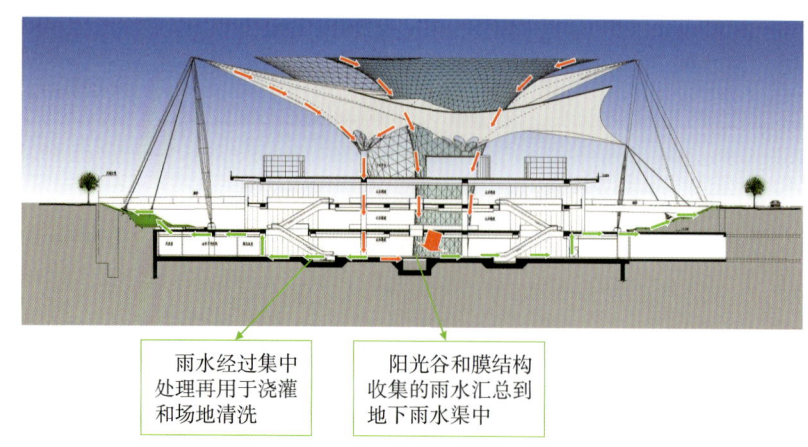

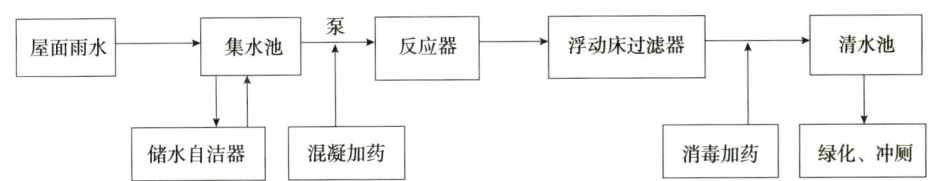

世博轴雨水处理工艺流程图

(2) 污水资源化，中水利用

中水利用在很多国家被采用，如日本每年将经处理的1.3亿立方米的再生水用于工业或绿化用。

北京2010年再生水利用量达6.8亿立方米，占总用水量的19%，但利用率仅为60%，"十二五"期间计划将全市再生水用到10亿立方米以上，利用率提高到75%。

天津市自2002年正式启动中水管网建设以来，已铺设了400多公里的中水管道。目前天津市已建成再生水厂8座，全年中水回用2500万吨左右，大部分用在大型的工业项目中，还有河西梅江居住区和南开水上公园周边一些小区用上了中水。

(3) 利用地下构造储水

建坝修水库必然会淹没耕地和村镇，从而动迁居民，引起库区的森林淹没和泥沙淤积以及下游河道冲刷，从而影响生态和环境，包括珍稀动植物。甚至会诱发地震和山体滑坡。对于欧美的"废坝运动"，要警惕有人别有用心怀有政治目的，他们"拆除"的大坝实际是拆除病坝和废坝，更学习他们代替建坝修水库储备水资源的科技新思路。

欧美发达国家现在已经基本放弃修建地表水库来储备水资源的传统做法，而是利用地下含水层，建立"水银行"来调节和缓解供水。包括地表入渗和井灌的人工补给是一种可行、费用低廉的解决供水的办法。

它一举多得：

①储备地下水，在水短缺的时候提供水量满足用水需求；

②控制由于地下水位下降引起的海水入侵和地面沉降；

③提高地下水位，减少地下水的抽取费用；

④维持河流的基流；

⑤通过土壤中的细菌作用、吸附作用和其他物理、化学作用改善水质；

⑥通过处理后的污水入渗来实现污水的循环再利用，以管理不断增加的大量污水；

⑦最后也是最重要的是保护了生态环境。

20世纪80年代开始，美国开展了钻孔补给含水层的恢复工程（ASR系统）。美国已运行的系统有18个，正在建设之中的ASR系统有40多个。瑞典、荷兰和德国的AR工程在总供水中占20%、15%和10%。

荷兰到1990年，地下水人工补给量达到1.8亿立方米/年。伦敦利用地下含水层人工补给后提供水资源90兆升/天。

在干旱和半干旱的中东国家，如约旦、科威特、摩洛哥和以色列，都在开展城市污水经处理后进行地下水补给工程。

阿曼从20世纪80年代初，开始在海岸平原和冲积干谷地区通过地下补给坝截获洪水，进行地下人工水补给工程。

神华神东矿区建成35座煤矿地下水库，储水总量约为3100万立方米，年供水量超过了6000万立方米，可供应神东矿区95%的用水，近三年因此而创造的经济效益超过30亿元。

科学利用地下储水空间，实施地下水资源战略储备，是实现水资源联合调蓄的关键。

北京的地下水储藏回补条件非常好，可建设永定河地下水库、潮白河地下水库、沟错河地下水库、温榆河地下水库及大石河地下水库等五大地下水库，作为2008年"后奥运时代"水资源应急储备和战略储备的空间，总库容可达47亿立方米。

山东省济南市从卧虎山水库经玉符河放水0.1亿立方米，通过强渗漏带渗入地下，并在24个强渗漏带建立了53个地表水拦截坝，拦截了南部山区每年白白流失的数亿立方米的地表水，使其渗入地下水库，其成功实施使消失多年的趵突泉焕发了新的生命。

五、城市地下空间开发利用是"转变城市发展方式"解决"城市病"的主要着力点，是"建设和谐宜居、美丽的环境友好型城市"的主要途径

（一）现状

城市交通拥堵、空气雾霾以及雨洪内涝之严重，已成为建设我国环境友好型和谐城市的最强不和谐音符。

过去注意高层建筑、大广场、宽马路城市"面子"建设的粗放发展模式酿成"里子"城下基础设施建设短板的恶果。

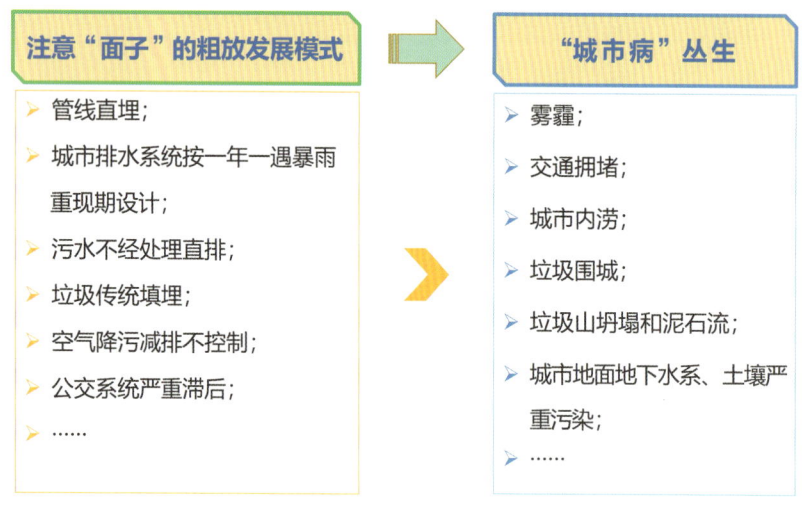

1. 紧迫性

2015年荷兰导航经营商TomTom发布了全球最拥堵城市排名前三十名中，中国大陆有十个城市上榜。在中国79.3%的城市居民每天上下班时感觉交通拥堵，33.6%的人感觉"非常堵"。

据中国交通部发布的数据显示，交通拥堵带来的经济损失占城市人口可支配收入的20%，相当于每年国内生产总值（GDP）损失5%~8%，每年达2500亿元。

中国科学院可持续发展战略研究组组长、首席科学家牛文元指出，中国15座大城市的居民每天上班比欧洲发达国家多消耗28.8亿分钟，每天因交通拥堵损失近10亿元。

2. 城市交通拥堵

据公安部交管局于2010年底发布的数据显示，至2010年9月底，全国667个城市中，约有三分之二

的城市交通在高峰时段出现拥堵。我国的许多大城市交通拥堵问题相当严重，已成常态且有加剧的趋势，一些二、三线城市也急速驶入"拥堵时代"。

北京交通发展研究中心2016年1月发布的《北京市交通运行分析报告(2015年)》提供的数据如下：2015年，北京交通拥堵较2014年有所加剧，日均拥堵时间(中度拥堵、严重拥堵)共计3小时，与2014年的1小时55分钟相比大大加长。

城市交通拥堵的新特点如下：

①全国城市交通进入"拥堵时代"：667个城市(包括二、三线城市)，约有三分之二的城市在高峰时段出现交通拥堵。

②交通拥堵由中心城区向郊区拓展，拥堵区域更大。

③拥堵时段由高峰向其他时段拓展：白天堵，晚上也堵。

④拥堵天数增多：工作日堵，周休日更堵。

3. 城市空气污染

《中华人民共和国国家环境分析》报告称，中国500个大型城市中，空气质量达到世卫组织推荐标准的不足5个。世界上污染最严重的10个城市之中，有7个位于中国。2013年1月席卷我国中东部的雾霾天气，在全国74个监测城市中，有33个城市的部分检测站点检测的空气质量指数超过300，即空气质量达到了严重污染，北京PM2.5浓度连续多天爆表。

根据环保局统计数据，2016年全国地级及以上城市空气质量总体呈改善趋势，但是进入11月份后，京津冀及周边地区连续发生三次大范围重污染天气，东北地区多个城市也出现了极端重污染天气过程；此外，与去年同期相比，全国338个地级及以上城市优良天数比例下降7.5个百分点，重度及以上污染天数比例上升0.3个百分点。

空气污染新特点如下：大力治理工业污染后，空气污染仍然很严重，特别是中国北方地区：11月27日至12月2日，12月8日至13日，北京地区PM达500，部分地区达900。沈阳地区PM达2500。

1943年著名的"洛杉矶雾霾"事件与1952年加州理工学院哈根斯米特教授的研究发现汽车尾气引起的光化学烟雾。

中国科学研究院大气灰霾追因与控制研究成果表明，北京的雾霾是"伦敦烟雾"和"洛杉矶雾霾"污染物的混合体：PM2.5来源中机动车尾气占1/4，燃煤和外来输送(包括沙尘、气溶胶)各占1/5。

陈吉宁部长："机动车已成为许多大中城市细颗粒物的主要来源，其中北京占31.3%，上海占29.3%，杭州占28%""最核心的问题是北京必须解决重污染车问题""第三个就是农村散煤问题，……，全国大概有2亿吨散煤"。

4. 城市雨洪内涝

2010年，住房和城乡建设部对351个城市进行专项调研结果就显示，仅2008—2010年间，全国62%的城市发生过城市内涝，内涝灾害超过3次以上的城市有137个。2010年我国有258座城市受淹，在发生过内涝的城市中，57个城市的最大积水时间超过12小时。

全国仍有340座城市(占总数53%)没有达到国家规定的防洪标准。特别是非农业人口150万人以上的34座特大城市中仅有7座(占总数21%)达到规定的防洪标准。

中国水利水电科学研究院防洪领域专家程晓陶曾根据2006—2014年的《中国水旱灾害公报》统计，这几年内，我国大陆每年遭受洪涝的城市都在百座以上。其中，2010年、2012年、2013年受淹城市分别高达258座、184座和243座，相应洪灾直接经济损失3745亿元、2674亿元和3168亿元。汛期"城市看海"几乎成为常态。

①北京"7·21"特大暴雨

2012年7月21日，北京遭遇暴雨及洪涝灾害，当地道路、桥梁、水利工程受损，民房多处倒塌，近 80 人遇难。全市因灾造成直接经济损失 116.4 亿元。

②浙江余姚城区被淹

2013 年 10 月，受台风"菲特"影响，余姚市过程雨量 496.4 毫米，罕见的雨情、水情，致使主城区 70% 以上地区受淹，交通瘫痪，全线停水、停电。山区公路交通全部中断。山区溪道、电站、灌溉等设施受损严重，平原河网的姚东浦塘全线漫堤。余姚市直接经济损失 69.91 亿元。

5. 垃圾围城

2009 年 2 月 15 日深圳某垃圾填埋场污泥坑上堆填垃圾过程中引发污泥外涌及下游堆体失稳事故，污泥从污泥坑下游垃圾围堰的薄弱部位挤出外涌，在下游面上冲开一道 1 米多宽的裂缝。污泥如火山熔岩般涌出，共持续了 4 小时，涌出污泥量达 4 万多立方米。

2015 年 12 月 20 日，深圳市光明新区红坳余泥渣土临时受纳场发生渣土堆填体失稳滑坡，造成特别重大滑坡事故。滑到下游的土方量达 275 万立方米，覆盖面积约 26 万平方米，下游堆积厚度达数米至十数米不等，造成 33 栋（间）建筑物被掩埋或不同程度损毁。截至 2016 年 1 月 6 日 12 时，77 人失联，其中 58 人遇难。

"马路拉链""空中蜘蛛网"、城市内涝、城市水系污染、地面变压器的随处可见、垃圾围城……使得城市变得脏、乱、差。

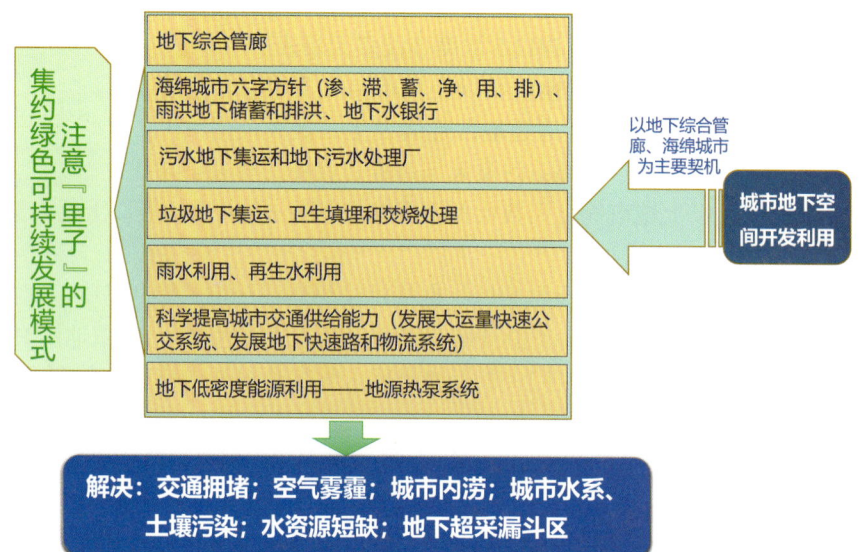

全国城市供水管网平均漏损水量达 25%。

全国城市道路 1% "开膛破肚"。

全国每年因施工产生的地下管线事故造成的直接经济损失近 50 亿元，间接经济损失约 400 亿元并造成人员重大伤亡。例如青岛东黄输油管泄露爆炸造成 62 人死亡，南京地下丙烯管道泄漏爆燃造成 22 人死亡。

（二）应对策略

随着以城市地下综合管廊为契机的城市地下空间开发利用，城市将会变得越来越秀美。

①综合管廊的建成，"马路拉链"消除了。

②电缆入廊，5000 千伏上海世博地下输变电工程建成，总建筑面积 57615 平方米，地下面积 55809 平方米，地面仅 1806 平方米，意味着"空中蜘蛛网"彻底消除；城市内再也不见架空高压线和地面变电器。

③污水入廊，污水处理，污水利用，解决了水污染。

④垃圾入廊，垃圾卫生填埋或焚烧，不再污染城市土壤和造成事故。

⑤地铁、地下快速路系统、地下物流系统建成，城市内见不到运货的货车，城市上不再交通拥堵，汽车尾气在地下收集处理，雾霾成为历史。

⑥修建雨洪入渗地下工程和雨洪地下调蓄设施，消除城市内涝现象和"看海"现象，雨洪资源得到利用。

1. 发展轨道交通，缓解城市交通拥堵，降低城市大气污染

中国城市轨道交通进入了空前发展的阶段。

（1）轨交综合体

成功范例：上海中山公园龙之梦。

通过多个中庭立体串联交通换乘。

规划特色：通过3个直接采光的中庭空间来高效组织地块南侧地铁2号线，地块西侧高架轻轨3号线、4号线，地块北侧公交车站以及地下2层出租车站的多向密集人流。

（2）形式常态化："轨交+综合体"形式比例稳步增长

综合体增量中轨交综合体占比过半并持续上升，二线城市尤为明显。随着轨交体系的持续铺开，"轨交+综合体"的形式被越来越广泛使用，逐渐显示出常态化的态势。

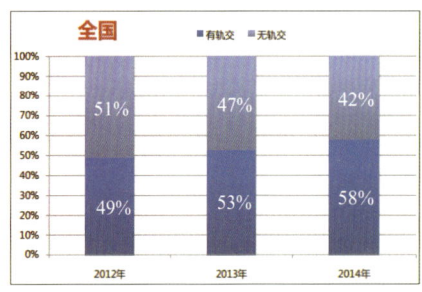 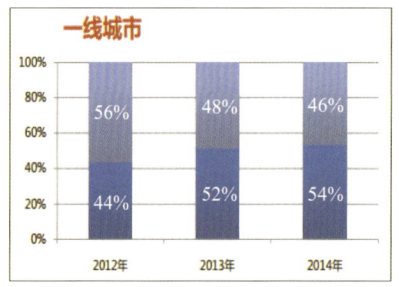 

（3）轨交物业开发融合

改变通道式连接，加强轨交综合体和轨交物业融合开发，提升价值，降低亏损，实现可持续。

若轨交参与物业全程开发运营，协同规划设置合理动线，将大幅提升价值。

分析：高价值项目一线城市居多，上海居首；经验开发商多集中于一线城市，意识先进；政府政策引导有助于优质产品打造。

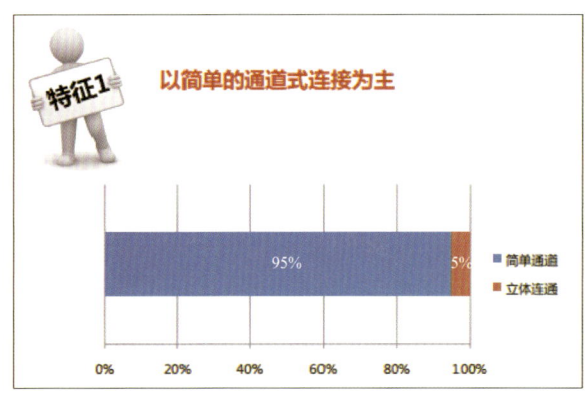 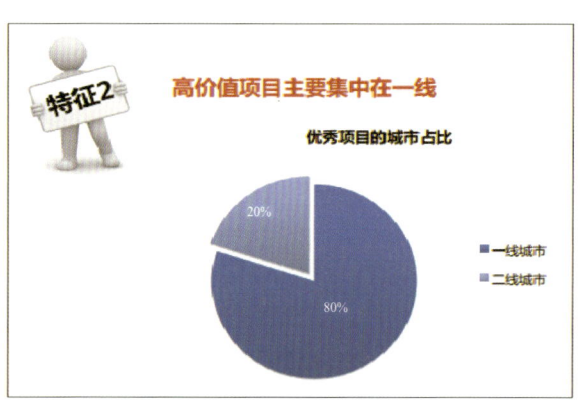

2. 创新转变交通方式，科学提高城市交通供给能力，降低尾气污染、城市内涝

除了抑制交通需求，最根本的是必须大力提高交通供给能力。即使在限购政策下，"存量"不减，"增量"攀升，车越来越多，交通需求也越来越膨胀。

对于北京来说，由于中心区已高强度开发，地面的土地资源已利用殆尽，受保护北京历史文化古城的特色和风貌的制约，中心城区不可能发展高架交通；可以说北京当前道路已无地可建；另一方面目前地面公共交通也已相当饱和，地面公交车排队进站，占用车道已达极限，所以即使"公交优先"也提高不了多少交通供给能力。

车多路不增必然拥堵。

大力发展以轨道交通为主的公共交通是解决交通拥堵的主要举措！但能否把"治堵"的希望完全寄托于轨道交通的未来发展上呢？

北京轨道交通已建成600多公里，但在上下班高峰时段，地铁上、下车拥挤之惨状已成谈虎色变。

环顾轨道交通相当完善的发达国家大城市，如伦敦、东京、新加坡、波士顿、马德里、莫斯科等，它们仍困扰于交通拥堵，正都纷纷寻找发展传统轨道交通以外的解决城市交通拥堵之良策，以应对小汽车进入家庭，机动交通快速增长之现实。因此我们更应考虑到中国特大城市多，城市人口密度特大的国情。

在20世纪末到21世纪初，国际上很多发达国家的特大城市开始把"治堵"目光转向了"地下快速路"和"地下物流系统"的建设——可以无限提高城市交通供给能力，降低尾气污染和城市内涝，一举三得。

（1）地下快速路

①美国波士顿1994年开始拆除高架路，10年间建成8~10车道的城市地下快速路系统。

②东京正在地下40米建设中央环状地下快速线，11公里的通过池袋、新宿、涩谷三大商业中心的新宿线已通车，正在建设品川线。

③新加坡已在城市中心区建成了15公里长的四车道环行快速线，承担了中心区交通量的40%。

④马来西亚吉隆坡建成12公里城市泄洪与快速路两用的城市地下路。

⑤马德里已建成36公里的快速线。

⑥悉尼建成4条共11.4公里的快速线。

⑦布里斯班建成两条6公里地下快速线。

⑧巴黎建成20公里地下快速线，并已完成了第2代城市地下快速路网设计。

⑨莫斯科在三环线中建成了3.6公里地下线。

（2）地下道路系统简介

①亚洲——东京

东京中央环状新宿线长11公里；埋深>40米；双向四车道（13米）；3大商业中心：池袋、新宿、涩谷；6个出入口；投资90亿美元。

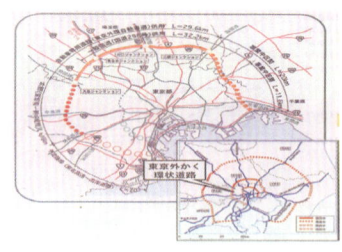
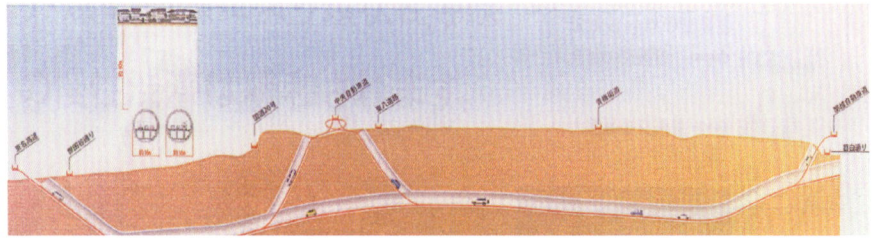

效果：

a. 节约时间：池袋→涩谷（50分→20分）。

b. 改善环境：每年减少2.5万吨 CO_2，160万吨 NO，16吨悬浮颗粒。

c. 降低成本：低于地面投资约 20%。

d. 国防建设。

东京外环线于 1966 年宣布为高架路方案，但由于沿线居民以噪声和污染严重为由强烈反对，于 1970 年宣布冻结。2003 年 1 月日本政府举行新闻发布会，宣布外环线改为建设地下 40 米深的高速路。由于地下无须动迁和征地费用，可比地面高架路节约建设费用 20%~30%，工期由 15 年缩短为 8 年。由于地下隧道内设置有可将汽车尾气中的烟尘和有害物质进行高科技过滤和分解的设施，排出的是无污染空气。

② 新加坡地下道路系统（SURS）

a. 长 15 公里。

b. 四车道，环形。

c. 造价 48 亿美元。

d. 承担城市中心区交通量的 40%。

日本东京外环品川地区联合开发

（三）发展城市核心区的地下车行系统

CBD 等核心区规划建设具有容积率高、地面支路网密、街坊面积小、道路红线窄等特点，核心区高密度高容积率的开发导致交通需求较大，导致区域支路上机非矛盾突出，机动车交通组织混乱，通行效率低下，安全隐患增大，有必要建设地下车行系统——地下车库联络道，贯通核心区地下车库，出入地块的车辆通过核心区外围的出入口集中进出地下车库，以减轻核心区地面支路网的压力，提升支路车辆及人非通行品质。

1. 北京金融街中心区

金融街地下环廊已建好 2.3 公里，总建筑面积 2.6 万平方米，可以连接金融街中心区 11 座大厦的地下车库，总计有 8000 多个车位。

2. 深圳前海地下空间规划

前海的地下空间规划建设在地下车行系统方面，将构建"地下快速路 + 地下环路 + 车库"的三级地下车行系统，海滨大道、妈湾跨海通道地下道路以分离过境交通为主，以服务区域到发交通为辅；听海路服务于区域到发交通，从而有效改善前海地面交通环境，缓解小街坊路网交通压力。

听海路及其地下道路是全国首创多点进出的城市地下道路，进出口多，与海滨大道相交节点为全地下互通立交。其中地面道路为双向六车道，北起双界河路，南至沿江高速地面道路，长约 2.16 公里，地下道路为双向四车道加集散车道。包含一座互通式立交（含 3 条匝道），4 个地下人行联络通道，2 个车行通道，4 条连接地面道路匝道，5 个预留接口。

（四）地下物流系统

即将城市货运逐步转移至地下，货运交通转入地下必将对治理交通拥堵做出重要贡献。

根据世界经合组织 2003 年统计，发达国家主要城市的货运占城市交通总量的 10%~15%。而在"世界工厂"和到处是建设工地的中国，其占城市交通总量则相应为 20%~30%。

地下货物运输系统，又称地下物流系统（ULS）是基于城内运输和城外运输的结合。城外货物通过各种运输手段到城市边缘的物流园区（CLP），经处理后由 CLP 通过 ULS 输送到各个终端。它以集装箱和货盘车为基本单元，以自动导向车（AGV）为运输工具。

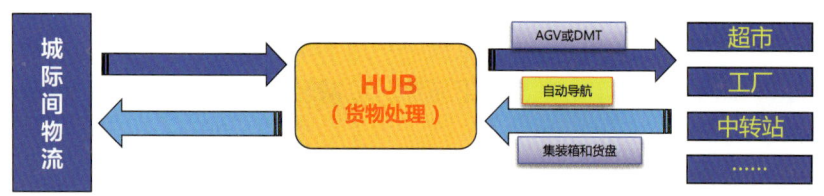

ULS 的环境效应显著。世界经济合作组织在 2003 年的《配送：21 世纪城市货运的挑战》报告中指出，发达国家主要城市的货运占城市交通总量的 10%~15%，而货运车辆对城市环境污染总量的贡献为 40%~60%。日本东京建设 300 公里地下物流系统评估报告中指出：一氧化氮和二氧化碳将分别减少 10% 和 18%，能源消耗减少 18%，运输速度提高 24%。

（1）针对港口区域，建议建设地下集装箱物流运输系统，创新采用新型集疏运模式，缓解交通拥堵、港城矛盾，实现节能减排。

传统的港口集疏运模式以公路集卡运输为主，需要穿越城市，客货交通相互影响，港口陆路集疏运交通需要大量穿越城市或占用城市道路资源，带来了交通拥堵、环境污染等问题。

既能支撑港区发展又能融合城市发展，解决港城矛盾，需要更集约化、可持续发展的集疏运新模式——地下集装箱物流系统。

首先将原先分散进入港口的货物集中在外围物流综合枢纽，再通过专用货运通道集约化转运到港口，实现高效率、规模化运输。可以有效避免集装箱卡车穿越城市，采用地下穿越形式，尽可能释放地面资源，不影响中心城交通、环境。

地下集装箱物流系统主要包括：

①荷兰地下物流系统（OLS-ASH）；
②纽约港和临近的新泽西港口（New Jersey）地下集装箱系统；
③比利时安特卫普（Antwerp）港地下集装箱运输系统；
④日本东京地下集装箱运输系统等。

以天津港为例，以集装箱货车为主的货运交通极大地影响了城市道路的服务水平和通行能力，导致港口城市交通综合环境日益恶化。2010年天津港有88%集装箱货运交通需要经过滨海核心区，道路客货运交通混行导致城市交通不畅，同时也降低了疏港交通的效率，高峰时段核心区内第九大街和泰达大街部分路段饱和度均超过1，货运比例超过0.8。

根据相关调研数据，2020年，天津港集装箱吞吐量将达2800万标准箱，若采用新型疏解模式即集装箱港口集疏运，2020年，专用地下物流通道可解决天津港55%~65%的集装箱运输，及1540万~1820万标准箱，能源消费量减少、二氧化碳排放减少将相当可观，同时还可以释放大量堆场土地。

（2）针对会展中心、机场等，建设区域性的地下物流运输系统。

传统会展中心的货运一般采用集卡运输，存在以下一些问题：

①必须设置足够货车轮候区，占用宝贵的土地资源；

②人流、货运重叠，交通组织难度高；

③布撤展时间短，对于分期大展，国内撤展时间1~2天，布展时间2~3天，货车大量提前到达保证按时进出，需要长时间展馆外路边等候；

④展会期间对城市交通冲击影响大。

未来在会展中心建设中应考虑必要的新型货运模式——地下物流系统，其系统由物流园区、运输隧道、展区三个子系统组成。首先会展货运车辆先集中到物流园区，集装箱货物被卸载后通过地下隧道转运至会展区域进行吊运拆箱。

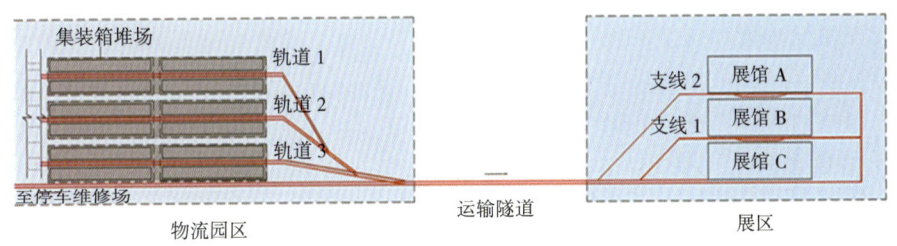

会展中心区域性的地下物流通道示意图

a. 采用地下物流方式解决展会的交通问题，能够较好地解决展会区域地面交通带来的巨大压力；

b. 减少碳排放，节省能源，改善区域周边环境；

c. 对会展综合体，通过有序组织拆卸箱，提高撤布展的效率；

d. 设置物流园区，除了早到货物的存储外，还可以用于调节货运需求的峰值；

e. 由于节省专门用于卸货转运的配套区，可显著提高土地的集约利用程度。

（3）针对城市区域，建设城市级的系统化网络化的地下物流运输系统，解决城市物流配送、垃圾运输等问题。

需求激增带来的地面车辆、车次数量剧增是导致城市交通拥堵的主要原因，其中对货物物流的需求增长占较大份额。2014年我国仅快递业务量已达140亿件，同比增长52%。如北京，按常规的车辆换算系数，货运车辆所占用的道路资源占40%。城市中大气污染物约60%来自机动车排放，机动车成为城市PM2.5的最大来源，交通也已成为城市噪声的最大污染源。据世界经合组织在《配送：21世纪城市货运挑战》报告中指出：货运对城市污染的占比达40%~60%，因为一辆重型货车污染排放量相当于100辆小汽车。

（五）跨江、跨海隧道

在发展地下高速路的同时，中国还正在发展跨越江河的水下隧道，已经在上海黄浦江下建设了13条水下越江隧道，宁波甬江下和广州珠江下各建造了两条城市公路隧道。此外，三条跨越长江的水下隧道（上海、南京和武汉）和厦门翔安海底隧道（项目全长9公里），青岛胶州湾海底隧道（线路全长6.17

公里，隧道长 5.55 公里）均已通车运营；正在建设南京和武汉多条跨长江隧道；跨越伶仃洋的港珠澳大桥海底隧道工程和厦门第二条跨海隧道以及两条跨越琼州、渤海的海底隧道正在论证中。

上海长江隧桥工程

穿越江河和海湾，应因地制宜发挥桥梁和隧道的各自优势，总体上把握"宜桥则桥，宜隧则隧，桥、隧并举"的指导思想。

交通是一项复杂的系统工程，不能只顾跨越江河、海湾的公铁交通，而损害穿梭江河、海湾的航运交通，必须统筹兼顾。桥梁由于桥高、桥孔宽度的限制，对航运总要产生影响。长江就是一例，自南京长江大桥建成起，黄金水道的通行能力就从万吨轮船降为 4 千吨，而撞船、撞桥事故不断发生。为减少撞桥事故，交通运输部规定船队过桥要解驳减速。

我国沿海多台风，很多地区又多雾，大桥易受台风、大雾、雨雪气候条件影响，不能全年通行；而隧道则可全天候运营。战时，交通枢纽是敌人轰炸的首选目标，桥梁易被发现和命中，斜拉桥和悬索桥一旦索塔被毁，整桥即坍塌，无法维修。而隧道则被发现和毁坏概率相对较低。所以，国外的跨越大江河和海湾的通道工程多"桥、隧"并举，如纽约的跨哈得逊河交通，桥 10 余座，而隧则 41 座。海湾，如美切什彼克湾、东京湾等海湾的主航道下都建海底隧道以保证航运。

1. 日本青函隧道

日本青函隧道就是当前世界上最长的一条隧道，因连接日本本州青森地区和北海道函馆地区而得名。青函隧道为双线隧道，全长为 53860 米，其中海底部分为 23300 米。青函隧道于 1964 年开始启动，历经 24 年建成，隧道海底段长 23.30 公里。最大水深 140 米，最小覆盖层厚 100 米。

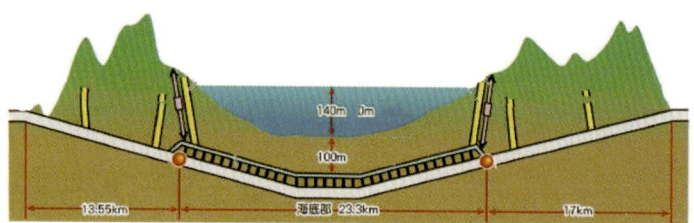

2. 直布罗陀海底隧道

直布罗陀海底隧道是在 20 世纪 70 年代末西班牙和摩洛哥两国在直布罗陀海峡共同规划、兴建一条连接欧洲和非洲的海底隧道，工程预计将花费 100 亿美元。目前，该隧道建设的可行性报告已经完成，根据设计方案，直布罗陀海峡隧道全长为 37.7 公里，其中 27.2 公里位于海底以下 300 米深处，隧道的起点和终点分别为西班牙的塔里法和摩洛哥北部的马拉巴塔角，隧道内分铁路和公路。

3. 马尔马拉隧道（博斯普鲁斯隧道）

马尔马拉隧道全长 13.6 公里，水下部分 1.4 公里，具有抗地震设计，总造价 40 亿美元，是世界上首条横跨欧洲与亚洲的海底隧道，2013 年 10 月 29 日正式启用，把土耳其伊斯坦布尔的亚洲和欧洲海岸连接起来。

4. 屯门隧道

香港屯门至赤蜡角的连接线隧道工程已经开始掘进，工程使用了一台直径达 17.6m 的盾构机，该盾构超过了西雅图 SR99 工程中直径为 17.45m 的 Bertha 号，成为目前世界直径最大的盾构机。

屯门隧道为一条 4.2 公里的双向四车道海底公路隧道，北起屯门西南第 40 区望后石接连屯门西绕道，经跨海高架桥及沉管海底隧道后到达港珠澳大桥香港口岸，再连接赤鱲角或大屿山。

（六）建设城市地下雨洪调蓄设施，解决城市内涝

1. 吉隆坡城市泄洪与公路两用隧道 Kuala Lumpur – SMART

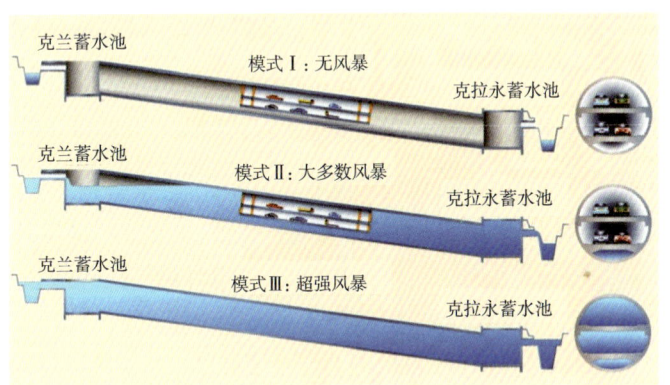

2. 日本雨洪地下调蓄设施

日本东京地下河川　　　　　　　日本横滨地下河

3. 芬兰 Paijanne 地下输水工程

世界上最长的引水隧道长 120 公里，是芬兰的由 Paijanne 湖至大赫尔辛基的输水隧道，截面面积为 13.8~18 平方米，流速 10 立方米/秒，位于地下 30~100 米深处。

4. 地下污水集、输和处理系统

即将污水处理设施放置在地下。这将使目前地面污水处理厂所占的土地得到开发利用，而且与它们邻近被污染不能被利用的土地也能得到开发。

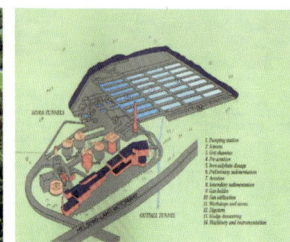

赫尔辛基污水处理系统　　　　　　　　　新加坡污水集、输和治理系统

5. 城市垃圾地下集、输、地下处理系统

中国广州金沙洲和亚运城都建设了真空垃圾收集系统。

上海世博园区内的垃圾隐形处理系统总覆盖面积约 0.5 平方公里，地下输送管线总长达 5.5 公里，主要服务于一轴四馆，永久性建筑及其周围区域，共有 51 个室外的智能垃圾投放口。

瑞典是在 20 世纪 60 年代建设并实施采用压缩空气吹运垃圾系统（PWT）的国家，PWT 系统建在 1700 人口的居民区内，该系统与收集、处理系统配套，投资在 3~4 年内得到回报。

发展地下垃圾处理系统节约了城市宝贵的土地资源，控制了对环境的噪声和恶臭的污染，减少了对环境的视觉负面影响。

（七）能源地下储存

能源是一个国家的经济命脉。为应付石油供应中断的突发事件，石油进口大国都制定了应急的石油储备目标，石油储备是稳定供求关系、平抑市场价格、应付突发事件、保障国家能源安全的重要手段。

能源（石油、天然气）储存的方式有陆上储罐、海上储罐和地下储存等几种方式，但地下储存为多数。

地下能源储库不易被发现和破坏，易于保卫，是一种平时和战时都较为安全、稳妥的储存方式，被称为"高度战略安全的储库"。

能源储库是敌打击的首选目标，南斯拉夫联盟共和国的陆上储罐几乎被敌全部摧毁。20 世纪 80 年代，我国黄岛陆上石油储罐被雷电击中引起大火，被烧毁殆尽。

1. 舟山储备油库

舟山国家石油储备基地库容为 500 万立方米，储备原油 398 万吨。

2. 镇海储备油库

镇海国家石油储备基地库容为 520 万立方米，储备原油 378 万吨。

3. 黄岛国家石油储备地下洞库

黄岛国家石油储备基地库容为 320 万立方米，储备原油 250 万吨。

地下储存有岩体（一般为花岗岩）中水封储存和盐岩中溶腔储存。由地质条件决定，前者被北欧诸国采用，后者广泛采用于美、德、法、俄、墨西哥、加拿大。

①盐岩的超低渗透特性与良好的蠕变行为，能够保证储库的密闭性；
②盐岩的力学性能稳定，能够适应储压的变化；
③盐岩溶于水的特性使盐岩洞库的施工（溶腔）较为容易和经济。

中国已在湖北应城、江苏淮安、河南平顶山和湖北潜江等地建成了盐岩储油库，规划中的江苏盐城的石油储库建成后可储油 600 万吨。

4. 长江三角洲地区的储气库工程

该矿区距西气东输干线仅为 30 公里，盐矿构造面积 60.5 平方公里，盐矿顶深 860~1140 米，盐矿厚度 70~220 米，盐矿上部黏土矿厚度及其密封能力比较大，顶板强度达到 30~60 兆帕，整个储库储气总量可达 9.8 亿立方米。

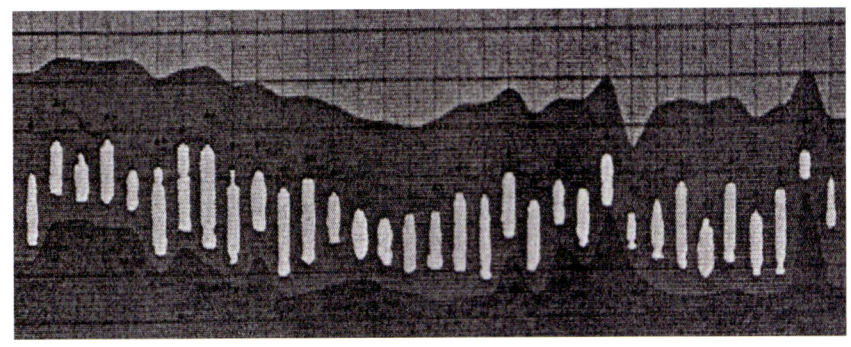

5. 核废料深地质处置

核废料具有放射性，深地质处置被认为是安全处置高放废物最现实可行的方法。建造和运营地下实验室并展开大规模存储实验也在进行中。

世界各国的主要地下储存场所

国家	机构名称	地点	废料类型	地质条件	深度	运营状态
加拿大	安大略电厂深地质储存所	安大略省	200000 立方米低及中低放射性废物	石灰岩	680 米	2011 年申
芬兰	VLJ	奥尔基洛托核电站	低及中低放射性废物	英云闪长岩	60~100 米	1992 年开始运营
芬兰		洛维萨	低及中低放射性废物	花岗岩	120 米	1998 年开始运营
芬兰	昂加洛乏燃料储存所	奥尔基洛托核电站	乏燃料	花岗岩	400 米	在建
德国	康拉德盐矿	下萨克森	303000 立方米低及中低放射性废物	积岩	800 米	在建
瑞典		福斯马克	乏燃料	花岗岩	450 米	2011 年申请执照
美国	废物分离中试厂		超铀元素核废料	盐床	655 米	1999 年开始运营

我国甘肃北山地区的核废料深地质储库以及深地实验室通过多年论证准备，已经进入正式建设阶段。

转变城市发展方式，积极推进城市地下空间的关键是工程建设方案比较指标由商务、技术两指标转变为三指标，增加环境效益。为此，研究环境效益货币化是必要的。

六、系统科学地编制好城市地下空间规划

（1）地下工程地质、水文地质情况要明；
（2）地下空间开发利用现状和问题要清；
（3）面对问题，通过研究，着力创新；
（4）地下地上空间规划要相互协调；
（5）涉及地下的各项规划要互相衔接、协调，尽力"促进'多规合一'"；
（6）地下空间横向要连通，竖向分层立体开发安排要合理，时序要科学；
（7）近期建设与中远期发展互相衔接形成逐步推进格局。

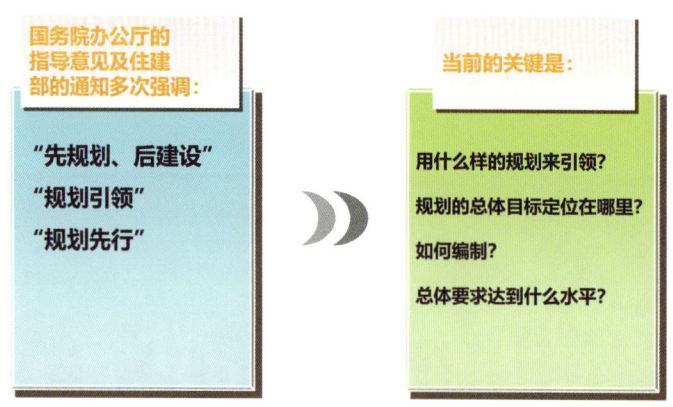

七、借鉴国外、境外先进经验，依法治国，制定完善的法规体系

"根本点是要提高城市地方人民代表大会的法治权威性"，国外先进经验是先立法、后建设，这样才能有序有力地推进建设。这是因为地下空间开发利用是一个跨行业、跨部门与单位的协同工程。过去我国城市地下空间开发利用综合管理水平低、问题多的病根在于分头管理的市政体制与条块分割的部门所

有制，破除分头管理与部门分割的利器是"法治"，依法执政首先要有法可依，光有指导意见不够，还需要对城市地下空间的所有权、规划权、建设权、管理权、经营权、使用权以及有偿使用费的收取原则等做出明确的完善的具体规定。

目前我国城市地下空间法规很不完善，亟待健全。我国现行法律中没有地下空间权的概念和相关规定，只有《中华人民共和国物权法》原则规定建设用地使用权可以在地表、地上和地下分别设立，地下空间最基础的民事权利并没有得到法律的确认。全国人大和国务院没有制定城市地下空间开发利用的法律法规，原建设部颁布的《城市地下空间开发利用管理规定》属于部门规章，许可权限和法律效力比较低，加之内容不完备、规定不具体，权威性和操作性都不够，在实践中难以有效地贯彻执行。

在这方面，我们可以借鉴德国的《城市建设法典》，日本的《大深度地下公共使用特别措施法》《共同沟法》，美国各州制定的空间法，瑞士的《瑞士民法典》，意大利的《民法典》等。

八、建立统一管理、分工协作、衔接有序的城市地下空间综合管理体制机制

（1）推进城市管理机构改革、创新城市工作体制机制。

目前我国城市地下空间管理体制不统一，亟待理顺。城市地下空间开发利用涉及发改委、国土、规划、建设、市政、交通、人防和房产等多个管理部门，而国家在体制上并没有明确一个综合管理机构对其进行统一管理，基本上是各部门依照相关职责对其实施分散管理，相互间缺乏统一协调机制，处于条块分割、各自为战的状态，形成多头管理与无人管理并存的局面。目前，许多城市对地下空间开发利用管理进行了有益探索，分别采取了以规划管理、建设管理、人防管理、安全管理和协调管理为主导的管理模式，但都没有起到统一管理的效果。

（2）必须对管理体制和管理措施进行系统完善。

管理法规是实施综合管理的基本依据，应明确规定管理体制和管理措施。管理体制是实践综合管理的执行主体，应在管理法规的法律授权或行政授权下采取各种管理措施。管理措施是实现综合管理的具体方法，应由管理机构依据管理法规，通过有效的管理机制，进行综合运用。

建设城市地下综合管廊转变城市发展方式

习近平总书记指出,城市面貌是城市的"面子",这些年有了很大的变化,但城市基础设施,特别是地下管网设施这些城市的"里子",目前仍然薄弱。

李克强总理也曾强调,"面子"是城市的风貌,而"里子"则是城市的良心,只有筑牢"里子",才能撑起"面子",这是城市建设的百年大计;"目前中国正处于城镇化快速发展时期,但我们的地下管廊建设严重滞后。加快这方面的建设,很有必要。"

国务院直接部署下,国务院办公厅连发文件具体指导城市地下综合管廊建设:《国务院办公厅关于加强城市地下管线建设管理的指导意见》《国务院办公厅关于推进城市地下综合管廊建设的指导意见》《中共中央 国务院关于进一步加强城市规划建设管理工作的若干意见》。

全国城市地下综合管廊建设正在热火朝天地推进,今年《政府工作报告》指出全年建设2000公里综合管廊。

为什么党中央、国务院这么高度重视城市地下综合管廊建设呢?

一、建设地下综合管廊对推进新型城镇化、补齐城市基础设施短板,实现城市发展方式转变的重大意义

(一)建设城市地下综合管廊是保障城市运营安全的重要环节

城市水、电力、电信、燃气、油等管线是城市的"生命线",生命线的安全是城市运营安全的基础。

然而,全国城市供水管线平均水量漏损率达25%;城市道路每开挖1m,城市规划额外的费用就多了1.4万元;全国拥有城市道路35.2万km、道路面积68.3亿m^2。1%"开膛剖肚",最直接的市政修补费用约6025万元。

城市"开膛破肚",造成停水断电断炊,中断通信、网络和电视,造成巨大经济损失,全国每年因施工产生的地下管线事故造成的直接经济损失近50亿元,间接经济损失约400亿元。

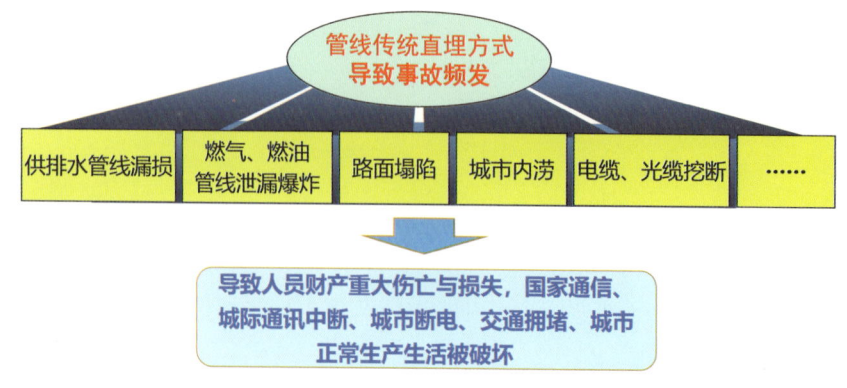

报告时间:2016年12月6日。

① 2010 年 7 月 28 日，位于江苏省南京市栖霞区迈皋桥街道的南京塑料四厂地块拆除工地发生地下丙烯管道泄漏爆燃事故，共造成 22 人死亡，120 人住院治疗，其中 14 人重伤，直接经济损失 4784 万元。

② 2013 年 11 月 22 日，位于山东省青岛经济技术开发区的中国石油化工股份有限公司管道储运分公司东黄输油管道泄漏原油进入市政排水暗渠，在形成密闭空间的暗渠内油气积聚遇火花发生爆炸，造成 62 人死亡、136 人受伤，直接经济损失 75172 万元。

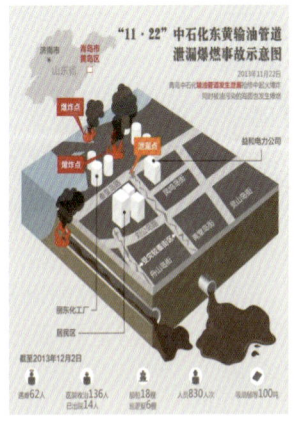 

③ 2013 年 11 月 24 日 16 时 30 分许，浙江省温州市鹿城区马鞍池东路与大南路交叉口附近一工地发生煤气管道爆炸事故，该事故造成 3 人受伤。

④ 2013 年 8 月 14 日，哈尔滨市辽阳街路面突然塌陷，4 人落入深坑，造成两死、两伤。事故原因是连续几场强降雨造成土质沉降，致使老旧排污管线断裂，泥沙灌入人防工程洞体，最终导致地面塌陷。

⑤ 2011 年 3 月 16 日上午，北京市海淀区四道口附近一处地热管道发生爆裂事故，事故将路面炸开直径数米的洞口，管道中热气大量外喷，浓烟达数十米，路面形成积水，有不少行人没能及时躲避，被水蒸气烫伤，至少有 7 人被送往医院治疗。

⑥ 2007 年 2 月 5 日凌晨 6：00 左右，南京汉中路牌楼巷与汉中路交叉路口北侧，正在施工的南京地铁 2 号线出现渗水塌陷，造成天然气管道断裂爆炸。事故导致附近 5000 多户居民停水、停电、停气，附近的金鹏大厦被爆燃的火苗"袭击"，事故没有造成人员伤亡。

⑦ 2012 年 4 月 1 日，一名女子在北京市北礼士路人行道上行走，突然落入一个热力管道渗漏形成的热水坑，不幸身亡。

⑧ 2013 年 5 月 20 日，深圳市龙岗区横岗街道华茂工业园由于地下排水箱涵老化发生地陷，导致 5 人死亡。

建设现代化的城市地下综合管廊可以加强管线检查维护，掌握管线管理信息，及时发现管线安全风险、消除安全隐患，避免发生管线事故。

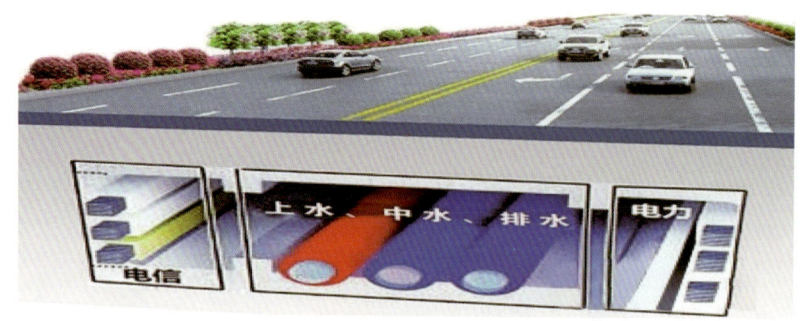

（二）建设城市地下综合管廊是我国城市发展方式由粗放发展向集约绿色可持续发展模式转变的关键契机

习近平总书记曾指出，"突出绿色、低碳、环保、可持续发展理念，最先进的理念、最高的标准、最好的质量""发扬'工匠'精神，精心推进，不留历史遗憾"进行城市规划设计"坚持世界眼光、国际标准、中国特色、高点定位，以创造历史、追求艺术的精神进行城市规划建设"。

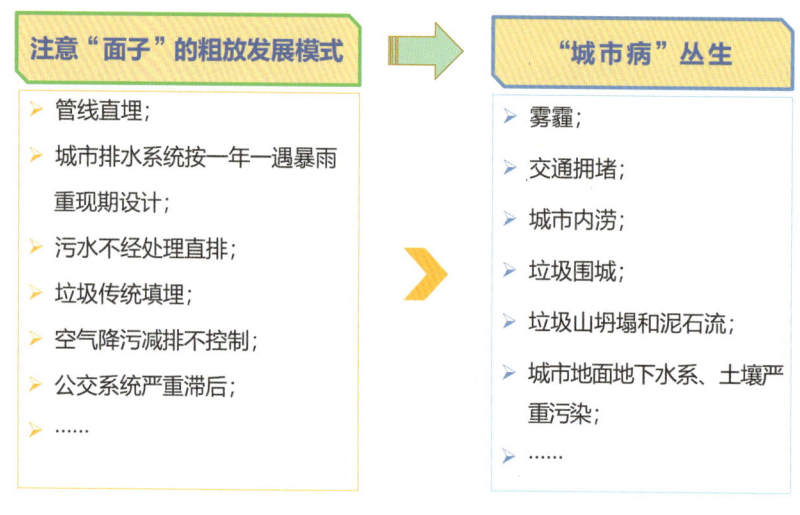

过去注意高层建筑、大广场、宽马路城市"面子"建设的粗放发展模式酿成城市基础设施"里子"建设短板的恶果。

2009年2月15日深圳某垃圾填埋场污泥坑上堆填垃圾过程中引发污泥外涌及下游堆体失稳事故，污泥从污泥坑下游垃圾围堰的薄弱部位挤出外涌，在下游面上冲开一道1 m多宽的裂缝。污泥如火山熔岩般涌出，共持续了4 h，涌出污泥量达4万 m^3。

2015年12月20日,深圳市光明新区红坳余泥渣土临时受纳场发生渣土堆填体失稳滑坡,造成特别重大滑坡事故。滑到下游的土方量达275万 m³,覆盖面积约26万 m²,下游堆积厚度达数米至十数米不等,造成33栋(间)建筑物被掩埋或不同程度损毁。截至2016年1月6日12时,77人失联,其中58人遇难。

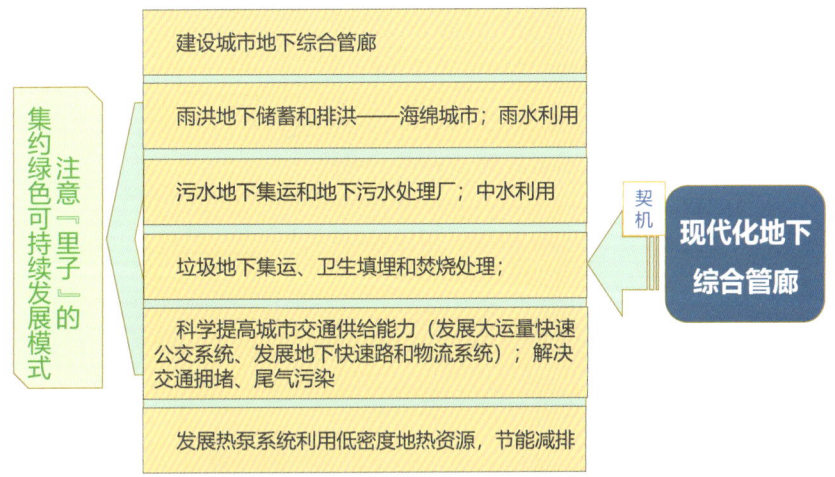

(三)地下综合管廊建设是拉动我国经济增长的一项重要措施

李克强曾指出中国的城镇化是最大的内需。其中,以地下综合管廊建设为龙头的补齐城市地下基础设施短板建设是"供给侧"结构性改革的一个重要方面。

城市建设是国家现代化建设的重要引擎,城市发展带动整个经济社会发展,补齐城市地下基础短板可以带领内需消费,从内部消化过剩产能,增加有效需求和有效供给(高质量的城市基础设施建设不会产生新的过剩),可以打造经济发展新动力。

城市地下综合管廊建设一次性建设成本较高,但后期效益十分显著。后期效益包括避免供水漏水;高压电缆、变压器入地而节约的土地利用;路面事故及其修复;管线维修等。以我国最大综合管廊珠海横琴综合管廊建设为例,全长33.4km,总投资22亿元,而横琴因建设地下管廊而节约的土地等产生的直接经济效益就超过80亿元。

住建部统计,按地下综合管廊全寿命周期计算,50年地下综合管廊与直埋管线相比,所需费用持平。超过50年,综合管廊比直埋管线所需费用低,到100年综合管廊与直埋管线相比低23%。

住建部陈政高部长指出:建设8000km的管廊,每公里造价1.2亿元,总投资近一万亿,还未包括间接拉动投资,消化过剩产能、拉动经济内需作用明显。

(四)城市地下综合管廊建设和管理是提升我国城市领导执政能力和执政水平的重要抓手

地下综合管廊是一个跨部门、跨领域的复杂工程。给水、排水、电力、电信、广电、燃气、供热、路灯、垃圾……各管线单位分属不同行政主管。

管线性质各异,有液体、气体,有强电、弱电,有管有缆,有防震的,有防爆的,具有消防、通风、供电、照明、监控和报警、标识等各种附属设施,涉及部门还有公安、人防、消防、园林,是一个跨行业、跨组织的协同工程。

过去城市地下综合管廊之所以存在"建而不入""建后难管"的现象,其病根在于分头管理的市政体制与条块分割的部门所有制。

破除分头管理与部门分割的关键,是推进城市管理体制机制的改革:落实2015年城市工作会议提出的"推进城市管理机构改革、创新城市工作体制机制""根本点是要提高城市地方人民代表大会的法治权威性";其次是提高市政领导"依法治政"的执政能力和执政水平,克服急功近利意识,处理好"面子"和"里子"的关系,秉持"执政为民"的信念,迎难而上,破除部门利益的束缚,处理各种矛盾。

欧洲为了破除"马路拉链",市议会通过立法不准马路开膛,台湾建设"共同管道"的经验是立法先行+"长官魄力与支持"。

(五)城市地下综合管廊建设是建设美丽城市,实现中国城市梦的迫切需要

"马路拉链""空中蜘蛛网"、城市内涝、城市水系污染、地面变压器的随处可见、垃圾围城……使得城市脏、乱、差。

以我国台湾为例(1990年):台北市挖路13000次(面积31万m²),占台北市面积1/32;高雄市挖路4000次(面积18万m²);台北县挖路5000次(面积23万m²),挖路造成交通堵塞、尘土飞扬、噪声不断、空气污染、市民怨声载道。

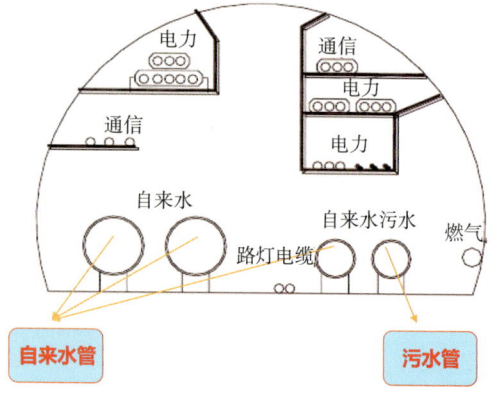

英国伦敦共同沟(1861年)

随着以城市地下综合管廊为契机的城市地下空间开发利用,城市将会变得越来越秀美。

①"马路拉链"消除了,污水、热力管道、燃力管入廊,马路坍塌、爆炸事故消除了。

②电缆入廊,5000千伏上海世博地下输变电工程建成,总建筑面积57615m²,地下55809m²,地面仅1806m²,意味着"空中蜘蛛网"彻底消除;城市内再也不见架空高压线和地面变电器以及电信电缆。

③污水入廊,新加坡全国7个地面污水处理站改造成4个地下污水处理站,城市再也见不到污水,闻不到臭气,中水可以利用。

④挪威垃圾入廊吹运至郊区高温焚烧发电,城市地面见不到垃圾。

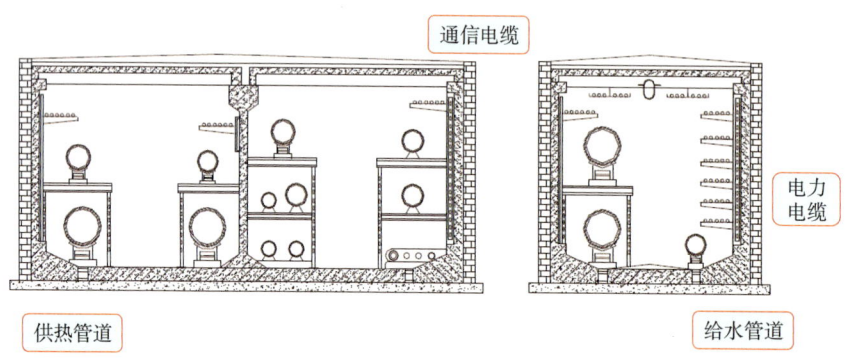

莫斯科共同沟(1933年)

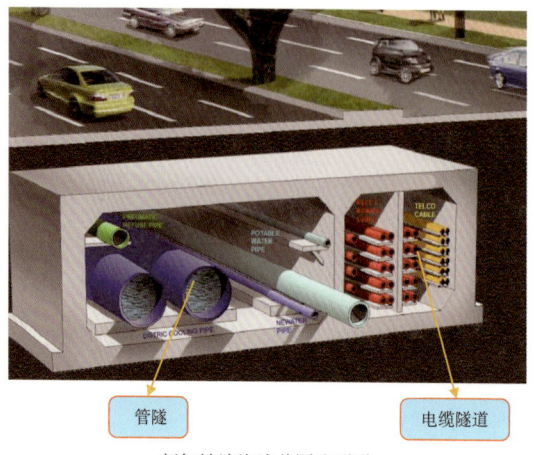

新加坡滨海湾共同沟隧道

⑤日本的燃气管道设置在一个封闭的管沟中，以防燃气泄漏引起的爆炸损害其他管线。

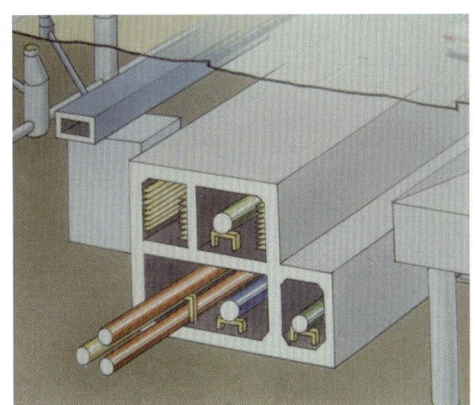
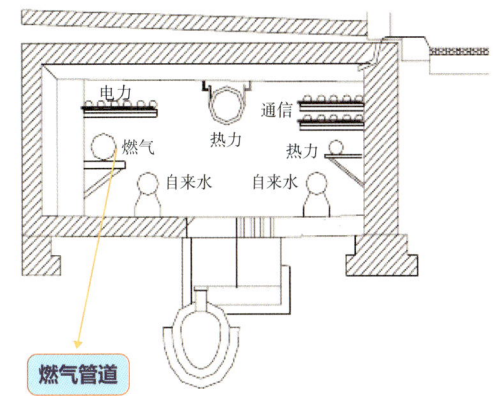

德国汉堡共同沟（1893年）

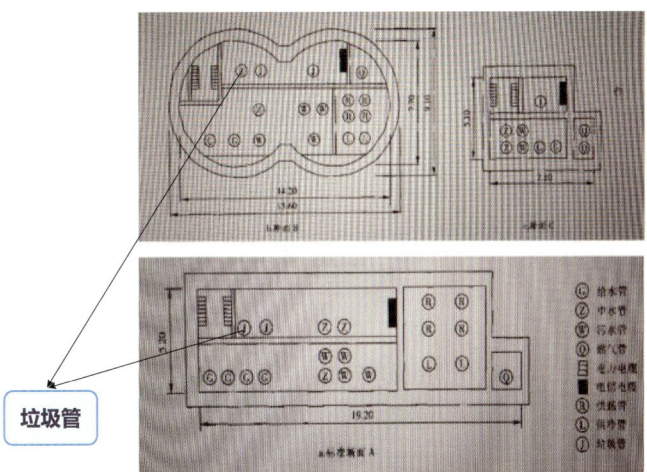

东京临海副都心共同沟

赫尔辛基污水处理系统

⑥地下快速路系统建成，新加坡已在城市中心区建成了15km长的四车道环行快速线，承担了中心区交通量的40%。

⑦地下物流系统建成，城市内见不到运货的货车，加上地铁、地下快速路系统建成，城市上不再交通拥堵，汽车尾气在地下收集处理，雾霾成为历史。

雨洪地下调蓄设施建成，城市内涝可以防止、雨洪资源得到利用。

日本东京地下河川

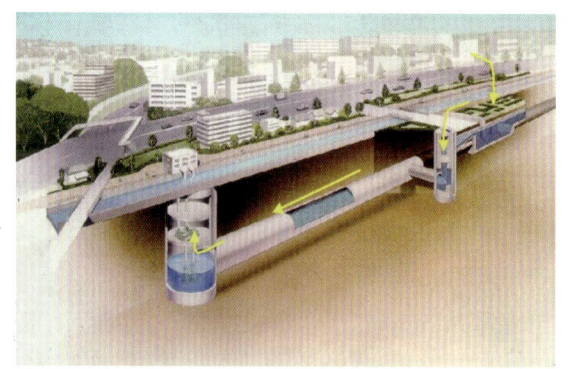

日本横滨地下河

⑧地源热泵系统建成，低密度地热资源得到利用，节能减排。

1. 地下结构与地温能开发一体化技术

地温能指的是地层中温度小于25℃一般层区的地层热能。它把传统空调器的冷凝器或蒸发器直接埋入地下，利用传热循环介质与大地进行热交换，从而提取地温能。

地下换热系统可埋设在地下结构基坑围护结构（地下连续墙、排桩）、基础底板下、桩基（钻孔灌注桩、预制桩、PHC桩等）内；还可埋设在新奥法施工的隧道衬砌内或以能源锚杆的形式埋设在其围岩中；也可埋设在地铁区间隧道内、地下输水管道内。

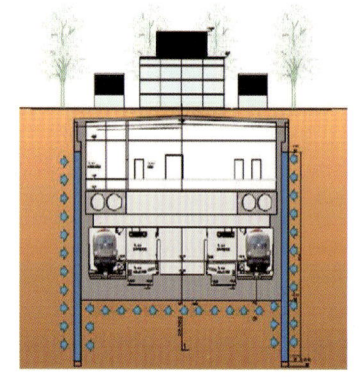

地铁车站的地温能提取示意图
（资料来源：Brandl, 2016）

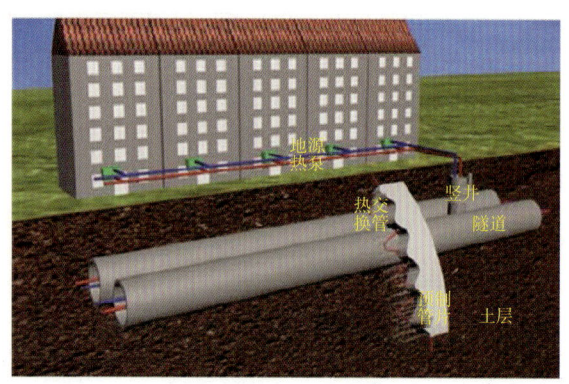

隧道结构内埋管换热器工作示意图

2. 利用低密度地热能源，节能减排

地源热泵系统利用地热进行节能，到达地下一定深度时（5m以下），四季的地层温度保持在一稳定值。低能耗的温度调节方式包括：①覆土结构或住宅；②地下气热管；③地源热泵系统；④活化桩。

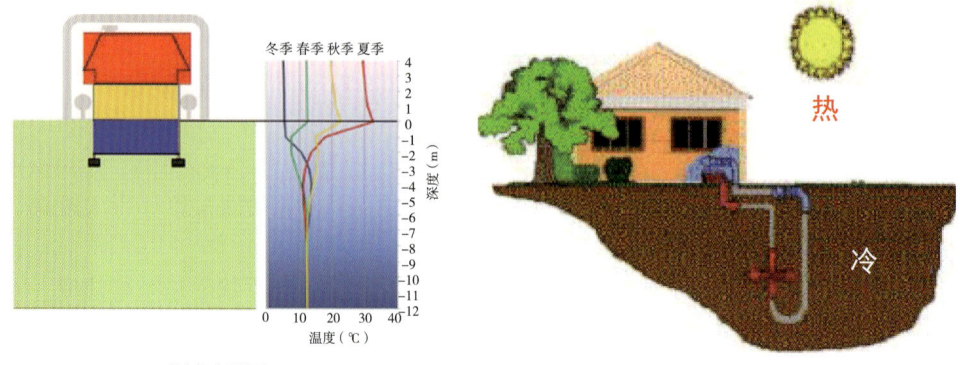

温度廓线图

3. 科罗拉多的地源热泵系统

安装地源热泵系统后，每个家庭可降低电力峰值需求、节约能源和减少 CO_2 排放 10%~30%。

4. 阿布扎比的地源热泵系统

阿布扎比的地源热泵系统在建筑物基坑桩基里面安装地源热泵管道系统。通过采用竖直的地源热泵线圈，来节约成本和节能。

二、国内外成功经验

（1）雨污分流：污水入综合管廊；雨洪不入廊，单独成系统。

街区的污水依重力汇入污水收集站后，由污水泵打入共同沟内的污水干管，经多次加压后被送至水处理厂，处理后达到中水水质标准，再由共同沟内的中水管道回输到城市各处，作景观、绿化、喷洗道路及冲厕等。

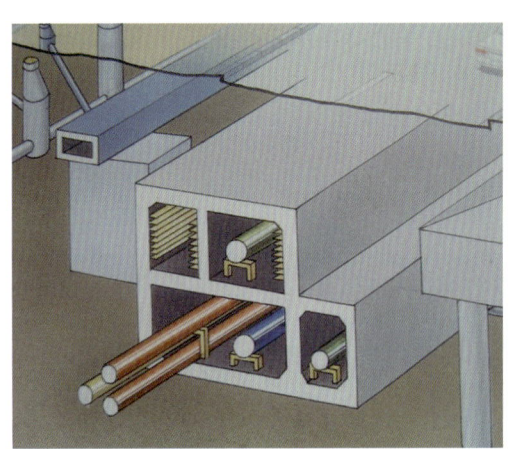

（2）大部分国家燃气入廊（英、法、德……），日本入廊单独设单沟。

日本：燃气管道设置在一个封闭的管沟中，以防燃气泄漏引起的爆炸损害其他管线。

（3）垃圾入廊，单独设管，真空吹送至垃圾焚烧厂，高温焚烧。

（4）管廊内保留一些待铺入廊管线的平台。

（5）组织先行，立法、规划紧跟。

以我国台湾为例：

1979 年 7 月 17 日，台北市政府第 570 次市政会议通过建设共同管道决议；

1980 年 2 月 15 日，成立工务局新工处共同管道科；

1989 年 6 月 14 日，共同管道法公布实施；

1990 年 12 月 28 日，公布共同管道发施行细则；

1990年12月19日，公布共同管道经费分摊办法；

1991年5月1日，公布共同管道系统上下空土地使用征收及补偿办法；

1992年4月23日，公布共同管道设计标准。

（6）综合管廊建设稳步推进、质量第一。

日本1963年通过共同沟措施法，于中央建设省下设26个共同管道科，在全国26个大都市全面推进共同沟工程，全国规划2000km，电线共同沟15000km，至1990年完成共同沟392km，进度"不算快"，预计今后五年内（至1997年）将陆续兴建168km共同沟与3000km的电线共同沟。

日本历经1995年阪神大地震，共同沟完好使用。

（7）与地铁、城市地下空间、快速路等城市基础设施建设整合。

如台北快速路共同管道的建设全长6.3km，其中2.7km与地铁建设整合；2.5km与地下街、地下车库整合建设；独立施工仅1.1km，从而大大降低造价，加快建设进度。再如美国波士顿拆高速路建地下路时，整合建设综合管廊。

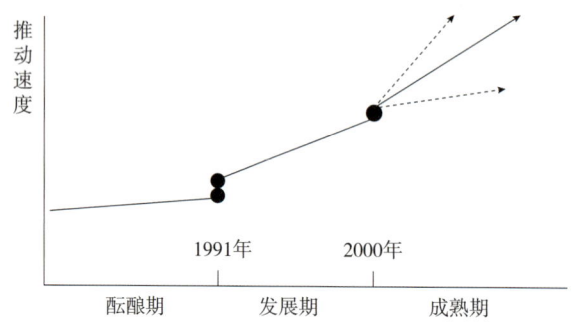

（8）高度信息化建设、管理。

"BIM技术与共同管道作业结合，以数位化整合式管理，辅助现场作业，提供统一的BIM数位资料库，使共同管道维护管理作业达到3D及电脑化管理模式"。

南港经贸园区共同管道Bentley BIM模型展示（未来3D模型）

 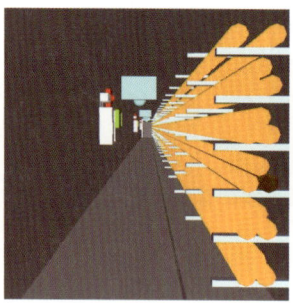  

三、加强顶层设计、完善相关政策，落实相关具体措施，推进地下综合管廊建设

（一）加强顶层设计、积极有序推进

组织先行：明确主管部门、建立协调机制；立法、规划跟进；制订路线图、时间表，积极有序推进工作部署。

(二)完善相关配套政策

(1)明确入廊要求

①城市规划区范围内的各类管线原则上应敷设于地下空间。

②已建设地下综合管廊的区域,该区域内的所有管线必须入廊。

③在地下综合管廊以外的位置新建管线的,规划部门不予许可审批,建设部门不予施工许可审批,市政道路部门不予掘路许可审批。

④既有管线应根据实际情况逐步有序迁移至地下综合管廊。

⑤各行业主管部门和有关企业要积极配合城市人民政府做好各自管线入廊工作。

(2)实行有偿使用

①入廊管线单位应向地下综合管廊建设运营单位交纳入廊费和日常维护费,具体收费标准要统筹考虑建设和运营、成本和收益的关系,由地下综合管廊建设运营单位与入廊管线单位根据市场化原则共同协商确定。

②入廊费主要根据地下综合管廊本体及附属设施建设成本,以及各入廊管线单独敷设和更新改造成本确定。

③日常维护费主要根据地下综合管廊本体及附属设施维修、更新等维护成本,以及管线占用地下综合管廊空间比例、对附属设施使用强度等因素合理确定。公益性文化企业的有线电视网入廊,有关收费标准可适当给予优惠,由发展改革委会同住房城乡建设部制定指导意见,引导规范供需双方协商确定地下综合管廊收费标准,形成合理的收费机制。

④在地下综合管廊运营初期不能通过收费弥补成本的,地方人民政府视情给予必要的财政补贴。

(3)加大政府投入

①中央财政要发挥"四两拨千斤"的作用,积极引导地下综合管廊建设,通过现有渠道统筹安排资金予以支持。

②地方各级人民政府要进一步加大地下综合管廊建设资金投入。省级人民政府要加强地下综合管廊建设资金的统筹,城市人民政府要在年度预算和建设计划中优先安排地下综合管廊项目,并纳入地方政府采购范围。有条件的城市人民政府可对地下综合管廊项目给予贷款贴息。

(4)完善融资支持

①将地下综合管廊建设作为国家重点支持的民生工程,充分发挥开发性金融作用,鼓励相关金融机构积极加大对地下综合管廊建设的信贷支持力度。

②鼓励银行业金融机构在风险可控、商业可持续的前提下,为地下综合管廊项目提供中长期信贷支持,积极开展特许经营权、收费权和购买服务协议预期收益等担保创新类贷款业务,加大对地下综合管廊项目的支持力度。

③将地下综合管廊建设列入专项金融债支持范围予以长期投资。

④支持符合条件的地下综合管廊建设运营企业发行企业债券和项目收益票据,专项用于地下综合管廊建设项目。

(三)落实具体措施

(1)借鉴国内外先进经验,制定完善的法规体系

2015年全国城市工作会议提出:"根本点是要提高城市地方人民代表大会的法治权威性"。

国外先进经验是先立法、后建设,这样才能有序有力推进建设。这是因为建设管理综合管廊是一个跨行业、跨部门与单位的协同工程。过去我国综合管廊"建而不入""建后难管"的病根在于分头管理的市政体制与条块分割的部门所有制,破除分头管理与部门分割的利器是"法治",依法执政首先要有法可依,光有指导意见不够,需要对综合管廊的所有权、规划权、建设权、管理权、经营权、使用权以

及入廊有偿使用费的收取原则等作出明确的完善的具体规定。

在这方面，我们可以借鉴德国的《城市建设法典》、日本的《关于建设共同沟的特别措施法》和台湾地区的《共同沟管道法》《共同管道法施行细则》《共同沟建设管线基金收支保管及运用办法》《共同沟建设及管理经费分摊办法》等现行法规；也可借鉴上海市的《上海市城市道路架线管理办法》和重庆市的《重庆市管线工程规划管理办法》等，进行完善和补充。

（2）系统科学地编制好城市综合管廊规划

又好又快、好字当头，编制好城市综合管廊规划是城市地下综合管廊建设的关键。

地下综合管廊是城市地下空间利用的一个有机组成部分，城市地下空间资源是宝贵而有限的，而且是不可逆的；城市地下空间利用应与城市地面、地上空间利用达到互补和互相协调。因此原则上城市地下综合管廊规划应与城市地下空间规划及城市建设总体规划同步协调进行。

条件不允许必须先行或单独规划的综合管廊规划，必须根据城市总体规划与现有和拟建的城市地下步行道系统、地下轨道交通、地下快速路系统以及为解决城市内涝问题而建设的城市的地面与地下"海绵"设施和贮排雨洪的城市地下河流、蓄排洪设施保持衔接即通过充分协调，达到统筹规划的编制。

综合管廊建设是百年大计，规划时还应充分考虑埋设深度，以及城市未来发展中对地下物流、垃圾集运和处理、雨水利用、中水输送乃至利用地热地冷资源实现城市供热供冷所需空间的规划要求，同时这也有利于提升综合管廊的经营性。

综合管廊应建设自动化监测系统和报警系统，变定期检查和维修为实时发现管线泄漏和超温现象及部位，及时或自动化维修。

为彻底解决城市内涝问题，雨洪的排泄系统建议参考日本经验，应包括雨洪调蓄设施，修建为城市地下河川或参考吉隆坡的经验，将暴雨排洪设施与城市地下快速路相整合，无雨时快速路通车，中小雨时，路面下排泄雨水，路面上仍通车，暴雨时，则快速路禁止通车仅作排洪。

应大力提倡综合管廊与地铁建设、地下街建设和地下快速路建设相整合，从而明显降低建设成本，减小社会干扰，缩短施工工期，避免重复建设、重复投资。

武汉地铁3号线宗关站，将电力管、排水管、通信管等管道集结在一起，与2、3号出入口工程整合建设，既解决了"城市蛛网"埋设随意性较大、分布不合理的问题，也可避免检修时对城市道路重复"开膛破肚"。

乌鲁木齐市在远景规划城市轨道交通线网中，也考虑了同步进行地下综合管廊建设。

编好综合管廊规划的前提和基础是充分掌握各类已有和需建管线的信息和各类地下空间信息，为此在编制规划前必须开展和完成城市地下管线普查工作。

（3）建立统一管理、分工协作、衔接有序的综合管廊管理体制

2015年全国城市工作会议提出："推进城市管理机构改革、创新城市工作体制机制"。为了克服条块分割、分头管理的弊端，达到统一协调，当前在尊重现有各部门职能分工的基础上，应尽快明确每个城市的综合管廊的综合管理机构。建议成立城市综合管廊管理委员会，承担〔2014〕27号文件所规定的政府对综合管廊的建设管理责任。由政府主要负责人担任主任，相关政府部门及管线主管部门担任委员，下设综合管理办公室，承担规划编制、项目审批、设计审查、安全和质量监管、地下管线信息共享、项目投融资、专家咨询等的综合协调事宜，制订地下综合管廊管理办法、指导运营单位完善管理制度，组织综合管廊建设运营单位与入廊管线单位签订协议，确保地下综合管理正常运行，指导有关政府部门做好突发事件和应急管理工作。

（4）地下综合管廊兼顾人防应到位，不越位、不错位

根据《中华人民共和国人民防空法》第十四条规定，城市的地下交通干线以及其他地下工程的建设，应当兼顾人民防空需要。因此，城市地下综合管廊建设必须兼顾人防需要。什么叫兼顾人防要求，就是设计时，要使管廊战时具有一定的抗毁能力，保证战时生命线的安全。

到位就是必须兼顾人防，从而既提高平时管廊的防灾减灾能力，提高战时的防空袭能力。但到位不能

越位，如果兼顾人防的要求过高，使其工程造价提高过大，就会影响综合管廊的建设，就叫越位。更不能错位，即改变了管廊的功能定位，如有的战时兼顾人防要求定位为战时管廊作人防掩蔽部使用，因防护单元的要求需在管廊内施作防爆隔断设施，从而既影响管线敷设，又提高了成本，这就是错位。

（5）创新投融资机制，推广 PPP 模式

创新投融资机制的关键是吸引社会资本参与综合管廊的建设与运营。除了用足投资补贴、特许经营权、收费权和购置服务预期收益等担保类贷款贴息和发行专项金融债等财政金融政策外，考虑到社会资本的逐利性是其基本属性之一，要吸引社会资本参与，还必须妥善解决如下问题。

①综合管廊集中统一运营，降低运营成本。目前国内已建综合管廊每公里运营成本达到了 100 万元以上，如此高的运营成本，不利于综合管廊产业的健康发展。降低运营成本，是综合管廊投资获取合理回报、降低管线单位负担的基础。

在综合管廊集中统一运营条件下可打造智慧管廊，充分运用物联网监控、BIM 信息管理等技术，实现集中远程控制，建立市场机制下的维护管理，从而大幅降低安全风险和运营维护成本。

②严格"风险分担"。即政府理应承担的风险不应由社会资本"共同承担"，PPP 模式中投资风险不应全由项目公司承担，政府更不应让社会资本捆绑打包更大的建设项目。

③风险分担的具体方案：政府承担政治风险、法律风险、规划与标准变更风险、土地获取风险、市场需求风险、决策风险、管线单位付费风险、以及因政府方原因造成的工期延误风险和成本超支风险等；社会资本承担融资风险、财务风险、技术风险、因社会资本方原因造成的工期延误和成本超支风险、运营管理风险等；双方共同承担利率变化、通货膨胀、不可抗力等风险。

④采取政府提供购买服务协议的方式解决经营收益不确定性问题。待相关标准出台后，可采取"使用者付费＋政府可行性缺口补贴"模式，即在管线单位入廊收费无法平衡项目合理投资回报及运营成本时，由政府提供可行性缺口补贴。如果运营收益能够平衡投资回报及运营成本时，政府不需要提供补贴。此时，由于相关收费标准明确，政府能够准确测算需提供的补贴。

⑤杜绝地方政府捆绑打包，"小马拉大车"，增大未来投资风险。部分地方政府利用国家对地方综合管廊建设的补贴资金，吸引社会资本投资建设过程中，要求社会资本捆绑打包其他更大的建设项目，用中央补贴资金拉动地方其他大项目建设的行为，造成投资风险的极大增加。

隧道岩爆监测预警

一、隧道岩爆监测预警可行性

隧道岩爆发生在工程活动范围内，工程不开挖不会发生岩爆（与地震不同）。

岩爆的孕育过程：岩体破裂→块体形成→块体弹射

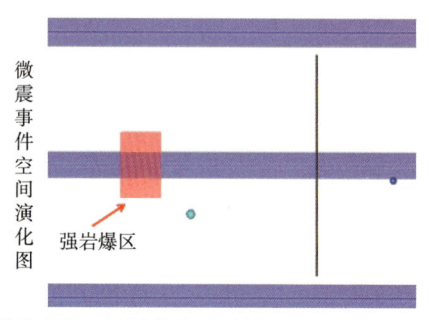

微震监测：监测岩体破裂及其能量释放的有效手段

在多数岩爆孕育过程中，微震事件及其能量的演化具有自相似性（时间、能量及空间分形特征），且有微震信息前兆特征。

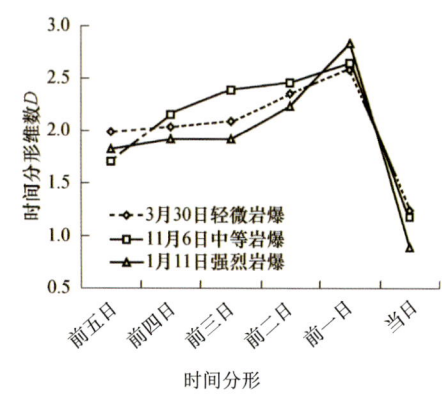

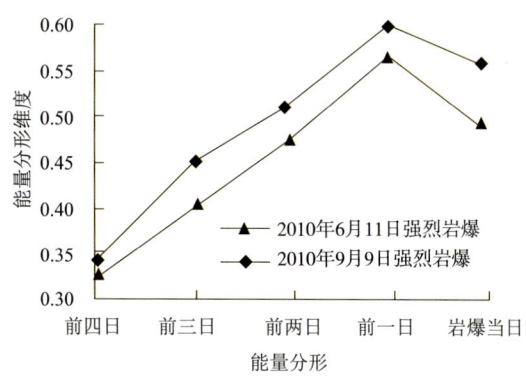

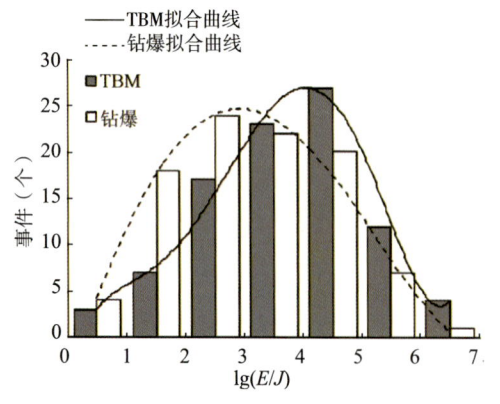

这表明，在大多数情况下，可以利用已监测到的微震活动性，在基于未来施工不变的情况下，对岩爆的区域和等级进行预警。

二、隧道岩爆监测预警技术与方法

岩爆孕育过程的预警方法：按施工方法分别预警 TBM 与钻爆法施工岩爆孕育存在显著差异：

报告时间：2016年8月18日。

① TBM 高等级岩爆孕育伴随低等级岩爆；
② TBM 同一区域常发生多次岩爆；
③ TBM 岩爆多发生于开挖过程中，钻爆法几小时到几天不等；
④ TBM 诱发的微震事件能量一般较大。

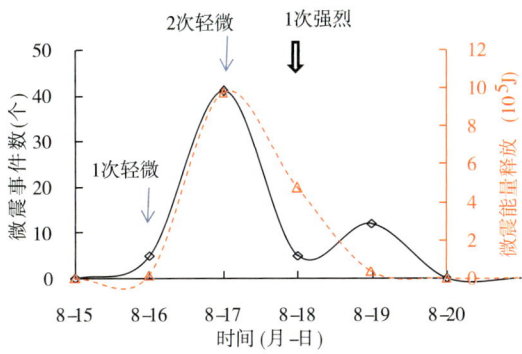
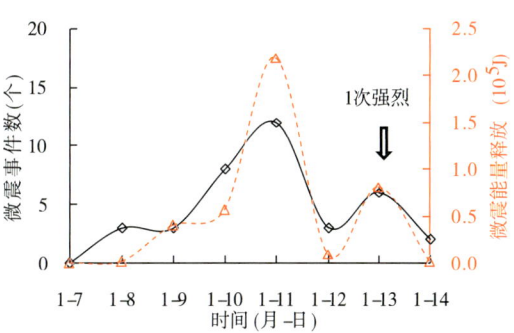

岩爆孕育过程的预警方法：再按岩爆类型进行定量动态预警。

$$P_i^{mr} = \sum_{j=1}^{6} w_j^{mr} P_{ji}^{mr}$$

式中，m 为施工方法，r 为岩爆类型，i 为等级岩爆发生概率。施工方法 m 分为钻爆法和 TBM 法；岩爆类型 r 分为应变型、应变—结构面滑移型、断裂滑移型。

钻爆法施工不同类型强烈岩爆的 6 变量 j 预警阈值

微震参数	应变型	应变-结构面滑移型
事件数（个）	49.7	32.1
释放能 [lg(E/J)]	6.3	5.8
视体积 [lg(V/J)]	5.0	4.9
事件率（个/天）	5.5	3.0
释放能速率 [lg(E/J)/天]	5.6	4.5
视体积率 [lg(V/J)/天]	4.1	3.9

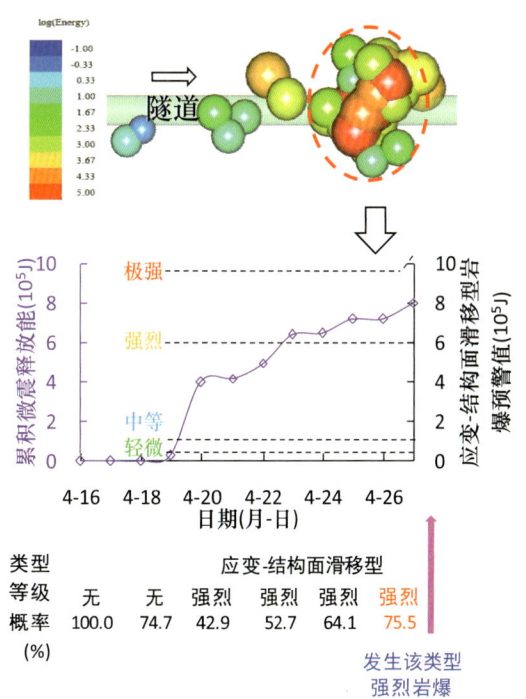

三、隧道岩爆孕育过程动态调控

1. TBM 开挖岩爆预警与调控
（1）动态调控：①开挖速率降低；②支护措施：增加锚杆。
（2）调控后：①微震活动性降低；②岩爆灾害减弱。

2. TBM 开挖岩爆预警与调控

（1）动态调控：①开挖速率：7m/d 降低到 5m/d；②支护措施：增加应力释放孔，TH 梁间距由 1.5m 减少到 1.0m。

（2）调控后：①微震活动性降低；②岩爆灾害减弱。

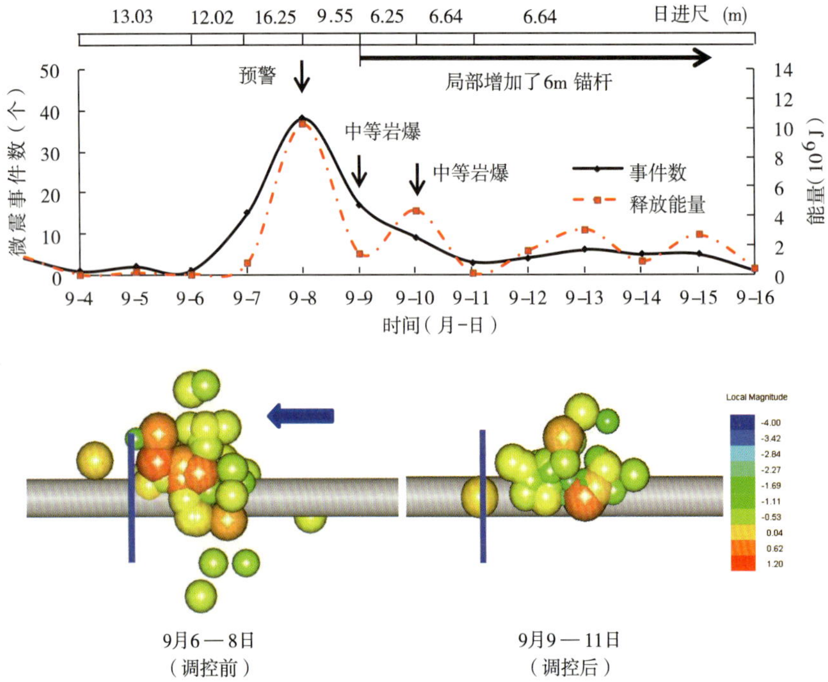

预警应变型强烈岩爆风险

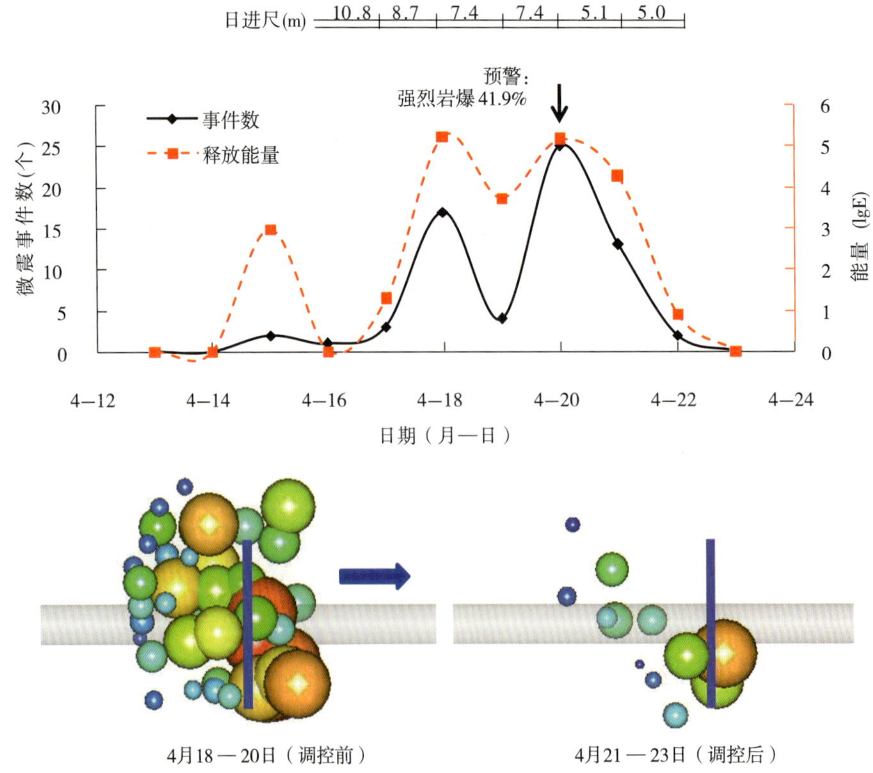

预警应变型强烈岩爆风险

3. 钻爆法开挖岩爆预警与调控

（1）动态调控。①开挖方式：27日停止B工作面掘进；②支护措施：增加系统锚杆。
（2）调控后。①微震活动性降低；②岩爆灾害减弱。

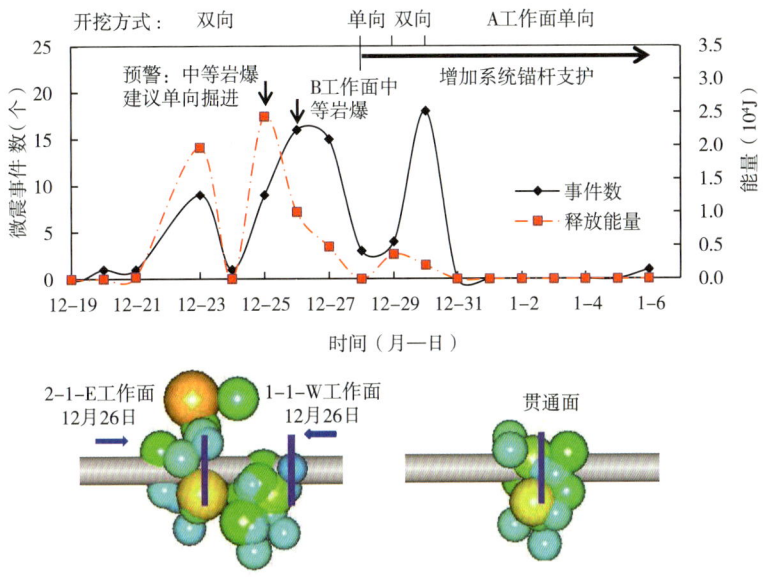

预警应变—结构面滑移型中等岩爆风险

四、隧道岩爆监测预警工程实践

1. 锦屏二级水电站深埋隧洞开挖过程岩爆监测预警与防控

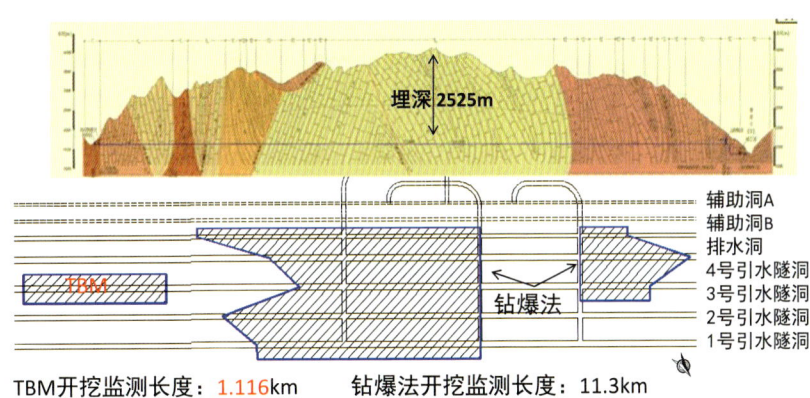

（1）2条交通洞：岩爆造成人员死亡和严重恐慌，施工队伍被迫更换多次，工期延误1年以上。
（2）排水洞："2009.11.28"极强岩爆，TBM被毁，7人丧生，停工半年，更改施工方案。在2009.11.28之前，50天掘进了40m。

相邻平行最大埋深洞段岩爆风险控制效果比较

对比项	技术应用前	技术应用后
岩爆长度占总长度比（%）	辅助洞A：20.2 辅助洞B：23.9	引水隧洞：平均7.3 排水洞：5.7
强烈岩爆次数	辅助洞A和B：平均10.5次	引水隧洞平均2.3次

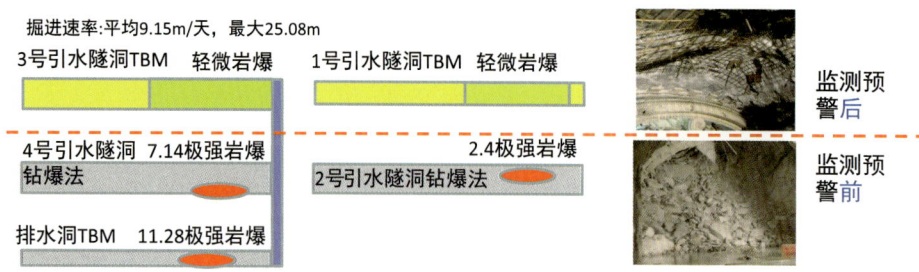

2. 锦屏地下实验室二期开挖全过程灾害监测预警与防控

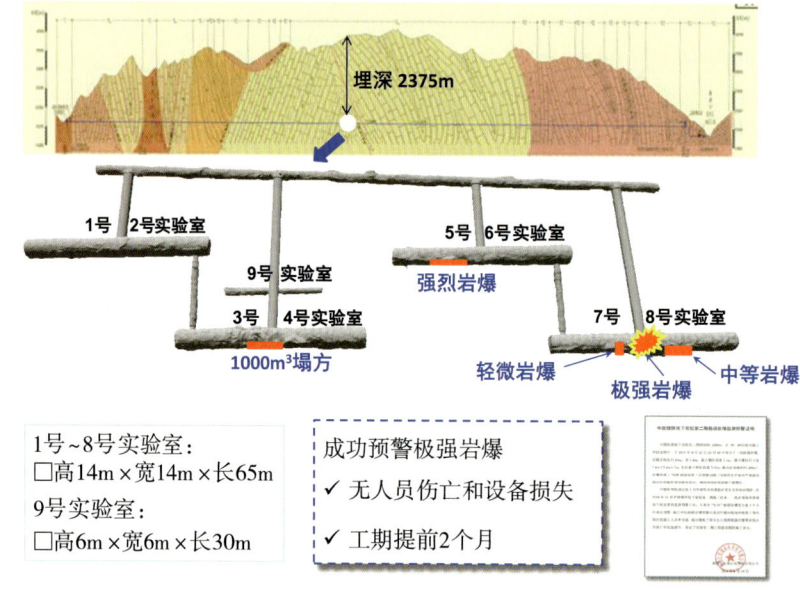

3. 巴基斯坦 NEELUM JHELUM 深埋隧道岩爆监测预警与控制

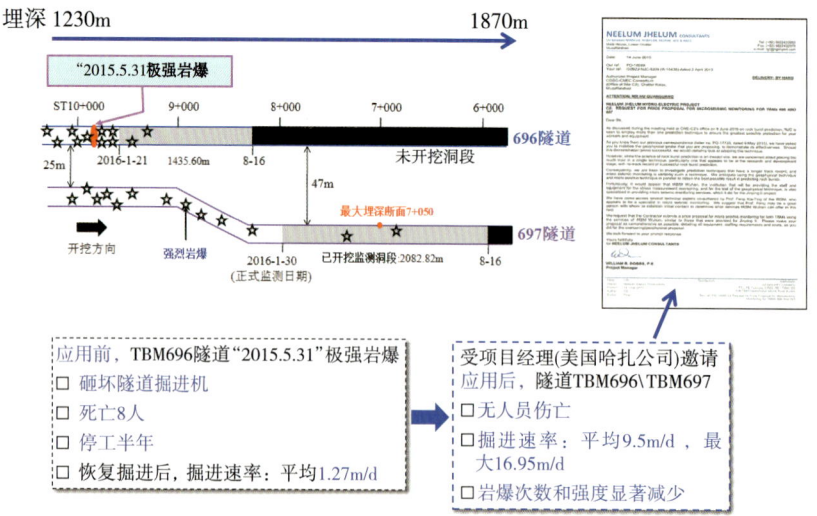

学 术 报 告——隧道及地下工程

地下工程建设安全面临的挑战与对策

一、地下工程建设安全的现状和分析

（一）建设安全现状

2008—2011 我国重特大事故统计

年份	事故性质	发生次数（次）	死亡人数（人）
2008	重大事故	86	1304
2008	特大事故	11	667
2009	重特大事故	67	1128
2010	重特大事故	74	1438
2011	重特大事故	59	897

（1）安全形势虽有好转，但事故发生率仍居高不下；
（2）土木工程领域事故伤亡人数仅低于道路交通事故和煤矿事故，居第三位；
（3）在全国重特大事故中，土木工程领域约占 10%。
近四年来地下工程事故及伤亡数没有明显的下降，总体上安全形势没有明显改善。

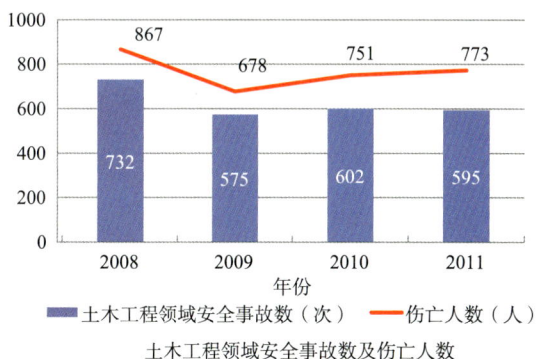

土木工程领域安全事故数及伤亡人数

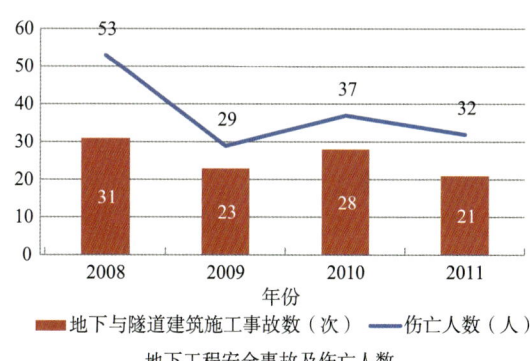

地下工程安全事故及伤亡人数

（二）事故分析

本文收集了 50 多起矿山法工程事故案例、20 多起浅埋暗挖法工程事故案例、40 多起盾构与 TBM 施工事故案例、100 多起明挖法基坑及地铁工程事故案例。

报告时间：2015年5月21日。

1. 事故类型分析

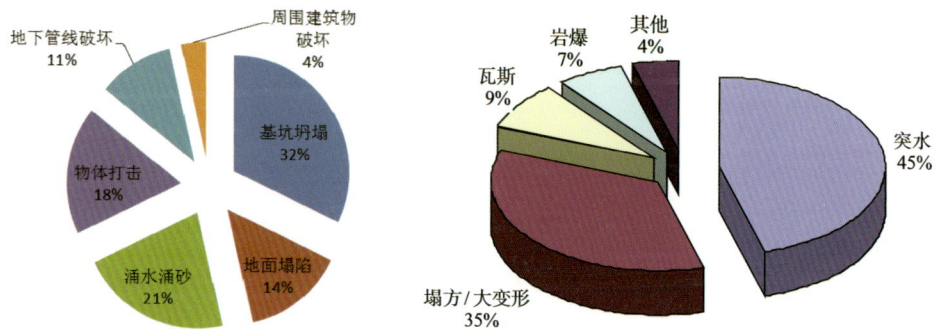

明挖法地下工程安全事故案例事故类型　　矿山法地下工程安全事故类型

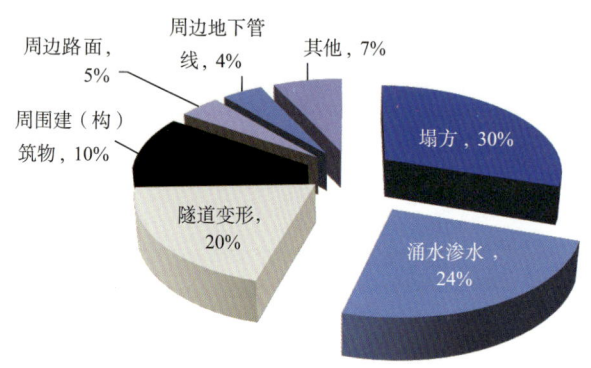

浅埋暗挖法地下工程安全事故类型

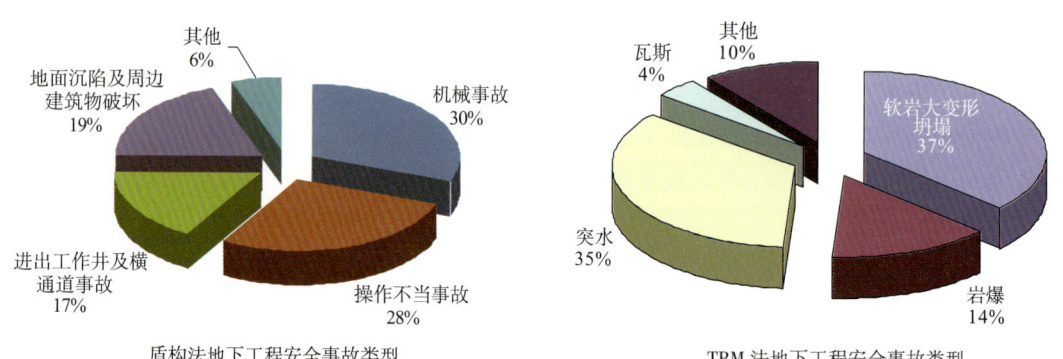

盾构法地下工程安全事故类型　　TBM法地下工程安全事故类型

2. 事故原因分析

四分之一（22%）完全为客观原因；三分之一多（37%）完全是主观原因；三分之一多（40%）既有客观原因，又有主观原因。

所有的主观原因可归为责任事故和人的因素：

（1）政府监管部门监管责任缺失，以罚代管、以包代管；

（2）施工方安全管理人员未能履行好安全管理职责，施工指挥人员违规指挥，施工人员违规操作，施工方案不完善；

（3）监理不严，责任心缺失；

（4）业主方招标时违规发包、转包、抢工期；设计方案存在缺陷，勘察资料不全等。

主观原因的深层次原因主要有：

（1）赶工期：政府、业主干预；

（2）工程造价低，安全措施投入少；

(3)用工制度造成的人员技术水平低，安全意识差，无证上岗人员多；
(4)工程招投标过程中存在违规现象，违规发包，层层转包，以包代管，以罚代管。

客观原因为绝大多数事故（坍塌、涌水涌砂、岩爆等）的客观原因是环境和气象，而工程地质环境、水文地质环境复杂是其第一主因。特别是突水突泥及岩爆难度最大，挑战最强。

二、地下工程安全管理面临的挑战和对策

（一）中国地下工程安全管理面临的挑战

地下工程建设规模大、发展快（城市地下空间开发利用；城市轨道交通建设；城市地下快速路建设），地下工程技术和管理力量难以充分保证，工期偏紧：
(1)前期工作（地质勘察和工程可行性研究）不充分；
(2)设计人员、施工管理人员青黄不接；
(3)对于地下工程的高风险特点认识不足；
(4)管理体制比较混乱、转包分包现象比较广泛；
(5)工程招投标制度不规范，项目管理水平参差不齐；
(6)工程造价偏低造成安全措施投入、风险管理的投入不到位。

（二）工程事故预防是可行的

事故发生前有预兆，预兆可以反映到监测数据中来。事故萌芽至险情发生一般历时 2~3 天。

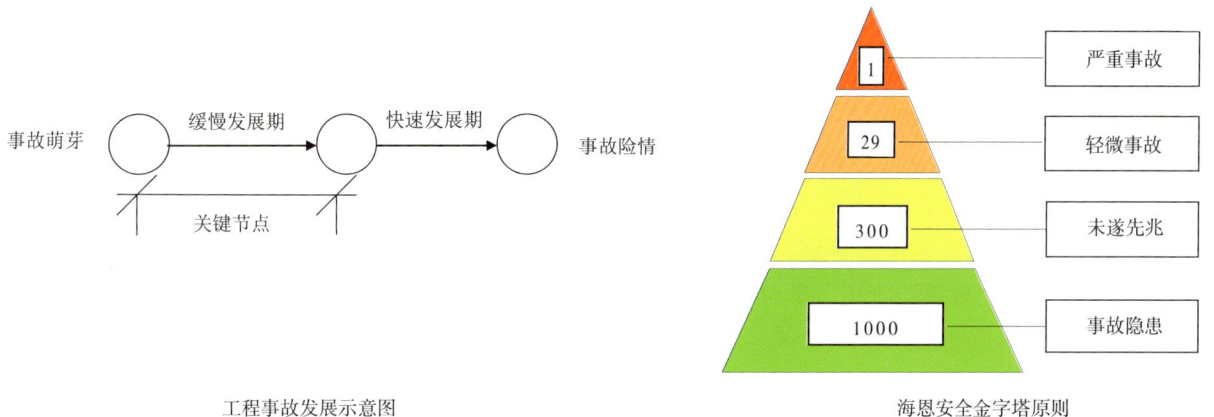

工程事故发展示意图　　　　　　　　海恩安全金字塔原则

1. 海因里希安全法则
通过 55 万次事故的统计分析，可知事故的原因。

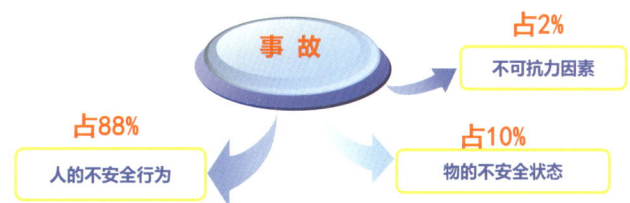

2. 约翰逊的工程变化——失误理论模型
通过失误理论模型，可知一系列的失误和变化才会引发事故的发生。

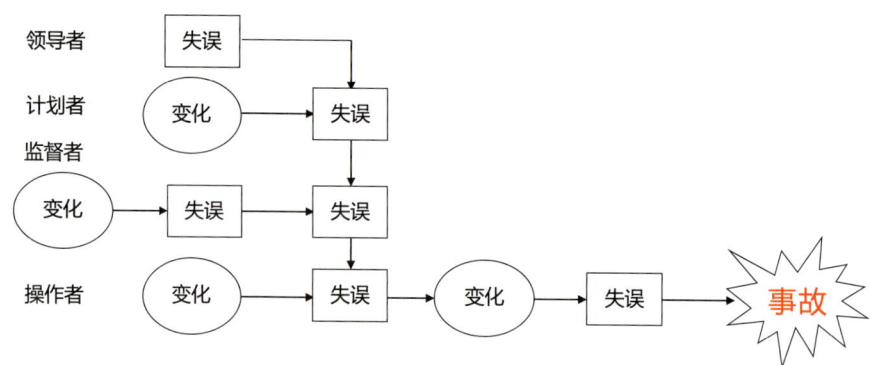

上述法则、理论和模型揭示了一个十分重要的事故预防原理：

(1) 要预防死亡重伤害事故，必须预防轻伤害事故。
(2) 预防轻伤害事故，必须预防无伤害虚惊事故。
(3) 预防无伤害虚惊事故，必须消除日常不安全行为和不安全状态。
(4) 能否消除日常不安全行为和不安全状态，则取决于日常管理是否到位，也就是常说的安全风险的细节管理，这是作为预防死亡重伤害事故的最重要的基础工作。

（三）地下工程建设安全管理应该实施的对策

1. 加强针对地下工程安全风险管理的法规建设

(1) 需要政府有关部门（住房和城乡建设部、交通运输部和水利部等）以及行业协会制订强有力的法规作指导和规范。
(2) 通过法规明确政府监管部门、业主、监理、设计和施工等各有关单位的安全风险管理责任，建立全面科学的安全风险管理体系。
(3) 通过法规强制实行重大工程立项前的安全风险分析、评估研究，凡安全风险评估为"严重"等级的工程方案应实行一票否决制，必须进行修改。
(4) 强制实行重大地下工程的设计和施工方案审核制。
(5) 通过法规规定在工程预算及定额中安全风险管理的相关费用，及其在整个地下工程建设费用中所占的合理比例，且为"专项提取"，不列入商务造价标。

2. 实施地下工程安全风险管理目的在"预防"、重在预防

(1) 提高地下工程风险意识，全面推广地下工程安全风险管理制度，将安全风险管理作为地下工程建设管理的必要组成部分。
(2) 安全风险管理主要包括风险因素辨识和分析、风险工程评估和风险监测和控制，内容涉及探明风险源、分析确定风险因素导致事故发生的概率和事故后果影响的大小，评估风险工程的等级以及建立监测系统、采取预防措施及预警应急方案以规避、转移和减缓风险等。
(3) 安全风险管理应贯穿地下工程建设的全过程，包括勘察阶段、方案设计阶段、初步设计阶段、施工设计阶段、施工阶段及工后阶段。

3. 建立地下工程安全监控中心（分中心）和第三方监测监理制度

(1) 安全监控中心是地下工程安全风险管理的信息枢纽，负责组织工程各部分、各阶段的安全风险监控、综合预警发布、信息管理和相关咨询工作，并进行监督管理。
(2) 对于安全风险专业咨询单位和第三方监测、监理单位的资质和监测、监理人员的考核、审查，政府有关部门应建立相应的严格管理制度。

4. 建立基于现代化信息化技术的地下工程建设安全风险管理信息系统

(1) 克服目前工程各方各自为政、人工流转信息不畅的弊端，利用现代化技术将工程有关技术信息和管理信息实时地传递到有关各方，使有关各方的"操作"在系统中"透明"，有利于信息共享、落实责任。

（2）利用地理地质信息系统，为风险管理提供地理、地质、工程和环境等信息支持；利用风险基础信息数据库，为风险管理提供风险工程信息，风险阈值，风险监测数据库（空域和时域），风险处置方案等；包含风险预警管理模块，为风险管理提供风险分析，预警发布，风险处置消警及历史预警查询等。

5. 加强地下工程安全风险管理理论以及重大事故预测预报和防治技术的研究

三、突水突泥地质灾害的挑战和对策

（一）突水突泥地质灾害对地下工程建设安全的挑战

中国是世界上岩溶分布最广的国家，可熔岩层分布面积达到国土总面积的1/3，长大隧道的建设将会遇到很多的岩溶和不良地质问题。

部分长大隧道突水突泥灾害详情表

隧道名称	地质灾害描述
大瑶山铁路隧道	施工期涌水量：4000～15000m^3/d，平导DK1994+213涌水造成竖井被淹、洞内机具被淹没达数月；正洞DK1994+600涌水淹没隧道200余米，水深1.4m，隧底淤积泥沙厚1m，中断施工长达1年之久；DK1994+636～DK1994+637处发生涌泥、涌沙80m^3，淹没轨道，造成短时行车中断；地表斑古坳地区生产生活用水枯竭；地表坍塌约413次
圆梁山铁路隧道	施工期涌水量：110000m^3/d（出口DK361+764）；DK354+450～DK354+510溶洞发生涌水为$9.6×10^4$～$1.656×10^5$ m^3/d，伴随涌沙涌泥，淤积长度130m，高度2.5m，涌沙量约1300m^3；DK354+879溶洞发生涌水，伴有涌泥涌沙（总量约6000m^3），最大涌水量达$7.2×10^4m^3$/d，造成人员伤亡事故，被迫采用迂回导坑通过；DK360+873涌砂，淹没导洞近200m；DK361+764处发生涌水随涌沙涌泥，涌水量为240000m^3，涌泥沙覆盖整个掌子面，淤积量约15000m^3
歌乐山铁路隧道	DK2+619.6发生涌水，涌水量14400m^3/d，涌水泥沙含量达20%～30%
宜万马鹿箐隧道	2006年1月位于湖北省恩施市屯堡乡的宜万铁路马鹿箐隧道发生涌水，死亡11人
宜万野三关隧道	2007年8月5日发生突水突泥事故，在30min内，突出15104方水、泥、石等物，导致大量机械被冲出，并造成人员伤亡，治理时间至少半年，淹没正洞500余米，掌子面附近200米全被大块孤石充满

渝怀线铁路圆梁山隧道、广渝线公路华蓥山隧道、渝怀线铁路武隆隧道、宜万线铁路野三关隧道、宜万线铁路马鹿箐隧道、衡广线铁路大瑶山隧道等均发生过严重的突水突泥事故。

宜万线野三关隧道突水

在国内外隧道特大事故中，突水事故在死亡人数和发生次数上均居于前列。据我国隧道施工的不完全统计，在水利水电、铁路、公路、矿山的隧（巷）道施工过程中遇到的突水、突泥等地质灾害所造成的停工时间约占总工期的30%。

随着长大隧道等地下工程埋深的增加，突水突泥带来的经济损失及人员伤亡呈增长趋势。

（二）突水突泥地质灾害的发生是有科学规律可循的

突水突泥的发生条件：含水构造的能量储存，含水的动水性能和能量释放以及围岩的稳定性。

突水突泥的影响因素，总体上可概括为工程因素和地质因素：工程因素包括开挖、爆破、注浆以及其他人为活动；地质因素主要包括地下水、岩组、地质结构面、岩层产状、沉积期和剥蚀面、地貌和降雨等七大因素。

（三）突水突泥地质灾害发生前是有预兆的，因此一般是可预测预报的

突水突泥地质灾害的前兆包括地质调查前兆、含水构造在地球物理勘探中的前兆以及临突水前的前兆。

1. 突水突泥地质灾害的地质调查前兆特征

（1）在地下工程施工中，通过观测钻孔出水情况以及对开挖揭露的围岩变化情况进行调查分析，可以获取一定的前兆信息，从而予以预报。

（2）在地下施工中，当接近不良地质或岩溶水时，可以发现一定的突水突泥灾害特征。

①断层破碎带前兆；②临近大型溶洞水体或暗河前兆；③岩溶管道前兆；④淤泥带的前兆；⑤临近人为坑洞积水的前兆。

2. 突水突泥地质灾害在地球物理勘探中的前兆特征

岩溶水体和不良地质体的存在是突水突泥地质灾害发生的前提，当地下工程开挖邻近它们时，它们在电场、磁场和地震波等地球物理信息场上有一定的响应特征，即表现为特殊的前兆信号：地质雷达波通过地下水、含水溶洞及富水介质等含水体后，高频成分被吸收，含水层的反射波的优势频率会降低；电磁波通过含水裂隙或富水破碎带时，其电导率远大于无水岩石，因此雷达波的反射强烈；雷达波遇含水层的反射波相位，相对于入射波，相位差 180°。

3. 突水突泥地质灾害在临突水前的前兆信息

模型试验和现场监测表明，当开挖面接近岩溶水体构造时，隔水层及其附近围岩内部经历了裂隙萌生、扩展、贯通的细观破裂过程和围岩渗水、局部涌水和整体失稳突水的宏观灾变过程。

（四）应对突水突泥地质灾害的对策

（1）突水突泥地质灾害事故的绝大部分在实行综合超前地质预报与预警机制后是可以避免的。

为了预防和避免突水突泥灾害事故，可以和必须实施超前地质预测预报，提高预报的准确性和精确度，必须实施综合超前地质预报，从而扬长避短、相互补充、相互印证，多角度、多参数地对施工前方的地质情况进行预报：宏观超前地质预报（工程地质法）、长距离（50~200m）超前地质预报（工程地质法、TSP探测等）、短距离（50m以内）超前地质预报（地质雷达、红外探水、瞬变电磁、超前钻探、超前导洞及经验法等）。根据综合地质超前预报报告制定相应的预警方案和处理措施。

（2）为了预防和避免突水突泥地质灾害事故，保证重大地下工程的建设安全和顺利进行，必须在岩溶地区实施地下工程安全风险管理制度。

①首先，在重大地下工程可研阶段实行工程安全风险管理立项研究，通过工程风险分析和风险评价决定工程是否批准立项。

②其次，在立项后工程实施过程中，首先通过地质分析确定所要预报工程各段的围岩情况进行风险等级评价和划分，对不同地质灾害风险等级的工程段实行不同的综合预报方案。

③灾害预测预警级别可分为四级：红色（特别严重，Ⅰ级）、橙色（严重，Ⅱ级）、黄色（较重，Ⅲ级）和蓝色（一般，Ⅳ级）。

④预警信息一经发布，必须严格按照预警级别及应急预案制定相应的工程预防措施并按措施进行施工。

近年来，山东大学在突水突泥地质灾害超前地质预报和预警的工程实践上其成功率已达 90%。

四、岩爆地质灾害的挑战和对策

（一）岩爆地质灾害对地下工程建设安全的挑战

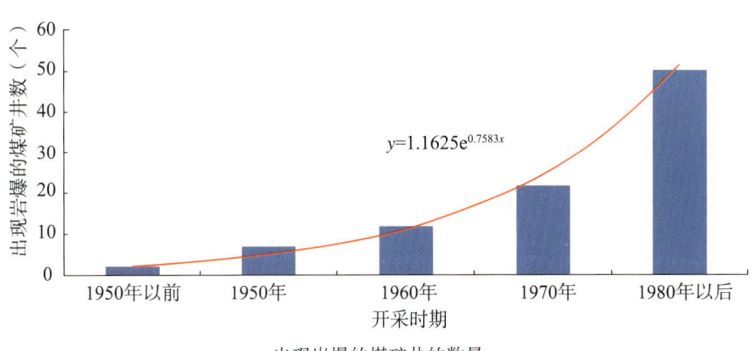

出现岩爆的煤矿井的数量

（1）发生岩爆灾害的煤矿井数量呈指数规律增长。

（2）金属矿的开采深度普遍大于煤矿采深，金属矿山的岩爆灾害更为严重突出。

我国发生岩爆的隧洞工程汇总（不完全统计）

工程名称	竣工年份	最大埋深（m）	岩爆等级及比例			岩爆次数（次）	岩爆段长度(m)	备注
			轻微（%）	中等（%）	强烈及极强（%）			
成昆铁路关村坝隧道	1966	1650	为主	少量	无			零星岩爆
二滩水电站左岸导流洞	1993	200	为主	少量	无	315		工程区位于深切河谷卸荷集中区域，最大主应力 σ_1 为26MPa，方位角 N34° E，倾角23°，因而以水平应力为主
岷江太平驿水电站引水隧洞	1993	600	为主	少量	少量	>400		
天生桥二级水电站引水隧洞	1996	800	70	29.5	0.5	30		比例依据岩爆次数统计
秦岭铁路隧道	1998	1615	59.3	34.3	6.4		1894	比例依据岩爆段长度统计
川藏公路二郎山隧道	2001	760	为主	少量	无	>200	1252	
重庆通渝隧道	2002	1050	91	7.8	1.2		655	比例依据岩爆段长度统计
重庆陆家岭隧道	2004	600	55.8	39.7	4.5	93		比例依据岩爆次数统计
禄布沟水电站进厂交通洞	2005	420				183		工程区位于深切河谷卸荷高应力集中区内，地应力方向沿着河谷边坡向与隧洞呈大角度相交
秦岭终南山特长公路隧道	2007	1600	61.7	25.6	12.7		2664	比例依据岩爆段长度
锦屏二级水电站引水隧洞	2011	2525	44.9	46.3	8.8	>750		比例依据岩爆次数统计
江边电站引水隧洞	2012	1678	46.4	50.4	3.2	>300		比例依据岩爆次数统计

随着地下工程埋深增加，岩爆灾害发生的频次和等级均明显增高，呈现岩爆等级向中级和强烈等级明显增加的趋势。

锦屏二级水电站引水隧洞工程更多次发生历史上罕见的极强岩爆。

锦屏二级水电站岩爆现场

（二）岩爆是可以预测的

岩爆地质灾害发生的主要机理已大致清晰，在一定程度上可以对岩爆进行安全风险分析意义上的预测。

从目前岩爆研究进展的成果看，其发生的主要（虽非全部）机理已逐渐清晰，因而可以对其进行大致的（虽非准确的）预测；我国大量已经进行的对岩爆监测预报的工程实践也表明，对岩爆的发生部位与强度进行大致的预报是可行的；缺乏对岩爆进行预测、预报的预警将是盲目的，因而也是事倍功半的。

1. 岩爆的定义和分类

将岩爆定义为高地应力地区由于地下工程开挖卸荷引起的围岩弹射性破裂的现象，而将无动力弹射现象的围岩脆性破坏归之于围岩静力稳定性丧失现象。

从其发生机制出发，基于围岩中有无结构面，可将岩爆类型分为应变型岩爆、地质构造型岩爆以及应变与构造杂交型岩爆。

此定义和分类将有利于对岩爆进行预测预报和采取预警措施和工程应急方案。

2. 均质岩体应变型岩爆的预测原理

对于围岩中只存在微缺陷，不存在断层、裂隙、节理等宏观裂纹的围岩称之为均质岩体，在这类岩体中发生的岩爆，一般为应变型岩爆。应变型岩爆的理论机制概括为：工程所在地区的初始高地应力在开挖卸荷引起的应力集中后，使得围岩中的局部地应力水平

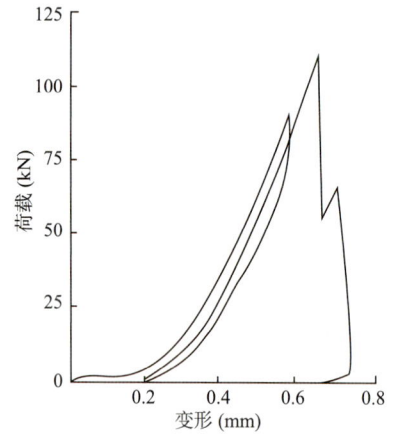

粉砂岩的峰值前加载曲线和荷载变形

超过岩体极限强度从而造成围岩的局部破坏，而当破坏岩体相应其围岩高应力水平的围岩弹性应变能超过围岩破坏所耗散的破坏能时，其剩余应变能因围岩的应力梯度转化成破坏岩体的动能，从而破坏岩体向临空面弹射形成岩爆。

当岩石加载至峰值强度后，外载做功 A_1 所产生的能量可分为弹性变形能 Φ_{SP} 和塑性变形能 Φ_{ST} 两部分。前者为岩石储存的变形能，后者则在加载过程中以热能等形式耗散掉；而峰值强度后曲线下降段曲线所围的面积则为岩石破坏所耗散的能量 A_2。发生岩爆的岩石动能即为剩余能量 $\Delta = \Phi_{SP} - A_2$。这只有在 A_2 较小，即 Δ 为正时才可能，即峰值后下降段曲线较陡才有可能，也即脆性破坏岩石才有可能。因为：$\Delta = \Phi_{SP} - A_2$；所以：$\Delta = A_1 - \Phi_{ST} - A_2$。

岩爆预测的四个条件：
(1) 获得相应围岩岩石的加－卸载过程的全应力－应变曲线；
(2) 计算出高地应力地区洞室开挖卸荷应力重分布后，非连续非协调变形岩体中的应力(应变)场；
(3) 计算出相应峰值强度前塑性变形能的微缺陷(裂纹)扩展以及次生裂纹稳定扩展所消耗的能量；
(4) 计算出相应于峰值强度下降段曲线岩石破坏能的次生裂纹失稳扩展合并成宏观裂纹的耗散能。

建立基于断裂力学和非欧非连续非协调变形介质解析计算模型或扩散有限元数值计算模型，根据该模型已可预测围岩中可能发生岩爆的危险区的部位、面积以及破坏岩块的弹射速度(即岩爆等级)。

3. 围岩中邻近洞室存在单独结构面时应变型岩爆的预测

基于工程实践，当邻近洞室的围岩节理裂隙发育到一定程度，地应力得到松弛，一般不易发生岩爆。但当邻近洞室仅存在单独结构面(裂隙)时，则另当别论，此时构造地应力松弛不充分，而裂隙位于开挖引起的应力重分布的应力集中区，导致原生裂隙滑动、启动出现次生裂隙，扩展到一定长度后，与洞壁发生相互作用，产生失稳扩展，形成破裂区，对于存在原生裂隙区的脆性岩石，其裂隙扩展、破裂消耗的能量更小，所以与无宏观裂隙的围岩相比，含单条裂隙的围岩洞室更易发生岩爆，岩爆的面积更大，岩爆的强度(弹性速度)更高，这已为预测计算所证实。

(三) 岩爆监测预报在理论上和工程实践上是可行的

岩爆的宏观破裂是由大量的微破裂的产生合并，集群导致的失稳破裂发展而成，就是岩爆的前兆；微破裂可以被微震监测器在其有效作用范围内所接收。微破裂产生、合并、集群导致的失稳破裂发展的规律，即是岩爆宏观破坏的规律，也是岩爆监测预报可行性的理论基础。

微震监测系统监测范围广，具有监测自动化、信息化、智能化和远程化的特点。微震监测系统的监测范围可以覆盖地下工程开挖卸荷引起的岩爆发生范围。中科院岩土力学所和大连理工大学的预报准确率均超过了85%。

(四) 对于岩爆地质灾害的对策

综合以上岩爆预测预报的理论研究和工程实践成果，应对岩爆地质灾害的对策包括：
(1) 在高地应力地区建设重大地下工程时，必须在原位实测与反演分析的基础上，掌握工程埋深区域的实际初始地应力的分布资料；在室内试验的基础上，获得工程开挖区域岩石的加卸载全过程的应力应变曲线。
(2) 进行工程开挖前的结构面(断层、裂隙等)的超前地质预报。
(3) 进行工程各开挖段的岩爆预测数值研究分析。
(4) 选择有经验的科技队伍，进行连续有效的微震监测，为了保证监测的连续有效，必须对现场监测设备及线缆进行有效的保护。
(5) 在综合对比分析超前地质预报的结构面信息、微震监测信息、岩爆预测数据的基础上，提出岩爆监测风险预报报告。
(6) 建立施工各方领导参加的对岩爆风险预报报告的会商制度(每天、每三天、每周)提出相应岩爆风险等级的防治策略及措施：降低开挖进尺的施工调整方案；增设超前钻孔爆破的应力释放孔；加强抗爆支护措施(增加喷层厚度、系统锚杆密度及长度、加大锚杆垫片的厚度及尺寸、安装等阻大变形抗爆锚杆等)。

世界地下物流系统研究动态与新进展

一、地下物流系统概念与研究动因

（一）地下物流系统概念

城市地下物流系统是指城市内部及城市间通过地下管道、隧道等运输固体货物的一种全新概念的自动运输和供应系统，将物流基地或园区的货物通过地下物流系统配送到各个终端，这些终端包括超市、工厂、中转站、甚至是小区等，是除传统的公路、铁路、航空及水路运输之外新型运输和供应系统。

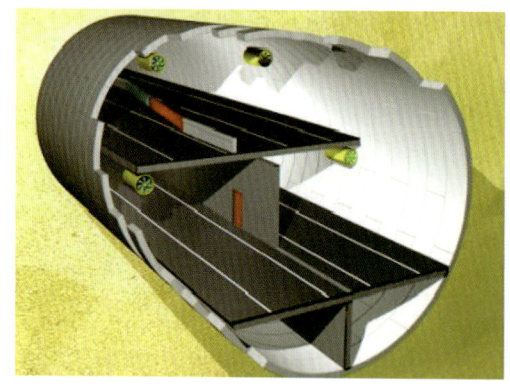

隧道（tube） 自动机车（Cargo mover）

管道（隧道）+ 自动运输和供应系统

（二）研究动因

1. 世界范围内城市交通拥堵解决思路

（1）拓展现有交通运输系统

报告时间：2011年9月20日。

长达25m的传统公路货车

长达33m的集装箱货车

美国的双层集装箱列车

运载能力达3.5万t的驳船

从长远的观点来看，对现有交通运输系统的扩展不可能满足未来集装箱集疏运快速增长的需要，也不是一种可持续的增长方式。

根据Downs定律，仅仅依靠单纯的增加道路供给并不能满足由于社会经济发展而产生的交通需求，道路交通的设施建设将会诱发潜在的交通需求，而这部分潜在的交通需求将很快占有新增的道路空间。

如欧洲在1975年至2002年，高速公路（highways）增长了65%，地区道路（regional roads）增加了15%，而交通强度分别增加了206%和90%，总的车辆公里数增加了144%，并且这种情形正在继续恶化。而对于公路双层集装箱货车或所谓的"Eco-Combis"（超过25m的货车），实施起来占地多，环境影响大，无论是目前或将来在欧洲都不会被批准。

（2）发展新运输方式

自动运输机车

 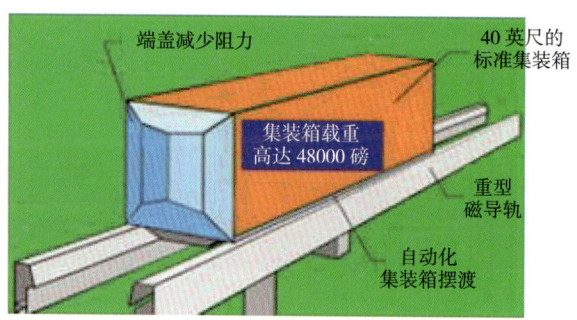

"悬挂"集装箱运输系统理念

发展新运输方式的最重要方向之一就是自动运输技术与地下隧道或管道相结合,即地下物流系统,它符合高效、低成本(节地节能)、环境友好等可持续发展原则。

因此,美国 Texas Transportation Institute(TTI)、Texas A&M University 与 Freight Pipeline Company,荷兰 Delft University of Technology,德国 Ruhr University Bochum,比利时 University of Antwerp 和日本等国家的相关科研机构等都不约而同提出并进行了地下物流系统的探索和研究。

2. 城市空间资源紧缺的约束

在美国南加利福尼亚的交通拥堵是一个"传奇"(legendary),仅洛杉矶每年由于交通拥堵所造成的损失达 110 亿美元。不断增加交通量使得该地区交通处于停顿状态。"即使当地政府、政治团体和社区领导能够同意通过拓展建设公路或铁路,地面已没有可供拓展的空间,甚至采取高架的形式也没有建设的可能性,因为高架方式仍然会有"足迹"(footprint)占用地面面积,而地面已没有可以利用的空间了。

荷兰的鹿特丹港口,在其战略规划《2020 港口展望》(《Port Vision 2020》)中,仅越来越重视环境,注重提高港口的竞争地位,投入了对环境友好的交通方式投入较大注意力,而且意识到"空间"("space")正在成为一个真正的奢侈品(a real luxury),空间不再首先考虑交通及港口发展方面的利用,海港内部及外部的空间越来越处于一种紧缺状态。

我国城市也开始面临空间紧缺问题,即使像深圳这样的新兴城市也同样存在问题,如《深圳 2030 城市发展策略》:

经过 25 年的快速发展,过于依赖物质资本积累和自然资源消耗的经济发展模式已经难以为继。城市不断增长的发展需求与资源有效供给不足(特别是空间资源和水资源短缺)的矛盾,成为制约深圳未来发展的主要"瓶颈"。

《北京交通发展纲要(2004—2020)》编制思路和要点:"城市土地和空间资源的严重短缺""更值得重视的问题是今后供给的能力将受到严重制约";"这样一个特殊的先天条件决定了北京市不可能持续大幅度扩充道路设施"。

3. 能源资源紧缺约束

从全球看,1962 年是石油发现的顶峰,也是世界石油储量增长的拐点。

美国能源部发布的世界和中国的已探明的矿物燃料资源可使用年限

项目	煤	天然气	石油
世界	221 年	80 年	39 年
中国	85 年	62 年	19 年

根据比利时研究，各种运输方式 1L 燃料运输 1km 的质量见下表。

运输方式	道路	铁路	水路	地下物流系统
质量	50t	97t	127t	≥200t

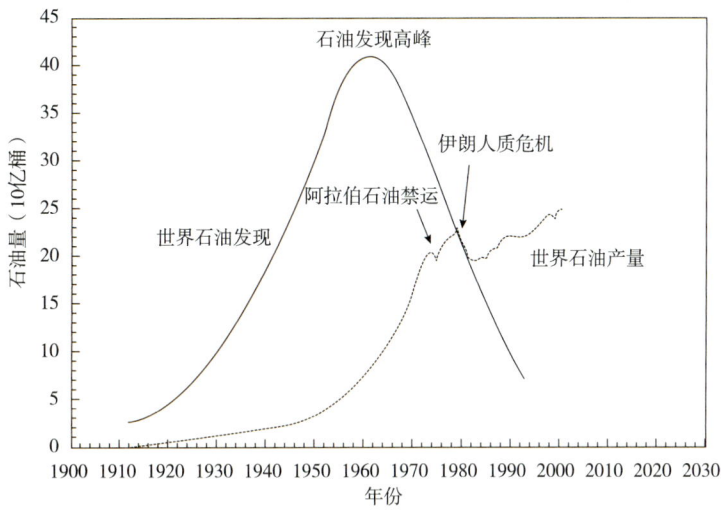

由上表可知，地下物流系统的能耗小于货车能耗的 25%。

根据美国研究，铁路与货车两种运输情况对比见下表。

方式	能耗	设施能力	费用	安全
铁路	455t·km/加仑	每年 2.16 亿 t/每轨道	2.7 美分/吨英里	0.61 次严重事故/吨英里；12.4 次事故/10 亿吨英里
货车	105t·km/加仑	每年 3780 万 t/每车道	5.0 美分/吨英里	1.45 次严重事故/吨英里；36.4 次事故/10 亿吨英里

注：1 英里 =1.6093km。

由上表可知，货车能耗约为铁路 26%，而地下物流系统（PCP）的能耗不足货车运输的 10%；地下物流系统的能耗为货车的 10%~25% 之间。

4. 面临的环境问题

世界经济合作组织（Organization for Economic Co-operation and Development, OECD）在 2003 年的《配送：21 世纪城市货运的挑战》报告中指出，发达国家主要城市的货运交通占城市交通总量 10%~15%，而货运车辆对城市环境污染则占污染总量的 40%~60%。

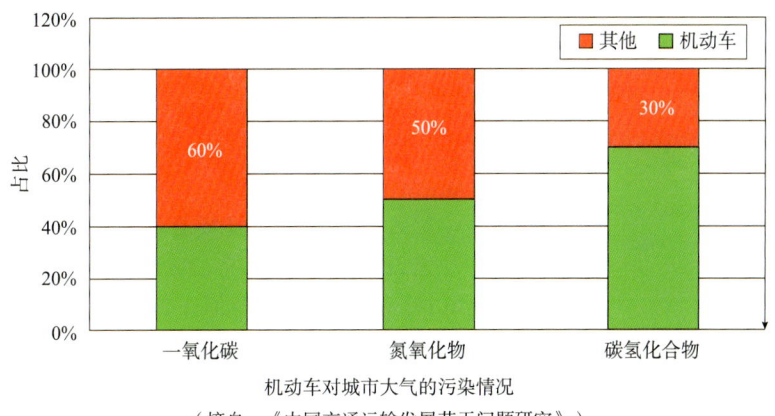

机动车对城市大气的污染情况

（摘自：《中国交通运输发展若干问题研究》）

二、各国地下物流系统的研究概况

截至目前,已召开6次国际地下物流会议,其中,1999年在美国、2000年在荷兰、2002年在德国、2005年在中国、2008年在美国、2010年在中国。

(一) 日本

从20世纪80年代开始,日本建设省下的公共工程研究院(The Public Works Research Institute of the Ministry of Construction)就开始致力于地下物流系统的研究。公共工程研究院发展了两用货车(DMT)。

两用货车以电力为能源,地下无人驾驶,也可在地面正常道路上有人驾驶。缓解地面上的交通堵塞和减少交通事故的发生率。

减少货车有毒气体排放,但是由于电池的限制,货车不能运载超过2t的货物。

1. 新项目——东京地下集装箱运输研究

根据第4次日本物流调查(the 4th Survey on Physical Distribution),仅在东京都市区内部的物流量是126万t/天。

"京都议定书(The Kyoto Protocol)"约束工业国家需减少温室气体的排放,已在2005年2月生效。以1990年作为基准,从2008年至2012年日本需减少6%的温室气体的排放。

2. 日本东京地下集装箱运输研究

建设地下配送系统(Underground Distribution System),确保可靠的运行,减小环境压力,消除港口周围的交通拥堵。

改善工作条件,减小交通事故的发生。

减少二氧化碳的排放,高效利用港口的土地,改善港口地区的景观。

其基本要求是：

(1) 可以运输 20 ft（1ft=0.3048m）的标准集装箱；

(2) 地下运输系统是无人驾驶的；

(3) 地下隧道将在地下 40m 或更深的层面上；

(4) 中间可与快速路相会合；

(5) 隧道是一个圆断面，其多余部分安排其他设施；

3. 日本东京地下集装箱运输研究

研究认为：该地下线路具有可行性，并符合《大深度地下空间法（Super Deep Underground Law）》要求。目前正在进行相关一系列研究，如信息设施、能够对大型灾害作出反应等。

（二）德国

在德国，道路作为一种传统的交通方式已成为交通的"瓶颈"。

每天大约有 10% 的高速公路（11000km）因为交通拥堵变成"停车场"，伴随的是 3300 万 L 汽油和 1.3 万 h 工作时间的浪费。

每年因交通堵塞而造成的损失达到 1000 亿欧元，这种对交通需求随着机动车数量的增长速度大于路网的增长速度在进一步加剧。

1998 年，受 Ministry of Education, Science and Research of North Rhine Westphalia 的支持，德国波鸿-鲁尔大学（Ruhr University Bochum）进行了一项各学科间的联合研究"地下的交通和供应系统（Transport and Supply Systems under the Earth）"，其目标是寻求一种通过管道运输货物的解决方案，这项研究的成果就是提出革新的 CargoCap 概念。

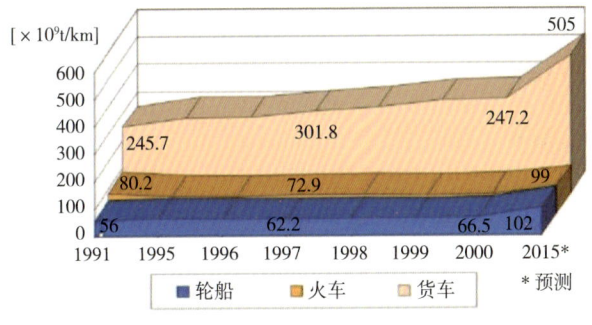

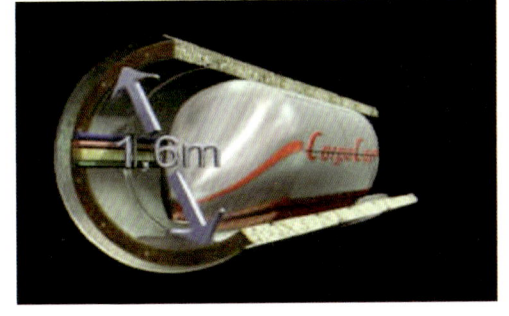

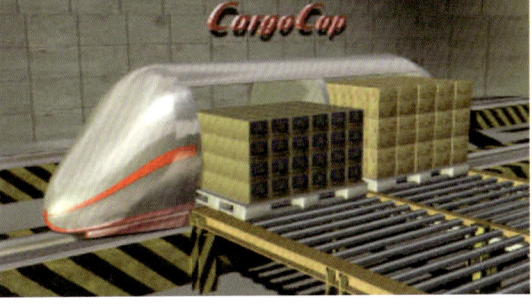

根据目前发展的 CargoCap 的概念，不需要任何的转运和重新包装，约有 2/3 的德国货物可直接适合 CargoCap 系统运输。

（三）美国

1991 年，National Science Foundation 在 University of Missouri-Columbia 成立了 Capsule Pipeline Research Center，标志着美国地下物流系统研究的开始。来自密苏里大学土木工程、机械工程、化学工程、电力工程、工业工程、采矿工程和法律等七个系的 15 名学者和 100 名学生参与这项研究工作。

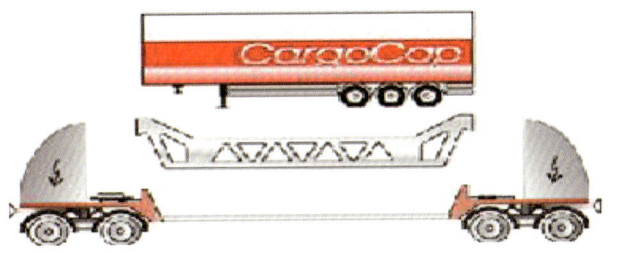

其资助资金主要来自美国国家科学基金会

(NSF), Department of Economic Development, Industry consortium,U.S., 以及 Department of Energy of Federal Energy Technology Center, 资金大约600万美元。

1994年,马萨诸塞州的Volpe国家交通系统中心提出一个报告将其定义为"管道运输(Tube Transportation)"。这个报告是为联邦高速管理部门落实《多式联运效力法案(ISTEA)》的6020条款而提出的。

1996年,美国土木工程师协会(ASCE)举办了一次国家级的关于管道研究的专题讨论会,会上提出了各种不同类型的管道舱物流系统。

圆形PCP系统

方形PCP系统

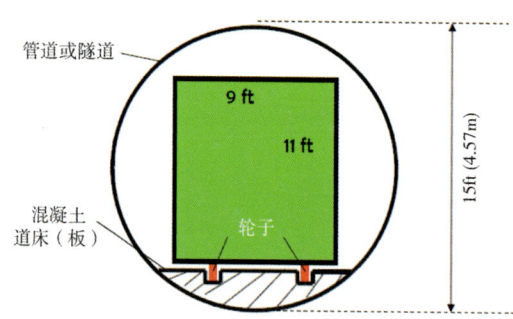

1997年,美国国家科学与技术理事会提交了一个关于"运输科学与技术战略"的报告。在这个报告中,管道舱物流系统被作为改善货物运输的方式之一。

1998年,美国国会通过了《交通效力法案》(TEA-21)。在这个法案中,国会向得克萨斯州交通研究院(TTI)批准112.5万美元的经费进行货物管道运输的研究。

(1) PCP系统集疏纽约港口集装箱可行性研究

根据标准箱尺寸,圆形隧道直径应该是15ft。

对于方形的导管需要9ft宽,11ft高。

每个capsule将是8.5ft宽,10ft高,42ft长,可以运输一个40ft长的集装箱,或2个20ft的标准箱。

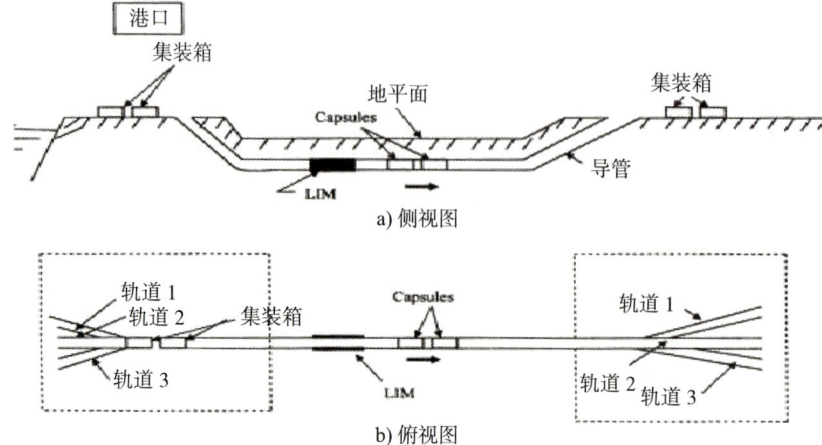

(2) 美国安全货物机车 (Safe Freight Shuttle)

自1998年，得克萨斯州交通研究院（Texas Transportation Institute –TTI）开始进行货物管道运输的研究。

通过近六年的研究，得克萨斯州交通研究院提出安全货物机车（Safe Freight Shuttle）新概念。

(3) 克利夫兰医疗中心地下 AGV 系统

克利夫兰医疗中心约227000平方英尺，现有81台自动导车（AGV）受中心各部门的触摸控制面板派遣，每天运行超过1000英里运送材料，极大地提升医疗中心的运行效率，并大幅削减开支。

（四）荷兰

1994年，荷兰政府为可持续技术发展纲要组织了几次会议，与会的专家学者提出了地下货物运输是未来可持续发展运输的潜在技术。

1996—1997年，进行了城市地区管道运输说明研究（Illustration process for tube transportation of freight within urban areas）。在这项研究中，提出了一种新物流概念：这个物流系统由地下管道自动运输网组成，联结商业区、居住区、工业区。装载的单位为1m³左右。

自动导向车（AGV）

1997年，荷兰成立IPOT（Interdepartemental Projectorganisatie Ondergronds Transport）组织，由来自荷兰交通部、公共事务与水管理部门的代表组成，对地下货物运输进行探索性研究。探索性研究表明：每年30%的国内货物运输（2.45亿吨货物）适合地下货物运输系统，并且具有良好的社会效益，如减少能源消耗、空气污染等。另一方面，建设这个新设施，需要很长时间才能完成，这个网络由4600km管道组成，其投资将达到250~500亿美元。

OLS-ASH工程可行性研究：Hoofddorp, Schiphol 和 Aalsmeer 这三个区域将会通过地下物流系统彼此连接，尤其可以缓解Aalsmeer花卉拍卖市场和Schiphol机场区域的道路拥堵。

（五）比利时

2002年，安特卫普集装箱集疏分担比例为60%，9%，31%（公路，铁路和水运），并且从长期来看，已不能适应其集装箱量的增长需要。

公路：安特卫普的环路（Antwerp Ring road）在高峰时段已经饱和。其环路全部贯通后（即所谓

Oosterweel link 计划在2010年或更早一点运营),将扩大其道路的通行能力,并在短期内缓解其交通拥堵,但从长期来看,是不可能是满足Deurganckdock码头集装箱集疏运的可持续方案。

铁路:当前铁路运输潜力也已经被充分利用,而其规划的Liefkenshoek铁路至少在2012年前完成。

水运:在将来可能没有足够的泊位(新码头全部运营后,至少每年有2万驳船在运输)。

传统集装箱集疏运不可能适应快速增长的集装箱运输需求,利用地下物流系统进行集装箱运输被提出。

安特卫普港采用垂直竖向进出口形式(vertical shaft entry)时,如果系统速度为7km/h,则地下物流系统每小时可运输100个自然箱(box),在该速度下,每天可运输5000个自然箱,每年达到180万个自然箱。如果速度提高到15~16km/h,每年可达到400万自然箱。

按20年寿命周期计算,运输每个集装箱少于30欧元。与传统运输方式相比,具有相当的竞争力。

安特卫普采用水平坡道进出口形式(horizontal ramp entry)时,自动导向车AGV形式,隧道内部无人驾驶,由计算机程序控制集装箱或货物的装卸,可为固定路线,也可无确定的路线,该系统在隧道外也可以利用。

其优点是灵活,可靠,要求的空间小而效率高。

其缺点是费用高;需它们自己电力供应,电池不能满足需要;因此,柴油电机也许是选择之一,因而需要通风系统。

轨道机车(rail Shuttle):无人驾驶,需要有专用轨道,由电驱动。由于其费用高,灵活性差,不在考虑的范围之内。

(六)澳大利亚

悉尼Botany海港:2025年前,集装箱吞吐量将保持5%~6%增长速度;85%的集疏运集中在海港40km以内区域;过度依赖集装箱货车。

提出隧道与PCP系统相结合,实现集装箱及大宗货物从海港至联合运输终端之间的运输。

(七)中国

在国家自然科学基金委、中国博士后基金、北京首规委、上海建交委等部门资助下:解放军理工大学、中国地质大学、上海市政工程设计研究总院等单位分别进行了地下物流系统研究,并分别纳入北京、上海等城市地下空间规划。

(1)北京地下物流规划研究

根据首规委与解放军理工大学初步研究:确定出未来北京地下物流系统网络可能节点,包括物流中心、近期配送中心和远期配送中心以及网络构想。

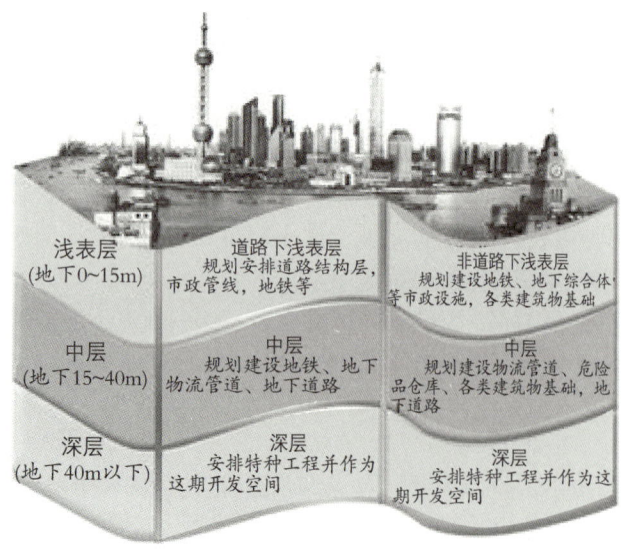

上海地下物流系统规划开发深度

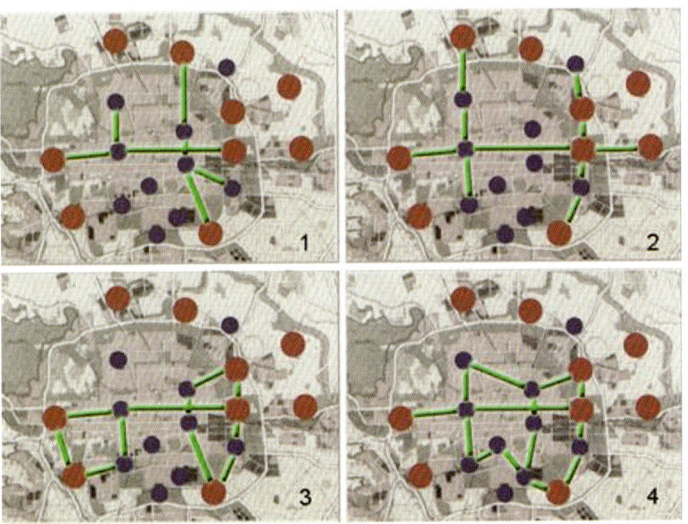

（2）上海地下集装箱运输系统基础研究

通过对上海港概况、集疏运系统现状及面临的挑战的分析，提出洋山港至芦潮港物流园区可以作为地下集装箱运输系统的应用可能线路，并作为东海二桥的替代方案备选。

通过对上海港概况、集疏运系统现状及面临的挑战的分析，提出洋山港内部，北部物流园区与前方集装箱港区作为地下集装箱运输系统的应用可能线路。

（3）上海市建设城市地下环境物流系统的规划和概念设计

将城市中心环卫规划用地或现有环卫设施用地转化为节点的形式，即根据城市具体环卫设施分布，通过分析选择出具有代表性的地区作为地下物流节点。此外，还需要根据城市的具体情况，结合地铁建设规划以及地面物流中心规划等，实现地下物流网络共享。

浦西中心城区地下环境物流网络布局方案

上海市地下环境物流网络布局方案

三、现状总体评价与应用前景

（一）研究现状总体评价

技术方面：目前各国发展的地下物流系统都是基于可获得的技术（自动卡车、自动机车、管道/舱体技术和磁浮等），但是它们都是在概念设计阶段，目前仍然没有已建好的设计原型或示范。

经济方面：根据各国的研究，不同地下物流系统几乎全部都可以获得经济可行性，如 PCP 系统、安全货物机车系统、CargoCap 系统等。

应用领域：规划考虑的是网络形式，其首先实施的一般是考虑从一个终端到另一个终端，点对点。

面临问题：建设这样一个新系统，耗资巨大，风险较大，是系统建设运行的最主要障碍。

（二）应用前景

日本尾岛俊雄教授和美国格兰尼教授提出，未来地下城市基本理念之一，是将地上城市中的城市运输和传输系统逐步转移到城市地下的不同层次里。释放出的地上空间，用作大片的自然植被和安全的步行，并且使地上逐步摆脱与传统运输相联系的环境干扰。重新恢复生机的地上环境将改善整体的社会和自然环境，并进一步增进城市的文明。

美国 Brad Swartzwelter 先生在《比飞机还快：美国长期运输问题解决方案》建议了一种未来的交通运输地下方案：地球上地表空间是供人们休养生息，享受万物精灵和大自然之美丽的。地下广阔的空间才是供人类开发交通运输通道的场所，那里在深度和广度上几乎不受限制，你可自由利用。地下通行权可自由获得，运输线路可以取得很直，很平，不会因地形变化、地面工程或天气条件而中断。直线是从一点到另一点最短的路径，在摩擦很低的环境中，直线运输线路的效率更显著。而在地面上建一条没有一点弯曲的运输通道极不现实。

像任何其他的前卫理念一样，地下物流系统也有赖于时间来调整和充实。虽然，急速扩张的城市面临着不断增长的复杂情况，但随着人们对它特点研究的深入，将会发现它确实是对复杂情况的富有远见的一种解决方案。

The evaluation of status quo and outlook for the underground logistics system (ULS) development in China

With the rapid development of economy and urbanization, there appears in Chinese cities, especially some metropolises severe traffic congestion, increasing traffic accidents and aggravating environmental pollution. One of the important reasons for the above problems is the rise in freight volume.

National freight volume increase from 1980 to 2020

Index	Volume	1980	2000	2020
Volume of freight	100 million tons	55	136	210
Freight turnover	100 million tons.km	12026	44452	84000

The national freight structure differs somewhat from that of urban freight transportation.

The proportion of different means of domestic goods transport

Year	Percentage of Freight transportation mode			
	railroad	road transit	waterway	pipeline
1980	20.6	70.4	7.1	1.9
1985	15.7	77.3	5.3	1.7
1990	15.7	75.3	7.4	1.6
1995	13.4	76.2	9.2	1.2
2000	13.1	76.5	9.0	1.4
2005	14.4	72.1	11.8	1.7

According to the freight transportation make up of some cities, road transit accounts for the largest proportion of urban freight.

Beijing, for example, among the 300 million tons goods, road transport took up 91.2%, railroad 8.5%, other means 0.3%

The number of automobiles in China is not so large, but they do occupy considerably large road resources.

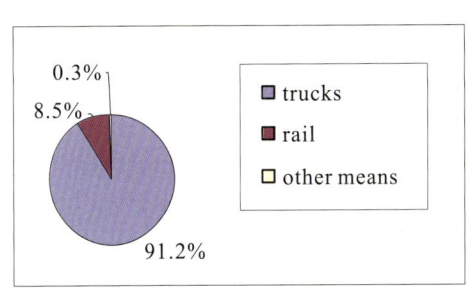

Proportion of Beijing's goods transport by trucks in 2000

Take Beijing for example. The proportions of various kinds of automobiles in 2004 are: cars 74.1%, buses 1.6%, light truck 5.5%, trucks 2.8%, the rest 16%. However, when all above vehicles is converted to PCU, automobiles for transporting goods assume 40%~50% of total vehicle-kilometer.

Under this situation, the underground logistics system has aroused the interest of Chinese scholars, who are exploring and considering moving freight transit underground to solve the above mentioned freight traffic problem.

At present, there are no internationally uniform concepts and standards with regard to ULS.

The present ULS developed in some countries

country	name	means of transportation
Netherlands	underground logistics system（ULS）; underground freight transportsystem（UFTS）	Automated Guided Vehicle（AGV）
Germany	Cargo cap	Cargo cap
Japan	underground freight transport system（UFTS）	Dual Mode Truck（DMT）
usA	Freight Transport by UndergroundPipeline	PCP, HCP or LCP

In China, ULS is generally accepted to refer to the above typical concepts of freight transportation systems that are wholly or partially underground. The application area can be inner-city or inter-city.

The present paper gives an introduction of the research, practice and academic activities of China's urban ULS, and attempts to put forth a prospect for its development.

1 The status quo of china urban ULS

China has paid attention to underground logistics since 2002. At first, introductions to the development of foreign underground logistics system constitute the major part of all efforts. The time sequence of related document and events is as follows:

In 2002 January, the New urban Underground Freight Transport System written by Yang Tao, Yang Donyuan and He Yongzhan was published in Urban Planning Overseas. The article introduced the method and result of cost & benefit evaluation if ULS is to be constructed in Tokyo's 11th District.

In 2002, The Exploration for Urban Pipeline Logistics written by Zhang Yaoping and Wang Daqing was published in Technical Economy in the 7th Issue, and it gives an introduction to ISUFT1999 and ISUFT 2000.

In 2003 September, The New Urban Underground Logistics System written by Nie Xiaofang and Tian Yuxin was published in Combined Transportation mainly deals with the concept of underground logistics system and its functions to a city.

Since 2004, China began to explore for ULS applying, including both research and practicing efforts.

1.1 Research on motivation of ULS in China

Studies of this group usually start with an introduction to the development of urban ULS overseas, and analyze the application feasibility and cut-in point of the system in China in accordance with our specific conditions. Some typical efforts are as follows:

Tentative Idea on Development of Urban Underground Pipeline for Rapid Physical Distribution System written by Prof. Ma Baosong et al was published in the first issue of Underground Space in 2004. The article discusses the present and history of ULS development in Japan, Germany, etc. and the possible role of such a system in Chinese cities.

The author of the present paper put forward in 2004 that ULS could be a new approach to solve the traffic congestion problems in metropolises (for example, Beijing and Shanghai). The idea was substantiated in Constructing Underground Expressway and Underground Logistic Systems in Metropolises of China, an article is published in the 4th Issue, Science & Technology Review. Furthermore, the author had the opportunity to analyze the direct motivation and constraining factors for their construction in China, and describe a future work for their development at the 4th ISUFT held in Shanghai in Oct. 2005.

At the 4th Chinese annual logistics convention (Nov. 2005) in Guangzhou, China, ULS arrested the attention of experts and scholars, who agreed that this is a new and promising area.

March 2005 witnessed the success of The Research on ULS Construction in Chinese Metropolises by the Research Center for Underground Space, PLA University of Science and Technology. This research illustrates the motivation, financial analysis, risk assessment and network of the system, taking into account the practical problems in developing China's traffic. With respect to the specific circumstances in Beijing, it also proposes the establishment of an ULS there, and draws some useful conclusions. For example, about 40%~50% of the capital's road resources will be taken up by freight traffic. And more than 75% of all goods there could be transported through ULS. Those fundamental findings provide a theoretical basis for the development of ULS in China.

As a result of the research, the book whose title is Introduction to Urban Underground Logistic System was published in Oct. 2007.

1.2 Research on planning and construction technology of ULS

The book Urban Underground Space Planning, written by Prof. Chen Zhilong, came out in Jan. 2005. It is the first attempt at ULS analysis in terms of planning. With planning guidelines, targets and content of the ULS network, it marks the commencement of research on planning for ULS in China.

As the chief solid part of the system, the urban ULS network performs the major functions such as system management and control, storage and safekeeping, as well as transport and distribution. It is necessary to conduct planning and design research before its development. In Mar. 2006, PLA University of Science and Technology completed the Planning and Design Research of Urban ULS, a preliminary systematic exploration into its planning and design, laying a foundation for subsequent studies and practices in China.

In 2005, the Research Center of Trenchless Technology, China University of Geosciences, carried out a basic research on a fast urban automatic logistics pipeline system. Sponsored by the National Natural Science foundation of China, the research studies and analyzes various non-excavation technology, including laying underground pipeline technologies such as vertical drilling, pipe jacking, and shield. It also approaches the means of underground logistics transport in China.

1.3 Practicing research-Beijing

The Plan For Underground Space Development and Utilization in Beijing's Central Urban Areas was completed in 2005, as a result of the joint effort by several research institutes. It draws a conclusion that ULS has a great potential, as it can alleviate the traffic congestion in the urban areas, especially the central parts.

A systematic research follows up in 2006, when Beijing Planning & Design Institute and the Research Center for Underground Space, PLA University of Science and Technology continued the study of developing underground logistics system in the capital city. The analysis concludes that the international logistics park at

Beijing capital international airport would be an appropriate choice as the basis or the jumping-off point of ULS. The intermediate-term construction could include the section between freight terminals and distribution stations, while the long-term aim could be the development of the underground freight corridors between Beijing City, Tianjin City and the major cities of Hebei Province.

1.4 Practicing research–Shanghai

The Concept Plan for Shanghai Underground Space was sanctioned by the municipal government at the beginning of 2005.

The Plan clearly provides that the deployment of all kinds of underground facilities should accord with the comprehensive plan of pipelines and achieve horizontal concordance. If the latter fails to be accomplished, stratification could be adopted.

The Plan puts forward the notion of building underground logistics pipeline in the mid-level.

Further, Shanghai began to address the feasibility of container transport through ULS. The city's port handling capacity reached 18080 thousand TEU in 2005. It is estimated that in 2010, its capacity will amount to 2.3 to 2.5 million TEU; in 2020, 3.6~3.8 TEU. The capacity of Yangshan Port accounts for half of that of Shanghai, hence the threat of congestion in container distribution.

At present, the Basic Research on Shanghai Underground Container Logistic System has started, which was supported by Shanghai General Municipal Engineering Design General Institute and China Postdoctoral Foundation, aiming to explore the feasibility of container transport via underground logistic system in Yangshan Port.

An Underground Waste Pneumatic Pipelines Transport System (UWPTS), will be put into action in World Expo Shanghai in 2010.

diameter: 50cm.

thickness: 6~25mm.

speed: 70-90km/h.

pressure: <40kPa.

underground depth: >1.25m service life: 25 years.

1.5 Practicing research—Guangzhou underground waste pipelines transport system

The plan for an Underground Waste Pipelines Transport System (UWPTS), had been put into action in the residential areas of Jin Shazhou, Guangzhou in 2006. Employing trenchless technology and powered by air, the system is made up of 4 groups of vacuum pipes, with a maximum daily disposal volume of 139 tons. The pipes are 35cm and 50cm in diameters, 6mm and 12mm in thickness respectively with service life of 30 to 60 years. The municipal government invested 150 million yuan, and the project is the first UWPTS in China, anticipated to be put into use in 2007, when 15 thousand households or 50 thousand residents will benefit from it.

In addition to Jin Shazhou's residential areas, the underground shops there will also install systems of this kind, according to the plan of the Pearl River New Town, Guangzhou. Both office buildings and stores have access to the vacuum pipeline, which collects waste to the central station.

Besides the above research and practices related to ULS, there are also a lot of other studies and exchange activities that have been accomplished or in full swing. For instance, the Ministry of Construction is compiling the Criteria for Urban Underground Space Plan, which aims to nail down the definitions of China's ULS and some related concepts.

Beijing Underground Space Planning, published in Sept. 2006 by Tsinghua University Press, makes a detailed description of studies of Beijing ULS.

Dr. Zhang Yaoping at Southwest Jiaotong University translated Brad Swartzwelter's Faster Than Jets: A Solution to America's Long-Term Transportation Problems. All the efforts stimulate and promote to a certain degree the development of research and practices in the field.

2 The status quo evaluation and outlook

2.1 The status quo evaluation

As a new type of sustainable freight system, ULS has drawn the attention of state and local governments. Metropolises such as Beijing and Shanghai have already begun studies and practices of the kind.

Technology: constructed design models or examples are yet to be established, even the concepts that fit Chinese situation are not built on. There is a lack of basic research and experiments, and a considerable gap between developed countries and us. The applied areas of ULS are mainly within a city, not between cities.

Finance: it is calculated that ULS in Chinese metropolises is financially feasible and risk-assistant. The ULS of Beijing, for example, if 30% (a modest estimation) of urban freight is conveyed through the system, the internal rate of return (indirect benefit not included) is 12.4% to 14.4% under normal condition, 7.9% and 8.4% at worst.

Difficulties: Constructing a ULS is both costly and risky, involving complicated and knotty planning, while the return of the investment lies only ahead after the line is in operation, which definitely takes a long time. Such is the chief hindrance the government is concerned about.

2.2 The prospect

The 21st Century has witnessed the rapid concentration of China's population in big cities, the frequent economic and cultural activities of urban people, and the rise in the number of household cars and trip, raising a higher demand for urban traffic.

A modern city needs to ensure enough traffic space, but the Chinese only have 8.27m² road area per person. It is unpractical for us to achieve the same quantity of 20 to 100 m² as that of the developed countries, whether in the light of land consumption or funds. In this situation, a major mode of solution to future traffic development could be exploiting underground space. The development and exploitation of underground space in terms of its traffic function and environmental function is the core of China's urban underground space construction in this century, and ULS is one important aspect.

Given the current circumstances, there is a wait before ULS can independently assume all the urban freight volume. One of the immediate tasks we face is to select a starting point, and to study the feasibility of such a system.

2.3 The starting point of a practical Chinese ULS: From metropolis center.

The extension and improvement of current traffic infrastructure alone cannot eradicate the urban diseases in China, especially those in metropolises: serious traffic jam, recurrent traffic accidents, and environmental problems such as noise and air pollution. With regard to the severe shortage of urban land and space, as well as the urgent need of historical and cultural heritage protection in some cities, we cannot possibly build more road facilities on a continual and large scale. Consequently freight accessibility and quality are limited.

Megacities such as Beijing and Shanghai, for example, have banned trucks during daytime hours in center areas. Therefore, in the light of sustainable urban development, an effective procedure for developing ULS is to shifting underground, begin with urban center, and move gradually to circumference of the city. For the populated city centers, urban ULS could be the first choice.

Conceivably, the development of ULS will be a long-term process in China. It could be an auxiliary to present or now-built ground traffic facilities, jointly constituting the urban logistics system. In the future, it can cooperate with traditional means of transportation, or even replace the truck-based logistics system.

3 Conclusions

Like any other avant-garde concept, it takes time for ULS to perfect and substantiate itself. However, as research on its properties goes deeper, it will surely end up as a foresighted approach to solve China's urban traffic problems.

The present situation and future prospect in application of tunneling machine in china underground engineering construction

1 Foreword

With the increasing requirements of excavating technology in underground engineering construction and the improvements of its scientific technology level, tunneling machine and its technology has been widely applied and swiftly developed in China.

According to different conditions of stratum and different practice of construction methods, the tunneling machines are, in general and our paper, categorized into three groups, i.e., shield machine, tunnel boring machine (TBM), and pipe jacking machine.

With the progress in large-scale construction of infrastructure facilities, great exploitation in west china, and the success in bidding for holding 2008 Olympic Game in Beijing, there exist a great need in China for constructing a large number of big and long tunnels in the fields of railway, highway, large-and medium-size hydropower stations, south-to-north water diversion projects and west-to-east gas transportation items in 21 century. And there are more and more construction of subways, municipal works (drainage and water supply pipelines and etc.), trans-river tunnels in the modern urban construction.

When we excavate these large number of underground tunnels, the application of tunneling machine is doubtless the best choice in handling the problems of speed, environmental protection and efficiency, for the tunneling machine entails shorter construction period which has great effects on economic efficiency and ecological environment, and the advantages which can solve the problem that the driving face of underground project is often greatly limited.

2 The development and application of tunneling machine in China's underground engineering construction and its achievements

2.1 The application of tunneling machine in China's underground engineering construction

2.1.1 The development and application of shield machine in China

In China the shield machine are mainly used in Shanghai with a special foundation of soft soil.

In 1963, the first China-made hand-digging shield machine with a diameter of 4.2m was tested in Pudong Dyke Bridge, Shanghai. It excavated a tunnel with a length of 63m.

In 1965, two China-made 5.8m-diameter mesh type extrusion shield machine was put into operation in the

报告时间：2007年5月25日。

experimental project of Hengshan Subway section to dig two tunnels with a total length of 1200m.

In 1966, the first China-made super large 10.2m-diameter mesh type shield machine, was successfully used in trans-river highway tunnel project under Huangpu River of Dapu Road, Shanghai, with the help of air pressure to stabilize the driving face.

In 1975 to 1979, mesh type shield machines were used to construct sewerage and seawater diversion tunnels with diameter from 3.5~4.4m in Shanghai, transporting soil by waterpower.

In 1980, Shanghai developed a 6.41m-diameter cutter disc type shield machine. For some technical problems, it was turned into the use of mesh type extrusion shield machine for excavating a tunnel with a length of 1230m.

In 1985, an 11.3m-diameter mesh type shield machine through hydraulic mechanical mucking excavating was used to build a round main of 1476m-long in Shanghai Yan'an Road Trans-River Tunnel Project. Since 1987, it has been widely used to build water supply and drainage tunnels at the bottom of river or sea.

In 1985, a 4.33m-diameter small reamer tray earth pressure shield machine was introduced from Japanese for the Shanghai Furongjiang Road Drainage Tunnel Project, with a tunneling length of 1450m.

In 1987, Shanghai Tunnel Engineering Co.Ltd. succeeded in developing the first 4.3m-diameter mud-adding type earth pressure balance shield machine in China. All parts and components were domestic products made in China. It was used in Shanghai South Railway Station Trans-River Cable Project with a total tunneling length of 534m.

In 1990, seven sets of 6.34m-diameter earth pressure balance shield machines, jointly manufactured by France FCB Company, Shanghai Tunnel Engineering Co.Ltd., Shanghai Tunnel Engineering Design Institute, Shanghai Shipbuilding Yard, were applied to interval tunnel in Shanghai No.1 Subway.

In 1993, a 6.34m-diameter earth pressure shield machine was produced and used in Nanjing Jiajiang Drainage Tunnel Project, which passed through silt stratum with a digging length of 1294m.

In 1996, Shanghai No.2 Subway used the above-mentioned seven earth pressure shield machines again, and introduced other two machines from France FMT Company to drive 24km of interval tunnel.

In 1996, three 6.14m-diameter shield machines from Japanese Kawasaki Heavies Corp. were selected to use in 8.82km-long interval tunnel of Guangzhou No.1 Subway Project.

In 1998, Shanghai bought a second-hand 7.65m-diameter hinged-type earth pressure balance shield machine for Shanghai Huangpu Sightseeing Tunnel Project from a foreign country, and excavating a tunnel of 644m length.

In 2000, Shanghai No.3 Subway used 10 earth pressure shield machines, which had been used in No.2 Subway, to drive interval tunnel.

In 2000, two German-made 6.14m-diameter compound earth pressure shield machines were selected to use in the 3423m-long interval tunnel, which runs in Guangzhou No.2 Subway.

In May 2001, the subway section from Sanyuanli to Yuexiu Park was tunneled with introduction of German-made 6.2m-diameter shield machine.

In June 2001, Beijing No.5 Subway began its construction with introduction of a German-made 6.14m-diameter earth pressure shield machine.

In June 2001, 6.34m-diameter earth pressure balance shield machine made by Japanese Mitsubishi Corp. was used in the interval tunnel of Nanjing No.1 Subway.

φ11.3m super large mesh type hydraulic mechanical shield machine used in Yan An Dong Road circular tunnel

φ6.34m-diameter EPB shield machine　　φ4.3m-diameter EPB shied machine　　φ7.65m articulated EPB shield machine used in Shanghai Huangpu Sightseeing tunnel

φ11.2m slurry shied used in duplicate line project of Yan AnDong Roadtunnel

2.1.2　The development and application of tunnel boring machine (TBM)

In 1966, Shanghai Waterworks Mechanical Plant designed and produced China's first 3.5m-diameter full-face TBM, which was put into operation in the water diversion tunnel project of Xierhe Hydropower Station in Yunnan Province.

In 1969, Guangzhou Municipal Mechanical and Electrical Engineering Bureau manufactured a 4.0m-diameter machine, which was used for trial operation in Guilin of Guangxi Province, and the total length of excavation was 245m.

During 1970, Coal Machinery Repairing Plant of Pingxiang Mining Bureau manufactured a 2.6m-diameter machine, which was used in laneway excavation project of Pingxiang Qingshan Mine.

In the same year, Xian Coal Machinery Plant tried and produced a 3.5m-diameter TBM too, which was put into used by Tongchuan Mining Bureau, with an excavation length of 669m.

In 1971, Guangzhou Municipal Mechanical and Electrical Engineering Bureau, Railway Corps, Fushun Mining Bureau and Shanghai Waterworks Machinery Plant tried and produced 2.5m-diameter, 5.5m-diameter, 3.8m-diameter and 5.9m-diameter TBMs respectively.

From 1977 to 1978, a SJ-58 TBM, produced by Shanghai Waterworks Mechanical Plant, was put into operation in the underwater tunnel of Xierhe Hydropower Station in Yunnan Province.

In 1980, Pingxiang Mining Bureau in Jiangxi Province manufactured a 30m-diameter machine, which was used in Pingxiang Mine, with a total excavation length of 735m.

In 1981, an improved SJ-58 TBM, primary produced by Shanghai Waterworks Mechanical Plant, was used to gig tunnel in the Project of diverting water from Lanhe River into Tianjiang, with a total excavation length of 1444.5m.

In 1983, the SJ-58 machine was put into use in Xinwangzhuang and Gurenzhuang Tunnel in Project. The excavation length of the later project was 478m.

In March 1983, the No.2 Bureau of Railway used a 2.5m-diameter machine, manufactured by Guangzhou Municipal Mechanical and Electrical Engineering Bureau, to dig tunnel in Guiding and Laoluopu Tunnel Project in Guizhou Province, with an excavation length of 263m.

In April 1983, Railway Corps put a homemade 5.0m-diameter machine into use in Laohutai Coal Mine, with a tunneling length of 287m.

In December 1983, the No.2 Engineering Bureau of Water and Electricity Ministry in China used a SJ-58 machine, produced by Shanghai Waterworks Mechanical Plant, in Luopoling Tunnel Project in Beijing, with 5.9m in diameter and 1250m in length.

In 1985, in construction of Guangxi Ronglin Tianshengqiao Hydroelectric Power Station, 10.08m-diameter TBM made by American Robins Manufacturing Company, was adopted.

In 1991, a 188-277 telescoping type (hinged shield machine) TBM, made by American Robins Manufacturing Company, was applied in the large-scale Datonghe-Qingwangchuan trans-drainage area irrigation project in Gansu Province to excavate the 30A tunnel. Its total length is 11.649km, and the diameter of the TBM was 5.53m.

In 1997, two sets of 8.8m-diameter TB880 TBMs, made by German Wirth Company, were used to construct No.1 Qingling Tunnel Line of Ankang Railway. Its total tunneling length was 18.46km.

In 1997, four sets of 4.88m-diameter double-shield full cross section TBMs were used to construct 88km length of tunnels in Shanxi Wanjiazhai Water Conversion Project. Two of these four TBMs were jointly manufactured by Shanghai Tunnel Co.Ltd. and American Atlas-Robins Co.Ltd.

In 1998, Shanghai Tunnel Engineering Co.Ltd. and other domestic units assembled and refit some import TBMs used in the Hydro-engineering of diverting water from Huanghe River into Shanxi Province, and supplied components for these machines.

In 1999, France FMT Company and Shanghai Heavy Machinery Plant cooperated to produce a 8.75m-diameter TBM for Hongkong.

In May 2001, the Huangtang Tunnel Project of Jiaojing Express Highway used s100 TBM, made by Jiamusi Lishi TBM Machinery Co.Ltd., to dig the central pilot. The tunnel's cross-section was 18.9 square meters and its length was 275m.

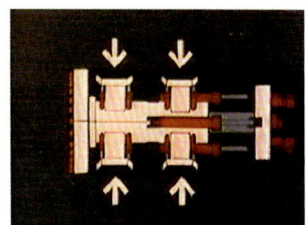

ϕ10.8m Robins TBM used in Guangxi Ronglin Tiansgengqian Hydroelectric power station

JS45 TBM used in Waterworks Tunnel of Fujian Longmentan second stage Hydroelectric Power Station

ϕ 8.8m TBM(WIRTH) used No.1 Qingling Tunnel Line of Ankang Railway

s100 TBM used in the Huangtang Tunnel Project of Jiaojing Express Highway

ϕ 4.8m full-section hare-rock TBM used in Shanxi Wanjiazai Water Conversion Project (south trunk line)

2.1.3 The development and application of pipe jacking machine in China

Around 1967, Shanghai succeeded in developing a small-caliber remote control earth pressure pipe jacking machine, with which it was unnecessary for workers to enter pipeline. The caliber of the machine ranged from 700mm to 1050mm. By the end of 1969, this kind of machines had constructed tunnels of more than 400m long in total.

Around 1978, Shanghai successfully developed an extrusion pipe jacking machine, which was especially suitable for soft soil and sullage clay.

Around 1984, Chinese cities like Beijing, Shanghai and Nanjing introduced advanced mechanical pipe jacking machines from abroad in succession.

In 1985, Shanghai Municipal Works Company introduced pipe jacking machine with 800m-inner diameter from Japanese Iseki Company.

In 1988, Shanghai developed China's first 2.72m multi-cutter earth pressure balance tunneler, which was put into operation in Hongcao Road Project and Fujian Road Project and etc. Up to now, the machine has completed a tunneling length of 4.6km in accumulative total.

During 1990s, Shanghai Tunnel Co.Ltd. manufactured more than 10 sets of 1.5~3.0m-diameter earth pressure balance pipe jacking machine with small cutter tray or large cutter tray, and their total tunneling length reached to 20km.

In 1992, Shanghai developed the country's first mud-adding 1440mm-earth pressure balance pipe jacking machine, which was used in the construction under the busy section of Shantou Jinsha East Road in Guangdong Province. This kind of tunneler now has been developed into serial products.

In 1998, the first 2.2m-diameter slurry pressure-added balance pipe jacking machine in China, made by Shanghai Tunnel Co.Ltd., was used in the Second Phase Construction of Trans-River inverted Pipeline of Sewage Treatment Project in Shanghai and its tunneling length was 1220m.

In May 1999, Shanghai Pudong Lujiazhui Subway Station used China's first 3.8m×3.8m rectangular compound cutter disc type earth pressure balance pipe jacking machine, which was made by Shanghai Tunnel Co.Ltd. It excavated a 120m-long tunnel and with which two 62m-long pedestrian underpasses were completed.

In 2000, Shanghai Fengxian Sewage Southward Drainage Project and Suzhou River Interflow Sewage Cut-off Project used ϕ600/800 micro pipe jacking machines to finish tunnel boring of over 400m.

Rectangular shied cutterhead Used in metro pedestrian passageway 3.8m×3.8m

800mm pipe jacking machineShanghai suzhou River Interflow Sewage Cut-off Project

2.2 Current manufacturing and application technology of tunneling machines in China

2.2.1 Shield machine

At present, many Chinese cities use closed type of shield machine to build infrastructure facilities and underground transportation facilities, particularly the use of Earth Pressure Balance Type of shield machine is very wide. With the construction practice, some definite experiences with controlling influence on surround environment in construction with shield machine, with technology of machine starting off and arriving at and technology of manufacturing reinforced concrete segment, some construction of tunnels which are super near each other and are super near to buildings that have been completed. Combining with Chinese condition enterprises have found the mode of repeated use of machine in many tunnel constructions.

Ever since 1990s, China has developed more than 10 sets of 3.8~6.34m-diameter earth pressure balance shield machine. Most of the machines are used to construct subway tunnel, water conversion and drainage tunnel, cable tunnel, and their technological standard is close to the world level, but manufacturing technologies of valves and hydraulic pumps in China have difference of standard from international advanced standard, external spares and parts are adopted in Chinese Shield machines.

2.2.2 Tunneling boring machine

As for TBM, since 1960s, China have developed more twenty TBM sets of three generations, but there is large difference of standard between domestic and oversea TBM sets. Mainly it is low reliability, short service life, low intact ratio and low availability. Statistical data shows that availability of domestic TBM sets is 15%. Reason of this difference is resulted to essential parts and back-integrated parts of TBM. Last years manufacture of TBM in China has been stopped.

2.2.3 Pipe jacking machine

In terms of 1.2~3m-diameter pipe jacking machine, China has developed, in succession, advanced backhoe pipe jacking machine, earth pressure balance pipe jacking machine and slurry pressure-added pipe jacking machine. China has completely had the ability to produce domestic machinery to replace the imported equipment. Shanghai developed 3.8m×3.8m rectangular pipe jacking machine, with which two 62m-long pedestrian underpasses were completed. As for miniature tunneling machines, China has also developed 600-800mm-diameter central-screw earth-excavating pipe jacking machine, tamping pipe jacking machine and horizontal orientation boring unit and etc.

2.3 Analysis of construction effects with tunneling machine on China's underground projects

2.3.1 Shield machine

(1) Speed

Speed	Place	
	Shanghai	Guangzhou
Average on day (m/day)	8	6
Maximum on day (m/day)	20	12
Average in month (m/month)	240	150~187
Maximum in month (m/month)	320	230

(2) Construction cost

In Shanghai: aggregate construction cost 30~35 thousands RMB/m;construction cost of Tunneling 10 thousands REM/m; lining cost 10 thousands RMB/m; lend cost of Shield machine 6 thousands RMB/m; water proof cost and so on 4~9 thousands RMB/m.

In Guangzhou: aggregate construction cost 35~45 thousands RMB/m.

(3) Environment Protection

The surface subsidence level is +1~-3mm in Shanghai, maximum subsidence is 3.7mm in Guangzhou, it is very safe to let shield machine to pass through under high-rise building, flood prevention walls, underground sidewalks.

(4) The stratum compatibility

At present, the shield machine used in China is suitable for use in weak and water-rich earth layer (in Shanghai), common earth layer (in Beijing) and shizolite rock stratum (in Guangzhou).

2.3.2 TBM

(1) Tunneling speed

Project name	Tunnel diameter (m)	Monthly tunneling rates (m/month)		Rock compressive strength
		Average	Highest	
Water Diversion project from Luan river to Tangshan: 1.Xingwang Zhuang tunnel(1981) 2.Guren Zhuang tunnel(1983)	5.8 5.8	90 100	201.5 213.4	80~130 60~100
Trans-Area Irrigation Project from Datong River to Qing (1991)	5.53	1100	1400 (1300)	26~133.7
Water tunnel of Tiansheng Qiao Hydro-electric Power (1985)	8.7~9.8	60	242	150
Water Diversion Project from Yellows River to: 1.Shanxi (south stream) 2.Master stream	4.88 4.88		1821 1200	
Qingling Railway tunnel	8.8		509	
Tao Hua Pu Railway tunnel	8.8		520	

（2）Construction cost

Project name	Construction cost of tunneling (yuan/m³)		Aggregate construction cost (yuan/m³)	
	TBM	B&B	TBM	B&B
From Luan to Tanshen Xingwang Zhuang tunnel	77.51	65.13	3756.39	6921.38
Guoren Zhuag			4335.82	7045.48
From Datong to Qing			6000	
From Yellow River to Shanxi			13000~20000	6000~10000
Qingling Railway tunnel			72000	32000
Tao Hua Pu Railway tunnel			45000	30000

Here I must point out that the large difference between the average and highest monthly tunneling rates is resulted to adverse geo-environment (Rock burst, dissolved cavern), to outmoded imported TBM and providing cutter not on time, and that non-consistency between construction cost and aggregate construction cost is resulted to low liming cost.

2.3.3 Pipe jacking mchine

Up to now, pipe jacking machine has been more and more popular in urban in China, and their application areas also become more and more extended. Pipe jacking machine was initially used in sewerage construction, and now it has been applied in pipeline construction of water supply pipe, gas pipe, power electric cable, communication cable and circulative water cooling system of power plants and so on. And it has included level oriented drilling machine, spiral drilling machine and has been developed into a non-excavating engineering technology based on pipe jacking machine. China has established the Non-Excavating Engineering Association.

3 Analysis of China's planned tunnels in first twenty years of 21st century

3.1 Railway tunnel projects

According to the Ministry of Railways' plan to improve 8 longitudinal and 8 latitudinal railway networks and the country's west exploitation strategic plan, the Chinese Government has planned a batch of new railroads for the future, including many railroads which run across multitudinous mountainous areas. Some of railroads have a large number of tunnels.

Railroads across mountainous areas are constructed and will be constructed in 2000—2010 ten years

Number	Railroads	Length（km）
1	Lasa–Geermu	810
2	Xian–Anyang–Xinyang	540
3	Chongqing–Huaihua	420
4	Guiyang–Huaihua–Changsha	630
5	Lanzhou–Baji–Xian	360
6	She nmu–Suide–Yanan–Xian	450
7	Daxian–Wanxian–Enshi–Zhicheng	390

continued

Number	Railroads	Length (km)
8	Zhongwei–Yanchi–Suide	450
9	Yibin–Yiliang–Liupanshan	240
10	Lanzhou–Guangyuan–Chongqing	480
11	Jinmen–Wuhan	180
12	Dali–Burma	300
13	Atush–Kirgizsan	180
14	Lasa–Baxin–Deyin–Dali–Laos	1620

Above mentioned 14 railroads at mountainous area have length 7000km, if length of tunnels is 30% of total length of railroads, length of tunnels amounts to 2100km. Moreover, Xian-Nanjing Railway just is beginning to construct, its western section is located at mountainous area, it has 74 tunnels, their length adds up to 77.6km, Fuzhou-Xiamen Railway and Wenzhou-Fuzhou Railway are located at southeast mountainous area, they have many tunnels.

3.2 Highway tunnel projects

The Chinese Government projects to finish five longitudinal and seven latitudinal highways by the year 2020, whose total length is 35960km. 3670km of these highways run through mountainous or half-mountainous regions, making up 20% of the total. During the highway constructions in these mountainous areas, roads surrounding mountains will be replaced with roads running across mountains, and the number of long and large road tunnels will be increased more, for instance, the length of Qinling Zhongnanshan super long highway tunnel under planning amounts to 18.4km.

Five longitudinal and seven latitudinal highways

	Name of highways	Total length (km)	Mountainous length (km)
Longitudinal highways	Tong tang–Sanya	5700	350
	Beijing–Fuzhou	2540	230
	Beijing–Zhuhai	2310	170
	Erlian ha ote	3610	580
	Chongqing–Zhanjiang	1430	320
Latitudial highways	Suifenhe–Manzhouli	1280	50
	Dandong–Lasa	4590	370
	Qingdao–Yinchuan	1610	210
	Lianyuanggang–Huoerduosi	3980	270
	Shanghai–Chengdu	2970	120
	Shanghai–Ruili	4090	820
	Hengyang–Kunming	1980	180

Eight province–level highways in west of China (no including inner Mongolia, Guangxi)

Name of highways	Total length (km)	Mountainouslength (km)	Controlling cities
Lanzhou–Mohan	2490	2490	Lanzhou, Chengdu, Xichang, Kunming, Mohan port
Baoji–Beihai	3400	2495	Baotou, Yanan, Xian, Chongqing, Guiyang, Beihai
Aletai–Hongqilafu	2990	2990	Aleta, Kela mayi, Ku erle, Hongqilafu port

Name of highways	Total length (km)	Mountainouslength (km)	Controlling cities
Yinchuan–Wuhan	1620	1120	Yinchuan, Xian, Xiangfan, Wuhan
Xian–Hefei	860	230	Xian, Nanyang, Xiangfan, Hefei
Changsha–Chongqing	1000	457	Changsha, Jishou, Chongqing
Xining–Kuerle	2190	2190	Xining, Jiayuguan, Dunhuang, Kuerle
Chengdu–Zhangmu	2897	2897	Chengdu, Batang, Lasa, Rikaze.Zhangmu port

3.3 Underground tunnel projects of hydropower stations

Between 2000—2020, China will build over 10 extra large and medium-size hydropower stations one after another, each will have several tunnels to be excavated for current directing, water diversion and flood discharging, whose total length will amount to 1100km. In the near future, China will have to excavate 180km of waterwork tunnels every year. These hydropower stations are: Xiluodu, Xiangjiaba, Baihetan, Wudongde, Jinping, Pubugou, Hutiaoxia, Guandi, Guanyinyan, Tongzilin, Zipingpu.

The west line project of Northward Water Conversion Project lies in mountainous plateau with an altitude of 3000~4000m. The topographical and geological stations are very complicated and the weather there is very cold. Water-surface elevation of water-piloting streams in the upper reaches of Yangtze River is lower than that of Yellow River's transmitting point, so high dykes (302m, 175m, 296m) and long tunnels need to be constructed in the west line project. The total length of tunnels is 653km (158km, 131km, 28.5 or 29.4km), the shortest tunnel is 30km long and the longest tunnel is 158km long.

The total length of tunnels at first stage of Project is 244km. The central line and east line of the North Water Conversion Project include two 9.6m-diameter under-river tunnels in depth of over 20m excavated at the bottom of the Yellow River near Henan's Zhenzhou and Shandong's Liaocheng.

The Dongshen large water diversion project in province Guangdong is from Taiyuan Pump station of Dong River to Dongguan, Shenzhen and Hongkong, diversion water is $24 \times 10^8 m^3$, length of channel is 51.7km. This project has many tunnels.

3.4 Subway interval tunnel

Subways and light railways, which are under construction or which are approved by government in the cities of Beijing, Tianjin, Shanghai, Guangzhou, Qingdao, Nanjing, Chongqing, Shenyang and Shenzhen reach 360km, the planned subways in 15 cities within 20 years 21st century are expected to reach 2280km.

3.5 City's common ditch

In order to change situation, that accident of utilities pipelines brings about a loss of 700 million RMB every year and open construction of pipelines brings harmful influence on urban transport and environment. Chinese cities like Shanghai, Beijing, Nanjing, have started a plan to put some utilities pipelines such as electricity, telecommunication, electric cable, telephone, coal gas, water supply and drainage pipeline into a same underground tunnel.

This synthesis pipelines corridor is called "common ditch", and it has become material carrier of various public utilities pipelines in modern cities. In combination with the development and construction of the Pudong New Area, Shanghai completed the first scaled Zhangyang Road common ditch and Chengdu completed the Dongfeng Road common ditch in 1994. Confronted with high-concentration of modern cities and high standardization of people's living level, demands for various kinds of supply facilities will be sharply increased. Preliminary estimation shows that by 2020, common ditchs in major Chinese cities will total 800km.

3.6 Trans–sea and trans–river

In the next decade, Shanghai will construct three over 1000m long highway and subway tunnels under Huangpu River. Two trans-river tunnel in Wuhan and highspeed railway tunnel in Nanjing under Yangtze River will be constructed.

Moreover, China completed feasibility study on the undersea tunnel in Qiongzhou Strait, plan of river bottom tunnel, which runs across the Yangtze River seaport and connects Shanghai, Chongming, and Qidong, has been demonstrated on meeting, about plan of the tunnel across the Taiwan Strait and plan of channel bridge in the south and tunnel in the north across Bohai Bay the symposiums have been held.

3.7 Many gas tunnels emerging in west–to–east gas transmission project

West –to –east gas transmission project is a project that spans longest, biggest investment, and has the most far-reaching influence in the history of China's pipeline construction.

It includes Talimu basin-Jiangshu, Zhejiang, Shanghai line, Qinghai-Lanzhou, Shanxi-Beijing line, Shichuan-Hunan, Hubei line. The main pipeline totals 4200km, whose diameter is 1118mm, gas pressure is 10Mpa. The technical problems in disposing these pipelines are complicated, and their construction has strict requirements. Especially when it runs across Luliangshan Mountain and Taihang Mountain, a number of mini-tunnels need excavating through the mountain bodies.

In a word, in China during first 20 years of the 21st century; the demand for subway, municipal works (e.g. sewage draining tunnel, and water supply tunnel), trans-river tunnels in modern urban constructions keep rising. Gross statistics of tunnel demand mentioned above shows. China will construct 6000km –long underground tunnels (railway, highway, waterworks and hydropower, oil and gas pipelines, subway, municipal tunnels and coal mining, etc.)during the period between 2000 and 2020, and averagely 300km each year.

4 Forecast on demand for tunneling machine during 2001—2010

With the vast market of underground space exploitation and application in Chinese cities and the increasing of non-excavation technology during 21st century, more than 20 cities have to build subways during the 10 years from 2000 to 2010, and shield machine will be the inevitable choice in constructing interval tunnels.

Nearly 40 shield machines of around Φ6m-diameter will be needed for construction of subways. Some of railway tunnels will be bored by TBM, and the demand for TBM of around ϕ8.6m during the next ten years will be about 6 sets, demand for TBM with inner diameter of around 8.6m will also be 6 sets. Some individual highway tunnels will use single-arm TBM to excavate central pilots, and the demand is about 10 sets.

Quite a great part of hydropower tunnels will use TBM, and about 20 sets of around ϕ4m-diameter TBM will be needed. Demand for ϕ2.44m TBM used in petrochemical gas tunnel construction is about 4 sets. In piping projects of other cities, demand for ϕ3~5m shield machine is about 100sets, and demand for ϕ0.6~3m pipe jacking machine is about 30 sets. Therefore demand of tunneling machine amounts to 195 sets, demand of pipe jacking machine-30 sets.

5 Some suggestions on application of tunneling machine in China's tunnel and enlargement of China's TBM market

According to China's current situation, solution of following problems is urgent for enlargement of China's tunneling machine market.

5.1 Localization of tunnel industrial system

China has a vast territory and its geological structure is complicated. It has a vast market of underground space development. In a long term view, a developing country like China can not all the time depend on imported equipment and talents to develop its underground space, but establishes localized tunnel industrial system, thus satisfying urgent demand of engineering and saving large amount of foreign exchange to propel the development of other sectors.

Localization of tunnel industrial system is essential measure to enlarge China's tunneler market also, because now in China project's construction cost of tunnel solution>project's construction cost of bridge solution; project's construction cost of solution with tunneling machine>project's construction cost of sunk tube; project's construction cost of solution with TBM>project's construction cost of with B&B. Reducing price of tunneling machine is main measure to enlarge application of tunneling machine, main measure to reduce price of tunneling machine is to manufacture it and its spares or parts in China.

Localization of tunneling machine don't get rid of abroad technology, expert and investment, on the contrary, welcome corporation with abroad enterprise to enlarge tunneling market together.

Moreover, it is needed to promote more Chinese civil engineers to know intimately application technology and management knowledge of tunneling machine. Only if more Chinese civil engineers know intimately construction technology with tunneling machine, construction solution with tunneling machine will be accepted among all competition tenders, thus tunneling machine market will be enlarged.

5.2 Repeated application of tunneling machine

Because now tunneling machine is very expensive equipment one of China's project do not have ability of full depreciation. Though there is no "universal TBM", in China's reality tunneling machine is repeatedly applied. According to repeated application of TBM, the TBM must be designed and manufactured as more multifunctional as possible, the reform technology and repair technology must be developed to satisfy requirement of tunneling market.

5.3 Raising wear ability and developing replacement technology

In China TBM mainly is applied to construct long tunnel, therefore it is necessary to raise wear ability and to develop replacement technology of cutting head and acting ring and other vulnerable parts.

5.4 Researches on sketch cross section shield machine

In construction of double-line tunnel with one-circle shield machine influence of succeeding tunnel to forerunning tunnel is easy to happen if distance between tunnels is super near, it is needed to apply double-circle or spectacle cross shield machine Three-circle cross section shield machine is applied to construct subway station. Rectangular or U-shaped tunnel are suitable for pedestrian underpass and underground common ditch, determinative factor to apply sketch cross section shield machine is cost of these machine, namely it is economic comparing with different solutions with different shield machines.

5.5 Technology of tube segment and popularization of ECL construction method

The characteristics of ECL construction method is that it uses fresh pressure-added concrete to prime in-situ at the end of tunneling shield machine, thus establishing lining which closely touches earth layer after curing. With the merits of long-distance speedy construction, small ground subsiding, no need of precipitation, good water-proof performance, high mechanization, labor-saving, safety and economy. ECL construction method is the most advanced shield construction in weak and water-bearing stratum. Its lining materials can be pressure-added concrete, pressure-added steel fiber concrete and reinforced pressure-added concrete.

5.6 The serialization of medium–and small–caliber pipe jacking machine and its popularization

Medium-caliber pipe jacking machine with 1.2~3.9m diameter is mainly used for digging water drainage, heat supply, electric cable and other public utilities pipelines in cities. China has made various types of pipe jacking machines. With more and more importance attached to urban traffic management and environment protection, open-trenched tunnel construction has been banned. But in China, most of this kind of machines currently in use are laggard hand digging pipe jacking machine.

In order to improve construction speed, to lower labor intensity and reduce effects on environment, our country must popularize the application of earth pressure balance type, slurry balance type and backhoe type of pipe jacking machines.

Mini-pipe jacking machine with a diameter smaller than 1km is widely used in urban public pipelines. Mini-pipe jacking machine is kind of remote control tunneling machine. We should study its corresponding surveying and guiding system, and monitoring system to improve its technological level of tunneling used in China's urban pipelines.

Finally must emphasize, now Chinese Government have begun to attach importance to localization of tunneling machine industry. In this year, work group for localization of tunneling machine industry has been established under Chinese Heavy Machine Society, work group for enlargement application of tunneling machine has been established under Chinese Rock mechanic and Engineering Society. Former director of Significant equipment Department of Machine industry Ministry Wang Jianye and Qian Qihu are appointed to the directors of these groups.

Chinese Second heavy machine group-company will be appointed to localization base of tunneling machine industry. Government will support localization work of group-company with policy and investment. Developing tunneling machine is placed to emphasis project of Ministry by Science and Technology Ministry. Work group for localization of tunneling machine industry and Chinese Second heavy machine group-company favorably receive heavy international famous experts, companies of manufacturing and designing TBM or consultant and construction company with TBM to develop tunneling machine market of China.

21世纪前期我国城市地下空间开发利用的战略及对策

一、研究的背景

实施城市的可持续发展就必须节约资源，保护环境，实现城市经济、建设与资源、环境的相互协调。

（一）开发利用地下空间是应对城市发展与土地资源稀缺的矛盾的主要出路

城市发展与土地资源稀缺的矛盾，是可持续城市化面临的最大挑战，出路在于集约使用城市土地资源，开发利用地下空间。我国城市土地利用的集约化程度处于国际上较低水平，城市发展沿用"摊煎饼"式的粗放经营模式，单位城市用地的国民经济总产值，北京（1989）仅为东京（1986）的6.1%。

按照城市化发展的相关分析，以目前人均城市用地100平方米的水平计算，到21世纪中叶，我国的城市发展将再占地1亿多亩。按人口平均，中国是耕地资源小国。人均仅有1.44亩，仅及世界人均值4.65亩的31%。极其有限的耕地资源将继续减少。城市人口的急剧发展与地域规模的限制已成为制约中国城市可持续发展的突出矛盾。因此，我国城市发展只能走土地资源集约化使用的发展模式。

纵观当今世界，很多发达国家和发展中国家已把对地下空间开发利用作为解决城市资源与环境危机的重要措施。1991年召开的城市地下空间国际学术会议通过了《东京宣言》，提出了"21世纪是人类开发利用地下空间的世纪"。向地下要土地、要空间已成为世界性的发展趋势，并以此作为衡量城市现代化的重要标志。城市地下空间是一个十分巨大而丰富的空间资源，如得到合理开发，将土地资源集约化使用，对于缓解城市中心区建筑高密度的效果是十分明显的。国际上有的学者预测：21世纪末将有三分之一的世界人口工作、生活在地下空间。

国外城市地下空间开发利用的经验是：把一切可转入地下的设施转入地下，城市发展的成功与否取决于地下空间是否得到了合理的开发利用。世界各国开发利用地下空间的实践表明，可转入地下的设施领域非常广泛，包括交通设施、市政基础设施、商业设施、文化娱乐体育设施、防灾设施、储存及生产设施、能源设施、研究实验设施、会议展览及图书馆设施。

（二）地下空间的开发利用是能源、水资源及其他资源持续利用和合理利用的重要途径

按人口平均，我国也是资源小国。我国人均能源占有量不到世界平均水平的一半，人均水资源为世界人均水平的1/4。因此，实现资源可持续利用有着重要意义。在这方面，地下空间的利用大有可为。

地下空间利用岩土良好的隔热性，地面以下5米的室内气温常年恒定。因此地下建筑比地面建筑要明显少消耗能量。

应当指出，地下空间可以合理利用能源，特别是可再生能源。太阳能是巨大的洁净可再生能源，但其来源随季节、昼夜有很大的不稳定性。太阳辐射热的主要部分一般仅在夏季得到，这就需要季节性贮存，在地下的水、岩石和土壤中贮存热量往往是最佳的甚至是唯一的选择。

报告时间：2007年5月25日。

地下空间的储热和储冷，由于岩、土的热稳定性与密闭性，热量或冷量损失小，不需要保温材料，利用岩石的自承能力，构筑简单，维护保养费大为降低。这就使天然能源或工业大量余热的利用富有成效。

我国水资源短缺问题日益明显。水资源在时空分布上很不均匀。在缺水的同时，又有大量淡水因为没有足够储存设施白白流向大海。我国若如挪威、芬兰等国那样，利用松散岩层、断层裂隙和岩洞以及疏干的地下含水层，或如日本那样在东京、横滨、名古屋以及札幌等建造人工地下河川、蓄水池和地下融雪槽，储存丰水季节中多余的大气降水供缺水季节使用，就可以部分解决水资源在时间上分布不均匀的问题。

地下空间还可为物资贮存和产品生产提供更为适宜的环境。地下空间独具的热稳定性和封闭性对贮存某些物资极为有利。

据统计资料分析：地下冷库的运行费用比地面的低25%~50%；在地下建造油库，不仅有利于减少火灾和爆炸危险，而且由于地下温度稳定，受大气影响较小，因而油料不易挥发和变质，可比地面油库节省20%~30%的管理费用；在处理好防潮防虫害基础上，利用地下温度稳定的优势建造地下粮库，也具明显的经济效益。

某些产品的生产对环境温湿度、清洁度、防微振、防电磁屏蔽提出更高要求，如在无线电技术生产和测试中，不仅要求高精度空气环境，而且常要求工作间不受外界电磁干扰。在地面建筑中如创造此类环境条件必须增加复杂的空调系统，配合各种高效过滤器并远离铁道、公路和其他工业生产震源，需要专门的电磁屏蔽装置以切断电磁波的干扰等。而在地下空间内则可利用岩、土良好的热稳定性和密闭性，大大减少空调费用，减少粉尘来源；利用岩土层的厚度和阻尼，使地面振动的波幅大大减少和使电磁波受到极大的削弱。

（三）开发利用地下空间是缓解城市发展中各种矛盾的有效措施

城市化的快速粗放发展的另一后果，是正在形成的"城市综合症"，集中表现在：交通阻塞；环境污染；生态恶化。

在解决"城市综合症"的诸多方面，地下空间的开发利用都将大有可为。

（1）发达国家解决城市"交通难"的主要措施是发展高效率的地下有轨公共交通，形成四通八达的地下交通网。

（2）解决"停车难"，在很多国家的现代化城市，其主要出路是修建地下停车库。

（3）改善城市的生态环境，减少大气污染，除发展地铁、轻轨等使用电能的公共交通网，减少尾气污染外，还要改变燃料能源结构。

（4）建设便于维修管理检查的多功能公用隧道"城市共同沟"，是城市现代化的标志，可以避免道路的反复开挖，以及施工对交通和城市居民生活的影响，特别是便于维护检查和拆换，减少事故，提高城市基础设施的抗灾能力。

（5）地下空间具有较强的抗灾特性。对地面上难以抗御的外部灾害，如战争空袭、地震、风暴、地面火灾等有较强的防御能力，提供灾害时的避难空间、储备防灾物资的防灾仓库、紧急饮用水仓库以及救灾安全通道。

二、国外城市地下空间开发的趋势

（1）综合开发；
（2）分层开发与深层开发；
（3）地下环境条件改善；

(4) 各种联合掘进机（TBM）和盾构将成为地下隧道快速开挖的主要趋势；
(5) 在钻爆法掘进中采用数字化掘进的趋势将加强；
(6) 地铁隧道断面将减小，地铁成本将降低；
(7) 微型隧道工程将加速发展；
(8) 市政公用隧道（共同沟）在二十一世纪将得到更广泛应用；
(9) 三S技术在地下空间开发中的作用将得到加强；
(10) 勘察、设计和施工的信息化整合；
(11) 地下空间的单体工程建筑形态将趋向功能化、艺术化。

三、我国城市地下空间开发的现状、问题及原因分析

（一）现状

目前地下空间开发利用的主要模式有：

(1) 结合地铁建设修建集商业、娱乐、地铁换乘等多功能为一体的地下综合体，与地面广场、汽车站、过街地道等有机结合。

(2) 地下过街通道（商场型），在市区交通拥挤的道路交叉口，修建过街地道为主，兼有商业和文娱设施的地下人行道系统，既缓解了地面交通的混乱状态，做到人车分流，又可获得可观的经济效益。是一种值得推广的模式。

(3) 站前广场的独立地下商场和车库（商场型）。在火车站等有良好的经济地理条件的地方建造的以方便旅客和市民购物为目的的地下商场，如沈阳站前广场地下综合体。有的地下商场下层设地下车库。

(4) 在城市中心区繁华地带，结合广场、绿化、道路，修建综合性商业设施，集商业、文化娱乐、停车及公共设施于一身，并逐步创造条件，向建设地下城发展。如上海人民广场地下商场、地下车库和香港街联合体。

(5) 在历史名城或城市的历史地段、风景名胜地区，为保护地面传统风貌和自然景观不受破坏，合理开发利用地下空间。如西安钟鼓楼地下广场。

(6) 高层建筑的地下室。一般高层建筑多采用箱形基础，有较大埋深，箱形基础本身的内部空间为建造高层建筑中的多层地下室提供了条件。将车库、设备用房和仓库等放在高层建筑的地下室中，是常规作法。

(7) 已建地下建筑、人防工程的改建利用是我国近年利用地下空间的一个主要方面，改建后的地下建筑常被用作商店、车库、仓库等。

（二）存在的主要问题

(1) 缺乏整体的发展战略和全面规划；
(2) 管理机构条块分割，不能形成合力；
(3) 城市地下空间开发利用无法可依；
(4) 固有观念和认识误区影响了城市地下空间开发的积极性；
(5) 没有广开渠道、多种形式解决资金来源；
(6) 平战结合处理不当，影响了地下空间开发的积极性；
(7) 已开发利用的地下空间功能较为单一，不同类型设施的开发比例不平衡；
(8) 地下空间的开发缺乏建筑师的参与。

四、开发地下空间的技术经济可行性

（一）城市经济发展的必然

其标志是人均国内生产总值（GDP）和地价的增长。纵观世界各国城市地下空间开发利用的发展历程，可以得出一个经验判断：即一个国家的人均国内生产总值在 500 美元以上时，开始出现城市地下空间的开发利用的需要；当人均 GDP 进入 500~2000 美元阶段时，城市地下空间的开发利用得到较为广泛的发展；当人均 GDP 在 2000 美元以上时，城市地下空间的开发利用将向高水平发展。

据中国社会科学院的宏观经济预测模型，我国人均 GDP 在 2000 年、2010 年、2020 年与 2030 年将分别为 818、1547、2543 与 3695 美元。按此发展水平，参照国际上关于城市地下空间开发的经验判断，可以认为我国城市地下空间开发目前已具有必需的经济基础，应当启动。21 世纪 20 到 30 年代就应向高水平发展。这样的判断是就全国而言的，有些城市经济发展水平远高于全国平均水平，这些城市地下空间开发的发展还应大大提前于全国进程。

我国人口众多，土地资源十分紧缺，一些大城市的人口压力、交通拥挤和环境污染的程度远高于发达国家。还有，我国城市化进程正在加快。北京、上海、广州等大城市近年来兴建了数以千计的高层建筑。如此大规模发展地面建筑的同时，不相应地开发地下空间，岂不是坐失良机吗？甚至造成不可逆转的地下空间资源的浪费，给今后地下空间的开发造成困难。所以，对有些城市来说，当务之急是制订城市地下空间开发的规划，然后随着城市建设的发展，结合地面建筑的发展逐步落实规划，使地下空间开发不失时机，健康、有序地发展。

（二）地下工程的造价有下降的空间

除地铁以外的地下工程与同类地面框架建筑的造价比较如下：

（1）1985 年前，两者的综合造价大致相当；

（2）1986—1990 年间，地下工程的综合造价明显低于地面同类工程造价，高层建筑的土建造价比地下工程平均高出 36.5%；

（3）1992 年后，地面高层框架建筑的土建部分造价略高出地下工程造价 1/3，多层商业建筑土建部分造价与地下建筑土建造价大致相当，而多层住宅建筑的土建造价约为人防工程土建造价的 1/2。

目前我国城市地铁造价偏高甚多，8 亿元/公里左右，使地铁的发展受到很大制约。

降低地铁工程造价的措施有：适当加大站间距，减少车站数量；缩小行车间隔，减小车站规模；设备以国产为主，不宜多进口，力争不进口；车站建筑装修从简；正确选择地铁隧道的平、纵断面；科学管理，避免资金浪费；坚持按建设程序办事，坚持工程招标。如果我们确实做到以上各项，就完全有可能把我国城市地铁造价控制在每公里 4~5 亿元之间。

（三）多种渠道筹资开发城市地下空间

我国过去城市地下空间的开发利用其主要构成为平战两用的人防工程，其经费主要靠政府的人防拨款和城市建筑的人防易地建设费，由于资金的限制，规模很难扩大。近年来，随着城市地下空间开发领域的扩大，情况已经发生了变化，其中有不少值得推广的经验。如：成都顺城街地下商业街；西安钟鼓楼广场地下综合体；大连市人民体育场外场改造地下工程；哈尔滨红博广场地下综合体。

这几个典型项目，各有不同的筹资方式，但共同的一点是少花或不花国家的钱，却把城市地下空间开发起来了。由于一方面城市浅层地下建筑工程造价比地面高层建筑造价低，另一方面，更主要的，在目前开发地下空间不需要支付或少支付土地费用的情况下，开发地下空间对房地产开发商投资具有很大的吸引力。

（四）我国已具备开发浅层开发的勘察、设计、施工和防灾的技术能力

一是我国的地下工程结构设计技术与地下施工技术发展较快，与国外先进水平差距不大。

二是我国对岩、土介质中结构分析的学术水平在有的方向已居国际先进水平。

三是我国的岩土工程勘察技术力量在学术素质上接近发达国家水平，规范标准初步建立，先进的岩土工程勘察手段已有很多成功应用。

四是我国多功能、综合性的城市地下空间规划尚处在起步阶段，缺乏经验，急需开展研究，制定各种技术规范，并研制相应的软件。

综上所述，我国地下空间的勘察、设计、施工和防灾虽尚存在若干薄弱环节，但已有一定的理论研究和工程实践基础，已具备开发浅层地下空间的技术水平和技术力量，对于复杂的、难度较大的地下空间开发任务，只要集中力量，也是可以完成的。但是，为迎接我国地下空间开发利用的新发展，还要在勘察、设计、施工及防灾技术等若干薄弱领域积极开展研究，消除空白领域，缩短与世界先进水平的差距。

五、目标与路径

（一）发展目标

在目前初步开发利用的基础上，采取积极稳妥的方针，加强立法和政策的制定，加强全面规划和管理，有序、合理、综合、高效地开发利用地下空间资源，力争我国城市地下空间的开发利用有一个较大规模的广泛发展，逐步达到国际先进水平，使其在城市现代化改造与建设中，发挥提高土地利用效率与节约土地资源、扩充基础设施容量、疏解中心区高密度人口，提高综合防灾能力、改善生态环境及保护历史文化景观等有效作用，成为贯彻可持续发展，建设生态节能型城市不可分割的重要组成部分。

（二）路径

充分开发地下10m以内浅层城市地下空间，适度开发地下30m以内或以下的次浅层、深层地下空间，根据特殊需求个别开发深层地下空间，逐步把一切可能转入地下的设施转入地下，全面改善城市地面环境和生态质量。具体可在以下方面着力：

（1）高效利用土地；
（2）进一步开发城市地下交通；
（3）市政基础设施统一规划，尽可能入地；
（4）进一步发展综合防灾设施；
（5）贮存能源、水资源及其他物资；
（6）把一部分高精度生产系统和研究实验系统设在地下。

六、保障措施和建议

（1）加强城市政府对地下空间开发利用的统一领导，建立城市地下空间开发利用委员会；
（2）加强宣传；
（3）尽早立法；
（4）政策优惠；
（5）远近兼顾，平战结合；

（6）全面规划，分步实施；

（7）积极稳妥，区别对待；

（8）严格审批，加强管理；

（9）加强科研，培养人才；

（10）发展我国地下施工掘进机械，地下施工监测控制设备和地铁及轻轨系统的车辆和信号设备产业，提供性能优良和低成本的国产设备。

"双碳"目标下的城市建设

一、"双碳"目标提出的极端重要性和紧迫性

联合国秘书长古特雷斯 2020 年曾指出:虽然病毒是目前最大的担忧,但气候变化仍然是当代面临的核心挑战。

全球 11000 名科学家联名签署提出:"所有这一切都是为我们即将到来的、可能是最糟糕的全球灾难—气候变化—发出的巨大警告信号。"到 21 世纪末,全球气温若上升 1.9℃,高海拔地区冰雪将减少 80%,水塔的减少将导致 19 亿人缺水,海平面上升 1 米左右,全球岛国淹没,将使 2.8 亿人失去家园,全球 35.7% 到 49.5% 的海滩将消失。

今年的情况甚为可忧,今年联合国埃及气候变化大会前全球已升温达 1.15℃,将很难达到巴黎大会确定的 1.5℃目标!埃及大会的一系列报告预示着地球已经靠近一场气候灾难:当下的行动力度导致全球气温将灾难性上升到 2.5℃,将抵到导致气候混乱不可逆转的临界点。以欧洲冰川为例,其消融速度"前所未有"。比利牛斯山维涅马勒峰自 2000 年以来,其冰川厚度已经减少了 40 米,比利牛斯山自 1850 年以来,冰川覆盖面积已经减少了 90%,2000 年统计的 44 个冰川,到 2022 年只剩下 23 个,平均每年消失一个,到 2050 年,科学家估计比利牛斯山将不再有冰川;阿尔卑斯山情况也是"灾难性"的,2022 年其冰川体量减少了 6%,是过去 20 年平均水平的 3 倍,全球变暖到 21 世纪会导致阿尔卑斯山区升温 3℃,届时,那里的冰川会全部消失!

联合国秘书长古特雷斯在今年埃及气候大会上指出:"我们正飞奔在走向地狱的快速路上。"

导致全球气候变暖的主要原因是二氧化碳等温室气体的排放,据估计,全球约三分之二的碳排放来自城市。

2020 年 9 月,习近平主席提出"双碳"目标,是我国积极参与应对气候变化全球治理的重要举措。党的二十大报告指出,推进生态优先、节约集约、绿色低碳发展。各地各行业聚焦"双碳"目标,加快促进经济社会全面绿色转型。

二、建设"公园城市",促进生态碳汇

(一)建设"公园城市"是生态碳汇的重要措施

习近平总书记 2018 年在成都首次提出"公园城市"新理念和城市发展新模式,指出"要突出公园城市特点,把生态价值考虑进去,努力打造新的增长极"。

生态碳汇是通过植被造林、植被恢复、森林管理等措施,利用植物光合作用吸收大气中的温室气体,并将其固定在植被和土壤中,从而减少温室气体在大气中的浓度。即发展绿水青山的生态建设。森林、湿地和绿色植被是世界上抵御气候变化的前沿防线。

《Science》杂志发表的一项研究预测,如果新种植 9 亿公顷树木——大约 5000 亿棵树苗——那么,它们在成熟时可以吸收 2050 亿吨碳。

报告时间:2022 年 12 月 14 日。

科学家估计：单一树种的林地每公顷吸收 12 吨 CO_2，混合树种的林地每公顷吸收 32 吨 CO_2。

浙江农业大学从竹笋生长、加工、包装、运输的全过程测算，其温室气体吸收和排放量：每亩笋基地每年可吸收 0.238 吨 CO_2，每公顷笋基地可吸收 3.57 吨 CO_2。

研究发现：混交树种林的树高、直径和生物量均显著高于单一树种林，平均增加量分别为 5.5%、6.8% 和 25.5%；这种增产效应主要是由种间互补作用导致的；随着混交树种的增加，增产效应呈增加趋势。

湿地碳汇是生态碳汇的另一重要手段，每平方公里湿地吸收 CO_2 量达 220 吨。这些数据表明，气候变化的应对之策应是推进绿色发展，实施生态大保护：建设绿色生态城市；保护世界原有森林和绿地；建设绿色国土，发展绿水青山。

中国"十四五"规划对绿色生态建设有 5 个约束性指标，其中森林覆盖率将由目前的 23.04% 提高到 24.1%；计划在这个十年末，每 10 座城市中有 7 座城市绿化率达到 40%；我国约有 170 个城市开展创建"森林城市"行动，参与城市年均新增造林面积约 1.3 万公顷。北京市"十三五"以来，已完成两轮百万亩造林任务！

《河北雄安新城规划纲要》规定，新城建筑空间仅占 30%，而 70% 的城市空间为绿蓝生态空间，新城今年已基本建成，森林覆盖面积达 32%；《长三角生态绿色一体化发展示范区总体方案》强调，一体化示范区要生态筑底、绿色发展，先行启动区蓝绿空间占比不低于 75%，长三角绿色一体化发展示范区将引领城市群的建设方向。

积极建设城市口袋公园、街边袖珍绿地、湿地公园。据 2022 年统计，全国累计建设口袋公园 2 万余个、湿地公园 1600 余处。

（二）"建设地上地下一体化、互相衔接、互相协调的立体化城市"是建设公园城市的重要途径

开发利用地下空间，节约地面土地资源，用以建设绿色生态，才能实现生态碳汇；开发利用地下空间，建设紧缩城市（Compact city），节省交通能源，助推减排。

2019 年底全国城市地下空间总面积约 22 亿平方米，新增地下空间建筑面积 2.57 亿平方米，占同期城市建筑面积 19%，全国已开通运营的铁路、公路、地铁隧道总里程达 42187 公里。

全国三线及以上城市面积约 149.02 万平方公里，对应的地下空间尚未得到充分开发。

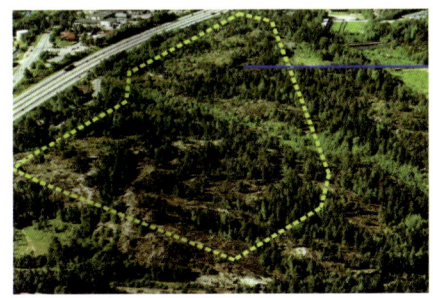
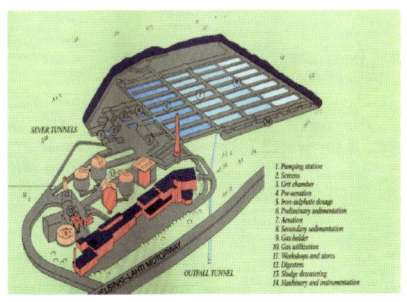

赫尔辛基污水处理系统

地面景观

天府新区第一污水处理厂地下厂区

波士顿拆除穿过市中心的六车道高架路，建设了 8~10 车道的地下高速路，原有的地面变成林荫路和街心公园，这样的结果是市区空气的一氧化碳浓度降低了 12%，增加了 260 英亩新的公园和开敞空间。

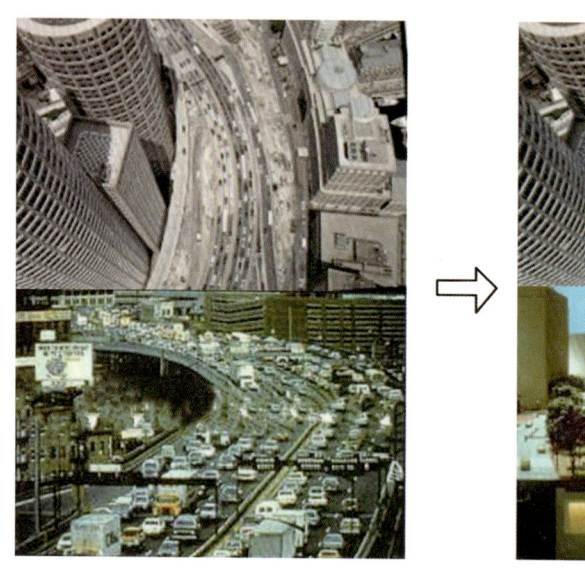

a) 改建前（地面高架路）　　　　　　　　b) 改建后（地下快速路）

波士顿道路入地改造

三、发展城市双碳建筑

2019 年全国建筑总量 644 亿 m^2，碳排放总量 22 亿吨，建筑碳排放量占比 30%~40%。
需求：到 2030 年非化石能源占比达到 25%；碳排放强度比 2005 年下降 65% 以上，占排放总量 22%。
途径：电气化；太阳能发电；地热供暖制冷；立体园林建筑。

（一）建筑设备电气化是关键

供暖、制冷、照明、烹饪、家用电器等均电气化。

（二）发展与建筑结合的光伏 BIPV（光储直流发电），电力自发自用，潜力巨大

芮城试点：每户投资 10~12 万元，屋顶光伏发电每年 2.2 万 $kW \cdot h$，自用外，可外送 1 万 $kW \cdot h$。
全国城乡建筑约 650 亿 m^2，可利用面积 250 亿 m^2，若利用 50%，可发电 2 万亿 $kW \cdot h$。
十年来光伏系统成本下降超过 90%，光伏电价已低于火电电价。同时，这一模式实现了电力产销结合、自产直销，电从身边来，增加城市能源自给率，安全性。
光伏发展新趋势：轻质化、柔性化太阳能电池。

①我国有40%的工业厂房和彩钢瓦屋顶缺乏承载能力，轻量化光伏组件能有效解决这一痛点。
②能够解决防水材料在屋顶安装无法穿透的问题。
③还适用于具有弧度的曲面屋顶。
典型应用：大型集中式光伏发电场 → 离散式分布式厂房和街区屋顶 → 城市新家居和新景观。

超薄柔性铜铟镓硒(CIGS)太阳能电池

最薄 / 总厚度≤35μm，相当于A4纸张1/3，传统晶硅电池的1/200厚度

最柔 / 电池可实现360°卷绕

最轻 / 最具重量能效比的电池

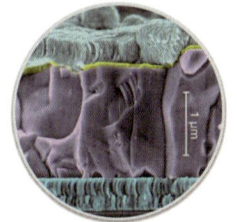

☑ 全光谱 弱光性好 转换率高　　☑ 3-5nm精度工艺，稳定可靠
☑ 独特封装 IP68防水 适应海边盐雾　　☑ –40～80℃ 纳米自洁

可与**移动电力、智慧城市、5G基站、特种军用、便携民用**等各行业完美结合，填补国家新型能源供给空白，应用场景无限。

典型案例-零能耗建筑系列

梦想小镇分布式光伏共享系统项目

- 地点：浙江·杭州未来科技城
- 组件：柔性/刚性CIGS组件
- 成果：项目单日发电量最高为7980kW·h；投产后年发电量125.23万度；为业主增加收益累计达150余万元

典型案例-零能耗街区系列

东湖区块（含北大街）零能耗街区项目

- 地点：临平区北大街及东湖街道区块
- 项目：自2017年第一座智慧候车厅落地以来，先后分4期建成了绿色智慧站台200余座，智慧合杆/智慧灯杆160余杆，通过分布式发电、分布式储能和分布式智慧应用的一体化系统方案，提供新能源+物联网的新型服务。
- 成果：项目首次在国内街区层级搭建了绿色+智慧的全新街区，应用产品包括绿色智慧灯杆、绿色智慧站台、智慧标牌等，打造面向未来的零能耗智慧街区

（三）利用地热供暖与制冷

利用地热为建筑供暖和制冷是经济适宜、生态减碳的选择，应当成为建筑能源的首选方案。

浅层地温能利用：能源桩、能源墙、能源隧道。

它把传统空调器的冷凝器或蒸发器直接埋入地下，利用传热循环介质与大地和江、湖水源进行热交换（热泵），从而提取地温能与水温能。

地下换热系统可埋设在地下结构基坑围护结构（地下连续墙、排桩内）、基础底板下、桩基（钻孔灌注桩、预制桩、PHC桩等）内；可埋设在新奥法施工的隧道衬砌内或以能源锚杆的形式埋设在其围岩中；也可埋设在地铁区间隧道内、地下输水管道内。

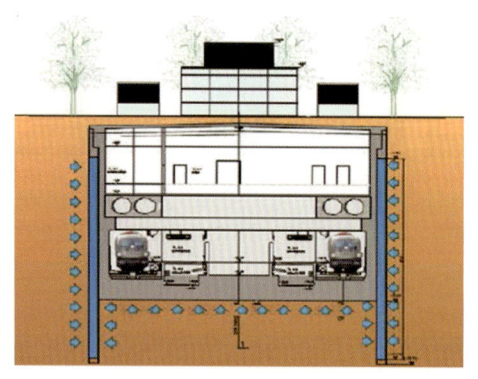

地铁车站的地温能提取示意图

我国地级以上城市规划区范围内浅层地热能资源年可采量折合标准煤7亿吨，可实现建筑物供暖制冷面积320亿 m^2。水热型地热能是指较深的地下水或蒸汽中蕴含的地热资源，主要蕴藏于温泉或人工钻井开采得到的地热流体中。我国水热型地热资源总量折合标准煤约1.25万亿吨，年可采量折合标准煤约19亿吨。干热岩是指埋藏于地球深部、温度高于180℃的异常高温岩体，可通过增强型地热系统进行开发。我国陆区地下3至10千米范围内干热岩资源量折合标准煤约856万亿吨。

在有条件的地方，积极开发浅中层温泉地热能（雄县）和深层干热岩地热能（长安信息大厦）。

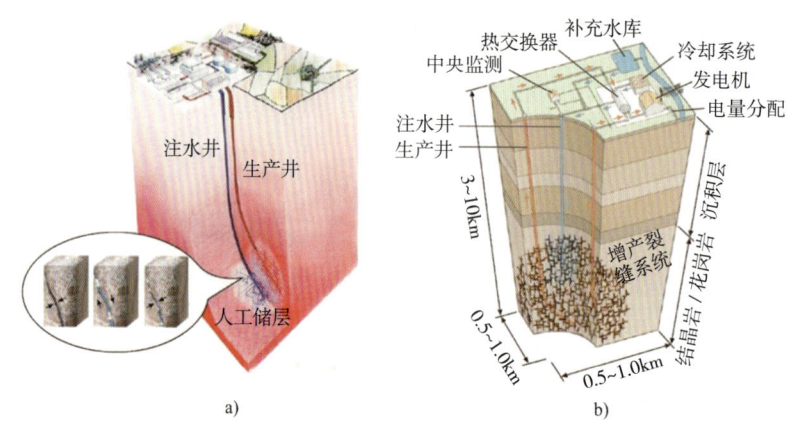

干热岩体中双井式增强型地热系统的概念模型

上海自然博物馆

上海自然博物馆占地面积约12000m^2，总建筑面积约为45086m^2，其中地上建筑面积12128m^2，地下建筑面积32958m^2。建筑总高度18m，地上三层，地下两层，采用地源热泵系统承担建筑部分夏季冷负荷以及冬季热负荷。夏季土壤换热器最大热负荷为1639千瓦，冬季土壤换热器最大热负荷为1178千瓦。

2019年8月，北京大兴国际机场地源热泵工程1号能源站开始启用，提取浅层地温能，实现大兴机场公共区域250万m^2的供热和制冷，每年节省天然气1735.89m^3，相当于21078吨煤，减少碳排放1.58万吨以上。

2017年底，我国地源热泵装机容量达2万兆瓦（2000万千瓦），实现供暖（制冷）面积5亿平方米，利用地热能折合1900多万吨煤。

当然立体园林建筑也可发挥综合减碳作用，不再赘述。

立体园林建筑可发挥综合减碳作用

四、发展城市双碳交通

化石燃料交通是碳排放大户，约占总排放量1/3。各国已在启动规划燃油车退出的路线图和时间表，我国海南岛已宣布，2030年禁驶化石燃料汽车。

途径：以电代油；以氢代油（氢燃料电池汽车、甲醇燃料）；光储一体化。

以电代油措施：电力驱动的轨道交通；电动汽车。

案例：2025年建成轨道上的长三角

8000公里高速铁路，长三角与相邻城市群3小时区际交通圈；1500公里城际铁路，三角区内相邻城市形成1~1.5小时城际交通圈；1000公里市域（郊）铁路，各都市圈内形成0.5~1小时通勤交通圈。

电动汽车产业化达到爆发节点；氢燃料电池进入成本下降的快车道。

中国电动车行动计划指出，2040年电动汽车保有量达到3亿辆，与之相应，发展电化学储能，城市要建设充电桩、充换电站、加氢站。

2020年底，电化学储能装机规模3269.2兆瓦，其中锂离子电池装机规模2902.4兆瓦。

"十四五"期间，中石化将规划布划布局5000座充换电站；今年全球首座全智能换电（池）站落户中国石化（朝英换电站：面积60m^2；单日可提供换电服务312次；自动停泊、自动换电、用户无需下车一键完成泊车换电业务）。

2025年中国拥有5万辆氢燃料电池汽车；每年可再生能源制氢（绿氢）10万~20万吨。日本宣布：今年三月试运行氢能源列车，2030年正式推广；2040年前氢燃料动力飞机投入实际使用；德国开通了"全球首秀"氢能列车，车顶安装了"氢燃料电池"，只产生蒸汽和冷凝水，时速80~120公里，最高时速140公里，续航里程1000公里（添加一次燃料），每年减少排放4400吨二氧化碳。

甲醇燃料又名液态阳光，是用甲醇形式实现太阳能和风能的储能、远距离运输及利用。光能、风能分解水→氢+CO_2→甲醇。甲醇汽车内燃机燃烧→水+氢。

光储一体化：基于地铁、地下物流系统的地下运行特性，依托地下空间修建储能系统减少占用地表空间，消纳谷价电力，储存盈余电量，响应电网削峰填谷，依托地表物业光伏发电，缓解电网供电压力。储能设施建于地下，降低所受外部环境的影响与降低储能故障事故对外部环境的影响。

综上所述，发展城市双碳交通主要途径有：建设轨道交通及其地下储能设施（飞轮、电池、光储一体）；新能源汽车（电动汽车、氢燃料电池汽车及其附属设施）。

五、积极稳步发展城市CCUS（Carbon Dioxide Capture Utilization and Storage）与减排

城市工业排放CO_2占总量三分之一，降排措施就是CCUS。

捕集：将化工、火电、钢铁、水泥、化肥等工厂化在能源利用过程中产生的CO_2进行分离和富集的过程。

例如上海石洞口电厂系统运行中每年捕集12万吨CO_2，相当于植树900万棵。

齐鲁石化——胜利油田百万吨级CCUS项目今年8月29日投入注气运行：齐鲁石化捕集提纯CO_2，然后运送至胜利油田进行驱油封存，首次实现二氧化碳长距离超临界压力管输送，做到制、输、用全过程全密闭。

利用：将CO_2注入地下，生产或强化能源、资源开采，主要用于提高石油、地热、深地层咸水等的采量。中石化华东油气分公司位于苏北平原，地跨11个县区，实现年注入大于10万吨的CO_2的驱油封存规模化应用，累计注入40万吨，累计增油13万吨（中石化南京化工公司的高浓度CO_2通过江苏发达的水系船运到油田）。

利用地下空间的封闭性、稳定性等优势，可实现CO_2的永久地下封存，也称是人工碳汇。目前测得全国煤矿采空区地下空间约138.4亿m^3，盐穴地下空间约1.3亿m^3，其他金属和非金属矿洞可利用空间约15亿m^3。部分改造后可用以封存CO_2。

封存：将捕采的CO_2以超临界流体形式封存于地质构造中还有枯竭油气田封存、海底咸水层封存、陆上咸水层封存。根据估算，中国可封存容量大，CO_2咸水层封存具有很大应用潜力：对于富集煤炭、石油资源，水资源短缺的西部地区，对东部沉降严重地区以及对南方深层卤水资源开采区具有很大吸引力和潜力。

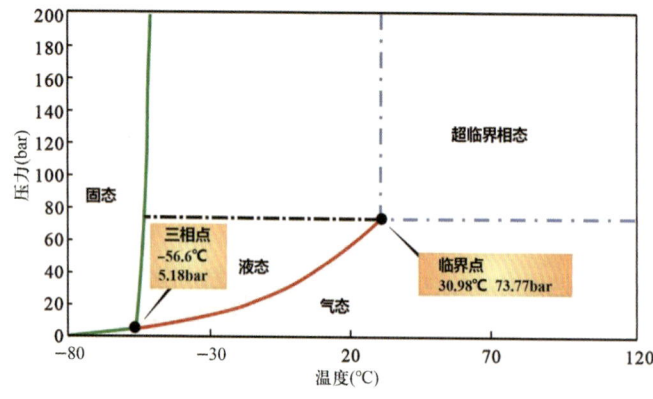 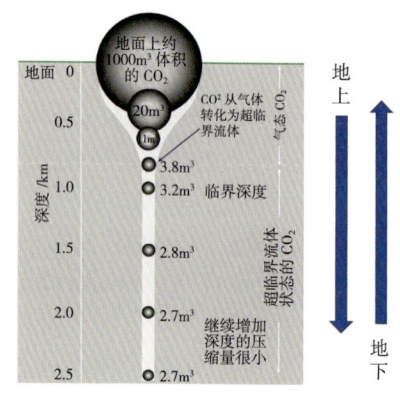

储能减碳：实现"双碳"目标最主要的CO_2的封存措施是转变化石能源为可再生的绿色能源，我国绿源的主力为光、风能，根据国家规划，到2030年风电和光伏的装机规模为12亿kW，实际上到今年已达10亿kW，风、光能具有波动性、随机性，从而影响电网的安全稳定运行，其对策是储能调节消纳，主要措施是电解水制氢（零碳燃料）储能。其方式有二：高纯度氢气的大容量、长周期的地下储存；

电化学储能，即锂离子电池等各种氢能电池。

纯氢直接储存的挑战：氢的密度较低，需要较高的存储压力，地下围岩可以提高存储稳定性；氢是一种小分子，具有很高的流动性和渗透性，除地下盐穴储库外，内衬式岩洞（LRC）储气库，由于选址灵活，库址易靠近能源消耗量大的城市，有利安全与减少运输成本。研究表明：比较而言，钢内衬效果良好，注气三天左右不发生泄漏，但其长期暴露在氢气中，特别在高温高压下，将发生氢致气泡、氢制开裂和氢蚀脆化等现象，从而降低其耐久性。所以目前我国尚无储氢实际工程，该领域工程技术亟须研究。

电化学（电池）储能：结合我国新能源汽车的发展迅猛，中国已并网投入运营了包括江苏镇江101/202MW·h储能电站在内的多个电化学储能系统。但其安全性亟待研究解决，全球共发生50余起安全事故，地下储存应为其可选择途径。

压缩空气储能：是指利用用电低谷期电能带动压缩机将空气压缩后存储于地下储气库，在用电高峰期将压缩空气释放带动发电机供应电能，它具有规模大、成本低、寿命长、储能周期不受限、清洁无污等优势，可实现电力系统调峰、安全备用，从而提高电力系统效率、安全性和经济性。

其他还有抽水蓄能储能、创新的重力储能技术（活塞式和水囊式）等。

国内首套百兆瓦（100/400MW·h）压缩空气电站已于今年在张家口顺利并网。最新技术是压缩空气至20个大气压，冷却到−160℃，使其变成液态，发电时再对液态空气加热使其膨胀，带动涡轮机发电，液态时体积为气态时的1/300，便于在城市地区使用。

"双碳"目标下的能源地下工程

一、前言

习近平总书记多次强调:应对气候变化不是别人要我们做,而是我们自己要做,是我国可持续发展的内在要求,是推动构成人类命运共同体的责任担当。

习近平主席在联合国大会上发言提出:"中国将提高国家自主贡献力度,采取更加有力的政策和措施,二氧化碳排放力争于 2030 年前达到峰值,努力争取 2060 年前实现碳中和。各国要树立创新、协调、绿色、开放、共享的新发展理念,抓住新一轮科技革命和产业变革的历史性机遇,推动疫情后世界经济'绿色复苏',汇聚起可持续发展的强大合力。"

地下空间作为一种具有广阔发展与应用前景的战略性国土资源,它本身是地热资源载体,且具有种种优势:受外界干扰小,具有封闭性;热与力学稳定性;对腐蚀、地震、台风、辐射、电磁干扰具有防护性。因此它可为多种碳减排、碳增汇技术提供充足且稳定的空间支撑,极具开发价值与潜力。

二、碳封存地下工程

碳封存包括地质利用与地质封存。

地质利用:将二氧化碳注入地下,生产或强化能源、资源的开采过程,不但封存了二氧化碳,还可用于提高石油、煤层气、地热、地层深部咸水、铀矿等资源采收率。

地质封存:将捕采的二氧化碳储存于地质构造中,实现与大气长期隔绝。主要划分为陆上咸水层封存、海底咸水层封存、枯竭油气田封存等。

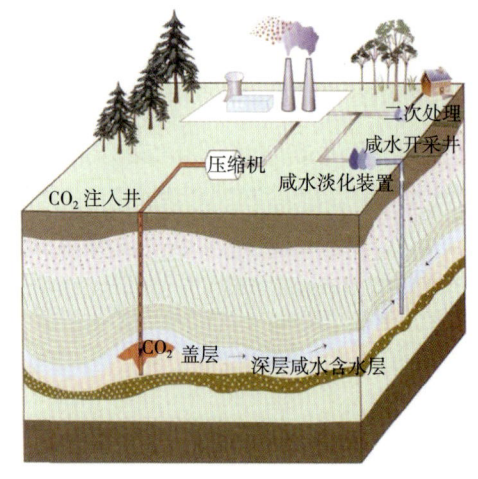

根据中国国情估算,中国可封存容量大(1.44×10^{11} t),CO_2 咸水层封存具有很大应用潜力:对于富集煤炭和石油资源、水资源短缺的西部地区,东部京津唐、苏锡常等沉降严重地区以及南方深层卤水资源开采区具有很大吸引力和潜力。

液相注入和超临界流体注入,使其溶解于咸水层中或与周围的岩石矿物产生作用形成新的矿物。

2019 年中国共有 18 个捕集项目在运行,二氧化碳捕集量约 170 万吨;12 个地质利用项目运行中,地质利用量约 100 万吨;化工利用量约 25 万吨、生物利用量约 6 万吨。

国内碳地质利用项目,例如中石化华东油气田 CCUS 苏北草舍泰州组低渗油藏先导项目实现年注入 CO_2 大于 10 万吨的驱油封存规模化应用,累计注入 CO_2 40 万吨,增油 13 万吨。船舶槽运 200 元/吨。

国内碳封存项目,例如国家能源投资集团有限责任公司神华深部咸水层二氧化碳地质封存示范工程,于 2011 年实施注入实验,截至 2015 年底,完成 30 万吨 CO_2 注入的目标。采用罐车运输,全流程成本 249 元/吨,驱油封存成本 120~800 元/吨。

报告时间:2021 年 12 月 24 日。

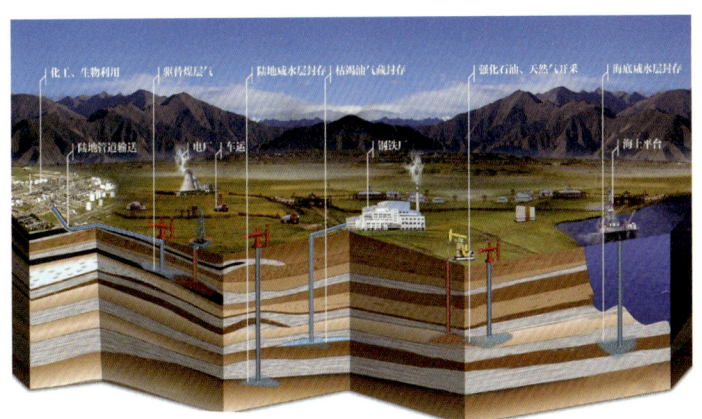

CCUS 技术类型示意图

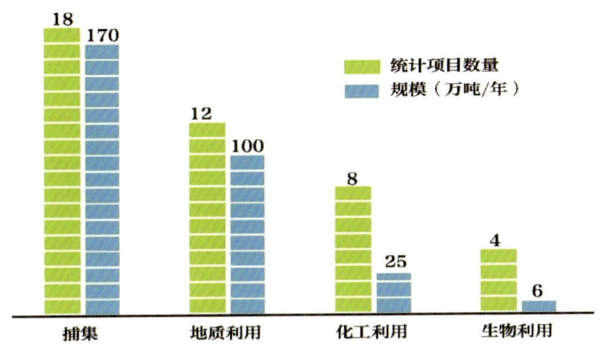

2019 年中国 CCUS 项目统计

注：①神华咸水层封存项目已于 2015 年结束注入，上图中不包含此项目。
②图上的化工利用、生物利用项目数量指统计数量，不代表全国实际水平，因统计原因，可能存在遗漏。
③化工利用的规模指的是全国规模，不代表本报告统计项目的利用量。数据来源于《中国 CCUS 发展路线图（2019）》。

三、地热能利用地下工程

浅层地温能利用工程包括：能源桩、能源墙、能源隧道。

它把传统空调器的冷凝器或蒸发器直接埋入地下，利用传热循环介质与大地和江湖水源进行热交换（热泵），从而提取地温能与水温能。

地下换热系统可埋设在地下结构基坑围护结构（地下连续墙、排桩）内、基础底板下、桩基（钻孔灌注桩、预制桩、PHC 桩等）内；可埋设在新奥法施工的隧道衬砌内或以能源锚杆的形式埋设在其围岩中；也可埋设在地铁区间隧道内、地下输水管道内。

利用地热为建筑供暖制冷是经济适宜的选择，应当成为建筑能源的首选方案。

根据现有资料估算表明，长三角地区 3 千米以浅的地热资源总量达 1200 亿吨标煤的热量，年可利用量相当于 1.7 亿吨标煤。可实现建筑物冬季供暖面积 24.8 亿 m^2，夏季制冷面积 14.5 亿 m^2。然而，目前长三角地区利用地热供暖和制冷的建筑面积仅有 0.57 亿 m^2，开发利用潜力巨大。

大兴机场建成我国最大的能源桩工程。

四、压缩空气地下储能工程

利用盐穴等地下洞室稳定的力学性能，储存大容量高压空气消纳谷价电力，建设蓄能电站（长三角地区金坛等地）。

五、地下抽水蓄能电站

在山顶建水库，山谷地下建电站，利用位势差蓄能消纳谷价电力。

六、基于地铁、地下物流系统的地下储能系统

基于地铁、地下物流系统的地下运行特性，依托地下空间修建地下储能系统减少地表空间，消纳谷价电力，储存盈余电量，响应电网削峰填谷，缓解电网供电能力。储能设施建于地下，降低所受外部环境的影响与储能故障事故对外部环境的影响。

七、零碳能源的地下储能工程

能源绿色转型的根本之策是以可再生清洁能源为主，辅以火力发电调节。可再生清洁能源（光伏能、风能、水力能）具有超远距逆负荷分布特性和出力的波动性、随机性。前者需要特高压传输网，中国已经基本建成。后者面临可再生能源消纳、电网安全稳定运行的挑战，其对策是储能调节消纳，无论是电磁储能、高压空气储能、热力储能、电解水制氢储能以及水力蓄能电站都需要开发利用地下空间，利用地下空间的封闭性、隔热性、电磁屏蔽性进行规模化储能。

主要包括制氢储能与电化学储能两类。电解水制氢储能是一类清洁低碳的制氢储能技术，利用电能制取高纯度氢气并进行大容量、长周期的储存。目前，较成熟的电解水制氢技术包括碱性电解技术、固体聚合物电解技术和固体氧化物电解技术。2021年我国正式投产了"国家级太阳能电解水制氢综合示范项目"，其电解水制氢装置产能可达每小时2万标方。电化学储能是指通过发生可逆化学反应储存或者释放电能的储能技术，根据化学物质的不同可以分为铅酸电池、液流电池、钠硫电池、锂离子电池等。目前，中国已并网投运了包括江苏镇江101MW/202MW·h储能电站在内的多个电化学储能系统。

制氢储能地下与化学电池储存地下的一系列技术亟待开发，包括相应的风险预防。

八、风、光、储一体化的地下工程

江苏海泰长江隧道正在研究依托风、光、储一体化设施建设国内第一条零碳隧道工程。

九、核能的地下高放核废物的地质储能

我国正在甘肃北山荒漠地区建设高放核废物深层地质储存库的地下实验室。

审读大百科轨道交通文稿时的思考

一、贯彻"以人民为中心"的思考

建设轨道交通的目的是以提高乘客的方便与舒适为目的。

设计规划地铁车站时应考虑：

（1）在站内设立货客电梯，这样可运送轮椅上的残疾人，失火时又用作备用的疏散通道。

（2）在站厅层设置公共厕所。

（3）上、下电动梯的范围从3.5米开始，提高乘客的流动速。

（4）通风照明系统和车站建筑形态要消除地下空间对乘客的负面影响。

（5）站厅、过厅的艺术形态、照明指示装置及广告板，创造易于记忆的形象，以便捷乘客的方向识别。

设计规划地铁换乘站时，应该达到的运行指标如下：

（1）线路每小时列车通行量达到40对，列车时间间隔1.5min。完善的供电系统、调度系统及列车运行的安全性允许通行能力在必要时再提高10%。

（2）道路设施，其平面及断面配置，现代化的行驶列车能保证速度达到每小时90~100km，而平时的行驶速度为每小时45~55km。

（3）站台参数必须保证计算的列车车厢数，以允许在高峰期乘客在最满载的车厢里有足够高的舒适度。

（4）出入口站厅及换乘设施规模的规划方案按乘客的乘行时间来讲，要保证良好的条件。

达到如此高的发车效率，重要的因素在于车库与折返线的设置：

（1）应在每一条地铁线上设置车库，线长大于20公里设第二个车库，线长大于40公里设第三个车库。在论证的基础上，当两条线上车厢为相同类型时，在新线的第一个十年期可由一个车库供两条线使用。

（2）地铁与车库必须以双轨支线相连。

（3）地铁线与它的一条或两条相交线路由单轨支线相连。

（4）在地铁线上，为了组织运营，行驶车辆的储放以及列车的回转，应每5~8km规定有一个独头线（折返线），在长度在20km的线路上，在其启动区段，应规定备有行驶车辆的技术服务点（ПТО），点内设有生产房间和卫生生活房间。超过20km长度时，下一个技术服务点配置在五年以上的地铁终点站后面，进一步的技术服务点数量由计算确定。

（5）在带折返线的车站后面有车库时，则不规定技术服务点，折返线的长度由计算确定，取决于行驶时最大的车厢数量。车辆的晚间存放应该规定在车库、线路的回转与存放折返线内。

二、噪声和振动的防护

（1）地铁设计时，应规定乘客与地铁员工、沿线居民建筑与其他对噪声和振动水平有要求的大楼对噪声和振动的防护。

报告时间：2020年7月2日。

（2）技术措施应保证在居民建筑中的振动水平不超过下表所指出的振动水平规范值（分贝）。

（3）由列车行驶引起的隧道衬砌振动水平的计算，振动由隧道至大楼的传播计算，以及防振动的构筑物的计算，参考规范 BCH-211-91。

（4）当隧道墙至居住大楼基础的最短距离达到 40m 时，地铁设计中不校核振动水平。

（5）配置在地铁运营线附近的居住建筑的设计应引入防振的结构与总体规划方案。这些方案的防振效果由计算方法和配置在基础点的仪器测量确定。

（6）为降低列车的噪声，在每一层路轨下设置柔软结构。

振动参数	平均几何概率（赫兹）					
	2	4	8	16	31.5	63
振速水平	69	63	57	57	57	57
振动加速度水平	15	15	15	21	27	33
振动位移水平	123	111	99	93	87	81

三、关于采用渡线箱和叉枝的问题

（一）为实现并行行驶

国外实践上相当流行的是，不同线路上的列车在一个车站的同一些道路上并行行驶，它是通过在车站的前后设置带指针式道岔的渡线箱来实现的。在这些情况下，一个车站就变成地铁线的换乘站，而没有任何规划和结构方案的改变。一条地铁线的列车利用另外一条线的部分长度实现并行行驶，似乎第一条地铁线有一个叉枝。俄罗斯的规范文件没有规定这样的并行，但是在 СНиП2.05.04—92 中也规定在个别情况下允许采用组织枝叉行驶。在运行实际中，也采用两条线的列车并行行驶，此时列车开始在一条线上行驶，然后通过联结的枝叉转入另一条线，实际上它是按特例采用的。

（二）在两条线路依次投入运行而双层分阶段修建的双线双层联合枢纽中的应用

随路线依次投入运行而二层分阶段修建的双线双层联合单拱换乘枢纽。

当上述联合枢纽两线中的一条线被长期推迟投入使用时，这时枢纽的部分投资被"冻结"了，还需要额外的费用以维护用于尚未投入使用的第二条线路的房间。为了消除上述已有的缺点，建议采用分阶段投入使用的结构施工方案，以便消除投资的"冻结"。

与上述双线双层联合枢纽不同的是：代替原来的装配式层间楼板的是预应力整体式钢筋混凝土拱形式的层间楼板方案。该面左侧显示的是第一条线路使用时的结构，当在上层使用同时修建第二条地铁线路时，在层间拱及侧支座的保护下，用联合机械开挖下层岩土，用梁式起重机安装仰拱的拱条砌块，借助千斤顶砌块及拱条分压向岩石。然后安装装配式钢筋混凝土构件组成的下层站台。

为了换乘的方便，必须相互交换上、下每一层道路的位置，使得乘客从一条线到另一条线的顺向换乘可在每一层中实现，而反向换乘（较少的乘客）利用层间电动梯来实现。

四、换乘枢纽

（一）概述

随着地铁网长度的增加，新的线路陆续投入使用，在地铁线的交叉点、地铁线和铁路的交叉点处会出现换乘枢纽。地铁网的发展及其加密，导致地铁换乘枢纽数量的增加。交叉于枢纽的线路数开始是两条，然后是三条、四条及更多。

如右图可见，转乘枢纽数量的增加与地铁线路数并不成正比，而以更快的速度增加。

在各地铁城市换乘枢纽总数量中，双线相交的最多占67.1%~88.3%(大阪为100%)，三线相交的其次，从11.7%~26.4%。四线相交的份额：在巴黎为7.1%（5个枢纽），伦敦为1.9%（1个），莫斯科为4.8%（1个）。仅仅在个别情况下遇见5线相交的枢纽（巴黎）及8线相交的枢纽（纽约）。

各地铁城市绝大多数换乘站的93%~100%，为两条地铁线或三条地铁线相交的换乘站。

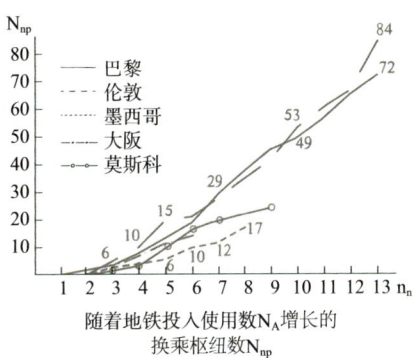

随着地铁投入使用数N_A增长的换乘枢纽数N_{np}

（二）评价换乘枢纽的原则和指标

建设地铁网的目的是提高乘客的方便与舒适。因此，评定换乘枢纽规划与结构方案有效性的最重要指标是节省换乘时间，以减少居民乘行的疲劳，增加乘客的个人自由时间。所以，应力求最大限度地减少换乘枢纽的各车站站台中心（车站纵、横轴交点）之间的距离。

为了增加城市土地的集约利用程度，合理地综合利用地下空间，减少地下开挖对岩土稳定性和地面沉陷的负面影响，特别在密集的城市建筑区和历史文化名胜区，应研究设计紧凑型换乘枢纽。

拟制地铁网的未来发展方案，应当紧密结合城市综合开发利用地下空间的规划。因此，研究设计同时综合开发地下空间的地铁枢纽的新的规划、结构和工艺方案及其有效的构筑方法与机械是紧迫的科学技术问题，对于大城市的建设具有重要意义。它应高水平规划地铁网本身，还应规划邻近枢纽的地下空间，并要考虑到建设城市新区或改造城市老中心区的前景。

（三）换乘枢纽的思考

（1）随着地铁网长度和地铁线路的增加，换乘枢纽的数量以较快的速度增长。为了及时地拟制大城市的地铁发展总方案以及综合开发地下空间的方案，分析研究新的、方便乘客的以及在规划与结构方面有效的地铁换乘枢纽是必须和紧迫的。

（2）探讨新的换乘枢纽方案，首先要合理地保证两线或三线的换乘，这在很大程度上适应了大城市地铁发展的趋势。

（3）按照舒适性、安全性、节省换乘时间、紧凑性来讲，联合换乘枢纽是最方便的，在联合换乘枢纽里，顺向与逆向的乘行站台相容于一个统一的结构内。

（4）对采用联合换乘枢纽有效性的综合评价，应该考虑到换乘枢纽固有的舒适性、换乘时间、由于站台的相容而使邻近换乘枢纽的区间的延长。不利用换乘枢纽的乘客探寻时间的增加、修建与运行成本费用以及其他因素。

（5）在其他相同条件下，应偏向于这些换乘枢纽，它们在较少的程度上在该枢纽处改变相交地铁的线路。

（四）换乘枢纽的推荐方案

（1）每一层上带有岛式站台的，双层三跨圆柱形车站形式的联合双线换乘枢纽。

优点：顺向换乘很方便，反向换乘也方便，在中厅中央由升高不大的楼梯或电动梯实施。与地面上站厅的联系可由电动梯实施。车站一端连接上层站台，车站另一端连接下层站。较紧凑，占地约5300m^2。

缺点：结构复杂，施工难度大，需要相应的联合盾构。

（2）华盛顿双线联合型换乘枢纽。

优点：换乘时间短，最多走2/3站台就可走到最近一组层间电动梯，换乘时间为30~120s；服务于换乘的电梯多而分散，从而可消除在一个位置处的状况；枢纽不要求地铁线的线路弯曲，枢纽允许在45°~90°范围内改变两个地铁线路的交角而没有任何恶化；枢纽到城市有4个或更多的出入口提供了大的城市服务面积。

缺点：枢纽占有的地下空间面积相当大，约为14000m²。上面与下面车站进出城市都必须利用中间电动梯到前厅后才能走向电动梯隧道。

（3）装配式层间楼板单拱结构双线双层联合枢纽。

该枢纽的上拱与仰拱的侧向支座是由混凝土浇筑于直径9.8m的辅助隧道，辅助隧道是由装配式板状钢筋的劲性钢筋混凝土结构衬砌的。支座中建造纵向通风通道，通过横向管道，排出来自行于车站上的列车的热空气。

路基及站台面下的承重结构为内部纵向布置的两排装配式钢筋混凝土圆柱梁结构。两层的站台由装配式钢筋混凝土构件组成。下层站台下面设置了服务间及电缆集合管道。上下层由小电动梯相连，为了降低列车的噪声，在每一层路轨下设置柔软结构。

在上层站台有三个直径4.8m的圆孔，在统一的艺术构思下将上、下层空间相连通。

五、车站的预制装配式结构

（一）深埋暗挖双层联合换乘车站的预制装配式结构

该枢纽的上拱与仰拱的侧向支座隧道是由区间盾构隧道扩挖而成，原有隧道的管片拆掉扩挖到外径9.8m的组合结构：外衬砌是装配式铸铁弧形板；内衬砌采用可置换模板和混凝土泵整体浇筑。

上拱外径11.2m，由12个厚70cm，没车站方向宽50cm的钢筋混凝土构件所组成。拱对岩体的压缩是由安放在拱顶的推力块体实现的，该推力块配有两个千斤顶。如在勇敢广场车站，所有的构件块在其端部都塞有厚度可变的弹塑性聚氯乙烯垫块，以保证按最小的偏心率中心地传递法向力。

仰拱内径15m，由13个构件块组成，它由两个带衬垫的钢筋楔形接头挤压紧而闭合，仰拱宽度1m。内部的装配式钢筋结构是作为上层站台和道路下的支撑结构，该结构成梁柱组合体形式，柱距4m，柱安设在纵向整体钢筋混凝土构件的"杯子"中，构件是混凝土浇筑在刚性基础界限内，刚性基础铺设在仰拱衬砌上。在柱的顶部安设装配式钢筋混凝土梁，这些梁连成连续桁梁。为了承接上层道路行驶列车的荷载，在路轨下设置了钢筋混凝土装配式梁，一端支撑在桁上，另一端支撑在整体式混凝土支柱的托架上。

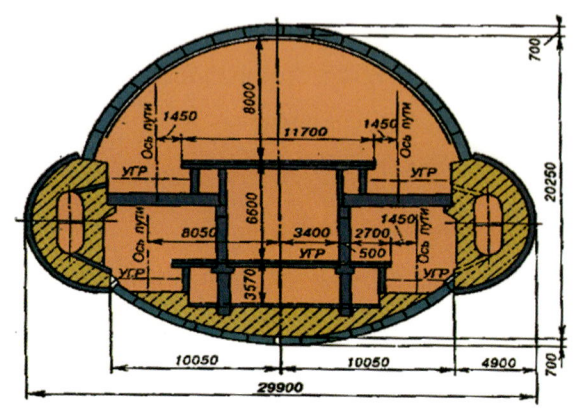
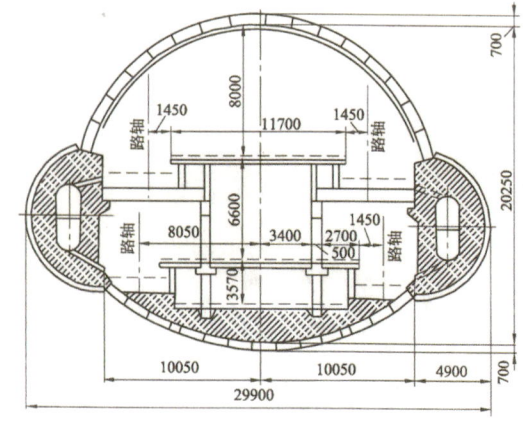

尺寸单位：mm

双层站台都由装配式钢筋混凝土构件修建的。上站台宽 11.7m，下站台被柱子占了一点宽度，宽 13.2m。下层站台下设置了服务房间和电缆总管，上层站台下仅有电缆总管。电动梯隧道以四组电动梯衔接上层，使车站与街道下的站厅相连。

在上层站台有三个直径 4.8m 的圆孔，在统一的艺术构思下将上、下层站台通过小电动梯相连通。

暗挖双层联合换乘预制装配式车站的优越性：

（1）能大大缩短乘客的换乘时间，包括反向换乘的乘客。

（2）与其他所有介绍的换乘枢纽相比，旅客通过能力最大。

（3）枢纽结构排除了在分离车站换乘枢纽中设置连通走廊的需要。

（4）要求用在枢纽维护保养上的运行费用最小。

（5）枢纽结构结合了深埋单拱车站所有表现得好的优点：上拱与仰拱压向岩体，使所建结构对岩体的影响最小并使地面沉降最小；车站修建采用工业化方法，外径 5.5m 的支座隧道用机械化盾构来掘进，拱用机械化的机组来修建，而基本核心岩石的开挖采用联合机和挖掘机。

（6）结构的紧凑性保证枢纽所占的地下空间面积最小。

（二）深埋暗挖单层车站的预制装配式结构

车站（右图）为钢筋混凝土块体装配而成的上拱和侧拱所构成的深埋单拱结构。直径 5.5m 的隧道作为拱脚，隧道是由钢筋混凝土铸铁块装配成的，部分整体浇筑混凝土。靠近拱连接处的混凝土用钢筋网配筋。上拱跨度为 18.4m，它是由互不相连的圆拱带拼成的，每个圆拱带宽度 0.5m，内径 9.8m。圆拱带由 12 个截面密致的块高 70cm 的标准块和一个闭合块所组成，这个闭合块带有两个拱的松紧千斤顶，在装配时压注特殊成分的水泥浆。标准块在其径向两端装有弹塑性的聚氯乙烯填块，这个块是用以减少标准块端的应力集中和降低法向力传递时的偏心。

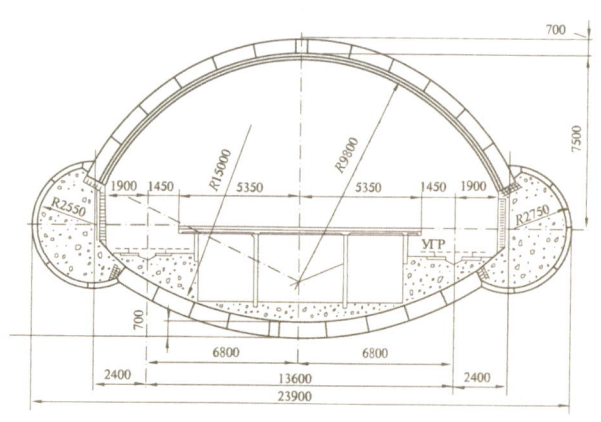

车站示意图（尺寸单位：mm）

仰拱也是由互不相连的宽 0.5m、内径 15m 的圆拱所组成。圆拱中 10 块构件，其中 9 块是致密截面的标准块，一个是带 1 个平面千斤顶的闭合块。圆拱的块体后面压注水泥砂浆。上拱与侧拱借助于千斤顶松紧使得车站结构衬砌完全压向岩层。

在支撑结构里面浇筑了刚性基础的整体混凝土，安装了站台和站台下小室的装配式结构以及防水的配筋水泥罩。

车站的修建是从支座隧道掘进、铺设铸铁弧形板开始的，随后修建单拱车站并贯穿支座隧道，翻修车站前后的邻近区间。

在贯穿的隧道中，借助移动式金属模板，用 6m 小巷灌筑拱支座混凝土，灌筑混凝土的速度稳定在 33m/月。

拱的修建是由专门的隧道上部切槽中的预制构件块安装机械进行的。岩石采用安放在 1m 小巷（两个拱下面）的平台上的破碎锤进行开挖。岩块由传送带抛到小车里，小车在支座隧道中的道路上移动。拱的构件块由安装器的绞盘从支座隧道提升到构件安装器上。构件块的安装在拱的中央由下而上、逐次从左至右实施。然后进行拱的松紧，第一个千斤顶用 735~785kN 的力，第二个千斤顶用 1960kN 的力。填缝和衬砌后的压注滞后于工作面 10~15m 处进行。施工是按连续工作周的流水图进行的。

与隧道上部切槽的开挖以及上部拱的安装平行但滞后 70~80m 进行核心岩石的两台阶开挖，岩石开挖是由带主动挖斗的电动挖土机进行的，岩块由皮带运输机供入小车。

仰拱的拱安装是采用电动梁式起重机完成的。装配好构件的拱在水泥浆第一次压注后，在滞后于第一个装配拱 3~4m 处，由千斤顶用 785~980kN 的力压紧。在滞后 15~20m 处敷设刚性基础的整浇混凝土，安装站台下房间及站台结构，然后进行防水罩、道路的安装工作、装饰及其他工作。

（三）明挖车站的预制装

概况：外轮廓净宽 20.5m，净高 17.45m；内轮廓净宽 19.1m，净高 15.35m；底板厚 800mm，侧墙厚 700mm，顶拱厚 800mm。

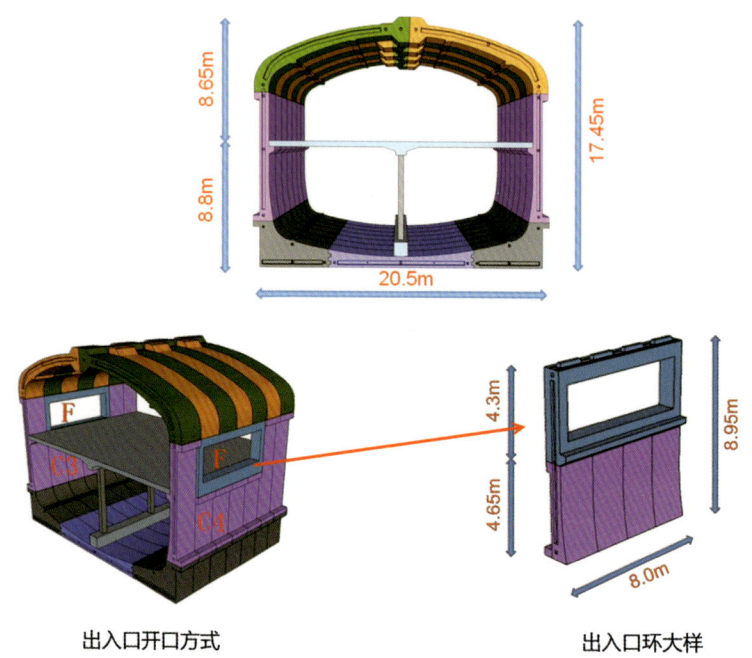

(1) 构件接头力学性能

采用的构件接头形式：榫槽灌浆接头，接头缝隙内灌注化学浆液，接头外部设置辅助加力钢棒。

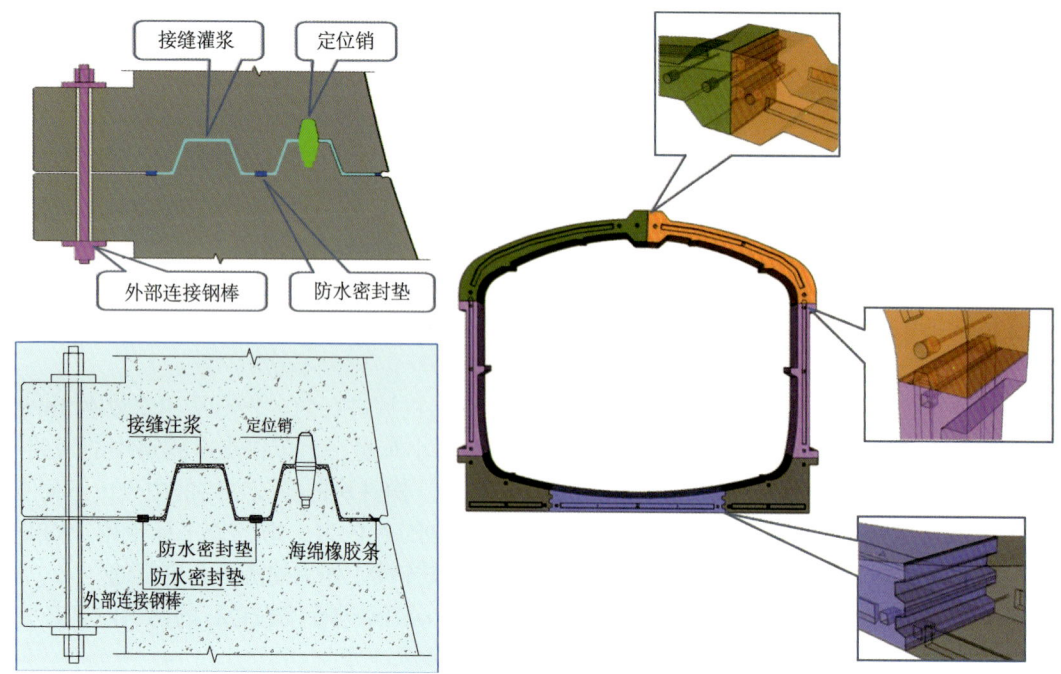

专门研制接头注浆材料,研发了专用注浆设备。

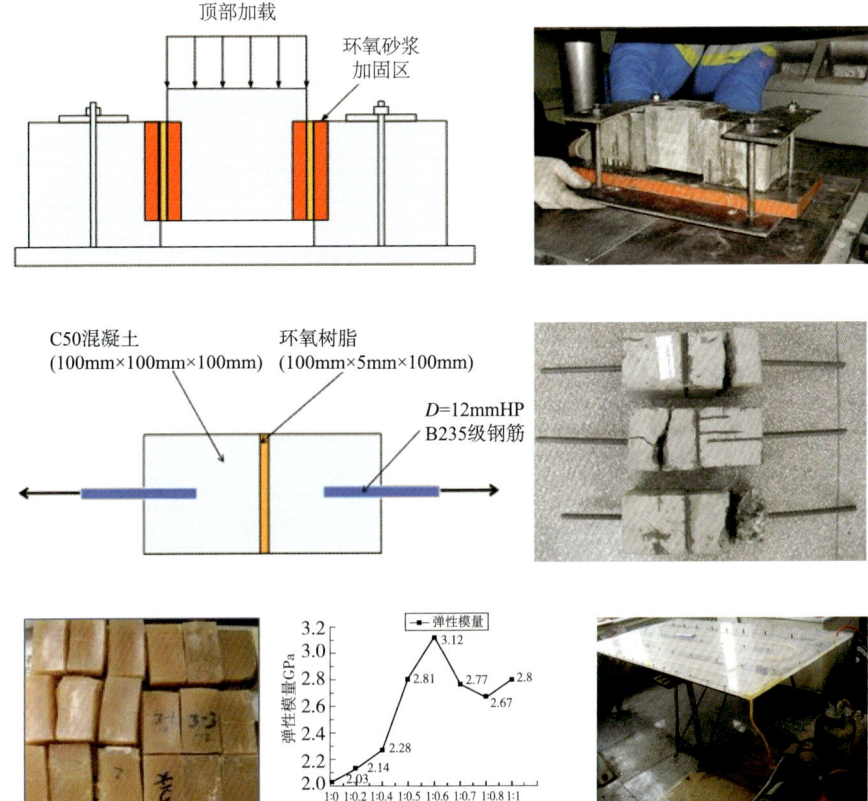

（2）结构防水

对装配式结构防水体系、接缝防水构造、接缝防水材料性能进行了研究,实际应用接头防水效果很好。

采用多道防线防水体系实现高性能防水,防范 80m 水头压力。

（3）预制装配技术的优势

①两者工程造价相当（高 10% 左右）。

②工期节省 4～6 个月（20%～30%）。

③施工人员每班由 130～150 人减少到 15～20 人。

④节省钢筋 800t、混凝土 20%、木材 800m³,施工废弃量减少 50%。

⑤施工场地减少约 1000m²。

⑥在施工安全性、噪声及空气污染等方面具有很强的优势。

（4）造价略高的原因

①原因一：采用无柱拱形结构,结构断面净高增加 3~4m。

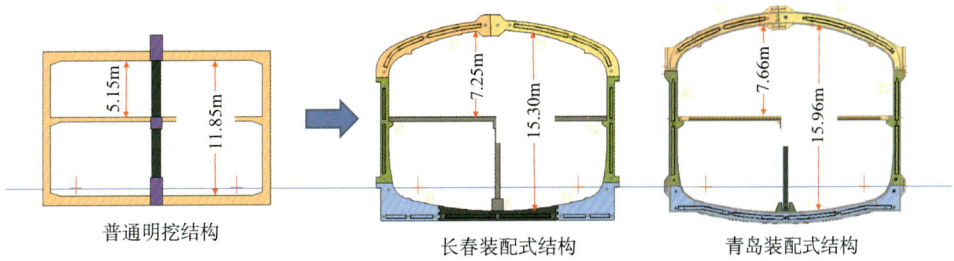

②原因二：因拼装作业要求，需在结构与基坑围护墙之间预留肥槽。

两侧各预留肥槽尺寸 200~600mm，导致基坑宽度加大。

肥槽后期用素混凝土回填，回填量 13.46m³/ 延米（双侧）。

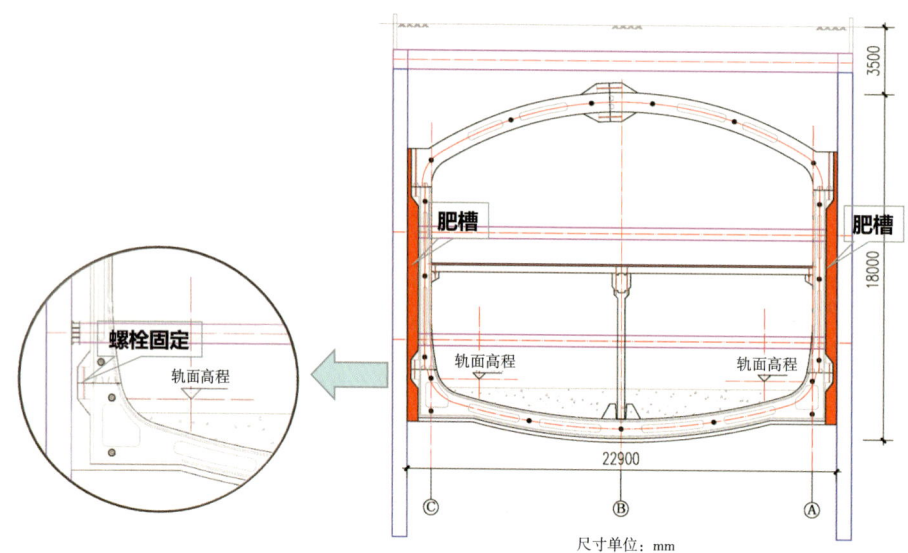

③原因三：预制混凝土结构单价高于现浇混凝土结构。

装配式地面建筑混凝土单价比较

楼　板　高54%

矩形梁　高69%

楼　梯　高57%

单位：元/m³

多种预制构件单价比较

序号	构件名称	单位	单价	备注
1	装配式车站构件	元/m³	3785	含利润及税金，到工地价格
2	预制管片	元/m³	2700	到工地价格
3	大型 U 形梁	元/m³	3556	按地铁全统定额测算，场地费按 400 元/m³ 增加，铜筋含量按 200kg/m³ 调整（实际钢含量大于 350kg/m³）
4	市政预制箱梁	元/m³	3425	按 2016 版市政消耗量定额预算，场地费按 300 元/m³ 增加，钢筋含量按 200kg/m³ 调整（实际钢含量大于 300kg/m³）
5	预制管廊	元/m³	2700	询价黄岛一家预制场，按含钢量 200kg/m³ 计算，含运输 100 元/m³

④原因四：目前装配式车站应用数量少，建厂成本摊销高

预制厂成本摊销直接影响装配式结构的造价，初步估算：

按 30 座车站摊销计算，每环构件摊销费用约 7.4 万元。

按 20 座车站摊销计算，每环构件摊销费用约 11.08 万元。

按 10 座车站摊销计算，每环构件摊销费用约 22.17 万元。

采用预制装配式结构建设车站的动因：工期紧张；一线施工人员紧缺；建设期环保要求高；严寒地区施工受气候影响大。

长春地铁西环城路站

长春地铁兴隆堡站

利用地下空间，助力发展绿色建筑与绿色城市

一、绿色发展的紧迫性

我们为什么要开发利用地下空间？我认为是由于地球正迫切地需要绿色发展，尤其是在资源和环境两个方面。首先，我想谈一谈我们正面临的最紧迫的问题，也就是关系到人类生存命运的问题。

去年联合国气候变化专门委员会发布了一个报告，称如果温室气体减排不利，北极的融冰将致使 21 世纪末海平面上升最少 60cm，比委员会 2013 年发布的报告高出 10cm。由于气温升高，全球的冰川退缩，我国的冰川也在明显退缩。山区的冰川在 21 世纪末将至少融化五分之一，阿尔卑斯山的冰山很可能会消失。随着海洋吸收二氧化碳的增加，海水酸性越来越强。21 世纪末，海洋动物总量将减少 13%，海洋渔业的捕获量将减少 24%。到 21 世纪末，海平面的上升将导致全球 2.8 亿人流离失所，上海、纽约等大城市将面临灾难性的洪水，北极 20% 的永冻土易受突然解冻的影响。由于气温升高，厄尔尼诺现象和拉尼娜现象导致全球气候变化极端异常，自然灾害频发。这些都将危及人类的生存空间，所以地下空间开发非常紧迫，也非常必要。

世界上的有识之士已经意识到这些风险，认为世界正在迎接绿色发展的时代。绿色发展时代的标志是什么呢？一些欧美国家，例如丹麦，要求在 2030 年停止使用燃料汽车、燃料柴油车。很多国家提出零碳排放，也就是实现碳排放为零的绿色建设和绿色城市。我国也发布了技术路线时间表和路线图，要求停止使用和销售化学燃料汽车，其中海南省是第一个发布的省份。

大家能很明显地感觉到，自十八大以来，中央把生态文明建设和绿色发展提到了一个新的战略高度，绿色已成为中国国家建设发展的五大理念之一。同时，中国的绿色发展正在进入世界绿色发展的先进行列。举三个例子：第一，在联合国发布的世界新增绿色植被中，中国占了四分之一，美国宇航局监测显示，中国是全球绿色增长的贡献者；第二，我国的可再生能源的开发，包括太阳能、风能等，已处于世界前列，中国是世界最大的太阳能电池板、风力涡轮机、电池和电动汽车的生产国、出口国和安装国；第三，我国清洁能源利用在全球清洁能源使用中占主要部分，例如自 2015 年，中国地热利用量、地源热泵装机容量、利用的热量就居世界第一，到 2020 年，我国地源热泵装机容量占全球总量的 50% 左右。所以中国绿色发展正在进入世界绿色发展的先进行列，将来有可能引领世界绿色发展。

二、绿色建筑与绿色城市概念的理解

为什么绿色发展的重点在城市？联合国统计显示，人类碳排放中城市占了 80% 多，所以城市绿色建设非常重要。什么是绿色建设？简单来说就是环保，就是节约能源、土地、水、材料和保护环境。什么叫绿色城市呢？绿色城市一方面是指绿色建筑规模化的必然结果，另外一方面，要加上其他的城市要素。所以，简单地用一个公式来说，绿色城市就是绿色建设加上绿色城市基础设施，再加上绿色的城市

报告时间：2020年4月27日。

自然环境，也就是蓝天、绿水、青山、净土，最后加上绿色的社会环境，包括犯罪率、人与人的和谐关系等。建设绿色城市的最终目标是促进城市可持续发展，提高城市的宜居水平。

三、地下空间在绿色城市发展中的重要作用

下面我将根据绿色城市的含义分别介绍地下空间在绿色城市发展中起到的重要作用。总体来说，地下空间的开发利用是发展绿色建设、绿色城市的一个主要途径和主要着力点。

（一）节约土地

绿色发展的紧迫性的一方面体现在资源的紧迫性上，而在中国，最重要的资源就是土地资源。中国节约土地的紧迫性和必要性体现在哪里呢？根据国务院的规定，中国的耕地红线是 18 亿亩，但是现有耕地是 18.58 亿亩，城市发展正面临无地可用的困境，所以要建设地下空间。

建设地下空间的潜力有多大？主要体现在两个方面：一个方面是紧密化和功能化已有建筑的土地，例如怎么建设小区的地下空间；另一方面是如何在绿地、公园、湖泊等没有建筑的地区下面建设地下空间。首先要把城市交通（地铁和其他轨道交通，地下快速路，越江和越海湾隧道等）转入地下，其次把其他一切可以转入地下的设施，特别是脏乱差、有毒的设施，例如停车场、污水处理厂、商场、餐饮等，尽可能地建于地下，提升土地利用效率，实现节约土地的要求。

据《中国城市地下空间发展蓝皮书（2018）》中的数据显示（图1），上海近五年地下空间开发增速有所降低，而深圳则实现了突破。蓝皮书中提到，2018 年北京地下空间开发规模为 9600 万 m²，上海 9400 万 m²，深圳 5200 万 m²。但是刚才那位专家在演讲中说上海如今已经突破一亿平方米了，深圳也已经达到六千多万了，说明近几年的增长速度是非常快的。所以，中国利用地下空间来节约土地的潜力是非常大的。

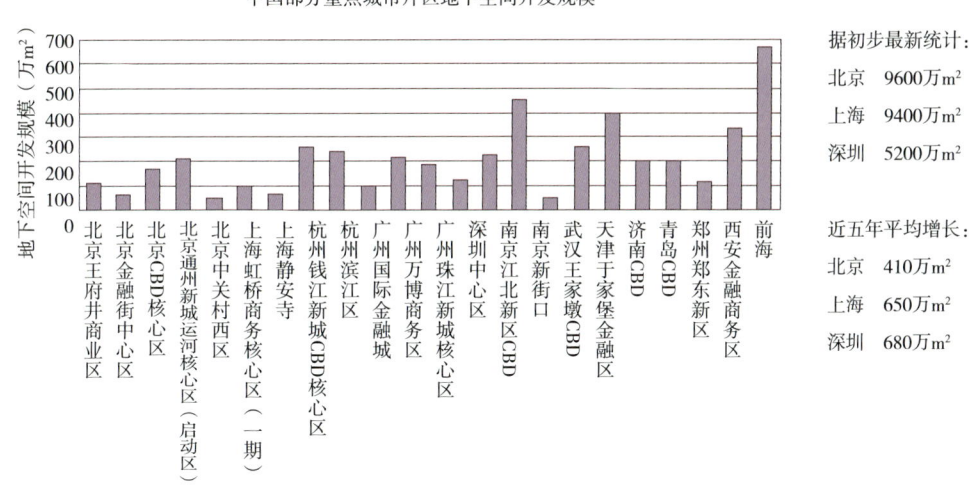

图 1　中国部分重点城市片区地下空间开发规模

（二）利用地热能

从能源的角度而言，可以从多方面节约能源，这里主要跟大家介绍一下中国利用地热能的情况。地热能的利用主要分成三个方面：地温能、浅中层地热能和深层地热能。

第一，地温能。地温能指的是比地层中温度小于 25℃ 的地层热能。地面上的温度会随着春夏秋冬四季变化而变化，但到达地下一定深度时，地下温度会保持在一个稳定值上，通俗来讲就是地下冬暖夏凉。举一个上海自然博物馆的例子，上海自然博物馆利用地源热泵系统承担部分夏季冷负荷和冬季热负荷，夏季的土壤换热器最大算热负荷为 1639kW，冬季 1178kW。此外，上海自然博物馆还采用了地下结

构内埋管热交换系统，每年可节省 111.7t 标准煤，减排二氧化碳 195.5t。而博物馆的建设成本在近两年便可收回。

最新的一个例子是北京大兴国际机场的地源热泵工程，其中 1 号能源站已于今年 8 月份启用，标志着国内最大的多能互补地源热泵系统工程启动运作。该系统每年提取浅层地温能 56.56 万 GJ，实现大兴机场 250 万 m^2 办公场地的供热和制冷，节约天然气 1735.69m^3，相当于节约 21078t 煤，减少碳排放 1.58 万 t 以上。有了这两个例子，希望全球致力于地下空间开发的专家将来能够想到地源热泵系统，并将其作为节省能源的首选解决方案。

中国现在地源热泵的使用处于什么水平？2017 年底我国地源热泵装机能量达到 2 万 MW，也就是 2000 万 kW，比三峡水电站的发电量还要多，可以实现供暖（制冷）面积 5 亿 m^2，利用地热能可折合 1900 多万 t 煤。从 2015 年起，我国地源热泵装机总量世界第一，2015 年占全球比重为 23.61%，2020 年上升到 50%。所以，中国的地下空间开发利用总体规模、水平都是居于世界前列的。

第二，浅中层地热能，也就是温泉。以河北省为例，2015 年河北省地热资源开采量突破了 1.1 亿 m^3，地热供暖面积达 6300 万 m^2。其中，雄县正在推广地热供暖模式，如今地热供暖面积已占城区集中供暖面积的 85%，每年可减少二氧化碳排放量 12 万 t。

第三，深层的地热能，也就是干热岩。我国陆域干热岩资源量是 856 万亿 t 标准煤。青海共和盆地 3705m 深钻获得 236℃的高温干热岩体，其干热岩质量非常高。此外，我国成功在陕西省实现了一个开发先例，也就是将干热岩用于供热的商业应用——长安信息大厦 2013 年共计 3.8 万 m^2 中应用干热岩供热，效果良好。如果按照 2% 的可开采资源量计算，我国可开采量为 17 万亿 t 标准煤，相当于 2016 年全国能源消费量（43.6 亿 t）的近 4000 倍。这意味着，中国如果成功开发干热岩，我们的能源将用之不尽。当然，目前我国还没有进行大规模开发，仍在加紧研究开发技术。因为传统的开发技术是由美国研发的，成本太高，另外，传统的开发方式为 EGS 技术，即深钻孔水裂干热岩技术，但现在专家正在研究这种技术是不是采热量有限，成本过高。

（三）节水

节水的方法之一是利用雨水。上海世博会期间进行过统计，世博轴自来水日用量是 2000m^3，但可利用雨水（绿色雨水）时段里，自来水日用量可减少近一半，自来水替代率为 40%~45%，所以雨水的利用量还是很大的。此外，西沙群岛上修建了可采集雨水达 14 万 t 的地下储水工程，收集的雨水经技术处理，已达到了国家饮用水卫生标准，从而结束了吃水靠大陆船运的历史；北京每年 6~9 月份降水中，可利用的雨水为 2.3 亿 m^3，相当于城区全年用水量的五分之一多。

另一种节水方法为利用中水。据统计，北京 2010 年再生水利用量达 6.8 亿 m^3，占总用水量的 19%，但目前利用率仅为 60%，"十二五"期间计划将全市再生水利用提高到 10 亿 m^3 以上，即利用率提高到 75%；天津自 2002 年正式启动中水管网建设以来，已铺设了 400 多公里的中水管道。目前天津已建成再生水厂 8 座，全年中水回用 2500 万 t 左右。我最近参观了成都天府的一家污水处理厂（图 2），发现其污水处理工艺比国际上的一些污水处理厂更先进。它已放弃传统工艺，而是采用生物降解工艺，进行封闭式处理，在厂内一点臭味也闻不到，同时通过渗透来分离水和泥，最后获得可再利用的水，也就是中水。

（四）绿色城市基础设施

绿色的城市基础设施是建设绿色城市的重点。过去我们采用粗放的发展方式，注意"面子"工程，建设高层建筑、大广场、宽马路等，这种发展模式造成城市"里子"（城市地下基础设施）建设出现短板的恶果，"城市病"丛生。由于现在要施行绿色发展模式，所以建设城市地下空间是"转变城市发展方式"、治理"城市病"和建设绿色城市的主要着力点。图 3 介绍了一些可持续发展的绿色基础设施项目，以及这些项目可解决的"城市病"。

图 2　成都天府地下污水处理厂

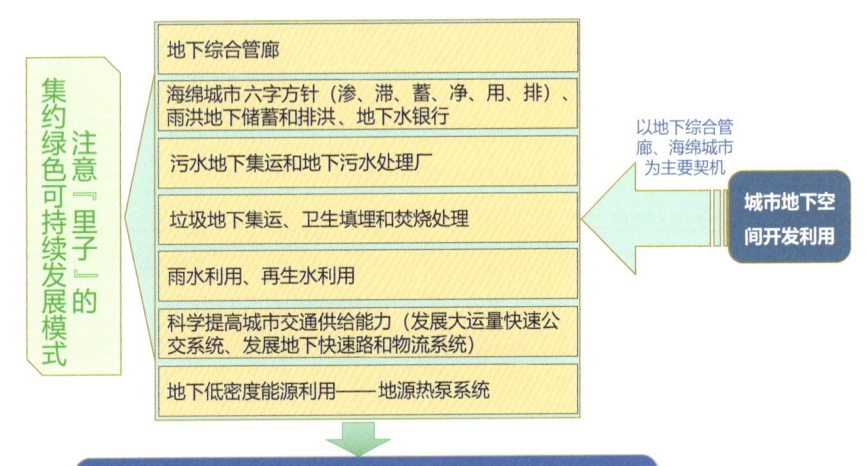

图 3　可持续发展的绿色基础设施项目及其可解决的"城市病"

（五）绿色客运

交通是碳排放的主要原因，所以我们要发展占地少、低能耗、无污染的绿色交通。城市客运交通以地下轨道交通为主，轨道交通是电驱动的，没有污染，也不会造成碳排放。未来的城际交通的发展趋势则是发展时速 600km 以上的地下高速磁悬浮，由于磁悬浮列车是在低真空的隧道里行驶，没有摩擦阻力，空气阻力也很少，因此能量消耗最低，当然也最绿色。

我非常同意，不能仅发展轨道交通，还要发展城市快速路，这是为什么呢？因为现在越来越多的家庭有了小汽车，人们上下班时可以乘坐公交交通，但是约会、旅游时还是希望开私家车，所以仅发展轨道交通尚不足以解决城市交通拥堵的问题。以日本为例，日本的地下轨道交通非常发达，虽然日本去年宣布不再建地铁了，但是日本地上交通拥堵问题仍未解决，地铁里的乘客也非常多，非常拥挤。可见轨道交通不能完全解决市民的出行问题，所以，世界上已开始进入建设地下快速路的阶段。

世界上有很多城市，如波士顿、东京、新加坡、马德里等，都在建地下快速路，国内也有很多城市在建设，上海就是比较积极的城市之一。特别要提到的是深圳前海地下路系统，这是一个"地下快速路 +

地下环路+地下车库"的三级地下车行系统，其中地下快速路解决过境交通，地下环路与地下车库互联互通可解决区域性到发交通问题。这一地下车行系统有效改善了前海地面交通环境，缓解了城市中心区和小街坊路网交通压力，是建设绿色交通的一个方案。

（六）未来的城市货运交通

现在，城市货运交通仍然是个大问题，因为货运交通是造成交通拥堵和城市污染的重要原因之一，所以世界上提出利用地下货运交通，也称地下物流系统。也就是说，所有的货物，无论是从城外运进来的货，还是从城内运出的货，都先通过各种运输手段到城市边缘的物流园区，然后通过地下物流系统送到各个终端。整个运输过程以集装箱和货盘车为基本单元，以自动导向车为运输工具，电力驱动，绿色环保。

我想重点讲两个中国的地下物流系统：港城地下物流系统和城市地下立体智慧物流系统。第一，为什么港城地下物流系统很重要？中国有很多排在世界前列的港口，比如上海港早在2012年就已经成为集装箱吞吐量世界第一的港口。但是远超规划的集装箱吞吐量也引起了交通拥堵、港口无序停靠、环境污染等问题，所以当务之急就是建设港城地下物流系统。

第二，为什么要建设城市内部的地下立体智慧物流？这是因为新零售推动城市物流爆发增长。中国2019年网上零售额比2018年增长了30%，超过1万亿美元，占全球网上零售总额的一半以上。此外，中国2014年快递包裹140亿件，2016年313亿件（全球是650亿件），2017年400亿，2018年490亿，可以说全球一半以上的快递包裹都是中国的。快递暴增促进了智慧物流，包括智慧化布局配送中心、数字化仓储、运输和配送系统和智慧化作业，如菜鸟全自动智慧仓储基地和菜鸟机器人。然而，智慧物流也有一些无法突破的瓶颈，即"最后一公里"快递员的地面配送，电子商务的地面配送加剧了城市交通拥堵和空气污染，使已经进入"拥堵时代"的中国无法承受，其中更为突出的是交通事故问题。据统计，2019上半年，上海市共发生快递、外卖行业各类交通事故325起，造成5人死亡。因此，我们需要将地面运输转移至地下，建设地下智慧物流系统。试想一下，这种地下物流系统一旦建成，系统末端配送可以与居民小区建筑运输管道物相连，最终发展成一个连接城市各居民楼或生活小区的地下管道物流运输网络，并达到高度智能化，人们购买任何商品都只需点一下鼠标，所购商品就像自来水一样通过地下管道很快地"流入"家中（小区的自动提货柜）。这就是我们所设想的未来绿色城市地下物流系统，并且这是可以实现的。世界上还没有国家成功建设这一地下物流系统，希望我们中国能率先实现。

（七）城市的绿色污水、雨洪蓄排系统

我们要建设雨污分流的深层隧道排水系统，在这方面新加坡值得我们学习。新加坡总面积仅为714.3km²，年平均降水量为2400mm左右。新加坡是全球少数几个采用雨污分流系统的国家之一。新加坡深层隧道充当的角色是将整个城市收集的污水输送至处理厂集中处理排放。我国上海、广州、武汉等城市也正在建设这种深层隧道排水系统。

（八）城市的绿色垃圾处理系统

在大力推广垃圾分类的基础上建设地下垃圾集运系统，实行垃圾焚烧，彻底解决困扰中国的垃圾围城现象，释放大批城市土地资源，改善城市环境。

四、地下空间绿色建造与运营

中国正通过智慧城市的建设大力推动地下空间的绿色运营，包括智慧管理、智慧监测和智慧检修三个部分。图4展示的是江苏南通的一个地下智能立体车库，这种停车场占地少，机械化程度高，很好的体现了地下空间绿色建造于运营的理念。以乌镇智能停车系统为模式的城市的智慧停车系统正在中国各地推广。

图 4　地下智能立体车库

综合管廊的智慧检修包括给水、排水、燃气、电力管线等的自动监测：漏水不漏？何处漏？漏损以后的官网的自动关闭以及机器人的自动维修。

实现绿色城市的愿景需要时间，但我们不求快只求好。相信通过若干个五年计划持之以恒的努力，我们一定能将地上城市中的城市运输与交通系统逐步转移到城市地下去，不但将旅客交通与货物运输转入地下，还包括垃圾、污水等的传输都转入地下，使地上逐步乃至彻底摆脱与传统运输相联系的环境干扰，尤其是交通拥堵与PM2.5超标的困扰，从而释放出的地上空间，用于大片的自然植被和安全的步行，这不是科学幻想，《欧洲地下建设战略研究议程》就已经明确提出，其2030年的远景目标是"将地面交由市民自由使用，将基础设施放入地下"。相信我们中国的也一定能够实现绿色发展和绿色城市建设。

绿色城镇与绿色城镇基础设施

一、城市建设面临的挑战与问题

（一）不可逆的全球发展趋势

1. 世界人口的加速增长

2050 年人口将达 97 亿（Alan Lewis）。

2. 城市化快速发展

全球每周有 140 万人迁到城市（联合国秘书长、联合国人居署执行主任）。

3. 碳排放、全球升温、气候变化异常对环境影响加剧

联合国气候变化专门委员会（IPCC）今年报告称，如果温室气体减排不利，北极的融冰将导致 21 世纪末海平面上升 60cm 至 1.1m，比 IPCC 2013 发布的报告高 10cm。

随着气温升高，全球冰川退缩，山区的冰川在 21 世纪末至少会有五分之一融化掉，阿尔卑斯山的冰川可能消失殆尽。我国冰川已消融面积 1628.94km²，占冰川总面积的 20%。

随着海洋吸收 CO_2 的增加，其酸性越来越强，威胁到珊瑚礁和渔业生存，到 21 世纪末，海洋动物的总生物量将减少 13%，渔业的捕获量将减少 24%。

到 21 世纪末，海平面上升可导致全球 2.8 亿人流离失所，上海、纽约等大城市将可能面临灾难性洪水，北极约 20% 的永冻土易受突然解冻的影响。

全球气温的上升引发的厄尔尼诺现象和拉尼娜现象，导致全球气候变化异常，自然灾害频发。

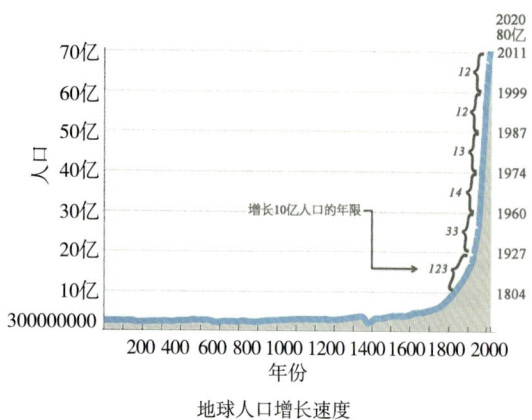

地球人口增长速度

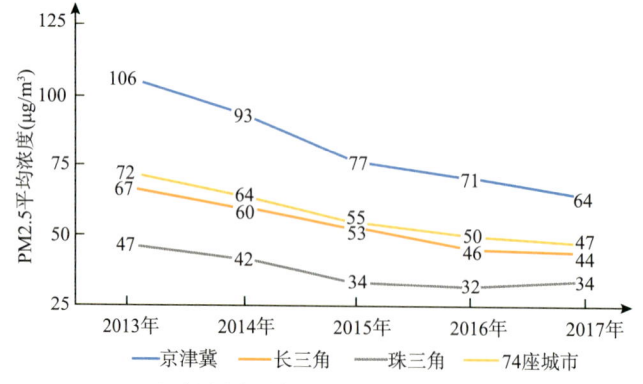

2013-2017 年全国重点区域及 74 座城市 PM2.5 平均浓度变化

报告时间：2019 年 12 月 10 日。

我国 PM2.5 达标具有一定的紧迫性。由图可知，2013 年至 2017 年，全国特别是京津冀地区空气质量有了明显提升。按此下降速度，京津冀地区 PM2.5 浓度应为 51μg/m³，但是世界卫生组织推荐的标准：第一阶段 35μg/m³；第二阶段 25μg/m³；第三阶段 15μg/m³。美国的标准是 15μg/m³，欧盟的标准是 15μg/m³。显然，我们的与世界的先进标准还存在较大差距。

（二）城市发展面临的挑战与应对措施

城市建设面临上述趋势产生的下述问题：城市无序蔓延、城市土地资源短缺、环境污染、交通拥堵、城市停车困难、能耗攀升、频繁的自然灾害、不足的城市基础设施。

联合国对全球城市化的挑战予以极大的关注，自 1993 年以来，联合国召开多次全球首脑会议讨论可持续发展和气候变化问题，制订了《可持续发展议程》《新城市议程》和《气候变化应对京都议定书》《气候变化巴黎议定书》，确定 10 月 31 日是世界城市日，提出建设"可持续、韧性和生态安全"城市，以应对快速城市化和自然和人为灾害。指出"快速城市化……促使自然灾害和人为灾害的风险增大，但是风险并非必然变成灾难的答案在于建设城市抵御风暴、洪水、地震、火灾、流行病和经济危机的能力"（联合国秘书长古特雷斯）。

二、应对挑战与问题的措施（城市地下空间开发利用全球峰会上海宣言 2019）

建设未来的可持续发展的、韧性的、生态安全的绿色城市的关键途径和主要着力点是开发利用城市地下空间，补齐并创新城市基础设施短板。

目前，ACUUS（国际地下空间联合研究中心）是地下空间利用领域唯一与联合国签署了实际合作行动的国际组织。联合国自然资源委员会于 1981 年 5 月把地下空间确定为一种重要的自然资源，联合国经济和社会理事会也于 1983 年通过了确定地下空间为重要资源的决议（理事会第 1983/58 号决议），要提高地下空间对城市可持续发展和韧性的贡献。

绿色城市基础设施：过去注意高层建筑、大广场、宽马路城市"面子"建设的粗放发展模式酿成"里子"城市地下基础设施建设短板的恶果："城市病"丛生。

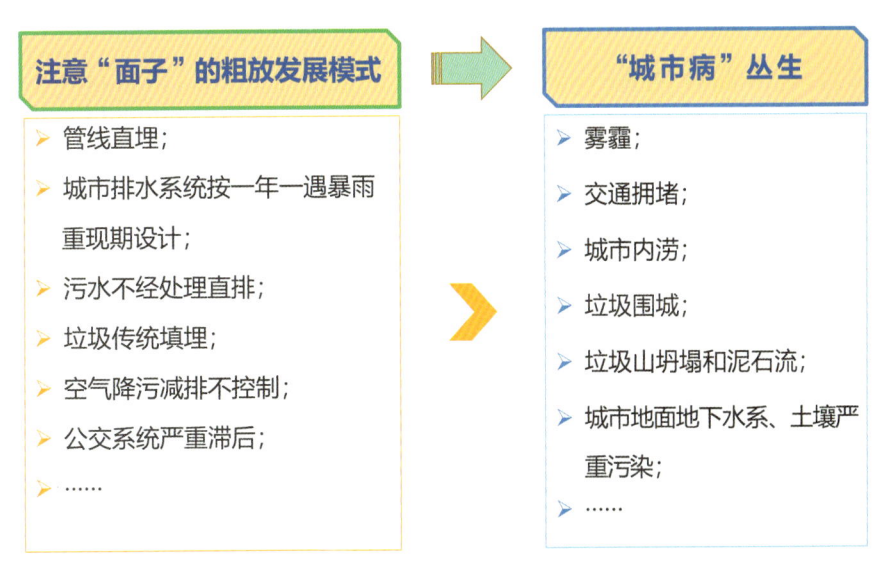

建设城市地下空间是"转变城市发展方式"，治理"城市病"，建设绿色城市的主要着力点。

1. 世界城市日上海宣言2019"更美好的城市,更美好的生活"的三个核心概念

(1) 基本城市服务实施于地下,释放出更多的地面土地实现城市绿色发展是城市地下空间对全球城市的关键贡献。

(2) 驱动城市地下基础设施面向韧性和生态安全性的创新。

(3) 城市决策者在对话与合作平台上的紧密联系使得城市可持续发展转型成为可能。

2. 核心概念衍生的会议六条政策性结论

基于这一共同认可,会议得出以下结论,并在此公开宣布它们为"上海宣言":

(1) 利用城市地下空间节省土地资源、提高土地利用效率,达到更好的城市土地利用。应鼓励配套式地下空间开发,预留珍贵且不可逆的城市地下空间资源,充分利用绿地的地下空间;利用地下空间实现混合用地模式,促进功能融合,提高设施利用效率;促使多专业规划设计协同一体化,加大地下空间综合结建力度。

(2) 利用城市地下空间缓解城市交通压力,达到更好的城市的连通性。应加大力度构建以轨道交通(地铁)为骨架的城市交通体系,科学合理建设地下车行道路,促进城市车行交通健康发展;优化重点区域的立体交通枢纽规划设计方案,形成高效的交通网络分级体系;推进中心城区地下专用物流系统的应用,平衡港城矛盾,并应对新兴互联网商业在货运交通方面给城市带来的巨大压力。

(3) 利用地下空间提高城市地块的步行连通性,以增强商业等城市活力达到更好的社会—经济活力。应加大重点区域地下步行系统建设,鼓励设施间的互联互通;并改善交通枢纽、核心商业区和园区景区地下空间建筑环境条件,构建具有吸引力的地下人行、商业和休憩空间。

(4) 利用地下空间增强城市综合抗灾能力,提高城市韧性。应扩充地下基础设施容量,利用地下专用及综合管廊,提升市政生命线系统的维护效率和安全性;并开展城市地下仓储建设,提升城市韧性,完善战略灾备能力;统一科学利用地下人防设施,开展地下避难场所建设,提高城市综合防灾能力。

(5) 利用城市地下空间,加强城市生态环境和保护城市生态环境,实现城市更好的绿色发展。在城市局部核心商务区或大型集中居住区,宜以地下快速路取代地面及高架快速路,改善中心城区地面景观及步行环境;促进市政设施地下化,提升土地价值,增加城市绿地面积。

(6) 利用地下空间减少环境污染,节约能源。应增大地下轨道交通比重,减少汽车尾气排放,并为地下道路中的汽车尾气的收集处理提供条件;鼓励地源热泵对于地温能的利用,降低能源消耗和碳排放,缓解空气污染与城市热岛效应。

三、未来绿色城镇基础设施

绿色城镇建设的指导思想:坚持"世界眼光、国际标准、中国特色、高点定位"的理念,建设"可持续发展、韧性和生态安全的"绿色城市。

习近平绿色发展思想是习近平新时代中国特色社会主义思想的主要组成部分:党的十八大以来,党中央提出了一系列治国理政新理念新思想新战略,把生态文明建设和绿色发展提到新的战略高度。绿色已列为中国国家建设发展的五大理念之中;中国的绿色发展正在进入世界绿色发展的先进行列:绿色植被;可再生能源的发展;清洁能源的利用。

绿色城镇基础设施特点:有效利用地下空间(少占地)、低能耗、低碳或零碳排放。

(一) 未来的城市供暖制冷系统

当前,首先要加速煤改电、煤改气,积极推进地热供暖制冷。

1. 地温能

地温能指的是地层中温度小于25℃的地层热能。利用低密度地热能源(地温能),可实现节能减排。

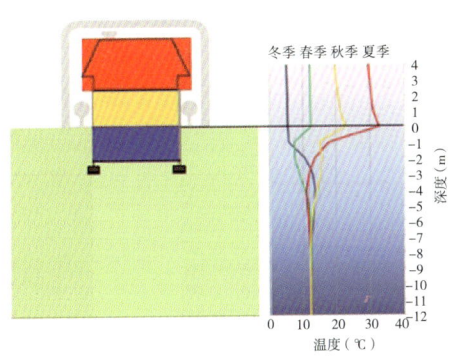
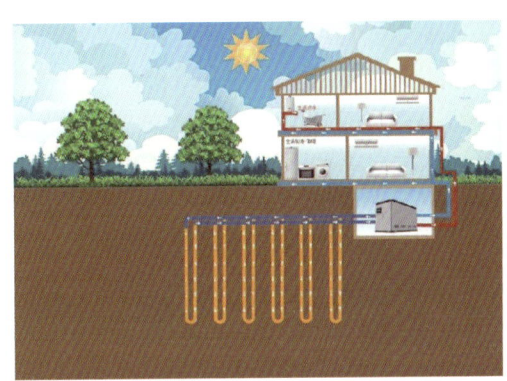

温度廓线图

到达地下一定深度时（5m以下），四季的地层温度保持在一稳定值。

该系统把传统空调器的冷凝器或蒸发器直接埋入地下，利用传热循环介质与大地进行热交换，从而提取地温能。

地下换热系统可埋设在地下结构基坑围护结构（地下连续墙、排桩）内、基础底板下、桩基（钻孔灌注桩、预制桩、PHC桩等）内；可埋设在新奥法施工的隧道衬砌内或以能源锚杆的形式埋设在其围岩中；也可埋设在地铁区间隧道内、地下输水管道内。

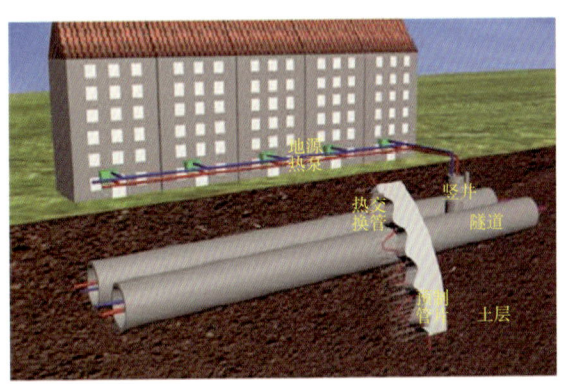

隧道结构内埋管换热器工作示意图

上海自然博物馆新馆位于上海市静安区，北京西路、石门二路处在建雕塑公园北部，北邻山海关路。占地面积约12000m²，总建筑面积约为45086m²，其中地上建筑面积12128m²，地下建筑面积32958m²。建筑总高度18m，地上三层，地下两层，采用地源热泵系统承担建筑部分夏季冷负荷以及冬季热负荷。夏季土壤换热器最大算热负荷为1639kW，冬季土壤换热器最大热负荷为1178kW。

上海自然博物馆

工程采用灌注桩与地下连续墙内埋管两种形式，其中，灌注桩埋内埋管 393 个，平均桩长为 45m，采用 W 形埋管，地下连续墙内埋管共 452 个，采用 W 形埋管。灌注桩及地下连续墙内埋管布置见图。

上海自然博物馆已于 2015 年投入运营，该馆采用了地下结构内埋管热交换系统，其初投资比传统的冷水机组 + 锅炉系统增加 210.2 万元，但该系统利用了地温能实现建筑制冷和供暖，年运行费用可节省 22.3 万元，动态回收期 11.98 年，该系统利用清洁的地温能，每年可节约 117.7t 标准煤，减排二氧化碳 195.5t。

上海中心大厦裙楼、上海世博轴和上海富士康大厦等国内重大工程均采用了地下结构内埋管热交换系统。

2019 年 8 月，北京大兴国际机场地源热泵工程 1 号能源站开始启用，标志着中国国内最大的多能互补地源热泵系统工程启动运作。大兴机场的方案是在机场公园湖区，按间距 5m 布设深 120~140m 钻孔 1 万多个，孔中安装 U 形热交换管，获取地温能，热泵是在能源中心的机组内完成的，大兴机场地源热泵系统每年提取浅层地温能 56.36 万 GJ，实现大兴机场公共区域 250 万 m^3 办公场地的供热和制冷，节省天然气 1735.89m^3，相当于 21078t 煤，减少碳排放 1.58 万 t 以上，地源热泵将成为建筑能源的首选解决方案。

（1）总共设置换热孔 10500 个，其中，南区 3120 个，北区 7380 个；

（2）地埋孔深度：南区 120m，北区 135 ~ 140m；

（3）孔间距 5m，局部区域根据地形情况，调整为 4.5m；

（4）地埋管分单元设置，北区设置 20 个单元，南区设置 8 个单元。

2017 年底，我国地源热泵装机容量达 2 万 MW（2000 万 kW），实现供暖（制冷）面积 5 亿 m^2，利用地热能折合 1900 多万 t 煤。2015 年起，地源热泵装机容量占世界第一，占全球比重 2015 年为 23.61%，2020 年将上升到 50%。

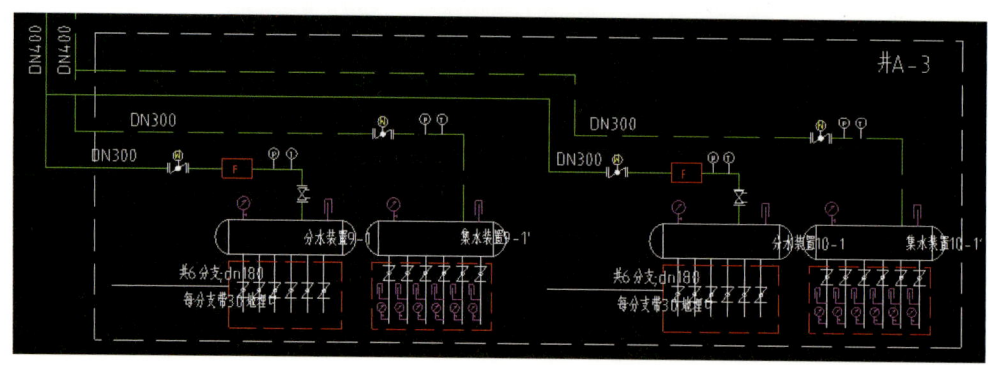

2. 浅中层地热能（温泉）

河北省地热能资源总量位居全国第二位，2015 年地热资源开采量突破了 1.1 亿 m^3，地热供暖面积达到了 6300 万 m^2。其中，雄县位于河北省保定地区，是国内首个通过地热供暖实现"无烟城"的县城，拥有享誉全国的"雄县模式"。雄县地热资源分布面积广，出水量大，水温高，现如今地热集中供暖面积已占城区集中供暖的 85%，覆盖县城 80% 以上的居民小区，每年可减少二氧化碳排放量 12 万 t。在收费方面，雄县居民地下水取暖收费为 16 元 /m^2，相较之前燃煤取暖 25 元 /m^2 的费用更为便宜。

雄县地热供暖工程方案：

2010 年，雄县采暖建筑面积 324.97 万 m^2。

采暖热负荷：158.78MW。

每眼新打地热井出水能力 150m^3/h，出水温度 70℃，按地热直接供暖方式产热能力 8.67MW，雄县工程新打 11 眼井，热负荷可达约 96MW（占 60%）。

该方式地板采暖的尾水排放温度为 20℃，尾水经水源地热热泵系统将低温热能水提升为高温热能水转到供热系统中，其提供的热负荷约 64MW（占 40%）。

由于地热水对金属腐蚀普遍存在，所以本工程采用热交换器换热，供给用户使用的为换热后的二次水。

为保证地热资源开采的可持续性：开采量必须小于可开采量；此外被冷却的地热水回落到地下距开采井足够距离的另一口井，即形成开采回落井对。

2005 年世界地热直接利用前 10 位国家

序号	国名	年产能（GW·h）
1	中国	12604.6
2	瑞典	10000.8
3	美国	8678.2
4	冰岛	6615.3
5	土耳其	5451.3
6	奥地利	2229.2
7	匈牙利	2205.7
8	意大利	2098.5
9	新西兰	1968.5
10	巴西	1839.7

3. 深层地热能（干热岩）

干热岩的研究最早始于 20 世纪 70 年代的美国，经过半个世纪的不断发展，干热岩在理论和实践方面都有了长足发展。美国、法国、德国、日本、澳大利亚等国最早开启 EGS 技术研发与工程建设，部分项目进入试验性运行发电阶段。据统计，全球在建及投运 EGS 工程已达 30 个，其中 14 个实现了运行发电，目前尚在运行的有 5 处，总装机容量为 12.2 MW，尚未实现规模化、商业化运行。

110 名科学家参与的美国 Fenton Hill 干热岩发电试验是世界上最大的试验项目，也是全球首例建立地下深部工业干热岩热储的增强型地热系统工程，17 年耗资 1.5 亿美元；4500m 的钻井深度，330℃的高温岩体，3600m 的热交换系统深度，世界上首座干热岩发电站于 1984 年建成，发电功率达 10MW，250MW/m² 的地热流值，如图所示。发电站因井壁损坏、设备缺陷和资金等诸多问题，于 2000 年被完全废弃。

我国干热岩资源勘查及 EGS 技术研究还处在起步阶段，目前开展了初步的资源勘查和相关技术研究，取得了一定的成果，但是还未建成我国第一个 EGS 示范研究场地。

全国陆域干热岩资源量为：856 万亿 t 标准煤。青海共和盆地 3705m 深钻获 236℃的高温干热岩体。

我国已成功在陕西省内进行了干热岩用于供热的商业应用——长安信息大厦 2013 年共计 3.8 万 m² 中应用干热岩供热，效果良好。

按照 2% 的可开采资源量计算，我国可开采量为 17 万亿 t 标准煤，相当于 2016 年全国能源消耗量（43.6 亿 t）的近 4000 倍。

至"十三五"末，地热能年利用量相当于替代化石能源 7000 万 t 标准煤，减排 CO_2 1.7 万亿 t！

（1）青海共和盆地

中国地质调查局和青海省国土资源厅共同组织实施的青海共和盆地干热岩勘查项目，在青海共和盆地成功施工 5 眼干热岩勘探孔，孔深为 3000~3705m，井底温度达 180~236℃，5 眼勘探孔均钻获干热岩体。其中 GR1 干热岩勘探孔，孔深 3705m，孔底温度达 236℃，孔内 3366m 以下深度平均地温梯度为 8.8℃/100m，根据国际标准，属高品质干热岩体，实现了我国干热岩勘查重大突破，同时采用地球物理、地球化学、放射性调查等综合技术手段圈定了面积达 3000km² 的干热岩体分布区。

（2）福建漳州地热田

2014年，中国地质调查局水环所联合闽南地质大队开展了漳州地区区域重力、重力剖面以及大地电磁测深等物探工作。通过结合大地热流、酸性岩体分布、居里面埋深以及区域构造特征，选定福建省漳州地区是我国干热岩潜在分布的有利区，并确定了钻探靶区。2015年5月21日，由中国地质调查局组织实施的我国首个干热岩科学钻探深井，在福建漳州龙海市东泗乡清泉林场开钻，目前钻探深度已达4000m，岩性为花岗岩，井底热储温度为109℃，勘探成果不及预期。

（3）河北献县干热岩钻探项目

2017年，河北省煤田地质局水文地质队在河北省献县实施钻探了GRY1井。该井隶属的干热岩预查项目为河北省地勘基金项目，主要设计干热岩参数孔一个，设计孔深4000m，并进行高温固井、抽水试验、压裂试验等科学试验工作。目前在4000m深度完井，完井层位雾迷山组，测得温度110℃左右，没有达到预计的干热岩钻探效果。

（二）未来的城市地下污水处理系统及地下垃圾集运及焚烧系统

成都天府新区第一污水处理厂总规划26万t/天，一期规模10万t/天，服务面积55km²，污水干管11.0km，系统收集区域内污水。污水处理地下完成，地面是集环保科普、休闲游憩于一体的活水公园及科普展厅（中小学生生态环境教育基地）。

（1）工艺

预处理——粗细格栅，去除污水中杂质和砂粒；

生化池——生物降解除臭，除磷，去氮，降解有机物；

MBR膜池——以渗透膜组件进行泥水分离；

紫外消毒——利用紫外线杀灭水中细菌；

脱泥——污泥离心脱水，含水率脱至80%。

厂区通道

膜池

生化池除臭系统

污水厂-51

（2）项目特点

资源节约型——全地下式结构，节约土地 320 亩；出水指标达标准水Ⅳ类标准，优于国家再生水标准，全部回用；80% 含水率以下污泥用于园林用肥。

环境友好型——地下处理消除了臭味、噪声对周边的影响，取得了良好的环境效益，解决了"邻避"问题。

（3）经济效益分析

建设城市——地下厂成本 6000~7000 元 / 吨水 >4000 元 / 吨水，比地面厂高 50% 左右。

运行成本——地下厂比地面厂高 30% 左右。

（4）总效益

节约土地 320 亩，价值 25.6 亿元，扣除增加造价，节约资金 22.6 亿元。为防止"邻避效应"，提升了周边土地价值，改善了环境。

中国广州金沙洲和亚运城都建设了真空垃圾收集系统。

上海世博园区内的垃圾隐形处理系统总覆盖面积约 0.5km^2，地下输送管线总长达 5.5km，主要服务于一轴四馆，永久性建筑及其周围区域，共有 51 个室外的智能垃圾投放口。

瑞典是在 20 世纪 60 年代建设并实施采用压缩空气吹运垃圾系统（PWT）的国家，PWT 系统建在 1700 人口的居民区内，该系统与收集、处理系统配套，投资在 3~4 年内得到回报。其他的例子，如设在香港岛西端大维山岩硐中 RTS。

发展地下垃圾处理系统节约了城市宝贵的土地资源，有很大的回报，控制了对环境的噪声和恶臭的污染，减少了对环境的视觉负面影响。

在大力推广垃圾分类与地下集运的基础上实行垃圾焚烧，彻底解决垃圾围城现象，释放大批城市土地资源，改善城市环境。

（三）未来城市的深层排水调蓄隧道系统

1. 深隧工程介绍

（1）排水深隧定义

深隧（Deep tunnel）是深层排水调蓄管道系统的简称，一般是指埋设在深层地下空间，用于调蓄、输送雨水或合流污水的、通常具有较大调蓄容量的系统工程。

日本东京排水系统　　　　　　　美国芝加哥蓄洪隧道工程　　　　　　法国巴黎合流排水系统

（2）深隧的功能

主要包括污染控制、洪涝控制和其他功能。

污染控制主要包括污水的输送，或者减少合流制溢流污染以及初期雨水污染。

洪涝控制主要包括提高排水标准、提高防汛标准、减少内涝灾害、输送过量雨水等。

2. 国内外工程案例

国际上深层调蓄隧道技术日趋成熟，成功案例有 20 余项，大部分深隧以单一功能为主。

序号	所在城市	隧道系统名称	工程规模	隧道主要功能
1	美国芝加哥	芝加哥隧道和水库方案	长175km，内径2.4~10.8m，深46~88m，1000万m³	控制水体污染
2	美国奥斯汀	沃勒河深层隧道	长1.7km，内径6.10~7.82m，深21.94m³	提高防洪标准
3	美国密尔沃基	密尔沃基隧道系统	长45.5km，内径10m，深100m，200万m³	控制水体污染
4	美国旧金山	旧金山输送调蓄系统	长3km，内径9m，埋深30m	控制水体污染
5	美国印第安纳	印第安纳波利斯深层隧道系统	长40km，内径5.5m，深75m，调蓄100万m³	控制水体污染
6	日本东京都	首都圈外围排水道	长6.3km，内径10.6m，深50m	缓解内涝
7	日本东京都	和田弥生干线	长4.5km，内径12.5m，深40m，调蓄量54万m³	缓解内涝
8	日本东京都	东京都古川地下调节池	长3.26km，内径7.5m，深36m，13.3万m³	缓解内涝
9	日本大阪府	寝室川南部地下河川	长11.2km，内径9.8~6.9m，深25m，调蓄量96万m³	缓解内涝，兼顾控制水体污染
10	日本横滨市	今井川地下河川	长2.0km，内径10.8m，深50m，17.8万m³	缓解内涝
11	日本神奈川县	矢上川地下调节池	长4.0km，内径7.9m，深55m，19.4万m³	缓解内涝
12	日本神奈川县	鹤见川地下河川	长4.0km，内径10m，深50m	缓解内涝
13	英国	泰晤士河隧道工程	长25km，内径6.5~7.2m，深30~65m	控制水体污染
14	马来西亚吉隆坡	SMART隧道	长9.7km，φ13.2m，调蓄量300万m³	解决内涝，缓解交通
15	法国巴黎	巴黎调蓄隧道和调蓄池	4条隧道，φ6~7m，长5.1km	缓解塞内涝
16	墨西哥城	东部深层排水隧道工程	长63km，φ7m，深200m	提高雨季过流能力
17	中国香港	荔枝角雨水排放隧道工程	长3.4km，内径4.9m，深40m	提高排水标准
18	中国香港	荃湾雨水排放隧道	长5.1km，φ6.5m，最大埋深约200m	提高排水标准
19	中国香港	净化海港计划污水隧道	长23.6km，埋深100m，内径0.9~3m	污水输送
20	中国广州	广州深层隧道排水系统	总长86.4km，近期东濠涌支段长1.7km，φ5.3m，埋深40m	控制水体污染，提高排水标准
21	中国深圳	前海南山排水深隧	长3.74km，φ4~6m，埋深45m	控制水体污染，缓解内涝

（1）洪涝控制型——日本东京首都圈外围排水道

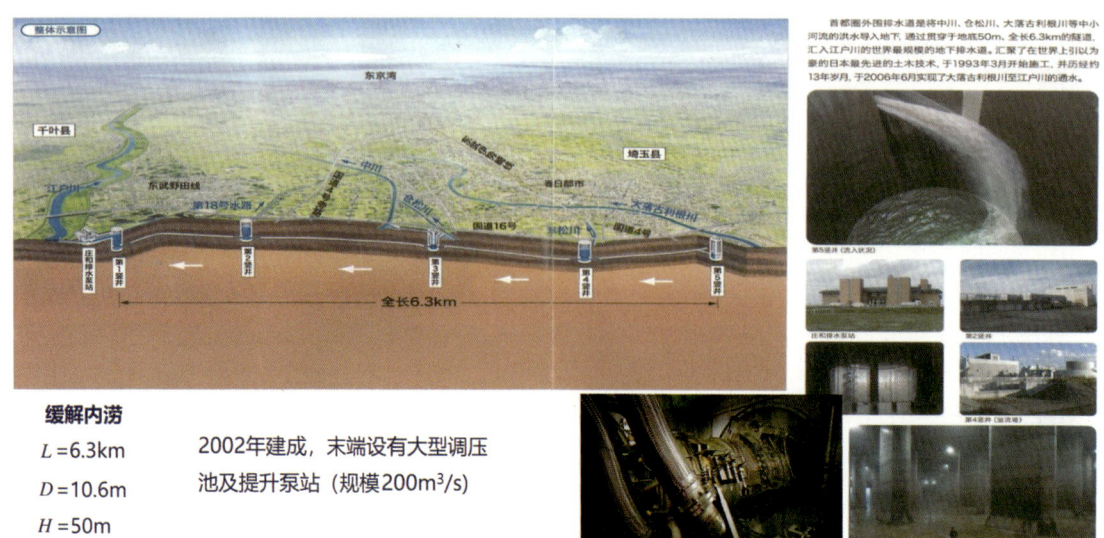

缓解内涝

$L = 6.3$ km
$D = 10.6$ m
$H = 50$ m

2002年建成，末端设有大型调压池及提升泵站（规模200m³/s）

（2）多功能型——马来西亚吉隆坡 SMART 工程

缓解内涝：晴天底排水，中上层交通；雨天三层均排水。

目前已建成，可将上游洪水转移至旁路隧道临时储存后排入郊外下游水库，减缓了河水倒灌及关键路段积水严重的现象。

（3）污染控制型——英国泰晤士河隧道工程

① Lee 隧道（Lee Tunnel）：

总长 6.9km，直径 7.2m，埋深 75~80m，建设耗时 6 年，2016 年 1 月投入运行，每年减排 40%（1600 万 t）溢流污水排放。

②潮路隧道（Thames Tideway Tunnel）：

总长 25km，直径 6.5~7.2m，埋深 30~65m，2016 年开工建设，计划 2024 年投入运行。

可收集伦敦泰晤士河 97% 的溢流污水，溢流次数由目前每年 60 次减少到 4 次，有效地改善了泰晤士河水体环境。

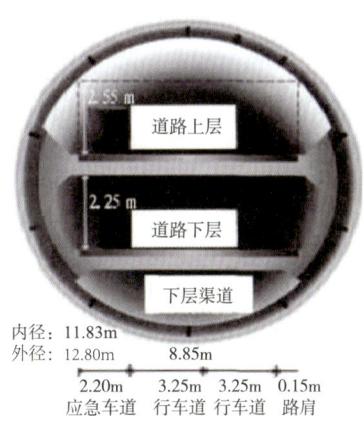

锚定式隧道管片密封垫

二次衬砌钢纤维衬砌厚 300mm

3. 苏州河项目建设的必要性

（1）满足排水标准提高需要

苏州河地区的涉及市政雨水排水系统共有 25 个，服务面积总计 57.92km²。

25 个系统中 19 个合流制，6 个分流制，均为强排系统，现状泵排流量 303.2m³/s。

25 个系统中除广肇系统为 2 年一遇外，其他均为 1 年一遇。在 5 年一遇标准下，区域内将近 22% 路段积水，平均深度达 25cm。

城镇类型	中心城区	非中心城区	中心城区的重要地区
超大城市和特大城市	3~5	2~3	5~10
大城市	2~5	2~3	5~10
中等城市和小城市	2~3	2~3	3~5

注：人口密集、内涝易发且经济条件较好的城镇，宜采用规定的上限。

（2）满足面源污染控制需要

苏州河经历一期、二期、三期 14 年的综合整治，取得了卓著的成果，但河道初雨污染仍然严重，河道水质在 V~劣 V 类间波动，逢雨即黑现象普遍存在。

初雨污染

2018 年苏州河主要监测断面水质情况（摘自市环保局官网）

断面	1月	2月	3月	4月	5月	6月	7月	8月	9月	10月	11月	12月
赵屯	V+	V+	V	V	V	V	V+	V	IV	V	IV	IV
北新泾	IV	V+	V+	V+	V+	V+	V+	V	IV	V	IV	V+
浙江路	III	IV	V+	III	V+	V+	V+	V+	V+	V+	IV	IV

4. 苏州河深隧概况

苏州河段深层调蓄管道系统工程主线工程

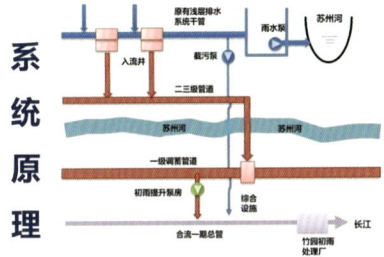

（四）未来城市交通和城际交通

当前的城市交通应发展以轮轨地铁为主的轨道交通。

1. 城市轨道交通

截至 2017 年底国内已有 35 个城市开通轨道交通，运营里程达 5033km。

未来城市交通应积极发展地下低速磁悬浮交通，磁浮没有地面摩阻力，能耗低，少占地，噪声少，电力驱动无二次污染。未来城际交通应发展地下低真空高速磁浮交通（>600km/h），无地面阻力，空气阻力极低，能耗最低、安全舒适，精确停车。

将来发展 1000km/h 的地下低真空高速磁浮交通，可取代高污染、高碳排放的喷气式客机。

2. 辅以城市快速路

（1）限购限行，利少弊多，不能持久。

（2）"存量"不减、"增量"攀升，中心城区地面已无地可建（路）。

（3）国外曾做过伦敦中心区地面交通供给能力的模拟测算，其结果是即使把伦敦中心区所有建筑架空，其下层地面全变成道路，也不能解决交通拥堵问题。

仅大力发展轨道交通不能从根本上解决交通拥堵。即使是轨道交通相当完善的发达国家大城市，如伦敦、东京、芝加哥、马德里、莫斯科等，它们也困扰于交通拥堵。

以东京为例，公共交通出行比例达到了 60%，地铁建设里程已达 300km（地铁集中于城中心区，仅占轨道交通的 1/10），仍然没有解决交通拥堵。伦敦也是，它的第一条地铁建成于 1863 年，至今天已建成 439km，但仍然要实行收取"交通拥堵费"以缓解拥堵之苦。

在 20 世纪末、21 世纪初，国际上很多发达国家的特大城市和有识之士把"治堵"目光转向了"地下快速路"和"地下物流系统"。

国内大城市的地下快速路建设也已初见端倪，并在缓解交通拥堵上发挥了明显作用。

①南京已建成玄武湖地下快速路、城东干道地下线和内环线地下路（占总长 1/2）。

②上海正在建设井字形通道方案，其中地下线 26km。

③深圳已建成港深通道的7km地下快速路。

④深圳前海的地下空间规划建设在地下车行系统方面，将构建"地下快速路+地下环路+地下车库"的三级地下车行系统，海滨大道、妈湾跨海通道地下道路以分离过境交通为主，以服务区域到发交通为辅；听海路服务于区域到发交通，从而有效改善前海地面交通环境，缓解城市中心区和小街坊路网交通压力。

这些地下快速路的天然优势是全天候，在暴风、雨雪和大雾等最易造成地面交通拥堵的情况下最能发挥作用。这在我国城市建设用地严格控制和大城市人均道路指标普遍偏低的情况下，非常适合我国地少人多的国情。

城市地下快速路系统以及地下物流系统的建设也必将为消除汽车尾气对城市空气的污染以及噪声污染作出重要贡献。例如采用如下方法消除尾气污染：除尘换气，氢气动力，电动汽车。

（五）未来的城市货运交通

科技创新转变城市交通运输方式还包括地下物流系统，以替代城市货车限行的行政措施，将城市货运逐步并最终转移至地下。

地下货物运输系统，又称地下物流系统（ULS）是基于城内运输和城外运输的结合。城外货物通过各种运输手段到城市边缘的物流园区（CLP），经处理后由CLP通过ULS输送到各个终端。它以集装箱和货盘车为基本单元，以自动导向车（AGV）为运输工具。

根据世界经合组织2003年统计，发达国家主要城市的货运占城市交通总量的10%~15%；在"世界工厂"和到处是建设工地的中国则为20%~30%；而在当今电子商务、快递和送快餐发达的中国，这个比率还要高。货运交通转入地下必将对缓解交通拥堵作出重要贡献。

1. 港城地下物流系统

传统的港口集疏运模式以公路集卡运输为主，需要穿越城市，占用城市道路资源，客货交通相互影响，带来了交通拥堵、环境污染等问题。需要更集约化、可持续发展的集疏运新模式——地下集装箱物流系统，解决港城矛盾。

将原先分散进入港口的货物，首先集中在外围物流综合枢纽，再通过地下专用货运通道集约化转运到港口，实现高效率、规模化运输。有效避免集装箱卡车穿越城市，释放地面资源，不影响中心城交通、环境。

以上海港为例，2009年以来，上海港集装箱吞吐量基本保持在20%~30%的增幅增长。2012年上海港集装箱吞吐量达到3253万标准箱，位列世界第一；2016年总量3713.3万TEU，超过所有欧洲港口集装箱吞吐量总和。根据《上海港总体规划》预测，2020年上海港集装箱吞吐量将达4000万标准箱。

上海港集装箱集疏运分析：2015年，上海港集装箱各集疏运方式中，公路集疏运占54.8%，水路集疏运占45.0%，铁路集疏运占0.15%。2020年，预测水水中转50%，公路集疏运占48%，铁路集疏运占2%。

（1）公路集疏运比例过高，方式单一。

①外环线和郊环线的集卡交通量较大，特别是外环线的东段和北段，以及郊环线北段，最高断面达6万~7万PCU/12。

②北部地区集卡流量集中，高等级道路供给不足。

③越江通道能力不足。

（2）集疏运比例不合理，公路运输比例过大引发交通问题。

①公路运输比例大，现有道路不足以提供通行能力，集卡为赶船期，只能提前前往港区附近。

②港区附近停车位不足，提前到达的集卡在路边无序停靠。

③无序停靠的载货汽车占用路面资源，进一步降低通行能力，造成恶性循环。

（3）集疏运比例不合理，公路运输比例过大由此引发环境问题。

①运输结构失衡，过度依赖公路运输，带来严重环境污染及能源消耗，使得上海港的快速发展对"节能减排"工作造成巨大的压力。

②据测算：2016 年，上海公路集装箱集疏运的能源消费量约 10.5 亿升柴油，折合标准煤约 131.25 万 t，排放二氧化碳约 286.47 万 t。

2. 城市地下立体智慧物流系统

诺贝尔经济奖得主斯蒂格利茨的预言正在成为现实：20 世纪信息技术的创新将在 21 世纪引发社会经济最伟大的变革。

互联网技术正在改革世界：互联网改造的第一个传统行业是商贸流通业，通过电子商务进入销售互联网时代，引发第四次零售业革命——新零售。

（1）新零售推动城市物流需求爆发增长

单位与居民物品物流总额近 10 年增长 20 多倍，是增长最快的领域。据罗兰贝格等机构分析，预计中国 2019 年的网上零售额将增长 30%，超过 1 万亿美元，占全球网上零售总额的一半以上。2019 年我国前九个月统计网上销售总额 7.8 万亿元，远超美国，居世界第一。

中国 2014 年快递包裹 140 亿件；2016 年 313 亿件（全球 650 亿）；2017 年 400.9 亿件；2018 年 490 亿件。

交通需求激增带来的地面车辆、车次数量剧增是导致城市交通拥堵的主要原因，其中对货物物流的需求增长占了较大份额。如北京，按常规的车辆换算系数，货运车辆所占用的道路资源占 40%。城市中大气污染物约 60% 来自机动车排放，机动车成为城市 PM2.5 的最大来源，交通也已成为城市噪声的最大污染源。

新零售成为城市交通事故的突出问题。2019 年上半年，上海市共发生快递、外卖行业各类交通事故 325 起，造成 5 人死亡。上述事故中快递人员 80% 以上是美团和饿了么等公司送快递和送餐的。2017 年全年，仅 117 起快递、外卖行业交通事故。

（2）新零售的核心是新物流，新零售呼唤新物流：智慧物流和地下物流

智慧物流配送智慧化平台——数据驱动，智慧布局。

例如通过构建综合评价模型、成本最优模型、站点数量最少模型等多维度模型，基于订单量、路区坐标等输入参数以及传站时间、配送半径等约束条件，采用遗传算法等智能算法进行求解，得出最优的站点数量、每个站点的坐标、平均派送半径等规划决策。在北京市，通过优化已经将 1000km 的距离缩短至 200~500km。

数字化运营——仓储管理系统、运输管理系统等	
仓储	智能存储 + 智能拣选 + 智能耗材推荐 +……
	WMS+ 算法 - 最优布局、定位、最短路径、设备调度……
运输	智能调度
	TMS+ 算法 - 动态规划……
配送	智能分拣 路径规划
	配送管理系统 + 算法 - 遗传、蚁群……

（3）智能化作业——菜鸟全自动智慧仓储基地

采用国内目前最好的"货找人"处理系统，消费者在电商平台下单后，仓内同时形成一个贴有货物信息条码的快递箱，扫码后箱子进入自动化轨道"寻找"仓内货物，到达相应货架后，拣货员只需按提示把身后的货物放入箱中即可。

（4）智能化作业——菜鸟机器人

菜鸟网络智慧仓内，单个AGV搬运机器人形同"扫地机器人"。在这里，超过百台这样的机器人。以消费者在猫超下单为例，下单之后，菜鸟网络惠阳智慧仓内的机器人便会接到指令。机器人会自动前往相应的货架，将货架拉到拣货员面前，由拣货员将市民购买的物品放置在购物箱内，随后进行打包配送。目前，机器人与拣货员搭配干活，一个拣货员一小时的拣货数量比传统拣货员多了三倍多。

（5）智慧物流无法突破的瓶颈指向地下物流

①受交通拥堵、交通管制和气候（雨、雪、台风）影响，配送时间和配送效率的要求无法满足；

②城市地面配送使交通拥堵的加剧，使已经进入"拥堵时代"的中国城市无法承受；

③在尾气污染已成为城市空气污染元凶的今天，物流的地面货车与摩托车城市配送对城市雾霾的加剧将使得社会无法承受；

④当前最后一公里的快递和外卖主要是摩托车配送，其突出问题是摩托车快递不遵守交通规则，违规行驶严重，影响了城市交通秩序，造成了交通事故，社会影响严重。

（6）城市的地下智慧物流配送系统

在城市，物流配送中心与地下物流系统枢纽相结合，或者地下物流系统的物流配送中心和大型零售企业结合在一起，也可以与城市商超结合，建立商超地下物流配送，实现网络相互衔接，客户在网上下订单以后，物流中心接到订单后，在地下进行客户货物的专业仓储、分拣、加工、包装、分割、组配、配送、交接、信息协同等基础作业或增值作业，通过地下管道物流智能运输系统和分拣配送系统进行运输或配送。

最终地下物流系统末端配送可以与居民小区建筑运输管道物相连，发展成一个连接城市各居民楼或生活小区的地下管道物流运输网络，并达到高度智能化。当这一地下物流系统建成后，人们购买任何商品都只需点一下鼠标。所购商品就像自来水一样通过地下管道很快地"流入"家中（小区的自动提货柜）。

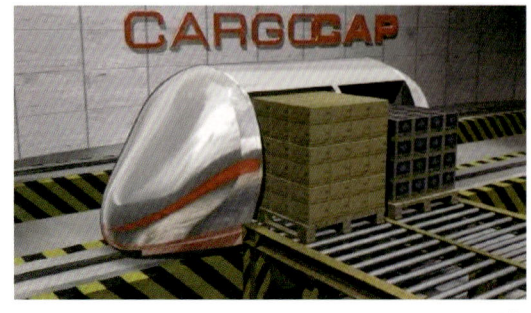 

地下物流系统示意图

(六)未来城市地下空间建设的机械化、自动化及智慧掘进与智慧运营

1. 新装备

(1) 地下隧道工程新装备——异形掘进机

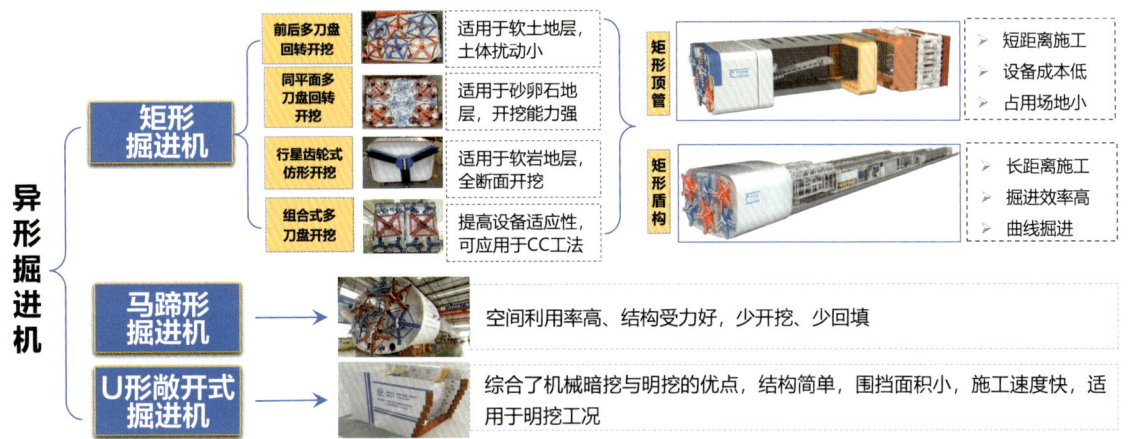

(2) 新维度掘进机——联络通道盾构

工法	敞口盾构法	顶管法		盾构法
图例				
工程案例	德国汉堡第四易北河救援通道	墨西哥 Emisor Oriente 污水隧道联络通道	日本大阪御筋堂综合管廊	香港屯门至赤鱲角连接线联络通道	宁波地铁三号线联络通道
穿越地层	云母、砂、淤泥、黏土、泥灰	砂、黏土	砂土、黏土	花岗岩、沉积岩、砂砾层	淤泥质黏土、粉土
加固情况	—	—	—	未加固	未加固
功效	—	11天掘进103m	—	20天	18天

(3) 新维度掘进机——斜井掘进机与竖井掘进机

斜井掘进机集斜井开挖、衬砌、出渣、运输、通风、排水等功能于一体,能够满足软岩、硬岩及复合地层等不同地质条件的斜井施。针对斜井施工中埋深大、水压高、下坡长、地层多变等问题,结合原盾构、TBM技术,重点突破,创新工法,适用于煤矿、水电引水等斜井施工,下行可完成 −10° 斜井施工,上行可完成 +45° 斜井施工,实现斜井安全、高效施工。

2. 新工法

（1）地下停车场施工技术

项目概况：本项目位于中铁工程装备集团有限公司的内部的停车场内，为地下一层结构，停车规模约 36.2m × 85.85m（宽 × 长），停车约 93 辆。

断面尺寸：单台 2.87m × 5.02m，组合后 5.74m × 5.02m；

覆土深度：3m；

隧道长度：7 × 61.58m；

地质概况：粉土；

项目难点：零间隙施工，轴线偏差、地表沉降要求高。

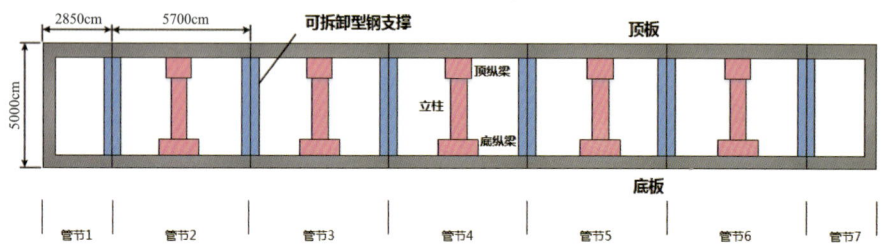 

地下停车场实景

工法与装备相结合，对解决城市停车难意义重大。

（2）口琴工法

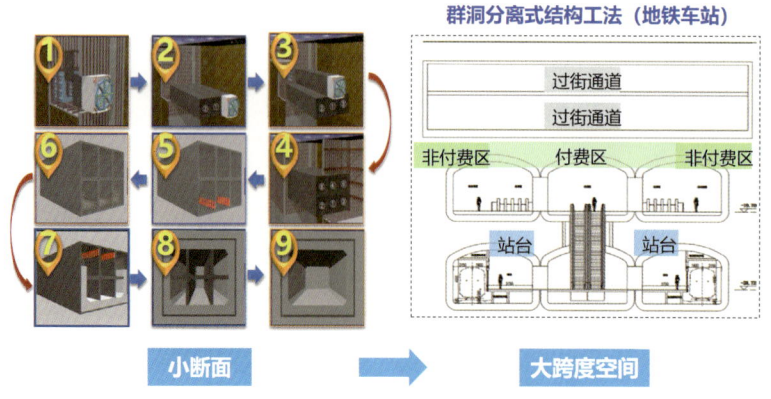

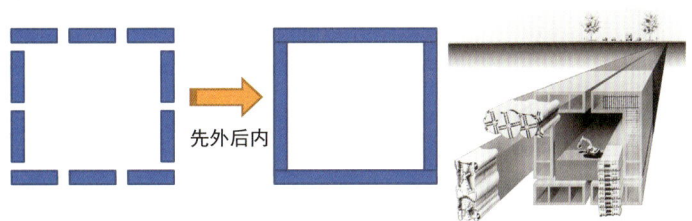

通过好多个小断面盾构机预先掘削隧道的外壳部位，然后相互之间予以连接，构筑成隧道的外壳部位的躯体之后，再掘削内部土砂、构筑起大断面的隧道工法。

3. 盾构产业智能化

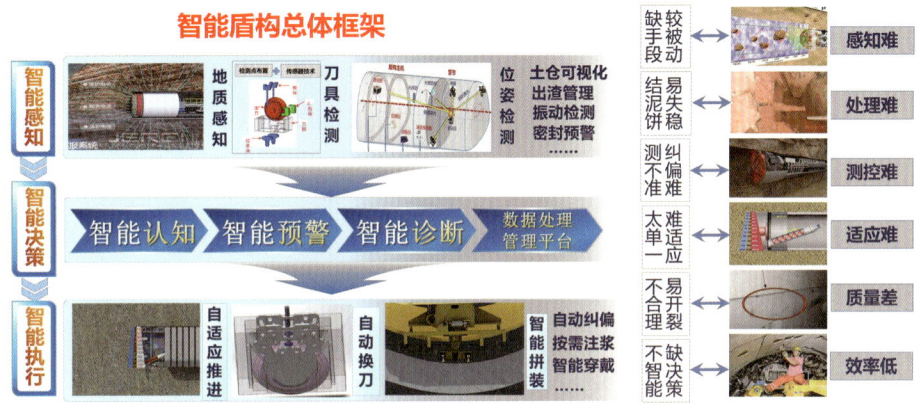

4. 超前预报地质感知系统

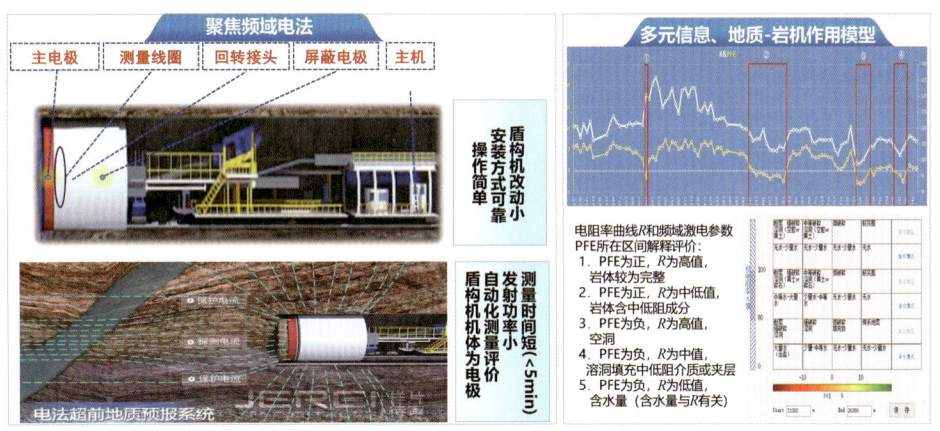

◆ 智慧管理　　◆ 智慧监测　　◆ 智慧检修

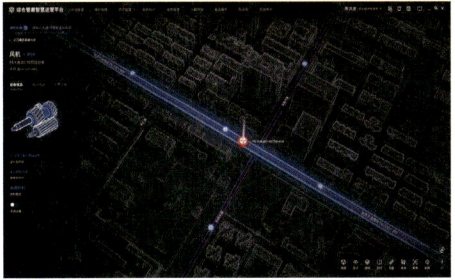

平台 GIS 视角

平台 BIM 视角

5. 城市的智慧停车系统

乌镇的智能停车系统和南通的智能地下立体车库为已建成居民小区紧密化和功能变化的模板。

江苏南通智能地下立体车库

四、结语

通过若干个五年计划持之以恒的努力,将地上城市中的城市运输与交通系统逐步转移到城市地下去。不但使旅客交通与货物运输转入地下,还将垃圾、污水等的传输都转入地下,使地上逐步乃至彻底摆脱与传统运输相联系的环境干扰,首先是交通拥堵与PM2.5超标的困扰,而释放出的地上空间将用作大片的自然植被和安全的步行,这不是科学幻想,《欧洲地下建设战略研究议程(Strategic Research Agenda For the European Underground Construction Sector)》就已经明确提出,其2030年的远景目标是"将地面交由市民自由使用,将基础设施放入地下"。中国的雄安新城规划正是按照美丽城市的理念编制,它的规划纲要中建设空间仅占30%,预留60%~70%的空间保证生态环境的健康。

| 学术报告——绿色建筑与智能建造

数字隧道和智慧隧道

——隧道建设信息化技术的发展方向

一、前言

隧道工程是高风险工程。极端复杂的不良地质条件是制约隧道建设安全高效的主要因素，并成为建设难点。建设人员不能正确应对与处理复杂不良地质条件酿成了地质灾害和工程事故，对隧道建设的安全、工期和成本造成严重危害。

暗河、溶洞、断层破碎带、孤石等不良地质体都会对钻爆法和 TBM/盾构掘进机隧道安全施工造成严重危害。

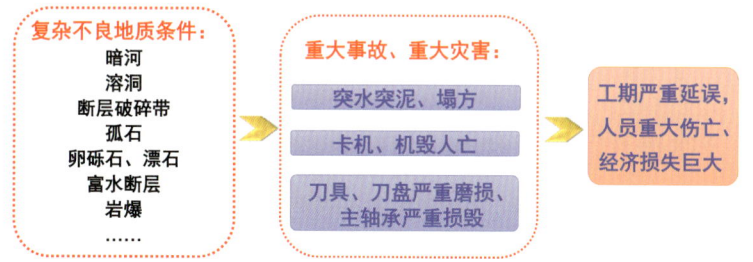

隧道工程之所以成为高风险工程，其关键问题是地下地质情况不掌握，事故发生机理不清楚，从信息学的角度来说，就是地下围岩环境信息采集和传递不及时，信息分析处理不完善，信息共享和利用不充分。因此隧道建设要实现安全高效的关键是提高隧道工程建设的信息化水平，逐步实现数字隧道和智慧隧道。

二、数字隧道

数字隧道是隧道工程信息化的初级阶段，是"物理隧道"（实体隧道）的虚拟对照体，以信息化手段对隧道建设过程中的勘察、设计、施工及监测等数据进行集中有效地管理。具体体现为数字隧道工程基

报告时间：2017年8月7日。

础平台，该平台是集数据信息存储、查询、三维可视化建模及虚拟浏览为一体的综合系统。其信息包括地形、地理的基础地理数据，包括工程地质、水文地质和环境地质的地质数据（二者共称为地层数据）、隧道主体的设计、施工及监测数据等。通过该系统实现隧道信息的分布式共享、协同和利用机制，最大限度地挖掘和发挥隧道数据的潜能和作用，并贯穿于隧道规划、勘察、设计、建设和运营的全过程，保障隧道的科学决策和管理。其基础是虚拟现实技术和 BIM 技术。

（一）虚拟现实技术

虚拟现实（Virtual Reality）技术是指以计算机三维图形处理技术为基础，配合强大的计算机语言开发环境，制作出可以模拟真实客观实体的技术。利用虚拟现实技术创建出逼真的高层建筑建构和三维隧道及其围岩环境，让管理者和建设者更直观地感知高层建筑和隧道及其环境的建设和运行情况。

（二）基于 BIM 技术的隧道建设

BIM（Building Information Modeling）技术的本质就是把数据转化为信息，通过数字信息仿真模拟建筑物（隧道）具有的真实信息，以三维虚拟现实技术建模，实现可视化的工程数据模型。

通过 BIM 技术，可以解决原二维平面（纸质设计图）不能解决的问题，从设计阶段开始为建设项目建立互相协调、内部一致的可运算三维信息模型，从而降低各个参与方进行项目管理的难度。BIM 不仅是一种模型工具，更是一个协同的工作流程，通过三维模型可视化演示，可以对拟建工程项目进行碰撞检测、施工进度模拟、工程质量的分析、安全风险的防控等。

例如，地铁管线综合施工中，因为要在非常有限的空间内安装完成通信、信号、综合监控以及通风、暖通、消防、电力等十几个专业系统，要合理布置这些系统与建筑、结构间的空间关系，碰撞冲突、返工修改等问题突出。另一方面，二维图纸的非可视化表达使施工前各专业的冲突问题无法得到解决。鉴于 BIM 技术的三维模型数字化、可视化、真实化的特点，采用 BIM 技术进行车站的施工碰撞研究，可以进行无间隙共享及无障碍交流，这使得项目的各个参与方能在整个项目周期有效地协同工作，并能有效地解决上述问题。

BIM 技术对于复杂的超高层结构（例如鸟巢钢结构和上海中心等）已基本上能满足其安全高效高质量的要求，但是对于高不确定性、高模糊性复杂地质环境中建设的隧道工程来说，BIM 技术远不能满足安全、高效建设的要求，还需要提升和发展。

（三）基于 BIM 技术提升隧道安全风险监控

为了降低高风险特殊区段地铁施工灾难性事故发生概率，需要实时感知安全风险，以便及时防范事故发生。进一步通过包含岩土地质、地下管线、周边建筑技术、机械人员、施工监测等在内的工程实体与施工工序对应的时间维的信息进行集成，形成地铁施工的 4D 模型。

下面以武汉地铁越江隧道联络通道施工风险控制为例，介绍其施工安全风险实时感知及预警系统。

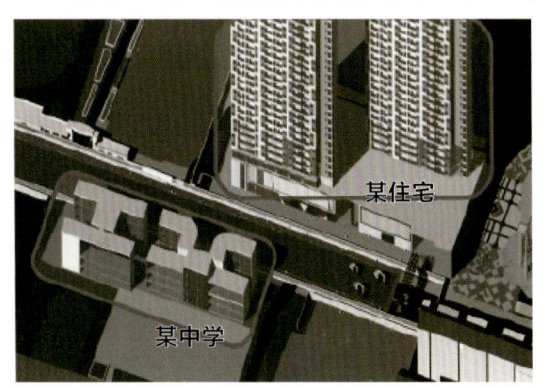

某地铁车站 BIM 模型

1. 安全风险实时感知系统

为了实现环境与结构的多物理量连续实时感知，提高风险特殊区段的地铁施工安全信息的采集和传输能力，首次将光纤光栅传感技术引入长江隧道联络通道冻结施工中，利用光纤光栅传感器具有耦合监测、高精度、自动连续、抗电磁干扰、不受水和潮气影响、可远距离传输等诸多优点，分别进行水平冻土、联络通道初期支护和既有隧道管片的温度-应变耦合监测，构建基于光纤光栅的联络通道施工多场耦合实时感知系统，实现整个施工周期内自动连续采集数据，实时分析与预警。

某地铁车站 4D 模型

对于前述的岩溶、断层和卵砾石复杂不良地质围岩环境，上述传感系统的感知系统显然不够，还需要超前地质预报系统。

实时感知系统由独立供电系统、数据存储分析系统、数据实时采集系统三大部分组成，并增设防尘防水保护系统。

从隧道左线向右线打入 5 根预埋传感器的冷冻管，共计 20 个传感器，最深处传感器位置距离管片外壁 10m，利用光纤光栅解析仪分析得到冻土温度变化规律。

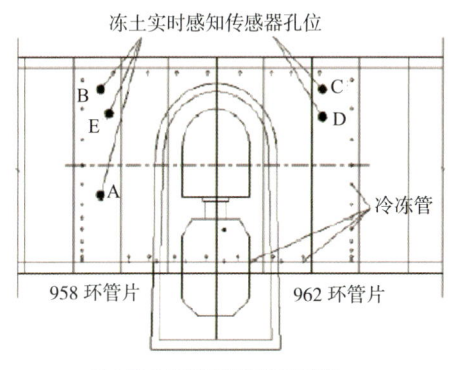

冻土安全实时感知布置平面图

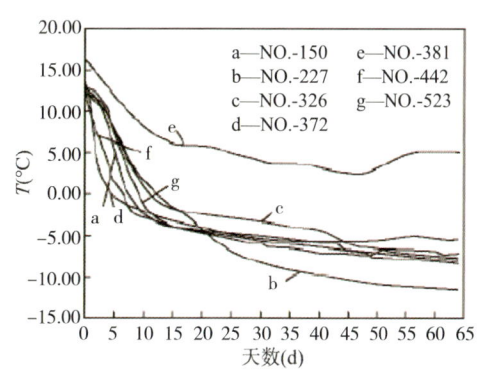

冻土温度变化曲线

2. 安全风险实时预警系统

根据海因里希事故连锁理论、轨迹交叉理论等事故致因理论，施工中物的不安全状态或人的不安全行为将导致安全事故的发生。因此，江底地铁隧道联络通道施工中，除了建立多场耦合实时感知系统以实时获取物的状态外，人的行为对于施工安全控制而言更为重要。在多场耦合实时感知系统的基础上，实现复杂环境下长大隧道中实时跟踪移动目标（射频识别系统），并将联络通道施工过程中环境、结构和人的安全信息综合起来进行安全分析判断，及时有效地发布预警信息，第一时间通知现场作业人员采取应急措施，实现安全知控一体化和实时化。

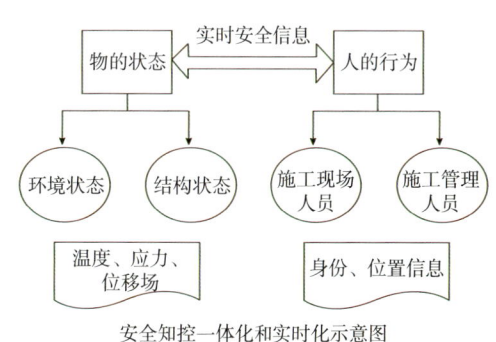

安全知控一体化和实时化示意图

地铁施工过程中产生大量的安全风险信息，有效的安全控制必须在充分、及时掌握（传输和分析处理）这些安全风险信息的条件下进行。因此，BIM、数据融合、物联网等现代前沿信息技术的应用是提升地铁工程施工安全控制水平重要的途径之一，它有利于提高地铁施工过程中安全信息的采集、传输、分析和挖掘能力，降低施工安全风险。

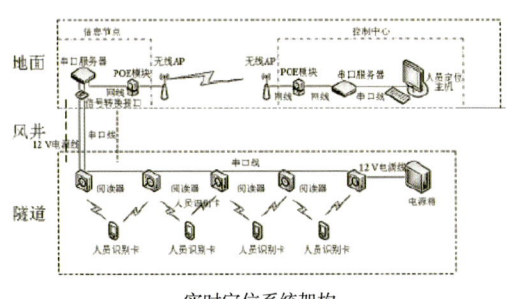
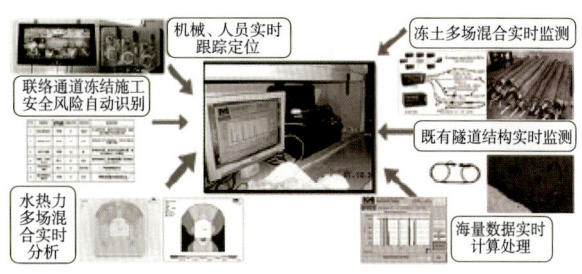

实时定位系统架构　　　　　　越江联络通道冻结施工的现场实时预警控制装置

三、数字隧道向智慧隧道（建设和运营维护）的发展

（一）智慧隧道

智慧隧道是为隧道工程建设和运营服务的隧道工程信息化的高级阶段。

什么是智慧和智能？

智慧是针对大系统和巨系统而言，例如城市是一个巨系统，包括人、自然和社会等的综合体，隧道也是一个大系统，包括隧道本体、隧道环境、隧道建设者和运营者的综合体。

而智能是对某项技术、某个功能和某种仪器设备而言，如智能手机、智能传感机等。

具体来说，智慧隧道的目的就是让作为隧道系统主体的隧道工程建设者和运营者更聪明。

首先，它通过互联网把无处不在的被植入隧道本体和周围地层中的智能化传感器和实时跟踪移动目标的 GPS 定位单元和无线射频识别单元以及地质超前预报系统连接起来形成物联网，实现对物理隧道（隧道本体和环境）和隧道建设者和设备的全面感知，通过互联网快速传输感知的信息。

其次，它利用云计算技术对感知信息进行数据融合和处理分析，实现网上"数字隧道"与物联网的融合。

最后，在运用大数据技术等智能分析处理后，发出对超前地质预报、设计方案和修改、施工方案的实施（如盾构推进的操作参数的确定等）、预警信息的发布、应急防治方案的实施等做出智能化响应和决策支持的指令，实现智慧服务。

概括来说，智慧隧道是对隧道建设（掘进和支护）和运营的机械、设备、围岩环境、人员和技术、材料、后勤保障等进行主动感知和认知、数据传输和融合、自动分析、智能化处理和智能化服务。

其本质是安全隧道、高效隧道、高质隧道、智能隧道。

隧道的数字化、信息化、自动化、智能化是智慧隧道的前提和基础。

（二）智慧隧道和数字隧道的区别和联系

数字隧道是物理隧道的虚拟对照体，两者是分离的。

智慧隧道通过物联网把数字隧道与物理隧道无缝连接在一起；是物联网与"数字隧道"的融合，智慧隧道是数字隧道功能的延伸、拓展和升华，是数字隧道的智能化。利用云计算对实时感知数据进行快速融合和协同处理，并在大数据技术所具备的关联分析能力、逻辑思维能力、自学习与自适应能力和行为决策能力的基础上提供智能化服务。

智慧隧道和数字隧道之间并无绝对界线。如上所述，其中间是可以过渡的，例如有 4D 数字隧道、N 维数字隧道……，相应有 4D 和 ND（B+SR）IM 技术。

（三）智慧隧道的"智"的体现

（1）透彻感知：无处不在的智能传感器，对隧道、环境、设备和人及其状态实现全面、综合的感知和对其运行状态的实时感测。

（2）全面互联：通过物联网将所有传感器全面连接，通过互联网实现感知数据的智能传输和存储。

（3）深度整合：物联网和互联网完全链接和融合，将多源异构数据整合为一致性的数据——隧道工程建设和运营全图。

（4）智能服务：在隧道智慧信息设施（网络、数据）基础上，利用云计算这种新的服务模式，构架一种新的能提供服务的系统结构，在大数据技术基础上，对海量感知数据进行并行处理、数据挖掘与知识发现，为隧道建设和运营提供各种不同层次、不同要求的高效率智能化服务。

（四）如何建设智慧隧道

关键是做好智慧隧道工程数据交换共享平台建设，这是一个智慧信息基础设施→智慧应用服务技术支持层→智慧应用服务层的推进过程。

需要有以下4方面的提升：

（1）三维可视化表达→统一时空基准的四维信息；
（2）"静态数据+周期性更新"→"实时获取+动态更新"；
（3）"有限服务"→"全面深度服务"；
（4）"事后分析+辅助决策"→"实时分析+智能数据挖掘+知识发现+实时决策"。

外国相类似的智慧工程实例有：

① 2006年，新加坡启动"智慧国家2015"计划，建立了针对交通堵塞预报的智慧城市系统。
② 2009年，韩国仁川实现了房间耗能的智能控制与通过网络监测病人状况的智慧城市应用。
③昆士兰建立了对桥的智能安全系统，通过装在桥上的传感器，确保桥的安全。
④爱尔兰"智慧港湾"利用桥梁上的传感器及渔民手机，监测水面上的漂浮物、水流，沟通渔民和餐厅的交易。
⑤瑞典斯德哥尔摩智慧城市系统，使汽车使用量降低25%，尾气排放降低8%~14%。

应用的前景：

①可避免2008年杭州地铁湘湖车站基坑野蛮开挖事故。
②可避免2003年上海4号线地铁冻结法事故。

四、开展基于大数据技术的TBM/盾构施工时的分析与控制研究

隧道建设时的工程事故和TBM/盾构的低效运行除了与客观不良地质条件有关，还与TBM/盾构的选型不当以及操控参数选择不合理有关，如何使选型与参数选择从经验上升为科学，大数据技术是一个可行的方向。

（一）大数据技术基础上的智能服务

大数据技术是指对数据规模大、结构复杂度高、关联度强的数据集进行处理与应用的处理技术。其涉及的技术内涵主要包括数据挖掘与关联分析技术、机器学习、模式识别、预测模型、时序分析以及可视化处理等。

（1）数据挖掘。数据挖掘是一个知识发现的过程，即从大量的数据中自动搜索隐藏在其中的知识或特殊关系信息的过程。

（2）关联分析。关联分析是用于从大量数据中分析各数据项之间有价值的相关关系。比如20世纪90年代美国沃尔玛超市基于销售数据关联分析将啤酒与尿布两个看上去没有关系的商品放在一起进行销售，获得了很好的收益。

（3）机器学习。机器学习主要研究如何使用电脑来模拟和实现人类学习时获取知识的过程，重构已

有知识，提升自身处理问题从而达到创新的能力。机器学习的最终目的是从数据中自动分析获取规律性知识，并用于对未知数据进行预测、判断和评估。机器学习与统计推断学联系密切。

（二）大数据技术对TBM、盾构施工的应用可行性分析

1. 大数据的概念和TBM/盾构施工数据具有天然的契合性

TBM/盾构施工数据是典型的大数据问题。TBM/盾构在施工掘进过程中连续自动采集数据并存储在数据库中。比如海瑞克盾构每2.5s会自动采集，每10s会自动存储一次数据。

对于一条长约25km的隧道来说，单就掘进参数而言，每台盾构机每分钟产生包括盾构扭矩、推力、转速、贯入度等掘进参数在内的6组掘进数据，推进2h，一条线路将产生约4.2×10^5组掘进数据，盾构/TBM掘进数据不仅包括自身的掘进参数（扭矩、推力、转速、贯入度等），还有地层变形数据包括应力、位移以及掘进地层地质数据，比如地层类型、土体物理力学参数、地下水位等，加上各监测点获取的数据，包括区间隧道上方、周边布置的监测点，以及大量管线、建筑物、道路布置的测点，数据形式多样，包括数值、文字、图片等资料。总的来说，盾构/TBM施工数据量巨大。盾构掘进参数之间、盾构掘进参数与地层变形之间相互关联，非常复杂，呈现出大量、多态、多源、多维的大数据特征。因此，盾构/TBM施工数据是典型的大数据应用范例。

2. 大数据技术在TBM/盾构施工中应用前景的分析

目前中国绝大多数正在建设地铁的城市中地铁工程都在进行盾构施工，山岭隧道和引水隧道的长大隧道建设中采用TBM施工的也越来越多，并都研发了施工信息管理系统，系统用以支持TBM/盾构的远程监控并通过多种传输方式将掘进数据发送至中心服务器并自动存储分析。因此TBM/盾构施工管理信息化技术的应用为大数据分析提供了数据基础。

通过大数据技术有效利用TBM/盾构施工中的大数据对优化TBM/盾构的设计，正确进行TBM/盾构的选型，提升TBM/盾构的掘进效率和防止施工时刀具刀盘磨损、卡机、突水突泥、工作面坍塌等重大事故的发生具有重要意义。

基于大数据技术开展TBM/盾构施工中的应用研究在国内刚刚起步，是一个富有探索性和挑战性的问题，尚未有成功的应用实例。例如选型中的扭矩确定、刀盘开口率的确定等：TBM/盾构的扭矩是其选型的重要参数，现在依据的是日本公式$T=\alpha D^2$，影响参数单一仅为盾构外径D，D大时，偏大，不能充分发挥切削能力，D小时，偏小，出现扭矩过载。理论研究因其复杂性很难，可以通过大数据技术对以往掘进施工数据进行统计挖掘，分析不同地质条件下不同类型盾构机掘进参数的变化规律，得到适用性强的经验公式，以优化国产盾构选型参数提供技术支持，再如如何选择与地层特性相适应的掘进参数，从而保证开挖面稳定、减小地层变形位移是盾构施工控制的难点。可以通过在不同地层内掘进时的掘进参数与地层变形参数进行统计挖掘，分析盾构掘进参数与地层变形之间复杂的规律性关系，为有效预测复杂条件下地层变形，防治地面隆起或坍塌事故提供依据。

建立数据联盟以及对TBM盾构施工海量数据的数据挖掘和关联分析技术的研究和提升是大数据技术应用的关键。

五、结语

随着物联网、物探技术、5G互联网的发展，结合多网融合技术、大数据智能分析技术、利用云计算或超级计算机实现对海量数据的处理和分析，完成对隧道系统中的人员、设备、掘进和支护的自动管理和控制，可以有效实现灾害预警预报、安全高效和绿色建设，最终实现无人隧道建设和运营。

若干重要建设工程中的哲学思考

研究的工程是狭义的工程，所谈的建设工程是指为建设一个新的客观实体而进行的一系列工程技术活动，是包括工程决策（工程方案的评价、比较与确定）、工程规划、工程勘察、工程设计、工程施工和工程运行等的集成。

工程建设的一系列工程技术活动相互联系、相互制约、相互作用，既受大系统制约，又反作用于大系统。这些相互联系、相互制约、相互作用又处于不断地运动、变化和发展中，表现为建设工程中进行每一工程技术活动的条件和建设工程总体所受制约的条件。

工程建设活动的核心环节即工程方案的评价、比较与决定，都依条件、空间和时间为转移。认识工程、研究工程都要从这些多种多样的联系中去进行，特别是研究其中的本质和必然的联系。

工程科学的任务就是揭示工程建设的规律性，也就是揭示决定工程建设发展趋势和基本过程的本质联系，它是如何在一定的条件下发生作用的？发生什么作用？又是如何反作用于工程所建的大系统？

唯物辩证法是帮助我们揭示各种事物复杂联系的显微镜和望远镜。工程哲学对于建设工程，特别是对于体现工程理念的工程决策的制定，具有重要的不可替代的意义和实践价值。它是杜绝各类"形象工程""政绩工程""豆腐渣"工程的重要保证，也是将粗放发展的工程理念转变为科学、正确的工程理念的重要杠杆。

一、从全局观点谈圆明园湖底的防渗工程

圆明园人工湖底用防渗膜修筑湖底防渗工程引发了岩土工程界的一场大争论，最终以决定拆除防渗膜、回填黏土和原湖底的底泥告终。这场争论的启示是：比较工程方案是从系统的全局还是局部来分析？

从局部观点看："利大于弊"。减少灌水次数，降低管理成本，对圆明园周围微观环境影响不大。

从全局观点看：超采地下水 $\xrightarrow{\text{只采不补}}$ 世界最大的华北地下水降落漏斗，引发地面沉降、地面开裂、海水入侵。

欧美先进地区建立"水银行" $\xrightarrow[\text{井灌入渗}]{\text{地表入渗}}$ 采补、平衡，维持地下水位。

产生的效益：

储备地下水，提供用水需求。

控制地下水位下降，防止海水入渗、地面沉降。

维持地下水位，减少地下水抽取费用。

通过土壤中的细菌作用和吸附作用，改善水质。

二级处理后，污水入渗再生。

维持河流基流，保护生态环境。

结论：

圆明园湖底防渗，不因其恶小而为之，只采不补，弊大于利。

圆明园湖底入渗，不因其善小而不为之，采补平衡，利大于弊。

报告时间：2011年7月24日。

二、建坝和反坝等工程理念:"一切都以地点、时间和条件为转移"

认识和处理问题,包括工程方案的制订,一切都依条件、时间和地点为转移。任何工程方案,既有利又有弊,利弊随工程实施的地点、时间和条件而变。因此,就要不断突破传统思维模式,开拓发展新的思路。

1. 建坝与反坝

在三峡地区建坝"利大于弊",在华北地区建坝应慎重处理。

时间、地点、条件的变化:经济发展、人口增加;工业、农业用水需求增加;城市化水平提高;森林、水面、绿地面积大为减少。

华北地区采补不平衡的原因:建坝拦水、蓄水、引水,缺乏流域上游(太行山、燕山)对流域下游地下水的补给;地下水缺乏平原地区的雨洪和雨水补给;城市化水平提高,城市不透水面积急剧增加。

导致的结果:不能维持华北地区河流基流,"有河则干";加剧华北平原大面积地下水降落漏斗。

因此,应改变传统思维模式、学习发展地下储水新思路(水银行):华北地区减少上游建水库、拦水、蓄水;大规模实施包括地表入渗、井灌入渗的雨洪补给和污水处理补给地下水的系统工程(ASR、AR 工程)。

2. 遇水架桥与桥隧并举

条件变化:架桥技术产生遇水架桥;地下开挖技术发展促进"桥隧并举"。

时间、空间变化:航运交通;气候影响;战时生存。

结论:桥隧并举。

3. 逢山开路与暗挖隧道

条件变化:炸药的发明产生逢山开路;TBM 的发展促进长大隧道的建设,可持续发展的战略。

时间、空间的变化:生态环境;线路长度、线形与交通性能;边坡稳定。

结论:建设长大隧道,少开挖、高边坡。

4. 高架路与地下交通

条件变化:交通需求发展,电子商务(自动化、信息化),环境意识与要求提高。

时间、地点的发展:城市环境;气候影响;经济价值。

结论:发展城市轨道交通、地下物流、地下快速路。

工程技术和一切事物一样,一切都在变化、运动和发展中。创新是永远的,只有创新才有生命力,应在所有的工程领域,贯彻创新思想,随着条件和需求(也是条件)的变化,不断开辟新技术途径。

三、工程方案的选择必须分析工程对象特性的主要矛盾和矛盾的主要方面

所有的工程领域,没有适用任何时间、地点、条件的亘古不变的工程技术途径和方案。任何工程方案都有其优越性和局限性,只有在适用条件和适用范围内,才可发挥其优越性,若超出其适用范围,就无法发挥其优越性,只能呈现其劣势的一面。

1. 长江水下隧道

修建方案的变更:

西气东输地质钻 —→ 盾构在深层开挖;

南京长江下隧道沉管法修建 —→ 盾构深层开挖方案;

武汉长江下隧道沉管法修建 —→ 盾构深层开挖方案。

变更的原因：三峡工程的建成以及长江上游干、支流上修建一系列水库以及有效的水土保持工程后，长江的泥沙特性和泥沙运动规律已经和将来都会发生重大变化，从而造成长江中下游的河势演变发生重大变化，不再遵循以往历史演变的规律，决定河床演变特性的是河床的冲淤这一主要矛盾，即：中下游冲淤平衡 ——→ 中下游冲刷为主，沉管浅埋 ——→ 盾构深挖。

2. 大城市机动车数量预测

预测失败：北京市机动车数量达到 200 万辆的来临时间比预测提前七年多。

失败的原因：预测的模型大多是依据历史情况分析，估计增长率，以此推算出未来年的机动车总量和交通需求量。这种模型只能适用于量变阶段，不能适用于关节点，即曲线的拐点，以及以后部分质变的新阶段。

机动车发展的质变：国际上预计当某地区人均生产总值大于 3000 美元，小汽车进入家庭。

四、从当代防护工程面临的挑战看工程技术发展与创新的动力

高技术进攻武器的发展向防护工程提出了严峻的挑战，面对挑战，有两种观点和态度：

一是"防护工程无用论"，无能为力，无所作为。

二是积极应对，取得防护技术的新发展、新进步。

从唯物辩证法来看，矛盾是一切事物包括工程技术发展的源泉和动力，矛盾着的对立面又统一、又斗争，推动着一切事物的运动、变化和发展。打击武器和防护工程构成一对矛盾，其双方的对立统一构成军事斗争的一个侧面，正是武器和防护工程的相互斗争、相互竞赛推动着它们的进步和发展。

进攻武器和防护工程对立矛盾的特点：

（1）武器的发展引起和规定了防护手段的发展。

（2）打击武器的发展超前于防护工程的发展。

（3）打击武器的威力越强大、越复杂、越多样化，防护手段研制越复杂、越艰巨。

（4）武器与防护工程的发展进步的主要形式表现为"扬弃"。

面对现代高技术武器的挑战，我们应该做的不是放弃、削弱，而是发展、完善我们的防护工程体系，发展中既包括对过去成功经验的继承，也包括对过时做法的抛弃，即哲学上所谓的"扬弃"。

中国岩石工程进展和规划

一、中国岩石工程研究进展

中国人民把简单的岩石力学知识用于工程建设可以追到很久。公元前613—591年，我国人民在安徽淠河上修建了历史上第一座拦河坝；公元前256—251年，在四川岷江修建了都江堰水利工程；公元前219年在广西开凿了沟通长江和珠江水系的灵渠，筑有砌石分水堰；公元前221—206年在北部山区修建了万里长城。到20世纪初，我国杰出的工程师詹天佑主持建成了北京—张家口铁路上一座长约1km的八达岭隧道。

中华人民共和国成立以后，随着水利、水电、采矿、交通、建筑、冶金以及国防等建设工程的发展，提出了大量亟待解决的岩石力学问题，从而使这门学科逐步发展壮大起来。从1958年开始，为适应长江三峡水利枢纽建设的需要，国家成立了三峡岩基研究组，集中了来自全国水利、水电、建工、矿冶及相关专业的18个单位100余名科技人员，下设大坝地基、地下结构、岩质边坡、动力特性、灌浆加固等5个专业组，在室内外开展了大量试验研究工作，如岩体流变试验、隧洞压水试验、地应力测试、振动爆破试验等。根据科研工作的需要，还成功研制出一批仪器设备，如岩石三轴仪大型电磁振动台、岩石扭转流变仪、液压钢枕等。三峡岩基组的工作，不仅对水工建设岩石力学的发展起重要作用，也为中国岩石力学和岩石工程的发展奠定了基础。

20世纪60—70年代，中国科技人员在若干重点水电工程（如葛洲坝、刘家峡）、矿山工程（如金川镍矿）、铁路工程（如成昆线）及国防工程建设中，作出了许多贡献。其中，针对葛洲坝水利枢纽进行的岩石力学研究最有代表性。

葛洲坝工程是中国于20世纪70年代在长江干流上自行设计、自行施工的第1座巨型水利枢纽，有"万里长江第1坝"之称。主要的工程地质问题是坝基中广泛分布有原生或构造软弱夹层，总计约80余层。为全面探索夹层的矿物、物理、力学、化学性质，除进行一系列宏观、微观分析外，着重进行了室内和野外抗剪、蠕变、松弛、振动爆破及抗力体试验。根据大量试验研究结果，运用流变力学原理，解决了软弱夹层及非连续、层状岩体结构的力学问题，并提出了有关的数学、力学模型及计算方法，为大坝设计和施工提供了重要依据。

第6个国民经济发展五年计划（1981—1985年）开始，中国全面系统地进行岩石力学与岩石工程前沿课题科研攻关。

"六五"期间，涉及岩石工程方面的国家科技攻关项目为：

（1）地下工程快速施工。

（2）复杂地基的勘测与处理技术研究。

"七五"期间，专门就三峡工程单独列项进行攻关的课题有：

（1）坝区及外围地壳稳定性研究。

（2）坝区高边坡和建基岩体工程问题研究。

（3）水库诱发地震研究。

（4）库岸稳定性研究。

（5）环境地质评价与预测。

报告时间：2007年5月25日。

（6）三峡工程高陡边坡开挖、加固技术研究。

"八五"期间岩石工程国家重点科技攻关涉及的课题有：

（1）高边坡稳定及处理技术研究：

①岩质高边坡稳定分析方法及软件系统；

②岩质高边坡开挖及加固措施研究；

③岩质高边坡勘测及监测技术方法研究。

（2）高拱坝坝体、库水和坝基相互作用动、静力分析研究。

（3）勘测关键技术及综合应用研究。

"九五"期间岩石工程国家科技攻关涉及的课题有：

（1）快速勘探技术的综合应用。

（2）300m级高拱坝坝基稳定研究。

（3）超大型地下洞室群合理布置及围岩稳定研究。

（4）坝址及近岸滑坡体稳定性研究。

①溪洛渡近坝滑坡体的稳定性分析和治理检测方案；

②小湾坝前堆积体稳定与加固方案研究。

近期开展的重点项目：

（1）灾害环境下重大工程安全性的基础研究。

（2）深部开采工程岩石力学关键问题。

通过科技攻关，有力地提高了我国岩石工程总体水平。

二、中国岩石工程地理地质环境特点

（一）地理特征

中国位于亚洲东部，太平洋西岸。陆地面积约960万km²，其中岛屿面积7万多平方公里；海域面积472.7km²，海洋平均深度961m，最大深度5559m。

"三山六水一分田"，山地面积占33%，高原面积占26%，盆地面积占19%，平原面积占12%，丘陵面积占10%。

西部主要是高山、高原和盆地；东部主要为丘陵和平原。

西部有四大高原：青藏高原、内蒙古高原、黄土高原、云贵高原。西边最高峰有世界最高大雄伟的喜马拉雅山脉，其珠穆朗玛峰海拔8848m，为世界最高峰。

西部四大盆地：塔里木盆地、准噶尔盆地、四川盆地。

东部三大平原：东北平原、华北平原、长江中下游平原。

东部丘陵较多：辽东丘陵、山东丘陵、东南丘陵等。

地势西高东低呈阶梯状分布，自西向东可分为三大地势阶梯：

第一阶梯：西部青藏高原有"世界屋脊"之称，平均海拔在4000m以上。

第二阶梯：青藏高原边缘以东和以北，是一系列宽广的高原和巨大的盆地，海拔1000~2000m。

第三阶梯：东部的丘陵和平原分布区，大部分海拔500m以下。

第三阶梯继续向海洋延伸，形成了近海的大陆架。

（二）大地构造位置

全球划分为六个刚性的岩石圈板块在软流层之上漂移，不断地运动（包括转动），我国处于欧亚板块。板块的变形主要集中在块体的边界，我国是欧亚大陆与印度板块和太平洋板块边界。

大陆可以分割成不同尺度的次级块体，变形主要集中在块体的边界，我国大陆上的一系列地震带就是典型的例子。

在印度板块和太平洋板块共同作用下，我国的构造地应力场、构造带的规模和性质变化复杂。其中欧亚板块与印度板块的陆－陆碰撞边界起控制作用。

（三）地形变和应力场分布

西部：新生代欧亚大陆和印度次大陆碰撞引发大规模地壳挤压缩短，岩石圈增厚和高原抬升。

东部：出现地幔隆起，大规模陆内伸展，岩石圈减薄，地壳沉降，盆地充填。

GPS观测到的中国大陆现今构造背景场的水平分量，自西向东成运动弧线散开状，西部大东部小。

（1）板块边界的作用力是中国大陆新生代和现今构造变形的主要动力源。

（2）变形呈弥散型（千公里以上），集中在不同次级块体的边界上，具有相对低的运动速率（小于10mm/a）。

（3）次级块体内部变形微弱或整体运动。

（4）边界带常常是地质、地球化学和地球物理的结构面，绝大多数大陆地震发生在活动带上。

西部地壳抬升，而东部地壳下降沿板块边界和板内活断层分布。

西部：新生代两大陆碰撞引发大规模地壳挤压缩短，岩石圈增厚和高原抬升。

东部：出现地幔隆起，大规模陆内伸展，岩石圈减薄，地壳沉降，盆地充填。新生代西部岩石圈汇聚与东部岩石圈离散之间是否存在耦合关系？

来自板块边界的作用力是中国大陆新生代和现今构造变形的主要动力源。

断裂的强活动性是高烈度地震发生和分布背景，构成高地壳应力环境。在西南部地区，区域应力量级可达10MPa；河谷集中应力量级可达30~55MPa；边坡应力量级可达22MPa。

构造活动性强烈而复杂的西南地区是我国水电工程的集中区域，岩石力学遇到了前所未有的挑战。

（1）独特的地质构造单元。

（2）青藏高原、黄土高原、岩溶地区。

（3）板块内部地震活动性强。

（4）地质演化与地质环境的多样化。

（5）各个地质时代不同类型地层俱全，构成了种类万千的岩土介质。

（6）宽阔的陆架、漫长的海岸带、典型的边缘海。

（7）流经多地貌单元的河流，从高原、高山到丘陵、平原、海洋各类地貌单元一应俱全。

（8）跨越多种气候带和独特的季风气候造就了复杂多样的地球表生态环境。

（9）高度的人为活动与悠久的历史记录。

三、中国岩石工程各领域成就和规划

（一）水利工程

我国水资源时空分布不均，长距离、跨流域洞水工程是21世纪我国水利建设一项重要而关键的举措。目前，以西线、中线和东线南水北调工程为主的跨地区，跨流域的远距离调水工程将是今后20~30年内水利建设的重点。它是迄今为止，世界上最大的水利工程。

首先启动的中线工程，主要解决京津及华北地区城市缺水问题，通过该工程将汉江的丹江口水库水引入北方。近期（2010年前后）总调水量400亿 m^3。这一工程涉及岩土工程的关键技术包括：

（1）丹江口大坝加高14.6m（相应增加库容116亿 m^3），还涉及左岸1223m土石坝地基处理和坝体加固。

（2）中线干渠全长1427.2km（其中输水总干渠以陶岔引水闸为调水渠首，经长江与淮河流域分水岭——方城垭口，沿唐白河平原北部和黄淮海平原西部边缘布置，大部分分渠段与京广线路平行），跨越88条河流，33次穿越铁路，133次与高等级公路交叉，其中穿越一部分膨胀土地区，如何防渗是极为重要课题。

（3）穿越黄河工程全长3450m，它位于郑州以西约30km处，包括南岸连接渠道，隧洞进口建筑物，穿黄盾构隧道，隧洞出口建筑物，北岸河滩明渠和北岸连接渠道。

（4）另外还有两条隧洞：一是穿新峰山隧洞，洞线由晏窑起经新贺庄、山张、三峰山、十里铺至颖河，全长4km，要穿过新老采空区，地下成洞条件差，施工困难大；二是潮河段切岗隧洞，南端起新郑市梨园，北端终于郑州毕河，全长18.794km，采用双洞，洞径12m，壁厚0.55m。经技术经济比较，决定采用混合式盾构法施工。

西线调水一期工程位于青藏高原东部边缘地带，工程处在海拔3500m左右，主要引大渡河支流的阿柯河、麻尔曲、杜柯河和雅砻江的支流泥曲、达曲之水补给黄河支流贾曲。由"五坝七洞一渠"组成。五坝即引水河流五座引水枢纽，坝高分别为115m、108m、104m、123m、63m。引水线路全长260km，其中隧洞长244km。线路通过的支流使隧洞自然分为七段，故为七洞，最长的洞段为73km。该工程是人类工程史上最长的隧洞输水工程；"一渠"是隧洞出口处一段长16km的明渠。

西线工程涉及的岩土工程问题更为复杂，如高寒高纬度、高海拔、深覆盖、复杂地质条件和高地震烈度地区进行深达千米的勘察钻孔；穿越世界屋脊——青藏高原的长隧洞；长年冻土中大直径的引水隧洞等，其设计和施工将是对岩土工程的巨大挑战。

东线工程从长江下流调水，向黄淮海平原东部和山东半岛补充水源，与中线和西线工程一起，共同解决北方地区水资源紧缺问题。

东线一期工程是利用江苏省江水北调工程，扩大规模，向北延伸，供水范围是苏北、皖东北、鲁西南、鲁北和山东半岛，并向天津市应急用水创造条件。调水线路从江苏扬州附近长江干流引水，有三江营和高港2个引水口门。线路总长1466km，其中长江至东平湖1045km，黄河以北173.5km，胶东输水干线239.65km穿黄河段7.85km。一期工程江苏段已于2002年底开工，预期2007年建成通水。

在黄河河底打一条穿黄隧道是东路一期工程的重点和难点。

各省市调水工程也有很多项目，如规划中云南滇中调水工程，从虎跳峡水库引水，利用洱海作反调节库，向大理、楚雄到昆明滇池调水，引水线路长408km，其中隧洞249km。新疆依犁河调水工程，包括长达40km引水隧洞。为解决关中水资源短缺的陕西调水工程，包括西线的红济石，中线的引汉济渭和东线的引乾济石三条跨流域调水线路，需兴建跨秦岭引水隧洞。甘肃省引洮工程包括长185.63km总干渠和16条长639.94km分干渠和1285km支渠。

（二）在水电方面

截至2004年9月，我国水电装机容量已突破1亿kW大关，这是电力建设史上一个重要里程碑，我国水电建设从几十万kW到5000万kW用了46年，而从5000万kW到1亿kW，仅用了9年，可见我国水电建设突飞猛进。按照规划，2020年水电建设要达到2.5亿kW，也就是说，未来15年水电将要新增装机容量1.5亿kW，平均每年增加1000万kW，这是一个巨大的任务。

全国水电资源分配情况见下图。

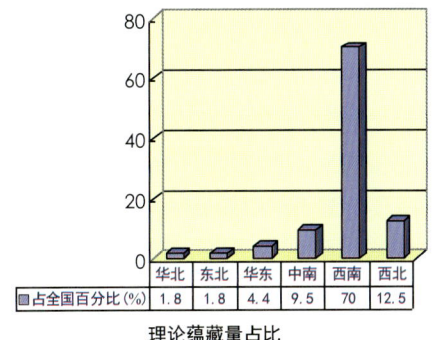

理论蕴藏量占比

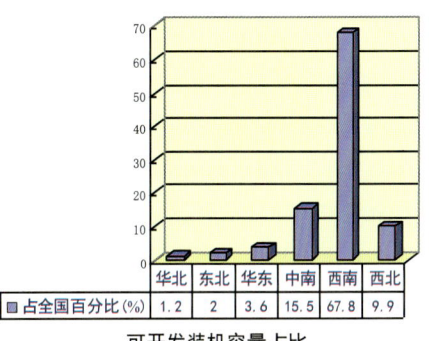

可开发装机容量占比

从上图可以看出，无论是理论蕴藏量还是可开发装机容量，西南地区都占 70%，今后整个水电建设重点肯定是西南地区。

下面介绍几个典型水电工程。

（1）三峡工程

这是举世瞩目的世界最大的水电工程，总装机容量为 1820 万 kW。三峡工程分三期完成，总工期为 17 年。

第一期（1993—1997 年）以实现大江截流为标志，是世界最大的截流工程，施工期为 5 年。

第二期（1998—2003 年）以蓄水、发电和永久船闸通航为标志，施工期为 6 年。

第三期（2004—2009 年），施工期为 6 年。

三峡工程的主体建筑：大坝、电站厂房、永久船闸、升船机。

①大坝：为混凝土重力坝，轴线全长 2309.5m，坝顶高 185m。其中 1600m 左岸大坝已经完工，还有 700m 右岸大坝，从 2004 年开始，每年长高 40m，到 2006 年和左岸坝平齐达到 185m，同时三峡水库蓄水将由 139m 抬高到 156m。

②电厂厂房：左岸电厂，目前已安装 11 台 70 万 kW 发电机组，2005 年还需安装 3 台 70 万 kW 发电机组，装机容量将达到 980 万 kW。右岸电厂，2007—2008 年要安装 12 台单机 70 万 kW 发电机组。到 2009 年三峡电厂总装机容量将达 1820 万 kW。

③永久船闸：为双线五级船闸，闸室段长 1600m，连同上、下游引航道船闸线路总长 6442m，闸室净高 56m，两闸室之间留有 60m 宽中隔离墩，人工开挖土石方量为 4166 万 m³。目前是世界上级数最多、总水头和级间输水水头最高的内河船闸。

④升船机：到 2009 年世界上最大的升船机工程将投入运行。

与一、二期工程相比，从 2004 年开始的三期工程更为艰巨复杂。具体工程要完成右岸大坝和电站厂房施工、地下厂房建设及设备安装、右岸 12 台机组安装，完成世界最大的垂直升船机的设计、制造和安装，完成临时船闸改冲沙闸和电源电站的建设，确保 2006 年汛后水库蓄水至高程 156m，大坝全线达到坝顶高程 185m。三期工程中主要技术难题是：世界最大的升船机设计制造国内外没有先例；水库蓄水后在环境与生态上出现的新问题需要解决；枢纽部分投入运行后，如何优化梯级枢纽调度，最大限度发挥防洪、发电、航运的综合效益等。

永久船闸通航　　　　　　　　　　　　永久船闸通航

（2）"三江"（金沙江、雅砻江、大渡河）水电资源开发

金沙江、雅砻江、大渡河三大江河水电资源可开发量分别为 7274 万 kW、2856 万 kW 和 2340 万 kW，资源量大且相对集中，其总量分别占西南地区和全国水电资源可开发量的 53.6% 和 32.9%，是全国乃至世界少有的水力资源"富矿"，是我国 12 大水电基地中的三大主要水电基地。

金沙江干流：分上、中、下三段，共规划 20 个梯级，总装机容量 7274 万 kW，其中 100 万 kW 以上的电站有 17 座，100m 以上高坝有 14 座，溪洛渡电站拱坝高 278m。

上段：8 个梯级电站，总装机容量 1416 万 kW。

中段：8个梯级电站，总装机容量2058万kW。

下段：4个梯级（乌东德、白鹤滩、溪洛渡、白家坝）电站，总装机容量3800万kW。

全国十二大水电基地

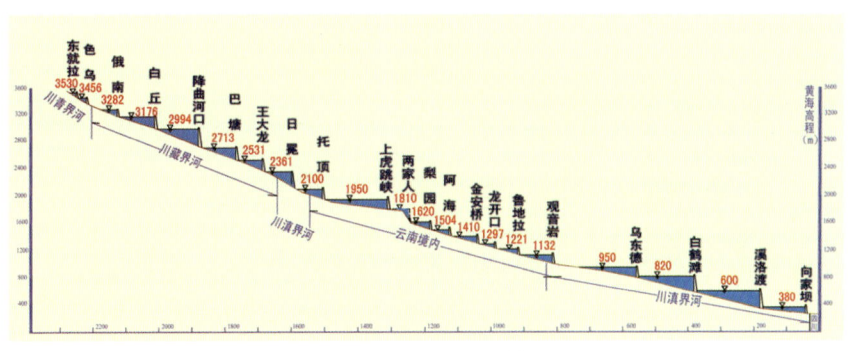

金沙江干流梯级电站纵剖面

雅砻江干流：共规划21级开发方案，总装机容量为2825万kW。其中100万kW以上的电站有10座，最大装机容量为锦屏二级电站为440万kW；100m以上高坝8座。锦屏一级电站坝高305m，已建成的二滩电站拱坝高240m。

雅砻江上游两河口以上规划10个梯级，装机容量325万kW；中游两河口至卡拉为6级开发，装机容量为1176万kW；下游卡拉至江口为5级开发，装机容量为1355万kW。

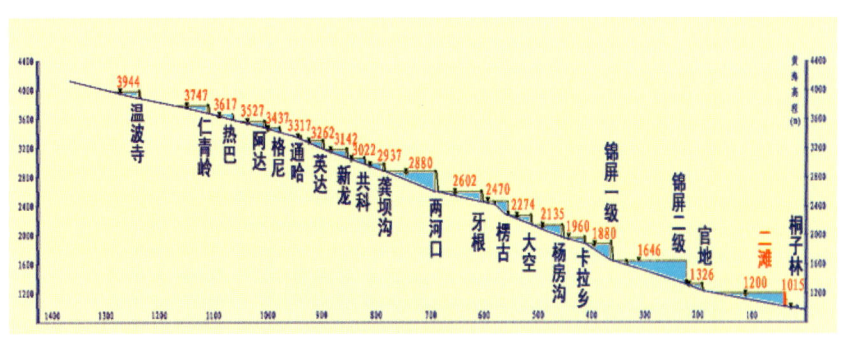

雅砻江干流梯级电站纵剖面图

大渡河干流：规划22个梯级，总装机容量为2340万kW，其中100万kW以上电站有7座，最大装机容量的瀑布沟电站为360万kW；100m以上高坝有12座，双江口电站坝高达300m。

双江口以上规划四个梯级，总装机容量为176万kW，双江口至铜街子河段采用"两库18级"开发，利用落差2087m，装机容量为2164万kW。

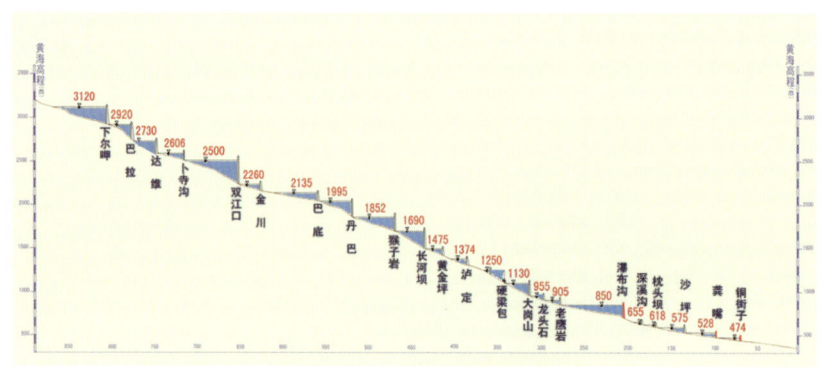

大渡河干流梯级电站纵剖面

到目前为止，金沙江开发利用率为零；雅砻江建成了二滩水电站装机容量为330万kW，开发利用率为11.6%；大渡河建成龚嘴（装机容量70万kW）和铜街子（装机容量60万kW）两级，开发利用率为5.6%。可见开发潜力巨大。

二滩水电站

三江地区位于青藏高原向四川盆地过渡地带，地势北高南低，地质构造十分复杂，活动断裂发育，地震活动频繁，垮山、滑坡、泥石流等地质现象时有发生。使得坝基岩体变形稳定、抗滑稳定和渗流稳定问题十分突出。另外，地应力偏高，对整个工程稳定也有不利影响。

由于受地形条件限制，引水发电建筑物往往布置在地下岩体中，如二滩、锦屏、瀑布沟、溪洛渡、太平峰等地下厂房空间尺度大，其中二滩地下厂房开挖跨度达30.7m，锦屏二级电站4条引水隧道各长18km，将雅砻江大河湾截弯取直穿越锦屏山，最大埋深为2600m。这些工程均存在不同程度岩爆、漏水、高温、有害气体等技术问题，施工难度极大。

另外，水电开发与环境生态保护的关系，已引起社会广泛注意，如何确保水电开发和环境保护相协调，是一项亟待解决的大问题。

雅砻江大河湾截弯取直

（3）大型抽水蓄能电站的建设

为了调节"峰"和"谷"，我国已建、在建及近期拟建大的大型抽水蓄能电站见下表。

我国已建、在建及近期拟建的大型抽水蓄能电站

电站名称	地点	河流	装机容量（万kW）	年发电量（亿kW·h）	水头（m）	建成年份（年）	电网	备注
潘家口	河北迁西	廉河	27	5.64	86	1992	京津唐	另有1台常规机组
十三陵	北京吕平	东沙河	80	12	477	1997	京津唐	
广州	广东从化	流溪河	240	23.8	536	2000	广东	
天荒坪	浙江安吉	大溪	180	31.6	667	2000	华东	
泰安	山东泰安	沱河	100	14	256		山东	在建中
桐柏	浙江天台	信史河	120	21.2	270		浙江	在建中
白山	吉林桦甸	第二松花江	30				华北	另有5台常规机组
琅琊山	安徽滁州	小沙河	60	8.56	145		安徽	在建中
宜兴	江苏宜兴		100	14.9	335		苏南	在建中
西龙池	山西五台	漳沱河	120	20.1	530		山西	在建中
宝泉	河南辉县	峪河	120	20.1	530		河南	前期已开工
张河湾	河北井陉	甘河	100	16.75	310		冀南	立项并开工
惠州	广东博罗		240		532		广东	已立项
白莲河一期	湖北罗田		60		195		湘北	已建成
盘龙	重庆綦江				500		重庆	申请中

广州抽水蓄能电站是我国第一座高水头、大容量的抽水蓄能电站，它位于广州从化市吕田镇境内，电站一期和二期总共安装8台300MW的蓄能机组，总装机容量达2400MW，是目前世界上最大的抽水蓄能电站。

（三）矿业工程

中国煤矿资源比较丰富，目前在能源结构中占主导地位，煤炭产量已达16.5亿t（占一次性能源构成的74%），与之相配套巷道掘进量每年为7000~8000km，据预测2020年煤炭产量为29亿t。与这配套的每年巷道掘进量大的在15000km以上。

中国金属矿采矿发展也很快。目前建成的县级以上国有金属矿山已达900余座。金属矿石年产量达到4亿t以上。中国已相继建成矿石生产能力300万t以上的金属露天矿18座，其中年生产能力在1000万t以上的金属露天矿有德兴铜矿、齐大山铁矿、水厂铁矿等。

随着浅部资源的日益减少，目前采矿工程最突出的问题是"深部开采"。中国煤炭在已探明储量中，埋深在1000m以下的为2.95万亿t，占煤炭总储量的53%。其中埋深在1000~1500m的煤炭储量为1.2万亿t，这一部分储量是中国21世纪后备能源保障。目前开采深度以每年8~12m速度增加，近年已有一批矿山进入深部开采，如沈阳采屯矿开采深度为1197m、开滦赵各庄矿开采深度为1159m、徐州张小楼矿开采深为1100m、北票冠山矿开采深度为1059m、新汶孙村矿开采深度为1055m、北京门头沟开采深度为1008m、长广矿开采深度为1000m。

兖州矿务局机械化开采（煤矿采场围岩）　　　　金川矿区1400中段采用铲运机出矿（现代化开采技术）

在金属矿开采方面，红透山铜矿目前开采深度已进入900~1100m，冬瓜山铜矿现已建成2条超1000m竖井来进行深部开采，弓长岭铁矿设计开拓水平750m，距地表达1000m，夹皮沟金矿二道沟坑口矿体延深至1050m，湘西金矿开拓38个中段，垂深超过850m。此外，还有寿王坟铜矿、凡口铅体矿、金川镍矿、乳山金矿等许多矿山都将进行深部开采。

预计在未来20年，我国很多煤矿开采深度将进入1000~1500m。同时，在今后10~20年内我国金属矿山开采深度也将进入1000~2000m。

深部开采与浅部开采的明显区别在于深部岩石所处的特殊环境，即"三高一扰动"的复杂力学环境。"三高"主要是指高地应力、高地温、高岩溶水压。

（1）高地应力

进入深部开采以后，仅重力引起的垂直原岩应力通常就超过工程岩体的抗压强度，而由于工程开挖所引起的应力集中水平则更是远大于工程岩体的强度。同时，据已有的地应力资料显示，深部岩体还存有构造应力场或残余构造应力场，造成在深部岩体中形成了异常高的地应力场。据南非地应力测定，高程在3500~5000m之间，地应力水平为95~135MPa。在如此高的应力状态下进行工程开挖，面临严峻挑战。

（2）高地温

根据量测，越往地下深处，地温越高。地温梯度一般在30~50℃/km不等。有些地区如断层附近或导热率高的异常局部地区，地温梯度有时高达200℃/km。地温可以使岩体热胀冷缩破碎，而且岩体内温度变化1℃可生产0.4~0.5MPa的地应力变化。岩体温度升高产生的地应力变化对工程岩体的力学特性会产生显著的影响。

（3）高岩溶水压

进入深部以后，随着地应力和地温的升高，同时将会伴随着岩溶水压的升高，在采深大于1000m的深部，其岩溶水压将高达7MPa，甚至更高。岩溶水压的升高，使得矿井突水灾害更为严重。

"一扰动"主要是指强烈的开采扰动。

面对"三高一扰动"的复杂环境，使深部开采遇到了岩爆，高温、突水、瓦斯爆炸、大变形等地质灾害，其成灾机理及控制技术是采矿工程领域面临的最大挑战性。

（四）铁路工程

1. 铁路建设

中国铁路目前营业里程已超过7万km，居世界第3位。根据规划，到2020年铁路营运里程将达到10万km，将修建2200座2270多公里长铁路隧道，最近2~3年开工的隧道长度将超过1000km。我们跨世纪的铁路工程除已建成的4纵4横线路外，还有若干重要的在建项目。如地质构造极为复杂的西藏地区，除青藏铁路外，滇藏线、川藏线等具有世界级难度的铁路工程也将在21世纪逐步修建起来。

在青藏铁路建设中"高寒缺氧，多年冻土，生态脆弱"三大世界性难题已取得突破性进展。全长

1100km的青藏铁路横跨可可西里和唐古拉山无人区，大部分地区氧气含量仅为海平面的50%，极端气温可达-40℃。青藏铁路还穿越550km常年冻土地区，冻土地区纬度低、海拔高、日照强烈，地质构造运动频繁，冻土复杂性和独特性举世无双。建设者在施工中采用热棒、片石通风路基、铺设保温板等多项措施，提高冻土路基稳定性；另外对高度极不稳定冻土区采取"以桥代路"。在冻土区打隧道时，技术人员创造性地研制了两台大型隧道空调机组，使隧道施工温度控制在-5~5℃之间，有效地防止了地下冰融化滑塌，并相继攻克了多项冻土施工难题，使隧道顺利贯通。另外青藏铁路用于环保投资大约为21亿元，占整个项目投资的8%。

2. 铁路隧道

1908年，由杰出的工程师詹天佑主持，在北京至张家口的铁路上用18个月的时间修建了长1091m的八达岭隧道，这是中国自力修建的第一座越岭隧道，在中国隧道修建史上写下了重要的一页。20世纪60年代以前，中国隧道修建采用矿山法施工，利用导坑先行，实行分部开挖以尽量减少对周围岩体的扰动。20世纪70—80年代，新奥法的应用，使隧道建设技术得到飞速发展，最具代表性的是长14.95km的大瑶山隧道，这也是我国修建长度10km以上的铁路隧道实践的开始。在这座隧道的施工中，采用凿岩台车、衬砌模板台车和高效能的装运工具等机具配套作业，实行全断面开挖。大瑶山隧道是我国山岭隧道采用重型机具综合机械化施工的开端，将隧道工程的修建技术和修建长大隧道的能力提高到一个新的阶段。20世纪90年代，秦岭铁路隧道修建时采用全断面隧道掘进机（TBM），使隧道修建技术再次迈上新台阶。目前正在修建的乌鞘岭隧道长20.1km，建成后将是中国最长的铁路隧道；在正在建设的青藏铁路线上，修建了世界上海拔最高的两座铁路隧道——风火山隧道和昆仑山隧道，隧道海拔接近5000m。

从上述发展过程可以看出，中国铁路隧道工程发展的趋势是越修越长，技术标准不断提高。从修建技术上看，中国铁路隧道工程在机械化程度、修建速度、长大隧道的修建能力等方面与国际水平相差无几，有些方面如穿过困难地质条件的能力、设计规范的水准等还处于国际领先水平。差距主要表现在隧道施工机械化设计、制造、配套能力较低。

各时期铁路隧道修建数量

时期	隧道数量（座）	隧道长度（km）
清朝（1988—1911年）	237	42.199
中华民国（1912—1949年）	427	113.881
中华人民共和国（1950—1959年）	1005	306.387
中华人民共和国（1960—1969年）	1113	660.302
中华人民共和国（1970—1979年）	1954	1034.668
中华人民共和国（1980—1989年）	319	198.602
中华人民共和国（1990—1999年）	1822	1310.721
中华人民共和国（2000—2003年）	530	539.845

各时期修建的长度大于3km的隧道数量和长度

时期	1900—1959年	1960—1969年	1970—1979年	1980—1989年	1990—1999年	2000—2003年
隧道数量（座）	5	20	36	13	80	40
隧道长度（km）	17.789	75.582	155.892	66.046	360.866	234.442

正在建设中的特长隧道

序号	隧道名称	线别	长度（m）	备注
1	乌鞘岭隧道	兰（州）武（威）二线	20050	两座单线
2	野三关隧道	宜（昌）万（州）线	13846	单线
3	堡镇隧道	宜（昌）万（州）线	11588	双线
4	齐岳山隧道	宜（昌）万（州）线	10463	单线

（五）公路建设

1. 公路建设情况

到 2003 年底，中国公路通车总里程为 181 万 km，其中高速公路近 3 万 km。同江至三亚、北京至珠海、连云港至霍尔果斯、上海至成都四条国道主干线基本贯通，从而实现了"五纵七横"国道主干线系统第一阶段建设目标。

到 2010 年，全国公路总里程达到 210~230km，全面建成"五纵七横"国道主干线，基本建成西部 8 条省际通道，高速公路连接 90% 人口在 20 万以上的城市，东部地区基本形成高速公路网，高速公路总里程达到 5 万 km。通公路的行政班车通达率达到 95% 以上。

到 2020 年，全国公路总里程达到 260~300km，其中高速公路里程达到 7 万 km。

（1）"五纵"主干线（总长约 15100km）

①同江至三亚国道主干线（约 5200km）。

②北京至珠海国道主干线（约 2400km）。

③重庆至湛江国道主干线（约 1400km）。

④北京至福州国道主干线（约 2500km）。

⑤二连浩特至河口国道主干线（3600km）。

（2）"七横"主干线（总长约 20400km）

①上海至成都国道主干线（约 2500km）。

②连云港至新疆霍尔果斯国道主干线（约 4400km）。

③黑龙江绥芬河至满洲里国道主干线（约 1300km）。

④丹东至拉萨国道主干线（约 4600km）。

⑤青岛至银川国道主干线（约 1600km）。

⑥上海至云南瑞丽国道主干线（4000km）。

⑦衡阳至昆明国道主干线（约 2000km）。

2. 公路隧道建设

我国是一个多山的国家，75% 左右的国土是山地或丘陵，为了根除道路病害，保护自然环境，在山区高等级公路建设中必须重视隧道方案，并努力提高公路隧道工程科学技术水平。

20 世纪 80 年代以后，隧道建设速度加快。近几年还建成了多座特长或宽体扁坦隧道，如中梁山隧道（3100m×2）、缙云山隧道（4116m×2）、大溪岭隧道（4116m×2）、二郎山隧道（4200m×2）、飞鸾岭隧道、潭峪沟隧道、中山门隧道、大宝山隧道、铁山坪隧道和真武山隧道等。目前，公路隧道的单洞长度越来越长，修建技术日趋复杂。如福建美菰岭隧道（5300m×2）、湖南雪峰山隧道（约 7000m×2）、四川泥巴山隧道（约 8000m×2）、陕西秦岭终南山隧道（约 18400m×2），以及沈大高速公路八车道路扁平大断面隧道等，都将遇到大量岩土力学方面的技术课题。

到 2002 年底，我国已建成公路隧道 1972 座，总里程 835km。从 1997 年起至今，公路隧道以每年 100km 的速度增长。

长 3000m 以上的公路隧道

序号	隧道名称	长度（m）	省份	车道 × 洞数	通风方式
1	秦岭终南山隧道	18300	陕西	2×2	送排式纵向通风
2	雪峰山隧道	6890	湖南	2×2	送排式纵向通风
3	秦岭Ⅰ号隧道	6144	陕西	2×2	上行送排式纵向通风 下行射流纵向通风
4	秦岭Ⅱ号隧道	6125	陕西	2×2	上行送排式纵向通风 下行射流纵向通风
5	美菰岭隧道	5560	福建	2×2	斜井单吸式通风
6	雁门关隧道	5200	山西	2×2	送排式纵向通风
7	秦岭Ⅲ号隧道	4930	陕西	2×2	射流纵向通风
8	华莹山隧道	4706	四川	2×2	射流纵向通风
9	鹧鸪山隧道	4423	四川	2×1	平导半横向通风
10	通渝隧道	4279	重庆	2×1	平导射流纵向通风
11	二郎山隧道	4160	四川	2×1	平导半横向通风
12	大溪—湖雾岭隧道	4116	浙江	2×2	送排式纵向通风
13	凉风垭隧道	4085	贵州	2×2	射流纵向通风
14	北碚隧道	4020	重庆	2×2	射流纵向通风
15	牛郎河隧道	3922	山西	2×2	射流纵向通风
16	云台山隧道	3800	江苏	2×1	射流纵向通风
17	猫狸岭隧道	3600	浙江	2×2	射流纵向通风
18	罗盘基隧道	3407.5	福建	2×2	射流纵向通风
19	潭峪沟隧道	3400	北京	3×1	射流纵向通风
20	赤岭隧道	3372.5	福建	2×2	射流纵向通风
21	大风垭口隧道	3290	云南	2×2	射流纵向通风
22	九顿坡隧道	3200	云南	2×2	射流纵向通风
23	飞鸾岭隧道	3167.5	福建	2×2	射流纵向通风
24	中梁山隧道	3165	重庆	2×2	上行竖井单吸式通风 下行射流纵向通风
25	清潭隧道	3020	广东	2×2	射流纵向通风
26	金鸡山隧道	3000	福建	2×2	射流纵向通风

近 10 年来，围绕隧道技术问题，取得一批高水平的研究成果。如依托于青海大坂山隧道的高海拔隧道防冻防水技术。青海省大坂山越岭段，隧道南口路面中心高程 3792.75m，是目前世界上已建成的海拔最高的公路隧道，年平均气温 –3.1℃，极端最低气温 –34℃，冬令期长达 218d，是典型的高海拔寒区隧道。设计中对防冻胀做了认真的考虑，全隧道设仰拱。设立了以防寒泄水洞为主通道的排水系统，设计衬砌保温层，并研究了稀土材料和硬质泡沫塑料在当地低温环境中的物理、力学性能、隔温效果以及泡沫塑料的施工工艺，保温层在大坂山隧道实践中取得成功。另外如依托于川藏公路二郎山隧道的高地应力治理技术，依托于重庆真武山隧道、深圳大梅沙隧道的大跨度扁平隧道设计施工技术，依托于福建京福高速公路隧道群的连体隧道、近间距隧道设计施工技术，以及依托于浙江大溪—湖雾岭隧道、猫狸岭隧道、景山隧道的竖井送排式纵向组合通风技术、逆光照明技术、总线监控技术等，这些研究成果强有力地支撑了公路隧道建设。

深圳大梅沙隧道（三车道断面，1250m×2）　　重庆北碚隧道（4320m×2）

江苏老山隧道环保型棚洞结构（$L=400$m）　　广东洋碰隧道（2650m×2）　　福州—泉州高速公路雷打石隧道进口

今后科研工作总体研究思路是，解决山区公路隧道修筑关键技术，重点是复合柔性衬砌设计技术、围岩稳定非确定性动态分析及合理的支护结构和施工方案、防排水技术及其新材料和新工艺、纵向通风、电吸尘装置的开发、隧道交通与环境控制技术等。

（六）石油天然气工程

在与石油天然气工程有关的岩石力学研究中，所涉及的地层深度大多在2000~8000m范围内，研究对象以沉积岩层为主体，岩石处于较高的围压（可达200MPa）、较高的温度（可达200℃）和较高的孔隙压力（可达200MPa）作用下。这与水电站的坝基设计、高边坡稳定、隧道和巷道的开挖及支护、建筑的桩基工程、地下洞室城市地铁建造等不超过1000m深度的地表或浅层问题不同，也不同于以火成岩和变质岩的研究主体，深度超过万米的下地壳、上地幔岩石物理力学问题。

20世纪60年代以来，随着我国大庆、胜利、辽河等油田的开发，岩石力学在石油工程中日益显示了其重要性。主要研究范围包括：

（1）深层地应力测量。
（2）深部地层环境下的岩石力学性质。
（3）岩石应力、渗透性的声学响应特性及岩石物理力学性质的地球物理解释。
（4）构造应力场的数值模拟及其在油气勘探与开发中的应用。
（5）深层岩石中天然裂缝的形态、分布和预测理论。
（6）固液耦合理论及在油藏工程中的应用。
（7）岩石破坏机理与井眼稳定问题。
（8）人工裂缝的起裂、扩展及水力压裂工程设计。
（9）弱固结地层的固相产出问题。
（10）地层错动、蠕变与套管损坏问题。

随着石油天然气勘探开发范围的不断扩大，深层岩石力学在边际油田开发、深水钻井、大位移井、多底井、鱼骨刺井等高难度油气资源开发工程中扮演着越来越重要的角色。石油工程中已建、在建及计划建设的、涉及岩石力学的重要工程很多，现简要介绍如下：

1. 西气东输工程

西气东输工程是中国管道史上管线最长、投资最大、影响深远的一项工程。管道全长4300km，管径1118mm。将塔里木盆地及陕甘宁气田的天然气经甘肃、宁夏、陕西、山西、河南、安徽、江苏东送到上海，初期年供气120亿m³。管线穿越涉及吕梁山、太行山一定数量的隧洞外，其关键工程是延水关黄河隧道工程。该工程位于陕晋大峡谷中部，全长518m，在黄河水下20多米处通过，地质条件复杂，技术难度较大。

2. 大位移井的普及

大位移钻井技术指在一个平台上开发周边距离较远的多个油田，从而减少平台建造数量和工程费用投资，是海上油田开发的重要方法之一，也是当今国际钻井技术的前沿。大位移井的井身结构设计、井壁稳定等关键技术均与岩石力学有着密不可分的关系，我国目前在海洋与陆地均有成熟的应用。

如大港油田建成我国陆上油田水平位移最大的深层大位移定向井港深69×1井（垂深4318.14m，水平位移3118.04m）。唐海油田建成北堡西3×1井，这是一口四开四段制双目标大位移定向井，设计斜深4189m，垂深2450m，位移3056.9m，最大井斜角为65.09°，井斜方位角288°，垂深与位移之比为1:1.25。该井从开钻到完钻，仅用了59天23.5小时，创造了同类井速度最快、周期最短、机械钻速最高的纪录。

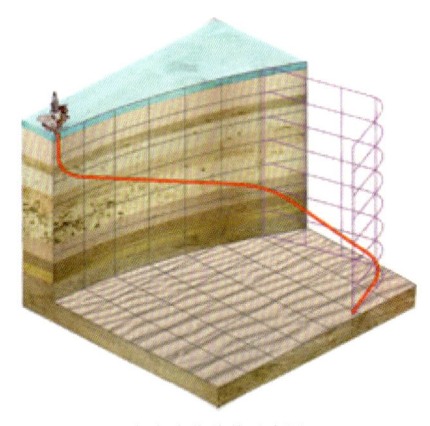

复杂大位移井示意图

3. 长庆油田水力压裂技术

长庆各类油藏普遍具有低渗、低压、低产的地质特点。水力压裂作为关键技术为油田经济、高效开发提供了有力技术支撑。低孔、低渗、低压、非均质性强、油水关系复杂是制约长庆低渗油田改造的难点。经多年研究及矿场试验，形成了从压裂地质研究—室内试验—压裂液支撑剂优化—优化设计及实施—压裂实时监测控制—压后评估完备的技术模式。技术水平上也由单项工艺发展到整体压裂技术和开发压裂技术。形成了以"安塞、靖安长6特低渗油层压裂改造技术""底水油藏储层改造技术""浅油层压裂改造技术"等为代表的主体技术，为近年长庆油田实现油气产量大发展起到了重要的促进作用，也为西气东输工程提供了有力的资源支撑。

4. 苏里格气田

苏里格气田位于内蒙古自治区鄂尔多斯市内的苏里格庙地区，勘探面积20000km²，已探明地质储量6025亿m³，最终可探明储量达7000亿m³，成为中国目前第一特大型气田，跃入世界级知名大气田行列。2000年8月26日，苏里格气田发现井苏6井喷出120.16万m³的高产工业气流。

该地区地表主要为沙漠覆盖，含气层为上古生界二叠系下石盒子组的盒8段及山西组的山1段，气藏主要受控于近南北向分布的大型河流、三角洲砂体带，是典型的岩性圈闭气藏，气层由多个单砂体横向复合叠置而成，基本属于低孔、低渗、低产、低丰度的大型气藏。

该气田从1999年开始进入大范围勘探，气田的开发目前也已进入实施阶段，包头、呼和浩特的天然气进城工程正在铺设管道。

岩石力学与石油天然气工程的结合，促进中国石油天然气勘探开发的发展。从20世纪60年代大庆油田的勘探到后来的胜利油田、辽河油田、渤海油田，直到近年来获得重要突破性进展的长庆油田和苏里格庙世界级整装大气田的发现，水平井开发技术、欠平衡钻井技术的运用，塔里木油田钻前井壁稳定技术和双阶梯水平井技术的运用，都与岩石力学的理论研究密不可分。

随着石油勘探开发技术的发展，石油天然气工程岩石力学面临新的挑战：

（1）深部地层岩石力学参数的预测、监测和检测精度的提高。

（2）高温高压条件下岩石变形与破坏的细观力学机理研究。

（3）多重孔隙地层中与应力相关的多相渗流油藏模拟技术研究。

（4）基于地震速度谱的高精度地层坍塌压力、破裂压力预测方法研究。

（5）非平面多水力裂缝的干扰、弯曲、转向的岩石力学判别准则及其在井水力压裂中的应用。
（6）地层漏失压力的力学机理与钻井安全泥浆密度的设计理论。
（7）深部地层岩石后效变形的有限变形理论与本构方程的建立。

石油天然气工程岩石力学的深入研究，不但会丰富岩石力学理论，而且更会极大地促进石油勘探开发新技术的发展。

（七）城市建设中地下工程

中国城市建设中地下空间利用在较长时期内发展速度缓慢，直到20世纪90年代才有了根本改变。特别是地下轨道交通建设从90年代以后如雨后春笋。

纵观中国城市轨道交通建设史，从1965年北京地铁一期工程开工至今，现已有北京、上海、天津、广州、长春、武汉、大连等地的城市地下轨道交通投入了运营，线路总长318km。除北京地铁1号线和2号线近40km外，其余都是20世纪90年代后修建的。目前在建的线路有10个城市17条线路共360km，共投资约1100亿元。另外，还有10个城市正在积极开展城市地下轨道交通建设的前期准备工作。据统计，到2010年前全国开通运营的城市地下轨道交通线路将达到1200km以上，今后每年平均建成的线路为180km。据规划，到2025年北京市轨道交通线路在600km以上；到2020年上海将建成540km轨道交通网络；南京已规划好的6条地铁总长度达300km；深圳未来25年规划建设地铁、轻轨365km；广州的未来地铁将超出本市，与佛山、南海等地区相连。

在经济发达地区，如珠江三角洲地区、长江三角洲地区、京津塘地区，正在酝酿建设城市间轨道交通前期工作。珠江三角洲步伐更快一些，该地区城际快速轨道交通分3个层次：一是，高速铁路作为轨道交通主骨架，由京广、京九、广深铁路，东部沿海铁路和西南铁路构成；二是，城际快速铁路以深港和广州两大都会为核心，沿深广和广珠两发展轴，形成连接珠江三角核心城市快速轨道交通系统；三是，城际区域线，即四通八达的客运轨道线，作为快速轨道交通的补充。目标是：中心城区居民大多数可以在500m范围内到达地铁站，特区内居民平行出行时间控制在30min内。

按规划，珠江三角城际轨道交通线总长将超过1000km。城市地铁及车站的科研、设计、施工、技术，由于城市土质大，地下水形态和运动规律复杂，地面建筑和地下管线密集，存在技术难度很大的岩土工程问题。经过近十几年的努力，中国在设计和施工等方面，如明挖法、盖挖法、沉埋法、盾构法都已达到国际先进水平，大跨度暗挖法和平顶直墙暗挖法还属于国际领先水平。

（八）越江跨海隧道工程

和建桥相比，越江跨海隧道可以不受气候变化的影响，做到全天候运营，具有稳定的运行能力；具有较强的抗地震能力；可一洞多用，同时安排城市给排水、电力、热力、电缆电视、光缆等通过；还可做到不拆迁或少拆迁；结构维护费低；设计承载能力大。因此，桥隧并举是更合理的选择。

在上海穿越黄浦江已建8条水底隧道，另外杭州、宁波、南京、武汉、广州、厦门也已建、在建或筹建越江跨海隧道。其工程情况见下表。

国内长大隧道一览表

工程名称	建造时间	工程概况	
		长度	隧道结构构造
打浦路隧道	1965—1971年	全长2.736km，其中盾构法隧道1.322km	黄浦江下的城市道路隧道。 单管隧道：双向两车道，设计车速40km/h。单层衬砌结构，装配式衬砌$\phi_{外}$10.0m，$\phi_{内}$8.8m，衬砌厚0.60m
延安东路北线隧道	1984—1989年	全长2.261km，其中盾构法隧道1.476km	黄浦江下的城市道路隧道。 单管隧道：双向两车道（3.75m×2），设计车速40km/h。单层衬砌结构，装配式衬砌$\phi_{外}$11.0m，$\phi_{内}$9.9m，衬砌厚0.55m

续上表

工程名称	建造时间	工程概况	
		长度	隧道结构构造
延安东路南线隧道	1994—1996年	盾构法隧道1.31km	黄浦江下的城市道路隧道。单管隧道：同向两车道（3.75m×2），设计车速50km/h。单层衬砌结构，装配式衬砌$\phi_{外}$11.0m，$\phi_{内}$9.9m，衬砌厚0.55m
上海外环隧道	2000—2003年	沉管法隧道，全长2.882km，江中沉管段长736m	黄浦江下的城市道路隧道。双向八车道公路隧道，江中沉管段全长736m，分为7节。管段横断面为三孔两管廊形式。边孔为三车道，中孔为两车道。管段宽43m，高9.55m，单节管段最长108m。管段设柔性接头，采用灌砂基础
大连路隧道	2001—2003年	全长2.567km，其中盾构法隧道1.27km	黄浦江下的城市道路隧道。双管隧道：每管同向两车道（3.75m×2），设计车速40km/h。单层衬砌结构，装配式衬砌$\phi_{外}$11.0m，$\phi_{内}$10.04m，衬砌厚0.48m
复兴东路隧道	2001—2004年	全长2.785km，其中盾构法隧道1.21km	黄浦江下的城市道路隧道。双管隧道：每管双层同向三车道，上层两根小车道，下层一根大车道加紧急停车带，设计车速40km/h。单层衬砌结构，装配式衬砌$\phi_{外}$11.0m，$\phi_{内}$10.04m，衬砌厚0.48m
翔殷路隧道	2003—2005年	全长2.3km，其中盾构法隧道1.5km	黄浦江下的城市道路隧道。双管隧道：每管同向两车道（3.75m+3.5m），设计车速80km/h。单层衬砌结构，装配式衬砌$\phi_{外}$11.36m，$\phi_{内}$10.4m，衬砌厚0.48m
上中路隧道	2003—2006年	全长2.8km，其中盾构法隧道1.21km	黄浦江下的城市道路隧道。双管隧道：每管双层同向四车道，车道宽度均为（3.75m+3.5m），设计车速80km/h。单层衬砌结构，装配式衬砌$\phi_{外}$14.5m，$\phi_{内}$13.3m，衬砌厚0.6m
崇明南港隧道	2005—2009年	全长8.95km，其中盾构法隧道7.47km	长江下的城市道路隧道。双管隧道：每管同向三车道（3.75m×2），设计车速80km/h。单层衬砌结构，装配式衬砌$\phi_{外}$15.0m，$\phi_{内}$13.7m，衬砌厚0.65m
宁波甬江隧道	1987—1995年	沉管法隧道，全长3.822km，江中沉管段长420m	甬江下的城市道路隧道。单管双车道结构，江中沉管段全长420m，分为5节。管段内壁宽10.20m，高6.5m，通行净高4.5m。设计车速60km/h。管段设柔性接头，采用碎石基础
宁波常洪隧道	1999—2001年	沉管法隧道，全长2.065km，江中沉管段长395m	甬江下的城市道路隧道。穿越甬江连接宁镇公路与江南公路，是我国首条桩基础沉管隧道。江中沉管段长395m，共分4节管段。双向四车道，通行净高5.0m。计算行车速度60m/h，管段外包宽22.8m，高8.45m
杭州西湖隧道	2002—2003年	水中明挖法施工。隧道分为西线、东线。西线隧道全长1258m，东线隧道全长1228m	西湖下的城市道路隧道。工程北起环城西路教场路交叉口，南出口布置在南山路，南进口布置在解放路延安路口，整个线路呈倒Y字形。工程设计车速为40km/h，为城市次干路，每条车道宽3.0m，通行净高3.2m；主体结构采用双孔双向四车道矩形箱涵结构形式
南京玄武湖隧道	2001—2003年	水中明挖法，全长2.66km	玄武湖下的城市道路隧道。工程连接新模范马路（规划中的纬三路）和新庄立交二期工程，穿芦席、中央路、明城墙，在玄武湖底向东绕过梁洲和翠洲。主线为双向六车道折板拱箱涵形式。工程设计行车速度主线为60km/h，匝道为40km/h。车道宽3.75m+3.5m+3.75m，通行限界净高≥4.5m
南京长江四桥越江隧道		全长约6.65km，盾构段长2.67km和1.67km	长江下的城市道路隧道。双向六车道高速公路隧道（3×3.75m）；设计车速80km/h；衬砌外径拟定为15.3m
厦门市东通道	在建	全长约9km	跨海的城市道路隧道。隧道横断面按双向六车道设计；行车道布置均为：0.5+3×3.75+0.5=12.25m；隧道建筑限界净高5m。计算行车速度主线80km/h，匝道40~50km/h。车辆荷载等级：汽车-超20级，挂车-120
武汉长江隧道	初步设计阶段	全长3.63km，其中盾构段长2.338km	长江下的城市道路隧道。双管隧道：每管同向两车道（3.75m+3.5m），设计车速50km/h。单层衬砌结构，装配式衬砌$\phi_{外}$11.20m，$\phi_{内}$10.2m，衬砌厚0.50m

续上表

工程名称	建造时间	工程概况	
		长度	隧道结构构造
广州市进生物岛隧道	初步设计结束，施工招标阶段	全长约1.1km，沉管段长约320m	珠江下的城市道路隧道。两孔一管廊，每孔双向两车道（3.75m+3.5m），设计车速50km/h，管段外包宽度23m，外包高度8.55m
广州市生物岛—大学城联络隧道	初步设计阶段	全长约1.6km，沉管段长约240m	

上海崇明越江通道工程采用"南隧北桥"方案，是目前世界上最大的桥隧工程。工程建成后，驱车从浦东五号沟至崇明陈家镇，不到30min。

受长江天堑阻隔，上海至今没有一条连接崇明岛进而沟通长江以北地区的陆路通道。目前，崇明岛和上海市区的交通联系仅靠轮渡，通行能力低。根据上海城市总体规划，崇明的开发将作为继浦东开发之后上海改革开放的又一战略重点；崇明越江通道的建设，将从根本上改变上海与崇明岛乃至长江江北地区交通联系不便的状况，有利于拓展上海市的城市空间，促进长江三角洲地区经济均衡发展，同时也可以满足长江南北两岸往来交通需求，完善干线公路网布局的需要。

崇明越江隧道部分将采用盾构法施工，工程所需的盾构将是世界上最大的盾构，国内外目前尚无成熟的工程实践经验。有关专家透露，经过深入研究和相应的科学试验，一系列国际先进技术将运用于这条隧道。据介绍，根据隧道通风要求，南港隧道在长江近长兴岛侧浅滩处将设置一座造型独特的通风井，届时施工人员将在江中修建一个直径为80m的人工岛，形成一道独特的景观。

中国第一条海底隧道——厦门东通道工程已于年内动工，预计2010年建成。

东通道总投资32.8亿元，项目全长9km，其中跨海主体工程长约6km，隧道最深在海平面下约70m。该隧道位于厦门岛东侧跨海连接大陆，建成后将与厦门北面的厦门大桥、西面的海沧大桥一起构成三条连接大陆的路网格局，形成厦门岛经济圈。东通道建成后，可从厦门大桥、海沧大桥吸引30%的进出岛车流量，同时形成金门—大嶝—东通道—厦门岛—海沧的黄金旅游线。据规划，通道按双向六车道设计，行车时速80km，两岸分别以互通立交与一级快速道路连接。

在最近的二三十年，中国正考虑建造5条跨海隧道，分别是：大连到烟台的渤海湾跨海隧道；由上海到宁波的杭州海湾工程；连接香港、澳门与广州、深圳和珠海的伶仃洋跨海工程（港珠澳大桥工程）；连接广东和海南两省的琼州海峡的跨海工程；连接福建和台湾的台湾海峡跨海工程。这些宏伟艰巨的工程中可以是桥隧结合，其中海底隧道将是主要的部分，有许多岩土工程的课题有待解决。

（九）中国高放核废料深埋处置中岩石力学研究进展

中国已和其他有核国家一样，经历几十年的核军工，已面临如何处置核废料的问题。除军工产生的核废料外，核电站也将产生核废料。我国已有9个核电机组在运营，另外2个机组也将建成。根据中国能源发展需求制定的中国核电发展规划，到2020年，中国核电的装机容量将达3600万kW，即相当于今后16年总共将建设28个核电机组。据此规划估计，到2010年我国将积累管道燃料达1000t，到2015年达2000t。2020年以前，每年将增加1000t乏废料。因此，寻求安全、有效、永久性处置核废料的课题也是我国较为紧迫的课题之一。

我国自1986年开始了放射性废料深埋处置库研究工作，主要研究工作有：

（1）经过全国筛选、对比，已初步选出甘肃北山地区为我国高放废物处置的最有利候选地区，对该区的评价工作正在进行。

（2）已在北京郊区筛选出2个地下实验室场址：昌平阳坊和怀柔油峪。

（3）经全国筛选，已选出内蒙古高庙子膨润土矿床作为我国高放废物处置库缓冲/回填材料的首选矿床。今后与工程屏障有关的研究、测试工作将围绕该矿床的膨润土进行。

(4) 已初步获得一批核素在花岗岩和膨润土上的吸附、扩散数据，进行了核素迁移的初步方法学研究。

(5) 我国的膨润土各种性能测试表明适宜作为高放废物处置库的缓冲/回填材料，在其中适当加入添加剂可以阻滞核素的迁移。

(6) 在锕系元素水溶液化学方面获得了一批试验数据。

(7) 地球化学方面，获得了铀矿床中pu239、I129和CI36的含量数据，进一步深入研究了超铀锕系元素在天然地质体和地下水中的迁移规律。

从1999年开始，在甘肃北山地区的三个重点地段开始了场址特性初步评价研究，并在其中两个地段首次施工了4口深钻孔，针对现场地质条件，中国开展了以下三方面的岩石力学研究工作：① 3.5km² 范围内的岩体裂隙的地质调查和三维岩体节理数值模拟；② 深部地应力测试研究和深部岩体渗透性研究；③ 深部岩石力学特性研究。

由于高放废物中含有强放射性、高毒性和长寿命放射性核素这一特殊性，因此，必须充分、彻底、可靠地永久隔离高放废物，使之不能进入生物圈和危害人类的生存，这也就要求"埋葬"高放废物的地下工程——地质处置库至少应有1万年的寿命。这一系列要求是目前任何工程所没有的。因此，安全处置高放废物在科学技术和工程上面临一系列重大挑战，中国正在制定规划建立"高放废物地质处置重大科研项目"迎接挑战，争取在十几年内有所突破。

（十）丝绸之路古遗址保护

1. 古遗址的情况

在古丝绸之路上，保存有大小石窟约600多处，目前有1处被列入世界遗产，即著名的敦煌莫高窟；被列入国家级保护单位的石窟18处。丝绸之路的石窟主要分布在甘肃省和新疆维吾尔自治区，其次是陕西省和宁夏回族自治区，青海省石窟较少。丝绸之路石窟，最著名的除上述甘肃敦煌的莫高窟外，还有甘肃天水市的麦积山石窟和永靖县的炳灵寺石窟；新疆维吾尔自治区拜城县的克孜尔石窟、库车县的库木吐拉石窟、吐鲁番市的柏孜克里石窟等；宁夏回族自治区固原区的须弥山石窟；陕西省的彬县大佛寺石窟等。

由于丝绸之路的大部分地域干旱少雨，因此保存下大量古代土建筑遗址。初步调查，丝绸之路上保存下的土建筑遗址约600处，被公布为国家级保护单位的土遗址约70多处。

土遗址中有相当一部分是指古代建筑被毁后遗存的土建筑部分，以我国新石器时代的两处人类居住遗址最著名：一处是西安东南郊的半坡遗址，另一处是甘肃省秦安县的大地湾遗址。

土遗址除上述的古建筑土遗址外，还有下列几种类型：

（1）古城

如新疆吐鲁番的交河古城、高昌古城和河北易县的战国古城遗址等。

（2）长城、关隘、烽燧及土塔等

这一类型的土遗址以敦煌西北的汉长城、阳关、玉门关及其附近的烽燧、土塔最具代表性。

（3）陵墓

宁夏回族自治区东部贺兰山下的西夏王陵最具有代表性。西夏王陵土遗址包括巨大的粉质沙土夯筑的坟土、陵院的墙体、土塔和部分木构建遗迹等。

（4）出土的坑、穴、窑、窖

这一类的土遗址有两个典型代表：一是陕西省临潼区的秦始皇兵马俑土坑，坑道是在生土层中挖建的；另一个土坑遗址是河南省三门峡的虢国墓地马坑。

2. 古遗址主要病害

（1）风化。

（2）风蚀。

（3）雨蚀。

（4）构造裂隙和卸荷裂隙发育。

（5）岩体坍塌。

（6）垦荒、修路人为破坏。

3. 古遗址加固保护

砂砾岩石窟和土遗址的加固保护，所采用的工程措施基本相同，如对危岩和危土体以锚固使其稳定；对岩体和土体裂隙进行灌浆处理；岩体表面和土体表面进行化学加固以防止风蚀、雨蚀破坏。但因砂砾岩体和土体有不同的物理力学及工程特性，加固使用的材料和工艺方法有所不同。

（1）裂隙灌浆。

裂隙灌浆是石窟加固和土遗址加固中一项很重要的工程措施。在石窟加固工程中，裂隙被灌浆充填密实后，它不仅可以防止大气降水沿裂隙入渗洞窟内破坏壁画，而且浆液结石体与锚索相结合，增加了岩土的完整性和结构的整体性，同时防止锚索锈蚀，以保证锚索的锚固强度。

过去没有找到适合于松散而强度很低的砂砾岩及土体裂隙的灌浆材料，经过多年的试验研究，研制出一种 PS-F 无机灌浆材料（PS：高模数硅酸钾；F：粉煤灰）。在榆林窟加固工程中，将 PS-F 用于岩体裂隙灌浆，取得很好的效果。

同样，我们在砂砾岩石窟岩体裂隙灌浆研究的基础上，经过不断试验，也为土遗址土体裂隙灌浆找到了一种适合的灌浆材料，即 PS-（F+C）（C：与遗址土体性质相近的黏土），将这种浆材用于玉门关的裂隙灌浆，也获得了很好的加固效果。

（2）表面和土体表面防风蚀、雨蚀加固

经过多年研究，研制出一种 PS 无机加固材料，这种加固剂与黏土作用后可以改变黏土的结构，使原来松散、晶态的黏土变成一种非晶态的胶凝体，大大提高土体的力学强度和耐水强度。风洞试验证明，经 PS 处理的遗址土体，其抗风蚀强度可提高十多倍，耐雨蚀强度也有非常明显的提高。在榆林窟加固工程中，用 PS 喷洒加固风化崖面、斜坡 9000m^2，取得很好的加固效果；在玉门关加固工程中，加固墙面 2500m^2；在河仓城加固工程中，加固墙面 3300m^2，都有很好的防风蚀、雨蚀效果。

四、结语

中国是岩石工程第一大国，不仅工程量极大，而且地质条件复杂，技术难度也是空前的。近年来，在勘测设计、施工、监测等方面，中国已积累了丰富的经验，也有许多教训，这些都是岩石工程领域的宝贵财富，我们愿意和世界各国岩石力学与岩石工程科技工作者共享。同时我们也欢迎世界各国岩石力学与岩石工程科技工作者参与中国岩石工程的建设，让我们携起手来，共同推进岩石力学与岩石工程发展，开辟岩石工程新纪元。

思想与观点

冀以尘雾之微补益山海，
萤烛末光增辉日月

Qian Qihu
Yuanshi
Wenji

钱七虎 院士 文集

建设科技强国迫切需要科学家精神

我国科技创新"三步走"战略要求,到2020年进入创新型国家行列,到2030年跻身创新型国家前列,到2050年建成世界科技强国。当前形势下弘扬科学家精神,最重要的是,和建设世界科技强国的伟大战略紧密结合起来。

一、追求真理,献身科学是科学家精神的精髓和核心

科学研究应该是科学家的事业,而非获得名利的手段。科学家精神是历代科学家在长期实践中形成和发展出来的。追求真理、献身科学的精神在古今中外科学家身上表现得充分淋漓。

医药学家李时珍冒着生命危险尝百草,为后世留下了宝贵遗产《本草纲目》;地理学家徐霞客,备尝艰辛,访遍山路陡峭、人迹罕至的名山大川;微生物学奠基人、化学家巴斯德为了验证狂犬病疫苗是否能安全地在人体上使用,以身试苗;居里夫人冒着生命危险,发现了有放射性的镭;"两弹一星"元勋邓稼先为了得到研究数据,冒险进入核试验区研究调查。这些科学家不为名利所动,不为个人的得失,甚至牺牲个人生命健康的精神,就是献身精神。

二、科学家精神拥有丰富的内涵

科学没有国界,但是科学家是有国家的、有祖国的。中外科学家都具有高度的爱国主义精神,具体讲,科学家精神有6个丰富内涵。

一是无私报国的爱国主义和服务人类的国际主义精神。

老一辈科学家把救国救民当作自己神圣的使命。我们的国家在站起来、强起来的过程中,无数科学家做出了突出贡献。许多在国外已经卓有成就的科学家,如钱学森、朱光亚、邓稼先等,怀揣着将中国建设成强大国家的爱国主义精神,响应祖国号召,归国报效。

"兵器工业之父"吴运铎,在抗日战争、解放战争中,为了制造军备,在兵工试验中,把眼睛炸瞎,手臂炸残,身上留下无数弹片伤痕。许多拥有吴运铎精神的科学家,把一切献给了党和国家。加拿大共产党员白求恩先是志愿到西班牙参加反法西斯斗争,后来到中国参加抗日革命,献出了生命。

二是严谨治学、追求真理的奉献精神。

著名生理学家巴甫洛夫关于严谨治学的一段话对我启发很大,"决不要企图掩饰自己知识上的缺陷,哪怕是用最大胆的猜度和假设作为借口来掩饰,不管这种肥皂泡的美丽色彩怎样使你们炫目,但肥皂泡是不免要破裂的,那时你们除了羞惭之外,一无所得的。"这就是说,严谨治学不能掩盖自身的缺陷,不能搞假。在治学时,一定要反对一知半解,反对夸夸其谈,这是严谨治学的要求。巴甫洛夫还要求学生要"循序渐进,循序渐进,循序渐进"。治学不能搞跳跃式,要一步一步来。不要故意隐瞒自己研究中的问题,在自己的文章中,要把问题暴露出来,让大家讨论。

追求真理,实事求是的求实精神,最难的是对自己实事求是。特别是对自己的学问,一定要自以为

本文原载于《科技导报》(2021年第39卷第10期)。

非,找出自己的缺点和问题,才能不断地进步、不断地取得新的成就。在学术上,"贵在自以为非,贵在不自以为是",这样学术才会不断地长进。

三是淡泊名利、潜心研究的奉献精神。

真正伟大的科学家都把名利看得很轻。爱因斯坦不止一次讲到,"我从来不把安逸和快乐看作是生活目的的本身,这种伦理基础,我叫它猪栏的理想。""人们所努力追求的庸俗的目标——财产、虚荣、奢侈的生活,我总觉得都是可鄙的。"

《爱迪生传》里提到,"一个人的价值,应当看他贡献什么,而不应当看他取得了什么。人只有献身于社会,才能找出那短暂而有风险的生命意义。"居里夫人也讲过,"人类也需要梦想者,这种人最醉心于一种事业的大公无私的发展,因而不能注意自身的物质利益。"在我看来,真正的科学家应当淡泊名利,有正确的财富观。

四是勇攀高峰、敢为人先的创新精神。

创新,是前人所没有做过的。创新没有近路,不可能短平快。马克思讲,科学上没有平坦的道路可走,只有不畏艰险、攀登陡峭山路的人,才能达到光辉的顶点。我们要搞创新,要取得大的成果,就要勇攀高峰,敢为人先,不怕困难。创新可能有失败、挫折,肯定也会有曲折,要不怕挫折,调查研究,找出问题所在,要克服困难,找出挫折的原因所在。

五是集智攻关、团结协作的协同精神。

科学技术是集体的事业,自然科学奖、技术发明奖,都是集体完成的成果。某个项目的科学理念、技术的途径,可能是由某一个科学家提出来的,但要实现科学的理念和技术的途径,要靠许多人计算、模拟、实验,没有团结协作的集体是干不出大事来的。以前也有科学家讲过,"脱离集体就是末路的开始。"一个团队如果有内讧,争排序、争名位,四分五裂,这个集体肯定没希望。

怎么能够团结协作?华罗庚讲过"人家帮我,永志不忘;我帮人家,莫记心上。"就是要感恩。一个集体要想气氛搞得好,就是要领头人、骨干,有名有利的事往后靠,吃苦耐劳的事往前赶,这个集体肯定好。

六是甘为人梯、奖掖后学的育人精神。

生命是有限的,事业是永恒的,真理是一条长河,无数真理的集合才形成了一个真理。牛顿讲过"如果说我看得远,不如说我是站在巨人的肩上。"特别是现在,建设科技强国需要几十年、需要几代人,不是靠一个人、一代人,而是靠几代人、成千上万的人,所以一定要有这样的精神。个人和集体的关系好像树枝和树干的关系,树枝离开了树干是要萎缩、枯死的,个人离开了集体也要萎缩的。

三、如何弘扬科学家精神,尤其是爱国和创新精神

一是弘扬科学家精神,每个科技工作者首先要把个人的理想与党和国家的建设科技创新型国家、建设科技强国的伟大战略紧密结合起来。

习近平总书记在两院院士大会上提出了建设科技创新型国家"三步走"的战略。"三步走"战略非常重要,科学家精神对于建设科技强国非常重要且迫切。面对复杂多变的国际形势,关键力量是科技。大国间的竞争,最后就是科技的竞争。我们建设科技强国、建设科技创新型国家已经到了非常迫切的领跑型的拐点,科技创新是应对世界大变局挑战的关键因素、关键力量,要认识到这一点。

中国制度、中国精神和中国效率我国建设创新型国家的3个独特优势。中国社会主义制度可以集中力量办大事;科学家精神是中国精神中不可或缺的,是当前科技创新的关键力量和因素;中国效率则归结为我们有马列主义世界观和方法论,我们懂得区分主要矛盾、次要矛盾,抓主要矛盾、矛盾的主要方面,所以效率高。这3个优势是别的国家没有的,3个因素不是相互独立,是相互交织的,缺一不可的。

二是弘扬科学家精神,要努力树立自己的远大理想,树立社会主义价值观、人生观和世界观。

科学家精神是党性在科学家身上的体现和反映,是革命人生观、马克思世界观在科学家身上的反映和体现。弘扬科学家精神就要打牢科学家精神的思想根基和道德根基,要树立远大理想以及正确的价值观和世界观。钱学森为什么讲他的学术成就归结到马克思主义对他的影响,那是因为他非常信仰马克思主义,他认为马克思主义的世界观、方法论对他的学术有很大的影响。

三是要和背离科学家精神、违反科研道德规范、突破科研诚信底线的行为做坚决的斗争。

Nature 上刊登的一篇评论文章中指出,中国的专利数量世界第一,但专利里很多垃圾专利,是为专利而专利,论文里也名堂百出,剽窃、伪造数据、伪造同行评议。中国贡献的科研文章占全球 8.2%,撤销论文的数量占到全球 24.2%,是贡献率的 3 倍,这值得引起我们的高度警惕,一定要群起而反对,在当前弘扬科学家精神,一定要坚决和作风、学风建设败坏行为做坚决的斗争。

四、如何培育科学家精神

科学家精神的培育就是要不断学习,最重要的是要学习英雄烈士、老一辈革命家、老一辈科学家等伟人的精神,帮助树立自己的远大理想,树立自己的革命人生观。伟人的一生是我们终生学习的榜样,伟人闪耀思想光芒的语言时刻铭记在心,就能指引自己不断地前行,伟人的光辉行为永远激励自己不忘初心,是我们前进不竭的动力。所以要不断学习、不断警惕、不断激励自己,才能培育好自己的科学家精神。

白鹤滩工程所取得的经验对地下工程建设有重要借鉴意义

【编者按】 白鹤滩水电站是世界上在建装机规模最大的水电站,是仅次于三峡工程的世界第二大水电站,坝址区工程地质条件复杂,面临着高地震烈度、高水头大流量泄洪消能、特高边坡开挖与治理、坝址区河谷地形不对称、坝基层间层内错动带发育和柱状节理玄武岩开挖卸荷变形控制、干热河谷混凝土温控防裂,以及世界上最大的地下电站洞室群开挖变形与稳定、巨型水轮发电机组制造安装等难题,堪称我国乃至世界上综合技术难度最大的水电工程。如何破解这些难题,推进我国水利水电技术进步,本刊采访了参与工程咨询和研究的钱七虎院士,请他分享观点和感受。

我首先谈谈对白鹤滩水电站工程的认识。白鹤滩水电站作为世界上规模最大的在建水电站工程,其建设有着重大的战略意义。现在世界已经进入绿色发展时代,《巴黎气候变化协定》的签署,形成全球应对气候变化的国家共识,欧洲一些国家政府已宣布禁止销售燃油汽车的时间表、路线图,海南省成为我国第一个提出燃油车禁售时间表的省份;德国、澳大利亚等一些国家已提出禁止采挖煤炭。

习近平总书记提出建设中国特色社会主义,绿色发展是特色社会主义很重要的组成部分。特别是党的十八大以来,已经把生态文明、绿色发展提到了新的战略高度,绿色发展进入"五大发展理念"中。长江大保护、三江大保护等列入生态文明建设重要内容之一。从这个背景来看,水电是可再生的绿色能源,开发水能资源,建设白鹤滩水电站意义重大。

从另一个方面来讲,白鹤滩水电站节能减排效益显著。在金沙江4个水电站中,白鹤滩水电站的规模是最大的,多年平均发电量600多亿kWh,约占4个水电站发电量的1/3,巨大的发电效益可平均每年减少标煤消耗量约2000万t,减少二氧化碳排放量约5000万t,节能减排效益显著。

白鹤滩水电站地下厂房工程规模巨大,地下厂房长438m,高88.7m,岩梁以上宽34m,岩梁以下宽31m,为目前世界上跨度最大的长廊形地下洞室。地下厂房开挖有两大难题:一是高地应力引起的顶拱变形问题,它的特点是水平主应力高于垂直主应力;二是层间层内错动带的问题,这是一个软弱沉积错动带,并且斜交地下厂房顶拱和边墙。我作为专家参与了咨询,当时也提出了一些意见,得到了采纳。

白鹤滩工程在组织管理创新上做得比较好。针对地下厂房复杂地质条件,白鹤滩工程建设部组织成立了由专家组、科研、建设、设计、监测、施工等单位共同组成的综合反馈分析体系,通过各方面的反馈,优化了设计,调整了开挖方案,对围岩支护进行动态设计,并与施工组织管理有机结合,成功解决了以上难题。

白鹤滩地下厂房开挖支护所形成的技术,对地下工程建设有着较大理论意义。目前,我国地下空间的开发向深度推进,在地下进行深部开挖就会碰到高地应力和岩爆问题,因此白鹤滩水电站工程所取得的经验对地下工程建设有着重要的借鉴意义和参考价值。

(记者:韦凤年)

本文原载于《中国水利》(2019年第18卷)。

又好又快稳步推进城市地下综合管廊建设的思考

国办发〔2014〕27号与国办发〔2015〕61号文件的发布对推进城市地下综合管廊建设具有里程碑式的意义与作用,文件原则上解决了管线入廊、管廊运行的经营以及管廊建设的融资问题。但是20世纪以来国内综合管廊先行城市上海、成都、北京等已建管廊建而不入、建后难管的教训与国外先进城市的综合管廊建设管理的成功经验相对照,要达到文件所提出"建成较为完善的城市地下管线体系,使地下管线建设管理水平能够适应经济社会发展需要"的目标,能够实现"培育一个投资额万亿级的综合管廊行业,产生一批专门搞管廊的企业"的良好前景尚需艰苦的努力,必须对综合管廊的管理法规、管理体制和管理措施进行系统完善。管理法规是实施严格管理的基本依据,在法规上明确规定管理体制和管理措施,管理体制是管理的执行主体,在法规的法律授权或行政授权下实施管理措施,管理措施是体现管理的具体方式方法,由管理机构依据法规,通过有效的管理机制综合运用。现具体分述如下。

一、借鉴国外先进经验,制定完善的法规体系

建设管理综合管廊是一个跨行业、跨部门与单位的协调工程。过去我国综合管廊"建而不入""建后难管"的病根在于分头管理的市政体制与条块分割的部门所有制,破除分头管理与部门分割的利器是"法治",依法执政首先要有法可依,光有指导意见不够,需要对综合管廊的所有权、规划权、建设权、管理权、经营权和使用权等作出明确、完善的具体规定。在这方面,我们可以借鉴德国的《城市建设法典》、日本的《关于建设共同沟的特别措施法》,以及我国台湾地区的《共同沟管道法》《共同沟管道法施行细则》《共同沟建设管线基金收支保管及运用办法》《共同沟建设及管理经费分摊办法》等现行法规,也可借鉴上海市的《上海市城市道路架线管理办法》和重庆市的《重庆市管线工程规划管理办法》等进行完善和补充。

二、系统科学地编制好城市综合管廊规划

地下综合管廊是城市地下空间利用的一个有机组成部分,城市地下空间资源是宝贵而有限的;城市地下空间利用应与城市地面、地上空间利用达到互补和互相协调。因此原则上城市地下综合管廊规划应与城市地下空间规划以及城市建设总体规划同步协调进行。条件不允许必须先行或单独规划的综合管廊规划,必须首先与现有和拟建的城市地下步行道系统、地下轨道交通、地下快速路系统以及为解决城市内涝问题而建设的城市地面与地下"海绵"设施和储雨排洪的城市地下河流、蓄洪设施充分协调、统筹规划地编制。综合管廊建设是百年大计,规划时应充分考虑城市未来发展中对地下物流、垃圾传输、净水输送乃至利用地热地冷资源实现城市供热供冷所需空间,同时也提升综合管廊的经营性。

现在关于燃气管线是否纳入综合管廊还有争议,争论的焦点主要在于燃气管线易燃易爆,危害综合管廊安全。综观以往的燃气管线事故,例如济南和南京等地发生的地下煤气管线爆燃爆炸事故,以及黄岛发生的地下输油管线爆炸等,都是由于燃气(油)管线检查维护管理不到位,导致燃气(油)泄漏引发的。将燃气管线统一纳入综合管廊后,可以实现对燃气管线的及时检查、维护和监控,从而能够避免燃气管线泄漏和爆炸事故的发生,因此将燃气管线纳入综合管廊当中也是合理可行的。

关于雨洪排泄管道入廊问题,首先城市要实行雨污分流,污水管道应纳入综合管廊中。而将雨洪排泄设施纳入管廊,则因雨洪管的设计至少要按百年一遇洪水进行,体量较大,导致雨洪排泄设施入廊很难,故不宜入廊。为彻底解决城市内涝问题,雨洪的排泄设施建议参考日本经验,修建为城市地下河川或参考吉隆坡的经验,将暴雨排洪设施与城市地下快速路相整合,无雨时快速路通车,中小雨时,路面下排泄雨水,路面上仍通车,暴雨时,则快速路禁止通车仅作排洪。

应大力提倡综合管廊与地铁建设、地下街建设和地下快速路建设相整合,从而明显降低建设成本,减小社会干扰,缩短施工工期,避免重复建设、重复投资。如台北东西快速道路共同沟的建设,全长6.3km,其中2.7km与地铁整合建设;2.5km与地下街、地下车库整合建设;独立施工的共同沟仅1.1km,从而大大地降低了建设总成本,有效地推进了共同沟的发展。武汉地铁三号线宗关站,将电力管、排水管、通信管等管道集结在一起,与2、3号出入口工程整合建设,既解决了"城市蛛网"埋设随意性较大、分布不合理的问题,也可避免检修时对城市道路重复"开膛破肚"。乌鲁木齐市在远景规划城市轨道交通线网中,也考虑了同步进行地下综合管廊建设。

三、建立统一管理、分工协作、衔接有序的综合管廊管理体制

为了克服条块分割、分头管理的弊端,达到统一协调,当前在尊重现有各部门职能分工的基础上,应尽快明确每个城市综合管廊的综合管廊机构。建议成立城市综合管廊管理委员会,承担〔2014〕27号文件所规定的政府对综合管廊的建设管理责任。由政府主要负责人担任主任,相关政府部门担任委员,下设综合管理办公室,承担规划编制、项目审批、设计审查、安全监管、信息共享、项目投融资、专家咨询等的综合协调事宜,制定地下综合管廊管理办法、指导运营单位完善管理制度,与入廊管线单位签订协议,确保地下综合管理正常运行,指导有关政府部门做好突发事件和应急管理工作。

四、地下综合管廊兼顾人防应到位不越位

根据《中华人民共和国人民防空法》第十四条规定,城市的地下交通干线以及其他地下工程的建设,应当兼顾人民防空需要。因此,城市地下综合管廊建设必须兼顾人防需要。

到位就是必须兼顾人防,从而提高平时管廊的防灾减灾能力,提高战时的防空袭能力。但到位不能越位,如果兼顾人防过高提高工程造价,就会影响综合管廊的建设,就是越位,战时兼顾的人防功能也应有所选择,如战时作人防掩蔽所使用,因防护单元的要求需在管廊内施作防爆隔断设施,从而既影响管线敷设,又提高了成本,这也是越位。综合管廊建于地下,提供了一定的防护能力,可理解为兼顾人防,如有必要,建议管廊战时可作为疏散通道使用。

五、创新投融资机制

创新投融资机制的关键是吸引社会资本参与综合管廊的建设与运营。考虑到社会资本的逐利性是其基本属性之一,要吸引社会资本参与,必须妥善解决如下问题:

1. 综合管廊集中统一运营,降低运营成本

目前国内已建综合管廊每公里运营成本达到了100万元以上,如此高的运营成本,不利于综合管廊产业的健康发展。降低运营成本,是综合管廊投资获取合理回报、降低管线单位负担的基础。

在综合管廊集中统一运营条件下可打造智慧管廊,充分运用物联网监控、BIM信息管理等技术,实现集中远程控制,建立市场机制下的维护管理,从而大幅降低运营维护成本。

2. 严格"风险分担"

政府理应承担的风险不应由社会资本"共同承担",PPP模式中投资风险不应全由项目公司承担,政府更不应让社会资本捆绑打包更大的建设项目。

风险分担的具体方案是:

(1)政府承担:政治风险、法律风险、规划与标准变更风险、土地获取风险、市场需求风险、决策风险、管线单位付费风险以及因政府方原因造成的工期延误风险和成本超支风险等。

(2)社会资本承担:融资风险、财务风险、技术风险、因社会资本方原因造成的工期延误和成本超支风险、运营管理风险等。

(3)双方共同承担:利率变化、通货膨胀、不可抗力等风险。

3. 合理确定项目公司(SPV)资本金比例和政府在项目公司中的持股比例

(1)项目公司资本金一般应不超过项目总投资的30%,在建设期可以采取增信措施,由社会资本方提供建设期担保解决建设融资问题。

(2)基于PPP项目要求,政府在项目公司中持股比例应当低于50%,同时为保证政府在项目公司拥有重大事项的控制权(超过三分之一持股比例),建议政府在项目公司的持股比例为34%~49%,社会资本在项目公司的持股比例为51%~66%。

4. 采取政府提供购买服务协议的方式解决经营收益不确定性问题

管廊收费机制和标准的制定需要顶层设计、高层协商,相关标准在没有出台前,综合管廊建设项目可视为非经营性项目,采取"政府付费"模式,由政府提供购买服务协议的方式解决经营收益不确定性的问题,建立基于"可用性+绩效付费"的考核机制。同时,综合管廊运营收取的费用全部归政府。

待相关标准出台后,可采取"使用者付费+政府可行性缺口补贴"模式,即在管线单位入廊收费无法平衡项目合理投资回报及运营成本时,由政府提供可行性缺口补贴。如果运营收益能够平衡投资回报及运营成本时,政府不需要提供补贴。此时,由于相关收费标准明确,政府能够准确测算需提供的补贴。

5. 杜绝地方政府捆绑打包"小马拉大车",增大未来投资风险

部分地方政府利用国家对地方综合管廊建设的补贴资金,吸引社会资本投资建设过程中,要求社会资本捆绑打包其他更大的建设项目,用中央补贴资金拉动地方其他大项目建设的行为,造成投资风险的极大增加。

21世纪,让我们向地下要空间

【导语】 看过忍者神龟吗?喜欢小乌龟们住的地下空间吗?那里不仅有宽大的房间,还能让汽车快速行驶。动画片中的一切,在21世纪也许会成为现实。当地面空间越来越稀缺,杭州鼓励大家往"下"发展。杭州市政府已正式下发《关于加强城市地下空间开发利用管理的若干意见》,这份将于今年1月21日起实施的《意见》提出,在符合城市规划和地下空间开发利用专项规划的前提下,鼓励开发建设利用地下空间的公共服务设施。

未来,我们怎样向地下要空间?人类可以生活在地底下吗?

1月7日,应本报与浙江省科协联手推出的科学会客厅之邀,中国工程院院士、解放军理工大学教授钱七虎来到杭州,为公众揭开神奇的"地下生活"。在钱院士口中,地下生活不仅可以给人类带来更广阔的空间,还可以节约能源、净化空气、减震避灾……

让我们随"科学会客厅"一起,走进钱七虎院士的神奇地下世界吧。

19世纪是桥的世纪,20世纪是高层建筑的世纪,21世纪将如何?

1991年在东京召开的地下空间国际学术会议通过的《东京宣言》指出:"二十一世纪是地下空间开发利用的世纪"。

20世纪末在巴西里约热内卢召开了联合国世界可持续发展大会,制定了世界的发展战略——可持续发展。

可持续发展也是我国的基本国策。

而可持续发展的一个重要方面,就是充分开发和利用地下空间。

这是因为,可持续性代表着,不仅要满足当代人的需求,经济、社会和资源、环境的协调发展,还要遵循科学发展观,构建资源节约型、环境友好型社会和城市,建设具有高防灾减灾水平的社会和城市,而不影响后代的需求。

解决这些问题的关键,在于运用先进的城乡规划理论、先进适用的科学技术来进行城乡的建设和运营。

这就离不开地下空间的开发和利用。

开发地下,可充分利用土地

据第六次人口普查的结果,全国总人口为13.397亿人,如此众多的人口需要大量耕地以保证粮食供应问题,国务院提出我国耕地面积红线是18亿亩。

国土资源部2010年数据显示,我国现有耕地为18.18亿亩。而按照联合国的统计,每一个进城的人约需要200~300平方米的生活空间。

改革开放以来,中国的城市化速率为每年接近1%,今后也将如此,每年约需要使用520万亩土地以满足城市化发展的需求,如果不考虑土地的复垦,我国的土地余量仅仅能支撑3~5年的城市化进程,城市建设将面临无地可用的困境。

因此,必须严格控制土地需求,科学合理地利用土地。

这就需要开发利用地下空间:把城市交通(地铁和轨道交通、地下快速路、越江和越海湾隧道)尽可

本文原载于《钱江晚报》,系2012年1月7日"科学会客厅"栏目实录。

能转入地下,把其他一切可以转入地下的设施(如停车库、污水处理厂、商场、餐饮、休闲、娱乐、健身等)尽可能建于地下,就可实现土地的多重利用,提高土地利用效率。

城市节地的一个主要方面在于宏观上努力实现土地的多重利用。土地的多重利用可以沿两个方向实施:城市无建筑区的土地额外利用;城市已建成区的紧密化和功能变化。

比如南京的玄武湖隧道,是国内最早在城市内湖泊下进行土地利用的例子;而杭州西湖的湖滨隧道,也是一个很好的例子。这样,能在无地可建路的情况下,开出一条新路。

北京的中关村西区,又是一个很好的例子:它采用立体交通系统,实现人车分流,各建筑物地上、地下均可贯通。

在上海,截至2010年底,已经建成地下工程总建筑面积为5200万平方米。有超过10座以上功能各异的"地下城"。大家很熟悉的虹桥综合交通枢纽,就是高速铁路、城际和城市轨道交通、公共汽车、出租车及航空港紧密衔接的国际一流的现代化大型综合交通枢纽;它的建筑面积约100万平方米,其中地下空间面积超过50万平方米。

还有杭州的钱江新城核心区的地下城,以波浪文化城和地铁1、2号线换乘站为骨干,形成地下城,地下空间总量达到200万平方米。

国外对于地下空间的利用,早在中国之前:法国巴黎卢浮宫的地下空间开发、瑞典斯德哥尔的摩伯尔瓦尔德地下音乐厅、挪威霍姆德亚地下运动厅和游泳池,都非常成功;尤其是美国明尼苏达大学的地下系馆,它的实验室采光获得了总统奖——他们通过光学系统,使人在地下实验室中,可以看到地面的景物,可以照到地面的阳光。

地下开发,可以节约能源

2010年,我国石油消耗量4.41亿吨,仅次于美国的10亿吨,位居世界第二位;交通领域所消耗原油约占原油总消耗量的68%,其中车用石油消费占57%左右。

与之对应的是,我国原油对外依存度正逐年上涨:2010年中国原油进口2.39亿吨,原油对外依存度达54%。国际能源机构(IEA)预测,2020年我国石油对外依存度将超过60%,2030年可能达到80%。

节约能耗,提上了日程。

在城市总耗能中,交通耗能和建筑能耗占据较高比例。

据住建部的统计,中国城市交通能耗约占总能耗的20%,并且正在逐年上升,正向30%迈进。

建筑能耗,包括建造能耗、生活能耗、采暖空调等,约占全社会总能耗的30%。按照目前的建筑能耗状况,到2020年我国建筑耗能的新增量相当于目前建筑总能耗的1.3倍。我国城市化进程如果按照发达国家发展模式,使人均建筑能耗接近发达国家的人均水平,需要消耗全球目前消耗的能源总量的1/4来满足中国建筑的用能要求。因此,必须大幅度降低建筑能耗,实现城市建设的可持续发展。

地下空间的开发利用可使得城市更紧凑,建成"紧密型"城市,城区的面积因而更小,城市出行对机动交通的依赖度降低,有利于推广步行和自行车出行,从而节约交通能耗。掩土住宅和新式窑洞的推广有助于节约建筑能耗。

利用地下空间,可以为可再生和洁净能源利用开辟一条有效途径——在地下岩、土介质中修建热水洞库、水库、压缩空气储库和超导磁储电库。德国已经在地下岩层中建成一个巨大的地下压缩空气储库;我国也正在江苏金坛地区论证建设这样的地下压缩空气储库。

美国、德国还正在研究开发地下永久非枯竭的清洁能源——深层干热岩发电,美国LosAlamos实验室在卡尔德拉的芬登山上建成了一个深层干热岩发电站,利用地下高温岩体发电。

充分开发利用地下空间,建设节能型城市还包括充分利用地下土壤、地下水的低密度天然能源作为冬季热源和夏季冷源,然后再由热泵机组向建筑物供热供冷,地源热泵技术就是这样一种利用地下可再生能源的既可供暖又可供冷的新型中央空调系统。这已在上海世博园中得到成功运用,节能效果显著,达到一半以上。

地下"水银行",缓解城市供水

国际上有人说,21世纪,是水的世纪。目前世界上有14亿人口缺乏安全清洁的饮用水。

中国是世界13个贫水国之一,人均水资源量仅为2251立方米,居世界第110位,是世界平均水平的1/4。据统计,我国城市每年缺水60亿立方米,在全国666座建制市中,有近400座城市缺水。

建设节水型城市,关键是要建立城市水系统的良性循环机制,实现水资源的可持续利用。在这方面外国的先进经验是:充分利用雨水;开发污水再生利用和建立地下"水银行"调节和缓解城市供水。这些经验的实施都需要开发利用地下空间。

欧美发达国家现在正在改变修建地表水库来储备水资源的传统做法,利用地下含水层建立"水银行"来调节和缓解供水。它除了可摆脱水库淹没地面迁移居民的困扰,还一举多得:储备地下水,在水短缺的时候提供水量满足用水需求;控制由于地下水位下降引起的海水入侵和地面沉降;提高地下水位,减少地下水的抽取费用;维持河流的基流;通过土壤中的细菌作用、吸附作用和其他物理、化学作用改善水质;通过处理后的污水入渗来实现污水的循环再利用,以管理不断增加的大量污水;最后也是最重要的是保护了生态环境。

收集雨水到地下也是扩大可用水资源的一种方法:比如日本、瑞士和牙买加等国,就利用屋顶收集雨水,并通过管道将雨水输送到地下储水库,日本雨水利用的供水量每年为0.05亿立方米,仅名古屋体育馆每年就积蓄雨水3.6万立方米;中国也在西沙群岛的永兴岛上修建了可采集雨水达14万吨的地下储水工程。通过技术处理,已达到了国家饮用水卫生标准,从而结束了吃水只能靠大陆船运的历史。

将污水处理设施放置在地下,是利用地下"水银行"的另一种方式,这将使目前地面污水处理厂所占的土地得到开发利用,而且使它们邻近被污染不能被利用的土地也得到开发。比如日本,每年就将经处理的1.3亿立方米的再生水用于工业或绿化。

建设地下工程,让环境更美好

国际上曾有专家预言:21世纪世界上两大最具影响的事件之一是中国的城市化。因为,中国的城市化与全球的资源和环境问题密切相关。

随着社会和经济的不断发展,环境质量必将受到越来越多的关注。我们应当看到对于环境、景观和适居性的质量意识不断增长的过程,从对环境质量要求的角度来看,地下空间的利用起着重要作用。

首先,可以保护地面自然环境。在芬兰,很多公路宁可以高工程成本修建暗挖隧道,也不以低工程经费开挖地面、修建边坡;在荷兰,在其"绿色心脏地带"阿姆斯特丹、鹿特丹、海牙、乌特列希等城市组成的具有重要观光和自然景观的大都市区,只建设地下工程以便将对其现有景观的破坏最小化,同时在某些程度上满足了空间增加的要求。

其次,可以改善城市环境。美国波士顿拆除了穿过市中心的六车道高架路,建设8~10车道的地下高速路,原有的地面变成林荫路和街心公园;市中栽植了2400棵乔木树,7000多棵灌木树;在海湾的景观岛上栽植了另外的2400棵乔木和26000棵灌木;增加了260英亩新的公园和开敞空间。这样做的结果,使市区空气的一氧化碳浓度降低了12%。

第三,可以通过发展地下交通、缓解城市交通拥挤、降低城市大气污染。比如2003年1月日本政府举行新闻发布会,宣布东京外环线改为建设地下40米深的高速路;由于地下无须动迁和征地费用,可比地面高架路节约建设费用20%~30%,工期由15年缩短为8年;由于地下隧道内设置了将汽车尾气中的烟尘和有害物质进行高科技过滤和分解的设施,有害气体可以得到集中处理。

第四,可以利用地下空间,建设城市地下市政管线公用隧道。将自来水管、排污管、供热管、电缆和通信线路放在市政管线公用隧道内,可缩短路线长度达30%;也易于检查,易于修理,减少管线破坏事故;减少对环境的影响,不影响地面土地的使用。在上海浦东的张相路和成都的东风路已经建成了两条地下市政管线公用隧道;在北京的中关村地区即将建设高标准的"共同沟"。

第五，可以在地下建设天然气输配工程，改变城市燃料结构，治理城市大气污染和酸雨区。

第六，地下可以建造能源（石油、天然气）的地下战略储存库。

最后，可以发展地下垃圾处理系统，消除垃圾围城现象。20世纪60年代，瑞典就建设并实施采用压缩空气吹运垃圾系统。发展地下垃圾处理系统节约了城市宝贵的土地资源，控制了对环境的噪声和恶臭的污染，减少了对环境的视觉负面影响。

而从防灾减灾角度来看，地下空间具有比地面建筑高得多的抗地震能力和抗风暴能力；东京的地下河川、地下集水设施建设表明可以用来缓解近年来困扰我国不少城市的由暴雨导致城市内涝的问题；但地下空间利用有很多火灾、洪灾的严重教训，所以在规划、设计和施工地下空间时，借鉴日本的经验，需要提前考虑这些问题。

建设地下工程，要确保环境友好

在建设地下工程过程中，一方面需要考虑到为了建设环境友好型社会与城市，促进地下空间的开发利用，一个最重要的办法是建立相应的法规，如欧洲很多城市做的那样。即工程方案的比较应"与时俱进"，通过法律规定提倡对不同方案进行社会经济、环境、技术的综合评估，以代替纯经济技术评估。

政府应积极制定严格的环境和安全方面的政策和法规，以进行环境影响的评估，从而保证不同实施方案的分析更加客观。

为此，应积极开展"环境效益与损害的货币化研究"，从而使社会的生态意识真正融入工程中。

【现场互动】

问：对于一些大型城市，比如杭州，地下能源利用应该如何进行？您有何建议？

钱七虎：干热电站，美国有一个专门的科研机构研究这个事，现在还不能推广到家庭应用，但是，地源热泵和江湖水源热泵，以及太阳能空调，很多城市应当积极推广。2014年，欧洲要建成绿色建筑，新建的住宅要实现零排放，其中采取一个措施就是推广家庭地源热泵和太阳能的利用，以抵消消耗的能源。在中国，我希望城市建设的领导机关，或者是科研院校单位，能够一起研究推广、呼吁这个事。

问：空间利用，是不是让人都住到地下去？

钱七虎：我说的空间利用，是让人们尽量把运动、娱乐、交通等转移到地下去；然后为人类在地面上的生活，腾出更多空间。人还是要生活在地面上，不要到地下，我们把一些设施放在地下是为了地面上更宜人。

高度重视地下工程安全
——专访中国工程院钱七虎院士

随城市地下空间开发利用规模不断增大,我国城市地下空间的总体规模和总量已居世界首位。我国已成为世界上隧道最多、建设发展最快的国家。国内已投入使用的各类隧道1.2万座以上,延长总计7000公里。其中铁路隧道7000多座,总长度4000多公里,公路隧道4000多座,总长度2000多公里。到2020年,我国将规划建设铁路隧道5000座,长度超9000公里。10年内建设155公里长的城市公路隧道,其中许多是长大和深埋隧道。目前我国建成的水下隧道有3座长江隧道、8座跨黄浦江隧道。

钱院士在演讲中介绍,中国许多特大和大城市将结合地铁建设、旧城改造和新区建设,建成众多大型城市地下综合体,特大城市的地下快速路的建设,也正越来越受到重视。

地下工程是一项高风险建设工程。地质环境复杂,基础信息不充分,施工技术复杂,不可预见因素多,社会环境影响大。我国地下工程建设事故频发,形势严峻,令人担忧。面对地下工程建设事故频发的严峻形势以及地下工程建设规模大、发展快的客观现实,必须更加坚持又好又快的科学发展观。"速度服从安全,速度服从质量,速度服从效益;实现确保安全前提下的高速度必须科学规划、科学勘察、科学设计、科学施工、科学管理;必要的管理力量、技术力量的充分投入是实现这'五个科学'的基础。"

地下工程建设规模大、发展快是安全形势严峻的客观原因。严峻的地下工程安全形势是实行安全风险管理必要和紧迫的客观需求。

"建立风险管理,对拟建和在建的城市地下工程项目,进行风险评估,继而进行风险控制十分必要。"风险管理制度应扩大到整个地下工程领域。地下工程的特点在于地下工程具有很高的风险性,但绝大多数事故是有征兆的,因此是可预防的,而且安全事故预防关键在人。

钱院士在演讲中介绍了目前国际上安全风险管理的发展趋势,钱院士说,安全风险管理已经成为大型项目发展中的一个例行程序;安全风险管理与工程项目管理日趋结合;迫切需要为安全风险管理制定强制性的法规、规范,特别是针对施工建设安全的法规、规范。为此,应掌握地下工程安全风险管理的现状及对地下工程安全风险管理必须树立的理念。

当前在中国,安全风险管理也越来越引起重视。2005年中国土木工程学会召开了中国第一次全国范围的地下工程安全风险分析研讨会,推动了地下工程安全风险研究的全面开展。安全风险管理的实际应用在近几年得到迅速发展,特别是在地铁建设方面,新建地铁项目大都进行了风险分析与评估。我国也在积极制定隧道和轨道交通工程项目安全风险管理的法规、规范等,例如《城市轨道交通建设项目管理规范》正在研究制定中。

钱院士认为,总体上,我国地下工程安全风险管理研究与实践已经得到了实质性进展,但还没有达到化解"风险管理化解地下工程建设之痛"的程度。其特点是区域、行业发展水平不平衡;主要开展的是风险分析和风险评估,并局限于定性分析或者半定量分析;对风险预警和控制方法还没有深入研究,仅实现监测功能,还不能"控",还没有整合成统一的安全风险管理系统;"事故处理"仍然是安全风险管理的工作重心,迫切需要完成由"事故处理"型到"事故预测"型的转变;缺乏坚实的地质地理信息系统和符合工

本文原载于《中国桥梁网》(2011年11月23日)。

| 思想与观点

程实际地质、水文地质和环境条件的风险阈值数据库系统。

安全风险管理的费用往往被首先挤压,导致一些施工单位在利益的驱使下,冒险施工,吃"风险饭","赌"不会发生事故的概率。工程安全风险管理队伍不够专业规范,管理水平参差不齐。国内对工程安全风险管理咨询评估的从业单位和人员还没有明确的资质管理制度。缺乏安全风险评估规定、安全评估单位资质管理办法、安全评估人员考核与管理办法等相关的法规。目前风险管理相关技术控制规范不够全面。因此,编制地下工程施工不同工法的技术规范,研制不同城市与不同工法适应不同岩土工程地质、水文地质、环境条件的风险阈值数据库系统已是当务之急。

目前,风险评估等工作虽已经纳入地铁建设的工作环节中,但由于缺乏国内工程安全风险统计资料,风险概率和后果计算依据不足,得到的定量风险评估数据缺乏足够的可信度,工程实际指导性还不强。

针对我国安全风险管理实践中存在的问题,钱院士提出了切实可行的建议:加强针对地下工程安全风险管理的法规建设工作;建立全面的安全风险管理体系,明确各方安全风险管理责任;大型工程立项前必须进行风险评估,对"严重"等级风险的工程一票否决;规定安全投入为"专项提取"及其占整个建设费用中的比例,并予审计、监管;强制实行设计和施工中的风险专项设计和方案。明确各方安全责任,实现动态施工,动态管理;提高地下工程风险意识,全面推广安全风险管理制度,将安全风险管理作为地下工程管理的一个必要组成部分;建立地下工程安全监控咨询和第三方监测制度;建立基于现代化信息技术的地下工程建设安全风险管理信息系统;结合中国地下工程建设实际情况,研究操作性强的安全风险评估和管理方法;加强地下工程安全风险管理理论以及重大事故预测预报和防治技术的研究。

钱院士的满头白发使人对他的第一印象更像个和蔼的老人,但是,只要听了他的演讲、与他有了简短的交流,就会发现,虽然已是古稀之年的老人,但他仍保留着军人硬朗、直爽的作风,科学家严谨、执着的精神,仍怀着对祖国发展与建设的殷切希望和发自内心的关注。钱院士的发言中,除了观点和理念的先进,更让我们体会到了老一辈科学家对我国科技发展、地下工程安全的深思与忧虑。

钱院士在接受记者采访时,对我国的工程技术水平给予了很高的评价:"从工程技术的角度来讲,我国的桥梁与隧道水平都已经走在了世界前列。桥梁方面,最大跨度的斜拉桥是中国建造的,第二大跨径的悬索桥也在中国;隧道方面,锦屏二级水电站等隧道工程也具有很高的水平,证明了我国在深埋长大隧道建设的能力。"

与此同时,钱院士对我们桥梁的整体水平也有着深刻而理性的分析:"从理论、实验方法以及装备等方面,我们与国际相比还是有一定距离的。比如隧道岩石力学理论大都还采用外国的理论,TBM 设备方面比较落后,目前大多数还需要引进,虽然我们自己也合作制造了一些设备,但是核心部件还是要靠国外。现在我们是桥梁大国,岩石工程的大国,但还不是强国。"

最后,钱院士表达了他对我国地下工程的信心和愿望:"只要上下一心重视和加强地下工程安全风险管理工作和重大事故预防的研究,我们相信,大量的安全事故是可以避免和减少的,降低地下工程灾害发生率是可以做到的!"

关于北京地下道路建设和地下空间开发相关问题的访谈

中国工程院院士、我国著名地下工程专家钱七虎是最早提出北京应建设地下快速路的专家之一。近日,就北京地下道路建设和地下空间开发相关问题,钱七虎院士接受了本报记者专访。

京华时报:北京如何解决交通拥堵问题?

钱七虎:解决交通拥堵问题是系统工程,抑制交通需求和增加交通供给要"双管齐下"。一方面,北京面积不能太大、人口不能太多,要通过"减肥"来降低出行需求;同时,北京新城、商业区规划要注重交通、医疗、教育等配套设施,避免出现"睡城"和由此引发的潮汐客流。另一方面,要通过挖掘地下空间潜能,发展地下快速路来增加交通供给。

京华时报:发展地下交通对北京缓解拥堵意义何在?

钱七虎:北京为缓堵已经想了、用了很多办法。随着经济社会的发展,以前小打小闹的治堵方法要换成大手笔、大措施才能突破瓶颈,必须向科技要出路,向地下空间要出路,因此产生了发展地下交通的解决方案。目前的科技水平,已经具备挖掘地下空间的能力,波士顿、马德里、东京、南京等城市的地下空间开发也为北京提供了成功范例。

京华时报:北京为何选择在东西二环建地下快速路?

钱七虎:因为东西二环路最堵,迫切需要增加道路供给来缓堵,但是北京二环内是古城,土地资源紧张、地面道路资源开发的很充分,已经无路可建了。而地下100米仍可开发利用,在东西二环修建地下快速路发展空间很大。

京华时报:北京建设地下快速路的条件如何?东西二环是否适合建设地下快速路?

钱七虎:从地质条件来看,北京优于上海、南京、广州等城市。在此前地铁建设的勘察过程中,我们已经发现北京的土质不是淤泥,而且砾石含量少,适合进行地下施工。从技术上来看,北京地下100米都是可以开发利用的,具体到东西二环地下快速路,施工前一定要进行科学勘察、科学规划。

京华时报:东西二环地下快速路建设在线位方面有何要求?目前规划部门制定了深埋、浅埋多套方案,何种更为科学?

钱七虎:线位和方案的选择是施工前的必要流程。结合北京实际,在地铁二号线下方施工的深埋方案以及偏离二环路的浅埋方案都是可行的,技术上没有不可以克服的地方。两种方案比较时,要考虑经济上是否合算,还要考虑文物保护和管网改造方面的因素。

本文原载于《京华时报》(2011年1月15日)。

京华时报：东西二环地下快速路建设如何解决文物保护、地下管网的问题？

钱七虎：地下快速路的建设需要科学勘察、科学规划、科学设计、科学施工。以现在的技术水平，完全可以掌握地下文物、地下管网的分布和状况，前期规划时，既可以制定避开文物、管网的施工方案，也可以制定文物搬迁、管网改造的方案。

京华时报：地下快速路一旦发生车祸、火灾，如何解决安全救援方面的问题？

钱七虎：地下安全不是不可逾越的鸿沟，国际上已有丰富的经验可以借鉴，可以通过增加地下标识物、安装地下检测设备、研究救援逃生方案等措施解决。总之，技术上没有不可以实现的。

京华时报：北京建设地下快速路的成本如何？

钱七虎：与地铁建设每公里平均造价6~7亿元相比，地下快速路省去了地铁车站、列车、信号系统等设备设施的费用，每公里造价估计约为3亿元，成本仅为地铁的一半。与地上道路相比，地下快速路可以省去拆迁成本。

京华时报：东西二环地下快速路的出入口应该设置在什么地方？此前有专家提出，出入口应该避开车流较多的路段，以免加剧拥堵。

钱七虎：如果出入口选择地面车流大的地方，容易加剧地面流量。但是，如果刻意引到车少的地方，说明这个地方不符合大多数人的出行需求，这种不便利将使得车主不愿选择地下快速路出行。因此，设置多少出入口、在什么地方设置都要遵循一定原则，在参考路网车流的基础上，要协调地面路网和出行需求的关系。

京华时报：除了东西二环路外，您认为哪些路段还需要开发地下空间来缓堵？

钱七虎：目前，钟楼鼓楼、后海、玉渊潭、北京大学、清华大学附近区域也处于无路可建的状况，可以通过发展地下路来缓堵。北京不但应建快速路，还应建地下物流系统，将运货车分流到地下，不给地面添堵。

京华时报：北京地下快速路建设以及地下空间开发未来前景如何？

钱七虎：北京的道路拥堵状况，已经到了必须开发地下空间来缓堵的地步。如果北京早几年开始建设地下路，可能现在就不会这么拥堵。总之，地下快速路的发展前景很好，发展空间很大，甚至可以说发展空间无限。

【幕后】 东西二环地下快速路将重启

早在2004年，中国工程院院士、我国著名地下工程专家钱七虎就建议，北京应大力挖掘地下空间的潜力，大力发展地下道路系统，包括建设东西二环地下快速路。

2005年7月，北京市政府批准《北京市中心城中心地区地下空间开发利用规划（2004—2020年）》，该规划提出了"四纵两横"的地下快速路系统方案，希望借此缓解二环、三环及长安街拥堵问题。东西二环地下快速路的建设首次提上官方议程。

钱七虎院士回忆，当时，包括东西二环快速路在内的地下快速路系统方案，得到了时任北京市市长王岐山、时任规划委主任陈刚等领导的支持，并经市长办公会议讨论通过。但随后多年，地下快速路建设工程一直未有进展，也没有开工建设的时间表。

钱七虎院士认为，有两大原因导致地下快速路建设暂时搁浅，除了当时北京筹办奥运会，建设任务较重，一时未能抽身投入人力、物力启动地下快速路建设这个原因外，更重要的原因在于阻力很大，专家反对的声音很多。一些专家提出了关于文物保护、水源保护、工程成本、建设周期等方面的意见。当时甚至有老专家当面问钱七虎,"你说故宫下能修地下路吗?"。对此,钱七虎用两个问题做出了回应,"为什么一定要在故宫下修?故宫下为什么一定不能修呢?",地质勘查、文物保护工作借助现代科学技术已经可以实现。

时隔5年,在2010年末的治堵新政发布前,钱七虎院士听闻,东西二环等地下快速路系统再次纳入官方视野,"北京的拥堵到了非治不可的地步,刘淇书记下决心了,必须向地下要空间,用高科技措施来破解发展瓶颈。"

(记者:邓 杭)

通过数字化向智慧建造迈进

【新京报讯】 2020年中国国际服务贸易交易会(简称"服贸会")即将开幕。日前,中国工程院院士、国家最高科学技术奖获得者钱七虎接受新京报记者专访表示,此次作为服贸会建筑服务领域主题论坛主讲嘉宾,他带来的关键词是"绿色发展""绿色建造"和"智慧建造"。

钱七虎指出,目前全球正面临气候变暖导致的生态环境恶化等多重挑战,应对之策是要推进绿色发展和绿色建造。而在工程建设领域,实现绿色建造的必然选择和最佳途径是智慧建造。"信息技术要继续深入地应用到建设工程中去,全面推广数字工程和智慧工程,未来还要向少人化、无人化的方向发展。"

谈绿色建造:"绿色发展呼吁建设向绿色建造转型"

新京报: 在20世纪90年代我国"城市病"问题并不突出的时候,您就率先提出开发城市地下空间。当时为何会提出这一想法?

钱七虎: 在20世纪90年代提出开发利用地下空间,可能对当时人均GDP刚超过1000美元的中国来说,是略为超前的,但在全世界范围内就不是了。

我作为一名学者,参加了一些国际会议,也接触到科学的前沿。当时我看了三本讨论全球生态环境的著作,《寂静的春天》《增长的极限》和《只有一个地球》。《寂静的春天》里讲述了由于大规模采用双对氯苯基三氯乙烷(DDT)等剧毒杀虫剂,造成了"没有鸟儿歌唱的树林""死亡的河流"的典型案例,说明"向大自然宣战""征服大自然"的不科学,而应与之和谐相处。《增长的极限》讲的是19世纪到20世纪全球生产力的发展、经济社会的发展取得了巨大成就,但也带来了严重的生态破坏、环境污染、"城市病"等问题。经济社会的发展要有能够适应资源和环境的客观限制,人类不能一味追求增长,不应该去想着战胜自然,要适应自然,与自然和谐相处。所有这些谈到的其实都是一个可持续发展的问题。

1992年,为了履行联合国环境与发展大会通过的《21世纪议程》,中国承诺将可持续发展纳入国家战略,建设可持续发展的城市和社会。虽然当时中国因为城市化水平和经济发展程度相对其他国家滞后,但是发达国家面临的那些问题在中国也将很快就显露出来。

当时我们向中国工程院提出了咨询课题"21世纪中国地下空间开发利用的战略和对策"。充分利用地下空间,特别是大力发展以地铁为骨架的轨道交通系统和集约可持续的城市基础设施,从而节省土地资源,提高土地利用效率,释放出大批地上空间用作绿色植被和生态空间,建设可持续城市。

可持续发展是一个长久的问题,现在进一步发展到绿色发展。在这次服贸会上,我也要讲到绿色发展、绿色建造的问题。

新京报: 在当下的经济社会发展背景下,为何需要呼吁"绿色建造"?

钱七虎: 现在可持续发展、绿色发展已经成为我们全社会包括国家领导人、科学家和大众的普遍共识,绿色发展已成为习近平新时代中国特色社会主义的一个重要组成部分。在20世纪,很多人还没有意

本文原载于《新京报》(2020年9月4日)。

识到气候变化、碳排放等可能引发的全球灾难。现在我们能够切身感受到了,比如,全球气温上升,洪水、风暴潮、台风、飓风等自然灾害增加。

在这次服贸会上,我也要给大家讲一讲在我们的工程建设领域,应该怎么样去应对未来全球气候变化的挑战。

为了适应绿色发展的要求,工程建设应该向"绿色建造"转型。建筑业是全球最大的原材料和能源消耗产业。目前,全球建筑运营能耗已占到总能耗的30%以上,加上建设过程中的能耗,这一指标接近50%。我国传统的粗放建造资源消耗量大、浪费现象严重。许多老旧工程拆除数量规模巨大,许多远未达到使用寿命期限的道路、桥梁、大楼被提前拆除,城市建筑垃圾和建筑物贡献的温室气体,与建筑有关的空气污染、光污染等,都在呼吁"绿色建造"。

谈智慧建造:"智慧建造是实现绿色建造的必然选择和最佳途径"

新京报:工程建设应该如何往"绿色建造"的方向发展?

钱七虎:首先是理念上的推动,大家要认识到绿色建造的重要性,它与人类命运共同体、与未来可能的全球灾难是什么样的关系,要重视并自觉地在实践中贯彻执行绿色发展的理念。其次是政策驱动,比如政府部门对单位和家庭不同等量的用水、用电采取不同的收费价格,鼓励大家节能;也可以通过标准驱动,例如制定绿色建筑标准等对建设项目进行准入,使得高耗能项目退出。

当然最重要的,还是要依靠科技的发展,发展大数据技术、人工智能技术、物联网、5G技术等,有了这些科技基础,建设才会更加高效节能、低碳排放,甚至零排放。这些技术应用到工程建设中,其实就是这次要谈到的另一个主题——智慧建造。智慧建造是实现绿色建造的必然选择和最佳途径。

新京报:在工程建设中,"智慧建造"有哪些体现?

钱七虎:首先是全面的透彻感知系统。拿我熟悉的地下工程、隧道工程来讲,地下工程是高风险的工程,因为地下存在很多不确定性,比如地下有没有水、水量多大,地下岩体是破碎的还是坚固的,破碎的岩体有没有可能导致塌方等。这些情况,人都是看不到感受不到的,要通过设备、传感器、信息化的设备去全面感知,摸清情况。

第二是通过物联网、互联网的全面互联实现感知信息(数据)的高速和实时传输。只能感知还不行,获取的信息一定要快速传输出去,如果当下获得的信息要过几天才能看到,只能进行事后分析,工程建设就不能实时地反馈和服务。有了互联网、物联网、5G技术后,信息传输非常快,可以即时地反映认知。

另一个是智慧平台的打造,技术人员要通过这个平台对反馈来的海量数据进行综合分析、处理、模拟,得出决策,从而及时发布安全预警和处理对策预案,这是非常有必要的。有了这些技术,工程建设的风险更低,施工人员更安全,同时也能最大限度地节省材料和减少环境破坏。

新京报:"智慧建造"中的"智慧"和"智能"的区别是什么?

钱七虎:我认为,智能是相对于简单的、局部的系统,是对某项技术、某个功能和某种设备单元而言的,比如智能手机、智能传感器等。

智慧是对复杂的大系统和巨系统而言,包括感知、认知、传输、处理、服务等多功能的系统。例如城市是一个巨系统,包括人、自然和社会等的综合体,所以有智慧城市。建设工程也是一个大系统,包括工程本体、工程环境、工程建设者和运营者的综合体,所以有智慧工程。在工程建造里面,智慧也代表着最高水平、最高技术含量的产物。

谈科技创新:"智慧工程未来要向无人化、少人化发展"

新京报:"智慧建造"以科技创新为基础。我国工程建设领域的科技创新,经历了怎样的发展历程?

钱七虎:我国工程领域的科技创新经历了从低到高,从局部到全面的发展历程。改革开放以来,工程建设经历了机械化和信息化的发展,比如地下工程原来都是采取钻爆法、人工打眼、人工放炮,后来可以大量应用机械台钻,多钻台车施工,现在可以采取数字化掘进。再如地铁隧道施工目前大都采用盾构法,使用全断面的隧道掘进机(TBM)施工。我们已经成功研发生产了直径15m的盾构掘进机,正在设计直径18.2m的,这是机械化的巨大进步。

信息化的发展体现在地下工程的地质探测工作中。建设地下工程,需要把地下的情况了解得很清楚,地下有没有水,石头有没有节理、裂隙,有没有断层,如何防止地下安全事故发生,这都是靠信息化设备来提升安全性能的。

新京报:未来工程领域的创新方向是什么?

钱七虎:今后,工程领域的进步要通过数字化向智慧方向迈进,比如数字工程的BIM技术在建筑工程中的应用。原来我们做设计是用图纸的,但图纸和工程实体是分离的,比如不同图纸上画出来的线可能是交叉的、矛盾的,到建设中就会出现很多问题。进入BIM时代后,技术人员可以在计算机里建立立体的可视化的工程模型。另外怎么样安装、怎么样开挖效率最高,通过这个模型都可以模拟得到结论。

当前科技创新最主要的标志,就是信息技术继续深入地应用到建设工程中去,也就是全面地推广数字工程和智慧工程。未来,数字工程和智慧工程还要向少人化、无人化的方向发展。

(记者:吴娇颖)

特大城市解决交通拥堵问题的思路与出路

一、问题的提出

2010年12月24日,北京市政府出台《北京市人民政府关于进一步推进首都交通科学发展,加大力度缓解交通拥堵工作的意见》(以下简称"治堵新政"),"治堵新政"中提到的优先发展公共交通、加快基础设施建设、调整重点区域停车收费价格等措施,非常必要。但从长远来看,这些措施仍难以有效解决北京市交通拥堵的难题。

城市交通拥堵本质是交通需求大于供给。正如住建部仇保兴副部长所言:"城市中心区有限的空间与日益增长的向心交通,特别是小汽车交通,形成了难以调和的供需矛盾"。

北京"治堵新政"实际上是采取交通需求管理的思路,从源头强行限制交通需求,并改变不同交通方式的分担率。如明年小客车上牌24万台的限购,提高城市中心区停车收费等政策,其实质是限制小汽车发展、大力发展公共交通的一种方式,其目标是到2015年,中心城公共交通出行比例达到50%左右,自行车出行比例保持在18%左右,小客车出行比例控制在25%以下。

大力发展公共交通能否从根本上解决城市交通拥堵问题呢?以日本为例,公共交通出行比例达到60%以上,仍然没有从根本上解决交通拥堵问题,并以牺牲交通个性化与舒适性为代价。目前北京的地铁拥挤也到一定程度,即使"十二五"期间实现"公共交通出行比例达到50%左右"、轨道交通"2号线、4号线、5号线、10号线高峰时段运力分别提高12.5%、50%、25%、50%"的目标,能否彻底解决交通拥堵问题仍存疑问。

事实上,特大城市的交通拥堵在国际上是一个普遍难题,小打小闹解决不了,需要大思路、大手笔,其出路在科技,出路在地下。

二、解决交通拥堵的思路

如前所述,城市交通拥堵的本质是交通需求大于供给。从长远来看,主要依托现有公共交通方式来同样无法特大城市解决交通拥堵,更无法满足人们对交通出行方式的个性化、舒适性的需求;在这样背景下,解决交通拥堵的大思路有两个:

(1)通过技术革命,发展新的交通运输方式,从根本上改变供需关系,达到供需平衡;

(2)在现有交通供需关系背景下,科学减小需求,并提高供给;从整体上时间与空间的重组,减小需求;拓展现有交通运输系统,提高供给,从而达到新共需平衡。

从目前来看,还没有成熟的新概念运输方式能在效率、个性化及舒适性等方面来替代地铁、小汽车等出行方式,其只能科学减小交通需求,提高供给的方式,使交通需求与供给达到平衡。

三、如何科学抑制交通需求?

1. 严格控制城市中心区的无序蔓延

从历史上来看,交通方式创新与城市空间布局模式是相互影响的。

1634年,伦敦出现雇佣马车,车速约4.8km/h;

1828年,伦敦开始有了公共马车,为了提高车速,保护道路,在路面铺上了石块。

1863年,世界上第一条地铁开通;

1870年,日本东京有了公共马车,十年后又出现了铁路马车。

1880年,伦敦人口已增至近600万人,城市中开始建设有轨电车线路,车速达到16km/h,线路可从市中心向外延伸8km。

由于交通工具的变化,车速和运力的提高,就有可能在一定的旅行时间内,将乘客运送更远的距离,引起城市规模的不断扩展。在现有交通工具(地铁、小汽车等),目前现代世界上几个特大城市的交通圈半径都在50km左右,但都面临着不同程度的交通拥堵问题。

目前,北京现有城市人口已达2000万人左右,机动车数量接近420万辆,由于多年来北京市区以原紫禁城为中心"摊大饼"式的城市发展格局从未改变,北京城区面积超过1000km^2,几乎相当于2个新加坡。

城市人口的增加和范围的扩大,使现有交通方式越来越不能满足对城市交通日益增长的需求,即北京空间范围已超过现有交通方式适合的距离,其空间不能再进行拓展,北京应编制适应现有交通方式的空间规划。

2. 倡导混合用地模式,从源头上减小新与中心区之间的工作出行量

北京自行车出行比例由2000年38.4%下降到2009年的18.1%;小汽车出行比例由2000年的23.2%上升到2009年的34.0%;表面的原因是依赖小汽车出行的交通习惯的因素,实质上与上述交通方式已不适应城市空间布局有很大关系。北京近年建成了天通苑、回龙观等"睡城"以及"科学城"等多个单一功能区,从源头上加剧了新城与中心区之间的工作出行量。

在未来,应转变城市空间的发展模式。在城市总体规划的指导下,加快新城建设,疏散中心区人口和工作岗位,倡导混合用地模式,形成就业和居住基本平衡、配套设施完善、具有较强人口反磁力的综合性新城。

四、科学提高供给

北京中心城区由于高强度开发,地面的土地已经很少,地面空间资源已利用殆尽,水平及向上拓展潜力已有限,可以说是无地可建道路。而且地面公共交通也已饱和,公交车排队进站,占用大量车道。即使能够修建,国外曾做过伦敦中心区的模拟测算,即使把伦敦中心区所有建筑架空,下层全变成道路,也不能解决交通拥堵问题。

因此,在地面空间已不能拓展的情况下,供给的出路在地下,甚至是唯一能够大规模提供交通供给的出路,并具有巨大开发潜力。以深圳市政府评估为例,仅地下30m以内,深圳市浅层地下空间开发量为4.11km^2,次浅层地下空间为9.1亿m^{2}的建筑面积。

从国际上看,向地下要交通空间是未来城市交通发展的主要模式之一。从以地下交通功能层次和地下环境功能层次为主的地下空间的开发利用,将是二十一世纪城市地下空间建设的核心内容。

北京"治堵新政"提到"地下"关键词共7处,如强调"地下环廊"使用、"建设地下停车场、地下快速通道工程"等,但从长远来看,甚至需要考虑"地下城市"理念。美国格兰尼教授和日本尾岛俊雄教授指出,地下城市基本理念之一,是将地上城市中的城市运输和传输系统逐步转移到城市地下的不同层次里。释放出的地上空间,用作大片的自然植被和安全的步行,并且使地上逐步摆脱与传统运输相联系的环境干扰。重新恢复生机的地上环境将改善整体的社会和自然环境,并进一步增进城市的文明。

《欧洲地下建设战略研究议程(Strategic Research Agenda for the European Underground Construction Sector)》明确指出,其2030年的远景目标是"将地面交由市民自由使用,将基础设施放入地下"。

美国 Brad Swartzwelter 先生在其《比飞机还快:美国长期运输问题解决方案》一书中的结语写道:地球上地表空间是供人们休养生息,享受万物精灵和大自然之美丽的。地下广阔的空间才是供人类开发交通运输通道的场所,那里在深度和广度上几乎不受限制,你可自由利用。地下通行权可自由获得,运输线路可以取得很直、很平,不会因地形变化而受到限制。

从目前技术看,地下 100m 以上都是可以建设的。如果一层交通空间不够,还可以建两层、三层,甚至多层,利用的空间还是相当大的。按上述理念,通过地铁、地下道路以及地下物流系统等方式,将北京地面客货运交通按规划逐步部分甚至全部分转为地下,可以预测,特大城市的交通将从根本得到改善。

五、建议

像任何其他的新兴事物一样,地下交通也有赖于时间来调整和充实。虽然,急速扩张的城市面临着不断增长的复杂情况,但随着人们对它特点研究的深入,发现它确实是对复杂情况的富有远见的一种解决方案。

20 世纪后期发展起来的移动通信和互联网带给人类无限的通信自由,彻底是解放交流沟通的"瓶颈";目前,"交通或物流",已经成为电子商务以及社会经济与人类生活方式进一步发展的"瓶颈"。在今天的信息技术(IT)时代之后,即将迎来一场运输革命,而地下交通网将成为这场革命的重要支撑。

从目前来看,地铁、地下道路甚至是地下物流系统,都是基于可获得的技术需要的是"决心、投资"。从城市的可持续发展的迫切需要出发,实施城市地下交通的可行程序是采纳日本西淳二教授的观点:把运输转移至地下中层区域应首先从城市的中心区开始,继而逐步向城市周边移动。

在发展地铁、公交优先、控制小汽车基础上,科学减小交通需求,提高供给,并与智能交通系统(ITS)和建立信息系统等政策和措施相协调,从现在坚持,必然达到既治"标"也治"本"目标,必有成效。

本文参考文献
[1] 钱七虎,郭东军. 城市地下物流系统导论[M]. 北京:人民交通出版社,2007.
[2] 北京市城市规划设计研究院. 城市土地使用与交通协调发展[M]. 北京:中国建筑工业出版社,2009.
[3] 吉迪恩·S·格兰尼. 城市地下空间设计[M]. 北京:中国建筑工业出版社,2005.
[4] 王玉北,陈志龙. 世界地下交通[M]. 南京:东南大学出版社,2010.
[5] 童林旭,祝文君. 城市地下空间资源评估与开发利用规划[M]. 北京:中国建筑工业出版社,2009.

又快又好地推进地下空间建设
——对话中国工程院钱七虎院士

【编者按】 地下空间是城市发展的战略性空间，是一种新型的国土资源。近百年来，在国际城市复兴和新城建设过程中，开发利用地下空间，通过空间形态竖向优化克服"城市病"，已成为城市发展的重要布局原则和成功模式。当前我国城市地下空间建设，正以世界罕见的规模和速度迅速展开，成绩巨大，但正因为规模大、战线多、速度快、时限短，反映暴露的问题值得我们认真研究思考。在2016中国（上海）地下空间开发大会上，本刊编辑带着一系列问题走访了著名地下工程专家、中国工程院钱七虎院士。

城市地下空间开发利用是"转变城市发展方式"解决"城市病"的主要着力点，是"建设和谐宜居、美丽城市"的主要途径。

记者：钱院士您好，这是您第二次参加地下空间大会，我们也看到您一直活跃在地下空间开发领域，给予我们很多意见和指导，在您看来，推动开发地下空间最主要的原因是什么？

钱七虎院士：首先，地下空间开发是城市可持续发展的重要途径。地下空间具有良好的隐蔽性和封闭性，有助于土地的多重利用和空间的复合开发，它不仅能内涵式地拓展空间、优化城市功能，也能少占用地面空间、保护地面景观、增加绿地，并减少尾气和浊气的直接排放。此外，地下冬暖夏凉的特性以及储能、储水、排水、管廊等设施和浅层地热的利用，有利于建设节能、节水、低碳的城市社会。其抵御空袭、地震、洪水等外部灾害的优势，则能提高城市安全性。

其次，利用地下空间科解决城市交通矛盾。交通是城市运行的基本支撑体系。大量建造的地铁、地下快速路、地下停车库、地下步行系统，有效扩大了交通空间供给，提高了交通效率。在解决交通矛盾的同时，也加强了交通与公共空间的水平连通和地上地下空间的立体协同，丰富和扩展了土地利用与交通系统的立体互动关系与范畴，对繁荣经济、保护环境、激发城市活力等，都产生了牵引作用和连锁效应。

第三，地下空间开发利用能够取得综合效益。地下空间的建设直接成本高于地面建设，但从节地、节能、环保等方面综合考虑，更有可行性。以荷兰某铁路工程为例，高架方案最经济，但影响环境，所以看造价不能只看直接造价。

项目	高架方案	隧道（掘开式）	隧道（TBM方法）
施工造价（万元）	308	494	536
土地/迁移/破坏（万元）	122	146	37
分项目1：直接造价（万元）	430	640	573
	100%	148%	133%
运营和维护周期（万元）	31	64	47
分项目2：生命造价（万元）	461	704	620
	100%	152%	134%
损害（换算成金钱）（万元）	133	168	5
分项目3：整个造价和损害（万元）	594	872	625
	100%	147%	102%

珠海横琴综合管廊全长33.4公里，总投资22亿元，因建设地下管廊而节约的土地，产生的直接经济效益就超过80亿元，其次还有供水管网漏损率降低产生的间接经济效益。

以上几点是比较明显的优势，因此开发利用地下空间是势在必行的。

记者：进入21世纪，我国城市地下空间开发快速增长，体系不断完善，特大城市地下空间开发总体规模和速度已居世界同类城市前列，对提高城市空间容量、缓解交通、保护地面环境做出了重要贡献。其中的一些成果您能为我们介绍一下吗？

钱七虎院士：地下交通设施建设取得巨大发展成就。目前，我国地铁建设运营里程居世界前列，有效缓解了交通拥挤。我国跨江河湖海隧道和城市地下快速路的建设也举世瞩目，消除了城市交通的空间屏障，保护了地面环境。除地下停车全面普及外，地下步行及过街系统、地下交通枢纽建设也受到关注。交通地下化有效拓展了城市交通资源，改善了拥堵，同时节约了地面用地，减少了尾气直接排放和景观破坏。

城市大型地下综合体建设项目多、规模大、水平高。特大城市结合地铁建设和旧城改造、新区开发进行地下空间开发，建设了北京中关村西区、上海世博轴、广州珠江新城、杭州钱江新城波浪文化城等融交通、商业、文化、娱乐、市政于一体的地下综合体，单体规模在数十万至数百万平方米之间。

内部环境质量更加舒适，内部采光、通风和视觉环境更人性化，有效改善了交通、购物和休闲环境，提高了城市生活质量，改善了地下空间形象，增强了地下空间的吸引力。

坚持世界眼光、国际标准、中国特色、高点定位，借鉴国外、境外先进经验，努力把城市建设成为人与人、人与自然和谐相处的美丽家园。

记者：日本、加拿大、美国、法国等老牌发达国家和中国香港、台湾地区，以及首尔、新加坡等城市，在地下空间开发过程中，形成了因地制宜、各具特色的利用模式，我们有哪些可以学习利用？

钱七虎院士：现在城市内涝问题很严重，国外很早提出了污水、雨水利用，建设水银行。通过建设雨洪地下调蓄措施，防止城市内涝；实施雨污分流，污水入综合管廊；雨洪不入廊，单独成系统。

日本地下河川的雨洪设施单独设置。在马来西亚吉隆坡市，城市快速路充作吸纳雨洪和排泄的通道，已解决雨涝之害。另外，日本、瑞士和牙买加等国家还利用屋顶收集雨水，并通过管子将雨水输送到地下储水库。日本雨水利用的供水量每年为0.05亿立方米。东京都在公园、校园、体育场、停车场等处的地下，修建了大量的雨水贮留池。凡是新建筑物，包括住宅楼，都要求设置雨水贮留设施。1989年开业的东京港区的野鸟公园，园内用水皆来自雨水，形成了湿地、芦苇荡、草地、树林等景点，成为东京地区的著名观光点之一。

欧美发达国家现在已经基本放弃修建地表水库来储备水资源的传统做法，而是利用地下含水层，建立"水银行"来调节和缓解供水。包括雨洪地表入渗和井灌的人工补给是一种可行、费用低廉的解决供水的办法。它一举多得：储备地下水，在水短缺的时候提供水量满足用水需求；控制由于地下水位下降引起的海水入侵和地面沉降；提高地下水位，减少地下水的抽取费用；维持河流的基流；通过土壤中的细菌作用、吸附作用和其他物理、化学作用改善水质；通过处理后的污水入渗来实现污水的循环再利用，以管理不断增加的大量污水；最后也是最重要的是保护了生态环境。

20世纪80年代，美国开展了钻孔不给含水层的回复工程ASR系统，美国已运行的系统有18个，正在建设之中的ASR系统有40多个，瑞典汉兰和德国的AR工程在总供水中占20%，我们是不是可以向他们学习？

其次，就是目前很热门的综合管廊问题。日本在综合管廊建设方面很先进，燃气和垃圾管道都纳入管廊，并且单独设管。瑞典是在20世纪60年代建设并实施采用压缩空气吹运垃圾系统（PWT）的国家，把垃圾用于焚烧发电！发展地下垃圾处理系统节约了城市宝贵的土地资源，有很大的回报，控制了对环境的噪声和恶臭的污染，减少了对环境的视觉负面影响。

充分开发利用地下空间，建设节能型城市还包括充分利用地下土壤、地下水的低密度天然能源作为冬季热源和夏季冷源,然后再由热泵机组向建筑物供热供冷,地源热泵技术就是这样一种利用地下可再生能源的既可供暖又可供冷的新型中央空调系统,它包括地下埋管式地源热泵和抽取地下水然后回灌的水源热泵。埋管式地源热泵技术目前在国外大面积推广,在欧美已得到普遍应用,是一种成熟可行的可持续发展的节能新技术。目前,地源热泵正进一步与太阳能结合,应对它的一些局限性。

这里我还想讲一下地下货物运输系统,大家可能对地下交通运输很熟悉了,地下货物运输系统,又称地下物流系统(ULS)是基于城内运输和城外运输的结合。城外货物通过各种运输手段到城市边缘的物流园区(CLP),经处理后由CLP通过ULS输送到各个终端。它以集装箱和货盘车为基本单元,以自动导向车(AGV)为运输工具。

ULS的环境效应显著。世界经济合作组织在2003年的《配送:21世纪城市货运的挑战》报告中指出,发达国家主要城市的货运占城市交通总量的10%~15%,而货运车辆对城市环境污染总量的贡献为40%~60%。日本东京建设300公里地下物流系统评估报告中:一氧化氮和二氧化碳将分别减少10%和18%,能源消耗减少18%,运输速度提高24%,所以这个系统也是我们以后可以发展的。

记者:总体上看,目前我国已经成为世界城市地下空间开发利用的大国,但不是强国,由于缺乏整体性的规划、综合性评估机制等原因,我国城市地下空间开发中还存在一些亟待解决的问题,这方面您怎么看?

钱七虎院士:我国在地下空间开发方面取得了一定的成果,但是还存在一些问题,主要表现在,形态上水平方向连通少、地上地下空间关联差,效果上未形成交通和空间的动态立体关系;一些地下空间规模偏大、开发时缺少条件预留、缺少综合规划等。

概括起来,造成上述问题的原因主要是:缺乏完善的地下空间法律、法规;缺乏有效的综合性评估机制、管理机制;地下空间规划与城市规划脱节,缺乏有效的规划控制等。

国外先进经验是先立法、后建设,这样才能有序有力推进建设。在这方面,我们可以借鉴德国的《城市建设法典》,日本的《大深度地下公共使用特别措施法》《共同沟法》,美国各州制定的空间法,瑞士的《瑞士民法典》,意大利的《民法典》等,我国台湾地区的《共同管道法》《大众捷运法》。

"突出绿色、低碳、环保、可持续发展理念,最先进的理念、最高的标准、最好的质量""发扬'工匠'精神,精心推进,不留历史遗憾"进行规划设计。

记者:城市地下空间资源是宝贵的有限的不可逆的资源,您也多次提到了"规划"的重要性,那么在规划地下空间开发利用的时候我们应该注意些什么?

钱七虎院士:当前,我国地下空间开发特别是地下综合管廊和海绵城市的建设,形势非常好,规模速度是空前的。但是正因为规模大、战线多、速度快、时间短,暴露的问题更值得我们认真研究思考。问题的核心是什么呢? 两个字,一个是好,一个是稳。怎么来研究解决这些问题呢?

第一个问题就是科学系统地编制好城市地下空间发展规划。先规划后建设,规划引领,规划先行,科学的规划,好的规划,因为规划科学是最大的效率,规划的失误是最大的失误,城市工作要把创造优良宜居环境作为中心目标,努力把城市建设成为人与人、人与自然和谐相处的美丽家园。

以"三最一突出"为原则,突出绿色、低碳、环保、可持续发展的理念,三最是最先进的理念,最高的标准,最好的质量。要发扬工匠精神,精心推进,不留下历史遗憾地进行规划设计。

具体来讲,我们怎么把规划做好呢? 制定地下空间规划,地下的工程地质、水文地质情况要明。不是所有的地下空间都要开发利用,这是不对的。有的地方可以开发,有的地方不宜开发,有的地方要控制开发,什么道理呢? 就是水文地质问题,这个要搞清楚,不搞清楚,必然会有失误。另外目前地下空间开发情况要清楚,目前开发得怎么样,有什么问题要清楚。比如综合管廊,目前有多少管线,它的现状问题在什么地方,要清楚。现在综合管廊建设反映的问题很多,施工麻烦,成本太高;出入口、逃生口、管线交叉和外面的接口很多。能否在规范方面、在研究阶段有所突破。比如说我们的管廊不是有人的,仅仅在维修的时候有些人,那么是不是一定需要进风口、排风口,是不是搞一些纵向排风就可以? 是不是逃生口可以不用那么多？这些都值得研究。

比如以前电力电缆，放到地下有没有问题。还有地上地下要协调，地下不光是综合管廊，不光是海绵城市建设，还要有轨道交通，还有地下通行道，还要有停车场，还要有围绕地下空间的站点等等，所以空间布置要合理协调，要有工匠精神，精心研究。要稳，不能贪快。我的体会一个是好，一个是稳，这就是我觉得最重要的两个点。

【后记】 当前我国城市化发展面临严峻挑战，土地紧张，绿色面积减少，交通拥堵，环境污染，纵观世界发达国家和我国城市建设的现状，向地下要空间要土地要资源，已经成为现代化城市发展的必然趋势。近年来，随着工业化城市化的进程推进，我国城市空间推进进入快速发展阶段，据不完全统计，地下空间开发与同期地面建筑竣工面积的比例，从约10%增加到15%。尤其是一些大城市，在轨道交通和地上地下综合建设的带动下，低下空间开发规模增长迅速，类型呈现多样化、深度化、复杂化的发展趋势。科学规划、有序建设，系统建设安全运营，综合利用地下空间是城市发展的必然趋势。

| 思想与观点

迎接气候变化的挑战，推进绿色建设、实施生态大保护

联合国政府间气候变化专门委员会（IPCC）于2019年9月发布报告称：如果温室气体减排不力，全球变暖导致全球冰库加快融化，21世纪末海平面将上升0.6～1.1m，比2013年发布的报告增加0.1m；海洋变暖，吸收CO_2量增加导致酸性提高，从而威胁鱼类的生存，21世纪末全世界渔业的收获量将减少24%；海洋变暖更催生厄尔尼诺现象和拉尼娜现象，导致极端气候频繁，台风飓风等风暴潮和洪涝灾害以及海平面上升将使全球2.8亿人失去家园，上海、纽约等大城市将面临灾难性洪水。

全球温室气体的排放有2/3来自城市，但是全球人口仍加速增长，从1999年60亿人到2050年将近100亿人，城市化正加速发展，联合国人类住区规划署报告称，全球每周有140万人迁到城市，这也导致温室气体排放加速。

全球生态环境的出路在哪里，对策在何方？

碳排放主要来自化石能源的燃烧。森林、湿地和绿色植被是世界上抵御气候变化的前沿防线，每年吸收的人为碳排放量高达总量的1/4。Science发表的一项研究预测，如果新种植9亿公顷树木——大约5000亿棵树苗——那么它们在成熟时可吸收2050亿吨碳，相当于工业革命以来人为碳排放量的2/3。这些数据表明，气候变化的应对之策应是推进绿色发展，实施生态大保护：建设绿色生态城市，特别是发展绿色能源和绿色交通；保护世界原有森林和绿地；建设绿色国土，发展绿水青山。

中国正在绿色发展和生态大保护的道路上快步前进。生态文明建设和绿色发展思想是习近平新时代中国特色社会主义思想的一个不可或缺的重要部分。党的十八大以来，党中央提出一系列治国理政新理念、新战略，其中生态文明建设和绿色发展提到了前未有的战略新高度，绿色已成为中国国家建设发展的五大理念之一。

中国已经进入世界绿色发展的先进行列：联合国报告，世界新增绿色植被的1/4是由中国创造的，塞罕坝就是联合国表彰的典型；全球能源转型地缘政治委员会发布报告称，中国成为可再生能源的太阳能电池板、风力涡轮机的最大生产国、出口国和安装国，以及清洁能源的电池和电动汽车的最大生产国、出口国和安装国；中国在可再生能源技术方面远超美国，居领先地位，占专利总数30%；中国有世界最大规模清洁能源的高铁，里程达3.1万km。据美国《纽约时报》网站刊登的美国前国务卿约翰·克里题为《不要让中国在绿色竞赛中取胜》的文章称，2016年，中国可再生能源投资1260亿美元，而美国仅400多亿美元。目前，我国在可再生能源地热利用的总量上已位居世界第一，浅层和中深层地热能供热建筑面积超过11亿m^2。

《河北雄安新区规划纲要》规定新城建设空间仅占30%，而60%～70%的城市空间为绿色生态空间，雄安新区将向世界指明未来城市的建设方向；《长三角生态绿色一体化发展示范区总体方案》强调一体化示范区要生态筑底、绿色发展，先行启动区蓝绿空间占比不低于75%，长三角绿色一体化发展示范区将引领城市群的建设方向。

如何实现上述目标？雄安新区和北京副中心通州区的建设控制性详细规划表明，必须充分利用地下空间，特别是大力发展以地铁为骨架的轨道交通系统和城区地下物流系统为特征的绿色交通和绿色城市基础设施，不但使市民交通和货物运输转入地下，还包括垃圾、污水等的集输运以及一切脏、乱、差的建筑

和设施都转入地下,从而节省土地资源,提高土地利用效率,释放出大批地上空间用作大片的绿色植被和生态空间。建设绿色城市,中国正在实施长江流域生态大保护、黄河流域生态大保护、三江源生态大保护、长三角生态绿色一体化发展,正在大片国土上发展绿水青山,在沙漠中建设绿洲,相信经过若干个五年计划,持之以恒,我们一定能实现绿色美丽中国的伟大梦想!

(作者:钱七虎)

依托中国的独特优势,加速迈向科技强国的伟大目标

目前在很多科技领域,中国只能说是"大国"而不是强国,正走在"由大变强"的路上。

以我从事的岩石力学与工程为例,中国的岩石力学与工程科技工作者的数量居世界第一,中国建成与在建的岩石工程的数量居世界第一,规模最大最复杂的工程也在中国。中国以前的大型岩土工程装备,例如开挖隧道的盾构机和TBM硬岩掘进机都是从国外引进,不能建长大深埋隧道,现在这些高端装备的国产化率和高端装备的市场占有率都有所提高,可以出口海外,长大深埋隧道也具备建设能力。经过10多年的努力,我担任主编的英文学术期刊 Journal of Rock Mechanics and Geotechnical Engineering(《岩石力学与工程学报》)2019年也被SCI收录了;中国的冯夏庭教授担任了国际岩石力学学会主席。

尽管如此,建成岩石力学与工程强国,还需要多方面攻坚。包括高端装备的核心部件研制,岩石工程的技术标准和试验方法的创新,岩石力学的基础理论的探索。此外,更需要在国际性的学会与学术期刊以及制定技术标准等平台上取得更大的话语权。

科技需求是科技创新的动力,中国强大的科技需求必将牵引中国科技创新的快速发展。以我从事的防护工程领域来说,川藏铁路建设提供了大量必须攻克的岩石力学与工程科技的难题;川藏铁路全线80%以上是长大深埋隧道,平均长度10km,平均埋深1000m,最大埋深大于2900m,这些位于印度板块与欧亚板块碰撞带高海拔地区的隧道建设,亟须解决包括岩爆、挤压大变形等一系列技术挑战。因此,正如国际岩石力学学会前主席英国工程院院士哈德逊所言,"21世纪的岩石工程与岩石力学在中国"。

长期以来,我国在应对各种挑战的同时,形成了独特优势:一是集中力量办大事的社会主义制度优势;二是中华民族家国情怀的优良传统与博大精深的中华文化;三是应用马克思主义世界观和方法论解难事的优势。这些优势必将在建设科技强国的道路上充分发挥出巨大作用。

2020年是中国建设世界科技强国"三步走"战略目标的重要时间节点,站在这个时间节点上,基于中国最强大的科技创新需求,凭借独特优势,只要我们科技工作者众志成城,充分贡献聪明才智,我们一定能实现建成领跑世界的科技强国的伟大目标!

在第四个全国科技工作者日来临之际,向全国科技工作者致以节日的祝福,让我们继承老一辈科学家"无私奉献、科技报国"的光荣传统,弘扬"科技为民、奋斗有我"的科学家精神和科技志愿服务精神,紧紧团结在以习近平同志为核心的党中央周围,为实现全面建成小康社会目标任务、建设创新型国家和世界科技强国不懈奋斗。

本文原载于《科技导报》(2020年38卷第10期)。

在第四个全国科技工作者日来临之际，向全国科技工作者致以节日祝福。让我们继承老一辈科学家"无私奉献、科技报国"的光荣传统，弘扬"科技为民、奋斗有我"的科学家精神和科技志愿服务精神，紧密团结在以习近平同志为核心的党中央周围，为实现全面建成小康社会目标任务、建设创新型国家和世界科技强国不懈奋斗。

钱七虎
二〇二〇年四月八日

钱七虎院士为《科技导报》"2020全国科技工作者日专刊"题词

（作者：钱七虎）

| 思想与观点

隧道与地下开发实现历史性"穿越"
——访中国工程院院士钱七虎

【编者按】"穿越"这个词2012年特别火,指的是新派文学或是热门电视连续剧,那里的穿越指的是时空。而我认为,"穿越"这个词用在隧道产业再贴切不过了。无论是高铁、还是城市轨道交通;无论是海底隧道、桥隧,还是山洞;无论是地上,还是地下……这些伟大的"穿越"正是由一条条隧道实现的!而进行隧道建设的人们就是名副其实的"穿越者"!

"隧贯山河,道通天下"。在崎岖的大山里运用科学和技术为我们打通一条通向目的地的坦途,"穿越者"在城市下面为人们开拓地下交通和生存空间,他们在一望无际的深邃海底运用高科技设备静静地开拓出一条通往海那一边的捷径……

当人类进入21世纪,隧道建设与地下开发的过程是怎样的?我国隧道产业在增强自主创新能力方面,"穿越者"进行了怎样的探索,完成了时代赋予的哪些新课题?日前,本刊记者专访了中国工程院院士钱七虎。

RT《轨道交通》:据了解,钱院士曾主持了"21世纪中国城市地下空间发展战略及对策"的国家咨询课题研究,请您谈谈我国未来城市地下空间发展战略及对策?

钱院士:**中国地下空间发展现状**:目前中国地下空间开发以前所未有的速度增长,城市(尤其是城市中心区)地下空间利用规模越来越大。目前我国北京、上海等特大城市地下空间建成面积已超过5000万平方米,深圳市地下空间现有开发量达到1900万平方米,杭州市地下空间现在开发量达到800万平方米;并每年都以建成百万级平方米的速度持续增长。据预测,到2020年,北京地下空间开发总面积将达到9000万平方米,中心城区的地下空间将达到地面建筑总面积的20%~30%;并分别在丰台丽泽商务区和通州新城规划了230万平方米和210万平方米的超大地下空间,将立体开发到5~6层。

近期发展战略重点建议:对于城镇土地资源合理利用与地下空间开发利用的必要性、重要性已毋庸置疑。为促进我国地下空间快速可持续健康发展,其近期发展战略重点建议如下:

(1)**建立综合管理机构,理顺城市地下空间管理体制**。针对地下空间的开发利用涉及国土资源、规划、建设、市政、交通、人防、园林、公安消防等各个行政管理部门实际,抓紧研究建立统一管理、分工协作、衔接有序的地下空间管理体制。在尊重现有各部门职能分工的基础上,尽快明确一个综合管理机构,承担综合管理的职能。

(2)**完善城市地下空间管理法规,明确产权**。尽快明确地下空间的使用权的性质、取得方式、转让方法及权属管理等民事基础权利。在《物权法》中要体现地下空间使用权的立法,并修订《土地管理法》《城市房地产管理法》等相关民法与之相衔接,明确地下空间等。按先单项立法,后综合立法原则,不断完善城市地下空间管理法规。

(3)**科技支撑战略**:多层次、多尺度对我国城市地下空间开发过程中面临的重点、难点任务进行联合攻关,建立系统的城镇土地资源合理利用与地下空间开发利用创新性技术体系,为我国城镇化的快速发展以及可持续发展战略与科学发展观的实施提供坚实的技术支撑。

(4)**试点示范**:按"图难由易"的原则,选择典型城市进行地下空间管理体制、法规和科技创新方面的

本文原载于RT《轨道交通》。

试点和示范,逐步展开,确保我国地下空间快速健康发展。

RT《轨道交通》:钱院士学识渊博,在土木工程、岩土工程、地下工程和爆破工程等多个领域造诣很深。您首次提出了我国工业经济防护研究的理论,现任国际城市地下空间联合研究中心亚洲区主任。请您谈谈目前国际城市地下空间开发的现状及发展趋势和策略?

钱院士:从世界范围来看,城市空间资源"硬"约束是地下空间开发利用的首要动因,深层化和城市与城市间交通的地下化是目前地下空间开发利用最重要的趋势。

(1)深层化。国际地下空间领域著名学者,美国宾夕法尼亚州立大学 Gideon S. Golany(格兰尼)教授认为:"从理论上,地下空间的利用可以拓展到几乎无限的深度。"目前地下工程施工技术进展(地下100米以上都可以建设),使得深层地下空间开发利用成为可能。其标志性事件有两个:一是2001年日本《大深度地下公共使用特别措施法》正式施行;另一个瑞士国家研究基金资助洛桑联邦理工(Aurele Parriax 教授)进行了"深层城市(Deep City)"课题研究。而事实上,苏联地铁在20世纪其建设深度已达数十米,日本东京2000年后新建的地铁深度已在40米以下。

(2)城市与城市间交通的地下化。包括两个方面,一是运"人":如地铁,地下道路等;在大城市建设大运量快速公共交通系统-地铁已成为共识,而美国史上最大的交通工程项目"桥改隧"工程:"bigdig",对世界范围内地下道路建设起到了强烈的示范效应;二是运"货":即地下物流系统,如现在正在使用的真空垃圾处理系统,以及正在研究发展的智能化程度更高的地下物流系统,如荷兰正在研究的地下 AGV 系统,德国的 Cargo Cap 系统等。此外,城市间交通的地下化也正在研究实践,如正在研究的瑞士地铁(Swiss Metro)系统,以及正在实施的日本磁悬浮列车中央新干线计划(其中100公里在大深度地下)等。

RT《轨道交通》:有人提出,地面太挤了我们住地下吧!除了地铁以外,还要建设城市地下快速路,甚至可以建造地下商场和住宅。对此,您有什么看法?

钱院士:首先,我反对人住在地下,即在地下修建住宅,有的院士提倡未来将有1/3的人口住在地下,这是错误的。这是因为地下的环境较地上为差,而且有的地下还有氡离子的放射影响。我主张城市交通,包括人流和物流转入地下。未来的城市转入地下,是指的将不宜放在地上的一切设施转入地下。

1. 修建地下道路可以节约城市日益紧缺的土地资源

我国拥有接近14亿的人口,如此多的人口需要大量耕地以保证粮食供应问题,国务院提出的我国耕地面积红线是18亿亩。根据国土资源部数据显示,我国2010年现有耕地为18.18亿亩。

按照联合国的统计,每一个城市人口需要200平方米以上城市用地为其提供生活和社会活动需要,中国的城镇化速率平均为1%(我国城市化率1980年为19%,2010年达到47%,预计2025年将达到59%),那么每年约需要使用520万亩土地以满足城市化发展的需求,现有土地余量仅仅能支撑3~5年的城市化进程,因此,必须严格控制土地需求,科学合理进行土地利用。城市交通用地是城市建设用地的大户,把城市交通尽可能转入地下,就可实现土地的多重利用,提高土地利用效率,实现节地的要求。

2. 修建城市地下快速路可以减少城市污染,改善城市环境

当前城市污染中,机动车尾气排放成为部分大中城市大气污染的主要来源。交通是温室效应气体 CO_2 及污染气体 CO 和 NO_X 的主要排放源,每年排放碳粒子4.75亿吨,机动车排放的污染物占城市大气污染物的比例高达40%~60%,并正以每年2%~3%的比例增加。汽车尾气已经成为危害城市居民身体健康的元凶。

修建地下快速路,可以将地下快速路进行封闭,不使尾气直接排放到大气中去,并采取合理措施和方法对汽车尾气进行集中处理,包括高科技处理为无害气体,从而可以大大减少对城市的污染。

国外已经有了较成熟的修建地下快速路的经验。美国波士顿1994年拆除原6车道的中央高架大道,代之以修建一条8~10车道的地下快速路和一条穿越波士顿港湾通向机场的8车道水底下隧道。工程包括总长7.8英里的匝道连接线和以单车道长度设计的161英里的道路,以及世界上最大的通风和废气处理系统等。工程竣工后,快速路拥堵时间将从现在的10小时缩短到一般城市的2小时,将可以降低城市12%的CO排放量,从而改善城市空气质量;俄罗斯莫斯科为了保护列福尔托地区众多历史文化遗迹和人文景观,在三环线上地下36米深处修建近4公里的快速路段;东京修建了连接池袋、新宿和涩谷三个商业中心的中央环状新宿线,其中埋深在40米以下段长11公里,环线完工后,由池袋到涩谷的时间由原来的50分钟缩减到20分钟,并且每年减少2.5万吨CO_2、160万吨NO、16万吨悬浮颗粒的排放,大大改善了周边环境。

国内的上海、北京、重庆、南京等也都开始了修建城市地下快速路的探索和实践。

3. 修建地下快速路可以实现人车分流,为地面创造更好的宜居环境

在城市修建地下快速路,可以很好地将人流和车流分离,将车流交通引入地下,为地面上生活的人创造了更加宜人的环境。

修建地下道路可以大大减小对地面环境和景观的破坏,保持地面景观和历史文化遗迹等,同时,地下道路的修建减少了地面土地的利用,减少了地面的道路需求,这样就减少了地面道路对城市景观的隔断和破坏,地面可以有更大的开敞空间,有更多的空间用来做绿地和活动广场等,使得城市环境更加宜居。

RT《轨道交通》:钱院士曾经主持和参加了国内多条地铁工程、水下隧道和海底隧道等重大工程的设计方案审查和评标工作;针对地铁工程地下空间开发利用方面,地铁建设是高风险的地下工程,安全事故频繁发生,这困扰了地铁建设者。如何建立安全风险管理体系,有效规避、降低施工安全风险?

钱院士:地铁工程是一项高风险工程,在施工过程中极易引发安全事故。这就要求我们的地铁建设者在工程建设过程中首先要树立风险意识,要有"如履薄冰、如临深渊"的态度来对待工程建设中的每一项工作。

研究表明,地铁的安全风险主要来源于我们对地下客观状态的认知的不确定性和人的不安全行为的不确定性。

我们知道,现有的勘察技术手段对地下的工程地质、水文地质和地下环境的探查存在一定的缺陷,不可能百分之百的搞清楚。加之在地下施工过程中由于施工扰动使得地下环境发生动态的变化。这些在客观上带来了施工的安全风险。这就要求我们在工程建设过程中引入风险管理,建立安全风险管理体系,包括对勘察、规划、设计、施工和工后全过程的风险管理,从技术和管理上高度重视地下工程的不确定性,在规划设计阶段要采取合理措施规避和降低施工风险,在施工阶段采用动态管理的思想,密切监视施工过程中工程和环境的变化,并及时调整技术措施来降低风险。在工后阶段要评估运营风险,并采取有效监控手段和技术措施来降低运营风险。

另一方面,由于工程建设过程中由人来实施的决策、设计、管理、操作方面的不安全行为,也给地铁施工带来了风险隐患。这其中,人的不安全行为是主要因素。因此,在建立安全风险管理体系过程中,要特别重视对人的管理。首先要建设安全风险文化,树立全员安全风险意识;其次要建立健全各种规章制度,利用制度来约束人的行为,防止错误的发生。

RT《轨道交通》:面对国内二线城市轨道交通蓬勃发展的势头,钱院士提出"建地铁科学管理不能光喊口号!我国二线城市建地铁起步晚更应将安全风险管理体系建设贯穿始终。"地铁建设风险点主要集中在哪些方面?相对于北京、上海、深圳等大型城市来讲,二线城市地铁建设应该如何吸收和借鉴以往地铁建设中所采用的先进技术?针对安全风险管理体系,二线城市如何提高轨道交通建设的安全风险管理水平?

钱院士:我国二线城市地铁建设相对一线城市起步较晚,建设规模也相对较小,但是安全风险形势一样严峻。地铁建设是一项利民工程,不能因为工程的安全风险而变成"害民工程"和"扰民工程"。一线城市在地铁建设过程中,有很多经验教训值得二线城市吸取,例如,要科学合理的安排工期,不能因为其他因素盲目加快工期,而带来巨大的安全风险。同时,一线城市在实践中逐渐摸索出的安全风险管理的好的经验,如采用第三方监测、开展风险评估工作、建立风险监控系统等,值得二线城市学习和借鉴。

二线城市地铁建设应根据本地的特点,建立合适自己的风险管理体系,切忌照搬照抄。首先,要合理规划工程规模,不能盲目"求大求全",而不顾城市承受能力带来安全风险。另外,二线城市要根据自身区域地质特点,及时总结经验,找到合适自己的技术手段来进行工程建设。例如,地铁工程大量使用盾构法,但是对硬岩地区使用盾构法则不适应,容易引起刀盘磨损、盾构卡壳等问题。最后,要建立适合本地情况的高效的风险管理组织体系,要灵活高效地开展风险管理工作,切忌为了安全管理而衍生出新的臃肿的"衙门"而人浮于事,起不到风险管理的作用。

二线城市在地铁建设中有"后发优势",我相信,只要不断学习和总结,一定能够不断提高轨道交通建设的安全风险管理水平,让地铁造福于人民。

(记者:刘苊萱)

"岩爆"可测时代即将到来

日前,第十三届全国岩石动力学会学术会议在重庆召开,近200名学者、专家参加了此次会议。防护工程专家、中国工程院院士钱七虎在会后接受《中国科学报》记者采访时表示,当前我国的岩石动力学已经取得丰硕成果,并在岩石爆炸预测预报方面取得重要进展。

钱七虎介绍说,作为岩土工程的一个分支学科,地震、爆炸、岩爆等都是岩石动力学所涉及、研究的内容。当前,岩土动力学可广泛应用在高层建设时土体爆炸压密、地震波预报、液燃气开采等方面。

钱七虎认为,岩石动力学学科的发展是由工程牵引的,而目前我国的大规模建设比较多,"工程带动学科发展,使得我们取得了丰硕成果"。

他举例说,现在我国已经能在2500米深的水下建水电站(锦屏二级水电站),在1000米深的地下建煤矿。在这样的压力下,岩石极易反弹,一开挖容易产生岩石爆炸(简称岩爆),易造成人员伤亡。世界上对防岩爆的研究很多,目前中国学者在岩爆预测预报方面取得了很大进展,"岩石要爆炸首先会裂开,也可能裂开了没有爆炸,但可以以此进行预测。大连理工大学和中科院共同进行的实验显示,我们的预报准确率在80%以上,这已经很了不起了"。

而在地震预报中,地震波会先于地震发生,可能只有几分钟,而这足以让人逃生。因此地震波在岩石中的传播研究与地震预报密切相关,非常复杂,"关于地震波在岩体中的传播,目前我们也取得了很多成果和进步。"钱七虎说,"在岩土的爆炸效应、定向大爆炸的技术等方面,我国也取得了一定的进展。"

然而和国外的相关研究相比,我国尚存在哪些不足?对此,钱七虎坦承,首先,岩爆预报的软件是我们自己设计的,但监测微震的仪器还依赖进口。"我们的预报水平比国外高,但硬件还需要从国外引进。"

"我们的大爆破规模比人家大,复杂程度比人家高。但液燃气的开采技术人家比我们先进。"钱七虎说道。

对于现在的年轻学者,尤其是研究岩土动力学的青年人,钱七虎提出了自己的期待:"一是要密切结合工程实际,要知道工程需要什么,为经济建设服务,为基础建设服务;二是瞄准学科发展前沿,我国是岩石工程大国,但还不是强国。我们要把理论科学搞上去。在大国向强国发展的路上,年轻人要为此多作贡献。"他还勉励年轻人要瞄准前沿,多作出具有科技创新性的成果、应用性的成果,"要乐于奉献,不为名利"。

谈到自己下一步的工作计划,钱七虎介绍说工作重点有两个,一个是岩石爆炸的动力效应;另一个则是岩石中的爆炸动力学,力图把岩爆的等级、范围计算出来,"岩爆可预报的时代快到了"。

(记者:彭科峰)

本文原载于《中国科学报》(2014年1月16日)。

利用地下空间建设"花园城市"

一、前言

"花园城市"的概念于1820年由著名的空想社会主义者欧文最先提出。1898年城市现代化运动发起人霍华德首次提出了"花园城市"的理论，主要内容包括：一是通过对城市和其周围5倍土地的统一规划，各种分散自发的城乡系统组合形成更有效的花园城市；二是大都市圈是一个星系结构，大都市圈的城市群相互之间应有快速公交和自由畅通的汽车道相互联系，各个城镇之间应由永久保护的农田、公园和保护的土地相互隔离；三是花园城市应以中央公园和花园广场为中心，四周发展经济文化社会康乐设施，环形花园大道与放射性的林荫大道作为内区和外区、街区与街区的隔离。其核心思想是使人们能够生活在既有良好的社会经济环境，又有美好的自然环境的新型城市之中。

随着时代的发展和可持续发展战略的提出，人们的环境意识在不断加强，又相继出现了"生态城市""山水城市"等概念，它们与"花园城市"的概念并不矛盾，而是相互融合的，都强调人与环境的和谐，从而达成城市的可持续发展。如在1971年在联合国教科文组织第16届会议上提出的"生态城市"，其发展目标是实现人—社会—自然的和谐，包含人与人和谐、人与自然和谐、自然系统和谐三方面内容其中追求自然系统和谐、人与自然和谐是基础和条件，实现人与人和谐才是"生态城市"的根本目的。于1990年由钱学森首先提出的"山水城市"，是从中国传统的山水自然观、天人合一的哲学观基础上提出的未来城市构想，可视为生态城市的一种形式，其目的是让人类与自然更接近一些，让生命有机地生长于大地环境，与自然界形成一种良性的生态平衡，为人类提供能提高人的能力、产生活力的场所。国际上一些专家预言：二十一世纪世界上两大最具影响的事件之一是中国的城市化。因为，中国的城市化与全球的资源和环境问题密切相关。在我国城市化进程中，按可持续发展的原则，提出了建设"花园城市"与"生态城市"，如北京申奥成功后，就提出要建设"生态城市"。但同时又面临人口多、土地少、交通挤、环境差、用水缺、基础薄等问题。因此，为拓展空间资源，应向地下要土地，大力开发利用城市地下空间，这不仅是城市发展的必然趋势，也是衡量城市现代化的重要标志。

二、走地下化发展的道路，通过土地的多重利用，增加人均绿地面积

联合国建议城市公共绿地应达到人均40m²时，这是基于城市生态环境的需要。《中国二十一世纪议程》提出：我国城市绿化覆盖率达到30%，人均公共绿地面积达到7m²。但据1990年统计，我国城市人均绿地面积只有5.29m²，上海市人均绿地面积仅1.9m²，与此标准相距甚远。与国外发达国家大城市相比，差距就更大，如伦敦人均22.8m²，巴黎人均25m²，莫斯科人均44m²，华盛顿人均40m²，世界"绿都"华沙人均70m²。为此，要想建设"花园城市"，就需要腾出更多的地面空间用于城市绿化。

我国的城市发展沿用的是"摊煎饼"式的粗放经营模式，这种城市蔓延的现象，侵占了大量本来十分有限的土地资源。在1986—1996十年间，全国31个特大城市城区实际占地规模扩大50%。按目前人均城市用地100平方米的水平计算，到21世纪中叶，我国的城市发展将再占地1亿多亩。而按人口平均，

本文原载于《中国建设信息》（2005年第3期）。

我国是耕地资源小国,人均只有1.44亩,仅及世界人均值4.65亩的31%。据中国科学院的研究报告,我国土地能供应17亿人口的粮食是以人均耕地基本维持目前水平为基础的。为此,我国实行了耕地预警制度,确保耕地数量动态平衡,城市新一轮总体规划原则上不再增加城市用地。因此,为解决这种供与需的矛盾,城市应走地下化发展的道路,通过土地的多重利用,增加人均绿地面积。这种土地的多重利用包括三方面:

(1)将适宜转入地下的地面建筑设施转入地下,腾出绿化用地。国外城市地下空间开发利用的一个重要观点是:把一切可转入地下的设施尽可能逐步转入地下。世界各国开发利用地下空间的实践表明,可转入地下的设施非常广泛,包括交通设施、市政基础设施、商业设施、文化娱乐体育设施、防灾设施、储存以及生产设施、能源设施、研究实验设施等。如美国的波士顿拆除穿过市中心的六车道高架路,改为建设8~10车道的地下高速路,原有的地面变成林荫路和街心公园。这样做的结果:市中栽植了2400棵乔木树、7000多棵灌木树;在海湾的景观岛上栽植了另外的2400棵乔木和26000棵灌木;增加了260英亩新的公园和开敞空间。

(2)在绿地集中地区下方,新建地下建筑设施。在不减少地面原有绿化种植的基础上,可在其地下开发适量的建筑设施,并可采用"玻璃连拱廊""采光宵顶"或"下沉式广场""下沉式小庭院"等建筑形式改善和美化地下空间环境,并将阳光、绿化引入地下,使地面丰富的绿化种植为地下空间所共享。如荷兰提出在其"绿色心脏地带"(阿姆斯特丹、鹿特丹、海牙、乌特列希等城市组成的具有重要观光和自然景观的大都市区),只建设地下工程以提供对其现有景观破坏最小化的机会,同时在某些程度上满足了空间增加的要求。

(3)注意在建设"花园城市"的同时,使城市更加紧密。现在国际上提倡的紧密式的城市结构,不但节省了土地资源,而且相对降低了对城区内交通的需求,使得高质量的服务更接近居民(例如高水平的文化艺术设施),更容易保护城市的文化历史景观。同时,这也是与我们耕地资源紧缺的国情是相符的。

三、发展地下轨道交通,缓解城市交通拥挤,降低汽车尾气污染

美国宾夕法尼亚大学城市规划专业高级教授格兰尼曾预言:未来的城市应向地下发展,首先是交通要转入地下,为什么?因为随着城市机动车数量的日益快速增长,交通阻塞、停车难和尾气污染已成为许多城市特别是大城市的突出问题。如北京市自20世纪90年代中期以来,机动车拥有量年均增长率超过10%,目前已达180多万辆,这使干道平均车速比10年前降低50%以上,而且正以每年递减2km/h的速度继续下降。市区183个路口中,严重阻塞的达60%,阻塞时间长达半个小时。同时,由于没有车位,汽车乱停占用道路,使交通压力更大。据统计,在我国500多座城市中,大气质量达到一级标准的不到1%,在世界20个污染大城市中,我国就有10个,而汽车尾气的大量排放,恰恰是城市大气污染的最主要原因,控制汽车污染,已成为根治城市大气污染的关键之一。地下轨道交通具有速度快、运量大、准时、污染小、占地面积少的特点。速度快:地铁运营速度达35km/h;城铁将高于35km/h。运量大:高峰时,地下轨道交通每小时通过客流达4~6万人,而公共汽车每小时通过客流只有1800人。准时:地下轨道交通由于全封闭,基本不堵车,能安全、准点到达。污染小:由于使用电力,基本不产生废气污染。占地面积少:由于处于地下,不占用地面有限的土地,为地面腾出了空间。因此,发展地下轨道交通,正在成为解决大城市地面交通拥挤、空气噪声污染等问题的主要交通方式。如加拿大蒙特利尔地下交通网是由东西两条地铁轴线、南北两条地铁轴线及环形地铁线和伸向城区中心地下的两条郊区火车道组成。实践表明:这种建设城市地铁交通网与地下通道、地下停车场、郊区火车相结合的交通体系是缓解城市交通压力、降低城市大气污染的有力措施。

我国继六十年代在北京、七十年代在天津、八十年代在上海兴建地铁后,目前正在北京、天津、上海、广州、深圳、南京、重庆、青岛、沈阳等大城市建设地铁,以代替地面交通。据统计,在21世纪的前20年

间,我国的地铁线路将超过 2000 公里,这种交通方式的变革将产生重要的效益。

同时,在城市中心区以及具有突出景观价值的区域,除发展旅客地下轨道交通外,还应发展地下货物运输系统,包括 UTP(管道中的集装箱运输)、ULS(地下后勤系统)和 PWT(气动废物运输),运送货物和废料。除了环境效益外,在一定的地区或这些货物通过管道和集合点组成的地下无人系统情况下,这些系统以其质量指标和运送的高效率在经济上具有竞争性。

为解决"停车难"的问题,主要出路是修建地下停车库。地下停车库的突出优点是容量大、用地少、布局接近服务对象。因此,在地下街、地下综合体的建设中,应当使停车场的面积保持适当的比例,特别是结合地铁车站修建地下库,方便换乘,减轻城市中心区的交通压力,既提高地铁的利用率,又减轻了由汽车造成的公害。风景秀丽的瑞士,为解决停车问题,在日内瓦湖底修建了五层的地下停车场,用他们的话说:虽工程巨大,但保护了环境,综合效费比高。

四、利用地下空间的特性,规划地下市政基础设施,提高城市环境质量

从建筑空间环境和使用功能特性角度看,地下空间具有温度稳定性(保温隔热)、隔离性(防风尘、隔噪声、减震等)、防护性和抗震性等特性。根据地下空间的这些特性,在规划市政基础设施时,应尽可能入地,从而提高城市环境质量。

1. 发展城市地下市政管线"共同沟"

将自来水管、排污管、供热管、电缆和通信线路等放在市政管线公用隧道内。我国城市的道路被戏称为"拉链式"道路,经常被开膛破肚,不仅影响交通,而且有碍景观。建设地下市政管线"共同沟",不仅可缩短路线长度多达 30%,而且具有易于检查、易于修理、消除管线破坏和破坏事故、减少对环境的影响、不影响地面土地的使用等优点。如在上海浦东的张杨路和成都的东风路已经建成了两条地下市政管线公用隧道;在北京的中关村地区即将建设高标准的"共同沟"。

2. 发展地下垃圾处理系统

我国年产垃圾 1.46 亿吨,处理仅占 2.3% 左右,其余在城郊堆积,存量高达 60 多亿吨,占地 30 多万亩,垃圾围城现象严重。我国许多城市正在发展垃圾处理系统,但都建在地面,不仅占用宝贵的土地,而且对四周环境造成了相当大的污染。国外先进城市发展的地下垃圾处理系统,不仅能节约城市宝贵的土地资源,有很大的回报;而且能控制对环境的噪声和恶臭的污染,减少对环境的视觉负面影响。瑞典在二十世纪六十年代建设并实施采用压缩空气吹运垃圾系统(PWT)。PWT 系统建在 1700 人口的居民区内,该系统与收集、处理系统配套,投资在 3~4 年内得到回报。香港也在岛西端大维山岩铜中建成处理垃圾的 RTS 系统,在其入口前,根本就没有垃圾堆成山、灰尘满天飞和臭气四周散的情况,消除了对环境的二次污染。

3. 发展地下储水系统

国际上有人说,21 世纪争水,是水的世纪。目前世界上有 14 亿人口缺乏安全清洁的饮用水。我国是世界 13 个贫水国之一,人均水资源量仅为 2251 立方米,居世界第 110 位,是世界平均水平的 1/4。据统计,我国城市每年缺水 60 亿立方米,在全国 666 座建制市中,有近 400 座城市缺水。利用地下构造储水,不仅储存空间大,而且能减少水的流失,降低水的污染。20 世纪 80 年代开始,美国开展了钻子蚌 L 给含水层的恢复工程(ASR 系统),纽约市的大型供水系统,完全布置在地下岩层中。瑞典、荷兰和德国的 ASR 工程在总供水中占 20%、15% 和 10%。同时,还要开展雨水利用和污水回用(建设中水道系统)的地下工程建设。如我国在西沙群岛上修建了可采集雨水达 14 万吨的地下贮水工程,通过技术处理,达到了国家饮用水卫生标准,结束了吃水靠大陆船运的历史。北京市准备在西山地区利用冲积扇岩层建立地下储水系统,不仅平时可收集雨水,解决用水紧张的问题;而且在发生山洪的时候,起到防洪的作用。将污水处理设施放置在地下,将使目前地面污水处理厂所占的土地得到开发利用,而且使它们邻近被污

染不能被利用的土地也得到开发。如芬兰赫尔辛基地下污水处理场设在未来居民区地下100万米深的岩洞中,可处理70万居民的生活污水和城市工业废水。中水利用在很多国家被采用,如日本每年将经处理的1.3亿立方米的再生水用于工业或绿化用。

五、加强软硬件建设,确保城市地下工程成为环境友好工程

促进建设环境友好工程的最重要因素是法规。应当提倡对不同方案进行社会经济、环境、技术评估,以代替纯经济技术评估。政府应积极制定严格的环境和安全方面的政策和法规,以进行环境影响的评估,从而保证不同实施方案的分析更加客观。

应当对地下工程的设计、施工方案提出地面稳定和地下水系保护的要求。这些要求可以通过负责的地球物理勘探,负责的岩土工程结构设计以及负责的工程的施工来完成。

为了建立"洞察地下"的信息系统,需要进行大量的原地和实验室的研究。现代信息技术可以整合基岩研究、岩土工程设计和工程的实际施工,使之一体化为统一过程。现代化的材料也应是耐久的。而设备的自动化和不断增长的广泛的计算机控制已经改进了效率、精度和开挖工程的安全,使得地下工程正在变得不但经济和有效,而且在环境上也是可持续发展的。因为不断提高的效率意味着更低的能源消耗、机械施工时更小的扰动和更少的有害物质的散发。

六、结语

建设"花园城市"是一项系统工程,开发利用城市地下空间实现土地的多重利用,使得城市结构更紧密;相对降低城区内交通的需求,提高城市空气质量,降低城市水污染,是改善城市环境、实现城市可持续发展、达成人与自然和谐统一的重要方面。地下空间是一个十分巨大而丰富的资源,21世纪将是人类开发利用地下空间的世纪,"花园城市"的建设要利用这个发展契机,在开发利用地下空间上有所作为。

(作者:钱七虎　陈晓强)

春风化雨　润物无声

——深切缅怀潘家铮院士

时光荏苒,我们敬爱的潘家铮院士驾鹤归天,与我们永别将近一个春秋了。这期间,斗转星移、风云变幻,但我们始终觉得逝者的灵魂和生者的怀念是永恒的。想到这里,不由得回忆起一些往事。见微知著、一叶知秋,试图从一个侧面追思潘院士的高尚人格和大家风范。

众所周知,潘院士是国内外著名的工程技术权威。半个多世纪以来,始终秉承"求是"学风,弘扬创新精神,在我国水利水电开发、建设中无私奉献,开拓进取,创造了非凡的业绩,做出了杰出的贡献,堪称水利水电工程界的泰斗。

在国际活动方面,潘院士也是有口皆碑,他长期担任中国大坝委员会主席,为祖国争得诸多荣誉。

历年来,潘院士的传奇人生和大家风范,通过中央电视台、《科技日报》等主流媒体的传播,广为人知,不需多加笔墨。

这里想补充提出的是潘院士在岩石力学与工程界的丰功伟绩。

事实上,潘院士从20世纪50年代初就开始对这门新兴学科开始深入研究。当时,他的一系列著作,如《建筑物的抗滑稳定和滑坡分析》等,都被广大同仁视为传世佳作。20世纪70年代末,潘院士与陈宗基先生共同组织国内知名专家倡议成立全国性一级学会——中国岩石力学与工程学会,开创成立我学会的嚆矢。1985年中国岩石力学与工程学会成立以后,潘院士先后担任副理事长、理事长(1991—1994),名誉理事长。近三十年来,他时刻关注着学会的成长、壮大,并为此付出了辛勤劳动,倾注了大量心血。历年来,我学会下属各级组织充分发挥跨学科、跨行业、人才荟萃、知识密集的优势,在国际、国内学术交流,科技咨询,人才培养等各个方面都取得骄人的业绩和举世瞩目的成就。特别是在国际活动各条战线上可以说是佳音不断,捷报频传。例如,2011年10月我国成功地举办了素有岩石力学界"奥林匹克"之称的第12届国际岩石力学大会,实现了几代人的梦想。在该次大会上,我国青年科学家冯夏庭教授就任国际岩石力学学会主席(2011—2015),成为该国际组织成立半个世纪以来首次当选的中国科学家。国际岩石力学学会主席团于2012年5月将终身最高荣誉称号——国际岩石力学学会会士(FellowISRM)头衔授予中国岩石力学与工程学会理事长钱七虎院士。再如,目前国际学会下设的15个专业委员会中,有5个专业委员会的主席由中国专家担任等等,不一一列举。毫无疑问,上述各项成就,无不与潘院士的关心与支持息息相关。

另外,还值得提出的是潘院士的团结协作、无私奉献精神,例如去年2月份当他得悉国务院南水北调办公室专家委员会要举行会议时,作为该委员会的主任委员,潘院士不顾长期卧床、重病缠身,婉言谢绝亲友的劝阻,毅然决然在陪同人员的护送下亲自赴会,并在会议上语重心长地发表了诸多指导性意见和建议,对新任专家委员会主任寄予厚望,对今后工作做了全面安排。当时距潘院士仙逝只有不到半年时间,他不顾个人安危,以超常的胆识和智慧毅然赴会,实在是一件令人万分敬佩、万分感动的壮举。这种"先天下之忧而忧,后天下之乐而乐"的崇高精神和行为永远值得我们学习。

大家还知道,潘院士一直关心青年科技人才的教育、培养。历年来,从潘院士门下脱颖而出的一批博士研究生,如今都成为国家建设的栋梁之材,以潘院士名义设立的"潘家铮奖",面向全国水利水电工程界,激励着无数杰出人才积极投身到科技创新事业。

此外,潘院士对杰出接班人的培养、提携也倾注了大量心血,在这方面给笔者印象最深刻的一个实例是中国工程院林元培院士在申报过程中的一段曲折历程。在2005年前的中国工程院第一、二轮投票选举时,林院士均未获通过。后来,潘院士慧眼识英雄,在全面、深入地了解了真实情况后力排某些争议,坚持原则,提出了客观、公平、公正、令人信服的真知灼见,致使在2005年的选举中林元培院士当选。

潘院士数十年如一日,勤勤恳恳,兢兢业业,礼贤下士,平易近人,生活俭朴,公正廉洁,从不以领导和专家自居。他一贯职业道德高尚,治学态度严谨,富有献身精神,无愧为大家学习的榜样。

潘院士的一生是革命的一生,战斗的一生,无私奉献的一生。

大音希声,大象无形。他既伟大又平凡的业绩将永远激励我们继承他的遗志,化悲痛为力量,沿着他开辟的道路奋力拼搏,再创辉煌。

衷心祝愿潘院士在天之灵永远安息!

(作者:钱七虎 傅冰骏)

城市交通拥堵、空气污染以及雨洪内涝的治本之策

中国城市交通拥堵、空气雾霾以及雨洪内涝之严重,已成为建设环境友好型城市的最强不和谐音符。习近平总书记曾指出"治理交通拥堵必须标本兼治"。如何寻求治本之策?什么是治本之策?必须在确定治理方针和措施前研究明确。世界发达国家城市治理的历史经验表明惰政思维、急功近利,只能治标不能治本。治本必须转变传统思维,要从产生交通拥堵、空气污染以及雨洪内涝的根本原因入手,立足大思路、大手笔,才能产生明显的大效果。

一、科学治理交通拥堵

城市交通拥堵的根本原因是交通需求与交通供给的失衡。治理它必须从科学抑制交通需求和科学提高交通供给两方面入手。目前采用的行政手段限购限行,不是科学抑制交通需求,因为私家车已是中国社会强劲的消费需求,限购限行不利于依靠消费内需发展经济,不利于城乡交流、发展农业提高农民收入。科学抑制城市交通需求首先要抑制城市规模,为此,一是抑制城市人口,人少车就少;二是限制城市面积,城市大市民出行不能依靠步行和自行车,必然激发机动交通需求,为此必须大力开发利用地下空间,实现土地的多重利用,建设"紧凑"型城市。如北京,为控制城市规模,必须树立科学的"首善"理念,"首善"是质量,不是数量,城市发展应科学定位,不应追求政治、经济、教育、文化、体育、艺术等所有方面都首善。要学习美国、法国、巴西、南非、韩国等国,把首都仅定位于行政,把其他如立法、司法、经济、文化、体育、高等教育、艺术等分散在其他城市,其城市"减肥"的效果将立竿见影。

科学抑制城市交通需求,还必须实施面向交通的城市规划模式。为此首先要倡导混合用地模式,不搞 CBD、金融街、工业开发区、大学城等单一功能布局,更不能在郊区大建"睡城",按功能分区必然导致跨区交通出行强度的激增,抑制了居民的出行量与出行距离,也就抑制了交通需求。其次,按照面向交通的城市规模模式,就应逐步实现由轨道交通相联系的城市主中心区与周边副中心区(卫星城)相结合的城市多中心规划布局,以遏制城市主中心区人口密度,同时引导人口与产业和城市功能同时分散至副中心(卫星城)这些卫星城发展独立的产业和经济基础以及较高水平的商业、医疗、教育和文体设施,使居民可以就近择业,不必每天通勤至中心城,让居民可以享受到不亚于主中心区的城市服务。

科学提高交通供给能力,首先是发展大容量快速轨道交通,以抑制私家车的出行强度,但是能否把"治堵"的希望完全寄托于轨道交通的发展上呢?北京轨道交通已建成 400 多公里,但上下班高峰时段地铁上下车拥堵之惨状已令市民谈虎色变,环顾发达国家的特大城市,其轨道交通已相当完善,它们仍困扰于交通拥堵,正纷纷寻找轨道交通以外之良策,以应对小汽车进入家庭、机动交通快速增长之现实。它们把"治堵"目光转向了"地下快速路"和"地下物流系统"的建设,这是因为在城市地面空间已不能拓展交通供给的情况下,唯一的出路在地下。地下快速路的天然优势是全天候通行,在暴风、风雪和大雾等最易造成地面交通拥堵的情况下最能发挥作用。

美国波士顿 1994 年开始拆除高架路,10 年间建成 8~10 车道的城市地下快速路系统;东京正在地下 40m 建设中央环状地下快速线,11km 的通过池袋、新宿、涩谷 3 大商业中心的新宿线已通车,正在建设品川线;新加坡、吉隆坡、马德里、布里斯班、巴黎、莫斯科等也已建成或正在建设地下快速路交通系统。

地下物流系统就是将城市货运转移至城市地下,从而对治理交通拥堵作出重要贡献。根据世界经济合作组织2003年统计,发达国家主要城市的货运占城市交通总量的10%~15%。而在"世界工厂"和到处是建设工地的中国其占城市交通总量则相应为20%~30%。

修建城市地下快速路和地下物流系统非常适合我国人多地少的国情,我国的特大城市,其城市规划土地余量已所剩无几,已步入无地可建路的窘状。

二、科学治理空气雾霾

城市空气污染的根本原因是工业污染和汽车尾气,而后者已超越前者成为城市$PM2_5$超标的主因。治理工业污染,可以关、停、并、转高能耗、高污染企业,实现"能源转型"与"产业升级"。只要决心大,这在社会主义国家的条件下,是能够而且相对较易实施的。困难的是汽车尾气的治理,无法收集处理数百万辆机动车排出的尾气。

建设地下轨道交通和地下物流系统,其运转工具为电力驱动,从根本上消除了尾气污染。至于地下快速路系统是隧道,其中行驶的机动车尾气易于收集和处理。波士顿和东京地下快速路的经验是:先将尾气污染的空气由引气机引流至处理区(换气站),再通过静电除尘、化学吸附、光催化、等离子等技术去除其烟尘等固体颗粒和使有害成分转化成无害成分,从而排出过滤和处理后的无害气体。

实践与研究表明,城市地下快速路系统以及地下物流系统的建设在解决交通拥堵的同时,必将为消除汽车尾气对城市空气的污染做出决定性贡献:波士顿修建地下快速路后其市区CO浓度降低了12%;东京11km新宿地下线的建成后,每年减少了3.4万t CO_2 排放,其中静电除尘装置可以过滤80%以上的颗粒物,低溶度脱硝装置可以过滤90%以上的NO_2;东京建设300km地下物流系统的评估报告指出,该地下物流系统建成后,东京市交通能耗减少18%,NO_x和CO_2浓度将分别减少10%和18%,货运速度提高24%。

三、治理城市雨洪内涝

中国城市雨洪内涝严重的根本原因在于,一方面客观上地球气候极端异常、暴雨强度和频度增加,另一方面由于城市化快速发展,城市面积快速增大,不透水地面占比极高,排洪系统难以适应。应对地球气候之异常,非一国一时之力所能胜任,我们能做的就是提高城市建设对暴雨雨洪的适应性措施:一方面通过透水铺装建设透水路面和透水地面、下凹式绿地和雨水花园、植草沟等措施尽可能扩大城市透水地面占比,以所谓的建设"海绵城市"来科学抑制城市的排洪需求,另一方面是科学提高城市排水系统的排洪能力。日本应对城市雨洪内涝的经验是在东京、横滨等特大城市修建地下河川和大容量的地下雨水调储池;马来西亚吉隆坡市则修建地下快速路,在特大暴雨情况下,关闭机动车通行,地下快速路充作排泄雨洪的通道;而在一般暴雨情况下,仅地下快速路路面下空间为排洪通道机动车照常行使。总之为应对极端气候的暴雨,城市排水系统的排洪能力必须大大增加,而我国城市排水管道的口径几十年一贯制必然会产生城市内涝。

综上所述,治理城市交通拥堵、空气雾霾和雨洪内涝的治本之策是多管齐下、综合治理,而关键是建设城市地下交通和排洪系统,这是一举三得的措施,集治理三大城市病于一役,我相信,通过若干个五年计划持之以恒地建设上述地下系统,不但使人员交通与货物运输转至地下,还包括雨洪、垃圾、污水等的传输都转入地下,使地面上彻底摆脱交通拥堵、PM2.5超标和内涝的困扰,而释放出的地上空间用作大片的自然植被和安全的步行,实现习近平总书记所设想的"要依托现有的山水脉络等独特风光,让城市融入大自然,让居民望得见山,看得见水,记得住乡愁"。

(作者:钱七虎)

城市化发展呼唤积极和科学开发利用城市地下空间

2010年3月25日,联合国经济与社会事务部人口司在纽约总部发布了《世界城市化展望2009年修正版》。报告显示:中国的城市化速度居世界第一,城市化比例在过去的30年间增长了1倍多,从1980年的19%跃升至2010年的47%,预计至2025年将达到59%,中国正经历着社会的重要转型。

随着我国城市化进程的迅速发展,一些问题也相伴产生。一是土地紧缺。耕地资源是一个国家最重要的战略资源之一,土地资源的可持续利用是我国实施可持续发展战略的基础。城市人口的急剧发展与地域规模的限制已成为中国城市发展的突出矛盾。我国耕地面积2007年为18.26亿亩,只约占国土面积的10%,人均耕地仅1.4亩,不到世界平均水平的1/2。按照城市化发展的相关分析,未来15年城市化水平每提高1%需要新增城市建设用地3459km^2。因此,未来几十年城市化进程中将面临18亿亩耕地的红线被突破的危险,也就是将凸显"土地安全"的问题。二是交通拥挤。近年来,我国许多大城市不同程度地受到交通拥挤的困扰,"出行难、乘车难"成为许多政府和市民普遍头疼的问题,交通拥挤已经成为制约城市社会和经济发展的瓶颈。以北京为例,2009年北京全市机动车保有量达401.9万辆;全年路网交通拥堵指数为5.41,已属轻度拥堵,非限行日时的晚高峰达到了中度拥堵,拥堵范围和常发拥堵路段数量不断增大,五环内路网的平均时速为24.7km(早高峰)和22.3km(晚高峰)。三是环境污染。根据世界银行等机构提供的数据,在全球污染最重的20个城市中,有16个城市位于中国,其中前10位中中国占7个。我国二氧化硫排放量世界第一,碳排放量世界第二,4亿城市人口呼吸不到新鲜空气,1/3的城市污染非常严重。2006年监测的559个城市中,空气质量为三级标准和劣于三级标准的城市已占28.5%和9.1%。30%的重点城市饮用水源地水质达不到Ⅲ类标准,流经城市90%的河段受到不同程度的污染,有50个城市的水环境功能区水质达标率低于50%。

出路在何方?适度发展城市的地下化。向地下要土地、要空间,改善城市环境,提高城市防灾抗毁能力,是全世界城市发展应对严峻挑战的必要趋势,并已成为衡量城市现代化的重要标志。

开发利用地下空间对扩大城市空间容量节约土地资源的潜力非常可观。据对北京城市中心地区324km^{2}范围内的地下空间资源调查结果,如以开发地下30m的浅层和次浅层地下空间计算,可供合理开发的有效利用的资源为1.19亿m^2,为现有建筑总量2.9亿m^{2}的41%。扣除北京市已开发的地下空间总量3000万m^{2}后,意味着在保持现有容积率和建筑密度的情况下,不需要扩大城市用地,就可以扩大城市空间总量的30%。

适度发展城市地下化,首先是把交通(客运和货运)以及停车库放到地下,其次是把城市基础设施,特别是诱发脏、乱、差的设施放到地下,如污水和垃圾的集运和处理等,还包括部分商场和餐饮、休闲和健身,如洗浴和游泳,以及部分需要恒温和无噪音的生产设施。由于把产生汽车尾气的交通以及产生污气、污液和噪声的设施放入地下,城市的环境就大会改善。如美国的波士顿,在21世纪初,拆除了全市的高架路,改为建成8~10车道的地下快速路,原有的地面变成了林荫大道和街心公园,城市增加了很多绿地面积和开放空间,包括增加了260英亩公园,城市空气质量大为改善,其中城市一氧化碳浓度降低了

本文原载于《科技导报》(2010年第28卷第10期)。

12%。部分地下商场、文化、休闲和娱乐健身设施还可采用"玻璃连拱廊""采光穹顶"或"下沉式广场""下沉式庭院"等建筑形式改善和美化地下环境,将阳光、绿化引入地下,使其为地下空间所共享。

城市面积越大,对机动交通的依赖度就越大,交通能耗就越多。开发利用地下空间,有助于城市"瘦身",建设成"紧凑型"的城市结构,从而相对增加市民的步行和自行车的出行比例,导致能耗大户——交通能耗的降低。开发利用地下空间,还可促进利用地下土壤、地下水的低密度能源作为夏季冷源和冬季热源,借助热泵向建筑物供热供冷。

雨水利用和污水的再利用也离不开地下空间的开发利用。利用地表入渗和井灌补给来实现地下含水构造贮水,特别是贮存雨洪。建立这种"水银行"来调节供水已是欧美先进城市行之有效的、低廉的解决城市部分供水的一种办法。除供水外,它还能控制由于地下水位下降引起的海水入侵和城市地面沉降;维持河流的基流,不致出现华北地区"有河则干"的现象;通过土壤的细菌作用、吸附作用及其他物理化学等作用改善水质,更重要的是保护生态环境。

以上一切表明了开发利用城市地下空间有助于建设资源节约型和环境友好型城市。

近年来,按照可持续发展的战略,我国城市地下空间开发利用作为城市建设和改造的重要组成部分,在原有人防工程的基础上,呈现出大规模、高速度、多层次、多类别的趋势。为适应城市地下空间"综合化、规模化、集约化、深层化和一体化"的发展需求,要真正使之成为建设资源节约型、环境友好型社会的重要途径,还必须十分注意科学合理地开发利用城市地下空间:一是要加强相关立法,形成完善的法律法规体系,为积极科学开发利用城市地下空间提供法律依据。二是要制定相关政策,积极引导、鼓励促进城市地下空间的科学开发利用。三是要搞好整体规划,使地上地下相协调、各地下系统相协调,整合城市空间资源。四是要统一管理体制,根除多头管理的弊端,形成城市地下空间开发利用管理的整体合力。五是要完善标准体系和加强管理,促进和达到精心勘察、精心设计、精心施工、精心管理,实现城市地下空间的施工安全、环境安全、功能优化和综合高效。

(作者:钱七虎　陈晓强)

在苏州地铁 11 号线开通仪式上的发言

(2023 年 6 月 24 日)

各位领导、各位嘉宾：

大家上午好！

很高兴和大家一起见证苏州地铁 11 号线的开通。11 号线西连苏州地铁 3 号线,东接上海地铁 11 号线,是苏州轨道交通具有里程碑意义的线路。到年底,苏州地铁 11 号线将与 3 号线贯通,同时通过双"11 号线"花桥站的无缝换乘,为苏州、昆山、上海之间的通勤提供了极大便利;从这个角度上看,地铁 11 号线将汇聚更多的资源和人才,进一步推动昆山城市的发展、推动苏州市域一体化发展。

建地铁就是建城市。前两年,昆山邀请我共同参加了地铁沿线开发研究,我发现在坐地铁 11 号线的时候,他们已经系统地梳理了沿线的水系,融入了海绵城市的理念,提高了排涝标准,同时积极整合沿线土地资源,开展地上地下空间的综合利用、以公共交通为导向的综合开发(TOD 综合开发),进行了公交线网调整和轨道沿线道路、立面恢复。在研究、建设过程中,昆山和瑾晖城市建设基金组建了地下空间研究院,着力引导城市、特别是地铁沿线用地集约化发展,在地下空间、安全生产方面进行了很多探索,具体工作呈现三方面特点:一是突出轨交项目的绿色智能建造。11 号线是苏州首条全过程应用 BIM 技术的轨道交通,在关键区间以智慧化手段实现盾构穿越风险源及沿线中高风险老旧房屋自动化监测、地下病害安全性检测,保障了盾构顺利贯通,也打开了轨交智慧建设新局面。二是突出地下空间的综合开发利用。通过玉山广场站、祖冲之路站等多个站点 TOD 综合开发、人防控制性详细规划、沿线地下筒仓式智能停车库规划建设、海绵示范建设等举措,打造地上地下一体化的站城综合体。三是突出数字赋能和韧性城市建设。扎实推进数字地下空间和生命线工程建设,一方面为智慧城市提供数字化的基础设施和数据资源,实现对轨道交通工程全寿命的数字化支撑以及对城市地下空间的高效利用和精细化管理,另一方面为韧性城市提供安全可靠的基础设施和风险防控体系,全力保障城市安全稳定运行。听说,下来昆山还要做 9 号线,要站在昆山城市高质量发展的高度,确定 9 号线在昆山的定位,尽早谋划线位、设站等事宜,做好 9 号线沿线土地综合利用规划和地上地下综合规划,我也愿意继续支持和配合,把昆山的地下空间和城市建设搞得越来越好。

苏州地铁 11 号线的开通,是长三角一体化、苏州市域一体化融合发展的一个缩影,也是地下空间开发利用的一次有益实践,作为家乡人,对家乡的快速发展感到由衷的高兴,衷心希望轨道交通给家乡父老带来更多便利,也衷心祝愿苏州轨道交通迎来更加辉煌的明天、苏州经济社会发展越来越好!

在中国城镇供热协会地下综合管廊运营维护专业委员会成立大会上的致辞

（2023年6月2日）

尊敬的各位领导、各位嘉宾：

大家好！

非常高兴受邀参加"中国城镇供热协会地下综合管廊运营维护专业委员会成立大会"，对专业委员会的成立表示热烈的祝贺！

早在20年前，针对我国长期存在的"重地上、轻地下"的现象，我们7位院士向中央提出了建设综合管廊、开发地下空间的建议，得到了中央领导同志的认可，也得到很好的实践。近10年来，党中央国务院从推动城市基础设施高质量发展的战略高度，部署推进综合管廊建设。

住房和城乡建设部等部门组织全国的力量，认真贯彻落实党中央国务院决策部署，因地制宜、统筹安排地下综合管廊建设。全国485个城市编制了综合管廊建设规划，累计建设管廊超过6000km，形成了世界上最大规模的综合管廊群体，其系统性和功能性不断增强，不仅保障了地下管线运营维护安全，还在统筹地上地下空间利用、解决"马路拉链"问题、助力稳投资稳增长等方面发挥了重要作用。促进了我国地下管线建设管理方式由粗放发展向高效、集约、绿色的可持续发展模式转变。

推进综合管廊建设需要全行业的鼎力协作。希望发挥专委会的资源整合优势，建立专家智库，发挥桥梁纽带作用，加强行业互助和合作，进一步深入总结推广可复制经验，充分调动政府、企业等各方积极性，为建立健全有利于综合管廊可持续发展的长效机制贡献经验和智慧。

我相信，专委会的成立将对我国因地制宜好字当头稳步有序推进城市地下综合管廊高质量建设发挥重要促进作用。预祝本次会议圆满成功！

谢谢大家！

在2021年中国国际服务贸易交易会智慧建造与绿色发展高峰论坛的发言

(2021年9月5日)

各位建筑业同行、各位嘉宾：

大家上午好！

很荣幸能够被邀请参加由北京市住建委、市规自委联合主办的2021年服贸会工程咨询与建筑服务专题，首先我对专题的顺利召开，表示热烈的祝贺和衷心的祝福。

2021年作为"十四五"的开局之年，在实现"双碳"目标的发展过程中，是一个重要的节点和里程碑。"碳达峰""碳中和"是为了应对全球气候变化，是人类面临的最大挑战。我国的《第十四个五年规划和2035年远景目标纲要》也为应对气候变化提出了明确要求：落实2030年应对气候变化国家自主贡献目标，制定2030年前碳排放达峰行动方案；锚定努力争取2060年前实现碳中和，采取更加有力的政策和措施。而"碳达峰""碳中和"主要有两个方面，一个是碳排放；一个是碳的负排放，也就是碳的吸收。

常规的三个碳排放大户为：工业、建筑、交通各占约1/3。一般而言，建筑的碳排放可以按建材生产、建材运输、建筑施工、建筑运营、建筑维修、建筑拆除、废弃物处理七个环节，构成全生命周期的排放量。多数研究认为，在一般情况下，建筑全生命期的碳排放中，运行阶段占据了最大比重，达到60%~80%；其次是建材生产的碳排放，20%~40%；而施工过程只占5%~10%，拆除阶段占比更低。面对严峻的碳排放攀升形势，为进一步加强建筑领域的绿色化和减碳力度，转变传统建造方式，大力发展绿色建筑，是实现降低建筑领域碳排放的重要举措，建筑领域除降低用能实现"碳中和"，更需要通过技术创新来实现绿色的发展，推动以建筑设计为主体的技术方法创新，推进空间节能和设备节能的融合，大幅降低供暖、空调、照明、电器等用能需求，促进部分时间、部分空间的低碳用能的理念落实，对减少建筑运行阶段的碳排放至关重要。

从长远看，要准确把握智慧建造是产业转型升级高质量发展的关键。今天就我结合"智慧建造、绿色发展"这个主题，与大家分享一下我对绿色发展与智慧建造的若干思考。

首先讲一讲绿色发展在世界和我们国家的地位和作用，我想是这几点，一是世界迎来了绿色发展的时代，绿色发展的思想是习近平新时代中国特色社会主义思想一个重要组成部分。二是绿色已成为中国国家建设发展、社会发展的五大理念之一，中国的绿色发展正在进入世界绿色发展的先进行列，而且可以肯定将来肯定位于世界发展的前列。

目前，全球正面临气候变暖导致的生态环境恶化等多重挑战，应对之策是要推进绿色发展和绿色建造。而在工程建设领域，实现绿色建造的必然选择和最佳途径是智慧建造。

一、工程建设应向"绿色建造"转型

1992年，为了履行联合国环境与大发展大会通过的《21世纪议程》，中国承诺将可持续发展纳入国家战略，建设可持续发展的城市和社会。当时我们向中国工程院提出了咨询课题《21世纪中国地下空间开发利用的战略和对策》，充分利用地下空间，特别是大力发展以地铁为骨架的轨道交通系统和集约可持续的城市基础设施，从而节省土地资源，提高土地利用效率，释出大批地上空间作绿色植被和生态空间，建设可持续城市。

建筑业是全球最大的原材料和能源消耗产业，目前全球建筑运营能耗已占到总能耗的30%以上，加上建设过程中的能耗，这一指标接近50%。我国传统的粗放建造资源消耗量大、浪费现象严重，也产生许多污染。

为了适应绿色发展的要求，工程建设应该向"绿色建造"转型。

二、智慧建造是实现绿色建造的必然选择和最佳途径

工程建设要往"绿色建造"的方向发展，首先是理念上的推动，大家要认识到绿色建造的重要性。其次是政策驱动，鼓励大家节能，也可以通过标准驱动，例如制定绿色建筑标准等对建设项目进行准入，使得高耗能项目退出。当然最重要的，还是依靠科技的发展，发展大数据技术、人工智能技术、物联网、第五代移动通信(5G)技术等，有了这些科技基础，建设才会更加高效节能、低碳排放，甚至零排放。

这些技术应用到工程建设中，就要涉及"智慧建造"，智慧建造是实现绿色建造的必然选择和最佳途径。

在工程建设中，"智慧建造"首先体现在全面的透彻感知系统。第二是通过物联网、互联网的全面互联实现感知信息(数据)的高速和实时传输。获取的信息一定要快速传输出去，如果当下获得的信息要过几天才能看到，工程建设就不能实时地反馈和服务。有了互联网、物联网、5G技术后，信息传输非常快，可以即时地反映认知。

另一个是智慧平台的打造，技术人员要通过这个平台对反馈来的海量数据进行综合分析、处理、模拟，得出决策，从而及时发布安全预警和处理对策预案。有了这些技术，工程建设的风险更低，施工人员更安全，同时也能最大限度地节省材料和减少环境破坏。

三、智慧工程未来要向少人化、无人化发展

我国工程领域的科技创新经历了从低到高，从局部到全面的发展历程。改革开放以来，工程建设经历了机械化和信息化的发展。比如地下工程原来都是采取钻爆法、人工打眼、人工放炮，后来可以大量应用机械台钻，多钻台车施工，现在可以采取数字化掘进，这是机械化的进步。

信息化的发展体现在地下工程的地质探测中，建地下工程，需要把地下的情况了解得很清楚，地下有没有水，有没有断层，如何防止地下安全事故发生，这都是靠信息化设备来提升安全性能的。

今后，工程领域的进步要通过数字化向智慧方向迈进，比如数字工程的建筑信息模型(BIM)技术在建筑工程中的应用，原来我们做设计是用图纸，但图纸和工程实体是分离的，进入BIM时代后，技术人员可以在计算机里建立可视化的工程模型。

当前科技创新最主要的标志，就是信息技术继续深入地应用到建设工程中去，也就是全面地推广数字工程和智慧工程。未来，数字工程和智慧工程还要向少人化、无人化的方向发展。

以上是我对绿色发展与智慧建造的一些思考，谢谢大家!

永远跟党走 永葆革命青春

——在中国工程院学习习近平总书记"七一"重要讲话精神报告会上的报告

（2021年7月13日）

工程院领导指示要我向大家汇报对习近平总书记建党百年讲话精神的学习领悟心得，心得谈不上，就讲一讲当面聆听讲话后的感想！

首先我为我党这一百年来所开辟的伟大道路、创造的伟大事业、取得的伟大成就所震撼！这是中华民族几千年历史上最恢宏的史诗：它开辟了一条在农村建立根据地，以农村包围城市的中国特色成功革命道路，推翻了压在中国人民头上的三座大山，使中国人民得到了解放；它开辟了一条以充满活力的社会主义市场经济体制为核心的、以改革开放为特征的中国特色社会主义建设道路，实现了总体小康，消灭了绝对贫困，奔向全面小康的历史性跨越！世界各国的史册，不乏繁荣的历史，不乏骄人的盛世，但鲜有哪个时代，哪个国家能像今天的中国，致力于"一个地区也不少"，"一个民族也不落"的人民全面福祉，全面建成小康社会，将是人类社会发展史上的伟大壮举。作为世界最大政党的我党，取得的丰功伟业，使得在苏联解体、东欧变色、苏共自行消亡的今天，成为世界人民仰望的马克思主义、社会主义和共产主义的坚强堡垒，成为世界人民心中的和平、正义、走向人类命运共同体的希望灯塔。我为我们党无比骄傲，我又为我65年来参加了党的队伍，参与了党的百年征程，在党的百年华诞之际向党交出了一个合格答卷，成为我党的优秀党员而自豪！

江山就是人民，人民就是江山。我党打江山，是为了人民的解放，打败侵略者和剥削者，我党守江山，就为了让人民过上好日子，从而守住人民的心。我是一名党员科学家，要牢记我党为人民服务的宗旨，我们攻坚克难搞科研，不是为了个人的名利，我们的科技创新要服从于人民、也就是服从于国家的需求，服务于国家经济社会发展也就是服务于人民。我从事防护工程专业，我们的科技工作就是为国家铸造坚不可摧的地下钢铁长城，为人民构筑绝对可靠的安全屏障，让妄想欺侮我们、逼迫我们、奴役我们的一切反动派在我们的钢铁长城面前碰得头破血流！

"敢于斗争，敢于胜利，是中国共产党不可战胜的强大精神力量，实现伟大梦想就是要顽强拼搏，不懈奋斗""世界上没有哪个党像我们的党这样，在百年征程上，遭遇过如此多艰难险阻，经历过如此多生死考验，付出过如此多壮烈牺牲"。面对困难、风险和挑战，只有不躲、不退、不让，敢于斗争、敢于牺牲、敢于胜利，拼搏奉献才能创造一个又一个的人间奇迹！在今天建成科技现代化、科技创新强国的道路上一定会遇到无数风险挑战，我们只有继承弘扬无私奉献的科学家精神，潜心顽强拼搏，坚韧不拔，才能跋山涉水为实现中国梦强国梦强军梦作出新的大贡献。

"中国共产党为什么能，中国社会主义为什么好，归根结底是因为马克思主义行"，对国家、对党、对每一个科技工作者，马克思主义都是认识世界，把握真理、追求真理、改造世界的强大思想武器，钱学森曾经深情的回忆，马克思主义认识论、方法论在他的科研工作中发挥了巨大作用。在我的科研进程中，马克思主义方法论在如何分析课题科研中的主要矛盾与矛盾的主要方面，如何认识理论和实践的关系，如何辩证地对待科研中的成功与失败、顺利与曲折等方面，起了很大的指导作用。我们科技工作者在实现建

注：考虑与前后文有重合之处，本文略做精简。

设科技强国的第二个百年奋斗目标的时候,应当自觉学习马克思主义理论,努力掌握其中蕴含的立场、观点、方法,做到知其言更知其义,知其然更知其所以然,努力往深里走、往实里走、往心里走,把学习成效转化为履行新时代使命的实际行动和实际效果。

再讲几句与大家共勉的话。

总书记在七一大会上向全世界庄严宣告:我们正在意气风发向着全面建成社会主义现代化强国的第二个百年奋斗目标迈进。百年新征程,世界正经历百年未有之大变局,国际环境错综复杂,国际形势充满风险和不确定性。形势逼人,挑战逼人!一代人有一代人的长征路,一代人有一代人的担当!实现第二个百年奋斗目标,关键在人才,核心是科技创新,"关键核心技术是要不来,买不来,讨不来的"。因此我认为,实现建设现代化强国,特别是科技强国、军事强国的第二个奋斗目标的责任和使命,首先落到当代科学家身上,实现关键核心科技的自立自强在当今突显极端的重要性和紧迫性。

靠什么实现关键科技的自立自强?回顾我国科技事业所取得的每一项重大成就,都离不开科学家无私报国、服务人民的高尚情怀和优秀品质。自力更生取得的"两弹一星"辉煌成就,靠的就是两弹一星精神!两弹一星元勋的绝大多数都像钱学森、朱光亚、邓稼先那样在国外已有很高的职位、优渥的待遇、学术上已有成就,但他们响应党的号召,为建设我们强大的祖国埋名隐姓、艰苦奋斗,弘扬的就是一个无私报国的科学家精神。科学家精神说到底就是党性在科学家身上的体现和反映,就是革命人生观、马克思主义世界观在科学家身上的体现和反映。我已年逾八十多岁,但我是一名承上启下、具有65年党龄的科学家。共产党员是所谓有特殊材料做成的人,我有责任继续践行科学家精神,让我余生继续发光发热。要学习钱学森,把毕生精力、目光始终关注到国家重大需求上;要学习华罗庚,勤奋工作到最后一刻,倒在讲台上。同志们,我今年84岁,已经退休了,但我觉得还有很多事情能做、想做、要做。所以我继续担任军事科学院首席专家,火箭军首席工程专家,空军工程顾问。我总觉得党给我的荣誉太多了——国家最高科学技术奖、全国道德模范、全国优秀党员,我必须活到老、学到老、革命到老,必须始终做到无须扬鞭自奋蹄,继续为国防建设用脑尽力,为国家铸就钢铁后盾,为经济和社会建设继续做贡献!我希望我们院士同志们,要继承弘扬好、实践好科学家精神,身体力行,用自身的榜样,教育青年科技工作者,把个人理想和国家重大需求相结合,排除名利束缚,严谨治学,潜心研究,直面问题,迎难而上,奋勇拼搏,敢闯科学"无人区"。要敢为人先,敢攀高峰,集智攻关,团结协作,勇当新兴交叉领域的开拓者,勇担关键科技领域的领跑者,抢占世界科技发展的制高点,为开启伟大新征程作出更大贡献!同志们,全国人民正把基础研究自立自强的希望寄托到科学院院士身上,把关键核心科技的自立自强希望寄托到我们工程院院士身上。重托在肩,让我们努力拼搏吧!我相信,在未来的新征程上,一定会涌现很多"两弹一星"元勋那样的新功臣,群星璀璨!

在中国科协学会党建工作指导委员会成立大会暨学会党建工作先进表彰大会上的发言

(2021年6月24日)

尊敬的怀进鹏书记、各位学会党委书记、同志们：

在喜迎建党百年华诞之际，科协党组决定组建学会党建工作指导委员会是中国科协党组坚决贯彻落实党中央重大战略决策部署的一项重要举措，也是进一步加强党对学会的全面领导，促进学会事业改革创新发展的有效方式。

中国科协组织召开本次大会，是对学会党建工作成效的一次总结，也是加强学会党建工作的又一次动员。指导委员会吸纳科学家作为委员，这是对科学家的激励和鼓舞，符合科协组织的特点，能够充分发挥科学家的示范引领作用，加强对学会党建重大共性问题和科技界重大事件的战略研究，怀书记邀请我来担任指导委员会的主任之一，我深感荣幸，又觉责任重大。在此我就弘扬科学家精神，加强学会党建工作的认识与大家探讨交流。

一、弘扬科学家精神是加强学会党建强化价值引领的重要内容

习近平总书记指出"科学无国界，科学家有祖国"，中国的现代化历程充满艰辛，正是在党的领导下，在中华民族伟大复兴事业的感召下，国家利益和人民利益的现实需要，成为一代又一代科学家投身科学事业的内在动力。我理解科学家精神的精髓和核心就是追求真理，献身科学，服务国家发展，是我国科技工作者在长期科学实践中积累的宝贵精神财富。科学家精神就是胸怀祖国、服务人民的爱国精神，勇攀高峰、敢为人先的创新精神，追求真理、严谨治学的求实精神，淡泊名利、潜心研究的奉献精神，集智攻关、团结协作的协同精神，甘为人梯、奖掖后学的育人精神。

新中国成立以来，我国科技事业所取得的每一个辉煌成就，都离不开科学家矢志报国、服务人民的高尚情怀和优秀品质。大家都在学习弘扬"两弹一星"精神，两弹的元勋绝大部分像钱学森、朱光亚、邓稼先等等都是在国外已经有很高的职位、待遇，学术上更是卓有成就，他们响应了党的号召，为建设我们强大的国家埋名隐姓、艰苦奋斗，弘扬的就是这么一个无私报国的爱国主义精神。说到底，科学家精神就是党性在科学家身上的体现和反映，是革命人生观、马克思主义世界观在科学家身上的反映和体现。当前，新一轮科技革命和产业变革突飞猛进，形势逼人、挑战逼人，我们要切实加强学会党建，引导广大科技工作者继承和弘扬这种科学家精神，坚决贯彻落实习近平总书记在两院院士大会、中国科协十大上的重要讲话精神要求。

二、引领学会事业全面融入国家发展大局是加强学会党建的重要任务

学会作为党领导下团结联系广大科技工作者的社会团体，党建工作应充分体现党的政治引领作用，

有效发挥政治功能,实现学会事业与国家发展大局血脉相连。一是坚定"四个自信",增强在党的领导下建成世界科技强国的信心和决心,引领广大科技工作者报效祖国,把人生理想融入为实现中华民族伟大复兴中国梦的奋斗中;二是始终心怀"国之大者",自觉从党和国家工作大局着眼,把论文写在祖国大地上,助力实现高水平科技自立自强;三是认真学习"四史",从党的百年奋斗历程中汲取奋进力量,弘扬科学家精神,坚定不移听党话、跟党走;四是坚持"四服务",积极参与全球科技治理,全面提升开放创新能力,推动构建人类命运共同体。

2016年以来我不再担任中国岩石力学与工程学会理事长,但对学会工作力所能及地予以关注和参与。近年来,学会以党建为引领,紧紧围绕党和国家的重大发展战略,认真落实创新驱动发展战略,努力实现高水平科技自立自强,推动科技创新成果不断涌现并转化为现实生产力。从三峡工程到溪洛渡、向家坝,再到乌东德、白鹤滩,为国家大型水电工程建设提供咨询服务;采用"直面问题、结合工程、深入研讨、商定对策"的形式,为川藏铁路建设难点和军民融合提建议、破难题;深度参与全球科技治理,积极开展"一带一路"国家科技交流。

三、激励广大科技工作者当好科技自立自强排头兵是加强学会党建的重要使命

习近平总书记在两院院士大会和中国科协十大上的重要讲话振奋人心,为科技界指明了前进方向。当前世界百年未有之大变局加速演进,国际环境错综复杂,新一轮科技革命和产业变革突飞猛进,科学研究范式正在发生深刻变革,学科交叉融合不断发展,科学技术和经济社会发展加速渗透融合,凸显了科技自立自强的极端重要性和空前的紧迫性。科技工作者面向世界科技前沿、面向经济主战场、面向国家重大需求、面向人民生命健康的责任更重,压力更大。

老一辈科学家都是把救国救民当作自己神圣的责任,在党的领导下科技事业取得了历史性成就。新形势下,我们要加强学会党建,激励新一代科技工作者发扬勇攀高峰、敢为人先的创新精神,从国家急迫需要和长远需求出发,树立敢为天下先的雄心壮志,直面问题,迎难而上,敢于探索科学"无人区",勇于挑战最前沿的科学问题,力争在重要科技领域成为领跑者、在新兴前沿交叉领域成为开拓者,抢占世界科技发展的制高点,为开局"十四五",开启新征程贡献力量。

涓涓不塞,是为江河;源源不断,是为奋斗;生生不息,是为中国。科协党组领导让我来担任委员会主任,这既是对我的信任,也是对我这名老党员的再一次考验,我一定发挥一名六十五年党龄的老党员、老科技工作者的积极作用,尽心履职,推动学会党建工作再开新局。

在"国土空间规划契机下地下空间的机遇与挑战学术研讨会暨《2020中国城市地下空间发展蓝皮书》发布会"上的致辞

(2020年12月25日)

尊敬的各位来宾、代表们：

上午好！2020年注定是不平凡的一年，这一年我们风雨同舟、共克时艰，我们砥砺前行，守望春天。在崭新的2021年即将到来之际，非常高兴能与各位相聚，共享地下空间的学术盛宴。在此，我对"国土空间规划契机下地下空间的机遇与挑战学术研讨会暨《2020中国城市地下空间发展蓝皮书》发布会"的召开表示热烈的祝贺，向百忙之中参会的各位领导，专家以及远道而来的各位学者、同行表示热烈的欢迎！

《中共中央、国务院关于建立国土空间规划体系并监督实施的若干意见》中明确提出，国土空间规划是国家空间发展的指南、可持续发展的空间蓝图，是各类开发保护建设活动的基本依据。本次会议正处于国土空间规划体制改革的关键时期。在这一契机下，城市地下空间当前的进展与影响、地下空间规划的转型与变革、学科和行业的推动与发展，成为我们今天思考的核心。

《中国城市地下空间发展蓝皮书》年度报告是我国城市地下空间开发利用的权威研究报告，自2015年以来，已连续发布五年，在国内引起了广泛响应并获得业界好评，被众多城市官方引录。此次，中国工程院战略咨询中心、中国岩石力学与工程学会地下空间分会、中国城市规划学会，共同发布"蓝皮书"，以三方的学术平台和传播媒介，冀获行业内更为广泛的瞩目，为城市可持续发展和空间资源合理利用提供指导，为我国地下空间行业发展做出更大贡献。

学术交流是科技创新之源，是促进学科发展和助力人才成长之境。虽然，经过几十年的不断探索和实践，地下空间规划领域逐步凝聚了一批高水平的学者和团队，收获了丰硕研究成果，在城乡规划行业内产生了重要的影响。但是，相比较其他专业方向，地下空间规划各方面尚处于起步阶段，在学术界、产业界的渗透力和影响力尚显不足，需要以更加专业的学术平台，推动学科发展，开展决策咨询，推进科学普及、托举人才成长。

当前，城市地下空间开发利用面临许多问题，我认为关键在政策与法规、规划与管理等顶层设计方面，本次会议围绕国土空间规划背景下的城市地下空间规划开展研讨，将有利于从城市地下空间规划层面推动地下空间的开发利用。鉴于此，为了促进我国城市地下空间规划体系和管理制度完善，促进现代科学技术的普及和推广，促进地下空间规划科技人才的成长和提高，我衷心地希望中国城市规划学会加强地下空间规划领域的学术研究与交流，为城市地下空间开发利用解决关键问题、提供技术支撑、作出引领示范，为可持续的城市建设做出积极贡献。

同时，我也希望所有地下空间规划的学者、研究人员能够把握机遇、砥砺前行，不断创新地下空间规划理论、服务城市发展实践，让城市地下空间成为中国城市发展的新动力、新源泉。

新契机赋予新使命，新使命开启新征程。最后，我要借此机会，向为本次大会付出辛勤劳动的专家、领导与嘉宾们表示最诚挚的谢意！并预祝本次大会圆满成功！

谢谢大家！

| 思想与观点

在江苏省高层次人才主题学习会上的发言

——让生命在科技报国中闪光

（2020年11月15日）

各位领导、同志们：

  大家好！非常高兴和大家分享我的一些感悟和体会。我本人是江苏昆山人，今天，能够在我们江苏省组织的这次主题学习会上发言，感到格外荣幸与亲切。我见证了国家的沧桑巨变，见证了我国国防建设取得的巨大成就，特别是亲身参与并见证了我国防护工程研究与建设从跟跑到并跑、再到有所领跑的全过程，感到很幸运也很幸福。习近平总书记深刻指出，"硬实力、软实力，归根到底要靠人才实力。谁拥有了一流创新人才、拥有了一流科学家，谁就能在科技创新中占优势。"参加我们今天这个会的都是全省的高层次人才，肩负着全省乃至全国科技创新、科技报国的重要使命。回顾我的人生历程，我想送给大家四句话。

  第一句话，爱党信党跟党走，是我一生中最正确、最坚定的选择。

  1937年10月，我出生在国歌所唱的"中华民族最危险的时候"，母亲在逃难途中的小船上生下了我。我的童年是在日寇的铁蹄下度过的。逃难途中，为了怕我啼哭引得日本兵发现，父亲差点把我捂死。7岁时，父亲因贫病离世，母亲靠摆小摊维持生计。每当回忆起童年，除了生活的艰辛，有两个场景时常浮现在我的脑海：一个是侵华日军将杀害的游击队员尸体放在学校操场上示众；另一个是美军在北京强奸了北大学生沈崇、在上海残暴打死一名三轮车车夫。亲眼看见侵略者、殖民者戕害同胞的残暴行径，让我深深感到，没有强大的祖国，人民不可能有幸福安宁的生活。

  中学时代，我就读于著名的上海中学，由于学习刻苦、成绩优异，当年上海代表团去朝鲜战场慰问时，把我的成绩单作为慰问品带给前线的志愿军战士。新中国成立后，依靠人民政府的助学金，我顺利完成了中学学业。中学毕业时，组织上本来要选送我到苏联留学。但我国急需军事人才，新成立不久的哈尔滨军事工程学院在上海中学招收优秀毕业生。作为优秀学员，组织就希望我到"哈军工"学习。出国留学还是在本国读军校？我当时就想，我这个乡下的穷孩子，能受到良好的教育，所有的一切都是党的培养，唯有献身党的事业，方能报答党的恩情。在初中抗美援朝时，我就报考军干校，因年龄太小、双眼红绿色弱，未能批准。这次我也决定放弃到苏联深造的机会，迈进了哈尔滨军事工程学院的大门，成为"哈军工"组建后招收的第三期学生。

  进入"哈军工"以后，当时防护工程专业选的人不多，因为要跟黄土铁铲打交道，大家都觉得太土。我是班上骨干，带头服从组织分配，选择了防护工程，那可以说是我人生中第一次与防护工程结缘。

  1960年，我被组织选派到莫斯科古比雪夫军事工程学院学习深造。有位老红军跟我说，我们国家还有人吃都吃不饱，国家得用多少金条才能送你们去留学啊。我深知机会来之不易。到了苏联后，我没有辜负党和国家的期望和重托，刻苦地钻研、学习，成为工学副博士。留学归国后，我克服各种困难，一门心思做学问、搞研究。后来，组织又考察选拔我当了院长、评了院士。

  现在回想起来，如果没有抗日战争的胜利，没有新中国的成立，没有国家的助学金支持，我早就会和哥哥姐姐一样失学、失业；如果没有组织的保送，我就不可能进入哈军工的大门；如果没有组织的选派，我也不可能去留学深造；如果没有组织的培养，我更不可能获得今天这些成就。所以我一直没有忘记党的恩情，没有忘记自己的初心，就是要回报党、回报组织。

获得国家最高科学技术奖后，我决定把国家奖给我的800万元奖金全部捐出，资助我国西部的贫困学生。之所以这么做，是因为我觉得，我们现在的幸福生活都是由烈士先辈流血牺牲奋斗换来的，他们把生命都献给了党和国家，我还有什么不能贡献呢？现在很多贫困学生需要帮助，如果他们能像我一样完成学业，成为一个对国家和社会有用的人才，那是一件非常有意义而又幸福的事。

今年，新冠肺炎疫情暴发后，我把江苏省配套奖励给我的800万中的650万捐给了武汉抗疫一线，其余的150万捐给了母校和学会。在危急关头，广大医务人员白衣为甲、逆行出征，舍生忘死挽救生命。今天，钟南山院士、张伯礼院士、张定宇博士、陈薇院士也都来到会场为我们作了报告，他们不顾个人安危、为党为国分忧的事迹让我们感动。我作为一名共产党员，能够贡献我的一点力量，也是在回报社会、回报党的恩情。

一路走来，我深刻认识到：没有中国共产党就没有中国人民的解放，也不会有中国人民的幸福生活。我常对老伴和身边同志说，没有党的培养就没有我的一切，爱党信党跟党走，是我一生中最正确、最坚定的选择。

第二句话，只有把个人的理想与国家和民族的前途命运紧密联系在一起，才能有所成就、实现价值。

爱因斯坦曾说，把安逸和享乐当作人生唯一目标，那不过是猪圈里的理想。我认为，一个人，他的注意力、关注点在哪里很重要。大家知道，"两弹一星"功勋奖章获得者朱光亚1950年回国前，在给留美同学的一封公开信中这样写道："让我们回去，把我们的血汗洒在祖国的大地上，灌溉出灿烂的花朵。"这是多么高尚的情怀啊！而另一位著名科学家钱学森一回国，就强烈建议国家搞原子弹、导弹。现在回过头来看，这又是多么具有战略意义的建议啊。一个人想什么，关注什么，和他的世界观、人生观是密切相关的。这就是老一辈科学家爱国奉献、科学报国的真实写照。

20世纪六七十年代，我国面临严峻的核威胁环境，但我们奉行的是积极防御战略，不打第一枪，那如何打好第二枪也就是保证"二次反击"呢？我当时就想，只有铸牢防护工程这面坚固的盾牌，才能确保我国首脑指挥工程和重要战略武器工程的安全。因为，这是我们国防安全的最后一道防线。从那时起，为祖国铸就坚不可摧的"地下钢铁长城"，就成了我毕生的追求。

20世纪70年代初，我受命进行某飞机洞库门设计。为了获得准确的实验数据，我赶赴核爆试验现场进行实地调查研究和收集数据。在现场发现，虽然核爆后的飞机洞库门没有被炸毁，飞机也没有受损，但是防护门出现严重变形，致使无法开启。为了解决那一问题，我率先引入当时世界上刚兴起的有限元计算理论，加班加点学习计算机语言，编制出电子计算机计算程序，翻译整理出十多万字的外文资料，通过当时中国最大的晶体管电子计算机计算，设计出了当时跨度最大、能抵抗抗力最高的核爆炸冲击波的机库大门。同时，我还出版了专著《有限元原理在工程结构计算中的应用》，并获得了1979年全国科学大会重大科技成果奖。

世间万物，相生相克，有矛必有盾。如果说核弹是军事斗争中锐利的"矛"，那么防护工程则是一面坚固的"盾"。我和团队时刻跟踪着新型进攻武器的发展，只要是进攻性武器这个"矛"发展一步，就琢磨和研究让我们的"盾"如何更坚固一层。从核空爆到核触地爆，再到核钻地爆；从普通爆炸弹到钻地弹，经过长达几十年的研究，我们攻克了一个个难关、突破了一系列技术难题，为我国战略工程装上了"金钟罩"。

各位领导、同志们，大国间的竞争，不仅体现在军事实力上，更体现在综合国力上。作为中国工程院院士，作为一名革命军人、一名科技工作者，既要关注如何增强国防实力、维护国家安全，也要关心提高综合国力、谋求人民幸福。1992年初，珠海特区建设机场，要炸平一座山，爆破总量1085万立方米，要求一次性爆破成功，一半的土石方定向投入大海，还要确保邻近1000米内的两处村庄安全。这样大的爆破，

| 思想与观点

世界还无先例,难度很大。我带领团队六下珠海,和大家一起研究方案和施工环节,当年12月成功实施爆破,创造了世界爆破史上的新纪录。

这些年,我还在长江隧道、西气东输工程、港珠澳大桥海底隧道、深中通道、能源地下储备、核废物深地处置等方面,贡献了我的学识和才智。我觉得,这是我作为院士的使命所系,也是我作为科学家的幸福所在。

我常对我的学生讲,一定要将个人命运与国家、民族的命运结合起来,哪些事情对国家和人民有利,我们的兴趣和爱好就要向哪里聚焦,这也是一名科学家应有的情怀和担当。事实上,一个人只有自觉把个人理想与党和国家的需要、民族的前途命运紧密联系在一起,才能有所成就、彰显价值,才能收获成功的事业和幸福的人生。

第三句话,科学是老老实实的学问,容不得一点点的马虎和心浮气躁。

根据我的人生体验,科技的成就或者人生的成功,有时候看来好像是一个偶然的机遇,但是我想说,机遇永远只垂青那些有准备的人,我们要做好迎接成功的准备。

成功的准备首先是科学基础的准备、基本功的准备,我回忆中学、大学学习知识的扎实功底帮了我的大忙。前面说过我中学学习的成绩单被送去慰问志愿军战士,大学基本五年全优,因为基础打得好,所以自学能力就强,能够在任务需要的时候,很快掌握空气动力学、有限元理论和计算机语言,能够独立编写出大型计算机程序并上机操作。而打好扎实的基础就要勤奋学习。我深信著名作家高尔基的名言:天才出自勤奋。我就读"哈军工"期间,哈尔滨松花江畔很漂亮,特别春天江水融化的美景,但我一次都没去看过,几乎把所有的时间和精力都投入到学习当中,六年假期只回过一次家,每个假期我都主动留校学习。所以才有了年年都被评为优秀学员和社会主义建设积极分子,也成了全年级唯一的全优毕业生。留苏4年,除了莫斯科没去过其他城市,到过红场,没有参观过列宁墓,因为怕排队占用过多时间,上街基本上也都是去书店或图书馆。

想成功还要有意志和品质的准备——不怕苦难、不怕挫折、不被干扰。马克思曾说:"科学上没有平坦的大道,只有不畏艰险沿着陡峭山路攀登的人,才有希望达到光辉的顶点。"我虽然没有达到顶点,但也克服了艰难登上了科技的高地。我记得在我设计飞机洞库大门时,没有学过有限元理论、计算机语言,没有编过电子计算机程序。在七机部五院、中科院计算机所的大型计算机上机计算时,只能是人家不上机的午饭时间和晚上,为此得了十二指肠溃疡、胃溃疡,随后又诱发了痔疮。这些困难我都克服了过去,坚持了下来。我也经历过失败,我设计的柔性大变形工事,在核试验中,工事中的狗被冲击波打死了,摄像机被打坏了,但我没有灰心,认真查找原因,发现是工事的门没有焊接好,焊缝脱落,门被打进去了。找到了原因,问题很好地解决了。

这么多年,无论外部环境如何动荡,我始终心无旁骛,安心教书育人,即使"文化大革命"时,我也潜心学习研究,这为我在防护工程领域的开疆拓土、创新引领打下了坚实基础。在1979年的全国性学术会议上,我一下子提交了8篇防护工程前沿领域的学术研究论文,震动了该领域同仁。由于成绩突出,1980年职称评审时,连助教、讲师都不是的我被直接评为副教授。回顾那段历史,我感到,遇到挫折不低头、碰到困难不退缩,让我没有停止科学研究,那是我人生积淀与进步的一个重要时期。

如何能做到不怕苦难、不怕挫折、不被干扰? 我认为,一个人只有树立了远大理想,才能有坚强的事业心;只有树立远大的理想,才能有巨大的动力,才能沉得下心气,耐得住寂寞,不被名利干扰,才能不断拼搏进取,始终走在科技的前沿。

如何才能树立远大的理想? 我的人生经历告诉我,烈士、英雄、伟人的事迹在我树立远大理想的过程中起到了巨大的作用。奥斯特洛夫斯基、卓娅和舒拉、吴运铎、刘胡兰、黄继光、董存瑞等一个个英雄人物激励我不断前进、不断成长。对我而言,烈士英雄一直活在自己心中,是他们的思想言行照亮了我永远前进的道路,让我永葆我革命的青春。

最后一句话，只有摆正个人和组织的关系，摆正个人和集体的关系，摆正个人和群众的关系，才能顺利前进。

去年1月8日，习近平总书记在国家科学技术奖励大会上，亲自为我颁发了国家最高科学技术奖奖章和证书。我深知，这份荣誉和褒奖，不属于我个人，它归功于党和政府对我们科技人员的热忱关心，属于防护工程领域的全体科技人员，归功于社会各界对科技创新的充分尊重，归功于领导和同志们对我的支持帮助。

科学技术研究是集体、团队的研究。我能参加核试验，完成科研任务，离不开组织的支持。我能完成跨度最大、抗力最高的飞机洞库大门设计任务，是团队和原济南空军设计所共同支持帮助的结果。我能当上教授、中国工程院首届院士，是已故我国著名老科学家张维、李国豪推荐提携的结果，我能获得这些国家级科研奖项，都是我的团队和合作团队集体奋斗的结晶。我能获得深部岩石力学的研究成果，是和一些专家教授共同研究完成的。我能连续三届担任中国岩石力学与工程学会理事长，连续四届担任防护工程学会理事长，是防护工程领域同志们信任、支持的结果，乃至我个人能当上原工程兵学院院长，当上全军爱国奉献优秀干部典型，也是学院广大干部大力支持的结果。

总结我一生，我深深体会到，一个人没有大家的支持，就不可能有什么进步和发展，作为科技人员，更是如此。而要得到别人的支持，就要首先支持别人；一个领导要得到群众的支持，就要树立和实践小平同志教导的"领导就是服务"的理念。只要大家在吃苦担当的时候能勇于往前冲，排名报奖的时候学会往后靠，就一定能获得他人的支持。著名数学家华罗庚曾说过："人家帮我，永志不忘；我帮人家，莫记心上"。我觉得任何事业都是集体的事业，不是个人单打独斗就能成功的。我们要把助人为乐作为人生准则，培养自己良好的团队精神，处理好个人和群众的关系、个人和集体的关系，团结力量干大事。

各位领导、同志们，我今年83岁了，已经退休了，但我觉得还有很多事情要做、想做、能做。在我有生之年，我将始终做到无须扬鞭自奋蹄，继续在防护工程领域潜心研究，带好学生、培养人才、关心团队建设，为国家铸就钢铁强盾，为经济和社会建设做出新的更大贡献。

最后，我想用《钢铁是怎样炼成的》中的一段话和大家共勉："人最宝贵的是生命。生命属于人只有一次。人的一生应当这样度过：当他回首往事的时候，不会因为碌碌无为、虚度年华而悔恨，也不会因为为人卑鄙、生活庸俗而愧疚。这样，在临终的时候，他就能够说："我的整个生命和全部精力都已献给了世界上最壮丽的事业——为人类的解放而斗争。"

衷心地希望大家大力弘扬科学家精神，立鸿鹄志、做奋斗者，也祝愿大家都能够在自己的领域有大的建树和作为，努力成为实现中国梦的栋梁之材！谢谢大家！

在习近平总书记关于乌东德水电站首批水轮机组发电亲切祝贺和重要指示的重大意义的认识座谈会上的发言

(2020年6月29日)

各位领导、各位同仁：

上午好！

很高兴参加座谈会，汇报我对总书记关于乌东德水电站首批水轮机组发电亲切祝贺和重要指示的重大意义的认识。习近平总书记指示中肯定了水力发电生态优先、绿色发展、造福人民的重要贡献，强调了在保护中发展、在发展中保护有序推进水能资源开发的技术路线。

众所周知，部分环保人士对于水力发电的发展持有异议，他们只看到建大坝、修水库对于局部生态环境的影响，我认为总书记是从马克思主义认识论出发，坚持辩证法、坚持一分为二的矛盾论，坚持局部服从全局抓主要矛盾，从全球的生态保护与全球的绿色发展出发来认识水力发电的利弊。

从全球的生态环境与人类面临的挑战来看，正如联合国秘书长最近指出："虽然病毒是目前最大的担忧，但气候变化仍然是当代面临的核心挑战"。联合国气候变化专门委员会报告：到21世纪末北极的融冰导致海平面上升60cm至1.1m，随着气温升高，全球冰川退缩，山区的冰川21世纪末1/5将融化掉，阿尔卑斯山的冰川将消失殆尽；海洋吸收CO_2导致酸性增加，海洋渔业捕获量将减少24%。*Nature*发表的科研报告指出：如果到21世纪末全球气温上升1.9℃，高海拔地区冰雪将减少80%，"水塔"的减少导致全球19亿人缺水，海平面上升导致全球2.8亿人流离失所，纽约、上海等临海城市面临灾难性洪水。而如果全球气温升高1.5℃，那么75%高海拔地区冰雪将得以保留。所以，出路在于全球合作治理控制温室气体碳排放、控制全球气温升高，这一切的关键是绿色发展，其核心是采用清洁能源和生态保护。

人类正在迎战新冠病毒大流行的人类灾难，而气候变化将是最糟糕的全球灾难，也是对人类命运共同体的更大挑战。中国在防控新冠病毒疫情中取得了不凡的成绩，在迎战新型冠状病毒全球大流行中作出了很大的贡献，我相信我们在迎战即将到来的气候变化大灾难、构建人类命运共同体中必将做出更大的贡献！因为，绿色发展是习近平中国特色社会主义思想的一个重要的组织部分。我们应当从绿色出发迎战全球气候变化，以构建人类命运共同体的高度来认识习近平总书记发出的"长江大保护""黄河大保护"、乌东德首批水轮机组发电的指示：生态优先、绿色发展、保护中开发、开发中保护的重大意义。

我们水电科技工作者、岩石力学与工程科技工作者应当认识到我们在全球生态保护与绿色发展中的重大责任与担当，拿水电发电来说，不但要开发中国一亿千瓦水力资源，建设西藏的水电站，还要走出国门，到一带一路沿线国家去，有序开发世界更多更好的清洁能源，为战胜气候变化大灾难、构建全人类命运共同体作出更大的新贡献。

在中国岩石力学与工程学会"弘扬科学家精神,加强作风和学风建设"主题宣讲会上的发言

(2020年5月28日)

广大的青年朋友们:

今天,我非常高兴以这种新形式与你们见面。在我们即将迎来第四个"全国科技工作者日"前期,冯夏庭理事长让我给青年朋友们讲讲,我也希望通过这个方式与青年朋友们谈谈我的想法。

我很高兴地听到冯夏庭理事长汇报说,学会目前各项工作都有了发展和进步,学会会员已超过2万人,越来越多的青年学者在学会发挥作用,除了青年工作委员会的组织以外,学会的青年托举人才,就是学会这几年依托中国科协"青年人才托举工程"项目,从全国选拔出来的32岁以下的优秀青年,目前已达到34名,做得非常好,这项工作是冯理事长倡导申报并直接领导的,我主持过一次评审工作,选拔竞争很激烈,可以说为学会培养未来的学术领军人才和工程大师储备了力量。这项工作充分证明学会对青年科技工作者的重视,希望越来越多的青年学者在做好科研工作的同时,参与到学会的工作中来。

今天宣讲的主题是"弘扬科学家精神,加强作风和学风建设",非常有必要,特别对我们青年朋友们,在事业起步之初,就要重视作风和学风建设。

首先说一说什么是科学家精神,为什么要大力弘扬科学家精神。"科学"是探索未知、发现真理、发展先进、改造世界、造福人类的学问。"科学家"是从事科学研究的人员,科学家精神是为真理和信仰献身的精神,是探索未知和奉献新知的精神,是科学家在长期科学实践活动中逐渐形成的。在中华民族的伟大进程中,经过一代代科学家的不懈努力,铸造了中国科学家独特的精神品质和鲜明的文化气质,铸就了中国科学家精神。中国科学家精神的主要内涵:一是胸怀祖国、服务人民的爱国精神;二是勇攀高峰、敢为人先的创新精神;三是追求真理、严谨治学的求实精神;四是淡泊名利、潜心研究的奉献精神;五是集智攻关、团结协作的协同精神;六是甘为人梯、奖掖后学的育人精神。

其次,我们谈一谈如何弘扬科学家精神。一代人有一代人的奋斗,一个时代有一个时代的担当。当前,我国正处于近代以来最好的发展时期,世界亦处于百年未有之大变局。新一轮科技革命和产业变革不仅打开了国际科技竞争的机会窗口,而且更进一步推动了科技与经济、科技与社会、科技与人的全面融合发展,国际社会在各种制度、体制、机制的不断蜕变中正呈现新的面貌,国际关系内涵发生了深刻变化,我们正前所未有地接近实现中华民族伟大复兴的目标,前所未有地走近世界舞台中央。

面对复杂多变的世界形势,我们能否在新科技革命的激烈竞争中勇立潮头,既取决于我们的发展战略、创新制度是否完善,也取决于我们的精神气质是否符合历史潮流和客观规律。"洋火洋油"的时代需要时刻铭记,"动车高铁、长江水下隧道、三峡大坝"的奇迹更要永远追忆。

正如党的十九大以来,党中央、国务院号召我们紧密团结在以习近平同志为核心的党中央周围,增强"四个意识"、坚定"四个自信"、做到"两个维护",在践行社会主义核心价值观的过程中,大力弘扬科学家精神,就是引导我们广大岩石力学与工程科技工作者,特别是青年科技工作者,不断激发创新发展活力,助力川藏铁路建设,服务国家重大工程。同时,加强学术诚信和学风建设,争做高尚品德的模范、良好科研作风和学风的标兵。下面就这个主题,我重点讲四点内容。

1. 要坚持党的政治引领，大力弘扬科学家精神

学会工作要始终坚持党的政治引领，将党建工作贯穿于学会事业改革发展的全过程，提高政治站位，强化政治引领，增强科技工作者对党的政治认同、思想认同、理论认同和情感认同。古人云：不积跬步无以至千里。青年学者应从小处做起，从年轻时期建立良好的习惯和正确的世界观。大力弘扬老一辈岩石力学与工程领域科学家爱国奉献、求真务实、协作创新、无私育人的高尚品德和奉献精神，争做科技强国建设的奉献者、科技创新的践行者、学术道德的垂范者、新人成长的引路者。

2. 坚守科研诚信底线，遵循科研道德规范

我最近注意到一些关于论文作假的报道。

事件回顾1：

2020年4月，国际学术出版集团施普林格（Springer）旗下期刊 *Multimedia Tools and Applications* 宣布批量撤销33篇论文。撤稿理由涉及五大原因：

duplicated from an unpublished manuscript（剽窃他人未发表手稿）

authorship manipulation（操纵作者身份）

an attempt to subvert the peer review process（试图颠覆同行评议发表系统）

substantial overlap most notably with the article cited（内容抄袭）

figure duplication without appropriate permission（图像未经允许不当复制）

此次的国际期刊批量撤稿事件涉及国内的多所高校，甚至多所知名高校赫然在列，其中不少论文还曾获得过国家自然科学基金的资助。

事件回顾2：

2019年，*Nature* 刊登了一篇评论文章，文章指出中国现有四百万名科研人员，这个研究人员的体量，给全球学术界带来的影响比任何国家都要多得多。"根据数据统计，仅在2008年，在Web of Science 上来自中国的论文收录数量就已经超过了英国，排名居世界第二位。到了2018年，中国的论文发表数量已达到了41.2万篇。然而在这些学术成果中，也存在着很多问题，比如伪造同行评审，或者论文中存在欺诈性内容。调查显示，截至2017年，由中国贡献的科研文章数量占全球的8.2%，但撤销的论文数量却惊人地占到了全球的24.2%。"

上述事例可以说已经达到事态非常严重的情况。因此，我们应牢固树立"诚信"意识，严于自律，严格按照对科研成果的创造性贡献大小据实署名和排序，反对无实质学术贡献者"挂名"，反对科研数据成果造假、反对抄袭剽窃科研成果、反对委托代写代发论文、反对庸俗化学术评价，这"四个反对"是我们的底线。

广大青年朋友们更应该在新时期做到严守科研道德规范和科研诚信底线，敢于同违背科研诚信、科研道德要求的现象作斗争，切实加强道德修养，努力做践行社会主义核心价值观的楷模、弘扬中华民族传统美德的典范。

3. 坚持实事求是，反对急功近利、浮夸浮躁之风

求真务实、严谨治学，坚决反对浮夸浮躁、急功近利，不人为夸大研究基础和学术价值，不向公众传播未经科学验证的现象和观点。杜绝违反国家有关保密的法律、法规和单位有关保密的规定，科研新成果在学术期刊或学术会议上发表前（有合同限制的除外），不先向媒体或公众发布，公布突破性科技成果和重大科研进展需经所在单位同意，推广转化科技成果不夸大技术价值和经济社会效益，不隐瞒技术风险，学术成果确实经得起同行评审、工程建设单位认可。

4. 崇尚学术民主，履行社会责任

坚持学术民主，倡导学术争鸣，尊重他人学术话语权，鼓励年轻科技工作者大胆提出自己的学术观点；主动走近公众，开展科普活动，传播科学知识和科学方法，弘扬科学精神与科学文化，引领更多青少年投身岩石力学与工程事业，在全社会形成热爱科学、崇尚科学家精神、注重科技创新的学风氛围。

年轻的岩石力学与工程科技工作者朋友们，最后，我想对你们说："雄关漫道真如铁，而今漫步从头

越"。我们要将"弘扬科学家精神"作为深化"不忘初心、牢记使命"主题教育的重要抓手,让我们紧密团结在以习近平同志为核心的党中央周围,坚持以习近平新时代中国特色社会主义思想为指导,以实际行动弘扬科学家精神,加强自身作风和学风建设,努力践行学风道德规范,引领科研诚信之风,坚持为科技工作者服务、为创新驱动发展服务、为提高全民素质服务、为党和政府科学决策服务的职责定位,团结协作,凝心聚力,改革创新,为创建世界一流学会而努力奋斗!

最后,在全国科技者工作日到来之际,祝各位科技工作者节日快乐!谢谢!

在超大直径全断面竖井掘进机下线仪式上的发言

(2020年1月12日)

各位领导、各位专家、女士们、先生们：

大家上午好！

今天，我们齐聚"工程机械之都"长沙，共同见证"全球首台超大直径全断面竖井掘进机"隆重下线，首先为铁建重工取得的成绩表示祝贺，也对各位对中国工程装备发展的关注表示感谢！

铁建重工自2007年成立以来，我来过几次，每一次都能感觉到铁建重工紧跟时代步伐、展现使命担当的情怀，每一次都能看到铁建重工推陈出新建造出一台台引领行业的精品。铁建重工经过短短十余年发展，现在已经成为行业知名的隧道掘进机制造企业，不仅在TBM研发能力上一直独领风骚，在竖井掘进机上的迈进步伐也铿锵有力。

我曾参与过铁建重工直径8m级全断面竖井掘进机的立项评审和成果鉴定，其在建井工法、装备技术等方面已达到国际领先水平。2019年9月，我再次受邀参加了今天这台"超级装备"的设计方案评审，让我惊讶的是，铁建重工仅用4个月时间便研制出了产品！当然，这不仅是铁建重工高效、优质的高端装备研发制造能力，更是由刘飞香带领的科研团队、制造团队多年来科技创新的深厚积淀！

同志们，现在世界各国纷纷布局地下空间建设，城镇建设层次化、资源开采深地化、军事防护地下化已成为趋势，而地下空间建设的核心是设备，施工设备进步才能推动地下工程建设进步。今天下线的超大直径全断面竖井掘进机在国防工程等重大地下工程建设中，具有非常高的战略价值和广阔的市场前景，也一直是世界地下工程装备领域试图攻克的重大技术难题。所以，今天的成功不仅仅是一个企业的成功，更体现了什么是中国创新、中国速度和中国高度！

再次祝贺铁建重工在大国重器的征途中谱写了华彩篇章，也祝愿这台"超级装备"在未来"超级工程"中征战成功！

祝大家身体健康、工作顺利、阖家幸福！谢谢！

在江苏省过江通道建设技术专家委员会成立大会上的发言

(2019年12月17日)

各位领导、各位专家、同志们:

今天我们相聚在美丽的古都南京,参加江苏省过江通道建设技术专家委员会成立大会。在此,我代表隧道专家组全体委员,向近年来江苏过江通道建设取得的进展和突破,表示热烈的祝贺!向江苏省人民政府及交通运输部门对在座专家的信任和重视,表示衷心的感谢!

江苏省地处长江下游,是万里长江经济带上经济体量最大的省份。为贯彻落实"交通强国"战略,更好地满足经济社会高质量发展和人民群众美好出行需要,近年来江苏省全面加快过江通道规划和建设步伐,在过江隧道建设领域取得了令人瞩目的成绩,南京长江隧道、扬子江隧道、南京地铁10号线等多座过江隧道顺利建成,为融合长江南北发展、促进长三角一体化、"一带一路"和长江经济带建设,发挥了显著的经济效益和社会效益。

在江苏新一轮过江通道建设中,江阴第二过江通道是当前最为迫切的重点项目之一。江阴第二过江通道是江苏省高速公路网规划S90高速公路的重要组成部分,南北分别连接江阴市和靖江市。修建这条过江通道,对分流江阴大桥交通压力,服务跨江联动发展,支撑扬子江城市群建设具有重大意义。经过比选,江阴第二通道拟采用隧道结构形式。江阴第二通道具有"大断面、长距离、高水压"的典型特征,是中国内地最大断面的盾构隧道,是全国长度排名第二的长江隧道,同时也是目前国内承受水压最高的水下隧道。工程建设技术难度大,安全风险高,工程管理复杂,工程建设面临巨大的挑战。

在江阴第二过江通道今后的设计、施工乃至运营过程中,隧道专家委员会将充分发挥专业特长,为工程建设定向把脉、精准施策,指导重大技术攻关,帮助解决各阶段遇到的重难点问题,助力建设单位在建设过程中形成一批创新成果,助力江阴第二过江通道顺利建成国内高水压大直径盾构隧道品质示范性工程。

我们相信,江苏省作为交通建国建设先行试点省份,在省委省政府的坚强领导下,在工程建设者的共同努力下,一定能够攻坚克难、创新突破,圆满实现江阴第二过江通道建设既定目标。我们期待,未来江苏省在过江隧道建设方面,不仅从数量规模上补齐短板,更能在建设方式和安全防控等方面进行积极的探索和创新,安全建造更多世界领先的过江隧道,为打造交通强国样板、展示交通强国形象作出江苏贡献。

谢谢大家!

在纪念于学馥先生百年诞辰大会上的发言

(2019年9月19日)

各位专家、各位来宾、各位朋友：

大家上午好！很抱歉！因有事冲突不能前来参加大会。

今天，我们怀着崇敬的心情，在这里隆重集会，纪念中国工程岩石力学的创始人、奠基人与学科前缘开拓者之一于学馥先生诞生一百周年。在此，以我个人的名义向于学馥先生致以深切的缅怀和追思，向于学馥先生的亲属致以诚挚的问候，向参加今天纪念活动的各位院士、同行及社会各界来宾、朋友们，表示衷心的感谢！

于学馥先生是我国采矿工程界第一批博士生导师（全国共2人），岩石力学与采矿工程专家，是中国工程岩石力学领域的创始人、奠基人与学科前沿开拓者之一，是北京科技大学矿业工程学科主要创建人之一，也是这一领域最早的教育家，为我国岩石力学和采矿工程、岩土工程等领域培养输送了大批高级科学技术人才（包括中国工程院院士2名）。在国际岩石力学界，于学馥先生首先提出"轴变论""岩石记忆"与不确定性分析科学决策理论，为我国现代岩石力学与采矿科学理论发展奠定了基础。

于学馥先生曾任国际岩石力学学会成员、国际岩石力学学会中国小组成员、中国岩石力学与工程学会常务理事、教育工作委员会主任委员、地下与地下空间岩石工程专业委员会副主任委员、《岩石力学与工程学报》主编、中国核学会铀矿冶学会理事等。

于学馥先生的一生是无私奉献的一生。今天我们纪念他，就是要学习他毫无保留地把自己的全部才智贡献给祖国人民的奉献精神。不确定性分析科学决策方法与科学研究新体系是于学馥前后经过了40余年的科学研究与生产实践中总结发展起来的。此外，在20世纪80年代初期，他在理论研究与工程实践中建立了一套基于非均质、非连续介质理论，以现代数学、力学和计算机数值模拟分析为主的地下围岩稳定分析理论。在金川资源综合利用的项目上集数十年工程试验与实践的经验，提炼升华了全新的地下空间开挖理论，在金川资源综合利用研究中作出了突出贡献。为此，于学馥先生1989年获国家科学技术进步特等奖。

各位来宾和朋友，让我们继承发扬于学馥先生高尚的精神和品格，乘着"创新2035"的东风，开拓创新，踏实工作，为祖国科技事业的创新进步和国家的繁荣强盛而努力奋斗！

谢谢大家！

在中国岩石力学与工程学会水下隧道工程技术分会成立大会上的发言

(2019年8月23日)

尊敬的各位领导、专家及同行们,大家好!

今天,我们相聚在美丽的"泉城"济南,共同参加"中国岩石力学与工程学会水下隧道工程技术分会"成立大会,及即将开始的"中国大盾构隧道智能建造高峰论坛",这是一次全国从事水下隧道科学技术工作者齐聚交流的盛会,代表中国水下隧道国家队的实力,我们在全球范围内立起了一座座水下隧道丰碑,如万里长江第一隧南京长江隧道、水下高铁大盾构典型工程广深港狮子洋隧道、万里长江第一廊苏通综合电力管廊工程等。今天,水下隧道工程技术分会的成立,以及中国大盾构隧道智能建造高峰论坛的召开,昭示着我们中国水下建设者不再沉默,我们将扛起世界水下隧道建设王牌军的大旗,引领世界水下隧道建设新潮流!值此机会,我作为一个从事水下隧道、从事地下工程科技工作的老者,向各位莅临揭牌仪式的各位领导、各位专家、各位代表,致以诚挚的问候,并表示衷心的感谢!

今年是新中国成立七十周年的大庆之年,数十年间我国水下隧道建造从无到有、从学习到领跑,积淀着一代代新中国建设者的辛勤和汗水,有不屈,有无畏,更有苦尽甘来的荣耀。近年来,随着国家经济一体化和大通道建设快速发展,我国水下隧道工程建设数量和规模不断提高,尤其是跨江、越海水下隧道工程,如雨后春笋般不断涌现,目前"渤海海峡跨海通道""琼州海峡跨海通道"等世界级跨海工程进入加速论证阶段,随着"复杂地质、超高水压、超大直径、超长距离"等技术难题的不断攻破,我国即将迎来海峡大通道建设的黄金时代,在座各位将都是我国水下隧道建设黄金时代的建设者、亲历者、荣耀者,共同推进我国水下隧道行业科技研发、设计、施工、装备制造、运营保养等领域取得长远进步。

为了更好地开发和利用地下空间,特别是建设好水下隧道工程,2018年底中国岩石力学与工程学会研究决定成立中国岩石力学与工程学会水下隧道工程技术分会,旨在构建交流共享平台、推动技术进步与成果转化,紧紧围绕未来水下隧道以至地下空间综合开发建造需求,推动深埋、大断面或大空间、长距离或超长距离、智能建造、绿色建造等领域技术水平发展,针对未来"新""难"技术挑战,我们要做好以下准备。

1. 掌握跨江越海水下隧道建造成套建造技术

未来,10km、20km乃至100km以上水下隧道工程将逐步实施,由于存在隧道单次掘进距离长、结构强度及耐久性要求高、隧道通风和竖井设计施工要求标准高、开挖设备整体性能要求提升等严峻挑战,隧道设计验算、施工开挖、施工通风、防灾减灾、运营保养等领域技术水平均亟待跨越式提高。

2. 智能感知掘进综合技术及成套装置亟待研发

目前,水下隧道施工受限于目前探测技术发展水平的制约,无法准确地探测出未知风险,加之海域、水域环境下水文地质、工程地质极其复杂,给工程实施带来极大的风险。超前物探、智能分析、三维重建、智能预测预警等,将在水下隧道掘进中的作用,将日趋重要。智能传感器、超前地质预报设备、智能拼装、智能换刀、智能换管等智能化装置亟待进行应用化研发制造,指导复杂水下隧道工程的潜行掘进。

3. 水下隧道大数据综合建造技术亟待整合

目前,世界都在抢占4G市场、超前研发5G技术,未来世界数字化特质将越来越显现,万物互联的时代很快到来！研发运用 BIM 技术、大数据、物联网、GIS、人工智能、5G、云计算等技术,深化传统土建、机械行业,需要与数字、智能、大数据进行深入融合,实现综合精细化、一体化规划、设计、施工与运维,全方位创新水下隧道领域及相邻行业的综合管理水平,推进"精细化、智能化、安全化"建造 2.0 版。

4. 超深、超大地下空间一体化建造技术

现代城市的快速发展,超大城市逐渐增多,人口急剧膨胀,许多城市都出现建设用地紧张、生存空间拥挤、交通阻塞等一系列问题,解决这些问题就需要我们加大超深、超大地下空间的开发利用。这就需要我们对深部岩体力学机理及勘探技术,地下超大空间开挖、支护技术,超深、超大地下空间开挖设备技术研发,超深、超大地下空间节能通风技术等关键技术的研发,通过学会搭台、协同发展等方式,共同促进我国地下空间的开发利用。

5. 新科技发展绿色综合建造技术

随着社会经济的发展,我国的环境污染问题越来越严重,高能耗、高污染、高排放的建筑行业已经无法适应当前国家、社会对建筑行业的要求。在大环境的影响下,人们对环保的要求越来越严格。在水下隧道设计、施工、运维中对绿色、健康、环保的要求越来越高。响应习近平总书记"绿水青山就是金山银山"的号召,需加大研究绿色设计、渣土绿色处理等相关技术,构建环境友好型社会,为我们的子孙后代留一片蓝天白云。

同志们,今天成立中国岩石力学与工程学会水下隧道工程技术分会,是我国推动水下隧道及地下工程快速、健康发展的关键时期,成立的一个服务行业、服务全国、服务广大技术工作者的一个高水平行业分会,希望各位专家、各位同行、各位领导多多支持学会工作,加强学术交流、加深技术合作、积极成功共享,共同将水下隧道工程技术分会这个平台建设好、使用好！最后,祝贺中国岩石力学与工程学会水下隧道工程技术分会成立大吉！预祝中国大盾构隧道智能建造高峰论坛取得圆满成功！祝各位领导、专家、与会代表在参会期间交流愉快！

谢谢！

在2019年盾构与掘进关键技术暨盾构再制造技术国际峰会的致辞

（2019年3月28日）

尊敬的各位领导、各位院士、专家、朋友们：

大家上午好！

我是中国工程院院士钱七虎，也是盾构及掘进技术国家重点实验室第二届学术委员会主任委员。今天非常高兴能够参加盾构及掘进技术国家重点实验室主办的2019年盾构与掘进关键技术暨盾构再制造技术国际峰会。我对本次峰会的成功举办表示祝贺！向莅临本次会议的各位领导、各位院士、专家和朋友们表示热烈的欢迎！

今年是新中国成立70周年，改革开放也刚刚过去40年，正是改革开放开启了科学的春天。改革开放使我国科技发展取得举世瞩目的伟大成就，科技整体能力持续提升，一些重要领域方向目前已跻身世界先进行列，为推动经济社会发展作出了重要贡献。面对复杂严峻的国际形势和艰巨繁重的改革发展稳定任务，在以习近平同志为核心的党中央坚强领导下，我国经济社会持续健康发展，创新驱动发展战略积极推进，科技创新再创佳绩。基础研究、前沿技术、高端装备、重大工程等领域取得一批重大成果，嫦娥四号探测器成功在月球背面着陆。人工智能、大数据、云计算等新技术与实体经济加快融合，新产业、新业态、新模式蓬勃发展，新动能持续快速成长，为稳增长保就业、推动高质量发展提供了有力支撑，成为当代中国创新发展的新标志。

纵观隧道施工技术的进步与发展，每一次技术进步都体现在机械化、智能化程度的不断提高。隧道施工主要经历了钢钎大锤法、手持风钻法、液压凿岩台车法和盾构TBM法等四个发展阶段。第一阶段主要依靠人力开挖，第二阶段使用手持风钻和小型矿车，使隧道修建从人力开挖发展到半机械化开挖，第三阶段使得隧道修建从半机械化开挖发展到机械化开挖，第四阶段盾构TBM法使得隧道修建从机械化开挖跨越到机械化智能化全断面施工。盾构TBM法具有快速、安全、环保、自动化等优点，与我国隧道及地下工程快速、安全、高质量发展的要求不谋而合。

在盾构TBM法施工技术应用前的漫长岁月中，隧道施工主要得益于中国四大发明之一的火药。那时，为了避免修建长隧道，必须尽可能地采用迂回展线来克服地形高差，使线路尽可能地依山傍水。宝成铁路上的34条隧道，最长的秦岭隧道也仅2363m，由于当时机械化程度低，中国修建这条长度在2km以上的隧道绝非易事，施工中首次使用了风动凿岩机和轨行式矿车。20世纪60年代中期修建的成昆铁路全长1085km，隧道占全线31%，其中关村坝隧道和沙木拉达隧道长度均超过6km，这批隧道采用了轻型机具，运用分部开挖法进行"小型机械化"施工，修建速度达到了每月"百米成洞"。中国修建长度10km以上隧道，是从代表中国隧道施工技术的专业化集团——中铁隧道工程局修建14.295km长的大瑶山隧道开始的，在这条双线铁路隧道施工中，采用了全液压凿岩台车，衬砌钢模板台车和装运机械等机械化成套装备。1999年8月，由中铁隧道局和中铁第十八局建设的中国当时最长的铁路隧道——秦岭隧道贯通，这条长18.457km的隧道使用了TBM法顺利建成，标志着中国隧道修建技术进入了世界先进行列。2018年，由中国中铁采用自主研制的世界首台马蹄形盾构修建的蒙华铁路白城隧道获世界隧道界最高奖（国际隧道协会技术创新项目奖），标志着异形盾构成功应用于山岭隧道。

| 思 想 与 观 点

我们在看到成绩的同时,仍然需要保持清醒,当前我国经济正处在转变发展方式、优化经济结构、转换增长动力的攻关期,保持经济平稳运行、促进高质量发展,必须更好发挥创新引领作用。要以习近平新时代中国特色社会主义思想为指导,把握世界新一轮科技革命和产业变革大势,紧扣重要战略机遇新内涵,深入实施创新驱动发展战略,加快创新型国家和世界科技强国建设,不断增强经济创新力和竞争力,让"中国盾构"在国际上不断展示中国人的自信!

最后,预祝本次峰会取得圆满成功,祝各位代表身体健康、工作顺利,谢谢大家!

在国家科学技术奖励大会上的发言

(2019年1月8日)

尊敬的习近平总书记、各位领导、同志们:

  我是中国人民解放军陆军工程大学的钱七虎。刚刚习近平总书记为我颁奖,这是党和国家的亲切关怀和深情鼓励。今天的成就,归功于以习近平同志为核心的党中央对科技事业的坚强领导,归功于军队和社会各界对科技创新的高度重视。这个奖属于防护工程乃至土木工程领域的所有科技工作者。在此,我向党中央、国务院表示衷心的感谢!向拼搏在科研一线的各位同仁致以崇高的敬意!我17岁被选拔保送到哈尔滨军事工程学院学习,从此穿上了军装,与防护工程结下了一生之缘。防护工程作为抵御核武器、常规武器打击的地下国防设施,是国家积极防御战略的重要基石,是确保国家核心力量安全的最后一道防线。新中国成立以来,一代代防护工程科技工作者自力更生、攻坚克难,解决了首脑指挥工程、战略武器洞库等一系列工程防护关键技术难题,构建了规模巨大、体系完备的"地下钢铁长城",为国家和国防安全作出了重要贡献。特别是党的十八大以来,我有幸亲历了我国防护工程科技水平从跟跑、并跑、再到部分领跑的飞跃发展,对此我感到无比的自豪和无上的光荣。

  作为党和军队培养的一名科技工作者,我深深懂得,没有强大的国防就没有人民的幸福安宁,没有科技的强盛就没有国家的强大,科学技术从来没有像今天这样深刻影响着国家的前途命运。科技工作者只有把个人理想与国家民族的命运紧密联系在一起,并为此奋斗一生,才能有所成就、彰显价值。这也是新时代科技工作者的使命所在、幸福所在。

  习近平总书记在庆祝改革开放40周年大会上指出,伟大梦想不是等得来、喊得来的,而是拼出来、干出来的。我们有幸赶上中华民族伟大复兴的新时代,党和国家对加快建设世界科技强国作出一系列战略部署,对广大科技工作者而言,是机遇,更是挑战。我们一定要紧密团结在以习近平同志为核心的党中央周围,以习近平新时代中国特色社会主义思想为指导,勇敢担负起新时代赋予的新使命,继承和发扬老一辈科学家无私奉献、科学报国的优良传统,想国家之所想,急国家之所急,瞄准前沿,迎难而上,把关系国家经济安全、国防安全和其他安全的关键核心技术,牢牢掌握在我们自己手中,不断增强自主创新能力,以实干兴邦的信心和决心勇攀科技高峰,为决胜全面建成小康社会、实现中华民族伟大复兴的中国梦而不懈奋斗!

  谢谢大家!

在第三届全国工程安全与防护学术会议上的发言

（2018年10月13日）

女士们、先生们：

工程安全与防护分会是中国岩石力学与工程学会一个比较年轻而活跃的分会，昨天，经过选举产生了第二届理事会。首先，我代表中国岩石力学与工程学会向新当选的各位理事、常务理事表示热烈的祝贺！感谢大家对学会工作的支持！同时，向做出辛勤工作的第一届理事会表示衷心的感谢！感谢他们为学会的发展和壮大付出了辛劳！

工程安全与防护分会成立四年来，坚持正确的办会方向，积极工作，做出了不少成绩，通过举办学术活动、开展工程咨询等活动不断宣传和扩大分会的影响。特别是在创新思路，积极发展会员和会员单位，积极筹措经费方面做出了较大的成绩。这与第一届理事会的努力工作是分不开的。应该说，工程安全与防护分会实现了良好的开局，为今后的发展打下了良好的基础。

但是，后面要面临的工作和任务也是相当艰巨的。我们大家有这样的责任和义务为把我国由岩石力学大国变成岩石力学强国这样一个目标而努力奋斗。这是目标，是挑战，也是学会发展的重要战略机遇！在大规模工程建设的背景下，工程安全与防护问题日益突出，我们工程安全与防护分会对促进该领域的发展有义不容辞的责任。为此，我提几点希望：

一是要断往开来，与时俱进，以改革为中心，推进学会工作新进展，迈上学会工作新台阶。学会工作要全面贯彻"三个代表"重要思想，全面落实科学发展观，在十八大精神指引下，坚持引导工程安全与防护领域的科技工作者实现科技进步和创新，围绕促进学科发展和为广大会员服务两个基本宗旨，全面推进学会的各项工作。

二是要与企业结合，与工程结合。工程安全问题是一个实践性问题，学会不仅要广大学术界学者的参加，更需要战斗在工程一线的企业管理者、技术工作者的积极参与。学会要发挥桥梁和纽带的作用，为政府决策当好"参谋"，为工程建设排忧解难，出谋划策，为科研学术的创新提供服务，使学会在国内工程安全与防护领域真正具有学术权威的影响和地位。

三是开展国际合作和交流，逐步树立学会在国际工程安全与防护领域的影响。要多渠道组织国内外、港澳地区和海峡两岸的学术交流和合作，积极参加国外相应学术活动，拓展国际活动空间。

四是要将着力培养青年人才作为重要任务。青年是学科发展的希望，学会要为青年人才的培养和脱颖而出发挥独特的作用，要在课题立项、论文发表、成果奖励方面为青年人才做好服务工作。

本次工程安全与防护学术大会确定了"地下工程安全与防护面临的新挑战"会议主题，这是近年来在工程安全与防护领域工程实践和科学研究成果的一次盛大交流，衷心祝愿全体参会人员能在此次会议中增进交流、开拓视野、收获友谊。

最后，祝大会取得圆满成功！祝各位专家、学者、朋友身体健康！

谢谢大家！

在中国岩石力学与工程学会工程安全与防护分会成立大会上的发言

(2008年5月27日)

前面几位领导关于工程安全与防护的意义讲了很多,讲得很好,我补充三点意见。

第一点,对我们工程安全防护分会的成立,对产生的分会的领导、理事长、副理事长、常务理事、理事表示热烈的祝贺。前面学会的副秘书长方祖烈教授介绍了我们学会的有关情况,我在此补充一点,就是我们学会是在前面几届理事会的领导下打好的基础上才取得的今天的成绩,我们学会的第一任理事长是已故的国际知名的岩土工程学家,也是我们全国人大常委会的委员陈中基院士,第二位理事长是国际原大坝委员会主席、中国工程院的副院长潘家铮院士,第三位理事长是我国著名的岩土工程学家、同济大学的孙钧院士,第四位是中国科学院地震研究所的国际知名的工程地质学家王思敬院士。有他们打下的良好基础,我们才取得了今天的成绩。

第二点,我简单地介绍一下我们学会为什么申请工程安全防护分会。大家都对5.12汶川大地震的惨烈程度,造成的人民生命财产的损失以及经济损失留下了深刻的印象。地震是不可抗拒的,不可避免的,但是地震造成的人民生命财产的损失是可以减少的。日本的地震专家说:地震中人的死亡不是地震震死的,绝大部分是由于建筑倒塌而砸死的。如果我们的烈度标准、设计标准是正确的,抗震隔震设计是科学合理的,建筑倒塌应该不是完全不可避免的。截至5月25日,此次汶川大地震已经死亡6万人了,如果将来工程安全防护搞得好,抗震隔震设计搞得好,烈度标准提高了,像日本那样,我们就不会死这么多人。另外,近年来,我们的工程安全防护事故不断,有煤矿的,有地铁的(北京、广州和南京),铁路隧道已经出现三次事故死人。这些都充分说明了工程安全防护的重要性和必要性,所以我们学会要申请成立工程安全防护分会。

第三点,我们有很多学会,比如,中国土木工程学会防护工程分会,防护工程学会也搞防护,我们这个学会也搞防护,还有地震学会,还有防灾减灾学会,学会这么多,那我们这个学会有什么特点呢?今后的交流空间和发展空间在哪里?我谈一点想法供分会的领导参考。我们这个分会是岩石力学与工程学会下的分会,应该搞岩土工程的安全与防护,也就是搞地下工程的安全防护,而不是搞地面建筑的安全防护。21世纪是地下空间开发利用的世纪,这方面的任务特别大,而且这方面的事故也特别多,除了这次地震,大部分的事故都是地下工程的安全事故。而且现在工程的抗震设计规范地面是有的,而地下的没有,需要研究,需要交流。我们说到地下空间,一般的观点是其抗震设计标准降低一级,这个是否科学,需要我们认真的研究和交流。我们的学会如果要搞工程安全,就要搞地下工程的抗震安全。去年济南暴雨,由于洪水死了100多人,因此抗洪是地下工程要考虑的问题,同时地下隧道的火灾的预防和救援也是一个很难解决的问题。因此我们这个分会有很大的工作和发展空间。我们学会与防护工程学会的区别在于,防护工程学会主要是军队单位、人防单位,是战事的防护,是军事工程的防护与研究,比如一些高抗的掩蔽部,一些战略武器的掩蔽工程,而我们研究的是民用工程的防护,包括反恐。这样两个学会的工作才不会重复。我认为,工程安全防护分会的成立具有重要的意义和紧迫性,它为我们岩土力学工程界从事工程安全防护的广大科技工作者、工程技术人员创造了许多互相熟悉、互相交流的平台。在我们国家未来的建设中一定会发挥有力的作用。

最后,祝我们工程安全防护分会在理事会的领导下,在全体理事、常务理事和全体会议支持者的共同努力下越办越好,谢谢大家!

| 思想与观点

在第七届中俄深部岩石动力学高层论坛开幕式上的发言

(2017年8月2日)

尊敬的任廷琦校长、阿巴林院士、宋振骐院士、蔡美峰院士、何满潮院士、中俄双方各位专家学者、女士们、先生们：

在这海风拂面的凉爽时节,我们共同迎来了第七届中俄深部岩石动力学高层论坛。我谨代表中国岩石力学与工程学会,向参加本次论坛的各位专家学者表示热烈的欢迎！向精心筹备此次论坛的山东科技大学表示衷心的感谢！

长期以来,中国岩石力学界始终致力于解决关键科学问题和技术难题,从而处于岩石力学与工程领域的先进行列,并与国际岩石力学界开展了广泛的学术交流,其中与俄罗斯科学院及相关科研院所有着长期交流合作的历史。自2011年在辽宁工程技术大学举办第一届中俄深部岩石动力学高层论坛以来,已在中俄双方成功举办了六届。在这六届论坛中,中俄双方的专家学者围绕深部岩石力学和岩石动力学问题进行了充分的学术交流,结下了深厚的友谊。本届为第七届论坛,本届论坛主要围绕深部岩石动力学理论、深部地下工程灾变监测预警与风险评价、深部工程监测、检测新设备与新技术等主题进行研讨。

参加本次会议的中俄双方代表共计××人,大部分是岩石力学界的高层学者、著名专家,为岩石力学的研究与发展做出了重要贡献。论坛期间共有××人将在大会上作报告,分享他们的研究成果。衷心地希望大家能够就论坛主题深入研究和广泛讨论,为提高岩石力学理论技术水平做出贡献。我相信在各位的努力下,本次会议一定会取得丰硕成果。

预祝大家在青岛这个美丽的城市度过几天美好的时光！

谢谢大家！

在中国城市建设科学发展论坛上的发言

(2017年7月24日)

尊敬的各位领导、各位来宾：

上午好！

去年10月，我受邀出席了中国城市建设科学发展论坛；时隔一年，我非常高兴地看到，在相关管理部门的关怀和指导下，同志们继往开来深耕管廊行业结出了硕果，代表我国综合管廊建设领域先进水平的"中国城市地下综合管廊产业联盟"宣告成立。这对加强综合管廊产业链各环节机构的密切合作，推进城市市政基础设施建设具有积极作用。在此我对联盟的成立表示热烈祝贺，并希望联盟在今后的工作中充分发挥上下联动、组织协调的功能，促进我国综合管廊产业和谐发展，实现"创新、共赢、合作、发展"目标。

一是加强地下管线建设改造，有序开展综合管廊建设，是保障城市运行"生命线"的重要举措，得到党和国家领导人的高度重视。

习近平总书记曾指出：城市面貌是城市的"面子"，这些年有了很大的变化，但城市基础设施，特别是地下管网设施这些城市的"里子"，目前仍然薄弱。李克强总理指出："面子"是城市的风貌，而"里子"则是城市的良心，只有筑牢"里子"，才能撑起"面子"，这是城市建设的百年大计；目前中国正处于城镇化快速发展时期，但我们的地下管廊建设严重滞后。加快这方面的建设，很有必要。

国务院、住房城乡建设部等国家机关先后多次专门下发文件，指导城市地下综合管廊建设。2017年5月颁布的《全国城市市政基础设施建设"十三五"规划》中提出，"十三五"期间，建设干线、支线地下综合管廊8000km以上。

二是党中央国务院高度重视城市地下综合管廊建设，因为其对推进新型城镇化、补齐城市基础设施短板，实现城市发展方式转变有重大意义。

建设综合管廊是保障城市运营安全的重要环节，是我国城市发展方式由粗放发展向集约绿色可持续发展模式转变的关键契机，是拉动我国经济增长的一项重要措施，是提升我国城市领导执政能力和执政水平的重要抓手。通过地下综合管廊建设，可实现人车分离，天蓝、地绿、水清，美丽城市、美丽农村的中国梦一定能实现。

三是推进地下综合管廊建设、做好相关工作，应加强顶层设计、完善相关政策，落实相关具体措施。

首先，应加强顶层设计，积极有序推进。明确主管部门、建立协调机制；保证立法、规划跟进；制定路线图、时间表，积极有序推进工作部署。

第二，应完善相关配套政策。明确入廊要求，实行有偿使用，加大政府投入，完善融资支持。

入廊管线单位应统筹考虑建设和运营、成本和收益的关系，向地下综合管廊建设运营单位交纳入廊费和日常维护费，实行管廊有偿使用。中央财政要发挥"四两拨千斤"的作用，积极引导地下综合管廊建设，通过现有渠道统筹安排资金予以支持。地方各级人民政府要进一步加大地下综合管廊建设资金投入。将地下综合管廊建设作为国家重点支持的民生工程,充分发挥开发性金融作用,鼓励相关金融机构积极加大对地下综合管廊建设的信贷支持力度。鼓励银行业金融机构提供中长期信贷支持,积极开展特许经营权、收费权和购买服务协议预期收益等担保创新类贷款业务。将地下综合管廊建设列入专项金融债支持范围予以长期投资。支持符合条件的地下综合管廊建设运营企业发行企业债券和项目收益票据,专项用于地下综合管廊建设项目。

第三,应针对管廊建设的实际,创新和落实相关具体措施。

(1)借鉴国外先进经验,制定完善的法规体系。国外先进经验是先立法、后建设,这样才能有序有力推进建设。我国应借鉴相关经验,对综合管廊的所有权、规划权、建设权、管理权、经营权、使用权以及入廊有偿使用费的收取原则等作出明确、完善的具体规定。

(2)系统科学地编制好城市综合管廊规划。又好又快、好字当头编制好城市综合管廊规划是城市地下综合管廊建设的关键。城市地下空间利用应与城市地面、地上空间利用达到互补和互相协调。原则上城市地下综合管廊规划应与城市地下空间规划以及城市建设总体规划同步协调进行。条件不允许必须先行或单独规划的综合管廊规划,必须通过充分协调,达到统筹规划的编制。

推进燃气管线、雨洪排泄管道、电力电缆等管线入廊的规划论证,大力提倡综合管廊与地铁建设、地下街建设和地下快速路建设相整合,从而明显降低建设成本,减小社会干扰,缩短施工工期,避免重复建设、重复投资。

(3)建立统一管理、分工协作、衔接有序的综合管廊管理体制。为了克服条块分割、分头管理的弊端,达到统一协调,当前在尊重现有各部门职能分工的基础上,应尽快明确每个城市综合管廊的综合管理机构。

(4)地下综合管廊兼顾人防应到位不越位、不错位;兼顾抗震,保证管线在震灾下的安全运行。

(5)创新投融资机制,推广PPP模式。创新投融资机制的关键是吸引社会资本参与综合管廊的建设与运营。除了用足投资补贴、特许经营权、收费权和购置服务预期收益等担保类贷款贴息和发行专项金融债等财政金融政策外,考虑到社会资本的逐利性是其基本属性之一,要吸引社会资本参与,还必须妥善解决以下问题。

综合管廊集中统一运营,降低运营成本。在综合管廊集中统一运营条件下可打造智慧管廊,充分运用物联网监控、BIM信息管理等技术,实现集中远程控制,建立市场机制下的维护管理,从而大幅降低安全风险和运营维护成本。

严格"风险分担",即政府理应承担的风险不应由社会资本"共同承担",PPP模式中投资风险不应全由项目公司承担,政府更不应让社会资本捆绑打包更大的建设项目。

采取政府提供购买服务协议的方式解决经营收益不确定性问题。待相关标准出台后,可采取"使用者付费+政府可行性缺口补贴"模式,即在管线单位入廊收费无法平衡项目合理投资回报及运营成本时,由政府提供可行性缺口补贴。

杜绝地方政府捆绑打包"小马拉大车",增大未来投资风险。部分地方政府利用国家对地方综合管廊建设的补贴资金,吸引社会资本投资建设过程中,要求社会资本捆绑打包其他更大的建设项目,用中央补贴资金拉动地方其他大项目建设的行为,造成投资风险的极大增加。

以上是我对推进城市地下综合管廊建设,转变城市发展方式的一些思考。我相信,在党和国家各级主管部门的部署、指导下,在广大管廊建设企业和工作者的共同努力下,在相关产业组织的推动与促进下,一定能实现《全国城市市政基础设施建设"十三五"规划》中提出的管廊建设8000km的中期目标,一定能实现入廊完备、收费合理、运行顺畅、保障充分、监管有力的综合管廊长效运行管理机制。谢谢大家!

在第十四次全国岩石力学与工程学术大会开幕式上的发言

(2016年12月13日)

第十四次全国岩石力学与工程学术大会今天在广州隆重开幕了。全国各地的岩石力学与工程领域的专家、学者齐聚广州,交流近年来在本领域取得的新理论、新方法、新技术,研讨岩石力学与工程发展战略,是岩石力学与岩石工程界的喜事和盛事。

学术交流是学会的"主业",是学会的生存之本,开展高质量的学术交流是展现学会能力和提高学会凝聚力、吸引力的重要标志,也是满足科技工作者需求的最重要方面,是学会参与科技创新、建设创新型国家的主要切入点。

近年来,学会学术交流活动数量稳步增长,质量和成效提高,"精品化"学术会议建设取得了明显成效。

两年一次的全国性大会,是我们学会最主要的标志性学术品牌活动。希望本次会议能上新水平、再上一个新的台阶。

2016年是"十三五"规划的开局之年,创新驱动已成为加快转变经济发展方式、增强经济发展内生动力和活力的国家发展战略。随着岩土工程的科技创新和新技术的涌现,以及重大工程项目建设难题的提出,为岩石力学与工程领域带来新机遇和挑战。特别是"一带一路"倡议实施,如何置身于创新驱动发展大战略中,做强岩石力学与工程学科,是历史赋予我们岩石力学与岩石工程领域广大科技工作者新的任务和挑战。

我相信,在与会代表共同努力下,我们这次会议一定会有丰硕的成果,取得圆满成功!

在苏通 GIL 综合管廊工程专家聘任座谈会上的发言

(2016 年 11 月 29 日)

各位专家、同志们：

上午好！

今天，国网江苏省电力公司在这里召开苏通 GIL 综合管廊工程专家座谈会，颁发专家聘书，包括隧道工程、GIL 安装、GIL 设备、GIL 运输等专业领域的 14 位专家受聘。很荣幸我受聘担任专家组组长，望在今后得到各位专家的支持。

苏通 GIL 综合管廊工程是目前世界上电压等级最高、输送容量最大、技术水平最高的超长距离 GIL 创新工程，该工程对于有效提高华东、江苏电网安全稳定水平，缓解大气污染、促进江苏乃至华东地区经济社会发展都具有重要意义。

工程将建设的越江隧道长达 5.5km，隧道结构最低点高程 −74.83m，水土压力超过 0.9MPa，是目前国内埋深最大、水压最高的隧道。隧道一次性长距离掘进超过 5400m，如何确保主驱动和盾尾密封的安全有效，是本工程成败的关键所在。管廊隧道采用泥水平衡盾构机施工，盾构隧道穿越标准贯入击数大于 50 的密实砂层长度约 3300m，砂层石英含量超过 70%，该地层对刀盘刀具的磨损巨大，江底换刀作业不可避免。如何选择合适的刀盘刀具设计，尽量减少换刀频次，并实现安全高效的换刀作业，是越江施工能否成功的重点，也是工期控制的难点。

苏通 GIL 综合管廊工程是世界上首个特高压 GIL 管廊工程，在电压等级和应用长度方面为世界之最，需要在座的专家为工程建设提供技术支撑。

各位专家造诣深厚，专业知识渊博，实践经验丰富，我衷心祝贺专家组的成立，衷心祝愿专家组能够把专家们紧紧团结在一起。坚持公正性和科学性，得出经得起国内外同行及科学家检验的结论。衷心希望各位专家多为苏通 GIL 综合管廊工程的建设献计献策，协助国网江苏省电力公司做好工程重大技术难题、重大施工方案、科技项目、关键性工程技术措施、优化设计等的咨询和论证工作，为这一世界级创新工程建设、推动特高压输电技术再上新台阶做出自己的贡献！

谢谢大家！

在东华理工大学建校60周年庆典大会上的致辞

(2016年10月16日)

尊敬的各位领导、各位来宾,亲爱的老师们、同学们、校友们:

大家好!

作为亲身经历东华理工大学蓬勃发展的一员,能够荣幸地参加建校60周年庆祝大会,从内心感到非常高兴。学校是一所老军工院校,自创办之日起,坚持国家利益至上、民族利益至上的宗旨,以培养服务国防科技工业人才为己任。历代东华理工人默默奉献、一心为国的精神深深地打动了我。自2002年我受江西省人民政府聘任为学校名誉校长以来,一直以主人翁身份关注和参与学校的发展,为学校的每一个进步和成绩感到由衷的高兴。今天,学校隆重举行60周年华诞庆典,这是东华理工大学发展历程中的又一重要里程碑,作为东华理工人,心里更是感到荣幸和自豪。

站在六十年新起点上,回首过去,东华理工人为了伟大的国防军工事业,怀揣着崇高的理想,满载着豪情壮志,从四面八方汇聚在培养核地学人才这一高扬旗帜下,担负着神圣使命,开始了不平凡的征程。虽也曾几起几落,但在全体教职员工的坚持下,在众多校友的热情支持下,即使在最困难的时期,也未曾放弃为国家、为核事业培养人才作贡献。尤其让我感到高兴的是,近些年来,学校的发展令人瞩目,成果丰硕,成功实现一系列新突破。一是学位点建设取得重大突破。获批服务国家特殊需求博士人才培养项目,学校拥有从本科到博士研究生完整的人才培养体系。二是实现了科研奖励的突破。陈焕文教授两次获得江西省自然科学一等奖,汤彬教授获得江西省科技进步一等奖。令大家非常期待的,国家科技进步奖已经公示了,汤彬教授获得二等奖,这是令全校师生员工欢欣鼓舞的一件喜事。三是实现了科研项目的突破。2012年陈焕文教授获批"国家杰出青年科学基金",填补了江西省自2004年以来的空白。获批国家自然科学基金重点项目,充分提升了科研实力。四是科研平台的突破。2015年,"质谱科学与仪器国际联合研究中心"成为江西省首家也是目前唯一一家国家级国际联合研究中心。五是实现了教育部科技创新团队计划的突破,提升了人才队伍层次。六是人才培养取得突破。让我们引以为豪的是,我们培养的学生有入选中科院院士的,有国家大型企事业单位的领导,有公司老总等,走出校门的校友,已经成长为中国国防军工事业和国家、地方经济建设的骨干之才,正带领更多的年轻校友,走向中华民族的复兴。

当前,国家战略发展的核心是建立创新型国家。创新的根基在于人才,人才的关键在于教育。习近平总书记指出:"教育是提高人民综合素质、促进人的全面发展的重要途径,是民族振兴、社会进步的重要基石,是对中华民族伟大复兴具有决定性意义的事业。"为此,我希望学校:

一、坚持特色,努力培养德才兼备的应用型创新人才。60年来学校为国家培养了十余万专门人才,积累了丰富的育人经验,形成了具有特色的人才培养模式。今后要继续发挥学校传统学科的优势,创新培养模式,提高人才培养质量。

二、深化合作,共谋双赢发展。要落实省部共建、军民融合、校地合作、校企合作等合作协议,建立起开放、集成、高效的协同创新机制,推动资源共享,联合攻关,共谋双赢发展。

三、加强科技创新,凝聚学校发展强大动力。科技创新是学校发展的内在动力、活力之源。六十年来,学校已经具备很好的科教资源和创新基础,要把科技创新摆在学校发展全局的核心位置,紧密结合国家战略需求,瞄准科技发展前沿,进一步提高自主创新能力,为建设创新型国家、实现创新驱动发展作出重要贡献。

我相信,在新的历史时期,东华理工大学一定能够传承历史使命,勇于担当,弘扬军工精神和民族优良传统文化,为发展中国国防军工事业和地方经济发展作出新的贡献。

谢谢大家!

在第四届 GeoChina 国际会议上的致辞

(2016 年 7 月 25 日)

尊敬的各位来宾、各位代表、各位朋友:

今天,我们在美丽的济南隆重举办第四届 GeoChina 国际会议,共同研讨交通基础设施建设相关学术问题。首先,请允许我代表中国岩石力学与工程学会、第四届 GeoChina 国际会议组委会,向与会的各位专家学者表示热烈的欢迎!向给予这次国际学术会议鼎力支持和辛勤劳动的各有关单位致以诚挚的谢意!

随着区域经济一体化的发展,跨区域互联互通基础设施的需求日益增长,全球交通基础设施建设正迎来一轮发展新机遇。交通基础设施建设,不仅有利于改善民生,提高民众的生活质量和便利程度,而且是驱动世界经济复苏和可持续增长的重要引擎。放眼全球,各国交通基建需求强劲。发展中国家在加速工业化和城市化发展中,需要加大基础设施投资建设。发达国家出于更新升级老化基础设施和刺激经济复苏的双重目的,也在陆续推出规模庞大的基础设施建设计划。

本次 GeoChina 国际会议,是交通基础建设领域的一次学术盛会。会议以极端气候下交通基础设施中创新科技的应用为主题,就国内外岩土工程、路面工程、桥梁工程、隧洞工程、铁道工程、大坝工程中的科学问题开展广泛的学术交流和研讨,必将进一步促进交通领域的国际交流与合作促进交通及相关学科快速发展。

中国岩石力学与工程学会是中国岩石力学与工程科技工作者的学术性群众团体。学会始终把提升学术水平和影响力放在重要位置,在国际上形成了系列多方位的交流格局。借此机会,向长期以来对中国岩石力学与工程学会给予热情帮助和大力支持的各位专家学者、各位朋友表示衷心的感谢!

最后,衷心地祝愿第四届 GeoChina 国际学术会议取得丰硕成果!祝愿各位朋友、嘉宾身体健康,工作顺利,万事如意!

在全国隧道及地下工程不良地质超前预报与突水突泥灾害防治学术会议上的发言

(2015年7月29日)

各位领导、来宾、女士们、先生们：

大家上午好！

今天我们在湖北恩施召开全国隧道及地下工程不良地质超前预报与突水突泥灾害防治学术会议，共同研讨"隧道突水突泥灾害预测预报预警与防治控制技术"问题。值此论坛开幕之际，首先，请允许我代表论坛组委会及中国岩石力学与工程学会，向参加论坛的所有来宾致以热烈的欢迎和衷心的感谢！

众所周知，近年来我国铁路、公路、水利水电、市政等领域隧道与地下工程的建设蓬勃发展。在交通领域，截止到2013年，我国[1]运营铁路隧道11074座，总长8938.78km，在建铁路隧道4206座，长度7795km，正规划铁路隧道4600余座，总长约10600km；2013年底，我国运营公路隧道11359座，总长9605km，到2020年公路隧道总里程将突破10000km。未来20年中，我国将新建近100万km各级公路，其中高速公路约6万km；新建3万km铁路，其中高速铁路超过1万km。在水利水电工程领域，我国有20多个世界级的大型水利水电工程和数十条跨流域调水工程正在建设。与之相配套，将建设数百条引水隧洞工程，如规划中的南水北调西线第一期工程主要由深埋长隧洞组成，总长264km，最长洞段为73km，最大埋深1150m；规划中的滇中引水工程输水线路总长超过600km，由超过60座隧洞构成，其关键性控制工程香炉山隧洞全长约60km，最大埋深超过1000m；已经开工的引汉济渭工程秦岭引水隧洞全长98.30km，最大埋深超过2000m。

今后，我国隧道建设埋深将越来越大，深部地质环境赋存的高地应力和高水压条件，导致施工过程中面临的突水突泥、塌方等地质灾害问题越来越严峻，隧道建设安全受到严重威胁。在我国交通隧道修建史上，渝怀铁路圆梁山隧道、宜万铁路野三关隧道等在建设过程中均发生过重大突水突泥灾害，造成了重大损失。水电领域的锦屏二级水电站4条引水隧洞，单洞长约17km，最大埋深达2525m，具有"高水压、大流量、高地应力"的显著特点。施工中遭遇多次遇到突水等灾害，水压超过10MPa，最大流量超过$7m^3/s$，最大地应力接近70MPa，严重影响了施工进度。

特别是随着"一带一路"倡议的深入实施，中国铁路、公路、水利建设将越来越多的走向世界。如何杜绝隧道施工灾害、保证建设安全关系着我国隧道建设在世界上的核心竞争力，是我们大家必须面对和解决的重大难题。本次学术会议的召开为隧道不良地质超前预报、突水突泥等灾害的预警和防控提供了一个很好的交流平台，将进一步促进我们攻克隧道灾害防治的重大难题。

机遇与挑战并存。近年来，我国科技工作者在突水突泥防治技术领域都取得了诸多成就。会议主办方山东大学李术才团队把湖北作为一个主战场，他们自2002年以来深入湖北工程建设一线，历经湖北沪蓉西高速公路、三峡翻坝高速公路、宜巴高速公路、保宜高速公路、鄂西高速公路等，理论研究与工程实践密切结合，承担了几十条高风险隧道的地质预报、灾害治理、安全监控等方面的工作，为保障建设安全做

---

[1] 本文统计数据不包括中国台湾地区数据，余同。

出了重要贡献。该团队通过十余年的科学研究和工程实践,在地下工程突水突泥灾害源定位定量预报及灾害控制方面取得了重要进展,获得了相关的国家科技进步二等奖2项。

在为我国隧道防灾减灾技术的进步感到欣慰的同时,我们也应该深刻认识到,在地下工程灾害防治技术领域,仍然有很长的路要走,面临的挑战也越来越多,越来越复杂。衷心希望各位专家学者能利用此次论坛充分分享和交流新知识、新技术和新理念,共同促进隧道与地下工程建设技术的发展,进一步推动该领域科技工作者的友谊和合作。

预祝此次学术大会取得圆满成功!

谢谢大家!

在"精细爆破学术研讨会"上的致辞

(2015年1月18日)

尊敬的各位院士、爆破界各位专家、各位领导：

大家早上好！

由中国工程院土木、水利与建筑工程学部和中国工程爆破协会主办，武汉市政建设集团、武汉爆破有限公司承办的"精细爆破学术研讨会"现在开始。按照中央提倡节俭办会的精神，我们今天的会议程序从简，莅临本会的领导不再一一致辞。由我代表中国工程院学部工作局阮宝君副局长、土水建学部领导、讨论会协同主席谢礼立院士、中国工程爆破协会理事长汪旭光院士和承办单位向莅临本次会议的院士、专家、领导表示衷心的感谢和热烈的欢迎。

工程爆破作为我国国民经济建设的重要支撑技术之一，已广泛应用于铁路、矿山、水利水电、城市建设和国防等领域，为我国的国民经济建设做出了重要贡献。进入21世纪以来，随着爆破基础理论研究的不断深入和相关学科的深度交叉融合，以及钻孔机具、爆破器材的快速进步，工程爆破技术得到了飞速的发展，在国民经济建设各领域应用的广度和深度逐渐增加。例如，为满足我国工业基础设施升级改造、城市改扩建和节能减排等工作快速推进的需要，建（构）筑物拆除爆破技术得到了广泛的应用，已逐步发展成为一门与工程建设相平行的工程技术。与此同时，工程爆破已发展至"精细爆破"新阶段，近年来，涌现了一大批堪称精细爆破典范的精品工程，取得了显著的经济、社会和环保效益。

今天，我们的研讨会将围绕精细爆破关键技术和发展方向展开研讨，将有2位中国工程院院士、10位工程爆破界专家做关于精细爆破的学术报告，内容涵盖了基础理论、爆破器材、设计施工关键技术以及爆破与相关学科的交叉融合等方面。报告结束后，将进行开放式讨论与交流，希望大家踊跃发言、积极交流，共同促进精细爆破的发展和工程爆破行业的进步。

作为工程院的系列学术会议之一，院部领导对本次会议给予了大力支持，体现了中国工程院对工程爆破行业的重视和关怀，中国工程爆破协会理事长汪旭光院士为本次会议的顺利召开了倾注了大量心血，会议承办方武汉市政建设集团为本次会议的召开做了大量的工作，在此一并表示衷心的感谢。

在汕头市苏埃通道工程专家委员会成立暨技术咨询会上的发言

(2014年11月26日)

尊敬的女士们、先生们：

非常高兴受邀来到美丽的花城——广州，参加汕头市苏埃通道工程专家委员会成立暨技术咨询会，也特别荣幸能够被聘为专家委员会的一员。在此我谨代表所有受聘专家向中信房地产股份有限公司，中信滨海新城工作领导小组，汕头市苏埃通道建设投资发展有限公司及各位领导，会议筹备、组织人员，参会人员，工作人员表示诚挚的感谢！

苏埃通道工程是连接汕头市南北两岸的重要交通工程，在促进濠江区、特别是滨海新城的建设和发展方面担当着至关重要的角色；对形成高效的中心城区对外交通的城市快速路网主骨架，完善汕头城市交通路网布局，提高城市运作效率，改善城市面貌都将会起到重要作用。

苏埃通道工程是目前世界最大直径盾构工程之一，为国内首座在8度抗震区采用超大直径盾构施工的海底隧道，对隧道结构的抗震性能提出了很高的要求。

苏埃通道工程地质条件复杂，既有流塑状淤泥，更有强度超过200MPa花岗岩基岩及花岗岩球状风化体，对施工设备和施工技术提出了前所未有的要求，是目前国内外最具挑战性的超大直径盾构工程。

世界的工程在亚洲，亚洲的工程在中国，中国的工程在交通。近20年来，中国的城市交通，城际交通，高速铁路等得到迅猛发展，盾构施工在其中扮演了重要的角色，也因此积累了在各类地质条件下施工的丰富经验，培养了大批的具有国际一流水平的施工队伍和优秀施工技术及管理人才。世界的盾构施工设备、设计、施工水平都因为中国而获得飞跃式进步。

我们坚信在汕头市人民政府和中信地产的领导下，在苏埃通道公司的严格管理下，中国的工程技术人员一定能够攻坚克难，高效、优质、安全地建成苏埃通道工程。我们全体专家委员会成员也将不遗余力地为苏埃通道工程献计献策。

最后，祝各位身体健康，工作愉快，家庭幸福，谢谢大家！

在"中国矿业科学协同创新联盟"成立大会上的发言

(2015年1月27日)

尊敬的宋军部长、各位领导和代表：

大家好！

今天，我们隆重召开"中国矿业科学协同创新联盟"成立大会，首先我代表中国岩石力学与工程学会，对出席和参加大会的各位领导和代表表示热烈的欢迎，对支持和参与"中国矿业科学协同创新联盟"事业的各理事单位和社会各界表示崇高的敬意和衷心的感谢！

党的十八大明确提出实施创新驱动发展战略，强调科技创新是提高社会生产力和综合国力的战略支撑，必须摆在国家发展全局的核心位置。

为了充分发挥科协及其所属学会在创新驱动发展中的重要作用，进一步推进学会为国家经济社会发展提供科技和人才支撑，中国科协于2014年10月10日，组织实施创新驱动助力工程，印发了《中国科协关于实施创新驱动助力工程的意见》。

中国岩石力学与工程学会认真学习《中国科协关于实施创新驱动助力工程的意见》，于去年11月27日，在北京召开了"中国矿业科学协同创新联盟"建设研讨会，会议研讨并确定了联盟章程、工作规划和组织机构管理办法，提出了联盟建设的目标，为今天联盟的成立奠定了基础。

中国矿业科学协同创新联盟是中国科学技术协会指导下的矿业领域创新驱动助力工程的重要组织，中国岩石力学与工程学会是联盟的组织与协调部门，"创新联盟"的成立对于提高我国矿业领域的整体技术水平，推动产业转型，带动经济发展，促进我国从"矿业大国"向"矿业强国"迈进将具有深远的战略意义和重要的现实意义。

创新联盟由创新载体（高校国家重点实验室、高新技术制造企业）和驱动载体（行业重点企业和行业大型企业）组成，在中国岩石力学与工程学会组织与协调的平台上，创新载体不断提供高新技术与装备，驱动载体利用此技术与装备创造出更高的经济和社会效益，两个载体相互交流，协同创新，良性互动，实现共赢。

创新联盟是我国矿业领域第一个协同创新战略联盟，是一个新生事物，希望大家多提出宝贵的意见，多支持联盟的发展，为创新联盟建设多作有益的贡献。

最后，祝各位领导和代表身体健康，工作顺利！

祝愿创新联盟办出特色，办出成效，越办越好！

谢谢大家！

中国科协第 80 期新观点新学说学术沙龙开幕词

(2013 年 8 月 22 日)

由中国科协主办,中国岩石力学与工程学会承办的"地球演化与全球变暖"学术沙龙现在开始。参加本次学术沙龙的除了岩石力学、岩土工程和矿业领域的专家外,主要的是地质领域的专家,包括构造地质、大地构造、地质成因、大地测量、地球化学、地球物理、工程地质、古生物、地震等领域的专家,大家都对"地球演化与全球变暖"的主题感兴趣。

2012 年初,我学会原副理事长唐春安教授提出了"地球演化与全球变暖"的全新观点。他认为:地球演化的历史就是一部地球热能在冷—热周期不断转换中渐进变衰的历史。地壳与地幔物质在热平衡条件下的不断相互转化,诱发了一系列以升温与冷却、膨胀与收缩、熔融与凝固为特征的地表变迁史,并在国内外首次提出了"地球演化与全球变暖"的新模型。2012 年 11 月 3 日,学会邀请了相关领域专家学者 30 余人参加的"地球演化与全球变暖新观点"研讨会,一方面听取了唐教授关于"地球演化与全球变暖"新模型的汇报,另一方面也是为本次学术沙龙进行了热身和准备。

科学的本质是批判,交流的本质是质疑。根据规定,除了主题发言以外,其他每人发言时间不超过 10 分钟,希望大家遵守,发言简明扼要,讲清观点,以便让更多的同志有机会发言。另外,因为是新观点的沙龙,是具有探索性的,所以允许不同观点和意见的交锋,但要讲清道理,我们鼓励有不同的观点,这样才能够把讨论引向深入,能够造成观点的碰撞,产生新的观点。"地球演化与全球变暖"沙龙活动,今天只是一个开端,希望以后在这个领域能够有更多的讨论,更多地争论,更有理性的探索并产生新的学术观点。

| 思想与观点

在人民交通出版社创建六十周年纪念活动上的发言

(2012年12月8日)

尊敬的杨传堂部长、各位领导、各位来宾：

大家下午好！

很荣幸，能作为人民交通出版社优秀作者的代表，参加人民交通出版社创建六十周年纪念活动。交通社是新中国成立后第一批成立的出版社，风雨兼程一个甲子，辛勤耕耘，硕果累累。今天，作为交通社的作者之一，有幸见证你们的荣光时刻，内心与你们一样激动。作为交通社众多作者的代表，在此，我想表达三层意思：

首先是祝贺！人民交通出版社成立六十年来，出版了一批高品质、有影响的图书，为人才培养、技术进步以及促进交通运输事业的发展做出了应有的贡献。近些年来，更是在激烈的市场竞争中取得了长足的发展，获得了"百佳出版社"的称号，长期保持在科技出版社第一梯队的行列，不容易！在这里，我也代表交通社的众多作者，向你们表示衷心的祝贺！

第二是感谢！正是像交通社这样一大批高品位的科技社出版的一大批高质量的图书，为我们科学界的研究奠定了坚实的基础。任何科学上、工程上的创见与发现，都是在传承前人成果的基础上取得的，而这些成果主要通过图书进行记载、总结。近年来，我国科学界取得了丰硕的成果，对此，你们出版社功不可没！站在读者的角度，我要感谢你们！

同时，站在作者的角度，我也要感谢出版社！你们是联系读者与作者的桥梁。你们更加清楚读者的需求，进行精心的策划，促使我们进行创作。一方面，促进我们科技工作者的研究成果系统化，更好地服务于广大读者；另一方面，促进了科技人才的成长。我自己的体会是，著书立说的过程，也是自己系统思考问题、收获提高的过程。

作为交通社的作者，我主编了一本《岩土工程师手册》。大家知道，这些年我主要侧重于一些前沿领域的研究，写一些理论与应用方面的专著，手册工具书的编写不是我的强项。我愿意写这本基础性手册的原因之一，就是被责任编辑的责任感打动了，被出版社的使命感打动了。我国岩土领域科技发展了这么久，取得了这么多的成果，的确缺少一个平台集中地予以展示。在组织编写过程中，也锻炼了一批岩土工程领域的中青年技术人员，现在很多其中的参编人员已经成为独当一面的专家，有的已经是某一细分领域的学术带头人。这本手册编写难度很大，前后用了四五年，在与责任编辑反复沟通协调的过程中，我切实地感受到了交通社对传播知识、传承文化的那份沉甸甸的责任感和使命感。出版后，这本手册也得到了大家的认可，获得了国家级"十一五"重点出版项目称号，在读者群体中形成了较好的口碑！相信今天在座的很多专家也是基于这种理念的认同、责任的共鸣，成了交通社的作者。

第三是祝愿！交通社六十年的发展，取得了丰硕的成果，也为自身进一步发展奠定了坚实的基础。希望交通社借文化大发展大繁荣的东风，坚持"立足交通、服务交通、服务社会"的办社宗旨，推动行业进步、文化发展，做好文化传承与积累，多出精品力作，无愧于我们这个伟大的时代！在走向百年老社的征程上，立意更高，视野更宽，发展更快！

最后，祝愿人民交通出版社的明天更美好！

谢谢大家！

在中国工程科技论坛上的发言

(2012年11月25日)

尊敬的白玉良秘书长、尊敬的副省长,各位来宾、女士们、先生们:

大家上午好!

今天我们在山东济南召开中国工程科技论坛,共同研讨"岩爆、突水突泥灾害预测预报预警与防治控制技术"问题。值此论坛开幕之际,首先,请允许我代表论坛组委会及中国岩石力学与工程学会,向参加论坛的所有来宾致以热烈的欢迎和衷心的感谢!

众所周知,近年来我国地下工程的建设蓬勃发展,交通隧道、水利水电以及矿山能源工程等,建设埋深越来越大,深部地质环境赋存的高地应力和高水压条件,导致施工过程中面临的岩爆和突水突泥问题越来越严峻。

在交通隧道修建史上,渝怀铁路圆梁山隧道、宜万铁路野三关隧道、马鹿箐隧道等在建设过程均发生过重大突水突泥灾害,造成了重大损失。未来20年中,我国将新建近100万km公路,交通路网将向西部纵深拓展;预计到2020年,我国铁路隧道建设将达到"两个一万",即隧道数量达到10000座、里程达到10000km。大量隧道在建设过程中所面临的挑战越来越多。正在修建的成兰铁路,线路将多处穿越活动断裂带和可溶岩区域,隧道最大埋深达1900m,最长达15km,工程建设难度极大。

在水利水电方面,"十二五"到"十三五"的10年建设投资将突破4万亿元,20多个世界级的大型水利水电工程正在修建。如锦屏二级水电站4条引水隧洞,单洞长约17km,最大埋深达2525m,具有"高水压、大流量、高地应力"的显著特点。施工中遭遇多次遇到岩爆和突水等灾害,水压超过10MPa,最大流量超过$7m^3/s$,最大地应力接近70MPa,严重影响了施工进度。

此外,在矿山能源方面,在近20年有250多对矿井因突水而淹没,经济损失高达350多亿元。目前开采深度超过1km的矿井越来越多,面临的冲击地压和突水等灾害问题也愈加突出。

然而,机遇与挑战并存。近年来,我国科技工作者在岩爆和突水突泥防治技术领域都取得了诸多成就。山东大学李术才团队通过近十年的科学研究和工程实践,在地下工程突水突泥灾害源定位定量预报及灾害控制方面取得了重要进展。中科院岩土所冯夏庭团队、大连理工大学唐春安团队和北京科技大学姜福兴团队在岩爆灾害防治方面取得了可喜的进步。同时,我们也应该深刻认识到,在地下工程灾害防治技术领域,仍然有很长的路要走。

本次论坛的召开为岩爆、突水突泥的预测预报预警和防控技术提供了一个很好的交流平台,将极大地促进重大地下工程建设的发展,进一步推动该领域科技工作者的友谊和合作。

衷心希望各位专家学者能利用此次论坛充分分享和交流新知识、新技术和新理念。同时也祝愿大家能够在泉城济南度过两天美好的时光。

预祝此次工程科技论坛取得圆满成功!

谢谢大家!

在中国人民解放军理工大学溯源碑揭幕仪式上的发言

(2011年12月13日)

各位领导、各位老师、各位校友,同志们:

大家上午好!

今天我们大家欢聚在哈军工[1]老五系(工兵工程系)楼前,郑重举行工程兵人"传承哈军工精神、继承哈军工传统"工兵工程系旧址纪念碑落成揭幕仪式,我谨代表哈军工工兵工程系的全体校友对关心和支持立碑的哈尔滨工程大学、哈军工北京校友会、解放军理工大学和工程兵工程学院等单位的领导和同志们表示衷心的感谢。

工兵工程系旧址纪念碑的落成是我们五系校友多年的心愿,今天实现了!大家非常高兴、非常激动!

忆往昔,在这里哈军工精神哺育我们成长,使我们立志献身国防。在这里,我们在老师和领导的精心培育下刻苦学习,求真务实,拼搏向上,开拓创新。学到了献身国防的本领,确立了全心全意为人民服务的人生观。

今天我们在这里树碑纪念,就是弘扬哈军工精神,传承哈军工传统,使更多后继学子忠于祖国、忠于人民,为了祖国的繁荣昌盛、民族复兴,更加努力地拼搏奋进,永远高举哈军工精神大旗。

我们中的多数虽已年过古稀,但传承哈军工精神的责任没有尽头。我们一定要志存高远,保持晚节,永远做一名真正的哈军工人。让年轻一代在理想信念的指引下青出于蓝而胜于蓝,使革命事业后浪推前浪,英雄辈出,后继有人。

最后再次对哈尔滨工程大学的领导和同志们对我们纪念立碑的大力协助和热情接待,对北京哈军工校友会领导远道赶来祝贺表示衷心感谢!

---

[1] 中国人民解放军军事工程学院,通称"哈尔滨军事工程学院",简称"哈军工"。

在中国岩石力学与工程学会岩爆机理探索沙龙上的发言

(2011年7月8日)

尊敬的各位专家、各位学者：

大家早上好，今天我们在此举行岩石力学与工程学会岩爆机理探索沙龙。我谨代表中国岩石力学与工程学会，向参加本次沙龙的各位专家学者表示热烈的欢迎！向精心筹备此次沙龙的学会秘书处工作人员表示衷心的感谢！

参加本次沙龙的各位代表都是国内岩石力学界岩爆研究领域的高层学者、著名专家，都曾在岩爆领域进行了深入的研究，作了大量的工作，取得了令人瞩目的成果。虽然以岩爆为主题的国际学术会议已经举行了十多次，但是岩爆机理的研究大多停留在定性解释的阶段，岩爆的预测预警也进行了大量工程实践的探索和研究，取得了不少成功的经验，但是尚未上升到系统的理论，甚至有专家断言岩爆的科学预报不可能实现。我国现阶段的岩石工程规模大、难度高，为世界所瞩目，无论是矿山工程，还是水电工程和交通工程，很多工程的开发进入了深部地下空间，不少工程都遇到了岩爆现象。近年来更发生了多起岩爆引起的重大事故，造成了人员伤亡和施工装备的严重破坏，严重阻碍了工程的顺利进行。岩石工程的实践已经向岩石工程技术工作者提出了严重的挑战。岩爆机理及其预测预报预警的研究已经成为我国岩石力学界必须致力解决的关键科学问题和技术难题。为了解决深地下矿产能源资源开发，西南巨大水能资源开发，长距离、大埋深交通隧道开挖，高放核废料地下处置等一系列深部岩石工程问题，我们岩石力学工作者特别是岩爆研究的技术人员，必须深入研究和弄清岩爆的机理，为深部岩石工程顺利开展提供科学理论和技术支持。

经中国科协批准，我们学会专门组织了这次以岩爆机理定量探索为主题的沙龙活动。此次沙龙主要围绕岩爆的孕育条件，岩爆的发生发展机理，岩爆的定性定量预测预报预警方法，岩爆对工程围岩的冲击破坏，岩爆条件下的巷道支护和围岩稳定性问题，岩爆防治、控制的技术措施以及岩爆相关的其他问题等主题进行研讨。衷心地希望与会专家学者展开广泛深入地交流，为提高岩爆机理定量研究的理论技术水平和岩爆控制的技术措施做出贡献。我相信在各位的努力下，本次沙龙一定会取得丰硕成果。

谢谢大家！

在第三届全国水工岩石力学学术会议上的致辞

(2010 年 8 月 28 日)

尊敬的各位代表、各位来宾：

中国岩石力学与工程学会主办、同济大学承办的第三届全国水工岩石力学学术会议于今天在这美丽的上海开幕了。出席本次会议的有来自全国各地的岩石力学工程领域的专家、学者和工程技术人员代表共计 200 余人。首先，我代表中国岩石力学与工程学会向出席大会的各位代表和来宾表示热烈的欢迎和衷心的感谢！向大会的承办单位同济大学以及组织委员会和秘书处的同志们所做的大量的会议筹备工作表示感谢！

中国岩石力学与工程学会是全国岩石力学与工程科技工作者的学术性群众团体，是中国科协的组成部分。学会前身是国际岩石力学学会中国小组，是 1978 年由中国科学院、外交部联合报请国务院批准成立的。在此基础上，1981 年成立了学会筹备组，经过 4 年多的努力，于 1985 年 6 月正式成立了中国岩石力学与工程学会。

学会经过近 30 年的发展，现在已经发展成为在国内具有重要影响力的科研学术组织。会员从成立时的 2000 余名发展到 15000 有余，约为国际岩石力学学会会员总数的两倍。学会在历届理事长的努力推动下，得到了广大科技工作者的积极参与和支持，围绕学科发展和工程贡献，卓有成效地开展国内外学术交流，并对重大工程建设项目提供了广泛的咨询和技术支持。

学会现在下设学术论文评审委员会、编辑工作委员会、教育工作委员会等 5 个工作委员会。此外，根据行业特点，学会共成立了地下工程分会、岩石动力学专业委员会、废物地下处置专业委员会、地面岩石工程专业委员会、隧道掘进机工程应用分会等 18 个专业委员会。学会期刊《岩石力学与工程学报》为 EI 核心收录期刊，力学类总被引频次和影响因子居国内前列，是中国科协优秀期刊和中国百种杰出学术期刊。

近几年来，我国在水利水电领域陆续开展了一些大型的工程项目。以溪洛渡、锦屏、白鹤滩等为代表的一批国家重大水电工程基础建设项目的实施，对水工岩石力学提出了一系列挑战性的研究课题，同时也为我国水工岩石力学的发展提供了一个机遇。前一段时间，国内发生了几起大的泥石流滑坡事故，造成了很大的人员伤亡。面对灾区地质地貌复杂，居住环境脆弱，如何预防泥石流等地质灾害的发生，我想也应该引起在座学者的关注。

今天的水工岩石力学大会已经是第三次了。第一、二次全国水工岩石力学学术会议分别于 2006 年 10 月、2008 年 10 月在郑州和武汉成功召开。已经成为水利水电工程岩石力学基础理论与工程应用研究的高层次学术交流盛会。

会议在为广大的专家、学者和工程技术人员提供了水工岩石力学和工程经验的交流平台，使与会学者能够在会议上精心交流学术成果和工程建设经验，对我国的水利水电工程事业起到推动作用。

本次会议的主题是："和谐地球上的水工岩石力学"，从会议通知发出之后，得到了全国广大科研工作者和工程技术人员的踊跃参与，会议共收到稿件 400 多篇，稿件内容范围涵盖了地质建模、数值模拟、围岩破坏、盾构开挖和边坡稳定性等岩石力学领域的各个方面。经过会议学术委员会的评审和推荐，共有 60 余篇优秀论文被《岩石力学与工程学报》正刊和增刊收录。

本次大会的议题主要有：

(1) 岩石、结构面力学和渗流性质;
(2) 岩体工程勘察、测试、探测与试验新技术、新方法、新设备;
(3) 岩体工程物理与数值模拟方法与工程应用;
(4) 岩体工程数字化技术与健康诊断理论;
(5) 高坝坝基设计理论与分析方法;
(6) 岩质边坡工程稳定性分析与治理;
(7) 大型地下洞室群设计与安全控制;
(8) 深埋长隧洞及 TBM 施工中的岩石力学问题;
(9) 岩体及坝基和边坡的动力特性及响应;
(10) 环境岩土工程问题。

根据会议日程安排,会议初步计划开三天,今天举行大会开幕式和安排特邀、大会报告,明天上午进行大会报告,下午分四个小组进行分组的汇报讨论,讨论完毕之后进行闭幕式。在会议期间还将召开《岩石力学与工程学报》七届四次编委会扩大会议以及测试专业委员会、岩体物理与数值模拟专业委员会的年会。会后 30 日对上海世博会进行参观和考察。

会议时间不长,但是希望大家能够积极发表意见,充分进行交流,共同努力,开好这次会议。

最后,我预祝此次会议取得圆满成功!

吾家吾国

钱七虎院士文集

人既发扬踔厉矣，则邦国亦以兴起

Qian Qihu
Yuanshi
Wenji

淡泊名利品自高

近日,"八一勋章"获得者、中国工程院院士钱七虎再次向"瑾晖慈善基金"捐赠善款100万元,用于关爱"事实无人抚养"和患白血病的困境儿童等。10年前,钱七虎从母亲和妻子名中各取一字做慈善,在家乡昆山设立"瑾晖慈善基金",多年来已累计捐赠约1168万元,惠及困难学生、老人780人次。淡泊名利品自高。钱七虎用实际行动亮明了自己的态度,也立起了共产党人的精神标杆。

正确的名利观,映照的是初心使命,体现的是责任担当。钱七虎把为党分忧、为国奉献、为民服务作为一生的坚守,怀揣许党报国信念,勇攀国防科技高峰,建立从浅埋工程到深埋工程防护、从单体工程到工程体系防护、从常规抗力到超高抗力防护等学术理论与技术体系,制定我国首部人防工程防护标准,创建我国防护工程人才培养体系,解决工程防护一系列关键技术难题,为我国防护工程发展作出巨大贡献。

淡泊名利的人,心中有"大我",在国家利益、人民利益和个人利益之间,总会不计个人得失,自觉把国家、人民利益放在前面。党的创始人之一李大钊,心中只有革命事业,牺牲时家中一贫如洗;县委书记的好榜样焦裕禄,"心中装着全体人民、唯独没有他自己";曾任云南省保山地委书记的杨善洲,退休后回到家乡义务植树造林,建成面积5.6万亩、价值3亿元的大亮山林场,2010年将林场经营管理权无偿移交国家……这些人之所以能不计个人得失、不改英雄本色,就是因为在他们心中,追求生命的意义与价值远比追名逐利更高尚,对党忠诚、为民造福远比汲汲于个人名利更可贵。

对共产党员来说,看重什么、看轻什么、坚守什么、舍弃什么,是一个关乎共产党人得失观、名利观的大问题。实践证明,把党和人民的事业看重一些、把个人得失名利看轻一些,方能保持平和心态,心无旁骛干工作,方能正心明道、怀德自重,始终同人民群众心心相印、生死相依、命运与共。面对利益取舍,自觉坚守"大我"舍弃"小我",不耿耿于一时之名,不汲汲于一事之利,始终保持共产党人的卓然风范和浩然正气,用一生回答好名利之问,一辈子追求"共产党员"这个光辉之"名"、实现"人民利益"这个崇高之"利",这才是共产党人应有的名利观。

个人名利淡如水,党的事业重如山。奋进新征程,建功新时代,树立正确的名利观、得失观是应有之义,也是必然要求。党员干部当端正价值追求,淡泊名利得失,把心思精力倾注在本职工作岗位上,把实现个人价值融入强国复兴伟大实践中。

(记者:王李彬)

本文原载于《中国纪检监察》(2023年5月26日)。

耿耿丹心　为国铸盾

【人物】"八一勋章"获得者钱七虎

【故事】

钱七虎，1937年10月生，中共党员，陆军工程大学教授，中国工程院首批院士。他投身现代防护工程理论开创性研究，带领团队成功研制出我国首套爆炸压力模拟器，建成爆炸冲击防灾减灾国家重点实验室。他创建了我国防护工程学科和人才培养体系，建立了我国城市人防工程毁伤理论模型和分析方法，制定了我国首部人防工程防护标准，曾取得国家科技进步一等奖、军队科技进步一等奖等，获得2018年度国家最高科学技术奖。前不久，钱七虎被授予"八一勋章"。

【点评】

"一个人活着是为了什么？"这是60多年前钱七虎就读于哈尔滨军事工程学院时，在一堂革命人生观教育课上开始思考的问题。钱七虎用超过一甲子的奋斗和实践给出了自己的答案："科技强军、为国铸盾，是我的毕生追求，也是我的幸福所在。"

防护工程被誉为一个国家的"地下钢铁长城"，也是钱七虎毕生钻研的课题。20世纪70年代初，戈壁深处一声巨响，荒漠升起蘑菇云……当人们欢呼庆贺时，一群身着防护服的科研人员迅速冲进核爆中心勘察爆炸现场，钱七虎便是其中一员。从荒凉戈壁到边防海岛，从浅埋工程到深埋工程防护，从常规抗力到超高抗力防护，在为国铸盾的道路上，钱七虎始终奔波在工程一线、奋战于专业前沿。几十年来，他和团队攻克了一个个难关、突破了一系列技术难题，为我国战略工程装上了"金钟罩"，擦亮了对党绝对忠诚、矢志强军报国的军人本色。

"哪些事情对国家和人民有利，科技工作者的兴趣和爱好就要向哪里聚焦。"秉持这样的信念，钱七虎把自己不知疲倦的忙碌身影和奔波足迹留在了一个个大国工程的建设现场。面对被称为"天下第一爆"的炮台山爆破，他迎难而上、不辱使命；面对当时施工风险最大的"万里长江第一隧"，他勇挑重担、攻坚克难，被授予"南京长江隧道工程建设一等功臣"；退休后，仍积极为川藏铁路建设、渤海湾海底隧道论证建言献策……钱七虎用驰而不息的奋斗生动诠释着这样的信念：只有把个人理想与党和国家的需要、民族的前途命运紧密联系在一起，才能有所成就、彰显价值。

今年高考后，贵州铜仁一名取得好成绩的考生说，最想感谢的人是钱七虎。正是因为钱七虎设立的瑾晖慈善基金的资助，他得以在学校安心学习，专心备考。在这位考生的家乡，获得瑾晖慈善基金资助的学生有400多名。从20世纪90年代起，钱七虎就将所获得的各种津贴奖金等用于公益慈善事业。无论是将国家最高科学技术奖800万元奖金全部捐助贫困学生，还是捐款650万元支持武汉抗击新冠肺炎疫情，无不传递着一个科学家心中有国、心中有民、心中有仁、心中有爱的动人情怀。

大国利剑舞长空，耿耿丹心筑长城。一批又一批像钱七虎这样的科技工作者，将个人的拼搏奋斗融进祖国的江河、民族的星空，让一个民族风雨无阻向前进的雄伟身姿傲然挺立，也让一个国家攀登世界科技高峰的征途无限光明。新时代新征程，广大科技工作者潜心科研、矢志报国，在科技创新的征途上不断前行，一定能为实现高水平科技自立自强再立新功。

（记者：邹　翔）

本文原载于《人民日报》（2022年8月3日）。

| 吾家吾国

科技强军、为国铸盾的防护工程专家

刚参加完某项目鉴定会,钱七虎院士又拎着行李箱,踏上出差的旅程。

很难想象,这是一位80多岁老人的生活。作为现代防护工程理论的奠基人、防护工程学科的创立者,在许多人早已退休的年纪,钱七虎还像"空中飞人"一般奔波。这样的生活,他并不觉得累,而是感觉"活得很充实,很有成就感,也有幸福感"。

"一个人活着是为了什么?"这是60多年前钱七虎在哈尔滨军事工程学院就读时,接受的第一堂革命人生观教育课。

奋斗一甲子,投身强国梦。参加工作以来,钱七虎始终坚守爱党、报国、强军的赤子情怀,战斗在大山深处、戈壁荒漠、边防海岛等工程一线,为我国防护工程发展作出了巨大贡献,用实际行动交出了自己的人生答卷:国之需要,我之理想。

初心,是时代镌刻在钱七虎心中永恒的烙印。20世纪70年代初,戈壁深处一声巨响,荒漠升起一片蘑菇云……当人们欢呼庆贺时,一群身着防护服的科研人员迅速冲进了核爆中心勘察爆炸现场,钱七虎便是其中一员。

那一年,钱七虎受命为空军设计飞机洞库防护门。为了找准原有设计方案存在的问题,他专门到核爆试验现场调查研究。

经过深入调查思考,钱七虎大胆决定采用刚刚兴起的有限单元法,但这涉及大量的工程结构计算。

当时,国内大型计算机设备紧缺,为了求取最科学的方案,他来回奔波于北京、上海,利用节假日和别人吃饭、睡觉的空隙打"时间差"蹭设备用。

两年后,他研究的当时我国跨度最大、抗力最高的地下飞机洞库防护门通过成果鉴定时,他却患上了严重的十二指肠溃疡和胃溃疡。

勇于攀登的人脚步从不停歇。紧接着,他又趁热打铁开始"有限单元法在工程结构计算中的应用"的研究攻关。长期的劳累又诱发了痔疮,但他还是坚持每天工作12小时以上,疼得不敢坐,就趴在床上写专著,该专著问世不久就获得全国科技大会重大科技成果奖。

"世间万物,相生相克,有矛必有盾。"当时,我国面临严峻的外部威胁,在钱七虎看来,如果说核武器是对付敌对军事力量锐利的"矛",那么防护工程则是一面坚固的"盾"。

随着侦察手段的不断更新、高技术武器与精确制导武器的相继涌现,防护工程在高度透明化的战场中,常常是"藏不了、抗不住",特别是世界军事强国开始研制精确制导钻地弹,给防护工程造成了巨大威胁。

"'矛'升级了,我们的'盾'就要及时升级。"面对一项项世界级国防工程的防护难题,钱七虎带领团队勇攀科技高峰,建立了从浅埋工程到深埋工程防护、从单体工程到工程体系防护、从常规抗力到超高抗力防护等学术理论与技术体系,制定了我国首部人防工程防护标准,解决了核武器和常规武器工程防护一系列关键技术难题。2019年,钱七虎获得我国科技领域最高奖项——2018年度国家最高科学技术奖。

师之大者,为国为民。作为多个国家重大工程的专家组成员,钱七虎在港珠澳大桥建设、雄安新区规划、南水北调工程、西气东输工程、能源地下储备等方面提出了切实可行的重大咨询建议。退休后,仍活跃在国家战略防护工程建设前沿,积极为川藏铁路建设、渤海湾海底隧道论证建言献策。

本文原载于《人民日报》(2022年8月1日)。

"一个人,只有把个人的理想与党和国家的需要、民族的前途命运紧密联系在一起,才能有所成就、彰显价值。"谈及自己的人生历程,钱七虎的目光中透露出坚定不移的信念。

"把更好的机会留给年轻人",这是钱七虎的育人理念。2009年,他主动放弃自己被提名的机会,推荐年轻学者冯夏庭出任国际岩石力学学会主席:"世界岩石力学研究中心在中国,冯夏庭年轻有能力、有梦想。"最终,冯夏庭成为目前为止担任该学会主席的唯一中国专家。

"一辈子搞了那么多课题,但我感觉培养人才始终是我最大的课题。"聊起未来,钱七虎充满对人才辈出的殷殷期盼。

数十年来,钱七虎培养的大批优秀人才成为防护工程领域的中坚力量。2019年,他将国家最高科学技术奖800万元奖金全部捐助贫困学生,帮助贫困学子圆了上学梦。

"虽然我已经80多岁了,但我总觉得还有很多事要做。为祖国铸就坚不可摧的'地下钢铁长城',是我的毕生追求,也是我的事业所在、幸福所在!"钱七虎表示,有生之年他将继续在为国铸盾的道路上不懈冲锋。

铸盾一生

采访钱老是在一个冬日的下午,缕缕阳光透过窗户洒在我们身上,让人倍感温暖。

那时,他刚刚获得 2018 年度国家最高科学技术奖。采访时,荧幕上那位满头白发的老人走上领奖台,从习近平主席手中接过获奖证书的画面还一直萦绕在我的脑海中。而此刻,他正和我们一起聊生活、聊理想、聊经历过的苦难,神采奕奕,和蔼可亲,就像一位邻家老爷爷。在他的娓娓讲述中,我们仿佛触摸到了那段激情燃烧的岁月……

他就是钱七虎,我国现代防护工程理论的奠基人、防护工程学科的创立者,中国工程院院士,原解放军理工大学教授。

一

1937 年 10 月,在江南水乡的乌篷船上,一个小生命呱呱坠地。

那一年,淞沪会战爆发,日本侵略者占领上海。血腥的战争逼得邻近的江苏昆山的老百姓流离失所。在那个风雨飘摇的年代,这个刚出生的小生命给逃难中的母亲带来了希望。家人期望他长得虎背熊腰、力大无比,又希望他有颗英雄虎胆、驱寇降魔,且因他在家中排行老七,故取名"七虎"。

在抗日战争的枪炮声中,钱七虎度过了穷苦的童年时期。社会动荡,7 岁丧父,家里子女全靠母亲摆小摊维持生计。新中国成立后,依靠政府的助学金,钱七虎完成了中学学业。强烈的新旧社会对比,在他心中深深埋下了矢志报党报国的种子。

1954 年 8 月,作为优秀中学毕业生,钱七虎高兴地迈进了哈尔滨军事工程学院的大门,成为哈军工组建后招收的第 3 期学生。

有人问钱七虎为什么会做出这样的选择,他深情地回答说:"我这个乡下的穷孩子,能受到良好的教育,所有的一切都源于党的培养,唯有献身党的事业,方能报答党的恩情。"

"一个人活着为了什么?"军校的第一堂革命人生观课让钱七虎至今记忆犹新。学校的预科主任教育他们说:是为了保护人民的生命财产安全,是为了保家卫国、建设钢铁长城。

"那是我第一次感受到军队的这种博大情怀,听了很振奋,记得很牢。"说话时,钱七虎的眼神中饱含激动与振奋,仿佛他又回到了那个熟悉的校园、那个青春的时代。

从哈军工毕业后不久,钱七虎又被选拔到苏联留学深造。

"我非常尊敬周总理,他一生严格要求自己,鞠躬尽瘁为人民,毫无私心。周总理一生为党为国的高贵品质一直深深地影响着我。"1961 年,周恩来总理赴苏联参加苏共二十二大会议,其间周总理接见了中国留学生,并合影留念,勉励大家努力学习,回国后更好地建设新中国。钱七虎就是其中的一位。至今,在钱七虎的办公室里,仍然摆放着周恩来总理的照片。看到电视剧《海棠依旧》里的场景,钱七虎常常禁不住潸然泪下。

1965 年,钱七虎学成回国。根据组织安排,钱七虎在原西安工程兵工程学院担任教员。从那时起,为国家铸就坚不可摧的"地下钢铁长城",就成了他毕生的事业追求。

本文原载于《解放军报》(2022 年 7 月 8 日)。

二

"世间万物,相生相克,有矛必有盾。"那个年代,我国面临严峻的核威胁环境。在钱七虎看来,如果说核武器是锐利的"矛",那么防护工程则是一面坚固的"盾"。

那是20世纪70年代初的一次核试验,随着戈壁深处的一声巨响,荒漠升起一片蘑菇云……当人们欢呼庆贺之时,一群身着防护服的科研人员迅速冲进了核爆中心勘察爆炸现场,钱七虎便是其中一员。

"我有任务,走了。"在那些不分昼夜、紧锣密鼓工作的日子里,钱七虎和很多从事绝密工程的科学家一样,临走时只能给家人留下这句话。

离家那一年,钱七虎受命进行空军飞机洞库门的设计。为了准确找出原有设计方案存在的问题和原因,他专门请求到核爆试验现场调查研究。

在核爆现场,他发现飞机洞库的防护门虽然没有被炸坏,里面的飞机也没有受损,但是防护门发生了严重变形,导致无法开启。

"门打不开、飞机出不去,就无法反击敌人。必须找出问题,进一步优化设计方案。"钱七虎首先想到的是改良传统手算模式,使用先进计算理论和设备。那个时候,有限单元法作为一种工程结构问题的数值分析方法刚刚兴起,他决定用这种方法解决飞机洞库门的计算问题,这在当时属国内首创。

当时先进的计算设备是晶体管计算机,自己单位还没有。辗转多方协调,钱七虎联系到国内少数几个拥有大型晶体管计算机的科研单位借用。刚开始,面对从未接触过的巨型计算设备和天书似的上机手册,整个团队一时间有些束手无策。此前,钱七虎自学了计算机基础理论,但从没有上机操作过。于是,他把自己关在房间里啃这本"天书"。两天后,当他再次站在团队成员面前,说的第一句话却是:"可以上机操作了!"两天时间,他不仅看懂了,而且准备着手编写大型防护结构的计算程序。

"当年,大型的计算机只有航天部门有,因为工作时间人家也要排算,我们的使用时间就是他们的吃饭时间、休息时间,就是在这样的工作条件下取得了一些科研成果。"采访中,钱七虎说起当年的艰难攻关,感慨万千。

终于,钱七虎利用有限单元法进行工程结构的计算,解决了大型防护门变形控制等设计难题。为了缩短防护门的启闭时间,他创新提出使用气动式升降门方案。钱七虎说:"气动试验做了几十次,用了整整一年时间。失败了总结一下,就接着准备下一次试验,每一次试验过程都是学习提高的过程。不知道的东西经过总结学习变成知道的,那真是十分愉快的事情。"

那段时间,他经常睡在办公室。历时2年多,钱七虎成功设计出当时国内跨度最大、抗力最高的飞机洞库防护门。那年,他38岁。

"矛"与"盾"总是在攻防对抗的进程中不断碰撞出新的"火花"。随着侦察手段的不断更新、高技术武器与精确制导武器的相继涌现,防护工程在高度透明化的战场中,常常是"藏不了、抗不住",特别是世界军事强国开始研制精确制导钻地弹,给防护工程造成了巨大威胁和挑战。

"防护工程是我们国家的地下钢铁长城,'矛'升级了,我们的'盾'就要及时升级。"钱七虎说。

功夫不负有心人。经过多年的研究,他攻克一个个难关,建立了从浅埋工程到深埋工程防护、从单体工程到工程体系防护、从常规抗力到超高抗力防护等学术理论与技术体系,制定我国首部人防工程防护标准,解决核武器和常规武器工程防护一系列关键技术难题,为我国防护工程发展作出巨大贡献。

三

"眼睛永远要向利国利民的方向聚焦。"回忆自己的人生经历时,钱七虎这句话掷地有声。

1992年,珠海机场开工动土。雄踞三灶岛南端的炮台山成为机场建设的天然障碍,炮台山爆破工程

| 吾家吾国

被列为重中之重。

1992年12月28日，珠海三灶岛。

那一天，在距炮台山3公里处的主席台上，坐着许多军地领导和媒体记者，现场的气氛宁静，又紧迫。作为这场爆破的主持者，他既充满信心，也有些许忐忑。他深知，这场爆破工程的难度是前所未有的：爆破总方量1085万立方米，50%的土石方要一次性定向抛至大海，50%的土石方要松动破碎；让半座山进入大海，需要上万吨炸药，还要确保山外600米和1000米两处村庄安全，房屋不得倒塌……

"是挑战，更是机遇！"在这之前一年，时任原解放军工程兵工程学院院长的钱七虎立下了军令状，"我们有最好的爆破工程专家和最过硬的团队，我胸有成竹！"

时针，指向13时50分。随着爆破工程现场总指挥一声令下，霎时间，1.2万吨炸药和数万支雷管在程控起爆器精确控制下，分成33批，在38秒内逐一起爆。

远远望去，炮台山在抖动中先后被切成3段，每段山体又一片一片被削抛向天空，掷入大海，宛如天女散花，又似蛟龙入海，蔚为壮观。

爆破成功，达到预期效果！一时间，这在我国爆破领域引起了轰动，军地领导、科研单位纷纷发来贺电。

炮台山的成功爆破，被媒体称为"亚洲第一爆"。这一爆不仅吹响了珠海特区改革前进的新号角，同时也开辟了我国爆破技术新的应用领域，为城市的快速建设发展增添了新动力。

"前进，就要走前人没有走过的路！"钱七虎从军60多年，为军事防护工程作出了杰出贡献。同时，他的科研触角还不止于此。他始终认为，作为中国工程院院士，有责任、有义务关心研究国家的建设发展，这也是一名科学家必须具备的情怀和担当。

时光进入新世纪。2002年，党和国家领导人召见了钱七虎等50位科学家。当时，钱七虎提出在长江上修建越江隧道。两年多后，南京长江隧道纳入南京"五桥一隧"总体规划。这让钱七虎兴奋不已。作为专家委员会主任，他也深知肩上责任之重。

南京长江隧道是当时已建的隧道中所经地质条件最复杂、技术难题最多、施工风险也最大的工程，被称为"万里长江第一隧"。如何进行隧道掘进，确定工程建设方案，是钱七虎首先要解决的问题。一开始，设计单位提出了用"沉管法"。钱七虎认为用"沉管法"存在安全隐患。

"由于三峡水电站的修建，泥沙含量将大为减少，长江中下游冲将大于淤。经过几十年、上百年江水冲刷，冲淤平衡被破坏，下游的管道就会露出江底，'沉管法'隐患太大！"钱七虎说道。

为此，他建议采用盾构机开掘隧道。盾构机在如此复杂的地质环境下施工，这在世界上尚属首次。

2008年8月，最担心的事还是发生了。当盾构机掘进到第659环时，因刀具、刀盘磨损严重，盾构机突然停止工作。

盾构机罢工，隧道施工搁浅。一夜之间，街头巷尾议论纷纷，远在某电站的钱七虎当即表示："工程绝不能报废，更不会'烂尾'，我们一定能解决。"

对已经从事科研数十年的钱七虎来说，继续隧道工程，除了积累多年的经验，还要有一种偏向虎山行的魄力。这种精神是撕开乌云的一道强光。

后来，在钱七虎建议下，磨损刀具更换，刀盘修复，国内厂家对刀具进行自主改良。改良后的刀具性能大幅增加，南京长江隧道掘进历经磨难，再次启程。

2010年5月28日，南京长江隧道在历经5年的建设之后全线通车运营。

作为多个国家重大工程的专家组成员，他在港珠澳大桥、南水北调、西气东输等工程中提出了切实可行的决策建议，并多次赴现场提出关键性难题的解决方案。当然，这其中也包括雄安的建设。

"雄安，这座中国未来之城的建设是国家的千年大计。建设之初就要考虑解决传统城市建设中交通拥堵、内涝等城市病问题，要多管齐下，既要'面子'，也要'里子'。"在一次雄安建设规划的相关会议上，钱七虎用全局长远的眼光审视这座即将崛起的新兴城市。

老骥伏枥，志在千里。耄耋之年的钱七虎仍以满腔热情履行着自己作为中国工程院院士的职责。

"居里夫人曾说,把安逸和享乐当作人生唯一目标,那不过是猪圈里的理想。如果一个人是追求个人名利,那你的关注点始终是在个人幸福的方向;如果你树立的是一个基于国家的价值观,你关注的就是国家、民族和人民。"钱七虎说。

四

走在校园的林荫道上,老院士总是一副和颜悦色的模样,教职员工见到他都主动打招呼,就连一些家属认识他的也能聊几句。这是钱七虎留给很多人的印象。

然而,很多学生提起师从钱七虎的经历,却觉得是"痛苦而有收获的煎熬",因为他们经常会受到导师的批评。对学生的论文,钱七虎总是不厌其烦,逐字逐句地推敲,提出修改意见。

"科学是老老实实的学问,来不得半点马虎。"钱七虎常常对他的学生们说。

陆军工程大学某实验室教授赵跃堂是钱七虎的得意弟子,他多次谈起自己博士论文修改答辩的故事。

"不能仅仅是为了拿到学位,你的科研成果是要运用到实践中的,理论和技术都要能立得住、经得起实践检验。"在导师的严格要求下,赵跃堂的论文先后做了4次较大调整。

1996年,当赵跃堂把自己的博士论文第7次修改稿交给导师钱七虎时,钱七虎正好动身去北京参加一个重要会议。半个月后,钱七虎开完会返回南京,赵跃堂拿到了自己的论文,发现上面写满了蝇头小字,从观点、公式到表格、数据,200多页的论文几乎每页都有增减和修改。

原来,答辩日期临近,钱七虎带着学生论文去开会。会议活动繁忙,直到晚上10点夜深人静以后,钱七虎才坐在台灯下,逐字逐句地修改推敲学生的论文,有几天一直改到凌晨。

教学科研生涯中,钱七虎始终把倾心育人、提携学生作为自己的责任。"把更好的机会留给年轻人!"科研学术中,钱七虎从不争名争利。在他的启发、指导和设计下,很多科研工作结出了累累硕果,但在获奖排名的时候,他总是推让。

熟悉钱七虎的人都知道,他有一项不成文的"规矩":每年都要举办几场特殊的"学术民主生活会"。会上,老中青科研人员"三代同堂",围绕学术前沿或科研难点展开研讨。在他带领的团队中,许多科研项目由年轻人担纲完成,许多专家已经成为该领域学科带头人和国家重点科研项目负责人。

就是靠这些一点一滴的努力,钱七虎创建了我国防护工程学人才培养体系。

钱七虎用自己的奖金和工资设立"瑾晖"基金,长期资助烈士子女和成绩优异的贫困学生,帮助不少贫困学生圆了上学梦。2019年,钱七虎将国家最高科学技术奖奖励的800万元奖金全部捐出,重点资助西部和少数民族的贫困学子。

"黄卷青灯伏书案,披繁星,戴骄阳。三尺讲台八千日,传师道,育儒将……"这是学生吕亚茹为导师钱七虎写的一首词,也是对他的一个评价。他说,自己很庆幸遇到了钱老师,让他找到了人生前进的方向。而钱老对自己的评价却是:"我一生做了很多课题、搞了很多研究,但培养人才始终是我最大的课题。"

(记者:王含丰 云利孝)

科研工作者要永远跟党走

今年是中国共产党建党100周年。长期以来,广大科研工作者无私奉献、顽强拼搏,以时不我待、只争朝夕的紧迫感和使命感,在科技领域创造了一个又一个的奇迹,一大批国之重器和重大工程上天入地、通江达海,为国家的繁荣富强奠定了坚实的基础。

什么是科研工作者的初心与使命?如何通过科学技术,筑牢钢铁城墙,为实现第二个百年奋斗目标添砖加瓦?日前,记者采访了中国工程院院士、国家最高科学技术奖获得者钱七虎。

记者:回顾中国共产党百年奋斗历程,作为一名科研工作者,您有怎样的感受?

钱七虎:江山就是人民,人民就是江山。中国共产党打江山,是为了人民的解放,打败侵略者和剥削者。我党守江山,是为了让人民过上好日子。作为一名党员科研工作者,要牢记我党为人民服务的宗旨。我们攻坚克难搞科研,不是为了个人的名利,我们的科技创新要服从于人民、也就是服从于国家的需求,服务于国家经济社会发展也就是服务于人民。

我长期从事国家的防护工程事业,目的就是给国家铸造坚不可摧的地下"钢铁长城",为人民构筑绝对可靠的安全屏障,让妄图欺侮、逼迫我们的人碰得头破血流。在建成科技现代化、科技创新强国的道路上,我们一定会遇到无数风险挑战,只有继承弘扬无私奉献的科学家精神,坚韧不拔、顽强拼搏,才能为实现中国梦强国梦作出新的贡献。

在我一生的科研工作中,马克思主义方法论在分析课题科研中的主要矛盾与矛盾的主要方面,如何认识理论和实践的关系,如何辩证地对待科研中的成功与失败,起了很大的指导作用。我觉得科研工作者,应当自觉学习马克思主义理论,努力掌握其中蕴含的立场、观点、方法,做到知其言更知其义,知其然更知其所以然。努力往深里走、往实里走、往心里走,把学习成效转化为履行新时代使命的实际行动和实际效果。

记者:您出生于战火纷飞的年代,在您的人生经历中,对中国共产党有怎样的感情?

钱七虎:爱党信党跟党走,是我一生中最正确的选择。1937年10月,我出生在国歌中所唱的"中华民族最危险的时候"。儿时的苦难经历,以及往后的人生历程,让我深深地感受到,没有中国共产党,就没有新中国、强大的中国,就没有14亿中国人民的幸福生活。这是我爱党信党跟党走的最根本原因。

如何跟党走?就是在关键时刻听从党的召唤,服从党的分配。1951年,朝鲜战争爆发,我响应党的号召,报名参加军干校,因年纪太小未被批准;高中毕业,我被选送到苏联留学,随后组织找我谈话,希望我能够去刚成立不久的哈军工,作为团支部书记,我毅然放弃留苏进了哈军工。当时选工程兵防护工程专业的人很少,大家觉得要跟黄土铁锹打交道太土。我作为班长,带头服从组织分配;本科毕业后,组织又选派我到苏联莫斯科军事工程学院读研究生,并最终获得副博士学位回国;在分配的时候,当时的工程兵政治部先通知我到北京某设计院报到,后来又说西安新成立的工程兵学院急需教员,我二话不说,马上先去西安报到;前年,在获得国家最高科学技术奖后,我决定把国家奖给我的800万元和江苏省配套奖给的800万元全部捐出,也是为了响应党的脱贫攻坚的号召,资助我国西部的贫困学子和抗击疫情的号召。

记者：您为我国的国防事业奋斗了一辈子，这是否是您的理想所在？

钱七虎：我觉得只有把个人的理想与国家和民族的前途命运紧密联系在一起，才能有所成就、实现价值。一个科学家，他的注意力、关注点在哪里很重要。"两弹一星"功勋奖章获得者朱光亚1950年回国前，在给留美同学的一封公开信中这样写道："让我们回去，把我们的血汗洒在祖国的大地上，灌溉出灿烂的花朵"。当年，钱学森先生一回国，就强烈建议国家搞原子弹、导弹。正是老一辈科研工作者的高瞻远瞩和家国情怀，才为中国的长治久安打下了牢固基础。

从走上工作岗位的那一刻起，为祖国铸就坚不可摧的地下"钢铁长城"，就成了我毕生的追求。只有铸牢防护工程这面坚固的盾牌，才能确保我国指挥工程和重要战略武器工程的安全，进而维护国家的安全喝稳定。

20世纪70年代初，我受命进行某地下飞机洞库门设计。为了获得准确的实验数据，我赶赴核爆试验现场进行实地调查研究和收集数据。在现场发现，虽然核爆后的飞机洞库门没有被炸毁，飞机也没有受损，但是防护门出现严重变形，导致无法开启。门打不开、飞机出不来，就无法反击敌人，必须进行改进。当时飞机洞库门设计采用简单手算的公式计算精度差。计算精确变形需要应用先进的理论在大型电子计算机上计算。与此同时，世界上刚兴起有限元计算理论，这对我来说完全是一个陌生领域。为了攻克难题，我加班加点学习有限元理论、计算机语言。当时，中国只有七机部五院，中科院计算机所才有大型晶体管计算机，我们就利用人家不上机的午饭和下班时间上机。经过团队艰苦卓绝的努力，我们终于编制出大型计算程序，设计出了当时跨度最大能抵抗核爆炸冲击波的抗力最高的机库大门，还出版了《有限元原理在工程结构计算中的应用》，并获得了1979年全国科学大会重大科技成果奖。

这些年，我还在长江隧道、南水北调工程、西气东输工程、港珠澳大桥海底隧道、能源地下储备、核废物深地质处置等方面，贡献了自己的学识和才智。我觉得，这是我作为院士的使命所系，也是我作为一名科研工作者的幸福所在。

记者：您认为做好科研工作，要树立怎样的意识？

钱七虎：科学技术研究是集体、团队的研究，只有摆正个人和组织的关系，摆正个人和集体的关系，摆正个人和群众的关系，才能顺利前进。我能参加核试验，完成科研任务，离不开组织的支持；我能完成跨度最大、抗力最高的飞机洞库大门设计任务，是团队和原济南空军设计所共同支持帮助的结果；我能当上教授、中国工程院首届院士，是已故我国著名老科学家张维、李国豪推荐、提携的结果。我能获得这些国家级科研奖项，都是我的团队和合作团队集体奋斗的结晶。

总结我一生，我深深体会到，一个人没有大家的支持，就不可能有什么进步和发展，作为科技人员，更是如此。科技事业不是个人单打独斗能成功的，靠的是和谐团结的团队，团队的领导要带领大家处理好个人和群众的关系、个人和集体的关系，这样团队才能凝聚力量干大事。

记者：当前，我国正向着建设社会主义现代化强国的目标迈进。您认为作为一名科研工作者，身上肩负着怎样的重担？

钱七虎：建设社会主义现代化强国，建设科技强国、军事强国的责任和使命，首先要落到当代科学家身上，实现关键核心科技的自立自强，在国际环境日益复杂的今天，显得尤为重要。

靠什么实现关键科技的自立自强？回顾我国科技事业所取得的每一项重大成就，都离不开科学家无私报国、服务人民的科学家精神。科学家精神说到底就是党性在科学家身上的体现和反映，就是马克思主义世界观在科学家身上的体现和反映。如今，我已年过八旬，但作为一名承上启下，拥有65年党龄的科学家，我有责任继续践行科学家精神，继续发光发热。

我觉得我还有很多事情能做、想做、要做。所以我继续担任军事科学院首席专家、火箭军首席工程专家、空军工程顾问等。我必须活到老、学到老、革命到老，必须始终做到扬鞭自奋蹄，继续为国防建设用脑尽力,为国家铸就钢铁后盾,为经济和社会建设继续做贡献。

我希望一代代的科研工作者,都要弘扬好、实践好科学家精神,把个人理想和国家重大需求相结合,排除名利束缚、严谨治学、潜心研究、直面问题、奋勇拼搏。要敢闯科学"无人区",敢为人先、集智攻关、团结协作,勇当新兴交叉领域的开拓者,勇担关键科技领域的领跑者,抢占世界科技发展的制高点,为开启伟大新征程作出更大贡献。重托在肩,让我们努力拼搏吧!

(记者:常　理)

科学家精神的核心是追求真理和献身科学

一、什么是科学家精神？

我理解科学家精神的精髓和核心是追求真理,献身科学。科学家精神是通过多少代科学家长期的实践形成和发展的,这在老一辈科学家身上表现得淋漓尽致。一些故事常在我的脑海中回想,激励着我。

居里夫人冒着放射危害的危险,发现了镭;"两弹一星"科学家邓稼先,在核试验过程中为了得到数据,冒着生命危险到现场研究调查;法国科学家巴斯德是微生物学的奠基人,狂犬病疫苗就是他研制的,疫苗研制出来以后,是他本人第一个去试这个疫苗,这就是献身精神。

我国古代一些科学家也是如此,地理学家徐霞客,足迹北至燕赵,南至人迹罕至的云贵两广。云贵地势陡峭,甚至没有交通,徐霞客彼时的艰辛可想而知;药物学家李时珍编撰的《本草纲目》里有500多种药物,大部分是他冒着生命健康的危险,尝百草总结出来,给后世留下了一份宝贵的遗产。

这些人不为名利所动,不计个人的得失,为追求真理甚至愿以生命为代价,这叫献身。"追求真理,献身科学"的科学家精神是在长期的实践里形成的,我们要搞科学,是当成事业,不是为了获得名利的手段。

二、科学家精神有丰富内涵

一是无私报国的爱国主义精神,服务人类的国际主义精神。科学没有国界,但科学家是有家国情怀的,老一辈科学家都是把救国救民当作神圣的己任。

"两弹一星"元勋钱学森、朱光亚、邓稼先等科学家,在国外已经有很高的职位了,为什么回国来?因为他们心怀爱国主义精神,响应了建设强大新中国的号召。

吴运铎是"兵器工业之父",抗日战争、解放战争时期,条件非常简陋,搞兵工要做试验、要做爆炸,把眼睛炸瞎了、手臂炸残了,身上无数弹片伤痕。这样的科学家,把一切献给了党、献给了国家。

加拿大共产党员白求恩先是到西班牙反对佛朗哥独裁政权,支持西班牙人民的反独裁战争,后来到了中国,献出了生命。共产党的宗旨是要为人类的解放而奋斗,服务人类。

二是严谨治学、追求真理的奉献精神。巴甫洛夫是国际上有名的生理学家,关于严谨治学他有一段话对我启发很大,"决不要企图掩饰自己知识上的缺陷,哪怕是用最大胆的猜度和假设作为借口来掩饰,不管这种肥皂泡的美丽色彩怎样使你们炫目,但肥皂泡是不免要破裂的,那时你们除了羞惭之外,一无所得。"我们治学一定要反对一知半解,反对半瓶子醋,夸夸其谈,这就是严谨治学的要求。实事求是难,难在对自己实事求是,特别是对自己的学问。治学一定要"自以为非",找出自己的缺点、找出自己的问题,才能不断地进步、不断地取得新的成就。

巴甫洛夫关于严谨治学还有一句名言,他对学生的要求是"循序渐进,循序渐进,循序渐进",讲了三遍。治学不能搞跳跃式,要一步一步来,我自己感觉我的中学学习、大学学习比较扎实,扎实的一个关键是不要跳过"拦路虎",学东西不能有一个问题就不管了,绕过去,那是不行的,这就不是严谨治学。自己写的文章,不能把问题隐含在里面,问题要暴露出来,大家讨论,不要故意隐瞒自己研究中的问题,这是自

本文整理自钱七虎在中国岩石力学与工程学会"弘扬科学家精神、加强作风和学风建设"主题宣讲会上的演讲。

己骗自己,骗读者。

三是淡泊名利、潜心研究的奉献精神。爱因斯坦不止一次讲到,"我从来不把安逸和快乐看作是生活目的的本身,这种伦理基础,我叫它猪栏的理想。"他认为猪的追求就是吃好睡好,如果人也是追求舒适和享乐,和猪没有两样,那是猪的理想。他还讲过,"人们所努力追求庸俗的目标——财产、虚荣、奢侈的生活,我总觉得都是可鄙的。"所以应当有正确的财富观,真正的科学家应当是淡泊名利的。

在为武汉捐款650万元以前,我捐了800万元,以前多少年的奖金加起来几十万、上百万元都捐过。钱干什么用?钱是为了生活。生活是为了什么?为了工作、为了贡献、为了做事业,现在生活足够了,国家给你这么多待遇,你要这么多财富干什么?

大家看看《爱迪生传》,书上说,"一个人的价值,应当看他贡献什么,而不应当看他取得了什么。人只有献身于社会,才能找出那短暂而有风险的生命意义。"居里夫人讲过,"人类也需要梦想者,这种人最醉心于一种事业的大公无私的发展,因而不能注意自身的物质利益。"更谈不上斤斤计较个人的事,所以这就是科学家。

四是勇攀高峰、敢为人先的创新精神。什么叫创新?创新就是前面没有人做过的。所以,创新没有捷径,不可能短平快。马克思讲得很好,"科学上没有平坦的道路可走,只有不畏艰险,攀登陡峭山路的人,才能达到光辉的顶点"。我们要搞创新,要想取得大的成果,就要勇攀高峰,敢为人先,不怕困难。因为是创新,所以可能有失败,可能有挫折。要不怕挫折,调查研究,找出问题所在,找出原因所在,克服困难,下次就成功了。

五是集智攻关、团结协作的协同精神。科学技术是集体的事业,先人讲过,"一朵鲜花是打扮不出美丽春天的"。也有科学家讲过,"脱离集体就是末路的开始"。一个团队如果有内讧,争排序,争名位,四分五裂,这个集体肯定没希望。

怎么能够团结协作?华罗庚讲过,"人家帮我,永志不忘;我帮人家,莫记心上。"就是要感恩,感人家的恩。一个集体里面怎么能够把气氛搞得好?就要求领头人、骨干将有名有利的事往后靠,吃苦耐劳的事往前赶,这个集体肯定好。所以,人际关系不好的同志,要想一想自己在什么地方做错了,是不是树立了团结协作的协同精神。这也是科学家精神的一个内涵。

六是甘为人梯、奖掖后学的育人精神。生命是有限的,事业是永恒的,真理是一条长河,无数真理的集合才形成了一个真理。现在,我们要建设科技强国,不是靠一个人、一代人,而是靠几代人的共同努力。所以,科学家一定要有这样的精神才行。

三、如何弘扬科学家精神?

在当前形势下,我对青年科技工作者弘扬科学家精神提一些要求,重点讲作风。

第一,弘扬科学家精神,要与党和国家的建设科技创新型国家、建设科技强国的伟大战略紧密结合起来。

中共中央、国务院《国家创新驱动发展战略纲要》提出了建设科技创新型国家"三步走"的战略——到2020年要进入科技创新型国家的行列(这个目标已经达到),到2030年要进入科技创新型国家的前列,到2050年,两个一百年的时候,要建成科技强国。

"三步走"非常重要,科学家精神对于建设科技强国非常重要、非常迫切。现在一些文章都在讲,疫情以后世界格局要有大变局、大变化,全球化、地缘政治都要变。大国间的竞争,经济战、外交战,最后是科技战。我们建设科技强国、建设科技创新型国家已经到了非常迫切的领跑型的拐点,科技创新是世界大变局的关键因素、关键力量,要认识到这一点。

我国建设科技创新型国家有三个独特优势。

一是中国制度。我们是真正的社会主义制度,集中力量办大事。

二是中国精神。中国的精神包含科学家精神,这是中国优秀的传统——家国情怀。中国精神在当前核心,就是科学家的精神。

三是中国效率。中国效率归结为我们有马列主义世界观和方法论。中国效率为什么这么高,因为我们懂得区分主要矛盾、次要矛盾,抓主要矛盾、矛盾的主要方面。

这三个因素不是相互独立,而是相互交织的,但是离开了精神,都没有了,缺一不可。所以,我感觉到,当前弘扬科学家精神要和建设科技强国结合起来,因为建设科技强国迫切地需要科学家精神。

第二,弘扬科学家精神,就要努力树立自己的远大理想,树立社会主义价值观、人生观和世界观。

说到底,科学家精神是党性在科学家身上的体现和反映,是革命人生观、马克思世界观在科学家身上的反映和体现。为什么有白求恩精神?因为白求恩是共产党员,他的党性就是为追求全人类的解放,所以他到中国来。爱因斯坦的价值观、世界观是人生价值不在于获取什么,而是要付出什么。他认为这样的人生是有价值的,这就是我们的社会主义价值观。

所以,弘扬科学家精神就要打牢科学家精神的思想根基和道德根基,要树立自己远大理想,树立自己正确的价值观、世界观。钱学森讲过,为什么他把自己的学术成就归结到马克思主义对他的影响,是因为他非常信仰马克思主义。他认为,马克思主义的世界观、方法论对他的学术有很大的影响。

第三,要和背离科学家精神、违反科研道德规范、突破科研诚信底线的行为做坚决的斗争。《Nature》刊登过一篇评论文章,中国的专利数量已经登上了世界第一,但是专利里有很多垃圾专利,为专利而专利。论文里名堂百出,剽窃、伪造数据,还有伪造同行的评审,名堂很多,同行评审是假的评审。有一组调查数据显示,截至2017年,中国贡献的科研文章数量占全球的8.2%,但撤销的论文数量却占到了全球的24.2%,几乎是3倍,这值得引起我们高度警惕。在当前弘扬科学家精神,一定要坚决和作风、学风建设败坏行为做坚决的斗争,不然长此以往,我们不但建不成科技强国,而且在世界上会落个骂名、臭名。

很多科学家的故事,离开了自己的勤奋和艰苦的奋斗,弄虚作假,在治学上掩饰自己的缺陷,肥皂泡终究是要破裂的。年轻人一定要吸取教训,假的终究是假的,科学是老老实实的学问,最后身败名裂,害了国家,也害了自己。

如何培育自己的科学家精神呢?根据我的体会,就是要学习学习再学习。学习什么?最重要的要学习伟人,这个"伟人"指英雄烈士,指老一辈革命家、老一辈科学家,这样才能帮助树立自己的远大理想,树立自己的革命人生观。是共产党员就会永葆自己的革命青春。伟人的一生是我终生学习的榜样,伟人闪耀思想光芒的语言时刻铭记在心,就能指引自己不断地前行,伟人的光辉行为永远激励自己不忘初心,是我前进不竭的动力。

在《钢铁是怎样炼成的》中保尔·柯察金有一句话,"人的一生,应当这样度过:当他回首往事时,不因虚度年华而悔恨,也不因碌碌无为而羞耻。""这样才能说,把伟大的生命献给了伟大的共产主义事业。"我想以此与大家共勉,不断学习、不断警惕、不断激励自己,才能培育好自己的科学家精神。

淡泊名利是科学家精神的重要内核

"爱党信党跟党走,是我一生中最正确、最坚定的选择。"获得"全国优秀共产党员"称号的中国工程院院士、国家最高科学技术奖获得者钱七虎已经在很多场合说过这句话。7月13日,在中国工程院举办的首场"学习习近平总书记'七一'重要讲话精神"报告会上,钱七虎再次向听众分享了自己的人生经历和政治感悟。

在钱七虎报告结束后,按流程由中国工程院院长李晓红继续作主旨演讲。李晓红首先讲了两个故事——两件钱七虎没有提及的事情。

2005年左右,时任重庆大学校长的李晓红决心提升该校土木工程学科的教研水平,准备聘请钱七虎为本校教授,聘期四年。"这是实打实的聘任,要谈很多指标。"李晓红解释说,"这些指标就是学科排名、人才引进和科研奖项。"我们当时开出了大概10万元的待遇,实话说不多,同时工作要求是很高的,要在几年之内完成很多指标。我们软磨硬泡才谈了下来,钱院士也提出了新的条件。

但是,钱七虎提出的新条件,并不是为自己提高待遇,而是要求重庆大学为土木工程学院追加至少一千万元的经费投入。

"作为学术大师,为了实现我们的小小协议,他兢兢业业,最终完成了我们提出的所有目标,专业排名也提升到了全国第九。"李晓红说。

"而他自己得到的钱,事后都捐出去了。"讲到这里,报告厅里响起了听众的掌声。

第二件事,是钱七虎捐赠所获奖金的幕后故事。

钱七虎淡泊名利,曾多次捐出自己的奖金。其中,钱七虎在2018年捐赠800万元国家最高科学技术奖奖金一事流传较广,也多次被媒体报道提及。

在报告会上,李晓红披露了一些细节。

原来,国家最高科学技术奖奖金发放的目的,是资助科学家的科研项目和家庭、个人生活,一般不用于捐赠。相关部门向李晓红说明了情况,李晓红又告知了钱七虎,但钱七虎并没有因此顺水推舟,撤销慈善计划,而是设立了基金,继续支持贫困学生上学。虽然形式做了调整,但是慈善的实质没有变,奉献的初心更是贯穿始终。

"他一生淡泊名利。"李晓红说。

在讲完这两个故事之后,李晓红以《立志千秋伟业 迈向伟大复兴》为题作了主旨演讲。在演讲当中,李晓红指出,科学家精神就是党性在科学家群体身上的体现。

报告会结束之后,钱七虎和科技日报记者谈起自己捐赠的故事,表示奉献比起享受对于自己来说更有价值。"我的孩子都是看到报道才知道我把奖金捐了出去。"钱七虎笑着说。

本文原载于《科技日报》(2021年7月15日)。

为武汉捐赠650万元,钱七虎:
烈士献出了生命,我有什么不能贡献?

"我是一个党员科学家,要牢记我党为人民服务的宗旨,我们攻坚克难搞科研,不是为了个人的名利,我们的科技创新要服从于人民、也就是服从于国家的需求。"

中国工程院今天(7月13日)举办"学习习近平总书记'七一'重要讲话精神"报告会,获得"全国优秀共产党员"称号的中国工程院院士、国家最高科学技术奖获得者钱七虎作报告,讲述了六十余年的科研经历。

今年84岁的钱七虎满头花白,但目光炯炯。"我从事的是防护工程专业,我们的科技工作就是为国家铸造坚不可摧的地下钢铁长城,为人民构筑绝对可靠的安全屏障。"

谈及为武汉抗击新冠肺炎疫情捐款650万元一事,他说钱对自己意义不大。"我如果追求奢侈的生活,对我这个共产党员来讲是可耻的。"

曾经在日本铁蹄下度过童年,坚定跟党走信念

"爱党信党跟党走,是我一生中最正确、最坚定的选择。"钱七虎回忆说,1937年10月,他出生在国歌中所唱的"中华民族最危险的时候",母亲在逃难途中生下了他。他的童年中有八年是在日本铁蹄下度过的,"日本鬼子将杀害的游击队员尸体放在小学操场上示众"的场景,他一直难忘。

新中国成立初期,毛主席发出了"一定要治好小虫病"的伟大号召,治好了他和江南水乡许多人的血吸虫病;他经历了志愿军在抗美援朝中的英勇作战迫使侵略者止步于三八线的历史。这一切使他深刻感到,没有党,就没有强大的祖国和人民的幸福生活。"这就是我信党跟党走,十四岁入团,十八岁入党的最本质的原因。"

不惧与黄土铁锹打交道,选择防护工程专业

如何跟党走?钱七虎说,就是在关键时刻听从党的召唤,服从党的分配。高中毕业时,钱七虎因学习成绩优秀,被计划选送到苏联留学。随后,成立不久的哈尔滨军事工程学院到他的母校上海中学招收优秀毕业生,组织上又谈话要报送他到"哈军工"。作为团支部书记,钱七虎毅然放弃留学机会进了哈军工。"当时选工程兵防护工程专业的人很少,大家觉得要跟黄土铁锹打交道太土。我是班长,所以带头服从组织分配。"

毕业后,组织选派钱七虎到苏联莫斯科军事工程学院读研究生。回国面临分配时,工程兵政治部先通知他到位于北京的工程兵科研设计院报到,休完假后,领导又找他谈话,说新成立的西安工程兵工程学院急需教员,迫切希望他去。

"当时我准备结婚,对象在北京工作,但我俩都是共产党员,我二话不说,马上先去西安报到,两个月后趁来北京参加首届全国防护工程学术会议发表论文的会议间隙结了婚。"

为祖国铸"地下钢铁长城"

为祖国铸就坚不可摧的"地下钢铁长城",是钱七虎毕生的追求。

本文原载于《新京报》(2021年7月14日)。

20世纪70年代初,钱七虎受命进行某地下飞机洞库门设计。为了获得准确的试验数据,他赶赴核爆试验现场进行实地调查研究和收集数据。在现场发现,虽然核爆后的飞机洞库门没有被炸毁,飞机也没有受损,但是防护门出现严重变形,致使无法开启。

当时飞机洞库门设计采用简单手算的公式,计算精度差,而计算精确变形需要应用先进的理论在大型电子计算机上计算。与此同时,世界上刚兴起有限元计算理论,钱七虎没有学过有限元理论、计算机语言,没有编写过大型程序,他只能加班加点学习,终于编制出大型计算程序。当时中国只有第七机械工业部第五研究院(七机部五院)、中科院计算机所才有大型晶体管计算机,他利用别人不上机的午饭和下班时间上机,时间久了得了十二指肠溃疡,又诱发了痔疮。

"这些困难我都克服了,坚持了下来。"最终,钱七虎设计出了当时跨度最大能抵抗核爆炸冲击波的抗力最高的机库大门,还出版了《有限元原理在工程结构计算中的应用》,上述成果获得了1979年全国科学大会重大科技成果奖。

20世纪六七十年代,我国面临严峻的核环境,他和团队时刻跟踪着新型进攻武器的发展。"只要是敌人核武器这个'矛'发展一步,就琢磨和研究让我们的'盾'如何更坚固一层。"从核空爆到核触地爆,再到核钻地爆;从普通爆炸弹到钻地弹,经过长达几十年的研究,他和团队攻克了一个个难关、突破了一系列技术难题,为我国战略工程装上了"金钟罩"。

关注民生工程,率先提出开发地下城市空间

钱七虎说,一个有担当的科学家,既要关注如何增强国防实力,也要关心国之重器的重大民生工程。

1992年初,珠海特区建设机场,要炸平一座山,爆炸总量1085万立方米,要求一次性爆炸成功,数万发雷管不能有哑爆,一半的土石方定向投入大海,一半的土石方必须松动破碎,而且要确保邻近1000米内的两处村庄安全。这样大的爆炸,世界还无先例,难度很大。钱七虎带领团队六下珠海,和大家一起研究设计方案和施工方案,于当年12月成功实施爆破,创造了世界爆破史上的新纪录。

自20世纪90年代末起,为预防和治理交通拥堵、空气污染、城市内涝等"城市病",他利用研究地下工程占有大量国内外学术资料的优势,率先提出开发利用地下城市空间、发展城市地下快速路、地下物流等创新观点,先后组织编制、主持和评审了全国20多个重点设防城市地下空间规划。

近些年,钱七虎在长江隧道、南水北调工程、西气东输工程、港珠澳大桥海底隧道、能源地下储备、核废物深地质处置等方面,贡献了学识和才智。

对话钱七虎:重大工程需要我,我就会去

新京报:建党百年之际获得"全国优秀共产党员"称号,您有什么样的感受?

钱七虎:我觉得党给我的荣誉太多了:国家最高科学技术奖、全国道德模范、全国优秀党员,我必须活到老、学到老、革命到老。我要给党交一个合格的答卷,无愧于这个称号。

新京报:2020年新冠肺炎疫情暴发后,你为武汉抗疫一线捐赠了650万元。当时是处于什么考虑?家人支持你吗?你在生活中是不是也很简朴?

钱七虎:烈士把头颅鲜血先给国家了,我还有什么不能贡献呢!

2018年,我获得国家最高科学技术奖,为了响应党的脱贫攻坚的号召,我把800万元奖金全部捐出,资助我国西部的贫困学生。2020年新冠肺炎疫情暴发后,我把江苏省配套奖励的800万元中的650万元捐给了武汉抗疫一线,其余的150万元分别捐给了母校上海中学和中国岩石力学与工程学会。

我们的生活比普通老百姓高了很多了,钱对我意义不大。我如果追求奢侈的生活,对我这个共产党员来讲是可耻的。

我们要把助人为乐作为人生准则。我成立瑾晖基金,"瑾晖"二字是我母亲和爱人名字中各取一个字,我做捐赠我爱人是支持的,她也是党员。基金每年资助家乡贫困学生两万元,孤寡老人一万元,我们

还救助过患艾滋病的孤儿。

新京报：您已经84岁,称自己还有很多事情要做、想做、能做,指的是什么？您工作的节奏是什么样的？

钱七虎：具有65年党龄的科学家,我有责任继续践行科学家精神,让余生继续发光发热。我虽然已经退休了,但继续担任军事科学院首席专家,火箭军首席工程专家。

这几天,我在团中央做报告,昨天参加了东六环工程盾构方案论证。重大工程特别是水下的大工程,肯定要叫我去,我基本上有空就去。这周四、周五,我还要去南通、成都等地出差,参加工程方案论证、作"在碳中和碳达峰目标下,地下空间怎么发挥作用"的报告。后面两个礼拜,我要把时间空出来治病。

新京报：地下空间建设,对于实现碳中和碳达峰有怎样的意义？中小城市和大城市地下空间发展有侧重吗？

钱七虎：简单来说,地面建筑搬到地下,地面就可以大量地植树造林,吸收二氧化碳。另外,地下空间还能为储能提供空间。

大城市人口集中,要先把交通转入地下空间,通过地铁、地下物流等减少碳排放。

(记者：张 璐)

老党员履新职

"涓涓不塞,是为江河;源源不断,是为奋斗;生生不息,是为中国。"6月25日,白发苍苍的钱七虎院士表达了自己将继续为党奋斗的决心:"我一定继续奋斗,发挥65年党龄的老党员、老科技工作者的积极作用。"

钱七虎院士曾获国家最高科学技术奖,他是在中国科学技术协会学会党建工作指导委员会成立大会暨学会党建工作先进表彰大会上吐露了这番心声。

中国科学技术协会是由同一学科的研究者组成的学术团体,目前所属的全国学会有200余个。在喜迎建党百年华诞之际,中国科学技术协会党组决定组建学会党建工作指导委员会,进一步加强党对学会的全面领导,由钱七虎担任该指导委员会双主任之一。

希望"再开新局面"

"这既是对我信任,也是对我这名老党员的再一次考验。"钱七虎说:"我将尽心履职,推动学会党建工作再开新局面。"在钱七虎看来,吸纳科学家作为委员,这是对科学家的激励和鼓舞,符合科协组织的特点,能够充分发挥科学家的示范引领作用,加强对学会党建重大共性问题和科技界重大事件的战略研究。

"大家都在学习弘扬'两弹一星'精神,两弹元勋绝大部分像钱学森、朱光亚、邓稼先等都是在国外已经有很高的职位、待遇,学术上更是卓有成就,他们响应党的号召,为建设我们强大的国家埋名隐姓、艰苦奋斗,弘扬的就是这么一个无私报国的爱国主义精神。"钱七虎说:"说到底,科学家精神就是党性在科学家身上的体现和反映,是革命人生观、马克思主义世界观在科学家身上的反映和体现。"钱七虎表示,当前,新一轮科技革命和产业变革突飞猛进,形势逼人、挑战逼人,应该切实加强学会党建,引导广大科技工作者继承和弘扬这种科学家精神。

再履职有更多期待

钱七虎曾担任过中国岩石力学与工程学会理事长,尽管2016年他已卸任,但对学会工作仍予以关注和参与。据他介绍,该学会以党建为引领,曾为三峡工程、溪洛渡、向家坝、乌东德、白鹤滩等国家大型水电工程建设提供咨询服务,并且曾为川藏铁路建设难点和军民融合提建议、破难题,还积极开展"一带一路"国家科技交流。

此次以老党员的身份再次履职,开展全国学会党建工作,钱七虎表示,将增强广大科技工作者在党的领导下建成世界科技强国的信心和决心,引领广大科技工作者报效祖国,把人生理想融入为实现中华民族伟大复兴中国梦的奋斗中。

"科技工作者要始终自觉从党和国家工作大局着眼,把论文写在祖国大地上,助力实现高水平科技自立自强。"他还希望广大科技工作者认真学习党史,从党的百年奋斗历程中汲取奋进力量,弘扬科学家精神,坚定不移听党话、跟党走。积极参与全球科技治理,全面提升开放创新能力,推动构建人类命运共同体。

本文原载于《光明日报》(2021年6月26日)。

将直面问题迎难而上

谈到当前的形势,钱七虎表示,当前国际环境错综复杂,新一轮科技革命和产业变革突飞猛进,凸显了科技自立自强的极端重要性和空前的紧迫性。"科技工作者面向世界科技前沿、面向经济主战场、面向国家重大需求、面向人民生命健康的责任更重,压力更大。"

钱七虎认为,面临这种形势,更要加强党建,激励新一代科技工作者发扬勇攀高峰、敢为人先的创新精神,从国家急迫需要和长远需求出发,树立敢为天下先的雄心壮志。

"老一辈科学家都是把救国救民当作自己神圣的责任,在党的领导下科技事业取得了历史性成就。"钱七虎说,当下更要"直面问题,迎难而上,敢于探索科学'无人区',勇于挑战最前沿的科学问题,力争成为领跑者、开拓者。"

(记者:詹 媛)

中国智慧建造必将走在世界前列

"今年新冠肺炎疫情防控和复工复产的速度,中国是全球最快的。我相信,未来智慧建造和智慧工程的发展,我们也会是全球最快的。"说这句话时,中国工程院院士、83岁的钱七虎语气坚定。

本届服贸会上,本报记者专访了钱七虎院士。这位为防护工程事业奉献了整整60年的专家表示,短短四十年,工程建设便完成了从人工向机械化的转变。未来,随着大数据、物联网和5G技术的发展,工程建设必将向智慧化、少人化乃至无人化的方向发展。

智慧建造:让工程有了"眼睛和耳朵"

"你知道吗?有研究报告表明,如果全球气温升高5℃,则有六成鱼类不能生存;但如果实现巴黎协定规定的气温升高1.5℃的话,这个数字将降低到一成。我们必须赶快行动起来!"谈起发展智慧建造的必要性,钱七虎先说起了全球气候变化的严峻形势。

他讲道,控制气温上升的关键是控制碳排放,特别是城市的碳排放。这其中,建筑恰巧是全球能源和原材料的消耗大户,因而也是碳排放的大户。目前,全球建筑运营能耗已占到总能耗的30%以上,若再加上建设过程中的能耗,这一指标接近50%。此外,城市建筑垃圾占到垃圾总量的30%~40%,与建筑有关的空气污染、光污染等占到综合环境总体污染的34%。"这一个个数字都表明,绿色发展呼唤绿色制造,而智慧建造又是绿色建造的必然选择和最佳途径。"

钱七虎解释道,就像人的五官具备感知功能一样,智慧建造让工程也有了"眼睛和耳朵"。"大数据、物联网和5G技术的发展,为智慧建造提供了技术支撑。只要我们能坚持不懈地运用这些技术,让智慧发展的理念深入人心,中国的智慧建造定会走在全球前列。"

未来城市建设:基础设施要转到地下去

"一座低碳高效的绿色城市;一座蓝绿交织的森林城市;一座自然生态的海绵城市。"《北京城市副中心控制性详细规划》中,描绘了千年之城的未来面貌。《河北雄安新区规划纲要》也提出,未来雄安新区的蓝绿空间占比要大于70%。

如何告别"混凝土的森林",真正在北京城市副中心和雄安新区的建设中实现蓝绿交织?钱七虎提出:充分开发利用地下空间,特别是大力发展以地铁为骨架的轨道交通系统和城区地下物流系统为特征的绿色城市基础设施。

"要将所有能转入地下的建筑都转到地下去。"钱七虎告诉记者,不但要让市民交通和货物运输转入地下,一切脏、乱、差的建筑和设施都可以转入地下,比如污水处理设施,从而节省土地资源,释放出大批地上空间用作大片的绿色植被和生态空间,建设绿色城市。这才是没有"城市病"的未来城市。

建筑工地:必将向少人化、无人化发展

从手工打眼、钻孔到使用数字化台车钻孔,从人工放线到激光扫描测量,从手工喷涂到自动化喷涂,过去四十年来建筑工地的发展变化,在钱七虎眼中可以用日新月异、翻天覆地两个词来形容。"如今,中

本文原载于《北京晚报》(2020年9月7日)。

国的机械化水平已是国际领先。"他举例,以"国之重器"盾构机来说,中国不仅自主研发出了直径15米的超大直径盾构机,还将设计更大直径的18.2米盾构机。

同时,随着建筑信息模型(BIM)技术的发展和应用,工地的数字化水平也越来越高。过去,设计人员设计的工程图纸和工程实体是分离的,设计图上的管线如果完全照搬到施工中,往往就会出现交叉、碰撞等许多问题。现在,有了BIM技术后,设计人员可以在计算机里模拟出立体化、可视化的工程模型,设计和施工不再是"两张皮"。

"目前,体量最大的工地在中国,规模最大的工地在中国,地质环境最复杂的工地也在中国。"在钱七虎的畅想中,随着信息技术的深入应用,建筑工地必将更加安全、高质和高效,并在未来走向少人化甚至无人化。

(记者:赵莹莹)

和平年代要树立忧患意识
科学家更要有责任担当

今年是中国人民抗日战争暨世界反法西斯战争胜利 75 周年，铭记历史，方能开创未来。"科技兴则民族兴，科技强则国家强。"习近平总书记多次强调科技对民族进步、国家发展的重要作用。

钱七虎院士是我国著名的防护工程学家，现代防护工程理论的奠基人、防护工程学科的创立者、防护工程科技创新的引领者，为我国防护工程各个时期的建设发展做出了杰出贡献。他倾尽心血为国铸就"地下钢铁长城"，将科技强军作为毕生事业。近日，钱七虎院士在接受人民网和科技部人才中心采访时表示，一定要在和平环境中树立忧患意识，科学家要有为中国建设科技创新型国家的责任感和担当精神。

"和平年代，一定要树立忧患意识"

人民网：今年是中国人民抗日战争暨世界反法西斯战争胜利 75 周年，现在虽然处于和平年代，但是世界格局仍然变化万千，您如何看待当前我国所处的国际环境？

钱七虎：首先，永远不要忘记和平环境是怎么来的。和平年代，一定要树立忧患意识。十四年抗战告诉我们，一个国家没有强大的国防，没有强大的军队，就要受人欺负。正是有强大的人民解放军保卫着我们国防，我们才保持着和平的发展环境。在"后疫情时代"，世界战略格局会重新调整。在这个变化过程中关键之一是科技创新，而科技创新的关键因素是科学家、人才。

人民网：在这种国际环境中，我们的科研工作者应当具备怎样的素质？

钱七虎：在这样的环境下，我们的科学家应当有政治敏感度，不仅要钻研科学技术，还要树立国家民族的认同感，树立为国家、为民族复兴而从事科技创新事业的责任感，这是首要的。身为科学家，在和平年代也同样要明确自己的责任担当，为中国建设科技创新型国家作出贡献。

"想要活出价值，就要干出一番对人类有贡献的事业"

人民网：您六十余载一直在国内从事防护工程研究和人才培养工作，对我国防护工程各个时期的建设发展做出了突出贡献。当初为何选择这一领域？

钱七虎：我是 1937 年出生的，那真是"中华民族到了最危险的时候"。我深深明白，国家弱、军队弱，人民就要遭难，国家就要沦亡。十七岁时我可以留学苏联成为一名水电站建设的工程师，但是我却接受了另一个选择——去中国人民解放军军事工程学院就读工程兵工程系，从此我和防护工程结下了不解之缘。回想起来，十七八岁时的选择，是正确的、无悔的选择。当站在人生的岔路口，面临多个选择的时候，要秉持一个原则——把个人的命运和国家事业的需要、人民的需要相结合。

人民网：在您看来是什么支撑着您坚持到现在？

钱七虎：居里夫人曾经讲过，一个人要活得有价值，就要能够干出一番对人类有贡献的事业。爱因斯

本文原载于《人民网——时政频道》（2020 年 7 月 27 日）。

坦讲过,每个人都有一定的理想,这种理想决定着他的努力和判断的方向,不应将安逸和快乐看作是生活目的本身。像爱因斯坦和居里夫人这样的科学家,不仅科学成就很令我崇拜,他们很多言论对我产生很大的教育意义,对我的人生观产生重要影响。

"青年人要为人类、为国家、为民族、为人民贡献自己的才能"

人民网:想要做好科学研究工作,您认为需要哪些品质?

钱七虎:机遇永远只是给那些有准备的人,我们要做好迎接成功的准备。所谓的准备,包括远大理想,不怕困难,不怕挫折的品质,以及不断锤炼的才能。

在这方面我有很多的体会。在科研工作中,有时我的身体状况并不是很好,但是都坚持下来了。另外,一些科研项目需要借用其他单位的大型计算机进行运算,只能在中午吃饭时、晚上睡觉时才有空闲时间供我们使用,条件是非常艰苦的。诸如此类的考验还有很多,只有克服困难,才能创造出成绩。

人民网:您对青年科研工作者有哪些期望?

钱七虎:首先,要树立远大的理想。现在有些年轻人毕业后首先想到的是待遇怎么样,说明缺少一个远大的理想。首先要想到的是把自己的才能贡献出来,为人类、为国家、为民族、为人民作贡献,这个要放在第一位。

第二,要学习伟人,学习老一辈无产阶级革命家,老一辈的科学家,通过他们的事迹激励自己,要铭记他们,这样才能继承和发扬无私奉献的科学报国的传统和精神。

第三,要勤奋学习,勤奋工作。我相信马克思所说的,天才就是勤奋。只有勤奋学习,才能积累自己的才华与能力;只有勤奋工作,才能创造一番成就,这是我想对青年人说的话。

(记者:王欲然　宋子节　钟　焯)

| 吾家吾国

一生一事,为国为民

——采访1954届校友钱七虎院士

钱七虎院士在我国防护工程领域辛勤工作60年,创建了我国防护工程学科,建成了国家重点学科、重点实验室和创新研究群体。他建立了从浅埋工程到深埋工程防护,从单体工程到工程体系防护,从常规抗力到超高抗力防护等理论与技术体系;解决了空中核爆、触地核爆、钻地核爆以及高技术常规武器侵彻爆炸等一系列工程防护技术难题。他提出的防护工程建设转型、建设超高抗力深地下防护工程、战略通道桥隧并举、能源地下储备等多项发展战略建议,被采纳实施。

钱七虎院士是我国现代防护工程理论的奠基人、防护工程学科的创立者、防护工程科技创新的引领者,对我国防护工程各个时期的建设发展做出了突出贡献。

我初中是敬业中学,在老西门附近的文庙街,是较好的市立学校。但我们都知道"上中"(上海中学)是最好的。所以当时初中毕业以后,我就报了两个学校,一个是"上中",另一个是国立高机(上海机械专科学校),这两所院校都是上海最好的、最难考的。两个我都考了,都录取了。"上中"先发榜录取了,所以我就进了"上中"。

追忆上中:似水年华

我上中学一共只有三年,但是我和我的上中同学、校友都感觉这三年在我们的人生经历里是影响很大,可以值得回忆的事情很多很多。现在很多年过去了——我是1951年到1954年,1954年离开——到现在已经是60多年了,但老师上课的一些形象现在还很亲切,讲的东西现在印象还是很深的。比如教我们立体几何的(老师)叫余元庆,他是很有名的,我牢记他课上讲的立体几何,讲三点决定一个平面,什么叫决定?可以且仅可以做一个平面,叫作决定,讲得十分形象,我们都记得很清楚。

数理化(数学、物理、化学)本来是很枯燥的一些学科,但是我们听"上中"老师的课感到享受,很有趣、很生动。比如我们生物老师叫褚祈,给我们讲生物的用进废退论,为什么长颈鹿的脖子很长,为什么猴子变人没有尾巴,讲得生动,非常有趣;我们的大代数老师是唐秀英,也是很有名的老师,她讲的大代数,我们都感觉到很享受;还有教物理的老师会举马拉车、车拉马的例子来讲作用力与反作用力。

可以说上海中学拥有的不是一个、几个很好的老师,而是一个优秀的老师群,这些老师大都是有留学经历的,教学已经达到了相当高的水平了,所以能讲得引人入胜。上海中学老师的教学,使得我们在科学的道路上走出了第一步,打好了扎实的基础,而且培养了对于数理化这些基础功课的热爱。

另外上海中学不是死读书的学校,在培养我们高尚的人生目标,培养我们优秀品质方面,也有很多值得回忆的地方。比如我们在上海中学的时候,学校请来了很多外面的专家来做报告,为我们培养树立高

本文采访时间:2020年7月23日。
口述:钱七虎
采访:杨嵘、常宇鑫、曹翔宇、陈云上、王一舟、邬旭恺、谢东豫、张文睿、方源、陈星皓
整理:方源、曹翔宇、陈云上、王一舟、邬旭恺、谢东豫、张文睿

尚的理想等等。

我举几个例子：一个是卓娅和舒拉的母亲，你们知道不知道他们是谁？卓娅，苏联卫国战争女英雄，游击队员，被德国人杀害。卓娅是那个时候人人皆知的苏联卫国战争的一个女英雄，就像刘胡兰一样。卓娅的弟弟叫舒拉，也是参加卫国战争牺牲了。他的母亲到我们上中做报告，讲她的女儿和儿子的英勇的行为，为国献身的壮举和他们从小的优秀品质，给我们很大的教育。使我们树立了志向，从青少年时代，要向英雄学习。还有一个报告，学校把李白烈士（《永不消逝的电波》中的人物原型）的夫人裘慧英请来，讲李白的事迹，他怎么在敌人的眼下给延安电台英勇发报、不怕牺牲，直到被别人逮捕的时候，他还是继续发报。另外一个报告人是黄宗英（电影演员），她是赵丹的夫人，黄宗英给我们做访苏代表团的故事，她介绍了苏联的很多情况。后来我到苏联学习，证明她讲的都是事实，全都是社会主义国家的特征，全民的教育，物价便宜，全民达到温饱，全民有休假等等。她讲了集体农庄，讲到苏联的今天就是祖国的明天，所以我们都向往要努力建设社会主义。

我之所以能取得一些成就，之所以能养成一些优秀品质，是因为上海中学时候的教育确实起了很大的影响。我一直让一些历史上的先烈英雄的形象活在心中，用他们的言行指导自己的行动。

上海中学也会举办一些社会活动——所以我讲不是"死读书"。在高中时我学习成绩很好，但是我并不只是读书的。我记得参加的两个活动很有意思：一个是农民夜校，我们高中学生晚上给农民上课，自愿的，我就参加了这项活动；还有参加纠察队，保卫上中，进行晚上巡逻。这些活动我感觉都很好。另外上中重视动手能力，我们学习了解剖鲫鱼，还有生物课学习嫁接，化学课学习怎么做肥皂？……这些都深刻地影响了我。

上海中学团的活动非常活跃，团员很多。比如说，有一次举办了一个团日活动，我们就如何树立远大理想展开讨论，我们是1951年到1954年在上中学习的，1953年，我们国家第一个5年计划开始了，苏联援助156项大型的工厂、工业要建设起来，所以我们就谈我们怎么树立伟大理想，要为祖国的建设贡献自己的一生。大家有说是要为祖国造飞机，要为中国造出大水电站，要为祖国炼出钢铁，所有人都热血沸腾。等我们1954年高中毕业，我们高三5班一个班五十几个同学，毕业后，很少人报上海的学校，只有几个人家里有些特殊原因，报上海交大，其他有报航空学院的，也有清华大学的，当然还有留苏的。大家都是抱着要成为工程师的理想去考大学。这个"团日"活动，给我最深刻的印象就是我们年轻人要树立远大的理想，为祖国的建设，为国家，为人民作贡献。

我们在上海中学时读了一些优秀的课外读物，比如说《钢铁是怎样炼成的》，讲的是保尔·柯察金的故事，书中有句名言：一个人回顾一生的时候，不要为自己碌碌无为而羞耻，为虚度年华而悔恨。所以我们都立志于建设我们中国最伟大的，也是人类最伟大的事业——共产主义事业，而奋斗。这些内容，大家都背得很熟。而在团日活动中，我们也会交流阅读的好书，比如吴运铎的《把一切献给党》，还有萧三写的《毛泽东的青少年时代》，我们不是死读书，课外读了很多好的作品，对我们的成长很有帮助。

上海中学能回忆的美好的东西有很多。因为我是初三入团的，后来高一到上中来，团支部选我当宣传委员，我们的团委会书记叫许俊生，组织部长叫陈忠，他们晚上自习时就办公，各个团支部有什么工作上的问题去找他们商量，非常活跃。所以这个团日活动组织得很好，很有教育意义。

体育锻炼，上中也搞得很好，当时叫劳卫制，即劳动卫国制锻炼，我们每天下午上完课后就按规定锻炼，这些生活习惯是很好的。还有大扫除，我还记得我大扫除后写了一篇作文，后来我们的语文老师（钱云）把我写的作文里一段在班上念给大家听。她说我有一句话让她印象很深："大扫除以后校园露出了少女般美丽的脸庞。"她说："为什么念这一段？这就是拟人化，写文章，要有拟人化，不能很枯燥。"

投身科学工程：个人的理想与国家的需要

黄宗英讲："苏联的今天就是祖国的明天。"所以我对于留学苏联很向往。但是后来哈军工（中国人民解放军军事工程学院）开始招生。当时党中央要培养军事工程技术人才，哈军工每年除了招收部队一些人，在各地中学里用保送的办法招收一批优秀的学生。所以后来学校又告诉我，要送我到哈军工去，当

| 吾家吾国

时到哈军工先保送后考试,就是说指定你到哈军工,但是要经过考试合格才能正式进入哈军工。我们这代青年有一个光荣的传统:服从组织分配。党的需要,就是我个人的需要,党分配到哪里,我就到哪里去。我们是共青团员,所以我们要服从安排,好好学习。比如说刘少奇写的《论共产党员的修养》,那个时候,我记得有一条:要做党的驯服的工具,要做我们国家革命事业的螺丝钉。所以团委书记叫我到哈军工,我就高高兴兴去了。当我被录取的时候,我们上中还给我家里寄了一份通知书,上面告诉我的母亲说,钱七虎被录取到了哈军工,是您的光荣,也是学校的光荣。

哈军工是6年制,一年预科,5年本科。预科阶段,学俄语、学军事,还有培养军人的一些基本的素质,然后预科完了才报志愿,报志愿的时候我报了三个志愿,那个时候学校里一共有5个系,一系是空军系,二系是炮兵系,三系是海军工程系,四系是装甲兵工程系,第五系就是工程兵工程技术系,每个系里有好多专业。报志愿的时候,不能报具体的专业,只能报系,从五个系中报三个系。当时大家都愿意报空军、海军,进了能穿空军海军服装的大盖帽,而二系、五系呢,这个帽子都是软的。尤其是五系(工程兵工程系),人们印象就是和黄土、洋锹、十字镐打交道,所以报的人少。但是我是班长,一个班二十几个人,班长要带头,所以我在三个系里报了工程兵工程系,然后就被录取了。就这样,我开始学国防工程研究,服从分配。

上甘岭战役里面有个很重要的就是坑道工事。飞机大炮别人比我们强,但是我们怎么能取得胜利?就别人飞机大炮轰炸、炮击的时候,我们的战士、我们的部队躲在坑道里面(躲过轰炸炮击),所以坑道起了很大的作用。以前是土办法造坑道,后来要抗原子弹氢弹爆炸,我们这个专业就是做计算、设计、建造,就是干这个事。

在哈军工我也是学习很好的,5年全优嘛,每年都是优秀学员,并且每年都是社会主义积极分子。所以后来我毕业的时候又被送去苏联去读研究生了,在苏联也是读的军事工程学院,当时就是副博士研究生(副博,实际上是候补博士)。英美硕士两年,苏联的副博士是三年。在苏联学习的时候,我刻苦学习,很少外出,红场我是去过的,但瞻仰列宁墓要排队,我舍不得时间,所以我在苏联学习4年,没有瞻仰列宁墓。

回国以后,我先被通知分配在北京工程兵国防工程设计研究院,后来,工程兵政治部告诉我了,说是变化了。为什么?因为哈军工后来建分院了,我们工程兵工程系就改为西安工程兵工程学院,后来总政就调我到西安去了。当时我马上要结婚了,我65年8月回国,计划10月份结婚。在这种情况下,我还是很愉快地服从分配。我的爱人也是党员,也服从,没有给我阻力。所以我就到了西安工程兵工程学院。

所以我说,要把国家和人民放在第一位,这不是空话,要体现在行动上。个人的志愿和国家的要求需要根本上是一致的,但是有的时候是并不一致的。这个时候就看你是把什么放在第一位:把个人放在第一位,还是把国家放在第一位?我们以前都是把国家放在第一位,于是就愉快地去工作了。

"文化大革命"期间,我继续研究专业。当时我还有一个任务就是教越南的留学生,我就给他们上课深造,其他的时间我就用来钻研以前在苏联留学时候积累的一些问题,收集资料,做一些理论研究、做些计算。所以"文化大革命"结束后我在防护工程全国学术会议上一下子能拿出8篇论文,数量大,而且质量有水平,引起了全国的一些科学家的重视,比如说同济大学原来的老校长李国豪,还有比如说清华大学副校长力学家张维。他们看了我的论文,对我特别赏识,所以后来我的职称要评定,推荐我当教授都是这些人。张维还推荐我当院士。

所以我就成为中国工程院成立以时的第一届的工程院院士,那么当了院士以后,是国家的院士,不能光干你本职工作。毕竟中国工程院是工程技术方面的最高的咨询机构和荣誉机构。所以你当了院士,你也要想为国家作贡献,发挥作用。

我的专业叫防护工程,防护工程是干在地下的。要抵抗原子弹、氢弹的爆炸冲击,就肯定要在地下要通过岩石岩体的抗力的作用才能抗原子弹的破坏,我们学防护工程,一方面要学它怎么抗原子弹、氢弹、炮弹炸弹的爆炸冲击怎么计算,同时还要学地下工程的那一套理论。所以我还能并积极地参加国家的重大工程的建设,包括长江隧道,包括黄河的隧道,包括青岛黄岛间的胶州湾隧道,包括厦门的,还包括港珠澳海底隧道,还有西气东输的地下工程都要参加,作为咨询审查这些工程,担任它们的专家委员会主任、委员,所以在这方面也发挥了一些作用,做了一些贡献。后来我在评国家最高科学技术奖的时候,组织上把我的这些方面发挥的作用、贡献都汇报上去,作为成就的一个方面。

在2002年中央通知我到北戴河休假,一共有50位,其中25位是国防科技的,共有五位代表指定发言,规定提一个建议,我就提了一个我们需要建、能建长江隧道。后来,国家就提出了建设长江隧道,当时是两条,一条武汉长江隧道,一条南京长江隧道。刚开始他们设计单位提出的方案,叫沉管隧道。后来我提出了这个方案不行,为什么沉管隧道不行?因为沉管是一种浅埋法——这里讲复杂了,就挖一个沟,放个管节,一节一节的钢筋混凝土管子放在下面埋起来,通汽车——这样很浅。我参加了三峡水库的攻关项目的审查,知道长江三峡建了以后,泥沙含量少了,冲刷大于回淤,冲淤不平衡,所以这个方案不行。我建议要做隧道埋得深的方案,就是盾构法。后来他们采纳我的建议,采用全断面开挖隧道的盾构法,这个盾构法就是在前面一个刀盘上装了很多刀,滚刀、刮刀在转,把那石头压转碎了,然后通过去。转碎的时候,这个刀的就磨损很厉害,出现了刀盘转不动等很多问题,我就和大家一起想办法,克服了这些困难,最后就建成了。所以南京长江隧道最后建成以后,南京市委市政府就公布嘉奖了建设功臣,我是建设功臣的第一位。

"盾构先锋":家国情怀

在上中学的时候,我们学习爱因斯坦、居里夫人他们事迹、言行。爱因斯坦讲过,一个人如果追求个人的安逸和享乐,从伦理学来讲,这种理想属于"猪圈的理想",因为猪就是吃饱就好,那么追求个人名利本质上是与此一样的。居里夫人、爱迪生他们的传记都说,一个人活着,要为世界做贡献,为人类做贡献。

每一个时代,每个人都要有自己的责任和担当。我们以前老一辈革命家、烈士,他们为了中华民族的解放,反抗帝国主义、反动派的压迫贡献了自己生命。他们完成了那个时代的使命,那么现在我们这个时代的使命,就是中华民族的伟大复兴,复兴首先需要我们贡献自己的才智,也需要我们贡献我们自己的才力。

特别是我,我以前念书的时候,也得到了国家资助的助学金。现在我们国家还有很多地区发展不平衡,很多边远地区的山区的青年还因为贫困走不出大山。所以我们这一代的责任就要帮助中华民族复兴,要帮助那些大山里的青年成才,为国做贡献。

当时我获得了国家最高科学技术奖,奖金800万元,我把这800万元拿出来,要资助边远地区、贫困地区的学生。这个事情实际上我20个世纪就开始了。我自己工资不是有剩余吗?所以我那个时候就资助贵州、青海、北京郊区、山东的这些青年学生。我的老家淀山湖镇,我每年也会拿出2万块钱资助学生,1万块钱给贫困的老人和孤寡老人,所有这些我很早就开始做了。爱因斯坦他们讲的不要追求个人享乐,我们有了钱,也不去追求豪华的汽车、豪华的别墅,这不是我们想要的。

钱是干什么呢?为了生活,为了工作,生活工作国家照顾得很好了,剩余的钱为国家为人民做点事。后来的江苏省配套奖励了我800万元,我是全都拿出来,其中100万元给中国岩石力学与工程学会设立钱七虎奖学金,50万元捐给了我的母校——上海中学,剩下的650万捐给了武汉疫情防控。

我有一个生活的准则,就是知足常乐,自己的一些生活需求满足了,所以就感到生活上没有什么还需要追求的,助人为乐,帮助人家解决问题很高兴。有些我帮助的青年学生,他们大专毕业以后,工作我要帮他们找,他们找到工作我感到很高兴,他们找对象我也帮他们找,解决了我很高兴,所以是真正尝到了快乐。就这样的心态好,所以身体也就好。

现在青年目标定得不高,就是为了追求个人的高的待遇、高的薪金。这样理想太低了。以前巴甫洛夫讲:"科学需要人一整个的生命。"马克思讲:"科学上是没有平坦大道,只有不畏艰险的山路的人才能攀登到光辉的顶点。"要有这样的目标,才能做出很大的贡献来,做出成就来。我7岁的时候父亲去世了,父亲临终前把我们子女叫到旁边,他对我讲了一句话,你要创一番事业,这是父亲说的话,和上面说的一个意思:我们要一个人活着一辈子要为世界做一点贡献,创一番事业。

采访感悟

想到这次的采访对象是"国家最高科学技术奖"获得者、享誉全国的钱七虎院士，我们不免有些紧张和忐忑。好在网络渠道提供的资料很多，我们反复修改了提纲、做好了前期准备。只是真正和钱院士展开对话后，我们才近距离感受到了钱院士的人格魅力和崇高的精神品质。他最后给予上中学子的六字寄语——"立志、勤奋、友善"，看似简单却富有深意。那包含的正是钱院士自己一路坎坷求索、终获成功的宝贵经验，也是最值得我们回味的人生财富。

1. 长风破浪会有时，直挂云帆济沧海——志向要大

在采访过程中，钱院士不止一次提到"上中的学生不是'死读书'，是要培养高尚的人生目标与品质"，正如他回忆的烈士家属和访苏代表团成员黄宗英的报告，钱院士不无感慨"它们使得烈士和英雄的形象始终活在我的心中，指导了我以后的行动"。原来这就是钱院士在上中形成的英雄情怀和高尚价值观；钱院士回忆了自己积极参加农民夜校和纠察队的经历，这些是钱院士全面发展能力、开拓视野的典范；钱院士还活跃于大量团活动和校园事务中。这些更让我们看到了一位践行上中传统、德智体美劳全面发展的上中学生楷模形象。可见，我们上中学子应当注重自身的多元、全面发展，形成广阔的视野和家国情怀、不畏牺牲不畏艰险的精神，才能不负优秀的前辈对我们的期望！

此外，钱院士着重回忆了在上中时的各位恩师，他反复赞扬老师们"教书育人成绩很大""他们的教学已经达到了相当高的水平了，所以引人入胜""讲得生动、非常有趣""数理化本来是很枯燥的一些科学内容，但是我们听上中老师的课感到享受"，诚然上中老师对钱院士产生了深远的影响，但我们也同样注意到了钱院士话语中流露出的对学科的热爱和学习的热情，学习对他从不是一种压力与任务，而是一种兴趣与修行。这将激励我们也在以后的学习过程中保有同等的动力和兴趣，享受上中的课堂过程与收获，并将它们转化为自身宝贵的财富。

最后，我们还感受到了钱院士"服从组织分配"的大局观和"为国为天下"理想与情怀。无论是响应国家培养先进军事人才的号召而放弃留苏，转入哈军工，还是身为哈军工班长带头报名冷门系"工程兵系"，都体现了钱院士在抉择面前以大局为先、以国家需要为先、以组织规定为先的高尚品德。

培养优秀的品德，树立对于学科的喜爱，以及拥有心怀天下的家国情怀——这是钱院士立下的志向，也应是我们每一个人的心之所向。毕竟，只有注视着星光，才终有一天能驶向那无垠的星海。

2. 三更灯火五更鸡，正是男儿读书时——为学要勤

在采访钱七虎院士前，我们就已知道他是一个十足的学霸，他自己也坦言到自己无论是在上中，哈军工还是苏联军事学院，成绩都是极好的，在上中时成绩单被作为慰问前线将士的礼物，在哈军工的五年成绩全优，均为优秀学员。而在这些光彩的背后，是院士的勤奋好学以及对专业始终如一的钻研精神。

钱七虎院士在赴苏留学的时光里，连最想去的列宁墓都没有去成，因为要排长队，而院士不舍得珍贵的学习时间，直到受邀参加国际会议时，才完成了这个多年的愿望。"文化大革命"时期，他也仍没有停下学习的步伐。这段时间里，他思考琢磨赴苏留学时期下的难题，在自己的专业领域下狠心钻研，"文化大革命"结束后便一下拿出8篇高质量的论文，引起了学术界的重视。除了军工之外，钱院士更是投身基础设施的建设，无论是长江隧道、港珠澳大桥，还是如今关于地下空间的开发和利用中，都能看到这位朝气蓬勃的先生勤劳的身影。就连在采访中，我们也深刻地感受到钱院士的忙碌与勤奋。

在这位已步入杖朝之年的先生身上，勤奋二字似乎依然闪烁着它独特的光芒与魅力。纵观历代人才，无一例外都具有"勤奋"的特点，而我们作为上中学子也更应具备这项能力。坚持，咬紧牙关的勤奋绝不仅限于学习，更应体现在生活里，如钱院士在"文化大革命"期间的埋头研究，忙碌于各项工程中，不愿让自己停下来变得懒惰，散漫。勤奋是一辈子的习惯，也应该是当我们毕业之际，这三年的生活留给我们的馈赠。

勤奋不一定能铸就成功，但成功的垫脚石里一定会有勤奋。钱院士的人生经历让我们明白，勤奋不是不懂变通，一条路走到黑，更应是一种处变不惊的定力，一种生活中应坚守的信仰。

最后借用钱院士告诉我们的马克思的话——"什么是勤奋？天才就是勤奋,勤奋出才干,勤奋的学习出才干,勤奋的工作出成就"——这不仅是钱院士本人一生的写照,更是历史上无数次应验的经验总结。

3. 千里家书只为墙,让它三尺又何妨——为人要善

钱院士身上还有一个让我们印象深刻的闪光点：为人友善,恃才不傲。他不无痛心地指出,现在的青年人,尤其是有才华的青年,因是独身女子的关系,大多骄纵,孤傲,难以相处。在团队相处合作的过程中,难以形成友善的团队协作关系,而这样的人在集体中越多,集体的团结度越低,越难完成集体的任务。因此钱院士总结认为,我们青年人应当做友善之人,从根源上促使集体的团结。

而钱院士本人,就是为人友善最好的例子。这不仅体现在生活中一点一滴的小事里,更体现在社会上救济贫困的大善上。从坚持捐赠贫困学生读书,到最近将650万元奖金捐给武汉抗疫前线,而说起为什么会无私地捐出这么多钱,钱院士提到生活讲究的是知足常乐与助人为乐,即满足于自己的一些生活待遇并能够力所能及地为他人提供帮助。如钱院士所言,当一个人做到了知足常乐时,他的心态必然是平和的,心态的平和带来的是给人以亲近的感受,而当一个人做到了助人为乐时,定然是众乐乐而非独乐乐。在众人心气平和而又能够互相帮助带来快乐的社会环境中,人人友善,集体的合作加强,社会的凝聚力也能够得到提升。

牺牲自己的享乐,换取国家的繁荣；拿出生活上的积蓄,换来更多人的笑脸：钱院士的大善为他赢得了名望,也指得我们每一个人去思考、学习。

从少年时的立志高远,到青年时的发愤图强,从中年时的埋首科研,到如今的矢志不渝,抹去繁华,我们看到的是一名共产党员,一名上中学子的初心。从钱院士身上,我们看到的是20世纪那些为国奉献了一生的科学家们的写照——他们用自己的智慧与汗水,将中国的科技实现了从零到一,从一到万的飞跃,他们载着旧时代的伤痛而来,开创了一个全新的时代,吉利如今的我们继往开来。

纵观钱院士的采访中,我们也捕捉到了不少批判性的智慧的光辉。他不无担忧地说,现在好学生都喜欢选金融,为了更好的薪金来选择专业。钱七虎的话让我们深思：我们作为上中人,作为实现中华民族伟大复兴的轴承,是选择为了更优越的生活活着,还是积极地将自己的价值实现与国家融合在一起。不错,每个人都有选择的权利,可作为上中人,我们理应有更高的眼界,更大的胸怀。

老一辈的龙门学子站在将来向我们眺望,我们也应继承他们的情怀：中国之复兴,舍我上中人,其谁！

| 吾家吾国

铸就共和国"地下钢铁长城"

各位听众朋友、各位网友：

大家好！非常高兴来到这里，和大家分享我的一些感悟和体会。我出生在战火纷飞的抗战时期，亲眼见证了国家的沧桑巨变，见证了我们国防和军队建设取得的巨大成就，特别是亲身参与并见证了我国防护工程研究与建设从跟跑到并跑、再到有所领跑的全过程，感到很幸运也很幸福。不少朋友和年轻战友与我探讨人生的意义，谈论军人的价值和知识分子的追求，还问我有什么成功的秘诀。回顾我的人生历程，可以用三句话来概括。

第一句话，献身党的事业，是我不变的初心。

1937年10月，母亲在逃难路上的小船上生下了我。为了怕我的啼哭引来日本兵，差一点把我捂死。7岁时，父亲因贫病离世，母亲靠摆小摊维持生计。说实话，如果没有抗日战争的胜利，没有1949年的全国解放，没有国家的助学金支持，我早就会和哥哥姐姐一样失学、失业。八年日寇铁蹄下的生活，四年国民党统治下的暴行，新旧社会的对比，使我深深懂得没有共产党就没有人民的解放和国家的独立，从此，坚定了我跟党走的信念，所以我14岁就入了团，18岁入党。1954年高中毕业哈军工需要学员，我放弃了留学的机会，转入哈军工，我服从了国家需要。当时防护工程专业没人选，因为要跟黄土铁铲打交道，但是我始终服从组织分配，让我学什么就学什么。大学六年我只回过一次家，年年都被评为优秀学员，是全年级唯一的全优毕业生。1960年，我被选派到苏联莫斯科古比雪夫军事工程学院学习深造。当时的工程兵政治部主任李大同（一位老红军），对我说，"我们国家还有人吃不饱，国家得用多少金条才能送你们去留学啊。"我始终牢记使命、刻苦学习，获得工学副博士学位。毕业回来，刚成立的西安工程兵工程学院需要教员，我又一次没有二话服从分配到西安工作。在1978年的全国科学大会上，我获得重大科技成果奖。

今年1月8日，在国家科学技术奖励大会上，习近平主席为我颁发了国家最高科学技术奖奖章和证书。我深知这份荣誉和褒奖不属于我个人，它归功于党和国家对科技人员的热忱关心，归功于我们所处的这个伟大时代。回顾自己八十余年的人生岁月，从亲身经历被侵略欺凌的旧社会，到满怀激情跨入中华民族伟大复兴的新时代，从切身感受国家翻天覆地的沧桑巨变，到无数次在个人志愿与组织需要时的无悔抉择，我深深感到，没有党的培养就没有我的一切，唯有献身党的事业，才能报答党的恩情。

第二句话，锻造"地下钢铁长城"，是我毕生的追求。

"世间万物，相生相克，有矛必有盾。"如果说核弹是军事斗争中锐利的"矛"，那么防护工程则是一面坚固的"盾"。20世纪六七十年代，国际冷战加剧，军备竞赛激烈，我国面临着严峻的核安全威胁。但我们奉行的是积极防御战略，就是不打第一枪，那如何打好第二枪也就是保证"二次反击"呢？关键就是要铸牢防护工程这面坚固"盾牌"，确保我国首脑指挥工程和战略武器防护工程的安全。从那时起，为国设计打不烂、炸不毁的"钢城坚盾"成了我一生未曾动摇的目标。20世纪70年代初，我受命担负飞机洞库门设计，专门赶赴核爆试验现场进行实地勘察和数据收集。我在核爆现场发现，虽然此前设计的飞机洞库门没有被炸坏，里面的飞机也没有受损，但是防护门出现严重变形，导致无法开启。门打不开、飞机出不来，就无法反击敌人、打击敌人。当时飞机洞库防护门的相关设计均采用手算的方式，计算精度差、效率低。我率先引入有限元计算方法,加班加点翻译整理出十多万字的外文资料,学习电子计算机语言,创新编写出内存容量最少的有限元计算程序,利用七机部五院当时中国最大的晶体管电子计算机中午和晚上的计算时间,设计出当时跨度最大、抗力最高,能抵抗原子弹爆炸压力最高的核爆炸冲击波的机库大门。

"矛"与"盾"总是在攻防对抗的进程中不断碰撞出新的"火花"。特别是世界军事强国开始研制精确制导钻地弹,给防护工程造成了巨大威胁和挑战。为此,我继续开展抗深钻地武器防护的系统研究。通过钻研已解密公开的苏联地下核试验等大量资料,经过近千次细致的推导计算,我创造性地提出了建设超抗力深地下防护工程的总体构想,并带领团队开始了艰难的跋涉。功夫不负有心人,经过长达十多年的研究,我和团队攻克了一个个难题,为抗钻地核武器防护工程的设计与建设提供了理论依据,为我国战略工程装上了"金钟罩"。

第三句话,谋求国富民强,是我最大的心愿。

当今世界,正处于大发展、大变革、大调整时期。大国间的竞争,不仅体现在军事实力上,更体现在综合国力上。作为革命军人、一名科技工作者,既要关注如何增强国防实力、维护国家安全,更要关心提高综合国力、谋求人民幸福。1992年初,珠海特区建设机场,要炸掉三灶岛炮台山一座山,12000吨炸药,爆破总量1085万立方米,要求一次性爆破成功,一半的土石方要求爆后定向投入大海,而且要确保邻近800米内的两处村庄安全。这样大的爆破,世界还无先例,难度很大,许多人都犹豫了、退缩了。我带领团队先后六下珠海,和大家一起研究爆破总体方案和装药坑道的每个施工环节,当年12月成功实施爆破,创造了世界爆破史上的新纪录。

我始终觉得,做科研工作不能仅仅着眼当下看得见的事情,更应该站在国家的全局进行前瞻思考。自20世纪90年代末起,国外一些大都市已经出现交通拥堵、空气污染、城市水涝等问题,地上空间利用远远跟不上人口的增长,而中国城市的规划建设者尚未充分意识到城市地下空间的开发利用在治理城市病和节约土地能源等资源方面的意义和作用。于是,我利用自己研究地下工程占有大量国内外学术资料的优势,开始从事相关研究工作,率先提出开发地下城市空间、发展地下交通、城市地下物流等创新观点,竭力为解决我国城市可持续发展问题出力献策。我先后组织编制了全国20多个重点设防城市的地下空间规划,主持了北京、深圳、南京、青岛等十几个城市地下空间规划的评审工作。这些年,我还在海底隧道、长江隧道、南水北调工程、西气东输工程、港珠澳大桥海底隧道、能源地下储备、核废物深地质处置等方面,贡献了我的学识和才智。我始终认为,这些都是我作为院士的使命所系,也是我作为科学家的幸福所在。我也常和我的学生讲,一定要将个人命运与国家、民族的命运结合起来,哪些事情对国家和人民有利,我们的兴趣和爱好就要向哪里聚焦。只有这样,才能收获成功的事业和幸福的人生。

各位网友,我今年82岁了,职务已经退休了,但我觉得思想还不能退休,工作不应退休,还有很多事情我可以做,需要我要做。在我有生之年,我将始终做到无须扬鞭自奋蹄,继续在防护工程领域和地下工程领域潜心研究,继续带好学生、培养人才、关心团队建设,为国家铸就钢铁强盾,为经济和社会建设做出新的更大贡献。

谢谢大家!

建设科技强国迫切需要工匠精神

我国科技创新确立了三步走战略:到2020年进入创新型国家行列,到2030年跻身创新型国家前列,到2050年建成世界科技强国。大家知道,建设科技强国靠人才,研制"两弹一星"的经验告诉我们:一切伟大,来自我们的大脑和双手的创造,建设科技强国不但需要大批具有科学家精神的科技人才、也需要大批具有大国工匠精神的技能人才。所以建设世界科技强国迫切需要工匠精神。今天我就和大家交流一下学习心得体现:什么是工匠精神?为什么要弘扬工匠精神?如何在新时代继承弘扬工匠精神?如何培育自己的工匠精神?

一、工匠精神的科学内涵

党的十八大以来,习近平总书记多次礼赞劳动创造,讴歌工匠精神。2020年11月24日,习近平总书记在全国劳动模范和先进工作者表彰大会上精辟阐释了工匠精神"执着专注、精益求精、一丝不苟、追求卓越"的科学内涵。

(一)执着专注

"执着专注"是一种踏踏实实做事的工作态度,是一种"术业有专攻"的人生追求。古有鲁班炉火纯青的高超技艺令人瞠目结舌,庖丁解牛的游刃有余令人啧啧称赞;今有徐立平精雕细琢,为我国的航天事业保驾护航;李刚蒙眼插线,方寸之间也能插接百条线路,领跑世界的"中国制造"。执着于自己所选择的事业,专注于自己所投身的领域。"干一行、爱一行",一旦定下目标,便风雨无阻,砥砺前行,为自己所热爱的事业奉献自己的光和热。热爱是最好的老师,爱干比能干更重要,热爱远超于对工作的责任感,以热爱作为引导,在心底埋下执着专注的种子,让精神之花在伟大实践的肥沃土壤里生根发芽。专注就是内心笃定而着眼于细节的耐心、执着、坚持的精神,这是一切"大国工匠"所必须具备的精神特质。

从中外实践经验来看,工匠精神都意味着一种执着,即一种几十年如一日的坚持与韧性。德国除了有人们耳熟能详的奔驰、宝马、奥迪、西门子等知名品牌之外,还有数以千计普通消费者没有听说过的中小企业,它们大部分"术业有专攻",一旦选定行业,就一门心思扎根下去,心无旁骛,在一个细分产品上不断积累优势,在各自领域成为"领头羊"。其实,在中国早就有"艺痴者技必良"的说法。古代工匠大多穷其一生只专注于做一件事,或几件内容相近的事情。《庄子》中记载的游刃有余的"庖丁解牛"、《核舟记》中记载的奇巧人王叔远等大抵如此。工匠以工艺专长造物,在专业的不断精进与突破中演绎着"能人所不能"的精湛技艺。实际上,得心应手的技能、巧夺天工的技术和出神入化的技艺,正是来源于专注。工匠始于学徒、技工,应学会干实事、细事、小事。选择了某个行当,就应沉潜下来,不能心猿意马。要在一个领域精雕细琢、精耕细作,必须坐得了冷板凳,耐得住寂寞,而这一切必须不忘初心、坚守理想。

(二)精益求精

"精益求精"出自《论语·学而》,原文是"如切如磋;如琢如磨"。它原本指的是君子的自我修养就像加工骨器,切了还要磋;就像加工玉器,琢了还得磨。新时代赋予其新的内涵,指的是一种追求工艺精湛的职业品质。"干一行,精一行",没有最好,只有更好。古有蔡伦造纸,不断推敲,反复实验,终于制出轻薄柔韧的纸张;而今神州十三号的成功发射离不开背后工作人员的精益求精,精雕细琢的工作态度。"干一行,精一行",成功绝非偶然,不断追求品质的卓越和一流的技艺,专注于自己的领域,坚持做到"术业有专攻"。精益就是精益求精,是从业者对每件产品、每道工序都凝神聚力、精益求精、追求极致的职业品质。所谓精益求精,是指已经做得很好了,还要求做得更好,"即使做一颗螺丝钉也要做到最好"。正如老子所说,"天下大事,必作于细"。能基业长青的企业,无不是精益求精才获得成功的。瑞士手表得以誉满天下、畅销世界、成为经典,靠的就是制表匠们对每一个零件、每一道工序、每一块手表都精心打磨、专心雕琢的精益精神。工匠们练就了炉火纯青之技,成为行业"绝活"的创始人、传承者。在技术创新的各领域,有那么一群"隐形冠军"企业,他们瞄准细分市场,集中力量进入并成为领先者。中铁二局第二工程有限公司隧道爆破高级技师彭祥华在软若豆腐般的岩层间精准爆破,误差控制远小于规定的最小误差;沪东中华造船集团公司焊工张冬伟手工焊缝长14公里厚0.7毫米的内胆,先修"心境"而后达"技境"。古人说"熟能生巧",正是因为不断精益求精,才能不断自我超越。

(三) 一丝不苟

"一丝不苟"是追求细节完美的工作作风,是一种做事严谨认真的精神品质。"失之毫厘谬以千里",小的失误会造成大的错误,细节的完美成就事业的成功,严谨的态度是影响成功的重要因素。"干一行,钻一行",小到螺丝钉,大到国之重器,都需要匠人们倾尽匠心,一丝不苟地慢慢打磨。以严谨的工作态度和求真务实的工作作风投身于自己所热爱的事业,从每个细节上积累优势,在每个细节上做足功夫,在平凡的岗位中创造出不平凡,以一流的精品去接受党和人民的检验。早在春秋时期,孔子就主张人在一生中始终要"执事敬""事思敬""修己以敬"。"执事敬",是指行事要严肃认真不怠慢;"事思敬",是指临事要专心致志不懈怠;"修己以敬",是指加强自身修养保持恭敬谦逊的态度。宋代大思想家朱熹将敬业解释为"专心致志,以事其业"。所以,工匠必须具有严谨的态度,必不能造伪器。每一项具体技术的研究开发与应用扩散,往往都具有严格的规程和标准,来不得半点马虎将就,走捷径搞变通是行不通的。就像拧螺丝这种简单技术,拧几圈回几圈以及施加多大扭矩都要遵守严格规定,否则就可能造成严重误差。"炮制虽繁必不敢省人工,品味虽贵必不敢减物力。"不放过任何一个细节,不忽视任何一个细微之处,一丝不苟、倾注匠心,才能创造出巧夺天工的精品。

(四) 追求卓越

"追求卓越"是工匠们不断进取,开拓创新的精神体现。它是一种不停滞不前,不满足于现状的工作状态。追求卓越是匠人的价值追求,将专业的理论知识与反复实验的耐心和勇气相结合,以技能报国的信念和与时俱进地创新品格铸就自己的工匠人生。大国工匠徐立平,精雕细琢火药药面,将误差从0.5毫米缩小到0.2毫米,从实践中不断吸取经验,设计发明出20多种药面整形刀具,并获得两项国家专利,为我国的航天事业贡献自己的毕生才智,无愧于大国工匠的称号。"变则通,通则久"工匠们与时俱进,吐故纳新,不断追求完美的境界,在尊重客观规律的前提下,充分发挥自己的主观能动性,投身于自己所热爱的事业。"工匠精神"强调执着、坚持、专注甚至是陶醉、痴迷,但绝不等同于因循守旧、拘泥一格的"匠气",其中包括着追求突破、追求革新的创新内蕴。这意味着,工匠必须把"匠心"融入生产的每个环节,既要对职业有敬畏、对质量够精准,又要富有追求突破、追求革新的创新活力。事实上,古往今来,热衷于创新和发明的工匠们一直是世界科技进步的重要推动力量。新中国成立初期,我国涌现出一大批优秀的工匠,如倪志福、郝建秀等,他们为社会主义建设事业做出了突出贡献。改革开放以来,"汉字激光照排系统之父"王选、"中国第一、全球第二的充电电池制造商"王传福、从事高铁研制生产的铁路工人和从事特高压、智能电网研究运行的电力工人等都是"工匠精神"的优秀传承者,他们让中国创新重新影响了世界。当技能达到一定水平后,根据边际报酬递减规律,持续以单一时间要素投入获取的技能提升增量会逐渐下降。此时就要通过改变要素投入组合,改变其他投入要素的量,或者投入其他要素,实现工艺改造和产品创新。在数字经济时代,技术的更新换代加快,技术生命周期呈现出新旧共性技术演绎创造性破坏的特征。正因此,弘扬工匠精神就要追求卓越、敢于创新,永不满足于现有水平,永不停滞于当前状态,而要向更高、更好、更强不断迈进。

二、为什么要弘扬工匠精神

在新时代大力弘扬工匠精神,对于推动我国建设高质量发展具有重要意义,对如期实现两个百年奋斗目标有着不可替代的关键作用。

(一) 弘扬工匠精神是推动人类进步的重要力量

人类的一切伟大与进步都是源于人类的大脑和双手,具体来说需要在科技快速迭代的推动下,通过高质量地控制和把握过程,才能高水平地实现进步的目标,而工匠精神是助推科技创新能力和高品质制造提升的重要动力,只有精益求精、追求卓越,才能从微创新、小创新发展到重大创新,才能从点创新到面创新,从局部创新到体系创新,才能创造出全国品牌和世界品牌的高质量水平!

(二) 弘扬工匠精神就是弘扬我国悠久历史文化积淀的优良传统

"不惰者,众善之师也。"我国自古就有尊崇和弘扬工匠精神的优良传统。《诗经》中的"如切如磋,如琢如磨",反映的就是古代工匠在切割、打磨、雕刻各类器物时精益求精、反复琢磨的工作态度。《庄子》中讲庖丁解牛游刃有余,"道也,进乎技矣"。新中国成立以来,一批又一批劳动者在党的领导下,始终坚持弘扬工匠精神,用奋斗创造了一个又一个"中国奇迹"。无论是红旗渠、南京长江大桥、"两弹一星",还是北斗导航和探月、探火、载人航天工程取得的辉煌成就,都展现出和包含着对工匠精神的继承与发扬。工匠精神是构成中国共产党人的精神谱系的重要组成部分,是中国共产党人红色基因的重要组成部分,已成为新时代引领社会风尚,构建新发展格局,实现中华民族伟大复兴"中国梦"的重要支撑。

(三) 弘扬工匠精神是实现制造业转型升级的重要基础

我国是世界制造业第一大国,在世界500多种主要工业产品中,我国220多种工业产品的产量位居世界第一。但总体而言,我国制造业大而不强,打造更多世界名牌产品,实现制造业转型升级迫在眉睫。在此,要做到坚持两个理念,一是大力推进名牌战略的实施,尽快改变我国产品物美价廉的生产模式,做到物优价也优,这就需要大量的独创和原创,掌握更多的自主知识产权。二是要坚持以人为本,人的质量决定产品质量,应加大对各类人才的培养力度,建立和完善广大干部群众终身学习和教育保障制度,搭建好各类人才的任用和成长平台。我们要通过弘扬工匠精神,培育众多劳动者追求完美、勇于创新的精神,为实施创新驱动发展战略、推动产业转型升级奠定坚实基础,加快建设制造强国,推动经济高质量发展。

(四) 技术工人队伍是支撑中国制造、中国创造的重要基础,对推动经济高质量发展具有重要作用

这就要求我们倡导尊崇工匠精神的社会风尚,为弘扬工匠精神营造良好社会氛围。在弘扬工匠精神的同时一定要坚持建立在自信基础上的创新和奉献,坚持持续打磨精湛的技术技艺。要形成相应体制机制,为劳模和技能人才发挥作用搭建宽广舞台,使他们在经济上有保障、发展上有空间、社会上有地位。特别是,大力弘扬工匠精神,对培养更多高素质技术技能人才、能工巧匠、大国工匠,为全面建设社会主义现代化国家、实现中华民族伟大复兴的中国梦提供有力人才和技能支撑。为此,我们要深刻理解工匠精神的时代价值和重要意义,立足本岗,大力弘扬工匠精神,谱写高质量发展的时代凯歌。

三、如何在新时代继承弘扬工匠精神

当前弘扬工匠精神面临新的形势。在革命年代和改革开放以前，革命热情、激情澎湃，传统的举国体制、贫富差距不大的环境为劳动者和技能工作者发挥主观能动性、践行工匠精神提供了外部约束和目标督促。当前一部分人先富起来、贫富差距明显、充满诱惑的市场经济环境和相对宽松自由的科研环境，对劳动者和技能工作者的精神自律、行为的自我约束提出了更高的要求，这就要求我们在不忘初心、坚守理想的基础上做到：

（一）应时而变、适需而变

大力弘扬劳模精神、劳动精神、工匠精神，首先是符合党和国家的事业需要。我国的劳模精神、劳动精神、工匠精神孕育于革命战争年代，形成于社会主义革命和建设时期，发展于改革开放新时期。如今，在激烈的国际竞争和国内产业转型升级压力下，"劳模精神、劳动精神、工匠精神"不断被赋予国家竞争综合实力的新内涵，早已不单单是练就炉火纯青技术的工人，抑或是成为自身行业"绝活"的创始人、传承者，更需要瞄准国家发展大局，紧盯方方面面的细分行业，特别在"卡脖子"的关键技术各领域，培育领先者、领跑者，为推动国家高质量发展储备人才力量。

（二）勤学苦练、深入钻研

大力弘扬劳模精神、劳动精神、工匠精神，必须落实到提高劳动者素质上。大国工匠们在专业的不断精进与突破中演绎着"能人所不能"的精湛技艺，凭借的正是勤学苦练、深入钻研。正如科技进步没有尽头一样，技术也没有尽头。只是当技能水平达到一定高度后，付出相同的时间、精力投入获取的技能提升增量会逐渐下降，这就需要引导广大劳动者不忘初心、坚守理想，坚定职业信条和行为操守，始终树立终身学习理念，密切关注行业、产业前沿知识和技术进展，不断精益求精，才能不断自我超越。

（三）勇于创新、敢为人先

大力弘扬劳模精神、劳动精神、工匠精神，就是永不满足于现有水平，永不停滞于当前状态，而要向更高、更好、更精的方向努力。近年来，大国工匠们凭借勇于创新、追求卓越的干劲，实现了一项项工艺革新、技术迭代，完成了一系列科技攻坚：全海深救人潜水器"奋斗者号"、全球首座中国自主三代核电站"华龙一号"、全球最大单口径射电望远镜"中国天眼"等一大批大国重器捷报频传。对成绩的最好纪念，就是创造新的成绩，不少研究周期长、攻关难度大的项目需要突破，尤其需要科研人员和技术人员耐得住寂寞、坐得住冷板凳，多出奇招妙招，努力"弯道超车"，为加快创新发展作出贡献。历史总会眷顾奋斗者，唯有大力弘扬劳模精神、劳动精神、工匠精神，通过不懈奋斗，复兴梦想才会走进现实。让我们以劳模精神、劳动精神、工匠精神为动力，以模范人物为榜样，不断为推动高质量发展、实施制造强国战略、全面建设社会主义现代化国家贡献智慧和力量。

四、如何培育自己的工匠精神

在弘扬执着专注、精益求精、一丝不苟、追求卓越工匠精神的道路上一定会像大国工匠那样遇到无数困难、艰险和干扰，怎样才能不避艰险不怕困难和排除干扰呢？

什么样的人才能不避艰险、不怕困难，排除干扰呢？只有具有远大理想和远大志向的人才能！为什么古人云：无私才能无畏！任弼时曾说过："世上无难事，只畏有心人。有心之人，即立志之坚者也，志坚则不畏事之不成！"因为只有树立了远大的理想，人生才有远大的目标，在前进的道路上才有责任感和担当心,有了担当,才能百折不挠力求作为,才有巨大的动力、魄力和毅力。只有树立了远大的理想和志向,才能无私,心底无私天地宽!无私才能无畏,才能具有巨大的动力、巨大的魄力、巨大的毅力,如同无产阶级革命家,伟大的科学家和大国工匠所表现出来的那样!相反,没有高尚的志向就会小人常戚戚,就会被名利欲望所引诱,这山望着那山高,坚持不了就走近道、向后转,当思想上的逃兵。

说到底工匠精神是党性在劳动者身上、士官身上的体现和反映,是革命人生观、马克思主义世界观在劳动者身上、士官身上的反映和体现。

如何树立自己的远大理想和志向,树立自己的革命人生观和马克思主义世界观?我的人生体会是:要让伟人、先烈、英雄和大国工匠活在自己的心中,让伟人、先烈、英雄、大国工匠的光辉思想言行照耀自己的人生道路,榜样的力量是无穷的!要自觉以他们为榜样,指导自己的思想行动!

最后再说几句共勉的话。

我已年逾八十多岁,但我是一名承上启下、具有66年党龄的科学家。共产党员是所谓有特殊材料做成的人,我有责任继续践行科学家精神,让我余生继续发光发热。要学习钱学森,把毕生精力、目光始终关注到国家重大需求上;要学习华罗庚,勤奋工作到最后一刻,倒在讲台上。

我虽已退休了,但我觉得还有很多事情能做、想做、要做。继续为国防建设用脑尽力,为国家铸就钢铁后盾,所以我继续担任军事科学院首席专家,火箭军首席工程专家,空军工程顾问。国家重大工程有需要,我就应召而去!为经济和社会建设继续做贡献!我总觉得党给我的荣誉太多了,我必须活到老、学到老、革命到老,必须始终做到无须扬鞭自奋蹄。

我希望同志们,要继承弘扬好、实践好工匠精神,身体力行,把个人理想和国家重大需求相结合,排除名利束缚,严谨治学,潜心研究,直面问题,迎难而上,奋勇拼搏,敢闯科学"无人区"。要敢为人先,敢攀高峰,集智攻关,团结协作,群策群力参与开拓新兴领域与交叉领域,群策群力抢占世界科技发展的制高点,为开启伟大新征程作出更大贡献!

大事记

1937 年 10 月 26 日 ▶ 出生于江苏省苏州市昆山市。

1948 年 – **1954** 年 ▶ 先后就读于上海敬业中学、上海中学。

1954 年 08 月 ▶ 接受组织安排，保送进入中国人民解放军工程学院（通称"哈尔滨军事工程学院"，简称"哈军工"）。

1956 年 04 月 ▶ 加入中国共产党。

1960 年 ▶ 从哈军工本科毕业，留校任教。

1961 年 09 月 ▶ 赴苏联古比雪夫军事工程学院留学。

1965 年 07 月 ▶ 完成了副博士学位论文《压缩波作用下地下织物–骨架结构单元的变形计算》，通过答辩。

1965 年 08 月 ▶ 到西安工程兵工程学院任教。

1965 年 10 月 ▶ 在第一届全国防护工程学术交流会上发表了回国后的第一篇论文《孔口防护中的空气动力学计算原理》；与妻子袁晖结婚。

1969 年 12 月 ▶ 调到刚组建不久的南京工程兵工程学校从事教学工作。

1972 年 09 月 ▶ 著作《武器效应与地下工程防护》由总字 340 部队（工程兵学校）出版。

1973 年 10 月 – **1975** 年 02 月 ▶ 采用有限元方法对空军 40 米跨钢防护门进行试验及理论分析，撰写了《40 米跨钢防护门试验及试验理论分析》学术报告。

1976 年 11 月 ▶ 著作《有限元法在工程结构计算中的应用》由空军后勤部和工程兵学校联合出版。

1981 年 05 月 ▶ 任南京工程兵工程学院训练部副部长；09 月 ▶ 著作《防护结构计算原理》由工程兵工程学院出版。

1982 年 09 月 ▶ 作为党代表参加"中国共产党第十二次全国代表大会"。

1983 年 09 月 ▶ 任南京工程兵工程学院院长。

1987 年 02 月 ▶ 主持的"团坚固防御阵地防护能力稳定性系统分析"获军队科学技术进步二等奖；08 月 ▶ "城市人民防空防护标准"获国家人防科学技术进步一等奖。

1988 年 ▶ 被授予中国人民解放军少将军衔。

1989 年 07 月 ▶ 获中国人民解放军三总部（总参谋部、总政治部、总后勤部）联合颁发的"全国优秀教学成果军队级一等奖"。

1990 年 03 月 ▶ "核爆条件下土中浅埋结构实用计算法"获军队科学技术进步二等奖；12 月 ▶ 被国家教育委员会和国家科学技术委员会联合授予"全国高等学校先进科技工作者"；"核爆条件下土中浅埋结构实用计算法"获国家科学技术进步三等奖；被国家人事部授予"中青年有突出贡献专家"。

1993 年 03 月 ▶ 当选"中国人民政治协商会议第八届全国委员会委员"，并连任第八、九、十届全国政协委员。其间，提交"关于编制城市地下空间规划的建议"和"关于加强隧道掘进机国产化的建议"等提案。

1994 年 05 月 ▶ 当选中国工程院院士（首批）。

1995 年 04 月 ▶ 获国务院颁发"政府特殊津贴"；10 月 ▶ 被中国人民解放军总政治部评为"爱国奉献优秀干部"。

1996 年 01 月 ▶ 著作《民防学》由国防工业出版社出版；07 月 ▶ 在"第四届全

国岩石力学与工程学术大会"作大会报告"岩体中爆炸与冲击作用研究进展";

08月 ▶ 获军队科技进步一、二等奖各一项;调任总参谋部科学技术委员会副主任。

1996年06月 – 2000年06月 ▶ 担任中国工程院土木、水利与建筑工程学部副主任。

1997年09月 ▶ "面向国防现代化建设,创一流防护工程学科"获军队教学成果一等奖;主持中国工程院咨询项目"21世纪中国城市地下空间开发利用战略及对策"。

1998年12月 ▶ "断裂构造对触地爆中远区地应力地运动参数的影响"获国家科学技术进步二等奖。

1999年06月 ▶ 参加由国家自然科学基金委员会、中国长江三峡工程开发总公司组织的"三峡水利枢纽工程几个关键问题的应用基础研究重大项目验收会";

07月 ▶ 获中国人民解放军四总部(总参谋部、总政治部、总后勤部、总装备部)联合颁发"中国人民解放军专业技术重大贡献奖";参加"宜昌至万州铁路特长复杂岩溶隧道工程"实地考察和咨询调研活动;**08月** ▶ 率专家团队对山西万家寨引黄工程南干线进行咨询。

2000年06月 – 2004年06月 ▶ 担任中国工程院工程管理学部副主任。

2001年01月 ▶ 主编《中国城市地下空间开发利用研究》(1~4)由中国建筑工业出版社出版。

2002年01月 ▶ 译著《俄罗斯地下铁道建设精要》由中国铁道出版社出版;**05月** ▶ 出席在意大利举办的"2002世界隧道及地下工程博览会暨学术交流会",并作报告"隧道掘进机在中国地下工程中的应用现状及发展前景"。

2003年03月 ▶ 在中国岩石力学与工程学会第五次会员代表大会上,当选中国岩石力学与工程学会第五届理事会理事长;**09月** ▶ 在南非约翰内斯堡参加"第十届国际岩石力学大会(10th Congress ISRM)",当选国际岩石力学学会副主席;完成中国工程院科研咨询项目"我国重要国民经济目标战时防护对策的研究"(2001–2003)。

2004 年 01 月 ▶ 与刘源张院士共同牵头完成科技部委托重大科技专项——《重要技术标准研究》的咨询评估，提交咨询评议总报告和 19 篇专项评估报告；02 月 ▶ 受聘于国务院南水北调工程建设委员会为"专家委员会委员"；05 月 ▶ 参加"SINOROCK 2004 国际岩石力学研讨会"，作特邀报告"非线性岩石力学的新进展(Some Key Problems Related to Non-linear Rock Mechanics at Great Depth)"；06 月 ▶ 参加国家自然科学基金"深部岩体力学基础研究与应用"重大项目启动会；09 月 ▶ 与潘自强等共同向中国工程院提交"关于我国高放废物地质处置科技工作的建议"；10 月 ▶ 率专家组对溪洛渡水电工程设计施工中的岩石力学问题进行咨询；著作《我国重要经济目标防护措施及对策》由解放军出版社出版。

2005 年 06 月 ▶ 受聘于南京市人民政府为"南京长江隧道工程政府特聘专家"；10 月 ▶ 参加"第五届全国工程结构安全防护学术会议"，作专题报告"战略防护工程面临的核钻地威胁及连续介质力学模型的不适用性"；12 月 ▶ 受聘为上海长江隧桥（崇明越江通道）工程技术专家。

2006 年 05 月 ▶ 承担中国工程院"装备制造业自主创新战略研究"咨询项目；组织开展锦屏水电枢纽"岩爆条件下隧道掘进"技术咨询。

2007 年 11 月 ▶ 在中国工程院和美国工程院联合举办的"城市反爆炸恐怖科学技术问题研讨会"上，作报告"重要目标防爆抗爆的主要技术措施（The Main Technical Measures on Preventing and Resisting Explosion of Important Objectives）"；12 月 ▶ 著作《地下空间科学开发与利用》由江苏科学技术出版社出版；完成中国工程院咨询项目"我国重大工程项目可行性研究的经验教训及改进方向研究"。

2008 年 06 月 ▶ 主持中国工程院重点咨询项目"城市地下空间综合管理研究"；

07月 ▶ 在"第十次全国岩石力学与工程学术大会"上，作特邀报告"分区破裂化现象研究的新进展及今后研究方向的思考"。

2009年04月 ▶ 在"纪念汶川地震一周年——抗震加盟、减灾专题学术讨论会"上，作特邀报告"隧道工程动力响应特性与汶川地震隧道震害分析及启示"；08月 ▶ "深地下大跨度战略工程抗直接打击防护关键技术研究"获军队科技进步一等奖；10月 ▶ 著作《高等防护结构计算理论》由江苏科学技术出版社出版；11月 ▶ 在锦屏一级水电站就"地下厂房及左岸边坡关键技术"组织专家咨询。

2010年02月 ▶ 受聘为"港珠澳大桥技术专家组专家"；03月 ▶ 著作《岩土中的冲击爆炸效应》由国防工业出版社出版；05月 ▶ 主持"锦屏辅助洞深埋长隧道工程关键技术研究及应用"研究成果鉴定会；为表彰对南京长江隧道工程建设做出的突出贡献，中共南京市委、市人民政府记其一等功；受聘于国家发展和改革委员会为"国家中长期科技基础设施编制专家"；08月 ▶ 向中国工程院提交"关于推进我国海上风电又好又快发展的思考和建议"；著作《岩土工程师手册》由人民交通出版社出版。

2011年01月 ▶ 受聘于南京市人民政府为"城市地下管线数字化建设顾问专家"；07月 ▶ 主持"中国科协第51期新观点新学说学术沙龙——岩爆机理探索"，并作专题报告"岩爆机理的简化分析及决定性参数的思考"；向中国工程院提交院士建议："关于我国城市轨道交通发展的若干建议""关于加强城市地下空间开发利用综合管理的建议""关于将核电站反应堆置于地下的建议"；10月 ▶ 作为大会主席，在北京主持召开"第12届国际岩石力学大会"；12月 ▶ "深地下防护工程建设理论与关键技术研究"获国家科技进步一等奖。

2012年12月 ▶ 作为论坛主席，在济南组织召开第152场中国工程科技论坛——"岩爆、突水突泥灾害预测预报预警与防治控制技术"。

2013 年 04 月 ▶ "中央军委给钱七虎记一等功庆功大会"在南京举行,会上宣读了中央军委主席习近平签署的通令,并颁发一等功证书和奖章;**09 月** ▶ 获何梁何利基金科学与技术进步奖;**11 月** ▶ 主持中国工程院咨询项目"深部地下空间开发对关键科学问题研究的工程技术挑战与展望"。

2014 年 05 月 ▶ 主持中国工程院重点咨询课题"深部地下空间开发对关键科学问题研究的工程技术挑战与展望";**11 月** ▶ 获国际地下空间联合研究中心会士奖(ACUUS Fellow)。

2016 年 09 月 ▶ 受聘于国家电网公司为"苏通 GIL 综合管廊工程专家组组长"。

2017 年 05 月 ▶ 在白鹤滩工程建设管理中心主持"白鹤滩水电站地下厂房洞室群岩石力学问题现场技术咨询会";**09 月** ▶ 受聘于中国大百科全书总编辑委员会为"《中国大百科全书》第三版总编辑委员会、土木工程学科主编";**11 月** ▶ 主持"FAST 开挖系统关键技术及安全性研究"成果鉴定会。

2018 年 01 月 ▶ 受聘于南京市政府为"南京长江隧道建设技术专家组组长"。

2019 年 01 月 ▶ 获 2018 年度国家最高科学技术奖;将国家最高科学技术奖的 800 万元奖金全部捐献,纳入此前成立的"瑾晖慈善基金",重点资助西部地区和少数民族的贫困学生;**09 月** ▶ 获得第七届全国道德模范"全国敬业奉献模范"。

2020 年 02 月 ▶ 向武汉市慈善总会捐款 650 万元;**07 月** ▶ 受聘为"川藏铁路建设地质安全风险评价专家委员会"副主任。

2021 年 06 月 ▶ 被中共中央授予"全国优秀共产党员"称号。

2022 年 07 月 ▶ 获颁"八一勋章"。